WORLD RADIO TV HANDBOOK

WRTH

THE DIRECTORY OF INTERNATIONAL BROADCASTING

1999

WR TH

VOLUME 53 - 1999

Editor:
David Bobbett

Contributing Editors:
Olle Alm
Bengt Ericson
Thord Knutsson
Tore Larsson

Special Contributors:
Mike Bird
George Jacobs MSEE

Editor Emeritus:
Jens M. Frost

Cover Design:
Bichard Boxall Design Associates

Production:
Nicholas Hardyman

Advertising Manager:
Beth Leinbach

Publisher:
Geoff Cowen

WRTH Publications Limited
PO Box 7373
Milton Keynes
MK 12 5ZL
UK

Fax: +44 (0) 1908 321030
E-mail: editor@wrth.demom.co.uk

© WRTH Publications Limited 1999

Sales Offices
USA, Canada & Mexico
Watson-Guptill Publications
1515 Broadway
New York
NY 10036
USA
Rest of the World
Windor Books International
11 The Boundary
Wheatley Road
Garsington
Oxford
OX44 9EJ
UK

ISBN 0–8230–5931–6
Printed and bound in Great Britain by Bath Press Limited

WORLD RADIO TV HANDBOOK

CONTENTS

Editorial

The WRTH has come a long way during its 53 year history, growing in terms of both content and size along the way, and this year sees a continuation of that trend. You will find an extra 32 pages of full colour in this edition - bringing the number of pages up to 640 and consolidating its position as being the largest and most comprehensive publication of its type.

RECENT DEVELOPMENTS
1998 proved to be one of the more eventful years in the history of WRTH with a change of editorial staff, a change of company and even a change of country. Following many years of valuable service with WRTH, Editor in Chief Andrew Sennitt decided to pursue a freelance career as a web author and broadcasting consultant, raising the question of how and where the book would be produced and of course, by who. In brief, a new company - WRTH Publications Ltd. - was formed in the United Kingdom and offices were established in Milton Keynes, some 80km north of London. Coming from a background of 25 years in telecommunications journalism, my own involvement began in the early stages of the project, culminating in the opening of the WRTH office in April 1998. One of my first tasks was to travel to the Netherlands to meet Andy Sennitt who - as I am sure anyone who knows him will confirm - I found to be an unassuming, highly professional journalist possessing a passionate and encylopaedic knowledge of international broadcasting matters. On a personal note, I would like to thank Andy for the unstinting effort, support and encouragement which he gave me over these critical early times - it is fair to say that not only was his contribution indispensable, it was also greatly appreciated.

Thanks are also due to our co-editors and contributors, all of whom made a tremendous effort to ensure that it really was 'business as usual' at WRTH and that we were able to produce the 1999 edition on time - the result of many hundreds of hours of work on their part.

DESIGN CHANGES
As regular readers will notice, there have been some changes made to the appearance of WRTH. For many years, most people have referred to the *World Radio TV Handbook* simply as WRTH and in recognition of this, we decided to re-design the cover in order to accommodate the WRTH acronym. In response to readers suggestions we have also added 'side-bars' to the directory section so as to make navigation in such a large book a little easier - now you can refer to the contents page, find the region of interest and simply flick forward to the appropriate section. We have also redesigned the features section of the book, where you will find equipment reviews, space weather, propagation predictions and antenna advice - all presented in a more colourful, attractive and accessible format.

REVIEWS
Also, you will find that we have re-assessed the way in which we conduct our equipment tests, incorporating the feedback which we have received from readers - the vast majority of whom were looking for a more 'user-friendly' approach to equipment testing. Essentially, most people said that they were looking for a more balanced, broader assessment, where outright technical performance would be only one of a number of factors. As a result we have assessed additional aspects such as ease of use, ergonomics, display legibility, quality of manual, standard of construction etc. which provides a more representative perspective of a radio as an entire package. Technological advances over recent years have produced alternative approaches to receiver design and this year we felt that it was time to take a look at such alternatives. As a result we have reviewed both PC radios and scanners in addition to the more traditional fare In the upper price bracket, we also were interested to know how the short-wave receiver of today compares with a similarly priced amateur radio transceiver having a general coverage receiver facility. Given that the amateur market is larger than that of the short-wave listener, we wanted to know if the economies of scale resulted in better value for money.

MAKE YOUR VOICE HEARD
We have already seen broadcasters in Australia, New Zealand and Norway, to name a few, having budgets slashed and services reduced or suspended. The broadcaster forum at the 1998 EDXC meeting in Sweden threw up two further trends which should worry the dedicated short-wave listener. Firstly, the replacement of international short-wave broadcasts with an Internet service - as in Estonia - and secondly, an increasing dependence upon satellite down feeds and local cable distribution agreements. The fact remains that the vast majority of the worlds population cannot afford these new technologies, which means that such policy changes actually shrink audience sizes where the programmes are needed most. Readers of the 'Space Weather' article will find that satellites are far from infallible during sunspot maxima and there is also the problem of how to listen to your favourite international broadcaster when you are not near a computer or satellite system. So what can we, as individuals, do about such situations? The answer is to write, E-mail or fax stations who are losing services and complain *loudly*. Professional broadcasters are just as opposed to cutbacks as their listeners, but often find their hands tied through being government employees - an avalanche of listeners' complaints can often tilt the balance.

COMPETITION
Finally, I would like to invite readers to enter our competition for a JRC NRD-545, winner of this year's WRTH 'Best semi-pro receiver' award. All you have to do is follow the instructions on the 1999 reader response card to enter the draw - the chances of winning are much better than most national lotteries, and you can even request further information from our advertisers. The draw will take place on the 1st of June 1998 and the lucky winner will be named on the Internet and in WRTH 2000. Good luck!

David Bobbett

Editor, WRTH

How to use WRTH

OPERATING TECHNIQUES

When operating their short-wave receivers, the majority of short-wave listeners tend to operate in one of two main modes, switching between them as and when they deem appropriate. One mode is employed when looking for a given station, or country of interest where the operator wishes to monitor the frequencies known to carry broadcasts from the desired 'target' area. The other approach, often employed by monitoring organisations, is simply to cruise a specific band, stopping on frequencies of interest and logging the appropriate details of each station which is encountered. Repeating this process at different times eventually provides a band occupancy profile which may be compared with similar 'band-scans' to establish details of new stations, closedowns and frequency changes.

ORGANISATION OF THE BOOK

WRTH has been designed in such a way that readers can operate in either or both of these modes by referring to the relevant section. The book consists of three main sections: **Features**, consisting of equipment reviews, solar activity reports, broadcasting predictions and informative radio-related articles: **Directory**, which is sub-divided into *national* and *international* radio services: and the **Reference** section where miscellaneous information covering wider aspects of broadcasting can be found.

TARGETING

When operating in the targeting mode, you can go to Contents on page 5, locate the appropriate region - Europe, Africa etc. and then use the 'side-bars' on the page edges to quickly find the required part of the book. Once in the correct region, the countries are listed alphabetically, so it is a simple matter of leafing forward until you find the desired country. A second option, if you are unsure of the correct region, is to use the country index at the back of the book, which will tell you the exact page number. This will also allow you to look at countries which are adjacent to your target so you may assess the propagation between you and that part of the world.

BAND-SCANNING

Should you prefer to use band-scanning, there are listings of both medium wave and international short-wave broadcasts available in the international section of the directory. These can also be useful for casual listening, but in either case can help to identify a station by frequency - whereupon further details can be obtained by using the country index to identify alternative, hopefully better frequencies for the station of interest.

BAND SELECTION

When choosing a band to monitor, it is well worthwhile referring to the table on page 49, which recommends the most suitable frequency bands for a variety of locations. It is however important to bear in mind that such tables are only guides and cannot be considered to be 100% reliable - ionospheric events can not only enhance propagation conditions but also operate to their detriment - nevertheless, such tables remain an excellent starting point and their use is strongly recommended.

RECEPTION REPORTS

WRTH recommends use of the simple SIO code using the following scale:

S=Signal Strength	I=Interference	O=Overall Merit
5=Excellent	5=None	5=Excellent
4=Good	4=Slight	4=Good
3=Fair	3=Moderate	3=Fair
2=Poor	2=Severe	2=Poor
1=Barely Audible	1=Extreme	1=Worthless

It is courteous to enclose return postage when writing to small domestic broadcasters. This can be in the form of an International Reply Coupon (IRC) available from major post offices. Some DX clubs buy IRC's in bulk and sell them to members. In all cases, when writing to radio stations you must write clearly. Remember - if the station cannot read your address, then you cannot expect to receive a reply!

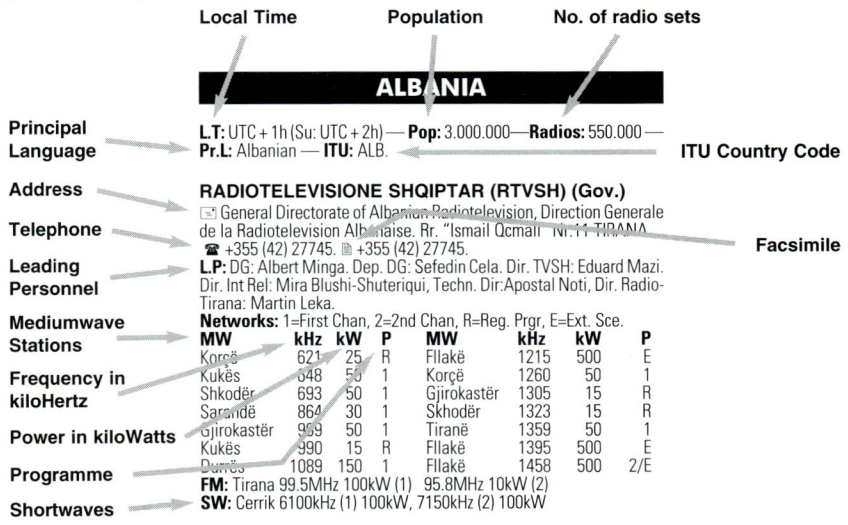

	Local Time	Population	No. of radio sets

ALBANIA

Principal	**L.T:** UTC + 1h (Su: UTC + 2h) — **Pop:** 3.000.000 — **Radios:** 550.000 —	
Language	**Pr.L:** Albanian — **ITU:** ALB.	ITU Country Code
Address	**RADIOTELEVISIONE SHQIPTAR (RTVSH) (Gov.)**	
	✉ General Directorate of Albanian Radiotelevision, Direction Generale	
Telephone	de la Radiotelevision Albanaise. Rr. "Ismail Qemali" Nr.11 TIRANA	
	☎ +355 (42) 27745. 🖷 +355 (42) 27745.	Facsimile
Leading	**L.P:** DG: Albert Minga. Dep. DG: Sefedin Cela. Dir. TVSH: Eduard Mazi.	
Personnel	Dir. Int Rel: Mira Blushi-Shuteriqui, Techn. Dir:Apostal Noti, Dir. Radio-Tirana: Martin Leka.	

Networks: 1=First Chan, 2=2nd Chan, R=Reg. Prgr, E=Ext. Sce.

MW	kHz	kW	P	MW	kHz	kW	P
Korçë	621	25	R	Fllakë	1215	500	E
Kukës	648	50	1	Korçë	1260	50	1
Shkodër	693	50	1	Gjirokastër	1305	15	R
Sarandë	864	30	1	Shkodër	1323	15	R
Gjirokastër	959	50	1	Tiranë	1359	50	1
Kukës	990	15	R	Fllakë	1395	500	E
Durrës	1089	150	1	Fllakë	1458	500	2/E

FM: Tirana 99.5MHz 100kW (1)   95.8MHz 10kW (2)
SW: Cerrik 6100kHz (1) 100kW, 7150kHz (2) 100kW

Mediumwave Stations — Frequency in kiloHertz — Power in kiloWatts — Programme — Shortwaves

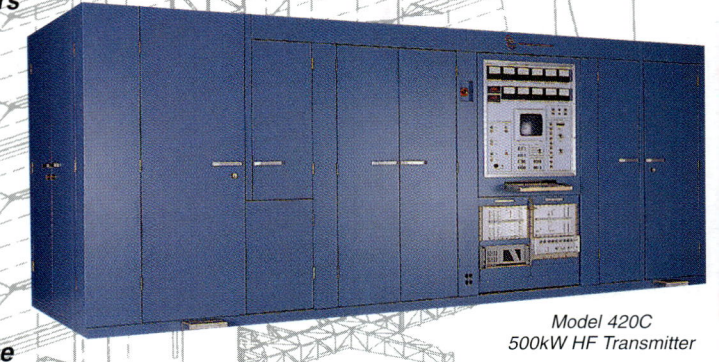

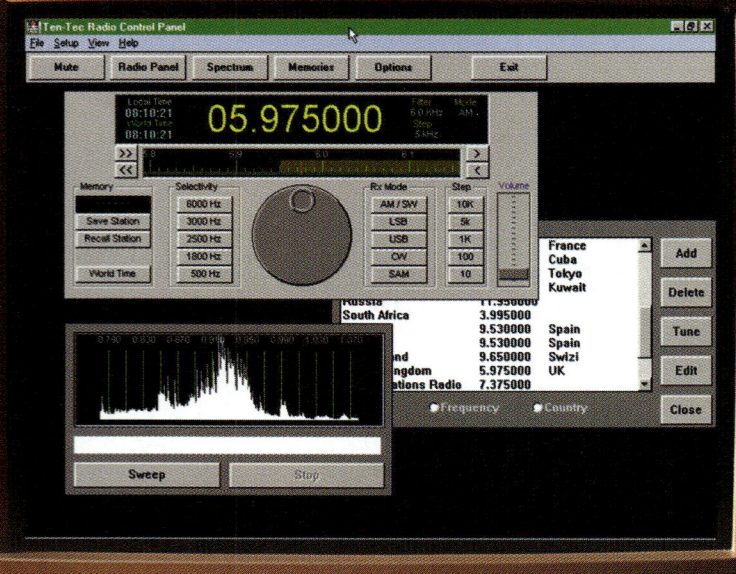

LISTEN TO THE WORLD ON YOUR PC

Worldwide shortwave listening is now only a mouse click away with **PC RADIO**. Designed for both the PC user who has never listened to shortwave and the experienced SWL who appreciates the powerful marriage of PC and shortwave listening.

Local AM broadcasting offers in-depth news, sports, and talk radio. Most countries broadcast around the globe on nearly a dozen international shortwave bands. Catch both world and local news from their viewpoint and a wealth of culture, politics, and music. Hams are spread out across nine bands, talking across town or half way around the world. You

will also hear military operations, commercial airlines, and CB. The manual includes a phenomenal beginner's guide written by respected author and columnist, Joe Carr. Launch the PC

RADIO, tune in an interesting station, and then put it in the background while you do other PC tasks. In fact, you could surf the Web and listen to shortwave at the same time!

Unlike conventional receivers, this is a revolutionary Digital Signal Processing, or "DSP", based design. This cutting edge, software-based technology dramatically reduces the number of individual electronic components inside. It's possible to provide features only dreamed of in previous receivers in this price class. Even the most experienced enthusiasts will marvel at the performance.

- tunes 100 kHz to 30 MHz
- needs only a serial port and one 1 meg of hard drive space
- runs on either Windows® 3.1 or 95
- no need to go inside your PC
- built-in, telescoping whip antenna

Call **1-800-833-7373** to request literature, or visit our web site at www.tentec.com.

$295*
factory-assembled only

- No-Risk 30-day Money-Back Guarantee**
- We accept VISA, Mastercard, and Discover
- Visit our home page at www.tentec.com

*Plus shipping and handling
**Customer pays shipping both ways.

CALL TODAY 1-800-833-7373
Monday - Friday
9:00 a.m. - 5:30 p.m. EST

You can reach us at:

1185 Dolly Parton Parkway
Sevierville, TN 37862
Office: (423) 453-7172
FAX: (423) 428-4483
e-mail: sales@tentec.com
Repair Dept.: (423) 428-0364 (8a - 4p EST)

TEN-TEC
MADE IN TENNESSEE

For those who like to build, we offer two other receivers in kit form. Request our T-KIT catalog covering these and many more budget-priced projects.

9-BAND SWL RECEIVER

Build in 8 hrs

"First radio kit" classic. Five transistor, 3 IC design. Tune both AM broadcast and SSB/CW from 1.8 - 22 MHz. Has Main and Fine tuning, Regen, RF gain, Volume. Use built-in speaker, your own, or stereo phones. Use 8 C cells or ext. 12 VDC. **1253..........................$59***

PORTABLE SWL RECEIVER

Enjoy quality shortwave listening comparable to factory built portables. Listen to AM broadcast as well as SSB/CW.

- 100 kHz - 30 MHz
- 15 memories
- 2.5 kHz and 100 kHz tuning steps
- Dual conversion, superheterodyne
- 13.8 VDC operation; AC wall transformer included

Build in 25 hrs

1254..............................$195*

WRTH Receiver Reviews 1999

For this edition we have placed the 14 receivers tested into four main groups. The traditional 'table-top' category designed principally for use with domestic AC power, the 'portable' and 'hand-held scanner' groups intended mainly for use from batteries, and the 'PC radio' which is designed to operate in conjunction with a personal computer.

A DIVERSE APPROACH

The rationale behind including the two latter groups is to establish whether 'non-traditional' approaches to receiver design have yet reached a stage of development where they become viable alternatives to conventional SW radios. Scanning receivers are very popular in those parts of the world where legislation does not prohibit their use and their very popularity (which brings with it relatively low cost) means that they become inevitable contenders for consideration for HF broadcast reception. As might be expected, there are some trade-offs for the wideband coverage of a typical 'scanner' – usually 500kHz or lower to at least 1GHz – as opposed to the much more restricted range of the traditional HF receiver. However, provided that one is not looking for the ultimate in dynamic range or expecting state-of-the-art IF and filter performance, these need not be limiting.

PC RADIO

The newest category is the 'PC radio'. Products in this group consist of radio receivers in the form of 'black boxes' whose control, display and memory functions are devolved to an IBM-compatible PC. There are many reasons why this is a good idea, and a few reasons why it can be a very bad one indeed if not carefully implemented. By virtue of the way in which it operates, the PC and its monitor are prolific sources of interference. Ensuring that this does not become an issue represents a demanding design task. That all three PC receivers produce such good performance in this area is a tribute to what must have been hard work on the part of those involved with producing them. In one form or another, especially with the advent of DAB and similar digital radio systems, there seems little doubt that more and more radios will make use of the enormous processing power and speed of the modern PC. Amongst other things, this probably represents the most cost-effective way to implement complex signal-processing algorithms and paves the way for relatively low-cost high-performance receivers which operate in the digital domain throughout.

AMATEUR RADIO EQUIPMENT

You may be wondering why we have included a typical HF amateur-radio transceiver (the Kenwood TS-570D) in the 'table-top' category, given that a relatively small proportion of WRTH readers hold a transmitting licence. There are two main reasons.
One is that modern amateur HF transceivers almost always incorporate 'general-coverage' receivers, and the relatively small and very crowded HF allocations in which very weak signals must be resolved in the presence of extremely strong ones make radio amateurs

very demanding customers. The other is that the amateur-radio market is far larger than what might be called the 'domestic' market for HF receivers and intuitively one might expect economies of scale to apply: the same order of receiver performance might cost the radio amateur less than it costs the SWL.
Given that the cost of the TS-570D is in the same general area as that of several 'table-top' models, and the widespread belief that amateur HF transceivers often perform better than the equivalently priced receivers, we felt it would be useful to make some comparisons.

ASSESSMENT CRITERIA

In the case of HF receivers, outright performance – sensitivity, the ability to receive weak signals in the presence of strong signals and so on – is undoubtedly important. But so are other factors, for example ergonomics tend to be given far less attention by designers, yet the argument that ease of use over long periods is the most important design factor of all is difficult to refute. As you will see, we have paid more attention to ergonomic issues than hitherto and we have also carefully scrutinised the manuals and other literature accompanying the receivers tested.
In recent years it has become fashionable to evaluate the strong-signal performance of radio receivers by measurement of parameters such as intercept points, phase noise, reciprocal mixing, 1dB compression point and so on. In our view it is important not to fall into the trap of thinking that because something can be measured, it is *ipso facto* important. It is certainly the case that strong-signal performance is important, but it is only one part of the overall picture.

MEASURING WHAT MATTERS

What really matters to the user of any radio receiver is whether some deficiency in it – or more correctly in the receiving system as a whole, including the antenna and other local factors – prevents reception of the required signal. In our view, reception failure hardly ever results from gross design deficiencies.
In the HF bands, the usable sensitivity of any receiving system is inevitably limited by external noise and whether there is any propagation between you and the part of the world you want to listen to - both factors being outside the user's control. A very important factor within the user's control, however, is how well the receiver is operated. It is a definite skill, which must be acquired over a period of time and kept sharp by practice.
So although we have made a very comprehensive suite of traditional measurements, and discussed them in the text when necessary, we have also carefully considered issues of operability and how well the receiver can be made to function as part of a receiving system with you at the centre. We have also given a good deal of thought to presentation of the measurement results in a way which will appeal to a wide audience.
We hope you enjoy this slightly different approach to evaluating receiver performance, and that you have as much pleasure from your eventual purchase as we did whilst reviewing and assessing.

for up to 18 SW, 18 FM and 9 MW stations is provided, and storage and recall are eminently straightforward. A feature called 'Auto Preset System', operating on MW and FM only, will perform a scan of a selected band and store stations in preset memories based on signal-strength order. Such a facility is useful for the foreign traveller and worked well on FM, but on MW the R881 tended not to be able to distinguish between a radio station and an interference source. For best results it was necessary to take considerable care with the siting of the radio before invoking the facility. The 'memory scan' feature worked rather better, sequentially presenting the stations stored in each preset for about seven seconds.

Roberts R881

US$125   £79   Euro110

OVERVIEW

The Roberts R881 shares with the Sony ICF SW-30 the distinction of being the lowest-cost radio in our tests. A degree of comparison is therefore inevitable. The R881 is slightly smaller than the ICF-SW30, measuring 164 x 100 x 35mm, but is marginally heavier at about 550g. This is mainly because it uses four AA cells as an internal power source rather than the three of the Sony. Visually the unit bears a close resemblance to the Sangean ATS-404 and is probably made in the same factory. The country of origin is not stated on the unit or anywhere in the packaging.

CONTROLS

Finished in a rather pleasant silver, the left-hand half of the R881's front panel is taken up with the speaker. The large liquid-crystal display is at upper right, with the power-on button next to it. This is not guarded in any way, although the 'lock' switch adjacent to the 9/10kHz MW step-size switch on the lower edge of the unit disables all the radio's controls. The keypad lies beneath the display, with up/down tuning buttons to the right and band select, scan and time-set buttons to the left. On the right-hand end of the unit are a rotary volume control and three-position tone switch: on the left are sockets for 6V DC power input, headphone output and a stereo/mono switch for FM reception. Roberts claims "approximately 16-20h of listening when used for 4h/day using alkaline batteries", but we found this rather optimistic and more like 10-12h was obtained under similar conditions. An AC adaptor must be purchased separately, although one very welcome accessory was a good-quality carrying case complete with earpiece in a separate partition. A generally well-written and illustrated manual and a frequency guide based on the WRTH also fit nicely into the case.

SPECIFICATION

The R881 covers 520/522-1710kHz on MW, fourteen short-wave bands between 120m and 11m and 87·5-108MHz for FM broadcasts. There is no long-wave coverage, which is an unfortunate omission. Pressing the 'SW Select' button sets the received frequency to the bottom edge of each band, and pressing the up-tuning button tunes the receiver up the band in steps of 5kHz. No other tuning steps are available. Direct frequency entry is also possible via the keypad but only frequencies entering in 0 or 5 can be input, the receiver forcing everything else to the nearest 10kHz. Memory storage

PERFORMANCE

In performance terms the R881 was reasonably sensitive at most frequencies (although a little less so than desirable on the higher bands) but neither its adjacent-channel rejection nor its image rejection were very good and the strong-signal performance was not outstanding either. Some difficulties were experienced with modulation hum when running the unit off external power supplies, including a Hewlett-Packard laboratory-grade regulated PSU; this could have been a sample fault. In its favour, the FM sensitivity and ability to cope with fading and flutter were excellent and the audio quality via stereo headphones was very good indeed; as with the Sony ICF-SW30, some very enjoyable concerts were enjoyed in this way during the test period. The clear display and switchable display backlight were also good features.

CONCLUSION

All in all, the Roberts R881 offers a lot of functionality for the money, above-average audio when signals are strong and steady, and what appeared to be quite acceptable constructional quality. The lack of fine tuning may occasionally be a nuisance, but at the price it is difficult to complain too loudly about this and the lack of SSB reception facilities. Comparisons with the Sony ICF-SW30 are interesting, with the latter offering a generally higher standard of performance and a long-wave band but considerably less overall functionality.

Rating table Roberts R881

Mechanical design	★★★
Constructional quality	★★★
Ergonomics	★★★
Sensitivity	★★★
Dynamic range	★★★
RF intermodulation	★★★
IF filters	★★★
IF performance	★★★
Audio quality	★★★★
Software	n/a
Manual	★★★
Versatility	★★★
VFM – absolute	★★★
VFM – comparative	★★★★
Overall rating	**★★★**

Key:
★ = Poor   ★★ = Fair   ★★★ = Average
★★★★ = Good   ★★★★★ = Excellent
VFM = Value for money

Sony ICF-SW30

US$125   £80   Euro115

OVERVIEW

At £79.95 in the UK, the Sony ICF-SW30 costs fractionally less than the Roberts R881 and is thus the lowest-cost radio in this year's group. It is a small but solidly-made-in-Japan portable which measures170 x 112 x 35mm and weighs about 450g; as such it is about twice the size and weight of the delightful ICF-SW100 but even so is not likely to be an obtrusive travelling companion.

The left-hand half of the ICF-SW30's front panel is taken up with the speaker, and the audio quality from the unit is really very good considering its size. The liquid-crystal display is at upper right, with the guarded power switch adjacent to it. A large up/down rocker switch is at lower right, and five station preset buttons are situated to its left. Ancillary buttons for various purposes are disposed around the display and along the bottom edge of the front panel.

The volume control and a speech/music tone switch are on the right-hand edge and a 4.5V DC power socket, a headphone socket and a sensitivity switch can be found on the left edge. Sony claims a 13-hour battery life on the basis of listening periods of 4h/day and we found this to be slightly pessimistic; fresh manganese-alkaline cells managed about 15h on this basis. The button-activated battery condition indicator was useful and reasonably accurate.

SPECIFICATION

Various versions of the ICF-SW30 are available but all cover 531-1602 or 530-1710kHz for medium-wave reception: a 9/10kHz switch concealed beneath the batteries adjusts the MW synthesizer steps as required. There are ten standard SW bands, which unfortunately do not include the 18·9-19·02MHz band or the tropical bands at 2·3-2·495 and 3·2-3·4MHz. An FM band starting at either 76 or 87·5MHz depending on the version and ending at 108MHz is provided, and some models apparently also cover the 153-279kHz LW band so it is worth checking to ensure that the model you buy is the correct version for your area. A pleasant surprise is that switching to FM and plugging a pair of stereo headphones into the socket provided reveals that the ICF-SW30 has a stereo decoder – a point not mentioned in the literature as far as we could see. Considerable use was made of this feature in the course of several rail journeys during the 1998 BBC Promenade Concerts season! You will have noticed that nowhere above was either a tuning knob or keypad mentioned, and this

brings us to what may for some users be the unit's biggest limitation. Tuning (in 1kHz or 5kHz steps) can only be carried out by the up/down rocker key, and frequencies cannot be entered directly. Admittedly there is a rather rudimentary scan facility, but it is slow and responds only to strong signals.

PERFORMANCE

Together with the fact that there are only five station presets per band, the immediate impression gen-

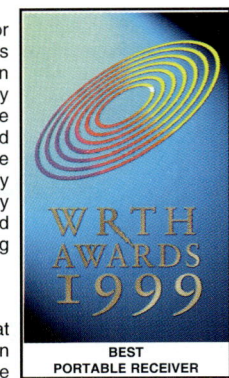

erated is that the ICF-SW30 is a radio for those looking for relatively good performance but with no ambitions to listen to anything other than a few standard stations per band. The performance is certainly there in large measure; despite its low cost the Sony receiver uses a double-conversion scheme for the HF bands and displays very good RF performance in almost all respects.

The selectivity is excellent, with the IF filters displaying remarkably good passband and stopband characteristics for a low-cost unit. Sensitivity is also very good, especially on the higher HF bands, and the RF intermodulation performance was considerably better than expected.

CONCLUSIONS

It may be that the inability to tune elsewhere than in defined bands, the lack of SSB reception facilities and the slow and awkward tuning system will rule the ICF-SW30 out of contention, even at its low price.

Lack of tropical-band and 15m coverage may also be an irritant. However, if your requirements run more along the lines of having reliable access to a few stations per band than winkling out weak stations on unusual frequencies, the Sony receiver's ease of use and excellent performance may make it the ideal travelling companion.

Rating table Sony ICF-SW30

Mechanical design	★★★★
Constructional quality	★★★★
Ergonomics	★★★
Sensitivity	★★★★
Dynamic range	★★★
RF intermodulation	★★★
IF filters	★★★★
IF performance	★★★★
Audio quality	★★★★
Software	n/a
Manual	★★★
Versatility	★★★
VFM – absolute	★★★★
VFM – comparative	★★★★
Overall rating	**★★★★**

Key:
★ = Poor   ★★ = Fair   ★★★ = Average
★★★★ = Good   ★★★★★ = Excellent
VFM = Value for money

Icom ICR-10

US$500   £260   Euro600

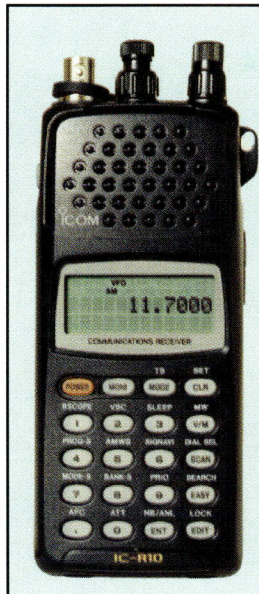

OVERVIEW

Like the Yupiteru MVT-9000, the Icom ICR-10 is one of a class of what might be referred to as 'hand-portable' receivers, as opposed to more classical 'portables' such as the Roberts or Sony examples tested. As with the Yupiteru family, much liked by the scanning fraternity, the ICR-10 visually resembles a hand-held transceiver or cellular telephone, measuring 145 x 60 x 32mm (excluding the 160mm antenna) and weighing 320g. The lockable keypad and display are front-mounted adjacent to the speaker, with the rotary volume/squelch and tuning controls placed next to the whip-type antenna. This latter connects via a BNC socket, implying that an external antenna can easily be connected if required. In normal use the ICR-10 runs from four rechargeable or primary AA cells. A socket on the side of the unit allows a 12V DC power supply to be connected, and a switch in the battery compartment permits rechargeable cells to be charged from the external source if required. This is a good feature, but care must be taken not to connect an external supply if non-rechargeable cells are in use unless you are quite sure that the recharge switch is off. For some reason no charger was supplied with our sample, although the manual (which incidentally is extremely comprehensive, well written and illustrated) implies that it should have been. The maximum current drain is about 200mA at full volume. Above the power socket is an interface socket for the optional CI-V level converter, which allows control of the receiver via an RS232C port.

SPECIFICATION

The ICR-10 covers 500kHz-1300MHz and is a triple superheterodyne, up-converting signal frequencies below 340MHz to a first IF of 429·1MHz and down-converting the range 340-1000MHz to 266·7MHz. Thereafter, the common second IF is 10·7MHz and the third IF is 455kHz. Wideband and narrowband FM, AM and SSB/CW can all be demodulated, and the tuning steps are 0·1, 0·5, 1, 5, 6·25, 8, 9, 10, 12·5, 15, 20, 25, 30, 50 and 100kHz. As if these were not enough, any other required step size between 0·1 and 999·99kHz (in 0·1kHz steps) can be programmed. In terms of memory, the ICR-10 has 1000 memory channels arranged in 15 banks of 50 for conventional storage together with two banks of 100 in which the results of auto-seek and 'program skip' scan modes are stored. Each bank and memory channel can be named, the former with up to 10 alphanumeric characters and the latter with eight. A variety of scan modes are available, and as a hand-held 'scanner' the unit works very well. The display is a large LCD with clear characters and a pleasant green backlight which automatically switches off after a short period to conserve battery life. Amongst its many functions are an S-meter, bandscope (on FM only, unfortunately) and memory/bank name indicator.

PERFORMANCE

It is no reflection on the abilities of the ICR-10 to observe that it is a better VHF/UHF scanning receiver than it is an HF broadcast receiver. Its chief limitation is that the only filter available on AM and narrowband FM has a specified -6dB point of "more than 15kHz" and the result was rather poor selectivity. On 6MHz at night, using just the whip antenna, moderately strong signals apparently took up several 5kHz channels. For example, a wanted station on 6195kHz was also audible on 6185, 6190, 6200 and 6205kHz. Strong stations, such as might be received with an external antenna connected, caused even more low-level interference several channels away. The ICR-10 was not short of sensitivity but its strong-signal performance was not very good, especially if external antennas were used. Under these conditions the attenuator can be expected to be in circuit for most of the time! In fairness, this is a common situation with hand-portables optimised for wide-range scanning.

CONCLUSIONS

All in all, within its limitations the ICR-10 did very well and (as would be expected from the 'wide' filter) produced very pleasant audio provided the signal was reasonably strong and that there were no other strong signals within about 20-30kHz of the wanted signal. It also proved to be a useful FM broadcast receiver, producing pleasant audio (non-stereo, alas) in either the internal speaker or an external headset. If good-quality reception of HF broadcasts is your main aim, you would undoubtedly do better to consider a more conventional portable unit. However, if you usually want to listen to other things but make occasional forays on to the broadcast bands, you may well find that the ICR-10 will fit the bill.

Rating table Icom ICR-10

Mechanical design	★★★★
Constructional quality	★★★★
Ergonomics	★★★
Sensitivity	★★★★
Dynamic range	★★★
RF intermodulation	★★★
IF filters	★★★
IF performance	★★
Audio quality	★★★
Software	n/a
Manual	★★★★
Versatility	★★★
VFM – absolute	★★★
VFM – comparative	★★★

Overall rating   ★★★

Key:
★ = Poor   ★★ = Fair   ★★★ = Average
★★★★ = Good   ★★★★★ = Excellent
VFM = Value for money

Yupiteru MVT-9000

US$455 £400 Euro550

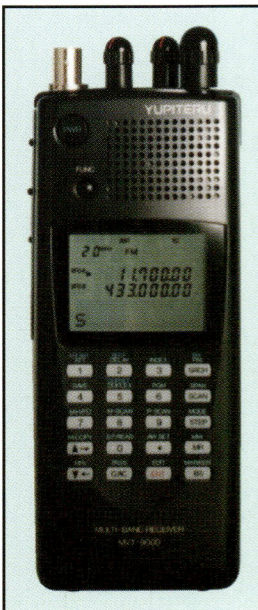

BEST
HANDHELD
SCANNER

OVERVIEW

The Yupiteru MVT-9000 is like the Icom ICR-10 insofar as it is a wide-range hand-portable receiver principally intended for scanning use but covering the MF and HF broadcast bands in addition. The Yupiteru company produces a comprehensive selection of scanning receivers: all have excellent reputations, and the MVT-9000 represents the top of the line. Measuring 173 x 65 x 40mm exclusive of the stubby whip antenna which adds another 160mm, the unit is rather larger than the ICR-10 and at 430g somewhat heavier.

The complexity of its functions is such that all 20 keys on the keypad have at least one level of shifted function and four have a second. Careful study of the 68-page manual is recommended before using the receiver, and – if the English-language version we saw is typical – some allowance will have to be made for the rather indifferent quality of the translation.

Visually, the unit resembles a hand-held transceiver. The keypad, dot-matrix liquid-crystal display and speaker dominate the front panel; the display has a timer-controlled backlight and is highly legible from all viewing angles. Power-on and shift buttons are next to the speaker, the latter acting as a 'shift' key. Volume, squelch and tuning knobs are on the top panel, next to the BNC antenna socket. Incidentally the MVT-9000 also has an internal ferrite-rod antenna for MF reception.

Three small buttons on the left-hand side control the display backlight, temporarily override the squelch setting and lock the keypad. On the right-hand side are sockets for the DC input and headphone output. Overall, the unit is very easy and intuitive to use despite what may seem initially to be a dauntingly comprehensive feature set, and all buttons and switches have delightfully light and positive actions.

SPECIFICATION

The MVT-9000 covers a very wide range. Frequencies between 530kHz and 2039MHz can be tuned, and the available modulation modes are narrow and wide FM, narrow and wide AM, SSB and CW. Tuning steps are 50, 100, 200 and 500Hz and 1, 5, 6·25, 8, 9, 10, 12·5, 15, 20, 25, 30, 50, 100 and 125kHz. There are two separate VFOs and both currently tuned frequencies are shown on the display; switching between them and between memory banks requires merely a press of one key. The memory structure is 20 banks of 50 channels per bank, with one of the latter serving as the storage space for the results of seek-and-store scanning. An additional ten channels are for priority-scan purposes. Banks can be coupled together in any sequence and scanned contiguously, and channels in a bank can be easily bypassed. The claimed scan rate is 30 per second, although tests suggested that this was a little optimistic. There is a useful band-scope function.

The four Ni-Cd cells supplied with our sample, together with a combined wall charger and power supply, gave about seven hours of listening, relatively short battery life seems characteristic of several Yupiteru receivers we have tested. The charging information in the manual was not considered very helpful.

PERFORMANCE

In performance terms the MVT-9000 was generally very good with sensitivity being excellent across the spectrum. The narrow AM filter was ideal for use on crowded HF bands, with its -6dB points at about 3·8kHz, and the wider AM filter (its -6dB points at 8·1kHz in our sample) was quite usable for strong stations with no competition on adjacent channels. Naturally the audio quality was better with the latter.

Strong-signal performance was not the MVT-9000's best point, as usual with this class of receiver, and it did not take at all kindly to being fed from large antennas. Like the Icom ICR-10, it is certainly not ideal for picking out weak HF broadcast signals adjacent to strong ones and indeed is not intended to be. However, some very good results were obtainable with care.

CONCLUSION

The MVT-9000 is a very high-performance scanning receiver primarily oriented towards the VHF/UHF listener. However, it has quite usable HF and VHF/FM broadcast reception facilities built in, and will prove a very good travelling companion for those wanting as much versatility and performance as possible.

Rating table Yupiteru MVT-9000

Mechanical design	★★★★
Constructional quality	★★★★
Ergonomics	★★★★
Sensitivity	★★★★
Dynamic range	★★★
RF intermodulation	★★★
IF filters	★★★
IF performance	★★★
Audio quality	★★★★
Software	n/a
Manual	★★★
Versatility	★★★★
VFM – absolute	★★★★
VFM – comparative	★★★★
Overall rating	**★★★★**

Key:
★ = Poor   ★★ = Fair   ★★★ = Average
★★★★ = Good   ★★★★★ = Excellent
VFM = Value for money

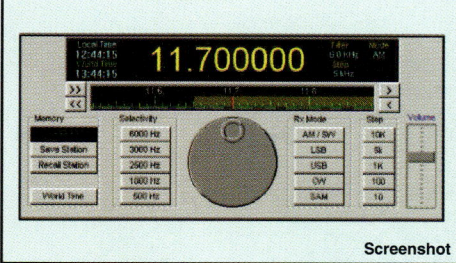

Screenshot

Ten-Tec RX320

US$300   £300   Euro360

OVERVIEW

Another PC-controlled radio to be assessed for this edition of the *WRTH* is the Ten-Tec RX-320. The name of this Tennessee-based company may not be well known to some short-wave listeners, but it is extremely familiar to almost every radio amateur. Its line of high-performance transceivers has been justly celebrated for the very high-grade receivers incorporated within them, and Ten-Tec's reputation for customer service is second to none. The RX-320 represents something of a new departure for the company, and we were very keen to examine the unit in view of what was widely rumoured to be class-leading performance at a low price. However, the realities of the publishing business – in particular the relentless pressure of deadlines – meant that only a very early pre-production sample could be supplied for assessment, and it was not clear how representative the finished version would be in terms of either hardware or software. For the very latest information we recommend that you check out the Ten-Tec web site (http://www.tentec.com).

HARDWARE

So what is the RX-320? Unfortunately no manual was supplied with our pre-production sample, so we may have missed some of the finer points. Be that as it may, the unit is a black box measuring 165 x 155 x 70mm and weighing just over 1kg. The front panel is blank apart from the Ten-Tec logo – there is not so much as an LED to indicate that the unit is powered – and all the action takes place on the rear panel. Here we find an input socket for the external power supply (a 120V version was supplied with our sample, which would not have taken kindly to being plugged into the UK mains supply), an on-off switch, the 9-pin D-type serial connector, line output and external speaker jacks, and the antenna input. For reasons best known to Ten-Tec, this latter is an RCA socket; we would much prefer to have seen something better, such as a BNC or TNC. A hole in the top panel also allows for a telescopic antenna to be directly connected. There is no internal speaker, the RX-320 being evidently intended for connection either to your computer's sound card via the 'line out' socket or to an external speaker via the 'external speaker' socket. Both worked well in tests.

SOFTWARE

The controlling software (Version 1.1 in our case, although later versions will undoubtedly be available) is supplied on floppy disc, and will run on either Windows 3.1 or Windows 95/98. No problems in installing and using the software on a 133MHz Pentium under Windows 98 were experienced, and Ten-Tec has even done the job properly by including an uninstall routine – would that all PC-radio manufacturers did the same! Clicking on the 'TTRCX' icon displays a receiver front panel with a number of large and clearly marked buttons; operation even without a manual seemed wholly intuitive and we were soon in business. The radio covers 100kHz-30MHz with step sizes of 10, 5 and 1kHz and 100/10Hz and caters for AM, SSB, CW and narrowband FM. A synchronous AM facility is also provided although it seemed to work rather strangely in our sample, only locking correctly on very strong signals.

The DSP-based IF filtering is 6, 3, 2·5, 1·8kHz and 500Hz, the latter being ideal for CW reception. Memory capacity appears to be limited only by the capacity of the host PC, and pressing the 'Memories' button on the toolbar showed that Ten-Tec had thoughtfully provided some broadcast-station names and frequencies to start the listener off.

PERFORMANCE

The performance of the RX-320 was generally very good. If anything, there was slightly too much sensitivity for use with large antennas, the receiver clearly having been designed with electrically short whip or rod-type aerials chiefly in mind. However, the IF widths proved almost ideal for HF use, and the shape factors are very good – as would be expected with DSP-derived filters. On some frequencies there were problems with 'hash' breakthrough, probably via the interface cable, and there was considerable interference on the 21MHz band which was evidently originating in the monitor. These are trivial issues which we would expect to have been fixed in production versions. Strong-signal performance was quite good, although the receiver proved a little intolerant of large antennas and some second- and third-order intermodulation was noted on the lower bands. An attenuator facility would have been useful.

CONCLUSION

Overall, the RX-320 is definitely worth considering if you are interested in a PC-controlled radio optimised principally for the HF bands.

Rating table Ten-Tec RX320

Mechanical design	★★★
Constructional quality	★★★
Ergonomics	★★★★
Sensitivity	★★★★
Dynamic range	★★★
RF intermodulation	★★★
IF filters	★★★★
IF performance	★★★★
Audio quality	★★★★
Software	★★★★
Manual	n/a
Versatility	★★★
VFM – absolute	★★★
VFM – comparative	★★★★
Overall rating	**★★★**

Key:
★ = Poor   ★★ = Fair   ★★★ = Average
★★★★ = Good   ★★★★★ = Excellent
VFM = Value for money

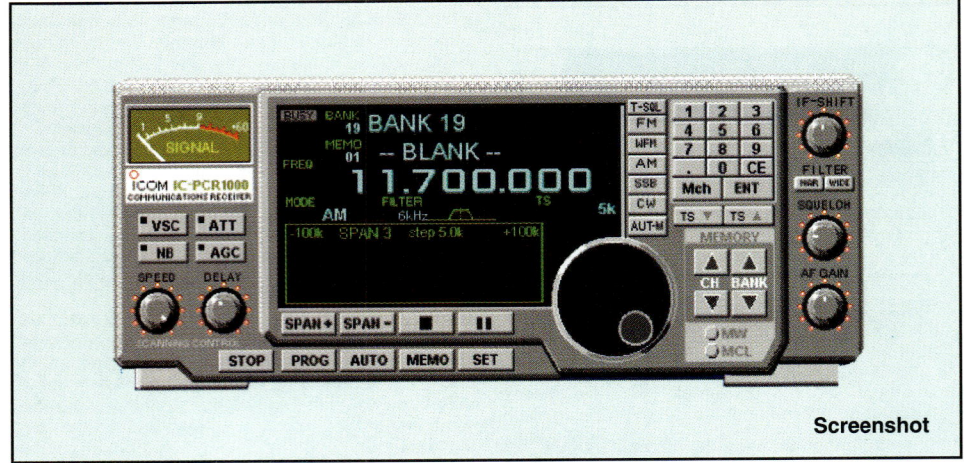

Screenshot

Icom IC-PCR1000

US$490   £320   Euro590

OVERVIEW

The Icom IC-PCR1000 is another of the 'new breed' of receivers whose control and memory functions are largely devolved to an IBM-compatible PC. Already enormously popular amongst the scanning fraternity, the IC-PCR1000's frequency range of 100kHz-1300MHz makes it an obvious candidate for purchase by those who want wide-range reception capabilities in a single unit. It also makes an interesting comparison with the WinRadio WR-1500e.

The receiver itself consists of the proverbial black box, measuring 200 x 128 x 30mm and weighing about 1kg. The only user control is the power switch, on what passes for the front panel. On the rear is an input socket for a 13·8V supply. Icom provides an AC adaptor as standard, although some 'field tests' suggested that the PCR-1000 is quite at home operating from any reasonably stable supply of approximately the correct value provided that it can source about 1A. A standard RS-232C socket is adjacent to the power connector. There is a jack for an optional external speaker, a ground connector and a BNC antenna socket. As might be expected, the quality of audio available from the tiny internal speaker located on the top of the IC-PCR1000 is nowhere near as good as can be obtained by plugging in something better, although it is adequate for monitoring purposes. If preferred, the received audio from the IC-PCR1000 can be sent to the computer's sound card and relayed through a multimedia system. One minor shortcoming of this potentially useful facility is that the PCR-1000 does not embody a suitable decoder for FM stereo broadcasts. A simple plug-in antenna is provided, although Icom quite rightly points out in the manual that something better is required for maximum performance.

HARDWARE

The receiver is essentially a triple superheterodyne with a first IF of 266·7MHz and a second IF of 10·7MHz. A third conversion to 455kHz takes place for non-wideband FM signals. The basic memory architecture is

20 bands of 50 memories, all of which are stored in the PC in the form of comma-delimited databases, but the ability to use multiple files means that there is no real limit on the amount of data which can be generated and used in conjunction with the radio apart from the size of the PC's hard drive. It would literally be possible to store millions of frequencies and associate a tuning step, mode, attenuator setting and label with each one. In respect of frequency storage and recall, the Icom software is generally capable and fairly user-friendly but displays some irritating limitations, especially from the point of view of the scanning enthusiast rather than the HF listener. Certain quite basic operations such as combining banks and scanning them, for example, are not possible. Composing the database in the first place is made more tedious than it need be by the non-compliance of the entry format with normal Windows conventions in certain respects. You cannot set global defaults for items such as step size or mode, or easily set a channel spacing for a particular band and apply it to a group of entries. For this and other reasons, it is a matter for some regret that Icom has chosen to regard its software as proprietary. History and hindsight strongly suggest that open software architectures almost always dominate the marketplace, especially when their originator is the first to market, and Icom has missed a trick here.

MODES & STEPS

The modulation modes covered by the PCR-1000 are wide and narrow FM, AM and SSB/CW. There are probably more available tuning steps on the Icom unit than in any other radio currently available, namely 1, 10, 20, 50, 100 and 500Hz, 1, 2·5, 5, 6·25, 9, 10, 12·5, 20, 25, 30, 50, 100 and 500kHz and 1MHz. It seems all the more surprising that the 8·33kHz step shortly to be introduced into portions of the 118-137MHz international aeronautical band was not available, but there was some speculation on the Internet at the time of our assessment

that a newer release of the operating software (our sample came with v1.3) would address this. The filter specification refers to the passband in terms of -6dB points. On SSB, AM and CW the narrowest filter is 2·8kHz (although the display actually indicates this as 3kHz), and for all these modes together with FM there is a 6kHz filter. A 15kHz filter is available for FM and AM, and there are also 50 and 230kHz filters for wideband FM. An 'IF shift' control is available in SSB mode.

SOFTWARE

Assuming you have at least a 486DX4 or (preferably) a 100MHz Pentium – ideally with at least one free serial port – and are running either Windows 3.1 or Windows 9x, the PCR-1000 is simplicity itself to set up. The software proved easy to install under Windows 98, and double-clicking on the desktop icon brought the receiver to life in a few seconds. There are three user-selectable interfaces, which are almost all other housekeeping functions are invoked from the toolbar. The simplest is called 'radio' and is somewhat representative of a simple tuner. Relatively little of the IC-PCR1000's potential functionality is available in this mode. The second option is the 'communications receiver', where the screen display looks more like the front panel of a fairly complex scanning receiver such as the Icom ICR-8500. The third is the 'component' screen, in which four stackable modules of a notional ideal receiver can be selected and grouped in any order.

One module could be considered as the tuning system, with a mouse-operated VFO knob, keypad, memories and display. A second handles the very bright and clear S-meter display and scan functions. The third controls mode, volume, step size, AFC, AGC and so on. The fourth introduces a 'bandscope' whereby activity within a maximum of ±200kHz of the currently tuned signal can be observed, although for some reason the audio is muted when in SSB or CW mode and the bandscope function is enabled. Any or all modules can be displayed or removed from the screen, and all functions are controllable via the mouse. Squelch and volume levels are controlled by dragging and dropping sliding bar-type elements, and a frequency is entered by clicking on the rather small numeric keypad adjacent to the display.

INSTRUCTION MANUAL

The twelve-page 'manual' is not a very comprehensive guide to the PCR-1000, and the help facility built in to the software and available from the toolbar is much more thorough. Indeed, it is best to regard the manual as merely a guide to getting started. As with most software, you learn at least as much about how it works by using it as by reading about it, and that of the PCR-1000 is no exception. The learning curve is a little steep and a few functions are somewhat counter-intuitive, but no particular difficulties should be expected.

PERFORMANCE

As far as performance is concerned, the IC-PCR1000 generally acquitted itself well. Using a 133MHz Pentium, the maximum scanning speed was about 18 channels per second and the various scan functions – which are generally similar to those on the ICR-8500 – appeared to work nicely. Sensitivity on the HF bands was adequate across most of the HF region, if perhaps a little lacking at the higher frequencies, and strong signal handling was reasonable if not outstanding.

Measurements suggested that the RFIM performance was such that the unit might not be comfortable with large antennas, and some second-order IMD from broadcast stations in the 7MHz band was noticed when tuning around 14MHz in the evening whilst using a multi-element Yagi. Judicious use of the 20dB attenuator usually restored order, however. Used in conjunction with a discone or a short random wire, the PCR1000 gave a good account of itself.

The filters were not quite ideal for HF broadcast reception; the 2·8kHz filter was a trifle narrow and the 6kHz filter (whose -6dB points in our sample were actually at 7·3kHz) was a little too wide for use on crowded bands. That said, the IC-PCR1000 proved to be very useful for general scanning and monitoring tasks and much less susceptible to strong-signalproblems than the WinRadio WR-1500e.

CONCLUSION

The overall verdict is that although the IC-PCR1000 does not display leading-edge performance, it is remarkably inexpensive for what it offers and in certain respects – such as versatility and almost unlimited storage capacity – it is in a category by itself.

There is no doubt that as time goes by, the marriage of radio receiver and PC will be a very fruitful one; on that basis the Icom IC-PCR1000 is the shape of things to come. One might perhaps hope for RDS, DAB, some DSP modes, a stereo decoder and filters better optimised for broadcast reception – perhaps with slightly friendlier and more flexible software – in the next version, but for now the IC-PCR1000 is well worth a look and something of a bargain at the price.

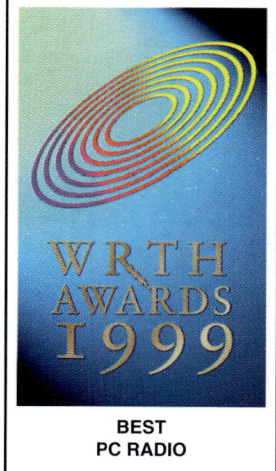

**BEST
PC RADIO**

Rating table Icom IC-PCR1000

Mechanical design	★★★★
Constructional quality	★★★★
Ergonomics	★★★★
Sensitivity	★★★★
Dynamic range	★★★★
RF intermodulation	★★★
IF filters	★★★
IF performance	★★★
Audio quality	★★★★
Software	★★★
Manual	★★★
Versatility	★★★★
VFM – absolute	★★★★
VFM – comparative	★★★★
Overall rating	**★★★★**

Key:
★ = Poor   ★★ = Fair   ★★★ = Average
★★★★ = Good   ★★★★★ = Excellent
VFM = Value for money

Screenshot

WinRadio 1500e

US$550   £400   Euro660

OVERVIEW

The earlier version of the WinRadio, essentially a wide-range receiver on an ISA-bus card for mounting inside an IBM-compatible PC, was assessed in the 1998 edition of *WRTH*. Since then the range of WinRadios has diversified into two main models – the WR-1000 and WR-1500 – each of which is manufactured in 'internal' and 'external' variants. As might be expected, the internal versions are ISA cards whereas the external versions are stand-alone boxes for connection to an RS-232 serial port. In this respect they are conceptually similar to the Icom IC-PCR1000 and some comparisons with that unit are inevitable.

The sample tested was an early production version of the WR-1500e, where the 'e' implies external. This is a beautifully finished grey box measuring 205 x 120 x 45mm and weighing an enormous 1·5kg. Clearly Messrs WinRadio build their machinery to last! The only user control is the power switch on the front panel, with an associated red LED. On the rear are standard D-type sockets for a PCMCIA card, the RS-232 connection, a BNC socket for the antenna, an RCA female socket for data output and a 12V DC input. A wall-adaptor power supply is provided, as is an assortment of cables and other accessories together with a very well-written and illustrated manual.

HARDWARE

The WR-1500e is notionally a triple superheterodyne covering 500kHz to 1500MHz, although the specification seems to imply that there are actually five intermediate frequencies (455kHz and 556·325, 249·125, 58·075 and 10·7MHz) of which a maximum of three are used depending on the tuned frequency. The -6dB points of the filters are given as 2·5kHz for SSB, 6kHz for AM, 17kHz for FM and 230kHz for wide FM.

These are quite similar to those in the IC-PCR1000 and are clearly oriented towards scanning-receiver use rather than HF broadcast reception. Tuning steps default to the smallest suitable increment for the chosen mode or can be user-defined. Reception modes are SSB/CW, AM, FM and wideband FM. The memory capacity per 'memory file' is 1000 and the total number of these is limited only by the size of the hard drive.

SOFTWARE

The software runs under Windows 3.1 or Windows 9x. Using a 133MHz Pentium with Windows 98, installation was very straightforward. Clicking on the shortcut displayed a large and clear screen resembling a control

panel, and operation is very simple and intuitive. There is insufficient space to describe all the unit's functionality in detail, but the software (our unit was supplied with v2.51) is really very good indeed. And as if that was not encouraging enough, developers are positively encouraged to produce their own software for the unit. Programming information is readily available from the WinRadio Web site (www.winradio.com) and unlike Icom seems to us to be very much the right approach.

PERFORMANCE

The WR-1500e was tried initially with a discone at 11m AGL; the test site is about 160m ASL in a rural location and with a good take-off in most directions. It was immediately clear that strong-signal performance was likely to be an issue since a signal on 116·8MHz was completely buried underneath what sounded like a combination of four or five local Band II broadcasting stations. Unfortunately this proved to be not untypical and at HF there were also frequent problems of complex RF intermodulation. For example, when listening to an amateur contact on 14·235MHz which was about S8 on the discone, a traffic list from a coastal radio station was about S6 together with some strong WEFAX from an unknown source.

Similar problems were experienced on almost all HF bands, and it was not deemed sensible to attempt tests with larger antennas than the discone. In fairness, WinRadio refers to this issue in its manual and suggests that using the 'local' function (an 18dB attenuator) will get rid of the intermodulation, but it did not always do so in practice and the degraded sensitivity which resulted was rather performance-limiting in certain bands. For HF broadcast reception under other than very favourable conditions, the AM filter was really too wide.

CONCLUSION

At present, the WR-1500e receives a qualified welcome. Its software is excellent – in our view more versatile and less idiosyncratic than that of the Icom IC-PCR1000 – but the basic receiver performance is evidently not quite there yet. Hopefully WinRadio will stay with its approach to integrating the PC and radio receiver and produce something really outstanding for next year.

Rating table WinRadio 1500e

Mechanical design	★★★★★
Constructional quality	★★★★
Ergonomics	★★★★
Sensitivity	★★★
Dynamic range	★★
RF intermodulation	★★
IF filters	★★
IF performance	★★
Audio quality	★★★
Software	★★★★★
Manual	★★★★
Versatility	★★★
VFM – absolute	★★★
VFM – comparative	★★★
Overall rating	**★★★**

Key:
★ = Poor   ★★ = Fair   ★★★ = Average
★★★★ = Good   ★★★★★ = Excellent
VFM = Value for money

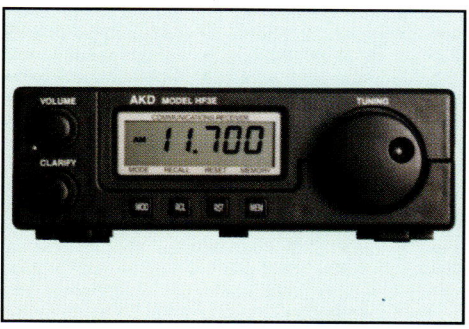

AKD HF3E

US$480   £299   Euro420

OVERVIEW

An earlier incarnation of the AKD HF-3E, the 'NASA HF-3', was reviewed in the 1998 edition of the *WRTH* and was observed to be a "very basic receiver" which at that time cost £160. The newer version is visually and operationally quite similar but embodies somewhat more in the way of functionality at the cost of an 87% increase in price. Briefly, the unit is a simple receiver covering 30kHz-30MHz in 1kHz steps and capable of demodulating AM, USB and LSB. Built in a clamshell-type plastic box, the HF-3E measures 185 x 180 x 60mm and weighs just under 1kg. It is intended for use with an external 12V regulated power supply, for which a socket is provided on the rear, and a suitable unit is supplied. There are only two rotary controls apart from the VFO, for volume and clarifier respectively. The latter varies the frequency of the second local oscillator to provide fine tuning. There are four possible tuning rates, selected by the speed at which the VFO knob is rotated, but no method of manually entering a frequency. Ten memories are provided and are selected for storage and recall by a combination of VFO knob rotation and pressing one of the four buttons beneath the display. The latter, incidentally, is extremely clear and the level and evenness of the backlighting is excellent. Other irritating ergonomic shortcomings mentioned in the earlier review are still present.

The original version of the HF-3E was criticized for using an RCA socket as the antenna input, and this has now been replaced by an SO239. This is probably the least worst type of socket for most users, although we take the view that in a rational world the entire UHF connector family would have been abolished many years ago and replaced by the N or BNC/TNC series. A more significant addition to the rear panel is a 9-way D-type socket for interfacing the unit with an IBM-compatible PC.

SOFTWARE

Suitable control software together with a Weatherfax decoder is provided on a disc supplied with the HF-3E. According to the manual (which by the way is written in a curiously archaic and authoritarian style, with some very strange grammar) the minimum system requirements are a 386SX with 640K of RAM, VGA graphics, 1·5MB of hard-drive space and a free serial port. As you will probably have gathered, these parameters imply that the software is not exactly state-of-the-art – and indeed the manual clearly specifies that

it is necessary to install and run the program under DOS. Unfortunately this advice proved to be literally true. Even using the full repertoire of tricks and tweaks known only to long-serving DOS acolytes, the software refused to run in a DOS window under Win95/98 on a 133MHz Pentium and it was necessary to have recourse to the IBM PS/2 (a 66MHz 486) which runs a variety of test and measurement software in the reviewer's laboratory. By modern standards the graphics were rather crude but the software allowed for easy storage and recall of up to 500 frequencies and direct frequency entry was easily possible. Four selectable scan widths between 200kHz and 2MHz were provided, and the spectral display plot worked quite well.

PERFORMANCE

The performance of the HF-3E was found to be very similar in most respects to that of the earlier HF-3. There was something of a shortfall in sensitivity, particularly at the higher frequencies, although in all except very quiet locations this will not prove limiting. The display on our sample was about 2kHz in error on SSB, which was irritating, and the combination of clarifier and VFO proved awkward to use when tuning SSB utility stations. As mentioned in the earlier review, the unit continues to display a degree of temperature sensitivity. Strong-signal performance was reasonable but not outstanding, and the 12dB attenuator proved to be necessary on several occasions. It is perhaps a little unfortunate that such a high value was chosen, especially given that the basic radio is not over-sensitive.

CONCLUSION

Although the HF-3E is easy to use and performs reasonably well, it strikes us as having an uncompetitive price-performance ratio and idiosyncratic ergonomics. For £300 you could buy one of several fully portable radios covering a far wider frequency range and with considerably more functionality. And although some of us have fond recollections of the era when 640K was a lot of memory and the C:\ prompt was part of everyday life, providing DOS-based control software in 1998 is taking nostalgia a little too far! Perhaps one for the technophobe.

Rating table AKD HF3E

Mechanical design	★★★
Constructional quality	★★★
Ergonomics	★★★
Sensitivity	★★★
Dynamic range	★★★
RF intermodulation	★★★
IF filters	★★★
IF performance	★★★
Audio quality	★★★
Software	★★
Manual	★★
Versatility	★★★
VFM – absolute	★★★
VFM – comparative	★★★
Overall rating	**★★★**

Key:
★ = Poor   ★★ = Fair   ★★★ = Average
★★★★ = Good   ★★★★★ = Excellent
VFM = Value for money

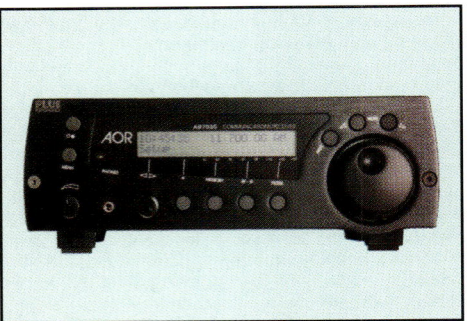

AOR 7030 Plus

US$1700   £950   Euro2040

OVERVIEW

When first introduced, the original AOR 7030 attracted much praise for its extremely high performance. It also attracted criticism for its very unusual menu-driven software-based control system. Originally reviewed in the 1997 *WRTH* and generating considerable controversy in respect of its strong-signal performance – which to at least one reader seemed to produce more heat than light – the 7030 is now available in 'Plus' form. This consists of a new module, the NB7030, which adds a tunable notch filter, an impulse noise blanker and an "enhanced CPU" and EEPROM. These latter endow the receiver with 400 memory channels (which can be tagged with a 14-character name), multiple timers and some new menu commands. Each memory can store frequency, mode, filter, PBS, AGC and squelch settings, and memory editing and searching facilities are also provided.

NEW FEATURES

The 7030 Plus is physically identical to the earlier model, measuring 238 x 227 x 95mm and weighing a hefty 2·2kg. The latter is largely a result of the fact that the mechanical designers did not stint on the use of substantial items of metalwork, and the overall finish is extremely good. The basic circuit architecture remains the same, namely a double superheterodyne with IFs of 45MHz and 455kHz which can be tuned between 0 and 32MHz in basic steps of 2·655Hz. The receiver caters for AM, SSB, CW, narrowband FM and some data modes. The new notch filter works exceedingly well. The measured notch depth was 56dB at 700Hz, which is excellent in itself, but the notch is also very narrow. The notch is manually tunable between 150Hz and 6kHz but there is also an auto-tracking mode which can either follow a varying heterodyne or be set up to find and null the first steady heterodyne encountered. The noise blanker also worked extremely well, dealing with a variety of impulsive interference very effectively. The only caveat was that its effectiveness tended to vary somewhat with signal strength.

OTHER ENHANCEMENTS

The other enhancements work nicely, but it must be said that they add even more complexity to what is already a complicated (and completely non-intuitive) menu-based control system. That said, what appears at first sight to be a dauntingly obscurantist method of operating a radio

becomes considerably easier with time and familiarity. In our view, the 7030 Plus is a prime example of a radio which would benefit enormously from an IBM-compatible PC interface and really good software. Happily, just such an interface is provided and the 'Data-Master' software mentioned in the original review is available from AOR for £129.00. The AOR 7030 Plus is a totally state-of-the-art HF receiver, whose specification puts that of almost everything else on the market to shame. If our sample is typical – and there is no reason to suppose it was not – some aspects of its performance almost certainly exceed the ability of the average RF test laboratory to measure and establish them. So much about this receiver is utterly right.

SENSITIVITY

The sensitivity is maintained right down to the lowest frequencies; the S-meter calibration is extremely linear; there are virtually no spurious responses; the audio quality is superb; and the AGC characteristics are close to perfection. A word of praise is also in order for the truly excellent manual, which is comprehensive without being obfuscatory and clear without being patronising. Other manufacturers would do well to consider how it was written and structured, and to resolve to try and emulate it. Our reservations are very few: a larger display with better contrast and a wider viewing angle would be very welcome, and we noted a mild tendency to what sounded like modulation hum (or perhaps stray coupling into a VCO) if the unit was sited too close to a transformer or power unit. One might also wonder whether an extra £150 for a noise blanker, a notch filter and some extra memories and a further £129 for the 'Data-Master' software represent rather steep premiums on the cost of an already expensive unit.

CONCLUSION

Whether the 7030 Plus justifies its relatively high price is always going to be a subjective judgement, especially in view of its rather idiosyncratic control system. It is not for the novice or the technophobe but for sheer performance allied to extreme versatility and excellent build quality, the AOR 7030 Plus remains one of the world's most desirable HF receivers.

Rating table AOR 7030 Plus

Mechanical design	★★★★★
Constructional quality	★★★★★
Ergonomics	★★★
Sensitivity	★★★★★
Dynamic range	★★★★★
RF intermodulation	★★★★★
IF filters	★★★★★
IF performance	★★★★★
Audio quality	★★★★★
Software	★★★★
Manual	★★★★★
Versatility	★★★★
VFM – absolute	★★★
VFM – comparative	★★★★
Overall rating	**★★★★★**

Key:
★ = Poor   ★★ = Fair   ★★★ = Average
★★★★ = Good   ★★★★★ = Excellent
VFM = Value for money

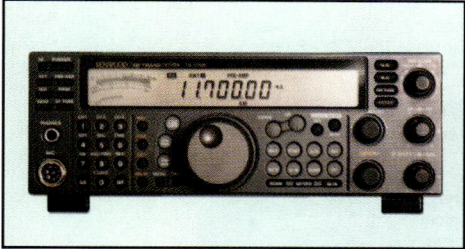

Kenwood TS-570D

US$1170   £1300   Euro1400

OVERVIEW

With the recent cessation of production of the R-5000, Kenwood no longer manufactures a stand-alone HF receiver. However, the Kenwood TS-570D is an HF amateur-band transceiver whose receiver covers 500kHz-30MHz. There are few applications for an HF receiver which are more difficult and demanding than hearing weak non-channelised amateur-band signals in the overcrowded and noisy conditions which prevail, especially during contest or DXpedition operations. This being so, it ought to be a safe bet that a receiver that meets such onerous requirements well should have little difficulty with HF broadcast reception. To establish whether this was the case, we assessed the TS-570D as a possible contender for the listener.

At 6·8kg the unit is rather heavier than might be expected for its size of 270 x 270 x 100mm, reflecting heavyweight internal construction and a good deal of internal screening. Even so, the TS-570D does not contain its own power supply and an external 13·8V power unit (capable of sourcing about 2·5A for reception but a good 20A for transmission) is required.The front panel is dominated by the large and clear display, evenly backlit in yellow, together with the smooth-running VFO knob beneath. The buttons and switches feel delightfully positive in use, with just the right combination of short travel, low breakout force and high tactile feedback.

SPECIFICATION

The unit is a triple superheterodyne, with first and second IFs of 73·05 and 8·83MHz: the 455kHz third IF is for narrowband FM only. It can handle AM, narrowband FM, USB/LSB, CW and FSK, with default tuning rates of 10Hz for the latter three and 100Hz for the rest. Pressing a button reduces these to 1Hz and 10Hz respectively. A small button adjacent to the VFO knob allows tuning in 9 or 10kHz steps, which was very useful for broadcast reception, and high-speed frequency shifts can be made with the DOWN and UP buttons to the right of the VFO. It is also possible to enter the required frequency directly with the keypad, although one minor anomaly here is that there is no decimal point and it is necessary to enter frequencies less than 10MHz with the appropriate leading zeros. Ten banks of 10 tunable memories are provided, with 90 being assignable to channels and ten to other functions such as the results of programmable scan. A 9-pin D-type provides an interface to an IBM-compatible PC, and the RCP 570 control software is freely available from the Kenwood web site and elsewhere. This ran very well on a 133MHz Pentium under Windows 95/98 and is notable for being freely customisable: it is possible to vary the 'look and feel' of the program within very wide limits. Several other shareware programs are available for the TS-570D including the very popular Rig-EQF and Log-EQF.

DIGITAL SIGNAL PROCESSING

As supplied, the IF filters are 2·2kHz for SSB. CW and FSK, 4kHz for AM and 12kHz for FM but a range of other filters is available. However, the *forte* of the TS-570D is its on-board digital signal processing (DSP). The HF and LF response of the filters can be independently adjusted by concentric knobs, making it possible to tailor the bandwidth precisely to that of the received signal. In addition, a 'beat cancel' function removes multiple heterodyne whistles from the selected bandwidth and worked extremely well. There is also a 'noise reduction' system which is a form of adaptive filter and has a rather peculiar but on the whole beneficial effect on noisy signals. To go with the DSP functions there is an IF shift facility, selectable AGC time constants, a noise blanker and a switchable preamp and attenuator. This latter has a fixed 20dB setting, which is arguably too high.

The TS-570D displayed a combination of excellent ergonomics and extremely convincing performance. Sensitivity was well judged and strong-signal handling was very good indeed, giving a high dynamic range. The IF filter system was simply delightful in use, giving away very little to the NRD-545 overall, and extended listening sessions left us with the impression of a receiver which was barely challenged by even the most demanding broadcast-reception conditions. The lack of synchronous AM detection was never an issue during testing, and the low level of audio distortion and quiet AF stages were of great benefit during extended periods of listening. And for those owners who are (or who might become) licensed amateurs, the transmitter worked extremely well too…!

CONCLUSION

The overall verdict is that the Kenwood TS-570D is very well worth consideration for those looking for a first-class HF receiver. We greatly enjoyed our time with the unit and were sorry to see it go. Incidentally, several UK dealers can carry out a (reversible) modification to disable the transmitter if required.

Rating table Kenwood TS-570D

Mechanical design	★★★★★
Constructional quality	★★★★★
Ergonomics	★★★★
Sensitivity	★★★★★
Dynamic range	★★★★★
RF intermodulation	★★★★★
IF filters	★★★★★
IF performance	★★★★
Audio quality	★★★★
Software	n/a
Manual	★★★★
Versatility	★★★★
VFM – absolute	★★★★
VFM – comparative	★★★★
Overall rating	**★★★★**

Key:
★ = Poor   ★★ = Fair   ★★★ = Average
★★★★ = Good   ★★★★★ = Excellent
VFM = Value for money

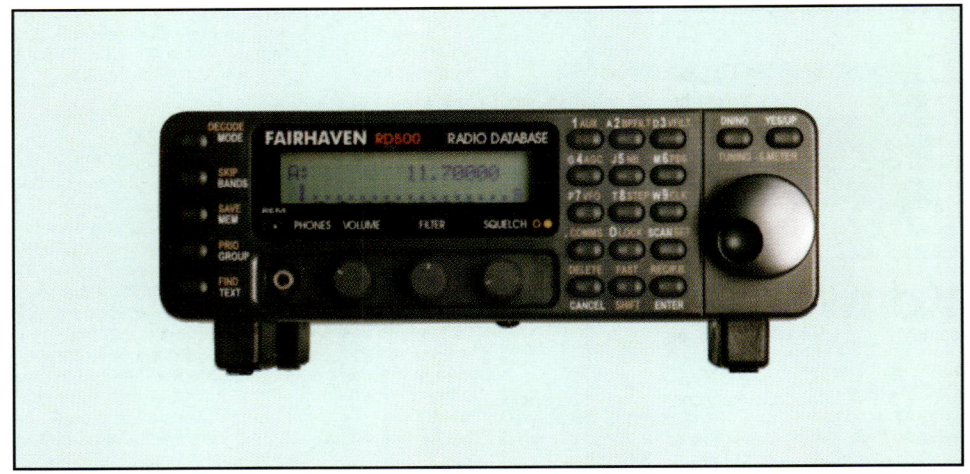

Fairhaven RD500

US$1280　£800　Euro1120

OVERVIEW

A rather unusual radio from Fairhaven Electronics of Derby, UK, the RD500 is a fascinating combination of receiver and computer. As a receiver, the unit rather resembles the AR7030 insofar as it makes extensive use of software-controlled menus for basic operation. It also contains a comprehensive integral frequency database of HF broadcasting stations. However, catering as it does for frequencies up to 1775MHz and having fairly limited IF filter options, the RD500 is perhaps best considered as a scanner which happens to embody MF/HF reception facilities rather than as a mainstream broadcast receiver. As we shall see, however, it embodies some distinctly unusual features.

FRONT PANEL

Physically, the RD500 is a nicely satin-finished black box measuring 205 x 195 x 65mm and weighing just under 1kg. From the front, the unit rather resembles the AR5000 insofar as both have quite small but decidedly well-populated front panels. The VFO control is on the extreme right of the panel, with a 5 x 3 keypad matrix to its left. The primary functions of ten of the keys are numeric entry, with decimal and enter keys in the usual places. However, all keys have shifted functions, as do a further group of five keys on the extreme left. Two more keys are positioned above the VFO knob. Ergonomically speaking, all keys have a combination of long travel, low breakout force and weak tactile feedback. In use they feel slightly 'soft' with rather indeterminate operating points. The keys of the remote controller which is supplied as standard have similar characteristics.

SPECIFICATION

The RD500 is a double superheterodyne with intermediate frequencies of 45MHz and 455kHz. The specification is quoted in terms of 0-36 and 48-1775MHz. The receiver is stated to be capable of reception of CW, SSB, AM, synchronous AM, narrow and wide FM (including stereo for the latter) and TV sound and video. The specification only refers to three IF filters, however, namely 2·4kHz for SSB, 6kHz for 'AM Wide' (there is no 'narrow' filter) and 12·5kHz for NBFM. Since the RD500 is certainly capable of producing excellent-quality stereo audio from FM broadcasting and television sound with the correct subcarrier offset, it clearly has some other IF filters as well! A passband shift control is provided, and tuning steps are 5Hz in CW, SSB and AMS; 100Hz in AM; and 20kHz in wide FM and TV modes.

REAR PANEL

On the rear panel are no less than four antenna sockets. Two of these are high-grade N-types and are used for 48-860MHz and 860-1750MHz respectively. The other two are SO239s, again of good quality (insofar as UHF connectors ever are) and these are for 'main' and 'whip' HF antennas respectively.

The manual explains that the 'whip' input is for high-impedance short antennas but that it "does not have the large signal-handling ability of the main input". A rear-panel switch selects between the two HF inputs, and in its centre position these are combined in antiphase. In certain circumstances this allows an effective and worthwhile degree of noise cancellation to be introduced via main and secondary antennas, and Fairhaven should be congratulated for providing such a simple and potentially powerful feature which is often neglected by other manufacturers.

Also to be found on the rear is a socket for a 13·8V input (the manual implies that a suitable mains PSU is supplied with the RD500 but none was with our sample, which was presumably an oversight); a DIN socket for a PC keyboard input and RS-232 output; another DIN socket for audio and cassette-recorder operation; two gold-plated RCA sockets for stereo audio output; and two lesser-quality RCA sockets for video and TV audio output. Finally there is a socket for use with an external loudspeaker.

DISPLAY

The display is a backlit dot-matrix type, which is inevitably on the small side. In the same way as that of the AR5000, it suffers somewhat from a shortage of contrast unless viewed almost head-on and also from decidedly restricted viewing angles. The RD500 can be tilted upwards towards the user by two hinged front feet, which assists somewhat in improving display legibility, and the resulting distance of the front panel above the surface on which the receiver is mounted is almost ideal for the average finger to reach the VFO knob. Inevitably the knob and its finger depression are on the small side, but VFO operation becomes quite simple and pleasant with practice. Three other rotary controls beneath the display are for volume, squelch and 'filter', the latter controlling a selectable notch or peak function.

INTERNAL SOFTWARE

Given that there are four rotary controls and 22 operating keys, you could be forgiven for suspecting that the RD500 is highly software-dependent – and you would be quite right. Although not perhaps as daunting as the system in the AR7030, access to much of the RD500's functionality is by means of menu options coupled with 'soft' keys. For example, pressing SHIFT and 1 gives access to the 'auxiliary menu' consisting of six functions: attenuator, preamplifier, cassette control, low-pass filter, AGC and an option referred to as "more", which provides access to the next part of the menu. Pressing SHIFT and 2 brings up a menu of options associated with CW filtering; low-pass, bandpass and so on. Pressing SHIFT and 7 gives access to no less than 26 VFOs, which can be used in a variety of different ways. Each is identified by a letter, and can be named with a text string. A variety of other functions are addressed in the same way. For example, pressing SHIFT and COMMS (the shifted function of the frequency decimal point) activates the RS-232 port, which according to the manual enables uploading and downloading of databases. Suitable software is provided, but unfortunately we could not persuade this to run despite much effort; both a 133MHz Pentium and a PS/2 486 locked up to such an extent that hard resetting via the power switch was the only recourse.

SKIMPY MANUAL

The information given in the manual is rather skimpy, and indeed the manual as a whole is nowhere near good enough for its job. It is heavily text-based, which is bad enough, but the text itself is frequently incoherent and reads very much as though it was thrown together by a junior engineer as an afterthought the day before the first production unit was to be shipped to the dealer! Frankly, a receiver as good as the RD500 deserves a manual and accompanying literature several orders of magnitude better than that which currently accompanies it. For example, what do you make of this?

"….the database system provides 234 groups (A1 to Z9) which can each hold between 1 and 999 records although the maximum total number of records is limited to 13290 with the standard 512K of RAM and 54681 with the maximum of 2MB, these figures are reduced when the memory is partitioned for sound recording."

Although we used the RD500 intensively for almost a month, we strongly suspect that it has a number of features we never fully explored because of the limitations of the manual. We never quite got to grips

with the scan modes, for example, and neither did we get to the bottom of some of the band functionality. Once we found our way past some of the more obscure ergonomics, however, the RD500 proved to be a very good receiver which was relatively easy to use in most aspects of everyday operation.

Sensitivity was about optimum for HF use, and there were few strong-signal problems unless very large arrays were used on the lower bands at night. Some second-order problems at 14MHz were very effectively dealt with by means of the 20dB attenuator, which was perhaps a little too heavy-handed; two separate 10dB steps would have been preferable, and should be easy to manage under software control. The peak and notch filters worked quite well but unfortunately the noise blanker did not, apparently having no effect whatsoever on a variety of different types of impulsive interference. This may have been a sample fault.

FILTERS

The 6kHz AM filter was too wide for reception of weak HF broadcast signals adjacent to strong ones, but switching to synchronous detection improved matters considerably and switching to LSB or USB improved them even more. The SSB filters were considered a little wide for single-sideband reception. A good test at the review locality is RAF VOLMET on 5450kHz, which in the evening requires some sharp filters with good shape factor if continuous splatter from adjacent channels is to be avoided. The RD500 was not quite as good as some other receivers in this respect, although it produced quite acceptable and intelligible results. Incidentally, the 6kHz AM filter is too narrow for multiply offset ground stations in the VHF airband and the 12·5kHz FM filter is also too narrow for certain applications.

CONCLUSION

Overall, we rather liked the Fairhaven RD500 even though we considered it badly let down by its manual. It is a singular vision of how radios of the future might be conceived; it almost seems to us like a brilliant one-off rather than a unit intended for large-scale production and use. Whilst not absolutely convinced by its ergonomics or its idiosyncratic approach to such things as bands and memories, we were still sorry to have to return it – which is arguably the best test of all of any radio!

Rating table Fairhaven RD500

Mechanical design	★★★★
Constructional quality	★★★★
Ergonomics	★★★
Sensitivity	★★★★
Dynamic range	★★★★
RF intermodulation	★★★★
IF filters	★★★
IF performance	★★★★
Audio quality	★★★★
Software	★
Manual	★
Versatility	★★★★
VFM – absolute	★★★
VFM – comparative	★★★
Overall rating	**★★★**

Key:
★ = Poor   ★★ = Fair   ★★★ = Average
★★★★ = Good   ★★★★★ = Excellent
VFM = Value for money

Icom ICR-8500

US$1500   £1550   Euro1800

OVERVIEW

The Icom ICR-8500 is second only to the NRD-545 in terms of price, costing £1,550 in the UK. Measuring 310 x 288 x 112mm and turning the scales at 7kg, the ICR-8500 is relatively large and heavy and conveys a reassuring feeling of solidity and strength. The frequency range covered by the European version is 100kHz-2GHz, although a footnote states "Specifications guaranteed 0·1-1000MHz and 1240-1300MHz". The North American version omits the 824-849 and 869-894MHz cellular telephone bands and a French variant omits the 87·5-108MHz FM broadcast band.

FACILITIES

Available modes are FM (wideband and narrowband), AM (wide, normal and narrow) USB/LSB and CW ("normal" and "narrow" with an optional FL-52A filter). The filter specifications are given in terms of their -6dB points and are "More than 150kHz" for wide FM, "More than 12kHz" for FM and wide AM, "More than 5·5kHz" for narrow FM and AM and "More than 2·2kHz" for narrow AM, SSB and CW. The FL-52A is quoted as "More than 500Hz". There are separate 50 Ohm coaxial antenna inputs for use above and below 30MHz – an SO239 for the latter and an N-type for the former – together with an RCA female connector forming a 500 Ohm input for random-wire HF antennas.

The mains power supply is an external item measuring 120 x 75 x 65mm and stated to be capable of supplying 16V DC at 1·5A, although the radio seems from the manual to be intended for use from a 13·8V supply and draws almost 2A DC with full volume set. The mains supply becomes decidedly warm in use, even with the receiver on standby, and the receiver itself also runs somewhat warm. However, in this respect the ICR-8500 is a great improvement on earlier Icom receivers such as the IC-R7000 with their integral and very hot-running power supplies. Whilst on the rear panel, there is also an RS-232C connector allowing direct control of the receiver via a PC, and a jack for the proprietary CI-V control interface. A 10·7MHz IF output is provided, as is a socket which can either output an AGC signal for the optional TV-R7100 television receiver adaptor or a non-de-emphasised audio signal for 9600bp data.

MEMORIES

That the ICR-8500 is intended to be considered principally as a scanning receiver is evident from its memory architecture. In essence it has 1,000 user-accessible memory channels arranged as 20 banks of 40, together with 100 further channels in which the results of a scan within user-selectable band limits are stored and another 100 in which unwanted frequencies to be skipped during a scan are memorized. A "free" bank provides temporary storage for new channels, which can be copied-and-pasted to other banks, and another 20 memories are used to store the upper and lower edges of user-selected bands for scanning. Eight-character names can be associated with all memories, and each bank can also have a five-character name. Each individual memory channel can have mode, tuning step and attenuator information associated with it. The main scanning modes are memory scan (all memories in a bank), "memory select scan" (for specified channels in a bank), "programmed scan" (which searches for signals within a specified range) and "auto memory write scan" (which operates in the same way as programmed scan but stores any frequencies on which signals are received into memories). There are also priority scan and modulation-specific scan modes. Scan speed and delay are adjustable. The specification states that the maximum scan speed is 40 channels per second, and this is certainly true when all channels in a bank are set to the same mode and filter width.

TUNING STEPS

There are no less than 13 tuning steps available, namely 10, 50 and 100Hz, 1, 2·5, 5, 9, 10, 12·5, 20, 25 and

100kHz and 1MHz. A programmable tuning step is also available, which can be set anywhere between 0·5 and 199·5kHz in 0·5kHz steps. The ICR-8500 was the only receiver in the tested group, other than the hand-portable Icom ICR-10 to provide this feature. With such a desirable and useful multiplicity of tuning steps it is perhaps a little unfortunate that 8·33kHz was not also provided, since portions of the 118-137MHz aeronautical band in Europe are shortly to be so channelised.

FRONT PANEL

The front panel is clean and uncluttered. The dominant feature is the yellow backlit LCD display, which is commendably clear and easy to read almost irrespective of viewing angle. The brightness of the backlighting can be toggled between high and low, and indeed a number of other receiver parameters can similarly be switched between two states (e.g. confirming 'beep' on or off, HF antenna connector 50 or 500 Ohm, etc.) in a software menu mode. For night use, even the dim setting of the backlight may be found to be slightly on the bright side. The S-meter at upper left is a little small by comparison with that in some earlier Icom receivers such as the IC-R7000. Next to it is the power on-off switch, and below this is a 'sleep' switch which can be used to turn the radio off after one of four preset periods. Three jacks for recorder activation, recorder audio and variable-level headphone output are beneath, and on their right are the three main rotary controls for AF gain, squelch and a concentric peak filter/IF shift function. The peak filter works quite well, but unfortunately there is no notch filter. The four mode switches above the rotary controls are marked WFM, FM, AM and SSB/CW and each toggles between the filter settings available for that mode.

DISPLAY

Icons indicating WIDE and NAR appear in the display when the appropriate filters are selected, but it is a little unfortunate that the bandwidth indication is replaced after a few seconds by the bank name. The SSB/CW button toggles between USB, LSB, CW and narrow CW. Some users might initially find the toggled filter settings awkward in use, but a little familiarity soon makes them feel quite natural. The lower row of buttons controls AGC speed (fast or slow; it cannot be switched off), noise blanking and AFC, and attenuation. This latter is available in 10, 20 or 30dB steps by combinations of buttons, and is an excellent feature. It is a fact of life that wideband receivers affordable by ordinary mortals tend to have considerably more limited spurious-free dynamic ranges than those covering more restricted bands, and the attenuator can be a very useful facility when trying to resolve weak signals adjacent to very strong ones.

ERGONOMICS

All the buttons in the IC-R8500 have low breakout force, moderate tactile feedback and well defined operating points, making them feel very pleasant in use. In this they are aided by slightly soft surfaces. The generally good 'feel' of the receiver is enhanced by the VFO knob, which has just the right amount of inertia, an adjustable brake and an indented rubber outer surface which is ideal for edge-of-the-finger rapid tuning. The deep indent into the front face is nicely sized for the average forefinger, and with the bail raised the front of the receiver is at the precise height necessary for fatigue-free operation of the VFO. The up/down tuning-step buttons fall naturally to hand at the right of the tuning knob. The keypad, at upper right, is easy to use for numeric frequency entry and also naming of banks and

memory channels, and the row of buttons beneath control the memory scan functions. All in all, the ergonomics of the IC-R8500 are ideal for extended periods of use once the inevitable complexities of such a versatile receiver have been mastered. The manual is generally helpful, being well written and illustrated and also offering a useful troubleshooting guide.

PERFORMANCE

The IC-R8500 is a triple superheterodyne with up-conversion to a first IF of 48·8MHz for frequencies below 30MHz; the image rejection figure of 60dB given in the specification is conservative, with our sample measuring considerably better. Specification sensitivities are given in terms of microvolts unreferenced to a signal-to-noise ratio or with any indication of whether they are intended as EMF or PD. In actual fact the IC-R8500 displays considerably more sensitivity below 30MHz than would be usable from a real-world antenna, and to some extent this is reflected in its modest strong-signal performance. The AM filters are also not ideal for HF broadcast reception. At a nominal 5·5kHz, the 'normal' AM filter is actually rather too wide whereas the 2·2kHz 'narrow' AM filter is somewhat too narrow for comfortable listening. The 'wide' filter, at 12kHz, is usable enough if you have a local medium-wave station and tend not to listen after dark but not much use for any other form of broadcast reception. Ideally Icom would make available an optional AM filter which is somewhere between the existing 'narrow' and 'normal' filters. Making use of the SSB filter is not very rewarding, and of course the absence of true synchronous detection does not help either. And as a last reservation, the built-in speaker does not do justice to the received audio, which via an external loudspeaker or headphones can be extremely good.

CONCLUSION

All in all, however, the Icom IC-R8500 is a fine receiver if you feel that you must have a wideband reception capability in a single unit. It is beautifully made, well presented and finished and a delight in use once the available functionality has been mastered. It has arguably a few more rough edges and omissions than are strictly desirable at the price, but you will probably not be disappointed if you buy one.

Rating table Icom ICR-8500

Mechanical design	★★★★
Constructional quality	★★★★
Ergonomics	★★★★
Sensitivity	★★★★
Dynamic range	★★★
RF intermodulation	★★★
IF filters	★★★
IF performance	★★★
Audio quality	★★★★
Software	n/a
Manual	★★★★
Versatility	★★★★
VFM – absolute	★★★
VFM – comparative	★★★★
Overall rating	**★★★★**

Key:
★ = Poor   ★★ = Fair   ★★★ = Average
★★★★ = Good   ★★★★★ = Excellent
VFM = Value for money

AOR AR5000+3

US$2895   £1550   Euro3475

OVERVIEW

The AOR company has a long history of producing innovative high-performance radios for a mass market. The AR2001 VHF/UHF scanner introduced in the early 1980s was a landmark in its time and many are still in use; its successors set the standard for wideband scanners for several years. Along with Yupiteru, AOR was one of the first manufacturers to introduce wide-coverage portable scanning receivers and some of these had quite respectable performance on HF. The AR7030 set new standards for HF reception and the AR3000 was one of the first 'tabletop' wide-range receivers. A few years ago AOR introduced the AR5000, which provided an almost incredible combination of functionality and performance for an admittedly quite high price. The original model is still available for £1325 but the 'AR 5000 + 3' embodies three factory-fitted modifications which enhance the performance (and unfortunately the cost) still further. At £1555 the AR5000 + 3 makes for an interesting comparison with the Icom ICR-8500which costs virtually the same at UK prices.

SPECIFICATION

The basic AR5000 is 260 x 220 x 100mm in size and weighs 3·5kg. Substantially constructed in metal casework, it is considerably smaller then the ICR-8500 and since much of the front panel is dominated by the display, the controls and keys are necessarily rather miniature. The VFO knob in particular is only about 35mm in diameter, and the finger recess is distinctly on

---

. . . rather spidery characters and limited contrast . . .

---

the small side. The display, backlit in a rather pleasant shade of green, has rather spidery characters and limited contrast at acute viewing angles. As seems to be almost universal nowadays, the AR5000 does not

contain its own power supply; a three-pin socket for an external 12-13V supply capable of sourcing about 1A is provided on the rear and a mains power unit is supplied with the receiver. This ran noticeably warm when operated from a 230V stabilised UPS. There is a large and clear S-meter at upper left, with power and function buttons adjacent. The main operating keys, all of which have shifted functions, are to the left of the VFO. The

---

. . . the audio quality belies its small size . . .

---

buttons have short travel, low breakout force and high tactile feedback, all of which gives them a positive if slightly 'hard' operating characteristic. The right-hand column of buttons is inconveniently close to the VFO knob, and left-handed users in particular would be likely to find the AR5000 something of a trial initially. The VFO has a lever-operated tension control adjacent to it, together with what the manual calls a "sub-dial" on its right. This can be programmed to carry out a number of different tasks, especially in changing parameters such as the IF bandwidth and the attenuator setting. It can also act as a subsidiary tuning control operating in channels rather than tuning steps. An interesting detail is that the speaker is mounted beneath the radio and its output is 'ducted' to the front via a suitably shaped plastic moulding. In consequence, the audio quality obtainable from the AR5000 belies its small size.

FACILITIES

The AR5000 is a triple superheterodyne with IFs of 622 and 10·7MHz and 455kHz. It covers the extremely wide frequency range of 10kHz-2600MHz and can demodulate AM, SSB, CW and narrowband FM. A 'numerically controlled oscillator' with basic tuning steps of 1Hz derives its reference from a TCXO and can be

programmed to provide any required tuning step size. If an external high-grade 10MHz reference is available, the AR5000 can make use of this via a rear-panel socket. As supplied, the nominal filter bandwidths are 3, 6, 15, 30, 110 and 220kHz although an optional Collins 500Hz mechanical filter (costing £70 in the UK) can be easily retrofitted by those needing better narrowband CW performance. In addition there are selectable audio

---

. . . if required the preselector can be manually controlled . . .

---

filters (50, 200, 300 and 400Hz high-pass and 3, 4, 6 and 12kHz low-pass) and selectable FM de-emphasis (off, 25, 50, 75 or 750mS). Either 10 or 20dB attenuation is available, as is an internal preamplifier for use below 230MHz, and automatic front-end preselector tracking is provided between 500kHz and 1000MHz. If required, the preselector can be manually controlled. There are five separate VFOs and 1000 memory channels, arranged as ten banks of 100 channels; there are also 20 search banks with an associated auto-seek and store function. Each memory bank and channel can be given an eight character alphanumeric name.

SCAN AND SEARCH FACILITIES

As would be expected, the scan and search functions are extremely comprehensive. The quoted scan speed is 25 channels per second but this can be speeded up to about 45 channels per second in a so-called "Cyberscan" mode in which the frequency or channel name are not displayed. Both scan modes worked well. An internal DTMF decoder is provided, and there is provision for an optional CTCSS decoder, a speech-inversion descrambler and various other items. There are many other excellent features, notably a sleep timer and alarms, an RS232 interface, built-in market-specific auto-mode bandplan and the ability to plug directly into an SDU5000 spectrum display. There are two antenna sockets, a front-panel accessory socket and a headphone jack. You will not be surprised to learn that the manual is a substantial and lengthy item of literature, and frequent reference to it will be necessary in the early stages of learning your way around the unit.

EXTRA 'PLUS' FEATURES

Apart from the surprising absence of a notch filter and any form of IF shift or passband tuning system, the AR5000 has possibly the most comprehensive features list of any commercially available receiver. However, adding the '+3' option endows it with synchronous AM reception capability, an AFC facility and a noise blanker. One might, of course, argue that a receiver which already costs in excess of £1300 should have all three facilities fitted, and that £230 is an excessive amount to

---

. . . the AFC has some particularly useful properties . . .

---

pay for them. Whatever conclusion you come to, it is certainly true that the new functions work well. The AFC has some particularly useful properties; it is capable of coping easily with the offsets sometimes encountered in reception of air-band ground stations, and in conjunction with the synchronous reception facility makes for very easy and pleasant tuning of HF broadcast stations. The

noise blanker also appeared to work very well indeed, even coping with a totally unsuppressed petrol-driven hedge-trimmer used a few metres from the antenna.

CONCLUSION

Both in the laboratory and in the course of extended listening sessions, the AR5000 + 3 acquitted itself honourably. Sensitivity was more than adequate and the strong-signal handling was generally very good, especially in view of the enormous frequency range covered. It is a very difficult design task to produce a wide-range receiver which on the one hand is sensitive enough to cope with the requirements of VHF and UHF reception with small antennas and on the other can handle the output of a large HF array, and in this respect the AR5000 + 3 is remarkably able. There were no difficulties even when using full-size dipole and Yagi antennas, and only very occasional problems of second-order IMD from 6/7MHz broadcast stations when listening on the 14MHz amateur band in the evening. The 'manual tracking' preselector facility proved useful on occasions, as did the attenuator. The 3kHz filter worked well in crowded bands or when listening to weak HF broadcast stations adjacent to strong ones, and the 6kHz filter was ideal for reception of strong stations in the clear. The audio filters were often helpful in obtaining the best possible results. Overall, the AR5000 + 3 is undeniably expensive but offers an outstanding and unusual combination of coverage, performance and

---

. . . better strong signal handling and superior filters . . .

---

functionality. Compared with the Icom ICR-8500 it offers considerably more features, better strong-signal handling, wider coverage and decidedly superior filters. However, the designers have crammed a lot of radio into a rather small space, and the unit displays some ergonomic shortcomings as a result. If you can live with these, and train your fingers to operate the receiver with the requisite delicacy, the AR5000 should be high on your list of receivers if you are in the market for a high-grade no compromise unit offering "DC to light" coverage.

Rating table AOR AR5000+3

Mechanical design	★★★★
Constructional quality	★★★★
Ergonomics	★★★
Sensitivity	★★★★
Dynamic range	★★★★
RF intermodulation	★★★★
IF filters	★★★★★
IF performance	★★★★
Audio quality	★★★★
Software	n/a
Manual	★★★★
Versatility	★★★★
VFM – absolute	★★★
VFM – comparative	★★★
Overall rating	**★★★★**

Key:
★ = Poor   ★★ = Fair   ★★★ = Average
★★★★ = Good   ★★★★★ = Excellent
VFM = Value for money

JRC NRD-545

US$1800   £1600   Euro2150

OVERVIEW

Marginally the most expensive receiver in our tested group, the NRD-545 is exceedingly impressive in all respects. Given that it costs just £50 more than both the Icom ICR-8500 and AR5000, comparisons seem natural and inevitable. However, to compare these units would be to attempt to compare apples and oranges. Whereas the ICR-8500 and AR5000 are both very capable wide-range receivers which to some extent trade-off performance and functionality for enormous frequency coverage, the NRD-545 is an utterly dedicated HF-band receiver optimised for the best possible performance between 100kHz and 30MHz. It embodies digital IF filtering and signal processing, a single-chip 1Hz-stepping DDS synthesizer, a powerful kit of tools for interference reduction including two noise blankers, a notch filter, continuously variable bandwidth control (which is quite delightful in use), passband shift and excellent facilities for ECSS reception. The front-end RF amplifier uses four parallel-connected JFETs and a quad-FET double-balanced mixer for good strong-signal handling. RTTY demodulation for 170, 425 and 850Hz shifts at baud rates between 37 and 75 is provided as standard, and the unit can be interfaced with an IBM-compatible PC. All in all, the features list is very impressive. Users requiring better than the existing 2ppm/hr frequency stability can install an optional extra CGD-197 TCXO, which will provide 0·5ppm, and coverage to 2GHz can be added by fitting the CHE-199 wideband converter. However, a note in the manual suggests that only AM, FM and WFM modes can be received above 30MHz.

SPECIFICATION

The NRD-545 measures 330 x 290 x 130mm and weighs just under 8kg. It is very solidly built and beautifully finished in all respects. The front panel is initially rather daunting, and careful reading of the densely written and well illustrated manual is recommended before switching on for the first time. The panel is dominated by the bright

and clear display, which uses multiple colours for better legibility; the frequency display is yellow whereas the electronic S-meter shows levels up to S9 in a pale blue and levels above that (to +60dB) in red. Current settings of parameters such as filters, mode and AGC are shown along the upper edge of the display. A keypad for direct frequency entry is located to the right of the display, and the large rotary VFO knob lies underneath. All the other controls are more or less symmetrically disposed about the front panel, and we especially liked the way the timer facilities were interconnected with the power on/off switch. Probably the only criticism one could make of the unit's ergonomics is that the finger hole in the VFO knob is a little small and sharp-edged and does not contain an insert which can remain fixed in the plane of the finger whilst the knob rotates. Fast tuning is perhaps best carried out by the up/down keys either side of the knob! One other very trivial complaint is that the display brightness can be set to one of two levels but the difference between them is slight and the 'dim' position is still too bright for use in low ambient light levels.

FILTERS

The NRD-545 is essentially a triple-conversion superheterodyne. The first IF is 70·455MHz, the second IF is 455kHz and the third is 20·22kHz. The minimum tuning-step size available is 1Hz but any other step between 10Hz and 100kHz is selectable. The digital IF filters are extremely versatile. In essence there are 'narrow', 'inter' (i.e. intermediate) and 'wide' settings which default to different values according to the selected mode. In wideband FM or FM modes, the 'wide' filter is the only one available. Bandwidths of the 'narrow' filter are 500Hz for CW and RTTY, 1·8kHz for SSB and 2·4kHz for AM. For the 'inter' the figures are 1kHz for CW/RTTY. 2·4kHz for SSB and 4·5kHz for AM. The 'wide' filter offers 2·4kHz for CW/RTTY, 2·7kHz for SSB and 6kHz for AM. But the default values – which for general listening work well – are only the starting-point

since the filter bandwidths are continuously adjustable with a single rotary control between 10Hz and 9·99kHz in 10 or 100Hz steps. The bandwidth is continuously displayed adjacent to the frequency, and the default value is reset each time the basic filter bandwidth is changed by means of the three buttons next to the VFO. In addition to the variable bandwidth is a notch filter with a good 46dB depth and variable over a ±2·5kHz range, allied to automatic notch tracking with a range of ±10kHz. There are also two noise blankers, continuously variable passband tuning with a range of ±2·3kHz and a 'noise reduction/beat canceller' facility. Given a little practice and familiarity with the receiver's operation, the overall impression is of tremendous IF performance allied to consummate ease of control, and the ease with which the IF filter parameters can be varied to suit the needs of the moment belies the enormous amount of work being performed by the NRD-545's circuitry. In this context it should be mentioned that the AGC release time can be continuously varied between 0·04 to 5·1 seconds in 20ms steps, and the time constant is displayed. Different constants can of course be selected for each mode, and the AGC can be turned off if required.

ATTENTUATOR

A 20dB attenuator is available. In the four weeks or so of the test period, the NRD-545 was used almost every day and was connected to a variety of antennas from a seven-element trapped array to assorted dipoles, long wires and a Beverage. It was very unusual for any strong-signal problems to be encountered and from memory the attenuator was used only once. Measurements suggested that the third-order intermodulation intercept point referred to a 1mV signal was +11dBm, which is an excellent figure, and that other important parameters also measured very well. Indeed, both the reciprocal-mixing and 1dB compression points were by some margin the best of the tested group. Second-order intermodulation to the 1mV point was an enormous +66dBm, which suggests that in practice it will simply never be an issue. In our view it is a mistake to place undue emphasis in receiver assessment on the occasional appearance of IMD products somewhere near the noise floor, which in no way affect reception of wanted signals.

MEMORY

There are 1000 channels of memory storage available, and the parameters stored in each one are frequency, mode, IF filter bandwidth, AGC, attenuator setting and tuning step. The channels are organised into twenty groups of 50, and it is possible to scan between any two channels. In addition, it is possible to search ten frequency bands which are initially factory-preset but can be programmed to individual requirements. Other ancillary functions include a versatile timer and a 'user setup' menu which allows some receiver parameters such as tuning step, S-meter indication, scan and sweep time, beep tone and so on to be set according to personal preference.

IN USE

In operational terms the NRD-545 was simply delightful, and monitoring sessions notionally for testing purposes became addictive periods of extended listening. A few minor ergonomic aspects of the receiver were mildly irritating at first but one rapidly became used to them; the scan and sweep functions in particular were a little awkward, and saving a frequency and its associated parameters to memory required rather a lot of button-pressing. But the combination of such superb performance and overall ease of use was thoroughly beguiling. The audio output of the NRD-545 was very low in distortion, and even lengthy periods of listening via a good-quality headset were not at all fatiguing. The combination of well-judged sensitivity, excellent strong-signal handling and a 'quiet' synthesiser allowed signals to stand out from the quiet background, in a way which is very characteristic of the best receivers, and the very well implemented noise blanking and passband tuning allowed even quite horrendous interference to be effectively dealt with. The only facility which did not seem to produce very convincing results was the noise reduction/beat canceller, but perhaps we needed more practice and experience with it. ECSS reception worked well, although it must be said that there were very few occasions when it was necessary to have recourse to it.

CONCLUSION

All in all, the JRC NRD-545 is one of the finest HF receivers ever to come our way and we enjoyed every moment of the test period. At £1600 it could not be classed as inexpensive, but if cost is a secondary consideration to all-out performance allied to ease of use there are very few other choices. Perhaps the most eloquent tribute we can give to the NRD-545 is that we were very sorry to have to return it!

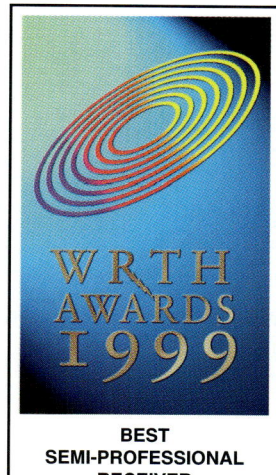

BEST SEMI-PROFESSIONAL RECEIVER

Rating table JRC NRD-545

Mechanical design	★★★★★
Constructional quality	★★★★★
Ergonomics	★★★★★
Sensitivity	★★★★★
Dynamic range	★★★★★
RF intermodulation	★★★★★
IF filters	★★★★★
IF performance	★★★★★
Audio quality	★★★★★
Software	★★★★
Manual	★★★★
Versatility	★★★★
VFM – absolute	★★★★
VFM – comparative	★★★★
Overall rating	**★★★★★**

Key:
★ = Poor   ★★ = Fair   ★★★ = Average
★★★★ = Good   ★★★★★ = Excellent
VFM = Value for money

REVIEWS ROUNDUP
1999

TABLE TOP

Of the traditional 'table-top' mains-powered receivers, there is no doubt that the JRC NRD-545 offers the finest overall performance. And so it should; it is the most expensive of the tested group and covers only the traditional HF spectrum, as opposed to the considerably extended ranges of receivers in comparable price brackets such as the Icom ICR-8500 and AOR 5000 + 3. Both the latter perform well, although the ICR-8500 would benefit from better IF filters and the AOR receiver does not offer a particularly attractive combination of price, performance and functionality. All in all, both strike us as somewhat expensive for what is on offer. The NRD-545 has quite superb filters, an excellent balance of sensitivity and strong-signal performance and is a delight to use; for the dedicated HF enthusiast who insists on something approaching state-of-the-art performance and functionality, the JRC unit is the best. The AOR 7030 Plus offers essentially equivalent performance in certain respects, although without the formidable DSP facilities of the NRD-545, and its idiosyncratic ergonomics will delight some potential purchasers and be the despair of others. With this receiver above all, 'try before you buy' is an essential precept. Something similar could be said of the Fairhaven RD500. Although ill-served by its thoroughly inadequate manual, the versatile RD500 represents a very individualistic vision of how a radio receiver of the future might be configured and in certain ways we liked it as much as any of the other units tested. In quite another category, unfortunately, the AKD HF-3 was the least enjoyable receiver we tested and was considered to be decidedly over-priced for its limited feature set and indifferent performance. There seemsto be little reason to consider this unit.

Also in the table-top category comes the Kenwood TS-570D amateur-band transceiver with wideband HF coverage. This unit (which as we went to press was rumoured to be replaced shortly by an augmented version at the same price) proved capable of excellent performance and in many respects it was as good as any other receiver on the market. Although lacking certain features which those principally interested in AM broadcast reception would wish to have, the unit's excellent DSP facilities, fine balance of sensitivity and strong-signal performance and excellent ergonomics make it well worth considering.

PORTABLES

As far as the two 'conventional' portables were concerned, the Sony ICF-SW30 was preferred to the Roberts R881 on several counts although the latter has a few features which would have been welcome on the Sony. However, the relatively low cost and reasonable performance of both units suggests that one would be relatively safe in buying either. In our view the markedly superior usability of the ICF-SW30 undoubtedly makes it the better choice, and its performance is better in several important respects as well. A word of praise is also in order for the excellent FM performance of the Sony receiver.

SCANNERS

The two hand-held scanners we tested are typical of their breed insofar as their circuitry and memory architectures are optimised for high-speed band-searching and scanning across VHF/UHF channels. Although it would be unfair to say that HF/MF broadcast reception capability is something of an afterthought in these units, it was clearly not a primary design goal; neither unit is comfortable with large antennas, and their IF filter parameters are not ideal for crowded HF bands. However, the Yupiteru MVT-9000 proved to be surprisingly usable for listening to the stronger HF stations, and its better filters and much more user-friendly ergonomics made it the 'better buy' of the scanning receivers. Further investigation of the Yupiteru receiver suggested that this level of performance resulted from the incorporation of a high-grade Murata 455kHz filter, reportedly similar in specification to that in the AOR 7030. The Icom ICR-10, which is admittedly considerably cheaper than the MVT-9000, is not really a contender for MF and HF broadcast listening although it has undoubted abilities as a scanning receiver. Icom equipments are often felt to have idiosycratic operating characteristics, and the ICR-10 is no exception.

PC RADIOS

The three 'PC radios' we tested were all interesting, and all had qualities which were highly desirable. That none of the three individually embodied all the desirable features probably reflects the fact that this market is nowhere near mature. The Icom IC-PCR1000 and WinRadio 1500e are evidently aimed more at the scanning-receiver enthusiast than the pure HF listener. Our early production sample of the WinRadio suggests that some strong-signal performance issues need to be tackled, but the software is excellent and the 'open' approach to software development adopted by the manufacturer is highly praiseworthy. The IC-PCR1000 is a fine receiver which again is not optimised for broadcast reception but makes a reasonable job of it. The software is rather limited in certain respects, and Icom chooses to regard it as proprietary. Even so, the relatively low cost of the unit makes it well worth consideration. Although limited to HF coverage, the Ten-Tec RX320 also looks promising and it will be very interesting to examine a full production version next year.

CONCLUSION

In consumer electronics it is traditional to end the conclusion of a group test with a which-was-best question. However, given the widely different abilities of the receivers in the tested group, not to mention the enormous 20:1 cost differential between the least and most expensive, such an exercise is not really possible here. I could find a good home for almost all of them in my radio room, with price of place given to the NRD-545 on sheer performance and functionality grounds, and probably manage without the WinRadio WR1500e in its present form, the Icom ICR-10, the Roberts R881 and the AKD HF-3E. Any or all of the rest would be welcome.

WRTH Receiver Guide 1999

Budget, Hand-held & Travel Portables

Maker	Model	Size	SEL	DR	OV	US$	£	Euro
Grundig	Yacht Boy 207	S	***	***	****	50	33	60
Grundig	Yacht Boy 305	S	***	***	****	130	-	155
Grundig	Yacht Boy 400	S	****	***	*****	170	120	205
Grundig	Yacht Boy 500	M	***	**	**	300	160	360
Icom	ICR-10	H	***	**	***	500	260	600
Panasonic	RFB-45	S	***	***	***	150	140	180
Philips	AE-3625	S	**	*	*	100	70	120
Philips	AE-3650	S	**	***	***	-	70	120
R Shack	DX-394	M	***	***	****	250	245	300
Sangean	ATS-404	S	***	***	***	125	80	115
Sangean	ATS-606	S	***	***	***	160	130	190
Sangean	ATS800A	S	***	***	***	80	-	95
Sangean	ATS808	S	****	***	***	120	120	145
Sony	ICF-SW10	S	***	***	***	50	60	60
Sony	ICF-SW12	S	**	**	*	90	60	110
Sony	ICF-SW30	S	****	***	****	125	80	115
Sony	ICF-SW40	S	****	***	*****	120	90	145
Sony	ICF-SW100E/S	S	****	***	****	360(S)	160	430
Sony	ICF-SW600	M	****	***	****	55	-	65
Sony	ICF-SW7600G	S	****	***	****	170	130	205
Sony	ICF-SW1000T	S	****	****	****	540	380	650
Yupiteru	MVT-9000	H	****	***	****	455	400	550

PC Radios, Serious Short-wave & Semi-pro Receivers

Maker	Model	Size	SEL	DR	OV	US$	£	Euro
AKD	Target HF-3	M	***	***	****	-	160	225
AKD	Target HF-3E	M	***	***	**	480	299	420
AOR	AR5000 + 3	M	***	***	***	2895	1550	3475
AOR	7030	M	*****	*****	*****	1480	730	1775
AOR	7030 Plus	M	*****	*****	****	1700	950	2050
Drake	SW-1	M	***	***	***	250	-	300
Drake	SW-2	M	****	****	****	500	500	600
Drake	SW-8	L	***	***	***	780	500	935
Drake	R8A	L	****	****	****	1060	995	1270
Fairhaven	RD500	M	***	***	***	1280	800	1120
Grundig	Satellit 700	L	****	***	***	-	330	460
Icom	R-72E	L	***	****	****	-	830	1160
Icom	ICR-8500	L	****	****	***	1500	1550	1800
Icom	ICR-9000	L	*****	*****	****	6200	-	7440
Icom	IC-PCR1000	C	***	***	****	490	320	590
JRC	NRD 345	L	****	****	***	800	800	960
JRC	NRD 535	L	****	*****	*****	1200	1750	1440
JRC	NRD 535D	L	*****	*****	****	1690	1995	2030
JRC	NRD 545	L	*****	*****	****	1800	1600	2150
Kenwood	R5000	L	****	****	****	1050	1000	1260
Kenwood	TS-570D	L	****	****	****	1170	1300	1400
Lowe	HF-150	M	***	****	*****	500	420	600
Lowe	HF-225	M	***	***	****	850	500	1020
Lowe	HF-250E	M	****	****	*****	1100	520	1320
Lowe	HF-250	M	***	***	***	1200	800	1440
Sangean	ATS-818C	L	***	***	****	150	150	180
Sangean	ATS-909	M	****	***	*****	270	120	170
Sony	ICF-2010	M	****	****	*****	350	-	420
Sony	ICF-SW55	M	****	***	****	350	260	420
Sony	ICF-77	L	****	****	****	470	360	565
Ten-Tec	RX320	C	***	***	***	300	300	360
Watkins	HF-1000	L	*****	*****	*****	3800	4500	4560
WinRadio	1500e	C	***	**	**	550	400	660
Yaesu	FRG-100B	L	****	****	****	590	450	710

KEY:
SEL = Selectivity. DR = Dynamic Range. OV = Overall Value. H = Hand-held, C = PC radio, S = Small, easily portable.
M = Medium, suitcase size. L = Large, table top use. * = Avoid ** = Poor *** = Fair **** = Good ***** = Excellent.
NOTES: Prices are approximate due to exchange rate fluctuations. Some models may be unavailable in certain markets.

Space Weather 1999

By Nancy Crooker, Patrick MacIntosh and Richard Thompson

Edited by Mike Bird

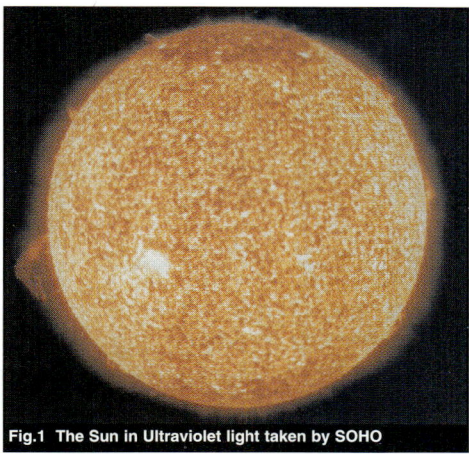

Fig.1 The Sun in Ultraviolet light taken by SOHO

RETROSPECTIVE

During the past year the Sun has been in the news many times. Back in April 1998, scientists from National Aeronautical Space Administration (NASA) and European Space Agency (ESA) announced they had discovered the Sun experiences tall, gyrating storms far larger than tornadoes on Earth. Apparently these storms appear near the Sun's north and south poles gusting up to 500,000 kilometres per second. In May scientists again working with NASA/ESA reported they had witnessed a solar flare producing seismic waves in the Sun's interior that closely resembled those created by earthquakes on our planet. According to their press release, researchers observed a flare generated solar quake that contained almost 40,000 times the energy released in the great earthquake that devastated San Francisco in 1906. The amount of energy released was enough to power the United States for 20 years at its current level of consumption, and was equivalent to an 11.3 magnitude earthquake.

June saw NASA/ESA scientists announce another solar happening. Apparently they had observed two comets plunging into the Sun's atmosphere in close succession. This unusual event on Earth's own star was followed by a possibly unrelated but dramatic ejection of solar gas and magnetic fields on the Sun's south-western limb. All of these observations came about due to the NASA/ESA SOHO spacecraft which was launched in December 1995. SOHO's payload includes 12 experiments which have helped the scientific community observe and understand the Sun in far greater detail than using ground-based observatories. Unfortunately, on 25th June NASA/ESA scientists suddenly lost contact with SOHO and have been trying in vain ever since to regain communication and control of the spacecraft.

While all of this is important and adds much to our knowledge of the Sun, the main story for readers of the WRTH is the march of Solar Cycle 23.

SOLAR ACTIVITY

This present cycle officially started in May 1996, since when it has slowly gathered momentum. Historically cycles have been measured by counting sunspots, those dark, cooler regions which appear in the Sun's photosphere. (See WRTH 1997)

Sunspots come and go on average every 11 years. When the sunspot count is low, the Sun produces less ultraviolet radiation which in turn makes for a weaker ionosphere. When the sunspot count is high, say around maximum, it produces more ultraviolet illumination which creates a stronger ionosphere. During the solar maximum period the ionosphere is capable of refracting shortwave signals up to and beyond 30 MHz.

The solar cycle contains four distinctive periods and they are Minimum, Rise (present), Maximum and Decline. The period between Minimum and Maximum is on average four and a half years, while the declining phase is longer, around six and a half years.

Presently we are witnessing the rise of Solar Cycle 23 (this is the 23rd cycle to be catalogued). The ionosphere is becoming stronger but the Sun is also exhibiting more activity such as solar flares and Coronal Mass Ejections (CMEs). These can lead to shortwave fadeouts and geomagnetic storms.

AUGUST 1997 - JULY 1998

The average smoothed sunspot number increased steadily throughout the year as shown in **Fig.2**. In August 1997 it was 25 and by December 1997 it had reached 39. Because the average sunspot number is computed over a 13 month period, the published figure is always 6 months in arrears. At the time of writing it is expected that the average sunspot number will have reached 85. This means that the sunspot number has not risen as fast as we first thought it would.

There were four periods during the past year when the Sun produced significant activity resulting in a disturbed geomagnetic field and/or a depressed ionosphere.

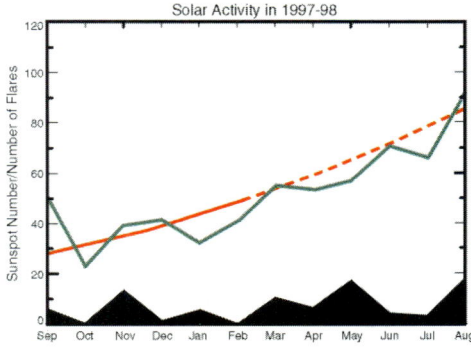

Fig.2 Solar Activity 1997 - 98

The first was in November 1997 when 10 M class and 3 X-class flares were observed. The magentic field ranged between active and storm thresholds on the 7th, 22nd and 23rd. This also resulted in a severely depressed ionosphere during the 8th, 23rd and 24th of November. During these three days up to 60% depressions to the ionosphere were observed.

March 1998 saw 10 low strength M class flares. The geomagnetic field reached storm thresholds on the 6th, 10th, 21st, 25th, 27th and 29th. The ionosphere became depressed on the 2nd, 6th, 11th and 16th.

During April, 4 M and 2 X-class flares were reported. The geomagnetic field was active or stormy on the 2nd, 3rd, 4th as well as the 29th and 30th. Up to 40% depressions to the ionosphere were observed on these days.

In May the Sun was once again very active. Between the 1st and the 10th 15 M and 2 X-class flares were reported, and on the 28th and 29th a further 2 M class events were also witnessed. Because of these flares the geomagnetic field became very disturbed on 3rd and 4th and then again on the 29th and 30th reaching major and minor storm thresholds respectively.

This catalogue of solar activity is proof of the fact that Solar Cycle 23 is progressing to the maximum phase. The comparison of indices for the last two years makes interesting reading.

SOLAR FLARES

Solar Flare Classification Energy (Watts/m^2) 1-8 Å
M-class flares $= 10^{-5} - 10^{-4}$
X-class flares $= > 10^{-4}$

September 1996 - August 1997
M class flares = 4   X-class flares = 0

September 1997 - August 1998
M class flares = 72   X-class flares = 11

Geomagnetic Activity (days disturbed)
Planetary Ap Index

0-7	=	Quiet
8-18	=	Unsettled
16-29	=	Active
30-49	=	Minor Storm
50-99	=	Major Storm
100-400	=	Severe Storm

	Sept 96 - Aug 97 (Final)	Sept 97 Aug 98 (Interim)
Quiet	202	180
Unsettled	114	123
Active	37	53
Minor Storm	11	13
Major Storm	1	6
Severe Storm	0	1

Month	Solar Flux Monthly Average	Sunspot Number Monthly Average	Sunspot Number Yearly Average	Ap Index (interim) Monthy Average	Flares >M1.0
Sept 97	96.2	51.3	28.4	9.8	6
Oct 97	85.0	23.3	31.9	10.5	0
Nov 97	99.5	39.3	35.1	10.5	13
Dec 97	98.8	41.5	39.1	4.5	1
Jan 98	93.5	32.3	43.8	7.5	5
Feb 98	93.5	40.7	48.9	7.9	0
Mar 98	109.3	54.8	53.7e	12.7	10
Apr 98	108.3	53.3	59.1e	9.8	6
May 98	106.6	56.9	65.5e	17.7	17
Jun 98	108.4	70.5	71.7e	9.8	4
Jul 98	114.0	66.2	78.6e	10.8	3
Aug 98	136.1	91.7	85.9e	18.2	18

Table 1. Monthly Values

THE STATE OF THE PRESENT SUNSPOT CYCLE

Each year we invite leading scientists working within the Spaceweather discipline to contribute their thoughts on the state of the science. This year we welcome back Patrick McIntosh. Patrick is one of the world's experts on solar features. Each day he scans the Sun interpreting its ever-changing state. He is also renowned for his sunspot group classification system which is used extensively. In the last few years, Patrick has run a solar advisory and warning service known as Heliosynoptics based in Boulder, Colorado.

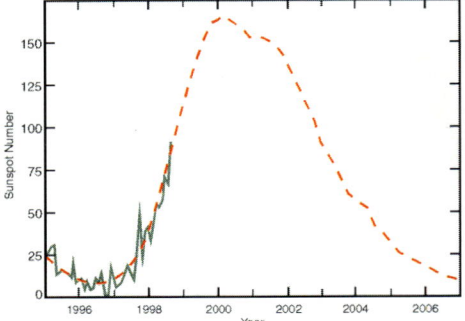

"The Sun is not the pure and featureless celestial object that Aristotle claimed it was. It has outbreaks of blemishes, sunspots, that are accompanied by violent eruptions that upset the Earth and interplanetary environments. The human race is more easily upset by solar eruptions now that we depend on satellite communications and other forms of high-technology. Even our changing climate may have a contribution coming from rising solar activity. It has become necessary to attempt a prediction of solar cycles and their eruptive events.

It is now two years since a minimum in sunspot activity (**Fig.3**) marked the boundary between one 11-year sunspot cycle and the next. This is the 23rd cycle since continuous sunspot observations began in 1749. The graph of the current cycle displays enough shape to compare with each of the earlier cycles. The fit of one curve to another tells how long until maximum activity and, perhaps, how high the cycle may be.

In early August 1998 the sunspot numbers increased after a 4-month pause in the rise of the cycle, doing so with the outbreak of a particularly large sunspot group with flare potential. The comparisons with earlier cycles suggests this is the beginning of the most rapid rate of rise. By November the maximum phase of Cycle 23 will begin and continue through 2002. The peak sunspot numbers should occur in the last quarter of 1999.

The amplitude of the sunspot maximum is more difficult to predict. The record of past cycles promised a reliable predictor in the observation that the odd-numbered cycles are often higher than the previous even-number cycle. This was true for 16 out of the 22 cycles, and all of the past 12 cycles. If this rule continues, it predicts Cycle 23 will be stronger than Cycle 22. The amplitude of Cycle 22 ranked third in the historic record; therefore, Cycle 23 should be extremely active. Persistence supports this prediction. Four of the past five cycles are the highest cycles on record.

The exceptions to the even-odd pairing of sunspot cycles warn that this pairing may be a coincidence. No one has offered a physical reason for this pairing, other

than noting that the magnetic poles of the Sun reverse every 11 years, making a full magnetic cycle a double sunspot cycle. That association still lacks any notion of a process that would relate the intensity of one cycle to the next. The recent pause in the rise of Cycle 23 is additional warning that the amplitude may be less than predicted by the even-odd paradigm.

Solar flares are recorded for the last five sunspot cycles. The times in the cycle when major flares first occur suggest how soon major events will occur again. The present cycle already produced large flares with serious high-energy particles in November 1997 and May 1998; however, these events pale in comparison with the greatest events of past cycles. The rapid rise expected in the next three months would bring this cycle to a level comparable to the point in Cycle 19 when the Sun produced an historic event. The great flare of February 1956 produced an x-ray burst so strong that electrical current was generated in the Earth's core, suddenly slowing Earth's rate of rotation by a fraction of a second! The comparisons between atomic clock and celestial determinations of time allowed detection of this minute, but awesome, effect. The 1956 event also produced a strong fluence of high-energy particles which would have been dangerous for both astronauts and spacecraft, had either existed at that time.

The most intense solar flares, particularly ones that generate high-energy protons, occur in sunspot groups with distinctive size and structure, making them more predictable than weaker flares. Unfortunately, this prediction is not possible until the sunspots form these distinctive structures. Often these appear only hours before the great flare occurs.

The regions which produce the most bothersome eruptive events, so-called super-regions, are relatively rare and they are not clustered around the time of sunspot cycle maximum. They can occur throughout the cycle, albeit most often in the last two years of the maximum phase. The number and magnitude of episodes of great activity differ from one cycle to another, and the rarity of these episodes makes research on their causes very slow. These sobering facts make it important to have vigilant patrols of the Sun and experienced observers who understand the distinctive characteristics of the most threatening of solar active regions.

The trouble with sunspots is they do not tell us about the other solar features that can create eruptions or change the solar output of electromagnetic radiation. Modern solar telescopes can record solar images in such a narrow range of wavelengths that the image represents light from a single atomic element. Images in the red light of hydrogen-alpha, the brightest hydrogen line in the visible solar spectrum, show a Sun with all its surface filled with amazing detail, see **Fig.4**. Activity is occurring in hydrogen whether there are sunspots or not, even in the period of sunspot cycle minimum.

The hydrogen-alpha (H-alpha) images have been taken regularly since 1908, and the modern flare patrol networks use this wavelength exclusively. These images show all solar flares in great detail, even those with very weak emissions. The images also reveal large-scale patterns of dark filaments and fine thread-like fibrils that permit the inference of magnetic-polarity boundaries over all the solar surface, right up to the polar regions of the sun. Maps of these boundaries have been constructed for each 27-day solar rotation for the past 34 years. Studies of solar global circulation are possible from these maps, and promise a physical understanding of solar activity that should improve solar predictions.

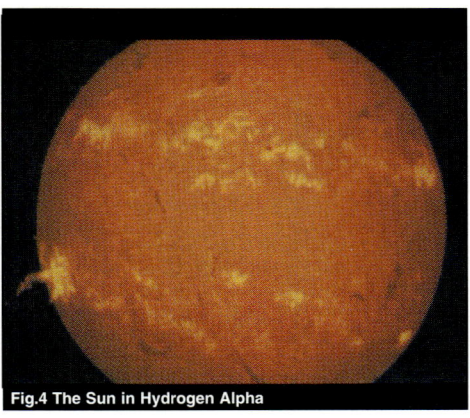

Fig.4 The Sun in Hydrogen Alpha

When the sites of eruptive activity are placed on these maps there are a number of relationships evident.

Some of the super-regions develop near large and growing coronal holes, dark openings in the solar corona first detected in X-ray and ultraviolet images obtained from space. Coronal holes are closely associated with the heliospheric current sheet. The surface manifestation of this sheet is a long magnetic-polarity boundary on global H-alpha maps. This boundary is uninterrupted in its convoluted path around the solar globe.

The most recent H-alpha maps show a rapid filling-in of quiet areas on the sun, replacing blank areas with active regions, filaments and moving patterns that suggest organized flows similar to "jet-streams" in Earth's atmosphere. We see, independent of sunspots, the increase in the solar cycle."

(Patrick MacIntosh)

SATELLITE PROBLEMS AHEAD?

Within the past year there have been warnings that the next solar maximum period might put some of the many satellites currently orbiting our planet at risk. All satellites operate within a hostile environment and are at the mercy of the Sun.

For example, in March and October 1989, the Sun became very active producing some of the largest solar flares ever witnessed. These flares in turn generated vast quantities of x-ray and proton radiation resulting in some of the biggest geomagnetic storms.

Throughout these two periods, shortwave communication was virtually useless, but on the positive side, we experienced some beautiful auroral displays. It was in this period that many satellites received enormous doses of radiation which in many cases caused considerable damage to their systems. For example, on 29th September 1989 the GOES 7 weather satellite received a dose of protons which degraded its circuitry as much as would normally have been expected in one year. Geomagnetic storms produce heat which causes the atmosphere to expand; this in turn creates drag and can lead satellites to descend off station to unexpected lower altitudes.

If these problems were not enough, a new self-inflicted complication appears to be about to make its presence felt. In May 1998, Professor David Thiele who is Director of the Radio Science Laboratory at Brisbane's Griffiths University, warned that satellite circuitry was becoming so small that it posed a risk in a high solar radiation environment.

"Electronic circuits which are not radiation hardened are more susceptible to the radiation flux form the sun. This is exacerbated by the shrinking unit size of the active devices in integrated circuits, from approximately 10 microns ten years ago, to less than 1 micron in 1998. Satellites launched since 1990 have not been subjected to the massive solar flux predicted to occur in the years 2000-2001, and so, at least to a certain extent, these effects cannot be predicted. The US government, however, is predicting a number of dire consequences for the years of solar maximum. These include:

* The possible downtime of the GPS network for accurate navigation is likely to be 20%.

* The likelihood of large numbers of communications satellites failing. Currently communications companies are privately taking stock of how to minimise the downtime resulting from the permanent disablement of satellites in their constellations.

It is essential that governments and private industry make some plans to accommodate the loss of satellite communications technology and navigation technology.....

There is also an urgent need to pay more attention to "space weather", because it is possible to change the configuration of satellite systems to minimise risk during a massive solar storm.

It is certainly my view that the onset of solar maximum is likely to have far more impact on Australian society than the so-called millenium bug which has provided a significant revenue stream for software programmers."

(David Thiele)

CORONAL MASS EJECTIONS

Over the past few years, this article has explored the makeup of the Sun and its major features (see WRTH 1997 and 1998).

This year, with the help of Nancy Crooker, we look at Coronal Mass Ejections or CME's as they're known. Nancy Crooker is a research professor at the Centre for Space Physics at Boston University, USA. She has authored nearly 100 papers for professional journals on the solar wind and magnetospheric research. Recently Nancy co-edited a monograph for the American Geophysical Union entitled 'Coronal Mass Ejections'.

Fig.5 The Sun's Corona in all it's glory

"Should a modern Joshua stop the moon in its tracks during a total solar eclipse so that for hours on end we could watch the hazy corona surrounding the sun, we would see an amazing phenomenon called a 'coronal mass ejection' - known as a CME for short (see **Fig.5**). We would be able to see a huge bubble of coronal material form and then leave the sun, heading out into space. In fact we would have witnessed the birth of a space storm.

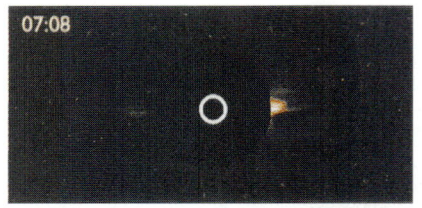

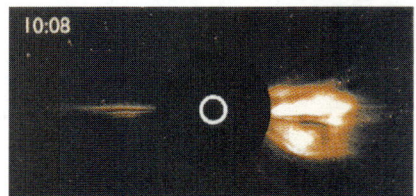

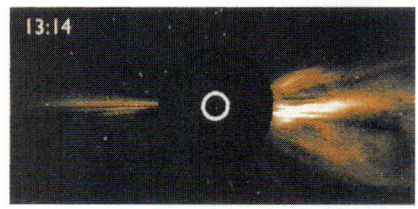

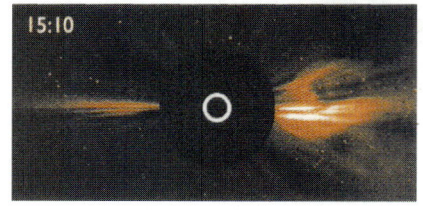

Fig.6 Outburst from the Sun - a coronal mass ejection

CMEs were recognized as a common solar phenomenon in the 1970s when a technological Joshua called a 'coronagraph' flying on an orbiting laboratory called 'Skylab' held an artificial moon fixed in front of the Sun and inaugurated an era of routine CME observations. A coronagraph (see **Fig.6**) simulates a total solar eclipse by blocking the bright light from the Sun with a disk. To see CMEs, a coronagraph must be above the atmosphere, where the sky is black, as it is at ground level during a solar eclipse. Blue sky is too bright to see the solar corona through it. What makes the sky bright and blue is sunlight scattered in all directions by the molecules that make up the atmosphere, with most scattering occurring at short wavelengths at the blue end of the visible spectrum. During a solar eclipse, the moon's shadow blocks not only the bright solar disk but also the sunlight that shines on the atmosphere, so that no scattering occurs. A coronagraph simulates the condition of no sunlight from the solar disk, but not the condition of no skylight. For that it must be placed above the atmosphere.

Since routine CME observations were inaugurated, it has taken scientists years more of research to establish that CMEs are space storms and to significantly improve means of predicting them. Understanding and predicting space storms is important to society because of our

rapidly increasing dependence upon technology that space storms affect. Space storms can cripple spacecraft, interrupt communications systems, and cause power failures. We now understand that the key storm ingredient in a CME is its magnetic field, which is coiled inside it. To appreciate why coiled magnetic fields bring stormy space weather, one needs to know something about fair weather in space, namely, the background flow called the "solar wind" and the configuration of its imbedded magnetic fields.

The story of the solar wind is fascinating in its own right. It begins with the fact that the solar corona, which is the Sun's upper atmosphere, is about 100 times hotter than the Sun's surface. In spite of this unexpected temperature difference, scientists in the 1950s, prior to the Space Age, believed that the solar atmosphere behaved in ways similar to Earth's atmosphere. Attempts to construct quantitative models incorporating the high coronal temperature, however, were met with seemingly unphysical requirements. In 1958, Eugene Parker found a solution that demanded a continually expanding solar atmosphere. His critics thought the solution was absurd. How could the Sun keep supplying material for its atmosphere? Undaunted, Parker took the solution at face value and predicted that the expanding corona would form a solar wind with speeds around 400 km/s. That same year, scientist Konstantin Gringauz convinced the Soviet authorities to add his transmitter to the Sputnik payload and created the beep heard 'round the world. The Space Age had begun. One year later, Gringauz' instrument on Lunik 2, the first spacecraft to travel beyond Earth's magnetic shield, the "magnetosphere," detected the solar wind predicted by Parker.

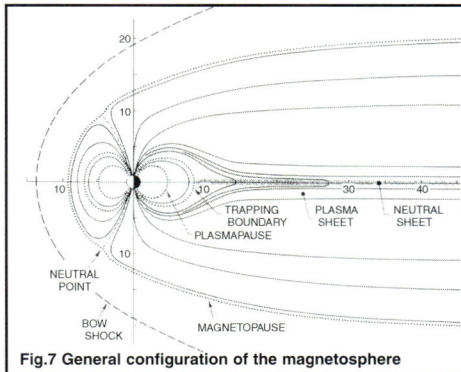

Fig.7 General configuration of the magnetosphere

magnetosphere is a bullet-shaped obstacle in the solar wind created by Earth's magnetic field, see **Fig.7**. At the nose of the bullet, where it faces into the wind, Earth's magnetic field points northward, perpendicular to the spiral field. Fields perpendicular to each other, for the most part, slip past each other. Antiparallel fields, on the other hand, connect when they meet. Thus any southward pointing fields in the solar wind link with Earth's field, and, like newly connected pipes, allow solar wind particles and energy to pour into the magnetosphere. The magnetosphere's response is a geomagnetic storm, complete with auroral displays and potential problems for spacecraft, power grids, and radio transmission. What heralds bad space weather for Earth, then, are southward fields in the solar wind; and the stronger they are, the worse the storm.

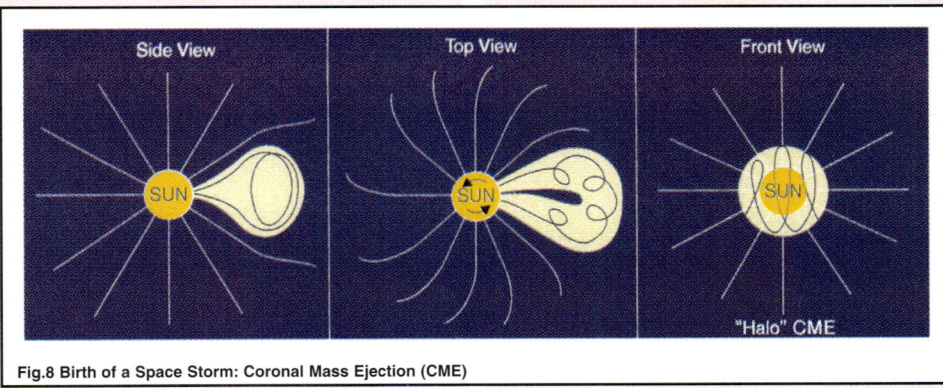

Fig.8 Birth of a Space Storm: Coronal Mass Ejection (CME)

We now know that interplanetary space is filled with the Sun's expanding atmosphere, the solar wind, which sets the fair weather backdrop for CMEs. Unlike weather on Earth, however, the important parameter is not the wind itself but its imbedded magnetic field, which comes from the sun. Parker recognized that the solar wind would draw the solar field out into space as if it were frozen to the flow. Further, he deduced that in the ecliptic plane, that is, the plane in which the planets rotate about the sun, the Sun's rotation about its own axis would cause the field lines to radiate away from the Sun in a spiral pattern like water from a rotating garden sprinkler. Subsequent spacecraft observations have proven Parker correct.

For Earth, the gently spiralling magnetic fields carried by the solar wind herald fair weather because Earth's magnetic shield, the magnetosphere, deflects them. The

What makes CMEs space storms (**Fig.8**), then, is their strong, coiled fields, which nearly guarantee some portions pointing southward. Moreover, for those CMEs that travel faster than the background wind, the background fields ahead of them become strong through compression, strengthening any southward fluctuations there.

The contrast between fair weather magnetic fields and CME fields is illustrated from three viewpoints in the schematic figure. The side view on the left is how CMEs were first observed, as a bubble coming off the limb of the sun. The solid lines indicate invisible magnetic field lines, which have been deduced from theory and from measurements in space. Those coming straight out of the Sun are the fair weather fields, while the CME shows an example of a storm field line. The storm line is coiled like a spring with its ends connected to the sun. The view

looks down the axis of the cylindrical space formed by the coil. Nested within are other coiled field lines, not shown. From this point of view, if Earth is off to the right, in the ecliptic plane, which cuts across the middle of the diagram, it is clear that the CME will bring southward (downward) fields either at the leading or trailing edge of the coil, depending upon the polarity of the coiled field line. In contrast, a fair weather field line in the ecliptic plane, such as the one off to the left, has no southward component.

The middle panel of the figure shows a view from above the solar system, looking down onto the ecliptic plane. The background field lines curve slightly to suggest the spiral pattern into which they will evolve, although this close to the Sun the curvature would hardly be noticeable in a picture drawn to scale. The CME field line from this perspective shows its coil configuration best. The coil is bent into a loop connected to the Sun at both ends. In three dimensions, the loop fills a bubble shaped like a light bulb.

The front view in the third panel illustrates a CME headed toward the observer. As in the first panel, the potential for southward fields is apparent here, too. Viewed from the front, a CME in a coronagraph looks like a uniform halo around the disk blocking the sun, hence the name, "halo CME." Halo CMEs have been difficult to detect because they are not as bright as CMEs viewed from the side. Recent technological advances, however, have made them routinely detectable with Guenter Brueckner's coronagraph on the SOHO spacecraft.

This recently developed ability to detect halo CMEs will be a boon for space weather predictions. In a retrospective study of data from December 1996 to June 1997, nine halo CMEs were detected, and all nine produced geomagnetic storms. Only three additional storms occurred which could not be associated with halo CMEs. This nearly one-to-one correspondence is remarkable in view of past prediction capabilities. Traditionally solar flares have been used to predict storms, and the failure rate has been high. Flares are brightening events on the Sun primarily at wavelengths shorter than visible light. Viewed in x-rays, for example, flares make the Sun look like a constantly sparkling jewel because they occur so frequently. The brightest flares have been used for storm predictions, and these often do accompany CMEs, hence their predictive capability. But many CMEs occur with no apparent associated flares or with flares so weak that they go undetected in standard monitoring techniques. Now that we can see the CMEs themselves, there is no need to depend upon such an unreliable predictor as a flare.

From an ethereal sight to a source of technological problems, CMEs continue to fascinate scientists. How they form, why they occur, and what makes some of them very fast are walls in our understanding that have yet to come tumbling down."

(Nancy Crooker)

An interesting story, and one I am sure we will hear more of in the coming years. Astute readers will note subtle differences in the contributors' interpretations of solar phenomena. Solar geophysics is a young, multifaceted science subject to individual perspective.

SOLAR WORKSHOP

As last year's article went to press, the world's leading solar cycle forecasters were invited by NASA to meet for the second year running to consider forecasts for Solar Cycle 23. The venue was the Sacramento Peak Solar Observatory in New Mexico. NASA and its space

program have a vested interest in predicting the present cycle. A large amplitude cycle can produce additional heating of the outer atmosphere causing it to expand. This can increase frictional drag on satellites and spacecraft. During the workshop, a few more published forecasts were compared with the findings of the previous year's gathering. After two weeks of deliberations, the panel of twelve experts saw no reason to change their published predictions for Solar Cycle 23. They are:

Solar Maximum Window

Jun 1999	**MARCH 2000**		Jan 2001
Parameter	Low end	Maximum	High end
Smoothed monthly sunspot number	130	160	190

SOLAR CYCLE 23

The primary guide to where we are in any Solar Cycle has historically been measured by counting sunspots. Richard Thompson is the solar forecaster at IPS Radio and Space Services and also one of the twelve who meet at the NASA workshops. Richard earned a considerable reputation during Solar Cycle 22 for his accurate forecasts.

"According to the traditional measure Solar Cycle 23 began in May 1996 (see **Fig.9**) when the smoothed sunspot number reached its lowest value (this date is called 'solar minimum'). The smoothed sunspot number then began to rise, although very slowly at first, and the new cycle was officially underway. However, the exact timing of solar minimum is very sensitive to any small burst of activity on the Sun which can shift it

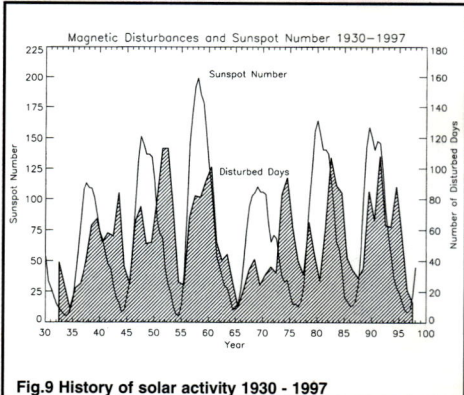

Magnetic Disturbances and Sunspot Number 1930-1997

Fig.9 History of solar activity 1930 - 1997

considerably. This may be the case for the onset of Cycle 23 because there are good reasons to believe that the actual minimum took place later in 1996. We can distinguish sunspot regions of the old and new solar cycles and we found that the Cycle 23 regions only became common right at the end of 1996, more than six months after the official start of the cycle. Also, the complete absence of sunspots is a good indicator of the low point of the cycle. For the last minimum, we experienced the longest sequence of such days days late in 1996, especially in September/October.

So if the real Cycle 23 was late for its official start, in more than two years since this time, the cycle has continued to promise much but deliver little. Predictions

of the amplitude of the cycle suggest that it will be of large amplitude reaching a peak smoothed sunspot number similar to that of the last two cycles (both of which were exceptional in comparison with historical cycles). Predictions suggest that this peak will take place in early-to-mid 2000 - yet another 'problem' for the new millennium.

By the end of 1996, the smoothed sunspot number had struggled to only a value of 10, up from its low of 8 in May 1996. By December 1997, the cycle reached 39, still unimpressive for a cycle almost 2 years old. Into 1998 the sunspot number showed signs of moving upwards much more rapidly, especially from March onwards. This appears to mark the rapid onset phase of the cycle when the trend becomes steeply upward.

The rise of Cycle 23 can be compared with that of other cycles. The smoothed sunspot number for Cycle 23 has risen by a total of 36 in the 20 months of available data since May 1996. In this respect Cycle 23 ranks very much in the middle of observed cycles - well behind Cycle 22 which at the same stage had sprinted upwards with a rise of 72. Naturally, many of these past cycles reached nowhere near the large peak amplitude predicted for Cycle 23; but this does not necessarily indicate that the cycle will not reach its predicted amplitude. For example, the rise of Cycle 23 is just behind that of Cycle 18 which went on to a peak of 152 in smoothed sunspot number.

If October 1996 is taken as the 'true' starting point of the Cycle then the rise of Cycle 23 since then is acutally one of the fastest and is certainly consistent with a large ultimate amplitude.

So what is the outlook for the next few years? Does the relatively slow rise of Cycle 23 suggest that it will not reach the heights predicted for it? At the time of writing it is too early to really answer. It is important to recall that many people predicted a huge cycle from the rise of Cycle 22. Cycle 22 was impressive in amplitude but nowhere near as much as its rise suggested.

My opinion is that the onset of Cycle 23 was just delayed and that it will still reach its predicted peak sunspot number of around 160. However, it would not be surprising if this peak is delayed until late 2000."

(Richard Thompson)

1999 - THE YEAR AHEAD

Next year will be an interesting time if not an exciting one for the shortwave listener. Solar Cycle 23 will continue on its merry way and hopefully the sunspot number will climb to predicted levels.

At the time of writing (August 1998) the Sun has once again woken up and given us a true tast of what is to come. In this period, 14 M and 4 X-class flares were observed. An X1 flare which was witnessed on 24 August produced vast quantities of X-ray and proton emissions that kicked off a polar cap absorption event a day or so later lasting for some four days. On the 26th, a geomagnetic storm started which continued until around midday on the 28th. During this period the magnetic field disturbance indicator the Planetary A (Ap) index reached 112 which corresponds to a Severe Storm, the first we have encountered in this cycle.

As the sunspot number increases so too will the ionosphere strengthen. IPS Radio and Space Services have computed that during the past year the available shortwave spectrum increased on average around 16% and during the next year it will increase further by another 23.5%. What this means is that the higher shortwave bands such as 15, 17, 21 and 26 Mhz will open earlier and close later in the day as well as being generally stronger. Broadcasters now have an opportunity to start using the high end of the shortwave spectrum which in turn will relieve much of the congestion in the lower bands. As always, the most likely period for disturbances to shortwave propagation is around the equinox period, when the solar wind and geomagnetic field coupling is at its strongest.

Good listening.

ACKNOWLEDGMENT

My thanks, as always, to IPS Radio & Space Services for their continued support of this article, in particular, Dr Richard Thompson, Garth Patterson and Patrick Phelan.

For further information contact

IPS Radio and Space Services

Street address:	Postal address:
Level 10	*PO Box 1386*
477 Pitt Street	*Haymarket*
Sydney	*NSW 1240*
NSW 2000	*Australia*
Australia	

General Enquiries
Tel: +61 2 9213 8000 Fax: +61 2 9213 8060
Duty Forecaster
Tel: +61 2 9213 8010
Prediction Services
Tel: +61 2 9213 8011
Daily Solar & Geophysical Report
Tel: +61 2 9213 8012

Internet Homepage
http://www.ips.gov.au/office@ips.gov.au
E-mail
office@ips.gov.au

Space Environment Centre
325 Broadway
Boulder
Colorado 80303
USA

General Enquiries
Tel: +1 303 497 7127
Space Weather Operations
Tel: +1 303 497 3171
Fax: +1 303 497 3137
Internet http://www.sec.noa.gov

Sunspots and the Solar Cycle sponsored by NOAA-Marshall Space Science Laboratory.
The Web site has an appropriate URL:
http://www.sunspotcycle.com/

PREDICTED SUNSPOT NUMBERS

Year	Jan	Feb	Mar	Apr	May	Jun	Jul	Aug	Sep	Oct	Nov	Dec
1999	118.9	124.8	129.9	134.3	139.4	144.6	148.4	151.6	155.7	159.2	161.3	161.9
2000	162.4	164.7	165.0	163.7	162.7	161.3	160.2	159.5	157.9	155.6	153.1	152.6

(Issued by IPS September 1998)

HF BROADCAST RECEPTION CONDITIONS EXPECTED DURING 1999

Prepared by George Jacobs, MSEE, Consulting Broadcast Engineer
P.O. Box 1714, Silver Spring, Maryland, USA 20915
e-mail <gja@gjainc.com>
Web Site <http://www.gjainc.com>

1999 - A GREAT YEAR!

Solar activity, upon which the propagation and reception of high frequency (HF) or shortwave broadcasts depend, is expected to reach peak intensity during 1999, at a level much higher than has been observed during the past ten years. The present 11- year cycle of sunspots is the 23rd since daily telescopic observations of the sun began late during the 18th century, since when sunspots have been used as a measure of solar activity.

Cycle 23 began during 1996, rose slowly during 1997, and increased considerably during 1998. It is expected to begin 1999 with a count greater than 110, and rise to a peak intensity whihc is likely to exceed 155 by year's end. This would be the highest level of solar activity to occur since 1989.

GOOD NEWS

This is good news for broadcast listeners on the HF bands, since it portends exceptionally good reception conditions during 1999. The greater the number of sunspots, the stronger is the ionosphere, that electrified region in the earth's upper atmosphere which reflects HF radio waves over great distances. For a more detailed discussion of the solar activity expected during 1999, refer to Mike Bird's *Space Weather* article which appears ealier in this edition.

Improved reception conditions are also expected during 1999 as a result of the continuing success of the High Frequency Coordinating Committee (HFCC). This international organization, comprised of approximately 50 HF broadcasting organizations, is responsible for coordinating more than 70% of the frequencies used by the world's shortwave broadcasting stations. This means less interference between cooperating stations, and clearer reception.

The combination of increased solar activity, leading to improved propagation conditions and the continued success of the HFCC should result in stronger, clearer reception on the HF bands during 1999.

INTERNET AND WORLDWIDE WEB

High Frequency broadcasting owes much of its continued popularity in this high tech era to its directness, immediacy, intimacy, free access, and its relatively low cost for transmission and reception. It is expected to continue to enjoy its unique and important role as a worldwide electronic market place for ideas, information and entertainment, to be freely admitted into one's home without an intermediate "gatekeeper" for well into the 21st Century.

But, a complementary medium for International Broadcasting is developing rapidly on the INTERNET. Despite the fact that a relatively high cost computer is required, and it must be linked to the INTERNET for a fee through a 'gatekeeper' or ISP, webcasting is becoming very popular. It is estimated that all major International Broadcasting stations from over 50 countries now place their broadcasts on the web, either

Fig.1 A sample page from the WRN website

in real-time on demand or in archives.

A 'super' webcasting site has been developed on the Internet by the World Radio Network (WRN) in London. WRN, formed in 1992 by a small group of BBC engineers, has developed an extensive worldwide international broadcast delivery system using satellites, terrestrial AM, FM, Cable and HF affiliate stations, as well as the Internet. Their website address (URL) is: <http://www.wrn.org> and the 24-hour site contains real-time on-demand audio broadcasts from thirty international broadcast stations, including the VOA, Deutsche Welle, Vatican Radio, Voice of Russia, Radio Netherlands, KBS Radio Korea, etc.

The site also maintains archives for up to three months of Kim Elliot's Communications World, Jonathan Marks' Media Network and Glenn Hauser's World of Radio.

TRANSMITTER DOCUMENTATION PROJECT

Over the past decade Ludo Maes of Belgium has assembled a wealth of information concerning HF broadcast transmitters in his Transmitter Documentation Project (TDP). He has collected technical information on over 100 different types of transmitters used for HF broadcasting, and he has a database locating almost every HF broadcast transmitter in use over the past 50 years! If it is related to shortwave broadcast transmitters, you will probably find the information on the TDP website at: <http://www.ping.be/tdp>.

Links are also maintained to HF broadcast station websites, to websites of manufacturers of shortwave transmitters, antennas and tubes, as well as links to HF stations that offer airtime for sale. There are also links to information about digital HF radio, and to several HF radio DX clubs. For readers of WRTH who are connected to the Internet, another website address (URL) that contains not only information about HF broadcasting, but also a wealth of data about propagation, the ionosphere, geomagnetism, sunspots

and solar activity provided by some of the world's leading research institutes is: <http://www.gjainc.com>.

To reach these websites requires a 486 PC, but preferably a Pentium, a connection to the Internet through a local Internet provider, appropriate software, a modem rated at 28.8kB, preferably 56kB, and Real Audio or equivalent sound file capability.

In conclusion, the last year of this millenium looks like being a great year for all those who are interested in broadcast reception on the HF bands, or indeed on the Internet.

George Jacobs (right) meeting recently in Brussels with Ludo Maes, creator of the Transmitter Documentation Project (TDP) and associated website, which contains data on almost every HF broadcast transmitter in the world.

Photo by Bea

1999 Reception Conditions

The following is a general summary of reception conditions expected in each high frequency broadcasting band during 1999.

26 MHz Band (11 Metres):

This band is usually completely dead for skywave propagation during more than half of a ten year sunspot cycle, but it suddenly springs to life during years of peak intensity. With the high level of sunspot counts expected during 1999, an increasing number of stations should make use of this band. Long distance reception, up to several thousands of miles, should be best during the daylight hours of the late fall, winter and early spring seasons. Signals will be strongest over north-south paths between the hemispheres during the equinoctial seasons, and strongest on east-west paths during the winter. Reception will be better in southern and tropical areas than at higher latitudes. During the summer months, very good reception should be possible over shorter distances, on the order of 800-1,200 miles, or 1,300-2,000 kilometers. Reception is expected to be virtually free of interference.

21 MHz Band (13 Metres):

This is another daytime   band which should see a considerable increase in the number of stations that will use it during 1999. It will be suitable for long distance reception over several thousands of miles from just after sunrise, through most of the day, to sundown. Conditions should be best during the fall, winter and spring seasons. During the summer months, reception should be good on north-south paths, but less so in easterly and westerly directions. Since most major broadcasters are expected to make increased use of this band during 1999, congestion and interference levels are likely to be a problem at times.

17 and 19 MHz Bands (16 Metres):

This is primarily another daytime band, with excellent reception expected throughout the year from shortly after sunrise, until sunset.. Stations should be heard as close as 1,000 miles (1,750 km), to as far away as several thousand miles. Reception of stations located in tropical or equatorial areas may also be possible well into the hours of darkness. Most major broadcasters are expected to make use of the 17 MHz band, increasing the possibility of congestion and interference. The new

19 MHz, used only by a single broadcaster in years past, is expected to see a considerable increase in occupancy during 1999.

15 MHz Band (19 Metres):

This is likely to be the most popular band for international broadcasting during 1999. Conditions are expected to be excellent for both medium and long distances of approximately 800 to several thousands of miles. Exceptionally strong signals during all seasons of the year are expected from shortly after sunrise, through the daylight hours, and well into the evening. During all but the winter months, excellent reception should also be possible during the hours of darkness, particularly from stations broadcasting from tropical and equatorial regions. Exceptional long distance reception, from stations up to half-way around the world, should be possible for an hour or two after local sunrise and again during the late afternoon and early evening hours. At times, strong signals from stations as near as several hundred miles to as far as the opposite side of the globe may occur simultaneously. With such favorable worldwide propagation conditions, this band very often will be congested and the interference level may be high, particularly during the late afternoon and early evening.

13 MHz Band (22 Metres):

Increased use is expected to be made of this band during 1999. Propagation conditions are similar to the 15MHz Band. The band should provide excellent reception during all seasons from shortly after sunrise to well past sundown. During the daylight hours reception should be best between distances of approximately 700 and 2,400 miles (1,200 to 4,000 km) and considerably beyond this range during the winter season. During the late afternoon and early evening hours reception should favor longer distances between approximately 1,000 miles (1,800 km) and up to half-way round the world. During the spring and summer months, expect reception of long distance stations well into the hours of darkness. Expect a high level of congestion, particularly during the late afternoon and early evening hours.

11 MHz Band (25 Metres):

This is expected to be an excellent band for medium distance reception during the daylight hours of all seasons, between approximately 500 and 1,500 miles

(800 to 2,400km). Longer distance reception, from stations up to several thousands of miles, should be possible for an hour or two after local sunrise, and again during the late afternoon and early evening. During the spring and summer months, medium and long distance reception should also be quite good throughout most of the hours of darkness. Heavy congestion is expected during the late afternoon and early evening, with considerable use being made of this band by both international and domestic broadcasters.

9 MHz Band (31 Metres):

This is a popular band throughout the entire solar cycle for both international and domestic HF broadcasting. Medium distance daytime reception ranges between approximately 400 and 1,200 miles (700 to 2,000 km), during all seasons. During the spring, summer and fall months longer distance reception, up to several thousands of miles, should also be possible during the hours of darkness, and until two to three hours after local sunrise. With increasing sunspot numbers this is also expected to be a good band for nighttime long distance reception during the winter months. A high level of congestion is likely, particularly during the late afternoon and early evening hours.

7 and 6 MHz Bands (41 and 49 Metres):

Both bands are expected to continue to be very popular throughout 1999 for daytime domestic broadcasting, and for short to medium distance international broadcasting. Excellent reception conditions should be possible for stations up to approximately 750 miles away (1,200km). As night time approaches, and through the hours of darkness until sunrise, reception should improve over increasingly greater distances up to several thousands of miles. High static levels during the late spring and summer months may often make long distance reception difficult at times. Since propagation conditions in these bands do not vary much throughout a solar cycle they are heavily used by international and domestic HF broadcast stations. A high level of congestion and interference is likely, particularly during the late afternoon and evening hours. Although the allocated 7 MHz band is not permitted in the Western Hemisphere, an increasing number of stations are expected to operate on a non-interference basis adjacent to it between 7.3 and 7.6MHz.

5, 4, 3 and 2MHz Bands (60, 75, 90 and 120 Metres)

All except the 4MHz band are used primarily in designated tropical areas for domestic broadcasting. The entire 4MHz band (3.9 to 4MHz) is available for domestic broadcasting in Asia, and the portion between 3.95 and 4MHz is available for use throughout Europe. These bands are not noticeably affected by changes in sunspot count, and they retain their propagational characteristics throughout a solar cycle. During the hours of daylight, reception should be possible over a range extending out to approximately 500 miles (800 km). After sunset, during the hours of darkness and the sunrise period, the range of reception should increase to a thousand miles or more, with the more powerful stations heard over greater distances. Very high static levels are likely to make reception difficult at times, particularly during the late spring, summer and early fall months. Because of nighttime congestion in the 6 and 7MHz bands, an increasing number of international stations are expected to use frequencies in or adjacent to the 5,4,3 and 2MHz bands on a non-interference basis during 1999.

For specific times when reception of HF stations is expected to be optimum on various bands during 1999, refer to the table of 'Most Suitable Megahertz Broadcasting Bands 1999' overleaf.

WARC-92 bands are to be allocated officially for use by HF broadcasting stations in 2007. Meanwhile an increasing number of stations are expected to use these bands in 1999 on a de facto non-interference basis, as well as other out-of-band frequencies on an non-interference basis adjacent to allocated HF broadcast bands.

HIGH FREQUENCY BROADCASTING BANDS

kHz	MHz		Metre Band
2,300-2,495	2		120 *
3,200-3,400	3		90 *
3,900-3,950	4		75 **#
3,950-4,000	4		75***#
4,750-5,060	5		60 *
5,730-5,900	6	NIB	49 @
5,900-5,950	6	WARC-92	49
5,950-6,200	6		49
6,200-6,295	6	NIB	49 @
6,890-6,990	7	NIB	41 @
7,100-7,300	7		41 #
7,300-7,350	7	RC-92	41
7,350-7,600	7	NIB	41 @
9,250-9,400	9	NIB	31 @
9,400-9,500	9	WARC-92	31
9,500-9,900	9		31
11,500-11,600	11	NIB	25 @
11,600-11,650	11	WARC-92	25
11,650-12,050	11		25
12,050-12,100	11	WARC-92	25
12,100-12,160	11	NIB	25@
13,570-13,600	13	WARC-92	22
13,600-13,800	13		22
13,800-13,870	13	WARC-92	22
15,030-15,100	15	NIB	19
15,100-15,600	15		19
15,600- 15,800	15	WARC-92	19
17,480-17,550	17	WARC-92	16
17,550-17,900	17		16
18,900-19,020	19	WARC-92	15
21,450-21850	21		13
25,670-26,100		25	11

* = Tropical bands, for broadcasting use only in designated tropical areas. ** = Regional band, used for broadcasting only in Asia. *** = Regional band, used for broadcasting in Asia and Europe. # = Not allocated for broadcasting in the Western Hemisphere.@ = Out-of-Band usage on a non-interference basis. All other bands are on a worldwide basis.

ABOUT THE AUTHOR

George Jacobs is truly a legend In the field of International Broadcasting. As an engineer, diplomat and journalist with a fierce belief in the free flow of information, his professional career spans a period of 58 years. He is the Dean of Contributing Editors to the WRTH, and this is the 37th consecutive year that his articles have appeared in this publication. He has received numerous awards, honors and Fellowships from his peers for his accomplishments and in 1997 the National Association of Broadcasters (USA) bestowed upon him its prestigious Lifetime Achievement Radio Engineering. His hobby is amateur radio and he is well known throughout the world by his distinctive call sign W3ASK.

Table of the Most Suitable Megahertz Broadcasting Bands 1999

Prepared by George Jacobs, MSEE, Consulting Broadcast Engineer
PO Box 1714, Silver Spring, Maryland, USA 20915
E-mail: gja@gjainc.com

TRANSMITTING STATION LOCATION

LISTENER'S AREA	LOCAL TIME	APPROX. UTC TIME	JAN./FEB. & NOV./DEC.								MAR./APR. & SEPT./OCT.								MAY–AUGUST							
			EUR/NAF	N.AM(E)	N.AM(W)	C/S.AM	C/S.AF	ME/S.AS	E.AS	AUS/NZ	EUR/NAF	N.AM(E)	N.AM(W)	C/S.AM	C/S.AF	ME/S.AS	E.AS	AUS/NZ	EUR/NAF	N.AM(E)	N.AM(W)	C/S.AM	C/S.AF	ME/S.AS	E.AS	AUS/NZ
EUROPE AND NORTH AFRICA	0000-0400	2300-0300	6	9	9	9	9	9	9	9	6	9	9	11	9	9	9	9	9	11	11	11	9	15	15	11
	0400-0800	0300-0700	6	9	9	9	9	9	9	9	6	9	9	11	11	9	11	11	9	11	11	11	11	11	15	9
	0800-1200	0700-1100	11	6	9	9	17	15	17	17	11	9	11	11	17	15	17	17	11	15	11	11	17	17	17	17
	1200-1600	1100-1500	11	17	9	17	17	17	17	17	11	15	11	17	21	15	17	17	11	15	15	17	17	15	17	17
	1600-2000	1500-1900	15	21	17	17	21	15	11	9	15	17	17	17	21	15	11	11	15	17	17	15	21	15	15	9
	2000-2400	1900-2300	6	11	15	11	11	7	9	9	9	15	15	15	11	9	9	9	9	15	15	15	15	9	11	9
NORTH AMERICA (EAST)	10PM-2AM	0300-0700	6	6	9	9	9	9	11	11	9	6	11	9	9	9	11	11	9	9	11	9	9	9	11	15
	2AM-6AM	0700-1100	9	6	9	9	11	9	9	9	11	6	9	6	9	11	11	11	11	6	9	9	11	11	11	11
	6AM-10AM	1100-1500	15	15	9	11	17	15	9	9	15	15	9	11	15	15	9	11	15	15	9	11	17	15	11	9
	10AM-2PM	1500-1900	21	15	15	17	21	17	11	11	21	15	15	17	21	17	9	15	17	15	15	17	21	15	15	15
	2PM-6PM	1900-2300	15	11	15	17	15	15	9	11	15	11	15	15	15	15	9	11	15	15	15	15	15	15	15	15
	6PM-10PM	2300-0300	9	9	11	9	11	9	15	15	9	6	15	9	11	9	17	17	11	9	15	9	11	11	17	17
NORTH AMERICA (WEST)	12M-4AM	0800-1200	9	6	9	6	9	11	9	11	9	6	9	11	9	9	9	11	11	9	9	6	9	11	11	11
	4AM-8AM	1200-1600	11	15	9	11	11	9	11	11	11	11	6	11	15	11	9	11	11	11	6	11	11	11	11	11
	4AM-12N	1600-2000	17	17	9	15	17	15	11	11	17	17	9	15	17	15	11	11	15	15	9	15	15	17	15	11
	12N-4PM	2000-2400	11	15	15	17	15	15	15	17	11	15	15	17	15	15	15	17	17	15	15	17	15	15	15	17
	4PM-8PM	0000-0400	9	11	15	11	11	15	17	17	9	15	15	11	11	15	17	17	11	15	15	11	11	15	15	17
	8PM-12M	0400-0800	6	9	9	9	9	11	15	15	9	9	9	9	9	11	15	15	11	9	11	9	9	11	15	15
CENTRAL AND SOUTH AMERICA	12M-4AM	0500-0900	9	9	11	6	9	9	11	15	9	11	11	6	9	9	11	15	11	11	11	6	9	11	11	15
	4AM-8AM	0900-1300	11	11	11	9	11	9	11	11	11	11	11	9	11	11	9	11	11	9	9	9	11	11	11	11
	8AM-12N	1300-1700	17	17	11	15	17	17	11	11	17	17	11	15	17	17	11	11	17	15	11	15	17	17	11	11
	12N-4PM	1700-2100	17	17	15	15	17	17	11	11	17	17	15	15	17	17	15	11	17	17	15	15	17	15	15	11
	4PM-8PM	2100-0100	15	15	17	15	15	15	15	17	15	15	17	15	15	15	15	11	15	15	17	15	15	15	15	11
	8PM-12M	0100-0500	9	11	15	9	11	11	15	17	11	11	15	6	11	11	17	17	11	15	15	6	11	15	15	17
CENTRAL AND SOUTH AFRICA	0000-0400	2200-0200	9	11	15	11	7	9	11	15	11	11	15	11	9	9	11	15	11	15	15	11	9	9	11	15
	0400-0800	0200-0600	9	9	9	9	9	11	11	17	9	9	11	9	9	11	11	17	11	9	15	9	9	11	11	17
	0800-1200	0600-1000	17	9	11	15	11	15	15	15	17	11	11	11	11	15	15	15	17	17	11	11	15	15	15	21
	1200-1600	1000-1400	17	15	17	15	15	15	15	11	21	17	15	21	15	15	15	11	17	17	15	17	15	15	17	21
	1600-2000	1400-1800	21	21	15	21	15	15	15	9	17	21	17	17	15	15	15	9	21	17	15	21	15	15	15	15
	2000-2400	1800-2200	11	15	11	15	11	11	11	15	11	17	17	17	11	11	15	15	17	15	17	11	15	15	15	15
MIDDLE EAST AND SOUTH ASIA	0000-0400	2100-0100	9	9	11	11	11	9	9	9	9	9	11	11	11	9	9	11	11	11	11	11	11	9	11	11
	0400-0800	0100-0500	9	7	11	11	11	9	15	15	9	9	11	11	9	9	15	15	11	11	11	11	9	9	15	15
	0800-1200	0500-0900	11	9	9	11	15	15	17	17	11	9	11	15	15	15	17	21	15	11	9	11	17	15	17	17
	1200-1600	0900-1300	17	9	9	11	17	15	15	11	17	11	9	11	17	15	15	11	17	15	9	11	17	15	15	11
	1600-2000	1300-1700	17	17	11	17	21	15	11	9	21	17	11	17	21	15	11	9	17	15	11	15	17	15	11	9
	2000-2400	1700-2100	9	15	11	17	17	9	11	9	15	15	11	17	17	9	11	9	15	15	11	17	11	9	11	9
EAST ASIA AND FAR EAST	0000-0400	1600-2000	9	11	11	11	11	9	6	11	11	15	15	15	9	9	9	11	15	15	15	15	11	11	9	11
	0400-0800	2000-2400	9	11	15	11	11	11	7	15	11	15	15	15	11	11	9	15	11	15	15	15	11	11	9	15
	0800-1200	0000-0400	15	17	15	15	15	15	9	15	11	15	17	15	15	15	15	17	11	15	15	15	15	15	15	17
	1200-1600	0400-0800	11	9	11	11	15	17	15	17	11	11	11	15	17	15	17	17	15	11	15	11	15	17	15	17
	1600-2000	0800-1200	17	9	9	11	17	15	17	17	17	9	11	11	17	15	17	17	17	11	11	11	17	17	15	17
	2000-2400	1200-1600	15	15	9	15	15	15	9	11	17	15	9	15	15	15	9	11	17	15	9	15	15	15	9	9
AUSTRALIA AND NEW ZEALAND	0000-0400	1400-1800	15	15	9	11	11	9	11	9	15	15	9	11	11	9	11	9	15	15	11	11	11	9	11	9
	0400-0800	1800-2200	9	15	11	11	11	9	11	9	9	15	11	11	11	9	11	9	11	15	11	11	11	11	11	9
	0800-1200	2200-0200	15	15	21	17	11	15	15	11	15	17	21	17	11	15	15	15	15	17	17	11	15	15	15	15
	1200-1600	0200-0600	11	11	15	15	15	15	15	15	11	15	15	15	15	15	17	15	11	11	15	15	15	17	15	15
	1600-2000	0600-1000	9	9	11	15	17	15	15	15	11	11	11	15	17	15	15	15	15	11	11	15	17	15	15	15
	2000-2400	1000-1400	17	9	9	11	11	11	15	9	17	9	9	11	15	11	11	9	17	15	15	15	11	11	11	9

Band Selections have been made taking into account both propagation conditions and station operating schedules.

Where the 11MHz band is shown as the most suitable, also check the 13MHz band,
where the 6MHz band is shown as the most suitable, also check the 7MHz band and vice versa.

Introduction to Antennas

by Joseph J Carr
PO Box 1099
Falls Church
VA 22041, USA
E-mail: carrjj@aol.com

A GOOD ANTENNA

There is an old bit of wisdom among radio enthusiasts: After buying a good receiver, put your money and efforts into a good antenna before buying any accessories that improve reception. That wisdom is as old as radio itself, but still holds true. Nothing improves your reception more than a good antenna.

Oddly, many newcomers to the radio hobbies will spend large amounts of money on a good receiver, and then insult it with a poor antenna. Sure, the quality of the receiver can make up for some problems, but a good antenna will improve the reception of most receivers quite nicely. We'll deal with the exception shortly.

THE FUNDAMENTAL PROBLEM

First, though, let's discuss what radio reception is all about. Your radio receiver has specifications for sensitivity and selectivity. The purpose of these specifications is to help the receiver pick up and demodulate radio signals. Both are based on the fundamental problem of radio reception: signal-to-noise ratio (S/N or SNR).

Although SNR is defined in detailed technical terms, the practical definition for the usual listener is that it is the ratio between the signal you want to hear, and everything else that the radio can pick up at the same time. The actual strength of the desired signal is not nearly as important as the ratio between its strength and the strengths of all the other stations, plus natural and man-made noise sources. Improve this ratio, and you will improve your listening pleasure. The sensitivity determines how well the receiver picks up weak signals, and the selectivity how it rejects signals not on the frequency of the desired signal. Both, in other words, enhance SNR but by different means.

PRACTICAL DESIGNS

Now to the antenna. Depending on its design, an antenna can do something to help either or both the 'S' and 'N' parts of the S/N ratio. It can help the 'S' side by increasing the strength of received signals. Directional antennas have gain, so will increase the levels of signal received from their direction of maximum sensitivity.

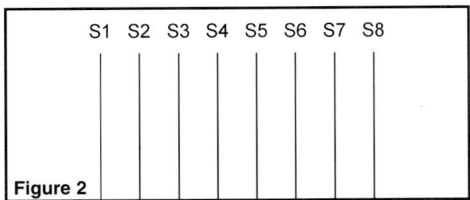

S1 S2 S3 S4 S5 S6 S7 S8

Figure 2

They also reduce the level of undesired signals and radio noise (both parts of the 'N' side) from all directions other than the direction of maximum sensitivity.

Let's take a look at a practical example. **Fig.1** shows a situation that is quite common on the shortwave bands. The dotted line around 'Your Radio' represents an antenna pattern that is equally sensitive in all directions. This pattern is more or less characteristic of the pattern of the small whip antenna that comes with portable receivers. For the sake of explanation we will assume that all signals (S1 through S8) appear at the antenna with equal strengths. The signal that we desire to receive is S1. Note in **Fig 2** that all signals, from all directions around the compass, appear with equal amplitude in the receiver. The result is a mess that assaults your ears with little or no hope of understanding, never mind enjoying, signal S1.

Now suppose we erect a simple bi-directional antenna such as a dipole. It produces a 'figure-8' pattern such as shown in **Fig.3**. There are two directions of maximum sensitivity (maxima) and two directions of very little sensitivity (minima or nulls). The sensitivity of the antenna for points between the maxima and the null is graded proportionally.

Fig.4 shows the relative signal strengths of S1 through S8 with the antenna of Fig.3. Note that S1, the desired signal, is positioned right at the maxima so has the strongest amplitude. Similarly, signal S4 is positioned to the null on the right side, so has the minimum amplitude. The overall result is a considerable reduction in the cacophony of signals that assault the receiver, and subsequently your ears. The listening situation portrayed in Figs.3 and 4 is greatly improved over that of Figs.1 and 2.

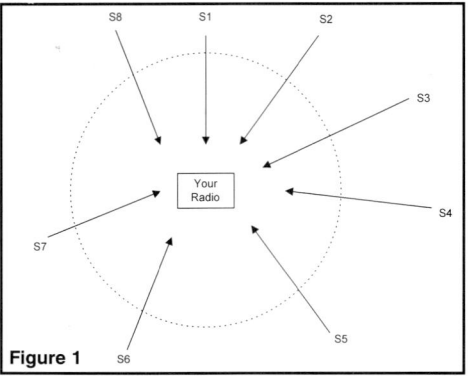

Figure 1

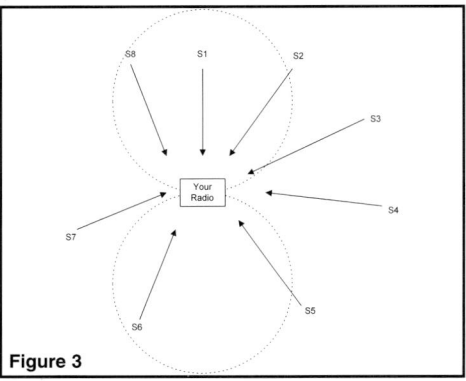

Figure 3

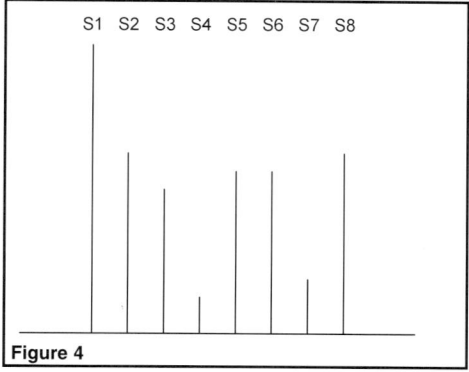

S1 S2 S3 S4 S5 S6 S7 S8

Figure 4

The antenna that produces the situation of Figs.3 and 4 could be a simple half-wavelength horizontal dipole. And it need not be expensive. Although you can buy the parts individually, you can also buy an antenna kit that contains all of the parts in one lot. Pricing shortwave dipole and other wire antenna kits while researching this article I found prices from US$20 to US$50, although it was difficult to see the difference in quality from lowest to highest priced offerings.

$$Length = \frac{72}{F_{MHz}} \text{ metres} \qquad [1]$$

ANTENNAS FOR YOU

Money spent on a good antenna for your receiver is money well spent. And admittedly some of the best antennas cost a lot of money and take up a lot of space. However, it is not necessary to spend huge quantities of money or own a Texas ranch to build an antenna that is considerably better than what you are presently using.

In the following section I will take a look at several antennas that are designed for small yards, or no yard, and for people with a tight budget. There are other antennas, to be sure, and they may even perform better than the selected antennas. The criteria for selection here, however, had to include limited space and money. Before beginning any antenna construction project be sure to read the Safety Note accompanying this article.

SIMPLE MARCONI ANTENNA

A Marconi antenna is unbalanced electrically with respect to ground, so it requires a good ground connection for proper operation.

Consisting of a single piece of wire 10 to 30 metres in length. The Marconi can be run directly from the high impedance antenna terminal on the receiver, but the most effective approach is shown in **Fig.5**. In this approach, the antenna is mounted between two end

supports, and is roughly horizontal to the ground. In general, the longer and higher the wire, the better the performance.

The antenna wire itself can be uninsulated. It is normally made with #14 stranded wire (or any close size). The wire is usually made of either hard drawn copper, or copper clad steel wire to improve its strength and durability.

The antenna is connected to the receiver through an insulated downlead. This wire can be almost any form of insulated hook-up wire, although sizes smaller than about #22 are not recommended (wire gauges get larger as the wire gets thinner). It is best to use #14 through #18 insulated, stranded wire for the download.

The download is connected to the antenna terminal on the back of the receiver. If there is an option, select the terminal marked 'high impedance' or 'hi-Z' (or some variant of these).

All antennas work better with a good ground. Do not use the electrical wiring ground in your house for the antenna ground. Besides the obvious danger if you are not experienced working with electrical systems, there is also the fact that electrical power grounds are terribly noisy, and will induce noise signals into the receiver that would not otherwise be there.

HALF WAVELENGTH DIPOLE

The pattern of the Marconi depends very much on frequency of operation. The nice clean figure-8 pattern shown in Fig.3 earlier can be realized on one band by using a half wavelength horizontal dipole shown in **Fig.6** overleaf. This antenna consists of two quarter wavelength sections fed at their junction from 75 Ohm coaxial cable. The length of each segment is a function of frequency derived from the formula: Where F_{MHz} is the centre frequency of the band of interest, expressed in megahertz (MHz).

The dipole will perform best as a directional antenna at the frequency used to design it, and close by frequencies. It will also operate directionally, but with different patterns, at odd multiples of the design frequency. Thus, an antenna designed for, say, 9.5MHz, will also perform well at 28.5MHz. At other frequencies, it will act much like a random length Marconi cut to approximately the same dimensions.

The dipole antenna operates bi-directionally, so reception off the ends is weak. When trying to suppress interfering signals in that direction, that feature is highly useful. But it's also a problem if the desired signal arrives from that direction. Some people solve the problem by making the antenna such that it can be rotated. That approach, however, greatly complicates the installation and construction of the antenna, and raises costs considerably. It is also not legal in many localities if the antenna tower is close to a property line.

Another approach is to rig two dipoles at right angles to each other. For some signals you would connect the North-South antenna, and for others the East-West antenna. This approach does, however, require a larger footprint on the ground. Some people's yard (e.g. my own) are longer in one direction than the other, so it's difficult to erect a bi-directional antenna without violating a neighbor's space.

THE QUAD LOOP ANTENNA

The main antenna problem for a very large number of shortwave

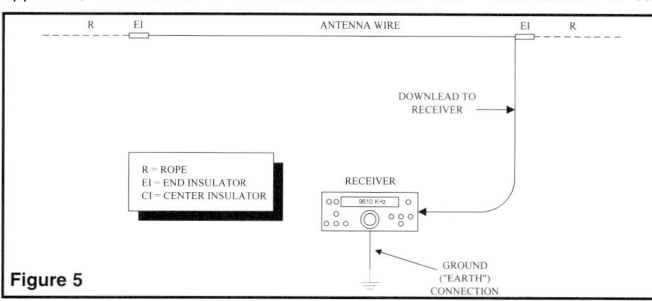

R   EI   ANTENNA WIRE   EI   R

DOWNLEAD TO RECEIVER

R = ROPE
EI = END INSULATOR
CI = CENTER INSULATOR

RECEIVER

9610 KHz

GROUND ("EARTH") CONNECTION

Figure 5

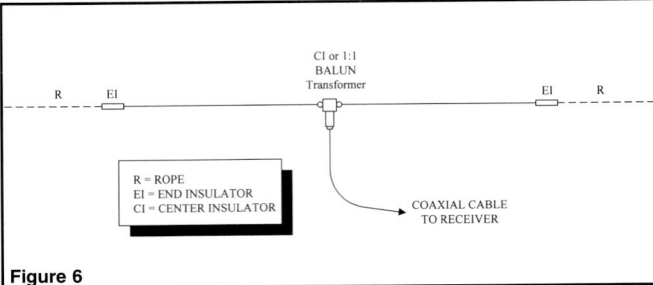

R = ROPE
EI = END INSULATOR
CI = CENTER INSULATOR

CI or 1:1 BALUN Transformer

COAXIAL CABLE TO RECEIVER

Figure 6

receiver owners is a small yard. Unfortunately, unless you have a kind neighbor and a liberal attitude on the part of local authorities, you will not be able to cross a property line with your antenna. In my locality, I cannot come within three metres of a property line, although I've noted the rule is not widely followed. A solution to that problem is the quad loop of **Fig. 7a**, with additional detail in **Fig.7b**.

The basic quad loop is a one wavelength square loop of wire, supported at the corners by end insulators and

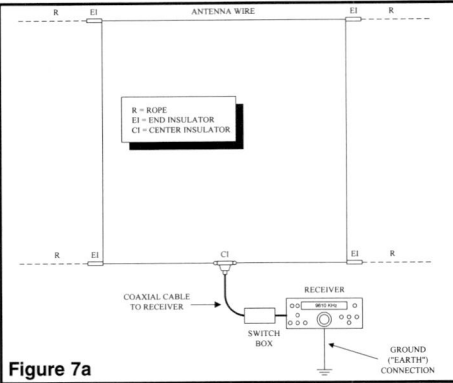

R = ROPE
EI = END INSULATOR
CI = CENTER INSULATOR

COAXIAL CABLE TO RECEIVER

RECEIVER

SWITCH BOX

GROUND ("EARTH") CONNECTION

Figure 7a

rope. Each side of the loop is quarter wavelength. The overall length is, therefore, $308/F_{MHz}$ metres, and each side is $77/F_{MHz}$ metres long. An antenna cut for the middle of the 25 metre band (say, 11.75MHz) will be 8-metres long on each side. The antenna is mounted such that the bottom end is about 2 metres from the surface of the Earth, making its overall height 10 metres. Such a height is relatively easily obtained with supports such as trees and masts on the roofs of houses.

The classic quad loop antenna is fed on the centre of one side with coaxial cable. The bottom edge is shown here, and is probably the most convenient. However, you can also use the top edge with equal results. The wire loop is broken at that point so that one side can be connected to the coax center conductor and the shield to the other side. This configuration offers a small footprint yet produces a figure-8 bi-directional pattern similar to a dipole that requires twice the horizontal space. The nulls are to the right and left in the version of Fig.7a, with the maxima being at right angles to the plane of the wire (i.e. in and out of the page).

The configuration quad loop in Fig.7a is designed for horizontally polarized signals. To optimize the antenna for vertically polarized signals connect the feedline to one of the vertical sides, rather than the horizontal side. The quad loop is bi-directional like a dipole, so can be used for directional listening. Some people place two

quad loops at right angles to each other, with separate feed lines, to be able to shift the direction of reception.

The twin loops idea is quite workable, but it might not meet the criterion of small yard antenna. It would, after all, take two 8-metre lengths. A better approach might be to use the switch box shown in Fig.7b. The box is not needed if you want to make a resonant loop, but if you want to make a bi-directional antenna with a straight loop, then it is possible to reach a compromise using the switch box approach. A switch is located in a shielded metal box. The switch is a double-pole, three position switch that will permit the antenna to be used as either an open loop or a closed loop.

If you are not particularly handy with electronic assembly tools, then you can still use this system. There is only a small difference between the patterns of the two open loops. Connect the antenna in the classical form, with the coaxial cable to the receiver. When you want to operate as an open loop, then unscrew the outer shield collar of the coaxial connector and back the connector out of the receiver antenna terminal. Leave the center conductor at least partially inserted into the antenna terminal. This method is not quite the same thing as the switch box method, but it works nicely.

The quad loop is a good antenna for a small yard, but is in no way a second rate antenna. People with large estates might also wish to consider this antenna. There are two advantages to the quad loop that must be considered. First, the antenna works as advertised close to the Earth's surface, while certain other antennas work according to the rules only at significant height off the surface. Second, the antenna is somewhat less sensitive to locally generated electrical noise. There is a tremendous amount of noise that finds its way into a radio receiver, so it is a bonus when the antenna is less sensitive to the noise.

Note that the classical quad loop design is square. Related antennas are triangular (delta loop) or rectangular. If a square loop is not feasible, then these other shapes can be used just as well.

INDOOR LOOPS

One of the problems faced by many people is that they are prohibited from installing any form of outdoor antenna. In the USA many townhouse, condominium and apartment house dwellings have such rules, I understand that similar situations exist in other countries as well.

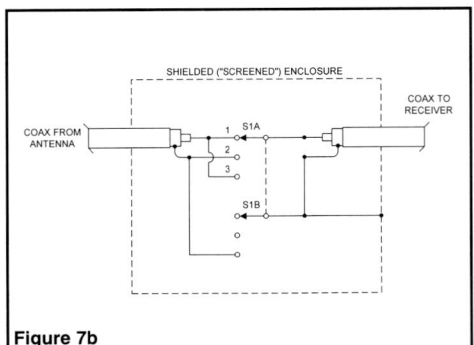

SHIELDED ("SCREENED") ENCLOSURE

COAX FROM ANTENNA

COAX TO RECEIVER

S1A

S1B

Figure 7b

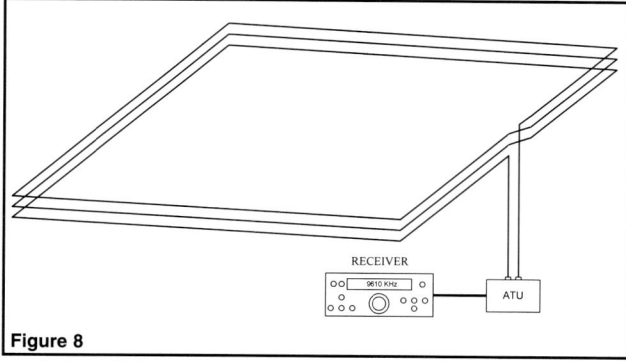

Figure 8

case of the Marconi, the length of the wire is usually a quarter wavelength, which is the length given by the equation [1].

In both antennas the wire is run vertically up a support which is insulated from the Earth, but can be rather close to it. The dipole version is fed with 72 Ohm coaxial cable in the center, while in the case of the Marconi it is fed at the base. The center conductor of 52 Ohm coaxial cable is connected to the antenna, while the coax shield is connected to a ground rod at the base of the antenna. Earlier I mentioned that not all radio receivers are improved by using an external antenna. Some portable receivers are an exception. Unless they have a terminal for an external antenna, then the radio may not have sufficient dynamic range for the higher signal levels that will be achieved. Some of these receivers may paradoxically produce lower output levels when the input signal level increases.

One solution to the problem is the quad loop discussed previously. Although the lowest frequency of operation will be adversely affected, a considerable quad loop antenna can be deployed indoors. In the USA room walls are typically 8 feet (2.44 metres) high. If the room is, say, 12ft. x 12ft. (3.66m x 3.66m), then you can install a loop antenna that is rectangular with an overall length of 12.2 metres. This is not an optimal length for the lower end of the shortwave bands, but it will still work quite nicely compared with lesser antennas.

Another approach that is open to some people is to string the antenna through the attic. The download or transmission line can be routed through the walls to the room where the radio receiver is located. I have successfully used dipole and random length Marconi antennas mounted in the attic on a number of occasions. Some people also find the loop antenna easy to install in this manner.

Alternatively, you can install the antenna on the roof outside of the building. The main issue in banning outdoor antennas is aesthetic, so If the neighbors can't see it, then they can't complain, can they? It is remarkably difficult to see a length of #18 wire at a distance, especially if the insulation is colored similarly to the roof - dark on dark, or light on light - colors don't have to match.

MIDDLE EAST ROOM LOOP

Fig.8 shows an antenna that a reader living in a Middle East country uses. He is in a country where using shortwave radio receivers is lega, but suspect, so he installed this antenna in one of the rooms of the apartment that he rents.

The wire is wrapped around the room where the walls meet the ceiling. A total of two to five continuous turns are used, for a total length of 50 to 100 metres. The turns can be held to the walls using various fasteners, but my correspondent used picture hangers. In a similar instalation in my own home, I used flower basket hangers, i.e. hooks set in the ceiling and the wire is dropped down a wall close to the receiver. It will work nicely straight into the receiver, but best performance is realized if an antenna tuning unit (ATU) is used between the antenna and the receiver input terminals. The ATU needs to be re-tuned when large changes in frequency are made, but the extra performance is worth the effort.

VERTICAL ANTENNAS

A vertical antenna is omnidirectional, so it will not improve the S/N situation very much, except for the S portion of the equation when compared with the small antenna on the radio. A vertical antenna can be made using either the dipole or Marconi designs and in the

GROUNDING

Radio antennas should be equipped with a ground (earth) connection for two reasons: 1) protection against lightning, and 2) improvement of reception. For most antennas, it is sufficient for the second goal to connect the GROUND or GND terminal on the rear panel of the receiver to a ground point with a thick wire.

Some manuals recommend use of a cold water pipe for grounding shortwave receivers. That is not always such a good idea. First, if you have a natural gas service in your house, then you may mistake a gas pipe for a water pipe, and that could be disastrous under the right circumstances. Second, even metal cold water pipes do not guarantee a good ground connection, and may be a noisy ground. Third, in modern construction cold water pipes are made of PVC so are not conductive. The best ground is a copper or copper-clad steel rod driven into the soil. Lengths of 1 metre (3-feet) to 2.5 metres (8 feet) are preferable but the longer the better - most radio or general electrical suppliers will stock suitable rods. Under *no circumstances* should you use the electrical power mains ground wire for grounding your receiver's antenna.

LIGHTNING

Lightning is a fact of life in our world. An antenna does not 'draw' lightning that would not strike that site anyway, but it does draw lightning in the sense of a lightning rod, so actually provides a protective function, even if a fairly poor one. Make no mistake about it: if lightning directly strikes the antenna it will destroy the receiver. However, direct strikes are extremely rare. Indeed, in forty years as a radio hobbyist and an electronics engineer, I've never heard a credible first-hand account of such a strike.

But that's not the end of it. It is possible for a lightning bolt to destroy your receiver even if it passes overhead or strikes nearby. The lightning produces high voltages in the antenna, and these do the devil's work in your circuitry. For that reason, and to provide a measure of protection in the case of a more direct strike, it is essential that all external antennas be fitted with a lightning arrester device appropriate for the type of transmission line or download used. The arrester is connected into the feedline from the antenna, and is grounded. It will pass radio signals from the antenna to the receiver with no ill effects, but will allow high voltages

associated with lightning to pass to ground. These devices can be obtained from radio supplies dealers and are quite low in cost for the protection they provide.

SAFETY GUIDELINES

Radio antennas are relatively safe to install, but they can become dangerous if you do not use common sense and follow certain precautions. It is not possible to foresee all of the situations that you might face when erecting an antenna. I would like to give you all possible warnings, but that is not even humanly possible. You take responsibility when installing an antenna. I can, however, give you some general safety guidelines. Knowledge of what you face, some hard nosed sound judgement, modulated by common sense, are the best tools on any antenna job.

One rule that is an absolute is that no antenna should ever be erected where the antenna, the feedline or any part thereof crosses over a power line. EVER! This is a 'no kidder' - don't do it! Power lines look insulated, but there are often small breaks or weakened spots that can bring the antenna into lethal contact with the live power line. Every year or so we hear about an SWL, scanner/monitor buff or ham radio operator being killed by tossing an antenna wire over a power line. Avoid making yourself into a sad statistic!

And the same rule applies to situations where the antenna can fall onto a power line if it falls down or breaks. You have to examine the situation with a critical eye to see if there is any possible way for that antenna, or its support structure, to fall onto a power line if it breaks in any way whatsoever. When a wire antenna breaks and is wind-whipped, it should not touch a power line. You also need to be sure that the antenna does not fall onto a neighbors property, or a public thoroughfare, where you might injure someone or someone's property and thus incur legal liability.

OTHER PRECAUTIONS

Another caution is that you be physically fit to do the work. While the on-the-ground portion of the work is not too strenuous, any climbing at all, even on ladders, can be taxing for some people. Antenna materials are deceptively lightweight on the ground, but when you get up on even a small ladder, they can be remarkably difficult to handle. If the wind blows even a few miles per hour, then the danger is magnified considerably. Antennas, even small wire antennas, have a bigger 'sail area' to wind than appears to be the case. Even a small wind can cause the antenna to whip around and become dangerous. Be Careful...and follow the 'buddy system' rule when installing antennas. In other words, always work with a buddy. Ask as many friends as are needed to safely do the job, and always have at least one assistant even when you think you can do it alone. Erecting a large antenna - and some small ones - without help is just plain stupid. At least have someone around who can call emergency medical services if you make a really bad mistake.

MATERIALS

Always use quality materials and use good work practices. Antennas, being potentially dangerous, should always have the best of both goods and workmanship in order to keep quality high. It is not just the electrical or radio reception workings that are important, but also the ability to stay up in the air and safe.

When planning the antenna job, keep in mind that pedestrian traffic in your yard could possibly affect the antenna system. Wires are difficult to see, and if an antenna wire is low enough to intersect someone's body, then it is possible to cause very serious injury to passersby. Even when the person is a trespasser, the courts may hold you liable for injuries caused by an inappropriately designed and installed antenna. Take care for safety not only of yourself, but of others.

LOCAL REGULATIONS

One necessary reminder is that your local government might have some interesting ideas - legal requirements actually - concerning your antenna installation. The electrical, mechanical and zoning codes must be observed. There is a great deal of similarity between local codes because most of them are adaptations from certain national standards. But there are enough differences that one needs to consult local authorities. Indeed, you may need a license or building permit to install the antenna in the first place.

Save all paperwork regarding your building permit, including inspection decals or papers, and the original drawings (with the local building inspector's stamps). If a casualty occurs, then your insurance company may elect to not pay off if you have violated an electrical, mechanical, building or zoning code. That clause may be overlooked by an enthusiastic antenna builder, but it could prove to be a costly oversight if something goes wrong.

ACCESSORIES

There is seemingly no end to the accessories that one can buy for a shortwave receiver. Perhaps the most common are antenna tuners, preamplifiers, preselectors and audio filters. All of these devices will improve reception some of the time, or even most of the time, but none of them are the universal answer to all problems. Indeed, use of some of them, either singly or in combination with others, may adversely affect rather than improve reception. In other words, there is no silver bullet!

In any event, with all the claims made for some accessories, one is tempted to think that some of them are much like the gasoline (petrol) saving devices that we often see advertised for automobiles. If we were to install every device advertised in even a single car buff magazine, then we would have to either risk rupture of the engorged fuel tank or stop every 100 kilometres to siphon off the excess gas that was produced! The lesson is that some devices may or may not work at all, some are useful under the right circumstances, and others work well by themselves but not in conjunction with one or more of the others. For example, two preamplifiers in cascade often have disastrous effects on the performance of your receiver, even if one preamp would be quite useful!

When evaluating a proposed device one must keep in mind the fundamental problem of radio reception: getting sufficient signal-to-noise ratio (S/N) to provide comfortable listening. In this context 'signal' means that particular radio signal that you want to hear, and 'noise' is everything else the receiver can pick up at the same time including other radio signals.

ANTENNA TUNERS

The antenna tuner, or antenna tuning unit (ATU) is an often under-estimated device. Consisting of a collection of inductors and capacitors, it is used to match (or tune) the impedance of the antenna feedpoint to the impedance of the receiver. This reduces losses in the system consequently helping to improve the S side of the S/N equation.

The ATU also serves as one additional tuning mechanism that will reduce unwanted signals. The lower the level of the unwanted signals reaching the receiver, the better. In this role the ATU serves as a species of RF filter. This use is probably most important to receiver owners because overcoming mismatched impedance losses results in less improvement than is commonly imagined.

Transmitter operators need to worry about the impedance match for other reasons entirely. Even relatively low power transmitters can be seriously damaged if the mismatch rises above certain limits, with expensive final stage components having to be replaced.

For the short wave listener there are, nevertheless, some cautions. First, if the antenna is already impedance matched, then the ATU might be a waste of money. Indeed, it may present less signal to the receiver because there is always an insertion loss to overcome due to plug to socket and internal component losses. The goal here is to reduce noise signals more than the desired signal, so that an improvement in the S/N ratio is obtained. Second, some receiver designs do not like tuned circuits being connected to their inputs unless they are designed to be impedance matched to that particular filter. As a result, anomalies in the ATU setting may result in a detrimental effect on receiver performance.

PREAMPLIFIERS

A preamplifier is a wideband RF amplifier that is connected between the antenna and the receiver's antenna input terminal. It boosts the signal level, so it must be a good thing, right? Not always. There are two problems.

First, the goal is to improve S/N, remember? If an amplifier boosts noise as much as signals, then the S/N is not improved. The preamplifier is a waste. The preamplifier should only be used in cases where it will improve the S/N ratio. In the medium wave and lower shortwave bands a preamplifier usually only improves things if the receiver is a bit marginal in the sensitivity department. Even then, the improvement is often not large. In the higher shortwave bands, and in the VHF/UHF scanner bands, the overall reception may be improved considerably if the preamplifier has a much lower noise figure than the receiver, especially when weak signals are sought.

Second, if the receiver lacks sufficient dynamic range, or if there are a large number of strong signals on the band, then the preamplifier might drive the receiver into nonlinearity. In this state the receiver might generate various species of spurious signals, or become desensitized.

A number of times I've seen people connect 60 dB preamplifiers to their receivers (60 dB is an increase of 1,000,000!). They are surprised when they find that signals appear at two or more spots on the dial, or that the overall level of the desired signal is lowered. Turning off the amplifier will actually improve matters! This phenomenon is why many high quality receivers include an attenuator that can be switched on and off. An attenuator is the opposite of a preamplifier, i.e. it causes a signal loss. If you are trying to receive an extremely strong signal, then overall performance may be improved by cranking in 6 to 30dB of attenuation, rather than by pre-amplification.

Generally speaking, although preamplifiers can be beneficial in certain, limited circumstances, modern receivers - especially those designed for the HF bands - rarely need them and perform better without them.

PRESELECTORS

A preselector is a tuned circuit, or filter, that will pass some desired band of frequencies and reject all others. The preselector may be passive, and causes a small insertion loss, or be amplified. It is often the case that preamplifiers also include preselectors. The idea is to limit the levels of interfering adjacent channel signals that could either desensitize the receiver, or generate spurious signals.

Preselectors come in bandpass and single frequency designs. A bandpass design is essentially a bandpass filter in series with the antenna. A single frequency design has a tuning knob that must be adjusted to the same frequency as the receiver. Otherwise, signal levels are decreased substantially.

A preselector is a very useful accessory in either of two situations: 1) if there are numerous high level interfering adjacent channel signals deteriorating receiver performance, or 2) if the receiver is a generally poor performer. In the latter case, the image rejection and front-end selectivity are the items that a preselector will improve.

AUDIO FILTER

Noise and adjacent channel interference can deteriorate the usability of a receiver considerably. An audio filter limits the bandwidth of the output, so can help reject unwanted signals. An internal IF filter, inside the receiver, is technically a better solution but may not be practical. Such filters must be installed by professional technicians, and are costly so it is preferable to order them with the radio when it is first being purchased. For the after-market case however, the audio filter is a viable alternative. These devices typically plug into an earphone or loudspeaker jack on the receiver. Some versions are passive, so require considerable audio power from the receiver, whilst others have their own internal amplifiers. The audio filter should have bandwidths matched to the type of reception, as the goal is to use only the bandwidth that passes the signal without distortion. A narrower than necessary filter will degrade the received audio and a wider than optimum filter will allow unwanted audio products to be heard.

Typically, an AM broadcast-band receiver will need to use 6 to 10kHz, while an AM shortwave receiver might be better off with 4 to 6kHz bandwidth. In the latter case we sacrifice a bit of fidelity for improved rejection of co-channel interference. For SSB reception, a 2.8kHz filter might be used, for RTTY a 1.8kHz filter, and for CW (Morse code) a 250 to 500Hz filter is appropriate.

CONCLUSION

The accessories you see advertised for your receiver can make a substantial difference in performance under the right circumstances. But there is still no silver bullet, and some care must be exercised in selecting and using these 'outboard' add-ons. Essentially, if the advert promises an improvement in performance which seems to be too good to be true - it probably *is* too good to be true!

Joseph J. Carr is the author of numerous books, and more than 700 articles on radio and electronics topics, including the 'Practical Antenna Handbook' 3rd Edition (McGraw-Hill), 'Antenna Toolkit w/CD-ROM' and 'Joe Carr's Receiving Antenna Handbook' (Universal Radio Research). Mister Carr can be reached at:

P.O. Box 1099, Falls Church
VA, 22041, USA
E-mail: carrjj@aol.com

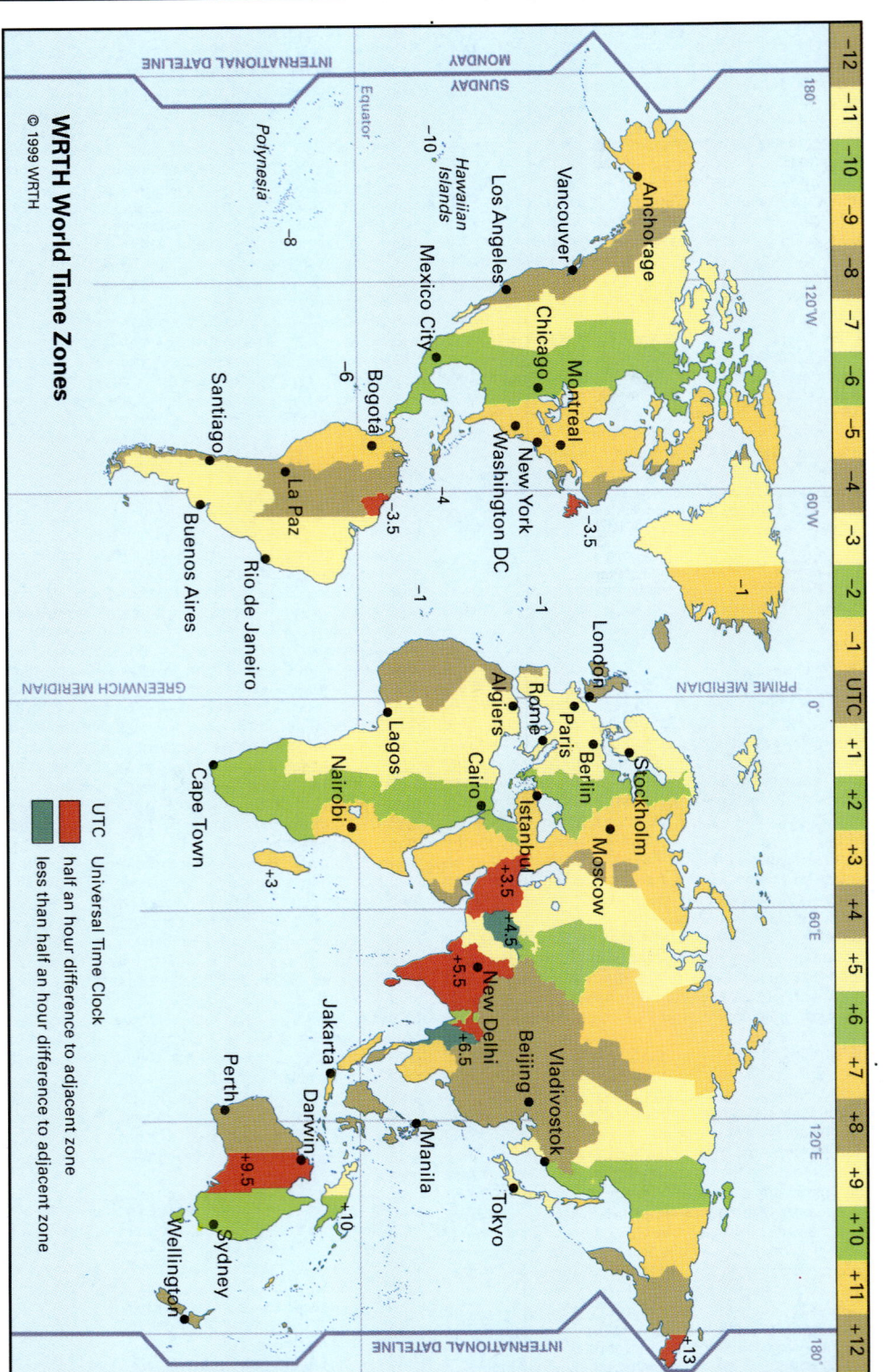

WRTH World Time Zones

© 1999 WRTH

UTC Universal Time Clock

half an hour difference to adjacent zone

less than half an hour difference to adjacent zone

NATIONAL RADIO
EUROPE

ALBANIA	AUSTRIA

ALBANIA

L.T: UTC + 1h (Su: UTC + 2h) — **Pop:** 3.000.000—**Radios:** 550.000 — **Pr.L:** Albanian — **ITU:** ALB.

RADIOTELEVISIONE SHQIPTAR (RTVSH) (Gov.)
✆ General Directorate of Albanian Radiotelevision, Direction Generale de la Radiotelevision Albanaise. Rr. "Ismail Qcmali" Nr.11 TIRANA ☎ +355 (42) 27745. 🖹 +355 (42) 27745.
L.P: DG: Albert Minga. Dep. DG: Sefedin Cela. Dir. TVSH: Eduard Mazi. Dir. Int Rel: Mira Blushi-Shuteriqui, Techn. Dir:Apostal Noti, Dir. Radio-Tirana: Martin Leka.
Networks: 1=First Chan, 2=2nd Chan, R=Reg. Prgr, E=Ext. Sce.

MW	kHz	kW	P	MW	kHz	kW	P
Korçë	621	25	R	Fllakë	1215	500	E
Kukës	648	50	1	Korçë	1260	50	1
Shkodër	693	50	1	Gjirokastër	1305	15	R
Sarandë	864	30	1	Skhodër	1323	15	R
Gjirokastër	909	50	1	Tiranë	1359	50	1
Kukës	990	15	R	Fllakë	1395	500	E
Durrës	1089	150	1	Fllakë	1458	500	2/E

FM: Tirana 99.5MHz 100kW (1) 95.8MHz 10kW (2)
SW: Cerrik 6100kHz (1) 100kW, 7150kHz (2) 100kW
First Channel: 0400-2300. **N:** 0500W, 0530 Sat & Sun, 0700, 0800, 0900, 1000, 1100, 1200, 1300, 1430, 1700, 1830, 2100, 2230.
2nd Channel: 1000-1500 — **Regional Prgrs:** rel. Tirana & own prgrs.
EXTERNAL SCE: see International Broadcasting section.

OTHER STATIONS
BBC World Service: Mnt. Dajti 103.9MHz - 10kW

TRANS WORLD RADIO (Rlg.)
see International Broadcasting section

ANDORRA

L.T: UTC + 1h (Su: UTC + 2h) — **Pop:** 65.000 — **Radios:** 10.000 — **Pr.L:** Catalan, French, Spanish — **E.C:** 50Hz, 220V — **ITU:** AND.

SERVEI DE TELECOMUNICACIONS D'ANDORRA
✆ Av. Meritxell 112, Andorra la Vella. ☎ +376 821021. 🖹 +376 864274. **E-mail:** sta.comer@andorra.ad
L.P: Admin Dir: Jaume Salvat. Tech. Dir: Joan Marc Lauga.
FM: (0.1kW): 102.0MHz France-Inter, 100.2MHz France-Musique, 103.8MHz France-Culture, 104.6MHz Catalunya R, 106.0MHz R4, 106.3MHz R. Barcelona FM, 106.8MHz RNE-1, 107.7MHz R3.

ORGANITZACIÓ DE RADIOTELEVISIÓ D'ANDORRA (ORTA) (Gov.)
✆ Avgda. Merixtell 9, 6Þ, Andorra la Vella. ☎ +33 628 865443. 🖹 +33 628 64999 — **L.P:** DG: Joan Muxella Botos

RADIO NACIONAL D'ANDORRA
✆Baixada del Moli, nÞ24, Andorra la Vella. ☎ +376 873777. 🖹 +376 864999 — **L.P:** DG: Enric Castellet
FM: 91.4MHz 2kW, 94.2MHz 0.5kW.
D.PRGR: 24h. Cultural & Social Information
ANDORA MÚSICA
FM: 97.0MHz 0.5kW
D.PRGR: 24h. 100% music.

RADIO VALIRA (Comm.)
✆ Avgda.Merixtell 9, 1st Floor, Andorra la Vella. ☎ +33 628 25461, 27619. **E-mail:** r_valira@andornet.ad
Web: http://www.andornet.ad/r_valira/
L.P: SM: Gualberto Osorio.
FM: Corroi 93.3MHz, Pas de la Casca 93.6MHz, 98.4MHz, Organya (Spain) 98.1/105.0MHz 2.5kW.
Prgr. 1 on 93.3/98.1MHz: W 0700-2000, Sun 0900-1300 in Catalan exc. **French** 1300-1330, **English** 1330-1400. Other times rel. Onda Cero (Spain) — **Prgr. 2** on 105.0MHz: times as Prgr.1.
V. by QSL-card.

AUSTRIA

L.T: UTC + 1h (Su: UTC + 2h) — **Pop:** 7.734.000 — **Radios:** 4.710.000 — **Pr.L:** German — **E.C:** 50Hz, 220V — **ITU:** AUT.

ÖSTERREICHISCHER RUNDFUNK (ORF)
✆ Würzburggasse 30, 1136 Wien. ☎ +43 (1) 878780. 🖹 +43 (1) 50108250.0. **Cable:** ORF Wien. ➀ 133601 orfz a.
Web: http://www.orf.at/home.htm
Blue Danube R: Argentinierstr. 30a, 1040 Wien. ☎ +43 (1) 50101 8901. 🖹 +43 (1) 50101 8900.
L.P: DG: Gerhard Weis. All other officeholders yet to be nominated

MW: Wien/Bissamberg 1476kHz 60kW.

FM (MHz)	Ö-1	Ö-Reg	Ö-3	BDR	kW
Bad Gleichen.		94.9			6
Bludenz 1	87.6	96.0h	98.8		4
Bregenz 1	93.3	98.2h	89.6	102.1	100
Bruck/Mur 1	87.6	93.2f	98.7	102.1	20
Graz 1	91.2	95.4f	89.2	101.7	67
Innsbruck 1	92.5	96.4g	88.5	101.4	50
Klagenfurt 1	92.8	97.8b	90.4	102.9	100/60
Kufstein	97.5	95.4g	103.9	99.9	5
Lienz	89.3	93.8b	99.3	101.0	2.6
		95.9g			2.6
Linz 1	97.5	95.2d	88.8	104.0	100
		90.1c			10
Mattersburg	89.0	96.2a	100.9		3
Rechnitz	90.6	93.5a	87.9	97.4	3/6/3
Salzburg	90.9	94.8e	99.0	104.6	100/2x75
		101.2d			8
St. Pölten	97.0	91.5c	89.4	98.8	100
Schärding	92.5	99.5d			4
Schladming 1	94.3	96.3f	101.3	103.3	4
Semmering	90.3	95.8c	88.2	92.4	10.2
Spittal/Drau 1	91.6	100.4b	87.9	103.6	3
Weitra	97.7	95.7c	98.1	101.4	2
Wolfsberg 1	96.7	94.5b	99.5	102.3	1.5
Wien 1	92.0	97.9c	99.9		100
		94.7a		103.8	2.4/5

+ more than 500 low power trs — BDR=Blue Danube R.
Österreich-1: 24h. **N:** 0500, 0600, 0800, 0900, 1000, 1100 (W), 1400, 1600, 1700, 2100, 2200, 2300. **Newsreel:** 0700, 1200 (MF), 1700 (Sun), 1800 (MF), 2200 (MF). **N. in English/French** (rel. Blue Danube Radio **Radio 1476:** special "Access R." prgrs 1700-2308 on 1476kHz (2200-2300 rel. R. Austria International). A joint project between ORF and the Polycollege in Wien.
Ö-2: Key to regional sces (studios in brackets): a) Burgenland (Elsenstadt), b) Kärnten (Klagenfurt), c) Niederösterreich (Wien), d) Oberösterreich (Linz), e) Salzburg (Salzburg), f) Steiermark (Graz), g) Tirol (Innsbruck), h) Vorarlberg (Dornbirn), i) Wien (Wien).
Regional addr: a) Buchgraben 51,7001 Eisenstadt — b) Sponheimerstr. 13, 9010 Klagenfurt — c) Argentinierstr. 30a, 1040 Wien — d) Europaplatz 3, 4010 Linz — e) Nonntaler-Haupstr. 49d, 5010 Salzburg — f) Marburgerstr. 20, 8042 Graz — g) Rennweg 14, 6010 Innsbruck — h) Höchsterstrasse 38, 6850 Dornbirn.
D.PRGR: 24h. **N:** W 0400, 0500, 0600, 0800, 0900, 1103, 1203, 1300, 1400, 1500, 1600, 1800, 1900, 2000, 2100, 2200, 2300 ; Sun 0600, 0700, 0900, 1203, 1400, 1500, 1600, 1700, 1900, 2000, 2100, 2200. **Local Prgrs:** Mon-Fri 0505-0700, 0803-0900, 0903-1030, 1145-1650, 1705-1800, 1903-2100 (Thurs); Sat 0505-0745, 1003-1030, 1145-1800, 1820-1900; Sun 0605-0755, 1200-1500, 1607-1800, 1820-1900 — **Ö-3:** 24h. **N:** on the h — **Blue Danube R (BDR):** 0600-2400 (rel. Ö-1 0000-0600). Prgrs in English, French & German. **N:** every h. on the half h. in English/German. **E-mail:** bdr-feats@orf.at **Web:** http://www.via.at/fobdr/
ANN: "Österreich 1", "Ö2 . . . (Wien, Niederösterreich, Tirol . . .)", "Ö3" — **IS:** Österreich 1: composition by Austrian composer Werner Pirchner. Ö2: Composition by Austrian Composer Bert Breit. Ö3: Electronic Music.

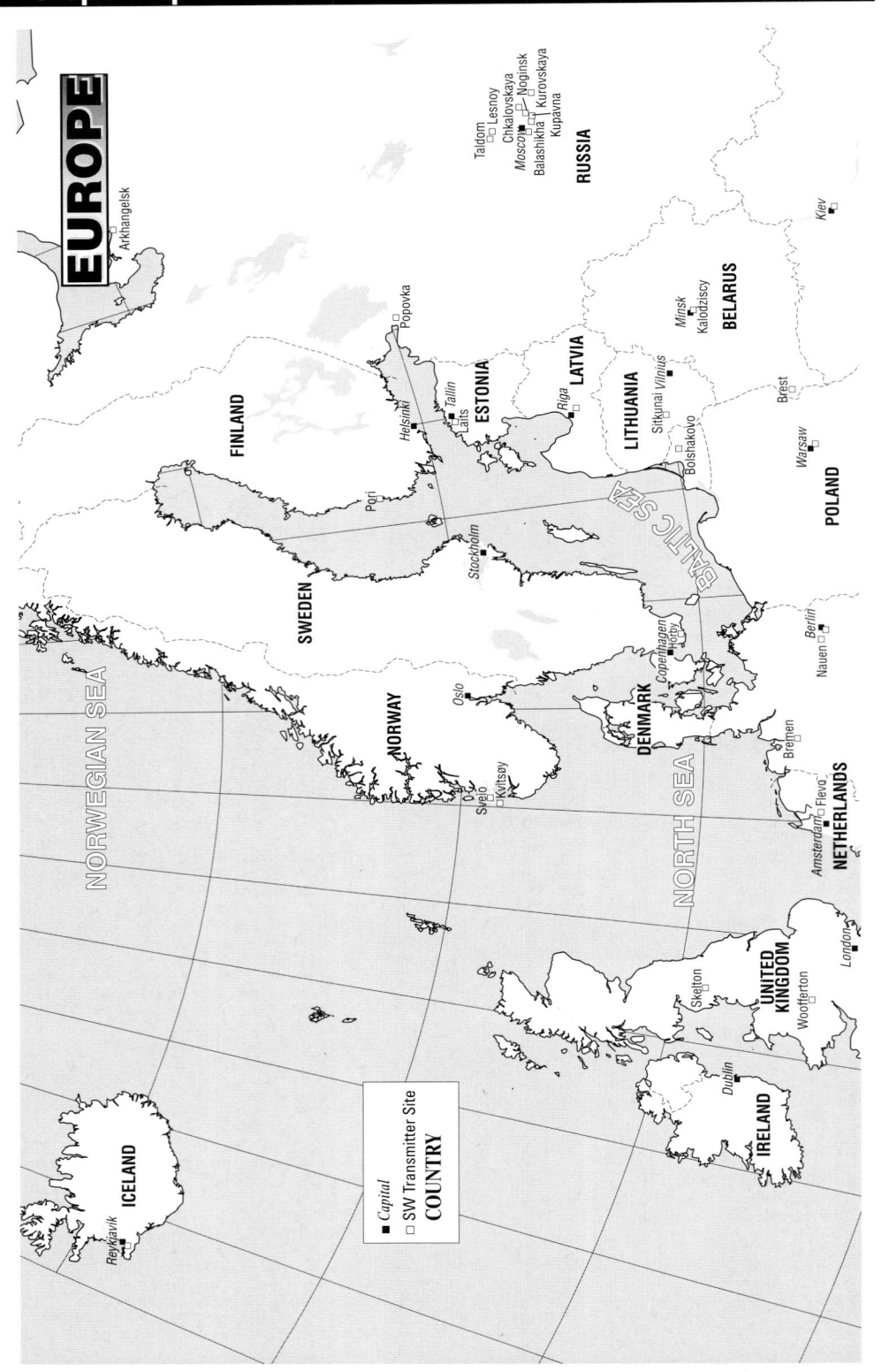

EUROPE

Arkhangelsk

RUSSIA
Taldom
Lesnoy
Chkalovskaya
Moscow
Balashikha
Noginsk
Kurovskaya
Kupavna

Kiev

Minsk
Kalodziscy
BELARUS

Brest

Popovka

FINLAND

ESTONIA
Tallin
Helsinki
Laits

LATVIA
Riga

LITHUANIA
Sitkunai Vilnius
Bolshakovo

Warsaw

POLAND

Pori

BALTIC SEA

Stockholm

SWEDEN

Berlin
Nauen

NORWAY
Oslo
Svejo
Kvitsoy

Copenhagen
Hoby
DENMARK

NORTH SEA

Bremen

NETHERLANDS
Amsterdam
Flevo

London

UNITED
KINGDOM
Skelton
Woofferton

NORWEGIAN SEA

Dublin
IRELAND

ICELAND
Reykjavik

■ *Capital*
□ SW Transmitter Site
COUNTRY

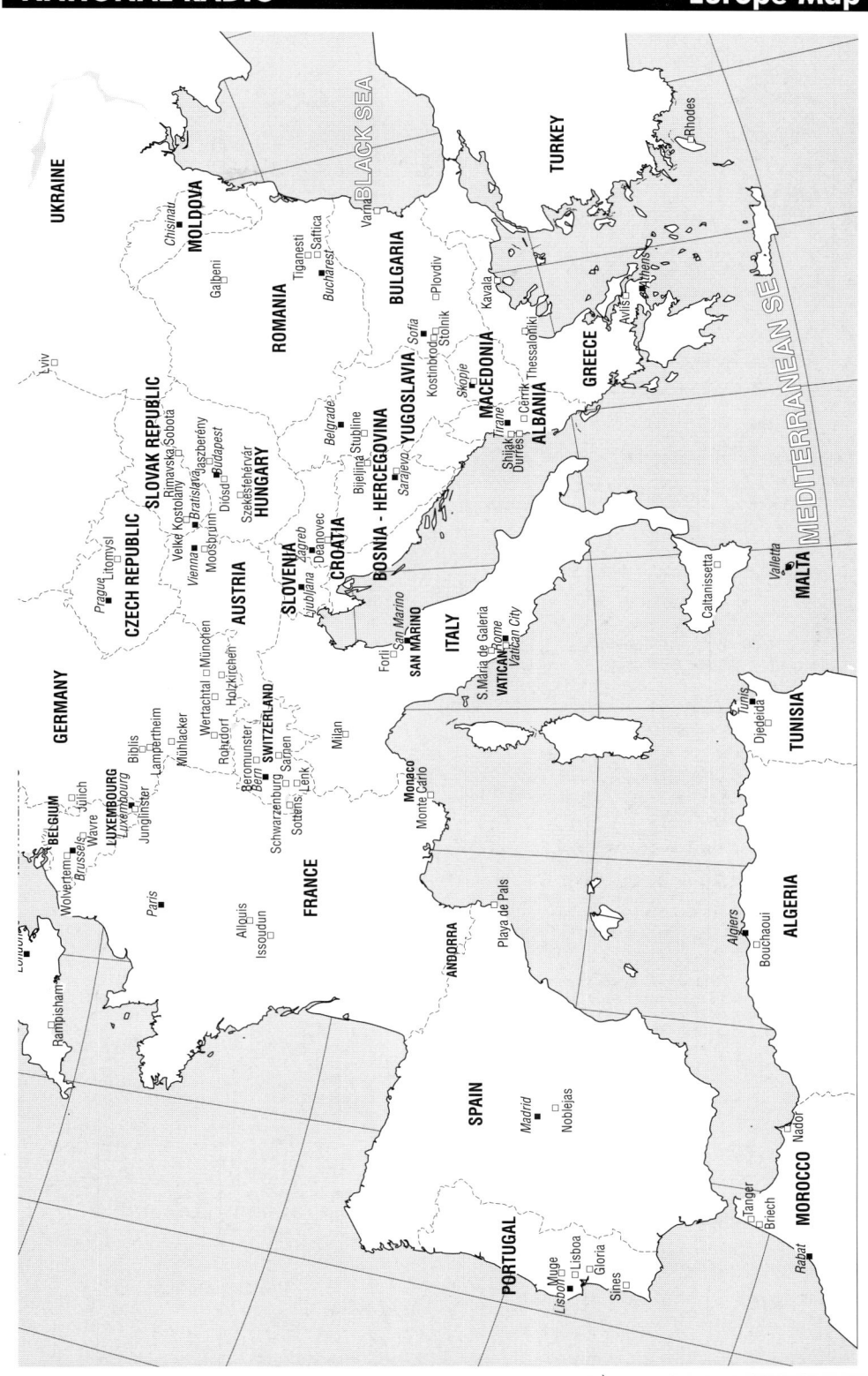

RADIO AUSTRIA INTERNATIONAL
see International Broadcasting section.

Private Stations
WIEN (Vienna): 88.6 L!ve, 88.6MHz – **Antenne 102,5 Wien,** 102.5MHz – **92,9 RTL,** 92.9MHz – **Energy 104,2,** 104.2MHz – **R. Stephansdom,** 107.3MHz – **Orange 93,3,** 93.3MHz. **NIEDERÖSTERREICH (Lower Austria): R. RPN,** 105.3MHz + 4rly – **R. PL1,** 100.8MHz – **Maria heute,** 104.7MHz **Antenne W4,** 100.8MHz + 3rly. **BURGENLAND: Servus Radio,** 103.4MHz + 2rly – **R. MORA,** 106.3MHz + 1rly. **OBERÖSTERREICH (Upper Austria): Life Radio,** 100.5MHz + 14rly – **Welle 1 92,6 Music Radio,** 92.6MHz – **FRO,** 105.0MHz – **Antenne Arcadia Innviertel,** 104.9MHz +1rly. **SALZBURG: R. Melody,** 101.8MHz + 14rly – **Welle 1 106,2 Music Radio,** 106.2MHz – **R. Arabella,** 107.4MHz – **Welle Alpin,** 107.3MHz + 3rly. **STEIERMARK (Styria): Antenne Steiermark,** 99.1MHz + 10rly – **107,5 das Grazer Radio,** 107.5MHz – **KÄRNTEN (Carinthia): Antenne Kärnten,** 104.9MHz + 4rly – **R. Wöthersee,** 95.2MHz – **Gute Laune,** 106.6MHz + 2rly – **R. Villach,** 101.6MHz – **Radiofreunde,** 106.3MHz – **R. Starlet,** 102.5MHz. **TIROL (Tyrol): Antenne Tirol,** 103.4MHz + 16rly – **Welle 1 106,5 Music Radio,** 106.5MHz – **SRI,** 105.1MHz. **TIROL (Osttirol): R. Osttirol,** 107.8MHz + 11rly – **Grizzly Radio,** 106.4MHz + 8rly. **VORARLBER: Antenne Vorarlberg,** 106.5MHz + 9rly – **R. Bregenz,** 95.9MHz + 1rly. **R. Edelweiss,** Bachlechnerstr. 46, 6020 Innsbruck: 101.1MHz – **R. M1,** Salzburgstr. 33, 6060 Hall im Tirol: 96.9/104.2/104.9MHz – **R. Zirog,** Bachlechnerstr. 46, 6020 Innsbruck: 103.4MHz.

AZORES (Portuguese)

L.T: UTC - 1h (Su: UTC) — **Pop:** 315.000 — **Radios:** 75.000 (est.) — **Pr.L:** Portuguese — **E.C:** 50Hz, 220/380V — **ITU:** AZR

Mediumwaves	kHz	kW	Times
4) Lajes	648	1	24h
1) Terceira	693	10/5	24h
1) Flores	828	1	24h
1) Barrosa	837	10/5	24h
2) Angra do Heroismo	909	10/1	24h (r. irr.)
1) Faial	1260	1	24h
5) Lajes	1503	0.05	24h (**English**)
3) Vila do Porto	1566	10	24h

FM (MHz): 1) Terceira 90.5/22kW, Faial 88.9/1kW, Pico da Barrosa 97.9/22kW, Terceira 90.5/22kW + 5 relay st's — 2) 101.2/0.4kW — 3) 94.9/1kW — 4) 93.5/0.125kW — 5) 96.1/0.1kW.

Addresses & other information: St's 2) and 3) are comm. 1) Radiodifusão Portuguesa, Centro Regional da R.D.P. – Açores, Rua Dr. Aristides Moreira da Mota 33, 9500 Ponta Delgada, S. Miguel. ☎ +351 (96) 22045. 🖷 +351 (96) 22039 — Dir: Carlos Melo Tavares. **ANN:** "Centro Regional dos Açores da RDP" — **IS:** Regional Hymn. **V.** by letter or QSL-card.
2) R. Clube de Angra (CSB80), Ap. 12, 9701 Angra do Heroísmo Codex, Terceira. ☎ +351 (95) 2310.— Dir. Tec: António Santos Caiado — **ANN:** "Aqui estação CSB80 do R. Club de Angra (A Voz da Terçeira)" — **IS:** Viola — **V.** by letter.
3) Estacão Emissora do Clube Asas do Atlântico (CSB81), Aeroporto de Santa Maria, 9580 Vila do Porto. ☎ +351 (96) 86182. — Pres & Dir. Tec: Manuel Pacheco de Carvalho — **ANN:** "Estação Emissora do Clube Asas do Atlântico" — **V.** by QSL-card or letter — **IS:** Opens and closes with Regional and National Hymn.
4) R. Lajes - Voz da Força Aerea Portuguesa (Portuguese Air Force), Zona Aerea dos Açores, 9760 Praia da Victoria. ☎ +351 (95) 542770/1/2. 🖷 +351 (95) 542200 — DG: Antonio Cunha. CE: Arnando Fernandez. PD/Inf. Dept: Carlos Barbosa — 24h (separate prgrs on MW & FM). **ANN:** MW: "Radio Lajes-a Voz da Força Aérea Portuguesa" — **IS:** Nat. Anthem at s/on & s/off — **V.** by letter. Rec. acc.
5) Station Chief, AFN Lajes Field (AB4), Detachment 6 AFBS, APO AE 09720-7755, USA (FM st. carries separate prgrs) — **ANN:** "Island AM/FM".

OTHER FM STATIONS
Antena Nove, Rua S. Joao 28B, Apart. 118, 9900 Horta Codex: 93.1MHz 0.5kW – **R. Pico,** Rua D. Jaime García Goulart, 9950 Madalena do Pico: 100.2MHz 0.5kW – **R. Lumena,** Bairro da

Conceicao, 9800 Velas S. Jorge: 107.2MHz 0.5kW – **R. Graciosa,** 9880 Santa Cruz de Graciosa: 107.8MHz 0.5kW – **R. Atlantida,** São Miguel: 99.3MHz — **R. Nova Cidade,** S. Miguel: 105.6MHz — **R. Horizonte:** Terceira 104.4 & 106.2MHz, S. Miguel 101.4 & 107.1MHz – **R. Açores,** S. Miguel: 99.3MHz.

BELARUS (Belorussia)

L.T: UTC + 2h (Su: UTC + 3h) — **Pop:** 10.442.000 — **Radios:** 3.150.000 — **Pr.L:** Belarusian, Russian — **E.C:** 50Hz 127/220V — **ITU:** BLR.

MINISTERSTVA SUVJAZI
(Ministry of Posts & Telecommunications)
✉ pr. F. Skaryny 10, 220050 Minsk. ☎ +375 (17) 2273861. 🖷 +375 (17) 2260848 — **L.P:** Minister: Uladzimir I. Goncharenka.

NACYJANALNAJA DZJARZA UNAJA TELERADYJOKAMPANIJA BELARUSI (National State TV & Radio Co. of the Rep. of Belarus, Gov.)
✉ vul. Makajonka 9, 220807 Minsk. ☎ +375 (17) 2649286. 🖷 +375 (17) 2648182. ☺ 252267 TV BY.
L.P: Chmn: Ryhor Kisel. Head of Tech. Dept: G. M. Grinevetsky.

BELARUSKAJE RADYJO (Gov.)
✉ vul. Cyrvonaja 4, 220807 Minsk. ☎ +375 (17) 2333922.
L.P: Gen. Dir: A. S. Ulasenka.

Rg	Stations	kHz	kW	H. of tr.	Prgr.
MA	Sasnovy	171	1000	0300-2200	R1
MA	Sasnovy	279	500	0400-2300	BR1, Reg
MA	Sasnovy	549	1000	0400-2300	RSM
MI	Minsk-Kalodziscy	873	150	0200-0000	RR
MA	Babrujsk	1008	50	0500-2200	BR2
HR	Hrodna	1008	7	0500-2200	BR2
HR	Slonim	1008	50	0500-2200	BR2
HO	Smiatanicy	1008	40	0500-2200	BR2
VI	Usacy	1008	50	0500-2200	BR2
BR	Brest	1026	7	0500-2200	BR2
HR	Heraniony	1026	5	0500-2200	BR2
MA	Mahiliou	1026	50	0500-2200	BR2
MI	Miadzel	1026	25	0500-2200	BR2
BR	Pinsk	1026	7	0500-2200	BR2
MI	Salihorsk	1026	5	0500-2200	BR2
MI	Minsk-Kalodziscy	1125	150	0400-2200	RSY
BR	Baranavicy	1125	10	0400-2200	RSY
MA	Sasnovy*	1170	1000	1600-2300	BR2, F
BR	Baranavicy	1197	5	0500-2200	BR2
VI	Braslau	1197	5	0500-2200	BR2
MI	Minsk-Kalodziscy	1197	50	0500-2200	BR2
VI	Viciebsk	1197	40	0500-2200	BR2
BR	Brest	1278	10	0400-2300	BR1
BR	Baranavicy	1566	5	0300-2300	RSM
BR	Brest	1566	7	0300-2300	RSM
HR	Heraniony	1566	5	0300-2300	RSM
MI	Miadziel	1566	10	0300-2300	RSM
BR	Pinsk	1566	7	0300-2300	RSM
MA	Mahiliou	5965	10	0400-1600	BR1
BR	Brest	6080	10	0400-1700	BR1
MI	Minsk-Kalodziscy	6115	75	0400-2300	BR1, Reg
HR	Hrodna	7140	5	0400-1700	BR1, Reg
MA	Mahiliou	7145	10	1600-2300	BR1
MI	Minsk-Kalodziscy	7210	75	0400-2300	BR1, Reg, F
HR	Hrodna	7265	5	0500-2200	BR2
BR	Brest	11690	10	1700-2300	BR1
MA	Mahiliou	15175	10	1700-2300	BR1, RSM
HR	Hrodna	17805	5	1700-2300	BR1

+) Minsk reg. prgr — *) directional 244Þ.

Rg	FM (MHz)	BR1+R	BR2$	R1	RSM$	RSY	MIR
BR	Brest	72.47	70.91	69.68		71.69	
BR	Brest	103.7					
BR	Pinsk	69.08	67.10	66.32		67.88	
HO	Brahin	68.30	69.92	67.37		69.11	
HO	Homiel	67.76	66.20	69.26		66.98	
HO	Homiel	101.1					
HO	Smjatanicy	67.22	70.28	68.00		66.44	
HO	Zlobin	69.68	71.81	71.03		68.45	
HR	Heraniony	71.54	68.39	69.26		72.32	
HR	Hrodna	66.98	68.90	66.20		67.76	

Rg	FM (MHz)	BR1+R	BR2$	R1	RSM$	RSY	MIR
HR	Hrodna	100.5					
HR	Slonim	69.44	67.34	68.12		66.56	
HR	Slonim	104.0					
HR	Smarhon	67.97*	68.12*				
MA	Asipovicy	68.96	71.69				
MA	Babrujsk	66.02	67.46			71.45	
MA	Kascjukovicy	66.47	68.03				
MA	Mahiliou	70.10	71.96	71.18		72.74	
MA	Mahiliou	100.9					
MI	Berazino		70.79				
MI	Miadziel	68.69	70.31	67.64		66.86	
MI	Miadziel	103.9					
MI	Minsk-Kalodziscy	71.33	70.43+	72.89	72.11		
MI	Minsk-Kalodziscy	103.7					107.1
MI	Salihorsk	70.22	72.23	69.56#	68.57		
MI	Salihorsk	100.3					
		102.8#					
VI	Braslau	69.08	71.99				
VI	Usacy	66.74	68.30	70.94		72.65	
VI	Vidzy	72.80*	74.00*				
VI	Viciebsk	70.67	72.26	69.92		71.48	
		105.5					

+ also carries Minsk musical prgr's 1300-2000. $ = stereo. *) Low power, others mostly 4kW (12-17kW ERP). # = F.PI.
D.PRGR: BR1: 0400-2300 in Belarusian/Russian — **BR2:** 0500-2200 in Belarusian, incl. RSM. relays 0500-0600, 1230-1300, 2000-2200 — F = Foreign Sce, cf. International Broadcasting section.
Relays from Russia: R1 = Radio 1 — RR = R. Rossii — RSY = R. St. Yunost — RSM = R. St. Mayak
MIR: TRK Mir Belarus, pr. Masherova 19, 220078 Minsk. 0900-1400 in Russian.
Regional Prgr's (within BR1): Mon-Fri 0430/0440-0500, 1600-1640.
Regional centres: (vobl. = voblast (province): BR) Bresckaja vobl: vul. Kujbysava 64, 224030 Brest — HO) Homielskaja vobl: vul. Puskina 8, 246050 Homiel — HR) Hrodzenskaja vobl: vul. Horkaga 85, 230015 Hrodna — MA) Mahiliouskaja vobl: vul. Persamajskaja 83, 212030 Mahiliou — MI) Minskaja vobl: Radyjostancyja Stalica (Capital Radio), ul. Saharova 42, 220034 Minsk. — VI) Viciebskaja vobl: vul. Kamunistycnaja 8, 210026 Viciebsk.

INDEPENDENT STATIONS

	MHz	kW*	Station, location, h. of tr.
1)	67.70	4	Avtokanal, Minsk (Kalodziscy): inactive
4)	68.84	4	Radyjo B.A., Minsk (Kalodziscy): 0400-2200
2)	101.2	4	R. Rossii-Nostalgie, Minsk: **F.PI.**
3)	102.1	1	R. ROKS-M, Minsk (Kalodziscy): 0300-2200
4)	103.4	1	Radyjo B.A., Minsk (Mahiliou): **F.PI.**
4)	104.6	4	Radyjo B.A., Minsk (Kalodziscy): 0400-2200

*) transmitter power (TRP).
Addresses:
1) P.O. Box 279, 220050 Minsk — 2) Minsk — 3) vul. Kamunistycnaja 6A, 220029 Minsk (studio) or vul. Storazhevaya 8-a, Minsk 220002. Rel. R. ROKS-Moskva + own ad's on the 7th & 50th min of each h — 4) vul. Ckalova 5, komn. 305, Minsk 220002 (office) or as 3) (studio).

BELGIUM

L.T: UTC + 1h (Su: UTC + 2h) — **Pop:** 10m — **Radio sets:** 5m — **Pr.L:** Dutch, French — **E.C:** a/c 50, 220V — **ITU:** BEL.

VLAAMSE RADIO EN TELEVISIEOMROEP (VRT)
(Public Sector Public Limited Company)
Dutch Language Network
Public Sce. grants by Flemish Council.
✉ Omroepcentrum, August Reyerslaan 52, B-1043 Brussels. ☎ +32 (2) 741 31 11. 🖷 +32 (2) 734 9351.
Web: http://www.vrt.be. **E-mail:** info@vrt.be
L.P: Man. Dir: Bert De Graeve. Adm. Gen: C. Goossens. Dir. Gen (Radio): Chris Cleeren. Dir. Gen. (Television): Piet Van Roe. Dir. of Production Dept: Marleen Bergen. Dir. of Prgr. Dept: Frans Ieven. Dir. Gen: H. van Roost. Tech. Manager: Marc Cabus. Head of Int. Rel: Mrs. Jaklien Caenberghs.

Mediumwaves	kHz	Prgr	kW
Waver-Overijse	540	2	150
O&W VI.	927	1	100
Kortrijk	1188	2	5

FM: (all freqs. MHz, all prgrs stereo)

Stations	1	2	3	4	5	kW
Veltem		88.7		94.8		1
Brussels		90.7			88.3	1
Egem O.-Vl.	95.7	98.6	90.4	102.1	101.5	50
Egem W.-Vl.		100.1				50
Genk	99.9	97.9	89.9	101.4	102.0	20/20/20
					93.0	40/40/3
Leuven	98.5	88.7		88.0	94.8	0.5
Schoten	94.2	97.5	96.4	100.9	89.0	20/20/3/40/20
Antwerpen	94.2		92.0	100.9	89.0	1
			96.4			
St-Pieters-L.		93.7	89.5	100.6		50/50/50/50/2
Gent				94.5		0.5
Brabant	91.7		89.5	100.6	97.0	
Limburg	99.9		89.9	101.4	102.0	
					93.0	
O&W Vl.	95.7		90.4	102.1	93.0	
Diest		92.4				
Dendermonde-A-T			90.7			

N: News prgrs. on radio: (not on Saturdays and Sundays)
D.PRGR: Night prgr on all frqs of Radio 1, Radio 2, Studio Brussels and Donna. Until 0200 also on MW 540 kHz and on MW 1188 kHz all night.
ANN: 1:"Radio Een", 2: "Radio Twee", 3: "Radio Drie", 4: "Studio Brussel", 5: "Radio Donna"

RADIO VLAANDEREN INTERNATIONAAL
See International Broadcasting section.

RADIO-TÉLÉVISION BELGE DE LA COMMUNAUTE FRANCAISE (R.T.B.F.)
French Language Network
Public sce. Grants by Parliament through French language Council.
✉ Cité de la Radio-Television, B-1044 Brussels. ☎ +32 (2) 737 21 11. **Cable:** RTBF Brussels. ☎ 63132. 🖷 +32 (2) 737 4357 **Web:** http://www.rtbf.be/.
L.P: Admin. Gen: R. Stéphane. Dir. Radio: Ph. Dasnoy. Tech. Dept: Chief Eng: F. Fantuzzi. Head Int. Rel: C. Vial.
Regional & Local Centres: Esplanade Anne Charlotte de Lorraine, 7000 Mons; Ave. Golenvaux 8, 5000 Namur; Passage de la Bourse, 6000 Charleroi; Rue de Verviers 203, 4821 Andrimont; Maison de la Culture, Parc des Expositions, 6700 Arlon; rue du Parc 17-21, 4020 Liège.

Mediumwaves	kHz	Prgr	kW		
Wavre	621	1	300		
FM stereo (MHz)	**1**	**2**	**3**	**4**	**kW**
Anderlues		92.3	96.6	99.1	50
Bruxelles	92.5		91.2	93.2	3
Léglise	87.6		94.1	96.4	10
Liège	96.4	90.5	99.5	95.6	1/3x50
Profondeville	102.7	92.8	98.3	90.8	10
Tournai	106.0	101.8	102.6	104.6	5
Verviers		103.0			2
Wavre	96.1	101.1			50/0.5

+ additional low power tr's.
Network 1 (Première): 0430 (Sat/Sun 0530)-2240. **N,Wrp:** 0430, 0500, 0530, 0600, 0630, 0700 and hourly to 1900, 2100, 2200.
Network 2 (Fréquence Wallonie): 0430 (Sat/Sun 0530)-2240. **N,Wrp:** 0430, 0500, 0800, 0900, 1000, 1200, 1400, 1500, 1800, 2100. **R. Chouette** (computerized music prgr.): 2240-0430/0530 on FM only.
Reg. Prgrs: W 0530-0800, 1200-1300, 1600-1800; Fri 1800-2100 — R. Hainaut (Mons) on 1125kHz + 92.3/101.8MHz — R. Liège on 1233kHz + 90.5/103.0/94.6/89.1/89.4MHz — R. Namur on 1305kHz + 97.3/92.8/89.3/91.5/89.4/90.2MHz — R. Bruxelles on 93.2MHz.
Local Prgrs: R. Verviers (Radiolène) W 0000-1200 on 103.0/94.6/89.1/89.4MHz.
Musique 3: 0558-2245. **N,Wrp:** 0600, 0630, 0700, 0800, 1200, 1800-2240.
Network 4 (Radio 21): 0430 (Sat/Sun 0530)-2240, then R. Chouette (see R.2).
Bruxelles Capitale: Mon-Fri on 99.3MHz 1kW (rel. on 93.2MHz 0530-0800).
ANN: "Vous écoutez R. Une, R. Deux, R. Trois ou Radio 21" — **IS:** Xylophone: Extract from melody: "Où peut on être mieux?" by André-Modeste Grétry.
V. by QSL-card. (Re. to Room 4U40). (Reg/Local centres: by letter).

BELGISCHES RUNDFUNK-UND FERNSEHZENTRUM DER DEUTSCHSPRACHIGEN GEMEINSCHAFT (BRF)
German Language Network (Public Sce.)
Grants by RDG-Rat (German speaking community council).

✉ Herbesthaler Str. 82, B-4700 Eupen. ☎ +32 (87) 59 4411/99. **Cable:** 49427 Brf eu b. **Web:** http://www.brf.be/brf/
Regional ✉ Blvd. Reyers 52, B-1040 Brussels – Malmedyer Str. 25, B-4780 St. Vith.
L.P: Pres. of Admin. Council: H. Ossemann; Dir. of BRF: H. Engels. Chief Editor: P. Maraite, CE: J. Schifflers.

STATIONS:	MHz	kW		MHz	kW
Lüttich	88.5	50	Brussel	95.2	2
Auel	92.2	0.16	Eupen	100.5	1
Recht	94.9	5	Recht	104.1	20

D.PRGR: 0500-2100. **N,Wrp:** 0500, 0530, 0600, 0630, 0700, 0800, 0900, 1000, 11000, 1130, 1300, 1400, 1500, 1600, 1700, 1730, 1900, 2000W. **ANN:** "Hier ist der Belgischer Rundfunk, Sendungen in der deutschen Sprache" — **V.** by QSL-card.

COMMERCIAL NETWORKS:

BEL-RTL
✉ rue Ariane 1, 1201 Brussels. ☎ +32 (2) 778 6878. 🖷 +32 (2) 778 6809.
Stations: Waals Brabant 100.2 MHz; Brussel 104.0 MHz; Brussel (C) 103.0 MHz; Charleroi 103.5 MHz; Famenne 101.6 MHz; Luik 103.6 MHz; Bergen 103.4 MHz; Namen 101.6 MHz; Nivelles 107.1 MHz; South Luxembourg 101.8 MHz.

RADIO CONTACT
✉ Avenue Télémaque 33, 1190 Brussels. ☎ +32 (2) 347 4880. 🖷 +32 (2) 347 5307 — Pres: Francis Lemaire.
Two networks serving Flanders and Wallonia respectively.

RADIO NOSTALGIE NETWORK (Belgian)
✉ RADIO Nostalgie (Belgium), Boulevard Ans Pach, 30-36, 1000 Bruxelles. ☎ +32 (02) 2270450 🖷 +32 (02) 2231455 — Dir: J-C Fyon

RADIO NOSTALGIE NETWORK (Dutch)
✉ RADIO Nostalgie, Transvaalstraat 9, 2600 Berchem. ☎ +32 (3) 2810100 🖷 +32 (3) 2810900 — PD: Max van Zanten.

RADIO NOSTALGIE NETWORK (French)
✉ Blvd. Ernest Melot 12, 5000 Namur. ☎ +32 (81) 220 044. 🖷 +32 (81) 241 254 — Dir: J. Fyon. PD: Pierre Joye.

PRIVATE FM STATIONS IN MAIN CITIES:
Brussels: R. Judaica 90.2 — R. Nostalgie 100.0 — R. Ciel 100.4 — R. Alma 101.9 — R. Contact (French) 102.2 — R. Contact Gold (Dutch) 102.5 — R. Contact (Dutch) 103.1 — R. Fantastique 103.4 — NRJ 103.7 — BEL-RTL 104.0 — Frequence L/Europe 2 104.3MHz — Fun R. 104.7MHz — R. Panik 105.4MHz — R. Contact Gold (French) 106.1 — Action FM 106.9 — R. Campus 107.2 — BFM Inforadio 107.6.
Antwerp: R. Antigoor 103.2 — R. Contact 102.5/104.8 — R. Expres 102.8 — R. Nostalgie 107.4/R. Totaal 104.8MHz.
Ghent: R. Contact 104.9 — R. Free 103.5 — R. Go 103.2 — R. Sis 102.8MHz.
Charleroi: Chérie FM 102. — R. Contact 104.0MHz.
Liège: Chérie FM 103.6 — R. Contact 100.9 — R. Nostalgie 107.5MHz.

AMERICAN FORCES NETWORK, SHAPE
✉ Box 7, 7010 SHAPE. (APO AE 09700). ☎ +32 (65) 44 41 21.
L.P: Officer-in-charge: Cpt. G. Martel. Broadc. Superv: SFC C. Kubicek. Chief Eng: René Libre.
STATIONS: Kleine Broegel 106.2MHz 0.1kW, Brussels 101.7MHz 0.9kW, SHAPE 104.2/106.5MHz 4kW, Florennes 107.7MHz 0.1kW, Chievres 107.9.
D.PRGR: 24h on 101.7/104.2/107.7MHz. Own prgrs Mon-Fri 0500-0800, 1400-1700; Sat 0800-1200. Other times rel. AFN Europe.
AFN-2: 24h easy listening stereo prgr. on 106.5MHz.
V. by QSL-card.

BRITISH FORCES BROADCASTING SERVICE
STATIONS: SHAPE 107.6MHz/Antwerp 107.4MHz 0.05kW.
D.PRGR: rel. BFBS Germany (see entry under "Germany").

BOSNIA-HERCEGOVINA

L.T: UTC + 1h (Su: UTC + 2h) — **Pop:** 3.265.000 — **Radios:** 840.000 — **Pr.L:** Serbian, Croatian — **E.C:** 50Hz, 220V — **ITU:** BIH.

RADIO TELEVIZIJA BOSNE I HERCEGOVINE

Bulevar M. Selimovica 4, 71000 Sarajevo. ☎ +387 (71) 646014 or 462886. ⊕ +387 (71) 645142 or 464061.
L.P: Ag. GM: Mrs. Amilia Omersoftic.

Mediumwaves	kHz	kW	Mediumwaves	kHz	kW
Sarajevo	612	600	Zavidovici	1503	1
Bihac	774	2	Mostar	1584	1

FM: Sarajevo 93.1MHz, Vlasic 97.0MHz 100kW + unknown locations on 89.3 & 95.6MHz.
SW: Sarajevo 7102.5kHz (USB + carrier).
D.PRGR: 24h. **Main N:** 0700, 0900, 1400MF, 1900 (audio of 1730 TV bulletin), 2100, 2300 — **Sarajevo 202:** local sce. on 93.1MHz.

RADIO MIR (operated by NATO forces)

MW: Cazin 1017kHz 1kW — **FM:** 106.2MHz. 18h daily.

INDEPENDENT STATIONS

Studio 99, Djure Djakovica 41, 71000 Sarajevo. ☎ +387 (71) 664550. ⊕ +387 (71) 664551 — **LP:** Editor-in-Chief: Adil Kulenovic. Prgr. Editor: Zoran Ilic — **D.PRGR:** 24h on 99.8MHz (rel. RFI 2000-2030).
R. Zid, Ise Jovanovica 8, 71000 Sarajevo. ☎ +387 (71) 443770, 443771, 470584. ⊕ (via USA): +1 (412) 339 4724. **E-mail:** radio_zid@zamir-sa.ztn.zer.de — **LP:** Chmn: Zdravko Grebo. Editor-in-Chief: Vlado Azinovic — **D.PRGR:** 24h.
R. Kameleon, Ulica Oktobarska 31, 75000 Tuzla. ☎ +387 (75) 238247, 235487, 231692. ⊕ +387 (75) 238247. **E-mail:** radio_kameleon@ zamir-tz.ztn.apc.org — **LP:** Chmn: Zlatko Berbic.
Azur FM, Sarajevo: 105.6MHz (music prgrs from R. France Int.).
R. Stari Grad, Sarajevo.
R. Vrhbosna, Sarajevo: serves the Croat community
Fateh (Islamic), Sarajevo. ☎ +387 (71) 532694 — Mgr: Mensud Basic.
FM: 105.9MHz.
Adventist World Radio: Sarajevo 105.2MHz 0.25kW.
A total of 40 st's are reportedly operating in this part of the country.

Herceg-Bosna

RADIO HERCEG-BOSNA

Mostar.
MW: 756kHz 10kW, 1071kHz 40kW.
Kupres R: Kupres 90.5MHz.
A total of 15 st's are reportedly operating in this part of the country.

Republika Srpska

SERBIAN RADIO-TELEVISION (SRT)

Pale — **L.P:** DG: Miroslav Tohol. Editor-in-Chief: Drago Vukovic.
FM: Kozara 92.7/95.3MHz 30kW + unknown locations on 95.8 & 102.3MHz.

SERBIAN RADIO-TELEVISION (SRT)

Banja Luka — **L.P:** Ag. DG: Radomir Neskovic.
FM: 91.8MHz 1kW.
NB: At editorial deadline this st. was operating independently of the st. in Pale.

29 other st's, mostly municipally owned, rel. news from SRT.
R. Big, Banja Luka
R. Krajina, Banja Luka.
R. Recnik, Kozara: 90.3/92.7MHz 30kW.
A total of 36 st's are reportedly operating in this part of the country.

AFN Bosnia (U.S. Mil.)

MW: 1143kHz — **FM:** Tuzla 100.1MHz, Sarajevo 104.5MHz.

BULGARIA

L.T: UTC + 2h (Su: UTC + 3h) — **Pop:** 8.963.000 — **Radios:** 3.920.000 — **Pr.L:** Bulgarian — **E.C:** 50Hz, 220V — **ITU:** BUL.

BÅLGARSKO NATIONALNO RADIO (Gov.)

4 Dragan Tzankov Str, 1421 Sofia. ☎ +359 (2) 661218. ⊕ +359 (2) 657298. **Cable:** Bulgarradio. ① 067 22557 radio bg.
Regional RTV Center, 2 kn. Dondukov-Korsakov Str, 4000 Plovdiv — 22 blv. Primorski, 9000 Varna — 56 blv. A.Velitchkov, 2700 Blagoevgrad — 75 N. Nikolov Str, 6000 Stara Zagora — 7 D. Voynikov Str, 9700 Shumen.
L.P: DG: Ivan Obretenov. Dep. DG's: Borislav Dzhamdziev, Rayna Konstantinova. Chief Secr: Boyan Bonev. Dir. of Prgrs: Rumyan Petrov (Horizont), Sonya Angelova (Christo Botev). Dir. Foreign Sce: Angel

Nedyalkov. Head of Int. Rel: Georgy Stoyanov. TD: Georgy Trenev.
Prgrs: 1 & 2=National 1st & 2nd Prgrs, FL=Foreign language prgr. from Varna, FS=Foreign Sce, T=Turkish language prgr. **Powers:** 2 x indicates that two tr's are used in parallel for part of the time only.

LW/MW	Prgr.	kHz	kW	H. of tr.
Sofia	1+P	261	60	24h
Vidin	2	576	500	0400-2200
Pleven	1	594	250	0400-2200
Dobrich	1	594	2 x 5	24h
Burgas	1	594	2 x 5	24h
Plovdiv	2+R	648	30	24h
Pirin	2	702	2 x 20	24h
Silistra	1	711	1	24h
Petrich	1+R	747	500	0400-2200
Shumen	1+T	747	2 x 5	24h
Salmanovo	1+R	747	10	24h
Sofia	1+C	774	50	24h
Suvorovo	1	774	5	24h
Shumen	2	828	500	0400-2200
Sofia	2	828	50	24h
Blagoevgrad	2	864	150	24h
Samuil	1+T	864	10	24h
St. Zagora	2+R	873	25	24h
Shumen	1+R	963	25	24h
Sofia	1	963	40	24h
Pirin	1	963	2 x 20	24h
Malko Tarnovo	1	963	5	24h
Varna	R+FL	981	150	24h
Kardzali	1+T	1017	50	24h
Vidin	1	1017	2 x 20	24h
Varna	1	1143	50	24h
St. Zagora	1	1161	500	0400-2200
Sofia	1	1161	2 x 20	24h
Targoviste	1+T	1161	2 x 5	24h
Dulovo	1+T	1161	2 x 5	24h
Vidin	FS+1	1224	500	24h
Smolyan	1	1287	1	24h
Kardzhali	2	1296	150	0400-2200
Pleven	2	1296	30	24h
Haskovo	2	1485	1	24h
Stolnik	1	7670	15	0400-2200
Plovdiv	1+FS	9850	250	24h

FM (MHz)	1	2	R	kW (ERP)
Belogradchik	67.8	67.1		6
Blagoevgrad	105.2	89.7s	105.2	1
Botev Mt.	70.88	72.44	71.66s	6
	100.9			10
Burgas	67.58	66.80	88.5*	6/6/1
	102.5s			10
Dobrich	71.42	65.84s	88.7s	3/3/1
G. Deltchev	72.92	71.36s	72.92	6
	100.3s			10
Jakoruda	101.7s			0.25
Kardzhali	67.4	65.84s		6
	105.6s			10
Kjustendil	67.76	66.98s	67.76	5
	102.1s			10
Petrochan	101.4s			10
Pleven	102.7s	100.2s		1
Plovdiv	94.0s	90.6	94.0	0.5
Ruse	67.22	69.32s		3
	103.0s			10
Silistra	71.78			0.6
	103.3s	90.3s	90.3s	0.5
Sliven	71.24	72.8		6
Smolijan	67.19	67.97		3
	101.6s	96.0s	103.1s	5/1/3
Sofia	68.06s	66.5s		10
	69.26s		69.26ss	10
	102.4s	92.8s		10
Shumen	71.0	72.56		6
	102.0s	87.6s	87.6s	10
			69.8s	6
St. Zagora		98.3s	88.3s	1
Varna	67.31	68.09	69.41s	3
	100.9s			5
			103.4s	1

s) stereo

Prgr. 1 ("Horizont"): 24h music and news.
Prgr. 2 ("Christo Botev"): 24h culture, education & music.
Regional Prgrs (R): Blagoevgrad: 0400-1700 on 747kHz + 67.76/
72.92MHz — **Plovdiv:** 0400-1700 on 648kHz + 71.66/94.0MHz —
Stara Zagora: 0400-1700 on 873kHz + 88.3MHz — **Shumen:** 0400-
1700 on 963kHz + 90.3/87.6MHz — **Varna:** 24h on 981kHz + 69.41/
88.5/88.7/103.4MHz.
Foreign Language Prgr of R. Varna (FL): May-Oct 0700-1000 &
1100-1230 in English, French and German.
Turkish (T): W 0430-0500 & 1030-1100, Sun 0600-0630 & 1030-1100
on 747/864/1017/1161kHz + 69.8MHz (Shumen).
Capital Sofia (C): W 1100-1200 on 69.26MHz.
Parliamentary Channel (P): times vary.

FOREIGN SERVICE: see International Broadcasting section.

Private Stations

MHz	Station	Location
87.7	R. Almatea	Padzardzik
87.8	R. 99	Shumen
87.9	Classic FM R.	Pleven
87.9	Darik R.	Vratsa
88.9	Darik R.	Gorna Orjahovitsa
89.3	Kanal KOM	Varna
90.2	R. Bella	Petrich
90.4	R. Treli	Stara Zagora
90.5	R. Glarus	Burgas
90.9	R. 99	Smoljan
91.2	R. Veselina	Jambol
91.6	R. Ogosta	Montana
91.9	Petra FM	Petrich
92.2	R. Express	Kjustendil
92.6	R. Atlantik	Galatea Varna
92.6	VOA Express	Varna
93.0	R. Nevrokop	Gotse Delchev
94.0	R. Viva	Sofia
94.3	R. Domino	Starae Zagora
94.7	R. Bimako	Sliven
94.9	R. FM Plus	Sofia
95.3	R. Haskovo	Haskovo
95.5	R. TNN	Plovdiv
95.7	R. Veselina	Stare Zagora
95.9	R. Oberon	Pazardzik
95.9	Classik FM R.	Varna
96.4	Kanal KOM	Plovdiv
96.4	R. Bravo	Varna
96.4	R. Gabrovo	Gabrovo
96.7	R. Tangra	Sofia
97.0	R. Atlantik	Plovdiv
97.0	VOA Express	Plovdiv
97.6	R. Maja	Blagoevgrad
97.6	R. Geya	Sandanski
97.6	R. Vitosha	Sofia (partly rel. VOA)
97.8	Classic FM R.	Vidin
97.7	R. Veselina	Plovdiv
97.7	VOA Express	Plovdiv
98.3	Darik R.	Sofia
98.7	R. Astra	Pleven
98.9	R. Aura	Blagoevgrad
98.9	Non Stop Musik	Vratsa
99.0	Tsche-el-Ka	Velingrad
99.0	R. Boomerang	Stara Zagora
99.3	Darik R.	Shabla
99.4	Darik R.	Jambol
99.7	R. Pazardzik	Pazardzik
100.1	Classic FM R.	Stara Zagora
100.5	R. Pristis	Ruse
100.6	R. Targoviste	Targoviste
101.7	R. Boomerang	Gabrovo
102.1	Classic FM R.	Haskovo
102.4	R. Veliko Tarnovo	Velike Tarnovo
103.7	Darik R.	Dobrich
103.7	Classic FM R.	Plovdiv
103.8	R. Strimon	Blagoevgrad
103.9	R. Maja	Burgas
103.9	VOA Express	Burgas
105.0	R. Express	Sofia
105.1	Classic FM R.	Burgas
105.4	R. Express	Plovdiv
105.4	R. Bravo	Shumen

MHz	Station	Location
105.7	Classic FM R.	Veliko Tarnovo
105.8	R. Melodia	Silistra
105.8	R. Tempo	Russe
106.0	R. 99	Sofia
106.2	Darik R.	Kjustendil
106.4	Kanal KOM	Sliven
106.5	Juzhen Briag	Burgas
106.5	R. Forte	Shumen
107.0	R. Express	Smoljan
107.7	Kanal KOM	Vidin
107.9	Sedem Dni	Sofia

FM relays of International Broadcasters:
BBC: Sofia 91.0MHz 1kW 24h
Deutsche Welle: 95.7MHz 1kW 24h
R. Free Europe: Pleven 87.9MHz, Sofia 89.1MHz, Varna 95.9MHz,
Vidin 97.8MHz, Stara Zagora 100.1MHz, Khaskovo 102.1MHz, Plovdiv
103.7MHz, Burgas 105.1MHz, V. Tirnovo 105.7MHz.

RFI BULGARIE (R. France Internationale)

🖃 V. Lesvki Str, 1000 Sofia. ☎ +359 (2) 810272. 🖷 +359 (2) 802741
— **LP:** SM: Ludmil Fotev. Editor-in-Chief: Sava Dinev.
FM: Sofia 103.6MHz. 24h.

CROATIA

L.T: UTC + 1h — **Pop:** 4.784.265 — **Radios:** 1.100.000 — **Pr.L:**
Croatian — **E.C:** 50Hz, 220V — **ITU:** HRB.

MINISTARTSVO POMORSTVA, PROMETA I VEZA
(Ministry of Maritime Affairs, Transport & Communications)
🖃 Prisavlje 14, Zagreb. ☎ +385 (1) 517000. 🖷 +385 (1) 615015.
① 21275 — **LP:** Minister: Ivica Mudrinic.

HRVATSKA RADIOTELEVIZIJA (HRT)
🖃 Dezmanova 10, 41000 Zagreb, Croatia. ☎ +385 (1) 6163691.
🖷 +385 (1) 6163692. ① 21326 HRT CRO.
Web: http://www.hrt.hr
LP: Gen. Man: Ivica Vrkic. Gen. Dir: Antun Vrdoljak. ☎ +38 (1) 272372.
Tech. Dir: Nikola Percin. 🖃 Prisavlje 3, Zagreb. ☎ +385 (1) 663663. 🖷
+385 (1) 6163308.

HRVATSKI RADIO (HR)
🖃 Prisavlje 3, Zagreb. ☎ +385 (1) 6163355. 🖷 +385 (1) 6163347.
① 21154 HR CRO — **LP:** Dir: Mrs Ivanka Lucev. ☎ +385 (1) 6163308.

MW	kHz	kW	Prgr.	H. of tr.
Osijek	594	10	1	24h
Koprivnica	765		L	0500-2100
Hvar	774	50/10	1	0500-0100
Buje	783	10	1	0500-0100
Vrbovec	828		L	1200-2300
Deanovac	1125	25/100	1	24h
Zadar	1134	600/1200	1	1400-0630
Tovarnik*	1143	300	1	
Nasice	1233		L	0600-1600
Cakovec	1305		L	0400-2100
Zabok	1350		L	0600-2300
Karlovac	1449	10/1	L	0600-2300
Dubrovnik	1485	1	L	
Brod	1584	1	L	24h
Osijek	1557	20	L	
Posega	1602		L	0630-2100
*) inactive				

FM	HR1	HR2	HR3	Local
Belje	93.3	98.1		102.8
Biokovo	89.7	98.9	-	102.0e
Blato na Korculi	87.7	92.8	-	97.2
Borinci	88.3	96.1		105.5
Brac	99.8		88.8	104.5e
Brezovica	90.1	95.0	107.7	
Brinje	91.6	99.2	-	
Brodski Stupnik	93.1	98.3		
Buje	91.3	103.7	93.2	96.4a
Cabar	89.2	90.7		
Celavac	95.1	98.1		103.0

THE WRTH 1999 QUESTIONAIRE

TELL US WHAT YOU THINK *and* **WIN** A SHORTWAVE RECEIVER!

This is your chance to tell us what you think of WRTH and a little bit about yourself so that we can shape the book to your needs and serve you even better in the years ahead. Complete entries will be eligible to enter a prize draw. The first prize is none other than the winner of the Best Semi-Pro Receiver 1999, the JRC NRD545. What a prize! Five runners-up will each get a copy of the 2000 edition of WRTH.

1. Name and address (use CAPITALS only please)

   Name

   Street

   City   State

   Zip or Postal code   Country

2. Gender: a. ❑ male   b. ❑ female
3. Age: a. ❑ 15–30   b. ❑ 31–40   c. ❑ 41–50   d. ❑ 50+
4. Which category best describes your situation?
   a. ❑ broadcasting professional   b. ❑ dedicated radio hobbyist/DXer   c. ❑ casual listener
5. How many times have you bought WRTH? a. ❑ 0   b. ❑ 1–3   c. ❑ 4–6   d. ❑ 6+
6. How often do you buy WRTH?
   a. ❑ every year   b. ❑ once every two years   c. ❑ less than once every two years
7. Are you a member of a DX club? a. ❑ Yes   b. ❑ No
8. What brand is your receiver? a. ❑ AOR   b. ❑ Drake   c. ❑ Grundig   d. ❑ JRC
   e. ❑ Kenwood   f. ❑ Lowe   g. ❑ Sangean   h. ❑ Sony   i. ❑ Yaesu   j. ❑ other
9. How much do you spend on radio equipment annually?
   a. ❑ less than $200   b. ❑ $200–$500   c. ❑ $500–$2000   d. ❑ $2000–$5000
10. Do you have plans to buy new equipment next year? a. ❑ Yes   b. ❑ No
11. Are you connected to the Internet, or do you expect to be in the next six months?
    a. ❑ Yes   b. ❑ No   c. ❑ No PC
12. If WRTH had a Website would you subscribe to a service with monthly news on international broadcasting and updates to broadcasting schedules? a. ❑ Yes   b. ❑ No
13. How much would you be willing to pay for this service as a monthly fee/per visit?
    a. ❑ nothing   b. ❑ $1–$5   c. ❑ $6–$10   d. ❑ $10+
14. Would you prefer national listings by region or by country? a. ❑ region   b. ❑ country
15. Would you like the TV listings to be with the national radio services? a. ❑ Yes   b. ❑ No
16. How many other people read your copy of WRTH? a. ❑ 0   b. ❑ 1   c. ❑ 2   d. ❑ 3+
17. WRTH is the only handbook with current winter broadcasting schedules at the time of publication. Is it important to you to have these winter schedules? a. ❑ Yes   b. ❑ No
18. Do you find all the information you need on radio listening and/or TV in WRTH?
    a. ❑ Yes   b. ❑ No
19. If no, what information should be added to WRTH? Or are there any changes you would like to see to the content of WRTH? _____

FREE INFORMATION FORM
WORLD RADIO TV HANDBOOK 1999

Dear WRTH reader,

To get **free information** from WRTH advertisers, check the box next to the advertiser's name you are interested in. Please select no more than five choices. Advertisers will send the information to you directly.
Please do not forget to fill in the other side of this card in order to have a chance to win the JRC NRD545. Return this card to arrive before 30 April 1999. Cards received after that date cannot be processed.

1. ❑ Adventist World Radio
2. ❑ AKG Acoustics
3. ❑ AOR (UK) Limited
4. ❑ Atlantic Ham Radio
5. ❑ Barry Electronics Corp.
6. ❑ BBC Monitoring
7. ❑ Continental Electronics
8. ❑ Cumbre DX

9. ❑ Danish Shortwave Club
10. ❑ Family Radio
11. ❑ Merlin Communications
12. ❑ Monitoring Times
13. ❑ Popular Communications
14. ❑ Radio Netherland
15. ❑ Radio Taipei
16. ❑ RF Systems

17. ❑ Shortwave Magazine
18. ❑ Sony
19. ❑ Ten-Tec
20. ❑ Thomcast
21. ❑ Universal Radio
22. ❑ Vacuum Tube Industries
23. ❑ Voice of America

(fold along this line and tape the card when mailing)

Place
Stamp
Here

WORLD RADIO TV HANDBOOK
P.O. Box 7373
MILTON KEYNES
MK12 5ZL
U.K.

FM	HR1	HR2	HR3	Local
Cres	87.9	95.6		97.9
Drenovci	92.1	104.4		102.4b
Duge Njive	92.9	100.5		
Fuzine	87.8	107.8		
Gerovo	88.7	98.6		
Gruda II	101.7	106.1		89.5
Hvar	93.7	97.0		100.2
Ilok	91.6			105.8
Ivancica	102.4	106.4	96.1	
Kalnik	90.8	105.8	107.8	
Knin				90.2
Komiza		95.6		93.3
Korcula	94.4	100.3		103.8
Koromacno	90.0	88.1		96.3
Krk			87.6	
Kupjacki Vrh	90.0	105.1	103.9	
Labinstica	91.3	96.1		101.0e
Lastovo	90.3	92.4		106.2
Licka Pljesivica	87.7	90.5		100.3
Limski kanal	90.2	102.6		93.9
Lopud	99.6			106.5
Mali Losinj	92.0	97.5		107.4
Martinscica	88.6	92.7		
Mirkovica	91.3	93.3		102.7
Murter	92.7	99.4	104.1	
Novigrad (Istra)	90.6	102.2		93.8
Obrovac	87.9	99.2		
Osijek II				102.4b
Pag	98.5	103.4		105.9
Papuk	94.9	106.8	97.7	
Peljesac	92.1	101.3		88.2
Porec	94.3	88.3		
Prezid	88.0	96.0		101.7
Promina	89.6	99.0		94.4
Psunj	97.3	99.7		102.0
Pula	91.4	102.1	94.2	100.0
Pulac				95.1
Rabac	88.4	98.2		
Rasa	88.7	92.1		103.8
Rovinj	92.4	103.6		
Savudrija	89.9	94.4		
Sibenik-Subicevac	94.0		102.3	
Slano	97.7	100.3		103.7
Slatina	96.6	100.7	91.1	
Slavonski Brod	91.3	105.1	107.9	
Sljeme	92.1	98.5	101.0	88.1d
Srb	88.4	91.9		
Srdj	88.9	98.5		105.0g
Starigrad Paklenica	99.0	89.0		
Stipanov Gric	102.3	97.5	89.7	
Straza	93.5	102.2	106.8	
Ugljan	91.6	87.6		101.8
Uljenje	95.1	103.0	105.6	
Umag	92.4	106.0		
Ucka	99.3	105.3		101.3a
				104.7c
Vela Luka	103.4	94.2		101.1
Velika Petka			102.4	
Vratnik	88.2	100.9		
VrgoracGomila	94.9	96.5		
Vrlika	87.7	96.9		105.8
Vrsar	89.0	107.2		100.0a
Zadar			107.3	
Zlatarevac	99.5	89.3		105.6

SW: see International Broadcasting Section
HR1: 24h. **N:** on the h. **N. in English:** MF 0705W, 0805 Sun, 0903 W, 1005 Sun, 1305, 1905, 2305 — **HR2:** 0600-2300 — **HR3:** 0700-2300.
Local Prgrs (freqs as above):
a) **R. Pula**, Riva 10, 52100 Pula. ☎ +385 (52) 22393. **Web:** http://www.hrt.hr/hr/hrpula
b) **R. Osijek**, Samacka 13, 3100 Osijek.
c) **R. Rijeka**, Korzo 24, 51000 Rijeka. ☎ +385 (51) 657 777. 🖹 +385 (51) 657 745. **E-mail:** radiori-hrt@alf.tel.hr
Web: http://www.multilink.hr/radiorijeka/
d) R.Sljeme (Zagreb), e) Split, f) Zadar, g) Dubrovnik.

ANN: "Ovdje Hrvatski Radio". E: "This is Croatian Radio".
IS: Tune to Dubrovnik's poem "Lovely, Dear, Sweet Liberty" played on celeste — **V.** by letter. Re. in SINPO code to Zelimir Klasan, HRT, Prisavlje 3, Zagreb, Croatia. ☎ +385 (1) 6163355. 🖹 +385 (1) 6163347. **E-Mail:** Zelimir.Klasan@hrt.hr

OTHER STATIONS

Station & Location	MHz
Obiteljski Radio, Zagreb	89.7
Radio Martin, Zagreb	90.3
4) R. 057, Zadar	91.0
1) R. Baranja, Osijek: 24h	91.2
8) R. Koprivnica	91.7
Radio Samobor, Zagreb	93.0
Radio Sesvete, Zagreb	93.2
Radio Cibona, Zagreb	93.6
2) R. Velika Gorica	94.9
3) R. Maestral, Pula	95.4
10) R. Marija, Zagreb	96.4
12) Gradski R, Varazdin	96.5
11) R. Dalmacija, Split	96.8
Plavi Radio, Zagreb	98.0
6) Gradski R, Osijek	99.1
9) Radio 101, Zagreb	94.3
Radio Zapresic, Zagreb	99.5
Radio Student, Zagreb	100.5
7) R. Brod, Slavonski Brod	101.3
Radio Samobor, Zagreb	102.1
Hrvatski Katolicki Radio, Zagreb	103.5
1) R. Baranja, Slavonia	104.4
Radio Cibona, Zagreb	104.5
5) R. Ritam, Sibenik	106.4

Addresses and other information
1) 31000 Osijek ☎ +385 (31) 121 336. **Web:** http://www.open.hr/com/radio_baranja/ — **L.P:** Dir: Karoly Janosi. Tech. Mgr: Mario Fuis. Mktg. Mgr: Dario Jagarinec.
2) Radio Turopolje d.o.o., Zagrebacka 3, 10410 Velika Gorica. **Web:** http://www.rvg.com
3) Narodni trg 9, 52100 Pula. ☎ +385 (52) 217 713. 🖹 +385 52 217 795. **E-mail:** maestral@gradpula.com
Web: http://www.gradpula.com/maestral/
4) C.F. Bianchia 2, 23000 Zadar. ☎ +385 (23) 314 057. 🖹 +385 (23) 314 457. **E-mail:** radio057@dalmacija.com
Web: http://www.dalmacija.com/radio057/ — **L.P:** Dir: Darko Smrkiniæ.
5) S. Radlac 24, 22000 Sibenik. ☎ +385 (22) 212 852 🖹 +385 (22) 212 858. **E-mail:** ritam@dalmacija.com.
Web: http://www.dalmacija.com/radioritam/ — **L.P:** Dir: Miso Bijelic
6) Trg. A. Slarcevica 7/1. 31000 Osijek. ☎ +385 (31) 168991. 🖹 +385 (31) 169876. **E-mail:** gradski-radio@gradski-radio.tel.hr. **Web:** http://www.gradski-radio.tel.hr/gradski-radio/
7) Slavonski Brod. **E-mail:** rtvsb@sb.tel.hr
Web: http://www.iridis.com/rtvsb/ — **L.P:** Mgr. & Chief Editor: Frano Piplovic.
8) Matije Gupca 2, 48000 Koprivnica. ☎ +385 (48) 625 011, 625 012. 🖹 +385 (48) 625 036. **E-mail:** rkc@radio-koprivnica.hr
Web: http://www.radio-koprivnica.hr
9) Zagreb. ☎ +385 (1) 42710. **Web:** http://pubwww.srce.hr/r101/
10) IX. Podbreje 35, 10020 Zagreb. ☎ +385 (1) 655 02 34. **E-mail:** radio-marija@zg.tel.hr **Web:** http://jagor.srce.hr/~bgrsic/rm/radio.htm.
11) Kralja Zvonimira 14/IX, Poŕtanski Pretinac 174, 21000 Split. ☎ +385 (21) 342910. 🖹 +385 (21) 551092. **Web:** http://www.rd.tel.hr/rd/
12) Zagrebacka 10, 42000 Varazdin. ☎🖹 +385 (42) 213100, 213103. **E-mail:** info@radio-042.hr. **Web:** http://www.radio-042.hr

AMERICAN FORCES NETWORK (US Mil.)
FM: Zagreb 99.0MHz, Sovanski Brod 103.5MHz (rel. AFN Bosnia).

CZECH REPUBLIC

L.T: UTC + 1h (Su: UTC + 2h) — **Pop:** 10.514.360 — **Radios:** 9.100.000 — **Pr.L:** Czech — **ITU:** CZE.

CESKÉ RADIOKOMUNIKACE, a.s.

⌨ U nákladového nádraží 4, 130 00 Praha 3. ☎ +420 (2) 6700 5111.
🖷 +420 (2) 691 9292 — **L.P:** Gen. Dir: Miroslav Curín.
Operates the TV and radio transmission facilities.

CESKÝ ROZHLAS

⌨ Czech Radio, 120 99 Praha 2, Vinohradská 12. ☎ +420 (2) 2409
4111. 🖷 +420 (2) 2422 2223. ☉ 121100. **Cable:** Radio Praha.
L.P: DG: Vlastimil Jezek. PD: Josef Havel. TD: Jaromír Holý. Dir. Foreign
Rel. & Prgr. Exchange: Jarmila Kopecková.

Long & Mediumwaves: Reg=region (see regional addresses)

Reg	Stations	kHz	kW	Prgr.
	Uherské Hradiště	270	750-	Radiozurnál + CRo 1
5)	Broumov	558	2	Reg. H. Kralové + CRo 2
2)	Lipno	558	1	Reg. C. Budejovice + CRo 2
3)	Tachov	558	1	Reg. Plzen + CRo 2
	Praha (Liblice)	639	1500	CRo 2
	Ostrava	639	30	CRo 2
	Praha-Zbraslav	792	40	Radiozurnál + CRo 1
	České Budejovice	846	7	Radiozurnál + CRo 1
7)	Ostrava-Svinov	846	2	Reg. Ostrava + CRo 2
	Strakonice	864	7	CRo 2
	Brno	900	25	CRo 2 (daytime only)
	Brno	954	200	CRo 2
	Plzen	954	40	CRo 2
	České Budejovice	954	30	CRo 2
	Liberec	954	3	CRo 2
	Jihlava	981	7	CRo 2
	Melnik	1287	400	R6 - RFE
	Litomysl	1287	300	R6 - RFE
	Ostrava	1287	2	R6 - RFE
	Karlovy Vary	1332	20	Reg. Plzen
	Moravské Budejovice	1332	50	CRo 2

FM (MHz)

Reg	Location	CRo 1	CRo 3	Reg./CRo 2	kW
6)	Brno	95.1	102.0	106.5	90/65
6)	Brno C.	92.6	90.4	93.1	5.75/0.2
2)	C. Budejovice	96.1	91.1	106.4	62
4)	Chomutov	98.9	96.3	103.1	10
3)	Domazlice	90.3	98.0	105.3	2
8)	Jeseník	94.3	91.2	106.8	18.8
6)	Jihlava	90.7	95.4	93.1	20
4)	Liberec	95.9	103.9	89.9	5.3
7)	Ostrava	101.4	104.8	107.3	50/45/2.8
5)	Pardubice	89.7	102.7	100.1	93/72
3)	Plzen	89.1	95.6	101.7	65
3)	Plzen C.	99.2	93.3	106.7	5
1)	Praha			100.7	16.5
1)	Praha C.	94.6	105.0	92.6	5

Reg	Location	CRo 1	CRo 3	Reg./CRo 2	kW
	Strakonice	97.0			1
7)	Trinec	92.1	101.9	105.3	1
5)	Trutnov	88.5	90.5	101.9	10/5
	Votice	93.1	103.2		95
4)	Ústí nad Labem	90.9	88.8	104.5	80
6)	Zlín	99.5	94.8	107.1	5.5

+ 8 low power sites not mentioned

CRo 1: 24h. **N:** 0700, 1200, 1800, 2300 & short news on the h — **CRo
2:** 24h — **CRo 3:** 24h. **N:** on the h — **CRo1:** 0700-1500 on MW freqs
— **Radio 6 - RFE:** 0500-2300. Combined prgr. of Radio Free Europe in
Czech & Slovak (12h daily), Radiozurnal (5h), VOA, BBC and Deutsche
Welle. **N:** on the h.

Regional Stations (own prgrs and relays of CRo 2): 1) Hybešova 10,
180 00 Praha 8: 0505 (SS 0800)-2300 — 2) U tri lvu 1, 370 29 České
Budejovice: 0500 (SS 0700)-2300 — 3) Nám Miru 4, 320 70 Plzen:
0600-2300 — 4) Na schodech 10, 400 91 Ústí nad Labem: 0500-2300.
0500 (Sat 0800, Sun 0900)-2300 — 5) Vrchlického 985, 501 82 Hradec
Králové: 0500 (Sat 0800, Sun 0900)-2300 — 6) Beethovenova 4, 657
42 Brno. **Web:** http://zocalo.net/cz/vbp-en/4/cro.html: 0500 -1700 —
7) Dr. Smerala 2, 729 91 Ostrava: 0400 (SS 0600)-2300 — 8) Olomouc.

FOREIGN SERVICE: See International Broadcasting section.

Major Private Stations/Networks

COUNTRY RADIO (Comm.)

⌨ Zenklova 34, 180 00 Praha 8. ☎ +420 (2) 826660. 🖷 +420 (2) 822145.
E-mail: country@ecn.cz **Web:** http://www.ecn.cz/country/

L.P: MD: Zdenek Petera. PD: Jan Sramek. CE: Miroslav Kasan.
MW: Praha 1062kHz 20/1kW — **FM:** Praha 89.5MHz.
D.PRGR: 24h.

RADIO ALFA (Comm.)

⌨ Na Porici 12, 110 00 Praha 1. ☎ +420 (2) 2487 2822. 🖷 +420 (2)
2487 2823 — **L.P:** MD: V. Kasik. PD: J. Havel. CE: K. Zyka.
FM: see freq. list below — **D.PRGR:** 24h.

FREKVENCE 1 (Comm.)

⌨ Nadrazni 56, 150 05 Praha 5. ☎ +420 (2) 24 57001900. 🖷 +420 (2)
57314186. **Email:** info@frekvence1.cz
Web: http://www.frekvence1.cz
L.P: Pres: Michael Fleischmann. MD: Vit Vojtech. PD: Ivo Mravinac.
FM: see list below.

EVROPA 2

⌨ Nádrazní 56,150 05 Praha 5. ☎ +420 (2) 57001808. 🖷 +420 (2)
57001807. **Email:** info@evropa2.cz/ **Web:** http://www.evropa2.cz/
L.P: Pres: Michael Fleischmann. MD: Vit Vojtech. PD: Milos Pokorny.
Mus. Dir: Roman Ondracek. **FM:** see list below.

FM PLUS

⌨ Box 40, Plzen 320 90 ☎ +420 (19) 276666, 7422222. 🖷 +420 (19)
7422221. **E-mail:** info@fmplus.cz/ **Web:** http://www.fmplus.cz/
L.P: Owner: Vaclav Jezek. Dir: Zbynek Suchy. PD: Vaclav Vobornik.
FM: see list below.

Listing of FM stations by frequency
(based on information provided by the Czechoslovak DX Club)

MHz	kW	Station	Location
87.6	70	R. Alfa	Brno
87.6	0.4	R. FM Plus	Beroun
87.8	1	R. Rokko	Praha
87.8	1	R. Cerná hora	Králíky
87.8	0.1	R. Krumlov	České Budejovice
88.0	1	Radioclub/R. Evropa 2	Ústí nad Labem
88.0	0.1	R. Faktor	Ceské Budejovice
88.1	10	R. Pohoda	Jeseník
88.1	0.18	R. FM Plus	Susice
88.1	1	R. Evropa 2	Liberec
88.2	3.7	R. Evropa 2	Praha
88.3	10	R. Kiss Hády 88 FM	Brno
88.3	0.2	R. Egrensis	Karlovy Vary
88.4	1	Eldorádio	Ceské Budejovice
88.6	0.5	R. Orion	Opava
88.9	10	R. Jih	Breclav
89.0	1	R. Práchen	Písek
89.0	45	R. Alfa	Ostrava
89.3	5	R. Presston	Benesov
89.3	0.2	R. Apollo	Bystrice pod Hostýnem
89.5	1.6	Country Radio	Praha
89.5	0.6	Radiohrad	Terlicko-Hradiste
89.6	7	AZ Rádio	Zlín
89.6	1	R. West/R. Evropa 2	Plzen
90.0	10	R. Kiss ProTon 90 FM	Plzen
90.0	0.08	R. Dragon	As
90.0	0.5	R. Dragon	Cheb
90.2	1	R. Styl	Kutná Hora
90.3	1	R. Limonádový Joe	Praha
90.3	3	Hit R. Publikum	Zlín
90.5	0.5	R. C. Budejovice/Evropa 2	Ceské Budejovice
90.6	0.6	R. Most	Chomutov
90.6	0.5	R. Proglas	Bystrice pod Hostýnem
91.0	70	R. Frekvence 1	Ostrava
91.3	0.5	R. Classic	Liberec
91.4	66	R. Alfa	Plzen
91.6	1	R. Decín	Decín
91.6	0.5	R. Life	Pardubice
91.7	2	R. Zlín	Zlín
91.7	0.20	R. Bonton	Benesov
91.8	0.50	R. umava	Kasperské Hory
91.9	1	R. 1	Praha
92.1	5	R. Alfa	Trutnov
92.3	1	R. Relax	Kladno
92.3	5	R. Haná	Pohorany
92.5	1	R. Egrensis	Mariánské Lázne
92.8	1	R. Sprint	Ostrava
92.8	1	R. Metuje	Náchod

MHz	kW	Station	Location
92.9	1	R. Delta	Mladá Boleslav
93.2	1	R. Egrensis	Cheb
MHz	kW	Station	Location
93.3	2	R. Proglas	Jeseník
93.4	26	R. Frekvence 1	Jihlava
93.5	5	R. Frekvence 1	Ústí nad Labem
93.6	1	R. Faktor	Písek
93.7	45	R. Hellax	Ostrava
93.7	5	R. City	Praha
93.8		R. Evropa 2	Karlovy Vary
93.9	80	R. OK	Pardubice
94.1	50	R. Frekvence 1	České Budejovice
94.3	10	R. Vysocina	Jihlava
94.4	0.8	R. Pro 94.4	Králíky
94.5	0.5	R. Duha/R. Evropa 2	Vrchlabí
94.7	0.2	R. Metuje	Vamberk
94.7	0.1	R. Metuje	Broumov
94.8	0.1	R. Bonton	Mladá Boleslav
94.8	0.5	R. Decín	Decín
95.0	95	R. Vox	Votice
95.2	0.5	R. North Music	Ústí nad Labem
95.2	0.8	R. Sumava	Klatovy
95.3	1	R. Vox-	Praha
95.5	0.2	R. Evropa 2	Zlín
95.6	0.1	AZ Rádio	Vsetín
95.6	0.2	R. Apollo	Olomouc
95.7	0.15	R. Profil	Hradec Králové
95.8	0.50	R. Most	Most
96.2	0.02	R. Cerná hora	Hradec Králové
96.2	1	R. Zlín	Uherský Brod
96.4	4	R. Orion	Ostrava
96.5	1	R. Prima	Sumperk
96.5	0.1	R. Most	Teplice
96.6	3.5	R. Alfa	Praha-mesto
96.9	1	R. Profil	Pardubice
96.9	0.3	R. Bonton	Ceská Lípa
97.1	0.8	R. Rubi	Pohorany
97.1	0.5	R. Euro K	Liberec
97.2	1	R. Zlatá Praha	Praha
97.2	0.1	R. FM Plus	Zelezná Ruda
97.4	50	R. Frekvence 1	Pardubice
97.4	0.2	R. Relax	Rakovník
97.7	12	R. Classic	Praha
97.9	1	R. Triangl	Liberec
98.1	1	R. Agara	Chomutov
98.1	1	R. Kiss 98 FM	Praha
98.3	1	R. Sprint	Trinec
98.7	1	R. Classic	Praha
98.7	1	R. Orion	Trinec
99.3		R. Evropa 2	Opava
99.3	10	R. Attack	Jeseník
99.3	1	R. Krumlov	Ceský Krumlov
99.5	1	R. Evropa 2	Hradec Králové
99.5	0.4	R. Práchen	Príbram
99.7	1	R. Bonton	Praha
99.7	1	R. Faktor 2	České Budejovice
99.7	1	R. Dragon	Karlovy Vary
99.8	1	R. Apollo	Valasské Mezirící
99.9	1	R. Crystal	Ceská Lípa
100.2	0.2	R. FM Plus	Praha
100.3	26	R. Alfa	Jihlava
100.6	2	R. Tep	Teplice
100.9	12.5	R. Alfa	Jeseník
101.1	1	R. Kiss Morava	Frýdek-Místek
101.3		R. Evropa 2	Plzen
101.4	20	R. Contact Liberec (RCL)	Liberec
101.8	1	R. Classic	Tábor
101.9	0.02	R. Apollo	Vsetín
102.0	50	R. Alfa	Ústí nad Labem
102.5	3.5	R. Frekvence 1	Praha-mesto
102.5	0.1	R. Zlín	Vsetín
102.8	0.4	R. Dragon	Cheb
102.9	50	R. Alfa	Ceské Budejovice
103.0	10	R. Krokodýl	Brno
103.4	1	R. Brno	Brno
103.4	1	R. OK	Hradec Králové
103.8		R. Frekvence 1	Klatovy
103.9	0.5	R. Orion	Valasské Mezirící
104.1	50	R. Frekvence 1	Plzen
104.2	1	R. Karyon	Znojmo
104.3	31.5	R. Faktor	Ceské Budejovice
104.3	20	R. Frekvence 1	Jeseník
104.5	50	R. Frekvence 1	Brno
104.7	10	R. Karolína	Plzen
105.3	1	R. Cerná hora	Trutnov
105.5	95	R. Lion/R. Evropa 2	Votice
105.7	1	R. Jizera	Mladá Boleslav
105.8	8	R. FM Plus	W. Bohemia
106.0	50	R. Alfa	Pardubice
106.1	1	R. FM Plus	Plzen
106.7	1	R. Evropa 2	Znojmo
107.2	0.2	R. Evropa 2/Radioclub	Ústí nad Labem
107.5	2.8	R. Proglas	Brno

RADIO FREE EUROPE/RADIO LIBERTY Inc.
See International Broadcasting section.

BBC World Sce: Praha 101.1MHz + 5 relay st's. Prgrs in English & Czech — **R. France Internationale:** Praha 99.3MHz 24h.

DENMARK

L.T: UTC + 1h (Su: UTC + 2h) - **Pop:** 5.250.000 - **Radios:** 5.200.000 - **Pr.L:**
Danish. - **EC:** 50Hz, 230/380V - **ITU:** DNK.

TELE DANMARK A/S
✉ Broadcast, Telegade 2, DK-2630 Taastrup. ☎+45 43344334. ▤+45 43711143.
Telecom A/S is responsible for the operation of the tr's carrying the prgrs. of Danmarks Radio.

DANMARKS RADIO (Gov.)
✉ Radio House, Rosenorns Alle 22, DK-1999 Frederiksberg C.
☎+ 45 35203040. ▤+45 35202644. **Listeners' Sce:** (+45 35203520.
Web: http://www.dr.dk - **E-mail:** webmaster@dr.dk
L.P: DG: Christian S. Nissen. Head of Inf. Dept: Soren Elmquist. Dir. Prgr. (radio): vacant. Head of Channel 1: Finn Slumstrup. Head of P2 Music: Steen Frederiksen. Head of Channel 3: Jesper Grunwald. Head of News and Actualities: vacant. Head of Regional Prgr: Tyge Dinsen. Head of Sport Dept. (radio): Henrik Brandt. Head of Tech: Alic Rosenlund.

LW/MW: Kalundborg 243kHz 300kW, 1062kHz 250kW.

FM (all stereo)	1	4	P2m	3	kW
Bornholm	96.2	99.3	99.3	90.0	30
Copenhagen	90.8	96.5*	102.3**	93.9	60
Hammeren	91.6	93.7	93.7	88.4	0.035
Hobro	94.9	96.7	96.7	92.0	0.06
Funen	89.0	96.8	96.8	92.6	60
Horsens		96.4	96.4		0.16
Mariager		89.1	89.1		0.1
No. Jutland	91.0	94.4	94.4	96.6	8
Odsherred	88.4	92.0	92.0	94.3	3
So. Jutland	95.1	99.9*	102.1**	97.2	60
So. Zealand	94.8	97.5	97.5	99.6	100
So. We. Jutland	88.7	99.0*	102.5**	92.3	10
		97.7*			10
Thisted	91.4	95.6*	101.3**	99.2	2
Vejle	95.5	94.0	94.0	90.7	10
Aalborg	93.3	98.1	98.1	89.7	30
Aarhus	88.1	95.9	95.9	91.7	60
We. Jutland	90.2	98.5*	100.3**	92.9	60

Digital Audio Broadcasting (DAB): Copenhagen (Margretheholm + Hove) and Holstebro (Mejrup) 227.360MHz 0.5kW.
1st Prgr. on 243kHz + FM1: MF 0500-2300, Sat 0600-2350, Sun 0600-2300.
N: 0500MF, 0600, 0700, 0800, 1102, 1600MF, 1730, 2200. Chimes of the Copenhagen Town Hall: 1100. Rel: 0710-0730 Mon-Sat, 0855-1005 Sunday. N. for Greenlanders in Danish: 1725MF. FM tr's rel. P2music at other times.
P2m = P2Music (classical/jazz) 1750-0500, except **=0000-2400.
4th Prgr: MF 0500-1730, Sat 0700-1100, Sun 0700-1000: News, entertainment and regional prgrs. N. on the hour. Rel. P2Music at other times. Fq's marked * carries a special "adult popular" musical prgr 1750-2300 MF, 0700-2300 Sat/Sun, and relays P3 from 2300.

Regional Prgrs:
Nordjyllands Radio, Fredrik Bajers Vej 9, DK-9220 Aalborg: on 94.4/96.7/98.1MHz
Radio Midt- & Vest, Vestergade 1, DK-7500 Holstebro: on 95.6/97.7/98.5MHz.
Østjyllands Radio, Olof Palmes Alle 10-12, DK-8200 Aarhus N: on 89.1/95.9 MHz.
Kanal 94, Karl Bjarnhofs Vej 2, DK-7120 Vejle: on 94.0MHz.
Radio Syd, H.P. Hanssensgade 11, DK-6220 Aabenraa: on 99.0/99.9MHz.
Radio Fyn, Lille Tornbjergvej 10, DK-5220 Odense S: on 96.8MHz.
Regionalen, Vadestedet 1, DK-4700 Naestved: on 92.0/ 97.5MHz.
Københavns Radio, Landskronagade 68, DK-2100 Copenhagen: on 96.5MHz.
Bornholms Radio, Aakirkebyvej 52, DK-3700 Roenne: on 93.7/99.3MHz.
3rd Prgr. on 1062 kHz + FM3: Popular music, news and sport: 24h. **N.** on the h. + 0529, 0629, 0729, 1529, 1629. N in **English**: 0731-0736 MF.
Special Prgrs. on 243/1062kHz: Wrp: 0445, 0745, 1045, 1645, 2145.
Navigational Warnings: 1700. **Fish prices:** 1025-1040 MF. **Gymnastics:** 0730-
0740MF. - **On 1062 kHz:** For Foreign Workers in Urdu, Southslavian, Turkish and Arabic: 1535-1605MF and 1605-1645 Tues-Fri in each language, 0800-1040 Sun. - **N.** from Greenland in Greenlandic & Danish:
1720-1730 MF. **Greenlandic:** 1750-1820 Thurs. **Faroese:** 1800-1830 Sat.
Radio Data System: RDS signals are broadcast on P1, P3 and P4 tr's, introducing RDS-TMC (Traffic Message Channel) late 1998.
ANN: DR P (or: Du lytter til P) et/to-musik/tre/fire (1st, 2nd, 3rd & 4th prgr).

FOREIGN SERVICE: see International Broadcasting section.

Private stations (local radio):
Approx. 240 organizations are operating low-powered (max. 160W ERP at 40 m. height) FM-tr's. Currently approx. 290 transmitters are on the air. A full listing of these st's, (plus DR, TV2, including all regional outlets + Greenland and Faroe Isl.), called Radio-TV Håndbogen can be obtained from Hartvig Media ApS, P.O. Box 112, DK-8900 Randers, at a cost of 75 DKK, 14 IRCs or equivalent.
Major stations in the main cities are as follows:
Copenhagen: The Voice, Magstræde 10B, 1204 Copenhagen K: 104.9MHz and Sirus 2 (5° East 11,938 GHz RC - 7.74 & 7.92 MHz) – R. Uptown, Box 200, 1006 Copenhagen K: 103.9 MHz – NRJ, Studiestræde 24, 1455 Copenhagen K: 95.0 MHz. – Danmarks Erhvervsradio, Børsbygningen, 1217 Copenhagen K: 90.4/96.1/106.6 MHz. – Station København, Hammershusgade 15, 2100 Copenhagen Ø: 102.9 MHz
Esbjerg: R. Victor, Danmarksgade 1, 6700 Esbjerg: 101.7/105.4MHz – R. Holsted, Vestergade 17, 6670 Holsted: 94.6/98.2/101.5/104.0/104.8/105.0/105.1/106.4/107.5/107.7MHz.
Horsens: Radio Horsens, Nørregade 42, 8700 Horsens: 91.1/97.9/104.4/104.8/106.4 MHz. Kolding: Radio Skala FM, Box 436, 6000 Kolding: 98.0/105.2 MHz
Nykøbing F.: Radio Sydhavsøerne, Tværgade 18, 4800 Nykøbing Falster: 87.8/90.6/107.8 MHz
Nykøbing M.: Radio Limfjord, Box 108, 7900 Nykøbing Mors: 104.7/106.9/107.8 MHz
Næstved: Radio Næstved, Fabriksvej 26, 4700 Næstved: 98.1/106.5/107.9 MHz
Odense:– The Voice, Brogade 6-8, 5000 Odense C: 195.1/106.0/107.6 MHz – Radio 3, Box 312, 5000 Odense C: 98.9 MHz – Radio 49, Box 317, 5000 Odense C: 103.5 MHz
Randers: R. ABC Box 174, 8900 Randers: 89.5/95.3/104.4/104.6/105.4/105.7/106.3/106.7/106.9/107.0 MHz – Radio Alfa, Box 174, 8900 Randers: 95.2/97.1/103.6/105.0/105.2/107.2/107.4/107.7/107.8 MHz a.o.
Silkeborg: Radio Silkeborg, Fredensgade 1, 8600 Silkeborg: 89.2/90.0/106.6/106.8/107.7 MHz
Vejle: VLR, Nyboesgade 35, 7100 Vejle: 104.1/104.5/105.9 MHz a.o.
Viborg: R. Viborg, Box 501, 8800 Viborg: 100.6/101.5/104.1/104.8/105.0/105.6/106.1/106.8 MHz.
Aabenraa: Radio Mojn, Box 44, 6200 Aabenraa: 102.6/104.0/104.5/105.1/106.2/107.6 MHz
Aalborg: ANR, Box 7089, 9200 Aalborg SV: 87.6/88.5/103.2/103.8/107.9 MHz a.o.
Aarhus: Aarhus Nærradio/R. Colombo, Søren Frichs Vej 42 B, 8230

Åbyhøj: 93.1/93.7/94.6/106.5MHz. **F.PI.:** Will change station name.

ESTONIA

L.T: UTC + 2h (Su: UTC + 3h) — **Pop:** 1.475.000 — **Pr.L:** Estonian, Russian — **ITU:** EST.

TRANSPORDI-JA SIDEMINISTEERIUM
(Ministry of Transports and Telecommunications)
✉ Viru 9, EE0101 Tallinn, Estonia. ☎ +372-639 7613. 🖷 +372 639 7606. — **L.P:** Raivo vare.

INSPECTION OF TELECOMMUNICATIONS
✉ Ädala 4 d, EE-0006 Tallinn. ☎ +372-639 9054. 🖷 +372-639 9055 & +372 (2) 771509. — **L.P:** Jüri Jõema.
This organization is reponsible for freq. management.

EESTI RINGHÄÄLINGU SAATEKESKUS
✉ Gonsiori 21, EE-0100 Tallinn. ☎ +372 (2) 2424945. 🖷 +372 (2) 6312476. — **L.P:** D.G. Rein Ruudi.
This organization is reponsible for the transmitter networks.

EESTI RAADIO
✉ Gonsiori 21, EE-0100 Tallinn. ☎ +372 (2) 2434115. 🖷 +372 (2) 2434457. ☉ 173271 radio ee. **Web:** http://www.er.ee/
L.P: Dir. Gen: Peeter Sookruus. Man. Dir: Hannes Valdma. Prgr. Dir: Mart Ummelas (ER1), Riina Sildos (ER2), n/a (ER3), Mari Velmet (ER4). Dir. Tec: Tiit Käsametsa. Head Of Foreign Rel: Ilona Hausmann.

Mediumwaves:	kHz	kW	Prgr
Laitse	612	100	*
Kohtla-Nõmme	612	20	*
Laitse	1035	150	*

*) All MW tr's inactive from July 1998.

FM (MHz)	ER1	ER2	ER3	ER4
Haapsalu	—	—	—	93.6+
Koeru	105.1++	102.6++	107.6++	93.4++
Kohtla-Nõmme	71.90	71.12	70.13	72.68
Kohtla-Nõmme	105.4	—	—	102.9
Narva	104.7+	—	—	102.3*
Narva	72.23*	73.01*	—	—
Orissaare	105.9	103.4	107.8	—
Orissaare	72.23	—	70.67	—
Pärnu	104.8	102.3+	107.3	94.8
Pärnu	72.23	—	—	—
Tallinn	104.1++	101.6++	106.6++	94.5
Tallinn-city	103.5*	—	—	—
Tartu	—	—	103.0*	94.4*
Valgjärve	106.1++	103.6++	—	—
Valgjärve	72.14	71.36	70.58	72.92
Viljandi	—	—	103.3+x	—

ERP approximately 10-15kW, exc. *) 100-150W, +) 1-2kW, ++) 25-40kW, x) **F.PI.**
D.PRGR: ER1 (Vikerraadio) in Estonian: 0330-2300.
ER2 (Raadio Kaks) in Estonian: 24h..
ER3 (Klassikaraadio) in Estonian: 0500-2100. Rel. VOA 1630-1700 & RFE 2000-2100 in Estonian.
ER4 (Raadio 4/Radio Chetyrye) in Russian: 24h on Tallinn 67.97/94.5, other FM tr's 0400 (SS 0500)-2200. 1730-1800 rel Deutsche Welle (Mon), BBC (Tues), R. Canada Int. (Wed), UNO-Radio (Thurs), VOA (Fri) in Russian. 2015-2100 Belorussian (Mon), Ukranian (Wed), Hebrew (Thurs).
ER Raadio Tallinn 103.5MHz: 0330-2205 (Sat 2300). Relays of BBC World Sce. 1000-1500, R. Sweden, R. Estonia, using ER1 as filler.

FOREIGN SERVICE: see International Broadcasting section.

INTERNET: all four channels of ER - Vikerraadio, Raadio 2, Klassikaraadio & Raasio 4 (in Russian) - can be listened to on the Internet in real time. News in Estonian, news in English & Esperanto are also on the Internet.
Web: http://www.er.ee/

PRIVATE AND INDEPENDENT STATIONS			
	MHz	kW*	FM
5)	69.62	0.1	Raadio Tartu, Tartu
6G)	88.3	0.8	Raadio Elmar, Tallinn
7)	88.8	0.6	Top Classics, Tallinn (**F.PI.**)

8)	88.9	3	Pereraadio, Rakvere
9)	89.0	6.0	Kuressaare Pereraadio, Kuressaare
10)	89.0	0.2	Tartu Pereraadio, Tartu
8)	89.4	1	Pereraadio, Haapsalu
8)	89.6	0.5	Tallinna Pereraadio, Tallinn
11)	90.0	0.8	Raadio Uuno, Märjamaa
7)	90.6	0.6	Sky Classics/Europa Plus, Tallinn
12)	91.0	3	Raadio B3, Pärnu (F.PI.)
12)	91.4	10	Raadio Pulss B3, Valgjärve
12)	91.5	0.3	Raadio Pulss B3, Tallinn
12)	91.6	1	Raadio Pulss B3, Koeru
7)	92.1	1	Top Raadio, Aegvildu (F.PI.)
12)	92.2	3	Raadio Pulss B3, Haapsalu
25)	92.5	0.1	Viru Raadio, Rakvere
5)	92.6	1	Raadio Tartu, Ardu (F.PI.)
7)	92.7	1	Pärnu Top Raadio, Pärnu (F.PI.)
7)	92.9	3	Top Raadio, Orissaare (F.PI.)
7)	93.0	0.3	Top Raadio, Märjamaa (F.PI.)
	95.4	0.1	Sky Pluss, Tallinn
13)	96.1	3	Panda/Viroona, Tamsalu
	96.3	1	Sky Pluss, Kuressaare
	96.3	0.8	Raadio 100, Kohtle Nõmme
14)	96.5	0.1	Raadio Vali, Põltsamaa
3)	96.6	10W	Saaremaa Raadio, Kuressaare
15)	96.6	0.3	Raadio Ruut, Valga
15a)	96.6	0.3	Mega FM, Tallinn
	96.8	1	Sky Pluss, Pärnu
	96.9	1	Sky Pluss, Haapsalu
6A)	97.2	0.8	Raadio Uuno, Tallinn
6A)	97.2	0.8	Raadio Uuno, Tartu
6A)	97.4	10	Raadio Uuno, Koeru
6A)	97.5	3	Raadio Uuno, Kuressaare
18)	97.8	1.3	Raadio Love, Tallinn
18)	97.9	1	Raadio Love, Viljandi
18)	98.0	0.3	Raadio Love, Märjamaa (F.PI.)
18)	98.1	1	Raadio Love, Paide
18)	98.3	1	Raadio Love, Pärnu
19)	98.4	0.8	Raadio Sky, Tallinn
18)	98.5	1	Raadio Love, Rakvere
18)	98.6	1	Raadio Love, Tartu
12)	99.0	3	Raadio Pulss B3, Pärnu
16)	99.0	1	Narva Päikeseraadio (F.PI.)
16)	99.2	1	Kohtla-Järve Päikeseraadio, Kohtla-Järve (F.PI.)
20)	99.3	0.3	Nõmme Raadio, Tallinn
21)	99.4	0.2	Tartu Päikeseraadio, Tartu
3)	99.6	1.0	Saaremaa Raadio, Orissaare
16)	99.7	3	Viljandi Päikeseraadio, Viljandi (F.PI.)
6A)	99.8	2	Raadio Uuno, Haapsalu
22)	99.9	0.8	Raadio Valga, Valga
6E)	99.9	3	Lääne Raadio, Haapsalu (F.PI.)
6B)	100.0	0.25	Radio 100, Tallinn
6C)	100.0	1.3	Radio 100, Narva
5)	100.2	0.6	Raadio Tartu, Tartu
23)	100.3	1.0	Pärnu Päikeseraadio, Pärnu
24)	100.4	0.5	Hiiu Raadio, Kärdla
25)	100.4	0.8	Viru Raadio, Jõhvi
26)	100.5	0.3	Järva Kuku Raadio, Paide
6D)	100.7	0.8	Raadio Kuku, Tallinn
22)	100.7	1.5	Kaguraadio, Põlva
4)	100.8	0.8	Mulgi Kuku, Viljandi
2)	100.9	2.5	Oma Raadio, Haapsalu
27)	101.0	0.5	Kuma Raadio, Paide
28)	101.2	0.8	Tartu Kuku Raadio, Tartu
29)	101.5	0.3	Raadio Kadi, Kuressaare
1)	101.7	0.8	Raadio Ring, Võru
6F)	101.9	3	Raadio Vooremaa, Jõgeva (F.PI.)
30)	103.1	1	Raadio 7, Tallinn
6G)	105.6	0.8	Raadio Elmar, Tartu

*) ERP.

Addresses and other information

NB: Many st's broadcast nonstop music nights. Some stations rel. ER prgr's when not broadcasting own prgr's.
1) Vee 6a, EE-2710 Võru. On 810kHz 1040-1100, 101.7MHz: 0700-1600. At other times rel. ER2 — 2) Ehitajate tee 3, EE-3170 Uuemõisa, Läänemaa — 3) PK 247, EE-3300 Kuressaare. W 1015-1100, at other times rel. ER2 — 4) Riia 55, EE-2900 Viljandi. 0600-0900, 1000-1100, 1300-1600, Sun 1300-1500; at other times rel. 6D) — 5) Laulupeo puiestee 25, EE-2400 Tartu. 24h, incl. rel. ER2/VOA Europe 2200-0400 — 6A, B, C, D, E, F, G) PK 3657, EE-0090 Tallinn. 24h. 6B, C in Russian. 6C mostly rel. 6B — 7) Regati puiestee 1-6K, EE-0019 Tallinn. 24h —

8) Lai 50, EE-0001 Tallinn. 24h — 9) Tallinna 45, EE-3300 Kuressaare. 0645-1600 — 10) Annemõisa 8, EE-2400 Tartu. 24h — 11) Märjamaa-Orgita, Rapla maakonna — 12) Ahtri 12, EE-0002 Tallinn — 13) Tamsalu. Rel R. Tartu & local prgr's — 14) Silla 2, EE-2353 Põltsamaa — 15) Sepa 2, EE-2500 Valga — 15a) Narva nmt 20, Tallinn — 16) c/o Mediainvest, Tallinn — 18) Koidula 17/ Poska 45 (5. korrus), EE-0001 Tallinn. 24h — 19) Akadeemia tee 24, EE-0026 Tallinn. 24h — 20) Nurme 40, EE-0016 Tallinn — 21) Riia 12, EE-2400 Tartu. 24h — 22) Kesk 16, EE-2600 Põlva. 24h in Estonian and Setu, incl. rel. ER2 — 23) Esplanaadi 10, EE-3600 Pärnu. 24h — 24) Leigri väljak 5, EE-3200 Kärdla. 24h incl. rel ER2 — 25) Tallinna 30, EE-2100 Rakvere; Oru Põik 3a, EE-2045 Jõhvi. 24h — 26) Ruubassaare tee 6A, EE-2820 Paide. 24h incl. rel. 6D) — 27) Prääma tee 11, EE-2820 Paide. 24h — 28) Opeta-ja 9a, EE-2400 Tartu. 24h incl. rel. 6D) — 29) PK 84, EE-3300 Kuressaare. 24h incl. rel. VOA — 30) PK 3396, EE-0090 Tallinn.

FAROE ISLANDS (Danish)

L.T: UTC (Su: UTC + 1h) — **Pop:** 43.000 — **Radios:** 21.000 — **Pr.L:** Faroese — **E.C:** 50Hz, 220V.

ÚTVARP FØROYA
✉ P.O. Box 328, FR-110 Tórshavn. ☎ +298 316566. 🖷 +298 310471. ① 81226. **E-mail:** uf@utvarp.olivant.fo **Web:** http://utvarp.olivant.fo **L.P:** SM: Jógvan Jespersen. PD: Jógvan Arge. TD: Hans Andor Johannesen.
MW: Akraberg 531kHz 200kW (usually operates at 100kW).
FM: Tórshavn 89.9MHz 31kW, Klaksvík 94.3MHz 41kW, Suduroy 97.5MHz 27kW + 10 st's low power.
D.PRGR: MF 0715-2000, Sat 0715-0100, Sun 1000-1900.
ANN: "Útvarp Føroya" — **IS:** Xylophone — **V.** by QSL-card.

FINLAND

L.T: UTC + 2h (Su: UTC + 3h) — **Pop:** 5m — **Radios:** 4.950.000 — **Pr.L:** Finnish, Swedish — **E.C:** 50Hz, 230/400V — **ITU:** FIN.

TELECOMMUNICATIONS ADMINISTRATION CENTRE
✉ P.O. Box 313, FIN-00181 Helsinki. ☎ +358 (9) 69661. 🖷 +358 (9) 6966410. ① 124545 thk fi. **E-mail:** info@thk.fi — **L.P:** CEO: Reijo Svensson. Dir. Radio Administration: Kari Koho. Dir. Telecommunications Administration & Standardization: Tapani Rantanen. Dir. of Television Licence Centre: Esko Kotilainen.

YLEISRADIO OY (Non-comm. Public Broadc. Sce.)
✉ Box 99, FIN-00024 YLEISRADIO. ☎ +358 (9) 14801. 🖷 +358 (9) 1480 3391. ① 124735 radio sf. **E-mail:** fbc@yle.fi
Web: http://www.yle.fi/
L.P: DG: Arne Wessberg. Dep. DG: & Dir. of Radio: Tapio Siikala. Dir Corporate Affairs: Jussi Tunturi. Dir. Finance: Martti Partanen. Dir. Radio 1: O. Alho. Dir. Radio 2: J. Haarma. Dir. Radio 3: R. Vanninen. Dir. Radio 4 (Swedish): Annika Nyberg-Frankenhaeuser. Dir. of Swedish Radio & TV: Ann Sandelin. Dir. Eng: Seppo Kalli. Head of N: J. Heino. Head of Int. Rel : Ulla-Kristiina Haarma. Int. Rel. (Radio): P.Rintakoski.

MW: Helsinki 558kHz 50kW (1/4), Pori 963kHz 600kW (1/4).

FM (MHz)	1	2	3	4	5	6	kW
Aavasaksa	87.9	89.8	94.7				3
Ahvenanmaa			100.3	104.9	93.1		10
Anjalankoski	88.5	92.8	96.9	99.5			30
Espoo	87.9	91.9	94.0	98.9	101.1		60
Eurajoki	87.7	92.0	94.8	99.4	103.0		30
Fiskars	90.9	93.1	97.0	102.5	99.7		3
Haapavesi	89.0	96.1	98.4				30
Hämeenlinna			99.2				0.5
Iisalmi	87.7	92.8	96.5				2
Inari	88.4	92.8	98.0			101.9	50
Joutseno	88.0	90.9	98.5				30
Jyväskylä	89.9	92.5		103.5			30
			87.6				3
Karigasniemi	89.5	93.4	96.8		100.8		2
Kerimäki	90.5	95.8	99.1				30
Kiihtelysvaara	88.4	94.9	97.2				5
Kilpisjärvi	88.0	90.9	96.0			103.7	0.1
Koli	90.2	93.4	99.6	102.4			30
Kruunupyy	91.4	94.0	97.6	99.7	102.7		60

Location							kW
Kuopio	91.6	93.9	98.1	100.2			50
Kuttanen	94.1	97.2	99.6		102.2		3
Lahti	93.2	95.5	97.9	100.6			50
Lammaskoski	88.5	91.4	98.7		101.2		5
Lapua	88.2	90.1	93.1	95.2	101.5		60
Mikkeli	88.9	92.1	94.6				30
Nuorgram	88.6	93.9	97.7		101.2		3
Nuvvus	88.1	90.2	94.4		101.7		0.05
Näätämä	87.6	89.9	97.4			103.3	0.3
Oulu	90.4	93.2	97.3	100.3			50
Pello	90.2	97.0	99.7				3
Pernaja	89.5	92.3	95.0	102.2	98.3		3
Pieksämäoki	89.4	95.3	97.4				2
Porvoo			90.3		91.4		0.5
Pihtipudas	88.6	91.1	97.0	100.8			50
Posio	87.6	91.5	98.6				30
Pyhötunturi	91.0	97.6	99.9				50
Pyhävuori	88.9	91.0	94.2	98.6	102.6		30
Rovaniemi	88.2	94.0	96.7	103.0			30
Ruka	90.7	92.8	95.1				3
Savitaipale			97.2				0.3
Taivalkoski	89.2	91.9	99.2				60
Tammela	89.2	91.2	96.0				5
Tampere	90.7	93.7	99.9	102.1			60
Tenola	89.0	94.1	95.8		100.5		0.02
Tervola	88.6	92.6	95.6				30
Turku	89.8	92.6	94.3	98.2	101.4		50
Utsjoki	90.7	93.1	99.4		102.6		2
Vaasa	87.8	89.6	94.8	97.3	101.0		1
Vuokatti	92.3	94.3	98.9				60
Vuotso	87.8	90.1	94.3		101.3		3
Yllös	92.2	95.3	98.1			103.8	60
Ähtäri	91.9	94.6	96.6				3

Additional transmitters

Location	MHz	kW	Programme
Ahvenanmaa	95.0	10	Swedish Radio P1
Ahvenanmaa	97.1	10	Swedish Radio P2
Ahvenanmaa	88.6	10	Swedish Radio P3
Ahvenanmaa	91.3	10	Åland Radio
Ahvenanmaa	102.3	10	Swedish Radio P4
Espoo	97.5	3	FM1
Espoo Capital	103.7	3	Ext prgrs
Haapavesi	101.9	3	Regional prgrs
Jyväskylä Local	99.3	30	Local prgrs
Kerimäoki	97.7	6	Regional prgrs
Kerimäoki	103.2	6	Regional prgrs
Kuopio	88.1	5	R. Kantti
Lahti	90.3	0.2	R. Masto
Pyhävuori	97.2	30	Regional prgrs
Taivalkoski	103.6	60	Regional prgrs
Turku	96.7	6	R. Aurora

FM1 "Yylen Ykkönen" (arts & music channel): 24h. **N:** 0400, 0500, 0600, 0900, 1100, 1400, 1700, 2000, 2200
FM2 "Radiomafia" (popular culture): 24h. **N:** on the h. Exc. 1500-2100. **N. in Latin:** Fri 0755, Thurs 1655
FM3 "Radio Suomi" (news and regional prgrs): 24h. **N:** on the h. **N. in English:** 2055. **N. in German:** 1955
FM4 "Radio Extrem" (Swedish language prgr for young people).
FM5 "Radio Vega" (Swedish language prgr for over 35's).
FM6 "Sámiradio" in Sami (⌨ Tammukkatie, 99870 Inari).

Regional & Local Prgrs
In Finnish on YLE 3: Ylen aikainen, Radiokatu 5, 00240 Helsinki: 94.0MHz – **R. Itä-Uusimaa,** Kaivokatu 40, 06100 Porvoo: 95.0MHz – **Ylen läntinen,** Karstuntie 4, 08101 Lohja: 97.0MHz – **Tampereen R,** Vuolteenkatu 20, 33100 Tampere: 99.9MHz – **Lahden R,** Aleksanterinkatu 8, 15110 Lahti: 97.9MHz – **R. Häme,** Viipurintie 4, 13200 Hämeenlinna: 96.0MHz – **Turun R,** Käsityöläiskatu 10, 20100 Turku: 94.3MHz – **Satakunnan R,** Mikonkatu 4 D, 28100 Pori: 94.8/97.2MHz – **R. Keski-Suomi,** Matarankatu 6, 40100 Jyväskylä: 87.6/97.0/99.3MHz – **R. Savo,** Kotkankallionkatu 12, 70600 Kuopio: 96.5/98.1MHz – **Etelä-Savon R,** Maaherrankatu 10, 50100 Mikkeli: 94.6/97.4/98.4MHz – **Pohjois-Karjalan R,** Siltakatu 16C, 80100 Joensuu: 97.2/97.7/99.6MHz – **Etelä-Karjalan R,** Kristiinankatu 11A, 53100 Lappeenranta: 97.2/98.5/103.2MHz – **Kymenlaakson R,** Salpausselänkatu 40, 45100 Kouvola/Takojantie 4, 48230 Kotka: 96.9MHz – **Pohjanmaan R,** Ahventie 20, 65200 Vaasa/Marttilantie 24, 60100 Seinäjoki: 93.1/94.2/94.8/96.6MHz – **R. Keski-**

Pohjanmaa, Torikatu 50, 67100 Kokkola: 97.6/101.9MHz – **Oulu-R,** 1MHz. Hanhitie 13, 90150 Oulu: 95.1/97.3/98.4/99.2MHz – **Kainuun R,** Lönnrotinkatu 14 B, 87100 Kajaani: 98.9/103.6MHz – **Lapin R,** Jorma Etontie 22, 96100 Rovaniemi: on 15 FM freqs – **R. Perämeri,** Sankarikatu 10, 94100 Kemi: 95.6/97.4MHz.
In Swedish on YLE 5: R. Mellannyland, Radiogatan 5, 00240 Helsingfors: 101.1MHz – **R. Östnyland,** Bruunsgatan 40, 06100 Borgå: 98.3MHz – **R. Västnyland,** Långgatan 13, 10600 Ekenäs: 99.7MHz – **R. Åboland,** Hantverkaregatan 10, 20100 Åbo: 101.4MHz – **R. Åland,** Ålanddsvägen 24, 22100 Mariehamn: 91.3MHz – **R. Österbotten,** Abborrevägen 20, 65200 Vasa: 101.0/101.5/102.6/102.7MHz.

FOREIGN SCE: see International Broadcasting section

Other Stations

	MHz	Station	Tr. location
1)	87.7	R. Albatrossi	Kotka
2)	87.9	Järviradio	Viitasaari
N1)	88.1	Classic FM	Hämeenlinna
8)	88.2	R. 88.2	Salo
16)	89.0	R. Inari	Inari
4)	89.0	Meriradio	Rymättylä
	89.0	R. Dei	Helsinki
11a)	89.4	R. Alex Oulu	Oulu
33)	89.4	R. Pori	Pori
2)	89.5	Järviradio	Alajärvi
41)	89.5	R. Salminen	Iisalmi
N2)	89.6	Kiss FM	Tampere
35)	89.7	R. RPM	Mikkelin Mlk
18)	89.7	R. Janne	Lammi
55)	90.1	Voima Asema	Hyvinkää
12)	90.5	R. Auran Aallot	Turku
N1)	90.8	Classic FM	Porvoo
20)	90.9	R. Kajaus	Hyrynsalmi
45)	91.1	R. Stadi	Helsinki
38)	91.5	R. Robin Hood	Turku
	MHz	**Station**	**Tr. location**
34)	92.0	R. Provinssi	Tornio
N1)	92.2	Classic FM	Tampere
29)	92.3	R. Paitapiiska	Kurikka
6)	92.7	Oikea Asema	Varkaus
22)	92.7	R. Kokkola	Kannus
37)	92.8	R. Rex	Liperi
N1)	92.9	Classic FM	Helsinki
49)	93.0	RadioTupla	Huittinen
36)	93.3	R. Ramona	Rauma
28)	93.5	R. Norppa	Lappeenranta
52)	93.8	SBC radio	Kouvola
19)	94.4	R. Jyväskylä	Jämsä
1)	94.4	R. Albatrossi	Hamina
13)	94.6	R. Botnia	Pietarsaari
55)	94.7	Voima Asema	Riihimäki
31)	94.8	R. Pooki	Raahe
N1)	94.8	Classic FM	Kuopio
10)	94.9	R. 99	Sysmä
27)	94.9	R. Moro	Ruovesi
11)	94.9	R. Alex	Helsinki
31)	95.0	R. Pooki	Haapajärvi
23)	95.1	R. Lännentie	Tammisaari
12)	95.3	R. Auran Aallot	Loimaa
27)	95.4	R. Moro	Vilppula
20)	95.4	R. Kajaus	Sotkamo
31)	95.5	R. Pooki	Ylivieska
16)	95.7	R. Inari	Inari
9)	95.7	R. 957	Tampere
20)	95.7	R. Kajaus	Kajaani
	95.9	R. Väst	Mariehamn
	96.2	Classic FM	Jyväskylä
14)	96.2	R. City	Helsinki
43)	96.3	R. Satahäme	Ikaalinen
52)	96.5	SBC Radio	Lappeenranta
23)	96.5	R. Lännentie	Lohja
51)	96.6	Rytmiradio Lahti	Lahti
6)	96.7	Oikea Asema	Kuopio
22)	96.7	R. Kokkola	Kokkola
35)	96.7	R. RPM	Savonlinna
32)	96.7	R. Pop	Valkeakoski

N1)	96.7	Classic FM	Vaasa
5)	96.8	NRJ Helsinki (Energy)	Helsinki
53)	96.9	Simpsiö 969	Lapua
36)	97.3	R. Ramona	Lappi
12)	97.5	R. Auran Aallot	Uusikaupunki
19)	97.7	R. Jyväskylä	Jyväskylä
18)	97.7	R. Janne	Forssa
29)	97.7	R. Paitapiiska	Teuva
46)	97.8	R. Toiska	Ähtäri
26)	98.4	R. Moreeni	Tampere
10)	98.6	R. 99	Lahti
21)	98.7	R. Keskil-Savo	Pieksämäki
N2)	98.7	Kiss FM	Turku
50)	98.8	Radioankkuri	Keminmaa
17)	99.0	R. Iniö	Iniö
27)	99.0	R. Moro	Virrat
2)	99.1	Järviradio	Vimpeli
47)	99.5	R. Vaasa	Vaasa
N1)	99.6	Classic FM	Oulu
15)	99.8	R. Foni	Porvoo
42)	100.1	R. Sata	Turku
19)	100.3	R. Jyväskylä	Mänttä
3)	100.3	Lähiradio	Helsinki
49)	100.4	Radiotupla	Nakkila
31)	100.5	R. Pooki	Nivala
27)	100.9	R. Alex	Tampere
19)	101.0	R. Jyväskylä	Vaajakoski
31)	101.0	R. Pooki	Kalajoki
34)	101.0	R. Provinssi	Tornio
N1)	101.0	Classic FM	Pori
39)	101.1	R. Roy	Rovaniemi
24)	101.2	R. Manta	Vammala
N2)	101.4	Kiss FM	Oulu
18)	101.7	R. Janne	Hämeenlinna
40)	101.8	R. Ruka	Kuusamo
N1)	101.9	Classic FM	Turku
41)	102.0	R. Salminen	Siilinjärvi
37)	102.2	R. Rex	Kitee
48)	102.3	R. Väst	Mariehamn
19)	102.3	R. Jyväskylä	Äänekoski
56)	102.4	Reissu R.	Helsinki
54)	102.8	U103 Golden	Tuusula
25)	103.1	R. Mega	Oulu
20)	103.2	R. Kajaus	Kuhmo
44)	103.3	R. Seinäjoki	Seinäjoki
30)	103.3	R. Pikku Pariisi	Haapavesi
37)	103.3	R. Rex	Juuka
7)	103.4	R. 103.4	Mäntsälä
22)	103.7	R. Kokkola	Kaustinen
27)	103.8	R. Moro	Orivesi
N3)	103.9	R. Nova	Turku
N3)	104.1	R. Nova	Haapavesi
N3)	104.3	R. Nova	Koli
N3)	104.4	R. Nova	Lahti
N2)	104.6	Kiss FM	Helsinki
N3)	104.7	R. Nova	Tampere
N3)	104.8	R. Nova	Oulu
N3)	105.1	R. Nova	Pihtipudas
N3)	105.5	R. Nova	Rovaniemi
N3)	105.7	R. Nova	Anjalankoski
N3)	105.7	R. Nova	Vuokatti
N3)	105.8	R. Nova	Jyväskylä
N3)	106.0	R. Nova	Eurajoki
N3)	106.2	R. Nova	Espoo
N3)	106.3	R. Nova	Mikkeli
N3)	106.5	R. Nova	Lapua
N3)	106.7	R. Nova	Kuopio
41)	107.1	R. Salminen	Vehmasjärvi
N3)	107.6	R. Nova	Pyhävuori
N3)	107.7	R. Nova	Kerimäki

National Networks **N1) CLASSIC FM (Comm.)**

World Trade Centre, Aleksanterink 17, P.O. Box 800, FIN-00101, Helsinki. ☎ +358 (9) 476 7800. 🖹 +358 (9) 476 78767 — **D.PRGR:** 24h.

N2) KISS FM (Comm.)

Tallberginkatu 1C, FIN-00180 Helsinki. ☎ +358 (9) 680 96501. 🖹 +358 (9) 680 96521. **Web:** http://kiss.fi/

N3) RADIO NOVA (Comm.)

Ilmalankatu 2 C 5.krs, P.O. Box 123, FIN-00241 Helsinki. ☎ +358 (9) 8848 8700. 🖹 +358 (9) 8848 8720. **Web:** http://www.radionova.fi/

Addresses and other information

1) Jussinpolku 1, 48400 Kotka. ☎ +358 (5) 215000. 🖹 +358 (5) 2264009. **E-mail:** albatros@planet.fi **Web:** http://albatros.planet.fi/

2) Kinaprcintie 11, 62900 Alajärvi. ☎ +358 (6) 5575895. 🖹 +358 (6) 5572899. **Web:** http://www.japo.fi/~jr/

3) Kumpulantie 7, 6.krs, 00520 Helsinki. ☎ +358 (9) 1482614. 🖹 +358 (9) 1482610. **Web:** http://www.kaapeli.fi:81/~kslvid/radio.htm

4) Sahantie 3, 21100 Naantali. ☎ +358 (2) 4353444. 🖹 +358 (2) 4353377.

5) Pursimiehenkatu 26, 00151 Helsinki. ☎ +358 (9) 47807400. 🖹 +358 (9) 6224090. **E-mail:** nrj@dlc.fi **Web:** http://www.dlc.fi/~nrj/

6) Koljonniemenkatu 2, PL 967, 70101 Kuopio. ☎ +358 (17) 2616400. 🖹 +358 (17) 2632344. **Web:** http://www.oa967.fi/

7) Koskenranta, PL 6, 04601 Mäntsälä. ☎ +358 (19) 6871573. 🖹 +358 (19) 6872133.

8) Vilhonkatu 11-13, PL 882, 24101 Salo. ☎ +358 (2) 777882. 🖹 +358 (2) 777883. **Web:** http://gnosite.oy.utu.fi/homepage/radio882/

9) Koskikeskus, PL 957, 33101 Tampere. ☎ +358 (3) 2130311. 🖹 +358 (3) 2133899. **Web:** http://www.radio957.fi/

10) Rautatienkatu 22, 15110 Lahti. ☎ +358 (3) 7832499. 🖹 +358 (3) 7832599.

11) Yrjönkatu 11 A 5, PL 125, 00101 Helsinki. ☎ +358 (9) 507 7949. 🖹 +358 (9) 507 8949. **Web:** http://www.alexpress.fi/alex/alexhes.html

11a) Kirkkokatu 31, 90100 Oulu. ☎ +358 (8) 311 4894. 🖹 +358 (8) 311 4893. **Web:** http://www.alexpress.fi/alex/alexoes.html

12) Valkoinen talo, Laivurinkatu 1, PL 320, 20101 Turku. ☎ +358 (2) 2341666. 🖹 +358 (2) 2357102.

13) Ristisuonraitti 1, 68600 Pietarsaari. ☎ +358 (6) 7233111. 🖹 +358 (6) 7233177.

14) Tallberginkatu 1 C, PL 291, 00181 Helsinki. ☎ +358 (9) 680960. 🖹 +358 (9) 68096296. **Web:** http://www.radiocity.fi/

15) Kaivokatu 29 A, PL 266, 06101 Porvoo. ☎ +358 (19) 583388. 🖹 +358 (19) 583331. **Web:** http://www.radiofoni.fi/

16) PL 50, 99871 Inari. ☎ +358 (16) 671326. 🖹 +358 (16) 671426.

17) 23390 Iniö. ☎ +358 (2) 4635261. 🖹 +358 (2) 4635229.

18) Palokunnankatu 21, PL 212, 13101 Hämeenlinna. ☎ +358 (3) 6537722. 🖹 +358 (3) 6537033.

19) Ilmarisenkatu 17, PL 480, 40101 Jyväskylä. ☎ +358 (14) 611900. 🖹 +358 (14) 620090. **Web:** http://www.radiojkl.fi/

20) Kauppakatu 228 A, PL 230, 87101 Kajaani. ☎ +358 (8) 629957. 🖹 +358 (8) 624114.

21) Häyrisentie 4, 76100 Pieksämäki. ☎ +358 (15) 483454. 🖹 +358 (15) 484987.

22) Kaarlelankatu 21, 67100 Kokkola. ☎ +358 (6) 8224122. 🖹 +358 (6) 8225926.

23) Kalevankatu 5, 08100 Lohja. ☎ +358 (19) 315675. 🖹 +358 (19) 315676. **Web:** http://www.lannentie.lohja.fi/

24) Hopunkatu 1, PL 101, 38201 Vammala. ☎ +358 (3) 5141416. 🖹 +358 (3) 5113097.

25) Asemakatu 18, PL 460, 90101 Oulu. ☎ +368 (8) 3112555. 🖹 +358 (8) 379744. **Web:** http://www.radiomega.fi/

26) PL 607, 33101 Tampere. ☎ +358 (3) 2156946. 🖹 +358 (3) 2157250. **Web:** http://www.uta.fi/moreeni/

27) Hallituskatu 16, PL 55, 33201 Tampere. ☎ +358 (3) 2666100. 🖹 +358 (3) 2666969. **Web:** http://www.alexpress.fi/moro/

28) Raatimiehenkatu 22, PL 19, 53101 Lappeenranta. ☎ +358 (5) 541 1175. **E-mail:** albatros@planet.fi **Web:** http://norppa.planet.fi/

29) Laulajantie 4, PL 50, 61301 Kurikka. ☎ +358 (6) 4502333. 🖹 +358 (6) 4502332. **Web:** http://www.kauhajoki.fi/paitapiiska

30) Haapaveden Opisto, Vanhatie 45, PL 62, 86601 Haapavesi. ☎ +358 (8) 450211. 🖹 +358 (8) 455 450445. **Web:** http://www.haapop.fi/

31) Kirkkokatu 43, 92101 Raahe. ☎ +358 (8) 221112. 🖹 +358 (8) 220366.

32) Valtakatu 9-11, Lokero 28, 37600 Valkeakoski. ☎ +358 (3) 5869000. 🖹 +358 (3) 5869020.

33) Yrjönkatu 15 A, PL 894, 28101 Pori. ☎ +358 (2) 6300100. 🖹 +358 (2) 6300140. **Web:** http://www.kolumbus.fi/radiopori/

34) Urheilukatu 6, 95400 Tornio. ☎ +358 (16) 480076. 🖹 +358 (16) 431487.

35) Maaherrankatu 20, PL 182, 50101 Mikkeli. ☎ +358 (15) 360897. 🖹 +358 (15) 360106.

36) Kaivopuistontie 1, PL 5, 26101 Rauma. ☎ +358 (2) 8336284. 🖹 +358 (2) 8336293. **Web:** http://stip.net/~ramona/

37) Torikatu 29 A, 2.krs, PL 928, 80101 Joensuu. ☎ +358 (13) 2489281. 🖹 +358 (13) 126804. **Web:** http://typhon.jpoy.fi/rex/

38) Hämeenkatu 15, 20500 Turku. ☎ +358 (2) 2500337. 🖹 +358 (2)

2500905. **Web:** http://www.atlasnet.fi/yritys/robinhood/
39) Koskikatu 10, 96200 Rovaniemi. ☎ +358 (16) 310100. 🖹 +358 (16) 318561.
40) Torangintaival 4, 93600 Kuusamo. ☎ +358 (8) 8506658. 🖹 +358 (8) 8506666.
41) Satamakatu 7, 74100 Iisalmi. ☎ +385 (17) 814771. 🖹 +385 (17) 814775.
42) Aurakatu 8, PL 1 20251 Turku. ☎ +358 (2) 2747101. 🖹 +358 (2) 2747110. **Web:** http://www.sata.fi/
43) Pilvilinnankatu 1, PL 67, 39501 Ikaalinen. ☎ +358 (3) 4587787. 🖹 +358 (3) 4586226.
44) Kalevankatu 10, PL 102, 60101 Seinäjoki. ☎ +358 (6) 4232900. 🖹 +358 (6) 4147666.
45) Kaisaniemenkatu 1 B a, PL 442, 00100 Helsinki. ☎ +358 (9) 680 911. 🖹 +358 (9) 6809 1223. **Web:** http://www.radiostadi.fi/
46) Ostolantie 11, PL 1001, 63701 Ähtäri. ☎ +358 (6) 5337978. 🖹 +358 (6) 5337979.
47) Hovioikeudenpuistikko 16, PL 371, 65101 Vaasa. ☎ +358 (6) 3175200. 🖹 +358 (6) 3126410. **Web:** http://www.walli.uwasa.fi/995/
48) Doppningvägen 4 B, 22100 Mariehamn. ☎ +358 (18) 22076. 🖹 +358 (18) 22079.
49) Tuomolankatu 44, 29200 Harjavalta. ☎ +358 (2) 6744939. 🖹 +358 (2) 6744040.
50) Sairaalakatu 2 C, PL 201, 94101 Kemi. ☎ +358 (16) 257600. 🖹 +358 (16) 257555.
51) Aleksanterinkatu 20, PL 85, 15141 Lahti. ☎ +358 (3) 7527911. 🖹 +358 (3) 7527912.
52) Ojaäyrääntie 2, PL 14, 45151 Kouvola. ☎ +358 (5) 3752938. 🖹 +358 (5) 3752101.
53) Sanomatie, PL 96, 62101 Lapua. ☎ +385 (6) 4387352. 🖹 +385 (6) 4331909. **Web:** http://www.travelnet.fi/969/
54) Ilmalantori 1 B, 00240 Helsinki. ☎ +385 (9) 876 1103. 🖹 +358 (9) 876 1027. **Web:** http://www.u103.com/
55) Salonkatu 11, PL 77, 05801 Hyvinkää. ☎ +385 (19) 431670. 🖹 +385 (19) 416955. **Web:** http://voima_asema.softavenue.fi/
56) Neljäs linja 17-19 E, PL 30, 00531 Helsinki. ☎ +358 (9) 701 8951. 🖹 +358 (9) 762 539.

FRANCE

L.T: UTC + 1h (Su: UTC + 2h) — **Pop:** 58.915.000 — **Radios:** 50.000.000 — **Pr.L:** French — **E.C:** 50Hz, 220/240V — **ITU:** F.

CONSEIL SUPÉRIEUR DE L'AUDIOVISUEL (C.S.A.)
✉ 39 Quai André Citroen, F-75015 Paris. ☎ +33 (0) 1 40 58 38 00.
LP: Pres: Hervé Bourges. **Web:** http://www.csa.fr
The CSA regulates TV and radio and issues broadcast licenses.

TÉLÉDIFFUSION DE FRANCE (TDF)
✉ 21/27 rue Barbès, 92542 Montrouge Cédex. ☎ +33 (0) 1 49 65 10 00. ℺ 250738F — **LP:** Prés. Dir. Gen: Bruno Chetaille.
Regional Centres: Paris Centre Nord: 21/23 rue de la Vanne, 92542 Montrouge — **Est (East):** 14 Route de Mirecourt, 54500 Vandoeuvre — **Centre Est (Centre East):** 44 Bd. Vivier Merle, B.P. 68, 69398 Lyon Cédex 3 — **Sud Est (South East):** 14 Bd. Edouard Herriot, 13271 Marseille Cédex 08 — **Sud Ouest (South West):** La Cépière, 31081 Toulouse Cédex — **Ouest (West):** B.P. 79, Ave. de Belle Fontaine, 35510 Cesson Sévigné.
TDF operates the radio & TV transmitters.

RADIO TÉLÉVISION FRANÇAISE D'OUTRE-MER (RFO)
✉ 5 Avenue Recteur Poincaré, 75016 Paris. ☎ +33 (0) 1 4215 7100. **Web:** http://www.rfo.fr — **LP:** Prés. DG:Jean-Marie Cavada.
RFO controls broadcasting in the French overseas territories.

RADIO FRANCE
✉ 116, Av. du Président Kennedy, F-75786 Paris Cedex 16.
☎ +33 (0) 1 42.30.22.22. 🖹 +33 (0) 1 4230.14.88. ℺ 200 002 RAND-FRAN F. **Web:** http://www.radio-france.fr
LP: Prés. DG: Michel Boyon.

HOME SERVICES
N=Networks: A=France-Inter, B=Radio Bleue, E=rel. Ext. Sce (RFI), L=rel. local st's at certain times.

MW	N	kHz	kW	MW	N	kHz	kW
Allouis	A	162	2000	Lille	B	1377	300
Paris	F	585	10	Ajaccio	B+L	1404	20
Lyon	B+L	603	300	Brest	B	1404	20
Rennes	B+L	711	300	Dijon	B	1404	5
Paris	E	738	5	Grenoble	B	1404	5
Limoges	B	792	300	Pau	B	1404	20
Nancy	B	837	200	Bastia	B+L	1494	20
Paris)	B	864	300	Bayonne	B+L	1494	4
Toulouse	B+L	945	300	Besançon	B	1494	5
Bordeaux	B	1206	300	Clermont-			
Marseille	B	1242	150	Ferrand	B	1494	20
Strasbourg	B+L	1278	300	Nice	B	1557	300

FM: C=France Inter (stereo), D=France-Culture (stereo), E=France-Musique (stereo), F=France Info (mono). RDS on all tr's.

Station	C	D	E	F	kW(ERP)
Abbeville	93.1	97.4	89.8	105.8	4
Ajaccio	92.4	97.6	88.0		5
Ajaccio La Punta	88.6	103.9	92.0	105.5	4
Alençon	93.0	88.0	91.0	105.5	10
Ales	87.6	96.1	98.6	105.1	1
Amiens	95.4	102.5	99.4	105.5	20
Amiens-Dury	92.6	97.0	89.3	105.5	2
Angers	93.2	91.4	97.4	105.5	8
Angoulême	92.4	87.6	95.1	105.5	2
Arcachon	87.7	97.0	91.0	105.5	1.2
Argenton	101.9	89.8	97.2		5
Arnay le Duc	94.6	90.2	100.3		3
Aurillac	94.5	98.0	91.9	105.5	7
Autun	88.1	97.3	94.1		10
Auxerres	99.5	89.5	92.8	105.5	5
Avignon	97.4	90.7	93.2	105.2	4
Bar le Duc	90.9	88.4	92.7	104.5	10
Bastia	95.9	89.2	93.9	105.5	5
Bayonne	89.0	96.1	92.7	105.5	40
Beaucaire				105.5	1
Beauvais				105.6	1

Station	C	D	E	F	kW(ERP)
Bergerac	92.3	94.0	97.1	105.5	50
Besançon (Town)	98.7	89.3	95.0	104.4	10
Besançon Lomont	90.0	97.7	92.9		22
Beziers				105.1	1
Bordeaux	89.7	97.7	93.5	106.0	6
Boulogne	103.3	99.9	89.4	105.5	1
Bourges	94.9	88.5	91.8	105.5	100
Brest	95.4	97.8	89.4	105.5	50
Briançon	91.5	97.8	89.5	105.4	1
Brignoles	106.7	104.0	105.5		1.5
Caen	99.6	91.5	95.6	105.5	50
Calais	104.7			105.6	1
Cannes				105.3	1
Carcassone	88.3	96.5	90.9	105.1	125
Castres				105.2	2
Chalon sur Saone				105.0	5
Chambery	93.5	90.5	98.6	105.1	8
Champagnole	88.5	91.7	98.3		1
Chartres	94.6	98.1	89.7	105.7	50
Chateauroux				105.5	1
Chaumont	96.5	90.4	93.3	105.5	10
Cherbourg	94.1	89.2	92.3	105.6	1
Cholet				105.9	1
Clermont-Ferrand	90.4	98.4	95.5	105.5	35
Compiegne				105.3	1
Corse(East)	96.8	92.3	103.5		17
Corte	98.2	91.0	94.8		1
Creil				105.6	1
Dijon	95.9	93.7	99.2	105.1	10
Dunquerque				99.0	1
Epinal	98.6	92.4	89.4	106.5	10
Evreux	87.7	98.9	102.0	105.5	1
Gap	98.3	88.5	95.3	105.5	5
Gex	94.4	96.7	89.6	101.1	20
Grenoble Chamrousse	99.4	88.2	91.8	105.1	2
Gueret	100.7	98.8	90.8	105.5	12
Hirson	94.4	99.7	97.2	105.3	5
Hyeres	91.6	97.5	94.5	107.1	1.5
Laval	95.1	88.3	92.1	105.5	5
La Rochelle				105.5	1.5
Le Havre	88.9	93.3	98.5	105.5	1
Le Mans	92.6	89.0	92.8	105.5	100
Le Puy	99.3	89.3	92.8	105.5	10

Station	C	D	E	F	kW(ERP)
Lesparre	92.4	90.3	95.1		1.5
Lille (Bouvigny)	103.7	98.0	88.7	105.2	400
Limoges	93.0	89.5	97.5	105.5	150
Longwy	98.1	88.3	91.0	104.3	5
Lyon Mont Pilat	99.8	88.8	92.4	103.4	50
Lyon (Town)	101.1	94.1	98.0	105.4	1
Mantes	95.0	92.4	97.1	105.6	5
Marseille	91.3	99.0	94.2	105.3	50
Marseille (Town)	91.7	98.6	94.7		1
Maubeuge				106.2	2
Melun				106.2	1
Mende	90.1	96.9	93.7		10
Menton	97.0	89.6	91.7	105.5	5
Metz	99.8	94.5	89.7	106.8	50
Mezieres	95.8	90.1	93.5	105.9	10
Millau	94.9	99.2	88.9	105.4	6
Mont de Marsan				105.5	6
Montargis	102.9	98.8	94.1	105.5	1
Montauban				105.7	1
Montereau				106.2	1
Montlieu la Garde	87.8	104.8	98.8		3
Montluçon				105.5	1
Montpellier	89.4	97.8	92.9		9
Montpellier (Town)				102.7	105.1	1
Morosaglia	97.1	88.8	93.4		1
Mulhouse	95.7	88.6	91.6	105.5	50
Nancy	96.9	88.7	91.7	105.9	10
Nantes	90.6	94.2	98.9	105.5	200
Neufchateau	96.3	100.3	91.5		1
Neufchatel-en-Bray	92.7	96.0	90.2		5
Nevers				105.5	1
Nice	100.2	101.9	92.2	105.7	100
Nice (town)			94.4	105.3	1
Nimes				105.1	5
Niort	99.4	96.4	91.1	105.5	200
Orleans	99.2	95.8	90.7	105.5	5
Paris (Tour Eiffel)	87.8	93.5	91.7	105.5	10
Paris (Romainville)		93.9	92.1		10
Parthenay	93.8	87.9	98.5	105.5	13
Pau				105.5	1
Perpignan	92.1	99.8	97.2	105.1	10
Porto Vecchio	96.8	90.8	98.9		1.5
Porto Vecchio (Punto di a Varra)	92.6	87.9	94.6		1.5
Privas	89.8	96.5	94.7	105.2	1.5
Reims	96.8	98.8	89.2	105.5	150
Rennes	93.5	98.3	89.9	105.5	100
Rouen	96.5	94.0	92.0	105.7	100
Saint Brieuc				105.5	1
St. Etienne	88.0	91.7	97.1	105.6	2
Saint-Nazaire	95.2	92.2	102.6	105.5	1.5
Saint-Raphael	96.3	88.7	99.6	105.4	10
Saint Quentin				105.6	1
Sarrebourg	93.1	99.4	90.3		10
Sens	96.3	98.5	93.8	94.3	10
Soissons				105.7	1
Strasbourg	97.3	87.7	95.0	104.4	50
Toulon	92.0	97.1	94.9	105.8	5
Toulouse (Bonhouré)	103.5	90.5	93.1	105.5	1
Toulouse (Pic du Midi)	87.9	95.7	91.5		56
Tours	99.9	97.8	92.2	105.5	10
Troyes	95.3	97.9	91.4	105.5	50
Ussel	96.0	88.2	99.7		10
Valence				105.4	3
Vannes	88.6	96.0	91.8	105.5	5
Verdun	92.1	99.3	94.2	106.3	5
Villebon sur Yvette	97.6	94.1	99.7		1
Villers-Cotterets	91.1	89.6	92.9		10
Vittel	98.2	89.0	94.0		5
Voiron	91.5	89.2	95.7	105.4	1

+ 1690 st's under 1kW not mentioned.

Additional FM-tr's used for Le Mouv'

Station	MHz	kW	Station	MHz	kW
Agen	99.4	0.5	Le Puy	101.1	0.5
Alençon	101.8	0.5	Mende	106.3	10
Angouleme	101.5	2	Montelimar	107.3	0.8
Bourgoin	93.1	0.1	Moulins	99.9	0.5

Chalon sur Saone	103.1	0.5	Niort	101.0	1
Chartres	97.3	4	Poitiers	87.6	0.5
Chatellerault	103.5	0.5	Toulouse	95.2	5
Evreux	89.5	1	Villeneuve sur Lot	100.0	0.2
Gap	101.8	0.25			

France-Inter: Network A on LW, C on FM; Allouis 162kHz: 24h exc. Tues 0005-0358. FM tr's: 24h.
N: Hourly, plus 0430, 0530, 0630, 0730, 2130.
R. Bleue (for older listeners): 0600-1800 on MW Network B, 107.1MHz 1kW (Paris) and 100.7MHz (Cannes) — **France-Culture:** 24h on FM (Network D) (stereo). **N:** 0630, 0800, 1130, 1730, 2130 — **France-Musique** (Network E in stereo): 24h. **N:** 0600, 0700, 0830, 1130, 1800, 2300 — **France Info** (Network F): 24h news and information — **Le Mouv':** 24h in stereo for young listeners. ✉ 61 boulevard Lazare Carnot, 31000 Toulouse. ☎ +33 (0) 62 307000.

Local Stations ("F.I.P.")

F.I.P. Bordeaux, 95 rue Judaïque, 33000 Bordeaux: 96.7MHz.
F.I.P. Lille, 66 rue de l'Hopital Militaire, 59000 Lille: 91.0MHz.
F.I.P. Lyon, 15 rue des Archers, 69292 Lyon Cedex 02: 87.8MHz.
F.I.P. Marseille, 2 rue Papère, 13001 Marseille: 96.4/96.8MHz.
F.I.P. Metz, 3 Place St. Martin, 57000 Metz: 98.5/98.8MHz.
F.I.P. Nantes, 1 rue de Alger, 44100 Nantes: 95.7/97.2MHz.
F.I.P. Côte d'Azur, B.P. 704, 06012 Nice: 94.2/94.8/101.1/103.8MHz.
F.I.P. Strasbourg, 4 rue Joseph Massol, 67080 Strasbourg Cédex: 92.3MHz.
F.I.P. Paris, 116 avenue du Président Kennedy, 75786 Paris Cédex 16: 585kHz, 105.1MHz.
D.PRGRS: on MW: 0700-1500 (Su: 1600). On FM: 24h (stereo). Prgrs consist of music and N.

Local Stations (R.F. = Radio France)

R.F. Alsace, 4 rue Joseph Massol, 67080 Strasbourg Cedex: 101.4/ 102.6MHz – **R.F. Armorique,** 14 ave. Janvier, 35000 Rennes: 101.3/ 103.1MHz – **R.F. Auxerre,** B.P. 101, 89002 Auxerre Cedex: 100.5/ 101.3MHz – **R.F. Belfort,** B.P. 439, 90000 Belfort: 106.8MHz – **R.F. Berry Sud,** 10 rue de la République, 36000 Chateauroux: 93.5/103.2MHz – **R.F. Besançon,** La Citadelle, 25027 Besançon Cedex: 101.4/ 102.8MHz – **R.F. Bordeaux Gironde,** B.P. 585, 33006 Bordeaux Cedex: 100.1/101.6/101.8MHz – **R.F. Bourgogne,** 29 rue Guillaume Tell, 21000 Dijon: 87.8/103.4/103.7MHz – **R.F. Bretagne Ouest,** B.P. 1119, 29101 Quimper: 93.0MHz – **R.F. Champagne,** B.P. 1094, 51054 Reims Cedex: 94.8/103.4MHz – **R.F. Cherbourg,** Place du Générale de Gaulle, 50100 Cherbourg: 100.7MHz – **R.F. Corse Frequenza Mora,** 4 rue Favalelli, 20200 Bastia: 88.2/97.0/100.0/100.5/101.7/101.8/ 104.6/105.4MHz – **R.F. Creuse,** B.P. 249, 23005 Gueret: 94.3MHz – **R.F. Drome,** B.P. 519, 26005 Valence: 87.9/98.4MHz – **R.F. Frequence Nord,** 14 rue Léon Trulin, 59002 Lille: 87.8/94.7/95.5/97.8/106.2MHz – **R.F. Herault,** B.P. 1256, 34011 Montpellier Cedex 1: 101.1MHz – **R.F. Isère,** 10 rue Etienne Forest, 38027 Grenoble Cedex: 99.1/101.8/102.8MHz – **R.F. Landes,** 13 place Jean Jaurès, 40000 Mont de Marsan: 98.8/100.5/103.4MHz – **R.F. La Rochelle,** B.P. 1134, 17008 La Rochelle: 101.1/103.9MHz – **R.F. Limoges,** B.P. 422, 87012 Limoges Cedex: 92.5/101.4/103.5MHz – **R.F. Loire Ocean,** 11 rue Flandres Dunkerque, 44053 Nantes Cedex 04: 88.1/96.0/101.8MHz – **R.F. Mayenne,** 41 ave. Robert Buron, 53000 Laval: 96.6MHz – **R.F. Melun,** 24 place Saint Jean, 77000 Melun: 92.3MHz – **R.F. Nancy Lorraine,** 14 route de Mirecourt, 54042 Nancy Cedex: 100.0/100.5/102.6/103.0MHz – **R.F. Nimes,** 12 ave. Carnot, 3000 Nimes: 90.2/91.6/104.9MHz – **R.F. Normandie Caen,** 75 rue Basse, 14000 Caen: 102.2/102.6MHz – **R.F. Normandie Rouen,** 45 blvd. des Belges, 76000 Rouen: 95.1/100.1/101.6MHz – **R.F. Orleans,** 8 rue d'Illiers, 45057 Orleans Cedex 1: 93.9/100.9/106.8MHz – **R.F. Pau Bearn,** 2 rue O'Quin, 64000 Pau: 93.2/102.5MHz – **R.F. Pays Basque,** 20 rue Orbe, 64116 Bayonne Cedex: 101.3MHz – **R.F. Pays de Savoie,** 45 place de la Brigade de Savoie, 73000 Chamberry: 95.2/103.9/106.1MHz – **R.F. Perigord,** B.P. 3033, 24003 Perigueux Cedex: 91.7/99.0MHz – **R.F. Picardie,** Rue de Maréchal de Lattre de Tassigny, 80000 Amiens: 100.2/100.6/101.3/102.8MHz – **R.F. Provence,** 28 rue Granet, 13100 Aix en Provence: 102.1/102.5/102.9/103.6MHz – **R.F. Puy de Dome,** B.P. 277, 63008 Clermont-Ferrand: 102.5MHz – **R.F. Roussillon,** 34 ave. du Générale de Gaulle, 66000 Perpignan: 101.6MHz – **R.F. Tours,** B.P. 3231, 37032 Tours Cedex: 105.0MHz – **R.F. Vaucluse,** B.P. 230, 84021 Avignon Cedex: 100.4MHz + 319 tr's less than 1kW not mentioned.

RADIO FRANCE INTERNATIONALE

Polish on Lille 1377 kHz: MF 1800-1830.

▣ R.F.I. Emissions pour les travailleurs étrangers immigrés, 116 Av. Prés. Kennedy, 75016 Paris.
FOREIGN SERVICE: see International Broadcasting section.

SUD RADIO (Comm.)
▣ 4 place Alfonse Jourdain, 31071 Toulouse Cédéx. ☎ +33 (0) 5 6163 2020. ▤ +33 05 6163 2064 — **MW:** Toulouse 819kHz 20kW.
D.PRGR: 0530-1730 (for MW).
FM: 40 st's (stereo), mainly 102-105MHz — **D.PRGR:** 24h.

EUROPE 1 (Comm.)
▣ 26 Bis rue François 1er, 75008 Paris. ☎ +33 (0) 1 4232 9000. ▤ +33 (0) 1 4723 1900
Web: http://www.europe-dev-intl.fr/ **LW:** see Germany —
FM: 186 st's (mono), mainly on 104.7MHz. — **D.PRGR:** 24h.

RTL (Comm.)
▣ 22 rue Bayard, 75008 Paris. ☎ +33 (0) 1 4070 4070.
Web: http://www.rtl.fr
LW: see Luxembourg — **FM:** 158 st's (stereo), mainly on 104.3MHz.
D.PRGR: 24h.

RADIO MONTE CARLO (Comm.)
▣ 16 boulevard Princesse Charlotte, MC-98090 Monaco. ☎ +377 9315 1617. ▤ +377 9315 1630 — **LW:** Roumoules 216kHz 1400kW.
FM: 133 st's (stereo), mainly between 104.3 & 105.1MHz.
D.PRGR: 24h.

OTHER STATIONS
As of Sept. 1998, a total of 3235 licenses (transmitters) were allocated to private commercial and non-commercial FM st's. Approx. 1650 st's are affiliated to one of the following private commercial networks. Prgrs are distributed via satellite (no. of affiliates is approx.).

MAIN NATIONAL NETWORKS
Chérie FM, 22 rue Boileau, 75116 Paris. ☎ +33 (0) 1 40 71 40 00. ▤ +33 (0) 1 40 71 40 40 (129 st's affiliated) — **Europe 2,** 26 bis rue Francois 1er, 75008 Paris. ☎ +33 (0) 1 47 23 10 63. ▤ +33 (0) 1 47 23 10 59. **Web:** http://www.europe2.fr (193 st's affiliated) — **Fun R.,** 143 ave. Charles de Gaulle, 92521 Neuilly sur Seine. ☎ +33 (0) 1 46 40 48 48. ▤ +33 (0) 1 47 47 48 22. **Web:** http://www.funradio.fr (195 st's affiliated) — **Montmartre FM,** 4 rue Euler, 75008 Paris. ☎ +33 (0) 1 53 67 67 67. ▤ +33 (0) 1 53 67 67 68 (55 st's affiliated) — **NRJ,** 22 rue Boileau, 75116 Paris. ☎ +33 (0) 1 40 71 40 00. ▤ +33 (0) 1 40 71 40 40. **Web:** http://www.nrj.fr (267 st's affiliated) — **R. Classique,** 12 bis place Henri Bergson, 75008 Paris. ☎ +33 (0) 1 40 08 50 00. ▤ +33 (0) 1 40 08 50 80 (51 st's affiliated) — **R. Nostalgie,** 11 rue Franquet, 75015 Paris. ☎ +33 (0) 1 53 68 80 00. ▤ +33 (0) 1 45 32 10 31 (190 st's affiliated) — **RFM,** 28 rue François 1er, 75008 Paris. ☎ +33 (0) 1 42 32 20 00. ▤ +33 (0) 1 42 32 20 01 (147 st's affiliated) — **Rire et Chansons,** 22 rue Boileau, 75116 Paris. ☎ +33 (0) 1 40 71 40 00. ▤ +33 (0) 1 40 71 40 40 (62 st's affiliated) — **RTL2,** 22 rue Bayard, 75008 Paris. ☎ +33 (0) 1 40 70 40 00. ▤ +33 (0) 1 40 70 43 50. **Web:** http://www.rtl.fr (104 st's affiliated) — **Skyrock,** 37 bis rue Greneta, 75002 Paris. ☎ +33 (0) 1 53 40 30 20 (92 st's affiliated) — **R.C.F.,** 7 place St. Irénée, 69005 Lyon. ☎ +33 (0) 4 72 38 20 22. ▤ +33 (0) 4 72 38 20 57. **Web:** http://www.cef.fr/rcf/ (143 st's affiliated). – **Beur FM,** B.P. 226, 75524 Paris Cedex 11. ☎ +33 (0) 1 4929 43 43. ▤ +33 (0) 1 48 06 36 45. (9 st's affiliated). — **BFM,** 43 rue Marius Jacotot, 92800 Pute Aux. ☎ +33 (0) 1 41 25 19 00. ▤ +33 (0) 1 41 25 19 50.
Web: http://www.bfmradio.com/ (16 st's affiliated).

GERMANY

L.T: UTC + 1h (Su: UTC + 2) — **Pop:** 80m — **Radio sets:** 150m —
Pr.L: German — **E.C:** a/c 50, 230V

BUNDESMINISTERIUM FÜR POST UND TELEKOMMUNIKATION (PTT)
▣ Heinrich-von-Stephan-Straße 1, 53175 Bonn. ☎ +49 (228) 14-0 ① 8861 101 bpm d ▤ +49 (228) 1488 72
L.P: Fed. Min. für Post und Telekommunikation: Wolfgang Bötsch

ARBEITSGEMEINSCHAFT DER ÖFFENTLICH-RECHTLICHEN RUNDFUNKANSTALTEN DER BUNDESREPUBLIK DEUTSCHLAND (ARD)
▣ Bertramstraße 8, 60320 Frankfurt. ☎ +49 (69) 59 06 07 ① 4111 27

▤ +49 (69) 155 2075 **Web:** http://www.ard.de
NB: The ARD is the umbrella organisation which represents German public right (öffentlich-rechtlichen) broadcasters. The chairmanship rotates amongst the members.

ARD members:
(Anstalten des öffentlichen Rechts)

Bayerischer Rundfunk (BR), 80300 Munich (letters only)
▣ Rundfunkplatz 1, 80300 Munich. ☎ +49 (89) 5900 01 ▤ +49 (89) 5900 2375. **Web:** http://www.br-online.de/
L.P: Dir: Albert Scharf; Press: Sybille Giel

Deutsche Welle
▣ Raderberggürtel 50, 50968 Cologne. ☎ +49 (221) 389 0. ▤ +49 (221) 389 3000 **Web:** http://www-dw.dwelle.de
L.P: Dir: Dieter Weirich; Press: Ralf Siepmann

Hessischer Rundfunk (HR), 60222 Frankfurt (letters only)
▣ Bretramstraße 8, 60320 Frankfurt. ☎ +49 (69) 155 1 ▤ +49 (69) 155 2900 **Web:** http://www.hr-online.de
L.P: Dir: Klaus Berg; Press: Miahael Dartsch

Mitteldeutscher Rundfunk (MDR),
▣ Kantstraße 71-73, Leipzig. ☎ +49 (341) 3000 ▤ +49 (341) 300 5544 **Web:** http://www.mdr.de
L.P: Dir: Udo Reiter; Press: Susan E. Knoll

Norddeutscher Rundfunk (NDR)
▣ Rothenbaumchaussee 132-134, 20149 Hamburg. ☎ +49 (40) 4156 0 ▤ +49 (40) 4476 02 **Web:** http://www.ndr.de
L.P: Dir: Jobst Plog; Press: Martin Gratzke

Ostdeutscher Rundfunk Brandenburg (ORB)
▣ August-Bebel-Sraße 26-53, 14482 Potsdam-Babelsberg. ☎ +49 (331) 965 10 ▤ +49 (331) 965 3571
L.P: Dir: Hansjürgen Rosenbauer; Press: Pia Stein

Radio Bremen (RB)
▣ Bürgermeister-Spitta-Allee 45, 28329 Bremen. ☎ +49 (421) 246 0 ▤ +49 (421) 246 1010 **Web:** http://www.radiobremen.de
L.P: Dir: Karl-Heinz Klostermeier; Press: Michael Glöckner

Saarländischer Rundfunk (SR), 66100 Saarbrücken (letters only)
▣ Funkhaus Halberg, 66121 Saarbrücken. ☎ +49 (681) 602 0 ▤ +49 (681) 602 3874 **Web:** http://www.sr-sb.de
L.P: Dir: Fritz Raff; Press: Rolf-Dieter Ganz

Sender Freies Berlin (SFB)
▣ Masurenallee 8-14, 14057 Berlin. ☎ +49 (30) 3031 0 ▤ +49 (30) 3015 062 **Web:**http://www.sfb-berlin.de
L.P: Dir: Horst Schättle; Press: Thomas Strätling

Südwestrundfunk (SWR),
SWR Stuttgart - ▣ Neckarstraße 230, D-70190 Stuttgart. ☎ +49 (711) 9290. **L.P:** DG: Prof. Peter Voss.
SWR Mainz - ▣ Am fort Gonsenheim 39, D-55122 Mainz. ☎ +49 (6131) 9290.
SWR Baden-Baden - ▣ Hans-Bredow-Str., D-76530 Baden-Baden. ☎ +49 (7121) 9290 **Web:** http://www.swr-online.de

Westdeutscher Rundfunk (WDR), 50600 Cologne (letters only)
▣ Appellhofplatz 1, 50667 Cologne. ☎ +49 (221) 2201 ▤ +49 (221) 2 20 48 00 **Web:** http://www.wdr.de
L.P: Dir: Fritz Pleitgen; Press: Rüdiger Oppers

D.PRGR: All ARD members (except DLF and DW) produce common programming, in turn, for nighttime broadcasts carried by most members. Foreign language programs are produced and carried by HR, NDR, RB, SFB and WDR.
Nachtexpress: (light music, news & traffic on the h.) 2130-0300 UTC; Radiowecker: 0300-0500 UTC; Nachtkonzert: (serious music) 2305-0500 UTC (Mon-Fri 0105-0400 produced by DLF Cologne and carried also on DLF's frqs); ARD-Popnacht: 2305-0400 (Fri&Sat 0500) UTC

STATIONS:
BAYERISCHER RUNDFUNK
(Anstalt des öffentlichen Rechts)
N.B: Bayerischer Rundfunk is the public broadcasting authority for Bavaria.

Mediumwaves: Hof 801 kHz0.2kW, München 801kHz 100 kW, Würzburg 801 kHz 0.2kW, Dillberg/Neumarkt 801kHz 50kW
D.PRGR: 1700-2020 (UTC) prgrs. for foreign workers
Shortwave: (G.C: 11.45E/48.15N) DMR24 6085 kHz 100kW
FM Stereo MHz:

	I	II	III	IV	V	kW
Bad Reichenhall	91.8	89.0OB	96.7	98.3	105.0	0.3
Bambg./Geisbg.	94.8	89.6F	99.8	102.9	97.4	25/5
Berchtesgaden/						
Schönau	90.4	99.6OB	96.9	94.1	106.4*	0.01/3

	I	II	III	IV	V	kW
Brotjacklrgl.	92.1	96.5NO	94.4	100.9	106.9	100/50
Burgsinn	92.2	90.0MF	99.3	99.0		0.01/0.2
Büttelberg/						
Frankenhöhe	91.4	88.2F	99.3	95.5	104.0	25/10/25
Coburg/						
Eckardtsberg	93.5	88.3F	99.2	97.7		5/0.5/5
Dillberg/	88.9	92.3F/	97.9	87.6	102.0	25
Neumarkt		104.5NO			5	
G.-Partnkrchn.	89.2	93.5OB	97.7	95.9	104.9	0.5/0.5
Gelbelstein	101.6	90.5B	97.6	88.0	106.1	25/10/10
Grünten/Allgäu	90.7	88.7S	95.8	101.0	106.9	100
Hohe Linie	95.0	93.0NO	99.6	97.0	105.0	25/5/25
Hohenpeißenbg.	92.8	94.2OB	99.2	100.4	106.3*	25
Hoher Bogen/						
Kötzting	96.8	91.6NO	94.7	88.3	104.4	50/5/25
Hühnerberg	91.9	96.1S	99.5	93.1	107.6	25/10/11
Kreuzberg/Rhön	98.3	93.1MF	96.3	107.9	105.3	100/50/60
Landshut-						
Altdorf	90.2	97.8NO	95.3	93.2	106.6	1
München	91.3	88.4M	97.3	103.2	90.0	25
Ochsenkopf/						
Fichtelgeb.	90.7	96.0F	99.4	102.3	107.1	100/100
Passau/						
Kühberg	87.7	93.2NO	90.4	95.6		1
Pfaffenberg	95.6	88.4MF	93.4	98.0	106.4	25/1/25
Reit im Winkl	91.0	87.90B	97.1	93.1	104.8	0.1
Tegernseer Tal	94.0	87.70B	99.7	97.9		0.5
Traunstein/						
Hochberg	98.0	91.5OB	95.9	97.0	107.1	5/0.5/5
Wendelstein	93.7	89.5OB	98.5	102.3	105.7	100
Würzburg	90.9	90.0MF	97.6	89.0	105.7	5/5/5/0.3
Weiler/Allgäu	87.7	100.3	99.4	—	106.0	0.1

+ 10 lps.

F=Mittel- and Oberfranken — NO=Niederbayern/Oberpfalz — M=München — MF=Mainfranken — S=Schwaben — OB=Oberbayern
D.PRGR: Bayern 1 + regional prgrs. on FM I, MW & SW (nights rel. ARD "Nachtexpress"); Bayern 2 Radio on FM II (nights rel. ARD "Nachtkonzert"); Bayern 3 on FM III (nights rel. SWF-3); Bayern 4 Klassik on FM IV (nights rel. ARD "Nachtkonzert"); B5 Aktuell on FM V & SW in summer.(2300-0500 rel. MDR-Info).
Ann: "Hier ist der Bayerische Rundfunk."; Bayern 3 "...Bayern Drei".
V: by QSL-card or letter.

DEUTSCHLANDRADIO

(Nationaler Hörfunk)
Deutschlandradio broadcasts in two regions, Cologne and Berlin. The station's format mainly concists of cultural, musical and political programming.Its headquarters are in Cologne. From Berlin most cultural and musical programs are provided, whereas from Cologne news and political programs are supplied under the old name Deutschlandfunk.
DeutschlandRadio Cologne "Deutschlandfunk"
Long- & Mediumwaves: Donebach (Odenwald) 153 kHz 500kW, Aholming 207 kHz 500kW, Thurnau/Bayreuth 549 kHz 100kW, Nordkirchen 549 kHz 100kW, Braunschweig/Königslutter 756 kHz 200kW, Ravensburg 756 kHz 100kW, Neumünster 1269 kHz 300kW, Heusweiler 1422kHz 600kW.
FM Stereo: Cottbus (City) 88.6 3kW, Hamburg 88.7 MHz 3kW, Bonn 89.1 MHz 5.0kW, Auerbach 94.5 MHz 3.kW, Leipzig 96.6 MHz 100kW, Chemnitz (Geyer) 97.0 MHz 100kW, Dresden (Wachwitz) 97.3 MHz100kW, Berlin 97.7 MHz100kW, Löbau 99.5 MHz 5kW, Brotjackriegel 100.1 MHz 100kW, Högl 100.3 MHz 15kW, Ochsenkopf 100.3 MHz 100kW, Witthoh 100.6 MHz 40kW, Aurich 101.8 MHz100kW, Bungsberg/Eutin 101.9 MHz 95kW, Kapaunberg 102.0 MHz 20kW, Lingen 102.0 MHz 15kW, Höhbeck 102.2 MHz 95kW, Nordhelle 102.7 MHz 20kW, Wesel 102.8 MHz 100kW, Flensburg 103.3 MHz 20kW, Rhön (Heidelstein) 103.3 MHz 100kW, Torfhaus/Harz 103.5 MHz 100kW, Garz 104.0 MHz 10kW, Bremerhaven 104.3 MHz

10kW, Saarburg 104.6 MHz 20kW, Blauen 105.1 MHz 10kW, Casekow 105.2 MHz 6.3 kW, Hornisgrinde 106.3 MHz 100kW, Rostock 106.5 MHz 1kW, Bremen 107.1 MHz 100kW, Helpterberg 96.5 MHz 100kW, Kassel 107.5 MHz (1kW), Frankfurt/Main 97.6 MHz (0.3kW), Helgoland 89.8 MHz (0.1kW), Bielefeld 95.5 MHz (0.1kW), Köln 89.9 MHz (0.03kW), Eschwege 100.6 MHz (0.5kW), Suhl 98.8 MHz (0.1kW), Erfurt 103.1 MHz (2kW), Wiesbaden 103.7 (0.5kW)
F.PL: Koblenz 99.8 MHz.(0.5kW) and Bad Kreuznach 106.5 MHz (0.05kW)
Ann: "Deutschlandfunk".

DeutschlandRadio Berlin
Long- & Mediumwaves: Oranienburg (Zehlerdorf) 177 kHz 500kW, Berlin-Britz 990 kHz 100kW.
Shortwave: Berlin-Britz 6005 kHz/100kW (24h)
FM Stereo: Völklingen 88.6 MHz 0.1kW, Hamburg 89.1 MHz 0.05kW, Hof 89.3 MHz 20kW, Berlin 89.6 MHz 100kW, Sankt Wendel 90.3 0.1kW, Cottbus 90.8 MHz 30kW, Frankfurt/Oder 92.7 MHz 5kW, Dresden 93.2 MHz 1kW, Sonneberg 94.2 MHz 100kW, Schwerin 95.3 MHz 100kW, Saarlouis 96.3 MHz 0.1kW, Marlow 96.7 30kW, München 96.8 MHz 0.1kW, Dequede 96.9 MHz 7kW, Helpterberg 97.1 MHz 30kW, Inselsberg 97.2 MHz 100kW, Brocken 97.4 MHz 100kW, Garding (planned) 101.7 MHz 0.5kW, Olsberg 106.1 MHz 10kW, Eifel-Bärbelkreuz 106.1 MHz 20kW, Cuxhaven 107.7 MHz 20kW, Lebach 107.9 MHz 0.1kW, Bremen 100.3 MHz (1kW), Bremerhaven 106.5 (0.5kW), Neumünster 107.8 MHz (0.1kW), Saarbrücken 107.5 MHz (0.5kW), Sulzbach 96.8 MHz (0.1kW), Bielefeld 106.2 MHz (0.1kW), Fulda 90.7 MHz (0.3kW), Nürnberg 105.6 MHz (0.1kW), Homburg 98.6 (0.2kW), Stuttgart 87.9 MHz (1kW), Leipzig 100.4 MHz (?kW),
F.PL: Trier 94.3 MHz (0.2kW), Idar-Oberstein 94.7 MHz (0.2kW), Kaiserslautern 96.9 MHz (0.5kW) and Karlsruhe 96.6 MHz (0.2kW)

HESSISCHER RUNDFUNK
(Anstalt des öffentlichen Rechts)
N.B: Der Hessische Rundfunk is the public broadcasting authority for Hesse.
Mediumwaves: Frankfurt 594 kHz 300kW, H. Meissner 594 kHz 200kW
FM Stereo MHz:

	I	II	III	IV l.s.	kW
Biedenkopf	91.0	99.6	87.6	104.3L/	100
	—	—	—	102.3K	5
Frankfurt	—	—	—	*90.4RM	1
	—	—	—	103.9B	
Gr. Feldberg	94.4	96.7	89.3	102.5RM	100
Habichtswald	—	—	101.2	103.2K	20
Hardberg	90.6	95.3	92.7	101.6B	50
H. Meissner	99.0	95.5	89.5	101.7K	100
Limburg	—	—	—	97.1L	0.2
Rhön/Heidelst.	104.8	—	106.2	107.3F	50/50/50
Rimberg	91.3	95.0	97.7	91.9F	50/20
Wiesbaden	—	—	101.4	—	0.5
Würzburg	88.1	97.4	89.7	103.8B	1

+ 6 lps.
*) Freqs used by Messeradio during trade fairs. — l.s.) local sts.
D.PRGR: HR1 on FM I (nights rel. ARD "Nachtexpress"); HR1-Plus on MW; HR2 on FM II (nights rel. ARD "Nachtkonzert"); HR3 on FM III (nights rel ARD "Popnacht"); HR4 on FM IV (nights rel. ARD "Nachtexpress") — L=Lahn, K=R. Kurhessen, F=R. Fulda, RM=R. Frankfurt, B=Radio Bergstraße.
Ann: "Hier ist der Hessische Rundfunk". **V:** by QSL-card.

MITTELDEUTSCHER RUNDFUNK
(Anstalt des öffentlichen Rechts)
N.B: MDR is the public broadcasting authority for Saxony, Saxony-Anhalt and Thuringia.
F.PL: MDR-Info on FM (lp. tx)

Mediumwaves:	kHz	kW	Prgr
Burg	531	20	MDR-Info
Leipzig-Wiederau	783	100	MDR-Info
Wachenbrunn	882	20	MDR-Info
Dresden-Wilsdruff	1044	20	MDR-Info
Reichenbach	1188	3	MDR-Info

FM Stereo MHz:

	I	II	III	IV	kW
Auerbach/Schöneck	88.7S	101.2	98.7	—	3/30/3
Brocken	94.6SA	91.5	107.8	—	100/100/16
Chemnitz	92.8S	89.8	87.7	—	100
Dequede	94.9SA	98.9	89.4	—	10
Dresden	92.2S	90.1	95.4	—	100

	I	II	III	IV	kW
Erfurt	94.4T	—	—	—	1
Halle	100.8SA	—	—	104.4	5
Hoyerswerda	*100.4S	—	—	—	30
Inselsberg	92.5T	90.2	87.9	—	100/60
Jena	88.2	—	—	—	1
Kapaunberg	96.1SA	—	107.4	—	20/20
Leipzig	93.9S	—	88.4	—	100
	106.5SA	90.4	—	—	23/100
Löbau	98.2S	91.8	96.2	—	5
Oschatz	101.8S	103.7	105.9	—	30
Saalfeld-Remda	103.6	105.6	100.7	—	60
Sonneberg	91.7T	96.9	95.2	—	100/20
Suhl	93.7T	91.1	89.8	—	1/0.1/0.2
Ronneburg	97.8T	100.9	103.9	—	10/30
Weimar	93.3T	—	—	—	5
Wittenberg	88.1SA	101.6	104.0	—	30/80/80

*) MDR1-Radio Sachsen together with Sorbischer Rundfunk. (Mon-Fri 0405-0700 and 1100-1200, Sat 0505-0800, Sun 1000-1300 UTC). Programs in Sorbisch (minority dialect).
D.PRGR: MDR-1 on FM I: 24h. (S=MDR1-Radio Sachsen, SA=MDR1-Radio Sachsen-Anhalt, T=MDR1-Radio Thüringen); MDR-Life on FM II; MDR-Kultur on FM III, at nights relays "ARD-Nachtkonzert".; MDR-Sputnik on FM IV; MDR-Info only on MW: 24 h.
V: by QSL-card or letter.

NORDDEUTSCHER RUNDFUNK

(Anstalt des öffentlichen Rechts)
N.B: The NDR isthe public broadcasting authority for Lower Saxony, Schleswig-Holstein, Mecklenburg-Vorpommern and Hamburg.
Mediumwaves: Helpterberg 657kHz 20kW, Lingen 792kHz 5kW, Hannover 828kHz 50/10kW, Flensburg 702kHz 7.5kW, Hamburg 972kHz 100kW, Putbus 729 kHz 5kW.
D.PRGR: NDR 4 prgr on MW, 1800-2120 UTC prgrs for foreign workers on MW.

FM Stereo MHz:

	I	II	III	IV	V	kW
Alfeld	87.8NS	93.6	96.5	91.1	—	0.05
Anklam	94.6MV	—	—	—	—	6.3
Aurich	95.8NS	98.1	90.0	96.4	92.7	25(3x)/10/1
Bad Doberan	—	—	—	—	103.7	5
Bad Pyrmont	88.6NS	92.6	95.7	98.5	—	0.05
Bd. Rothenfelde	—	—	—	97.9	91.2	0.3/0.1
Barth	87.6MV	—	—	—	95.0	0.4/0.3
Braunschweig	—	—	—	—	100.3	15
Bremen(3)	—	—	—	95.0	—	1
Bremerhaven(3)	—	—	—	98.9	—	0.5
Braunlage	—	—	—	93.6	96.1	0.02
Bungsberg	97.8N	91.9	89.9	96.6	—	50(3x)
Cuxhaven	105.4NS	97.9	94.6	93.1	91.6	20/10/0.3/
	98.4 (0.3kW)					10/10
Dannenberg	91.2NS	96.4	93.3	90.7	94.0	25/25/10/3/1
Flensburg	89.6N	93.2	96.1	87.7	—	25(3x)/10
Goslar	88.2NS	93.7	95.1	96.0	96.5	0.1
Göttingen	88.5NS	94.1	96.8	99.9/	95.9	5/5/0.5/5/1
				107.1		0.5
Greifswald	101.0MV	—	—	—	—	0.16
Hamburg	89.5N/	—	—	—	94.2	10/1
	90.3H	87.6	99.2	92.3	—	80(3x)/5
Hannover	*90.9NS	96.2	98.7	—	92.6	15/3/15/0.5
Hann. Münden	88.2NS	96.1	90.8	92.9	94.8	0.05
Harz/Torfhaus	98.0NS	92.1	89.9	99.5	—	100(3x)/50
Heide	90.5N	96.3	99.4	87.9	94.9	15(4x)/0.5
Helgoland	88.9N	93.4	97.0	92.5	—	0.01
Helpterberg	90.5MV	99.1	96.0	101.8	—	100
	94.2MV (6.3kW)					
Heringsdorf	97.6MV	94.0	102.7	100.5	92.3	1
Holzminden	92.7NS	96.0	98.4	88.6	—	0.5
Kiel	91.3N	98.5	95.7	99.7	94.5	15/15/1/0.4/15
Klütz/Hamberge	—	—	—	—	103.4	5
Lauenburg	94.7N	—	—	96.8	99.8	0.3
Lingen	92.8NS	97.8	90.2	88.9	96.6	15(3x)/0.2/
						0.5
Lübeck	93.1N	90.7	88.0	95.9	94.0	0.5(3x)/0.1/
						0.5
Malchin	—	—	—	103.5	94.5	1
Marlow	91.0MV	93.5	88.2	102.8	—	100/30/100/
						100
Mölln	104.5N	—	—	—	—	20

	I	II	III	IV	V	kW
Neubrandenbg.	—	—	—	—	89.5	1
Neumünster	106.4N	—	90.8	98.7		20/1/0.5
Osnabrück	92.4NS	89.2	98.8	87.6	96.4	8(3x)/0.2
Pasewalk	93.7MV	—	—	—	—	2.5
Ribnitz-D.	—	—	—	—	99.4	0.3
Röbel	88.5MV	107.0	94.7	100.4	97.4	10/24/10/4/4
Rosengarten	103.2NS	—	—	—	—	20
Rostock	—	—	—	—	88.9	2
Rügen	102.5MV	99.8	91.5	88.6	95.5	50/10(4x)
Schwerin	92.8MV	98.5	89.2	105.3	99.5	30/100/30/
						100/2
Stadthagen	100.8NS	102.6	104.4	—	—	25
Steinkimmen	91.1NS	99.8	94.4	98.6	92.9	1
Stralsund	92.1MV	—	—	—	—	0.4
Sylt	90.9N	98.7	94.3	92.7	95.6	5
Überherrn	90.1MV	—	—	—	—	4
Visselhövede	91.8NS	95.9	87.8	98.4	90.1	5/5/2/5/1
Wolgast	89.0MV	—	—	—	93.2	0.4/0.3

NS="NDR 1-Radio Niedersachsen" — N="NDR 1Welle Nord" — H="NDR-Hamburg-Welle 90.3" — MV="NDR-1 Radio Mecklenburg-Vorpommern
(3)=Radio Bremen transmitter sites partially used by NDR
*) Freqs used by NDR-Messewelle only during Hannover trade fair.
-**N-Joy Radio**, Bebelallee 1, 22299 Hamburg. ☎ +49 (40) 480560.
-**NDR-Messewelle**, Rudolf-von-Bensingen-Ufer 22, 30169 Hannover. ☎ +49 (511) 88621
D.PRGR: NDR 1 ("Welle Nord", "R. Niedersachsen", "Radio Mecklenburg-Vorpommern" and "NDR Hamburg-Welle 90.3") on FM I at nights relay ARD "Nachtexpress"; (NDR Hamburg-Welle on 90.3 at nights rel. NDR 2); NDR 2 on FM II; NDR 3 on FM III (at nights rel. ARD "Nachtkonzert"); NDR 4 on FM IV & MW (At nights rel. ARD "Nachtexpress"). N-Joy Radio = 24 hrs. youth prgr on FM V.
Ann: "Hier ist NDR (1,2,3,4)", "N-Joy Radio" **V:** by QSL-card.
F.PL: New program "N-Rock".

OSTDEUTSCHER RUNDFUNK BRANDENBURG

(Anstalt des öffentlichen Rechts)
N.B: ORB is the public broadcasting authority for Brandenburg and represents 3 sts on FM only: Antenne Brandenburg, Radio Brandenburg & Radio Fritz.
-**Antenne Brandenburg**, Puschkinallee 4, 14469 Potsdam. ☎ +49 (331) 3200
-**Radio Brandenburg & Fritz**, August-Bebel-Straße 26-53, Potsdam. ☎ +49 (331) 965 10.
FM Stereo MHz:

	I	II	III	IV	kW
Belzig	106.2	99.3	91.9	100.2	100/30/30/30
Berlin-A'platz	99.7	95.8	102.6	93.1	100/100/80/25
Cottbus	98.6	95.1	103.2	104.4	1000/30/100/60
Calau	—	—	—	93.4+	30
Casekow	91.1	106.1	100.1	104.4	20/60/50/10
Forst	103.9	—	—	—	6
Guben	100.9	—	—	—	5
Dequede	91.4	—	—	—	4
Frankfurt/O	87.6	89.1	101.5	96.8	10/5/30/5
Prenzlau	99.4	—	—	—	0.5
Pritzwalk	106.6	99.9	103.1	91.7	100/10/60/10
Zehlendorf	90.8	—	91.8/	—	1.3
			107.9		5

+)Mon-Fri 0405-0700 + 1100-1200, Sa 0505-0800, SU 1000-1400 on FM 93.4 program in Sorbisch (minority language), produced by MDR in Bautzen and ORB in Cottbus. Other times relays "Radio B- Zwei".
D.PRGR: Antenne Brandenburg on FM I; Radio Brandenburg on FM II; Radio Fritz on FM III, Info-Radio on FM IV (except 93.4 (R.S.2).
Ann: "Antenne Brandenburg" (jingle), "Hier ist Radio Brandenburg/Fritz" (jingle)
V: by QSL-card.

RADIO BREMEN

(Anstalt des öffenlichen Rechts)
N.B: Radio Bremen is the public broadcasting authority for Bremen.
Mediumwave: Bremen 936kHz 100kW (prgrs for foreign workers)
Shortwave: Bremen 6190kHz 50kW
FM Stereo MHz:

	I	II	III	IV	kW
Bremen	93.8	88.3	96.7	101.2	100/100/50/100
Bremerhaven	89.3	92.1	95.4	100.8	25/25/6.5/6.5

D.PRGR: RB-1 on FM I, MW & SW (nights rel. SWF 3); RB-2 on FM II (nights rel. ARD "Nachtkonzert"); Radio Bremen Melodie on FM III

(nights rel. ARD "Nachtexpress"); RB-4 on FM IV (nights rel. ARD "Popnacht"). I=entertainment; II=cultural; III=Light music; IV=youth prgrs.
Ann: "Hier ist Radio Bremen Eins-Hansawelle"; "...Radio Bremen Zwei/Melodie/Vier". **V:** by QSL-card

SAARLÄNDISCHER RUNDFUNK
(Anstalt des öffentlichen Rechts)
N.B: Saarländischer Rundfunk is the public broadcasting authority for Saarland.
-Offener Kanal, Eschberger Weg 65, W-6600 Saarbrücken. ☎ +49 (681) 812026
FM Stereo MHz:

	I	II	III	IV	kW
Bliestal	92.3	98.0	89.1	105.0	5
Göttelborner H.	88.0	91.3	95.5	103.7	100/40
Merzig	89.3	92.1	98.0	105.1	0.1
Mettlach	98.6	88.5	96.0	106.1	0.1
Moseltal	91.9	88.6	96.1	106.2	5
Saarbrücken (H)	94.2 (SR-1 & Stadtr. Saarbrücken)				0.5

D.PRGR: SR 1, Europawelle Saar on FM I only(nights rel. ARD "Popnacht"); SR 2 on FM II and partially on FM IV (nights rel. ARD "Nachtkonzert"); SR 3 Saarwelle on FM III (nights rel. ARD "Nachtexpress"); SR 4 & "Offener Kanal" (Mon-Fri 1400-1800, Sat 1100-1700), other times rel. SR 2 on FM IV.
V: by QSL-card

SENDER FREIES BERLIN
(Anstalt des öffentlichen Rechts)
N.B: SFB is the public broadcasting authority for Berlin.
Mediumwaves: Berlin-Spandau 567 kHz 50kW
FM Stereo MHz:

	I	II	III	IV	V	VI	kW
Berlin	88.8	92.4	96.3	106.8	93.1	*104.1	80(3x)/1/25

*) Freq. used by "Fritz Love Radio" only during the Love Parade Weekend in July (in 1998: Fri 10- Sun 12 July)
D.PRGR: "Berlin 88'8" on FM I; "Radio B Zwei" on FM II and 567 kHz; "SFB3" on FM III, "SFB 4 Multikulti" on FM IV. (German & foreign languages), "Inforadio" on FM V.

SÜDWESTRUNDFUNK
(Anstalt des öffentlichen Rechts)
N.B: Süddeutscher Rundfunk & Südwestfunkhave merged to create Südwestrundfunk. SWR commenced broadcasting as of September 1st. 1998.
D.PRGR: SWR 1, SWR 2, SWR 3, SWR 4.
STATIONS: All frequencies formerly used by Süddeutscher Rundfunk and Südwestfunk.

Baden-Württemberg
Mediumwaves: Baden-Baden 1485kHz 1kW SWR4, Bd. Mergenth. 711kHz 3kW SWR1, Bodenseesender 666kHz 150kW SWR4, Freiburg 828kHz 10kW SWR4, Heidelberg 711kHz 5kW SWR1, Heilbronn 711kHz, 5kW SWR1, Mühlacker 576kHz 200kW SWR1, Rheinsender 1017kHz 200kW SWR1, Ulm 711kHz 5kW SWR1.
Shortwave: Mühlacker 6030 kHz 20kW SWR3, Rohrdorf 7265kHz 20kW SWR3.
FM Stereo MHz:

	I	II	III	IV	Reg.
Aalen	95.1	91.1	98.1	96.9	UL
Bd. Mergenth.	87.8	93.2	99.7	105.5	HN
Baden-Baden	90.9	98.9	94.1	88.5	KA
Blaubeuren	89.6	91.5	98.9	96.4	UL
Blauen	89.2	92.6	97.0	—	
Brandenkopf	95.4	—	99.7	97.6	OG
Buchen	91.9	97.1	94.1	107.5	MA
Creglingen	89.8	92.5	97.2	94.9	HN
Donautal	93.7	—	92.8	87.6	RV
Eberbach	97.2	87.8	94.2	—	
Elzach	—	—	—	101.9	FR
Enztal	92.5	98.0	94.3	—	
Eyachtal	—	—	—	99.5	TU
Feldberg	89.8	97.9	93.8	104.0	FR
Freiburg	107.0	91.1	99.2	—	
Freudenberg	90.3	97.2	94.9	91.6	HN
Geislingen	90.3	88.5	95.5	107.9	S
Giengen/Brenz	89.9	87.8	91.5	97.7	UL
Grenzach-Whln.	97.2	96.4	92.3	—	

	I	II	III	IV	
Grünten	98.7	—	103.0	—	
Heidelberg	97.8	88.8	99.9	104.1	MA
Heidelberg	91.5	—	—	—	#
Heidenheim	87.6	99.1	97.6	89.8	UL
Heilbronn	—	—	—	99.5	HN
Hohe Möhr	87.6	100.2	96.8	—	
Hornisgrinde	93.5	96.2	98.4	94.0	OG
Ipf	—	—	—	94.1	UL
Langenbrand	92.9	—	89.5	—	
Laufenburg	94.0	91.4	90.1	—	
Mötzingen	90.5	—	97.2	87.6	S
Mühlacker	—	—	—	95.7	KA
Murgtal	89.1	—	97.5	—	
Nagoldtal	97.4	90.4	99.5	—	
Oberes Kinzigtal	90.5	—	94.5	99.2	OG
Pforzheim	—	99.5	—	87.6	KA
Raichberg	88.3	91.8	94.3	107.3	TU
Rauenberg	95.1	91.2	92.4	—	
Schliengen	—	—	—	96.6	FR
Schramberg	89.4	98.8	91.2	—	
Schussental	99.0/107.2	—	87.9	—	
Schw. Gmünd	—	—	—	100.9	UL
Sigmaringen	—	—	—	101.2	RV
St. Chrishona	87.9	92.0	89.5	98.3	LO
Strasbourg	—	—	—	88.9	OG
Stuttgart	94.7	105.7	92.2	90.1	S
Stuttgart-Fkh.	99.6	93.1	90.8	—	
Ulm	92.6	89.2	97.4	94.5	UL
Vaihingen	—	104.7	—	—	
Villingen	—	—	—	91.1	FR
Vogtsburg	—	—	—	100.7	FR
Waldburg	—	94.9	—	91.2	RV
Waldenburg	98.8	93.8	96.5	106.6	HN
Wannenberg	95.1	92.8	98.5	87.7	LO
Wattkopf	90.3	—	—	97.0	KA
Weinheim	97.1	93.1	99.5	100.7	MA
Wertheim	96.9	91.8	94.6	101.2	HN
Wiesensteig	90.6	97.2	99.3	—	
Witthoh	92.4	90.4	97.1	89.0	RV*
Wittigbachtal	92.4	—	94.1	90.4	HN

= Programmes for foreign nationals 1900-2230. * = R. Breisgau 1300 to 1400. KA = Baden Radio; RV = Bodensee Radio; HN = Franken Radio; MA = Kurpfalz Radio; OG = Ortenau Radio; FR = Radio Breisgau; UL = Schwaben Radio; TU = Radio Tübingen; S = Radio Stuttgart; LO = Hochrhein Radio.

Rheinland-Pfalz
Mediumwaves: Messkirch Rohrd 666 kHz 150kW, Mainz Wolfsheim1017 kHz 200kW(F.PL: down to 100kW), Freiburg Lehen 828RB kHz 10kW, Baden-Baden 1485 kHz 1kW
Shortwave: Rohrdorf 7265 kHz 20kW
FM Stereo MHz:

	I	II	III	IV	Reg.
Alf-Bullay	99.5	95.1	92.6	—	
Bad Marienberg	89.8	95.4	92.8	106.3	KO
Bad Marienberg	—	—	91.3	—	#
Bleialf	88.3	99.7	98.9	94.6	TR
Bornberg	90.8	93.9	97.5	99.6	KL
Donnersberg	99.1	92.0	101.1	105.6	KL
Eifel	91.1	—	98.5	93.6	TR
Haardtkopf	97.7	93.0	90.0	107.1	TR
Hohe Wurzel	—	—	—	107.9	MZ
Idar-Oberstein	—	—	—	106.4	TR
Kettrichhof	100.8	—	107.2	104.2	KL
Kirn	96.4	90.4	93.3	106.1	TR
Koblenz	96.1	94.0	91.6	107.4	KO
Koblenz	—	—	—	99.4	#
Lahntal	88.4	93.4	98.2	—	
Linz	92.4	—	94.8	97.4	KO
Mainz	87.7	103.2	93.7	91.4	MZ
Nahetal	88.5	95.1	98.1	—	
Nierstein	—	—	87.9	92.9	MZ
Oberes Ahrtal	88.5	—	99.4	93.3	KO
Rheinsender	—	—	—	94.9	MZ
Saarburg	99.2	93.8	90.6	101.2	TR
Trier	94.9	89.4	98.2	98.8	TR
Weinbiet	89.9	102.2	—	95.9	LU

= Programmes for foreign nationals 1900-2230. KO = Radio Koblenz, KL = Radio Kaiserslautern, LU = Radio Ludwigshafen, MZ = Radio Mainz, TR = Radio Trier.

WESTDEUTSCHER RUNDFUNK

(Anstalt des öffentlichen Rechts)
N.B: The WDR is the public broadcasting authority for North Rhine-Westphalia.
Mediumwaves: Langenberg 720 kHz 200kW (at daytime only), Bonn 774 kHz 5kW
FM Stereo MHz:

	I	II	III	IV	V	kW
Aachen-Stolb.	106.4	100.8	95.9	93.9	101.9A	20
Bonn (Venusb.)	102.4	100.4	93.1	90.7	88.0K	50/10/35
Ederkopf	107.2	101.8	105.2	100.7	95.8S	15
Kleve	103.7	93.3	97.3	101.7	99.7D	2
Köln	—	—	98.6	—	—	1
Langenberg	106.7	99.2	95.1	101.3	88.8D	85/100
	—	103.3E	—	—	—	100
Münster	107.9	94.2	89.7	100.0	92.0M	25
Nordhelle	104.7	93.5	98.1	103.8	90.3S	35
Olsberg	107.0	98.6	100.9	104.1	98.6S	10
Teutoburger W.	105.5	93.2	97.0	100.5	90.6B	100
Wittgenstein	—	92.3	88.7	—	—	15
Wuppertal	—	99.8	—	—	—	1

+ 27 lps.

D.PRGR: "Eins Live" on FM I (nights rel.WDR-2); WDR 2 on FM II & MW (720 & 774kHz); WDR 3 on FM III (nights rel. ARD "Nachtkonzert"); WDR 4 on FM IV (nights rel. ARD "Nachtexpress"); "WDR Radio 5" on FM (nights rel. ARD "Nachkonzert")
WDR 2 regional prgrs: A=Aachen — B=Bielefeld — D=Düsseldorf — E=Essen — K=Köln — M=Münster — S=Siegen — W=Wuppertal)
Ann: "Hier ist Eins Live/WDR 2/WDR 3/WDR 4/WDR Radio 5
V: by QSL-card.

COMMERCIAL RADIO STATIONS

There are no nationwide commercial radio networks in Germany. The Federal Republic of Germany comprises 16 so called "Bundesländer" (states). Each state has its own authority, the so called "Landesmedienanstalt", which is responsible for the allocation of radio licenses to commercial radio stations. In general either one or two statewide high-power (up to 100kW) commercial radio stations, or a number of low power (± 0.1 kW) local commercial radio stations are allowed. Only in Bavaria and Saxony both types of commercial radio exist. Statewide commercial radio stations operate in: Schleswig-Holstein — Hamburg — Mecklenburg-Vorpommern — Lower Saxony — Bremen — Brandenburg — Saxony-Anhalt — Berlin — Saxony — Hesse — Rhineland-Palatinate — Saarland — Bavaria — Thuringia. Local commercial radio stations operate in: Saxony — North Rhine-Westphalia — Baden-Württemberg — Bavaria.
N.B: Many sts. are also relayed via satellite. For detailed information see the Satellite and TV Guide, another Watson-Guptill publication.

BREMEN (City)

-Radio 107.1, Erste Schlachtpforte 1, 28195 Bremen. ☎ +49 (421) 335660. 🖷 +49 (421) 3356677.
STATIONS: Bremen 107.1 MHz (100kW & Bremerhaven 104.3 MHz (10kW) **D.PRGR:** Mon-Fri 0700-0900 & 1300-1430, Sat & Sun 0700-1000 (other times rel. DLF)
F.PL: New freqs. for "Radio 107.1" (24 hrs): Bremen 100.3 (1kW), Bremerhaven 106.2 (0.5kW)
-Offener Kanal Bremen(Public), Findorffstraße 22-24, 28215 Bremen. ☎ +49 (421) 350 1010/350 1030.
-Offener Kanal Bremerhaven (Public), Hafenstraße 156, 27576 Bremerhaven. ☎ 49 (471) 954 9595
D.PRGR: Tue 0900-1100 & 1300-2000, Wed 1300-2000. Thu & Fri 0900-2000, Sat 1300-1700. Thu 1600-2000 rel. Radio 46 Stadtstudio Walle. Mon 0900-1100 & 1300-1800 rel. Offener Kanal Bremer Umland. Other times relays Radio Bremen 2.
STATIONS: Bremen 92.5 MHz (0.3kW), Bremerhaven 90.7 MHz (0.2kW)

SCHLESWIG-HOLSTEIN (State)

-Radio Schleswig-Holstein (RSH), Funkhaus Wittland, 24109 Kiel. ☎ +49 (431) 58 700 🖷 +49 (431) 588 646
-delta radio, Werftstraße 214, 24143 Kiel. ☎ +49 (431) 70200
-Nordostseeradio (NORA), Im Saal 2, 24145 Kiel. ☎ +49 (431) 717273 🖷 +49 (431) 7177803

STATIONS:

Loc.	RSH	delta radio	NORA	R. Küstenrock	kW
Eutin (Bunsberg)	100.2	104.1	106.2	—	50/0.25
Flensburg	101.4	105.6	88.5	106.5	20/0.16/0.5
Garding	—	—	94.1	91.7	0.5/0.5
Hamburg	100.0	93.4	—	—	0.1/2
Hamburg	102.0	107.7	—	—	0.1/0.1
Heide	103.8	100.4	—	—	15
Helgoland	100.0	103.5	101.6	—	0.05/0.5
Itzehoe	—	—	104.9	92.7	1/0.5
Kaltenkirchen	102.9	107.4	101.1	—	20/100
Kiel	102.4	105.9	97.9	97.4	15/0.1/0.16
Lauenburg	102.5	105.6	97.4	—	1/0.3
Lübeck	101.5	107.9	91.5	93.6	20/1/0.5/1
Neumünster	—	—	88.9	—	0.5
Niebüll	—	—	107.2	—	
Schleswig	—	—	92.4	100.8	1
Sylt (Westerland)	102.8	104.8	89.1	89.8	5/1/0.5
Rendsburg	—	—	93.6	92.9	0.5/0.5

D.PRGR: Radio Schleswig-Holstein also rel. regional prgrs. from Flensburg, Kiel, Heide, Lübeck, Pinneberg and Hamburg.
-Offener Kanal Lübeck (Public), Kanalstraße 42-48, 23552 Lübeck. ☎ +49 (451) 705002
D.PRGR: Mon-Sat 1300-2100 UTC
STATIONS: Berkenthin (near Ratzenburg) 98.8 MHz (0.5kW)
-Offener Kanal Westküste (Public).
D.PRGR: daily 1500-1800 UTC
STATIONS: Berkenthin (near Ratzenburg) 98.8 MHz 0.5kW, Garding 97.6 MHz (0.5kW).
F.PL: New commercial radio station for Schleswig-Holstein on MW 612 kHz named "Power 612". **D.Prgr:** 0500-1800 UTC
-Regattaradio, 105.75 MHz (0.025kW) only during "International Rose Regatta" in the Baltic Sea.

HAMBURG (City)

-Magic FM, Spaldingstraße 218, 20097 Hamburg. ☎ +49 (40) 32 73 30.
-Offener Kanal (public), Stresemannstraße 375, 22761 Hamburg. ☎ +49 (40) 89 81 51.
-Energy Hamburg, Winterhuder Marktplatz 6-7, 22299 Hamburg. ☎ +49 (40) 4800190
-Klassik Radio, Brandstwiete 4, 20457 Hamburg. ☎ +49 (40) 3005050 🖷 +49 (40) 30050511.
-Radio Hamburg, Speersort 10, 20095 Hamburg. ☎ +49 (40) 339 71 40.
-AlsterRadio, Rödingsmarkt 29, 20459 Hamburg. ☎ +49 (40) 370907 15
STATIONS: Magic FM on 95.0 & 88.1 (0.1kW); Offener Kanal on 96.0 (0.16kW); Energy Hamburg on 97.1 (0.1kW) & 100.9 (0.1kW); Klassik Radio on 98.1 (0.1kW); Radio Hamburg on Hamburg 103.6 (80kW), Hamburg 104.0 (0.20kW) & Cuxhaven 88.5 (2kW); AlsterRadio on 106.8 (40kW) & Cuxhaven on 93.6 (2kW); Radio FSK on 93.0 (0.05kW), RSH on 100.0 & 102.0 (0.1kW); delta radio on 93.4 (2kW) & 107.7 (0.1kW)
-Radio FSK, Schulterblatt 23c, 20457 Hamburg. ☎ +49 (40) 434324

MECKLENBURG-VORPOMMERN (State)

-Antenne MV, Funkhaus Plate, 19086 Plate. ☎ +49 (3861) 55000
STATIONS: Heringsdorf 105.4 MHz 10kW, Röbel 93.8 MHz 4kW, Helpterberg 103.8 MHz 100kW, Schwerin 101.3 MHz 100kW, Marlow 100.8 MHz 100kW, Rostock 97.3 MHz, Putbus/Rügen 105.1 MHz 50kW, Güstrov 107.7 (1.3kW), Klützer Winkel 105.8 (0.1kW)
-Ostseewelle, Kröpeliner Straße 83, 18055 Rostock. ☎ +49 (381) 491340. 🖷 +49 (381) 4913419
STATIONS: Marlow 104.8 (100kW); Helpterberg 105.8 (100kW); Heringsdorf 103.3 (2kW); Schwerin 107.3 (100kW).
-Offener Kanal Neubrandenburg, Treptower Straße 9, 17033 Neubrandenburg.
STATION: Neubrandenburg 87.8
F.PL: New commercial radio station for Mecklenburg-Vorpommern on MW Schwerin 576 kHz (250kW) & Greifswald 1017 (5kW) called "Mega Radio MV" on MW 999 kHz (Schwerin) and on 558 kHz (Neubrandenburg, Rügen and Rostock).

LOWER SAXONY (State)

-Airport Radio, Flughafen Han over, Abflugebene B, Flughafenstraße, 30855 Langenhagen. ☎ +49 (511) 9770. **D.Prgr:** 0400-0700 UTC prgrs for airline passengers.
-Pfarrfunk Breitenberg, c/o Georg Borghard, Lindenweg 2, 37115 Duderstadt-Breitenberg

STATION: 98.4 MHz 6W.
-ffn, Stiftstraße 8, 30159 Hannover. ☎ +49 (511) 1666 0. 🖷 +49 (511) 1666 110
-Antenne Das Radio, Goseriede 9, 30159 Hannover. ☎ +49 (511) 91 180.
STATIONS:

Loc.	ffn	Antenne	kW
Aurich	103.1B/O	104.9OL	25/2
Barsinghausen	101.9HAN	103.8H	25
Braunschweig	103.1BS	106.9BS	15/0.16
Cuxhaven	102.6B/O	104.6	20/2
Dannenberg	102.7H/L	106.1	25
Göttingen	102.8GÖ	106.0GÖ/93.4	5
Hann. Münden	100.7GÖ	106.7GÖ	0.5
Holzminden	102.2GÖ	105.7GÖ	0.5
Lingen	101.5OS	104.3OS	15
Lüneburg	—	95.5	1
Osnabrück	103.4OS	105.9OS	10
Rosengarten	100.6H/L	105.1	20
Steinkimmen	102.3BO	105.7OL	100
Torfhaus	102.4BS	106.3	100
Visselhövede	101.7H/L	104.2	10

D.PRGR: Radio ffn broadcasts loc. prgrs. from Hamburg, Lüneburg (H/L), Oldenburg & Bremen (B/O), Osnabrück (OS), Hannover (HAN), Braunschweig (BS), Göttingen (GÖ).

Local Radio Stations in Lower Saxony (non-commercial):
-Radio Aktiv Hameln, Hastenbeckerweg 8 (TGZ), 31785 Hameln. ☎ +49 (5151) 55155
-OK Emsland, c/o Volkshochschule Lingen, Postfach 2144, 49791 Lingen. ☎ +49 (591) 9120210.
-OK Oldenburg, Bahnhofstraße 11, 26112 Oldenburg. ☎ +49 (441) 26444.
-OK Osnabrück, Lohstraße 45a, 49074 Osnabrück. ☎ +49 (541) 23489
-Radio OKerwelle, Leopoldstraße 6/7, 38100 Braunschweig. ☎ +49 (531) 244410.
-Radio Flora, Wlhelm-Blum-Straße 12, 30451 Hannover. ☎ +49 (511) 21979 0.
-Radio Jade, Kieler Straße 63, 26382 Wilhelmshaven. ☎ +49 (4421) 44566/42215.
-Radio Velzen, Brauereistraße 32, 29525 Velzen. ☎ +49 (581) 2227
-Radio Take Off, RTO, Postfach 1410, 49644 Cloppenburg.
STATIONS:

Loc.	freq.	kW	Station
Bad Pyrmont	94.8	0.05	R. Aktiv Hameln
Braunschweig	104.6	0.5	R. OKerwelle
Bremen	92.5	0.3	Bremer Umland
Bremerhaven	90.7	0.2	OK Wesermündung
Cuxhaven	88.5	2	Radio Hamburg
Cuxhaven	93.6	2	AlsterRadio
Cloppenburg	99.3	1	R. Take Off
Göttingen	107.1	1	StadtRadio
Hameln	100.0	0.3	R. Aktiv Hameln
Hannover	106.5	0.3	R. Flora/Stadtradio
Lingen	95.6	1	Ems-Vechte-Welle
Lüneburg	95.5	1	Radio Velzen
Oldenburg	106.5	1	OK Oldenburg
Osnabrück	104.8	1	OK Osnabrück
Uelzen	88.0	1	Radio Uelzen
Wilhelmshaven	87.8	1	Radio Jade

BERLIN (City)
-Berliner Rundfunk, Leipziger Straße 62, 10117 Berlin. ☎ +49 (30) 201910
-r.s.2, Voltastraße 5, 13355 Berlin. ☎ +49 (30) 467030
-Spreeradio, Schloßplatz 7, 10178 Berlin. ☎ +49 (30) 201 99600.
-Radio Hundert,6, Katharina-Heinroth-Ufer 1, 10787 Berlin. ☎ +49 (30) 25403143.
-Energy 103.4, Potsdamer Straße 88, 10785 Berlin. ☎ +49 (30) 254350.
-104.6 RTL, Kurfürstendamm 207-208, 10719 Berlin. ☎ +49 (30) 884840.
-Kiss-FM, Voltastraße 5, 13355 Berlin. ☎+49 (30) 4649090
-Jazz Radio Berlin, Pestalozzistraße 105, 10625 Berlin. ☎ +49 (30) 3124550
-NewsTalk 93.6, Kurfürstendamm 207-208, 10719 Berlin. ☎ +49 (30) 88484552. 🖷 +49 (30) 88484551
-Radio Paradiso, Am Kleinen Wannsee 5, 14109 Berlin. ☎ +49 (30) 8069200

STATIONS: VOA Express/Uniradio BB on 87.9 (1kW); Berliner Rundfunk on Berlin 91.4 (100kW), Cottbus 102.2 (3kW), Angermünde 100.9 (1kW). Frankfurt/Oder 104.2 20kW; r.s.2 on Berlin 94.3 (50kW), Angermünde 107.3 (12kW), Frankfurt/Oder 94.7 (3kW); Spremberg 105.3 (4kW), Cottbus 95.6 (1kW) and Lübben 100.1 (3kW); Spreeradio on 105.5 MHz (1.5kW); Hundert,6 on 100.6 (10kW); Energy 103.4 on 103.4 (50kW); 104.6 RTL on104.6 (8kW); Radio Paradiso on 98.2 (8kW); Kiss-FM on 98.8 (3kW); Jazz Radio Berlin on FM 101.9 (1kW); NewsTalk 93.6 on 93.6 (3kW); Klassik radio on 101.3 (5kW), RTL Radio on MW 891 kHz (5kW), Radio Makaria on 94.8 (3kW).

BRANDENBURG (State)
-BB Radio, Friedrich-Engels-Straße 37, 14482 Potsdam. ☎ +49 (331) 74787-0
STATIONS: Cottbus (Calau) 107.2 MHz 100kW, Berlin 107.5 MHz 13kW, Frankfurt /Oder 107.8 MHz 30kW, Angermünde 102.1 MHz 10kW, Pritzwalk 104.3 MHz 94kW, Potsdam 97.2 MHz (1kW), Zehlendorf 95.4 (1.3kW).

SAXONY-ANHALT (State)
-Radio Brocken, Brachwitzer Straße 16, 06118 Halle/Saale. ☎ +49 (345) 5258201
-Radio SAW (Sachsen-Anhalt-Welle), Hansapark 1, 39116 Magdeburg. ☎ +49 (391)6300. 🖷 +49 (391) 6309
STATIONS:

Loc.	R. Brocken	R. SAW	kW
Brocken	89.0	101.4	60/100
Dequede	101.0	95.6	60/1.5
Halle	93.5	103.3	10/5
Leipzig	—	104.9	90
Magdeburg	105.7	100.1	4/3
Dessau	90.6	92.6	2
Naumburg	98.8		10
Wittenberg	102.3	98.4	4/5

SAXONY (State)
-Antenne Sachsen, Breitscheidstraße 40, 01237 Dresden. ☎ +49 (351) 25770. 🖷 +49 (351) 2577180
-Radio PSR, Delitzscher Straße 97, 04129 Leipzig. ☎ +49 (341) 5636600.
-Radio Blau, Steinstraße 18, 04275 Leipzig. ☎ +49 (341) 3010097
-Radio T, Zwickauer Straß 152, 09116 Chemnitz. ☎ +49 9371) 4884165
-coloRadio, Schandauer Straße 64. 01277 Dresden. ☎ +49 (351) 3463207
-Radioropa-Infowelle Sachsen, Reichpietschstraße 23, 04317 Leipzig. ☎ +49 (341) 696060
-Radio Mephisto, Internes PF 0099, Augustusplatz 2, 04109 Leipzig. ☎ +49 (341) 9737960

STATIONS:
Radioropa on Burg 261 kHz (50kW)

Loc.	A. Sachsen	R. PSR	Radioropa	kW
Leipzig	106.9	102.9	97.6*	100/100/2
Dresden	105.2	102.4	89.2	100/100/1
Chemnitz	105.4	100.0	91.0**	100/100/3
Löbau	105.6	101.0	—	30
Oschatz	104.7	98.0	—	3
A/Schöneck	106.0	92.0	—	30/5
Auerbach	—	—	88.2	?
Fichtelberg	—	—	107.7	?
Hoyerswerda	—	—	102.8	?
Reichenbach	—	—	100.5	?

*) together with Radio Mephisto and Radio Blau
**) together with Radio T.
OTHERS: Energy Chemnitz 97.5 on 97.5MHz 3kW (Chemnitz); Radio Chemnitz on 102.1 MHz 3kW (Chemnitz); Energy Dresden 100.2 on 100.2 MHz 5kW (Dresden); Radio Leipzig 91.3 on 91.3 MHz 4kW (Leipzig); Energy Leipzig 99.8 on 99.8 MHz 4kW (Leipzig); Radio Lausitz on 107.6 MHz 30kW (Löbau); Radio Dresden 103.5 on 103.5 (2kW); Energy Zwickau on 98.2 0.1kW; Radio Zwickau on 96.2 MHz (0.5kW).

NORTH RHINE-WESTPHALIA (State)
N.B: There are 45 loc. radio stations occupying approx. 114 freqs. in the FM-Band with a power of 10W-4kW. All stations use a common supporting program formula ("Mantelprogram" or "Rahmenprogram"), but always with local ID's, some local news and advertising.
Most important sts. are:
-Radio Köln, Mediapark 5, 50670 Köln. ☎ +49 (221) 951990.

-Antenne Düsseldorf, Kaistraße 7-9, 40221 Düsseldorf. ☎ +49 (211) 9301010
-Radio Duisburg, Ruhrorter Straße 187, 47119 Duisburg. ☎ +49 (203) 800890
-Radio Essen, Sachsenstraße 36, 45128 Essen. ☎ +49 (201) 245850
-Radio Bonn/Rhein-Sieg, Kennedybrücke 4, 53225 Bonn. ☎ +49 (228) 400 710
-Radio FiV (Funk im Vest), Schaumburgstraße 14, 45657 Recklinghausen. ☎ +49 (2361) 9460.
-Radio 91.2, Karl-Zahn-Straße 11, 44141 Dortmund. ☎ +49 (231) 957735
-Radio Berg, Friedrich-Ebert-Straße, 51429 Bergisch Gladbach. ☎ +49 (2204) 844000.
-Welle Niederrhein, Uerdinger Straße 543, 47800 Krefeld. ☎ +49 (2151) 50602
-Radio Neandertal, Elberfelder Straße 81, 40804 Mettmann. ☎ +49 (2104) 919010.
STATIONS:

Loc.	MHz	kW	St.
Bonn	98.9	0.1	R. Bonn/Rhein-Sieg
Dorsten	105.2	0.1	Radio FiV
Dortmund	91.2	0.2	Radio 91.2
Duisburg	92.2	0.1	R. Duisburg
Düsseldorf	104.2	1	A. Düsseldorf
Essen	102.2	0.3	Radio Essen
Haltern	95.6	0.1	Radio FiV
Herchen	10.7.9	0.1	R. Bonn/Rhein-Sieg
Krefeld	87.7	0.2	Welle Niederrhein
Köln	99.7	0.5	R. Berg
Köln	107.1	0.5	R. Köln
Langenberg	97.6	4	Radio Neandertal
Lindlar	105.2	4	Radio berg
Much	94.2	0.1	R. Bonn/Rhein-Sieg
Opladen	96.9	0.5	Radio berg
Recklinghausen	94.6	0.1	Radio FiV
Siegburg	91.2	0.2	R. Bonn/Rhein-Sieg
Viersen	102.5	0.3	Welle Niederrhein
Waldbröl	105.7	1	Radio Berg

-evosonic radio, Follerstraße 2, 50676 Köln. ☎ +49 (221) 92134210. 🖹 +49 (221) 9213 4211 (only via satellite)

THURINGIA (State)
-Antenne Thüringen, Belvederer Allee 25, 99425 Weimar. ☎ +49 (3643) 5525 52.
STATIONS: 102.2 100kW (Inselsberg); 102.7 60kW (Sonneberg); 102.5 30kW (Ronneburg); 107.6 60kW (Remda); 101.3 (Suhl); 90.9 MHz 1kW (Jena); 103.9 MHz 5kW (Dingelstädt); 100.2 MHz 3kW (Erfurt); 104.7 MHz 3kW (Kulpenberg), 107.2 (Weimar).
-Landeswelle Thüringen, Gorkistraße 9, 99084 Erfurt. ☎ +49 (351) 2222120
STATIONS: Suhl/Ringberg 88.6 MHz 1kW, Gotha 88.8 MHz 0.1kW, Ronneburg/Gera 94.9 MHz 1kW, Saalfeld/Remda 95.7 MHz 10kW, Kulpenberg 96.8 MHz 3kW, Apolda 97.8 MHz 1kW, Rudolstadt 99.7 MHz 0.1kW, Sangerhausen 101.1 MHz 0.1kW, Weimar/Ettersberg 101.9 MHz 1kW, Inselsberg 104.2 MHz 10kW, Nordhausen 104.3 MHz 0.1kW, Mühlhausen 105.8 MHz 0.1kW, Sonneber-Stadt 105.8 MHz 1kW, Jena-Oßmaritz 106.1 MHz 1kW, Eisenach 106.5 MHz 1 kW, Sonneberg/Bleßberg 106.7 MHz 60kW.
-Star*Sat Radio, Berkaer Straße 37a, 99837 Dippach. ☎ +49 (36922) 21226 (only via satellite)

HESSE (State)
-Hit Radio FFH, Graf-Vollrath-Weg 6, 60489 Frankfurt. ☎ +49 (69) 78 97 90.
STATIONS: Feldberg 105.9 (100kW); Hoher Meißner 105.1 (100kW); Rhön 100.9 (50kW); Rimberg 103.0 (50kW); Dillenburg 100.0 (30kW); Driedorf 106.8 (30kW); Kassel 103.7 (20kW); Krehberg 105.0 (20kW); Vogelsberg 104.8 (20); Korbach 107.7 (20kW); Wiesbaden 102.0 (0.1kW)
-Planet Radio, Graff-Vollrath-Weg 6, 60477 Frankfurt. ☎ +49 (69) 9783000
STATIONS: Frankfurt 100.2 (1kW), Kassel 104.6 (0.5kW), Dieburg 90.1 (0.8kW), Marburg 101.0 (?), Korbach 94.0 (?)
-Local stations (non-commercial): Frankfurt 97.1 R. X-MIX; Wiesbaden 92.5 K2R/Rheinwell; Rüsselsheim 90.9 K2R/Rheinwelle; Darmstadt 103.4 RadaR Radio; Marburg 90.1 R. Unerhört; Kassel 88.9 Freies R. Kassel; Eschwege 99.7 Das! Regional Radio.
-Evangeliums Rundfunk (ERF), Berliner Ring 62, 35576 Wetzlar. ☎ +49 (6441) 9570.
Wetzlar started broadcasting on Mainflingen 1539 kHz MW in 1996 (formerly used by Deutschlandfunk). **Web:** http://www.erf.de
D.PRGR: 0600-230 & 2100-0400 ERF2, other times ERF1

RHINELAND-PALATINATE (State)
-Radio RPR, Turmstraße 8, 67059 Ludwigshafen. ☎ +49 (621) 59 00 00.
STATIONS:

Loc.	RPR 1	RPR 2	kW
Ahrweiler	103.5	104.9	30
Bad Dürkheim	98.1	—	5
Bad Kreuznach	—	104.8	0.2
Bad Marienberg	102.9	—	25
Bornberg	103.1	107.6	25
Eifel	102.1	106.6	20/9.5
Haardtkopf	100.1	—	50
Idar-Oberstein	100.3	—	1
Kalmit	103.6	106.7	25
Kettrichhof	104.7	—	2.5
Koblenz	101.5	104.0	40
Ober-Olm	100.6	104.5	20
Saarburg	102.6	—	20
Trier	102.9	106.4	0.1
Zweibrücken	103.3	106.6	1/0.1

D.PRGR: RPR1: pop music and reg. prgrs. from Koblenz, Mainz, Trier & Ludwigshafen; RPR-2: light music.
-Radio Donnersberg, Postfach 1323, 67286 Kirchheimbolanden.
STATION: 97.1 (0.02kW).
-Radio Nürburgring, 87.7 MHz (0.05kW) during motor races from Nürburg.
-Radio Hockenheimring, 106.1 MHz (0.05kW) during motor races from Hockenheim.
-Energy, Kaiser-Wilhelm-Straße 34, 67059 Ludwigshafen. ☎ +49 (621) 5902644
STATIONS: 88.3 MHz 0.25kW (Koblenz), 87.6 MHz 0.2kW (Idar-Oberstein), 88.3 MHz 0.1kW (Bad Kreuznach), Kaiserslautern 98.1 0.16kW, 105.8 0.1kW (Trier).
F.PL: Nürburg 87.7 (0.05kW), Linz 96.9 (0.2kW), Daun 92.2 (0.08kW)
-Radio Kampanile, Turmstraße 8, 67059 Ludwigshafen. ☎ +49 (621) 591980 (only via satellite)

SAARLAND (State)
-Radio SALÜ, Richard-Wagner-Straße 58-60, 66111 Saarbrücken. ☎ +49 (681) 39 127.
-Antenne Saar, Richard-Wagener-Straße 58-60, 66111 Saarbrücken. ☎ +49 (681) 9377 0
STATIONS: Homburg 98.6 (0.2kW), Lebach 107.9 (0.1kW), Saarbrücken 107.5 (0.1kW), Saarlouis 96.3 (0.1kW), Sankt Wendel 90.3 (0.1kW), Salzburg 96.8 (0.1kW), Völklingen 88.6 (0.1kW)
D.PRGR: 0600-1000 UTC, other times rel. Deutschland Radio
STATIONS: Heusweiler II/Schoksberg 101.7 (100kW); Bliestal 100.0 (5kW); Perl 100.3 (5kW); Merzig 103.0 (0.1kW); Mettlach 104.2 (0.1kW)
D.PRGR: 24 h. hit radio.

BADEN-WÜRTTEMBERG (State)
N.B: There are 3 regional & 15 local commercial radio stations on 72 freqs. (0.005-100kW) in Baden-Württemberg. Also broadcasting are 8 public non-commercial radio stations.
Regional Radio Stations:
-Antenne, Plieninger Straße 150, 70567 Stuttgart. ☎ +49 (711) 720 585 0.
STATIONS: Stuttgart 101.3 (75kW); Raichberg 103.4 (100kW); Langenburg 100.1 (50kW); Langenbrand 105.2 (20kW); Heilbronn 89.1 (0.5kW); Bad Mergentheim 101.2 (0.005kW); Bad Urach 89.3 (0.025kW).
-Radio 7, Gaisenbergstraße 29, 89073 Ulm. ☎ +49 (731) 1477 0.
STATIONS: Aalen 103.7 (50kW); Iberger Kugel 105.0 (50kW); Witthoh 102.5 (40kW); Ulm 101.8 (10kW); Geislingen 100.3 (5kW); Schussental 96.9 (0.1kW); Villingen-Schwenningen 101.2 (0.1)
-Radio Regenbogen, Dudenstraße 12-26, 68167 Mannheim. ☎ +49 (621) 33750
STATIONS: Hornisgrinde 100.4 (80kW), Heidelberg 102.8 (50kW), Blauen 101.1 (8.4kW)
Local Radio Stations
-Radio Ton (Tauber-Odenwald-Neckar), Allee 2, 74072 Heilbronn. ☎ +49 (7131) 6500.
STATIONS: Heilbronn 103.2 (25kW); Bad Mergentheim 103.5 (20kW); Schwäbisch Hall 102.6 (0.5kW); Wertheim 104.7 (0.1kW); Crailsheim 104.8 (0.1kW); Künzelsau 96.0 (0.1kW)
-Welle Fidelitas, Am Sandfeld 13, 76149 Karlsruhe. ☎ +49 (721) 97060
STATIONS: Karlsruhe 101.8 (25kW), Mühlacker 100.7 (20kW), Baden-Baden 100.9 (0.8kW), Bruchsal 107.3 (0.1kW)
-Sunshine Live, Scheffelstraße 55, 68723 Schwetzingen. ☎ +49 (6202) 2820.

STATIONS: Mudau 102.1 (25kW), Heidelberg 106.1 (1kW), Weinheim 107.7 (0.1kW), Wiesloch 107.1 (0.1kW), Mosbach 107.9 (0.1kW)
-Radio KOMMA 1, Bahnhofstraße 65, 73430 Aalen. ☎ +49 (7361) 57270.
STATIONS: Aalen 107.1 (20kW), Heidenheim 104.2 (0.1kW)
-Radio Neckarburg, August-Schuhmacher Straße 10-12, 78664 Esbronn-Mariazell. ☎ +49 (7403) 8000
STATIONS: Rottweil 102.0 (0.3kW), Villingen 93.1 (1kW).
-Radio Seefunk, Konzilstraße 1, 78462 Konstanz. ☎ +49 (7531) 28650.
STATIONS: St. Chrischona/Rührberg 103.1 (5kW), Wannenberg 107.0 (3kW), Konstanz 101.8 (0.5kW), Weingarten 102.6 (0.1kW), Singen 105.3 (0.1kW), Friedrichshafen 99.3(0.1kW), Waldshut-Tiengen 105.4 (0.1kW), Laufenburg 102.4 (0.1kW), Lörrach 104.3 (0.1kW).
-Radio FR 1, Sasbacher Sraße 12, 79111 Freiburg. ☎ +49 (761) 452660.
STATIONS: Blauen 106.0 (8.4kW), Freiburg 94.7 (0.5kW); Hochfirst 106.6 (0.130kW).
-Radio 7 melody, Leipzigstraße 26, 88400 Biberach. ☎ +49 (7351) 50400.
STATIONS: Ulm 105.9 (2.5kW), Biberach 103.9 (0.4kW), Riedlingen 106.2 (0.5kW)
-Radio Ohr, Hauptstraße 83a, 77652 Offenburg. ☎ +49 (781) 5043000.
STATIONS: Wilstätt 104.9 (5kW), Lahr 107.4 (5kW), Bühl 103.8 (0.8kW), Haslach 93.0 (0.1kW)
-Radio BB (Böblingen), Planierstraße 11/1, 71063 Sindelfingen. ☎ +49 (7031) 69030.
STATIONS: Sindelfingen 104.3 (1kW), Herrenberg 89.0 (1kW), Leonberg 106.9 (0.1kW), Calw 103.0 (0.1kW), Nagold 102.7 (0.1kW)
-Neckar Alb Radio, Silberburgstraße 50, 72764 Reutlingen. ☎ +49 (7121)16600.
STATIONS: Reutlingen 104.8 (1kW), Tübingen 100.9 (1kW), Ermstal 99.0 (0.1kW), Hechingen 100.1 (0.1kW)
-RMB-Radio (Rems-Murr-Bürgerradio), Anton-Schmidt-Straße 36, 71332 Waiblingen. ☎ +49 (7151) 959660
STATIONS: Backnang 101.8 (1kW), Ludwigsburg 103.9 (0.3kW), Waiblingen 104.9 (0.1kW), Schorndorf 104.6 (0.1kW), Schönbühl 104.5 (2kW)
-Radio 7 ES, Zeppelinstraße 116, 73730 Esslingen. ☎ +49 (711) 939390.
STATIONS: Esslingen 97.5 (0.5kW), Kirchheim 100.8 (0.1kW); Nürtingen 106.8 (1kW).
-Antenne Filstal, Lange Straße 36, 73033 Göppingen. ☎ +49 (7161) 673020. (often rel. Antenne from Stuttgart nights and partly during daytime)
STATIONS: Geislingen 105.4 (1kW), Göppingen 103.0 (0.1kW)
-Stadtradio 107.7, Königstraße 2, 70173 Stuttgart. ☎ +49 (711) 16326110.
STATIONS: Stuttgart 107.7 (1kW); Rottweil 102.0 (0.3 kW); Villingen 93.1 (1kW)
-Radio Dreyeckland (pub., non-commercial), Adlerstraße 12, 79098 Freiburg. ☎ +49 (761) 30407
STATIONS: Freiburg 102.3 (1kW).
-Querfunk Karlsruhe (pub., non-commercial), Steinstraße 23, 76133 Karlsruhe. ☎ +49 (721)387858
D.PRGR: Mon-Thu 1100-1600 & 2100-0600 UTC, Fri-Sun: 24hr.
STATION: Karlsruhe 104.8 (1kW)
-Radio aus Bruchsal, IFM, Karlsruher Straße 20, 76646 Bruchsal. ☎ +49 (7251) 91230
D.PRGR: Mon-Fri 0600-1100 UTC
STATION: Karlsruhe 104.8 (1kW)
-Lernradio (radio for students), Staatliche Hochschule für Musik, Wolfartsweierer Straße 7a, 76131 Karlsruhe. ☎ +49 (721) 662970
STATION: Karlsruhe 104.8 (1kW)
-Freies Radio Stuttgart, Falbenhennenstraße 11, 70180 Stuttgart. ☎ +49 (711) 6400442
-Radio Kormista, Öffinger Straße 7, 70736 Fellbach. ☎ +49 (7195) 57743
STATION: Stuttgart 97.2 (0.1kW)
-Freies Radio "Wüste Welle", Hechinger Straße 203, 72072 Tübingen. ☎ +49 (7071) 760337
-Uni Welle, Brunnenstraße 30, 72074 Tübingen. ☎ +49 (7071) 2972514
-radio helle welle, Kanzleistraße 19, 72764 Reutlingen. ☎ +49 (7121) 321272
STATION: Tübingen 96.6 (1kW)
Radio StHörfunk, Lange Straße 13, 74523 Schäbisch Hall. ☎ +49 (791) 973333
STATION: Schwäbisch Hall 97.5 (0.1kW)
-Freies Radio Freudenstadt, Forststraße 23, 72250 Freudenstadt. ☎ +49 (7441) 88221

STATIONS: Freudenstadt 100.0 (0.1kW); Baiersbronn 104.1 (0.1kW)
-Radio "Kanal Ratte", Bahnhofstraße 3, 79650 Schopfheim. ☎ +49 (7622) 62788.
STATION: Hohe Möhr 104.5 (0.5kW)
-Radio Free FM, Söflingerstraße 206, 89077 Ulm. ☎ +49 (731) 9386284.
STATION: Ulm 102.6 (1kW)
RCG 102.6, Schützenstraße 20, 89231 Neu-Ulm. ☎ +49 (731) 82492
STATION: Ulm 102.6 (1kW)

BAVARIA (State)
-Antenne Bayern, Münchener Straße 20, 85774 Unterföhring. ☎ +49 (89) 992770
-Klassik Radio, Brandstwiete 4, 20457 Hamburg. ☎ +49 (40) 3005050
STATIONS:

Loc.	A. Bayern	Klassik R.	kW
Augsburg	—	92.2	0.3
Bad Reichenhall	88.2	—	0.3
Balderschwang	97.3	—	0.02
Bamberg	101.1	—	16
Berchtesgaden	107.9	—	0.1
Brotjacklriegl	103.5	—	100
Burgbernheim	101.5	—	25
Coburg	103.8	—	5
Dillberg	100.6	—	25
Gelbelsee	100.2	—	25
Grünten	104.4	—	10
Herzogstand	102.0	—	0.1
Hochries	107.7	—	50
Hohenpeißenberg	103.8	—	25
Hoher Bogen	101.9	—	50
Inntal	104.2	—	0.05 (F.PI)
Landshut	104.1	—	1
Lindau	99.0	—	0.5
München	101.3	107.2	0.3
Nürnberg	—	105.1	0.9
Ochsenkopf	103.2	—	100
Passau	102.1	—	1
Pfaffenberg	103.0	—	25
Regensburg	103.0	91.1	25
Reit im Winkl	101.6	—	0.1
Rhön	101.9	—	100
Tegernseer Tal	101.1	—	0.5
Traunstein	103.7	—	5
Unterringingen	103.3	—	25
Untersberg	105.3	—	1 (F.PI)
Weiter	106.0	—	0.1
Würzburg	104.4	92.1	0.16/5
Zugspitze	102.7	—	2

N.B: In addition approx. 50 loc. sts are broadcasting on approx. 100 freqs. (all lps).

MUNICH (City)
-89 Hit FM, Münchner Straße 23, 85767 Unterföhring. ☎ +49 (89) 89898989.
-Relax FM, Titurelstraße 9, 81925 München. ☎ +49 (89) 9833 60
-AFK Radio München, c/o BLR, Einsteinstraße 172, 81675 München. ☎ +49 (89) 4550 4413
-Radio 2 Day, Postfach 701563, 81315 München. ☎ +49 (89) 7237 750
-Radio Arabella & Charivari Radio, Paul-Heyse-Straße 2-4, 80336 München. ☎ +49 (89) 5447000, 5141 000
-Energy München, Pestalozzistraße 23, 80469 München. ☎ +49 (89) 393933
-Radio Gong 96.3, Franz-Joseph-Straße 14, 80801 München. ☎ +49 (89) 3816 6111
STATIONS: Relax FM on 92.4 MHz, ENERGY München on 93.3 MHz; Radio 2 Day on 89.0; Radio Arabella on 105.2 MHz; Charivari R. on 95.5; R. Gong 2000 on 96.3; "M-94.5" on 94.5 MHz.
N.B: 89.0 and 92.4 are used by different stations: 89.0 0500-1100 Radio 2 Day, 1100-1700 89 Hit FM, 1700-2300 Radio 2 Day, 2300-0500 89 Hit FM; 92.4 0500-1500 Relax FM, 1500-1700 Radio Feierwerk, 1700-2000 Radio Lora, 2000-0500 Relax FM

NÜRNBERG (City)
-Hit-Radio N 1, Senefelderstraße 7, 90409 Nürnberg. ☎ +49 (911) 5191240
-Radio F, Senefelderstraße 7, 90409 Nürnberg. ☎ +49 (911) 5191390
-Energy 95.8, Äußere Brucker Straße 51, 91052 Erlangen. ☎ +49 (9131) 80900
-Radio Charivari, Senefelderstraße 7, 90409 Nürnberg. ☎ +49 (911) 5191290

STATIONS: Hit Radio N1 on 92.9; Radio F on 94.5; Energy 95.8 (2300-1500 UTC) & Radio Z (1500-2300) on 95.8; Radio Gong on 97.1; R. Charivari on 98.6.

AUGSBURG (City)

-Radio Kö., Konrad-Adenauer-Allee 11, 86150 Augsburg. ☎ +49 (821) 5020 10
-Radio Fantasy, Maximilianstraße 41, 86150 Augsburg. ☎ +49 (821) 1510 30
-Radio RT Eins, Curt-Frenzel-Straße 4, 86133 Augsburg. ☎ +49 (821) 7774 202
STATIONS: R. Kö on 87.9; R. Fantasy on 93.4, R. RT Eins on 96.7

FOREIGN SERVICES

DEUTSCHE WELLE (DW)
See International Broadcasting Section.

Others:

VOICE OF AMERICA (WERTACHTAL)
STATION: (G.C: 10.41E/48.05N): 500kW.
For frequencies and schedule see International Broadcasting section.

VOICE OF AMERICA (MUNICH)
MEDIUMWAVE: 1197kHz 300kW
For SW frequencies and schedule see International Broadcasting section.

VOICE OF AMERICA BERLIN
STATION: Berlin FM 87.9 1kW
D.PRGR: 1700-1600 daily VOA Express

VOICE OF RUSSIA RELAY
STATION: Berlin MW 693 kHz (5kW), Wachenbrunn 1323 kHz (1000kW 0600-1900), (150kW other times)
N.B: Full schedule in International Broadcasting section.

BBC BERLIN RELAY
◨ see UK
STATION: FM 90.2 MHz (50kW)
D.PRGR: Relays of BBC World Service, German Service and English service by radio. For more details see BBC schedules in the International Broadcasting section.

AMERICAN FORCES NETWORK EUROPE (AFN)
◨ (local studios); APO 09757-4310, Frankfurt a/M.; APO 09742-4317, Kaiserslautern; APO 09696-4313 Nürnberg; APO 09036-4318, Würzburg.

Mediumwaves	kHz	kW		kHz	kW
Frankfurt a/M	873	150	Heidelberg	1143	1
Stuttgart	1143	10	Mönchengldb	1143	1

+ 11 sts. 0.3kW on 1143 kHz (Bad Kissingen, Bamberg, Bitburg, Schweinfurt, Würzburg, Wildflecken) and 6 sts. on 1485 kHz (0.3kW) (Ansbach, Augsburg, Berchtesgaden).

FM:	MHz	kW		MHz	kW
Frankfurt*	98.7	60	Stuttgart*	102.3♭	100
Augsburg	100.0!	0.01	Bremerhaven	107.9	0.3
Kaiserslautern	100.2	7	Bonn	107.6	0.08
Garmisch P.	90.3	0.05	Memmingen	92.2	0.05
Kalkar	106.1	1	Nürnberg	107.4	0.9
Pirmasens	103.0	0.5	Illesheim	101.0	0.4
Würzburg	104.9	0.16	Heidelberg	104.6	0.1
Bitburg	105.1	0.1	Bad Aibling	97.7	0.1
Mannheim	107.3	0.05			

♭) stereo — *) separate prgrs.and Z-FM — !) also rel. Z. FM
D.PRGR in English: 24h. — **N:** every h. on the h. (except 0400). major newscasts (30 min) 0500, 0600, 1700, 2100. — local prgrs: Mon-Fri 0405-0500, 0505-0600, 0705-0800, 0905-1055, 1205-1400, 1405-1700; Sat 0705-1100. FM separate prgrs. are mostly easy listening.
N.B: AFN Bavaria is the new ID for a group of AFN stations in southern Germany. HQ is Vilsek. ☎ +49 (9662) 831490. Z-FM rel. Unistar Adult Rock'n Roll daily from 2300-0400 (Sat-0700, Sun-1030) & Sat 1700-2000 live via satellite from the US (AFRTS)
Ann: "This is the AFN (Bavaria), The American Forces Network Europe". **V:** by QSL-card.

BRITISH FORCES BROADCASTING SERVICE
N.B: A division of the Service Sound and Vision Corp.
◨ Wentworth Barracks, Liststraße, 32049 Herford. ☎ +49 (5221) 88 090

BFBS 1st Network
FM	MHz	kW		MHz	kW
Drachenberg	93.0	80	Langenberg	96.5	30
Bonn	97.8	0.5	Visselhöv.	97.6	30
Bielefeld	103.0	70	Hameln	99.3	0.1
Kiel	88.4	0.03	Nordhorn	106.5	0.1

BFBS 2nd Network
FM	MHz	kW		MHz	kW
Bielefeld	101.6	0.3	Osnabrück	106.3	0.1
Münster	102.2	0.3	Lippstadt	105.0	0.16
Bergen-Hohne	104.8	0.16	Rheindalen	104.3	0.3
Fallingbostel	95.2	0.05			

+2 sts in Belgium & 2 sts. in The Netherlands listed separately.
D.PRGR: 24h in English. **ANN:** "BFBS Germany".
V: by QSL-card

EUROPE 1 (Independent Comm.)
◨ Europäische Rundfunk und Fernsehen AG, Europa Nr. 1, Postfach 111, 66011 Saarbrücken. ☎ +49 (681) 30782
STATIONS: Felsberg 183kHz 2000kW.
D.PRGR: 0400-2200 (Sat-Sun 24h). **N:** hourly & 0530, 0630, 0730, 0830, 2200.
Ann: F: "Europe 1" **V:** by QSL-card. Re. to: Europe 1, 66713 Saarlouis, P.O. Box 1365, Germany.

RADIO FRANCE INTERNATIONALE
◨ see France
STATIONS: Berlin 106.0 MHz (1kW)
D.PRGR: 24h.

L.T: UTC + 1h (Su: UTC + 2h) — **Pop:** 30.000 — **Radios:** 17.200 — **Pr.L:** English, Spanish — **E.C:** 50Hz, 240V — **ITU:** GIB.

GIBRALTAR BROADCASTING CORP. (Part Comm.)
◨ Broadcasting House, 18 So. Barrack Rd, Gibraltar. ☎ +350 79760. 🖷 +350 78673. **Web:** http://www.gibnet.com/gbc/, http://www.gib-nynex.gi/buss/gbc
L.P: GM: George Valarino. Senior Eng: John Tewkesbury. Senior Producer: Joe Adamberry. Senior Presenter: Richard Cartwright. News Editor: Stephen Neish.
MW: 1458kHz 0.5/2kW.
FM: 92.6MHz 1kW, 91.35MHz 0.2kW (stereo).
Local sce. in English: 0600-1300 & 1500-1800. **Spanish:** 1300-1500. **Rel. BBC World Sce:** 1800-0600.
ANN: "Radio Gibraltar" — **V.** by QSL-card. Rec. acc.

Rock FM: 106.9MHz 0.3kW.

BRITISH FORCES BROADC. SCE. GIBRALTAR
◨ BFBS Gibraltar, BFPO 52. ☎ +350 54211. 🖷 +350 54443 — **L.P:** SM: Tony Davis. Sen. Eng: Julian Brower. PD: Roger Woods — **FM:** 89.4/99.5MHz 0.2kW, 93.5/97.8MHz 1kW.
BFBS 1 on 93.5/97.8MHz & **BFBS 2** on 89.4/99.5MHz: 24h. **Local Prgrs:** 0700-0900, 1200-1400, 1600-1800, 2200-2300. **Rel BBC World Sce:** 0000(SS 2300)-0500.
ANN: "BFBS Community Radio on the Rock, BFBS 1 FM".

L.T: UTC + 2h (Su: UTC + 3h) — **Pop:** 10.500.000 — **Radios:** 4.200.000 — **Pr.L:** Greek — **E.C:** 50Hz, 220V — **ITU:** GRC.

ELLINIKI RADIOPHONIA TELEORASSI S.A. (ERT S.A.)
◨ 432 Messoghion Ave, 153 42 Aghia Paraskevi, Athens. ☎ +30 (1) 600 6900. 🖷 +30 (1) 639 0652. ☏ 216066.
L.P: Pres. & MD: Eugenios Giannakopolous.

GREEK RADIO – ERA
◨ P.O. Box 60019, 153 42 Aghia Paraskevi, Athens. ☎ +30 (1) 639 6700. 🖷 +30 (1) 638 0323 — **L.P:** DG: Apostolos Kossonas. Dir. Local

R: Michalis Messinis. Head of ERA 1: Christoforos Kontaxis. Head of ERA 2: Takis Psaridis. Head of ERA 3: Giorgos Tsangaris. Head of ERA 4: Makis Papazissis. Head of ERA 5: Filotas Gianottas. Dir. of Int. Rel: Evi Demiri.

Mediumwaves	kHz	kW	Prgr.
Athens	666	15	ERA 3
Athens1	729	150	ERA 1
Ioannina	765	10	R + ERA 1, 2, 4
Zakynthos	927	50	R + ERA 1, 2, 4
Larissa	945	5	R + ERA 1, 2, 4
Heraklion	954	10	R + ERA 1, 2, 4
Athens 4	981	200	ERA4
Corfu Isl.	1008	50	R + ERA 1, 2, 4
Thessaloniki 1	1044	150	M1
Orestiada 2	1080	10	R + ERA 1, 2, 4
Thessaloniki 2	1179	50	M2 + ERA 2,4
Florina	1278	10	R + ERA 1, 2, 4
Tripolis	1314	10	R + ERA 1, 2, 4
Pyrgos	1350	4	R + ERA 1, 2, 4
Athens2	1386	50	ERA2
Komotini	1404	50	R + ERA 1, 2, 4
Patra	1485	1	R + ERA 1, 2, 4
Volos	1485	1	R + ERA 1, 2, 4
Orestiada1	1485	1	ERA 1
Rhodes Isl.	1494	50	R + ERA 1, 2, 4
Chania	1512	50	R + ERA 1, 2, 4
Serres	1584	1	R + ERA 1, 2, 4
Kavala	1602	1	R + ERA 1, 2, 4
Kozani	1602	1	R + ERA 1, 2, 4
Samos	1602	1	ERA 1

Prgrs: 1=ERA 1, 2=ERA 2, 3=ERA 3, 4=ERA SPOR, M1=Macedonia, M2=Macedonia 2, R=Regional prgrs.

FM(MHz)	ERA1	ERA2	ERA3	ERA4	kW(ERP)
Alexandroupolis	89.8	91.8			3
Athens	91.6	93.6	95.6	100.9	100
	105.8	103.7	107.7	101.8	100
Chania	92.6	94.6			3
Corinth	97.6	99.9			3
Volos	92.8	94.8	96.8		10
Heraklion	94.4	96.4			3
Santorini	96.9	98.9			10/10/35
Ioannina	97.8	99.8			3
Kavala	89.2	91.2			10
Kalamata	92.2				3
Kastoria	88.6	90.6			10
Kerkyra Isl.	91.8	93.8			3
Kephallonia Isl.	96.9	98.9			10
Mytilini	92.3	94.3			3/3/35
Rhodos Isl.	88.4	90.4			3
Thessaloniki	88.0	90.0	92.0+		10
Tripolis	88.3	90.3			10

+) also carrying regional prgrs, *) stereo.

D.PRGR: ERA 1, ERA 2,. ERA 3, ERA 4 (ERA SPOR): all 24h.

MACEDONIA RADIO STATION
✉ 2 Angelaki Str, 54621 Thessaloniki. ☎ +30 (31) 244979. 🖷 +30 (31) 236370.
SW: Thessaloniki (G.C: 23.00E/40.36N): 3 x 35kW
kHz: 6245, 9935, 11595
Macedonia 1 on 1044kHz + FM 102.0MHz, 0600-2305 on 9935kHz (Eu) and 11595kHz (ME), 1500-2200 on 6245kHz (Eu).
Macedonia 2 on 1179kHz + 95.8MHz.
ANN: "Elliniki Radiophonia, Radiophonikos Stathmos Makedonias"

ERA regional programmes on FM
Aegaeon: Thira Isl. 93.3MHz 35kW, Lesvos Isl. 104.4MHz 35kW (both stereo) – **Chania:** 104.0MHz 35kW – **Corfu:** 99.3MHz – **Florina:** 102.3MHz – **Heraklion:** 97.5MHz 3kW – **Ioannina:** 102.1MHz – **Kalamata:** 94.2MHz 3kW – **Kavala:** 96.3MHz – **Komotini:** 98.1MHz – **Kozani** 100.2MHz – **Larissa:** 98.3MHz – **Mytilini:** 104.4MHz – **Orestiada:** 103.5MHz – **Patra:** 92.5MHz – **Pyrgos:** 100.0MHz – **Rhodes:** 92.7MHz – **Santorini:** 93.3MHz – **Serres:** 101.5MHz – **Tripoli:** 95.3MHz – **Volos:** 101.2MHz 35kW – **Zakynthos:** 95.2MHz.

THE VOICE OF GREECE (ERA 5th Prgr.)
see International Broadcasting section.

Private stations in Major Cities
Athens
98.4FM Stereo, Liossion 22, 104 38 Athens: 98.4MHz – Antenna 97.1 FM Stereo, Kifissias 10-12, 151 25 Athens: 97.1/102.9MHz – Capital Gold: 96.5MHz – Cool FM Athens, Kifissias 64, 151 25 Athens: 98.7MHz – Ellada FM, 438 Aharnon Ave, 111 43 Athens: 88.3MHz – Flash 96.1 FM Stereo, Kiffissias 64, 105 25 Athens: 95.8/96.1MHz – Galaxy 92FM: 91.8/92.1MHz – Jeronimo Groovy (JGRS), P.O. Box 65, 154 10 Athens: 88.9MHz – Kiss 909 FM: 43 Mitropoleos Center, 151 24 Athens: 90.9MHz – Klik FM, Fragoklissias 7, 151 25 Athens: 88.0MHz – Lampsi FM: 92.4MHz – Melodia FM 100, Philellinon 24, 105 57 Athens: 100.0MHz – Nitro R: 102.4MHz – Piraeus Ch. 1, Euripidou 79, 185 35 Piraeus: 90.6 MHz – Planet FM: 104.5MHz – R. Gold. **E-mail:** radiogold@hol.gr. **Web:** http://www.radiogold.gr/radiogold/: 105.0MHz – Rock FM: 96.8MHz – Rodo FM: 94.4MHz – Sfera FM: 102.1MHz – Skai 100.4 FM Stereo, 2 Falireos, Ethnarchou Makariou Str, 185 47 Piraeus: 100.4/100.7MHz – Sport FM: 94.6MHz.

Thessaloniki: Antenna FM: 97.5Mhz (see Athens) – Cosmoradio: 95.3MHz – Flash FM: 99.7MHz (see Athens) – FM 100: 100.0MHz – FM 101: 101.0MHz – Jeronimo Groovy: 93.4MHz (see Athens) – Kanali 1: 88.0MHz – Kiss FM: 87.6MHz (see Athens) – Panorama 984: 98.4MHz – R. Paratiritis: 93.0MHz – R. Radio: 90.8MHz – Stal FM: 88.5MHz (see Athens) – Star FM: 97.1/103.7MHz – Top FM: 101.4MHz – R. Thessaloniki: 94.5MHz.

Patra: Radio 1: 89.4MHz – R. Gamma: 94.0NMHz – Kanali 11: 95.3MHz – Max FM: 97.1MHz – R. Patra: 90.5MHz.

Iraklio: Iraklio FM: 98.4MHz – R. Kastro: 91.0MHz – R. Kriti: 101.5MHz – R. Kymata: 96.1MHz.

Larissa: R. Europe: 96.3MHz – Ihos FM: 90.9MHz – Larissa FM: 90.5MHz.

Volos: Akroama FM: 98.3MHz – R. Delta: 104.6MHz – R. Volos: 95.8MHz

BBC World Sce: Athens 107.1MHz.

ARMED FORCES RADIO & TV SERVICE
(United States Air Force)
✉ Det 4, Air Force European Broadc. Squadron (AFEBS), APO AE 09846, USA — SM: MSgt. Scott W. Kirby.
FM: Gournes 106.1MHz: 24h — **ANN:** "This is EBS Iraklion".

VOICE OF AMERICA RADIO STATION KAVALA
MW: 792kHz 500kW — **SW** (G.C: 24.50E/40.53N): 10 x 250 kW.
For SW frequencies and schedule see International Broadcasting section.

VOICE OF AMERICA RADIO STATION RHODES
MW: 1260kHz 500kW — **SW** (G.C: 28.10E/36.15N): 2 x 50kW.
For SW frequencies and schedule see International Broadcasting section.

HUNGARY

L.T: UTC + 1h (Su: UTC + 2h) — **Pop:** 10.325.030 — **Radios:** 6.250.000 — **Pr.L:** Hungarian — **E.C:** 50Hz, 220V — **ITU:** HNG.

MAGYAR RÁDIÓ (Gov. Semi-Comm.)
✉ Bródy Sándor u. 5-7, H-1800 Budapest. ☎ +36 (1) 1388388. 🖷 36 (1) 1388910 or 1388943. ① (22) 5158. **Calypso 873:** ☎ +36 (1) 2691993. **Web:** Kossuth Rádió: http://www.kossuth.enet.hu – Petöfi Rádió: http://www.petofi.enet.hu/
L.P: Pres: István Hajdu. Vice-Pres:Miklòs Papp. MD: Laszlo Csucs. Dir. Tech: Jeno Rednai. Head of Foreign Rel. Dept: László Ovári.
Regional stations: a) 7621 Pécs, Szent Mór u. 1. ☎ +36 (72) 210666. 🖷 +36 (72) 210424 – b) 5000 Szolnok, Kolozsvári útca. 2. ☎ +36 (56) 421133. 🖷 +36 (56) 425610 – c) 4400 Nyíregyháza, Szent István út. 42. ☎ +36 (42) 410611. 🖷 +36 (42) 410472 – d) 9017 Györ, nagy Imre útca 28. ☎ +36 (96) 412722 – e) 3527 Miskolc, Bajcsy-Zsilinszky útca 15. ☎ +36 (46) 346666. 🖷 +36 (96) 341688 – f) 6720 Szeged, Stefánia útca 7. ☎ +36 (62) 475725. 🖷 +36 (62) 474777 – g) 4024 Debrecen, Piacz. u. 28/c. ☎ +36 (52) 412311. 🖷 +36 (52) 413358.

Mediumwaves	kHz	kW	Prgr (+ regional)
Solt	540	2000/1000	1
Lakihegy	873	20	h
Pécs	873	20	2 + a
Mosonmagyaróvár	1116	5	5 + d
Szolnok	1188	135/5	2
Szombathely	1188	25	2 + d
Nyíregyháza	1251	25	2 + c
Siófok	1251	135	2

Mediumwaves	kHz	kW	Prgr (+ regional)
Marceli	1251	500/1000	2
Lakihegy	1341	150/300	2
Pécs	1350	10	2
Györ	1350	5	2 + d
Szolnok	1350	5	2 + b

Prgrs: 1=Kossuth (1st prgr); 2=Petöfi (2nd prgr); 3=Bartók (3rd prgr) stereo, a-h) carries regional prgrs (see below)
SW: Székesfehérvár (G.C: 18.24E/47.10N): 6025kHz 100kW (1). a-g) carries regional prgrs (see below).
FM (MHz): stations in italics are **F.Pl.** at time of editing.

	1	2	3	kW
Budapest	67.40	66.62	69.38	10
		94.8	105.3	10
Csávoly		89.4		3
Debrecen		89.00	106.6	1
Györ	72.86d		67.04	3
		93.1	106.8	3
Kabhegy	72.98d	71.42	70.64	10
		93.9	105.0	10
Kékes	71.21e	72.77	70.10	3
		102.7	90.7	3
Kiskörös		95.1	105.9	3
Komádi	66.14	66.92	68.24	3
		96.7		3
Miskolc	66.80e	66.02	68.48	3
		102.3	107.5	3
Nagykanizsa	71.03a	69.98	68.36	10
		94.3	104.7	10
Pécs	71.81a	67.19	67.97	10
			107.6	10
Sopron	72.86	70.40	72.08	10
		99.5	107.9	10
Szeged	94.9f	104.6	105.7	0.5/1/1
Szentes	66.29f	67.85	68.72	3
		98.8	107.3	3/10
Tokaj	71.33e	72.11	70.43	10
		92.7	105.5	10
Úzd		90.3	106.9	3
Vasvár		98.2	106.9	3

Prgr. 1 (Kossuth Rádió): 0330 (Sun 0500)-2315. **N:** on the h. **Rel. BBC Hungarian Sce.** on Kossuth OIRT-FM: 2200-2230. **Rel. VOA Hungarian Sce.** on Kossuth OIRT-FM: 2000-2100. **Slovak:** 1730-1800. **Romanian:** 1800-1830. **German:** 1830-1900 (also on Pécs 873kHz). **Croatian:** 1900-1930 (also on Pécs 873kHz). **Serbian:** 1930-2000 (also on Pécs 873kHz).
Prgr. 2 (Petöfi Rádió): 24h on FM, 0330 (Sun 0500)-2330 on MW. **N:** 0330 and half hourly to 0700, 0800, 1000, 1200, 1400, 1600, 1800, 2000, 2200.
Prgr. 3 (Bartók): 0500-2310. **N:** 0500, 0530, 0600, 0700, 1100, 1500, 1800, 2230.
Reg. Prgrs:
a) Pécs: on 873kHz: MF 0455-2000, Sat 0800-1100, Sun 0800-0900. **Rel. BBC Hungarian:** MF 0645-0700. **Prgrs. for national minorities: Serbian:** 1100-1230. **Croatian:** 1230-1400. **German:** W 1700-1830, Sun 1400-1500 — on FM: MF 0455-1100, Sat 0800-1000, Sun 0800-0900.
b) Szolnok: MF 1055-1700, Sat 0700-1100, Sun 0700-1700.
c) Nyíregyháza: MF 0500-0800 & 1600-1800, SS 0800-1000. **Rel. BBC Hungarian:** MF 0645-0700.
d) Györ: MF 0355-0400, 0412-0500, 0512-0600, 0612-0700, 0712-0800, 1600-1800. Sat 0800-1000. Sun 0730-0900. **Slovene:** Sun 0835-0900.
e) **Miskolc:** on 1116kHz: Mon-Thurs 0455-0500, 0510-0800 & 1400-1700. Fri 0455-0500, 0510-0600, 0610-0700, 0710-1700. SS 0455-0500 & 1120-1700. **Slovak:** Sun 0740-0900 — on FM: Mon-Thurs 0555-0600, 0610-0700, 0710-0800. Fri as 1116kHz (SS rel. Kossuth R. all day).
f) Szeged: on 66.29MHz: MF 0455-0500, 0512-0600, 0612-0800, 1200-1500. Sat 0800-1000 & 1200-1500. Sun 0730-0900 & 1200-1500 — on 94.9MHz: MF 0455-0500, 0512-0600, 0612-0900, 1200-1800. SS 0655-1100 & 1200-1500. **Prgrs. for national minorities** (both freqs): **Slovak:** 1200-1330. **Romanian:** 1330-1500.

g) **Debrecen** on 103.5MHz: MF 0455-0500, 0512-0600, 0612-0800. Sat 0655-0700, 0705-0900, 0910-1100. Sun 0655-0900, 0910-1100. **Rel. BBC Hungarian:** 0800-0810 — **Cívis R.** (comm.): Sun-Fri 1500-1800.
h) **Rádió Calypso:** 0500-1800 (Sun 1500).
ANN: "Kossuth Radio, Budapest", "Petöfi Radio, Budapest", Bartók Radio, Budapest".

FOREIGN SERVICE: see International Broadcasting Section

OTHER STATIONS

ORSZÁGOS RÁDIÓ ÉS TELEVÍSIÓ TESTÜLET (ORTT) (National Radio and Television Board)
✉ 1088 Budapest, Reviczy utca 5. ☎ +36 91) 2672590.

SZABAD RÁDIÓK MAGYAROSZÁGI SZERVEZETE (SZARÁMASZTER) (Organization of Free Hungarian Radio Stations)
✉ 1066 Budapest Ó utca 11. II/6. ☎ +36 91) 1111855, 3111855.

HELYI RÁDIÓK ORSZÁGOS SZERVEZETE (National Organization of Local Radio Stations)
✉ 6000 Kecskemét Deák, ferenc tér 3. ☎ +36 (76) 480880. 🖷 +36 (76) 481328.

RADIO DANUBIUS (Comm.)
NB: this station is curently operated by Magyar Radio, but will be transferred to private ownership.
✉ Bródy Sándor Utca 40, Budapest 1088. ☎ +36 (1) 1387840. 🖷 +36 (1) 1388925. **Web:** http://www.danubius.hu
LP: MD: Gbor Gyorgy.

FM:	MHz	kW	FM:	MHz	kW
Szeged	94.9	1	Budapest	103.3	15
Miskolc	98.3	3	Györ	101.4	3
Kabhegy	100.5	30	Tokaj	103.5	10
Komádi	101.6	3	Kékes	104.7	3
Sopron	102.0	30	Pecs	105.5	10

D.PRGR: 24h.

OTHER STATIONS

MW	kHz	kW	Name & Location	Times
1)	810	15	Juventus R, Siofok	24h
2)	1485	1/0.25	Mohács Rádió, Mohács	0700-1900
12)	1485	1/	Rádió Hat, Debrecen	0700-1900
3)	1602	1	Somogy Rádió, Kaposvár	MF 0500-2200
13)	1602	1	Kaposvár Rádió, Kaposvár	MF 2200-0500
5)	1602	1	Komádi Rádió, Komádi	0700-1900
11)	1602	1	Rádió Egerszeg	0700-1900

FM (stations with power of less than 250 watts not mentioned)

	MHz	kW	Name & Location	Times
6)	88.8	0.25	Rádió Jam, Veszprém	0500-1900
7)	88.9	0.25	Start Rádió, Békéscsaba	0045-1245
1)	89.5	10	Juventus R, Budapest	24h
14)	91.1	0.25	Duna Rádió, Szentendre	0500-1700
15)			Pilis Rádió, Szentendre	1700-0500
16)	91.9	0.25	Inforum Rádió, Vác	0000-1200
17)			Dunakanyar Rádió, Vác	1200-2400
1)	93.3	10	Juventus R, Kabhegy	0500-2300
18)	94.1	0.25	Rádió Ikva, Sopron	0100-0900
19)			Rádió Sopron, Sopron	1700-0100
20)	95.2	0.25	Cervinus Rádió	24h
21)	96.3	0.25	City Rádió	0900-2100
22)	96.4	0.284	Rádió Torökszentmiklós	0500-1100
				1600-2100
23)	96.9	0.25	TOP Rádió, Kazincbarcika	2300-1100
24)	97.8	0.25	Radír Rádió, Tatabány	0500-1700
25)	99.2	0.25	Vorosmarty Rádió Székesfehérvár	1000-1700
26)			Reflex Rádió, Székesfehérvár	0400-1000 / 1700-2100
27)			Club Rádió, Székesfehérvár	2100-0400
28)	100.0	0.25	Kalocsai Rádió, Kalocsa	see below
29)			Pulzus Rádió, Kalosca	see below
30)	100.2	0.25	Média 6 Rádió, Szeged	1600-0400
31)			Tisza Rádió, Szeged	1000-1600
32)			Rádió 88, Szeged	1600-0400
33)	100.3	0.25	Rádió 1, Budapest	24h
34)	100.4	0.25	Focus Rádió, Salgótarján	MF 0700-2300 / SS 1100-2300

MHz	kW	Name & Location	Times
8)	102.1	7	R. Bridge, Budapest	24h
9)	103.0	0.25	T-Rádió, Miskolc	2100-0900
35)			Nonstop Rádió, Miskolc	0900-2100
36)	104.0	0.25	Csaba Rádió, Békéscsaba	1100-0300
37)			Alfold Rádió, Békéscsaba	0300-1100
10)	104.4	5	Dráva Rádió, Pécs	0500-1700

Addresses and other information
1) 1075 Budapest, Rubah Sebesén útca 19-21. ☎ +36 (1) 2679831. **Web:** http://www.juventus.hu
2) 7700 Mohács,Bacáus u. 5. ☎ +36 (69) 311447. 📠 +36 (69) 31733.
3) 7400 Kaposvár, Dombóvári út 1. ☎ +36 (82) 422444. 📠 +36 (82) 410500.
5) Hósök tere 4, 4138 Komádi. ☎ +36 (54) 438013.
6) 8200 Veszprém, Zrinyi út. 3. ☎📠 +36 (88) 422944, 425056. **Web:** http://www.netropolisz.net/jam
7) 5600 Békéscsaba, Derkovics sor 2 (IFIHÁZ). ☎📠 +36 (66) 442555.
8) 1016 Budapest, Naphegy tér 8. ☎ +36 (1) 2010198. 📠 +36 (1) 2025252. **Web:** http://www.netropolisz.net/bridge
9) Klapka U. 42, 3524 Miskolc. ☎ +36 (46) 4121645. 📠 +36 (46) 411802. **Web:** http://www.uni-miskolc.hu:8080/ ~www/english/szk/ magyar/szili/radio.html
10) 7622 Pécs, Nagy Lajos Király út 9. ☎ +36 (72) 212121. 📠 +36 (72) 313744. **Web:** http://www.dravanet.hu/dravaradio/
11) 8900 Zalaegerszeg, Köztársaság útja 1/A. ☎ +36 (92) 325400. 📠 +36 (92) 325810.
12) 4025 Debrecen, Wesselényi út 25.
13) 7400 Kaposvár, Kossuth Lajost ut. 6. ☎ +36 (82) 424445. 📠 +36 (82) 424452.
14) 2000 Szentendre, Pátriárka út. 7. ☎ +36 (26) 31900. 📠 +36 (26) 319801.
15) 2000 Szentendre, Szentlásloi út 13/A (✉ P.O. Box 297). ☎ +36 (26) 314151. 📠 +36 (26) 319090.
16) 2600 Vác, Galcsek ut. 8-10. ☎ +36 (27) 311322. 📠 +36 (27) 311405.
17) 2600 Vác, Fekete út. 65. ☎ +36 (27) 316248. 📠 +36 (27) 319225.
18) 9400 Sopron, Szent Gyorgy út. 16. ☎ +36 (99) 340240. 📠 +36 (99) 341441.
19) 9400 Sopron, Lovér korút 74. ☎ +36 (99) 341900. 📠 +36 (99) 341800.
20) 5540 Szarvas, Kossuth tér 3. ☎📠 +36 (66) 313036.
21) 3524 Miskolc, Klapka Gyorgy út. 42. ☎ +36 (46) 431400, 431300. 📠 +36 (46) 431131.
22) 5200, Torókszentmiklós, Kossuth tér 6. ☎ +36 (56) 390676. 📠 +36 (56) 390405.
23) 3700 Kazincbarcika, Jószerencsét út. 10. ☎ +36 (48) 311774. 📠 +36 (48) 312411.
24) 2800 Tatabány, Fó tér 36. ☎ +36 (34) 310536. 📠 +36 (34) 310624.
25) 8000 Székesfehhérvár, Szabadságharcos út. 59. ☎📠 +36 (22) 316011.
26) 8000 Székesfehhérvár, Rákóczi út 3. (+36 (22) 348888. 📠 +36 (22) 349999.
27) 8000 Székesfehhérvár, Rákóczi út 3. (+36 (22) 34877. 📠 +36 (22) 349999.
28) 6300 Kalocsa, Szent István Király út. 2-4. ☎📠 +36 (78) 461955. Odd weeks Tues-Fri 1200-1700, even weeks Tues-Sat 0700-1200.
29) 6300 kalosca, Egres út. 9. ☎📠 +36 (78) 462488. Mon 0700-1700, odd weeks Tues-Fri 0700-1200, even weeks Tues-Fri 1200-1700.
30) 6724 Szeged, Feltámadás utca 36. ☎ +36 (62) 421601. 📠 +36 (62) 317113.
31) 6753 Szeged, budai nagy Antal utca 36. ☎ +36 (62) 495495.
32) 6725 Szeged, Szabad satjó út 30. ☎ +36 (61) 421353. 📠 +36 (61) 421253.
33) 1087 Budapest, Kerepesi út 27A. ☎ +36 (1) 3335321. 📠 +36 (1) 3335323. **Web:** http://www.datanet.hu/bri/radio1
34) 3100 Salgótarján, P.O. Box 162. ☎ +36 (32) 310735.
35) 3530 Miskolc, Kisavas Alsósor 22. ☎ +36 (46) 412025. 📠 +36 (46) 354476.
36) 5600 Békéscsaba, Teleki utca 5. ☎ +36 (66) 441112. 📠 +36 (66) 446056. **Web:** http://www.sziksci.hu/comps/csabaradio
37) 5600 Békéscsaba, Irányi utca 4-6. ☎ +36 (66) 326798. 📠 +36 (66) 442666.

ICELAND

L.T: UTC — **Pop:** 273.407 — **Radios:** 197.000 — **Pr.L:** Icelandic — **E.C:** 50Hz, 220V — **ITU:** ISL.

RÍKISÚTVARPID (National Broadcasting Service)
✉ Efstaleiti 1, 150 Reykjavík. ☎ +354 515 3000. 📠 +354 515 3010. **E-mail:** isradio@ruv.is **Web:** http://www.ruv.is

L.P: DG: Markús Örn Antonsson. Chmn. Prgr. Council: Guunlaugur Saevar Gunnlaugsson. Dir. Admin: Hörður Vilhjálmsson. Dir. Radio: Dóra Ingvadóttir. PD: Margaret Oddsdóttir (Ch. 1), Óscar Ingólfsson (Ch. 2). Dir. N: Kari Jónasson. Head of Int. Rel: Gudrún Eyberg. Head of Features and magazine programmes: Þorgerður K. Gunnarsdóttir.

MW	kHz	kW	MW	kHz	kW
Gufuskálar	189	300	Eidar	207	100
Raufarhöfn	1485	0.02			

F.PI: a) on 189kHz will replace b) on 207kHz.
+ 2 low power tr's on 1485kHz.

FM: (MHz) stereo	1	2	kW
Gagnheidi	88.8	87.7	8.8/11.8
Girdisholt	92.9		3.5
Háfell	93.8	98.7	14/34
Hegranes	90.6	98.8	3.1/5
Hnjúkar	89.1	95.5	6.0/6.2
Hvítabjarnarhóll	95.1	90.3	0.8/1
Höfn	90.3	104.8	1.2/1
Skálafell	92.4	99.9	24
Stykkishólmur	88.0	96.3	2.5
Thódólsholt		88.3	2.5
Vadlaheidi	91.6	96.5R	9.3
Vatnsendi	93.5	90.1R	2.5
Vidarfjall	88.1	96.1	3.3
Vestmannaeyar	97.1	88.1	17/24

+ 43 st's below 1kW — R) carries regional prgrs.
Ch. 1 on LW/MW/FM1: 0645 (Sun 0800)-0100. **N:** 0700W, 0800, 0900, 1000, 1100, 1220, 1400, 1500, 1600, 1700, 1900, 2200. **N. in English:** 0855 (June-August). **Night Prgr:** 0130-0605 (SS 0730). **N:** on the h —
Ch. 2 on FM2: 24h.
Regional Prgrs: Útvarp Nordurland (Akureyri): MF 0810-0830 & 1835-1900. Útvarp Austurland (Egilsstadir) & Svaedisútvarp Vestfjarda (Isafjördur): Wed/Thurs/Fri 1835-1900.
SW: see International Broadcasting section.
ANN: "Utvarp Reykjavík" (P1), "Rás tvö" (P2) — **V.** by QSL-card.

Private Stations
Adalstödin, P.O. Box 46, 121 Reykjavík: Reykjavík 90.9MHz, Akureyri/So.Iceland 103.2MHz, **X** (2nd Prgr): Reykjavík 97.7MHz — **FM 95.7,** P.O. Box 9057, 129 Reykjavík: 95.7MHz — **Stjarnan,** Sigtun 7, 107 Reykjavík: 102.2MHz — **Top Bylgjan,** Lynghálsi 5, 110 Reykjavík: 100.9MHz.

RADIO ALPHA & OMEGA (Rlg.)
See International Broadcasting section.

NAVY BROADCASTING SERVICE (U.S. Mil.)
✉ Officer in Charge, NAVMEDIACEN Broadcasting Det Keflavík, PSC 1003, P.O. Box 25, FPO, AE 09728-0325, USA (or Box 25, NATO Base, IS-235 Keflavíkurflugvollur).
L.P: SM: LT E.C.Zeigler. CE: S.Jonsson.
MW: Keflavík 1530kHz 0.25kW — **D.PRGR. in English:** 24h.
Local Prgrs: MF 0005-0200, 0605-0800, 1105-1300, 1305-1400 (Fri) 1605-1800. Sat 0005-0200, 0805-1000. Sun 0005-0200, 2305-2400.
ANN: "AM 1530", "Power 104FM" — **V.** by letter.

IRELAND

L.T: UTC (Su: UTC +1h) — **Pop:** 3.585.713 — **Radios:** 2.150.000 — **Pr.L:** Irish, English — **E.C:** 50Hz, 220/380V — **ITU:** IRL.

RADIO TELEFÍS ÉIREANN (Statutory Corporation)
✉ Donnybrook, Dublin 4. ☎ +353 (1) 208 3111. 📠 +353 (1) 208 3080.
Cable: Broadcasting.
Web: http://www.rte.ie **2FM Web:** http://www.2fm.ie
L.P: Chmn: Farrel Corcoran. DG: Bob Collins. MD Organisation & Development: Liam Miller. MD Commercial: Conor Sexton. Dir. of Finance: Gerry O'Brien. Dir. of Radio: Helen Shaw. Dir. of News: Ed Mulhall. Dir. of Public Affairs: Kevin Healy. Dir. of Corporate Affairs: Tom Quinn.
Raidió Na Gaeltachta, Casla, Conamara, Co. na Gailmhe. ☎ +353 (91) 506677. 📠 +353 (91) 506666. **Web:** http://www.rte.ie/radio/rnag — **L.P:** MD: Pól Ó Gallchóir. PD: Seosamh Ó Braonáin. CE: Cyril Ryan.
RTE Radio Cork, Fr. Mathew Street, Cork. ☎ +353 (21) 805805. 📠 +353 (21) 273829. **Web:** http://www.iol.ie/media/rtecork/
Networks: 1=R1, 2=2FM, 3=FM3 Classical Radio, 4=Raidió Na Gaeltachta.

Mediumwaves	kHz	kW	N	Mediumwaves	kHz	kW	N
Casla	540	2	4	Ballydavid	828	1	4
Tullamore	567	500	1	Derrybeg	963	10	4
Athlone	612	100	2	Cork	1278	10	2
Mediumwaves	**kHz**	**kW**	**N**	**Mediumwaves**	**kHz**	**kW**	**N**
Cork	729	10	1/2	Dublin	1278	10	2

SW: R1 "News at 6.30" is rebroadcast via WWCR, Nashville, USA at 1930 (Sat 2000, Sun 2100), repeated the following day at 1000 (SS 1100). See International Broadcasting section for latest frequencies. RTE also broadcasts special events on SW by hiring airtime from various European broadcasters.

FM (MHz)	1	2	3/4	kW
Achill	89.3	91.5	93.7	2
Aranmore	89.6	91.8	94.0	3
Athlone	89.9	92.1		0.05
Cahirciveen	89.5	91.7	93.9	2
Castlebar	89.8	92.0	94.2	3
Castletownbere	88.3	90.5	92.7	1
Clermont Carn	95.2	97.0	102.7	40
Clifden	89.5	91.7	93.9	3
Clonmel	88.3	90.5	92.7	1
Cork City	89.2L	91.4	93.6	5
Crosshaven	88.2	90.4	92.6	3
Derryberg			93.0	1
Dungarvan	88.5	90.7	92.9	3
Fanad	89.8	92.0	94.2	4
Holywell Hill	89.2	91.4	93.6	2
Kinsale			93.4	0.01
Kippure	89.1	91.3	93.5	40
Knockmoyle	88.4	90.6	92.8	1
Limerick City	89.4	91.6	93.8	2.5
Maghera	88.8	91.0	93.2	160
Monaghan	89.9	91.1	93.3	3
Moville	88.3	90.5	92.9	10
Mt. Leinster	89.6	91.8	94.0	100
Mullaghanish	90.0	92.2	94.4	160
Schull/Skibbereen	89.1	91.3	93.5	5
Suir Valley	89.0	91.2	93.4	3
Three Rock	88.5	90.7	92.9	10
Truskmore	88.2	90.4	92.6	80

L) also carries Cork Local R.

Radio 1: 24h in English & Irish on 567kHz + FM (also via Astra Satellite transponder 22 audio subcarrier 7.56MHz). **N. in English:** on the h 0600-0100 + 0630, 0730W, 0830MF, 1830.
2FM: 24h in English on 612/1278kHz + FM2. **N:** on the h.
FM3: 0630 (SS 0730)-0800 & 1930 (Sun 1945)-0045.
Raidió Na Gaeltachta: 0800-1930 (Sun 1945) in Irish on FM3 tr's.
RTE Radio Cork: MF 1130-1700, Sat 1000-1700, Sun 1000-1300 & 1445-1800 on 729kHz + 89.2MHz. Other times rel Radio 1.
IS: Radio 1: Musical Box playing 4 bars of Irish air "O'Donnell Abu".
V. by QSL-card.

ATLANTIC 252 (Comm.)

✉ Mornington House, Summerhill Rd, Trim, Co. Meath. ☎ +353 (46) 36655. 🖷 +353 (46) 36688. ✉ in U.K: 74 Newman Str, London W1P 3LA ☎ +44 (171) 637 5252. 🖷 +44 (171) 637 3925. **E-mail:** programming@atlantic252.com or engineering@ atlantic252.com
Web: http://www.atlantic252.com
LP: MD: Travis Baxter. SM: Cathryn Geraghty. Prgr. Dir: Al Dunne.
LW: Clarkestown 252kHz 500/100kW.
D.PRGR: 24h (100kW 1900-0600). Music prgrs aimed at the 15-24 age group — **V.** by QSL-card. Re. to London addr.

INDEPENDENT RADIO & TV COMMISSION (IRTC)

✉ Marine House, Clanwilliam Court, Dublin 2. ☎ +353 (1) 676 0966. 🖷 +353 (1) 676 0948. **Web:** http://www.irtc.ie
L.P: Chairman: Niall Stokes. Chief Exec: Michael O'Keeffe

RADIO IRELAND (Comm.)

✉ Radio Ireland House, 124 Upper Abbey Str, Dublin 1. ☎ + 353 (1) 804 9000. 🖷 + 353 (1) 804 9099.
LP: MD: Dick Hill. Head of News/Current Affairs: Conor Kavanagh. Head of Sales & Mktg: Dave Hammond.

FM	MHz	kW	FM	MHz	kW
Truskmore	100.0	250	Holywell Hill	101.0	12
Crosshaven	100.0		Spur Hill	101.0	10
Knockmoyle	100.2		Achill	101.1	
Three Rock	100.3	12	Clonmel	100.1	
Cappoquin	100.3		Woodcock Hill	101.2	
Maghera	100.6	320	Mt. Leinster	101.4	400

FM	MHz	kW	FM	MHz	kW
Monaghan	100.7		Castlebar	101.6	
Mullaghanish	100.8.	320	Fanad	101.6	
Suir Valley	100.8		Mullaghanish	101.8	
Kippure	100.9	100	Clermont Carn	105.5	80

D.PRGR: 24h. **N:** on the h, also on the half h at peak times.

LOCAL STATIONS

St's 1) - 21) are comm. — **H.of tr:** 24h exc. where indicated.

	MHz	kW	Name, tr. location & times
10)	94.7	0.05	Highland R, Ballybofey
18)	94.8	2	Northern Sound, Slieve Glah
	MHz	**kW**	**Name, tr. location & times**
1)	94.9	5	East Coast R, Windgates
2)	95.0	4	Limerick 2000, Limerick
16)	95.1	5	WLR FM, Faha
10)	95.2	1	Highland R, Arranmore
20)	95.2	0.25	R. Kerry, Dingle: 0700-0100
5)	95.4	0.02	MWR FM, Westport: 0700-0100
7)	95.5		Clare FM, Kilrush: 0700-0100
4)	95.5		LM FM, Co. Meath
3)	95.6		South East R.
14)	95.7	0.05	Shannonside 104FM, Boyle
4)	95.8	5	LM FM, Mt. Oriel: 0630-0100
9)	95.8	0.1	Galway Bay FM, Galway
7)	95.9		Clare FM, Ea. Clare
8)	96.0		R. Kilkenny, Corbally Wood: 0700-0200
5)	96.1	5	MWR FM,Kiltimagh: 0700-0100
20)	96.2	3	R. Kerry, Cahirciveen
1)	96.2		East Coast R, Bray
3)	96.2	0.025	South East R, Gorey
18)	96.3	5	Northern Sound, Corcaghan
21)	96.3	5	North West R, Mt. Charles: 0700-0100
17)	96.4	2	Cork's 96 FM, Cork C.
7)	96.4	1	Clare FM, Maghera: 0700-0100
3)	96.4		South East R.
8)	96.6	5	R. Kilkenny, Johnswell: 0700-0200
17)	96.7	0.02	Cork's 96 FM, Monatrea (Youghal)
9)	96.8		Galway Bay FM, Abbeyknockmoy
20)	97.0	20	R. Kerry, Mullaghanish: 0700-0100
11)	97.1	5	Tipp FM, Woodruff
5)	97.1	1.5	MWR FM, Achill: 0700-0100
5)	97.2	0.02	MWR FM, Ballina: 0700-0100
12)	97.3	1	CKR, Kileshin
9)	97.4		Galway Bay FM, Ballinasloe
16)	97.5	5	WLR FM, Carrickpherish
18)	97.5	0.1	Northern Sound, Carrickmacross
20)	97.6	1	R. Kerry, Knockanore: 0700-0100
12)	97.6	1	CKR, Red Gap
13)	98.1	5	98 FM, Dublin
3)	99.2	2	South East R, Mt. Leinster
6)	102.1		Midlands Radio 3, Athlone
23)	102.2		Raidió Na Life, Dublin
21)	102.5	2	North West R, Truskmore: 0700-0100
1)	102.9	4	East Coast R, Ballyguille
10)	103.3	5	Highland R, Brunfoot
17)	103.3	5	103 FM County Sound West, Nowen Hill
18)	103.4	0.05	Northern Sound, Kingscourt
6)	103.5	1.25	Midlands Radio 3, Slieve Bloom
17)	103.7	2.5	103 FM County Sound North, Mt. Hillery
22)	103.8		Anna Livia 103 FM, Dublin: 1700 (Sat 0800, Sun 0900)-0200
11)	103.9	1.6	Tipp FM, Devils Bit
14)	104.1	5	Shannonside 104FM, Slieve Bawn
1)	104.4		East Coast R, Arklow
15)	104.4	5	FM 104, Dublin
19)	104.8	5	Tipperary Mid-West R, Dangardargan: 0900-2400
4)	104.9	0.02	LM FM, Dundalk: 0630-0200
21)	105.0	5	North West R, Oughtdarnid: 0700-0100
8)	105.0		R. Kilkenny
12)	107.4		CKR FM

Addresses

1) 9 Prince of Wales Terrace, Quinsboro Rd, Bray, Co. Wicklow. **E-mail:** ecoast@indigo.ie — 2) Mount Kennett House, Henry Str, Limerick — 3) Custom House Quay, Wexford Town, Co. Wexford. **E-mail:** wexford@iol.ie — 4) Boyne Shopping Centre, Drogheda, Co. Louth — 5) Abbey Str, Ballyhaunis, Co. Mayo — 6) The Mall, William Str,

Tullamore, Co. Offaly. **E-mail:** mr3@iol.ie **Web:** http://www.iol.ie/mr3/ — 7) The Abbeyfield Centre, Francis Str, Ennis, Co. Clare. **E-mail:** clarefm@iol.ie **Web:** http://www.iol.ie/~clarefm/ — 8) 56 Hebron Rd, Kilkenny — 9) Unit 3, Sandy Rd, Galway — 10) Pine Hill, Letterkenny, Co. Donegal — 11) Whitbridge House, Old Waterford Rd, Clonmel, Co. Tipperary — 12) Lismard House, Tullow Str, Carlow or ACC House, 51 So. Main Str, Naas, Co Kildare — 13) South Block, The Malt House, Grand Canal Quay, Dublin 2.. **E-mail:** 98fm@iol.ie **Web:** http://ireland.iol.ie/ resource/98fm/ — 14) Minard House, Sligo Rd, Longford or Castle Str, Roscommon — 15) Hume House, Ballsbridge, Dublin 4. **E-mail:** fm104@clubi.ie **Web:** http://www.clubi.ie/fm104/ — 16) The Radio Centre, George's Str, Waterford. **E-mail:** wlrfm@iol.ie **Web:** http://www.iol.ie/~wlrfm/ — 17) Broadcasting House, Patrick's Place, Cork. **Web:** http://www.cis.ie:80/ marketplace/96fm/ — 17) Broadcasting House, Patrick's Place, Cork — 18) 33 Glaslough Str, Monaghan, Co. Monaghan or Archcourt, Townhall Str, Cavan — 19) St. Michael's Str, Tipperary. **E-mail:** tmw@iol.ie — 20) Maine Str, Tralee, Co. Kerry — 21) Market Yard, Sligo, Co. Sligo. **E-mail:** nwr@iol.ie **Web:** http://www.nwconnect.ie/NW_Radio/ — 22) 3 Grafton Str, Dublin 2 (non-comm, public sce. st.). **Web:** http://slarti.ucd.ie/annalivia/ — 23) 46 Stráid Chill Dara, Baile Átha Cliath 2 (Irish language st).

Community st's: 10 st's in operation as of August 1997.

Unlicensed stations: A number of unlicensed stations are operating on MW and FM. Details of these stations can be found on the World Wide Web at http://www.clubi.ie/hibernia/irish_radio_database/

ITALY

L.T: UTC + 1h (Su: UTC + 2h) — **Pop:** 57.880.000 — **Radios:** 45.350.000 — **Pr.L:** Italian — **E.C:** 50Hz, Voltages vary — **ITU:** I.

RAI-RADIOTELEVISIONE ITALIANA
(State Service by Public act to RAI)
✉ Viale Mazzini 14, 00195 Roma. ☎ +39 (6) 38781. 🖷 +39 (6) 3226070. **Web:** http://www.rai.it
Tech. Dept: Via Cernaia 33, 10121 Torino. 🖷 +39 (11) 539429. ① 221035 RAI TOI.
LP: Chmn: Enzo Siciliano. GM: Franco Iseppi. Dir. Radio Coordinator: Pietro Vecchione. Dir. Radio Prgrs: Stefano Gigotti. Dir. Radio News: Marcello Sorgi. Dir. Radio-International: Roberto Morrione. Head of PR: Carlo Sartori.
Regional Centres: Abruzzo: Via de Amicis 27, 65123 Pescara — **Alto Adige:** Piazza Mazzini 23, 39100 Bolzano/Bozen — **Basilicata:** Viale de Basento, 85100 Potenza — **Calabria:** Strada Statale 19 bis, 87100 Cosenza — **Campania:** Via Marconi 9, 80125 Napoli — **Emilia-Romagna:** Viale della Fiera 13, 40127 Bologna — **Friuli-Venezia-Giulia:** Via Fabio Severo 7, 34133 Trieste — **Liguria:** Corso Europa 125, 16132 Genova — **Lombardia:** Corso Sempione 27, 20145 Milano — **Marche:** Piazza della Repubblica 1, 60121 Ancona — **Molise:** Viale Principe di Piemonte 59, 86100 Campobasso — **Piemonte:** Via Verdi 16, 10124 Torino — **Puglia:** Via Dalmazia 104, 70125 Bari — **Sardegna:** Viale Bonaria 124, 09100 Cagliari — **Sicilia:** Viale Strasburgo 19, 90146 Palermo — **Toscana:** largo Alcide de Gasperi 1, 50136 Firenze — **Trentino:** Via Perini 141, 38100 Trento — **Umbria:** Via L. Masi 2, 06100 Perugia — **Veneto:** Palazzo Labia, Campo S. Geremia 275, 30121 Venezia.

HOME SERVICES

LW/MW	kHz	kW	LW/MW	kHz	kW
Caltanissetta	189	10	Sassari	1143	10
Caltanissetta	567	20	Messina	1143	6
Bologna	567	20	San Remo	1188	6
Sassari	567	10	La Spezia	1296	2
Salento	567	6	Udine	1296	2
Aosta	567	2	Genova	1305	12
Napoli	657	120	Pisa	1305	2
Firenze	657	100	Campobasso	1314	2
Torino	657	50	Cantanzaro	1314	1
Bolzano	657	25	Matera	1314	2
Venezia	657	20	Ancona	1314	10
Potenza	693	20	Roma	1332	600
Milano	693	100	Bari	1332	50
Trieste	819	20	Pescara	1332	25
Roma	846	1200	Palermo	1332	10

LW/MW	kHz	kW	LW/MW	kHz	kW
Taranto	873	2	Cagliari	1368	2
Trapani	936	10	Venezia	1368	20
Venezia	936	20	Napoli	1368	10
Genova	936	10	Milano	1368	12
Trieste	981	10	Torino	1368	6
Potenza	990	10	Messina	1368	2
Torino	999	20	Catania	1368	2
Perugia	999	20	Firenze	1368	2
Rimini	999	6	Sassari	1368	2
Capo Vaticano	999	2	Trento	1368	2
Milano	1035	50	Bari	1368	1
Napoli	1035	20	Foggia	1431	2
Trieste	1035	10	Pesaro	1431	2
Firenze	1035	12	Taranto	1431	2
Pescara	1035	6	Squinzano	1449	50
Salento	1035	6	Catania	1449	6
Caltanissetta	1035	2	12 st's	1449	2
Oristano	1035	2	La Spezia	1449	2
Cagliari	1062	25	6 st's	1449	0.15
Squinzano	1062	25	5 st's	1485	2
Ancona	1062	10	L'Aquila	1485	1
Catania	1062	2	2 st's	1485	1
Verona	1062	2	3 st's	1485	0.15
Pisa	1062	2	Palermo	1512	2
Trento	1062	2	Genova	1575	50
Roma	1107	6	Perugia	1575	20
Bologna	1116	60	6 st's	1575	2
Pisa	1116	25	3 st's	1575	1
Cuneo	1116	20	1 st's	1575	0.15
Palermo	1116	10	7 st's	1602	2
Aosta	1116	2			

MW/LW daily transmission hours: All stations transmit from 0600 to 2400, except for Milano 900kHz-600kW and Roma 846kHz-600/1200kW transmitting 24h.

SW: Caltanissetta (G.C: 14.04E/37.30N): 6060kHz 25kW, 7175/9515kHz 5kW.

FM (MHz)	R1	R2	R3	S	kW(ERP
Bertinoro (FO)	90.8	93.4	99.6		30
Bologna (BO)	89.5	91.7	93.9		60
Bolzano (BZ)	91.5	93.7	97.1	99.6	14
Ca' del Vento	92.1	96.5	98.5		40
Canepina (VT)			93.7		12
Capo Spartivento (RC)	95.6	97.6	99.7		10
Col Visentin (BL)	91.1	93.1	95.5		30
Crotone (CZ)	94.9	97.9	99.9		10
Firenze (FI)	87.8	91.1	98.4		10
Friscano (PN)	88.4	90.5	94.1		40
Gambarie (RC)	95.3	97.3	99.3		40
			103.9		40
Genova (GE)	89.5	91.9	95.1		80
Golfo di Policastro (SA)	88.5	90.5	92.5		10
Golfo di Salerno (SA)	95.1	97.1	99.1		20
Gorizia (GO)	89.5	92.3	94.6	98.3	10
Martina Franca (TA)	89.1	91.1	93.1		100
Milano (MI)	90.6	93.7	99.4	102.2	60
Monte Argentario (GR)	90.1	92.1	94.3		70
		89.0			16
Monte Beigua (SV)	91.5	94.6	98.9		40
Monte Caccia (BA)	94.6	96.7	99.2		100
Monte Cammarata (AG)	91.1	95.9	99.9		100
Monte Canate (PR)		95.9			24
Monte Cavo (RM)	87.6	91.2	98.4		80
Monte Conero (AN)	88.3	90.3	92.3		100
Monte Faito (NA)	94.1	96.3	98.1		100
Monte Favone (FR)	88.8	90.9	92.9		30
Monte Lauro (SR)	94.7	96.7	98.7		100
Monte Limbara (SS)	88.9	95.3	99.3		60
Monte Luco (SI)	88.1	92.5	96.2		30
Monte Nerone (PS)	94.7	96.6	98.7		100
Monte Peglia (TR)	98.7	97.7	99.7		60
		88.3			30
Monte Penice (PV)	94.2	97.4	99.9		120
		103.0			120
Monte Pierfaone (PZ)	88.1	90.1	92.1		45
Monte Sambuco (PG)	88.6	90.7	93.5		100
		100.7			100
Monte Scuro (CS)	88.5	90.5	92.5		30
Monte Serpeddi (CA)	90.7	92.7	96.3		70
Monte Serra (PI)	88.5	90.5	92.9		70

FM (MHz)	R1	R2	R3	S	kW(ERP)
Monte Soro (ME)	89.9	91.9	93.9		30
Monte Subasio (PG)	89.3	91.4	93.5		30
Monte Venda (PD)	88.1	89.0	89.9		160
Monte Vergine (AV)	87.9	90.3	92.3		20
Martina Franca	89.1	91.1	93.1		100
Napoli Camaldoli (NA)	89.3	91.3	93.3	103.9	12
Nova Siri		89.6			10
Palermo M. Pellegrino (PA)	94.9	96.9	98.9		40
Pecara S. Silvestro (PE)	89.2	94.3	96.4		70
Pomarico (MT)	88.7	92.7	95.7		10
Punta Badde Urbara (OR)	91.3	93.3	97.3		70
Roma M. Mario (RM)	89.7	91.7	93.7	100.3	100
Roseto Capo Spulico (CS)	94.4	96.5	98.5		10
Salento Turrisi (LE)	90.7	95.5	97.5		60
San Cerbone (FI)	95.3	97.3	99.3		12
San Zeno di Montagna (VR)	93.2	96.5	98.5		10
Selva Piana (BS)	88.4	90.3	92.4		20
Torino Eremo (TO)	92.1	95.6	98.2	101.8	100
Trapani Erice (TP)	88.4	90.5	92.5		60
Trieste M. Belvedere (TS)	91.5	93.6	95.8	103.9	30
Udine (UD)	94.9	97.2	99.8		60
Velletri (RM)	88.7	90.7	92.7		15

R1=Radiouno, R2=Radiodue, R3=Radiotre, S=Special Prgrs.
+ over 2450 st's below 10kW not mentioned.
Radiouno (Prgr. 1): 0500-2300 on 567, 657, 819, 900, 990, 1062, 1296, 1332, 1575, 6060, 9515kHz + FM. **N:** 0600, 0700, 0900, 1000, 1100, 1200, 1400, 1600, 1800, 2100. **Night Prgrs: "Misteri della Notte":** 2300-0500 on FM. **"Notturno Italiano":** 2230-0500 on 846/900/6060kHz. **N. in Italian:** 2300, 2330, 0103-0403 hourly. **N. in English:** 0006-0406 hourly. **N. in French:** 0009-0409 hourly. **N. in German:** 0012-0412 hourly. **Regional Prgrs:** 0620-0630.
Radiodue (Prgr. 2): 0500-2300 on 189, 693, 846, 936, 999, 1035, 1116, 1143, 1314, 1449, 1485, 7175kHz + FM. **N:** 0530, 0630, 0730, 0930, 1130, 1230, 1430, 1530, 1530, 1730, 1830, 2130. **Night Prgr: "Stereonotte":** 2300-0500 on FM. **Regional Prgrs:** 1110-1130.
Radiotre (Prgr. 3): 0500-2300 on 1107, 1305, 1368, 1512, 1602kHz + FM. **N:** 0545, 0745, 1245, 1745. **Night Prgr: "R. Tre Notte Classica":** 2300-0500 on FM. **Regional Prgrs:** Venezia: W 0620-0630, 1110-1130, 1430-1530.

SPECIAL PRGRS

RAI 2 ISO Radio: 24h sce. for motorway users on 103.3MHz FM.
Sender Bozen (Bolzano): Prgrs in **German** on 1602kHz (Bolzano/Brunico/Merano/Bressanone 1kW) + FM (46 tr's of 1kW or less). **D.PRGR:** 0630 (Sun 0700)-2200. **N.** 0715 (W), 0800 (Sun), 1000 (W), 1100, 1200, 1300, 1700 (W), 1930.
Regional Prgr. in Slovene: Trieste 981kHz 10kW + 103.9MHz 20kW (and 22 additional FM-tr's). **D.PRGR:** 0600 (Sun 0700)-1830. **N:** W 0600, 0700, 0900, 1200, 1300, 1600, 1800; Sun 0700, 1200, 1300, 1800. **N. in German:** 0900 (W).
N. in Arabic (for Mediterranean area): W 1430-1445 on Caltanissetta 567kHz + SW.
ANN: Home Sce: "Radiouno", "Radiodue", "Radiotre" as appropriate. Night Prgr: "RAI—Radiotelevisione Italiana stazioni a onda media di Roma kHz 846 pari a m.355, e di Milano kHz 900 pari a m.333, e stazione ad onda corta di kHz 6060 pari a m.49.50 RAI International "Notturno Italiano"

FOREIGN SERVICE: see International Broadcasting section.

RUNDFUNK ANSTALT SÜDTIROL (RAS)

✉ Europaallee 164/A, 39100 Bozen. ☎ +39 (471) 932933.
L.P: Pres: Helmuth Hendrich. MD: Klaus Gruber. Dir. Tec: Helmuth Schäfer. RAS is a public body of the autonomous region of Southern Tyrol whose purpose is to relay TV and radio from Germany, Austria and Switzerland to the German-speaking population.

FM	RAS 1	RAS 2	RAS 3	kW
Kronplatz	100.7	103.0	104.7	2
Meransen	101.3	103.9	107.3	1
Obervinschgau	100.5	103.0	106.1	0.6
Penegal	103.3	100.3	104.7	2
Perdonig	101.8	104.0	106.0	1
Plose	99.8	102.0	105.6	1
Vinschgau	101.1	102.9	105.0	2

+ low power st's.
RAS 1: rel. OE-3 (Austria)— **RAS-2:** rel. OE-R (Austria)— **RAS-3:** rel. OE-1 (Austria)

RADIO TIROL GmbH

✉ Postfach 26/Aichweg 4, 39019 Dorf Tirol.
☎ +39 (473) 93656. ① 401406. 🖷 +39 (473) 93663.

L.P: GM: Dr. Gerald Fleischmann.

FM (Stereo)	MHz	kW		MHz	kW
Taufers			Zirog-Alm	101.5	67
Münstertal	99.3	8.5		105.3	
Innichen	103.7	1.0			

+ low power relays.
D.PRGR. in German: 24h. **N:** W 0600, 0700, 0900, 1130, 1800; **Wrp:** W 0535, 0615, 0705, 0905, 1215, 1815; Sun 0700, 1130.

PRIVATE STATIONS

This section was compiled by our collaborator Dario Monferini. Only st's r. with MW/SW broadcasts and FM networks are listed. A number of other st's are heard irr. There are approx 2000 FM st's.

	MW	kW	Name, location and h. of tr.
1)	1548	1	R. Star Veneto, Vicenza, rel. of R. Cuore: 0700-1700v
2)	1584	2.5	R. Sanluchino, Bologna: 24h
3)	1584	2	R. Studio X, Momigno: 24h

	SW	kW	Name, location and h. of tr.
4)	3985	0.5	R. Mariquita/Alpen Adria, Treviso: irr.
5)	6260v	0.5	R. Mistero Gosh Planet, Torino: irr.
6)	6295v	0.5	Good Fun R., Treviso: irr.
7)	7140	5	R. Maria Network, relay Spoleto: 24h
8)	7306	0.5	R. Europe, Pioltello (USB): 0800-1400v
9)	7475	0.5	R. Strike, Palermo: irr.
10)	7500	1	R. Internazionale, Padova: irr.

Addresses and other information
1) c/o PROGRES Veneto, Via Turazza 48/c, IT-35128 Padova. ☎ +39 (49) 8070755. 2 +39 (49) 774800. – **FM:** 93MHz 5kW. **V.** by letter. Rp.
2) Via San Luca 35, IT-40135 Bologna. ☎ +39 (51) 434525. 🖷 +39 (51) 435651 – SM: S. Piancastelli. **FM:** 104.7/100.4MHz 5kW.
3) Via Mammianese 687, IT-51030 Momigno (PT). ☎ +39 (572) 894019 – SM: L. Betti. **FM:** 87.5MHz kW. **V.** by letter. Rp. Re. to PLAY-DX, Via Davanzati 8, IT-20158 Milano. Rp.
4) c/o G.A.M.T., P. O. Box 3, Succ.10, IT-31100 Treviso. **V.** by QSL-card. Rp.
5) c/o P.O. Box 220 342, DE-42373 Wuppertal, Germany – SM: B. Pecciatto. **V.** by letter. Rp.
6) c/o G.A.M.T., P. O. Box 3, Succ.10, IT-31100 Treviso. **V.** by QSL-card. Rp.
7) Via Turati 7, IT-22100 Eraba (LC). ☎ +39 (31) 610610. 🖷 +39 (31) 611288. SM: E. Ferrario. **V.** by letter. Rp.
8) c/o PLAY-DX, Via Davanzati, IT-20158 Milano. 🖷 +39 (2) 8645 0149. **E-mail:** PLAYDX@hotmail.com – SM: A. Bertini. **V.** by QSL-card. Rp.
DX-prgr: Sun 0815-0845 irr.
9) c/o R. Scaglione, P. O. Box 119, Succ.34, IT-90144 Palermo. **V.** by QSL-card. Rp.
11) c/o E-mail: r7500@hotmai.com – SM: G. Belloni. **V.** by letter. Rp. Re. to PLAY-DX, Via Davanzati 8, IT-20158 Milano.

FM NETWORKS

1) **Capital FM Network** (110 st's), c/o Forum di Assago, Milano Fiori, I-20090 Assago (MI). ☎ +39 (2) 488501. 🖷 +39 (2) 48850220. SM: C. Cecchetto. **V.** by letter. Rp.
2) **Circuito Marconi** (talk radio) (30 st's), Via San Antonio 5, I-20122 Milano. ☎ +39 (2) 588 03682. 🖷 +39 (2) 583 03618. SM: Don A. Cattaneo. V. by letter. Rp.
3) **C.N.R. Channel News Radio** (50 st's), Via C. Simonetta 10, I-20122 Milano. ☎ +39 (2) 5810 2782. 🖷 +39 (2) 5810 2463. **Web:** http://worknet.sii.it/cnr/ SM: G. Bozzo. **V.** by letter. Rp.
4) Company Network (40 st's), Via Salata 58, I-35027 Noventa Padovana (PD). ☎ +39 (49) 901616. 🖷 +39 (49) 9801666. SM: C. Rampazzo. **V.** by letter.
5) **Italia Network** (50 st's), Via dei Giudei 21-39, I-40050 Funo di Argelato (BO). ☎ +39 (51) 6646208. 🖷 +39 (51) 6646780. SM: A. Gandolfi.
6) **Italia Radio** (30 st's), Piazza del Gesù 47, I-00186 Roma. ☎ +39 (6) 6791412, 6796539. 🖷 +39 (6) 6781936. SM: C.Fotia.
7) **Italia Vera** (20 st's), Via Bovio 6, I-20159 Milano. ☎ +39 (2) 6900 6593. 🖷 39 (2) 6901 0206. **Web:** http://www.energy.it/italiavera. SM: E. Palazzolo. **V.** by letter. Rp.
8) **Latte e Miele l'Italiana** (30 st's), Via Dozza 54, I-40013 Castelmaggiore (BO). ☎ + 39 (51) 713919. 🖷 +39 (51) 715626. SM: F. Mignani.
9) **Magic One-o-One Network** (35 st's), Via Locatelli 6, I-20124 Milano. ☎ +39 (2) 6698 2551. 🖷 +39 (2) 6704900.**Web:** http://www.galactica.it/101. SM: A.Villa. **V.** by QSL-card. Rp. Re. to

D.Monferini, Via Davanzati 8, I-20158 Milano.
10) **Popolare Network** (40 st's), Via Stradella 5, I-20129 Milano. ☎ +39 (2) 2952 4141. 🗎 +39 (2) 2940 5506. Web: http://www.altair.it/ XXV/ SM: P. Scaramucci. **V.** by letter. Rp.
11) **R. Cuore & R. Cuore 2** (110 st's), Via Giovanni da Verrazzano, I-56038 Ponsacco (PI). ☎ +39 (587) 733880. 🗎 +39 (587) 733861. SM: L. Bessi.
12) **R. Deejay** (110 st's), Via Massena 2, I-20145 Milano. ☎ +39 (2) 345711. 🗎 +39 (2) 312629.**Web:** http://www.deejay.it SM: Linus. **V.** by letter.
13) **R. Dimensione Suono Network** (60 st's), Viale Mazzini 119, I-00195 Roma. ☎ +39 (6) 3728488. 🗎 +39 (6) 3725363. **Web:** http://www.f;ashmet.it/rds SM: E. Montefusco. **V.** by letter. Rp.
14) **Kiss Kiss FM** (40 st's), Via Sgambati 61, I-80131 Napoli. ☎ +39 (81) 541212, 5462700. 🗎 +39 (81) 5467789. SM: G.Simioli. **V.** by letter. Rp.
15) **R. Italia solo Musica Italiana & R. Italia Anni 60,** (110 st's), Via F. Casati 2, I-20124 Milano. ☎ +39 (2) 2951 2955. 🗎 +39 (2) 2951 7463. SM: M. Volanti. **V.** by letter. Rp.
16) **R. Maria Network Europa** (590 st's), Via Turati 7, I-22036 Erba (LC). ☎ +39 (31) 610610. 🗎 +39 (31) 611288. **V.** by letter. Rp.
17) **R. MonteCarlo FM-Italia** (30 st's), Via Turati 40, I-20121 Milano. ☎ +39 (2) 6596116. 🗎 +39 (2) 6592272. Re. to Monaco office (Services Techniques B. Poizet. ☎ +377 93 151960). **V.** by letter.
18) **R. Radicale** (110 st's), Via Principe Amedeo 2, I-00185 Roma. ☎ +39 (6) 4880541. 🗎 +39 (6) 4880196. SM: P. Vigevano. **V.** by letter. Rp.
19) **Rete 105 The Radio** (110 st's), P. O. Box 1448, I-20100 Milano. ☎ +39 (2) 6596116. 🗎 +39 (2) 6592272. SM: A. Hazan. **V.** by letter. Re. to Servizio Tecnico M. Cavestro.
20) **Rete Italia** (50 st's), Via F. Casati 2, I-20124 Milano. ☎ +39 (2) 2951 2955. ☎ +39 (2) 2951 7463.
21) **Rete 8 Network** (20 st's), Via Reni 37, I-21100 Varese. ☎ +39 (332) 287888. 🗎 +39 (332) 235332. **V.** by letter. Rp.
22) **RTL Hit Radio S.r.l.** (110 st's), Via Suardi 42, I-24040 Arcene (BG). ☎ +39 (35) 879294-8. 🗎 +39 (35) 878012. SM: L. Suraci. **V.** by letter. Rp.

ADVENTIST WORLD RADIO EUROPE (Rlg.)
See Internationai Broadcasting section.

NEXUS – INTERNATIONAL BROADCASTING ASSOCIATION
See International Broadcasting section.

SOUTHERN EUROPEAN BROADC. SCE. (U.S. Mil.)
✉ APO New York, NY 09221, USA.
L.P: Commander Bernard L. Miles, MAJ, SC.
1st Prgr. on 106.0 MHz. **Key Station:** Vicenza (1kW); other st's: Aviano/Livorno/Verona/San Vito (all 0.25kW), Napoli (1kW), Mt. Virgine/Rimini/Decimomannu/La Maddelena (all 0.05kW).
2nd Prgr. on 107.0MHz (rel. AFN Germany). **Key Station:** Vicenza (1kW); other st's; Aviano/Livorno/Sigonella/ La Maddelena (all 0.25kW), Napoli (1kW), San Vito (0.05kW).

LATVIA

LT: UTC + 2h (Su: UTC + 3h) — **Pop:** 2.475.000 — **Radios:** 1.396.000 — **Pr.L:** Latvian (official), Russian — **E.C:** 50Hz, 220V — ITU: LVA.

LATVIJAS VALSTS RADIO UN TELEVĪZIJAS CENTRS (Gov.) (Latvian State Radio and TV Centre)
✉ Elizabetes ielā 41/43, LV-1010 Rīga. ☎ +371 (7) 228687. 🗎 +371 (7) 333886. ① 161318 LVRTC LV — **L.P:** Dir: Māris Pauders.
LVRTC is responsible for state transmission networks.

NACIONĀLĀ RADIO UN TELEVĪZIJAS PADOME (Public) (National Radio and TV Council)
✉ Smilsu ielā 1/3, LV-1939 Rīga. ☎ +371 (7) 221848. 🗎 +371 (7) 220448 — **L.P:** Chmn: Ojārs Rubenis.
NRTP is the radio and TV broadcasting regulatory and licensing body.

LATVIJAS RADIO (Gov.)
✉ Doma laukumā 8, LV-1505 Rīga. ☎ +371 (7) 206722. 🗎 +371 (7) 206709. ① 161266 RADIO LV. **Web:** http://www.radio.org.lv - Prgr1 Real Audio.
L.P: DG: Dzintris Kolāts. PD: Dārija Juskevica.

MW	kHz	kW	Prgr
Rīga	945	150	1
Cesvaine	1350	50	1
Kuldīga	1350	50	1
Rēzekne	1422	25	1
Valmiera	1422	25	1
Daugavpils	1539	7	1
Liepāja	1539	7	1
Ventspils	1602	0.2	1

FM (MHz)	FM1	FM2	FM3	kW (ERP)
Cesvaine	67.85	105		12
Daugavpils	69.92	100.7		30
Dundaga	91.1	106.7		6/4
Jēkabpils	66.04			0.6
Kuldīga	66.92	101.9		8
Liepāja	72.32	101.0		4.5/10
Livāni	67.28			0.2
Rēzekne	71.12	101.6		12
Rīga	$P65.71			36
Rīga	$107.7	$91.5	$103.7	3.5/20/5
Valmiera	68.27	66.68		20
Ventspils	71.84	103.0		0.1

$ = stereo. P = Pilot tone instead of OIRT system.

Prgr 1 (national): On MW1 + FM1 tr's 0400-2300 in Latvian. **N, wrp, comm:** 0700W, 0705MF, 0728MF, 0758 (Sun), 0800MF, 0805MF, 0828MF, 0900W, 0905MF, 0938MF, 0950MF, 1000 (Wed, SS), 1005 (Sun), 1015 (Wed), 1130 (Mon, Tues, Thurs, Fri), 1200 (Mon, Tues, Thurs-Sun), 1205 (Mon, Tues, Thurs, Fri, Sun), 1300 (Mon, Tues, Thurs, Fri), 1305 (Mon, Tues, Thurs, Fri), 1400SS, 1500MF, 1505MF, 1600, 1605MF, 1700MF, 1705 (Mon-Wed, Fri), 1800, 1900 2000MF, 2200W.
Rel. international broadcasters: RFE Latvian 1700-1800 (FM1 exc. 107.7MHz) & 1900-2000 (107.7MHz), R. Sweden Latvian 2125-2140, VOA Latvian 2140-2200MF — Parliament session reports: Thurs 0700-(time varies) in Latvian on FM1 exc. 107.7MHz.
Prgr 2 (ethnic): On FM2 tr's 0430(SS0530)-2200 in Latvian, Russian and others. **N, wrp in Russian:** 0532 (Sun), 0600 (Sun), 0700 (Sun), 0727W, 0800 (Sun), 0825MF, 0925MF, 0955 (Sun). Comm. in Russian: 0820MF. **Prgr's for national minorities and cultural communities:** 0930 (Sat 0945)-1000. **Programmes for the younger generation** (Radio 2): 1000-2100 in Latvian. **N:** on the h.
Prgr "Klasika" (cultural): On FM3 tr's 0500-2300 in Latvian. Serious music and entertainment. **N, wrp:** (rel. Prgr 1) 0700MF, 0900MF, 1200, 1400, 1600, 1800.
ANN: "Raida Rīga" or "Runā Rīga", "Latvijas Radio pirmā/otrā programma" (prgr 1/2).
IS: Melody from folk song "Pūt, vējiņi!"

FOREIGN SERVICE: see International Broadcasting section.

OTHER STATIONS

FM	MHz	kW (ERP)	Name, location
1)	67.60	0.07m	R. Merkūrijs, Balvi
2)	68.93	0.4	Super FM, Rīga
3)	70.21	0.1	R. Trîs, Cēsis
4)	70.50	0.1	R. Sigulda, Sigulda
5)	70.63	0.2	R. Miks's, Iecava
7)	71.63	1.7m	Alise Plus, Daugavpils
8)	72.97	0.7	R. Rīgai, Rīga
9)	73.08	0.4	R. Ef-Ei, Rēzekne
10)	89.2	3	R. SWH+
11)	90.3	0.5	R. Mazsalaca, Mazsalaca
12)	99.1		R. Era, Cēsis
14)	99.5	4	R. "Bizness & Baltija", Rīga
15)	100.0	0.5	Rīga (F.P.l.)
16)	100.5	0.9	BBC, Rīga
17)	100.9	0.2	R. Saules Iela, Cēsis
7)	101.6	0.25	Alise Plus, Daugavpils
18)	101.8	4.5	Latvijas Kristī Radio, Rīga
19)	101.9	0.4	R. BČN, Ventspils
20)	102.2	3.5	R. SWH, Talsi
21)	102.4	0.8	R. Imanta, Valmiera
22)	102.7	9	R. Mix FM 102.7, Rīga
2)	104.3	4	Super FM, Rīga
18)	104.6	0.8	Latvijas Kristī Radio, Liepāja
3)	104.7	1	R. Trîs, Cēsis
18)	104.7	0.7	Latvijas Kristī Radio, Talsi
20)	105.2	4	R. SWH, Rīga
23)	105.7	4	R. Amadeus, Rīga

FM	MHz	kW (ERP)	Name, location
24)	105.8	0.2	R. FM 102, Liepāja
8)	106.2	3	R. Rīgai, Rīga
25)	106.4	10	Kurzemes R, Kuldīga
26)	107.1	1.6	R. Liepāja, Liepāja
27)	107.2	4	R. Skonto, Rīga
28)	107.4	4	R. Skala, Kuldīga
29)	107.9	0.4	Saldus R, Saldus

All sts commercial exc. 19). FM: m = mono, others stereo.
St's using only the OIRT band will move to the CCIR band, but freq's not yet fixed.

Addresses and other information:
1) Bērzpils ielā 2a, LV-4501 Balvi. E-mail: merkurijs@ridzene.lv 24h in Latvian, Russian, rel. VOA Express 2200-0445 — 2) Bezdelīgu ielā 12, LV-1007 Rīga. 24h in Latvian — 3) Valnu ielā 5, LV-4101 Cēsis. 24h in Latvian — 4) L. Paegles ielā 3, LV-2150 Sigulda. 24h in Latvian, non-stop music overnight — 5) Rīgas ielā 18, LV-3913 Iecava. 0600-1600 in Latvian (**F.PI**: 24h) — 7) Raiņa ielā 27, LV-5400 Daugavpils. E-mail: ineta@alise.latg.lv 24h in Russian, Latvian — 8) Televīzijas centrs, Zausalas krastmalā 3, LV-1509 Rīga. 24h in Latvian, nonstop music overnight — 9) Atbrīvošanas aleja 98, LV-4600 Rēzekne. **E-mail:** efei@mailbox.riga.lv 24h in Russian, Latvian — 10) Addr: As 20). **Web:** http://www.radio.svh.lv/~radio/swhplus.html 24h in Russian — 11) Avotu ielā 13, LV-4215 Mazsalaca. 0600-2100 in Latvian — 12) Irregular operation at editorial deadline — 14) Balasta dambī3, LV-1081 Rīga. 24h in Latvian — 15) Rīga — 16) Addr: See UK. 24h BBC World Service — 17) Saules ielā 8a, LV-4100 Cēsis. 24h in Latvian — 18) Lācplēsa ielā 37, LV-1011 Rīga. 24h in Latvian, Russian, E — 19) Fabrikas ielā 2, LV-3602 Ventspils — 20) Skanstes ielā 13, LV-1013 Rīga. Reg. st: Talsu raj. majas "Cumulas", Talsi. **Web:** http://www.radio.svh.lv/~radio/swh.html 24h in Latvian — 21) Tērbatas ielā 1, LV-4201 Valmiera. **E-mail:** radio@valmiera.lanet.lv 24h in Latvian, Russian, E, rel. VOA in Latvian, DW in Russian — 22) L. Nometnu ielā 62, LV-1002 Rīga. 24h in Russian — 23) Addr: As 20). **Web:** http://www.radio.svh.lv/~radio/amadeus.html 24h in Latvian (classic music) — 24) Peldu ielā 1/5, 4. stāvā, LV-3400 Liepāja. 0530-2100 in Latvian, 2100-0530 VOA Express — 25) Pilsētas laukumā 4, LV-3300 Kuldīga. 24h in Latvian, incl. rel. Latvijas R. 1st prgr — 26) Avotu ielā 10, LV-3400 Liepāja. 24h in Latvian — 27) Elizabetes ielā 75, LV-1011 Rīga. **E-mail:** ppskonto@mail.vernet.lv 24h in Latvian — 28) Pilsētas laukumā 3, 2. stāvā, LV-3300 Kuldīga. 0600-2200 in Latvian — 29) Brīvības ielā 16, LV-3800 Saldus. 0400-2200 in Latvian.

LIECHTENSTEIN

L.T: UTC + 1h (Su: UTC +2h) — **Pop:** 31.000 — **Pr.L:** Allemanian (a German dialect), German — **E.C:** 50Hz, 220V — **ITU:** LIE.

RADIO L (Comm.)
✉ Dorfstr. 24, 9495 Triesen. ☎ +41 (75) 399 1313. 🗎 +41 (75) 399 1399. **E-mail:** radiol@radiol.lol.li **Web:** http://www.lol.li/RADIOL/
L.P: GM: Roman Banzer.
FM: Vaduz 96.9MHz 0.2kW, Nendeln 103.7MHz 0.1kW, Rüthi 89.2MHz 0.8kW, Trübbach 106.0MHz 0.2kW.
D.PRGR: 24h — **V.** by letter. Rp.

LITHUANIA

L.T: UTC + 2h (Su: UTC +3h) — **Pop:** 3.725.000 — **Radios:** 1.420.000 — **Pr.L:** Lithuanian (official), Russian, Polish — **E.C:** 50Hz, 220V — **ITU:** LTU.

RYSIU IR INFORMATIKOS MINISTERIJA (Ministry of Communications and Informatics)
✉ Vilniaus 33, LT-2001 Vilnius. ☎ +370 (2) 620448. 🗎 +370 (2) 225070. ☺ 261166 rim lt. **Web:** http://www.is.lt/RIM/engl/rimhome.HTML
L.P: Minister: Rimantas Pleikys. Dir. Foreign Rel. Dept: Arūnas Luksas.

LIETUVOS RADIJAS IR TELEVIZIJA (Gov.)
✉ Konarskio 43, LT-2674, Vilnius MTP. ☎ +370 (2) 633182. 🗎 +370 (2) 263282. ☺ 261151 litvtr lt. **Web:** http://www.lrtv.lt
L.P: DG: Arvydas Ilginis. TD: Juozas Vilciauskas.
Parent organization for Lietuvos Radijas, Lietuvos Televizija and LRT Kauno Programu Direkcija.

LIETUVOS RADIJAS (Gov.)
✉ Konarskio 49, LT-2674 Vilnius MTP. ☎ +370 (2) 634471. 🗎 +370 (2) 263282. ☺ 261151 litvr lt. **Web:** http://www.lrtv.lt/lr.htm
L.P: Dir: Mrs. Laima Grumadienė.

MW	kHz	kW	P		MW	kHz	kW	P
Vilnius	612	50	1		Vilnius	666	500	1
Klaipéda	612	25	1					

SW: Sitkûnai (G.C: 23.49E/55.02N): 9710kHz 50kW 259Þ.

FM stereo (MHz)	1	2	3	4	kW (ERP)
Alytus			103.7		0.46
Bubiai	71.15	70.07	103.4	100.9	12/12/1/1
Druskininkai			107.6	102.5	0.46
Juragiai	70.64	72.98	107.6	106.0	12/12/1/1
Klaipéda	67.13	66.35	105.3	102.6	12/12/1/1
Neringa			106.8	103.3	0.46
Pazagieniai			107.5	105.3	1/0.46
Tauragé	66.89	68.51	107.4	105.7	9/8/0.46/0.46
Vilnius	69.80	72.59	102.6	105.1	12/12/1/1
Viesintos	68.03	66.47		100.4	12/12/0.46

Prgr. 1: 24h on FM1 & FM3, 0400-2200 on 612/666kHz, 1000-1100 (SS 1000-1200) on 9710kHz. **N:** 0300MF, 0400-1200 on the h, 1400, 1500 (exc.Sat), 1700, 1900, 2100. **Russian:** 1300-1400, 0000-0030 (Mon 0000-0040). **Polish:** 1600-1630. **N. in Belorussian:** 1905-1915. **Night Prgr:** 0000-0400. **DX-prgr:** Sun 1349, Mon 0230.

Prgr. 2: 0500-2200 (exc. relays, cf. below) on FM2, 1300-2200 (exc. relays) on FM4, 1100-1200MF on 9710kHz. N: 0500, 0600, 0700 (exc. Sat), 0800, 1000, 1200 (Sun), 1300, 1400, 1500 (exc.Sat), 1600, 200MF, 2100. **Russian:** 1700-1730. **Byelorussian:** Sat 0930-1000. **Yiddish:** 1st & 3rd Sun 0930-1000. **Ukrainian:** 2nd & 4th Sat 1230-1300. **Tartar/Russian:** 1st & 3rd Sat 1230-1300.

Relays: FS (R. Vilnius): 2000-2030 & 2200-2300 on Prgr.1. **VOA in Lithuanian:** MF 2030-2100 on Prgr.1. **R Free Europe in Lithuanian:** 1800-1900 on Prgr.1. **Vatican Radio in Lithuanian:** 0505-0525 (pre-recorded) on Prgr.2. **Blagovest** (Belgium) **in Russian:** 1930-2000 on Prgr. 2.

ANN: 1/2/Night Prgr: "Lietuvos Radijo pirmoji/antroji/ Nakties programa" — **IS:** 10 bar orchestral melody.
PUB: "Kalba Vilnius", weekly in Lithuanian, "Programmy TeleRadio", weekly in Russian — **V.** by letter for DX program only. Rec.acc. IRC's appreciated. Re. to DX-Editor S.Zilionis.

FOREIGN SERVICE: see International Broadcasting section.

LRT KAUNO PROGRAMU DIREKCIJA (Gov.)
✉ Daukanto 28a, LT-3000 Kaunas. ☎ +370 (7) 227596. 🗎 +370 (7) 205566 — **L.P:** Dir: R. Garnys — **D.PRGR:** prepares certain prgrs for Lietuvos Radijas — **V:** does not verify.

PRIVATE COMMERCIAL STATIONS

ALYTAUS RADIJAS
✉ Pramonés 1a, LT-4580 Alytus. ☎ +370 (35) 57711. 🗎 +370 (35) 25656 — **L.P:** Editor-in-Chief: Liudas Ramanauskas.
FM: Alytus 70.04MHz 30W ERP, 99.0MHz 3kW ERP — **D.PRGR:** 24h.
V. by letter.

AUKSTAITIJOS RADIJAS
✉ Laisvés a. 1, LT-5300 Panevézys. ☎ +370 (5) 465775, 466285. 🗎 +370 (5) 465775. **E-mail:** ar@omnitel.net — **L.P:** Dir: Algirdas Satas.
FM: Panevézys 106.9MHz 2kW ERP — **D.PRGR:** 24h.

BUMSAS
✉ Kauno 3-108, LT-5802 Klaipéda. ☎ +370 (6) 320909. 🗎 +370 (6) 322020 — **L.P:** Dir: Dalius Noreika.
FM: Klaipéda 91.4MHz 2kW ERP — **D.PRGR:** 24h.

ELEKTRÉNU "VERSMÉS" RADIJAS
✉ Versmés vid. m-la, Saulés 30, LT-4061 Elektrénai. ☎ +370 (37) 35443. **E-mail:** root@vers.elek.soros.lt
Web: http://193.219.86.34/MokVizit/4/tele.htm
FM: Elektrénai 70.31MHz 0.2kW ERP.

KAPSAI
✉ P. Armino g., LT-4520 Marijampolé. ☎ +370 (43) 54512. — **L.P:** Dir: Benjaminas Masalaitis. PD: Raimundas Maruskevicius.
FM: Marijampolé 100.2MHz 2kW ERP — **D.PRGR:** 24h.

KAUNO FONAS
✉ Radastu 2, LT-3000 Kaunas. ☎ +370 (7) 733675. 🗎 +370 (7) 730946. **L.P:** Dir: Arûnas Gruseckas.
FM stereo: Kaunas 105.4MHz 0.5kW ERP.
D.PRGR: 24h. **E:** Sun 1630-1700 — **V.** by letter.

LAISVOJI BANGA

Naugarduko 51, LT-2006 Vilnius. ☎🖹 +370 (2) 263836.
E-mail: lbanga@auste.elnet.lt. **Web:** http://www1.omnitel.net/lb/
L.P: Dir: Ceslovas Burba. TD: Vytautas Bartkus.

FM (MHz)	1	kW (ERP)
Juragiai	104.5	3
Klaipéda	104.1	3
Pazagieniai	104.8	0.8
Vilnius	104.7	3

D.PRGR: 24h — **ANN:** "Cia radijo stotis Laisvoji Banga" – **V.** by letter.

LALUNA

M.Mazvydo al. 11, LT-5800 Klaipéda. ☎🖹+370 (6) 343232. **E-mail:** laluna@klaipeda.omnitel.net. **Web:** http://www1.omnitel.net/laluna/
L.P: Dir: Ruslanas Aleksandravicius.
FM: Klaipéda 100.8MHz 2.5kW ERP — **D.PRGR:** 24h.

M-1

P.d. 1747, LT-2300 Vilnius. ☎+370 (2) 428370, 705805. 🖹 +370 (2) 421327, 429152. **E-mail:** m-1@m-1.lt. **Web:** http://www.m-1.lt
L.P: DG: Hubertas Grusnys.

FM stereo (MHz)	1	2	3	4	kW (ERP)
Alytus	106.0				2
Bubiai	106.3	72.71	100.5	71.93	2/7/2/3.5
Ignalina	105.9				2
Juragiai	106.6	72.20	97.6	71.45	2/7/2/3.5
Klaipéda	106.5	69.11	98.3	67.91	2/7/2/3.5
Marijampolé	106.3				2
Pazagieniai	106.0				2
Raseiniai	106.4				2
Tauragé	106.2				2
Tryskiai	106.0				2
Utena	106.3		98.7		2/2
Vilnius	106.8	71.81	106.2	71.03	1/7/2/3.5
Viesintos		68.24		70.52	1/7/3.5

Prgr. 1 ("M-1"): 24h on FM1, 0500-2400 on 71.81/72.20MHz, 0500-2200 on 68.24/68.39/69.11/72.71 MHz.
Prgr. 2 ("M-1 Plius"): 0500-2400 on FM3 & FM4. **Rel. VOA in Lithuanian:** 0500-0510.
ANN: M-1: "M-vienas" (jingle). M-1 Plius: "Radijo stotis, radijo sto-tis, radijo stotis M-vienas-plius" (jingle).

MAZEIKIU AIDAS

Sodu 13-93, LT-5500 Mazeikiai. ☎+370 (93) 65095, 65600. 🖹 +370 (93) 65600 — **L.P:** Dir: Alvyda Purauskyté.
FM: Mazeikiai 100.4MHz 6kW ERP.
D.PRGR: 24h — **ANN:** "Radijo stotis Mazeikiu aidas".

NEVÉZIO RADIJAS

P.d. 107, LT-5304 Panevézys. ☎+370 (5) 435854. 🖹+370 (5) 424624.
L.P: Dir: Audrius Rudys. TD: Rolandas Meiliūnas.
FM stereo: Pazagieniai 103.0MHz 3kW ERP, Birzai 103.3 1kW ERP.
D.PRGR: MF 1100-1500 in **Lithuanian.**

PUKAS

P.d. 1283, LT-3002 Kaunas. ☎+370 (7) 747444. 🖹+370 (7) 207632.
L.P: Dir: Arvydas Dimsa.
FM stereo: Kaunas 104.9MHz 2kW ERP, 70.22MHz 2kW ERP.
D.PRGR: 24h. **N:** 0630, 0730. **ANN:** "Radijo stotis Pûkas".
V. by letter.

RADIOCENTRAS

Laisvés pr. 60, LT-2056 Vilnius. ☎+370 (2) 429463. 🖹+370 (2) 429073, 429463. **E-mail:** radiocentras@mail.tipas.lt
L.P: DG: Gintautas Babravicius. PD: Gintaras Ruplénas. TD: Artûras Mironcikas.

FM-stereo (MHz)	1	2	kW (ERP)
Alytus	101.1		2
Juragiai	107.1	67.67	2/2
Klaipéda	101.5	67.52	2/2
Marijampolé	101.8		2
Pazagieniai	101.9	67.55	2/2
Raseiniai	99.9		2
Siauliai	101.7	70.40	2/2
Tauragé	102.7		2
Tryskiai	101.8		2
Utena	101.0		2

FM-stereo

FM-stereo (MHz)	1	2	kW (ERP)
Vilnius	101.5	67.94	2/2
Vilnius *	105.6		0.74

*) rel. **VOA Express.**
D.PRGR: 24h. **Rel. VOA in Lithuanian:** 1800-1810.
V. does not verify.

RADIOLA

V.Mykolaicio-Putino 5, LT-2001 Vilnius. ☎ & 🖹 +370 (2) 251458. 🖹 +370 (2) . — **L.P:** Dir: Stanislovas Kairys.
FM: Vilnius 99.7MHz 2kW ERP.
D.PRGR: tests only — **F.Pl:** 24h, music.

RATEKONA

P.d. 3300, LT-2013 Vilnius. ☎+370 (2) 656621. 🖹 +370 (2) 227454.
E-mail: ratekona@is.lt — **L.P:** Dir: Sigitas Zilionis.
FM: Vilnius 99.3MHz 500W ERP. **D.PRGR:** 24h. **Rel. RAI International in Lithuanian** 0445-0505, **Universal Life** (rlg.) in **German/English:** Fri 1700-1830.

SAULÉS RADIJAS

Dvaro 88, LT-5400 Siauliai. ☎+370 (1) 422431. 🖹+370 (1) 432816.
E-mail: src@siauliai.aiva.lt.
L.P: Editor-in-Chief: Valentinas Didzgalvis.
FM: Bubiai 102.5MHz 1kW ERP.
D.PRGR: 24h in Lithuanian. **ANN:** "Saulés Radijas".

TAU

Savanoriu 31, LT-3000 Kaunas. ☎+370 (7) 228207, 774000. 🖹 +370 (7) 206548. **E-mail:** merica@kaunas.omnitel.net
L.P: Dir: Algirdas Kepezinskas.
FM: Juragiai 102.9MHz 2kW ERP. **D.PRGR:** 24h.

TITANIKA

Sporto 6, LT-3000 Kaunas. ☎+370 (7) 203404. 🖹 +370 (7) 208222.
E-mail: titanika@kaunas.sav.lt
Web: http://www.kaunas.sav.lt/titanika/
FM stereo: Kaunas 105.9MHz 65W ERP.
D.PRGR: 24h in Lithuanian.

ULTRA VIRES

P.d. 2094, LT-3000 Kaunas. ☎🖹 +370 (7) 734734 — **L.P:** Dir: Evaldas Vaiciūnas. PD: Ausrys Krisciūnas. TD: Giedrius Lipnickas.

FM stereo	MHz	kW(ERP)	FM stereo	MHz	kW(ERP)
Alytus	103.3	3	Pazagieniai*	103.0	3
Birzai*	103.3	1	Tauragé	103.1	3
Bubiai	103.9	3	Utena	103.4	3
Juragiai	103.5	6	Vilnius	103.1	6
Klaipéda	103.7	6	Vilnius	69.05	12

*) rel. **Nevezio Radijas** MF 1100-1500.
D. PRGR: 24h — **V.** by letter.

VENTUS

Montuotoju 2, LT-5500 Mazeikiai. ☎+370 (93) 74477. 🖹 +370 (93) 73503. — **L.P:** DG: Giedrius Stelmokas.
FM: Mazeikiai 105.6MHz — **D.PRGR:** 24h.

ZEMAITIJOS RADIJAS

Mazeikiu 18, LT-5610 Telsiai. ☎+370 (94) 51403.
L.P: DG: Remigijus Valauskas.
FM: Telsiai 95.4MHz 1kW ERP — **D.PRGR:** 24h.

ZNAD WILII

Laisvés pr. 60, LT-2056 Vilnius. ☎+370 (2) 429457.
🖹 +370 (2) 429465. **E-mail:** znad.wilii@mail.tipas.lt
L.P: DG: Konstanty Wincel. PD: Robert Rauluszewicz.
FM: Vilnius 103.8MHz 4kW ERP, 73.34MHz 12kW ERP.
D.PRGR: 24h in **Polish** exc. **Russian** Mon 2000-2130, Tues & Thurs 1430-1500, Fri 2000-2100, Sun 1200-1300. **Rel. Polskie R** (Poland): **Polish:** W 1630-1700, **Lithuanian:** 2130-2200, **E:** Sun 0430-0500 — **ANN:** "Radio Znad Wilii" — **V.** by letter.**French** exc. **Russian** 1400-1500, 1900-2000 and **Polish** 0545-0600, 1700-1800, 2200-2300.
BBC World Service: Vilnius 100.1MHz 1kW
Radio France Internationale: Vilnius 98.3MHz 1kW ERP: 24h

LUXEMBOURG

L.T: UTC + 1h (Su: UTC + 2h) — **Pop:** 4C0.000 — **Radios:** 240.000 — **Pr.L:** Luxembourgish, French, German — **E.C:** 50Hz, 110/220V — **ITU:** LUX.

CLT MULTI MEDIA (Comm.)
✉ 45 blvd. Pierre Frieden, L-2850 Luxembourg. ☎ +352 4214 22175. 🖷 +352 4214 22756. **Web:** http://www.cltmulti.com
L.P: Pres./DG: Gaston Thorn. Asst. DG: Jules Felton. Dir. Radio: Remy Sautter. Head of Radio & TV Prgrs: Alain Berwick.
Luxembourg Sce: RTL Radio Lëtzebuerg: ☎ +352 4214 23. 🖷 +352 4214 22737. **Web:** http://www.rtl.lu — **L.P:** Chief Editor: Roby Rauchs. PD: Fernand Mathes.
German Sce: RTL Radio - die Grössten Oldies: ☎ +352 4214 23500. 🖷 +352 4214 22738. **Web:** http://www.rtlradio.com — **L.P:** MD: Bernt von zur Muehlen. PD: Holger Richter.
French Sce: RTL: 22 rue Bayard, F-75008 Paris. ☎ +33 (1) 4070 4070. 🖷 +33 (1) 4070 4272. **RTL2:** ☎+33 (1) 4070 4000. 🖷 +33 (1) 4070 4350. **Web:** http://www.rtl2.fr — **L.P:** CEO: Jacques Rigaud. VP & PD: Philippe Labro. Prgr. Mgr RTL2: Frédéric Jouve.
LW/MW: Junglinster 234kHz 2000kW, Marnach 1440kHz 1200kW.
FM: Marnach 88.9/97.0/107.0MHz 100kW, 92.5MHz 50kW. Dudelange 93.3MHz 100kW. Also FM relays in France & Germany.
F.PL: 107.7MHz 100kW.
RTL Radio Lëtzebuerg in English/German/Luxembourgish: 24h on 92.5MHz (2000-0500 rel. German Sce.)
RTL: 24h on 234kHz + FM network in France.
RTL2: 24h on FM network in France
RTL Radio - die Grössten Oldies: 24h on 88.9MHz + satellite and via tr's in Germany.
Italian: 2000-2030 on 1440kHz (produced by RAI, Italy).
V. by QSL-card.

OTHER STATIONS
Den Neie Radio, P.O. Box 1522, L-1015 Luxembourg: 102.9/104.2MHz — **Eldoradio,** P. O. Box 1344, L-1013 Luxembourg. **E-mail:** eldoradio@vo.lu **Web:** http://eldoradio.vo.lu/eldoradio/ 24h on 105.0/107.2MHz — **Honnert.7** (non-comm.): 100.7MHz. **Web:** http://www.100komma7.lu/ — **R. Ara,** P. O. Box 266, L-2012 Luxembourg. **Web:** http://www.ara.lu/ 103.3/105.2MHz — **R. Latina,** 2 rue Astrid, L-1143 Luxembourg: 101.2/103.1MHz — **R. LRB,** B.P. 8, L-3201 Bettembourg. **E-mail:** lrb@digicron.com **Web:** http://www.gms.lu/lrb/ 24h on 103.9MHz — **Radio WAKY Power FM 107,** P. O. Box 70, L-5801 Hesperange, Luxembourg: 24h on 107.0MHz in English & Luxembourgish — **Sunshine Radio,** 29 ave. de la Financerie, L-1510 Luxembourg: 24h on 102.2MHz in English & Luxembourgish.

MACEDONIA

L.T: UTC + 1h (Su: UTC + 2h) — **Pop:** 2.200.000 — **Radios:** 350.000 — **Pr.L:** Macedonian — **E.C:** 50Hz, 220V — **ITU:** MKD.

MAKEDONSKA RADIOTELEVIZIJA (Gov.)
✉ Goce Delcev bb, 91000 Skopje. ☎🖷 +389 (1) 112578. 🖷 (Int. Rel. Dept.) +389 (1) 225212 ① 597 51 157 MFT — **L.P:** DG: T. Ilievski.

Mediumwaves	kHz	kW	Mediumwaves	kHz	kW
Strumica	567	10	Bitola	1323	10
Skopje 1	810	1000	Delcevo	1323	10
Gevgelija	936	10	Debar	1485	1
Ohrid 1	1242	5	Kicevo	1485	1
Kriva Palanka	1197	1	Berovo	1485	1
Skopje 2/3	1314	100	Debar	1602	1
Ohrid 2/3	1314	10	Kicevo	1602	1
Gostivar	1323	10	Resen	1602	1

FM (MHz)	MR 1	R. 2000	R. Kultura	kW
Belasica	91.5	97.8		10
Boskija	95.3	98.1		10
Crni Vrh	97.3	94.1		100
Golak			102.7	10
Pelister	92.3	96.1		20
Mali Vlaj	93.3	91.0	95.7	10
		97.7		10
Sveti Erazmo	89.9			10
Turtel	93.3	90.5		50
Vodno	87.8	89.2	98.9	10

+ st's less than 10kW.

MR 1: 24h in Macedonian — **R. 2000:** prgrs. produced by 32 local st's in Macedonian, Albanian and Turkish — **R. Kultura:** 1900-2200 — **R. Biljana:** prgrs. for Macedonians abroad.

Local Stations	kHz	kW	Local Stations	kHz	kW
Stip	639	1	Kratovo	1386	1
Struga	720	1	Probistip	1431	1
Tetovo	738	1	Kavadarci	1485	1
Kumanovo	945	2	Sveti Nikole	1530	1
Skopje	1026	5	Negotino	1548	1
Kocani	1287	1	Prilep	1584	1
Titov Veles	1341	1	Radovis	1584	1

+ 5 st's less than 1kW.
F.PI: Macedonian radio news via Internet.

Other stations: R. Libertas (Comm.): Skopje — **Super R,** Ohrid: regional st. — **Kanal-77,** Stip:independent st.

BBC World Sce: Skopje 104.7MHz 0.5kW.

AMERICAN FORCES NETWORK (US Mil.)
FM: Skopje 98.5/99.5MHz (rel. AFN Bosnia).

MALTA

L.T: UTC + 1h (Su: UTC + 2h) — **Pop:** 377.988 — **Radios:** 95.000 — **Pr.L:** English, Maltese — **E.C:** 50Hz, 240V — **ITU:** MLT.

MALTA BROADCASTING AUTHORITY (Regulatory Authority)
✉ Mile-end Rd, Hamrun. ☎ +356 247908, 221281. 🖷 +356 240855. **Cable:** Broadcasts Malta.
L.P: Chief Exec. & Head of Int. Rel: Antoine J. Ellul. Dir. of Prgrs & Head of Public Affairs: Harry Zammit Cordina. Head of Production: Ms. Josanne Cassar. Tech. Mgr: Brian Tilcock.

PUBLIC BROADCASTING SERVICES LTD.
✉ Box 82, Valletta. ☎ +356 225051. ① MW1443. 🖷 +356 244601 — **L.P:** Head of Radio: J.Inguanez. CE: A.V. Mallia.
MW: Bizbizja 999kHz 5kW — **FM:** 93.7MHz 4kW (stereo).
R. Malta 1: 0500-2200 on 999kHz. **N:** 0600, 0700, 0900, 1100, 1400, 1700, 2130 — **R. Malta 2:** 0500-2200 on 93.7MHz. **N:** 0600, 0900, 1100, 1600. **BBC N:** 0800.

Commercial Stations:
Bay R, St. Julians. **E-mail:** elg_lb@maltanet.omnes.net. **Web:** http://prometheus.megabyte.net/~eden/bayradio/home.html — 89.7MHz,
Super One R, 288, Qasam Industrijali, Marsa. ☎ +356 226634. 🖷 +356 231472. **E-mail:** radionet@keyworld.net **Web:** http://www.super1.com/ — 92.7MHz.
Radju 99, Valletta. ☎ 356 235397 **Web:** http://www.geocities.com/Heartland/Plains/1699/index.html — 99.5MHz.
Live FM (Floriana) 100.2MHz
R. 101, San Gwann — 101.0MHz
Island Sound R, Floriana — 101.8MHz
R. Calypso, Gozo — 102.3MHz
RTK, Blata-I-Badja — 103.0MHz
Radju ta' l-Universita', Valletta. **Web:** http://dream.vol.net.mt/uni-rad/ — 103.7MHz
Smash R, Paola — 104.6MHz.

VOICE OF THE MEDITERRANEAN (Gov.)
See International Broadcasting section.

MOLDOVA (Moldavia)

L.T: UTC + 2h (Su: UTC + 3h) — **Pop:** 4.362.900 — **Radios:** 1.556.000 — **Pr.L:** Moldavian, Russian, Ukrainian, Gagauz — **E.C:** a/c 50Hz, 127/220V (no) — **ITU:** MDA.

MINISTERUL COMUNICATIILOR SI INFORMATICII
(Ministry of Communications and Informatics)
✉ bul. Stefan cel Mare 134, MD-2012 Chisinau. ☎+373 (2) 221001.

⌔ +373 (2) 241553. Ⓢ 163227 MSMLD.
L.P: Minister: Ion Casian.

INSPECTORATUL DE STAT PENTRU FRECUENTE RADIO
(State Inspection for Radio Frecuencies)
⌔ str. Drumul Viilor 28/2, MD-2028 Chisinau. ☏+373 (2) 735392.
L.P: Dir: Tudor Ciclici.
This organization is responsible for frequency allocations.

CONSILIUL COORDONATOR AL AUDIOVIZUALULUI
(Radio & TV Coordinating Council, public body)
⌔ Bd. Stefan cel Mare 73, cam. 340, MD-2001 Chisinau. ☏ +373
(2) 224095 or 222534.
L.P: Chmn: Alexei Ciubasenko. Secr. Stefan Casian.
CCAV is the radio & TV broadcasting regulatory and licensing body.

COMPANIA DE STAT "TELERADIO-MOLDOVA" (Gov.)
⌔ Sos. Hancesti 64, MD-2028 Chisinau.
L.P: Chmn: Tudor Olaru.

RADIO MOLDOVA
⌔ str. Miorita 1, MD-2028 Chisinau.
☏ +373 (2) 721388. ⌔ +373 (2) 723537. Ⓢ 163130 RADIO SU.
L.P: Gen. Dir: Constantin Rotaru.

MW	kHz	kW	Prgr.
Chisinau	873	75	RM1
Chisinau	1449	30	RM2
Cahul	1494	30	RM1
Edinet	1494	30	RM1
Chisinau	1593	5	RM1*

*) also carries RM night prgr (see below). RM1 also rel. via Iasi,
Romania, on 1053kHz (1000kW) MF2100-2200.

FM (MHz)	RM1	RM2	BBC+	RFE/RL++	kW (ERP)
Balti	66.68	68.24$	68.24		17
Cahul	67.37	69.14	69.14		17
Causeni	71.24*	72.80$			17
Cimislia	103.5				10
Chisinau	100.5				8
Edinet	70.31	67.46$		70.31	17
Rezina	72.35				4
Rezina	103.3				10
Straseni	72.02*	66.80$	68.48	67.58	17
Ungheni	68.00	69.53$		69.53	17

$ = stereo. +) rel. BBC prgr's in Romanian, Russian, E, Ukrainian 0330-
0630, 1200-1230, 1600-2100. ++) rel. RFE/RL prgr's in Romanian,
Russian 1600-2200. *) also carries RM Night Prgr.
RM1 (Programul Unu): 0400-2200 in Moldavian, exc. **Russian:**
0700-0715, 0900-0915, 1300-1315, 1600 (Tues, Thurs 1515)-1615, MF
1900 (2nd & 4th Wed 1830, last Mon 1930)-2000, **Ukrainian**
("**Vidrodzhennya**"): Thurs 1830-1900, **Bulgarian:** 1st & 3rd Fri
1830-1900, **Gagauz:** 2nd & 4th Fri 1830-1900. **Gipsy:** Last Tues
1830-1900, **Hebrew/Yiddish:** 1st & 3rd Wed 1830-1900.
RM2 (Luceafarul): 0400-2200 in Moldavian & limited Russian. Rel
RM1: 1000-1100 (SS1030). Prgr "Welcome to Moldova": 0602-0615
in **English**, 0615-0630 in **French**, 0630-0700 in **Spanish**.
RM Night Prgr (Radio Nocturn): 2200-0100, 0200-0400 on
1593kHz, 72.02/71.24MHz in Moldavian, incl. R. Moldova Int. prgrs in
English, French, Spanish
ANN: RM1/RM2/Night Prgr: Moldavian: "Aici Chisinau, Radio
Moldova, programul unu/doi, Luceafarul/nocturn". Russian: "Govorit
Kishineu, Radio Moldovy".

FOREIGN SERVICE: See International Broadcasting section.

INDEPENDENT STATIONS

	MHz	kW*	Station, location
1)	66.41	0.1	R. Poli-Disc, Chisinau
2)	69.44	0.1	R. Micul Samaritean, Chisinau

	MHz	kW*	Station, location
3)	71.00	0.1	Europa Plus Moldova, Chisinau
2)	71.39	0.1	R. Micul Samaritean, Drochia
4)	72.71	0.1	R. Unda Libera, Chisinau
5)	73.28	0.1	R. Molda, Chisinau
	100.1	1	F.PI, Calaras
	100.1	1	F.PI, Cornesti
2)	100.7	1	R. Micul Samaritean, Balti
8)	100.9	0.1	R. Contact, Chisinau
	101.3	1	F.PI, Hancesti
2)	101.5	1	R. Micul Samaritean, Causeni
	101.9	1	F.PI, Taraclia
	102.1	1	F.PI, Comrat
	102.5	0.5	F.PI, Glodeni
	102.6	1	F.PI, Nisporeni
4)	102.7	0.1	R. Unda Libera, Chisinau
6)	103.2	0.1	R. d'Or, Chisinau
2)	103.2	1	R. Micul Samaritean, Taraclia (F.PI.)
	103.5	1	F.PI, Balti
1)	103.7	1	Russkoye R, Chisinau
2)	103.8	0.1	R. Micul Samaritean, Edinet
	104.0	0.6	F.PI, Bender
2)	104.1	0.1	R. Micul Samaritean, Chisinau
	104.6	1	F.PI, Ceadar-Lunga
	105.0	1	F.PI, Dubasari
	105.2	0.1	F.PI, Chisinau
	105.7	1	F.PI, Nisporeni
7)	105.9	0.1	R. Nova, Chisinau
3)	106.4	0.1	Europa Plus Moldova, Chisinau
	106.6	1	F.PI, Comrat
	106.8	1	F.PI, Floresti
	106.8		F.PI, Calaras
	106.9	0.1	F.PI, Chisinau
2)	107.0	4	R. Micul Samaritean, Ungheni
2)	107.7	1	R. Micul Samaritean, Cahul
	107.9	0.1	F.PI, Chisinau

*) Transmitter power = TRP. H. of tr: 24h in Moldavian, Russian.
Addresses and other information
1) sos. Hancesti 59/1, MD-2028 Chisinau — 2) str. Bucuresti 68, MD-
2012 Chisinau — 3) str. Alecu Russo 1, et. 16, MD-2068 Chisinau —
4) Universitatea de Stat, str. Mateevici 60, MD-2014 Chisinau — 5)
str. Alecu Russo 7/1, ap. 1, MD-2068 Chisinau — 6) Casa Lumea
Deschisa, bul Stefan cel Mare 180, et. 14, MD-2004 Chisinau — 7)
str. Bucuresti 68, cam. 724, MD-2012 Chisinau.

FM relays of international broadcasters:
BBC: Causeni 101.5MHz 1kW: 0330-0630, 1200-1230, 1600-2100.
R. France Internationale: Chisinau 102.3MHz 1kW 24h.

PRIDNESTROVYE

GOSUDARSTVENNYY KOMITET PO TELEVIDENIYU, RADIOVESHCHANIYU I PECHATI
(State Committee for TV, Radio & Press)
ADDR: Verkhovnyy Sovet, ul. 25 Oktyabrya 45, 278000 Tiraspol. ☏
+373 (33) 35167.
L.P: Chmn: Boris Akulov.

RADIO PRIDNESTROVYE
ADDR: ul. Rozy Lyuksemburg 10, 278000 Tiraspol (editorial) or
NGDXM-"RP", P.O. Box 3297, MD-2044 Chisinau (foreign mail). ☏
+373 (33) 35570 or 32417. ☏/⌔ +373 (33) 32245.
L.P: Editor-in-Chief: Aleksandr Radchenko. Asst. Editor-in-Chief:
Lyubov Belinskaya.

Station	kHz	kW	H. of tr. Prgr.	
Maiac	1467	150	0400-1800	RP, RR

R. Pridnestrovye (RP): 0500-0600, 1700-1830 in Russian exc.

Moldavian 0545-0600, 1745-1800, Ukrainian 0530-0545, 1730-1745.
R. Rossii (RR): 0400-0500, 0600-1700 relay from Moscow, Russia.
ANN: RP: Russian: "Govorit Tiraspol, stolitsa Pridnestrovskoy Moldavskoy Respubliki. V efire Radio Pridnestrovya". Moldavian: "Aici Tiraspol, Postul de Radio Nistrean". "Hovorit Tiraspol. Vas vitae ukrayinska redaktsyya Radio Pridnestrovya".
V. by QSL-card. Re to Chisinau addr. **F.PI:** Resumption of FS, reactivation of 234/594kHz (Maiac 1000kW).

RELAY SERVICES

MW: Maiac 999kHz 500kW, 1467kHz 150kW.
SW: Maiac (near Grigoriopol, G.C: 47.17N/29.24E): 5x1000kW.
Relays of V. of Russia, BBC, Deutsche Welle, TWR. Cf. these stations for details.

MONACO

L.T: UTC + 1h (Su: UTC + 2h) — **Pop:** 30.521 — **Radios:** 30.000 — **Pr.L:** French — **E.C:** 50Hz, 220V — **ITU:** MCO.

RADIO MONTE CARLO (Comm.)

✉ 16 Blvd. Princesse Charlotte, MC-98080 Monaco Cédex. ☎ +377 151617, 151780. 🖷 +377 151703. ① 469926. ✉ **in Italy:** Via Turati 40, 20121 Milano. ☎ +39 (2) 2900 1636. 🖷 +39 (2) 655 1451.
L.P: Pres: C.C.Solamito. **French Sce:** MD: Jean Noel Tassez. PD: Jean Pierre Foucoult. **Italian Sce:** MD: Alberto Hazan. PD: Francesco Migiliozzi.

Long & Mediumwaves H.of tr.	kHz	kW	Prgr.	
Roumoules (France)	216	1400	RMC French	0400-0005
Monaco	702	40	RMC French	0400-1200
Roumoules (France)	1467	1000	RMC French	0500-1845

FM (all sces. 24h) Location	MHz	kW	Prgr.
Mt. Angel (France)	90.3	50	Montmartre FM
Monaco	93.5		Nostalgie FM
Monaco	93.8		Nostalgie FM
Monaco	95.4	0.05	R. Maria (Italian)
Mt. Angel (France)	98.5	50	RMC French
Monaco	98.8	1	RMC French
Mt. Angel (France)	102.7	50	R. Classique
Monaco	102.9	1	R. Classique
Monaco	103.3		RMC French
Monaco	106.8	1	RMC Italian

RIVIERA RADIO (Comm.)

✉ 16 Blvd. Princesse Charlotte, 98000 Monaco. ☎ +377 93254906. 🖷 +377 93304245. **E-mail:** rivieraradio@monaco.mc
Web: http://www.riviera-radio.com/
L.P: MD: Randall Kehrig.
FM: Monaco 106.3MHz, San Remo/Saint Tropez 106.5MHz.
D.PRGR. in English: MF 0500-2200, Sat 0600-2200, Sun 0700-2100.
Bloomberg Business Report: MF 0730, 1430, 1730.
Rel. BBCWS overnight + Sat 1430-1730, Sun 1100-1130

TRANS WORLD RADIO (Rlg.)

See International Broadcasting section.

THE NETHERLANDS

L.T: UTC + 1h (Su: UTC + 2h) — **Pop:** 15m — **Radio Sets:** 12m — **Pr.L:** Dutch — **E.C:** 50hz, 220V — **ITU:** HOL

N.B: domestic frequencies will be reorganized. Final details not available at deadline.

NEDERLANDSE OMROEP STICHTING (NOS)

✉ Sumatralaan 45, 1217 GP Hilversum, P.O. Box 26600, 1202 JT Hilversum, The Netherlands. ☎ +31 (35) 6779222 (TV/Radio) 🖷 +31 (35) 6772649 ① 43287 (TV/Radio)
L.P: Dir. Radio and TV: Ruurd Bierman; Hd. Comm: F. de Vries
Board of Management: G.J. Wolffensperger, H. van Beers, P.B.A. Dirks

NEDERLANDSE PROGRAMMA STICHTING (NPS)

✉ P.O. Box 29000, 1202 MA Hilversum. ☎ +31 (35) 6779333. 🖷 +31 (35) 6774517 — **L.P:** Dir: W.J.M. van Beusekom

NB: The Dutch prgrs. are provided by the NOS, NPS and seven broadcasting organizations:
WWW (for all public broadcasters): http://www.omroep.nl
-Algemene Omroepvereniging AVRO, 's Gravelandseweg 52, 1217 ET Hilversum, Postbus 2 1200 JA Hilversum. ☎ +31 (35) 6717911 🖷 +31 (35) 6717439
-Vereniging Evangelische Omroep EO, Oude Amersfoortseweg 79a, 1213 AC Hilversum, Postbus 21000, 1202 BB Hilversum. ☎ +31 (35) 6882411 🖷 +31 (35) 6882685
-Katholieke Radio Omroep KRO, Emmastraat 52, 1213 AL Hilversum, Postbus 23000, 1202 EA Hilversum. ☎ +31 (35) 6713911 🖷 +31 (35) 6237345
-Nederlandse Christelijke Radio Vereniging NCRV, Bergweg 30, 1217 GN Hilversum, Postbus 25000, 1202 HB Hilversum. ☎ +31 (35) 6719911 🖷 +31 (35) 6719285
-TROS, Lage Naarderweg 45-47, 1217 GN Hilversum, Postbus 28450, 1202 LL Hilversum. ☎ +31 (35) 6715715 🖷 +31 (35) 6715236
-Omroepvereniging VARA, Heuvellaan 50, 1217 JN Hilversum, Postbus 175, 1200 AD Hilversum. ☎ +31 (35) 6711911 🖷 +31 (35) 6711333
-Omroepvereniging VPRO, 's Gravelandseweg 63-73, 1217 EJ Hilversum, Postbus 11, 1200 JC Hilversum. ☎ +31 (35) 6712911 🖷 +31 (35) 6712254

Mediumwaves: Loc.	kHz	kW	Prgr.
Flevoland	747	400	Radio 1
Flevoland	1008	400	Radio 5
Hulsberg	891	20	Radio 5
Hulsberg	1251	10	Radio 1

FM: Location	1	2	3	4	kW
Arnhem	104.1	—	—	—	50
Goes	104.4	87.9	95.0	99.8	50/15
Haarlem	94.3	—	—	—	10
Hulsberg	105.3	92.1	103.9	98.7	10
Loon op Zand	98.2	—	—	—	40
Lopik		92.6	96.8	98.9	100
Markelo	104.6	91.4	96.2	98.4	100
Roermond	104.8	88.2	90.9	94.5	100
Rotterdam	90.2	—	—	—	10
Smilde	101.1	88.0	91.8	94.8	100
Wieringermeer	101.6	87.7	89.8	92.2	100

D.PRGR: Radio 1: 0600-1800 (rel. Radio 2 1800-0600) N: on the h.; Radio 2: 24h. N: on th h.; Radio 3: 0600(Sun 0700)-2300 N: on the h.; Radio 4: 0600(Sun 0700)-2300 N: on the h.; Radio 5: 0800-2130 (Sat 2200, Sun 2140, Mon 2100) N: various languages.
ANN: "Dit is de VARA..", "Dit is de VPRO..". etc. **V:** by QSL-card.

STICHTING ETHER RECLAME STER

✉ Laapersveld 70, 1213 VB Hilversum, Postbus 344, 1200 AH Hilversum. ☎ +31 (35) 6725500 🖷 +31 (35) 6218940
NB: The STER is the only organization that provides commercials to the public broadcasting organisations, who are not allowed to carry any commercials themselves.

Regional Stations

Omroep Brabant, Heggeranklaan 1, Postbus 108, 5600 AC Eindhoven. ☎ +31 (40) 2116262 — **D.Prgr:** 53 hrs per week.
Omroep Flevoland, Meentweg 1, Postbus 567, 8200 AN Lelystad. ☎ +31 (320) 242221 — **D.Prgr:** 76 hrs per week.
Omrop Fryslân, Groeneweg 1, Postbus 642, 8901 BK Leeuwarden. ☎ +31 (58) 299 7799 — **D.Prgr:** 80 hrs per week.
Omroep Gelderland, Rosendaalseweg 704, Postbus 747, 6800 AS Arnhem. ☎ +31 (26) 371 3713 — **D.Prgr:** 57 hrs per week.
Omroep Limburg, Bankastraat 3, Postbus 94, 6200 AB Maastricht. ☎ +31 (43) 346 7777 — **D.Prgr:** 71 hrs per week.
Omroep Zeeland, Kanaalstraat 64, Postbus 1090, 4388 ZH Oost-Souburg. ☎ +31 (118)466355 — **D.Prgr:** 19 hrs, 28 min per hour.
Radio Drenthe, Beilerstraat 24, Postbus 999, 9400 AZ Assen. ☎ +31 (592) 338080 — **D.Prgr:** 45 hrs, 17 min per week.
Radio Noord, Martinikerkhof 23, Postbus 30101, 9700 RP Groningen. ☎ +31 (50) 318 3456 — **D.Prgr:** 57 hrs, 54 min per week.
Radio Noord-Holland, Sloterkade 133, 1058 HM Amsterdam — ☎+31(20)5121212 — **D.Prgr:** 68 hrs per week.
Radio Oost, Hazenweg 25, 7556 BM, Postbus 1000, 7550 BA Hengelo. ☎ +31 (74) 245 6456 — **D.Prgr:** 90hrs per week.

Radio Rijnmond, Delftsestraat 21, Postbus 1515, 3000 BM Rotterdam. ☎ +31 (10) 4364436 — **D.Prgr:** 80 hrs per week.
Radio Utrecht, Laan van Beek en Rooyen 45, Postbus 666, 3700 AR Zeist. ☎ +31 (030) 6935566 — **D.Prgr:** 59 hrs, 58 min per week.
Radio West, Van Vredenburghweg 71, Postbus 1220, 2280 CE Rijswijk. ☎ +31 (70) 3078888 — **D.Prgr:** 69 hrs, 23 min per week.

STATIONS:

Location	MHz	kW	Pol.	St.
Amsterdam	95.2	1	V	Radio Noord Holland
Arnhem	102.4	0.1	V	Omroep Gelderland
Den Haag	88.4	5	V	Radio West
Goes	101.9	50	V	Omroep Zeeland
Haarlem	97.6	1	V	Radio Noord Holland
Hilversum	93.1	0.1	V	Radio Noord Holland
Hoogezand	97.5	15	V	Radio Noord
Hulsberg	95.3	10	H	Omroep Limburg
Irnsum	88.6	15	V	Omrop Fryslân
Lelystad	102.1	5	V	Omroep Flevoland
Loon op Zand	91.9	2	V	Omroep Brabant
Lopik	100.1			Radio Utrecht
Losser	89.4	10	V	Radio Oost
Markelo	95.6	5	H	Radio Oost
Megen	89.1	5	V	Omroep Gelderland
Megen 2	95.8	5	V	Omroep Brabant
Mierlo	87.6	1	V	Omroep Brabant
Philippine	97.8	10	V	Omroep Zeeland
Roermond	100.3	100	H	Omroep Limburg
Roosendaal	95.4	1	V	Omroep Brabant
Rotterdam	93.4	10	V	Radio Rijnmond
Ruurlo	90.4	10	V	Omroep Gelderland
Smilde	90.8	4	H	Radio Drenthe
Ugchelen	103.5	20	V	Omroep Gelderland
Wieringermeer	93.9	10	H	Radio Noord Holland
Zwollerkerspel	99.4	25	V	Radio Oost

+ 264 local stations.

Commercial Stations

RADIO 10 GOLD
▭ Vijzelgracht 55, Amsterdam. ☎ +31 (20) 4203 203
L.P: GM: A. Ch. Feiner; PD: T.B. Muller.
STATION: 103.0 MHz (Lelystad)

HITRADIO VERONICA
▭ Laapersveld 75, 1213 VB Hilversum, Postbus 1234, 1200 BE Hilversum. ☎ +31 (35) 6716716 ▤ +31 (35) 6249771
L.P: GM: U. Glorie.
STATIONS: 88.1 MHz (Hilversum), 93.0 MHz (Dishoek), 93.4 MHz (Maastricht), 103.8 MHz (Rotterdam).

JAZZ RADIO
▭Vondelstraat 13, 1054 GC Amsterdam, P.O. Box 74045, 1070 BA Amsterdam.
☎ +31 (20) 5158 10. ▤ +31 (20) 515811
L.P: GM: S. van Mierlo
STATIONS: 88.6 MHz (Eindhoven), 89.5 MHz (Utrecht), 90.1 MHz (Den Bosch), 91.1 MHz (Alkmaar), 91.5 MHz (Oostburg).

RADIO NOORDZEE NATIONAAL
▭ Flevolaan 41, 1411 KC Naarden. ☎ +31 (35) 6958440/958599. ▤ +31 (35) 6946173
L.P: M: J. Flink.
STATIONS: 102.7 MHz 50kW (Rotterdam), 88.9 MHz 2kW (Amsterdam), 99.1 MHz (Hoogezand), 102.2 MHz (Gennep), 102.3 MHz (Ruurlo), 102.5 (Wieringermeer).

CLASSIC FM
▭ Ambachtsmark 3, P.O. Box 50073, 1305 AB Almere. ☎ +31 (36) 5471666. ▤ +31 (36) 5349797 — **L.P:** GM: Paul Beerkens
STATIONS: 90.7 MHz (Gouda), 93.7 MHz (Losser).

SKY RADIO
▭ Naarderpoort 2, 1411 MA Naarden. ☎ +31 (35) 6991007. ▤ +31 (35) 6991008
L.P: GM: Ton Lathouwers, MD: Vranz van Maaren
STATIONS: 100.7 MHz 100kW (Lopik)

RADIO 538
▭ Graaf Wichmanlaan 46, 1405 HB Bussum. ☎ 31 (35) 6948538. ▤ +31 (35) 6942696

LP: GM: Erik de Zwart, PD: Bart van Leeuwen
STATIONS: 89.0 MHz (Groningen), 93.2 MHz (Irnsum), 93.6 MHz (Zwollekerspe), 101.2 (Diemen).

RADIO NEDERLAND WERELDOMROEP (RNW)
N.B: For details and schedule see International Broadcasting Section

OTHER STATIONS

AMERICAN FORCES NETWORK
STATIONS: Brunssum 89.2MHz 1kW, Volkel 93.6MHz 0.05kW.
D.PRGR: 24h.

BRITISH FORCES BROADCASTING SERVICE
STATIONS: Maastricht 87.7MHz; Hoensbroek 90.2 MHz 0.05 kW
D.PRGR: Relays BFBS prgrs Germany.

CANADIAN FORCES NETWORK-BRUNSSUM
STATION: Brunssum 91.5 MHz 0.1 kW (stereo)
D.PRGR: 24h rel. R. Canada Int. + local prgrs.

RADIO KERK VAN BLOEMENDAAL
▭ Vijverweg 14, Postbus 26, 2060 AA Bloemendaal. ☎ 31 (23) 5250471 — **L.P:** Secr: P.D. Amels; TD: N. Andrea
STATION: 1116kHz 0.5kW
D.PRGR: Sun & Christian Holidays 0800-1600
V: by QSL-card.

NORWAY

L.T: UTC + 1h (Su: UTC + 2h) — **Pop:** 4.337.000 — **Radios:** 3.342.000 — **Pr.L:** Norwegian — **EC:** 50Hz, 230V — **ITU:** NOR.

POST OG TELETILSYNET
Norwegian Post and Telecommunications Authority
▭ P. O. Box 447-Sentrum, N-0104 Oslo. ☎ +47 22 824600. ▤ +47 22 824640. ① 79 544 NTRAN.
L.P: Sen. Exec. Officer, Freq. Admin. Dept. (Planning): Olav Mo Grimdalen. (**E-mail:** olavmo@online.no).

NORSK RIKSKRINGKASTING
(non-comm. enterprise operated by an independent public foundation)
▭ N-0340 Oslo. ☎ +47 2304 7000. ☎**Int. Rel:** +47 2304 8833). ▤ +47 2245 7440. ①76820. **Cable:** STUDIO Oslo.
E-mail: info@nrk.no
P1/P3: ▭ N-7005 Trondheim. ☎ +47 7388 1400.
Web: http://www.nrk.no
L.P: DG (Radio & TV): Einar Førde. Dep. DG (Radio & TV): Olav Nilssen. Tech. Dir: Geir Jan Sundal. Dir. Radio: Tor Fuglevik. Dir. TV: Hans-Tore Bjerkaas. Dir. Finance: Kjersti Nordtveit. Dir. Admin. Einar Li. Dir.Regional Activities & News: Tom Berntsen. Dir. Resources: Arild Hellgren. Head Int. Relations: Kjell Lokvam

LW/MW	kHz	kW		LW/MW	kHz	kW
Vadsø	153/216	100 F.PI		Vadsø	702	20
Vigra	630	100		Kvitsøy	1314	1200
Bodø	675	10				

FM	P1	P2	P3	P4+	kW
Alta	89.7	88.5/94.6ʰ	91.3	101.0	3.5
Bagn	91.7	95.3	88.0	102.1*	35/7*
Bergen	89.1	94.8	99.0	102.5*	46
Bjerkreim	94.2	98.7	91.8	101.0*	60/6*
Bokn	93.5	97.3	91.1	102.8	120
Bremanger	93.6	98.1	91.3	103.4*	46/9.2*
Førde	92.8	88.7	97.1	102.0*	12/2.4*
Gamlemsveten	91.9	96.3	90.0	102.8	50
Gausta	89.5	96.4	99.7	103.1*	55/5.5*
	101.0				6.75
Greipstad	88.8	92.6	97.0	100.1	57.5
Grong	91.9	96.6	88.9	102.0*	95/9.5*
Gulen	88.0	94.5	97.6	101.4*	39/7.8*
Hadsel	92.4	99.3ʰ	94.5	101.4*	30/6*
Halden	94.1	89.1	101.5	106.1	72.5
Hammerfest	96.6	91.6/87.7ʰ	93.6	101.9*	24/4.2*
Harstad		88.4			
Hasvik	90.1	99.0ʰ	94.9		2.2
Hemnes	88.5	99.8	96.1	104.2*	36/7.2*

FM	P1	P2	P3	P4+	kW
Hovdefjell	87.8	93.7	96.0	103.6*	25/5*
Iskuras	88.7	96.1Þ	92.0		2.2
Jetta	95.9	99.5	91.1	101.6*	85/8.5*
Kappfjell	95.5	99.4	93.4		1.25
Karasjok	87.9	94.7Þ	98.4*	104.5	1.5/0.06*
Kautokeino	90.3	93.8Þ	99.2		35
Kistefjell	91.8	95.7Þ	99.8	103.1*	44/4.4*
Kongsberg	91.3	95.5	97.8	102.5	60
Kongsvinger	89.8	93.9	96.1	107.2*	33/6.6*
	96.1				11
Kopparen	88.3	94.5	96.0	102.4*	40/8*
Lyngdal	97.6	88.3	95.0	102.0*	50/10*
Lyngen	93.3	97.5Þ			4.2
Lønahorgi	93.3	88.3	96.7	100.6	48/4.8*
Melhus	92.4	97.2	99.1	101.1	60
Mosvik	90.9	98.4	93.4	103.8	33
Narvik	88.8	96.7/98.9Þ	91.1	101.1*	90/9*
Nordfjordeid	89.4	99.3	92.3	101.2*	12/2.4*
Nordkapp	89.2	95.4Þ	98.2	102.5*	15/3*
Oslo	88.7	100.0	93.5	103.9	90
Reinsfjell	89.1	95.1	90.7	100.2*	24/4.8*
Salten	93.3	95.5	89.8	100.4*	48/9.8*
Skien	88.2	92.3	100.4	105.2	80
	90.3				7.25
Sogndal	91.5	95.1	98.7	103.9*	24/2.4*
Steigen	90.3	97.8Þ	93.9	102.1*	70/7*
Stord	96.0	99.6	92.6	101.8*	60/6*
Store Jaekkir	99.9	97.3Þ	90.9	103.3	1.15
Tana	92.5	97.0c	91.1		24
Trolltind	88.2	94.0Þ	90.5	101.7*	50/5*
Tromsø	92.2	99.1	102.6	104.0	90
Tron	98.3	88.6	94.3	102.5*	24/4.8*
Varanger	88.1	91.8Þ	102.9	105.8*	30/6*
Vega	89.3	95.2	98.2	102.8*	55/5.5*
Vadsø		96.0			
Varanger		99.0			

+ more than 1600 low power tr's — +) Private comm. st. (see below)
— Þ) carries Lappish sce. All tr's are stereo.
P1: 24h on FM.**N:** MF on the h. also 0530, 0630, 1130, 1530, 1630.
Sat on the h. also 0630, 1130, 1530. Sun on the h. also 1130, 1530.
Regional Prgrs: MF 0504-0530,0536-0559, 0604-0629, 0706-0800, 0903-0906, 1003-1100, 1103-1130, 1303-1306, 1503-1530, 1533-1600, 1603-1630. Sat 0705-0708, 0903-0906, 1003-1100, 1103-1130.
P2: 24h (cultural prgr.) on FM. **N:** MF on the hour 0500-2300 except 1900 and 2000. Also 0530, 0630, 0730, 1130, 1530. Sat: on the h. 0500-2300 except 0900, 1600, 2000, 2100. Also 0630, 1130, 1530. Sun: on the h. 0500-2300 except 0900. **Night prgr:** Notturno rly (EBU classical channel) 2305-0500 .**P3:** 24h (youth channel). Joint sce. with 1st Prgr. 2305-0500.
P3: 24h youth channel on FM. Rly P1 2300-0500. **N:** MF on the h. also 0530, 0630, 0730, 0830. Sat :on the h. except 1700, 1900, 2100, 2200. Sun: on the h. except 2200.
NRK Alltid Klassik: Oslo 91.9MHz 0.9kW, Bergen 98.2 (running DAB tests): 24h. classical music, fully digit:sed station.
NRK Alltid Nyheter: 24h all-news sce. on Oslo 93.0MHz, Stavanger 93.0MHz, Kristiansand 94.0MHz, Bergen 93.8MHz, Trondheim 94.9MHz, Tromsø 89.8MHz, Bodø 96.8MHz. Also on satellite via Intelsat 707 and TV NRK-1 outside normal hours.
Europakanalen (Special programmes on MW): misxture of P1, P2, P3 and R.Norway International.
ANN: 1st Prgr: "P1". 2nd Prgr: "P2". 3rd Prgr: "NRK Petre". Lappish: "Datlae Sámeradio, Kárássjagás".

REGIONAL SERVICES

D.PRGR: MF 0504-0530, 0536-0559, 0604-0629, 0706-0800, 0903-0906, 1003-1100, 1103-1130, 1303-1306, 1503-1330, 1533-1600,1603-1630. Sat 0705-0708, 0903-0906, 1003-1100, 1103-1130..
NRK Buskerud, Postboks 7030, 3007Nedre Eiker: 91.3/99.7MHz.
E-mail: buskerud@nrk.no
NRK Finnmark, Postboks 613, 9801 Vadsø: 702kHz + 87.9/88.1/ 89.2/89.7/90.3/92.5/96.6MHz. **E-mail:** nrk.finnmark@nrk.no
NRK Hedmark, Postboks 216, 2401 Elverum: 87.6/89.8/98.3MHz. **E-mail:** hedmark@nrk.no
NRK Hordaland, 6002, 5020 Bergen: 89.1/93.3/ 96.0MHz. **E-mail:** nrk.hordaland@nrk.no
NRK Møre og Romsdal, 6025 Ålesund: 630kHz + 89.1/91.9MHz. **E-mail:** more.og.romsdal@nrk.no
NRK Nordland, Postboks 303, 8001 Bodø: 675kHz + 88.5/88.8/ 89.3/90.3/92.4/93.3MHz. **E-mail:** nrk.nordland@nrk.no

NRK Nord-Trøndelag, Postboks 2055, 7701 Steinkjer: 90.9/91.9MHz. **E-mail:** nord.trondelag@nrk.no
NRK Oppland, Postboks 273, 2601 Lillehammer: 90.4/95.9/97.1MHz. **E-mail:** oppland@nrk.no
NRK Rogaland, Postboks 614 Madla, 4040 Hafrsfjord: 93.5/94.2MHz.
E-mail: rogaland@nrk.no
NRK Sogn og Fjordane, Postboks 100, 6801 Førde: 88.0/89.4/91.5/ 92.8/93.6MHz. **E-mail:** sogn.og.fjordane@nrk.no
NRK Sørlandet, Tangen 29, N-4610 Kristiansand. + Postboks 2000, Posebyen, 4602 Kristiansand: 87.8/88.8/97.6MHz.
E-mail: sorlandet@nrk.no
NRK Sør-Trøndelag, Tyholt, 7005 Trondheim: 88.3/92.4MHz.
E-mail: sor.trondelag@nrk.no
NRK Telemark, Postboks 284, 3901 Porsgrunn: 88.2/89.5MHz.
E-mail: telemark@nrk.no
NRK Troms, Krognesveien 29, 9001 Tromsø: 88.2/ 91.8MHz.
E-mail: troms@nrk.no
NRK Vestfold, Postboks 700, 3101 Tønsberg: 94.1MHz.
E-mail: nrk.vestfold@nrk.no
NRK Østfold, Postboks 1138, 1631 Gamle Fredrikstad: 94.8MHz.
E-mail: ostfold@nrk.no
NRK Østlandssendingen, Postboks 4555 Torshov, 0404 Oslo: 88.7/91.9MHz. **E-mail:** ostlandssendingen@nrk.no

SAMERADIO (special programmes in Lappish)
✉ Postboks 183, 9730 Karasjok. ☎ +47 (784) 69200 🖷 +47 (784) 69223. **LP:** Nils Johan Haetta.
Stations: see main FM list — **D.Prgr:** P1: Sun 2130-2200 (in Norwegian from Sami community). P2: MF: 1103-1130, also in northern region plus Oslo FM 90.2MHz MF 0630-0800, 1300-1630, (Fri. 1230-1700), Sat 1700-1800
V. Reception reports on long- and MW-trs must be on reg. transmissions to be verified.

RADIO NORWAY INTERNATIONAL
see International Broadcasting section.

P4 – RADIO HELE NORGE (Comm.)
✉ Postboks 414, 2601 Lillehammer. ☎ +47 61 26 26 60. 🖷 +47 6126 29 20. **Oslo office:** Karl Johansgt. 27, 0159 Oslo. ☎ +47 22 42 44 44. 🖷 +47 22 42 0510. **Tromsø office:** Strandgt. 5, 9008 Tromsø. ☎ +47 77 61 12 80. **E-mail:** p4@p4.no **Web:** http://www.p4.no
FM: see P4 network (above).
D.PRGR: 24h (music prgrs aimed at 25-50 year age group). **N:** on the h. 0500-2300, also 0530, 0630, 0730, 1530, 1630.

RADIO 1 NORGE (Comm.),
✉ Gjerdrumsvei 12, 0486 Oslo. ☎ +47 220 23300. 🖷 +47 223 33102. Main office for network with stns. In Oslo 102.0MHz, Bergen 105.2, 105.3, 105.8, 106.4, 106.7 & 107.1MHz, Trondheim 100.5, 104.2, & 106.6MHz and Stavanger 102.2MHz.
More than 400 low power FM commercial st's are in operation.

LOCAL FM STATIONS

AFRTS (U.S. Mil.)
FM: Lifljel 101.5MHz — **D.PRGR:** 24h.

SVALBARD (SPITSBERGEN) (Norwegian Territory)
LT: UTC +1h — **Pr.L:** Norwegian — **E.C:** a/c 50, 230V.

NRK SVALBARD
+ 9170 Longyearbyen.
MW: Longyearbyen (G.C: 25.24E/78.14N): 1485kHz 1kW.
FM: Svea 89.05MHz 15W, Longyearbyen 92.4MHz 15W.
D.PRGR: Rel. NRK Prgr. 1. During regional prgrs, rel. NRK Troms in morning & NRK Østlandssendinga at 0902 & 1404. Also rel. Wrp. at 1100-1115 (as 1314kHz) & R. Norway Int. at 1600-1700.
V. by letter.

POLAND

LT: UTC+1h (Su: UTC+2h) — **Pop:** 30m — **Radios:** 16.300.000 — **Pr.L:** Polish — **E.C:** 50 Hz, 220V — **ITU:** POL.

POLSKIE RADIO S.A.
✉ Al.Niepodleglosci 77/85, 00-977 Warszawa. ☎ +48 (2) 6459212. 🖷 +48 (22) 444119. ① 614825. **Web:** http://www.radio.com.pl/

L.P: Pres.of the Board: Krzysztof Michalski. Head of Int.Rel: Hanna Dabrowska. Dir. Ch1: Jacek Fuglewicz. Dir. Ch2: Edward Pallasz. Dir. Ch3: Pawel Zegadlowicz. Dir. Ch4 (R. Bis): Krystyna Kępska-Michalska.

LW/MW	kHz	kW	Prgr.	MW	kHz	kW	Prgr.
Raszyn	225	600	PR1	Koszêcin	1080	375	PR4
Zorawina	1206	200	PR4	Tuszyn	1305	60	PR4
Przebedowo	738	300	PR4	Koszalin	1206	10	PR4
Warszewo	1260	10	PR4	Sowlany	1305	60	PR4
Wola Raztow.	819	300	PR4	Bozy Dar	1260	60	PR4
Boguchwala	1305	60	PR4	Kraków	1368	20	PR4

Networks: 1 = Polskie Radio 1, 4 = Radio Bis.

FM (MHz)	PR1	PR2	PR3	PR4	Reg	kW (TRP)
Baranówka (Rzeszów)	–	–	92.0	88.0	90.5m	3/4/0.2
Bialystok	–	–	96.0	92.3	100.2a	1/1/3
Bogatynia	102.8	–	–	–	–	10
Chelmce (Kalisz)	–	–	96.4	–	95.6j	3
Chelmska Góra (Koszalin)	–	–	97.4	105.3	103.1f	10/1/20
Choragwica (Kraków)	–	67.67	66.89	–	68.75g	10
	–	–	99.4	–	101.6g	10
Chrzelice (Koftanów)	–	70.31	66.77	–	72.89l	10
	–	–	90.3	88.3	103.2l	2/10/1
Chwaszczyno (Gdansk)	–	70.31	66.29	–	67.85c	10
	–	–	99.9	95.7	103.7c	1/1/20
Czarna Góra (Klodzko)	–	67.64	69.74	–	72.44r	3/1/1
			89.2			
	–	–	–	97.6	96.0r	1/3
Częstochowa-Bleszno	–	67.79	66.23	–	68.96d	6/6/5
Gizycko	–	–	–	–	99.6a	1
Gologóra (Koszalin)	–	69.92	66.95	–	67.73f	10
Góra Chelmiec (Walbrzych)	–	–	–	–	95.5r	10
Góra Parkowa (Kudowa Zdrój)	–	68.51	65.90	–	73.83r	0.05
			99.3			
Gorzów Wielkopolski	–	–	99.1	–	–	1
Gubalowka (Zakopane)	–	70.31	71.45	–	73.85g	0.05
Jelenia Góra	–	92.5	94.0	–	96.7r	10/20/10
Jemiolów (Sulecin)	–	72.50	71.72	–	69.14s	10
	–	–	94.1	89.9	103.0s	4/5/5
Kolowo (Szczecin)	–	68.78	66.74	–	67.52o	10
	96.3	–	100.3	102.3	92.0o	3x10/20
Krynica	–	–	–	–	102.1g	3
Krynice (Bialystok)	–	70.01	72.02	–	72.80a	10
Krzemieniucha (Suwalki)	–	68.60	71.12	–	72.68a	5/6/10
	–	–	96.6	–	–	10
Lódz	–	–	–	–	73.43i	0.5
	–	–	103.8	91.4	99.2i	5/3/5
Losice (Biala Podlaska)	–	70.22	66.41	–	68.03p	6
	–	–	90.5	88.3	103.4h	5/6/6
Lublin-Bozy Dar	–	69.92	71.81	–	72.59h	10
Myslowice-Kosztowy	–	67.55	65.99	–	68.33d	10
	–	–	99.7	97.9	102.2d	10
Nowa Karczma (Luban)	–	69.56	–	–	67.46r	6
	–	–	91.5	99.0	103.6r	10/10/30
Olsztyn-Pieczewo	–	69.56	67.25	–	70.79k	10
	–	–	99.1	97.3	103.2k	10/0.5/10
Opole	–	–	–	–	103.2l	1
			88.3	72.89l		
Ostrolêka	–	–	98.5	96.3	100.8p	0.5/2/0.3
Plaski (Lublin)	–	–	104.2	90.8	102.2h	20/10/20
Plock-Rachocin	–	68.72	70.97	71.45	72.53p	6
	–	–	96.1	92.2	101.9p	10/1/5
Poznan-Platkowo	–	–	–	–	102.7j	0.5
Prehyba (Nowy Sacz)	–	–	–	–	88.0g	0.5
Raszyn (Warszawa)	65.75	69.20	71.45	–	67.94p	10
Rusinowo (Walcz)	–	69.38	72.02	–	72.80f	10/6/6
	–	–	90.9	101.9	103.6j	10/6/3
Skórowo (Lêbork)	–	–	–	–	91.1f	5
Skrzyczne (Wisla)	–	–	100.8	91.5	103.0d	1/3/2
Sleza (Sobotka)	–	70.67	72.11	–	71.33r	10

FM (MHz)	PR1	PR2	PR3	PR4	Reg	kW (TRP)
	–	–	100.2	98.8	102.3r	10
Slupsk	–	–	–	–	95.3f+n	3
Sniezne Kotly (Jelenia Góra)	–	71.72	68.78	–	73.70r	0.05
Srem	–	69.74	66.56	–	67.40j	10
	–	–	96.4	92.3	100.9j	5
Sucha Góra (Rzeszów)	–	68.24	65.90	–	67.46m	10
Swiety Krzyz (Nowa Slupia)	–	70.49	72.71	–	71.15e	10
	–	–	96.2	92.3	101.4e	10/0.4/5.5
Swinoujscie	107.7	–	–	–	106.3o	2
Tarnawatka (Zamosc)	–	69.38	66.68	–	67.61h	6
	–	–	91.3	87.6	–	5/6
Tatarska Góra (Przemysl)	–	68.60	71.69	–	72.41m	0.05
	–	–	99.6	–	–	1
Toporzyk (Lobez)	–	–	–	–	88.1f	5
Trzebnica	87.7	–	–	–	89.8r	20
Trzeciewiec (Bydgoszcz)	–	68.96	71.84	–	72.62b	5/5/6
	–	–	102.1	97.6	100.1b	4/1/10)
Warszawa	92.0	–	98.8	102.4	101.0p	0.5/3/2/1
Zolwieniec (Konin)	–	–	103.3	87.7	91.9j	1/1/2
Zygry (Zdunska Wola)	–	68.51	72.23	–	73.01i	10

PR1: 24h (news & music). **N:** on the h. N. in English & German: 1000 — **PR2:** 24h (classical music) — **PR3:** 24h (youth channel) — **PR4** (R. Bis): 0500-2300 (SS 0120) (educational & musical prgrs).
ANN: "Tu Polskie Radio Warszawa w programie pierwszym (1)/ drugim (2)/ trzecim (3).
V. by QSL-card. **F.PI:** 5th prgr.

REGIONAL SERVICES (24h): PR = Polskie Radio
PR Bialystok (a): ul.Swierkowa 1, 15-328 Bialystok – **PR Pomorza i Kujaw** [PR PiK] (b): ul.Gdanska 48/50, 85-006 Bydgoszcz – **PR Gdansk** (c): ul.Uphagena 1, 80-237 Gdansk – **PR Katowice** (d): ul.J.Ligonia 29, 40-953 Katowice – **PR Kielce** (e): ul.Radiowa 4, 25-317 Kielce – **PR Koszalin** (f): ul.Pilsudskiego 43/49, 75-502 Koszalin. Local sce ("PR Miejskie Koszalin"): on 103.1MHz – **PR Kraków** (g): ul.Szlak 71, 31-153 Kraków. **Web:** http://www.radio-krakow.pl/ – **PR Lublin** (h): ul.Obronców Pokoju 2, 20-030 Lublin – **PR Lódz** (i): ul.Narutowicza 130, 90-146 Lódz – **PR Merkury** (j): ul.Berwinskiego 5, 60-765 Poznan – **PR Olsztyn** (k): ul.Radiowa 24, 10-206 Olsztyn – **PR Pro FM** (l): ul.Strzelców Bytomskich 8 (or ul.Piastowska 20), 45-084 Opole. **E-mail:** pro_fm@radio.opole.pl **Web:** http://www.radio.opole.pl/ – **PR Rzeszów** (m): ul.Zamkowa 3, 35-032 Rzeszów – **PR Slupsk** (n): ul.Sienkiewicza 20, 76-200 Slupsk – **PR Szczecin** (o): ul. Niedzialkowskiego 24, 71-410 Szczecin – **PR Dla Ciebe** (p): ul.Mysliwiecka 3/5/7, 00-977 Warszawa – **PR Wroclaw** (r): ul.Karkonoska 10, 53-015 Wroclaw – **PR Zachod** (s): ul.Kukulcza 1, 65-472 Zielona Góra.

FOREIGN SERVICE: see International Broadcasting section.

NATIONAL COMMERCIAL NETWORKS

RADIO ZET
✉ ul.Piekna 66-a, 00-672 Warszawa. ☎ +48 (22) 6227676. 🖷 +48 (22) 6273300. **Web:** http://www.radiozet.com.pl/
L.P: Dir: Dorota Zawadowska-Wojciechowska.

FM (MHz)	OIRT	CCIR	kW (TRP)
Baranówka (Rzeszów)	72.23	89.9	0.1/1
Bialystok	65.99	107.3	0.4/1
Bielsko-Biala	71.03	–	0.1
Bydgoszcz	71.21	92.1	3.6/3.6
Chelmska Góra (Koszalin)	–	107.4	3.5
Chwaszczyno (Gdansk)	71.69	105.0	0.25/2
Czarna Góra (Klodzko)	–	103.8	1
Częstochowa	71.96	103.4	0.5/1
Cziuchow	–	107.0	2
Elblag	71.36	104.2	0.2/0.5
Gdynia	71.69	–	0.5
Gizycko	–	104.0	2
Góra Chelmiec (Walbrzych)	–	97.2	2
Gorzów Wielkopolski	–	99.6	0.5
Gubalowka (Zakopane)	–	106.3	2
Jawor (Baligorod)	–	103.1	1
Katowice	72.44	102.8	1.8/3.9
Kielce	–	90.4	0.2
Kolobrzeg	–	104.2	
Kraków	71.96	104.1	0.3/1

FM (MHz)	OIRT	CCIR	kW (TRP)
Krzemieniucha (Suwalki)	–	101.4	2
Nowa Karczma (Luban)	–	89.4	10
Lublin	71.03	107.0	0.3/1.9
Lódz	71.63	90.1	0.4/3.6
Modlimowo		92.9	
Nysa		98.6	
Olsztyn-Pieczewo	73.04	107.7	3.6
Opole	68.21	92.2	0.3/1.2
Plock-Rachocin	–	97.3	1
Poznan-Platkowo	72.32	103.4	0.3
Prehyba (Nowy Sacz)	–	97.8	2
Radom	–	88.7	3
Rosan	–	102.8	0.3
Rusinowo (Walcz)	–	97.9	3
Siedlce	–	91.3	0.3
Skórowo (Lêbork)	–	96.6	2
Skrzyczne (Wisla)	–	95.7	1
Slupsk	71.45	88.5	0.4/1
Sniezne Kotly (Jelenia Góra)	–	104.2	5
Srem	–	97.0	2
Szczecin/Kolowo		95.2	
Szczecin-Zelechowo	–	91.2	0.1
Sucha Góra (Rzeszów)	–	104.9	4
Swinoujscie	–	91.8	3
Tatarska Góra (Przemysl)	–	107.9	2
Torun	73.10	–	3.6
Trzeciewiec (Bydgoszcz)	–	95.6	5
Warszawa	67.00	107.5	0.4/6
Wielun	–	88.6	0.3
Wilkanowo		107.0	
Wroclaw	68.84	93.6	0.3/1
Zamosc	–	100.7	3
Zelechowo (Chojna)	–	95.2	1
Zielona Góra-Wilkanow	–	88.3	2
Zolwieniec (Konin)	–	107.1	1

RADIO MUZYKA FAKTY (RMF FM)

Al.Waszyngtona, Kopiec Kosciuszki, 30-204 Kraków. ☎ +48 (12) 219696. 🖷 +48 (12) 217895. **Web:** http://www.rmf.pl/
L.P: Dir: Jolanta Wisniewska.

FM (MHz)	OIRT	CCIR	kW (TRP)
Bialystok	–	91.1	0.2
Bielsko-Biala	72.77	89.2	0.1/0.3
Chelmce (Kalisz)	–	98.0	10
Chelmska Góra (Koszalin)	–	104.9	5
Choragwica (Kraków)	70.06	96.0	10/5
Chwaszczyno (Gdansk)	73.52	98.4	1/3.5
Ciechanów	–	94.3	1
Czarna Góra (Klodzko)	–	101.6	3
Cêstochowa-Bleszno	70.88	92.4	1/0.1
Elblag	–	101.2	0.3
Gizycko	–	102.0	5
Góra Chelmiec (Walbrzych)	–	102.9	3
Gorzów Wielkopolski	–	106.7	2
Gubalowka (Zakopane)	72.86	101.8	0.3/3
Jawor (Baligrod)	–	101.1	3
Jemiolów (Sulecin)	–	106.4	20
Kolowo (Szczecin)	–	106.7	10
Koszalin	–	89.3	5
Krzemieniucha (Suwalki)	–	89.0	10
Legnica	–	96.1	4
Lódz	70.10	93.5	0.3/3
Lomza	–	91.5	0.5
Losice (Biala Podlaska)	–	91.9	10
Lublin	67.19	–	0.5
Nowa Klarczma (Luban)	–	93.8	1
Olsztyn-Pieczewo	72.20	89.40	0.5/0.4
Olsztyn-Pieczewo	–	95.3	2
Opole	71.06	95.3	0.4/1
Ostrolêka	–	91.5	0.3
Pila-Roldo	70.85	–	0.3
Plaski (Lublin)	–	89.3	5
Plock-Rachocin	–	94.3	2
Poznan-Platkowo	73.52	–	0.5
Prehyba (Nowy Sacz)	–	103.2	5
Radom	–	100.3	1
Rusinowo (Walcz)	–	96.6	2
Siemianowice Slaskie-Mytkow	71.75	93.0	0.8/10
Skórovo (Lêbork)	–	103.4	1

FM (MHz)	OIRT	CCIR	kW (TRP)
Sniezne Kotly (Jelenia Góra)	–	100.8	25
Srem	–	94.6	4
Swiety Krzyz (Nowa Slupia)	–	88.2	5
Swinoujscie	–	101.2	1
Tarnawatka (Zamosc)	–	107.7	6
Tatarska Góra (Przemysl)	–	103.4	3
Toporzyk (Lobez)	–	91.3	20
Torun	70.46	–	1
Trzeciewiec (Bydgoszcz)	70.46	93.3	0.4/0.5
Tyczyn (Rzeszów)	70.88	100.1	0.5/5
Warszawa	66.17	91.0	0.5
Wodzislaw Slaski	–	90.0	0.1
Wroclaw	68.09	92.9	0.5/2
Zawada (Tarnów)	70.06	95.4	5
Zolwiniec (Konin)	–	98.9	5

RADIO MARYJA

📧 ul. Zwirki i Wigury 80, 87-100 Torun. ☎ +48 (56) 552361. 🖷 +48 (56) 552362. **Web:** http://www.logon.bydgoszcz.pl/goscie/mateusz/rm/
L.P: Dir: Tadeusz Rydzyk.

FM (MHz)	OIRT	CCIR	kW (TRP)
Barlinek	–	107.2	2
Biala Podlaska	67.40	–	0.5
Bialystok	67.55	104.7	0.1/3
Bielsko-Biala	72.17	88.4	1
Bogatynia	–	100.3	2
Braniewo	–	106.2	1
Brzesko		98.7	
Bydgoszcz	67.61	88.5	0.3
Chelmce (Kalisz)	70.16	103.1	0.2/1
Chelmno (Swiecle nad Wisla)	70.67	104.0	0.3/1
Chelmska Góra (Koszalin)	70.55	102.6	0.3/2
Ciechanów	66.74	–	0.5
Czersk	–	104.1	0.5
Dêbno	–	98.8	4
Dabrowa Górnicza-Golonog	–	103.3	1
Drawsko Pomorskie	–	104.7	1
Elblag	69.11	102.3	0.5/3
Elk	–	102.6	3
Gdansk	72.29	88.9	
Gdynia	–	102.3	0.3
Glogów	–	100.6	1
Góra Chelmiec (Walbrzych)	–	107.4	1
Gorzow Wielkopolski	–	68.12	
Grojec	–	99.8	
Gryfice	–	102.9	
Hajnówka	–	102.0	5
Hrubieszów	–	107.5	0.3
Inowroclaw	66.17	–	0.3
Jemiolów (Sulecin)	–	98.4	0.5
Kalwaria Zebrzydowska	–	94.3	0.3
Kazimierz Dolny	–	89.9	0.2
Kêdzierzyn-Kozle	–	97.7	0.3
Kielce	67.34	102.7	0.1/0.3
Klodawa	71.24	–	0.3
Klodzko	–	106.3	1.5
Kolno	–	104.0	1
Kolobrzeg	–	94.4	0.3
Konin	68.00	105.1	0.5/1
Koszêcin (Lubliniec)	–	103.7	4
Kraków	71.36	90.6	0.3
Krasnik Fabryczny	–	98.0	0.5
Krosno	71.90	–	
Krotoszyn	72.59	–	
Kutno	69.47	–	0.2
Kwidzyn	–	107.4	0.3
Laziska Górne (Tychy)	–	107.6	3
Lebork	–	92.7	
Legnica	66.08	–	0.3
Leszno	68.39	–	0.3
Letnica	–	90.3	0.5
Lezajsk	–	106.3	1
Lêbork	–	92.7	1
Lidzbark (Ciechanów)	69.08	–	0.3
Lidzbark Warminski	68.84	–	0.3
Lódz	69.44	87.9	0.5/0.4
Lomza	73.10	103.6	1
Lowicz	–	103.5	5
Lubaczów	–	105.1	0.3

FM (MHz)	OIRT	CCIR	kW (TRP)
Lubaczowa	71.45	99.9	0.1/1
Luban	–	95.2	
Lublin	67.85	100.3	0.3/0.5
Lubon Wielki (Rabka)	–	100.7	2
Malawa	–	103.8	2
Miejsce Piastowe (Krosno)	–	104.5	1
Mikolajki	–	88.4	0.3
Nysa	–	100.4	1
Olkusz	–	104.6	0.1
Olsztyn-Pieczewo	71.60	102.2	0.3/1
Opoczno	–	95.4	1.5
Opole	–	89.2	
Orneta	–	94.5	0.3
Ostrów Mazowiecka	–	100.4	1
Paslêk	72.29	–	0.3
Pila	–	100.4	0.1
Piotrków Trybunalski	–	94.7	0.2
Plonsk	71.72	–	0.3
Poznan-Piatkowo	65.90	97.7	0.2/1
Przemysl	68.00	98.4	0.2/1
Radom	67.40	–	0.1
Rzeszów	71.51	100.9	0.5/0.1
Sadowe (Ostrów Wlkp.)	69.29	–	0.2
Sieradz	67.49	–	0.3
Sleza (Sobotka)	–	88.9	5
Slupsk	70.85	102.0	1
Srem	–	106.8	5
Stalowa Wola	–	104.4	0.2
Stargard Szczec	–	97.4	
Starograd Gdanski	–	87.6	
Suwalki	–	105.5	2
Szczecin	–	101.6	1
Szczytno	–	88.1	0.3
Szpetal Górny (Wloclawek)	–	100.9	1
Swinoujscie	–	87.7	1
Toporzyk	–	100.6	
Torun	66.41	100.6	0.3/5
Warszawa	73.70	89.1	10/1
Wegorzewo	–	100.2	
Wlodawa	–	100.6	1
Wlostów (Ostrowiec Sietokrz.)	–	100.9	0.3
Wloszczowa	–	90.2	0.2
Wojakowa	71.63	–	0.2
Wolsztyn	–	98.7	0.1
Wroclaw	66.05	94.5	0.5/0.4
Wysoka Wies	–	100.4	1
Zagan	–	101.2	1
Zamosc	70.34	90.1	0.5/1
Zbrosza Duza	70.82	–	0.2
Zlotów	71.42	101.1	0.3

OTHER STATIONS

D.PRGR: 24h unless otherwise stated

MW	kHz	kW	Station, location, schedule
1)	1602	1	R. Ton, Radom: 0400-2200
2)	65.90	0.3	Classic FM, Lódz
3)	65.90	0.3	KR. Ciechanów, Ciechanów
4)	65.90	0.3	Vigor FM, Slupsk (Kobylnica)
5)	65.96	0.2	KR. As, Szczecin
6)	65.99	0.2	KR. Plock, Plock
7)	65.99	0.3	R. Puls, Lublin
8)	66.08	0.1	Radio 5, Suwalki
9)	66.08	0.3	R. Centrum Kultury i Stzuki, Kalisz
10)	66.08	0.5	Rekord FM, Radom
11)	66.14	0.5	R. O`le, Opole
12)	66.17	0.3	R. Maks, Tarnów (Zawada): 0400-2200
13)	66.17	0.3	R. Wanda, Kraków
	66.23		Katolickie R. Ciechanow, Ciechanow
14)	66.44	0.3	R. Leliwa, Tarnobrzeg: 0400-2200
15)	66.50	0.2	R. City, Slupsk
16)	66.50	0.3	R. Oko, Ostrolêka
17)	66.59	0.5	R. Fama, Kielce
18)	66.68	0.3	KR. Emaus, Lódz
19)	66.71	0.3	R. Vox, Bydgoszcz
20)	66.80	0.3	R. Wloclawek,Wloclawek/Zawisle
21)	66.89	0.5	R. Hot, Przemysl (Tatarska Góra)
	66.98		R. Opatow, Opatow
22)	67.00	1	Radio MR-FM, Jelenia Góra (Góra Szybowcowa)

FM	MHz	kW*	Station, location, schedule
23)	67.01	0.1	KR. Fiat, Czêstochowa: 0500-2000
24)	67.07	5	KR. Plus, Gdansk (Chwaszczyno)
25)	67.07	0.5	R. Kormoran, Wagorzewo
26)	67.10	0.2	Plockie R. Puls, Plock
17)	67.34	0.1	R. Fama, Sochaczew
27)	67.37	0.4	R. Park, Wysoka
28)	67.46	0.3	Radio PM, Kwidzyn
29)	67.58	0.3	R. Sud, Kêpno: 0500-2100
30)	67.70	0.2	Plockie R. Boss, Plock
31)	67.82	1	KR. Legnica, Legnica
32)	67.85	0.3	R. Inowroclaw, Inowroclaw
33)	67.98	0.1	R. Opatow, Opatow: 1300-1800
34)	68.15	0.3	R. Gra, Torun
35)	68.24	0.3	R. Ilawa, Ilawa
36)	68.24	0.5	R. Weekend, Chojnice
37)	68.30	0.1	Radio El, Bydgoszcz
	68.60		R. Hit FM, Stalowa Wola
38)	68.63	0.5	R. Eska Nord, Gdynia (Gdansk)
39)	68.84	0.3	R. Pro Kolor, Opole
	69.05		R. Gama, Bochnia
40)	69.23	0.2	R. Pólnoc, Koszalin
31)	69.38	0.3	KR. Legnica, Legnica (Jelenia Góra)
41)	69.38	1	R. Top, Katowice-Koszutka
38)	69.44	0.1	R. Eska Wroclaw, Wroclaw
42)	69.44	0.2	R. Ziemii Wielunskiei, Wielun
43)	69.59	0.5	Hit FM, Mielec: 0400-2200
44)	69.59	0.5	R. Radom, Radom
45)	69.62	0.3	R. Fon, Czêstochowa
46)	69.65	0.3	KR. Dobra Nowina, Tarnów: 0400-2200
47)	69.68	0.3	R. ARnet, Gdansk
	69.74		R. MTM FM, Starachowice
48)	69.77	0.05	R. Bielsko, Bilesko-Biala
49)	69.80	0.3	R. Legnica, Legnica
50)	69.80	0.5	R. WaWa, Warszawa
51)	69.83	0.1	R. Las Vegas, Ciechocinek: 0500-2100
52)	69.92	1	R. Flash Reporterzy, Gliwice (Zabrze)
53)	69.92	0.1	R. Piotrków, Piotrków Trybunalski
54)	70.04	0.3	R. Torun, Torun
55)	70.19	0.1	Akademickie R. Centrum, Rzeszów: 1100-2300
56)	70.19	0.3	R. Warmia-Mazury (WAMA), Olsztyn
	70.22		R. Jowisz, Jelenia Gora
	70.25		R. Pila, Pila
89)	70.25		R. Gaga, Belchatow
57)	70.30	0.3	KR. Gorzów, Gorzów Wielkopolski
58)	70.34	0.2	KR. Archidiecezjalna Rozglosnia Radiowa, Poznan: 1800-2000
59)	70.40	1	R. Delta, Bielsko-Biala
	70.40		Nasze R., Mecka Wola
	70.52		R. 5, Elk
	70.55		R. Hit, Wloclawek
60)	70.70	0.5	KR. Warszawa, Warszawa (Miedzeszyn): 0430-2100
4)	70.75	0.3	Vigor FM, Slupsk (r.)
61)	70.76	0.3	KR. Mariackie Kraków, Kraków: 0400-2200
62)	70.82	0.2	KR. Zbrosza Duza, Jasieniewo (Zbrosza Duza)
63)	70.85	0.3	R. Manhattan, Lódz
	70.94		R. Vox FM, Feliksowka
64)	71.00	0.5	R. Jowisz, Jezow Sudecki (Góra Szybocowa)
65)	71.03	0.5	KR. Puls, Gliwice (r.)
66)	71.03	0.3	R. Echo, Nowy Sacz (Wysokie)
67)	71.09	0.6	KR. Glos Pelplin, Pelplin: 0400-1000, 1455-2010
	71.09		Ja R. Jarocin, Jarocin
68)	71.12	0.3	R. Go, Gorzów Wielkopolski
47)	71.15	0.3	R. ARnet, Gdansk
69)	71.15	0.1	R. Kolobrzeg, Kolobrzeg
70)	71.24	0.3	R. Akadera, Bialystok
	71.27		R. Mazury, Ostroda
71)	71.54	0.5	R. Konin, Konin
36)	71.57	0.5	R. Weekend, Chojnice (r.)
	71.57		R. Leliwa, Machow
	71.60		R. Maryka, Olsztyn
17)	71.69	0.3	R. Fama, Opole
17)	71.80	0.1	R. Fama, Sochaczew (r.)
72)	71.96	0.3	KR. Jednosc, Kielce
73)	71.99	0.1	KR. Radomskie Ave, Radom: 0400-2000
52)	72.20	0.5	Flash FM, Gliwice
74)	72.25	0.1	R. Alex, Zakopane (Nowy Targ)

FM	MHz	kW*	Station, location, schedule
74)	72.26	0.5	R. Alex, Zakopane (Gubalowka)
75)	72.32	0.3	KR. Bis, Elblag
76)	72.38	0.5	R. Kolor, Warszawa
77)	72.38	0.4	R. Reja, Szczecinek
65)	72.44	0.5	KR. Puls, Gliwice
78)	72.50	0.3	R. City, Częstochowa
60)	72.53	6	KR. Warszawa, Warszawa (Plock-Rachocin)
79)	72.56	0.5	R. Alfa, Kraków (Węgrzce)
80)	72.62	0.1	R. Mega FM, "Radio FM", Pszczyna
81)	72.65	0.2	KR. FM-Lipiany, Lipiany: 0500-1600
37)	72.74	0.5	Radio EL, Elblag
82)	72.80	0.3	KR. Jasna Góra, Częstochowa: 0400-2100
79)	72.80	1	R. Alfa, Kraków
38)	72.92	0.5	R. Eska Nord, Gdynia
83)	72.92	5	R. S-Poznan, Poznan
84)	73.20	0.1	R. ABC, Szczecin
38)	73.20	10	R. Eska, Warszawa (Kaweczyn)
80)	73.20	0.1	R. Plesino, Pszczyna: 0300-2200 (SS 24h) (r.)
85)	73.22	0.3	R. Elka, Leszno
86)	73.25	0.5	R. Pomoze, Bydgoszcz
8)	73.28	0.5	Radio 5, Suwalki
38)	73.28	0.5	R. Eska, Lulbin
87)	73.70	0.3	R. Bartelik, Olsztyn
88)	73.71	0.3	R. Kiks, Lódz
	73.73		R. Fama, Gniezno, Gniezno
17)	87.6	0.2	R. Fama, Opole (Kielce)
61)	87.8	3	KR. Mariackie Kraków, Kraków: 0400-2200
81)	87.9	0.5	KR. FM-Lipiany, Lipiany: 0500-1600
90)	87.9	2	KR. Lublin, Lublin-Łozy Dar: 0400-2200
59)	87.9	1.6	R. Delta, Bielsko-Biala
	88.1		R. Fan, Knurow
	88.3		R. Dobra Nowyna, Krynica
	88.4		R. Jazz FM, Poznan
42)	88.6	0.2	R. Ziemii Wielunskiej, Wielun
34)	88.8	1	R. Gra, Torun
	88.80		Student R. Lodz, Lodz
5)	88.9	3	KR. As, Szczecin
55)	89.0	0.3	Akademickie R. Zentrum, Rzeszów
91)	89.2	0.3	R. Bit, Bialystok
	89.30		KR. Puls, Bytom
89)	89.4	0.2	R. GaGa, Belchatów
92)	89.5	1	KR. Archidiecezji Gnieznienskiej (KR. Wojciech), Gniezno
2)	89.6	0.3	Classic FM, Lódz
93)	89.8	1	Radio 44, Swinoujscie
50)	89.8	0.3	R. WaWa, Warszawa
	89.8		Archidiecezjalna Rozglosnia, Poznan/Piatkowo
87)	89.9	0.3	R. Bartelik, Olsztyn
94)	90.0	0.1	Radio 90 FM, Wodzislaw Slaski
69)	90.2	0.1	R. Kolobrzeg, Kolobrzeg
56)	90.2	0.3	R. Warmia-Mazury (WAMA), Olsztyn (Ilawa)
76)	90.4	0.3	R. Kolor Wroclaw, Wroclaw
56)	90.5	0.3	R. Warmia-Mazury (WAMA), Olsztyn
	90.6		R. Frem, Zgorzelec
	90.6		RMI FM, Poznan
95)	90.7	3	KR. FM-Gryfice, Gryfice
47)	90.7	0.3	R. ARnet, Gdansk
71)	90.7	2	R. Konin, Konin
17)	90.8	0.3	R. Fama, Opole
32)	90.8	0.3	R. Inowroclaw, Inowroclaw
	91.1		R. Mercury, Chelmce
	91.2		Katolickie R. Radom, Radom
96)	91.3	0.5	Radio M, Wroclaw
67)	91.4	0.5	KR. Glos Pelplin, Pelplin: 0400-1000, 1455-2010
4)	91.5	3	Vigor FM, Slupsk
	91.7		Lubus R. Nadzeja, Zielona Gora
97)	91.8	0.1	R. Art-Press, Lublin
97)	91.8	0.5	Radio Top 91.8 FM, Lublin
98)	91.8	0.5	Inforadio, Walbrzych
98)	91.8	0.3	R. Harcowka, Walbrzych
99)	92.0	0.2	KR. Rodzina, Wroclaw: 0400-2000
100)	92.0	0.7	Rozglosnia Harcerska, Warszawa (Gdansk)
80)	92.3	0.3	R. Plesino, Pszczyna: 0300-2200 (SS 24h)
13)	92.5	0.3	R. Wanda, Kraków
37)	92.6	0.3	Radio El, Elblag
51)	92.8	0.2	R. Las Vegas, Ciechocinek: 0500-2100
83)	93.0	4	R. S-Poznan, Poznan
101)	93.2	0.5	R. Goleniów, Goleniów

FM	MHz	kW*	Station, location, schedule
102)	93.4	0.3	R. Sopot, Sopot
103)	93.5	0.3	R. Obywatelskie, Poznan
	93.6		R. Barys, Siemianowice
33)	93.7	0.1	R. Opatow, Opatow: 1300-1800
	93.7		R. Mariacke Krakow, Mogilany
66)	93.8	0.3	R. Echo, Nowy Sacz
68)	93.8	3	R. Go, Gorzów Wielkopolski
7)	93.8	0.2	R. Puls, Lublin
	94.0		R. Art, Warszawa
	94.4		R. Jowisz, Zary
41)	94.5	0.5	R. Top, Katowice-Koszutka
23)	94.7	2	KR. Fiat, Częstochowa: 0500-2000
28)	94.8	0.3	Radio PM, Kwidzyn
31)	94.9	10	KR. Legnica, Legnica (Chrosnica-Dziwiszow)
	94.9		R. Fama, Sochaczew
104)	95.1	0.5	Radio SBB Rodlo, Bytom
38)	95.1	0.3	R. Eska Wroclaw, Wroclaw
	95.2		Inforadio, Gdansk Jaskowa Kopa
49)	95.7	0.3	R. Legnica, Legnica
105)	95.8	0.3	R. Mazowsze, Lomianki (Nowy Dwor Mazowiecki)
	95.8		Inforadio, Wroclaw
40)	95.9	0.5	R. Pólnoc, Koszalin
	96.0		R. Index, Zielona Gora
65)	96.1	2	KR. Puls, Gliwice
37)	96.2	0.3	Radio El, Bydgoszcz
38)	96.4	0.3	R. Eska Nord, Gdynia (Gdansk)
100)	96.4	0.6	Rozglosnia Harcerska, Warszawa (Rzeszów)
106)	96.4	1	R. Sudety, Dzierzoniów: 0500-1100, 1400-1800
107)	96.5	0.5	KR. Józef, Warszawa: 0400-2200
78)	96.6	0.3	R. City, Częstochowa
108)	96.7	0.1	R. Lan, Proszowice (Koniusza): 0400-2200
54)	96.7	0.2	R. Torun, Torun
109)	96.8	0.5	R. Jarocin, Jarocin
110)	96.9	0.1	R. Plama, Szczecin
	97.0		R. Katowice, Raciborc
36)	97.1	2	R. Weekend, Chojnice
111)	97.1	5	R. Zielona Góra, Zielona Góra
	97.4		Inforadio, Katowice
	97.4		Inforadio, Lodz
58)	97.7	0.5	KR. Archidiecezjalna Rozglosnia Radiowa, Poznan
112)	97.7	0.5	R. Blue, Wieliczka
	97.7		Inforadio, Warszawa
	97.7		Inforadio, Poznan/Piatkowo
	97.7		R. 9, Slocina
	97.80		Inforadio, Gdynia Oksywie
113)	97.9	0.2	Pomorska Stacja Radiowa - PSR, Szczecin
88)	97.9	0.3	R. Kiks, Lódz
72)	98.0	1	KR. Jednosc, Keilce
114)	98.1	0.5	R. Jutrzenka, Warszawa
12)	98.1	0.2	R. Maks, Zawada (Tarnów)
115)	98.2	0.2	R. Centrum, Lublin
53)	98.2	0.1	R. Piotrków, Piotrków Trybunalski
	98.2		R. Ave Maryja, Jaroslaw
14)	98.3	0.2	R. Leliwa, Tarnobrzeg: 0400-2200
84)	98.4	0.1	R. ABC, Szczecin
85)	98.5	0.3	R. Elka, Leszno
116)	98.6	0.3	Studenckie Radio Afera, Poznan: 0400-2200
	98.6		R. Niepokalanow, Lodz
77)	99.0	0.5	R. Reja, Szczecinek
117)	99.1	0.7	R. Rezonans, Sosnowiec
	99.3		Inforadio, Szczecin
36)	99.3	2	R. Weekend, Chojnice (r.)
118)	99.4	5	Radio Muzyczno-Informacyjne (RMI), Poznan (Srem)
63)	99.8	0.2	R. Manhattan, Lódz
	100.1		R. Pogoda, Warszawa
119)	100.2	0.3	R. Fan, Poznan
120)	100.3	1.5	R. Vanessa, Racibórz
18)	100.4	1	KR. Emaus, Lódz
	100.4		R. Pila, Pila
121)	100.5	0.3	R. Akademickie Kraków, Kraków: 0400-2200
82)	100.6	5	KR. Jasna Góra, Częstochowa: 0400-2100
57)	100.7	2	KR. Gorzów, Gorzów Wielkopolski
39)	100.7	0.2	R. Pro Kolor, Opole
15)	100.9	0.2	R. City, Slupsk
38)	100.9	0.5	R. Eska, Lublin

FM	MHz	kW*	Station, location, schedule
122)	101.1	2	BRW 101.1FM, Walbrzych
122)	101.1	2	R. Walbrzych, Walbrzych
9)	101.1	5	R. Centrum Kultury i Stzuki, Kalisz
100)	101.1	0.4	Rozglosnia Harcerska, Warszawa (Gdynia)
46)	101.2	5	KR. Dobra Nowina, Tarnów (Wysokie): 0400-2200
19)	101.2	0.3	R. Vox, Bydgoszcz
100)	101.5	0.3	Rozglosnia Harcerska, Warszawa
	101.5		R. Mazury, Ostroda
100)	101.6	0.3	Rozglosnia Harcerska, Warszawa (Poznan)
24)	101.7	5	KR. Plus, Gdansk (Chwaszczyno)
123)	101.7	20	KR. Podlasia, Siedlce (Losice): 0400-2100
29)	101.7	0.5	R. Sud, Kêpno: 0500-2100
27)	101.8	0.3	R. Park, Kedzerzyn-Kozle
38)	102.0	0.6	R. Eska, Warszawa (Kaweczyn)
	102.0		R. Flash, Gliwice
	102.1		R. MTM FM, Starachowice
75)	102.3	1	KR. Bis, Elblag
43)	102.4	1	Hit FM, Mielec: 0400-2200
79)	102.4	0.6	R. Alfa, Kraków (Wêgrzce)
124)	102.6	2	KR. Diecezji Koszalinsko-Kolobrzeskiej, Koszalin
45)	102.6	0.3	R. Fon, Czêstochowa
	102.6		KR. Legnica, Polkowice
125)	102.7	2	KR. Niepokalanów, Teresin (Kampinos): 0400-2100
	102.7		R. Mariacke Krakow, Rabka
	102.9		Inforadio, Krakow
76)	103.0	0.5	R. Kolor, Warszawa
126)	103.1	1	KR. Diecezji Kaliskiej, Kalisz (Chelmce)
127)	103.5	5	KR. Victoria, Lowicz: 0400-2200
112)	103.5	0.4	R. Blue, Biskupice
86)	103.5	0.3	R. Pomoze, Bydgoszcz
	103.5		Katolickie R. Lowicz Via, Lowicz
46)	103.6	1	KR. Dobra Nowina, Tarnów (Lichwin): 0400-2200
	103.6		R. ESKA, Lublin
	103.6		R. Mercury, Pila / Walcz-Rusinowo
128)	103.7	0.5	R. Classik, Warszawa
29)	103.8	2	KR. Diecezji Rzeszówskiej "Via", Krasne (Rzeszów): 1400-1700
3)	103.9	0.3	KR. Ciechanów, Ciechanów
130)	103.9	0.3	R. Darlowo, Darlowo: 0400-2200
131)	104.1	1	Radio 100, Pila
11)	104.1	0.3	R. O`le, Opole
6)	104.3	1	KR. Plock, Plock
	104.3		Katolickie R. Lipiany, Lipiany
	104.4		R. Wa-Wa, Gda Jaskowa Kopa
	104.8		R. Wa-Ma, Mragowo
74)	105.2	0.05	R. Alex, Zakopane (Kasprowy Wierch)
	105.0		R. Flash, unknown location
	105.5		R. Flash, Bedzin
132)	105.6	0.3	R. Pila, Pila
	105.6		R. Wa-Wa, Gdynia Oksywie
133)	106.1	0.3	R. Klakson, Wroclaw
10)	106.2	0.2	Rekord FM, Radom
64)	106.2	3	R. Jowicz, Jezow Sudecki (Góra Szybowcowa)
	106.2		KR, Warszawa/Szczecin
52)	106.4	1.5	R. Flash, Gliwice (Zabrze)
126)	106.4	5	R. Katolickie, Kalisz (Chelmce)
	106.4		R. Wa-Wa, Krakow
22)	106.7	3	Radio MR-FM, Jelenia Góra (Góra Szybowcowa)
48)	106.7	0.1	R. Bielsko, Bielsko-Biala (Skrzyczne)
38)	106.7	0.6	R. Eska Nord, Gdynia
	106.8		R. Jazz, Warszawa/Kaweczyn
100)	106.9	0.3	Rozglosnia Harcerska, Warszawa (Wroclaw)
44)	106.9	2	R. Radom, Radom
25)	107.0	5	R. Kormoran, Wegorzewo
134)	107.6	1	KR. Archidiecezji Katowickiej, Laziska Górne
135)	107.9	2	KR. Góra Swietej Anny, Opole (Wysoka): Mon-Sat 0900-1600

*) transmitter power. **NB:** KR = Katolckie Radio (Catholic R.).
Addresses:
1) ul. Przytycka 2, 26-610 Radom – 2) ul. Traugutta 25, 90-113 Lódz – 3) ul. Sciegiennego 18, 06-400 Ciechanów – 4) ul. Poznanskiego 1, 76-200 Slupsk – 5) ul. Paska 35-a, 71-622 Szczecin – 6) ul. Zdunska 1, 09-402 Plock – 7) ul. Sklodowskiej-Curie 5, 20-029 Lublin – 8) ul. Nowomiejska 3, 16-400 Suwalki – 9) ul. Lazienna 6, 62-800 Kalisz –

10) ul. ˜eromskiego 88, 26-600 Radom – 11) ul. Zielonogórka 3, 45-955 Opole – 12) ul. Nowy Swiat 3, 33-100 Tarnów – 13) ul. Fatimska 13, 31-831 Kraków – 14) Al.Niepodleglosci 9, 39-400 Tarnobrzeg – 15) ul. Sienkiewicza 20, 76-200 Slupsk – 16) ul. Kopernika 15, 07-400 Ostrolêka – 17) HQ: ul. Harcerska 15, 45-158 Opole. Reg.st's: ul. Piotrkowska 25-510 Kielce; ul. Niemcewicza 1, 96-500 Sochaczew – 18) ul. Skorupki 1-a, 90-458 Lódz – 19) ul. Kujawska 117, 85-152 Bydgoszcz – 20) ul. Królewicka 45/49, 87-800 Wloclawek – 21) Przemysl – 22) Pl.Wyszynskiego 45/1, 58-500 Jelenia Góra – 23) ul. BrzeZnicka 59, 42-400 Czêstochowa – 24) ul. Sawalska 46, 80-215 Gdansk. **Web:** http://www.poland-biz.net/clients/radio-plus/ – 25) ul. M.Kopernika 21, 11-600 Wegorzewo – 26) ul. Otolinska 25, 09-400 Plock – 27) Kedzierzyn-Kozle – 28) ul. ˜irowa 2-a, 82-500 Kwidzyn – 29) Al.Marcinkowskiego 12-k, 63-300 Kêpno – 30) ul. Jachowicza 2, 09-402 Plock – 31) ul. Zielona 5/6, 59-220 Legnica – 32) ul. Chobrego 75, 88-100 Inowroclaw – 33) ul. Partyzantów 13, 27-500 Opatow – 34) ul. Poznanska 152, 87-100 Torun – 35) ul. Ostródzka 53, 14-200 Ilawa – 36) ul. Jana Pawla II 1, 89-600 Chojnice – 37) HQ: ul. 1 Maja 2, 82-300 Elblag. Reg.st: ul. Sobieskiego 1, 85-959 Bydgoszcz – 38) HQ: Al.Jerozolimskie 125/127, 02-017 Warszawa. **Web:** http://www.pol.pl/witryna._p/eska/ Reg.st's: Al.˜eromskiego 26, 81-364 Gdynia. **Web:** http://www.poland-biz.net/clients/eska/; ul. Powstancow Slaskich 95, 53-332 Wroclaw – 39) ul. Ozimska 19/ 916, 45-057 Opole – 40) ul. Zwyciêstwa 137, 75-604 Koszalin – 41) ul. Jesinowa 9-a, 40-146 Katowice – 42) ul. Kilinskiego 23, 98-300 Wielun – 43) ul. Kusocinskiego 4, 39-300 Mielec – 44) ul. Jagiellonski 15, 26-600 Radom – 45) ul. Nowowiejskiego 26, 42-217 Czêstochowa. **Web:** http://www.matinf.pcz.czest.pl/~qbak/fon.html – 46) ul. Bema 14, 33-100 Tarnów – 47) ul. Wyspianskiego 9, 80-434 Gdansk – 48) ul. Cieszynska 317, 43-303 Bielsko-Biala. **Web:** http://host1.bielbit.bielsko.pl/radiob/ – 49) addr as 31) – 50) ul. Nowolipki 9-b, 00-151 Warszawa – 51) ul. Nieszawska 21, 87-720 Ciechocinek – 52) ul. Pszczynska 89, 44-100 Gliwice. **Web:** http://www.flash.gliwice.pl/ ~flash – 53) ul. Jagiellonska 7 (or Al.3 Maja 4), 97-300 Piotrków-Trybunalski – 54) ul. Szosa Lubiska 2/18, 870100 Torun – 45) ul. Akademicka 6, 35-084 Rzeszów – 56) ul. Dabrowszczaków 39, 10-542 Olsztyn – 57) ul. Kazimierza Jagiellonczyka 8, 66-400 Gorzow Wlkp – 58) Os.Jagielly 105, 60-681 Poznan – 59) ul. Browarna 2, 43-300 Bielsko-Biala – 60) ul. Szafirowa 58, 04-954 Warszawa – 61) ul. Zakopanska 86, 30-418 Kraków – 62) ul. Zbrosza Duza 45, 05-604 Jasieniec – 63) ul. Pilsudskiego 7, 90-368 Lódz. **Web:** http://www.doskomp.lodz.pl/manhat.htm – 64) Jezow Sudecki. **Web:** http://www.cpu-zeto.com.pl/Os/radio-jowisz/ – 65) ul. Glowackiego 3, 44-100 Gliwice – 66) ul. Limanowskiego , 33-300 Nowy Sacz – 67) ul. B.Dominika 11, 83-130 Pelplin – 68) ul. Podmiejska 23, 66-400 Gorzow Wlkp – 69) ul. IV Dyw. Piechoty 60, 78-100 Kolobrzeg – 70) ul. Zwierzynieckia 4, 15-333 Bialystok – 71) ul. 1 Maja 13, 62-510 Konin – 72) ul. Jana Pawla II 4, 25-013 Kielce – 73) ul. Mlynska 23/25, 26-612 Radom – 74) ul. Krupówki 48, 34-500 Zakopane – 75) ul. Saperów 20, 82-300 Elblag – 76) HQ: ul. Nabrutta 41/43, 02-536 Warszawa. Reg.branch: ul. Wita Stwosza 2, 50-148 Wroclaw. **Web:** http://www.math.uni.wroc.pl:80/~bkuba/kolor.html – 77) ul. Fabryczna 13, 78-400 Szczecinek – 78) ul. Dekabrystow 41, Czêstochowa. **Web:** http://etcetera.iie.ae.wroc.pl/han/Media/City/ – 79) ul. Wêgrzyce 206, 32-086 Kraków, or Al.3 Maja 9, 30-062 Kraków – 80) ul. Kilinskiego 5, 43-200 Pszczyna – 81) ul. Okrzei 5, 72-240 Lipiany – 82) ul. Kordeckiego 2, 42-225 Czêstochowa – 83) ul. Piekary 14/15, 61-823 Poznan. **Web:** http://www.capella.ae.poznan.pl/~radios – 84) ul. Mazowiecka 14, 70-526 Szczecin. **Web:** http://www.szczecin.pl/abcradio/ – 85) ul. Spóldzielcza 6, 64-100 Leszno – 86) ul. Kaszubska 25, 85-048 Bydgoszcz – 87) Olsztyn – 88) ul. Piotrkowska 77, 90-423 Lódz – 89) ul. Czaplinecak 44-b, 97-400 Belchatów – 90) ul. Jana Pawla II 11, 20-535 Lublin – 91) Bialystok – 92) ul. Cicha 3, 62-200 Gniezno – 93) Swinoujscie – 94) ul. Mlodziezowa 67-b, 44-373 Wodzislaw Slaski – 95) ul. 3 Maja 17, 72-300 Gryfice – 96) Wroclaw – 97) ul. Raabego 2-a, 20-030 Lublin – 98) ul. Chrobrego 45, 58-300 Walbrzych – 99) ul. Katedralna 4, 50-328 Wroclaw – 100) ul. Konopnickiej 6, 00-491 Warszawa – 101) ul. Grunwaldzka 4, 72-100 Goleniów – 102) Sopot – 103) ul. 28 Czerwca 1956 231/239, 61-485 Poznan – 104) ul. Mickiewicza 68, 41-902 Bytom. **Web:** http://www.silesia.pik-net.pl/rodlo – 105) ul. Kosciuszki 6, 05-092 Lomianki – 106) ul. Pocztowa 5, 58-200 Dzierzoniów – 107) Skwer Wyszynskiego 9, 01-015 Warszawa – 108) ul. Rynek 18, 32-100 Proszowice – 109) ul. Slubianki 21, 63-200 Jarocin – 110) ul. Dunska 25, 71-795 Szczecin – 111) ul. Kukulcza 1, 65-472 Zielona Góra – 112) ul. Lazany 162-a, 32-041 Biskupice – 113) ul. Pl.Rodla 8, 70-419 Szczecin – 114) Warszawa – 115) ul. Radziszewskiego 16, 20-031 Lublin – 116) ul. Sw.Rocha 11-a, 61-142 Poznan – 117) ul. Bêdzinska 39, 41-200 Sosnowiec – 118) ul. Sw.Marcin 58/64, 61-807 Poznan – 119) ul. Dozynkowa 9-g, 61-662 Poznan – 120) ul. Starowiejska 75, 47-400 Racibórz – 121) ul. Rastafinskiego 8, 30-073 Kraków – 122) ul.

Wysockiego 45, 58-304 Walbrzych – 123) ul. Swirskiego 56, 08-100 Siedlce – 124) ul. Staszica 38, 75-452 Koszalin – 125) ul. Niepokalanów, 96-515 Teresin – 126) Kalisz – 127) ul. Seminaryjna 6, 99-400 Lowicz – 128) Warszawa – 129) Malawa 515, 36-007 Krasne – 130) ul. WOP 12, 76-150 Darlowo – 131) Pila – 132) ul. Dabrowskiego 8, 64-920 Pila – 133) ul. Pl.Sw.Macieja 21, 50-244 Wroclaw – 134) ul. Koscielna 4-a, 43-170 Laziska Górne – 135) Pl.Katedralny 4, 45-005 Opole.

PORTUGAL

L.T: UTC (Su: UTC + 1h) — **Pop:** 10.676.872 — **Radios:** 2.220.000 — **Pr.L:** Portuguese — **E.C:** 50Hz, 220V — **ITU:** POR.

RADIODIFUSÃO PORTUGUESA EP (RDP)

✉ Av. Eng. Duarte Pacheco 6, 1000 Lisboa. ☎ +351 (1) 3871109. 🖷 +351 (1) 692298, 657988 (**Int. Rel:** +351 (1) 3871402). ① 64774 **E-mail:** rdp@telepac.pt **Web:** http://www.rdp.pt/rdp/
Antena 1/Antena 2: Rua do Quelhas 2, 1200 Lisboa. ☎ +351 (1) 3960181. 🖷 +351 (1) 602170 (RDP1), +351 (1) 3878057 (RDP2) — **News Dept:** Rua do Quelhos 10, 1200 Lisboa. ☎ +351 (1) 3960181. 🖷 +351 (1) 609889.
RDP/Açores & RDP/Madeira: see respective land listings.
Admin Board: Pres: Arlindo de Carvalho. Vice-Pres: Teresa Nunes. Members: Sérgio Azevedo, José Marques de Freitas, Inácio Moraes Mendes.
L.P: Dir. Human Resources: Guilherme Barbosa. TD: Francisco Mascarenhas. Dir. Exploitation Sces: Celso ce Albuquerque. Dir. Legal Dept: Júlio César Pereira. PD: Pedro Castelo. Dir. RDP2: José Manuel Nunes. Dir. N: Pedro Cid. Dir. Marketing & Communications: Luis Ochoa. Dir. Finance: Ferro de Carvalho. Dir. Computing: António Barriga. Reg. Dir (No.): Dialino Estevas. Reg. Dir. (So.): Jorge Amorim. Head of Int. Rel: Antonio Ribeiro.

RDP Onda Média	kHz	kW	RDP Onda Média	kHz	kW
Chaves	630	1	Montemor-o-Novo	630	10
Miranda do Douro	630	1	Bragança	666	1
Covilhã	666	1	Faro	720	10
Lisboa	666	135	Guarda	720	10
Valença	666	10	Mirandela	720	10
Vila Real	666	10	Porto	720	100
Viseu	666	10	Lamego	756L	
Castelho Branco	720	10	Portalegre	1287	1
Elvas	720	10	L) local prgrs.		

FM (MHz)	1	2	3	L	kW
Alcoutim	97.6				
Banática	99.4				
Bornes	92.8	91.1	102.1		5
Braga	91.3	88.0	103.0		
Bragança	96.4	98.2	104.2		6
Coimbra			94.9		
Elvas - Vila Boím	103.8				
Faro	97.6	93.4	100.7	100.7b	5
Gardunha	96.4	93.9	101.3		6
Guarda	94.7	88.4	100.6		5
Leiria	98.7	104.2	106.4		
Lisboa	95.7	94.4	100.3		100
Lousã	87.9	89.3	102.2	102.2c	50
Marão	95.2	99.8	101.5	101.5a	6
Mendro	87.7	91.1	102.4		5
Minhéu	94.9	88.0	104.7		6
Monchique	88.9	91.5	101.9	101.9b	30
Monsanto (Lisbon)	95.7	94.4	100.3		
Montejunto	98.3	88.7	105.2		10
Muro	88.3	94.6	102.0		5
Portalegre	97.9	92.9	102.8		5
Porto	96.7	92.5	100.4		80
Santarem	98.8				
Santiago do Cacém	99.2	90.7			6
Valença	98.2	89.6	104.0		5
Viseu			101.8		

D. PRGR: 24h. 1=Antena 1 (general prgrs), 2=Antena 2 (serious music), 3=Antena 3 (music for 15-45 age group.
Local stations: a) Casa do Douro (Sun 0800-0900), b) RDP Algarve (MF 0800-0900, 1700-1900), c) RDP Centro (MF 1000-1200 & 1500-1900, SS 1400-2000).

RDP INTERNACIONAL – RADIO PORTUGAL
see International Broadcasting section.

RÁDIO COMERCIAL (Priv.)

✉ Rua Sampaia e Pina 26, 1000 Lisboa. ☎ +351 (1) 387 2071. 🖷 +351 (1) 689551. **E-mail:** administracao@radiocomercial.pt, programas@radiocomercial.pt, tecnica@radiocomercial.pt, geral@radiocomercial.pt **Web:** http://www.radiocomercial.pt/
L.P: MD: Toão David Nunaf. PD: Rui Paco. CE: Carlos Silva.

MW	kHz	kW	MW	kHz	kW
Faro/Albufeira	558	10	Porto	1170	1
Aveiro	783		VilaReal/Lamego	1170	10
Mirandela	783	10	Valença	1170	10
Castelo Branco	828	1	Bragança	1170	1
Covilhã	828	1	Elvas	1332	1
Guarda	828	1	Chaves	1575	1
Coimbra/Leiria	828		Braga	1575	1
Viseu	828	1	Miranda do Douro	1575	1
Lisboa/Setúbal	1035	100			

FM (MHz)	R. Comercial	R. Nostalgia	kW
Alcácer do Sal	92.3		6
Braga	99.2		5
Bragança	93.9		
Chaves	91.9		
Coimbra	90.8		
Évora	92.0	106.4	
Faro	96.1	106.1	5
Fundão	98.2		
Guarda	96.1		5
Lagos	88.1	107.5	
Lamego	88.7		
Leiria	99.8	96.4	
Lisboa	97.4	104.3	100
Portalegre	98.9	106.7	6
Porto	97.7	100.8	40
Santiago do Cacém		107.1	
Valença	99.0		5
Vila Real	88.9		

D.PRGR: 24h (all three networks).
Main N. on AM network: 0630, 0730,0830, 1730, 1830, 1930.

RÁDIO RENASCENÇA (Rlg. Comm.)

✉ Rua Ivens 14, 1294 Lisboa Codex. ☎ +351 (1) 347 5270. 🖷 +351 (1) 342 2658.
L.P: MD: Oliveria Pires. PD: Pedro Tojal. CE: João Ramos.

MW	kHz	kW	MW	kHz	kW
Braga	576	10	Coimbrà	981	10
Muge	594	100	Guarda	981	1
Vila Moura	891	10	Chaves	1251L	10
Evora	927L	1	Porto	1251L	10
Seixal	963L	10	Castelo Branco	1251	1
Bragança	981	1	Viseu	1251L	10
Vila Real	981	1	L) incl. local prgrs.		

FM (MHz)	Ch.1	RFM	kW
Arrabida	105.8	89.9	10
Bornes	89.6	101.1	5/8
Bragança		99.5	1
Foia	98.6	104.9	28
Gardunha	103.4	99.5	8
Guarda		90.2	1
Lamego	98.6	106.2	10
Leiria/Fatima	95.0		10
Lisboa	103.4	93.2	100
Lousã	106.0	91.7	100
Marofa	94.2		6
Mendro	96.5	100.9	5/14
Minhéu	89.8	102.6	5/8
Montejunto	90.2	106.8	40
Muro	90.4	103.4	10
S.Mamede	95.3	101.1	8/4
S.Miguel	103.8	89.6	4/6
Valença	100.0	95.4	8
Valongo	93.7	104.1	100

D.PRGR: 24h (Ch.1 on MW/FM, RFM on FM only).

EXTERNAL SERVICE: see International Broadcasting section.

Main FM Networks
CORREIO DE MANHA RÁDIO (Comm.)
✉ Rua Tierno G. Amoreias, Torre 3, sal. 706, 1000 Lisboa. ☎ +351 (1) 387 0044. 🖷 +351 (1) 658876 — **L.P:** MD: Rui Pego.
FM (MHz): Lisboa 104.3, Faro 107.5, Leiria 96.4, Setubal 107.1

FM RADICAL (Comm.)
✉ Avenida de Ceuta 1, 2 piso, 1300 Lisboa. ☎ +351 (1) 364 2786. 🖷 +351 (1) 363 5863. **E-Mail:** fmradical@mail.telepac.pt **Web:** http://www.fmradical.pt/ — **L.P:** Dir: Paulo Alves Guerra.
FM (MHz): Lisboa 92.4, Porto 90.0. Coimbra 98.4, Faro 90.9

TSF RÁDIO JORNAL (Comm.)
✉ Av.Eng.D. Pacheco, Edif.Amoreiras 2-P-6, 1000 Lisboa. ☎ +351 (1) 387 0406. 🖷 +351 (1) 388 2791 — **L.P:** MD: D. Borges.
FM (MHz): Beja 93.7, Coimbra 90.0, Far 99.1, Lisboa 89.5, Porto 105.8.

Other Stations
FM stations in Lisbon (MHz): 87.6 R. Seixal – 88.0 R. Ocidente – 89.1 R. Leziria – 89.5 TSF – 90.4 R. Paris Lisboa (rel. R. France Internationale + 8h daily local prgs) – 90.9 Popular FM – 91.2 R. Clube de Sintra – 91.4 R. Iris – 92.0 R. Nova Antena – 92.4 FM Radical – 92.8 R. Horizonte Tejo – 93.2 RFM – 93.7 Lights FM – 94.4 Antena 2 – 95.0 R. Miramar – 95.3 Latina FM. 95.7 Antena 1 – 96.2 Super FM – 96.6 R. Nova/RCP – 97.8 R. Voz de Almada – 98.1 R. Marginal – 98.7 R. Baía – 100.3 Antena 3 – 100.8 R. Capital – 101.1 Memória FM – 101.9 Estação Orbital – 102.2 R. PAL – 102.6 R. Comercial da Linha – 103.9 R. Santiago – 104.8 R. Echo – 105.4 R. Clube de Cascais – 107.2 R. Cidade – 107.7 R. Nossa.

RADIO ALTITUDE
✉ Rua Batalha Reis, 6300 Guarda — **MW:** 1584kHz 1kW — **FM:** 107.7MHz — **D.PRGR:** 24h.

SOCIEDADE DE RÁDIORETRANSMISSÃO Ldª
DEUTSCHE WELLE RELAY STATION
RADIO TRANS EUROPE
See International Broadcasting Section.

ROMANIA

L.T: UTC + 2h (Su: UTC + 3h) — **Pop:** 22.680.000 — **Radios:** 4.500.000 — **Pr.L:** Romanian — **E.C:** 50Hz, 220V — **ITU:** ROU.

CONSILIUL NAŢIONAL AL AUDIOVIZUALULUI (PUB.) (National Audio-Visual Council)
✉ Bd. Libertataţii 14, sector 5, 70060 Bucureşti. ☎ +40 (1) 3126004, 4100357. 🖷 +40 (1) 3124634
L.P: Gen.Dir: Mariana Jurian.
CNA is the autonomous public regulatory & licensing authority.

SOCIETATEA ROMÂNA DE RADIODIFUZIUNE
✉ str. General Berthelot 60-62, RO-70747 Bucureşti. ☎ +40 (1) 650355, 6159350. 🖷 +40 (1) 2232612. ➀ 11252 — **L.P:** Pres: Tudor Catineanu. Dir. Foreign Rel: Titus Vajeu. T.D: Ilie Dragan.

LW/MW	kHz	kW	P
Bod (Braşov)	153	1200	1
Petroşani	531	14	1
Târgu Jiu	558	400	2/R
Bod (Braşov)	567	50	1
Satu Mare	567	50	1
Botoşani	603	50	1
Bucureşti	603	50	2
Oradea	603	14	1
Turnu Severin	603	14	1
Timişoara	630	400	1/R
Voineşti (Prahova)	630	400	2/5
Sighet (Maramureş)	711	50	2
Baia Mare	720	7	1
Isacea (Tulcea)	720	14	1
Sinaia	720	14	1
Boldur (Lugoj)	756	400	2+
Tâncăbeşti (Bucureşti)	855	1500	1
Cluj	909	200	4/R
Timişoara	909	50	1
Miercurea Ciuc	945	14	1
Urucani (Iaşi)	1053	1000	1/R*
Cluj	1152	950	1
Galbeni (Bacau)	1178	200	1
Vascau (Reşita)	1179	7	1
Bod (Braşov)	1197	14	4/R
Craiova	1314	7	1
Timişoara	1314	30	2/4
Valul Traian (Constanţa)	1314	14	1
Târgu Mureş	1323	7	4/R
Galaţi	1332	50	1

LW/MW	kHz	kW	P
Sibiu	1404	7	2
Sighet (Maramureş)	1404	50	1
Olaneşti (R. Vâlcea)	1422	7	1
Agigea (Constanţa)	1458	50	2/5/R
Mahmudia (Tulcea)	1530	14	2
Saveni (Radauţi)	1530	14	1
Harghita	1593	14	4/R
Ion Corvin (Constanţa)	1593	14	2
Oradea	1593	2	4/R
Sibiu	1593	7	4/R

+) also carries VOA 0445-0500, 0530-0600, 2300-2330. *) also rel. R. Moldova prgrs MF 2100-2200.

FM (MHz) (Powers are TRP)	1	2	3	kW
Alba (Bihor)			72.98	6
Balota (Turnu Severin)		69.68	72.71	6
Baneasa (Constanţa)		66.92	71.12	6
Bârlad		66.36	72.32	6
Bucegi (Ploieşti)		72.92	71.30	10/12
Cerbu (Novaci)		70.04	71.90	6/7
Comaneşti (Bacau)		67.01	68.87	6
Constanţa		67.79	70.01	6
Constanţa		100.1*		6
Cozia (R. Vâlcea)		72.68	67.25	6
Feleac (Cluj)		68.36	66.76	6
Gheorghieni (Harghita)		65.96	68.60	6
Heniu (Bistriţa)		72.44	69.47	6
Herastrau (Bucureşti)	72.08	70.40+	68.24	6
Herastrau (Bucureşti)	101.3		98.30	0.1/0.5
Magura Boiului (Deva)		70.64	72.20	6
Magura Odobeşti (Focşani)		70.22	67.88	6
Mangalia	70.52			0.3
Mogoşa (Baia Mare)		66.56	68.12	6
Oradea		71.00	69.86	6
Parang (Petroşani)		71.06	72.80	6
Piatra Neamţ		71.06	66.17	6
Pietraria (Iaşi)		69.92	71.84	6
Rarau (Câmpulung Moldovenesc)		70.85	72.20	6
Saveni (Botoşani)			71.42	6
Semenic (Reşiţa)		68.44	72.365	6
Sibiu		66.44	69.35	6
Siria (Arad)		72.56	70.79	6
Suceava		72.98	70.61	6
Timişoara		71.72	69.65	6
Topolog (Tulcea)		68.72	71.96	6
Vacareni (Tulcea)		69.65	70.76	6
Varatec (Maramureş)		69.11	67.34	6
Zalau		67.67	65.96	6

*) also carries prgrs. 5/Reg, +) also carries prgrs 4/Reg.
Radio România Actualitaţi (1): 24h. **N:** on the h – **Radio România Cultural (2):** 0300-2200 – **Radio România Tineret (3):** 0600-2300, Sat 2300-0600 – **Program "Maghiar-German" (4):** M-Sat 1200-1400, Sun 0800-0830 in Hungarian/German – **Program "Antena Satelor" (5):** Mon-Sat 0354-0700, 1700-2000. **Program "Studiouri Teritoriale" (Reg): Cluj:** str. Donath nr. 160, 3400 Cluj-Napoca. On 909/1593 (Oradea + Sibiu) kHz: 0500 (Sun 0600)-1300, 1600-2000 – **Constanţa:** Vila nr 1, 8741 Mamaia. On 1458kHz, 100.1MHz: 0700-1700, **R. Vacanţa (R. Holidays):** May-Sept. 0800-1000, 1700-1900 in Romanian, E, F, G, Russian – **Craiova:** str. Ştirbei Voda nr. 3, 1100 Craiova. On 558kHz: W0400-2200, Sun 0500-1100 - **Iaşi:** str. Lascar Catargi nr. 44, 6600 Iaşi. On 1053kHz: 0500 (SS0600)-2000 – **Timişoara:** str. Pestalozzi nr. 14-a, 1900 Timişoara. On 630kHz: 0500-2000 – **Târgu Mureş:** b-dul 1 Decembrie 1918 nr. 109, 4300 Târgu Mureş. On 1197/1323/1593 (Harghita) kHz: 0500-1200, 1400-2000.

RADIO ROMÂNIA INTERNATIONAL
See International Broadcasting section.

Private stations

	MHz	kW	Name and location
1)	65.84	0.05	Univers FM, Braila
2)	65.90	0.1	R. Top '91, Suceava
3)	65.95	0.1	R. Arad, Arad
4)	66.02	0.05	T5 ABC, Haţeg
5)	66.14	0.2	R. Etalon, Râmnicu Vâlcea
6)	66.35	0.01	R. Star, Braşov (Predeal) (rel. RFE 1700-1800)
7)	66.56	0.05	R. Sica, Alba Iulia (Hunedoara)
8)	66.74	0.05	R. Dolly-do Braila, Galaţi (Braila) (rel. RFE 1700-1800)
9)	66.92	0.08	R. Argus, Tulcea
10)	66.92	0.1	R. Târgu Jiu, Tg. Jiu

	MHz	kW	Name and location
11)	67.01	0.1	R. Sibiu, Sibiu
12)	67.10	0.1	R. Campus, Urziceni (Slobozia)
7)	67.40	0.1	R. Sica, Alba Iulia (DEva)
13)	67.34	0.1	R. Alpin, Braşov
14)	67.43	0.1	R. Europea Nova, Bucureşti
	67.58		Strasgeny (rel. RFE 1600-2000)
	67.64		Braila
15)	67.67	0.1	R. Tîmpa, Braşov
6)	67.67	0.1	R. Star, Braşov (Fagaras) (rel. RFE 1800-1900)
16)	67.73	0.1	Unison R, Bârlad (Vaslui)
17)	67.79	1	R. Star B, Bacau
18)	67.80	0.05	FUN R, Bucureşti
19)	67.80	0.05	Radio Z, Bucureşti (rel. RFE 1600-1900)
20)	67.85	0.1	R. Cinemar, Sibiu
21)	68.24	0.1	R. Eveniment, Sibiu
22)	68.24	0.1	Radio 1, Braşov
23)	68.24	0.1	Vocea Speranţei, Bucureşti (Braşov)
24)	68.26	0.1	R. Transilvania L.B.M, Oradea (Zalau)
25)	68.36	0.5	R. Tineret, Focşani
26)	68.60	0.08	Radio D, Târgovişte
27)	68.70	0.05	R. Media Buzau, Buzau
28)	68.70	0.1	R. Sonic, Cluj-Napoca
29)	68.70	0.2	R. Tinerama, Bucureşti
30)	68.72	0.1	R. Activ, Sibiu (rel. RFE 0600-0700, 1600-2000)
31)	68.75	0.05	R. Galaţi, Galaţi
24)	68.90	0.1	R. Transilvania L.B.M, Oradea (Satu Mare)
32)	68.96	0.5	R. Doina FM, Constanţa
14)	69.11	0.08	R. Europa Nova, Lugoj
24)	69.11	0.1	R. Transilvania L.B.M, Oradea (Bistriţa)
33)	69.26	0.08	R. Prahova, Ploieşti
	69.53		Ungeny (rel. RFE 1600-2000)
34)	69.80	0.05	R. Bruzau, Bruzau
35)	69.80	0.08	Uniplus R, Bucureşti
36)	70.04	0.1	Radio XXI, Bucureşti (Alba Iulia)
	70.31		Edinetz (rel. RFE 1600-2000)
37)	70.46	0.05	R. Terra, Piatra Neamţ
38)	70.76	0.1	R. Vox-T, Iaşi
39)	71.21	0.2	R. Vâlcea 1, Vâlcea (rel. RFE 1700-1900)
24)	71.39		R. Transilvania L.B.M. Oradea (Turda)
40)	71.54	0.1	R. Nord 22, Suceava
41)	71.81	0.1	R. Sonvest, Oradea
20)	71.84	0.2	R. Cinemar, Sibiu (Baia Mare)
42)	72.44	0.3	R. Nord-Est, Iaşi
43)	72.68	0.4	R. Romantic, Bucureşti
35)	72.68	0.1	Uniplus R, Bucureşti (Târgu Mureş)
44)	72.80	0.1	R. Giurgiu, Giurgiu
14)	72.92	0.2	R. Europa Nova, Iaşi
12)	87.7	0.1	R. Campus, Urziceni (Slobozia)
45)	87.8	0.08	PRO FM, Bucureşti (Costineşti)
46)	87.8	0.1	R. Braşov, Braşov (rel. RFE 1700-1800)
47)	88.1	0.1	Mediaş 725, Mediaş
7)	88.1	0.1	R. Sica, Alba Iulia
35)	88.3	0.1	Uniplus R, Bucureşti
48)	88.3	0.1	Vocea Evangheliei, Bucureşti (Cluj-Napoca)
49)	88.5	0.08	R. Contact, Bucureşti (Baia Mare)
35)	88.7	0.1	Uniplus R, Bucureşti
50)	88.7	0.1	R. Vest, Timişoara
48)	88.7	0.1	Vocea Evangheliei, Bucureşti (Tmişoara)
46)	88.8	0.1	R. Braşov, Braşov (Fagaras) (rel. RFE 1700-1800)
15)	88.8	0.1	R Tîmpa, Braşov (Fagaras)
16)	88.8	0.1	Unison R, Bârlad
51)	89.0	0.05	R. Metronom-GX, Râmnicu Vâlcea
52)	89.3	0.1	R. Deva, Deva
35)	89.4	0.08	Uniplus R, Bucureşti (Constanţa)
23)	89.4	0.08	Vocea Speranţei, Bucureşti (Constanţa)
53)	89.6	0.1	R. Alfa, Bacau
14)	89.7	0.1	R. Europa Nova, Timişoara
23)	89.7	0.1	Vocea Speranţei, Bucureşti (Timişoara)
35)	89.7	0.2	Uniplus R, Bucureşti (Botoşani)
49)	89.8	0.1	R. Contact, Bucureşti (Cluj-Napoca)
49)	90.2	0.1	R. Contact (Bucureşti (Târgu Mureş)
14)	90.2	0.05	R. Europa Nova, Lugoj
54)	90.7	0.1	Studioul TVC, Piteşti
	91.0		Vaslui (rel. RFE 0500-0700, 1600-1900)
49)	91.1	0.08	R. Contact, Bucureşti (Constanţa)
49)	91.8	0.1	R. Contact, Bucureşti (Sibiu)
49)	91.9	0.1	R. Contact, Bucureşti (Iaşi)
55)	92.1	0.08	R. Galaxia, Baia Mare
45)	92.1	0.075	PRO FM, Bucureşti (Arad)
48)	92.1	0.1	Vocea Evangheliei, Bucureşti (Oradea)
56)	92.7	0.2	Radio 2M+, Bucureşti
57)	92.8	0.1	C.D. Radio Napoca, Cluj-Napoca
49)	92.8	0.08	R. Contact, Bucureşti (Ploieşti)
58)	93.1	0.08	R. Meridian, Botoşani
59)	93.5	1	R. Şcoala "Delta", Bucureşti
60)	94.0	0.1	R. Samtel, Satu Mare
24)	94.2		R. Transilvania L.B.M, Oradea (Alba Iulia)
61)	94.2	0.5	R. Total, Bucureşti
48)	94.2	0.5	Vocea Evangheliei, Bucureşti
23)	94.2	0.5	Vocea Speranţei, Bucureşti
48)	94.2	0.1	Vocea Evangheliei, Bucureşti (Sibiu)
62)	94.9	0.1	R. Hit, Iaşi
63)	95.0	0.1	Studioul de Radio-TV Sicam, Târgu Secuiesc
49)	96.1	0.05	R. Contact, Bucureşti
45)	96.5	0.08	PRO-FM, Bucureşti (Ploieşti)
36)	97.7	0.1	Radio XXI, Bucureşti (Alba Iulia)
12)	98.0	0.08	R. Campus, Urziceni (Buzaau)
64)	98.5	0.08	R. Minisat, Târgovişte
53)	98.8	0.1	R. Alfa, Bacau (Braila)
53)	99.0	0.05	R. Alfa, Bacau (Galaţi)
12)	99.0	0.05	R. Campus, Urziceni
35)	99.0	0.1	Uniplus R, Bucureşti (Suceava)
48)	99.0	0.1	Vocea Evangheliei, Bucureşti (Suceava)
2)	99.2	0.05	R. Top '91, Suceava, Radauţi
36)	100.2	0.5	Radio XXI, Bucureşti
65)	100.5	0.08	R. Galaxy, Dr. Tr. Severin
66)	101.1	0.05	Sky FM Stereo, Constanţa
67)	101.1	0.5	Voces Campi, Calaraşi
43)	101.7	0.3	R. Romantic, Bucureşti
68)	101.7	0.08	Radio S.O.S, Ploieşti
45)	102.8	0.3	PRO FM, Bucureşti
69)	103.3		R. Bârlad. Bârlad
70)	103.6	0.1	R. Korion, Prudeni (Craiova)
9)	FM		R. Argus, Cluj-Napoca
71)	FM		R Teneş, Cluj-Napoca
72)	FM		R. Galaţi, Galaţi

Location within brackets = tr location when different from the studio location

Addresses:

1) str. Victoriei, bl. 8, ap. 48, 6100 Braila – 2) str Ştefan cel Mare 20-a. 5800 Suceava – 3) str. Aurel Vlaicu, Piaţa Agroalimentara, et. 1, 2900 Arad – 4) Piaţa Inirii 1, bl. P/33, Haţeg – 5) spl. Independenţei 9, bl. 10, sc. A, et. 2, sp. 8, 1000 Râmnicu Vâlcea – 6) str. N. Balcescu 18, 2200 Braşov – 7) str. Republicii 14, Alba Iulia – 8) str. Mazepa II, bl. LC8, sc. 1, ap. 10, 6200 Galaţi – 9) str. Garii 2, bl. 1, ap. 1-a, 8800 Tulcea – 10) str. Eroilor 36, Târgu Jiu – 11) Aleea Infanteriştilor, bl. 4, ap. 24, 2400 Sibiu – 12) Calea Bucureşti 67, bl. 314, et. 3, ap. 16, Urziceni – 13) Bd. 15 Noiembrie 47, 2200 Braşov – 14) HQ: str. Dr. Lister 6, sector 5, Bucureşti. Reg. st's: str. Cronicar Mustea 17. 6600 Iaşi – str. Ion Creanga 5-9, 1800 Lugoj – Bd. Revoluţiei 8, 1900 Timişoara – 15) str. Radunicii 3, bl. 4, sc. B, ap. 8, 2200 Braşov – 16) str. 1 Decembrie 2, Bârlad – 17) str. 9 Mai 56, 5500 Bacau – 18) str. Capitan Vijelie 6, sector 5, Bucureşti – 19) Calea Victoriei 133-135, sector 1, Bucureşti – 20) str. Rahova 14, ap. 18, 2400 Sibiu – 21) str. Piaţa Mica 22, 2400 Sibiu – 22) bd. 15 Noiembrie 47, 2200 Braşov – 23) str. Argas 8, sector 2, Bucureşti – 24) str. Crişan 14, 3700 Oradea – 25) str. 1 Decembrie 1918, 32, Focşani – 26) str. Gr. Alexandrescu bl. E8, parter, 0200 Târguvişte – 27) str. D. Gherea, bl.6, ap. 8, 5100 Buzau – 28) str. Dimbrovita 77, 3400 Cluj-Napoca – 29) str. Ion Câmpineanu 20, et. 5, sector 1, Bucureşti – 30) str. Miron Costin 18, 2400 Sibiu – 31) str. Basarabiei 18, 6200 Galaţi – 32) str. 1 Decembrie 118, bl L30, sc. C, ap. 39, 8700 Constanţa – 33) str. Româna 124, 2000 Ploieşti – 34) str. Unirii, bl. E3, et. 1, ap. 40, 5100 Bazau – 35) str. Carol I, 46, sector 2, Bucureşti – 36) Intr. Filioara 3-5, sector 1, Bucureşti – 37) Piaţa Petrodava 1, Hotel Central, et. 12, ap. 1201, 5600 Piatra Neamţ – 38) str. Arcu 1, sc. A, et. 1, ap. 2, 6600 Iaşi – 39) str. Marşal Ion Antonescu 11, bl. M, sc. A, et. 3, ap. 5, 1000 Râmnicu Vâlcea – 40) Bd. 1 Decembrie 1918, 10, et. 5, 5800 Suceava – 41) str. Vrancei 4, 3700 Oradea – 42) str. Smîrdan 5, 6600 Iaşi – 43) şos. Bucureşti-Ploieşti 25- 27, sector 1, Bucureşti – 44) Tineretului, bl. C200CF, sc. B, ap. 40, 8375 Giurgiu – 45) addr. as 19) – 46) str. Stancii 2, 2200 Braşov – 47) str. Ulmului 53, 3125 Mediaş – 48) str. Orzani 84, sector 2, Bucureşti – 49) spi. Independenţei 202-a, et. 12, sector 6, Bucureşti – 50) Bd. M. Viteazu 1, 1900 Timişoara – 51) str.Ştirbei Voda 54, 1000 Râmnicu Vâlcea – 52) str. Lajerului 2, sector 2, Bucureşti – 53) HQ: str. Razboeni, bl. 26, 1000 Râmnicu Vâlcea – Reg. st: str. Saturn 32, bl. 16, et. 10, ap. 153, 6200 Galaţi – 54) str. Triviale, bl. 47, sc. B, et. 3, ap. 15, 0300 Piteşti – 55) Bd. Unirii 14- a, Hotel Sport, cam. 109, 4800 Baia Mare – 56) str. Hristo Botev 8. sector 3, Bucureşti – 57) str. Tache Ionescu 99, ap. 11, 3400 Cluj-Napoca

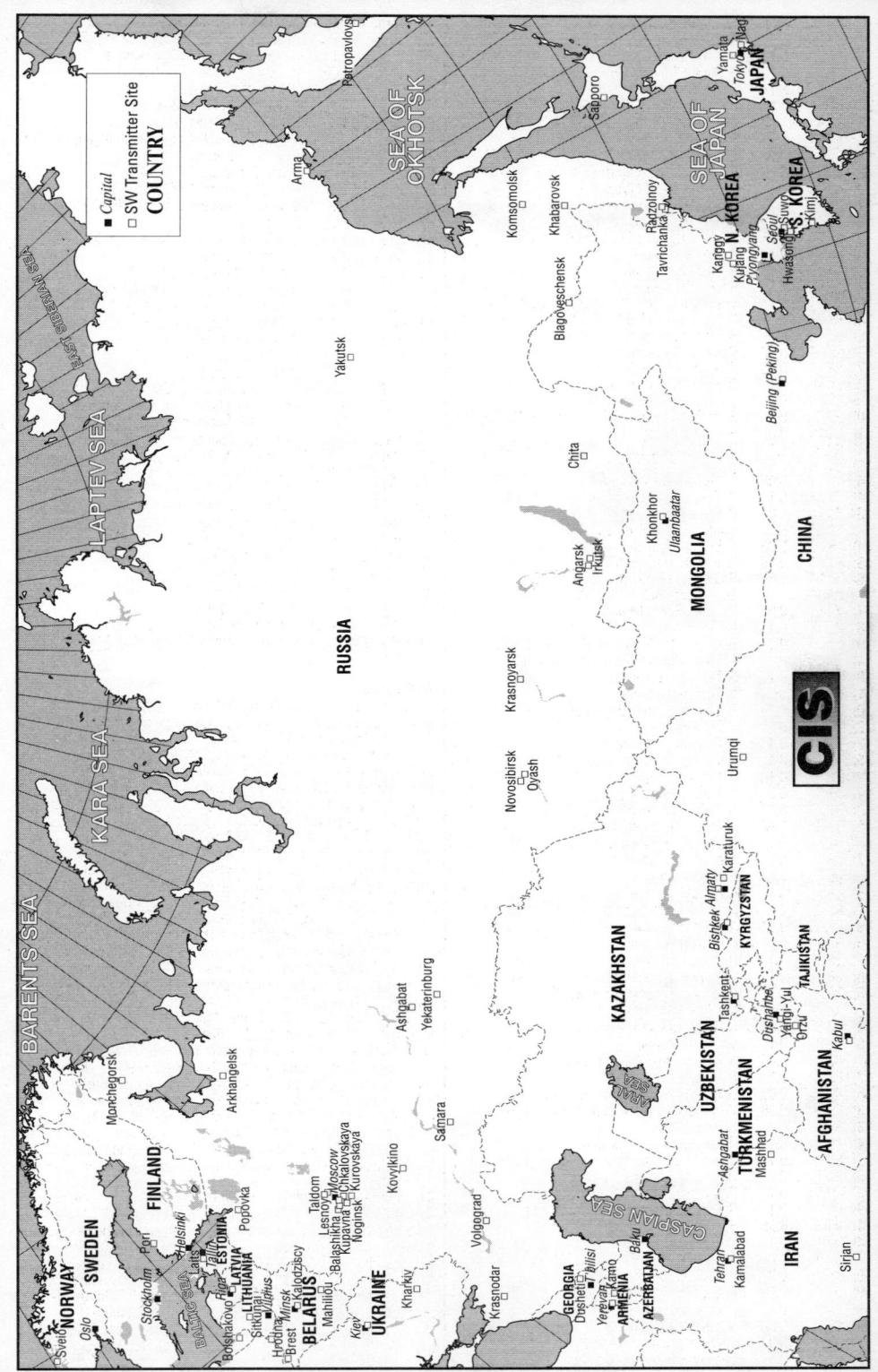

■ *Capital*
□ SW Transmitter Site
COUNTRY

EAST SIBERIAN SEA
LAPTEV SEA
KARA SEA
BARENTS SEA
BALTIC SEA
SEA OF OKHOTSK
SEA OF JAPAN
ARAL SEA
CASPIAN SEA

Petropavlovsk
Arma
Komsomolsk
Khabarovsk
Radzolnoy
Tavrichanka
Yakutsk
Blagoveschensk
Chita
Khonkhor
Ulaanbaatar
Angarsk
Irkutsk
Krasnoyarsk
Novosibirsk
Oyash
Urumqi

RUSSIA
MONGOLIA
CHINA

Yamata
Tokyo Nag
JAPAN
Kanggye
Kujang N. KOREA
Seoul S. KOREA
Pyongyang Hwaŭng Suwŏn Kimj
Sapporo
Beijing (Peking)

CIS

Karaturuk
Bishkek Almaty
Karatal
KYRGYZSTAN
TAJIKISTAN
Tashkent
Dūshanbe
Yangi-Yul
Orzu
Kabul

KAZAKHSTAN
UZBEKISTAN
AFGHANISTAN

Ashgabat
Yekaterinburg
Ashgabat
TURKMENISTAN
Mashhad
Kamalabad
Tehran
IRAN
Baku
Sirjan

Arkhangelsk
Monchegorsk
Samara
Volggograd
Ashgabat

SWEDEN
NORWAY
FINLAND
ESTONIA
LATVIA
LITHUANIA
BELARUS
UKRAINE
GEORGIA
ARMENIA
AZERBAIJAN

Oslo
Svelo
Son
Stockholm
Lais
Helsinki
Tallinn
Bolshakovo Riga
Vln
Sikilda
Hrodna
Brest
Minsk
Mahilioŭ
Kiev
Kharkiv
Krasnodar
Kovylkino
Taldom
Lesnoy
Moscow
Balashikha
Kupavna
Noginsk
Chkalovskaya
Kurovskaya
Popovka
Dosheti
Tbilisi
Yerevan
Gamo
Baku

105

– 58) Bd. Camil Ressu 56, sector 3, Bucureşti – 59) IPB, Fac. Electronica-telecom, Bd. A. Poporului 1-3, Bucureşti – 60) Piaţa 25 Octombrie 3, 3900 Satu Mare – 61) str. A.I. Cuza 22, sector 1, Bucureşti – 62) Bd. Copou 3-5, 6600 Iaşi – 63) str. Dealului 8, bl. 11, sc. B, ap. 8, 4000 Sfîntu-Gheorghe – 64) Bd. Independenţei 55, bl. 32, sc. B, et. 4, ap. 37, 0200 Târgovişte – 65) str. Crişan 25, 1500 Dr. Tr. Severin – 66) str. 1 Decembrie 1918, bl. L30, sc. C, ap. 39, 8700 Constanţa – 67) Be. 1 Mai, bl. A11/5, sc. C, ap. 7-8, 8500 Calaraşi – 68) str. Industriei bl. 16-b, sc. B, ap. 32, 2000 Ploieşti – 69) Bârlad – 70) str. Zavideni Gara 5, Prundeni – 71) Cluj-Napoca – 72) Galaţi

RUSSIA

L.T: Moscow: UTC + 3h (Su: UTC + 4h); See world time table for other zones — **Pop:** 147.200.000 — **Radios:** N/a — **Pr.L:** Russian, numerous minority languages — **E.C:** a/c 50Hz, 127/220V (no) — **ITU:** RUS

MINISTERSTVO SVYAZI
(Ministry of Communications)
ul. Tverskaya 5, 103375 Moskva. ☎ –7 (095) 2927070. **Fax:** +7 (095) 2972128. ① 113748 MSSR RU.
LP: Minister: Vladimir B. Bulgak.

MAIN CENTRE FOR CONTROL OF BROADCASTING NETWORKS
ul. Nikolskaya 7, 103012 Moskva. ☎ +7 (095) 9212501. ☏ +7 (095) 9216024. ① 412912 GCUB — Chief: Anatoliy T. Titov.

RUSSIAN FEDERAL SERVICE FOR TV AND BROADCASTING (FSRTR)
ul. Pyatnitskaya 25, 113326 Moskva. ☎ +7 (095) 2336603. ☏ +7 (095) 2332893.
L.P: Head: N/a. Deputy head: Vladimir Makoveyev.
FSRTR is responsible for the licensing of radio and TV stations and for the administration and coordination as well as financing of the Russia-wide and regional state TV and radio broadcasting organizations.

RADIO 1 KULTURA (Gov.)
ul. Malaya Nikitskaya 24, 121812 Moskva. ☎ +7 (095) 2179962. ☏ +7 (095) 2151314.
L.P: Gen. Dir: Yevgeniy A. Khoroshevtsev.

RADIOSTANTSIYA MAYAK (Gov.)
ul. Pyatnitskaya 25, 113326 Moskva. ☎ +7 (095) 2179340. ☏ +7 (095) 2156956.
L.P: Gen. Dir: Vladimir Povolyayev.

RADIOSTANTSIYA YUNOST (Gov.)
ul. Pyatnitskaya 25, 113326 Moskva.
☎ +7 (095) 2336767. ☏ +7 (095) 2336244.
L.P: Gen. Dir: Yevgeniy V. Pavlov.

RADIO ORFEY (Gov.)
ul. Kachalova 24, 121069 Moskva.
☎ +7 (095) 2220127. ☏ +7 (095) ...

VSEROSSIYSKAYA GOSUDARSTVENNAYA TELERADIOKOMPANIYA (VGTRK, Gov.) 'RADIO ROSSII'
Yamskogo polya 5-ya ul. 19/21, 125124 Moskva. ☎ +7 (095) 2514050. ☏ +7 (095) 2144767.
L.P: Chmn: Eduard Sagalayev. Gen. Dir: Anatoliy G. Lysenko.

RUSSIAN STATE RADIO BROADCASTING COMPANY 'VOICE OF RUSSIA' (Gov.)

ul. Pyatnitskaya 25, 113326 Moskva. ☎ +7 (095) 2337860.
LP: Chmn: Armen Oganesyan.
This company produces the Voice of Russia/Golos Rossii external service programmes.

HOME SERVICES
Abbreviations: R-1, RO-1/RO-4 = Radio 1 prgr versions — RSM = R. St. Mayak — RSY = R. St. Yunost — RO = R. Orfey — RR = R. Rossii — Reg = Regional prgr's — VOR = V. of Russia prgrs for local listeners — F = Foreign sce's — E = European part of Russia — S = Siberia — FE = Far East — v = variable times, cf. addr. section.

LONG & MEDIUMWAVES (Rg = Region)

Rg	Location	kHz	kW	H. of tr.	Prgr.
MO	Taldom, E	153	300*	0230-0000	RSY
KH	Komsomolsk, FE	153	1200	2000-1600	RR, Reg
BA	Ufa, E	162	300	0100-2000	RO-4, Reg
TM	Norilsk, S	162	150	2200-1800	RR, Reg
MO	Elektrostal, E	171	1250	0400-2100	R-1
KA	Bolshakovo, E	171	1200	0400-2100	R-1
KO	Syktyvkar, E	171	150	0400-2100	R-1
KD	Tbiliskoye, E	171	1200	0400-2100	R-1
MU	Murmansk, E	171	150	0400-2100	R-1
NK	Oyash, S	171	250	2200-1800	RR, Reg
SA	Yakutsk, S	171	150	2000-1600	RR, Reg
CH	Chita, S	180	150	1800-1600	RSM
KM	Petropavlovsk, FE	180	150	1800-1400	RR, Reg
AM	Belogorsk, S	189	1200	2000-1600	RR, Reg
MO	Noginsk, E	198	150	0230-2300	RSM
SP	Olgino, E	198	150	0000-2300	RSM
BA	Ufa, E	198	150	0000-2300	RSM
IR	Angarsk, S	198	250	2300-1600	RSM
AM	Tynda, S	207	150	2000-1900	RSM
KN	Krasnoyarsk, S	216	150	2200-1800	RR, Reg
YV	Birobidzhan, FE	216	30	2000-1600	RR, Reg
KY	Surgut, S	225	1000	0000-2000	RR, Reg
SP	Krasnyy Bor, E	234	1200	0400-2100	R-1
AR	Arkhangelsk, E	234	500	0400-2100	R-1
SM	Syzran, E	234	2000	0400-2100	R-1
IR	Angarsk, S	234	1000/300	2200-1800	RR, Reg
MA	Arman, FE	234	1000	1800-1400	RR, Reg
PM	Razdolnoye, FE	243	1200	2000-1600	RR, Reg
TS	Kazan, E	252	150	0400-2100	R-1, Reg
MO	Taldom, E	261	2500	0200-2200	RR
CH	Chita, S	261	1200	2000-1600	RR, Reg
OR	Orenburg, S	270	50	0000-2000	RR, Reg
NK	Novosibirsk, S	270	150	2200-1800	RR, Reg
KH	Khabarovsk, FE	270	150	2100-1600	RO-2
YE	Yekaterinburg, S	279	150	0000-2000	RR, Reg
AT	Gorno-Altaysk, S	279	50	2200-1800	RR, Reg
BU	Selenginsk, S	279	150	2158-2100	RR, Reg
SL	Yuzhno-Sakhalinsk, FE	279	1000	1800-1400	RR, Reg
CV	Cheboksary, E	531	30	0145-2200	RSM, Reg
OR	Orenburg, E	540	50	2300-2200	RSM
MO	Kurovskaya, E	549	1000*	0300-0000	RSM
SP	Krasnyy Bor, E	549	1200	0400-2300	RSM
KA	Kaliningrad, E	549	50	0400-2300	RSM
KO	Syktyvkar, E	549	150	0400-2300	RSM
RO	Novocherkassk, E	549	50	0400-2300	RSM
SA	Yakutsk, S	549	50	2000-1900	RSM
AM	Svobodnyy, S	549	150	2000-1900	RSM
MA	Magadan, FE	549	50	2000-1800	RSM
PM	Tavrichanka, FE	549	500	2000-1400	RSM
BU	Ulan-Ude, S	558	25	0700-0600	RSM
VO	Volgograd, E	567	1000	0200-2200	RR, Reg
TU	Kyzyl, S	567	150	2200-1800	RR, Reg

Rg	Location	kHz	kW	H. of tr.	Prgr.
KB	Nalchik, E	576	25	0100-2100	RSM, Reg
KB	Tegenekli, E	576	1	0100-2100	RSM
AS	Astrakhan, E	576	50	0200-0100	RSM
TY	Tyumen, S	576	100	0000-2000	RR, Reg
NK	Oyash, S	576	1000	0000-1200	RSM
IR	Angarsk, S	576	250	2300-1600	RSM
KH	Khabarovsk, FE	576	150	2000-1400	RSM
KM	Petropavlovsk, FE	576	150	1800-1300	RSM
PR	Perm, E	585	150	0000-2000	RR, Reg
AM	Belogorsk, S	585	1200		F
UD	Izhevsk, E	594	40	0200-2200	RR, Reg
SO	Vladikavkaz, E	594	25	0200-2200	RSM, Reg
KY	Surgut, S	594	1000	2345-2300	RSM
KN	Krasnoyarsk, S	594	150	2200-2100	RSM
YN	Salekhard, S	603	50	0000-2000	RR, Reg
AM	Belogorsk, S	603	30	2100-2200	RSM
AM	Skovorodino, S	603	30	2100-2200	RSM
MO	Kurkino, E	612	40	0300-2200+	VOR
KT	Petrozavodsk, E	612	150	24h	RSM
TM	Norilsk, S	612	25	2200-2000	RSM
KO	Syktyvkar, E	621	50	0200-2200	RR, Reg
DA	Makhachkala, E	621	50	0200-2200	RR, Reg
DA	Kochubey, E	621	5	0200-2200	RR, Reg
SA	Yakutsk, S	621	25	2100-1600	RO-2
KH	Khabarovsk, FE	621	50	2000-1300	RR, Reg
SR	Saratov, E	630	42	0400-2100	R-1, Reg
KH	Komsomolsk, FE	630	500	1100-1400	F
OM	Omsk, S	639	150	0000-2000	RR, Reg
PM	Ussuriysk, FE	648	1000	2000-1900	F
MU	Murmansk, E	657	150	0200-2200	RR, Reg
CC	Groznyy, E	657	50		
KD	Sochi, E	666	25	0200-2200	RSM
SA	Aldan, S	666	7	2100-1600	RO-2
KH	Komsomolsk, FE	666	150	2000-1400	RSM
KO	Ukhta, E	675	5	0400-2100	R-1
NK	Oyash, S	675	250	0100-2000	RSY
BA	Ufa, E	693	150	0000-2000	RR
CK	Anadyr, FE	693	25	1800-1400	RR, Reg
IR	Bratsk, S	702	7	2200-1700	RSM
NE	Naryan-Mar, E	711	7	0200-2200	RR, Reg
KH	Okhotsk, FE	711	7	2100-1500	RSM, Reg
SL	Yuzhno-Sakhalinsk, FE	720	1000	1100-1400	F
				1800-0900	RSM
MO	Moskva, E	738	5	0300-2100	RO
CB	Chelyabinsk, S	738	40	0000-2000	RSM, Reg
KK	Palana, FE	738	25	1800-1400	RR, Reg
	Unknown, FE	756		1900-1400	RO-1
KT	Petrozavodsk, E	765	150	0200-2200	RR, Reg
VN	Somovo, E	774	30	0200-0100	RSM, Reg
PM	Tavrichanka, FE	783	500		
AS	Astrakhan, E	792	50	0400-2100	R-1, Reg
RK	Abakan, S	792	25	2200-1800	RR, Reg
SL	Aleksandrovsk-Sakhalinskiy, FE	792	50	1800-1400	RR, Reg
SP	Krasnyy Bor, E	801	600	0300-2200	Reg
CH	Chita, S	801	1200	2100-1600	RO-2, F
VO	Volgograd, E	810	500	0100-2400	RSM
YE	Yekaterinburg, S	810	500	0100-2400	RSM
PM	Razdolnoye, FE	810	150	2100-1000	RR, Reg
NN	Nizhniy Novgorod, E	828	50	0200-0100	RSM, Reg
DA	Unknown, E	828			?, Reg
TU	Kyzyl, S	828	150	1900-1800	RSM
MO	Noginsk, E	846	150	0300-1900v	Reg
KG	Elista, E	846	42	0200-2200	RR, Reg
PE	Kamenka, E	855	50	0200-2200	RR, Reg
KM	Petropavlovsk, FE	855	20	1900-1400	RO-1
AM	Blagoveshchensk, S	864	25	2100-1600	RO-2
MO	Elektrostal, E	873	1200	0200-2200	RR
SP	Olgino, E	873	150	0200-2200	RR
SM	Novosemeykino, E	873	2000	0200-2200	RR
KA	Kaliningrad, E	873	50	0300-2200	RR
BE	Novyy Oskol, E	873	1	0300-2200	RSY
BE	Volokonovka, E	873	1	0200-0000	RSY
	Unknown, FE	873		2000-1600	RR
ST	Stavropol, E	882	30	0200-2200	RSM, Reg
SL	Aleksandrovsk, FE	882	10	1900-1400	RO-1
TY	Tyumen, E	891	50	0100-2000	RO-4
ME	Sovietskiy, E	900	25	0155-2200	RSM, Reg
	Unknown, So. E	909		0200-2200	RSM
PM	Dalnerechensk, FE	909	5	2100-1600	RO-2
AR	Arkhangelsk, E	918	150	0200-2200	RR, Reg
DA	Makhachkala, E	918	50	0200-0100	RSM, Reg
KU	Makushino, E	918	5	0000-2000	RSM
KU	Shadrinsk, S	918	7	0000-2000	RSM
KU	Shumikha, S	918	5	0000-2000	RSM
KH	Komsomolsk, FE	927	25	2100-1600	RO-2
OR	Matveyevka, E	936	5	0000-2300	RR, Reg
KN	Krasnoyarsk, S	936	5	2200-1800	RSY
PM	Dalnegorsk, FE	936	5	2100-1600	RO-2
RO	Rostov-na-Donu, E	945	150	0200-2200	RR, Reg
BU	Guzino-Ozersk, S	963	1	2158-2000	RR, Reg
BU	Zakamensk, S	963	20	2158-2000	RR, Reg
SL	Yuzhno-Sakhalinsk, FE	972	30	1900-1400	RO-1
KN	Krasnoyarsk, S	981	10	2300-1800	RO-3
CB	Yuryuzan, S	990	1	0000-2200	RSM
YV	Birobidzhan, FE	999	7.5	24h	RSM
KD	Tuapse, E	1008	25	0200-2200	RSM
KM	Petropavlovsk, FE	1008	50	1800-1400	RSY, Reg
KH	Khabarovsk, FE	1008	25	2000-1600	RSY
AR	Arkhangelsk, E	1026	5	0200-2200	RSM
AR	Nyandoma, E	1026	7	0200-2200	RSM
AR	Porog, E	1026	5	0200-2200	RSM
AR	Urdoma, E	1026	5	0200-2200	RSM
AR	Vazhskiy, E	1026	5	0400-2100	R-1
NK	Oyash, S	1026	1000		
AR	Kotlas, E	1044	7	0400-2100	R-1
AM	Belogorsk, S	1053	25	2100-1600	RO-2
MD	Saransk, E	1062	25	0200-2200	RR, Reg
KM	Ust-Kamchatsk, FE	1062	1	1800-1700	RSM
AM	Zeya, S	1071	7	2100-2000	RSM
MU	Murmansk, E	1080	30	0300-2200	RSY
IR	Angarsk, S	1080	1000	1230-1600	F
KD	Tbilisskoye, E	1089	1200	0200-1600	RR, Reg
				1600-2300	F, RSM
KK	Tilichiki, FE	1089	5	1800-1400	RR, Reg
VG	Vologda, E	1098	7	0200-2200	RR, Reg
VG	Chagoda, E	1098	7	0200-2200	RR, Reg
VG	Nikolsk, E	1098	5	0200-2200	RR, Reg
VG	Totma, E	1098	7	0200-2200	RR, Reg
VG	Ustyuzhna, E	1098	7	0200-2200	RR, Reg
AR	Arkhangelsk, E	1107	50	0200-0100	RSY
SM	Samara, E	1107	150	0200-1600v	Reg
KB	Nalchik, E	1107	25	0400-2100	R-1, Reg
KB	Tegenekli, E	1107	1	0200-0000	R-1, Reg
KA	Kaliningrad, E	1116	30	0200-2200	RSY, Reg
KD	Sochi, E	1116	30	0200-2200	RR
YV	Birobidzhan, FE	1116	25	2100-1500	RSY
SP	Olgino, E	1125	150	0300-2100	RO
MU	Murmansk, E	1134	75	0300-0200	RSM
MD	Saransk, E	1134	30	0200-0000	RSM
PS	Velikiye Luki, E	1134	7	0300-2200	RSM

Rg	Location	kHz	kW	H. of tr.	Prgr.	Rg	Location	kHz	kW	H. of tr.	Prgr.
RO	Salsk, E	1134	5	0400-2100	R-1	SA	Sangar, S	1413		2000-1600	RR
RO	Veshenskaya, E	1134	5	0400-2100	R-1	SM	Samara, E	1422	300	0300-2200	RSY
RO	Volgodonsk, E	1134	5	0400-2100	R-1	CH	Kuanda, S	1422	5	0000-2000	RR
BR	Shvedchiki, E	1134	7	0200-2200	RSM	SA	Neryungri, S	1431	7	2000-1600	RSM
KA	Bolshakovo, E	1143	150	0500-0000	RSM, F	SL	Nogliki, FE	1431	0.2	1900-1400	RO-1
SM	Mekhzavod, E	1143	150	0025-2300	RSM	AL	Kosh-Agach, S	1440	5	2200-1800	RR, Reg
IR	Tayshet, S	1143	7	2200-1700	RSM	AL	Ust-Koksa, S	1440	5	2200-1800	RR, Reg
MO	Kupavna, E	1152	150	0300-2100	RO	TV	Kushalino, E	1449	20	1100-2100	RSY
KY	Khanty-Mansiysk, S	1152	20	0100-2100	RR, Reg	AR	Plesetsk, E	1449	7	0200-2200	RSM
KH	Komsomolsk, FE	1152	50	2000-1400	RSM, Reg	MU	Gremikha, E	1449	7	0200-2200	RSM
VO	Volgograd, E	1161	75	0300-2100	RO	MU	Kandalaksha, E	1449	1	0200-2200	RSM
KD	Tbilisskoye, E	1170	1200	1500-2100	F	MU	Kirovsk, E	1449	5	0200-2200	RSM
IR	Angarsk, S	1170	250	2300-1800	RO-3	MU	Monchegorsk, E	1449	42	0200-2200	RSM
KY	Berezovo, S	1197	5	0000-2000	RR, Reg	MU	Nikel, E	1449	1	0200-2200	RSM
KA	Bolshakovo, E	1215	600			MU	Umba, E	1449	7	0200-2200	RSM
IR	Ust-Kut, S	1224	7	2200-1700	RSM	BR	Unecha, E	1449	7	0200-2200	RSM
AR	Velsk, E	1233	25	0400-2100	R-1	BE	Valuyki, E	1458	5	0300-2200	RSY
KC	Cherkessk, E	1251	7	0200-2200	RSM, Reg	KV	Kirov, E	1458	7	0300-2200	RSY
KC	Urup, E	1251	1	0200-2200	RR, Reg	KV	Pinyug, E	1458	7	0200-2200	RSM
ST	Letnyaya Stavka, E	1251	5	0200-0000	RSM	KP	Kudymkar, E	1458	7	0100-2300	RR, Reg
ST	Neftekumsk, E	1251	1	0200-0000	RSM	OM	Tara, S	1458	5	0100-2000	RO-4
UD	Izhevsk, E	1251	7	0200-2200	RSY	TL	Aleksin, E	1476	1	0400-2100	R-1
PM	Ussuriysk, FE	1251	600	2100-1700	RO-2, F	AL	Onguday, S	1476	20	2200-1800	RR, Reg
NN	Nizhniy Novgorod, E	1278	30	0200-2200	RSY	PM	Tavrichanka, FE	1476	1000	2100-1600	RSY
SR	Balakovo, E	1278	5	0200-2200	RSY	PR	Oktyabrskiy, E	1485	1	0000-2000	RR
SR	Balashov, E	1278	5	0200-2200	RSY	CB	Kartaly, S	1485	5	0000-2200	RSM, Reg
SR	Saratov, E	1278	30	0200-2200	RSY	TY	Gaz Sale, S	1485	1	0000-2000	RR
SR	Yershov, E	1278	5	0200-2200	RSY	YN	Tazovskiy, S	1485	1	0000-2000	RR
UD	Balezino, E	1278	5	0200-2200	RSY	YN	Krasnoselkup, S	1485	1	0000-2000	RR
VO	Dubovka, E	1278	50	0200-2200	RSY	SA	Cherskiy, S	1485	1	2000-1800	RSM
RO	Novocherkassk, E	1278	50	0200-2200	RSY	SA	Tomtor, S	1485	0.2	2000-1800	RSM
BU	Barguzin, S	1278	20	2158-2000	RR, Reg	SA	Ust-Maya, S	1485	1	2000-1800	RSM
BU	Bagdarin, S	1278	5	2158-2000	RR	SA	Yakutsk, S	1485	1		
BU	Severobaykalsk, S	1278	7	2158-2000	RR, Reg	KK	Kamenskoye, FE	1485	1	1800-1400	RR, Reg
KL	Sukhinichi, E	1287	5	0200-2200	RSY	PM	Chkalovskoye, FE	1485	1	2100-1600	RO-2
CC	Groznyy, E	1287	7			PM	Plastun, FE	1485	1	2000-1800	RSM
KO	Syktyvkar, E	1287	10	0200-2200	RSY	SP	Krasnyy Bor, E	1494	1200	1600-2200	F
BA	Yazykovo, E	1287	50	0200-2200	RSY	BE	Belgorod, E	1503	7	0300-2200	RSY
NK	Novosibirsk, S	1287	75			YN	Salekhard, S	1503	5	2300-2000	RSM
BU	Kyakhta, S	1287	5	2158-2000	RR, Reg	BE	Staryy Oskol, E	1512	5	0300-2200	RSY
TU	Shagonar, S	1287	7	2200-1800	RR, Reg	KD	Sochi, E	1512	7	0400-2100	R-1
TU	Chadan, S	1287	7	2200-1800	RR, Reg	CV	Ibresi, E	1512	7	0200-0000	RSY, Reg
MO	Kupavna, E	1305	300	0400-2100	R-1	PR	Chaykovskiy, E	1512	5	0000-2000	RR, Reg
YE	Serov, S	1305	7	24h	RSM	MU	Zapolyarnyy, E	1521	7	0000-2300	RSM
AL	Onguday, S	1350	5	2200-2000	RSM	TS	Kazan, E	1521	20	0100-2200	RSM
AL	Choya, S	1350	1	2200-2000	RSM	KN	Boguchany, S	1521	5	2200-1800	RR
AL	Shebalino, S	1350	1	2200-2000	RSM	CH	Krasnyy Chikoy, S	1530	5	1900-1500	RSM
AL	Ust-Kan, S	1350	5	2200-1800	RR, Reg	KD	Sochi, E	1539	5	0300-2200	RSY
AL	Ust-Ulagan, S	1350	5	2200-1800	RR, Reg	SL	Aleksandrovsk-Sakhalinskiy, FE	1548	10	1800-1400	RSM
MO	Chkalovskaya, E	1359	150			SL	Yuzhno-Sakhalinsk, FE	1575	7	1800-1600	RSY, Reg
PR	Perm, E	1359	40	0100-0000	RSM		Unknown, So. E	1584		0200-2200	RR
IR	Ust-Ilimsk, S	1359	7	2200-1700	RSM	OR	Pleshanovo, E	1584	1	0000-2000	RR, Reg
YE	Yekaterinburg, S	1377	50	0100-2000	RSY	DA	Khunzakh, E	1584	7	0200-2200	RR
PM	Tavrichanka, FE	1377	75	2000-1500	RSY	BU	Taksimo, S	1584	1	2200-1800	RR
SL	Okha, FE	1377	5	1800-1400	RR, Reg	SA	Aykhal, S	1584	5	2000-1600	RSM
KA	Bolshakovo, E	1386	1200	1600-2000	F	SA	Ust-Nera, S	1584	1	2000-1800	RSM
KL	Obninsk, E	1386	5	0300-2200	RSY	SA	Yakutsk, S	1584	1	24h	RSM
	Unknown, E	1386		0300-2200	RSY	KM	Klyuchi, FE	1584	1	1800-1400	RR, Reg
KT	Petrozavodsk, E	1395	50	0300-2200	RSY	KK	Tigil, FE	1584	1	1800-1400	RR, Reg
OR	Buguruslan, S	1395	5	0000-2000	RR, Reg	IR	Irkutsk, S	1593	50	2300-1700	RSY
YA	Dubki, E	1395	1	0200-2200	RSY	KP	Gayny, E	1602	5	0000-2000	RR, Reg
YA	Lyubim, E	1395	1	0200-2200	RSY	BU	Novo-Ilinsk, S	1602	1	2200-1800	RR
YA	Volga, E	1395	5	0200-2200	RSY	BU	Ust-Barguzin, S	1602	1	2200-1800	RR
ST	Stavropol, E	1413	30	0200-2200	RSY	SL	Kurilsk, FE	1602	1	24h	RSM
VO	Dubovka, E	1413	20	0400-2100	R-1	SL	Severo-Kurilsk, FE	1602	1	24h	RSM
VO	Volgograd, E	1413	7	0400-2100	R-1						

*) half power during the night. +) shared with independent stations.

SHORTWAVES

Domestic Services on Shortwave

Rg	St. name, location	kHz	kW	H. of tr.	Prgr
EV	Tura, S	4040	5	2200-1800	RR, Reg
SA	Yakutsk, S	4395	15	2000-1600	RSY
BA	Yazykovo, E	4485	17	0000-2000	RR, Reg
KK	Palana, FE	4520	2	2000-0800v	Reg
KY	Khanty-Mansiysk, S	4520	5	0000-2000	RSM
BU	Ulan-Ude, S	4795	50		RR, Reg
KY	Khanty-Mansiysk, S	4820	50	0000-2000	RR, Reg
SA	Yakutsk, S	4825	50	2000-1600	RR, Reg
CH	Chita	4860	80	2000-1600	RR
TY	Tyumen, S	4895	50	0000-2000	RR, Reg
PM	Tavrichanka, FE	5015	100		
PR	Perm, E	5290	5	0000-1600	RR, Reg
KN	Krasnoyarsk, S	5290	50	2200-1800	RR, Reg
MU	Monchegorsk, E	5930	40	0200-2200	RR, Reg
MA	Arman, FE	5940	100	1800-1400	RR, Reg
AM	Blagoveshchensk, S	6060	5	2000-1600	RR, Reg
KY	Khanty-Mansiysk, S	6095	2.7	0000-1600	RR, Reg
AR	Arkhangelsk, E	6160	40	0200-2200	RR, Reg
SA	Yakutsk, S	7140	50	2000-1600	RR, Reg
ME	Yoshkar-Ola, E	7200	5	0200-1600	RR, Reg
SA	Yakutsk, S	7200	100	2000-1600	RR, Reg
KH	Khabarovsk, FE	7210	100	2000-1600	RR, Reg
MA	Arman, FE	7320	100	1800-1400	RR, Reg
SA	Yakutsk, S	7345	50	2000-1600	RR, Reg
MA	Arman, FE	9530	100	1800-1400	RR, Reg
MA	Arman, FE	9600	100	1800-1400	RR, Reg
SL	Yuzhno-Sakhalinsk, FE	11840	15	1800-1400	RR, Reg
ME	Yoshkar-Ola, E	15165	5	1605-2300	RR, Reg

The national services are also using high power transmitters in the 16-49m bands. Times and frequencies vary over the year.

NATIONAL PRGRS

Radio 1 (R-1): Aired in a 'European' edition 0400-2100 for time zones UTC +2, +3, +4 and four time-shifted 'Orbita' versions. Radio 1 (Radio Odin) is a general style 'full service' prgr.

Radiostantsiya Mayak (RSM): Light music, sports. On the air continuously. IS and news, weather or other information every hour at :00, :20, :40.

Radiostantsiya Yunost-Molodezhniy kanal (RSY): The youth channel, with music and other prgr's for the younger generation. On the air continuously. European block: 0300-2200.

Radio Orfey (RO): Classical music, drama. On the air to European Russia 0300-2100.

R-1, RSY and RSM are rebroadcast to a varying extent in several other CIS countries.

Radio Rossii: A general style 'full service' prgr with a heavy news emphasis. There are five time shifted versions. The European version is on the air 0200-2300.

REGIONAL SERVICES (Gov.)

In most cases regional services produced by provincial state radio broadcasting companies (Gosudarstvennaya teleradiokompaniya — GTRK) are broadcast for a few hours a day on tr's shared with national services (usually R. Rossii, in some cases R-1, RSM or RSY). A number of provinces have no regional services on AM and use FM and wired networks only.

Addresses and other information

Regional services by province:

FM: Key FM freq's (usually 17kW ERP) carrying regional or local sces are given. A freq. with no location shown normally refers to the provincial capital.

NB: Several provincial prgr companies also run prgrs or prgr channels with a name similar to the names of independent sts.

Abbreviations: Resp. = respublika (republic); obl. = oblast (province); ul. = ulitsa (street).

AB) Aginskiy Buryatskiy avt. okrug: 674460 Aginskoye. **Reg:** On 70.93MHz in Buryat/Russian: Mon 2230-2320, Tues/Fri 2300-2330, Wed/Thurs 2300-2345, Tues/Wed 0410-0500, Sat 2300-2400, 0300-0400.

AD) Resp. Adygeya: GTRK Respubliki Adygeya, ul. Zhukovskogo 24, 352700 Maykop. **Reg:** On 1089kHz, 69.08MHz: W 0510-0600, 1310-1400, Mon/Tues, Thurs-Sat 1110-1200, Sun 0510-0600. On 1089kHz + 49 or 41 mb SW: Mon/Fri 1800-1900.

AL) Resp. Altay: GTRK 'Gornyy Altay', pr. Kommunisticheskiy 37, 659700 Altaysk. On 1350/1440/1476kHz, 67.22MHz: MF 2300-2400, 1100-1200.

AM) Amurskaya obl: GTRK 'Amur', per. Svyatitelya Innokentiya 15, 675000 Blagoveshchensk. **Reg:** On 189kHz, (MHz) Belogorsk 66.26, Skovorodino 67.22, Zeya 68.24, Shimanovsk 66.92, Progress 66.74, Tynda 70.50, Svobodnyy n/a: MF 2000-0000, 0300-0400, 0900-1000, Sat 2200-2300, Sun 0010-0100. Prgr Zeya on 72.4MHz Sat/Sun 0100-0400.

AR) Arkhangelskaya obl: GTRK 'Pomorye', ul. Popova 2, 163000 Arkhangelsk. **Reg:** On 918/6160kHz, 66.08MHz + 8 FM relays: MF 0400-0500 & 1520-1600, Sat 0500-0600, 1100-1130 & 1300-1310. On 68.60MHz Sun-Fri 0700-0800 (Thurs 0900).

AS) Astrakhanskaya obl: GTRK 'Lotos', ul. Lyakhova 4, 414040 Astrakhan. **Reg:** On 792kHz, 66.02MHz + 68.57MHz Chernyy Yar, 70.16MHz Tambovka: MF 0320-0400, 0900-1000, 1520-1600, Sat/Sun 0500-0600, 0900-1000, 1500-1700. On 67.58MHz: Thurs 1630-1730.

AT) Altayskiy kray: GTRK 'Altay', Zmeinogorskiy trakt 27-A, 656020 Barnaul. **Reg:** On 279kHz, 68.60MHz + (MHz) Biysk 70.40, Rubtsovsk 69.68, Kamen na Obi 70.31, Blagoveshchenka 66.95, Mamontovo 67.16, Ust-Kalmanka 67.85, Zarinsk 69.53: MF 0100-0200, 0410-0500, 1300-1400, Sat 0200-0500, Sun 0500-0600. **German:** Sat 0425-0455. On 68.60MHz: Sun 0400-0500.

BA) Resp. Bashkortostan: GTRK 'Bashkortostan', ul. Gafuri 9, 450076 Ufa. **Reg:** On 162/693/4485kHz, 68.30MHz in Russian/Bashkirian: MF 0200-0300, 0700-0900, 1300-1500. Sat 0200-0300, 0400-0900, Sun 0300-0400. On 66.68MHz: Radio-2: 0500-2200.

BE) Belgorodskaya obl: GTRK 'Belgorod', ul. Shchertsa 8, 308000 Belgorod. **Reg:** On (MHz) Belgorod 70.16, Staryy Oskol 71.09, Valuyki 66.80: MF 0310-0400, 0510-0600, 1600-1700, Sat 0500-0600, 0700-0800, 1110-1200, Sun 0510-0600.

BR) Bryanskaya obl: GTRK 'Bryansk', ul. Stanke Dimitrova 77, 241033 Bryansk. **Reg:** On (MHz) Bryansk 67.58, Shvedchiki 70.04, Unecha 70.55: W 0410-0500, Tues-Sun 1120-1200, Sat 1500-1600, Sun/Fri 1520-1600. 68.78MHz: 1700-1800.

BU) Resp. Buryatiya: Buryatskoye GTRK, Erbanova 7, 670000 Ulan-Ude. **Reg:** On 279/963/1278kHz, Ulan-Ude 69.74MHz, Kyakhta 70.16MHz in Russian/Buryat: D 2158-2200 & 2210-2300; MF 2320-0000 & 1120-1200; Sun 0430-0500; Sat (varying) 0100-0200, 1000-1100. On 71.84MHz: Tues/Thurs/Fri 0200-0400, Sat 0200-0300.

CB) Chelyabinskaya obl: Chelyabinskoye GTRK, ul. Ordzhonikidze 54-B, 454000 Chelyabinsk. **Reg:** On 738/1485kHz, 71.96MHz + Magnitogorsk 71.81, Yuryuzan 67.25, Kartaly 66.65, Zlatoust 70.16, Kyshtym 67.13, Stepnoye 68.36, Mezhozernoye n/a: MF 0110-0300, 0400-0500, 1310-1400, Tues-Thurs, SS 0500-0600, Sat 0300-0400. On 69.86MHz: Music prgr.

CC) Chechenskaya Resp. Itchkeriya: GTRK 'Vaynakh', ul. Dzerzhinskogo, 364021 Groznyy. **Reg:** On 657kHz, 67.37MHz in Russian, Chechen.

CH) Chitinskaya obl: Chitinskaya GTRK, ul. Kostyushko-Grigorovicha 27, 672090 Chita. **Reg:** In Russian/Buryat on 261/4860kHz, 66.32MHz + (MHz) Orlovskiy 71.93, Khada-Bulak 71.66, Kholbon 69.80, Krasnokamensk 70.67: 1959-2003, MF 2110-2200, D 2210-2300, MF 1000-1100, 1120-1200, Wed/Fri 1210-1300, Sat/Sun 2320-2400,

0010-0100, 0110-0200. On 67.92MHz: 0500-0800 'Dauriya'.

CK) Chukotskiy avt. okrug: GTRK 'Chukotka', ul. Lenina 18, 686710 Anadyr. **Reg:** On 692kHz (4030kHz inactive) in Russian/Chukchi/Eskimo: D 1825-1900, MF 1940-2000, 2015-2100, Sat 2115-2145, Thurs 0115-0145, MF 0700-0800, Mon/Wed/Fri 0815-0900 (also on 5940/9600kHz via Armar, MA). Also rel. GTRK 'Magadan', MA.

CV) Chuvashskaya Resp: Chuvashskaya GTRK, ul. Nikolayeva 4, 428000 Cheboksary-20. **Reg:** On 531/1512kHz, 67.04MHz + Ibresi 70.85MHz in Russian/Chuvash: MF 0320-0537, 0910-1000, 1500-1600 (Wed 1700), Sat 0320-0500, 1400-1700, Sun 0400-0600, 1400-1700.

DA) Resp. Dagestan: GTRK 'Dagestan', ul. Magomed Gadzhieva 188, 367032 Makhachkala. **Reg:** On 621/828/918/1584kHz, 68.87MHz + (MHz) Kochubey 67.04, Gergebil 69.95 (separate prgr's). Russian: D 0340-0400, Sat 0800-0900 & 1215-1300, Sun 0525-0600 & 0820-0900, Wed 0815-0900, 1500-1540/1600. At 0315 and other times in Avar, Azeri, Chechen, Dargin, Kumyk, Lak, Lezgin, Nogay, Tabasaran, Tat.

EV) Evenkiyskiy avt. okrug: Evenkiyskaya GTRK 'Kheglen', ul. 50 let Oktyabrya 28, 663370 Tura. **Reg:** On 4040kHz in Russian, Evenkish: D 0100-0200, 0500-0515.

IN) Ingushskaya Resp: GTRK 'Ingushetiya', 247521 Nazran. Details not available.

IR) Irkutskaya obl: Irkutskaya GTRK, ul. Gorkogo 15, 664000 Irkutsk. **Reg:** On 234kHz, 70.31MHz + 16 additional FM st's: MF 2210-2300, W 2320 (Sat 2302)-0010, Sun 0100-0210, Sat 0200-0300, MF 0400-0410 & 1100-1200. Local stereo prgr on 72.02MHz 0800-1030.

IV) Ivanovskaya obl: Ivanovskaya GTRK, ul. Varentsovoy 24, 153647 Ivanovo. **Reg:** On Ivanovo 71.25MHz, Rodniki 70.13MHz: MF 0320-0400, 1120-1200, 1520-1600, SS 0720-0800, 1120-1200. On Ivanovo 72.77MHz: 1800-1900: Local musical prgr.

KA) Kaliningradskaya obl: GTRK 'Yantar', ul. Klinicheskaya 16, 236016 Kaliningrad. **Reg:** On 1116kHz, 66.02/102.5MHz + Veselovka 65.90MHz: MF 0510-0600, 0610-0700, 1500-1600. Sat/Sun 0600-0700, 1300-1400. At other times on FM only.

KB) Kabardino-Balkarskaya Resp: GTRK 'Kabbalkteleradio', ul. Nogmova 38, 360000 Nalchik. **Reg:** On 1107kHz, 70.52MHz: In Russian, Balkar, Kabardin 0400-0700, 1400-1600. On 576kHz: 0400-0600, 1300-1600. On 72.23MHz: 0600-0700. On 1089kHz + 49 or 41 mb SW: Wed/Thurs/Sun 1830-1900.

KC) Karachayevo-Cherkesskaya Resp: GTRK 'Karachayevo-Cherkesiya', 357100 Cherkessk. **Reg:** On 1251kHz, 72.11 MHz in Russian, Abazin, Cherkess, Karachai, Nogai: 0330-0400, 0615-0700, 0715-0800 (Tues-Sun), 0815-0900 (Thurs/Fri), 1300-1400, 1415-1500 (Mon/Fri), 1645-1745 (Mon/Wed). Also rel. ST. On 70.31MHz music prgr.

KD) Krasnodarskiy kray: GTRK 'Kuban', ul. Radio 5, 350630 Krasnodar. **Reg:** On 1089/1116kHz, 102.5MHz + (MHz) Tbilisskoye 66.20, Sochi 71.93, Kanevskaya 68.36: W 0400-0500, Sun 0350-0400, 0510-0600, Sat 0700-0800, Tues-Fri 0800-0810, Sun 0810-0900, MF 1500-1600. **Radio Sochi:** ul. Teatralnaya 11a, 354000 Sochi. On 71.93MHz: MF 1045-1100, 1600-1615, Tues-Sat 0445-0500, Sat 1200-1300.

KE) Kemerovskaya obl: GTRK 'Kuzbass', Krasnoarmeyskaya ul. 137-A, 650099 Kemorovo. **Reg:** On 66.56MHz: Mon 0000-0110, Tues-Fri 0000-0100, Mon-Wed, Fri 0510-0600, MF 1120-1200, Tues-Thurs 1300-1500, Sat 0110-0200, 1000-1100.

KG) Resp. Kalmykiya-Khalmg Tangch: Kalmytskaya GTRK, ul. M. Gorkogo 34, 358000 Elista. **Reg:** In Russian/Kalmyk on 846kHz, 67.28MHz + (MHz) Sadovoye 66.95, Utta 68.24, Ulan-Kholl 69.59: W 0320-0400, 0410-0500, MF 1600-1700, Mon 1200-1210, Tues-Fri 1100-1120, Sat 1100-1200, 1700-1800. Stereo prgr on 69.14MHz: Tues-Fri 1800-1900, Sat/Sun 1700-1900.

KH) Khabarovskiy kray: GTRK 'Dalnyevostok', pl. Slavy, 682632 Khabarovsk. **Reg: 1st prgr:** On 153/216/621/7210kHz, 70.28MHz + (MHz) Birobidzhan 66.32, Komsomolsk 68.72, Chegdomyn 70.16,

Bidzhan 70.07: 2010-2100, MF 0200-0300 & 0930-1100. R. St. Tikhiy Okean: Cf. PM. **2nd prgr:** On 711/999/1152kHz: 2200-0400. Prgr. 'Amur': 72.02MHz: 0400-0900. Komsomolsk local prgr: On 68.72MHz: MF 0930-0945.

KK) Koryakskiy avt. okrug: Koryakskaya GTRK, ul. Obukhova 4, 684620 Palana. **Reg:** On 738/1089/1485/1584/4520kHz in Russian, Koryak: D 2000-2030, Tues-Thurs 2145-2200, Tues-Wed 0115-0145, Sat 0700-0800.

KL) Kaluzhskaya obl: GTRK 'Kaluga', Pole Svobody 40A, 248021 Kaluga. **Reg:** On 66.23MHz: MF 0410-0500, 1500-1600, Sat 1100-1200. On 68.60MHz: 1800-1900.

KM) Kamchatskaya obl: GTRK 'Kamchatka', ul Sovetskaya 62 (or: Nikolskaya sopka), 683000 Petropavlovsk-Kamchatskiy. **Reg:** On 180/1584kHz, 69.68MHz + Sobolevo 72.08MHz: W1830-2000 (Sat 1900), Sat/Sun 2030-2100, Fri 0000-0100, MF 0720-0800. Prgr for fishermen 'Kamchatka rybatskaya': Tues/Thurs/Sat 2000-2100 on 2 freq's in the 31, 25 or 21mb (a.o. 7255, 7300, 12050kHz).

KN) Krasnoyarskiy kray: Krasnoyarskaya GTRK 'Tsentr Rossii', ul. Mechnikova 44a, 660028 Krasnoyarsk. **Reg:** On 216/5290kHz, 67.31MHz + (MHz) Achinsk 72.14, Solyanka 68.84, Uzhur 69.56 (rel. by TM 162kHz, RK 792kHz): MF 2320-0100, & 1120-1300, SS 0000-0100, Sun 1100-1300. On 69.68MHz: MF 1200-1400, Sat/Sun 0200-0400.

KO) Resp. Komi: GTRK 'Komi-gor', Oktyabrskiy pr. 164, 167610 Syktyvkar. **Reg:** On 621kHz, 66.80MHz + 9 FM relay st's: Russian, Komi: W MF 0320-0500, 1520-1600, Sat 0320-0600, Sun 0510-0600.

KP) Komi-Permyatskiy avt. okrug: GTRK 'Komi Permyatskaya', ul. Volodarskogo 18, 617240 Kudymkar. **Reg:** On 1458/1602kHz, 67.19MHz: MF 0230-0300 & 1410-1500 Sat 0200-0300, Sun 0230-0300. Also rel. PR.

KR) Kurskaya obl: GTRK 'Kursk', ul. Sovetskaya 32, 305016 Kursk. **Reg:** Kursk 69.71MHz, Kshen 70.85MHz. Sched. not available.

KS) Kostromskaya obl: Kostromskaya GTRK, ul. Nikitskaya 10, 156005 Kostroma. **Reg:** On (MHz) Kostroma 69.85, Sharia 67.10, Galich 66.74: MF 0400-0500, Sun-Fri 1600-1710, SS 0600-0700. On 72.56MHz: Thurs 1800-1900, Sun 1200-1300. Other details not available.

KT) Resp. Kareliya (Karjalan tasavalta): GTRK 'Kareliya', ul. Frunze 20, 185630 Petrozavodsk. **Reg:** On 765kHz, 70.52MHz + (MHz) Sortavala 67.13, Medvezhyegorsk 72.47, Nadvoitsy 66.29, Muezerskiy 70.55, Kostomuksha 70.07 in Russian, Finnish, Karelian, Vepsian: MF 0400-0600, 1120-1200, 1520-1600, Sat 0500-0700, 1200-1300, 1400-1500, Sun 0600-0700, 0800-0900. Finnish: Wed 0445-0500, W 1100-1115 (Wed, Fri 1145), MF 1520-1535, Sat 1200-1300, Sun 1300-1400.

KU) Kurganskaya obl: Kurganskaya GTRK, ul. Sovetskaya 105, 640018 Kurgan. **Reg:** On 71.87MHz + (MHz) Shumikha 66.89, Makushino 68.48, Shadrinsk 69.22: W 0315-0330 & 0410-0440, MF 1300-1400.

KV) Kirovskaya obl: GTRK 'Vyatka', ul. Uritskogo 34, 610002 Kirov **Reg:** On (MHz) Kirov 66.44, Shmelevo 67.22, Urzhum 69.14, Pinyug 70.55, Kirs 66.86: Mon 0610-0700, MF 0320-0400, 1520-1600, Tues-Fri 0640-0700, Sat 0510-0600, Sun 0610-0700. On 68.00/69.14MHz 1630-1730.

KY) Khanty-Mansiyskiy avt. Khanty-Mansiyskaya GTRK 'Yugoriya', okrug: ul. Mira 7, 626200 Khanty-Mansiysk. **Reg (R. Yugry):** On 225/1152/1197/4820/4520(irr.)kHz, 66.00MHz + Surgut 68.84MHz in Russian, Khanty, Mansi: Tues-Fri 0200-0230, Mon 0210-0230, W 0800-0900, Sat 0400-0500. Also rel. Tyumen, TY.

LI) Lipetskaya obl: GTRK 'Lipetsk', pl. Plekhanova 1, 398050 Lipetsk. **Reg:** On 66.53MHz: MF 0320-0500, 1315-1600, SS 1300-1500. Prgr 'Lipetskiy Radiokanal'.

MA) Magadanskaya obl: GTRK 'Magadan', ul. Kommuny 8/12, 685000 Magadan-13. **Reg:** On 234/5930/7320/ 9530/9600kHz: D 0010-0100, MF 2000-2100 & 0115-0200 & 0320-0400 & 0720-0800v, Sat/Sun 2020-2100, Sat 0111-0120(0200). 0815-0900 rel. R. St. Tikhiy Okean on 234kHz and available SW (D). Chukchi/Eskimo prgr from GTRK 'Chukotka', CK on 5940/9600kHz: MF 0720-0800.

MD) Resp. Mordoviya: GTRK 'Mordoviya', ul. Dokuchayeva 29, 430000 Saransk. **Reg:** On 1062kHz, 66.68MHz + (MHz) Atyuryevo 71.43, Ardatov 69.53 in Russian/Mordovian: MF 0040-0500, 0900-1100, 1520-1600, Sat 1300-1400, Sun 0400-0500, 1400-1500.

ME) Resp. Mariy El: GTRK 'Mariy El', ul. Osipenko 50, 424014 Yoshkar-Ola. **Reg:** On 900/7200kHz, 70.34MHz in Russian/Mari: W 0320 (Sat 0300)-0500, D 0410-0500 (Sun 0600), Sat 0720-0800, MF 1500-1600, Sat 1410-1500, Sun 1410-1600. On 71.96MHz: Thurs 0800-0900.

MO) Moskva & Moskovskaya obl: GTRK 'Moskva', ul. Akademika Koroleva 19, 127427 Moskva. **Reg:** For Moskovskaya obl. ('R. Podmoskovye'), ul. Pyatnitskaya 25, 113326 Moskva. On 846kHz/66.44MHz: D 0300-0400, MF 0800-0900, Mon-Thurs 1600-1700, Fri/Sat 1400-1500, Sat 0530-0600, Sun 1000-1100. For Moskva city ('Radiokompania Moskva') on 846kHz, 66.44/69.80MHz: Mon-Sat 0400-0500, MF 0600-0700, Mon-Thurs 1400-1600, Fri 1500-1900, Sat 1200-1300. **R-1** on 171/1305/1359kHz, 72.92MHz. **Radio-1-Retro** alternative musical sce. on 1359kHz, 72.92MHz: 0400-0700, 1000-1430, 1800-2100. **R. St. Yunost** on 153kHz, 68.84MHz. **R. St. Mayak** on 198/549kHz, 67.22MHz. **R. Orfey** on 738/1152kHz, 72.14MHz. **R. Rossii** on 261/873kHz, 66.44MHz. **Prgr's for foreigners** (V. of Russia relays) on 612/918/kHz (variable times).

MU) Murmanskaya obl: GTRK 'Murman', sopka Varnichnaya, 183042 Murmansk. **Reg:** On 657/5930kHz, 67.22MHz: MF 0400-0500, 0800-0810, 1520-1700, Mon-Sat 0910-1000, Sat 0610-0700, 0720-0800, Sun 0700-0710, 1000-1100, 1200-1300, D 2010-2100.

NE) Nenetskiy avt. okrug: Nenetskaya GTRK, ul. Smidovicha 19, 164700 Naryan-Mar. **Reg:** On 711kHz: 0330-0400.

NK) Novosibirskaya obl: GTRK 'Novosibirsk', ul. Vertkovskaya 10, 630048 Novosibirsk. **Reg:** On 270kHz, 67.10MHz + (MHz) Tatarsk 69.50, Kuybyshev 69.68, Dovolnoye 69.38, Bagan 68.36: Mon-Sat 2330-0000, Tues-Fri 0010-0100, Sat/Sun 0215-0300 & 0330-0400, Sat 0415-0500 & 0715-0800 & 0900-1000, Sun 0930-1000, MF 1015-1100, Sun-Fri 1115-1145. Stereo prgr on 69.28MHz 1100-1500.

NN) Nizhegorodskaya obl: GTRK 'Nizhniy Novgorod', ul. Belinskogo 9-A, 603600 Nizhniy Novgorod, GSP-228. **Reg:** On 828kHz, 67.95MHz + (MHz) Arzamas 69.95, Shakhunya 71.69, Krasnyye Baki 70.64, Sergach 67.16, Vyksa 71.09, Lukoyanov 72.52: MF 0320-0350, 0410-0500 0800-0810, 1520-1600, Sat 0510-0520, 1010-1100, 1120-1200, 1210-1300.

NO) Novgorodskaya obl: GTRK 'Slaviya', ul. Nikolskaya 16, 173610 Novgorod. **Reg:** On (MHz) Novgorod 67.25, Borovichi 67.07, Zaluchye 71.93. Sched not avaliable. On 72.95MHz: MF 1300-1400, Sat/Sun 1800-1900. Radio Nova on 70.61/103.7MHz.

OL) Orlovskaya obl: ul. Orlovskaya GTRK, 7 noyabrya 43, 302028 Orel. **Reg:** On 70.31MHz Orel, 67.19MHz Libny. Sched. not available. On 69.76MHz: 0400-0500. On 72.83MHz: Wed/Fri 0500-1100, Sat/Sun 0700-1400. On 72.05MHz Orel, 68.66MHz Libny: 1000-1100.

OM) Omskaya obl: GTRK 'Irtysh', pr. Mira 2, 644008 Omsk. **Reg:** On 639kHz, 69.74MHz + (MHz) Isilkul 66.50, Tara 66.77, Cherlak 66.86, Ust-Ishim 67.04, Khutora 70.43, Nazyvayevsk 67.28: 0100-0200, 0700-0720, 1300-1400, Sat 0410-0600, Sun 0300-0600. On 71.84MHz: MF: 0200-0300, SS 0300-0500.

OR) Orenburgskaya obl: GTRK 'Orenburg', Televizionniy per. 3, 460024 Orenburg. **Reg:** On 270/935/1395/1584kHz, 66.02MHz + (MHz) Orsk 66.92, Yazniy 69.71, Buzuluk 66.62, Kubandyk 70.04 in Russian, Chuvash, Tatar: MF 0110-0300, 0700-0900, Mon 1400-1410, Mon/Thurs/Fri 1520-1600, Tues/Wed 1400-1600, Sat 0400-0900, 1100-1120, 1300-1520, 1600-1700, Sun 0320-0700, 1310-1600. On 71.84MHz: Fri 1530-1800, Sat 1300-1530, Sun 1530-1730.

PE) Penzenskaya obl: Penzenskaya GTRK, ul. Lermontova 39, 440602 Penza. **Reg:** On 855kHz, 70.67MHz + (MHz) Pachelma 66.80, Blagodatka 61.07, Meshcherskoye 66.98: W 0400-0500, Sat 0530-0600, Sat/Sun 0530-0600, MF 1520-1600.

PM) Primorskiy kray: Tikhookeanskaya GTRK 'Vladivostok', ul. Uborevitsa 20-A, 690000 Vladivostok-Central. **Reg:** 1st prgr on 234kHz, 5015kHz (part time), 71.84MHz + (MHz) Arsenyev 66.86,

Dalnegorsk 70.04, Dalnerechensk 69.32, Novozhatkovo 72.20, Chkalovskoye 70.40: 1954-2000, MF 2030-2200, Sat/Sun 2058-2315 & 0115-0200, Sat 0230-0400, Tues-Fri 0300-0400, 0815-0900, 0930-1010 (MF 1030), Mon/Wed/Thurs 1045-1100, Fri 1100-1115. 2nd prgr on 810/5015 (from 0700) kHz: 0045-0100, 0500-0515, Wed/Thurs 0700-0715, Mon/Tues 0700-0815, Fri 0700-1400, Sat/Sun 0715-1400, other days 0930-1400. **Radiostantsiya Tikhiy Okean (R. St. Pacific Ocean):** 0815-0900, 1900-1945 on various LW/MW/SW freq's in eastern Russia, locally also 1215 on 810/5015kHz. Segment in **English** Sat 0850-0853.

PR) Permskaya obl: Permskaya GTRK, ul. Tekhnicheskaya 7, 614070 Perm. **Reg:** On 585/1458/1512/1602/5290kHz, 66.02MHz + (MHz) Berezniki 71.87, Barda 67.10, Chusovoy 70.67, Kudymkar 67.10, Kungur 66.65: Sun-Thurs 0058-0100, MF 0110-0230, 0320-0400, 0600-0610, 0900-0910, 1300-1400, Wed 1610-1700, Sat 0200-0300, 0800-0900, Sun 0410-0500, 0610-0700.

PS) Pskovskaya obl: GTRK 'Pskov', Rizhskiy pr. 71, 180024 Pskov. On (MHz) Pskov 66.05, Velikiye Luki 67.25, Glubokoye 71.27, Dedovichi 69.86. MF 0400-0500, 1500-1600, Sat 0600-0900.

RK) Resp. Khakasiya: ul. GTRK Respubliki Khakasiya, Rozy Luksemburg 94, 662000 Abakan. **Reg:** On 792kHz, 66.89MHz + (MHz) Novomikhailovka 67.19, Shira 70.28: In Russian, Khakass: MF 0000-0100 & 1120-1200, Sat 0110-0200 & 0320-0400, Sun 0320-0400.

RO) Rostovskaya obl: GTRK 'Don-TR', ul. 1-a Barrikadnaya 18, 344101 Rostov-na-Donu. **Reg:** On 945kHz, 72.65MHz + (MHz) Salsk 66.86, Kamensk 70.28, Volgodonsk 70.13, Morozovsk 68.99, Veshenskaya 70.67: D 0510-0600, MF 0410-0500, 1500-1600, Fri/Sat 0610-0700, Sun 0610-0700, 1510-1600, 1610-1700. On 71.39MHz: MF 1400-1500, Mon-Thurs 0500-0600, Fri 1500-1600.

RY) Ryazanskaya obl: GTRK 'Oka', ul. Skomoroshinskaya 20, 390006 Ryazan. **Reg:** On Ryazan 67.87MHz, Mosolovo 71.66MHz: MF 0340-0350, 0400-0500, Mon-Thurs 1520-1600, Sat 0600-0700, Sun 0600-0700. On 68.42/103.5MHz 'R. Kontakt': 1000-1500, 1700-2000.

SA) Resp. Sakha (Yakutiya): GTRK Respubliki Sakha, Ordzhonikidze 48, 677007 Yakutsk. **Reg:** In Russian/Yakut on 171kHz + SW + (MHz) Yakutsk 70.40, Aldan 67.28, Neryungri 68.24: D 2200-2300, 2320-0000, MF 0210-0610, 1010-1100, SS 0210-0630.

SL) Sakhalinskaya obl: GTRK 'Sakhalin', Komsomolskaya 209, 693000 Yuzhno-Sakhalinsk. **Reg:** On 279/792/1377kHz, 67.64/106MHz + (MHz) Poronaysk 69.92, Uglegorsk n/a: 2000-2100, 0200-0300, 0800-0815, 1120-1210. R. St. Tikhiy Okean (from Vladivostok): 0815-0900. Korean on 279kHz: 0730-0800.

SM) Samarskaya obl: GTRK 'Samara', ul. Sovetskoy Armii 205, 443011 Samara. **Reg:** On 1107kHz, 70.31MHz + (MHz) Zhigukevsk 67.31, Sergiyevsk 66.71: MF 0300-0600, 1400-1700, SS 0600-1000. In Erzya: Sun 0730-0800. On 71.55MHz (Zhigulevsk): Sched. not available.

SN) Smolenskaya obl: GTRK 'Smolensk', ul. Nakhimova 1, 214025 Smolensk. **Reg:** On (MHz) Smolensk 68.54, Smogiri 68.96, Vyazma 69.20, Roslavl 70.91: MF 0415-0500 & 1500-1540, Sat 0700-0800/0900 & 1215-1315, Sun 0700-0800/0900, Thurs 1700-1800. On 66.35MHz: MF: 1400-1500, Sat/Sun 1115-1215.

SO) Resp. Severnaya Osetaya: GTRK 'Osetiya', Osetinskaya gorka 2, 362007 Vladikavkaz. **Reg:** In Osetian/ Russian on 594kHz, 71.24MHz: W 0430-0500 & 0525-0600 (Mon 0545), Sun 0800-0900 & 1000-1200, Sat 0945-1100, Wed 1300-1400, MF 1500-1540/1600.

SP) Sankt-Peterburg & Leningradskaya obl: GTRK 'Peterburg — 5-y kanal', ul. Italyanskaya 27, 191011 Sankt-Peterburg. **Reg:** On 801kHz, 69.47MHz: 0300-1500. **R. Klassika:** On 69.47MHz: 1500-2200.

SR) Saratovskaya obl: GTRK 'Saratov', 2-a Sadovaya 7, 410017 Saratov. **Reg:** On 630kHz, 71.09MHz + (MHz) Yershov 66.80, Aleksandrov Gay 69.68, Balakovo 70.52, Balashov 70.16, Perelyub 70.55: MF 0330-0400, 0810-0815, 1500-1540; Tues-Sat 0525-0545; Sat 0435-0500, 1215-1300, 1500-1600; Sun 1645-1730.

ST) Stavropolskiy kray: Stavropolskaya GTRK, ul. Artema 35-A, 355000 Stavropol. **Reg:** On 882/1251kHz, 69.53MHz + (MHz) Pyatigorsk 67.19, Neftekumsk 71.90: W 0400-0500, 1500-1600, Sun

0600-0700, 1500-1600.

TA) Tambovskaya obl: Tambovskaya GTRK, ul. Michurinskaya 8-a, 392720 Tambov. **Reg:** On 71.00MHz. W 0359-0400, 1500-1600, MF 0400-0510, Sat 0459-0500, 0510-0610, 0700-0720, 1710-1810. Musical prgr on 71.78MHz: Mon/Wed/Fri/Sun 1300-1400.

TL) Tulskaya obl: GTRK 'Tula', Staronikitskaya ul 1, 300600 Tula. **Reg:** On Tula 71.15MHz, Novomoskovsk 71.40MHz: MF 0410-0500, 1500-1600. Sat 1100-1200. 'Radiomayak-Tula' on 70.07MHz SS0900-1000.

TM) Taymyrskiy (Dolgano-Nenetskiy) avt. okrug: GTRK 'Taymyr', ul. Sovetskaya 26, 663210 Dudinka. **Reg:** On 162kHz, 69.68MHz: MF 0030-0100, 0610-0700, SS 0510-0600. In Dolgan, Nenets: MF 1210-1300. Also rel. KN.

TO) Tomskaya obl: Tomskaya GTRK, ul. Batenkova 3, 634032 Tomsk. **Reg:** On 171/1602kHz, 67.22MHz + (MHz) Kolpashevo 68.87, Strezhevoy 66.78: 0010-0100, 0110-0200 (Sat 0300), 0610-0700, 1120-1200.

TS) Resp. Tatarstan: GTRK 'Tatarstan', ul. M. Gorkogo 15, 420015 Kazan. **Reg:** In Tatar/Russian/Chuvash on 252/6115//7185/11905/11985kHz + FM (MHz): Kazan 68.50, Leninogorsk 68.63, Naberezhnyye Chelny 69.32, Bilyarsk 70.13, Nizhnekamensk 72.29: MF 0300-0500, 1520-1800, Tues-Fri 0900-1015, Sat 0300-1115, Sun 0300-1315. On 66.92MHz: Music prgr: 1700-2000. **ANN:** Tatar: 'Kazan söyli'.

TU) Resp. Tyva: GTRK Respubliki Tyva, ul. Gornaya 31, 667003 Kyzyl. **Reg:** On 567/1287kHz, Kyzyl 67.10MHz, Shagonar 70.64MHz: W 0015-0100 & 0115-0200, Sun 0040-0100 & 0125-0200, Sat 0340-0400, MF 1115-1200. On 69.08MHz: 1200-1300.

TV) Tverskaya obl: GTRK 'Tver', ul. Vagzhanova 9, 170000 Tver. **Reg:** On (MHz) Tver 68.48, Selizharovo 69.68, Maksatikha 71.24: MF 0410-0500, 0510-0530 & 1520-1600, Thurs 1610-1700, SS 0610-0700. **R. Tver Express:** On 71.09MHz 1000-1200

TY) Tyumenskaya obl: GTRK 'Region-Tyumen', Permyakova 6, 625013 Tyumen. **Reg:** On 225/576/1152/1197/1485/4820/4895/6095kHz + (MHz) Tyumen 71.66, Gagarino 66.89, Surgut 68.84, Tobolsk 70.04, Shabanovo 70.55, Nizhne-Vartovsk 72.56: MF 0110-0200, 0230-0300, 0600-0610, 1000-1010, 1400-1410, SS 0300-0500, Sat 1300-1400, Sun 0700-0800. On 69.56MHz: 1530-1630. Tatar: Sat 1300-1400.

UB) Ust-Ordynskiy Buryatskiy avt. okrug: 666110 Ust-Ordynskiy. On 234kHz (Irkutsk) in Buryat: 2245-2250, 1145-1155.

UD) Udmurtskaya Resp: GTRK 'Udmurtiya', ul. Komunarov 216, 426004 Izhevsk. **Reg:** In Russian/Udmurt on 594kHz, 68.06MHz + Balezino 70.94MHz in Udmurt, Russian: MF 0200-0300, D 0320-0400, SS 0500-0610, MF 0810-0900, SS 0500-0610, Sun 0610-0700, Sat 0700-0800, Sun 0720-0800, M 1500-1740, Tues 1410-1620, Wed 1500-1600, Thurs 1410-1610, Fri 1410-1640. Local prgr's 'R. Ingur' on 70.40MHz MF0410-0700, 1100-1300, SS 0900-1300.

UL) Ulyanovskaya obl: GTRK 'Volga', ul. Simbirskaya 5, 432030 Ulyanovsk. On (MHz) Ulyanovsk 71.00/72.56, Dimitrovgrad 67.19/69.23, Novospasskoye 67.07/68.99, Veshkayma ??/71.99: **Reg.** prgr's on 1st freq's Mon-Sat 0320-0410, 1520-1610, Sun 0610-0700. Stereo prgr on 2nd freq's: D 1800-1900.

VG) Vologodskaya obl: Vologodskaya GTRK, ul. Predtechenskaya 32, 160000 Vologda. **Reg:** On 1098kHz, 70.43MHz + (MHz) Cherepovets 66.38, Totma 71.20, Lipin Bor 69.65, Yakutino 66.86, Sludno 66.77: MF 0410-0500, Mon/Fri 1500-1600, Tues-Thurs 1510-1600, SS 0600-0700.

VL) Vladimirskaya obl: GTRK 'Vladimir', Vorontsovskiy 4, 600000 Vladimir. **Reg:** On 69.47MHz (Sudogda): W 0400-0500, MF 1600-1700, Sat 1200-1300, Sun 0500-0700..

VN) Voronezhskaya obl: Voronezhskaya GTRK, ul. Karl Marksa 114, 394625 Voronezh. **Reg:** On 774kHz, 72.17MHz + (MHz) Boguchar 71.90, Borisoglebsk 70.82, Bobrov 67.04: MF 0320-0345 (Fri 0400), 1520-1600, Sat 1520-1600, 1710-1800, Sun 0610-0700, 1710-1800. On 69.83MHz: Sat 1000-1400, Sun 0900-1200.

VO) Volgogradskaya obl: GTRK 'Volgograd-TRV', ul. Mira 9, 400066 Volgograd. **Reg:** On 567kHz, 70.43MHz + (MHz) Kletskaya 70.94,

Mikhailovka 66.83, Kamyshin 67.28, Chilekovo 69.44: MF 0410-0500, 1500-1600, Thurs 1610-1700, Sat/Sun 0500-0800.

YA) Yaroslavskaya obl: GTRK 'Yaroslavl', ul. Bogdanovicha 20, 150014 Yaroslavl. **Reg:** On 68.66MHz + (MHz) Dubki 69.56, Volga 70.88: MF 0400-0500, 1500-1600, Sat 0400-0500, SS 1410-1700.

YE) Sverdlovskaya obl: Sverdlovskaya GTRK, ul. Lunacharskogo 212, 620219 Yekaterinburg. **Reg:** On 279kHz, 71.06MHz + (MHz) Baranča 69.29, Serov 69.65, Zaykovo 66.83, Afanasyevskoye 70.43, Nizhniye Sergi 67.01, Azanka 70.43: MF 0110-0300, 0910-1000, 1310-1400, Sat 0410-0720, Sun 0410-0700.

YN) Yamalo-Nenetskiy avt. okrug: GTRK 'Yamal', ul. Lambinykh 3, 626600 Salekhard. **Reg:** On 603kHz: W 0200-0215, MF 0800-0900 & 1000-1010, Mon-Thurs 1415-1445; Sat 1100-1200, 1300-1400.

YV) Yevreyskaya avt. obl: GTRK 'Bira', ul. Sovetskaya 13, 682290 Birobidzhan. **Reg:** On 216kHz, 66.32MHz in Russian, Yiddish: Mon 2115, Sun 2145, Sat 2315-2400, 0140-0300, Wed 0515, Fri 0950, and others. Also rel. KH.

VOICE OF TATARSTAN

OTHER STATIONS

INDEPENDENT PRIVATE PRGR PRODUCERS
A number of companies produce prgrammes using government owned studios and/or transmitters. In some cases various prgr producers share transmitter time. Several of these production companies are joint ventures with the government owned radio. Due to financial difficulties changes among these stations are very frequent.

INDEPENDENT STATIONS
Companies which possess their own production and transmitting equipment.

M = Moskva — SP = Sankt-Peterburg — Yek = Yekaterinburg — d:o = using the same transmitter as the previous station — RS = Radiostantsiya.
The tr location is shown within brackets if different from the studio location.

	kHz	kW	Station, location, schedule
51)	270	150	Transibirskaya Studiya, Novosibirsk: MF 0000 0050, 0720-0800, Sun 0800-0900 RS Slovo, d:o: 0300-0700, 1400-1800
2)	612	40	R. Alef, M (Kurkino): Thurs 1700-1800
3)			Islamskaya Volna, d:o: Tues-Fri 1600-1700
3)			R. Kala Aturaya, d:o: Sat 1600-1700
3)			RS Radonezh, d:o: 1800-2000
52)	612	10	R. VBC, Vladivostok: 24h
150)	621	50	R. Vostok, Khabarovsk: 0700-0830 (Sun 0800)
53)			R. "A", d:o: 2100-1200
54)			R. Prospekt, d:o: 1200-
55)	666	10	R. Maksimum, Yek: 2300-1600
	675		Obshchestvennoye R. Primorya, Vladivostok: 2100-0900
56)	684	10	Peterburgskiy Meridian, SP (Olgino)
224)			Pravoslavnoye R, d:o: Sat 1000-1300, D 1500-1800
3)			R. Radonezh, d:o: 1800-2000
—	693	10	Deutsche Welle, M: 0300-2300
57)	693	0.2	ARK "Narodno-Trudovy Soyuz", Astrakhan: 1800-2100
58)	738		R. Novaya Volna, Vladivostok: 2000-1500
59)	738		R. Raduga, Izhevsk
60)	756	10	R. Titan, Ufa: 0200-1600
61)	783	500	Russkoye R. Lenna, Vladivostok (Tavrichanka): 2000-1400

kHz	kW	Station, location, schedule
5)	810	R. Meditsina dlya vas, M: 0400-1400
62)	810 7.5	Avtoradio, Krasnoyarsk: 24h
220)	810 10	AST-Radio, Irkutsk: 24h
63)	810 150	R. DVS, Vladivostok (Razdolnoye): 2100-1100
64)	828 10	Eldoradio-Retro, SP: 0600-1500
—		R. Liberty, d:o: 0400-0600, 1500-1900, 2000-2100
—		VOA, d:o: 1900-2000
3)	846 150	RS Radonezh, M (Noginsk): 1800-2000
6)		R. Rezonans, d:o: Sun-Fri 1200-1400, Sat 1700-1800
66)	855 20	Radio 3, Petropavlovsk-Kamchatskiy: W 1800-2100
67)	909 10	Studiya Gorod, Yek: 0100-0200, 1200-1500
		R. Liberty, d:o: 0200-0900, 1500-1800
15)		RFI, d:o 1800-2000 & SS1300-1500
3)	918 75	R. Svobodnaya Rossiya, M (Balashikha): MF 1600-1900
—		D:o: Relays of BBC, VOA, R. Liberty
68)	954	FEBC, Vladivostok (?): 1100-1200
7)	963 15	R. Krishnaloka, M: 24h
8)	990	RS Slavyanka, M (Povarovo): MF 1400-1600
6)	1017 500	R. Rezonans, M (Lesnoy): W 1400-1800
3)	1044 20	RS Nadezhda, M (Kurkino): 0300-2200
3)	1053 10	RS Nadezhda, M (SP): 0300-2200
209)	1053	R. Menovyy Dvor, Orenburg
N1)	1053 20	Europa Plus Yenisey, Krasnoyarsk: 2300-1700
	1053	RS Novyy Gorod, Krasnoyarsk
9)	1071 10	RS RIA-Radio, M: MF 0600-1000, MF 1200-1600
69)	1080 30	69th Parallel, Murmansk: MF 0400-1700, SS 0600 (Sun 0700)-1100
70)	1089 20	R. Teos, SP: 0500-2100
3)	1089 1200	Islamskaya Volna, M (Tbilisskaya): Tues-Fri 1600-1700
2)		R. Alef, d:o: Thurs 1700-1800
3)		R. Kala Aturaya, d:o: Sat 1600-1700
—	1098 10	R. Liberty, Moskva: 0300-2100
71)	1098 75	R. Vladivostok, Vladivostok: 2100-1400
100)	1107	R. Sem iz Samary, Samara
10)	1116 10	Khristyanskiy Tserkovnyy Obshchestvennyy Kanal, M: 0400-2100
72)	1116 30	Avtoradio, Perm: 0100-1800
N5)	1116 50	Russkoye R, Chita
70)	1134 20	R. Teos, SP (Kurkino): 0500-2100
73)	1134 7	R. Invar, Khabarovsk
3)	1152 150	R. Svobodnaya Rossiya, M (Kupavna): MF 1600-1900
3)	1170 1200	Islamskaya volna, M (Tbilisskaya): Tues-Fri 1600-1700
2)		R. Alef, d:o: Thurs 1700-1800
3)		R. Kala Aturaya, d:o: Sat 1600-1700
—	1188 10	Deutsche Welle, SP: 0300-2300
68)	1188 10	FEBC, Khabarovsk: 0500-1200
74)	1224	R. Evraziya, Elista: 0400-2100 (Sat 2200)
12)	1233 150	Armyanskoye R, M (Chkalovskaya): MF 0700-1000
3)	1233	Narodnoye R, d:o: 0500-0645, 1015-1600
75)	1242 10	Angara R, Angarsk: 0000-1400
76)	1251 1	R. Rus, Stavropol (Neftekumsk)
77)	1251 600	R. Muzykalnaya Klassika, Vladivostok (Ussuriysk)
13)	1260 10	BBC WS, M: 0330-2130
13)	1260 10	BBC WS, SP: 0330-2130
13)	1260 10	BBC WS, Yek: 0100-2130
78)	1278 75	R. Novosibirsk, Novosibirsk: MF 0000-0100, SS 0100-0200

kHz	kW	Station, location, schedule
79)	1296 10	Delovaya Volna, Vladivostok: 2330-0100
68)		FEBC, d:o: 0800-1100
80)	1323 10	Krishnaloka, SP 0400-1800
81)	1332 1	R. Dipol-Patrul, Tyumen: 24h
82)	1332	R. Russkaya Volna, Krasnoyarsk
83)	1332 10	RS Volna Baykala, Irkutsk: 24h
84)	1350 10	R. Vtoroy Kanal, Ufa: 0300-1500
3)	1359 150	R. Svobodnaya Rossiya, M (Chkalovskaya): MF 0400-0700, 1000-1300, 1600-2100
84)	1395 50	R. Pervyy Kanal, Ufa: 0100-1900
14)	1413 40	Kamerton R, M (Balashikha): W 0500-1000, 1200-1600
15)	1440 10	RFI, M: 0100-1500, 1600-0000
15)	1440 10	RFI, SP: 0300-2100
85)	1449 0.2	PO "Irtysh", Omsk
86)	1449 5	RS Lena, Yakutsk: 0300-1000
87)	1467	R. Vedo, Volgograd: 0500-2130
88)	1476 1000	R. Studiya O'Key, Vladivostok (Tavrichanka): 2000-1200v
16)	1485 20	R. Tsentr, M (Kurkino): 0400-0600, 1400-2100
17)	1485	R. Sadko, M: MF 0700-0900
5)	1485	R. Meditsina dlya vas, d:o:
89)	1485	Avtorskoye R, Omsk: 0000-2200
90)	1485	CCR, Saratov: 0600-1100, 1400-2100
91)	1485 0.2	RS Vostok, Irkutsk: 2300-1600
3)	1494 1200	R. Kala Aturaya, M (Krasnyy Bor) Sat 1600-1700
3)		Islamskaya Volna, d:o: Tues-Fri 1600-1700
92)	1503	R. Ussuriysk, Ussuriysk: 2300-1400
18)	1539 150	Delovaya Volna, M (Lesnoy): 1400-2200
93)	1557 2	INTA-Radio, Irkutsk: 24h
94)	1557	R. L-Tsentr, Stavropol
95)	1557	Radio 1557, Vladivostok: -1300
N1)	1557	Europa Plus, Vladivostok
96)	1566	R. Kanal 3, Barnaul: inactive
97)	1566 0.2	R. Salyut, Omsk
N1)	1584	Europa Plus, Omsk (Russkaya Polyana): 24h
98)	1584	R. ITN, Iskitim
99)	1593	Russkoye R. Barguzin, Angarsk: 24h
N5)	1602	Russkoye R, M (Tyumen)

Shortwaves (freqs variable during the year)

86)		RS Lena, Yakutsk on 5920kHz 2kW, 6125kHz 5kW: 0300-1000.
3)		R. Kala Aturaya, M (M, Tbilisskaya): Sat 1700-1800
3)		Islamskaya Volna, M (M, Tbilisskaya): Mon-Fri 1600-1700.

FM	MHz kW*	Station, location
N3)	65.9	R. Rossii-Nostalgie, Miass
101)	65.96	R. Aleks, Rostov-na-Donu
102)	66.02 4	R. Nika, Murmansk
N1)	66.14 1	Europa Plus, Kemerovo
84)	66.17	R. Vtoroy Kanal, Ufa
)	66.17	R. Roksana, Ufa
72)	66.20 1	Avtoradio, Perm (Bereznyki)
197)	66.23	Nashe Radio, Voronezh
103)	66.26 1	R. Niks, Cheboksary
N2)	66.26	R. ROKS, Azov
104)	66.41 0.2	R. Sfera, Pyatigorsk
105)	66.41 1	R. Puls, Azov
N4)	66.62	R. 101, Yekaterinburg
106)	66.74 4	Intervolna, Chelyabinsk
107)	66.80	R. Kora, Orenburg
108)	66.80 0.35	R. Passazh, Kazan
109)	66.83 0.5	Nord-Vest R, Pskov
100)	66.83 0.1	R. Sem iz Samary, Samara
N1)	66.86 1	Europa Plus, Yuzhno-Sakhalinsk
19)	66.86 7.5	R. Maksimum, M

FM	MHz kW*	Station, location
N5)	67.04 0.125	Russkoye R, Tyumen
110)	67.10 0.1	R. Mirazh, Rostov-na-Donu
72)	67.10 1	Avtoradio, Perm (Solikamsk)
111)	67.19 1	R. Provintsiya, Pyatigorsk (Stavropol)
112)	67.25	R. Pilot, Pskov (Velikiye Luki)
N3)	67.46	R. Rossii-Nostalgie, Yek
N5)	67.50 2	Russkoye R, Ufa
N4)	67.55	R. 101-Kontinental, Chelyabinsk
N1)	67.58 0.3	Europa Plus, Pskov
113)	67.61 30W	R. Megapolis, Samara
114)	67.61 1	R. Novaya Volna, Volgograd (Mikhailovka)
115)	67.64	R. Universitet, Barnaul
N3)	67.70 1	R. Rossii-Nostalgie, Voronezh
116)	67.91 0.1	Radio "N", Novocherkassk
20)	68.00 1	Avtoradio, Moskva
N1)	68.09 1	Europa Plus, Penza
N1)	68.09 1.5	Europa Plus, Tolyatti
117)	68.10 0.25	R. 103, Rostov-na-Donu
N5)	68.12 4	Russkoye R, Kemerovo
118)	68.20	R. Provintsiya, Rostov-na-Donu
N3)	68.24 1	R. Rossii-Nostalgie, SP
N1)	68.36 1	Europa Plus, Kaliningrad
119)	68.36 0.1	R. Reyting, Obninsk
223)	68.36	Formula Da, Naberezhnyye Chelny
N1)	68.39 1	Europa Plus, Kurgan
120)	68.39 1	R. Si, Yekaterinburg
121)	68.42 1	R. Ekho, Ryazan
N3)	68.45 0.2	R. Rossii-Nostalgie, Rostov-na-Donu
122)	68.48	R. Oktava, Kursk
123)	68.48 1	R. Sibiri, Novosibirsk
124)	68.51 0.1	SBC, Samara
N1)	68.57 1	Europa Plus, Voronezh
N1)	68.60 1	Europa Plus, Saratov
N5)	68.60 1	Russkoye R, Omsk
125)	68.66 30W	R. Mariya, SP
126)	68.75 2	R. Novaya Volna-2, Chelyabinsk (inact?)
98)	68.84	R. ITN, Iskitim
127)	68.87	R. Novaya Provintsiya, Rostov-na-Donu
128)	68.90	Mestnoye vremya, Krasnoyarsk
N1)	69.02 1	Europa Plus, Tula
N1)	69.0	Europa Plus, Oktyabrskiy
N5)	69.02	Russkoye R, Tambov
193)	69.02	R. Dzhem, Yekaterinburg
	69.05 1	R. Gardarika, SP
N1)	69.11 1	Europa Plus, Barnaul
N5)	69.26	Russkoye R, Moskva
129)	69.26	Music R, Perm
130)	69.27	R. Novaya Volna, Novosibirsk
131)	69.27	R. Lyubimiy Gorod, Novosibirsk
N2)	69.29	R. ROKS, M (Ivanovo)
N5)	69.32	Russkoye R, Arkhangelsk
N6)	69.38 1	R. Modern, Petrozavodsk
132)	69.38	R. Os, Vladikavkaz
N6)	69.38 1	R. Modern, Petrozavodsk
212)	69.44	R. Ekho Rostova, Rostov-na-Donu
N1)	69.50 4	Europa Plus Baykal, Irkutsk (Shelekhov)
133)	69.50 0.1	R. Sibir, Tomsk (Asino)
N1)	69.56	Europa Plus, Velikiye Luki
N1)	69.59 2	Europa Plus, Volgograd
N3)	69.65 4	R. Rossii-Nostalgie, Chelyabinsk
84)	69.68 1	R. Pervyy Kanal, Ufa
134)	69.68	Radiochannel Pelikan, Vladivostok
111)	69.71 0.1	R. Provintsiya, Pyatigorsk (Kislovodsk)
N1)	69.74 1	Europa Plus, Naberezhnyye Chelny
N1)	69.74 1	Europa Plus, Ulyanovsk
N1)	69.80 15	Europa Plus, Moskva

FM	MHz kW*	Station, location
133)	69.80 35W	R. Sibir, Tomsk (Strezhevoy)
135)	69.89 0.1	R. Telekom, Azov (inact)
136)	69.9	R. Pyataya Vershina, Pyatigorsk
N5)	69.92	Russkoye R, Chita
72)	70.04 2	Avtoradio, Perm
N1)	70.07 1	Europa Plus, Novokuznetsk
N1)	70.13	Europa Plus, Tver
137)	70.16 2	R. Randevyu, Nizhniy Novgorod
N3)	70.19 1	R. Rossii-Nostalgie, M (Balashikha)
138)	70.25 1	R. Piligrim, Ulyanovsk
N1)	70.37 1	Europa Plus, Yugorsk
N5)	70.43 1	Russkoye R, Ulan-Ude
139)	70.52 1	R. Trek, Yekaterinburg
140)	70.52 0.1	R. Tron, Murmansk
141)	70.52	R. Kray, Barnaul
142)	70.55 0.8	R. Antenna-7, Omsk (F.Pl.)
143)	70.64 30W	R. Avgust, Tolyatti
N3)	70.67	R. Rossii-Nostalgie, Bryansk
144)	70.82 4	R. Maksimum-Perm, Perm
N1)	70.85	Europa Plus Ural, Orenburg (Orsk)
145)	70.88	R. NTN, Novosibirsk
N5)	70.91	Russkoye R, Zlatoust
146)	71.00 1	Radio 71, Kostroma
147)	71.03	R. LUX, Magnitogorsk
148)	71.09 0.1	R. 45, Pyatigorsk
N2)	71.15	R. ROKS, Sochi
191)	71.18	R. Assonans, Kursk
N1)	71.24 1	Europa Plus, Khabarovsk
149)	71.24 10	R. Baltika, SP (Olgino)
N1)	71.27 1	Europa Plus, Samara
21)	71.30 10	M-Radio, Moskva
150)	71.30	R. Serebryannyy Dozhd, Ufa
151)	71.30	R. Viktoria, Yakutsk
N3)	71.31	R. Rossii-Nostalgie, Kemerovo
152)	71.45	Studiya 2, Nalchik
153)	71.45	R. Premier, Penza
)	71.46	Petrozavodsk
N1)	71.54 1	Europa Plus, Smolensk
96)	71.57	Kanal 3, Barnaul
176)	71.60	R. Rus, Stavropol
154)	71.66 15	R. 1-Petrograd, SP
155)	71.72 4	Em-Si (MC) Radio, Kaluga
N5)	71.78 1	Russkoye R, Ulyanovsk
156)	71.84 1	R. Relaks, Kirov
87)	71.89	R. Vedo. Volgograd
N1)	71.93 1	Europa Plus, Surgut
N3)	71.96	R. Rossii-Nostalgie, Chelyabinsk
112)	71.99	R. Pilot, Pskov
133)	72.00 0.2	R. Sibir, Tomsk
59)	72.02 1	R. Raduga, Izhevsk
157)	72.11 1	R. Baltik Plus, Kaliningrad
N5)	72.11	Russkoye R, Orenburg
)	72.14 5	Moskva
N6)	72.14 5	R. Modern, SP
143)	72.17	R. Avgust, Tolyatti (Dimitrovgrad)
N1)	72.26	Europa Plus, Arkhangelsk
N1)	72.26 1	Europa Plus, Astrakhan
158)	72.29	Populyarnoye R, Chita
N5)	72.41	Russkoye R, Stavropol
N3)	72.44	R. Rossii-Nostalgie, Perm
N1)	72.44 1	Europa Plus, Novosibirsk
81)	72.44 2	R. Dipol-Patrul, Tyumen
158)	72.44 1	R. Samara-Maksimum, Samara (Tolyatti)
N5)	72.59 1	Russkoye R, Magnitogorsk
160)	72.60	Studiya M, Vladimir
161)	72.65 1	R. Oskol, Staryy Oskol

FM	MHz	kW*	Station, location
162)	72.65		R. Russkiy Vityaz, Barnaul
N1)	72.68	1	Europa Plus, SP
N5)	72.74	4	Russkoye R, Chelyabinsk
N3)	72.74		R. Rossii-Nostalgie, Zlatoust
N1)	72.80	4	Europa Plus, Lipetsk
N1)	72.80	1	Europa Plus, Vyborg
N1)	72.80	1	Europa Plus, Izhevsk
163)	72.83	1	R. Vostok, Orel
164)	72.83	1	R. Express, Orel
N5)	72.83	1	Russkoye R, Samara
N5)	72.89		Russkoye R, Cherkessk
N1)	73.01	2	Europa Plus, Nizhniy Novgorod
164)	73.10	1	Eldoradio-Retro, SP
165)	73.13	0.25	R. Elis, Yaroslavl
N1)	73.16		Europa Plus, Perm
N1)	73.16	1	Europa Plus Ural, Orenburg
166)	73.16	0.1	R. Alternativa, Omsk
167)	73.16	0.2	R. BAS, Kaliningrad
N5)	73.19	1	Russkoye R, Kirov
168)	73.19	0.2	R. Tsentr, Ulyanovsk
169)	73.23		R. Express, Penza
N1)	73.25	1	Europa Plus, Vladimir
170)	73.25		R. Afontovo, Krasnoyarsk
171)	73.28		R. Dulkyn, Kazan
N5)	73.28	0.1	Russkoye R, Krasnoyarsk
N3)	73.31		R. Rossii-Nostalgie, Kemerovo
N1)	73.37	30W	Europa Plus, Dimitrovgrad
N1)	73.40	1	Europa Plus, Yek
172)	73.40		Avtoradio, Izhevsk
11)	73.40	0.1	R. Provintsiya, Pyatigorsk
24)	73.40	4	R. 7, Moskva
173)	73.43		R. Tonik, Saratov
174)	73.43		Radio S, Saratov
175)	73.49	1	Kurs Radio, Kursk
176)	73.52		R. Zet, Chelyabinsk
N5)	73.55		Russkoye R, Voronezh
N3)	73.55		R. Rossii-Nostalgie, Tyumen
N3)	73.61		R. Rossii-Nostalgie, Tver
158)	73.61	1	R. Samara-Maksimum, Samara
N4)	73.61		R. 101, Tambov
N1)	73.64	1	Europa Plus, Stavropol
114)	73.67	1	R. Novaya Volna, Volgograd
N3)	73.73		R. Rossii-Nostalgie, Arkhangelsk
11)	73.82	10	Ekho Moskvy, Moskva
177)	73.82	1	R. Maksimum, SP
N5)	73.85		Russkoye R, Kurgan
N1)	73.94	1	Europa Plus, Yaroslavl
N1)	73.94	1	Europa Plus, Omsk
N3)	73.97		R. Rossii-Nostalgie, Kursk
178)	73.97	0.25	BIM-Radio-2, Kazan
179)	73.97	0.25	R. Kuray, Kazan
180)	73.97		R. Mariya, Kirov
181)	73.97		Stil FM, Yek
	90.1	5	R. Varyag, SP (F.Pl.)
11)	90.2		Ekho Moskvy, Moskva
222)	90.6		R. 90.6, SP
20)	91.4		Avtoradio, Moskva
11)	91.5		Ekho Moskvy, M (SP)
87)	100.0		R. Vedo, Volgograd
182)	100.0		Studiya Enn, Novosibirsk
N6)	100.05		R. Modern, Perm
22)	100.1		R. Serebryannyy Dozhd, M
183)	100.1	0.1	R. Shok, Kaliningrad
N1)	100.3	1	Europa Plus, Voronezh
184)	100.3		R. Gorod, Krasnoyarsk
N1)	100.4	1	Europa Plus, Petrozavodsk
N3)	100.4		R. Rossii-Nostalgie, Yek
N4)	100.4	1	R. 101-Kontinental, Chelyabinsk
N3)	100.55		R. Rossii-Nostalgie, M
N1)	100.5	10	Europa Plus, SP
N3)	100.5		R. Rossii-Nostalgie, Kazan
N5)	100.6		Russkoye R, Tver
N1)	100.6	1	Europa Plus, Volgograd
129)	100.75		Music R, Perm
121)	100.7	1	R. Ekho, Ryazan
212)	100.7		R. Ekho Rostova, Rostov-na-Donu
145)	100.7		R. NTN, Novosibirsk
167)	100.8	0.2	R. BAS, Kaliningrad
3)	100.8		RS Nadezhda, Krasnodar
N4)	100.95		R. 101, Nizhniy Novgorod
N2)	100.9	0.1	R. ROKS, M (Anapa)
154)	100.9		R. 1-Petrograd, SP (Krasnyy Bor)
N3)	101.1	0.25	R. Rossii-Nostalgie, Yaroslavl
N4)	101.2	10	R. 101, Moskva
N1)	101.2	5	Europa Plus, Yekaterinburg
N3)	101.2		R. Rossii-Nostalgie, Kemerovo
185)	101.2	0.1	R. OVERSUN, Rostov-na-Donu
N1)	101.3		Europa Plus, Lipetsk
186)	101.3	0.25	R. Tatarus, Kazan
64)	101.4	10	Eldoradio, SP (Olgino)
220)	101.4		AST-Radio, Irkutsk
52)	101.4		R. VBC, Vladivostok
N3)	101.5		R. Rossii-Nostalgie, Samara
N3)	101.5		R. Rossii-Nostalgie, Perm
N5)	101.5		Russkoye R, Saratov
N1)	101.6	1	Europa Plus, Chelyabinsk
187)	101.6	0.1	R. Rostov, Rostov-na-Donu
84)	101.6	1	R. Pervyy Kanal, Ufa
188)	101.6		R. 101, Khanty-Mansiysk
N1)	101.7	1	Europa Plus, Ulyanovsk
22)	101.7		R. NSN, Moskva
N2)	101.7	0.1	R. ROKS, M (Temryuk)
N5)	101.8	1	Russkoye R, Penza
N5)	101.8		Russkoye R, Stavropol
N1)	101.8	0.1	Europa Plus Germes, Tyumen
N1)	101.9	1	Europa Plus, Naberezhnyy Chelny
N1)	101.9	0.1	Europa Plus, Novorossiysk
N1)	101.9	1	Europa Plus, Omsk
N6)	101.9	5	R. Modern, Tula
190)	101.9		R. Nika, Sochi
N2)	101.9	0.1	R. ROKS, M (Gelendzhik)
N4)	101.9		R. 101, Nizhniy Novgorod
N3)	102.0		R. Rossii-Nostalgie, Arkhangelsk
114)	102.0		R. Novaya Volna, Volgograd
N2)	102.0	10	R. ROKS, SP
N4)	102.0		R. 101, Yek
109)	102.1	1	Nord-Vest R, Pskov
191)	102.1		R. Assonans, Kursk
174)	102.1		Radio S, Saratov
168)	102.1	1	R. Samara-Maksimum, Samara (Tolyatti)
N5)	102.1	1	Russkoye R, Tomsk
N5)	102.1		Russkoye R, Cherepovets
N3)	102.1	1	R. Rossii-Nostalgie, Irkutsk
)	102.1		Moskva
N1)	102.2	1	Europa Plus, Kirov
N1)	102.2	0.2	Europa Plus, Rostov-na-Donu
N1)	102.2		Europa Plus, Krasnodar
N5)	102.2	1	Russkoye R, Astrakhan
143)	102.3		R. Avgust, Tolyatti
N5)	102.3		Russkoye R, Chelyabinsk
	102.4	5	R. Gardarika, SP
192)	102.4		R. Stily, Vladimir

FM	MHz kW*	Station, location
141)	102.4	R. Kray, Barnaul
N2)	102.4 1	R. ROKS, M (Novorossiysk)
N1)	102.5 1	Europa Plus, Cheboksary
N1)	102.5 1	Europa Plus, Yuzhno-Sakhalinsk
3)	102.5 3	Otkrytoye R, M (Balashikha)
193)	102.5	R. Dzhem, Yekaterinburg
140)	102.5	R. Tron, Murmansk
N3)	102.5 0.2	R. Rossii-Nostalgie, Kaliningrad
N3)	102.5	R. Rossii-Nostalgie, Sochi
N3)	102.5	R. Rossii-Nostalgie, Ivanovo
194)	102.5	R. Bulgar, Ufa
)	102.5 1	R. Zodiak, Omsk (F.Pl.)
N5)	102.6	Russkoye R, Yaroslavl
195)	102.6	R. Mir, Novosibirsk
116)	102.6 0.1	Radio "N", Novocherkassk
72)	102.7	Avtoradio, Perm
N1)	102.7 1	Europa Plus, Tolyatti
196)	102.7 1	M-Radio, Tver
N5)	102.7	Russkoye R, Belgorod
N6)	102.7 5	R. Modern, Krasnodar
N1)	102.7 1	Europa Plus, Astrakhan
178)	102.8 0.25	BIM-Radio, Kazan
N1)	102.8 1	Europa Plus, Arkhangelsk
N1)	102.8 1	Europa Plus, Ulan-Ude
197)	102.8	Nashe Radio, Voronezh
177)	102.8 10	R. Maksimum, SP (Olgino)
106)	102.9 1	Intervolna, Chelyabinsk
N5)	102.9	Russkoye R, Nizhniy Novgorod
124)	102.9 1	SBC, Samara
N5)	102.9	Russkoye R, Barnau
N1)	103.0 1	Europa Plus, Izhevsk
N6)	103.0 1	R. Modern, Murmansk
23)	103.0 5	M (Balashikha)
117)	103.0 0.1	R. 103, Rostov-na-Donu
84)	103.0	R. Vtoroy Kanal, Ufa
)	103.0	R. Roksana, Ufa
198)	103.0	R. Ekofond, Nizhniy Tagil
N6)	103.0 1	R. Modern, Omsk
N5)	103.0 0.1	Russkoye R, Novokuznetsk
N5)	103.1 0.1	Russkoye R, Pyatigorsk
N2)	103.1 1	R. ROKS, Volgograd
N5)	103.1	Russkoye R, Sochi
151)	103.1	R. Viktoria, Yakutsk
N1)	103.2	Europa Plus, Kurgan
	103.2	R. Iks, Krasnodar
N1)	103.2 1	Europa Plus, Novosibirsk
N1)	103.2 1	Europa Plus, Ryazan
144)	103.2 5	R. Maksimum-Perm, Perm
139)	103.2	Stil FM, Yek
N4)	103.2	R. 101, Tolyatti
N2)	103.3 1	R. ROKS, Azov
200)	103.3	R. Vizavi, Tula
201)	103.3 1	Puls R, Kazan
202)	103.3	R. Novaya Volna, Ulan-Ude
N3)	103.3 1	R. Rossii-Nostalgie, Irkutsk (Shelekhov)
105)	103.3	R. Puls, Azov
158)	103.3	Populyarnoye R, Chita
28)	103.4	R. Sport, M
137)	103.4 5	R. Randevyu, Nizhniy Novgorod
N5)	103.4	Russkoye R, Tambov
N4)	103.4	Radio 101, Voronezh
169)	103.5	69th Parallel, Murmansk
N1)	103.5	Europa Plus Vavilon, Belgorod
N1)	103.5 1	Europa Plus, Saratov
203)	103.5 1	Radio 3, Omsk
150)	103.5	Supervolna, Ufa

FM	MHz kW*	Station, location
204)	103.5	R. Vavilon, Belgorod
205)	103.5	R. Lukomorye, Petropavlovsk-Kamchatskiy
N1)	103.6 2	Europa Plus, Stavropol
N1)	103.6	Europa Plus, Ivanovo
113)	103.6 5	R. Megapolis, Samara
19)	103.7 10	R. Maksimum, M
N3)	103.7 0.1	R. Rossii-Nostalgie, Rostov-na-Donu
120)	103.7 5	R. Si, Yekaterinburg
N1)	103.8 4	Europa Plus Baykal, Irkutsk
N1)	103.8	Europa Plus Yenisey, Krasnoyarsk
N1)	103.8 1	Europa Plus, Penza
N5	103.9 2	Russkoye R, Novosibirsk
N1)	104.0 1	Europa Plus, Yugorsk
N6)	104.0 10	R. Modern, SP (Olgino)
206)	104.0	R. Novaya Rossiya, Novorossiysk
N3)	104.0	R. Rossii-Nostalgie, Tolyatti
207)	104.0	R. Volga, Kazan
60)	104.0	R. Titan, Ufa
208)	104.1 0.2	Donskaya Volna, Rostov-na-Donu
65)	104.1	R. Mediana, Perm
176)	104.1 1	R. Zet, Chelyabinsk
N1)	104.2	Europa Plus, Sochi (inact?)
N3)	104.2	R. Rossii-Nostalgie, Magadan
3)	104.2	RS Nadezhda, M (Balashikha)
58)	104.2	R. Novaya Volna, Vladivostok
N5)	104.3	Russkoye R, Voronezh
158)	104.3 1	R. Samara-Maksimum, Samara
N5)	104.4 5	Russkoye R, SP
209)	104.4	Golos Tuly (VOT R.), Tula
N1)	104.4	Europa Plus, Sochi
N1)	104.5	R. Europa Plus, Kaliningrad
210)	104.5	R. Magnat, Volgograd
110)	104.5 0.1	R. Mirazh, Rostov-na-Donu (inact?)
59)	104.5 1	R. Raduga, Izhevsk
	104.5	R. Felix, Dzerzhinsk
N5)	104.5	Russkoye R, Ufa
N5)	104.6	Russkoye R, Pskov
133)	104.6 0.2	R. Sibir, Tomsk
N1)	104.7 1	Europa Plus, Perm
N6)	104.7 1	R. Modern, Arkhangelsk
N5)	104.7	Russkoye R, Petrozavodsk
)	104.7	Severodvinsk
24)	104.7 7.5	Radio 7, Moskva
108)	104.7 1	R. Passazh, Kazan
120)	104.7	R. Si, Yek (Kamensk-Uralskiy)
149)	104.8 5	R. Baltika, SP
N3)	104.8	R. Rossii-Nostalgie, Saratov
N3)	104.8	R. Rossii-Nostalgie, Yoshkar-Ola
N1)	104.9 1	Europa Plus, Barnaul
	105.0	R. Pilot, Yekaterinburg
211)	105.0	Neru Plus, Murmansk
N3)	105.0	R. Rossii-Nostalgie, Ufa
N1)	105.1 1	Europa Plus, Yaroslavl
N5)	105.1 1	Russkoye R, Rostov-na-Donu
94)	105.1	R. L-Tsentr, Stavropol
N1)	105.2 1	Europa Plus, Kemerovo
18)	105.2	Delovaya volna, Moskva
157)	105.2 1	R. Baltik Plus, Kaliningrad
N2)	105.3	R. ROKS, Krasnodar
N3)	105.3 5	R. Rossii-Nostalgie, SP
171)	105.3	R. Dulkyn, Kazan
N4)	105.3	R. 101, Izhevsk
N5)	105.4	Russkoye R, Ryazan
N1)	105.6 1	Europa Plus, Khabarovsk
N5)	105.6 1	Russkoye R, Volgograd
99)	105.6	Russkoye R. Barguzin, Angarsk

FM	MHz kW*	Station, location
N6)	105.7 0.5	R. Modern, Petrozavodsk
N5)	105.7 4	Russkoye R, M
N5)	105.7	Russkoye R, Tolyatti
N5)	105.7	Russkoye R, Yek
213)	105.7 2	Radio 2, Novosibirsk
214)	105.8	R. Premier, Kaliningrad
N4)	105.9 1.2	R. 101, SP
N1)	106.0 1	Europa Plus, Ufa
N1)	106.0 1	Europa Plus, Vyborg
101)	106.0	R. Aleks, Rostov-na-Donu
	106.0	Novoye R, Krasnodar
N1)	106.1 1	Europa Plus, Samara
172)	106.1	Avtoradio, Izhevsk
N6)	106.2 4	R. Modern, Novgorod
N1)	106.2 10	Europa Plus, Moskva
180)	106.2	R. Mariya, Kirov
139)	106.2 4	Radio Trek, Yekaterinburg
215)	106.2	Narodnaya TVR-kompaniya, Khabarovsk
N1)	106.2	Europa Plus, Volgograd
216)	106.3 5	R. Rekord, SP
217)	106.4	R. Avtos, Angarsk
N1)	106.8	Europa Plus, Kazan
25)	106.8	R. Stantsiya, Moskva
218)	107.0 5	R. Port-FM, SP
219)	107.1	R. Pik, Irkutsk
N5)	107.3	Russkoye R, Kazan
26)	107.4	R. Hit FM, Moskva
67)	107.6	Studiya Gorod, Yek
105)	107.7	R. Puls, Azov
	107.7	Volnaya Kuban, Krasnodar
27)	107.8	Militseskaya volna, Moskva

*) Transmitter power (TRP).

Addresses:
N1) **HQ:** ul. Akademika Koroleva 19, 127427 Moskva. **E-mail:** public@europaplus.msk.su **Web:** http://www.infoart.ru *Regional stations:* pr. Pobedy 290, k. 306, 454106 Chelyabinsk — pr. Dimitrova 8, 433510 Dimitrovgrad — ul. Narvskaya 37/39, 236019 Kaliningrad — ul. K. Marksa 127, ofis 316, 610000 Kirov — pr. Davidkovskiy 3;5;40, 156016 Kostroma — ul. Dekabristov 26, 660017 Krasnoyarsk — ul. Koli Myagotina 56-a, 640001 Kurgan — pr. Pobedy 8, 398001 Lipetsk — ul. Belinskogo 9-a, 603600 N. Novgorod — P.O. Box 296, 630090 Novosibirsk — ul. Uchebnaya 83, 644024 Omsk — ul. Kiseleva 36, 460024 Orenburg or ul. Ostakhova 13, 460000 Orenburg — ul. Engelsa 12, 185000 Petrozavodsk — per. Internatsionalnyy 1, 180024 Pskov — ul. Varfolomeyeva 259, 344010 Rostov-na-Donu — ul. Televizionnaya 4, 390000 Ryazan — ul. Rabochaya 7, 443010 Samara — ul. Professora Popova 47, 197346 Sankt-Peterburg — ul. Volskaya 3, 410002 Saratov — P.O. Box 35, 666020 Shelekhov — ul. Kominterna 21, 355000 Stavropol — ul. Energetikov 14, 626400 Surgut — ul. Novozavodskaya 57, Tolyatti — pr. Lenina 108-a, 300026 Tula — Smolenskiy per. 29, 170000 Tver — ul. Permyakova 3-a, 625013 Tyumen — ul. Lenina 27, 670000 Ulan-Ude — ul. Goncharova 23, 432021 Ulyanovsk — ul. 3-go Internatsionalnaya 9, 600000 Vladimir — ul. Parkhomenko 47-b, 400050 Volgograd — ul. Ordzhonikidze 25, 394636 Voronezh — ul. Chkalova 2, 150005 Yaroslavl — ul. Zavidskaya 12, 620028 Yekaterinburg — ul. Komsomolskaya 213-a, 693016 Yuzhno-Sakhalinsk. Others not available.
N2) **HQ:** Khomutovskiy tipik 5-a, Sadovoye koltso, 103064 Moskva. *Regional stations:* ul. Privokzalnaya 10, 346740 Azov — pr. Lenina 2, 185000 Petrozavodsk — ul. Dzerzhinskogo 57, komn. 1201, 357000 Pyatigorsk — ul. Chapygina 4, 197022 Sankt-Peterburg — ul. Parkhomenko 43-b, 400066 Volgograd. Others not available.
N3) **HQ:** ul. Pyatnitskaya 25, 113326 Moskva. **Web:** http://www.nostalgie.ru *Regional stations:* ul. Ordzhonikidze 54-b, 454000 Chelyabinsk — P.O. Box 87, 630016 Novosibirsk — ul. Skomoroshinskaya 20, 390006 Ryazan — c/o press agency "Tsefey", nab. Fontanki 46, 191025 Sankt-Peterburg — ul. Sevenovskaya 22, 170000 Tver — pr. Revolyutsii 1-a, 394000 Voronezh — ul. Shchapova 20, 150054 Yaroslavl. Others not available.
N4) **HQ:** ul. Pyatnitskaya 25, 113326 Moskva. **E-mail:** radio101@dol.ru **Web:** http://www.101.ru *Regional stations:* ul.

Ordzhonikidze 50, 454000 Chelyabinsk. Others not available.
N5) **HQ:** ul. Kazakova 16, 103064 Moskva. **E-mail:** rusradio@asvt.ru **Web:** http://www.russia.net/rusradio *Regional stations:* ul. Chapygina 4, Hotel 'Druzhba', 5th floor, 197138 Sankt-Peterburg — ul. Frunze 100, 308000 Belgorod. Others not available.
N6) **HQ:** ul Malaya Posadskaya 8, 197046 Sankt-Peterburg. *Regional stations:* c/o R. City, pl. Lenina 2, 185000 Petrozavodsk. Others not available.
1) ul. Lesnaya 41, 101514 Moskva (editorial), or 3) (studio) — 2) P.O. Box 72, 123154 Moskva — 3) ul. Pyatnitskaya 25, 113326 Moskva — 4) 2-ya Baumanskaya ul. 9/23, korp. 18, 107005 Moskva — 5) P.O. Box 112, 121019 Moskva — 6) ul. Akademika Koroleva 19, 127427 Moskva — 7) Khoroshevskoye shosse 8, kor. 3, 125284 Moskva — 8) Kolymazhniy per. 14, 103160 Moskva, or Room 717 at 3) — 9) Zubovskiy bul. 1, 119021 Moskva — 10) P.O. Box 73, 125422 Moskva — 11) ul. Novyy Arbat 11, 121803 Moskva — 12) ul. Akademika Koroleva 12, 127427 Moskva — 13) P.O. Box 100, 113525 Moskva — 14) ul. Akademika Koroleva 15, korp. 3, 127427 Moskva — 15) ul. Nikolo-Yamskaya 1, 109189 Moskva — 16) ul. Nikolskaya 7, 103012 Moskva — 17) ul M. Nikitskaya 24, Moskva — 18) Moskva — 19) ul. Tverskaya 16/2, 103829 Moskva — 20) addr. as 10) — 21) ul. Ak. Koroleva 15, 127427 Moskva — 22) ul. Demyana Bednogo 24, 123308 Moskva — 23) Moskva — 24) ul. Leninskiy pr. 2a, 117936 Moskva — 25) P.O. Box 1068, 103064 Moskva — 26) Moskva — 27) Golutvinskiy per 10/8, 113186 Moskva. — 28) ul. B. Tatarskaya d. 35 kor. 4, 113184 Moskva — 51) ul. Vertkovskaya 10, 630048 Novosibirsk — 52) Vladivostok-91 — 53) ul. Lenina 4, Khabarovsk — 54) P.O. Box 1353, 680042 Khabarovsk — 55) ul. Chebysheva 6, komn. 521, 620219 Yekaterinburg — 56) ul. Dekabristov 34, 190000 Sankt-Peterburg — 57) Astrakhan — 58) Vladivostok — 59) P.O. Box 5037, 426057 Izhevsk — 60) P.O. Box 450005 Ufa — 61) Vladivostok — 62) ul. Voronova 12-180, 660112 Krasnoyarsk — 63) Vladivostok — 64) P.O. Box 101, 193144 Sankt-Peterburg — 65) ul Tekhnicheskaya 7, 614070 Perm — 66) Petropavlovsk-Kamchatskiy — 67) pr. Lenina 24-a, 620038 Yekaterinburg — 68) P.O. Box 2128, 680020 Khabarovsk — 69) per. Rusanova 7, 183042 Murmansk — 70) P.O. Box 171, 194356 Sankt-Peterburg — 71) ul. Uborevicha 20-a, 690670 Vladivostok — 72) bul. Gagarina 80, 614007 Perm — 73) ul. Gamarinka 51, kv. 53, 680030 Khabarovsk — 74) ul. Nometo Achirova 4, Elista — 75) P.O. Box 5016, 665841 Angarsk — 76) ul. Artema 35-A, 355000 Stavropol — 77) Vladivostok — 78) P.O. Box 104, 630015 Novosibirsk — 79) Vladivostok — 80) Sankt-Peterburg — 81) ul. Volodarskogo 38, 625000 Tyumen — 82) Krasnoyarsk — 83) P.O. Box 86, 665003 Irkutsk — 84) ul. Mingazheva 160, 450005 Ufa — 85) Omsk — 86) ul. Semena Dezhneva 75/4, 677002 Yakutsk — 87) P.O. Box 1940, 400123 Volgograd — 88) Vladivostok — 89) Omsk — 90) Saratov — 91) ul. Sergeyeva 3, 664004 Irkutsk — 92) ul Kirova 28, 692525 Ussuriysk — 93) ul. Chkalova 33, 664020 Irkutsk — 94) Stavropol — 95) Vladivostok — 96) Barnaul — 97) Omsk — 98) Iskitim — 99) P.O. Box 3471, 665839 Angarsk — 100) ul. Gagarina 6-a, 443079 Samara — 101) Rostov-na-Donu — 102) ul Zoyi Kosmodemyanskoy 16-35, 183005 Murmansk — 103) Cheboksary — 104) Dom Sovetov, pl. Lenina, komn. 111, 357530 Pyatigorsk — 105) ul. Privokzalnaya 10, 346740 Azov — 106) ul. Ordzhonikidze 35, 454091 Chelyabinsk — 107) Orenburg — 108) P.O. Box 122, 420503 Kazan — 109) P.O. Box 10, Glavpochtamt, 180000 Pskov — 110) pr. Budennovskiy 50, 344011 Rostov-na-Donu — 111) P.O. Box 164, 137500 Pyatigorsk — 112) Pskov — 113) ul. Aerodromnaya 13, 443070 Samara — 114) ul. Komsomolskaya 8, 400066 Volgograd — 115) Barnaul — 116) Baklanovskiy pr. 118, ofis 413, 346412 Novocherkassk — 117) pr. Narodnogo Opolcheniya 2, 344017 Rostov-na-Donu — 118) Rostov-na-Donu — 119) pr. Lenina 127, ofis 408, 249022 Obninsk — 120) P.O. Box 707, 620055 Yekaterinburg — 121) ul. Televizionnaya 6, 390011 Ryazan — 122) Kursk — 123) P.O. Box 372, 630048 Novosibirsk-48 — 124) ul. Sovetskoy Armii 217, 443011 Samara — 125) P.O. Box 731, 190068 Sankt-Peterburg — 126) ul. Vorovskogo 6, 454091 Chelyabinsk — 127) Rostov-na-Donu — 128) Krasnoyarsk — 129) addr. as 144) — 130) P.O. Box 372, 630048 Novosibirsk — 131) Novosibirsk — 132) Vladikavkaz — 133) P.O. Box 86, 634050 Tomsk — 134) Vladivostok — 135) ul. Promyshlennaya 5, 346740 Azov — 136) Pyatigorsk — 137) ul. Belinskogo 9-a, 603600 Nizhnyy Novgorod — 138) Penzenskiy pr. 20, 432067 Ulyanovsk — 139) ul. Volodarskogo 9, 620063 Yekaterinburg — 140) pr. Geroyev Severomortsev 2, 183038 Murmansk — 141) Barnaul — 142) Omsk — 143) P.O. Box 26, 445050 Tolyatti — 144) P.O. Box 4158, 614000 Perm — 145) ul. Rimskogo-Korsakova 9, 630048 Novosibirsk — 146) Kostroma — 147) Magnitogorsk — 148) ul. Zavodskaya 1, 357562 Pyatigorsk — 149) Kamennoostrovskiy pr. 67, 197022 Sankt-Peterburg — 150) Ufa — 151) Yakutsk — 152) Nalchik — 153) Penza — 154) P.O. Box 29,

197136 Sankt-Peterburg — 155) ul. Pukhova 52, 248010 Kaluga — 156) P.O. Box 2728, 610031 Kirov — 157) Sovetskiy pr. 43, 236000 Kaliningrad — 158) Chita. — 158) ul. Novosadovaya 3, ofis 702, 443002 Samara — 160) Vladimir — 161) ul. Lenina 11-7, Staryy Oskol — 162) Barnaul — 163) ul. Oktyabrskaya 27, 302036 Orel — 164) ul. 7 Noyabrya 43, 302028 Orel — 165) ul. Sobinova 15/14, 150000 Yaroslavl — 166) Omsk — 167) ul. Ozernaya 4/8-103, 236029 Kaliningrad — 168) ul. Sovetskaya 6, Ulyanovsk — 169) Penza — 170) P.O. Box 6180, 660017 Krasnoyarsk — 171) Kazan — 172) Izhevsk — 173) Saratov — 174) P.O. Box 882, 410027 Saratov — 175) Kursk — 176) Chelyabinsk — 177) ul. Mokhovaya 17, 191028 Sankt-Peterburg — 178) P.O. Box 16, 4205C3 Kazan — 179) Kazan — 180) P.O. Box 2796, 610031 Kirov — 181) Yekaterinburg — 182) Novosibirsk — 183) ul. Karla Marksa 18, 236000 Kaliningrad — 184) ul. Dubinskogo 6, ofis 33, Krasnoyarsk — 185) Rostov-na-Donu — 186) P.O. Box 10, 420015 Kazan — 187) ul. Bolshaya Sadovaya 82, ofis 1, 344007 Rostov-na-Donu — 188) Khanty-Mansiysk — 190) ul. Tonnelnaya 16, 354000 Sochi — 191) Kursk — 192) Vladimir — 193) Yekaterinburg — 194) Ufa — 195) ul. Gorskaya 16, 630032 Novosibirsk — 196) ul. Sevenovskaya 22, 170000 Tver — 197) Voronezh — 198) Nizhniy Tagil — 200) ul. Turgenevskaya 69, 300000 Tula — 201) ul. Dekabristov 2, 420066 Kazan — 202) Ulan Ude — 203) Omsk — 204) ul. Michurina 56, komn. 206, 308007 Belgorod — 205) Petropavlovsk-Kamchatskiy — 206) Novorossiysk — 207) Kazan — 208) P.O. Box 2218, 344038 Rostov-na-Donu — 209) P.O. Box 461, 300000 Tula — 210) Volgograd — 211) Murmansk — 212) Rostov-na-Donu — 213) Novosobirsk — 214) Kaliningrad — 215) Khabarovsk — 216) ul. Instrumentalnaya 6. 197022 Sankt-Peterburg — 217) P.O. Box 3485, 665939 Angarsk — 218) Sankt-Peterburg — 219) Irkutsk — 220) P.O. Box 1377, С64000 Irkutsk — 221) Televizyonniy per. 3, 460024 Orenburg-98 — 222) ul. Chapygina

SAN MARINO (Republic of)

L.T. UTC + 1h (Su: UTC + 2h) — **Pop.** 25.000 — **Radios:** 12.600 — **Pr.L:** Italian — **ITU:** SMR.

RADIO TITANO (Priv.)
Relay R. Cuore (see under Italy Private Networks).
FM: 97MHz 10kW (stereo) — **D.PRGR:** 24h

SAN MARINO RTV (Gov.)
Palazzo dei Congressi, RSM 47031, Rep. San Marino.
☎ +39 549 882000. 🖷 +39 549 882060 — **L.P:** Dir. S. Zavoli.
FM: 102.7MHz 10kW, 103.2MHz 2kW — **D.PRGR:** 24h
donwave– 209) P.O. Box 461, 300000 Tula – 210) Volgograd – 211) Murmansk – 212) Rostov-na-Donu – 213) Novosobirsk – 214) Kaliningrad – 215) Khabarovsk – 216) ul. Instrumentalnaya 6, 197022 Sankt-Peterburg – 217) Angarsk – 218) Sankt-Peterburg – 219) Irkutsk – 220) Irkutsk.

SLOVAKIA

L.T: UTC + 1h (Su: UTC + 2h) — **Pop:** 5m — **Radios:** 2.895.000 — **Pr.L:** Slovak — **E.C:** 50Hz, 220V — **ITU:** SVK.

SLOVENSKÝ ROZHLAS
Mýtna 1, 812 90 Bratislava. ☎ +42 (7) 493174, 496488. 🖷 +42 (7) 498923. **Cable:** Radiobratislava. ① 93352.
Web: http://www.slovakradio.sk
L.P: DG: vacant. PD: Vladimir Holan.

Long & Mediumwaves	kHz	kW	Prgr.
Rimavská Sobota	567	20	S4
Zilina	567	14	S4
Orava	621	14	S4
Banská Bystrica	702	200	S1
Bratislava	792	10	S4
Skacin	864	1	S4
Poprad	900	14	S4
Kosice	927	10	S4
Nitra	927	50	S4
Bratislava C.	1017	5	S4
Rimavská Sobota	1017	50	S4
Banská Bystrica	1035	7	S4

Long & Mediumwaves	kHz	kW	Prgr.
Presov	1071	40	S4
Nitra	1098	750	S1**FM** (MHz)

S1		S2	S3	kW
Banská Bystrica	90.1	102.0	101.5	30
Banská Stiavnica			102.60	
Bardejov			101.7	2.5
Bratislava	96.6	99.3	89.3	60
Kosice-Dubnik	96.6	101.3	100.3	30
Kosice C.		101.3		1
Lucenec			103.6	3
Námestovo		100.4	102.4	3.8
Nitra			102.2	1
Nové Mesto	103.2		100.7	1
Poprad	92.2	96.9	104.3	30
Roznava			105.9	
Ruzomberck	102.1	103.8	100.6	5
Stúrovo			103.7	3.2
Trencín			101.2	5.9
Zilina	97.2	103.5	100.1	5

+ relay st's less than 1kW.
S1=Slovensko 1, S2=Slovensko 2, S3=Slovensko 3 (Rock FM R.), H=prgrs for Hungarian minority, R=prgrs for Ruthenian minority, U=prgrs for Ukrainian minority.
S1=Slovensko 1: 24h(national prgr, news). Main N: 0600W, 1100W, 1700D, 2310D.
S2=R. Devín: 0600-2400 (cultural prgrs).
S3=Rock FM Radio: 24h (rock music). **N:** on the h. (🖃 P.O. Box 90, 81790 Bratislava 15. ☎ +42 (7) 494695.493935, 494694).
S4=Regional Prgrs for Hungarian, Ruthenian & Ukrainian minority.
Regional Prgrs:
1) Broadcasting for We. Slovakia, Mýtna 1, 812 90 Bratislava.
2) Broadcasting for Bratislava, Mýtna 1,.812 90 Bratislava.
3) Studio Banská Bystrica, Prof. Sáru 1.975 68 Banská Bystrica.
4) Studio Kosice, Moyzesova 7,. 041 61 Kosice.
5) Bastová 25, Studio Presov, 081 68 Presov.

FOREIGN SERVICE: see International Broadcasting section.

Private Stations

ASOCIÁCIA NEZÁVISLÝCH ROZHLASOVÝCH STAVIC (Association of Independent Radio Stations)
🖃 Stúrova 9, 811 02 Bratislava. ☎ +42 (7) 323 065.

MHz	kW	Station	Location
90.8	10	R. Forte	Trnava
91.1	1	R. Twist	Liptovský Mikulás
92.7	F.PI.	R. Mystic	Nové Zámky
93.3	0.08	Hit R.	Trencín
93.3	1	R. Lumen	Banská Stiavnica
94.2	30	R. Rebeca	Poprad
94.3	89	Fun R.	Bratislava
94.5	0.08	R. Zilina	Zilina
95.2	10	R. N	Nitra
95.9	10	R. Kiks	Humenné
96.0		R. Plus	Lucenec
97.0	F.PI.	R. Levice	Nové Zámky
97.6	F.PI.	R. Nitra	Banská Bystrica
97.8	10	R. Lumen	Trencín
98.5	10	R. Koliba	Nové Mesto nad Váhom
98.6	F.PI.	R. Východ	Kosice
99.2	2	Fun R.	Zilina
99.4		R. Forte	Trnava
100.3	F.PI.	R. Nitra	Bratislava
100.9	10	R. Koliba	Poprad
101.8	89	R. Twist	Bratislava
101.9	0.5	R. TOP	Kosice
102.0	1	R. Beta	Bojnice
102.5	1	R. Tatry	Poprad
102.8	1	R. Local FM	Komárno
102.9	2	R. Lumen	Vysoké Tatry
102.9	0.1	R. Lumen	Banská Bystrica
102.9	0.06	Fun R.	Kosice-mesto
103.3	1.5	R. Kiks	Michalovce
103.7	1.5	R. Flash	Presov
104.0	95	Fun R.	Banská Bystrica
104.1	0.5	R. Kiks	Presov

MHz	kW	Station	Location
104.2	0.1	R. Kiks	Domasa
104.4	5	R. RMC	Bratislava-mesto
104.5	0.12	R. Kiks	Kosice
104.5	1	R. Lumen	Banská Bystrica
104.6	0.75	R. Frontinus	Zilina
104.8	10	R. Koliba	Bratislava
105.1	10	R. Twist	Zvolen
105.5	5.8	R. DCA	Trencín
106.0	50	R. Koliba	Banská Bystrica
106.3	3	R. Lumen	Lucenec
106.6	2	R. Twist	Zvolen
106.6	3	R. Ragtime	Bratislava
106.9	2.5	R. Rebeca	Zilina

RADIO FREE EUROPE

⌕ & **L.P:** see main entry in International Broadcasting section.
MW: 1287kHz: Velké Kostolany 400kW, Poprad 14kW.
D.PRGR: 0600-0700MF, 1900-2200 in Slovak.

BBC World Sce: Bratislava 93.8MHz 3kW, Kosice 103.2MHz 0.04kW,
Banská Bystrica 105.4MHz 0.08kW: 24h in English/Slovak.

ADVENTIST WORLD RADIO (Rlg.)

⌕ AWR Europe, P.O. Box 100252, D-64202 Darmstadt, Germany —
SW: see R. Slovakia International.
For frequencies and schedule see International Broadcasting section

SLOVENIA

L.T: UTC + 1h (Su: UTC + 2h) — **Pop:** 2m — **Radios:** 620.000 (house-
holds) — **Pr.L:** Slovene — **E.C:** 50Hz, 220V — **ITU:** SVN.

RADIOTELEVIZIJA SLOVENIJA (Gov.)

⌕ Kolodvorska Ulica 2-4, SI-1550 Ljubljana. ☎ +386 (61) 1311333.
🖷 +386 (61) 1319171. ① 32283 **Web:** http://www.rtvs.si/
L.P: DG: J. Knez, a.i.
Regional Centers: RTV Koper-Capodistria, Ulica O.F. 12, Koper.
☎ +386 (66) 38533. 🖷 +386 (66) 34402 — RTV Studio Maribor, Ilichova
33, Maribor. ☎ +386 (62) 101244. 🖷 +386 (62) 101455.

MW	kHz	kW	Prgr.	MW	kHz	kW	Prgr.
Beli kriz	549	20	1	B. Kriz	1170	300	C
Maribor	558	20	1+L	Ptuj	1485	1	1+L
Cerkno	594	1	L	Radlje	1485	1	1+L
?	630		C	Brezice	1584	1	1+L
Nemcavi	648	10	L	Kovor	1584	1	2+L
Domzale	918	600	1	Ormoz	1602	1	1+L

C=R. Capodistria in Italian

FM (MHz)	Slo 1	Slo 2	Slo 3	Reg.
Ajdovscina-Planina	89.5	98.6		
Beli Kriz	92.0	94.1		104.3k
Bizeljsko	92.7	96.2	101.5	
Boc				90.4m
Breginj	95.7	97.6		
Cerkno-Lajse		90.9		
Crnonelj-Plesivica	89.0	96.5	99.0	
Fara		99.0		
Grahovo	93.5	90.0		
Idrija I		90.9		
Idrija II-Gradisce	88.9	90.0		
Jezersko	87.7	97.2		
Kanin	89.8	91.6	93.8	
Kocevje - Deklosca Gora		92.6		
Kovor	100.7	92.4		
Kozje	89.6			
Kranjska Gora - Brvogi	94.7	96.8		
Krim	88.0	93.5	96.5	
Krvavec	91.8	98.9	102.0	
Kuk	96.4	88.7		100.6k
Kum	94.1	99.9	103.9	
Lendava				103.3h
Mozirje	87.6	94.7	96.2	
Nanos	92.9	95.3	105.7	88.6k
Pec	100.1			
Pecarovci				87.6h
Plesivec	90.0	92.4	101.4	
Pohorje	88.5	96.9	105.3	93.1m

FM (MHz)	Slo 1	Slo 2	Slo 3	Reg.
				102.8mi
Skalnica				100.3k
Strari trg ob Kolpi	88.6	94.0		
Tinjan	89.3	94.6		107.6k
Trdinov Vrh	90.9	97.6	100.6	
Trenta-Lomovje	89.1	92.2		
Trenta-Skala	94.7	101.1		
Trzic 2	96.9			
Vogel	93.7	97.6		

Reg. st's: c=Onda Blu in Italian, h=R. Murska Sobota in Hungarian,
k=Modri val, m=R. Maribor, mi=R. Maribor Int,
R. Slovenija 1: 24h (Mon 2300-Tues 0300 on FM only). **N. in English
& German:** 2130.
R. Slovenija 2: 0430 (Sun 0500)-2300. Pop + entertainment.
R. Slovenija 3: 1500-2300. Serious music, educational.

RADIO KOPER - CAPODISTRIA

⌕ Ulica OF 15, Koper. ☎ +386 (66) 38861. 🖷 +386 (66) 391051 (Slovene
Dept.) +386 (66) 391050 (Italian Dept.). **E-mail:** Radio.koper@irts.si
Web: http://www.irts.si/radio-kp/
⌕ Rejèeva ulica 6, Nova Gorica. ☎+386 (65) 24312.
L.P: Editor-in-Chief (Slovene): Leon Horvatic. Editor-in-Chief (Italian):
Bruno Fonda.
Modri val in Slovenian: 0500-0600 & 1200-1700 on 549kHz + FM
Nanos 88.6,Kuk 100.6,Skalnica 100.3,Tinjan 107.6,Koper 96.4 and Beli
Kriz 104.3 MHz. Other times rel. Slovenija 1.
Onda Blu in Italian: 0500-1900 on 1170kHz + 88.4/98.9/103.1MHz
(also r. on 630kHz), 1900-0500 rel. R. Maribor Int. (FM only).

LOCAL/PRIVATE STATIONS

	kHz	Station	Tr. location
1)	594	R. Cerkno	Cerkno
43)	648	Pomurski Madzarski Radio	Nemcavci
29)	1485	R. Ptuj	Beljski vrh
6)	1584	R. Posavje	Sremice
5)	1602	R. Ormoz	Ormoz
	MHz	Station	Location
43)	87.6	Pomurski Madzarski Radio	Pecarovci
2)	87.7	R. Salomon	Bl.Dobrava
3)	87.9	R. Capris	Poljane
4)	88.4	R. Poslovni Kanal	Krvavec
5)	88.9	R. Ormoz	Ormoz
6)	88.9	R. Posavje	Brezice
7)	88.9	R. Velenje	Velenje
42)	88.9	T. Trzic	Trzic
8)	89.3	R. Student	Lj.- Sance
9)	89.5	R. Radlje	Radlje ob Dravi
10)	89.8	R. Trbovlje	Jesenice
11)	89.8	R. Zeleni Val	Spodnja Slivnica
14)	89.8	R. Ziri	Koprivnik
12)	90.3	R. Univox	Kokoska
39)	90.5	R. Val	Sveta gora
1)	90.9	R. Cerkno	Cerkno
13)	91.1	Notranjski Radio-Logatec	Sekirica
45)	91.2	Radio Ognjisce	Tinjan
14)	91.2	R. Ziri	Lubnik
15)	91.5	R. Ton	Sl.Bistrica
11)	93.1	R. Zeleni Val	Polzevo
16)	93.4	R. Morje	Markov hrib
41)	93.7	Radio Smarje pri Jelsah	Boc
17)	94.6	R. Murski Val	Pecarovci
18)	94.6	R. Sraka	Trdinov Vrh
34)	94.8	R. Nova	Planina
19)	94.9	R. Veseljak	Ljubljana- Sance
42)	95.0	T. Trzic	Kovor
20)	95.1	R. Celje	Boc
21)	95.6	R. Hit	Dobeno
22)	95.7	Radio & Radio	Pecarovci
20)	95.9	R. Celje	Malic
6)	95.9	R. Posavje	Sremic
44)	95.9	Mariborski R. Student	Meljski hrib
23)	96.0	R. Triglav	Ravne nad Valvazorjem
24)	96.4	R. Slovenske Gorice	Lenart
14)	96.4	R. Ziri	Miklav - Selca
25)	96.7	R. Sevnica	Laze
1)	97.2	R. Cerkno	Idrija I
26)	97.2	R. Koroski R.	Plesivec
27)	97.3	R. Kranj	Smarjetna Gora
10)	98.1	R. Trbovlje	Kum

MHz	Station	Location
28) 98.2	R. 94	Precna Reber
29) 98.2	R. Ptuj	Ptuj
1) 99.5	R. Cerkno	Gradisèe
30) 99.5	R. Gama	Ljubljana - Sance
31) 99.5	R. Robin	Nova Gorica
31) 100.0	R. Robin	Trstelj
32) 100.2	R. Glas Ljubljane	Krim
20) 100.3	R. Celje	Celje - Golovec
33) 100.6	R. City	Meljski Hrib
23) 101.1	R. Triglav	Vogel
23) 101.5	R. Triglav	Brvogi
2) 101.6	R. Salomon	Ljubljana- Sance
1) 102.5	R. Cerkno	Cerkno I - Lajse
35) 103.0	R. Studio D	Trdinov Vrh
43) 103.3	Pomurski Madzarski radio	Nemcavci
36) 103.4	R. Dur	Smarna Gora
28) 104.1	R. 94	Slivnica
29) 104.3	R. Ptuj	Beljski Vrh
45) 104.5	Radio Ognjisce	Krvavec
32) 104.8	R. Glas Ljubljane	Boc
3) 105.1	R. Capris	Slavnik
32) 105.1	R. Glas Ljubljane	Ljubljana
37) 105.1	R. Kobarid	Stol
25) 105.2	R. Sevnica	Sevnica-Grad
30) 106.4	R. Gama	Krim
38) 106.6	R. Krka	Trdinov Vrh
39) 106.9	R. Val	Tinjan
13) 107.0	Notranjski Radio-Logatec	Trije Kralji
45) 107.3	R. Ognjisce	Boc
12) 107.5	R. Univox	Lovski Vrh
45) 107.5	R. Ognjisce	Skalnica Sv. Gora
40) 107.8	R. Alfa	Plesivec (shared time)
7) 107.8	R. Velenje	Plesivec (shared time)

Addresses & other information
1) Platiseva 39, 5282 Cerkno.
2) 1532 Ljubljana.
3) Ulica 15.maja, 6000 Koper.
4) Trzaska 55, 1000 Ljubljana.
5) Kolodvorska 9, 2270 Ormoz.
7) Fojtova 10, 3320 Velenje.
8) Cesta 27. aprila 31, 1000 Ljubljana. **E-mail:** rs@radiostudent.si
Web: http://www.radiostudent.si/
9) Koroska cesta 2, 2360 Radlje ob Dravi.
10) Trg svobode 11a, 1420 Trbovlje.
11) Spodnja Slivnica 16, 1290 Grosuplje.
12) Rozna ulica 39, 1330 Koèevje.
13) Trzaska 148, 1370 Logatec.
14) Poljanska 2, 4220 Skofja Loka.
15) Trg svobode 28, 2310 Slovenska Bistrica.
16) Polje 5i, p.p. 96, 6310 Izola.
17) Ul. Arhitekta Novaka 13, 9000 Murska Sobota.
18) Valanticevo 17, 8000 Novo mesto.
19) Salomonov oglasnik d.o.o., 1532 Ljubljana.
20) Presernova 19, 3000 Celje.
21) Ljubljanska 36, 1203 Domzale.
22) Ul. Staneta Rozmana 1c, 9000 Murska Sobota.
23) Cufarjev trg 4, 4270 Jesenice.
24) Trg osvoboditve 5, 2230 Lenart (partly relays BBC)
25) Heroja Maroka 14, 8290 Sevnica.
26) Meskova 21, 2380 Slovenj Gradec.
27) Slovenski trg 1, 4000 Kranj.
28) Kolodvorska 5a, 7230 Postojna.
29) Raiceva 6, 2250 Ptuj.
30) Stegne 21c, 1101 Ljubljana.
31) Tolminskih puntarjev 12, 5000 Nova Gorica.
32) Cesta 24. junija 23, 1000 Ljubljana.
33) Slovenska ul. 35, 2000 Maribor. **Web:** http://www.radiocity.si
34) Gorisca 17, 5270 Ajdovscina.
35) Seidlova 27, 8000 Novo mesto.
36) Zerjalova 8, 1210 Ljubljana - Sentvid.
37) Idrsko 87, 5222 Kobarid.
38) Adamiceva 2, 8000 Novo mesto.
Web: http://www.insert.si/radio-krka/
40) Cankarjeva 1, 2380 Slovenj Gradec.
41) Askercev trg 21, p.p. 4, 3240 Smarje.
42) Balos 4, 4290 Trzic.
43) Partizanska 120, 9220 Lendava.
44) Gosposvetska 87b, 2000 Maribor.
45) Stula 23, 1210 Ljubljana - Sentvid.

SPAIN

L.T: UTC + lh (Su: UTC + 2h) - Pop: 39.500.000 - Radios: 12m - **Pr.L:** Spanish, Catalan, Valencian, Galician, Basque - **E.C:** 50Hz, 127/220V - **ITU:** E.

MINISTERIO DE FOMENTO
Secretaría General de Comunicaciones
✉ Alcalá 50, Palacio de Comunicaciones, 28071 Madrid.

RADIO NACIONAL DE ESPANA
✉ Casa de la Radio, Prado del Rey, 28223 Madrid. ☎ +34 91-3461000. 📠 +34 91-3461249.
L.P: Dir.RNE: Diego Armario, Assistant Dir: Beatriz Pécker, Deputy Dir.of Int.Rel: Alfonso Gallego, CE: Eladio Gutiérrez. Dir.R.1: Miguel Vila, Dir.R.2: Adolfo Gross, Dir.R.3: José Luis Ramos, Dir.R.5: Pedro Roncal. **Web:** http://www.rtve.es.

HOME SERVICE
Mediumwaves: R.1 and R.5.

	kHz	kW	Net	Rg	Location
GA02)	531	25	R.5	GA	Pontevedra
AS01)	531	20	R.5	AS	Oviedo°
AN02)	531	10	R.5	AN	Córdoba
NA01)	531	10	R.5	NA	Pamplona°
VA01)	558	50	R.5	VA	Valencia°
GA01)	558	25	R.5	GA	La Coruña°
PV02)	558	10	R.5	PV	San Sebastián
MU01)	567	50	R.5	MU	Murcia°
CM02)	567	10	R.5	CM	Socuéllamos (rel.of Ciudad Real)
AN03)	567	5	R.5	AN	Marbella
CA01)	576	100	R.5	CA	Barcelona°
MA01)	585	600	R.1	MA	Madrid°
AN01)	603	50	R.5	AN	Sevilla°
CL02)	603	5	R.5	CL	Palencia
CA02)	612	10	R.1	CA	Lleida
PV01)	612	10	R.1	PV	Vitoria°
AN04)	621	10	R.1	AN	Jaén
CL03)	621	10	R.1	CL	Avila
BA01)	621	50	R.1	BA	Palma de Mallorca°
GA01)	639	300	R.1	GA	La Coruña°
AR01)	639	50	R.1	AR	Zaragoza°
PV03)	639	50	R.1	PV	Bilbao
AN05)	639	20	R.1	AN	Almería
CM03)	639	10	R.1	CM	Albacete
EX02)	648	50	R.1	EX	Badajoz
MA01)	657	50	R.5	MA	Madrid°
AN01)	684	600	R.1	AN	Sevilla°
CM01)	693	20	R.1	CM	Toledo°
CA01)	693	5	R.1	CA	Tortosa (rel.of Barcelona)
AS01)	729	100	R.1	AS	Oviedo°
AN06)	729	20	R.1	AN	Málaga
RI01)	729	20	R.1	RI	Logroño°
CL01)	729	20	R.1	CL	Valladolid°
CM04)	729	10	R.1	CM	Cuenca
VA02)	729	10	R.1	VA	Alicante
CA01)	738	600	R.1	CA	Barcelona°
AN07)	747	10	R.5	AN	Cádiz
VA01)	774	100	R.1	VA	Valencia°
EX01)	774	50	R.1	EX	Cáceres°
PV02)	774	25	R.1	PV	San Sebastián
GA03)	774	20	R.1	GA	Orense
CL04)	774	10	R.1	CL	León
CL05)	774	10	R.1	CL	Soria
AN08)	774	10	R.1	AN	Granada
AN09)	774	10	R.1	AN	La Línea
CM05)	801	20	R.1	CM	Ciudad Real
GA04)	801	20	R.1	GA	Lugo
CA04)	801	10	R.1	CA	Girona
CL06)	801	10	R.1	CL	Burgos
VA03)	801	10	R.1	VA	Castellón
CL07)	801	10	R.1	CL	Zamora
MU01)	855	300	R.1	MU	Murcia°
CT01)	855	50	R.1	CT	Santander°
GA02)	855	20	R.1	GA	Pontevedra
CL08)	855	10	R.1	CL	Ponferrada
CL09)	855	10	R.1	CL	Salamanca

	kHz	kW	Net	Rg	Location
AN10)	855	10	R.1	AN	Huelva
AR02)	855	10	R.1	AR	Teruel
NA01)	855	10	R.1	NA	Pamplona°
CA03)	855	10	R.1	CA	Tarragona
AN03)	855	5	R.1	AN	Marbella
CM02)	864	5	R.1	CM	Socuellamos (rel.of Toledo)
BA01)	909	10	R.5	BA	Palma de Mallorca°
AR01)	936	50	R.5	AR	Zaragoza°
CL01)	936	20	R.5	CL	Valladolid°
VA02)	936	10	R.5	VA	Alicante
AN01)	972	5	R.1	AN	Cabra (rel.of Sevilla)
GA05)	972	5	R.1	GA	Monforte de Lemos
AN08)	1017	10	R.5	AN	Granada
CL06)	1017	10	R.5	CL	Burgos
GA04)	1098	20	R.5	GA	Lugo
AN05)	1098	20	R.5	AN	Almería
CL03)	1098	10	R.5	CL	Avila
AN10)	1098	10	R.5	AN	Huelva
RI01)	1107	25	R.5	RI	Logroño°
CT01)	1107	20	R.5	CT	Santander°
EX01)	1107	20	R.5	EX	Cáceres°
CL08)	1107	10	R.5	CL	Ponferrada (rel.of León)
AR02)	1107	10	R.5	AR	Teruel
CM01)	1125	20	R.5	CM	Toledo°
VA03)	1125	10	R.5	VA	Castellón
PV01)	1125	10	R.5	PV	Vitoria°
CL05)	1125	10	R.5	CL	Soria
EX02)	1125	10	R.5	EX	Badajoz
AN06)	1152	20	R.5	AN	Málaga
CL07)	1152	10	R.5	CL	Zamora
CA02)	1152	10	R.5	CA	Lleida
MU02)	1152	10	R.5	MU	Cartagena
CM03)	1152	10	R.5	CM	Albacete
GA03)	1305	25	R.5	GA	Orense
PV03)	1305	25	R.5	PV	Bilbao
CM05)	1305	20	R.5	CM	Ciudad Real
CL04)	1305	10	R.5	CL	León
CM04)	1314	10	R.5	CM	Cuenca
CL09)	1314	10	R.5	CL	Salamanca
CA03)	1314	10	R.5	CA	Tarragona
MA01)	1359	600	MA		Madrid°° (irr.)
AN04)	1413	10	R.5	AN	Jaén
CA04)	1413	5	R.5	CA	Girona
GA06)	1413	5	R.5	GA	Vigo
GA05)	1503	2	R.5	GA	Monforte de Lemos (rel.of Lugo)

°= regional key station
°°= operating from sunset to sunrise//Madrid 585 kHz except 1930-2200=cultural prgrs//R.2 FM network.

FM:

Andalucia

	R.1	R.2	R.3	R.4	R.5	kW
AN02) Cabra	95.1					1
AN01) Guadalcanal		90.6				5
AN07) Jerez	103.5	94.5	96.7	106.3		10
AN02) Lagar de la Cruz	92.2	97.5	98.6			10
AN06) Los Montes		99.2	104.0		92.5	1
AN11) Marbella					87.6	1
AN06) Mijas	106.6	98.1	99.8		88.0	10
AN08) Parapanda	103.0	91.1	93.9			5
AN05) Pechina	100.9	92.4	94.9			5
AN10) Punta Umbria	95.2	92.6	99.0			5
AN06) Ronda	106.1					5
AN08) San Miguel	104.2	96.4	94.4		98.5	1
AN07) San Roque	105.2					1
AN04) Sierra Almadén	105.4	90.0	96.0			10
AN08) Sierra Lújar	96.7	90.4	94.2			5
AN07) Tajo	105.0	94.0	103.1			5
AN01) Valencina	91.2	93.7	98.8		90.0	5

Aragon

	R.1	R.2	R.3	R.4	R.5	kW
AR02) Alcañiz	89.5					1
AR03) Arguis	103.9	88.1	101.5		92.8	5
AR03) Barbastro	89.6					1
AR01) Caspe	90.2					1
AR01) Cuarte Torrero	94.5			100.0		5
AR03) Fraga	95.0					1
AR01) Inogés	89.4					5
AR03) Jaca				98.7		1
AR02) Javalambre		90.0	93.9			1
AR01) La Muela	104.4	90.9	96.3			10
AR02) Teruel	104.7	89.2	94.5		95.6	1

Asturias

FM:	R.1	R.2	R.3	R.4	R.5	kW
AS01) Avilés	100.0					1
AS01) Boal	93.2	97.8	88.2		90.5	1
AS01) Cangas de Narcea		97.2	99.0	106.2	89.8	1
AS01) Cangas de Onis	88.8	90.1	95.7		101.7	1
AS01) Gamoniteiro	102.5	92.2	94.4		104.4	10
AS02) Gijón	99.2				89.9	5
AS01) Ibias	95.8	98.7	102.9		105.1	1
AS01) Llanes	106.1				97.3	1
AS01) Los Oscos	89.7	96.6	105.7		100.0	1
AS01) Luarca	96.8	93.8	100.3			1
AS01) Mieres					101.8	1
AS01) Oviedo	89.4	96.0	90.3			1
AS01) Sama de Langreo					88.7	1

Baleares

	R.1	R.2	R.3	R.4	R.5	kW
BA01) Alfabia	90.1	87.9	92.3		104.5	10
BA01) Ibiza	101.6	104.0	105.7			1
BA01) Monte Toro	94.6	97.1	105.8		100.4	1
BA01) Pollensa		95.4	97.4			1

Catalunya

	R.1	R.2	R.3	R.4	R.5	kW
CA02) Alpicat	94.6	89.2	97.8	87.9		10
CA02) Baquéira	92.2			93.3		1
CA02) Bosost	94.4					1
CA01) Collsuspina		97.9	103.1	104.7		1
CA01) Igualada	89.4	90.9	105.1	106.9		1
CA03) Monte Caro	104.3	96.6	99.6	90.7		5
CA01) Montserrat	94.3	99.0		103.8		2
CA03) Musara		91.5	94.5	88.8		5
CA01) San Pedro Ribas		95.2	97.5	106.3		1
CA04) Rocacorba	93.3	91.1	95.9	106.2		5
CA02) Soriguera	99.9			90.6		1
CA01) Tibidabo	88.3	93.0	98.7	100.8		20
CA03) Ulldecona	95.0					4
CA02) Viella	90.0			102.6		1

Castilla y León

	R.1	R.2	R.3	R.4	R.5	kW
CL06) Aranda Duero	90.0	92.7	101.6			1
CL03) Arenas S.Pedro	102.4	90.0				1
CL03) Avila	87.6	92.0	97.8		102.4	1
CL09) Béjar	99.9	101.6	104.7			1
CL07) Benavente	87.8					1
CL06) Burgos	93.6	90.3	91.2		106.6	1
CL04) Castropodame	103.3	93.0	99.9		105.9	5
CL02) Cervera	88.6					1
CL09) El Cabaco	102.9	92.4	95.4			5
CL02) Guardo	89.8					3
CL04) León	87.1	91.1	89.3		102.2	1
CL06) Miranda de Ebro	89.7					1
CL02) Palencia	91.8	101.0	97.6		88.0	1
CL07) Pbla de Samabria	93.6	103.5	100.3		91.9	1
CL09) Salamanca	94.5	88.1	91.4		102.2	1
CL10) Segovia	97.0				91.5	1
CL05) Soria	89.7	91.5	94.3		104.6	5
CL01) Valladolid	97.3	93.1	92.2		95.1	5
CL04) Villablino		89.0	91.4			1
CL06) Villadiego		102.3	103.3			1
CL04) Villafranca		89.7	97.5			1
CL07) Zamora	101.8	96.7	98.5		88.8	5

Castilla-La Mancha

	R.1	R.2	R.3	R.4	R.5	kW
CM03) Almansa	91.2					1
CM03) Chinchilla	91.8	93.6	99.0		106.3	5
CM05) Ciudad Real	95.7	92.8	106.8		88.8	1
CM04) Cuenca	105.6	93.0	92.0		96.1	1
CM06) Guadalajara	103.7	93.5	96.9		102.1	1
CM05) La Mancha		89.8	94.5			10
CM05) Socuéllamos	88.4					1
CM07) Talavera	97.8	105.5	94.7		89.4	5
CM01) Toledo	102.0	103.9	106.4		99.9	1

Cantabria

	R.1	R.2	R.3	R.4	R.5	kW
CT01) Liérganes	96.9	93.0	102.9		105.0	10
CT02) Torrelavega	99.5	97.9			89.4	1

Extremadura

	R.1	R.2	R.3	R.4	R.5	kW
EX02) Badajoz	94.9	90.1	92.2		106.0	1
EX01) Cáceres	95.1	101.7	93.7		88.2	1
EX02) La Luneta					106.0	1
EX02) Mérida					101.3	1
EX01) Montánchez	105.3	97.7	99.3			5
EX01) Plasencia	88.6		93.3		104.4	1

Galicia

	R.1	R.2	R.3	R.4	R.5	kW
GA01) Bailadora	100.4	91.6	94.5		95.8	10
GA02) Domaio	90.1	92.1	97.4			5

FM:		R.1	R.2	R.3	R.4	R.5	kW
GA06)	El Castro					96.0	1
GA04)	Monforte					88.8	1
GA03)	Monte Meda	102.8	91.2	94.3			5
GA02)	M.Tomba					104.3	1
GA04)	Páramo	101.7	88.2	99.6		92.8	5
GA07)	Santiago	103.1	98.1	99.0		93.7	5
GA03)	Verin	90.7	98.4	106.4			1
GA03)	Vilar	100.6	97.2	99.4		95.1	5
GA04)	Xistral	89.5					1
Madrid							
MA01)	Navacerrada	104.9	98.8	95.8			30
MA01)	Torrespaña	88.2	96.5	93.2		90.3	10
Murcia							
MU01)	Carrascoy	101.7	98.2	96.0		92.1	10
MU02)	Cartagena		94.5	97.5		103.5	1
MU01)	Jumilla	89.1	93.1	100.1			1
MU01)	Yecla	88.8	93.4	103.7			1
Navarra							
NA01)	Estella	89.0	101.2			90.9	1
NA01)	Gorramendi	88.3	99.0	100.6		103.9	1
NA01)	Ibañeta	89.6	93.8	103.4		101.9	1
NA01)	Isaba	90.3	95.1	103.0		91.8	1
NA01)	Leire	88.6	90.5	99.6		101.0	1
NA01)	Monreal	106.1	97.5	93.0		95.7	5
NA01)	San Miguel Aralar		96.7	100.0			
102.7	1						
NA01)	Tudela	100.9	102.2	91.3		88.3	1
País Vasco							
PV03)	Archanda	100,7	90.6	99.2		96.3	5
PV02)	Azcoitia	88.7	104.9	106.9			1
PV02)	Beasain	100.2	98.4	94.9			1
PV02)	Eibar	92.9	98.7	95.9			1
PV02)	Jaizquibel	104.7	90.0	92.1			10
PV02)	Monte Ulia	87.6				93.3	1
PV03)	Oiz	106.4	105.3	102.1			5
PV01)	San Leon					93.3	1
PV03)	Sollube	105.9	93.9	95.4			5
PV02)	Tolosa	101.9	98.8	96.0			1
PV01)	Vitoria	92.5	96.9	99.5			1
La Rioja							
RI01)	Calahorra	87.6				105.4	1
RI01)	Logroño	95.4	98.2	101.4		97.2	1
RI01)	San León	102.0					1
Comunidad Valenciana							
VA02)	Aitana	104.8	88.6	99.7			10
VA02)	Alcoy	95.8	92.3	91.1			1
VA03)	Benicasim	89.3	90.3	92.8		95.5	5
VA04)	Crevillente	92.5	105.8			96.9	1
VA02)	Elda	93.9					1
VA01)	Monduber	97.4	99.3	100.1			5
VA01)	Monte Picayo	89.8	106.6	95.1		88.2	10
VA01)	Onteniente	100.7					1
VA02)	Santa Pola	92.5	100.1	94.3		104.2	5
VA02)	Santa Pola					105.8	5
VA01)	Utiel	98.1	96.6			87.9	1
VA02)	Villena	90.7					1

R.1: (MW and FM): 24h. **N:** On the h. Regional prgrs originating from key stn of each region: MF 0555-0600, 0655-0700, 1005-1010, 1210-1300, 1605-1610; SS 1105-1200.
R.2: (FM): Classical music & cultural prgrs: 24h.
R.3: (FM): Young people's music prgr: 24h. **N:** 0000, 0300, 0500, 0700, 0900, 1100, 1300, 1500, 1800.
R.4: (FM): Regional network in Catalunya: 24h.in Catalán.
R.5: (MW and FM): 'Todo Noticias'-All news: 24h. Relays R.1 0100-0600.

+ for RNE regional key stns:
AN Andalucia: Edif.RTVE, Parque del Alamillo, 41092 Sevilla
AR Aragon: José Luís Albareda 1-3, 50004 Zaragoza.
AS Asturias: Melquiades Alvarez 9, 33002 Oviedo.
BA Baleares: Aragón 26, 07006 Palma de Mallorca.
CA Catalunya: Paseo de Grácia 1, 08007 Barcelona.
CL Castilla y León: García Morato 27-29, 47307 Valladolid.
CM Castilla-La Mancha: Plaza de San Cristóbal s/n, 45002 Toledo.
CT Cantabria: Polígono de Raos s/n, 39609 Camargo (Santander).
EX Extremadura: Ronda de San Francisco s/n, 10003 Cáceres.
GA Galicia: Paseo Méndez Nuñez 12, 15006 La Coruña.
MA Madrid: Casa de la Radio, Prado del Rey, 28223 Madrid.
MU Murcia: La Olma 27-29, 30005 Murcia.
NA Navarra: Aoiz 117, 31004 Pamplona.

PV País Vasco: Plaza de Simón Bolívar 13, 01003 Vitoria.
RI La Rioja: Av.de Portugal 2 (or: Ap.247), 26001 Logroño.
VA Comunidad Valenciana: Av.Colón 13, 46004 Valencia.

OTHER STATIONS
(Nationwide networks)

(SER) SOCIEDAD ESPANOLA DE RADIODIFUSION (Pr.)

🖃 Gran Vía 32, 28013 Madrid. ☎ +34 91-3470700. 🖷 +34 91-3470709.
L.P: Pres: Jesús de Polanco Gutiérrez. DG: Augusto Delkader Teig. Dir.Tec: Agustín Ruíz de Aguirre. Dir.Publ.Rel: Conchita Migoya Calvo-Sotelo.
FM st's ID as Cadena 40.

(COPE) CADENA DE ONDAS POPULARES ESPANOLAS (Pr.)

🖃 Valenzuela 1, 28014 Madrid. ☎ +34 91-3090000. 🖷 +34 91-5317517.
L.P: DG: Eugenio Galdón Brugarolas. Dir.Tec: Antonio Baena García.
Web: http://www.cope.es.
FM st's ID as Cadena 100. Radio Voz and Ultima Hora Radio (FM) belong to COPE.

(OCR) ONDA CERO RADIO (Pr.)

🖃 Pintor Rosales 76, 28008 Madrid. ☎ +34 91-5386300. 🖷 +34 91-5386323.
L.P: DG: D.Antonio Martinez Henajeros. Dir.Comm: Raúl Domingo de Blas.

(D) CADENA DIAL (Pr.)

🖃 Gran Vía 32, 28013 Madrid. ☎ +34 91-3470880. 🖷 +34 91-3470709. Belongs to SER.

(M80) M-80 RADIO (Pr.)

🖃 Gran Vía 32, 28013 Madrid. ☎ +34 91-3470805. 🖷 +34 91-5228692.
L.P: Gen.Mgr: Sandro D'Angeli, Musical. Mgr: Santiago Alcanda. Belongs to SER.

(A3) ANTENA 3 (Pr.)

🖃 Gran Vía 32, 28013 Madrid. ☎ +34 91-3470808-09, 🖷 +34 91-5228693.
L.P: Dir: Jorge de Anton, Red.Jefe: Miguel Sánchez. Sinfo Radio (classical music); Radiolé (Spanish music). Belongs to SER.

(Regional networks)

(CR) CORPORACIO CATALANA DE RADIO I TELEVISIO (Catalan Autonomous Gov.)

🖃 Av. Diagonal 614-616, 08021 Barcelona. ☎ +34 93-2019911. 🖷 +34 93-2006224. Web: http://www.catradio.es.
Ch.1: Catalunya Radio; Ch.2: Catalunya Música: Ch.3: RAC Radio Associació de Catalunya); Ch.4: Catalunya Informació.
L.P: Mgr: Lluís Olivia i Vazquez.

(EI) EUSKO IRRATI TELEBISTA
(Basque Autonomous Gov.)

🖃 Euskadi Gaztea (FM in Basque and Sp.) Miramón 172, 20014 San Sebastián. ☎ +34 943-423630. 🖷 +34 943-468236.
L.P: Dir: Julian Beloki Guerra.
🖃 Euskadi Irratia (MW and FM in Basque) Miramón 172, 20014 San Sebastián. ☎ +34 943-423630. 🖷 +34 943-468236.
L.P: Dir: Julian Beloki Guerra.
🖃 R.Euskadi (MW and FM in Sp.) Gran Vía 85, 48011 Bilbao. ☎ +34 94-4280800. 🖷 +34 94-4425177.
L.P: Dir: José Maria Iriondo Unanue.
🖃 R.Vitoria (MW and FM in Sp.) Pasaje Postas 32, 01001 Vitoria. ☎ +34 945-144500. 🖷 +34 945-133828.

L.P: Dir: José Ramon Diez Unzueta.

(RG) RADIO GALEGA - RADIO TELEVISION GALICIA
(Galician Autonomous Gov.)

✉ Casa de la Radio, San Marcos, 15820 Santiago de Compostela. ☎ +34 981-540940. 🖹 +34 981-540949. **Web:** http://www.crtvg.es. **LP:** DG: Arturo Maneiro Vila.

(R9) RADIO TELEVISION VALENCIANA
(Valencian Autonomous Gov.)

✉ Av. Blasco Ibañez 134, 46022 Valencia. ☎ +34 96-3721011. 🖹 +34 96-3728513.

(CS) CANAL SUR RADIO

✉ Carr.San Juan de Aznalfarache km 1.300, 41920 Sevilla. ☎ +34 95-5607600. 🖹 +34 95-5607845. **Web:** http://cica.es.
Canal Sur 1 (International music and Spanish pop and rock music).
Canal Sur 2 (Andalucian and Spanish music, news and sports).

	kHz	kW	Net	Rg	Station and location	FM
CA07)	540	50	OCR	CA	Onda Cero R., Barcelona	
CA08)	666	50	SER	CA	R.Barcelona, Barcelona	93.9
MU05)	711	25	COPE	MU	R.Popular, Murcia	89.7
PV07)	756	25	EI	PV	R.Euskadi, Bilbao	91.7
CA09)	783	50	COPE	CA	R.Miramar, Barcelona	
AN15)	792	50	SER	AN	R.Sevilla, Sevilla	97.1
MA05)	810	50	SER	MA	R.Madrid, Madrid	93.9
PV08)	819	10	EI	PV	R.Euskadi, San Sebastián	96.5
AN16)	837	50	COPE	AN	R.Popular, Sevilla	99.6
CL15)	837	10	COPE	CL	R.Popular, Burgos	95.5
GA09)	837	2	COPE	GA	R.Popular, El Ferrol	88.7
BA05)	837	2	COPE	BA	R.Popular, Ibiza	89.1
AR05)	873	25	SER	AR	R.Zaragoza, Zaragoza	95.3
GA10)	873	10	SER	GA	R.Galicia, Stgo de Comp.	90.6
CA10)	882	20		CA	Catalunya O. M., Barcelona	
AN17)	882	5	COPE	AN	R.Popular, Málaga	89.4
AS05)	882	2	COPE	AS	R.Popular, Gijón	103.6
CL16)	882	5	COPE	CL	R.Popular, Valladolid	88.5
VA07)	882	2	COPE	VA	R.Popular, Alicante	95.6
PV10)	900	25		PV	R.Popular, Bilbao	101.5
GA11)	900	5	COPE	GA	R.Popular, Vigo	87.8
AN18)	900	5	COPE	AN	R.Popular, Granada	88.2
EX05)	900	5	COPE	EX	R.Popular, Cáceres	88.2
MA06)	918	50		MA	R.Intercontinental, Madrid	95.1
MA07)	954	50		MA	R.España, Madrid	97.2
PV09)	963	10	EI	PV	R.Euskadi, Vitoria	90.9
PV11)	990	10	SER	PV	R.Bilbao, Bilbao	89.5
AN19)	990	5	SER	AN	R.Cádiz, Cádiz	89.4
MA08)	999	50	COPE	MA	R.Popular, Madrid	99.5
EX06)	1008	5	SER	EX	R.Extremadura, Badajoz	96.9
CA11)	1008	5	SER	CA	R.Girona, Girona	88.1
VA08)	1008	5	SER	VA	R.Alicante, Alicante	91.0
CA12)	1026	10	SER	CA	R.Reus, Reus	101.4
AS06)	1026	5	SER	AS	R.Asturias, Oviedo	97.5
GA12)	1026	5	SER	GA	R.Vigo, Vigo	99.4
CL17)	1026	2	SER	CL	R.Salamanca, Salamanca	96.9
AN20)	1026	2	SER	AN	R.Jaén, Jaén	96.9
AN21)	1026	2	SER	AN	R.Jerez, J. de la Frontera	97.8
GA17)	1026	2	SER	GA	R.Lugo, Lugo (1287 kHz)	
PV12)	1044	10	SER	PV	R.San Sebastián, San Seb	97.2
CL18)	1044	5	SER	CL	R.Valladolid, Valladolid	90.9
AR06)	1053	25	COPE	AR	R.Popular, Zaragoza	97.9
VA09)	1053	5	COPE	VA	R.Popular Castellón, Vila R	91.7
PV07)	1071	50	EI	PV	Euskadi Irratia, Bilbao	88.9
BA06)	1080	5	SER	BA	R.Mallorca, P. de Mallorca	94.1
AN22)	1080	10	SER	AN	R.Granada, Granada	95.4
AR07)	1080	5	SER	AR	R.Huesca, Huesca	96.9

	kHz	kW	Net	Rg	Station and location	FM
GA13)	1080	5	SER	GA	R.Coruña, La Coruña	91.0
CM11)	1080	5	OCR	CM	Onda Cero R., Toledo	100.8
CA13)	1116	5	SER	CA	R.Tarrasa, Tarassa	95.5
GA14)	1116	5	SER	GA	R.Pontevedra, Pontevedra	89.1
CM12)	1116	5	SER	CM	R.Albacete, Albacete	89.6
CM13)	1134	5	COPE	CM	R.Popular, Puertollano	97.5
BA07)	1134	5	COPE	BA	R.Pop Menorca, Ciutadella	89.6
AN23)	1134	2	COPE	AN	R.Pop Jerez de la Frontera	92.4
NA05)	1134	2		NA	R.Popular, Pamplona	87.9
CL19)	1134	2	COPE	CL	R.Popular, Salamanca	90.0
CL20)	1134	2	COPE	CL	R.Popular, Astorga	87.6
AN24)	1143	5	COPE	AN	R.Popular, Jaén	88.8
AS07)	1143	5	COPE	AS	R.Popular, Oviedo	92.8
GA15)	1143	2	COPE	GA	R.Popular, Orense	92.4
CA14)	1143	2	COPE	CA	R.Popular, Reus	89.7
PV08)	1161	50	EI	PV	Euskadi Irratia, San Seb.	94.4
VA10)	1179	50	SER	VA	R.Valencia, Valencia	94.2
RI05)	1179	2	SER	RI	R.Rioja, Logroño	91.7
PV09)	1197	10	EI	PV	Euskadi Irratia, Vitoria	95.0
CL21)	1215	5	COPE	CL	R.Popular, León	93.3
CT05)	1215	5	COPE	CT	R.Popular, Santander	88.4
AN25)	1215	2	COPE	AN	R.Popular, Córdoba	87.6
MU06)	1215	2	COPE	MU	R.Popular, Lorca	93.5
PV13)	1224	10		PV	R.Popular, San Sebastián	88.5
CM14)	1224	2	COPE	CM	R.Popular, Albacete	95.4
AN26)	1224	2	COPE	AN	R.Popular, Huelva	91.9
GA16)	1224	2	COPE	GA	R.Popular, Lugo	90.0
CA15)	1224	2	COPE	CA	R.Popular, Lérida	96.0
BA08)	1224	2	COPE	BA	R.Popular, Palma de Mallorca	97.6
AN27)	1224	2	COPE	AN	R.Popular, Almería	97.1
MU07)	1260	5	SER	MU	R.Murcia, Murcia	91.3
AN28)	1260	5	SER	AN	R.Algeciras, Algeciras	95.7
CA16)	1269	10	COPE	CA	R.Popular, Figueres	89.4
CM15)	1269	10	COPE	CM	R.Popular, Ciudad Real	93.6
CL22)	1269	5	COPE	CL	R.Popular, Zamora	94.9
EX07)	1269	5	COPE	EX	R.Popular, Badajoz	89.1
CA17)	1287	10	SER	CA	R.Lérida, Lérida	92.6
CL23)	1287	2	SER	CL	R.Castilla, Burgos	89.1
GA17)	1287	2	SER	GA	R.Lugo, Lugo (1026 kHz)	91.8
VA11)	1296	20	COPE	VA	R.Popular, Valencia	99.0
CL24)	1341	2	SER	CL	R.León, León	88.2
AN29)	1341	2	OCR	AN	Onda Cero R., Almería	93.8
CM16)	1341	2	OCR	CM	Onda Cero R., Ciudad Real	92.1
VA12)	1485	2	SER	VA	R.Alcoy, Alcoy	96.3
CT06)	1485	5	SER	CT	R.Santander, Santander	90.9
CA18)	1485	5	OCR	CA	Onda Cero R., Vilanova	96.3
AN30)	1485	2	OCR	AN	Onda Cero R., Antequera	96.3
VA13)	1521	0	SER	VA	R.Castellón, Castelló	94.8
VA14)	1539	6	SER	VA	R.Elche, Elche	94.8
CA19)	1539	5	SER	CA	R.Manresa, Manresa	91.8
AN31)	1575	5	SER	AN	R.Córdoba, Córdoba	96.6
NA06)	1575	5	SER	NA	R.Pamplona, Pamplona	92.2
VA15)	1584	2	SER	VA	R.Gandía, Gandía	96.5
GA18)	1584	5	SER	GA	R.Orense, Orense	87.6
CL25)	1584	2	SER	CL	R.Zamora, Zamora	89.8
AN32)	1602	5	SER	AN	R.Linares, Linares	94.9
MU08)	1602	2	SER	MU	R.Cartagena, Cartagena	102.3
VA16)	1602	2	SER	VA	R.Onteniente, Onten.	95.3+97.2
CL26)	1602	2	SER	CL	R.Segovia, Segovia	93.6
PV09)	1602	10	EI	PV	R.Vitoria, Vitoria	104.1

STATIONS OPERATING ON FM
(St's less than 5 kW omitted)

	MHz	kW	Net	Rg	Station and location
CA25)	87.6	20	CR	CA	Catalunya Informació, Soriguera
AN35)	87.7	6	OCR	AN	Onda Musical, Jerez de la Frontera
AN36)	87.9	5	SER	AN	R.Morón, Morón de la Frontera

	MHz	kW	Net	Rg	Station and location
CA25)	87.9	5	CR	CA	RAC 105, Collsuspina
CA25)	88.0	25	CR	CA	RAC 105, La Mussara
AN37)	88.3	5	CS	AN	Canal Sur 2, Algeciras
CA25)	88.4	25	CR	CA	Catalunya R., Montcaro
CA25)	88.6	20	CR	CA	Catalunya Música, Soriguera
CA25)	88.9	25	CR	CA	RAC 105, Rocacorba
PV07)	88.9	20	EI	PV	Euskadi Irratia, Bilbao
MA05)	89.0	20	M80	MA	M80 Madrid, Madrid
CA07)	89.1	8	OCR	CA	R.Salud, Barcelona
VA21)	89.2	8	OCR	VA	Onda Cero R., Alicante
NA11)	89.3	6	OCR	NA	Onda Cero R., Pamplona
BA11)	89.5	8	OCR	BA	Onda Musical, Palma de Mallorca
EX11)	89.5	6	OCR	EX	Onda Cero R., Cáceres
AN38)	90.1	8	OCR	AN	Onda Musical, Málaga
AN35)	90.3	6	OCR	AN	Onda Cero R., Jerez de la Frontera
EX12)	90.4	5	OCR	EX	Onda Cero R., Mérida
CA08)	90.5	8	M80	CA	M80 Barcelona, Barcelona
CL31)	90.5	5	SER	CL	R.Miranda, Miranda de Ebro
CA25)	90.8	10	CR	CA	Catalunya Música, Macanet
AN39)	90.8	5	M80	AN	M80 Puerto, El Puerto de S.M.
PV09)	90.9	20	EI	PV	R.Euskadi, Vitoria
MA11)	91.0	30		MA	Onda Mini, Madrid
AS06)	91.1	6	D	AS	Dial Asturias, Oviedo
PV07)	91.2	20	EI	PV	Euskadi Gaztea, Bilbao
EX13)	91.4	10	SER	EX	R.Plasencia, Plasencia
AN40)	91.4	8	OCR	AN	Onda Cero Melodía, Córdoba
PV21)	91.5	8	OCR	PV	Onda Cero R., San Sebastián
EX14)	91.5	5		EX	R.Guadiana, Badajoz
PV07)	91.7	20	EI	PV	R.Euskadi, Bilbao
MA05)	91.7	10	D	MA	Dial Madrid, Madrid
AN41)	91.7	5	CS	AN	Canal Sur 1, Málaga
CA25)	91.9	80	CR	CA	Catalunya Música, Alpicat
CT11)	91.9	6	OCR	CT	Onda Cero R., Santander
AR11)	92.0	40	A3	AR	Sinfo R.Zaragoza, Zaragoza
CM16)	92.1	5	OCR	CM	Onda Musical, Ciudad Real
MA12)	92.1	5		MA	R.Sensación, Torrejón de Ardoz
CA25)	92.2	10	CR	CA	Catalunya Música, Falsel
EX15)	92.2	5		EX	R.Guadiana, Mérida
MA05)	92.4	14	A3	MA	Radiolé, Madrid
GA21)	92.6	8		GA	R.Voz, La Coruña
CM21)	92.7	6	OCR	CM	Onda Cero R., Albacete
AN22)	92.8	8	SER	AN	R.Granaca 2, Granada
GA22)	93.1	5		GA	R.Voz, Pontevedra
GA13)	93.4	8	SER	GA	R.Coruña 2, La Coruña
AR05)	93.5	8	SER	AR	R.Zaragoza 2, Zaragoza
CA25)	93.8	10	CR	CA	Catalunya R., Falsel
AN29)	93.8	5	OCR	AN	R.Luz/Onda Musical, Almería
AN42)	94.0	20	CS	AN	Canal Sur 2, Huelva
GA23)	94.0	6	OCR	GA	Onda Cero R., Vigo
CA25)	94.3	10	CR	CA	Catalunya R., Flix
PV08)	94.4	20	EI	PV	Euskadi Irratia, San Sebastián
CA25)	94.5	5	CR	CA	Catalunya Informació, Collsuspina
PV07)	94.7	20	EI	PV	Euskadi Gaztea, Bilbao
AN15)	94.8	40	M80	AN	M80 Sevilla, Sevilla
CA26)	94.9	10		CA	R.Tiempo/R.Top 40, Barcelona
PV09)	95.0	20	EI	PV	Euskadi Irratia, Vitoria
GA24)	95.0	5		GA	R.Compostela/R.Top 40, S. de Comp.
AN43)	95.1	5	CS	AN	Canal Sur 2, Granada
MA06)	95.1	5		MA	R.Inter Economia, Madrid
CA25)	95.4	20	CR	CA	Catalunya R., Soriguera
AN44)	95.4	8	OCR	AN	Onda Cero R., Cádiz
AN40)	95.6	8	OCR	AN	Onda Musical, Córdoba
AN45)	95.9	40	OCR	AN	Onda Cero R., Sevilla
VA10)	96.1	10	M80	VA	M80 Valencia, Valencia
CL32)	96.2	6	OCR	CL	Onda Musical, Salamanca
AN46)	96.2	20	CS	AN	Canal Sur 1, Cadiz
AN47)	96.3	5	OCR	AN	Onda Musical, Antequera
PV08)	96.5	20	EI	PV	R.Euskadi, San Sebastián
CA25)	96.5	10	CR	CA	RAC 105, Montserrat
VA22)	96.5	5	R9	VA	R.Nou, Alicante
CA25)	96.7	25	CR	CA	Catalunya Música, Rocacorba
MU11)	96.7	6	OCR	MU	Onda Cero R., Cartegena
VA23)	96.9	40	OCR	VA	Onda Musical, Valencia
CA08)	96.9	5	SER	CA	R.Barcelona 2, Barcelona
CA25)	97.0	80	CR	CA	Catalunya Informació, Alpicat
EX16)	97.0	6		EX	R.Estudio, Cáceres
AR05)	97.1	5	D	AR	Dial Zaragoza, Zaragoza
MA07)	97.2	100		MA	R.España Top 40, Madrid
CA25)	97.3	10	CR	CA	Catalunya R., Montserrat
CA25)	97.3	10	CR	CA	Catalunya Música, Flix
CL32)	97.6	6	OCR	CL	Onda Cero R., Salamanca
VA11)	97.7	5		VA	97 Punto 7, Valencia
PV10)	97.8	5	COPE	PV	R.Correo, Bilbao
AN48)	97.9	25	CS	AN	Canal Sur 1, Jaén
MA13)	98.0	13	OCR	MA	Onda Cero R., Madrid
AN49)	98.1	5	SER	AN	R.Huelva, Huelva
CA25)	98.3	10	CR	CA	Catalunya Informació, Montserrat
VA10)	98.4	5	D	VA	Dial Mediterraneo, Valencia
AN50)	98.5	80	CS	AN	Canal Sur 1, Jerez de la Frontera
AN42)	98.5	20	CS	AN	Canal Sur 1, Huelva
CA25)	98.5	10	CR	CA	Catalunya Informació, Montcaro
CT12)	98.5	6	OCR	CT	Onda Musical, Santander
CM22)	98.5	5	OCR	CM	Onda Cero R., Talavera de la Reina
AR11)	98.6	40	M80	AR	M80 Zaragoza, Zaragoza
AN42)	98.6	20	CS	AN	Canal Sur 2, Huelva
MU12)	98.8	6		MU	R.Voz, Cartagena
BA12)	98.8	5		BA	Ultima Hora R., Palma de Mallorca
NA12)	99.2	6		NA	R.Top 40, Pamplona
MU13)	99.3	8	OCR	MU	Onda Musical, Murcia
CA25)	99.4	10	CR	CA	Catalunya Informació, Macanet
CA08)	99.4	10	D	CA	Dial Barcelona, Barcelona
CL33)	99.4	8	OCR	CL	Onda Musical, Valladolid
AN51)	99.5	8	OCR	AN	Onda Musical, Granada
CA25)	99.7	5	CR	CA	Catalunya R., Collsuspina
AN19)	99.9	8	D	AN	Dial Bahía, Cádiz
BA13)	99.9	8		BA	R.Balear Ciutat, Palma de Mallorca
CA09)	100.0	8	COPE	CA	Cadena 100, Barcelona
AN52)	100.2	5	SER	AN	R.Gaviota, Adra.
AN45)	100.3	40	OCR	AN	Onda Musical, Sevilla
CA25)	100.3	25	CR	CA	Catalunya R., La Mussara
AN53)	100.3	8		AN	R.Priego, Priego de Córdoba
AN54)	100.4	6	A3	AN	Sinfo R., Málaga
PV22)	100.4	5	SER	PV	Cadena 40, Vitoria
CA25)	100.5	5	CR	CA	Catalunya Música, Collsuspina
AN48)	100.6	5	CS	AN	Canal Sur 2, Jaén
CA25)	100.7	80	CR	CA	Catalunya R., Alpicat
CM11)	100.8	5	OCR	CM	R.Tajo/Onda Musical, Toledo
AN55)	100.8	5	CS	AN	Canal Sur 2, Córdoba
GA25)	100.9	30	RG	GA	R.Galega, Xesteiras
AN54)	101.1	5	M80	AN	M80 Málaga, Málaga
VA23)	101.2	20	OCR	VA	Onda Cero R., Valencia
AN56)	101.2	5	OCR	AN	Onda Cero R., Huelva
VA22)	101.2	5	R9	VA	R.Nou, Benidorm
MA14)	101.3	100		MA	Onda Madrid, Madrid
AN55)	101.3	60	CS	AN	Canal Sur 1, Córdoba
BA14)	101.4	5		BA	R.Club 25 Balear Inca, Inca
CA25)	101.5	80	CR	CA	Catalunya Música, Collserola
AN15)	101.5	40	A3	AN	Sinfo R., Sevilla
CA09)	102.0	10	COPE	CA	R.Popular, Barcelona
VA24)	102.2	10	R9	VA	R.Nou, Sagunto
CA25)	102.2	70	CR	CA	Catalunya R., Rocacorba
AN42)	102.2	30	CS	AN	Canal Sur 1, Huelva

MHz	kW	Net	Rg	Station and location
GA25) 102.3	40	RG	GA	R.Galega, Domaio
CA25) 102.4	10	CR	CA	Catalunya Música, Montserrat
AN57) 102.5	70	CS	AN	Canal Sur 1, Almería
CA25) 102.5	30	CR	CA	Catalunya Música, Montcaro
AN22) 102.5	8	D	AN	Dial Granada, Granada
AN58) 102.5	5		AN	R.Carolina, La Carolina
PV23) 102.6	15	COPE	PV	Bizkaia Irratia, Bilbao
MA13) 102.7	100	OCR	MA	Onda Musical, Madrid
GA26) 102.7	8	OCR	GA	Onda Cero R., La Coruña
CA25) 102.8	10	CR	CA	Catalunya R., Collserola
CA25) 102.8	10	CR	CA	RAC 105, Macanet
BA15) 102.8	5	SER	BA	R.Ibiza, Ibiza
VA22) 103.0	10	R9	VA	R.Nou, Aitana
PV07) 103.2	20	EI	PV	R.Euskadi, Bilbao
PV08) 103.5	20	EI	PV	Euskadi Gaztea, San Sebastián
VA24) 103.5	10	R9	VA	R.Nou, Monduber
VA24) 103.5	5	R9	VA	R.Nou, Benicàssim
MA15) 103.5	5		MA	R.Voz, Madrid
AN55) 103.6	60	CS	AN	Canal Sur 2, Córdoba
AN59) 103.9	60	CS	AN	Canal Sur 1, Sevilla
GA25) 103.9	40	RG	GA	R.Galega, Bailadora
MU07) 103.9	8	D	MU	Dial Murcia, Murcia
PV09) 104.1	20	EI	PV	R.Vitoria, Vitoria
AN60) 104.1	6	A3	AN	Radiolé, Almería
AN43) 104.2	8	CS	AN	Canal Sur 1, Granada
CA08) 104.2	8	A3	CA	Sinfo R., Barcelona
PV07) 104.4	20	EI	PV	Euskadi Irratia, Bilbao
AN42) 104.5	30	CS	AN	Canal Sur 2, Huelva
CA25) 104.5	25	CR	CA	Catalunya Informació, La Mussara
AN41) 104.6	60	CS	AN	Canal Sur 2, Málaga
AN50) 104.8	80	CS	AN	Canal Sur 2, Jerez de la Frontera
GA25) 104.8	40	RG	GA	R.Galega, Monte Meda
AN57) 104.8	5	CS	AN	Canal Sur 2, Almería
AN43) 104.9	60	CS	AN	Canal Sur 2, Granada
CA25) 104.9	10	CR	AN	RAC 105, Montcaro
CA25) 104.9	10	CR	CA	RAC 105, Falsel
PV09) 104.9	8	EI	PV	Euskadi Irratia, Vitoria
CA25) 105.0	5	CR	CA	RAC 105, Collserola
AN59) 105.1	5	CS	AN	Canal Sur 2, Sevilla
GA14) 105.1	5	M80	GA	M80 Pontevedra, Pontevedra
CL33) 105.2	8	OCR	CL	Onda Cero R., Valladolid
AN43) 105.3	8	CS	AN	Canal Sur 1, Granada
MA05) 105.4	100	SER	MA	R.Madrid 2, Madrid
CA25) 105.4	40	CR	CA	Catalunya Música, La Mussara
CA25) 105.5	20	CR	CA	RAC 105, Soriguera
CA25) 105.5	10	CR	CA	Catalunya R., Macanet
AN55) 105.5	5	CS	AN	Canal Sur 1, Córdoba
CL34) 105.5	6	OCR	CL	Onda Cero R., Burgos
AR12) 105.8	40	OCR	AR	R.Aragón/Onda Musical, Zaragoza
AN41) 105.8	8	CS	AN	Canal Sur 1, Málaga
MA14) 106.0	30		MA	Onda Madrid, Madrid
CA25) 106.1	80	CR	CA	RAC 105, Alpicat
BA12) 106.1	8	RS	BA	Ultima Hora R., Palma de Mallorca
GA27) 106.1	8		GA	R.Voz, Santiago de Compostela
CA25) 106.3	10	CR	CA	RAC 105, Flix
CA27) 106.6	5		CA	R.Estel, Barcelona
CA25) 106.8	5	CR	CA	RAC 105, Alfábia
AN61) 106.9	50		AN	R.América, Sevilla
EX17) 107.2	5		EX	R.Azuaga, Azuaga

Addresses and other information:
AN00) Andalucia
AN01) Edif.RTVE, Parque del Alamillo, 41092 Sevilla.
AN02) Góngora 3, 14002 Córdoba.
AN03) Av.Ricardo Soriano 11, 29600 Marbella.
AN04) Av.de Granada 57, 23001 Jaén.

AN05) Hermanos Machado 23, 04004 Almería.
AN06) Av.de la Aurora 40, 29006 Málaga.
AN07) Av.de Andalucía 67, 11007 Cádiz.
AN08) Plaza Carretas 5, 18009 Granada.
AN09) Real 24, 11300 La Línea de la Concepción.
AN10) La Fuente 4, 21004 Huelva.
AN11) Ricardo Soriano 11, 29600 Marbella.
AN15) Rafael González Abreu 6, 41001 Sevilla.
AN16) Rioja 4, 41001 Sevilla.
AN17) Linaje 2, 29001 Málaga.
AN18) Gran Vía de Colón 28, 18001 Granada.
AN19) Paseo Marítimo 1, Edif.Reina Victoria, 11010 Cádiz.
AN20) Obispo Aguilar 1, 23001 Jaén.
AN21) Guadalete 12, 11403 Jerez de la Frontera.
AN22) Santa Paula 2 (or: Ap.158), 18001 Granada.
AN23) San Agustín 11 (or: Ap.364), 11401 Jerez de la Frontera.
AN24) Av.Madrid 68, 23008 Jaén.
AN25) Plaza Cardenal Toledo 4, 14001 Córdoba.
AN26) José María Amo 2, 21001 Huelva.
AN27) Padre Luque 11, 04001 Almería.
AN28) José Antonio 7, 11201 Algeciras.
AN29) Av.Federico García Lorca 105, 04005 Almería.
AN30) San Agustín 4, 29200 Antequera.
AN31) García Lovera 3, 14002 Córdoba.
AN32) Plaza Ramón y Cajal 8, 23700 Linares.
AN35) Caballeros 2-2, 11402 Jerez de la Frontera.
AN36) Pozo Nuevo 60-2, 41530 Morón de la Frontera.
AN37) Dr.Ramón P.Rodriguez 36, 11203 Algeciras.
AN38) Comp.Lhemberg Ruíz 4, Edif.Galaxia, 29007 Málaga.
AN39) Misericordia 10, 11500 Puerto de Santa María.
AN40) Barroso 4, 14003 Córdoba.
AN41) Carr.de Cádiz 307, Av.Velázquez, 29004 Málaga.
AN42) Plaza de San Pedro 3 y 4, 21001 Huelva.
AN43) Recogidas 24, 18002 Granada.
AN44) Plaza de Asdrúbal 16, 11008 Cádiz.
AN45) Miguel de Mañara 16, 41004 Sevilla.
AN46) Plaza de España 15, 11006 Cádiz.
AN47) Rodaljarros 8, 29200 Antequera.
AN48) Av.del Ejército Español 6, 23007 Jaén.
AN49) Mendez Nuñez 15-5-6, 21001 Huelva.
AN50) Corredera 53, 11402 Jerez de la Frontera.
AN51) Paseo de Ronda 101, 18003 Granada.
AN52) Albéniz 11, 04770 Adra.
AN53) Av.América 8, Cana 22D, 14800 Priego de Córdoba.
AN54) Palestina 1, 29007 Málaga.
AN55) Av.del Gran Capitán 2-4, 14008 Córdoba.
AN56) Arquitecto Pérez Carasa 14-16, 21001 Huelva.
AN57) Residencial Oliveros, 04004 Almería.
AN58) Casa Carolina, 23200 La Carolina.
AN59) Carr.San Juan de Aznalfarache km 1.300, 41920 Sevilla.
AN60) Av.Cabo de Gata 2, 04007 Almería.
AN61) Placentines 2, 41004 Sevilla.

AR00) Aragon
AR01) José Luís Albareda 1-3, 50004 Zaragoza.
AR02) Nueva 1, 44001 Teruel.
AR03) José Gil Caves 1, 22005 Huesca.
AR05) Paseo de la Constitución 21, 50001 Zaragoza.
AR06) Paseo de Sagasta 50 (or: Ap.42), 50006 Zaragoza.
AR07) Loreto 2, 22003 Huesca.
AR11) Coso 46, 50004 Zaragoza.
AR12) Paseo Echegaray y Caballero 76, 50003 Zaragoza.

AS00) Asturias
AS01) Melquiades Alvarez 9, 33002 Oviedo.
AS02) Plaza del Instituto 3, 33201 Gijón.
AS05) Carr.de la Costa 87 (or: Ap.235), 33205 Gijón.

AS06) Asturias 19, 33007 Oviedo.
AS07) Prado Picón 16, 33008 Oviedo.

BA00) Baleares
BA01) Aragón 26, 07006 Palma de Mallorca.
BA05) Felipe II N° 28, 07800 Ibiza.
BA06) Son Moix s/n, 07011 Palma de Mallorca.
BA07) Av.Negrete 3, 07760 Ciutadela.
BA08) Av.Rey Jaime III N° 18, 07012 Palma de Mallorca.
BA11) Av.Alejandro Roselló 23-7, 07002 Palma de Mallorca.
BA12) Passeig Mallorca 32, 07012 Palma de Mallorca.
BA13) Menacor 171, 07007 Palma de Mallorca.
BA14) Llorenc Villalonga 25, 07300 Inca.
BA15) Av.de la Paz s/n, 07800 Ibiza.

CA00) Catalunya
CA01) Paseo de Grácia 1, 08007 Barcelona.
CA02) Carrer Lluis Companys 1, 25003 Lérida.
CA03) Rambla Nova 23, 43003 Tarragona.
CA04) Gran Vía Jaume I N° 60, 17001 Girona.
CA07) La Ramblas 94-98, 08002 Barcelona.
CA08) Caspe 6, 08010 Barcelona.
CA09) Av.Diagonal 297, Entre suelo, 08013 Barcelona.
CA10) Gran Vía 643, 08010 Barcelona.
CA11) Joan Miró 2, 17001 Girona.
CA12) Pintor Tomás Bergadá 3, 43204 Reus.
CA13) Gütemberg 3, 08224 Tarrasa. FM: R.Club 25.
CA14) Noguera 54-56, Edif.Boule, 43201 Reus.
CA15) Academia 17, 25002 Lérida.
CA16) Sant Llatzer 21, 17600 Figueres.
CA17) Marqués de Vila Antónia 5, 25007 Lérida.
CA18) Rambla Josef Vidal Pascual 10, 08800 Vilanova i La Geltrú.
CA19) Plana de l'Om 2, 08240 Manresa.
CA25) Av.Diagonal 614-616, 08021 Barcelona.
CA26) Via Laietana 33, 08003 Barcelona.
CA27) Av.Diagonal 460, 08006 Barcelona.

CL00) Castilla y León
CL01) García Morato 27-29, 47007 Valladolid.
CL02) Becerro de Bengoa 7, 34002 Palencia.
CL03) Santa Clara 2, 05001 Avila.
CL04) Ordoño II N° 28, 24001 León.
CL05) Campo 5, 42001 Soria.
CL06) Barrio Gimeno 11-13-15, 09001 Burgos.
CL07) Av.de Requejo 21, 49012 Zamora.
CL08) Ave María 11, 24400 Ponferrada.
CL09) Plaza de Colón 4, 37001 Salamanca.
CL10) Paseo Ezequiel Gonzales 24, 40002 Segovia.
CL15) Av.del Cid 8, 09005 Burgos.
CL16) Duque de la Victoria 23, 47001 Valladolid.
CL17) Arco 16-20 (or: Ap.211), 37002 Salamanca.
CL18) Montero Calvo 7, 47001 Valladolid.
CL19) Sol Oriente 11-15, 37001 Salamanca.
CL20) Hermanos La Salle 2, 24700 Astorga.
CL21) Lope de Vega 1, 24002 León.
CL22) Plaza Fernández Duró 3 (or: Ap.42), 49001 Zamora.
CL23) Conde Jordana 1, 09004 Burgos.
CL24) Villafranca 6, 24001 León.
CL25) Calle Santa Ana 6, 49006 Zamora.
CL26) Plaza Cirilo Rodríguez 2, 40001 Segovia.
CL31) Vitoria 24, 09200 Miranda de Ebro.
CL32) Paseo de Canalejas 68-72, 37001 Salamanca.
CL33) Santiago 14, 47001 Valladolid.
CL34) Vitoria 29, 09004 Burgos.

CM00) Castilla-La Mancha
CM01) Plaza de San Cristóbal s/n, 45002 Toledo.
CM02) Ramiro Ledesma 8, 13630 Socuéllamos.

CM03) Nuestra Señora de Araceli 1, 02001 Albacete.
CM04) Camino de la Resinera 1, 16003 Cuenca.
CM05) Ronda del Carmen s/n (or: Ap.150), 13002 Ciudad Real.
CM06) Plaza de Consejo, Centro Civico, 19001 Guadalajara.
CM07) Ronda del Canillo 35, 45600 Talavera de la Reina.
CM11) Plaza de la Merced 1, 45002 Toledo.
CM12) Concepción 25, 02002 Albacete.
CM13) Alejandro Prieto 2, 13500 Puertollano.
CM14) Tesifonte Gallego 7, 02002 Albacete.
CM15) Pasaje San Isidro 3, 13001 Ciudad Real.
CM16) Av.del Rey Santo 8, Edif.Europa, 13001 Ciudad Real.
CM21) Av.de la Estación 5, 02001 Albacete.
CM22) Av. del Principe 25, 45600 Talavera de la Reina.

CT00) Cantabria
CT01) Polígono de Raos s/n, 39609 Camargo (Santander).
CT02) Av.del Besaya 1 (or: Ap.46), 39300 Torrelavega.
CT05) Rualasal 5, 39001 Santander.
CT06) Pasaje de Peña 2, 39008 Santander.
CT11) Fernandez de Isla 14B, 39008 Santander.
CT12) Cuesta 4 entreplanta, 39002 Santander.

EX00) Extremadura
EX01) Ronda de San Francisco s/n, 10005 Cáceres.
EX02) Plaza de España 5, 06002 Badajoz.
EX05) Av.de España 15, 10002 Cáceres.
EX06) Ramón Albarrán 2, 06002 Badajoz.
EX07) Menacho 12, 06001 Badajoz.
EX11) Av.de España 9-6, 10004 Cáceres.
EX12) Av.de Portugal s/n, Ctro Comercial El Foro, 06800 Mérida.
EX13) Santa Isabel 4, 10600 Plasencia.
EX14) Felipe Checa 15, 06001 Badajoz.
EX15) Plaza Santa María 2, 06800 Mérida.
EX16) Luis Alvarez Lancero 8, 10001 Cáceres.
EX17) Plaza de la Merced 1, 06920 Azuaga.

GA00) Galicia
GA01) Paseo Méndes Nuñez 12, 15006 La Coruña.
GA02) Lepanto 7, 36001 Pontevedra.
GA03) Rua de Progreso 115 (or: Ap.268), 32003 Orense.
GA04) Orense 59-63 (or: Ap.73), 27004 Lugo.
GA05) Plaza de España 4, 27400 Monforte de Lemos.
GA06) Av.García Barbón 36, 36201 Vigo.
GA07) San Marcos s/n, Edif.TVE, 15780 Santiago de Compostela.
GA09) Plaza de España 5-6, 15403 El Ferrol.
GA10) San Pedro de Mezonzo 3, 15701 Santiago de Compostela.
GA11) Principe 57, 36202 Vigo.
GA12) Policarpo Sanz 36, 36202 Vigo.
GA13) Plaza de Orense 3, 15004 La Coruña.
GA14) Daniel de la Sota 5, 36001 Pontevedra.
GA15) Rua de Progreso 89, 32003 Orense.
GA16) Rua de Valiño s/n, 27001 Lugo.
GA17) Plaza de Santo Domingo 3, 27001 Lugo.
GA18) Rúa do Paseo 30 (or: Ap.1017), 32003 Orense.
GA21) Concepción Arenal 11-13, 15006 La Coruña.
GA22) Sagasta 1, 36001 Pontevedra.
GA23) Plaza de Compostela 17, 36201 Vigo.
GA24) Horreo 17, 15760 Santiago de Compostela.
GA25) Casa de la Radio, San Marcos, 15820 Santiago de Compostela.
GA26) Ronda D'Outeiro, 15009 La Coruña.
GA27) Rua de Villar 37, 15705 Santiago de Compostela.

MA00) Madrid
MA01) Casa de la Radio, Prado del Rey, 28223 Madrid.
MA05) Gran Vía 32, 28013 Madrid.
MA06) Modesto Lafuente 42, 28003 Madrid.
MA07) Manuel Silvela 9, 28010 Madrid.

MA08) Valenzuela 1, 28014 Madrid.
MA11) Gran Vía 31, 28013 Madrid.
MA12) Ferrocarril 15, 28850 Torrejón de Ardoz.
MA13) Pintor Rosales 76, 28008 Madrid.
MA14) García de Paredes 65, 28010 Madrid.
MA15) Bolivia 31, 28016 Madrid.

MU00)Murcia
MU01) La Olma 27-29, 30005 Murcia.
MU02) Paseo Alfonso XIII N° 51, 30203 Cartagena.
MU05) Arco de Santo Domingo 2, Edif.Fontanar, 30001 Murcia.
MU06) Av.Juan Carlos I N° 63, 30800 Lorca.
MU07) Radio Murcia 4, 30001 Murcia.
MU08) Real 82, 30201 Cartagena.
MU11) Edif.Mediterráneo, Puerta Murcia 11, 30201 Cartagena.
MU12) Alameda de San Antón 9, 30280 Cartagena.
MU13) Madre de Dios 15, 30004 Murcia.

NA00) Navarra
NA01) Aoiz 117, 31004 Pamplona.
NA05) Amaya 2-B, 31002 Pamplona.
NA06) Yangüas y Miranda 17, 31002 Pamplona.
NA11) Plaza del Castillo 43, 31001 Pamplona.
NA12) Plaza del Castillo 20, 31001 Pamplona.

PV00) País Vasco
PV01) Plaza de Simón Bolívar 13, 01003 Vitoria.
PV02) Easo 12, 20006 San Sebastián.
PV03) Licenciado Poza 55, 48013 Bilbao.
PV07) Gran Vía 85, 48011 Bilbao.
PV08) ETB-Miramón, 20014 San Sebastián.
PV09) Pasaje Postas 32, 01001 Vitoria.
PV10) Plaza Sagrado Corazón 5, 48011 Bilbao.
PV11) Epalza 8, 48007 Bilbao.
PV12) Av.de la Libertad 27, 20004 San Sebastián.
PV13) Garibal 19, 20004 San Sebastián.
PV21) Loyola 1, 20004 San Sebastián.
PV22) General Alava 10-6 Depto 9, 01005 Vitoria.
PV23) Fontecha y Salazar 9-5, 48007 Bilbao.

RI00) La Rioja
RI01) Av.de Portugal 2 (or: Ap.247), 26001 Logroño.
RI05) Av.de Portugal 12 (or: Ap.149), 26001 Logroño

VA00) Comunidad Valenciana
VA01) Colón 13, 46004 Valencia.
VA02) Angel Lozano 18, 03001 Alicante.
VA03) Paseo de Ribalta 5, 12001 Castelló.
VA04) Av.Genera Primo de Ribera 37, 03202 Elche.
VA07) Rambla de Méndez Nuñez 45, 03002 Alicante.
VA08) Calderón de la Barca 26, 03004 Alicante.
VA09) Av.Francisco Tárrega 69, 12540 Villa Real.
VA10) Don Juan de Austria 3, 46002 Valencia.
VA11) Pasaje Dr.Serra 2, 46004 Valencia.
VA12) La Cordeta 4, 03801 Alcoy.
VA13) Moyano 5, 12002 Castelló.
VA14) Dr.Caro 43, 03201 Elche.
VA15) Alcala del Olmo 15 y Loreto 38, 46700 Gandía.
VA16) Ereta 2A (or: Ap.84), 46870 Onteniente.
VA21) Av.Maissonnave 19-21, 03003 Alicante.
VA22) Av.Aguilera 1, 03007 Alicante.
VA23) Cirilo Amorós 27, 46004 Valencia.
VA24) Av.Blasco Ibañez 134, 46022 Valencia.

AMERICAN FORCES RADIO & TV SERVICE (Mil.)
92.1   Morón de la Frontera. ☞ Base Aerea USAF, 41530 Morón de la Frontera.

100.2   Torrejón de Ardoz. ☞ Base Aerea USAF, 28850 Torrejón de Ardoz.
102.5   Rota. ☞ Base Naval, 11520 Rota.
D.PRGR: All st's 24h.

SWEDEN

LT: UTC + 1h (Su: UTC + 2h) — **Pop:** 8.975.000 — **Radios:** 7.450.000 — **Pr.L:** Swedish — **E.C:** 50Hz, 220V — **ITU:** S.

TERACOM
Has the responsibility for the distribution of the prgrs. produced by Sveriges Radio (Swedish Broadcasting Corporation) and by most of the commercial radio stations and community radio associations.
HQ: Medborgarplatsen 3, Stockholm (☞ Box 17666, SE-118 92 Stockholm). ☎ +46 (8) 671200. 🖷 +46 (8) 6712001. **LP:** Pres. & CEO: Valdemar Persson.Chmn. Board of Directors: Gösta Gunnarson.

SVERIGES RADIO AB (Swedish Broadcasting Corporation) (Non-comm. broadcasting service)
HQ: Radiohuset, Oxenstiernsgatan 20, Stockholm. (☞ SE-10510 Stockholm). ☎ + 46 (8) 784 0000. 🖷 + 46 (8) 662 6992.
Web: http://www.sr.se
LP: Chmn. Board of Governors: Olle Wästberg. GD: Lisa Soederberg. Dep. GD: Jan Engdahl. Admin. Dir: Tomas Roxström. Controllers: Ewonne Winblad (P1), Christina Mattsson (P2), Mats Åkerlund (P3), Kjerstin Oscarson (P4). Contr. Int. Rel: Carina Nilsson. Legal Advisor: Gunhild Frylen.

MW: Sölvesborg 1179kHz 600/10kW: carries Home Sce. 1st Prgr. from s/on until 1600 with 10kW, Foreign Sce. 1600-0030 with 600kW.
SW: see International Broadcasting section.

FM (MHz)	Area►	1	2	3	4	kW
Arvidsjaur	24)	89.4	94.2	97.1	100.6	60
Bollnäs	20)	88.4	91.7	96.0	103.8	60
Borlänge	19)	89.4	93.0	97.7	101.3	60
Borås	25)	88.5	94.6	97.9	102.9	10
Bäckefors	14)	92.7	96.8	99.1	102.2	60
Emmaboda	6)	93.0	96.7	99.7	101.8	60
	7)				95.6	60
Filipstad	16)	88.5	90.1	98.8	103.2	2.5
Finnveden	5)	90.6	94.2	99.9	103.4	30
Gällivare	24)	88.3	94.9	98.5	100.9	60
Gävle	20)	88.1	97.4	99.8	102.0	60
Göteborg	13)	89.3	96.3	99.4	101.9	60
Halmstad	12)	87.7	91.2	95.4	97.3	60
	10)				102.6	3
Helsingborg	11)	89.8	95.7	98.4	103.2	6
Hudiksvall/Forsa	20)	87.6	90.2	93.8	100.7	60
Hörby	10)	88.8	92.4	97.0	101.4	60
	11)				89.5	5
Jönköping	5)	91.6	93.7	97.1	100.8	3
Kalix	24)	91.3	93.6	97.9	100.2	60
Karlshamn	9)	90.3	93.4	98.3	100.4	15
Karlskrona	9)	89.1	95.0	97.7	100.7	10
Karlstad	16)	90.5	94.2	96.5	103.5	15
Kiruna	24)	89.1	92.7	96.4	102.7	60
Kisa	4)	90.5	92.4	96.9	103.6	30
Lycksele	23)	92.9	95.4	98.7	103.3	60
Malmö	11)	87.9	93.3	98.0	102.0	6
Mora	19)	92.2	96.7	99.0	101.0	60
Motala	4)	91.1	94.0	98.2	101.2	20
Norrköping	3)	90.0	93.5	98.7	102.3	60
	4)				94.8	60
Nässjö	5)	89.6	92.1	99.0	102.1	60
Pajala	24)	90.8	93.0	95.9	100.2	60
Skellefteå	23)	93.8	96.3	100.0	103.9	10
Skövde	15)	88.9	95.1	97.5	100.3	60
Sollefteå	21)	89.3	93.5	98.1	101.2	60
Stockholm*	1)	92.4	96.2	99.3	103.3	60
	1)				93.8[1]	0.9
Storuman	23)	87.6	91.2	99.0	102.5	60
Sundsvall	21)	92.7	96.9	99.2	102.8	60
Sunne	16)	90.9	94.5	98.5	101.8	60
Sveg	22)	90.6	94.9	97.9	102.2	60
Trollhättan	14)	91.9	95.7	99.8	103.7	3

FM (MHz)	Areaᵇ	1	2	3	4	kW
Tåsjö	22)	89.9	94.7	97.5	100.8	60
	23)				88.2	60
Uddevalla	14)	89.9	93.1	97.2	103.3	10
Uppsala	2)	90.3	93.3	96.6	102.5	20
Varberg	12)	90.4	93.6	98.8	103.8	10
Visby	8)	87.6	94.1	97.2	100.2	60
Vislanda	6)	88.0	90.6	94.7	101.0	20
Väddö	1)				94.7	3
Vännäs	23)	88.5	92.1	95.8	103.6	60
Västervik	7)	8.3	91.8	96.0	102.7	60
Västerås	18)	90.7	95.8	98.0	100.5	60
Ånge	21)	93.2	95.6	99.6	103.1	60
	22)				94.5	60
Älvsbyn	24)	90.6	94.5	99.4	102.9	60
Örebro	17)	87.9	91.5	99.6	102.8	60
Örnsköldsvik	21)	90.8	94.4	97.8	100.1	60
Östersund	22)	87.9	91.5	94.0	100.4	60
Östhammar	2)	89.1	92.8	95.5	101.6	60
Överkalix	24)	88.9	91.7	99.0	103.2	3

+ 350 low power tr's.

Þ) Local radio area, ¹⁾Immigrant languages
*) Additional tr on 89.6MHz, 0.7kW, relays Foreign Service, also Finnish prgrs (see P2).

First Prgr. (P1) (news & spoken word): MF 0430-001, SS0425-2325. **N:** MF 0430, 0500, 0530, 0600, 0630, 0700, 0800, 0900, 1000, 1130, 1300, 1400, 1500, 1545, 1645, 1800, 1900, 2000, 2100, 2200. **Wrp** (incl. forecast for Swedish waters): 0455, 0705, 1200, 1455, 2050 — **Time Signal:** D 1059-1100.
Rel. of 1st programme on SW: See Foreign Sce. schedule.
Second Prgr (P2): serious music, Sami, Finnish and prgrs for immigrants. **Serious music:** MF 2300-0655, 1100-1430, 1700-1800, 1820-2400. Sat 2300-0600, 0600, 0800-1200, 1400-1500, 1700-2300, Sun 2300-0600. 0800-1430, 1700-2300. **Sami:** MF 0655-0730, 1200-1230. **Prgrs in Finnish:** 0730-0800, 1500-1600, 1300-1400 (Sat), 0600-0800 (Sun), 1505-1600 (Sun). **N. in Finnish:** 0455, 0730, 1500, 1555, 1730. **Prgrs for Immigrants:** MF 1000-1100, 1600-1700. SS 1600-1700. **Educational Prgrs:** MF 0800-1000, Sat 0600-0730.
Stockholm Classic FM over P2 in Stockholm and Malmö areas: serious Music from Second Prgr. 0000-0600, 0700-2400 music.
Third Prgr (P3): light music, entertainment, current affairs, news for people up to the age of 35/40: 24h. **N:** on the hour.
Fourth Prgr (P4): Regional network (frequencies as above, addresses given below).

ANN: Nat. Prgr. "Sveriges Radio" and the service f. inst. "Sveriges Radio P1".

Addresses of Regional Centres
1) SR Stockholm, Pipersgatan 45, 107 45 Stockholm – 2) SR Uppland, Box 1552, 751 45 Uppsala – 3) SR Sörmland, Box 641, 631 08 Eskilstuna – 4) SR Östergötland, Box 500, 601 17 Norrköping – 5) SR Jönköping, Bäckalyckevägen 14, 551 92 Jönköping – 6) SR Kronoberg, Box 62, 351 03 Växjö – 7) SR Kalmar, Box 805, 391 28 Kalmar – 8) SR Gotland, Box 1324, 621 24 Visby – 9) SR Blekinge, Box 305, 371 25 Karlskrona – 10) SR Kristianstad, Box 505, 291 25 Kristianstad – 11) SR Malmöhus, Baltzarsgatan 16, 211 01 Malmö – 12) SR Halland, Box 133, 301 04 Halmstad – 13) SR Göteborg, Delsjövägen, 405 13 Göteborg – 14) SR Väst, Box 654, 451 24 Uddevalla – 15) SR Västmanland, Box 850, 721 22 Västerås – 16) SR Skaraborg, Kyrkogatan 9, 541 24 Skövde – 17) SR Värmland, Box 98, 651 03 Karlstad – 18) SR Örebro, Box 1800, 701 18 Örebro – 19) SR Dalarna, Box 123, 791 23 Falun – 20) SR Gävleborg, Box 545, 801 07 Gävle – 21) SR Västernorrland, Krönvägen 18, 851 80 Sundsvall – 22) SR Jämtland, Lingonvägen 7 B, 831 62 Östersund – 23) SR Västerbotten, Mariehemsvägen 4, 906 15 Umeå – 24) SR Norrbotten, Nyagatan 3, 971 71 Luleå – 25) SR Sjuhärad, Box 27, 501 02 Borås.

FOREIGN SERVICE: see International Broadcasting section.

SVERIGES UTBILDNINGSRADIO AB
(Swedish Educational Broadcasting Company)
(Non-Comm. Public Broadcasting Service).

SE-115 80 Stockholm. Visitors: Tegeluddsvägen 23, Stockholm. ☎ + 46 (8) 784 00 00 (Managing director, administration, pre-school, school programmes, adult education, University). Stockholmsvägen 30, Stocksund. ☎ + 46 (8) 850420 (Engineering and studios). **Cable:** Broadcast — **L.P:** Chmn. Board of Governors: K. Mattson. MD: LK. Hansson. Int. Rel: Fred FleicheSR
FM: See Swedish Radio.

COMMERCIAL RADIO

Main Networks

RADIO RIX
Box 17820, 118 94 Stockholm. ☎ +46 (8) 562 720 00. ᵮ +46 (8) 562 720 82. **E-mail:** radio.rix@zradio.se
Web: http://www.rix.se/
FM (MHz): Borås 107.1, Eskilstuna 104.5, Gävle 104.9, Göteborg 105.9, Halmstad 104.2, Helsingborg/Ängelholm 107.6, Hudiksvall 105.1, Jönköping 106.0, Karlstad 104.4, Kristianstad 107.3, Lycksele 107.3, Luleå 105.6, Malmö 106.7, Nyköping 107.7, Skellefteå 92.4, Stockholm 101.9, Sundsvall/Härnösand 105.5, Trollhättan/ Uddevalla/Vänersborg 105.0, Umeå 104.2, Varberg 106.5, Västerås 106.1, Växjö 104.3, Åre 96.9, Örebro 106.3, Örnsköldsvik 104.8, Östersund 104.0

ENERGY
P.O. Box 4345, 104 67 Stockholm.
FM (MHz): Stockholm 105.1, Göteborg 105.2,Malmö 105.2.

MIX MEGAPOL
P.O. Box 3126, 103 62 Stoclkholm. **Web:** http://www.megapol.se
FM (MHz): Stockholm 104.3, Eskilstuna 107.3, Gävle/Sandviken 106.7, Göteborg 107.8, Linköping & Vasterås 106.9, Malmö 106.1, Skövde 106.4, Uddevalla 104.2, Örebro 104.7.

Other Stations in Main Cities

Stockholm
Golden Hits, Nybrokajen 7, 4tr, 111 48 Stockholm: 100.8MHz. **E-mail:** goldenhits@goldenhits.se **Web:** http://www.goldenhits.se
Bandit 105.5, Hammarby Kajvag 18, 102 30 Stockholm: 105.5MHz. **Web:** http://www.bandit.se/
R. City, P.O. Box 27852, 115 93 Stockholm: 105.9MHz. **E-mail:** mailbox@radiocity.se **Web:** http://www.sto.ims.se:80/city/
Classic FM, P.O. Box 10750, 121 29 Stockholm: 107.5MHz 1kW. **Web:** http://www.classicnet.co.uk/ sweden/

Göteborg
City 107, P.O. Box 11335, 404 27 Göteborg: 101.1MHz

Helsingborg
Stella 106,. Box 22273, 250 24 Helsingborg: 94.9/99.2MHz.

Malmö
City 107, Box 4417, 203 15 Malmö: 102.6MHz.

Norrköping
East FM 106.5, St. Persg. 19, 601 86 Norrköping: 106.5MHz.
Gold 105, St. Persg. 19, 601 86 Norrköping: 104.9MHz.

COMMUNITY RADIO

NÄRRADIONÄMNDEN
Box 16334, SE-103 26 Stockholm. ☎ + 46 (8) 237210.
STATIONS (FM): Independent broadcasting for local populations. Any non-commercial organization, whose main activity is other than broadcasting, may obtain a permit for community radio broadcasting. The transmitters are made available at a nominal fee. The number of tr's is 182, with powers between 5 and 300W ERP.

SWITZERLAND

L.T: UTC + 1h (Su: UTC + 2h) — **Pop:** 7.150.000 — **Radios:** 5.600.000 — **Pr.L:** German, Swiss German dialects, French, Italian, Rumantsch — **EC:** 50Hz, 220V — **ITU:** SUI.

SWISS BROADCASTING CORPORATION (SBC)
(A non-profit-making Company responsible for the radio & tv programme services).

⌨ Schweizerische Radio-und Fernsehgesellschaft (SRG); Société suisse de radiodiffusion et télévision (SSR); Società Svizzera di Radiotelevisione (SSR); Giacomettistrasse 3, (P.O. Box 610), CH-3000 Bern 15. **Cable:** Radif. ☎ +41 (31) 3509111. ▤ +41 (31) 3509256. ℺ 911 590 ssr ch.
Web: http://www.srg-ssr.ch/
L.P: Pres. SRG: Eric Lehmann. DG: Antonio Riva. Secr. Gen & Dir. Legal Dept: Beat Durrer. Dir. Finance: François Landgraf. Dir. Eng: Daniel Kramer. Dir. Human Resources: Raymond Zumsteg. Television Affairs: Tiziana Mona. Radio Affairs: Félix Bollmann. Dir. Communication & Marketing: Roy Oppenheim. Press Officer: Dr. Oswald Sigg.

GERMAN LANGUAGE NETWORK
Radio der deutschen und der raeto romanischen Schweiz: Schweizer Radio (DRS).
⌨ Radiodirektion SR DRS, Novarastrasse 2, Postfach 4024 Basel. ☎ +41 (61) 365 3484. ▤ +41 (61) 365 3483.
E-mail: drs@schweizerradiodrs.ch
Web: http://www.srg-ssr.ch/SRDRS/index.html
Reg. Studios: Pargau/Solothurn: Zohnhofstrasse 88, 5001 Aarau — Innerschweiz: Inseliquai 8, 6002 Luzern — Ostschweiz: Rorschacherstr. 150, 9006 St. Gallen — **Radio studios SR DRS:** Schwarztorstrasse 21 Postfach, 3000 Bern 14 — Novarastrasse 2, Postach 4024 Basel — Brunnenhofstrasse 22, Postfach, 8042 Zürich – Bahnhofstrasse 88, Postfach, 5001 Aarau – Inseliquai 8, Postfach 4069, 6002 Luzern – Rorschacherstrasse 150, Postfach 719, 9006 St. Gallen
L.P: MD: Andreas Blum. PD's: DRS-1: Heinrich von Grueningen. DRS-2: Arthur Godel. DRS-3: Andreas Schefer.
MW (DRS 1/2): Beromünster 531kHz 500kW (0500-2210).

FM (MHz)	DRS 1	DRS 2	DRS 3	DRS-R	kW
Biel-Magglingen	90.2	92.2	107.3		1.0
Feschel	88.2	90.3	101.5		2.2
Froburg	96.0	98.7	91.3		1.2
Gebidem	89.4	93.9	103.9		1
Gotschnagrat	95.2	97.8	102.9	89.4	1.2
Grono	94.8				1.0
Lenzerheide	92.0	96.6	107.4	88.2	1.5
Monte Ceneri-Passo	96.9				1.1
Monte S. Salvatore	96.3				17.0
Niederhorn	93.6	97.2	105.8		35
Rigi	90.9	96.6	103.8		30
Ruschein	96.4	100.4	102.0	91.5	1.2
St. Chrischona	90.6	99.0	103.6		47
Säntis	101.5	95.4	105.6		58
Solothurn	89.7	98.0			1.5
Uetilberg	94.6	106.7	97.5		1.8

+ additional st's less than 1kW — DRS-R=Rumantsch (see below).

1st Prgr: 24h. **N:** on the h (exc. 1100SS, 1200, 1800, 2000. Also 0430, 0530, 0630, 1730). **Local Broadc. in German:** MF 0553-0600, 1103-1112, 1630-1700 — **2nd Prgr:** 24h. **N:** 0500, 0600, 0700, 0800W, 0900W, 1000, 1130, 1300SS, 1500W, 1900W, 2100MF, 2200Sun, 2300. **N. in Rumantsch:** 1820Sun, 1850W — **3rd Prgr:** 24h — **ANN:** "Schweizer Radio DRS 1", "DRS 2" or "DRS 3".— **V.** by QSL-card.

RUMANTSCH LANGUAGE RADIO (RR)
⌨ R. Rumantsch, Theaterweg 1, CH-7002 Chur. ☎ +41 (81) 229566 (from April 1 1996 252 9566). ▤ +41 (81) 2523501.
Web: http://www.srg-ssr.ch/RTR/
L.P: Dir: Chasper Stupan.
FM: see DRS-R network (above) — **D.PRGR:** MF 0600-1230 & 1700-2000, SS 0800-1400 & 1700-2100.

FRENCH LANGUAGE NETWORK
Radio Suisse Romande (RSR)
⌨ Ave. du Temple 40, CH-1010 Lausanne. ☎ +41 (21) 3181111. ▤ +41 (21) 6523719. **Web:** http://www.rsr.ch/
L.P: MD RSR:Gérald Sapey. PD's: RSR1: Jacques Donzel. RSR2: François Page. RSR3: Blaise Duc.Option Musique: Jacques Boffard.
RSR, Studio de. Genève: Bd Carl-Vogt 86, 1205 Genève.
MW: (RSR 1/2): Sottens 765kHz 600kW, Savièse 1485kHz 1kW.

FM (MHz)	RSR 1	RRS 2	RSR 3	kW
Bantiger	95.1	99.3		12
Biel-Magglingen	96.4	99.7	103.0	1

FM (MHz)	RSR 1	RRS 2	RSR 3	kW
La Chaux-de-Fonds	92.3	96.3	103.4	1.5
La Dôle	94.8	91.2	100.1	45
Feschel	91.4	96.1	107.4	2.2
Gibloux	89.4			1.7
Grono	97.5			1.0
Haute-Nendaz	94.4	96.5	106.0	2.5
Les Ordons	94.2	99.6	104.8	10
Monte Ceneri-Passo	105.3			1.1
Mont-Pélerin	95.3	98.5	104.7	1
Monte S. Salvatore	104.0			17.0
Mont Salève (F)	105.6	102.5	100.7	1.0
Premier	96.2	97.8		1
Ravoire	94.0	95.7	100.5	1.2

+ low power st's less than 1kW.

1st Prgr: 24h. **N:** On the h. (exc. 2100, 2200). Also 0530, 0630, 1130, 2130 — **2nd Prgr:** 24h. **N:** 0500, 0600, 0700, 0800, 1130, 1600, 1800, 2130, 2300. **N. in Rumantsch:** 2150 — **3rd Prgr:** 24h non-stop music and N. **ANN:** "Radio Suisse Romande". Prgr. 1: "RSR-La Première". Prgr. 2: "RSR-Espace 2". Prgr. 3: "Couleur 3" — **V.** by QSL-card.

ITALIAN LANGUAGE NETWORK
Radiotelevisione Svizzera di Lingua Italiana (RSI)
⌨ Casella postale, CH-6903 Lugano-Besso. ☎ +41 (91) 803 93 13. ▤ +41 (91) 803 93 14. **Web:** http://www.rtsi.ch/
L.P: Regional Radio & TV Dir: Marco Blaser. PD: RSRI-1: Jacky Marti. RSI-2: Carlo Piccardi. RSI-3: Angelo Passora.
MW (RSI 1/2/3): Monte Ceneri-Cima 558kHz 300kW.

FM (MHz)	RSI I	RSI 2	RSI 3	DRS 1	RSR1	kW
Cardada	91.3	97.8	103.3	99.6	104.3	0.6
Grono	88.7	90.9	102.3	94.8	95.7	1
M. Ceneri-Passo	89.4	93.5	107.4	96.9	105.3	1.1
M. Morello	88.8	93.0	104.5	98.8	87.8	0.75
M. San Salvatore	88.1	91.5	106.0	96.3	104.0	17
Pizzo Matro	88.4	90.0	105.9	92.4	103.9	0.37

+ 22 low power st's.

RSI-1: 24h. **N:** hourly (not 1300, 1900) + 1230, 1830— **RSI-2:** 24h. **N:** 0600, 0800Sun, 0900Sat, 1000MF, 1030Sun, 1100Sat, 1500W, 1515Sun, 1630MF, 2230MF, 2300, 2400.
RSI- 3: non-stop music and N.
ANN: "R. Svizzera di lingua Italiana Rete 1", "Rete 2" or "Rete 3".

SWISS RADIO INTERNATIONAL
See International Broadcasting section.

ICRC RADIO
⌨ Red Cross Broadcasting Service, 19 Ave de la Paix, 1202 Geneva. ☎ +41 (22) 7346001. ▤ +41 (22) 733 2057. ℺ 414 226 CCR CH.
L.P: Ag. Head of Communication Dept: Michèle Mercie. Head of Press Div: Carlos Bauverd. Editor-in-Charge: Patrick Piper.
ANN: E: "This is Geneva—Red Cross Broadcasting Service". F: "Ici Genève—Service de Radiodiffusion du Comité International de la Croix-Rouge". Sp: "Aqui Ginebra — Servicio de Radiodifusión del Comité Internacional de la Cruz Roja". G: "Hier ist der Rundfunkdienst des Internationalen Komitees vom Roten Kreuz in Genf". P: "Aqui Genebra, Servico de Radiodifusão do Comite International da Cruz Vermelha" — **V.** by QSL-card. IRC appreciated. (B).
For frequencies and schedule see International Broadcasting section.

UNITED NATIONS RADIO
⌨ Nations Unies, Palais de Nations, CH-1211 Geneva 10. ☎ +41 (22) 917-1248/49. ▤ +41 (22) 917-0031. **Cable:** UNATIONS, GENEVE. ℺ 41 29 62.
L.P: Dir. Inf: T. Gastaut; Chief, Radio/TV: P. Klee.
United Nations radio programmes can be heard on radio stations throughout the world. The UN radio studios in the Palais des Nations, Geneva are used by broadcast journalists who feed their stories by radio/telephone circuits to their broadcast organizations. UN-produced news stories and interviews are also fed to the International Broadcast Centre at UN HQ New York for inclusion in UN Radio programmes produced there.

WORLD RADIO GENEVA (Comm.)
⌨ WRG-FM SA, 1 passage de la Radio, 1205 Geneva. ☎.+41 (22) 809 5040. ▤ +41 (22) 809 5045 — **FM:** Geneva 88.4MHz

D.PRGR. in English: 24h. **Music & N:** 0530-0900, 1530-1800.

Other Private Stations in main cities

Basel: R. Basilisk 94.5MHz — **Bern:** R. Extra Bern 97.7MHz, R. Förderband Bern 96.7MHz — **Geneva:** R. Cité 91.8MHz, R. Lac 97.6MHz — **Zürich:** Alternatives Lokal-Radio Zürich 104.5MHz, R. 24 102.8MHz, R. Z 100.9MHz, R. Zürisee 90.1MHz.

UKRAINE

L.T: UTC + 2h (Su: UTC + 3h) — **Pop:** 51.270.000 — **Radios:** 18.000.000 — **Pr.L:** Ukrainian, Russian — **E.C:** a/c 50Hz, 127/220V — **ITU:** UKR.

MINISTERSTVO ZVYAZKU TA INFORMATIKI
(Ministry of Communications and Informatics)
✉ vul. Khreshchatyk 22, 252001 Kyyiv. ☎ +380 (44) 2262140. 📠 +380 (44) 2286141. ① 132419 IRIS UA.
L.P: Minister: V. P. Efrimov.

DERZHAVNA TELERADIOMOVNA KOMPANIYA UKRAYINI,
Ukrainian State TV & Radio Company (Gov.)
✉ vul. Khreshchatyk 26, 252001 Kyyiv. **L.P:** Chmn: Zynoviy Kulyk.

NATSIONALNA RADIOKOMPANIYA UKRAYINI — UKRAYINSKE RADIO (Gov.)
✉ vul. Khreshchatyk 26, 252001 Kyyiv. ☎ +380 (44) 2291285. 📠 +380 (44) 2991170, 2994557. ① 131163 TON.

Rg	Stations	kHz	kW	H. of tr.	Prgr.
LV	Lviv-Krasne	171	1000	0327-0100	UR1, F
KY	Kyyiv-Brovary	207	500	0327-2300	UR1
DO	Mariupol	549	7	0245-2200	UR2
KR	Kerch	549	5	0245-2200	UR2
KY	Kyyiv-Brovary	549	150	0100-0000	UR2
MY	Mykolayiv-Luch	549	500	0100-0000	UR2
VI	Vinnytsya-Zarvantsi	549	50	0245-2300	UR2
ZK	Mizhhirya	549	7	0329-2300	UR3
ZK	Tyachiv	549	7	0329-2300	UR3
KY	Kyyiv	612	5	0300-2300	BBC
KR	Oktyabrske	648	150	0327-2300	UR1/Reg
CV	Chernivtsi	657	25	0329-2300	UR3/F
KA	Kharkiv	675	5	0300-2300	BBC
ZK	Uzhhorod	675	25	0329-2300	UR3
DO	Dokuchayevsk	711	50	0327-2300	UR1/Reg
OD	Petrivka	765	50	0327-2300	UR1/Reg
KY	Kyyiv-Brovary	783	150	0329-2300	UR3
KA	Kharkiv	810	7	0329-2300	UR3
VO	Lutsk-Pidhaytsi	810	5	0329-2300	UR3
ZP	Tokmak	810	25	0329-2300	UR3
CV	Novodnistrovsk	819	8	0400-2200	Reg
CV	Chernivtsi	837	30	0400-2200	Reg
KA	Kharkiv	837	150	0245-2200	UR2
DN	Dnipropetrovsk	873	10	0400-0700	RL
				1800-2100	RL
KM	Volochisk	873	25	as above	RL
KR	Simferopol	873	1	as above	RL
KY	Kyyiv-Brovary	873	50	as above	RL
VI	Vinnytsya-Zarvantsi	873	7	as above	RL
ZK	Uzhhorod	891	150	0327-2300	UR1/Reg
LV	Lviv-Krasne	936	1000	1730-2300	
				0327-0800a	UR1
LU	Starobilsk	936	3	0327-2300	UR1/Reg/L
MY	Mykolayiv-Luch	972	500	0327-2300b	UR1
CV	Chernivtsi	1071	10		Reg
DN	Dnipropetrovsk	1071	50	0327-2300	UR1/Reg
LU	Luhansk	1134	5	0329-2300	UR3
CH	Kholmy	1242	5	0245-2200	UR2
DO	Dokuchayevsk	1242	50	0245-2200	UR2
KM	Volochisk	1242	50	0245-2200	UR2
KR	Oktyabrske	1242	50	0245-2200	UR2/Ind
LU	Starobilsk	1242	7	0245-2200	UR2
OD	Odesa	1242	30	0245-2200	UR2
ZK	Tyachiv	1242	50	0245-2200	UR2
KA	Kharkiv	1260	50	as 873kHz	RL
KR	Sevastopol	1278	7	0329-2300	UR3/RL
OD	Petrivka	1278	150	as 873kHz	RL
OD	Odesa	1332	7	1000-2000	Reg

Rg	Stations	kHz	kW	H. of tr.	Prgr.
CV	Novodnistrovsk	1350	8	0400-2200	Reg
CV	Putila	1350	1	0245-2200	UR2
LU	Starobilsk	1350	3	0329-2300	UR3
MY	Mykolayiv	1350	7	0327-2300	Local
DO	Dokuchayevsk	1359	50	0400-2100	RL/Reg
KR	Simferopol	1359	1	0329-2300	UR3
CV	Chernivtsi	1377	50	0245-2200	UR2
KH	Tsybuleve	1377	7	0245-2200	UR2
MY	Mykolayiv	1377	7	0245-2200	UR2
VO	Lutsk-Pidhaytsi	1377	50	0245-2200	UR2
ZK	Uzhhorod	1377	50	0245-2200	UR2
ZP	Tokmak	1377	50	0245-2245	UR2
KR	Oktyabrske	1395	50	0329-2300	UR3
DN	Dnipropetrovsk	1404	30	0245-2200	UR2
IF	Strymba	1404	7	0245-2200	UR2
KE	Vasylivka	1404	7	0245-2200	UR2
LU	Luhansk	1404	25	0245-2200	UR2
LV	Lviv-Murovane	1404	50	0245-2200	UR2
OD	Izmayil	1404	25	0245-2200	UR2
SU	Shostka	1404	5	0245-2200	UR2
VO	Kovel	1404	7	0245-2200	UR2
ZK	Mizhhirya	1404	7	0245-2200	UR2
MY	Mykolayiv-Luch	1431	500	0329-2300	UR3/F
CV	Chernivtsi	1449	50	0327-2300	UR1/Reg
DO	Mariupol	1458	3	0329-2300	UR3
KR	Yalta	1467	0.1	0329-2300	UR3
LV	Lviv-Murovane	1476	30	as 873kHz	RL
KA	Kharkiv	1485	7	0327-2300	UR1/Reg
KR	Simferopol	1485	7	0327-2300	UR1/Reg
KY	Kyyiv	1485	7	0327-2300	UR1/Reg
KY	Kyyiv	1512	10	0100-0000	F
SU	Trostyanets	1530	5	0329-2300	UR3
VI	Vinnytsya-Zarvantsi	1530	30	0327-2300	UR1/Reg
CV	Chernivtsi	1557	8	0327-2300	UR1/Reg
LU	Starobilsk	1566	1	0329-2300	RSM/L
OD	Odesa	1566	7	0327-2300	UR1/Reg
KR	Botanichne	1575	0.1	0100-0000	RSM
DO	Mariupol	1584	1	0327-2300	UR1/Reg
IF	Verkhovyna	1584	1	0327-2300	UR1/Reg
KR	Simferopol	1584	1	0245-2200	UR2
KR	Parkove	1584	0.1	0400-2300	RSM
KM	Polonne	1593	1	0329-2300	UR3
KR	Sevastopol	1602	7	0245-2200	UR2/Reg
MY	Mykolayiv	1602	5	0327-2300	UR1/Reg
ZK	Uzhhorod	1602	1	0300-2300	BBC

a) directional 232 deg. b) directional 355 deg.

FM: 4kW TRP approximately = 17kW ERP.

Rg	Location	UR1+R	UR2$	UR3	Local	kW (TRP)
CH	Chernihiv	69.47	71.57	70.79		4
CH	Pryluky	71.00	72.56			4
CK	Cherkasy	70.64	72.98			4
CK	Buky	67.88	66.32			4
CV	Chernivtsi	69.26	67.19	67.94	66.41	4
CV	Novodnistrovsk	69.59	68.39			4
DN	Dnipropetrovsk	68.36	66.74	67.52	70.37	4
DN	Kryvyy Rih	71.63	69.56			1
DN	Nikopol	68.38				1
DN	Mezhova		F.PI.	F.PI.		0.1
DN	Pavlohrad	71.15	72.71			0.1
DN	Pershotravensk		70.07	71.93		0.1
DO	Vilnohirsk	F.PI.	F.PI.			0.1
DO	Donetsk	69.77	70.97	72.53	71.75	4
DO	Mariupol	67.34	69.44			4
DO	Kramatorsk	69.41	67.28			4
IF	Ivano-Frankivsk	71.24	72.80			4
KA	Izyum	72.08	70.46	72.86		4
KA	Kharkiv	67.13	67.91	66.35		4
KA	Kupyansk		69.71	70.70		0.1
KA	Velykyy Burluk		69.02	67.73		0.1
KA	Zacheplivka	71.57				0.1
KE	Kherson	71.90	70.04			4
KE	Vasylivka	69.23	67.10			4
KH	Kirovohrad	66.98*	68.84*			4
KM	Khmelnytskyy	67.70	70.46		104.6	4/4/1
KM	Polonne	68.90	71.09			4
KR	Oktyabrske	66.68	68.24			4
KY	Kyyiv-Brovary	68.45	71.33	72.86		4
KY	Kyyiv	105.0+	105.0+			4

Rg	Location	UR1+R	UR2$	UR3	Local	kW (TRP)
LU	Luhansk	68.75	70.83		—	4
LU	Rovenky	69.08	67.73		—	0.1
LU	Starobilsk	69.65	71.66		—	4
LV	Lviv-Murovane	67.10	68.96		—	4
MY	Mykolayiv	69.80	71.78	72.56	—	4
MY	Pervomaysk	69.92	68.03		—	4
OD	Odesa	70.52	72.14		—	4
OD	Izmayil	72.53			—	1
OD	Kamyanske	66.59	68.15		—	4
OD	Kotovsk	67.25	69.35		—	4
OD	Zhovten	68.99	67.07		—	4
PO	Poltava	73.88	71.21		101.9	0.1
PO	Chishata			71.79		0.1
PO	Iskivtsi	73.49	65.93			0.1
PO	Hrebinka	69.35	73.79			0.1
PO	Krasnohorivka	66.08	68.60	66.86	67.64	0.1
PO	Kremenchuk	71.21	73.79		107.7	0.1
PO	Lokhvytsya	73.88				0.1
RI	Antopil	66.56	69.77	71.21	68.21	0.1
SU	Sumy		67.28	69.02		0.1
SU	Bilopillya	66.50	68.00			4
SU	Shostka	67.49	65.93			4
SU	Trostyanets	68.75	69.92			4
TE	Ternopil	69.86				4
VI	Vinnytsya-Zarv.	71.69	68.57	72.47	70.91	4
VI	Vinnytsya-Zarv.				103.7	5
VI	Bershad	71.93	70.10			4
VI	Pishchanka	F.Pl.	F.Pl.			0.1
VI	Rosava	F.Pl.	F.Pl.			0.1
VI	Tomashpil	F.Pl.	F.Pl.			0.1
VO	Lutsk (Pidhaytsi)	70.82	72.38		107.3	0.1/0.1/1
VO	Kovel	66.02	67.73		68.48	4
VO	Kovel				103.9*	1
ZH	Zhytomyr	71.90	70.04	72.68	71.12	4
ZP	Zaporizhzhya	72.29*	70.73*	69.92		1
ZP	Berdyansk	73.01	68.57			0.1
ZP	Komysh-Zorya	67.07	68.99			3
ZP	Melitopol	66.14	68.72	67.70		3
ZP	Prymorsk	69.17	68.30			0.1
ZP	Tokmak	71.06	71.84			0.1

*) freq's reported to be interchanged. +) shared by UR1 & UR2.

D.PRGR: UR1 (Persha prohrama): 0327(SS0357)-2300 in Ukrainian (news & information). N: 0600, 0800, 0900, 1300, 1400, 1600, 1900, 2150, 2250. N. in Russian: 0700-0710, 1000-1010 — **UR2 (Promin):** 24h in Ukrainian and limited Russian (music & entertainment) — **UR3 (Tretya prohrama, R. Muz):** 0329(SS0359)-2300 in Ukrainian (cultural prgrs), incl. Russian 1500-1600, 2200-2300. Rel. R. Canada Int. in Ukrainian: 1700-1800. Rel. VOA Ukrainian 2100-2200 — **R1** = Radio Odin relay (cf. Russia) — **RSM** = R. St. Mayak relay (cf. Russia) — **RL** = R. Liberty relay (cf. Czech Rep.); addr. in Kyiv: vul. Lenina 8 — **F** = Foreign Sce (see below).

Regional services (within UR1, mostly on FM and wired networks): Mon/Wed/Fri 0430-0455, Tues/Thurs 0445-0455, D 0610-0630, 1340-1400, 1610-1700, 1800-1830. For FM freq's of reg. sce's refer to the above list. Some provincial companies also run separate metropolitan or regional prgr channels.

Addresses and other information.
Obl. = oblast (province).
CH) Chernihivska obl: vul. Uritskogo 36 (or: vul. Lenina 38-a), 250000 Chernihiv.
CK) Cherkaska obl: vul. Baydy Vishnevetskogo 35/1, 257236 Cherkasy.
CV) Chernivetska obl: vul. Holovna 91, 274000 Chernivtsi. Reg. prgr "Bukovina" on 819/837kHz, 66.41MHz 0700-2100, incl. Romanian 1300-1400. Reg. 2nd prgr on 1071/1350kHz.
DN) Dnipropetrovska obl: vul. Ivana Sirko 43 (or RTV, vul. Telebachennya 3), 320000 Dnipropetrovsk. Reg on 1071kHz + FM. "R. Mryya": 0700-1600 on 70.37MHz.
DO) Donetska obl: vul. Kuybysheva 61 340000 Donetsk (or: vul. Artema 131, 340015 Donetsk). 1st reg. prgr on 711kHz + FM. 2nd reg prgr on 1359kHz, 71.75MHz 0800-1700. Mariupol local prgr "R. Priazovya" 1800-1830 on 1584kHz, 67.34MHz.
IF) Ivano-Frankivska obl: vul. Sichovykh Streltsiv 30-a, 284000 Ivano-Frankivsk. Reg. on 1584kHz + FM.
KA) Kharkivska obl: vul. Chernyshevskogo 22, 310000 Kharkiv (or: pl. Svobody, 310002 Kharkiv). Reg. on 1485kHz + FM.
KE) Khersonska obl: vul. Perekopska 10, 325000 Kherson.
KH) Kirovohradska obl: Dom Sovetov, pl. Kirova 1, 316022 Kirovohrad (or: RTV, .vul. Pashutinska 13, 316000 Kirovohrad.
KM) Khmelnytska obl: vul. Vladimirska 92, 280000 Khmelnytskyy. Reg. 1st prgr on 873kHz + FM. Reg 2nd prgr "R. Podillya Tsentr" on 104.6MHz 0500-1700.
KR) Respublika Krym (Crimea): Krymskoye Radio, ul. Karayimskaya 6, 333011 Simferopol. Reg. on 648/1485/1602kHz + FM in Russian, Ukrainian, Armenian. Prgr of Dzhankoy studio in Krymo-Tatar: Sun 1610-1700. Yalta studio: nab. imeni Lenina 31, 334200 Yalta.
KY) Kyyivska obl: vul. Khreshchatyk 26, 252001 Kyyiv. Reg. prgr "Kyyivshchina" on FM.
LU) Luhanska obl: vul. Demekhina 25, 384055 Luhansk. Luhansk reg. on FM. Starobilsk reg. on 936/1566kHz + FM.
LV) Lvivska obl: vul. Knyazya Romana 6, 290005 Lviv.
MY) Mykolayivska obl: pr. Lenina 24-b, 327029 Mykolayiv. Reg. prgr "Buzka Khvylya" on FM. "Radiokanal Nikolayev" on 1350kHz: Tues-Fri 0700-1500, SS 0745-1400 in Russian.
OD) Odeska obl: vul Troyitska 43-b, 270011 Odesa. Reg. prgr "Chernomorskiy Mayak" on 765kHz + FM in Ukrainian & Russian, incl. Romanian Sun 1610-1700, Bulgarian/Gagauz Sun 1800-1830. Prgr for seamen "Ridna Havan" Sat 1200. "Radiokanal Odesa-2" on 1332kHz: 1000-2000 in Ukrainian, Russian, incl. reg. prgr relays.
PO) Poltavska obl: vul. Rozy Lyuksemburg 1, 314000 Poltava. Local on 67.64MHz 0800-1900.
RI) Rivnenska obl: vul. Kotlyarevskogo 20-A, 266028 Rivne. Prgr "Nova Khvylya" on 68.21MHz 1200-1600.
SU) Sumska obl: vul. Petropavlovska 125, 244021 Sumy.
TE) Ternopilska obl: bul. Tarasa Shevchenko 17, 282000 Ternopil.
VI) Vinnytska obl: vul. Teatralna (Ckalova) 15, 287100 Vinnytsya. 2nd prgr on 70.91MHz planned.
VO) Volynska obl: vul. Slovatskogo 9 (or: vul. Vinnichenko 17), 263000 Lutsk. "R. Lutsk" on 68.48/103.9/107.3MHz 0500-1100, 1600.2100 incl reg. prgr relays.
ZH) Zhytomyrska obl: vul. Lyubarska 1-a, 262000 Zhytomyr. Polish 3rd Fri 1800-1830.
ZK) Zakarpatska obl: Kyyivska nab. 18, 294018 Uzhhorod. Reg. on 891kHz.
ZP) Zaporizka obl: vul. Matrosova 24, 332057 Zaporizhzhya.

INDEPENDENT STATIONS

MW	kHz	kW	Station, location, h. of tr.
1)	612	10	Novoye R, Kharkiv: r. inactive
2)	873	10	R. Miks, Dnipropetrovsk: 0700-1400
61)	873	7	R. Vita, Vinnytsya: 0700-1800
3)	1215	7	R. DI, Dnipropetrovsk: 0600-1600
4)	1242	50	R. Mazhor, Simferopol (Oktyabrske): 0400-1800
5)	1278	150	R. Top, Odesa (Petrivka): 0700-1600
6)	1476	30	R. Nezalezhnist, Lviv: Tues/Fri 1500-1600
6)	1476	30	Polskie Radio Lwów, Lviv Sun 0900-1000
7)	1476	30	R. Vilne Slovo z Ukrayiny, Lviv: 2100-2200
8)	1476	7	R. Briz, Sevastopol
9)	1539	7	Oniks R, Kharkiv 24h
4)	1584	0.1	R. Mazhor, Simferopol (Parkove): SS 1800-2200
10)	1584	0.1	R. VM, Chernihiv: SS 0700-1600
11)	1602	1	R. Skyway, Luhansk

FM	MHz	kW	Station, location, h. of tr. (24h or as below)
12)	66.26	1	Lvivska Khvylya, Lviv
13)	66.86		R. Towak, Poltava: 0600-0800
14)	67.46	1	R. ROKS-Krym, Sevastopol (Simferopol)
15)	67.52	1	R. Sistema, Kryvyy Rih: 0500-2100
16)	67.70	1	R. ROKS-U, Kyyiv
17)	67.82	1.5	R. Lyuks, Lviv
18)	68.00	0.1	R. Aleks, Zaporizhzhya
19)	68.15	0.1	Radio 107.5, Cherkasy
4)	68.24	4	R. Mazhor, Simferopol
8)	68.36	0.6	R. Top, Odesa (F.Pl.)
20)	68.84	2	Izyum (F.Pl.)
21)	69.68	1	Gala R, Kyyiv
	69.68	1	Biznes R, Kyyiv: Sun 0500-0900
22)	69.80		R. Favorit, Kharkiv
23)	69.95	1	R. Premier, Dnipropetrovsk
24)	69.97	1	R. Yutar, Kyyiv (Chernivtsi)
11)	69.98	1	R. Skyway, Luhansk
25)	70.40	4	R. Nart, Kyyiv: 0500-2100
14)	70.40	1	R. ROKS-Krym, Sevastopol (Yalta)
26)	70.76	2	R. Simon, Kharkiv
27)	71.00	4	RadioSet, Mykolayiv: 0500-2200
7)	71.36	1	R. Vilne Slovo z Ukrayiny, Lviv (inact?)
28)	71.45	0.1	R. Saga, Zaporizhzhya: inactive
29)	71.69	4	Trans-M-Radio, Vinnytsya: Fri 1630-1700

FM	MHz	kW	Station, location, h. of tr. (24h or as below)
29)	71.93	4	Trans-M-Radio, Vinnytsya: Fri 1630-1700
30)	72.08	1	Biznes R, Kyiv
8)	72.20		R. Briz, Sevastopol
31)	72.25	0.1	R. Ankor, Khmelnitskyy
32)	73.13	0.3	R. Megapolis, Sevastopol (inact?)
33)	73.37	2	Europa Plus, Odesa
56)	73.40	0.2	Magic R, Dnipropetrovsk
34)	73.58		R. Da, Donetsk
35)	73.64	1	Kyyivski Vedomosti. Kyyiv
4)	73.79	4	R. 50, Kharkiv
21)	100.0	5	Gala R, Kyyiv
7)	100.0	1	R. Vilne Slovo z Ukrayiny, Lviv
37)	100.4	0.1	R. Megapolis, Luhansk
38)	100.5	1.5	R. Super Nova, Kyy v
1)	100.5	1	Master R, Kharkiv
	100.5		Mega R, Donetsk
	100.8	1	Novoye R, Zaporizhzhya
	100.8	1	R. Universitet, Zaporizhzhya
12)	100.8	1	Lvivska Khvylya, Lv v
39)	100.9	5	R. Kontinent, Kyyiv
40)	101.5	10	Music R, Kyyiv
41)	101.5		R. Rossii-Nostalgie, Dnipropetrovsk
42)	101.5		R. M, Dnipropetrovsk
43)	101.7	0.1	Visma R, Kremenchuk
44)	101.8		Boychuk-Studiya, Ivano-Frankivsk
45)	101.8	0.1	R. Velykyy Luh, Zaporizhzhya
24)	101.8	2	R. Yutar, Kyyiv (Odesa): 0700-2300
46)	101.8	0.1	Poltava
47)	101.9	1	R. Lider, Kyyiv
8)	102.0		R. Briz, Sevastopol
23)	102.0	2	R. Premier, Dnipropetrovsk
48)	102.1	0.1	R. Niko, Chernivtsi (Khmelnytskyy)
49)	102.1		R. Klass, Donetsk
24)	102.1	5	R. Yutar, Kyyiv (Mykolayiv):0700-2300
50)	102.2	0.1	R. Yuzhnye Prostory, Melitopol
51)	102.3		Trans-M-Radio, Simferopol
52)	102.5	1.4	Prosto R, Odesa (Kviyu)
33)	102.5	0.5	Europa Plus, Dnipropetrovsk
3)	102.5	0.5	R. DI, Dnipropetrovsk (F.Pl.)
15)	102.5	1	R. Sistema, Krivyy Rih: 0500-2100
24)	102.6	1.5	R. Yutar, Kyyiv (Kharkiv)
	102.8		R. Kontinent, Sevastopol: 0500-1700
52)	102.9	0.1	R. Voyazh, Luhansk
33)	102.9	0.05	Europa Plus, Odesa (Reni)
9)	103.0		Oniks R, Kharkiv
32)	103.0		R. Megapolis, Sevastopol (inact?)
17)	103.1	5	R. Lyuks, Lviv (Kyyiv)
53)	103.3		Classic R, Dnipropetrovsk
16)	103.5	1	R. ROKS-U, Kyyiv
54)	103.5	0.5	R. Da, Donetsk
24)	103.7	1	R. Yutar, Kyyiv (Zaporizhzhya)
18)	103.7	0.1	R. Aleks, Zaporyzhzhya
5)	103.8	2	R. Top, Odesa (F.Pl.)
4)	104.0	1	R. Niko, Chernivtsi
55)	104.0	4	R. Hit, Kyyiv
56)	104.0		Magic R, Dnipropetrovsk
33)	104.0	0.2	Europa Plus. Odesa (Izmayil)
57)	104.3		R. Zakhidnyy Polyus, Ivano-Frankivsk
41)	104.3		R. Rossii-Nostalgie, Nikopol
33)	104.5	1	Europa Plus, Zaporizhzhya
14)	104.5	1	R. ROKS-Krym, Sevastopol
14)	104.5	1	R. ROKS-Krym, Sevastopol (Yevpatoriya)
41)	104.5		R. Rossii-Nostalgie, Melitopol
52)	104.6	1.2	Prosto R, Odesa (Mykolayiv)
17)	104.7	1.5	R. Lyuks, Lviv
59)	104.8	0.5	R. Kontinent, Luhansk
24)	105.0	0.5	R. Yutar, Kyyiv (Cherkasy): 0700-2300
15)	105.2		R. Sistema, Kryvyy Rih: 0500-2100
	105.2		R. Fora, Kharkiv
52)	105.3	5	Prosto R, Odesa
24)	105.3	1.5	R. Yutar, Kyyiv (Ternopil)
58)	105.5		R. Aktivnost, Kyyiv
4)	105.7	1	R. 50, Kharkiv
60)	105.8		Avtoradio, Dnipropetrovsk
61)	105.9	0.1	R. Vita, Vinnytsya
35)	106.0	1	Kyyivski Vedomosti, Kyyiv 0500.2100
33)	106.0	2	Europa Plus, Odesa
48)	106.1	1	R. Niko, Chernivtsi (Ternopil)
28)	106.2	0.1	R. Saga, Zaporizhzhya

FM	MHz	kW	Station, location, h. of tr. (24h or as below)
41)	106.5	10	R. Yutar, Kyyiv: 0700-2300
62)	106.6	2	R. Glas, Odesa
26)	106.6	1	R. Simon, Kharkiv
63)	106.7	1	Biznes-Kanal, Lviv: 1400-1500
6)	106.7	1	R. Nezalezhnist, Lviv: 1100-1400, 1500-1900
64)	106.8	0.1	R. Poltava Plus, Poltava
33)	106.8		R. Europa Plus, Donetsk
3)	106.8	0.5	R. DI, Dnipropetrovsk
11)	106.9	1	R. Skyway, Luhansk
33)	107.0	1	Europa Plus, Kyyiv
65)	107.1	1	R. Kontakt, Khmelnitskyy 0500-2100
2)	107.3	1	R. Miks, Dnipropetrovsk
66)	107.5		R. De Vira, Kyyiv
33)	107.5	1	R. Europa Plus, Odesa
67)	107.5	0.1	R. Rossii-Nostalgie, Zaporizhzhya
19)	107.5	0.1	Radio 107.5, Cherkasy
17)	107.7		R. Lyuks, Lviv (Zhytomyr)
33)	107.8		R. Europa Plus, Sevastopol
4)	107.8		R. Mazhor, Simferopol
47)	107.9	1	Nashe Radio, Kyyiv
74)	FM		R. Diana-Master, Donetsk
75)	FM		R. Orfey, Mariupol
76)	FM		R. Lavensari, Mariupol

FM powers usually are TRP.

Addresses, schedules:
1) vul. Sumska 11, 310057 Kharkiv — 2) P.O. Box 26, 320106 Dnipropetrovsk — 3) vul. Prohrebnyaka 25-a, 320000 Dnipropetrovsk — 4) ul. Karayimskaya 6, 333001 Simferopol — 5) ul. Kanatnaya 134, 270011 Odessa. On MW rel. R. Glas + own ads. F.PI: FM sce — 6) vul. Knyazya Romana 12, 290005 Lviv — 7) pr. Tarasa Shevchenko 13, 290005 Lviv — 8) Sevastopol — 9) vul. Chernyshevskogo 15, 310057 Kharkiv — 10) P.O. Box 1343, GPO, 250000 Chernihiv — 11) Khudozhestvennyy kombinat. 3-y etazh, vul. Sent-Etienovskaya 40, 384000 Luhansk (office) or vul. Demyokhina 25, 384055 Luhansk (studio). Also rel. VOA Russian/Ukrainian — 12) Box 8552, 290058 Lviv. In Ukrainian, Russian, Polish, incl. VOA & BBC relays — 13) Poltava — 14) P.O. Box 39, 335023 Sevastopol — 15) vul. Televiziyna 3-a, 335087 Kryvyy Rih — 16) vul. Khreshchatyk 2, 252001 Kyyiv — 17) pl. Galitska 5, 290008 Lviv. Branch st: vul. Proreznaya 27, Kyyiv — 18) vul. 8 Marta 48, 332057 Zaporizhzhya — 19) vul. Oktyabrskaya (Kotovskogo) 25, 257000 Cherkasy — 20) Izyum — 21) vul. Saksaganskogo 91, 252001 Kyyiv — 22) Kharkiv — 23) vul. Ivana Sirko 43, 320000 Dnipropetrovsk — 24) vul. Artyoma 37, 252053 Kyyiv. Reg. branches: ul. Marshala Govorova 4, 270063 Odesa; vul. Palekhi 2, 257000 Cherkasy; vul. Rabochaya 2-a, 327021 Mykolyiv — 25) pr. 50 let Oktyabrya 2-b, 252148 Kyyiv — 26) vul. Derevyanko 1-a, 310103 Kharkiv — 27) vul. Ochakovska 1-a, 327036 Mykolayiv (studio) or vul. Artilleriyska 7, 327030 Mykolayiv (office) — 28) Dom Byta "Yubileynyy", pl. Pushkina, 332000 Zaporizhzhya — 29) vul. Maksimovicha 3-a, 286036 Vinnytsya — 30) Kyyiv — 31) Khmelnitskyy — 32) P.O. Box 284, 335011 Sevastopol — 33) vul. Institutskaya 28, 252079 Kyyiv. Reg. st's: TRK Skifiya, vul. Gogolya 15, 320044 Dnipropetrovsk; ul. Kanatnaya 83, komn. 420, 270107 Odesa; vul. Pobedy 131-a, 332057 Zaporizhzhya — 34) Donetsk — 35) vul. Degtyariavska 31, 252680 Kyyiv. Incl. rel BBC Ukrainian — 37) vul. Lermontova 2, 384000 Luhansk — 38) vul. Khmelnytskogo 10, kv. 39, 252000 Kyyiv — 39) P.O. Box 34, 252001 Kyyiv — 40) vul. Dorogozhitska 10, 252000 Kyyiv — 41) TRK "Lend", pr. Kirova 111-b, 320026 Dnipropetrovsk — 42) Dnipropetrovsk — 43) Kremenchuk — 44) Ivano-Frankivsk — 45) vul. Matrosova 8-a, 330057 Zaporizhzhya — 46) Poltava — 47) P.O. Box 16, Kyyiv 254119 — 48) TRK "Niko-PR", vul. Holovna 36, 274000 Chernivtsi — 49) Donetsk — 50) pr. Bogdana Khmelnytskogo 66-a, Melitopol — 51) ul. Geroyev Sevastopolya 6, 333047 Simferopol — 52) ul. Tereshkovoy 15, etazh 11, 270076 Odesa — 53) Dnipropetrovsk — 54) vul. Khirurgicheskaya 22, 340010 Donetsk — 55) vul. Nagornaya 24, 252 116 Kyyiv — 56) Dnipropetrovsk — 57) Ivano-Frankivsk — 58) Kyyiv — 59) vul. Kotsyubinskogo 2, Luhansk 384000 — 60) Dnipropetrovsk — 61) vul. Edelshteina 8, 286000 Vinnytsya — 62) ul. Kanatnaya 83, 270107 Odesa — 63) vul. Volodymyra Velykogo 5-a, 290026 Lviv — 64) Poltava — 65) Khmelnitskyy — 66) pr. Pobedy 63-b, Kyyiv — 67) vul. Sedovaya 12, 332057 Zaporyzhzhya — 74) Donetsk — 75) Mariupol — 76) Mariupol — 77) kv. Dimitrova 24, 384009 Luhansk.

UNITED KINGDOM

L.T: UTC (Su: UTC + 1h) — **Pop:** 58m — **Radios:** 65.400.000 — **Pr.L:** English — **E.C:** 50Hz, 240V — **ITU:** G.

THE BRITISH BROADCASTING CORPORATION

The BBC is an independent body created by a Royal Charter and operated under a licence and agreement with the Minister exercising functions under the Wireless Telegraphy Acts.

☞ Broadcasting House, London W1A 1AA. ☎ +44 (171) 580 4468. 🖷 +44 (171) 637 1630. ① 265781 BBC HQ G. **Cable:** "Broadcasts, London". **Web:** http://www.bbc.co.uk

Board of Governers: Chmn: Sir Christopher Bland. Vice-Chmn: Lord Cocks of Hartcliffe. National Governor for Northern Ireland: Sir Kenneth Bloomfield KCB. National Governor for Wales: Dr Gwyn Jones. National Governor for Scotland: The Rev Norman Drummond. Governors: Bill Jordan CBE, Lord Nicholas Gordon Lennox KCMG KCVO, Margaret Spurr OBE, Janet Cohen, Sir David Scholey CBE, Richard Eyre CBE, Adrian White CBE.

Board of Management: DG: John Birt. Deputy DG and Chief Exec. BBC Worldwide: Bob Phillis. Chief Exec. BBC Broadcast: Will Wyatt. Chief Exec. BBC Production: Ronald Neil. Chief Exec. BBC News: Tony Hall. Chief Exec. BBC Resources: Rod Lynch. Dir. of Personnel: Margaret Salmon. Dir. of Finance and IT: Rodney Baker-Bates. Dir. of Policy and Planning: Patricia Hodgson. Dir. of Corporate Affairs: Colin Browne. MD World Sce: Sam Younger. Dir. of Education: Jane Drabble. Dir. of Regional Broadcasting: Mark Byford. Director of Radio: Matthew Bannister. Dir. of Television: Michael Jackson. Dir. of Programmes: Alan Yentob.

Radio 4 UK

	kHz	kW	Radio 4 UK	kHz	kW
Burghead	198	50	Londonderry	720	0.25
Droitwich	198	500	Redruth	756	2
Westerglen	198	50	Enniskillen	774	1
Newcastle	603	2	Plymouth	774	1+
Lisnagarvey	720	10	Redmoss	1449	2
London (Lots Rd.)	720	0.75	Carlisle	1485	1
Barrow	693	1	Droitwich	693	150
Bexhill	693	1	Enniskillen	693	1
Brighton	693	1	Folkestone	693	1
Burghead	693	25	Postwick	693	10
Redmoss	693	1	Lisnagarvey	909	10
Stagshaw	693	50	Londonderry	909	1
Start Point	693	50	Moorside Edge	909	200
Bournemouth	909	0.25	Redruth	909	2
Brookmans Park	909	150	Westerglen	909	50
Clevedon	909	50	Whitehaven	909	1
Exeter	909	1	Tywyn	990	1
Fareham	909	1			

FM (all stereo)

England, Isle of Man, Channel Isl.

England, Isle of Man, Channel Isl.	R1	R2	R3	R4	kW ERP
Barnstaple	98.1	88.5	90.7	92.9	1
Beacon Hill North	98.4	88.7	90.9	93.1	1
Belmont	98.3	88.8	90.9	93.1	16
Bilsdale	98.6	89.0	91.2	93.4	5
Bow Brickhill	98.2	88.6	90.8	93.0	10
Bristol	98.9	89.3	91.5	93.7	1.3
Chatton	99.7	90.1	92.3	94.5	5.6
Crystal Palace	98.5	88.8	91.0	93.2	4
Douglas I.O.M.	98.0	88.4	90.6	92.8	11
Guildford	97.7	88.1	90.3	92.5	3
Hemdean	99.4	89.8	92.0	94.2	1
Holme Moss	98.9	89.3	91.5	93.7	250
Keighley	98.5	88.9	91.1	93.3	1
Les Platons (C.I.)	97.1	89.6	91.1	94.8	16
Manningtree	97.7	88.1	90.3	92.5	5
Morecambe Bay	99.6	90.0	92.2	94.4	10
North Hessary Tor	97.7	88.1	90.3	92.5	160
Oxford	99.1	89.5	91.7	93.9	46
Pendle Forest	97.8	90.2	92.6	94.6	1
Peterborough	99.7	90.1	92.3	94.5	40
Pontop Pike	98.1	88.5	90.7	92.9	134
Redruth	99.3	89.7	91.9	94.1	25
Ridge Hill	98.2	88.6	90.8	93.0	10
Rowridge	98.2	88.5	90.7	92.9	250
Sandale	97.7	88.1	90.3	92.5	250
Stanton Moor	99.4	89.8	92.0	94.2	1.2
Sutton Coldfield	97.9	88.3	90.5	92.7	250
Swingate (Dover)	99.1	90.0	92.4	94.4	11
Tacolneston	99.3	89.7	91.9	94.1	250
Wenvoe	99.5	89.9	92.1	94.3	250
Winter Hill	98.2	88.6	90.8	93.0	4
Woolmoor	99.6	90.2	92.2	94.4	4.5
Wrotham	98.8*	89.1	91.3	93.5	125*/250

+ low power tr's less than 1kW.

Radio 1: Pop music incl. dance and rap, live concerts, social action campaigns, comedy: 24h. **N:** MF on the half h + 0600, 0700, 0800, 1700, 1800. Sat on the half h. Sun on the half h. exc. 1730, 1830, 1930.
Radio 2: Popular music, entertainment, comedy, the arts: 24h. **N:** MF on the h (exc. 2000 Fri), also 0530, 0630, 0730, 0830. Sat on the h exc. 2000 & 2100, also 0630, 0730. Sun on the h exc. 2000, also 0730.
Radio 3: Classical music drama,.documentaries, poetry, school prgrs: 24h. **N:** MF 0700, 0800, 1300, 1800, 1900, 0030; Sat 0900, 1300, 0100. Sun 0900, 1300, 0030.
Radio 4: News, documentaries, drama, entertainment, and cricket on LW in season: 0555-0040 (approx). **N:** MF 0600-0900 every half h, 1000-1900 hourly, 2200, 0000; Sat 0600, 0700, 0730, 0800, 0830, 0900, 1000, 1100, 1300, 1400, 1600, 1800, 2200, 0000. Sun 0600, 0700, 0800, 0900, 1300, 1700, 1800, 2200, 0000. **Radio 5 Live:** News & sport: 24h. **N:** on the h. and half h.

BBC Local Radio

FM: stereo exc. m=mono — **NB:** Some MW-tr's may be taken out of service to accommodate new commercial radio sces.

	Station	kHz	kW	MHz	kW
1)	R. Cornwall (Redruth)	630	2	103.9	17
1)	Isles of Scilly (r)			96.0	0.06
31)	3 Counties R. (Luton)	630	0.2	103.8	0.5
1)	Bodmin (r)	657	0.5		
1)	Caradon Hill (r)			95.2	4.3
2)	R. York	666	0.5	103.7	2
2)	Woolmoor (r)			104.3	0.5
33)	BBC Essex	729	0.2		
37)	BBC H&W, Worcester	738	0.037	94.7	2
37)	Kidderminster (r)			104.6	0.5
37)	Hereford (r)			104.0	2
3)	R. Cumbria	756	1	95.6	15
33)	Chelmsford (r)	765	0.5	103.5	12
4)	R. Leeds	774	0.5	92.4	5.6
4)	Luddenden (r)			95.3	0.083
4)	Wharfedale (r)			95.3	0.04
4)	Beercroft Hill (r)			103.9	0.1
4)	Keighley (r)			102.7	1
14)	R. Kent	774	0.7	96.7	8.7
14)	Swingate (r)			104.2	10
14)	Folkestone (r)			97.6	0.1
5)	Barnstaple (r)	801	2		
6)	Asian Network (r)	828	0.2		
7)	R. Cumbria South	837	1.5	96.1	3.2
7)	Kendal (r)			95.2	0.1
7)	Windermere (r)			104.2	0.065
8)	Asian Network (r)	837	0.4		
	R. Leicester			104.9	8
5)	R. Devon	855	1	103.4	15
5)	Okehampton (r)			96.0	0.07
9)	R. Lancashire	855	1.5	95.5	1.6
9)	Winter Hill (r)			103.9	1.8
10)	R. Norfolk	855	1.5	95.1	5.7
10)	We. Norfolk (r)	873	0.3	104.4m	4.2
5)	Exeter (r)	990	1	95.8	0.4
11)	BBC Solent	999	1	96.1	10
11)	Dorset (r)			103.8	0.5
12)	R. Cambridgeshire	1026	0.5	96.0	1
13)	R. Jersey	1026	1	88.8	3.8
15)	R. Sheffield	1035	1	104.1	4.4
15)	Sheffield (r)			88.6	0.16
15)	Chesterfield (r)			94.7	0.4
17)	R. Derby	1116	1.2	104.5	5.4
17)	Stanton Moor (r)			95.3	1.2
17)	Derby (r)			94.2	0.01
18)	R. Guernsey	1116	0.5	93.2	1
31)	3 Counties R (Bedford)	1161	0.1	95.5	1
	Milton Keynes			104.5	2.2
19)	Bexhill (r)	1161	1		
2)	Scarborough (r)	1260	0.5	95.5	0.25
29a)	Somerset Sound	1323	0.63		
39)	Wiltshire Sound	1332	0.3	104.3	0.6
11)	Bournemouth (r)	1359	0.85		
19)	Reigate (r)	1368	0.5	104.0	3.8
20)	R. Lincolnshire	1368	2	94.9	6
20)	Grantham (r)			104.7	0.035
39)	Swindon (r)	1368	0.1	103.6	0.5
39)	Salisbury (r)			103.5	1
39)	Marlborough (r)			104.9	0.1
12)	Peterborough (r)	1449	0.15	95.7	5.1

	Station	kHz	kW	MHz	kW
6)	Asian Network	1458	5		
	R. WM			95.6	11
6)	Meriden (r)			94.8	2.2
6)	Lark Stoke (r)			103.7	1.4
6)	Nuneaton (r)			104.0	0.05
3)	Whitehaven (r)	1458	0.5	104.1	
21)	R. Newcastle	1458	2	95.4	10
21)	Fenham (r)			104.4	0.042
21)	Chatton (r)			96.0	5.6
7)	Torbay (r)	1458	2	94.8	0.675
19)	Southern Counties R	1485	1	95.3	1.2
19)	Heathfield (r)			104.5	10
19)	Horsham (r)			95.1	0.04
19)	Newhaven (r)			95.0	0.1
19)	Chichester (r)			104.8	2
19)	Guildford (r)			104.6	3
24)	R. Humberside	1485	2	95.9	9.6
26)	R. Merseyside	1485	1.2	95.8	8.2
27)	R. Stoke-on-Trent	1503	1	94.6	6.1
	Stafford (r)			104.1	0.75
33)	Southend (r)	1530	0.15	95.3	1.2
29)	R. Bristol	1548	5	95.5	9
29)	Bath (r)			104.6	0.082
29)	Bristol (r)			94.9	0.95
9)	Lancaster (r)	1557	0.25	104.5	2.1
28)	BBC Nottingham	1584	1	95.5	2
28)	Mapperley Ridge (r)			103.8	1
32)	R. Shropshire	1584	0.5	96.0	4.8
32)	Ludlow (r)			95.0	0.01
14)	Rusthall (r)	1602	0.25		

STATIONS ON FM ONLY

	Station	MHz	kW
25)	BBC Thames Valley R., Henley (r)	94.6	0.25
22)	GLR	94.9	4
30)	BBC Cleveland	95.0	5
34)	Stroud (r)	95.0	0.1
23)	GMR	95.1	5.6
25)	BBC Thames Valley R.	95.2	5.8
25)	BBC Thames Valley R., Windsor	95.4	0.5
38)	R. Suffolk, Lowestoft	95.5	0.05
34)	Cirencester (r)	95.8	0.08
30)	Whitby (r)	95.8	0.1
16)	Geddington (r)	103.6	0.8
38)	R. Suffolk, Manningtree	103.9	5
25)	BBC Thames Valley R., Hannington	104.1	4
16)	BBC Northampton	104.2	4
25)	BBC Thames Valley R., Reading	104.4	1
38)	R. Suffolk, Gt. Barton	104.6	2
34)	R. Gloucestershire	104.7	2

Addresses

1) Phoenix Wharf, Truro, Cornwall TR1 1UA
2) 20 Bootham Row, York YO3 7BR.
Web: http://www.bbc-ne-cu.co.uk/yo.htm
3) Ametwell Str, Carlisle, Cumbria CA3 8BB.
Web: http://www.bbc-ne-cu.co.uk/cu.htm
4) Broadcasting House, Woodhouse Lane, Leeds LS2 9PN
5) Broadcasting House, Seymour Rd, Mannamead, Plymouth PL3 5BD
5a) rel. 5)
6) P. O. Box 206, Birmingham B5 7QQ
7) rel. R. Cumbria
8) Epic House, Charles Str, Leicester LE1 3SH
9) 20-26 Darwen Str, Blackburn, Lancs. BB2 2EA
10) Norfolk Tower, Surrey Str, Norwich NR1 3PA. **E-mail:**
norfolk@bbc.co.uk **Web:** http://www.bbc.co.uk/radio_norfolk/
11) Broadcasting House, Havelock Rd, Southampton SO14 7PW.
Web: http://www.bbc.co.uk/south_today/solent.htm
12) Broadcasting House, 104 Hills Rd, Cambridge CB2 1LQ
13) 18 Parade Rd, St. Helier, Jersey
15) Ashdell Grove, 60 Westbourne Rd, Sheffield S10 2QU
16) P. O. Box 1107, Abington Str, Northampton NN1 2BE.
E-mail: northampton@nc.bbc.co.uk
17) P.O. Box 269, 56 St. Helens Str, Derby DE1 3HY
18) Commerce House, Les Banques, St. Peter Port, Guernsey GY2 4QQ–
19) Broadcasting House, Guildford GU2 5AP
20) P.O. Box 219, Newport, Lincoln LN1 3XY.
Web: http://www.bbclincs.demon.co.uk
21) Broadcasting Centre, Barrack Rd, Newcastle-upon-Tyne NE99 1RN.

Web: http://www.bbc-ne-cu.co.uk/ne.htm
22) P. O. Box 94.9, 35C Marylebone High Str, London W1A 4LG
23) New Broadcasting House, Oxford Rd, Manchester M60 1SJ
24) 9 Chapel Str, Hull HU1 3NU. **E-mail:** radio.humberside@bbc.co.uk
25) 269 Banbury Rd, Oxford OX2 7DW; Thomas House, 42A Portman
Rd, Reading RG3 1NB
26) 55 Paradise Str, Liverpool L1 3BP
27) Conway House, Cheapside, Hanley, Stoke-on-Trent, Staffs ST1 1JJ
28) York House, Mansfield Rd, Nottingham NG1 3JB
29) P. O. Box 194, Bristol BS99 7QT
30) P. O. Box 95, Middlesborough, Cleveland TS1 5DG.
Web: http://www.bbc-ne-cu.co.uk/cl.htm
31) P.O. Box 476, Hastings Str, Luton LU1 5BA
32) P. O. Box 397, Shrewsbury SY1 3TT
33) P. O. Box 765, Chelmsford, Essex CM2 9XB.
E-mail: Essex@nca.bbc.co.uk
34) London Rd, Gloucester GL1 1SW
37) Hylton Rd, Worcester WR2 5WW; 43 Broad Str, Hereford HR4 9HH
38) Broadc. House, St. Matthews Str, Ipswich IP1 3EP
39) Broadcasting House, Prospect Place, Swindon SN1 3RW.
D.PRGR: St's generally carry local prgrs from 0600 to 1800 or 1900,
regional prgrs until 2400, then BBCWS overnight.
V. by QSL-card or letter. Reports to Engineer-in-Charge.

BBC SCOTLAND

✉ Broadcasting House, Queen Margaret Drive, Glasgow G12 8DG.
☎ +44 (141) 338 2000. 🖷 +44 (141) 334 0614.
Web: http://www.bbc.co.uk/scotland/
R. Scotland: 810kHz: Burghead 100kW, Westerglen 100kW, Redmoss
5kW.
R. Aberdeen: 990kHz 1kW — **BBC Solway:** 585kHz 2kW.
FM: stereo exc. m=mono

Location	1FM	R2	R3	R4	RS/L	kW
Ashkirk	98.7	89.1	91.3	103.9	93.5f	50
Ben Gullipen	98.3	88.7	90.9	104.9	93.1	1
Black Hill	99.5	89.9	92.1		94.3	250
					95.8	200
Bressay	97.9	88.3	90.5	94.9	92.7abc	43
Clettraval	97.7	88.1	90.3	95.1	92.5	2
Daliburgh	98.9	89.3	91.5	95.9	93.7	1
Darvel	99.1	89.5	91.7	104.3	93.9	10
Durris	99.0	89.4	91.6	95.9	93.8a	2.1
Eitshal	99.4	89.8	92.0	95.1	94.2d	2
Forfar	97.9	88.3	90.5	94.9	92.7	17
Fort William	98.9	89.3	91.5	95.9	93.7d	3
Glengorm	99.1	89.5	91.7	96.1	93.9d	4.6
Keelylang Hill	98.9	89.3	91.5	95.9	93.7ab	41
Meldrum	98.3	88.7	90.9	95.3	93.1a	150
Melvaig	98.7	89.1	91.3	95.7	93.5d	50
					103.9	50
Oban	98.5	88.9	91.1	95.3	93.3d	3.6
Rosemarkie	99.2	89.6	91.8	103.6	94.0d	20
Rumster Forest	99.7	90.1	92.3	104.5	94.5d	10
Sandale	97.7	88.1	90.3	92.5	94.7e	250
Skriaig	98.1	88.5	90.7	94.8	92.9d	30
					104.7	30
So. Knapdale	98.9	89.3	91.5	95.6	93.7	1.1
					104.8m	1.1

+ low power tr's less than 1kW.
RS/L=R. Scotland + alternative & local prgrs – a) RS: Aberdeen – b)
RS: Orkney – c) RS: Shetland – d) RS: Inverness – e) RS: Dumfries –
f) RS: Selkirk.
D.PRGR: 0600 (SS 0700)-2400. Relay BBC World Sce. overnight.

Local Services (MF only). Freqs as above.

a) Beachgrove Terrace, Aberdeen AB9 2ZT. 0655-0700, 0750-0800,
1255-1300, 1655-1700. **Web:** http://www.bbc.co.uk/aberdeen/
index.html – b) Commercial Assurance Bldg, Castle Str, Kirkwall,
Orkney KW15 1DF: 0730-0800, 1815-1900 – c) Brentham House,
Lerwick, Shetland ZE1 0LR: 0730-0800, 1730-1800, 1815-1930 – d) 7
Culduthel Rd, Inverness IV2 4AD. 0655-0700, 0750-0800, 1255-1300,
1655-1700 – e) Elmbank, Lovers Walk, Dumfries DG1 1NZ: 0655-0700,
0750-0800, 1255-1300, 1655-1700 – f) Old Municipal Bldg, High Str,
Selkirk TD7 4BU: 0655-0700, 0750-0800, 1255-1300, 1655-1700.

Radio Nan Gaidheal

✉ 7 Culduthel Rd, Inverness IV2 4AD. ☎ +44 (1463) 720720.

FM	MHz	kW	FM	MHz	kW
Clettraval	93.7	2	Eitshal	104.3	2
Glengorm	103.5	4.6	Oban	104.6	3.6
Malvaig	103.9	50	Skriaig	104.7	30
Daliburgh	104.2	1	So. Knapdale	104.8	1.1
Fort William	104.2	3	Rosemarkie	104.9	20

BBC WALES

⌨ Broadcasting House, Llantrisant Rd, Llandaff, Cardiff CF5 2YQ.
☎ +44 (1222) 572 888. 🖥 +44 (1222) 552 973 – The Old School House,
Glan Afon Road, Mold, Clwyd CH7 1DA. ☎ +44 (1352) 700367 –
Broadcasting House, Meirion Road, Bangor, Gwynedd LL57 3BY. ☎ +44
(1248) 370880 – Broadcasting House, 32 Alexandra Road, Swansea
SA1 5DT. ☎ +44 (1792) 654986.
Web: http://www.bbc.wales.com
R. Wales: 882kHz: Penmon 10kW, Tywyn 5kW, Washford 100kW,
Forden 1kW — Llandrindod Wells 1125kHz 1kW.

FM stereo	R1	R2	R3	R4	R..Cymru	kW (ERP)
Blaenplwyf	98.3	88.7	90.9	104.0	93.1	250
Carmel	98.0	88.4	90.6	104.6	92.8	3.2
Haverfordwest	98.9	89.3	91.5	104.9	93.7	20
Llanddona	99.4	89.8	92.0	103.6	94.2	21
Llandrindod						
Wells	98.7	89.1	91.3	103.8	93.5	2.8
Llangollen	98.5	88.9	91.1	93.3	104.3	15.6
Wenvoe	99.5	89.9	92.1	94.3	96.8	250

+ low power tr's less than 1kW.
R. Wales on 882/1125kHz: 0600-2400 (Sat 2300, Sun 2130). Other
times rel. R4 or World Sce.— **R. Cymru** on FM: 0600-2400 (Sat 1945).
Other times rel. BBC World Sce.

BBC NORTHERN IRELAND

⌨ Broadcasting House, Ormeau Avenue, Belfast BT2 8HQ. ☎ +44
(1232) 338 000. 🖥 +44 (1232) 338 048.
MW: Enniskillen 873kHz 1kW, Lisnagarvey 1341kHz 100kW.

FM stereo	R1	R2	R3	R4	Radio Ulster	kW ERP
Brougher Mountain	99.0	89.4	91.6	95.6	93.8	9.8
Camlough	98.3	88.7	90.9	104.6	93.1	2
Divis	99.7	90.1	92.3	96.0*	94.5*	250/125*
Limavady	99.2	89.6	91.8	94.0	95.4	3.4
Londonderry	98.3	88.7	90.9		93.1h	31
				94.9		10

+ low power tr's less than 1kW — h) R. Foyle (see below).
R. Ulster: 0630 (Sat 0755, Sun 0740)-2400. Other times rel. BBC World
Sce.— **V.** by QSL-card.

BBC Radio Foyle

⌨ 8 Northland Rd, Londonderry BT48 7NE. ☎ +44 (1504) 262 244.
MW: Londonderry 792kHz 1kW — **FM:** 93.1MHz 31kW.
D.PRGR: MF 0730-1730, Sat 1330-1630, 1400-1700 (own prgrs. and
relay BBC R. Ulster).

CASTLE TRANSMISSION INTERNATIONAI

⌨ P.O. Box 98, Warwick CV34 6TN. ☎ +44 (1926) 411212. 🖥 +44
(1926) 450 812. **E-mail:** trans.bus.dev@bbc.co.uk
Web: http://www.ctxi.com/
Castle Transmission International operates all the BBC's domestic
transmitter sites.

BBC WORLD SERVICE

See International Broadcasting section.

THE RADIO AUTHORITY (Regulatory Authority)

⌨ Holbrook House, 14 Great Queen Str, London WC2B 5DG. ☎ +44
(171) 430 2724. 🖥 +44 (171) 405 7062.
L.P: Chief Exec: Tony Stoller. Press & Inf. Officer: Tracey Mullins.

COMMERCIAL RADIO COMPANIES ASSOCIATION (CRCA)

⌨ 77 Shaftesbury Ave, London W1V 7AD. ☎ +44 (171) 306 2603.
🖥 +44 (171) 470 0062.
CRCA (formerly the Association of Independent Radio Companies
(AIRC) is the trade body for UK commercial radio. It represents commercial radio to Government, the Radio Authority, Copyright Societies
and other organizations concerned with radio.

NATIONAL COMMERCIAL STATIONS

CLASSIC FM

⌨ Academic House, 24/28 Oval Rd, London NW1 7DQ. ☎ +44 (171)
284 3000. 🖥 +44 (171) 713 2630.
E-mail: programmes@classicfm.co.uk *or* engineers@classicfm.co.uk
Web: http://www.classicnet.co.uk/
L.P: MD: J.Spearman. PD: Michael Bukht.

FM	MHz	kW		MHz	kW
Sandale	99.9	250	Blaen Plwyf	101.1	20
No.Hessary Tor	100.0	160	Oxford	101.3	46
Sutton Coldfield	100.1	250	Fremont Point	101.3	16
Angus	100.1	12	Darvel	101.3	10
Douglas I.O.M.	100.2	0.8	Kilvey Hill	101.3	1
Rowridge	100.3	250	Swansea	101.3	0.92
Pontop Pike	100.3	150	Inverness	101.4	10
Milton Keynes	100.4	10	Tacolneston	101.5	250
Ridge Hill	100.4	10	Redruth	101.5	17
Meldrum	100.5	150	Rowridge	101.6	46
Londonderry	100.5	31	Gt. Ormes Head	101.6	2.5
Presely	100.5	20	Black Hill	101.7	250
Belmont	100.5	7	Wenvoe	101.7	250
Crystal Palace	100.6	2	Sheffield	101.7	0.3
Arfon	100.7	20	Morecambe Bay	101.8	10
Wrotham	100.9	250	Dover	101.8	7
Selkirk	100.9	10	Divis	101.9	250
Newcastle	101.0	0.05	Peterborough	101.9	40
Holme Moss	101.1	250	Brighton	101.9	0.5

D.PRGR: 24h. **N:** on the h.

TALK RADIO UK

⌨ 74 Newman Str, London W1P 3LA. ☎ +44 (171) 343 2222. 🖥 +44
(171) 343 2200.
L.P: GM: Paul Robinson. MD: Travis Baxter. PD: Jason Bryant.

Mediumwaves	kHz	kW	Mediumwaves	kHz	kW
Droitwich	1053	100	Newcastle	1071	1
Dumfries	1053	10	Wallasey	1071	0.5
Postwick	1053	10	Brookmans Park	1089	100
Rusthall	1053	4	Dartford Tunnel	1089	0.004
Brighton	1053	2	Moorside Edge	1089	100
Bournemouth	1053	1	Washford	1089	50
Dundee	1053	1	Westerglen	1089	50
Exeter	1053	1	Lisnagarvey	1089	10
Hull	1053	1	Redmoss	1089	2
Inverness	1053	1	Redruth	1089	2
Londonderry	1053	1	Boston	1107	2
Lydd	1053	1	Gatwick Airport	1107	2
Plymouth	1053	1	Lydd	1107	2
Stockton	1053	1	Fareham	1107	1
Clipstone	1071	1	Torbay	1107	1

D.PRGR: 24h

VIRGIN RADIO

⌨ 1 Golden Square, London W1R 4DJ. ☎ +44 (171) 434 1215.
🖥 +44 (171) 434 1197. **Web:** http://www.virginradio.co.uk/
L.P: Chief Exec: David Campbell. MD: John Pearson. PD: Mark Story.

Mediumwaves	kHz	kW	Mediumwaves	kHz	kW
Cheltenham	1197	1	Aberdeen	1215	2
Northampton	1197	0.5	Plymouth	1215	1
Nottingham	1197	0.5	Dartford Tunnel	1215	0.004
Gatwick	1197	0.2	Washford	1215	50
Hoo (Kent)	1197	1	Lisnagarvey	1215	10
Oxford	1197	0.5	Fareham	1215	1
Wallasey	1197	0.5	Norwich	1215	1
Bournemouth	1197	0.5	Hull	1215	0.3
Torbay	1197	1	Manningtree	1224	0.5
Cambridge	1197	0.35	Reading	1233	0.2
Moorside Edge	1215	100	Swindon	1233	0.16
Droitwich	1215	50	Dundee	1242	1
Newcastle	1215	2	Stockton	1242	0.5
Brighton	1215	1	Sheffield	1242	0.5
Brookmans Park	1215	50	Guildford	1242	0.5
Westerglen	1215	50			

FM: London 105.8MHz 2kW (separate prgrs for a few h. per day).
D.PRGR: 24h (Rock music). **N:** on the h — **ANN:** "Virgin 1215".
FM Sce (Virgin FM): rel. AM sce. with some local programming.

REGIONAL COMMERCIAL STATIONS

Name	MHz	kW	Tr. site	Area
3) Century R.	96.2	0.05	Fenham	local coverage

	Name	MHz	kW	Tr. site	Area
3)	Century R.	96.4	0.1	Hexham	local coverage
5)	Galaxy 101	97.2	0.15	Pur Down	Severn Estuary
1)	Scot FM	100.3	20	Black Hill	Glasgow
2)	Jazz FM	100.4	5	Winter Hill	No. We. England
3)	Century R.	100.7	9	Bilsdale	Cleveland
4)	Heart FM	100.7	11.4	Sutton Coldfield	We. Midlands
5)	Galaxy 101	101.0	40	Mendip	Bristol & So. Wales
1)	Scot FM	101.1	10	Craigkelly	Edinburgh
3)	Century R.	101.8	9	Burnhope	Newcastle
8)		105.2	10	F.Pl.	Isle of Wight
7)		105.4		F.Pl.	No. We. England
6)	Vibe FM	105.6	1	Madingley	We. Norfolk
8)		105.8	1	F.Pl.	Poole
9)	R. 106 FM	106.0	8	Copt Oak	Ea. Midlands
6)	Vibe FM	106.1	4	Stoke Holy Cross	Ea. Norfolk
6)	Vibe FM	106.4	20	Mendlesham	Suffolk
6)	Vibe FM	107.7	0.2	Gunthorpe	Peterborough

Addresses and other information

1) Number 1 Shed, Albert Quay, Leith EH6 7DN. ☎ +44 (131) 554 6677. ▤ +44 (131) 554 2266. **E-mail:** scotfm@easynet.co.uk — **L.P:** MD: Bob Christie.
2) The World Trade Centre, Exchange Quay, Manchester M5 3EJ. ☎ +44 (161) 877 1004. ▤ +44 (161) 877 1005. **Web:** http://www.jazzfm.co.uk — **L.P:** Chmn:Richard Wheatly. SM: Mike Henfield (owned by Golden Rose Communications plc).
3) P.O. Box 100, Gateshead NE8 2YY. ☎ +44 (191) 477 6666. ▤ +44 (191) 477 5660 — **L.P:** MD: John Myers. PD: John Simons.
4) 1 The Square, 111 Broad Str, Edgbaston, Birmingham B15 1AS. ☎ +44 (121) 626 1007. ▤ +44 (121) 696 1007 — **L.P:** MD: Philip Riley. PD: Paul Fairburn. (owned by Chrysalis Group plc).
5) Millenium House, 26 baldwin Str, Bristol BS1 1SE. ☎ +44 (117) 901 0101. ▤ +44 (117) 901 4666. **E-mail:** galaxy@dial.pipex.com. **Web:** http://dspace.dial.pipex.com/town/plaza/hc30/ — **L.P:** Chief Exec: Steve Parkinson. PD: Tristan Bolitho. (owned by Chrysalis Group plc)
6) Radio House, 19-20 Clifftown Rd, Southend-on-Sea, Essex SS1 1SX.
9) 103) City Link, Nottingham NG2 4DP.
Web: http://www.radio106fm.co.uk/

LOCAL COMMERCIAL STATIONS

MW	kHz	kW	Name or Slogan	Location
30)	558	0.8	Spectrum Int. R.	London
43c)	603	0.1	Invicta SuperGold	Littlebourne
35a)	603	0.1	The Cheltenham R. St.	Cheltenham
2)	666	0.34	Gemini AM	Exeter
47)	756	0.63	R. Maldwyn	Powys
4e)	774	0.14	Classic Gold 774	Gloucester
4b)	792	0.2	Classic Gold 792	Bedford
20b)	828	0.12	Magic 828	Leeds
4b)	828	0.2	Classic Gold 828	Dunstable
7g)	828	0.27	Classic Gold 828	Bournemouth
76)	828	0.8	Townland R.	Cookstown
6)	855	0.15	Sunshine 855.	Ludlow
7c)	936	0.18	Brunel Classic Gold	We. Wilts
90)	936	1	Yorkshire Dales R.	Hawes
7p)	945	0.2	Gem AM	Derby
43d)	945	0.7	South Coast R.	Bexhill
8)	954	0.16	R. Wyvern	Hereford
2)	954	0.32	Gemini AM	Torquay
69)	963	1	Viva 963	Southall
77)	963	0.8	Asian Sound R.	Blackburn
7j)	990	0.09	WABC	Wolverhampton
27c)	990	0.25	Magic AM	Doncaster
7p)	999	0.25	Gem AM	Nottingham
20c)	999	0.8	Red Rose Gold	Preston
7j)	1017	2	WABC	Shrewsbury
12)	1026	1.7	Downtown R.	Belfast
23c)	1035	0.78	NorthSound Two	Aberdeen
14)	1035	0.32	West Sound R.	Ayr
62)	1035	1	Country 1035	London
15)	1107	1.5	Moray Firth R.	Inverness
87)	1116		Valley Radio	Welsh valleys
16)	1152	23.5	LBC 1152	London
23d)	1152	3.6	Clyde 2	Glasgow

MW	kHz	kW	Name or Slogan	Location
43b)	1152	3	Xtra AM	Birmingham
7s)	1152	0.83	Amber R.	Norwich
20a)	1152	1.5	Piccadilly Gold	Manchester
20g)	1152	1.8	GNR	Newcastle
1)	1152	0.32	Plymouth Sound AM	Plymouth
20i)	1161	0.35	Magic 1161	Hull
23a)	1161	1.4	R. Tay AM	Dundee
7c)	1161	0.16	Brunel Classic Gold	Swindon
27b)	1170	0.58	Swansea Sound	Swansea
20g)	1170	0.32	GNR	Stockton
18)	1170	0.25	ElevenSeventy AM	High Wycombe
27)	1170	0.2	Signal Two	Stoke-on-Trent
43d)	1170	0.12	South Coast R.	Portsmouth
7t)	1170	0.28	Amber R.	Ipswich
43c)	1242	0.32	Invicta SuperGold	Maidstone
35d)	1242	0.5	Isle of Wight R.	Isle of Wight
7u)	1251	0.76	Amber R.	Bury St.Edmunds
81)	1260	0.29	Sabras Sound	Leicester
7a)	1260	1.6	Brunel Classic Gold	Bristol
11)	1260	0.64	Marcher Gold	Wrexham
27c)	1278	0.43	Easy Magic 1278	Bradford
63)	1296	6.4	Radio XL	Birmingham
20e)	1305	0.2	Touch R.	Newport
20h)	1305	0.15	Magic AM	Barnsley
71)	1305	0.5.	Premier R.	Ewell
71)	1305	0.5	Premier R.	Enfield
43e)	1323	0.5	South Coast R.	Brighton
7h)	1332	0.6	Classic Gold 1332	Peterborough
71)	1332	1	Premier R.	Bow (London)
7b)	1359	0.27	Classic Gold 1359	Coventry
20e)	1359	0.2	Touch R.	Cardiff
10)	1359	0.26	The Breeze	Chelmsford
77)	1377	0.1	Asian Sound R.	Rochdale
71)	1413	0.5	Premier R.	Heathrow
71)	1413	0.5	Premier R.	Dartford
90)	1413	0.1	Yorkshire Dales R.	Skipton
7e)	1431	0.14	Classic Gold 1431	Reading
10)	1431	0.48	The Breeze	Rayleigh
24)	1458	50	Sunrise R.	London
38b)	1458	5	1458 Lite AM	Manchester
42)	1476	1	County Sound	Guildford
7e)	1485	1	Classic Gold	Newbury
38c)	1521	0.64	Fame 1521	Reigate
85)	1521		R. 1521	Craigavon
21d)	1530	0.74	Easy Magic 1530	Huddersfield
8)	1530	0.52	R. Wyvern	Worcester
43a)	1548	97.5	Capital Gold	London
20h)	1548	0.74	Magic AM	Sheffield
45)	1548	4.4	Magic 1548	Liverpool
23)	1548	2.2	Max AM	Edinburgh
4c)	1557	0.76	Classic Gold 1557	Northampton
43d)	1557	0.5	South Coast R.	Southampton
9)	1557	0.125	Mellow 1557	Clacton
23a)	1584	0.21	R. Tay AM	Perth
35)	1584	0.04	KCBC R.	Kettering
35)	1584	0.01	KCBC R.	Corby
58)	1584	0.2	London Turkish R. LTR	Haringey
28)	93.7	0.01	Island FM	Alderney
43a)	95.8	4	Capital FM	London
43c)	95.9	0.27	Invicta FM	Thanet
43c)	96.1	0.25	Invicta FM	Ashford
19d)	96.1	0.5	SGR-FM	Colchester
7p)	96.2	1	Trent FM	Nottingham
17)	96.2	4	SIBC	Bressay
52)	96.2	2.5	Lantern R.	Bideford
73)	96.2	0.15	KFM	Tonbridge
60)	96.2	1	Mix 96	Aylesbury
39a)	96.2	0.625	Yorkshire Coast R.	Scarborough
40)	96.2	0.08	Kix 96	Coventry
20b)	96.3	0.6	96.3 Aire FM	Leeds

FM	MHz	ERP	Name or slogan	Location	FM	MHz	ERP	Name or slogan	Location
7a)	96.3	2	GWR FM	Bristol	43d)	97.5	0.85	Ocean FM	Portsmouth
10)	96.3	1.03	Essex FM	Southend	27c)	97.5	0.5	The Pulse	Bradford
38)	96.3	0.2	Q96.3	Paisley	38a)	97.5	0.03	R. Mercury	Horsham
32)	96.3	1.25	Coast FM	Colwyn Bay	23g)	97.5	0.01	R. Borders	Berwick
2)	96.4	1	Gemini FM	Torquay	54)	97.5	0.115	Heartland FM	Pitlochry
12)	96.4	2	Downtown R.	Limavady	4a)	97.6	1	Chiltern FM	Dunstable
43b)	96.4	9.8	BRMB FM	Birmingham	8)	97.6	0.78	R. Wyvern	Hereford
27b)	96.4	0.625	Sound Wave 96.4	Swansea	23)	97.6	0.2	Forth FM	Bathgate
7u)	96.4	2	SGR-FM	Bury St.Edmunds	23c)	97.6	0.025	Northsound One	Balgownie
42)	96.4	3	The Eagle	Guildford	5c)	100.0	2	Kiss 100 FM	London
23a)	96.4	0.35	Tay FM	Perth	99)	100.4		Medway FM	Rainham
55)	96.4	1.25	CFM	Carlisle	103)	101.2		Waves R.	Peterhead
97)	96.4		Neptune R.	Folkestone	59)	101.6		Wey Valley Radio	Four Marks
7)	96.5	0.06	GWR FM	Marlborough	73)	101.6	0.1	KFM	Sevenoaks
7p)	96.5	0.05	Trent FM	Mansfield	9a)	101.7	0.1	Ten 17	Harlow
50)	96.5	0.1	R. Wave	Blackpool	5)	102.0	0.5	Kiss 102.	Manchester
23f)	96.5		South West Sound	Stranraer	43e)	102.0	0.2	Southern FM	Hastings
12)	96.6	10	Downtown R.	Enniskillen	35b)	102.0	1.25	Spire FM	Salisbury
4c)	96.6	4	Northants FM	Northampton	59)	102.0	0.09	Wey Valley Radio	Alton
1)	96.6	0.04	Plymouth Sound FM	Tavistock	83)	102.0	1.5	The Bear	Stratford
20j)	96.6	8.9	TFM	Stockton	67)	102.1	1.25	No. Ea. Community R.	Meldrum
53)	96.6	0.4	R. Ceredigion	We. Wales	7)	102.2	0.5	GWR FM	We. Wilts
66)	96.6	0.045	Nevis FM	Fort William	33)	102.2	2	Jazz FM 102.2	London
10a)	96.6	0.25	Oasis FM	St. Albans	17)	102.2	0.01	SIBC	Lerwick
15)	96.6	0.6	Moray Firth R. + Speysound	Cairngorm	36)	102.2	6.4	Lincs FM	Lincoln
82)	96.6	1	Spirit FM	Chichester	26)	102.2	2.5	Pirate FM 102	Liskeard
23e)	96.7	2.2	West Sound R.	Ayr	31a)	102.2	0.04	Choice FM	Birmingham
43d)	96.7	0.5	Ocean FM	Winchester	55)	102.2	0.8	CFM	Workington
45)	96.7	8.2	R. City 96.7	Liverpool	96)	102.2	0.02	Lochbroom FM	Ullapool
46)	96.7	0.55	BCR	Belfast	7g)	102.3	1	2CR FM	Bournemouth
91)	96.7	2.5	KLFM	King's Lynn	56)	102.3	0.04	The Bay	Windermere
36)	96.7	0.05	Lincs FM	Grantham	82)	102.3	0.5	Spirit FM	Littlehampton
8)	96.7	0.05	R. Wyvern	Kidderminster	39)	102.3	0.44	Minster FM	Thirsk
15)	96.7		Moray Firth R. + Kinniard R.	Fraserburgh	66)	102.3	0.8	Nevis FM	Mallaig
23g)	96.8	5	R. Borders	Selkirk	43e)	102.4	8.2	Southern FM	Eastbourne
20i)	96.9	10	Viking FM	Hull	4e)	102.4	2	Severn Sound FM	Gloucester
4b)	96.9	0.89	B97 FM	Bedford	12)	102.4	10	Downtown R.	Londonderry
23c)	96.9	10	NorthSound One	Aberdeen	7s)	102.4	3.3	R. Broadland 102	Norwich
43e)	96.9	0.1	Southern FM	Newhaven	93)	102.4	1	Wish FM	Billinge Hill
27)	96.9	0.1	Signal One	Stafford	66)	102.4		Nevis FM (F.P.I.)	Glencoe
31)	96.9	0.1	Choice FM	So. London	23d)	102.5	15	Clyde 1	Glasgow
56)	96.9	3.2	The Bay	Lancaster	27c)	102.5	2	The Pulse	Halifax
43c)	97.0	0.5	Invicta FM	Dover	15)	102.5	1	Moray Firth R/Caithness FM	Thurso
2)	97.0	1	Gemini FM	Exeter	55)	102.5	0.1	CFM	Penrith
1)	97.0	1	Plymouth Sound FM	Plymouth	27)	102.6	4	Signal One	Stoke-on-Trent
7b)	97.0	2	Mercia FM	Coventry	10)	102.6	2	Essex FM	Chelmsford
7e)	97.0	0.5	2-Ten FM	Reading	49)	102.6	9	Fox FM	Oxford
23d)	97.0	0.05	Clyde 1	Vale of Leven	20g)	102.6	0.2	Metro R.	Alnwick
20g)	97.1	10	Metro FM	Newcastle	51)	102.6	4	Orchard FM	Taunton
7t)	97.1	4	SGR-FM	Ipswich	67)	102.6	0.3	No. Ea. Community R.	Kildrummy
11)	97.1	0.2	MFM	Wrexham	7h)	102.7	4	Hereward FM	Peterborough
42)	97.1	0.025	Delta R.	Haslemere	38a)	102.7	3.6	R. Mercury FM	Reigate
51)	97.1	0.4	Orchard FM	Yeovil	43c)	102.8	1	Invicta FM	Dunkirk
7j)	97.2	2	Beacon R.	Wolverhampton	7q)	102.8	0.6	Ram FM	Derby
7c)	97.2	0.72	GWR FM	Swindon	8)	102.8	1	R. Wyvern	Malvern
23f)	97.2	1	South West Sound	Dumfries	23a)	102.8	5.25	Tay FM	Dundee
61)	97.2	0.625	Wessex FM	Dorchester	26)	102.8	11.25	Pirate FM 102	Redruth
65)	97.2	0.2	Stray FM	Harrogate	15)	102.8	1	Moray Firth R./KCR	Keith
16)	97.3	4	News Direct 97.3	London	20h)	102.9	0.5	Hallam FM	Barnsley
23)	97.3	6.9	Forth FM	Edinburgh	7e)	102.9	4	2-Ten FM	Basingstoke
15)	97.3	2.8	Moray Firth R.	Inverness	7b)	102.9	0.05	Mercia FM	Leamington Spa
20d)	97.4	1.9	Rock FM	Preston	13)	102.9	7	Q 102.9	Londonderry
37)	97.4	3.2	Cool FM	Belfast	20a)	103.0	4	Piccadilly Key 103 FM	Manchester
20e)	97.4	0.5	Red Dragon FM	Newport	4e)	103.0	0.1	Severn Sound	Stroud
20h)	97.4	0.4	Hallam FM	Sheffield	7a)	103.0	0.08	GWR FM	Bath
49)	97.4	0.3	Fox FM	Banbury	7k)	103.0	1	Q103 FM	Cambridge
7k)	97.4	0.05	Q103 FM	Newmarket	20g)	103.0	0.05	Metro FM	Fenham
53)	97.4	0.4	R. Ceredigion	Penwaun	2)	103.0	1	Gemini FM	Ea. Devon
35c)	97.5	0.4	Gold Radio	Shaftesbury	23c)	103.0	0.175	NorthSound One.	Peterhead
23e)	97.5	0.15	West Sound	Girvan	)	103.0		F.PI.	Caernarfon

FM	MHz	ERP	Name or slogan	Location
23f)	103.0		South West Sound	Kirkcudbright
7j)	103.1	1	Beacon R.	Shrewsbury
43c)	103.1	4	Invicta FM	Maidstone
23b)	103.1	0.5	Central FM	Stirling
12)	103.1	0.5	Downtown R.	Camlough
23g)	103.1	0.03	R. Borders	Peebles
39a)	103.1	0.08	Yorkshire Coast R.	Whitby
20e)	103.2	2	Red Dragon FM	Cardiff
43d)	103.2	2	Power FM	Southampton
24a)	103.2	0.5	Sunrise FM	Bradford
56)	103.2	0.1	The Bay	Kendal
80)	103.2	0.4	Alpha 103.2	Darlington
20g)	103.2	0.1	Metro FM	Hexham
67)	103.2	0.3	No. Ea. Community R.	Colpy
34)	103.3	0.05	London Greek R.	Haringey
53)	103.3	5.8	R. Ceredigion	Blaen Plwyf
4d)	103.3	2.2	Horizon	Milton Keynes
23d)	103.3	0.1	Clyde 1	Firth of Clyde
89)	103.3	0.19	Oban FM	Oban
20h)	103.4	1.5	Hallam FM	Doncaster
11)	103.4	1.4	MFM	Wrexham
23g)	103.4	0.5	R. Borders	Eyemouth
29)	103.4	0.08	Sun FM	Sunderland
55)	103.4	0.2	CFM	Whitehaven
88)	103.4	2	The Beach	Lowestoft
43e)	103.5	0.9	Southern FM	Brighton
41)	103.7	3.8	Channel 103 FM	Jersey
22)	103.8	0.16	RTM	Thamesmead
39)	104.7	2.5	Minster FM	York
28)	104.7	1.25	Island FM	Guernsey
27a)	104.9	0.2	Signal 105	Stockport
102)	104.9	2	XFM	London (C. Palace)
5a)	105.1	2.58	Kiss 105	Emley Moore
48)	105.4	4	Melody FM	London
7r)	105.4	6	Leicester Sound FM	Billesdon
5a)	105.6	0.25	Kiss 105	Sheffield
5a)	105.6	0.5	Kiss 105	Bradford
72)	105.8	2	Virgin FM	London
5a)	105.8	9.4	Kiss 105	Humberside
100)	106.0		CTFM	Canterbury
70)	106.2	4	Heart 106.2	London
57)	106.6	0.25	Star FM	Slough
74)	106.6	0.025	Channel Travel R. (French)	Folkestone
97)	106.8		Neptune R.	Dover
101)	107.2		Thanet Local R.	Isle of Thanet
99)	107.5		Easy 107	Eastbourne
74)	107.6	0.025	Channel Travel R.	Folkestone
94)	107.7		The Wolf	Wolverhampton
105)	107.8	0.1	Thames FM	Kingston
92)	107.9	0.1	Oxygen FM	Oxford
98)	107.9		Dune FM	Southport
99)	107.9	0.1	Medway FM	Bluebell Hill
104)	107.9		Huddersfield FM (**F.PI.**)	Huddersfield

Addresses & other information

H. of tr: Most st's operate 24h. Stations belonging to groups general-ly have some shared programming, especially overnight, but local ID's may still be heard (eg. before N.on the h.)
1) Earl's Acre, Alma Road, Plymouth PL3 4HX.
2) Hawthorne House, Exeter Business Park, Exeter EX1 3QS.
4a) Chiltern Rd, Dunstable LU6 1HQ
4b) Broadcast Centre, Goldington Rd, Bedford MK40 3LS
4c) The Broadcast Centre, 19021 St. Edminds Rd, Northampton NN1 5DY
4d) Broadcast Centre, Crown Hill, Milton Keynes MK8 0AB
4e) Broadcast Centre, Southgate Str, Gloucester GL1 2DQ.
5) Kiss House, P.O. Box 102, Manchester M60 1GJ. **E-mail:** BigShout@kiss102.u-net.com **Web:** http://www.u-net.com/kiss102/
5a) 2a St. Joseph's Well, Hanover Way, Leecs LS3 1AB.
Web: http://www.kiss105.co.uk

5c) Kiss House, 80 Holloway Rd, London N7 8JG.
Web: http://www.kiss100.com
6) Sunshine House, Waterside, Ludlow, Shropshire SY8 1GS.
7a) P.O. Box 2000, Bristol BS99 7SN. **Web:** http://gwrfm.avonibp.co.uk.
7b) Hertford Place, Coventry CV1 3TT
7c) P.O. Box 2000, Swindon SN4 7EX
7d) 3a North Str, Tavistock PL19 4AN
7e) P.O. Box 210, Reading RG31 7RZ. **Web:** http://www.2-tenfm.co.uk
7g) 5 Southcote Rd, Bournemouth BH1 3LR.
Web: http://www.classicgold828.co.uk, http://www.2crfm.co.uk
7h) P.O. Box 2020, Queensgate Centre, Peterborough PE1 1LL
7j) 267 Tettenhall Rd, Wolverhampton WV6 0DQ
7k) The Vision Park, Chivers Way, Histon, Cambridge CB4 4WW
7p) 29-31 Castle Gate, Nottingham NG1 7AP
7q) The Market Place, Derby DE1 3AA
7r) Granville House, Granville Rd, Leicester LE1 7RW
7s) St.George's Plain, 47-49 Colegate, Norwich NR3 1D.
MW: E-mail: prog@amber.radio.co.uk or eng@amber.radio.co.uk.
Web: http://www.amber.radio.co.uk.
FM: E-mail: prog@broadland102.co.uk or eng@broadland102.co.uk.
Web: www.broadland102.co.uk.
Audio: http://norfolkweb.co.uk/audio/nir.htm
7t) Radio House, Alpha Business Park, White House Rd, Ipswich IP1 5LT. **MW: Web, E-mail:** as 7s. **FM: E-mail:** prog@sgrfm.co.uk or eng@sgrfm.co.uk **Web:** www.sgrfm.co.uk
7u) P.O. Box 250, Bury St Edmunds IP33 1AD. **E-mail, Web:** as 7s
7v) Abbeygate Two, 9 Whitewell Rd, Colchester CO2 7DE. **E-mail, Web:** as 7s
8) 5/6 Barbourne Terrace, Worcester WR1 3JZ.
9) 2 St. Johns Wynd, Culver Centre, Colchester, Essex CO1 1WQ.
E-mail: ten17@netforce.net
9a) Latton Bush Business Centre, Southern Way, Harlow, Essex CM18 7BU **Web:** http://www.ten17.netforce.net/
10) Radio House, Cliftown Rd, Southend-on-Sea, Essex SS1 1SX.
E-mail: breeze@netforce.net, essexfm@netforce.net. **Web:** http://www.breeze.netforce.net http://www.essexfm.netforce.net
10a) 7 Hatfield Rd, St. Albans AL1 3RS.
11) The Studios, Mold Rd, Wrexham, Clwyd LL11 4AF.
Web: http://www.mfmradio.co.uk/
12) Newtonards, Co. Down BT23 4ES. **Web:** http://downtown.co.uk
13) The Riveside Suite, The Old Waterside Railway Station, Duke Str, Waterside, Londonderry, BT46 1DH.
15) P.O. Box 271, Inverness IV3 6SF. Keith Community R. (KCR) evenings on 102.8MHz.
16) 200 Gray's Inn Rd, London WC1X 8XZ.
Web: http://www.lbc.co.uk, http://www.newsdirect973fm.co.uk
17) Market Str, Lerwick, Shetland ZE1 0JN.
18) P.O. Box 1170, High Wycombe, Bucks HP13 6YT.
Web: http://ds.dial.pipex.com/town/plaza/raj16/
20a) 127-131 The Piazza, Piccadilly Plaza, Manchester, M1 4AW
20b) P.O. Box 2000, 51 Burley Rd, Leeds LS3 1LR.
Web: http://www.airefm.co.uk/
20c) P.O. Box 999, St.Paul's Square, Preston PR1 1XR.
Web: http://www.redrose.demon.co.uk/999.htm
20d) P.O. Box 974, St.Paul's Square, Preston PR1 1XS.
Web: http://www.redrose.demon.co.uk/rock.htm
20e) P.O. Box 99, Cardiff CF1 5YJ. **Web:** http://www.rdfm.co.uk
20g) Radio House, Long Rigg, Swalwell, Newcastle-upon-Tyne NE99 1B. **E-mail:** onair@metrofm.ace.co.uk
20h) 900 Herries Rd, Sheffield S6 1RH.
Web: http://www.hallamfm.co.uk
20i) Commercial Road, Hull HU1 2SG
21ghi) **Web:** http://www.millhouse.co.uk/gygold/tune.htm
20j) Yale Crescent, Thornaby, Stockton-on-Tees TS17 6AA.
22) Harrow Manor Way, Thamesmead South, London SE2 9XH.
23) Forth House, Forth Str, Edinburgh EH1 3LF.
Web: http://www.almac.co.uk/personal/sctwww/
23a) 6 North Isla St, Dundee DD3 7JQ
23b) Stirling Enterprise Park, Kerse Rd, Stirling FK7 7YJ.
23c) 45 King's Gate, Aberdeen, AB2 6BL
23d) Clydebank Business Park, Clydebank, Glasgow G81 2RX

23e) Radio House, 54 Holmston Road, Ayr KA7 3BE
23f) Campbell House, Bankend Rd, Dumfries DG1 4TH
23g) Tweedside Park, Galashiels, Selkirkshire TD1 3TD.
24) Sunrise House, Sunrise Rd, Southall UB2 4AU
24a) 30 Chapel Str, Little Germany, Bradford BD1 5DN.
26) Carn Brea Studios, Wilson Way, Redruth, Cornwall TR15 3XX.
Web: http://www.cableol.net/ukrd/pirate1.html
27) Studio 257, 67-73 Stoke Rd, Stoke-on-Trent ST4 2SR.
E-mail: SignalOne@SignalRadio.netcentral.co.uk *or*
SignalGold@SignalRadio.netcentral.co.uk
Web: http://www.netcentral.co.uk/signal/one.html *or*
http://www.netcentral.co.uk/signal/gold.html
27a) 1st Flr, Regent House, Heaton Lane, Stockport SK4 1BX. Web:
http://www.netcentral.co.uk/signal/cheshire.html
Web: http://www.netcentral.co.uk/signal/cheshire.html
27b) Victoria Road, Gowerton, Swansea SA4 3AB
27c) Forster Square, Bradford BD1 5NE. **E-mail:** general@pulse.co.uk
28) 12 Westerbrook, St. Sampson, Guernsey GY2 4QQ.
E-mail: islandfm@itl.net
29) 39 Holmeside, Sunderland.
E-mail: general@suncity.co.uk **Web:** http://www.suncity.co.uk/
30) 204/206 Queenstown Rd, Battersea, London SW8 3XA.
E-mail: ethnic@ethnic.demon.co.uk
31) 16-18 Trinity Gardens, Brixton, London SW8 3NR
31a) 95 Broad Str, Birmingham B15 1AU.
32) The Studios, 41 Conwy Rd, Colwyn Bay, Clwyd LL28 5AB
33) 26/27 Castlereagh Str, London W1H 6DJ.
Web: http://www.jazzfm.co.uk
34) Florentia Village, Vale Rd, London N4 1TD.
35) P.O. Box 1584, Robinson Close,.Telford Industrial Estate, Kettering
NN16 8PU
35a) Churchill Studios, Churchill Rd, Cheltenham GL53 7EP
35b) City Hall Studios, Malthouse Lane, Salisbury SP2 7QQ
35c) Longmead, Shaftesbury, Dorset SP7 8QQ. **Web:** http://our-
world.compuserve.com:80/homepages/bkingsley/goldrad.htm
35d) Dodnor Park, Newport, Isle of Wight PO30 5XE.
E-mail: iow@musicradio.com
36) Witham Park, Waterside South, Lincoln LN5 7JN.
37) P.O. Box 974, Belfast BT1 1RT.
E-mail: music@coolfm.co.uk **Web:** http://www.coolfm.co.uk
38) 6 Lady Lane, Paisley PA1 2LG
38a) The Friary, Guildford GU1 4YX. **Web:** http://www. cableol.
net/ukrd/county1.html *or* http://www.cableol.net/ ukrd/eagle1.html
38b) P.O. Box 1458, Trafford Park, Manchester M17 1FL.
E-mail: mail@fortune1458.u-net.com
38c) Broadfield House, Brighton Rd, Crawley RH11 9TT
Web: http://www.invicta.co.uk/mercury/fame/
39) P.O. Box 123, Dunnington, York YO1 5ZX
39a) P.O. Box 962, Scarborough YO12 5YX.
E-mail: ycr@yorkshirecoast.co.uk
Web: http://www.yorkshirecoast.co.uk/ycr/
40) Ringway House, Hill Str, Coventry CV1 4AN. **E-mail:** kix96@mark-
keen.demon.co.uk **Web:** http://www.indiscreet.com/kix96/
41) 6 Tunnel Str, St. Helier, Jersey JE2 4LU.
E-mail: radio@103fm.itl.net **Web:** http://www.103fm.itl.net
42) 65 Weyhill, Haslemere GU27 1HN. **Web:** http://www.cableol.net/
ukrd/county1.html http://www.cableol.net/ukrd/delta1.html
43a) 29-30 Leicester Square, London WC2H 7LE.
Web: http://www.capitalfm.co.uk http://www.capitalgold.co.uk
43b) Radio House, Aston Rd.North, Birmingham B6 4BX
43c) Radio House, John Wilson Business Park, Whitstable CT5 3QX
43d) Radio House, Whittle Ave, Segensworth West, Fareham PO15
5SH –43e) P.O. Box 2000, Brighton BN41 2SS.
45) 8-10 Stanley Str, Liverpool L1 6AF.
Web: http://www.connect. org.uk/merseyworld/cityFM/
46) Russel Court Bldg, Claremont Str, Lisburn Rd, Belfast BT9 6JX.
47) The Studios, The Park, Newtown, Powys SY16 2NZ.
48) 180 Brompton Rd, London SW3 1HF.
Web: http://www.netlink. co.uk/users/melnet/
49) Brush House, Pony Rd, Cowley, Oxford OX4 2XR.
E-mail: fox@foxfm.co.uk **Web:** http://www.foxfm.co.uk/

50) 965 Mowbray Drive, Blackpool FY3 7JR.
Web: http://www. radiowave.net/
51) Haygrove House, Shoreditch, Taunton TA3 7BT. **E-mail:**
orchard@orchardfm.co.uk **Web:** http://www.orchardfm.co.uk/
52) The Light House, 17 Market Place, Bideford EX39 2DR.
Web: http://bigweb.castlelink.co.uk/north_devon/lantern/
53) Yr Hen Ysgol Gymraeg, Fford Alexandra, Aberysytwyth, Dyfed SY23
1LF.
54) Atholl Curling Rink, Lower Oakfield, Pitlochry, PH16 5DS.
55) P.O. Box 964, Carlisle CA1 3NG.
56) P.O. Box 969, St. George's Quay, Lancaster LA1 3LD
57) The Observatory Shopping Centre, Slough SL1 1LH.
Web: http://www.cableol.net/ukrd/star.html
58) 185 High Rd, Wood Green, London N22 6BA.
59) Prospect Place, Mill Lane, Alton, Hants GU34 2SY.
Web: http://www.cableol.net/ukrd/weyv1.html
60) 11 Bourbon Str, Aylesbury, Bucks HP20 2PZ.
E-mail: mix@mix96.demon.co.uk
61) Radio House, Trinity Str, Dorchester DT1 1DJ.
62) 7, Hurlingham Business Park, Sullivan Rd, London SW6 3DU.
63) KMS House, Bradford Str, Birmingham B12 0JD.
65) P.O. Box 972, Station Parade, Harrogate HG1 5YF.
66) Inverlochry, Fort William, Inverness-shire PH33 6LU.
67) Town House, Kintore, Aberdeenshire AB51 0US.
69) Golden Rose House, 26-27 Castlereagh Str, London W1H 6DJ.
70) P.O. Box 1062, London W10 6SP.
71) Glen House, Stag Place, London SW1E 5AG.
Web: http://www.wordnet.co.uk/premier.html
72) See national st. Virgin Radio (above)
73) 1 East Str, Tonbridge, Kent TN9 1AR.
E-mail: kfm@cis.compuserve.com
Web: http://www.cityscape.co.uk/users/bm22/kfm.html
74) P.O. Box 2000, Folkestone, Kent CT18 8XY.
76) 2c Park Ave, Cookstown, Co. Tyrone BT80 8AH.
77) Globe House, Southall Str, Manchester M3 1LG.
E-mail: nigel@xtml.u-net.com
Web: http://www.u-net.com/manchester/radio/asiansound/
79) P.O. Box 828, Cookstown, Co. Tyrone, BT80 9LQ.
80) Radio House, 11 Woodland Rd, Darlington DL3 7BJ.
81) Radio House, 63 Melton Rd, Leicester LE4 6PN.
82) Dukes Court, Bognor Regis. Chichester PO19 2FX.
83) P.O. Box 102, Stratford-upon-Avon CV37 9GF
85) Carn Business Park, Craigavon, Co. Armargh BT63 5RH.
86) P.O. Box 255, Stoke-on-Trent, ST4 8YY.
E-mail: 100664.1057@compuserve.com
Web: http://www.dungeon.com/~sherwood/home.html
87) P.O. Box 1116, Ebbw Vale, NP3 5YJ.
88) P.O. Box 103.4, Lowestoft NR32 2TL.
Web: http://www.thebeach. co.uk
89) McLeod Units, Lochavullin Estate, Oban, Argyll.
90) YDR House, Gargrave Rd, Skipton BD23 1YD.
91) P.O. Box 77, King's Lynn, Norfolk PE30 1NN.
92) Suite 41, the Westgate Centre, Oxford OX1 1PD. **E-mail:** mail@oxy-
gen.demon.co.uk **Web:** http://www.oxygen.demon.co.uk
93) Orrell Lodge, Orrell Rd, Wigan WN5 8HJ.
94) 51-53 Queen Str, Wolverhampton Wv1 3BU.
95) London. **Web:** http://www.xfm.co.uk
96) 24 Argyll Str, Ullapool. Local prgrs: MF 0700-0900, 1700-
2400/0100. Also weekends. Other times rel. Virgin Radio.
97) Branscombe, Upper Str, Kingsdown, Deal CT14 8BH.
98) P.O. Box 1006, Southport PR9 7RP.
99) Berkley House, 186 High Str, Rochester ME1 1EY.
100) Canterbury.
101) Isle of Thanet.
102) **Web:** http://www.xfm.co.uk/
103) Peterhead (on air from Dec 1st 1997).
104) Huddersfield.
105) So. We. London.
V. Re. direct to individual stations.

Restricted Service Licences

Licences are granted for low power special event stations operating for up to 28 days (occ. longer) on MW or FM.

MANX RADIO (Comm.)

✉ P.O. Box 1368, Douglas, Isle of Man IM99 1SW. ☎ +44 (1624) 682610. 📠 +44 (1624) 682604. **E-mail:** postbox@manxradio.com
Web: http://www.manxradio.com/
L.P: GM: Stewart Watterson. PD: George Ferguson. CE: Ewan Leeming.
MW: 1368kHz 20kW
FM: 89.0/97.2/103.7MHz.
D.PRGR: 0630-0100. Separate prgrs on FM: MF 0730-0830, Sun 0900-1900
"Radio TT", a special sce. for the Manx TT motorcycle races, will be on the air June 1st-15th 1998 on 1368kHz at 0600-0100 daily. May include some prgrs in French or German.
V. by QSL-card.

BRITISH FORCES BROADCASTING SCE. LONDON

(a division of the Services Sound & Vision Corp.)
✉ Bridge House, North Wharf Rd, London W2 1LA. ☎ +44 (171) 724 1234. 📠 +44 (171) 706 1582. ① 25704 BFBS G. **E-mail:** info@bfbs.com
Web: http://www.bfbs.co.uk/
L.P: Dir. of Broadc: Peter McDonagh. Contr: Charly Lowndes. PD: Chris Russel. CE: Ian Martin.
STATIONS: see under Belize, Brunei, Cyprus, Falkland Islands, Germany, Gibraltar.

VOICE OF AMERICA RELAY STATION

See International Broadcasting section.

VATICAN CITY STATE

L.T: UTC + 1h (Su: UTC + 2h) — **Pop:** 1.000— **ITU:** CVA.

RADIO VATICANA (Rlg.)

✉ Vatican Radio, I-00120 Vatican City. ☎ +39 (06) 6988 3551. (**Int. Rel:** ☎ +39 (06) 6988 3237. ① 2023 VA. **Cable:** RADVAT. **E-mail:** relint@vatiradio.va
Web: http://www.vatican.va - http://www.wrn.org/vatican-radio/
L.P: DG: Rev. Pasquale Borgomeo S.I. PD: Rev. Federico Lombardi S.I. TD: Rev. Eugenio Matis S.I. CE: Pier Vincenzo Giudici. Head of Int. Rel: Mrs. Solange de Maillardoz.
MW: 526kHz 5kW, 1530kHz 300/600kW, 1611kHz 100kW.
FM: 93.3/96.3/103.8/105.0MHz 10kW (stereo).
SW: S.M. Galeria, Italy (G.C: 12.19E/42.03N): 100/250/500kW — Vatican City (G.C: 12.27E/41.54N): 1 x 80kW, 1 x 10kW.
ANN: Before all transmissions: Latin: "Laudetur Jesus Christus" (Praised be Jesus Christ), repeated in the language of the broadcast, then station identification — **IS:** "Christus Vincit" — **V.** by QSL-card. For SW frequencies and schedule see International Broadcasting section

YUGOSLAVIA (Federal Republic of)

L.T: UTC + 1h (Su: UTC + 2h) — **Pop:** 14.000.000 — **Radios:** 2.692.310 — **Pr.L:** Serbian — **E.C:** 50Hz, 220V — **ITU:** YUG.

UDRUZENJE JUGOSLOVENSKIH RADIOTELEVIZIJA d.O.O. (JRT)

✉ Generala Zdanova 28, 11000 Beograd. ☎ +381 (11) 330194. 📠 +381 (11) 334380. **Cable:** Jurate, Beograd. ① 71048 or 12158 yu jurate, Beograd — **L.P:** MD: S. Dumić.
JRT comprises Radiotelevizija Srbije, Radiotelevizija Crne Gore, and the External Service (R. Yugoslavija):

SERBIA & MONTENEGRO

RADIOTELEVIZIJA SRBIJE

✉ Takovska 10, 11000 Beograd. ☎ +381 (11) 340911. 📠 +381 (11) 341630. ① 11884 — **L.P:** DG: M. Vucelic.
R. Beograd: Hilendarska 2, 11000 Beograd. ☎ +381 (11) 324 8888 —

L.P: Dir: Milivoje Pavlovic.

Mediumwaves	kHz	kW		kHz	kW
Bosilegrad	675	10/5	Kladovo	999	10/5d
Dimitrovgrad 1	675	10/5	Bosilegrad 2	999	10b
Beograd	684	2000	Beograd 2/3	1008	200
Negotin	693	10/5d	Aleksinac	1008	200b
Nis	711	100a	Novi Pazar	1062	1a
Medvedja	765	1d	Vranje	1296	10d
Jagodina	1440	20/10b	Tutin	1485	1a
Bujanovac	1467	1a	Zajecar	1485	1d
Crna Trava	1485	1d	Priboj	1485	1a
Kladovo	1485	1a	Leskovac	1602	1a
Pirot	1485	1a	Negotin	1602	1e
Trgoviste	1485	1a			

SW: Stubline (G.C: 20.09E/44.34N): 7200kHz 100kW (rel. Beograd 1 0600-2305).

FM Stereo (MHz)	I	II/III	IV	L	kW
Avala	95.3	97.6	101.4	104.0	80
Besna Kobila	91.7	95.3	105.6		15
Crni Vrh	89.7	99.3	101.0		25
Crveni Cot	94.5	96.5	101.8	103.2	80
Deli Jovan	87.7	94.9			2
Jastrebac	96.9	89.3	103.5	106.0	100
Kopaonik	93.7	90.9	102.1		50/25
Mokra Gora	91.5	100.1	103.2		15
Nis	99.5			97.9	3.5/0.7
Ovcar	88.1	90.1	101.6		25
Pirot	91.3	98.5			1
Tornik	90.6	97.5	100.2		15
Tupiznica	92.5	96.1	100.4	105.3	40
Vrsac	95.7				30

+ additional low power st's not mentioned.
a) Beograd 1 + reg. prgr — b) rel. Beograd 2/3 — c) rel. Beograd 202 + reg. prgr — d) rel. Beograd 1 — e) rel. Beograd 2/3 + reg. prgr — L) local st's.

R. Beograd 1: 24h. **N:** W 0303, 0330, 0400, 0430, 0500, 0540, 0700, 0800, 0900, 1000, 1100, 1200, 1300, 1400, 1600, 1700, 1830, 2100, 2200, 2300; Sun 0430, 0500, 0530, 0600, 0700, 0800, 0900, 1000, 1100, 1200, 1400, 1600, 1830, 2000, 2200, 2300.
R. Beograd 2: W 0400-1900 (Sun 0600-1900). **N:** W 1130, 1230, 1330, 1500, 1550, 1850. Sun 0630, 0730, 0930, 1055, 1130, 1330, 1730, 1850 — **R. Beograd 3:** 1900-2300 (Serious prgr.) — **R. Beograd 202:** 0400-2400 on 1503kHz, 104.0MHz + FM IV (0000-0400 rel. R. Beograd 1) — **Stereorama:** 0700-1700SS on FM IV. Other times rel. Beograd 202 — **Music R. 101,** Beograd: 24h on 98.5MHz + FM III — **Youth R.,** Beograd: 24h on 92.5MHz 1kW — **R. Politika:** Makedonska 29, 11000 Beograd: 24h on 105.2MHz 3kW + Rudnik 91.5MHz 6kW. **F.P.I.** 6 FM tr's to cover most of Serbia — **Ju Radio:** Hilandarska 2, 11000 Beograd — 24h on 100.4MHz 1kW (operated by R. Yugoslavija) — **R. Index/University R,** Beograd: 24h on 88.9MHz 1kW. **Web:** http://www.radioindex.co.yu

Local st's	kHz	kW	MHz
Uzice	531	10	92.0
Vranje	531	1	96.5
Soko Banja	639	1	90.5
Lazarevac	648	1	100.9
Sabac	702	10	106.6
Krusevac	738	10	92.2
Arandjelovac	792	1	98.9
Studio "B" 2	900	2	94.9
Smederevo	945	1	96.1
Bor	981	10	93.2
Cacak	981	10	92.8
Pozarevac	990	1	90.1
Kragujevac	1026	10	106.8
Jagodina	1062	10	97.3
Novi Pazar	1062	1	90.0
Vrnjacka Banja	1170	1	93.2
Majdanpek	1206	1	96.7
Mladenovac	1215	1c	90.8
Kraljevo	1242	10	106.6
Loznica	1296	10	104.8

Local st's	kHz	kW	MHz
Zajecar	1341	10	98.1
Studio "B" 1	1350	10	100.8
Valjevo	1368	10	88.6
Kladovo	1458	1	89.9
Priboj	1485	1	87.6
Smed. Palanka	1566	1	88.3
Prijepolje	1584	1	95.9
Leskovac	1602	1	99.0
Sjenica	1602	1	

FOREIGN SERVICE

RADIO YUGOSLAVIA

see International Broadcasting section.
Local/Private FM st's in Beograd:
Studio "B", P.O. Box 10, 11001 Beograd, Masarikova 5. **E-mail:** srdjan.kusovac@zamir-bg.ztn.zer.de **Prgr. 1** on 1350kHz +100.8MHz: 24h. **Prgr. 2** on 900kHz +94.9/106.4MHz: 24h. **Prgr. 3** on 99.1/107.2MHz 1600-2300 – **R. "Pingvin":** Autoput 2, 11070 N. Beograd: 24h on 90.9MHz 1kW – R. Vozd 88.4 – R. Index 88.9 – R. Bonton 89.5 – P. Pingvin 90.9 – R. Pink 91.3 – R. B-92 92.5. – R. Krik 93.3 – R. Studio B 2 94.9 – R. Beograd 1 95.3 – R. Beograd 2/3 97.6 – R. 101 98.5 – R. Studio B 3 99.1 – R. Grom 99.4 – YU Radio 100.4 – R. Studio B 1 100.8 – R. 101 101.4 – R. Kosava 102.2 – R. Novosti 102.6 – R. Golf 103.6 – R. 202 104.0 – R. Papagaj 104.4 – R. S 104.7 – R. Politika 105.2 – R. Vozd 105.7 – R. Ponos 106.0 – R. Studio B 3 106.4 – R. Studio B 1 107.2 – R. Roda 107.9MHz.
Other local st's on FM: Bajina Basta 87.6 – R. Nemanja, Cuprija 87.6 – R. Srpski Venac, Bujanovci 88.5 – Kosjeric 88.5 – R. Dabar, Pozarevac 88.8 – R. M, Cacak 88.9 – Gornij Milanovac 88.9 – R. 34, Kragujevac 88.9 – R. Zanuki, Pirot 88.9 – R. Rec Naroda, Pozarevac 90.1 – R. Osa, Pirot 90.3 – Velika Plana 90.4 – OK Studio, Krusevac 91.3 – R. Politika, Rudnik 91.5 – R. Sunce, Arandjelovac 91.9 – R. Elektronik, Knjazevac 92.1 – R. Bum 93, Pozarevac 93.4 – R. Bubamara, Svrljig 93.5 – Valjevo 93.9 – R. Papagaj, Gornji Milanovac 94.3 – Knjazevac 94.5 – R. Maestral, Vranje 95.5 – R. Belami, Nis 95.6 – R. M 31, Uzice 95.6 – Pirot 95.8 – R. 48, Trstenik 95.8 – R. Kis, Vranje 95.9 – R. Maki, Jagodina 96.5 – R. S, Smederevo 97.7 – Paracin 97.8 – R. Jugohol, Cacak 98.3 – R. B 90, Negotin 99.6 – R. Duga Sky, Pozarevac 100.2 – R. Sabor, Cuprija 100.6 – Nezavisni RTV, Aleksinac 101.3 – R. O16, Leskovac 101.6 – R. Resava, Svilajnac 101.9 – R. Globus, Kraljevo 102.5 – R. Fast, Nis 102.7 – Sremcica 102.9 – R. As, Sabac 103.7 – R. Sunce, Nis 106.1 – R. Mars Valjevo 106.2 – Omea Radio, Loznica 107.0 – R. Melos, Kraljevo 107.1 – Krusevac 107.2 – R. F, Leskovac 107.4 – Cuprija 107.8.

RADIOTELEVIZIJA CRNE GORE

⌧ Cetinjski put bb, 81000 Podgorica. ☎ +381 (81) 41800. ↻ 61133—**L.P:** DG: Z. Joković.

Mediumwaves	kHz	kW	Mediumwaves	kHz	kW
Pljevlja Grad	567	10L	Niksić	1458	1b
Niksić	594	10L	Cetinje	1485	1L
Bar	738	5a	Plav	1485	1a
Podgorica	882	600/	Pluzine	1485	1a
		300a	Zabljak	1485	1a
Bijelo Polje	882	10a	Ulcinj	1503	10a
Herceg Novi	882	10a	Bar	1548	1b
Pljevlja	882	10a	Rozaje	1584	1a
Rozaje	882	10a	Budva	1602	1a
Ulcinj	882	1a	Ivangrad	1602	1a
Bar	1026	1L	Pljevlja	1602	1a
Ulcinj Grad	1161	1L	Tivat	1602	1a
Herceg Novi	1305	1L	Savnik	1602	1a

R. Podgorica 1: 24h (a) – L: local prgr 0800-1500.
R. Podgorica 2: 24h (b).

FM (MHz)	I	II	III	L	kW
Lovćen	94.9	98.0	101.0	89.8	54
Bjelasica	92.1	99.3			54
Velji Grad	89.6	99.7			10
Mozura			102.6		1
Durmitor	91.3				1
Sudjina Glava	88.0	98.9			10

+ 11 tr. sites less than 1kW.

Local/private st's: R. Montenegro, Podgorica 87.6MHz 0.3kW – R. Cetinje – R. Antena M.

VOJVODINA (Autonomous Province)

RADIO NOVI SAD

⌧ Zarka Zrenjanina 3, 21000 Novi Sad. ☎ +381 (21) 611588. 🖷 +381 (21) 26624. ↻ 14127.

Mediumwaves	kHz	kW	Mediumwaves	kHz	kW
Sombor	837	50y	Novi Sad 1	1269	750
Novi Sad 2	1107	150			

FM (MHz)	I	II	M	kW
Novi Sad	87.7	90.5	92.9(g)	50/50/10
			94.1(f)	10
			97.2(h)	10
Subotica	99.3	92.5		10/10
Pancevo	99.6			1
Vrsac			91.7(f)	10

y) rel. Novi Sad 1 – M) prgr's for national minorities: Romanian (f), Rossinian (g), Slovak (h).
R. Novi Sad 1: 24h in Serbian
R. Novi Sad 2: 0400-2305 in Hungarian, Romanian, Rossinian. Slovak.

Local st's	kHz	kW	FM
Sombor	666	10	90.9
Stara Pazova 2	783	0.2	107.4
Ruma	936	1	102.7
Temerin	1044	1	93.5
Subotica	1089	10	91.5
Kovin	1161	0.2	88.5
Backa Topola	1170	0.2	97.8
Sid	1323	1	89.1
Vrbas-Kula	1359	2	
Kovacica	1395	0.2	93.2
Beocin	1431	0.2	97.8
Srbobran	1449	2	102.6
Zrenjanin	1467	2	103.6
Novi Sad L	1485	2	93.7
Apatin	1494	0.2	98.7
Odzaci	1539	0.2	89.7
Stara Pazova 1	1548	1	91.5
Bac	1575	0.2	
Pancevo	1584	1	92.1
Indjija	1593	0.2	96.0
Vrsac	1602	1	98.1

Local/private st's on FM: R. Yusaco, Novi Sad 88.6 – Kula 89.2 – Kikinda 89.3 – Backi Petrovac 91.4 – R. M, Sremska Mitrovica 91.9/106.2 – Backa Palanka 95.1/99.1 – Vrbas 95.5 – Futog 99.5 – R. 100, Novi Sad 100.6 – R. Spektar, Sombor 101.3 – R. Politika, Novi Sad 102.5 – Becej 105.4 – R. Pink, Novi Sad 107.0 – R. Safir, Pancevo 107.7MHz.

KOSOVO-METOHIJA (Autonomous Province)RADIO

PRISTINA

⌧ Marsala Tita bb, 38000 Pristina. ☎ +381 (13) 826255. 🖷 +381 (13) 832073. ↻ 18134 — **L.P:** DG: Adem Gashi.

Mediumwaves	kHz	kW	Mediumwaves	kHz	kW
Pristina 3	549	100	Pristina 2	1512	10
Pristina 1	1413	100			

FM (MHz)	1	2	3	kW
Pristina	91.9	100.6	99.2	1/1/1
Goles	95.7		97.7	30/30
Cviljen	88.9	92.4	94.5	30/30/30

R. Pristina 1: 0400-2305 in Albanian.
R. Pristina 2: 0800-1700 in Albanian & Turkish.
R. Pristina 3: 0400-2305 in Serbian.

Local sts	kHz	kW	MHz	Local sts	kHz	kW	MHz
Djakovica	936	10	95.2	Prizren	1377	10	94.0
Gnjilane	954	10	95.1	Peć	1539	10	94.1
Mitrovica	1035	10	91.1	Urosevac	1584	2	
Kacanik	1179	1					

AFRICA

ALGERIA

L.T: UTC + 1h — **Pop:** 30.509.393 — **Radios:** 3.500.000 — **Pr.L:** French, Arabic, Berber dialects — **E.C:** 50Hz, 220V — **ITU:** ALG.

ENTREPRISE NATIONALE DE RADIODIFFUSION SONORE

✉ 21 Blvd. des Martyrs, Alger 16000. ☎ +213 (2) 590700. 🖷 +213 (2) 605814. ✑ 65265 DZ — **L.P:** DG: Lamine Bechichi. Dir. Tec: Sadek Laskri. Dir. Coop: Abdelkader Lalmi.
Ch.: 1=Arabic, 2=Berber, 3=French, 4=International Sce, L=Local.

LW/MW	kHz	kW	Ch.
Bechar	153	1000	1
Ouargla	198	1000	1
Tipaza	252	1500*	3/4
Ain-El-Beida	531	600*	1
Les Trembles	549	600*	1
Bechar	576	400	1
Tindouf	666	5	1
Reggane	693	5	1
Ain-El-Hammam	693	5	2
Ain-Amenas	738	5	1
Djanet	783	5	1
Ben-Abbas	837	5	1
Ghardaia	873	5	1
Alger	891	600*	1
Tamanrasset	909	5	1
Timimoun	927	5	1
Alger	981	600*	1
Hassi-Messaoud	1026	5	1
Adrar	1089	5	1
Ouargla	1098	5	1
Ain-Salah	1161	5	1
El-Golea	1287	5	1
Constantine	1305	20	3
Alger	1422	50	3
Constantine	1422	10	C

*) half power used 1800-0600 — **FM:** 94.2/94.7MHz (Ch. 3).
SW: see International Broadcasting section.
Ch. 1 in Arabic: 24h. **N:** on the h — **Ch. 2 in Berber:** 0600-0100. **N:** on the h — **Ch. 3 in French:** 0500-0100 on 1305kHz, 0500-2000 & 2100-0100 on 252kHz, 0500-1600 & 2200-0100 on 1422kHz. **N:** on the h — **Cultural R (C):** 1600-2200 on 1422kHz.
INTERNATIONAL SCE: see International Broadcasting section.
ANN: A: "Huna El Djazair, idha'atu-El Djoumhouriya El Djazairia". C: "al-Idha'atu al-Thaqafiyah". F: "Alger chêine 3, Radiodiffusion Algerienne" — **IS:** Oriental Lute (Ud) — **V.** irr. by QSL-card. Rp.

ANGOLA (People's Republic)

L.T: UTC + 1h — **Pop:** 10.900.979 — **Radios:** 450.000 — **Pr.L:** Portuguese + 10 ethnic — **E.C:** 50Hz, 220V — **ITU:** AGL.

RÁDIO NACIONAL DE ANGOLA (Gov.)

✉ Rua Comandante Gika (C.P. 1329), Luanda. ☎ +244 (2) 323172, 321258. 🖷 +244 (2) 324647, 391234. ✑ 0991 3121 emissora an
L.P: DG: Agostinho Vieira Lopes. PD: Arlindo Macedo. ID: Luis Fernando. TD: Filipe D. Sungu Miguel. AD: Rui Vieira Lopes.

Location	kHz	kW	Prgr.	H. of tr.
Mulenvos	702	10	3	1000-2400
Cazenga	944	25	2	0430-1900
Cazenga	944	1	E	1900-2300
Mulenvos	1088	100	N	1800-0500
Mulenvos	1088	25	N	0500-1800
Mulenvos	1367	100	N	1600-0400
Cazenga	3355	10	2	2300-0900
			E	1900-2300
Mulenvos	3375v	10	N	24h
Mulenvos	4950v	10	N	24h
Cazenga	7215	10	2	0430-1900
Mulenvos	7245	100	N	24h
Mulenvos	9535	100	2	0430-1900
			E	1900-2300
Mulenvos	9720	100	N	0400-1600
Mulenvos	11955	100	N	0500-1800

FM: Luanda (4kW): 93.5MHz (N - inactive), 96.5MHz (RFM Stéreo), 101.4MHz (2 + Ext. Sce.).

Emissão Nacional (N) in Portuguese: 24h. **N:** on the h.
Antenna 2 in Portuguese & ethnic languages: 0430-1900. **N:** rel Emissão Nacional — **RFM Stéreo** (musical): 1000-2400.
EXTERNAL SCE: see International Broadcasting section.
ANN: "De Luanda, capital da República Popular de Angola transmite a Rádio Nacional de Angola" — **IS:** Vibraphone — **V.** by QSL-card.

PROVINCIAL STATIONS (EP=Emissora Províncial)

Mediumwaves	kHz	kW	H. of tr.
2) EP do Huambo	1010	1	0300-2200
4) EP do Kuanza-Sul	1110	1	alt. to 1115
4) EP do Kuanza-Sul	1115	10	0500-2300
5) EP do Kuando-Kubango	1148	10	0500-1200, 1600-2100
3) EP do Zaire	1152	10	0400-1200, 1600-2100
2) EP do Huambo	1160	10/1	0300-2200
6) EP de Malanje	1188	10/1	0355-2200
7) EP de Benguela	1200	1	24h
8) EP da Huíla	1232	10	0500-2300
9) EP da Lunda-Sul	1241	5	1830-2200
10) EP do Kuanza Norte	1259	10/1	0400-1300, 1600-2200
11) EP do Uige	1260	10	0300-2200
8) EP da Huíla	1313	1	0500-2300
12) EP do Namibe	1313	1	0500-2200
1) EP do Lobito	1403	1	0400-2200
13) EP do Bié (alt. 1386)	1404	10	0355-2200
14) EP da Lunda-Norte	1440	1/10	0500-2300
7) EP de Benguela	1502	10	0500-2300
15) EP de Cabinda	1570	10	0500-2300
16) EP do Moxico	1586	10	0500-2300

Shortwaves (all freqs variable)			
5) EP do Kuando-Kubango	4780	5	0500-0800, 1700-2300
8) EP da Huíla	4820	3	0400-2300
15) EP da Cabinda	4970	5	0400-2100
12) EP do Namibe	5015	1	r. 1845-2300
7) EP de Benguela	5043	1	0400-2200
7) EP de Benguela	6150	1	0500-2300
1) EP do Lobito	7150	1	0400-2200
8) EP da Huíla	7350		
1) EP do Lobito	11815		

FM (MHz): 1) 104.9/106.9 — 2) 92 & 95 — 4) 104.5

Addresses & other information

1) C.P. 56, Lobito. Dir: Ribeiro dos Santos Tadeu. PD: André Moisés. CE: António Mendes Filipe. Dir.Admin: Francisco Jerónimo Chilombo. **V.** by QSL-card. Rec.acc — 2) C.P. 125, Huambo. Dir: Nelson de Jesus Ferreira dos Santos. PD: José Manuel Lucas. TD: Francisco António Monteiro. Dir. Admin: Maria Isabel Faria Serrão. **F.PL:** 50/10kW – 3) M'Banza Kongo. Dir: Pedro Sirdes — 4) C.P. 10, Sumbe. Dir: Sebastião Daniel Neto. PD: Sabino Nicolau Noy Jr. Tech. Asst's: Carlos Diniz de Resende Costa, Joaquim Fernando Manuel – 5) C.P. 36, Menongue. Dir: Miguel Alves Perrera – 6) C.P. 83, Malanje. Dir: Carlos Matariquiriri – 7) C.P. 19, Benguela. Dir: Carlos A.A.Gregório. Dir. Tec: António Mendes Felipe. PD: Lilas Andre Orlov. Dir. Admin: Manuel Luis Morais. **V.** by letter – 8) C.P. 111, Lubango. Dir: Celso do Amaral – 9) C.P. 116, Saurimo. Dir: Zaqueu Dias Kaferro – 10) C.P. 174, N'dalatando. Dir: Augusto Luis Macosso – 11) C.P. 140, Uige – 12) C.P. 174, Namibe. Dir: Joaquim dos Santos – 13) C.P. 33, Kuito. Dir: Cordeiro Chimo – 14) C.P. 1247, Luachimo. Dir: Abilio Melo de Macedo – 15) Cabinda. Dir: Carlos da Cruz Luis – 16) C.P. 74, Luena. Dir: Paulo Cahílo – 19) EP Bengo, Caxito. Dir: A.S.Helen (no details of freq.or schedule received).
N: All st's rel. Luanda at 0600, 1100, 1200, 1900 and/or 2200. Many st's also rel. headlines.

RÁDIO ECLÉSIA (Rlg.)

✉ Rua Commandante Bula, 118 CP 3579, Luanda. ☎ +244 (2) 343041. 🖷 +244 (2) 343093. **L.P:** DG: D.Benidito Roberto. DE: Aristides Neiva. **E-mail:** reclesia@ebonet.net. **FM:** 97.5MHz, 5kW. 0600-2130.

LUANDA ANTENA COMERCIAL (Comm.)

✉ Luanda. **E-mail:** lac@ebonet.net

ASCENSION ISLAND (British)

L.T: UTC — **Pop:** 1.800 — **Radios:** 1.000 — **Pr.L:** English — **E.C:** 50Hz, 220V — **ITU:** ASC.

ASCENSION RADIO (USAF)

✉ Ascension Radio Station, Ascension AAF, P.O. Box 4235, Patrick

AFB, FL 32925-0235, USA.
MW: ZD8VR 1602kHz 0.1kW (1kW authorized) — **FM:** 95.1MHz 40W, 98.7MHz 0.4kW (stereo) — **Prgr. 1:** 24h on 1602kHz (95.1MHz inactive) — **Prgr. 2:** 24h on 98.7MHz (instrumental).

BBC ATLANTIC RELAY STATION
✉ English Bay, Ascension Island, So. Atlantic.
Local Sce: MW: 1485kHz 0.5kW — **FM:** 93.2MHz 15W (24h relay of BBCWS in English plus occ. local prgrs).
SW (G.C. 14.23W/7.54S): 6 x 250kW tr's.
See International Broadcasting section for further details.

VOICE OF AMERICA RELAY STATION
See International Broadcasting section.

BENIN

L.T: UTC + 1h — **Pop:** 5.450.000 — **Radios:** 400.000 — **Pr.L:** French + 18 ethnic — **E.C:** 50Hz, 220V — **ITU:** BEN.

OFFICE DE RADIODIFFUSION ET TÉLÉVISION DU BENIN (Gov.)
✉ B. P. 366, Cotonou. ☎ +229 301096, 300481. ☺ ORTB 5132 CTNOU - 5208 A.B.P — **L.P:** Dir: Emile Désiré Ologoudou. Chief Tech. Sces: Anastase Adjoko.
Regional ✉ B.P. 128, Parakou. ☎ +229 610773 or 610774.
MW: Cotonou 1475.1kHz 50/20kW, Parakou 963kHz.

SW	kHz	kW	Times
Cotonou	4870v	30	MF 0500-0845, 1200-1400**, 1600-2300 SS 0600-2300 (**Wed 2300)
Parakou	5025v	10	MF 0500-0800, 1100-1400, 1700-2200 SS 0700-2300
Parakou+	7190	10	MF 0500-0845, 1200-1400**, 1600-2300 SS 0600-2300 (**Wed 2300)
Cotonou++	7210.2	30	MF 0500-0845, 1200-1400**, 1600-2300 SS 0600-2300 (**Wed 2300)

+) alt. to 5025v, ++) alt. to 4870v
Cotonou: FM: Cotonou 88.4/94.7/98.2MHz 0.05kW.
N. in French: 0615MF, 0800SS, 1200SS, 1215MF, 1930, 2115.
Atlantique FM: Mon/Tues/Thurs/Fri 0900-1055 & 1430-1555, Wed 1000-1155 on 98.2MHz.
Parakou: regional prgrs in French + ethnic languages.
ANN: "Ici R. Bénin, Office de Radiodiffusion et Télévision du Bénin, émettant de Cotonou". "Ici Parakou, Office de Radiodiffusion et Télévision du Bénin, station regionale" — **IS:** Bénin Tam-Tam.
V. by QSL-card. Rp. Rec. acc. (Parakou by own QSL for direct rx.)

R. Rurale (community st's set up by the development agency ACTT): st's are operating in Tanguiéta, Ouesse, Lalo, Banikoara & Ouaké.

Privately owned stations: A few st's have been licensed, with at least 19 applications pending. No further information received.

R. France Internationale: Cotonou 90.0MHz.

BOTSWANA

L.T: UTC + 2h — **Pop:** 1.493.000 — **Radios:** 600.000 — **Pr.L:** English, Setswana — **E.C:** 50Hz, 220V — **ITU:** BOT.

RADIO BOTSWANA (Gov.)
✉ Private Bag 0060, Gaborone. ☎ +267 352541. 🖶 +267 371588.
E-mail: Rbeng@info.bw
L.P: Dir. of Inf. & Broadc: Ted Makgekenene. CE: Habuji Sosome. Principal Broadc. Officers: Moreri Gabakgore (Culture & Entertainment), Oshinka Tsiang (N. & Current Affairs). Principal Broadc. Engineers: Kingsley Reetsang (studios), Daniel Manyake (transmitters). Educ. Broadc. Officer: Mrs Queen Pilane. Head of Training: Antony Masete.
G.C: Sebele: 25.58E/24.34S

Location	kHz	H. of tr.	Location	kHz	H. of tr.
Maun	531	0430-2400	Sebele	3356	0430-0800
Selebi-Phikwe	621	0430-2400	Sebele	3356	1730-2400
Mopipi	648	0430-2400	Sebele	4830	0430-2400
Gantsi	873	F.PI.	Sebele	7255	0430-2000
Sebele	972	0430-2400	Sebele	9595	0800-1730
Jwaneng	1071	0430-2400			
Mahalapye	1215	0430-2400			

FM: Gaborone 89.9 & 103.0MHz 5kW + 13 additional st's.

National Sce. on SW/MW/FM stereo: 0430-2400. **N. in English:** 0710,1310,1810,2110.
Commercial Sce: on FM (currently only covers Gaborone and surrounding areas).
ANN: E: "This is R. Botswana broadcasting from Gaborone" or "You're listening to R. Botswana, the station at the heart of the nation" or "This is the General Service of R.Botswana". Setswana: "Ke Seromamowa sa Botswana mo Gaborone".
IS: Cow Bells. Imitation of farm animals at s/on. National Anthem at opening and closing — **V.** by letter.

VOICE OF AMERICA RELAY STATION
See International Broadcasting section for details.

BURKINA FASO

L.T: UTC — **Pop:** 11.320.000 — **Radios:** 512.500 — **Pr.L:** French + 16 ethnic — **E.C:** 50Hz, 220V — **ITU:** BFA.

RADIODIFFUSION NATIONALE DU BURKINA (Gov.)
✉ 03 B.P. 7029, Ouagadougou 03. ☎ +226 324055, 324697, 3243999. 🖶 +226 310441. ☺ (Min. of Inf. & Culture) 5132UV.
L.P: Dir: Lézin Didier Zongo. Head of Tr. Centre: Marcel Teho.
STATIONS: Ouagadougou (G.C: 01.31W/12.22N):

kHz	kW	Times
747	100	0530-0900, 0900-1200SS, 1200-2400
1341	1	standby tr.
4815	50	0530-0800, 1700-2400
7230	50	0800-0900, 0900-1200SS, 1200-1700
9515	50	alt. to 7230

*) Standby tr — **FM:** 92.0MHz 0.02kW.
N. in French: 0630MF, 1000SS + Thurs, 1245 (regional), 1300, 1900, 2200. **N. in English:** 1920 (approx).
ANN: "R. Burkina" — **IS:** Balafon — **V.** by letter. Rp.

REGIONAL STATIONS
Radio Bobo, B.P. 392, Bobo-Dioulasso. ☎ +226 991158 — **L.P:** SM: Alphousseini Bassolet — **MW:** 1008kHz 10kW — **FM:** 92.0MHz 0.02kW — MF 0600-0800, 1200-1400, 1600-2400. SS 0800-2400.
Radio Gaoua, Gaoua. ☎ +226 870348, 870198 — **L.P:** SM: François Xavier Sanon.

RADIO CANAL ARC-EN-CIEL (Comm.)
✉ Ouagadougou. ☎ +226 324545, 324141.
FM: 96.6MHz — **L.P:** PD: Gnama Pako Drabo.

RADIO HORIZON FM STEREO – FRÉQUENCE MAGIQUE (Comm.)
✉ P.O. Box 2710, Ouagadougou. ☎ +226 312858, 308547. 🖶 +226 312858 — Bobo-Dioulasso. ☎ +226 972727, 970666.
L.P: DG: Moustapha Laabli Thiombano — **FM:** Tenkodogo 97.6MHz, Banfora 98.0MHz, Koudougou 98.7MHz, Ouayigouya 100.4MHz, Bobo-Dioulasso 102.7MHz Ouagadougou 104.4MHz.
D.PRGR: 24h in French, English and ethnic languages.

RADIO PULSAR 94.8 FM (Priv./Comm.)
✉ P.O. Box 5976 Ouagadougou 01. ☎🖶 +226 314199.
L.P: GM: Ouedraogo Ousmane.
FM: Ouagadougou 94.8MHz 0.4kW. **F.PI:** tr. in Bobo Diolassou.
D.PRGR. in French/English/ethnic: 0500-2400.

GOSPEL DEVELOPMENT RADIO (Rlg.)
✉ Emanagoungue Bldg, Ouagadougou.
FM: 93.4MHz 0.1kW + 2 relay st's.

Other Private Stations
La Voix du Paysan, Ouahigouya 97.0MHz
R. Energie: Kaya 92.9MHz, Yako 94.9MHz, Fada N'Gourma 98.8MHz.

R. Rurale (community st's set up by the development agency ACTT): st's are operating in Diapaga, Djibasso, Gassan, Kongoussi, Orodara & Poura.

R. France Int.: Ouagadougou 94.0MHz, Bobo-Dioulasso 99.40MHz. Aldo partly rel. via R. Energie & La Voix du Paysan (see above).
Africa No. 1: Ouagadougou 90.3MHz.

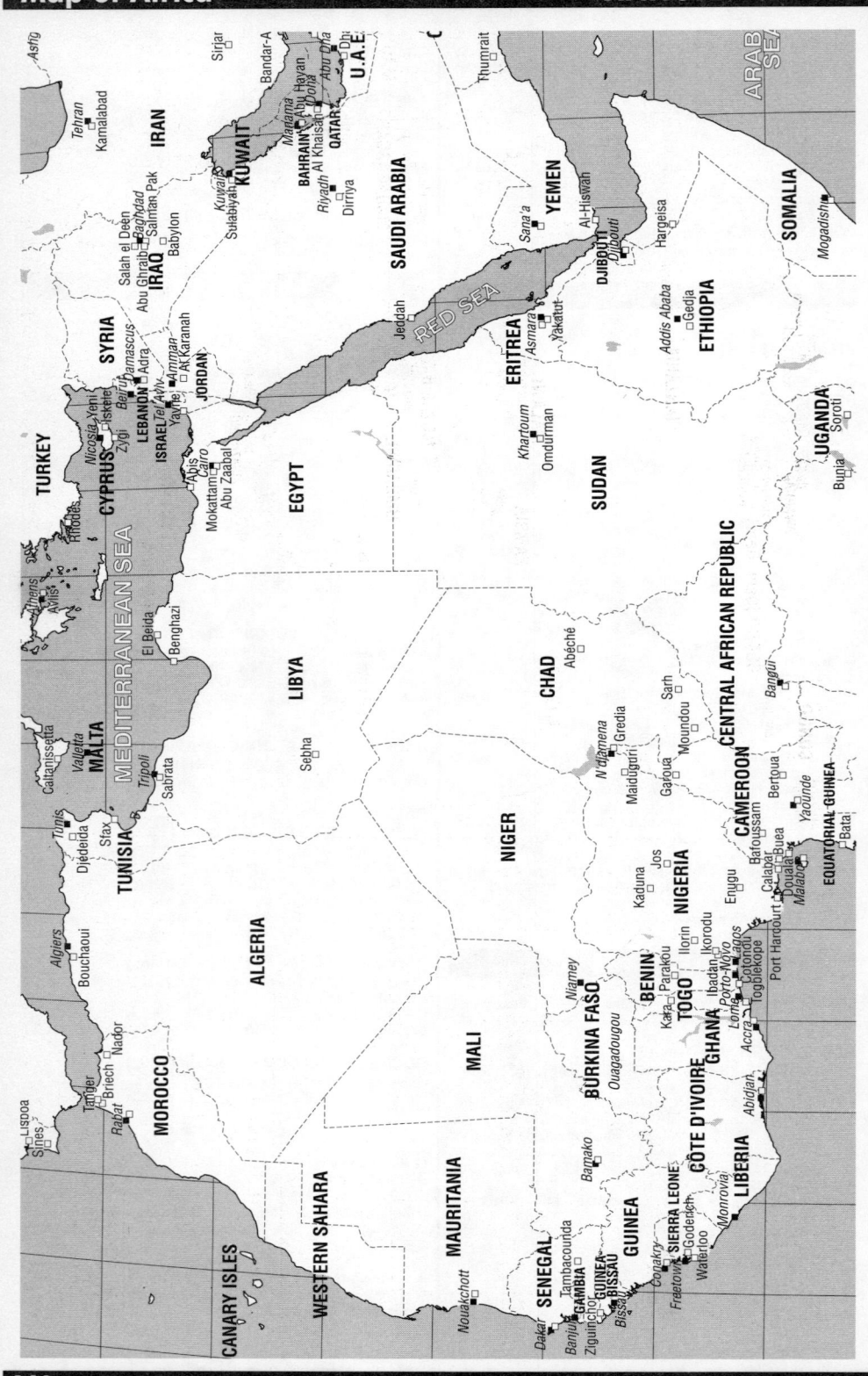

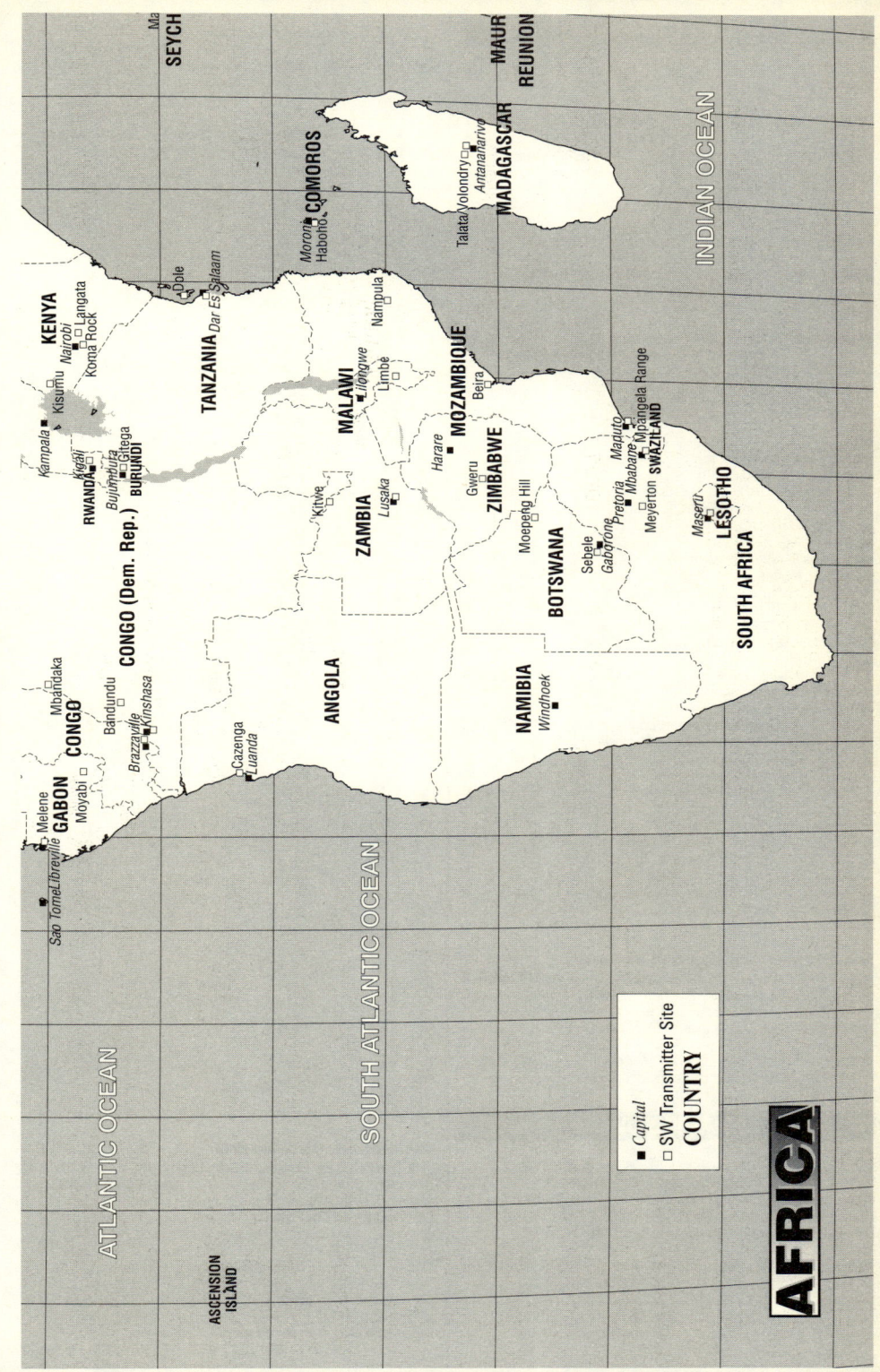

AFRICA

- ■ Capital
- □ SW Transmitter Site
- **COUNTRY**

Map labels:

SEYCH

MAUR

REUNION

MADAGASCAR

INDIAN OCEAN

COMOROS — Moroni, Haboho

Talata/Volondry, Antananarivo

KENYA — Nairobi, Langata, Koma Rock, Kisumu

Obie

TANZANIA — Dar Es Salaam

Nampula

MALAWI — Lilongwe, Limbe

MOZAMBIQUE — Beira

Kampala

RWANDA — Kigali

BURUNDI — Bujumbura, Gitega

Kitwe

ZAMBIA — Lusaka

Harare

Gweru

ZIMBABWE — Moepeng Hill

Maputo — Mpangela Range

Mbabane, SWAZILAND

Pretoria, Meyerton

Sebele, Gaborone

BOTSWANA

Maseru, LESOTHO

SOUTH AFRICA

CONGO (Dem. Rep.)

Mbandaka

Bandundu, Kinshasa

Brazzaville, Moyabi

CONGO

GABON — Libreville, Melene

Sao Tome

ANGOLA — Cazenga, Luanda

NAMIBIA — Windhoek

ATLANTIC OCEAN

SOUTH ATLANTIC OCEAN

ASCENSION ISLAND

CAMEROON

L.T: UTC + 1h — **Pop:** 14.740.000 — **Radios:** 6m — **Pr.L:** French, English, ethnic — **E.C:** 50Hz, 220V — **ITU:** CME.

CAMEROON RADIO TELEVISION (C.R.T.V.) (Gov.)

B.P. 1634, Yaoundé. ☎ +237 214088. ▤ +237 204340. **Web:** http://www.ditof.cm/crtv. **L.P:** GM: Prof. Gervais Mendo Ze. Dir. of Admin & Finance: Therese Owoundi . Dir. of Inf: Antoine Marie Ngono. Dir. of Prgrs: Daouda Moutchangou. Asst. Dir. of Radio Prgrs: Peter Essoka. Tech. Dir: Eyebe Tanga. Asst. Tech. Dir. (Radio): Hilaire Thima. Editor-in-Chief Radio N: Michel Ndjock Abanda. Head of CRTV Training Centre: Marie Rose Nzie. Head of Comm. Div: Madeleine Bonou.

Mediumwaves	kHz	kW	H. of tr.
2)Douala	1106	5	0430-2300
7)Bamenda	1152	300/1	0430-2300

Shortwaves	kHz	kW	H. of tr.
3)Buea	3970v	4	0430-0730, 1630-2300
1)Yaounde	4850	100	0430-0730, 1630-2400
4)Garoua	5010v	100/4	0430-0730, 1630-2300 (irr.)
3)Buea	6005	4	0730-1630
1)Yaounde	6060	100	0730-1630
4)Garoua	7240	100	0730-1630

FM: 1)Yaounde 88.8/94.0/101.9MHz 10/10/1kW — 2)Doula 89.2/91.3/94.5/104.9MHz 10/10/3/10kW — 3)Buea 94.5/98.6MHz — 4) Garoua 97.7/101.2MHz 10/10kW — 5)Bertoua 89.8/92.9MHz 10/10kW — 6)Bafoussam 91.1/93.5MHz 10/10kW — 7)Bamenda 89.4/ 92.5MHz 10/10kW — 8)Ebolowa 96.0MHz 10kW — Ngaoundere 102.5MHz 10kW.

Addresses & other information

1) as above. **National Sce.** on 899/4850kHz + 88.8MHz. **N. in French:** 0530MF, 1200MF, 1900. **N. in English:** 0500MF, 0600Sun, 1400, 1800, 2100. **ANN:** E: "This is Yaoundé, the national station of CRTV". F: "Ici Yaoundé, poste nationale de la CRTV" — **Provincial Sce.** on 101.9MHz — **Urban Sce. in French** on 94.0MHz (rel. National Sce. 0600-1700).

2) B.P. 986, Douala — **ANN:** E: "This is Douala, the Littoral provincial st. of CRTV. "Ici Douala, CRTV, station provinciale de la Cameroon Radio Television du Sud-Ouest".

3) P.M.B, Buea. **Local N. in French/English:** 0610, 0830, 1615, 1850, 2130 — **ANN:** F: "Ici Buea, station provinciale de la Cameroon Radio-Television, du Sud-Ouest". E: "This is Buea, the So.We. provincial st. of the Cameroon Radio-Television".

4) B.P. 103, Garoua. **Local N: English:** 1330. **French:** 0600 — **ANN:** E: "This is Garoua, provincial st. of CRTV". F: "Ici Garoua, station provinciale de la C.R.T.V. du nord".

5) P.O. Box 230, Bertoua, Eastern Province. Provincial Sce: 0430-0910 & 1100-2300 on 92.9MHz. **Local N. in English:** 1715. Rel National Sce: 0430-2400 on 89.8MHz — **ANN:** E: "This is Bertoua, the Eastern provincial st. of CRTV, Cameroon Radio & Television Corporation". F: "Ici Bertoua, station provinciale de la CRTV".

6) B.P. 970, Bafoussam, Western Province — **ANN:** E: "This is Bafoussam, the Western provincial st. of R. Cameroon". F: "Ici Bafoussam, station provinciale de R.Cameroon".

7) B.P. 4049, Bamenda, North-Western Province. **Local N: English:** 0515. 1705, 2130. **French:** 0720, 1840 — **ANN:** E: "This is Bamenda, the North Western provincial st. of R. Cameroon". F: "Ici Bamenda, station provinciale de la Cameroun Radio Télévision, CRTV".

8) Ebolowa.

V: Yaoundé & Buea reply direct. Re. for all other st's to James Achanyi Fontem, B.P. 1460, Douala, Literal Province. Rp.

CANARY ISLANDS

L.T: UTC (Su: UTC + 1h) — **Pop:** 1.700.000 — **Radios:** 600.000 (est.) — **Pr.L:** Spanish — **E.C:** 50Hz, 110/220V — **ITU:** CNR. **Abbreviations:** GC=Gran Canaria, GCF=Fuerteventura, GCL=Lanzarote, TF=Tenerife, TFP=Isla de la Palma, TFG=Isla de la Gomera. (For network abbreviations please refer to Spain).

MW	kHz	kW	Net	Location	Island
1)	621	100	RNE-1	Santa Cruz	TF
1)	720	20	RNE-5	Santa Cruz	TF
1)	747	20	RNE-5	Las Palmas	GC
4)	837	10	COPE	Las Palmas	GC
5)	882	20	COPE	La Laguna	TF
6)	1008	10	SER	Las Palmas	GC
7)	1098	2	RNE-5	Santa Cruz	TFP

MW	kHz	kW	Net	Location	Island
9)	1179	20	SER	Santa Cruz	TF
8)	1269	20	ECCA	Las Palmas	GC

FM:	MHz	kW	Net	Location	Island
33)	87.8		DIAL	La Laguna	TF
7)	89.6	2	RNE-5	Santa Cruz	TFP
28)	89.7	4	A3	Arrecife	GCL
1)	90.0	60	RNE-3	Izaña	TF
8)	90.4	2	ECCA	Las Palmas	GC
12)	90.7	4	SER	Arrecife	GCL
13)	91.0	8	A3	Santa Cruz	TF
36)	91.2		OCR X	Fuenteventura	GC
33)	91.6	2	D	Los Realejos	TF
4)	91.8	1	C100	Las Palmas	GC
8)	91.8	1	ECCA	Santa Cruz	TF
1)	92.3	60	RNE-1	Izaña	TF
1)	92.8	50	RNE-1	La Isleta	GC
8)	93.0	1	ECCA	Pto del Rosario	GCF
9)	93.2	25	SER	La Laguna	TF
9)	93.3	2.5	SER	La Gomera	TCF
1)	93.9	10	RNE-5	Finca España	TF
37)	94.0	8	OCR C	Santa Cruz	TF
14)	94.4	8	R.80	Las Palmas	GC
32)	94.4	2	OCR	Tacoronte	TF
5)	95.1	50	RNE-2	La Isleta	GC
5)	95.1	2	COPE	Santa Cruz	TFP
15)	95.3	4	SER	Maspalomas	GC
6)	95.8	8	TOP	Las Palmas	GC
1)	96.2	60	RNE-2	Izaña	TF
16)	96.8	5	M	Arrecife	GCL
5)	97.1	2	COPE	La Laguna	TF
17)	97.5		L	Guía	GC
1)	97.7		RNE-5	La Isleta	GC
5)	98.4	5	RNE-5	El Paso	TFP
1)	98.5	50	RNE-3	La Isleta	GC
35)	98.6	2	A3	Llanos de Aridane	TFP
18)	99.1	2	RM	Pto. la Cruz	TF
8)	99.5	2	ECCA	Santa Cruz	TFP
38)	99.9	8	OCR C	Las Palmas	GC
19)	99.8	2	SER	La Orotava	TF
31)	100.0	1	OCR	Adeje	TF
20)	100.0	5	IND	Arrecife	GCL
13)	100.1	8	R.80	Santa Cruz	TF
11)	100.3	2	SER	Telde	GC
9)	101.1		SER	Santa Cruz	TF
22)	101.4	2	RM	Las Palmas	GC
40)	101.6		SER	Santa Cruz	TFP
1)	102.1	10	RNE-2	Las Mesas	TF
23)	102.2		IND	Pto. del Rosario	GCF
10)	102.4	1	A3	Playa del Inglés	GC
1)	102.7	5	RNE-1	El Paso	TFP
1)	103.0	3.7	SER	Las Palmas	GC
25)	104.0	4	OCR X	Arrecife	GCL
26)	104.1	2	DIAL	Llanos de Aridane	TF
1)	104.5	5	RNE-2	El Paso	TFP
29)	105.0	1	RL	Pto. del Rosario	GC
1)	105.7	10	RNE-3	Las Mesas	TF
27)	106.0		L	Las Palmas	GC
1)	106.1	1	RNE-3	El Paso	TFP
30)	106.8	8	OCR	Las Palmas	GL
39)	107.3		R. Isora		
34)	107.5		COPE	Pto. del Rosario	GC

Addresses and other information

1) **R. Nacional de España,** Aptdo. 1233, Santa Cruz de Tenerife 38001. ☎ +34 (922) 288162. **R.1:** 24h on 621kHz. **N:** On the h. **R.2:** (classical music) 24h. **R.3:** 24h. **R.5:** 24h.

4) **R. Popular de Las Palmas,** Av. Escaleritas 60, Las Palmas 35011. ☎ +34 (928) 255840. Dir: José Antonio Baeza Betancor. **D.Prgr:** 0500-0100 (occ. 24h). **N.** summaries in English, German & Swedish between 1330 & 1830, produced by **Canary Tourist R,** Paseo de Chil 117, 35014 Las Palmas. Dir: Xavier Palin. **FM:** 91.8 R. Insular.

5) **R. Popular de Tenerife,** Obispo Rey Redondo 49, La Laguna 38201. ☎ +34 (922) 261211. Dir: José Silverio Pérez. **D.Prgr:** 0700-2400.

6) **R. Las Palmas,** C/ Profesor Lozano nº5, Urb. El Cebadal, Las Palmas de Gran Canaria. ☎ +34 (922) 462052/53. **E-mail:**

topradio@canari.step.es **Web:** http://www.step.es/top/ — Dir: Andrés Roda Díaz. **D.Prgr:** 0815-0100. **FM:** R. Gran Canaria.

7) **R. Nacional de España,** Plaza de la Constitución 4, Santa Cruz de la Palma 38700. ☎ +34 (922) 288400. Dir: Julio Marante Díaz. **D.Prgr:** 0800-2300.

8) **R. ECCA,** Av. Mesa y López 38 (or: Aptdo. 994), Las Palmas 35007. ☎ +34 (928) 275454. Dir: Luis Espina Cepeda. **D.Prgr:** 0655-2300 (Sun 2100). **FM 91.8:** Lope de Vega 5, 38700 Sta, Cruzde Tenerife. **FM 93.0:** Plaza de España, Pto. del Rosario, Fuerteventura. **FM 99.5:** Av. Francisco de Abreu 6, Santa Cruz de la Palma 38700.

9) **R. Club Tenerife,** Av. de Anaga 35, Santa Cruz de Tenerife 38001. ☎ +34 (922) 270404. Dir: Francisco Padrón García. **D.Prgr:** 24h.

10) **Antena 3, R. Canarias Sur,** Av. Tirajana 1, Playa del Inglés 35290. ☎ +34 (928) 764884. **D. Prgr.** in Spanish 24h. exc: W. 0910-1000, 1700-1800, 1910-2100 **German,** 1000-1100 **Dutch,** 1800-1900 **Swedish**.

11) **SER Las Palmas,** C/ General Blanes, s/n, Edificio Maphre 4ª plta, Las Palmas de Gran Canaria 35008. ☎ +34 (928) 463007.

12) **R. Lanzarote,** Fred Olsen 14, 2Þ, Apdo. 234, 35500 Arrecife. ☎ +34 (928) 811517. Dir: Agustín Cruz. **D. Prgr.** in Spanish 0700-2400 exc: Mon-Fri 1530-1630 **German,** 1830-2000 **English**.

13) **Antena 3/R. 80 Serie Oro Tenerife,** R.J. Hamilton 14, Santa Cruz de Tenerife 38001. ☎ +34 (922) 240227. Dir: Gabriel Mesa Barrera. **D. Prgr:** 24h.

14) **R. 80 Serie Oro,** José Franchy Roca 5, Las Palmas 35007. ☎ +34 (928) 279911. Dir: Gabriel Mesa Barrera. **D.Prgr:** 24h. **DX Prgrs** produced by José Luis Martín: Sat 1300-1330 "El Mundo de las Ondas", Mon 2030-2100 "Radioaficionados", Mon 1205-1220 "Historia de la Radio".

15) **R. Maspalomas,** Morro Besudo s/n, Maspalomas 35100. ☎ +34 (922) 765853. Dir: Juan José Roda. **D.Prgr.** in Spanish 24h. exc: W. 0830-0900, 1705-1800 **German,** 0905-0930 **English,** 0930-1000 **Swedish**.

16) **R. Onda Insular,** García Escámez 159, Arrecife 35580. ☎ +34 (928) 815750. Dir: Armando de León Expósito. **D.Prgr:** 0700-2400.

17) **R. Guía,** 35450 Santa María de Guía.

18) **R. Minuto,** Pl. de los Reyes Católicos 3, Pto. la Cruz 38400. ☎ +34 (922) 380012. Dir: Ignacio Baute España.

19) **R. Club Sur,** Av. Marítima s/n, Playa de Américas 38660. **D.Prgr:** 0800-0100 (Fri 0300, Sat 0200).

20) **R. Tourist Lanzarote,** Av. Fred Olsen 14 (or: Aptdo. 260), Arrecife 35580. Prgr. Dir: Barbara Graf. **D.Prgr:** 0655-2300. **N.** in English on the h., in German on the halfhour. **German** 0740-0930, 1530-1630, **Scandinavian** 1630-1730, **English** 0930-1100, 1800-2000.

22) **R. Minuto,** Paseo de Chil 1013, Las Palmas 35010. ☎ +34 (928) 268900.

23) **R. Fuerteventua,** Cervantes 122, 35600 Pto. del Rosario.

25) **R. 106,** Aptdo. 28, Arrecife 35580 (Lanzarote).

26) **Dial Llanos de Aridane,** Av. Anaga 35, 38001 Sta. Cruz de Tenerife.

27) **Guiniguada R.,** Ap. 67, 35080 Las Palmas, Las Palmas de Gran Canaria.

28) **R. Volcán,** Luis Morote 20, 35500 Arrecife (Lanzarote).

29) **R. Archipielago,** La Juventud 31, 35600 Fuerteventura.

30) **Onda Cero Color,** León y Castillo 41, 35005 Las Palmas.

31) **Onda Cero R.,** C.C. San Eugenio Adeje, 38660 Playa Las Américas.

32) **Onda Cero, R. 21,** Crtra/ Gral. Norte 29, 38350 Tacoronte.

33) **Dial Norte,** Av.Av. Anaga 35, 38001 Sta. Cruz de Tenerife. ☎ +34 (922) 270408.

34) **R. ECCA Pto. del Rosario,** Dr. Fleming s/n, 35600 Pto. del Rosario.

35) **Dial Tenerife,** Av. Anaga 35, 38001 Sta. Cruz de Tenerife.

36) **Canal 28 Fuenteventura,** Calle Segura 29, 3Þ, 35600 Pto. del Rosario. ☎ +34 (928) 530255.

37) Salamanca 5, 38006 Sta. Cruz de Tenerife.

38) Canal 28 Las Palmas, Dr. Juan Dominguez Pérez 12, 1Þ. 35008 Las Palmas.

39) Apto de correo 76, Guia de Isora, 0 38680. ☎ +34 (228) 50302. **Web:** http://tool-box.com/rc/owa/freeweb.page?id=1038768

40) **R. La Palma,** Santa Cruz de la Palma

CAPE VERDE (Republic of)

L.T: UTC - 1h — **Pop:** 476.133 — **Radios:** 57.000 — **Pr.L:** Portuguese, Crioulo — **E.C:** 50Hz, 220V — **ITU:** CPV.

RÁDIO NACIONAL DE CABO VERDE (Gov.)

✉ Praca Alexandre Albuquerque, Praia, Santiago. ☎ +238 615755/6/7. 📠 +238 615754.

C.P. 29, Mindelo, S. Vicente. ☎ +238 311513, 311731. 📠 +238 311006.
C.P. 40, Espargos, Ilha do Sal. ☎ +238 411333. 📠 +238 411444.

L.P: Dir: Manuela Fonseca. PD: Giordano Custodio. Dir. Inf: Salome Monteiro. Dir. Tec: Francisco Lopes Monteiro.
FM: Santiago: M. Tchota 91.6MHz 0.5kW, Praia 98.1MHz 0.1kW
S. Vicente: Mindelo 95.6MHz 0.1kW, M. Verde 97.6MHz 0.1kW.
Sal: M. Curral 89.7MHz 0.25kW (+ 9 low power relays).
D.PRGR: 24h.

RÁDIO NOVA — EMISSORA CRISTÃ/CABO VERDE

✉ C.P. 426, Mindelo, S. Vicente. ☎ +238 317819 (office), 311480 (studios). 📠 +238 311475 — **L.P:** Dir: Antonio Fidalgo Barros.
FM: Santiago 101.6MHz 0.25kWm S. Vicente 104.3MHz 0.5kW + 3 relay st's — **D.PRGR:** 14h daily.

R. France Internationale: Santiago 99.3MHz 1kW, S. Vicente 100.7MHz 0.25kW.
RDP (Portugal): Santiago 105.2MHz 1kW, S. Vicente 93.9MHz 1kW + 2 low power relays.

CENTRAL AFRICAN REPUBLIC

L.T: UTC + 1h — **Pop:** 3.250.000 — **Radios:** 180.000 — **Pr.L:** French, Sango, ethnic — **E.C:** 50Hz, 220V — **ITU:** CAF.

RADIODIFFUSION-TÉLÉVISION CENTRAFRICAINE (Gov.)

✉ B.P. 940, Bangui. ☎ +236 (61) 2588 or 1650.
L.P: MD: Paul Service. PD: Pascal Songolema. TD: Michael Bata.
MW: 1440kHz 20/50kW — **FM:** 87.78/90.2MHz.
SW (G.C: 18.35E/04.21N) 100kW:
kHz: 5033.5v (0430-0700 & 1630-2100), 7220 (0700-1630).
D.PRGR: 0430-2300.
N. in French: 0500, 0600, 0700, 0800, 1300, 1800.
ANN: F: "Ici Bangui, Station Nationale de la Radiodiffusion- Télévision Centrafricaine" — **IS:** Repeated piano chord. Opens and closes with National Anthem — **V.** by QSL-card.

R. Rurale (community st's set up by the development agency ACTT): st's are operating in Bouar, Nola, Berberati & Bambari.

Africa No. 1: Bangui 94.5MHz (see main entry under Gabon).
R. France Internationale: Bangui 99.8MHz.

CHAD

L.T: UTC + 1h — **Pop:** 5.450.000 — **Radios:** 1.310.000 — **Pr.L:** French, Arabic, 8 ethnic — **E.C:** 50Hz, 220V — **ITU:** TCD.

RADIODIFFUSION NATIONALE TCHADIENNE (Gov.)

✉ B.P. 892, N'Djamena. ☎ +235 516071. **Cable:** Radio Tchad.
L.P: Dir. RNT: Khamis Togoï. S/Dir. of Tech. Sces: Raphael Mbaissané.
R. Rurale (prgrs for rural listeners): S/Dir: Amadou Batouré. ☎ +235 515238.
MW: 840kHz 20kW — **FM:** 94.05MHz 0.1kW.
SW (G.C: 15.03E/12.08N): 100kW:

kHz	Times
4904.5	0425-0730, 1600-s/off)
6165	0430-0600, 1600-s/off
7120	MF 0600-0730 & 1025-1600, SS 0600-1600

D.PRGR: MF 0425-0730 & 1025-2200, Sat 0425-2300, Sun 0425-2200.
N. in French: 0500, 1300, 1330, 1830, 1900, 2100.
ANN: "Ici N'Djamena, Radiodiffusion Nationale Tchadienne".
IS: Balafon — **V.** by QSL-card.

REGIONAL STATIONS

R. Moundou, B.P. 122, Moundou. ☎ +235 691322 — **L.P:** Dir: Dimanangar Djainta — **SW:** 5286.2kHz 5kW (irr.). **FM:** 94.05MHz 450/200W. **D.PRGR:** 0500-0800, 1400-1830 — **ANN:** "Ici Moundou, station régionale de la Radiodiffusion Nationale Tchadienne, émettant de Moundou" — **V.** by letter.
R. Sarh, B.P. 270, Sarh. ☎ +235 681361, 681422 — **L.P:** Dir: Biana Fouda Nactouandi — **MW:** 850kHz 1kW — **D.PRGR:** 1500-1800.
R. Abéché, B.P. 105, Abéché. ☎ +235 698149 — **L.P:** Dir: Sanoussi Saïd — **SW:** 5900kHz 1kW — **D.PRGR:** 0500-0600, 1630-1800. **ANN:** "Ici Radio Abéché, Station Régionale de la Radiodiffusion Nationale Tchadienne émettant d'Abéché".

Africa No. 1: N'Djamena 103.0MHz (see main entry under Gabon).
R. France Internationale: N'Djamena 100.2MHz.

COMOROS (Federal Islamic Rep.)

L.T: UTC + 3h — **Pr.L:** French, Comorian, Arabic — **Pop:** 630.000 — **Radios:** 61.000 — **E.C:** 50Hz, 220V — **ITU:** COM.

RADIO COMORO (Gov.)

BP 452, Moroni, Grand Comoro. ☎ +269 (73) 2531. ⓔ +269 (73) 0303. **LP:** Tech. Dir: Abdullar Radjab. **MW:** Haboho 1089kHz 40kW. **SW:** 3331kHz 60kW. **(F.PI:** SW-tr. to be reactivated.) **FM:** Moroni, R.Studio 1 101.2MHz. Nkazi, R.Nkazi 107.0MHz. 6 x 1kW, 3 x 0.5kW tr's. **D.PRGR:** 0300-1900. **N. in French:** 1000, 1700. **ANN:** F: "Ici Radio Comoro" — **V.** by letter. Rp. Rec.acc.

R. Dziyalandze: Anjouan 90.0MHz (rel. RFI 1700-1030).

R. France Internationale: 103.0MHz.

CONGO (Dem. Rep. of)

L.T: Kinshasha/Mbandaka UTC + 1h, Lubumbashi/Kisangani/ Bukavu/Mbuji-Mayi/Kananga: UTC + 2h — **Pop:** 48.399.104 — **Radios:** 3.480.000 — **Pr.L:** French, ethnic — **E.C:** 50Hz, 220V — **ITU:** ZAI.

RTNC - RADIO-TÉLÉVISION NATIONALE CONGOLAISE (Gov.)

Station Nationale, B.P. 3164 or B.P. 3171, Kinshasa-Gombe. ☎ +243 (12) 23171-5.

Mediumwaves	kHz	kW
7) Matadi	1160	50
5) Mbandaka	1268	2
1) Kinshasa	1448	2
Shortwaves	**kHz**	**kW**
1) Kinshasa	15245	100

FM: 1) 100MHz 10kW, 92.0MHz 3kW — 3) 88.04MHz 1.5kW — 4) 93.3MHz 1.5kW/90MHz 0.05kW — 6) 89MHz 50kW/92MHz 1kW. **Addresses and other information** 1) B.P. 3171, Kinshasa-Gombe. 24h in French/Swahili/Lingala/ Tshiluba/Kikongo. **N. in French:** 0100, 0300, 0500, 0600, 0700, 0900, 1000, 1130, 1300, 1400, 1500, 1800, 2000, 2200, 2300 — also rel. by other sts. **ANN:** "RTNC, Radio-Télévision Nationale Congolaise, emet- tant de Kinshasa" **V.** by letter — 2) B.P. 7296, Lubumbashi. W 0300- 0700, 1000-1245, 1500-1830v; Sun 0300-1830v. **V.** by letter — 3) B.P. 475, Bukavu. W 0300-0700, 1000-1245, 1500-1830; Sun 0300-1830. **Regional N. in French:** 0430, 0630, 1030, 1630. **V.** by letter — 4) B.P. 1061, Mbandaka. W 0500-0800, 1000-1200, 1600-2100; Sat 0500- 0800, 1000-2200; Sun 0500-2200. **N.** 0430 1030 1630. **V.** by letter. Rec. acc. — 5) B.P. 1232 Mbuji-Mayi — 6) B.P. 708, Kananga, Kasai Occidental. — 7) B.P. 704, Matadi — 8) B.P. 1745, Kisangani. 0400-

LA VOIX DU PEUPLE

B.P. 373, Bunia. **kHz:** *3390.2, 5066.4, 7150* **FM:** 91.4/95/98MHz 0.02kW. **D.PRGR:** 0400-0800, 1000-1700

R. Raga (priv.): Kinshasa 90.5MHz. **R. Moto (Community st.):** B.P. 9, Kivu. Dir: Fr. Joseph Delvordre — **FM:** 103MHz 1.2kW. **R. France Internationale:** Kinshasa 93.2MHz.

CONGO (Republic of the)

L.T: UTC + 1h — **Pop:** 2.525.000 — **Radios:** 240.000 — **Pr.L:** French, Lingala, Kikongo — **E.C:** 50Hz, 220V — **ITU:** COG.

RADIODIFFUSION-TÉLÉVISION CONGOLAISE (Gov.)

B.P. 2241, Brazzaville. ☎ +242 812473. **Cable:** Radio-Congo. **Regional** B.P. 1063, Point Noire — **LP:** Pres: Jean Gilbert Foutou. DG: Gilbert-David Mutakala. PD: Jean Pierre Nkouka. **MW:** Brazzaville 1476kHz 20kW — **FM:** Pt. Noire 98.95/96.4MHz. **SW:** Brazzaville (G.C: 15.18E/04.15S) 100kW.

kHz	H. of tr.	kHz	H. of tr.
4765	0400-1100, 1700-2130	6014.7	1700-2130
5985	0400-0700, 1700-2130	9610	1100-1700

National Network: 0400-2130. **Main N. in French:** 0500, 0700, 0800, 1200, 1330, 1500, 1830, 2100. **Regional Network** (Pt. Noire): 0445-0700, 1030-1330, 1445-2000. **ANN:** "Radiodiffusion Nationale Congolaise", "R. Congo" or "RNC". **IS:** Zansi solo. Opens and closes with National Anthem. **V.** by letter. Rp. Rec.acc.

R. Rurale (community st's set up by the development agency ACTT): st's are operating in Sembé, Étoumbi, Nkayi & Mossendjo.

R. France Internationale: Brazzaville 93.2MHz. **Africa No. 1:** Brazzaville 90.3MHz, Pt. Noire 93.5MHz (see main entry under Gabon).

CÔTE D'IVOIRE

L.T: UTC — **Pop:** 14.555.000 — **Radios:** 1.600.000 — **Pr.L:** French, Diola, 12 ethnic — **E.C:** 50Hz, 220/380V — **ITU:** CTI.

RADIODIFFUSION TÉLÉVISION IVOIRIENNE (Gov.)

01 B.P. 8091, Abidjan 01. ☎ +225 331179. ⓔ +225 224974. **Cable:** Diraci 22635. ⓣ 0983 22777. **LP:** DG: Gnozié Outtara. TD: A. Amangoua. **MW:** Abidjan 1494kHz 20kW, Bondoukou 1240kHz 1kW. **FM:** Abidjan 88.9/92.0/95.5/96.0MHz + 9 relays. **SW:** Abidjan (G.C: 03.57W/05.21N):

kHz	kW	kHz	kW	kHz	kW
6015	500	7215	20	11920	500

Chaîne Une: 0459-2400 on 6015/7215kHz + 88.9MHz, 0459-0759 + 1829-2400 on 1494kHz + 88.9/92.0/95.5MHz. **Main N:** 0530, 0615, 0700, 0745, 1245, 1900, 2200. **Fréquence Deux:** 24h on 1240/11920kHz, 0759-1829 on 1494kHz + 96.0MHz, 0000-1829 on 92.0MHz. **Main N:** 1330. **Rel. Chaîne Une:** 0455-0800. **English:** 1833-1930. **N:** 1845. **ANN:** "R. Côte d'Ivoire, Chaîne Une" or "Fréquence Deux" — **IS:** s/on with clock chimes & National Anthem — **V.** by QSL-card. Rp.

R. Rurale (community st's set up by the development agency ACTT): st's are operating in San Pédro, Bin-Houyé, Tengréla & Bouna.

R. France Internationale: Abidjan 97.6MHz. **Société de Radiodiffusion de Côte d'Ivoire:** Abidjan 91.1MHz (rel. Africa No. 1, Gabon). **BBC World Sce. in French:** Abidjan 94.3MHz.

RADIO NOSTALGIE (Priv.)

1 Avenue Chardy, 01 B.P. 157 Abidjan 01. ☎ +225 211052. ⓔ +225 225926 **Web:** http://www.africaonline.co.ci/AfricaOnline/societes/nostalgie/ **LP:** Ahmed Bakayoko — **FM:** Abidjan 101.1MHz.

CANAL PLUS HORIZON (Priv.)

Tour Alpha 2000, 3rd Floor, 01 B.P. 1132 Abidjan 01. ☎ +225 219999. ⓔ +225 214590 — **LP:** GM: Jack Barbier de Crozes

Radio Espoir & Radio JAM: no details received.

DJIBOUTI

L.T: UTC + 3h — **Pop:** 575.000 — **Radios:** 35.000 — **Pr.L:** French, Somali, Afar, Arabic — **E.C:** 50Hz, 220V — **ITU:** DJI.

RADIODIFFUSION-TÉLÉVISION DE DJIBOUTI (RTD)

B.P. 97, Djibouti. ☎ +253 352294. ⓔ +253 356502. ⓣ 5863 DJ. **LP:** DG: Mohamed Djama Aden. Dir. Tec: Areh Houssein Ragueh. PD: Omar said Bileh. Dir. Inf: Mohamed Daher Hassan. **MW:** 1170/1539kHz 20kW — **FM:** 91.0MHz.

National Network in Afar/Arabic/Somali: 0300-0700 & 0900-2200 (Fri 0300-2200) on 1539kHz + FM — **International Network in French:** 0300-0430 & 0900-1730 (Fri 0300-1730) on 1170kHz. **V.** by QSL-card or letter. Rp. Rec.acc. **R. France Internationale:** 92.0MHz.

EGYPT (Arab Republic of)

L.T: UTC + 2h — **Pop:** 62m — **Radios:** 16.450.000 — **Pr.L:** Arabic — **E.C:** 50Hz, 220V — **ITU:** EGY.

EGYPTIAN RADIO & TV UNION (Gov.)
P.O. Box 11511, 1186 Cairo. ☎ +20 (2) 5787120, 5757155, 770355. **Cable:** CIBROTEV. ☉ 92152 (KARADUN)
Web: http://www.sis.gov.eg/vidaudio/html/audiofm.htm
L.P: Pres. ERTU: Amin Bassiouni. Chmn. Eng. Sector: Ibrahim A. Ibrahim. Chmn. Sound Prgr. Sector: Farouk Susha. Chmn. of TV Prgr. Sector: Suhair El Attriby.

MW	kHz	kW	P	Times
Cairo	558	40	8k	1600-2000
Barnis	603	100	1	0400-2200
Batrah	621	1000	6a	24h
El Kharga	702	10	2	0800-1500 (not Fri 0900-1100)
			4	2000-2200
			8h	0400-0800, 0900-1100Fri, 1500-2000
Asswan	702	10	2	0700 (Fri 1130)-1600
			4	0200-0500, 2000-2200
			8j	0500-0700 (Fri 1130), 1500-1900
Tanta	711	100	2	0400-2400
Qena	756	10	2	0700 (Fri 1130)-1600
			4	0200-0500, 2000-2200
			8j	0500-0700 (Fri 1130), 1500-1900
Abis	774	500	5	0400-2330
Batra	819	1000	1	24h
Santah	864	500	4	24h
Matruh	882	10	1	1100-0700 (Fri 24h)
Hurghada	918	10	2	0500-1600
			4	0200-0500, 2000-2200
			8k	1600-2000
Bawiti	918	10	1	24h
Salum	936	10	1	24h
Assiut	981	10	4	0200-0500, 2000-2200
			8a	0500-2000
Baris	981	1	1	0300-2200
Abu Simbil	981	1	1	24h
El Arish	1008	100	6a	0800-1000, 2100-0600
			6c	0600-0800, 1200-1500
			7	1500-2100
El Fayoun	1008	10	8a	0500-2000
Cairo	1071	100	8b	0350-2200
El Minya	1080	10	1	0400-2200
Luxor	1080	10	1	0400-2200
Batrah	1107	600	6a	1800-0500
			6b	0500-0600, 1500-1800
			6c	0600-1000, 1200-1600
Sohag	1143	10	1	0400-2200
Tanta	1161	100	8d	0400-0700 (Fri 1100), 1200-1900 (Thurs 2100)
Quena	1179	10	1	0400-2200
Ras Gharib	1188	10	1	0400-2200
Alexandria	1197	25	8c	0400-2200
Asswan	1278	10	1	0400-2200
Assiut	1305	10	1	0400-2200
Hurghada	1314	10	1	0400-2200
Nag Hamadi	1314	1	1	0400-2200
Abu Simbil	1314	1	2	0700 (Fri 1130)-1600
			4	0200-0500, 2000-2200
			8j	0500-0700 (Fri 1130), 1500-1900
Cairo	1341	100	3	1600-2400
Bawiti	1341	10	2	0500-1600
			4	0200-0500, 2000-2200
			8k	1600-2000
Idfu	1341	10	1	0400-2200
Siwa	1341	10	1	24h
Quseir	1350	10	1	0400-2200
Batra	1359	450	3	1600-2400
El Kharga	1368	10	1	0400-2200
El Farafra	1368	1	2	0500-1600
			4	0200-0500, 2000-2200
			8k	1600-2000

MW	kHz	kW	P	Times
El Dakhla	1386	10	1	0400-2200
Luxor	1386	10	2	0700 (Fri 1130)-1600
			4	0200-0500, 2000-2200
			8j	0500-0700 (Fri 1130), 1500-1900
Ras Gharib	1422	10	2	0500-1600
			4	0200-0500, 2000-2200
			8i	0500-0700, 1300-2000
Salum	1422	10	2	0700-1300
			4	0200-0500, 2000-2200
Sohag	1476	10	2	0700 (Fri 1130)-1600
			4	0200-0500, 2000-2200
			8j	0500-0700 (Fri 1130), 1500-1900
El Minya	1476	10	4	0200-0500, 2000-2200
			8a	0500-2000
El Tur	1485	1	2	0900 (Fri 1100)-1500
			4	2000-2200
			8f	0500-0900 (Fri 1100), 1500-2000
El Arish	1503	25	4	2000-2200
			8e	0400-1900
Quseir	1575	10	2	0500-1600
			4	0200-0500, 2000-2200
			8k	1600-2000
Idfu	1584	10	2	0700 (Fri 1130)-1600
			4	0200-0500, 2000-2200
			8j	0500-0700 (Fri 1130), 1500-1900
Baris	1584	1	2	0800-1500 (not Fri 0900-1100)
			4	2000-2200
			8h	0400-0800, 0900-1100Fri, 1500-2000
Port Said	1584	1	8g	0400-0800 (Fri 1100), 1300-2200 (Thurs 2300)
Matruh	1593	10	2	0700-1300
			4	0200-0500, 2000-2200
			8i	0500-0700, 1300-2000
El Dakhla	1602	10	2	0800-1500 (not Fri 0900-1100)
			4	2000-2200
			8h	0500-0800, 0900-1100Fri, 1500-2000
Nag Hamadi	1602	10	2	0700 (Fri 1130)-1600
			4	0200-0500, 2000-2200
			8i	0500-0700, 1300-2000
Siwa	1602	10	2	0700-1300
			4	0200-0500, 2000-2200

Prgrs: 1=General Prgr, 2=Youth & Sports Prgr, 3=Cultural Prgr, 4=Holy Koran Prgr, 5=Middle East Commercial Prgr, 6a=Voice of the Arabs, 6b=Wady El Nile Prgr, 6c=Palestine Prgr, 7=Hebrew Prgr, 8a-k=Local Prgrs: a) North Upper Egypt, b) Greater Cairo, c) Alexandria, d) Mid Delta, e) No. Sinai, f) So. Sinai, g) El Canal, h) Wady El Gadid, i) Matruh. j) South Upper Egypt, k) Educational Prgr.

FM (MHz)	1	2	3	4	kW
Cairo	98.8M	89.5C	95.4E	108.0Y	20
Alexandria	90.1M	97.0E			20
Aswan	89.0M	92.1E	95.3K		10
Luxor	90.0M	96.3E	103.1K		18
El Minya	87.9M	94.2E	103.1K		11
Ismalia	93.5E	96.7R			50
Suez	88.1M	91.2E	97.7R		10
Nuweiba	89.5M	92.6G	95.8R		10
El Arish	90.9M	94.1E	97.4R		80
Sharm El Sheikh	88.0M	91.1G	94.3R		7
Assiut	89.1M	92.6E	95.8R		10
Mahalla	93.1M	99.6E	89.2K		160
Matruh	95.8M	92.6E	99.1K	102.6R	10
El Tur	92.5M	95.7G	99.0R		9.1

Prgrs: C=Cultural Prgr, E=European Prgr, G=General Prgr, K=Koran, M=Musical Prgr, R=Regional Prgr.

SW: Abu Zaabal (G.C: 31.22E/30.16N): 100/250kW, Abis (G.C: 30.05E/31.10N): 250kW (+ 1 x 500kW currently being installed), Mokattam (G.C: 31.15E/ 30.03N): 50/100kW.
Further details in International Broadcasting Section

ANN: General Prgr: "Idha'atu jumhuriya misr al'arabbiya min al-qahira". Voice of the Arabs: "Saut al-'arab, min al-qahira". Quran prgr: "Idha'atu-I-Quran min al-qahira" — **V.** by QSL-card. Re.to Box 1186, Cairo. Schedules available on request.

EQUATORIAL GUINEA

L.T: UTC + 1h — **Pop:** 453.803 — **Radios:** 200.000 — **Pr.L:** Spanish, ethnic — **E.C:** 50Hz, 220/380V — **ITU:** GNE.

RADIO NACIONAL DE GUINEA ECUATORIAL
✉ Barrio Comandachina. ☎ +240 (8) 2592. 🖷 +240 (8) 2093 — Av. 3 de Agosto 90, Malabo. ☎ +240 (9) 2260. 🖷 +240 (9) 2097 & 3122 (Tech. Dept.).
L.P: DG of Information: José Mbá Obama Bendomo. Dir. Tech. Radio & TV: Hermengildo Moliko Chele. **Bata:** Dir: Sebastian Eló Aseko. PD: Esteban Mangué Mbá. **Malabo:** Dir: Juan Eyene Ekua. PD: Román Manuel Mañe.

Location	kHz	kW	Times
Bata	4926	50	0430-2200 (alt. to 5003.5v)
Bata	5005v	50	0430-2200 (alt. to 4926)
Malabo	6250v	10	0600-0800, 1100-2200

N: 0600, 1415, 2100 (simultaneously on both networks).
FM: Bata 98.0MHz.
V: Bata by QSL-card & on air at 2130, Malabo by card or letter.

RADIO AFRICA & RADIO EAST AFRICA (Comm.)
See International Broadcasting section.

R. France Internationale: Malabo 97.5MHz in French/Spanish.

ERITREA (State of)

L.T: UTC + 3h — **Pop:** 4.640.000 — **Pr.L:** Afar, Amharic, Kunama, Tigre, Tigrigna — **E.C:** 50Hz, 127/220V — **ITU:** ERI.

ERITREAN POST & TELECOMMUNICATIONS AUTHORITY (Gov.)
✉ P.O. Box 234, Asmara. ☎ +291 (1) 112868. 🖷 +291 (1) 1209328.
L.P: GM (Telecommunications Sces): Seyoum Gebechristos. Head of Freq. Management: Woldeslassie Musse.

VOICE OF THE BROAD MASSES OF ERITREA (Gov.)
✉ Ministry of Information, Radio Division, P.O. Box 872, Asmara (Technical Branch, P.O. Box 243, Asmara).
L.P: Minister of Inf: Beraki Gebre Selasie.
MW: Asmara 945kHz 50kW (Prgr.1), 837kHz 10kW (Prgr.2).
SW: Selea Daro: 2 x 100kW — Asmara: (G.C: 28.50E/15.15N): 1 x 10kW.

kHz	Prgr.	kHz	Prgr.	kHz	Prgr.	kHz	Prgr.
945	1	6190	1	837	2	7175	2

Prgr. 1 Tigriyna 0330-0430, 0930-1030, 1700-1830. Tigre 0430-0530, 1600-1700. Kunama 1500-1600.
Prgr. 2 in Arabic: 0330-0430, 0930-1030, 1700-1800. Afar 1600-1700. Bile (Mo,We) 1500 - 1530. Saho (Tu,Fr) 1500-1530. Amharic 0430-0500, 1530-1600.
ANN: Amharic:"Yeh be Asmera ketema yemigegne yesifiw yeritrea hezeb demts yeamarigna agelgilot new". Arabic: "Huna Asmara, Idha'at Sawt al-Jamahir al-Iritriyyah". Tigrigna: "Ezi kab Asmara Zemehalalef Medeber Radio Demtsi Hafash Eritrea Eyu".
V. by QSL-card. Rp — **F.PL:** Ext. Sce. in English.

ETHIOPIA

L.T: UTC + 3h — **Pop:** 61.330.000 — **Radios:** 9m — **Pr.L:** Amharic, Somali, Tigre, Oromo — **E.C:** 50Hz, 127/220V — **ITU:** ETH.

RADIO ETHIOPIA (Gov.)
✉ P.O. Box 1020, Addis Ababa. ☎ +251 (1) 21011, 21063. **Cable:** Ethiocast. **Ext. Sce:** P. O. Box 654, Addis Ababa; ☎ +251 (1) 711111.
L.P: SM: Kasa Miloko. CE: Kebede Gobena. Head of International Sce: Moges Taffese.

Mediumwaves	kHz	kW	N	Mediumwaves	kHz	kW	N
Metu	684	100	N	Addis Ababa	990	1	E
Harar	855	100	N	3 st's	1485	1	N
Addis Ababa	873	100	N				

SW: Gedja (G.C: 38.38E/08.47N): 100kW.
kHz: 5990, 7110, 9705
N) National Sce: MF 0330-0600, 0900-1000, 1030-1100, 1500-2000. Sat 0300-0600, 1000-1400, 1500-2000. Sun 0400-0900, 1000-1400, 1500-2000. **English:** MF 1030-1100.

E) Ext. Sce: see International Broadcasting section.
ANN: Nat. Sce: "Yeh Ye-Ethiopia Radio Naw". E: "This is R. Ethiopia broadcasting in English".
IS: Electronic keyboard — **V.** by letter (IRC's acc.).

VOICE OF PEACE: see International Broadcasting section.

GABON

L.T: UTC + 1h — **Pop:** 1.300.000 — **Radios:** 155.000 — **Pr.L:** French, Fang, Bopounou, Obamba, Djebi — **E.C:** 50Hz, 220V — **ITU:** GAB.

RADIODIFFUSION TÉLÉVISION GABONAISE (Gov.)
✉ B.P. 10150, Libreville. ☎ +241 732784 (**Chaîne 1:** +241 732025). ① RTG 5342 GO – B.P. 270, Franceville. ☎ +241 677218 – B.P. 230, Port Gentil. ☎ +241 552511 – B.P. 459, Oyem. ☎ +241 986004 – B.P. 280, Makokou. ☎ +241 903019.
L.P: DG RTG-1:John Joseph Mbourou. DG RTG-2: Luc Mvoumba. Asst. DG's: Radio: Gilles Terence Nzoghe. TV: Jules César Lekogho. Tech: Claude Nganga. Provincial St's: Robert Aloli.

MW	kHz	kW	N	Times
Oyem	549	20	2	0430-0630, 1030-1430, 1600-2300
Epila*	990	20	2	0430-0700, 1030-1430, 1600-2300
Oyem	1206	1	1	0455-2300
Port Gentil	1520	1	1	0455-2300
Tchibanga	1575b	20	2	0430-0630, 1020-1400, 1600-2130/2145
Melen	1554	20	1	0455-2300

*) studio in Franceville
FM (MHz): Franceville 87.86, Libreville 87.7/96.54, Makokou 100.5, Oyem 87.94, Pt. Gentil 88.03, Tchibanga 91.04.
SW: Libreville (G.C: 13.31E/01.40S) 100kW

kHz	Times	kHz	Times
4777	0455-0800, 1500-2300	7270	0800-1500

1=RTG-1 in French. **N:** 0500, 0530, 0600, 0700, 0800, 0900, 1000, 1200, 1400, 1500, 1700, 1800, 1830, 2100.
2=RTG-2 (provincial network) in French & ethnic languages.
FM Prgr: 0500-2305 on FM only.
ANN: "Ici Libreville, vouz écoutez Radio Gabon, châine 1" (or "Ici ...(town)...station provinciale") de la Radiodiffusion-Télévision Gabonaise". Or Franceville: "Ici Radio Masuku". Makokou: "Ici Radio 6". Tchibanga: "Ici Radio Massanga" — **IS:** Indigenous instruments. Opens and closes with National Anthem — **V.** by QSL-folder.

AFRICA No. 1 (Comm.)
✉ B.P. 1, Libreville. ☎ +241 760001. 🖷 +241 742133. ① 5558 GO. (✉ **in France:** Eurafripub, 27 rue Guersant, F-75017 Paris or B.P. 944, F-75829 Paris Cédéx 17. ☎ +33 (1) 4574 8383. 🖷 +33 (1) 4574 1769) —
L.P: Pres: Louis Barthélémy Mapangou. Dir Délegué: Michel Koumbangoye. Pierre Devoluy. Dir. Tec: Gaston Ombolo Ki-Obo. Dir. Prgrs. & Adv: Augustin Letamba. Dir. Inf: Jean Valère Mbina Mandza.
FM: Libreville 94.5MHz + relays in Burkina Faso, Central African Rep, Chad, Congo, Côte d'Ivoire, France, Mali, Niger, Senegal, Togo, Zaire.
Main N: 0530, 0630, 0730, 1115W, 1215, 1700, 1830, 2200.
SW: Moyabi-Moanda (G.C: 13.31E/01.40S): 5 x 500kW tr's.
Further details in International Broadcasting section.

Radio Generation Nouvelle — Dir: General Jean-Boniface Assele (no further details known).
R. Soleil, R. Mandarine: no further details known.
R. France Internationale: Libreville 104.0MHz.

GAMBIA

L.T: UTC — **Pop:** 1.083.526 — **Radios:** 500.000 — **Pr.L:** English, Fula, Mandinka, Wollof, Jola, Serer, Serahule, Manjago, Karonimka — **E.C:** 50Hz, 220/240V — **ITU:** GMB.

GAMBIA RADIO & TELEVISION SERVICES (Gov.)
✉ Mile 7, Banjul. ☎ +220 495101/497419. 🖷 +220 495102.
L.P: Dir. of Broadc. Sces (Radio & TV): Tombong Saidy. Prgr. Mgr. (radio): Mrs. Sara Grey Johnson. Eng. Mgr. (radio): Yankuba Toure.
MW: Bonto 648kHz 50kW, Basse 747kHz 10kW — **FM:** 91.4MHz.
D.PRGR: Mon-Thurs 0555-1400, 1655-2400. Fri-Sun 0555-2400. **N. in English:** (SS)0700, 1300, 1800, 1900,2200.
ANN: "You are tuned to Gambia Radio & Television Services from Banjul".
IS: Cora (harp) — **V.** by letter.

RADIO SYD (Comm.)

✉ P.O.Box 279/280, Banjul. ☎ +220 221870. 🖳 +220 226490. **Cable:** SYD — **L.P:** GM: Miss Constance Wadner Enhörning.
MW: Banjul 909kHz 2.5kW.
D.PRGR: 0600-0200. **N:** rel. R. Gambia.
Swedish (Nov 15-April 15): 0800-0830, 2000-2030 — **V.** by letter.

Citizen FM (Comm.), Banjul — **FM:** 105.7MHz (relays some BBC prgrs).
R. 1 FM, Kairaba Ave, Serrekunda: 24h on 102.0MHz 0.5kW.

GHANA

L.T: UTC — **Pop:** 19.444.200 — **Radios:** 12.500.000 — **Pr.L:** English, Hausa, French + ethnic — **E.C:** 50Hz, 230V — **ITU:** GHA.

GHANA BROADCASTING CORPORATION
(Autonomous Statutory Body)

✉ P. O. Box 1633, Accra. ☎ +233 (21) 221161. 🖳 +233 (21) 221153. ☏ 2114 GBC GH E-mail: Gtv@ncs.com.gh
L.P: DG: Dr. Kofi Frimpong. Dir, Special Duties: Anna Sai. Dir. of Admin: Seth K. Sosu. Ag. Dir. of Radio: Cris Tackie. Dir. of Eng: K.D. Frimpong. Ag. Dir. Finance: Albert Ammafio. Dir. Corporate Affairs: Mrs. Maud Blankson-Mills. Ag. Business Dir: Rev. George Lomotey. Pub. Rel. Officer: Kweku Boham.
SW: Accra (G.C: 00.10E/05.31N):

kHz	kW	Prgr.	Times
3366	50	Radio 2	0530-0915, 1700-2400
4915	50	Radio 1	0530-0915, 1230-2400
6130	50	Radio 2	1230-2400

Radio 1 (multilingual): as above. **N. in English:** 0600, 0700, 0900, 1100SS, 1300, 1400, 1800, 1900MF, 2000, 2200, 0000. 0045.
Radio 2 in English (commercial): as above. **N:** as R1. **Schools Prgr:** 0915-1040MF — **FM:** 0400-0200.
ANN: "This is Radio 1/2" — **IS:** Sign on with drum beat. **V.** by QSL-card. Re. to Dir. of Engineering.
FM (MHz)

Location	MHz	Name	Region
Bolgatanga	89.7	URA Radio	Upper East
Ho	91.05	Volta Star	Volta
Tamale	91.2	R.Savanna	Northern
Kumasi	92.1	Garden City R.	Ashanti
Cape Coast	92.5	R.Central	Central
Wa	93.9	Upper West R.	Upper West
Sekondi-Takoradi	94.7	Twin City R.	Western
Dormaa Ahenkro	94.9	Dormaa Ahenkro	Brong Ahafo
Accra	95.7	GAR	Greater Accra
Apam	96.5	Apam FM	Central
Swedru	98.6	Swedru	Central

	MHz	Private Stations Name and location	H. of tr.
1)	90.4	Gold FM, Lartehbiokorshie	0530-2400
2)	91.9	Vibe FM, Osu	0530-2400
3)	95.774	GAR FM, Kanda	0530-2400
4)	99.7	Joy FM, Kokomiemie	24h
5)	106.3	Groove FM, La	0530-2400
6)	102.3	Choice FM, Roman Ridge	0430-2400

Addresses
1) Link Rd, Lartehbiokorshie.
2) Former Cave du Roy Night Club Bldg, near Green Leaf Hotel, Osu.
3) Broadcasting House, Kanda.
4) Nar SSB, Kokomiemie.
5) Near Trade Fair Center, La.

R. France Internationale: Accra 90.0MHz.

GUINEA (Republic)

L.T: UTC — **Pop:** 7.038.478 — **Radios:** 230.000 — **Pr.L:** French, Fulah, Maninké, Soussou — **E.C:** 50Hz, 220/380V — **ITU:** GUI.

RADIODIFFUSION TÉLÉVISION GUINÉENNE (Gov.)

✉ B. P. 391, Conakry. ☎ +224 442205. ☏ 22341 RTG.
L.P: Dir. National: Justin Morel Jr. Head of Public Rel: Momo Toure.

MW	kHz	kW	MW	kHz	kW
Kankan	603	25	Labé	1386	50
Conakry	1295	1	Conakry*	1404	200

*)r. inactive
SW: Conakry (G.C: 13.40W/09.32N):

kHz	kW	H. of tr.
4900	18	W 0555-0800 & 1200-2400, Sun 0800-2400
6155+	50	W 0555-0800 & 1200-2400, Sun 0800-2400
7125.5v	18	W 0555-0800 & 1200-2400, Sun 0800-2400

+) irr. Also irr. heard on 9650 and 15310kHz.
FM: Conakry 88.55MHz 2.5kW + 7 regional st's.
D.PRGR. in French & ethnic languages: W 0555-0800 & 1200-2400, Sun 0800-2400. **N: French:** 0645W, 0800W, 0915Sun, 1245W, 1615, 1800, 1900, 1945, 2200, 2350. **English:** 1845 (irr.).
R. Rurale: regional sce. on 1386kHz.
ANN: F: "R. Conakry", "R. Guineé", "R. Rurale, station de la Moyenne-Guinée — **IS:** Guitar — **V.** rarely by letter. Re.to Head of Public Rel.

R. Rurale (community st's set up by the development agency ACTT): st's are operating in Boké, Mamou and Kissidougou.

GUINEA-BISSAU (Republic of)

L.T: UTC — **Pop:** 1.098.000 — **Radios:** 40.000 — **Pr.L:** Portuguese — **E.C:** 50Hz, 220/380V — **ITU:** GNB.

RADIODIFUSÃO NACIONAL (Gov.)

✉ C. P. 191, Bissau — **MW:** 1034kHz 5kW — **FM:** 92.9MHz.
D.PRGR: 0600-0800, 1200-2400. **N:** 0700, 2215.
ANN: "Escutar a Radiodifusão Nacional de Republica de Guiné-Bissau" — **V.** by QSL-card. Rp.

Radio Bombolom FM (Priv.): incl. relays of BBC Portuguese sce.
R. France Internationale: Bissau 94.7MHz.

KENYA

L.T: UTC + 3h — **Pop:** 29.681.599 — **Radios:** 5m — **Pr.L:** English, Swahili, Hindustani, Ethnic — **E.C:** 50Hz, 240V — **ITU:** KEN.

KENYA BROADCASTING CORPORATION (KBC)

✉ Box 30456, Harry Thuku Rd, Nairobi. ☎ +254 (2) 334567. 🖳 +254 (2) 220675. **Web:** http://www.africaonline.co.ke/AfricaOnline/netradio.html
Local ✉ Box 799, Nyeri — Box 1585, Nakuru — Box 1327, Meru — Box 3287, Kibirichia — Box 90200, Mombasa — Box 844, Kisumu.
L.P: MD: Simeon Anabwani. Radio Prgrs. Mgr: Mrs. Eulalia Namai. Editor-in-Chief: Kipserem Maritim. Asst. Mgr. Tech Sces (Radio): Daniel Githua. Asst. Mgr. Tech. Sces (Research): Daniel Obam. Chief, Int. & Pub. Rel: Chris Opiyo. Mktg. Mgr: Emmanuel Oywero.

MW	kHz	kW	Sce.	MW	kHz	kW	Sce.
Voi	540	100	N	Nyamninia+	954	100	G
Kisumu	558	20	W	Voi	981	100	G
Garissa	567	50	N	Nyeri	1017	20	N
Ngong++	612	100	N	Malindi	1044	100	N
Garissa	639	50	G	Maralal	1107	100	N
Marsabit	675	50	N	Kitale	1134	50	G
Marania+++	702	100	N	Wajir	1152	50	N
Ngong++	747	100	G	Marsabit	1233	50	G
Nyamninia+	846	100	N	Ngong++	1268	20	C
Kitale	882	50	N	Wajir	1305	50	G
Marania+++	900	100	G	Nyeri	1368	20	G
Malindi	927	100	N	Maralal	1386	100	G

*) temporarily inactive.
+) near Kisumu, ++) near Nairobi, +++) near Meru.
SW: Langata (G.C: 36.47E/01.21S): 2 x 100kW, 2 x 10kW, 2x 5kW — Koma Rock (G.C: 37.09E/01.16S): 2 x 250 kW, 1 x 20kW.

kHz	Sce.	Times
4885	E	0200-0700, 1330 (Sun 1400)-1745
	N	1715-1910
	G	1910-2110
4915	C	W 0230-1930, Sun 0700-1700
4935	G	0200 (Sun 0230)-0630, 1300-2010 (SS 2110)
6075	N	0200-0625
6150	N	1325-2110
7140	N	0625-1325

FM (MHz)	N	G	E	C
Meru	90.4	103.5		
Ce. Nairobi	89.9	89.7	89.3	89.5
Greater Nairobi	92.9	95.6		
Malindi	90.3	93.6		
Mombasa	104.4	100.8		
We. Kenya	88.6	91.5		
Unknown location	99.5	97.3		

National Sce. (N) in Swahili: 0200-2110.
General Sce. (G) in English: 0200 (Sun 0230)-2010 (SS 2110). **N:** 0215, hourly 0300-1900, 2100.
Central Sce. (C) in Hindi/ethnic langs: MF 0300-1100 & 1300-1905.
Western Sce. (W) in ethnic langs: MF 0300-1905 & Sat 1100-1905.
Eastern Sce. (E) in Somali & ethnic langs: MF 0900-1905.
Metro 101.9FM: Nairobi 101.9MHz 24h.
ANN: E: "This is KBC". Swahili: "Hii ni KBC".
IS: Flute & drum melody. National Anthem at s/on and s/off.
V. by QSL-Aerogramme.

CAPITAL FM (Comm.)
19th flr, Lohnro House,Standard St, (P.O. Box 74933), Nairobi. ☎ +254 (2) 210020. 🖷 +254 (2) 332349.
L.P: PD: Phil Matthews (**E-mail:** lynxah1@users.africaonline.co.ke).
FM: Nairobi 98.4MHz

LESOTHO

L.T: UTC + 2h — **Pop:** 2.142.401 — **Radios:** 1.100.000 — **Pr.L:** SeSotho, English — **E.C:** 50Hz, 220V — **ITU:** LSO.

LESOTHO NATIONAL BROADCASTING SERVICE (Gov.)
P. O. Box 552, Maseru 100. ☎ +266 323561. 🖷 +266 310003.
⏱ 4340 LENA 10 — **L.P:** Principal Secr, Min. of Inf. & Broadc: Ms. Mpine Tente. Dir. Broadc: Molahlehi Letlotlo. Ag. Dir. of Prgrs: Mamonyane Matsaba. Ch. Eng: Lebohang Monnapula. Studio Eng: Motlatsi Monyane. Tr. Eng: Basia Maraisane.
MW: Maseru 891kHz 50kW — **FM:** Maseru 93.9/99.8MHz.
SW: Lancer's Gap (G.C: 27.32E/29.19S): 4800kHz 100kW.
D.PRGR: 0300-2100. **N. in English:** 0500, 1130, 1600.
ANN: E: "This is Radio Lesotho" or "This is the Lesotho National Broadcasting Service, Maseru". SeSotho: "Se-ea-le-moea sa Lesotho, Maseru" — **IS:** National Anthem at s/on and s/off.
V. by QSL-card. Rec.acc.

LIBERIA

L.T: UTC — **Pop:** 2.950.000 — **Radios:** 600.000 — **Pr.L:** English + 18 ethnic — **E.C:** 60Hz, 120V — **ITU:** LBR.

LIBERIA BROADCASTING CORPORATION
(supports ECOMOG peace-keeping forces)
P.O. Box 10-594, 1000 Monrovia 10. ☎ +231 224984, 221036, 222758, 222647. **Cable:** Broadcasts.
L.P: DG: Jesse B. Karnley. Dep. Dir. of Broadc: James Morlu.
FM: 91.7MHz 5kW — **SW:** 7275kHz 10kW.
D.PRGR on FM: MF 0550-1100 & 1355-2400, Sat 0550-2400, Sun 0700-2400. **N. in English:** 0700W, 0900MF, 1200SS, 1400, 1600MF, 1900, 2100, 2300 — **D.PRGR on 7275kHz:** 0615 (0655SS)-1000, 1355-1730. **N. in English:** 0700W, 0900W, 1400, 1600MF.
ANN: "This is ELBC-FM 91.7, the voice of peace, harmony and reconciliation in Monrovia".

LIBERIAN COMMUNICATION NETWORK
P.O. Box 1103, 1000 Monrovia 10. ☎ +231 226963. 🖷 +231 226003.
Web: http://www.afric-network.fr/afric/liberia/liberia.html
L.P: Dep. Mgr: James Kassoyen. Engineer-in-Charge: Isaac P. Davis.
FM: Kiss-FM 89.9MHz.
SW: See International Broadcasting section.

RADIO MONROVIA 98 FM (Comm.)
P.O. Box 10-3501, Monrovia. ☎ +231 225301. 🖷 +231 222743.
L.P: Pres: Charles A. Snetter Jr. Business Mgr: Emmanuel Kuyateh.
FM: 98.0MHz 1.35kW — **D.PRGR:** 0600-2400. **N:** rel. VOA.

D.C. 101.1 FM (Comm.)
Ducor Broadc. Corp, P.O. Box 1312, Monrovia. ☎ +231 226464.
L.P: Pres: Fred Bass-Golokeh. PD: Martin Brown.

FM: 101.1MHz — **D.PRGR:** 0600-2400.

RADIO VERITAS (Catholic. Rlg.)
P.O. Box 3569, Monrovia. ☎ +231 221658.
FM: ELCM Monrovia 97.8MHz 5kW.
D.PRGR: 0500-1100, 1800-2300.

STAR RADIO
Monrovia — GM: George Bennett.
☎ +231 226820, 226176, 227390. 🖷 +231-227360. **E-mail:** libe@atge.automail.com. Funded by Fondation Hirondelle, 3, Rue Traversière, CH 1018-Lausanne, Switzerland. ☎ +41 (21) 6472805. 🖷 +41 (21) 6474469. **E-mail:** fondhi@atge.automail.com.
FM: Monrovia 104MHz — **SW:** 4 x 10kW tr's.
D.PRGR: 0500-2000. **N. in English:** 0630. 0730, 0757, 1830,1930, 1957.
F.PL: SW sce. on 3400 and 5890kHz.

LIBYAN ARAB JAMAHIRIYA

L.T: UTC + 1h (Su: UTC + 2h) — **Pop:** 5.250.000 — **Radios:** 1m — **Pr.L:** Arabic — **E.C:** a/c 50, 127/230V — **ITU:** LBY.

LIBYAN JAMAHIRIYA BROADCASTING
P.O. Box 9333 (**Ext. Sce:** Box 4677), Souq al Jama, Tripoli. ☎ +218 (21) 603191/5. ⏱ 20010. **Cable:** Ida Shaabia Tripoli.

MW	kHz	kW	N	MW	kHz	kW	N
Tobruk	648	300	V	Tripoli	1053	50	H
Benghazi	675	100	H	Ajedabia	1080	40	H
Ghadames	711	50	V	Jalo	1080	20	H
Jefren	711	50	V	El Beida	1125	500	V
Sebha	711	50	V+S	Tripoli	1251	500	V
Sirte	792	20	H	Tripoli	1404	20	Q
Sebha	828	300	V	Misurata	1449	20	H
Ghiagboub	909	20	H	Brak	1485	1	F
Kufra	909	10	H				

N=Network: H=Home Sce, V=V. of the Great Homeland, Q=Holy Quran Prgr, F=Foreign Language Prgr, E=Ext. Sce. S=Sebha Local R.
FM: Tripoli 94.0/104.0MHz, El Beida 87.9MHz, Tobruk 98.4MHz.
SW: Tripoli/Sabrata (G.C: 13.11E/32.45N): 500kW — Sebha (G.C: 14.50E/25.52N): 100kW —) Benghazi (G.C: 20.04E/32.08N): 100kW. Full schedule in International Broadcasting section.
Home Service: 0500-2350. **N:** 0600, 0900, 1200, 1330, 1600, 1700, 1900, 2130, 2300 — **Voice of the Great Homeland:** 1115-1315, 1745-0500. **N:** 1230, 1815, 2115, 0015, 0215, 0430 — **Holy Quran Prgr:** 0500v-1900v on 1404kHz, 0500v-1400 on 1485kHz.
Foreign Language Prgr: French 1400-1600, **English** 1600-1800.
Tripoli Radio of the Arabs on 104.0MHz: 0700-1100 & 1600-1800.
EXTERNAL SERVICE: See International Broadcasting section.
ANN: H: "Idha'at al jamahiriya al-'arrabiya al-libyya ash-sha'abiyya al-sitirakkiya". V: "Idha'at saout al-watan al'arabbiy al-Kabir".
IS: Prgrs open and close with National Anthem — **V.** by QSL-card.

MADAGASCAR (Dem. Rep.)

L.T: UTC + 3h — **Pop:** 14.350.000 — **Radios:** 2.300.000 — **Pr.L:** Malagasy, French — **E.C:** 50Hz,110/220V — **ITU:** MDG.

RADIO-TELEVISION MALAGASY (Gov.)
B.P. 1202, Antananarivo. ☎ +261 (2) 22381 or 21745.
L.P: DG: Mamy Rafenomanantsoa — **MW:** Antananarivo 630kHz 150/75kW, 1394/1502kHz 4kW. Relay st's (1kW) on 1260, 1278, 1323, 1394, 1502, 1554, 1562 and 1602kHz + 12 st's 0.25kW.
FM: Antananarivo 99/100MHz, Toamasina & Ambositra 99MHz.
SW (G.C: 47.35E/18.50S):

kHz	kW	N	H. of tr.
3288v	100	1	0256-1600
		2	1600-2000
5009.5v	100	1	0256-0500, 1400-1900 (SS 1500-2100)
6135.4	30	1	0500-1500

Prgr. 1 in Malagasy: MF 0200-1900, SS 0300-2100 on MW (exc. 1502kHz) and SW as above — **Frequence 2 in French:** 1600 (Sat 1400, Sun 0900)-2000 on 1502kHz + FM. **N:** 1000Sun, 1600MF. **English N:** 1745 — **ANN:** 1: "Radio Madagasikara ity".
V. by QSL-card. Rp. Re. in French.

RADIO TELEVISION ANALAMANGA (Priv.)
B.P. 7547, Ampesiloha. ☎🖷 +261 (2) 62804.
L.P: Owner: Ottacio Ermini.

FM: freq unknown — **SW:** 5950kHz (unlicensed).
D.PRGR: 24h.

TSIOKA VAO (Priv.)
✉ B.P. 315, Tana. ☎ +261 (2) 21749.
L.P: Dir: Detkou Dedonnais.
FM: freq unknown — **SW:** 6075kHz (unlicensed).
D.PRGR: 0300-1900

R. France Internationale: Antananarivo 96 0MHz.
Alliance Française: Antananarivo 92.0MHz, Antsiranana 98.0MHz (rel. RFI 0300-2100).

RADIO NETHERLANDS RELAY STATION
See International Broadcasting section.

MADEIRA (Portuguese)

L.T: UTC — **Pop:** 300.000 — **Radios:** 200.000 (est.) — **Pr.L:** Portuguese — **E.C:** 50Hz, 220/380V — **ITU:** MDR.

CENTRO REGIONAL RDP/MADEIRA (RDP)
✉ Rua Tenente Coronel Sarmento 15, 9000 Funchal. ☎ +351 91 229155. 🖷 +351 91 230753. ✆ 72111 RDPMAD.
L.P: Dir: João Afonso de Almeida.
MW (Canal 1): Porto Santo 531kHz 10kW, Pico de Areeiro 603kHz 10kW, Ponta do Pargo 1125kHz 1kW, Funchal 1332kHz 1kW.
FM (MHz):

Main Coverage area	Canal 1	Super FM	kW
Achadas da Cruz	104.3	105.0	11/21
Areeiro e Zonas Altas da Madeira	95.5	94.1	150/250
Costa Norte e Porto Santo	100.5	96.5	110/210
Do Caniço à Ponta do Sol	96.7	94.8	11/21
Funchal	104.6	89.8	11/21
Machico	93.1	90.8	10/20
Paúl da Serra	101.9	93.3	11/21
Ponta do Pargo	90.2	94.6	11/21
Ribeira Brava		103.1	21
Santa Cruz		91.3	21

D.PRGR: 24h (both networks).
ANN: Canal 1: "RDP Madeira, Rádio Provincial".
V. by QSL-card via RDP Lisboa, Portugal.

POSTO EMISSOR DE RADIODIFUSÂO DO FUNCHAL (Comm.)
✉ Rua Ponte de São Lázaro 3, 9000 Funchal. ☎ +351 91 230393. 🖷 +351 91 221797 — **MW:** Funchal 1530kHz 10kW, Santana 1017kHz 1kW — **FM:** Funchal 92.0MHz 2kW (stereo).
D.PRGR: 0600-2400 (Sat 0100) on MW, 24h on FM (different prgrs).

ESTAÇÂO RÁDIO DE MADEIRA (Comm.)
✉ Caixa Postal 450, 9006 Funchal Codex. ☎ +351 91 766066. 🖷 +351 91 764395 — **L.P:** Dir: Teresa Sardinha.
MW: 1485kHz 1kW — **D.PRGR:** 0600-2400.
M Rádio: ☎ +351 91 763315 (✉ 🖷 as above) — **L.P:** Dir: Nuno Portela Ribeiro — **FM:** Funchal 96.0MHz 0.4kW..
Rádio Turista: D.PRGR. on 1485kHz: W 0830-0900 Danish/Norwegian/Swedish, 0900-0930 German, 1745-1830 English, 1830-1900 French. ☎ +351 91 229663.

OTHER STATIONS
R. Zarco (Machico) 89.6MHz 1kW — **R. Palmeira** (Sta. Cruz) 96.1MHz 0.5kW — **R. Girão** (Estreito Camara de Lobitos) 98.8MHz 1kW — **R. Sol** (Ponta do Sol) 103.7MHz 0.5kW — **R. Clube** (Funchal) 106.8MHz 0.4kW — **R. Brava** (Rio Brava) 98.4MHz 0.5kW.

MALAWI

L.T: UTC + 2h — **Pop:** 10.602.795 — **Radios:** 1.060.000 — **Pr.L:** Chichewa, English — **E.C:** 50Hz, 230V — **ITU:** MWI.

MALAWI BROADCASTING CORPORATION (MBC) (Statutory Body, part-comm.)
✉ P. O. Box 30133, Chichiri, Blantyre 3. ☎ +265 671222. 🖷 +265 671257 or 671353. ✆ 0904 44425 Wailesi MI. **E-mail**: dgmbc@malawi.net
L.P: Ag. DG: Sam Gunde. Dir. of N. & Current Affairs: Molland Nkhata. Dir. of Eng: Joseph Chikagwa. Dir. of Prgrs: Matheas Manyeka (R.1), Brighton Matewere (R.2). Ag. Dir. Corporate: Owen Maunde. Financial

Contr: Dafton Chitseko. Eng. Consultant: Phillip Chinseu. Contr. of Tr's: Edwin Lungu. Contr. of N. & Current Affairs: Don Chimera. Contr. of Studios: Fletcher Njowe. Contr. of Business Affairs: Aureliano Kakowa.

Mediumwaves	kHz	kW	Mediumwaves	kHz	kW
Mangochi	540	10	Bangula	810	10
Karonga	558	10	Nkhota Kota	1107	1
Lilongwe	594	25	Chitipa	1277	1
Mzuzu	675	50	Kasungu	1278	1
Blantyre	756	25			

FM: Radio 1 Blantyre 95.4MHz + 15 rlys. Radio 2 Blantyre 92.2MHz, Lilongwe 91.5MHz, Mzuzu 91.3MHz, 24h.**SW:** Limbe (G.C: 35.02E/15.42S): 1 x 100kW, 1 x 50kW, 1 x 20kW.

kHz	kW	Times	kHz	kW	Times
3380	50	0255-0800	7130	50	1400-2210
5995	10	Standby			

D.PRGR: 0255-2210. **N. in English:** 0300, 0500W, 0600Sun, 0700W, 0800Sun, 0900W, 1000Sun, 1030MF, 1100SS, 1200MF, 1400MF, 1600, 1800, 2000W, 2100, 2200. **International Service:** 1600-1800.
MBC Radio 2 on FM: no details received.
ANN: E "This is the MBC" or "This is the International Sce. of the MBC". Chichewa: "Kuno ndi ku Interanashonolo Sevesi ya MBC".
IS: 0255 Cock crow and rapid drum beat. Piksífon just before 1600 and 1700 — **V.** by letter.

MALI (Rep. of)

L.T: UTC — **Pop:** 10.211.672 — **Radios:** 1.600.000 — **Pr.L:** French + ethnic langs — **E.C:** 50Hz, 220V — **ITU:** MLI.

OFFICE DE RADIODIFFUSION TÉLÉVISION MALIENNE (ORTM)
✉ B.P. 171, Bamako. ☎ +223 (22) 2019 or 2474. 🖷 +223 (22) 4205.
Cable: Radio-Mali — **L.P:** DG: Abdoulaye Sidibe. DG Adj: Sidiki Konate. CE: Mahamadou Sow.
MW: Bamako 540kHz 50kW — **FM:** 87.6/91.6MHz
SW: Bamako (G.C: 08.01W/12.39N) 1 x 100kW, 6 x 50kW.

kHz	Times
540	0555 (Sun 0655)-2400
4783	0555 (Sun 0655)-0758, 1757-2400
4835	0555 (Sun 0655)-0758, 1757-2400
5995	0555 (Sun 0655)-0758, 1757-2400
7185	0758-1757
7284.4	0758-1757
9635	0758-1757
11960	0758-1757

N. in French: 0600, 0700MF, 0800MF, 0900, 1000MF, 1300, 1500, 1800, 1945, 2230. **UN Radio Prgr:** Sat 2130-2200. **English:** Sun 1840-1900
Radio 2 on FM only: no details available.
IS: Guitar — **V.** by QSL-card. Rec.acc.

R. Bamakan (Community St.), Nouveau Marche de Mdine, Bamako. ☎ +223 (22) 2760 — PD: Mohamamadou Cisse. **FM:** 100MHz.

Other FM Stations in Bamako: R. Patriote 88.1MHz, FR3 93.8MHz, R. Tabaley 94.4MHz, R. Quintan 94.7MHz, Chaîne 2 95.8MHz, R. Binkan 97.1MHz, R.Liberté 97.7MHz, R.Klédu 101.2MHz, R.Kaira 104.4MHz, R. Voix de l'Islam 107.4MHz.

R. Rurale (community st's set up by the development agency ACTT): st's are operating in Niono, Kadiolo, Bandiagara & Kidal.

R. France Internationale: Bamako 98.5MHz.
Africa No. 1: Bamako 102.0MHz (see main entry under Gabon).

CHINA RADIO INTERNATIONAL RELAY STATION
See International broadcasting Section.

MAURITANIA (Islamic Republic of)

L.T: UTC — **Pop:** 2.250.000 — **Radios:** 1m — **Pr.L:** Arabic, Poular, Soninké, Wolof, French — **E.C:** 50Hz, 220/380V — **ITU:** MTN.

RADIO MAURITANIE (Gov.)

✉ B.P. 200, Nouakchott. ☎ +222 252164, 252679, 252101. 🖷 +222 251264. ① 515 MTN — **L.P:** DG:Abdallahi Ould Mohamedou.
MW: Nouakchott 1349kHz 20kW.
SW (G.C: 15.57W/18.07N): 4845v/7245kHz 100kW.
FM: Nouakchott 98.01MHz, Nouadhibou 93.0MHz.
D.PRGR: Sat-Thurs 0630-1600 & 1800-0100 on FM, 0630-0830, 1200-1600 & 1800-0100 on 1349kHz, 0630-0830 & 1800-0100 on 4845kHz, 1200-1600 on 7245kHz. Fri 0800-0100 on 1349kHz + FM, 0800-1700 on 7245kHz, 1700-0100 on 4845kHz. **N: Arabic:** 0700, 1200, 1500, 2030, 2230. **French:** 0730, 0930 (Fri), 1430, 1930.
ANN: F: "R.Mauritanie émettant de Nouakchott". A: "Huna Nouakchott iza'at gumhuriyati I-Islamiyya al mauritaniyya".
IS: Mauritanian guitar — **V.** by QSL-card or letter.

R. France Internationale: Nouakchott 93.3MHz.

MAURITIUS

L.T: UTC + 4h — **Pop:** 1.157.429 — **Radios:** 400.000 — **Pr.L:** English, French, 6 Indian languages, Chinese — **E.C:** 50Hz, 240V — **ITU:** MAU.

MAURITIUS BROADCASTING CORPORATION (Part-comm.)

✉ P. O. Box 48, Curepipe. ☎ +230 675 5001. 🖷 +230 675 7332. **E-mail:** mbc@bow.intnet.mu. **Web:** http://xensei.com/users/mbc/
L.P: DG: Trilock Dwarka. CE Radio: C Mootoosamy. CE Finance and Admin: C Jaunbocus. Dir. Corporate Affairs: Atma Bumma. Dir. Human Resources): Hamid Seelarbokus.
MW: 684/819kHz 10kW — Rodrigues 1206kHz 10kW, 666kHz 1kW.
SW: F.PI: reactivation of SW-tr's on 4855 & 9710kHz at 1300-0200.

FM (MHz)	Sugar FM	FM Planet	FM Sansar
Ce. Plateau/No.	97.3	90.8	94.0
No./No.We./P.Louis	91.7	94.9	98.2
So.	89.3	92.4	95.6

RM1 on 684/1206kHz: 24h. **English:** 1430-1030. **N:** 0400, 1440 (BBC).
N. in French: 0300, 0830, 1230, 1430.
RM2 on 819/666kHz: 24h. **English:** 1030-0300.
FM Sces: all 24h.
ANN: E: "This is the Mauritius Broadcasting Corporation". F: 'Ici MBC, Office de Radiodiffusion-Télévision de Maurice".
V. by QSL-card. Rec.acc. (exc. BBC relays).

BBC WORLD SERVICE: 1575kHz 1kW.

MAYOTTE (French Territory)

L.T: UTC + 3h — **Pop:** 112.000 — **Radios:** 50.000 — **Pr.L:** French, Mahorian — **E.C:** 50Hz, 220V — **ITU:** MYT.

RADIO-TÉLÉVISION FRANÇAISE D'OUTRE-MER (RFO-MAYOTTE)

✉ B.P. 103, F-97610 Pamandzi, Ile de Mayotte. ☎ +269 601017. 🖷 +269 601852. ① 915822. **Web:** http://www.rfo.fr
L.P: St. Dir: Robert Xavier. Dir. Tec: Serge Sulpice-Timothee.
MW: Pamandzi 1458kHz 100kW — **FM:** Dzaoudzi 91MHz 100W, Lima Combani 92MHz 500W — **D.PRGR:** 24h. **N. in French:** 0330, 0430, 0600, 0700*, 0800*, 0930, 1000, 1100*, 1600, 1700.*) France-Inter.
Rel. France-Inter: 1915-0300 (Fri/Sat 0400).
ANN: "Vous êtes à l'écoute de RFO-Mayotte".
IS: Melody on guitar — **V.** by letter.

Europe 2: Boueni 90.2MHz, Mamoudzou 99.1MHz, Pamandzi 97.7MHz.

MOROCCO

L.T: UTC — **Pop:** 27.866.000 — **Radios:** 5.100.000 — **Pr.L:** Arabic, French, Spanish, English, Berber, Hassania — **E.C:** 50Hz, 127/220V — **ITU:** MRC.

RADIODIFFUSION-TÉLÉVISION MAROCAINE (Gov.)

✉ 1, Rue El Brihi (or B.P. 1042), 10000 Rabat. ☎ +212 (7) 709613. 🖷 +212 (7) 703208. ① 0407 dt rtm 36687 m.

Web: http://www.maroc.net/rc/
Regional ✉ B.P. 459, Laayoune.
L.P: DG: Mohamed Tricha. Dir. Radio: Abderrahman Achour. Dir. TV: Mohamed Issari. Dir. Tec: Jamal Eddine Tanane. Dir. Ext. Rel: Ali M'Barek. Dir .Finance & Admin: Mehdi Bouzekri.

LW/MW	kHz	kW	N	LW/MW	kHz	kW	N
Azilal	207	800	A	Agadir	774	50	C
Sidi Bennour*	540	600	A	Rabat	819	25	A
Oujda	594	100	A+B	Oujda	828	100	C
Sebaa-Aioun*	612	300	A	Errachidia	864	15	A
Sebaa-Aioun	702	140	C	Safi	909	5	B
Laayoune*	711	600	A	Agadir	936	600	A

LW/MW	kHz	kW	N	LW/MW	kHz	kW	N
Ad-Dakhla*	999	10	A	Agadir	1197	20	B
Rabat	1026	1	B	Tanger	1233	200	R
Tanger*	1053	600	A	Safi	1325	5	C
Casablanca	1080	1	A	Casablanca	1485	1	C
Casablanca	1188	1	B	Marrakech	1593	1	C

*) carries regional prgrs.
SW: Tanger (G.C:05.50W/35.45N): 100/50kW.

kHz	Area	P	Times	kHz	Area	P	Times
11920	EEu/ME	A	0000-0500	15345	ME	A	0900-2200
15335	WEu	A	1100-1500	17595	Eu/Af	A	1400-1700
			2200-2400	17815	Eu/Af	B	1700-1900

FM (MHz)	Prgr. A	Prgr. B	kW
Boukhouali		93.4	10
Cap-Spartel	88.7	91.8	20
Casablanca		90.0	39
Meghrez	89.8	99.4	12
Oukaimeden	91.7	98.8	31
Rabat		87.9	39
Zerhoun	98.4	95.1	10

Netw. A in Arabic: 24h. **N:** on the h.
Regional Prgrs (exc.Sun): **Tetuan:** 0805-0900 & 1000-1200 on 1053kHz. **Tangier:** 1500-1600 & 1620-1800 on 1053kHz. **Oujda:** 1200-1300 & 1330-1500 on 594kHz. **Marrakech:** 1500-1600 & 1620-1800 on 540kHz. **Casablanca:** 1500-1600 & 1620-1800 on 1080kHz. **Agadir:** 1500-1600 & 1620-1800 on 936kHz. **Laayoune:** Fri/Sat 2100-2200 in **Spanish** on 711kHz. **Dhakla** on 999kHz: times not known.
Int. Netw. B: 0600-0100. **English:** W 1000-1200, Sun 1400-1500v.
Spanish: 0900-1000. **N:** 0915, 0930. Other times in **French.**
Netw. C in Berber/Arabic: 0600-0100 (incl. relays of Netw. A).
ANN: Arabic: "Huna Ribat, idha'atu-I-mamlaka al Maghribiyya". French: "Ici Rabat, Radiodiffusion Télévision Marocaine". Berber: "Dahab Rbad Lidaa Attalfaza Li mamlaka Lmaghrib".
V. by QSL-card. Rec.acc. Rp.

RADIO MÉDITERRANÉE INTERNATIONALE (Gov. Comm.)

✉ 3 et 5, rue Emsallah (B.P. 2055), Tanger. ☎🖷 +212 (9) 936363. ① RADMED 33711M. (✉ **in France:** 78 av.Raymond Poincaré, 75116 Paris) — **LW:** Nador 171kHz 2000kW.
MW: Sebaa-Aioun 1044kHz 300kW — **SW:** Nador (G.C: 02.55W/34.58N): 9575kHz 250kW — **FM:**Agadir 104.6, Casablanca 99.6, Fès 101.4, Marrakech 105.3, Nador 94.0/99.9, Meknes 105.5, Oujda 102.9, Rabat 97.5, Tanger 95.6, Tetouan 103.7MHz.
D.PRGR. in French/Arabic: 0500-0100. **N. in French:** 0630, 0730, 0830, 1230, 1700, 1930, 2200. **N. in Arabic:** 0600, 0700, 0800, 1200, 2000, 2300 — **ANN:** "Médi 1" — **V.** by letter.

VOICE OF AMERICA RELAY STATION

SW: Briech (G.C: 35.34N/05.58W): 10 x 500kW tr's.
Further details in International Broadcasting section.

Ceuta (Spanish)

RADIO NACIONAL DE ESPAÑA, S.A.

✉ Beatrice de Silva 12, E-11701 Ceuta. ☎ +34 (956) 522203. 🖷 +34 (956) 519067 — **FM:** 105.2MHz 1kW (rel.RNE R1).

RADIO CEUTA (Comm.)

✉ Ap.180, Real 90, E-11701 Ceuta. ☎ +34 (956) 511820. 🖷 +34 (956) 516820 — **MW:** 1584kHz 5kW 24h (rel. SER Madrid 2200-0500) — **FM:** 88.5MHz (music), 96.2MHz (R. Sol).

COPE CEUTA (Comm.)

✉ Sargento Mena 8, 1º izq, E-11701 Ceuta. ☎ +34 (956) 511122. 🖷 +34 (956) 517603 — **FM:** 89.8MHz.

OCR CONVENCIAL (Comm.)

✉ Grupos Alfa 4, 3º inz, E-11701 Ceuta. ☎ +34 (956) 617886. 🖷 +34 (956) 517004 — **FM:** 101.4MHz 3kW.

Melilla (Spanish)

RADIO NACIONAL DE ESPAÑA, S.A.

✉ Ap. de Correos 222, E-29801 Melilla. ☎ +34 (956) 681907. 🖷 +34 (956) 687332 — **L.P:** Dir: Pedro A.Medina Barrenechea.
MW: 972kHz 5kW (R1).
FM (0.3kW): 97.7 MHz (R1), 105.3MHz (R2), 107.6MHz (R3).

R.MELILLA (Comm.)

✉ C/Muelle Puerto Ribera, 18 , E-52005 Melilla. ☎ +34 (95) 2681708. 🖷 +34 (95) 2681753 — Dir: Gaspar Diaz Cerdá.
MW: EAJ21 1485kHz 4kW — 24h.

RADIO SINFO MELILLA (Comm.)

✉ Edif. Melilla, Urbaniz Rusadir, E-29801 Melilla. ☎ +34 (956) 688840. 🖷 +34 (956) 674317 — **FM:** 101.2MHz 0.2kW.

MOZAMBIQUE

L.T: UTC + 2h — **Pop:** 19.720.165 — **Radios:** 5m — **Pr.L:** Portuguese, ethnic — **E.C:** 50Hz, 220/380V — **ITU:** MOZ.

RÁDIO MOÇAMBIQUE (Gov. Comm.)

✉ C.P. 2000, Maputo. ☎ +258 (1) 431679. 🖷 +258 (1) 421816.
L.P: Chmn: Dr. Manuel Fernando Vetorano. TD: Eduardo Rufino de Matos. Inf. Dir: Ms. Maria Orlanda M. Mendes. Dir. of Prgrs: Daude Amade. Comm. Dir: João B. de Sousa. Head of Int. Rel: Marcos Mulezdera.

MW	kHz	kW	N	MW	kHz	kW	N
4) Inhambane	557	5	P	Maputo	1008	50	I
Maputo	738	50	N	Maputo	1079	50	C
5) Lichinga	782	5	P 8)	Tete	1160	5	P
3) Chimoio	836	5	P 6)	Quelimane	1179	50	P
1) Beira	872	50	P	Nampula	1223	5	P
Maputo	918	50	C/E 6)	Quelimane	1295	10	P
1) Beira	935	1	P 7)	Pemba	1493	5	P

FM: Maputo 98.01MHz (C), 97.9MHz (E), Beira 91.6MHz 10kW (N), 97.9MHz (C/E).
SW (G.C: Maputo 32.28E/25.57S, Beira 34.58E/19.40S):

Station	kHz	kW	N	H. of tr.
Maputo	3210	10	N	0250-0510, 1600-2210 (irr.)
1) Beira	3277	100	P	0250-0500, 1500-2100 (irr.)
Maputo	3338	10	N	0250-0530, 1600-2100
Beira	3370v	10	P	0250-0500, 1500-2210
Maputo	4855v	20	I/E	0250-2210
Maputo	4925v	7.5	I	0250-0430, 1500-2210
Maputo	5824.6		C	1530-2100 (Sun 2300)
Maputo	5845		c	r. 1400
Maputo	5932	10	I	0250-0430, 1500-2200
1) Beira	6025v	10	P	0250-2205 (irr.)
Maputo	6111v	10	N	0250-0700, 1600-2210
Maputo	9619	120	N/E	0400-2210
1) Beira	9637	100	P	0250-2255
Maputo	11812	120	E	1100-1135
Maputo	15291.8	120	N	0830-1205

Emissão Nacional (N) in Portuguese: 0250-2210. **N:** 0300-0900 hourly, 1030, 1200-1600 hourly, 1730. 1900, 2100, 2200.
Emissão Interprovincial de Maputo e Gaza (I) in Tsonga & Portuguese: 0250-2200. **N. in Portuguese:** 0530, 1700.
Emissão "C" (R. Cidade): 0800-2210 (Sun 2300). **Rel Ext. Sce. in English:** 1100-1135, 1800-1900.

Provincial Stations (P):

These st's broadcast in Portuguese and the ethnic languages of the province.
1) **Delegação de Beira,** C.P. 1942, Beira. ☎ +258 (3) 324674. Dir: Valentim Daniel — **"A" Prgr.** in Portuguese/ethnic languages on 872/935/3370/6025kHz, **"B" Prgr.** in ethnic languages/Swahili on 3280/9637kHz.
2) **Emissor Provincial de Nampula,** C.P. 93, Nampula. ☎ +258

213171. Dir: Gabriel Sauzande Jeque — 0250-2215
3) **Emissor Provincial de Manica,** C.P. 390, Chimoio. ☎ +258 (51) 22563. Dir: Maria Luísa Meneses — MF 0250-0515, 0955-1215, 1455-2010. Sat 0250-0515, 0955-2210. Sun 0250-2010.
4) **Emissor Provincial de Inhambane,** C.P. 196, Imhambane. ☎ +258 2456. Dir: Carlito José — MF 0400-0515, 1000-1215, 1600-2000. Sat 0400-0515, 1000-2000. Sun 0400-0700, 1000-2000.
5) **Emissor Provincial do Niassa,** C.P. 171, Lichinga. ☎ 258 2225. Dir: Américo Viana — 0250-2000.
6) **Emissor Provincial de Zambézia,** C.P. 333, Quelimane. ☎ +258 3024. Dir: António Barros — MF 0250-0515, 1000-2115. Sat 0250-0515, 1000-2215. Sun 0250-2155.
7) **Emissor Provincial de Cabo Delgado,** C.P. 45, Pemba. ☎ +258 2313. Dir: Carlitos José — MF 0400-0515, 1000-1215, 1600-2000. Sat 0400-0515, 1000-2000. Sun 0400-0700, 1000-2000.
8) **Emissor Provincial de Tete,** C.P. 384, Tete. ☎ +258 23174. Dir: Arsénio Palha — MF 0250-0515, 0955-2000. Sat 0250-0515, 0955-2215. Sun 0250-2000.
9) **Delegação de Xai-Xai,** C.P. 130, Xai-Xai. ☎ +258 22352. Dir: Gaspar Zunguene.

EXT. SCE: see International Broadcasting section.
ANN: "Dos seus estúdios em Maputo escutam a Rádio Moçambique, Emissão Nacional/Emissão C", "Rádio Moçambique, dos seus estúdios em Maputo transmite a Emissão Interprovincial de Maputo e Gaza". **E:** "This is the Ext. Sce. of R. Mozambique" or "This is R. Maputo, the Ext. Sce. of R. Mozambique".
IS: Mbira (indigenous xylophone) and spoken ID in Portuguese, English, French & Swahili. Opens and closes with National .Anthem. 4 note gong between prgrs — **V.** by QSL-card. Re. for most provincial st's via Maputo. Nampula verifies direct.

RADIODIFUSÃO PORTUGUESA

Beira 94.8MHz 5kW (rel. RDP-Africa).

R. France Internationale: Maputo 105.0MHz.

NAMIBIA

L.T: UTC + 2h — **Pop:** 1.827.915 — **Radios:** 230.000 — **Pr.L:** English (official), Afrikaans, German, Oshiwambo, Otjiherero, Tswana, Damara, Nama, Kwangali, Mbukushu, Gcirku, Lozi — **E.C:** 50Hz, 220V — **ITU:** NMB.

NAMIBIAN BROADCASTING CORPORATION

✉ P.O. Box 321, Windhoek 9000. ☎ +264 (61) 291 9111, 291 3111. 🖷 + 264 (61) 217760. ① 622. **E-mail:** nbcra@iwwn.com.n. **Web:** http://www.oneworld.org/cba/nbc.htm
L.P: DG: Dr Ben Mulongeni. Sen. Contr. Prgrs: Gabriel Haindaka. Sen. Contr. Auxiliary & Support Sces: Vitura Kavari. Contr. Mktg: Cyril Lowe. Contr. Tech. Sces: Maarten Venter. Head of Training & Development: Jimmy Amupala.

Mediumwaves	kHz	kW	Mediumwaves	kHz	kW
Tsumeb	594a	100	Gobabis	747b	100

a) Oshiwambo Sce. — b) Otjiherero Sce./Tirelo ya Setswana.
SW: Windhoek (G.C: 17.13E/22.33S): 2 x 100kW.

kHz	Prgr.	H. of tr.
3270	2	1700-0700
3290	1	1600-0800
4930	2	0700-1700
6060	1	0800-1600

Prgr. 1=Otjiherero Sce., Damara/Nama Sce, Prgr. 2=Afrikaans, German. Exact times of relays for each sce. are given below.
FM: 31 medium & low power st's.
National Sces 1 & 2: 0400-2300 on 21 FM-tr's. On all freq's MF 0400-0800 & 1900-2300, and on all Windhoek-based sces MF 1500-1600. **All Night Sce:** 2300-0400 on all freq's.
Oshiwambo Sce: 0800 (SS 0400)-1900 on 594kHz + 10 FM-tr's.
✉ Box 123, Oshakati. **IS:** Call of the Fish Eagle and wooden pestles — **Otjiherero Sce. & Tirelo ya Setswana:** MF 0800-1500 & 1600-1900, SS 0400-1900 on 747kHz + 7 FM tr's — **Damara/ Nama Sce:** MF 0800-1500 & 1600-1900, SS 0400-1900 on 18 FM-tr's — **Damara/Nama & Otjiherero:** 0800-1600 on 6060kHz, 1600-2000 on 3270kHz — **German Sce:** MF 0800-1500 & 1600-1900, SS 0400-1900 on 14 FM-tr's — **Afrikaans Sce:** MF 0800-1500 & 1600-1900, SS 0400-1900 on 22 FM-tr's — **German/Afrikaans:** 0600-1600 on 6175kHz, 0400-0600 & 1600-2300 on 3290kHz — **Rukavango Sce:**

0800 (Sat/Sun 0400)-1900 on 7 FM-tr's. ✉ Box 1000, Rundu —
Caprivi Sce: 0800 (Sat/Sun 0330)-1900 on 1 FM-tr. ✉ Box 1066, Katima Mulilo.
ANN: All Night Sce: "Here is the National Sce. of the NBC Nationwide".
G: "Hier ist das Deutsche Hörfunkprogramm der NBC". Dam/Nam: "Nes ge Damara/Nama Gowab loabas NBC's disa". A: "Dit is die Afrikaanse diens van die NBC". Otjiher: "Indji oradio ja Namibia moru-pa rueraka Otjiherero" — **IS:** at s/on: National Anthem with choir/orchestra — **V.** by QSL-card. Re. to NBC Shortwave Section, P.O. Box 321, Windhoek, Namibia.

RADIO 99 (Comm.)
✉ P.O. Box 11849, Windhoek. ☎ +264 (61) 223634, 225182.
🖷 +264 (61) 230964 — **FM:** Windhoek 99MHz + 4 rlys.

RADIO ENERGY (Comm.)
✉ P.O. Box 11720, Windhoek. ☎ +264 (61) 223863.
🖷 +264 (61) 230964 — **FM:** Windhoek 100MHz + 1 rly.

KATUTURA COMMUNITY RADIO (KCR)
✉ Katutura — **FM:** 106.2MHz — **D.PRGR:** 10h daily.

L.T: UTC + 1h — **Pop:** 9.015.000 — **Radios:** 440.000 — **Pr.L:** French + ethnic — **E.C:** 50Hz, 220/380V — **ITU:** NGR.

LA VOIX DU SAHEL (Gov.)
✉ B.P. 309, Niamey. ☎ +227 723155. ☼ 5229 NI.
L.P: DG: M.Diallo. Dir. R: A.Khamed. TD: Y.A.Tidjani. Dir. Admin: Y.Ibrahim. Editor-in-Chief: Adamou Oumarou.

MW	kHz	kW	MW	kHz	kW
Niamey	1125	20	Difa	1484	0.1
Tahoua	1215	0.1	Niamey	1575	1
Meninsoroua	1331	20			

FM: Niamey 90.2/93.6MHz + 13 relay st's.
SW: Niamey (G.C: 02.06E/13.30N):

kHz	kW	H. of tr.
3260v	4	0430-0700, 1700-2203 (Sat 1630-2300)
5020v	30/100	0430-0700, 1700-2203 (Sat 1630-2300)
7155	20	0700-2203
9705	100	1100-1400

D.PRGR: 0430-1400 & 1700-2203 (Sat 1630-2300). **N. in French:** 0545, 1200, 1900. **N. in English:** 2000 (Sun).
N.B: splits into two local language networks at 1745-1830.
ANN: "Ici la Voix du Sahel" — **V.** by QSL-card.

R. France Internationale: Niamey 96.2MHz.
Africa No. 1: Niamey 103.0MHz (see main entry under Gabon).

L.T: UTC + 1h — **Pop:** 122.840.000 — **Radios:** 17.200.000 — **Pr.L:** English, Yoruba, Hausa, Igbo — **E.C:** 50Hz, 230V — **ITU:** NIG.

NATIONAL BROADCASTING COMMISSION
✉ P.O. Box 55021, Lagos. ☎ +234 (1) 2647867. 🖷 +234 (1) 2647868.
L.P: DG: Dr. A. Tom Adaba.

NIGERIAN TELECOMMUNICATIONS LIMITED
✉ Dir. of Telecommunications, International Relations Division, Nigerian Telecommunications Limited, Private Mail Bag 12557, Lagos.
Cable: Dintrel.

MW	Location	kHz	kW	MW	Location	kHz	kW
2)	Akure	531	50	17)	Enugu	828	25
3)	Sokoto	540	50	21)	Azare	846	10
14)	Tukun Tawa	549	50	10)	Katsina Ala	846	10
4)	Alaho	567	50	22)	Port Harcourt	854	10
6)	Moniya	576	25	8)	Kaduna	882	25
6)	Abakaliki	585	50	18)	Abeokuta	900	25
17)	Enugu	585		19)	Fwagwa Lada	909	50
5)	Jaji	594	100	20)	Ikeja	918	50
18)	Ibesi	603	30	10)	Makurdi	918	50
16)	Ilorin	612	50	1)	Ikorodu	936	10
28)	Akwa	621	20	27)	Birnin Kebbi	945	10

MW	Location	kHz	kW	MW	Location	kHz	kW
6)	Enugu	621	10	6)	Enugu	954	10
9)	Maiduguri	630	50	25)	Katsina	972	50
22)	Port Harcourt	630	50	35)	Otite	972	10
8)	Kaduna	639	25	12)	New Bussa	981	10
29)	Warri	639	5	22)	Port Harcourt	981	10
4)	Ibadan	657	100	21)	Bauchi	990	50
7)	Benin	666	50	20)	Ikeja	990	10
15)	Obmomasho	675	25	15)	Ibadan	1008	10
30)	Damaturu	684	20	12)	Kontagora	1008	10
35)	Ayangba	693	10	34)	Osogbo	1008	10
12)	Suleja	695	10	11)	Yola	1017	10
31)	Wukari	702	25	33)	Dutse	1026	25
13)	Owerri	720	50	6)	Onitsha	1062	10
11)	Bali	722	20	6)	Nsukka	1071	
14)	Jogana	729	25	12)	Minna	1080	50
9)	Damagun	756	25	1)	Sogunle	1089	20
15)	Ibadan	756	100	11)	Yola	1089	50
12)	Minna	756	50	5)	Jaji	1107	25
11)	Bali	774	10	10)	Ogoja	1134	10
31)	Wukari	774	10	12)	Bida	1143	10
35)	Ochaja	783	50	18)	Abeokuta	1170	25
15)	Aha	792	25	11)	Yola	1206	10
30)	Damaturu	801	20	24)	Jos	1224	50
27)	Zuru	801	1	31)	Jalingo	1269	10
9)	Maiduguri	1296	20	29)	Warri	1397	10
24)	Jos	1314	10	21)	Gombe	1404	10
34)	Iwo	1359	10	8)	Kaduna	1416	25
23)	Calabar	1368	50	11)	Yola	1440	58
35)	Egbe	1390	10	1)	Sogunle	1458	1
26)	Abak	1395	10	12)	Mokwa	1476	10

Shortwaves	kHz	kW	Shortwaves	kHz	kW
1) Lagos	3326	50	4) Ibadan	6050	50
17) Enugu	3970	100	5) Kaduna	v6090	250/50
5) Kaduna	4770	50	5) Kaduna	7275	100
1) Lagos	4990	50	1) Lagos	7285	50
17) Enugu	6025	10	5) Kaduna	9570	50

FM: (MHz) 1) 92.936/97.666 – 3) 96.375 – 4) 93.54 – 5) 96.1 – 6) 96.1 – 7) 95.7 – 8) 90.892 – 9) 94.5 – 10) 95.0 – 11) 95.774 – 12) 91.02 – 13) 94.4 stereo – 14) 88.98/89.324/96.892 – 15) 95.8/98.5 – 16) 99.0 – 17) 92.85 – 18) 90.356 – 20) 89.5 – 21) 94.5 – 22) 99.128 – 23) 92.7 stereo – 24) 88.636/90.5 – 26) 90.258 stereo – 28) 88.5 – 29) 88.636/97.924 – 31) 90.614 – 32) 88.1 – 34) 89.76.

FEDERAL RADIO CORPORATION OF NIGERIA (Statutory Corporation).
✉ Radio House, P.M.B. 452, Garki, Abuja. ☎ +234 (09) 2346318, 2346319, 2346487. 🖷 + 234 (09) 2346486. Director General: Alhaji Abdurrahman Micika; Director, Finance and Supplies: Mr. Richard Isaiah; Director Personnel Management and Administration: Alhaji Usman Shettima; Director, Technical Services: Engineer P.V.I. Okonye; Director, News and Programmes: Mr. Tajudeen Akanbi; Secretary/Legal Adviser to the Corporation: Alhaji Mustapha Kurama; Head, Public Affairs Department: Mr. Olasupo Olaleye; Head, Training School and Manpower Development: Mr. Kola Ilori.
Addresses and other information
1) FRCN Lagos National Station, Broadcasting House, P.M.B. 12504, Ikoyi, Lagos. ☎+234 (1) 2690340. **L.P.:** Exec. Dir.: Atilade Atoyebi.
Radio Nigeria 1 (RN-1) in English: 0430-2300 on 1089/4990kHz, 0430-1000 & 1700-2300 on 3326kHz, 1000-1600 on 7285kHz, 1089kHz. **N:** On the h. **NB:** Nigerian N. at 0600, 1500 & 2100 is relayed by all FRCN stations and most state stations.
Radio Nigeria 2 (RN-2) in English: 0500-2300 on 1458kHz/97.666MHz (stereo). **N:** on the half h
Radio Nigeria 3 (RN-3) in Pidgin/English/Yoruba/Hausa/Igbo on 92.936MHz (stereo): 0430-2300.
— **IS:** Talking Drum. (Gangan). Opens and closes with the Nat. Anthem and spoken pledge.
2) Ondo Radio Vision Corp, P.M.B. 709, Akure, Ondo State — **GM:** Ade Adekanmbi. 0400-2300 in English/Yoruba.
3) Sokoto State Broadc. Corp, Private Mail Bag 2156, Sokoto, Sokoto State — **GM:** Alh. Sambo Bello. 0430-2315 in English/Hausa. **N:** 1100, 1430, 1900. **V.** by letter.
4) FRCN Ibadan National Station, Broadcasting House, Oba, P.M.B. 5003, Ibadan, Oyo State. ☎ +234 (2) 2413930, 2400660. 🖷 + 234 (2) 2413930. — Exec. Dir: S.O. Ajani. 0430-2305 in English/Yoruba/Edo/Igala/Urhobo. **V.** by letter.
5) FRCN Kaduna National Station, Dahiru Modibo House, 7 Yakubu Gowon Way, P.O. Box 250, Kaduna, Kaduna State. ☎ +234(62) 241069,

241070, 241072. ▤ + 234 (62) 241071 — Exec. Dir.: Alhaji Muhammad Ardo. Ch 1 in Hausa: 0400-2305 on 594/ 6090/7275 kHz. Ch 2 in English/Hausa/Fulfulde/Kanuri/Nupe: 0430-2305 on 1107/4770/9570kHz. **N.** on the h. **English N.** (both ch's): 1100, 1600, 1700. Ch 3 in English: 0430-2305 on 96.1MHz.

6) Enugu State Broadc. Service (ESBS), P.M.B. 01600, Enugu, Enugu State — MD: A. Okafor — **D.Prgr:** in English/Igbo on 585kHz: 0430-2300. **N:** on the h — **Prgr. 2** on 1062kHz: 0500-1800 (Sat/Sun 2400). **N. in English:** 0530, 1430, 2000 (rel. 1st Prgr). Relay R. Nigeria N. on all freq's: 0600, 1500. **Prgr. 3** on FM: ("Sunrise 96"): 0500-2100.

7) Edo Broadc. Sce, P.M.B. 1012, Benin City, Edo State — GM: Amokpae Egunibar. 0400-2305 in English + 12 local languages. **N:** 0530, 0600, 1200, 1500, 2100. **Ann:** "BBS Radio".

8) Kaduna State Media Corp., P.M.B. 2013, Kaduna, Kaduna State — MD: Alh. Muh. Bello Tukur. 0430-2305 in English/Hausa **Ann:** "This is the Kaduna State Media Corp, Kaduna".

9) Borno Radio Corp., P.M.B. 1020, Maiduguri, Borno State — GM: Kaith A. Gazali. 0400-2305 in English/Hausa/Kanuri. **Ann:** "Radio Sce. of BRTV broadcasting from Maiduguri".

10) R. Benue, P.M.B. 102202, Makurdi, Benue State — GM: I. Edime — **Prgr. 1:** 0430-2305 in English/Hausa + 7 local languages. **Prgr. 2** on **FM:** 0500-2105. **N:** 0800, 1100, 1600. **Ann:** "This is R. Benue, Makurdi".

11) Adamawa Broadc. Corp, P.M.B. 2123, Yola, Adamawa State — Administrator: Nyaka Mindapa. 0430-2300 in English/Hausa + 6 Nigerian languages. **N:** on the h + 0530, 1445. **Ann:** "This is GBC Yola, your No. 1 Radio Station".

12) Broadc. Corp. of Niger State, P.M.B. 88, Minna, Niger State — GM: Alh. Abdullahi Paiko. 0430-2330 on 756kHz, 0530-1900 on 1143kHz in English/Hausa.

13) Imo Broadc. Corp, P.O. Box 329, Owerri, Imo State — DG: Theophilus C. Okeke. **Prgr. 1:** 0425-2305 on MW. **Prgr. 2:** 0440-2305 on FM. **English:** 0430-0630, 1100-1830, 2100-2300 (Sat/Sun 0100). **N.** 0500, 0530, 1100, 1300, 1400, 1600, 1700, 2100, 2200, 2300. **Distance Education Broadcast:** Mon-Fri 1630-2200, Sat 1930-2130. **Rel. R. Nigeria:** 0600, 1200, 1500. Other times in Igbo. **Ann:** "This is Imo Broadc. Corp, Owerri".

14) Kano Broadc. Corp, P.M.B. 3014, Kano, Kano State — MD: Alh. Hassan Sulaiman. 0430-2320 in English/Hausa. **Prgr 2:** on FM: 0550-2320. — **Ann:** "This is KBC, the Kano State Radio Corp".

15) Broadc. Corp. of Oyo State, PMB 1, Ile Akede, Basorun, Ibadan, Oyo State — GM: J. Fademi. **Prgr. 1:** 0400-2200 in English/Yoruba. **Prgr 2:** on FM: 0700-2100. **Ann:** "R. Oyo Ibadan".

16) Kwara State Broadc. Corp., P.M.B. 1345, Ilorin, Kwara State — GM: J. Oni. 0400-2305 in English/Yoruba/Nupe/Barba. **N. in English:** 0500 (State news), 0600, 1200, 1400 (State news), 1500, 1900, 2100. **Ann:** "This is R. Kwara, Ilorin". **V.** by letter.

17) FRCN Enugu National Station, Broadcasting House, Onitsha Road, Enugu, Enugu State. ☎ +234 (42) 254371, 254400, 234137. ▤ + 234 (42) 255354 — Exec. Dir.: Andy Uchenna Anorado. 0430-2300 on 828kHz in English/Igbo (0430-1030 and 1445-2030 on 3970kHz); 0430-2300 on 6025kHz + FM (separate prgrs) in English/Igbo/Tiv/Efik and Izon. **V.** by letter. **F.PI:** 3 new SW-tr's.

18) Ogun State Broadc. Corp, P.M.B. 2084, Abeokuta, Ogun State — GM: Bashorun Tunde Elegbede. CE: A.A. Majiyagbe. Ch. 1 on MW, Ch. 2 on FM: 0400-2400 in English/Yoruba/Egun. **Ann:** "Nation's Model Station", "Ogun-Radio", "Voice from the Rocks".

19) FRCN Abuja National Station, Broadcasting House, Gwagwalada, Abuja Federal Capital Territory. ☎ +234 (9) 8821040, 8821065. — Head/Dir.: Mallam Shuaibu D. Ibrahim. 0430-2300 in English/Hausa/Igbo/ Yoruba and local languages. **Ann:** "The Capital R, the Voice of Unity".

20) Lagos State Broadc. Corp, P.M.B. 21035, Ikeja, Lagos State — GM: Engr. Ilesanmi H. Idowu. 0430-0005 in English/Yoruba/Ogu. **Ann:** "This is R. Lagos, Ikeja".

21) Bauchi Radio Corp, P.M.B. 0133, Bauchi, Bauchi State — GM: Alh. Hassan Adamu Shira. 0430-2300 on MW, 0500-1700 in Fulfulde (separate prgrs) on 94.5MHz. **F.PI:** 24h.

22) Rivers State Broadc. Corp P.M.B. 5170, Port Harcourt, Rivers State — GM: Gloria B. Fiofori. **Prgr. 1:** 0450-2310 on 630kHz, in English/Ikwerre/Kalabari/Khana/Kolokoma/Ekpeye/Okrika/Gokana/Nembe/Pidgin. **Prgr. 2:** 0450-2310 on FM. **Ann:** "This is R. Rivers".

23) Cross River State Broadc. Corp, P.M.B. 1035, Calabar, Cross Rivers State. GM: Joseph Okpa Oru. Prgrs in English/Efik/Ejagham/ Bekwara: 0430-2315 on 1368kHz, 0500-1200 & 1400-2030 on 1134kHz. **N:** on the h. **Ann:** "This is Cross River Radio, Calabar". **V.** by letter.

24) Plateau Radio Television Corp., P.M.B. 2043, Jos, Plateau State — Administrator: James Dimka. Ch. 1 on MW: 0500-2300 in English/Hausa/Ndago/Ngas/Berom/Eggon/Tarok/Goemai/Afizere/Mwaghavul. Ch. 2 on FM Stereo: 0500-2300. **Ann:** "This is Radio Plateau 1 AM", "This is Radio Plateau 2, 90.5 FM Stereo".

25) Katsina State Broadc. Corp., P.M.B. 2011, Katsina, Katsina State — GM: I. Hamisu. 0430-2300 in English/Hausa. **N. in English:** 0530, 0600, 1100, 1400, 1500, 1800, 2100. **Ann:** "This is Katsina State R."

26) Akwa Ibom Broadc. Corp., P.M.B. 1122, Uyo, Akwa Ibom State — GM: Rev. F. Mbaba. 0500-2300.

27) Kebbi Broadc. Corp., Broadc. House, Kalgo Rd, Birnin Kebbi, Kebbi State — GM: Alhaji Muhammadu Dan Kano. 0500-2300 in English/Hausa/Fulfulde/Dakarci/Zabarmanci. **Ann:** "You are tuned to Birnin Kebbi and listening to the services of the Kebbi Broadcasting Corporation broadcasting on 945 kHz, 317 medium wave".

28) Anambra Broadc. Sce, P.M.B. 5070, Awka, Anambra State — MD: Mrs. Edna Izuorah. 0500-2300.

29) Delta State Broadc. Sce, Broadc. House P.M.B. 5032, Asaba, Delta State — GM: Donald Obveredjo. 0500-2300.

30) Yobe Broadc. Corp, P.M.B. 1044, Damaturu, Yobe State — GM: Alh. Goni Fika. 0500-2300.

31) Taraba State Broadc. Sce, P.M.B. 1038, Jalingo, Taraba State — GM: Mall. Jibrin Tafida. 0500-2300.

32) Broadc. Corp. of Abia State, Broadc. House, Umuahia, Abia State — DG: Engr. Chiso Ekeke. 0500-2300.

33) Jigawa Broadc. Corp, Broadc. House, Dutse, Jigawa State — 0500-2300.

34) Osun State Broadc. Corp., P.M.B. 4425, Osogbo, Osun State — GM: Kayode Ajibade. 0500-2300 in English/Yoruba.

35) Kogi Broadc. Corp, 1 Mount Patti Rd, Lokoja, Kogi State — GM: Alhaji Abu Onaji. 0500-2300.

ANN: Exc. where stated, FRCN st's identify as "R. Nigeria" followed by location.

INTERNATIONAL SCE: see International Broadcasting section.

Ray Power 100 FM, DAAR Communications Ltd., Communication Village, Ilapo Village, Alagbado, Off Lagos-Abeokuta Expressway, Lagos — **L.P:** Chief Exec: Chief Raymond Dokpesi. GM: Olusesan-Olusola. **FM:** Lagos 100.5MHz. 24h.

Minaj Radio, Minaj Holdings Ltd., Ivie House (2nd Floor), 4/6 Ajose Adeogun Str, Victoria Island, Lagos — **L.P:** Exec. Chmn: Mike Ajegbo. **FM:** Obosi 88.5MHz.

REUNION

L.T: UTC + 4h — **Pop:** 642.000 — **Radios:** 170.000 — **Pr.L:** French — **E.C:** 50Hz, 220V — **ITU:** REU.

RADIO-TÉLÉVISION FRANÇAISE D'OUTRE MER (RFO)

✉ RFO-Réunion, 1 rue Jean-Chatel, F-97716 St. Denis Messag Cédex 09. ☎ +262 406767. ▤ +262 216484.
L.P: Regional Dir: Jean-Philippe Roussy. PD: Roland Mallett. Eng: Hubert Coulmont.
MW: St. Pierre 666kHz 20kW, St. André 729kHz 20kW

FM	MHz	FM	MHz
Sainte-Rose	87.9	Saint-Joseph	90.7
Saint-Pierre	89.0	Salazie	90.8
Saint-Denis Ville	89.2	Le Port	92.6
Piton Textor	89.6	Saint-Phillipe	92.6
Saint-Benoît	89.6	Silaos	93.8
Sainte-Suzanne	90.0	Saint-Denis Montagne	96.6
Saint-Leu	90.7		

D.PRGR: 0100-1900 (Sat 2000) .
ANN: "Société Nationale de Radio-Télévision Française d'Outre Mer, Station de la Réunion".
IS: "Séga & Maloya" (Réunion Folklore) — **V.** by QSL-card. Rp.

FRANCE-INTER: 97.1/98.8MHz + 7 relay st's.

PRIVATE STATIONS: Over 40 st's broadcast locally on FM.

RWANDA (Republic of)

L.T: UTC + 2h — **Pop:** 5.5m — **Pr.L:** Kinyarwanda, Swahili, French, English — **E.C:** 50Hz, 220V — **ITU:** RRW.

RADIO RWANDA

✉ B.P. 83, Kigali. ☎ +250 76665. ▤ +250 76185.
L.P: Dir. Broadcasting: Mweusi Karake. Dir Prgrs: Louise Kayibanda.

Ag. Ch.Editor: Faustin Karangira, Ch.Tech: Charles Nahayo.
SW: Kigali (G.C: 30.04E/01.58S): 6055kHz 50kW.
FM:
Channel I: 89.9/96.5/99.3/100.4/100.7MHz 15hrs MF, 18hrs W/E.
Channel II: 90.6/90.7/101.1/106.1MHz ,15hrs MF, 18hrs W/E.
93.7MHz BBC rebroadcast 24h.

DEUTSCHE WELLE RELAY STATION KIGALI
See International Broadcasting section.

SÃO TOMÉ E PRINCÍPE

L.T: UTC — **Pop:** 151.752 — **Radios:** 31.000 — **Pr.L:** Portuguese —
E.C: 50Hz, 220V — **ITU:** STP.

RÁDIO NACIONAL DE SÃO TOMÉ E PRINCÍPE
✉ Avenida Marginal 12 de Julho (C P. 44), São Tomé. ☎ +239 22875.
Cable: Rádio Nacional. **Telex:** 0967 217 radio st
L.P: Dir: Adelino Lucas dos Santos.
MW: 945kHz 20kW — **FM:** 91.5 (stereo)/95.5/99.5MHz.
D.PRGR: 0530-2300. **N:** 0600, 0700, 1000, 1200, 1400, 1600, 1745,
1930, 2200 — **ANN:** "Aqui São Tomé, Capital da República
Democratica de S. Tomé e Príncipe, transmite a Rádio Nacional.
IS: 1 note Gong, guitar — **V.** by letter. Rp.

R. France Internationale: on 102.8MHz in French & Portuguese.

VOICE OF AMERICA RELAY STATION
MW: Pinheira 1530kHz 600kW.
SW: 4 x 500kW tr's.
Further details in International Broadcasting section.

SENEGAL

L.T: UTC — **Pop:** 8.715.000 — **Radios:** 850.000 — **Pr.L:** French, ethnic — **E.C:** 50Hz, 220V — **ITU:** SEN.

SOCIÉTÉ NATIONALE DE RADIODIFFUSION TÉLÉVISION SÉNÉGALAISE (RTS)
✉ Triangle Sud (B.P. 1765), Dakar. ☎ +221 217801 or 236349.
📠 +221 223490. **Cable:** Radio Sénégal Dakar. ① 21818 — **L.P:** DG:
Guila Thiam. Dir. Radio: Ibrahim Sane. Dir. Tech. Sces: Joseph Nesseim.

Mediumwaves	kHz	kW	Sce.
1) Dakar	765	300/200	N
Podor	810	1	N
Matam	963	1	N
4) Ziguinchor	1224	20/1	R
5) Kaolack	1287	4	R
1) Dakar	1305	20	I
Linguére	1323	1	N
2) Saint-Louis	1368	20	R
3) Tambacounda	1503	20	R
1) Dakar	1539	10	N
Shortwaves	**kHz**	**kW**	**Sce.**
1) Dakar	4890v	100	N (inactive)
1) Dakar	7170	100	N

FM: Dakar 94.5MHz (N) — **G.C:** Dakar 17.26W/14.39N.
Chaîne Nationale (N): 0558 (Sun 0700)-0100. **N. in French:** 0600MF,
0630MF, 0700SS, 0800-2200 hourly (exc. 1200, 1600, 1700), 2355.
Chaîne Inter (I): MF 0558-1405 & 1758-0005, SS 0658-0005. **N:**
French: on the h. **Portuguese/Arabic/English:** MF 2005-2040.
Regional Centres: 2) B.P. 375, Saint-Louis. ☎ +221 611519. **ANN:**
"La Voix du Nord" — 3) Tambacounda. ☎ +221 811175. 0600-2200
— 4) B.P. 173, Ziguinchor. ☎ +221 911048. 0600-0800, 1200-1400,
1800-2400 — 5) B.P 321, Kaolack. ☎ +221 412265. 0600-0800, 1200-
1400, 1800-2230.
ANN: N: "Radiodiffusion Télévision du Sénégal émettant de Dakar".
I: "Sénégal-Inter" — **IS:** Melody on "Cora" (local harp).
V. by QSL-card. Rp.

SUD FM (Comm.)
✉ Immeuble Fahd, Bld. Djily Mbaye x rue Macodou Ndiaye (5ème
étage), Dakar. ☎ +221 225393, 224205. 📠 +221 225290.
Web: http://www.metissacana.sn/sud/prog.html
L.P: DG: Cherif Elvalide Seye. Editor-in-Chief: Oumar Diouf Fall.

FM (MHz)	MHz	kW	FM (MHz)	MHz	kW
Saint Louis	93.2	2	Ziguinchor	95.6	2
Kaolack	94.2	2	Matam	98.4	2
Diourbel	94.6	2	Dakar	98.5	3
Kolda	95.4	0.4M	Thiès	102.2	0.2

M) Mono. Others stereo.
Africa No. 1: Dakar 102.0MHz (see main entry under Gabon).
R. France Internationale: Dakar 92.0MHz, Kaolack 91.5MHz, Saint-
Louis 99.7MHz.
BBC World Sce: Dakar 105.6MHz (24h).

SEYCHELLES

L.T: UTC + 4h — **Pop:** 74.490 — **Radios:** 50.000 — **Pr.L:** Creole,
English, French — **E.C:** 50Hz, 240V — **ITU:** SEY.

SEYCHELLES BROADCASTING CORPORATION
✉ Box 321, Hermitage, Mahé. ☎ +248 224161. 📠 +248 225641. ①
2315 INFOTV SEZ. **E-mail:** — **L.P:** MD: Ibrahim Afif. Mgr. Admin &
Personnel: Ms Johan Ernesta. Prgr. Mgr.(AM Radio): Miss Marguerite
Hermitte. Prgr. Mgr.(FM Radio): Sarah Carpin. Financial Controller:
Parakrama Nanyakkara.
MW: ZCQ3 1368kHz 50kW.
D.PRGR: MF 0200-0930 & 1100-1800, SS 0200-1800. **N: English:**
0300, 0600, 0900, 1500. **French:** 0330, 0700, 1300, 1700. **Creole:**
0230, 0500, 0800, 1600.
ANN: E: "This is SBC Radio". F: "Ici la Radio SBC". C: "Isi Radyo SBC"
— **IS:** Instrumental music — **V.** by QSL-card.

R. France Internationale: Victoria 103.8MHz, Anse Soleil 102.8MHz.

FAR EAST BROADCASTING ASSOCIATION (FEBA)
BBC INDIAN OCEAN RELAY STATION
Further details in International Broadcasting Section.

SIERRA LEONE

L.T: UTC — **Pop:** 5.138.090 — **Radios:** 1m — **Pr.L:** English, ethnic
— **E.C:** 50Hz, 230/240V — **ITU:** SRL.

SIERRA LEONE BROADCASTING SERVICE (Gov.)
✉ New England, Freetown. ☎ +232 (22) 240123. 📠 +232 (22) 240922.
Cable: Broadcasts, Freetown. ① 3334 TADTEX SL.
L.P: Ag. DG: Mrs. Gina Banda-Thomas. Ag. Head of Prgrs (radio): Denis
Smith. Ag. Dir. Eng. Sces: F.B. Bunduka. Ag. CE: B.D.H. Taylor. Ag.
Contr. Eng. Sces: A.K. Sheriff. Project Eng: Steve Konteh.
MW: Goderich 1206kHz 50/10kW — **FM:** Freetown 99.9MHz.
SW: Goderich (G.C: 13.14W/08.30N): 3316kHz 10kW
National Sce: on MW/SW: 0558-2400. **N. in English:** 0700, 0800,
1900, 2000, 2100, 2200 — **FM Sce.** for Freetown: 24h.
ANN: "This is the SLBS in Freetown" or "You are tuned to Freetown",
IS: 5-note chime, military band — **V.** by QSL-card. Rp.

Other Stations
FM 94, c/o Bintumani Hotel, Aberdeen: 94MHz — **KISS 104,** Bo:
104MHz — **FM 93,** c/o Grace Brethren Church, Freetown: 93MHz —
Handicapped Radio, Freetown: 96.6MHz — **FM 96:** Freetown
96.0MHz — **Sky FM:** Freetown 106.0MHz (incl. relays of VOA) — **FM**
(freq unknown) c/o Wesleyan Mission, Makeni.

R. France Internationale: Freetown 89.9MHz.

SOMALIA

L.T: UTC + 3h — **Pop:** 11.387.872 — **Radios:** 300.000 — **Pr.L:** Somali
— **E.C:** 50Hz, 230V — **ITU:** SOM.

RADIO MOGADISHU, Voice of the Somali Republic
(in support of Interim President Ali Mahdi Muhammad)
SW: 6822vkHz — **D.PRGR:** 0400-0500 (Fri 0600), 1000-1100, 1300-
1430, 1600-1800. **N. in Arabic:** 1730.
ANN: "Halkani wa Radio Moqdisho, Odka Jamhuriyada Somalida".

RADIO MOGADISHU, Voice of the Masses of the Somali Republic
(in support of Saddam Aidid)

L.P: Chmn: Farah Hasanm Ayoboqore. Vice-Chmn: Abukar Haji Muhammad Gobdon. Secr. & Coordinator: Abd al-Karim Muhammad Kariyeh — **SW:** 6890vkHz.
D.PRGR: 0330-0500 (Fri 0400-0600), 0900-1300, 1500-1900.
ANN: "Halkani wa Radio Moqdisho, Odka Sha'abka ee Jamhuriyada Soomaaliyeed".

RADIO MOGADISHU, Voice of Somali Pacification
(in support of Osman Ato)
L.P: Head of Prgrs: Abdi Muhammad Isma'il.
SW: 6732vkHz.
D.PRGR: 0300-0500, 0930-1200, 1500-1900. **N. in English:** 1845.
ANN: S: "Halkani wa Radio Moqdisho, Odka Nabadeynta Soomaaliyeed". E: "R. Mogadishu, V. of the Somali Peace Processing"

HOLY QURAN RADIO
(in support of fundamental Islamic organization Ahlu Sunnag Waljama)
SW: 6545v kHz — **D.PRGR:** 1500 (Fri 1600)-1800.

NORTHERN SOMALIA

RADIO HARGEISA
P.O. Box 14, Hargeisa — **SW** (G.C: 44.03E/09.33N): 7540vkHz 1kW.
D.PRGR: 1000-1230, 1600-1800.
ANN: Halkani wa Radio Hargaisa, Codka Jamhuriyada Somaliland

Radio Borama: acc. to Reporters Sans Frontieres, a station of this name is also operating. No further details known.

SOUTH AFRICA (Republic of)

L.T: UTC + 2h — **Pop:** 44m — **Radio sets:** 7.500.000 — **Pr.L:** English, Afrikaans, No. & So. Sotho, Zulu, Tswana, Xhosa, Venda, Tsonga — **E.C:** 50Hz, 220/380V + d/c — **ITU:** AFS.

SOUTH AFRICAN BROADCASTING CORPORATION (SABC)
Private Bag XI, Auckland Park 2006. ☎ +27 (11) 714 9111. +27 (11) 714 4086. **Web:** http://www.sabc.co.za **Regional** P.O. Box 2551, Cape Town 8000 – P.O. Box 1588, Durban 4000 – P.O. Box 563, Bloemfontein 9300 – P.O. Box 912600, Silverton 0127 – P.O. Box 1040, Port Elizabeth 6000 – P.O. Box 395, Pietersburg 0700 – P.O. Box 2724, Nelspruit 1200.
L.P: Group Chief Exec: Zwelakhe Sisulu. Chief Exec. TV: Molefe Mokgatle. Chief Exec. Radio: Rev. Hawu Mbatha. Chief Exec. Human Resources: Ntombi Langa-Royds. Chief Exec. Business: Talib Sadik. Group Functions Co-Ordinator: Leslie Xinwa. Group Mgr. Communications: Enoch Sitole. Sen. GM. Strategic Planning: Solly Mokoetle. GM. Africa Enterprises: Madala Mphahlele. GM Community Radio: Nico de Kock. GM Metropolitan Radio: Koos Radebe. GM Regional Radio: Charlotte Mampane. Exec. Editor Channel Africa: Hans-Dieter Winkens. Mgr. Human Resources (Radio): Busi Mlotshwa. Editor-in-Chief Radio N: Barney Mthombothi. Mgr. Educ. Radio: Fakir Hassen.

HOME SERVICES (Comm.)

Mediumwaves:
Location	Svc	kHz	Location	Svc	kHz
Klipheuwel	Cape Talk	567	Komga	R.Xhosa	846
Meyerton	R.Metro	576	Ga-Rankuwa	Mmabatho R.	1098
Meyerton	Pulpit/Kansel	657	Welgedacht	R.Swazi	1287
Klipheuwel	Punt Media	729	Welgedacht	R.Ndebele	1404
Ga-Rankuwa	R.702	702			

Northern Province
FM (MHz)	Grense	SAFM	R. Sonder R. 2000	5 FM
Blouberg	102.3	105.9	-	95.5
Hoedspruit	102.0	105.6	98.5#	-
Louis Trichardt	100.7	104.3	97.2#	-
Nylstroom	102.9	106.5	-	-
Potgietersrus	101.4	105.0	97.9#	-
Thabazimbi	101.9	105.5	98.4#	-
Tzaneen	102.6	106.2	107.7#	-

NW Province
FM (MHz)	Grense	SAFM	R. Sonder R. 2000	5 FM
Christiana	103.6	107.2	-	-
Enzelsberg	101.6	105.2	-	-
Groot Marico	102.3	105.9	-	-
Klerksdorp	101.2	104.8	97.7#	-

FM (MHz)	Grense	SAFM	R. 2000	5 FM
Piet Plessis	102.8	106.4	-	-
Pomfret	101.1	104.7	-	-
Rustenburg	100.7	104.3	97.2#	-
Schweize-Renke	103.1	106.7	99.6#	-
Zeerust	102.6	106.2	99.1#	-

Gauteng — R. Sonder
FM (MHz)	Grense	SAFM	R. 2000	HVST
Heidelburg	100.8	104.4	97.3#	94.0
Johannesburg	101.5	105.1	99.7#	94.5
Menlo Park	102.1	105.7	98.6#	-
Pretoria	101.0	104.6	97.5#	-
Welverdiend	102.0	105.6	98.5#	95.5

Mpumalanga — R. Sonder
FM (MHz)	Grense	SAFM	R. 2000	JAKR
Carolina	103.0	106.6	-	96.2
Davel	103.5	107.1	100.0#	96.7
Dullstroom	100.8	104.4	-	94.0
Lydenburg	102.8	106.4	-	96.0
Middelburg	101.8	105.4	-	95.0
Nelspruit	102.5	106.1	99.0#	95.7
Piet Retief	102.1	105.7	-	95.3
Sabie	104.2	107.9	-	97.1
Volksrust	102.6	106.2	-	95.8

Northern Cape — R. Sonder
FM (MHz)	Grense	SAFM	R. 2000	5 FM
Alexander Bay	102.2#	105.8#	98.7#	92.2#
Calvinia	101.5	105.1	-	-
Carnavon	102.5	106.1	-	-
Colesburg	103.8	107.5	-	-
De Aar	102.0	105.6	-	-
Douglas	102.9	106.5	-	-
Faans Grove	103.0	106.6	-	-
Garies	100.7#	104.3#	-	-
Kimberley	101.0	104.6	97.5#	91.0
Kuruman	102.4	106.0	-	-
Pofadder	102.8@	106.4@	-	-
Prieska	100.8	104.4	-	-
Sringbok	101.6@	105.2	-	-
Upington	101.7	105.3	-	-
Victoria West	101.1	104.7	-	-
Williston	103.2#	-	-	-

Free State — R. Sonder
FM (MHz)	Grense	SAFM	R. 2000	5 FM
Bethlehem	101.9	105.5	98.4#	-
Bloemfontein	103.0	106.6	99.5#	91.6
Boesmanskop	101.2	104.8	-	-
Ficksburg	103.7	107.3	-	-
Kroonstad	103.4	107.0	99.9#	93.4
Ladybrand	102.1	105.7	-	-
Petrus Steyn	102.3	105.9	98.8#	-
Senekal	101.1	104.7	97.6#	-
Springfontein	102.6	106.2	99.1#	-
Theunissen	102.5	106.1	99.0#	92.5
Witsieshoek	101.3	104.9	-	-

Kwazulu Natal — R. Sonder
FM (MHz)	Grense	SAFM	R. 2000	5 FM
Donnybrook	102.7	106.3	99.2#	-
Durban	100.8	104.4	97.3#	89.9
Durban North	102.5	106.1	99.0#	103.8
Eshowe	103.4	107.0	99.9#	-
Glencoe	103.1	106.7	99.6#	-
Greytown	101.7	105.3	98.2#	-
Kokstad	101.0	104.6	-	-
Ladysmith	101.0	104.6	97.5#	-
Matatiele	101.5	105.1	-	-
Mooi River	102.2	105.8	98.7#	-
Nongoma	102.9	106.5	99.4#	-
Pietermaritzburg	101.4	105.0	97.9#	100.3
Port Shepstone	101.3	104.9	97.8#	-
The Bluff	102.0	105.6	98.5#	107.4
Ubombo	102.4	106.0	98.9#	-
Vryheid	101.2	104.8	97.7#	-

Western Cape — R. Sonder
FM (MHz)	Grense	SAFM	R. 2000	5 FM
Beaufort West	100.7@	104.3@	-	-
Constantiaberg	102.1	105.7	98.6#	89.0
Ceres	103.7	107.3	-	-
Franschhoek	100.7	104.3	97.2	-
George	101.7	105.3	98.2#	91.7
Grabouw	101.7	105.3	-	-
Hermanus	100.8	104.4	97.3#	-

Western Cape FM (MHz)	R. Sonder Grense	SAFM	R. 2000	5 FM
Hex River	102.0	105.6	-	-
Hout Bay	100.9	104.5	97.4#	87.8
Kleinmond	104.2	107.9	-	-
Knysna	102.2	105.8	98.7#	92.2
Ladismith	101.4	105.0	-	-
Matjiesfontein	102.8	106.4	-	-
Montagu	104.2	107.9	-	-
Napier	102.4	106.0	-	-
Oudtshoorn	102.6	106.2	99.1#	92.6
Pearl	101.6	105.2	98.1#	88.5
Piketberg	101.1	104.7	97.6#	-
Plettenberg	100.8	104.4	-	-
Riversdale	100.9	104.5	-	-
Sea Point	103.5	107.1	100.0#	90.4
Simonstown	100.7	104.3	97.2#	87.6
Stellenbosch	100.9	104.5	97.4#	87.8
Table Mountain	102.6	106.2	99.1#	89.9
Tygerberg	103.0	106.6	99.5#	88.2
Uniondale	103.4	107.0	-	-
Vanrhynsdorp	103.4	107.0	-	-
Villiersdorp	103.3	106.9	99.8#	-

Eastern Cape FM (MHz)	R. Sonder Grense	SAFM	R. 2000	JAKR
Aliwal North	101.7	105.3	-	-
Andrieskraal	103.2	106.8	-	-
Barkly East	100.9	104.5	-	-
Bedford	100.8	104.4	-	-
Burgersdorp	103.9	107.6	-	-
Butterworth	101.1	104.7	-	-
Cala	103.4	107.0	-	-
Cradock	102.7	106.3	-	-
East London	101.6	105.2	98.1#	88.5
Elliot	101.4	105.0	-	-
Graaff-Reinet	103.3	106.9	-	-
Grahamstown	103.5	107.1	100.0#	90.4
Hankey	101.0	104.6	-	-
Kareedouw	102.9	106.5	-	-
King Williams Town	103.0	106.6	-	-
Mount Ayliff	103.2	106.8	99.7#	-
Noupoort	101.4	105.0	-	-
Patensie	101.5	105.0	-	-
Parsons Hill (PE)	101.0	104.6	-	-
Paul Sauer Dam	103.6	107.2	-	-
Port Elizabeth	102.3	105.9	98.8#	89.2
Port St.Johns	103.7	107.3	100.2#	-
Queenstown	102.2	105.8	98.7#	-
Suurberg	101.8	105.4	-	-
Ugie	102.6	106.2	-	-
Umtata	102.0	105.6	98.5#	-
Willowmore	101.2	104.8	-	-

= mono no RDS, @ = mono RDS, HVST = Highveld Stereo, JAKR = Jakarander Stereo

NATIONAL SHORTWAVE SERVICES

SW: Meyerton (G.C: 28.08E/26.35S): 4 x 100kW tr's + 1 standby tr.

kHz	Sce.	H. of Tr.	kHz	Sce.	H. of Tr.
3320	RSG	1700-2300	6000	RSG	0435-0655
	R. 2000	2300-0300	7185	RSG	0700-1655

RSG=R. Sonder Grense

COMMERCIAL AND COMMUNITY RADIO

5 FM in English: 24h. **N:** 0400W, 0500, 0600, 1100, 1500, 1630MF.
Headlines (MF): 0430, 0530, 0630, 1430, 1530, 1630.
R. Metro, P.O. Box 93116, Johannesburg 2000 — **MW:** Meyerton 576kHz 50kW. **FM:** Johannesburg 96.4MHz, Durban 93.0MHz, Durban No. 107.9MHz, Cape Town 93.0MHz, Pretoria 92.4MHz + 7 relays. 24h in **English** on 576kHz. **N:** 0400 & 0500 (Mon-Fri), 0600, 1100, 1400 (Mon-Fri), 1600. Rel. on all FM/MW st's: 2100-0300 (exc. R. Xhosa 2200-0200) — **ANN:** "This is R. Metro 5-7-6 stereo", the Action st.", "The Sound of the City".
R. 2000: 0300-2200.
R. Good Hope Stereo: (Cape Town): 95.3 MHz 3kW (stereo) + 30 relays.
R. Lotus (Durban): 87.7MHz 10kW + 7 relays.

METROPOLITAN RADIO

SAFM in English: 0300-1930 on FM only. **N:** 0330 (W), 0400 (W), 0500, 0600, 0900, 1100, 1400 & 1500 (Mon-Fri), 1600, 1700 (W), 1900.

Radio Sonder Grense in Afrikaans: 0300-1930 on FM. **SW:** as above.
Umhlobo Wenene FM Stereo, P.O. Box 36, King William's Town 5600 — FM: King Williams Town 93.0MHz 3kW + 42 relays.
Ukhozi FM, P.O. Box 4559, Johannesburg 2000; P.O. Box 1588, Durban 4000 — **FM:** Johannesburg 91.5MHz 10kW + 19 relays.
Lesedi, P.O. Box 1970, Johannesburg 2000 — **FM:** Bloemfontein 89.9/Johannesburg 88.4/Kroonstad 90.3MHz 10kW + 13 relays.

REGIONAL RADIO

Motsweding FM, P.O. Box 28151, Sunnyside 0132 — **FM:** Bloemfontein 93.0MHz 10kW + 16 relays.
Munghana Lonene FM (Pietersburg): **FM:**
Thobela FM, P.O. Box 1869, Pietersburg 0700
Phalaphala FM, Pietersburg — **FM:** Johannesburg 87.9MHz 3kW + 4 relays.
Ligwalagwala FM, P.O. Box 962, Pretoria 0001 — **FM:** Nelspruit 92.5MHz + 7 relays.
Ikwekwezi FM, P.O. Box 962, Pretoria 0001 —**FM:** Middelburg 91.8MHz 3kW + 5 relays.

CHANNEL AFRICA

See International Broadcasting section.

OTHER STATIONS

INDEPENDENT BROADCASTING AUTHORITY (IBA)

✉ 26 Baker Str, Rosebank, Johannesburg 2196 or Private Bag X31 Parklands 2121. ☎ +27 11 447 6180/7. 🖷 +27 11 447 6186/9. **E-mail:** theiba@wn.apc.org **Web:** http://wn.apc.org/iba/
LP: Co-chairpersons: Ms Felleng Sekha.The IBA is charged with the promotion of a diverse range of sound and television broadcasting services on a national, regional and local level which, when viewed collectively, caters for all languages and cultural groups and provides entertainment, education and information.

CAPE TALK (Comm.)

✉ Private Bag X567, Vlaeberg, Western Cape. ☎ +27 (21) 488 1500. 🖷 +27 (21) 488 1600. **LP:** M. Wills. **E-mail:** 567@capetalk.co.za.
MW: Cape Town 567kHz, 50kW.
✉ PO Box 211, Greenpoint, Western Cape. ☎ +27 (21) 406 8900. 🖷 +27 (21) 406 8960. **LP:** T.Ntokwana **E-mail:** p4@p4radio.co.zaradio.
FM: Cape Town 104.9MHz, 2.5kW.

PUNT OP MEDIUM GOLF (Comm.)

✉ PO Box 3662, Tyger Valley, W. Cape. ☎ +27 (21)418 9737. 🖷 +27 (21) 418 9748. **LP:** D.van Lill. **E-mail:** puntgesels@punt.co.za
MW: Cape Town 729kHz, 50kW.
✉ Private Bag X201, Midrand, Gauteng. ☎ +27 (11) 805 5490. 🖷 +27 (11) 805 5495. **LP:** D.du Toit. **E-mail:** puntgesels@punt.co.za
MW: Cape Town 1332kHz, 50kW

Y-FM

✉ PO Box 94244, Yeoville, Gauteng. ☎ +27 (11) 624 3417. 🖷 +27 (11) 618 4206. **LP:** R.Abrahams. **E-mail:** yfm@yfm.co.za.
FM: Johannesburg 99.2MHz, 5kW.

CLASSIC FM

✉ PO Box 782, Auckland Park, Gauteng. ☎ +27 (11) 408 5235. 🖷 +27 (11) 408 5249. **LP:** E.de Vos. **E-mail:** classicfm@icon.co.za.
FM: Johannesburg 102.7MHz

RADIO KFM

✉ Private Bag X945, Cape Town. ☎ +27 (21) 418 7000. 🖷 +27 (21) 418 8647.**LP:** Kevin Savage. **E-mail:** hod@kfm.co.za
FM: Cape Town 94.5MHz + 23 relays.

RADIO ALGOA/BRFM

✉ PO Box 180, Port Elizabeth, Eastern Cape. ☎ +27 (41) 339 497. 🖷 +27 (41) 339 279. **LP:** M.Vincent. **Web:** http://algoa@radioalgoa.
FM: Port Elizabeth 95.5MHz + 19 relays.

RADIO ORANJE

✉ PO Box 7117, Bloemfontein, Free State. ☎ +27 (51) 505 0900. 🖷 +27 (51) 505 0905. **LP:** A.Grobbelaar. **E-mail:** ofm@internext.co.za.
FM: Bloemfontein 96.2MHz + 16 relays.

HIGHVELD STEREO

✉ PO Box 3438, Rivonia, Gauteng. ☎ +27 (11) 883 1883. 🖷 +27 (11) 282 3652. **LP:** M.Freid. **E-mail:** 94.7@highveld.co.za
FM: Johannesburg 94.7MHz + 3 relays.

EAST COAST RADIO

✉ Private Bag X9495, Durban, Kwazulu-Natal. ☎ +27 (31) 204 9495. 🖷 +27 (31) 207 9415. **LP**: D.Macleod. **Web**: http://ecr@dbn.lia.net.
FM: Durban 94.0MHz + 16 relays.

RADIO P4

✉ PO Box 4995, Durban. ☎ +27 (31) 30579. 🖷 +27 (31) 30579.
LP: Z.Mapipa. **FM**: Durban 99.5MHz.

RADIO JACARANDA

✉ PO Box 11961, Centurion. ☎ +27 (12) 673 9100. 🖷 +27 (12) 657 0105.
LP: W.Engelbrecht. **FM**: Menlo Park 95.3MHz + 16relays.

Community Stations

August 1998 - 70 community stations holding 1 year temporary licences licenced by IBA .

L.T: UTC — **Pop**: 5.644 — **Radios**: 2.500 — **Only L**: English — **E.C**: 50Hz, 240V — **ITU**: SHN.

ST. HELENA GOVERNMENT BROADCASTING SCE.

✉ The Castle, Jamestown, Island of St. Helena, So. Atlantic Ocean. ☎ +290 4669. 🖷 +290 4542. ① 4202 GOVT HL.
Web: http://www.sthelena.se/ — **LP**: SM: Tony Leo.
MW (G.C: 15.55S/05.32W): 1548kHz 1/0.1kW.
SW: uses SSB tr. of Cable & Wireless one day per year (usually in October) for special DX test on 11092.5kHz at 1900-2300 UTC. Check the Web site for details of the 1998 transmission.
D.PRGR. on MW: 0900-1500MF rel.BBC prgrs, 1400-2330D local prgrs incl. **BBC N**. 2000 & 2200 — **ANN**: "Radio St. Helena" — **IS**: Trumpet call 11 seconds repeated at 5 second intervals for 64 seconds followed by "Life on the Ocean Waves" — **V.** by QSL-card. Rp.

L.T: UTC + 2h — **Pop**: 32.294.201 — **Radios**: 5.755.000 — **Pr.L**: Arabic — **E.C**: 50Hz, 240V — **ITU**: SDN.

SUDAN NATIONAL BROADCASTING CORP.

✉ P.O. Box 572, Omdurman ☎ +249 (11) 53151. **Cable**: IZZA.
Regional ✉ Box 532, Khartoum; Box 126, Juba — **LP**: DG: Salah al-Din al-Fadhil Usud. DG Eng. & Tech. Sces: Abbas Sidig.

Location	kHz	kW	Prgr.	Times
Nyala	540*	50		
Omdurman	576	7	G	0300-0600
Khartoum	576	100	G	0300-0600
			K	1500-1700
El Obeid	639*	5		
Juba	639*	200		
Kassala	666*	5		
Omdurman	765	50	G	0300-0715, 1200-1500
			E	1530-1630, 1700-1900
			2	1900-2200
Atbara	783*	5		
Dongola	819*	5		
Medani	873*	5		
Khartoum	963	100	G	0300-1330, 1600-2200
			H	1330-1600
Reba	1296	1500	G	0300-0500, 1400-1700
			V	1700-1900
			G	1900-2200
Omdurman	7200	100	G	0300-0830 & 1900-2200
Omdurman	9024v	20	G	0300-0600, 2100-2200
			N	1245-1530
			E	1530-1630, 1700-1900

*) listed but unconfirmed.
Prgrs: G=General Prgr, 2=Second Prgr, E=External Sce, H=Holy Quran Sce, K=Khartoum State Radio, N=National Unity Radio, V=Voice of the Nation (from Khartoum)
FM: Omdurman 91.0MHz.
SW: Omdurman (C.C: 32.28E/15.30N): 2 x 120kW, 1 x 100kW, 2 x 20kW, 1 x 7.5kW, 2 x 2kW, 1 x 1kW.
EXT. SCE: see International Broadcasting Section.
ANN: "Huna Omdurman, idha'atu-l-gümhuriya as-Sudan ad-dimuqratiya" — **IS**: Omdurman: Sudanese music. Opens and closes with National Anthem.

L.T: UTC + 2h — **Pop**: 1.063.736 — **Radios**: 500.000 — **Pr.L**: English, SiSwati — **E.C**: 60Hz, 220V — **ITU**: SWZ.

SWAZILAND BROADCASTING SCE. (Gov. Comm.)

✉ P. O. Box 338, Mbabane. ☎ +268 47263/4/5. 🖷 +268 42774.
L.P: Dir: T.Makama. Dep. Dir: Miss Thembie Mkoko. Principal Prgrs. Officer: V.Maziya. Prgr. Coordinator: Timothy Shongwe. Tr. Engineer: Stan M. Motsa. Principal Inf. Officer: Jerome Dlamini. PR Officer: Percy Simelane.
MW: Sidvokodvo 954kHz 50/30kW (stereo).
FM: 93.02/95.6MHz 10kW + 4 low power relays (all stereo).
D.PRGR: 0255-2000. **English** (approx): 0255-0630, 1400-2000.
N: 0400, 0500, 1600, 1700, 1900 (VOA). **Rel. BBC African Sce**. 1710-1745MF — **ANN**: "This is the English sce. of R. Swaziland 954 AM stereo". "Lona ngu Mawakato waka Ngwane".
IS: at s/on, Cilongo (Swazi instrument). English Sce: cock crow, fanfare, spoken ID, instrumental theme.

SWAZILAND COMMERCIAL RADIO (Pty) Ltd.

✉ P. O. Box 1586, Alberton 1450, So. Africa. ☎ +27 (11) 434 4333. 🖷 +27 (11) 434 4777. **E-mail**:cicade@icon.co.za
L.P: MD: Mr A.de Andrade. PD: Helder dos Santos.
MW: Sandlane 1377kHz 50kW. 1400-2100 R.Cidade Portuguese
SW: Mbabane (G.C: 30.48E/26.38S): 6155 (alt:9704)kHz 100kW.
English to So. Africa: MF 1700-2030, Sat 1700-2000, Sun 0500-0600 & 1700-2045.
ANN: "This is Swazi Radio" — **V.** by letter. Re. to Chief Eng.

COMMUNITIES BROADCASTING SCES. cc (Comm.)

✉ P.O. Box 15535, Doornfontein 2028, So. Africa. ☎ + 27 (11) 434 1431/2. 🖷 +27 (11) 434 1867 — **L.P**: PD: D.Bettancourt.
SW: Mbabane (G.C: 30.48E/26.38S): 6155kHz 100kW.
FM: 99/100MHz.
Portuguese ("R. Cidade" ✉ 81 Main St, Kenilworth, Johannesburg 2190, So. Africa): MF 0700-1000, Sat 0900-1200, Sun 1200-1500.
Portuguese/English ("R. Cidade Jovem"): Sat 1200-1600. **Italian** ("Transmissione Italiana"): Sat 0600-0900. **Greek** ("Hellenic Heart Beat"): Sun 1100-1200.
ANN: P: "Radio Cidade...ques ouve, gosta".

TRANS WORLD RADIO (Rlg.)

MW: 1170kHz 50kW.
Further details in International Broadcasting section.

L.T: UTC + 3h — **Pop**: 30.953.173 — **Radios**: 4.500.000 — **Pr.L**: Swahili, English — **E.C**: 50Hz, 240V — **ITU**: TZA.

RADIO TANZANIA - DAR ES SALAAM (Gov. Comm.)

✉ Nyerere Rd, P. O. Box 9191, Dar es Salaam. ☎ +255 (51) 860760. 🖷 +255 (51) 865577.
L.P: Dir. of Broadc: Abdul Ngarawa. Contr. of Prgrs: Mrs. Edda Sanga. Ag. Ch.Ed.Mr E.Kahuranga. Head of English Sce. & Int. Rel. Unit: Ms.P.Nyamungumi. CE: Taha Usi. Dep. CE: Emmanuel Mangula.
Networks: G=General Sce, R=Regional Sces, E=English Sce.

Mediumwaves	kHz	kW	Sce.
Dar es Salaam	531	10	G
Dodoma	603	100/10	G/R
Mbeya	621	50/10	G/R
Nachingwea	648	100/10	G/R
Dar es Salaam	657	2 x 50	G
Kigoma	711	100/10	G/R
Mwanza	721	50/10	G/R
Dar es Salaam	837	1	G/E (standby tr.)
Songea	990	100/10	G/R
Dar es Salaam	1035	10	E
Arusha	1215	50/10	G/R

SW: Dar es Salaam (G.C:39.14E/06.50S): 2 x 100kW, 2 x 50kW, 4 x 10kW.

kHz	kW	Sce.	Times
5050	10	G	0500-1000, 1615-2400
	10	E	0500-1000, 1615-2400
5985	100	G	0500-1000, 1615-2400
6105	100	G	0500-1000, 1615-2400
7165	100	G	1000-1615
7280	100	G	1000-1615
9530	100	G	1000-1615

FM (G): Dar es Salaam 92.5MHz, Dodoma 87.7MHz, Kigoma 88.4MHz, Masasi 92.3/105MHz, Songea 98.7MHz — (G): Dar es Salaam 94.6MHz — (E) Dar es Salaam 89.9MHz.
General Sce. in Swahili: 0500-2400. **N:** on the hr.
Regional Sce. in Swahili: 1300-1600.
External Sce. in English: see International Broadcasting section.
ANN: G: "Hii ni Radio Tanzania Dar es Salaam".
IS: Celeste — **V.** by QSL-card.

PRIVATE STATIONS

R. Kwizera, N'Gara: 97.9MHz (incl. prgrs from R. France Int.).
R. One, P.O. Box 4374, Dar es Salaam. ☒ +255 (51) 75914, 73980, 73998. ☒ +255 51 75915. **Telex:** 41133 IPP–TZ — Dir: Reginald Mengi. CE: M. Mazana — 1440kHz + 99.6MHz.
R. Tumaini, P.O. Box 167, Dar es Salaam: 96.3MHz. Operated by the Catholic Church. Prgrs incl. relays of R. France Int, Deutsche Welle, R. Tanzania.

ZANZIBAR & PEMBA

SAUTI YA TANZANIA ZANZIBAR (Gov.)

☒ P.O. Box 1178, Zanzibar, Tanzania. ☎ +255 (54) 31088/9. ① 57207 STZ — Dir. of Broadc: Yusuf Omar Chunda. Chief Eng: Abdulrahman M.Said — **MW:** Chumbum 585kHz 10kW.
SW: Dole (G.C: 39.14E/06.05S): 6014.6kHz 50kHz.
D.PRGR: 0300-0500 on 585kHz, 1100-2000 on 585/6014.6kHz.
N. in Swahili: 0900, 1300, 1500, 1645.
ANN: "Hii ni Sauti ya Tanzania Zanzibar".
IS: Indigenous xylophone — **V.** rarely by letter.

TOGO

L.T: UTC — **Pop:** 4.236.000 — **Radios:** 720.000 — **Pr.L:** French, ethnic — **E.C:** 50Hz, 127/220V — **ITU:** TGO.

RADIODIFFUSION TOGOLAISE (Gov.)

☒ B.P. 434, Lomé. ☎ +228 212493. **Cable:** Radio Togo.
L.P: Dir: Bawa Semedo. CE: Dozdi Soares.

kHz	kW	H. of tr.
1394	20/1	0455-0005
5047	50/100	0455-0903, 1608-0005
7265	50/100	0904-1607

FM: 88.3/92.7/96.8/99.4MHz.
Main N. in French: 0600, 0700, 1000, 1100, 1230, 1600, 1900, 2200.
N. in English: 1005, 1945 (varies 1945-2015).
ANN: "Radiodiffusion Togolaise Chaîne Inter", "Radio Togo", "Radio Lomé" — **IS:** Soft tempo chime. National Anthem at s/on and s/off.
V. by QSL-card Re. in F, E. Rec.acc.

RADIO KARA (Regional st.)

☒ B.P. 21, Kara. ☎ +228 606060.
L.P: Dir: Kao Pérézi. CE: Tete Anani.

kHz	kW	H. of tr.
1502	10	0525-0905, 1200-1435, 1625-2105
3222v	10	0525-0905, 1625-2305
6155	10	1200-1435

FM: 91.4/91.9/94.6/99.2MHz.
N. in French: 0535, 0700, 0900, 1205, 1300, 1430, 1700, 1930, 2130, 2300 — **ANN:** "Radiofiffusion Kara" — **V.** by QSL-card.

R. France Internationale: Lomé 91.5MHz.

TRISTAN DA CUNHA

L.T: UTC — **Pop:** 313 — **Only L:** English — **E.C:** 50Hz, 220V — **ITU:** TRC.

TRISTAN BROADCASTING SERVICE (Gov.)

☒ The Administrator, Tristan da Cunha, So. Atlantic via Cape Town, So. Africa — Dir. of Broadc: A. Swain — **FM:** 93.5MHz 0.015kW — **D.PRGR:** Mon/Wed/Fri 1630-1800, Sun 1000-1200.

TUNISIA

L.T: UTC + 1h — **Pop:** 8.800.000 — **Radios:** 1.700.000 — **Pr.L:** Arabic — **E.C:** 50Hz, 115/220/380V — **ITU:** TUN.

ÉTABLISSEMENT DE LA RADIODIFFUSION-TÉLÉVISION TUNISIENNE (ERTT)

☒ 71, ave. de la Liberté, Tunis 1002. ☎ +216 (1) 287300. ☒ +216 (1) 785146. ① 14560 or 18270 TN **E-mail:** info@radiotunis.com **Web:** http://www.radiotunis.com
L.P: DG: Abdelhafidh Herguem. Dir. R. Monastir: Abdelkader Aguir. Dir. Ext. Rel: Mokhtar Rassaâ.
Regional stations: R. Gafsa, Avenue Habib Bourguiba, 2100 Gafsa. ☎+216 (6) 226461. ☒ +216 (6) 226 260 — **R. le Kef,** Rue Monji Slim, 7100 Le Kef. ☎+216 (8) 222000. ☒ +216 (8) 222395 — **R. Monastir,** 5019 Monastir. ☎+216 (3) 462900. ☒ +216 (3) 460909 — **R. Sfax,** Route de Menzel Chaker, km 1.5, 3058 Sfax. ☎+216 (4) 240 655. ☒ + 216 (4) 245 971.

MW	kHz	kW	Prgr.	Times
Gafsa	585	350	N	0400-2400
Sousse	603	10	R	0550-1715
Tunis	630	600	N	24h
Sfax	720	100	R	0430-2300
Remada	882	1	N	0400-2400
Tunis	963	200	I	0500 (Mon 0900)-2300
Sfax	1566	1200	N	1600-2400

SW: Sfax (G.C: 10.53E/34.48N): 3 x 100kW — Djedeida (G.C: 09.57E/36.50N): 1 x 50kW.

kHz	Prgr.	Times
7280	N	1710-2330
7475	N	0400-0600, 1905-2330
11730	N	0600-1710
12005	N	0500-0600, 1710-2330
15450	N	0500-1710
17500	N	0600-1710

FM (MHz)	N	I	Y	kW
Ain Draham	90.3	96.6		5.6
Biadha	95.0	101.8		40
Gorrâa	89.1	95.4		3.8
Kasserine	89.6	99.2	89.6	49
Ghraba (Sfax)	93.0	99.5		60
Tunis	88.6	98.2	88.6	1.2
Zaghouan	94.0	96.5	96.5	20
	92.0			
Zarzis	93.9	97.2		72.5

National Network in Arabic (N): 24h. **N.** on the h — **Tunis International R. (I):** French: s/on-1302, 1600-2300. **N:** 0900 (exc. Mon), 1200, 1300, 1400, 1500, 1630, 2130. **English:** 1302-1400. **N:** 1330. **German:** 1402-1500. **N:** 1430 (local), 1435 (rel. Deutsche Welle). **Italian:** 1502-1600. **N:** 1530 — **Youth Radio Station (Y):** W 0500-0800, 1100-1400, 1700-2300. Sun 0500-2300..
ANN: F: "Ici Radio Tunisia Internationale"; Arabic: "Idha'atu-l-gumhuriya at-tunisiyya" — **V.** by letter (Re. in E, F to Dir. Gen. des Telecommunications. Rec. acc.).

UGANDA

L.T: UTC + 3h — **Pop:** 29.924.406 — **Radios:** 10m — **Pr.L:** English, Swahili, ethnic — **E.C:** 50Hz, 240V — **ITU:** UGA.

RADIO UGANDA (Gov.)

☒ P.O. Box 7142 or 2038, Kampala. ☎ +256 (41) 257256. ☒ + 256 (41) 256888. ① 61084 UG. **Cable:** Knowledge Kampala.
L.P: Director of Broadc: Vacant. *Commissioners:* Radio Broadc: Jack Turyamwijuka. Eng: Tom Okurut Anyoka. Educ. B'casting: Herbert Bisamunyu. Inf: Dr. Justin Okullu-Mura. Ag. Contr. of Prgrs (Radio): Charles Byekwaso. Ag. Chief N. Editor (Radio): Sam Muwanguzi. Ag. Principal Eng. (Radio): Leopold B. Lubega. Ag. Chief Commercial Mgr. (Radio): Paul Olungi. Head. of Public Rel: Mrs. Florence Sewanyana.

STATIONS (G.C: Kampala 32.36E/00.20N):
Channels: B=Blue, R=Red, G=Green (FM only).

MW	kHz	kW	Ch.	MW	kHz	kW	Ch.
Mawagga	576	100	B	Arua	1485t	10	B
Kampala	639	50	B	Butebo	729	100	R
Bobi	810	100	B	Kampala	909	20	R
Kabale	999	100	B				

SW	kHz	kW	Ch.	Times
Kampala	4976	10	R	0300 (SS 0345)-0600, 1300-2100
Kampala	5026	20	B	0300-1200MF, 1300-2100 MF
				0345-2100SS
Kampala	7110	20	R	0600-1200 (SS 1430)
Kampala	7195v	10	R	0600-1200 (SS 1430)

t) freq. tentative — **FM** (MHz): Kampala 98.0/93.7 (G), 97.2 (R).
National Prgr. (Blue Channel) in English, ethnic: MF 0300-1200 & 1300-2100, SS 0345-2100. **N. in English:** 0400, 0600Sun, 0700W, 1100 Sat, 1300Sat, 1400 (exc. Sat), 1900.
Regional Prgr. (Red Channel) in English, Swahili and ethnic: MF 0300-1200 & 1300-2105, SS 0300-2105. **N. in English:** 0400, 0500MF, 0600Sun, 0700 (exc. Sat), 1000, 1100, 1200Sat, 1300Sat, 1400 (exc.Sat), 1700, 1900, 2000.
FM Sce. (Green Channel): 0300-2100 (rel. Red Channel 0600-2100).
Regional st's: 8h daily in regional languages. **N:** rel. Kampala.
ANN: E: "This is Radio Uganda, Kampala".
IS: Local Xylophone — **V.** by QSL-card. Rec. acc.

Central Broadc. Sce: ✉ Box 12760,Kampala. Kampala 88.8MHz. M prgrs in Luganda for Kingdom of Buganda.
Capital Radio: ✉ Box 7638, Kampala — MD: Patrick Quarcoo.
E-mail: capital@imul.com. Kampala 91.3MHz, Mbale 90.6MHz, Mbarara 88.7MHz. (incl. relays of BBCWS). **F.Pl:** additional relays, Pan-African satellite sce.
R. Maria: Mbarara (we. Uganda): non-comm, religious prgrs.
R. One: ✉Box 4589, Kampala. 90.0MHz.
R. Sanyu: ✉ Katto Plaza, Nkurumah Rd, Kampala. Kampala 88.2MHz (incl. relays of Voice of America).
R. Simba: No details known.
R. Tooro: ✉ Box 2203, Kampala. Kampala 100.5, Fort-Portal 101.0, Mbarara 95.0MHz, Mubende 97.5MHz.
Top Radio: Kampala 104.9MHz

R. France Internationale: Kampala 93.7MHz
BBC World Sce: Kampala 101.3MHz 24h.

ZAMBIA (Republic)

L.T: UTC + 2h — **Pop:** 10.231.670 — **Radios:** 1.300.000 — **Pr.L:** English, ethnic — **E.C:** 50Hz, 220/240V — **ITU:** ZMB.

ZAMBIA NATIONAL BROADCASTING CORP.
(Independent Statutory Body, Comm.)
✉ Broadcasting House, P. O. Box 50015, Lusaka. ☎ +260 (1) 254989. 250524, 251881. ▤ +260 (1) 254317. **Cable:** Broadcasts Ridgeway. ① (0902) 41221 ZA — P. O. Box 20748, Kitwe. ☎ +260 (2) 223555.
Cable: Broadcasts Kitwe.
L.P: DG: Duncan H. Mbazima. Dir. Eng: Felix K. Chabala. Dir. of Prgrs: Frank Mutubila. Dir. of Finance: Leonard H. Mwenda. Contr. Radio (Local Languages): Mwansa Kapeya. Contr. Radio (English Sces): Kenneth Maduma. Regional Contr: Cornelius H. Mapulanga. Public Rel. Mgr: Keith M. Nalumango. Contr. N. & Current Affairs: Wellington Kalwisha.

MW	kHz	kW	Sce.	MW	kHz	kW	Sce.
Kitwe	549	100	R2	Kasama	882	10	R2
Kasama	567	10	R1	Solwezi	909	10	R1
Mongu	603	10	R1	Mansa	918	10	R1
Lusaka	630	500	R2	Livingstone	927	10	R2
Chipata	666	10	R1	Kitwe	1071	100	R1
Livingstone	729	10	R1	Kabwe	1124	1	R1
Lusaka	819	200	R1	Solweizi	1125	10	R2

MW	kHz	kW	Sce.	MW	kHz	kW	Sce.
Mongu	828	10	R2	Kabwe	1157	1	R2
Chipata	855	10	R2	Mansa	1269	10	R2

FM: R2: Lusaka 88.2MHz. R4: Kabwe/Kitwe 92.2, Lusaka 92.6, Ndola/Livingstone 95.5MHz.

SW: Lusaka (G.C:28.15E/15.30S): 1 x 150kW, 3 x 50kW.

kHz	Sce.	Times
4910	R1	0245-0530, 1430-2205
6165	R2	0250-2210
7220	R1	0500-1430

Radio One in Zambian languages: 0245-2205. **N. in English** (rel. R2): 0500, 1115, 1800 — **Radio Two in English:** 0250-2210. **N:** 0400, 0500, 0600W, 0800W, 1000W, 11015MF, 1400MF, 1600W, 1800, 2000, 2100 — **Radio Four in English:** 0300-2200 in stereo.
ANN: R2: "This is Radio Two of ZNBC Radio and Television Network broadcasting from Lusaka". — **IS:** "Call of the Fish Eagle".
V. by QSL-card. Rp.

CHRISTIAN VOICE (Rlg.)
✉ Private Bag E606, Lusaka. ☎ +260 (1) 274251. ▤ +260 (1) 274526.
E-mail: cvoice@zamnet.zm (✉ **in UK:** Christian Vision, Ryder Str.,West Bromwich, West Midlands B70 0EJ. ☎ +44 (121) 5226087. ▤ +44 (121) 5226083. **E-mail:** 100131.3711@compuserve.com).
L.P: SM: Philip Haggar. Tr. Engineer: Andrew Flynn.
SW: 1 x 100kW tr.

kHz	(alt. kHz)	Area	Times
3330	(4965)	Af	0400-0700, 1600-2200
6065	(7250)	Af	0700-1600

FM: Lusaka 106.2MHz.
D.PRGR. in English: 0400-2030. **Main N:** 1100, 1600, 2000.

RADIO PHOENIX (Priv.)
✉ Pvt. Bag E702, Lusaka. +260 (1) 226652, 226841, 224210, 224211.
▤ +260 (1) 226839 — **L.P:** MD: Errol Hickey.
F.Pl: a 300kW MW-tr. has been installed at Kabwe and is awaiting a frequency allocation.

ZIMBABWE

L.T: UTC + 2h — **Pop:** 11.745.485 — **Radios:** 1.300.000 — **Pr.L:** English, Shona, Ndebele, Chewa — **E.C:** 50Hz, 220V — **ITU:** ZWE.

ZIMBABWE BROADCASTING CORPORATION
(Independent Statutory Body, Comm.)
✉ P .O. Box HG 444, Highlands, Harare. ☎ +263 (4) 498610. ▤ +263 (4) 498613. ▤ (**Engineering:**) +263 (4) 498608. **Cable:** Broadcasts. ① 24175 ZBCHOV ZW. **Radio 2 Sce. (Mbare):** P.O. Box 9048, Harare. **Radio 2 Sce. (Bulawayo):** P.O. Box 2279, Bulawayo. ☎ +263 (9) 71811.**Cable:** Broadcasts Bulawayo. ① 3345 ZW.
L.P: DG: Edward Moyo. Dir. Mktg, Admin. & Finance: Onias Z. Gumbo. Dir. Eng. & Tech. Sces: Elliot Muchimbiri. Dir. Prgrs, N. & Current Affairs: Thomas Mandigora. Dir. Human Resources: Victor Mhizha-Murira.

FM (MHz)	R1	R2	R3	R4	kW
Harare	92.8	96.0	99.3	102.8	10
Mutare	102.2	89.1	95.4	98.7	3
Bulawayo	90.0	96.3	99.6	103.1	10
Gweru	90.7	93.9	97.2	100.7	5
Kadoma	88.5	94.8	98.1	101.6	10

+ relay st's at 17 locations.
SW: Gweru (G.C: 29.51E/19.26S): 4 x 100kW, 1 x 20kW, 1 x 10kW.
kHz: 3306, 3396, 4828, 5012 5975, 6045, 7285
R1 (mainly English): 0300-2200. **N:** 0400, 0500, 0600, 0800MF, 1000W, 1100, 1400MF, 1600, 1800, 2000, 2100.
R2 in Shona/Ndebele/English: 0300-2200. Rel. on SW: ?-1630 on 5012kHz.
R3 in English: 0300-2200. **Rel. on SW:** 0300-0600 & 1800-2200 on 3306kHz, 0300-0600 on 3396kHz, 0300-0400 & 1630-2000 on 3306kHz, **N:** on the h.
R4 (Educational): MF 0800-2000. **Rel. on SW:** 0800-1630 on 5975/6045 (alt. 7285)kHz, 1630-2000 on 3306/4828kHz. **N. in English:** 1130, 1800.
ANN: E: "This is the Zimbabwe Broadcasting Corporation" (or "Voice of Zimbabwe") R1, 2, 3 or 4". R3 also "Zimbabwe 3". R4 also "You're tuned to R4, the educational st. of the Zimbabwe Broadc. Corp."
Other stations:R. Icengelo (no further details received).

NEAR & MIDDLE EAST

AFGHANISTAN

L.T: UTC + 4½h — **Pop:** 31.876.539 — **Radios:** 1.670.000 — **Pr.L:** Pushtu, Dari — **E.C:** 50Hz, 220V — **ITU:** AFG.

RADIO AFGHANISTAN

✉ P.O. Box 544, Ansari Wat, Kabul. ☎ +93 25241. ✆ 24288 (AFGRTV AF) Kabul. **Cable:** RADIOTAF Kabul
L.P: Pres, Radio and TV, Min. of Communications: Habibullah Jaghouri.
MW: Kabul 1278/1600kHz 100kW.
Other tr's (inactive at time of editing): Taluqan 648kHz 10kW, Baghlan 936kHz 10kW, Farah 945kHz 10kW, Kabul 1107kHz 1000kW, Mazar i Sharif 1206kHz 400kW.
SW: Yakatut 7200kHz 100kW.
Pushtu/Dari: 0130-0335 (Sat-Thurs), 0335-0800 (Fri) & 1400-1630. **N:** 0145, 0230, 0320, 1330, 1430, 1530, 1600.
Ext. Sce: Urdu 1630-1645 & **English** 1645-1700.
ANN: Pushtu: "Da R. Afghanistan Kabul Dai". Dari: "Injá Radyu Afghanistan Kabul ast". — **IS:** A record on a national flute.
V. by QSL-card. Rp.

ARMENIA

L.T: UTC + 4h — **Pop:** 3.600.000 — **Radios:** 785.000 — **Pr.L:** Armenian (official), Russian — **E.C:** a/c 50Hz 220V — **ITU:** ARM.

HAYASDANI HANRABEDOUTYAN GABI NAKHARAROUTYOUN (Ministry of Communications of the Rep. of Armenia)

✉ 1 Independence Square, 375010 Yerevan. ☎ +374 (2) 526632. 🖷 +374 (2) 538 645 — **L.P:** Minister: Grigor A. Poghpatyan.

HAYASTANI DZAYN (Gov.)

✉ 5 Alek Manoukyan, 375025 Yerevan. ☎ +374 (2) 558010. 🖷 +374 (2) 551513. **Telex:** 243260 HORZIB AM.
L.P: DG: Armen Amiryan. DG of Yerevan Studio: Tigran Khzmalyan.

Stations	kHz	kW	H. of tr.	Prgr.
Kamo	234	500	0300-2200	R1
Yerevan	252	150	0200-0000	AR1
Yerevan	702	150	2300-2200	RSM
Kamo	1314	1000		RSY, F
Sisian	1377			AR1
Yerevan	1395		0200-2200	AR1/2
Kamo	1458	30	0100-0000	RSM
Yerevan	1566	5	0200-2200	RR

NB: Powers and locations are tentative only.
FM: Yerevan (17kW ERP): AR1: 66.22MHz; AR2/RSM: 69.77MHz; AR3/R1: 72.56MHz; no other details available.
AR1 in Armenian, exc. Russian W0620-0700, Kurdish 1330-1400.
AR2 (Tsiatsan): 1300-2000 in Armenian; at other times rel. AR1.
AR3: relay of AR1 & F.
F = Foreign Sce (see International Broadcasting section).
Relays from Russia: R1 = Radio 1 — RSM = R. St. Mayak — RSY R. St. Yunost — RR = R. Rossii. Cf. Russia.
ANN: Armenian: "Khosum e Yerevana".

OTHER STATIONS

Hai-FM: Yerevan. ☎🖷 +374 (2) 56000. **Web:** http://www.arminco.com/homepages/haifm/ — **SM:** Anahit Tarkhanian. 105.5MHz 10kW.
Radio Lasto: Yerevan 106.5MHz.

R. France Internationale: Yerevan 102.5MHz.

AZERBAIJAN

L.T: UTC + 4h (Su: UTC + 5h) — **Pop:** 7.570.000 — **Radios:** 1.174.000 — **Pr.L:** Azerbaijani (official), Russian — **E.C:** 50Hz, 220V — **ITU:** AZE.

AZÄRBAYCAN RESPUBLIKASI RABITÄ NAZIRLIYI (Ministry of Communications of Azerbaijan Rep.)

✉ Azärbaycan pr. 33, 370139 Baki. ☎ +994 (12) 930004. 🖷 +994 (12) 934480. ✆ 142105 AINAM AI — **L.P:** Minister: Nadar Ahmedov.

TELERADIO - AZÄRBAYCAN RABITÄ NAZIRLIYI TELERADIO ISTEHSALAT BIRLIYI (TELERADIO - Radio & TV Association of the Ministry of Communications of Azerbaijan)

✉ Azärbaycan pr. 33, 370139 Baki. ☎ +994 (12) 98 8066. 🖷 +994 (12) 98 3325. ✆ 142472 DALGA AI — **L.P:** Dir: Vagif Musayev.

AZÄRBAYCAN RADIOTELEVIZIYA ŞIRKÄTI (Radio & TV Co. of Azerbaijan, Gov.)

✉ Mehdi Hüseyin küç. 1, 370011 Baki. ☎ Chmn: +994 (12) 923807. Dir. Radio: +994 (12) 928727. **Infoline:** +994 (12) 929099. 🖷 +994 (12) 395452, 392777. ✆ 142214 TEMBR AI.
L.P: Chmn: Nizami Hudiev. Dep. Dir: Movlud S. Balahisiev. Dir. Radio: Zejnal Zejnalov. Dir. AZR2: Iljas Adygezalov.

Stations	kHz	kW	H. of tr.	Prgr.
Gäncä	218	500	0200-2000	AZR1
Gäncä	549	50	0200-2100	AZR2
Baki	612	50	0300-0200	RSM
Pirsaqat	801	150	0200-2000	AZR1
Baki	891	30	0200-2000	AZR1
Baki	1089	7	0700-1500	Local
Pirsaqat	1296	125	see below	RL, F
Baki	1476	20	0200-2100	AZR2
Pirsaqat	1476	150	0200-2100	AZR2
Baki	1530	7	0200-2100	RL

FM (MHz)	AZR1	AZR2	AZR2 (CCIR)
Akstafa	67.37	68.87	
Astara	71.06	72.62	
Baki	69.53	68.24	90.00
Danaçi	71.84	69.92	
Duzdag	69.20	67.40	
Jebrayl	66.74	68.36	
Geokçay	67.31	69.11	
Gäncä	69.92	71.84	
Kuba	70.34	71.96	
Lenkoran	69.53	67.43	
Lerik	66.44	68.00	
Ordubad	71.27	72.83	
Şarur	69.53	70.79	
Şuşa	72.20	70.64	

All FM tr's are 17kW ERP. All OIRT FM tr's are due to be replaced by new tr's in the CCIR FM band.
AZR1: 0200-2100 in Azerbaijani.
AZR2 (Araz): 0200-2100 in Azerbaijani, French, German (see FS), Russian (0700, 1500), Armenian (1600), Lezgin, Talish. Also includes BBC relays.
F = Foreign Sce: see International Broadcasting section.
RSM = R. St. Mayak relay (cf. Russia).
RL = R. Liberty relay (cf. International Broadcasting Section): 0400-0500, 1600-1700, 1900-2000 in Azerbaijani, 1700-1900 in Russian.

Regional Services

Qaraba g Rep. (currently under separatist admin.): Azerbaijani-Qaraba g Radio, Stepanakert/Xankändi r. on 1377kHz in Armenian/Russian.
Naxçivan: Details not available. Includes Şarur & Ordubad tr sites.

OTHER STATIONS

FM	Station	Location	MHz	kW	Times
1)	R. ANS-GM	Baki	102.0	1.5	24h
2)	R. Sara	Baki	104.0	1	24h
3)	RFI	Baki	105.0	0.5	24h

In Qarabag Rep: R Hmayk, Stepanakert/Xankändi on FM in Armenian/Russian. Further details not available.
Addresses and other information:
1) Matbuat pr, kvartaly 504, kacme 1128, 370005 Baki. Also rel. BBC, VOA Express.
2) Salatin Askarov küç. 85, E. 17, 370001 Baki.
3) cf. France.
Foreign broadcasters relayed in Azerbaijan:
Golos Rossii: adress as for Azerbaijan RTV. Dir: Islam Gumiev.
VOA: Istigialiyat küç. 53, 370005 Baki. Dir: Rafael Gusejnov.
R. Liberty: Istigialiyat küç. 15, 370005 Baki. Dir: Elmira Ahmetova.
BBC: Azatlyk pr. 91, 370000 Baki. Dir: Eljan Aliev.

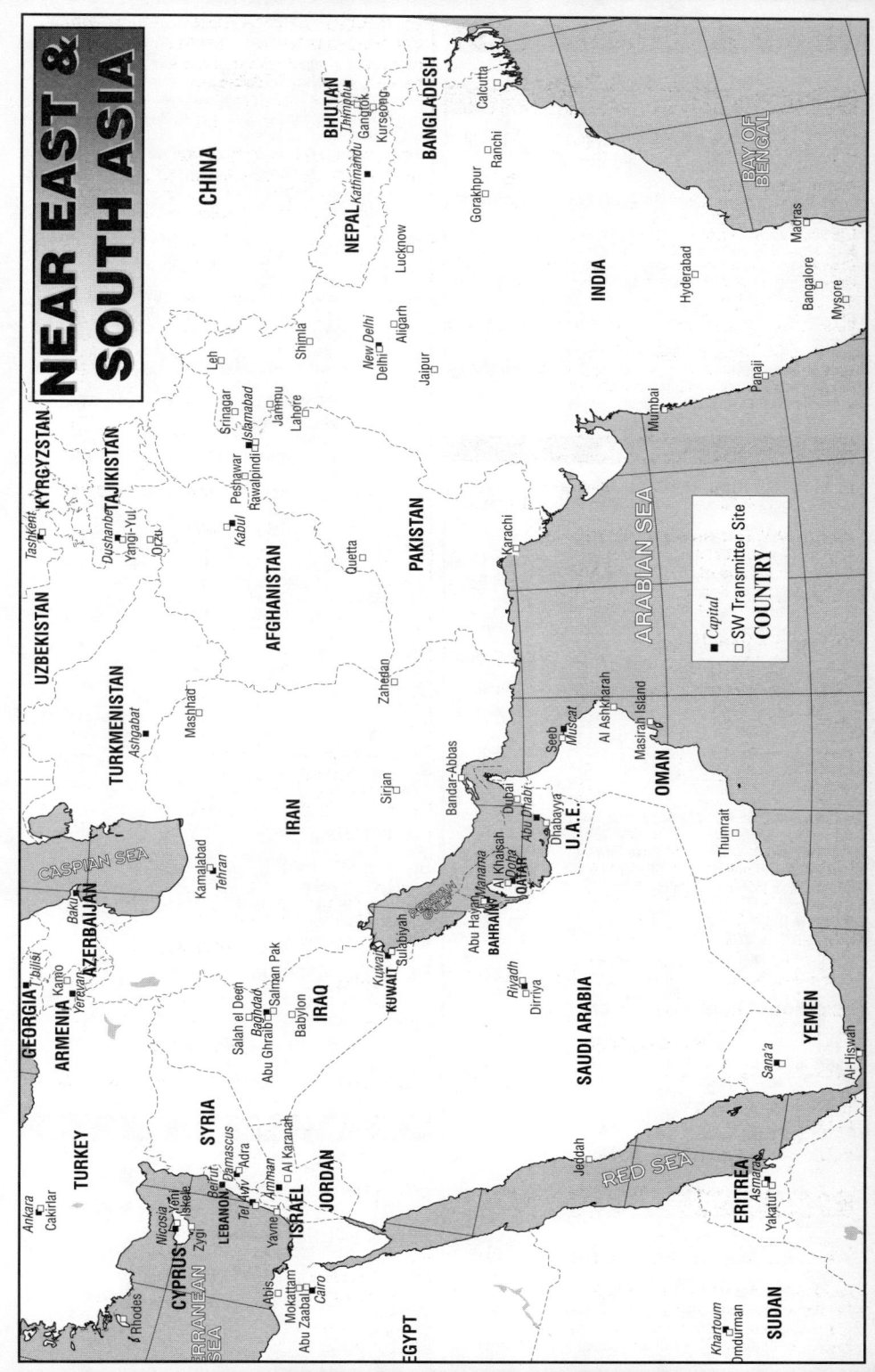

NEAR EAST & SOUTH ASIA

CHINA

KYRGYZSTAN

TAJIKISTAN

UZBEKISTAN

Tashkent

Dushanbe

Yangi-Yul

Otzu

BHUTAN

Thimphu

Gangtok

Kurseong

BANGLADESH

Calcutta

Ranchi

Gorakhpur

Lucknow

NEPAL

Kathmandu

Shimla

New Delhi

Delhi

Aligarh

Jaipur

INDIA

Hyderabad

Bangalore

Mysore

Panaji

BAY OF BENGAL

Madras

Leh

Srinagar

Islamabad

Jammu

Peshawar

Rawalpindi

Lahore

Kabul

AFGHANISTAN

Quetta

PAKISTAN

Zahedan

Karachi

TURKMENISTAN

Ashgabat

Mashhad

IRAN

Kamalabad

Tehran

Sirjan

Bandar-Abbas

Seeb

Muscat

Al Ashkharah

Masirah Island

OMAN

Thumrait

ARABIAN SEA

Dubai

Abu Dhabi

Dihabayya

U.A.E.

Seeb

| Capital
□ SW Transmitter Site
COUNTRY

CASPIAN SEA

Baku

AZERBAIJAN

GEORGIA

Tbilisi

ARMENIA

Kars

Yerevan

Salah el Deen

Baghdad

Abu Ghraib

Babylon

Salman Pak

IRAQ

Kuwait

KUWAIT

Sulaibiyah

Abu Hayir

Manama

BAHRAIN

Al Khaisah

Doha

QATAR

PERSIAN GULF

Riyadh

Dirriya

SAUDI ARABIA

YEMEN

Sana'a

Al-Hiswah

Ankara

Cakirlar

TURKEY

SYRIA

Beirut

Damascus

Adra

LEBANON

Tel Aviv

Amman

Al Karanah

Yavne

JORDAN

ISRAEL

Nicosia

Zygi

Yeni Iskele

CYPRUS

Rhodes

MEDITERRANEAN SEA

Abis

Mokattam

Cairo

Abu Zaabal

EGYPT

Jeddah

RED SEA

ERITREA

Asmara

Yakutut

SUDAN

Khartoum

Omdurman

BAHRAIN

L.T: UTC + 3h — **Pop:** 590.000 — **Radios:** 320.000 — **Pr.L:** Arabic — **E.C:** 50Hz, 230V — **ITU:** BHR.

BAHRAIN RADIO & TELEVISION CORP. (Gov.)

✉ P.O. Box 1705, Manama.
Arabic Sce: ☎ +973 781888, 629009. 🖷 +973 681544.
✉ PO Box 194, Manama.
English Sce (R. Bahrain): ☎ +973 681903, 629085. 🖷 +973 780911.
✉ PO Box 702, Manama.
Tech. Sces: 🖷 + 973 681544.
L.P: CE: Khalil Thawadi. Dir.of Broadc: Abdul Rahman Abdulla. Ag. Head of R. Bahrain: Salah Khalid.
MW: 612/801kHz 100kW, 1584kHz 1kW.
FM: 90.9/93.3MHz 3kW, 96.5MHz 2.5kW, 101.0MHz 0.5kW.
SW: Ras Hayan (G.C: 50.37E/26.02N): 6010/9745kHz 60kW. (SW-tr's reported inactive).
Arabic: 24h on 612/801kHz + FM 90.9/93.3MHz. **N:** on the h.
R. Bahrain in English: 24h on 1584kHz + 96.5/101.0MHz.**N:**on the h.
ANN: A: "Huna al Bahrain". E: "Radio Bahrain".
IS: Local composition on guitar and violin — **V.** by QSL-card.

CYPRUS

L.T: UTC + 2h (Su: UTC + 3h) — **Pop:** 756.000 — **Radios:** 210.000 — **Pr.L:** Greek, Turkish, Armenian — **E.C:** 50Hz, 240V — **ITU:** CYP.

CYPRUS BROADCASTING CORPORATION

✉ P.O. Box 4824, 1397 Nicosia. ☎ +357 (2) 422231. 🖷 +357 (2) 314050. ① 0605 2333 cybc cy. **Cable:** Broadcasts. **E-mail:** rik@cybc.com.cy
Web: http://www.cybc.com.cy
L.P: Chmn: Marios Eliades. PD: Panos Ioannides. Dir. Tech Sces: Andreas Michaelides. Head of Radio Prgrs: Kyriacos Charalambides. Head of Pub. & Int. Rel: Nayia Roussou.

MW	kHz	kW	Ch.	MW	kHz	kW	Ch.
Paphos	558	10	1	Paphos	918	10	3
Nicosia	603	100	3	Nicosia	963	100	1
Limassol	693	10	1	Limassol	1044	10	3
FM (MHz)	**Ch. 1**			**Ch. 2**		**Ch. 3**	
Larnaca	90.2			92.4		96.0	
Mt. Olympos	97.2			91.1		94.8	
Paphos	93.3			96.5		99.8	
Paralimni	91.4			94.2		97.9	

Ch. 1 in Greek: 24h. **Main N:** 0430, 0500, 0530, 0600, 0800, 1000, 1130, 1300, 1400, 1500, 1600, 1700, 1900, 2000, 2100, 2200.
Ch. 2: Turkish: 0400-1500 incl. **N. in English:** 1130. **Armenian:** 1500-1600. **Multilingual** (English, French, German & Russian for visitors to Cyprus): 1600-1800. **English:** 1800-2200. **N:** 1800.
Ch. 3 in Greek: 24h.
Ext. Sce. in Greek: see International Broadcasting section.
ANN: Greek: "Radiofonikon Idryma Kyprou". Turkish: "Burasi Kibris Radyo Yayin Korporasyonu" — **IS:** "Avkoritssa" (guitar).
V. by QSL-card. Rec.acc.

RADIO MONTE CARLO MIDDLE EAST (Comm.)

✉ B.P. 2026, Nicosia. ① 2162 SOMERA C.
L.P: SM: A.Pavlides — **MW:** Cape Greco 1233kHz 600kW.
D.PRGR. in French/Arabic: 0500-2040. **N. in Arabic:** on the h, also 0530 & 1730. **N. in French:** 1000, 1830.
V. Re. to B.P. 128, Monte Carlo.

TRANS WORLD RADIO (Rlg.)

MW: Cape Greco 1233kHz 600kW — **Arabic:** 0400-0420, 2140-2330 (Sun 2245). **Armenian:** 0420-0435. **Persian:** 0330-0400.
Other details: see International Broadcasting section (Monaco).

Other stations: Mega R, Larnaca 88.4 – Anastasis, Nicosia 88.6 – Skyline, Limassol 88.7 – Cosmos, Lakatamia 89.0 –R. Napa 90.9 – Kokkinochoria, Paralimni 91.5 – Niata, Larnaca 92.0 – R. Paphos 92.5 – Astra 92.8 – Zenon, Larnaca 93.5 – R. Eraklis, Nicosia 93.6 – Super 93.9 – R. Klik, Nicosia 94.3 – Kiniras, Paphos 95.6 – Kronos, Nicosia 95.7 – R. Zenith, Vavatsinia 96.4 – Cannel 7, Nicosia 98.4 – Channel 986, Limassol 98.6 – R. FM, Larnaca 99.0 – Proton 99.3 – R. Top, Nicosia 99.9 – Pyrkos, Limassol 100.2 – Gastro, Larnaca 100.3 –

Yialloussa, Paphos 100.5 – Super 100.7 – Logos,.Troodos 101.1 – Logos, Vavatsinia 101.8 – Epilogi,.Paphos 102.0 – Asteras, Limassol 102.2 – Magic, Larnaca 102.4 – Logos, Paphos 102.4 – Tymbou, Larnaca 102.7 – Fredrick, Nicosia 103.0 – Famagusta, Nicosia 103.6 – R. Kamares, Larnaca 103.9 – Astromeritis, Nicosia 104.0 – Epistrofi, Nicosia 104.3 – Aris, Paphos 105.0 – Astra, Larnaca 105.1 – Disastasi, Nicosia 105.4 – R. City, Limassol 105.7 – Proton, Larnaca 105.7 – Aphrodite, Larnaca 106.0 – R. One, Vavatsinia 106.0 – R. Contact, Peristeron 106.3 – Temptation, Nicosia 106.6 – R. Grammar, Nicosia 107.0 – R. Lykos, Nicosia 107.3 – R. Elios, Vavatsinia 107.6 – Pentadactilos, Nicosia 107.9MHz.

BRITISH FORCES BROADCASTING SCE. CYPRUS

✉ BFBS Akrotiri, BFPO 57. ☎ +357 474 8518. 🖷 +357 527 8580.
L.P: GM: Charles Foster. Eng.Mgr: J.Dunlop.

Location	Ch.1 (MHz)	Ch.2 (MHz)	kW
Akrotiri	92.1	89.9	25
Dhekelia	99.6	95.3	25
Nicosia	99.6	95.3	1.5

D.PRGR: 24h — **ANN:** "Forces FM" — **V.** by QSL-folder.

BRITISH EAST MEDITERRANEAN RELAY STATION

See International Broadcasting section.

NORTHERN CYPRUS

BAYRAK RADIO & TELEVISION CORPORATION

✉ Yeni Organize Sanayi Bolgesi, Lefkosa via Mersin 10, Turkey. ☎ +90 392 228 5555. ① 57264 brtk. **E-mail:** brt@cc.emu.edu.tr
Web: http://brt.emu.edu.tr
L.P: DG: Ismet Kotak. Head of Transmission Dept: Mustafa Tosun. Head of Admin: Süleyman Türem. Head of Prgrs: Hüseyin Cobanoglu. Head of Sales: Mehmet Kircailiar. Head of Int. N: Huriye Dimililer.
MW: Yeni Iskele 1098kHz 100kW, 1494kHz 10kW — Lefkosa 1070kHz 5kW.
SW: Yeni Iskele (G.C: 33.55E/35.17N): 6150 (alt.6160)kHz 7.5kW

FM (stereo)	1	2	3
Kantara	98.1	87.8	90.6
Nicosia	102.0	105.0	92.0

Ch. 1 in Turkish: 0400-2200 on 1089kHz + FM. **N:** 0500, 0600, 0800, 1000, 1330, 1500, 1730, 2330 — **Ch. 2 (Bayrak International) in Greek/English/Arabic/German:** 0400-2200 on 1494/6150kHz + FM.
N. in English: 1215, 1730 — **Ch. 3:** 24h on FM
ANN: "This is Bayrak International, the Voice of the Turkish Republic of Northern Cyprus" — **V.** by QSL-folder.

KIBRIS FM (Priv.)

✉ Dr. Fazil Kucuk Boulevard, Yeni Sanayi Bolgesi, Lefkosa. ☎ +90 (392) 228 1922. 🖷 +90 (392) 228 1934.
L.P: Head of Broadc. Exec. Council: Asil Nadir. Dir: Mehmet Ali Akpinar.
FM: Selvili Tep 103.4MHz 10kW, Quantara 100.2MHz 2.5kW.
D.PRGR: 0400-2100.
F.PI: prgrs in foreign languages.

FIRST FM & INTERFIRST FM (Priv.)

✉ Lefkosa — **FM:** 105.6MHz.

R. EMU (Priv.)

✉ G. Magusa. **E-mail:**radioemu@cc.emu.edu.tr
Web: http://www.emu.edu.tr/~radioemu
FM: 106.5MHz — **D.PRGR:** 0600-1900

GEORGIA

L.T: UTC + 4h — **Pop:** 5.370.000 — **Radios:** N/a — **Pr.L:** Georgian (official), Russian, Abkhazian, Ossetian — **E.C:** 50Hz, 127/220V — **ITU:** GEO.

MINISTRY OF COMMUNICATIONS

✉ 9th April Street 2, 380008 Tbilisi. ☎ +995 (32) 999528. 🖷 +995 (32) 934419. ① 212911 PTT GI — **L.P:** Minister: Fridon Injia.

STATE DEPARTMENT OF TELEVISION AND RADIO

✉ 68 M. Kostava Str, 380015 Tbilisi. ☎ +995 (32) 368166. **E-mail:** gtvr@iberiapac.ge — **L.P:** Chmn: Archil Gogelia.

GEORGIAN RADIO (Gov.)

✉ 68 M. Kostava Street, 380015 Tbilisi. ☎ +995 (32) 368362. 🖷 +995 (32) 368665. ➲ 212106 RADIO GI.
L.P: Head of Radio: Vakhtang D. Nanitashvili.

Stations	kHz	kW	H. of tr.	Prgr.
Tbilisi	189	500	0100-2300	GR1
Batumi (r)	693		0100-2100	GR1, Reg
Tbilisi	810	200	0100-0000	RSM
Tbilisi	1044	200	0200-2300	GR2, RSM
Tskhinvali	1323	30	0200-2200	RSM, Reg
Sokhumi	1350	50	0100-2100	GR1, Reg
Batumi	1422	30	0100-2100	GR1, Reg
Dusheti	4875	100	0200-1900	GR2, RSM
Dusheti	5040	50	0100-2300	GR1
Sokhumi	9495	5		

NB: Powers are tentative only.

FM (MHz)	GR1	GR2	R1	RSM
Batumi	+67.46	65.90		
Gori	67.04	68.69		
Kutaisi	72.50	70.94		
Sokhumi	+68.45	66.05		
Tbilisi	71.60	69.59	70.82	72.38

All FM tr's are 17kW ERP. +) carries regional prgrs at times (see below).
GR1: 0100-2300 in Georgian/Russian. **N. in Russian:** 1530-1545. **N. in English:** 1930-1935 — **GR2:** 0200-2300, own prgr's and rel. GR1.
Relays from Russia: R1 = Radio 1 — RSM = R. St. Mayak.

Regional Services

Abkhaz Rep.*: State Radio-TV Co, Aydghilar str, 384900 Sokhumi, Abkhazia. 0400-0500 on 1350/9495kHz, 1510-1600 (not Sun) on 1350kHz in Abkhaz/Georgian.
Adzharian Rep.*: Adzharian R, pr. Stalina 57, 384500 Batumi, Adzharia.
South Ossetian R: Tskhinvali 1323kHz.
*) Autonomous provinces, currently under separatist administration.

FOREIGN SERVICE: see International Broadcasting section.

IRAN

L.T: UTC + 3½h (Su: UTC + 4½h) — **Pop:** 69.167.670 — **Radios:** 13m — **Pr.L:** Farsi (Persian) — **E.C:** 50Hz, 220V — **ITU:** IRN.

ISLAMIC REPUBLIC OF IRAN BROADCASTING (IRIB) (Gov.)

✉ P.O. Box 19395-333, Tehran. (Int. Tech. Affairs: P.O. Box 15875-4344, Tehran). ☎ +98 (21) 21961. 🖷 +98 (21) 2045056, 2041051. ➲ 213910. **Cable:** IRIB — P.O. Box 555, Mashhad.
E-mail: irib@dci.iran.com (*Research Center:* iribrec@dci.iran.com)
Web: http://194.126.32.20/irib/
L.P: Pres: Dr. Ali Larijani. Ag. Pres. & Head of Eng: Mehdi Tabeshian. Vice-Pres. Admin. & Financial Affairs: Ali Kordan. Vice-Pres. Political Affairs: H. Mohamadi. Vice-Pres. Tech. Development & Planning: A. Aghamohammadi. Vice-Pres. Tech. Operation & Maintenance: H. Tabatabai. Vice-Pres. Ext. Prgr. Affairs: J. Sarafraz. Vice-Pres. Training Affairs: Shokrollah Heidari. Vice-Pres. Sound Prgr. Affairs: Mr. Mohajerani. Vice-Pres. TV Prgr. Affairs: Mr. Pournejati. Vice-Pres. Int. Rel: A. Ghasemzade. Dir. Int. Tech. Affairs & Tech. Liaison Officer: R. Moazzami. Dir of Bureau of Broadc. Union & Int. Rel: J. Mottaghi.
a) daytime, b) nighttime.

Mediumwaves	kHz	kW	Mediumwaves	kHz	kW
Iranshahr	531	20	Abadan	1080	750
Mashhad	540	200	Zabol	1098	200
Sirjan	549	400	Bandar Lengeh	1116	125
Gheslagh	558	1000	Bojnurd	1134	20
Abadan	576	750	Yasuj	1143	20
Tehran area	582	100	Tabriz	1152	100
Shiraz	594	400	Abadan	1160	10
Zahedan	603	100	Damghan	1170	2
Birjand	621	20	Tehran	1188	100
Bonab	639	400	Chalus	1197	20
Dasht sepid			Moghan	1197	20
Mashhad	684	100	Kerman	1242	10
Tayebad	720	400	Kermanshah	1278	200

Mediumwaves	kHz	kW	Mediumwaves	kHz	kW
Tabas	738	10	Bushehr	1305	20
Bandar-e-Torkaman	747a	400	Ardabil	1314	20
Sirjan	747	150	Jolfa	1323	20
Chahbahar	765	600	Tehran	1332	100
Arak	774	100	Bam	1341	1
Saravan	783	10	Bandar-Abbas	1350	10
Sari	819	20	Sari	1368	20
Esfahan	837	200	Ahwaz	1386	400
Mahabad	882	10	Kiashahr	1404	800
Yasuj	891	20	Kermanshah	1422	100
Tehran	900	50	Esfahan	1431	200
Jiroft	918	20	Bandar-e-Torkaman	1449b	800
Dorud	927	10	Marivan	1476	10
Urumiyeh	936	50	Abadeh	1485	10
Shahr-e-kord	936	10	Sabzevar	1485	1
Sanandaj	945	20	Izeh	1485	1
Birjand	963	20	Bushehr	1503	100
Ilam	972	100	Kiashahr	1521	100
Shiraz	v990	400	Yazd	1530	20
Semnan	1008	20	Sanandaj	1548	20
Tabriz	1026	100	Ardabil	1557	50
Yazd	1035	50	Bandar Abbas	1566	100
Khorramabad	1053	20	Zanjan	1602	2
Kerman	1062	20			

FM in Tehran: 91.3/93.9/97.7/100.7/106.7MHz 193kW.
1st Prgr: 24h in Farsi on MW + 93.9MHz. Relayed in part by regional st's. Also relayed on SW (see International Broadcasting section).
N: on the half h. **Main N.** at 0230, 0430, 1030, 1630, 2030.
Tehran Provincial Prgr: 24h on 1188kHz.
R. Khabar (News Radio): 0200-1730 on 900kHz + 104.7MHz.
Quran Prgr: 0200-0500 & 1330-1630 on 1332kHz + 106.7MHz. Additional Quran Prgrs from Ahvaz on 1080kHz & Bushehr on 1305kHz.
R. Javan (Youth R.): 13 daily on 1548kHz + FM 91.3MHz.
Foreign Language Prgrs on 100.7MHz: **Arabic:** 0330-0730, 0930-1130, 1700-2130. **Armenian:** 1530-1630. **Bengali:** 1430-1530. **English:** 0030-0130, 1130-1230. **French:** 2230-2330. **Persian** (V. of Ashena): 2130-2230, 0230-0330. **Pushtu:** 1230-1330. **Urdu:** 1330-1430. **Serbian/Croatian:** 1630-1700. **Spanish:** 0130-0230. On 106.7MHz: **English:** 2130-2230. **Russian:** 1930-2030. **Spanish:** 2030-2130.
Regional Stations: There are studios in 39 centres producing prgrs in Farsi and local languages, including some Ext. Sce. prgrs. These st's also relay some prgrs from Tehran. **Regional N.** usually at 1330, 1430 or 1500.

EXTERNAL SERVICE: see International Broadcasting section.

IRAQ

L.T: UTC + 3h — **Pop:** 23.034.378 — **Radios:** 3.700.000 — **Pr.L:** Arabic, Kurdish, Turkmen, Assyrian — **E.C:** 50Hz, 220V — **ITU:** IRQ.

REPUBLIC OF IRAQ RADIO (Gov.)

✉ Iraqi Broadcasting & TV Establishment, Salihiya, Baghdad. ☎ +964 (1) 884 4412/3. ➲ 212246 (IDAAH IK) Baghdad. **Cable:** IDAAH Baghdad, Iraq — **L.P:** DG: Sabah Yaseen. Dir. Pub. Rel: Shamil N. Sarsam. Dir. of Eng: Abdullah Al-Musrif. Dir. Mother of Battles R: Adib Nasir. Dir. FM Sce: Ghassan al-Mani'.

MW	kHz	kW	Prgr.	Times
Rutba	558	300	GS	0258-0025
Nineva	603	300	GS	0258-1350
Kirkuk	630	20	GS	0258-1350+
Tanaf	684	1000	GS	*
Basrah	693	600	V	0300-1700
			M	1700-2000
			V	2000-2100
Baghdad	756	300	GS	0258-0025
Nasiriya	846	300	GS	0258-0025
Baghdad	909	300	*	
Unk. location	963		GS	0258-0025
Babylon	1035	1000	GS	0258-0025
Unk. location	1044		GS	0258-0025
Unk. location	1084		Y	1630-2130
Rutba	1116	300	*	
Nineva	1206	300	K	0315-2130

MW	kHz	kW	Prgr.	Times
Nasiriya	1224	300	*	
Kirkuk	1360	120	*	
Tanaf	1377	1000	M	1700-2000
			GS	2000-0025
Missan	1431	1000	*	
Arbat	1539	1000	*	

*) inactive, +) ann. but not confirmed.

NB: Many transmitters are operating irregularly or with reduced power.

FM: Baghdad 91.1/91.9/92.5/93.3/95.0MHz + 8 additional tr's.

Networks: GS=General Sce, FS=Foreign Sce, K=Kurdish, M=Mother of Battles Radio (see Ext. Sce), V=V. of the Masses, Y=V. of Youth

SW: Abu Ghraib (G.C: 44.15E/33.19N): 6 x 250kW — Babylon (G.C: 44.30E/32.30N): 4 x 500kW — Salah el Deen (G.C: 44.10E/33.58N); 16 x 500kW — Salman Pack (G.C: 44.35E/33.09N: 4 x 100kW, 18 x 50kW, 8 x 10kW.

NB: Since the Gulf War of 1990, most SW tr's remain inactive.

kHz	Prgr.	Times
3935	M	1700-2000
4615	GS	0258-0025 (irr.)
6560	K	0315-2130
7150	M	2000-2300
9715	V	0615-0900
	M	1400-1600
9755	M	1900-2300
11292	M	0550-0900, 1900-2300
11785v	M	0900-1200
15340	GS	0258-0025

General Sce. in Arabic: 0258-0025. **Main N:** 0400, 1130, 1500, 1700, 1900, 2100 — **V. of the Masses in Arabic:** 0300-2100 — **Kurdish:** 0300-2300 on 1197/6560kHz.

ANN: GS: "Idha'at Jumhuriyat al-'Iraq min Baghdad". Y: "Sawt as-Shabab min Dar al-Salam". V: "Sawt al Jamaheer min Baghdad". Mother of Battles R: Idha'at Umm al-Ma'arik".

RADIO IRAQ INTERNATIONAL
See International Broadcasting section.

<hr/>

ISRAEL

L.T: UTC + 2h (Su: + 3h) — **Pop:** 5.674.961 — **Radios:** 2.250.000 — **Pr.L:** Hebrew, Arabic — **E.C:** 50Hz, 230/400V — **ITU:** ISR.

ISRAEL BROADCASTING AUTHORITY
✉ P.O. Box 28080, Jerusalem 91280. ☎ +972 (2) 240124. 📠 +972 (2) 257034.
L.P: DG: Mordechai Kirshenbaum. (☎ +972 (2) 252905).

KOL ISRAEL – THE VOICE OF ISRAEL
Main Studios: Heleni Hamalka 21, P.O. Box 1082, Jerusalem 91010. ☎ +972 (2) 248715. 📠 +972 (2) 253282. ✆ 25263. **Foreign Sce.** 📠 +972 (2) 302327. **News & Actuality Studios:** Torah M'Zion 15, Jerusalem. ☎ +972 (2) 383150 or +972 (2) 383173. **Cable:** Kolisrael.
Web: http://www.artificia.com/html/news.cgi
L.P: Dir. & PD: Amnon Nadav. CE: David Cohen. Dir. Israel Radio International: Shmuel Ben-Zvi.

MW	kHz	kW	Prgr.	Times
Jerusalem	531	100	C	0400 (Sat 0500)-2305
Tel Aviv	576	200	A	0400 (Sat 0500)-2205
Tel Aviv	657	200	B	24h
Tel Aviv	738	1200	D	0400-2215
Safad	846	5	B	24h
Beit Hillel	882	1	B	24h
Eilat	927	1	B	24h
Haifa	927	50	R	0500-2200
Jerusalem	954	100	R	0500-2200
Tel Aviv	1026	50	D	0400-2215
Jerusalem	1080	5	B	24h
Haifa	1206	50	B	24h
Jerusalem	1458	10	A	0400 (Sat 0500)-2205
Eilat	1458	10	A	0400 (Sat 0500)-2205
Haifa	1575	5	X	0500-1700

FM (MHz)	B	C	D	X	M	kW
Beersheva	103.8	101.4			88.8	100
Eitanim	95.5	97.8			91.3	40
Haifa	103.7	105.5			100.2	100
Jerusalem			90.3	88.2		1

+ additional tr's 1kW or less.
Prgrs: A = **Multilingual prgrs. for immigrants** incl. **English:** 0500-

0515, 1100-1130, 1500-1502, 1600-1630. **French:** 0515-0530, 1130-1200, 1502-1505. **N. in Hebrew:** rel. Prgr. B.– **B** = **Commercial Sce. N:** on the h – **C** = **Commercial Sce. N:** rel. Prgr. B – **D** = **Arabic Prgr.** – **X**=**Traffic Prgr. N:** as Prgr. B – **R**= **Immigrant Absorption Network** in Russian, Amharic, Yiddish, Hebrew, Ladino, Romanian, Spanish,. Moghrabi, Bukharian, Georgian – **M** = **Voice of Music:** 0400 (Sat 0500)-2205 in stereo

ANN: A: "Kolyisrael Mi Yerushalayim b'Reshet Aleph". B: "Kolyisrael b'Reshet Beth". C: "Kolyisrael b'Reshet Gimei". D: "Saout el Israil min Urshhalem el Kuds". M: "Kol Hamusica". X: "Boker Tov Hayom Yom".

ISRAEL RADIO INTERNATIONAL
See International Broadcasting section.

GALEI ZAHAL (Israel Defence Forces Radio)
✉ Military Post Office Box 01005, Israel. ☎ +972 (3) 569 4276. 📠 +972 (3) 814697. **E-mail:** galatz@glz.co.il **Web:** http://www.glz.co.il
L.P: Dir: M. Shionsky. Dir. Eng: S.Kasif.

MW	kHz	kW		kHz	kW
Beersheba	1224	20	Mitspe Ramon	1395	
Tel Aviv	1287	100	Shivta	1368	20
Haifa	1305	20	Rosh-pina	1368	50
Eilat	1305	5	Jerusalem	1404	10
FM (stereo)	**MHz**	**kW**		**MHz**	**kW**
Bnei-Brak	91.8+		Beith-Shean	99.8	3.5
Tel Aviv	92.3	3.5	Haifa	102.3	40
Jerusalem	93.9	20		104.0	
Eilat	93.9	0.3	N. Negev	104.1+	1
Abugosh	96.6	20		106.4+	

+) Galgalatz, others carry Main Prgr.
Main Prgr: 24h. **N:** on the h — **Galgalatz** (traffic reports and music): 0500-2200 (Fri 2400). Tr's rel. Main Prgr. overnight.
ANN: Main Prgr: "Galei Zahal, Shidure Tsva Hagana Le'Yisrael".
V. by QSL-card.

Private Stations
R. Shouth, Beersheva: 97.0MHz – **Haifa R.,** Haifa: 97.5MHz – **R. Jersulaem,** Jerusalem: 101.0MHz – **V. of the Red Sea,** P.O. Box 2148, 8121 Eilat. ☎ +972 (7) 340630. 📠 +972 (7) 340590. 101.0MHz 1kW – **R. Nonstop,** Ramat Gan: 103.0MHz.

West Bank & Gaza
Pop: West Bank 1.463.498, Gaza 929.469 — **Pr.L:** Arabic

PALESTINIAN BROADCASTING CORPORATION
Voice of Palestine and Palestinian TV
✉ P.O. Box 4052, Gaza. 📠 +972 (7) 823809.
✉ Jericho. ☎ +972 (2) 921220, 921221, 921222.
Web: http://www.bailasan.com/pinc/
L.P: Dir: Radwan Abu Hayash. Tech. Coordinator: Hisham Makki.
MW: Ramallah 675kHz.
D.PRGR. in Arabic: 0400-2300. **N:** on the h. **Local N:** 1100 (exc. Fri).
English: Mon/Wed 1505-1550. **N. in Hebrew:** Mon/Wed 1550.
ANN: "Sawt Filastin".

Private Stations: numerous local stations r. in operation, but no details received.

<hr/>

JORDAN

L.T: UTC + 2h (April 1-Oct 1: UTC + 3h) — **Pop:** 4.440.618 — **Radios:** 980.000 — **Pr.L:** Arabic, English — **E.C:** 50Hz, 220V — **ITU:** JOR.

JORDAN RADIO & TELEVISION CORP. (JRTV) (Gov.)
✉ P.O. Box 1041 or 2333, Amman. ☎ +962 (6) 77311/9. 📠 +962 (6) 751503 (Admin) or 788115 (Eng.). ✆ 23544 (JTVINT JO) International Relations or 24213 (RTVENG JO) Technical. **E-mail:** rj@jrtv.gov.jo (prgrs), eng@jrtv.gov.jo (techical), general@jrtv.gov.jo (general).
Web: http://www.jrtv.com
L.P: DG: Ihsan Ramzi. Dir. Radio: Hashem Khresat. Dir. Eng: Fawzi Saleh. Dir. Eng (Radio): Yousef Al-Areeny. Dir. Int. Rel: Mrs. Fatima Massri. Comm. and Mktg. Dir: Waleed Sinawi. Dir. Production: Nidal Dalqamouni. Dir. Arabic Radio Prgrs: Mrs. Firyal Zamakhshari. Dir. Foreign Radio Prgrs: Jawad Zada.

Mediumwaves	kHz	kW	Prgr.	Times
Al Karanah	207	600	Arabic	24h
Amman	612	200	Arabic	24h
Ajlun	801	2000	Arabic	24h
Amman	855	10	English	0500-2200

Aqaba	1485	5	Arabic	24h	
Al Karanah	1494	1000	Arabic	24h	
FM (MHz)	**Arabic**	**English**	**French**	**Quran**	**kW**
Ajlun		90.9	90.0		5
Amman	99.0	96.3		93.1	0.6/120/10
	105.0				3
Aqaba		98.7			3
Irpit	95.4				0.5

Arabic: 24h. **N:** on the h. (not 0400, 1100 Fri, 1300, 1700).
English: 0400-2400. **N:** 0500, 0800, 1000 (not Fri), 1100MF, 1200, 1400, 1600, 1700, 1900, 2000, 2100 — **French:** 1800-2400 — **Quran Prgr:** 0500-2200.

SW: Al Karanah (G.C: 36.26E/31.44N): 3 x 500kW tr's.
For frequencies and schedule see International Broadcasting section.

ANN: Arabic: "Huna 'amman, Idha'atu-l-mamlaka al-urduniyya al-hashimiyya". E: "This is R. Jordan, broadc. from Amman".
V. by QSL-card. Rec. acc. (Re. to P.O. Box 909, Amman).

BBC WORLD SERVICE
FM: Amman 103.1MHz 5kW: 0330-2115 (rel. BBC Arabic sce.).

KUWAIT

L.T: UTC + 3h — **Pop:** 2.255.227 — **Radios:** 1.1m — **Pr.L:** Arabic —
E.C: 50Hz, 240V — **ITU:** KWT.

MINISTRY OF INFORMATION (Gov.)
✉ P.O. Box 193 Safat, 13002 Safat. ☎ +965 2415301.
📠 +965 2434511. **Cable:** ALIRSHAD.
Web: http://www.moninfo.kw.gov

RADIO OF THE STATE OF KUWAIT (Gov.)
✉ P.O. Box 397 Safat, 13004 Safat. ☎ +965 2423774. 📠 +965 2456660, 2415498, 2415946. **Cable:** ALIRSHAD. ① (496) MI 46285 KT. **E-mail:** kwtfreq@ncc.moc.kw. **Web:** http://www.radiokuwait.org
FM Super Station: E-mail: fm997@ncc.moc.kw
Web: www.moc.kw/users/fm997
L.P: Minister: Sheikh Saud Nasir Al-Saud Al-Sabah. Under-Secr: Faisal Mohamed Al-Hajji. Asst. Under Secretaries: Broadc. Affairs: Dr. Abdulazeez Al-Mansour. Admin & Finance: Faisal Al-Malek Al-Sabah. Culture, Press & Documentation Affairs: Salman Al-Dawood Al-Sabah. Eng. Affairs: Jawad A. Al-Mazeedi. Int. Media: Mrs. Amal Al-Hamad. TV Affairs: Khalid Abdullah Al-Anezi. News Affairs & Political Prgrs: Mohamed Hamed A. Al-Qahtani. Tech. Adviser: Yacoub Y. Dashty. Dir. Eng.(TV): Abdulazeez Al-Baghli. Dir. Eng. Communications: Bader F. Al-Mazeedi. Dir. Eng.(Radio): Maher N. Al-Mutawa.

MW	kHz	kW	Prgr.	Times
	540	600	Main Arabic	24h
	630	10	Holy Quran	24h
	963	20	Main Arabic	24h
	1134	10	Main Arabic	24h
	1269	100	2nd Arabic music	24h
	1341	10	Holy Quran	0200-0700, 1700-2300
			2nd Arabic	0700-1700

F.Pl: tr's on 1485 and 1602kHz.

FM	MHz	Prgr.	Times
	87.9	Arabic Music	24h
	89.5	Main Arabic	24h
	90.5	Arabic Music	24h
	92.5	Easy FM (**English**)	24h
	95.0	Arabic Music	24h
	97.5	Holy Quran	0200-0700, 1700-0200
		2nd Arabic	0700-1700
	99.7	FM Super Station (**English**)	24h
	98.9	Main Arabic	24h
	100.5	TV sound (Prgr. 1/4)	24h
	103.7	2nd Arabic Music	24h
	105.9	Arabic Music	24h

SW: see International Broadcasting section.
Main Arabic Prgr: N: 0300-2100 hourly exc. 0900, 1100, 1600, 2000.
N. in English: 0400, 0600, 0900, 1200, 1500, 1800.
ANN: A: "Huna al-Kuwayt". E: "This is Radio Kuwait".
IS: old Kuwaiti tune on clarinet — **V.** by QSL-folder. Rec.acc.

VOICE OF AMERICA RELAY STATION
see International Broadcasting section.

LEBANON

L.T: UTC + 2h (May 1-Sept 30: UTC + 3h) — **Pop:** 3.939.470 — **Radios:** 2.247.000 — **Pr.L:** Arabic, French, English, Armenian — **E.C:** 50Hz, 110/190V — **ITU:** LBN.

RADIO LIBAN (Gov.)
✉ Ministry of Information, Beirut. ☎ +961 (1) 346880. **Cable:** Radio Liban. ① Minform 29786 — **L.P:** Dir: M. Sobhi Eid. Chief, Tech. Dept: Micheal Samaha. Chief, Prgr. Dept: Waheed Jalal. Chief, Public Rel: Faouzi Fehmy.
MW: Amchit 837kHz 100kW, 990kHz 10kW — **FM:** Beirut 96.2/100.2/106.8MHz, We. Biqa Valley 88.3MHz, Sidon 106.0MHz.
1st Prgr. in Arabic: 24h on all freqs exc. 96.2MHz. **Main N:** 0530, 1130, 1630. Summaries on the h. exc. 0600, 1200. **N. in French** (rel. Prgr. 2): 0505, 1105, 1606.
2nd Prgr. in French/Armenian/ English: 0445-2200 on 96.2MHz.
N. in French: 0505, 1105, 1606. **Rel. R. France Int:** 12h daily.
ANN: A: "Iza'at Loubnan min Beirut". F: "Ici Radio Liban émettant de Beyrouth" — **IS:** Opening notes from the Lebanese National anthem played on guitar — **V.** by QSL-card.
F.Pl: A major refurbishment is under way, which includes several high power SW-tr's for overseas broacasting.

OTHER STATIONS

MW	kHz	kW	Name & status
9)	585	100	R. Free & Unified Lebanon, Ehden
4)	680		Voice of Hope (Christian/commercial)
23)	684		Voice of the Oppressed, Ba'labakk (Hezbollah)

MW	kHz	kW	Name & status
1)	756		Voice of the South, Marjayoun
2)	873	80	R. Voice of Lebanon (Phalangist/commercial)
15)	902		R. Artiste, Aley
11)	930		Voice of Arab Lebanon, Saïda
4)	945	50	Voice of Hope (Christian/commercial)
20)	963	100	R. Free Lebanon, Zuk Musbih (Mil.)
14)	1000		Voice of Islam (Hezbollah)
12)	1014		Islamic Unification R. (Sunni Moslem)
21)	1050		Voice of the National Resistance
13)	1080	50	Voice of the Mountain (Progressive Socialist)
8)	1134		Voice of the Homeland (Conservative Sunni Moslem/commercial)
3)	1188		Voice of the People (Communist)
9)	1310	10	R. Free & Unified Lebanon, Ehden
10)	1368		The Voice of Truth/The Lebanese Radio
1)	1485	5	Voice of the South, Marjayoun

SW	kHz	kW	Name & H. of tr.
4)	6280	10	Voice of Hope, So. Lebanon: 24h
2)	6550		R. Voice of Lebanon: 24h
4)	9960	25	Voice of Hope, So. Lebanon: 2100-1330 *
4)	9965	25	Voice of Hope, So. Lebanon: 1330-2100 *

*) inactive at editorial deadline due to fire at transmitter site.

FM	MHz	kW	Name & status
18)	91.0		R. Orient, Beirut (Commercial)
6)	92.0		R. of the Light, Beirut (Chi-ite)
22)	92.4	10/2.5	La Une (commercial)
2)	93.2		R. Voice of Lebanon (Phalangist/commercial)
3)	93.5		Voice of the People (Communist)
15)	97.2		R.Artiste, Aley
17)	98.6		The Voice of Love, Ghadir (Christian)
19)	99.2		RML 99, Jounieh (commercial)
16)	99.5		New Lebanon Radio, Ea. Beirut (Priv.)
1)	99.9	0.08	Voice of the South, Marjayoun
17)	100.5		The Voice of Love, Dora (Christian)
19)	100.5		Hit FM, Beirut (commercial)
19)	101.0	30	R. Mont Liban, Jounieh (commercial)
5)	102.1		Magic 102, Jounieh (commercial)
12)	103.5		Islamic Unification R. (Sunni Moslem)
1)	103.6		Voice of the People (Communist)
7)	104.3		R. Jeunesse, Hadeth (commercial)
4)	104.5		Voice of Hope (Christian/commercial)
10)	104.8		The Voice of Truth/The Lebanese Radio
4)	105.1		Voice of Hope (Christian/commercial)
22)	105.5	10/1	Radio One (commercial)
13)	107.8		Voice of the Mountain (Progressive Socialist)

Addresses & other information
1) Marjayoun — 0500-1900. **ANN:** "Saout al Djenoub".
2) B.P. 165271, Ashrafieh, Beirut. ☎ +961 (1) 338400. ① VDL 23203

LE — **L.P:** DG: Sheikh Simon Wehbé el Khazen. CE's: William Merheb, Maroun Salameh — 0400-2330 (Sat 0100) **N: French:** 0800, 1300, 1715. **English:** 0900, 1200, 1700. **ANN:** A: "Huna Saout Loubnan, saout al-Hurriyah wa al-Karamah". F: "Ici La Voix du Liban". E: "This is the Voice of Lebanon" — **IS:** Opens with River Kwai March and Lebanese National Anthem. Closes with River Kwai March and Katäeb Party Anthem — **V.** by letter.

3) Saïda — DG: Hanna Saleh. 0400-2300. **ANN:** "Saout al Shab".
4) P.O. Box 3379, Limassol, Cyprus. ☎ + 972 (6) 959174 (English), +972 (6) 959889 (Arabic). ▤ + 972 (6) 997827 (English), +972 (6) 959889 (Arabic). (**USA** ✉ Voice of Hope, P.O. Box 100,. Simi Valley, CA 93062) — **L.P:** CEO: George Otis. Pres: John Tayloe. Asst. to the Pres: Paul Johnson. GM: Gary Hull — **ANN:** E: "High Adventure Ministries, the Voice of Hope for the whole of the Middle East". A: "Saut al Amal le Kul Sharq al Awsat" — **V.** by QSL-card or letter. Full prgr. and frequency schedule in International broadcasting section.
5) Jounieh. ☎ +961 (9) 914102. **English:** MF 0400-2300. **French:** Sat 24h, Sun 0500-2300.
6) Beirut — 0600-2200. **ANN:** "Iza'at Al-Ncur".
7) Hadeth. ☎ +961 (1) 461515 — **D.PRGR in French:** 0400-2300.
8) Mazraa, We. Beirut. ☎ +961 (1) 312784 — DG: Mohammed Kraem. 0400-2200. **N. in French:** 1125. **ANN:** "Saout al-Watan".
9) Zghorta, No. Lebanon — Dir: Ramez El Khazen. 0500-2200. **ANN:** "Iza'at Loubnan El Hor Wal Mouwahad".
10) Antélias. ☎ +961 (1) 406571. 0400-2300. **N: French:** 1200, 1730. **English:** 1215, 1745. **ANN:** "Saout Al-Haqq" or "Iza'at Al-Loubnaniyah".
11) Saïda (or Mazraa, We. Beirut). ☎ +961 (1) 314291 — Dir: Salah Bsirty. 0400-2100. **N: French:** 1230, 1900. **English:** 1915. **ANN:** "Saout Loubnan al-Arabiy".
12) Tripoli — 0600-0900, 1100-1500. **ANN:** "Iza'at El Tawhid Al Islami".
13) Saïda — DG: Ghazi Aridi. 0400-2200. **N. in French:** 1105, 1300, 1505.
14) 1400-1600. **ANN:** "Nida al Islam - Saout al Moustada Fin".
15) Ahl Al-Fan Radio, Aley. ☎ +961 (5) 551605 — 0430-2100. **ANN:** "Iza'at Aal Al-Fan".
16) rue Mar Elias, Ea. Beirut — **D.PRGR. in English:** 0400-2300.
17) Ghadir. ☎ +961 (9) 918090; Dora. ☎ +961 (1) 898514 — **D.PRGR:** 0400-2200. **ANN:** Saout Al-Farah".
18) Beirut — Dir: Hani Hammoud. 0400-2400. Relay st's in France (Paris, Lyon & Annemasse), Syria (Damascus), Comoros (Moroni).
19) RML Group, rue Fouad Chehab-Fassouh, 10th & 11th flr, P.O. Box 166000, Beirut. ☎ +961 (1) 202500. ▤ + 961 (1) 423121. **E-mail:** rml@dm.net.lb — Mgr: Jihad El Murr.
20) Adonis. ☎ +961 (9) 911888 — DG: M. Chawki Abousleiman. **D.PRGR:** 0400-2300. **N. in French:** 0930, 1330, 1730. **N. in English:** 0830, 1230, 1530, 1830.
21) Bir Hassan, We. Beirut. ☎ +961 (1) 831608 — 0400-2200. **ANN:** "Saout al Mukawama al-Wataniya".
22) Zakhem Bldg., Beit Mery El Metn. ☎ +961 (4) 961281, 961294, 961645, 972818. (via USA: +1 212 478 4237 or +1 212 444 8923) — GM: Raymond Gaspar. PD: (R. One): Abdc Kahhaleh, (La Une): Solo Feghaly. **Web:** http://www.inco.com.lb/lbci.com.lb/
23) Muqaddass Shaykh Habib al-Ibrahim Str, Ba'labakk. ☎ +961 (8) 370284, +961 (8) 370973.

OMAN (Sultanate of)

L.T: UTC + 4h — **Pop:** 2.100.000 — **Radios:** 900.000 — **Pr.L:** Arabic — **E.C:** 50Hz, 220V — **ITU:** OMA.

RADIO SULTANATE OF OMAN (Gov.)
✉ Ministry of Information, P.O. Box 600, 113 Muscat. ☎ +968 603222. ▤ +968 603812. ☏ 5265 INFORM ON. **Cable:** Information.
Web: http://www.oman-tv.gov.om
L.P: DG: Ali Bin Abdallah Al Mujeni. DG (Tech. Affairs): H.Y.Al-Kindy. CE (Radio): Abood Al-Sawafi. Head of Maintenance: Rashid Haroon Al Jabry. PD: Hamdan S.Al-Humeidy. Dir. Foreign Sce: Fatiyah Al Hinai.

Mediumwaves	kHz	kW	Mediumwaves	kHz	kW
Haima	576	?	Muscat	1242	200
Salalah	738	100			

SW: S=Seeb (G.C: 58.10E/23.40N), T=Thumrait (G.C: 53.56E/ 17.38N): 50/100kW.
FM: 88.1/90.4/91.7/94.3/97.4/99.1/99.9/101.5/107.3MHz.
Arabic: 0200-2130 on MW + 88.1/91.7/97.4/99.1/101.5/107.3MHz. Rel. on SW: see International Broadcasting section.
N: 0300, 0700 (Fri 0830), 1300, 1600, 1700, 1900, 2000.
English: 0300-1800 on Muscat 90.4 & Salalah 94.3MHz.
N: 0330, 1030, 1430.

ANN: A: "Idha'atu Saltanat Oman min Muscat." E: "This is the English Service of Radio Sultanate of Oman from Muscat".
V. by QSL-folder.

BBC EASTERN RELAY STATION
✉ P.O.Box 6898, 112 Ruwi, Sultanate of Oman.
MW: Masirah Island 720/1413kHz 750kW.
SW: Masirah Island (G.C: 58.53E/20.36N): 4 x 100kW tr's.
V. by letter. Re. to Tr. Mgr.
For SW frequencies and schedule see International Broadcasting section.
F.PI: This st. will be de-commissioned in 2001. A new tr. site is under construction at Al-Ashkharah. It will house 4 x 300kW SW and 2 x 600kW MW-tr's.

QATAR

L.T: UTC + 3h — **Pop:** 605.000 — **Radios:** 180.000 — **Pr.L:** Arabic — **E.C:** 50Hz, 240V — **ITU:** QAT.

MINISTRY OF INFORMATION & CULTURE
✉ P.O. Box 1836, Doha. ▤ +974 831447. **Cable:** Information Doha. ☏ 4229 INFOR DH.
L.P: Minister of Inf: Dr.Hamad Abdul Aziz Al Kawari. Under-Secr: Hamad Bin Tamer Al-Thani. Dir. of Tech. Office: Hassan Al-Mass.

QATAR BROADCASTING SERVICE (Gov.)
✉ QBS, P.O. Box 1414, Doha. ☎ +974 864111. ▤ +974 822888. **Cable:** Broadcasting. ☏ 4597 Q Radio DH.
L.P: Asst. Under-Secr. for Radio & TV: Abdul Rahman Saif Al-Madhadi (☎ +974 864823). Dir. of Broadc: Mubarak Jaham Al-Kawari (☎ +974 864805). Dir. Eng. Radio & TV: Hussain A.Jaffar. (☎ +974 864518).

Mediumwaves	kHz	kW	Prgr.	H. of tr.
Al Khaisah	675	100	A	0245-2130
Al Arish	954	1500**	A	0245-2130
Al Khaisah	999	50	S/U	1200-1500, 1600-1730
Al Khaisah	1233	100	E/F	0300-1830
Al Khaisah	1485	1	S	1200-1500
Doha	1602	1	S/U	1200-1500,1600-1730

) F.PI: 2000kW.
FM: Doha 97.5MHz (E) — Al-Jumailiyah 90.8MHz (A), 100.8MHz (F), 102.6MHz (E).
SW: Al Khaisah (G.C: 51.25E/25.25N): 1 x 500kW, 1 x 250kW, 1 x 100kW.
Arabic Prgr. (A): 0245-2130. **N:** 0330, 0430, 0530, 0700, 0900, 1030, 1200, 1300, 1530, 1700, 1800, 1900 (exc. Fri/Sat), 2000, 2100.
English Prgr. (E): 0300-2130 on FM, 0300-1000 on 1233kHz.
N: 0330, 0430, 0530, 1000, 1700 — **French Prgr. (F):** 1100-1400 — **Shaabi Prgr. (S):** 1200-1500 (Folklore prgrs. in local dialect) — **Urdu Prgr. (U):** 1600-1730.
ANN: A: "Idha'at at Qatar Min Al-Doha". E: "This is the Qatar Broadcasting Station, Doha".
IS: "Arabic melody on string instrument (23 notes) — **V.** by QSL-card.
For SW frequencies and schedule see International Broadcasting section.

SAUDI ARABIA

L.T: UTC + 3h — **Pop:** 17.500.000 — **Radios:** 3.800.000 — **Pr.L:** Arabic — **E.C:** 50 or 60Hz, 127 or 220V (variable) — **ITU:** ARS.

MINISTRY OF P.T.T., Riyadh
L.P: Minister of P.T.T: H.E. Ali Al-Gohany.

BROADCASTING SERVICE OF THE KINGDOM OF SAUDI ARABIA (Gov.)
✉ Ministry of Information, Riyadh.
L.P: Minister of Inf: H.E. Dr. Fouad Al-Farsy. Dep. Minister for Inf. Affairs: Abdullah Al-Jasir. Actg. Asst. Dep. Minister for Eng. Affairs: Mohamed Sahli (✉ P.O. Box 61718, Riyadh 11575). ☎ +966 (1) 4425170. ▤ +966 (1) 4041692. ☏ 401040 SJ.

Mediumwaves	kHz	kW	H. of Tr. & Prgr.
Qurayyat	549	2000	0300-1500 (A), 1500-1700 (CI)
Gizan	549	1	0300-1500 (A), 1500-1800 (CI) 1800-2300 (A)
Rafha	549	20	0300-1500 (A), 1500-1700 (CI) 1700-2300 (A)
Jeddah	558	1	as Dammam 1098kHz

Mediumwaves	kHz	kW	H. of Tr. & Prgr.
Abha	567	5	0300-2100 (H)
Afif	567	15	0300-2100 (H)
Gizan	567	1	0300-2100 (H)
Riyadh	585	1200	0300-1500 (A), 1500-1700 (CI) 1700-2300 (A)
Duba	594	2000	0300-1500 (A), 1500-1700 (CI)
Hail	612	5	0300-2100 (H)
Khamasin	612	15	0300-1500 (A), 1500-1700 (CI) 1700-2300 (A)
Gizan	630	20	0300-2100 (H)
Najran	630	10	0300-2100 (H)
Jeddah	648	2000	0300-1500 (A), 1500-1700 (CI) 1700-2300 (A)
Rafha	657	20	0300-2100 (H)
Abha	675	5	0300-1500 (A), 1500-2100 (A2)
Afif	675	20	0300-1500 (A), 1500-1700 (CI) 1700-2300 (A)
Riyadh	684	10	0300-2100 (A2)
Jeddah	684	50	0300-2100 (A2)
Duba	702	40	0300-2100 (A2)
Najran	747	1	as 585kHz
Qurayyat	765	10	0300-2100 (H)
Khamasin	765	20	0300-2100 (H)
Dammam	783	100	0300-2100 (A2)
Jeddah	792	50	0300-2100 (H)
Medinah	828	20	as 585kHz
Dammam	864	100	0300-2100 (H)
Qassim	873	10	0300-1500 (A), 1500-2100 (H)
Dammam	882	100	0100-0300 (CI), 0300-2100 (H)
Qurayyat	900	1000	as 585kHz
Riyadh	936	5	0100-0300 (CI), 0300-2100 (H)
Hail	945	5	as 585kHz
Madinah	981	20	0100-0300 (CI), 0300-2100 (H)
Tabuk	999	20	0300-2100 (H)
Qurayyat	1098	10	0300-2100 (A2)
Dammam	1098	5	0500-0800 (M), 0800-1300 (F) 1300-1400 (M), 1400-2100 (F)
Makkah	1287	5	0300-2100 (H)
Taif	1305	1	0300-2100 (A2)
Riyadh	1422	20	as Dammam 1098kHz
Dammam	1440	1600	as 585Hz
Jeddah	1485	1	1500-2100 (F)
Jeddah	1512	1000	0100-0300 (CI), 1300-2100 (A)
Duba	1521	2000	as 585kHz

Prgrs: A=General Arabic Prgr (from Riyadh), A2=Second Arabic Prgr (from Jeddah), H=Holy Quran (from Riyadh), CI=Call of Islam (from Makkah), F=Foreign Language Prgr, M=Music Prgr.
FM: Riyadh 91.2MHz (A/CI), 97.7MHz (M/F), 100.0MHz (H) — Jeddah 93.2MHz (A/CI), 98.0MHz (M/F) + 20 st's.
General Prgr. in Arabic: as above. **N:** 0400, 0600, 0800, 1030, 1130, 1300 (local), 1430, 1630, 1800, 2000, 2230.
Second Prgr: N: 0330, 0500, 0700, 0900, 1000, 1130 (rel. Gen. Prgr), 1400, 1500, 1700, 1900, 2030. **Call of Islam:** 0200, 1630.
Foreign Language Prgrs: English: 1000-1300, 1600-2100. **N:** 1830, 2045. **French:** 0800-1000, 1400-1600.
SW: Jeddah (G.C: 39.10E/21.32N) — Riyadh (G.C: 46.23E/24.30N): 50/500kW.
For frequencies and schedule see International Broadcasting section.
ANN: Arabic: "Ithaa till Mamlakah till Arabiyah al-Saudiyah". E: "This is the Broadcasting Sce. of the Kingdom of Saudi Arabia".
INT-SIG: 'Ud' (oriental lute). Opens and closes with National Anthem — **V.** by QSL-card. (Re. to Freq. Mgr.)**ARAMCO RADIO (Priv.)**

Serving staff of Saudi Aramco.
✉ Mgr. Media Prod. & Op's Dept., Bldg. 3030, LIP. Dhahran 31311.
☎ +966 (3) 876 6026. 🖷 +966 (3) 876 1608. ① ASAO SJ (DHAMEDIA).
LP: Coordinator, Media Broadc. Div: M.A. Al-Atani.
Studio 1 FM: Udhailiyah 88.8MHz, Dhahran 91.4MHz, Safaniya & Tanajib 103.9MHz — **Studio 2 FM:** Udhailiya 91.9MHz, Dhahran 101.4MHz, Safaniya & Tanajib 107.9MHz.
D.PRGR: 24h. **N:** 0945, 1030, 1530. Time Signal: on the h.

SYRIAN ARAB REPUBLIC

L.T: UTC + 2h (April-Nov UTC + 3h) — **Pop:** 14.500.000 — **Radios:** 3m — **Pr.L:** Arabic — **E.C:** 50Hz, 115/200V — **ITU:** SYR.

ORGANISME DE LA RADIO-TÉLÉVISION ARAB SYRIENNE (Gov.)

✉ Ommayad Square, Damascus. ☎ +963 (11) 228053, 228668. 2 +963 (11) 720700. ① SY 411138 GD
LP: DG: Khudr Omran. Dir. Eng: M.Bara. Dir. Public Rel: Mrs. Awafet Haffar. Dir. Admin & Finance: Zuheir Breidi.

Mediumwaves	kHz	kW	Prgr.
Adra	567	1000	1
?	612		1
Homs	617v	300	1
Sabboura	666	600	1
Sarakeb	747	100	1
Tartus	783	600	1
Deir el Zawr	827	100	1
Kharabo	877.3	10	2
Al-Hassake	918	200	1
Deir el Zawr	954	60	2
Tartus	1071	60	2
Al-Hassake	1125	200	2
Aleppo	1314	10	2
Homs	1485	10	1

FM: Damascus 92.1/99.0MHz (1), 97.1MHz (2).
Main Prgr (1): 0325-0045. **N:** 0415, 0515, 0915, 1115, 1215, 1515, 1615, 1815, 1915, 2115, 0015
Voice of the People (2): 0400-1600. **N:** 0600, 0800, 0900, 1100, 1400.

EXTERNAL SERVICE: see International Broadcasting section.

TURKEY

L.T: UTC + 2h (Su: UTC + 3h) — **Pop:** 62.531.000 — **Radios:** 8.800.000 — **Pr.L:** Turkish — **E.C:** 50Hz, 220/380V — **ITU:** TUR.

TÜRKIYE RADYO-TELEVIZYON KURUMU (TRT) (Turkish Radio-Television Corporation)

✉ TRT Sitesi, OR-AN.06450 Ankara. ☎ +90 (312) 4901730. 🖷 +90 (312) 4901733. **Cable:** TRT Ankara. ① 46826 trintr.
TRT Radio Dept: TRT Sitesi Kat: 9/B, OR-AN, 06450 Ankara. ☎ +90 (312) 4901797. 🖷 +90 (312) 4905636.
Regional Addr: TRT Ankara Radyosu, Atatürk Bulvari, Sihhiye, Ankara — TRT Istanbul Radyosu, Harbiye, Istanbul —TRT Ismir Radyosu, Kültür Parki, Izmir — TRT Çukurova Bölge Müdürlügü, 33130 Mersin.
L.P: DG: Prof. Dr. Tayfun Akgüner. Dep. DG (Eng.): Vural Tekeli. Dep. DG (Prgr.): Altan Kinal. Dep. DG (Admin): Abdullah Yazici. Dep. DG (Finance): Halinur Özçinar. Legal Adviser: Akin Besiroglu. Secr. Gen: Ziya Arikan. Dep. Gen. Secr: Halid Ertugrul. Head of Int. Rel: Güleser Güler. Head of Foreign Sce: Savas Kiratli. Head of TV Dept: Cetin Izbul. Head of Radio Dept: Bengi Baytur. Int. Tech. Rel: A.Dilek Cenkçiler.
STATIONS: *) Ankara, **) Istanbul.

Location	N	kHz	kW	Location	N	kHz	kW
Agri	2	162d	1000	Catalca**	2	702	150
Polatli*	2	180	1200	Gaziantep	1+R	765	600
Etimesgut*	1	198	120	Antalya	1+R	891	600
Van	1+R	225	600	Izmir	1	927	200
Erzurum	1+R	243	200	Trabzon	1+R	954	300
Denizli	1	558	600	Mundanya**	1	1017	1200
Malatya	1+R	594	600	Diyarbakir	1+R	1062	300
Çukurova	1	630d	300				

N=Network 1, 2, & 3 — R=carries Regional.prgrs at times.

FM-tr's in Main Cities:

Location	TRT-2	TRT-3	TRT-FM	kW
Istanbul	94.5	88.2	91.4	628/30/30
Izmir	94.7	88.0	91.2	30
Eskisehir		94.4	96.8	30
Ankara	100.3	91.2	88.0	628/30/30

Location	TRT-2	TRT-3	TRT-FM	kW
Bursa		97.6	95.0	30
Adana		89.2	92.5	30
Gaziantep		95.2	97.6	30
Konya		92.4	95.8	30
Kayseri		99.2	97.2	30
Diyarbakir		88.4	95.5	30

TRT-1: 0300-2300. **Main N:** 0530, 1100, 1700, 2100 — **TRT-2:** 0500-2300. **N:** 0500, 0530, 0700, 0900, 1100, 1300, 1500, 1700, 2100, 2255 — **TRT-3:** 24h. **N:** 0500, 0700, 1000, 1200, 1500, 1700, 2000, 2255. **N. in English/French/German** (3 min's each): 0703, 1003, 1203, 1503, 1703, 2003. **Tourist Prgrs: English:** 1715SS. **French:** 1515SS. **German:** 2015SS — **TRT-FM:** 24h — **OAP Radio:** 0300-1030MF on 1062kHz — **Regional st's:** Antalya/ Erzurum/Trabzon 0300-2300, Diyarbakir 1030 (0300SS)-2300 — **Tourist Station:** FM-stereo prgrs in **English, French, German & Greek** for tourists at 0530-1045 & 1630-2000: Izmir 100.5, Antalya 100.6, Marmaris 101.0, Pamukkale 101.0, Kusadasi 101.9, Nevsehir 103.0, Bodrum 97.4, Kalkan 105.9, Istanbul 101.6MHz (all 5kW). **N:** 0630, 0830, 1030, 1630 & 1930.
ANN: "Burasi TRT-1 (or Radio Bir), Burasi TRT-2 (or Radio Iki), Burasi TRT-3 (or Radio Üç)."

VOICE OF TURKEY (Foreign Sce.)

See International Broadcasting section.

OTHER STATIONS

	kHz	kW	Times
6)	1200		0800-1000. 1200-1400 (Sept-May)
3)	6325.1		0600-1700
2)	6900	5	0500-0950. 1200-1645
6)	7101.5	0.75	0800-1000. 1200-1400 (Sept-May)
1)	7370	10?	0550-1700 (occ. 24h)

FM	MHz	kW	Station/times
15)	88.6		Radio 2019, Ankara: 24h
35)	89.9		Show Radyo, Istanbul
28)	89.0		Lokum FM, Istanbul
12)	89.1		Number One FM, Sanlýurfa
17)	89.3		Alem FM, Istanbul
11)	90.2		R. Dunya, Bakanlýklar
12)	90.5		Number One FM, Izmir
12)	90.5		Number One FM, Antalya
12)	90.5		Number One FM. Bodrum
37)	90.8		Super FM, Istanbul
15)	91.5		Radio 2019, Izmir: 24h
23)	91.8		Hedef Radyo, Ankara
27)	92.0		Kral FM, Istanbul
8)	92.7		Yaprak FM, Gaizantep.
36)	93.0		Sky Radio, Izmir
4)	93.5		Super Stereo, Marmaris.

FM	MHz	kW	Station/times
3)	94.1		Istanbul Polis Radyosu: 0600-1700 (stereo)
1)	94.5	0.05	Türkiye Polis Radyosu: 1600-2200 (stereo)
20)	94.5		R. Blue, Istanbul
16)	94.7		Radyo5, Istanbul
14)	94.9		Acik Radio, Istanbul
15)	95.3		Radio 2019, Marmaris: 24h
22)	95.3		Radio Foreks/BBC, Istanbul
1)	95.5		Türkiye Polis Radyosu: 24h
32)	96.0		Radyo Ostim, Ankara
15)	96.2		Radio 2019, Istanbul: 24h
13)	96.9		R. Night & Day, Ankara
30)	97.2		Metro FM, Istanbul
15)	97.4		Radio 2019, Antalya: 24h
19)	98.4		Best FM, Istanbul
29)	98.9		Radyo Mega, Balikesir
21)	99.0		Fenerbahce FM, Istanbul
7)	99.5		Capital R, Ankara
33)	100.0		Power FM, Istanbul
8)	100.1		Yaprak FM, Gaizantep.
25)	100.5		Joy FM, Istanbul
24)	101.0		FM Izmir 101, Izmir
26)	101.1		Kent FM, Istanbul
12)	101.5		Number One FM, Ankara
12)	102.5		Number One FM, Istanbul
12)	102.7		Number One FM, Alanya
31)	103.1		Radyo ODTU, Ankara
34)	103.5		RadyoAktif, Izmir
18)	103.7		Radyo Anadolu, Ankara: 24h
5)	103.8	0.25	Istanbul Technical University R
38)	104.4		TGRT FM, Ankara
12)	106.5		Baskent Radyo TV, Bakanlýklar
10)	106.7		R. Bilkent, Bilkent
9)	107.9	0.25	R. Bogazici,Bebek

Addresses and other information

1) T.C. Içisleri Bakanliĝi, Emniyet Genel Müdürlüĝü, Ankara. **Prgrs:** Turkish music & short ann. **ANN:** "Burasi Türkiye Polis Radyosu." **V.** by QSL-card.
2) T.C. Çevre Bakanligi, Devlet Meteoroloji Isleri Genel Müdürlügü, 06120, Kalaba-Ankara. ☎ +90 (312) 359 75 45. ▤ +90 (312) 359 34 30. Dir: Faysal Geyik. **Prgrs:** Turkish music, weather rpts on the h. **ANN:** Burasi "Meteorolojinin Sesi Radyosu". **V.** by letter. Rp. Rec. acc.
3) Gamlica, Istanbul. **ANN:** "Burasi Istanbul Polis Radyosu".
4) Marmaris.
5) Istanbul. **Web:** http://titan.ehb.itu.edu.tr/~ituradyo/.
6) Cinarli Anadolu Teknik ye Endustri Meslek Lushsi Deneme Radyosu, Cinarli, 35110 Izmir. **English:** Thurs 1000-1030. **V.** by QSL-card. Rp.
7) P.K. 57 Asagi Ayranci, 06540 Ankara.
Web: http://www.capitalradio.com.tr
8) Fevzi Cakmak Bulvari 48, Gaizantep.
9) Bogazici Universitesi P.K.:2, 80815 Bebek - Istanbul.
E-mail: radyo@marconi.ee.boun.edu.tr
Web: http://marconi.ee. boun.edu.tr/~radyo/.
10) Bilkent University Main Campus, Engineering Building, 6th Floor, 06533 Bilkent/ Ankara. **E-mail:** radio@bilkent.edu.tr
Web: http://www.bilkent.edu.tr/ ~radio/.
11) R. Dunya, Atatürk Bulvarý No:137/7 Bakanlýklar 66640.
E-mail: radyo@dunya.com.tr **Web:** http://www.dunya.com.tr/.
12) Akay Cad. 25/7 Bakanlýklar / Ankara.
Web: http://www.NumberOne.com.tr/
13) Ankara. **Web:** http://www.radio.optima.com.tr.
14) Istanbul. **E-mail:** acikradyo@medyatext.com
Web: http://www.medyatext.com/acikradyo/

15) Istanbul. **Web:** http://www.radio2019.com.tr
16) Istanbul. **Web:** http://radyo5.superonline.com
17) Ergenekon Cad. Se-tat Tic. Merkezi No:100 Ferikoy-Istanbul.
Web: http://www.medyatext.com/alemfm
18) Ankara. **Web:** http://www.metu.edu.tr/AA/radyo.htm
19) Ankara. **E-mail:** info@bestfm.com.tr
Web: http://www.bilkent.edu.tr/~radio (+ nationwide FM relays)
20) Istanbul. **Web:** http://www.star.com.tr/radyo/blue.htm
21) Kusdili Caddesi İnsaatçilar Han, No. 8-6 Kadiköy - İstanbul.
Web: http://www.turkiyeonline.com/FbFm99
22) Istanbul. **Web:** http://www.auguste.com/planet/xn/foreks.html
23) Ziyabey Caddesi 3. Sokak No. 15/5 (06620) Balgat/Ankara (P.K. 74
Yenisehir/Ankara). **Web:** http://www.wec-net.com.tr/Hedef/
24) Izmir. **Web:** http://www.fm-izmir101.com.tr/
25) Istanbul. **Web:** http://www.star.com.tr/radyo/joy.htm
26) Istanbul. **Web:** http://www.kentfm.com.tr/
27) Istanbul. **Web:** http://www.star.com.tr/radyo/kral.htm
28) Istanbul. **Web:** http://www.star.com.tr/radyo/lokum.htm
29) Balikesir. **Web:** http://www.medyatext.com/radyomega/
30) Istanbul. **Web:** http://www.star.com.tr/radyo/metro.htm
31) Ankara. **E-mail:** radio@metu.edu.tr
Web: http://www.radio.metu.edu.tr/index2.html
32) Ankara. **Web:** http://www.ostim.com.tr/ostim/radyo/
33) Istanbul. **Web:** http://www.powerfm.com.tr
34) Izmir. **E-mail:** bilgi@radyoaktif.com.tr
Web: http://www.raksnet.com.tr/radyoaktif/
35) Istanbul. **Web:** http://www.medyatext.com/showradyo/
36) Izmir. **Web:** http://www.skyradio.com.tr
37) Istanbul. **Web:** http://www.star.com.tr/radyo/super.htm
38) Ankara. **Web:** http://www.tgrt.com.tr/TgrtFm/MenuTgrtFm.htm

UNITED ARAB EMIRATES

L.T: UTC + 4h — **Pop:** 2.822.000 — **Radios:** 490.000 — **Pr.L:** Arabic
— **E.C:** 50Hz, 220V — **ITU:** UAE.

MINISTRY OF INFORMATION & CULTURE
✉ P.O. Box 17, Abu Dhabi. ☎ +971 (2) 453000. 🖷 +971 (2) 452130.
☉ 94922283 INFOR EM.
L.P: Minister of Inf. & Culture: Khalfan M. Al-Roumi.

RADIO OF THE UNITED ARAB EMIRATES (Gov.)
✉ P.O. Box 63, Abu Dhabi. ☎ +971 (2) 451000. 🖷 +971 (2) 451155.
L.P: DG: Abdul Wahab al Radhwan. Contr. of Prgrs: Abdul Hadi al Mobarak.
Chief Eng (Tr's): Rushdi Hattab. Dir Foreign Prgrs: Ms. Aida Hamza.

MW	kHz	kW	Prgr.	Times
Sadiyat	657	2 x 50	M	0200-0800, 1800-2200
Sadiyat	729	2 x 750	M	0200-2200
MW	**kHz**	**kW**	**Prgr.**	**Times**
Maqtaa	810	50	F	0800 (Fri 0700)-1800
al Ain	828	1	M	0200-2200
Fujairah	972	2 x 50	M	0200-2200
Dabiya	1314	2 x 1000	M	0200-2200
Sharjah	1575	50	L	1130-1500

L) Local prgr.
FM: 100.5MHz 5kW, 98.0MHz 2kW, 88.6/94.8/98.9MHz 1kW.
Main Prgr in Arabic: 0200-2200. **N:** 0630, 0930, 1230, 1400, 1430.
Holy Quran Sce: 1200-1500 on 88.6/94.8/98.0MHz.
English (Capital FM): 0300-2000 on 100.5MHz.
Foreign Language Sce. (F): 0700-0800 (Fri) Bengali, 0800-1100
English, 1100-1300 French, 1300-1600 Urdu, 1600-1800 Pilipino.

SW (G.C: 54.17E/24.23N): 500kW — +) Makta (G.C: 54.34E/ 24.21N):
120kW.

For frequencies and schedule see International Broadcasting section.
ANN: A: "Idha'atu-l-imarat al-arabiyya al-muttahida min Abu Zabiy".
V. by QSL-folder. R. to Dir, Foreign Prgrs.

U.A.E. RADIO & TELEVISION – DUBAI (Gov.)
✉ P.O. Box 1695, Dubai. ☎ +971 (4) 370255. 🖷 +971 (4) 379275.
☉ 95845605 BRCAST EM — **L.P:** Dir. of Inf: Sheikh Hasher al
Maktoum. Ag. DG: Ahmed Saeed al Gaoud. Contr. Radio: Hassan
Ahmed. Contr. of Eng: Abdul Rehman Al Ali.

kHz	kW	kHz	kW	kHz	kW
1107	10	1251	600	1481	1500

FM: 92.5MHz 10kW.
Arabic: 0300-2100. **N:** 0315, 0500, 1000, 1300, 1700, 1900, 2030.
English: 24h on 92.5MHz. **N:** 0330, 0530, 1030, 1330, 1630.

SW (G.C: 55.16E/25.14N): 1 x 500kW, 3 x 300kW tr's.
For frequencies and schedule see International Broadcasting section.
ANN: A: "Idha'at al imarat al Arabiyyah al Mutahhida min Dubai". E:
"This is United Arab Emirates Radio in Dubai" — **V.** by QSL-card.

RAS AL KHAIMAH BROADCASTING STATION (Gov.)
✉ P.O. Box 141, Ras al Khaimah. ☎ +971 (7) 851151. 🖷 +971 (7)
353441.
MW: 1152kHz 50kW, 1175kHz 10kW — **FM:** 95.3MHz.
D.PRGR: 0200-1900. **Urdu:** 0830-1000. **Malayalam:** 1000-1100.
ANN: "Idha'at al imarat al Arabiyyah al Mutahhida min Ras al Khaima"
— **V.** by letter.

UMM AL QIWAIN BROADCASTING STATION (Gov.)
✉ P.O. Box 444, Umm al Qiwain. ☎ +971 (6) 666044. 🖷 +971 (6)
666055. ☉ 944/95869711 EM — **L.P:** DG: Ali Jassim Ahmed. CE:
Saeed el Naqueeb — **MW:** 846kHz 20kW.
D.PRGR: 0200-1900. **Sinhalese:** 0800-0900. **Urdu:** 0900-1055.
Prgrs.in Hindi, Malayalam, Filipino.

YEMEN (Republic of)

L.T: UTC + 3h — **Pop:** 14m — **Radios:** 665.000 — **Pr.L:** Arabic —
E.C: 50Hz, 230V — **ITU:** YEM.

YEMEN RADIO & TV CORPORATION (Gov.)
✉ P.O. Box 2182, Sana'a. ☎ +967 (1) 231181. 🖷 +967 (1) 230761.
☉ 2645 YARTV —P.O. Box 1222, Aden.
Networks: A=Arabic Netw, L=Local prgrs.
G.C: Sana'a 44.11E/15.22N.

Mediumwaves	kHz	kW	N.	Times
Mocha	711	200	A	0255-2130
Mukalla	756		L	24h
Sana'a	837	30	A	0255-2130
Sana'a	1008v	600	A	0255-2130
Taiz	1080	30	A	0600-1105
Hudaydah	1125		A	0700-0915, 1045-1630,
				1300-1600

Shortwaves: Sana'a (G.C: 44.11E/15.22N).

kHz	kW	N.	Times
5950	300	A	0255-2130
9780.2	50	A	0255-2130

+) no transmission break on Fri — **FM:** Sana'a 92.6/93.5MHz.

N. in Arabic: 0330, 0430, 0600, 0700, 0800, 0900, 1200-2100 hourly.
English: 1800-1900 (repeated next day 0600-0700) on 711/9780.2kHz
+ FM 93.5MHz. **N:** 1800, 1830.
ANN: "Idha'at-l-jumhuriya al-yamanniya min Sana'a/Aden". E: "The
Republic of Yemen Radio broadcasting from Sana'a".
V. by QSL-card or letter. Rp.

ASIA

BANGLADESH

L.T: UTC + 6h — **Pop:** 137.000.000 — **Radios:** 8m — **Pr.L:** Bengali — **E.C:** 50Hz, 220/440V — **ITU:** BGD.— **IDD:** 880.

BANGLADESH BETAR(Gov.)
National Broadcasting Authority, NBA House, 121 Kazi Nazrul Islam Ave, Dhaka-1000. ☎ +880 (2) 865294. 📠 +880 (2) 862021. **Cable:** Broadcast. ① 642228 NBA BJ.
L.P: DG: M.I. Chowdbury. CE: Syed Abdus Shaker. DG (Prgrs): Ashfaqur Rahman Khan. Dir. of Monitoring: M. Shahib Ali. Dir. Ext. Sces: Dilruba Begum. Dir. Liaison: Quazi Mahmudur Raham. Station Engineer (Research & Rec.): Monoranjan Das

Stations	kHz	kW	Times
Khulna	558	100	0000-0430, 0600-0930, 1000-1740
Dhaka-B	630	100	0000-0230, 0430-1740, 1800-2100
Dhaka-A	693	1000	0000-0430, 0830-1740
Rajshahi	846	10	0000-0430, 0600-0930, 1000-1740
Chittagong	873	10	0000-0430, 0600-0930, 1000-1740
Sylhet	963	20	0000-0430, 1000-1740
Thakurgaon	999	10	1200-1740
Rangpur	1053	20	0000-0430, 1000-1740
Rajshahi	1080	100	0000-0430, 0600-0930, 1000-1740
Rangamati	1161	10	0000-0430
Dhaka-C	1170	10	0800-1100
Comilla	1413	10	1200-1740
Shavar	4880	100	0000-0505, 1200-1610
Shavar	15520	100	1200-1740

G.C: Shavar: 90.12E/23.27N.
FM: Dhaka 100.0/Sylhet 101.5/Chittagong 102.5/Rajshahi 102.0/Rangpur 105.5/Khulna 106.5MHz (all 2kW) + 6 st's less than 1kW.
N. in English: 0200, 0700, 1100, 1530, 1805. **N. in Bengali:** 0100, 0300, 0400, 0500, 0600, 0800, 0900, 1000, 1200, 1430, 1700, 1800.
OVERSEAS SERVICE: see International Broadcasting section.

BHUTAN (Kingdom of)

L.T: UTC + 6h — **Pop:** 1.650.000 — **Radios:** 23.000 — **Pr.L:** Dzongha, Sharchopkha, Lhotsam, English — **E.C:** 50Hz, 220V — **ITU:** BTN.
BHUTAN BROADCASTING SERVICE – BBS (Gov.)
P .O. Box 101, Thimpu. ☎ +975 23070/1/2. 📠 +975 23073. ① 212 Infobro TPU BT.
L.P: Dir. of Inf: Tashi Phuntshog. Joint Dir: Phub Tshering. Prgr. Officer: Tashi Dhendup. St. Eng: Sonam Tobgay.
SW: 5030/6030/9615kHz 50/10kW — **FM:** 96/98MHz.
D.PRGR: W 1000-1500 on 5030kHz. **Tibetan:** Sun 0400-1000 on 6030 or 9615kHz. **English:** W 1415-1500 (**N:** 1415. **UN Radio Prgr:** Mon/Thurs 1430-1445. (**NB:** In summer, st. closes at 1400. The English Prgr. and UN Radio Prgr. are therefore aired 1h earlier).
V. by QSL-card. 15 min prgr details req. Rp. (2 IRC's).

BRITISH INDIAN OCEAN TERRITORY

L.T: UTC + 5h — **Pr.L:** English — **Pop:** variable (US & British military personnel). The original population of approx. 3000 were repatriated to Mauritius — **E.C:** 60Hz, 110/220V — **ITU:** BIO.

ARMED FORCES RADIO AND TELEVISION SERVICE (U.S. Mil.)
Naval Media Center Detachment-Diego Garcia, PSC 466 Box 14, FPO, AP 96595-0014. ☎ +246 370 3680/3685 📠 +246 370 3681
E-mail: Dgar@mediacen.navy.mil
MW: Island Talk, 1485kHz 200W, news, sports & talk
FM: Power 99, 99.1MHz 200W, weekdays 0600-1400, rock & roll, live DJ. **Island Variety,** 101.9MHz 200W, mixture of rock, alternative, urban & country. — **D.PRGR:** 24h — **V.** by letter.

BRUNEI DARUSSALAM

L.T: UTC + 8h — **Pop:** 316.000 — **Radios:** 125.000 — **Pr.L:** Malay, English, Chinese, Gurkha — **E.C:** 50Hz, 240V — **ITU:** BRU.

RADIO TALIVISHEN BRUNEI (Gov.)
Jalan Elizabeth II, Bandar Seri Bagawan 2042, Negara Brunei Darussalam. ☎ +673 (2) 243111. 📠 +673 (2) 241882. **E-mail:** rtbeng@brunet.bn **Web:** http://www.brunet.bn/gov/rtb/rtbmain.htm
L.P: Dir: Pengiran Dato Paduka Haji Badaruddin bin Pengiran Haji Ghani. Dep. Dir: Haji Md Yusof Bin Hj Abd Rahman. Sen. Admin. Officer: Pengiran Ali bin Pg. Haji Othman. Ag. Supt. Eng: Lim Sam Lee. Head of Prgrs: Pengiran Datin Hajah Normah Pengiran Daud. Head of Int. & Pub. Rel. & CBA/ COMBROAD Liaison Officer: Pengiran Hajah Fatimah binti PSJ PHA Momin. Head of N. & Current Affairs: Awang Johari bin Achee. Prgr. Prod. Mgr. (Radio): Hj Ramlee Hj Abd Said.

MW	kHz	kW	P	Times
Tutong	594	200d	RN	2030-1600
Serasa	675	200	RN	2030-1600
S. Hanching	711	20	E	2230-0030, 0300-0800, 1200-1400
			C	0030-0300, 0800-1100

P=Prgr: C=R. Pilihan in Chinese, **E**=R. Pilihan in **English**, RN=Radio Nasional in Malay, RPe=R. Pelangi (prgrs for young people), RH=R. Harmoni (music sce.).

FM (MHz)	RN	RPe	E/C	RH	kW
Andulau	93.8	91.0	96.9	93.1	5
Bukit Subok	92.8	91.4	95.9	94.1	5/0.5/5

N. in English: 2245, 2330, 0430, 1315.
IS: Synthesized chimes — **V.** by QSL-card.

BRITISH FORCES BROADCASTING SERVICE
BFPO 2 — 24h on 92.0MHz 1kW & 89.05MHz 0.05kW.

CAMBODIA (Kingdom of)

L.T: UTC + 7h — **Pop:** 10m — **Radios:** 1.500.000 — **Pr.L:** Cambodian (Khmère) — **E.C:** 50Hz, 120/208/220V — **ITU:** CBG.

NATIONAL RADIO OF CAMBODIA
106, Street Preah Kossamak, Phnomh Penh. ☎ +855 (2) 3369 or 2869. 📠 +855 (23) 27319.

Mediumwaves	kHz	kW	Mediumwaves	kHz	kW
Steung Treng	585	20	Phnom Penh	1300	1
Phnom Penh	740	150	Phnom Penh	1360	1
Phnom Penh	918	120	Battambang	1453	20
Sihanoukville	1255	20			

FM: Phnom Penh 103MHz.
SW (G.C: 104.51E/11.34N): 4907vkHz 50kW, 6090vkHz 15kW.
National Sce. on 740/918kHz + SW: 2230-0700, 1100-1500.
Local radio on 585/1255/1300/1453kHz + 103MHz.
Overseas Sce. on 1360kHz: 0000-0115, 1200-1315. 15 mins each in English, French, Thai, Laotian & Vietnamese.
ANN: HS: "Vithyu Cheat Kampuchea" — **V.** by letter.

Private stations: 2 st's in Phnom Penh on 90 and 99MHz Prgr. in **English** on 99MHz: 0600-0800.
R. France Internationale: Phnom Penh 92.0MHz.

CHINA (People's Rep. of)

L.T: UTC + 8h (exc. Xinjiang Uighur Aut. Reg: UTC + 6h). — **Pop:** 1,236,260,000 — **Radios:** 215,950,000 — **Pr.L:** Chinese, Amoy, Cantonese, Chaozhou, Hakka, Kazakh, Korean, Mongolian, Tibetan, Uighur, Zhuang, a.o. — **E.C:** a/c 50, 220V - **ITU:** CHN.

MINISTRY OF INFORMATION AND INDUSTRIES
Xi Chang'an Jie, Beijing 100031. **L.P:** Minister: Wu Jichuan.

THE STATE ADMINISTRATION OF RADIO, FILM AND TELEVISION (SARFT)(Gov.)
2 Fuxingmenwai Dajie, Beijing 100866 or P.O.Box 4501, Beijing. ☎+86 (10) 6851 3409. 📠 +86 (10) 6851 2174. ① 222236 RTPRC(CN)Beijing. **Cable:** RTPRC BEIJING. **L.P:** Dir: Tian Congming.

Official P.R.C Abbreviations: The 31 regions of People's Republic of China, with their abbreviations and names in Pinyin (Chinese Phonetic Alphabet) version followed by the old spelling in (brackets):
AH: Anhui (Anhwei) - BJ: Beijing M. (Peking) - CQ: Chongqing M. (Chungking) - FJ: Fujian (Fukien) - GD: Guangdong (Kwangtung) - GS: Gansu (Kansu) - GX: Guangxi Zhuang A.R. (Kwangsi) - GZ: Guizhou (Kweichow) - HAN: Hainan (Hainan) - HB: Hubei (Hupeh) - HEB: Hebei

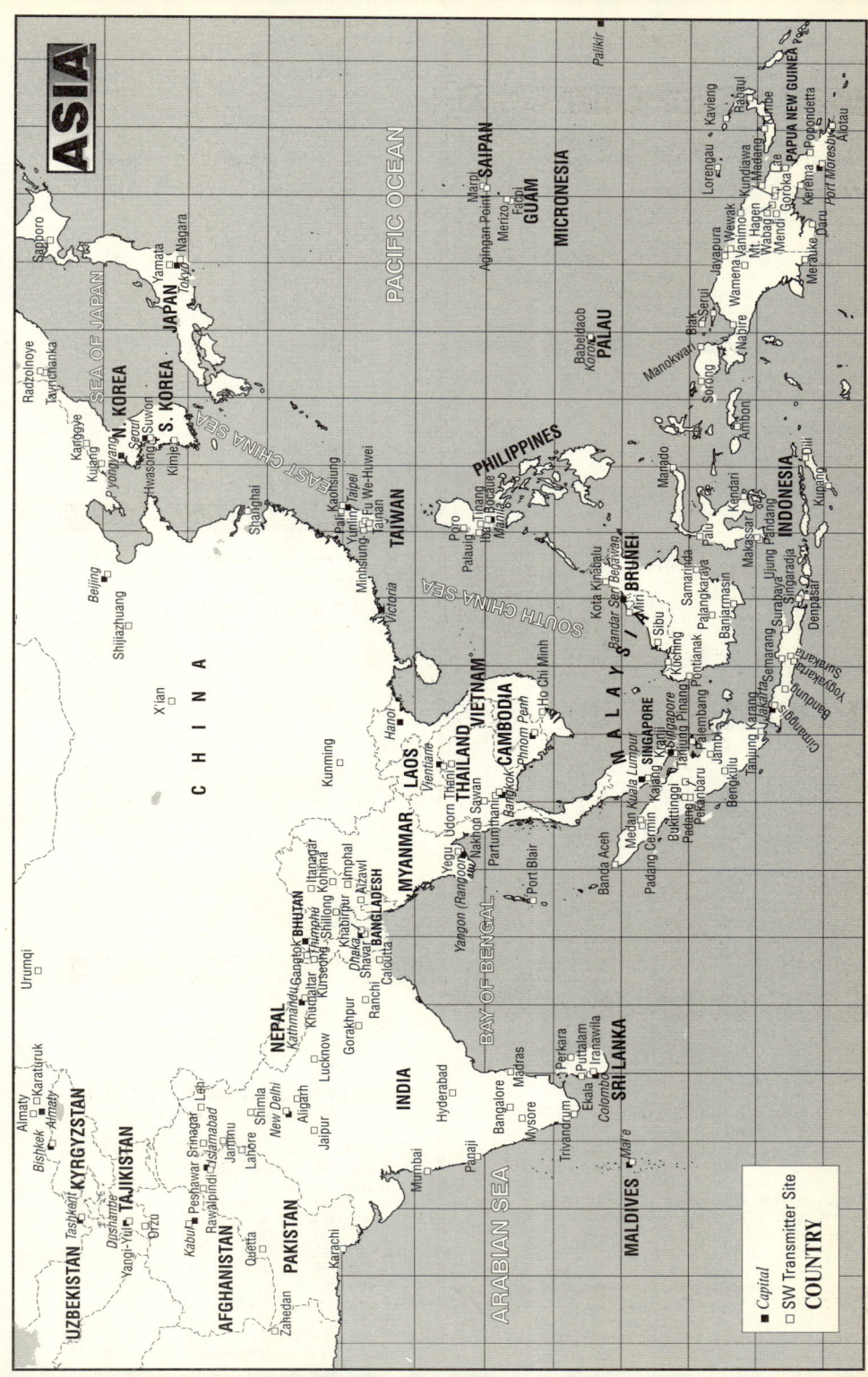

ASIA

PACIFIC OCEAN

Sapporo
Radzolnoye
Tayrchanka
SEA OF JAPAN
Yamata
Tokyo
Nagara

JAPAN

N. KOREA
S. KOREA
Kanggye
Kujang
Pyongyang
Hwasong
Seoul
Suwon
Kimje
EAST CHINA SEA

Beijing
Shijiazhuang
Shanghai

X'ian

C H I N A

Urumqi

Kunming

Hanoi
Vientiane
LAOS
VIETNAM

MYANMAR
Yegu (Rangoon)
Yangon (Rangoon)

THAILAND
Udorn Thani
Nakhon Sawan
Partumthani
Bangkok
Ban Dung

CAMBODIA
Phnom Penh

Ho Chi Minh

SOUTH CHINA SEA

SAIPAN
Mairi
Agingan Point
Merizo
Fadai
GUAM

MICRONESIA

Babeldaob
Koror
PALAU

PHILIPPINES

TAIWAN
Kaohsiung
Taipei
Yunlin
We-Huwei
Tainan
Minsfung
Fu
Victoria

Paro
Palauig
Iba Bocaue
Manila
Tinang

PAPUA NEW GUINEA

Almaty
Karatyuruk
Bishkek
Almaty
KYRGYZSTAN
Tashkent
UZBEKISTAN
Dushanbe
Yangi-Yul
TAJIKISTAN

AFGHANISTAN
Kabul
Peshawar
Rawalpindi
Islamabad
Srinagar
Jammu
Leh
Lahore
Shimla
New Delhi
Aligarh
Jaipur

Quetta
PAKISTAN
Zahedan
Karachi

NEPAL
Kathmandu
Gangtok
BHUTAN
Khudatar
Thimphu
Kursedng
Shillong
Kohima
Itanagar
Imphal
Aizawl

Gorakhpur
Lucknow
Khabi Pur
Dhaka
Shawal
BANGLADESH
Ranchi
Calcutta

INDIA
Hyderabad

Mumbai

Panaji

Bangalore
Perkara
Mysore
Ekala
Puttalam
Iranawila
Colombo
Mait e
SRI LANKA
Trivandrum
Madras

ARABIAN SEA

BAY OF BENGAL
Port Blair

MALDIVES

MALAYSIA
Kota Kinabalu
BRUNEI
Bandar Seri Begawan
Miri
Sibu
Kuching

Medan
Kuala Lumpur
Kajang
Kajji
SINGAPORE
Singapore
Tanjung Pinang
Bukittinggi
Padang Cermin
Pekanbaru
Jambi
Palembang
Pontianak
Bengkulu
Tanjung Karang
Padang

INDONESIA

Banda Aceh

Marado
Palu
Samarinda
Makassar
Kendari
Pangkajaja
Banjarmasin

Ujung Pandang
Surabaya
Singaradja
Denpasar
Yogyakarta
Surakarta
Semarang
Bandung
Cumanggis

Manokwari
Serui
Biak
Sorong
Nabire
Ambon

Jayapura
Wewak
Vanimo
Mt. Hagen
Wabag
Goroka
Mendi
Wamena

Dili
Kupang

Lorengau
Kavieng
Rabaul
Kundiawa
Madang
Kerema
Daru
Merauke

Popondetta
Port Moresby
Alotau

Palikir

■	Capital
□	SW Transmitter Site
	COUNTRY

(Hopeh) - HEN: Henan (Honan) - HL: Heilongjiang (Heilungkiang) - HN: Hunan (Hunan) - JL: Jilin (Kirin) - JS: Jiangsu (Kiangsu) - JX: Jiangxi (Kiangsi) - LN: Liaoning (Liaoning) - NM: Nei Menggu A.R. (Inner Mongolia) - NX: Ningxia Hui A.R. (Ningsia) - QH: Qinghai (Tsinghai) - SC: Sichuan (Szechwan) - SD: Shandong (Shentung) - SH: Shanghai M. (Shanghai) - SN: Shaanxi (Shensi) - SX: Shanxi (Shansi) - TJ: Tianjin M. (Tientsin) - XJ: Xinjiang Uighur A.R. (Sinkiang) - XZ: Xizang A.R. (Tibet) - YN: Yunnan (Yunnan) - ZJ: Zhejiang (Chekiang).

Regional Services: Add "Renmin Guangbo Diantai" (People's Broadcasting Station) to the station name shown in the table below to obtain the full name in Standard Chinese.
Abbreviations: 1 = 1st prgr, 2 = 2nd prgr, 3 = 3rd prgr; E = Economic st/ch, ED = Educational st/ch, FL = Foreign Language st, G = General st, L = Literary st/ch, M = Music st/ch, N = News st/ch, T = Traffic st/ch; EBS = Economic Broadcasting Station, IBS = Information Broadcasting Station, LBS = Literary Broadcasting Station, MBS = Music Broadcasting Station, TBS = Traffic Broadcasting Station.

Languages: Standard Chinese (Putonghua), based on the Beijing dialect, is used in broadcasts throughout China. Various dialects and minority languages are included in the relevant regional services and in broadcasts to Taiwan.

Abbreviations: Ch = Standard Chinese, Ko = Korean, Kz = Kazakh, Mo = Mongolian, Tb = Tibetan, Ug = Uighur.

Mediumwaves:

	kHz	kW	Station	Location
ZJ1)	531	10	Zhejiang	ZJ
1)	540		CNR 1	
1)	549	100	CNR Taiwan 1	FJ
NM5)	549	10	Chifeng-Ch	NM
NM12)	549	10	Alxa-Ch	NM, Bayanhot
EN2)	549	10	Zhengzhou	HEN
FJ1)	558	50	Fujian FJ	
XJ1)	558	20	Xinjiang-Ug	XJ, Urumqi
NM3)	558	10	Baotou	NM
1)	567		CNR 1	
TJ1)	567	20	Tianjin T	TJ
ZJ1)	576v		Zhejiang	ZJ
YN1)	576	10	Yunnan G	YN
JL14)	576	1	Yanji	JL
EN4)	576	1	Luoyang	HEN
EN13)	576		Nanyang	HEN
JS1)	585	10	Jiangsu EBS	JS, Nanjing
5)	585		Southeast BC	FJ, Fuzhou
JX5)	585	1	Xinyu	JX
JX1)	594	10	Jiangxi EBS	JX, Nanchang
SD1)	594	100	Shandong E	SD, Jinan
XZ1)	594	300	Xizang-Tb	XZ, Lhasa
100)	603		CRI DS	GD, Zhuhai
BJ1)	603	10	Beijing ED	BJ
SH1)	603		Shanghai E	SH
EN1)	603		Henan IBS	HEN, Zhengzhou
GD1)	603		Guangdong ED	GD, Guangzhou
GZ1)	603		Guizhou EBS	GZ, Guiyang
SN1)	603		Shaanxi EBS	SN, Xi'an
SX4A)	603		Datong 2	SX
NM8)	603	50	Hulun Buir-Ch	NM, Hailar
NM10)	603	10	Ordos BS-Mo	NM, Ulanhot
JS8)	603		Nantong EBS	JS
ZJ4)	603	10	V.O.Health	ZJ, Ningbo
SD5)	603		Zaozhuang EBS	SD
XJ3A)	603		Urumqi EBS	XJ
XJ9)	603	1	Ili-Ug	XJ, Yining
LN1)	612	10	Liaoning	LN, Dandong
FJ1)	612	100	Fujian	FJ
SC1)	612	10	Sichuan 1	SC
HL1)	621	200	Heilongjiang	HL, Harbin
HB9)	621		Yichang	HB
QH4)	621	20	Haixi	CH, Da Qaidam
1)	630		CNR 2	
NX1)	630	10	Ningxia EBS	NX, Yinchuan
1)	639		CNR 1	
SH1)	648		Shanghai T	SH
GD1)	648	150	Guangdong Sat.	GD, Guangzhou
AH3)	648	1	Huainan	AH
EN1)	657	300	Henan	HEN, Zhengzhou
JL8)	657	1	Baishan	JL
ZJ6)	657	1	Jiaxing	ZJ
101)	666		China Huayi BC	FJ, Fuzhou
QH1)	666	200	Qinghai 1	QH, Xining
LN8)	666	2	Jinzhou	LN
JL5)	666		Siping	JL
HL10)	666	10	Jiamusi	HL
ZJ9)	666		Jinhua	ZJ
	(Nom 675kHz)			
AH2)	666	10	Hefei	AH
SD10)	666		Jining N	SD
YN3)	666	1	Dongchuan	YN
NM1)	675	150	Nei Menggu-Ch	NM, Hohhot
XJ1)	675		Xinjiang-Ch	XJ
YN1)	675	1	Gejiu	YN
1)	684		CNR Taiwan 2	FJ
GS1)	684	10	Gansu 1	GS
XJ1)	684		Xinjiang-Ch	XJ
EB8)	684	10	Tangshan	HEB
LN5)	684	10	Fushun	LN
HL9)	684	10	Mudanjiang	HL
ZJ11)	684	7.5	Zhoushan	ZJ
AH14)	684		Suxian Diqu	AH, Suzhou
SN1)	693	200	Shaanxi Sat.	SN, Xianyang
LN16)	700		Chaoyang	LN, Lingyuan
JS1)	702	150	Jiangsu G	JS, Nanjing
XJ1)	702		Xinjiang-Ch	XJ
NM6)	702	10	Ulanqab-Ch	NM, Jining
NM15)	702	1	Manzhouli	NM
SC10)	702	1	Neijiang	SC
YN5)	702	10	Honghe	YN, Gejiu
QH1)	711	10	Qinghai 1	QH, Golmud
ZJ10)	711		Quzhou	ZJ
ZJ12)	711		Lishui	ZJ
AH16)	711		Lu'an	AH
EN2)	711	10	Zhengzhou EBS	HEN
SC6)	711		Panzhihua	SC
SC9)	711	1	Mianyang	SC
GZ3)	711		Liupanshui	GZ
1)	720		CNR 2	
SC7)	720	1	Deyang	SC
JX1)	729	150	Jiangxi 1	JX, Nanchang
EN14)	729		Shangqiu	HEN
JL1)	738	150	Jilin N	JL, Changchun
HN1)	738	150	Hunan Sat.	HN, Changsha
XJ1)	738	10	Xinjiang-Ch	XJ, Urumqi
ZJ8)	738		Shaoxing	ZJ
100)	747		CRI DS	AH, Hefei
SC1)	747		Sichuan IBS	SC, Chengdu
SN1)	747		Shaanxi LBS	SN, Xi'an
NM4)	747	1	Wuhai	NM
LN5)	747		Fushun M	LN
JS6A)	747		Yancheng EBS	JS
JS10)	747		Changzhou TBS	JS
FJ6)	747		Longyan	FJ
JX8)	747		Ganzhou Diqu BS	JX, Ganzhou
SD17)	747		Binzhou Diqu	SD, Binzhou
SC13)	747		Nanchong	SC
YN6)	747		Xishuangbanna	YN, Jinghong
LN16)	750		Chaoyang	LN, Jianping
GX6)	750		Beihai	GX
SC6)	750		Luzhou	SC
1)	756		CNR 1	
1)	765v		CNR Taiwan 1	FJ
NM1)	765	10	Nei Menggu-Ch	NM
GZ1)	765	10	Guizhou Sat.	GZ
GD8)	765	10	Shaoguan	GD
HB1)	774	100	Hubei	HB, Wuhan
EB1)	783	100	Hebei	HEB, Baoding
SH2)	792	50	Dongfang BS	SH
GX1)	792	100	Guangxi Sat.	GX, Nanning
NM10)	792		Ordos BS-Ch	NM, Otog
LN2)	792	10	Shenyang N	LN
SC3)	792		Chengdu 1	SC
GS4)	792	1	Jiayuguan	GS
100)	801		CRI DS	TJ, Tianjin
100)	801		CRI DS	AH, Hefei
100)	801		CRI DS	SD, Zibo/Weifang
GD2)	801	50	Zhujiang EBS	GD, Maoming
EB8)	801	10	Tangshan EBS	HEB
JS2)	801		Nanjing TBS	JS
JS3)	801		Xuzhou LBS	JS
JS5)	801		Huaiyin	JS
JS7)	801		Yangzhou EBS	JS

ID	Freq	Power	Name	Region
ZJ5)	801		Wenzhou E	ZJ
SD8)	801		Yantai E	SD
SD10)	801		Jining E	SD
SD12)	801		Linyi T	SD
EN8)	801		Xinxiang	HEN
NX2)	801	10	Yinchuan	NX
ZJ1)	810	200	Zhejiang	ZJ, Hangzhou
XJ1)	810		Xinjiang-Ug	XJ
LN16)	810	1	Chaoyang	LN
JL6)	810	1	Liaoyuan	JL
SN2)	810	10	Xi'an N	SN
XJ3)	810		Urumqi-Ch	XJ
SX1)	819	200	Shanxi	SX, Taiyuan
XJ8)	819		Kuytun-Kz	XJ
BJ1)	828	50	Beijing	BJ
EN1)	828	10	Henan	HEN
GD1)	828	50	Guangdong Sat.	GD, Heyuan
EN15)	828v		Zhoukou	HEN
HB8)	828		Jingzhou	HB
HB17)	828v	1	Xiantao	HB
1)	837		CNR 1	
XJ2)	837		Tianshan EBS	XJ, Urumqi
HL2)	837	10	Harbin	HL
2)	846		V.O.Strait	FJ
3)	846		V.O.Jinling	JS, Nanjing
EB10)	846		Cangzhou	HEB
SX1)	846	20	Shanxi	SX, Changzhi
JL1)	846		Jilin EBS	JL, Changchun
EN1)	846		Henan EBS	HEN
GX1)	846		Guangxi EBS	GX, Nanning
XZ1)	846		Xizang-Ch	XZ, Lhasa
XJ1)	846		Xinjiang-Ug	XJ
LN13)	846		Fuxin Mo BS	LN
JS10)	846	10	Changzhou	JS
SD7)	846		Weifang TBS	SD
EN6)	846		Jiaozuo	HEN
1)	855		CNR 2	
ZJ1)	864		Zhejiang	ZJ
AH1)	864	50	Anhui EBS	AH, Hefei
HL1)	873	200	Heilongjiang Ko	HL, Harbin
GS1)	873	50	Gansu 1	GS, Lanzhou
EB13)	873v		Xinji	HEB
SD12)	873		Linyi	SD
EN3)	873v	1	Kaifeng	HEN
HB3)	873	10	Wuhan	HB
XJ7)	873		Changji	XJ
FJ1)	882	100	Fujian	FJ, Fuzhou
EB2)	882		Shijiazhuang 1	HEB
NM2)	882	10	Hohhot	NM
LN2)	882	10	Shenyang EBS	LN
LN3)	882	10	Dalian	LN
EN9)	882		Anyang	HEN
QH3)	882	10	Yushu	QH
XJ4)	882		Karamay-Ch	XJ
XJ9)	882	1	Ili-Kz	XJ, Yining
NX1)	891	100	Ningxia 1	NX, Yinchuan
LN7)	891	50	Dandong EBS	LN
4)	900		V.O.Pujiang	SH
SX2)	900		SX Changcheng	SX, Taiyuan
HL1)	900	50	Heilongjiang	HL, Bei'an/Jiamusi
EN1)	900	10	Henan TBS	HEN, Zhengzhou
HN1)	900		Hunan EBS	HN, Changsha
GD1)	900		Guangdong ED	GD, Shenzhen
SN1)	900	10	Shaanxi SE BS	SN, Xi'an
NM17)	900	1	Dongsheng	NM
LN18)	900		Haicheng	LN
JL3)	900	10	Changchun	JL
JS2)	900		Nanjing EBS	JS
JS9)	900		Zhenjiang EBS	JS
JS11)	900		Wuxi L	JS
ZJ7)	900		Huzhou EBS	ZJ
EN12)	900		Luohe	HEN
GD6)	900		Zhuhai	GD
YN9)	900		Dehong	YN, Luxi
SN4)	900		Baoji EBS	SN
1)	909	100	CNR Taiwan 2	FJ
TJ1)	909	50	Tianjin N	TJ
SC1)	909	10	Sichuan 1	SC, Zigong
QH1)	909	10	Qinghai 1	QH
XJ1)	909	10	Xinjiang-Mo	XJ, Urumqi
JL7)	909	10	Tonghua	JL
HL8)	909	5	Yichun	HL
SD1)	918	10	Shandong	SD, Yantai
GX2)	918	1	Nanning	GX
1)	927		CNR Taiwan 2	FJ
BJ1)	927	50	Beijing T	BJ
LN1)	927	12.5	Liaoning IBS	LN, Shenyang
JS14)	927	1	Changshu 1	JS
JX1)	927		Jiangxi IBS	JX, Nanchang
GD3)	927		Yangcheng TBS	GD, Guangzhou
GZ1)	927	200	Guizhou Sat.	GZ, Kaili
NM7)	927		Xilingol-Mo	NM, Xilinhot
JL4)	927	10	Jilin-shi	JL
JL18)	927		Hunchun	JL
HN7)	927		Yueyang	HN
ZJ1)	930		Zhejiang LBS	ZJ
AH1)	936	200	Anhui 1	AH, Hefei
NM10)	936	10	Ordos BS-Ch	NM, Dongsheng
1)	945		CNR 2	
HA1)	954	30	Hainan	HAN, Haikou
SC1)	954		Sichuan GB	SC, Chengdu
XJ1)	954		Xinjiang-Kz	XJ
EB12)	954	1	Hengshui	HEB
LN4)	954	10	Anshan	LN
ZJ2)	954	10	Hangzhou	ZJ
SC6)	954		Luzhou	SC
GS2)	954		Lanzhou	GS
LN1)	963	50	Liaoning	LN, Dalian
ZJ1)	963	10	Zhejiang	ZJ
XJ1)	963		Xinjiang-Kz	XJ
EN1)	972	150	Henan EBS	HEN, Zhengzhou
XJ1)	972		Xinjiang-Kz	XJ
HL2)	972	10	Harbin EBS	HL
1)	981		CNR 1	
SH1)	990	50	Shanghai 1	SH
YN1)	990	20	Yunnan G	YN, Kunming
XJ1)	990		Xinjiang-Kz	XJ
EB9)	990		Qinhuangdao	HEB
LN1)	999	200	Liaoning EBS	LN, Shenyang
GD1)	999	7.5	Guangdong Health	GD, Guangzhou
SC1)	999		Sichuan EBS	SC, Chengdu
XJ1)	999		Xinjiang-Ch	XJ
GZ2)	999	10	Guiyang	GZ
1)	1008	10	CNR 1	YN, Jinping
GD1)	1008		Guangdong ED	GD, Jiangmen
SN1)	1008	50	Shaanxi Sat.	SN, Hanzhong/Yan'an
JS2)	1008		Nanjing N	JS
JS11)	1008		Wuxi T	JS
SD7)	1008		Weifang E	SD
EN2)	1008		Zhengzhou LBS	HEN
HN7)	1008v	1	Yueyang EBS	HN
1)	1017		CNR 1	
100)	1017	100	CRI	JL, Changchun
GD1)	1017		Guangdong Sat.	GD, Shaoguan/Shantou
BJ1)	1026	50	Beijing E	BJ
GZ1)	1026	150	Guizhou Sat.	GZ, Jianhe
LN10)	1026	1	Yingkou	LN
JS6)	1026		Yancheng	JS
JS16)	1026		Yizheng	JS
JX4)	1026	1	Pingxiang	JX
1)	1035	50	CNR 1	
100)	1044	300	CRI	JS, Changzhou
XJ1)	1044	10	Xinjiang-Ug	XJ
YN8)	1044	1	Dali	YN
ZJ1)	1050		Zhejiang LBS	ZJ
EB18)	1050		Shahe	HEB
EB1)	1053		Hebei T	HEB, Shijiazhuang
JS1)	1053		Jiangsu L	JS, Nanjing
JL11)	1053	7.5	Yanbian-Ch	JL, Yanji
AH2)	1053		Hefei T	AH
SD2)	1053	10	Jinan	SD
EN3)	1053		Kaifeng EBS	HEN
YN4)	1053	1	Wenshan	YN
GD2)	1062	150	Zhujiang EBS	GD, Guangzhou
HL11)	1062		Qitaihe	HL
TJ1)	1071	50	Tianjin EBS	TJ
ZJ1)	1071	10	Zhejiang FBS	ZJ, Hangzhou
FJ1)	1071v		Fujian N	FJ
GX1)	1071		Guangxi Sat.	GX
LN4)	1071	2	Anshan EBS	LN

ID	Freq	Power	Station	Location
ZJ7)	1071		Huzhou	ZJ
SN4)	1071	10	Baoji	SN
XJ3)	1071		Urumqi-Ug	XJ
100)	1080	200	CRI	YN, Kunming
ZJ1)	1080v		Zhejiang	ZJ
XJ1)	1080		Xinjiang-Ug	XJ
HL7)	1080		Daqing	HL
JS12)	1080	10	Suzhou	JS
GD7)	1080	5	Shantou N	GD
CQ1)	1080	10	Chongqing ED	CQ
FJ5)	1084v		Quanzhou	FJ
1)	1089		CNR Taiwan 2	FJ
LN1)	1089	200	Liaoning	LN, Shenyang
FJ1)	1089		Fujian LBS	FJ, Fuzhou
HN3)	1089	1	Zhuzhou	HN
JS9)	1098		Zhenjiang LBS	JS
JS15)	1098		Zhangjiagang	JS
ZJ11)	1098		Zhoushan EBS	ZJ
AH7)	1098v		Bengbu EBS	AH
EN15)	1098		Zhoukou EBS	HEN
GD19)	1098	5	Maoming	GD
XJ6)	1098		Hami-Ug	XJ
AH13A)	1100		Fuyang EBS	AH
SD12)	1100v		Linyi E	SD
JL1)	1107		Jilin N	JL
HA1)	1107	10	Hainan	HAN, Tongshi
XJ1)	1107	20	Xinjiang-Kz	XJ, Urumqi
HL5)	1107	1	Shuangyashan	HL
AH6)	1107	1	Tongling	AH
FJ3)	1107v		Xiamen N	FJ
EN7)	1107		Hebi	HEN
1)	1116	120	CNR 2	HL
1)	1116		CNR Taiwan 1	FJ
SC1)	1116	50	Sichuan 1	SC
AH13)	1116	1	Fuyang	AH
EB1)	1125		Hebei EBS	HEB, Shijiazhuang
HB4)	1125v		Changjiang EBS	HB, Wuhan
ZJ1)	1134v		Zhejiang	ZJ
GD18)	1134	10	Zhanjiang	GD
SN3)	1134	10	Tongchuan	SN
GS8)	1134	1	Yumen	GS
XJ9)	1134	1	Ili-Ch	XJ, Yining
ZJ1)	1143v		Zhejiang	ZJ
EN1)	1143		Henan L	HEN, Zhengzhou
QH1)	1143		Qinghai EBS	QH, Xining
EB16)	1143	1	Dingzhou	HEB
NM14)	1143	1	Hailar	NM
LN5)	1143		Fushun E	LN
LN14)	1143		Liaoyang EBS	LN
JL16)	1143		Tumen	JL
HL10)	1143		Jiamusi EBS	HL
JS2)	1143		Nanjing TBS	JS
JS3)	1143		Xuzhou EBS	JS
JS10)	1143		Changzhou EBS	JS
ZJ6)	1143		Jiaxing EBS	ZJ
AH15)	1143v		Chaohu	AH
SD4)	1143	1	Zibo	SD
GS3)	1143	10	Tianshui	GS
SC14)	1147		Yibin	SC
HN1)	1152		Hunan Sat.	HN, Hengyang
NM9)	1152	7.5	Jirem-Ch	NM, Tongliao
NM11)	1152	7.5	Bayannur	NM, Linhe
LN3)	1152	2	Dalian EBS	LN
1)	1161		CNR 1	
GX1)	1161		Guangxi EBS	GX, Beihai
JS11)	1161		Wuxi N	JS
SD7)	1161	10	Weifang	SD
SD5)	1170		Zaozhuang	SD
SD14)	1170		Qingzhou	SD
HN8)	1170		Changde	HN
AH13)	1175v		Lu'an	AH
HB2)	1179		Chutian BS N	HB, Wuhan
NM6)	1179	10	Ulanqab-Mo	NM, Jining
JS7)	1179		Yangzhou N	JS
XJ4)	1179		Karamay-Ug	XJ
100)	1188	300	CRI	YN
EB4)	1188		Xingtai	HEB
EB14)	1188	1	Botou	HEB
NM13)	1188	10	Hinggan-Ch	NM, Ulanhot
SH1)	1197		Shanghai L	SH
HL3)	1197	10	Qiqihar	HL
FJ2)	1197		Fuzhou 2	FJ
100)	1206	300	CRI	YN
GD1)	1206		Guangdong ED	GD, Zhaoqing
EB3)	1206		Handan 2	HEB
JL11)	1206	150	Yanbian-Ko	JL, Yanji
SD9)	1206v		Weihai	SD
1)	1215		CNR 2	
1)	1215		CNR 7	GD, Zhuhai
XJ2)	1215		Tianshan EBS	XJ, Urumqi
1)	1224		CNR Taiwan 2	FJ
GX1)	1224		Guangxi EBS	GX
EB11)	1224	1	Langfang	HEB
JS9)	1224		Zhenjiang N	JS
HN1)	1233		Hunan Sat.	HN, Yueyang
YN1)	1242		Yunnan EBS	YN, Kunming
LN9)	1242	1	Huludao	LN
HB20)	1242		Macheng	HB
ZJ10)	1250		Quzhou L	ZJ
AH8)	1250		Ma'anshan	AH
100)	1251		CRI DS	BJ
QH1)	1251	100	Qinghai-Tb	QH, Xining
JS11)	1251		Wuxi EBS	JS
ZJ8A)	1251		Shaoxing FM BS	ZJ, Shaoxing
ZJ13)	1251v		Xiaoshan	ZJ
SD3)	1251		Qingdao EBS	SD
SD13)	1251		Liaocheng Diqu	SD
EN10)	1251		Puyang	HEN
LN1)	1260		Liaoning	LN
XZ2)	1260	1	Shannan	XZ, Nedong
EB15)	1265v		Renqiu	HEB
100)	1269		CRI	YN
2)	1269	200	V.O.Strait	FJ
SX1)	1269		Shanxi	SX
JL17)	1269		Dunhua	JL
JS3)	1269		Xuzhou	JS
EB1)	1278	100	Hebei	HEB, Shijiazhuang
HL13)	1278		Daxing'anling	HL, Jagdaqi
FJ3)	1278	10	Xiamen N	FJ
JX2)	1278		Nanchang	JX
SD11)	1278		Rizhao	SD
1)	1287		CNR 1	
ZJ1)	1287		Zhejiang	ZJ
NX1)	1287		Ningxia 1	NX
LN12)	1287		Fuxin	LN
EN11)	1287		Xuchang	HEN
YN7)	1287	10	Chuxiong A.P.	YN, Chuxiong
100)	1296	300	CRI	YN
SH1)	1296	20	Shanghai FL	SH
EB19)	1296		Qinghe	HEB
LN6)	1296	10	Benxi	LN
SC9)	1296		Suining	SC
SN5)	1296	1	Xianyang	SN
1)	1305		CNR 2	
100)	1314	100	CRI	NM
JS1)	1314	10	Jiangsu G	JS, Suzhou
CQ1)	1314	15	Chongqing	CQ
SD8)	1314		Yantai	SD
HB6)	1314		Xiangfan	HB
100)	1323	100	CRI	JL, Jilin-shi
100)	1323	100	CRI	XZ
SN1)	1323	10	Shaanxi TBS	SN, Xi'an
LN17)	1323	1	Wafangdian	LN
JL10)	1323	10	Baicheng	JL
ZJ4)	1323	20	Ningbo	ZJ
HN2)	1323	10	V.O.Xingsha	HN, Changsha
EN1)	1332	10	Henan	HEN
JL3)	1332	2	Changchun EBS	JL
FJ2)	1332	10	Fuzhou 1	FJ
GS5)	1332	7.5	Gannan	GS, Hezuo
100)	1341	100	CRI	GD, Guangzhou
HL1)	1341	10	Heilongjiang	HL, Shuangyashan
LN2)	1341	10	Shenyang TBS	LN
JS13)	1341	5	Taizhou	JS
SD15)	1341	1	Qufu	SD
HB16)	1341		Yingcheng	HB
HB21)	1341	1	Puqi	HB
JX1)	1350v		Jiangxi 1	JX
NM9)	1350	50	Jirem-Mo	NM, Tongliao
YN2)	1350	50	Kunming 1	YN

	kHz	kW	Station	Location
1)	1359		CNR 1	
FJ1)	1368v		Fujian N	FJ
HL6)	1368		Jixi	HL
HB15)	1368		Guangshui	HB
2)	1377	200	V.O.Strait	FJ
EB5)	1377		Baoding N	HEB
AH11)	1377		Chuzhou	AH
SD3)	1377	10	Qingdao	SD
NX5)	1377	1	Qingtongxia	NX
TJ1)	1386	20	Tianjin L	TJ
HB1)	1386	1	Ezhou	HB
HB18)	1386	1	Shishou	HB
GX3)	1386	5	Liuzhou	GX
NM1)	1395		Nei Menggu-Mo	NM
AH1)	1395	10	Anhui 1	AH
NM7)	1395		Xilingol-Ch	NM, Xilinhot
ZJ1)	1404		Zhejiang	ZJ
HB1)	1404		Hubei	HB
LN7)	1404	10	Dandong	LN
JS1)	1413		Jiangsu G	JS, Wuxi
XJ1)	1413	100	Xinjiang-Ug	XJ
LN15)	1413		Tieling	LN
HL4)	1413		Hegang	HL
NX4)	1413		Wuzhong	NX
SH1)	1422	20	Shanghai E	SH
SX3)	1422	10	Taiyuan	SX
SC4)	1422	10	Zigong	SC
EB2)	1431		Shijiazhuang 2	HEB
NM8)	1431	200	Hulun Buir-Mo	NM, Hailar
JL9)	1431		Songyuan	JL
AH4)	1431	1	Huaibei	AH
AH10)	1431v	1	Huangshan	AH
HB19)	1431	1	Danjiangkou	HB
GX1)	1440		Guangxi Sat.	GX
NM5)	1440	50	Chifeng-Mo	NM
NM12)	1440	10	Alxa-Mo	NM, Bayanhot
LN19)	1440		Zhuanghe	LN
JS8)	1440		Nantong	JS
FJ4)	1440v		Putian	FJ
JX1)	1449	20	Jiangxi 1	JX
JS3)	1449		Xuzhou	JS
SD6)	1449		Dongying	SD
NM1)	1458	150	Nei Menggu-Mo	NM, Hohhot
JS4)	1458	5	Lianyuangang	JS
FJ1)	1467v	50	Fujian EBS	FJ, Fuzhou
NM13)	1467	10	Hinggan-Mo	NM, Ulanhot
JX3)	1467	1	Jingdezhen	JX
1)	1476	200	CNR 2	HL, Shuangyashan
ZJ1)	1476		Zhejiang	ZJ
JL19)	1476	1	Qian Gorlos	JL
HL9)	1476		Mudanjiang EBS	HL
HB5)	1476	10	Huangshi	HB
HB14)	1476v	1	Laohekou	HB
SC11)	1476		Leshan	SC
QH2)	1476	1	Xining	QH
XJ5)	1476		Shihezi	XJ
SD1)	1485		Shandong	SD
GS1)	1485		Gansu 1	GS
EB3)	1485	1	Handan 1	HEB
SX5)	1485		Yangquan	SX
LN11)	1485	1	Panjin	LN
JL12)	1485	1	Gongzhuling	JL
ZJ6)	1485		Jiaxing	ZJ
AH7)	1485	1	Bengbu N	AH
JX6)	1485	1	Jiujiang	JX
EN5)	1485	1	Pingdingshan	HEN
HB7)	1485	1	Shiyan	HB
GX4)	1485	1	Guilin	GX
GX5)	1485		Wuzhou	GX
SC3)	1485	1	Chengdu 2	SC
YN5)	1485	1	Honghe	YN, Jinping
YN10)	1485		Chuxiong	YN
XJ6)	1485		Hami-Ch	XJ
XJ8)	1485		Kuytun-Ch	XJ
XJ1)	1494	20	Xinjiang-Ch	XJ
NM16)	1494		Tongliao	NM
AH5)	1494v		Wuhu	AH
ZJ1)	1503		Zhejiang	ZJ
LN14)	1503	1	Liaoyang	LN
HN4)	1503	1	Xiangtan	HN
NM5)	1512		Chifeng-Ch	NM
GS6)	1512	10	Linxia	GS
100)	1521	500	CRI	XJ
HL1)	1521	1	Heilongjiang	HL, Jingbohu
SN1)	1521	1	Shaanxi Sat.	SN, Shangzhou
JL14)	1521	1	Taonan	JL
JS5)	1521		Huaiyin EBS	JS
JS7)	1521		Yangzhou EBS	JS
JS12)	1521		Suzhou EBS	JS
HN9)	1521	1	Jinshi	HN
GD20)	1521	1	Zhaoqing 1	GD
YN5)	1521		Honghe	YN
JL1)	1530		Jilin N	JL, Yanji
ZJ1)	1530	50	V.O. Health	ZJ, Hangzhou
HN5)	1530	10	Hengyang	HN
SD1)	1548	200	Shandong	SD, Jinan
HN6)	1548		Shaoyang	HN
EB10)	1557		Cangzhou	HEB
EB17)	1557	1	Nangong	HEB
EB6)	1566		Zhangjiakou	HEB
SX7)	1566		Jincheng	SX
SD16)	1566		Liaocheng	SD
HB12)	1566	1	Xiaogan	HB
GS7)	1566		Pingliang	GS
100)	1575	2	CRI DS	LN, Dalian
ZJ5)	1575	7.5	Wenzhou	ZJ
EB7)	1584	1	Chengde	HEB
SX4)	1584		Datong 1	SX
SX6)	1584	1	Changzhi	SX
SX8)	1584		Yuci	SX
JL13)	1584	1	Meihekou	JL
ZJ7)	1584		Huzhou	ZJ
ZJ14)	1584		Rui'an	ZJ
AH8)	1584v	1	Ma'anshan	AH
AH9)	1584v	1	Anqing	AH
JX7)	1584		Yingtan	JX
HB13)	1584	1	Suizhou	HB
HL1)	1593	10	Heilongjiang	HL
XJ1)	1593	100	Xinjiang-Mo	XJ, Urumqi
HL12)	1593	1	Suihua	HL
JS1)	1602	1	Jiangsu G	JS, Hongze
HB10)	1602		Jingmen	HB

Shortwaves: *) inactive.

	kHz	kW	Station	Location	Times
FJ1)	2340		Fujian	Fuzhou	2120-2315, 1030-1600
ZJ5)	2415	1	Wenzhou		-1330
JX1)	2445		Jiangxi 1	Nanchang	-1300
YN1)	*2460		Yunnan G	Kunming	
ZJ1)	*2475	20	Zhejiang	Hangzhou	
XJ1)	2560	15	Xinjiang-Ch	Urumqi	as 738kHz
2)	*3200		V.O.Strait		
1)	*3220	50	CNR 1		Winter only
GZ1)	*3260	10	Guizhou Sat.	Guiyang	
4)	3280		V.O.Pujiang	Shanghai	0255-0500, 0955-1400
1)	*3375		CNR 1		Winter only
NM8)	3900		Hulun Buir-Ch	Hailar	as 603kHz
2)	*3900		V.O.Strait		Winter only
HB1)	*3940v	10	Hubei	Wuhan	
QH1)	3950		Qinghai 1	Xining	as 666kHz
XJ1)	*3960	50	Xinjiang-Ch	Urumqi	Nov-Apr only
XJ1)	*3990	50	Xinjiang-Ug	Urumqi	Nov-Apr only
NM1)	4000	50	Nei Menggu-Ch	Hohhot	2150-0105, 0950-1515
XZ1)	4035		Xizang-Tb	Lhasa	as 594kHz
1)	4190		CNR Minor.		2130-2300, 1230-1300
XJ1)	*4220	15	Xinjiang-Mo	Urumqi	
XJ1)	4330		Xinjiang-Kz	Urumqi	as 1107kHz
1)	4460		CNR 1		2000-2300, 1330-1735
XJ1)	4500		Xinjiang-Ch	Urumqi	as 738kHz
NM1)	4525	50	Nei Menggu-Mo	Hohhot	2150-0105, 0950-1505
NM1)	4620	50	Nei Menggu-Ch	Hohhot	2150-0105, 0950-1515
XJ1)	4735		Xinjiang-Ug	Urumqi	as 558kHz
NM8)	4750		Hulun Buir-Mo	Hailar	as 1431kHz

Code	Freq	kW	Station	City	Notes / Times
XZ1)	4750		Xizang-Ch	Lhasa	as 846kHz
YN1)	*4760		Yunnan G	Kunming	
NM1)	4785	50	Nei Menggu-Mo	Hohhot	2150-0105, 0950-1505
ZJ1)	*4785	20	Zhejiang	Hangzhou	
XZ1)	4820		Xizang-Tb	Lhasa	as 594kHz
101)	*4830		China Huayi BC	Fuzhou	Winter Only
HL1)	4840		Heilongjiang	Harbin	as 621kHz
1)	4850		CNR 2		2100-2230, 1300-1605
GS1)	4865		Gansu 1	Lanzhou	as 873kHz
2)	4900		V.O.Strait		2055-2400, 1300-1800
GX1)	*4915		Guangxi Sat.	Nanning	2200-?, 0958-1600
YN5)	4930v		Honghe	Gejiu	as 702kHz
2)	4940		V.O.Strait		2055-2400, 0855-1800
4)	4950		V.O.Pujiang	Shanghai	0255-0500, 0955-1400
NM7)	4950		Xilingol-Mo	Xilinhot	as 927kHz
XJ1)	4970		Xinjiang-Kz	Urumqi	as 1107kHz
FJ1)	4975		Fujian	Fuzhou	as 882kHz
XJ1)	4980		Xinjiang-Mo	Urumqi	as 909kHz
HN1)	4990	10	Hunan Sat.	Xiangtan	as 738kHz
GX1)	*5010		Guangxi EBS	Nanning	2200-?, 0925-1400
JX1)	5020		Jiangxi 1	Nanchang	as 729kHz
XZ1)	5020v		Xizang-Tb	Lhasa	as 594kHz
1)	5030		CNR 1		2000-2330, 1230-1735
FJ1)	5040		Fujian	Fuzhou	2215-1030
2)	5050		V.O.Strait		2055-2400, 1400-1800
6)	5050	15	Guangxi FBS	Nanning	2300-0100, 1000-1600
XJ1)	5060		Xinjiang-Mo	Urumqi	as 909kHz
1)	5075		CNR 2		2100-2330, 1200-1605
4)	*5075		V.O.Pujiang	Shanghai	
1)	*5090		CNR Taiwan 2		Winter only
1)	*5125		CNR Taiwan 1		Winter only
1)	5163	50	CNR 2		2100-2230, 1430-1605
XZ1)	5240		Xizang-Ch	Lhasa	as 846kHz
1)	5320		CNR 1		2000-2200, 1330-1735
1)	5420		CNR Minor.		2130-2300, 1230-1300
3)	5860	50	V.O.Jinling	Nanjing	1200-1600
1)	5880		CNR 1		2000-0100, 1100-1735
SC1)	5900		Sichuan GB	Chengdu	-1400
1)	5915		CNR 1		2000-0100
GX1)	5920		Guangxi 3	Nanning	2300-0135, 1000-1235
XZ1)	*5935v	50	Xizang-Ch	Lhasa	
HL1)	5950	50	Heilongjiang Ko	Harbin	as 873kHz
XZ1)	5950	50	Xizang-Tb	Lhasa	as 594kHz
1)	5955	50	CNR 1		2000-0100
YN1)	5960	50	Yunnan ED	Kunming	0030-0400, 1045-1200
GS5)	5970	1	Gannan	Hezuo	as 1332kHz
XZ1)	5995	50	Xizang-Tb	Lhasa	as 594kHz
1)	*6015	50	CNR Taiwan 1		Winter only
NM12)	6025v	15	Alxa-Ch	Bayanhot	as 549kHz
1)	6030		CNR 1		2000-1735
XZ1)	6030		Xizang-Ch	Lhasa	as 846kHz
7)	6035	50	Yunnan BS	Kunming	2230-0130, 0945-1300
JX1)	6040	50	Jiangxi 1		2100-0530, 0950-1505
NM1)	6045	50	Nei Menggu-Ch	Hohhot	2150-0105, 0950-1515
SC1)	6060v	50	Sichuan 1	Xichang	as 621kHz
1)	6110		CNR 1		2000-2400, 1200-1735
2)	6115	50	V.O.Strait		2055-2400, 0855-1800
1)	6125	50	CNR 1		2000-0100, 1100-1735
SN1)	6176	15	Shaanxi Sat.	Xi'an	0100-0720
101)	6185		China Huayi BC	Fuzhou	0255-0700, 0855-1600
NM1)	6195	50	Nei Menggu-Mo	Hohhot	2150-0105, 0950-1505
XZ1)	6200		Xizang-Tb	Lhasa	as 594kHz
QH1)	6260		Qinghai 1	Xining	as 666kHz
QH1)	6500	50	Qinghai-Tb	Xining	as 1251kHz
1)	6750		CNR 1		2000-0100, 1100-1735
1)	*6790		CNR Taiwan 2		Winter only
1)	6890		CNR 2		2100-2400, 1200-1605
YN1)	6937v		Yunnan Minor.	Kunming	2225-0045, 0355-0545, 1100-1500
NM1)	7105	50	Nei Menggu-Ch	Hohhot	0120-0550
XZ1)	7110	50	Xizang-Tb	Lhasa	as 594kHz
1)	7140		CNR 2		2100-2300, 1130-1605
XZ1)	7170	50	Xizang-Ch	Lhasa	as 846kHz
XZ1)	7195	50	Xizang-Tb	Lhasa	as 594kHz
XJ1)	7195	50	Xinjiang-Ug	Urumqi	May-Oct only
YN1)	*7210	15	Yunnan G	Kunming	
3)	7215	50	V.O.Jinling	Nanjing	0000-0200
1)	7220		CNR 1		2155-0100
SC1)	7225	15	Sichuan 1	Chengdu	0940-
NM1)	7270	50	Nei Menggu-Mo	Hohhot	0120-0550
GZ1)	7275	7.5	Guizhou Sat.	Guiyang	
2)	7280	50	V.O.Strait		0855-1400
1)	7290		CNR 1		2000-2300, 1200-1735
XZ1)	*7290		Xizang-Tb	Lhasa	
1)	7315		CNR 2		2100-2300
1)	7345		CNR 1		2000-2330, 1230-1735
HL1)	7350		Heilongjiang		2205-0540, 0840-1130
XJ1)	7385	50	Xinjiang-Ch	Urumqi	May-Oct only
1)	7504		CNR 1		2000-1735
1)	*7620		CNR Taiwan 1		Winter only
1)	7770		CNR 2		2100-1605
1)	7935		CNR 1		2000-0130, 1000-1735
1)	8566		CNR Minor.		0500-0530, 1000-1100
1)	9064		CNR 2		2100-1605
1)	9080	50	CNR 1		2000-0030, 1100-1735
1)	9170		CNR Taiwan 2		2055-0105, 1300-1805
1)	9290		CNR 1		2300-1330
1)	9340		CNR 1		2330-1230
1)	9380		CNR Taiwan 1		1200-0005
1)	9500		CNR 2/Minor.		2100-1605
2)	9505	50	V.O.Strait		0855-1300
XZ1)	*9510		Xizang-Ch	Lhasa	
NM1)	9520	50	Nei Menggu-Ch	Hohhot	0120-0550
XJ1)	9560	50	Xinjiang-Ch	Urumqi	0430-0730
XJ1)	9595	50	Xinjiang-Ug	Urumqi	0430-0730
4)	9705		V.O.Pujiang	Shanghai	0955-1400
NM1)	9750	50	Nei Menggu-Mo	Hohhot	0120-0550
1)	9755	50	CNR 2		2100-2330, 1330-1605
1)	9775		CNR 2		2100-0100, 1100-1605
QH1)	9780		Qinghai 1	Xining	0355-0630
1)	9800		CNR 1		1100-1735
1)	9810		CNR 2		2230-1430
1)	9830		CNR 1		2000-2330, 1230-1735
1)	9845		CNR 1		2000-2400, 1130-1735
1)	9920		CNR Minor.		0500-0530, 1000-1100
1)	10260		CNR 1/Minor.		2000-0130, 0600-1500
1)	11000		CNR Taiwan 2		2055-0105, 0355-1805

1) 11040		CNR 2		0000-1200
1) 11100		CNR Taiwan 1		0055-0615, 0955-0005
1) 11375		CNR Minor.		0000-0130, 0530-0700, 1100-1230, 1300-1500
1) 11505		CNR 2		2230-1300
1) 11610		CNR 2		2300-1130
1) 11630		CNR 1/Minor.		2000-0200, 0600-1500
1) 11710		CNR 2		2100-2400, 1400-1605
1) 11740	50	CNR 2/Minor.		2300-1605
1) 11800		CNR 2		2100-1605
1) 11925		CNR 1		2000-2400, 1200-1735
1) 11960		CNR 1		0000-1200
1) 11935		CNR Taiwan 1		0055-0615, 0955-0005
XZ1) 11950v	15	Xizang-Tb	Lhasa	as 594kHz
1) 12030		CNR 1		2000-1735
1) 12080		CNR 2/Minor.		2330-1330
1) 12120		CNR 1		2200-1330
1) 13610		CNR 1		1055-1500
1) 15390	50	CNR 1		2330-1230
1) 15480		CNR 1		2300-1200
1) 15500		CNR 2		2330-1200
1) 15550		CNR 1		2330-1230
1) 15570		CNR 2		0030-1400
1) 15670		CNR Minor.		0000-0130, 0530-0700, 1100-1230, 1300-1500
1) 15710		CNR Taiwan 1		0055-0615, 0955-1200
1) 15880		CNR Taiwan 2		0355-1300
1) 17550		CNR 1		0000-1130
1) 17580		CNR 1		0000-1200
1) 17605		CNR 1		0030-1100
1) 17700	50	CNR 2		0100-1100

Addresses and other information:

1) CHINA NATIONAL RADIO (CNR)
✉ 2 Fuxingmenwai Dajie, Beijing 100866. ☎ +86 (10) 6851 5522. Directors Office: +86 (10) 6851 2435. Audience Relations Dept: +86 (10) 6852 2624.
1st Prgr: 2000-1735 (exc. Tues 0600-0855) on MW/SW/FM.
2nd Prgr: 2100-1605 (exc. Wed/Fri 0600-0955) on MW/SW/FM.
3rd Prgr "Leisure FM": 2155-1600 (exc. Tues 0600-0855) on FM only.

Minorities Prgr (4th Prgr) on SW - +)relayed by regional sts.

Kazakh

0100-0130	XJ	15670, 11630, 11375, 10260
0630-0700	XJ	15670, 11630, 11375, 10260
1200-1230	XJ	15670, 11630, 11375, 10260
1400-1430+	XJ	15670, 11630, 11375, 10260

Korean

2130-2230+	JL/HL	5420, 4190
1000-1100+	JL/HL	9920, 8566

Mongolian

2230-2300+	NM	5420, 4190
0500-0530+	NM	9920, 8566
1230-1300+	NM	5420, 4190
1430-1500+	XJ	15670, 11630, 11375, 10260

Tibetan

0000-0030	XZ/QH	15670, 12080, 11740, 11710, 11375, 9500
0530-0600	XZ/QH	15670, 15570, 12080, 11740, 11375, 9500
1100-1130	XZ/QH	15670, 15570, 12080, 11740, 11375, 9500
1300-1330+	XZ/QH	15670, 15570, 12080, 11740, 11375, 9500

Uighur

0030-0100	XJ	15670, 11630, 11375, 10260
0600-0630	XJ	15670, 11630, 11375, 10260
1130-1200+	XJ	15670, 11630, 11375, 10260
1330-1400+	XJ	15670, 11630, 11375, 10260

Taiwan Sce 1st Prgr (5th Prgr) in Chinese, Amoy and Hakka: 0055-0615, 0955-0005 on MW/SW/FM.
Taiwan Sce 2nd Prgr (6th Prgr) in Chinese, Amoy and Hakka: 2055-0105, 0355-1805 (exc. Tues 0605-0955) on MW/SW/FM.
V.O. Old China (7th Prgr) for the Zhujiang Delta: 2100-1805 (exc. Tues 0600-0853) on 1215kHz/87.8/104.9MHz.

BROADCASTS TO TAIWAN

2) Voice of the Strait (Haixia zhi Sheng), Xindian, Fuzhou or P.O.Box 187, Fuzhou, Fujian 350012. Operated by the People's Liberation Army of China - on MW/SW/90.6MHz in Chinese and Amoy. 2055-2400, 0855-1800.
3) Voice of Jinling (Jinling zhi Sheng), P.O.Box 268, Nanjing, Jiangsu 210002 - 0000-0200 on 7215kHz, 1200-1600 on 5860kHz. Also transmitted on 846kHz at 2155-2400, 0200-0400, 0955-1200.
4) Voice of Pujiang (Pujiang zhi Sheng), 1376 Hongqiao Lu, Shanghai or P.O.Box 3064, Shanghai 200051 - 0255-0430, 0955-1400.
5) China Southeast Broadcasting Company, 2 Gutian Lu, Fuzhou, Fujian 350001 - on 585kHz/97.6/106.2MHz and via Satellite in Ch and Amoy. 0955-1600.

BROADCASTS TO VIETNAM

6) Guangxi Foreign Broadc. St, 75 Minzu Dadao, Nanning, Guangxi 530022 - 2300-0100, 1000-1200, 1400-1600 in Vietnamese, 1200-1400 in Cantonese.
7) Yunnan Broadc. St, Renmin Xilu, Kunming, Yunnan 650031 - 2230-2320, 0030-0120, 1030-1100, 1130-1200, 1230-1300 in Vietnamese, 2320-2330, 0120-0130, 1000-1030, 1100-1130, 1200-1230 in Ch. Rel. CRI Vietnamese prgr. 2330-0030.

BEIJING MUNICIPALITY

BJ1) 14 Jianguomenwai Dajie, Chaoyang Qu, Beijing 100022 - on 828kHz and via Satellite 2130-1600 (exc. Thurs 0600-0800).
Economic St. on 1026kHz 2155(Sun 2225)-1500.
Traffic St. on 927kHz 2255-1400 (exc. Thurs 0600-0900).
Educational St. on 103.9MHz 2225(SS 2245)-1500 (exc. Tues 0500-0900).
Educational Special Sce. on 603kHz 2130-1600 (exc. Thurs 0600-0900).
Literary St: FM stereo on 87.6MHz 2155-1600 (exc. Tues 0500-0900).
Music St: FM stereo on 97.4MHz 24h.

TIANJIN MUNICIPALITY

TJ1) 143 Weijin Lu, Heping Qu, Tianjin 300070.
News Ch. on 909kHz 2150-0600, 0725-1600.
Economic Ch. on 1071kHz 2155-0530, 0700(Wed 0725)-1600.
Literary Ch. on 1386kHz 2225-0600, 0655(Wed 0725)-1600.
Traffic Ch. on 567kHz 2225-1600 (exc. Wed 0600-0800).
Educational Ch. on 106.8MHz 2155-0530, 0655(Wed 0830)-1540.
Music Ch: FM stereo on 99.0MHz 2155-1800 (exc. Wed 0600-0830).

HEBEI PROVINCE

EB1) 4 Yuhua Zhonglu, Shijiazhuang, Hebei 050012 - 2130-0630(Tues 0530), 0825-1525.
Traffic Music Ch: on 1053kHz/91.8/99.2MHz.
Hebei EBS: on 1125KHz/FM 2130-1600 (exc. Tues 0530-0900).
EB2) 33 Tiyu Zhong Dajie, Shijiazhuang, Hebei 050021.
1st Prgr. on 882kHz 2105-0035, 0320-0550, 0920-1440.
2nd Prgr. on 1431kHz/100.0MHz 2220-0600, 0855-1435.
Shijiazhuang LBS: FM stereo on 106.7MHz.
EB3) 246 Renmin Lu, Handan, Hebei 056002.
1st Prgr. on 1485kHz/101.0MHz 2155-?, ?-1305.
2nd Prgr. on 1206kHz/102.8MHz 2215-?, ?-1230.
EB4) Dahuoquan Lu, Xingtai, Hebei 054000 - 2220-2400, 0255-0535, 1020-?.
EB5) 30 Wusi Donglu, Baoding, Hebei 071000.
News Ch. on 1377kHz 2220-0600, 0950-1500.
EB6) 17 Jianguo Lu, Qiaodong Qu, Zhangjiakou, Hebei 075000 - 2200-0555, 0925-1410.
EB7) 17 Xiaotonggou Nanpo, Chengde, Hebei 067000.
1st Prgr. on 1584kHz/89.1/93.8MHz 2155-0600, 0940-1400.
2nd Prgr. on 97.6MHz 2155-0200, 0350-0530, 0950-1400.
EB8) 1 Guangda Jie, Tangshan, Hebei 063000 - 2155-0055, 0345-0625, 0935-1435.
Tangshan EBS: on 801kHz/95.5MHz 2200-?, ?-1430.
Tangshan Traffic Literary BS: on 102.0MHz.
EB9) 23 Wenhua Beilu, Haigang Qu, Qinhuangdao, Hebei 066000 - 2200-1600.
EB10) 12 Jiefang Xilu, Cangzhou, Hebei 061001 - 2225-0040, 0320-0530, 0855-1330.

EB11) 8 Yongfeng Dao, Langfang, Hebei 102800 - 2210-0100, 0350-0615, 0905-1325.
EB12) 49 Hongqi Dajie, Hengshui, Hebei 053000 - 2225-0535, 0825-1400.
EB13) Xinghua Lu, Xinji, Hebei 052360 - 2225-2355, 0255-0500, 1025-1250.
EB14) 393 Xiguan Xijie, Botou, Hebei 062150 - 2225-2355, 0345-0450, 1025-1235.
EB15) 6 Yuhua Zhonglu, Renqiu, Hebei 062550 - 2225-0005, 0325-?, 0925-1230.
EB16) Zhongshan Xilu, Dingzhou, Hebei 073000.
EB17) Muchang Jie, Nangong, Hebei 051830 - 2225-0045, 1005-1400.
EB18) Yingxin Jie, Shahe, Hebei 054100.
EB19) Sanyang Dongjie, Qinghe, Hebei 054800 - 2210-?, 0910-1230.

SHANXI PROVINCE

SX1) 318 Yingze Dajie, Taiyuan, Shanxi 030001 - on MW/FM 2155-0050, 0255-0650, 0905-1530(SS 1600).
Traffic Sce: FM stereo on 89.3MHz.
V.O. Health: on 94.0MHz 2220-1700.
Shanxi LBS: FM stereo on 101.5MHz 2200-1500.
SX2) Shanxi Changcheng BS, 318 Yingze Dajie, Taiyuan, Shanxi 030001 - on 900kHz/95.8MHz. 24h (exc. Tues 0600-0900).
SX3) 2 Yifen Jie, Taiyuan, Shanxi 030024 - W2150-0030, Sun0010-0630, W0150-0615, D0900-1510 (Sun 1440).
Taiyuan EBS: FM stereo on 104.4MHz 2200(Sun 2220)-1600.
Taiyuan TBS: on 107.0MHz.
SX4) 1 Xinjian Beilu, Datong, Shanxi 037006.
1st St. on 1584kHz W2145-0010, Sun2215-0205, W0245-0540, Sun0350-0630, D0955-1415.
SX4A) Yuhe Beilu, Datong, Shanxi 037044.
2nd St. on 603kHz 2210-?, ?-0950.
SX5) 10 Bei Dajie, Yangquan, Shanxi 045000 - 2150-2400, 0300-0535, 0955-1330.
SX6) 124 Yingxiong Lu, Changzhi, Shanxi 046000 - 2155-0050, 0355-0630, 0855-1400.
SX7) Xiaozhuang, Jincheng, Shanxi 048000 - 2200-?, 1025-1330.
SX8) 37 Taishan Miao Jie, Yuci, Shanxi 030600 - 2220-0015, 0355-0605, 0955-1400.

NEI MENGGU AUTONOMOUS REGION

NM1) 19 Xinhua Dajie, Hohhot, Nei Menggu 010058.
Ch Prgr. on MW/FM and via Satellite 2150-0550, 0950-1515.
Mo Prgr. on MW/FM and via Satellite 2150-0550, 0950-1505.
FM stereo: on 99.0MHz 0325-0530.
NM2) 56 Gongyuan Xilu, Hohhot, Nei Menggu 010035 - 2200-0130, 0150-0530, 0830-1400.
NM3) 31 Wulan Dao, Kundulun Qu, Baotou, Nei Menggu 014010 - 2145-0500, 0930-1500.
NM4) Haila Beilu, Wuhai, Nei Menggu 016000 - W2225-0025, Sun0025-0515, W0325-0520, D1025-1305(SS 1405).
NM5) Gangtie Jie, Hongshan Qu, Chifeng, Nei Menggu 024001.
Ch Prgr. on 549/1512kHz/FM 2120-0030, 0255-0605, 1000-1500.
Mo Prgr. on 1440kHz/FM 2120-2310, 0320-0600, 0955-1400.
FM stereo: on 105.7MHz.
NM6) 9 Qiaoxi Shahe Lu, Jining, Nei Menggu 012000.
Ch Prgr. on 702kHz 2150-0100, 0340-0540, 005-1500.
Mo Prgr. on 1179kHz 2125-0100, 0330-0600, 1010-1505.
NM7) Xilin Dajie, Xilinhot, Nei Menggu 026000.
Ch Prgr. on 1395kHz 2225-2400, 0255-0550, 0955-1305.
Mo Prgr. on 927/4950kHz 2220-0035, 0240-0450, 1025-1335.
NM8) 11 Shengli Dajie, Hailar, Nei Menggu 021008.
Ch Prgr. on 603/3900kHz 2150-0010, 0330-0500, 0920-1400.
Mo Prgr. on 1431/4750kHz 2130-2400, 0355-0600, 0925-1500.
NM9) 28 Huolinhe Dajie, Tongliao, Nei Menggu 028001.
Ch Prgr. on 1152kHz 2140-0005, 0320-0530, 0955-1330.
Mo Prgr. on 1350kHz 2120-2335, 0245-0530, 0955-1330.
NM10) Ordos BS, Dongsheng, Nei Menggu 017000.
Ch Prgr. on 936/792kHz 2220-0020, 0320-0600, 1000-1320.
Mo Prgr. on 603kHz.
NM11) 26 Xinhua Xijie, Linhe, Nei Menggu 015000.
NM12) 1 Elute Lu, Bayanhot Zhen, Alxa Zuoqi, Nei Menggu 750306.
Ch Prgr. on 549/6025kHz 2220-0020, 0350-0600, 0955-1305.
Mo Prgr. on 1440kHz 2220-0020, 0340-0530, 1050-1345.
NM13) Hinggan Beilu, Ulanhot, Nei Menggu 137400.
Ch Prgr. on 1188kHz 2155-2400, 0955-1400.
Mo Prgr. on 1467kHz.
NM14) 37 Xi Jianguo Jie, Hailar, Nei Menggu 021000.
NM15) Dianshi Jie, Manzhouli, Nei Menggu 021400.
NM16) Mingren Dajie, Tongliao, Nei Menggu 028001 - 2145-2355.

NM17) Sujiaqu, Dongsheng, Nei Menggu 017000.

LIAONING PROVINCE

LN1) 10 Guangrong Jie, Heping Qu, Shenyang, Liaoning 110003 - on MW/FM and via Satellite 2050-1500 (exc. Tues 0540-0855).
Liaoning EBS: on 999kHz/FM 2125-1500 (exc. Tues 0540-0855).
Liaoning IBS: on 927kHz/97.5MHz 2150-1600 (exc. Tues 0540-0855).
Liaoning Stereo LBS: on 95.9MHz 2155-1600 (exc. Tues 0540-0855).
LN2) 9 Heping Nan Dajie, Heping Qu, Shenyang, Liaoning 110003.
News St. on 792kHz 2100-1600.
Shenyang EBS: on 882kHz W2100-1400, Sun2300-1500.
Shenyang TBS: on 1341kHz 2225-1600(SS 1300) (exc. Thurs 0500-0855).
LN2A) Shenyang Stereo LBS, 89 Nan Sanhao Jie, Heping Qu, Shenyang, Liaoning 110003 - FM stereo on 92.1MHz 2255-1430.
LN3) 162 Minquan Jie, Shahekou Qu, Dalian, Liaoning 116022 - 2055-1500.
Dalian EBS: on 1152kHz/FM 2125-1600 (exc. Tues 0530-0900).
Dalian LBS: FM stereo on 105.7MHz 2125-1430 (exc. Tues 0530-0855).
LN4) 3, 219 Lu, Tiedong Qu, Anshan, Liaoning 114001 - 2055-1400.
Anshan EBS: on 1071kHz/89.7MHz.
LN5) 2 Hunhe Beilu, Xinfu Qu, Fushun, Liaoning 113006 - 2110-0215, 0350-0615, 0915-1325.
Economic Traffic Ch. on 1143kHz/106.1MHz 2155-0130, 0355-0610, 0925-1230.
Music Ch. on 747kHz/100.6MHz 2230-0600, 0800-1300.
LN6) 15 Tiyu Lu, Mingshan Qu, Benxi, Liaoning 117000 - 2125-0100, 0330-0530, 0855-1400.
LN7) 1 Shanshang Jie, Zhenxing Qu, Dandong, Liaoning 118000 - 2120-0510, 0850-1300.
Dandong EBS: on 891kHz/101.7MHz 2155-0030, 0300-0530, 0955-1220.
LN8) 4-3 Beijing Lu, Jinzhou, Liaoning 121001 - 2125-0030, 0315-0540, 0820-1430.
Jinzhou EBS: on 97.7MHz.
Jinzhou FM Stereo BS: on 100.3MHz.
LN9) Xinggong Jie, Lianshan Qu, Huludao, Liaoning 121500 - 2130-0500, 0930-1230.
LN10) 20 Jiankang Li, Bohai Dajie, Yingkou, Liaoning 115000 - 2120-0600, 0825-1400.
LN11) Liaohe Lu, Shuangtai Qu, Panjin, Liaoning 124000 - 2115-2350, 0355-0550, 0855-1335.
LN12) Zhonghua Lu, Haizhou Qu, Fuxin, Liaoning 123000 - 2115-0155, 0330-0540, 0925-1400.
LN13) Fuxin Mongolian BS, 52-2 Yan'an Lu, Haizhou Qu, Fuxin, Liaoning 123000 - 2155-0015, 0325-0530, 1040-1300 in Mo.
LN14) 59 Qingnian Dajie, Liaoyang, Liaoning 111000 - 2130-0005, 0150-0515, 0845-1305.
Liaoyang EBS: on 1143kHz/102.0MHz 2155-0530, 0750-1300.
LN15) 45 Gongren Jie, Yinzhou Qu, Tieling, Liaoning 112000 - 2125-0045, 0300-0610, 0900-1300.
LN16) 3-11 Xinhua Lu, Chaoyang, Liaoning 122000 - 2145-2355, 0345-0610, 1015-1445.
LN17) 3-4 Delin Jie, Wafangdian, Liaoning 116300 - 2135-0500, 0810-1235.
LN18) 15 Huancheng Xilu, Haicheng, Liaoning 114200 - 2120-?, 0315-0505, 0850-1230.
LN19) Xin Kailu Dajie, Zhuanghe, Liaoning 116400 - 2100-0100, 0855-1200.

JILIN PROVINCE

JL1) 2 Xi'an Dalu, Changchun, Jilin 130051.
News Ch. on MW/FM 2050-0600(Tues 0500/Sun 0730), 0820-1500.
Popular Life Ch. on 846kHz/95.3MHz 2200-1600.
Traffic Literary Ch. on 103.8MHz.
Northeast Asia Music Ch: FM stereo on 92.7MHz. 24h.
Jilin IBS: on 101.9MHz 2300-0550, 0855-1100.
JL3) Jia 2, Jinshui Lu, Changchun, Jilin 130061 - W2115-0700 (Tues 0530), Sun2115-1450, W0830-1450.
Changchun EBS: on 1332kHz/96.8MHz 2150-0700, 0830-1400.
JL4) 2 Jiangnan Dajie, Jilin-shi, Jilin 132011 - 2110-1430.
Jilin-shi IBS: on 90.3MHz.
JL5) 56 Nan Xinhua Dajie, Siping, Jilin 136000 - 2135-0030, 0300-0600, 0855-1300.
JL6) 20 Hebin Lu, Longshan Qu, Liaoyuan, Jilin 136200.
JL7) 24 Cuiquan Lu, Longquan Jie, Tonghua, Jilin 134001 - 2150-0530, 0920-1310.
JL8) 36 Hunjiang Dajie, Baishan, Jilin 134302 - 2125-0540, 0930-1240.
Traffic Literary St: FM stereo on 107.7MHz.
JL9) 71 Linjiang Lu, Fuyu Qu, Songyuan, Jilin 131200.

JL10) 99 Zhongxing Dong Dalu, Baicheng, Jilin 137000 - 2115-0100, 0945-1430.
JL11) 166 Juzi Jie, Yanji, Jilin 133000.
Ch Prgr. on 1053kHz 2155-0040, 0305-0525, 0900-1300.
Ko Prgr. on 1206kHz/103.0MHz 2100-0045, 0315-0530, 0900-1225.
JL12) Gongyuan Donglu, Gongzhuling, Jilin 136100.
JL13) 70 Henan Jie, Meihekou, Jilin 135000.
JL14) 29 Gushu Nanjie, Taonan, Jilin 137100.
JL15) 7 Yongle Jie, Yanji, Jilin 133000 - 2130-0200, 0300-0730, 0900-1200 in Ch and Ko.
FM stereo on 93.5MHz 0200-0300, 0730-0900, 1200-1230.
JL16) 12 Xiangshang Jie, Tumen, Jilin 133100 - 2125-2355, 0325-0500, 0855-1230 in Ch and Ko.
JL17) 1 Xinhua Xilu, Dunhua, Jilin 133700 - 2140-2330, 0320-0525, 0925-1245 in Ch and Ko.
JL18) 233 Hunchun Lu, Hunchun, Jilin 133300 - 2145-0025, 0300-0520, 0940-1200 in Ch and Ko.
JL19) Yucai Jie, Qian Gorlos, Jilin 131100 - 2125-2330, 0325-0500, 0955-1230 in Ch and Mo.

HEILONGJIANG PROVINCE
HL1) 181 Zhongshan Lu, Nangang Qu, Harbin, Heilongjiang 150001 - on MW/SW/FM and via Satellite 2055-0540(Fri 0510), 0855-1500. Ko prgr: 1400-1500.
Economic Sce: on 95.4MHz.
Heilongjiang Korean BS: on MW/SW/FM in Ko. W2100-2300, W0200-0400, Sun0000-0400.
Traffic Sce: on 99.8MHz 2300(Sun 0100)-0600, 0830-1230.
Music Sce: FM stereo on 104.5MHz 2200-0400.
HL2) 1 Huashan Lu, Xiangfang Qu, Harbin, Heilongjiang 150036 - 2055-0520, 0855-1445.
Harbin EBS: on 972kHz 2155-0505.
Harbin LBS: FM stereo on 98.4MHz 0100(SS 0000)-0500, 0900-1200.
HL3) 1 Shifu Lu, Qiqihar, Heilongjiang 161005 - 2125-0010, 0340-0610, 0855-1335.
HL4) Bama Lu, Xiangyang Qu, Hegang, Heilongjiang 154100 - 2055-2320, 0325-?, 0840-1430.
HL5) 74 Xinxing Dajie, Shuangyashan, Heilongjiang 155100 - 2050-2400, 0325-0500, 0850-1415.
HL6) 11 Diantai Lu, Jixi, Heilongjiang 158100 - 2120-0600, 0850-1350.
HL7) Jia 1, Dongfeng Lu, Daqing, Heilongjiang 163311 - 2125-0105, 0250-?, 0925-1430.
Daqing TBS: on 95.0MHz.
HL8) Tonghe Lu, Yichun Qu, Yichun, Heilongjiang 153000 - 2050-0040, 0325-0520, 0855-1330.
HL9) 138 Taiping Lu, Mudanjiang, Heilongjiang 157000 - 2100-0030, 0315-0620, 0935-1400.
Mudanjiang EBS: on 1476kHz/91.6MHz 2100-0600, 0900-1600.
HL10) 17 Shunhe Lu, Jiamusi, Heilongjiang 154002 - 2055-0100, 0250-0530, 0855-1300.
Jiamusi EBS: on 1143kHz/95.0MHz 2155-0500, 0855-1300.
Jiamusi Traffic LBS: FM stereo on 98.0MHz MF2300-0600, SS2355-0400, D0855-1200.
HL11) Shanhu Dajie, Taoshan Qu, Qitaihe, Heilongjiang 154600.
HL12) 83 Bei Ersi Lu, Suihua, Heilongjiang 152054.
HL13) 47-2 Xing'an Dajie, Jagdaqi Zhen, Heilongjiang 165000 - 2120-0155, 0255-0535, 0825-1430.

SHANGHAI MUNICIPALITY
SH1) 1376 Hongqiao Lu, Shanghai 200002.
1st Prgr. on 990kHz/93.4MHz 2058-1600 (exc. Mon 0500-0858).
Market Economic Ch. on 603/1422kHz 2155-1300.
Literary St. on 1197kHz in Ch and Shanghai dialect. 2158-1400 (exc. Wed 0500-0758).
Foreign Language St. "Shanghai Calling" on 1296kHz Sun2128-0300, W2158-0600,
Sun0358-1200, W0800-1500, Sun1258-1500.
Traffic Information St. on 648kHz/98.7MHz 2258-1300.
Music 1st St: FM stereo on 103.7MHz 2258-1600 (exc. Thurs 0600-0858).
Music 2nd St: FM stereo on 105.7MHz 0058-1700 (exc. Tues 0600-0858).
SH2) Shanghai Dongfang BS, 1376 Hongqiao Lu, Shanghai 200002 - on 792kHz
24h. FM stereo on 101.7MHz 2200-1600.
Financial St: on 97.7MHz 2200-1000.

JIANGSU PROVINCE
JS1) 8 Xi Citang Xiang, Zhongshan Donglu, Nanjing, Jiangsu 210002.
General St. on MW/FM and via Satellite 2055-1600 (exc. Tues/Thurs 0600-0855).
Music St: FM stereo on 89.7MHz 2200-1700.
Literary St: on 1053kHz/97.5MHz and via Satellite 2200-1600.
Jiangsu EBS: on 585kHz/101.1MHz and via Satellite 2200-1600. Rel. CNR 3: 1500-1600.
JS2) 50 Yanling Xiang, Nanjing, Jiangsu 210002.
News St. on 1008kHz 2130-1600.
Nanjing EBS: on 900kHz/104.3MHz. 24h.
Nanjing TBS: on 801/1143kHz 2200-1500.
Nanjing MBS: FM stereo on 105.8MHz 2200-1800.
JS3) 223 Zhongshan Nanlu, Xuzhou, Jiangsu 221003 - 2130-0115, 0255-0640, 0900-1430.
Xuzhou EBS: on 1143kHz/99.6MHz 2150-0600, 0950-1405.
Xuzhou LBS: on 801kHz/93.0MHz 2200-0545, 0900-1230.
JS4) 221 Jiefang Xilu, Xinpu, Lianyungang, Jiangsu 222003 - 2140-1510.
Lianyungang EBS: on 90.2MHz.
JS5) 6 Dazhi Lu, Huaiyin, Jiangsu 223001 - 2130-?, 0925-1350.
Huaiyin EBS: on 1521kHz/105.0MHz 2150-?.
JS6) 4 Shengyuan Lu, Yancheng, Jiangsu 224001 - 2155-0020, 0255-0520, 0855-1500.
JS6A) Yancheng EBS, 88 Huanghai Zhonglu, Yancheng, Jiangsu 224002 - on 747kHz/105.3MHz.
JS7) 8 Meiling Donglu, Yangzhou, Jiangsu 225002.
News St. on 1179kHz/98.5MHz 2150-1500 (exc. Tues 0600-0930).
Yangzhou EBS: on 801/1521kHz/94.9MHz 2130-1530.
JS8) 120 Renmin Zhonglu, Nantong, Jiangsu 226001 - 2135-0520, 0825-1440.
Nantong EBS: on 603kHz/103.0MHz 2150-0130, 0300-0500, 1000-1400.
JS9) 94 Zhongshan Xilu, Zhenjiang, Jiangsu 212001.
News St. on 1224kHz 2150-0010, 0215-0520, 0825-1430.
Zhenjiang EBS: on 900kHz/99.4MHz 2200-?, 0800-1450.
Zhenjiang LBS: on 1098kHz/96.3MHz 2200-1600.
JS10) 30 Damiao Nong, Changzhou, Jiangsu 213003 - 2135-0030, 0150-0505, 0750-1400.
V.O. Lanling: FM stereo on 100.1MHz 2155-?.
Changzhou EBS: on 1143kHz/105.2MHz 2200-1435.
Changzhou TBS: on 747kHz/90.0MHz 2200-1400.
JS11) 4 Hubin Lu, Wuxi, Jiangsu 214061.
News St. on 1161kHz 2130-1530.
Economic St. on 1251kHz/104.0MHz 2200-1600.
Traffic St. on 1008kHz/98.7MHz 2200-1300.
Literary St. on 900kHz/91.4MHz 2200-1600.
JS12) 4 Gongyuan Lu, Suzhou, Jiangsu 215006 - on 1080kHz in Ch and Suzhou dialect. ?-1600.
FM stereo: on 96.5MHz W0258-0555, Sun0058-0725, D0858-1140.
Suzhou EBS: on 1521kHz/104.8MHz 2155-1500.
JS13) 20 Qingnian Lu, Taizhou, Jiangsu 225300 - on 1341kHz/97.3MHz 2145-0530, 0645-1400.
JS14) Hubin Qiao Nan, Yushan Zhen, Changshu, Jiangsu 215500.
1st Prgr. on 927kHz/100.8MHz 2125-2350, 0155-0555, 0825-1340.
3rd Prgr: FM stereo on 102.7MHz 0340-0450, 0855-1110.
JS15) Chenjiachang Nong, Yangshe Zhen, Zhangjiagang, Jiangsu 215600 - 2115-0035, 0150-0625, 0845-1230.
JS16) 43 Gongnong Lu, Yizheng, Jiangsu 211400 - 2120-?.

ZHEJIANG PROVINCE
ZJ1) 111 Moganshan Lu, Hangzhou, Zhejiang 310005 - on MW/SW/FM and via Satellite 2130-1600 (exc. Tues 0500-0858).
V.O. the Health: AM stereo on 1530kHz 2200-1600.
V.O. the Traffic: on 93.0MHz 2255-1400.
Zhejiang EBS: FM stereo on 95.0MHz and via Satellite 2200-1600.
Zhejiang LBS: on 930/1050kHz/99.6MHz 24h (exc. SS1600-2200).
Zhejiang Foreign Language BS: on 1071kHz 2255-0730(Tues 0500), 0855-1500.
Rel. CRI DS: 2300-0300, MF1300-1500.
ZJ2) 86 Moganshan Lu, Hangzhou, Zhejiang 310005 - 2100-1600.
ZJ2A) Hangzhou EBS, 58 Tiyuchang Lu, Hangzhou, Zhejiang 310004 - FM stereo on 91.9MHz 2200-1600.
ZJ3) V.O. Xihu, 86 Moganshan Lu, Hangzhou, Zhejiang 310005 - FM stereo on 105.4MHz. 24h.
ZJ4) 3 Guangji Xiang, Jiefang Nanlu, Ningbo, Zhejiang 315010 - 2130-0505, 0755-1505.
Ningbo EBS: on 102.9MHz 2200-1600.
V.O. the Health: on 603kHz.
ZJ5) 19 Xianxue Qian, Lucheng Qu, Wenzhou, Zhejiang 325000 - on 1575/2415kHz/94.9MHz in Ch and Wenzhou dialect. 2155-1600 (exc. Tues 0610-0900).
Economic St: on 801kHz/92.7MHz 2200-1605 (exc. Tues 0600-0855).

ZJ6) 6 Dongsheng Lu, Jiaxing, Zhejiang 314001 - 2200-?, 0850-1230.
Suburban St. on 88.2MHz 0225-?, 0920-1235.
Jiaxing EBS: on 1143kHz/92.2MHz 2200-1300 (exc. Tues 0400-0900).
ZJ7) 40 Dongmen Mao'anqian, Huzhou, Zhejiang 313000 - 2210-?, 0915-1345.
Literary St: FM Stereo on 98.5MHz.
Huzhou EBS: on 900kHz.
ZJ8) 2 Banbi Nong, Fushan Zhijie, Shaoxing, Zhejiang 312000 - 2155-0105, 0245-0525, 0910-1335.
ZJ8A) Shaoxing FM BS (V.O. Yuezhou), Da Xiaochang, Shaoxing, Zhejiang 312000 - 2230-1500.
ZJ9) 30 Xinhua Jie, Jinhua, Zhejiang 321000 - 2145-0440, 0800-1320.
Jinhua EBS: on 101.4MHz.
ZJ10) 35 Nanjie, Quzhou, Zhejiang 324000 - 2200-1400 (exc. Tues 0500-0900).
Literary St: on 1250kHz/97.5MHz 2255-1600 (exc. Fri 0500-0900).
ZJ11) 137 Changguo Lu, Dinghai Qu, Zhoushan, Zhejiang 316000 - 2135-0530, 0955-1500.
Zhoushan EBS: on 1098kHz/97.8MHz 2200-?, 0900-1400.
Zhoushan FM BS: FM stereo on 91.0MHz 2140-0530, 0830-1500.
ZJ12) 149 Zhongdong Lu, Lishui, Zhejiang 323000.
ZJ13) 99 Renmin Lu, Chengxiang Zhen, Xiaoshan, Zhejiang 311200 - 2200-1400.
ZJ14) Xishan, Rui'an, Zhejiang 325200 - ?-1305.

ANHUI PROVINCE
AH1) 69 Lujiang Lu, Hefei, Anhui 230065.
1st Prgr. on MW/FM and via Satellite 2120-1550.
2nd Prgr: FM stereo on 98.8MHz. Rel. CNR 3: 0400-0600.
Anhui EBS: on 864kHz/89.5MHz and via Satellite 2200-0600, 0900-1500.
AH2) 100 Fuyang Lu, Hefei, Anhui 230001 - 2200-0530, 0900-1500.
Hefei TBS: on 1053kHz/102.6MHz 2300-0705(SS 0505), 0900-1500.
Hefei LBS: FM stereo on 87.6MHz 2300-0530(SS 0500), 0700-1400.
AH3) Dongshan, Huainan, Anhui 232001 - 2120-0010, 0255-0510, 0855-1300.
AH4) Huaihai Donglu, Xiangshan Qu, Huaibei, Anhui 235000 - 2200-2350, 0320-0450, 0900-1230.
AH5) 61 Laodong Lu, Wuhu, Anhui 241000 - 2158-0020, 0155-0510, 0850-1430.
Wuhu EBS: on 96.3MHz.
AH6) Yi'an Beilu, Tongling, Anhui 244000 - 2120-0005, 0155-0510, 0855-1305.
AH7) 150 Zhongrong Jie, Bengbu, Anhui 233000.
News St. on 1485kHz 2200-0005, 0235-0500, 0825-1255.
Bengbu EBS: on 1098kHz/104.2MHz 2150-1430.
AH8) Gongyuan Xincun, Hubei Lu, Ma'anshan, Anhui 243000 - 2150-0500, 0850-1400.
FM stereo: on 88.7MHz 0355-0520, 0945-1100.
AH9) 35 Guanyue Miao Jie, Anqing, Anhui 246004 - 2200-0530, 0930-1300.
AH10) 14 Huangshan Donglu, Tunxi Qu, Huangshan, Anhui 245000 - 2200-0025, 0310-0510, 0925-1335.
AH11) 37 Langxie Lu, Chuzhou, Anhui 239000 - 2155-?, ?-1135.
AH12) Wandong EBS, 37 Langxie Lu, Chuzhou, Anhui 239000 - on 85.8/95.0MHz.
AH13) 149 Renmin Donglu, Fuyang, Anhui 236014 - 2155-0610, 0855-1505.
AH13A) 73 Yingzhou Lu, Fuyang, Anhui 236001.
AH14) 123 Huaihai Zhonglu, Suzhou, Anhui 234000 - 2155-0035, 0330-0605, 0955-1235.
AH15) Dongfeng Lu, Chaohu, Anhui 238000 - 2135-?, 0245-0500, 0815-1125.
AH16) Gaocheng Lu, Lu'an, Anhui 237006 - ?-1235.

FUJIAN PROVINCE
FJ1) 2 Gutian Lu, Fuzhou, Fujian 350001 - on MW/SW/FM and via Satellite in
Ch and Amoy 2150-1600 (exc. Tues/Fri 0600-0855).
Fujian EBS: on MW/FM in Ch and Amoy 2200-1600 (exc. Tues 0600-0900).
Music Traffic Ch: on FM stereo 89.8/97.6/98.7/103.6MHz. 24h.
FJ2) 16 Shengmiao Lu, Gulou Qu, Fuzhou, Fujian 350001.
1st St. on 1332kHz/94.0MHz in Ch and Fuzhou dialect. W2135-2400, Sun2135-1500, W0155-0600, W0855-1500.
2nd St: on 1197kHz/89.3MHz. 24h (exc. Thurs 0600-1000).
FJ3) 8 Huyuan Lu, Xiamen, Fujian 361003.
News St. on 1278/1107kHz 2140-0600, 0825-1600 in Ch and Amoy.
Economic St. on 105.2MHz 2200-1600 (exc. Tues 0600-0900).
Music St: FM stereo on 90.9MHz. 24h.

FJ4) 2 Nanmenwai, Chengxiang Qu, Putian, Fujian 351100 - 2200-1430 in Ch and Puxian dialect.
FJ5) Puxi, Citong Nanlu, Quanzhou, Fujian 362000 - 2210-1605.
FJ6) 62 Heping Lu, Longyan, Fujian 364000 - 2158-1605 (exc. Tues 0600-1000).

JIANGXI PROVINCE
JX1) 111 Hongdu Zhong Dadao, Nanchang, Jiangxi 330046.
1st Prgr. on MW/SW/FM and via Satellite 2100-1505 (exc. Tues 0600-0855).
Literary Ch. (2nd Prgr): FM stereo on 103.4MHz 2200-1600.
Jiangxi EBS: on 594kHz/106.5MHz 2200-0600, 0900-1500.
Jiangxi IBS: on 927kHz/96.9MHz 2200-2400, 0300-0600, 0900-1100.
JX2) 401 Minde Lu, Nanchang, Jiangxi 330008 - 2130-0005, 0250-0530, 0825-1515.
JX3) 23 Lianshe Beilu, Jingdezhen, Jiangxi 333000 - 2205-0500, 0940-1500.
JX4) 1 Yingbin Lu, Pingxiang, Jiangxi 337055 - 2135-0025, 0325-0535, 0935-1425.
JX5) Ganxin Lukou, Xinyu, Jiangxi 336525 - 2125-2400, 0325-0530, 0925-1310.
JX6) 14 Jiaotong Lu, Jiujiang, Jiangxi 332000 - 2155-?.
JX7) 22 Shengli Xilu, Yingtan, Jiangxi 335000 - 2200-0015, 0325-0535, 0920-1300.
JX8) Ganzhou Diqu BS, 76 Hongqi Dadao, Ganzhou, Jiangxi 341000.

SHANDONG PROVINCE
SD1) 81 Jing 10 Lu, Lixia Qu, Jinan, Shandong 250001 - on MW/FM 2125-0610, 0845-1530.
Economic St. on 594kHz/FM 2155-0600, 0955-1600.
Educational St. on 105.0MHz 2155-0500, 1025-1500.
Literary St. on FM 0225-1600 (exc. Tues 0400-0825). Rel. CNR 3: 1400-1600.
SD2) 32 Jing 11 Lu, Lixia Qu, Jinan, Shandong 250014 - 2150-0540, 0850-1505.
Jinan EBS: on 103.1MHz.
SD3) 30 Shanxian Lu, Qingdao, Shandong 266002.
News St. on 1377kHz/107.6MHz 2155-1600.
Qingdao EBS: on 1251kHz/102.9MHz 2200-1600.
Qingdao LBS: FM stereo on 96.4MHz 2200-1600.
SD4) 10 Xi 2 Xiang, Liuquan Lu, Zhangdian Qu, Zibo, Shandong 255008 - 2150-0635, 0925-1405.
Zibo LBS: FM stereo on 100.0MHz 2145-1600.
SD5) 6 Xinsheng Lu, Zaozhuang, Shandong 277137 - 2050-0100, 0310-0600, 0925-1400.
Zaozhuang EBS: on 603kHz/100.3MHz.
SD6) 30 Shenghua Lu, Dongying, Shandong 257031 - 2155-0115, 0320-0620, 0955-1435.
Dongying EBS: on 105.3MHz.
SD7) 248 Dongfeng Dongjie, Weifang, Shandong 261041.
News St. on 1161kHz/100.2MHz 2155-0130, 0315-0545, 0855-1500.
Economic St. on 1008kHz/93.3MHz 2200-0200, 0300-0800, 1000-1500.
Traffic Music St. on 846kHz/95.9MHz.
SD8) 50 Qingnian Lu, Yantai, Shandong 264000 - 2125-0530, 0855-1430.
Economic Literary St: on 801kHz/105.9MHz.
SD9) 99 Guzhai Donglu, Weihai, Shandong 264200 - W2130-0600, Sun2100-0540, W0915-1500, Sun0915-1320.
Literary St: FM stereo on 96.1MHz 2155-0430, 0625-1200.
SD10) Changqing Lu, Jining, Shandong 272137.
News St. on 666kHz/104.2MHz. Economic St. on 801kHz/ 99.3MHz.
Literary St. on 101.8MHz.
SD11) 19 Wanghai Lu, Rizhao, Shandong 276800.
SD12) Jiefang Lu, Linyi, Shandong 270003 - 2125-0105, 0225-0605, 0925-1405.
Economic St. on 1100kHz/89.9MHz 2155-0530, 0855-1405.
Traffic Music St. on 801kHz/101.0MHz 2125-?, 0955-1420.
SD13) Beijiao, Liaocheng, Shandong 252051 - 2200-2400, 0300-0530, 0955-1400.
SD14) 21 Fangongting Xilu, Qingzhou, Shandong 262500 - 2145-?, 0325-0445, 0955-1350.
SD15) Gulou Beijie, Qufu, Shandong 273100 - 2215-0025, 0355-0530, 1045-1330.
Literary St: FM stereo on 98.4MHz.
SD16) 204 Wenhua Lu, Liaocheng, Shandong 252000 - 2155-0035, 0325-0600, 1025-1330.
SD17) 584 Huanghe 6 Lu, Binzhou, Shandong 256611.

HENAN PROVINCE
EN1) 2 Jing 5 Lu, Zhengzhou, Henan 450003 - on MW/FM and via

Satellite 2105-0050, 0330-1510.
Literary St. on 1143kHz/90.0MHz 2200-1500.
Henan EBS: on 846/972kHz/103.2MHz and via Satellite 2155-1600.
Henan TBS: on 900kHz/104.1MHz 2225-1500.
Henan IBS: on 603kHz 2200-1600.
EN2) Nanduan, Songshan Lu, Zhengzhou, Henan 450052 - 2150-0115, 0300-0555, 0900-1415.
Zhengzhou EBS: on 711kHz/92.6MHz 2225-0635, 0855-1600.
Zhengzhou LBS: on 1008kHz/94.0MHz 0900-1500.
EN3) 27 Gulou Jie, Kaifeng, Henan 475000 - 2155-0105, 0325-0640, 0955-1605.
Kaifeng EBS: on 1053kHz/100.2MHz.
EN4) Fu 67, Jiudu Lu, Luoyang, Henan 471000 - 2158-0100, 0158-0535, 0858-1510.
Luoyang EBS: FM stereo on 92.2MHz 0230-0530, 0925-1235.
EN5) Zhongduan, Kuanggong Lu, Pingdingshan, Henan 467000 - 2150-0005, 0355-0530, 1025-1405.
EN6) 43 Jiefang Zhonglu, Jiaozuo, Henan 454002 - ?-0030, 0345-?.
Jiaozuo EBS: FM stereo on 89.4MHz.
EN7) Zhongduan, Chunlei Lu, Hebi, Henan 458000 - 2155-2400, 0350-0535, 0955-1330.
EN8) Renmin Lu, Xinxiang, Henan 453000 - 2200-0215, 0330-0535, 0850-1335.
EN9) 21 Jiefang Lu, Anyang, Henan 455000 - 2155-0025, 0320-0515, 0900-1345.
EN10) Zhongyuan Lu, Puyang, Henan 457001 - 0925-?.
EN11) 72 Balong Lu, Xiao Nanhai, Xuchang, Henan 461000 - 2200-0045, 0355-0545, 0910-1330.
EN12) 1 Dongfeng Xiang, Luohe, Henan 462000 - 2140-0020, 0325-0530, 0920-1325.
EN13) Wolonggang Lu, Nanyang, Henan 473081 - 2225-0040, 0355-0525, 1050-1415.
EN14) Jiadao Jie, Shangqiu, Henan 476000 - 2205-0150, 0355-0555, 0840-1330.
EN15) Dongduan, Jianshe Lu, Zhoukou, Henan 466001 - 2155-0200, 0355-0730, 0955-1400.
Zhoukou EBS: on 1098kHz/89.3MHz 0925-1405.

HUBEI PROVINCE
HB1) 563 Jiefang Dadao, Hankou, Wuhan, Hubei 430022 - on MW/FM 2120-1615 (exc. Tues 0615-0850).
Economic St. on 99.8MHz 2200-1600.
Traffic Music St. on 107.8MHz 2155-1600 (exc. Tues 0615-0850).
Hubei LBS: FM stereo on 103.8MHz 2150-1600.
HB2) Chutian BS, 563 Jiefang Dadao, Hankou, Wuhan, Hubei 430022.
News Netw. on 1179kHz 2125-1630 (exc. Tues 0630-0855).
Life Netw. on 92.7MHz 2125-1600 (exc. Tues 0630-0855).
Music Netw. on 105.8MHz. 24h (exc. Tues 0630-0855).
HB3) Echengdun Xiqu, Jiang'an Qu, Wuhan, Hubei 430015 - 2159-0630(Wed 0530), 0929-1500.
Wuhan LBS: FM stereo on 101.8MHz 0055-0800, 1000-1605.
HB4) Jianshe Dadao, Hankou, Wuhan, Hubei 430015 - 2300-0600, 1000-1400.
HB5) 38 Wuhan Lu, Huangshi, Hubei 435000 - 2150-0500, 0855-1400.
HB6) 177 Zhongshan Houjie, Xiangfan, Hubei 441000 - 2120-0005, 0255-0540, 0955-1500.
HB7) Liuyan Shan, Shiyan, Hubei 442000 - 2200-?, 0855-1310.
HB8) 32 Yuanlin Lu, Jingzhou, Hubei 434000 - 2150-2400, 0255-0545, 0920-1500.
HB9) Yiling Lu, Yichang, Hubei 443000 - 2220-0015, 0315-0540, 0945-1300.
Yichang EBS: on 100.6MHz MF2220-0730, SS0050-0600, D0850-1200.
HB10) 13 Taying 1 Lu, Jingmen, Hubei 448000 - 2155-0035, 0325-0600, 0925-1330.
HB11) 23 Jianshe Xijie, Ezhou, Hubei 436000 - 2155-0530, 1000-1420.
HB12) 106 Changzheng Lu, Xiaogan, Hubei 432100 - 2220-0120, 0355-0535, 0957-1355.
HB13) 8 Lieshan Dadao, Suizhou, Hubei 441300 - 2130-2330, 0330-0500, 1030-1305.
HB14) 6 Xuefu Lu, Laohekou, Hubei 441800 - 2155-2400, ?-1255.
HB15) 8 Fuqian Jie, Guangshui, Hubei 432700 - 2155-0055, 0355-0515, 0955-1255.
HB16) Jianshe Jie, Yingcheng, Hubei 432400.
HB17) 117 Mianyang Dadao, Xiantao, Hubei 433000 - ?-1255.
HB18) 2 Shannan Xiaoqu, Shishou, Hubei 434400 - 2200-0005, 0955-1235.
HB19) 4 Renmin Lu, Danjiangkou, Hubei 441900 - 2200-2355, 0955-1155.
HB20) 1 Xinjian Jie, Macheng, Hubei 436100.
HB21) 50 Chunchuan Daqiao Lu, Puqi, Hubei 437300 - 0950-1340.

HUNAN PROVINCE
HN1) 27 Yuhua Lu, Changsha, Hunan 410007.
Satellite Sce. on MW/SW/FM and via Satellite 2125-1600 (exc. Tues/Fri 0600-0855).
Hunan EBS: on 900kHz/90.1/91.0MHz 2200-1600 (exc. Tues/Fri 0600-0800).
Hunan TBS: on 91.8MHz.
Hunan LBS: FM stereo on 97.5MHz 2225-0600, 0925-1600.
HN2) 5 Wuyi Xilu, Changsha, Hunan 410005.
V.O. Xingsha: on 1323kHz/105.0MHz 2145-0635, 0800-1600(Sun 1410).
HN3) Hejiatu, Jianshe Zhonglu, Zhuzhou, Hunan 412000 - 2155-0025, 0125-0535, W0925-1220, Sun1030-1300.
HN4) 9 Shaoshan Xilu, Xiangtan, Hunan 411100 - 2155-0015, 0325-0555, 0925-1430.
HN5) 154 Xianfeng Lu, Hengyang, Hunan 421001 - 2155-0100, 0325-0500, 0925-1405.
HN6) Zhangshu Long, Baoqing Xilu, Shaoyang, Hunan 422000 - 2150-0030, 0355-0530, 0930-1415.
HN7) 4 Jiefang Lu, Yueyang, Hunan 414000 - 2200-0015, 0330-0530, 0925-1340.
Yueyang EBS: on 1023kHz ?-1500.
HN8) Gaoshan Jie, Changde, Hunan 415000 - 2200-2355, 0400-0500, 0925-1100.
HN9) Renmin Lu, Jinshi, Hunan 415400.

GUANGDONG PROVINCE
GD1) 686 Renmin Beilu, Guangzhou, Guangdong 510012.
Satellite Sce. on MW/FM and via Satellite 2200-1900 (exc. Tues 0015-0355).
V.O. the City: FM stereo on 103.6MHz 2200-1800.
V.O. the Health: on 999kHz/93.6MHz 24h (exc. Mon 0500-1000).
Educational St. on 603/90/1008/1206kHz 0200-0600, 0900-1600.
Music St: FM stereo on 99.3/93.9MHz 2200-1800 (exc. Wed 0700-0900).
GD2) Zhujiang EBS, 686 Renmin Beilu, Guangzhou, Guangdong 510012 - on 1062/801kHz/FM in Cantonese 24h.
GD3) Yangcheng TBS, 686 Renmin Beilu, Guangzhou, Guangdong 510012 - on 927kHz/105.2MHz 24h.
GD4) Guangzhou PBS, 23 Xiatang Xilu, Guangzhou, Guangdong 510045.
1st St: FM stereo on 96.2MHz in Cantonese. 24h (exc. Sun 1600-2200).
Golden Ch. (2nd St): FM stereo on 102.7MHz in Cantonese. 24h.
GD5) Shenzhen BS, Yijing Lu, Shenzhen, Guangdong 518001.
1st St: FM stereo on 89.8MHz in Ch and Cantonese. 2155-1700.
2nd St: FM stereo on 97.1MHz in Ch and Cantonese. 2225-?, 0300-1700.
GD6) 8 Kangning Lu, Xiangzhou Qu, Zhuhai, Guangdong 519000 - on 900kHz/95.1MHz in Ch and Cantonese. 2225-1700.
GD7) Chaoshan Lu, Shantou, Guangdong 515031.
News St. on 1080kHz/99.3MHz in Ch and Chaozhou dialect. 2200-0600, 0900-1600.
Economic St: FM stereo on 102.5MHz 2300-1600.
GD8) 28 Huimin Lu, Shaoguan, Guangdong 512026 - on 765kHz/105.7MHz in Ch and Cantonese.
2220-0040, 0255-0805, 0900-1430.
GD9) Heyuan PBS, Yanjiang Lu, Yuancheng Zhen, Heyuan, Guangdong 517001 - FM stereo on 104.0MHz in Ch and Cantonese.
GD10) Meizhou PBS, Dong Jiaochang, Meizhou, Guangdong 514011 - FM stereo on 97.8MHz in Ch and Hakka. 2155-0100, 0255-0630, 0855-1310.
GD11) Huizhou PBS, 1 Zhongshan Gongyuan, Huizhou, Guangdong 516001 - FM stereo on 103.4MHz in Ch and Cantonese. 2230-0545, 0830-1230.
GD12) Shanwei PBS, 12 Chezhan Xilu, Shanwei, Guangdong 516611 - FM stereo on 90.0MHz in Ch and Hakka. 0940-?.
GD13) Dongguan PBS, 35 Xizheng Lu, Cheng Qu, Dongguan, Guangdong 511700 - FM stereo on 106.9MHz in Cantonese. ?-0140, 0935-1235.
GD14) Zhongshan BS, Xingzhong Dao, Shiqi Zhen, Zhongshan, Guangdong 528403 - FM stereo on 96.7MHz in Cantonese. 2225-0605, 0755-1800.
GD15) Jiangmen PBS, 19 Jianshe Lu, Jiangmen, Guangdong 529000 - FM stereo on 100.2MHz in Cantonese. 2200-1600.
GD16) Foshan PBS, 98 Yong'an Lu, Foshan, Guangdong 528000.
Zhen'ai Ch: FM stereo on 94.6MHz in Cantonese. 24h.
Qianse Ch: FM stereo on 98.5MHz in Cantonese. 24h.
GD17) Yangjiang PBS, Wangliaoling, Yangjiang, Guangdong 529500 - FM stereo on 97.2MHz in Ch and Cantonese.
GD18) 93 Yuejin Lu, Chikan Qu, Zhanjiang, Guangdong 524038 - 2200-1530 in Ch, Cantonese and Leizhou dialect.
Zhanjiang EBS: FM stereo on 104.6MHz 2200-0600, 0800-1600.
GD19) 11 Binnan Lu, Hedong Qu, Maoming, Guangdong 525000 - 2230-

1500 in Ch and Cantonese.
GD20) 41 Ganyuan Beilu, Zhaoqing, Guangdong 526040.
1st St: on 1521kHz/92.9MHz in Cantonese. 2200-1600.
Music Ch. (2nd St): FM stereo on 90.9MHz in Cantonese.
GD21) Qingyuan PBS, 30 Xianfeng Lu, Qingyuan, Guangdong 511500 - FM stereo on 88.7/96.7MHz in Ch and Cantonese. 2225-1405.

GUANGXI ZHUANG AUTONOMOUS REGION
GX1) 75 Minzu Dadao, Nanning, Guangxi 530022.
Satellite Sce. on MW/SW/FM and via Satellite 2200-1600.
3rd Prgr. on 5920kHz/93.0MHz 2300-0135, 1000-1235.
Literary St: FM stereo on 95.0MHz 0055-0500(exc. Tues), 0930-1035. Rel. CNR 3: 0100-0500(exc. Tues).
Guangxi EBS: on MW/SW/FM in Ch and Guangxi dialect. 2200-1600.
GX2) 7 Chaoyang Lu, Nanning, Guangxi 530012 - on 918kHz/101.4MHz in Ch and Guangxi dialect. 2200-0530, 0900-1500.
GX3) 22 Wuyi Lu, Liuzhou, Guangxi 545001 - on 1386kHz/92.5MHz in Ch and Liuzhou dialect. 24h.
Economic St. on 99.1MHz.
GX4) Anxinzhou, Guilin, Guangxi 541002 - 2225-0525(Sun 0350), W0910-1400, Sun1025-1410.
FM stereo: on 97.7MHz.
GX5) 1-1 Dieshan Li, Dieshan 2 Lu, Wuzhou, Guangxi 543002 - 2200-1600 (exc. Mon) in Ch and Guangxi dialect.
Economic St: on 107.5MHz 2255-1100 (exc. Thurs).
GX6) Guizhou Zhonglu, Beihai, Guangxi 536000 - on 750kHz/97.1MHz in Ch and Guangxi dialect. 2150-2335, 0330-0440, 0950-1330.
Beihai EBS: on 93.5MHz 2200-0030, 0300-0600, 0930-1600.

HAINAN PROVINCE
HA1) 1 Haixiu Dadao, Haikou, Hainan 570206 - on 954/1107kHz/91.0MHz in Ch and Hainan dialect. 2200-1530 (exc. Tues 0610-0920).
Economic Ch: FM stereo on 99.0/103.8MHz 2200-1600.
HA2) Haikou PBS, 31 Longhua Lu, Haikou, Hainan 570102 - on 95.4MHz in Ch and Hainan dialect. 0930-1500.
HA3) Sanya PBS, 4 Jiaoyu Xiang, Hedong 1 Lu, Sanya, Hainan 572000 - FM: 104.6MHz.

CHONGQING MUNICIPALITY
CQ1) 159 Zhongshan 3 Lu, Yuzhong Qu, Chongqing 630015 - on 1314kHz/105.5MHz 2200-1600.
Chongqing EBS: FM stereo on 101.5MHz 2220-1600.
Chongqing Commercial BS: FM stereo on 95.5MHz. 24h.
Chongqing Educational BS: on 1080kHz 0225-0600, 0755-1100.

SICHUAN PROVINCE
SC1) 119-1 Hongxing Zhonglu, Chengdu, Sichuan 610017.
1st Prgr. on MW/SW/FM 2155-1605.
V.O. the Golden Bridge (2nd Prgr): on MW/SW/FM in Ch, Tb and Yi. 2230-1600.
Sichuan EBS: on 999kHz/103.7MHz 2200-1700 (exc. Tues 0600-0900).
Sichuan IBS: on 747kHz 2230-1200.
SC2) Minjiang Music BS, 119-1 Hongxing Zhonglu, Chengdu, Sichuan 610017 - FM stereo on 89.4MHz. 24h.
SC3) 3 Duan, Dong 1 Huanlu, Chengdu, Sichuan 610051.
1st Prgr. on 792kHz/91.4MHz 2200(Sun 2225)-0700, 0940(Sun0945)-1500.
2nd Prgr. on 1485kHz 2155-0030, 0425-0600, 1055-1400.
Economic St: FM stereo on 105.6MHz.
SC4) 87 Tanmulin Jie, Zigong, Sichuan 643000 - 2120-0035, 0320-0530, 0920-1430.
SC5) Linjiang Lu, Panzhihua, Sichuan 617000 - 2225-0210, 0355-0635, 0810-1355.
SC6) Sanxing Jie, Luzhou, Sichuan 646000 - 2215-0205, 0325-0600, 0955-1300.
SC7) 63, Nan 1 Duan, Taishan Nanlu, Deyang, Sichuan 618000.
SC8) 4 Hongxing Nanlu, Mianyang, Sichuan 621000 - 2200-1500.
SC9) Dabei Jie, Suining, Sichuan 529000.
SC10) 33, 1 Xiang, Xianglong Lu, Neijiang, Sichuan 641000.
SC11) Haitang Lu, Leshan, Sichuan 614000.
SC12) Jiazhou EBS, 40 Dingdong Jie, Leshan, Sichuan 614000 - FM stereo on 96.8MHz 2200-1405 (exc. Wed 0500-1000).
SC13) 22 Fujiang Lu, Nanchong, Sichuan 637000 - 2155-0015, 0355-0520, 1155-1415.
SC14) 28 Renmin Lu, Yibin, Sichuan 644000.

GUIZHOU PROVINCE
GZ1) 259 Qingyun Lu, Guiyang, Guizhou 550002.

Satellite Sce. on MW/SW/FM and via Satellite 2150-0600, 0850-1605.
3rd Prgr: FM stereo on 91.6MHz 0150-0900, 1150-1505. Rel. CNR 3: W0230-0430, Sun0230-0530, W0600-0900.
Guizhou EBS: on 603kHz/106.2MHz 2210-0600, 0800-1610.
GZ2) 285 Qingyun Lu, Guiyang, Guizhou 550002.
Guiyang Traffic Literary BS: on 102.7MHz.
GZ3) 9 Minghu Lu, Zhongshan Qu, Liupanshui, Guizhou 553000 - 2225-0030, 0325-0530, 0955-1330.

YUNNAN PROVINCE
YN1) 73 Renmin Xilu, Kunming, Yunnan 650031.
General St. on MW/FM and via Satellite 2200-0100, 0255-0700(Tues 0600), 0930-1600.
Minority Language St. on 6937kHz in Lahu, Jingpo, Lisu, Dehong Dai and Xishuangbanna Dai. 2225-0045, 0355-0545, 1100-1500.
Educational St. on 5960kHz W0030-0400, D1045-1200.
Literary St: FM stereo on 103.4MHz 1155-1600.
Music St: FM stereo on 97.0/100.0MHz 2300-1600 (exc. Tues 0600-0900).
Yunnan EBS: on 1242kHz 2220-1100(Tues 0600).
YN2) 27-1 Xinwen Lu, Kunming, Yunnan 650032.
1st Prgr. on 1350kHz/100.8MHz 2225-1600 (exc. Tues 0400-0800).
2nd Prgr: FM stereo on 102.8MHz. 24h.
YN3) Xincun, Dongchuan, Yunnan 654100.
YN4) Zhudianpo, Wenshan Xian, Yunnan 663000 - on 1053kHz/96.0MHz in Ch, Zhuang, Miao and Yao. 2225-0100, 0225-0500, 0955-1400.
YN5) 32 Jianshe Donglu, Gejiu, Yunnan 661400 - on MW/SW/FM in Ch, Hani and Yi. 2225-0030, 0225-0505, 0855-1400.
YN6) 10 Jinghong Donglu, Jinghong, Yunnan 666100 - 2210-0100, 0250-0600, 1030-1540 in Ch, Xishuangbanna Dai and Hani.
YN7) 38 Lucheng Beilu, Chuxiong, Yunnan 675000 - 2230-2335, 0400-0500, 1000-1100.
YN8) 160 Cangshan Lu, Xiaguan Zhen, Dali, Yunnan 671000.
YN9) 31 Yingjian Lu, Mangshi Zhen, Luxi Xian, Yunnan 678400 - 2230-0110, 0330-0700, 1030-1530 in Ch, Dehong Dai, Jingpo and Zaiwa.
YN10) Lucheng Beilu, Chuxiong, Yunnan 675000 - W2225-2400, Sun2325-0200, D0325-0600, D0955-1405.
YN11) Zhongshan Lu, Gejiu, Yunnan 661400.

XIZANG AUTONOMOUS REGION
XZ1) 180 Beijing Zhonglu, Lhasa, Xizang 850000.
Ch Prgr. on MW/SW and via Satellite W2300-0235, Sun2300-0750, W0355-0750(Tues 0650), D0950-1545.
Tb Prgr. on MW/SW and via Satellite W2300-0235, Sun2300-0750, W0355-0750(Tues 0650), D0950-1545.
Economic St: FM stereo on 98.0MHz 0000-0200, 0430-0630, 1000-1200.
XZ2) 25 Nedong Lu, Zetang Zhen, Nedong, Xizang 856000 - 2335-0135, 0405-0535, 1005-1340 in Ch and Tb.

SHAANXI PROVINCE
SN1) 119 Bei Dajie, Xi'an, Shaanxi 710003.
Satellite Sce. on MW/SW/FM and via Satellite 2155-0720(Tues 0600), 0950-1505.
Shaanxi EBS: on 603kHz 2230-0600(Fri 0100), 0920-1500.
Shaanxi Scientific Educational BS: on 900kHz 2230-0600(Fri 0100), 0920-1500.
Shaanxi LBS: on 747kHz/98.8MHz 2258-1500. Rel. CNR 3: Sun2330-0600, W0400-0600, D0930-1500.
Shaanxi Traffic Music BS: on 1323kHz.
SN2) 4 Xushimiao Jie, Xi'an, Shaanxi 710003.
News St. on 810kHz 2150-1600.
Economic St: FM stereo on 104.3MHz 24h.
Music St: FM stereo on 93.1MHz 24h.
SN3) Miaopu Lu, Hongqi Jie, Tongchuan, Shaanxi 727000 - 2210-0015, 0330-0515, 0915-1405.
SN4) 47 Hongqi Lu, Baoji, Shaanxi 721000 - 2145-2400, 0325-0610, 0930-1500.
Baoji EBS: on 900kHz/FM 2155-0600, 0955-1400.
SN5) Bei'an Cun, Weiyang Xilu, Xianyang, Shaanxi 712000 - 2210-0025, 0345-0630, 0850-1110.

GANSU PROVINCE
GS1) 226 Donggang Xilu, Lanzhou, Gansu 730000.
1st Prgr. on MW/SW/FM 2150-1600 (exc. Tues 0600-0855).
2nd Prgr (City FM): on 102.2/106.6MHz 2229-1830 (exc. Tues 0600-0855).
3rd Prgr (Traffic Sce): on 103.5/104.8MHz 2250-1500 (exc. Tues 0600-0855).

GS1A) Gansu TBS, 226 Donggang Xilu, Lanzhou, Gansu 730000 - on 103.5/104.8MHz 2255-1450.
GS2) 34 Qingyang Lu, Lanzhou, Gansu 730030 - W2220-0255, Sun0000-0530, W0355-0855, 1000(Sun 0920)-1300.
GS3) 11-5 Huancheng Zhonglu, Qincheng Qu, Tianshui, Gansu 741000.
GS4) Wuyi Nanlu, Jiayuguan, Gansu 735100.
GS5) 49 Renmin Xijie, Hezuo Zhen, Xiahe, Gansu 747000 - on 1332/5970kHz/97.2MHz in Ch and Tb. 2220-0030(Sun0040), 0420-0620, 0950-1330(Sat/Sun 1415).
GS6) 109 Tuanjie Lu, Linxia, Gansu 731100 - 2255-0130(Sun 0230).
GS7) 29 Hongqi Jie, Pingliang, Gansu 744000.
GS8) Gongyuan Lu, Zhongping Qu, Yumen, Gansu 735200.

QINGHAI PROVINCE

QH1) 96 Kunlun Lu, Xining, Qinghai 810001.
1st Prgr. on MW/SW and via Satellite W2220-0030, Sun2225-0630, W0355-0630, D0925-1505.
3rd Prgr: FM stereo on 97.2MHz 1155-1500. Rel. CNR 3: 1200-1400.
Tb Prgr. on 1251/6500kHz and via Satellite 2225-0055, 0355-0545, 1055-1540.
Qinghai EBS: on 1143kHz 2225-0030, 0410-0640, 1005-1350(Sun 1405).
QH2) 105 Nan Dajie, Xining, Qinghai 810000 - 2225-0530(Tues 0015), 1025-1330(Sun/Mon 1400).
QH3) 21 Hongwei Lu, Jiegu Zhen, Yushu Xian, Qinghai 815000.
QH4) 24 Renmin Lu, Delingha, Qinghai 817000 - 1115-1130 in Ch, 1230-1315 in Mo.

NINGXIA HUI AUTONOMOUS REGION

NX1) 16 Beijing Donglu, Xinshi Qu, Yinchuan, Ningxia 750021.
1st Prgr. on MW/FM 2210-0130(Sun 0230), 0355-0620, 0855-1405.
2nd Prgr. FM stereo on 104.3MHz 0355-0545, 1025-1400. Rel. CNR 3: 1200-1400.
Ningxia EBS: on 630kHz 2230-0105, 0330-0630, 1130-1500.
NX2) 19 Zhongshan Beijie, Yinchuan, Ningxia 750004 - W2225-0145, Sun0000-0500, W0355-0530, D1025-1310(Sun 1330).
NX3) Chaoyang Dongjie, Dawukou Qu, Shizuishan, Ningxia 753000.
NX4) Yumin Dongjie, Wuzhong, Ningxia 751100.
NX5) Wenhua Jie, Xiaoba Zhen, Qingtongxia, Ningxia 751600.

XINJIANG UIGHUR AUTONOMOUS REGION

XJ1) 84 Tuanjie Lu, Urumqi, Xinjiang 830044.
Ch Prgr. on MW/SW 2300-0210(Sun 0330), 0430-0730, 1030-1650.
Ug Prgr. on MW/SW 2300-0200, 0330-0730, 1030-1650.
Kz Prgr. on MW/SW 0000-0230, 0530-0700, 1200-1650.
Mo Prgr. on MW/SW 2300-0230, 0500-0700, 1130-1650. Kirghiz prgr: 2300-2400, 1130-1230.
Music St: FM stereo on 90.9MHz W0000-0200, Sun0230-0530, Sun1030-1230, W1130-1330.
XJ2) Tianshan EBS, 84 Tuanjie Lu, Urumqi, Xinjiang 830044 - on 837/1215kHz/92.9MHz W0000-0210, Sun0200-1005, W0330-0700, W1000-1220.
XJ3) 28 Xinmin Lu, Urumqi, Xinjiang 830002.
Ch Prgr. on 810kHz 2345-0230, 0510-0900, 1050-1615.
Ug Prgr. on 1071kHz 2350-0210, 0500-0730(Sun 1005), 1130-1545.
FM stereo: on 100.7MHz.
XJ3A) 11 Nanchang Lu, Urumqi, Xinjiang 830002 - on 603kHz/97.4MHz 24h.
XJ4) 42 Tianshan Xilu, Karamay, Xinjiang 834000.
Ch Prgr. on 882kHz 0025-0240, 0600-0800, 1155-1505.
Ug Prgr. on 1179kHz.
XJ5) Bei 3 Lu, Shihezi, Xinjiang 832000.
XJ6) 2 Hongxing Xilu, Hami, Xinjiang 839000.
Ch Prgr. on 1485kHz W2310-0215, Sun2355-0450, W0500-0710, Sun1150-1450, W1155-1515.
Ug Prgr. on 1098kHz.
XJ7) Beijing Beilu, Changji, Xinjiang 831100 - W2335-0235, Sun0155-0540, W0500-0730, Sun1115-1350, W1135-1430.
XJ8) Korla Donglu, Kuytun, Xinjiang 833200.
Ch Prgr. on MW/SW W2355-0230, Sun0025-0335, Sun0528-0720, W0558-0740, D1123-1425.
Kz Prgr. on 819kHz.
XJ9) 13 Sidalin Jie, Yining, Xinjiang 835000.
Ch Prgr. on 1134kHz 2350-0200, 0550-0700, 1150-1600.
Ug Prgr. on 603kHz 2350-0200, 0550-0700, 1150-1600.
Kz Prgr. on 882kHz 2350-0200, 0550-0700, 1220-1500.

EXTERNAL SERVICES

100) CHINA RADIO INTERNATIONAL (CRI)
(Zhongguo Guoji Guangbo Diantai)

✉Jia 16, Shijingshan Lu, Shijingshan Qu, Beijing 100039. ☎ +86 (10) 6889 1001. CE's Office: +86 (10) 6889 1533. Audience Rel. +86 (10) 6889 1149. Foreign Affairs Office: +86 (10) 6889 1110. **Web:** http://www.cri.cngb.com/mainchn.htm

DOMESTIC SERVICE

Prgrs in English/French/German/Spanish/Japanese: News, current affairs, features, music & weather for visitors & resident foreigners.
Beijing (1251kHz + 91.5MHz): English 2300-0500, 0900-1200, 1300-1500; French 0500-0600; German 1500-1530; Spanish 2200-2300; Japanese 1200-1300.
Tianjin (801kHz): English 2300-0100, 0400-0500, 1300-1500.
Shanghai (1296kHz): English W2300-2400, Sun2300-0500.
Guangzhou (603kHz): English 2300-2400; (1584kHz): English 1400-1500.
Xi'an (1323kHz): English 1300-1400, 2300-0100; Japanese 1200-1300.
Lanzhou (98.2MHz): English 0900-1200, 1300-1500, 2300-0500; Japanese 1200-1300.
Jinan (105.0MHz): English 0000-0100, 0200-0400; Japanese 1200-1300.
Jingdezhen (102.0MHz): English 0900-1200, 1300-1500, 2300-0500.
Wuhan (105.8MHz): English 1300-1500.
Zibo (801kHz): English 0000-0100, 1300-1500.
Weifang (801kHz): English 1100-1200, 1300-1500, 2300-0100.
Qingdao (846KHz): English 1100-1200, 1300-1500, 2300-0100.
Hefei (747/801kHz + 92.4MHz): Japanese 1200-1300.
Hangzhou (1071kHz): English 2300-0300.
Special Prgr:
Zhuhai (603kHz + 107.1MHz): English 2200-2400, 0200-0400, 0600-0900, 1300-1700;
Chinese 0000-0200, 0900-1100; Cantonese 0400-0600, 1100-1300.
Dalian (1575kHz + 89.1MHz): Chinese 2125-1300; English 1300-1500.

OVERSEAS SERVICE

Mediumwave key frqs(kHz): 1017(JL), 1044(JS), 1080/1188/1206/1269/1296(YN), 1116(HL), 1314(NM), 1323(JL/XZ), 1341(GD), 1521(XJ).
Shortwave: Baoding, HEB(G.C: 115.44E/38.39N) 240/120kW; Beijing, BJ(G.C:116.27E/39.57N) 240/120/50kW; Jinhua, ZJ(G.C:119.39E/28.07N) 500kW; Kunming, YN 120/50kW; Shijiazhuang, HEB (G.C:114.28E/38.04N) 500kW; Xi'an, SN(G.C: 108.54E/34.12N) 300/150kW. Further trs in border areas are used at times.
Overseas relay st's: CRI exchanges transmitting facilities with R. Canada Int, R. France Int, R. Exterior de Espana, Swiss Radio Int and V.O. Russia, and also uses relay facilities in Brazil and Mali.
Relays: B=Brazil, C=Canada, E=Spain, F=France, G=Fr. Guiana, M=Mali, R=Russia, S=Switzerland.

Albanian	Area	kHz
1900-1927	Eu	9965 7385
1930-1957	Eu	9965 7405 4960
2100-2127	Eu	6150F
Amoy		
1400-1457	SEAs	11685 11650 9715 7335
Arabic		
1600-1657	ME	17580R 9440 7405
1830-1927	ME	15530M 13685M 12035R 7480 7315
2100-2157	ME	11515 9685 9440 7260
Bengali		
1500-1527	SAs	11825 8660 7335 1206
1530-1557	SAs	11825 8660 7335 1206
Bulgarian		
1830-1857	Eu	9860 7385 6020
2030-2057	Eu	9860 6150F
Burmese		
1130-1157	SEAs	11825 9880 8260 1269 1188
1300-1327	SEAs	11780 9880 8260 1269
1330-1357	SEAs	11780 9880 8260 1269
Cambodian		
1030-1127	SEAs	9870 9440 8260 1080
1200-1257	SEAs	9870 9440 8260 1080
1400-1457	SEAs	9440 8260
Cantonese		
1000-1057	SPac	11915 11715 11650
1100-1157	SEAs	15440 11945 11715 11685 7335
1200-1257	NAm	11965
1700-1757	SAs/SEAf	9900 7180
1900-1957	Eu	11945 9730
2330-2357	SEAs	15400 12065 12015 11685 9870

		9457 9440 8260 7335 7190 6140 5250
Chaozhou		
0030-0057	SEAs	15400 12065 12015 11685 9870 9457 9440 8260 7335 7190 6140 5250
Chinese		
0200-0257	SAm/NAm	15435 9690E
0300-0357	NAm	9730G
0900-0957	NEAs/SPac	15440 15180 12015 11945 11875 11685 11650 11515 9945 9480 6590 6010 5250
1000-1057	NEAs	15440 12015 11945 11685 6590 6010 5250
1200-1257	NEAs/SPac	15260 11945 11875 11685 9365
1300-1357	NEAs/SPac	15260 11945 11875 11685 9440 9365
1500-1557	SAs/EAf	11980 9715 9457 8260 7110
1730-1827	WAs/NAf/Eu	9820 9600 7800 7335 7315 7110 5250
2000-2027	Eu	6165S
2000-2057	WAs/NAf/Eu	11650 11515 9730 9710 9685 9600 7660 7185
2230-2257	NAm	15500M 11975M 11650 9535 7160
2230-2327	NEAs/SEAs	15400 12065 12015 11685 9870 9457 9440 8260 7335 7190 6140 5250
2300-2357	NAm	11975M
Czech		
1900-1927	Eu	7375 6020
1930-1957	Eu	7375 7305F
English		
0300-0357	NAm	9690E
0400-0457	NAm	9730G 9560C
0900-0957	SPac	11755 9890 9785
1000-1057	SPac	11755 9890 9785
1200-1257	ENAm	9565
	SEAs/SPac	15480 11980 11675 11660 9945 9715 7385 6950 1341
1300-1357	SEAs/SPac	15180 11980 11675 11660 9945 7405 7385 1341
1400-1457	WNAm	7405
	SAs	11825 9700 9535 7260
1500-1557	WNAm	9785
	SAs	7160
1600-1657	EAf/SAf	9620 9565
1700-1757	EAf/SAf	11910 9745 9570 7405 5220
1900-1957	WAf/NAf	11515 9600 9440
2000-2057	EAf/SAf	15500M 11975M 7160
	WAf/NAf	9440
	Eu	9920 6950 5220
2100-2127	EAf/SAf	15500M 11975M 9535 7180
	Eu	3985S
2100-2157	Eu	9920 6950 5220
2200-2257	Eu	9880R
Esperanto		
1100-1127	NEAs	9535 7170
1300-1327	SEAs	11840 11600
2000-2027	Eu	9965 7405 4960
2230-2257	SAm	9860 6950
Filipino/English		
1130-1157	EAs	11700 8660 1341
1200-1227	EAs	12110 11700 8660
1430-1457	EAs	11445 9880 8660 1341
French		
1830-1927	Af/Eu	9820 7800 7335 7110 5250
1930-2027	Af/Eu	9820 7800 7335 7315 7110 5250
2030-2057	Eu	3985S
2030-2127	Af/Eu	12010R 9820 7800 7335 7110 5250
2130-2227	Af/Eu	15500M 11975M 9820 7800 7335 7110 5250
German		
1800-1857	Eu	9920 6950 5220
1900-1957	Eu	9920 6950 5220
2000-2027	Eu	3985S
Hakka		
0000-0027	SEAs	15400 12065 12015 11685 9870 9457 9440 8260 7335 7190 6140 5250
1600-1657	SEAf	11825 9900

Hausa		
1730-1757	Af	15125M 11970M 9760 7235
1800-1827	Af	15125M 11970M 7405 7235 4960
Hindi		
1500-1557	SAs	11675 9920 7590 1323
1600-1657	SAs	11980 11675 9920 9745 8660 1269
Hungarian		
1800-1827	Eu	9860 7385 6020
2030-2057	Eu	9365 6020
2130-2157	Eu	6150F
Indonesian		
0830-0927	SEAs	17680 15600 15135 12055 8660
1030-1127	SEAs	15135 12055 11445 8660
1330-1427	SEAs	15135 11445 8660 1341
Italian		
1830-1857	Eu	9965 7405 4960
2030-2057	Eu	9965 7385
2100-2127	Eu	9965 9365
2130-2157	Eu	3985S
Japanese		
0930-1527	NEAs	9855 7190 1044
Korean		
1100-1257	NEAs	5965 1017
1300-1457	NEAs	5965 1017
Lao		
1230-1327	SEAs	7360 6590 6010
1430-1527	SEAs	7360 6590 6010 1080
Malay		
0930-1027	SEAs	17680 15135 12055 8660
1230-1327	SEAs	15135 11445 8660
Mongolian		
1100-1157	CAs	5850 5145
1200-1257	CAs	5850 5145 4883 4815 1314
1400-1457	CAs	4883 4815
Nepalese		
1500-1527	SAs	9535 7150 1269
1530-1557	SAs	9535 7150 6590 1269
Pashto		
1500-1527	ME	11515 7480
1530-1557	ME	11515 7480 5220
Persian		
1500-1527	ME	11660 9440
1800-1827	ME	15595R 11515 9440 7215
1830-1857	ME	11575 9785 9440 7215
Polish		
2000-2027	Eu	7375 6150F 6020
2030-2057	Eu	7405 7375 4960
Portuguese		
0030-0057	SAm	11515 7435
1900-1927	Af	9535 7180
1930-1957	Af/Eu	15500M 11975M 9785 7385 7150
2130-2157	Eu	6165S
2200-2227	SAm	7435 6950
Romanian		
1900-1927	Eu	9860 7305F
1930-1957	Eu	9860 6020
Russian		
0000-0057	Siberia	9725 7110
0300-0357	CAs	15435 11755 11515
1000-1057	FE/Siberia	11980 9725 7820 7110 5145 1521 1323 963
1100-1157	FE/Siberia	9725 7820 7110 1521 1323 1314 963
1200-1257	Siberia	1521 1323 963
1300-1357	E.Rus/CAs	7820 5850 5145 4883 4815 1521 1323 1314 963
1400-1457	Siberia	1521 1323 963
1500-1557	Eu/C.Rus	11650 9730 7375 4883 4815 1521 963
1600-1657	Eu/CAs	11650 11515 9965 9860 9730 9655 7375 1521
1700-1757	Eu/CAs	11650 9860 9730 9695 9655 9365 7375 6950 1521 1323
1800-1857	Eu	11685 9730 9695 9655 9535 9365 7375 1521
1900-1957	Eu	11685 9695 9655 9365 1521 1323
2000-2057	Eu	11915 9695 9655 1323
2300-2357	Siberia	9725 7110
Serbian/Croatian		
2000-2027	Eu	9860 9365 7385

2030-2057	Eu	6165S
2100-2127	Eu	9860 7405 4960

Sinhalese

1400-1427	SAs	11980 11780 7120
1430-1457	SAs	11980 11780 7120

Spanish

0000-0057	SAm	11650 7160
0100-0157	SAm	11650 9665M 9565 5250
0200-0257	SAm	11650 9665M
0300-0357	SAm	11765B 9560C
2100-2127	Eu	6165S
2100-2157	Eu	7360 6020
2200-2257	Eu	7360 6020
2300-2357	SAm	11650 7160

Swahili

1600-1627	Af	11600 9457 5250
1630-1657	Af	11600 9457 5250
1700-1727	Af	9457 7260 5250

Tamil

1400-1427	SAs	11575 9457
1430-1457	SAs	11575 9457

Thai

1130-1227	SEAs	9785 7360 6590 6010 1080
1330-1427	SEAs	7360 6590 6010

Turkish

1400-1427	WAs	11515 9480
1600-1627	WAs	11685 9715 8260 7480
1900-1927	WAs	9785 9685 7405 4960

Urdu

1600-1627	SAs	9945 7590 7160 5220 1323
1630-1657	SAs	9945 7590 7160 5220 1323

Vietnamese

1100-1157	SEAs	9550 7245 5250 1296
1200-1257	SEAs	9550 7245 5250 1296
1300-1357	SEAs	9550 7245 5250 1296
1400-1457	SEAs	9550 7245 5250 1296
1500-1557	SEAs	9550 7245
1600-1657	SEAs	7360 6590 6010

ANN: Arabic: "Idha'at as-Sin ad-Duwaliyah". Chinese: "Zhongguo guoji guangbo diantai." English: "This is China Radio International, CRI, broadcasting from Beijing". Esperanto: "C^ina Radio Internacia parolas en Pekino". German: "Hier ist Radio China International". Indonesian: "Inilah Radio CRI, China Radio Internasional". Japanese: "Kochirawa Chugoku Kokusai Hosokyoku desu". Malay: "Inilah Radio Antarabangsa China, dalam bahasa Melayu". Mongolian: "Hyatadyn Olon Ulsyn Radio". Spanish: "Esta es Radio Internacional de China". Vietnamese: "Day la dai phat thanh quoc te Trung quoc".
INT-SIG: First 4 bars of "The East is Red".

101) CHINA HUAYI BROADCASTING COMPANY (Priv.)

✉ P.O. Box 251, Fuzhou, Fujian 350001.
D.PRGR: on 666/4830/6185kHz + 99.6/107.1MHz for Hong Kong, Macao and Southeast Asia. 0255-0700(Tues 0600), 0855-1600.

CHINA (Republic of) (Taiwan)

LT: UTC + 8h — **Pop:** 21,400,000 — **Radio sets:** 8,620,000 — **Pr.L:** Mandarin(Chinese), Amoy, Hakka — **E.C:** a/c 60,110/220V — **ITU:** TWN.

BROADCASTING CORPORATION OF CHINA (BCC)

(Priv. enterprise under Gov. contract).
✉ 53 Jen'ai Road, Sec. 3, Taipei 10647. ☎ + 886 (2) 27710151. 🖷 +886 (2) 27519277. **Web:**http://www.bcc.com.tw **e-mail:**pr@bcc.com.tw
L.P: Pres: Lee Ching-ping, Dir.Eng.Dep.: Lee Muh-tsun.

Mediumwaves: Call: BE— N=Network; N=News Netw, H=Hakka Channnel

	Call	Location	kHz	kW	N
10)	G51	Ilan	630	10	3
1)	D34	Taipei	657	20	N
4)	D58	Taichung	720	10	N
6)	D92	Tainan	783	10	2
8)	D28	Taitung	819	10	N

	Call	Location	kHz	kW	N
4)	D43	Taichung	837	10	2
9)	D27	Hualien	855	10	N
7)	D25	Kaohsiung	864	10	N
2)	G77	Hsinchu	882	10	N
3)	G78	Miaoli	891	3.5	3
7)	D79	Kaohsiung	909	10	2
1)	D55	Taipei	954	20	2
8)	D88	Taitung	1008	10	2
2)	D53	Hsinchu	1017	10	2
5)	D26	Chia-i	1035	10	2
1)	D85	Taipei	1062	10	3
4)	D23	Taichung	1062	10	3
11)	D72	Yuli	1116	3.5	N
12)	D68	Puli	1152	1	2
10)	D86	Ilan	1161	10	2
3)	D89	Miaoli	1161	10	2
9)	D32	Hualien	1188	10	2
6)	D47	Tainan	1296	10	N
5)	D63	Chia-i	1350	10	N
11)	D74	Yuli	1386	3.5	2
2)	D87	Hsinchu	1386	3.5	3
10)	D65	Ilan	1404	10	N
3)	D54	Miaoli	1413	10	N
8)	D80	Taitung	1413	3.5	3
12)	D67	Puli	1413	1	N
7)	D48	Kaohsiung	1449	10	3
1)	D57	Taipei	1458	10	H
9)	D81	Huanlien	1467	10	3
5)	D82	Chia-i	1467	10	3
6)	D78	Tainan	1539	10	3

FM: *) relay station. P=Pop Netw, A=Amoy Language Netw.

	Location	P	A	Music	Local	kW
1)	Taipei	103.3	105.9	96.3		35/10
3)	Huoyenshan	102.9	101.5	96.1		10/0.1
4)	Taichung	102.1	106.9	96.3		35/10
5)	Chentoshan	103.1	104.3	96.1		10/0.1
7)	Kaohsiung	103.3	105.9	96.3	107.7	35/10
8)	Taitung	102.1	106.9	96.3		5/2.5
9)	Hualien	102.1	106.9	96.3		5/2.5
10)	Ilan	102.1	102.9	96.1		5/2.5
11)	*Yuli	103.3	105.7			2.5/1
12)	*Puli	107.3				0.1

News Netw (1st Netw): 24h in Mandarin.
2nd Netw: 24h in Amoy.
3rd Netw: 24h in Mandarin.
Hakka Channnel: 2300-1500 in Hakka.
FM Pop Netw: 24h in Mandarin.
FM Amoy Language Netw: 24h in Amoy.
Music Netw: 24h in Mandarin.

Local Prgrs
✉ 2) 9th Flr, 3, 55 Tungkuang Rd,Hsinchu 300 — 3) 1008-78 Chungshan Rd, Kaomiao Li, Miaoli 360 — 4) 35th Flr,758,Chung ming South Rd,nan,Taichung — 5) 121 Ufeng South Rd, Chia-i 600 — 6) 19th Flr, 5, 248 Yunghua 2nd Rd, Anping, Tainan 708 — 7) 24th Flr, 1 ,91 Chung shan 2nd Rd, Kaohsiung 806.
Web:http://w3.wownet.net/~coolfish/ks_index.htm — 8) 52-23 Kuilin South Rd, Taitung 930 — 9) 25 Shuiyuan Str, Hualien 950 — 10) 2 Kuchie Rd, Chuangyuan Village, Ilan. — 11) Yuli — 12) Puli.

ANN: Chinese: "Chungkuo Kuangpo Kungssu, Hsinwen/Liuhsing Wang" (News/Pop network) or "Chungkuo Kuangpo Kungssu (location) Kuangpo Tientai", Amoy: "Tiyon Gok Kon Po Kon Sih, (location) Kon Po Den Tai" -

CENTRAL BROADCASTING SYSTEM (CBS)

✉ 55 Pei'an Rd, Tachih, Taipei 104. ☎+ 886(2)25918161,🖷 +886(2)25850741. **Dir:** Huang Sze-chuan **e-mail** rtim@cbs.org.tw.
Web: http://www.cbs.org.tw, http://www.cbs-taipei.org.tw

M=Mandarin Program, N=News Netw, O=Overseas

MW	kHz	kW	N	MK	kHz	kW	N
Fangliao	585	1200	O	Lukang	1008	600	N
Lukang	603	1000	N/O	Kouhu	1098	300	N
Minhsiung	747	250	M/O	Minhsiung	1206	100	N/O
Kouhu	900	300	M	Changchih	1521	1200	M
Changchih	927	1200	M				

Shortwaves

kHz	kW	N	kHz	kW	N	kHz	kW	N
3335	100	M	9610	250	N/O	11840	100	M
6040	100	N/O	9630	300	N/O	11860	100	O
6085	300	M	9690	100	N/O	11905	100	O
6180	100	N	9765	250	O	11915	100	O
7105	300	M/N	9955	250	O	11970	100	M
7130	50	O	11550	100/250/300	O	15125	250	M
7250	100	N	11725	100	N	15270	100	O
7285	100	O	11745	250	O	15345	100	O
7445	100	O	11775	300	M			
9280	100	M/O	11825	100	O			

Mandarin Program
Taiwan Area

Kaoshiong, Pintung	1000-2200	927,1521kHz
Yunlin, Chiayi, Tainan	1600-0000	747kHz
	0400-0000	900kHz
	1000-0100	1098kHz
Taipei	0400-1000	6085kHz
	1000-1900	7105kHz
	0400-0800	11775kHz

Mainland China Area

North-eastern China	0200-1000	11970kHz
Northern China	1000-1900	6085kHz
	0000-0500	9280kHz
	0400-0800	11775kHz
	0400-0500	11840kHz
	0400-1100	900kHz
	1000-2200	1521kHz
Central China	0900-0000	3335kHz
	1000-1900	7105kHz
	1500-0000	747kHz
	1000-2200	927kHz
	1000-0100	1008kHz
Southern China	0400-1000	6085kHz
Southern Asia	0900-1800	15125kHz
	2200-0300	15125kHz

News Network
Taiwan Area

Changhua, Taichang, South Miaoli	1000-2200	603kHz
	1000-1800	1008kHz
	2200-0600	1008kHz
Yunlin, Chiayi Southeast	1600-2200	1206kHz
	0600-1600	1422kHz
	2200-0300	1422kHz

Mainland China Area

Northern China	0800-0400	7250kHz
	1000-1900	7250kHz
	2200-0600	9610kHz
Central China	1000-1900	11775kHz
	2200-0200	7105kHz
Southern China	1100-1900	9630, 6180kHz
	1400-1900	9690, 6040kHz
	2200-0000	9690, 6040kHz
	2200-0500	11725kHz
	0700-1600	11725kHz

KUANGHUA BROADCASTING STATION
P.O.Box 89, Taipei. ☎ (2) 26030429.🖷 (2) 26030433

MW	kHz	kW	MW	kHz	kW
Hsinfeng	711	250	Kinmen	1107	10
Kuanyin	801	250	Kuanyin	1251	100
Kuanyin	846	250	Matsu	1341	20
Matsu	963	10	Chingshui	1431	10
Hsinfeng	981	250	Taipei	1440	10
Kuanyin	1053	100	Kinmen	1467	50

D.PRGR: 0655-0100 in Chinese.
ANN: "Kuanghua chih Sheng, Kuanghua Kuangpo Tientai."

CBS-Radio Taipei International
55 Pei'an Rd, Tachih, Taipei 104. ☎ + 886(2)25918161.🖷 +886(2)25850741

MW:
Fangliao: 585kHz 1200kW,Lukang: 603kHz 1000kW,Minhsiung: 747kHz 250kW 1206kHz 100kW.

SW:
Annan(G.C 120.38E/23.11N:4 x 250kW-Huwei 1 x 300kW, 4 x 100kW-Kouhu 3 x 100kW
-Minhsiung(G.C 120.25E/23.33N: 1 x 50kW-Paochung(G.C 120.18E/23.43N: 5 x 100kW
-Tanshui: 3 x 300kW -*)relay via WYFR,USA
kHz:
5950*, 6040, 7130, 7285, 7445, 9280, 9610, 9630, 9680*, 9690, 9765, 9955, 9985*, 11550, 11740*, 11745, 11825, 11860, 11905, 11915, 15130*, 15215*, 15270, 15345, 15600*, 15695*, 15715*, 17750*, 17805*

	Area	kHz
Amoy		
0800-1200	CHN	603
0900-1500	CHN	747
0200-0400	SEAs	11550, 11915
0800-0900	Pac	11745
1000-1100	SEAs	7130, 11550, 11745, 15345
1300-1400	SEAs	7130, 11745, 11860
Arabic		
2000-2100	Af/ME	9630,11550
Cantonese		
0000-0600	CHN	6040, 9690
0400-0500	Pac	11550
0500-0600	SEAs	11825, 11915, 15270, 15345
0800-0900	SEAs	7445
1000-1100	SEAs	11915
1000-1100	Pac	9610
1100-1200	SEAs	15270
1300-1400	SEAs	9765, 15345
English		
0200-0300	NAm	11740*
0200-0400	NAm,WAm	5950*
	CAm	9680*
	SEAs	11825, 15345
0700-0800	NAm	5950*
1200-1300	Pac	9610
2200-2300	Eu	15600*, 17750*
French		
2000-2100	Eu	15715*
2100-2200	Eu	9610
German		
0600-0700	Eu	9985*
1800-1900	Eu	9955
2100-2200	Eu	15600*,17750*
Hakka		
0000-0600	CHN	1206
0900-1500	CHN	1206
0000-0200	SEAs	11745, 15345
1400-1500	SEAs	9765, 15345
1500-1600	Pac	9610
Indonesian		
0800-0900	SEAs	11550, 11915, 7130
1100-1200	SEAs	11550
1200-1300	SEAs	15345
Japanese		
1100-1200	FE	7130, 11745
1400-1500	FE	7130, 11745
0100-0200	FE	11745, 15270
Mandarin		
(Perspective Program)		
0415-0500	SEAs	15345
	SEAs	11550, 11825, 7130
	WAm, EAm	5950*
	CAm	9680*
1215-1300	SEAs	15270
	FE	11745
1915-2000	Af/ME	9955
(Pop Network Program)		
0115-0200	SAm	11825, 15215*
0715-0800	SEAs	7130
0915-1000	SEAs	7445
	SEAs	11550
	SEAs	11915
	Pac	9610
	FE	11745
1915-2000	Eu	11760*, 9985*
Mongolian		
1000-1100	CHN	11905
Spanish		

0400-0500	CAm	11740*
0600-0700	WAm	5950*
2300-2400	SAm`	15130*, 17805*
2000-2100	Eu	15715*
2100-2200	Eu	9610
Russian		
1300-1400	Eu	15695*
1700-1800	Eu	9955
Thai		
0600-0700	SEAs	15270, 15345
1400-0015	SEAs	15270, 11860
2200-2300	SEAs	585, 7445
Tibetan		
1100-1200	CHN	11905
Uighur		
1200-1300	CHN	11905
Vietnamese		
1500-1600	SEAs	11745, 15345
2300-2400	SEAs	11745, 15345

Voice of Asia

Indonesian		
1200-1300	SEAs	585, 7445, 11860
1600-1700	SEAs	585, 7445, 11745
Mandarin		
1300-1500	So.CHN	7285
1500-1900	No.CHN	9280
1750-2300	SEAs	585, 7445
Thai		
1500-1600	SEAs	585, 7445, 15270
2300-2400	SEAs	585, 7445, 15270

OTHER PUBLIC & COMMERCIAL ST'S
Call: BE— T=Traffic Prgr, S=AM stereo (C-QUAM system).

	Call	Station	Location	kHz	kW
2)	H7	Fu Hsing	Taipei 1	558	1
5)	M2	ICRT	Taipei	576S	10
2)	H2	Fu Hsing	Taipei 2	594	10
2a)	H38	Fu Hsing	Taichung	594	5
2b)	H44	Fu Hsing	Kaohsiung	594	10
6a)		Taiwan	Tahsi	621	1
6c)		Taiwan	Sungling	630	1
1b)	V59	Cheng Sheng	Taichung 2	657	10
9)	C22	Han Sheng	Taipei	684	10
2a)	H39	Fu Hsing	Taichung	693	10
2c)	H82	Fu Hsing	Hualien	693	1
9)	C25	Han Sheng	Taoyuan	693	10
9g)	C32	Han Sheng	Tainan	693	10
8b)	P24	Ching Cha	Taichung	702	10
11)	E43	Shih Hsin	Taipei	729	3
31)	L2	Taiwan Yuyeh	Penghu	738	100
12)		Shengli	Makung	756	1
6b)	V94	Taiwan	Taichung	774	10
13)	V88	Hsien Sheng	Taoyuan	774	10
9d)	C33	Han Sheng	Hualien	792	10
14)	V79	Keelung	Keelung	792	1
7)		Chien kuo	Hsinhua	801	1
6a)		Taiwan	Kuanhsi	810	1
15)	V54	Kuo Sheng	Changhua	810	10
8d)	P28	Ching Cha	Kaohsiung T	819	5
12)	V56	Shengli	Tainan 1	837	1
2b)	H56	Fu Hsing	Kaohsiung	846	10
9f)	C38	Han Sheng	Penghu	846	10
1a)	V72	Cheng Sheng	Chia-i	855	5
17)	V24	Min Pen	Taipei 2	855	1
18)	V52	Chung Hua	Taichung	864	10
19a)		Feng Ming	Penghu	882	1
2)	H3	Fu Hsing	Taipei 1	909	10
2c)	H79	Fu Hsing	Hualien	909	1
9a)	C42	Kuo Kuang	Taipei	936	10
7)	V85	Chien Kuo	Hsinying	954	10
20)		Yen Sheng	Wuho	954	1
21b)		Tien Sheng	Tunglo	954	1
c)	V84	Taiwan	Chunghsing	963	10
9)	V68	Feng Ming	Kaohsiung 2	981	1
1b)	V58	Cheng Sheng	Taichung 1	990	10
8e)	P38	Ching Cha	Ilan	990	1
8f)	P34	Ching Cha	Hualien	990	1
1f)	V60	Cheng Sheng	Kaohsiung	1008	5
2)	V92	Tien Nan	Taipei	1008	1
21a)		Tien Sheng	Yuanli	1026	1
23)	V51	Chung Hua	Sanchung 2	1026	1
3)	V98	Cheng Kung	Kaohsiung	1044	1
20)	V64	Yen Sheng	Hualien	1044	5
1d)	V82	Cheng Sheng	Ilan	1062	5
24)	V74	Min Li	Pingtung	1062	5
25)	V96	Tien Sheng	Tainan	1071	1
2)	H5	Fu Hsing	Taipei 1	1089	5
2c)	H80	Fu Hsing	Hualien	1089	10
9)	C31	Han Sheng	Yunlin	1089	10
16)	G28	Kaohsiung	Kaohsiung	1089	10
2a)	H34	Fu Hsing	Taichung	1116	10
8a)	P26	Ching Cha	Hsinchu	1116	5
8d)	P25	Ching Cha	Kaohsiung	1116	10
9)	C22	Han Sheng	Taipei	1116	10
9)	C30	Han Sheng	Ilan	1116	10
1c)	V36	Cheng Sheng	Yunlin	1125	5
8g)	P40	Ching Cha	Taitung	1125	1
26)	G26	Taipei	Taipei	1134	10
31)	L3	Taiwan Yuyeh	Penghu	1143	100
28)	V70	Hua Sheng	Taipei 1	1152	5
19)	V67	Feng Ming	Kaohsiung 1	1161	1
1)	V35	Cheng Sheng	Taipei	1170	5
15)		Kuo Sheng	Erhlin	1179	2.5
6)	V46	Taiwan	Taipei 2	1188	10
12)	V57	Shengli	Tainan 2	1188	1
6a)	V62	Taiwan	Hsinchu	1206	1
28)	V71	Hua Sheng	Taipei 2	1224	1
23)		Chung Hua	Juifang	1233	1
9c)	C29	Han Sheng	Kaohsiung	1251	10
1a)		Cheng Sheng	Putzu	1260	1
8)	P22	Ching Cha	Taipei	1260	10
1e)	V37	Cheng Sheng	Taitung	1269	1
9f)	C44	Han Sheng	Penghu	1269	10
9b)	C27	Han Sheng	Taichung	1287	10
24)		Min Li	Fangliao	1287	1
17)	V23	Min Pen	Taipei 1	1296	1
8c)	P32	Ching Cha	Hsinying	1314	5
21)	V76	Tien Sheng	Chunan	1314	10
6)	V45	Taiwan	Taipei 1	1323	1
6c)		Taiwan	Puli	1332	1
9e)	C36	Han Sheng	Tsoying	1332	10
23)	V50	Chung Hua	Sanchung 1	1350	2.5
9d)	C40	Han Sheng	Hualien	1359	5
1f)		Cheng Sheng	Chishan	1377	1
2)	H9	Fu Hsing	Taipei 2	1377	1
1c)		Cheng Sheng	Peikang	1386	1
30)	V78	Yi Shih	Keelung	1404	5
7)		Chien Kuo	Kuanyin	1422	1
2b)	H57	Fu Hsing	Kaohsiung	1476	10
4)	E32	Chiao Yu	Taipei	1494	10
4a)	E34	Chiao Yu	Changhua	1494	5
2a)	H36	Fu Hsing	Taichung	1512	5
2c)	H84	Fu Hsing	Hualien	1512	
8)	P27	Ching Cha	Taipei T	1512	10
27)		Pai Yun	Kaohsiung	1593	1
31a)		Taiwan Yuyeh	Ilan	1593	1

FM:

Call	MHz	kW		Call	MHz	kW
32)		92.1		5b)		100.1	30
10)		92.1		4e)	E39	100.3	1
38a)		92.3		5)	M3	100.7	30
33)		92.7		5a)		100.7	30
d)	P42	93.1	25	48)	M26	100.7	3
26)	G25	93.1	30	8f)	P35	101.3	1.5
55)		93.3		8g)	P39	101.3	1
56)		93.5		9g)	C28	101.3	35
57)		93.7		4)	E33	101.7	10
34)		93.7		4b)	E36	101.7	5
8)	P41	94.3	10	64)		102.5	
8b)	P43	94.3	30	49)	M27	102.5	1
8f)	P44	94.3	5	4a)	E35	103.5	5
8g)	P45	94.3	5	4f)	E40	103.5	10
16)	G29	94.3	35	4d)	E38	103.5	1
35)		96.7		4c)	E37	103.7	1
58)		96.9		4g)	E41	103.9	3
59)		96.9		32a)		103.9	
36)		97.1		1)	M22	104.1	3

No.	Call	Freq	Pwr	No.	Call	Freq	Pwr
37)		97.3		9)	C26	104.5	35
60)		97.3		9d)	C35	104.5	3
38)		97.5		8)	P29	104.9	10
39)		97.7		8c)	P31	104.9	3.5
61)		97.9		8d)	P32	104.9	25
40)	M23	98.1	3	8e)	P30	104.9	35
41)		98.3		9)	C39	105.3	3
38b)		98.3		50)		105.5	
62)		98.7		51)	M29	105.5	1
42)	M31	98.9	3	8g)	P29	106.5	1.5
4h)		99.1	4	9)	C24	106.5	35
43)		99.1		52)	M28	107.1	3
44)		99.3		9)	C34	107.3	35
45)	M24	99.5	3	29)		107.3	
38c)		99.5		9c)	C37	107.3	3
46)		99.7		53)	M25	107.7	3
63)		99.7		54)		107.7	
47)	M30	99.9	3	2a)	I44	107.8	30

NB: also some other Low power Community FM st, 69st's.

Address and other information:

1) Cheng Sheng Broadc. Co, 7th Flr, 66-1, Sec. 1, Chungching So. Rd, Taipei 100. 24h (Sat 2100-Sun 1800) on 1170kHz, 24h on 104.1MHz, - 1a) 17 Chuiyang Rd, Chia-i 600. 24h - 1b) 760, Sec. 2, Chunghsing Rd, Tali Village, Taichung 412. 1st prgr on 990kHz, 2nd prgr on 657kHz, both 24h(Sat 2100-Sun 1800) - 1c) 10 Shuiyuan Rd, Huwei Town, Yunlin 632. - 1d) 45 Chienchun Rd, Ilan 260. 24h(Sat 2200-Sun 1800). - 1e) 21, Lane 380, Hsinsheng Rd, Taitung 950. 2100(Sat 2125)-1700(Sun 1500). - 1f) 838 Chengching Rd, Niaosung Village, Kaohsiung 833. 24h(Sat 2155-Sun 1800). **Web:** http://www.csbc.com.tw

2) Fu Hsing Broadc. Co, 5, Lane 280, Sec. 5, Chungshan No. Rd, Taipei (operated by Dept. of the Interior).24h. **ANN:** "Fu Hsing Kuangpo Tientai, (location) Tai". -2a) 81 Chun she li chung tai Rd,Nantun,Taichung 408.(FM: 40-2 Shui ching Str, Shui yuan, Hsin she Village,Taichung 426).-2b) 819 Cheng ching Rd, Niaosung Village,Kaohsiung 833. -2c) 111 Peipin Str, Hualien,970.

3) Cheng Kung Broadc. St, 63 Chunghua 3rd Rd, Kaohsiung. 24h (Sat 2100-Sun 1400).

4) Chiao Yu (Educational) Broadc. System, 41 Nanhai Rd, Taipei 107.24h.
ANN: "Chiao Yu chih Sheng, Chiao Yu Kuangpo Tientai". - 4a) 5-1, Hukang Rd, Changhua 500. - 4b) 380 Kuangtung 3rd Rd, Kaohsiung 806. - 4c) 10-5 Linyuan Rd, Hualien 970. - 4d) 30, Lane 76, Shengli Str, Taitung 950. - 4e) Yuli (relay st.) - 4f) Ilan (relay st.)- 4g) Miaoli.(relay st.) -4h) Penghu.(relay st.). **Web:** http://www.ebs.gov.tw, Hualian **Web:** http://203.71.143.1

5) International Community Radio Taipei, 8 Chungyung 2nd Rd, Shantze Hou, Yangming Shan, Taipei 113. 24h in English (separate prgrs on MW and FM at times). N: on the h. - 5a) Kaohsiung - 5b) Taichung. **Web:** http://www.icrt.com.tw

6) Taiwan Broadc. Co, 4th Flr, 89, Shui yuan Rd, Chung cheng,Taipei 107. 1st prgr on 1323kHz, 2nd prgr on 1188kHz, both 24h(Sat2100-Sun1800). - 6a) 2, Lane 506, Kaofeng Rd, Hsinchu 300. 24h. - 6b) 42-1, Sec. 2, Nantun Rd, Taichung 408. 24h - 6c) 258-1, Fen tsao, Tsaotun Town, Nantou 542. 24h.

7) Chien Kuo Broadc. St, 78 Chienkuo Str, Peiliao Town, Hsinying. 2050(Sat 2150)-1700(Sun 1600).

8) Ching Cha Broadc. St (Public Radio System), 17 Kuangchou Str, Taipei 100. 24h on MW/FM. Taipei Traffic Prgr: 24h on 1512kHz + 94.3MHz. - 8a) 1-1, Chia hsing Rd, Chupei 302. - 8b) 99 Poai Str, Taichung 408. Traffic Prgr on 94.3MHz. - 8c) 49 Minchih Rd, Hsinying 730. - 8d) 246 Chungcheng 4th Rd, Kaohsiung 801. Traffic Prgr on 819kHz + 93.1MHz. - 8e) 110 Wusha Rd, Ilan 260. - 8f) 21-2 Fuchien Rd, Hualien 970. Traffic Prgr on 94.3MHz. - 8g) 191-3, Lane 719, Sec. 1, Chunghua Rd, Taitung 950. Traffic prgr on 94.3MHz. **Web:** http://www.prs.gov.tw

9) Han Sheng Broadc. St, 5 Flr, 3, Sec. 1, Hsin-i Rd, Taipei 100 (military broadc. st). MW, FM both 24h. - 9a) Kuo Kuang Broadc. St, 122, Sec. 1, Chungching So. Rd, Taipei 100. - 9b) 178 Chenhsing Rd, Taichung 401. - 9c) 248,Chung cheng 4th Rd, Kaohsiung 801. - 9d) 643, Chung cheng Rd, Hualien 970. - 9e) 40 Mingte New Village, Tsoying 813. - 9f) Chukuangying, Makong, Penghu 880. - 9g) Fuhsing Rd, Yongkang Village, Tainan 710.

10)Chin Sheng Broadc. St, 206, Kuanghua 1st Rd, Ling ya, Kaohsiung 802. 24h 11) Shih Hsin (World News) Broadc. St, 1, Lane 17, Sec. 1, Mushan Rd, Taipei 116. 2155-1400(Sun 0100-1300).

12) Shengli chih Sheng (Voice of Victory) Broadc. Co, 22, Sec. 1, Chienkang Rd, Tainan 700. 1st Prgr. on 837kHz, 2nd Prgr. on 1188kHz, both 2130(Sat 2220)-1825(Sun 1300). **ANN:** "Tainan Shengli chih Sheng Kuangpo Tientai".

13) Hsien Sheng Broadc. Co, 8, Lane 448, Chungshan Rd, Taoyuan 330. 24h (Sat 2300-Sun 1700).

14) Keelung Broadc. St, 12 Flr, 13 Chungsu Rd, Keelung. 2055-1810(Sun 1705).

15) Kuo Sheng Broadc. Co, 35 Wenchuan Rd, Pakuashan, Changhua 500. 2155(Sat 2200)-1800(Sun 1600). Erhlin relay st

16) Kaohsiung Broadc. St, 90 Hsinchiang Rd, Kaohsiung 804 (operated by Kaohsiung City Council). 2155-1600.

17) Min Pen Broadc. Co, 6 Flr, 325, Sec. 3, Huanho So. Rd, Taipei 109. 1st Prgr on 1296kHz, 2nd Prgr on 855kHz, both 24h.

18) Chung Sheng Broadc. Co, 134 Kuangfu Rd, Taichung 400. 2040-1800.

19) Feng Ming Broadc. Co, 492 Chiuju 2nd Rd, Kaohsiung 807. 1st Prgr on 1161kHz, 2nd Prgr on 981kHz, both 24h. - 19a) Chentieh Hsien, Li 38, Makung, Penghu.

20) Yen Sheng Broadc. St, 31, Sec. 1, Nanpin Rd, Tungchang, Chi'an Village, Hualien 873. 24h(sat2200-Sun1600).

21) Tien Sheng Broadc. St, 285 Kungyi Rd, Chunan 350. 2030(Sat 2200)-1645(Sun 1600). - 21a) 8 Kouya, Chungshue Rd, Yuanli. - 21b) 38-1, Tsotung, Wenhua Str, Tunglo.

22) Tien Nan Broadc. St, 31, Sec. 2, Hangchou So. Rd, Taipei 106. 24h.

23) Chung Hua (China) Broadc. Co, 15, Lane 1, Shuangyuan Str, Sanchung, Taipei 241. 1st Prgr on 1350kHz, 2nd prgr on 1026kHz, both 2150(Sat 2250)-1815(Sun 1600). - **ANN:** "Chung Hua Kuangpo Tientai Ti I/Erh Tai".

24) Min Li Broadc. St, 57-20 Minsheng Rd, Pingtung 900. 24h.

25) Tien Sheng Broadc. St, 15th Flr, 149, Sec. 1, Linsen Rd, Tainan 700. 2130(Sat 2220)-1825(Sun 1755).

26) Taipei Broadc. St, 62-2, sec. 3, Chungshan No. Rd, Taipei 104 (operated by Taipei City Council). 1134kHz , 93.1MHz. (0100- Non schedule).

27) Paiyun Broadc. Co, 7th Flr,1091,Yu chueng Rd, Kushan,Kaohsiung 804. 2100-1630. **Web:** http://www.sbs.org.tw/pai-yun/new.htm

28) Hua Sheng Broadc. Co, 18 Huansheng Str, Shihlin, Taipei 111. 1st Prgr on 1152kHz, 2nd Prgr on 1224kHz, both 24h.

29)Lan Yang FM Bradc,St, 12th Flr, 186 ,Sec3, Chung cheng Rd,. Wuchien Village,Ilan. 24h.

30) Yi Shih Braodc. St, 75 Paisan Str, Chitu, Keelung. 2130(Sat 2145)-1710(Sun 1610). **ANN:** "Keelung Yi Shih Kuangpo Tientai".

31) Taiwan Chu Yuyeh (Taiwan Area Fishery) Broadc. St, 5 Yukang No. 2nd Rd, Kaohsiung 806. 24h. Weather rpt. at every h. **ANN:** "Taiwan Chu Yuyeh Kuangpo Tientai". - 31a) Ilan(relay st.).

32) Fei tieh(UFO) Braodc. Co(UFO Netw), 25th Flr, 102, Sec2 ,Lossufou R d , Taipei.24h.**Web:**http://xsite.cynex.com.tw/xsite/ads/adpages/ufo/ufo1.htm,http://207.31.80.7/amebic/ -32a)UFO Netw Nan Taiwan Sheng 20th Flr, 12, Poai 3th Rd, Tso ying, Kaohsiung. **Web:**http://www.taconet.com.tw/ufo921/ UFO Netw:Taichung,Changhua,Nantou district 89.9MHz,Miaoli 91.3MHz,Yunlin Chia-I district 90.5MHz,Ilan 89.9MHz,Penghu 89.7MHz.

33) Ya chou(Asia) Braodc. St, 22th Flr, 2, 102, Chung ping Rd, Taoyuan 330. 24h. Taipei,Taoyuan district 92.7MHz,Hsinchu,Miaoli district 92.3MHz. **Web:**http://www.asiafm.com.tw

34) Sheng tu Braodc. Co, 233, Feng tsao,Tsaotun Town,Nantou 542. **Web:**http://www.fm937.com.tw

35) Huan yu Braodc. Co, 7th Flr,254,Kuang fou Rd,Hsinchu 300. 24h. **Web:**http://www.turc967.com.tw

36) Ilan chih sheng(Voice of Ilan) chung shan Braodc. Co(East Taiwan Super Netw), 12th Flr,289-3, Kung cheng Rd, Lo tung Town,Ilan 265. 24h.

37) Luse Heping Taiwan Wenhua (Greenpeace TaiwanCulture) Broadc.St, 25th Flr, 2, 97, Sec.4, Chunghsing Rd, Sanchung, Taipei 241. 24h.**Web:**http://www.greenpeace.com.tw

38) Kuaile (Happy) Broadc. St(Happy Netw), 8th Flr, 3, 63 Santo 4th Rd, Lingya, Kaohsiung 802. 24h(Sat 2200-Sun 1800). **Web:**http://www.happyradio.com.tw -38a) Chai-i(Chia le Bradc St) 16th Flr , 1, 193, Hsiao ya Rd, Chai-i. - 38b) Hualien(Huan le Bradc St) - 38c)Taitung(Lan yu Bradc St). Happy Netw: Taichung 89.5MHz,Taipei 89.3MHz, Penghu 91.3MHz.

39) Hao chia ting(Family) Braodc. Co, 37th Flr, 789, Chung ming So. Rd So., Taichung 402. 24h. 40) Taiwan Chuan min Broadc.St, Room No.607, 6th Flr, 88, Sec.2, Chung hsiao tung Rd, Taipei. 24h.

41) Kangtu Broadc.St(Kangtu Netw), 43th Flr , 1, 80 Mintsu 1st Rd, Kaohsiung 807. 24h.Kangtu Netw:Pingtung 89.3MHz.

42)Yes989 Jen-jen FM Broadc.St,20th Flr, 173 Chengkung Rd, Sanchung, Taipei 241. 2100-1800. **Web:**http://www.yes989.com.tw

43) Ta chien Broadc.St, 9th Flr,83, Hsueh shih Rd, No., Taichung 404. **Web:**http://www.superfm99-1.com.tw

44) Hsin sheng FM Broadc.St, 20th Flr, 1, 295, Sec 2, Kuang fou Rd, Hsinchu 300.

45) Shennung (Farmer Radio) Broadc.Co, 10th Flr, 234 Peiping Rd, Huwei, Yunlin 632. 24h.
46) Taipei Ai le Broadc.Co, 7th Flr,111-1, Tung hsing Str, Hsin i ,Taipei 110. 24h. **Web:**http://www.prtmusic.com.tw
47) Tachung Broadc.Co(Tachung Netw Kiss-FM), 34th Flr, 2, 6 Minchuan 2nd Rd, Kaohsiung 806. 24h. **Web:**http://www.kiss.com.tw Tachung Netw: Miaoli 98.3MHz, Nantou 99.7MHz, Tainan, Chai-i district 97.1MHz.
48) Taichung Broadc.Co. FM St, 21Flr, 345 Chungkang Rd, Taichung. 24h. **Web:**http://www.lucky7.com.tw
49) Kutu Broadc.Co, 15Flr, 1, 77 Sec.2, Chunghuatung Rd, Tainan. 2155-1800. **Web:**http://www.fm1025.com.tw
50)Tung shan he FM Broadc.St, 13th Flr, 452-1, Chun ching Rd, Lotung Village, Ilan.
51) Chuankuo Broadc.Co. Taichung FM St, 8Flr, 1-18, Chungkang Rd, Taichung 407. 24h. **Web:**http://www.taichungnet.com.tw
52) Taoyuan Broadc.St, 9Flr, 859, Sec.1, Chunghua, Chungli, Taoyuan 320. 24h.
53) Hualien FM Broadc.St(East Taiwan Super Netw), 55 Chunghsing Rd, Hualien 970. 2200-1600.
54) Taipei chih Yin (Sound of Taipei) Broadc.Co. 10th Flr, B, 15-1, Sec. 1, Hanchou So. Rd, Taipei. 24h. **Web:**http://www.vot.com.tw Music St(Hit fm):Taipei 91.7MHz,Taichung 91.5MHz, Kaohsiung 90.1MHz.
55)Yun chia Bradc. St, 9th Flr, 443, Jenai Rd, Chai-i. 24h. **Web:**http://www.chiayis.com.tw/933/fm933.htm
56)Hsin kechia Bradc.St, 16th Flr,411, Huannan Rd,Ping chen City,Taoyuan. **Web:**http://www.showtower.com.tw/~limkl/radio.html
57)Pao tao kechia Bradc.St, 17th Flr, 91, Sec2, Lossufu Rd,Taipei. Taipei&Ilan 93.7MHz, Taoyuan 98.7MHz,Hsinchu&Miaoli 102.5MHz. **Web:**http://www.taiwanese.com/org/hakradio
58)Tien tien Bradc.St, 42th Flr, 760, Chung ming South Rd, nan, Taichung.24h. **Web:**http://www.sky969.com.tw
59)Chu jen(boss)Bradc.St, 17th Flr, 155, Foujen Rd, Ling ya, Kaohsiung 802. 24h.
60)Ai yu Bradc.St, 824, Sec 3, Chin ma Rd, Changhua. 24h.
61)Touch Bradc.St(Touch netw), 21th Flr, 2, 425, Chunghua Rd, Yungkang City,Tainan. Touch Netw:Pingtung 92.5MHz.
62)Changhua FM Bradc.St, 8-2 Hukang Rd,Changhua. 24h. **Web:**http://www.fm987.com.tw
63)Chin ma chih sheng Bradc.St, 3rd Flr, 28 Huantao No.Rd, Chincheng Town, Kinmen. 2100-1430.
64)Pei tai chih sheng Bradc.St(East Taiwan Super Netw), P.O.Box500, Keelung. 24h. **Web:**http://w7.dj.net.tw/~fm102.5

CHRISTMAS ISLAND (Australian)

L.T: UTC + 7h — **Pop:** 2.500 — **Pr.L:** English, Malay, Cantonese, Mandarin — **E.C:** 50Hz, 240V — **ITU:** CHR.

CHRISTMAS ISLAND COMMUNITY RADIO SERVICE
✉ Admin of the Territory of Christmas Island, Indian Ocean 6798, Australia. ☎ +6724 (4) 8316. ▤ +6724 (4) 8315. **Technical Services:** ☎ +672 (4) 8499. ▤ +672 (4) 8496 — **MW:** VLU2 1422kHz 0.5kW. **FM:** VLU2-FM 102.1MHz 0.015kW.

D.PRGR: continuous music & local prgrs. The tr. is on the air continuously for shipping, aircraft and weather information.
ANN: "The Power Station" — **V.** by letter.

COCOS (Keeling) ISLANDS (Australian)

L.T: UTC + 6½h — **Pop:** 600 — **Radios:** 200 — **Pr.L:** English, Cocos Malay — **E.C:** 50Hz, 220V — **ITU:** ICO.

RADIO VKW (Gov.)
✉ P.O. Box 70, Cocos (Keeling) Islands, Indian Ocean 6799, Australia — **L.P:** SM: S. O'Neill. TD: K. Beard. PD: Ms. A.Parker.

MW: 1404kHz 0.1kW — **D.PRGR:** 24h. Rel. BBCWS 1300-1730. Rel. R.Australia 1730-2400 — **V.** by QSL-card.

HONG KONG (China)

L.T: UTC +8h — **Pop:** 6,310,000 — **Radio sets:** 3,700,000 — **Pr.L:** Cantonese, English — **E.C:** a/c 50 200, 220v — **ITU:** HKG

RADIO TELEVISION HONG KONG (Gov.)
✉ Broadcasting House, 30 Broadcast Drive, P.O.Box 70200, Kowloon Central Post Office, Hong Kong. ☎ +852 2339 6300. **Cable:** "Broadcasts". ⟳ 45568. ▤ +852 2338 0279. **E-mail:** www@rthk.org.hk, radio1@rthk.org.hk, radio2@rthk.org.hk, radio3@rthk.org.hk, radio4@rthk.org.hk, radio5@rthk.org.hk, putonghua@rthk.org.hk. Web: http://www.rthk.org.hk
L.P: Dir. of Broadc: Miss Man-yee Cheung, ISO, JP. Dep. Dir. of Broadc. Chu Pui-hing. Asst. Dir. of Broadc. (Radio): Shiu-Lo-sin. Asst. Dir. of Broadc. (PATV): NG Sek-fai. Depertmental Secretary: B. S. Pannu.
Mediumwaves: P. ch=Putonghua Channel

kHz	Netw.	Location	kW
567	Radio 3	Golden Hill	20
621	P. Ch	Golden Hill	20
675	Radio 6	Peng Chau	10
783	Radio 5	Golden Hill	20
1584	Radio 3	Chung Hom Kok	0.1

FM

MHz	Netw.	kW	Tr. Location	Target Div.
92.6	Radio 1	1	Mt.Gough	Kowloon
92.9	Radio 1	0.1	Golden Hill	Tsuen Wan
93.2	Radio 1	0.5	Cloudy Hill	Fan Ling
93.4	Radio 1	0.5	Castle Peak	Tuen Mun
93.5	Radio 1	0.1	Beacon Hill	Shatin
93.6	Radio 1	0.3	Lamma Island	HK Island south
94.4	Radio 1	0.5	Kowloon Peak	HK Island north,Sai Kung
94.8	Radio 2	1	Mt.Gough	Kowloon
95.3	Radio 2	0.5	Cloudy Hill	Fan Ling
95.6	Radio 2	0.1	Golden Hill	Tsuen Wan
96.0	Radio 2	0.3	Lamma Island	HK Island south
96.3	Radio 2	0.1	Beacon Hill	Shatin
96.4	Radio 2	0.5	Castle Peak	Tuen Mun
96.9	Radio 2	0.5	Kowloon Peak	HK Island north,Sai Kung
97.6	Radio 4	1	Mt.Gough	Kowloon
97.8	Radio 4	0.5	Cloudy Hill	Fan Ling
97.9	Radio 3		Mt.Nicholson	Jardine's Lookout
98.1	Radio 4	0.1	Beacon Hill	Shatin
98.2	Radio 4	0.3	Lamma Island	HK Island south
98.4	Radio 4	0.1	Golden Hill	Tsuen Wan
98.7	Radio 4	0.5	Castle Peak	Tuen Mun
98.9	Radio 4	0.5	Kowloon Peak	HK Island north,Sai Kung
100.9	P. Ch		Happy Valley	Happy Valley
106.8	Radio 5		Tuen Mun	Tuen Mun
106.8	Radio 3		Chung Hom Kok	HK Island south

RTHK Radio 1 in Cantonese/Chinese: 24h. N. half-hourly 2230-1430. Rel. Radio 2 1830-2200.
RTHK Radio 2 in Cantonese: 24h. **N.** half-hourly 2230-1430, hourly 1500-2200.
RTHK Radio 3 in English: 24h AM-stereo **N.** hourly.
RTHK Radio 4 in English/Cantonese: 24h **N.** hourly. Rel. Radio 3 1600-2200.
RTHK Radio 5 in Cantonese/Chinese: 24h AM-stereo. Rel. Radio 1 1400-1530 and 2300-0200, and Radio 2 1530-2300.
RTHK Radio 6 Relay BBCWS English: 24h.
RTHK Putonghua Channel in Chinese: 24h **N.** hourly. Rel. Radio 5 0800-1000, Radio 1 1800-1830 and Radio 2 1830-0200.
V. by QSL-card

HONG KONG COMMERCIAL BROADC. CO. LTD
✉ 3 Broadcast Drive, KCPO Box 73000, Hong Kong.
☎ +852 2336 5111. ▤ +852 2338 0021. **Cable:** Radioads.
L.P: Chmn G. Ho. Man.Dir. Winnie Yu.
Mediumwave

kHz	Location	Prgr.	kW
864	Peng Chau	Quote AM	10

FM

MHz	Netw.	kW	Tr. Location	Target Div.
88.1	CR1	1	Mt.Gough	Kowloon
88.3	CR1	0.5	Cloudy Hill	Fan Ling
88.6	CR1	0.5	Castle Peak	Tuen Mun
88.9	CR1	0.1	Golden Hill	Tsuen Wan
89.1	CR1	0.3	Lamma Island	HK Island south
89.2	CR1	0.1	Beacon Hill	Shatin
89.5	CR1	0.5	Kowloon Peak	HK Island north, Sai Kung
90.3	CR2	1	Mt.Gough	Kowloon
90.7	CR2	0.5	Cloudy Hill	Fan Ling
90.9	CR2	0.1	Golden Hill	Tsuen Wan
91.1	CR2	0.1	Beacon Hill	Shatin
91.2	CR2	0.5	Castle Peak	Tuen Mun

91.6   CR2   0.3   Lamma Island   HK Island south
92.1   CR2   0.5   Kowloon Peak   HK Island north, Sai Kung
HKCR CR1 in Cantonese. 24h N: half-hourly. Rel. Ch2 1800-2200, 0200-0400(Sun).
HKCR CR2 in Cantonese. 24h N: hourly.
HKCR Quote AM in English. 24h all music.
V. by letter

METRO BROADCAST CORPORATION LTD.
Site 11, Basement 1 Whampoa Gardens Hunghom, Kowloon, Hong Kong. ☎ +852 2364 9333 ▤ +852 2364 6577. **E-mail:** tech@metroradio.com.hk. **Web:** http://www.metroradio.com.hk

Mediumwave

kHz	Netw.	kW	Location
1044	Metro Plus	10	Peng Chau

FM

MHz	Netw.	kW	Tr. Location	Target Div.
99.7	Hit Radio	1	Mt.Gough	Kowloon
100.0	Hit Radio	0.5	Cloudy Hill	Fan Ling
100.4	Hit Radio	0.5	Castle Peak	Tuen Mun
100.5	Hit Radio	0.1	Beacon Hill	Shatin
101.6	Hit Radio	0.1	Golden Hill	Tsuen Wan
101.8	Hit Radio	0.5	Kowloon Peak	HK Island north, Sai Kung
102.1	Hit Radio	0.3	Lamma Island	HK Island south
102.4	FM Select	0.1	Beacon Hill	Shatin
102.5	FM Select	0.5	Castle Peak	Tuen Mun
104.0	FM Select	1	Mt.Gough	Kowloon
104.5	FM Select	0.3	Lamma Island	HK Island south
104.7	FM Select	0.5	Cloudy Hill	Fan ling
105.5	FM Select	0.1	Golden Hill	Tsuen Wan
106.3	FM Select	0.5	Kowloon Peak	HK Island north, Sai Kung

Metro Plus in English/Mandarin(Standard Chinese) 24h. AM-Stereo Music, news and Information. **Rel. China R. Int. :** 2200-2300(Sun-Thu) in English, 2300-2400(Sun-Thu) in Chinese, 0500-0600(Mon-Fri) in Cantonese.
Metro Hit Radio in Cantonese 24h. Music, conversation and information.
Metro FM Select in English/Cantonese 24h. Easy listening and many categories music with news hourly. Rel. Hit Radio 1800-2200.

INDIA

L.T: UTC + 5½h — **Pop:** 916.000.000 — **Radios:** 111.000.000 — **Pr.L:** Assamia, Bangla, Dogri, English, Gujarati, Hindi, Kannada, Kashmiri, Marathi, Malayalam, Nepali, Oriya, Punjabi, Sanskrit, Sindhi, Tamil, Telugu & Urdu — **E.C:** 50Hz (+ some d/c), 230/400V — **ITU:** IND

MINISTRY OF INFORMATION & BROADCASTING
Main Secretariat, A-Wing, Shastri Bhawan, New Delhi-110 001. ➀ 81-31-66349 mib in. **Cable:** Information.
L.P: Minister for Inf.& Broadc: K.P.Singh Deo. Inf.& Broadc.Secr: R.Bhargava.

AKASHVANI (Gov, semi-comm.)
Administration: Directorate General of All India Radio, Akashvani Bhawan, 1 Sansad Marg, New Delhi-110 001. ☎ +91 (11) 3710006 (pabx). ➀ 81-31-65585avdg in. **E-mail:** air@kode.net
Web: http://www.allindiaradio.org. **Audio:** http://air.kode.net
L.P: DG: S. Kapoor. Dep. DG: M.D. Gaikwad. Eng-in-Chief: H.M. Joshi. Dir. of Freq. Planning: A K Bhatnagar (**E-mail:** faair@giasdl01.vsnl.net.in). Asst. Dir. (Freq. Assignments): S.A.S. Adibi (in charge of processing reception reports).
Programming: Broadcasting House, 1 Sansad Marg, New Delhi-110 001. ☎ +91 (11) 3715411 (pabx). **Cable:** Akashvani — DG (News): B. Bhalla. Dir. Transcription & Prgr. Exchange Sces: G. Raghuram.
Commercial Sce: Vividh Bharati Sce, AIR, 101 M.K.Rd, Mumbai-400 020. ☎ +91 (22) 219415 or 291594. ➀ E1-11-85414 cbvu in. **Cable:** Vividh Bharati — Dep.Dir: V. Chhabra.
Audience Research: Audience Research Unit, AIR, 2nd Floor, PTI Building, Sansad Marg, New Delhi-110 001. ☎ +91 (11) 3710033. **Cable:** Listener — Dir: S.K. Khatri.
National Channel: AIR, Gate 22, Jawaharlal Nehru Stadium, Lodhi Road, New Delhi-110 003. ➀ 81-31-65560 nchjln in — Dir: G.Gururaj. St.Eng: S.K. Sharma.
Research & Development: Office of the Chief Eng R & D, All India Radio, 14-B Ring Road, Indraprasthta Estate, New Delhi-110 002. ☎ +91 (11) 3311762. ➀ 81-31-65478 cerd in. **Cable:** Airsearch — Chief Eng: A.S. Guin.
Monitoring: International Monitoring St., All India Radio, Dr. K.S.

Krishnan Rd, Todapur, New Delhi-110 097. ☎ +91 (11) 581461 — Asst. Research Eng (Frequency Planning): D.P. Chhabrah.
Central Monitoring St., AIR, Ayanagar, New Delhi-110 047. ☎ +91 (11) 666306. **Cable:** Airmonitor — Dir: S. Haider.

Regional Headquarters:
North Zone: Jamnagar House, Shahjahan Road, New Delhi-110 011 — **East Zone:** 4th Floor, Akashvani Bhawan, Eden Gardens, Calcutta-700 001 — **North-East Zone:** AIR Chandmari, Gowahati-781 003, Assam — **West Zone:** AIR, Old CGO Building, 101 M.K.Road, (P.O. Box 11452), Mumbai-400 020, Maharashtra — **South Zone:** AIR, Silingi Bldg, 38/39 Greams Road, Thousand Lights, Chennai-600 006.

MW	kHz	kW	Zone	MW	kHz	kW	Zone
Jodhpur A	531	100	N	Jammu B	1089	1	N
Aizawl	540	20	NE	Udipi	1089	20	S
Ranchi	549	100	A	Gulbarga	1107	10	S
Mumbai B	558	100	N	Srinagar A	1116	200	N
Dibrugarh	567	300	NE	Udaipur	1125	10	N
Alappuzha	576	100	S	Mogra	1134e	1000	E
Nagpur A	585	100	W	Ratnagiri	1143	20	W
Chinsurah	594e	1000	E	Rohtak	1143	20	N
Ajmer	603	200	N	Ranchi B	1152c	1	E
Bangalore A	612	200	S	Thiru'puram A	1161	10	S
Patna A	621	100	E	Hyderabad C	1170c	1	S
Trichur	630	100	S	Rewa	1179	20	N
Kohima	639	50	NE	Mumbai C	1188c	50	W
Indore A	648	100	N	Tirunelveli	1197	10	S
Calcutta A	657	100	E	Jodhpur B	1197c	1	N
Delhi B	666	100	N	Shillong	1197	1	NE
Chhattarpur	675	20	N	Bhawani-			
Itanagar	675	100	NE	patnam	1206	200	E
Bangalore B	675c	1	S	Puducheri	1215	1	S
Kozhikode A	684	10	S	Delhi	1215	20n	N
Port Blair	684	20	S	Calcutta D	1224	10	E
Jalandhar B	702e	300	N	Srinagar D	1224c	1	N
Siliguri	711	200	E	Bhopal B	1233	1	N
Chennai A	720	200	S	Tura	1233	20	NE
Gowahati A	729	100	NE	Varanasi	1242	100	N
Hyderabad A	738	50	S	Sangli	1251	20	W
Lucknow A	747	300	N	Ambikapur	1260	20	N
Jagdalpur	756	20	N	Madurai	1269	10	S
Dharwar	765	200	S	Agartala	1269	20	NE
Shimla	774	100	N	Jaipur B	1269c	1	N
Chennai C	783c	20	S	Lucknow C	1278c	10	N
Pune A	792	100	W	Panaji A	1287	100	W
Jabalpur	801	200	N	Darbhanga	1296	10	E
Rajkot A	810	300	W	Parbhani	1305	10	W
Delhi A	819	200	N	Bhuj	1314	10	W
Silchar	828	10	NE	Cuttack B	1314c	1	E
Vijayawada A	837	100	S	Calcutta C	1323c	20	E
Ahmedabad A	846	200	W	Tezu	1332	10	NE
Shillong	864	100	NE	Mangalore	1332	1	S
Jalandhar A	873	300	N	Kohima	1341	1	NE
Imphal	882	50	NE	Dharwar B	1350c	1	S
Rampur	891	10	N	Jalandhar C	1350c	1	N
Tiru'palli B	891c	1	S	Bhadravathi	1359	20	S
Cuddapah	900	100	S	Delhi C	1368c	10	N
Gorakhpur	909	100	N	Hyderabad B	1377	10	S
Suratgarh	918	300	N	Gwalior	1386	10	N
Visakapatnam	927	100	S	Chennai B	1395	10	S
Tiru'palli A	936	100	S	Bikaner	1395	20	N
Sambalpur	945	20	N	Gangtok	1404	20	E
Najibabad	954	100	N	Rajkot B	1422c	1	W
Jalgaon	963	20	W	Chandigarh	1431c	1	N
Cuttack A	972	100	E	Kozhikode B	1431c	1	S
Raipur	981	100	N	Port Blair	1440	1	S
Jammu A	990	300	N	Ahmedabad B	1440c	1	W
Almora	999	1	N	Kurseong	1440	1	NE
Coimbatore	999	10	S	Kanpur	1449c	1	N
Calcutta B	1008	100	E	Barmer	1458	20	N
Mysore	1017	1	S	Bhagalpur	1458	10	E
Delhi D	1017	20	N	Jeypore	1467	100	N
Allahabad	1026	20	N	Jaipur A	1476	1	N
Gowahati B	1035	10	NE	Adilabad	1485	1	S
Mumbai A	1044	100	W	Allahabad B	1485	1	N
Tuticorin	1053e	200	S	Vadodara	1485c	1	W
Leh	1053	10	N	Bhopal	1485	1	NE
Pasighat	1062	10	NE	Thiru'puram B	1494c	1	S
Rajkot	1071e	1000	W	Srinagar C	1503	1	N

MW	kHz	kW	Zone	MW	kHz	kW	Zone
Vijayawada B	1503c	1	S	Mathura	1584	1	
Tawang	1521	10	NE	Bhopal A	1593	10	N
Aurangabad	1521	1	W	Nagercoil	1602	1	S
Agra	1530	10	N	Nagpur B	1602c	1	W
Panaji B	1539c	20	W	Sholapur	1602	1	W
Nagpur	1566n	1000	W	Patna B	1602c	1	E
Jamshedpur	1584	1	E	Pune B	1602c	1	W
Keonjhar	1584	1	E	Varanasi	1602	1	NE
Kota	1584	1	N	Ottacumand	1602	1	S
Kavarathi	1584	1	S	Uttarkashi	1602	1	N
Indore B	1584c	1	N				

c) carries Vividh Bharati, e) carries Ext.Sce. n) carries National Channel.

NB: Ajmer 603kHz relays Jaipur A 1476kHz, Alappuzha 576kHz relays Thiruvananthapuram A 1161kHz. Udipi 1089 relays Mangalore 1332.
D.PRGR: 0000-1841 (with breaks, varies from st. to st.). May be extended for sports coverage and special events.
National Channel: 1325-0040 in Hindi and English.

FM Stations

MHz	kW	Location	Reg.	Times
100.1	6	Kothagudem	S	1230-1730
100.2	3	Haflong	NE	1130-1600
100.2	6	Patiala	N	1130-1530
100.2	6	Shivpuri	N	1130-1600
100.2	6	Poonch	N	
100.3	3	Jammu	N	1130-1600
100.4	6	Bareilly	N	
100.4	6	Keraikal	S	
100.5	6	Dhule	W	1130-1600
100.5	10	Hospet	S	1130-1600
100.6	6	Behrampur	E	1130-1600
100.6	6	Nagpur	W	test
100.7	6	Churu	N	1130-1600
100.7	6	Raigarh	N	1130-1600
101.0c	6	Pune	W	0100-0430, 0630-1730
101.1	6	Ahmednagar	W	1130-1600
101.1	6	Bathinda	N	1130-1530
101.1	6	Nanded	W	1130-1600
101.1	6	Surat	W	1130-1600
101.2	6	Khandwa	N	1130-1600
101.3	6	Balaghat	N	1130-1600
101.5	6	Sawai Madhopur	N	1130-1600
101.3	6	Banswara	N	1130-1600
101.3	3	Bolangir	E	1130-1600
101.3	6	Cuttak	E	
101.4	6	Idukki (Devikulam)	S	1130-1600
101.4	6	Kurukshetra	N	1130-1600
101.4	6	Nasik	W	
101.5	6	Kannur	S	1230-1600
101.5	6	Merkapuram	S	1230-1600
101.6	3	Indore	N	1130-1600
101.7	6	Anantapur	S	1130-1600
101.7	6	Chaibassa	E	1130-1600
101.8	6	Hamirpur (HP)	N	1130-1600
101.8	10	Jaisalmer	N	1130-1600
101.9	6	Faizabad	N	1130-1600
101.9	6	Lunglea	E	
102.0	6	Shahdol	N	1130-1600
102.1	6	Hamirpur (UP)	N	1130-1600
102.1	3	Hazaribagh	E	1130-1600
102.1	6	Raichur	S	1130-1600
102.2	6	Chindwara	N	1130-1600
102.2	6	Godhra	W	1130-1600
102.2	6	Hassan	S	1130-1600
102.2	6	Murshidabad	E	1200-1500
102.3	3	Guna	W	1130-1600
102.3	6	Karwar	S	1130-1600
102.3	6	Kochi	S	1230-1600
102.3	6	Purnia	E	1130-1600
102.3	3	Daman	S	
102.4	6	Akola	W	1130-1600
102.4	6	Kurnool	S	1130-1600
102.5	3	Patna	E	1130-1600
102.6	10	Delhi	N	0125-0430, 0736-1730#
102.6	6	Chitradurga	S	1230-1600
102.6	5	Sagar	W	1130-1600
102.6	6	Rourkela	E	
102.7	10	Jalandhar	N	1130-1600

MHz	kW	Location	Reg.	Times
102.7	6	Kolhapur	W	1130-1600
102.7	6	Nowgong	NE	
102.7	6	Obra	N	1130-1600
102.7	6	Yavatmal	W	1130-1600
102.8c	6	Hyderabad	S	0100-0430, 0630-1730
102.9	6	Baripada	E	1130-1600
102.9	6	Beed	W	1130-1600
102.9	6	Chittorgarh	N	1130-1600
103.0	6	Chanderpur	W	1130-1600
103.0	10	Daltonganj	E	1130-1600
103.0	6	Jhansi	N	1130-1600
103.1	6	Alwar	N	1130-1600
103.1	6	Betul	W	1130-1600
103.1	6	Satara	W	1130-1600
103.1	6	Mercara	S	1130-1600
103.2	6	Bilaspur	N	1130-1600
103.2	6	Jhalawar	N	113-1600
103.2	3	Kailasahar	NE	1130-1600
103.2	6	Nizamabad	S	1230-1600
103.2	3	Tirupathi	S	1230-1600
103.4	10	Dharamsala	N	1130-1600
103.4	10	Jorhat	NE	1130-1600
103.4	6	Sasaram	E	1130-1600
103.5	3	Bhopal	N	1130-1600
103.5	10	Warrangal	S	1130-1600
103.7	6	Nagaur	N	1130-1600
103.7	3	Belonia	NE	1130-1600
105.4	6	Panaji	W	1130-1600
106.4	6	Kathua	N	1230-1600
107.1s	10	Mumbai	W	0125-0430, 0736-1730
107.1	10	Calcutta	E	1230-1730
107.1s	10	Chennai	S	0125-0430, 0736-1730
107.2	10	Kasauli	N	1130-1600

c) carries Vividh Bharati sce. — s) stereo.

Regional Domestic SW st's

Station	kHz	kW	Times
Shimla	3223	50	0045-0200, 1300-1710
Lucknow	3245	50	1415-1740
Ranchi	3305	10	0045-0200, 1130-1740
Bhopal	3315	50	0025-0215, 1200-1740
Jaipur	3345	50	0025-0215, 1430-1740
Kurseong	3355	50	1130-1740
Delhi	3365	20	1230-1840
Gangtok	3390	50	0025-0330, 1200-1730
Leh	4760	50	0125-0430, 1130-1600
Port Blair	4760	20	0000-0300, 1030-1630
Imphal	4775	50	0025-0215, 1030-1730
Hyderabad	4800	50	0025-0215, 1145-1745
Calcutta	4820	50	0025-0215, 1230-1830
Mumbai	4840	50	0015-0400, 1230-1730
Kohima	4850	50	0000-0415, 0930-1630
Delhi	4860	50	0025-0400
Lucknow	4880	50	0025-0400, 1130-1415
Kurseong	4895	20	0100-0345, 1130-1740
Jaipur	4910	50	0230-0415, 1130-1415
Chennai	4920	50	0015-0245, 1200-1740
Shimla	4930	50	0045-0200, 1300-1741
Guwahati	4940	50	0015-0400, 1200-1700
Srinagar	4950	50	0025-0200, 1130-1745
Shillong	4970	50	0000-0400, 1045-1700
Itanagar	4990	50	0025-0400, 1030-1630
Thiru'puram	5010	20	0025-0215, 1145-1740
Aizawl	5050	10	0030-0400, 1200-1630
Leh	6000	50	0530-1100
Shimla	6020	50	0215-0400, 0700-1000, 1130-1230
Kohima	6065	50	0630-0915
Gangtok	6085	50	0700-0930
Srinagar	6110	50	0215-0430, 0630-1115
Ranchi	6140	10	0215-1000
Itanagar	6150	50	1030-1630
Delhi	6190	50	0730-1100
Shillong	6195	50	1330-1600
Lucknow	7105	50	0630-1000
Delhi	7110		0215-0225, 0228-0355/0400
Port Blair	7115	20	0315-0345, 0700-0850
Jaipur	7120	50	0700-1000
Shillong	7130	50	0645-0930
Hyderabad	7140	50	0228-0430, 0610-0915

Station	kHz	kW	Times
Imphal	7150	50	0230-0445, 0700-0945
Chennai	7160	50	0300-0415, 0605-1145
Bhopal	7180	50	0228-0405, 0700-0945
Calcutta	7210	50	0230-0400, 0730-1015
Kurseong	7230	50	0630-1000
Mumbai	7240	50	0530-1045
Guwahati	7280	50	0630-0940, 0945-1215
Thiru'puram	7290	20	0230-0400, 0630-0930
Aizawl	7295	10	0700-0930
Delhi	9565	50	1330-1420, 1430-1440, 1445-1615/1630/1700/1730v, 1730-1740
IDelhi	9685	50	1330-1445, 1450-1630/1700/1730v. 1730-1740
Delhi	10330	50	0025-0430, 0700 (Sun & Hol 0630)-1200, 1245-1730
Khampur	11620	250	1130-1140
Aligarh	11710	250	1115-1140
Delhi	11830	100	0030-0040, 0125-0225, 0228-0355/0400, 1115-1140
Delhi	15135	50	0125-0205, 0228-0355/0400
Delhi	15185	50	0730-0740, 0800-0850, 1115-1140
Delhi	15260	50	0700-0810, 0820-0850
Aligarh	17865	250	1220-1245

V. by QSL-card. Re.to Asst. Dir. Eng.(Freq. Assignments) at New Delhi addr.

Addresses of Regional Stations

Adilabad-504 001, Andhra Pradesh – Palace Compound, North Gate, **Agartala**-799 001, Tripura – Vivabhnagar, **Agra**-282 001, Uttar Pradesh – P.O.Box 4040, Navarangpura, Ashram Rd, **Ahmedabad**-380 009, Gujarat – **Ahmednagar**- 414 001, Maharashtra – Radio Tila, Tuikhuahtlang, **Aizawl**-796 001, Mizoram – 21/10 Vaishali Nagar, (P.O.Box 135), **Ajmer**-305 001, Rajasthan – **Akola**-444 001, Maharashtra – P.O.Box 4623, Shree Ram Temple Trust Bldg, near Cullen Bridge, **Alappuzha**-688 012, Kerala – Z-9 Dayanand Marg, (Thernhill Road), **Allahabad**-211 001, Uttar Pradesh – **Almora**-263 601, Kumaon District, Uttar Pradesh – **Alwar**-301 001, Rajasthan – Kumar Palace, **Ambikapur**-497 001, Surguja District, Madhya Pradesh – **Ananatapur**-515 001,Andhra Pradesh – Jalna Rd, **Aurangabad**-431 005, Maharashtra – **Balaghat**-481 001, Madhya Pradesh – P.O.Box 5096, Raj Bhavan Rd, (Cubbon Road), **Bangalore**-560 001, Karnataka – **Banswara**-327 001. Rajasthan – **Baripada**-757 001, Mayurbhanj District, Orissa – **Barmer**-344 001, Rajasthan – Bareilly-243 001, Uttar Pradesh, **Beed**-431 122, Maharashtra – Behrampur-760 001, Orissa – **Belonia**-799155, Tripura – **Betul**-460 001, Madhya Pradesh – Post Bag 108, J.P.S.Colony, Paper Town, **Bhadravati**-577 302, Karnataka – Port Campus, **Bhagalpur**-812 001, Bihar – **Bhawanipatna**-766 001, Orissa – **Bathinda**-151 001, Punjab – Akashvani Bhawan, Shamla Hills, **Bhopal**-462 002, Madhya Pradesh – **Bhuj**-370 001, Kachchh District, Gujarat – **Bikaner**-334 001, Rajasthan – Nutan Colony, **Bilaspur**-495 001, Madhya Pradesh – **Bolangir**-767 001, Orissa – Broadcasting House, P.O.Box 13034, Backbay Reclamation, **Mumbai**-400 020, Maharashtra – Akashvani Bhavan, Eden Gardens, (G.P.O. Box 696), **Calcutta**-700 001, West Bengal – Tungri Maidan, Chaibasa-833 201,Singhbhum District, Bihar – **Chanderpur**-442 401, Maharashtra – Sector-19B, **Chandigarh**-160 019, Union Territory – Kamrajar Salai, Mylapore, **Chennai**-600 004, Tamil Nadu – **Chhatarpur**-471 001, Madhya Pradesh – **Chindwara**-480 001, Madhya Pradesh – **Chitradurga**-577 501, Karnataka – **Churu**-331 001, Rajasthan – Trichy Rd, Ramanathapuram, (P.O. Box 7101), **Comibatore**-641 045, Tamil Nadu – Cooperative Colony, (P.O. Box 42), **Cuddapah**-516 001, Andhra Pradesh – Madhupur House, Bakshi Bazar, Cantonement Rd, **Cuttack**-753 001, Orissa – **Daltonganj**-822 101, Bihar – **Darbhanga**-846 004, Bihar – **Devikulam**-685 613, Kerala – **Dharamsala**-176 215, Himachal Pradesh – **Dhule**-424 001, Maharashtra – Saptapur, **Dharwad**-580 008, Karnataka – **Dibrugarh**-786 001, Assam – **Faizabad**-224 001, Uttar Pradesh – Old MLA hostel, **Gangtok**-737 101, Sikkim – **Godhra**-389 001, Gujarat – Post Bag 26, Town Hall, **Gorakhpur**-273 001, Uttar Pradesh – Aiwan-e-Shahi, Municipal Garden, **Gulbarga**-585 103, Karnataka – **Guna**-473 001, Madhya Pradesh – P.O.Box 28, Chandmari, **Gowahati**-781 003, Assam – Gandhi Rd, **Gwalior**-474 002, Madhya Pradesh – **Haflong**-788 819, Assam – **Hamirpur**-210 301, Uttar Pradesh – **Hamirpur**-177 001, Himachal Pradesh – **Hassan**-573 201, Karnataka – **Hazaribagh**-825 301, Bihar – **Hospet**-583 201, Karnataka – Rocklands, Saifabad, **Hyderabad**-500 004, Andhra Pradesh – **Idukki**-685 602, Kerala – Palau Rd, **Imphal**-795 001, Manipur – Malwa House, Residency Area,

Indore-452 001, Madhya Pradesh – Naharlagun, **Itanagar**-791 110, Arunachal Pradesh – 373 Napier Town, **Jabalpur**-482 001,Madhya Pradesh – Collectorate Rd, **Jagdalpur**-494 001 – 5 Park House, Mirza Ismail Rd, **Jaipur**-302 001, Rajasthan – **Jaisalmer**-345 001, Rajasthan – **Jalandhar**-144 001, Punjab – P.O. Box 52, Jilha Peth, **Jalgaon**-425 001, Maharashtra – Radio Kashmir, Begum Haveli, Old Palace Rd, **Jammu**-180 001, Jammu & Kashmir – Adityapur, Gamharia Rd, **Jamshedpur**-831 013, Singhbhum District, Bihar – **Jhalawar**-326 001, Rajasthan – Jeypore-764 005, Koraput District, Orissa – **Jhansi**-284 001, Uttar Pradesh – **Jodhpur**-342 001,Rajasthan – **Jorhat**-785 001, Assam – **Kannur**-670 001, Kerala – **Kanpur**-208 001, Uttar Pradesh – **Kailashsahar**-799 277, Tripura – **Karwar**-581 301, Karnataka – **Kasauli**-173 204, Solan District, Himachal Pradesh – **Kathua**-184 104, Jammu & Kashmir – **Khandwa**-450 001, Nimar District, Madhya Pradesh – **Keonjhar**-758 001, Keonjhargarh District, Orissa – B.M.C., Thrikakara, **Kochi**-682 021, Ernakulam District, Kerala – **Kohima**-797 001, Nagaland – behind Government Printing Press, **Kolhapur**-416 003, Maharashtra – Jhalawar Rd, **Kota**-324 001 – P.O. Ramavaram, **Kothagudem**-507 118, Khammam District, Andhra Pradesh – P.O. Box 118, Beach Rd, **Kozhikode**-673 032, Kerala – **Kurnool**-518 001, Andhra Pradesh – Mehta Club Bldg, **Kurseong**-734 203, Darjeeling District, West Bengal –**Kurushetra**-132 118, Haryana – Radio Kashmir, **Leh**-194 101, Ladakh District, Jammu & Kashmir – 18 Vidhan Sabha Marg, **Lucknow**-226 001, Uttar Pradesh – P.O. Box 49, Lady Doak College Rd, **Madurai**-625 002, Tamil Nadu – Kadri Hills, **Mangalore**-575 004, Dakshin Kanara District, Karnataka – Vrindavan Rd, P.O. Gaytri Tapobhumi, **Mathura**-281 003, Uttar Pradesh – **Mercara**- ?, Karnataka – **Merkapuram**-523 316, Andhra Pradesh – **Murshidabad**-742 101, West Bengal – Yadavagiri, **Mysore**-570 020, Karnataka – Basni Rd, **Nagaur**-341 001, Rajasthan – Konam, **Nagercoil**-629 004, Kanya Kumari District, Tamil Nadu – Civil Lines, Palam Rd, **Nagpur**-440 001, Maharashtra – Kotwali Rd, **Najibabad**-246 763, Bijnor District, Uttar Pradesh – **Nanded**-431 601, Maharashtra – **Nizamabad**-503 012, Andhra Pradesh – **Obra**- 314 401, Rajasthan – P.O. Box 220, Altinho, **Panaji**-403 001, Goa – Jamakar Colony, Nawa Mondha, **Parbhani**-431 401, Maharashtra – **Pasighat**-791 102, East Siang District, Arunachal Pradesh – Phase-I, Urban Estate, Rajpura Rd, **Patiala**-147 001, Punjab – Chhaju Bagh, **Patna**-800 001, Bihar – Dilanipur, **Port Blair**-744 102, Andaman & Nicobar Islands, Union Territory – 24 Coubert Avenue (Cours Chabrol), Gorimedu, **Pudducheri**-605 001, Union Territory – **Purnia**-854 302, Bihar – University Rd, Shivaji Nagar, **Pune**-411 005, Maharashtra – **Raichur**-584101, Karnataka – **Raigarh**-496 001, Madhya Pradesh – Kamla Nehru Marg, Civil Lines, (P.O. Box 97), **Raipur**-492 001, Madhya Pradesh – Opposite Race Course, Sitaram Pandit Marg, (Post Bag 155), **Rajkot**-360 001, Gujarat – **Rampur**-244 901, Uttar Pradesh – 6 Ratu Rd, **Ranchi**-834 001, Bihar – Thiba Palace Rd, **Ratnagiri**-415 612, Maharashtra – 6 Civil Lines, **Rewa**-486 001, Madhya Pradesh – Subhash Rd, **Rohtak**-124 001, Haryana – **Sagar**-470 001, Madhya Pradesh – P.O.Box 3, Kuchery Rd, **Sambalpur**-768 001, Orissa – Market Yard, Kolhapur Rd, **Sangli**-416 416, Maharashtra –**Sasaram**-821 301, Rohtas District, Bihar – **Satara**-415 001, Maharashtra – **Sawai Madhopur**-322 001, Rajasthan – **Shahdol**-484 001, Madhya Pradesh – Assembly Hall, (P.O. Box 14), **Shillong**-793 001, Meghalaya – Choura Maidan, **Shimla**-171 004, Himachal Pradesh – **Shivpuri**-473 551, Madhya Pradesh – **Sholapur**-413 006, Maharashtra – Trunk Rd, **Silchar**-788 001, Cachar District, Assam – 2 Mile Sevoke Rd, **Siliguri**-734 401, Darjeeling District, West Bengal – Radio Kashmir, Sherwani Rd, **Srinagar**-190 001, Jammu & Kashmir – **Suratgarh**-335 804, Sriganganagar District, Rajasthan – **Tawang**-790 104, Arunachal Pradesh – **Tezu**-792 001, Lohit District, Arunachal Pradesh – Ramavarampuram, **Thrissoor**-680 631, Kerala – P.O.Box 54, 288 Promenade Rd, **Tiruchirapalli**-620 001, Tamil Nadu – Schafter Hall, High Ground Rd, Near Palai Bus stand, Samathanapuram, Palayamkottai, **Tirunelveli**-627 006, Tamil Nadu – **Tirupathi**-517 501, Andhra Pradesh – P.O. Box 403, Bhakti Vilas, Vazuthacaud, **Thiruvananthapuram**-695 014, Kerala – Lower Chandmari, **Tura**-794 001, Meghalaya – **Udagamandalam**-643 001, Tamil Nadu – Chetak Circle, **Udaipur**-313 001, Rajasthan – Bhramavar/**Udipi**-576 213, Dakshina Kanara District, Karnataka – Makarpura Rd, **Vadodara**-390 009, Gujarat – Mahmoorganj, **Varanasi**-221 010, Uttar Pradesh – New Bldg, Bandar Rd, Punnammathota, **Vijayawada**-520 010, Andhra Pradesh – Sripuram Junction, **Visakapatnam**-530 003, Andhra Pradesh – **Warrangal**-506 002, Andhra Pradesh – **Yavatmal**- 445 001, Maharashtra.

EXTERNAL SERVICES: see International Broadcasting section.

INDONESIA

L.T: We. Indonesia (Jawa, Sumatra, We. & Ce. Kalimantan): UTC + 7h; Ce. Indonesia (So. & Ea. Kalimantan, Sulawesi, Bali, Nusa Tenggara): UTC + 8h; Ea. Indonesia (Maluku, Irian Jaya): UTC + 9h — **Pr.L:** Bahasa Indonesia (Indonesian) — **Pop:** 197m — **Radios:** 26.000.000 — **E.C:** 50Hz, 127/220V — **ITU:** INS.

DIRECTORATE GENERAL OF RADIO, TV & FILM
✉ Jalan Merdeka Barat 9, Jakarta. **Cable:** Radiogen.
L.P: Dep. Dir: M. Arsyad Subik.

DIRECTORATE OF RADIO
Department of Information, Jalan Merdeka Barat 4/5, Jakarta.
☏ Deppen 44349
L.P: Dep. Dir. for Technical Development: Mr. Sunendra.

FEDERATION OF INDONESIAN NATIONAL COMMERCIAL BROADCASTERS
Persatuan Radio Siaran Swasta Nasional Indonesia (PRSSNI)
✉ Pengurus Pusat, Persatuan Radio Siaran Swasta Nasional Indonesia, Jln K. H. Mas Mansyur 25A, Jakarta Pusat 10230.
L.P: Chmn: Siti Hardiyanti Rukmana.

Mediumwaves: v)=varies; r)=reported; u)=unconfirmed but official information, not completely monitored; *) r. inactive. Commercial st's broadcast between 2200 and 1700 only, with breaks in mid-morning and mid-afternoon.

	kHz	kW	Station
1)	540	2/10	RRI Bandung
5)	540	0.25	R. Dirgahayu, Barabai
8)	540		R. Voice of Yarden, Ambon
375)	540		R. Dei Marganusa, Metro
10)	558	0.25	R. Manggis B.S., Jambi
12)	558		R. Swara Angkasa Megah, Pandeglang
14)	558	0.25	R. Suara Jalesveva Juana Sakti, Pati
88)	558		R. Diantara Vita Kharisma (DVK), Kebumen
118)	558	0.25	R. Suara Cakrawala, Telukbetung
9)	567	0.25	RSPDKDT2 Semarang, Ungaran
37)	567		R. Kencana Nada Suara, Way Jepara
11)	576		R. Citra Prima Mahardika, Pringsewu
19)	585	50	RRI Surabaya
42)	594		R. Suara Pramudya Lestari (Puri), Sukadana
507)	603	0.5	RDKDT2 Sumedang
13)	603		R. Citra Airlangga, Kediri
347)	603		R. PTDI Suara Kenanga Citra Indah, Purworejo
21)	612		R. Dwikarya 69, Bandung
22)	612	0.25	R. Gema Bahari Selatan, Rangkasbitung
23)	612	0.22	R. Linggarjati Utama (Rasilima), Kuningan
24)	612	0.25	R.Suara Banyumas Aslia (Subali), Purwokerto
26)	612		R. Angkasa Bahana Citra (A.B.C.), Surakarta
28)	612	0.5	R. Aneka Rama Ria (AR), Denpasar
31)	612		R. Favorite, Ambon
355)	612		R. Bahurekso Sakti (BOS), Weleri
48)	612	0.25	R. Pahlawan Budi Sakti, Serang
49)	612		R. Geswara Pamanukan (GSP), Pamanukan
69)	612		R. Swara Barabai, Barabai
33)	630	50	RRI Ujung Pandang
30)	630		R. Saba Putra, Pringsewu
500)	630		R. Merapi Indah, Muntilan
36)	648	0.25	R. Gema Bunda Kandung (RBK), Panjang
39)	648	0.25	R. Gita Kanari Ria, Purwakarta
40)	648		R. Santo Bernadus D.S, Pekalongan
43)	648	0.25	R. Nirwana Lestari, Banjarmasin
46)	648	0.25	R. Montini Jaya, Manado
47)	648		R. Merpati, Ambon
110)	648		R. Suara Palagan Semesta Sehati (SPS), Ambarawa
162)	648		R. Bestari. Sukabumi
366)	648	0.5	RKPDT2 Lamongan
499)	648	1/5	RRI Dili
501)	648	0.25	R. Gema Merdeka, Denpasar
508)	648		R. Roro Djonggrang B.S., Prambanan
622)	648		R. Wijaya Adikusuma, Cilacap
7)	648		R. Batanghari Permai (BHP), Muarabulian
498)	655	0.25	RSPD Kotamadya Magelang
54)	666	0.25	R. Ramakusala, Surakarta

	kHz	kW	Station
55)	666	0.25	R. Borobudur Siebra Patria, Semarang
73)	666		R. Gita Nada Tebu Ireng, Jombang
86)	666		R. Swara Populer, Purwakarta
90)	666		R. Bali Perkasa, Denpasar
509)	666		R. Mercidiona, Bandung
50)	666		R. Den Bang, Bd. Jaya
574)	666	2	RDKDT2 Pandeglang
595)	675		RSPDT2 Kotamadya Pekalongan (Suara Kota Batik)
3)	684		R. Swara Pakusarakan Pratika, Sawangan
105)	684		R. C.M.B, Purwokerto
117)	684		R. Wisata Panataran (Wita), Blitar
139)	684		R. Swara Kijang Berantai (Kiber) Perkasa, Sambas
174)	684		R. Patrol Radika, Bandar Lampung
279)	684		R. Purnama Nada, Kandangan
35)	684		R. Swara Gorontalo Permai, Gorontalo
495)	690	0.5	RKP Kotamadya DT2 Pasuruan (R. Gema Suropati)
150)	693		RSPDK Kediri (R. Canda Bhirawa)
529)	693		RRI Sungai Liat
585)	693	0.25	RSPDKDT2 Purworejo (R. Suara Irama)
591)	693		RSPKDT2 Jepara (R. Kartini)
63)	702		R. Tjandra Buana Suara, Cianjur
510)	702		R. Swara Zenith Angkasa, Salatiga
267)	702	0.25	R. Siera Alfa Lima, Bantul
208)	702	0.25	R. Antar Nusa, Jakarta
336)	702		R. Ekatama Swara Gajahmada (El Gama), Banjarmasin
634)	702		R. Gloria Paramita, Lumajang
176)	702		R. Barisan Nauli, Sidikalang
192)	702		R. Besakih Rasisonia, Amlapura
243)	710	0.5	RKPKDT2 Serang
307)	710	0.5	RSPDKDT2 Cilacap
575)	711	0.5	RP Kotamadya DT2 Mojokerto (R. Gelora Mojopahit)
57)	711	0.5	RSPDKDT2 Wonogiri
553)	711	0.25	RSPDKDT2 Banjarnegara
572)	711	0.5	RSPDT2 Blora
58)	711	0.25	R. Rajawali Terbang, Bandar Lampung
15)	720		R. Purnamasidi, Wonosobo
59)	720		Suara Simpati Angkasa (Susia), Pinrang
67)	720	10	RRI Ambon
125)	720	0.1	R. Telerama, Banjarmasin
275)	720	0.25	R. Lusiana Namberwan, Semarang
513)	720	0.25	R.Suara Kinijani, Tabanan
544)	720		RKPDT1 Jawa Timur, Surabaya
627)	720		R. Gagah Sehat Berbobot (Gasebo), Majenang
633)	720		R. Gita Swara Alfina, Pemalang
64)	720		R. Suara Mitra Bayu Buana, Belitang
196)	720		R. Bhara Kharisma Suryajaya, Talangpadang
225)	729		R. Swara Mega Mustika, Tanjungkarang
70)	738	0.25	R. Andalas Raya Bestari, Padang
72)	738		R. Nada Kencana Agung, Bandung
74)	738	0.25	R. Galuh Surya Kencana, Tasikmalaya
75)	738		R. Swara Prima Sonata, Cirebon
76)	738		R. Konservatori, Surakarta
78)	738	0.45	R. Rinjani Permai, Cakranegara
79)	738	0.25	R. Kharisma Nada Rasisonia (La Fuzsy), Banjarmasin
597)	738		R. Suara Pamekasan Indah, Pamekasan
200)	738		R. Bharata Bhakti Nusa, Tangerang
201)	738		R. Swara Melati Gramedia, Mempawah
81)	738		R. Rina Bestari, Rantepao
82)	738	0.25	R. Sound of Love, Manado
183)	738	1	RRI Jember
600)	738		R. Trisara Kencana, Cianjur
27)	738		R. Duta Paramita, Metro
66)	738		R. Aditya Nada Jaya, Indralaya
546)	740		RSPDT2 Batanghari, Muarabulian
204)	743	0.3	RPDT2 Sumba Barat, Waikabubak
527)	747	5	RRI Bengkulu
639)	747	0.25	RSPDT2 Kudus (Suara Kudus)
212)	747		R. Wijaya Asthanaria, Kotabumi
373)	747	0.5	RKPDT2 Pasuruan (R. Untung Suropati)
306)	756	0.5	RKPDKDT2 Magetan (R. Magetan Indah)
646)	756	0.5	RKPKDT2 Mojokerto
83)	756	2/10	RRI Purwokerto

	kHz	kW	Station
84)	756		R. Venus Nusantara, Ujung Pandang
320)	756		R. El Gangga, Bekasi
32)	756		R. Barata Adya Swara, Bandar Lampung
213)	765		RKPD Kotamadya Blitar
257)	765	0.2	RRI Banjarmasin
541)	765	0.25	RSPDT2 Boyolali
68)	774		R. PTDI Walisongo, Pekalongan
87)	774		R. Antariksa Radang IV, Surabaya
602)	774		R. Bima Sakti, Keburmen
576)	774	0.5	RDK Tangerang
89)	774	0.25	R. Continental Emerald, Manado
136)	774		R. Yusyan Media, Sumedang
491)	774	0.2	R. Leonardus Buana Suara, Salatiga
34)	774		R. Ramayana Wiratama, Metro
215)	774		R. Citrah Anugerah, Negara
221)	774	0.35	RSPD Kotamadya Payakumbuh
378)	783	0.25	RSPDT2 Pati (Suara Pati)
91)	785	0.5	RKPDT2 Kotamadya Bandung
126)	786		RSPDT2 Tegal, Slawi
6)	792		R. Padaidi Fadaelo Sipatuo Sipatokkong, Ujung Pandang
108)	792		R. Swara Graha Jelita, Surakarta
60)	792		R. Cempaka Angkasa, Ciamis
124)	792		R. Andhika Lugas Suara (Andalus), Malang
604)	792		R. Suara Dwi Amanda, Gadingrejo
38)	792		R. Bayu Sakti, Purwokerto
226)	792		R. Suara Ria Jaya Sentosa (SRJS), Butaraja
228)	792		R. Romantika, Bondowoso
230)	792		R. Tirilolok, Kupang
538)	792		R. Cipta Bentara Swara (CBS), Magelang
372)	792	0.5	RKPDK Jombang
141)	792		R. Gema Warga Karya, Bandung
155)	792		R. Mitra Idola Kita, Pancor
92)	801	10	RRI Semarang
101)	801	1	RRI Medan
93)	810	7.5	RRI Merauke
94)	810	0.25	R. Warastra Bewara Swara, Palembang
95)	810	0.25	R. Romeo and Juliet Rasisonia, Tanjungkarang
97)	810		R. Indraswara Cakrawala Nada, Majalengka
98)	810		R. Gema Kuripan, Amuntai
605)	810		R. Suara Karamita, Pamekasan
100)	810	0.25	R. Mersy Maesaan Waya, Manado
514)	810		R. Nusantara Jaya (ROS), Jakarta
577)	810		RSPDK Bandung
234)	810		R. Suara Maung Sakti, Banjarnegara
235)	810		R. Saburai Alam Permai, Liwa
169)	810		R. Suara Tanjung Berjaya, Tanjungbalai
606)	819		R. Pancabayu Madugondo (Suara RPM), Sukoharjo
101)	819		RRI Medan *
103)	828	0.25	R. Leidya Swara Utama, Bandung
104)	828	0.25	R. Prabu Kiansantang, Tasikmalaya
427)	828	0.25	R. Cendrawasih (Suara Semarang), Semarang
107)	828	0.25	R. Swara Citra Suhada Jaya (RCS), Pekalongan
109)	828	0.25	R. Persatuan, Bantul
111)	828		R. Tritara Yaksa (TT-77), Malang
114)	828		R. Swara Christy Ria, Ujung Pandang
115)	828	0.25	R. Alkhairaat, Manado
116)	828	0.25	R. Suara Yudha, Denpasar
41)	828		R. Suara Tiara Indah, Kotabumi
616)	828		R. Angkasa Media, Kadipatan
236)	828		R. Gema Bukitasam (RGBA), Tanjung Enim
238)	828		R. Gita Segara, Bangkalan
71)	828		R.S.B.S., Purbalingga
170)	828		R. Adhika Pariwara, Pelabuhanratu
338)	837		R. Mayangkara Ria, Blitar
120)	846	0.25	Swara Fortune Indah, Bandung
121)	846		R. Suara Galunggung Giri Sakti, Tasikmalaya
122)	846	0.25	R. Immanuel, Surakarta
123)	846		R. Miniwati Pesona Indah, Surabaya
127)	846		R. Swara Caraka Ria, Semarang
609)	846		R. Swara Anggada Senatama, Purbalingga
497)	846	0.5	RKPDT2 Ponorogo (R. Suara Ponorogo)
253)	846		R. Menara Buana Suara Indah, Sukabumi
262)	846		R. Suara Tegal Agung Raya (Star), Tegal
263)	846		R. Swara Citra Survanada, Amuntai
317)	850	1	RRI Bogor
101)	855	50	RRI Medan (Padang Cermin)
128)	855	2/10	RRI Mataram
51)	855		R. Surya Gita Paramarta (SGP), Labuhan Maringgai
129)	864	2/10	RRI Cirebon
16)	864		R. Mataram Buana Suara, Yogyakarta
132)	864		R. Arief, Payakumbuh
134)	864	0.25	R. Mercy Jaya Raya, Tanjungkarang
135)	864		R. Assyafiyah, Jakarta
137)	864		R. Menara III, Surabaya
138)	864		R. Iskanada Mustika (Sylvania), Banjarmasin
45)	864		Suara AsAdiyah, Sengkang
322)	864	0.25	RSPD Kotamadya Salatiga
607)	864		R. Suara Karangbolong, Gombong
265)	864		R. Swara Trans Wahana Makmur, Kota Lais
269)	864		R. Suara Anggada Senatama (SAS), Banjarsari
484)	864		R. Prominda Dirgantara, Pontianak
272)	870		RKPD Kotamadya Probolinggo
610)	873	0.5	RSPDKDT2 Sragen (R. Suara Buana Asri)
52)	873		R. Sari Bunga Sadari (SBS), Sidomulyo
540)	873	0.25	RSPDK Batang
273)	873		R. Poliyama Indah, Gorontalo
578)	879	0.5	RKPKDT2 Sidoarjo
142)	882	0.25	R. Chandrika Widya Swara (CBS), Bandung
143)	882	0.3	R. Nusantara Bharata Citra (NBC), Garut
144)	882	0.25	R. Chandra Rasisonia, Banjarmasin
145)	882	0.5	R. Bambapuang, Pangkajene
61)	882		R. Ragam Tunas Lampung (Ratula), Kotabumi
172)	882		RRI Kendari
274)	882		R. Pelangi Nusantara, Jakarta
280)	882		R. Swara Kranggan Persada, Temanggung
281)	882		R. Gema Bahari Selatan (GBS), Malingping
146)	886	0.5	RRI Serui
147)	891	10	RRI Malang
77)	900		R. Pratama Mahardika, Gedong Tataan
612)	900	0.25	R. Gema Nugraha, Sungai Penuh
148)	900		R. Elkartika Angkasa Niaga, Padang
149)	900	0.25	R. Jelita Bahanswara, Bukittinggi
151)	900		R. Andalan Muda, Wates
660)	900	0.25	R. Safari Bina Budaya, Jakarta
152)	900		R. M.C.A., Surabaya
647)	900		R. Paramita Jaya, Tulungagung
289)	900		R. Nugraha Top, Palu
300)	900		R. Gematara Batakan, Pelaihari
302)	900		R. Permata Swaratama, Boyolali
389)	900		RPKDT2 Cianjur
463)	900	0.25	R. Gema Megantara Pratama, Tabanan
459)	900	0.25	R. Wahana Bewara Suara, Cirebon
191)	900		R. Aries Sanggau Perkasa, Sanggau
556)	905	0.25	RPDKDT2 Sumbawa
473)	909	0.25	RKPDKDT2 Pamekasan (Suara Dian Lestari)
567)	909		RRI Sorong
153)	918	0.075	R. Citra Remaja Maju Jaya, Langsa
154)	918		P.T.D.I., Jakarta
156)	918		R. Gema Nury (El Nury), Bogor
157)	918		R. Swara Pitaloka, Ciamis
232)	918		R. Dhirgantara, Negara
96)	918		R. Gelora Pertiwi, Medan *
53)	918	0.25	R. Chandra Kusuma, Pekalongan
579)	918	0.5	RKPKDT2 Gresik
614)	918		R. Suara Semeru Permai, Lumajang
62)	918		R. Suara Selomanik (RSS), Banjarnegara
309)	918		R. Balistik, Kupang
311)	925		RDK Subang (R. Benteng Pancasila - Benpas)
615)	925	0.25	R. Benpas, Bogor
158)	927	25	RRI Pekanbaru
65)	927		R. Mandiri, Kotabumi
313)	927	0.25	RKPDT2 Sampang (R. Suara Bahari)
677)	927		R. Primanada, Gisting
140)	936		R. Liga Perdana, Balikpapan
159)	936		R. Pasopati Andalan, Semarang
160)	936	0.25	R. Puspa Dwi Swara Cipta (P2SC), Jakarta
161)	936		R. Budaya Sari, Bandung
163)	936	0.25	R. Irama Adinada, Surakarta
291)	936		R. Widya Bhakti, Magelang
166)	936		Suara Mahameru, Kediri
167)	936		R. Suara Fiskarama, Bondowoso

	kHz	kW	Station		kHz	kW	Station
168)	936	0.25	R. Gemini Perkasa, Mataram	171)	1044		R. Sikamoni, Medan
169)	936		R. Refalado, Banjarmasin	211)	1044	10	RRI Biak
314)	936		R. Tanjung Puri Perkasa, Tabalong	629)	1044		R. Arena Duta Swara, Trenggalek
244)	936		R. Swara Dermagaria Persada Cakrawala, Sekadau	630)	1044		R. Duta Angkasa, Pangandaran
				214)	1044	0.25	R. Purnayudha, Sukabumi
197)	945		RKPDT2 Blitar	216)	1044	0.25	R. Raka, Tegal
250)	945		R. Tuah Swara Murni, Lubukpakam	217)	1044	0.25	P.T.D.I. Medari-Sleman
678)	945		R. Swara Buana Asri, Wonosobo	218)	1044		R. Camar, Surabaya
44)	954		R. Gandaria (Anging Mamiri), Ujung Pandang	219)	1044		Suara Akbar, Jember
172)	954	10	RRI Kendari	220)	1044	0.25	R. Cakra Swara Perkasa, Singaraja
173)	954	0.15	R. Garuda Kenten Jaya, Palembang	229)	1044		R. Arya Bomantara, Singkawang
177)	954	0.25	R. Sena Bahana Cakrawala, Sukabumi	352)	1044		R. Soreram Indah, Pekanbaru
618)	954		R. Gita Nusantara Perkasa (Studio 99), Purbalingga	354)	1044		R. Swara Tapin Raya Bastari (Swatara), Rantau
179)	954	0.25	R. Sahara, Kendal	357)	1044		R. Adyemaja, Lhokseumawe
180)	954	0.25	P. T. D. I. Kota Perak, Yogyakarta	359)	1044		R. Lima Swara Mandiri (Purnayudha), Bekasi
181)	954	0.25	R. El Bayu, Gresik	361)	1044		R. Bimareksa Dirgantara, Sanggau
182)	954	0.25	R. Gema Persada, Banjarmasin	363)	1044		R. Mega Primadona Nada Perkasa, Pangkalanbuun
4)	954		R. Nada Berlian, Jambi	436)	1044		R. Suara Ria Santana, Bengkulu
131)	954		R. Batur, Bangli	551)	1050		RPDT2 Aceh Timur, Langsa
318)	954		R. Ewangga, Kuningan	222)	1053	10	RRI Jayapura
319)	954		R. Suara Terunajaya, Pemeungpeuk	186)	1053	1	RRI Surakarta
183)	963	10	RRI Jember	223)	1062	0.25	Suara Subuh, Padang
619)	963		R. Kelana Sumbangsihku (Kasihku), Bumiayu	224)	1062	0.25	R. Gema Mutiara, Palembang
620)	963		R. Suara Lorosae, Dili	227)	1062		R. Maestro, Bandung
185)	966		RPDK Deli Serdang, Medan *	231)	1062		R. Sodiak, Surabaya
186)	972	50	RRI Surakarta	233)	1062	0.25	R. Suta Remaja, Mataram
621)	972		R. Megantara Bhinneka, Nganjuk	632)	1062		RKPDT2 Ngawi
188)	972	0.25	R. Diah Rosanti, Pontianak	99)	1062		R. Swara Sentosa Pratama, Ujung Pandang
189)	972		R. Swara Citra Esa Enang, Langowan	516)	1062		R. PTDI Kalimasadha Sakti, Semarang
580)	972	0.25	R. Suara Alas Roban (ARO), Batang	506)	1062	0.5	RPKDT2 Tasikmalaya
323)	972		R. Suara Harmoni, Situbondo	365)	1062		RPDK Kapuas, Kuala Kapuas
260)	972		R. Suara Adya Samudra (ADS), Bekasi	370)	1062		R. Gema Mutiara, Plaju
382)	972		R. Cempaka Asri, Bulukumba	376)	1071		R. Gita Pesona Swara Pandaran, Sampit
371)	981		R. Antares, Garut	586)	1071	0.25	RSPDT2 Pemalang (Suara Widuri), Pemalang
452)	981	0.25	RSPDKDT2 Magelang	237)	1080	2/10	RRI Singaraja
394)	981		R. Angraini Kalamaira, Siabat	239)	1080	0.25	R. Mancasuara, Bandung
193)	990		R. Sion, Tomohon	241)	1080	0.25	R. Carolina Arjuno, Surabaya
430)	990	0.25	R. Suara Calvary, Klungkung	528)	1080	1	RRI Gorontalo
431)	990		R. Bahana Nirmala, Martapura	381)	1080		R. Karya Pancaran Swara Media (Karysma), Boyolali
623)	990		R. Citra Wanodya Angkasa, Jombang				
25)	990	0.25	R. Kharisma Indah Swara (KIS), Semarang	383)	1080		R. Idola Nada, Manggala
324)	990		R. Gema Suara Gloria (GBS), Kupang	386)	1080		R. Katalina, Sigli
325)	990		R. Nias Mitra Dharma, Gunungsitoli	395)	1080		R. Swara Mahkota Polareksa, Sintang
331)	990		R. Cakra Donya Multi Swara, Lhokseumawe	441)	1080		R. Pariwara Swara Desa, Indrapura
341)	990		R. Gita Lestari, Brebes	479)	1080		R. Bahana Nusantara, Ampah
401)	990		R. Suara Sowerigading, Wonomulyo	245)	1097	0.3	RSPDKT2 Kendal
195)	999	1/150	RRI Jakarta	246)	1098	2/10	RRI Jambi
421)	1000		R. Kharisma Lombok Perkasa, Aikmel	247)	1098	0.5/10	RRI Sumenep
525)	1001	0.3	RRI Bukittinggi	248)	1098		R. Swara No Name, Palembang
625)	1005		RPDT2 Kutai, Tenggarong	249)	1098		R. Media Mahasiswa Tarumanegara, Jakarta
626)	1008	0.25	R. Adyafiri, Watansoppeng	261)	1098		R. Diaros Duta Swara, Singkawang
198)	1008	10	RRI Madiun	617)	1098		R. Serenada Gita Lestari, Slawi
106)	1008		R. Pelangi Cakrawala Nusantara, Subang	398)	1098		R. Gelora Indah Swara (IS), Wonogiri
342)	1008		R. Pentas Taruna Sriwijaya, Sumatera Selatan	400)	1098		R. Suara Kelana, Curup
343)	1008		R. Delta Pawan Indah, Ketapang	252)	1107	1/10	RRI Yogyakarta
350)	1008		R. Suara Amarta Sakti, Pekalongan	581)	1107	0.25	R. Bintang Niaga, Kisaran
433)	1008		R. Citra Tebingtinggi Idola Nada, Tebingtinggi	130)	1107	1/5	RRI Kupang
				255)	1116	0.25	R. Suzana Jaya Bhakti, Surabaya
199)	1017		R. Kardopa, Medan	532)	1116	0.25	RPDT2 Kotamadya Binjai
146)	1026	5	RRI Serui	582)	1116	0.25	Suara Batang, Batang
628)	1024	0.25	RPDT2 Belu, Atambua	158)	1116	0.3	RRI Pekanbaru
2)	1026		R. Gema Cakrawala Utama, Kuala Simpang	402)	1116		R. Indah Sragen Asri, Sragen
202)	1026		R. Fortuna, Sukabumi	403)	1116		R. Adhika Swara, Tangerang
203)	1026		R. Menawan Ceria Indonesia Rasitada (Mersi), Semarang	661)	1116		R. Swara Gipsi Pratita Cakrawala, Langsa
				256)	1120v		R. Andalas IX, Banda Aceh
205)	1026	0.25	R. El Victor, Surabaya	523)	1125	0.25	RPDT2 Luwu, Palopo
206)	1026	0.25	R. Taurus Adiswara, Kediri	315)	1125		RRI Pontianak
207)	1026	0.25	R. Swara Indonusa Perkasa, Gianyar	560)	1125	1	RRI Palu
208)	1026		R. Siaga Indah Marista (Santanimo), Banjarmasin	631)	1125	0.25	RSPDKDT2 Kebumen
				405)	1127	0.2	RPDT2 Kotamadya Medan *
406)	1026	0.25	R. Lita Sari, Bandung	257)	1134	25/1	RRI Banjarmasin
494)	1029	0.5	RPKDT2 Ciamis	258)	1134	0.1	RSPD Kotamadya Tegal
542)	1034		RPDT2 Ngada, Bajawa	259)	1134		R. Alpha Romeo, Banda Aceh
194)	1035		RSPDT2 Temanggung	388)	1134		R. Kauman, Bogor
209)	1035	1/5	RRI Tanjungkarang	411)	1134		R. Duta Nusantara Suara Ponorogo, Poncrogo
210)	1035	0.5	RKPDT2 Malang				
560)	1035		RRI Palu	413)	1134		R. Swara Delanggu, Delanggu

	kHz	kW	Station
423)	1134		R. Harau Megantara Angkasa, Tanjung Pati
665)	1134		R. Kazuma Bawana Swara, Lhokseumawe
264)	1135		R. Ria Cindelaras, Indramayu
666)	1140		R. Siaran Niaga dan Budaya Mhanda, Meulaboh
462)	1143	0.25	R. Pariwisata Senaputra, Malang
472)	1143		RKPDT2 Jember
285)	1145		R. Daerah Perdajaya Bebas Sabang, Sabang
268)	1152		R. Alvacos, Medan *
270)	1152	0.25	R. Enes Duabelas Ulu, Palembang
271)	1152	0.25	R. Suara Wajar, Pahoman
276)	1152	0.35	Suara Istana, Yogyakarta
277)	1152		R. Yasmara, Surabaya
278)	1152	0.185	R. Wijang Songko, Kediri
458)	1152		R. Swara Maya Nada, Bandung
424)	1152		R. Swara Wangi Timur, Banyuwangi
572)	1152		R. Swara Mersidiona, Cimone
258)	1161	0.1	RSPD Kotamadya Tegal
92)	1170	50	RRI Semarang
283)	1170	0.25	R. Ganesha Nada, Bandung
284)	1170	0.25	R. Swara Rama Lokantara, Serang
351)	1170		R. Rajawali, Surabaya
635)	1170		R. Swara Irama Kusuma Sena, Bogor
437)	1170		R. Swara Megapola Nada Indah (Megaphone), Sigli
442)	1170		R. Lariang Indah, Mamuju
286)	1179	2/10	RRI Padang
519)	1179		R. Kotamadya Cirebon (R. El Thema), Cirebon
290)	1188		R. Swara Selabintana Permai (SSP), Sukabumi
158)	1188	0.3	RRI Pekanbaru
292)	1188	0.25	R. Arma Sebelas, Yogyakarta
293)	1188		R. Swara Perak Jaya PTDI, Surabaya
294)	1188		R. Kutilang, Malang
295)	1188	0.25	R. La Barong, Singaraja
299)	1188		R. Dalka Mega Swara, Meulaboh
518)	1188		R. Gema Nirwana, Samarinda
443)	1188	0.25	R. Namora Swara Pratama, Curup
445)	1188		R. Suara Ayukarya Banjaran Adiwerna (RSA-Abadi), Tegal
448)	1188	0.25	R. Swara Carano Batirai Indah, Batusangkar
128)	1194	0.5	RRI Mataram
317)	1195	1	RRI Bogor
548)	1197	0.5	RSPDT2 Labuhan Batu, Rantau Prapat
533)	1197	1/5	RRI Palangkaraya
636)	1200		RSPDT2 Indragiri Hulu, Rengat
549)	1205		RPKDT2 Lebak, Rangkasbitung
456)	1205	0.2	RPDT2 Simalungun, Pematang Siantar
19)	1206	1	RRI Surabaya
20)	1206		R. El Philia, Telukbetung
296)	1206		R. Indah Bahagia Ceria (I.B.C.), Semarang
297)	1206	10/0.5	RRI Denpasar
298)	1206	0.25	RSPKDT2 Purbalingga
301)	1206		R. Suara Dikara Bawana (Dirgan Bravo), Padang
524)	1206	1	RRI Manado
583)	1206	0.25	RSPKDT2 Pekalongan (R. Kota Santri)
584)	1206		R. Suara Mitra, Tanjung Gading
251)	1206		R. Histori Gita Jaya, Karawang
667)	1206		R. Geunta Suara, Geudong
668)	1206		R. Suara Langkat Tanjung Persada, Tanjungpura
304)	1211	0.5	RKPDT2 Bondowoso
195)	1215	1	RRI Jakarta
305)	1215	10/0.5	RRI Samarinda
308)	1224	0.25	R. Prima Elita, Palembang
310)	1224	0.25	R. Buana Jaya, Tasikmalaya
496)	1224	0.5	RKPDT2 Bojonegoro
321)	1224		Suara Tawang Alun, Banyuwangi
482)	1224		R. Istana Blitar
119)	1224		R. Keluarga Cihanjuang (KC10), Indramayu
457)	1224		RSPDT2 Banyumas, Purwokerto
460)	1224		R. Prima Ukir Utama, Jepara
477)	1224		R. Indraswara, Tebingtinggi
480)	1233		RKPDT2 Sumenep (R. Dinamika Suara Pariwisata)
315)	1233	0.2/1/5	RRI Pontianak
316)	1233v	0.3	RPD Kotamadya Sabang
317)	1235	0.5	RRI Bogor
165)	1242		R. La Victor, Surabaya
481)	1242		R. Duta Suara Garuda Sakti, Blora
488)	1242		R. Gema Remaja, Kuningan
184)	1242		R. Airlangga, Sukabumi
520)	1242		R. Swara Rugeri, Garut
527)	1242	1	RRI Bengkulu
669)	1242		R. Suara Sibolga Indah Sempurna, Sibolga
317)	1250	1	RRI Bogor
128)	1251	1	RRI Mataram
326)	1251	10	RRI Banda Aceh
237)	1251	1	RRI Singaraja
492)	1251	0.5	RKPKDT2 Probolinggo
327)	1260	0.25	R. Khamasutra, Medan
328)	1260	0.25	R. Lokawisesa, Jakarta
329)	1260	0.25	R. Famor, Bandung
332)	1260		R. PTDI Suara Kaliwungu Dirgantara, Kaliwungu
333)	1260		R. Gabriel, Madiun
334)	1260		R. R.D.A. 45, Bangil
335)	1260		Suara Mandala, Banyuwangi
337)	1260		R. Rajawali Sakti, Balikpapan
164)	1260		R. Citra Angkasa Ikhsaniya (RCA), Tegal
515)	1260		R. Swara Jupti Indah, Sibolga
517)	1260		R. Panca Pesona Jaya (Papeja), Lubuklinggau
339)	1278		RPDT2 Lombok Timur, Selong
340)	1278v		R. Thosiba (TOSS), Banda Aceh
344)	1278		R. Irama Kusuma Sena, Bogor
345)	1278	0.25	R. Mandalika Rasiswana, Jepara
346)	1278		R. Citra Ikhsaniya, Tegal
348)	1278		R. Mercury Madesuk, Surabaya
521)	1278	0.25	RSPDK Grobogan, Purwodadi
662)	1278		R. Ringan Mutiara, Kendari
530)	1285		R. Gema Kahayan, Pangkalanbuun
349)	1287	0.5/50	RRI Palembang
637)	1287		RSPDKDT2 Karanganyar (Swara Intanpari)
555)	1290	0.2	RPDT2 Pematang Siantar
353)	1296	0.5	R. Duta Gita Bhyomantara Shinta Rama, Cakranegara
19)	1296		RRI Surabaya
557)	1296		R. R.B.B, Rembang
558)	1296		R. Suara Kelandka, Palopo
670)	1296		R. Begita, Kabanjahe
671)	1296		R. Merak Jaya, Muarateweh
672)	1296		R. Gapilar Rasisonia, Solok
587)	1304	0.5	RKPDKDT2 Nganjuk
534)	1305	0.25	RDKDT2 Sukabumi (Programa Dua)
673)	1305		R. Kerinci Giri Swara (KGS), Sungaipenuh
17)	1314		R. Rosa, Ujung Pandang
133)	1314		R. Gelora Ramona, Palembang
356)	1314		R. Mutiara, Bandung
358)	1314		R. Ratna Palupi, Karawang
360)	1314		Suara Sion Perdana, Karanganyar
362)	1314		R. Bintoro Karya, Demak
175)	1314		R. Suara Ronggohadi, Lamongan
561)	1314		R. Gema Sritanjung Mediatama (GSM), Jatibarang
562)	1314		R. Suara Asahan, Kisaran
564)	1314		R. Shinta Wahana, Bengkulu
663)	1314		R. Buana Sutra, Kendari
502)	1316	0.5	RKPDK Tulungagung
569)	1324		RDK Lampung Selatan, Kalianda
380)	1332		RKPD Kotamadya Surabaya (R. Gelora Surabaya)
195)	1332	10	RRI Jakarta
369)	1332		R. Ellusia, Pekanbaru
638)	1332		R. Swara Tugas Actari, Ciamis
566)	1332	1	RRI Wamena
504)	1332	0.5	RKPDKDT2 Lumajang
641)	1332		R. Suara Bhakti Nusantara, Bau-Bau
570)	1332		R. Atachi Brosser, Palu
571)	1332		R. Granada Tara Indah, Kuala Kapuas
287)	1341	1/5	RRI Tanjung Pinang
18)	1350		R. Suara Mesra, Parepare
364)	1350		Suara Kartika, Jember
374)	1350	0.25	R. LAmour Citra Budaya Sriwijaya, Palembang
377)	1350		R. Puspa Jaya, Bojonegoro
559)	1350	0.75	RPDT2 Tapanuli Utara, Balige
640)	1350		R. Airlangga, Sukabumi

	kHz	kW	Station
588)	1350	0.5	RKPDT2 Situbondo
573)	1351		R. Delijaya, Tebingtinggi
33)	1359	1	RRI Ujung Pandang
679)	1359		RKPDT2 Bangkalan (Suara Bangkalan Ceria)
545)	1363	0.35	RPDK Kulonprogo (R. Suara Indrakila), Wates
589)	1365	0.5	RKPDKDT2 Pacitan
379)	1365	0.5	RKPKDT2 Trenggalek
412)	1368		R. Sonya Manis, Bireuen
590)	1368	0.5	RPKDT2 Cirebon
85)	1368		R. Hay Citra Ria, Palangkaraya
266)	1368		R. Siaran Ichthus, Semarang
471)	1368	0.25	RKPDK Tuban (R. Gelora Ronggolawe)
603)	1368		R. Kharisma Swararia, Balige
594)	1377	0.25	RSPDK Sukoharjo (R. Sukoharjo Makmur)
385)	1385	0.25	RPKDT2 Majalengka
384)	1386		R. Karya Prima 70 (Karima 70), Banda Aceh
387)	1386	0.25	R. Cendra Wasih Pusat, Jakarta
391)	1386	0.25	R. Cakra Bhuwana, Malang
392)	1386		R. Mudhita Buana B.S., Pontianak
608)	1386		R. Dhara Perbawa Swara Pada, Pariaman
674)	1386		R. Swara Maya Prastha, Poso
1)	1398	50	RRI Bandung
592)	1400	0.5	RKPDKDT2 Banyuwangi (R. Swara Blambangan)
393)	1404		R. Simponi, Medan *
396)	1404	0.25	R. Bahtera Yudha, Surabaya
397)	1404		R. Moderato, Madiun
415)	1404		Suara Ramayana Jelita, Palu
611)	1404		R. Puspa Irama, Belitang Oku
522)	1405		R. Swara Nava Ria Gemilang, Palangkaraya
526)	1413		RRI Sibolga
399)	1422		R. El Em Bahama, Padangpanjang
643)	1422		R. Perkasa Muda Agung (PMA), Kraksaan
644)	1422		R. Suara Purwodadi Bersemi (Pursemi), Purwodadi
432)	1422		R.Suara Manusia Indah, Pontianak
593)	1422	0.25	R. Citra Kisarannada, Kisaran
624)	1422		R. Karastina, Tanjung Pinang
675)	1422		R. Dian Kusuma Jaya, Lubuklinggau
29)	1440	0.25	R. Telstar, Ujung Pandang
404)	1440	0.25	RPDT2 Brebes
407)	1440	0.25	R. Muria, Kudus
408)	1440		R. Dian Sindoro Suara Semesta (DSS), Temanggung
409)	1440	0.25	R. Nada Kemala Jaya, Sumenep
410)	1440	0.25	R. Gema Surya, Ponorogo
537)	1440		RPDT2 Bengkalis
563)	1440	0.25	R. Indra Kencana, Tembilahan
418)	1440		Suara Totabuan Ria, Kotamobagu
645)	1440		R. Budaya Karo, Kabanjahe
503)	1442	0.5	RDT2 Indramayu
550)	1442	0.5	R. Pesona Lematang, Lahat
596)	1450	0.2	RPDT2 Kampar, Bangkinang
254)	1458		R. Swara Cakrawala Sangkuriang, Bandung
414)	1458	0.25	R. Salvatore, Surabaya
651)	1458		R. Sritanjung Setia, Rogojampi
648)	1460		R. Suara Tunjung Nyaho, Palangkaraya
475)	1465		R. Kencana Perkasa, Pematang Siantar
652)	1465		R. Winayaka, Medan *
453)	1475	1	RKDT2 Karawang
451)	1476	0.25	R. Siaran Niaga Hiukencana (RHK), Semarang
653)	1476		R. R.H.K, Semarang
417)	1485	0.5	RKPDK Madiun
419)	1485		R. Media Mahasiswa Universitas Sumatera Utara (USU), Medan *
420)	1485	0.25	R. Bimantara, Bukittinggi
422)	1485	0.25	R. Dian Irama, Jambi
425)	1485		R. Angkatan Bersenjata (R. ABRI Suara Jakarta), Jakarta
426)	1485		R. Sindang Kasih, Cirebon
286)	1485	0.3	RRI Padang
428)	1485		R. Esti Mada Cita (EMC), Yogyakarta
429)	1485		R. Angkasa Jaya, Probolinggo
434)	1485	0.25	R. Swara Angela Permai, Manado
435)	1494	0.25	R. Toddo Puli (Topsi), Palu
642)	1494		R. Blora Sakti (RBS), Cepu
654)	1494		R. Al Rona Bahana, Padangsidempuan
416)	1503	0.5	RKPDT2 Kotamadya Kediri (R. Gema Kediri)

	kHz	kW	Station
438)	1503		R. Swara Lebak Ria, Rangkasbitung
439)	1503	0.25	R. Radiks 99 Suara Kebahagiaan, Semarang
444)	1503		R. Alnora, Medan
446)	1512	0.25	R. Swara Mitra Dirgantara (Ramona Jelita), Balikpapan
447)	1521		R. Dirgantara, Langsa
449)	1521		R. Cendrawasih Karya Murni, Pematangsiantar
450)	1521	0.1	RSPDK Klaten
531)	1521	0.3	RSPDKDT2 Wonosobo
595)	1521	0.25	RSPDT2 Kotamadya Pekalongan (Suara Kota Batik)
649)	1521		R. Antares, Sidoarjo
676)	1530		R. Swara Rhamagong, Kupang
454)	1539	0.2	R. Niaga dan Budaya Simalungun, Pematangsiantar
455)	1539	0.25	R. Esti Elita, Pekanbaru
461)	1539		R. Cakrawala Bhakti, Surabaya
464)	1539	0.25	R. Berita Harapan, Manado
367)	1539		R. Rhodisko Rasisonia, Banda Aceh
552)	1545	0.75	RKPKDT2 Bekasi
470)	1550	0.06	R. Programa Hiburan dan Informasi DT2 Indragiri Hilir, Tembilahan
524)	1551	0.1	RRI Manado
467)	1557		R. Miraka Yunior, Bogor
468)	1557	0.25	R. Fantasy 70, Jatiwangi
474)	1557	0.25	RSPDT2 Demak (Suara Kota Wali)
655)	1557		R. Al Masilah, Jakarta
656)	1563	0.3	RPDT2 Garut
469)	1566		R. Swara Primadona Mahardika, Cikampek
657)	1566		R. Pesona Bahari, Weleri
466)	1584		R. Pangkalan Remaja Derap Bhakti (PRDB), Samarinda
476)	1584		R. Dian Erata, Padangpanjang
483)	1584		Swara Al Karomah Pratama, Martapura
485)	1584	0.25	R. Hex, Balikpapan
554)	1584		R. Sumber Kasih, Manado
680)	1584		R. Gema Bhakti Yudha Seroja, Bekasi
547)	1593		R. Khusus Informasi Pertanian (RKIP), Surabaya
486)	1602		R. Lintas Triaga Angkasa (Lita), Bukittinggi
487)	1602	0.15	R. Cynthia Rama, Pekanbaru
489)	1602		R. Paksi, Bandung
490)	1602	0.25	R. Swadaya Cempaka 23, Karawang
493)	1602		R. Karya Dharma, Manado

Shortwaves: r=reported; v=varies; *=inactive.

	kHz	kW	Station
33)	2490v	2	RRI Ujung Pandang. 2055-0130, 0400-0800
658)	2580		RSPDT2 Timor Tengah Selatan, Soe: 2155-2400, 0900-1400
664)	2695	0.5	RPDT2 Ende: 0930-1455
542)	2899		RPDT2 Ngada, Bajawa: 0900-1500
543)	2960	0.3	RPDT2 Manggarai, Ruteng. 2145-2300, 0900-1425
178)	3115v		RSPDT2 Halmahera Tengah, Soasio: 2300-0300, 0830-1430
1)	3204	10	RRI Bandung *
524)	3215	10	RRI Manado: 2100-0200, 0800-1410
128)	3223	5	RRI Mataram *
287)	3225	10	RRI Tanjung Pinang: 2200-0100, 1000-1215
525)	3232	10	RRI Bukittinggi: 2200-0300, 1030-1705
257)	3250	20	RRI Banjarmasin: 2100-0300, 0900-1500
527)	3265	10	RRI Bengkulu: 2100-0300, 0900-1600
528)	3265v	10	RRI Gorontalo: 2100-0015, 0800-1300
305)	3295		RRI Samarinda: 0900-1500 irr.
533)	3325	10	RRI Palangkaraya: 2200-0100, 0800-1600
535)	3345	10	RRI Ternate: 2000-0030 (Sun 2330), 0750-1520
247)	3355	0.6	RRI Sumenep: 1700-2215
246)	3355	1	RRI Jambi: 0100-0900, 1700-2200
101)	3375	20	RRI Medan *
130)	3385	10	RRI Kupang: 2100-0155, 0900-1520 irr.
209)	3395	10	RRI Tanjung Karang: 2130-0230, 0830-1705 (Sun 2130)
650)	3542		RSPDT2 Sumba Timur, Waingapu: 1000-1515
190)	3578	0.4	RSPDT2 Maluku *
659)	3630v		RPDT2 Buol, Tolitoli: 1000-1355 irr.

kHz	kW	Station
93)	3905	10	RRI Merauke: 0700-1400, 1900-2200
326)	3905	50	RRI Banda Aceh: 2215-0100, 0930-1350 irr.
92)	3934	5/10	RRI Semarang *
209)	3946	2.5	RRI Tanjung Karang: 0230-0830, 1515-2200 (rel. Prgr. Nasional 1700-2200)
286)	3960	1	RRI Padang: 1645-2200
560)	3960	10	RRI Palu: 0900-1600, 2130-0100 irr.
315)	3976		RRI Pontianak: 2200-0100, 1000-1605
511)	3987	1	RRI Manokwari *
172)	3996		RRI Kendari: 0015-0750
315)	3996		RRI Pontianak: 1600-2050
172)	4000	5	RRI Kendari: 2030-0015, 0750-1600
286)	4000		RRI Padang (alt. to 4003)
286)	4003	10	RRI Padang: 2200-0300, 0900-1700
146)	4606	1	RRI Serui: 2000-2315, 0925-1530
547)	4697	2.5	R. Khusus Informasi Pertanian (RKIP), Surabaya: 2200-0200. 0500-0700, 0900-1515
33)	4719	50	RRI Ujung Pandang (irr. standby tr. operates only when tr. on 4750 breaks down)
33)	4753	20	RRI Ujung Pandang: 2055-0130, 0745-1605
101)	4766	20	RRI Medan *
195)	4777	7.5	RRI Jakarta (Kebayoran): 1045-2315
565)	4789	1	RRI Fak-Fak: 2000-0100, 0700-1000
67)	4845	10	RRI Ambon: 2000-0015, 0330-0600, 0800-1130
566)	4866	0.3	RRI Wamena *
566)	4871	0.05	RRI Wamena: alt. to 4867 *
567)	4875	10	RRI Sorong: 0800-1430 irr.
525)	4911		RRI Bukittinggi: 1030-1705, 2200-0300 irr.
287)	4920	1	RRI Tanjung Pinang: 1215-1705
246)	4925	10	RRI Jambi: alt. to 4927
246)	4927	7.5	RRI Jambi: 2200-0100 (Sun 0700), 0900-1700
186)	4931.6	10	RRI Surakarta: 1045-2200 (Sat only)
158)	5040		RRI Pekanbaru: 2200-0315, 0930-1705
252)	5046	20	RRI Yogyakarta *
568)	5055		RRI Nabire: 2000-0200, 0800-1215 irr.
252)	5059		RRI Yogyakarta: 0215-0305, 1155-1515
599)	5691.2	0.5	RPDT2 Berau, Tanjungredeb: 1000-1410 irr.
524)	5987.7	10	RRI Manado: 0200-0300
257)	6025		RRI Banjarmasin: alt. to 6031
257)	6031	10	RRI Banjarmasin: 0300-0900, 1500-2100
222)	6070	20	RRI Jayapura. 0700-0100 (**English:** Sun 2115-2130)
305)	6134	1	RRI Samarinda: 0200-0815
211)	6153		RRI Biak: 0730-1500, 2000-2400
511)	6189	0.75	RRI Manokwari. 2025-2400, 0225-0625, 0800-1400
252)	7099	7.5	RRI Yogyakarta: 2200-0215
146)	7173	0.5	RRI Serui: 2315-0925
565)	7231	0.5	RRI Fak-Fak: 0100-0700, 1000-1500
560)	7235		RRI Palu: 0100-0855
195)	9525	250	V. of Indonesia (Cimanggis): 0030-0400, 0800-1300, 1730-2100
33)	9552	7.5	RRI Ujung Pandang: 0000-0830
195)	9565	250	RRI Jakarta (Bonto Sunggu): 2100-1605
195)	9630	250	RRI Jakarta (Bonto Sunggu): 2100-1605
195)	9680	100	RRI Jakarta (Cimanggis). 2200-0200, 0900-1300
315)	9705	50	RRI Pontianak: 0315-0700
567)	9743	10	RRI Sorong: 0000-0755
195)	11750	250	RRI Jakarta (Bonto Sunggu): 2200-0200, 0900-1300
195)	11785	250	RRI Jakarta (Cimanggis): 0400-1605
195)	11885	250	RRI Jakarta (Bonto Sunggu): 2200-0200, 0900-1300
195)	15125	250	RRI Jakarta (Cimanggis): 2200-0800
195)	15150	250	RRI Jakarta (Cimanggis): 2200-0200, 0900-1300

FM: See address list for RRI FM freqs. There are 245 known non-RRI st's broadcasting on FM only, which are not included.

HOME SERVICES

RADIO REPUBLIK INDONESIA (Gov.)
✉ Jln Merdeka Barat 4-5, P.O. Box 356, Jakarta.
Central Station: Jakarta 195)
National Prgr. 1: (Programa Nasional Satu) on 999, 4777, 9565, 9630, 11785, 15125kHz + FM 93.2/101.4MHz: 24h. **N:** on the h.

National Prgr. 2 (Programa Nasional Dua) on 9680, 11750, 11885, 15150kHz: 2200-0200, 0900-1300 (may relay National Prgr. 1 at other times).
Metropolitan Prgr. 1(Programa Ibukota 1): 2300-1700 on 1332kHz + FM 99.8MHz. **English:** 0515-0600 (exc. Fri).
Metropolitan Prgr. 2 (Programa Ibukota 2): 2200-1800 on 1215kHz + FM 105.1MHz.

FOREIGN SERVICE

THE VOICE OF INDONESIA
✉ P.O. Box 1157, Jakarta 10001.
SW: Jakarta (Cimanggis): 9525kHz 250kW. **FM:** Jakarta 103.0MHz.
IS: Rayuan Pulau Kelapa (the song of the Coconut Islands) played on a Hammond Organ, Flute, Vibe and Piano.
V. by QSL-card or letter. Listeners reports are welcomed.
PUB: Weekly: RRI Radio dan Televisi (Home Prgr.)
For language schedule see International Broadcasting section.

Commercial Stations: Permitted power: 0.5kW. Permitted D. Prgr: 2200-1700.
Radio Angkatan Bersenjata/Radio Angkatan Udara (Military Stations).
Provincial Gov. Station: Radio Khusus Pemerintah Daerah Tingkat Satu.
District Stations: Radio Khusus Pemerintah Daerah Tingkat II (Dua). Sometimes "Khusus" (special) is deleted. "Kabupaten" is occ. used instead of, or as well as, Daerah Tingkat Dua. Both mean "District".
Municipal Stations: Only intended for particular towns; Radio Khusus Pemerintah Daerah (e.g. Kotamadya Surabaya).
Further Radio Pemerintah Daerah (RPD) stations are known to exist, the majority of which operate on MW frequencies. However, outside Java and Bali, there are several such stations which still use SW. Here is a province list of those Kabupaten (with capital town indicated in brackets where different from the Kabupaten name) and Kotamadya (KM) whose station frequencies were not available at the time of editing. Station headings (e.g. RPDT2) will vary from station to station.
D.I. Aceh. Aceh Barat (Meulaboh), Aceh Selatan (Tapaktuan).
Sumatera Utara. Langkat (Binjai), Pulau Nias (Gunungsitoli), KM Tanjung Balai, R. Informasi Pertanian (Tg. Morawa), Karo (Kabanjahe), Tanapuli Selatan (Padangsidempuan).
Jambi. Kerinci (Sungaipenuh).
Sumatera Selatan: Belitung (Tanjungpandan).
Jawa Barat. Bogor, KM Bogor, KM Sukabumi, Kuningan.
Jawa Timur. KM Madiun.
Bali. Jembrana (Negara), Karangasem (Amlapura).
Nusa Tenggara Barat. Lombok Tengah (Praya), Bima (Raba).
Nusa Tenggara Timur. Timor Tengah Utara (Kefamenanu).
Sulawesi Utara. Gorontalo (Limboto), Sangir Talaud (Tahuna), KM Bitung, Bolaang Mongondow (Kotamobagu).
Sulawesi Tengah. Donggala, Banggai (Luwuk), Poso.
Sulawesi Selatan. Soppeng (Watangsoppeng).
Kalimantan Timur. Bulungun (Tg. Selor).
Kalimantan Selatan. Hulu Sungai Tengah (Barabai), Hulu Sungai Utara (Amuntai), Tabalong (Tanjung), KM Banjarbaru, Hulu Sungai Selatan (Kandangan).
Kalimantan Tengah. Kotawaringin Barat (Pangkalanbuun).
Kalimantan Barat. Kapuas Hulu (Putussibau), Pontianak (Mempawah), Sambas (Singkawang), Sintang, Ketapang, Sanggau.
Maluku: Maluku Tenggara (Tual).

Indonesian Station Headings:
RDK	Radio Daerah Kabupaten
RKDT2	Radio Kabupaten Daerah Tingkat Dua
RKPD	Radio Khusus Pemerintah Daerah
RKPDT2	Radio Khusus Pemerintah Daerah Tingkat Dua
RKPKDT2	Radio Khusus Pemerintah Kabupaten Daerah Tingkat Dua
RKPDK	Radio Khusus Pemerintah Daerah Kabupaten
RKPDKT2	Radio Khusus Pemerintah Daerah Kabupaten Tingkat Dua
RPD	Radio Pemerintah Daerah
RPDT1	Radio Pemerintah Daerah Tingkat Satu
RPDK	Radio Pemerintah Daerah Kabupaten
RPDKDT2	Radio Pemerintah Daerah Kabupaten Daerah Tingkat Dua
RPDT2	Radio Pemerintah Daerah Tingkat Dua
RPKDT2	Radio Pemerintah Kabupaten Daerah Tingkat Dua

RRI   Radio Republik Indonesia
RS   Radio Siaran
Note: There are many other possible station headings, however, these can be determined by using the appropriate letters/words.
Addresses (Jln = Jalan)

1) Jln Diponegoro 61, Kotak Pos 1055, Bandung 40010, Jawa Barat. **FM:** 96.0/102.0/104.0MHz.
2) Jln Mawar 106, Karang Baru - Kuala Simpang, D.I. Aceh.
3) Jln Raya Bojongsari 17, Sawangan - Bogor 16516, Jawa Barat.
4) Jln Sumantri Bojonegoro 74, Jambi 36121, Propinsi Jambi.
5) Jln H. Sibli Imansyah 3, Barabai 71314, Kalimantan Selatan.
6) Jln Buru 34, P.O. Box 1479, Ujung Pandang 90171, Sulawesi Selatan.
7) Jln Sultan Thoha 1, Komplek Airpanas, Muarabulian 36610, Jambi.
8) Jln Kenanga 15, Ambon, Maluku.
9) Jln Brigjen. Slamet Riyadi, Ungaran.
Semarang 50511, Jawa Tengah.
10) Jln Tan Malaka 201C, Kp. Manggis - Jambi 36134, Propinsi Jambi.
11) Jln Jend. A. Yani 7, Pringsewu, Lampung.
12) Jln Raya Serang Km. 2, Kesambi - Pandeglang 42213, Jawa Barat.
13) Jln Dr. Soetomo 13, Pare - Kediri, Jawa Timur.
14) Jln Sunan Ngerang 2A, Juwana - Pati, Jawa Tengah.
15) Jln Dieng 1A, Wonosobo 56311, Jawa Tengah.
16) Jln Tegalgendu 12, P.O. Box 1247, Kotagede-Yogyakarta, D.I. Yogyakarta.
17) Jln Sungai Limboto 42, Ujung Pandang, Sulawesi Selatan.
18) Jln Sultan Hasanuddin 14, Parepare 91114, Sulawesi Selatan.
19) Jln Pemuda 82-90, Kotak Pos 239, Surabaya 60271, Jawa Timur. **FM:** 95.4/97.6MHz.
20) Jln Dr. Cipto Mangunkusumo 24, Telukbetung-Bandar Lampung, Lampung.
21) Jln Parakan Asri 17, Bandung, Jawa Barat.
22) Jln Raya Km. 4, Kaduagung Barat - Rangkasbitung 42311, Jawa Barat.
23) Jln Radio 9, Cirendang - Kuningan, Jawa Barat.
24) Jln Margantara Tanjung, P.O. Box 45, Purwokerto, Jawa Tengah.
25) Jln Karanganyar Gunung 254-255, Semarang 50255, Jawa Tengah.
26) Jln Kapt. Mulyadi 117, Surakarta 57113, Jawa Tengah.
27) Jln Jend. Sudirman 41A, Metro 34114, Lampung.
28) Jln Ciung Wenara, Komp. Nitimandala Renon, Denpasar, Bali.
29) Jln Ali Malaka 26, Ujung Pandang, Sulawesi Selatan.
30) Jln K.H. Ghalib Raya 416, Pringsewu 35373, Lampung.
31) Jln Kemakmuran 26, Ambon, Maluku.
32) Bandar Lampung, Lampung.
33) Jln Riburane 3, Kotak Pas 103, Ujung Pandang 90111, Sulawesi Selatan. **FM:** 97.6/99.8MHz.
34) Jln Jend. Sudirman Khairbras 1, Banjar Agung - Metro 34114, Lampung.
35) Jln Teuku Umar 46, Gorontalo 96115, Sulawesi Utara.
36) Jln Sukarno-Hatta Bypass 32, P.O. Box 91, Panjang - Bandar Lampung 35001, Lampung.
37) Jln Raya, Way Jepara, Lampung 38) Jln R.A. Wirya Atmaja 28, Purwokerto, Jawa Tengah.
39) Jln Pahlawan 313/41, P.O. Box 3, Purwakarta 41115, Jawa Barat.
40) Jln Barito 4, Pekalongan 51116, Jawa Tengah.
41) Jln Raya Candimas 561, Kotabumi, Lampung.
42) Jln Ir. Sukarno 320, Sukadana, Lampung 43) Jln Kol. Sugiono 72, Banjarmasin 70233, Kalimantan Selatan.
44) Jln Buru 28, Kotak Pos 45, Ujung Pandang 90171, Sulawesi Selatan.
45) Jln Mesjid Raya 100, Sengkang - Wayo, Sulawesi Selatan.
46) Jln Santo Yoseph 17A, Kleak - Manado 95115, Sulawesi Utara.
47) Jln Dr. Siwabessy SK 12/30, Wainutu - Ambon 97115, Maluku.
48) Jln K.H. Abdul Hadi 106, Serang 42117, Jawa Barat.
49) Jln Ion Martasasmita 24, Pamanukan - Subang 41254, Jawa Barat.
50) Jln Proklamator Raya 163, Bd. Jaya-Lampung Tengah, Lampung.
51) Jln. Raya Sribawono Panjang 576, Sribawono, Labuhan Maringgai 34198, Lampung.
52) Jln Raya Hamka 576, Sidomulyo - Kalianda, Lampung.
53) Jln Wonopringgo 414, Pekalongan, Jawa Tengah.
54) Jln Purworejo VI/10, Surakarta, Jawa Tengah.
55) Jln Jeruk Raya 27, Semarang 50249, Jawa Tengah.
57) Komplek Perluasan Kota, Lln. Plongkowati, Wonogiri, Jawa Tengah.
58) Jln Terusan Enim/Wai-Ketibung 25, Bandar Lampung 35214, Lampung.
59) Jln Sudirman 73, Pinrang, Sulawesi Selatan.
60) Jln Batulawang 1, Banjar - Ciamis 46133, Jawa Barat.
61) Jln Pahlawan 6, Kotabumi, Lampung.
62) Jln D.I. Panjaitan 3, Banjarnegara 53415, Jawa Tengah.
63) Jln Raya Jakarta Km. 3, Warang Batu - Cianjur 43251, Jawa Barat.
64) Jln Kapasan 112, Belitang - OKU 32182, Sumatera Selatan.
65) Kotabumi, Lampung.

66) Jln Simpang Tiga Tanjung Seteko, Indralaya - OKI 30662, Sumatera Selatan.
67) Jln Jendral Akhmad Yani 1, Ambon, Maluku. **FM:** 90.0/105.5/107.0MHz.
68) Jln Gajah Mada 5, Pekalongan 51118, Jawa Tengah.
69) Jln Abdul Muis Ridhani 5, Barabai, Kalimantan Selatan.
70) Jln Ratulangi 23, Padang, Sumatera Barat.
71) Jln Overste Isdiman 22, Purbalingga 53313, Jawa Tengah.
72) Jln Mohammad Toha 146, Bandung 40243, Jawa Barat.
73) Jln Irian Jaya IV/10, Tebu Ireng - Jombang, Jawa Timur 74) Jln Cagak Gobras 3, Tasikmalaya, Jawa Barat.
75) Jln Pangeran Drajat, Kesambi - Cirebon 45133, Jawa Barat.
76) Jln K.H. Agus Salim 22, Surakarta 57147, Jawa Tengah.
77) Jln Pemuda 521, Gedong Tataan- Lampung Selatan, Lampung.
78) Jln Sriwijaya 12, Cakranegara, Mataram, Lombok, Nusa Tenggara Barat.
79) Jln Manggis 31, Banjarmasin, Kalimantan Selatan.
80) Jln Rajawali 16, Ujung Pandang, Sulawesi Selatan.
81) Jln Kemakmuran 4, Rantepao, Sulawesi Selatan.
82) Jln Arie Lasut, Manado, Sulawesi Utara.
83) Jln Jendral Soedirman 427, Kotak Pos 5, Purwokerto 53116, Jawa Tengah. **FM:** 94.0/97.6/98.8/104.0MHz.
84) Jln Nusantara 261A, Ujung Pandang 90171, Sulawesi Selatan.
85) Jln Haji Ikap 47, Palangkaraya, Kalimantan Tengah.
86) Jln Jend. A. Yani 64, Purwakarta, Jawa Barat 87) Jln Anggrek 4, Surabaya, Jawa Timur.
88) Jln Kutoarjo 60, Kebumen, Jawa Tengah.
89) Jln Nusantara II/50, Manado, Sulawesi Utara.
90) Jln Veteran 4, Denpasar, Bali.
91) Jln Jendral A. Yani 296, Bandung, Jawa Barat.
92) Jln Ahmad Yani 144-146, P.O. Box 1073, Semarang 50241, Jawa Tengah. **FM:** 89.0/95.1MHz.
93) Jln Jendral A. Yani, P.O. Box 11, Merauke 99601, Irian Jaya. **FM:** 90.0/105.0MHz.
94) Jln Letjen. Bambang Utoyo 113, RT 13A, 3 Ilir, Palembang 30116, Sumatera Selatan.
95) Jln H.O.S. Cokroaminoto 10, Tanjungkarang-Bandar Lampung, Lampung.
96) Jln Cirebon 3, Belawan-Medan, Sumatera Utara.
97) Jln Pramuka 10, Majalengka 45418, Jawa Barat.
98) Jln Candi Agung 20/III, Amuntai 71418, Kalimantan Selatan.
99) Jln Macan 21, Ujung Pandang 90132, Sulawesi Selatan.
100) Jln S. Parman 161, Manado 95121, Sulawesi Utara.
101) Jln Letkol Martinus Lubis 5, Medan 20232, Sumatera Utara. **FM:** 88.0/90.6/92.4/93.5/97.8/105.0MHz.
102) Jln H. Peeng 9, Batusari - Kebun Jeruk, Jakarta Barat 13520.
103) Jln Siliwangi 5, Bandung 40132, Jawa Barat.
104) Jln Bojong Tengah 2B, Tasikmalaya, Jawa Barat.
105) Jln Baturaden 433 Km. 8, Purwokerto, Jawa Tengah.
106) Jln Letjend. S. Parman 45, Subang 41215, Jawa Barat.
107) Jln Kertijayan 234, Pekalongan 51171, Jawa Tengah.
108) Jln Bhayangkara 49, Surakarta, Jawa Tengah.
109) Jln Jend. A. Yani 22, Bantul, D.I. Yogyakarta.
110) Ambarawa, Jawa Tengah.
111) Jln Dr. Sutomo 26, Malang, Jawa Timur.
112) Jln Belitung Darat 11, Banjarmasin 70128, Kalimantan Selatan.
113) Jln Jendral Achmad Yani XX/118, Sampit, Kalimantan Tengah.
114) Jln Samiun 17, Ujung Pandang, Sulawesi Selatan.
115) Jln Hasanuddin 20, Manado 95235, Sulawesi Utara.
116) Jln Hayam Wuruk 78A, Denpasar 80235, Bali.
117) Komplek Taman Wisata Panataran, Blitar, Jawa Timur.
118) Jln Mongonsidi 73A, Telukbetung - Bandar Lampung, Lampung.
119) Perumahan Jati Indah, Blok 4, Jln Melati 4, Jatibarang - Indramayu 45273, Jawa Barat.
120) Jln L.R.E. Martadinata 229, Bandung, Jawa Barat.
121) Jln Raya Timur 12, Singaparna-Tasikmalaya, Jawa Barat.
122) Jln D.I. Panjaitan 3, Surakarta, Jawa Tengah.
123) Jln Dharmahusada Indah Barat, Blok A/75, Surabaya, Jawa Timur.
124) Jln Simpang Tlogomas II/14, Malang, Jawa Timur.
125) Jln Bali 23B, Banjarmasin 70114, Kalimantan Selatan.
126) Slawi - Tegal, Jawa Tengah.
127) Jln Kawi V/1, Semarang, Jawa Tengah.
128) Jln Langko 83, P.O. Box 2, Ampenan-Mataram 83114, Lombok, Nusa Tenggara Barat. **FM:** 89.1/104.0MHz.
129) Jln Brigjen. Dharsono/By Pass, Cirebon 45132, Jawa Barat. **FM:** 107.3MHz.
130) Jln Tompello 8, Kupang, Timor, Nusa Tenggara Timur. **FM:** 90.6/93.5/102.0MHz.
131) Jln Merdeka 99, Bangli 80614, Bali.
132) Jln Nusantara 14, Payakumbuh 26210, Sumatera Barat.
133) Jln Temon 259, 26 Ilir, Palembang, Sumatera Selatan.

134) Jln Mesuji 1, Pahoman - Telukbetung, Bandar Lampung, Lampung.
135) Jln Bali Matraman 17, Jakarta Selatan.
136) Jln Raya Cimalaka 152, Cimaaka - Sumedang, Jawa Barat.
137) Jln Simolawang I/96, Surabaya, Jawa Timur.
138) Jln Jend. A. Yani Km. 4 No. 321B, Banjarmasin 70113, Kalimantan Selatan.
139) Jln Raya Sambas Bukitluwing 1, Sambas 79162, Kalimantan Barat.
140) Jln Pembangunan 9 R.E. Martadinata, Balikpapan, Kalimantan Timur.
141) Jln Angkasa 1, Bandung, Jawa Barat.
142) Jln Paria 8, Bandung 40262, Jawa Barat.
143) Jln Pembangunan 7, Garut 44151, Jawa Barat.
144) Jln Piere Tendean 50, Banjarmasin, Kalimantan Selatan.
145) Jln Andi Naboang 1, Pangkajene, Sulawesi Selatan.
146) Jln Pattimura, Serui, Irian Jaya. **FM:** 97.5MHz.
147) Jln Candi Panggung 58, Malang, Jawa Timur. **FM:** 93.5/102.0/105.1MHz.
148) Jln Sisingamangaraja 1, Padang, Sumatera Barat.
149) Jln Jambu Air 18, P.O. Box 42, Bukittinggi, Sumatera Barat.
150) Jln Pang. Besar Sudirman 141, Kediri, Jawa Timur.
151) Jln Puntodewo, Gadingan - Wates, Kulon Progo, D.I. Yogyakarta.
152) Jln Gadung III/2, Surabaya, Jawa Timur.
153) Jln T. Nya' Arief 2, Langsa, D.I. Aceh.
154) Jln Tebet Timur Dalam I-N/249, Jakarta, Selatan.
155) Jln Jend. Sudirman 10, Pancor - Selong 83611, Lombok, Nusa Tenggara Barat.
156) Jln Raya Kedung Halang 2, Waru Jambu - Bogor 16710, Jawa Barat.
157) Jln Jend. Sudirman 265, Ciamis 46211, Jawa Barat.
158) Jln Candi Panggung 58, Mojolangu-Malang 65142, Jawa Timur. **FM:** 102.0/105.0 MHz.
159) Jln Satria Selatan III/H 262, Semarang, Jawa Tengah.
160) Jln Dakota V/1, Kemayoran, Jakarta 10630.
161) Jln Babakan 85, Majalaya - Bandung, Jawa Barat.
162) Jln Raya Siliwangi 311, Cicurug - Sukabumi 43159, Jawa Barat.
163) Jln Honggowongso Panularan 2/VI , Surakarta 57149, Jawa Tengah.
164) Jln Kapt. Sudiboyo 46, Lt. 2, Tegal 52113, Jawa Tengah.
165) Jln Adityawarman 81, Surabaya 60242, Jawa Timur.
166) Jln Banjaran II/29, Kediri 64124, Jawa Timur.
167) Jln Veteran 6B, Bondowoso 68211, Jawa Timur.
168) Jln Bung Karno 22, Mataram, Lombok, Nusa Tenggara Barat.
169) Jln M.T. Haryono 64, Tanjungbalai 21311, Sumatera Utara.
170) Jln Siliwangi 103, Pelabuhanratu - Sukabumi 43164, Jawa Barat.
171) Jln Monginsidi 89B, Medan 20157, Sumatera Utara.
172) Jln Laute Mandonga 44, Kotak Pos 7, Kendari 93111, Sulawesi Tenggara. **FM:** 91.2/103.0/107.1MHz.
173) Jln Dr. M. Isa 38, 8 Ilir, Palembang, Sumatera Selatan.
174) Jln Teuku Umar 65A, Rajabasa - Bandar Lampung, Lampung.
175) Jln Raya Bedahan 17, Babat - Lamongan, Jawa Timur.
176) Jln Dr. F. L. Tobing 59, Sidikalang 22212, Sumatera Utara.
177) Jln Perintis Kemerdekaan 86, Cibadak-Sukabumi, Jawa Barat.
178) Jln A. Malawat, Soasio 97812, Tidore, Maluku.
179) Jln Makam Pahlawan Kusumajati, P.O. Box 2, Kendal, Jawa Tengah.
180) Jln Mentaok Raya 9, Kotagede - Yogyakarta 55173, D.I. Yogyakarta.
181) Jln Aipda Karel Sasuit Tubun 15, Gresik 61114, Jawa Timur.
182) Jln Brigjen. H. Hasan Basri 64, Banjarmasin, Kalimantan Selatan.
183) Jln D.I. Panjaitan 61, Jember, Jawa Timur. **FM:** 91.0/104.0MHz.
184) Jln Pasundan 81, Sukabumi, Jawa Barat.
185) Jln Brig. Jendral Katamso 43, Lubuk Pakam-Medan, Sumatera Utara.
186) Jln Abdul Rahman Saleh 51 (or: Kotakpos No. 40), Surakarta, Jawa Tengah. **FM:** 102.0/105.0MHz.
187) Jln C. Asem Baris Raya 26, Jakarta.
188) Jln Nurali 30, Pontianak 78111, Kalimantan Barat.
189) Jln Desa Koyamas I/36, Langowan, Sulawesi Utara.
190) Jln Pattimura, Masohi, Seram, Maluku.
191) Jln Kom. Yos Sudarso 9, Sanggau 78582, Kalimantan Barat.
192) Jln Diponegoro 99, Amlapura, Bali.
193) Jln Bukit Inspirasi, P.O. Box 10, Tomohon-Minahasa, Sulawesi Utara.
194) Jln Jenderal Achmad Yani 32, Temanggung, Jawa Tengah.
195) Tromolpos 157, Jln Medan Merdeka Barat 4-5, Jakarta, 10001.
196) Jln Tanggamus 548, Talangpadang 35377, Lampung 197) Jln Sadanco Supriadi 40, Blitar, Jawa Timur.
198) Jln Mayjen. Panjaitan 10, Madiun, Jawa Timur. **FM:** 98.5/99.0/104.0MHz.
199) Jln Iskandar Muda 117A, Medan 20119 Sumatera Utara.

200) Jln Raya Cipondoh (K.H. Hasyim Ashari) 82, Tangerang 15140, Jawa Barat.
201) Jln D. Menambon 738, Mempawah 78912, Kalimantan Barat.
202) Jln Manggis I/20, Sukabumi, Jawa Barat.
203) Jln Dr. Cipto Mangunkusomo 175, Lt. 2, Semarang 50125, Jawa Tengah.
204) Waikabubak, Sumba, Nusa Tenggara Timur 205) Jln Darmokali 19, Surabaya 60241, Jawa Timur.
206) Jln Joyoboyo 77, Kediri 64132, Jawa Timur.
207) Jln Raya Gianyar 100x, Sukawati-Gianyar, Bali.
208) Jln Kayutangi II Jalur I/88, Banjarmasin 70110, Kalimantan Selatan.
209) Jln Gatot Subroto 26, Kotakpos 24, Pahoman-Bandar Lampung 35213, Lampung. **FM:** 90.9/93.0/ 98.0MHz.
210) Jln Raya Dilem 1, Kepanjen - Malang 65163, Jawa Timur.
211) Jln Achmad Yani, Kotakpos 505, Biak, Irian Jaya. **FM:** 93.7/107.6MHz.
212) Jln Jend. Sudirman 107, Kotabumi 35911, Lampung.
213) Jln Merdeka, Blitar 66110, Jawa Timur.
214) Jln Raya Tipar 16, Sukabumi, Jawa Barat.
215) Jln Ngurah Rai 141, Negara, Bali.
216) Jln Tentara Pelajar 52, Tegal 52122, Jawa Tengah.
217) Jln Bhayangkara 81, Medari-Sleman 55515, D.I. Yogyakarta.
218) Jln Dukuh Kupang Barat XVII/15, Surabaya, Jawa Timur.
219) Jln Kartini 25, Jember 68137, Jawa Timur.
220) Jln Jendral Sudirman 59, Singaraja, Bali.
221) Jln Jend. Sudirman 18, Payakumbuh 26211, Sumatera Barat.
222) Jln Tasangkapura 23, Jayapura, Irian Jaya. **FM:** 90.0/93.6/97.6MHz.
223) Jln Pontianak 22, UKT - Padang, Sumatera Barat.
224) Jln D.I. Panjaitan 3/41, Plaju - Palembang, Sumatera Selatan.
225) Jln Imam Bonjol, Gg. Pelita 45, Tanjungkarang, Bandar Lampung, Lampung.
226) Jln Dr. Moh. Hatta 644A, Baturaja - OKU 32110, Sumatera Selatan.
227) Jln Lengkong Besar, Bandung, Jawa Barat.
228) Jln Sucipto Yudohusodo 29, Bondowoso, Jawa Timur.
229) Jln Pelita 34, Singkawang 78123, Kalimantan Barat.
230) Jln M.H. Thamrin, Oepoi - Kupang 85362, Timor, Nusa Tenggara Timur.
231) Jln Gatot Kaca 45, Negara, Bali.
232) Jln Selaparang 60, Mataram, Lombok, Nusa Tenggara Barat.
234) Jln Letjend. S. Parman 28, Banjarnegara, Jawa Tengah.
235) Jln Raden Intan Way Mangaku, Liwa, Lampung.
236) Komplek PTBA, Jln Parigi 1 Plar A, Tanjung Enim 31716, Sumatera Selatan.
237) Jln Gajah Mada 144, Tromolpos 153, Singaraja 81113, Bali. **FM:** 93.8/103.7MHz.
238) Jln Kartini 2, Bangkalan 69110, Jawa Timur.
239) Jln Lengkong Besar 14, Bandung, Jawa Barat.
240) Jln Suryakencana 91, Sukabumi, Jawa Barat.
241) Jln Ngagel Jaya Utara IV/21, Surabaya 60283, Jawa Timur.
242) Jln Ciliwung 32A, Blitar 66116, Jawa Timur.
243) Jln K.H. Abdul Fatah Hasan 9B, Serang, Jawa Barat.
244) Jln Kawak 26, Sekadau 78582, Kalimantan Barat.
245) Jln Kyai Gembyang 7, Kendal, Jawa Tengah.
246) Jln Jenderal A. Yani 5, Telanaipura-Jambi 36122, Propinsi Jambi. **FM:** 88.8/103.7MHz.
247) Jln Urip Sumoharjo 26, Sumenep 69411, Madura, Jawa Timur. **FM:** 94.4/101.0MHz.
248) Jln Puncak Sekuning 462, Palembang, Sumatera Selatan.
249) Jln Letjen. S. Parman 1, Jakarta Barat.
250) Jln Galang 9, Lubukpakam 20510, Sumatera Utara.
251) Jln K.H.A. Dahlan 1, Karawang, Jawa Barat.
252) Jln Amat Jazuli 4, Tromolpos 18, Yogyakarta 55224, D.I. Yogyakarta. **FM:** 91.2/101.3/103.0/107.3MHz.
253) Jln Suryakencana 91, Sukabumi 43113, Jawa Barat 254) Jln Sukajadi Belakang 227, Bandung 40153, Jawa Barat.
255) Jln Taman Apsari 7, Surabaya 60271, Jawa Timur.
256) Jln Panglima Polim SK 3/2, Banda Aceh, D.I. Aceh.
257) Jln Jenderal A. Yani Km. 3.5, No. 234, Kotak Pos 117, Banjarmasin, Kalimantan Selatan. **FM:** 95.7/97.6/105.5MHz.
258) Jln Pemuda 4, "Balaikota", Tegal, Jawa Tengah.
259) Jln Tgk. Cik Ditiro 2, Banda Aceh, D.I. Aceh.
260) Jln Sukatani Raya 21, Bekasi 17630, Jawa Barat.
261) Jln Gunung Puteng 22, Singkawang, Kalimantan Barat.
262) Jln Sudarsono Km. 12, Tegal, Jawa Tengah.
263) Jln Nomor Umar 300, Amuntai 71410, Kalimantan Selatan.
264) Jln Olahraga 21, Indramayu, Jawa Barat.
265) Jln Raya Utama 6, Kota Lais 38653, Bengkulu.
266) Jln Muga Dalam IV/9-11,Tromolpos 200,Semarang,Jawa Tengah.

267) Jln Kesejahteraan Sosial 63, Wonosewu, Kel. Ngestiharjo, Kec. Kasihan, Bantul 55780, D.I. Yogyakarta.
268) Jln Sei Bohorok 8, Medan, Sumatera Utara.
269) Jln Raya Barat 98, Banjarsari - Ciamis 46383, Jawa Barat.
270) Jln K.H.A. Azhari 136, 12 Ulu, Palembang, Sumatera Selatan.
271) Jln Cendana 26, Rawalaut, Pahoman - Bandar Lampung 35127, Lampung.
272) Jln Suroyo 27, Probolinggo 67210, Jawa Timur.
273) Jln Raya Limboto 58, Gorontalo 96250, Sulawesi Utara.
274) Komplek Taman Mini Indonesia Indah, Pondok Gede, Jakarta.
275) Jln Raung 7, Candi Baru - Semarang, Jawa Tengah.
276) Jln Puro Pakwalaman, Yogyakarta, D.I. Yogyakarta.
277) Jln Amir Hamzah 18, Surabaya 60241, Jawa Timur.
278) Jln Kilisuci 42, Kediri 64132, Jawa Timur.
279) Jln Pahlawan 33, Kandangan 71211, Kalimantan Selatan.
280) Jln Kanjengen C-308, Kranggan - Temanggung 56271, Jawa Tengah.
281) Jln Raya Saketi, Malingping - Lebak 42391, Jawa Barat.
283) Jln Siliwangi 41, Bandung 40131, Jawa Barat.
284) Jln May. Syafei 66B, Serang, Jawa Barat.
285) Jln Diponegoro, Sabang, Weh, D.I. Aceh.
286) Jln Jendral Sudirman 12, Kotakpos 77, Padang 25121, Sumatera Barat. **FM:** 90.9/93.5/103.7MHz.
287) Jln Ahmad Yani, Tanjung Pinang, Riau. **FM:** 97.6MHz.
288) Jln Petojo Utara 7, Jakarta Pusat.
289) Jln Kijang II/12, Biro Puli - Palu 94114, Sulawesi Tengah.
290) Jln Selabintana 146, P.O. Box 59, Sukabumi, Jawa Barat.
291) Jln Pahlawan 134A, Magelang 56116, Jawa Tengah.
292) Jln K.H.A. Dahlan 3, P.O. Box 105, Yogyakarta, D.I. Yogyakarta.
293) Jln Teluk Aru 68, Surabaya 60165, Jawa Timur.
294) Jln Mondoroko 2, Singosari - Malang 65153, Jawa Timur.
295) Jln Jend. A. Yani 123, Singaraja, Bali.
296) Jln Pasir Mas Selatan B-61, Tanah Mas - Semarang 50143, Jawa Tengah.
297) Jln Hayam Wuruk, P.O. Box 31, Denpasar 80001, Bali. **FM:** 88.8/92.0/93.5.100.9MHz.
298) Jln Dipokusumo, Purbalingga, Jawa Tengah.
299) Jln Teuku G.K. Dipunding 15, Meulaboh, D.I. Aceh.
300) Jln Kemakmuran 8, Peleihari - Tanah Laut, Kalimantan Selatan.
301) Jln W.R. Monginsidi 4B, Lantai 2, Padang 25117, Sumatera Barat.
302) Jln Lintar Solo - Boyolali Km. 12, Boyolali 57373, P.O. Box 17, Kartosuro 57560, Jawa Tengah.
303) Jln Gembol Wetan 131, Bandung, Jawa Barat.
304) Jln Letnan Karsono 47, Bondowoso, Jawa Timur.
305) Jln M. Yamin 8, P.O. Box 45, Samarinda, Kalimantan Timur. **FM:** 88.8/93.5MHz.
306) Jln Basuki Rakhmat Timur, Magetan 63314, Jawa Timur.
307) Jln Jendral Soedirman 16A, Cilacap, Jawa Tengah.
308) Jln Veteran 495/9H, Palembang, Sumatera Selatan.
309) Jln Nusa Indah 21, Oepura - Kupang 85117, Timor, Nusa Tenggara Timur.
310) Jln Raya Sukamantri 107, Ciawi - Tasikmalaya, Jawa Barat.
311) Subang, Jawa Barat.
312) Andhika Bldg. Lt. III, Jln Simpang Dukuh 38-40, Surabaya 60275, Jawa Timur.
313) Jln Trunojoyo 20, Sampang, Jawa Timur.
314) Tanjung, Tabalong, Kalimantan Selatan.
315) Jln Jendral Sudirman 7, Kotak Pos 6, Pontianak 78111, Kalimantan Barat. **FM:** 90.3/102.0MHz.
316) Jln Diponegoro 53, Sabang, Weh, D.I. Aceh.
317) Jln Pangrango 34, P.O. Box 232, Bogor, Jawa Barat. **FM:** 91.1/94.0MHz.
318) Jln Raya Siliwangi 101, Ciawi Gebang - Kuningan 45591, Jawa Barat.
319) Jln Satria 222, Pemeungpeuk - Garut 44175, Jawa Barat.
320) Jln Serma Marzuki 30, Bekasi Selatan, Jawa Barat.
321) Jln Jember 17, Genteng-Banyuwangi 68465, Jawa Timur.
322) Jln Pemuda 3, P.O. Box 43, Salatiga 50711, Jawa Tengah.
323) Jln. W.R. Supratman 29, Situbondo, Jawa Timur.
324) Jln Untung Suropati 2B, Kupang 85119, Timor, Nusa Tenggara Timur.
325) Jin Diponegoro 69, Gungungsitoli, Nias, Sumatera Utara.
326) Jln Sultan Iskandar Muda 13, P.O. Box 112, Banda Aceh 23423, D.I. Aceh. **FM:** 88.5/93.0/97.6/103.7MHz.
327) Jln Hokki 21, Medan, Sumatera Utara.
328) Jln Srengseng Raya 45, Kebon Jeruk, Jakarta Barat.
329) Jln Gamelan 28, Bandung, Jawa Barat.
330) Jln S. Parman 45, Subang, Jawa Barat.
331) Jln Pase 10, P.O. Box 127, Lhokseumawe 24314, D.I. Aceh.
332) Jln Raya Kramat 1, Kaliwungu - Kendal 51372, Jawa Tengah.

333) Jln Pesanggrahan V Taman, Madiun 63131, Jawa Timur.
334) Jln Kauman Dawur 226, Bangil-Pasuruan 67153, Jawa Timur.
335) Jln Jaksa Agung Suprapto 35, Banyuwangi 68416, Jawa Timur.
336) Jln May. Jendral Sutoyo 5, Komplek Pondok Indah 26, Banjarmasin, Kalimantan Selatan.
337) Jln Jend. A. Yani 82, Balikpapan, Kalimantan Timur.
338) Jln Ciliwung 32A, Blitar 66116, Jawa Timur.
339) Jln Prof. Muh. Yamin S.H. 3, Selong 83622, Lombok, Nusa Tenggara Barat.
340) Jln Taman Siswa 86, Banda Aceh, D.I. Aceh.
341) Jln Pesantren 19, Ketanggungan - Brebes 52263, Jawa Tengah.
342) Jln R.A. Kartini 138, Prabumulih 31320, Sumatera Selatan.
343) Jln Basuki Rahmat 22, Ketapang 78812, Kalimantan Barat.
344) Jln Lebak Pasar 17, Bogor, Jawa Barat.
345) Jln Kol. Sugiyono 288, Jepara, Jawa Tengah.
346) Jln Waru 14, Tegal, Jawa Tengah.
347) Jln Brigjen. Katamso 132, Purworejo 54115, Jawa Tengah.
348) Jln Citandui 14, Surabaya 60241, Jawa Timur.
349) Jln Radio 2 Km. 4, Palembang 30128, Sumatera Selatan. **FM:** 91.8/93.5/97.2MHz.
350) Jln Tirta Raya 169, Pekalongan 51110, Jawa Tengah.
351) Jln Kacapiring 5, Surabaya 60272, Jawa Timur.
352) Pekanbaru, Riau.
353) Jln Miru 72, Cakranegara-Mataram, Lombok, Nusa Tenggara Barat.
354) Jln Brigjend. Hasan Basri 15, Rantau - Tapin 71111, Kalimantan Selatan.
355) Jln Pegadaian 108, Weleri - Kendal 51355, Jawah Tengah.
356) Jln Cikamiri 7, Cisadea - Bandung, Jawa Barat.
357) Jln Inpres 1, Hagu Tengah - Lhokseumawe 24351, D.I. Aceh.
358) Jln Ki Hajar Dewantara 69, P.O. Box 15, Karawang, Jawa Barat.
359) Jln Cendana 70, Bekasi 17100, Jawa Barat.
360) Jln Dr. Muwardi 47, Badranasri - Karanganyar 57712, Jawa Tengah.
361) Jln Sutan Syahrir 12, Sanggau 78513, Kalimantan Barat.
362) Jln Kyai Jebat 1, Demak, Jawa Tengah.
363) Jln Pangeran Antasari 105, Pangkalanbuun 74114, Kalimantan Tengah.
364) Jln Kartini 12, Jember 68137, Jawa Timur.
365) Jln Jend. A. Yani, Kuala Kapuas, Kalimantan Tengah.
366) Jln Kombes Pol M. Duryat 20, Lamongan 62217, Jawa Timur.
367) Jln Pemancar, Banda Aceh, D.I. Aceh.
368) Jln Durian 82, Pekanbaru, Riau.
370) Jln D.I. Panjaitan 3/41, Plaju 30265, Sumatera Selatan.
371) Jln Merdeka 92A, Garut, Jawa Barat.
372) Jln K.H. Wakhid Hasyim 133, Jombang 61419, Jawa Timur.
373) Jln P.B. Sudirman 52, Pasuruan 67110, Jawa Timur.
374) Jln Jend. Sudirman 1025F/RT.33, Palembang, Sumatera-Selatan.
375) Jln Jend. A. Yani 57, Metro 24111, Lampung.
376) Jln M.T. Haryono I/32, Sampit 74322, Kalimantan Tengah.
377) Jln J. A. Suprapto 85, Bojonegoro 62118, Jawa Timur.
378) Jln Dr. Wahidin 1, Pati, Jawa Tengah.
379) Jln K.H. Wakhid Hasyim 1, Trenggalek 66311, Jawa Timur.
380) Humas Gelora 10 Nopember, Jln Tambaksari, Surabaya 60136, Jawa Timur.
381) Jln Perintis Kemerdekaan, Siswodipuran - Boyolali 57310, Jawa Tengah.
382) Jln Labu 8, Bulukumba 92510, Sulawesi Selatan.
383) Jln Raya Tulung Bawang 148, Unit III, Manggala 34596, Lampung.
384) Jln Perdagangan 12, Lantai 2, Banda Aceh 23242, D.I. Aceh.
385) Jln Jawa Timur, Majalengka, Jawa Barat.
386) Jln Rukun 19, Sigli 24110, D.I. Aceh.
387) Jln Batu Ceper V/52, Jakarta Pusat 10120.
388) Jln R.E. Abdullah 3, Gunung Batu - Bogor 16610, Jawa Barat.
389) Jln Suroso 46, Cianjur 43214, Jawa Barat.
390) Sumatra Bldg. Lt. IV, Jln Sumatra 31 G-H, Surabaya 60281, Jawa Timur.
391) Jln Simpang Aluminium 12, Malang 65100, Jawa Timur.
392) Jln Jendral Urip, Gg. Kutilang 72, Pontianak 78111, Kalimantan Barat.
393) Jln Medan Area Selatan 820, Gg. Sairin, Medan, Sumatera Utara.
394) Jln Damai 4, Siabat - Langkat 20851, Sumatera Utara.
395) Jln M.T. Haryono 144, Sintang 78614, Kalimantan Barat.
396) Jln Bedagung 6, Surabaya 60265, Jawa Timur.
397) Jln Mayjen. Sungkono 137, Madiun, Jawa Timur.
398) Jln Kol. Sugiyono 18, P.O. Box 144, Wonogiri 57612, Jawa Tengah.
399) Jln Prof. M. Yamin SH 4, Padangpanjang 27116, Sumatera Barat.
400) Jln Nusirwan 375A, Curup 39112, Bengkulu.
401) Jln Brawijaya 2, Wonomulyo - Polmas 91352, Sulawesi Selatan.
402) Jln Raya Sukowati 530, Sragen 57215, Jawa Tengah.

403) Jln A. Yani 7, Tangerang, Jawa Barat.
404) Jln Veteran 6, Brebes, Jawa Tengah.
405) Medan, Sumatera Utara.
406) Jln Budhi 42, Cilember - Cimahi 40175, Jawa Barat.
407) Jln Johar 109, Kudus, Jawa Tengah.
408) Jln Kartini 34, Temanggung 56216, Jawa Tengah.
409) Jln K.H. Mansyur 65A, Sumenep 69411, Madura, Jawa Timur.
410) Jln Prof. Dr. M. Yamin 67, Ponorogo, Jawa Timur.
411) Jln Citarum 2A, Ponorogo 63410, Jawa Timur.
412) Jln Pabrik Padi 43, Bireuen 24201, D.I. Aceh.
413) Jln Raya, Delanggu - Klaten 57471, Jawa Tengah.
414) Jln Raya Darmo Permai Utara 73-80, Surabaya, Jawa Timur.
415) Jln Hasanuddin II/42C, Lt. 2, Palu 94112, Sulawesi Tengah.
416) Jln Jendral Basuki Rakhmad 15, Kediri 64123, Jawa Timur.
417) Jln Mayjend. Sungkono, Madiun, Jawa Timur.
418) Jln Teuku Umar 155, Kotamobagu, Sulawesi Utara.
419) Jln Kompl. Universitas Sumatera Utara, Medan, Sumatera Utara.
420) Jln Sanjai, Wisma Inkobra, Bukittinggi, Sumatera Barat.
421) Jln Koperasi (KLP Sinar Rinjani), Aikmel 83653, Lombok, Nusa Tenggara Barat.
422) Jln Prof. Dr. M. Yamin S.H. 19, Jambi 36135, Propinsi Jambi.
423) Jln Raya Negara Km. 7, Tanjung Pati 26271, Sumatera Barat.
424) Jln Blambangan 61, RT 03/RW 01, Banyuwangi 68410, Jawa Timur.
425) Jln Cipinang Cempedak I/51, Polonia Jatinegara, Jakarta.
426) Jln Pilang 463, Cirebon 45153, Jawa Barat.
427) Jln Banyumanik Raya 44, Semarang 50236, Jawa Tengah.
428) Jln Perintis Kemerdekaan 8, Yogyakarta 55161, D.I. Yogyakarta.
429) Jln Sukarno-Hatta 45, Probolinggo, Jawa Timur.
430) Jln Flamboyan 55, Klungkung, Bali.
431) Jln Barintik 35, P.O. Box 48, Martapura, Kalimantan Selatan.
432) Jln A. Rahman Saleh, Pontianak, Kalimantan Barat.
433) Jln Imam Bonjol 16, Tebingtinggi 20610, Sumatera Utara.
434) Jln Mianggas III, Manado, Sulawesi Utara.
435) Jln Setia Budi 55, Palu, Sulawesi Tengah.
436) Jln K.H. Ach. Dahlan 5, Bengkulu 38117, Propinsi Bengkulu.
437) Jln Garut 7, Sigli 24110, D.I. Aceh.
438) Jln K.H. Syam'un 2, Rangkasbitung 42311, Jawa Barat.
439) Jln Abdulrachman Saleh 514, Manyaran - Semarang 50147, Jawa Tengah.
441) Jln Jend. Sudirman 265, Indrapura - Asahan 21256, Sumatera Utara.
442) Jln Pasar Sentral 48, Mamuju, Sulawesi Selatan.
443) Jln D.I. Panjaitan 99, Curup - Bengkulu 39118, Propinsi Bengkulu.
444) Jln Ngalengko 15A, Medan 20236, Sumatera Utara.
445) Jln Raya Banjaran 34B, Adiwerna - Tegal 52194, Jawa Tengah.
446) Jln A. Yani 50, Balikpapan 76123, Kalimantan Timur.
447) Jln Rantau 19, Langsa, D.I. Aceh.
448) Jln Simpurut 34, Batusangkar 27211, Sumatera Barat.
449) Jln Simbolon 5, Pematangsiantar 21115, Sumatera Utara.
450) Jln Pemuda Tengah 56, Kotak Pos 113, Klaten, Jawa Tengah.
451) Jln H. Kimar III/5, Semarang 50249, Jawa Tengah.
452) Jln Pemuda Pucungrejo, Muntilan, Magelang, Jawa Tengah.
453) Jln Brigpol. Nasuha 2, Karawang, Jawa Barat.
454) Jln Bola Kaki 31, Pematangsiantar 21115, Sumatera Utara.
455) Jln Teratai 17, Pekanbaru, Riau.
456) Pematangsiantar, Sumatera Utara.
457) Purwokerto, Jawa Tengah.
458) Jln Kaum Kidul 415, Soreang - Bandung 40910, Jawa Barat.
459) Jln Dr. Wahidin S. IV/4, Cirebon, Jawa Barat.
460) Jln Raya Taunan Km. 6, Jepara, Jawa Tengah.
461) Jln Krembangan Bhakti 86, Surabaya 60176, Jawa Timur.
462) Jln Kahuripan 91, Malang 65119, Jawa Timur.
463) Kompleks Taman Sekar Kav. A-34, Jln Kartini, Kediri - Tabanan 82171, Bali.
464) Jln B.W. Lapian 38, Manado, Sulawesi Utara.
465) Jln Sunan Kudus 194, Kudus, Jawa Tengah.
466) Jln Argamulya Dalam III/32, Samarinda 75110, Kalimantan Timur.
467) Jln Raya Puncak 70, Cipayung - Bogor, Jawa Barat.
468) Jln Raya Timur 74, Jatiwangi-Majalengka, Jawa Barat.
469) Jln Siswa 56, Cikampek - Karawang 41373, Jawa Barat.
470) Jln Veteran 5, Tembilahan 29211, Riau.
471) Jln Dr. Wahidin Sudirohujodo 31C, Tuban 63210, Jawa Timur.
472) Jln Sudirman 1, Jember 68118, Jawa Timur.
473) Jln Pamong Praja 3, Pamekasan, Jawa Timur.
474) Jln Sultan Patah 3, Demak, Jawa Tengah.
475) Jln Kartini 137, Pematang Siantar, Sumatera Utara.
476) Jln Sukarno-Hatta 20, Padangpanjang, Sumatera Barat.
477) Jln Raya Medan 32, Tebingtinggi 20610, Sumatera Utara.
479) Jln Pongsongteleng 47, Ampah 73652, Kalimantan Tengah.

480) Jln Dr. Cipto, Sumenep 69410, Jawa Timur.
481) Jln Raya Jepon 147, Blora, Jawa Tengah.
482) Jln Raya Garum 7, Blitar, Jawa Timur.
483) Jln Demang Lehman RT. III/II. 5C, Pasayangan Utara - Martapura 70619, Kalimantan Selatan.
484) Jln Akcaya III/7 No. 32A, Pontianak 78121, Kalimantan Barat.
485) Jln Sentosa I/6, Balikpapan, Kalimantan Timur.
486) Jln Veteran 7, Bukittinggi, Sumatera Barat.
487) Jln Melati 46, P.O. Box 1006, Pekanbaru, Riau.
488) Jln Raya Waduk Darma 48, Kuningan 45500, Jawa Barat.
489) Jln Jend. A. Yani 662, Bandung, Jawa Barat.
490) Jln Cempaka 5, Karawang, Jawa Barat.
491) Jln Kemuning 30, P.O. Box 48, Salatiga 50724, Jawa Tengah.
492) Jln Pang. Besar Sudirman 2, Probolonggo, Jawa Timur.
493) Jln Toar 62, Manado, Sulawesi Utara.
494) Jln Ir. H. Juanda 128, Ciamis 46211, Jawa Barat.
495) Jln Pahlawan 20, Pasuruan 67126, Jawa Timur.
496) Jln AKBP M. Soeroko 11, Bojonegoro 62111, Jawa Timur.
497) Jln Kabupaten, Ponorogo 63413, Jawa Timur.
498) Jln Kartini 4, Magelang, Jawa Tengah.
499) Jln Kaikoli, Dili, Timor Timur. **FM:** 88.2/93.0/105.0MHz.
500) Jln Raya Gulon, Salam, Gulon - Magelang, Jawah Tengah.
501) Jln Raya Supratman 90, Denpasar, Bali.
502) Jln Timur Aloon-Aloon, Tulungagung, Jawa Timur.
503) Jln Olah Raga Komplek B.T.N., Indramayu 45213, Jawa Barat.
504) Jln W.R. Supratman 27, Lumajang 67310, Jawa Timur.
505) Jln A. Yani 1, Raba 84100, Sumbawa, Nusa Tenggara Barat.
506) Jln Dadaha 17, Tasikmalaya, Jawa Barat.
507) Jln P. Geusan Ulun 125, Sumedang, Jawa Barat.
508) Jln Pamukti Baru 9, Prambanan - Klaten 57454, Jawa Tengah.
509) Jln Cikutra 27, Cibeunying-Bandung, Jawa Barat.
510) Jln Osa Maliki 29, P.O. Box 57, Salatiga, Jawa Tengah.
511) Jln Merdeka 68, Manokwari, Irian Jaya. **FM:** 93.5MHz.
512) Jln Diponegoro 99, Amlapura, Bali.
513) Jln Durian 6, Tabanan 82113, Bali.
514) Jln Tebet Barat Raya 19, Jakarta Selatan.
515) Jln Letjend. Suprapto 101, Sibolga 22510, Sumatera Utara.
516) Jln Raya Pedurungan Kidul 24, Semarang, Jawa Tengah.
517) Jln Dempo 368, Lubuklinggau 31622, Sumatera Selatan.
518) Jln Ruhui Rahayu 45, Samarinda, Kalimantan Timur.
519) Jln Kalijaga 14, Cirebon 45110, Jawa Barat.
520) Jln Guntur 154, Garut, Jawa Barat.
521) Jln D.I. Panjaitan 47, Purwodadi, Jawa Tengah.
522) Jln Pangeran Diponegoro 22T, Palangkaraya, Kalimantan Tengah.
523) Jln Mangga 1, Palopo 91921, Sulawesi Selatan.
524) Jln Radio 1, Tikala Ares. Manado 95124, Sulawesi Utara. **FM:** 89.1/97.2/102.0MHz.
525) Jln Prof. Muhammad Yamin 199, Kuning, Bukittinggi, Sumatera Barat. **FM:** 93.0/97.7/103.0MHz.
526) Jln Ade Irma Suryani Nasution 5, Sibolga, Sumatera Utara. **FM:** 93.0/108.0MHz.
527) Jln Let. Jend. S. Parman 31, Kotak Pos 13, Bengkulu, Propinsi Bengkulu. **FM:** 90.6/93.0/97.0/105.5MHz.
528) Jln Jenderal Sudirman, Gorontalo, Sulawesi Utara. **FM:** 93.6/102.0MHz.
529) Sungai Liat, Bangka, Sumatera Selatan.
530) Jln Diponegoro 17, Pangkalanbuun 74114, Kalimantan Tengah.
531) Komplek Kabupaten Wonosobo, Jln Merdeka 1, Wonosobo, Jawa Tengah.
532) Jln Ismail 5A, Binjai, Sumatera Utara.
533) Jln M. Husni Thamrin 1, Palangkaraya 73112, Kalimantan Tengah. **FM:** 89.4/93.0MHz.
534) Komplek Asrama Haji, Cisalak - Sukabumi, Jawa Barat.
535) Jln Sultan Khairun, Ternate, Maluku. **FM:** 93.5/102.0/ 103.5MHz.
536) Jln Keramika, Tanjungpandan, Belitung, Sumatera Selatan.
537) Jln Jendral Ahmat Yani 74, Bengkalis, Riau.
538) Jln Pahlawan 99, Magelang, Jawa Tengah.
539) Jln Alun-Alun Utara, Blora, Jawa Tengah.
540) Jln Kartini Raya, Batang 51210, Jawa Tengah.
541) Jln Pandanaran 5, Boyolali 57311, Jawa Tengah.
542) Jln Soekarno-Hatta, Bajawa, Flores, Nusa Tenggara Timur.
543) Ruteng, Flores, Nusa Tenggara Timur.
544) Jln Pahlawan 110, Surabaya, Jawa Timur.
545) Jln Tamtama 3, Wates, D.I. Yogyakarta.
546) Jln Gajah Mada, Muarabulian 36610, Jambi.
547) Jln Jendral Achmad Yani 156, Kotakpos 247, Wonocolo - Surabaya, Jawa Timur.
548) Jln W.R. Supratman 37, Rantau Prapat, Sumatera Utara.
549) Jln Putih Derus 6, Rangkasbitung, Jawa Barat.
550) Jln Raya Bandar Agung 4, Lahat 31414, Sumatera Selatan.
551) Langsa, D.I. Aceh.

JAPAN

L.T: UTC + 9h — **Pop:** 126,280,000 — **Radio Sets:** 100,000,000 —
Pr.L: Japanese - **EC:** a/c 50 + 60Hz, 100V. - **ITU:** J.

BROADCASTING BUREAU
MINISTRY OF POSTS & TELECOMMUNICATIONS
✉ 3-2, Kasumigaseki 1-chome, Chiyoda-ku, Tokyo 100-8798. ☎
+81(3504)4411. **Web:** http://www.mpt.go.jp/outline/broad.html
L.P: Minister: S. Noda, DG of Broadcasting Bureau: M.Shinagawa.

NIPPON HOSO KYOKAI(NHK)
(The Japan Broadcasting Corporation)
✉ 2-1, Jinnan 2-chome, Shibuya-ku, Tokyo 150-8001. ☎ +81(3)3465-
1111. 🖷 +81(3)3469-8110. **Cable:** RADIONHK. ① J34179 RADJAPAN.
Web: http://www.nhk.or.jp
L.P: Chmn. (Board of Governors): H. Suda. Pres:K. Ebisawa, Exec. Vice-
Pres: H. Kanno. Gen. MD & Exec. Dir. Gen. Eng:T. Hasegawa, N. Kohno,
MD's M.Arase, K. Ishiwata, H. Sakai, M. Wada, Y. Ohba, T. Matsuo,
K. Tabata, Y.Haga.

STATIONS: Mediumwaves:
1)= Radio one; 2)=Radio two. Call: JO-.

Name & Prgr.	Call	kHz	kW
Morioka 1	QG	531	10
Yamagata 1	JG	540	5
Miyazaki 1	MG	540	5
Kitakyushu 1	SK	540	1
Okinawa 1	AP	549	10
Sapporo 1	IK	567	100
Kagoshima 1	HG	576	10
Kushiro 1	PG	585	10
Tokyo 1	AK	594	300
Okayama 1	KK	603	5
Obihiro 1	OG	603	5
Fukuoka 1	LK	612	100
Asahikawa 1	CG	621	3
Kyoto 1	OK	621	1
Oita 1	IP	639	5
Shizuoka 2	PB	639	10
Toyama 1	IG	648	5
Osaka 1	BK	666	100
Hakodate 1	VK	675	5
Yamaguchi 1	UG	675	5
Nagasaki 1	AG	684	5
Tokyo 2	AB	693	500
Kitami 2	KD	702	10
Hiroshima 2	FB	702	10
Nagoya 1	CK	729	50
Sapporo 2	IB	747	500
Kumamoto 1	GK	756	10
Akita 2	UB	774	500
Nagano 1	NK	819	5
Osaka 2	BB	828	300
Niigata 1	QK	837	10
Kumamoto 2	GB	873	500
Shizuoka 1	PK	882	10
Sendai 1	HK	891	20
Nagoya 2	CB	909	10
Kofu 1	KG	927	5
Fukui 1	FG	927	5
Muroran 1	IQ	945	3
Hikone 1	QP	945	1
Tokushima 1	XK	945	5
Aomori 1	TG	963	5
Matsuyama 1	ZK	963	5
Saga 1	SP	963	1
Kochi 1	RK	990	10
Hamamatsu 1	DG	999	1
Fukuoka 2	LB	1017	50
Toyama 2	IC	1035	1
Takamatsu 2	HD	1035	1
Hiroshima 1	FK	1071	20
Sendai 2	HB	1089	10
Obihiro 2	OC	1125	1
Muroran 2	IZ	1125	1
Tottori 2	LC	1125	1
Okinawa 2	AD	1125	10
Kushiro 2	PC	1152	10
Kochi 2	RB	1152	10

Name & Prgr.	Call	kHz	kW
Kitami 1	KP	1188	10
Kanazawa 1	JK	1224	10
Matsue 1	TK	1296	10
Fukushima 1	FP	1323	1
Tottori 1	LG	1368	1
Takamatsu 1	HP	1368	5
Yamaguchi 2	UC	1377	5
Nagasaki 2	AC	1377	1
Morioka 2	QC	1386	10
Kanazawa 2	JB	1386	10
Okayama 2	KB	1386	5
Kagoshima 2	HC	1386	10
Hakodate 2	VB	1467	1
Nagano 2	NB	1467	1
Oita 2	ID	1467	1
Miyazaki 2	MC	1467	1
Akita 1	UK	1503	10
Matsuyama 2	ZB	1512	5
Yamagata 2	JC	1521	1
Aomori 2	TC	1521	1
Hamamatsu 2	DC	1521	1
Fukui 2	FC	1521	1
Niigata 2	QB	1593	10
Matsue 2	TB	1593	10
Asahikawa 2	CC	1602	1
Kofu 2	KC	1602	1
Kitakyushu 2	SB	1602	1
Fukushima 2	FD	1602	1

SHORTWAVES

Location	kHz	kW	Mode	H. of tr
Fukuoka 1	3262.75	0.6	SSB	0800-1300
Osaka 2	3377.25	0.3	SSB	0800-1300
Tokyo 1	3611.25	0.9	SSB	0800-1300
Sapporo 1	3970	0.6	DSB	1300-1500
Nagoya 1	3973.75	0.3	SSB	2000-0030
				0400-1300
Osaka 2	5431.75	0.3	SSB	2030-0300
				0500-0730
Sapporo 1	6005	0.6	DSB	2000-0030
				0400-1230
Nagoya 1	6008.75	0.3	SSB	0100-0330
Fukuoka 1	6133.75	0.6	SSB	2000-0400
Tokyo 1	6178.75	0.9	SSB	2000-0030
Osaka 2	9184.75	0.3	SSB	0330-0430
Sapporo 1	9535	0.6	DSB	0100-0330
Fukuoka 1	9538.75	0.6	SSB	0430-0730
Tokyo 1	9553.75	0.9	SSB	0100-0730

These SWs are transmitted from standby/emergency trs. H. of tr. will
be changed in an emergency.

FM-Stations: Call: JO(call)-FM. St's below 1kW are not mentioned.

Location	Call	MHz	kW
Utsunomiya	BP	80.3	1
Chiba	MP	80.7	5
Toyama	IG	81.5	1
Maebashi	TP	81.6	1
Tsu	NP	81.8	3
Yokohama	GP	81.9	5
Yamagata	JG	82.1	1
Kanazawa	JK	82.2	1
Niigata	QK	82.3	1
Tokyo	AK	82.5	10
Nagoya	CK	82.5	10
Sendai	HK	82.5	5
Kyoto	OK	82.8	1
Morioka	QG	83.1	1
Mito	EP	83.2	1
Fukui	FG	83.4	1
Tokushima	XK	83.4	1
Gifu	OP	83.6	1
Otsu	QP	84.0	1
Fukuoka	LK	84.8	3
Urawa	LP	85.1	5
Sapporo	IK	85.2	5
Fukushima	FP	85.3	1
Kumamoto	GK	85.4	1
Kofu	KG	85.6	1
Kagoshima	HG	85.6	1

Location	Call	MHz	kW
Aomori	TG	86.0	3
Takamatsu	HP	86.0	1
Akita	UK	86.7	3
Matsuyama	ZK	87.7	1
Osaka	BK	88.1	10
Okinawa	AP	88.1	1
Hiroshima	FK	88.3	1
Okayama	KK	88.7	1
Shizuoka	PK	88.8	1
Oita	IP	88.9	1

+approx MW 250 and FM 480 st's.

☞ **of regional key st's:** Osaka: 3-43, Bamba-cho, Chuo-ku, Osaka 540-8501 - Nagoya: 13-3, Higashisakura 1-chome, Higashi-ku, Nagoya 461-8710 - Hiroshima: 11-10, Otemachi 2-chome, Naka-ku, Hiroshima 730-8672 - Fukuoka: 1-10, Ropponmatsu 1-chome, Chuo-ku, Fukuoka 810-8577 - Sendai: 11-1, Nishiki-machi 1-chome, Aoba-ku, Sendai 980-8435 - Sapporo: 1-1, Odori Nishi, Chuo-ku, Sapporo 060-8703 - Matsuyama: 5, Horinouchi, Matsuyama 790-8501.

R. One (General prgr): 24h(2nd & 4th Mon: 2000-1600) - **N:** W every h, also 2030, 2140, 2230, 2330, 0030, 0130, 0230, 0430, 0530, 0630, 0730, 0830, 1130. Sun: every h exc. 0000, . **Regional N:** W 2125, 2215(Sat: 2210), 2240, 2355, 0155, 0310, 0455, 0555, 0655, 0755, 0855, 0950, 1045(Sat: 1010), 1255, 1405(Sat: 1455). Sun 2110, 2255, 2355, 0310, 0555, 0855, 0950, 1010, 1255, 1405 - **Wrp:** W 2055, 2155, 0055, 0250, 0355, 1055, 1155 & after reg. N. Sun: 2055,2155, 0155, 0250, 0455, 0655, 0755, 1158 & after reg. N. **Local Prgrs** (the amount of local prgr's varies between the st's): 2055wrp/2125(W)N/wrp/inf, 2155wrp/inf, 2210(Sat & Sun)N/wrp/inf, 2215(Mon-Fri)N/wrp/inf, 2235(Sun)inf, 2240(W)N/wrp/inf, 2255(Sun)N/wrp/inf, 2318(W)inf/wrp, 2333(W)inf, 2355N/wrp/inf, 0028(W)inf, 0055(W)wrp/inf, 0128inf, 0155N/wrp/inf(Sun: wrp/inf), 0228inf, 0250wrp/inf, 0310N/wrp, 0355(W)wrp/inf, 0428inf, 0455N/wrp/inf(Sun: wrp/inf), 0528inf, 0555N/wrp/inf, 0628inf, 0655N/wrp/inf(Sun: wrp/inf), 0728inf, 0755N/wrp/inf(Sun: wrp/inf), 0826(W)inf, 0828(Sun)inf, 0855N/wrp/inf, 0928(W)inf, 0950N/wrp/inf, 1010(Sat & Sun)N/wrp, 1028(Thu-Sat)inf, 1045(Mon-Fri)N/wrp/inf, 1055(Sat)wrp/inf, 1058(Sun)inf, 1128(W)inf, 1155(W)wrp/inf, 1158(Sat & Sun)inf, 1255N/wrp/inf, 1410N/wrp, 1428wrp. **IS:** Original music played by Celesta.
ANN: "Daiichi Hoso". Local ID's with call letters, network & location given by studio st's just before (W): 2000, 0300, 1000. (Sun): 2000, 0000, 0300, 1000, 1200.

R. Two (Educational prgr): 2030-1500 - **Foreign language N**(rel. NHK World - R. Japan): **Chinese:** 0400-0410. **Korean:** 0410-0420. **English:** 0500-0515(Sat. & Sun: 0500-0510), 1400-1415(Sat. & Sun: 1400-1410). **Portuguese:** 0900-0910 - L.L: Basic English(1st step): W 2100-2115, 0515-0530, 0925-0940 - Basic English (2nd step): W 2115-2130, 0530-0545, 0940-0955 - Basic English (3rd step): W 2130-2145, 0545-0600, 0955-1010 - English Conversation(1st step): W 2145-2200, 0610-0625, 1010-1025 - English Conversation(2nd step): W 0625-0645, 1025-1045, 1320-1340 - Easy English for Businessmen: W 1340-1400, Sun 0200-0400 - German: W 2200-2220, 0340-0400 - French: W 2240-2300, 0420-0440 - Russian: W 2350-0010, 0740-0800 - Spanish: W 2300-2320, 0440-0500 - Chinese: W 2320-2340, 1435-1455 - Korean: W 2220-2240, 0720-0740 - Italian: W 0030-0050, 1415-1435. **Wrp:** 0010, 0700, 1300 (all 20 mins.). **IS:** Original music played by Celesta. Nat. Anthem at s/on on national holidays & s/off. **ANN:** "Daini Hoso". Local ID's on certain st's (as 1st Netw) just before 2030, 2300, 0030, 0720, 1320, 1500.

FM Netw: 24h (off the air Sun 1600-2000). Local ID's just before 2100, 0000(Sun), 0100(Sat), 0200(exc Sat. & Sun), 0900(Sat), 1000(Sun) & 1601(Sun).
V: Regional key st's have no organised QSL service. However, many local st's verify by QSL card or letter for DX reports. Rp. Pub: NHK Year Book (Japanese) and NHK Update(English).

FOREIGN SERVICE (NHK WORLD - R. JAPAN)
See International Broadcasting section

THE NATIONAL ASSOCIATION OF COMMERCIAL BROADCASTERS IN JAPAN
✉ 3-23, Kioi-cho, Chiyoda-ku, Tokyo 102-8577. ☎+81(3)5213 7711. 📠 +81(3)5213 7703. **Cable:** Mimporen Tokyo.
L.P: Pres: S. Ujiie, Vice-Presidents: N. Wakamatsu, M. Saito, M. Hashimoto, Exec. Dir: A. Sakai, Sec. Gen: T. Tamagawa.

Mediumwaves: Call: JO(call). ID: Company initials usually used as st. identification. S=AM Stereo(C-QUAM System)

Call	kHz	kW	ID	Name, location and h. of tr.
1)CR	558	20d	AMK	AM Kobe, Kobe: 2000-1800
36)WN	639	5	STV	Sapporo TV Hoso, Hakodate
2)DF	684	5	IBC	Iwate Hoso, Morioka
2)	684	1	IBC	Iwate Hoso, Ofunato
36)WN	639	5	STV	Sapporo TV Hoso, Hakodate
3)IL	720	1	KBC	Kyushu Asahi Hoso, Kitakyushu
4)LR	738	5	KNB	Kita Nihon Hoso, Toyama
4)	738	1	KNB	Kita Nihon Hoso, Takaoka
5)RR	738	10	RBC	Ryukyu Hoso, Naha
6)JF	765	5	YBS	Yamanashi Hoso, Kofu
15)PF	765	5	KRY	Yamaguchi Hoso, Tokuyama
7)XR	864	10	ROK	R. Okinawa, Naha: 2000-1800
8)SO	864	1	SBC	Shin'etsu Hoso, Matsumoto
9)HE	864	3	HBC	Hokkaido Hoso, Asahikawa
9)QF	864	3	HBC	Hokkaido Hoso, Muroran
10)PR	864	5d	FBC	Fukui Hoso, Fukui
48)XN	864	1	CRT	Tochigi Hoso, Nasu: 2100-2000
36)WS	882	3	STV	Sapporo TV Hoso, Kushiro
36)	882	1	STV	Sapporo TV Hoso, Esashi
9)HO	900	1	HBC	Hokkaido Hoso, Hakodate
11)HF	900	5d	BSS	San'in Hoso, Yonago: 24h
12)ZR	900	5d	RKC	Kochi Hoso, Kochi
36)VX	909	5	STV	Sapporo TV Hoso, Abashiri
13)EF	918	5	YBC	Yamagata Hoso, Yamagata
13)	918	1	YBC	Yamagata Hoso, Tsuruoka/Yonezawa
15)PM	918	1	KRY	Yamaguchi Hoso, Shimonoseki
15)PN	918	1	KRY	Yamaguchi Hoso, Iwakuni
14)TR	936	5	ABS	Akita Hoso, Akita
21)NF	936	5	MRT	Miyazaki Hoso, Miyazaki
21)	936	1	MRT	Miyazaki Hoso, 4 relay st's
16)KR	954S	100	TBS	Tokyo Hoso, Tokyo
17)NR	1008S	50	ABC	Asahi Hoso, Osaka
19)AR	1053S	50	CBC	Chubu Nippon Hoso, Nagoya
36)WM	1071	5	STV	Sapporo TV Hoso, Obihiro
8)SR	1098	5	SBC	Shin'etsu Hoso, Nagano
8)SW	1098	1	SBC	Shin'etsu Hoso, Iida
20)MF	1098	1	NBC	Nagasaki Hoso, Sasebo
22)GF	1098	5	OBS	Oita Hoso, Oita
23)WO	1098	5	RFC	R. Fukushima, Koriyama
24)CF	1107	20d	MBC	Minami Nihon Hoso, Kagoshima
24)	1107	5	MBC	Minami Nihon Hoso, Akune/Okuchi
26)MR	1107	5	MRO	Hokuriku Hoso, Kanazawa
26)	1107	1	MRO	Hokuriku Hoso, Nanao
27)AF	1116	5	RNB	Nankai Hoso, Matsuyama
27)AL	1116	1	RNB	Nankai Hoso, Niihama
27)AM	1116	1	RNB	Nankai Hoso, Uwajima
28)DR	1116	5	BSN	Niigata Hoso, Niigata
29)QR	1134S	100	NCB	Bunka Hoso, Tokyo
25)BR	1143	20	KBS	KBS Kyoto, Kyoto
33)OR	1179S	50	MBS	Mainichi Hoso, Osaka
12)	1197	1	RKC	Kochi Hoso, Nakamura
30)FO	1197	1	RKB	RKB Mainichi Hoso, Kitakyushu
31)BF	1197S	10d	RKK	Kumamoto Hoso, Kumamoto
31)	1197	1	RKK	Kumamoto Hoso, Aso/Goshonoura/Hitoyoshi
32)YF	1197	5	IBS	Ibaraki Hoso, Mito: 2100-1530
36)WL	1197	3	STV	Sapporo TV Hoso, Asahikawa
36)	1197	1	STV	Sapporo TV Hoso, Nayoro/Wakkanai
25)BO	1215	5	KBS	KBS Kyoto, Maizuru
25)BW	1215	1	KBS	KBS Shiga, Hikone
20)UR	1233	5	NBC	Nagasaki Hoso, Nagasaki
34)GR	1233	5d	RAB	Aomori Hoso, Aomori
35)LF	1242S	100	NBS	Nippon Hoso, Tokyo
37)IR	1260	20	TBC	Tohoku Hoso, Sendai
9)HW	1269	5	HBC	Hokkaido Hoso, Obihiro
9)FM	1269	1	HBC	Hokkaido Hoso, Esashi
38)JR	1269	5	JRT	Shikoku Hoso, Tokushima
38)	1269	1	JRT	Shikoku Hoso, Ikeda
30)FR	1278	50	RKB	RKB Mainichi Hoso, Fukuoka
9)HR	1287S	50	HBC	Hokkaido Hoso, Sapporo
39)UF	1314S	50	OBC	R. Osaka, Osaka
40)SF	1332S	50		Tokai R. Hoso, Nagoya
41)ER	1350S	20	RCC	Chugoku Hoso, Hiroshima
9)TS	1368	1	HBC	Hokkaido Hoso, Wakkanai
1)	1395	1	AMK	AM Kobe, Toyooka
23)WE	1395	1	RFC	R. Fukushima, Wakamatsu

9)QL	1404	5	HBC	Hokkaido Hoso, Kushiro
42)VR	1404	10	SBS	Shizuoka Hoso, Shizuoka
42)VO	1404	1d	SBS	Shizuoka Hoso, Hamamatsu
3)IF	1413S	50	KBC	Kyushu Asahi Hoso, Fukuoka
43)RF	1422	50d	RF	"RF"R. Nippon, Yokohama
11)HL	1431	1	BSS	San'in Hoso, Tottori: as 900kHz
11)	1431	1d	BSS	San'in Hoso, Izumo: as 900kHz
20)	1431	1	NBC	Nagasaki Hoso, Fukue
23)WW	1431	1	RFC	R. Fukushima, Iwaki
44)VF	1431	5	WBS	Wakayama Hoso, Wakayama
45)ZF	1431	5	GBS	Gifu Hoso, Gifu: (off the air Sun 1500-2000)
36)WF	1440	50	STV	Sapporo TV Hoso, Sapporo
36)	1440	3	STV	Sapporo TV Hoso, Muroran
36)	1440	1	STV	Sapporo TV Hoso, Tomakomai
9)QM	1449	5	HBC	Hokkaido Hoso, Abashiri
46)KF	1449	5	RNC	Nishi Nippon Hoso, Takamatsu(off the air Sun (1630-2030)
46)	1449	1	RNC	Nishi Nippon Hoso, Marugame(off the air Sun (1630-2030)
20)UO	1458	1	NBC	Nagasaki Hoso, Saga
23)WR	1458	1	RFC	R. Fukushima, Fukushima
32)YL	1458	1	IBS	Ibaraki Hoso, Tsuchiura: 2100-1500
32)	1458	1	IBS	Ibaraki Hoso, Sekijo: as above
41)	1458	1	RCC	Chugoku Hoso, Shobara
15)PL	1485	1	KRY	Yamaguchi Hoso, Hagi
34)GO	1485	1	RAB	Aomori Hoso, Hachinohe
9)TL	1494	1	HBC	Hokkaido Hoso, Nayoro
47)YR	1494S	10	RSK	Sanyo Hoso, Okayama
47)	1494	1	RSK	Sanyo Hoso, 5 rel. st's
28)DO	1530	1	BSN	Niigata Hoso, Joetsu
48)XF	1530	5	CRT	Tochigi Hoso, Utsunomiya: 2100-2000(Sun1500)
41)EO	1530	1	RCC	Chugoku Hoso, Fukuyama
41)	1530	1	RCC	Chugoku Hoso, Mihara

Relay stns below 1kW(approx 125 st`s) not mentioned.

Addresses and other information:

Schedules: 24h unless otherwise indicated. Most of 24h st's are off the air for 2 to 5 hours until 2000 on Sun except Mainichi Hoso and Nippon Hoso. All other days a network prgr is aired 1600 or 1800 to 2000 on most st's. Network prgr's also may be b'cast at other times of the day.

V: All st's verify by QSL-card. Rec. acc. Rp.

1)R. Kansai Co., Ltd., 5-7, Higashi Kawasaki-cho 1-chome, Chuo-ku, Kobe 650-8080. **Web:** http://www.memenet.or.jp/amkobe

2)Iwate Broadc. Co., Ltd., 6-1, Shike-cho, Morioka 020-8566. **Web:** http://www.nnettown.or.jp/IBC

3)Kyushu Asahi Broadc. Corp., 1-1, Nagahama 1-chome, Chuo-ku, Fukuoka 810-8571. **Web:** http://www.kbc.co.jp

4)Kita-nihon Broadc. Co., Ltd.,10-18 Ushijima-machi, Toyama 930-8585

5)Ryukyu Broadc. Corp., 3-1, Kumoji 2-chome, Naha 900-8711. **Web:** http://www.cosmos.ne.jp/rbc

6)Yamanashi Broadc. System, 6-10, Kitaguchi 2-chome, Kofu 400-8566. **Web:** http://www.sannichi-ybs.co.jp/YBS

7)R. Okinawa Broadc. Co., Ltd., 4-8, Nishi 1-chome, Naha 900-8604. An independent audio service via satellite also exists. **Web:** http://www.cosmos.ne.jp/rok

8)Shin-etsu Broadc. Co., Ltd., 21-24, Yoshida 1-chome, Nagano 381-8585. **Web:** http://www.sbc-nagano.co.jp

9)Hokkaido Broadc. Co., Ltd., 2, Nishi 5-chome, kita 1-jo, Chuo-ku, Sapporo 060-8501. **Web:** http://www.hbc.co.jp

10)Fukui Broadc. Co., Ltd., 5-105, Itagaki, Fukui 918-8677

11)Broadc. System of San-in, 1-71, Nishi-Fukubara 1-chome, Yonago 683-8670. **Web:** http://www.sanin-v.com/bss

12)Kochi Broadc. Co., Ltd., 2-15, Hon-machi 3-chome, Kochi 780-8550.**Web:** http://www.rkc-kochi.co.jp

13)Yamagata Broadc. Co., Ltd., 2-5, Hatago-machi, Yamagata 990-8555.**Web:** http://www.yamagata-japan.com/YBC

14)Akita Broadc. System, 9-42, Sanno 7-chome, Akita 010-8611.**Web:** http://www.akita-abs.co.jp

15)Yamaguchi Broadc. Co., Ltd., Koen-Ku, Tokuyama 745-8686. **Web:** http://www.kry.co.jp

16)Tokyo Broadc. System, Inc., 3-6, Akasaka 5-chome, Minato-ku, Tokyo 107-8006. **Web:** http://www.tbs.co.jp

17)Asahi Broadc. Corp., 2-48, Oyodominami 2-chome, Kita-ku, Osaka 530-8501. **Web:** http://www.asahi.co.jp

18)Chubu-Nippon Broadc. Co., Ltd., 2-8, Shinsakae 1-chome, Naka-ku, Nagoya 460-8405. **Web:** http://www.cbc-nagoya.co.jp

19)Nagasaki Broadc. Co., Ltd., 1-35, Uwamachi, Nagasaki 850-0054.

20)Nagasaki Broadc. Co., Ltd., 1-35, Uwamachi, Nagasaki 850-0054. **Web:** http://www.nbc-nagasaki.co.jp

21)Miyazaki Broadc. Co., Ltd., 6-7, Nishi 4-chome, Tachibana-dori, Miyazaki 880-8639. **Web:** http://m-surf.or.jp/mrt

22)Oita Broadc. System, 1-1, Imazuru 3-chome, Oita 870-8620. **Web:** http://www.obstv.co.jp

23)R. Fukushima Broadc. Co., Ltd., 13-17, Otamachi, Fukushima 960-8655. **Web:** http://www.rfc.co.jp

24)Minaminihon Broadc. Co., Ltd., 5-25, Korai-cho, Kagoshima 890-8570. **Web:** http://www.minc.or.jp/MBC

25)Kyoto Broadc. System Co., Ltd., Kamichojamachi, Karasumadori, Kamigyo-ku, Kyoto 602-8583. **Web:** http://www.kbs-kyoto.co.jp

26)Hokuriku Broadc. Co., Ltd., 2-1, Honda-machi 3-chome, Kanazawa 920-8560. **Web:** http://www.tbs.co.jp/jnn/mro

27)Nankai Broadc. Co., Ltd., 6-24 Dogohimata, Matsuyama 790-8510. **Web:** http://www.rnb.co.jp

28)Broadc. System of Niigata, Inc., 3-18, Kawagishi-cho, Niigata 951-8655. **Web:** http://www.bsn-niigata.co.jp

29)Nippon Cultural Broadc., Inc., 1-5, Wakaba, Shinjuku-ku, Tokyo 160-8002. **Web:** http://www.joqr.co.jp

30)RKB Mainichi Broadc. Corp., 3-8, Momojihama 2-chome, Sawara-ku, Fukuoka 814-8585. **Web:** http://www.bcc-net.co.jp/rkb

31)Kumamoto Broadc. Co., Ltd., 30, Yamasaki-machi, Kumamoto 860-8611. **Web:** http://www.rkk.co.jp

32)Ibaraki Broadc. System, 2084 Senba-cho, Mito 310-8505. **Web:** http://www.sunshine.or.jp/~ibshodo1

33)Mainichi Broadc. System, Inc., 17-1, Chayamachi, Kita-ku, Osaka 530-8304. **Web:** http://mbs.co.jp

34)Aomori Broadc. Corp., 8-1, Matsumori 1-chome, Aomori 030-0965. **Web:** http://www.rab.co.jp

35)Nippon Broadc. System, Inc., 4-8, Daiba 2-chome, Minato-ku, Tokyo 137-8686. **Web:** http://www.fujisankei-g.co.jp/jolf

36)The Sapporo Television Broadc. Co., Ltd., 1-1, Nishi 8-chome, kita 1-jo, Chuo-ku, Sapporo 060-8705. **Web:** http://www.dosanko.co.jp/stv

37)Tohoku Broadc. Co., Ltd., 26-1, Kasumi-cho, Yagiyama, Taihaku-ku, Sendai 982-0831. **Web:** http://www.tbc-sendai.co.jp

38)Shikoku Broadc. Co., Ltd., 5-2, Nakatokushima-cho 2-chome, Tokushima 770-8573. **Web:** http://www.jrt.co.jp http://www.inter.co.jp/TVNET-JAPAN/japan_index/TV/JRT

39)Osaka Broadc. Corp., 2-4, Benten 1-chome, Minato-ku, Osaka 552-8501. vhttp://www.obc1314.co.jp

40)Tokai Radio Broadc. Co., Ltd., 14-27, Higashisakura 1-chome, Higashi-ku, Nagoya 461-8503. **Web:** http://www.tokairadio.co.jp

41)Chugoku Broadc. Co., Ltd., 21-3, Moto-machi, Naka-ku, Hiroshima 730-8504. **Web:** http://www.rcc-hiroshima.co.jp

42)Shizuoka Broadc. System, 1-1, Toro 3-chome, Shizuoka 422-8033. **Web:** http://www.sbs-np.co.jp http://www.tbs.co.jp/jnn/sbs

43)RF Radio Nippon Co., Ltd., 5-85, Choja-machi, Naka-ku, Yokohama 231-8611. **Web:** http://jorf1422.excite.co.jp/jorf1422

44)Wakayama Broadc. System, 3-3, Minato-honmachi, Wakayama 640-8577. **Web:** http://www.wbs.co.jp

45)Gifu Broadc. System, 9, Imako-machi, Gifu 500-8588. **Web:** http://www.jic-gifu.or.jp/np/radio_tv/radio/radio

46)Nishi-nippon Broadc. Co., Ltd., 8-15, Marunouchi, Takamatsu 760-8575.**Web:** http://www.rnc.co.jp

47)Sanyo Broadc. Co., Ltd., 1-3, Marunouchi 2-chome, Okayama 700-8580. **Web:** http://www.rsk.co.jp

48)Tochigi Broadc. Co., Ltd., 12-11 Honcho, Utsunomiya 320-8601

V: All st's verify by QSL-card Rec. acc. Rp.

NIHON SHORT-WAVE BROADC. CO., LTD.
(Radio Tampa)

✉ 9-15, Akasaka 1-chome, Minato-ku, Tokyo 107-8373. ☎ +81(3)3583 8151. 🖷 +81(3)3583 7441. **Web:** http://www.tampa.co.jp

Call	kHz	kW	Prgr.
JOZ	3925	50	1
JOZ4	*3925	10	1
JOZ5	3945	10	2
JOZ2	6055	50	1
JOZ6	6115	50	2
JOZ3	9595	50	1
JOZ7	9760	50	2

There are satellite channels for both prgrs. *) Nemuro; others Nagara, Chiba (east of Tokyo).

1st Prgr: 2030(seasonally 2020)-1500(1510 Fri and Sun, 1550 Mon) on 3925/6055/9595 kHz; as above except 2300-0750 on 3925 kHz(Nemuro).

2nd Prgr: 2300-1300 on 3945 kHz, followed by 1st Prgr at 1300-1500. 2300-1000 6115 kHz, 2300-0800 9760kHz.

IS: Slow tempo chime with Japanese instrument "Koto" at sign on and sign off — **V.** by QSL card.

COMMERCIAL FM STATIONS

(Japanese FM-channels: 76-90MHz): Call JO(call)-FM

	Call	MHz	kW	Name, Location
1)	QU	76.1	1	FM Iwate, Morioka, Iwate
2)	LU	76.1	1	FM Fukui, Fukui
3)	DW	76.1	10	Inter FM, Tokyo
48)	FW	76.1	1	Love FM, Fukuoka
4)	SV	76.4	1	R. Berry, Utsunomiya, Tochigi
5)	AW	76.5	10	FM CO-CO-LO, Osaka
49)	UV	77.0	1	E-Radio, Otsu, Shiga
6)	JU	77.1	5	FM Sendai, Sendai, Miyagi
7)	SU	77.4	1	FM Naka Kyushu, Kumamoto
8)	VU	77.4	0.5	FM San'in, Matsue, Shimane
9)	XU	77.5	1	FM Radio Niigata, Niigata
46)		77.6	1	Kiss-FM, Himeji, Hyogo
10)	QV	77.8	10	ZIP FM, Nagoya, Aichi
11)	NV	77.9	0.5	FM Saga, Saga
12)	GV	78.0	5	BAY-FM, Chiba
13)	GU	78.2	1	Hiroshima FM, Hiroshima
14)	YU	78.6	1	FM Kagawa, Takamatsu, Kagawa
15)	RV	78.7	3	CROSS FM, Fukuoka
16)	NU	78.9	3	FM Mie, Tsu, Mie
17)	KU	79.2	1	K-MIX, Shizuoka
18)	UU	79.2	1	FM Yamaguchi, Yamaguchi
19)	HU	79.5	1	Smile-FM, Nagasaki
20)	DV	79.5	5	NACK 5, Urawa, Saitama
21)	EU	79.7	1	FM Ehime, Matsuyama
22)	ZU	79.7	1	FM Nagano, (Matsumoto), Nagano
23)	OV	79.8	1	FM Kagoshima, Kagoshima
24)	WU	80.0	1	FM Aomori, Aomori
25)	AU	80.0	10	Tokyo FM, Tokyo
26)	FV	80.2	10	FM 802, Osaka
27)	FU	80.4	5	AIR-G', Sapporo, Hokkaido
28)	EV	80.4	1	FM Yamagata, Yamagata
29)	HV	80.5	1	FM Ishikawa, Kanazawa
30)	CU	80.7	10	FM Aichi, Nagoya
31)	MV	80.7	1	FM Tokushima, Tokushima
32)	DU	80.7	3	FM Fukuoka, Fukuoka
33)	AV	81.3	10	J-WAVE, Tokyo
34)	LV	81.6	0.5	FM Kochi, Kochi
47)	TV	81.8	1	Fukushima FM, Fukushima
35)	PV	82.5	5	FM North Wave, Sapporo, Hokkaido
36)	OU	82.7	1	FM Toyama, Toyama
37)	PU	82.8	3	FM Akita, Akita
38)	CV	83.0	1	FM Fuji, Kofu, Yamanashi
39)	MU	83.2	1	FM Miyazaki, Miyazaki
40)	TU	84.7	5	FM Yokohama, Kanagawa
41)	BU	85.1	10	FM Osaka, Osaka
42)	RU	86.3	1	FM Gunma, Maebashi, Gunma
9)		86.5	1	FM Radio Niigata, Yamato, Niigata
8)		86.6	1	FM San-in, Hamada, Shimane
43)	IU	87.3	1	FM Okinawa, Naha, Okinawa
44)	JV	88.0	1	FM Oita, Oita
45)	KV	89.4	3	Alpha-STATION, Kyoto
46)	IV	89.9	1	Kiss-FM, Kobe, Hyogo

Relay stns below 1kW not mentioned.

H. of tr: 24h unless otherwise indicated. Most of 24h st's are off the air for 2 to 5 hours until 1900, 2000 or 2100 on Sun.

Addresses and other information

1)Iwate FM Broadc. Co., Ltd., 8-17, Morioka-Ekimaedoori, Morioka 020-8512. **Web:** http://www.fm-iwate.co.jp

2)Fukui FM Broac. Co., Ltd., 1-1, Miyuki 1-chome, Fukui 910-8553. **Web:** http://www.ffg.co.jp/FM

3)FM Inter-wave Inc., 5-4, Shibaura 4-chome, Minato-ku, Tokyo 108-3070 - Prgr. in English & foreign languages. A satellite audio service also exists. **Web:** http://www.interfm.co.jp

4)FM Tochigi Co., Ltd., 1-19, Ichijo 3-chome, Utsunomiya 320-8550

5)Kansai Intermedia Corp., 14-16, Nanko Kita 1-chome, Suminoe-ku, Osaka 559-0034. 2100-1500 foreign language prgr. in English, Chinese, Korean, etc.**Web:** http://www.cocolo.co.jp

6)Sendai FM Broadc., Inc., 10-28. Honcho 2-chome, Aoba-ku, Sendai 980-8420

7)FM Nakakyushu Broadc. Co., Ltd., 5-50, Chibajomachi, Kumamoto 860-0001

8)San-in FM Broadc. Co., Ltd., 383, Tono-machi, Matsue 690-8508.

9)FM Radio Niigata Co., Ltd., 1-1, Yachiyo 2-chome, Niigata 950-8581.**Web:** http://www.fm-niigata.co.jp

10)FM Nagoya Inc., 20-17, Marunouchi 3-chome, Naka-ku, Nagoya

460-8705. **Web:** http://zip-fm.co.jp

11)FM Saga Co., Ltd., 286-5, Fukuro, Honjo-Machi, Saga 840-0023. 2100-1800(Sun 1600)

12)FM Sound Chiba Co., Ltd., 11-1, Chuo 1-chome, Chuo-ku, Chiba 260-8625. **Web:** http://www.bayfm.co.jp

13)Hiroshima FM Broadc. Co., Ltd., 8-2, Minamimachi 1-chome, Minami-ku, Hiroshima 734-8511. **Web:** http://www.hiroshima-fm.co.jp

14)FM Kagawa Broadc. Co., Ltd., 4-23, Saiho-cho 1-chome, Takamatsu 760-8584. **Web:** http://www.fnc.co.jp/fm

15)FM Kyushu Co., Ltd., 9-11, Furusenba-machi, Kokurakita-ku, Kitakyushu 802-8570. **Web:** http: www.nishinippon.co.jp/cross-fm

16)Mie FM Broadc. Co., Ltd., 1043-1, Kannonji-cho, Tsu 514-8505

17)Shizuoka FM. Broadc. Co., Ltd., 133-24, Tokiwa-cho, Hamamatsu 430-8575. **Web:** http://www2.shizuokanet.or.jp/usr/kmix

18)FM Yamaguchi Broadc. Co., Ltd., 3-31, Midori-cho, Yamaguchi 753-8521. **Web:** http://www.fmy.co.jp

19)FM Nagasaki Co., Ltd., 5-5, Sakae-machi, Nagasaki 850-8550. 2100-1800(Sat. 2000, Sun. 1600).

20)FM Saitama, 16-2, Tokiwa 4-chome, Urawa 336-8579. **Web:** http://www.nack5.co.jp

21)FM Ehime Broadc. Co., Ltd., 10-7, Takehara-machi 1-chome, Matsuyama 790-8565. 2057-1803(Fri 1603, Sat 1703, Sun 1503). **Web:** http://www.joeuf.co.jp

22)Nagano FM Broadc. Co., Ltd., 13-5, Honjo 1-chome, Matsumoto 390-8520

23)FM Kagoshima Co., Ltd., 1-38, Higashisengoku-cho, Kagoshima 892-8579

24)Aomori FM Broadc. Co., Ltd., 7-19, Tsutsumi-machi 1-chome, Aomori 030-0812. **Web:** http:www.infoaomori.ne.jp/ENET/afb

25)Tokyo FM Broadc. Co., Ltd., 1-7 Kojimachi, Chiyoda-ku, Tokyo 102-8080. **Web:** http://www.tfm.co.jp

26)FM 802 Co., Ltd., Kita 2-6, Tenjinbashi 2-chome, Kita-ku, Osaka 530-8580. **Web:** http://www.fm802.co.jp

27)FM Hokkaido Broadc. Co., Ltd., 1, Nishi 2-chome, kita 1-jo, Chuo-ku, Sapporo 060-8532. **Web:** http://www.air-g.co.jp

28)Yamagata FM Broadc. Co., Ltd., 14-69, Matsuyama 3-chome, Yamagata 990-8543. **Web:** http://www.dewa.or.jp/boy-fm

29)FM Ishikawa Broadc. Co., Ltd., 1-45, Hikoso-machi 2-chome, Kanazawa 920-8605. **Web:** http://www.fmishikawa.co.jp

30)FM Aichi Broadc. Co., Ltd., 15-18, Chiyoda 2-chome, Naka-ku, Nagoya 460-8388. **Web:** http://www.nds-g.co.jp/fm_aichi

31)FM Tokushima Broadc. Co., Ltd., 1-6, Saiwai-cho, Tokushima 770-0847

32)Fukuoka FM Broadc. Co., Ltd., 1-82, Watanabe-dori 2-chome, Chuo-ku, Fukuoka 810-8575. **Web:** http://fmfukuoka.co.jp

33)FM Japan Ltd., 17-30, Nishiazabu 4-chome, Minato-ku, Tokyo 106-8088. A satellite audio service also exists. **Web:** http://www.j-wave.co.jp

34)FM Kochi Broadc. Co., Ltd., 3-21, Hon-machi 3-chome, Kochi 780-8532

35)FM North Wave Co., Ltd., 3-1, Nishi 4-chome, Kita 7-jo, Kita-Ku, Sapporo 060-8577. **Web:** http://demia.aaapc.co.jp/825

36)Toyama FM Broadc. Co., Ltd., 8-10, Marunouchi 1-chome, Toyama 930-8567

37)Akita FM Broadc. Ltd., 7-10, Yabase-Honcho 3-chome, Akita 010-0973

38)FM Fuji Co., Ltd., 7-23, Marunouchi 2-chome, Kofu 400-8550. 2000-1900(Sat 1600. Sun.1630). **Web:** http://www.yin.or.jp/fmfuji

39)Miyazaki FM Broadc. Co., Ltd., 2-78 Gion, Miyazaki 880-8583

40)Yokohama FM Broadc. Co., Ltd., 2-1, Minato-Mirai 2-chome, Nishi-ku, Yokohama 220-8110. **Web:** http://www.fmyokohama.co.jp

41)FM Osaka Co. Ltd., 2-4, Nakanoshima 3-chome, Kita-ku, Osaka 530-8285. **Web:** http://www.fmosaka.co.jp

42)Gunma FM Broadc. Co., Ltd., 4-8, Wakamiyacho 1-chome, Maebashi 371-8533. **Web:** http://www.gunmanet.or.jp/fm-gunma

43)FM Okinawa Broadc. Corp., 40, Kowan, Urasoe, Okinawa 901-2525

44)FM Oita Broadc., Co., Ltd., 17-19, Higashi kasuga-machi, Oita 870-8558

45)FM Kyoto, Inc., 98, Matsumoto-cho, Kamigamo, Kita-ku, Kyoto 603-8588. **Web:** http://www.fm-kyoto.co.jp

46)Hyogo FM Radio Broadc. Co., Ltd., 5-4 Hatoba-machi, Chuo-ku, Kobe 650-8589. **Web:** http://www.kiss-fm.co.jp

47)FM Fukushima Inc., 6-6, Sakae-machi, Fukushima 960-8031

48)Kyushu International FM Inc., 5-35, Tenjin 2-chome, Chuo-ku, Fukuoka 810-8565. Prgr. in English, Chinese and Korean. **Web:** http://www.lovefm.com

49)FM Shiga Co., Ltd., 19-10, Nishinosyo, Otsu 520-0818. **Web:** http://www.fm-shiga.com

UNIVERSITY BROADCASTING STATION

✉ Hoso Daigaku, 2-11, Wakaba, Mihama-ku, Chiba 261-8586. **Web:** http://www.u-air.ac.jp/hp
JOUD-FM 77.1MHz 10kW, Tokyo, 78.8MHz 1kW Maebashi.
D. PRGR: 2100-1500 - **Pub:** Broadcasts texts for distance university studies.

AMERICAN FORCES NETWORK (AFN)

The network serves the members of the US forces. The st's in Japan broadcast by authority of Commander, US Forces, Japan, in cooperation with the Japanese Radio Regulatory Bureau. St's are linked by land line and microwave.

✉ **AFN Tokyo**, Detachment 10, Air Force Broadcasting Service, Unit 5091, APO AP 96326-5091. ☎ +81(425)52 2510 ext 52374. **E-mail:** eagle810@pbs.pa.af.mil. **Web:** http://www.yokota.af.mil/afrts/index
✉ Other stns` **AFN Okinawa** - Det 11, AFBS, Unit 5154, APO AP 96368-5154. **Web:** http://www.okr.usmc.mil/offduty/fen **AFN Misawa** - Det12, AFBS, Unit 5033, APO AP 96319-5033. **Web:** http://www.misawa.af.mil/orgs/afn/main **AFN Iwakuni** - Det13, AFBS, PSC561, Box 1875, FPO AP96310-1875. **AFN Sasebo** - Det14, AFBS, PSC476, Box 8, FPO AP96322-2900.

Mediumwaves	kHz	kW	Mediumwaves	kHz	kW
Okinawa	648	10	Misawa	1575	0.25
Tokyo	810	50	Iwakuni	1575	1
Sasebo	1575	0.25			

FM: Okinawa 89.1MHz 6kW.
D. PRGR: 24h. **N:** on the h.
Local live prg's: Mon-Fri 2100-2400, 0100-0400, 0500-0900.
ANN: "This is the Armed Forces Radio and Television Service." For Tokyo station: "This is Eagle 810".
V. by QSL card - **PUB:** Master schedule quarterly.

KAZAKSTAN

L.T: UTC + 6h (eastern), 5h (western), 4h (Batys Qazaqstan province) (Su: UTC + 7h/6h/5h) — **Pop:** 16.677.000 — **Radios:** 4.180.000 — **Pr.L:** Kazak (official), Russian — **E.C:** a/c 50Hz, 127/220V — **ITU:** KAZ.

BAYLANIS MINISTIRLIKI
(Ministry of Telecommunications)

✉ Bogenbay Batira köçesi 134, 480000 Almaty. ☎ +7 (3272) 623194. ▤ +7 (3272) 637210. ① 251358 FAD.
L.P: Minister: Igor V. Ulyanov.

REPUBLICAN TV & RADIO BROADCASTING CORP. OF KAZAKSTAN (Gov.)

✉ Zheltoksan Str. 175 A, 480013 Almaty.
L.P: Pres: Ashirbek Kopishev.

KAZAK RADIO (Gov.)

✉ Zheltoksan Str. 175 A, 480013 Almaty. ☎ +7 (3272) 635629 (Int. Rel.). ▤ +7 (3272) 631207. ① 251114 TRVD.
L.P: Gen. Dir: A.N. Midike.

LW/MW	kHz	kW	Prgr.
Almaty	180	250	KR1
Aqtöbe	180	150	KR1
Türkistan (1)	180	50	KR1
Almaty	243	1000	KR2
Qaraghandy	243	1000	KR2
Almaty	549	1000	
Qostanay	684	30	KR1/Reg
Kökşetau	747		KR1/Reg
Qaraghandy	747	20	KR1/Reg
Oral (12)	747	50	KR1/Reg
Türkistan (1)	783	150	KR2/Reg
Arqalyq (9)	801	30	KR1/Reg
Öskemen (7)	801	30	KR1/Reg
Jambyl	846	50	KR1/Reg
Karaturuk (6)	900	150	RSY/F
Qaraghandy	936	1000	
Kaçiry (2)	945	50	KR1/Reg
Qyzylorda	954		KR1/Reg
Aqmola	972	30	KR1/Reg
Aqtöbe	981		KR2/Reg
Jilaydy (4)	1008	50	KR1/Reg
Öskemen (7)	1071	100	KR2
Almaty	1098	150	KR2/Reg
Şardara (1)	1098		KR2

LW/MW	kHz	kW	Prgr.
Jetisay (1)	1098		KR2
Qaragayly (3)	1098		KR2
Leninskoye (1)	1098		KR2
Şingoja (4)	1098		KR2
Jilaydy (4)	1098		KR2
Abay (1)	1188		KR2
Ayaköz (4)	1188		KR2
Kökşetau	1188		KR2
Yuzhnyy (4)	1188		KR2
Aqmola	1197		KR2
Unknown	1197		KR2
Almaty	1197		RUI
Jambyl	1224		KR2
Jezdy (10)	1224		KR2
Çelkar (11)	1224		KR2
Atyrau	1323	20	KR1/Reg
Aqtau (8)	1341	25	KR1/Reg
Almaty	1341	30	VOA/RL
Panfilov (5)	1395		KR2
Unknown	1422		KR2
Atasu (10)	1440	5	KR2
Bakanas(6)	1440		KR2
Qostanay	1440		RUI
Qyzylorda	1440		KR2
Arqalyq (9)	1458		KR2
Buran (7)	1458		KR2
Unknown	1494		RUI
Qostanay	1539		KR2
Andreyevka (5)	1548		KR2/Reg
Jansügirov (5)	1548		KR2/Reg
Lepsi (5)	1548		KR2/Reg
Tekeli (5)	1548		KR2/Reg
Antonovka (5)	1557		KR1/Reg
Kapal (5)	1557		KR1/Reg
Uç-Aral (5)	1557		KR1/Reg

Regions: (1) Şymkent; (2) Pavlodar; (3) Qaraghandy; (4) Semey; (5) Taldyqorghan; (6) Almaty; (7) Şyghys Qazakstan; (8) Mangghystau; (9) Torghay; (10) Jezkazghan; (11) Aqtöbe; (12) Batys Kazakstan; others = city name.

Shortwaves	kHz	kW	H. of tr.	Prgr.
Almaty	4545	20	0000-1900	KR1
Almaty	5970	100	0100-1800	KR2
Almaty	6155	100	1230-1700	F
Almaty	6160	200	2330-0200	F
Almaty	6180	100	0000-1900	KR1
Almaty	6230	100	1230-1700	F
			2330-0200	F
Almaty	7255	100		F
Almaty	9505	100	0100-1800	KR2
Almaty	11950	100	0000-1900	KR1

KR2 rel. via Kiev, Ukraine: 2 x 100kW, freq's variable.
NB: Powers are estimates only — **FM:** Information not available.

KR1 in Kazakh/Russian 0000-1830(1900SS); German: Sat 0900-0956.
KR2 (Şalkar) 0100-1800 in Kazakh incl. prgr's in Uyghur, German.
KR3 (FM only): Details not available.
RSY = R. St. Yunost relay (cf. Russia).
RL = R. Liberty relay (cf. Czech Rep.) 0000-0100, 0200-0300, 0500-0700, 1400-1600 in Russian; 0100-0200, 1300-1400 in Kazakh.
VOA = Voice of America relay (cf. USA).
RUI = R. Ukraine International relay (cf. Ukraine).
Regional Prgrs: 0115-0200 (ea. regions), 0215/0230-0300 (we. regions), 1300-1400.
✉ of regional centres: Moskva köç. 49, 473032 Aqmola — Lenin köç. 43, 463000 Aqtöbe — pr. Gornyakov 36, 459830 Arqalyq — pl. Abaya 25, 465017 Atyrau — Nekrasov köç. 15, 486024 Şymkent — Kumjat köç, 484000 Jambyl — Svyaz köç. 14, 470061 Qaraghandy — Puşkin köç. 54, 458000 Qostanay — Vokzal köç. 21, 637026 Pavlodar — Şugayev köç. 157, 490018 Semey — Kirov köç. 432, 488009 Taldy-Qorghan — pr. Lenina 204, 417000 Oral — Stakhanov köç. 70, Öskemen.

FOREIGN SERVICE: see International Broadcasting section.

OTHER STATIONS

MW	kHz	Station, location
1)	1494	Beat 16, Almaty

FM	MHz	Station, location
2)	66.00	Radio M, Almaty
3)	66.26	R. Terra, Qaraghandy: 0130.1600

FM	MHz	Station, location
4)	67.82	R. Rossii-Nostalgie, Qaraghandy
5)	68.96	Radio 69, Qaraghandy: See below
6)	69.70	R. Rik, Almaty
7)	70.70	R. Maksimum, Almaty
8)	71.51	Radio 31, Almaty
9)	73.31	R. Maks, Şimkent
4)	73.60	R. Rossii-Nostalgie, Aktöbe
10)	91.7	R. Totem, Almaty
9)	101.0	R. Maks, Jambyl
9)	102.2	R. Maks, Almaty
8)	103.5	Radio 31, Almaty
4)	103.8	R. Rossii-Nostalgie, Aktöbe
11)	104.0	R. Akbar, Almaty
12)	104.7	Radio MBC, Almaty
13)	106.0	Radio NS, Almaty

3) Mir blvd. 22/2, Qaraghandy — 4) cf. Russia — 5) ul. Voynov Internatsionalistov 14, 470061 Qaraghandy.

KOREA (Democratic People's Republic of)

L.T: UTC + 9h — **Pop:** 23,920,000 — **Radio sets:** 4,700,000 — **Pr.L:** Korean —
E.C: a/c 60Hz, 100/200/220V — **ITU:** KRE.

THE RADIO AND TELEVISION BROADCASTING COMMITEE OF THE DEMOCRATIC PEOPLE'S REPUBLIC OF KOREA

✉ Jonsung-dong, Moranbong District, Pyongyang. ☎ +850(2)816035; 🖷 +850(2)812100; ☉ 5508 RTKP; **Cable:** KPT PYONGYANG.
L.P: Chairman: Jong Ha Chol. Vice Chairman: Ri Pong Hui, Cha Sung Su, Han Kwang Hak, Kim Kwang Ho, Ro Sok Kyu, Kim Song Guk, Maeng Kyong Ho, Ri Hyong Jom, Jo Hyon Hwa, Kim Won Chol, Ri Yong Do.

KOREAN CENTRAL BROADCASTING STATION (Joson Jung-ang Pangsong)

✉ Jonsung-dong, Moranbong District, Pyongyang. ☎ +850(2)812301. 🖷 +850(2)814418.
C = Central Broadcast from Pyongyang, R = Regional Sce, E = rel. Ext. Sce.

Mediumwaves

	kHz	kW	Prgr		kHz	kW	Prgr
Chongjin	702	50	C/R	Wonsan	882	250	C/R
Wiwon	720	500	C/R	Hwangju	927	50	C/R
Hyesan	765	50	C/R	Hamhung	999	250	C/R
Pyongyang	819	500	C	Haeju	1080	1500	C/R
Sinuiju	864	250	C/R	Pyongyang	1368	2	E

Shortwavess (all freqs variable):

	kHz	Prgr		kHz	Prgr
Sariwon	2350	C/R	Kanggye	3960	C/R
Pyongyang	2850	C	Wonsan	3970	C/R
Hamhung	3220	C/R	Kanggye	6100	C
Pyongsong	3350	C/R	Pyongyang	9665	C
Hyesan	3920	C/R	Kanggye	11680	C
Chongjin	3940	C/R			

FM: Kaesong 102.3MHz.
D.PRGR in Korean: 2000-1800 on all freqs exc. 1080 (2000-1500), 6100 (2000-0630 & 1500-1800) and 9665 (2000-0930 &1730-1800).
N: 2100, 2200, 0100, 0300, 0600, 0800, 1100, 1200, 1300. Regional Prgrs: W0500-0600.
Rel. Pyongyang Broadc. St: 1500-1800 on 720/864kHz. 1500-2000 on 702kHz/102.3MHz. 1800-2000 on 3220/3940/3970kHz.
ANN: "Joson Jung-ang Pangsong-imnida". Reg. Prgrs: "(location) Pangsong-imnida". IS: Song of General Kim II Sung. Opening & closing music: Nat. Anthem.
V: does not normally verify.

PYONGYANG BROADCASTING STATION (Pyongyang Pangsong)

✉ Pyongyang.
P = Pyongyang Broadc. St, E = Ext. Sce.

Mediumwaves

	kHz	kW	Prgr		kHz	kW	Prgr
Chongjin	621	500	P/E	Sepo	729	50	P
Kangnam*	657	1500	P	Hwadae	801	500	P
Samgo	684	250	P	Sangwon	855	500	P

*) Key st. serving Pyongyang area.

Shortwaves (all freqs variable): Pyongyang 3250/3320/6250kHz, Kanggye 6400kHz.
FM: Pyongyang 89.1, 89.5, 90.3 , 91.1, 91.9, 92.9, 95.1, 95.9, 96.7, 97.3, 99.5, 101.1, 101.9, 104.5, 106.5, 107.3MHz - Kaesong 90.7, 91.4, 91.9, 98.2, 104.1, 104.5, 104.9, 105.9, 107.1MHz
D.PRGR. in Korean: 2100-2030 on 657/729/801/855kHz; 2100-1800 on 684/6400kHz; 2100-1900 on 801/3320kHz; 2100-0930 & 1500-1900 on 6250kHz; 2100-2300 & 0100-0700 on 621/3250kHz; 1500-2030 on 3250kHz; 1500-1800 on 621 kHz. **N:** 2200, 2300, 0100, 0300, 0600, 0800, 1100, 1200, 1300. Rel. the Voice of National Salvation (clandestine st.): 1030-1100(Sun.)
ANN: "Pyongyang Pangsong-imnida".
IS: Song of General Kim II Sung. Opening & closing music: Nat. Anthem.
V: does not normally verify.

PYONGYANG FM BROADCASTING STATION (Pyongyang FM Pangsong)

Stations:

	MHz	kW		MHz	kW
Pyongsong	90.1	2	Komdok	102.1	1
Kaesong	92.5	2	Sariwon	103.0	2
Kanggye	93.3	5	Haeju	103.7	10
Hyesan	93.8	2	Pyongyang	105.2	20
Wongsan	95.1	5	Chongjin	105.5	10
Heaju	97.8	10	Hamhung	106.1	20
Sinuiju	101.3	5	Nampo	107.2	2

D.PRGR: 0700-2000, 2100-2400 (National holidays: 2100-2030) (music, drama and novel).
ANN: "Pyongyang FM Pangsong-imnida".
IS: Song of Dear Leader Comrade Kim Jong II. Opening music: Pyongyang Is My Heart.

EXTERNAL SERVICE

✉ R. Pyongyang, Pyongyang
Mediumwave: Chongjin 621 kHz 500kW - Pyongyang 1368kHz 2kW.
Shortwaves: Kanggye (G.C: 126.36E/40.58N): 5x200kW, Kujang (G.C: 125.05E/40.05N): 5x200kW, Pyongyang (G.C: 125.33E/39.05N): 10x200kW.
kHz: 3250, 3560+, 4405+, 6070, 6125, 6520, 6575, 7200, 7580, 9325, 9345, 9600, 9640, 9650, 9975, 11335, 11700, 11735, 11740, 11845, 13650, 13760, 13790, 15130, 15180, 15230, 15340, 17735
+)These are standby/emergency feeder tr's.

Arabic

1500-1600*	ME/Af	6520, 9600
1800-1900	ME/Af	6520, 9600, 9975
2000-2100	ME/Af	6520, 9600, 9975

Chinese

0700-0800	CHN	4405, 6125, 7200, 9345
0900-1000*	CHN	4405, 6125, 7200, 9345
1100-1200	CHN	4405, 6125, 7200, 9345
1300-1400	CHN	4405, 6125, 7200, 9345

English

0000-0100	SEAs/LAm	3560, 11845, 13650, 15230
0500-0600*	Eu	3560, 11740, 13790
1100-1200	ME/Af	3560, 9640, 9975
1100-1200	SEAs/LAm	11335, 13650, 15230
1500-1600	ME/Af	3560, 9640, 9975
1500-1600	SEAs/LAm	11335, 11735, 13650
1800-1900	Eu	4405, 6575, 9345
1800-1900	NAm	11700, 13760
1900-2000	ME/Af	6520, 9600, 9975
2100-2200	Eu	4405, 6575, 9345
2100-2200	NAm	11700, 13760
2300-0000	NAm	4405, 11335, 11700, 13760, 15130

French

0400-0500*	Eu	3560, 11740, 13790
0500-0600	SEAs/LAm	13650, 15180
0500-0600	NAm	15340, 17735
1200-1300*	ME/Af	3560, 9640, 9975
1200-1300	SEAs/LAm	11335, 13650, 15230
1500-1600	Eu	4405, 6575, 9345
2000-2100	Eu	4405, 6575, 9345
2000-2100	NAm	11700, 13760
2100-2200	ME/Af	6520, 9600, 9975

German

1800-1900	Eu	3560, 9325, 13790
2000-2100	Eu	3560, 9325, 13790
2100-2200*	Eu	3560, 9325, 13790

Japanese

0000-0100	J	621, 3250, 7580, 9650
0700-0800	J	621, 3250, 6070, 7580, 9650
0800-0900*	J	621, 3250, 6070, 7580, 9650
0900-1000	J	621, 3250, 6070, 7580, 9650
1100-1200	J	621, 3250, 6070, 6520, 7580
1200-1300	J	621, 3250, 6070, 6520, 7580
1300-1400	J	621, 3250, 6070, 6520, 7580
2300-2400*	J	621, 3250, 7580, 9650

Korean

0400-0600	CHN	4405, 7200, 9345
0600-0650*	Eu	11740, 13790
0600-0650	SEAs/LAm	13650, 15180
0600-0650	ME/Af	15340, 17735
0800-0900	CHN	4405, 6125, 7200, 9345
1200-1300	CHN	4405, 6125, 7200, 9345
1400-1450*	ME/Af	3560, 9640, 9975
1400-1450	SEAs/LAm	11335, 13650, 15230
1600-1650	ME/Af	3560, 6520, 9600, 9640, 9975
1600-1650	Eu	4405, 6575, 9345
1600-1650	SEAs/LAm	11335, 11735, 13650
2200-2250	Eu	4405, 6575, 9345, 11700, 13760

Korean/Russian

0800-0900	RUS	3560 ,6575, 9975, 11740, 13790
1700-1800	RUS	6520, 9600, 9975

Russian

0700-0800*	RUS	3560 ,6575, 9975, 11740, 13790
0900-1000	RUS	3560 ,6575, 9975, 11740, 13790
1700-1800	Eu	3560, 9325, 13790
1900-2000	Eu	3560, 9325, 13790

Spanish

0000-0100*	NAm	4405, 11335, 11700, 13760, 15130
0400-0500	SEAs/LAm	13650, 15180, 15340, 17735
1300-1400*	ME/Af	3560, 9640, 9975
1300-1400	SEAs/LAm	11335, 13650, 15230
1700-1800	Eu	4405, 6575, 9345
1700-1800	NAm	11700, 13760
1900-2000	Eu	4405, 6575, 9345
1900-2000	NAm	11700, 13760
2300-0000	SEAs/LAm	3560, 11845, 13650, 15230

*) Transmissions are also aired on 1368kHz for Pyongyang area.

IS: Song of General Kim Il Sung. Opening & closing music for Korean sce: Nat. Anthem.

ANN: Korean: "Joson Jung-ang Pangsong-imnida" or "Pyongyang Pangsong-imnida". Arabic: "Idaat Jumhuriat Kuria Democratiat At Shabia min Pyongyang". Chinese: "Pingrang Guangbo Diantai". English: "This is Radio Pyongyang". French: "Ici Radio Pyongyang". German: "Hier ist Radio Pyongyang". Japanese: "Kochirawa Pyongyang Hoso desu". Russian: "Gavarit Penyan". Spanish: "Aqui Radio Pyongyang".

V. by QSL card - PUB: "Radio Pyongyang" upon request.

KOREA (Republic of)

L.T: UTC + 9h — **Pop:** 45,060,000 — **Radio sets:** 42m — **Pr.L:** Korean — **E.C:** a/c 60, 100/220V — **ITU:**KOR.

KOREAN BROADCASTING SYSTEM(KBS)
(Hanguk Pangsong Kongsa)(Public Corporation)
✉ 18, Yo-ui-do-dong, Yongdungp'o-gu, Seoul 150-790. ☎+82(2)781 2410. 🖷 +82(2)761 2499. ① KBSKB K24599. **Cable:** KBS, Seoul. **Web:** http://www.kbs.co.kr
L.P: Pres: Park, Kwon-Sang. Vice Pres: Lee, Hyung-Mo. Man. Dir.(Corporate Planning): Lee, Sang-Won. Man. Dir.(N. & Sports): Chon, Byung-Chae. Man. Dir.(TV): Kang, Dae-Young. Man. Dir.(R.): Kim, Jong-Il. Man. Dir.(Broadc. Operations & Eng.): Park, Hyun-Jong. Exec. Dir.(Public Rel.): Nam, Sun-Hyun. Dir.Technical Planning. Div: Chung, Woon-Sung. Dir. Int. Rel. Div: Cha, Myong-Hee.
Mediumwaves: N1= KBS R. One, N2 = KBS R. Two, L1 = First Liberty Prgr, L2 = Second Liberty Prgr., E = also used for Ext. sce., **N** = Netw. or local st. area, + = Regional key St, 2 = rel N2 exc. for local prgr's (other local st's take N1), *) Seoul area, Call: HL(call).

N	Location	Call	kHz	kW
13)	Changsu	SN	540	1
20)	Chomch'on	SC	540	1
11)	Hongsong	CZ	540	10
16)	Changhung	SM	540	1
20)	Taegu+2	QH	558	250
13)	Chonju+	KF	567	100
18)	Sunchon+2	KZ	576	1

N	Location	Call	kHz	kW
21)	Yongju	AG	594	10
N2)	Namyang*	SA	603	500
8)	T'aebaek+	SJ	621	10
26)	Sogwip'o	CF	621	10
9)	Yongdong	AY	621	1
3)	Inje	SE	630	5
19)	Yosu+	CY	630	10
N2)	Kaebong*	-	639	50
16)	Posong	SL	648	1
3)	Ch'unch'on+	KM	657	50
14)	Kunsan+2	AS	675	10
N1)	Sorae*	KA	711	500
20)	Taegu+	KG	738	100
16)	Kwangju+	KH	747	100
N1)	Yoju*	-	756	100
6)	Yongwol	CV	783	10
3)	Yanggu	SY	846	5
4)	Kangnung+	KR	864	100
11)	Taejon+	KI	882	20
2E)	Pusan+	KB	891	250
20)	Kumi	QY	909	10
N1)	Yonch'on*	-	918	50
12)	Puyo	QA	927	10
25)	Hadong	SU	927	1
7)	Hongch'on	QD	927	1
23)	Ch'angwon+2	KD	936	10
9)	Poun	QW	945	10
26)	Cheju+	KS	963	10
21)	Andong	CR	963	10
L1)	Tangjin*	CA	972	1500
5)	Sokch'o 2	CS	1008	50
3)	Hwach'on	CG	1026	10
25)	Koch'ang	KW	1026	1
22)	P'ohang+	CP	1035	10
4)	Samch'ok	CI	1044	10
10)	Chech'on	CD	1044	10
9)	Ch'ongju+	KQ	1062	50
10)	Ch'ungju+	CH	1089	10
25)	Chinju+	CJ	1098	20
L1)	Hwasong*	-	1134	500
7)	Wonju+	CW	1152	10
L2E)	Kimje*	SR	1170	500
6)	Chongson	SW	1206	1
21)	Ch'ongsong	QR	1206	1
16)	Kwangju+2	AA	1224	20
6)	P'yongch'ang	CC	1233	1
21)	Yongyang	QG	1233	1
15)	Namwon+	KL	1260	10
N1)	Yangju	-	1269	10
14)	Kurye	SI	1269	1
23)	Hapch'on	QV	1278	1
22)	Uljin	SV	1305	10
16)	Yong-gwang	QJ	1323	1
22)	Ullung	CU	1323	1
N1)	Kimp'o*	-	1341	10
13)	Muju	KO	1368	1
N1)	Ch'olwon*	CO	1395	10
24)	Ulsan+	QB	1449	10
25)	Hamyang	SH	1458	1
21)	Ponghwa	SD	1458	1
17)	Mokp'o+	KN	1467	50
18)	Kohung	QU	1485	1
12)	Kongju+	QS	1485	1
20)	Kimch'on	SK	1503	1
8)	Kosan	QC	1539	1
28)	Tanyang	DK	1584	1
25)	Sanch'ong	-	1584	1
11)	Kumsan	QZ	1584	1
6)	Sabuk	QE	1602	1

NB: Liberty st's and FM-st's do not use call letters (even if assigned). Other st's without call letters use the calls from their main st's.

Shortwaves: Hwasong (G.C. 126.47E/37.13N).

kHz	kW	Prgr.
3930	10	N1 + L
6135	10	L2 + R. Korea Int.
6015	100	L1

FM: (MHz) Reg = region in MW section. I-Standard F(R. One)M; II-KBS FM One; III a = FM Two, b = R. Two. *) are also RDS and SCA(Sarange Sori Pangsong)

Reg.	Location	I	II	III	kW
1)	Namsan		93.1	89.1a	-/10/10
1)	Kwanak-san	97.3*			10
1)	Yongmun-san	90.3*			1
2)	Yongdo		92.7	97.1b	-/5/3
3)	Hwa-ak-san	99.5*	91.1		5/5
4)	Kwaebang-san	98.9*	89.1		1/5
7)	Paegun-san		89.5		-/3
7)	T'aegi-san	95.5*			1
8)	Hambaek-san	93.7*	97.3		1/3
9)	Shikchang-san		102.1		-/3
9)	Huksong-san	89.9*			1
9)	Uam-san	89.3	94.1		1/1
9)	Kayop-san	90.9*			1
10)	Kayop-san		100.3		-/3
11)	Kyeryong-san	94.7*	98.5		1/5
13)	Mo-ak-san	96.9*	100.7		5/5
15)	Nogodan	88.3*	92.9		1/3
16)	Mudung-san	90.5*	92.3		5/5
17)	Yang-ul-san		98.3		-/1
19)	Mang-un-san	95.7*	94.5		1/3
20)	P'algong-san	101.3*	89.7		5/5
21)	Ilwol-san	92.1*			1
21)	Hakka-san		88.1		-/3
22)	Chohang-san	95.9*	93.5		1/3
23)	Pulmo-san	91.7*	93.9		1/1
24)	Muryong-san	90.7*	101.9		1/3
25)	Kamak-san		102.5		-/3
25)	Mangjin-san		89.3		-/1
26)	Kyonwolak	99.1*	96.3		5/3
26)	Sammaebong		99.9		-/3

+ low power relay st.

KBS R. One (KBS Che-il Radio, HLKA): 0300-0100 on 3930kHz, others 24h.
Non-commercial nationwide sce. Key freq's 711/756/3930kHz, 90.3/97.3MHz.
Also rel. by Standard FM St's and most reg. st's. All R. One St's carry a separate Liberty prgr 1600-1900. Reg. st's may broadcast local prgr's at designated times. **N:** hourly 2000-1500. **Local N:** 2205(Sun), 2210(W), 0000, 0310(Sun), 0315(W), 0600(Sun), 0605(W), 0800(Sun), 0805(W), 1100.
KBS R. Two (KBS Che-i Radio, HLSA): 2000-1800 (558kHz to 1500). Commercial. Key freq's 603/639kHz. Reg. st's may broadcast local prgr's at designated times. **N:** hourly 2000-1400. Local N: 2300, 0400, 0700, 1200. Liberty prgr 1700-1800.
LIBERTY Prgr. (Sahoe Kyoyuk Pangsong)(Social educ. sce.): A sce. for ethnic
Koreans living outside of the Rep. of Korea. Two separate sce's + prgr's on
KBS R. One and R. Two.
First Liberty Prgr. (Sahoe Kyoyuk Che-il Pangsong, HLCA): 0400-2400 on 972/1134/6015kHz, 1500-1800 on 558kHz. **N:** 0400(W), 0500(W), 0600, 0700, 0800, 0900, 1000, 1100, 1200, 1500(W), 2300(W).
Second Liberty Prgr. (Sahoe Kyoyuk Che-i Pangsong, HLSR): 1000-0400 on 1170kHz, 1000-2400 on 6135kHz. Relay First Liberty Prgr 1400-2400 and takes R. Korea Int. 1000-1400.
KBS FM One (KBS Che-il FM Pangsong, HLQK-FM): 24h. Mainly Korean traditional and western classical music.
KBS FM Two (KBS Che-i FM Pangsong, HLKC-FM): 24h. Mainly Korean and western popular and light classical music.
NB: Regional FM One st's air a combination of their own local prgr's, FM One and FM Two prgr's.
ANN: N1: "AM Ch'ilbaek-sib-il(711) kHz, FM Kusib-ch'il-chom-sam(97.3) MHz, Che-il Radiomnida. HLKA". N2: "KBS Che-i Radiomnida. HLSA". Liberty prgr: "Taehanminguk Seoul-eso Ponaedurinun KBS Sahoe Kyoyuk Pangsong-imnida".

🖃 Regional Key Stations
2) 63, Namch'on-dong, Suyong-gu, Pusan 608-790 - 3) 86-1, Nagwon-dong, Ch'unch'on 200-100 - 4) 62-5, Yong-gang-dong, Kangnung 210-070 - 5) 306-2, Tongmyong-dong, Sokch'o 217-020 - 6) 893-1, Yonghung-7-ri, Yongwol-up, Yongwol-gun 230-800 - 7) 79-1, Won-dong, Wonju 220-060 - 8) 82-2, Hwangji-1-dong, T'aebaek 235-011 - 9) 604-26, Sajik-dong, Ch'ongju 360-070 - 10) 417, Munhwa-dong, Ch'ungju 380-790 - 11) 15-3, Mok-dong, Chung-gu, Taejon 301-070 - 12) 571-2, Shingwan-dong, Konju 314-040 - 13) 523-3, Kumam-2-dong, Tokchin-gu, Chonju 560-180 - 14) 796, Naun-1-dong, Kunsan 572-020 - 15) 500, Hyanggyo-dong, Namwon 590-020 - 16)177-39, Sa-dong, Nam-gu, Kwangju 502-790 - 17) 1188-3, Yongdang-dong, Mokp'o

530-360 - 18) 91-3, Sokhyon-dong, Sunch'on 540-100 - 19) 8-7, Kwanmun-dong, Yosu 550-070 - 20) 100, Shinch'on-3-dong, Tong-gu, Taegu 701-790 - 21) 666, T'aehwa-dong, Andong 760-790 - 22) 655, Sangdo-dong, Nam-gu, P'ohang 790-790 - 23) 97-1, Shinwol-dong, Ch'angwon 641-790 - 24) 416-17, Tal-dong, Nam-gu, Ulsan 680-790 - 25) 13-22, Shinan-dong, Chinju 660-790 - 26) 302-2, Yon-dong, Cheju 690-170.

Local identifications: Within local prgrs. N1: just before the h. at 2000, 2100, 2300-0100, 0300, 0500, 0700, 0800, 0900, 1100, 1300-1600. N2: just before the h. 2000-1600.

RADIO KOREA INTERNATIONAL (Radio Hanguk)
🖃 18, Yo-ui-do-dong, Yongdungp'o-gu, Seoul 150-790. ☎ +82(2)781 3710; 🖷 +82(2)781 3799. **LP:** DG: Kim, Sun-Ok.
MEDIUMWAVES: See Home Sce.
SHORTWAVES: Kimje (G.C: 126.50E/35.50N): 3x100 kW, 3x250kW(+1x100kW reserve) - H) Hwasong (G.C: 126.47E/37.13N): 2x10kW, 2x100kW(+1x10 kW reserve) - S) via Sackville, Canada - B) via Skelton, United Kingdom.
kHz: 3970B, 5975H, 6135H, 6145B, 6145S, 6150, 6480, 7275, 7285, 7550, 9515, 9535B, 9570, 9580, 9640, 9650S, 9870, 9875B, 11715S, 11725, 13670, 15575 ND)Omnidirectional.

Arabic

1600-1700	6150, 7275	CHI
1900-2000	6480, 7550, 15575	Eu
	9515	ME/Af

Chinese

1130-1230	9640	ND
	6055	CHI
1300-1400	1170, 5975, 6135	ND
	7285	CHI
2200-2300	5975	ND
	7275	CHI
	9640	SEAs

English

0200-0300	7275	CHI
	11725, 11810	SAm
	15575	NAm
0800-0900	9570	AUS
	13670	Eu
1030-1100	11715S	SAm
1200-1300	7285	CHI
1230-1300	9570, 13670	SEAs
	6055	CHI
	9640	ND
1600-1700	5975	ND
	9515, 9870	ME/Af
1900-2000	5975	ND
	7275	CHI
2100-2130	3970B	Eu
	6480	Eu
2100-2200	15575	Eu

French

1700-1800	7275	CHI
	9515, 9870	ME/Af
1900-2000	6145B	Eu
	9870	ME/Af

German

1800-1900	6480	Eu
	7275	CHI
2000-2100	6145B, 7550, 15575	Eu

Indonesian

1130-1230	9570, 13670	SEAs
1400-1500	13670	SEAs
2300-2400	7275	CHI
	9640	SEAs

Japanese

0000-0100	9640	SEAs
	11810	SAm
0800-0900	5975, 7275, 9640	ND
1200-1300	1170, 5975, 6135	ND
1400-1500	5975, 7275	ND

Korean

0100-0200	7275	CHI
0300-0400	7275	CHI
	11725, 11810	SAm
	15575	NAm
0700-0800	7550, 9535B	Eu
0900-1000	7550	SAm

Time	Freq	Target
0900-1100	5975, 7275	ND
	9570	AUS
	13670	Eu
1000-1100	1170, 6135	ND
1100-1130	6145S, 9650S	NAm(E)
	9640	ND
	9580	SAm
1300-1400	9640	ND
	13670	SEAs
1700-1900	5975	ND
	7550, 15575	Eu
2100-2200	5975	ND
	7275	CHI
	9640	SEAs
2300-0100	5975	ND
	15575	NAm

Russian

1100-1200	1170, 5975, 6135, 7275	ND
1500-1600	9515	ME/Af
1800-1900	9875B	Eu
2000-2100	5975	ND

Spanish

0100-0200	11725, 11810	SAm
	15575	NAm
1000-1030	11715S	SAm
1000-1100	7550, 9580	SAm
1800-1900	9515, 9870	ME/Af
2000-2100	6480	Eu
	7275	CHI
	9870	ME/Af
2200-2300	6150	Eu

IS: Korean children's song "Tar-a Tar-a Palgun Tar-a(Oh, bright moon)" played by a glockenspiel. Original music "Dawn" composed by Kim, Hee-jo with KBS symphony orchestra. - **V.** by QSL card.
ANN: Korean: "Yoginun Taehanminguk Seoul-eso Ponaedurinun Hanguk Pasong Konsa, KBS-e Kukche Pangsong, Radio Hangug-imni-da". English: "This is Radio Korea International of the KBS". Japanese: "Kochirawa Rajio Kankoku, KBS-no Kokusai Hoso desu".
PUB: Prgr. schedules, pennant, folder, RKI Newsletter(each lang.), KBS Handbook (English), KBS information Sce.

EDUCATION BROADCASTING SYSTEM(EBS)
(Kyoyuk Pangsong)
Programming by Ministry of Education.
✉ 92-6, Umyon-dong, Soch'o-gu, Seoul. ☎ +82(2)572 5021, 572 5121, 522 8020. **Web:** http://www.ebs.co.kr
L.P: Pres: Park, Heung-Soo.
Call letters HLQL used for all stations.

Sce. area	Tr.location	MHz	kW
Ch'ungju	Kayop-san	104.1	5
Ch'angwon	Pulmo-san	104.3	5
Seoul	Kwanak-san	104.5	5
Chinju	Kamak-san	104.7	3
Kangnung	Kwaebang-san	104.9	3
Wonju	Paegun-san	104.9	3
Sogwip'o	Sammaebang	104.9	3
Taegu	P'algong-san	105.1	5
Kwangju	Mudung-san	105.3	5
Taejon	Kyeryong-san	105.7	5
Ulsan	Muryong-san	105.9	3
Yosu	Mang-un-san	106.3	1
Ch'unch'on	Hwa-ak-san	106.5	5
P'ohang	Chohang-san	106.7	3
Chonju	Mo-ak-san	106.9	5
T'aebaek	Hambaek-san	107.1	3
Cheju	Kyonwolak	107.3	3
Namwom	Nogodan	107.5	3
Andong	Hakka-san	107.7	3
Pusan	Yongdo	107.7	3
Ch'ongju	Shikchang-san	107.9	3

+ low power relay st's.
ANN: "EBS, Kyoyuk Pangsong-imnida. HLQL".

MUNHWA BROADC. CORP.(MBC)
(Munhwa Pangsong)
Nationwide Commercial Network.
✉ 31, Yo-ui-do-dong. Yongdungp'o-gu, Seoul 150-728. ☎ +82(2)784 2000. ① MBCNEWS K22203. **Cable:** MBCHLKV SEOUL - **Web:** http://www.mbc.co.kr

L.P: Pres: Duk-Ryul Lee. Exec. Man. Dir: Sung-Hee Kim. Auditor: Myung-Suk Lee. Man. Dir: Keung-Hee Lee (programming), Sang-Yul Lee (N. & sports), Yun-Hun Lee (production), Soo-Ryang Kim (operation & engineering), Kang-Jung Kim (corporate planning). Dir. Int. Rel: Hea-Myung Shin.
Mediumwaves: Call: HL(call).

	Call	kHz	kW	St. and h. of tr.
1)	CQ	765	10	Taejon MBC:1955-1700
2)	AJ	774	10	Cheju MBC:1955-1700
3)	AN	774	10	Ch'unch'on MBC:1955-1700
4)	CT	810	20	Taegu MBC: 1955-1700
5)	CN	819	20	Kwangju MBC: 1955-1700
6)	AU	846	10	Ulsan MBC: 1955-1600
7)	CX	855	10	Chonju MBC: 1955-1700
8)	KV	900	50	Seoul MBC: 24h.
9)	AP	990	10	Masan MBC: 1955-1600
10)	AW	1017	10	Andong MBC:1955-1700
11)	AT	1080	10	Yosu MBC:1955-1700
12)	AV	1107	10	P'ohang MBC:1955-1700
13)	KU	1161	20	Pusan MBC:1955-1700
14)	AK	1215	10	Chinju MBC:1955-1700
15)	SB	1242	10	Wonju MBC:1955-1600
16)	AF	1287	10	Kangnung MBC: 1955-1700
17)	AX	1287	10	Ch'ongju MBC: 1955-1630
18)	AO	1332	10	Ch'ongju MBC: 1955-1700
19)	AQ	1386	10	Samch'ok MBC:1955-1600
20)	AM	1386	10	Mokp'o MBC: 1955-1700

FM Stations

Location Studio (Transmitter)	Music FM MHz	kW	General FM MHz	kW
8) Seoul (Namsan)	91.9	10		
Seoul (Kwanak-san)	95.9	10		
13) Pusan (Yongdo)	88.9	5	95.9	3
4) Taegu (P'algong-san)	95.3	5	96.5	5
5) Kwangju (Mudung-san)	91.5	5	93.9	5
Kwangju (Nogodan)	95.1	3		
1) Taejon (Shikchang-san)	97.5	5	92.5	1
7) Chonju (Miruk-san)	99.1	5		
Chonju (Nogodan)	101.7	3		
9) Masan (Pulmo-san)	100.5	1		
3) Ch'unch'on (Pong-ui-san)	94.9	1		
17) Ch'ongju (Uam-san)	99.7	1		
2) Cheju (Kyonwol-ak)	90.1	3		
Cheju (Sammae-bong)	102.9	3	97.1	1
6) Ulsan (Muryong-san)	98.7	3		
16) Kangnung (Kwaebang-san)	94.3	5		
14) Chinju (Mangjin-san)	97.7	1		
14) Chinju (Kamak-san)	96.1	3		
14) Chinju (Samch'onp'o)			91.1	
20) Mokp'o (Yang-ul-san)	102.3	1		
11) Yosu (Kubong-san)	98.3	2		
Yosu (Ch'onhwang-san)	100.7	1		
10) Andong (Hakka-san)	91.3	3	100.1	3
15) Wonju (Paegun-san)	98.9	3	92.7	
18) Ch'ongju (Kayop-san)	88.7	3	96.1	1
Ch'ongju (Chech'on)	94.7	0.5		
Ch'ongju (Tanyang)	94.1	0.5		
19) Samch'ok (Hambaek-san)	98.1	3	101.5	1
Samch'ok (Ponghwang-san)	99.1			
12) P'ohang (Chohang-san)	97.9	3		

NB: General FM St's simulcast with the MW St in the same city. A separate sce. is provided to the Music FM St's. All regional st's broadcast a combination of a feed from Seoul and their own local prgr. General FM st's follow the same schedule as their corresponding MW outlet. Music FM of Seoul MBC sched: 24h.
ANN: "(location) Munhwa Pangsong-imnida. (call letters)" or "Munhwa Pangsong-imnida" or "MBC". Seoul: "Chungp'a Kubaek(900) kHz, P'yojun FM Kushib-o-chom-gu(95.9) MHz Munhwa Pangsong-imnida."
✉ Add "(location) Munhwa Broadc. Corp." to addr. 1)381-171, Sonhwa-1-dong, Chung-gu, Taejon 301-728 - 2)321-22, Yon-dong, Cheju-shi, Cheju-do 690-170 - 3) 238-3, Samch'on-dong, Ch'unch'on-shi, Kangwon-do 200-200 - 4)1, Pomo-dong, Susong-gu, Taegu 706-728 - 5) 300, Wolsan-dong, Nam-gu, Kwangju 502-728 - 6) San 7-1, Haksong-dong, Chung-gu, Ulsan-shi, Kyongsang Nam-do 681-728 - 7) 151-9, 2-ga, Chunghwasan-dong, Wansan-gu, Chonju-shi, Cholla Buk-do 560-728 - 8) National addr. - 9) 525-1, Yangdok-dong, Hoewon-gu, Masan-shi, Kyongsang Nam-do 630-490 - 10) 709-1, T'aehwa-dong, Andong-shi, Kyongsang Buk-do 760-290 - 11)101-1, Munsu-dong, Yosu-shi, Cholla Nam-do 550-728 - 12) 907-4, Taejam-dong, P'ohang-shi, Kyongsang Buk-do 790-728 - 13) 316-2, Millak-dong, Suyong-gu, Pusan

613-728 - 14) 47, P'yong-an-dong, Chinju-shi, Kyongsang Nam-do 660-728 - 15)1023-70, Haksong-1-dong, Wonju-shi, Kangwon-do 220-031 - 16) 1091-6, P'onam-2-dong, Kangnung-shi, Kangwon-do 210-112 - 17) 261-30, Uam-dong, Sangdang-gu, Ch'ongju-shi, Ch'ungch'ong Buk-do 360-728 - 18) 680, Ho-am-dong, Changwon-shi, Kangwon-do 245-090 380-130 - 19)111, Kalch'on-dong, Samch'ok-shi, Kangwon-do 245-090 - 20)1096-1, Yongdang-dong, Mokp'o-shi, Cholla Nam-do 530-728.
V. Many st's will verify directly. Pusan and Masan have QSL cards. Re. directly to individual st's.

CHRISTIAN BROADCASTING SYSTEM(CBS)
(Kidokkyo Pangsong)
Mediumwaves: Call: HL(call).

Call	kHz	kW	St.and h.of tr.
1)KY	837	50	*CBS Seoul: 1847-1703
2)CL	999	10	CBS Kwangju: 1947-1603
3)KT	1251	10	CBS Taegu: 1947-1603
4)CM	1314	10	CBS Chonbuk: 1947-1603
5)KP	1404	10	CBS Pusan: 1947-1603

*key st.
FM: 6) CBS Ch'ongju HLAC-FM 91.5MHz 3kW 1947-1603.
7) CBS Ch'unch'on HLDC-FM 93.7MHz 3kw 1947-1603.
1) CBS-FM Seoul HLKY-FM 93.9MHz 7kW 24h(Music FM).
+low power relay st's.
Addresses and other information:
1) 917-1, Mok-dong, Yangch'on-gu, Seoul 158-701 ☎ +82(2)650 7000. **Web:** http://www.cbs.co.kr - **LP:** Pres: Kwon, Ho-kyong. Exec. Man. Dir: Chung, Doo-Jin. Dir of Eng: Kim, Hang-Jin. **ANN:** "Pitkwa Pogum-e sori, P'albaek-samship-ch'il(837) kHz, CBS AM-imnida. HLKY."
2) 721-2, Kumho-dong, So-gu, Kwangju 500-020 - Dir: Cho, Chun-Taek.
3) 3-7, Chimsan-2-dong, Puk-ku, Taegu 702-703 - Dir: Park, Byong-hwa.
4) 86-8, Namjung-dong-1-ga, Iksan-shi, Cholla Puk-do 570-101 - Dir: Park, In-sok.
5) 1248-4, Onch'on-3-dong, Tongnae-gu, Pusan 607-063 - Dir: An, Yong-min.
6) 1408, Pongmyong-dong, Ch'ongju-shi, Ch'ungch'ong Puk-do 360-300 - Dir: Choi, Kon-jak.
7) 174-3, Ungyo-dong, Ch'unch'on-shi, Kangwon-do 200-080 - Dir: Kim, Yong-Han.
ANN: St s 2)-7) "(freq. and location) Kidokkyo Pangsong-imnida. (call letters)"
F.Pl: Standard FM in Seoul, Kwangju, Taegu, Chonbuk and Pusan. General FM in Taejon.
V: All St's verify reports for their own prgr's by QSL card.

SEOUL BROADCASTING SYSTEM(SBS)
(Seoul Pangsong)
✉10-2, Yo-ui-do-dong, Yongdungp'o-gu, Seoul 150-010. ☎ +82(2)786 0792. ☎+82(2) 786 0785. **Web:** http://www.sbs.co.kr
LP: Pres: Yoon, Hyok-Ki. Exec. Man. Dir: Park, Joon-Young (TV). Dir: Song, Do-Gyun (programming), Song, Suk-Hyung (N. & sports), Kim, Soo-Woong (administration). Dir. Tech: Lim, Young-Gyu.
MW: HLSQ Koyang (near Seoul) 792kHz 50kW (AM stereo; C-QUAM System).
D.Prgr: 2000-1700 (commercial).
Music FM: 107.7MHz HLSQ-FM 10kW: 24h.
ANN: "Chungp'a Ch'ilbaek-kusib-i(792) kHz, Yorobun-e SBS Seoul Pangsong-imnida. HLSQ.", "FM Paek-ch'il-chom-ch'il(107.7) MHz, yorobune SBS POWER FM imnida"
F.Pl: Standard FM in Seoul.

FAR EAST BROADCASTING CO., KOREA (Rlg.)

Mediumwaves	kHz	kW
1) HLKX, Seoul	1188	100
2) HLAZ, Cheju	1566	250/100
FM	**MHz**	**kW**
3) HLAD-FM, Taejon	93.3	3
4) HLDD-FM, Ch'angwon	98.1	3

+ lowpower relay st's.
Addresses and other information
1) Far East Broadc. Co.(Kuktong Pangsong), P.O. Box 88, Map'o , Seoul 121-707. ☎ +82(2) 320 0114; ☐ +82(2) 333 2627 **Web:** http://www.febc.co.kr - Pres: Dr. Billy Kim. Vice Pres: Ju, Kwang Jo. Exec. Man. Dir: Min, San-Woong. Man. Dir: Kim, Yong-Ho. Dir. Tec: Yoo, Jae-Ok.
2000-1700. **Korean:** 2000-1100. **English:** 1100-1230. **Chinese:** 1230-1700.
ANN: Korean "Kuwon-e Kippun Soshigul Jonhanun Yorobun-e Kuktong Pangsong-imnida". English: "This is HLKX Radio broadcasting with

100,000 watts of power on 1188kHz" - Fl: by contributions & free will offerings.
V: All St's by QSL card or letter. Rec. acc. (B).
2) Asia Broadc. Co.(Asea Pangsong), P.O. Box 1566, Cheju. - Dir: Hwang, Young-Jin. - 1000-0800. Korean: 1700-0800. Chinese: 1000-1230, 1345-1630.
Japanese: 1230-1345. Russian: 1630-1700 - ANN: Korean: "Kuwon-e Kippun Soshigul Jonhanun Yorobun-e Asea Pangsong-imnida". Japanese: "Kochirawa Kirisutokyo Hosokyoku FEBC desu".
3) Taejon Kuktong Pangsong, 233-15, Chijok-dong, Yusong-gu, Taejon. ☎ +82(42)825 9330. Man. Dir: Kim, Moon-Joon. D.PRGR: 2000-1500.
4) Ch'angwon Kuktong Pangsong, 11th floor, Ch'angwon Department Store, 73-2, Yongho-dong, Ch'angwon-shi, Kyongsang Nam-do. Te: +82(551) 81 7740. - Exec. Man. Dir: Hwang, Young-Il. D.PRGR: 2000-1500.

RADIO PACIS(PBC)(P'yonghwa Pangsong)
Endowment by the Catholic Church.
Stations:
1) Seoul HLQP-FM 105.3MHz 5kW: 1957-1702.
2) Kwangju HLDL-FM 99.9MHz: 3kW: 1957-1702.
3) Teagu HLDK-FM 93.1MHz: 3kW: 1957-1702.
✉1) 2-3, 1-ga, Ch'o-dong, Chung-gu, Seoul 100-031. ☎ +82(2) 270 2114; ☐+82(2) 278 4972. **LP:** Pres: Park, Shin-On. Exec. MD: Lee, Chul-Won. MD: Lee, Hae-Wook. Prgr. Dir: Lee, Ki-Hun (R.), Kim, Min-Soo (TV). Dir. of Eng: Kang, Hong-Shin. **Web:** http://www.pbc.co.kr
2) 3-5, 3-ga, Kumnam-ro, Tong-gu, Kwang-ju.
3) 71, 2-ga, Kyesan-dong, Chung-gu, Taegu.Ann: "Chong-hap FM Paego-chom-sam(105.3) MHz, Malgun sori, Palgun Sesang, PBC P'yonghwa Pangsong-imnida. HLQP."

BUDDHIST BROADCASTING SYSTEM(BBS)
(Pulgyo Pangsong)
Owned and operated by the Buddhist Church.
Stations:
1) Seoul HLSG-FM 101.9MHz 5kW: 2000-1700.
2) Kwangju HLDB-FM 89.7MHz 3kW: 2000-1700.
3) Pusan HLDA-FM 89.9MHz 3kW: 2000-1700.
4) Taegu HLDI-FM 94.5MHz 3kW: 2000-1700.
5) Ch'ongju HLDJ-FM 96.7MHz 3kW: 2000-1700.
ADDR:☐ 704 5114. ☐ 704 1878/9. **LP:** Pres: Sung, Nak-Seung. Exec. MD: Kwon, In-Hyun. Dir. of Eng: Paik, Nak-Chan.
2) Taesaeng Bldg, 78-2, Im-dong, Puk-gu, Kwangju. 500-010. Dir: Lee, Sang-Jin.
3) Posaeng Bldg, 833-13, Pomil-dong, Tong-gu, Pusan 601-060. Dir: Ryu, Jin-Soo.
4) Chingak Hoegwan, 156-1, Maebong-dong, Chung-gu, Taegu 700-430. Dir: Shin, Kwang-Soo.
5) 1695, Yong-am-dong, Sangdang-gu, Ch'ongju-shi, Ch'ungch'ong Buk-do 360-190. Dir: Kim, Kyung-Han.
Ann: 1) "FM Paeg-il-chom-gu(101.9) MHz, Kkae-ch'im-e Sori, Nanunun Kippum, BBS Pulgyo Pangsong-imnida. HLSG."

TRAFFIC BROADCASTING SYSTEM(TBS)
(Kyot'ong Pangsong)
Municipal Station. This station is operated by the Seoul Municipal Traffic Broadcast Headquarters to provide traffic information and education of the citizens of Seoul and surroundings.
✉ 3-8, Yejang-dong, Chung-gu, Seoul. ☎ +82(2) 311 5114.
LP: GM: Choi, In-Hwan. Dir. Tec: Yu, Pok-Il. Prgr. Dir: Kim, Sung-Gil.
Station: HLST-FM 95.1MHz 5kW: 2000-1700.
ANN: "Yollin Sori, Yollin Sesang. FM Kushib-o-chom-il(95.1) MHz, Kyot'ong Pangsong-imnida. HLST."

TRAFFIC BROADCASTING NETWORK(TBN)
(Kyot'ong Pangsong)
✉ 1) 665-2, Ssang-am-dong, Kwangsan-gu, Kwangju 506-303. ☎ +82(62)970 1114. 2) 580-8, Taeyon-3-dong, Nam-gu, Pusan 608-023. ☎ +82(51)625 9944.
Stations:
1) Kwangju 97.3MHz HLDM-FM 3kW: 2000-1700.
2) Pusan 94.9MHz HLDN-FM 3kW: 2100-1500.

PUSAN BROADCASTING CORP.(PSB)
(Pusan Pangsong)
✉ 603-8, Yonsan-4-dong, Yonje-gu, Pusan 611-084. ☎ +82(51)850 9000. **Web:** http://www. psb.co.kr
Station: HLSG-FM 99.9MHz 3kW: 24h.
ANN: "Kuship-ku-chom-ku(99.9) MHz, PSB-FM Bluewave-mnida.HLDG."

KYRGYZSTAN

L.T: UTC + 5h (Su: UTC + 6h) — **Pop:** 4.521.000 — **Radios:** 825.000 — **Pr.L:** Kyrgyz, Russian — **E.C:** 50Hz, 127/220V — **ITU:** KGZ.

BAYLANIŞ MINISTRLIGI
(Ministry of Communications)

✉ pr. Çuy 96, 720000 Bişkek. ☎ +7 (3312) 262034. 🖷 +7 (3312) 288362. ① 351334 PTB KH.
L.P: Minister: Abdyzhapar T. Tagayev.

MAMLEKETTIK BAYLANIŞ INSPEKTSIYASI
(Government Inspection of Communications)

✉ Sovet köçesi 7b, 720005 Bişkek. ☎ + 7 (3312) 422312. 🖷 +7 (3312) 477631 — **TD:** N. Nikolayev.

KYRGYZSTAN STATE RADIO AND TV BROADCASTING CO. (Gov.)

✉ Molodaya Gvardiya 63, 720300 Bishkek.

KYRGYZ RADIO (Gov.)

✉ Molodaya Gvardiya 63, 720300 Bishkek. ☎ +7 (3312) 253404. 🖷 +7 (3312) 257930. ① 245173 RADIO KH.
L.P: Gen. Dir: Tugelbay Kazakov.

Stations	kHz	kW	H. of tr.	Prgr.
Bişkek	198	150	0100-1800	RO4
Oş	576	40	0000-2000	
Bişkek	612	150	2300-2200	
Bişkek	882	500	0000-1900	KGR1
Bişkek	1278	150	0000-2000	RSY
Bişkek	1323	30	see below	RL
Jojomel	1404		0000-1900	KGR1
Haidarkan	1404		0000-1900	KGR1
Orgoçor	1404		0000-1900	KGR1
Talas	1404		0000-1900	KGR1
Çolpon-Ata	1404	1	0000-2000	
Şulyukta	1431		0000-1900	KGR1
Kanyş-Kiya	1431		0000-1900	KGR1
Jalal-Abad	1431		0000-1900	KGR1
Naryn	1431		0000-1900	KGR1
Bişkek	1467	30	0200-1600	KGR2
Batken	1467		0200-1600	KGR2
Kara-Kul	1467		0200-1600	KGR2
Oş	1467		0200-1600	KGR2
Taş-Kumyr	1530	5	0000-2000	
Bişkek	4010	100	0000-1900	KGR1
Bişkek	4050	100	0000-1200	KGR2+

Frequency usage not recently confirmed. +) Private stations reported on this freq.
NB: Powers and locations are tentative only.

FM (MHz)	RL	FM (MHz)	RL
Alay-Kuu	69.08	Oş	71.69
Bişkek	66.38	Sulyukta	70.88
Dedyumel	71.69	Talas	72.08
Gulça	69.68	Terek-Say	70.64
Kara-Kul	66.86	Vostochnaya	67.04
Naryn	72.38	Yuzhnaya-2	71.93
Orgoçor	67.82		

ERP in general appr. 8kW. No further details available.
KGR1 in Kyrgyz/Russian. N. in English: 0010-0015v. Dungan: W 1615-1700. German: Sat 1530-1615.
KGR2 in Kyrgyz, Russian and others.
RO4 = R. Odin Orbita 4 relay (cf. Russia).
RSY = R. St. Yunost relay (cf. Russia).
RL = R. Liberty (cf. Czech Rep.) 1800-1900 in Russian, 1300-1400, 1600-1700 in Kyrgyz. Also relays of R. France Internationale.
ANN: Kyrgyz: "Bişkekden söylöbüz" — **V.** by letter (irr.).

PRIVATE STATIONS

Radio Almaz: 6-38/1-28, 720060 Bişkek. On 1431kHz: 0145-0400, 1000-1230.
Radio Pyramid: Molodaya Gvardiya 59, 720300 Bişkek. On 1431kHz: 0400-1000, 1230-1500.
Radio Max CN: Bişkek. On 106.0MHz.
Several further stations are operating on FM, but no details available.

LAOS (People's Democratic Republic)

L.T: UTC + 7h — **Pop:** 5.245.852 — **Radios:** 575.000 — **Pr.L:** Lao Soung, Lao Theung — **E.C:** 50Hz, 220V — **ITU:** LAO.

RADIO NATIONALE LAO (Gov.)

✉ B. P. 310, Vientiane. ☎ +856 (21) 212432. 🖷 +856 (21) 212430.
L.P: DG: Bounthan Inthaxay. Dep. DG: Keungkham Vilayasith. Tech. MD: D. Sisombath.
S=Sce: N=National, R=Regional, C=City, I=International

Location	kHz	kW	S	H. of tr.
Vientiane	580	150	N	2200-0300, 0400-0730. 0930-1530
LP	705		R	2225-0130, 0355-0600, 1025-1400
Muang Hay	800	1	R	2300-0200, 1100-1400
Houa Phanh	1000		R	1255-1425
Vientiane	1030	10	I	2330-0030, 0500-0630, 1130-1400
Vientiane	1190		C	0030-1200v
Phonsavan	1215	1	R	2230-0100, 1000-1300
Pakse	1350	10	R	2300-0200, 0400-0600, 1000-1400
Savannakhet	1430	3	R	2225-0100, 0425-0600, 1025-1400
Xam Nua	4690	1	R	?-0000, 1000-1100
Vientiane	6130	25	N	2200-0300, 0400-0730, 0930-1530
LP	6975	1	R	2225-0130, 0355-0600, 1015-1400
Vientiane	7145	50	I	inactive

NB: LP=Louang Prabang — SW freq's vary.
FM: Vientiane 97.5MHz 0.05kW (0200-0400, 0700-0900).
National Sce: N: 2300, 0000, 0500, 1200, 1400. **Hmong:** 2200-2230, 0630-0700. **Khmu:** 2230-2300, 0700-0730. **English/French LL:** 1300-1330.
International Sce: Cambodian: 0000-0030, 1230-1300. **English:** 0600-0630, 1330-1400. **French:** 0530-0600, 1300-1330. **Thai:** 0500-0530, 1130-1200. **Vietnamese:** 2330-2400, 1200-1230.
ANN: National Sce: "Thini Withayu heng Sat, krachaisiang chak Wianchan, nakhong-lwang khong sathanalat pasatipatai pasason Lao". Vientiane Capital Sce: "Vitthayou Krachaistang Nakhonluang Vientiane". Regional st's : "Thini sathani vithayu krachaisiang heng Luang Prabang", "Thini vithayu krachaisiang heng sat Houa Phan", "Thini vithayu krachaisiang heng sat Ke Thai otaa Pakse", "Thini vithayu krachaisang heng sat kueng Savannakhet", "Thini vithayu krachaisiang heng Sat Oudomxay".
IS: (National Sce): Music on Khéne (mouth organ) & Solo (bamboo instrument) — **V.** by QSL-card. IRC's not acc.

MACAU

L.T: UTC+8h, **Pop:** 490,000, **Radio sets:** 250,000, **Pr.L:** Portuguese, Cantonese
E.C: a/c 50.220V, **ITU:** MAC

TELEDIFUSAO DE MACAU, SARL (Priv.comm.)

✉ Avenida Dr. Rodrigo Rodrigues, No. 223-225, Edif. Nam Kwong 7 Andar, Macau. ☎ +853 522978 or 335888. 🖷 +853 343199.
D.PRGR in **Portguese** 24h: 98.0MHz(stereo) 2.5kW, in **Cantonese** 24h: 100.7MHz (stereo) 2.5kW.

MALAYSIA (Federation of)

L.T: UTC + 8h — **Pop:** 21.078.922 — **Radios:** 9.800.000 — **Pr.L:** Bahasa Malaysia, English, Chinese, Tamil — **E.C:** 50Hz, 240V — **ITU:** MLA.

PENINSULAR MALAYSIA

RADIO TELEVISION MALAYSIA – RTM (Gov.)

✉ Dept. of Broadcasting, Angkasapuri, Bukit Putra, 50614 Kuala Lumpur. ☎ +60 (3) 2825333. 🖷 +60 (3) 2824735. ① MA 30283 Kuala Lumpur. **Cable:** Broadcasts Kuala Lumpur. **E-mail:** rtm@asiaconnect.com.my **Web:** http://www.asiaconnect.com.my/rtm-net
L.P: DG: Dato' Jafar Kamin. Dep. DG. (Prgr.): Dato' Salleh Pateh Akhir. Dep. DG. (Eng): Mohd. Nor Abdullah. MD: Mohd. Helan Abu. Dir. of Radio: Madzhi Johari. Dir. of Eng. (Radio & TV): Mohd. Nor Abdullah. Dep. Dir. of Eng. (Radio): Ms. Aminah Din. Dir. N. & Current Affairs: Ismail Mustapha. Head of Int. Rel: Ms. Nawiyah Che' Lah.
Mediumwaves: JB=Johore Bharu, KL=Kuala Lumpur (Kajang), KB=Kota Bharu, KT=Kuala Terengganu.

Location	kHz	kW	Sce	Location	kHz	kW	Sce
Tronoh	576	100	R3	KT	765	10	R1
JB	576	50	R1	Kuantan	810	10	R1
Kajang	594	200	R1	Melaka	1008	10	R1
Segamat	621	100	R1	Mersing	1053	20	R1/3
Kuala Lipis	648	20	R1	Grik	1089	20	R5/6

Grik	657	20	R1	Dungun	1260	5	R1/3
Penang	666	10	R1	Mersing	1314	20	R5/6
KB	702	20	R4				

FM: KL 98.3MHz (R1), 95.3MHz (R2), 100.9MHz (R. KL), 102.3MHz (R. KL), 100.1MHz (R4), 96.3MHz (R5), 106.7MHz (R6) + relay st's at 13 locations.
SW: Kajang (G.C: 101.46E/03.01N): 9 x 100kW, 1 x 50kW.

kHz	kW	Sce	Times
4845	100	R6	Mon-Thurs 0300-1070, 0800-1700, 2100-0100 Fri/Sun 2100-1500 Sat 0515-1500
5965	100	R1	1100-0900 (irr. 24h)
6025	100	R5	2200-1600 (irr. 24h)
7295	100	R4	

Radio 1 in Bahasa Malaysia: 24h — **Radio Muzik in FM-Stereo** 24h — **R. Kuala Lumpur in Bahasa Malaysia:** 2200-1600 — **Radio 4 in English:** 24h. **N:** 2300, 0530, 1030, 1200, 1330, 1500 — **Radio 5 in Chinese:** 24h — **Radio 6 in Tamil:** 2100-1600.
ANN: R1: "Radio Satu RTM". R2: "Radio Dua". R3: "Radio Tiga (name of state capital)." R5: "Malaixiya guangbo diantai diwutai". R6: "Teman Setia Anda".

OVERSEAS SERVICE: See International Broadcasting section.

RADIO PENERANGAN (Information Radio) (Gov.)
Kuala Lumpur area (freq. unknown): 2200-0600. **F.Pl:** 18h daily.

TIME HIGHWAY RADIO - THR (Comm.)
✉ 20th Flr, Plaza Berjaya, 12 Jalan Imbi, 55100 Kuala Lumpur. ☎ +60 (3) 2433088. **Web:** http://thr.time.com.my
L.P: Chief Exec: Hisham Rahman.

Coverage Area	MHz	Coverage Area	MHz
Ce. Perak	97.9	So. Perak	102.0
K. Lumpur/Selangor	99.3	Kedah	102.7
So. Penang	99.5	No. Penang	102.7
Melaka	99.7	Johor/Singapore	103.7
Langkawi	101.9	No. Perak	104.6
No. Sembilan	101.5		

D.PRGR. in English: 24h.

BEST 104 (Comm.)
✉ P.O. Box 1, Tamam Seri Terbrau, Johore Bharu. ☎ +60 (7) 311011.
FM: Segamat 94.8MHz, Johore Bharu 104.1MHz.
D.PRGR: 2200-1700 (Malay & English music).

RADIO MUZIK
✉ Tingkat 2, Wisma Radio Angkasapuri, Kuala Lumpur 50740. ☎ +60 (3) 285 7288.

SABAH
L.T: UTC + 8h — **Pr.L:** Bahasa Malaysia, English, Kadazan, Dusun, Murut, Chinese, Bajau — **E.C:** a/c 50, 230/400V.

RADIO TELEVISION MALAYSIA – KOTA KINABALU (Gov.)
✉ 88614, Kota Kinabalu. ☎ +60 (88) 52711. **Cable:** Broadcasts, Sabah. ✆ MA 80061 Kota Kinabalu.
STATIONS: KK=Kota Kinabalu (Tuaran) (G.C: 116.14E/06.12N), LD=Lahad Datu, SM=Suara Malaysia (Overseas Sce).
Languages: BM=Bahasa Malaysia, E=English/local dialects.

Location	kHz	kW	L	Location	kHz	kW	L
Tenom	567	10	BM	LD	828	10	E
KK	603	10	BM	Tawau	927	10	E
LD	675	10	BM	Sandakan	1080	10	E
KK	693	10	E	Kudat	1197	10	E
Tawau	747	10	BM	KK	1476	600	SM
Tenom	774	10	E	KK	4970	10	BM
Sandakan	783	10	BM	KK	5980	10	E
Kudat	801	10	BM				

FM: Layang-Layang 93.5MHz 10kW (E), 97.5MHz 10kW (BM).
Bahasa Malaysia: 2000-1600 — **English/local dialects:** 2200-0100 & 1000-1600. **Local N. in English:** 2300.
ANN: Bahasa Malaysia: "Inilah R. Malaysia Kota Kinabalu".
IS: Opens and closes with National Anthem.
V. by QSL-card. Re. to Publicity Officer.

SARAWAK
L.T: UTC + 8h — **Pr.L:** Bahasa Malaysia, English, Chinese and 8 local dialects — **E.C:** a/c 50, 230/400V.

RADIO TELEVISION MALAYSIA – SARAWAK (Gov.)
✉ Broadcasting House, Jalan P.Ramlee, 93614 Kuching. ☎ +60 (82) 248422. 🖷 +60 (82) 241914. **Cable:** Broadcast Kuching. ✆ MA70084.
STATIONS: Kuching (G.C: 110.20E/01.33N), Sibu (G.C: 111.49E/02.18N), Limbang (G.C: 115.00E/04.45N), Miri (G.C: 113.59E/ 04.23N), Sri Aman (G.C: 111.28E/01.13N).

Location	kHz	kW	Netw.	Location	kHz	kW	Netw.
Kuching	549	20	Y	Miri	1206	20	G
Miri	576	20	Y	Miri	3385	10	G
Sibu	621	20	R	Kuching	4835	10	Y
Limbang	648	20	Y	Kuching	4895	10	G
Kuching	729	20	Y	Sibu	5005	10	G
Miri	819	20	R	Kuching	5030	10	B
Kuching	846	10	G	Sibu	6050	10	G
Limbang	873	20	R+G	Miri	6060	10	G
Sibu	909	20	Y	Kuching	7130	10	B
Kuching	954	10	B	Kuching	7145	10	Y
Sri Aman	1044	20	Y	Kuching	7160	10	R
Sibu	1062	20	G	Kuching	7270	10	G
Sri Aman	1161	20	R+G				

FM: details not received. All networks operate 24h on FM.
Red Netw: W2200-0700 & 1030-1600, Sun 2200-1600 in **Chinese** exc. **English:** 2330-0100, 0500-0700, 1030-1200, 1400-1600. **Local N. in English:** 2330, 0530, 1030,1400.
Yellow Netw. (Y) in Bahasa Malaysia: 2200-1600.
Green Netw. (G) in Iban: W 2200-2300, 0400 0500, 1000-1500. Sun 2200-0500, 1000-1500.
Blue Netw. (B) in Bidayuh: W 2200-2400, 0100-0330 (schools prgr), 0400-0530, 1030-1500. Sun 2200-0200, 0400-0530, 1030-1500.
NB: All networks have regional prgrs at times.
ANN: Bahasa Malaysia: "Inilah R. Malaysia (location)".
IS: A musical phrase (played on a native instrument, the Sape), alternating between A and F, followed by the National Anthem.
V. by QSL-card (Miri/Sibu reply direct).

MALDIVES (Republic of)

L.T: UTC + 5h — **Pop:** 290.392 — **Radios:** 29.556 — **Pr.L:** Dhivehi (Maldivian) — **ITU:** MLD.

VOICE OF MALDIVES (Gov.)
✉ "Moonlight Higun", Male' 20-06. ☎ +960 325577. 🖷 +960 328357. ✆ 66120 ADU MF.
L.P: DG (Broadc.): Ibrahim Manik. DG (Eng): Maizan Ahmad Maniku.
MW: 1449kHz 10kW (standby tr: 1458kHz 5kW).
FM: 103.8MHz 20W. **F.Pl:** 89.0 & 99.0MHz 1kW.
SW: 5998.5kHz 1kW (occasional operation, using reduced power).
D.PRGR. on 1449kHz: 0025-1015, 1200-1800 (during Ramadan st. operates 0025-1800 continuously). **N:** 0200, 0900, 1200, 1400, 1700. **English:** 1200-1400. **N:** 1800.
R. Eke on 103.8MHz: 1500-1900. **N:** 1700 (rel MW sce).
ANN: MW: "Mee Dhivehi Raajjeyge Adu" — **V.** by QSL-card.
F.Pl: 10kW SW-tr. Regular SW sce (possibly in 11MHz band).

MONGOLIA

L.T: UTC + 8h (Su: UTC + 9h) — **Pop:** 2.691.643 — **Radios:** 280.000 — **Pr.L:** Mongolian — **E.C:** 50Hz, 220V — **ITU:** MNG.

MONGOL RADIO AND TELEVISION (Gov.)
✉ Huvisgalyn zam 3, Ulaanbaatar 11. ☎ +976 (1) 23520, 28978, 29766. 🖷 +976 (1) 327234. ✆ 223 RTV MH. **Cable:** Mongolian Radio.
E-mail: mrtv@magicnet.mn
Web: http://www.magicnet.mn/monradio/
L.P: Chmn: Byambajavyn Uvgenhuu. Head Int. Dept: Tuvdennyam Neeneegin. Protocol Officer Int. Dept: Mrs. Amarsaihan Elbrus.

LW/MW	kHz	kW	MW	kHz	kW
Ulaanbaatar	164	100	Choibalsan	882	750
Ulgii	209	60	Huvsgul	882	150
Dalanzadgad	209	150	Altai	990	150
Choibalsan	209	150	Ulaanbaatar	990+	500
Altai	227	150	Sainshand	1233	75
Murun	882	150	Dalanzadgad	1350	150

+) also carries Ext.Sce.

SW	kHz	kW	SW	kHz	kW
Dalanzadgad	3960v	12	Ulaanbaatar	4850	50

Uliosxai	3960v	1	Sainshand	4865v	12
Ulaanbaatar	4078v	50	Ulangoom	4865v	5
Ulgii	4750v	12	Murun	4895v	12
Altai	4828	12	Choibalsan	4995v	12

FM: 70.52/72.15MHz.
D.PRGR: 2200-1600. **Kazakh:** 1330-1500 on 4750kHz (not confirmed recently) — **Radiostation Tsataan Shonkor (priv.):** Wed 1400-1500 on 209/227/4080/4850kHz.

EXTERNAL SERVICE: see International Broadcasting section.

MYANMAR (Union of)

L.T: UTC + 6½h — **Pop:** 46m — **Radios:** 3.300.000 — **Pr.L:** Bamar, English, Kechin, Kaya, Kayin (Po & Sakaw), Chin, Mon, Rakhihe Shan — **E.C:** 50Hz, 230V — **ITU:** BRM.

MYANMA TV AND RADIO DEPARTMENT (Gov.)

✉ Pyay Rd, Kamayut-11041, Yangon (**Postal** ✉ GPO Box 1432, Yangon-11181). ☎ +95 (1) 37122. 🖷 +95 (1) 30211. **Cable:** Mytero. ☼ 21360 BBSXZK BM.
L.P: DG: U. Kyi Lwin. Dir. (Admin): U Khin Maung Htay. Dir. (Broadc.): U Ko Ko Htwe. CE: U Tin Wan. Dir (TV): U Phone Myint.
Transmitter site: Yangon (Yegu) (G.C: 96.10E/16.52N).

kHz	kW	P	Times
576	200	M	0030-0230 (SS 0730), 0330-0730, 0930-1800
4725	50	I	0930-1430/1530
5986v	50	M	1430-1600
7185	50	M	0030-0230
9730	M	M	0330 (SS 0230)-0730

P=Prgr: I=Indigenous, M=Main Prgr (incl. **English**).
FM: Yangon 98.0MHz 1.0kW, 102.0MHz 0.3kW, 104.0MHz 0.5kW. **English:** 0200-0230, 0700-0730, 1430-1600. **N:** 0200, 0700, 1445.
ANN: (0030: "Min Galar Nan Net Khin Bor Shin) "Thaw Ta Shin Myar Min Galar Ah Paung Ne Kha Naung Bar Zay Lo Hnoke Khun Hset Tha Ga Ra Wa Pyu Laik Par Dae Shin". E: "This is R. Myanmar, Yangon" — **IS:** Myanma Orchestral Music.
V. by letter if IRC is enclosed. Re. to DG, Myanma TV & Radio Dept.

DEFENCE FORCES BROADCASTING UNIT (Mil.)

✉ Taunggyi, Shan State — **SW:** 6570kHz 10kW.
D.PRGR. in Bamar and minority langs: 0200-0330, 0530-0630, 1030-1330 — **V.** Re. must be in Bamar.

MYAWDDY RADIO STATION (Mil.)

MW: 1440kHz — **SW:** 5973kHz.
D.PRGR: 1100-1330, 1430-1530.

NEPAL

L.T: UTC + 5¾h — **Pop:** 23.168.895 — **Radios:** 625.000 — **Pr.L:** Nepali, English — **E.C:** 50Hz, 220V — **ITU:** NPL.

RADIO NEPAL (Semi-Gov, Comm.)

✉ Radio Broadcasting Service, P.O. Box 634, Singha Durbar, Kathmandu. ☎ +977 (1) 233910 or 225467. 🖷 +977 (1) 221952. **Cable:** Radionepal. ☼ 2590 RNEPAL NP. **E-mail:** rne@rne.wlink.com.np
Web: http://www.catmando.com/news/radio-nepal/radionp.htm
L.P: Exec. Dir: Shailendra Raj Sharma. Dep. Exec. Dir: M. P. Adhikari.

MW	kHz	kW	MW	kHz	kW
Surkhet	576	100	Kathmandu	792	100
Dhankuta	648	100	Dipayal	810	10
Pokhara	684	100	Bardibas	1143	10

FM: Khumaltar 100MHz 1kW.
SW: Khumaltar (G.C: 85.12E/27.42N): 3 x 100kW.
kHz: 3230 (winter), 5005, 7165 (summer)
D.PRGR on MW/SW: 0015-0615 & 0715-1830 (Sat 0015-1830). **N. in English:** 0215, 0720, 1415. **Regional Prgrs:** 0400-0530 & 1235-1300.
Rel. BBCWS: 1715-1730 Hindi, 1730-1745 Urdu, 1745-1830 English.
Radio Nepal FM-Kathmandu: 6h daily on 100MHz.
ANN: Nepali: "Yo Radio Nepal Ho". Ext. Sce: "This is the Voice of Radio Nepal".
IS: Instruments used are conch shell, violin, piano and jal tarang.
V. by QSL-card.

R. SAGARMATHA (Priv.)

✉ Kathmandu 102.4MHz 0.1kW.
Kantipur FM 100: 0130-1615 in **English**.
Other stns include Image Channel, ✉ Music Nepal, Classic & Good Night.

PAKISTAN

L.T: UTC + 5h — **Pop:** 136.658.040 — **Radios:** 10.200.000 — **Pr.L:** Urdu — **E.C:** 50Hz, 220V — **ITU:** PAK.

PAKISTAN BROADCASTING CORPORATION (Gov.)

✉ Broadcasting House, Constitution Avenue, Islamabad 4400. ☎ +92 (51) 9214947. 🖷 +92 (51) 216657. ☼ 5816 PBCNO PK.
L.P: DG: Muhammad Abbas. Finance Dir: Ijaz Ahmed. Dir. Admin: Zaman Shah Rashdi. Dir. News: Nazir Ahmed Bokhari. Dir. Prgrs: Inayat Ullah Baloch. Dir. Eng: Mehdi Raza Abidi. Head of Training & Overseas Liaison/CBA Liaison Officer: Rashid Ali.
R=Region: N=No. We. Frontier Province & Northern tribal areas, F=Federal District of Islamabad, P=Punjab, S=Sind, B=Baluchistan.

Mediumwaves	kHz	kW	R	H. of tr.
Peshawar	540	300	N	0045-0405, 0600-1808
Khuzdar	567	300	N	1155-1808
Islamabad	585	1000	F	0045-0605, 0800-1900
Lahore-I	630	100	P	0045-0405 (Fri 0820), 0800 (Fri 1000)-1900
Karachi-II	639	10	S	0315-0545, 1300-1535
Peshawar II	729	100	N	1230-1705
Quetta-I	756	150	B	0045-0805, 1000-1810
Karachi-I	828	100	S	0045-0405, 0600-1900
Quetta-II	855	10	B	0200-0400, 0600-1810
Khairpur	927	100	S	0045-0605, 1000-1808
Hyderabad-I	1008	120	S	0045-0605, 0600-1808
Multan	1035	120	P	0045-0605, 0600-1808
Lahore-II	1080	50	P	1230-1705
Hyderabad-II	1098	50	S	1230-1705
Rawalpindi	1152	10	P	0045-0405, 0600-1808
Loralai	1251	10		
Bahawalpur	1341	10	P	0850-1808
Dera Ismail Khan	1404	10	N	0855-1600
Zob	1449	10		
Faisalabad	1476	10	P	0045-0820
Gilgit	1512	10	N	1000-1700
Skardu	1557	10	N	1000-1700
Turbat	1584	0.25	B	1300-1810
Sibbi	1584	0.25	B	0755-1108
Chitral	1584	0.25	N	1050-1515
Abbottabad	1602	0.25	N	0845-1415

D.PRGR: as above. **N. in English** (National): 0300, 0500, 0800, 1100, 1300, 1600.

Shortwaves	kHz	kW	Times
Rawalpindi	4780v	100	0045-0405, 1430-1810
Islamabad	4815v	100	0045-0245
Islamabad	4895v	100	0045-0405, 1430-1900
Islamabad	4915	100	0015-0215
Quetta	5025v	10	0045-0405 (Fri 0345), 1200-1805
Rawalpindi	5055v	100	1430-1530
Islamabad	5085	100	0045-0505, 1430-1805

Shortwaves	kHz	kW	Times
Islamabad	5825v	100	1700-1900 .
Islamabad	6060	100	0330-0430, 0045-0545, 1300-1600
Islamabad	6070	100	0100-0105, 0200-0215, 1300-1600
Peshawar	6250	10	1100-1400
Rawalpindi	7110	100	0600-1115
Islamabad	7125	100	1410-1430
Quetta	7170	10	0400-0820 (Fri), 0600 (Fri 1000)-1145
Islamabad (L)	7230	10	1600-1630, 1700-1800
Islamabad (L)	7270	10	0230-0245, 0300-0310, 0317-0320, 0400-0404, 0500-0504, 0600-0604, 0608-0612
Islamabad	9645	100	0612-0616
Islamabad (L)	9645	10	0900-1000, 1230-1304, 1311-1318, 1400-1410, 1500-1530

(L) Link services

EXTERNAL SERVICES: see International Broadcasting section.

Capital FM (Priv.): Islamabad 100MHz (stereo): 24h music and audience participation shows

AZAD KASHMIR RADIO (Gov.)

✉ Muzaffarabad, Azad Kashmir, via Pakistan (high power tr's).
L.P: Dep.Controller (Eng): Syed Ahmed.
STATIONS: Muzaffarabad 792kHz 150kW, 3665vkHz 1kW —
Rawalpindi (G.C: 73.00E/33.30N): 4790/7265kHz 100kW.
D.PRGR: 0045-0445 & 1000-1808 on 792/3664kHz, 0045-0435, 1230-
1330 & 1345-1815 on 4790kHz, 0800 (Fri 0930)-1415 on 7265kHz.
N. in English: 1600.
ANN: 792/3664kHz: "Yeh Azad Kashmir Radio Muzaffarabad Hay".
Other freqs: "Yeh Azad Kashmir Radio Trarkhal Hay".
IS: Azad Kashmir anthem at open and close — **V.** by letter. Rp.

PHILIPPINES (Republic of the)

L.T: UTC + 8h — **Pop:** 69.988.000 — **Radios:** 8.300.000 — **Pr.L:**
Philippine dialects, Cebuano Ilocano, Bicolano, English — **E.C:** 60Hz,
220/110V — **ITU:** PHL.

General Notes: Station identification is made using English alphabet.
Generally, station identifications are on the hour and half hour, and consist of callsign, frequency, and station location, and usually, system
or network name. Frequency is expressed in Spanish language, English
language, or Tagalog (Pilipino) language numerals. When known, individual station addresses should be used in preference to system or network HQ, or KBP repr. Following cities and towns comprise
Metropolitan Manila (Metro-Manila), a specially created province;
Makati, Malabon, Mandaluyong, Manila, Navotas, Pasay City, Pasig,
Quezon City, San Juan, Valenzuela, and Caloocan City.

NB: Cities may be referred to by their simple name, e.g. Baguio
City/Baguio, Naga City/Naga, Zamboanga City/Zamboanga. However,
Quezon City is **always** referred to by its full name.

Callsign assignments: DU = Shortwave only; DW = Luzon; DX =
Mindanao and Sulu; DY = Visayas and Palawan; DZ = Luzon.

NATIONAL TELECOMMUNICATIONS COMMISSION
(Department of Transportation and
Communications).

✉ 865 Vibal Bldg, Edsa Corner Times Str, Quezon City.
L.P: Commissioner: Josefina Lichauco. Dep. Commissioners: Aloysius
R. Santos, Florentino L. Ampil. Chief Broadcast Sce. Dept: Carlos D.
Saliuan Jr.

KAPISANAN NG MGA BRODKASTER SA PILIPINAS
(KBP)
(Association of Broadcasters in the Philippines).

✉ 6th Flr, LTA Bldg. 118 Perea Str, Legazpi Village, Makati, Metro-
Manila. ☎ +63 (2) 815 1990/1/2/3. 🖷 +63 (2) 815 1989.
L.P: Chmn: Jose E. Escaner, Jr. Pres: Mrs. Maloli E. Manastas. Chmn.
Metro-Manila Chapter: Mario Garcia.
Member st's: all st's exc. 1), 11), 32), 47), 87), 90), 97x), 99), 100), 102),
103), 105), 109), 111), 113), 114), 115), 117).

PHILIPPINE FEDERATION OF CATHOLIC
BROADCASTERS

✉ 2307 Pedro Gil Str, Santa Ana, (P.O. Box 3169), Manila 1099. ☎
+63 58-4828. **Cable:** Reuter Jesuitas Manila.
L.P: Pres: His Excellency Most Rev. Jesus Y. Varela. Chmn: Fr. James
Reuter S.J. (14 member st's).

MEDIUMWAVES: For locations, cf. MW Frequency List.
H. of tr: mostly 2100-1600. Variations, where known, are given in the
Address section.

Addresses and other information
FJE Group is formed by Manila Broadcasting Company, Pacific
Broadcasting System, Philippine Broadcasting Corporation and Cebu
Broadcasting Company.

1) Allied Broadc. Center Inc, 17th Flr, Unit C, Strata 200 Bldg, Emerald
Ave, Ortigas Complex, Pasig, Metro Manila – GM: Atty. E.V. Rosales
— 2) Audiovisual Communicators Inc, 17th Floor, Strata 200, Emerald
Ave, Ortigas Complex, Pasig, Metro Manila – GM: A.V. Barreiro Jr —
3) Super R., Global Media Arts Inc. GMA Broadcast Center, Phase V
Alta Tierra Village, Quintin Salas, Jaro, Iloilo C. 5000 — 4) Guzman
Institute of Technology, Sta. Cruz, Manila — 5) Bayanihan Broadc.
Corp, 120 J. Arellano Str, San Juan, Metro Manila – Pres. & GM: Fr.
F. Lucas SVD. 2100-1500 — 6) Beta Broadc. System Inc, Rm 211,
Margarita Bldg, J.P. Rizal, Makati, Metro Manila – GM: Ms. L. Dela
Cruz — 7) Bicol Broadc. Systems Inc, MSI Bldg, 5233 Farenheit Str,
Palanan, Makati, Metro Manila – Pres: L. Villafuerte. 2000-1500 — 8)
Bohol Chronicle Radio, c/o Suite 301, Legaspi Towers 200, Paseo de
Roxas, Legaspi Village, Makati, Metro Manila – GM: Atty. Z. Dejaresco
Jr — 9) Silangan Broadc. Corp., 55 Magallanes Str, Surigao C., 8400
Surigao del Norte – GM: Miguel Cinches. 2100-1400 — 10) Cadiz
Radio & TV Network Inc, c/o Radio St. DYAG, Cadiz C., Negros
Occidental – GM: Loel Katalbas — 11) Capitol Broadc. Center, 317
Roosevelt Ave, San Francisco del Monte Quezon C. – St. Mgr: Joey
Luison Jr. 2100-1500 — 12) Catholic Welfare Organization, 2307
Pedro Gil Str, St. Ana, Metro Manila – GM: Sr. S. Manapol, SPC —
12x) Radio Diwa, World Broadc. Corp., Burayan, San José, Tacloban
C. – Mgr.: Fr. Joseph Suson — 13) Catholic Welfare Organization,
Clergy House, Borongan, Western Samar —14) Christian Broadc. Sce,
Bo. Maligaya Bldg 2, 887 E. de los Santos Ave, Quezon C. – St. Mgr:
Nel C. Santiago. CE: A.H. Ceralde. 2100-1600 — 15) Bombo Radyo
Philippines. 2046 Florete Bldg., Edison cor. Nobel St., Makati, Metro
Manila. – DXMF 576: Bombo R. Broadc. Center, San Pedro St., Davao
C. – DZVV 585: Tamag, Vigan, Ilocos Sur – DYWR 594: Bombo R.
Broadc. Center, Sto. Nino cor. Imelda Ave., Tacloban C. DZVR 711: Brgy.
48 K, Kabungnaan Airport Ave., Laoag C. – DZSO 720: Pennsylvania Ave.,
San Fernando, La Union – DXPD 792: Bombo R. Broadc. Center, North
Diversion Road, Barangay Banale, Pagadian C. – DXES 801: Bombo R.
Broadc. Center, Brgy. Bula, General Santos C. – DZNC 801: Bombo R.
Broadc. Center, Barrio Menante II, Cauayan, Isabela – DYFM 837: Sky
City Tower, Mapa St., Iloilo C. – DZGR 891: Bombo R. Broadc. Center,
Población (Bagumbayan), Taguegarao, Cagayan – DYOW 900: Bombo
R. Broadc. Center, Arnaldo Blvd., Roxas C. – DZLG 927: Bombo R.
Broadc. Center, Tahao Road near Central City, Subdivision Gate,
Legaspi C. – DXBR 981: Bombo R. Broadc. Center, Aruj Ville Subd., Brgy.
Libertad, Butuan C. – DXMC 1026: Bombo R. Broadc. Center, General
Santos Drive, Koronadal, South Cotabato – DZWX 1035: Bombo R.
Broadc. Center, 14 Lourdes Subdivision, Bagauio C. – DZNG 1044:
Bombo R. Broadc. Center, Diversion Road, Brgy. Tabuko, Naga C. –
DYIN 1107: Oyo Torong St., cor. C. Laserna Ext. St., Kalibo, Aklan –
DZWN 1125: Bombo R. Broadc. Center, Maramba Bankers' Village,
Bonuan Catacdang, 2400 Dagypan C. SM: Minerva A. Caburnay –
DXLX 1188: Tambo, Barangay Hinaplon, Iligan C. – DXIF 1188: Bombo
R. Broadc. Center, Corrales Ave., Cagayan de Oro C. – DYWB 1269: CBS
Dev. Corp. Bldg., Lacson St., Mandalagan, Bacolod C. – DZVX 1269: J.
Pimentel St., Daet, Camarines Norte — 16) Cotabato Television Corp,
ROF Enterprises, Regional Complex, Cotabato C. – GM: Ramon Flores
— 17) Crusaders Broadc. System Inc, 209 E. de la Paz Str,
Mandaluyong, Metro Manila 3119 – GM: Atty. F. Borja. 2100-1500 —
18) Delta Broadc. System, Matthew St., Multinational Village,
Paranaque, Metro Manila – GM: G.Z. Velarde. St. Mgr: O.R. Talain.
2100-1600 — 19) Digos Broadc. System, Associated Labor Unions,
Elliptical Rd cor. Maharlika St., Diliman, Quezon C. – SM: J. Cabatingan
— 20) Communication Unit, University Extension, Central Mindanao
University, Musuan, Bukidnon – St. Mgr: Prof. Liberty Uego-Josue. CE:
Rainier A. Escarlos. 2100-0100 & 0900-1200 — 21) Eagle Broadc.
Corp, Maligaya Bldg 2, No. 887 E. delos Santos Ave, Quezon C. – GM:
Art V. de Guzman II. 2000-1600 — 21x) East Visayas Bctg, RYGE
Specialist, 3rd Flr, UniversalRE Bldg, 106 Paseo de Roxas, Makati, MM
– St. Mgr: Florencio Villamor — 22) FBS Radio Network Inc, 18th Flr,
Philcomcen Bldg, Ortigas Ave, Pasig, Metro Manila – Pres: Mrs. L. Vera.
2000-1600 — 23) Fairwaves Broadc. Network, Rm 601, Allied Bank
Bldg, Q. Paredes Str, Binondo, Metro Manila – GM: Atty. R.S.
Encarnacion — 24) **Far East Broadc. Co.** (see end of section) — 25)
Filipinas Broadc. Network, Rm. 306, Legaspi Towers 200, Paseo de
Roxas, Makati, Metro Manila – GM: Mrs. D. C. Gozum — 26) First
United Broadc. Corp, Zamboanga C., Zamboanga del Sur — 27) GMA
Network Inc., EDSA cor. Timog Ave. Diliman, Quezon C. 2000-1600 —
28) Tirad Pass Broadc., Candon, Ilocos Sur — 29) Trans-Radio Broadc.
Corp., Suite 608, Pacific Bank Bldg, Ayala Ave, Makati, Metro Manila
– Pres: E.A. Tuason — 29x) Good News Sorsogon Foundation Inc., Rizal
Str, Sorsogon – St. Mgr: Rev. Fr. Gerard V. Deyeza. 2100-1500 — 30)
Hypersonic Broadc. Center, 341 3rd Flr, J&T Bldg, Magsaysay Blvd, Sta.
Mesa, Metro Manila – Officer-in-Charge: Hernando E. Baldo — 31)
Insular Broadc. System, 46 Aguilar Str, Proj. 7, Quezon C — 32)
Banahaw Broadc. Corp., Broadcast City Complex, Capitol Hills,
Diliman, Quezon C. 2000-1600 —33) Kumintang Broadc. System, Suite
800, LPL Towers, 112 Legaspi Str, Legaspi Village, Makati, Metro
Manila – GM: Antonio Leviste — 34) Satellite Broadc. Inc., NLTI Bldg,
San Fernando, La Unión — 35) Mabuhay Broadc. System Inc.,
Centerpoint Condominium, Dona Julia Vargas Ave., Ortigas Center,
pasig C. 2000-1600 — 36) Magiliw Community Broadc. Co. Inc, 74 F.
Blumentritt Str, Mandaluyong, Metro Manila – GM: R. Estrella Jr —

37) People's Television – Channel 4, Media Center, Bohol Ave, Quezon C. – Op's Dir: R. Dalogdog — 38) Pacific Broadc. System Inc., Santiago, Isabela — 39) Manila Broadc. Co, 4th Flr, FJE Bldg, 105 Esteban Str, Legaspi Village, Makati C, Metro Manila — ☎+63 (2) 815-9131. ▤+63 (2) 810-9362. 24h.— 40) Interactive Broadc. Media Inc., 23 E. Rodriguez Sr. Blvd., Quezon C., Metro Manila – ☎+63 (2) 732-9121/9126. ▤+63 (2) 732-9127. 2000-1600 — 41) Masagana Broadc. Corp, 3rd Flr, Admin. Bldg, Araullo Univ, Bitas, Cabanatuan C. – Own: Rolan C. Esteban — 42) Masbate Community Broadc. Company Inc, 13 Madelaine Str, Parkway Village, SFDM, Quezon C. – GM: M.E. Manalastas — 43) Nueva Ecija Provincial Govt., Cabanatuan C., Nueva Ecija — 44) Dawnbreaker's Foundation Inc., Bulac, Talavera, Nueva Ecija — 45) Mountain Province Broadc. Corp, P.O. Box 156, Baguio C., 2600 Benguet – GM: Fr. Hugo Delbaere. SM: D. Pineda. 2000-1500 — 46) Mountain View College, c/o L. Cruz Compound, 29 Sta. Queteria, Caloocan C., Metro Manila – GM: R.B. Abordo Jr — 47) Nation Broadc. Corp, Jacinta Bldg, 914 Pasay Rd, Makati, Metro Manila – GM: A.L. Yabut. 2000-1500 — 48) Western Phil. Broadc. Corp., Puerto Princesa C., Palawan — 49) Newsounds Broadc. Network Inc, Ground Flr, Florete Bldg, 2406 Nobel cor. Edison Str, Makati, Metro Manila – GM: E. Billones – Bombo Radyo DXIF: Corrales Ave, 9000 Cagayan de Oro C. – SM: Albino B. Quinloc — 50) Notre Dame Broadc. Corp, Rm. 302, Gemini Bldg, Sep. G.J. Puyat, Makati, Metro Manila – GM: Fr. V.P. Quioque, OMI — 51) Ormoc Broadc. Co. Inc, Rms 402/403, State Condominium I, Salcedo Str, Legaspi Village, Makati, Metro Manila – GM: C.S. Canoy — 52) Palawan Broadc. Corp, No. 6, Rd 1, Highway Hills, Mandaluyong, Metro Manila – Exec. Vice Pres: Ms. L.I. Ilustre — 53) Pedro N. Roa Broadc. System, 4th Flr, National Life Insurance Bldg, Ayala Ave, Makati, Metro Manila – GM: R.N. Cui — 54) People's Broadc. Network, 64-C Cordillera Str, Quezon C. – GM: J.D. Bayona — 55) DWIZ, 5th Floor, Dominga Bldg., Pasong Tamo Cor. Dela Roza St., Makati, Metro Manila — 56) Philippine Broadc. Corp, 3rd Flr, FJE Bldg, Esteban Str, Legaspi Village, Makati, Metro Manila – Pres: E.D. Buyco — 57) Philippine Radio Corp, 10th Flr, National Life Insurance Bldg, Ayala Ave, Makati, Metro Manila – GM: Jose E. Escaner, Jr. CE: Floriano M. Gloria. 2100-1400 — 58) R. Veritas, 20th Floor, "The Centerpoint", Doña Juliana Vargas Ave/Corner Garnet Str.,Ortigas Complex, Pasig, 1600 Metro-Manila. ☎+63 (2) 635-2512. ▤+63 (2) 633-5453. 24h. — **FM:** DWCS 103.5MHz — 59) RCPI Broadc. Network, RCPI Bldg, E. delos Santos Ave, Quezon C. – GM: O.Q. Pajar — 60) RMC Broadc. Co. Inc, Madapo Hills, 8000 Davao C. – GM: Dr. Evelyn A. Magno. 1900-1400 — 61) Radio Corp. of the Philippines, 10th Flr, National Life Bldg, Ayala Ave, Makati, Metro Manila – Head Tec.Op's: J. Escaner. 2100-1500 — 62) Radio Inc, 10th Flr, National Life Insurance Bldg, Ayala Ave, Makati C, Metro Manila – GM: Jose E. Escaner, Jr. CE: Floriano M. Gloria. 2000-1500 — 63) R. Mindanao Network Inc, 15th Flr, Philcomcen Bldg., Ortigas Ave., Pasig, Metro Manila – GM: H.R. Canoy. 24h — 64) R. Philippines Network Inc, Broadcast C., Capitol Hills, Old Balara, Quezon C. – Pres. & GM: Jose Mari Gonzales. Eng. Mgr: Antonio T. Sorriano — 65) R. St. DXGE, Radio Spectrum Inc, Rm. 601, Campos Rueda Bldg, Tindao Str, Makati, Metro Manila – GM: Ms. E. Cruz-Rafols — 66) R. St. DZHH (non-comm.), Philippine Air Force, Villamor Air Base, Pasay C. —SM: Maj. V.A. Baes. 2100-1600 — 67) R. St. DZLB (non-comm.), UP-Los Banos, College Laguna 4031 – SM: G.L. Paje — 68) R. St. DZUP (non-comm.), Univ. of the Philippines, UP, Diliman, Quezon C. – GM: Ms. C.L. Lazaro. 1100-1300 — 69) R. Pilipino Corp, 12th Flr, National Life Insurance Bldg, Ayala Ave, Makati, Metro Manila – GM: J. Escaner Jr — 70) Ragde, Vicente & Sons, Suite 402/403, State Condominium I, Salcedo Str, Legaspi Village, Makati, Metro Manila – GM: Ret. Gen. H.M. Isleta — 71) Rajah Broadc. Network Inc, J&T Bldg., 3894 Ramon Magsaysay Blvd., Santa Mesa, Manila – GM: A. Santonia. 2100-1700 — 72) Southern Broadc. Network, 3881 E. Vallejo Str. Santol, Sta. Mesa, Metro Manila – GM: Mrs. C.B. Pacquing — 73) Southern Philippines Mass Communications Center, (UMBN) CRS, Camp Navarro, Calarian, Zamboanga C., Zamboanga del Sur – SM: Col. Buenaventura Ramos — 74) Subic Broadc. Corp, 1 Kasarinlan Rd., Olongapo C. 2200 – Pres. & GM: J. Gordon Jr. 1930-1400. **F.PI:** 5kW — 75) Sulu Tawi-Tawi Broadc. Corp. Inc., Suite 4, 3rd Flr, Celta Bldg, Carino Cor. South Superhighway, Makati, Metro Manila – SM: Sra. Ma. V. Adre, OND — 76) Tagbilaran Broadc. Corp, 301 Cityland III Condominium, Cor. Esteban & Makati Ave, Makati, MM – GM: N.C. Abacajan — 77) Times Broadc. Corp, 5th Flr, Union Bank Bldg. II, Pasay Rd, Makati, Metro Manila – GM: Alex V. Sy — 78) Conamor Broadc. Corp., Ibabang Dupay, Lucena C., Quezon — 79) United Broadc. Network, 11th Flr, FEMS-Tower I, Zobel Roxas Cor. South Superhighway, Makati, Metro Manila – GM: J. Hodreal — 80) Universal Broadc. Corp., Sunshine Village, Esperos Str, Tacloban C., Leyte — 81) Universal Broadc. System, DYXT Bldg., Luna Str, Tagbilaran C. 6300, Bohol – GM: R. T. Gatal — 82) Univ. of Mindanao Broadc. Network, Torres Arcade, 1002-

J, Pasay Rd. cor. Makati Ave, Makati, Metro Manila – Pres: G. Torres. 2100-1600 — 83) 2nd Flr, Univ. of San Agustin, Jalandoni Str., Iloilo C. 5901, Iloilo – GM: Miss Marivic Z. Hermano. CE: Elizer H. Arsenal. 2100-1600 — 84) Vanguard R. Network, 4th Floor, Diego Building, Maharlika Highway, Cabanatuann C. 2100.☎+63 (44) 463-2112. ▤+63(44)463-2373.— 85) Victory Broadc. System Inc, 3rd Flr, Jacinta Bldg, 914 Pasay Rd, Makati, Metro Manila – GM: J. Yabut — 86) Vimcontu Broadc. System, Broadcast Production & Training Center, ALU-Vimcontu Welfare Center, Port Area, Cebu C. – SM: C.M. Remonde — 87) Bureau of Broadc. Sces (Phill. Broadc. Sce.), Philippine Information Agency Bldg., Visayas Ave., Del Monte, Quezon C. – Dir: Francisco M. Batacan Jr, CE: Jose Q. Borromeo. DXBN 792 radyo na Bayan (SuperRadio): Do-ongan, Butuan C. – SM: Mansueto G. Catubay Sr. — 88) Penafrancia Broadc. Corp., Naga C., Camarines Sur — 89) Zambales Broadc. & Development Corp, DZOR Bldg, 1683 Rizal Avenue, Olongapo C., Zambales – Pres: Dr. H.S. Ruiz — 90) Caceres Broadc. Corp., Diversion Rd, Tabuco, Naga C., Camarines Sur — 91) ABS-CBN Broadc. Network Inc, Chronicle Bldg, Ortigas Ave, Pasig, Metro Manila – Mgr: L. Balquiedra. **Web:** http://www.abs-cbn.com/entertainment/radshows/dzmm/ – DYAB 1512: P. del Roasario corner Leon Kilat Str, Cebu C, Cebu – PM: Dante Luzon — 92) DZRA Broadc. Corp., Talisay,.Camarines Norte — 93) Catanduanes State College, Virac, Catanduanes — 94) Dept. of National Defence, Armed Forces of the Philippines, Camp Aguinaldo, Quezon C. – SM: Orly Punzalan. 2100-1600 — 95)Pacific Broadc. System Inc., Laoag C., Ilocos Norte — 96) Abra Community Broad. Corp, Bl. Arnold Janssen Communications Center, Zamora Str, Cor. Rizal Str, Bangued, Abra 2800 – GM: A. Bello — 97) Filipinas Broadc. Assoc, Baruyan, San Jose, Tacloban C. – GM: Fr. Mar Alingasa, SVD. 2200-1300 — 98) Ultrasonic Broadc. System, 145 Panay Ave, cor. Sgt. Esguerra St., Quezon C., Metro Manila. 2100-1600 — 99) Solid North Broadcasting, Vigan, Ilocos Sur — 100) GV Broadc. System, Angeles C., Pampanga — 101) ZOE Broadc. Network, Calamba, Laguna — 102) Ribbon Broadc. Netw, 5th Flr, LCC Bldg, Lipa C., Batangas — 103) Edelwina F. Pena, Puerto Princesa, Palawan — 104) OMARCO (Ben Viduya), Calapan, Oriental Mindoro — 105) Romblon Broadc. Netw, Odiongan, Romblon — 106) Muslim Mindanao Dev. Multi-Purp. Coop., Marawi C., Lanao del Sur — 107) Rinconada Broadc. Corp, UNEP Compound, San Roque, Iriga C., Camarines Sur — 108) R. Sorsogon Netw, Don Luis Lee Bldg, Plaza Bonifacio Sorsogon, Sorsogon — 109) UMBN, Ormoc C., Ormoc — 110) Franciscan Broadc. Corp, Bishop's Res., Dumaguete C., 6200 Negros Oriental – GM: Merlin T. Logronio, P.C. 2100-1630 — 111) Eastern Broadc. System, 67 Dr. Alejos Dr, Quezon C. — 112) R.T. Broadc. Specialists Philippines, Rm. 9, Poblete Bldg, 17 Sen. Gil J. Puyat Ave, Makati, MM – Pres. & GM: R.L. Tan — 113) DXZB/TV-13 Cooperative for Sce, Campaner Str, Zamboanga C., Zamboanga del Sur — 114) Ruta Broadc. System, C.M. Recto Str, Malaybalay, Bukidnon — 115) Islamic Dawah Broadc. Netw, Marawi C., Lanao del Sur — 116) Mindanao Broadc. Co. Inc, 7th Flr, Topman Center Bldg, Ayala Ave, Makati, MM — 117) Assoc. of Islamic Development, Banale Dist., Pagadian C., Zamboanga del Sur — 118) Cebu Broadc.Co., Bo. Tangue, Talisay, Cebu C., Cebu — 119) Intercontinental Broadc. Corp., Iligan C., Lanao del Norte — 120) National Council of Churches, Siliman Univ., Dumaguete, Negros

Call	kHz	kW		Call	kHz	kW
33) DZBR	531	5		87) DZFM	603	10
124) DXGH	531	5		56) DWSP	612	5
45) DZWT	540	10		63) DYHP	612	10
62) DYRB	540	10		87) DWCM	612	5
12) DXHM	549	5		64) DZTG	621	5
12) DYAF	549	1		63) DXDC	621	10
87) DWRP	549	10		87) DZFT	621	5
75) DXIM	549	1		10) DYAG	630	5
63) DZXL	558	40		91) DZMM	630	50
87) DWRP	558	10		64) DZRL	639	1
38) DXCH	567	5		63) DXKR	639	5
87) DYCA	567	10		38) DWRH	648	10
47) DZYZ	576	1		48) DWPS	648	5
15) DXMF	576	10		118) DYRC	648	5
87) DZMQ	576	5		114) DXMB	648	5
87) DYMR	576	10		103) DWRM	648	5
118) DZSR	576	5		84) DZXC	657	1
15) DZVV	585	5		12) DXDD	657	5
87) DYCI	585	10		34) DZLU	657	1
12) DXCP	585	5		57) DWRN	657	5
12) DXDB	594	10		63) DYVR	657	5
15) DYWR	594	10		111) DYFL	657	1
27) DZBB	594	20		39) DZRH	666	50
63) DXPR	603	10		87) DXRP	666	10
7) DWLV	603	10		32) DWLW	675	5

Call	kHz	kW
64) DYKC	675	5
75) DXCD	675	1
87) DYES	675	1
32) DWGW	684	1
41) DZCI	684	1
39) DYEZ	684	10
25) DZCV	684	5
64) DXDX	693	1
18) DXDN	693	1
39) DYKH	693	1
39) DYHP	693	10
63) DXBC	693	5
24) DZAS	702	50
47) DXRD	711	5
15) DZVR	711	5
21x) DYBR	711	5
47) DZYI	711	5
63) DXIC	711	5
15) DZSO	720	5
5) DZJO	720	5
39) DYOK	720	10
63) DXMY	729	5
87) DZPE	729	10
87) DWPE	729	10
54) DZGB	729	5
53) DXOR	729	5
87) DZRB	738	40
63) DYHB	747	10
50) DXND	747	5
39) DZJC	747	10
87) DWGC	756	10
87) DZGC	756	10
32) DWNW	756	5
61) DXJM	756	2
6) DWHL	756	1
47) DYCB	765	5
47) DZYT	765	5
52) DYPR	765	10
62) DXGS	765	5
87) DXSO	774	10
40) DWWW	774	25
56) DZNL	783	5
42) DYME	783	5
60) DXRA	783	10
15) DXPD	792	5
51) DYRR	792	5
100) DWGV	792	5
87) DXBN	792	10
15) DXES	801	10
110) DYWC	801	5
47) DXBL	801	1
15) DZNC	801	10
82) DXMZ	801	5
30) DWFA	801	1
12) DYKA	801	5
71) DZRJ	810	10
39) DYVL	819	10
82) DXMC	819	5
73) DXSC	819	1
47) DWRI	819	5
129) DXUM	819	10
30) DWZR	828	5
69) DZTC	828	1
63) DXCC	828	10
23) DZXE	837	5
47) DXRE	837	5
15) DYFM	837	10
58) DZNN	846	50
25) DZGE	855	10
39) DXGO	855	
38) DXDH	855	5
26) DXLA	855	5
37) DZPE	855	10
2) DXCT	855	1
118) DXZH	855	5
119) DXWG	855	1
47) DWSI	864	5
47) DZSP	864	
18) DWSF	864	1
123)DYHH	864	10
96) DZPA	873	5
25) DZRC	873	5
47) DXRB	873	5
47) DXRT	873	5
87) DXJS	873	10
55) DWIZ	882	50
50) DXMS	882	10
87) DYJR	882	10
87) DXRG	882	1
12) DWWM	891	5
15) dzgr	891	5
90) DWHQ	891	5
120) DYSR	891	10
15) DYOW	900	5
43) DWNE	900	5
82) DXRZ	900	5
72) DXSS	900	10
12) DZEA	909	5
86) DYLA	909	5
87) DZSR	918	50
75) DXMM	927	5
15) dzlg	927	5
99) DWRS	927	5
128) DXCA	927	5
61) DZXT	936	1
63) DYCC	936	1
64) DYKW	936	1
82) DXDN	936	5
87) DXIM	936	10
1) DYRO	945	5
47) DXRO	945	5
14) DZEM	954	40
80) DYMM	954	2.5
87) DWFB	954	10
87) DWSB	954	10
107) DZAL	954	5
47) DXRI	954	1
12) DZNS	963	5
47) DXYZ	963	5
55) DYMF	963	10
82) DXMO	972	5
78) DWTI	972	5
118) DXKH	972	5
118) DYSM	972	1
15) DXBR	981	10
47) DZRD	981	5
32) DYBQ	981	10
56) DWMT	981	5
57) DWRS	981	5
69) DXOW	981	10
63) DXDR	981	5
28) DWRT	990	15
95) DZMT	990	5
124) DYTH	990	5
131)	990	5
63) DXHP	999	1
27) DYSS	999	5
87) DXPT	999	1
87) DZEQ	999	1
12) DWBS	1008	5
64) DXXX	1008	10
74) DWGO	1008	2.5
32) DWDW	1017	10
1) DYRP	1017	10
65) DXGE	1017	5
9) DXSN	1017	5
87) DWLC	1017	10
130) DXRR	1017	10
30) DWXX	1026	5
15) DXMC	1026	5
82) DXMI	1026	1
47) DZAM	1026	25
15) DZWX	1035	10
12) DXCP	1035	2.5
49) DXWB	1035	5
69) DYRL	1035	10
109)	1035	5
15) DZNG	1044	10
69) DXCO	1044	5
49) DZDR	1044	10
112) DXLL	1044	5
19) DXML	1044	1
64) DXKD	1053	1
83) DYSA	1053	5
21) DZEL	1053	5
21) DZEC	1062	40
24) DXKI	1062	5
81) DYXT	1071	1
64) DXKT	1071	5
107) DZAL	1071	5
12) DWAM	1080	1
21) DWIN	1080	5
26) DXRH	1080	5
32) DYSJ	1080	1
118) DYBH	1080	1
64) DXKS	1080	1
62) DWRL	1080	5
106) DWCL	1080	1
30) DYHR	1089	1
82) DXCM	1098	5
17) DWAD	1098	10
47) DXCL	1098	5
15) DYIN	1107	5
27) DXBB	1107	5
49) DXMF	1107	10
87) DXRK	1107	5
122) DWDY	1107	10
24) DXAS	1116	5
67) DZLB	1116	5
76) DYTR	1116	1
57) DYRM	1125	1
24) DWAS	1125	5
27) DXGM	1125	5
15) DZWN	1125	10
126) DXGL	1125	10
57) DYRM	1134	1
87) DZPT	1134	1
82) DXMV	1134	5
94) DWDD	1134	10
42) DYCM	1152	5
63) DXMD	1152	5
54) DZMD	1161	5
8) DYRD	1161	5
56) DWCM	1161	5
63) DYKR	1161	1
82) DXDS	1161	1
87) DZCA	1170	10
87) DXMR	1170	10
87) DZRP	1170	50
49) DYCX	1179	5
115) DXMT	1179	10
59) DYRV	1188	1
15) DXIF	1188	5
61) DZLT	1188	5
15) DXLX	1188	5
84) DZXO	1188	5
24) DXFE	1197	5
1) DYRH	1197	5
16) DWBA	1197	5
90)	1197	10
32) DWAN	1206	10
63) DXRS	1206	1
12) DYRF	1215	10
21) DXED	1224	10
87) DWBF	1224	5
87) DZAG	1224	2.5
87) DXSM	1224	1
39) DWSR	1224	5
24) DYVS	1233	5
58) DWRV	1233	5
22) DWBL	1242	20
77) DXSY	1242	5
113) DXZB	1242	5
32) DYRG	1251	1
38) DXPR	1251	5
54) DZMS	1251	2.5
36) DWMC	1260	10
123) DYDD	1260	10
15) DYWB	1269	10
15) DZVX	1269	5
32) DXAM	1278	10
87) DZRM	1278	10
32) DYJJ	1287	1
39) DZZH	1287	5
108) DZRS	1287	1
1) DWPR	1296	5
1) DWLQ	1296	5
91) DXAB	1296	10
21) DYFX	1305	10
18) DWXI	1314	10
3) DYSI	1323	10
87) DZRK	1323	1
106) DXAD	1323	5
64) DZKI	1332	1
64) DWKI	1332	1
80) DYBB	1332	5
85) DWAY	1332	5
57) DZYS	1341	5
35) DZXQ	1350	10
87) DZER	1350	5
57) DZYR	1359	5
87) DYSL	1359	10
64) DXKO	1368	5
47) DWTT	1368	5
64) DZBS	1368	5
64) DXKP	1377	1
13) DYJC	1386	5
46) DXCR	1386	5
12) DZVT	1395	5
39) DYCH	1395	10
84) DWMG	1395	1
93) DZRA	1395	1
64) DYKB	1404	10
12x) DYDW	1413	10
97) DYXW	1413	5
79) DWBC	1422	15
89) DZOR	1422	1
20) DXMU	1422	5
70) DYRS	1431	1
39) DWPH	1440	10
104) DZOM	1440	1
125)DXSI	1440	0.01
116) DXSA	1449	5
121) DYAC	1449	5
24) DWRF	1458	10
101) DZOE	1458	1
123) DYZZ	1458	10
12) DXVP	1467	5
28) DZTP	1467	10
69) DZYA	1476	1
102) DWRB	1476	1
63) DYKR	1485	1
39) DYDH	1485	5
62) DXOC	1494	1
98) DWSS	1494	10
91) DYAB	1512	15
105) DYCR	1512	5
11) DZME	1530	25
127) DXDV	1530	10
57) DZYM	1539	1
12) DYDM	1548	5
117) DXID	1566	10
66) DZHH	1566	10
1) DYAY	1584	1
44) DZDF	1584	1
68) DZUP	1602	10
4) DWGI	1674	0.25
132) DZBF	1674	5

Occidental — 121) Ce. Visayas College of Agriculture, VIZCA Baybay, Leyte, Leyte — 122) Northeastern Broadc. Service, Ground Floor, Isabela Hotel, Mirante Uno, Cauayan Isabela — 123) SIAM Corp., Bantay Radyo, Mactan Business Center, Airport Road, Lapulapu C.,

Metro Cebu. — 124) Pacific Broadc. System, Davao C., Davao — 125) Southern Institute of Technology, Cagayan de Oro C., Misamis Oriental — 126) PEC Broadc. Corp, Butuan C., Agusan del Norte — 127) Vismin Radio & TV Broadc. Netw., Butuan C., Agusan del Norte — 128) Office of the Governor, San Francisco, Agusan del Sur — 129) Mt. Apo Science Foundation, Matina, Davao C., Davao — 130) Gertrudes Natividad. McArthur Highway, Matina, Davao C., Davao — 131) RBS, Cotobato C., Cotobato — 132) Municipality of Marikina, Marikina, Metro Manila.
FM: Approx. 290 FM-st's are operating.

RADYO PILIPINAS

✉ Philippine Broadc. Sce, 4th Flr. Media Center, Visayas Ave, Diliman, Quezon C., Metro Manila 1103. ☎ +63 (2) 924 2267. 🗎 +63 (2) 924 2745 — **L.P:** Dir's: Rafael Dante A. Cruz, Ben B. Tabisaura. Ag. Chief Eng. Div: Armando C. Remedo.
SW: Tinang (see Voice of America Philippines).
V. by QSL-card. Rp. (2 IRC's). Rec. acc.
For frequencies and schedule see International Broadcasting section.

RADIO VERITAS ASIA
Philippine Radio Education & Information Center
✉ P.O. Box 2642, Quezon C. 1166. ☎ +63 (2) 900012. 🗎 +63 (2) 907436. **Cable:** 64420 VERTAS PN. ① 632-900014.
L.P: Chmn: H.E.J.Cardinal Sin. Mgr: Erlinda G.So. Head of Prgrs: Msgr. Pietro Yan Tai. Tech. Consultant: Fr. H.Delbaere. Prgr. Consultant: Fr. J. Desautels. Tech. Dir: F.L.Kiguchi. Freq. Planning: R.S. Alonzo. Head of PR: Cleote Labindao.
SW: Palauig, Zambales (G.C: 119.50E/15.28N): 3 x 250kW tr's.
ANN: "This is R. Veritas Asia, Quezon C., Philippines".
IS: "O Via, Vita, Veritas" — **V.** by QSL-card.
For frequencies and schedule see International Broadcasting section.

FAR EAST BROADCASTING CO. (Rlg.)
✉ Box 1, 0560 Valenzuela, Metro-Manila. ☎ +63 (2) 361 1010. 🗎 +63 (2) 359490. **Cable:** FEBCOM MANILA. ① 40048 FEBCOM PM.
E-mail: febcomphil@febc.jmf.org.ph. **Web:** http://www.febc.org
L.P: MD: Efren M. Pallorina. Head of Prgrs: Peter McIntyre. CE: Romualdo M. Lintag. Head of PR: Priscilla R. Calica.
PHILIPPINE SCE. in English and Tagalog: 2130-1600 on DZAS 702kHz, 2100-2320 (Sat 2200-2330) & 0800-1100 on 3345kHz, 2200-2315 on 6030kHz.
Regional AM st's: DXKI 1062kHz (Marbel, So. Catabato) 2100-0400 & 0900-1400, DXAS 1116kHz (Zamboanga) 2100-0600 & 0900-1400, DWAS 1125kHz (Daraga, Albay) 2200-0500 & 0900-1300, DXFE 1197kHz (Davao C.) 2100-0500 & 0900-1300, DYVS 1233kHz (Bacolod, C.) 2200-0500 & 0900-1300, DWRF 1458 kHz (Iba, Zambales) 2200-0500 & 0900-1100.
FM: Manila 98.7MHz 60kW (2200-1600), Cebu 98.7MHz 9kW (2100-0500, 0900-1600).
SW: Bocaue, Bulacan (G.C: 120.55E/14.48N), Iba, Zambales (G.C: 119.57E/15.22N).
ANN: "Broadcasting from the Philippines, this is FEBC Radio International, The Sound Alternative"2200-2315
IS: "We have heard the Joyful Sound".
V. by QSL-card (B). Rp. preferred (3 IRCs for Airmail).
PUB: Prgr. schedule free on request. Signal (quarterly).
For frequencies and schedule see International broadcasting section.

FAR EAST NETWORK (AFRTS)
✉ & **L.P:** See Far East Network Japan.
STATIONS: Subic Bay 1251kHz 0.25kW, FM 95.1MHz 0.6kW.
D.PRGR: 24h.

VOICE OF AMERICA RADIO STATION PHILIPPINES
MW: Poro 1143kHz 1000kW
SW: see International Broadcasting section.

SINGAPORE

L.T: UTC + 8h — **Pop:** 2.864.000 — **Radios:** 822.000 — **Pr.L:** Malay, Mandarin, Tamil, English — **E.C:** 50Hz, 230V — **ITU:** SNG.

RADIO CORPORATION OF SINGAPORE (RCS)
✉ Caldecott Hill, Andrew Road, Singapore 1129. ☎ +65 251 8622. 🗎 +65 254 8062. **Web:** http://rcs.com.sg/
L.P: Chief Exec. Officer: Anthony Chia. VP (Radio): Chua Foo Yong. Asst. VP for English Programming: Florence Lian. Public Rel. Exec: Priscilla Yim Stacey.
SW: (G.C: 103.51E/01.24N): 6 x 250kW, 1 x 100kW.

kHz	kW	MHz	Network	Format	Lang.	H. of tr.
6000a	250	95.8	95.8 FM	N./Info	Chinese	2200-1800
		93.3	93.3 FM	CHR	Chinese	24h
		97.2	Love 97.2	AC	Chinese	24h
6155a	250	90.5	One FM	N./Info	English	24h
		92.4	FM 92.4	Easy, c&w	English	2200-1600
		95.0	Class 95	AC	English	24h
		98.7	Perfect 10	CHR	English	24h
7170b	100	96.8	Olikkalanjiam	Full sce.	Tamil	2100-1800
7250c	250	94.2	Warna	Full sce.	Malay.	2045-1800
		89.7	Ria	CHR	Malay	2200-1800

SW freqs operate as follows: a) 2300-1100 & 1400-1700, b) 2200-1700, c) 2200-1100 & 1400-1700
V. by QSL-card. Rp. Rec.acc.

RADIO SINGAPORE INTERNATIONAL (RSI)
See International Broadcasting section.

NTUC VOICE CO-OPERATIVE SOCIETY LTD.
Operated by National Trade Union Congress
✉ 510 Thomson Rd, #01-00 SLF Bldg, Singapore 1129. ☎ +65 353 6100 **Web:** http://www.heartfm913.org.sg/
FM: 91.3/100.3MHz (stereo).
R. Heart 91.3: 24h in English exc. 1500-1800 Malay.
100.3MHz: 24h in Mandarin exc. 1500-1800 Tamil. **Japanese:** Sun 0900-0945.

SAFRA RADIO
Operated by SRCC Pte Ltd.
✉ Defence Technology Towers, Tower B, #12-04, Depot Rd,.Singapore 0410. ☎ +65 373 1924. 🗎 +65 278 3039. **E-mail:** power98@letter-box.com
Web: http://power98.com.sg/ or http://www.fm883.com.sg/
L.P: Finance & Admin. Mgr: Vivien Leong — **FM:** 88.3/98.0MHz.
English: "Power 98": 2100-1800 on 98.0MHz.
Mandarin "Dongli 883": 2100-1800 on 88.3MHz.

BBC FAR EASTERN RELAY STATION
✉ P.O. Box 434, Singapore. ☎ +65 2601511. 🗎 +65 669 0834
FM: 88.9MHz 4kW (24h rel. of BBCWS in English).
SW: see International Broadcasting section.

SRI LANKA

L.T: UTC + 6h — **Pop:** 19.114.383 — **Radios:** 20m — **Pr.L:** Sinhala, Tamil, English — **E.C:** 50Hz, 230V — **ITU:** CLN.

SRI LANKA BROADCASTING CORPORATION
(Public Corporation)
✉ P.O. Box 574, Independence Square, Colombo 7. ☎ +94 (1) 696329, 584673, 697491. 🗎 +94 (1) 695488. **Cable:** Broadcast, Colombo. ① 21408 SLABCOR CE. **E-mail:** brzcast@sri.lanka.net
L.P: DG: M. P. J. Peiries. Exec. Working Dir: M.L.M.A. Farook. Advisor to the Corporation: Dr. Vijaya Corea. Add. DG: Newton Gunaratane. Dep. DG (Training, Audience Research & Educ.): V.A. Thirugnanasuntharam. Dep. DG (Eng.): H.P.A.L. Pinto. Ag. DG (Finance): N.P.W. Perera. Dir. Foreign Rel: M.J.M. Ashroff. Dir. Sinhala: Palitha Perera. Ag. Dir. Tamil/Nat. Sce: V.N. Mathialagan. Dir. English Sce: Nihal Bhareti. Dir. Muslim Sce: Z.L.M. Mohamed. Dir. N./Ag. Dir. Educ, Sce: Somapala Perera.
Home Sces: A=Sinhala National Sce, B=Sinhala Commercial Sce, C=Tamil Commercial Sce, D=English Commercial Sce, R=Regional Sce, S=Education & Sports Sce.

MW	kHz	kW	Sce.	MW	kHz	kW	Sce.
Ambewela	531	40	B/R	Ratnapura	729	10	A
Diyagama	558	20	C	Kantale	747	20	B
Senkadagala	567	10	A/R	Anuradhapura	774	10	A
Kantale	585	20	C	Maho	801	20	R
Ambewela	648	20	A	Kandy	819	20	B
Weeraketiya	594	20	A	Ampara	855	20	B
Ratnapura	603	10	A	Diyagama	873	20	D
Diyagama	621	20	B	Ampara	972	20	C
Maho	639	50	A	Galle	1026	10	A
Weeraketiya	675	20	B/R	Mahiyangana	1485	1	B
Ampara	693	20	A	Mahiyangana	1602	1	A
Diyagama	702	25	A				

SW: Ekala (G.C: 79.54E/07.06N): 10kW

kHz	Sce	H. of tr.
4870	B	0000-0200, 1000-1630
4902	A	2255-0700, 1025-1700 (Full Moon days 2255)
4940	D	0025-0200, 1100-1700
4970	S	Special Events
5020	C	2300-0200, 1000-1000
6050	S	Special Events
6075	A	0200-0700
6130	D	0200-1000
6150	B	0200-1000
6185	C	0200-1000

FM (MHz)	A	B	C	D	R	S
Colombo	98.3	88.4	105.65	93.3	91.7	101.35
Deniyaya	99.6	97.6		90.8	107.2	102.6
Haputale					105.4	
Hunasgiriya					102.0	
Karagahatenna	107.6	92.75	104.45	99.6	102.4	
Radella	97.0	103.5	87.5	100.2	89.8	89.7
Senkadagala					97.6	
Palali (Jaffna)					102.0	

Nat. Sce. in Sinhala (A): 2300-0130 & 1100-1700 on MW + FM, SW as above — **Comm. Sce. in Sinhala (B):** 2330-1600(SS 1700).
Comm. Sce. in Tamil (C): 2300-1700 — **Comm. Sce. in English (D):** 0025-1700 — **FM. Sce:** 24h.
Sri Lanka FM Savaya: 24h on 90.8MHz (Hanguranketa), 91.2MHz (Colombo), 94.4MHz (Radella) & 95MHz (Karagahatenna).
Vishva Sravani Vikashanaya: 24h on 90.MHz (Karagahatenna), 92.8MHz (Deniyaya), 95.6MHz (Colombo).
Regional Sces:
Rajarata Sevaya, Anuradhapura: 2300-0230 (SS 0700) & 1000-1530 on 801kHz + FM 102.4MHz.
Kandurata Savaya, Kandy: 2300-0230 (SS 0700) & 1000-1530 on 531/567kHz + 89.80/102.0MHz.
Ruhunu Sevaya, Matara: 2300-0200 (Sun 0700) 1000 (SS 1030)-1500 (SS 1630) on 675kHz + 105.4/107.2MHz.
ANN: A: "Me Sri Lanka Guwan Viduli Sansthave Welanda Sevaya". B: "Me Sri Lanka Guwan Viduli Sansthava Swadeshiya Sevaya". C: "Illangar Oliparappu Kootuthapanam Tamil Sevai". D: "This is the Sri Lanka Broadcasting Corporation".

EXTERNAL SERVICE: see International Broadcasting section.

TNL RADIO (Comm.)
Head Office: 5B Tower Bldg., Station Rd., Colombo 4. ☎+94 (1) 584107, 584871. **Comm. Office:** 9D Tower Bldg., 25 Station Rd, Colombo 4. ☎ +94 (1) 501681. 🖷 +94 (1) 501683.
Web: http://www.lanka.net/tnl/
L.P: Chmn. & MD: Shan Wickremesinghe. Comm. Dir: Ms. Ishini Wickremesionghe — **FM:** Colombo 90.0/101.7MHz 5kW.
D.PRGR. in English: 24h.

MBC NETWORKS (PVT) Ltd. (Comm.)
🖃 109 wnd Floor Colllettes Bldg., Rt. Hon. D.S. Senanayake Mw., Colombo 8. ☎ +94 (1) 689234-6. Station: Depanama. ☎ +94 (1) 851543-5.
Yes FM in English: 24h on Colombo 89.5/101.0MHz and Kandy 88.2MHz all 1kW (stereo).
Sirasa FM in Sinhala (🖃 P.O. Box 25, Depanama, Pannipitiya): 24h on 88.9/105.9/106.1/106.5MHz all 1kW (stereo).

COLOMBO COMMUNICATIONS (PVT) Ltd. (Comm.)
🖃 2/9 2nd Floor, Liberty Plaza, 250 R.A. de Mel Mw., Colombo 3. ☎ +94 (1) 577924-7, 330718-9. 🖷 +94 (1) 577929.
Capital Radio in English: 24h on 100.4MHz 1kW (stereo).
Savana in Sinhala: 24h on 99MHz 1kW (stereo).

TRANS WORLD RADIO (Rlg.)
🖃 P.O. Box 364, 91 Wijerama Mawatha, Colombo 7. ☎ +94 (1) 685235/6/7. 🖷 + 94 (1) 685245. **Cable:** Votan.
L.P: Dir. of Op's: Mark Blosser. CE: Darryl van Dyken.
Prgr. 🖃 Trans World Radio-India, Box 4407, L-15, Green Park, New Delhi 110 016. ☎ +91 (11) 662058. 🖷 +91 (11) 686 8049.
L.P: Regional Dir: Dr. N.Emil Jebasingh.
MW: Puttalam 882kHz 400kW.
D.PRGR: see International Broadcasting section.

RADIO JAPAN RELAY STATION
DEUTSCHE WELLE RELAY STATION
VOICE OF AMERICA RELAY STATION
see International Broadcasting section.

L.T: UTC + 5h — **Pop:** 5.945.000 — **Radios:** 854.000 — **Pr.L:** Tajik (official), Uzbek, Russian — **E.C:** 50Hz, 127/220V — **ITU:** TJK.

VAZORATI ALOKA (Ministry of Communications)
🖃 pr. Rudaki 57, 734025 Dushanbe. ☎ +7 (3772) 232 284. 🖷 +7 (2772) 279 806. ① 201119 MC.
L.P: Minister: N.N. Mukhitdinov.

STATE TV-RADIO BROADCASTING CO. OF THE TAJIK REPUBLIC (Gov.)
🖃 Chapayev Str. 31, 734025 Dushanbe.
L.P: Head of Radio & TV: Mirbobo Mirrakhimov. Dep. Chmn: G. Makhmudov.

TAJIK RADIO (Gov.)
🖃 Chapayev Str. 31, 734025 Dushanbe. ☎ +7 (3772) 276569. ① 201392 TELE.

LW/MW	kHz	kW	H. of tr.	Prgr.
Yangi-Yul	252	150	0030-1830	TR1
Dushanbe	549	40		
Orzu	648	1000		F
Orzu	702	150	0000-2200	RR
Orzu	972	1000	0030-1900	TR2
Yangi-Yul	1143	150	0030-1900	TR2, F
Orzu	1161	40	0030-1830	TR1
Dushanbe	1323		0030-1830	TR1
Dushanbe	1502	7	0100-2000	RO4
Yangi-Yul	4635	50	0030-1830	TR1
Yangi-Yul	5800	100	1630-1800	TR1
Yangi-Yul	7245	100	0030-1900	TR2, F
Yangi-Yul	7510	100	1630-1800	TR1
Yangi-Yul	9905	100	0130-0430	TR1
			1630-1800	TR1

FM (MHz)	TR1	TR3	kW (ERP)
Dushanbe	70.64	72.20	17
Qзrghonteppa	67.88	66.32	17
Khujand	72.56	69.80	17

TR1 in Tajik, Russian, Uzbek — **TR2** in Tajik, Russian: 0500-1400 "Payk-i 'Ajam", at other times FS — **TR3** music sce. in Tajik, Russian. 0700-1600 rel. R. St. Mayak (cf. Russia) — **RO4** = rel. R. Odin Orbita 4 (cf. Russia) — **RR** = rel. R. Rossii (cf. Russia) — **F** = Foreign Sce (see International Broadcasting section).

L.T: UTC + 7h — **Pop:** 59.233.000 — **Radios:** 10m — **Pr.L:** Thai — **E.C:** a/c 50, 220/380V — **ITU:** THA.

RADIO AND TELEVISION EXECUTIVE COMMITTEE (RTEC)
Constituted under the Broadcasting and TV Rule 1975, this body consists of 17 representatives from 14 Government agencies. It controls Administrative, Legal, Technical and Programming aspects of broadcasting in Thailand.
🖃 **Programme, Administration and Law Section:** Div. of RTEC Works, Gov. Public Rel. Dept, Rajchadamnern Klang Rd, Phra Nakhon Region, Bangkok 10200. **Technical Section:** Radio Frequency Management Office, Post & Telegraph Dept, Soi Sai Lom, Phaholyothin Rd, Saam Sen Nai, Phraya Thai Region, Bangkok 10400. ☎ +66 (2) 710151-60. 🖷 +66 (2) 713514. **Cable:** Telepost Bangkok. ① 82503 POSTEL TH.
L.P: Dir. of Int. Sces Div: Somchit Chularat.
STATIONS: All st's are operated by Gov. agencies and/or under the supervision of the Gov.
MW: (N=Nakhon, U=Ubon).(E)=Educational Sce, (N)=News Sce.

	kHz	kW	Location		kHz	kW	Location
1)	531	50	Maha Sarakham	1)	558	50	Songkhla (E)
13)	540	10	Bangkok	10)	558	10	Chaiyaphum
1)	549	10	Krabi	43)	576	20	Bangkok
1)	549	100	Lampang (E)	8)	585	10	Phrae
1)	549	10	Mukdahan	31)	585	20	Prachin Buri

	kHz	kW	Location		kHz	kW	Location
5)	585	10	Chumphon	1)	1026	10	Yala
8)	585	10	Phitsanulok	6)	1035	20	Bangkok
5)	594	10	Khon Kaen	15)	1044	10	Khon Kaen
40)	603	20	Bangkok	32)	1053	10	N. Sawan
20)	612	20	Lop Buri	23)	1053	10	Lampang
1)	621	100	Khon Kaen (E)	12)	1053	10	N. Si Thammarat
22)	630	10	Bangkok	22)	1062	20	Bangkok
1)	639	20	N. Si Thammarat	5)	1080	10	N. Sawan
1)	639	10	Phuket	5)	1080	10	Yala
1)	648	55	Khon Kaen	6)	1089	10	Udon Thani
10)	657	10	Samut Sakhon	7)	1098	10	Songkhla
11)	657	1	N. Ratchasima	1)	1098	10	Tak
8)	666	10	Tak	4)	1107	20	Chon Buri
17)	675	20	Chiang Mai	17)	1116	20	Samut Sakhon
4)	684	20	Bangkok	1)	1125	50	Chanthaburi
24)	693	10	Saraburi	1)	1134	10	Lampang
41)	702	10	Lop Buri	1)	1134	10	Kanchanaburi
19)	711	20	Chiang Mai	1)	1134	10	Phangnga
19)	711	20	Bangkok	39)	1143	25	Bangkok
1)	711	20	U. Ratchathani (E)	18)	1152	20	Chiang Mai
1)	729	50	N. Ratchasima	18)	1152	20	Khon Kaen
5)	738	10	Chiang Mai	12)	1152	1	Songkhla
18)	738	20	Bangkok	4)	1161	20	N. Phanom
11)	747	1	Chaiyaphum	4)	1161	10	Rayong
15)	747	10	Surin	25)	1179		Bangkok
14)	747	1	Prachin Buri	4)	1179	20	Chanthaburi
12)	747	1	Surat Thani	19)	1179	10	Chiang Rai
5)	747	10	Songkhla	15)	1188	10	Sakon Nakhon
11)	747	10	Udon Thani	8)	1188	10	Phitsanulok
9)	756	50	Narathivat	14)	1188	10	Prachin Buri
6)	765	20	Lampang	27)	1197	20	Bangkok
25)	774	10	Udon Thani	15)	1215	10	Phrae
20)	783	20	Lop Buri	11)	1215	10	U. Ratchathani
13)	792	20	Bangkok	1)	1215	50	Surat Thani
11)	792	10	Kalasin	1)	1224	10	Chiang Rai
2)	801	10	Chiang Rai	2)	1224	10	Bangkok
2)	801	5	U. Ratchathani	8)	1233	1	Uttaradit
2)	801	5	N. Ratchasima	5)	1233	10	Udon Thani
2)	801	5	Lop Buri	14)	1233	10	Prachuap Khiri
2)	801	1	Prachuap KhiriKhan	8)	1242	10	Lampang
1)	810	20	Nong Khai	26)	1242	1	Phayao
1)	810	7	Khon Kaen	8)	1242	10	Phetchabun
1)	810	10	Trang	10)	1242	10	Roi Et
1)	819	10	Bangkok	1)	1242	50	Surat Thani (E)
1)	819	10	Satun	2)	1251	20	Bangkok
8)	828	10	Sukhothai	1)	1260	50	Chiang Rai
13)	828	20	Rayong	17)	1269	20	Songkhla
11)	828	10	Khon Kaen	21)	1278	25	Bangkok
9)	837	50	Sakon Nakhon	5)	1287	10	Chiang Rai
1)	846	10	Bangkok	8)	1287	10	Uttaradit
1)	864	10	Tak	5)	1287	10	U. Ratchathani
3)	864	10	Bangkok	5)	1287	10	N. Si Thammarat
1)	873	10	Phatthalung	1)	1296	10	Pattani
1)	873	10	Phetchabun	13)	1305	10	Bangkok
1)	891	1000	Saraburi	17)	1314	20	Khon Kaen
1)	909	50	Surin	2)	1323	10	Chiang Mai
7)	918	10	Chiang Mai	2)	1323	1	N. Pathom
1)	918	1	Chanthaburi (E)	2)	1323	10	Surat Thani
1)	918	10	Bangkok	2)	1323	10	Songkhla
8)	936	10	Chiang Rai	33)	1332	10	Bangkok
11)	936	10	N. Ratchasima	1)	1341	10	Loei
12)	936	10	N. Si Thammarat	1)	1341	50	U. Ratchathani
12)	936	1.5	Pattani	1)	1341	10	Phangnga
1)	936	50	N. Sawan (E)	23)	1350	10	Lampang
2)	954	5	Chanthaburi	5)	1350	10	Trang
2)	954	5	N. Sawan	16)	1359	10	Bangkok
2)	954	5	Phitsanulok	1)	1368	50	Nan
2)	954	5	Udon Thani	1)	1368	1	Chumphon
2)	954	1.1	Maha Sarakham	26)	1377	10	Phitsanulok
2)	954	5	Bangkok	3)	1386	20	Pathum Thani
5)	963	10	N. Ratchasima	1)	1395	25	Chiang Rai
1)	963	50	Krabi	8)	1404	10	Phichit
9)	972	50	Phetchabun	1)	1404	50	Songkhla
1)	981	5	Mae Hong Son	10)	1404	10	Yasothon
1)	981	10	N. Phanom	6)	1413	10	Tak
3)	981	10	Bangkok	7)	1422	10	Phitsanulok
1)	981	50	Yala	4)	1431	20	Songkhla
7)	990	10	N. Ratchasima	8)	1440	1	Nan
16)	1008	10	Bangkok	11)	1440	10	N. Phanom
1)	1026	50	Phitsanulok	5)	1440	10	Samut Sakhon

	kHz	kW	Location		kHz	kW	Location
20)	1449	5	Chumphon	9)	1521	25	Bangkok
10)	1458	10	Si Sa Ket	5)	1530	10	Uttaradit
4)	1458	20	Phuket	14)	1530	10	Chanthaburi
1)	1467	100	Bangkok (E)	13)	1539	10	Udon Thani
1)	1476	100	Lamphun	1)	1539	10	Ratchaburi
1)	1485	1	Surat Thani	16)	1548	10	Bangkok
12)	1485	1	Trang	24)	1557	10	Phetchabun
12)	1485	1	Krabi	37)	1575	1000	Bangkok
1)	1485	1	Yala	3)	1584	10	Chiang Mai
12)	1485	1	Ratchaburi	14)	1584	1	Ratchaburi
39)	1494	25	Bangkok	1)	1584	1	Trad
11)	1503	10	Sakon Nakhon	12)	1584	1	Surat Thani
10)	1503	10	Surat Thani	12)	1584	1	Phatthalung
15)	1512	10	Phayao	1)	1593	1	Ratchaburi (N)
14)	1512	10	Suphan Buri	1)	1593	10	Ranong
11)	1512	10	Surin	1)	1602	1	Buriram

FM: The number of FM-st's operated by each st. is indicated in the address section.

SW: Pathumthani (G.C: 100.43E/14.03N)

	kHz	kW		kHz	kW		kHz	kW
1)	4830v	10	33)	6149.8r		1)	9655	50
1)	6070	10	1)	7115	10	1)	11905	100

Other information: Abbreviated short names, e.g. Thor Thor Thor, serve as station ann. if preceded by "Thini..." ("Here is..."). Full name, given in brackets, is usually preceded by "Thini, Sathani Withayu Krachaisiang..." ("Here is R. St...").

Numerals, preceded by "Thii" ("number") are: 0 soon, 1 nung, 2 song, 3 saam, 4 sii, 5 ha, 6 hock, 7 jed, 8 pat, 9 kao, 10 sip.

RADIO THAILAND (Gov.)

Gov. Public Rel. Dept, Rajchadamnern Klang Rd, Phra Nakhon Region, Bangkok 10200. ☎ +66 (2) 771814, 771840. (**Studio** 236 Vibhavadi Rangsit Superhighway, Din Daeng, Huai Khwang, Bangkok 10400).

L.P: DG, Govt. Pub. Rel. Dept: Bangern Musikapong. Dep. DG's: Tavach Meksawan, Arun Ngamdee. Dir, Radio Thailand: Somphong Visuttipat. Dir. Head Home Sce: Mrs. Chalermsri Huncharoen. Head Ext. Sce: Ms.Amporn.

Prgr. 1: 2200-1730 on 891/4830/6070/7115kHz + FM 92.5MHz. **N:** On the h. exc. 0500, 0600 & 1100. Also 0530 & 1545 — **Prgr. 2:** 2300-0400 & 0700-1400 on 846kHz. **N:** rel. Prgr. 1 — **Prgr. 3:** 2300-1700 on 819kHz (traffic and public affairs information). **N:** Relay Prgr. 1 — **FM prgrs:** "R. FM 88" on 88.0MHz, Music Prgrs on 95.5MHz ("Gold FM"), 105.5MHz ("Smooth FM") & 107.75MHz (KISS-FM. Rel. Foreign Sce. on 107.0MHz. **English** (rel. Foreign Sce.): 0000-0100, 0500-0530, 1200-1230, 1300-1330 on 95.5/105.0MHz. **Educational Sce:** 2200-1640 on all E tr's — **Provincial Stations:** See MW/FM Frequency List and "Other Stations" (below).

EXTERNAL SERVICE: see International Broadcasting section.

OTHER STATIONS

FM: Total no. of tr's indicated in brackets at end of each entry.

1) Sathani Withayu Krachaisiang Hang Prathet Thai, Sor. Wor. Tor. (Radio Thailand). HQ addr: See above; **Provincial address-es:** Most st's can be reached by quoting "Sor. Wor. Tor" and the location given in the freq. list, followed by the phrase "Muang District", and finally the city, which is generally the same as in the freq. list. Exceptions are the following: **Lamphun:** The tr. is in Lamphun, but the studios are in Chiang Mai; **Phangnga:** The addr. for 1134kHz is Takuapa District, Phangnga 82110; **Saraburi:** Studios in Bangkok. The 1000kW tr. addr. is: Rim Klong Hog Wa, Moo 4, Nong Rong, Nong Care, Saraburi 18140; **Tak:** The addr. for 1098kHz is Mae-Sot Tak Rd, Mae Sot District, Tak 63110; **Yala:** The addr. for 1026kHz is Betong District, Yala 95110. (73).

2) Tor. Or. (Thahaan Akart), Royal Thai Air Force, Directorate of Communications & Electronics, Bangkok Airport, Wip avadee Rungsit Rd, Don Muang, Bang Khen region, Bangkok 10900. (15).

3) Por. Chor. Sor. (Krom Pracha Samphun), Govt. Public Rel. Dept, Rajchadamnern Ave, Phra Nakhon region, Bangkok 10200. (no FM tr's).

4) Sor. Tor. Ror. (Siang Chak Thahaan Rua), The Voice of the Navy, Naval Communications Dept, Phra Ratchawang Derm Thonburi, Wat Arun, Bangkok Yai Region, Bangkok 10600 2200-1900. (13).

5) Wor. Por. Tho. (Withayu Pracham Thin), Communications Division, Signals Dept, Royal Thai Army. (6).

6) Nung. Por. Nor. (Krom Paisanee Thoralek), Post & Telegraph Dept, Sam Sen Nai, Phraya Thai Region, Bangkok 10400. ☎ 2710151-60 2200-1700 (5).

7) **Sor. Wor. Por.** (Sathani Withayu Pitaksantirat), Police Broadc. Sces, 2nd Communication Div., Directorate of Police Communications, Police Dept, Bang Khen Region, Bangkok 10900. (31).

8) **Tor. Por. Saam (3)** (Kongthap Pak Tii Saam), The 3rd Army Area, Somdej Phranarasuan Maharat Camp, Muang District, Phitsanulok 65000. (1).

9) **Wor. Sor. Kor. Ror. Por. Klang** (Withayu Krachaisiang, Kong Amnuay Kam Klang Raksa Kham Prod Phai (Hang Chat), Central Security Division Radio Station, Radio Broadcasting Division, Headquarters of Supreme Command, Sanam Sua Pa, Rajchadamnern Nok Ave, Dusit Region, Bangkok 10300. (5).

10) **Chor. Sor.** (Krom Chaye Thahaan Süesarn), Broadcasting Control Section, Radio and TV Division, Signals Dept, Royal Thai Army, Phra Ram V Rd, Saphan Daeng, Bangsüe, Dusit Region, Bangkok 10300. (2).

11) **Tor. Por. Song** (Kongthap Pak Tii Song), The 2nd Army Area, Suranaree Camp, Po Klang, Muang District, Nakon Ratchasima 30000. (8).

12) **Tor. Por. Sii** (Kongthap Pak Tii Sii) The 4th Army Area, Vachiravadh Camp, Rajchadamnern Rd, Pak Poon, Muang District, Nakhon Si Thammarat 80000 2200-1600. (6).

13) **Yan Kraw,** The 4th Cavalry Batallion (Armoured Unit, Royal Guard), Bangkrabüe Rd, Bangsüe, Dusit Region, Bangkok 10300. (On 792kHz, prgrs produced by Buddhism Diffusion Center, Apitham Mahathat College Foundation). (2).

14) **Tor. Por. Nung.** (Kongthap Pak Thii Nung), The 1st Army Area, Suan Missakawan, Rajchadamnern Nok Ave, Dusit Region, Bangkok 10300. (6).

15) **Kor. Wor. Sor.** (Kitkarn Withayu Krachai Siang), Radio Broadc. & TV Division, Signals Department, Royal Thai Army, Phra Ram V Rd, Saphan Daeng, Bangsüe, Dusit Region, Bangkok 10300. (no FM-tr's).

16) **Phol Nung Ror. Or.** (Kongphol Thii Nung Raksa Phra Ong), The 1st Infantry Division (Royal Guard) HQ, Phitsanulok Rd, Dusit Region, Bangkok 10300. (2).

17) **Mor. Kor.** (Mahavitthayalai Kasetsart), Extension & Training Center, Kasetsart University, Lad Yao, Bang Khen Region, Bangkok 10900. ☎ 5792294/5790537. (no FM-tr's).

18) **Ror. Dor.** (Kromkarn Raksa Dindan), Territorial Defence Dept, Suan Chaochet, Sanamchai Rd, Wat Po, Phra Nakhon Region, Bangkok 10200. (1).

19) **Wor. Por. Tor.** (Withayu Kromkarn Phalang Ngan Thahaan), Defence Energy Dept., Phetchabari Rd, Pratumwan Region, Bangkok 10500. (7).

20) **Wor. Sor. Por.** (Withayu Soonkarn Thahaan Pin Yai), Artillery Center, Phaholyothin Camp, Koke Kratiem, Muang District, Lop Buri 15160. (no FM-tr's).

21) **Kho. Sor. Tor. Bor.** (Kromkarn Khong Song Thahaan), Army Transportation Dept., Bang-Krabüe Rd, Dusit Region, Bangkok 10300. (1).

22) **Mor. Tor. Bor. Nung** (Monthol Thahaan Bok Thii Nung), HQ of the 1st Army Region, at the approach of Kesa Komon, Amnuay Songkram Rd, Dusit Region, Bangkok 10300 2100-1900. (no FM-tr's).

23) **Mor. Tor. Bor. Jed** (Monthol Thahaan Bok Thii Jed), The 7th Army Area HQ, Surasak Montree Camp, Phichai, Muang District, Lampang 52000. (no FM-tr's).

24) **Siang Adison,** The Voice of Adison Camp, Cavalry Center, Adison Camp, Muang District, Saraburi 18000. (no FM-tr's).

25) Siam Computer AD, 60/809 Ladprao Rd, Soi 53, Bangkok 10230. **FM:** Phuket 95.0, Roi Et 95.5, Udon Thani 105.75MHz.

26) **Wor. Phol. Sii** (Withayu Kong Phol Thii Sii) HQ of the 4th Infantry Division, Somdej Phranarasuan Maharat, Muang District, Phitsanulok 65000. (no FM-tr's).

27) **Wor. Sor. Sor.** (Withayu Suksa), Center for Educational Innovation and Technology, Dept. of Non-Formal Education, Min. of Education, Rajchadamnern Nok Ave, Dusit Region, Bangkok 10300. (1).

28) **Mor. Chor.** (Mahavitthayalai Chiang Mai) Mass Communication Branch, Faculty of Humanities, Chiang Mai University, the approach of Doi Suthep, Huay Krew Rd, Suthep, Muang District, Chiang Mai 50000. (1).

29) **Mor. Kho.** (Mahavitthayalai Khon Kaen), Khon Kaen University, Naimuang, Muang District, Khon Kaen 40000. (1).

30) **Mor. Sor.** (Mahavitthayalai Songkhla Nakarin), Songkhla Nakarin University, Had Yai Campus, Khohong, Hat Yai District, Songkhla 90110. (2).

31) **Mor. Tor. Bor. Song** (Monthol Thahaanbock Thii Song), The 2nd Army Region, Chak Krapong Camp, Dong Phra Ram, Muang District, Prachinburi 25000. (1).

32) **Mor. Tor. Bor. Sii** (Monthol Thahaanbock Thii Sii), HQ of the 4th Army Region, Jiraprawat Camp, Muang District, Nakhon Sawan 60000. (no FM-tr's).

33) **Or. Sor.** (Amporn Sathan), Amporn Sathan Throne, Dusit Palace (Chitra Lada Rahothan Palace), Rajchavithee Rd, Chitra Lada, Dusit Region, Bangkok 10300 Tues Sat 0330-0500 & 0900-1200, Sun 0230-0500 (no bc Mon) — **V.** by QSL-Folder. (1).

34) **Thai TV Si Chong Saam** (Thai TV Color Ch. 3 Radio St.), HS-TV3, Lak Song, Nong Khem District, Phetcha Kasem Rd, (Km No.18), Bangkok 10160. (1).

35) **TV Si Chong Jed** (TV7 Color Radio St.), 998/1 Soi Ruam Sirimitra,

Phaholyothin Rd, Saam Sen Nai, Phraya Thai Region, Bangkok 10400. ☎ 2781255. **Cable:** Bebetevee. (1).

36) **Withayu Chula** (Chulalongkorn University Radio St.), Audio-visual Center, Secretariat, Chulalongkorn University, Angreedonang Rd, Nang Mai, Pratumwan Region, Bangkok 10500. **Web:** http://www.chula.ac.th/radio/ — **FM:** 101.5MHz 5kW (1). 2300-1700.

37) **Wor. Or. Sor.** (Withayu Asia Seri), Voice of Free Asia (see below).

38) **Wor. Tor. Or.** (Withayalai Technology Lae Archeva Suska), Vocational and Technology College, Technological Engineering Faculty, Sii Sao Thewes Campus, Sam Sen Rd, Sii Sao Thewes, Dusit Region, Bangkok 10300. (1).

39) **Tor. Tor. Tor.** (Mass Media Organization of Thailand), 222, Mot Building, 6th Floor, Asok-Dindaeng Rd, Huayfang Region, Bangkok 10310. **N.** (R. Thailand): 2300-2330, 0530-0545, 1200-1230, 1300-1330, 1400-1410; **Army N.** 2345-2400, **Navy N.** 1245-1300, **Police N.** 1030-1100. (98).

40) **Por. Tor. Or.** (Kongphol Thahaan Punyai Torsue Akart Yan), Anti-Aircraft Artillery Division, Kiak Kay Junction, Thahaan Rd, Bangsüe, Dusit Region, Bangkok 10300. (1)

41) **Chor. Tor. Ror.** (Changwat Thahaanbock Lopburi), HQ of Lopburi Army Province, Chak Chupson, Muang District, Lopburi 15000. 43) **Tor. Chor. Dor.** (Tamruat Trawen Chaidan), HQ of Border Patrol Police Division, Phaholyothin Rd, Saam Sen Nai, Phraya Thai Region, Bangkok 10400. (5).

44) **Manager Radio,** 2 Trok Rongmai Chao-Fah Road, Bangkok 10200. **E-mail:** kumpee@mozart.inet.co.th **Web:** http://goldsite.com/Radio/Manager/. **FM:** 97.5 MHz. **English** (rel. BBC World Sce.): 1700-2230.

45) A Time Media/Radio Concept, CMIC Tower B 20-21 fl, 209/1 Rd, Soi 21, Bangkok 10110. **FM:** 88.0/91.5/93.5/1`04.5MHz.

46) Arun Radio Media, 99/137-8 Tessabansongkrot Rd, Ladyao, Bangkok 10900. **FM:** so. Thailand (no details available).

47) Ben Advertising, Silom-Surawongse Condo. 15 fl, 43/326 Soi Anumanrajdhon, Surawongse Rd, Bangrak, Bangkok 10500. **FM:** 101.5MHz.

48) Boom Radio, Bangkok. **FM:** 90.0MHz.

49) Fatima Broadcasting International, 91/7 Soi Areesampan 1, Phayathi Rd, Bangkok 10400. **FM:** 96.5MHz.

50) K.C.S. Advertising, 40/1-3 Sukhumvit Rd, Soi 3, Bangkok 10110. **FM:** 89.5MHz.

51) Media Plus, Monterey Tower 18 fl, 2170 Petchburi Rd, Bangkok 1031. **FM:** 96.0/98.0/99.5/103.0/107.0MHz.

52) Music Network & Siam Arts Entertainment, 110/2 Rama Vi rd, Phaythai, bangkok 10400. **FM:** 99.0MHz.

53) PN Promotion, 125/87 Ladprao Rd, Soi 87, Bangkapi, Bankok 10310. **FM:** 98.5MHz.

54) Pirate Radio, 573 Ramkamhang Rd, Soi 39, Bangkok 10310. **FM:** 89.0MHz.

55) R.S. Promotion 1992, Chetchotisak Bldg, 419/1 Ladprao Rd, Soi 15, Jatujak, Bangkok 10900. **FM:** 88.5/93.0/94.0/106.0MHz.

56) Ratana-Urai, 2229/23 Ramkamhang Rd, Bangkok 10240. **FM:** 103.5MHz.

57) Siam Broadcasting, Bangkok. **FM:** 94.5MHz.

58) Studio 107 Thailand, Jayanama Bldg, 61/223-225 Rama IX Rd, Bangkok 10310. **FM:** 97.0MHz.

59) Universal United, 296 St. Louis 3, Sathorn Rd, Yannawa, Bankok 10120. **FM:** 100.5MHz.

60) Watch Dog, Imperial Queen's Park Hotel, Room 6710672, 119 Sukhumvit Rd, Soi 22, Bangkok 10110. **FM:** 101.0MHz.

61) U & I Corporation, Vanit 2 Bldg, 1 fl, 1126 New Petchburi Rd, Bangkok 10400. **FM:** 95.5/105.0/105.5MHz.

N.B: All st's are required to relay **N.** from R. Thailand at 0000, 0530, 1200 & 1300 daily.

Reports: Geographical locations given refer to provinces ("Changwat"). St's are not always located in the provincial capitals but in a local district ("Amphur"). However, a report to a provincial station is known to reach its destination even if the "amphur" is unknown.

VOICE OF FREE ASIA (Wor. Or. Sor.)

✉ P.O. Box 2-131, Rajdamnoen, Bangkok 10200.
L.P: Dir: Anucha Osathanond.
STATION: Bangkok 1575kHz 1000kW (tr. operated by VOA).
D.PRGR: Mon-Fri: 1030-1130, 1500-1530 & 2230-2400. (1030-1100 Thai, 1100-1130 Malay, 1500-1530 **English,** 2230-2300 Vietnamese, 2300-2330 Cambodian, 2330-2400 Laotian).
INT-SIG: Thai National Anthem — **ANN:** "This is the Voice of Free Asia" — **V.** by QSL-card and letter.

VOICE OF AMERICA RELAY STATION
BBC RELAY STATION
RADIO FRANCE INTERNATIONALE RELAY STATION
See International Broadcasting section.

TURKMENISTAN

L.T: UTC + 5h — **Pop:** 4.574.000 — **Radios:** 850.000 — **Pr.L:** Turkmen (official), Russian — **E.C:** 50Hz, 127/220V — **ITU:** TKM.

ARAGATNAŞIK MINISTRLIK
(Ministry of Communications)
Jitmikov köçesi 36, 744000 Aşgabat. ☎ +7 (3632) 252153. 🖷 +7 (3632) 290420. ① 228111.
L.P: Minister: Amanmurad D. Cummiyew.

TÜRKMENISTANYŇ MILLI TELERADIOKOMPANIYASY (TMT, Gov.)
(National TV and Radio Co. of Turkmenistan)
Mollanepes köçesi 3, 744000 Aşgabat. ☎ +7 (3632) 251515.
L.P: Head of company: Isa Orazmedov. First Deputy Head of company: Abdyllah S. Yakubov. Dir. Tech: J.M. Pashayew.
Tech. Dept: "TMT-Tolkun", Gaudanskoye shosse, 2km, 744013 Aşgabat-13. ☎ +7 (3632) 290557, 290554 – **L.P:** Dir. Tech: Zaur Alekperow.

TURKMEN RADIO (Gov.)
Mahtumkuli köçesi 89, 744000 Aşgabat. ☎ +7 (3632) 251515. 🖷 +7 (3632) 251421 (manual). ① 228125 RADIO.

Stations	kHz	kW	H. of tr.	Prgr.
Aşgabat	153	500		
Aşgabat	279	150	0100-0000	TMR1
Aşgabat	576	150		
Aşgabat	675	150		
Ek-Arça (1)	720	1	0100-0000	TMR1
Gyzyletrek (1)	720	1	0100-0000	TMR1
Çärýew (2)	765	30		
Çärýew (2)	927	50	0000-2300	TMR1, Reg
Türkmenbaşi (1)	927	5		
Bekdaş (1)	927	1	0100-2000	RO4
Guşgy (3)	945	20		
Türkmenba.şi (1)	1080	7	0100-0000	TMR1
Guşgy (3)	1080	5	0100-0000	TMR1
Aşgabat	1125	1000	0100-1230	RO4
Syrtagta (3)	1233	40	0100-0000	TMR1
Bekdaş (1)	1287		(new)	
Aşgabat	4930	50	0050-2030	RSM, TMR2
Aşgabat	5015	100	0100-0000	TMR1

(1) Balkan province. (2) = Leban province. (3) Mary province.

FM (MHz)	TMR1	RSM*	kW
Arlan		72.20	17
Aşgabat		69.68	17
Bayramly		70.27	17
Çärýew	66.85	68.77	17
Kalininsk		67.22	17
Nebitdag		72.02	0.5
Tecent		72.14	17

*) RSM tr's may be inactice or carry another service. Further details not available.
TMR1: 0100-0000 in Turkmen, Russian — **TMR2:** 0300-0900, 1200-1900 in Turkmen, Russian, other times rel. RSM — **RO4** = R. Odin Orbita 4 (cf. Russia) — **RSM** = R. St. Mayak (cf. Russia).
Regional Prgr's (R): Çärýew: 0500-0530 & 1500-1600 on 927kHz + FM (also r. on 279kHz).

FOREIGN SERVICE: see International Broadcasting section.

UZBEKISTAN

L.T: UTC + 5h — **Pop:** 23.206.000 — **Radios:** 3.677.000 — **Pr.L:** Uzbek (official), Russian — **E.C:** 50Hz, 127/220V — **ITU:** UZB.

VAZIRLIGI ALOKA (Ministry of Communications)
Aleksey Tolstoy küçä 1, 700000 Toşkent. ☎ +7 (3712) 336645. 🖷 +7 (3712) 442603. ① 116108 PTB.
L.P: Minister: Komlcon R. Rakhimov.

TELEVISION AND RADIO COMPANY OF UZBEKISTAN (Gov.)
Khorazm küçä 49, 700047 Toşkent.
L.P: Chmn: Erkin K. Khaitboev.

UZBEK RADIO (Gov.)
Khorazm küçä 49, 700047 Toşkent. ☎ +7 (3712) 441210. 🖷 +7 (3712) 440021. ① 116062 EFIR.

Stations	kHz	kW	H. of tr.	Prgr.
Toşkent (11)	162	150	0000-2100	UZR1
Nukus (12)	549	5		
Vobkent (2)	576	7		
Toşkent (11)	576	50		
Samarkand (8)	648		0000-2100	UZR1/Reg
Toşkent (11)	666	30	0000-2300	UZR2
Samarkand (8)	666	1		
Koson (6)	675	20		
Kün girod (12)	675	5	0000-2200	RO4
Zarafşon (2)	675		0000-2000	UZR1
Buhoro (2)	711	25	0000-2300	UZR2
Urgonç (4)	711		0000-2300	UZR2
Piskent (11)	756	50	0000-2200	RO4
Samarkand (8)	756	5	0000-2300	RO4
Andijon (1)	1062	50	0000-2300	UZR2
Nukus (12)	1062		0000-2300	UZR2
Toşkent (11)	1062		0000-2100	UZR1/Reg
Nukus (12)	1260	100	0000-2100	UZR1/Reg
Namangan (7)	1269	5	0000-2200	RO4
Zarafşon (2)	1269	50	0000-2200	RO4
Koson (6)	1323		0000-2300	UZR2
Dangara (3)	1323	5		
Muborak (7)	1332		0000-2100	UZR1
Jizzah (5)	1485	1	0000-2200	RO4
Nurobod (8)	1485	1	0000-2300	RO4
Muynak (12)	1485	1		
Samarkand (8)	1539		0000-2300	UZR2/Reg
Muynak (12)	1584	1	0000-2200	RO4
Jar-Kür gon (9)	**1593		0000-2300	UZR2
Far gona (3)	4510		0930-1000	Reg
			1200-1230	Reg
Toşkent (11)	4850	50	0000-2300	UZR2
Toşkent (11)	5995	100	0000-2100	UZR1
Toşkent (11)	9540	100	2100-2300	UZR2
Toşkent (11)	9545	100	1300-1700	UZR2
			2100-2300	UZR2
Toşkent (11)	15165	100	0500-1230	UZR2
Toşkent (11)	+15200	100	0430-1400	UZR2

**) r. v1583-1585kHz. +) Wi 15330kHz. SW freq's subject to seasonal variations and breaks for FS prgr's — A major network rearrangement has been reported, but no details currently available.
Regions: (1) Andijon (2) Buhoro, (3) Far gona (4) Khorazm, (5) Jizzah, (6) Kaşkadaryo, (7) Namangan, (8) Samarkand, (9) Sur gondarya, (10) Sirdaryo, (11) Toşkent (Tashkent); (12) Rep. of Korakalpoqhiston (Karakalpakstan).

FM (MHz)	UZR1	UZR2	MR1	RSM
Toşkent	67.19	67.97	66.41	69.23

ERP 17kW. Full details of other stations not available.
UZR1 in Uzbek, Russian
UZR2 (R. Mash'al): 2300-1900
UZR3 (R. Dustlik): in Uzbek, Russian, Kazakh, Tajik, Crimean Tatar.
RO4 = R. Odin Orbita 4 relay (cf. Russia).
RSM = R. St. Mayak relay (cf. Russia). Re. inactive.

Regional stations

Sovet küçä 26, 710000 **Andijon.**
Buhoro: On 70.70MHz 1300-1330 in Uzbek, Russian, Tajik
Kalinin küçä 28, 712000 **Far gona**: On 4510kHz in Uzbek, Russian, Tajik, Tatar

Karakalpak R, **Nukus**: On 1260kHz in Karakalpak, Uzbek, Russian 0215-0300, 0700-0800 (Sun), 1230-1300 (Sun), 1330-1400, 1415-1445 (Sun 1500)
Samarkand: On 648/1539kHz, 67.34MHz 1300-1400 in Russian.
Toşkent: Provincial & city prgr's on 1062kHz, 67.26MHz: 0000-2000.

FOREIGN SERVICE: see International Broadcasting section.

VIETNAM

L.T: UTC + 7h — **Pop:** 77.805.880 — **Radios:** 7m — **Pr.L:** Vietnamese, ethnic — **E.C:** 50Hz, 120/127/230V — **ITU:** VTN.

DAI TIENG NÓI VIÊT NAM
(RADIO THE VOICE OF VIETNAM) (Gov.)
✉ 58 Quan Su Str, Hanoi. ☎ +84 (4) 254953. 📠 +84 (4) 255765.
Web: http://www.ioit.ac.vn/tieng_noi_vn/tnvn.html
LP: DG: Phan Quang. Head of Int. Rel: Dinh The Loc. Officer in charge of Technical Section: Mrs. Nguyen Thi Le Quan.

Mediumwaves	kHz	N	Mediumwaves	kHz	N
Mê Tri 1	549	2	Nghê An	783	1+P
Lam Dong	550	P	7) Son La	820	P
Quan Tre	558	2	Tien Giang	820	P
4) Ha Giang	570	P	Quang Binh	846	2+P
Mê Tri 1	570	2	Than Hoa	850	P
Dong Dé	572	2	6) Lai Cau	861	P
Khanh Hoa	580	P	Vin Phu	899	1
Tay Ninh	580	P	Ha Tinh	900	P
Yên Bai	580	P	Can Tho	900	2
Nhahtrang	580	P	Lang Son	914	P
Da Nangh	600	P	Cao Bang	918	1
Quang Nam	600	P	Ben Tre	930	P
Gia Lai	630	P	1) Hanoi	945	2
Dong Hoi	630	1	Vinh Long	950	P
An Nhon	648	1	8) Cao Bang	970	P
Ha Bac	650	P	Kiên Giang	970	P
3) Ho Chi Minh C	655	R	Song Be	970	P
Quan Tre	657	1	12) Kon Tum	980	P
Dong De	666	2	Quang tri	999	2+P
1) Hanoi	675	1	Hanoi	1008	FS
Dac Lac	693	1+P	Tra Vinh	1053	P
An Hai	694	1	An Giang	1070	P
Quang Ninh	700	P	Dac Lac	1089	2
An Hai	702	2	Binh Thuan	1100	P
Dong Thap	710	P	Minh Hai	1120	P
Dong Nai	720	P	Ha Tay	1180	P
Thua Thien Hué	720	P	Hai Hung	1195	P
An Nhon	738	1	Soc Trang	1200	P
Quan Tre	747	2	Hanoi	1242	FS
14) Long An	760	P	Thai Binh	1250	P
Binh Thuan	765	1	Nam Ha	1280	P
Quang Ngai	774	1+P	Ba Ria Vung tau	1500	P
13) Can Tho	780	P			

STATIONS: C=City — **Networks:** 1=National Network 1, 2=National Network 2, P=Provincial, FS=Foreign Language Sce.

Shortwaves: Freqs vary widely. Latest r. freq. in brackets.

Station	kHz	Times
6) Lai Cau 1	4215	1050-1215
Lam Dong	4675	1130-1230
5) Lao Cai	4677	1000-1500
7) R. Son La 1	4739.5	2200-2300, 1200-1400
Tuyen Quang	4740	1200-1350
11) R. TV Gia Lai	4722.5	2200-2300, 1030-1400
Dac Lac	4800	
12) R. TV Kontum	4800	1045-1155
Than Hoa	4880	r. inactive
1) Nat. Netw. 2	4960	2200-1600
7) R. Son La	4965	0300-0400, 1200-1400
Binh Thuan	5000	
R. Yên Bai	5000	
1) Dia Tieng Noi	5030	0500-0600, 1200-1230
(Hmong Sce.)		1300-1345, 2200-2230
Quang Tri BS	5050	1000-1200
Nghê An	5200	
5) Lao Cai	5597	1000-1500
1) Nat. Netw. 1	5925	2200-1600 (Fri 1700)
1) Hmong Sce	6165	0500-0600, 1200-1230
		1300-1345, 2200-2230
7) R. Son La 2	6300 (r.6329v)	2200-2300, 1200-1400
6) R. Lai Cau 2	6395	0400-0600, 1200-1330
8) R. Cao Bang	6530	r.0325, 1200-1400
10) Yên Bai	6541.8	1200-
5) R. Lao Cai	6702	1000-1500
9) RTV Bac Thai	7153	0300-0500, ?-1430
4) Ha Giang BS	7300	2330-2400, 1200-1400
1) Nat. Netw. 1	10060	2300-1600 (Fri 1700)
1) Nat. Netw. 2	12035	2300-1600

FM: Hanoi 100MHz — Phu Yen 96MHz — Ninh Binh 97.6MHz — Ho Chi Minh C. 99.9/104.5MHz — Ninh Thuan 100MHz — Dac Lac 103MHz.

Prgrs. from Hanoi

National Netw. 1: 2200-1600 (Fri 1700). **N:** 2200, 0100, 0200, 0400, 0800, 1300, 1545 — **National Netw. 2:** 2200-1600. **N:** 2200, 1545.
Hmong Sce on SW: 2200-2230, 0500-0530, 1200-1230, 1300-1345. **N:** 2200, 2300 (rel. Netw.1), 0500, 1100 (rel. Netw. 1), 1200, 1300.
ANN: "Da la tieng noi Viet Nam, phat thanh tu Hanoi, thu do nuoc cong hoa xa hoi chu ghia Viet Nam".
Ethnic Language Prgrs:
Ede on 693kHz : 2200-2230, 0430-0500, 1130-1200.
Bana on 693kHz: 2230-2300, 0400-0430, 1200-1230.
So. Khmer on 747kHz: 2200-2245, 0400-0445, 1300-1345.
Giarai on 1089kHz: 2200-2230, 0430-0500, 1130-1200.
News & Music Prgr. on FM: 24h. on 100MHz (Hanoi) & 104.5MHz (Ho Chi Minh C.). **N. in English:** 0200, 0800, 1400. **N. in French:** 0130, 0730, 1330. **N. in Russian:** 0330, 0930, 1530.
Prgrs. for foreigners in Hanoi on 1242kHz: **English:** 2330, 1000, 1230, 1330. **French:** 1300. **Russian:** 1130. **Spanish:** 1100. **Japanese:** 1200, 1400. **Cantonese:** 2230. **Thai:** 2300, 1500. **Khmer:** 0000. **Lao:** 0030. **Indonesian:** 0930, 1030, 1430.

Provincial Stations

These st's broadcast in Vietnamese and the main ethnic languages of their region: 3) 7 Xo Viet Nge Tinh Str, Ho Chi Minh C. — 4) Ha Giang, Ha Tuyen Province — 5) Lao Cai, Lao Cai Province — 6) Lai Cau Province — 7) Son La Province — 8) Cao Bang Province — 9) Tainquyen, Bac Tai Province — 10) Yên Bai Province — 11) Playcu, Gia Lai Province — 12) Kontum Province — 13) Can Tho, Hua Giang Province. — 14) Tanan, Long An Province.

FOREIGN SERVICE: see International Broadcasting section

PACIFIC

L.T: Victoria, New South Wales, Queensland, Tasmania: UTC + 10h (Vic, NSW, Tas: Oct-March UTC + 11h); South Australia: UTC + 9½h (Oct-March UTC + 10½h); Western Australia: UTC + 8h; Northern Territory: UTC + 9½h — **Pop:** 18.230.000 — **Radios:** 21m — **Pr.L:** English — **E.C:** 50Hz, 240V — **ITU:** AUS.

AUSTRALIAN BROADCASTING CORP. (ABC)

HQ: Ultimo Centre, 700 Harris Str, Ultimo, NSW 2007.(✉ GPO Box 9994, Sydney NSW 2001) ☎ +61 (02) 9333 1500. 📠 +61 (02) 9333 5305.
L.P: Steve Ford, Project Officer, National Communications Unit (Radio), **E-mail:** ford.steve@a2.abc.net.au).

MW: N = R. National, R = Regional R, M = Metropolitan Sce, P = Parliamentary & News Netw. **Call letters:** 2 = NSW (exc. Canberra = A.C.T.), 3 = Victoria, 4 = Queensland, 5 = So. Australia, 6 = We. Australia, 7=Tasmania, 8=Northern Territory.

Call	kHz	kW	Netw.	Location
1) 6DL	531	10	R	Dalwallinu
2) 4QL	540	10	R	Longreach
3) 2CR	549	50	R	Orange
4) 6WA	558	50	R	Wagin (Minding)
5) 4JK	567	10d	R	Julia Creek
1) 6PU	567	0.1	R	Paraburdoo
1) 6TP	567	0.1	R	Tom Price
1) 6MN	567	0.1	R	Newman
1) 6PN	567	0.1	R	Pannawonica
6) 2RN	576	50	N	Sydney
7) 6PB	585	10	P	Perth
7) 7RN	585	10	N	Hobart
8) 3WV	594	50	R	Horsham (Dooen)
6) 2RN	603	10d	N	Nowra
2) 4CH	603	10d	R	Charleville
10) 6PH	603	2	R	Port Hedland
11) 4QR	612	50	M	Brisbane
7) 6RN	612	10	N	Dalwallinu
12) 3RN	621	50	N	Melbourne
6) 2PB	630	10	P	Sydney
5) 4QN	630	50	R	Townsville (Brandon)
13) 6AL	630	5	R	Albany
9) 7RN	630	0.4	N	Queenstown
14) 4MS	639	1	R	Mossman
15) 5CK	639	10	R	Port Pirie (Crystal Brook)
16) 8RN	639	2	N	Katherine
17) 2NU	648	10	R	Manilla
18) 6GF	648	2	R	Kalgoorlie
3) 2BY	657	10d	R	Byrock
16) 8RN	657	2	N	Darwin
19) 2CN	666	5	M	Canberra ACT
20) 2CO	675	10	R	Corowa
1) 6BE	675	5	R	Broome
21) 2KP	684	10	R	Smithtown
22) 6BS	684	4	R	Busselton
16) 8RN	684	1	N	Tennant Creek
15) 5SY	693	2d	R	Streaky Bay
6) 2BL	702	50	M	Sydney
7) 6KP	702	10	R	Karratha
23) 4QW	711	10d	R	St.George
24) 7NT	711	10d	R	Launceston (Kelso)
25) 2ML	720	0.4	R	Murwillumbah
17) 2RN	720	0.05	N	Armidale
26) 3MT	720	2d	R	Omeo
14) 4AT	720	4	R	Atherton (Yungaburra)
7) 6WF	720	50	M	Perth
27) 5RN	729	50	N	Adelaide
9) 7PB	729	2	P	Hobart
25) 2NR	738	50	R	Lawrence
7) 6MJ	738	5d	R	Manjimup
23) 4QS	747	10	R	Toowoomba (Dalby)
16) 8JB	747	0.2	R	Jabiru
21) 2TR	756	5d	R	Taree
12) 3RN	756	10d	N	Wangaratta
12) 3LO	774	50	M	Melbourne

Call	kHz	kW	Netw.	Location
28) 8AL	783	2	R	Alice Springs
11) 4RN	792	25	N	Brisbane
14) 4QY	801	2	R	Cairns
29) 2BA	810	10	R	Bega
7) 6RN	810	10	N	Perth
17) 2GL	819	10	R	Glen Innes
1) 6KW	819	5	R	Kununurra
26) 3GI	828	10	R	Sale (Longford)
7) 6GN	828	10	R	Geraldton
30) 4RK	837	10	R	Rockhampton (Gracemore)
18) 6ED	837	0.1	R	Esperance
19) 2RN	846	10	N	Canberra
1) 6CA	846	2.5	R	Carnarvon
31) 4QB	855	10d	R	Pialba
31) 4QO	855	10	R	Eidsvold
7) 6DB	873	2	R	Derby
27) 5AN	891	50	M	Adelaide
11) 4PB	936	10	P	Brisbane
9) 7ZR	936	10d	M	Hobart
27) 5PB	972	2	P	Adelaide
12) 3RN	990	0.25	N	Albury-Wodonga
16) 8GO	990	0.5	R	Gove (Nhulunbuy)
32) 2NB	999	2d	R	Broken Hill
1) 6WH	1017	0.1	R	Wyndham
12) 3PB	1026	5	P	Melbourne
33) 2UH	1044	2d	R	Muswellbrook
14) 4WP	1044	0.5	R	Weipa
7) 6BR	1044	1	R	Bridgetown
14) 4TI	1062	2	R	Thursday Island
35) 5RN	1062	2.5	N	Renmark
6) 2RN	1098	0.2	N	Goulburn
34) 5PA	1161	10d	R	Naracoorte
24) 7FG	1161	1d	R	Fingal
7) 6XM	1188	2	R	Exmouth
7) 6NM	1215	0.5	R	Northam
6) 2NC	1233	10	M	Newcastle
4) 6RN	1296	10	N	Wagin
35) 5MV	1305	2	R	Renmark
3) 2LG	1395	0.2	N	Lithgow
6) 2RN	1431	2	N	Wollongong
6) 2PB	1440	2	P	Canberra
6) 2PB	1458	2	P	Newcastle
6) 2RN	1485	0.1	N	Wilcannia
5) 4HU	1485	0.05	R	Hughenden
15) 5LN	1485	0.2	R	Port Lincoln
32) 2RN	1512	10	N	Newcastle
8) 3RN	1548	5	N	Horsham (from Sep.97)
36) 4QD	1548	50	R	Emerald
31) 4GM	1566	0.2	R	Gympie
32) 2WA	1584	0.1	R	Wilcannia
34) 5MG	1584	0.2	R	Mt.Gambier
15) 5WM	1584	0.05	R	Woomera
24) 7SH	1584	0.05	R	St. Helens
29) 2CP	1602	0.05	R	Cooma
8) 3WL	1602	0.25	R	Warrnambool
15) 5LC	1602	0.1	R	Leigh Creek South

FM stations (Transmitters of greater than 1kW)
Networks: N=Radio National, R/M=Regional or Metropolitan Network, FM=Fine Music Network, JJJ=Triple J Network (alternative).

	Area	State	N	R/M	FM	JJJ
27)	Adelaide	SA			103.9	105.5
27)	Adel. Foothills	SA			97.5	95.9
28)	Alice Springs	NT	99.7		97.9	94.9
17)	Armidale	NSW		101.9	103.5	101.1
26)	Bairnsdale	Vic	106.3			
12)	Ballarat	Vic		107.9	105.5	107.1
6)	Batemans Bay	NSW	105.1	103.5	101.9	
29)	Bega/Cooma	NSW	100.9		99.3	100.1
12)	Bendigo	Vic		91.1	92.7	90.3
6)	Bombala	NSW		94.1		
6)	Bourke	NSW	101.1			
11)	Brisbane	Qld			106.1	107.7
32)	Broken Hill	NSW	102.9		103.7	102.1
1)	Broome	WA	102.7			
7)	Bunbury	WA	(F.PI)		93.3	94.1(F.PI)
14)	Cairns	Qld	105.1	106.7	105.9	107.5
14)	Cairns North	Qld	93.9	95.5	94.7	97.1

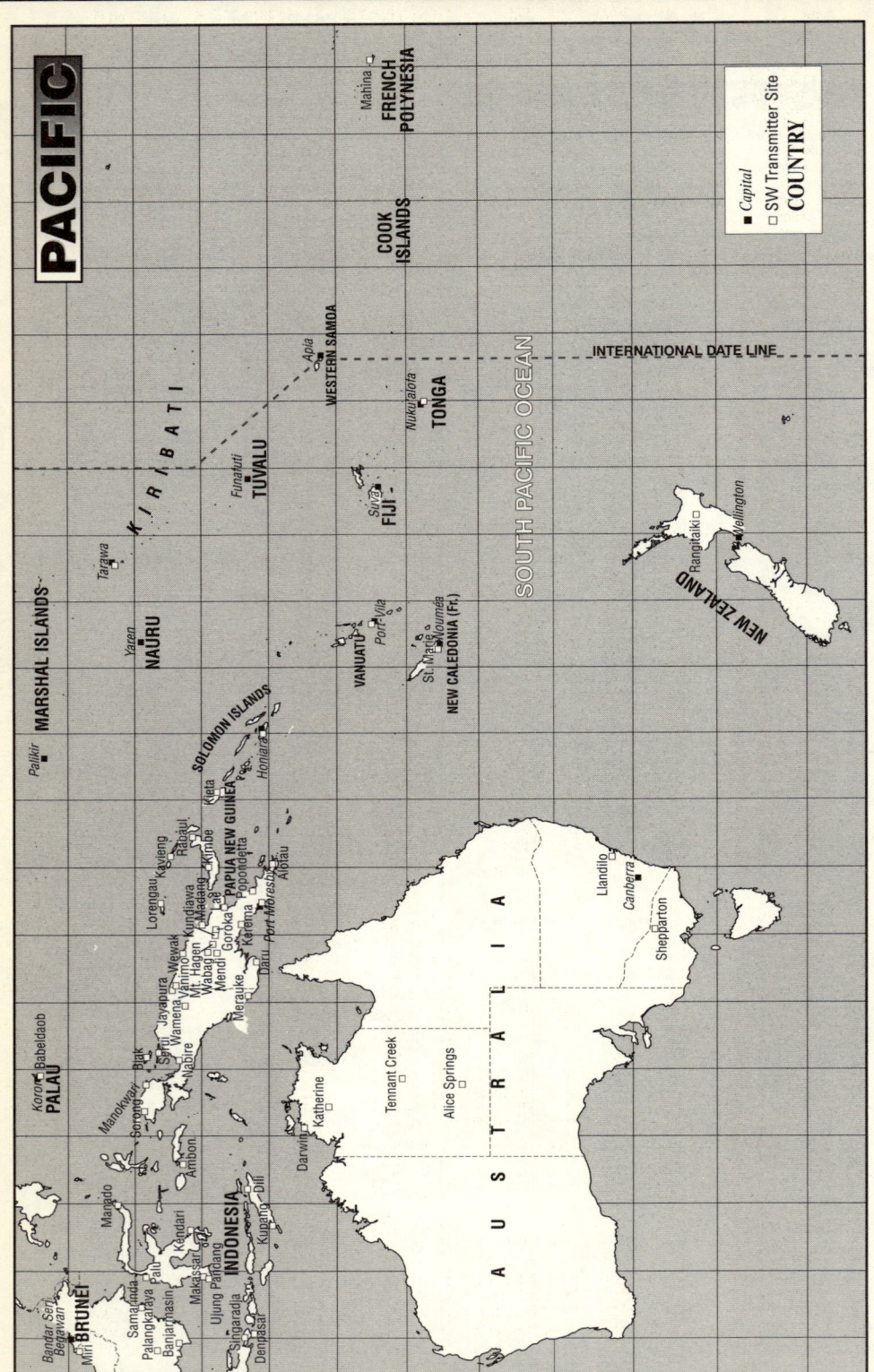

PACIFIC

Mahina □
FRENCH POLYNESIA

COOK ISLANDS

Legend:
- ■ *Capital*
- □ *SW Transmitter Site*
- **COUNTRY**

INTERNATIONAL DATE LINE

Apia ■
WESTERN SAMOA

Nukuʻalofa ■
TONGA

K I R I B A T I

Funafuti ■
TUVALU

Suva ■
FIJI

SOUTH PACIFIC OCEAN

Rangitaiki □
Wellington ■

NEW ZEALAND

Tarawa □

Yaren ■
NAURU

MARSHAL ISLANDS

Port Vila ■
VANUATU

St. Marie
Nouméa ■
NEW CALEDONIA (Fr.)

Palikir ■

SOLOMON ISLANDS

Kieta □
Honiara ■

Lorengau Kavieng
Rabaul
Kundiawa Kimbe
Vanimo
Mt. Hagen Lae
Wabag Kerema Popondetta
Goroka
Mendi
Wamena Madang
Jayapura
PAPUA NEW GUINEA
Daru Port Moresby
Merauke
Aitape

Sorong Biak Serui
Nabire
Manokwari

Koror ■ Babeldaob
PALAU

Llandilo
Canberra ■
Shepparton

A U S T R A L I A

Tennant Creek

Alice Springs

Darwin

Katherine

Manado

Samarinda
Palangkaraya Palu
Banjarmasin
Makassar Kendari
Ambon

Ujung Pandang
Singaradja Dili
Denpasar Kupang
INDONESIA

Bandar Seri Begawan ■
Miri **BRUNEI**

	Area	State	N	R/M	FM	JJJ
19)	Canberra	ACT			102.3	101.5
7)	Cen.Agricultural	WA			98.9	98.1
35)	Central East	SA			105.1	101.9(F.PI)
3)	Cen.Tablelands	NSW	104.3		102.7	101.9
6)	Cen. Wst. Slopes	NSW	107.9	107.1	105.5	102.3
6)	Coffs Harbour	NSW	99.5	92.3	97.9	91.5
23)	Darling Downs	Qld	105.7		107.3	104.1
16)	Darwin	NT		105.7	107.3	103.3
6)	Deniliquin	NSW	99.3			
36)	Emerald	Qld	93.9			
18)	Esperance	WA	106.3		104.7	
7)	Geraldton	WA	99.7		94.9	98.9
17)	Glen Innes	NSW	105.1			
11)	Gold Coast	Qld	90.1	91.7	88.5	97.7
6)	Goulburn Valley	Vic		97.7	96.1	94.5
31)	Gympie	Qld	96.9	95.3	93.7	
6)	Hay	NSW	88.9	88.1		
9)	Hobart	Tas			93.9	92.9
6)	Illawara	NSW		97.3	95.7	98.9
6)	Jerilderie	NSW	94.1			
18)	Kalgoorlie	WA	97.1		95.5	98.7
16)	Katherine	NT		106.1		
34)	Keith	SA	96.9			
9)	King Island	Tas		88.5		
26)	Latrobe Valley	Vic		100.7	101.5	96.7
24)	Launceston	Tas	94.1		93.3	102.1
9)	Lileah	Tas	89.7	91.3		
2)	Mackay	Qld	102.7	101.1	97.9	99.5
21)	Manning River	NSW	97.1	95.5	98.7	96.3
11)	Meandarra	Qld	104.3			
12)	Melbourne	Vic			105.9	107.5
12)	Mildura	Vic	105.9	104.3	102.7	101.1
11)	Monto	Qld	101.9		103.5(F.PI)	
11)	Moranbah	Qld	106.5	104.9		
34)	Mount Gambier	SA	103.3		104.1	102.5
12)	Murray Valley	Vic		102.1	103.7	105.3
6)	Murrumbidgee	NSW	98.9	100.5	97.3	96.5(F.PI)
	Irrigation Area					
11)	Nambour	Qld		90.3	88.7	89.5
33)	Newcastle	NSW			106.1	102.1
12)	Nhill	Vic	98.9			
27)	Oodnadatta	SA		95.3		
7)	Perth	WA			97.7	99.3
10)	Port Hedland	WA	95.7			
12)	Portland	Vic	98.5	96.9	88.1(F.PI)	
6)	Richmond	NSW	96.9	94.5	95.3	96.1
	and Tweed					
30)	Rockhampton	Qld	103.1		106.3	104.7
13)	S. Agricultural	WA	96.9		94.5	92.9
11)	South'n Downs	Qld	106.5	104.9	101.7	103.3
15)	Spencer Gulf N	SA	106.7		104.3	103.5(F.PI)
24)	St. Helens	Tas		96.1		
15)	Streaky Bay	SA	100.9			
6)	SW Slopes &	NSW	89.1	89.9	88.3	90.7(F.PI)
	East Riverina					
6)	Sydney	NSW			92.9	105.7
17)	Tamworth	NSW	93.9		103.1	94.7
16)	Tennant Creek	NT		106.1		
5)	Townsville	Qld	104.7		101.5	105.5
15)	Tumby Bay	SA	101.9			
6)	Upper Hunter	NSW		105.7		
12)	Upper Murray	NSW		106.5	104.1	103.3
6)	Upper Namoi	NSW	100.7	99.1	96.7	99.3
6)	Wagga Wagga	NSW	104.3	102.7	105.9	101.1(F.PI)
8)	Warrnambool	Vic	101.7		92.1(F.PI)	
8)	Western Vic.	Vic		90.9	92.5	94.9(F.PI)
11)	Wide Bay	Qld	100.9	100.1	98.5	99.3
11)	Winton	Qld	107.9			
27)	Wirrulla	SA	107.3			
27)	Wudinna	SA	107.7			

ABC regional addresses:
1) PO Box 211 Geraldton WA – 2) PO Box 318 Longreach Qld 4730 – 3) PO Box 863 Orange NSW 2600 – 4) PO Box 242 Bunbury WA 6230 – 5) PO Box 694 Townsville Qld 4810 – 6) PO Box 487 Sydney NSW 2001 – 7) GPO Box 190D Perth WA 6001 – 8) PO Box 506 Horsham Vic 3400 – 9) GPO Box 9994, Hobart Tas 7001 – 10) PO Box 387 Port Hedland WA 6721 – 11) GPO Box 293 Brisbane Qld 4001 – 12) GPR Box 1686 Melbourne Vic 3001 – 13) PO Box 489 Albany WA 6330 –

14) PO Box 932 Cairns Qld 4870 – 15) PO Box 289 Port Pirie SA 5540 – 16) PO Box 633 Darwin NT 0800 – 17) PO Box 558 Tamworth NSW 2340 – 18) PO Box 125 Kalgoorlie WA 6430 – 19) GPO Box 365 Canberra ACT 2601 – 20) PO Box 321 Albury NSW 2640 – 21) PO Box 76 West Kempsey NSW 2440 – 22) PO Box 242 Bunbury WA 6230 – 23) PO Box 358 Toowoomba Qld 4350 – 24) PO Box 201 Launceston Tas 7250 – 25) PO Box 435 Grafton NSW 2460 – 26) PO Box 330 Sale Vic 3850 – 27) GPO Box 1419H Adelaide SA 5001 – 28) PO Box 1144 Alice Springs NT 0871 – 29) PO Box 336 Bega NSW 2550 – 30) PO Box 911 Rockhampton Qld 4700 – 31) PO Box 376 Maryborough Qld 4650 – 32) PO Box 315 Broken Hill NSW 2880 – 33) 47 Newcomen St, Newcastle NSW 2300 – 34) PO Box 448 Mount Gambier SA 5290 – 35) PO Box 20 Renmark SA 5341 – 36) Telecom Radio Centre, Emerald Qld 4720.

RADIO AUSTRALIA (Overseas Sce.)

See International Broadcasting section.

NORTHERN TERRITORY SHORTWAVE SERVICE

⌨ Box 9994, Darwin, NT 0801 (**CAAMA Radio:** P.O. Box 2924, Alice Springs, NT 0871) — all tr's 50kW.
VL8A Alice Springs: 2310kHz (0830-2130), 4835kHz (2130-0830), 3230kHz (alt. freq) — **VL8T Tennant Creek:** 2325kHz (0830-2130), 4910kHz (2130-0830), 3315kHz (alt. freq) — **VL8K Katherine:** 2485kHz (0830-2130), 5025kHz (2130-0830), 3370kHz (alt.freq). Programming has been known to run over designated times.
D. PRGR. in English & Aboriginal languages: 24h. Prgrs produced by the Central Australian Aboriginal Media Association (CAAMA): VL8A/VL8T: MF 2300-0230, 0340-0730, 0940-1300. Prgrs produced by Top End Aboriginal Bush Broadc. Assoc. (TEABBA): VL8K: MF 2045-2330, 0730-0830. Rel. of ABC Alice Springs.

PRIVATELY-OWNED BROADCASTING STATIONS

FEDERATION OF AUSTRALIAN RADIO BROADCASTERS

⌨ P.O. Box 299, St. Leonards, NSW 2065.☎ +61 (2) 9906 5944. 🖹 +61 (2) 9906 5128. ☼ 25161. **Cable:** FARB, Sydney.
L.P: Federal Dir: Anthony M. King.
Abbreviations: N-1: News on the hour. N-2: News on the half hour. N-3: News on the hour and half hour. The numeral preceding the call letters indicates the state: 2=New South Wales, 3=Victoria, 4=Queensland, 5=South Australia, 6=Western Australia; 7=Tasmania, 8=Northern Territory.
News: Additional newscasts are often carried during breakfast and drive times — t=translator (relays main station).

	Call	kHz	kW		Call	kHz	kW
1)	2MC	531	5(d)	19)	6AM	864	2
2)	3GG	531	5(d)	21)	2GB	873	5
3)	4KZ	531	5(d)	22)	3YB	882	2(d)
4)	7SD	540	5(d)	24)	4BH	882	5(d)
5)	4AM	558	5(d)	23)	6PR	882	2
6)	4GY	558	5(d)	34)	4RR	891	5(d)
7)	7BU	558	2(d)	25)	2LM	900	5(d)
8)	2BH	567	0.5	107)	2LT	900	5(d)
133)	2CS	639	5(d)	168)	6BY	900	2
33)	4CC(t)	666	2.5	27)	7AD	900	2
103)	4LM	666	2	28)	8HA	900	2
164)	6LN	666	1	29)	2XL	918	2
105)	3EE	693	5(d)	153)	4VL	918	2
9)	4KQ	693	5(d)	31)	6NA	918	2
3)	4KZ(t)	693	0.5	32)	3UZ	927	5
103)	4LM(t)	693	0.5	33)	4CC	927	5(d)
164)	6TZ	693	2	69)	4HI(t)	945	1
131)	6SE	747	5	36)	2UE	954	5
10)	2EC	765	3.5(d)	17)	4CA(t)	954	0.5
73)	4GC(t)	765	0.5	38)	2RG	963	5
134)	5CC	765	5(d)	37)	4WK	963	5(d)
88)	6KA(t)	765	0.1	93)	5SE	963	5(d)
147)	8HOT(t)	765	0.5	86)	2DU(t)	972	0.3
11)	4TO	774	5(d)	39)	2MW	972	5
153)	4VL(t)	783	2 (F.PI)	112)	2NM	981	5(d)
13)	6VA	783	2	41)	3HA	981	2
157)	2RF	801	5	42)	6KG	981	2
14)	5RM	801	2	43)	4RO	990	5(d)
73)	4GC	828	1	45)	2ST	999	5(d)
16)	7XS	837	0.5(d)	46)	4TAB	1008	5(d)
17)	4CA	846	5(d)	48)	7TAB	1008	5(d)
18)	4GR	864	2	49)	2KY	1017	5

	Call	kHz	kW			Call	kHz	kW
52)	6NW	1026	2		90)	3AW	1278	20(d)
53)	5CS	1044	2		91)	2TM	1287	2
54)	2CA	1053	5(d)		162)	2ZW	1314	5(d)
55)	3CV	1071	5		94)	3BA	1314	5(d)
56)	4SB	1071	2		40)	5DN	1323	2(d)
31)	6WB	1071	2		98)	3SH	1332	2
57)	2MO	1080	2		99)	4BU	1332	5(d)
48)	7TAB	1080	5(d)		162)	2HH	1341	5
59)	2GZ	1089	10(d)		101)	3GL	1341	
60)	3WM	1089	5(d)		102)	2LF	1350	5(d)
61)	4LG	1098	2		37)	4WK(t)	1359	0.3
62)	6MD	1098	2		104)	2GN	1368	2
63)	7LA	1098	5(d)		105)	3MP	1377	5(d)
167)	3BM	1116	2		106)	5AA	1395	5(d)
65)	4BC	1116	5(d)		108)	2PK	1404	2
135)	6MM	1116	2		156)	3XY	1422	5
113)	5MU	1125	5(d)		5)	4AM(t)	1422	2
66)	2AD	1134	2(d)		111)	2MG	1449	5(d)
67)	3CS	1134	5(d)		114)	3MA	1467	2.5
31)	6CI	1134	2		115)	4ZR	1476	2
68)	2HD	1143	2		116)	2AY	1494	2
69)	4HI	1143	5(d)		117)	2BS	1503	5(d)
70)	2WG	1152	2		90)	3AK	1503	20(d)
71)	4MB	1161	2		119)	2QN	1521	2
72)	2CH	1170	5		120)	2VM	1530	2
75)	2NZ	1188	2		157)	2RF	1539	1
80)	2CC	1206	5(d)		106)	5TAB	1539	5
78)	2GF	1206	5(d)		121)	2RE	1557	2
161)	6TAB	1206	2		122)	3NE	1566	5(d)
45)	2ST(t)	1215	0.35		157)	2RF	1575	0.5
69)	4HI(t)	1215	0.1		10)	2EC(t)	1584	0.5
82)	3TR	1242	5(d)		33)	4CC(t)	1584	0.5
85)	4AK	1242	2		153)	4VL(t)	1584	0.2
84)	5AU	1242	2(d)		157)	3RG	1593	5(d)
163)	8TAB	1242	2		165)	8ADR	1611	(F.PI)
86)	2DU	1251	2		162)	2AM	1620	0.4
87)	3SR	1260	2		165)	4ADR	1620	0.4
88)	6KA	1260	1		165)	6ADR	1620	(F.PI)
89)	2SM	1269	5		165)	5ADR	1638	(F.PI)

FM-Stations (1kW and higher)

	FM Call	MHz	kW			FMCall	MHz	kW
154)	4SEA	90.9	16		81)	2UUS	101.7	20(d)
130)	5SSA (t)	91.1	1		153)	4VL(t)	101.7	5 (F.PI)
29)	2XL(t)	92.1	1		20)	7HHO	101.7	23
155)	4GLD	92.5	16		126)	3FOX	101.9	35
86)	2DU	92.7	(F.PI)		122)	3NNN	102.1	25
15)	4SSS	92.7	5		1)	2ROX	102.3	5
91)	2TTT	92.9	10		40)	5AD	102.3	15(d)
94)	3BA	92.9	10		119)	2MOR	102.5	5
44)	6PPM	92.9	20		45)	2ST(t)	102.7	40
70)	2WZD	93.1	50		109)	2KKO	102.9	10
35)	3BBO	93.5	60		94)	3BBA	103.1	5(d)
101)	3BAY	93.9	35		34)	4RAM	103.1	65(d)
99)	4RUM	93.9	2		17)	4HOT	103.5	5(d)
79)	6JKY	94.5	40(d)		86)	2DU	103.7	(F.PI)
45)	2WSK	94.9	25		163)	8TAB	103.7	1
101)	3CAT	95.5	55		128)	2DAY	104.1	35
12)	2ONE	96.1	5		74)	3KKZ	104.3	10
125)	6MMM	96.1	20		127)	4MMM	104.5	6(d)
29)	2XL(t)	96.3	2		138)	2ROC	104.7	10
95)	2UUL	96.5	5(d)		77)	5MMM	104.7	5(d)
153)	4VL(t)	96.5	1 (F.PI)		147)	8HOT	104.7	5(F.PI)
142)	6GGG	96.5	5		116)	2AAY	104.9	50
40)	5AD(t)	96.7			129)	2MMM	104.9	35
118)	3SUN	96.9	50		1)	2ROX	105.1	10
5)	4AM(t)	97.7	3		124)	3MMM	105.1	35
112)	2VLY	98.1	10(d)		139)	2NEW	105.3	20
123)	2WIN	98.1	25(d)		92)	4BBB	105.3	6
142)	6BAY	98.1	5		115)	4ZR(t)	105.3	0.5
120)	2NOW	98.3	10		10)	2EC(t)	105.9	10
29)	2XL(t)	98.7	1		137)	1CBR	106.3	10
114)	(3MA)	99.5	10		64)	2WFM	106.5	20
147)	8HOT	100.1	5		45)	2ST(t)	106.7	1
121)	2RE	100.3	2		100)	2XXX	106.9	10
51)	4MKY	100.3	50		141)	4QFM	106.9	16(d)
25)	2ZZZ	100.9	20(d)		130)	5SSA	107.1	7(d)
144)	7TTT	100.9	18		97)	2GGO	107.7	10(d)
50)	3TTT	101.1	35		51)	4MKY(t)	107.9	5
140)	2CFM	101.3	10(d)		34)	4RAM(t)	107.9	1(d)

Addresses and other information (ARN:Australian Radio Network)

NB: The term midnight-to-dawn refers to local time. Exact hours vary from st. to st.

1) Midcoast B'casters Pty Ltd, P.O. Box 1611, Port Macquarie NSW 2444 – 24h (N. and overnight, rel. Sky Radio). Supplementary st. on 102.3 and 105.1. **Web:** http://www.midcoast.com.au/users/2mc/

2) Votraint no. 691 Pty Ltd, P.O. Box 531, Traralgon, Vic. 3844. 24h (N-2) Rel. Arnsat overnight

3) Coastal B'casters Pty Ltd, P.O. Box 19, Innisfail, Qld. 4860 **E-mail:** zedamfm@kz.com.au – 24h (N-1: Sky Radio). Rel. 2UE 9:00am to noon and SKY3 from 6:00pm to 6:00am. Translators: Tully 693kHz 0.5kW, Dunk Island 88.5MHz 0.5kW.

4) North East Tasm. R. B'casters Pty. Ltd, P.O. Box 262, Devenport, Tas. 7310 – 24h (N-1).Part of Tasmanian Broadcasting Network

5) Your Station 4AM, P.O. Box 177, Mareeba, Qld 4880 – 24h. (N-1:Sky Radio) Translators: Port Douglas 1422 kHz, Weipa 97.7 MHz.

6) Gympie Noosa B'casters, P.O. Box 187, Noosa, Qld 4567. 24h (N-1). ID's as "Radio 56".

7) Burnie Broadc. Sce. Pty. Ltd, P.O. Box 120, Burnie, Tas. 7320 – 24h (N-1: Sky Radio)

8) Far West Radio Pty. Ltd, P.O. Box 699, Broken Hill, NSW 2880 – – 24h (N-3.) Supplementary st. on 106.9 – 24h

9) Radio 4KQ, P.O. Box 693, Stones Corner, Qld 4120 – 24h (N-1)

10) Ea. Coast Radio 2EC, P.O. Box 471, Bega, NSW 2550 – 24h. Translators: 1584=Narooma, 105.9Mhz = Batemans Bay.

11) Transmedia Holdings Pty Ltd, P.O. Box 986, Townsville, Qld 4810. **E-mail:** fourto@ultra.net.au **Web:** http://www.ozemail.com.au/~aschter – 24h (N-1:Sky Radio). Rel 4CA 846, 4AM 558, 4LM 666 for football telecasts

12) Hayden Nepean Broadc. Pty. Ltd, P.O. Box 145, Penrith, NSW 2750 – 24h (N-1:Sky Radio)

13) Belcap Investments Pty Ltd, P.O. Box 293, Albany, WA 6330 – 24h. (N-1)

14) Radio 5RM, 19 Vaughan Tce, Berri SA 5343 **E-mail:** fiverm@riverland.net. au. **Web:** http://www.riverland.net.au/~fiverm/ – 24h (N-1:Macquarie). Rel Sky Radio 6:00pm to 6:00am. **F.PI:** FM license

15) Sunshine Coast B'casters Ltd, P.O. Box 828, Nambour, Qld 4560 – 24h (N-1)

16) Tasmanian Broadcasting Network Pty Ltd, P.O. Box 315, Queenstown, Tas 7467 – 24h (N-3:Sky Radio). Translators at Strahan 105.1 MHz 25w & Rosebery 107.1 MHz 0.3kW

17) Greater Cairns Radio, P.O. Box 846 (for 4CA)/1035 (for 4HOT), Cairns, Qld 4870. **E-mail:** 4ca@iig.com.au and hotfm@iig. com.au – 24h (N-1:Sky Radio)

18) Gold Radio Sces Pty. Ltd, P.O. Box 111, Toowoomba, Qld 4350 – 24h (N-1)

19) Radio 6AM Northam Pty. Ltd, P.O. Box 265, Northam, WA 6401 – 24h

20) Comm. B'casters Pty. Ltd, GPO Box 542F, Hobart, Tas 7001 – 24h (N-3)

21) Harbour Radio. Ltd. GPO Box 4290, Sydney 2001 – 24h (N-3)

22) Regional Communications Pty Ltd, P.O. Box 485, Warrnambool, Vic. 3280 – 24h

23) Western Broadc. Sces Pty. Ltd, P.O. Box 6072, Hay Str. East, Perth, W.A. 6000 – 24h (N-1)

24) Old Radio Ltd, GPO Box 882, Brisbane, Qld 4001 – 24h (N-1:Macquarie)

25) Richmond River B'casters Pty. Ltd, P.O. Box 44, Lismore, NSW 2480 – 24h. (N-1:Sky Radio). 2LM 900 relays Radio 97: 7:00pm to 6:00am

26) Austereo Ltd., P.O. Box 33, Tuart Hill, WA 6060. 2200-1000 (N-3). Rel. 6VA 1000 (Sun 2200)-1600. 6IX– 24h (N-1). 6WB: Ellrale Pty Ltd, P.O. Box 148, Bunbury, WA 6230. **F.PI:** 6GL move to FM stereo, with new call sign

27) Northern Tas B'casters Pty. Ltd., P.O. Box 262, Devonport, Tas 7310 – 24h

28) Alice Springs Comm. B'casters Pty. Ltd, P.O. Box 2106, Alice Springs 0871 – 24h (N-1). Rel. sky R. translator at Yularaon 100.5MHz with 100w. Supplementary st. 8SUN on 96.9 MHz with 300w at Alice Springs – 24h

29) Cooma B'casters Pty. Ltd, P.O. Box 651, Cooma, NSW 2630 – 24h (N-1:Sky Radio) Relays 2UE 9:00am to 10:00am and AUSTEREO 4:00pm to 6:00pm. Translators: Thredbo 92.1 MHz 1kW, Jindabyne 96.3 MHz 2kW and Perisher 98.7 MHz 1kW

31) Radio West, PO Box 148 Bunbury WA 6230. Relays Radio West Network Program overnight, usually includes local ID's on the hour/half hour for the individual stations

32) 3UZ Pty Ltd, P.O. Box 927, Carlton, Vic 3053 – 24h (N-1) ID's as "Sport 927"

33) Regional Broadcasters Australia Pty. Ltd, P.O. Box 420, Gladstone, Qld 4680 – 24h. (N-1). Translator at Rockhampton on 1584 with 500w and at Biloela on 666KHz with 2.5 Kw

34) 4RAM/4RR, P.O. Box 986, Townsville, Qld 4810 (FM E-mail: hotfm@ultra.net.au) – 24h (N-1:Sky Radio). 4RR: Racing format, programming 8:00am-midnight, also relays 4TAB 1008. 4RAM: Translator at Mt Stuart 107.9 MHz 1kW, ID's as "103.1 Hot FM"

35) P.O. Box 108, Golden Square, Vic. 3555 – 24h (N-1)

36) R. 2UE Sydney Pty. Ltd. P.O. Box 950, North Sydney, NSW 2059 – 24h (N-3)

37) Amalgamated Marketing Pty. Ltd, P.O. Box 403, Toowoomba, Qld 4350 – 24h (N-1:Sky Radio) Rel 2TM 1287 7:00pm to 6:00am. Translator: Toowoomba 1359 kHz 0.3 kW

38) Radio 2RG Pty Ltd, P.O. Box 1005, Griffith, NSW 2680 – 24h (N-1:Sky Radio). Rel 2UE 9:00am to 10:00am M-F and Sky Radio 7:00pm to 5:30am.

39) Tweed R & Broadc. Co. Pty. Ltd, P.O. Box 97, Coolangatta, Qld 4225 – 24h (N-1). Ids as "Radio 97"

40) Southern State Broadcasters Pty Ltd, PO Box 5 Nth Adelaide SA 5006. Web: http://www.5dn.com.au/ – 24h (N-1:Macquarie)

41) Western District (3HA) B. Casting, P.O. Box 981, Hamilton, Vic 3300 – 24h (N-1:Southern Cross) Rel ACE Network 9:00am to 12:30pm and 1:00pm to 6:00am

42) Radio Station 6KG, P.O. Box 440, Kalgoorlie, WA 6430 – 24h (N-1)

43) Rockhampton Broadc. Co. Pty. Ltd, P.O. Box 159, Rockhampton, Qld 4700 – 24h (N-1)

44) Consolidated Broadc. Syst (WA) Pty. Ltd, P.O. Box 157, Subiaco, WA 6008 – 24h (N-1)

45) South Coast & Tablelands Broadc. Pty. Ltd, P.O. Box 540, Nowra 2540 – 24h (N-1). Translators: Bowrall 1215kHz, Uladulla 106.7MHz .Supplementary St. on 94.9 MHz. (N-1)

46) Radio 4TAB, P.O. Box 275, Albion, Qld 4010. Operating 6:00am to midnight (N-3: 6:00am to 8:00am M-F from 4KQ). Racing format

47) Great Northern B'casters Ltd, P.O. Box 141, Geraldton, WA 6530. W 2200-1500, Sun 2300-1400 (N-1). F.PI: – 24h

48) Tas Radio Pty. Ltd, G.P.O. Box 572F, Hobart, Tas 7001 – 24h on 1008 kHz & 1080 kHz, 11:00am to 11:00pm M-Sa on 87.6 MHz 1W narrowcast throughout Queenstown, Strahan, Zeehan, Roseberry, Tullah, Stanley& Smithton (N-1:Sky Radio). Racing format. Rel 2UE 9:00am to noon M-F

49) 2KY B'casters Pty. Ltd, 20 Wentworth St, Parramatta, NSW 2150 – 24h (N-3) Provides relays to over 100 NSW stations carrying racing: 9:00am-11:00pm M-F, 5:00am-11:00pm Sa & 5:00am to noon Su

50) Double T Radio P/L, Private Bag 1011, Richmond Vic. 3121 – 24h (N-1) Relays programming on 3GG 531 & 2AY 1494

51) Barrier Reef B'casting Pty Ltd, P.O. Box 183, Mackay, Qld 4740 – 24h (N-1). Airlie Beach on 94.7MHz. Bowen on 107.9MHz

52) Northwest Radio Pty. Ltd, 11 Court Place, South Hedland, WA 6722 – 24h. Rel. st. 44 1600-2130, rel. st. 88 0100-0200 & 0400-1600

53) Coast and Country Broadcasting Services Pty Ltd, P.O. Box 481, Pt. Pirie, SA 5540 – 24h (N-1:Sky Radio/Macquarie)

54) Austereo Ltd, G.P.O. Box 163, Canberra City, ACT 2601. Web: http://www.2ca.village.com.au/ – 24h. (N-3).

55) Radio 3CV Pty Ltd,PO Box 108 Goldern Sq. Bendigo Vict. 3555. 24h (N-1:Sky Radio)

56) Forsby Pty Ltd P.O. Box 305, Kingaroy, Qld 4610 – 24h (N-1: Australian Radio Network) ID's as "1071AM" and "Classic Gold" Rel ARNSAT 10:00pm to 6:00am

57) 2MO Gunnedah Pty. Ltd, P.O. Box 62, Gunnedah 2380

59) Radio 2GZ Pty Ltd, P.O. Box 1785, Orange, NSW 2800 – 24h. (N-1:Sky Radio) ID's as "Radio 1089" Rel Sky Radio 10:00pm to 5:30am

60) Associated Comm. Enterprises (3WM) Pty. Ltd, P.O. Box 606, Horsham, Vic 3400 – 24h. (N-1)

61) Glowweir Pty Ltd, P.O. Box 20, Longreach, Qld 4730

62) Mid-Districts Radio Pty. Ltd, P.O. Box 264, Merredin, WA 6415. (N-1)

63) 7LA Radio Pty. Ltd, P.O. Box 635G, Launceston, Tas 7250 – 24h (N-3)

64) Commonwealth Broadc. Corp. Pty. Ltd, P.O. Box 1107, Neutral Bay NSW 2089 – 24h (N-1:ARN). ID's as "Mix 106.5 FM"

65) Radio 4BC Pty Ltd, P.O. Box 10116, Adelaide St, Brisbane, Qld 4001 – 24h (N-1:Sky Radio)

66) New England B'casters Pty. Ltd, P.O. Box 270, Armidale, NSW 2350. E-mail: 2AD@mpx.com.au – 24h (N-1: Sky Radio) Rel Sky Radio 7:00pm to 5:00am

67) Ace Radio Broadcasters Pty Ltd, P.O. Box 63, Colac, Vic. 3250 – 24h (N-1). Rel. R. Active syndication midnight-to-dawn prgr

68) 2HD B'casters Pty. Ltd, P.O. Box 19, Hunter Region Mail Exchange, NSW 2310 – 24h (N-3)

69) Queensland Regional B'casting Pty. Ltd, P.O. Box 267, Emerald, Qld 4720 – 24h. (N-1:Sky Radio). Translators: 945 kHz 1kW, 1215 kHz 0.1 kW, 88.1 MHz 30W, 92.5MHz 10W, 98.2 MHz 0.1 kW, 102.1 MHz

0.25 kW. Rel 4AM 558 kHz, 4ZR 1476 kHz, 4CC 927 kHz 1:00pm to 2:00pm M-F and other times

70) Riverina B'casters (holdings) Pty. Ltd, P.O. Box 480, Wagga Wagga, NSW 2650. (N-1). Translator at Tumut on 107.9 MHz with 10w. Supplementary St. on 93.1 MHz. Both stns – 24h

71) Maryborough Broadc. Co. Pty. Ltd, P.O. Box 143, Maryborough, Qld 4650 – 24h. (N-1)

72) Radio 2CH Pty Ltd. P.O. Box 1170, Nth Sydney, NSW 2059 – 24h (N-1:Macquarie). F.PI: increase power

73) Radio 4GC Gold City Radio, P.O. Box 381, Charters Towers, Qld 4820 – 24h. Rel Sky Radio 9:00am to noon. Translator: Hughenden 765 kHz 0.5kW

74) GOLD FM, Private Bag 1043, Richmond Vic. 3121 – 24h ID's as "Gold FM"

75) Northern B'casters Pty. Ltd, P.O. Box 770, Inverell, NSW 2360 – 24h. (N-3)

76) Gold Coast Radio Broadc. Co. Pty. Ltd, P.M.B. 925, Gold Coast Mail Center Qld. 4217 – 24h (N-1)

77) Adelaide FM Radio Pty Ltd, GPO Box 1047, Unley, SA 5061 – 24h (N-1: AUSTEREO). Translator in Adelaide city on 98.3 MHz 0.5kW

78) Westlawn Investment Co. Pty. Ltd, P.O. Box 276, Grafton, NSW 2460 – 24h. (N-1:Sky Radio). Rel 2UE 9:00am to noon M-F

79) 6JKY Pty Ltd, P.O. Box 945, Subiaco, WA 6008 – 24h (N-1: BBC)

80) Capital City B'casters Pty. Ltd, P.O. Box 1499, Canberra, ACT 2601 – 24h (N-1). Rel. Arnsat at various times

81) Wesgo Communications Pty. Ltd, P.O. Box 234, Seven Hills, NSW 2147 – 24h (N-1:ARN) F.PI: Macquarie news & ARNSAT network pgm

82) Southern Cross Radio, P.O. Box 1242, Traralgon, Vic. 3844 – 24h (N-1)

84) 5AU B'casters Pty. Ltd, P.O. Box 496, Port Augusta, SA 5700 – 24h (N-1)

85) Radio 4AK Pty Ltd, P.O. Box 403, Toowoomba, Qld 4350 – 24h (N-1: Sky) Rel 2ZZZ 100.9 MHz 7:00pm to 6:00am

86) Western B'casters Pty. Ltd. P.O. Box 1221, Dubbo, NSW 2830. E-mail: 2du@lisp.com.au – 24h. (N-1:Sky Radio) Rel Radio 97 overnight. F.PI: FM station "ZOO FM" Dubbo 92.7 MHz, Cobar 103.7 MHz

87) Goulburn Valley B'casters Pty Ltd, P.O. Box 1260, Shepparton, Vic 3630 – 24h. Rel. Arnsat via satellite

88) Northwest Radio Pty. Ltd, P.O. Box 153, Karratha, WA 6714. (N-1)

89) Kick Media, P.O. Box 1270, North Sydney, NSW 2059 E-mail: contact@kick-am.com.au. Web: http://www.kick-am.com.au/ – 24h (N-1: Sky Radio) ID's as "Kick AM"

90) 3AW Southern Cross Radio Pty Ltd, GPO Box 369F, Melbourne 3001. Web: http://www.3aw.com.au/

91) Tamworth Radio Development Co, P.O. Box 497, Tamworth, NSW 2340 – 24h (N-1). Supplementray stn. on 92.9 MHz

92) B105FM Pty Ltd, P.O. Box 105, Albion, Qld 4010 – 24h (N-1: AUSTEREO) ID's as "B105"

93) Southern State B'casters Pty. Ltd, P.O. Box 500, Mt. Gambier, SA 5290 – 24h (N-1)

94) Radio Ballarat Pty. Ltd, P.O. Box 360, Ballarat, Vic 3350 – 24h. (N-1: from 3AW 1278)

95) Associated Communication Enterprises, P.O. Box 1234, Wollongong, NSW 2500. E-mail: mike@w151.aone.net.au – 24h (N-1: Sky)

97) Wesgo Communications Pty. Ltd, P.O. Box 564, Gosford, NSW 2250 – 24h (N-1:ARN)

98) Associated Communications Enterprises (3SH) Pty. Ltd., P.O. Box 504, Swan Hill, Vic 3585 – 24h (N-1:from 3AW 1278). Rel 2UE 9:00am to 10:00am M-F and ACE Radio Network from 3WM 1089 10:00am to 6:00am. F.PI: FM service in mid 1997 on 107.7 MHz

99) Bundaberg B'casters Pty. Ltd, P.O. Box 1059, Bundaberg, Qld 4670 – – 24h (N-1)

100) Radio Newcastle Pty Ltd, P.O. Box 97, Charlestown, NSW 2290 – 24h (N-1)

101) K-ROCK, P.O. Box 9550, Geelong, Vic 3220. E-mail: krock@slanreach.au – 24h (N-1). ID's as "K-Rock"

102) Young B'casters Pty. Ltd, P.O. Box 31, Young, NSW 2594 – 24h (N-1). Rel. Sky Radio overnight.

103) North Queensland Broadc. Co. Pty. Ltd, P.O. Box 780, Mount Isa, Qld 4825 (N-1). Relays to 4GC 828. Translator: Cloncurry . Supplementary FM license at Mt. Isa – 24h. (N-1)

104) Radio 2GN Pty. Ltd, P.O. Box 115, Goulburn, NSW 2580 – 24h (N-1: Sky Radio)

105) Greater West Television, P.O. Box 75, Frankston, Vic. 3199. E-mail: magic@magic.com.au – 24h (N-1:from 3AW 1278). 3EE ID's as "Magic"

106) Festival City B'casters Pty. Ltd, Pulteny Court, Adelaide SA 5000

– 24h (N-1)

107) Midwest Radio Network, Mailbag 90, Lithgow, NSW 2790. **E-mail:** 2lt@lisp.com.au – 24h. (N-1: Sky Radio). Relays 2UE 954 9:00am to 11:00am and 6:00pm to 7:00pm weekdays. (for QSL'ing purposes) c/o John Wright, 31 Chamberlain Ave, Carringbah NSW 2228.

108) Parkes Broadc. Co. Pty. Ltd, P.O. Box 295, Parkes, NSW 2870 – 24h. (N-1:Sky Radio) Rel Radio 97 7:00pm to 6:00am M-F & 2TM 1287 noon to 6:00am Sa/Su. **F.PI:** AM & FM translators

109) Radio 2KO Newcastle P/L, P.O. Box 606, Charlestown, NSW 2290 – 24h. (N-1)

111) Mudgee Broadc. Co. Pty. Ltd, P.O. Box 17, Mudgee, NSW 2850 – 24h

112) Radio Hunter Valley Pty. Ltd, 100 Bridge St., Muswelbrook, NSW 2333 – 24h (N-1:Sky Radio) 2NM 981 relays 2UE 954 9:00am to 10:00am M-F. 2VLY 98.1 ID's as "Power FM"

113) Murray Bridge Broadc. Co. Ltd, P.O. Box 470, Murray Bridge, SA 5253 – 24h (N-1:Macquarie). Serves Murray Bridge, The Coorong and Meningie. **F.PI:** FM license

114) Promote Pty Ltd, P.O. Box 539, Mildura, Vic 3500 – 24h (N-1:Sky Radio). Rel Sky Network 6:00pm to 6:00am on AM and FM. 3MA 99.5 ID's as "Today's Music 99.5FM"

115) Regional Broadcasters Australia Pty. Ltd, P.O. Box 22, Roma, Qld 4455 – 24h. (N-1:Sky Radio) Relays 4CC 927 2:00pm to 6:00am

116) Radio Albury Wodonga Ltd, P.O. Box 670, Albury, NSW 2640. **Web:** http://www.albury.net.au/radio.albury.wodonga/2ay. html – 24h (N-1). Supplementary stn. on FM

117) Bathurst B'casters Pty. Ltd, P.O. Box 310, Bathurst, NSW 2795. **E-mail:** stereo@2bs.ix.net.au or 2bs@csu. edu.au. **Web:** http://www.2bs.ix.net.au/ – 24h (N-1:Macquarie) **F.PI:** FM service on 99.3 MHz

118) P.O. Box 195, Shepparton, Vic. 3630

119) Rich Rivers Broadc. Pty. Ltd, P.O. Box 312, Deniliquin, NSW 2710 – 24h. (N-1:Sky Radio and 3AW 1278) 2MOR 102.5 ID's as "Classic Rock 102.5"

120) Moree Broadc. & Development Co. Ltd, P.O. Box 389, Moree, NSW 2400 – 24h. (N-1). Supplementary license on 98.3 MHz – 24h. Translator on 88.7 MHz with 250w – 24hr. (N-1)

121) Manning Valley Broadc. Pty. Ltd, P.O. Box 275, Taree, NSW 2430 – 24h (N-1). Translator: Gloucester 100.1MHz and Forster on 100.3 MHz

122) North East B'casters Pty. Ltd, P.O. Box 449, Wangaratta, Vic 3677 – 24h (N-1:on 3NE from 3AW 1278). 3NE Translators: Mt. Hotham 89.3MHz 0.02kW, Mt. Buffalo 105.3MHz 0.2kW, Mt. Beauty 90.3MHz 10w. 3NNN ID's as "Edge FM"

123) WIN Radio Pty Ltd, Locked Bag 6198 Sth Coast Mail Centre NSW 2521 – 24h (N-1: ARN) ID's as "98FM"

124) Melbourne F.M. Radio Pty. Ltd, GPO Box 105, Melbourne, Vic (N-1) – 24h

125) 96FM Stereo New Broadcasting Ltd, 111 Wellington Str, East Perth, WA 6004 – 24h. (N-1)

126) Austereo Ltd, P.O. Box 1019, St. Kilda, Vic. 3182 – 24h (N-1)

127) FM 104 Pty. Ltd, GPO Box 1041, Brisbane, Qld 4001 – 24h. (N-1:AUSTEREO)

128) 2DAY-FM Level 2, P.O. Box 920, Crows Nest, NSW 2065. **Web:** http://www.2dayfm.com.au/ (N-1)

129) The Triple M Broadc. Co. Pty. Ltd, GPO Box 442, Sydney, NSW 2001 – 24h (N-1). **Web:** http://www.geko.com.au/triplem/

130) SA FM, P.O. Box 1071, Unley, SA 5061.24h (N-1) Translator South Tce, Adelaide on 91.1 MHz 1kW. ID's as "SAFM"

131) Esperance B'Casters Pty. Ltd, P.O. Box 527, Esperance, WA 6450 – 24h. N-1. Rel. 6PPM-FM 1000-2200

132) Carnarvon Comm. B'casters Pty. Ltd, P.O. Box 665, Carnarvon, WA 6701. 2200-1500 (N-1). Translator: Exmouth

133) Comm. Radio Coffs Harbour Ltd, P.O. Box 1234, Coffs Harbour, NSW 2450 – 24h (N-1: Sky Radio). Rel Sky 9:00pm to 5:30am. Rp

134) Coast & Country Broadc. Sces Pty. Ltd, P.O. Box 483, Port Lincoln, SA 5606 – 24h. (N-1:local & from 5AD/5DN).

135) West Coast Radio Pty. Ltd, P.O. Box 688, Mandurah, WA 6210 – 24h (N-1). Rel. Arnsat overnight.

136) Austereo Ltd, GPO Box 163, Canberra, ACT 2601 – 24h. Belongs to 54). **F.PI:** translator for Tuggeranong area

137) Canberra FM 106.3, P.O. Box 106, Dickson, ACT 2602 – 24h. (N-1:ARN) ID's as "Mix 106.3"

138) Austereo Ltd, GPO Box 163, Canberra, A.C.T. 2601 – 24h (N-1)

139) Newcastle FM P/L – 24h (N-1). Relays Arnsat overnight

140) Radio 2CFM, PO Box 2101, Gosford, NSW 2250 – 24h. (N-1:ARN)

141) Ipswich & West Moreston B'casting Corp. Pty Ltd, P.O. Box 7, Ipswich, Qld 4305 – 24h (N-1:Macquarie) ID's as "Mix 106.9 QFM"

142) Geraldton FM Pty. Ltd, PO Box 128 Geraldton, WA 6530. 24 h (N-

1:Sky Radio)

144) So. Tasmania FM Stereo Pty. Ltd, G.P.O. Box 1800, Hobart, Tas. 7001 – 24h (N-1:local & ARNSAT)

147) P.O. Box 2510, Darwin, NT 0801 – 24h (N-1) Translators: Katherine 765 kHz 0.5 kW. **F.PI:** FM service in Darwin on 104.7 MHz 5 kW

153) Outback Radio 4VL, PO Box 84, Charleville, Qld 4470 – 24h (N-1:Sky Radio) ID's as "Outback Radio". Relays Sky Radio overnight. Translator: Cunnamulla 1584 kHz 0.2 kW. **F.PI:** 918 kHz increase power to 10kW, Cunnamulla freq change to 783 kHz 2 kW, FM service on 101.7 MHz 5kW from Charleville and 96.5 MHz 1 kW from Cunnamulla

154) 4SEA FM Pty Ltd, PO Box 5910 Gold Coast Mail Centre Bundall Qld 4217 – 24hr. (N-1)

155) Gold Coast FM Pty Ltd. Private Bag 925 Gold Coast Mail Centre Qld 4215 – 24h. (N-1).

156) Radio Hellas, 264-272 Rosslyn st West Melbourne Victoria 3003 – 24h Greek format

157) Radio Italia, (for QSL'ing purposes) c/o John Wright, 31 Chamberlain Ave, Carringbah NSW 2228 – 24h. (N-1:local and RAI Italy) Italian and sports programming. Rp

161) TAB Headquarters, 14 Hassler Rd, Osborne Park WA 6017 – 24h. Racing format. Rel racing from 2KY 1017, 3UZ 927, 4TAB 1008 and 5TAB 1539. Rel to Broome and Port Hedland on 87.6 MHz and to Dampier, Derby,Karratha & Northam on 88.0 MHz(sometimes 2KY, 3UZ, 4TAB & 5TAB.)

162) Tourist Radio PO Box 202 Caringbah NSW 2229. **E-mail:** radio@mpx.com.au – 24h. 2ZW 1314 & 2HH 1341 has relays from 2KY 1017 9:00am to 10:00pm Mo-Sa. 2AM 1620 & FM has relays from 2HH/2ZW overnight. For QSL'ing purposes: c/o John Wright 31 Chamberlain Ave, Caringbah NSW 2228. Rp. **F.PI:** 2AM 1620 freq. change, relays of Satellite World Radio Network

163) GPO Box 3170 Darwin NT 0801 – 24h. translator at Alice Springs on 959 KHz with 250w, Groote Eylandt on 103.7 MHz with 100w, Jabiru on 103.7 MHz with 25w, Katherine on 103.7 MHz with 1Kw, Nhulunbuy on 103.7 MHz with 50w, Pine Creek on 103.7 MHz with 25w and Tennant Creek on 103.7 MHz with 100w

164) Triple Six LN, PO Box 951 Carnarvon WA 6701 – 24h

165) Australian Dance Radio, PO Box 609 Nerang, Qld 4211 – 24h. Techno Music format incorporating commercial sponsorship. **F.PI:** Transmitters in Brisbane, Caloundra & Toowoomba all 400 watts on 1620 kHz. Transmitters in Adelaide on 1638, Perth on 1620 and Darwin on 1611

167) Radio 3BM, 33 Peel St, West Melbourne Vic 3003 – 24h. Italian Format

168) Radio 6BY, Gommes Lane, Yornup, PO Box 666, Bridgetown WA 6255. 24h. Relays Radio West overnight.

COMMUNITY BROADCASTING ASSOCIATION OF AUSTRALIA

Suite One, Level Three, 44-54 Botany Rd. Alexandria, NSW 2015. ☎ +61 (2) 9310 2999. 🖷 +61 (2) 9319 4545.
PRN: Public Radio Network, CBAA: Community Broadcasting Association of Australia. CBAA provides ComRadSat.

PUBLIC BROADCASTING STATIONS

MW	Call	kHz	kW	MW	Call	kHz	kW
1)	5UV	531	0.5	6)	4EB	1053	0.5
2)	2WEB	585	5(d)	47)	1RPH	1125	2(d)
68)	6WR	693	2.5	101)	6AR	1170	2
3)	3CR	855	2(d)	46)	3RPH	1179	5
55)	7RPH	864	2	48)	5RPH	1197	2
4)	6NR	927	2(d)	53)	2RPH	1224	5
56)	6RPH	990	2(d)	45)	4RPH	1296	5
5)	2XX	1008	0.3				

F.PI: Additional stations in the expanded AM band. These need not have regular two/three letter callsigns.

FM	Call	MHz	kW	FM	Call	MHz	kW
42)	3MFM	88.1	1	66)	6EBA	95.3	10
50)	4CCR	89.1	1	64)	2LVR	97.9	0.5
25)	4CRB	89.3	5	57)	2OOO	98.5	1
22)	3CCC	89.5	5	44)	3ONE	98.5	9
37)	2ARM	92.1	1	33)	6SON	98.5	10
58)	6RTR	92.1	5	60)	4AAA	98.9	6
8)	6UVS	92.1	5	36)	3RPC	99.3	2
7)	7THE	92.1	10	73)	5MBS	100.5	2
9)	2MCE	92.3	1.5	40)	4CBL	101.1	2(d)
38)	3ZZZ	92.3	10	29)	2GLA	101.5	5
10)	2NCR	92.9	3	14)	4ZZZ	102.1	10

FMCall		MHz	kW	FM	Call	MHz	kW
11)	5EBI	92.9	3	68)	6WR(t)	102.1	1
13)	5DDD	93.7	4	15)	2MBS	102.5	5
9)	2MCE(t)	94.7	1	17)	3RRR	102.7	55
16)	4DDB	102.7	2	31)	5TCB(t)	104.5	1
11)	5EBI	103.1	10(F.PI)	39)	3GCR	104.7	1.5
18)	2CBA	103.2	5	27)	7WAY	105.3	2
19)	7HFC	103.3	3	51)	2UUU	105.5	1
59)	3MBR	103.5	3(d)	67)	2NVR	105.9	0.2
12)	3MBS	103.5	10	30)	3MFM	106.1	1.5
21)	2NUR	103.7	3(d)	31)	5TCB	106.1	2
20)	4MBS	103.7	7.5	65)	7DBS	106.1	5
41)	7LTN	103.7	3	32)	3PBS	106.7	14
26)	1SSS	103.9	10	49)	2VOX	106.9	1
22)	3CCC(t)	103.9		62)	4KIG	107.1	10
61)	4TTT	103.9	1	52)	2REM	107.3	1
43)	2CHY	104.1	1	24)	2SER	107.3	9(d)
23)	8TOP	104.1	10	63)	5RAM	107.9	5(d)
34)	2BOB	104.5	1.8				

+ majority of stations below 1kW are not mentioned.

Addresses and other information

1) R. 5UV, Univ. of Adelaide, 228 North Terrace, Adelaide, SA 5000. **E-mail:** jlangdon@radio5uv.adelaide.edu.au – 24h. Education, classical music, jazz. Rel. BBCWS 12:30am to 6:00am. Member of Public Radio News Network.

2) Western Region Educational Broadc. Co. Ltd, P.O. Box 426, Bourke, NSW 2840 – 24h.

3) Community R. Federation Ltd, P.O. Box 277, Collingwood, Vic 3066 – 24h. Various foreign languages. **F.PI:** 2.5 kW(d).

4) Curtin Univ of Technology, GPO Box U1987, Perth, WA 6001. **E-mail:** nhill@c.c.curtin.edu.au – 24h. Rel. CBAA Network at times and BBCWS overnight. **F.PI:** Move to FM.

5) Community Radio 2XX Inc, P.O. Box 4, Canberra City, ACT 2600 – SM: E. O'Brien. TD: K. Vennenon – 23 foreign langs. **F.PI:** FM sce.

6) Ethnic Broadcasting Assoc. of Qld, P.O. Box 7300, East Brisbane, Qld 4169 – 24h. Various language prgrs.

7) Hobart FM Inc, P.O. Box 1324, GPO Hobart 7001 – 24h. **F.PI:** Move to 104.9 MHz.

8) Universities Radio Ltd, c/o Univ of Western Australia, Nedlands, WA 6009 – Rel. BBC-WS News. Pgrs in Italian, French and Russian.

9) Charles Sturt Univ., Mitchell Bathurst, NSW 2795 **E-mail:** 2mce@csu.edu.au – 24h. (N-1:PRN). Relays BBCWS 2:00am to 6:00am M-F. Filipino Sun 0900-1000. Translator at Orange on 94.7MHz 0.5kW.

10) Southern Cross Univ, P.O. Box 5123, East Lismore, NSW 2480. **E-mail:** fm-2ncr@scu.edu.au – 24h. PRN, BBC and Macquarie news.

11) Ethnic Broadcasters Inc, 10 Byron Place, Adelaide, S.A. 5000 – 24h. Ethnic format (46 languages). BBC news at noon. **F.PI:** Increase power to 10kW and move to 103.1MHz, satellite prgrs from Deutsche Welle.

12) Music Broadc. Society of Victoria Ltd, 146 Cotham Rd, Kew, Vic 3101 – 24h.

13) Progressive Music Broadc. Assoc, 43 Franklin Str, Adelaide, SA 5000 – 24h

14) Creative B.Casters Ltd, P.O. Box 509, Toowong, Qld 4066 – 24h.

15) Music Broadc. Society of NSW Co-op Ltd, 76 Candos Str, St. Leonards – 24h.

16) Darling Downs Broadc. Society, P.O. Box 400, Toowoomba, Qld 4350. **E-mail:** ddbfm@peg.apc.org – 24h. PRN and BBC news. Relays BBCWS midnight to 6:00am.

17) Triple R.B.Casters Ltd, P.O. Box 304, Fitzroy, Vic 3065 – 24h. **F.PI:** power increase.

18) Christian Broadc. Assoc, P.O. Box 54, Five Dock, NSW 2046 – 24h. Christian format. (N-1:Macquarie, BBC).

19) Hope Foundation Communications Inc, P.O. Box 1033, Newtown, Tas 7008. **E-mail:** hfc@trump.net.au – 24h.

20) Music Broadc. Society of Q'land, 384 Old Cleveland Rd, Coorparoo, Qld 4151 – 24h. **F.PI:** power increase.

21) Newcastle Community R, 2NUR-FM, Univ. of Newcastle, NSW 2308. **E-mail:** radio-2nur@newcastle.edu.au – 24h.

22) Radio 3CCC, 120 McCrae St, Bendigo Vic. 3550 — 23) TOP FM, Darwin NT 0909. **E-mail:** carole. miller@ntu.edu.au –24h. (N-1:Sky Radio). Relays 2UE 8:30am to 11:00am. **F.PI:** Tx at Adelaide River, Jabiru and Batchelor NT.

24) Sydney Educational Broadc. Ltd, P.O. Box 123, Broadway, NSW 2007. **E-mail:** radio.2ser-fm@2ser.uts.edu.au – 24h.

25) Gold Coast Christian & Community Broadc. Assoc, P.O. Box 86, Burleigh Heads, Qld 4220 – 24h. 5 Foreign language prgrs.

26) Canberra & District Racing & Sporting B'Casters Ltd, P.O. Box 24, Hackett, ACT 2602 – 24h. Sport & racing format. Relays 2KY

selectively 11:30am to 11:00pm and BBCWS overnight.

27) Launceston Christian B'Casters Inc, P.O. Box 1111, Launceston, Tas 7250.

29) 2GLA-FM, P.O. Box 1015, Foster NSW 2428 – 24h. **F.PI:** 2 translators

31) Tatiana Community FM B'casters Inc, P.O. Box 396, Border-town, SA 5268 – 24h.

32) Progressive Broadcasting Service Co-op Ltd, P.O. Box 210, St. Kilda Vic. 3182 – 24h.

33) Good News B'casters Inc, P.O. Box 430, Morley, WA 6062 – 24h.

34) Manning Media Co-op Society Ltd, P.O. Box 400, Taree, NSW 2430 – 24h. Various foreign lang prgrs, relays BBCWS overnight Mon-Wed and houly BBC news Sa/Su. **F.PI:** Translator and increased power.

35) Eurobodalla Access Radio Inc, P.O. Box 86, Moruya, NSW 2537 – 1900-1400. German/French. Fri 0200-0400. **F.PI:** 2kW & translator at Narooma.

36) Radio Portland Corp. Inc, P.O. Box 450, Portland, Vic. 3055 .

37) P.O. Box 707, Armidale, NSW 2350 – 24h (N-1). Foreign language prgrs.

38) Ethnic Public Broadc. Assoc. of Victoria, P.O. Box 1106, Collingwood, Vic. 3066 – 24h. Ethnic format, relays BBCWS midnight to 6:00am.

39) Gippsland Community Radio, P.O. Box 579, Morwell, Vic. 3840 – 24h. Various foreign language prgrs.

40) Radio Logan Inc, P.O. Box 2101, Logan City, Qld 4114 – 24h. (N-1) Various music types plus, Spanish Mon 8:00pm to 10:00pm. **F.PI:** FM Translator at Mt Tamborine.

41) Launceston Community FM Group Inc, P.O. Box 1501, Launceston, Tas 7250 – Foreign languages.

42) South Gippsland FM Radio Inc, P.O. Box 144, Inverloch, Vic. 3996 – 24h. (repeaters on 89.1 & 89.5MHz). **F.PI:** increase power.

43) Community Media CHY Ltd., P.O. Box J233, Coffs Harbour Jetty, NSW 2450 – 24h. (N-1:Sky Radio) relays 2KY when required. **F.PI:** increase power to 4.5 kW.

44) Goulburn Valley Community Radio Inc, P.O. Box 6824, Shepparton, Vic. 3632 – 24h. (N-1) Relays VCRN along with 3CCC 89.5 & 3BBB midnight to 6:00am and BBCWS noon to 12:30pm.

45) Queensland Radio for the Print-Handicapped Ltd, P.O. Box 146, Roma Str, Brisbane, Qld 4003 – 24h (N-3). Rel. BBCWS overnight. ID's as "Information Radio".

46) Assoc. for the Blind, 454 Glenferrie Rd, Kooyong 3144 – 24h. Relays BBCWS overnight.

47) Print-Handicapped Radio of ACT Inc, Barton Highway, Gungahlin, ACT 2912 – relays BBCWS Mon-Fri 1:00pm to 6:00pm, Sa noon to Su 9:30am.

48) Radio 5RPH, 231 Morphett str. Adelaide SA 5000 – 24h. Relays BBCWS overnight.

49) Illawarra Community FM Broadcasters. P.O. Box 1663 South Coast Mail Centre. NSW 2521. — Relays BBCWS overnight.

50) Cairns FM Broadc. Society, PO Box 300, Manunda, Cairns, Qld 4870 – 24h. 5 foreign langs, rel. BBCWS.

51) Shoalhaven Community R., P.O. Box 884, Nowra, NSW 2541.

52) P.O. Box 1079, Lavington, NSW 2641.

53) R. for the Print-Handicapped (NSW) Co-op Ltd, 2/252 Illawarra Rd, Marrickville NSW 2204 – 24h.

54) Townsville Community Broadc. Co. Ltd, P.O. Box 1033, Townsville, Qld. 4810 – 24h. 17 foreign language prgrs.

55) Radio 7RPH Broadcasting Services for Handicapped Inc. 136 Davey st Hobart Tas. 7000 – 24h. Information and reading service format. Relays BBCWS 11:00pm to 10:00am Mon-Sat, 24h Sunday.

56) Foundation for Information Radio, P.O. Box 101, Victoria Park WA 6100 – 24h. Information and print reading services. Rel selected programs from ComRadSat and relays BBCWS 11:00pm to 6:00am. **F.PI:** expansion to regional WA.

57) P.O. Box 624, Leichhardt, NSW 2040.

58) Arts Radio Ltd, P.O. Box 949, Nedlands, WA 6009 – 24h. (N-1:Pub.Rad.Net) Rel selected CBAA programs **F.PI:** increase power to 10 kW.

59) Mallee Community & Educ. B'casters Co-Op Ltd, P.O. Box 139, Murrayville, Vic. 3512 – BBC news.

60) Brisbane Indigenous Media Assoc, P.O. Box 6229, Fairfield Gardens, Qld 4103.

61) Townsville Community Broadc. Co. Ltd, P.O. Box 1033, Townsville, Qld 4810 – 24h.

62) Townsville Aboriginal & Island Mutual Assoc, MSO Box 5483, Townsville, Qld 4810 – 24h.

63) Alta Mira Fm, PO Box 1079 Kensington Rd, Kensington Park SA 5068. **E-mail:** altamira@academy.net.au – 24h. (N-1:Sky Radio) Christian based format, ID's as "Alta Mira FM" **F.PI:** increase power to 10kW.

64) Radio 2LVR FM-97.9, P.O. Box 618, Parkes Road, Forbes, NSW 2871 – Uses Sky Radio news. **F.PI:** 24h.

65) Coastal FM Inc, P.O. Box 333, Wynyard, Tas. 7325 – 24h. PRN news, Relays BBCWS Midnight to 6:30am. ID's as "Coastal FM" **F.PI:** 2 translators.

66) 6EBA-FM World Radio. 20 View St Nth Perth WA 6006 – 24h. Foreign languages.

67) 2NVR PO Box 69 Bowraville NSW 2449. 24 h. (N-1:BBC and PRN) **F.PI:** Mobile broadcasting.

68) Radio Station 6WR. PO Box 162 Kununurra WA 6743. **E-mail:** radio@perth.dialix.oz.au – 24h (N-1:CBAA) Aboriginal programs from National Indigenous Radio Service.

69) Business Community Radio, PO Box 132, East Brunswick, Vic. 3057 – Format: Spanish, Macedonian, Chinese, Vietnamese and business information in English. ID's as "Radio 1629AM" **F.PI:** power increase to 0.4 kW.

70) Radio Lebanon, PO Box 3, Thornleigh NSW 2120 – 24h (N-3) Format: Arabic pgm. **F.PI:** increased power.

71) Greek Radio, PO Box 512, Kingsgrove NSW 2208 – 24h. **F.PI:** New antenna and transmitters interstate.

72) The Access, PO Box 781, Nerang Qld 4211 – 24h Format: Dance, pop and rock music. ID's as "The Access" **F.PI:** To operate on further AM freq plus FM and SW, plans to relay other networks.

73) Good Music Round the Clock 5MBS, Reply Paid 33, PO Box 3020, Rundle Mall, Adelaide SA

SPECIAL BROADCASTING SERVICE (SBS)

✉ Locked Bag 028, Crows Nest, NSW 2065. ☎ +61 (02) 9430 2828. 📠 +61 (02) 9430 3700 – **LP:** Head of Radio: Mr Quang Luu.

	Call	kHz	kW	MHz	Location	Service.
2)	3EA	1224	5d		Melbourne	Melbourne 1. 24hrs
2)	3SBS		10	93.1	Melbourne	Melbourne 2. 24hrs
1)	2EA	1107	5		Sydney	Sydney 1. 24hrs
1)	2SBS		20	97.7	Sydney	Sydney 2. 24hrs
1)	2EA	1485	0.15		Wollongong	National Prgr. 24hrs
1)	2EA	1413	5		Newcastle	National Prgr. 24hrs
1)	4SBS		45	93.3	Brisbane	National Prgr. 24hrs
1)	5SBS(t)		1	95.1	Adelaide Hills	National Prgr. 24hrs
1)	6SBS		50	96.9	Perth	National Prgr. 24hrs
1)	8SBS		10	100.9	Darwin	National Prgr. 24hrs
1)	5SBS		20	106.3	Adelaide	National Prgr. 24hrs
1)	7SBS		35	105.7	Hobart	National Prgr. 24hrs
1)	1SBS		50	105.5	Canberra	National Prgr. 24hrs

Addresses and other information
1) Locked Bag 028, Crows Nest, NSW 2065.
2) P.O. Box 294, South Melbourne, Vic 3205.

COOK ISLANDS

L.T: UTC - 10½h (Su: UTC - 9½h) — **Pop:** 19.000 — **Radios:** 10.000 — **Pr.L:** English, Maori — **E.C:** 50Hz, 220V — **ITU:** CKH.

COOK ISLANDS BROADCASTING SERVICE

✉ P.O. Box 126, Avarua, Rarotonga. ☎ +682 29460. 📠 +682 21907. **Cable:** Cooksbroad — **LP:** C.E.O: Emile Karua.
MW: 630kHz 5kW (0.5kW reserve).
D.PRGR: 1600 (Sun 1700)-1000 (occ. 24h). **National N:** 1600W, 1645W, 2245, 0400, 0600. **N. in English** (rel. R. NZ or ABC): 0100, 0200, 0300, 0500. — **ANN:** "This is Radio Cook Islands calling".
IS: Symphony of Drums — **V.** by QSL-card. Rp.

KIA ORANA COUNTRY RADIO (Comm.)

✉ P.O. Box 521, Avarua, Rarotonga. ☎ +682 23203.
LP: Pres. & GM: David Schmidt.
FM: Penrhyn 95.3MHz 1kW, Rarotonga 103.3MHz 1kW.
D.PRGR: MF 1600-0900, Sat 1600-1000, Sun 1700-0800. **N:** on the h (rel. R. NZ or ABC) — **V.** by letter.

EASTER ISLAND (Chilean Dependency)

L.T: UTC - 6h — **Pop:** 2.800 — **Pr.L:** Spanish, Rapa Nui — E.C: 50Hz, 220V (Hotel Hangaroa: 60Hz, 220V) — **ITU:** PAQ.

RADIO MANUKENA
(operated by Chilean Air Force Volunteers)

✉ Aeropuerto Mataveri, Isla de Pascua, Chile. ☎ +5639 223271

MW: 580kHz 0.4kW — **FM:** 101.8MHz.

Chilean Navy Station: 98.5MHz FM stereo.

FIJI

L.T: UTC + 12h — **Pop:** 784.000 — **Radios:** 450.000 — **Pr.L:** English, Fijian, Hindustani — **E.C:** 50Hz, 240V — **ITU:** FJI.

FIJI BROADCASTING COMMISSION
(Independent Statutory Body)

✉ P.O. Box 334, Suva. ☎ +679 314333. 📠 +679 313606. **Cable:** BROADCOM SUVA. ② 2142 (FBC FJ SUV) Suva.
✉ P.O. Box 606, Lautoka. ☎ +679 62121. 📠 +679 65855.
✉ P.O. Box 1241, Labasa. ☎ +679 82888. 📠 +679 63450.
L.P: Chmn: Olota Tokavunisese. GM: Barry Ferber. Mgr. Eng. & Tech. Resources: Ram Deo Raj.

MW	kHz	kW	N	MW	kHz	kW	N
Suva	558	20	1	Lautoka	990	5	3
Lautoka	639	10	1	Suva	1089	1	3
Labasa	684	10	1	?	1152		1
Suva	774	20	2	Sigatoka	1206	2.5	2
Labasa	810	10	2	Rakiraki	1323	2.5	1
Lautoka	891	10	2	Rakiraki	1467	2.5	2
Sigatoka	927	2.5	1				

FM: Suva 101.2MHz (R. Fiji 1), 103.6MHz (R. Fiji 2)
Networks on FM only:

Location	RFG	B-FM	98FM	FM104	VoH
Lautoka	100.0		98.4	104.0	107.2
Nadi	89.4	91.0	88.6	94.2	91.8
Sigatoka	100.6	103.8	98.2	103.0	107.0
Suva	100.4	102.0	98.0	104.4	107.6
Tavua	94.6	91.4	89.0	93.8	92.2

1 = **R. Fiji 1** in Fijian & English, 2 = **R. Fiji 2** in Hindustani, 3 = **R. Fiji 3** in English, RFG = **R. Fiji Gold** in English, B-FM = **Bula FM** in Fijian, **98FM** in Hindi, **FM104** in English, VoH = **Voice of Hope** (Rlg.) in English. **BBC N. in English** (R. Fiji Gold): 0700.
IS: Fijian Lali (Log-Drum) — **V.** by QSL-card. Rp.

COMMUNICATIONS FIJI (Comm.)

✉ 231 Waimanu Rd, Suva (✉ Private Mail Bag, Suva). ☎ +679 314766. 📠 +679 303748. ② 2496 SUVA FM.
E-mail: commfiji@is.com.fj
✉ 101 Vitogo Parade, Lautoka. ☎ +679 64966 📠 +679 64996.
✉ P.O. Box 503, Labasa. ☎ +679 812791. 📠 +679 812177.
L.P: SM: William Parkinson. GM: John Eller. CE: N. Prasad. PD (Hindi): A. Diwarkar.

FM (MHz)	FM96	Navtarang	Viti FM
Ba	99.2	101.6	103.2
Lautoka	95.4	97.4	99.6
Sigatoka	99.0	102.2	107.8
Suva	96.0	98.8	102.8

FM 96 in English, **Navtarang** in Hindi, **Viti FM** in Fijian: all 24h.

RADIO PASIFIK FM 88.8

✉ USP Students' Association, Univ. of the So. Pacific, P.O. Box 1168, Suva. ☎ +679 313900 📠 +679 312591. **E-mail:** schuster_a@usp.ac.fj
L.P: GM: Alfred Schuster.
FM: 88.8MHz — **D.PRGR:** MF 1800-2000, 2200-0200, 0400-0800.

RADIO LIGHT FM 106 (Priv.)

✉ P.O. Box 319, Pacific Harbour, Fiji. ✉📠 +679 450007.
L.P: SM: Arnie Dykes — **FM:** Deuba 106.0Mhz 0.2kW
D.PRGR: Gospel music.

GUAM

L.T: UTC + 10h — **Pop:** 150.000 — **Radios:** 273.600 — **Pr.L:** English, Chamorro, Filipino — **E.C:** 60Hz, 110/220V — **ITU:** GUM.

	MW	kHz	kW		FM	MHz	kW
2)	KGUM	567	10	5)	KSDA-FM	91.9	3.8
1)	KUAM	612	10/1	1)	KUAM-FM	93.9	5.2
6)	KTWG	801	10	3)	KSTO	95.5	25
	FM	**MHz**	**kW**	2)	KZGZ	97.5	40
7)	KPRG	89.3	6.6	4)	KOKU	100.3	5
8)	KOLG	90.9	5.7				

Addresses and other information

1) Box 368, Agana, Guam 96910. **Web:** http://virtual.guam.net/KUAM/ — Pres. & GM: Joey Calvo. Asst. SM: James Castro. CE: Richard Garman. PD (AM): Lynda Evangelista. **E-mail:** lyndae@isla61.com — PD (FM): Kelly Muna. **E-mail:** kellykel@94jamz.com — KUAM-AM: "Isla 61" Chamorro Music st. w/Tagalog music 2h daily MF. KUAM-FM: "94 Jamz". (10kW 1400-0800, 1kW 0800-1400).

2) Sorensen Pacific Broadcasting Inc, P.O. Box GM, Agana, Guam 96932. **Web:** http://www.radiopacific.com — Chmn: Rex Sorensen. Pres (also GM/PD KGUM): Jon Anderson. VP: Kathleen Sorensen. SM/PD KZGZ: Albert Juan. KGUM: 24h news/talk in English. KZGZ: 24h CHR. **N.** on the h (rel. CBS). **ANN:** "Newstalk 57 AM", "Power 98 FM".

3) KSTereO FM 95.5, Inter-Island Communications, 4th Flr, Bank of Hawaii Bldg, Agana (or P.O. Box 20249, GMF 96921). **N:** rel. AP.

4) Hit Radio 100, 530 We. O'Brien, Agana, Guam 96910 — GM: E.Galito. PD: R.Gibson. Sales Mgr: Al Linton. 24h CHR. **N.** at :55.

5) Joy 92, 290 Chalan Palasyo, Agana Heights, Guam 96919. **Web:** http://www.tagnet.org/ksda/ — SM: John Geli. **N:** rel CNN Radio N.

6) see Trans World Radio Pacific (below).

7) Guam Educational Radio Foundation, Univ. of Guam, Bldg. 13, Dean's Circle, UOG Station, Mangilao 96923 (NPR afiiliate).

8) Light 91, Catholic Educational R, Archdiocese Agana, P.O. Box DZ, Agana 96910 — GM: Fr. David Quitugua.

ADVENTIST WORLD RADIO - ASIA (Rlg.)

See International Broadcasting section.

TRANS WORLD RADIO PACIFIC (Rlg.)

✉ P.O. Box CC, Agana,.Guam 96910-8980. ☎ +1 (671) 477 9701. 🖷 +1 (671) 477 2838. **E-mail:** ktwg@twr.hafa.net.gu **Web:** http://www.guam.net/pub/twr/ktwgguam.htm **L.P:** MD: Harry Bettig. PD: Jim Elliott. . CE: Robert Chick. Head of PR: Glenn Scheyhing.

MW: KTWG 801kHz 10kW : 2000-1300 in English exc. International Hour Mon-Fri 1100-1200 (Mon Tagalog, Tues Mandarin, Wed Korean, Thurs Chamorro, Fri Japanese).

SW: see International Broadcasting section.

IS: "We've a Story to Tell the Nations" played on an organ.

V. by QSL-card. 3 IRC's for airmail reply, one for surface mail. Rec. acc.

PUB: Frequency schedule on request.

HAWAII (U.S. State)

L.T: UTC–10h — **Pop:** 1.187.000 — **Radio sets:** 1m (est) — **Pr.L:** English, Japanese, Filipino — **E.C:** 60, mainly 115 V — **ITU:** HWA.

HAWAIIAN ASSOCIATION OF BROADCASTERS, INC.

✉ P.O. Box 22112, Honolulu HI 96823-2112 - **Pres:** Mark Haworth.

Mediumwaves: all sts comm. exc. *) non-comm.

	Call	kHz	kW	h. of tr.
1)	KMVI	550	5	24h (exc. Mon 1000-1400)
2)	KQNG	570	1	24h
3)	KSSK	590	7.5	24h
4)	KIPA	620	10	24h
4S)	KIPA-1	620	10	24h
5)	KHNR	650	10	24h
6)	KPUA	670	50	24h
7)	KQMQ	690	10	24h
8)	KUAI	720	5	Mon-Thurs 1500-1000, Fri/Sat 1500-1100, Sun 1500-0900
5)	KGU	760	10	24h
4)	KKON	790	5	24h
9)	KHVH	830	10	24h
10)	KHLO	850	5	24h
11)	KAIM	870	50	1500-1000
1)	KNUI	900	5	24h (exc. Mon 1000-1430)
12)	KJPN	940	10	M-F 1600-0600, SS 1800-0400
3)	KIKI	990	5	24h
13)	KLHT	1040	7.5	24h
14)	KAHU	1060	1	1530-0800
15)	KWAI	1080	5	24h
16)	KAOI	1110	5	24h
17)	K...	1130	10/6	F.Pl.
18)	KOHO	1170	5/4.8	silent
19)	KZOO	1210	1	1530-1000
20)	KNDI	1270	5	24h
21)	KIFO*	1380	6.2	24h
22)	KCCN	1420	5	24h
23)	KULA	1460	5	24h
24)	KUMU	1500	10	24h
15)	KISA	1540	5	24h
25)	KUAU	1570	1/0.5	1700-0300

	MHz	kW	Call		MHz	kW	Call
26)	93.5	72	KPOA	32)	99.9	51	KAYI
27)	95.9	1.13	KSRF	26)	101.1	72	KLHI-FM
28)	96.9	100	KFMN	27)	103.3	85	KAUI
29)	98.1	100	KAWV	33)	104.7	29	KONI
30)	98.9	100	KAWT	34)	105.5	6	KPMW
31)	99.5	100	KORL	35)	106.9	5.5	KWYI

Stations on FM only

Sts below 1kW not mentioned. Others: see below.

Addresses and other information:

Abbreviations: See under U.S.A. commercial sts.

Addresses: Add state abbreviation HI between location and zip code.

1) 311 Ano Str, Kahului, Maui 96732-1304 — GM: Pamelan Tsutsui. PD: Cliff Arquette. CE: Earl Tolley. KNUI: ADC, oldies, Hawaiian, KMVI: AS — **FM:** KNUI-FM 99.9MHz 100kW, KNUQ Paauilo 103.7MHz 100kW, KMVI-FM Pukalani 98.3MHz 100kW.

2) 4271 Hale Nani St, Lihue, Kauai 96766 — GM: Rodney T. Sanchez. CE: E. Nearman. Oldies, rel. FM (CHR). **FM:** KQNG-FM 93.5MHz 100kW.

3) 1505 Dillingham Blvd, Suite 208, Honolulu 96817-4905 — GM: Robert Longwell. PD: Dave Lancaster. CE: Dale Machado. KSSK: ADC/OLD, KIKI: C&W.. **FM:** KSSK-FM Waipahu 92.3MHz 100kW, KUCD Pearl City 101.9MHz 100kW.

4) 688 Kinoole Str, #112, Hilo 96720-3868 — GM: Hugh E. Gordon. PD: Victor Vierra. KIPA: Variety. KKON: OLD. **FM:** KHWI 100.3MHz 74kW, KAOE 92.7MHz 2,2kW, KPVS 95.9MHz 27kW, KLUA Kailua-Kona 93.9 32kW, KAOY Kealakekua 101.5MHz 7kW. (KONA lic. to Kealakekua.)

4S) Kalaoa, 10kW, Naalehu, 5kW, both synchronized operation with 4).

5) 560 N. Nimitz Hwy, #114B, Honolulu 96817. GM: Alan Zee. PD: John Williams. CE: Clayton Caughill. KHNR: All news, KGU: Sports.

6) 1145 Kilauea Ave, Hilo 96720-4203 — GM: Chris Leonard. CE: Glenn Ziegler. News, talk, sports. Japanese Mon-Sat 1500-1600. **FM:** KWXX 94.7MHz 100kW, KNWB 97.1MHz 40kW.

½7) 711 Kapiolani Blvd, Suite 1193, Honolulu 96813-5282 — GM: Bernie Armstrong. PD: Kimo Akane. CE: Ray Baca. CHR. **FM:** KQMQ 93.1MHz 54kW, KPOI-FM 97.5MHz 83kW, KHUL Waipahu 102.7MHz 61kW.

8) Box 720, Eleele, Kauai 96705-0720 — GM: William G. Dahle. PD: R. Deroos. CE: Michael Friedlander. ADC, Hawaiian (N-1).

9) 345 Queen Str, Suite 601, Honolulu 96813-4793 — GM: Robert Longwell. PD: Rick Hamada. CE: Jerry Varoujean. N/t. **FM:** KIKI-FM 93.9MHz 100kW. KKLV 98.5MHz 51kW.

10) 913 Kanoelehua Ave, Hilo 96720-5116 — GM: Phillip Brewer. OLD. **FM:** KKBG Hilo 97.9MHz 35kW, KLEO Kahalu'u-Kona 106.1MHz 7.3kW, KKOA Volcano 107.7MHz 18kW.

11) 3555 Harding Ave, Honolulu 96816-2491 — GM: Del Gibbs. PD: Jack Waters. CE: Ken Wooley. Rlg — **FM:** KAIM-FM 95.5MHz 99kW. (MW tr. in Molokai.)

12) 711 Kapiolani Blvd, #750, Honolulu 96813-5282 — GM: Ikuko Tomita. Japanese.

13) 1190 Nu'uanu Ave, Honolulu 96817-5122 — GM: Jim Neuman. PD: Debi Layne. CE: Ralph Wilson. Rlg.

14) Box 4727, Hilo 96720-0727 — GM: Frederick Baker, Jr. PD: Haunani Baker. CE: Jeremy Storm. Hawaiian.

15) 100 N. Beretania St, #401, Honolulu 96817 — GM: Sam Wagenvoord. PD: Barry Wagenvoord. CE: Ralph Wilson. Talk.

16) Box 38, Kahului, Maui 96732-0038 — Pres. & GM: John Detz. PD: Jack Gist. CE: Alex Kowalski. N/t. **FM:** KAOI-FM Wailuku 95.1MHz 100kW, KDLX Makawao 94.3MHz 3kW.

17) Honolulu.

18) Honolulu.

19) 250 Ward Ave, #209, Honolulu 96814-4066 — GM: Dave Furya. PD: Robin Furuya. CE: Ralph Wilson. Japanese, Rlg.

20) 1734 S. King Str, Honolulu 96814-2042 — GM: Leona Jona. PD: Harvey Weinstein. CE: Ralph Wilson. Variety.

21) 738 Kaheka St, Honolulu 96814-3726. **E-mail:** hpr@lava.net — GM/PD: Anna Kosof. CE: Jeff Ilardi. News, talk. **FM:** KIPO-FM Honolulu 89.3MHz 3.2kW, KKUA Wailuku 90.7MHz 7kW, KHPR Honolulu 88.1MHz 45kW, KANO Hilo 91.1MHz 100kW (F.Pl.). (KIPO lic. to Pearl

City.)
22) 900 Fort Street Mall, #400, Honolulu 96813-3797. **E-mail:** kccn@aloha.net **WWW:** http://www.keystroke.net/~kccn — GM: Michael W. Kelly. CE: Ernie Nearman. Sports. **FM:** KCCN-FM 100.3MHz 100kW, KINE-FM 105.1MHz 100kW.
23) 970 N. Kalaheo Ave, Suite C-107, Kailua, Oahu 96734 — ADC. **FM:** KRTR Kailua 96.3MHz 75kW, KGMZ Aiea 107.9MHz 100kW, KXME Kaneohe 104.3MHz 75kW.
24) 765 Amaha St, #206, Honolulu 96814 — GM: Jeff J. Coelho. SM: George Rudolph. CE: Ernie Nearman. Adult Standards. **FM:** KUMU-FM 94.7MHz 100kW.
25) Box 565, Kuau, Maui 96779 — Owner: Richard Miller. N/t, sports. (Lic. to Haiku.)
26) 505 Front St, #215, Lahaina, Maui 96761 — 27) KAUI: Kekaha, KSRF: Poipu (both U.C.) — 28) Box 1566, Lihue, Kauai 96766 — 29) Lihue-Kauai (U.C.) — 30) Princeville (U.C.) — 31) 735 Iwilei Rd, #140, Honolulu 96817 — 32) Princeville (U.C.) — 33) 10 Wailea Ekolu Pl, #1705, Kihei, Maui 96753 — 34) 230 Hana Hwy, Ste. 2, Kahului, Maui 96732 — 35) Box 6540, Kamuela, The Big Island 96743.

KIRIBATI

L.T: UTC + 12h — **Pop:** 84.121 — **Radios:** 6.050 — **Pr.L:** I-Kiribati, English — **E.C:** 50Hz, 240V — **ITU:** KIR.

RADIO KIRIBATI (A division of the Broadcasting and Publications Authority)

✉ P.O. Box 78, Bairiki, Tarawa. ☎ +686 21187. 🖷 +686 21096. ☼ K77024 (TELCBAI). **Cable:** INFORMATION TARAWA.
L.P: Mgr: Bill Reiher. Prgr. Organiser: Atiota Bauro. Engineer: Tooto Kabwebwenibeia.
MW: 846kHz 10kW.
SW: Tarawa (G.C: 172.56E/01.21N): 9810kHz 1kW (0530-0930 only).
FM: 98MHz 0.1kW (temporarily inactive)
D.PRGR in I-kiribati (90%)/English (10%): 1800-2000, 0000-0130, 0530-0930. **N. in English:** 0600 (rel. BBC).
ANN: "This is Radio Kiribati, the national broadcasting service of Kiribati in the Central Pacific". "Aio Bwanaan Kiribati te botaki ni kanako bwanaa i bukin Kiribati i nukan te Betebeke".
V. by QSL-card. Enclose US$1 for postage. IRC's not acc.

LORD HOWE ISLAND (Australian)

L.T: UTC + 10½h (UTC + 11h until March 1 and from Oct 25) — **Pr.L:** English — **E.C:** 50Hz, 240V.

RADIO LORD HOWE ISLAND

✉ c/o Post Office, Lord Howe Island, NSW 2898, Australia. ☎ +61 (65) 632123 — **L.P:** SM: Gary Millman.
FM: Signal Point 100.0MHz 30W.
D.PRGR: Thurs 0530-0830. Other times vary seasonally. Rel. 2CS Coffs Harbour when no local announcer available — **V.** by letter. Rp.

MARSHALL ISLANDS (Rep. of)

L.T: UTC + 12h (Kwajalein: UTC - 12h) — **Pop:** 62.915 — **Pr.L:** English, Marshallese — **E.C:** 60Hz, 110/220V — **ITU:** MHL.

MARSHALL ISLANDS BROADC. CO. (Gov. Comm.)

✉ Dept. of Interior & Outer Island Affairs, Majuro, Marshall Isl. 96960. ☎ +692 (29) 3240. 🖷 +692 (29) 3413
L.P: Chief Info Specialist: Billy Sawej.
MW: V7AD 1098kHz 5kW: 1900(Sun 2000)-1000 — **V.** by letter.
ANN: "Radio Majuro".

MICRONESIA HEATWAVE (Comm.)

✉ P.O. Box 1, Majuro 96960. ☎ +692 625 3250.
L.P: SM: Arden Sorimle — **MW:** B7RR 1557kHz 10kW.
D.PRGR: 24h (mostly music with no ann.).

V7AA (Rlg.), P.O. Drawer H, Majuro 96960-1008. ☎ +692 (625) 3141 🖷 +692 (625) 4690 — **Marshall Broadcasting Co,** P.O. Box 19, Majuro 96960. ☎ +692 (29) 3383, 3210 – 97.9MHz — **FM 96,** P. O Box, Majuro 96960. ☎ +692 (625) 4999. 🖷 +692 (625) 3699.

CENTRAL PACIFIC NETWORK (AFRTS - U.S. Mil.)

✉ Box 23, APO San Francisco, CA 96555, USA.
MW: Kwajalein 1224kHz 1kW.
FM: 99.9/101.1/102.1MHz 1kW.
D.PRGR: 24h (on FM: 1 automated and 2 satellite-fed sces.). **N:** hourly via satellite — **ANN:** "The Central Pacific Network".

MICRONESIA (Federated States of)

L.T: Yap, Chuuk UTC + 10h; Pohnpei UTC + 11h; Kosrae UTC + 12h — **Pop:** 125.000 — **Radios:** 70.000 — **Pr.L:** Yapese, Trukese, Ponapean, Kosraean, English — **E.C:** 60Hz, 110/220V — **ITU:** FSM.

FEDERATED STATES OF MICRONESIA BROADCASTING SERVICE (Gov.)

✉ P.O. Box 34, Palikir Station, Pohnpei FSM 96941, Eastern Caroline Islands. ☎ +691 320 2548. 🖷 +691 320 4356. ☼ 6807 (FSMGOV FM) Kolonia. **Cable:** FSM Goverment Pohnpei.
L.P: Special Asst. to Pres. for Information Office: Mrs. Terry Gamabruw. ABU Liaison Officer: Terry G. Thinom. ABU Tech. Liaison Officer: Elieser Rospel.

MW	Call	kHz	kW	H.of tr.*
1)	V6A*	1350		
2)	V6AH	1449	10	1900-1300
3)	V6AI	1494	10	2000-1400
4)	V6AJ	1584	1	1800-1200
5)	V6AK	1593	5	2000-1400

*) all st's operate 24h during adverse weather conditions.
FM: 6) V6AF 104.0MHz
Addresses and other information
1) Moen, Chuuk State, FSM 96942. Rlg. (Baptist) — 2) P .O. Box 1086, Kolonia, Pohnpei State, FSM 96941. **ANN:** "Met Station V6AH nan Pohnpei" — 3) P.O. Box 119, Colonia, Yap State, FSM 96943. **ANN:** "Pary e radio station V6AI nu Waab" — 4) P.O. Box 147, Tofol, Kosrae State, FSM 96944. **ANN:** "Painge station V6AJ, fwin an Kosrae" — 5) P. O Box 206, Weno, Chuuk State 96942. **ANN:** "Ach nenien appio V6AK lon Chuuk" — 5) c/o Calvary Baptist Church, Drawer H, Kolonia, Pohnpei, FSM 96941 — 6) Pohnpei FSM 96941.

MIDWAY ISLANDS (U.S. Territory)

L.T: UTC - 11h — **Pop:** 453 (military personnel) — **Only L:** English — **E.C:** 60Hz, 110/220V — **ITU:** MDW.

NAVY BROACASTING SERVICE (AFRTS)

✉ Det. 27, U.S. NAF Box 39, San Francisco, CA 96614, USA.
FM: 94.0MHz 50W — **D.PRGR:** 24h. **N:** on the h.

NAURU (Republic of)

L.T: UTC + 12h — **Pop:** 10.559 — **Radios:** 10.000 — **Pr.L:** English, Nauruan — **E.C:** 50Hz, 110/240V — **ITU:** NRU.

NAURU BROADCASTING SERVICE (Gov.)

✉ Information and Broadcasting Sces, Chief Secretary's Dept, Rep. of Nauru, Ce. Pacific. ☎ +674 4443109. 🖷 +674 4443195.
L.P: SM: Miss Rin Tsitisi. TD: Malcom Aroi.
MW: 1323kHz 0.2kW — **D.PRGR:** 1900-1130 — **V.** by letter.

NEW CALEDONIA (French)

L.T: UTC + 11h — **Pop:** 186.000 — **Radios:** 92.000 — **Pr.L:** French — **E.C:** 50Hz, 220V — **ITU:** NCL.

RADIODIFFUSION FRANCAISE D'OUTRE MER (RFO)

✉ Mt Coffyn, B.P. G3, Noumea. ☎ +687 274327. 🖷 +687 281252. ☼ RTOM3052NM.
L.P: Reg. Dir: Wallés Kotra. PD: Louis Palmieri.
MW: 666kHz 20kW — **FM:** Noumea 89.9/99.3MHz. + 17 relay st's.
D.PRGR: 24h in French. **Rel. France-Inter:** 1100-1900.
ANN: "Bonjour, vous êtes à l'ecoute de RFO Nouvelle-Caledonie". **France-Inter** is relayed 24h on 90.2MHz in Noumea.

Noumea Radio Joker 2000 (N.R.J. 2000), 41/43 rue Sebastopol, B.P. 179, Noumea. ☎ +687 272584, 263434. 🖷 +687 281627 — **FM:** Noumea 93.5MHz — **24h.**

Radio Rythme Bleu (RRB), B.P. 578 Noumea. ☎ +687 254646. 🖷 +687 284928 — **FM:** Noumea 100.4 MHz + 12 relay st's — 24h (rel. Europe 1).
Radio Djido, 29 r Mar Juin Ht Mgta, B.P. 1671, Noumea. ☎ +687 253515. 🖷 +687 253433 — **FM:** Noumea 97.4 MHz — **24h.**

NEW ZEALAND

L.T: UTC + 12h (Oct-March UTC + 13h) — **Pop:** 3.350.000 — **Radios:** 3.100.000 — **Pr.L:** English, Maori — **E.C:** 50Hz, 230V — **ITU:** NZL.

RADIO NEW ZEALAND (State-owned enterprise)

🖃 P.O. Box 2092, Wellington. ☎ +64 (4) 4741555. 🖷 +64 (4) 4741440. **Cable:** Radionet. ① NZ 30131. **Web:** http://www.rnz.co.nz/
L.P: Chief Exec: Nigel Milan. GM Eng. & Op's: Norm Collison. GM Metro St's: Ms. Lynne Clifton. GM Provincial Sts: Carl Terry. GM RNZ News: Ray Lilley. GM Public Radio (non-comm.): John Craig

Key: A=Access Radio, C=Concert Prog, F=Southern Star (Over 50's Network), G=R. Rhema (rlg.), M=ZM Net, N=National Radio, P=Private, S=Sports Network, T= R. Pacific Talk Radio. Z=RNZ Community Net. 1:=local hours.

MW STATIONS:

National Radio Net: 24h. Stations often have local sports particularly rugby games. **N:** 1200, 1300, 1400, 1600, 1700, 1800, 1830, 1900, 1930, 2000, 2030, 2100, 2200, 2300, 0000, 0030, 0100, 0200, 0300, 0400, 0500, 0600, 0800, 0900, 1000, 1100. **Regional news:** 1845, 2005, 0034, 0630, Maori news: 1841, 0505. **Main weather:** 1730, 1838, 1915, 0038, 1015. **"Pacific Islands Magazine":** Mon-Fri 0700-0730 with news in Samoan, Tongan, Cook Islands Maori, Niuean & Tokelauan, with a 12-minute magazine in a different language each week-day.

Concert FM: 1900-1200. **N:** 1800, 2000 (BBC), 2100, 2300, 0100, 0300 (BBC), 0500, 1100 (BBC).
Sports Roundup: carried by Net S stations.

Local Addresses of RNZ stations: see address section below

RADIO NEW ZEALAND INTERNATIONAL
See International Broadcasting section.

NB: the listings below are composite listings of RNZ & private st's.

MW	Call	kHz	kW	N	Name, location & h. of tr.
166)		531		P	Pacific Islands R, Auckland
40)		531	2	P	R. Central, Alexandra. # 4X0 1206 l: 1800-0600, Suns 2100-0000
42)	1XC	540	5	G	R. Rhema Tauranga
42)	2XV	540	4	G	R. Rhema, New Plymouth
42)		549		G	R. Rhema, Kaitaia (**F.PI.**)
2)	2YA	567	50	N	National R. Wellington
46)	2XR	585	2	P	R. Ngati Porou Ruatoria
42)		585		G	R. Rhema, Blenheim (**F.PI.**)
42)	3XL	594	2	G	R. Rhema, Timaru
42)		594		G	R. Rhema, Wanganui
41)	1XO	603	5	P	Irirangi Maori Radio, Auckland
42)	3XG	612	2	G	R. Rhema, Christchurch
42)	4XG	621	2	G	R. Rhema, Dunedin
2)	2YZ	630	20	N	National R, Napier
5)	4YW	639	2	N	National R, Alexandra
42)	2XC	648	5	G	R. Rhema, Napier
2)	2YC	657	60	S	Wellington
2)	3YA	675	20	N	National R, Christchurch
56)	1XP	702	5	T	R. Pacific, Auckland
79)	2XP	711	5	T	R. Pacific, Wellington
2)	4YZ	720	20	N	National R, Invercargill
2)	1YP	729	2	N	National R, Tokoroa
126)	4XX	729	0.9	P	Ranfurly R, Ranfurly
2)	1YA	756	10	N	National R, Auckland
124)	2XT	765	2.5	P	R. Kahungunu, Napier
139)	2YB	783	10	A	Access R, Wellington
175)	2YB	783	10	P	Samoan Capital R, Wellington
8)	1XSR	792	0.4	S	The Sports Network, Hamilton
42)	2XL	801	1	G	R. Rhema, Nelson 1: Sun-Thurs 1900-2100 or MF 0430-0530

MW	Call	kHz	kW	N	Name, location & h. of tr.
110)	1XU	810	2	A	Access R, Auckland
2)	4YA	810	20	N	National R, Dunedin
2)	1YZ	819	10	N	National R, Rotorua
43)	2XS	828	2	P	Classic Hits XS, Palmerston North
2)	1YX	837	2	N	National R, Whangarei
2)	1YX	837	2	N	National R, Kaitaia
12)	2ZD	846	2	Z	R. Wairarapa, Masterton
42)	1XH	855	2	G	R. Rhema, Hamilton
3)	4ZA	864	10	Z	News-Talk ZB, Invercargill
13)	3ZE	873	0.5	Z	Ashburton # 3ZB 1098 1: Mon-Fri 1800-0200, SS -0600
2)	1YC	882	10	S	Auckland
44)	2XW	891	5	P	The Breeze, Wellington
2)	4YC	900	10	S	Dunedin
42)		900		G	R. Rhema, Whangarei (**F.PI.**)
2)	2XD	909	0.8	S	Sports Roundup, Napier
2)	3YT	918		N	National R, Timaru
14)	2ZA	927	2	Z	Palmerston North
15)	2ZG	945	2	Z	Gisborne
145)	1XW	954	2	T	R. Pacific, Hamilton
2)	3YC	963	10	S	Christchurch
42)	2XG	972	5	G	R. Rhema, Wellington
2)	1YE	981	2	N	National R, Kaikohe
105)		990	1	P	BBC World Sce, Auckland
91)		990	1	P	Fifeshire Classics, Nelson
16)	1ZD	1008	10	Z	Easy B.O.P. AM, Tauranga
11)	1ZN	1026	2	Z	R. Northland, Whangarei
11)	1ZK	1026	2	Z	R. Northland, Kaitaia
8)	4XSR	1026	1	S	The Sports Network, Invercargill
27)	2ZB	1035	20	Z	Wellington
5)	4ZB	1044	10	Z	Dunedin
17)	2ZP	1053	2	Z	R. Taranaki, New Plymouth
2)	2YE	1071		N	National R, Masterton
28)	1ZB	1080	10	—	Newstalk 1ZB, Auckland
6)	3ZB	1098	10	Z	Christchurch
1)		1107	0.1	A	Coromandel Access Radio, Waihi
2)	2YX	1116	2	N	National R, Nelson
8)		1125	0.2	S	The Sports Network, Dunedin
8)		1125		S	The Sports Network, Napier
2)	4YQ	1134	2	N	National R, Queenstown
2)	1YW	1143	2	N	National R, Hamilton
20)	3ZC	1152	2	Z	R. Caroline,Timaru
70)	2XM	1161	5	P	Te Upoko o Te Ika, Wellington
30)	1ZW	1170	0.2	Z	R. Waitomo Te Kuiti # ZH-FM 98.6 1: 1800-
120)		1170		A	Access R., Invercargill (**F.PI.**)
40)	4XE	1179	0.1	P	Resort R, Wanaka
137)		1179	1	P	Ruia Mai, Auckland
2)	1YR	1188	0.4	N	National R, Rotorua
21)	2ZW	1197	2	Z	River City R, Wanganui
40)	4XO	1206	2	P	Otago R, Dunedin
130)	1XHC	1206	0.5	A	Access Community R, Hamilton
11)	1ZE	1215	2	Z	Kaikohe # 1ZN 1026
47)	4XF	1224	2	P	R. Foveaux, Invercargill
42)		1233	0.6	F	"Southern Star" Wellington
48)	1XX	1242	2	P	Whakatane
48)	1XX	1242	0.1	P	Galatea (synchro with 1XX)
29)	1XG	1251	5	G	R. Rhema, Auckland
64)	3XA	1260	2	P	C93 FM, Christchurch
18)	2ZT	1269	0.4	Z	R. Nelson Takaka # 2ZN 1341
4)	2ZC	1278	2	Z	Bay City R, Napier
7)	3ZW	1287	2	Z	R. Scenicland, Westport # 3ZA 747
19)	1ZH	1296	2.5	Z	ZH-AM, Hamilton
50)	4XD	1305	2	P	R. Dunedin, Dunedin
2)	2YW	1314	2	N	National R, Gisborne
8)		1332	5	S	The Sports Network, Auckland
18)	2ZN	1341	2	Z	R. Nelson, Nelson
40)	4XC	1359	1	P	Resort R, Queenstown
52)	1XT	1368	1	P	Village R, Tauranga Suns only 2100-0500
53)	2XX	1377	2	P	R. Horowhenua. Levin
162)		1386		P	R. Tarana, Auckland
23)	4ZW	1395	2	Z	R. Waitaki, Oamaru
42)	4XL	1404	2.5	G	R. Rhema, Invercargill
24)	1ZO	1413	2	Z	R. Forestland, Tokoroa # 1TPO 97.5 1: 1800-
108)	3XP	1413	0.1	P	R. Ferrymead, Christchurch Sun. 2100-0600

MW	Call	kHz	kW	N	Name, location & h. of tr.
111)	4XK	1422	0.1	P	R. Kingston, Kingston
60)	2XKC	1431	2	A	R. Kidnappers, Hastings
112)	1XK	1440	0.2	P	Te Reo o Tauranga Moana Tauranga
126)		1440	0.1	P	Goldfields R, Lawrence
2)	2YM	1449	2	N	National R, Palmerston North
2)	3YW	1458	2	N	National R, Westport
88)	1XD	1476	5	P	The Point, Auckland Airport
2)	1YT	1494	2.5	N	National R, Taupo
8)		1503		S	The Sports Network, Christchurch
8)		1503		S	The Sports Network, Wellington
25)	1ZU	1512	1	Z	King Country R, Taumaranui # 1TPO 97.5 1: Mon-Fri 1800-0030, Sat 1900-0000
9)		1512	0.3	A	"Coast Access Radio" Kapiti
87)	1XTR	1521		P	Classic Good Time Oldies, Tauranga
2)	2YP	1530	2	N	National R, New Plymouth
153)		1530	1/ 0.3	P	The Wireless Station, Hastings
26)	2ZE	1539	1	Z	R. Marlborough, Blenheim
89)	1XN	1548	0.99	P	Classic Country, Rotorua
31)	2ZH	1557	2	Z	R. Taranaki, Hawera # 2ZP 1053 1: MF 1800-2200, Sat 2000-2300
113)	4XS	1575	2.5	A	Hills AM, Dunedin
26)	2ZF	1584	0.4	Z	R. Marlborough, Picton # 2ZE 1539
42)	1XCB	1593	2.5	F	Southern Star, Auckland
57)	2XA	1602	1	P	Radio Reading Sce, Levin: W 2030-1000, Sun 0600-0900

SW	Call	kHz	kW	N	Name, location & h. of tr.
57)	ZLXA	3935	1	P	2030-1000 # 2XA 1602
57)	ZLXA	5960	1	P	2030-0600v # 2XA 1602
57)	ZLXA	7290	1	P	alt. to 7290

FM	MHz	kW	N	Name, location & h. of tr.
137)	88.6	1	P	Mai FM, Bastion Point, Auckland 64
2)	89.0	40	C	Concert FM, Palmerston North
2)	89.0		C	Concert FM, Horokaka
73)	89.0	0.1	P	R. Active, Wellington 1900-1300, Weekends 24h
82)	89.0	0.1	P	Contact FM, Hamilton 1900-1300, weekends 24h
71)	89.0	1	P	R. Te Arawa, Rotorua
138)	89.0		P	C FM, Paeroa
42)	89.0		G	R. Rhema, Wanaka (F.PI.)
56)	89.1		T	R. Pacific, Hokitika
138)	89.1	0.02	P	C FM, Whanagmata
56)	89.1	0.2	T	R. Pacific,.Greymouth
6)	89.2		M	ZM-FM, Sumner #3ZZM 91.3
47)	89.2		P	Fouveaux R, Invercargill
101)	89.3	0.1	P	Hitz 89 FM, Masterton
83)	89.3		P	89 FM, Gisborne
67)	89.3		P	Hot FM, Paeroa Range
28)	89.4	40	Z	Newstalk ZB, Auckland
27)	89.4	0.3	Z	B 90 FM, Kapiti
28)	89.4	3	Z	ZB-FM, Dunedin
4)	89.5		P	Bay City R, Mt. Threave, Napier
59)	89.6	10	P	K-Double-C FM, Hikurangi
21)	89.6	0.1	Z	River City FM, Wanganui
2)	89.7	40	C	Concert FM, Christchurch
160)	89.8	50	P	The Breeze, Hamilton
54)	89.8	0.1	P	Te Reo Irirangi o Rangitane, Palmerston North
28)	89.8		Z	R. Nelson, Nelson
95)	89.8	0.2	P	More FM, Colonial Knob, Porirua
56)	89.9		P	R. Pacific, Mt. Studholme (Timaru)
17)	90.0	3	Z	R. Taranaki, New Plymouth
7)	90.0		Z	R. Scenicland, Westport
2)	90.0		C	Concert FM, Invercargill
33)	90.0	0.3	Z	B 90 FM, Wellington
56)	90.1		T	R. Pacific, Westport
128)	90.1		P	Baptist R,. Wairoa
53)	90.2		P	2XX FM, Forest Heights
56)	90.2		T	R. Pacific, Auckland
59)	90.3	0.1	P	K Double C FM, Whangarei
2)	90.3		C	Concert FM, Tihiotonga, Rotorua
106)	90.3		P	I-90 FM, Napier
40)	90.3		P	R. Central, Alexandra
28)	90.4		Z	Classic Hits 90 FM, Motueka
48)	90.5	10	P	Triple X FM Whakatane

FM	MHz	kW	N	Name, location & h. of tr.
94)	90.5	0.05	P	Tahu FM, Christchurch
7)	90.5		Z	Scenicland FM, Franz Josef
138)	90.6	0.02	P	Coromandel FM, Mercury Bay
14)	90.6	40		Q 91 FM Palmerston North
97)	90.6	0.5	P	Rukawa FM, Rucu
42)	90.7		G	R. Rhema, New Plymouth (F.PI.)
93)	90.8		P	Tautoko FM, Hikurangi
17)	90.8	1	P	90 FM Rahotu
2)	90.9	50		ZM-FM Wellington
71)	90.9	0.1	P	97 FM, Paeroa Range
75)	91.0	0.9	P	R. One Dunedin 1900-0200, Weekends 24h
62)	91.0	50	P	The Breeze, Auckland
90)	91.0		P	Bay Rock, Tauranga
2)	91.1	3	C	Concert FM Napier
7)	91.1		P	Scenicland FM, Greymouth
181)	91.1	0.1	P	ZB FM Lawrence
59)	91.2	5	P	R. J-Wave, Queenstown (Japanese)
141)	91.2		P	Iwi FM Waipuna
6)	91.3	40		ZM-FM Christchurch
2)	91.4	40	C	Concert FM Hamilton
2)	91.4	3	C	Concert FM Grampians, Nelson
146)	91.4	1	P	Country FM Palmerston No.
2)	91.6	3	C	Concert FM Mt. Taranaki (New Plymouth)
142)	91.7	0.5	P	The Box. Mt. Kau Kau
151)	91.7		P	Turanganui a Kiwa, Gisborne
173)	91.8		P	More FM, Auckland
56)	91.9	0.25	T	R. Pacific, Rotorua
178)	91.9	4	P	More FM, Napier
145)	91.9		P	Maniapoto FM, Okahukura
143)	92.0		P	Q 92 FM, Queenstown
144)	92.1		P	R. Weka Waitangi, Chatham Is.
56)	92.1	0.5	T	R. Pacific, Whakatane
96)	92.1	34.5	P	More FM Christchurch
141)	92.1		P	Iwi FM, Raetihi
80)	92.2	1	P	Nga Iwi FM, Waihi
172)	92.3		P	Most FM, New Plymouth
59)	92.4	10	P	K Double C FM Hobson # 1KCC 89.3
22)	92.4	0.25	Z	R. Lakeland, Karangahape #1TPO 96.7
65)	92.4	1	P	Energy FM Rahotu
2)	92.5	40	C	Concert FM Wellington
13)	92.5		Z	3ZE FM, Ashburton
170)	92.5	0.2	P	Classic Rock, Tauranga
2)	92.6	40	C	Concert FM Auckland
2)	92.6	25	C	Concert FM Dunedin
63)	92.7	2.5	P	Hot 93 FM Hastings
42)	92.7	0.01	G	R. Rhema, Ohura
145)	92.7		P	Maniapoto FM, Pio Pio
93)	92.8	0.2	P	Tautoko FM, Maungataniwha
102)	92.8	0.25	P	Star FM, Wanganui
85)	92.8	0.25	P	Q 92 FM, Deer Park, Queenstown
64)	92.9	40	P	C93 FM Christchurch
74)	92.9	0.3	P	Sounds FM, Blenheim # 91)
147)	92.9		P	Big River R., Balclutha
107)	93.0		P	Rock 93 FM Hamilton
91)	93.0	3	P	Fifesfire FM Nelson
48)	93.0	0.05	P	Triple X FM Ohope # 1XXX 90.5
7)	93.1		Z	Scenicland FM, Paparoa
65)	93.2	3	P	Energy FM New Plymouth
78)	93.2		P	Port FM, Mt. Studholme
46)	93.3		P	R. Ngati Porou
90)	93.4	25	P	Coastline FM Tauranga
169)	93.4	1	P	Kool FM Auckland
40)	93.4		P	Rox FM Dunedin
104)	93.4		N	National Radio, Wanaka
66)	93.5	0.1	P	Kiss FM, Taupo
32)	93.5	0.4		ZM-FM Wellington # 2ZZM 90.9
179)	93.5		P	Central FM Waipukurau
42)	93.5		G	R. Rhema, Rotorua
60)	93.5		P	R. Kidnappers, Hastings
77)	93.6	0.5	P	Kaitaia Community R., Kaitaia
59)	93.7	0.1	P	K Double C FM Russell # 1KCC 89.6
56)	93.7		T	R. Pacific Christchurch
147)	93.7	0.04	P	Big River R. Kuriwao
148)	93.8		P	Gulf FM Waiheke Is.
56)	93.8	5	P	R. Pacific, Wharite
160)	93.8	2	P	Blue Skies FM, Hamilton
2)	93.9		C	Concert FM Edgecumbe
7)	93.9		Z	Scenicland FM, Greymouth

FM	MHz	kW	N	Name, location & h. of tr.
78)	93.9		P	Port FM, Twizel # 97.9
56)	94.0		T	R. Pacific, Invercargill
56)	94.0		T	R. Pacific, Whangarei
65)	94.0	0.3	P	Energy FM Pukeiti
138)	94.0	0.2	P	Coromandel FM, Pauanui/Tairua
44)	94.1	0.5	P	The Breeze Wellington
91)	94.1		P	Fifeshire FM Murchison
68)	94.2		P	FM Country, Auckland
150)	94.2		P	Tahi FM Taupo
181)	94.3		P	ZB FM Blue Spur
142)	94.3	0.3	P	The Box, Forest Heights
134)	94.3		P	The Fish, Taupo
114)	94.4	1	P	Hiku o Te Ika Kaitaia
180)	94.4		P	Ski FM Whakapapa (ski season)
153)	94.4		P	94 FM Country, Wanganui
42)	94.4	0.5	G	R. Rhema, Queenstown (F.PI.)
74)	94.5		P	Sounds FM Picton
45)	94.5		P	I 94.5 FM, Christchurch
64)	94.5	20	P	Easy Listening I-94.5, Christchurch
131)	94.6		P	FM Country, Hamilton
100)	94.7		P	Kiks FM, Lower Hutt
42)	94.7		G	R. Rhema, Greymouth
183)	94.7		P	Channel Z, Wellington
149)	94.8	0.6	P	Te Korimako o Taranaki New Plymouth
56)	94.8		T	R. Pacific Gisborne
165)	94.9	1	P	Fox FM, Ashburton
69)	95.0	5	P	B-FM. Auckland
16)	95.0		Z	R. B.O.P. Tauranga
53)	95.1		P	95 FM Levin
42)	95.1	0.1	G	R. Rhema Taupo
53)	95.1		P	95 FM Waikanae # 90.2
56)	95.1		T	R. Pacific Greymouth
56)	95.2		T	R. Pacific, Northland
2)	95.3	0.2	C	Concert FM Mt. Pearce, Banks Peninsula
96)	95.3	0.2	P	More FM Southshore
164)	95.6		P	Rock 95.6FM Mt. Egmont
117)	95.8	0.1	P	Coast FM Orema
84)	95.4	0.1	P	Te Reo Irirangi o Tainui Ngaruawahia
2)	95.4		C	Concert FM, Wanaka
127)	95.4	0.3	P	Mosgiel 95 FM, Mosgiel
151)	95.5		P	Turanganui a Kiwa, Wharekeoai
2)	95.6	0.6	C	Concert FM Wellington
56)	95.6		T	R. Pacific, Northland
97)	95.7		P	Ruakawa FM, Tokoroa
152)	95.8	2	P	Star FM/Peak FM Raetihi
137)	95.8		P	Mai FM, Manukau
5)	95.8	0.5	M	ZM FM, Dunedin
2)	95.8	0.02	N	National R., Twizel
137)	95.8		P	Mai FM, Manukau
67)	95.9	0.1	P	Lake City FM Rotorua
4)	95.9	3.0	Z	Greatest Hits 96 FM Hastings
56)	95.9		T	R. Pacific Alexandra
2)	96.0		C	Concert FM, Kaitaia
56)	96.1		T	R. Pacific Tauranga
173)	96.1		P	NZ Broadc. School Christchurch
58)	96.1		P	Rock 96 FM, Wellington
49)	96.1		P	Easy FM, Blenheim
88)	96.1		P	The Point, Auckland
138)	96.2	1	P	Coromandel FM Thames
143)	96.2	0.2	P	Q 92 FM, Mid Dome, we. Southland
182)	96.2		P	R. Dargaville, Dargaville
133)	96.2	1	P	B 96 FM, Gisborne
118)	96.3		P	Kawhia Moana, Kawhia
42)	96.3	0.63	G	R. Rhema Palmerston North
153)	96.3		P	Port FM, Otematata # 91.3
154)	96.3	1	P	Pirate FM Wellington
120)	96.4		A	Access R., Invercargill
65)	96.4		P	Energy FM, Oakura
145)	96.5	0.05	P	Maniapoto FM, Te Kawa
91)	96.5		P	Fifeshire FM, Westport
91)	96.6		P	Fifeshire FM Takaka # 93.0
155)	96.6		P	Mercury FM Coromandel
56)	96.6		T	R. Pacific Dunedin
180)	96.6		P	Ski FM Turoa (ski season only)
140)	96.6		P	The Wave, Whangamata
179)	96.6		P	Central FM, Tourere
22)	96.7	0.6	Z	R. Lakeland Taupo
71)	96.7	0.1	P	Whanau FM, Mercury Bay
42)	96.7		G	R. Rhema, Levin
85)	96.7		P	Q 92 FM, Alexandra
2)	96.8	0.1	C	Concert FM Whangarei
85)	96.8	0.5	P	Q 92 FM, Coronet Peak # 92.8
156)	96.9	0.1	P	Atiawa FM, Lower Hutt
61)	96.9	3.5	A	Plains R. Christchurch
26)	96.9		Z	R. Marlborough Marlborough # 2ZE 1539
132)	97.0	50	P	I 97 FM Hamilton
2)	97.2		C	Concert FM Whakapunake, Gisborne
157)	97.2		P	Tuwaretoa FM Mt Pihanga, Turangi
119)	97.2		P	Te Ika Whenua Murupara
153)	97.3		P	Port FM, Omarama #91.3
28)	97.4	40	M	Classic Hits 97FM Auckland
40)	97.4	1.2	P	4XO-FM Dunedin
10)	97.5	0.4	P	Geyserland FM. Rotorua
42)	97.5	0.1	G	R. Rhema, Taumarunui
56)	97.5		T	R. Pacific Napier
142)	97.5		P	The Quake, Kapiti
2)	97.5		C	Concert FM,Alexandra
56)	97.6		T	R. Pacific Nelson
158)	97.6	0.1	P	97.6 FM, Kaitaia
159)	97.7			Blenheim
176)	97.7		P	Bay Rock, Tauranga
6)	97.7	7.5	Z	B 98 FM, Christchurch
42)	97.8		G	R. Rhema, Greymouth
160)	97.8		P	Buzzard FM, Hamilton
78)	97.9		P	Port FM, Timaru
65)	98.0	2	P	Easy 98 FM
44)	98.1	0.1	P	Towai (#The Breeze FM 94.1)
161)	98.1		P	Hot FM, ?
98)	98.1		T	R. Pacific Masterton
46)	98.1		P	Radio Ngati Porou
51)	98.2		P	I 98 FM Auckland
2)	98.3	0.6	C	Kapiti Coast, Wellington
2)	98.3	0.2	C	Taupo
76)	98.3	0.5	P	R. U, Christchurch
162)	98.4		P	Tumeke FM, Whakatane
2)	98.4		C	Concert FM, Queenstown
2)	98.4		C	Concert FM, Hikurangi
49)	98.5		P	Easy FM, Picton
183)	98.5		P	Channel Z, Wellington
19)	98.6	40	Z	ZH-FM, Hamilton
91)	98.6		P	Fifeshire Classics, Nelson
43)	98.6	0.3	P	Magic 98 FM, Wharite
6)	98.7		M	Sumner
42)	98.7		G	R. Rhema, Whangarei
20)	98.7		Z	Classic Hits 99 FM, Timaru
3)	98.8	5	Z	4ZA-FM, Invercargill
2)	98.8		C	Concert FM, Russell
95)	98.9	0.25	P	More FM, Towai (Wellington)
165)	98.9	0.3	P	Fox FM, Ashburton
2)	99.0	0.1	C	Wanganui
5)	99.0	0.25	C	Highcliffe (#Dunedin 89.0)
55)	99.0	40	P	R. Hauraki, Auckland
99)	99.1		P	R. Foxton, Foxton
152)	99.1	1.5	P	Heartbeat, Taumarunui
161)	99.1		P	Hot FM, Rotorua
181)	99.1	0.005	P	ZB FM Blue Spur
42)	99.1		G	R. Rhema Mt.Erin
56)	99.1		T	R. Pacific, Taupo
2)	99.2	2.5	C	Masterton
40)	99.2		P	Resort R., Queenstown
140)	99.2		P	The Wave, Pauanui
163)	99.3		P	Life FM, Christchurch
48)	99.3		P	Bay Rock, Whakatane
72)	99.4	0.1	P	R. Massey, Palmerston No. 1900-0100 (weekends 24h)
131)	99.4		P	Pare Hauraki, Paeroa
92)	99.4		P	Te Reo Irirangi o Maniapoto Te Kuiti
80)	99.4	5W	P	Nga Iwi FM, Waihi
2)	99.4	0.02	C	Concert FM Saddle Hill, Dunedin
102)	99.4	1	P	Star/Peak FM Taihape
179)	99.4		P	Central FM, Dannevirke
145)	99.4		P	Maniapoto FM, Te Kuiti
135)	99.4		P	Fresh FM, Motueka
40)	99.4		P	Resort R., Wanaka
2)	99.5	10	C	Mt. Studholme, Timaru
121)	99.5	0.1	P	Ngati Hine, Whangarei
95)	99.6		P	More FM, Kapiti
17)	99.6	0.02	P	90 FM Oakura

FM	MHz	kW	N	Name, location & h. of tr.
9)	99.7	0.04	Z	Classic Rock 96 FM, Wairoa
90)	99.8		P	Coastline FM, Tauranga
136)	99.8		P	Today FM, Auckland
42)	99.8		G	R. Rhema, Hawera (**F.PI.**)
2)	99.9	0.8	C	Sumner/Redcliffs, Christchurch
106)	99.9		P	Extreme 100, Napier
55)	100.0		P	Hauraki FM, Auckland
123)	100.0		P	100 FM, Whanganui
95)	100.0	50	P	More FM, Wellington
122)	100.0		P	Whitestone FM, Oamaru
55)	100.0		P	R. Hauraki FM, Hamilton
40)	100.0		P	Resort Radio, Queenstown
93)	100.		P	Tautoko FM, Okaihau

Addresses & other information

INDEPENDENT BROADCASTERS ASSOCIATION (N.Z.) INC.

✉ P.O. Box 3762, Auckland. ☎ +64 (9) 378 0788. 🖷 +64 (9) 378 8180 — **LP:** Exec. Dir: B.G. Impey. Secr: Janine Bliss.

1) c/o 6 Victoria St, Waihi – 2) P.O. Box 123, Wellington – 3) P.O. Box 802, Invercargill – 4) P.O. Box 241, Napier – 5) P.O. Box 888, Dunedin – 6) P.O. Box 1484, Christchurch – 7) P.O. Box 378, Greymouth – 8) P.O. Box 2396, Wellington – 9) c/o 3 Horopito Place, Kapiti – 10) P.O. Box 1147, Rotorua – 11) P.O. Box 845, Whangarei – 12) P.O. Box 220, Masterton – 13) P.O. Box 465, Ashburton – 14) P.O. Box 1045, Palmerston North – 15) P.O. Box 1040, Gisborne – 16) P.O. Box 642, Tauranga – 17) P.O. Box 141, New Plymouth – 18) P.O. Box 43, Nelson – 19) P.O. Box 489, Hamilton – 20) P.O. Box 275, Timaru – 21) P.O. Box 632, Wanganui – 22) P.O. Box 740, Taupo – 23) P.O. Box 426, Oamaru – 24) P.O. Box 272, Tokoroa – 25) P.O. Box 383, Taumarunui – 26) P.O. Box 225, Blenheim – 27) Box 300, Wellington – 28) P.O. Box 3526, Auckland – 30) P.O. Box 276, Te Kuiti – 31) P.O. Box 341, Hawera – 32) P.O. Box 1991, Wellington – 40) R. Otago Ltd., Private Bag, Dunedin. (4XA: P.O. Box 143, Alexandra: 4XC: P.O. Box 224, Queenstown) – 41) Aotearoa R., P.O. Box 97-254, SAMC, Auckland – 42) R. Rhema, Private Bag 92-636, Auckland – 43) Manawatu R. Co., P.O. Box 446, Palmerston North – 44) Capital City R. Ltd., P.O. Box 11-441, Wellington – 46) R. Ngati Porou, P.O. Box 55, Ruatoria – 47) Foveaux R. Ltd., P.O. Box 1740, Invercargill – 48) R. Bay of Plenty Ltd., P.O. Box 383, Whakatane – 49) Seymour Str, Blenheim – 50) Otago R. Assoc., P.O. Box 404, Dunedin – 51) R.i Ltd., Private Bag, Auckland – 52) Village R., P.O. Box 597, Tauranga. Station is completely voluntary and operates under the auspices of the Tauranga District Museum as part of the Historic Village complex – 53) R. Horowhenua Ltd., P.O. Box 132, Paraparaumu – 54) Box 1341, Palmerston North – 55) Hauraki Enterprises Ltd., P.O. Box 1480, Auckland – 56) R. Pacific, Private Bag, Ponsonby – 57) NZ R. for the Print Disabled Inc, P.O. Box 360, Levin. Operated by volunteers. St. Dir: Allen Little. Studio Mgr: Ash Bell. Prgr. Supervisor/QSL Mgr: Brian Stokoe. V. by QSL-card.Rp or SAE req. – 58) 96 FM, Wellington – 59) Northland F.M. R. Ltd., P.O. Box 100, Whangarei – 60) Box 143, Alexandra – 61) Canterbury Communications Trust, P.O. Box 22297, Christchurch – 62) Metropolitan F.M., P.O. Box 33-644, Takapuna, Auckland 9 – 63) R. Hawkes Bay, P.O. Box 193, Hastings – 64) Canterbury F.M. Ltd., P.O. Box 4750, Christchurch 1 – 65) Energy F.M., P.O. Box 869, New Plymouth – 66) Kiss F.M., P.O. Box 393, Taupo – 67) Central FM Corp, P.O. Box 92, Rotorua – 68) Private Bag, Albany, Auckland – 69) Private Bag, Auckland – 70) 182 Wakefield Str, Wellington – 71) Te Arawa Maori Trust Board, P.O. Box 883, Rotorua – 72) Private Bag, Palmerston North – 73) Box 11-971, Wellington – 74) P.O. Box 930, Blenheim – 75) Otago Univ. Students Assoc., P.O. Box 1436, Dunedin – 76) Canterbury Univ. Students Assoc., Ilam Rd., Ilam, Christchurch – 77) Box 81, Kaitaia – 78) P.O. Box 635, Timaru – 79) P.O. Box 11-850, Wellington – 80) Box 135, Paeroa – 81) Box 224, Queenstown – 82) Private Bag 3059, Hamilton – 83) P.O. Box 230, Gisborne – 84) P.O. Box 208, Ngaruawahia – 85) Box 92, Queenstown – 86) Box 8880, Auckland – 87) Box 134, Tauranga – 88) Box 73-020, Auckland Airport, Auckland – 89) P.O. Box 1007, Rotorua – 90) P.O. Box 2429, Tauranga. **Web:** http://www.enternet.co.nz/ coastline/ – 91) P.O. Box 907, Nelson – 92) Box 15-213, Hamilton – 93) Mangamauku Bridge, RD2, Okaihu – 94) Box 15-115, Christchurch – 95) Box 27-000, Wellington – 96) P.O. Box 25-209, Christchurch – 97) P.O. Box 842, Tokoroa – 98) P.O. Box 209, Turangi – 99) 55 Main St., Foxton – 100) Box 2123, Foxton – 101) Box 811, Masterton – 102) Box 920., Wanganui – 103) Barnetts Radio & TV, Wanaka – 104) 2nd Floor, 68 Victoria Ave, Wanganui – 105) Box 1434, Auckland – 106) Box 105, Napier – 107) P.O. Box 19-293, Hamilton – 108) 269 Bridal Path Rd., Christchurch – 109) Box 980, Invercargill – 110) P.O. Box 5609, Wellington – 111) c/o

NZ Post, Kingston – 112) P.O. Box 382, Tauranga – 113) P.O. Box 2142, South Dunedin – 114) P.O. Box 458, Kaitaia – 115) Box 680. Hastings – 116) P.O. Box 2090, Whakatane – 117) Box 47-376, Auckland – 118) P.O. Box 41, Kawhia – 119) P.O. Box 98, Murupara – 120) Box 1, Invercargill – 121) P.O. Box 1127, Whangerei – 122) Box 12. Oamaru – 123) P.O. Box 430, Wanganui – 124) P.O. Box 7010, Taradale – 125) c/o Roy Gillions, 24 Ross Str, Lawrence – 126) 3 Charlemont St., Ranfurly – 127) 1 Gordon Rd, Mosgiel – 128) Box 427, Wairoa – 129) P.O. Box 2197, Auckland 1015 – 130) P.O. Box 15-213, Hamilton – 131) Box 9540, Hamilton – 132) P.O. Box 19-298, Hamilton – 133) Box 230, Gisborne – 134) Box 1131, Taupo – 135) Motueka Rec Centre, Old Wharf Rd., Motueka – 136) P.O. Box 92, Auckland – 137) P.O. Box 68-886, Newton, Auckland 1 – 138) Box 962, Thames – 139) Box 2396, Wellington – 140) 710 Port Rd, Whangamata – 141) Liverpool Str, Wanganui – 142) Box 10-399, Wellington – 143) P.O. Box 622, Queenstown – 144) Box 92, Waitangi. – 145) Te Reo Irirangi o Maniapoto Te Kuiti – 146) Box 754, Palmerston North – 147) 12 John Str, Balclutha – 148) Box 25, Oneroa. – 149) Box 4232, Ngamotu, Palmerston North – 150) Box 87, Turangi – 151) 29 Carnarvon Str, Gisborne – 152) Box 77, Taumarunui – 153) Box 1162, Hastings – 154) Box 9969, Wellington – 155) Box 16, Whitianga – 156) Waiwhetu Marae, Riverside Dr, Lower Hutt – 157) Box 292, Gore – 159) Box 883, Wairau – 160) Box 9540, Hamilton – 161) Box 2139, Rotorua – 162) Box 68-100, Newton, Auckland – 163) Box 8379, Christchurch – 164) Box 202, New Plymouth – 165) Box 521, Ashburton 8300 – 166) Box 11-320, Ellerslie – 167) Private Bag 4750, Christchurch – 168) P.O. Box 8116, Symonds Street, Auckland 1035 – 169) Box 68-393, Auckland – 170) Suite 3, Westpac Plaza, Grey Str, Tauranga – 171) Box 54-052, Wellington – 173) Box 22-095, Christchurch – 175) Bowen House, 32 Bowen Str, Wellington – 176) Bay Rock, Tauranga – 177) Box 72, Papakura – 178) Box 155, Napier – 179) Northumberland Str, Waipukurau – 180) P.O. Box 1369, Wellington – 181) c/o Roy Gillions, 24 Ross Str, Lawrence – 182) Box 362, Dargaville – 183) P.O. Box 9113, Wellington.

NIUE ISLAND

L.T: UTC - 11h — **Pop:** 2.532 — **Radios:** 1.000 (est.) — **Pr.L:** Niuean, English — **E.C:** 50Hz, 230V — **ITU:** NIU.

BROADCASTING CORPORATION OF NIUE (BCN)

✉ P.O. Box 67, Alofi, Niue Isl. ☎ +683 4026. 🖷 +683 4217.
L.P: Chmn: Henry Eveni. GM: Hima Douglas. CE: Trevor Tiakia.
MW: ZK2ZN Alofi 594kHz 0.25kW — **FM:** 91MHz 0.1kW.
D.PRGR: 1730-2000, 2230-0030, 0500-0830. **N:** on the h.
ANN: "This is Radio Sunshine" — **V.** by letter.

NORFOLK ISLAND (Australian)

L.T: UTC + 11½h — **Pop:** 1.880 — **Radios:** 2000 — **Pr.L:** English, Pitcairn Norfolk — **E.C:** 50Hz, 220V — **ITU:** NFK.

NORFOLK ISLAND BROADCASTING SCE. (Gov.)

✉ P.O. Box 456, Norfolk Island 2899, Australia. ☎ +6723 2137. 🖷 +6723 3298 — **L.P:** Broadc. Mgr: Margaret Meadows.
FM: 89.9/93.9/95.9MHz 0.05kW.
Local Prgrs: 1830-0930 on 93.9MHz. **Rel. ABC Regional R:** 24h on 95.9MHz. **Rel. ABC Fine Music:** 24h on 89.9MHz.

NORTHERN MARIANA ISLANDS (U.S. Commonwealth)

L.T: UTC + 10h — **Pop:** 23.500 — **Radios:** 10.500 (est.) — **Pr.L:** English, Chamorro, Carolinian, Filipino — **E.C:** a/c 60, 110V — **ITU:** MRA.

INTER ISLAND COMMUNICATIONS INC. (Comm.)

✉ P.O. Box 914, Saipan, CM 96950. ☎ +670 234 7239. 🖷 +670 234 0447.
L.P: GM: Hans W. Mickelson. PD: Ken Warnick. CE: Angel Ocampo. N. Dir: Ken Phillips.
MW: Chalan Kiya: KCNM 1053kHz 1kW.
FM: KZMI 103.9MHz 3.2kW (stereo).
D.PRGR: 24h. **N:** KCNM: AP Network N. on the h — **ANN:** "KCNM (or) KZMI, Your Music Station" — **V.** by QSL-card & letter.

KRSI, PPP 413, Box 10000, Saipan 96950. ☎ +670 233 9801/2. +670 233 8200 — **L.P:** GM: Patrick Williams — **FM:** 97.9MHz 4.5kW — **ANN:** "Hot 98", "Saipan's Roots, Rock and Reggae Station".

KPXP, PPP 415, Box 10000, Saipan, MP 96950. ☎ +670 235 7996. ▤ +670 235 7998. **Web:** http://www.radiopacific.com/p99/
L.P: CEO: Rex Sorensen. SM: Jeanne Borger.
FM: 99.5MHz 6.5kW — **ANN:** "Power 99".

FAR EAST BROADCASTING CO. (Rlg.)

▤ Box 209, Saipan, CM 96950. ☎ +670 234 6520 (KSAI), +670 322 9088 (KFBS). ▤ +670 322 3060.
L.P: Dir: Chris Slabaugh. PD: Frank Gray. CE: Bob Springer. Head of Public Rel: Bob Stiles.
MW: Susupe (G.C: 145.50E/15.08N): KSAI 936kHz 10kW.
D.PRGR. in English: 2000-1200. **N.** on the h.
SW: see International Broadcasting section.
V. by QSL-card. Rp. (2 IRC's). Rec. acc.

CHRISTIAN SCIENCE PUBLISHING SOCIETY - SHORTWAVE BROADCASTS (Rlg.)

See International Broadcasting section.

PALAU

L.T: UTC + 9h — **Pop:** 17.285 — **Radios:** 9.000 (est.) — **Pr.L:** Palauan, English — **E.C:** 60Hz, 115/230V — **ITU:** PAL.

WSZB BROADCASTING STATION

▤ Box 279, Koror State, Republic of Palau 96940. ☎ +680 488 2417. ▤ +680 488 1932. ① 728890 — **L.P:** SM: Salustiano Albert.
MW: WSZB Malakal Island 1584kHz 5kW.
D.PRGR: W 2130-1500, Sun 2200-1400. **N:** on the h.
ANN: "Radio Palau".

KHBN (Rlg.)

See International Broadcasting section.

PAPUA NEW GUINEA

L.T: UTC + 10h — **Pop:** 4.597.955 — **Radios:** 234.000 — **Pr.L:** English, Melanesian Pidgin, Hiri Motu + 30 ethnic languages — **E.C:** 50Hz, 240V — **ITU:** PNG.

PAPUA NEW GUINEA NATIONAL BROADCASTING CORPORATION

▤ P.O. Box 1359, Boroko, N.C.D. ☎ +675 3253022. ▤ +675 3255403. ① 0703 ne 22348.
L.P: MD & Chief Exec: Renagi R. Lohia, CBE. Dep. MD: Tonny Boski. Mgr. Inf. Prgrs: Ephraim Tammy. Mgr. Kundu Sces: Don Penias. Mgr. Karai Sces: Iga Kila. Contr. Tech. Sces: John Waingut. Head, Op's & Maintenance: Bob Kabewa.
STATIONS: N=Karai Sce. (National), P=Kundu Sce. (Provincial).

kHz	kW	N	Station & h. of tr.
1) 585	10	N	Port Moresby: 24h
1) 675	2	N	Lae: 24h
1) 675	2	N	Wewak: 24h
1) 810	2	N	Rabaul: 24h
1) 864	2	N	Madang: 24h
1) 900	2	N	Goroka: 24h
20) 1494	10	P	NBC Wabag:MF 2200-0700
8) 1593	10	P	NBC Vanimo: MF 2200-0700
19) 2410	10	P	R. Enga: 2000-2200, 0700-1300
3) 3205	10	P	R. West Sepik: 2000-2200, 0700-1300
9) 3220	10	P	R. Morobe: 2000-2200, 0700-1300
2) 3235	10	P	R. West New Britain: 2000-2200, 0700-1300
10) 3245	10	P	R. Gulf: 2000-2200, 0700-1300
11) 3260	10	P	R. Madang: 2000-2200, 0700-1300
12) 3275	10	P	R. Southern Highlands: 2000-2200, 0700-1300
13) 3290	10	P	R. Central: 2000-2200, 0700-1300
14) 3305	10	P	R. Western: 2000-2200, 0700-1300
18) 3315	10	P	R. Manus: 2000-2200, 0700-1300
15) 3325	10	P	R. North Solomons: 2000-2200, 0700-1300
16) 3335	10	P	R. East Sepik: 2000-2200, 0700-1300
7) 3345	10	P	R. Northern: 2000-2200, 0700-1300
3) 3355	10	P	R. Simbu: 2000-2200, 0700-1300
17) 3365	10	P	R. Milne Bay: 2000-2200, 0700-1300
6) 3375	10	P	R. Western Highlands: 2000-2200, 0700-1300
1) 3385	10	P	R. East New Britain: 1900-2200, 0700-1200
4) 3395	10	P	R. Eastern Highlands: 2000-2200, 0700-1300
5) 3905	10	P	R. New Ireland: 2000-2200, 0700-1300
1) 4890	100	N	Port Moresby: 24h
6) 5965	10	N	NBC Mt. Hagen: MF 2200-0700
17) 6040	10	N	NBC Alotau: MF 2200-0700
14) 6080	10	N	NBC Daru: MF 2200-0700
16) 6140	10	N	NBC Wewak: MF 2200-0700
1) 9520	100	N	Port Moresby: 2200-0700
1) 9675	100	N	Port Moresby: 2200-0700
1) 11880	10	N	Port Moresby: 2200-0700

Commercial Netw: R. Kalang, P.O. Box 1534, Pt. Moresby. ☎ +675 3259796. ▤ +675 3251747 — **L.P:** MD: John Malisa. CE: V. Kuppusamy — **H. of tr:** 1900-1400 — **FM:** Lae 100.3MHz 1kW, Rabaul 100.8MHz 1kW + low power st's on 100.2/100.3/100.4/100.5/100.8/101.0/102.0MHz.

Karai Service: Addr. & **L.P:** see above. **N. in English:** 2000W, 2045, 2145W, 2245Sun, 0230, 0330 (not. Sat), 0600MF, 0800 (not Sat), 0900, 1100, 1300. **V.** by QSL-card. Rp. Specific prgr. details, time, frequency, date req. for verifications.

Kundu (Provincial) Stations

2) P.O. Box 412, Kimbe. Mgr: Ruben Bale. **ANN:** "Maus bilong Tavur" – 3) P.O. Box 228, Kundiawa. Mgr: Luck Umbo. **ANN:** "Karai bilong Mambu" – 4) P.O. Box 311, Goroka. Mgr: Tonko Nonao. **ANN:** "Karai bilong Kumul" – 5) P.O. Box 140, Kavieng. Mgr: O. Malatanai. **ANN:** "Maus bilong Mai Mai" – 6) P.O. Box 311, Mount Hagen. Mgr: Winterford Suharupa. **ANN:** "Nek bilong Tarangau" – 7) P.O. Box 137, Popondetta. Mgr: Gerald Didymus. **ANN:** "Voice of Oro" – 8) P.O. Box 37, Vanimo. Mgr: Elias Rathley. **ANN:** "Maus bilong Sandaun" – 9) P.O. Box 1262, Lae. Mgr: Peter Manau. **ANN:** "Maus bilong Kundu" – 10) P.O. Box 36, Kerema. Mgr: Robin Wainetti. **ANN:** "Voice of the Seagull" – 11) P.O. Box 2138, Madang. Provincial Prgr. Mgr: Geo Gebading. **ANN:** "Maus bilong Garamut" – 12) P.O. Box 104, Mendi. Mgr: Andrew Meles. **ANN:** "Karai bilong Muruk" – 13) P.O. Box 1359, Boroko. Mgr: Akuss Matiki. **ANN:** "Kibi ena Gadona" – 14) P.O. Box 23, Daru. Mgr: Geo Gedabing. **ANN:** "Voice of the Sunrise" – 15) P.O. Box 35, Kieta. Mgr: Demas Kumaina. **ANN:** "Maus bilong Sunkamap" – 16) P.O. Box 65, Wewak. Mgr: Tonny Waine. **ANN:** "Nek bilong Sepik" – 17) P.O. Box 111, Alotau. Mgr: Daniel Mailau. **ANN:** "Voice of Kula" – 18) P.O. Box 505, Lorengau. Mgr: John Mandrakamo. **ANN:** "Maus bilong Chauka" – 19) P.O. Box 196, Wabag. Mgr: Roberto Papuvo. **ANN:** "Karai bilong Miok" – **V.** Most st's verify direct.

NAU FM & YUMI FM (Comm.)

▤ P.O. Box 744, Pt. Moresby, NCD. ☎ +675 3201996. ▤ +675 3201995. **Web:** http://www.datec.com.au/naufm/
L.P: GM (Nau FM): Mark Rogers. GM (Yumi FM): Justin Kili. CE: Alwin Agonia.

FM (MHz)	Nau FM	Yumi FM	kW
Goroka	96.1		1
Lae	96.3	93.7	1
Madang	96.3	93.7	1
Mt. Hagen	96.9	93.5	1
Port Moresby	96.5	93.1	1
Rabaul		93.9	1

D.PRGR: 24h (Nau FM aimed at young people, Yumi FM for an older audience).

POLYNESIA (French)

L.T: UTC - 10h — **Pop:** 215.000 — **Radio sets:** 105.000 — **Pr.L:** French, Tahitian — **E.C:** 60Hz, 220V — **ITU:** OCE.

SOCIÉTÉ NATIONALE DE RADIO TÉLÉVISION FRANÇAISE D'OUTRE MER (RFO)

▤ B. P. 125, Papeete, Tahiti. ☎ +689 430551. ▤ +689 413155. **Cable:** 291FP FROMRPT. ① 070200.
Web: http://www.tahiti-explorer.com/rfo.html
L.P: Dir: Claude Ruben . Editor-in-Chief: Patrick Durand-Gaillard. Dir. of Prgrs: Jean-Raymond Bodin.
MW: Mahina 738kHz 20kW — **SW:** Mahina (G.C: 149.00W/ 17.00S): 6135kHz 4kW, 11827/15168vkHz 20kW.
FM: Papeete 89.0MHz + 4 st's.

D.PRGR: 24h. Local N. in French: 1730. **Rel. France-Inter:** on the h. (N.) and during the evening/night hours.
ANN: F: "Ici Tahiti, Société Nationale de Radio-Télévision Française d'Outre Mer". Tahitian: "O Tahiti teie te RFO".
IS: Tahitian flute (vivo) and drums — **V.** by QSL-card.

PRIVATE STATIONS

Kiss FM, B.P. 4552, Papeete — **Radio 1,** B.P. 3601, Fare Ute, Papeete — **R. Maohi,** Maison des Jeunes de Pirae — **R. Papeete,** Ex Magasin Arupa, Papeete — **R. Soleil,** Qtr du Commerce, Papeete — **R. Tahiti-FM,** Mahina — **R. Tahiti Api,** Pirae — **R. Te Reo o Tefana,** B.P. 13069, Punaauia — **R. Tiare,** Fare Ute, Papeete.

SAMOA

L.T: UTC - 11h — **Pop:** 224.601 — **Radios:** 75.000 — **Pr.L:** Samoan, English — **E.C:** 50Hz, 230/410V — **ITU:** SMO.

SAMOAN BROADCASTING SERVICE – 2AP (Gov. Comm.)
📧 P.O. Box 1868, Mulinu'u, Apia. ☎ +685 21420. 📠 +685 21072.
L.P: DG: Kolotita Stowers Ah Kau. CE: Lua Nafoi. Prgr. Mgr: John Solofa. Tech. Supt: Lua Nafoi. Sales Mgr: Patrick Mamaia.
MW: Apia 540/747kHz 10kW.

Ch. 1 in Samoan/English: 1600 (Sun 2200)-1000 on 540kHz.
World N: 1800W, 1900W, 1930W. **Local N:** 1630W, 1730W, 1830W, 0730 — **Ch. 2 in English:** Sun 1800-0500 on 747kHz.
ANN: E: "Samoa Broadcasting Service, 2AP". "This is Channel 2 of the Samoan Broadcasting Sce" — **IS:** Gong — **V.** by letter.

RADIO POLYNESIA LTD. (Comm.)
📧 P.O. Box 762, Apia. ☎ +685 25150. 📠 +685 25147.
FM: 98.1/99.9*MHz — **D.PRGR:** 24h — **ANN:** "Magic 98 FM".
*) 99.9MHz does not operate on Sundays.

RADIO GRACELAND (Rlg.)
📧 Apia — **FM:** 90.1/106.1MHz.
D.PRGR: Sun 1800-1100 (gospel music).

SAMOA (American)

L.T: UTC - 11h — **Pop:** 53.000 — **Radio sets:** 20.000 (est.) — **Pr.L:** Samoan, English — **E.C:** 60Hz, 120V — **ITU:** SMA.

WVUV, Box 4894, Pago Pago, AS 96799, USA. ☎ +684 688 7397 — **MW:** 648kHz 10kW 24h.
Samoa Technologies Inc, Box 793, Pago Pago 96799. ☎ +684 633 7000. 📠 +684 633 5727 — **FM:** KSBS-FM 92.1MHz 15kW 18h.

SOLOMON ISLANDS

L.T: UTC + 11h — **Pop:** 441.325 — **Radios:** 45.000 — **Pr.L:** Pidgin, English — **E.C:** 50Hz, 240V — **ITU:** SLM.

SOLOMON ISLANDS BROADCASTING CORPORATION (Statutory Authority, Comm.)
📧 P.O. Box 654, Honiara. ☎ +677 20051. 📠 +677 23159. **Cable:** Broadcast Honiara. ① SIBC HQ66406.
L.P: GM: James Kilua. PD: David Palapu. CE:Cornelius Rathamana. Head of N: Dykes Angiki. Head of Finance & Admin: Duddley Marau.
MW: Honiara 1035kHz 10kW(d), Gizo 945kHz 10kW, Lata 1386kHz 5kW — **SW:** (G.C: 160.03E/09.25S): 5020/9545kHz 10kW.
D.PRGR: 1900-1130.
N. in English: 2000, 2200 (R. Australia), 0130W (local), 0200 (R. Australia), 0500W (local), 0600 (BBC), 0730 (local), 1000 (R. Australia), 1100 (local).
ANN: "This is the SIBC, Radio Hapi Isles".
IS: Drum and Bamboo Pipes.
V. by QSL-card. Rp. Re. must contain prgr. details.

TONGA (Kingdom of)

L.T: UTC + 13h — **Pop:** 108.090 — **Radios:** 40.000 — **Pr.L:** Tongan, English — **E.C:** 50Hz, 240V — **ITU:** TON.

TONGA BROADCASTING COMMISSION (Independent Statutory Board, part-comm.)
📧 P.O. Box 36, Nuku'alofa. ☎ +676 23295. 📠 +676 24417. ① 66225 Gentel TS — **L.P:** GM: Tavake Fusimalohi. Dep. GM: Ahongalu Fusimalohi. Ag. CE: Sioeli Maka Tohi. Editor: Miss Nanisc Fifita. Contr. Sales & Mktg: Lavinia Vikilani.
MW: A3Z 1017kHz 10kW — **FM:** 97.2MHz 0.1kW (R. Tonga 2).
SW: 5030kHz 1kW (inactive but due to be reactivated soon).
R. Tonga 1 on 1017kHz: 1750-1200. **N. in English:** 1800 (BBC), 1900 (ABC), 0000 (ABC or RNZ), 0700 (local), 0715 (ABC).
R. Tonga 2 on 97.2MHz: 1750-1200.

93FM (Rlg.)
📧 United Christian Broadcasters, Box 478, Nuku'alofa. ☎📠 +676 23076 — **FM:** A3R 93.1MHz 0.2kW (mostly music).

TUVALU

L.T: UTC + 12h — **Pop:** 11.000 — **Radios:** 3.000 — **Pr.L:** English, Tuvaluan — **E.C:** 50Hz, 240V (Funafuti only) — **ITU:** TUV.

RADIO TUVALU (Gov.)
📧 Private Mail Bag, Vaiaku, Funafuti. ☎ +688 20139. 📠 + 688 20732. ① 4800 TUV. **Cable:** Broadcast Tuvalu — **L.P:** Broadc. & Inf. Officer: Pusinelli Laafai. Head of Tech. Sces: John Sammons. Prgr. Producer: Mrs. Falahea Haleti. N. Editor: Faauoa Maani.
MW: Funafuti T2U2 621kHz 5kW — **D.PRGR:** 1852-2000, 2325-0100, 0550-0900W. **N. in English:** 1910, 0710.
ANN: E: "This is Radio Tuvalu" — **V.** by letter.

VANUATU (Republic of)

L.T: UTC+11h (Oct-March UTC + 12h) — **Pop:** 185.472 — **Radios:** 55.000 — **Pr.L:** Bislama, English, French — **ITU:** VUT.

VANUATU BROADCASTING AND TELEVISION CORPORATION (VBTC)
📧 P.M.B. 049, Port Vila. ☎ +678 22999, 23051. 📠 +678 22026. ① 1046.
L.P: Head of Prgrs: Abong Thompson.
MW: Emten Lagoon 1125kHz 10kW, Santo 1179kHz 2kW.
SW: Emten Lagoon 3945/4960kHz 10/2.5kW
FM: Santo 98.5MHz, Port Vila (separate prgrs) 98.0MHz.
N. B. Stations closed temporarily (Oct 98).
R. Vanuatu AM: 1900-1115 (Sun 1000). **On 3945kHz:** 1900-2300, 0600-1115 (Sun 1000). **On 4960kHz:** 2200-0700. **N. in English:** 0100W, 0900MF. **N. in French:** 0100W, 1200MF. **Rel. N. from foreign broadcasters:** 2000 (RA English), 2100 (RFI or VOA French), 2200 (exc. Fri/Sat. BBC or RA or RNZI English), 2300 (exc. Fri/Sat. RA or BBC English), 0000MF (RFI French).
N.B. Services reduced to 6 hours AM +4 hours PM.
ANN: "Radio Vanuatu", "Yu stap haren naoia Radio Vanuatu"
Nambawan FM: 1900-1100.

WALLIS & FUTUNA (French)

L.T: UTC + 12h — **Pop:** 18.128 — **Pr.L:** French, Wallisian — **ITU:** WAL.

RADIODIFFUSION FRANÇAISE D'OUTRE MER (RFO)
📧 B. P. 102, 97911 Mata-Utu, Iles de Wallis et Futuna (par Nouméa, Nouvelle-Calédonie). ☎ +681 722020. 📠 +681 722346 — **L.P:** SM: Joseph Blasco (☎ +681 722419). Head of Inf: Bernard Joyeux. ☎ +681 722818. 📠 +681 722713).
MW: Mata-Utu 1188kHz 2kW — **D.PRGR:** 1800-1000 — **ANN:** "Bonjour, vous êtes sur RFO Wallis et Futuna" — **V.** by letter.

NORTH AMERICA

L.T: UTC – 9h (Westernmost Aleutian Is. – 10h) (cf. World Time Table) — **Pop:** 609.300 — **Pr.L:** English — **E.C:** a/c 60, 120/240V — **ITU:** ALS.

ALASKA BROADCASTERS ASSOCIATION

✉ c/o KIAK, Box 73410, Fairbanks 99707 — Pres: Peter Van Nort.

	Call	kHz	kW	h. of tr.
1)	KUHB	540	2.5	F.Pl.
2)	KTZN	550	5	24h
3)	KVOK	560	1	24h
4)	KRSA	580	5	24h
5)	KHAR	590	5	24h
6)	KGTL	620	5	W 1400-0906
7)	KJNO	630	5/1	1400-0900 (Fri/Sat 24h)
8)	KIAM	630	5/3.1	1600-0800
9)	KYUK*	640	10	1500 (Sun 1700)-0930 (Sat 1030)
2)	KENI	650	50	24h
11)	KFAR	660	10	24h
12)	KDLG*	670	10	1500-0900 (June-Aug: 24h) (Sat, Sun 1600-1100)
14)	KBYR	700	10	24h
15)	KOTZ*	720	10	1500-0900
16)	KFQD	750	50	24h
17)	KCHU*	770	9.75	1500-0900
18)	KNOM*	780	25/14	1455-1110
19)	KCAM	790	5	24h
20)	KINY	800	10/7.8	24h
21)	KCBF	820	10	24h
22)	KSDP*	830	1	as 12)
23)	KABN	840	8	**
24)	KICY	850	50	1445-0930
25)	KSKO*	870	10	24h
26)	KBBI*	890	10	1430 (Sat/Sun 1500)-0900
27)	KZPA*	900	5	1400-0800
28)	KIYU*	910	5	24h
29)	KSRM	920	5	24h
30)	KTKN	930	5/1	1400-0900
31)	KNSA*	930	2.5	as 12)
32)	KSWD	950	1	**
33)	KIAK	970	5	24h
34)	KZXX	980	1	24h
35)	KAXX	1020	10	24h
36)	KASH	1080	10	24h
29)	KSLD	1140	10	24h
37)	KJNP	1170	50/23	1400-1100
38)	KVAK	1230	1	24h
39)	KIFW	1230	1	1500-0900
40)	KFSH	1240	1	F.Pl.
41)	KLAM	1450	0.25	24h

*) non-comm — **) silent or irregular operation at editorial deadline.
Stations below 250W not mentioned.

Addresses and other information:

Abbreviations: See under U.S.A. commercial st's.
Addresses: Add state abbreviation AK between city and zip code.
1) Pribilof School District, St. Paul 99660. **FM:** KUHB-FM 91.9MHz 3kW. Mostly rel. 12).
2) 800 E. Diamond Blvd, Ste. #3-320, Anchorage 99515 — GM: Gary Donavan (KTZN), Andrew Lohman (KENI). PD: Wayne Maloney (KENI), Ray Knight (KASH-FM). CE: Van Craft. KTZN: Sports. KENI: News/talk. KBFX: OLD. KASH-FM: C&W. **FM:** KBFX 100.5MHz 25kW, KASH-FM 107.5MHz 100kW.
3) 1227 Mill Bay Rd, Kodiak 99615 — GM: Andrew Tierney. OLD, talk. **FM:** KRXX 101.1MHz 3kW.
4) Box 650, Petersburg 99833 — GM: Andrew Mazzella. CE: Daniel Zachary. Variety.
5) 11259 Tower Rd, Anchorage 99515 — GM: Don Nordin. KHAR: AS. KEAG, KKRO: OLD. KBRJ: C&W. **FM:** KEAG: 97.3MHz 100kW, KBRJ 104.1MHz 55kW, KKRO 102.1MHz 23kW.
6) Box 109, Homer 99603 — GM/CE: David F. Becker. PD: Tim White. AS. **FM:** KWVV-FM 103.5MHz 100kW, KPEN-FM 101.7MHz

25kW.
7) 3161 Channel Drive, #2, Juneau 99801 — GM: Steve Rhyner. PD: Dorothy Michaels. CE: Brian Romeijn. Oldies based ADC. **FM:** KTKU 105.1MHz 4kW.
8) Box 00474, Nenana 99760 — GM: Robert Eldridge. PD: Kelvin Schubert. CE: Timothy Zook. Rlg.
9) Pouch 468, Bethel 99559 — GM: John McDonald. PD: Kate Hamilton. CE: B. Humelsine. Variety. Bilingual English/Yupik.
11) 1060 Aspen Str, Fairbanks 99709 — GM: Terry Walley. PD: Perry Walley. CE: Chuck Beck. KFAR: N/t. KWLF, KUWL: ADC. **FM:** KWLF 98.1kHz 28kW, KUWL 103.9MHz 2.9kW.
12) Box 670, Dillingham 99576 — GM, CE: Rob Carpenter. PD: Michelle M. Abrams. Variety.
13) Box 109, Barrow 99723 — Mgr: Don Rinker. PD: S. Hamlin. CE: Charles Lakaytis. Variety 20% Eskimo. **FM:** KBRW-FM 91.9MHz 0.89kW.
14) 1007 W. 32nd, Anchorage 99503 — GM: Rob Gottstein. PD: Bob Dean. CE: Duane Milsap. KBYR: Talk, sports. KNIK-FM: C%W. **FM:** KNIK-FM 105.3MHz 45kW.
15) Box 78, Kotzebue 99752 — GM: Suzy Erlich. PD: Karen Sherman. CE: Pierre Lonewolf. Variety. Bilingual English/Inupiaq.
16) 9200 Lake Otis Parkway, Anchorage 99507 — GM: Scott Smith. PD: Michael Rogers. CE: Jay White. KFQD: N/t. KWHL: Rock. KMXS: ADC. **FM:** KWHL 106.5MHz 100kW, KMXS 103.1MHz 27kW.
17) Box 467, Valdez 99686 — GM: Shanna Simmons. PD: Greg Williams. Variety. Partly relays KUAC 89.9MHz Fairbanks.
18) Box 988, Nome 99762 — GM: Tom A. Busch. PD: Katy Clark. CE: Tim Cochran. Variety. **FM:** 96.1MHz 0.1kW.
19) Box 249, Glennallen 99588 — GM, PD: Scott Yahr. CE: Dan J. Zachary. 50% C&W, 50% rlg.
20) 1107 8th St. W, #2, Juneau 99801 — GM: Dennis W. Egan. PD: Paul Ryder. CE: Charles Gray. ADC. **FM:** KSUP 106.3MHz 10kW.
21) 3528 International Way, Fairbanks 99701. **WWW:** http://www.kcbf.com — GM: Jerry Bever. PD: Bill Holzheimer. CE: John Antonuk. OLD. **FM:** KXLR 95.9MHz 25kW.
22) Box 328, Sand Point 99661 — Mgr: Roger L. Daniels. Variety. Mostly rel. 12).
23) Long Island.
24) Box 820, Nome 99762 — GM: John McBride. PD: Daniel A. Smith. CE: Tom Guilliam. ADC, rlg. **Russian** 0800-0930, **Eskimo** MF0230-0300. **FM:** 100.3MHz 0.084W.
25) Box 70, McGrath 99627 — GM: Amie Hind. CE: Ab Ross. Variety.
26) 3913 Kachemak Way, Homer 99603 — GM: Dave Hamock. PD: Kathy Steberl. CE: S. Morton. Variety, community radio.
27) Box 50, Fort Yukon 99740 — SM: Dorothy Carroll. Variety. Mostly rel. KUAC-FM 89.9MHz Fairbanks.
28) Box 165, Galena 99741 — GM: Robert C. Sommer. PD: Scott Campbell. Variety (mostly rel. 25).
29) HC 2 Box 852, Soldotna 99669. **E-mail:** ksrm@ptialaska.net — GM: John Davis. PD: Tom Farrell. CE: Richard Zook. KSRM: Talk, ADC, KSLD: OLD. **FM:** KWHQ 100.1 MHz 3kW. KKIS-FM 96.5MHz 10kW.
30) 526 Stedman St, Ketchikan 99901 — GM: Kent Colby. CE: Ken Eckland. ADC, talk. **FM:** KGTW 106.7 4kW.
31) #32 Airport Heights, Unalakleet 99684 — GM: Henry Ivanoff. Mostly rel. 12).
32) Seward 99664.
33) Box 73410, Fairbanks 99707 — GM: Peter Van Nort. CE: C. Beck. KIAK: N/t. KIAK-FM: C&W. KAKQ-FM: ADC. **FM:** KIAK-FM 102.5MHz 26kW, KAKQ-FM 101.1MHz 25kW.
34) 6672 Kenai Spur Hwy, Kenai 99611 — GM: Jim Wenstrom. PD: Paula Richardson. Oldies.
35) 9200 Lake Otis Parkway, Anchorage 99507 — Sports. (Lic. to Eagle River)
36) 3601 C Street, #290, Anchorage 99503 — Classical. **FM:** KLEF 98.1MHz 25kW.
37) Box 56359, North Pole 99705 — Pres: Donald L. Nelson. GM: Roger Skold, PD: Beverly Olson. CE: Reginald Swedberg. Rlg. **Eskimo/Indian:** Sun 0600-0900. **FM:** KJNP-FM 100.3MHz 25kW.
38) Box 367, Valdez 99686 — GM: Laurie Prax. ADC.
39) 611 Lake St, Sitka 99835 — GM: Bobbi Rusk. CE: Brian Romeijn. ADC. **FM:** KSBZ 103.1MHz 3kW.
40) Seward 99664.
41) Box 60, Cordova 99574 — GM/PD: J.R. Lewis. CE: Van Craft. Eclectic mx.

ALASKAN FORCES RADIO NETWORK (AFRN)

Netw. Addr: Air Force Arctic Broadcasting Squadron, Attn: Station Manager, Elmendorf AFB, Alaska 99506-5000.

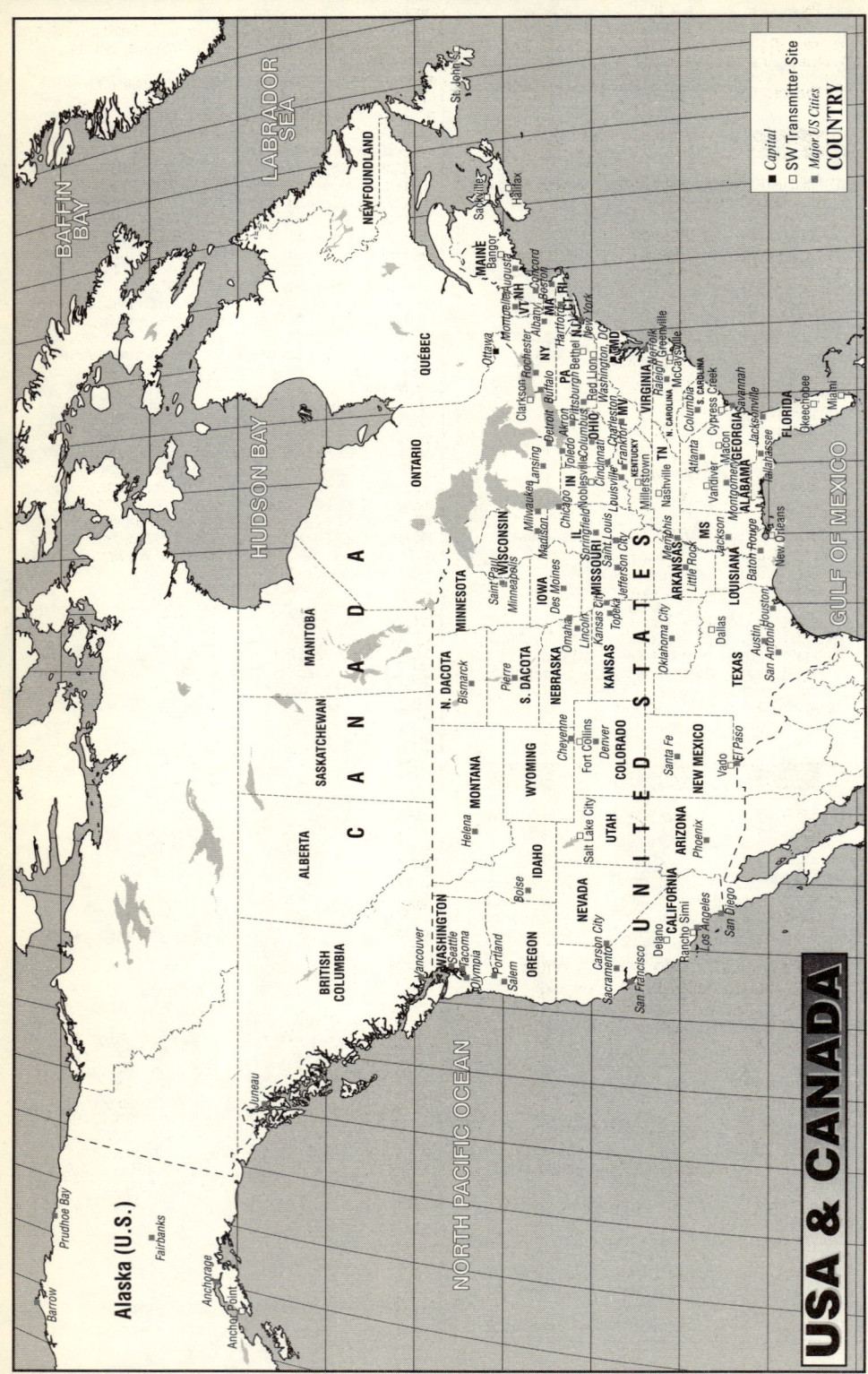

Addr of AFRN sts with own studios: OLGF, AFABS, Attn: Station Manager, APO Seattle, WA 98733, USA. (Fort Greely).
FM: Fort Greely 90.5/93.5MHz 0.3kW, Adak 101.0/103.0MHz 0.25kW. Local prgrs + AFRTS network.

KNLS — The New Life St. (Rlg.)

BERMUDA

L.T: UTC - 4h (Su: UTC - 3h) — **Pop:** 61.100 — **Radios:** 80.000 — **Pr.L:** English — **E.C:** 60Hz, 115/230V — **ITU:** BER.

ZBM-ZFB RADIO & TELEVISION (Comm.)
✉ Box HM 452, Hamilton HM BX (**Studios:** Fort Hill Rd, Devonshire). ☎ +1 (441) 295 2828. 🖷 +1 (441) 295 4282. ✆ 3702 ZBM-BA. **E-mail:** zbmtv@ibl.bm
L.P: Ag. GM: M.R. Fletcher. CE: D. Ingham. Prgr. Mgr: B. Lodge.
MW (1kW): ZFB-AM 1230kHz — ZBM-2 1340kHz.
FM: ZBM-FM 89.1MHz 15kW (stereo), ZFB-FM 94.9MHz 1kW.
ZBM: W 1000-0300, Sun 1100-0200. **N, Wrp:** W 1025, 1055, 1125, hourly 1200-0300, also 1608. Sun 1100, 1200, 1300, 1400, 1455, 1615, hourly 1700-0200.
ZFB: 1000 (Sun 1100)-0400. **N:** W 1055, 1130 and every h to 2230, 0230. Sun 1330, 1630, 2230, 0230. **Portuguese:** W 0930-1030, Mon 2330-0100, Fri 2300-0100.
ZBM-FM: 24h music prgr. **N:** 1100, 1200, 1600, 2200.
ZFB-FM: 24h music prgr. **N:** 1200, 1700, 2400.
V. by QSL-card or letter. Rp. Rec. acc.

DEFONTES BROADCASTING CO. LTD. (Comm.)
✉ P.O. Box 1450 (**studios:** 94 Reed Str.) Hamilton 5. ☎ +1 (441) 295 1450. 🖷 +1 (441) 295 1658. **E-mail:** vsbnews@ibl.bm.
L.P: Pres: Kenneth deFontes. SM: D.E. Browne. N. Dir: M. De Leon. TD: Ed. Tucker.
MW: 1160/1280/1450kHz 1kW — **FM:** 106.1MHz 2.5kW.
VSB1 on 1450kHz: 24h (country music). **N:** 1000, 1045, 1115, 1145, 1215, 1245, 1415, 1615, 1645, 1715, 1915, 2115, 2215, 0015 — **VSB2** on 1280kHz: 24h (religious) — **VSB3** on 1160kHz: 24h. Rel. BBC World Sce 1530-1500. Other times tourist info & special prgrs — **VSB-FM** on 106.1MHz: 24h (adult contemporary).
V. by QSL-card or letter.

CANADA

L.T: See World Time Table — **Pop:** 29.784.000 — **Radios:** 22.600.000 — **Pr.L:** English, French — **E.C:** 60Hz, 115/230V (generally). — **ITU:** CAN.

Abbreviations: NF=Newfoundland, NS=Nova Scotia, PEI=Prince Edward Isl, NB=New Brunswick, PQ=Quebec, ON=Ontario, MB=Manitoba, SK=Saskatchewan, AB=Alberta, BC=British Columbia, NWT=North West Territories, Lab=Labrador, YT=Yukon Territory.

CANADIAN BROADCASTING CORPORATION SOCIÉTÉ RADIO-CANADA (Publicly owned)
✉ Head Office: P.O. Box 8478, Ottawa, Ontario K1G 3J5. ☎ +1 (613) 724-1200. 🖷 +1 (613) 738-6749. **Cable:** Broadcasts. ✆ 053-4260.
Web: http://www.cbc.ca
L.P: Pres. and Chief Exec. Officer, Perrin Beatty. Chmn, Board of Directors: Guylaine Saucier. Sen. Vice-Pres, Resources: Louise Tremblay. Sen. Vice-Pres, Media: Vacant. Sen. Advisor to the Pres. and CEO: Michael McEwen. Vice-Pres, Internal Audit: Robert Hertzog. Vice-Pres, Gen. Counsel and Corp. Secr: Gerald Flaherty, Q.C. Vice-Pres, Human Resources: George C. B. Smith. Exec. Dir, Media Accountability: Donna Logan. Sen. Dir. of Corporate Communications and Public Affairs: Charlotte O'Dea.
English Networks: ✉ P.O. Box 500, Station "A", Toronto, ON M5W 1E6. ☎ +1 (416) 975-3311. **Cable:** Broadcast. ✆ 062-17796.
Web: http://www.radio.cbc.ca/
L.P: Vice-Pres, English Radio: Harold Redekopp. Head, CBC Newsworld, Slawko Klymkiw. Exec. Dir, Media Operations: Michael Harris. Sen. Dir, Media and Public Rel: Tom Curzon. Sen. Dir, Broadcast Communications: Diane Kenyon.
French Networks: ✉ P.O. Box 6000, Montreal, PQ H3C 3A8. ☎ +1 (514) 597-5970. **Cable:** Radcan. ✆ 05-267417.

Web: http://www.radio-canada.com/
L.P: Vice-Pres, French Radio: Marcel Pépin. DG, Communications: Raymond Guay. Dir. of Public Rel: Micheline Savoie. Exec. Dir: RDI, Renaud Gilbert
CBC Engineering: 7925 Côte St. Luc Rd., Montreal, PQ H4W 1R5. ☎ +1 (514) 485-1301. ✆ 055-66437.
L.P: Sen. Dir, Eng: Brian D. Baldry

English Network (CBC): =also on Shortwave.

	Call	kHz	kW			Call	kHz	kW
1)	CBT	540	10		14)	CBZ	970	10
2)	CBK	540	50		15)	CBY	990	10
3)	CFWH	570	5/1		16)	CBW	990	50/46
4)	CBN	640	10		17)	CBR	1010	50
5)	CBU	690	50		18)	CBA	1070	50
6)	CBL	740	50		19)	CBI	1140	10
7)	CBX	740	50		21)	CFFB	1230	1
10)	CFPR	860	10		23)	CFYK	1340	2.5
11)	CHAK	860	1		24)	CBG	1400	1
13)	CBM*	940	50		25)	CBE	1550	10

*) to be discontinued, replaced by FM 88.5MHz from fall 1998.
CBC English mono distributed through 416 rebroadcasters.
Shortwaves:
4) CKZN 6160kHz 0.3kW: 0930-0500 (rel. CBN)
5) CKZU 6160kHz 0.5kW: 1400 (Sun 1500)-0905 (rel. CBU).
FM: CBM (Montreal) 88.5MHz (from spring 1998), CFGB-FM (Happy Valley/Goose Bay) 89.5MHz 4.5kW, CBHA-FM (Halifax) 860MHz 90.5kW, CBCT-FM (Charlottetown) 96.1MHz 100kW, CBD-FM (Saint John) 91.3MHz 100kW, CBVE-FM (Quebec) 104.7MHz 100kW, CBO-FM (Ottawa) 92.0MHz 84kW, CBCS-FM (Sudbury) 99.9MHz 50kW, CBQT-FM (Thunder Bay) 88.3MHz 10/5kW, CBWK-FM (Thompson) 100.9MHz 0.094kW, CBKA-FM (La Ronge) 105.9MHz 0.08kW, CBQR-FM (Kivalik) 105.1MHz 100.2kW, CBTK-FM (Kelowna) 88.9MHz 4.7kW, CBYG-FM (Prince George) 91.5MHz 100kW.

FM (stereo):

Call	MHz	Location
CBX-FM	90.9	Edmonton
CBBC-FM	91.7	Lethbridge
CBU-FM-1	92.1	Victoria
CBWS-FM	92.7	Brandon
CBBK-FM	92.9	Kingston
CBM-FM	93.5	Montreal
CBL-FM	94.1	Toronto
CBA-FM	95.5	Moncton
CBK-FM	96.9	Regina
CBW-FM	98.3	Winnipeg
CBE-FM	98.9	Windsor
CBBL-FM	100.5	Toronto
CBZ-FM	101.5	Fredericton
CBQ-FM	101.7	Thunder Bay
CBR-FM	102.1	Calgary
CBH-FM	102.7	Halifax
CBOQ-FM	103.3	Ottawa
CBBP-FM	103.9	Peterborough
CBU-FM-2	105.1	Metchosin/Sooke
CBI-FM	105.1	Sydney
CBKS-FM	105.5	Saskatoon
CBU-FM	105.7	Vancouver
CBN-FM	106.9	St. John's

French Network (SRC):

	Call	kHz	kW			Call	kHz	kW
28)	CBGA-1	540	10		2)	CBKF-2	860	10
25)	CBEF	540	2.5/5		26)	CJBR	900	10
7)	CHFA	680	10		27)	CBV	980	50
13)	CBF	690	50		16)	CKSB	1050	10
2)	CBKF-1	690	5		28)	CBGA	1250	10/5
6)	CJBC	860	50		29)	CBJ	1580	50

CBC French radio is also distributed through 159 rebroadcasters.
FM: CBKF-FM (Regina) 97.7MHz 13.7kW, CBON-FM (Sudbury) 98.1MHz 50kW, CBSI-FM (Sept Iles) 98.1MHz 96.7kW, CBAF-FM-5 (Halifax) 92.3MHz 100kW, CBOF-FM 90.7MHz 84kW, CBAF-FM 88.5MHz 22kW.

FM (stereo):

Call	MHz	Location
CJBC-FM	90.3	Toronto
CBF-FM-2	90.7	Sherbrooke
CBV-FM	95.3	Quebec
CBUF-FM	97.7	Vancouver
CBAL-FM	98.3	Moncton
CBF-FM	100.7	Montreal

Call			Freq	Location
CBJ-FM			100.9	Chicoutimi
CJBR-FM			101.5	Rimouski
CBOX-FM			102.5	Ottawa
CBF-FM-1			104.3	Trois Rivières

Addresses:
1) P.O. Box 218, Grand Falls, NF, A2A 2J7.
Web: http://www.radio.cbc.ca/radio/regional/Newfoundland/
2) 2440 Broad Str, Regina, SK, S4P 4A1
Web: http://www.radio.cbc.ca/radio/regional/Saskatchewan/
3) 3103 — 3rd Avenue, Whitehorse, YT, Y1A 1E5.
E-mail: cbc@yknet.yk.ca
Web: http://www.radio.cbc.ca/radio/regional/North/Yukon/
4) P.O. Box 12010, St. A, St. Johns, NF, A1B 3T8.
Web: http://www.radio.cbc.ca/radio/regional/Newfoundland/
5) P.O. Box 4600, Vancouver, BC, V6B 4A2.
Web: http://www.radio.cbc.ca/radio/regional/BC/
6) P.O. Box 500, Station "A", Toronto, ON, M5W 1E6
Web: http://www.radio.cbc.ca/radio/regional/Ontario/
7) P.O. Box 555, Edmonton, AB, T5J 2P4.
Web: http://www.radio.cbc.ca/radio/regional/edmonton/
10) 222 3rd Ave. W, Prince Rupert, BC, V8J 1L1.
Web: http://www.radio.cbc.ca/radio/regional/BC/
11) Mackenzie Rd, Bag Sce. No. 8, Inuvik, NWT, X0E 0T0.
13) P.O. Box 6000, Montreal, PQ, H3C 3A8.
Web: http://www.radio.cbc.ca/radio/regional/Quebec/
14) P.O. Box 2200, Fredericton, NB, E3B 5G4. **Web:**
http://www.radio.cbc.ca/radio/regional/Maritimes/Fredericton/
15) P.O. Box 610, Corner Brook, NF, A2H 6G1.
Web: http://www.radio.cbc.ca/radio/regional/Newfoundland/
16) P.O. Box 160, Winnipeg, MB, R3C 2H1.
Web: http://www.radio.cbc.ca/radio/regional/Manitoba/
17) P.O. Box 2640, Calgary, AB, T2P 2M7.
Web: http://www.radio.cbc.ca/radio/regional/calgary/
18) P.O. Box 950, Moncton, NB, E1C 8N8. **Web:**
http://www.radio.cbc.ca/radio/regional/Maritimes/Moncton/
19) 285 Alexander Str, Sydney, NS, B1S 2E8 . **E-mail** infoam@atcon.
com **Web:** http://www.radio.cbc.ca/radio/regional/Maritimes/
Sydney/
21) P.O. Box 490, Iqaluit, NWT, X0A 0H0.
23) P.O. Box 160, Yellowknife, NWT, X1A 2N2.
24) P.O. Box 369, Gander, NF, A1V 1W7.
Web: http://www.radio.cbc.ca/radio/regional/Newfoundland/
25) P.O. Box 1609, Windsor, ON, N9A 1K7.
E-mail: rporter@windsor.cbc.ca
Web: http://WWW.Radio.CBC.CA/radio/regional/Ontario/Windsor/
26) 273 St. Jean Baptiste Ouest, Rimouski, PQ, G5L 4J8.
27) P.O. Box 10400, Ste-Foy, Quebec, PQ, G1V 2X2.
28) P.O. Box 2000, Matane, PQ, G4W 3P7.
29) P.O. Box 790, Chicoutimi, PQ, G7H 5E7.
V. Reports for sts 1), 15), 24) should be sent to 4). The QSL policy varies
from st. to st.

CBC NORTH - QUEBEC
SW Sce. in English/French/Inuktitut/Cree: 1155-0610 on 9625kHz via
Sackville transmitter site of R. Canada International.

RADIO CANADA INTERNATIONAL
see International Broadcasting section

PRIVATELY OWNED STATIONS
Mediumwaves: Stations less than 100W not mentioned. *=also
Shortwave, f=mainly French prgrs, e=mainly English prgrs,
b=bilingual E/F, m=multilingual/ethnic. +=inactive or F.PI. c=will
change to FM.

	Call	kHz	kW		Call	kHz	kW
101R)	CKHL	530	1/0.25e	471)	CKGL	570	10e
464)	CIAO	530	1/0.25m	601)	CJEM	c570	5/1f
17)	CFJC	550	25/5e	900)	CFCB	570	1e
1)	CKPG	550	10e	6)	CKXR	580	10/1e
400)	CHNO	550	50/10e	100)	CKUA	580	10e
500)	CHLN	550	10/5f	301)	CKY	580	50e
2)	CHTK	560	1/0.25e	403)	CFRA	580	50/10e
3)	CKNL	560	1e	404)	CKPR	580	5/1e
401)	CJKL	560	5e	405)	CKAP	580	10/1e
402)	CFOS	560	7.5/1e	406)	CKWW	580	0.5e
501)	CKCN	c560	10/5f	502)	CHLC	580	10/5f
901D)	CHVO	560	5e	800)	CJFX	580	25e
4A)	CKWL	570	1e	7)	CFTK	590	1e
5)	CKEK	570	10/1e	302)	CFAR	590	10/1e
212)	CKSW	570	10e	407)	CJCL	590	50e

	Call	kHz	kW		Call	kHz	kW
503)	CKRS	590	25/10f	114)	CHQT	880	50e
608A)	CJCW	590	1/0.25e	30)	CJDC	890	10e
901)	VOCM	590	20e	16)	CJVI	900	10e
8)	CKBD	600	10e	203)	CKBI	900	10e
200)	CJWW	600	25/8e	400)	CHYC	900	10/1f
408)	CKAT	600	10/5e	424)	CHML	900	50e
504)	CIQC	*600	10/5e	514)	CJER	900	1b
801)	CKCL	600	10/1e	516)	CKTS	900	10e
9)	CHNL	610	25/5e	540A)	CKVD	c900	10/2.5f
90)	CKRW	610	1e	803)	CKDH	900	1e
101)	CKYL	610	10e	106)	CKDQ	910	50e
303)	CHTM	610	1e	517)	CHRL	910	10f
409)	CKTB	610	10/5e	4)	CKCQ	920	10/1e
506)	CFLO	610	1/0.25f	308)	CFRY	920	25/15e
505)	CHNC	610	10/5f	428)	CKNX	920	10/1e
11)	CJCI	620	10e	604)	CJCJ	920	10/1e
201)	CKCK	620	10e	802)	CJCH	920	25e
410)	CKOY	620	10/5f	107)	CJCA	930	10e
502)	CFRP	620	1f	418A)	CKNS	930	10e
901B)	CKCM	620	10e	605)	CFBC	930	50e
12)	CKOV	630	5/1e	903)	CJYQ	930	50e
102)	CHED	630	50e	18)	CJIB	940	10e
411)	CFCO	630	10/1e	204)	CJGX	940	50/10e
412)	CJET	630	10e	309)	CFAM	950	10e
507)	CHLT	630	10/5f	606)	CKNB	950	10/1b
700)	CFCY	630	10e	805)	CHER	950	10e
422)	CFOB	640	1/0.5e	108)	CFAC	960	50e
452)	CHOG	640	25e	454)	CIAM	c960	1e
38)	CISL	650	10/9e	430)	CFFX	960	10/5e
208)	CKOM	650	10e	806)	CHNS	*960	10e
901c)	CKGA	650	5e	109)	CJYR	970	10e
128)	CFFR	660	50e	19)	CKNW	980	50e
903DR)	CKXB	670	10/5e	205)	CKRM	980	10/5e
305)	CJOB	680	10/5e	431)	CFPL	980	10/5e
415)	CFTR	680	50e	432)	CKRU	980	10/7.5e
903C)	CKXG	680	10e	519)	CKGM	990	50e
105)	CKRD	700	50/25e	807)	CKBW	1000	10e
416)	CJRN	710	10/5e	555)	CFLP	1000	10f
509)	CKVM	710	10/1f	903D)	CKXD	1010	1e
901E)	CKVO	710	10e	433)	CFRB	*1010	50e
701)	CHTN	720	10/7.5e	129C)	CKVH	1020	1/0.4e
13)	CKLG	730	50e	15)	CKST	1040	50e
306)	CKDM	730	10/5e	37)	CICF	1050	10/4e
510)	CKAC	730	50f	206)	CJNB	1050	10e
901A)	CHCM	740	10e	434)	CHUM	1050	50e
211)	CJVR	750	25e	111)	CKMX	*1060	50e
414)	CKGB	750	10/5e	20)	CFAX	1070	10e
24A)	CFLD	760	1/0.5e	436)	CHOK	1070	10e
25)	CKQR	c760	20/5e	112)	CKSA	1080	10e
104)	CHQR	770	50e	438)	CKKW	1090	10e
14)	CKOR	+780	10/2.8e	448)	CKTY	1110	10/1e
802)	CFDR	780	50/15e	21)	CKWX	1130	50e
103)	CFCW	790	50e	122)	CHRB	1140	50/46e
419)	CIGM	790	50e	22)	CKFR	1150	10e
602)	CFAN	790	5/1e	310)	CKX	1150	50/10e
900R)	CFNW	790	1e	439)	CJRC	1150	50/5f
14)	CKOR	800	10/0.5e	440)	CKOC	1150	50e
202)	CHAB	800	10e	505R)	CHGM	1150	5f
406)	CKLW	800	50e	116)	CKGY	1170	50/15
423)	CJBQ	800	10e	207)	CFSL	1190	10/5e
425)	CKDR	800	1/0.7e	23)	CKXM	1200	50e
511)	CHRC	800	50f	465)	CJBZ	1200	50e
512)	CJAD	800	50/10e	129A)	CKWA	1210	1e
902)	VOWR	800	10/2.5e	210A)	CFYM	1210	1e
307)	CKJS	810	10m	905)	VOAR	1210	10e
603)	CJVA	810	10f	117)	CJOC	1220	10/5e
450)	CHAM	820	10/5e	309A)	CJRB	1220	10e
127A)	CKKY	830	10/3.5e	442)	CHSC	1220	10e
466)	CFJR	830	5/1e	443)	CJRL	1220	5/1e
4B)	CKBX	840	1/0.5e	444)	CJSS	1220	1e
110)	CJXX	840	25/10e	522)	CKSM	1220	10/2.5f
27A)	CKMA	850	10e	556)	CFVM	1220	10/5f
129B)	CKBA	850	1e	608)	CKCW	1220	25e
513)	CKVL	850	50/35	7A)	CKTK	1230	1e
6R)	CKIR	870	1/0.25e	9A)	CJNL	1230	1e
24)	CFBV	870	1/0.25e	109A)	CIYR	1230	1e
900B)	CFSX	870	0.5e	118)	CJOK	c1230	1e
10R)	CKKC	880	1/0.25e	401A)	CJTT	c1230	1e
312)	CKLQ	880	10e	410R)	CHYK	1230	1/0.6f

Call	kHz	kW		Call	kHz	kW
524R) CJLP	1230	1f		810) CJLS	1340	5/4e
523) CHVD	1230	10/1f		900E) CFLW	1340	0.25e
514R) CJSA	1230	1f		907) CKHV	1340	1e
900C) CFGN	1230	0.25e		36A) CKCI	1350	10e
900D) CFLN	1230	1e		456) CKDO	1350	10/5e
1A) CKMK	1240	1e		812A) CKAD	1350	1e
5A) CFEK	1240	1e		538) CJVL	1360	10/5f
14A) CJOR	1240	1e		609) CKBC	1360	10e
26) CJAV	1240	1e		129) CFOK	1370	10e
27B) CKGO	1240	0.25e		458) CKLC	1380	10e
33B) CFNI	1240	1e		459) CKPC	1380	10e
206A) CJNS	1240	1e		524B) CFDA	1380	10f
302A) CJAR	1240	1/0.25e		126) CJCY	c1390	10e
423A) CJNH	1240	1e		5R) CJEK	1400	1e
427) CJWA	1240	1e		6B) CKGR	1400	1e
447) CJCS	1240	1e		14D) CIOR	1400	1e
526) CFLM	1240	1f		9R) CHNL-1	1400	1e
527) CJMD	1240	1f		106B) CKSQ	1400	1e
541A) CJAF	1240	1f		429A) CKCB	1400	1e
702) CJRW	1240	1e		514R) CKSJ	1400	1f
901BR)CKIM	1240	1/0.5e		524C) CKFL	1400	1f
309B) CHSM	1250	10e		540) CKRN	c1400	1f
449) CHWO	1250	10e		541) CJFP	1400	10/5f
119) CFRN	1260	50e		32) CFUN	1410	50e
610) CIHI	1260	10e		462) CKSL	1410	10e
27) CHWK	1270	10e		811) CIGO	1410	10e
120) CHAT	1270	10e		463) CKPT	1420	10/5e
423B) CJTN	1270	10e		812B) CKGY	1420	1e
530) CFGT	1270	10/5f		470) CHTK	1430	50m
808) CJCB	1270	10e		33) CFCP	c1440	1e
33C) CHQB	1280	1e		121) CKJR	1440	10e
207A) CJSL	1280	10e		14C) CHOR	1450	1e
543) CFMB	1280	50m		109R) CKYR	1450	0.5e
311) CIFX	1290	1e		467) CHUC	1450	8/1e
451) CJBK	1290	10e		541B) CHRT	1450	1f
533) CHRM	1290	10f		556R) C...	+1450	1f
209) CJME	1300	10e		812C) CFAB	1450	1e
127) CHLW	1310	10e		468) CJOY	1460	10e
453) CIWW	1310	50e		546) CKRB	1460	10/5f
29) CHMB	1320	50m		34) CJVB	1470	50m
449A) CJMR	1320	10e		469) CHOW	c1470	10e
809) CKEC	1320	25e		549) CHRD	c1480	10f
210) CJYM	1330	10e		14AR) CJOR-1	1490	1e
524) CKLD	1330	10f		33A) CFWB	1490	1e
6A) CKCR	1340	1e		117A) CJPR	1490	1e
9R) CINL	1340	1e		212A) CJSN	1490	1e
11A) CIVH	1340	1e		402R) CFPS	1490	1e
25A) CKGF	1340	1e		506A) CKLO	1490	1/0.25f
10B) CFKC	1340	0.25e		601R) CKMV	1490	1f
106A) CIBQ	1340	1e		812) CKEN	1490	1e
108AR)CFHC-1	c1340	1e		35) CKAY	1500	10/1e
127B) CJCM	1340	1e		472) CKOT	1510	10e
410R) CHOX	1340	1f		473) CHIN	1540	50/30m
524A) CJAN	1340	1f		36) CKEG	1570	10e
527R) CFED	1340	0.25f		113) CKTA	1570	10/5e
540C) CHAD	c1340	1f		309C) CKMW	1570	10e

Shortwaves

Call	kHz	kW	Relays
504) CFCX	6005	0.5	CIQC, Montreal: Inactive
111) CFVP	6030	0.1	CKMX, Calgary: Inactive
433) CFRX	6070	1	CFRB, Toronto: 24h
806) CHNX	6130	0.5	CHNS, Halifax: 24h

Addresses and other information:

Abbreviations: See under U.S.A. comm. st's.

Station Networks: There are several regional networks. The st's of a network often transmit a common prgr. during part of the broadcast day. The main st has been given a pure number, under which details common to all st's or pertaining to the main st only are found, while the other st's have been given the same number with an individual capital letter added, under which details pertaining only to this st are found. R. means a pure relay st.

Programme Networks:

French: Radiomédia, Télémédia, Radiomutuel.

English: Broadcast News, Standard Broadcast News, Telemedia Network, Western Information Network, Pelmorex Radio Network. These networks provide news and/or other prgrs.

CBC Relays: Certain st's are required to carry CBC prgrs for a few hours a day in areas with inadequate coverage by CBC's own tr's.

The number of such st's is steadily decreasing.

Verifications: Many stations verify correct reception reports, especially for exceptional distances. The details of the verification policies often change and thus are omitted here.

British Columbia

1) 1220 6th Ave, Prince George, BC V2L 3M8. **E-mail:** ckpgmail@ckpg.bc.ca — GM: G. Leighton. PD: D. Holien. CE: M. Fountain. 24h on 550/1240kHz.

1A) Box 1210, Mackenzie, BC V0J 2C0 — Mgr: J.D. McKenzie. Own prgrs: 1500-1900 Mon-Fri.

2) 346 Stiles Place, Prince Rupert, BC V8J 3S5 — Mgr. & PD: R. Langridge. CE: Harry Nuttma. 24h.

3) 10532 Alaska Rd, Fort St. John, BC V1T 1B3 — PD: M. Bodnar. 24h.

4) 160 Front Str, Quesnel, BC V2J 2K1 — Mgr: T. Cawdell. PD: Wayne Leslie. CE: Floyd Lust. 24h on 570/840/920kHz. **German:** Wed 0105-0130. **Italian:** Thurs 0105-0130.

4A) 83 S. First Ave, Williams Lake, BC V2G 1H4 — Mgr: Karen LeComte. Own prgrs 1400-1800.

4B) Box 939, 100 Mile House, BC V0K 2E0 — Own prgrs: Mon-Fri 1500-2100. Sat 1500-1700, 2000-2100.

5) 19-9th Ave. S, Cranbrook, BC V1C 2L9 — GM: Gary D. Cavers. PD: Rod Schween. CE: R. Schween. 24h on 570/1240kHz.. (CJEK=Sparwood repeater)

5A) Box 1170, Fernie, BC V0B 1M0 — Mgr: H. Ashmore. Own prgrs W 1500-2000, 2300-0200.

6) Box 69, Salmon Arm, BC V1E 4N2 — Mgr: H.F. Davidson. PD: P. Scott. CE: G. Hoffos. 24h on 580/870/1340/1400kHz. Ann: "Columbia Shuswap R." 6R) = Invermere.

6A) Box 1420, Revelstoke, BC V0E 2S0 — Mgr: T. Biggs. Own prgrs Mon-Fri 1400-2000, Sat 1400-1700.

6B) Box 1403, Golden, BC V0A 1H0. Own prgrs: As 6A). Mgr: S. Sterdan.

7) 4625 Lazelle Ave, Terrace, BC V8G 1S4. **E-mail:** info@osg.net **WWW:** http://www.osg.net — GM: T. Christie. PD: D. Graham. CE: N. Saele. 24h on 590/1230kHz.

7A) 350 City Centre, Kitimat V8C 1T6 — Mgr: G. Ball. Local prgrs: Mon-Fri 1400-1800, 2300-0100.

8) 1401 West 8th Ave, Vancouver, BC V6H 1C9 — GM: Gerry Siemens. PD: Marc Patric. CE: A. Jackson. 24h.

9) 611 Lansdowne St, Kamloops, BC V2C 1Y6 — GM: Robbie Dunn. PD. Jim Reynolds. CE: Dave Coulter. 24h on 610/1230/1340/1400kHz. 1340=Ashcroft/Cache Creek, BC, 1400=Clearwater, BC.

9A) Box 1630, Merritt, BC V0K 2B0 — GM: Elizabeth Laird. PD: Andrew Laird. Own prgrs Mon-Sat 1400-1800, 2000-2100.

10) CJAT Radio, 1560 2nd Ave, Trail, BC V1R 1M4 — GM: Karl Johnston. PD: A. White. CE: S. Schnell. 24h on FM 95.7. Ann: "KBS". CKKC=Nelson, BC.

10B) Box 310, Creston, BC V0B 1G0 — Own prgrs: 1200-0600.

11) 1940-3rd Ave, Prince George, BC V2M 1G7. **E-mail:** cjci@solutions-4u.com **WWW:** http://www.cjci.com — Pres: Ron A. East. GM: Terry Shepherd. PD: D. Coogan. CE: Dave Allendahl. 24h on 620/1340kHz.

11A) Box 1370, Vanderhoof, BC V0J 3A0 — PD: Tom Bulmer. Own prgrs Mon-Fri 1400-1700, 2300-0200.

12) 3805 Lakeshore Rd, Kelowna, BC V1W 3K6. **E-mail:** ckov@awinc.com — GM: Dean Cooper. PD: Rob Bye. CE: Lorne Gagnon. 24h.

13) 1006 Richards Str, Vancouver, BC V6B 1S8 — GM: C. Pandoff. PD: K. Gorman. CE: G. Jardine. 24h.

14) 33 Carmi Ave, Penticton, BC V2A 3G4 — Mgr: R. Clark. PD: R. Morgan. CE: H. Nutma. 24h on 800/1240/1450/1490kHz. Ann: "OR Network."

14A) Box 539, Osoyoos, BC V0H 1V0 — Mgr: R. Bedard. Own prgrs W 1400-0200 (Sat 2200) (CJOR-1=Oliver repeater).

14C) Box 1170, Summerland, BC V0H 1Z0 — Own prgrs 1400-0200 (Sat 2200).

14D) Box 1400, Princeton, V0X 1W0 — Mgr: L.M. Currie. Asst. Mgr: W.G. Elliott. CE: R. Dale. Own prgrs Mon-Fri 1400-1800, 2300-0200, Sat. 1400-1800.

15) Unit 100, 856 Homer St, Vancouver, BC V6B 2W5. **E-mail:** comments@am1040.com **WWW:** http://www.am1040.com — GM: R. Dixon. PD: D. Geddes. CE: Ed Jurak. 24h.

16) Box 900, Victoria, BC V8W 2S2 — GM: Kim Hesketh. PD: A. Simpson. CE: Cliff Smedley. 24h.

17) 460 Pemberton Terrace, Kamloops, BC V2C 1T5 — GM: Rick W. Arnish. PD: Doug Collins. CE: Kris Swamy. 24h.

18) 3313-32nd Ave, Vernon, BC V1T 2E1 — GM: P. Nicol. PD: Don

Weglo. CE: D. Schindel. 24h.

19) #2000-700 West Georgia St, Vancouver, BC V7Y 1K9. **E-mail:** cknw@wic.ca **WWW:** http://www.cknw.com — GM: Rod Gunn. PD: Tom Plasteras. CE: David Glasstetter. 24h.

20) 825 Broughton Str, Victoria, BC V8W 1E5. **E-mail:** cfax@islandnet.com **WWW:** http://www.cfax1070.com — GM: Mel Cooper. PD: Terry Spence. CE: Bud Goes. 24h.

21) 2440 Ash St, Vancouver, BC V5Z 4J6 — GM: Chuck McCoy. PD: D. Boute. CE: Jack Wiebe. 24h.

22) 2419 Hwy. 97N, Kelowna, BC V1X 4J2 — GM: M. Tindall. PD: B. Yaremus. CE: George J. Young. 24h.

23) 3795 Carey Rd, 2nd Floor, Victoria, V8Z 6T8. **E-mail:** radio@100.3theq.com **WWW:** http://www.100.3theq.com — GM: Dan McAllister. PD: J. Shields. CE: Bob Calder. 24h.

24) Box 335, Smithers, BC V0J 2N0. GM: T.A. Collison. PD: Chris Collins. CE: D. Alendal. 24h on 760/870kHz.

24A) Box 600, Burns Lake, BC V0J 1E0 — Own prgrs 1700-2000.

25) 525-11th Ave, Castlegar, BC V1N 1J6. **WWW:** http://www.ckgr.com — GM: Dennis Gerein. PD: Dan Harrington. CE: Kevin McKinnon. 24h on 760/1340kHz.

25A) Box 1570, Grand Forks, BC V0H 1H0 — PD: Gary Dorosz.

26) 2970 3rd Ave, Port Aberni, BC V9Y 7N4. **E-mail:** cjav@cedar.alberni.net **WWW:** http://www.alberni-net/~cjav/cjav/htm — GM: C. Gibson. PD: C. Talbot. CE: Leo Wouters. 24h.

27) Box 386, Chilliwack, BC V2P 6J7. **WWW:** http://www.fraservalley.com — Mgr: W.J. Coombes. PD: B. McArthur. CE: Arnie Schmidt. 24h on 1240/1270kHz. Ann: "Fraser Valley R."

27A) 2722 Allwood Str, Abbotsford, BC V2T 3R8. **E-mail:** max@fraservalley.com — Mgr: P. Alpen.

27B) Box 1600, Hope, BC V0X 1L0 — Mgr: T. DeSorcy. Own prgrs W1400-2100 (Sat 2000).

29) #100-1200 West 73rd Ave, Vancouver, BC V6P 6G5. **E-mail:** chmb@am1320.com **WWW:** http://www.am1320.com — GM: Wayne H. Lee. PD: Hanson Lau. 24h, mostly in Chinese.

30) 901-102nd Ave, Dawson Creek, BC V1G 2B6. **E-mail:** cjdcam@pris.bc.ca **WWW:** http://www.pris.bc.ca/cjdc_tel/ — R. Clark. CE: Glen Foisy. 24h.

32) 300-380 W. 2nd Ave, Vancouver, BC V5Y 1C8. **E-mail:** cfunmail@cfun.com **WWW:** http://www.cfun.com — GM: Paul Ski. PD: Neil Gallagher. CE: Dave Youell. 24h.

33) 1625-A McPhee Rd, Courtenay, BC V9N 3A6 — GM: Greg Phelps. PD: Murray Collins. CE: George J. Young. 24h on 1240/1280/1440/1490kHz.

33A) 909 Ironwood St, Campbell River, BC V9W 3E5 — Mgr: B. Langston.

33B) Box 1240, Port Hardy, BC V0N 2P0 — Mgr: J. James. 24h.

33C) 6816 Courtenay Str, Powell River, BC V8A 1X1 — PD: Derek Bouchard. 24h.

34) 814 Richards Str, Ste. 101, Vancouver, BC V6B 3A7 — Mgr: P. Wong. PD: B. Lo. 24h, in E and 21 other languages. Chinese: W0400-1700, English 2000-0200. and within foreign language prgrs.

35) #205-2700 Beverly St, Duncan, BC V9L 5C7 — Pres: Dick Drew. GM, PD: D. Hamilton. 24h.

36) 4550 Wellington Rd, Nanaimo, BC V9T 2H3 — GM: Bob Adshead. PD: Mike O'Brien. CE: Bob Calder. 24h.

36A) Box 1370, Parksville, BC V0R 2S0.

37) 2800-31st, Vernon, BC V1T 5H4. **WWW:** http://www.mix105.com — GM: Michael Tindall. PD: G. White. CE: R. Dale. 24h.

38) 20-11151 Horseshoe Way, Richmond, BC V7A 4S5. **E-mail:** zinfo@z95.com **WWW:** http://www.z95.com — GM: Gary Russell. PD: Eric Samuels. CE: Barry Johnston. 24h.

Yukon Territory

90) 203-4103 4th Ave, Whitehorse, YT Y1A 1H6 — GM: S. Rai. PD: M. Carlucci. CE: Bruce Whittington. 24h.

Alberta

100) 4th Flr, 10526 Jasper Ave, Edmonton, AB T5J 1Z7. **E-mail:** ckua@oanet.com **WWW:** http://www.ckua.org — GM: Ken Davis. PD: David Ward. CE: Larry King. Sun-Fri 1300(Sun 1400)-0700, Sat 1400-0900.

101) Bag 300, Peace River, AB T8S 1T5 — GM: Terry J. Babiy. PD: W. Kozak. CE: J. Sebryk. 24h. 530=High Level, AB.

102) 5204-84th Str, Edmonton, AB T6E 5N8 — GM: Doug Rutherford. PD: Dave Jamieson. CE: Tom Davies. 24h.

103) 4752-99th Str, Edmonton, AB T6E 5H5 — GM: A. Anderson. PD: J.R. Greening. CE: Lyndy Olson. 24h.

104) 1900, 125-9th Ave S,E, Calgary, AB T2G 0P6 — GM: Doug Rutherford. PD: J. Donald. CE: Mel B. Hoyme. 24h.

105) P.O. Bag 5700, Red Deer, AB T4N 6V5 — GM: W. H. Yuill. PD: J. Hall. CE: Steve McPherson. 24h.

106) Box 1480, Drumheller, AB T0J 0Y0 — GM: B. Hepp. PD: P. Larsen. CE: L. Jesperson. 24h on 910/1340/1400kHz.

106A) Box 180, Brooks, AB T1R 1B3 — Own prgrs Mon-Fri 1300-2000, Sat 1400-1900.

106B) 4703-58 Str, Stettler, AB T0C 2L1 — Mgr: F. Bonham. Own prgrs: Mon-Fri 1300-0100, Sat 1400-1900).

107) 10250-108 Str, Edmonton, AB T5J 2X3 — GM: Doug Main. PD: Dale Wolfe. CE: Ray Semenoff. 24h.

108) 3320-17th Ave SW, Calgary, AB T3E 6X6 — GM: Kevin McKanna. PD: Dawn Buffam. CE: Bob MacDonald. 24h on 960/1340/1450kHz.

109) Box 6600, Edson, AB T7E 1T9 — GM: Mel Lazarenko. PD: Dave Schuck. CE: L. Jesperson. 24h on 970/1230kHz.

109A) 118 Athabasca Ave, Hinton, AB, T7V 2A5. Own prgr's 1700-2200.

110) #202, 9817-101 Ave, Grande Prairie, AB T8V 0X6 — GM: Ken Truhn. PD.: Ken Norman. CE: Keith Henriksen. 24h.

111) P.O. Box 2750, Stn. "M", Calgary, AB T2P 4P8 — GM: Pat Holiday. CE: Ken Pasolli. 24h.

112) 5026-50th Str, Lloydminster, AB T9V 1P3 — Mgr: Ken Ruptash. PD: N. Bergen. CE: R. Green. 24h.

113) 401 Mayor Magrath Dr. S, Lethbridge, AB T1J 3L8 — GM: Georgina Kneitel. PD: Marv Gunderson. CE: Tyler Everitt. 24h. Location: Tabor, AB.

114) 10550-102th Str, #200, Edmonton, AB T5H 2T3. **E-mail:** cisn@compusmart.ab.ca — GM: R.M. Lang. PD: B. Spitzer. CE: D. Mattice. 24h.

115) 804-16th Ave. S.W, Calgary, AB T2R 0S9 — GM: L. Friesen. PD: J. Jackson. CE: V. Mooers. 24h.

116) Bag 5339, Red Deer, AB T4N 6W1 — GM: Ron Thompson. PD: Boyd Leader. CE: Cliffort Wheeler. 24h.

117) Box 820, Lethbridge, AB T1J 3Z9 — Mgr: Brent Seely. PD: C. Fay-Davies. CE: Kris Rodts. 24h on 1220/1490kHz.

117A) Box 840, Blaimore, AB T0K 0E0 — Mgr: Daryl Ferguson. Own prgrs: W 1300-1600 (Sat 2000).

118) 9912 Franklin Ave, Ft. McMurray, AB T9H 2K5 — GM: Kelly Boyd. PD: Russell Thomas. CE: Larry Howell. 24h.

119) #100-18520 Stony Plain Rd, Edmonton, AB T5S 2E2 — GM: Marty Forbes. PD: S. Moore. CE: Bruce Bedford. 24h.

120) Box 1270, Medicine Hat, AB T1A 7H5 — Mgr: D. Dietrich. PD: Jay Hitchen. CE: J. Simmons. 24h.

121) 5220-51st Ave, Wetaskiwin, AB T9A 3E2 — Mgr: Dave Lynes. CE: L. Jesperson. 24h.

122) #11-5th Ave S.E, High River, AB T1V 1G2 — Mgr: Lyndon Friesen. PD: Keith Leask. CE: Vernon Mooers. 24h.

126) 2nd Flr, 457-3rd Str. S.E, Medicine Hat, AB T1A 0G8 — GM: I. Heid. PD: John Carter. CE: Roy Findley. 24h.

127) #201, 4341-50th Ave, St. Paul, AB T0A 3A3 — Mgr: B. Labrie. PD: P. Larsen. CE: L. Jesperson. 24h on 830/1310/1340kHz. Partly in netw. with st. 129).

127A) #2, 1037 2nd Ave, Wainwright, AB T9W 1K7 — PD: S. Granigan.

127B) P.O. Box 433, Cold Lake, AB T9M 1R5 — Mgr: Roger Thorpe.

128) 2723-37 Ave. NE, Station M, Calgary, AB T1Y 5R8 — GM: Vince Cownden. PD: T. Voth. CE: K. Crook. 24h.

129) Box 1800, Westlock, AB T0G 2L0 — Mgr: W. Betts. PD: P. Larsen. CE: L. Jesperson. 24h.

129A) Box 2470, Slave Lake, AB T0G 2A0 — Mgr: Royal Watson. Local prgrs: Mon-Fri 1400-2000 2200-0200, Sat 1400-1800.

129B) Box 1800, Athabasca, AB T0G 0B0 — Mgr: Doug Hardy. Own prgrs: 1400-0600.

129C) Box 2219, High Prairie, AB T0G 1E0 — Mgr: D. Lynes.

Saskatchewan

200) 345-4th Ave. S, Saskatoon, SK S7K 5S5. **E-mail:** cjww.radio@sk.sympatico.ca **WWW:** http://www.saskatoon.sk.ca — GM: Vic Dubois. PD: R. Kitter. 24h.

201) Box 6200, Regina, SK S4P 3H7. **E-mail:** ckck.radio@sasknet.sk.ca — GM: J. Huschi. PD: M. Olstrom. CE: D. Ellis. 24h.

202) 1704 Main St. N, Moose Jaw, SK S6H 4P5 — GM: W. Michelson. PD: Barrie Vice. CE: R. McLaren. 24h.

203) Box 900, Prince Albert, SK S6V 7R4 — GM: Jim Scarrow. PD: Neil Headrick. CE: Dale Zimmerman. 24h.

204) 120 Smith Str. E, Yorkton, SK S3N 3V3. **E-Mail:** cjgx@sask.sympatico.ca — GM: Lyle J. Walsh. PD: Brad Bazin. CE:

Bryan Mierau. 24h.
205) 2060 Halifax St, Regina, SK S4P 1T7 — GM: John Huschi. PD: Willy Cole. CE: M. David Ellis. 24h.
206) Box 1460, North Battleford, SK S9A 2Z5 — GM: D. Dekker. PD: Doug Harrison. CE: Dave Senft. 24h on 1050/1240kHz.
206A) 225 Centre Street, Meadow Lake, SK S0M 1V0 — PD: K. Schiller. CE: K. Fischer. Own prgrs: W1500-1800, 1930-2400.
207) Box 340, Weyburn, SK S4H 2K2 — GM: L. Pappel. PD: S. Armstrong. CE: Jim Hutchings. 24h on 1190/1280kHz.
207A) Box 1280, Estevan, SK S4A 2H8 — Mgr: L. McGillivray.
208) Box 65000, Saskatoon, SK S7H 0W3 — Mgr: Pam Carley. PD: Greg Harrison. CE: Ken Fisher. 24h.
209) #210-2401 Saskatchewan Dr, Regina, SK S4P 4H8 — GM: M. Zaplitny. CE: David M. Senft. 24h.
210) Box 490, Rosetown, SK S0L 2V0 — GM: D. Dyck. PD: B. James. CE: Al Pippin. 1200-0700 (Sun 0600).
210A) Kindersley, SK (all correspondence to Rosetown).
211) Box 750, Melfort, SK S0E 1A0. **E-mail:** cjvr@sk.sympatico.ca **WWW:** http://www.saskstar.sk.ca/cjvr/ — GM: Gary Fitz. PD: Bill Wood. CE: Bayne Opseth. 24h.
212) 134 Central Ave. N, Swift Current, SK S9H 0L1 — GM: L. Friesen. PD: G. Schutte. CE: D. Funk. 24h on 570/1490kHz.
212A) Box 1176, Shaunavon, SK S0N 2M0 — Mgr: Joe Gregoire. Own prgrs: Mon-Fri 1330-1800.

Manitoba

301) Polo Park, 1440 Rapelje Ave, Winnipeg, MB R3G 0L7 — GM: Ron Kizney. PD: Frank Capozzolo. CE: George Buzunis. 24h.
302) Box 430, Flin Flon, MB R8A 1N3 — GM: Doug T. O'Brien. PD: Dave Baker. CE: Danny Parker. 24h on 590/1240kHz.
302A) Box 2980, The Pas, MB R9A 1R7. **E-mail:** cj1240@mts-net — Mgr: Darren L. Sayer.
303) 201 Hayes Rd, Thompson, MB R8N 1M5 — Mgr: Tom O'Brien. PD: Ron Krane. CE: Danny Parker. 24h. Cree: M-F 0100-0300.
305) 930 Portage Ave, Winnipeg, MB R3G 0P8. **WWW:** http://www.manitobanow.com — GM: Garth Buchko. PD: Ken Kilcullen. CE: Jack Hoeppner. 24h.
306) 27-3rd Ave. N.E, Dauphin, MB R7N 0Y5 — GM: Linus J. Westberg. PD: Bruce LePerre. CE: Peter Nykolaishen. 24h.
307) 520 Corydon Ave, Winnipeg, MB R3L 0P1. **E-mail:** ckjs@magic.mb.ca — GM Tony Carta. PD: G. Gigliotti. CE: Don Trueman. 24h.
308) 1500 Saskatchewan Ave. W, Portage la Prairie, MB R1N 0N6 — Pres. & CE: R.D. Hughes. GM: Red Hughes. PD: Bev Edmondson. 24h.
309) R. Southern Manitoba, Box 950, Altona, MB R0G 0B0 — GM: D. Wiebe. PD: Al Friesen. CE: Laverne Siemens. 24h on 950/1220/1250kHz.
309A) Box 1220, Boissevain, MB R0K 0E0.
309B) Box 1250, Steinbach, MB R0A 2A0.
309C) Box 1570, Winkler, MB R0G 2X0 — PD: Deb Kauenhofen.
310) 2940 Victoria Ave, Brandon, MB R7B 0N2 — GM: Alan Cruise. PD: H. Adams. CE: Paul Weger. 24h.
311) 1445 Pembina Hwy, Winnipeg, MB R3T 5C2 — GM: B. Stone. PD: H. Kroeger. CE: Lorne Anderson. 24h.
312) 624-14th Str. E, Brandon, MB R7A 7E1 — GM: D. Kille. PD: S. Antaya. CE: R. Landry. 24h.

Ontario

400) 295 Victoria Str, Sudbury, ON P3C 1K5 — CE: Alan Aysto. CHNO: GM: Don Shafer. PD: Bruce Lindsay. CHYC: Linda Miller. PD: Josee Perreault. CHYC 1100-0600 on 900kHz, CHNO 24h on 550kHz.
401) Box 430, Kirkland Lake, ON P2N 3J4 — GM./PD: Robin Connelly. CE: Don Elvidge. 24h on 560/1230kHz.
401A) Box 1058, New Liskeard, ON P0J 1P0. **E-mail:** cjtt@nt.net **WWW:** http://www.nt.net/1230_cjtt/radio.htm — GM/PD: Mike Perras. CE: Don Elvidge.
402) 270 9th St. E, Owen Sound, ON N4K 1N7. **E-mail:** bayshore@radioowensound.com **WWW:** http://www.radioowensound.com — Mgr: Ross Kentner. PD: Madelyn Hamilton. CE: R. Coyne. 24h on 560/1490kHz. (CFPS=Port Elgin repeater).
403) 1900 Walkley Rd, Ottawa, ON K1H 8P4. **E-mail:** mailbox@koolcfra.com **WWW:** http://www.cfra.com — GM/PD: Mark Maheu. CE: Harrie Jones. 24h.
404) 87 N. Hill Str, Thunder Bay, ON P7A 5V6 — GM: H. Fraser Dougall. PD: Rob Brown. CE: M. Volbracht. 24h.
405) 52. Riverside Dr, Kapuskasing, ON P5N 1X5 — Mgr: D. Todd. PD: Dave Palmer. CE: B. Boulianne. 1100-0500.
406) 1640 Ouellette Ave, Windsor, ON N8X 1L1 — GM: Eric

Proksch, PD: K. Chinnery (CKLW), W. Duff (CKWW). CE: Jim Valvasori. 24h.
407) 40 Holly Str, 9th Floor, Toronto M4S 3C3. **E-mail:** thefan@telemedia.org — GM: Dave Ackhurst. PD: Nelson Millman. CE: Wally Lennox. 24h.
408) Box 3000, North Bay, ON P1B 8K8 — GM: Rick Doughty. PD: P. McKeown. CE: C. Senyi. 24h.
409) P.O. Box 610, St. Catharines, ON L2R 6X7. **WWW:** http://www.htzfm.com — GM: Jamie O'Brien. PD: T. Denis. CE: S. Sloane. 24h.
410) Box 1340, Timmins, ON P4N 7J8 — GM: D. Bouchard. CE: D. Elvidge. 24h on 620/1230/1340kHz. (CHOH=Hearst, ON. CHYK=Kapuskasing, ON).
411) Box 100 Chatham, ON N7M 3H3. **E-mail:** cksy/fm@ciaccess.com — GM: Carl Veroba. PD: George Brooks. CE: Ron Wilken. 24h.
412) Box 630, Smiths Falls, ON K7A 4T4 — GM: Gary Perrin. PD: Brian Perkin. CE: Wayne Henwood. 24h.
414) Box 1046, Timmins, ON P4N 7H8 — GM: Bruce Lindsay. PD: Dave McLaughlin. CE: Jim Profit. 24h.
415) 36 Victoria St, Toronto, ON M5C 1H3 — GM: Sandy Sanderson. PD: Paul Fisher. CE: Kirk Nesbitt. 24h (all news).
416) Box 710, Niagara Falls, ON L2E 6X7. **E-mail:** rock@theplanet.com — GM: E. Lewis. PD: B. Auchterlony. CE: B. McDougall. 24h.
418A) 46 Mead Blvd, Espanola, ON P0P 1C0 — 24h. Own prgrs: M-F 1100-1500.
419) 880 Lasalle Blvd, Sudbury, ON P3A 1X5. **E-mail:** cigm@neilnet.com — GM: R. Doughty. PD: Jim Hamm. CE: C. Tryon. 24h.
422) 242 Scott St, Fort Frances, ON P9A 1G7 — GM: Don Kay. CE: John Sinclair. 24h.
423) Box 488, Belleville, ON K8N 5B2 — GM: W.A. Morton. PD: Peter Tho,pson. CE: Doug K. Vanderwater. 24h. on 800/1240kHz.
423A) Box 1240, Bancroft, ON K0L 1C0 — Mgr: Bob Rowbotham. CE: Larry Wortley. Own prgrs Mon-Fri 1400-1500, 1800-2000, 2300-2400. Sat 1410-1530, 1730-1900. Sun 1330-1500.
423B) Box 9, Trenton, ON K8V 5R1 — Mgr: R.S. Rowbotham. CE: M. Coffey. 24h.
424) 875 Main Str. West, Hamilton, ON L8S 4R1 — GM: Don Luzzi. PD: Darryl Hartwick. CE: Ted Townsend. 24h.
425) Box 580, Dryden, ON P8N 2Z3 — Mgr: Bruce Walchuk. 24h.
427) 53 Broadway Ave, Wava, ON. POS 1K0 — Relays CJQM-FM, addr: 642 Great Northern Rd, Sault Ste Marie, ON P6B 4Z9.
428) 215 Carling Terrace, Wingham, ON N0G 2W0 — GM: J. Kippen. PD: Dan Gall. 1100(Sun 1300)-0408.
429) Box 950, Barrie, ON L4M 4V1 — GM: D. Coulson. PD: J. Walther. CE: J. Evans. 24h on 101.1MHz/1400kHz.
429A) 1400 Highway 26 East, Collingwood, ON L9Y 4W2. Mgr: T. Aikins. PD: J. Nichols.
430) 479 Counter Str, Kingston, ON K7M 7J3 — GM/PD: Lorne Matthews. CE: Larry Cameron. 24h.
431) Box 2580, London, ON N6A 4H3 — GM: Rick A. Moss. PD: Brian Nuttall. CE: Andrew Bingle. 24h.
432) Box 4150, Peterborough, ON K9J 6Z9 — GM: Mike Ferguson. PD: Malcolm Sinclair. CE: P. Ward. 24h.
433) 2 St. Clair Ave West, Toronto, ON M4V 1L6 — GM: Gary Slaight. PD: Bob Mackowycz. CE: David Simon.. CFRX special ID twice an hour on 6070kHz — **V.** by QSL-card. Re. to ODXA, P.O. Box 161, Willowoak, ON M2N 5S8.
434) 1331 Yonge St, Toronto, ON M4T 1Y1. **E-mail:** webmaster@1050chum.com **WWW:** http://www.1050chum.com — GM: Jim Waters. PD: Brad Phillips. CE: Bruce Carnegie. 24h.
436) Box 1070, Sarnia, ON N7T 7K5. **E-mail:** chok@icis.on.ca **WWW:** http://www.chok.com — GM: Wayne Steele. PD: Paul Godfrey. CE: D. Connors. 24h.
438) 255 King St. N, Waterloo, ON N2J 4V2 — GM: Linda Benoit. PD: Paul Cugliari. CE: Allan Douglas. 24h. (Lic. to Kitchener, ON.)
439) 22 rue St-Louis, Gatineau, PQ J8T 2R9 — GM: Jacques Papin. CE: Pierre Sylvestre. 24h. (Lic. to Ottawa, ON.)
440) 883 Upper Wentworth St, Hamilton, ON L9A 4Y6. **E-mail:** ckoc@radiocorp.com **WWW:** http://www.radiocorp.ca — GM: Jim MacLeod. PD: Nevin Grant. CE: Dave Barry. 24h.
442) 36 Queenston St, St. Catharines, ON L2R 2Y9 — GM: Doug Setterington. PD: Ted Yates. CE: B. McDougall. 24h.
443) 128 Main St. S, Kenora, ON P9N 1S9 — Mgr: H. Syrja — PD: H. Fawcett. 24h.
444) Box 969, Cornwall, ON K6H 5V1. **E-mail:** cjssradio@cnwl.isg.net — GM: Paul Vincent. PD: John Bolton. CE: T. Smith. 24h.

447) Box 904, Stratford, ON N5A 6W3 — GM: S. Rae. PD: M. Philbin. 24h.
448) 1415 London Road, Sarnia, ON N7S 1P6 — GM: Wayne Steele. PD: R. Dann. CE: Brian Hinz. 24h.
449) 284 Church St, Oakville, ON L6J 7N2 — Pres, PD: Michael H. Caine. CE: S. Headon. 24h.
449A) Box 1320, Port Credit Postal Sta, Mississauga, ON L5G 4M3.
450) 151 York Blvd, Hamilton, ON L8R 3M2 — GM: Lyndon Friesen. PD: Joel Christie. 24h.
451) 743 Wellington Road S, London, ON N6C 4R5 — GM: Don Chamberlain. PD: R. Walters. CE: Jeffrey Guy. 24h.
452) 5255 Yonge St, Suite 1400, North York, ON M2N 6P4. **WWW:** http://www.talk640.com — GM: Don Luzzi. PD: C. Cardinal. CE: Rob Enders. 24h. (Lic. to Richmond Hill, ON.)
453) #1900-112 Kent Str, Ottawa, ON K1P 6J1 — GM: Scott Parsons. CE: Terence McDougall. 24h.
454) 46 Main Str, Cambridge, ON N1R 1V4 — GM: P. Osborne. PD: Ron Fitzpatrick. CE: Joe Mignacca. 24h.
456) 1200 Airport Rd, 2nd Fl, Oshawa, ON L1J 8P5 — GM: J.M. Leroy. PD: Shawn Turner. 24h.
458) Box 1380, Kingston, ON K7L 4Y5 — GM: Robert A. Wood. CE: Terry Kelly. 24h.
459) 571 West St, Brantford, ON N3T 5P8 — GM: Richard Buchanan. PD: Ted Lehman. CE: Stewart Bailey. 24h.
462) Box 1410, London, ON N6A 5J2 — GM: Jim Webb. PD: Bob O'Brien. CE: William Tofflemire. 24h.
463) Box 177, Peterborough, ON K9J 6Y8. **E-mail:** radio@ckpt.com **WWW:** http://www.ckpt.com — GM: Jim Blundell. PD: Rick Ringer. CE: Ed Crompton. 24h.
464) 50 Kennedy Rd. South, Suite 20, Brampton, ON L6W 3R7 — Pres: Angelo Cremisio. GM: Rick Sargent. 24h.
465) 1575 Carling Ave, Ottawa ON K1Z 7M3 — GM: Dianne M. Wilson. PD: Darcy Magee. CE: Marc Germain. 24h.
466) Box 666, Brockville, ON K6V 5V9. **E-mail:** cfjr@cfjr.brockville.com **WWW:** http://www.cfjr.brockville.com — GM: Robert Wood. PD: Greg Hinton. CE: T. Kelly. 24h (exc. Mon 0600-1100).
467) Box 520, Cobourg, ON K9A 4L3. **E-mail:** chuc@eagle.ca — GM: Don T. Conway. PD: Don Martin. CE: John Ton. 24h.
468) 75 Speedvale Ave, Guelph, ON N1E 6M3 — GM: Guus Hazelaar. PD: Larry Mellott. CE: Joe Mignacca. 24h.
469) Regional Rd. 23 West, Welland, ON L3B 5R6 — GM: Suzanne Rochan Burnett. PD: Pete Morena. CE: J. Gurney. 24h.
470) 1 Yonge St, #2500, Toronto, ON M5E 1E5.
471) 305 King Str. W, Kitchener, ON N2G 4E4 — GM: Wolfgang von Raesfeld. PD: George Gordon. CE: M. McCabe. 24h.
472) Box 10, Tillsonburg, ON N4G 4H3 — S:M: John D. Lamers Jr. PD: Doug Cooper. CE: R. Lamers. Daytime only (Jan. 1300-2215, July 1000-0100). Hungarian/Flemish/German: Sun 1800-2100.
473) 622 College Str, Toronto, ON M6G 1B6. **E-mail:** chin@istar.ca **WWW:** http://www.chinradio.com — Pres: J.B. Lombardi. GM: Dario Amaral. CE: Shahnaz Bhatti. 24h.

Quebec
500) 3550 Boul. Royal, Trois-Rivières, PQ G9A 5G8 — PD: P. DeMondehare. CE: D. Goudreault. 24h.
501) 437 Arnaud, Sept-Îles, PQ G4R 3B3 — GM: D. Marquis. PD: Patrick Sirois. CE: E. Piuze. 24h.
502) c/o CHLC, 399 rue de Puyjalon, Baie Comeau, PQ G5C 2Z7 — L.P. (CHLC 97.1MHz): GM: Camille St. Pierre. 24h. Location: Forestville, P.Q.
503) C.P. 59, Jonquière, PQ G7X 7V8 — Pres: Denis Langlois. CE: Daniel Dufour. 24h.
504) 211 Gordon Ave, 3rd Flr, Verdun, PQ H4G 2R2 — GM: Brian Kenemy. PD: Bob Linney. CE: K. Bickerdike. 24h. (Lic. to Montreal, PQ.)
505) C.P. 610, New Carlisle, PQ G0C 1Z0 — GM: Réginald Poirier. PD: Mario Loubier. CE: Arthur Houde. 24h. CHGM=Gaspé, PQ.
506) 332 de la Madone, Mont-Laurier, PQ J9L 1R9 — GM. & PD: Alain Desjardins. CE: Marcel Bellemare. 24h on 610/1490kHz. Nights rel. st. 510)
506A) 41 Principale, L'Annonciation, PQ J0T 1T0 — Own prgrs 1100-0500.
507) 4020 blvd. de Portland, Sherbrooke, PQ J1L 2V6 — GM: Michel Fortin. PD: Claire Francoeur. CE: Michel Laroche. 24h.
509) C.P. 3000, Ville-Marie, PQ J0Z 3W0 — GM. & PD:: Yvon Larivière. CE: Yves Hamel. 24h.
510) 1411 Peel Str, bureau 300, Montreal, PQ H3A 3L5 — GM: R. Briere. PD: A. Chevalier. CE: R. Dubois. 24h.
511) 2136 Chemin Ste-Foy, Ste-Foy, PQ G1V 1R8 — GM: A. Fleury.

PD: M. Paquin. CE: J. Ruel. 24h.
512) 1411 rue du Fort, Montreal, PQ H3H 2R1 — GM: Rob Braide. PD: Steve Kowch. CE: M. Kavanagh. 24h.
513) 211 Gordon Ave, Verdun, PQ H4G 2R2 — GM: Pierre Arcand. PD: Robert G. Hynes. CE: Réal Terrault. 24h.
514) 783, rue Labelle, St-Jérôme, PQ J7Z 5M4 — Mgr: Pierre Labonté. PD: Steeve Laplante.. CJSA=Ste-Agathe, CKSJ=St-Jovite repeaters. Not recently confirmed.
516) 901 Galt Str. East, Sherbrooke, PQ J1G 1Y6 — Reportedly simulcasting CJAD-800kHz or silent.
517) 568, boul. St-Joseph, Roberval, PQ G8H 2K6 — Mgr: Marc-André Levesque. PD: Louis Arcand. CE: Stepane Villeneuve. 24h (N. :15, :30)
519) 1310 Greene Ave, Montreal, PQ H3Z 2B5 — GM: Lee Hambleton. PD: I. McLean. CE: Denis Dion. 24h.
522) C.P. 695, Shawinigan, PQ G9N 6V9 — GM: Marineau Real. PD: Gille Forcier. CE: J. Fournier. 1030-0500.
523) 1975 boul. Wallberg, Dolbeau, PQ G8L 1J5 — Mgr: R. Leclerc. PD: M. Bouchard. CE: G. Gosselin. 24h, also rel. 517).
524) C.P. 69, Thetford Mines, PQ G6G 5S3 — GM/PD: Andrée Wright. CE: Renaud Cloutier. 24h on 1230/1330/1340/1380/1400kHz. Network st's have local prgrs Mon-Fri 1200-1430, 1600-1800, 2200-2230. Sat 1300-1830 (CJLP = Disraeli, studio closed).
524A) 185 rue Du Roi, bureau 301, Asbestos, PQ J1T 1S4 — GM: Marie-Paule Drouin. CE: R. Cloutier. 24h, own prgrs: 1200-1800, 2100-2230.
524B) C.P. 490, Victoriaville, PQ G6P 6T3 — GM. & PD: Robert Daneau. CE: Renaud Cloutier. 24h.
524C) 5088 rue Frontenac, Lac Mégantic, PQ G6B 1H3 — GM: Rémi-Mario Mayette. PD: Michel Brochu. CE: Jacques Vachon. 24h.
526) C.P. 850, La Tuque, PQ G9X 3P6 — GM: Réjean Leclerc. PD: Reine MacDonald. CE: Yves Brassard. 24h (N. at :15, :45)
527) 455-3e rue, Chibougamau, PQ G8P 1N6 — GM. M. Loubier. PD: G. Fournier. CE: Y. Gauthier. Prgr. on 1240/1340kHz. (CFED=Chapais).
530) 200-460 Place Sacre-Coeur O, Alma, PQ G8B 1L9 — GM: G. Pedneault. PD: M. Larouche. CE: C. Gagnon. 24h.
533) 800 ave. du Phare Ouest, Matane, PQ G4W 1V7. **E-mail:** chrmam@quebectel.com — GM: Kenneth Gagné. PD: Charles LePage. CE: R. Comeau. 24h (N. at :45).
538) 1360 rue Notre-Dame Sud, Ste-Marie-de-Beauce, PQ G6E 2W9 — GM: M. Jalbert. PD: Raymond Maheux. CE: Gaston Guay. 24h. Own prgrs: 1100-2200. Other times rel CIRO-FM and 546).
540) 380 ave. Murdoch, Rouyn-Noranda, PQ J9X 5C2. — GM: Denis Bouchard. PD: J.-P. Proulx. CE: C. Imbeau. 24h on 900/1240/1340/1400kHz. Netw. st's broadcast own prgrs W exc. 2300-1045 (Sat 1100) (N. at :45).
540A) 1729 3e Ave, Val d'Or, PQ J9P 1W3 — Mgr: Guy Laporte. CE: G. Beaulieu.
540C) 751-1ère Ave Ouest, Amos, PQ J9T 1V7 — Mgr: R. Audet. 24h.
541) 64 Hôtel de Ville, Rivière-du-Loup, PQ G5R 1L5 — GM: Guy Simard. PD: D. St-Pierre. 1045-0500 on 1240/1400/1450kHz.
541A) Cabano, PQ. Addr. etc. as 541).
541B) Pohénégamook, PQ. Addr. etc. as 541).
543) R. Montréal, 35 York St, Westmount, PQ H3Z 2Z5. **E-mail:** admin@cfmb.ca **WWW:** http://www.cfmb.ca — Pres. & PD: Andrzej Mielewczyk. PD: Walter Centa. CE: Luigi Valente. 24h in Italian and several other languages.
546) C.P. 100, St. George-de-Beauce, PQ G5Y 5C4 — GM: M. Jalbert. PD: Raymond Maheux. CE: Gaston Guay. 24h.
549) 2070 St-Georges, Drummondville, PQ J2C 5G6 — GM: R. Desmarais. CE: M. Courmoyer. 1130-0500 (N. at :15, :45).
555) 875 boul St-Germain, Rimouski, PQ G5L 3T9 — GM: Bertrand Bellavance. PD: Pierre Lindley. CE: J. Fournier. 24h.
556) C.P. 1840, Amqui, PQ G0J 1B0 — GM: Adalbert Lévesque. PD: Frédéric Bureau. CE: Jacques Beaudoin. 24h. (1450=Causapscal, PQ.)

New Brunswick
601) Box 188, Edmundston, NB E3V 3K8 — GM: Claude Boucher. CE: M. Soucie. Prgr (French and 5% English): 1000(Sat 1100, Sun 1200)-0330 (Sat 0400, Sun 0300) on 570/1490kHz. CKMV=Grand Sault, NB.
602) Box 338, Newcastle, NB E1V 3M4. **E-mail:** cfan@nb.sympatico.ca **WWW:** http://www.cfan.com — Mgr: Jim MacMullin. PD: E. Grey. CE: E. Rogers. 24h.
603) 93 blvd. St-Pierre Est, Caraquet, NB E0B 1K0 — Mgr: Armend Roussy. CE: René Lanteigne. 24h.

604) Box 920, Woodstock, NB E0J 2B0 — GM: Pat Brennan. PD: Rick McGuire. CE: Dick Cleveland. 24h.
605) 199 Chesley Dr, St. John, NB E2K 4S9. **E-mail:** info@radio.fundy.ca **WWW:** http://www.radio.fundy.ca — GM: George Ferguson. PD: Roxanne Kirkpatrick. CE: R. Vautour. 24h.
606) Box 340, Campbellton, NB E3N 3G7. **E-mail:** cknb@nb.sympatico.ca — GM: Claude Arsenault. PD: John Kennedy. CE: Miller Sound. 24h.
608) 1000 St. George Blvd, Moncton, NB E1C 2E1 — GM: Dan Roman. PD: Steve Jones. CE: Bob Oke. 24h on 590/1220kHz.
608A) Box 5900, Sussex, NB E0E 1P0. **E-mail:** cjcw@nbnet.nb.ca — GM: Dave Boone. PD: Dennis Vautour. CE: Miller Sound. Own prgrs Mon-Sat 1000-2200, Sun 1200-1600.
609) Box 1360, Bathurst, NB E2A 4J1 — GM: Jim Duncan. PD: P. Assaff. CE: Dick Cleveland. 24h. F: M-F 1800-1900.
610) 206 Rookwood Ave, Fredericton, NB E3B 2M2 — GM: John F. Eddy. PD: Tom Blizzard. CE: Richard Cleveland. 24h.

Prince Edward Island
700) Box 1060, Charlottetown, PEI C1A 7M7 — GM: Frank Lewis. PD: Jim Ferguson. CE: W. Corney. 24h.
701) Box 7200, Charlottetown, PEI C1A 8V7 — GM: Frank Lewis. PD: Gerard Murphy. CE: G. Mills. 24h.
702) 763 Water Str. East, Summerside, PEI C1N 4J3. **E-mail:** cjrw@atcon.com **WWW:** http://www.icondata.com/cjrw/index.htm — GM: Paul M. Schurman. CE: S. Harvey. 1000-0415.

Nova Scotia
800) Box 5800, Antigonish, NS B2G 2R9. **E-mail:** cjfx@atcom.com — Mgr: David MacLean. PD: Peter S. McCully. CE: Scott MacLeod. 24h.
801) 187 Industrial Ave, Truro, NS B2N 6V3. **E-mail:** ckcl@ckcl.ca **WWW:** http://www.ckcl.ca — GM: Dan Cormier. PD: M. Allard. CE: Ken Condon. 24h.
802) 2900 Agricola St, Halifax, NS B3J 2Z4 — GM: William Bodnarchuk. PD: Terry Williams. CE: Walter Labucki. 24h.
803) Box 670, Amherst, NS B4H 4B8. **E-mail:** am90@atcon.com **WWW:** http://www.ckdh.com — GM: Gary Crowell. PD: David March. CE: Bob Oke. 24h.
805) 188 Charlotte St, Sydney, NS B1P 1C5 — GM: Sean Russell. PD: Dan Barton. 24h.
806) Box 400, Halifax, NS B3J 2R2. **E-mail:** chns@ns.sympatico.ca — GM: Gary Barker. PD: Troy Michaels. CE: Wayne Harvey. 24h.
807) 215 Dominion Str, Bridgewater, NS B4V 2G8. **E-mail:** ckbw@ckbw.com **WWW:** http://www.ckbw.com — GM: Michael Prud'homme. PD: Gary Richards. CE: F. Grayney. 24h.
808) Box 1270, Sydney, NS B1P 6K2 — GM: Don L. Brown. PD: Donnie Graham. CE: Roy MacIntosh. 24h.
809) Box 519, New Glasgow, NS B2H 5E7 — CEO: Doug B. Freeman. PD: Rod Mackey. CE: P. Lann. 24h.
810) 328 Main Str. #201, Yarmouth, NS B5A 1E4. **E-mail:** cjls@atcon.com — GM: Grant P. Wyman. PD: J. Gramen. CE: Mike Evans. 24h.
811) Box 1410, Port Hawkesbury, NS B0E 2V0. **E-mail:** bob@cigo.com **WWW:** http://www.cigo.com — GM: Bob MacEachern. PD: Paul Knott. CE: Paul Williams. 24h.
812) Box 310, Kentville, NS B4N 1H5. **E-mail:** avr@glinx.com **WWW:** http://www.glinx.com — GM: Neil MacMullen. PD: Dianne Best-Redden. CE: H. A. David Morrison. 24h. Netw. st's carry local prgrs Mon-Fri 1300-1600.
812A) Box 550, Middleton, NS B0P 1J0 — Mgr: Rod Reeves.
812B) Box 1420, Digby, NS B0V 1A0 — Mgr: Karen Corey.
812C) Box 278, Windsor, NS B0N 2T0 — Mgr: Rod DeVillier.

Newfoundland
900) Box 570, Corner Brook, NF A2H 6H5 — GM: Roger Humber. PD: David Bouzane. CE: C. Poole. 24h on 570/750/870/1230/1340kHz. (CFNW Port au Choix repeater).
900B) 30 Oregon Dr, Stephenville, NF A2N 2X9. **E-mail:** cfcb870@hotmail.com — Mgr: Gerry Murphy. PD: Larry Bennett. 24h. Own prgrs 0930 (Sat 1030, Sun 1630)-2130.
900C) Box 1230, Port aux Basques, NF A0M 1C0 — Mgr: G.N. Critchell. PD: R. Edwards. CE: J. Goosney. 24h. Own prgrs: Mon-Fri 1230-2130.
900D) Box 4000, Stn "C", Goose Bay, Lab, NF A0P 1C0 — Mgr: Paul Saunders. 1000 (Sun 1200)-0500. Own prgrs: Mon-Sat 1100-2130 (studio in Corner Brook).
900E) Box 6000, Wabush, Lab, NF A0R 1B0 — Mgr: Gary Peckham. CE: R. Power. Own prgrs: 1100-2130 (studio in Corner Brook).
901) Box 8-590, Station A, St. John's, NF A1B 3P5. **WWW:**

http://www.vocm.com — GM: John Murphy. PD: Tom Ormsby. CE: Reg McCausland. 24h on 560/590/620/710/730/740/1240kHz.
901A) Box 560, Marystown, NF A0E 2M0 — S.M: Russ J. Murphy. PD: Gary Myles. CE: Harold Steele. Own prgrs: 0930-2130, rel.
901D) 2130-0330.
901B) Box 620, Grand Falls, NF A2A 2K2 — Mgr. & PD: Jim Coady. CE: H.G. Feltham. Own prgrs: 0930-2100. (CKIM=Baie Verte).
901C) Box 650 (TCH), Gander, NF A1V 1X2 — Mgr: Paul Stride. CE: Reg McCausland. Own prgrs: 0930-2130.
901D) #1 CHVO Drive, Carbonear, NF A1Y 1A2. GM: John Harvey. CE: Harold Steele.
901E) Clarenville, NF — Own prgrs 0930-2130, rel. 901D) 2130-0330.
902) Box 7430, St. John's, NF A1E 3Y5 — GM. & PD: John C. Tessier. CE: Reg McCausland. 24h. 30% rlg, non-comm.
903) P.O. Box 8010, Station A, St. John's NF A1B 3M7 — GM: Hilary. Montbourquette. PD: A. Newman. CE: Ken Condon. 24h. CJYQ rel. on 1010kHz. CKIX-FM rel. on 670/680/1010kHz.
903C) Box 810, Grand Falls, NF A2A 2M4 — Mgr: Dave Hillier. PD./CE: Richard King. Own prgrs 0930-approx. 2000.
903D) 78 Elizabeth Dr, Gander, NF A1V 1G8 — Mgr: Kenton Dunphy. CE: Ken Condon. Own prgrs 0930-2130. (CKXB=Musgravetown).
905) Box 2520, Mt. Pearl, NF A1N 4M7 — GM/PD: Cameron Beierle. CE: Brian Matthews. 24h. Rlg (Seventh-Day Adventist Church).

L.T: UTC - 3h (exc. Pituffik/Thule area: UTC - 4h and exc. Ittoqqortoormiit/Danmarkshavn UTC - 1h). Su: add one hour, exc. Thule area — **Pop:** 55.900 — **Radios:** 22.000 — **Pr.L:** Greenlandic, Danish — **E.C:** 50Hz, 220V. — **ITU:** GRL.

KALAALLIT NUNAATA RADIOA - KNR (Gov. Comm.)
☑ P.O. Box 1007, DK-3900 Nuuk. ☎ +299 321172. 🖷 +299 324703. **Web:** http://www.knr.gl
Regional addr: Kujataata Radioa (So. Greenland), Box 158, DK-3920 Qaqortoq — Aavannaata Radioa (No. Greenland), Box 223, DK-3952 Ilulissat.
LP: Dir: Peter Frederik Rosing . Head of programming: Elisabeth Lyberth.

Mediumwaves	kHz	kW	Mediumwaves	kHz	kW
Nuuk	570	5	Upernavik	810	5
Qeqertarsuaq	650	5	Uummannaq	900	5
Simiutaq	720	10			

FM *) stereo	MHz	W	FM	MHz	W
Nuuk *	90.5/95.5/96.0	50	Aasiaat	95.5	100
Sisimiut	95.0	100	Ilulissat	96.0	50
Kangerlussuaq	96.0	10	Tasillaq	96.0	50
Uummannaq	95.0	50	Tasillaq Dye Four	98.7	500

+ 60 additional st's 50W or less.
D.PRGR: MF 0930-0030, Sat 1000-0300, Sun 1100-0030. Relays DR P1, Denmark, daily from 0500.
ANN: "Kallaallit-Nunaata Radioa", "Grønlands Radio" — **IS:** "Sunnia Kalippoq" (The Whaleboat "Sonja" drags whale) played on celeste.
V. by QSL-card.

DR P1, Denmark. Sat. relay 24h: Nuuk 98.0 MHz 100W

Private stations (local radio):
Nuuk: Nuummiut Tusaataat, Box 1462, 3900 Nuuk: 93.0 MHz
Aasiaat: Tusaat Aasiaat, Box 20, 3950 Aasiaat: 93.0 MHz
Ilulissat: Ilulissat Radio-at 99 MHz, Box 1004, 3952 Ilulissat: 99.0 MHz
Kangerlussuaq: KLR Kangerlussuaq Lokal Radio, Box 37, Kangerlussuaq: 98.5 MHz
Maniitsoq: Maniitsup Tusaataa Akisuasoq, Box 29, 3912 Maniitsoq: 90.5/93.0 MHz
Narsarmiijit Radio: Narsarmiijit B.461, 3922 Nanortalik: 102.0 MHz
Paamiut: Paamiut Tusaataat, Anders Petersenip Aqq. B-95, 3940 Paamiut: 93.0 MHz
Pituffik/Thule Air Base: Radio OZ520, Den Danske Radio, SPE, Box 139, Thule Air Base, 3970 Pituffik: 97.1 MHz
Qaanag: Qaanaaq Radiunga, Box 57, 3971 Qaanag: 93.5 MHz
Qaqortoq: Qaqortup Tusaataa, Box 40, 3920 Qaqortoq: 93.0 MHz
Sisimiut: Sisimiut Tusaataat, Box 312, 3911 Sisimiut: 91.0/93.0/98.0 MHz

Tasiilaq: Tasiilap Tusaalaa, Ittimiini B.883, 3913 Tasiilaq: 93.0 MHz
Uummannaq: Tusaat Uummannaq, Box 195, 3961 Uummannaq: 98.2 MHz

MEXICO

LT: UTC -6h (Su: UTC -5h) in all states exc. UTC -7h (Su: -6h) in Baja California Sur, Sonora, Sinaloa and Nayarit; UTC -8h (Su: UTC -7h) in Baja California — **Pop:** 96.000.000 — **Radios:** 21.000.000 — **Pr.L:** Spanish — **E.C:** 60hz, 110/220V — **ITU:** MEX.

SECRETARIA DE COMUNICACIONES Y TRANSPORTES
Dirección General de Telecomunicaciones
✉ L.Cárdenas 567-11° Ala Norte, Narvarte, 03020 México. ☎ +52 (5) 5199734.
L.P: Dir.Gen: Ing.Enrique Luengas H.
Departamento de Frecuencias
✉ Av.Niño Perdido y Cumbres de Acultzingo, México 12, D.F. ☎ +52 (5) 5303060/9.
L.P: Dir: Ing.Luis Valencia P.

DIRECCION GENERAL DE CONCESIONES Y PERMISOS DE TELECOMUNICACIONES
Departamento de Asignación de Frecuencias
✉ Unidad Contel Sga-2, Av.de las Telecomunicaciones s/n, Ixtapalapa, 09310 México, D.F. - ☎ +52 (5) 6920077.
L.P: Dir: Ing.Sergio Cervantes.

CAMARA NACIONAL DE LA INDUSTRIA DE RADIO Y TELEVISION DE MEXICO
✉ Av.Horacio 1013, Col.Polanco Reforma, 11550 México, D.F. ☎ +52 (5) 7269909.
L.P: Pres: Raúl Aréchiga Espinosa. Gte: Lic. César Hernández Espejo.

Mediumwaves: Call XE-,° = also on shortwave, * = inactive, (r) = repeater, v = varying fq.The letters preceding the st.number indicate the State. Addresses are listed by State in alphabetical order.

	Call	kHz	kW	Name and h.of tr.
BC19)	BACH	540	25	X-Bach Stereo 5-40, Tijuana: 24h (Parallel to KNOB 540kHz, Costa Mesa, CA, USA)
CS22)	MIT	540	5/1	XEMIT, Comitán: 1100-0700
SL01)	WA	540	150	KeBuena WA/Grande Cadena W, San Luis Potosí: 24h
CH01)	PL	550	5	La Super Estación, Cd. Cuauhtémoc: 24h
GR01)	ACD	550	1	R.Lobo, Acapulco: 1200-0600
JL23)	ZK	550	1/0.25	R.Artena, Tepatitlán de Morelos: 1200-0300
OX01)	HLL	550	1	R.Mar, Salina Cruz: 1200-0500
OX22)	UC	*550	1/0.25	R.Felicidad, Tehuantepec
VE01)	KL	550	1/0.25	Primerísima 550, Jalapa: 1300-0400
YU01)	QW	550	2/0.35	QW La Poderosa, Mérida: 1130-0100
CL10)	MZA	560	5	La Buena Onda del Pacífico, Manzanillo: 1200-0500
DF01)	OC	560	5	R.Chapultepec, México: 1200-0700
DG05)	SRD	560	5d	La Ke Suave, Santiago Papasquiaro: 1200-0400
SO37)	YO	560	1	R.Lobo/R.5-60, Huatabampo: 1300-0500
ZC12)	XZ	560	5	R.Cañon, Zacatecas: 1200-0600
CO01)	TJ	570	1/0.25	La Tropical de Oro, Torreón: 1200-0400
JL29)	KZX	570	5	KZX Tejano 57, Cd. Guzmán: 1300-0100
MI31)	LQ	570	1	R.Imagen, Morelia: 1200-0600
NL02)	BJB	570	5	La Estación del Barrilito, Monterrey: 24h
OX02)	OA	570	5	R.Oaxaca, Oaxaca: 1100-0400
PU13)	VJP	570	1	Espacio 570, Xicotepec de Juárez: 1100-0300
SO26)	UK	570	1	UK AM Stereo, Caborca: 1400-0600
TB01)	VX	570	2.5	R.Fórmula/La Grande de Tabasco, Villahermosa: 1000-0300
YU09)	ME	570	1	R.Valladolid, Valladolid: 1200-0400
CH08)	FI	580	5/0.25	R.Mexicana, Chihuahua: 24h
CO02)	MU	580	5/2.5	La Rancherita del Aire, Piedras Negras: 24h

CS01)	UE	580	1	LV de Chiapas, Tuxtla Gutiérrez: 1100-0600
JL02)	AV	580	10/1	Canal 58/R.Guadalajara, Guadalajara: 1200-0700
QE08)	UAQ	580	2	R.Universidad, Querétaro: 1200-0600
QR01)	YI	580	1/0.25	R.Festival, Cancún: 1100-0600
SO06)	HO	580	1	R.Centro, Cd.Obregón: 24h
TM01)	HP	580	1	R.Exitos, Cd.Victoria: 1200-0400
VE02)	DZ	580	1	Canal 58, Córdoba: 1200-0400
CS23)	ZZZ	590	5/1	R.Z, Tapachula: 1100-0500
DF02)	PH	590	10/5	Sabrosita 590, México: 24h
DG01)	E	590	1	R.Festival, Durango: 24h
SO01)	HD	590	1d	Globo 590, Hermosillo: 1200-0700
TM02)	FD	590	5	La Super F-D, Reynosa: 1100-0500
CO01)	DN	600	1	La Mexicana, Torreón: 1200-0500
CS10)	OB	600	5	La Máquina Musical, Pichucalco: 1100-0500
CS24)	OCH	600	10/1	XEOCH, Ococingo
GJ10)	GTO	600	10	R.Cañon, León: 1200-0100 (rep.on 590)
GR02)	BB	600	5	La Comadre, Acapulco: 1200-0400
JL29)	KZX	600	5	Tejano 600, Cd.Guzmán: 1200-0600
MI02)	TA	600	0.5	R.Sensación, Zitácuaro: 1200-0400
NL03)	MN	600	1	La Regiomontana, Monterrey: 24h
SL16)	CV	600	5/1	La Gran Compañía, Cd.Valles: 1200-0500
SN20)	HW	600	5	Digital 600, Rosario: 1200-0700
YU02)	Z	600	5	R.600, Mérida: 1200-0500
CO03)	BX	610	5/0.5	La Primera Estación, Sabinas: 1100-0500
MI03)	UF	610	5	R.Variedades, Uruapan: 1200-0800
OX03)	KZ	610	1	LV del Istmo, Tehuantepec: 1200-0600
SN02)	GS	610	1	R.Guasave, Guasave: 1230-0600
VE01)	JA	610	1	R.Fiesta, Jalapa: 1200-0500
ZC01)	EL	610	5/2.5	Super Canal 610, Fresnillo: 1200-0600
CH03)	BU	620	5	La Norteñita, Chihuahua: 1200-0700
DF03)	NK	620	10/5	R.6-20, México: 1200-0600
DG02)	CK	620	1/0.5	Digital 620, Durango: 1200-0500
NA01)	OO	620	5/1	R.Triunfadora, Tepic: 24h
SL14)	WZ	620	2.5	R.6-20, San Luis Potosí: 1200-0430
TB05)	HGR	620	1	Globo 620, Villahermosa: 1200-0600
TM30)	GH	620	1	Bonita, Reynosa: 24h
GR02)	ACA	630	5	R.ACIR, Acapulco: 24h
JL02)	JB	630	5	R.Gobierno, Guadalajara: 1300-0700
NL03)	FB	630	20	Radar FB, Monterrey: 24h
SN03)	OPE	630	5/0.25	R.Hits, Mazatlán: 24h
SO02)	FX	630	1	La Profesionales de la Radio, Guaymas: 24h
TM27)	ERO	630	1	R.Tamaulipas, Altamira: 1200-0400
VE03)	FU	630	10/0.75	LV Amiga de la Cuenca, Cosamaloapan: 1200-0400
CH27)	JUA	640	5	Super Estelar, Cd.Juárez: 1200-0200
CH29)	HHI	640	10	R.Uno, Hidalgo del Parral: 24h
CS03)	WM	640	1d	R.6-40, San Cristóbal de las Casas: 1200-0300
HG04)	NQ	640	20/2.5	NQ La Vista/LV de la Provincia de la Provincia, Tulancingo: 24h
TM19)	TAM	640	1	Fiesta Mexicana, Cd.Victoria: 24h
ZC09)	YQ	640	10	R.Fresnillo, Fresnillo: 1200-0700
GR21)	CHH	650	5	La Explosiva, Chilpancingo: 1200-0300
JL03)	EJ	650	10	La Grande, Puerto Vallarta: 24h
MI04)	ZM	650	5/1	La Expresión de Zamora, Zamora: 1200-0600
SN04)	TNT	650	2.5	R.65/La Ley, Los Mochis: 24h
SO29)	VSS	650	2.5	Notiflash, Hermosillo: 1200-0200
TB15)	VLL	650		R.Felicidad, Villahermosa: 24h
YU11)	VG	650	1	Extasis, Mérida: 1200-0400
BS06)	JJC	660		XEJJC, San José del Cabo
CH04)	ACB	660	5/3	ACB AM Numero Uno, Cd.Delicias: 1200-0600
DF04)	DTL	660	50	La Candela 660, México: 24h
DG06)	WX	660	1/0.5	R.Mexicana, Durango: 1200-0600
NL02)	FZ	660	10	FZ Metrópoli, Monterrey: 24h
OX04)	YG	660	1	R.Consentida, Matías Romero: 1200-0100
QR09)	CPR	660	1	R.Chan Santa Cruz-LV de los Mayas, Felipe Carillo Puerto
TM20)	AR	660	5	La Mexicana, Tampico: 24h
ZC02)	EY	660	20d	La Consentida, Jalpa: 1100-0600
CO04)	TOR	670	1	R.Ranchito, Torreón: 1200-0600
JL01)	IS	670	5d	La Rancherita Consentida, Cd.Guzmán:

				1200-0600
QE01)	QG	670	1	Poder 6-70, Querétaro: 1200-0600
VE38)	SIC	670	0.5	R.Festival, Córdoba: 24h
CH08)	FO	680	1d	Fiesta Norteña, Chihuahua: 1200-0100
CS04)	KQ	680	5/0.5	R.Soconusco, Tapachula: 1100-0400
GJ02)	LG	680	10	LG La Grande, León: 24h
GR24)	CHG	680	1	Fiesta Mexicana, Chilpancingo: 1200-0600
PU01)	FJ	680	1	R.Teziutlán, Teziutlán: 1200-0400
SN05)	ORO	680	1	La Tremenda, Guasave: 1200-0600
SO30)	SON	680	1	R.ACIR, Hermosillo: 24h
BC01)	TRA	690	77.5	Sport 6-90, Tijuana: 24h (English, programmed from San Diego)
CL03)	CS	690	5	La Grande, Manzanillo: 24h
DF05)	N	690	20/5	Ondas del Lago, México: 1200-0600
MI10)	XL	690	1d	R.Láser, Pátzcuaro: 1200-0400
NL07)	RG	690	5	La Deportiva 6-90, Monterrey: 1200-0600
SN19)	ST	690	10	R.Sensación, Mazatlán: 1300-0500
VE07)	UY	690	2.5	R.Sensación, Coatzacoalcos: 24h
ZC03)	MA	690	10	La Onda Musical del Altiplano, Fresnillo: 24h
CA09)	PUJ	700	1	LV del Corazón de la Montaña, X'pujil
CH25)	GD	700	1	La Poderosa, Hidalgo del Parral: 1200-0600
JL04)	DKR	700	1	R.Red, Guadalajara: 24h (r:1110)
VE02)	VC	700	2.5	Canal 70, Córdoba: 1300-0100
CH05)	DP	710	10/1	La Ranchera de Cuauhtémoc, Cd.Cuauhtémoc: 2300-0700
CL01)	RL	710	1	La Super R-L, Colima: 1100-0600
CS05)	ON	710	3	R.Mexicana, Tuxtla Gutiérrez: 24h
DF04)	MP	710	10	R.7-10, Mexico: 24h
GR02)	MAR	710	1	R.Felicidad, Acapulco: 1200-0300
NA02)	RK	710	1/0.25	R.Korita, Tepic: 1200-0800
OX05)	RPO	710	5	R.Variedades, Oaxaca: 24h
SL10)	SMR	710	1	R.Fórmula, San Luis Potosí: 1200-0600
SN01)	BL	710	5	Fiesta Mexicana, Culiacán: 24h
SO03)	PS	710	1	R.Amistad, Empalme: 1300-0700
TB02)	KV	710	1	R.Sensación, Villahermosa: 24h
TM10)	OLA	710	1	La Reina Grupera, Tampico: 1200-0400
YU01)	YK	710	5/0.25	R.Mágica, Mérida: 1200-0400
CO19)	DE	720	5/0.25	Audio Digital 720, Saltillo: 24h
JL05)	QZ	720	1d	Ritmo 7-20, San Juan de los Lagos: 1300-0100
SN19)	VU	720	1	720 Super Radio, Mazatlán: 1300-0700
VE37)	AVR	720	5	R.Fórmula, Veracruz: 1100-0600
BC21)	EBC	730		R.ACIR, Ensenada: 1400-0800
BS01)	LBC	730		XELBC, Loreto
CO05)	PQ	730	1/0.1	La Sa Sa Sabrosita, Cd.Muzquiz: 1200-0400
DF06)	X	730	100	La X de México/Sintonía humana, México: 24h
JL30)	GDL	730	0.5	La Tremenda 730, Zapotlanejo: 1300-0100
AG02)	LTZ	740	1	El Planeta, Aguascalientes: 1200-0400
CO28)	QN	740	5	R.Capullo 7-40, Torreón: 1200-0600
JL25)	VAY	740	1/0.5	La Mexicana, Puerto Vallarta: 1200-0600
SN13)	HS	740	5	Audio A, Los Mochis: 24h
YU10)	PET	740	10	LV de los Mayas, Peto: 1100-2400 (SS -2000)
CH21)	OH	750	1d	Chiquitita, Camargo: 24h
GR22)	KOK	750	5	R.Sensación, Acapulco: 1130-0130
MI29)	IP	750	10	La Poderosa, Uruapán: 24h
NA09)	JMN	750	5	LV de los Cuatro Pueblos, Jesús María: 1200-1930
OX06)	CORO	750	5	R.Loma, Loma Bonita: 1200-0400
SL17)	RASA	750		La música que llego para Quedarse, San Luis Potosí: 1200-0600 (rel.of XENK 620 kHz)
SN23)	CSI	750	1	R.Mexicana, Culiacán: 24h
VE31)	TI	750	1	La Fabulosa, Tempoal: 1200-0400
CS07)	RA	750	1	R.Chiapas, San Cristóbal las Casas
DF07)	ABC	760	50/5	R.ABC, México: 24h
DG03)	DGO	760	5	La Super Grupera, Durango: 1100-0500
JL15)	ZZ	760	10/0.25	R.Gallito, Guadalajara: 1230-0600
MI25)	LAC	760	5	R.Azul, Cd.Lázaro Cárdenas: 1100-

				0600
SL10)	EQ	760	0.25d	La Pantera, San Luis Potosí: 1200-0100
SO07)	EB	760	1	La Rancherita, Cd.Obregón: 1200-0800
SO17)	NY	760	5/0.1	R.Geny, Nogales: 24h
YU01)	YW	760	2.5/0.5	La Tropiranchera, Mérida: 1200-0600
CH07)	HB	770	1d	R.Jilguerita, San Francisco del Oro: 1300-0100
GR20)	SUR	770	1.5	R.Chilapa, Chilapa: 1200-0200
MI05)	ML	770	1d	R.Tropical, Apatzingán: 1200-0500
SN04)	FTA	770	5	Fiesta 77, Los Mochis: 1300-0700
VE47)	QRV	770	5	La Costeñita, Veracruz: 1200-0600
ZC04)	IH	770	10/5	La Nueva Dimensión en Radio, Fresnillo: 24h
CS02)	TS	780	5	R.Tapachula, Tapachula: 1100-0500
GJ22)	ZN	780	0.5d	Stereo Digital, Celaya: 1200-0100
GR04)	XY	780	1d	LV del Balsas, Cd.Altamirano: 1200-0430
JL07)	LD	780	5d	R.Costa, Autlán: 1200-0500
OX15)	GLO	780	5	LV de la Sierra Guelatao de Juárez: 1145-2230 (SS 1300-2100)
TM20)	MTS	780	1	R.Fórmula, Tampico: 1200-0600
TM25)	SFT	780	5/1	La Bronca Norteña, San Fernando: 1200-0600
AG01)	BI	790	5	R.B-I, Aguascalientes: 24h
BC06)	SU	790	1	R.7-90, Mexicali: 1400-0200
BS01)	NT	790	5/0.75	La Paz, La Paz: 1300-0700
CH02)	RPC	790	5/0.25	R.Ranchito, Chihuahua: 24h
CO01)	GZ	790	1d	La Pantera, Torreón: 1200-0500
DF08)	RC	790	50	El Fonógrafo, México: 24h
JL26)	GAJ	790	1	R.790, Guadalajara: 1300-0600
TB03)	VA	790	10/0.25	R.Hogar, Villahermosa: 1000-0600
TM03)	FE	790	1/0.5	R.Fiesta, Nuevo Laredo: 1230-0600
VE11)	COV	790	1	R.Lobo, Poza Rica: 1200-0600
YU06)	UP	790	1	La Bonita de Oriente, Tizimín: 1200-0500
BC03)	MMM	800	0.5	R.80/Onda Ranchera, Tijuana: 1300-0800
CH12)	ROK	800	150	R.Cañon, Cd.Juárez: 24h
CS13)	UI	800	1	R.Comitán, Comitán: 1200-0400
GJ19)	GX	800	1	R.Cañon, San Luis de la Paz: 1300-0300
GR09)	ZV	800	3	LV de la Montaña, Tlapa de Comonfort: 1200-0100 (SS -2030)
JL27)	AN	800	1d	R.Alegría/R.Ocotlán, Ocotlán: 1200-0300
VE09)	QT	800	5	La Poderosa, Veracruz: 1200-0600
CL05)	MAX	810	3/0.25	Radiomax, Armería: 1100-0400
CO06)	IM	810	1	R.Capital, Saltillo: 1200-0600
CS04)	OE	810	1	R.Amistad, Tapachula: 1100-0400
CS06)	IN	810	1	LV del Valle, Cintalapa: 1100-0500
GJ12)	EMM	810	1	R.Salmantina, Salamanca: 1300-0400
GR18)	AGR	810	1/0.15	R.Fórmula, Acapulco: 24h
NA03)	UX	810	10/0.25	La Legendaria, Tuxpán: 1300-0100
QR04)	RB	810	5/1	Sol Stereo, Cozumel: 1200-0100
SO04)	RSV	810	5d	R.Alegría, Cd.Obregón: 1300-0200
TM04)	RI	810	1	Realmente Música, Reynosa: 24h
TM05)	FW	810	50/35	R.Estrella, Tampico: 1200-0400
TX01)	HT	810	2.5	R.Huamantla, Huamantla: 1200-0600
YU03)	MQ	°810	2	R.Fórmula, Mérida: 1230-0600
ZC05)	ZC	810	1/0.25	R.Felicidad, Río Grande: 1200-0500
BC07)	MVS	820	3.5d	R.Lobo, Mexicali: 24h
CA01)	ESC	820	0.75d	R.Escárcega, Escárcega: 1200-2400
CH09)	SB	820	1d	R.Mexicana, Santa Bárbara: 1100-0200
DG06)	DRD	820	1/0.5	La Poderosa, Durango: 1200-0600
JL15)	BA	820	10/1	La Consentida, Guadalajara: 24h
SN14)	UDO	820	1	R.Universidad de Occidente, Los Mochis: 1300-0500
VE02)	KG	820	2.5	Golden Hits, Córdoba: 1300-0100
DF17)	LA	830	10	Buena Música desde México, México: 24h
MI06)	PUR	830	1	LV de los Purépechas, Cheran: 1300-0100 (SS -2100)
NL13)	LN	830	3	R.Linares/La Llegadora, Linares: 1200-0600
SN25)	VQ	830	5	La Grande de Sinaloa, Culiacán: 1300-0200
SO28)	DR	830	1/0.25	La Marinera, Guaymas: 1100-0700
ZC12)	LK	830	10/0.5	R.Sensación, Zacatecas: 1200-0600
CA10)	CUC	840	0.5	Casa de la Cultura de Campeche,

Campeche

Code	ID	Freq	Power	Station
CS01)	IO	840	10d	R.Capital, Tuxtla Gutiérrez: 1200-0500
GJ04)	FG	840	5d	La Pachanga, Celaya: 1200-0600
JL16)	XXX	840	5	Fiesta Mexicana/Fiesta Digital, Tamazula: 1200-0600
NA13)	TEY	840	1	R.Sensación, Tepic: 1300-0500
OX07)	ACC	840	1	LV del Puerto, Puerto Escondido: 1300-0100
OX16)	OJN	840	5	LV de la Chinantla, San Lúcas Ojitlán: 1200-2400
PU09)	FS	840	1	R.Matamoros, Izúcar de Matamoros: 1100-0500
TM18)	MY	840	1	Tropical, Cd.Mante: 1200-0100
VE18)	PV	840	2.5	La Fiera Grupera, Papantla: 1200-0600
BC04)	ZF	850	0.5d	La Rancherita Consentida, Mexicali: 1300-0300
CH03)	M	850	5/0.5	R.Exitos, Chihuahua: 1200-0500
CO29)	ZR	850	1/0.25	La Formula 85, Zaragoza: 1200-0400
JL08)	MIA	850	1d	R.ACIR, Guadalajara: 1200-0600
QE02)	JAQ	850	1	R.Felicidad, Jalpan: 1200-0600
SO05)	US	850	1d	R.Universidad de Sonora, Hermosillo: 1200-0620
VE04)	TQ	850	10	La Q Orizabeña, Orizaba: 1300-0900
AG05)	PLA	860	5	R.Mexicana, Aguascalientes: 1100-0600
BC05)	MO	860	5	R.86/La Poderosa AM Estéreo, Tijuana: 24h
CH10)	ZOL	860	1	R.Noticias 860, Cd.Juárez: 24h
CL02)	AL	860	5	R.Mundo, Manzanillo: 1100-0700
CS08)	DB	860	5	DB La Máquina Musical, Tonalá: 1100-0500
DF10)	UN	860	50	R.UNAM-Universidad Nacional Autónoma de México, México: 1300-0700
DG03)	DU	860	5	D-U la que le gusta, Durango: 1100-0500
NL04)	NL	860	5	R.Recuerdo, Monterrey: 1200-0600
QR06)	CTL	860	5	R.Chetumal, Chetumal: 2300-0700
QR07)	CCN	860	5	R.Caribe, Cancún: 1100-0500
SN07)	NW	v860	1	La Radio de la Ciudad, Culiacán: 1300-0600
TB04)	ZX	860	1/0.15	Voz del Usumacinta, Tenosique: 1200-0200
TM05)	TW	860	10	R.Fiesta/La Sabrosona, Tampico: 1200-0100
CH18)	TAR	870	10	LV de la Sierra Tarahumara, Guachochi: 1200-0100
GJ05)	AMO	870	1/0.5	R.Amor, Irapuato: 1200-0600
GR14)	GRO	870	1	R.Guerrero, Acapulco: 1200-0700
MI01)	LY	870	1d	R.Moderna, Morelia: 1200-0300
PU06)	NG	870	1	R.Nueva Generación, Huauchinango: 24h
SN19)	FIL	870	1	R.Maz, Mazatlán: 1300-0700
CH08)	V	880	5	R.Fórmula, Chihuahua: 24h
CO16)	TC	880	10	R.Mayrán, Torreón: 1200-0600
GR10)	IG	880	1	La Super G/RCN La Grande de Iguala, Iguala: 1200-0600
JL09)	AAA	880	20/1	La Triple A/Noticias 880, Guadalajara: 24h
SL04)	EM	880	1d	R.Alegría, Río Verde: 1300-0100
SN18)	PNK	880	5d	Super Canal 88, Los Mochis: 1100-0600
TB10)	QQQ	880	3	Super Q, Villahermosa: 1100-0500
VE19)	YV	880	5	R.Variedades, Huatusco: 1200-0100
CS18)	FRT	890	20	R.Frontera, Comitán: 1100-0500
GJ11)	AK	890	5	R.Consentida, Acámbaro: 1200-0400
NA13)	PNA	890	1	R.Joya, Tepic: 1300-0700
OX19)	POR	890	5	R.8-90/LV de la Amistad, Putla de Guerrero: 1200-0130
SN01)	NZ	890	10/0.5	Fiesta Mexicana, Culiacán: 24h
VE24)	BY	890	1	R.Tuxpam, Tuxpam: 1200-0400
YU03)	PY	890	0.5	La Tropical Ardiente, Mérida: 1150-0500
ZC06)	PC	890	5/1	Sonido Estrella, Zacatecas: 24h
CS19)	TAK	890	1	Super K-90, Tapachula: 1100-0100
DF06)	W	900	250	LV de la América Latina/La W/La Gran Cadena W, México: 24h
VE05)	WB	900	50	LV de la América Latina/La W/La Gran Cadena W, Veracruz (r:900)
NL05)	OK	900	1	El Radiazo Mexicano, Monterrey: 24h
BC06)	AO	910	1	R.Mexicana, Mexicali: 24h
GJ06)	ACN	910	5	R.Metropoli, León: 24h
NA01)	NAY	910	10	La Buena Onda del Pacífico, Tepic: 24h
PU02)	OL	910	1	R.Impacto, Teziutlán: 1200-0400
BC09)	SDD	920	2.5	Onda 92/R.Mil, Ensenada: 1400-0800
CA11)	TEB	920	1.5	XETEB, Campeche: 1200-0600
CH03)	QD	920	1	La Divertida, Chihuahua: 1200-0700
CO04)	TAA	920	5	R.Exitos, Torreón: 24h
CO07)	OP	920	1	R.Estelar 920, Monclova: 1200-0600
CO08)	MJ	920	1	R.Exitos, Piedras Negras: 1200-0500
CS05)	VV	920	10	R.Diversión, Tuxtla Gutiérrez: 1200-0600
GJ04)	RE	920	5	La Comadre, Celaya: 1300-0130
JL15)	LT	920	10/0.25	R.Escucha, Guadalajara: 24h
MI27)	LCM	920	1	XELCM, Cd.Lázaro Cárdenas: 24h
OX24)	PNX	920	1	R.Costa, Pinotepa Nacional: 1200-0600
PU14)	ZAR	920	1	Que Bonita 9-20, Puebla: 24h
SL19)	BM	920	5	La Ranchera, San Luis Potosí: 24h
SN08)	CQ	920	5	La Rancherita de Culiacán, Culiacán: 24h
SO01)	BH	920	5/1	R.Interactiva, Hermosillo: 1300-0800
BS04)	VSD	930	1	La Señal del Progreso, Cd. Constitución: 1300-0700
CL09)	TTT	930	1	Caliente AM, Colima: 1200-0400
CO17)	SHT	930		Super Hits, Saltillo: 1200-0200
CS12)	MK	930	5	R.Mexicana, Huixtla: 1100-0100
OX17)	TLA	930	0.7	LV de la Mixteca, Tlaxiaco: 1200-0030 (SS -2030)
VE06)	U	°930	5	La U de Veracruz, Veracruz: 1130-0700
YU07)	UL	930	2.5	La Barracuda/Tu Nueva Onda, Progreso: 1200-0500
ZC03)	QS	930	10/5	Romance en Radio, Fresnillo: 24h
BC07)	WV	940	1	Fiesta Mexicana, Mexicali: 24h
DF06)	Q	940	50	La TropiQ, México: 24h
CS17)	REC	940	1	R.9-40, Reforma: 1200-2400
TM07)	RKS	940	1d	R.Cañon, Reynosa: 1200-0400
AG05)	CAA	950	1	La Doble A/R.Exitos, Aguascalientes: 1200-0600
BC08)	KAM	950	10/5	R.California, Tijuana: 24h
CA02)	MAB	950	1	Canal Internacional, Cd.del Carmen: 1200-0400
CH34)	FA	950	1	R.Rama/La Poderosa, Chihuahua: 24h
CO09)	YJ	950	10	La Nueva Imagen, Nueva Rosita: 24h
GJ24)	CEL	950	5	R.Lobo, Celaya: 24h
JL22)	MEX	950	5	950 La Mexicana, Cd.Guzmán: 1200-0600
NA07)	ZE	950	2.5	La Cotorra de Tu Radio, Santiago Ixcuintla: 1200-0200
NL06)	RN	950	1d	R.Naranjera, Monterrey: 1150-0600
SN09)	ORF	950	1	R.Exitos, Los Mochis: 1200-0700
SO12)	PB	950	10	R.Amor, Hermosillo: 1300-0800
TM20)	TO	950	1	R.Lobo, Tampico: 24h
CH11)	CC	960	0.25	Presencia en Radio, Cd.Camargo: 1300-0400
CO10)	KS	960	1	R.Exitos, Saltillo: 1100-0600
CS25)	TAP	960	5	XETAP, Tapachula: 1100-0100
GR05)	UQ	960	1/0.5	R.Variedades, Zihuatanejo: 1200-0300
GR12)	XC	960	1	Super Mil, Taxco: 1200-0600
JL10)	HK	960	1/0.25	LV de Guadalajara, Guadalajara: 24h
MI07)	MM	960	1	R.ACIR, Morelia: 24h
QR02)	ROO	960	5	La Pirata del Caribe, Chetumal: 1200-0500
SL03)	CZ	960	5	La Z/CZ, San Luis Potosí: 1200-0600
SO38)	IQ	960	1/0.5	R.Felicidad, Cd.Obregón: 24h
TM08)	K	960	5/1	La Estación Grande de Nuevo, Laredo: 1155-0400
VE07)	GB	960	1	R.Fiesta, Coatzacoalcos: 24h
VE08)	OZ	960	1	R.Festival, Jalapa: 1200-0600
CH30)	J	970	10/5	La J/R.Mexicana, Cd.Juárez: 24h
CO11)	MF	970	1	La Tremenda, Monclova: 24h
CS04)	TAP	970	5	La Tropical de Oro, Tapachula: 1100-0100 (rep.on 890 kHz)
DF11)	RFR	970	50	R.Fórmula/R.Noticias, México: 24h
GJ08)	UG	970	0.5	R.Universidad de Guanajuato, Guanajuato: 1300-0200
MI08)	CJ	970	1	R.Apatzingán, Apatzingán: 1200-0400
SN10)	VOX	970	5/0.25	Fiesta Mexicana/Fiesta Digital, Mazatlán: 24h
SO08)	EZ	970	5	La Super Z, Caborca: 1400-0600

ID	Call	Freq	Power	Station
TB05)	VT	970	5/2.5	Stereorey, Villahermosa: 24h
TM01)	BJ	970	1	R.9-70, Cd.Victoria: 1200-0600
TM09)	O	970	1	R.Gallito, Matamoros: 1200-0600
YU03)	MH	°970	5	Candela Tropicalmente, Mérida: 1200-0500
ZC13)	ZAZ	970	10	De Mil Amores 9-70, Zacatecas: 24h
CH24)	JK	980	1	R.Amistad, Cd.Delicias: 1300-0500
CO12)	NR	980	5/0.5	La N-R, Nueva Rosita: 1155-0600
MI09)	LC	980	5	Dual Stereo, La Piedad: 1230-0500
NA04)	XT	980	1	R.Fiesta, Tepic: 1300-0700
SL06)	FF	980	1	XEFF, Matehuala: 1200-0400
SO09)	FQ	980	1	LV de la Ciudad del Cobre, Cananea: 1200-0700
SO10)	KE	980	1	KE-98 Acción Digital, Navojoa: 24h
TM10)	TU	980	10	R.Tampico, Tampico: 1200-0600
VE03)	QO	980	5	R.Romance, Cosamaloapan: 1200-0400
BC04)	CL	990	5	La L de Mexicali, Mexicali: 24h
BS02)	HZ	990	1/0.25	R.9-90, La Paz: 1300-0700
CH13)	ER	990	5	R.Fórmula, Cd.Cuauhtémoc: 1300-0600
CS09)	TG	990	20	L Grande del Sureste, Tuxtla Gutiérrez: 24h
JL01)	BC	990	1	La Buena Onda, Cd.Guzmán: 1200-0600
MI28)	ATM	990	1	A Toda Máquina, Morelia: 1200-0900
NL15)	T	990	150	La T Grande de Monterrey, Monterrey: 24h
OX08)	IU	990	2.5	Estéreo Crystal, Oaxaca: 1200-0400
YU04)	UM	990	1	R.99, Valladolid: 24h
ZC10)	FP	990	10	R.Alegría, Jalpa: 24h
CH33)	FV	1000	1	La Rancherita, Cd.Juárez: 1200-0300
CS02)	TAC	1000	1d	Audio Mil de Chiapas, Tapachula: 1200-0400
DF02)	OY	°1000	50	R.Mil "Tradición y Excelencia en Radio", México: 24h
GJ10)	RZ	1000	1	Heavy Radio, León: 1200-0600
SN11)	MIL	1000	1	La Comadre, Los Mochis: 1200-0700
SN24)	MMS	1000	1	R.Ranchito, Mazatlán: 1300-0700
TM24)	NLT	1000	1	Laredo Radio, Nuevo Laredo: 1200-0600
VE30)	CSV	1000	1	La Tremenda, Coatzacoales: 1200-0600
YU01)	MYL	1000	5/0.25	Super MYL, Mérida: 1200-0400 (rep.on 1060 kHz)
BC22)	DX	1010	2	R.Variedades, Ensenada: 1400-0800
CH14)	LO	1010	1	R.Lobo, Chihuahua: 1200-0600
CH15)	TX	1010	1/0.25	Voz de la Sierra Madre Occidental, Nuevo Casas Grandes: 1200-0400
CO01)	VK	1010	5	Sonovida, Torreón: 1200-0500
CO13)	KD	1010	0.5	K de Oro, Cd.Acuña: 1200-0600
JL15)	HL	1010	50/5	La Poderosa, Guadalajara: 24h
SN12)	WS	1010	5d	R.Capital, Culiacán: 1300-0500
SO11)	XN	1010	0.5	R.Ures, Ures: 1300-0500
VE09)	FM	1010	1	La Máquina Tropical, Veracruz: 1200-0600
CL09)	VE	1020	1	La Comadre, Colima: 1200-0400
NA13)	PIC	1020	1	R.Hits, Tepic: 1300-0700
OX14)	OU	1020	0.5d	R.Joya, Huajuapan de León: 1300-0100
OX23)	YN	1020	1	Sonovida, Oaxaca: 24h
QE06)	KH	1020	1	R.Centro, Querétaro: 24h
QR03)	WO	1020	1d	R.Chetumal, Chetumal: 1200-0400
VE14)	PR	1020	5	R.Imagen, Poza Rica: 1200-0600 (rep.on 1480 kHz)
VE29)	GF	1020	2	R.Fiesta, Gutiérrez Zamora: 1200-0300
CA05)	BCC	1030	1	La Gaviota Musical del Golfo, Cd.del Carmen: 1200-0500
CH30)	YC	1030	1	R.Sensación, Cd.Juárez: 24h (rep.on 1460 kHz)
CS15)	VFS	1030	4	LV de la Frontera Sur, Las Margaritas: 1200-2200
DF08)	QR	1030	50	R.Centro, México: 24h
GR18)	VP	1030	1	R.Fiesta, Acapulco: 1200-0200
JL11)	LJ	1030	5/1	R.Central, Lagos de Moreno: 1300-0400
MI24)	GQ	1030	1	R.Variedades, Los Reyes: 1200-0200
SN13)	MPM	1030	10	R.Fama, Los Mochis: 1300-0700
TM31)	PAV	1030	1	Extasis Digital, Tampico: 1200-0600
VE42)	ZON	1030	1	LV de la Sierra Zongolica, Zongolica: 1200-2400 (rep.on 1360)
CH08)	HES	1040	1	La Poderosa, Chihuahua: 24h
CS01)	TRE	*1040	1	R.Palenque 10-40, Palenque
GJ26)	SAG	1040	1	Radiorama 1040, Irapuato: 1300-0100
JL09)	BBB	1040	10/1	X-E Triple B, Guadalajara: 24h
OX20)	TLX	1040	1	R.Tlaxiaco, Tlaxiaco: 1200-0400
SO42)	GYS	1040	1	Super Banda, Guaymas: 24h
VE01)	GR	1040	1d	R.Favorita, Jalapa: 1300-0100
AG06)	DC	1050	1	La Nueva 10-50, Aguascalientes
BC06)	D	1050	10	La Gran D, Mexicali: 24h
BS05)	BCS	1050	10/1	R.Cultural Surcalifornia, La Paz: 1300-0500
GR19)	ZUM	1050		R.Poder, Chilpancingo: 1200-0400
MI29)	URM	1050	1	R.Mexicana, Uruapán: 24h
NL08)	G	1050	100	La Ranchera de Monterrey, Monterrey: 24h
QR10)	QOO	1050	35/2.5	R.Pirata, Cancún: 24h
TB14)	TAB	1050	5/0.5	R.Juvenil, Villahermosa: 1200-0400
VE25)	JF	1050	5	R.Variedades, Tierra Blanca: 1200-0200
DF12)	EP	°1060	100/20	R.Educación, México: 24h
CA02)	IT	1070	1	R.Hit, Cd.del Carmen: 1200-0400
GR16)	AGS	1070		XEAGS, Acapulco
JL12)	SP	1070	5/1	R.Juventud Unica, Guadalajara: 24h
PU03)	GY	1070	1/0.25	R.Lobo, Tehuacán: 1200-0700
SL10)	EI	1070	1	R.Mexicana, San Luis Potosí: 24h
SL13)	ANT	1070	5	LV de las Huastecas, Tancanhuitz de Santos: 1145-0030 (SS 1300-2100)
SO18)	OBS	1070	1	R.Mexicana, Cd.Obregón: 24h
VE10)	MI	1070	1	La M Grande, Minatitlán: 1100-0600
CH31)	DT	1080	1	La Divertida, Cd.Cuauhtémoc: 1300-0100
CL04)	UU	1080	1	R.Variedades, Colima: 1200-0600
CO26)	SAC	1080	2	La Tremenda Tropical, Saltillo: 1200-0600
GJ05)	CN	1080	1	Lasser Hits, Irapuato: 1200-0600
ME07)	TUL	1080	5	Sistema XEGEM "R.Mexiquense", Tultitlán: 1200-0600 (r:1600)
QR11)	CAQ	1080	50	R.Fórmula, Cancún: 24h (rep.on 740)
SO31)	DY	1080	5	R.Gallo, San Luis Río Colorado: 1200-0700
VE11)	XK	1080	10	1080 R.Mundo, Poza Rica: 1200-0600
BC11)	PRS	1090	50	R.Express/La Gigante, Tijuana: 24h
JL13)	LB	1090	5	R.La Barca, La Barca: 1200-0200
NL07)	AU	1090	2.5	R.AU, Monterrey: 1200-0600
PU16)	HR	1090	1	R.ACIR, Puebla: 1200-0600
QE07)	XE	1090	1/0.2	R.Fórmula, Querétaro: 1200-0600
TM11)	WL	1090	1	Nostalgia, Nuevo Laredo: 1200-0200
VE12)	MCA	1090	5d	La Emisora de las Huastecas, Pánuco: 1200-0100
VE13)	IL	1090	1	La Comadre, Veracruz: 24h
YU01)	FC	1090	10/0.25	Estéreo Romantica, Mérida: 1200-0400
GJ09)	BV	1100	5	R.Alegría, Moroleón: 1200-0600
SL15)	PO	1100	1	R.Voz, San Luis Potosí: 24h
SO43)	NAS	1100	1	La Norteñita, Navojoa
ZC11)	TGO	1100	5/0.5	R.Alegría Digital, Tlaltenango: 1200-0600
CH03)	ES	1110	1/0.5	R.Alegría, Chihuahua: 1300-0500
CH33)	WR	1110	1d	R.Hits, Cd.Juárez: 1200-0200
DF08)	RED	1110	50/25	R.Red, México: 24h
GJ25)	LEO	1110	5d	La Rancherita, León: 1300-0100
JL25)	PVJ	1110	1	R.Vallarta, Puerto Vallarta: 1200-0600
SO32)	VS	1110	1d	R.Festival, Hermosillo: 24h
TB06)	ACM	1110	1d	R.Exitos, Cárdenas: 1200-0600
TM02)	OQ	1110	1d	La Pantera Cumbiambera, Reynosa: 1300-0100
BC23)	MX	1120	1	Sonido de Mexicali, Mexicali: 1400-0800
CO15)	IK	1120	1	La Super Invasora, Piedras Negras: 1200-0600 (rep.on 1360 kHz)
JL14)	UNO	1120	1/0.5	R.Centro, Guadalajara: 24h
OX09)	ZB	1120	1/0.1	La Tremenda, Oaxaca: 1200-0100
QE04)	GV	1120	0.5d	R.ACIR, Villa del Pueblito: 24h
SL05)	TR	1120	1d	R.Panorámica, Cd.Valles: 1200-0200
YU08)	RUY	1120	1/0.25	R.Universidad, Mérida: 1200-0600
GR15)	CHG	1130	1	R.Fiesta, Chilpancingo: 1200-0600
ME02)	TOL	1130	10	R.Lobo, Toluca: 24h
MI11)	FN	1130	1d	R.Moderna, Uruapan: 1300-0700
NA05)	LUP	1130	1	R.Lupita, Compostela: 1300-0100

SN27)	MOS	1130	1	Super Banda, Los Mochis: 24h
SO36)	HN	1130	1d	R.Fiesta, Nogales: 1200-0200
SO44)	EETCH	1130	1	XEETCH, Etchojoa
VE08)	ZL	1130	10	R.Centro, Jalapa: 1200-0400
GJ23)	XF	1140	5d	R.ACIR, León: 24h
MI22)	LIA	1140	1	La Tremenda, Morelia: 1200-0400
NL02)	MR	1140	50	La Quebradita, Monterrey: 24h
PU04)	TE	1140	1d	R.ACIR, Tehuacán: 1300-0100
BC06)	RM	1150	1	Baja R.Mexicali, Mexicali: 24h
CH16)	JS	1150	1	JS Digital, Hidalgo del Parral: 1245-0500
CO14)	BF	1150	1	R.Mexicana, San Pedro: 1200-0100
DF08)	CMQ	1150	20	Formato 21, México: 24h
JL06)	AD	1150	20	R.Metrópoli, Guadalajara: 24h
OX10)	XP	1150	5d	XEXP, Tuxtepec: 1200-0100
QE01)	QUE	1150	1	R.Querétaro, Querétaro
SN16)	UAS	1150	5/1	R.Universidad, Culiacán: 1300-0200
SO22)	SO	1150	5	La Poderosa, Cd.Obregón: 1200-0800
TB07)	RTM	1150	1/0.5	R.Variedades, Macuspana: 1200-0500
VE43)	TVR	1150	1.5	Azul 11-50, Tuxpam: 1100-0500
YU01)	RRF	1150	5/0.5	R.Juvenil, Mérida: 1200-0200
GJ11)	VW	1160	2.5	R.Sensación, Acámbaro: 1300-0130
MI12)	IW	1160	1d	Canal Juvenil, Uruapan: 1300-0400
SL12)	GI	1160	1	Reyna de las Huastecas, Tamazunchale: 1200-0200
VE16)	BE	1160	5/0.25	R.Perote, Perote: 1200-0300
AG01)	UVA	1170	5	La Rancherita, Aguascalientes: 1200-0700
CO11)	MDA	1170	1	Estéreo Digital, Monclova: 1300-0600
GJ17)	JE	1170	1/0.5	R.XEJE, Dolores Hidalgo: 1200-0400
JL19)	JTF	1170	1d	Prisma Musical, Zacoalco de Torres: 1300-0100 (rep.on 1450)
ME01)	RLK	1170	1	Super Stereo Miled, Atlacomulco: 24h
MI30)	LP	1170	0.25	R.Pia, La Piedad: 1300-0300
PU05)	CD	1170	2	LV de la Confianza: 1200-0600
SO32)	FEM	1170	1	R.Mil, Hermosillo: 1200-0600
SO33)	IB	1170	1	R.Sensación, Caborca: 1300-0700
TM07)	RT	1170	5d	La Gigante, Reynosa: 1200-0200
VE46)	ZS	1170	0.5	R.Hit 11-70, Coatzacoalcos: 1200-0600
BS05)	UBS	1180	1	R.Universidad Autonoma de Baja California Sur, La Paz
CH35)	DCH	1180	1	La Romántica, Cd.Delicias: 1200-0600
DF14)	FR	1180	10	Oxido 11-80, México: 24h
GJ05)	YA	1180	1	La Picosa, Irapuato: 1200-0600
OX11)	AH	1180	0.5	R.Hit, Juchitán: 1130-0630
BC07)	MBC	1190	0.5	La Picosita, Mexicali: 1400-0800
CH30)	PZ	1190	1d	R.Norteña, Cd.Juárez: 1200-0200
JL15)	WK	1190	50	La W de Guadalajara, Guadalajara: 24h
MI13)	SOL	1190	1d	R.Sol, Cd.Hidalgo: 1200-0400
M001)	JPA	1190	5	La Grande 11-90, Cuernavaca: 1200-0600
NL08)	CT	1190	1d	Nucléo Deportes 11-90, Monterrey: 1200-0800
TB12)	RV	1190	10	R.Villa, Villahermosa: 1100-0500
TM20)	TOT	1190	5	La Poderosa, Tampico: 1200-0600
VE32)	PP	1190	5	Super Mil Con El Poder de la Música, Orizaba: 1200-0600
AG06)	AGA	1200	1	La Bonita, Aguascalientes: 1200-0400
ME02)	QY	1200	1	Que Bonita 1200, Toluca: 24h
MI14)	ZI	1200	1	Canal Festivo 120, Zacapu: 1200-0100
SN06)	WT	1200	1	La Nueva WT, Culiacán: 1200-0300
SO32)	YF	1200	1	R.Fórmula, Hermosillo: 24h
VE11)	PW	1200	1/0.3	La Tremenda, Poza Rica: 1200-0600
CL07)	BCO	1210	50/10	R.Occidente/La Poderosa Voz de Colima, Colima: 24h
VE27)	BD	1210	10	R.Centro/R.Punto, Jalapa: 1200-0600
DF04)	B	1220	100	La B Grande, México: 24h
JL26)	ZAJ	1220	2.5d	Sono Ritmo 12-20, Guadalajara: 1200-0200
PU20)	TCP	1230	1	La Poderosa, Tehuacan
SN12)	EX	1230	1	R.ACIR, Culiacán: 24h
TB11)	TBH	1230	1/0.5	R.Tabasco, Villahermosa
VE15)	ID	1230	2.5	R.Alamo, Alamo: 1200-0400
CH32)	WG	1240	1	Mágico 12-40, Cd.Juárez: 1200-0600
CH17)	BN	1240	1/0.25	Radiola, Cd.Delicias: 1300-0400
CO15)	VM	1240	1	R.Sensación, Piedras Negras: 1100-0600
CS01)	LM	1240	1	R.Chiapas, Tuxtla Gutiérrez: 1200-0500
HG08)	RD	1240	1	R.Lobo, Pachuca: 1200-0500
MI15)	RPA	1240	1	R.Ranchito, Morelias: 1200-0300
NA06)	SI	1240	1	R.Afirmación, Santiago Ixcuintla: 1300-0500
NL04)	IZ	1240	1	La Potranquita, Monterrey: 1200-0600
OX25)	CE	1240	2.5	R.Hit 12-40, Oaxaca: 1130-0600
SO36)	CG	1240	1	R.Mexicana, Nogales: 1300-0700
SO14)	BQ	1240	1/0.25	Estelar 12-40/R.Mexicana, Guaymas: 1300-0500
TM10)	S	1240	1	Energía 12-40, Tampico: 1200-0400
VE04)	OV	1240	2.5	La Picosa, Orizaba: 1200-0700
CH19)	AT	1250	1	R.Onda, Hidalgo del Parral: 1200-0500
CO03)	SC	1250	1	R.1250, Sabinas: 1300-0100
CO30)	SJ	1250	5	R.Saltillo, Saltillo: 24h
CS11)	MG	1250	1/0.25	La Poderosa, Arriaga: 1200-0400
GR06)	PI	1250	2.5/0.5	La Consentida, Chilpancingo: 1200-0200
JL04)	DK	1250	5	Radiorama DK, Guadalajara: 1200-0600
ME07)	TEJ	1250	0.5	Sistema XEGEM "R.Mexiquense", Tejupilco: 1200-0600 (r:1600)
PU17)	ZT	1250	1	R.Trebuna 12-50, Puebla: 24h
QE05)	JX	1250	1	R.Lobo, Querétaro: 1200-0600
SO32)	DL	1250	5	R.Ambiente, Hermosillo: 24h
VE21)	TF	°1250	1	LV de Veracruz, Veracruz: 1130-0600
AG03)	YZ	1260	1	R.Variedades, Aguascalientes: 24h
CH20)	OG	1260	1	R.Ranchito, Ojinaga: 1200-0400
CO31)	WGR	1260	1	Estéreo Vida, Monclova: 24h
DF14)	L	1260	50	R.ACIR, México: 24h
GJ13)	ZH	1260	0.5d	La Estación que se oye, Salamanca: 1300-0500
JL24)	JY	1260	10	R.Sistema del Suroeste/LV del Valle, Autlán: 1200-0300
MI04)	QL	1260	1	Sonido Brillante, Zamora: 1200-0600
NL13)	R	1260	1	R. Linares, Linares: 1200-0400
OX18)	JAM	1260	5	LV de la Costa Chica, Santiago Jamiltepec
SL18)	XR	1260	5/1	R.Mensajera, Cd.Valles: 1200-0300
SN26)	SA	1260	5	R.Hits, Culiacán: 24h
SO15)	MW	1260	1	R.San Luis, San Luis: 24h
VE22)	MTV	1260	0.25	R.Lobo, Minatitlán: 1200-0100
VE48)	TBV	1260	1	La Poderosa, Tierra Blanca: 1200-0500
BC12)	AZ	1270	5	R.Zeta 13, Tijuana: 24h
CO04)	WN	1270	0.5	R.Variedades, Gómez Palacio: 1200-0600
DG04)	HD	1270	0.5d	Universidad Juárez del Estado de Durango, Durango
GJ01)	RPL	1270	10	La Poderosa, León: 24h
HG05)	QH	1270	1	R.Sinfonia, Ixmiquilpán: 1200-0100
MI16)	ZU	1270	0.5	R.Fiesta, Zacapu: 1200-0400
OX09)	AX	1270	5/0.5	Express 12.7, Oaxaca: 1200-0400
SO16)	GL	1270	1	La Pionera del Radio, Navojoa: 1200-0700
TB16)	VHT	1270	1	Romántica 12-70, Villahermosa: 24h
TM10)	RRT	1270	2	R.Ranchito, Cd.Madero: 24h
VE20)	RRR	1270	1/0.5	R.Ritmo, Poza Rica: 1200-0300
CA03)	CAM	1280	2.5	LV de las Murallas, Campeche: 1155-0500
CH03)	BW	1280	1/0.5	Canal 12-80, Chihuahua: 24h
CS12)	KY	1280	1	Dimensión 12-80, Huixtla: 1100-0300
GJ14)	SQ	1280	0.5	R.San Miguel. San Miguel de Allende: 1200-0400
JL28)	BON	1280	1	R.Morena, Guadalajara: 24h
NL04)	AW	1280	10	La Macroestación, Monterrey: 24h
PU17)	EG	1280	1	R.Fórmula 12-80 AM, Puebla: 24h
TM12)	TUT	1280	1	R.Tamaulipas, Tula: 1200-0400
VE38)	AG	1280	2	R.Capital, Córdoba: 24h
BC20)	QIN	1290	2.5	LV del Valle, San Quintín
CA04)	TH	1290	0.25	R.Palizada, Palizada: 1200-2400
DF09)	DA	1290	10	R.13, México: 24h
GJ03)	FAC	1290	5	La Mera Mera, Salvatierra: 1230-0130
MI17)	IX	1290	1	La Pantera, Suhuayo: 1300-0200
SL04)	IY	1290	1d	R.Juventud, Río Verde: 1200-0200
SN10)	NX	1290	1	La Poderosa NX, Mazatlán: 24h
SO18)	AP	v1290	1	R.Sensacion, Cd.Obregón: 24h
VE33)	VZ	1290	1	R.Sensación, Acayacan: 1200-0600
CH22)	SW	1300	1d	R.Madera, Cd.Madera: 1300-0400
CH30)	P	1300	1	R.13, Cd.Juárez: 24h
GJ15)	XV	1300	1	Stereo R.13, León: 1200-2400

MI07)	KW	1300	1	R.Festival, Morelia: 1200-0400
SN15)	JL	1300	1	R.Guamuchil, Guamuchil: 1200-0500
SO13)	XW	1300	1	R.Norteña, Nogales: 1300-0700
TM13)	LE	1300	1	R.13, Tampico: 1200-0600
VE23)	HU	1300	1	R.Tropical, Martinez de la Torre: 1200-0600
BC13)	C	1310	1	R.Enciso, Tijuana: 24h
CH23)	RU	1310	1/0.25	R.Universidad, Chihuahua
CO01)	BP	1310	10	R.Sensación, Torreón: 1200-0500
GR07)	HJ	1310	0.25d	R.Petatlán, Petatlán: 1200-0100
GR23)	GRT	1310	1	R.Guerrero, Taxco
JL06)	TIA	1310	10	R.Contacto, Guadalajara: 24h
NL02)	VB	1310	5	R.Alegría, Monterrey: 1200-0600
PU14)	HIT	1310	5	La Tremenda, Puebla: 24h
QE09)	HY	1310	1	R.Mexicana, Querétaro: 1200-0200
SO19)	FH	1310	1	R.Plan de Agua Prieta, Agua Prieta: 1400-0300
TM14)	AM	1310	5	La M Grande, Matamoros: 24h
VE06)	HV	1310	1	R.Trópico, Veracruz: 1200-0700
VE44)	TRC	1310	1	R.Gazeta de la Tierra Norte, Tepetzintla
AG07)	NM	1320	1	R.13-20, Aguascalientes
BS03)	SR	1320	0.5	R.Cachanía, Santa Rosalia: 1300-0700
CH06)	JZ	1320	1d	R.Variedades, Cd.Jimenez: 1300-0200
CO25)	CPN	1320	10	R.1320, Monclova: 24h
DF08)	JP	1320	20	R.Variedades, México: 24h
HG02)	CY	1320	1/0.25	R.Diversión, Huejutla: 1200-0100
MI18)	NI	1320	1	R.Festival, Uruapán: 1200-0800
OX10)	UH	1320	1	R.U-H, Tuxtepec: 1200-0400
SN03)	RJ	1320	5	La RJ/R.Mazatlán, Mazatlán: 24h
CL08)	MAC	1330		La Turquesita, Manzanillo: 1100-0600
CO18)	WQ	1330	1	La Superestación, Monclova: 24h
CO27)	AJ	1330	0.5	La Revolución del Radio, Saltillo: 24h
GJ16)	BO	1330	5	R.Variedades, Irapuato: 1030-0600
PU07)	EV	1330	0.5d	Alma Musical, Icúcar de Matamoros: 1200-0100
TM10)	RP	1330	1	R.Principal, Tampico: 1200-0600
VE23)	UZ	1330	1	R.Veracruzana, Martínez de la Torre: 1200-0600
BC24)	AA	1340	1	R.Idolos, Mexicali: 24h
CH20)	RCH	1340	0.25	R.Exitos, Ojinaga: 1400-0200
CO13)	DH	1340	1	La Tremenda Tropical, Cd.Acuña: 1200-0800
GR22)	CI	1340	1	La Nueva 13-40, Acapulco: 1200-1000
HG03)	QB	1340	1	La Divertida, Tulancingo: 1200-0600
JL04)	DKT	1340	1	R.Ranchito, Guadalajara: 1200-0600
MI05)	APM	1340	1	R.Romántica, Apatzingán: 1200-0300
MI19)	CR	1340	1	R.Variedades, Morelia: 1200-0300
MO01)	ASM	1340	1	R.Uno, Cuernavaca: 1200-0600
NL02)	NV	1340	1	La Sabrosita, Monterrey: 24h
OX12)	PX	1340	5/2	LV del Angel, Puerto Angel: 1200-0200
PU08)	LU	1340	5	R.Esmeralda, Cd.Serdán: 1300-0100
SL02)	SL	1340	2	R.Centro, San Luis Potosí: 1200-0600
SN17)	QE	*1340	1d	La Perla Camaronera, Escuinapa: 1100-0100
SN18)	CW	1340	1/0.2	Sonido Zeta, Los Mochis: 1200-0400
SO40)	OS	1340	1	R.Variedades, Cd.Obregón: 1300-0500
TB08)	YR	1340	1	R.13, Teapa
TM01)	RPV	1340	1	R.Festival, Cd.Victoria: 1300-0200
TM14)	MT	1340	1	La Tremenda Tropical, Matamoros: 1200-0300
TM15)	BK	1340	1	Super Grupera, Nuevo Laredo: 1155-0600 (Sat -0800, Sun -0200)
VE17)	OM	1340	1	R.ACIR, Coatzacoalcos: 24h
CO04)	TB	1350	5/0.5	R.Laguna, Torreón: 24h
CS16)	CAH	1350	5	R.Chiapas/LV Soconusco, Cacahoatán: 1100-0700
DF04)	QK	1350	5	La Hora Exacta, México: 24h
PU15)	CTZ	1350	2.5	LV de la Sierra Norte, Cuetzalán del Progreso: 1200-0100
SO20)	TM	1350	1d	El Heraldo de la Frontera, Naco: 1300-0200
TM16)	ZD	1350	0.25	La Doña de la Frontera, Camargo: 1200-0400
CH02)	DI	1360	1	R.Sensación, Chihuahua: 24h
CS21)	UD	1360	5	Florida Radio, Tuxtla Gutiérrez: 1100-0700
GJ07)	Y	1360	1/0.25	R.Celaya, Celaya: 1100-0600
GR08)	KF	1360	1	La K-F, Iguala: 1155-0500
VE26)	DQ	1360	1	R.Alegría, San Andrés Tuxtla: 1200-0600

ZC08)	XM	1360	1/0.1	R.Jerez, Jerez de García Salinas: 1200-0400
BC06)	HG	1370	0.5	R.Norteña, Mexicali: 1400-0200
CA06)	A	1370	1	La Grande A, Campeche: 1200-0600
DG07)	RPU	1370	1	Sonidoz, Durango: 24h
JL08)	PJ	1370	5	R.Capital, Guadalajara: 24h
MI20)	SV	1370	0.5d	R.Universidad Michoacana de San Nicolas de Hidalgo, Morelia
NL14)	MON	1370	10	R.Fórmula, Monterrey: 1200-0600
PU19)	PA	1370	10	La Super Fiera, Puebla: 1200-0600
SO36)	HF	1370	5	R.Arizona, Nogales: 1400-0800
TM17)	GNK	1370	0.25d	R.Mexicana, Nuevo Laredo: 1200-0130
BC14)	KT	1380	1	R.Variedades, Tecate: 24h
CO01)	RS	1380	1	R.Sinfonía, Torreón: 1300-0500
CO20)	VD	1380	1/0.1	R.Allende, Allende: 1200-0400
DF13)	CO	1380	10	Romántica 13-80, México: 24h
TM01)	GW	1380	5	Super Tropical, Cd.Victoria: 1200-0400
VE27)	TP	1380	5	R.Sensación. Jalapa: 1300-0100
CL06)	TY	1390	5/1	R.Tecomán, Tecomán: 1100-0300
GJ06)	RW	1390	10	Sonido 13-90, León: 24h
HG06)	ZG	1390	0.25	R.Mezquital y Huasteca Hidalguense, Ixmiquilpán: 1300-0100
MO03)	CTA	1390	5	R.Cuautla, Cuautla
SO21)	QC	1390	1d	La Reyna del Mar, Puerto Peñasco: 1200-0100
TM18)	XO	1390	5/0.25	La Super Buena, Cd.Mante: 1200-0400
TM02)	OR	1390	1	R.Gallito, Reynosa: 1230-0600
VE24)	TL	1390	5/1	LV de la Huasteca, Tuxpam: 1200-0600
AG02)	AC	1400	5	La A Grande, Aguascalientes: 1200-0400
BC15)	PF	1400	1	La Ranherita, Ensenada: 24h
GR03)	KJ	1400	1/0.25	R.Alegría, Acapulco: 1200-0400
ME03)	XI	1400	1	La I de Ixtapan, Ixtapan de la Sal: 1200-0600
MI21)	OJ	1400	1	R.Horizonte, Cd.Lázaro Cárdenas: 1200-0400
MI22)	I	1400	1/0.25	R.Morelia, Morelia: 24h
NA08)	LH	1400	1d	La Gardenia Musical, Acaponeta: 1300-0100
OX21)	UBJ	1400	0.5	R.Universidad Benito Juárez, Oaxaca
QE03)	VI	1400	0.5	Energia 14, San Juan del Río: 1200-0500
SL06)	WU	1400	0.25	XEWU, Matehuala: 1300-0100
SO23)	AB	1400	1	R.Santana, Santa Ana: 1400-0500
CO01)	YD	1410	1/0.1	R.Madero, Torreón: 1200-0400
DF02)	BS	1410	10	R.Sinfonola, México: 24h
JL17)	KB	1410	25/10	Canal 14-10, Guadalajara: 24h
NL09)	SH	1410	5	R.Sabinas, Cd.Sabinas: 1200-0600
SL11)	IR	1410	5	Estelar 14-10, Cd.Valles: 24h
SN13)	CF	1410	1/0.5	La Mexicana, Los Mochis: 24h
TM11)	AS	1410	1	La Tamaulipeca, Nuevo Laredo: 24h
BC16)	XX	1420	1	Super XX, Tijuana: 24h
CH33)	F	1420	5	La Fabulosa, Cd.Juárez: 24h
GJ05)	WE	1420	10/1	R.Irapuato, Irapuato: 1100-0500
HG01)	PK	1420	1	Que Bonita, Pachua: 1200-0600
JL18)	KMX	1420	1d	La Super X, Sayula: 1200-2400
MO04)	WF	1420	0.5	Fiesta Mexicana, Cuernavaca: 24h
NL03)	H	1420	5	La Tremenda Tropical, Monterrey: 24h
PU10)	WJ	1420	1/0.25	R.Popular, Tehuacán: 1200-0600
TM21)	EW	1420	1	LV del Bajo Bravo, Matamoros: 1155-0600
VE10)	JV	1420	1/0.25	La Super Joven, Minatitlan: 1200-0300
CA06)	RAC	1430	1	La Juvenil, Campeche: 1200-0600
CL09)	COC	1430	1	La Querida, Colima: 1200-0400
OX13)	TEKA	1430	1	R.Exitos, Ixtepec: 1200-0200
SO24)	OX	1430	5	O-X Canal 14-30, Cd.Obregón: 24h
TM22)	WD	1430	2	La Grande, Cd.Miguel Alemán: 1155-0400
TX02)	TT	1430	10	R.Tlaxcala, Tlaxcala: 1200-0600
VE06)	LL	1430	1	R.Onda, Veracruz: 1130-0700
DF16)	EST	1440	10	Cambio 14-40, México: 24h
JL09)	CCC	1440	10/1	Fiesta Mexicana, Guadalajara: 1300-0100
TB13)	NAC	1440	0.5	LV de los Chontales, Nacajuca
BC17)	SS	1450	1	R.Fiesta, Ensenada: 1400-0800
CH26)	ARE	1450	1/0.25	R.Pegüis, Ojinaga: 1200-0500
GR11)	RY	1450	1/0.1	R.Arcelia, Arcelia: 1200-0300

JL21)	ED	1450	1/0.25	ED Contigo, Ameca: 1200-0400
MI17)	GC	1450	1	R.Impacto, Sahuayo: 1300-0100
NA10)	TD	1450	1	Canal de la Juventud, Tecuala: 1200-2400
NL04)	JM	1450	1	La Tropicalísima, Monterrey: 1200-0600
PU16)	PUE	1450	1	Ella AM, Puebla: 1200-0600
QE05)	NA	1450	1/0.5	R.Alegría, Querétaro: 1200-0400
SL07)	IE	1450	5	R.Alegría, Matehuala: 1155-0405
SN18)	CU	1450	1	La Rancherita, Los Mochis: 24h
SO25)	DJ	1450	1	R.Clave, Magdalena: 1300-0300
SO41)	CB	1450	1	R.Ranchito, San Luis: 1300-0700
TM18)	CM	1450	1/0.5	Música Romántica, Cd.Mante: 1200-0400
TM29)	VH	1450	1	R.Galáctica, Matamoros: 1200-0600
VE20)	JD	1450	1	R.Tropicana, Poza Rica: 24h
VE22)	KM	1450	1/0.5	R.Mina/Onda 14-50, Minatitlán: 1100-0500
GR14)	GRA	1460	1	14-60 La Estación Familiar, Acapulco: 1200-0600
JL20)	HE	1460	1d	LV de Jalisco, Atotonilco el Alto: 1300-0100
MI02)	LX	1460	10	R.Mexicana, Zitácuaro: 1200-0300
OX05)	KC	1460	5/1	Stereo Exitos, Oaxaca: 1200-0800
QR05)	CPQ	1460	1	La Estrella Maya, Felipe Carrillo Puerto: 1300-0100
SL08)	XQ	°1460	0.25	R.Universidad de San Luis Potosí, San Luis Potosí
SO40)	HX	1460	1	La Consentida, Cd.Obregón: 1200-0400
VE08)	JH	1460	1	R.ACIR, Jalapa: 24h
BC05)	RCN	1470	5	RCN-R.Cadena Nacional, Tijuana: 1300-0800
CA08)	BAL	1470	2.5	R.Voz Maya de México, Bécal: 1200-0600
DF11)	AI	1470	20	Radio A-I, México: 24h
DG03)	CAV	1470	1	Estéreo Imagen, Durango: 1200-0500
HG07)	IND	1470	1	LV Sierra Hidalguense, Tlanchinol: 1200-0200
SN21)	ACE	1470	1/0.25	Stereo Exitos/La Gran Máquina Musical, Mazatlán: 24h
TM23)	HI	1470	3/0.25	El Heraldo Internacional, Cd.Miguel Alemán: 1155-0500
CO21)	XU	1480	1	R.Variedades, Monclova: 1200-0600
CH28)	HM	1480	1d	HM Radio, Cd.Delicias: 1300-0100
JL10)	ZJ	1480	1	XEZJ El 1480 AM, Guadalajara: 24h
NL07)	TKR	1480	1	La TKR/Rancherita y Regional, Monterrey: 1200-0600
QR08)	CCQ	1480	1	R.Sensación, Cancún: 1200-0300
SO10)	NS	1480	1	R.Ambiente, Navojoa: 1200-0700
TM28)	VIC	1480	5	R.Tamaulipas, Cd.Victoria: 1200-0800
AG03)	RO	1490	1	La Inolvidable, Aguascalientes: 24h
CH30)	CJC	1490	1	La Pantera, Cd.Juárez: 1200-0400
ME05)	CH	1490	1	R.Capital, Toluca: 1100-0500
MI04)	GT	1490	1	R.Alegría, Zamora: 1200-0600
MI23)	KN	1490	2.5	R.Variedades, Huetamo: 1200-0300
NA11)	SK	1490	0.25	La Super K, Cd.Ruiz: 1200-0300
PU16)	POP	1490	1	La Comadre, Puebla: 1200-0600
SO27)	AQ	1490	1	R.Juventud, Agua Prieta: 1500-0600
TM29)	MS	1490	1	R.Mexicana, Matamoros: 1200-0200
VE28)	YT	1490	1	R.Teocelo, Teocelo: 1200-0200
CO22)	JQ	1500	0.4	R.Parras, Parras: 1100-0100
DF11)	DF	1500	50	1500 AM, México: 24h
GJ20)	FL	1500	0.25d	R.Santa Fé de Guanajuato, Guanajuato: 1300-0100
VE34)	GN	1500	1	R.Gigante, Piedras Negras: 1200-0600
GJ04)	OF	1510	5d	Stereo Carnaval, Cortázar: 1200-0600
HG09)	HUI	1510	0.25	R.Huichapán, Huichapán
NL11)	QI	1510		Sistema R.Nuevo Leon, Monterrey
VE45)	JPM	1510	0.5	R.Veracruz, La Antigua Veracruz: 1200-0300
AG04)	UAA	1520	0.25d	Universidad Autonomica de Aguascalientes, Aguascalientes
CH30)	JCC	1520	1	R.Paso del Norte, Cd.Juárez: 24h
CO23)	VUC	1520	1d	Mi Compañera, Zaragoza: 1200-0200
ME07)	ATL	1520	1	Sistema XEGEM "R.Mexiquense" Atlacomulco: 1200-0600 (r:1600)
MO02)	ART	1520	2	La Señal de las Estrellas, Jojutla: 1200-0400
SO34)	EH	1520	1d	R.Exitos, San Luis: 1200-0600
TB09)	ZQ	1520	3/0.1	R.Futurama, Huimanguillo: 1200-0130
TM18)	YP	1520	1d	La Juvenil, El Limón: 1150-0100
VE35)	VO	1520	1d	R.San Rafael, San Rafael: 1300-0100
DF13)	UR	1530	20	La Poderosa, México: 24h
GJ21)	SD	1530	1	R.Actualidades, Silao: 1300-0400
CS14)	VF	1540	10/1	R.Variedades, Villaflores: 1200-0400
GJ07)	NC	1540	1/0.25	R.Voz Musical, Celaya: 1200-0500
NL15)	STN	1540	5	R.Red, Monterrey: 24h (r:1110)
PU12)	RTP	1540	1	La Musiquera, San Martín Texmelucán: 1200-0500
SO29)	HOS	1540	5	La Poderosa, Hermosillo: 24h
BC03)	BG	1550	1	R.15-50/La B-G, Tijuana: 24h
ME04)	XOO	1550	1	La O de Oro, El Oro
MI26)	REL	1550	0.5	R.Michoacán, Morelia
TM11)	NU	1550	5d	La Rancherita, Nuevo Laredo: 1200-0100
VE36)	RUV	1550	10	R.Universidad Veracruzana, Jalapa: 1100-0700
CA07)	SE	1560	5d	LV de Campeche, Champotón: 1200-2400
CH33)	JPV	1560	1d	La Nueva Radio Viva, Cd. Juárez: 1200-0200
DF08)	FAJ	1560	20/10	La Consentida, México: 24h
GJ12)	MAS	1560	1	La Estación de los Exitos, Salamanca: 1200-0500
NA12)	RIO	1560	5d	R.Triunfadora, Ixtlán del Río: 1300-0100
NL10)	DD	1560	0.5	D-D R.Amistad, Montemorelos: 24h
SL09)	ZW	1560	1d	R.Diversión, Cerritos
CO24)	RF	1570	50	La Poderosa, Cd.Acuña: 24h
SO41)	LBL	1570	0.5d	R.Centro, San Luis: 1300-0100
GJ07)	AF	1580	1/0.25	R.Luz, Celaya: 24h
GR13)	LI	1580	1	LV del Sur, Chilpancingo: 1200-0400
ME06)	VAB	1580	0.25	Super Stereo Miled, Valle del Bravo: 24h
SO35)	DM	1580	50	La Grande de Sonora/R.ACIR 15-80, Hermosillo: 1200-0700
BC10)	HC	1590	1	R.Bahía, Ensenada: 1400-0700
CH24)	BZ	1590	1	R.ACIR, Cd.Delicias: 24h
DF14)	VOZ	1590	20	Bonita AM, México: 24h
GJ26)	IRG	1590	1	La Campirana, Irapuato
NL03)	ACH	1590	5	R.Centro, Monterrey: 24h
TM02)	FD	1590	5/0.25	La Super F-D, Reynosa
VE39)	PT	1590	1	R.Misantla, Misantla: 1200-0200
BC02)	KTT	1600	1	La Tremenda, Tecate: 1400-0800
CO04)	LZ	1600	1	R.Consentida, Torreón: 1200-0600
CO13)	AE	1600	1	La Super Juvenil, Cd.Acuña: 1300-0700
CS20)	TUG	1600	1	Sonovida, Tuxtla Gutiérrez: 1200-0600
ME07)	GEM	1600	5	Sistema XEGEM "R.Mexiquense", Metepec: 1200-0600
ME08)	UACH	1610	0.03	R.Universidad Atónoma de Chapingo, Chipancingo: 2000-2400

Shortwaves:

VE40)	JN	2390	0.5	R.Huayacocotla, Huayacocotla: 1230-1600, 2100-0100
DF18)	RTA	4800	5	R.Transcontinental de América, México: irr.
NL12)	UJ	5980	0.5	LV Internacional de la Provincia Mexicana, Linares: 1230-0600
DF04)	RMX	5985	10	R.México Internacional, México: 1200-1600, 1800-2300
DF02)	OI	6010	1	R.Mil "La Estación de la Ciudad", México: 24h
VE06)	UW	*6020	0.25	La U de Veracruz, Veracruz
SL08)	XQ	*6045	0.25	R.Universidad de San Luis Potosí, San Luis Potosí
TM06)	CMT	*6090	1	Música Romántica, Cd.Mante
YU03)	QM	6105	0.25	Tus Panteras, Mérida: irr.
DF12)	PPM	6185	5	R.Educación, México: 0000-1200
VE21)	FT	*9545	0.25	LV de Veracruz, Veracruz
DF10)	YU	*9600	1	R.UNAM-Universidad Nacional Autónoma de México
DF04)	RMX	9705	5	R.México Internacional, México: 1200-1600, 1800-0500

For further R.México International information, see International Broadcasting section.

State abbreviations: AG = Aguascalientes; BC = Baja California; BS = Baja California Sur; CA = Campeche; CH = Chihuahua; CL = Colima; CO = Coahuila; CS = Chiapas; DF = Distrito Federal; DG = Durango; GJ = Guanajuato; GR = Guerrero; HG = Hidalgo; ME = Estado de México; MI = Michoacán; MO = Morelos; NA = Nayarit; NL = Nuevo León; OX = Oaxaca; PU = Puebla; QE = Querétaro; QR = Quintana; SL = San Luis Potosí; SN = Sinaloa; SO = Sonora; TB = Tabasco; TM = Tamaulipas; TX = Tlaxcala; VE = Veracruz; YU = Yucatán; ZC = Zacatecas.

N.B: These abbreviations are not officially recognized by the Mexican Post Office. Letters should therefor carry the abbreviations in brackets or full State name.

Addresses and other information:

AG00) AGUASCALIENTES (Ags.)
AG01) Ap.173, 20000 Aguascalientes - DG: Ing.A.Rivas G.
AG02) Av.Las Americas y Valparaíso, Fracc.La Fuente, 20000 Aguascalientes - DG: Lic.A.Morales P.
AG03) Morelos 222, 20000 Aguascalientes - DG: Alfredo Rivas G.
AG04) Universidad Autónomica de Aguascalientes, 20000 Aguascalientes.
AG05) Madero 333-501, Col.Centro, 20000 Aguascalientes.
AG06) República del Uruguay 205-C, Col.Centro, 20000 Aguascalientes.
AG07) Av.Adolfo López Mateos 426 Pnte, 20000 Aguascalientes.

BC00) BAJA CALIFORNIA NORTE (B.C.)
BC01) Ap.100, 22000 Tijuana (or: c/o Noble Broadcasting of San Diego, 4891 Pacific Highway, San Diego, CA 92110, USA) -Gte: Ing.Eduardo Liaño G. **Web:**http://xtrasports.com/.
BC02) Blvd.Benito Juárez 500, Local 2-B, Plaza Cuchuma, 21450 Tecate - DG: Ismael Chavarin R.
BC03) Av.de los Olivos 305, 22410 Tijuana - DG: Leopoldo Quiroz.
BC04) Av.Alfareros 1301, 21010 Mexicali (or: Box 1014, Calexico, CA 92231, USA) - DG: Silvia Lacarra de Vildosola.
BC05) Gral.Manuel Márquez de León 950, Zona Río, 22320 Tijuana (or: 713 Broadway, Suite "F", Chula Vista, CA 91910, USA) - DG: Ing.Luis Carlos Astiazaran O.
BC06) Av.Calafia 519, Centro Cívico, 21000 Mexicali - Dir: Jesús Ruiz Muñoz.
BC07) Prolong.Alfareros 253, Centro Cívico, 21000 Mexicali (or: Box 2897, Calexico, CA 92231, USA) - DG: Manuel Hurtado Simoni.
BC08) Carr. Escenica Tijuana-Ensenada km 22.5, 22440 Tijuana - DG: Joaquín Pasquel M.
BC09) Calle 3a y Floresta 1323-15, 22830 Ensenada - DG: Gustavo Lopez E.
BC10) Ap.777, 22800 Ensenada - DG: Alfredo Cañas M.
BC11) Blvd.Agua Caliente 10535-506, Fracc.Chapultepec. 22420 Tijuana (or: Box 5413, Chula Vista, CA 91912, USA) - Gte: Ing.J.Wilkins G.
BC12) Baja California 1310, Zona Norte, 22100 Tijuana (or: Box 430233, San Ysidro, CA 92073, USA) - DG: Javier Sánchez M.
BC13) Ap.23, 22000 Tijuana - DG: Gloria Enciso P.
BC14) Ap.19, 21400 Tecate - DG: Fco.J.Fimbres Durazno.
BC15) Ap.123, 22800 Ensenada - DG: José Enrique J.
BC16) Carlos Rovirosa 142, 22420 Tijuana - DG: Enrique Luterot V.
BC17) Ap.287, 22800 Ensenada - DG: José Marquez Muñoz de Cote.
BC19) Blvd.Lázaro Cárdenas 10183, Desp.201, 22450 Tijuana - DG: Torres Gustavo López.
BC20) Ap.217, 22930 San Quintín - Gte: Gabriel Neri Cornejo.
BC21) Matamoros 8, Fracc.Bahía, 22880 Ensenada - DG: Gonzalo Morales.
BC22) Ap.526, 22800 Ensenada - DG: Susana Vargas Villareal.
BC23) Lázaro Cárdenas, Esq.Colegio Militar, Centro Comercio, Villa Fontana, Loc 33 y 34, 21180 Mexicali - DG: Aldo Gonzales S.
BC24) Ap.1405, 21000 Mexicali - DG: Raul Reynoso Nuño.

BS00) BAJA CALIFORNIA SUR (B.C.S.)
BS01) Ap.105, 23010 La Paz - DG: Maria Guadalupe Lucero A.
BS02) Hidalgo 314-B, Centro, 23000 La Paz - DG: Raúl Antonio Arechiga Espinoza.
BS03) Av.Las Flores 1, 23920 Santa Rosalía - DG: Ernesto Robles C.
BS04) Ap.279, 23600 Cd.Constitución - DG: Ismael Tonche P.
BS05) Ap.19-B, 23010 La Paz.
BS06) San José del Cabo. Subdir.Gen: Ing.Fernando Martínez M.
BS07) Loreto.

CA00) CAMPECHE (Camp.)
CA01) Calle 44 y 21 s/n, 24350 Escárcega - Gte: R.Palma P.
CA02) Calle 22 N° 131, 24100 Cd.del Carmen - DG: Mario A.Boeta B.
CA03) Ap.33, 24000 Campeche - DG: Raul Ramos N.
CA04) Ap.22, 24200 Palizada - DG: R.Palma P.
CA05) Calle 30 N° 86-A, 24100 Cd.del Carmen - DG: Lic.José Guillermo Lliteras E.
CA06) Tamaulipas 15, Col.Santa Ana, 24050 Campeche - DG: Raul Palma Perez.
CA07) Carr.Cd.del Carmen-Champotón km 1, 24400 Champotón - DG: R.Palma P.
CA08) Ap.1, 24930 Bécal - DG: C.Reynaldo Catzín C.
CA09) X'pujil.
CA10) 24000 Campeche.
CA11) Prol.Calle 53, Esq.Av.16 de Septiembre s/n, 24000 Campeche - Ger: Lic.Fernando Solis Mejía.

CH00) CHIHUAHUA (Chih.)
CH01) Ap.412, 31500 Cd.Cuauhtémoc - DG: A.Moreno S.
CH02) Libertad 1306, 31000 Chihuahua - DG: Agustín Caldera M.
CH03) Calle Novena 513, 31000 Chihuahua - DG: R.Salmón de Boone.
CH04) Ap.50, 33000 Cd.Delicias - Gte: Roberto Díaz G.
CH05) Ap.710, 31500 Cd.Cuauhtémoc - DG: Israel Beltrán M.
CH06) Allende 613, 33980 Cd.Jiménez - Gte: R.Montes A.
CH07) Ap.8, 33500 San Francisco del Oro - Gte: A.Gutiérrez M.
CH08) Julián Carrillo 705-A, 31000 Chihuahua - DG: David Yberi Russek.
CH09) Coronado 71, 33580 Santa Bárbara - DG: Domingo Salayandia N.
CH10) Av.16 de Septiembre Ote 337-302, 32000 Cd.Juárez - DG: Benigno Olivas A.
CH11) Av.Benito Juárez 106, 33700 Cd.Camargo - Gte: H.Piñera L.
CH12) Ap.266, 32380 Cd.Juárez - DG: Antonio Carmona I.
CH13) Ap.177, 31500 Cd.Cuauhtémoc - DG: Alberto Ramos L.
CH14) Cuauhtémoc 2000, 31020 Chihuahua - DG: E. Lopez de la Rocha.
CH15) Ap.59, 31700 Nuevo Casas Grandes - DG: Fernando Salgado.
CH16) Ap.125, 33800 Hidalgo del Parral - DG: Gildardo Valles D.
CH17) Ap.222, 33000 Cd.Delicias - Gte: H.L.Pérez B.
CH18) Guachochi - DG: Victor Martínez Juárez.
CH19) Ap.33, 33800 Hidalgo del Parral - DG: A.Gutiérrez M.
CH20) Ap.29, 32880 Ojinaga - DG: Jorge Hernandez C.
CH21) Ap.44, 33700 Cd.Camargo - DG: Ernesto Salayandia G.
CH22) Ap.38, 31940 Cd.Madera - DG: A.Muñoz E.
CH23) Universidad de Chihuahua, 31000 Chihuahua - Dir: A.Varona T.
CH24) Ap.250, 33000 Cd.Delicias - DG: Jaime Narvaez G.
CH25) Ap.190, 33800 Hidalgo del Parral - DG: María del Carmen Paez de Salayandia.
CH26) Juárez y 2a 201, 32881 Ojinaga (or: Box 276, Presido, TX 79845, USA) - DG: Lic.Alfredo Rohana E.
CH27) Priv.Agustín Lara 37, 32000 Cd.Juárez - DG: Fco. Antonio Gonzales S.
CH28) Av.del Parque Sur 6, 33000 Cd.Delicias - DG: H.L.Pérez B.
CH29) Blvd.Ortíz Mena 52, P3, 33800 Hidalgo del Parral - DG: Arnoldo A.Rodríguez R.
CH30) Prof.Eliza Dosamentes 400, Fracc.Los Colorines, 32380 Cd.Huárez - DG: Lázaro Megret.
CH31) Agustín Melgar 602, Niños Heroes, 31500 Cd.Cuauhtémoc - DG: R.Salmon de Boone.
CH32) Ap.157, 32380 Cd.Juárez - DG: Lazaro Megret.
CH33) Ap.70, 32380 Cd.Juárez - DG: Benigno Olivas A.
CH34) Ap.9, 31000 Chihuahua - DG: Adrian Pereda Gómez.
CH35) Av.2a Norte N° 309, Local 107, Col.Centro, 33000 Cd.Delicias - DG: Arnoldo Rodríguez Reyes.

CL00 COLIMA (Col.)
CL01) Ap.200, 28000 Colima - DG: Roberto F.Levy.
CL02) Av.México 51-A, Esq.Juárez, 28200 Manzanillo - DG: Sra. Lucía Tene R.
CL03) Ingenieros 14, Esq.Av.México, 28200 Manzanillo - DG: A.Romo B.
CL04) Ignacio Sandoval 13, 28000 Colima - DG: Enrique Padilla V.
CL05) Allende 408-101, 28100 Tecomán - DG: Roberto F.Levy.
CL06) Ap.2, Tecomán - Gte: V.M.Martínez Escamilla.
CL07) Ap.2-1690, Suc.A, 28950 Colima - DG: Lic.Laura Sánchez Menchero.
CL08) Lote 4, Manzana B, Parque Industrial Fondeport, 28200 Manzanillo - DG: Lic.Jesus Granadas R.
CL09) Calzada La Armonía 270, La Armonía, 28020 Colima - DG: Roberto F.Levy.
CL10) Blvd.Costera Miguel de la Madrid km 10, Plaza Galería, Local 4, 28200 Manzanillo - DG: Ernesto Rodriguez M.

CO00) COAHUILA (Coah.)

CO01) Blvd.González de la Vega 195, 27000 Torreón - DG: Carlos Enrique F. - DG XEBP/XETJ/XEYD: Lic.Enrique Jan C.
CO02) Ap.3, 26000 Piedras Negras (or: Box 196, Eagle Pass, TX 78853-0196, USA) - DG: Claudio M.Bres M.
CO03) Ap.60, 26700 Sabinas - DG: R.Salmon de Boone.
CO04) Priv.Eulogio Ortiz y Pamanes, Col.Ampl.Los Angeles, 27140 Torreón - DG: José de León F.
CO05) Ap.74, 26340 Cd.Múzquiz - DG: Hugo H.Martínez T.
CO06) Piedras Negras 1812, 25280 Saltillo - DG: Emilio Lopez S.
CO07) Ap.285, 25700 Monclova - DG: Ing.Salvador Apud.
CO08) Ap.57, 26000 Piedras Negras - Gte: Arq.F.Elizondo C.
CO09) Reforma y América 9, 26850 Nueva Rosita - DG: Mario Alberto Franco Lopez.
CO10) Gral.Manuel Pérez Trevino Pte 839, 25000 Saltillo - Dir: J.M.López C.
CO11) Ap.108, 25700 Monclova - DG: Lic.Rolando Gonzalez T.
CO12) Ap.32, 26850 Nueva Rosita - DG: Daniel Boone M.
CO13) Ap.10, 26200 Cd.Acuña - DG: Sergio Benanidez G.
CO14) Ap.78, 27000 San Pedro - Gte: C.Enríquez F.
CO15) Av. Carranza 1104, Col.Roma, 26000 Piedras Negras (or: Box 1261, Eagle Pass, TX 78852, USA) - DG: Braulio Zavala Z.
CO16) Acuña 276 Sur, P2, 27000 Torreón - DG: Luis de la Rosa Montellano.
CO17) Escheverria 1017-204 Norte, Fracc.Cumbres, 25270 Saltillo - DG: Juan Alberto Jaubert L.
CO18) Ap.89, 25700 Monclova - Gte: Norberto R.Carrizales L.
CO19) Ap.57, (or: Ap.170) 25280 Saltillo - DG: Sra.Mª Magdalena Tafich de León.
CO20) Juárez 1400 Sur, 26530 Allende - Gte: H.Moreno V.
CO21) Ap.91, 25600 Monclova - DG: J.F.Elizondo V.
CO22) Fco.I.Madero 501 Pte, 27980 Parras - DG: Lorenzo Zuñiga M.
CO23) Zaragoza 505 Sur, 26450 Zaragoza - DG: Tomas García J.
CO24) Hidalgo 349 Pte, 26200 Cd.Acuña - Dir: Lic.Eduardo Sanchez Hernandes.
CO25) Venustiano Carranza 612, 25700 Monclova - DG: Alberto Gauthreau.
CO26) Colima, Esq.San Juan de Sabinas 483, República, 25280 Saltillo - DG: Lic.Delia M.Villareal R.
CO27) Periferico Luis Esheverria Norte 1017-203, 25270 Saltillo - DG: Juan Alberto Jaubert L.
CO28) Av.Morelos 1320-204, Edif.Monterrey, 27000 Torreón - DG: Gerardo Vargas de Santiago.
CO29) Ap.17, 26450 Zaragoza - DG: Tomas García J.
CO30) Ap.27, 25230 Saltillo - DG: Lic.Jorge Ruíz S.
CO31) Ap.181, 25750 Monclova - DG: Lic.Rolando Gonzales T.

CS00 CHIAPAS (Chis.)
CS01) Ap.59, 29000 Tuxtla Gutiérrez - DG:R.Angel Camejo R.
CS02) Ap.57, 30700 Tapachula - DG: Amín Simán Habib.
CS03) Ap.74, 29250 San Cristóbal de las Casas - DG: Fco. José Narváez R.
CS04) Ap.76, 30700 Tapachula - DG: Kenny Ordaz E.
CS05) Av.Central Pte 554-4, 29000 Tuxtla Gutiérrez DG: Simon Valanci B.
CS06) Ap.60, 30400 Cintalapa - DG: Fco.Siman Valanci B.
CS07) 29200 San Cristóbal las Casas.
CS08) Ap.19, 30500 Tonalá - DG: Amín Simán Habib.
CS09) Blvd.Belisario Dominguez 4810, 29000 Tuxtla Gutiérrez - DG: Fco.Simán Stefan.
CS10) Carr.Pichucalco-Teapa km 2, 29520 Pichucalco - DG: Amín Simán Habib.
CS11) Ap.28, 30450 Arriaga - DG: José Fco.Rueda M.
CS12) Av.Central Norte 8, 30640 Huixtla - DG: Hector Cruz Angel.
CS13) Ap.99, 30000 Comitán - DG: Pedro Aguilar M.
CS14) 1a Av.Norte Pte N° 53, 30470 Villaflores - DG: Oscar Fonseca Alfaro.
CS15) Las Margaritas - DG: Carlos Romo Zapata.
CS16) Cacahoatán.
CS17) Carr.Boca del Limón km 2.5, 29500 Reforma - DG: José J. Lopez D.
CS18) Primera Calle Norte Pte 7, 30000 Comitán - DG: Arturo Palacios P.
CS19) Primera Av.Sur 2, 30700 Tapachula - DG: Guillermo Flores Salvador.
CS20) Blvd.Belisario Dominguez km 1081, 29000 Tuxtla Gutierrez - DG: R.Angel Camejo R.
CS21) Ap.91, 29000 Tuxtla Gutiérrez - DG: Fco.Simán S.
CS22) Av.Chichimá 405, 30000 Comitán - DG: Lic.Marco Escobar G.
CS23) 7a Oriente N° 42, 30700 Tapachula.
CS24) Gobierno del Estado de Chiapas, Ococingo.

CS25) 1a Av.Sur N° 2, 30700 Tapachula - DG: Jaime Fernández A.

DF00) DISTRITO FEDERAL (D.F.)
DF01) Radio Chapultepec, Av.Chapultepec 473, P7, Col.Juárez, 06600 México — ☎+52 (5) 2116738 — DG: Oscar Obregón Mazón.
DF02) Núcleo Radio Mil, Insurgentes sur 1870, Col.Florida, 01030 México — ☎ +52 (5) 6626060, 6630590 — 🖷 +52 (5) 6630739 — Web:http://www.nrm.com.mx/ — Pres: Lic. E.Guillermo Salas Peyró. Radio Mil sw: Ap.21-1000, 04021 México.
DF03) Radiodifusoras Asociadas, Durango 341, Col.Roma, 06700 México — ☎ +52 (5) 2114911, 2114939 — 🖷 +52 (5) 2862774 — Web:http://www.rasa.com.mx/ — E-mail: info@rasa.com.mx. — DG: Lic.José Laris Rodríguez.
DF04) Instituto Mexicano de la Radio, Mayorazgo 83, Col.Xoco, 03330 México — ☎ +52 (5) 6281700 — 🖷 +52 (5) 6281693 — Web:http://www.telecommex.com/imer/ — E-mail: imerte04@telecommex.com — DG: CP Carlos Lara Sumano. Radio México Internacional, Ap.21-300, 04021 México — ☎ +52 (5) 5345210 - 🖷 +52 (5) 5241758 — E-mail: imer@mpsnet.com.mx: — Ger: Martín Rizo Gavira. See also Intrnational Broadcasting section.
DF05) Radio Sistema Mexicano, Insurgentes sur 1377, P4, Insurgentes Mixcoac, 03920 México — ☎ +52 (5) 5982488, 🖷+52 (5) 6110401 — E-mail: ondaslag@mpsnet.com.mx — DG: Tere Vale.
DF06) Sistema Radiopolis, Calzada de Tlalpan 3000, Col. Espartaco, 04870 México — ☎ +52 (5) 3272000 — Web: http://www.tele-visa.com.mx/radio: — Pres: Lic.Ricardo Rocha.
DF07) México Radio, Gómez Farias 51, Col.San Rafael, 06470 México — ☎ +52 (5) 7056746, 7052275 — 🖷+52 (5) 5350295 — DG: Ing.Antonio Aguilar Darriba.
DF08) Grupo Radio Centro, Av.Constituyentes 1154, Col.Lomas Altas, 11590 México — ☎ +52 (5) 7284800-10 — Pres: Mª Esther Gómez de Aguirre.
DF09) Radio S.A., Rodolfo Emerson 412, Col.Chapultepec Morales, 11570 México — ☎ +52 (5) 2035577 — 🖷 +52 (5) 5452078 — Pres: Lic.Carlos Quiñones Armendáriz.
DF10) Universidad Nacional Autónoma de México, Adolfo Prieto 133, Col.del Valle, 03100 México — ☎+52 (5)5239350, 5239356 — Web: http://www.unam.mx/radiounam/ - E-mail: radiounam@www.unam.mx — DG: Lic.Malena Mijares Fernández.
DF11) Organización R.Fórmula, Privada de Horacio 10, Col. Polanco, 11560 México — ☎ +52 (5) 2821020, 2821411 — Web: http://www.radioformula.com.mx — E-mail: radioformula@super-net.com.mx — Pres: Lic.Rogerio Azcárraga Madero.
DF12) Radio Educación, Angel Urraza 622, Col.del Valle, 03100 México — ☎ +52 (5) 5596169 — 🖷 +52 (5) 5756566 — DG: Luis Ernesto Pi Orozco. Radio Educación sw: Ap.21-940, 04021 México.
DF13) Radiorama, Paseo de la Reforma 56, P5, Col.Tabacalera, 06030 México — ☎ +52 (5) 5660299. 5660471 — 🖷 +52 (5) 5661454 — Web: http://www.radiorama.com.mx — DG: Lic.José Luis Chavero Reséndiz.
DF14) Grupo ACIR, Pirineos 770, Lomas de Chapultepec, 11000 México — ☎ +52 (5) 2590001, 5400810 — 🖷 +52 (5) 5404106 — DG: Lic.Antonio Ibarra Fariña.
DF15) Grupo ACIR, Blvd.de los Virreyes 1030, Lomas de Chapultepec, 11000 México — ☎ +52 (5) 5201956, 5202499.
DF16) Grupo 7 División Radio, Montecito 59, Col.Nápoles, 03810 México — ☎ +52 (5) 6824370, 6826767 — 🖷 +52 (5) 5364873 — E-mail: cambio@spin.com.mx — Pres: Francisco Javier Sánchez C.
DF17) MVS Radio, Trabajadores Sociales 309, Col.Magdalena Atlazalpa, 09410 México — ☎ +52 (5) 7220900 — 🖷+52 (5) 7220909 — Pres: Lic.José Luis Fernández Herrera.
DF18) Radio Transcontinental de América, Torre Latinoamericana, P37, 06007 México (or: Ap.653, 06002 México) — DG: Roberto Najera Martínez.

FM in México City: DF08) 88.1 Red FM — DF15) 88.9 Azul 89 — DF02) 89.7 Morena FM — DF17) 90.5 Pulsar FM — 90.9 Ibero — DF08) 91.3 Alfa — DF08) 92.1 La Z — DF06) 92.9 La Ke Buena — DF08) 93.7 Stereo Joya — DF04) 94.5 Opus 94 — DF14) 95.3 Inolvidable — 95.7 El Politécnico en Radio — DF10) 96.1 UNAM - DF06) 96.9 "W" — DF08) 97.7 Energy —DF17) 98.5 Radioactivo — DF15) 99.3 Digital 99 — DF02) 100.1 Stereo Cien — DF02) 100.9 Cien punto nueve — DF06) 101.7 Vox — DF17) 102.5 Stererey — DF11) 103.3 Fórmula 103 — DF11) 104.1 Uno — DF17) 104.9 FM Globo — DF04) 105.7 Orbita — DF15) 106.5 Mix 106 — DF08) 107.3 Universal.

DG00) DURANGO (Dgo.)
DG01) Ap.174, 34000 Durango - DG: Ricardo León Garza L.
DG02) Ap.405, 34000 Durango - DG: Luís Guillermo Manifacio T.
DG03) Negrete 405-B Oriente, 34000 Durango - DG: A.Armas H.

DG04) Universidad Juárez del Estado de Durango, 34270 Durango.
DG05) Fco.I.Madero y Heroico Colegio Militar s/n, 34600 Santiago Papasquiaro - Gte: Ricardo L.Garza.
DG06) Manuel Rangel 100, P3, 34270 Durango - DG: Lic.Julio Muciño Pereda.
DG07) Ap.115, 34080 Durango - DG: Carlos Ernesto Galvez R.

GJ00) GUANAJUATO (Gto.)
GJ01) Ap.424, 37160 León - DG: José Rogelio Esquerra A.
GJ02) Ap.301, 37160 León - Gte: J.Torres M.del Campo.
GJ03) Ap.3, 38900 Salvatierra - DG: Felipe Arizaga C.
GJ04) Blvd.López Mateos Ote 1117, 38070 Celaya - DG: Lic.Fdo Olivares Ramos.
GJ05) Morelos 110, 36500 Irapuato - DG: Ing.Marco António H.Contreras S.
GJ06) Ap.392, 37480 León - Gte: D.Pérez V.
GJ07) Ap.528, 38090 Celaya - DG: Jesus San Martín B.
GJ08) Universidad de Guanajuato, Ap.359, 36000 Guanajuato - DG: Enrique Ayala Negrete.
GJ09) Elodia Ledezma 658, 38890 Moroleón - DG: G.F.Ortiz M.
GJ10) Ap.311, 37370 León - DG: D.Castrejón T.
GJ11) Allende 16, 38600 Acámbaro - Gte: J.C.Sandoval A.
GJ12) Juárez 100, 36700 Salamanca - DG: Ing.Marco António H.Contreras S.
GJ13) Ap.24, 36700 Salamanca - DG: Sergio Rodríguez M.
GJ14) Ap.4, 37700 San Miguel de Allende - Gte: José Manuel Zavala Z.
GJ15) Ap.13, 37000 León - DG: Manuel Vázquez M.
GJ16) Ap.72, 36500 Irapuato - Gte: A.Martínez D.
GJ17) Ap.43, 37800 Dolores Hidalgo - Gte: Sergio Reyna Lopez.
GJ19) Ap.67, 37900 San Luis de la Paz - Gte: M.González S.
GJ20) Municipio Libre 8, 36080 Guanajuato - DG: M.A.Hernández L.
GJ21) Ap.60, 36100 Silao - Gte: J.Sánchez R.
GJ22) Ap.528, 38000 Celaya - DG: Alfonso Aizcorbe C.
GJ23) Ap.562, 37530 León - DG: D.Castrejón T.
GJ24) Priv.Venustiano Carranza 119-101, 38000 Celaya - DG: Alfonso Aizcorbe C.
GJ25) Ap.642, 37160 León - DG: J.Torres del Campo.
GJ26) Av.Guerrero y Francisco Sarabia, Centro Plaza Magna, Local 3-B, 36500 Irapuato - DG: Luis de Alba Padilla.

GR00) GUERRERO (Gro.)
GR01) Calle de la Paz 190-6, Edif.Nick, 39300 Acapulco - DG: Lic.Andres Martinez V.
GR02) Ap.60, 39390 Acapulco - Gte: Lic.José Luis Baños G.
GR03) Ap.21, 39300 Acapulco - DG: Lic.Andres Martinez V.
GR04) Prol.Pungarabato pte, La Cosita, 40660 Cd.Altamirano - DG: Rafael García V.
GR05) Ignacio Altamirano N°1, Esq.5 de Mayo, 40880 Zihuatanejo - DG: M.Morales Vallejo.
GR06) Ap.2, 39170 Tixtla - DG: Manuel Radilla R.
GR07) Ap.31, 40830 Petatlán - DG: Jesús Orta Z.
GR08) Ap.2, 40000 Iguala - DG: J.M.Garrido H.
GR09) Tlapa de Comfort.
GR10) Ap.52, 40000 Iguala - Gte: Sergio Fajardo C.
GR11) Entronque Carr.Igualada-Cd.Altamirano, 40500 Arcelia - DG: Rafael García V.
GR12) Cerro de la Bermeja s/n, 40200 Taxco - DG: José Bustamente S.
GR13) Ap.40, 39000 Chilpancingo - DG: Edgardo Lozada R.
GR14) Monteblanco 37, Fracc.Hornos-Insurgentes, 39350 Acapulco - DG: Arturo Solís.
GR15) Manzana 1, Fracc.Eduardo Neri, 39000 Chilpancingo - DG: Florentino Martinez V.
GR16) Av.La Suiza 19, Fracc.Las Playas, 39390 Acapulco - DG: Alberto Guilbot S.
GR18) Carr.Escenica 100, Playa El Guitarron, 39368 Acapulco - DG: Lic.Saul Estrada.
GR19) Morelos 6-3, 39000 Chilpancingo - DG: Antonio Mancilla J.
GR20) Calle Sur 305, 41100 Chilapa - DG: Lic.Sergio Fajardo C.
GR21) Av.Guerrero 10-B, Desp.2. 39000 Chilpancingo - DG: Lic. Luis Carlos Mendiola C.
GR22) Av.del Tanque y Ponciano Arriaga, Lázaro Cárdenas, 39740 Acapulco - DG: Lic.Guillermo Montes G.
GR23) 40200 Taxco.
GR24) Miguel Alemán 29, Col.Centro, 39000 Chilpancingo - DG: Lic.Florentino Martínez Vargas.

HG00 HIDALGO (Hgo.)
HG01) Ap.123, 42000 Pachuca - DG: Juan Manuel Larrieta E.
HG02) Ap.35, 43000 Huejutla - Gte: J.A.Reyes F.
HG03) Hidalgo Ote.209, 43600 Tulancingo - DG: Heberto Seguva O.
HG04) Plaza de la Constitución y Manuel F.Soto, 43600 Tulancingo - DG: L.Castelán L.
HG05) Carr.a Cardonal km 2.689, Barrio de San Nicolas, 42300 Ixmiquilpán - DG: Roberto Martinez O.
HG06) Félipe Angeles s/n, 42300 Ixmiquilpán.
HG07) Tlanchinol.
HG08) Plaza Juárez 103, 42000 Pachuca - DG: Juan Manuel Larrieta E.
HG09) Chávez Macotela 7A, 42400 Huichapan.

JL00) JALISCO (Jal.)
JL01) Ap.60, 49000 Cd.Guzmán - DG: Rodolfo Hernandez R.
JL02) Av.Vallarta 1458, 44140 Guadalajara - DG: Manuel Lopez A.
JL03) Paseo de las Gaviotas 198, Fracc.Las Gaviotas, 48328 Puerto Vallarta - DG: Sra Gloria Angélica Carrillo G.
JL04) Lorenzana 884, 45040 Guadalajara - DG: Carlos Guerrero D.
JL05) Carr.Tampico-Barra de Navidad km 695, 44100 San Juan de los Lagos - DG: José Ismael Alvarado R.
JL06) Av.México 3150, Fracc.Monraz, 44670 Guadalajara - DG: Roberto Ruvalcaba Barba.
JL07) Ap.7, 48900 Autlán - DG: J.A.Perales G.
JL08) Av.Lázaro Cárdenas 2820, Jardines del Bosque, 44520 Guadalajara - DG: Luís Suarez Huizar.
JL09) Av.Mariano Otero 3405, Fracc.Verde Valle, 45060 Guadalajara - DG: Ing.Jesús Orozco Godinez.
JL10) Vidrio 2056, 44100 Guadalajara - DG: Luis Carlos Fregoso Mendoza.
JL11) Ap.32, 47400 Lagos de Moreno - DG: Ing.Carlos Sanchez S.
JL12) Pablo Casal 567, Prados Providencia, 44670 Guadalajara - DG: Lic.Fco.Vidrio G.
JL13) Morelos 27, 47910 La Barca - Gte: Arturo Laris R.
JL14) Hidalgo 2055, Sector Hidalgo, 44500 Guadalajara - DG: Nicandro Tavares C. JL15) Lerdo de Tejada 2186, Col.Americana, 44140 Guadalajara -
DG: Nicandro Tavares C.
JL16) Ramón Corona 54-2, 49650 Tamazula - DG: Fco. Uribe C.
JL17) Ap.1-526, 44100 Guadalajara - DG: Ing.Daniel Martínez O.
JL18) Ap.36, 49300 Sayula - DG: J.A.Aguilar H.
JL19) Fco.I.Madero 77, 45750 Zacoalco de Torres - DG: José Toscano Figueroa.
JL20) Zaragoza 59 Altos, 47750 Atotonilco El Alto - DG: S.Rubio P.
JL21) Ap.16, 46060 Ameca - Gte: A.Navarro A.
JL22) Primero de Mayo 126-8, 49000 Cd.Guzmán - DG: Juan Fco. Aguilar G.
JL23) Hidalgo 78-4, 47600 Tepatitlán de Morelos - DG: J.Ismael Alvarado.
JL24) Ap.5, 48900 Autlán - DG: Luís Felipe Villaseñor M.
JL25) Manuel M.Diéguez 314, P2, 48380 Puerto Vallarta - DG: C.P.Juan Fco. Aguilar G.
JL26) Niños Héroes N° 1555-602, Edif.Plaza Tolsa, 44100 Guadalajara - DG: Enrique Pereda G.
JL27) Monterrey 190, Fracc.Camino Real, 47820 Ocotlán - DG: Alejandro Rubio Beltrán.
JL28) Av.México 3150, Fracc.Monraz, 44670 Guadalajara - DG: Alejandro Diaz R.
JL29) Moctezuma 68, 49000 Cd.Guzmán - DG: Samuel Diaz M.
JL30) Alvaro Obregón 96, Loma Dorada, 45430 Zapotlanejo - DG: Ruben Hernandez C.

ME00) ESTADO DE MEXICO (Edo.Méx.)
ME01) Carr.Panamericana km 24, 50450 Atlacomulco - DG: G.Libien S.
ME02) Paseo Tollocan Pnte 300, Col.Universidad, 50130 Toluca - Pres: Enrique Bernal Servin.
ME03) José María Morelos 948, 51900 Ixtapan de la Sal - DG: Rafael Salazar D.
ME04) Independencia 19, El Oro - DG: R.Ordorica C.
ME05) Ap.37, 50000 Toluca - DG: C.P.Victor X.Guadarrama P.
ME06) Independencia 506, 51200 Valle del Bravo - DG: Gabriela Libien S.
ME07) Av.Estado de México Km 1, Metepec - DG: Ing.J.Muñoz E.
ME08) Carr.México-Texcoco km 38.5, 56230 Chipancingo.

MI00) MICHOACAN (Mich.)
MI01) Aqua 78, Prados del Campestre, 58297 Morelia - DG: G.Libien S.
MI02) Ap.50, 61500 Zitácuaro - DG: M.Jiménez R.
MI03) Ap.6, 60000 Uruapan - DG: Ing. J.Luis G.Treviño N.
MI04) Av.5 de Mayo 501 Sur, Jardines de Catedral, 59670 Zamora - DG: Arturo Laris Rodríguez.
MI05) Av.Constitución de 1814 Norte 10 Altos, 60600 Apatzingán - DG:

Sra.Maria del Carmen Solis.
MI06) Cheran.
MI07) Laguna de Parras 630, Col.Ventura Puente, 58020 Morelia - DG: Raymundo Rodríguez Macias.
MI08) Av.Constitución de 1814 Norte 2 Altos, 60600 Apatzingán - DG: Armando Guzman G.
MI09) Ap.10, 59300 La Piedad - DG: H.Guízar A.
MI10) Ap.244, 61600 Pátzcuaro - DG: J.Becerra C.
MI11) Ap.132, 60000 Uruapan - Gte: J.Lira R.
MI12) Mazatlán 30, 60050 Uruapan - Gte: F.de Jesús Flores L.
MI13) Altos del Mercado Municipal F.Zapata, 61100 Cd.Hidalgo - DG: Jaime Robledo C.
MI14) Ap.65, 58600 Zacapu - DG: Juan Ortiz A.
MI15) Av.Madero Pte 644, 58000 Morelia - Gte: J.Gpe.Muñoz M.
MI16) Ap.50, 58600 Zacapu - DG: Alfonso Fugeman C.
MI17) Ap.60, 59000 Sahuayo - DG: José Raul Nava C.
MI18) Cupatitzio 42-1, 60000 Uruapan - DG: Jaime Pardo H.
MI19) Ap.275, 58020 Morelia - DG: J.M.Treviño N.
MI20) Universidad de San Nicolás, 58000 Morelia.
MI21) Av.Río Balsas 7, 60950 Cd.Lázaro Cárdenas - DG: José Manuel Virrueta S.
MI22) 20 de Noviembre 358, 58000 Morelia - DG: E.Salgado B.
MI23) Madero Norte 15, 61940 Huetamo - DG: Jorge M.Treviño R.
MI24) Ap.45, 60300 Los Reyes - DG: Lic.Eduardo Treviño Nuñez.
MI25) Ap.430, 60950 Cd.Lázaro Cárdenas - DG: Fdo.Escarcega V.
MI26) 58000 Morelia.
MI27) Carr.Lázaro Cárdenas-La Mira, 5 de Mayo, 60990 Lázaro Cárdenas - DG: Rogelio Ortega C.
MI28) Agua 78, Prados del Campestre, 58297 Morelia - DG: Lic. Arturo Herrera C.
MI29) Macarena 32, Inhuambo, 60130 Uruapan - DG: Ramón García Luis Ponce de León.
MI30) Ap.73, 59300 La Piedad - DG: Ernestina Ayala T.
MI31) Ap.136, 58297 Morelia - DG: Lic.Arturo Herrera C.

MO00) MORELOS (Mor.)
MO01) Ap.868, 62000 Cuernavaca - DG: Lic.Miguel Angel Castillo F.
MO02) Plaza Yuliana, P2, 62900 Jojutla - Gte: A.E.Martínez S.
MO03) 62746 Cuautla (alt.address: Hidalgo 105, 62220 Ocotepec, Cuernavaca).
MO04) Ap.720, 62000 Cuernavaca - DG: Andres Martínez Vargas.

NA00) NAYARIT (Nay.)
NA01) Ap.296, 63060 Tepic - DG: J.de Jesús Macías C.
NA02) Ap.95, 63060 Tepic - DG: C.P.Arnulfo Nuñez Iglesias.
NA03) Independencia 1330 Ote, 63200 Tuxpan - DG: G.Herena M.
NA04) Ap.68, 63060 Tepic - DG: Mario M.Bertrand P.
NA05) Ap.82, 63700 Compostela - Gte: Julio Mondragón G.
NA06) Ap.22, 63310 Santiago Ixcuintla - Gte: Julio Mondragón G.
NA07) Ap.4, 63310 Santiago Ixcuintla - DG: J.de Jesús Macías C.
NA08) Ap.49, 63440 Acaponeta - DG: Pedro Aguiar Villegas.
NA09) Jesús María.
NA10) Ap.7, 63440 Tecuala - DG: José Alberto A.
NA11) Puebla 3, 63600 Cd.Ruíz - DG: Arnulfo Nuñez I.
NA12) Ap.33, 63940 Ixtlán del Río - DG: José de Jesus Macías C.
NA13) Av.Insurgentes Sur 1046, El Rodeo, 63060 Tepic - DG: C.P.Arnulfo Nuñez Iglesias.

NL00) NUEVO LEON (N.L.)
NL02) Ap.2747, 64700 Monterrey - DG: Alberto Estrada Torres.
NL03) Juan Ignacio Ramón 506 Oriente, P20, Condomino del Norte, 64000 Monterrey - DG: Javier Aguilera de Alba.
NL04) Ap.628, 64700 Monterrey - DG: Fco.A.Gonzáles.
NL05) Ap.1111, 64700 Monterrey - DG: Emilio Lopez S.
NL06) Privada Rhin 647, 64000 Monterrey - DG: J.A.Gómez F.
NL07) Ap.1430, 64700 Monterrey - DG: Fco.A.González.
NL08) Ap.118, 64000 Monterrey - DG: Lic.Teofilo Bichara Z.
NL09) Ap.38, 65290 Cd.Sabinas Hidalgo - DG: Lic.Armando Ríos Leal.
NL10) Ap.45, 67500 Montemorelos - DG: A.G.Pezzino P.
NL11) 64000 Monterrey.
NL12) Ap.62, 67700 Linares - DG: Marcelo Becerra Gonzalez.
NL13) Ap.81, 67700 Linares - DG: Lic.Alberto Granados P.
NL15) Paricutín 316 Sur, Roma, 64000 Monterrey - DG: José Maria Diaz A.

OX00) OAXACA (Oax.)
OX01) Ap.46, 70600 Salina Cruz - DG: H.López Lena Jr.
OX02) Ap.175, 68000 Oaxaca - DG: W.Hernández M.
OX03) Ap.21, 70760 Tehuantepec - DG: Norma Angélica Santos G.
OX04) Ap.66, 70300 Matías Romero - DG: H.López L.

OX05) Bustamante 112, P1, Centro, 68000 Oaxaca - DG: Miguel A.Gonzales A.
OX06) 16 de Septiembre 23, Loma Bonita - DG: Sergio Palazeta P.
OX07) Domicilio Conocido, 71980 Puerto Escondido - DG: António Jalil V.
OX08) Jazmines 907, 68050 Oaxaca - DG: Manuel de la Lanza E.
OX09) Ap.145, 68000 Oaxaca - DG: Alberto Marquez R.
OX10) Ap.48, 68300 Tuxtepec - DG: María Eugenia Rodríguez N.
OX11) Ap.60, 70030 Juchitán - DG: H.López L.Jr.
OX12) Ap.35, 70900 Puerto Angel - DG: Ing.Julio Jalil.
OX13) Carr.Ixtepec-Juchitan km 2, 70110 Ixtepec - DG: H.López L.
OX14) Ap.48, 69000 Huajuapan de León - DG: Manuel de Jesus Siordia T.
OX15) Guelato de Juárez.
OX16) San Lúcas Ojitlán.
OX17) Tlaxiaco - DG: Benjamin Muratella.
OX18) Santiago Jamiltepec - DG: Teresa Peréz Díaz.
OX19) Morelos 6-2, 71000 Putla de Guerrero - DG: Codina Nestor Jimenez L.
OX20) Colón 12, 69800 Tlaxiaco - DG: J.A.Fugemann Chang.
OX21) Universidad Benito Juárez, 68000 Oaxaca.
OX22) 70760 Tehuantepec.
OX23) Macedonio Alcala 915, 68000 Oaxaca - DG: W.Hernández M.
OX24) Av.Alfonso Perez Gasca 504, 71600 Pinotepa Nacional - DG: David Jimenez L.
OX25) Ap.229, 68000 Oaxaca - DG: Lic.Fco.Pliego G.

PU00) PUEBLA (Pue.)
PU01) Allende 507, 73800 Teziutlán - DG: J.Sánchez T.
PU02) Ap.176, 73800 Teziutlán - DG: Juan Angel Dommarco A.
PU03) Ap.84, 75700 Tehuacán - DG: Francisco Sanchez M.
PU04) Primera Norte 101-6, 75700 Tehuacán - DG: Juan Manuel Olguín S.
PU05) Av.Juárez 2105, La Paz, 72160 Puebla - DG: J.Grajales S.
PU06) Ap.54, 73160 Huauchinango - DG: M.R.Rojano S.
PU07) Ap.49, 74400 Izúcar de Matamoros - DG: Sra.A.Rodríguez E.
PU08) Ap.21, 75520 Cd.Serdán - DG: A.Muñoz Bautista.
PU09) Zaragoza 31-A, 74400 Izucár de Matamoros - DG: C.P.Fidel Balbuena Sánchez.
PU10) Ap.44, 75700 Tehuacán - DG: Francisco Sánchez T.
PU12) Ap.4, 74000 San Martín Texmelucan - DG: R.Franco P.
PU13) Reforma N° 120, 73080 Xicotepec de Juárez - DG: Alejandro Wong L.
PU14) 15 Pte.1306, Santiago, 72000 Puebla - DG: Guillermo Craig.
PU15) Privada de Miguel Álvarado, 73560 Cuetzalán - DG: Lic. Angel Diez Mendoza.
PU16) Av.15 de Mayo 2939, Frac.Las Hadas, 72070 Puebla - DG: Lic.Rafael Cañedo B.
PU17) Calle Tres Sur 107-3, 72000 Puebla - DG: J.Grajales S.
PU18) Calle Tres Sur 107-5, 72000 Puebla - DG: Ramon Bojalil B.
PU19) Calle 25 Sur 304-4, La Paz, 72160 Puebla - DG: Lic. Salvador J. Martinez y Duarte.
PU20) 1 Sur 108, Desp.307, Col.Centro, 75700 Tehuacan - Ger: Gerardo Cuevas Rodríguez.

QE00) QUERETARO (Qro.)
QE01) Fray Sebastián de Aparicio 28, Esq.F.Pedro de Gante, 76030 Querétaro. DG: Ing.Marco António H.Contreras S.
QE02) Carr.San Juan del Rio-Xilitla km 181, Col.San José, 76340 Jalpan - DG: Mariano Ugalde García.
QE03) Av.Juárez 38 Pte, 76800 San Juan del Río - DG: E.Morales G.
QE04) Paseo del Prado 102-401, 76039 Villa del Pueblito - DG: E.Lara-Martinez L.
QE05) Zaragoza 15 Pte, Centro, 76000 Querétaro - DG: Emilio Nassar Rodríguez.
QE06) Av.Carrizal 28-F2, Fracc.Ampliación Carrizal, 76030 Querétaro - DG: José Luis Rodríguez A.
QE07) Av.Tecnológico Sur 2, Local 106, Col.Héroes, 76010 Querétaro - DG: Ing.Carlos Caballero M.
QE08) 76000 Querétaro.
QE09) Av.Tecnológico 100, Desp.306-307, Edif.Tec 100, 76000 Querétaro - DG: Raúl Estrada Tsuru.

QR00) QUINTANA ROO (Q.Roo.)
QR01) Ap.506, 77500 Cancún - DG: A.Mateos B.
QR02) Ap.96, 77010 Chetumal - DG: Ranier Aviles S.
QR03) Prol.Av.Heroes 680, 77000 Chetumal - DG: Maria Cristina Perea M.
QR04) Ap.299, 77600 Cozumel - DG: Luis A.Pavia M.
QR05) Ap.13, 77200 Felipe Carillo Puerto - DG: Eduardo Maldonado B.

QR06) Av.Miguel Hidalgo 201, 77000 Chetumal - DG: Eduardo Aguilar Barragan.
QR07) Av.Uxmal s/n, 77500 Cancún - DG: Lic.Sergio E.Cardenas E.
QR08) Rg.92, Manzana 62, Lotes 21-23, Z-7, 77500 Cancún - DG: Lic.Gaston Alegre L.
QR09) Av.Lázaro Cárdenas 46, 77200 Felipe Carillo Puerto.
QR10) Durazno 22, S.M.2-A, Zma 4, 77500 Cancún - DG: Jacobo Alvarez Lima.
QR11) Nader 27, Desp.1043, 77500 Cancún - DG: Benjamin Castro H.

SL00) SAN LUIS POTOSI (S.L.P.)
SL01) Ignacio Montes de Oca 345, Col.Tequisquiapan, 78250 San Luis Potosí - Ger: Laurentino Escamillas.
SL02) Capitan Caldera 315, Col.Tequisquiapan, 78250 San Luis Potosí - DG: Lic.Mª Elena Aguirre de Solis.
SL03) Los Bravo 445 Altos, 78000 San Luis Potosí - DG: Sra Hilda Briones R.
SL04) Hidalgo 7-A, 79600 Río Verde - DG: M.I.Martínez S.
SL05) Ap.160, 79050 Cd.Valles - DG: Lucía Lastra C.
SL06) Ap.80, 78700 Matehuala - DG: Guillermo Benitez C.
SL07) Ap.18, 78700 Matehuala - DG: N.Rueda L.
SL08) Universidad de San Luis Potosí, Ap.456, 78001 San Luis Potosí - Gte: M.Carillo G.
SL09) Martín de Turubiartes 500, 79400 Cerritos - DG: S.Ramírez H.
SL10) Av.Himno Nacional 1951, Fracc.Tangamanga, 78269 San Luis Potosí - DG: Ing.Salvador Cazares P.
SL11) Av.Hidalgo 1, Zona Centro, 79000 Cd.Valles - DG: Belem Altomirano I.
SL12) Privada Pemex 3. Barrio San Rafael, 79960 Tamazunchale - DG: Juan Roberto Reyna L.
SL13) Tancanhuitz de Santos.
SL14) Fausto Nieto 220, 78000 San Luis Potosí - DG: Ing.Sergio Ramirez H.
SL15) Venustiano Carranza 460, 78000 San Luis Potosí - DG: Carlos S.Camacho.
SL16) Ap.36, 79000 Cd.Valles - DG: Rafael Castro Torres.
SL17) Carranza 1408-interior, 78250 San Luis Potosí - DG: Ing. Carlos M.
SL18) Ap.231, 79090 Cd.Valles - DG: Rafael Castro Torres.
SL19) Capitan Caldera 420, Col.Tequisquiapan, 78250 San Luis Potosí - DG: César Augusto Garcés A.

SN00) SINALOA (Sin.)
SN01) Av.Insurgentes Sur N° 334, P2, 80129 Culiacán - DG: Alfonso Millan.
SN02) Ap.61, 81000 Guasave - DG: C.Chávez L.
SN03) Ap.60, 82010 Mazatlán - DG: O.Pérez G.
SN04) Ap.965, 81200 Los Mochis - DG: R.Chávez L.
SN05) Ap.68, 81000 Guasave - DG: Carlos Quiñones A.
SN06) Calz.Insurgentes 334 Sur, P2, 80000 Culiacán - DG: Emilio Salazar C.
SN07) Malecon y Bravo, 80000 Culiacán - DG: Héctor Ramos O.
SN08) Ap.233, 80000 Culiacán - DG: Jesus Diaz N.
SN09) Ap.1, 81200 Los Mochis - DG: Homero Esquerra B.
SN10) Ap.148, 82000 Mazatlán - DG: Julio Ruiz C.
SN11) Juárez 353 Ote, 81200 Los Mochis - DG: Manuel Perez M.
SN12) Ap.120, 80000 Culiacán - DG: D.L.Carillo B.
SN13) Aquiles Serdán 860 Pte, 81200 Los Mochis - DG: Ing. Manuel F.Perez M.
SN14) Universidad de Occidente, Blvd.Macario Gaxiola y Carr. Internacional, 81200 Los Mochis - DG: Lic.R.Irma Peñuelas C.
SN15) Escobedo y Morelos Altos, 81400 Guamuchil - DG: Hector Ramos R.
SN16) Universidad de Sinaloa, Rosales 284 Pte, 80000 Culiacán.
SN17) Hidalgo 408, 82400 Escuinapa - DG: Francisco M.Ramos.
SN18) Hidalgo 775 Pte, 81200 Los Mochis - DG: Orlando Navarro Z.
SN19) Av.Miguel Aleman 619 Ote, 82000 Mazatlán - DG: Esteban Muciño P.
SN20) Ap.35, 82800 Rosario - DG: M.Morales S.
SN21) Ap.422, 82000 Mazatlán - DG: Dan Pablo Xibillé P.
SN23) Insurgentes 334 Sur, Centro Sinaloa, 80129 Culiacán - DG: Pedro F.Guzman M.
SN24) Av.del Mar 80, 82010 Mazatlán - DG: Clemente Valdez V.
SN25) Ap.146, 80000 Culiacán - DG: D.L.Carillo B.
SN26) Ap.113, 80129 Culiacán - DG: Jesus Diaz N.
SN27) Sinaloa 442 Pte, Col.Centro, 81200 Los Mochis - DG: Ing.Javier Carrizales.

SO00) SONORA (Son.)
SO01) Ap.1519, 83000 Hermosillo - DG: J.Ignacio Miranda R.

SO02) Ap.630, 85480 Guaymas - DG: Ing.Fernando María A.
SO03) Ap.58, 85340 Empalme - DG: Rafael Leree.
SO04) Ap.75, 85160 Cd.Obregón - DG: Alicia Leyva Montoya.
SO05) Ap.1817, 83000 Hermosillo - DG: Lic.Emma Lourdes Lopez V.
SO06) Blvd.Rodolfo Elias Calles 252 Ote, 85000 Cd.Obregón - DG: Guillermo Macias V.
SO07) Sinaloa Sur N° 408, 85000 Cd.Obregón - DG: Lic.David Saldana C.
SO08) Av.Morelos y Obregón, Col.Centro, 83600 Caborca - DG: J.de Jesús P.
SO09) Ap.213, 84620 Cananea - DG: Carlos Lara S.
SO10) Ap.226, 85800 Navojoa - DG: J.S.Terminel U.
SO11) Ap.6, 84900 Ures - DG: Francisco Vidal E.
SO12) Heriberto Aja 96, 83000 Hermosillo - DG: Carlos Aparicio R.
SO13) Ap.199, 84000 Nogales - DG: Ing.Justino Muciño.
SO14) Ap.371, 85400 Guaymas - DG: A.A.Padilla.
SO15) Ap.203, 83400 San Luis Río Colorado - DG: Orlando Navarro Z.
SO16) Ap.9, 85800 Navojoa - DG: R.Gómez B.
SO17) Ap.256, 84000 Nogales (or: Box 1472, Nogales, AZ 85628, USA) - DG: Mª del Carmen Guzmán de Ojeda.
SO18) Guerrero y California, Plaza Tutuli, Local E17, 85000 Cd.Obregón - DG: Hector Pegueras P.
SO19) Ap.28, 84200 Agua Prieta - DG: H.Rivera E.
SO20) Ap.7, 84180 Naco - DG: J.M.Franco M.
SO21) Ap.66, 83550 Pto Peñasco - DG: Miguel Angel Tanori.
SO22) Miguel Alemán 668 Norte, 85000 Cd. Obregón - DG: Eduardo Montoya P.
SO23) Ap.44, 84600 Santa Ana - DG: Mario Rochin D.
SO24) Ap.158, 85000 Cd.Obregón - DG: Luis Felipe García de León M.
SO25) Ap.63, 84160 Magdalena - DG: Mario Rochin D.
SO26) Morelos y Calle Obregón, 83600 Caborca - DG: B.de Jesús P.
SO27) Ap.28, 84200 Agua Prieta - DG: Norberto Rivera T.
SO28) Ap.62, 85400 Guaymas - DG: Rafael Leree.
SO29) Rosales y Elias Calles, Desp.4007, Edif.Cremi Pitic, 83000 Hermosillo - DG: Gustavo Astiazaran R.
SO30) Ap.485, 83000 Hermosillo - DG: Fausto Soto S.
SO31) Ap.148, 83400 San Luis Río Colorado - DG: Renato Brassea E.
SO32) Heriberto Aja 96 y Nayarit, 83000 Hermosillo - DG: Carlos Aparicio R.
SO33) Av.13 de Julio 5, 83600 Caborca - DG: Sofia Muciño P.
SO34) Ap.333, 83400 San Luis Río Colorado - DG: Orlando Navarro Z.
SO35) Ap.285, 83000 Hermosillo - DG: Fausto Soto S.
SO36) Ap.199, 84000 Nogales - DG: Justino Muciño P.
SO37) Juarez 33 Ote, 85900 Huatabampo - DG: Guillermo Macias V.
SO38) Ap.75, 85000 Cd.Obregón - DG: Lic.David Saldaña C.
SO40) Veracruz 230 Sur Altos, 85000 Cd.Obregón - DG: Guillermo Macias V.
SO41) Ap.44, 83400 San Luis Río Colorado - DG: Orlando Navarro Z.
SO42) Edif.Leo, Abelardo Rodríguez 180, Desp.45, Col.Centro, 85400 Guaymas - DG: Ing.Javier Carrizales Luna.
SO43) 85800 Navojoa.
SO44) Etchojoa.

TB00) TABASCO (Tab.)
TB01) Paseo de la Ceiba 102, P1, Col.3 de Mayo, 86190 Villahermosa - DG: Moises Dagdug Lutzow.
TB02) J.Alvarez 301, 86000 Villahermosa - DG: E.Lodoza G.
TB03) Ap.270, 86000 Villahermosa - DG: V.Gómez B.
TB04) Calle 28 N° 117, 86900 Tenosique - DG: Jorge Rene Dominguez P.
TB05) N.Saenz 217, 86000 Villahermosa - DG: Lic.Gerardo Gaudiano Peralta.
TB06) Leandro Adriano y Rogelio Ruiz Rojas, 86500 Cárdenas - DG: A.Cálderon L.
TB07) Lerdo de Tejada 217, 86700 Macuspana - DG: M. Huerta M.
TB08) José Julián Dueñas 201, Parque Hidalgo, 86800 Teapa.
TB09) Allende 61, 86400 Huimanguillo - DG: C.Zerecero D.
TB10) Ap.397, 86000 Villahermosa - DG: Amín Simón Habib.
TB11) 86000 Villahermosa.
TB12) Sánchez Mármol 408, 86000 Villahermosa - DG: Oscar Bravo S.
TB13) Nacajuca.
TB14) Paseo Usumacinta, Esq.Ayuntamiento s/n, 86080 Villahermosa - DG: V.Gómez B.
TB15) Calle de la Ceiba 102-3, Primero de Mayo, 86190 Villahermosa - DG: Jorge Dias de Sandi.
TB16) Consitución 1011, Zona Centro, 86000 Villahermosa - Ger: María Inés Gómez.

TM00) TAMAULIPAS (Tamps.)
TM01) Gaspar de la Garza 170 Sur, 87000 Cd.Victoria - DG: B.Zurita M.

TM02) Ap.134, 88500 Reynosa - DG: Srta Naime Salem G.
TM03) Ap.4, 88000 Nuevo Laredo - DG: R.Esparza G.
TM04) Ap.246, 88500 Reynosa - DG: Antonio Gallegos G.
TM05) Ap.79, 89160 Tampico - DG: Ing.Victor J.Flores T.
TM06) Ap.79, 89901 Cd.Mante
TM07) Ap.52, 88500 Reynosa - Dir: Ismael Ramírez Acosta.
TM08) Ap.99, 88000 Nuevo Laredo - DG: Lic.E.Villareal M.
TM09) Ap.735, 87300 Matamoros (or: Box 1708, Brownsville, TX 78522, USA) - DG: Srta Naime Salem Gonzales.
TM10) Ap.414, 89150 Tampico - DG: José Antonio Purón Acevedo.
TM11) Gonzales y Mendoza 747, Col.Centro, 88000 Nuevo Laredo (or: 1510 Calle del Norte, Suite 2, Laredo, TX 78041, USA) DG: Carolina Irenne Villareal de Noguez.
TM12) Diego Acuña, Cd.Tula - DG: Lic.Teodoro Medina L.
TM13) Aquiles Seldan 119 Sur, 89000 Tampico - DG: José Manuel Benitez S.
TM14) Ap.540, 87300 Matamoros - DG: Gerardo Lopez C.
TM15) Ap.232, 88000 Nuevo Laredo - DG: Armando Cortes Delgado.
TM16) Ap.20, 88440 Cd.Camargo - DG: R.Hinojosa G.
TM17) Ap.110, 88000 Nuevo Laredo - DG: Carolina Villareal O.
TM18) Av.Juárez 703 Ote, 89800 Cd.Mante - DG: Sergio Braña A.
TM19) Calle 9 y 10, Blvd.P.Balboa 805-6, 87000 Cd.Victoria - DG: Ismael Vera V.
TM20) Benito Juárez 506-A, Col.Tolteca, 89160 Tampico - DG: Lic.Jorge de Zamacona B.
TM21) Ap.465, 87350 Matamoros - DG: J.Cárdenas G.
TM22) Ap.13, 83000 Cd.Miguel Alemán - DG: R.Hinojosa G.
TM23) Ap.1, 83000 Cd.Miguel Alemán - DG: Dr.Gustavo Gomez P.
TM24) Morelos 2513, Juarez, 88209 Nuevo Laredo - Mgr: Lic. Noé Cuéllar G.
TM25) Zaragoza 85, 87600 San Fernando - DG: Juan Fco.Aguilar G.
TM27) Altamira Calle Principal de Estereos, Carr.Tampico-Gonzalez, Altamira - DG: Lic.Teodoro Medina L.
TM28) Calle 8 y Cuauthémoc 125, Col.Pedro Sosa, 87120 Cd. Victoria - DG: Lic.Teodoro Medina L.
TM29) Ap.134, 87300 Matamoros - DG: Antonio Gallegos E.
TM30) Lázaro Cárdenas 210, Local 19,20 y 21, Col.Centro, 88500 Reynosa - Ger: Lic.Eduardo Echeverría Porras.
TM31) Valentín Gómez Farias 407, Col.Otomí, 89150 Tampico - DG: Carlos G.Cortés García.

TX00) TLAXCALA (Tlax.)
TX01) Av.Juárez Norte 203, 90500 Huamantla - DG: Raul Romero Rivera.
TX02) Calle Uno 420, 90070 Tlaxcala - DG: A.Hernández C.

VE00) VERACRUZ (Ver.)
VE01) Plaza Crystal, Loc.26, 91150 Jalapa - DG: Carlos Ferraez C.
VE02) Av.Tres 425, 94500 Córdoba - DG: Alonso Dominguez D.
VE03) Ap.18, 95400 Cosamaloapan - DG: Arnulfo Aguirre Salamanca.
VE04) Av.Oriente 6 No 261-210, 94300 Orizaba - DG: Alejandro Dominguez F.
VE05) (See DF06).
VE06) Melchor Ocampo 119, P7, Edif.Pazos, 91700 Veracruz - DG: Fco.Girón M.
VE07) Ap.13, 96400 Coatzacoalcos - DG: Jaime Fernandez A.
VE08) Pasaje Revolución 5, 91000 Jalapa - DG: Profr.Dionisio Castillo Z.
VE09) B.Franklin 4, 91700 Veracruz - DG: Alberto Ferraez C.
VE10) Juárez 100, 96700 Minatitlán - DG: I.Lazaro D.
VE11) Blvd.Adolfo Ruiz y Heriberto Kehoe, Obrera, 93260 Poza Rica - DG: E.Rojano S.
VE12) Ignacio de la Llave 38, 92000 Pánuco - DG: Jorge Lopez L.
VE13) Av.Salvador Díaz Mirón 2625, Esq.Heroico, Col.Militar, 91700 Veracruz - DG: Bernardo Sanchez.
VE14) Ap.4, 93300 Poza Rica - DG: Lic.Antonio Cassameni C.
VE15) Esq.Comunicación, Gabino Gonzales, 92730 Alamo - DG: Ana Luisa Sardo C.
VE16) Humboldt Sur 36, 91270 Perote - DG: Raul C y Luis Manuel Molina.
VE17) Rodríguez Malpica 1414, 96400 Coatzacoalcos - DG: Lic.Omar Bejár Gómez.
VE18) Ap.15, 93400 Papantla - DG: Ing.Enrique Rojano S.
VE19) Av.4 Pte.421-17, 94100 Huatusco - DG: José Luis Oliva M.
VE20) Ap.4, 93300 Poza Rica - DG: Antonio Cassameni G.
VE21) Bravo 1163 N° 201, 91700 Veracruz - DG: Juan de Dios Rodríguez Díaz.
VE22) Eulalio Vela 15, 96700 Minatitlán - DG: María E.González.
VE23) Ap.80, 93600 Martinez de la Torre - DG: Eliza Sainz Vda.de Manterola.

VE24) Av.Juárez 13, P4, 92800 Tuxpam. DG: Lic.Calixto Almazan F.
VE25) Libertad y Morelos 301, 95100 Tierra Blanca - DG: G.Haaz D.
VE26) Constitución 7, 95700 San Andrés Tuxtla - DG: I.Fariña G.
VE27) Plaza Crystal, Loc.20, 91150 Jalapa - DG: Lic. José L.Oliva M.
VE28) Ap.15, 91615 Teocelo - DG: Aaron Mora Ugarte.
VE29) Av.Manuel Avila Camacho 11, Col.Centro, 93550 Gutiérrez Zamora - DG: Luis Felipe Pintado C.
VE30) Ursulo Galván 403, Esq.A.Serdán, 96480 Coatzacoales DG: Romano D'Jacobis G.
VE31) Ap.1, 92060 Tempoal - DG: José António Purón Acevedo.
VE32) Ap.18, 94300 Orizaba - DG: Srta Irma Ortega A.
VE33) Ap.26, 96000 Acayucan - DG: Raymundo Martinez D.
VE34) Av.Libertad 201, 95220 Piedras Negras - DG: María Mora de Alarcon.
VE35) Carr.Nacional 38, 93620 San Rafael - DG: José Manuel M.
VE36) Universidad Veracruzana, Ap.629, 91000 Jalapa - Gte: Rafael Méndez A.
VE37) Jimenez Sur 4306, Entre Alvarado y Perez Abasc, Pascual Ortiz Rubio, 91750 Veracruz - DG: Yolanda Dominguez.
VE38) Calle 8 N° 119, Entre Av. 1 y 3, 94500 Córdoba - DG: Irma Ortega A.
VE39) Zaragoza 105, 93820 Misantla - DG: Eliza Sainz Vda.de Manterola.
VE40) Ap.13, 92600 Huayacocotla - DG: Francisco Ramos Solido. Prgrs in Sp., Otomi, Nahua and Tepehua.
VE41) Fernando Siliceo 801, 91970 Veracruz - DG: Arturo Zorilla M.
VE42) Zongolica.
VE43) Banderas 4, 92800 Tuxpam - DG: Emilio Velasco D.
VE44) Tepetzintla.
VE45) La Antigua Veracruz.
VE46) Av.Vicente Guerrero Sur 202, 96400 Coatzacoalcos - DG: Ing. José María Cassiano.
VE47) Ap.75, 91970 Veracruz - DG: Lic.Arturo Zorilla M.
VE48) Carr.Federal a Cd.Alemán km 38.5, Col.Pemex, 95180 Tierra Blanca - DG: Luis Ponce de León.

YU00) YUCATAN (Yuc.)
YU01) Calle 62 N° 465, Entre 53 y 55, 97000 Mérida - DG: Luís A.Rivas A.
YU02) Ap.152, 97001 Mérida - DG: Sra Cristina Cantillo J.
YU03) Ap.217, 97001 Mérida - DG: Jorge Carlos Iglesias B.
YU04) Ap.21, 97780 Valladolid - DG: Mario Gonzales P.
YU05) Calle 56 N° 447, 97000 Mérida - DG: R.Rodríguez R.
YU06) Ap.5, 97700 Tizimín - DG: José Laris R.
YU07) Ap.78, 97320 Progreso - DG: José Laris R.
YU08) Universidad Autonoma de Yucatán, Ap.63-B, 97000 Mérida - DG: Irving Berlin Villafaña F.
YU09) Calle 42 N° 194-A, Entre 35 y 37, 97000 Valladolid - DG: DG: Luís A.Rivas A.
YU10) Hacienda Aranjuea, 97930 Peto - DG: Manuel Angel Diez Mendoza.
YU11) Calle 60 N° 451, Entre 49 y 51, 97000 Mérida - DG: Jorge Carlos Iglesias B.

ZC00) ZACATECAS (Zac.)
ZC01) Carr.Panamericana km 724.6, 99030 Fresnillo - DG: Jaime Torres Gallegos.
ZC02) Esq.Heroe de Nacozari y Gaspar López 715, Jardines de la Fuentes, 99600 Jalpa (or: Av.Universidad 1001, P6, Desp.614, Edif.Plaza Bosques, 20127 Aguascalientes, Ags) - DG: A. Rodríguez Z.
ZC03) Ap.41, 99000 Fresnillo - DG: Jesus Bonilla E.
ZS04) Belisario Dominguez 303, 99000 Fresnillo - DG: José Antonio Casas T.
ZC05) Dr.Gilberto Delgadillo 18-3, 98400 Río Grande - DG: E.Llamas Saucedo.
ZC06) Calle Julián Aguirre 110, Col.Lomas de la Soledad, 98040 Zacatecas - DG: J.Jesús Jáquez Acuña.
ZC08) Ap.198, 99300 Jerez de García Salinas - DG: Jesus Avila Femat.
ZC09) Ap.324, 99000 Fresnillo - DG: Rafael Murillo M.
ZC10) Ocampo 22, Centro, 99600 Jalpa - DG: G.Díaz A.
ZC11) Josefa Ortíz de Dominguez 51, P3, 99700 Tlaltenango - DG: Everardo Garza G.
ZC12) Ap.78, 98000 Zacatecas - DG: Rafael Murillo M.
ZC13) Estudiante y Fernando Villalpando s/n, Int.1, 98000 Zacatecas - DG: Arnoldo Rodríguez Zermeño.

ST. PIERRE ET MIQUELON (French)

L.T: UTC - 3h (Su: UTC - 2h) — **Pop:** 6.432 — **Radios:** 3.000 (est.) —

Pr.L: French — **E.C:** 50Hz, 220V — **ITU:** SPM.

RADIO TÉLÉVISION FRANÇAISE D'OUTRE MER (RFO)
☒ B.P. 4227, F-97500 St. Pierre et Miquelon. ☎ +508 413824.
① FRTSPM 020428 QN — **L.P:** Dir: Joseph Edern. Dir. Tec: Daniel Beugin. Head of N: Jacques Barret.
MW: St. Pierre 1375kHz 20kW — **FM:** St. Pierre 97.9MHz 10W, 99.9MHz 0.5kW — Miquelon 98.9MHz 50W.
D.PRGR: 0930-0230. **Rel. France-Inter:** 0230-0930. **N (local):** 1000, 1530, 2200. **(Rel. France-Inter:)** 1100, 1200, 1300, 1400, 1800, 1900 — **ANN:** "Ici RFO, Station de Saint-Pierre et Miquelon".
IS: La Marseillaise — **V.** by QSL-card. Rec. acc.

R. Atlantique: 102.1MHz (rel. R. France Internationale).

UNITED STATES OF AMERICA

LT: See World Time Table — **Pop:** 267.840.000 — **Radios** 520.000.000 — **Pr.L:** English — **E.C:** 60Hz, 110V — **ITU:** USA.

THE FEDERAL COMMUNICATIONS COMMISSION
(FCC) Government licensing agency for broadcast stations.
☒ 1919 M Street, N.W., Washington, D.C. 20554. ☎ +1 (202) 632-6460. ① 7108220160. ▤ +1 (202) 653-5402.
Web: http://www.fcc.gov — **L.P:** Chmn: Reed Hunt.
The FCC is an independent Federal agency composed of five Commissioners appointed by the President with the consent of the Senate. One of its major activities is the general regulation of broadcasting, visual as well as aural. This regulation may be divided into three phases: 1) The allocation of spectrum space to the different types of broadcast services; 2) consideration of applications to build and operate individual stations; 3) regulation of their operations.
Call letters: International agreement provides for the identification of the radio station by the first letter or first two letters of its assigned call signal and for this purpose apportions the alphabet among different nations. USA nations use the initial letters K, N and W exclusively, and part of the A series. For broadcast stations calls beginning with K are assigned to stations west of the Mississippi River and in the territories and possessions, while W is assigned to broadcast stations east of the Mississippi. Calls consist of four letters, to which FM or TV may be added with a hyphen for FM or TV stations.
Note: A few exceptions with stations east of the Mississippi using a "K" callsign and stations west of the Mississippi using a "W" callsign will be noted. These are old callsigns that were assigned before the geographical division was introduced and are retained by special permission. Similarly some very old callsigns using only three letters may be retained by the stations that once were assigned these callsigns.
Stations: More than 12,000 st's are operating (MW + FM).
Major networks providing AM station programming

ABC RADIO NETWORKS
☒825 7th Ave, 4th fl, New York, NY 10019. ☎ +1 (212) 456-1777. ▤ +1 (212) 456-1899 — Pres: David Kantor — News services, sports. ☒ 13725 Montfort Dr, Dallas, TX 75240. ☎ +1 (972) 991-9200. ▤ +1 (972) 991-1071 — News, features, Satellite Music Network.
Web: http://www.abcradionet.com/

CBS RADIO NETWORKS
(CNNRadio, CNN News, NBC Radio Network, Westwood One)
☒ 1675 Broadway, New York, N.Y. 10019. ☎ +1 (212) 641-2000. ▤ +1 (212) 247-0393.

JONES SATELLITE NETWORK
☒ 8250 South Akron Street, #205, Englewood, CO 80112. ☎ +1 (303) 784-8700. ▤ +1 (303) 784-8612 — Music formats.

NATIONAL PUBLIC RADIO
Non-commercial nationwide system.
☒ 635 Massachusetts Ave, NW, Washington, D.C. 20001. ☎ +1 (202) 414-2000. ▤ +1 (202) 424-3329. **Web:** http://www.npr.org — **L.P:** Pres: Douglas J. Bennet — News, features.

USA RADIO NETWORK
☒ 2290 Springlake Rd, Suite 107, Dallas, TX 75234. ☎ +1 (972) 484-3900. ▤ +1 (972) 241-6826 — News, features, sports.

Web: http://www.night.net/usasched.html

Other National Organizations

NATIONAL ASSOCIATION OF BROADCASTERS
☒ 1771 N St. NW, Washington, D.C. 20036. ☎ +1 (202) 429-5300. ▤ +1 (202) 429-5343.

COMMERCIAL STATIONS
Especially in multi-station markets, radio stations concentrate their programming to appeal to a given segment of the population or a given listening taste. Many stations devote their entire broadcast day to news and/or talk programs. Others specialize in hit music (adult contemporary, top 40), country music, oldies (e.g. hits of the fifties and sixties), big bands/standards, black (urban contemporary, jazz, rhythm & blues), religious services and inspirational music, classical music, ethnic programs (e.g. programs entirely in Spanish).
Today satellites are widely used for the distribution of news/talk and musical program. Numerous such networks are in operation. Stations making extensive use of network programming may have only one local identification per hour, usually on top of the hour. The former "clear channel" stations today have a protected area extending to 700 miles. Outside this area the frequencies are also used by other stations. A few stations have been granted temporary licenses for increased powers to combat interference from neighboring countries. Many daytime stations may now operate after local sunset using low or very low powers. With the large decrease in AM listening in favour of FM an increasing number of AM stations go off the air for a longer or shorter period due to economical difficulties. The latest development is that stations on the so called regional channels, previously limited to 5kW power, may now apply for up to 50kW limited only by the required protection of other stations. Relaxed ownership rules have allowed groups of co-owned stations to form in larger markets with the group stations often broadcasting from a common studio address.
Mediumwaves: Stations less than 10kW are not mentioned. For a more comprehensive list of stations please refer to the North America medium wave table of the back of this book.
*=Presunrise Service Authority. A daytime station with a PSA may operate from 0600 local time to local sunrise with low power when local sunrise occurs later than 0600.
Expanded MW band: The MW band is being expanded to 1700kHz. A number of stations are now using this band and many more construction permits have been granted.
Call letters: On stations relaying an FM sister station or carrying certain network programs other call letters than those listed below may frequently be heard. All stations are required to announce their actual call letters and city of licence once per hour as close as possible to the top of the hour.
Stereo: Some stations are operating in AM stereo using the Motorola C-QUAM system.
Obs: Calendar days refer to local time.
Powers: Where a second power is mentioned this refers to early morning/late afternoon operation for daytime stations. For full time stations this appears as the middle one of three power levels.
Transmission hours: For full time stations not operating 24h the number of daily transmission hours is shown in the frequency list with full details given in the address section. D=local daytime only.
N.B: Hawaii and Alaska are listed separately — St. 353) Gov.

Call 540kHz	kW	H. of tr.		Call	kW	H. of tr.
			14)	WWJZ 540kHz	50/1	24h
1)	KNOB	25/0.3	24h	15)	WFNC	10/1	24h
2)	KIEZ	10/0.5		16)	WGOC	10/0.81	24h
3)	WQTM	50	24h	17)	WCRV	50/0.48	24h
	550kHz				650kHz		
4)	WDUN	10/2.5	18h	18)	KSTE	25/10	
5)	WASG	25/0.14	18h	19)	WNMT	10/1	24h
	570kHz			20)	WSM	50	24h
6)	WMCA	50/30	24h	21)	KMTI	10/0.9	24h
	580kHz				660kHz		
7)	WTCM	15/0.8	24h	22)	WDLT	10/0.85	
	610kHz			23)	KTNN	50	24h
8)	WIOD	10	24h	24)	KGDP	50/7	18h
	620kHz			25)	WBHR	10/0.25	24h
9)	WTMJ	50/10	24h	26)	WFAN	50	24h
	640kHz			27)	KZTU	10/0.07	
10)	KFI	50	24h	28)	WESC	50/10	D*
11	)WGST	50/1	24h	29)	KSKY	10/0.66	24h
12)	WNNZ	50/1	24h	30)	KAPS	10/1	24h
13)	KGVW	10/1	24h		670kHz		

Call	kW	H. of tr.
31) KLTT	50/1.4	
32) WWFE	50/1	24h
33) KBOI	50	24h
34) WMAQ	50	24h
35) WVNS	20/5	D
680kHz		
36) KNBR	50	24h
37) WCNN	50/15	24h
38) WCBM	10/5	24h
39) WRKO	50	24h
40) WDBC	10/1	20h
41) WPTF	50	24h
42) WJCE	10/5	24h
43) KKYX	50/10	24h
44) WCAW	10/0.25	24h
690 kHz		
45) WJOX	50/0.5	24h
46) WOKV	50/10	24h
47) KGGF	10/5	19h
48) WTIX	10/5	24h
49) KHEY	10	24h
50) WZAP	10/0.01	18h
700kHz		
51) KMJY	10/1	16h
52) WLW	50	24h
53) KGRV	25/0.5	24h
53a) KSEV	15/1	20h
54) KWLW	50/1	24h
710kHz		
55) KUET	50/4.2	F.PI.
56) KFIA	25/1	
57) KDIS	50/10	24h
58) KNUS	25/2.7	24h
59) WAQI	50	24h
60) KEEL	50/5	24h
61) KCMO	10/5	24h
62) WOR	50	24h
63) KXMR	10/4	F.PI.
64) KGNC	10	24h
65) WFNR	10	D*
66) KIRO	50	24h
67) WDSM	10/5	24h
720kHz		
68) WRZN	10/0.25	18h
69) WMXY	10	D
70) WGN	50	24h
71) KDWN	50	24h
71a) WQTH	50/0.5	F.PI.
72) WGCR	10/3	D
73) KSAH	10/0.89	24h
730kHz		
74) KBSU	15/0.5	18h
740kHz		
75) WMSP	10/0.34	
76) KBRT	10/0.11	D*
77) KCBS	50	24h
78) WWNZ	50	24h
79) WGSM	25/0.04	24h
80) WPAQ	10/1	D
81) KRMG	50/25	24h
82) KTRH	50	24h
750kHz		
83) WSB	50	24h
84) KERR	50/1	24h
85) KMMJ	10	D*
86) KXL	50/20	24h
87) KAMA	10/1	24h
88) KOAL	10/6.8	24h
760kHz		
89) KMTL	10	D
90) KFMB	50/5	24h
91) KTLK	50/1	24h
92) WBDN	10/5	24h
93) WVNE	25/8.6	D
94) WJR	50	24h
95) WCHP	35	D
96) KTKR	50/1	24h
770kHz		
97) WVNN	10/0.25	24h

Call	kW	H. of tr.
98) KCBC	50/1	24h
99) WWCN	10/1	24h
100) KATL	10/1	24h
101) KKOB	50	24h
102) WABC	50	24h
103) KPBC	10/1	24h
66) KNWX	50/5	24h
780kHz		
104) WBBM	50	24h
105) KKOH	50	24h
106) WWOL	10	D
790kHz		
107) WQXI	28/1	24h
810kHz		
108) WNSI	50/0.5	20h
109) KGO	50	24h
110) WSJC	50/0.5	
111) WHB	50/5	24h
112) WGY	50	24h
113) KBHB	25	D
114) KTBI	50/23	D*
820kHz		
115) WZTM	50/1	19h
116) WNYC	10/1	16h
117) WBAP	50	24h
118) WGGM	10/1	24h
119) KGNW	50/5	24h
830kHz		
120) KFLT	50/1	24h
121) KNCO	25/10	19h
122) KPLS	50/20	24h
123) WCCO	50	24h
124) KOTC	10	D
125) WXII	50/10	24h
126) KUYO	10	D
840kHz		
127) WBHY	10	D
128) WHGH	10	D
129) WHAS	50	24h
130) KXNT	50/25	24h
131) WCTG	50	D
132) KMAX	10/0.28	F.PI.
850kHz		
133) WYDE	9/1	24h
134) KAHS	25/1	24h
91) KOA	50	24h
39) WEEI	50	24h
136) WWJC	10	D
137) WQST	10	D
138) WRBZ	10/5	24h
139) WRMR	50/5	24h
140) WJAC	10	24h
141) WJBZ	50	D*
142) KEYH	10/0.18	D*
143) WTAR	50/25	24h
144) KHHO	10/1	24h
860kHz		
145) KTRB	50/10	24h
146) KKOW	10/5	24h
27) KPAM	50/10	
148) WTEL	10/0.5	D+
149) KCNR	10/3/0.224h	
150) WOAY	10/0.01	16h
870kHz		
151) WQRX	10/4.7	D
152) KIEV	20/3	24h
153) WWL	50	24h
154) WLAM	10/1	24h
155) WKAR	10	D
156) KPRM	25/1	17h
157) KLSQ	10/1	24h
158) WPWT	10	D
159) KFLD	10/0.25	24h
880kHz		
160) KGHT	50/0.22	
161) KKMC	10/1	24h
162) KJJR	10/0.5	24h
163) KRVN	50	24h
163a) KHAC	10/0.43	16h

Call	kW	H. of tr.
164) WCBS	50	24h
165) WRFD	23/6.1	D
53a) KJOJ	10/1	24h
167) KIXI	50/10	24h
168) WMEQ	10/0.21	24h
890kHz		
169) WLS	50	24h
170) WBPS	25/3.4	24h
171) WQIS	10	D
172) WBAJ	10.4/0.9D	F.PI.
173) KVOZ	10/1	24h
174) KDXU	10	24h
175) WKNV	10	D F.PI.
900kHz		
176) WJWL	10/1	24h
177) KTIS	25/0.3	24h
940kHz		
145) KFRE	50	24h
179) WINZ	50	24h
180) WMAC	50/10	24h
181) KXTK	10/5	24h
182) WYLD	10/0.5	24h
183) WCPC	50/0.25	17h
184) KXUX	10/0.06	17h
185) KBRE	10/0.04	24h
186) WKGM	10/3.1	24h
950kHz		
186a) WWJ	12/50	24h
187) KJR	50	24h
970kHz		
188) WJMX	10/3	24h
980kHz		
189) KFWB	50	24h
190) KCTY	10/0.48	24h
191) WTEM	50/5	24h
990kHz		
192) KTKT	10/1	24h
193) WHOO	50/0.25	24h
194) WEEB	10/0.03	24h
195) WZZD	50/10	24h
196) WVSC	50/0.1	18h
197) WALE	50/5	19h
141) WNOX	10	24h
199) KWAM	10/0.45	24h
200) KWFT	10/1	24h
1000kHz		
201) WDJL	10/5	D
202) WMVP	50	24h
203) KKIM	10/0.04	
204) KXRB	10	D*
205) KOMO	50	24h
1010kHz		
207) KIQI	10/1.5	
208) KSIR	25/0.28	
209) WQYK	50/5	24h
211) WIOJ	10/0.14	24h
212) WGUN	50/0.08	19h
213) WMOX	10/1	19h
214) KXEN	50/0.5	
215) WINS	50	24h
216) WFGW	50/0.5	D
217) KBBW	10/2.5	17h
218) KTUR	50/3/0.224h	
1020kHz		
219) KTNQ	50	24h
220) WJEP	10	D
221) WNTK	10	D
222) KCKN	50	20h
223) KDKA	50	24h
224) WRIX	10/3	D
1030kHz		
225) KEVT	10/1	
226) KFAY	10/1	24h
226a) KIOQ	50/1	F.PI.
227) WONQ	10/1.7	24h
228) WWGB	50	D
229) WBZ	50	24h
230) WCTS	50/1	19h
231) WNOW	10	D

Call	kW	H. of tr.
232) WFTK	50	D
233) KLLU	10/0.63	24h
235) WSFZ	50/10/1	24h
236) KCTA	50	D+
237a) WBGS	10/0.29	D
237) KMAS	10/1	24h
239) KTWO	50	24h
1040kHz		
240) WJNO	25/1.1	24h
241) WHO	50	24h
242) WSGH	10/2.5	D
243) WSKE	10/4	D
244) WQBB	10/3	D
1050kHz		
245) KTRJ	10	D
36) KTCT	50/10	24h
247) WTKA	10/0.5	24h
248) KMTA	10	D
249) WEVD	50	24h
1060kHz		
251) KLMO	10	D*
252) WAMT	10/5	24h
253) KBGN	10	D*
254) WLNO	50/5	24h
256) KYW	50	24h
257) KGFX	10/1	24h
258) KBNB	10	D*
259) KFNA	10	D*
260) KIJN	10	D
261) KKDS	10/1	16h
1070kHz		
262) WAPI	50/5	24h
263) KNX	50	24h
264) WFRF	10	D*
265) WIBC	50/10	24h
266) KFDI	50	24h
268) WNCT	10	24h
269) WKOK	10/1	24h
270) WHYZ	50	D*
271) WFLI	50/2.5	24h
199) WDIA	50/5	24h
273) KENR	10	24h
274) WIWS	10	D*
275) WTSO	10/5	24h
1080kHz		
276) KSCO	10/5	19h
277) WTIC	50	24h
278) WVCG	50/20	24h
279) WFIV	10	D
280) WFTD	10	D*
281) KVNI	10/1	24h
129) WKJK	10/1	24h
283) KOTK	50/10	24h
284) WPGR	50/25	D*
285) KRLD	50	24h
286) KRPX	10/5	D*
1090kHz		
287) KAAY	50	24h
288) KAJK	10	D
289) KMXA	50/0.5	24h
291) WBAL	50	24h
292) WKCV	10/1.8	D
293) KRPM	50	24h
1100kHz		
294) KCCF	50/1	
295) KFAX	50	24h
296) KNZZ	50/10	19h
297) WCGA	10	D
298) WHLI	10	D
299) WTAM	50	24h
300) KDRY	11/1	18h
1110kHz		
301) WBCA	10/2.5	D
303) KRLA	50/20	24h
304) WTIS	10	D*
305) WJML	10	D*
306) KFAB	50	24h
307) WBT	50	24h
308) KBND	10/5	24h

#	Call	kW	H. of tr.
309)	WCKO	50	D
1120kHz			
311)	WUST	20/3	D
312)	WBNM	10/2.5	D*
313)	KMOX	50	24h
314)	KPNW	50	24h
315)	KANN	10/1	24h
1130kHz			
316)	KSDO	10	24h
60)	KWKH	50	24h
319)	WDFN	50/10	24h
320)	KFAN	50/30	24h
321)	WBBR	50	24h
63)	KBMR	10	D
322)	KTMR	10	D
323)	WISN	50/10	24h
1140kHz			
324)	KCMJ	10/2.5	24h
325)	KHTK	50	24h
59)	WQBA	50/10	24h
327)	KGEM	10	24h
130)	KSFN	10/2.5	24h
330)	KSOO	10/5	24h
331)	WBXR	15/7.5	D
332)	WRVA	50	24h
333)	KZMQ	10	D
1150kHz			
134)	KXTA	50/44	24h
333a)	KSEN	10/5	24h
1160kHz			
334)	WERD	10/0.16	24h
335)	WSCR	50/4.2	24h
336)	WSKW	10/0.73	24h
337)	WVNJ	20/2.5	18h
338)	WJFJ	10/0.5	24h
339)	WCCS	10/1	24h
340)	WAMB	50/1	24h
341)	KENS	10/1	24h
342)	KSL	50	24h
1170kHz			
343)	WACV	10/1	24h
344)	KLOK	50/5	24h
345)	KCBQ	50/1.5	24h
347)	KVOO	50	24h
348)	WLGO	10/2.5	D
349)	KPUG	10/5	24h
350)	WWVA	50	24h
1180kHz			
351)	KYET	10/0.25	16h
352)	KERI	50/10	24h
353)	VOA	50	17.5h
354)	WJNT	50/0.5	24h
355)	KOFI	50/10	24h
356)	KOIL	25/1	24h
357)	WHAM	50	24h
358)	WMYT	10	D
359)	WHJM	10/2.6	D
360)	KGOL	50/1	17h
1190kHz			
361)	KORG	11/1.4	24h
362)	WGKA	10	D
363)	WOWO	50/9.8	24h
364)	WANN	10	D
365)	KXKS	10/0.02	D*
366)	WLIB	10/30	
367)	KEX	50	24h
368)	KLUV	50/5	24h
369)	WBDY	10	D
1200kHz			
372)	KOQI	25/1	F.Pl.
373)	WTLQ	10/1	24h
374)	WLXX	10/2.5	24h
376)	WKOX	50	24h
377)	WCHB	50/0.7	
378)	WGNY	10/2.5	24h
379)	WSML	10/1	24h
380)	KFNW	10/0.7	24h
381)	WRKK	10/0.25	24h
340)	WQDQ	10/0.5	24h

#	Call	kW	H. of tr.
96)	WOAI	50	24h
1210kHz			
382)	WQLS	10/5	D
383)	KQTL	10/1	24h
385)	KPRZ	20/5	24h
386)	WNMA	25/2.5	24h
387)	WDGR	10/2.5	D
388)	WSKR	10/1	19h
389)	WJZZ	50/2.5	D
390)	KGYN	10	24h
391)	WPHT	50	24h
392)	WGSF	10/0.25	24h
393)	KUBR	10/5	
394)	KONY	10/0.25	24h
395)	KBSG	27.5/10	24h
396)	KREW	10/1	16h
397)	KLDI	10/1	
1220kHz			
299)	WKNR	50	24h
1250kHz			
400)	WREN	15/5	
1260kHz			
401)	WWJQ	10/1	
401a)	WMIH	10/5	24h
1280kHz			
401b)	WADO	50/5	24h
402)	KDYL	10/0.6	24h
403)	WNAM	20/5	24h
1290kHz			
403a)	WBZT	10/4.9	24h
1300kHz			
404)	KAPL	20/5	19h
405)	WNQM	50/5	24h
119)	KKOL	35/16	24h
1310kHz			
407)	KMKY	20	24h
1320kHz			
408)	WDER	10/1	16h
1430kHz			
409)	KEZW	10/5	24h
410)	KLO	10/5	24h
1460kHz			
411)	KDON	10	
412)	KENO	10/0.65	24h
1480kHz			
413)	WHBC	15/5	24h
1500kHz			
414)	KSJX	10/5	24h
415)	WTOP	50	24h
416)	WLQV	50/5	24h
417)	KSTP	50	24h
1510kHz			
418)	KFNN	22/0.1	24h
419)	KIRV	10	D*
420)	KMSL	10/1	24h
421)	KDKO	10/1.3	24h
422)	WWJY	10/5	18h
423)	WNRB	50	24h
424)	KCTE	10/1	D*
424a)	WOLF	50/1	F.Pl.
425)	WLAC	50	24h
426)	KLLB	10	D
427)	KGA	50	24h
428)	WAUK	10	D*
1520kHz			
429)	KTRO	10/1	24h
430)	KDYS	10/0.5	24h
431)	WGAM	10	D
432)	WQWQ	10/1	
433)	KOLM	10/0.8	
434)	WWKB	50	24h
435)	KOMA	50	24h
436)	KKSN	50/15	24h
1530kHz			
437)	KHPY	50/6	D
18)	KFBK	50	24h
439)	WOBS	50	D
440)	WTTI	10	D
441)	WRPM	10/1	D*
442)	WRTP	10	D

#	Call	kW	H. of tr.
52)	WSAI	50	24h
444)	KQQA	10/0.99	D*
445)	KGBT	50/10	24h
1540kHz			
446)	KASA	10	D*
447)	KCTD	50/10	24h
448)	KMBY	10	24h
449)	KXEL	50	24h
450)	WDCD	50	24h
451)	WOGR	10	D*
452)	WNWR	50	D*
453)	WTBI	10/1	D
454)	KZMP	35/0.89	19h
455)	WREJ	10	D*
1550kHz			
456)	WLOR	50	D
457)	KUAT	50	D
458)	KYCY	50/10	24h
459)	KQXI	10/4.75	
460)	WAMA	10/0.13	18h
461)	WAZX	50/0.5	24h
462)	KVKI	10/0.5	18h
463)	WNTN	10	D
464)	KQWB	10/5	24h
465)	WBSC	10/5	19h
466)	WKQV	10/0.5	24h
467)	KMRI	10/0.5	24h
468)	WKBA	10	D
469)	KNTR	50/10	18h
470)	KSVY	10/2.5	16h
471)	KVAN	10	24h
1560kHz			
472)	KNZR	25/10	24h
473)	WRHC	45/4.4	24h
474)	WPAD	10/5	24h
476)	WQEW	50	24h
477)	WAGL	50	D*
478)	KKAA	10/5	20h
1580kHz			
479)	KCWW	50	24h
480)	KBLA	50	24h
481)	KWYD	10	D
482)	WTCL	10	D
483)	WSRF	10/5	24h
484)	WTKT	10	D
485)	WPGC	50/0.4	
486)	KDZZ	10/0.5	24h
487)	WLIM	55/0.5	16h
488)	WCTJ	10	D
1600kHz			
489)	WWRL	25/5	24h
1610kHz			
490)	K...	10/1	F.Pl.
1620kHz			
491)	WPHG	10/1	24h
492)	KSMH	10/1	F.Pl.
493)	KBLI	10/1	F.Pl.
494)	WJVA	10/1	F.Pl.
356)	KAZP	10/1	F.Pl.
496)	KNNT	10/1	F.Pl.
497)	K...	10/1	F.Pl.

#	Call	kW	H. of tr.
498)	WAZG	10/1	F.Pl.
499)	KAZW	10/1	F.Pl.
501)	KYIZ	10/1	24h
1630kHz			
502)	KBEG	10/1	F.Pl.
503)	WAWX	10/1	F.Pl.
504)	KCJK	10/1	F.Pl.
505)	KBCM	10/1	F.Pl.
506)	KKWY	10/1	F.Pl.
1640kHz			
507)	KDIA	10/1	24h
508)	WLHJ	10/1	F.Pl.
509)	KBFQ	10/1	F.Pl.
510)	KKJY	10/1	24h
512)	K...	10/1	F.Pl.
513)	WKSH	10/1	16h
1650kHz			
514)	KHFS	10/1	F.Pl.
515)	KKTR	10/0.93	24h
516)	K...	10/1	F.Pl.
517)	WAZJ	10/1	F.Pl.
518)	K...	10/1	F.Pl.
519)	K...	8.5/0.85	F.Pl.
520)	WHKT	10/1	F.Pl.
1660kHz			
521)	KAXW	10/1	F.Pl.
522)	WMIB	10/1	F.Pl.
400)	K...	10/1	F.Pl.
524)	W...	10/1	F.Pl.
525)	WBAH	10/1	24h
526)	W...	10/1	F.Pl.
464)	KQJD	10/1	F.Pl.
528)	KAXY	10/1	F.Pl.
529)	KXOL	10/1	
1670kHz			
530)	KAZT	10/1	F.Pl.
531)	WNML	10/1	24h
532)	WAWR	10/1	F.Pl.
533)	WTDY	10/1	24h
1680kHz			
534)	KAVT	10/1	F.Pl.
535)	K...	10/1	F.Pl.
536)	W...	10/1	F.Pl.
537)	WTTM	10/1	F.Pl.
119)	KAZJ	10/1	F.Pl.
1690kHz			
539)	KSXX	10/1	F.Pl.
459)	KAYK	10/1	24h
540)	WAZD	10/1	F.Pl.
541)	WHTE	10/1	F.Pl.
542)	WMDM	10/1	24h
1700kHz			
543)	W...	10/1	F.Pl.
386)	WCMQ	10/1	24h
544)	KBGG	10/1	24h
545)	WAYU	10/1	F.Pl.
546)	KCHT	10/1	F.Pl.
547)	KQXX	10/1	F.Pl.
548)	KTBK	10/1	F.Pl.
549)	WEZI	10/1	F.Pl.

Addresses and other information

GM=General Manager. PD=Program Director. CE=Chief Engineer. SM=Station Manager. Op's=Operations. ADC=Adult contemporary music. AS=Adult Standards. BB=Big Bands. BM=Beautiful Music. C&W=Country & Western music. CWM=Modern country music. CHR=Contemporary hit radio (Top 40). MoR=Middle-of-the-Road music. MYL=Music of Your Life (popular music of the 20's, 30's, 40's). NOS=Nostalgia. N/t=News, talk. OLD=Oldies (Popular music of the 50's, 60's, 70's). UC=Urban contemporary. Cont=Contemporary. Pop=Popular.

Verifications: The details of the verification policies often vary and thus are omitted here. Many stations will verify correct reports of reception over exceptional distances. Rp in the form of mint stamps and a self adhesive return adress label will be appreciated by most stations.

NB: In addresses N/S/W/E = North/South/West/East, St. = Street. 1) 7807 Girard Ave, #200, La Jolla, CA —— Classical music. (Lic. to

Costa Mesa, CA.)
2) 651 Cannery Row, Monterey, CA 93940 — Talk. (Lic. to Carmel Valley, CA.)
3) 2500 Maitland Center Pkw, #401, Maitland, FL 32751 — Sports. (Lic. to Pine Hills, FL.)
4) P.O. Box 10, Gainesville, GA 30503 — N/t.
??5) 1318 S. Main St, Atmore, AL 36502 — GM: Nathan Martin. C&W.
6) 201 Rt. 17 N, Ste. 601, Rutherford, NJ 07070 — GM: Joe Davis. PD: Carl Miller. Rlg. (Lic. to New York, NY.)
7) P.O. Box 472, Traverse City, MI 49685. **Web:** http://www.wtcmradio.com — GM: Ross Biederman. PD: Jack O'Malley. CE: Jim Sofonia. News, sports.
8) 194 NW 187th St, Miami, FL 33169 — N/t.
9) 720 E. Capitol Dr, Milwaukee, WI 53212 — GM: Jon Schweitzer. CE: Randy Price. N/t, sports.
10) 610 S. Ardmore Ave, Los Angeles, CA 90005. **E-mail:** KFlam640@aol.com — GM: Howard Neal. PD: George Oliva. CE: Marvin Collins. Talk.
11) 1819 Peachtree Rd. NE, #700, Atlanta, GA 30309 — GM: Pat McDonnell. PD: Nancy Zantak. CE: Mike Lawing. N/t.
12) 1500 Main St, Springfield, MA 01115 — GM: Curt Hahn. PD: Greg Johnson. CE: Bill Weeks. News/sports. (Lic. to Westfield, MA.)
13) 2050 Amsterdam Rd, Belgrade, MT 59714 — GM: Mark Brashear. CE: Dale A. Heidner. Rlg.
14) P.O. Box 640, Mt. Holly, NJ 08060 — AS.
15) P.O. Box 35297, Fayetteville, NC 28303 — GM: John Dawson. PD: Jeff Thompson. CE: Terry Jordan. N/t.
16) 640 Radio Way, #286-B, Blountville, TN 37617 — GM, CE: Mitch Sandidge. C&W (N. at :55).
17) 4990 Poplar Ave, Memphis, TN 38117 — GM: Mrs. Sunny Caldwell. CE: Dave Beveridge. Rlg talk. (Lic. to Collierville, TN)
18) 1440 Ethan Way, #200, Sacramento, CA 95825 — GM: Mark McCoy. PD: Ken Kohl. CE: Mark Stennett. KFBK: N/t. KSTE: Talk. (KSTE licensed to Rancho Cordova, CA.)
19) P.O. Box 1060, Hibbing, MN 55746 — CE: Danny Klaysmat. C&W. Lic. to Nashwauk, MN.)
20) 2644 McGavock Pike, Nashville, TN 37214 — GM: Bob Meyer. PD: Kyle Cantrell. CE: Hugh Hickerson. C&W (N: :55, :30).
21) P.O. Box 40, Manti, UT 84642 — Mgr: Douglas Barton. C&W.
22) 1204 Dauphin St, Mobile, AL 36604 — Black Gospel. (Lic. to Fairhope, AL.)
23) P.O. Box 2569, Window Rock, AZ 86515 — GM: Tazbah McCullah. PD: Michael D. Moore. CE: Abram Meria. Navajo 1200-1900, C&W.
24) 3070 Skyway Dr, # 501, Santa Maria, CA 93455. **E-mail:** eaglerock@aol.com — GM: Sherwood Patterson. PD: Dan Franklyn. Rlg. 1300-0700. (Lic. to Orcutt, CA.)
25) P.O. Box 366, Sauk Rapids, MN 56379 — GM: Herb Hoppe. PD: Scott Kloehn. CE: Terry J.J. Brehn. Childrens prgrs.
26) 3412 36th St, Astoria, NY 11106 — GM: Joel Hollander. PD: Mark Chernoff. CE: E. Knapp. Sports. (Lic. to New York, NY.)
27) 10209 S.E. Division St, #100, Portland, OR 97266 — Rlg. (KPAM lic. to Troutdale, OR, KZTU lic. to Eugene, OR.)
28) P.O. Box 660, Greenville, SC 29602 — GM: Allen Power. PD: Jeff Blake. CE: Don Gowens. C&W, sports.
29) 4144 N. Central Expressway, #266, Dallas, TX 75204 — GM: Ted Sauceman. PD: Kelly Williams. CE: Mike Vanhooser. Rlg. (Lic. to Balch Springs, TX.)
30) P.O. Box 70, Mt. Vernon, WA 98273 — GM: James Keane. C&W.
31) 2150 W. 29th Ave, #300, Denver, CO 80211 — Rlg. (Lic. to Commerce City, CO.)
32) 330 SW 27th Ave, #207, Miami, FL 33135 — GM: Jorge A. Rodríguez. PD: Emilio Milian, Sr. CE: "Eddy" Rodríguez. Spanish "R. Fé".
33) P.O. Box 1280, St, Boise, ID 83701. **E-mail:** kboi@worldnet.com — GM: Bob Rosenthal. CE: Willis Frahm. News, ADC.
34) 455 N. Cityfront Plaza, 6th Flr, Chicago, IL 60611 — GM: Weezie Kramer. PD: Chris Witting. CE: Gregory Scott Davis. News, sports.
35) 410 Briar Hill Rd, Norfolk, VA 23502 — Op's Mgr: Pat Murphy. News. (Lic to Claremont, VA.)
36) 55 Hawthorne St, #1100, San Francisco, CA 94105 — GM: Tony Salvadore. PD: Bob Agnew. CE: Bill Ruck. KNBR: Sports, talk. KTCT: Sports. (KTCT lic. to San Mateo, CA.)
37) 1601 W. Peachtree St. NE, Atlanta, GA 30309 — GM: Marc Morgan. CE: Bob Mayben. News. (Lic. to North Atlanta, GA.)
38) 11 Music Fair Rd, Owings Mills, MD 21117 — GM: Nicholas Mangione, Jr.. PD: Sean Casey. CE: Dwight Weller. Talk. (Lic. to

Baltimore, MD.)
39) 116 Huntington Ave, Boston, MA 02116 — GM: Brad Murray. PD: Al Mayers. CE: James F. Robinson. WRKO: Talk. WEEI: Sports.
40) 604 Ludington St, Escanaba, MI 49829 — GM: Alice Sabuco. PD: Kevin Scannell. CE: Bob Haslow. Full sce. light ADC. 1000-0600.
41) 3012 Highwoods Blvd, #200, Raleigh, NC 27604 — GM: George King. PD: Bryce Wilson. CE: Gary Leibish. N/t.
42) 5904 Ridgeway Ctr. Parkway, Memphis, TN 38120 — GM: Curt Peterson. PD: Tim Kirkland. CE: Bob Mayben. UC-OLD.
43) 8122 Datapoint Dr, #500, San Antonio, TX 78229 — GM: Ben Reed. PD: Carl Becker. CE: Paul K. Reynolds. C&W-OLD.
44) 1111 Virginia Street East, Charleston, WV 25301 — GM: Daler Miller. AS.
45) 244 Goodwin Crest Dr, #300, Birmingham, AL 35209 — GM: Davis H. Hawkins. PD: William Jenkins, Jr. CE: Frank Giardina. All sports.
46) 6869 Lenox Ave, Jacksonville, FL 32205 — GM: Mark Schwartz. Talk.
47) P.O. Box 1087, Coffeyville, KS 67337 — GM: John Leonard. CE: Bob Cauthon. N/t, sports. 1100 (Sun 1200)-0603.
48) 3313 Kingman St, Metairie, LA 70006 — GM: Robert Namer. CE: Barry Chickini. Talk. (Lic. to New Orleans, LA.)
49) 2419 N. Piedras, El Paso, TX 79930 — GM: Bill Struck. CE: Patrick Parks. Sports.
50) P.O. Box 369, Bristol, VA 24203 — GM: R. Al Morris. PD: Tommy Tester. CE: John Faniola. Rlg.
51) P.O. Box 1740, Old Town, ID 83822. GM's: Jim and Helen Stargel. CWM days, CHR nights. (1400-0600). (Lic. to Newport, WA.)
52) 1111 St. Gregory St, Cincinnati, OH 45202 — GM: J.D. Martin (WLW), Peter Zolnowski (WSAI). CE: Al Kenyon. WLW: Talk, sports, C&W. WSAI: AS, rlg.
53) 196 S.E. Main St, Winston, OR 97496 — GM: John Scott Welch. CE: Paul Brown. Rlg.
53a) 11767 Katy Freeway, #1170, Houston, TX 77079 — GM: Dan Patrick. N/t, sports, OLD. (KSEV lic. to Tomball, TX, KJOJ lic. to Conroe, TX.)
54) 2801 S. Decker Lake Dr, Salt Lake City, UT 84119 — GM: Rick Porter. CE: Troy German. Easy Listening. (Lic. to North Salt Lake.)
55) Black Canyon City, AZ.
56) 1425 River Park Dr, #520, Sacramento 95815, CA — GM: Joe Cruz. Rlg. (Lic. to Carmichael, CA.)
57) 3321 S. La Cienaga Blvd, Los Angeles, CA 90016 — GM: Bill Sommers. PD: Len Weiner. CE: Steve Blodgett. Childrens prgr's.
58) 3131 S. Vaughn Way, # 601, Aurora, CO 80014 — GM: Caroline Bernhardt. PD: Mason Lewis. N/t. (Lic. to Denver, CO.)
59) 2828 Coral Way, #102, Miami, FL 33145 — GM: Claudia Puig. CE: C. Fernández. Spanish n/t. WAQI = "R. Mambi", WQBA = "La Cubanísima".
60) 6341 Westport Ave, Shreveport, LA 71129 — GM: Bill Fry. CE: Armando Gonzales. KEEL: N/t, KWKH N/t, sports, C&W.
61) 4935 Belinder Rd, Westwood, KS 66205 — GM: Bob Zuroweste. PD: Brian Wilson. CE: Richard Myers. Talk. (Lic. to Kansas City, KS.)
62) 1440 Broadway, 24th Flr, New York, NY 10018. **Web:** http://www.wor710.com — GM: Robert Bruno. PD: David Bernstein. N/t.
63) 3500 E. Rosser, Bismarck, ND 58501 — GM: Alvin L. Anderson. PD: Charlie Williams. C&W (KBMR).
64) P.O. Box 710, Amarillo, TX 79189 — GM: Bob Russell. PD: Bob Reed. CE: John Harrold. N/t.
65) 485 Tower Rd, Christiansburg, VA 24073 — GM: Karen Travis. PD: Bob Travis. CE: J.J. Largen. N/t. (Lic. to Blacksburg, VA.)
66) 1820 Eastlake Ave, Seattle, WA 98102. **Web:** http://www.kiro710.com — GM: Richard Carlson. PD: Tom Clendening. CE: Buzz Anderson. KIRO: News, sports, info. KNWX: News.
67) 715 E. Central Entrance, Duluth, MN 55811 — GM: Debra Messer. CE: P. Maki. Talk, sports. (Lic. to Superior, WI.)
68) 3988 N. Roscoe Rd, Hernando, FL 34442 — GM: Patricia Brinker. PD: Steven Cox. CE: David Terry. BBD. 1100-0500.
69) P.O. Box 1114, La Grange, GA 30241 — GM: Richard Smith. P.D: Stacy Powers. Urban cont. (Lic. to Hogansville, GA.)
70) 435 N. Michigan Ave, Chicago, IL 60611. **Web:** http://www.wgnradio.com — GM: Steve Carver. PD: Mary June Rose. CE: James J. Carollo. N/t.
71) 1 Main St, Las Vegas, NV 89101 — Mgr: Claire Reis. PD: A.J. Williams. CE: M. Messina. N/t.
71a) Hanover, NH.
72) P.O. Box 720, Pisgah Forest, NC 28768 — GM: Randy C. Barton. PD: Larry Spears. CE: Jobie Sprinkle. Rlg.

73) 1777 N.E. Loop 410, #803, San Antonio, TX 78217 — GM: Dennis G. Roberts. PD: Oscar Rios. CE: Bill Haberer. Spanish CHR (R. Festival). (Lic. to Universal City, TX.)

74) 1910 University Dr, Boise, ID 83725 — GM. Jim Paluzzi. CE: Ralph Hogan. Jazz, Spanish.

75) P.O. Box 4999, Montgomery, AL 36103 — GM: Christy Patrick. CE: Larry Wilkins. Sports.

76) 3183-D Airway Ave, Costa Mesa, CA 92626 — GM/PD: Ed Personius. CE: Gary Bloodworth. Rlg talk. (Lic. to Avalon, CA.)

77) 1 Embarcadero Ctr, #3200, San Francisco, CA 94111 — GM: Frank Oxarart. News Dir: Ed Cavagnaro. CE: Shingo Kamada. All news.

78) 2500 Maitland Ctr. Parkway, #401, Maitland, FL 32751 — GM: Jenny Sue Rhoades. News, sports. (Lic. to Orlando, FL.)

79) 900 Walt Whitman Rd, Melville, NY 11747 — GM: Ron Gold. Childrens prgrs.

80) P.O. Box 907, Mt. Airy, NC 27030 — GM. & PD: Kelly Epperson. CE: Ralph D. Epperson. Bluegrass.

81) 7136 S.Yale, #500, Tulsa, OK 74136 — GM: Chuck Browning. PD: Michael Delgiorno. CE: Wayne Smith. N/t.

82) 510 Lovett Blvd, Houston, TX 77006 — GM: Laura Morris. SM: Bill van Rysdam. CE: Errol R. Coker. News, sports, talk.

83) 1601 W. Peachtree St. NE, Atlanta, GA 30309 — GM: Marc W. Morgan. PD: Greg Moceri. CE: Ron Wilson. N/t.

84) 581 N. Reservoir Rd, Polson, MT 59860 — GM: Dennis Anderson. C&W.

85) P.O. Box 4907, Grand Island, NE 68802 — GM: Shaun Schleis. Spanish.

86) 0234 S.W. Bancroft St, Portland, OR 97201 — GM: Tom McNamara. PD: Mike Dirkx. CE: Larry Wilson. N/t, sports.

87) 2211 E. Missouri Ave, #S-300, El Paso, TX 79903 — GM Kathy Clark. CE: David Stuart. Spanish "Puro tejano".

88) P.O. Box 875, Price, UT 84501 — GM, CE: Thomas B. Anderson. PD: Keith Mason. Talk, sports.

89) P.O. Box 6460, N. Little Rock, AR 72124 — Owner: George V. Domerese. CE: R. Loewy. Rlg. (Lic. to Sherwood, AR.)

90) 7677 Engineer Rd, San Diego, CA 92111 — GM: Bob Bolinger. PD: M. Larson. CE: John D. Weigand. N/t.

91) 1300 Lawrence St, Ste. 1300, Denver, CO 80204 — GM: Lee Larsen. PD: Robin Bertolucci. CE: Jan Chadwell. KTLK: Talk. KOA: N/t. (KTLK lic. to Thornton, CO.)

92) 13577 Feather Sound Dr. #680, Clearwater, FL 34622 — Spanish "Mega 760". (Lic. to Brandon, FL.)

93) 70 James St, #140, Worcester, MA 01603. **E-mail:** wvne@aol.com **Web:** http://www.gocin.com/lifechangingradio — GM: William A. Blount. PD: Steve Tuzeneu. CE: Lincoln W. Hubbard. Rlg. (Lic. to Leicester, MA.)

94) 2100 Fisher Bldg, Detroit, MI 48202. **E-mail:** wjrradio@aol.com **Web:** http://www.760wjr.com — GM: Michael Fezzey. PD: Al Mayers. CE: Ed Buterbaugh. N/t, C&W.

95) P.O. Box 888, Champlain, NY 12919 — GM: Teri Billiter. PD: Brandi Lloyd. CE: Mike Mathieu. Rlg. French: Sun-Fri 1900-2100, 2200-2230.

96) 6222 NW Interstate 10, San Antonio, TX 78201. **Web:** http://www.woai.com — GM: Betty Kocurek. PD: Andrew Ashwood. CE: Dan Walthers. Sports.

97) P.O. Box 389, Athens, AL 35612 — GM: W. Dunnavant. PD: Dave Stone. CE: C. Sampieri. N/t.

98) 10948 Cleveland Ave, Oakdale, CA 95361 — GM: Donald Crawford. Op's Dir: Rich Woodruff. Rlg. (Lic. to Riverbank, CA.)

99) 20125 S. Tamiani Trail, Estero, FL 33928 — GM: Bruce Beasley. CE: Richard Gallo. Talk, sports. (Lic. to North Fort Myers, FL.)

100) P.O. Box 700, Miles City, MT 59301 — GM, PD. & CE: Donald L. Richard. ADC.

101) 500 4th St. NW, Albuquerque, NM 87102 — GM: Paul Ehlis. PD: Bob Shomper. Contract CE: Mike Langner. Talk.

102) 2 Penn Plaza, 17th Fl, New York, NY 10121 — GM: Mitch Dolan. PD: John Mainelli. CE: Bill Krause. Talk.

103) 3201 Royalty Row, Irving, TX 75062 — GM: Paul Niven. PD: Don Evans. CE: Mike Vanhooser. Rlg. (Lic. to Garland, TX.)

104) 630 N. McClurg Court, Chicago, IL 60611 — GM: Rod Zimmerman. News Dir./PD: Chris Berry. CE: Mark Williams. All news.

105) 595 E. Plumb Lane, Reno, NV 89502 — GM: Leonard Smart. PD: Dan Mason. CE: Martin Stabbert. N/t.

106) 1381 W. Main St, Forest City, NC 28043. **Web:** http://www.wwol780.com — SM: Rev. Wade H. Huntley. PD: Ray Davis. CE: Julius Blanton. Rlg.

107) 3340 Peachtree Rd. NE, Atlanta, GA 30326 – sports.

108) P.O. Box 1118, Jacksonville, AL 36365 — Talk.

109) 900 Front St, San Francisco, CA 94111 — GM: M. Luckoff. Op's Dir: J. Swanson. CE: Bruce Schirmer. N/t.

110) Magee, MS 39111. Silent.

111) 1600 Genessee, #925, Kansas City, MO 64102 — GM: Mike Carter. CE: Chris Ostrander. C&W, Farm.

112) One Washington Sq, Albany, NY 12205. **E-mail:** wgy@aol.com **Web:** http://www.wgy.com — GM: Michael T. Whalen. PD: Tom Parker. CE: Bob Blanchard. Talk. (Lic. to Schenectady, NY.)

113) P.O. Box 99, Sturgis, SD 57785 — GM: Dana Caldwell. PD: Dean Kinney. CE: Steve Naeve. C&W-OLD.

114) P.O. Box 734, Ephrata, WA 98823 — Rlg.

115) 11300 4th St. N, #318, St. Petersburg, FL 33716 — CE: Shannon Murdoch. Sports. 1000-0500. (Lic. to Largo, FL.)

116) 1 Centre St, New York, NY 10007 — Dir: Arthur Cohen. CE: Steve Cellum. N/t. 1000-0300.

117) 2221 Lamar Blvd, #400, Arlington, TX 76006 — Pres: John Hare. PD: Ted Stecker. CE: Clay Steely. Talk, C&W. (Lic. to Fort Worth, TX.)

118) P.O. Box 676, Chester, VA 23831 — GM: Paul Scott. CE: Jeff Loughridge. Rlg.

119) 2815 2nd Ave, #550, Seattle, WA 98121 — GM: Richard M. Ulrich. PD: Kevin Manna. CE: James Dalke. Rlg.

120) P.O. Box 36868, Tucson, AZ 85740. **E-mail:** kflt@flr.org **Web:** http://www.flr.org — GM: Lee Escobedo. PD: Dave Ficere. CE: Randy Howard. Rlg.

121) 1255 E. Main St, #A, Grass Valley, CA 95945 – N/t.

122) 1592 N. Batavia St, #1, Orange, CA 92667 — Variety.

123) 625 2nd Ave. S, Minneapolis, MN 55402 — GM: James Gustafson. PD: Jon Quick. CE: J. Miller. Talk, sports.

124) P.O. Box 271, Kennett, MO 63857 — GM: Jeff Wheeler. PD: Mike Scallorns. CE: Larry Anthony. C&W.

125) P.O. Box 11847, Winston-Salem, NC 27106 — News. (Lic. to Eden, NC.)

126) P.O. Box 50607, Evansville, WY 82605 — GM: Steve Stumbo. PD: Aaron M. Remington. Rlg.

127) P.O. Box 1328, Mobile, AL 36633 — GM: Wilbur Goforth. PD: Stephen Goforth. CE: Steve Riggs. Rlg talk.

128) P.O. Box 2218, Thomasville, GA 31799 — GM: Sheila Baeti. PD: James Sherman. CE: James Devane. Black Gospel.

129) P.O. Box 1084, Louisville, KY 40201 — GM: Mark Thomas. PD: Rick Belcher. CE: Charles Strickland. WHAS: Talk. WKJK: NOS.

130) P.O. Box 14805, Las Vegas, NV 89114. **Web:** http://www.kluc.com — GM: Tom Humm. PD: Alan Eisenson (KSFN), Cat Thomas (KXNT). CE: Tracy Teagarden. KXNT: N/t, KSFN: Sports. (Lic. to North Las Vegas.)

131) P.O. Box 23840, Columbia, SC 29223 — Talk.

132) P.O. Box 710, Colfax, WA 99111 — GM: Robert Hauser. Talk.

133) 244 Goodwin Crest Dr, #G-126, Birmingham, AL 35209 — GM: Jerry Voyles. CE: Jim Gray. Childrens prgrs.

134) 3400 Riverside Dr, #800, Burbank, CA 91505 — CE: Mike Callaghan. Sports. (KAHS lic. to Thousand Oaks, CA. KXTA lic. to Los Angeles, CA.)

136) 1120 E. McCuen St, Duluth, MN 55808 — GM. & PD: Ted Elm. CE: John Talcott. Gospel.

137) 18844 Hwy. 80 E, Forest, MS 39074 — Rlg.

138) 5000 Falls of The Nuese, #308, Raleigh, NC 27609 — GM: Brian Maloney. PD: John Low. CE: Jon Hardee. Talk, sports.

139) One Radio Lane, Cleveland, OH 44114 — GM: Chris Maduri. PD: Jim Davis. CE: Ted Alexander. MYL.

140) 109 Plaza Dr, Johnstown, PA 15905 — GM: Michael F. Brosig, Sr. CE: James J. Boxler. OLD.

141) 4711 Old Kingston Pike, Knoxville, TN 37919 — GM: Bobby Denton. CE: Milton Jones. WIOL: N/t, WNOX: Talk, sports.

142) 1980 Post Oak Blvd, #1500, Houston, TX 77036 — GM: Tom Castro. CE: Robert Tindle. Spanish "La Ranchera".

143) 999 Waterside Dr, #500, Norfolk, VA 23510 — GM: Robert Sinclair. PD: Tony Macrini. CE: Joe Hardin. Talk, sports.

144) 950 Pacific Ave, #1200, Tacoma, WA 98402 — GM: Michele Grosenick. Sports.

145) 5087 E. McKinley Ave, Fresno, CA 93727 — GM: John F. Carpenter. PD (KFRE): Ed Monson. CE: Steve Dresser. KTRB: News. KFRE: Talk.

146) 1162 E. Hwy 126, Pittsburg, KS 66762 — GM: Lance Sayler. PD: Bob Capps. CE: T. Fast. C&W.

148) 555 City Line Ave, #300, Bala-Cynwyd, PA 19004 — GM: Raul G. Lahee. PD: Rafael Grullon. CE: Bill Stallman. Spanish "R. Tropical". 1200-2400.

149) 434 Bearcat Dr, Salt Lake City, UT 84115 — GM: Pete Benedetti. CE: Ritchie Bauer. Childrens prgr's.

150) P.O. Box 140, Oak Hill, WV 25901 — GM: Gene Ellison. PD:

Steve Bush. CE: Mark Smith. Rlg. 1100-0300.

151) P.O. Box 309, Valley Head, AL 35989 — GM: Evan Stone. CE: Timothy H. Dobson. ADC/EZL.

152) 5900 San Fernando Road, Glendale, CA 91202 — GM: Fred S. Beaton. PD: Dick Sinclair. CE: Hal Williams. N/t, rlg.

153) 1450 Poydras St, #440, New Orleans, LA 70112 — GM: Johnny Andrews. PD: Dave McNamara. CE: Robert L. Dunn. N/t, C&W.

154) 912 Washington St, Auburn, ME 04210 — GM: Ron Frizell. PD: Mac Dickson. CE: Bob Perry. AS. (Lic. to Gorham, ME.)

155) 283 Comm. Arts Bldg, Michigan State University, East Lansing, MI 48824-1212 — Mgr: Steve K. Meuche. PD: Curt Gilleo. CE: Harold Beer. N/t.

156) P.O. Box 49, Park Rapids, MN 56470 — GM: Ed P. De LaHunt Jr. C&W, talk.

157) 6767 W. Tropicana Ave, #102, Las Vegas, NV 89103 — GM: José Valle. PD: Roberto Ibarra. CE: Gordy Alsum. Mexican "La Super Q". (Lic. to Laughlin, NV.)

158) P.O. Box 2061, Bristol, TN 37621 — GM: Kenneth Hill. Talk. (Lic. to Colonial Heights, TN.)

159) 2621 W. A St, Pasco, WA 99301 — GM: Kevin Souhrada. PD: Tim O'Rourke. CE: Ronald S. Sweatte. Sports.

160) 1000 Warden Rd, North Little Rock, AR 72120 — GM: Jim Schmidt. CE: David Jones. Rlg. (Lic. to Sheridan, AR.)

161) 8 E. Alisal St, Ste. 501, Salinas, CA 93901 — GM, PD: John Dick. CE: Ronald Warren. Rlg. Spanish: W0000-0200, Sun 0300-0800. (Lic. to Gonzales, CA.)

162) Box 5409, Kalispell, MT 59903 — GM: Jim Paulson. Talk. (Lic. to Whitefish, MT.)

163) P.O. Box 880, Lexington, NE 68850. **E-mail:** krvnam@krvn.com **Web:** http://www.krvn.com — GM: Eric Brown. PD: Craig Larson. CE: Vern Killion. C&W.

163a) P.O. Box 9090, Window Rock, AZ 86515 — Rlg, Ethnic. (Lic. to Tse Bonito, NM.)

164) 51 W. 52nd St, New York, NY 10019 — GM: Dan Griffin. News Dir: Harvey Nagler. CE: Ed Schwartz. All news.

165) 8101 N. High St, #360, Columbus, OH 43235. — GM: Dan Craig. PD: Glenn Moore. Rlg.

167) 12011 NE 1st, #206, Bellevue, WA 98005 — GM: Marc Caye. CE: George Bisso. Standards. (Lic. to Mercer Island, WA.)

168) P.O. Box 880, Menomonie, WI 54751 — Mgr: Rick Hencley. N/t, C&W.

169) 190 N. State St, Chicago, IL 60601 — GM: Zemira Jones. PD: Drew Hayes, CE: Warren Schultz. Talk.

170) 197 8th St, Suite 500, Charlestown Navy Yard, MA 02129 — GM: Maury Warshauer. CE: Grady Motes. Sports, ethnical, brokered.

171) 51 Victory Rd, Laurel, MS 39440 — GM: Pete Carpey. CE: Glen A. Musgrove. AS.

172) Blythewood, SC. New, not on air.

173) P.O. Box 252, McAllen, TX 78501 — GM: Eloy Bernal. CE: Art Trevino. Spanish rlg. (Lic. to Laredo, TX.)

174) 750 W. Ridgeview Dr, St. George, UT 84770 — GM: Don Shelline. PD: Dave Hart. CE: Jeff Ward. N/t, sports.

175) P.O. Box 889, Blacksburg, VA 24063 — Rlg. (Lic. To Fairlawn, VA.)

176) 701 N. DuPont Hwy, Georgetown, DE 19947 — GM: Dan Lankford. PD: Cepth Michaels. CE: Terry Dalton. AS.

177) 3003 N. Snelling Ave, St. Paul, MN 55113 — GM: Don Rupp. CE: Rod Thannum. Rlg. (Lic. to Minneapolis, MN.)

179) 194 NW 187 St, Miami, FL 33169. **E-mail:** news@940winz.com **Web:** http://www.940winz.com — GM: Ronna Woulfe. PD: Peter Bolger. CE: George Butch. News.

180) P.O. Box 900, Macon, GA 31202 — GM: Douglas Grimm. PD: Don King. CE: W. Sowell. N/t, sports.

181) 1416 Locust, Des Moines, IA 50305 — GM: Phil Hoover. CE: J. Kosobucki. Talk.

182) 2228 Gravier St, New Orleans, LA 70119 — GM: Earnest James. PD: Steven Ross. CE: J.P. Robillard. UC, Gospel.

183) 1189 Hwy. 15 N, Houston, MS 38851 — GM: Robin H. Mathis. PD: Melanie Munlin. CE: Olen Booth. Rlg. 1100-0410.

184) P.O. Box 5068, Bend, OR 97708 — GM: Norm Louvau. PD: Kit Carson. CE: James Boyd. AS. 1100-0400.

185) P.O. Box 858, Cedar City, UT 84721 — GM: Art Challis. OLD.

186) P.O. Box 339, Smithfield, VA 23451 — Mgr: Larry W. Cobb. CE: H. Wood. Cont. Rlg.

186a) 16550 West 9 Mile Rd, Southfield, MI 48086 — News. (Lic. to Detroit, MI.)

187) 190 Queen Anne Ave. N. #100, Seattle, WA 98109 — GM: Michael O'Shea. Sports.

188) P.O. Box 103000, Florence, SC 29501 — GM: Philip Abdelnor. Sports, talk.

189) 6230 Yucca St, Los Angeles, CA 90028 — GM: Roger Nadel. CE: Richard A. Rudman. News.

190) 517 S. Main St, #201, Salinas, CA 93901. **E-mail:** radsuprema@aol.com — GM: Robert Dahlstrom. PD: Vicente Romero. Spanish "La Mexicana".

191) 11300 Rockville Pike, #707, Rockville, MD 20852 — Sports. (Lic. to Washington, DC.)

192) 1920 W. Copper Pl, Tucson, AZ 85745 — GM: Jim Cooley. PD: Larry Miles. CE: Rob McDonald. News.

193) 200 S. Orange Ave, #2240, Orlando, FL 32801. **Web:** http://www.whtq.com — GM: Debbie Morel. PD: J. T. Stevens. CE: Steve Flutier. AS.

194) P.O. Box 1855, Southern Pines, NC 28388 — GM: Richard McCarthy. CE: Jim Davis. N/t.

195) 117 Ridge Pike, Lafayette Hill, PA 19444 — GM: Russ Whitnar. PD: Carl Dean. CE: Stu Engelke. Cont. Christian. (Lic. to Philadelphia, PA.)

196) 2046 Husband Rd, Somerset, PA 15501 — GM: Michael Brosig. CE: Harold Showman. OLD.

197) 1185 N. Main St, Providence, RI 02904 — GM: Francis Battaglia. Talk, ethnic. 1100-0500. (Lic. to Greenville, RI.)

199) 112 Union Ave, Memphis, TN 38103 — GM: Bruce Demps. PD (WDIA): R. O'Jay. CE: A. Pendleton. KWAM: Gospel. WDIA: UC.

200) P.O. Box 787, Wichita Falls, TX 76307 — Silent.

201) 6420 Springfield Rd, Huntsville, AL 35806 — GM: Walter Peavy. CE: Carl Sampieri. Rlg.

202) 875 N. Michigan Ave, #3750, Chicago, IL 60611 — GM: John Cravens. SM: Greg Solk. CE: Tom Knauss. Sports, talk.

203) 300 San Mateo NE, #1000, Albuquerque, NM 87108 — GM: Jamie Randall. PD: John Lehman. CE: Bill Pace. Cont. Christian.

204) 3205 S. Meadow, Sioux Falls, SD 57106 — GM: Don Jacobs. PD: Charlie Walker. CE: Tony Randall. CWM.

205) 1809 7th Ave, #200, Seattle, WA 98101 — GM: J. Shannon Sweatte. PD: Paul Duckworth. CE: George Anderson. N/t.

207) 2601 Mission St, San Francisco, CA 94110 — GM: Richard Ferdinand. CE: John Scherer. Spanish talk "R. Centro".

208) 231 Main St, Ft. Morgan, CO 80701 — GM: Kevin Shaffer. N/t, farm. (Lic. to Brush, CO.)

209) 9450 Koger Blvd, St. Petersburg, FL 33702 — GM: Jay Miller. PD: Beecher Martin. CE: Fred Berry. N/t. (Lic. to Seffner, FL.)

211) 14286-19 Beach Blvd, #343, Jacksonville Beach, FL 32250 — GM: Charles McHan. PD: Amy McHan. CE: Jerry Smith. AC/OLD.

212) 2901 Mountain Industrial Blvd, Tucker, GA 30084 — GM: Fred J. Webb. PD: Chris Edmonds. CE: Bill Loudermilk. Rlg, health. 1100-0500. (Lic. to Atlanta, GA.)

213) P.O. Box 5184, Meridian, MS 39302 — GM: Eddie Smith. PD: Noel Adcock. CE: Jim Johnson. Talk, sports. 1100-0600.

214) P.O. Box 8085, Mitchell, IL 62040 — GM. & PD: Dan Allen. CE: Robert B. Niekamp. Rlg. (Lic. to Festus-St. Louis, MO.)

215) 888 7th Ave, 11th Flr, New York, NY 10106 — GM: Scott Herman. PD: Mark Mason. CE: Mark Olkowski. All news.

216) P.O. Box 159, Black Mountain, NC 28711 — GM: Edna Edwards. PD: Colin O'Brien. CE: Tim Neese. Rlg.

217) 1019 Washington St, Waco, TX 76701 — GM & PD: Steve Williams. CE: Dave Fricker. Rlg. 1200-0500.

218) 353 E. 200 S, #200, Salt Lake City, UT 84104 — Pres: Fred Bond. All news. (Lic. to Tooele, UT.)

219) 1645 N. Vine St, #200, Los Angeles, CA 90028 — GM: Richard Heftel. PD: David Gleason. CE: G. Ogonowski. Spanish talk.

220) P.O. Box 90, Thomasville, GA 31799. **Web:** http://www.lifelineministries.com — GM: Jimmy Keyton. PD: Bob Ferrell. CE: Clyde Scott. Cont. Rlg. (Lic. to Ochlocknee, GA.)

221) P.O. Box 2295, New London, NH 03257. **E-mail:** wntk1020@aol.com **Web:** http://www.wntk.com — GM: Robert L. Vinikoor. PD: Sheila Vinikoor. CE: Russ McAllister. N/t. (Lic. to Newport, NH.)

222) P.O. Box 670, Roswell, NM 88202 — GM: John Dunn. PD: Wayne Kirpatrick. CE: Donald Niccum. Light ADC, talk, rlg. Spanish 0530-0700.

223) One Gateway Center, Pittsburgh, PA 15222 — GM: Brian Whittemore. PD: Diane Cridland. CE: Mel Check. N/t.

224) Watson Village, Anderson, SC 29624 — GM: John Woodson. CE: Bob Bierman. Rlg. (Lic. to Homeland Park, SC.)

225) 5115 S. 12th Ave, Tucson, AZ 85706 — GM: Mosés Hernández. Spanish, rlg. (Lic. to Cortaro, AZ.)

226) P.O. Box 878, Fayetteville, AR 72702 — GM: Brett Hash. CE: Zeb Huffmaster. Talk, sports. (Lic. to Farmington, AR.)

226a) Folsom, CA.

227) 1033 Semoran Blvd, Ste. 253, Casselberry, FL 32707 — GM & CE: Paul Gamache. PD: Agustin Galarza. Spanish. (Lic. to Oviedo, FL.)

228) 5210 Auth Rd, 6th Fl, Suitland, MD 20746 — Rlg (Lic. to Indian Head, MD.)

229) 1170 Soldiers Field Rd, Boston, MA 02134 — GM: Ted Jordan. PD: Peter Casey. CE: Mark Mauvilian. N/t.

230) 1250 W. Broadway, Minneapolis, MN 55411. **E-mail:** DW9944@aol.com — GM. & CE: Dennis Whitehead. Rlg. 1130-0600.

231) P.O. Box 23509, Charlotte, NC 28227 — GM: Russ Jones. CE Mike Miranda. Rlg, Spanish. (Lic. to Mint Hill, NC.)

232) 707 Leon Street, Durham, NC 27704 — GM: Larry Cobb. CE: J.D. Watson. Rlg. (Lic. to Wake Forest, NC.)

233) P.O. Box 168, Reedsport, OR 97467 — GM: Robert Ratter. CE: Robin O'Kelley. C&W.

235) 3801 Premier Ave, Memphis, TN 38181 — GM: John Rainey. CE: Jerry Campbell. Sports.

236) 1302 S. Brownlee, Corpus Christi, TX 78404 — GM: Bill York. PD: John Boudresu. CE: Paul Easter. Rlg.

237) P.O. Box 760, Shelton, WA 98584 — GM: H.S. Greenberg. PD: Bill Meyer. CE: Tom Trotzer. ADC.

237a) 303 8th St, Point Pleasant, WV 25550 — Rlg.

239) 150 N. Nichols Ave, Casper, WY 82601 — GM: Robert D. Price. PD: Bruce King. CE: Alan R. Esterline. N/t, C&W.

240) 3071 Continental Dr, West Palm Beach, FL 33407 — GM: Steve Lapa. CE: Rick Rieke. N/t. (Lic. to Boynton Beach, FL.)

241) 1801 Grand Ave, Des Moines, IA 50309 — GM: Mark Halverson. PD: Van Harden. CE: K. Erickson. N/t, C&W.

242) 1647 Bloomtown Rd, East Bend, NC 27018 — GM: Rodney Baucom. Rlg. (Lic. to Lewisville, NC.)

243) P.O. Box 187 Everett, PA 15537 — GM: Sandra King. PD: Martin M. Bakner. CE: R. Resconsin. CWM.

244) 1114 W. Clinch Ave, Ste. 1A, Knoxville, TN 37916 — GM: James Staley. PD: Eddy Roy. CE: Bob Wallace — N/t. (Lic. to Powell, TN.)

245) P.O. Box 9775, Bakersfield, CA 93389 — Classical. (Lic. to Frazier Park, CA.)

247) 24 Frank Lloyd Wright Ave, Ann Arbor, MI 48108 — GM: Dave Paulus. PD: Dean Erskine. CE: John Grevers. N/t, sports.

248) P.O. Box 1426, Miles City, MT 59301 — GM: Kevin Senger. CE: Bob Parker. OLD.

249) 333 7th Ave, New York, NY 10001 — GM: Thomas Bird. PD: Jose Northover. CE: Glenn Blewis. Talk.

251) P.O. Box 799, Longmont, CO 80501 — GM: Wm. G. Stewart. PD: Jerry Steffen. CE: Ron Stephens. C&W.

252) 6305 State Road 46, Mims, FL 32754 — GM: Greg Sherlock. Standards, sports. (Lic. to Titusville, FL.)

253) 3303 E. Chicago St, Caldwell, ID 83605 — GM. & CE: N. Wilson. Rlg.

254) 401 Whitney Ave, #160, Gretna, LA 70056 — GM: Gayril Gibson. CE: Doug Booth, Jr. Rlg. (Lic. to New Orleans, LA.)

256) 101 S. Independence Mall E, Philadelphia, PA 19106 — GM: Roy Shapiro. PD: Mark Helms. CE: Janet Kowalczyk. All news.

257) 214 W. Pleasant Dr, Pierre, SD 57501 — GM: Mark Swendsen. PD: Charlie Hale. CE: C. Hesle. C&W, farm.

258) 2105 Anrhony Dr, Tyler, TX 75701 — GM: Sans Hawkins. CE: Wayne Shultice. News. (Lic. to Gilmer, TX.)

259) El Paso, TX. Silent.

260) P.O. Box 458, Farwell, TX 79325 — GM: Gil W. Patschke. Rlg.

261) P.O. Box 57760, Salt Lake City, UT 84157 — GM: Ralph J. Carlson. PD: Steve Carlson. CE: Kenneth Meyer. AS. 1300-0500.

262) 244 Goodwin Crest Dr, #300, Birmingham, AL 35209 — CE: Frank Giardina. News.

263) 6121 Sunset Blvd, Los Angeles, CA 90028 — GM: George Nicholaw. News Dir: Robert Sims. CE: Rick Sietsema. All news.

264) 1310 Paul Russell Rd, Tallahassee, FL 32301 — GM: Scott Beigle. CE: John Mathews. Rlg.

265) 9292 N. Meridian, Indianapolis, IN 46260 — GM: Tom Severino. CE: Ralph Beaty. Talk.

266) P.O. Box 1402, Wichita, KS 67201 — GM: Mike C. Oatman. PD: John Speer. CE: David A. Easley. C&W.

268) P.O. Box 7167, Greenville, NC 27835 — GM: Bruce Simel. PD: W. Baker. CE: Roy McCoury. OLD.

269) P.O. Box 1070, Sunbury, PA 17801 — GM: Joseph A. McGranaghan. PD: Jack Richards. CE: John W. Keller Jr. Talk, sports.

270) 200 N. Highway 25 Bypass, Greenville, SC 29617 — Talk. (Lic. to Sans Souci, SC.)

271) 621 O'Grady Dr, Chattanooga, TN 37419 — GM: Paul White.

PD: Jim Hill. CE: Joe Poteet. Rlg. (Lic. to Lookout Mountain, TN.)

273) 6161 Savoy, #1200, Houston, TX 77036 — GM: Gordon Marcy. CE: John Shadle. Talk, ethnical.

274) P.O. Box 1037, Beaver, WV 25813 — OLD. (Lic. to Beckley, WV.)

275) 5721 Tokay Blvd, Madison, WI 53719 — GM: Jeff Tyler. PD: Dan Masucci. CE: Jeff Zigler. NOS.

276) 2300 Portola Dr, Santa Cruz, CA 95062 — GM: Michael Olson. PD: Rosemary A.L. Chalmers. Talk. 1300-0800.

277) 1 Financial Plaza, Hartford, CT 06103 — GM: Suzanne McDonald. CE: Garnet Drakiotes. N/t.

278) 2828 W. Flagler St, Miami, FL 33135 — GM: Matthew Rowe. PD: Roberto Hernández. CE: Ron Streeter. Mon-Fri mostly Sp. rlg; Sat/Sun Carribbean. (Lic. to Coral Gables.)

279) 1080 Country Blvd, Kissimmee, FL 34741 — GM: Edward Allmon. CE: L. Campbell. Spanish.

280) 774 Roswell St, Marietta, GA 30060 — GM: Rocky Payne. CE: George Pass. Rlg.

281) 101 Lakeside Ave, Coeur d'Alene, ID 83814 — GM: Bruce Deming. PD: Dick Haugen. CE: Steve Franco. ADC full service.

283) 931 SW King Ave, Portland, OR 97205 — GM: Ron Carter. CE: Christopher Cullen. Talk.

284) 960 Penn Ave, #300, Pittsburgh, PA 15222 — GM: Phil Austin. CE: Cliff Bryson. Gospel.

285) 1080 Ballpark Way, Arlington TX 76011 — GM: Jerry Bobo. PD: Richard Walker. Dir. of Eng: Eric J. Disen. N/t. (Lic. to Dallas, TX.)

286) 163 E. 100 N, Price, UT 84501 — GM: Mike Halloran. Sr. PD: Claudine Vu. CE: Jim M. Dart, Jr. OLD.

287) 7123 I-30, #1, Little Rock, AR 72209 — GM, PD: Dianne McArthur. CE: F. McDonald. Rlg.

288) 337 W. 15th St, Eureka, CA 95501 — GM: M. Keith Allgood. CE: M. Householter. ADC, OLD. (Lic. to Fortuna, CA.)

289) 5660 Greenwood Plaza Blvd, #400, Englewood, CO 80111 — GM: Mike Murphy. CE: John Vigil. Spanish "R. Tricolor". (Lic. to Aurora, CO.)

291) 3800 Hooper Ave, Baltimore, MD 21211 — GM: Edward Kiernan. PD: J. Beauchamp. CE: Hank Volpe. N/t.

292) P.O. Box 2061, Bristol, TN 37621 — Rlg. (Lic. to Kingsport, TN.)

293) 3131 Elliott Ave, #750, Seattle, WA 98121 — GM: Fred Schumacher. CE: Robert Trimble. ADC.

294) 5227 North 7th St, #1100, Phoenix, AZ 85014 — Brokered talk. (Lic. to Cave Creek, AZ.)

295) 39138 Fremont Blvd, 3rd Flr, Fremont, CA 94538. **Web:** http://www.kfax.com — GM: Ron Walters. PD: Craig Roberts. Rlg.

296) 1360 E. Sherwood, Grand Junction, CO 81501 — GM: Jim TerLouw. PD: Christopher Newby. CE: Dwight C. Morgan. N/t. 1225-0710.

297) 1014 Providence Church Road, White Oak, GA 31568 — Talk. (Lic. to Woodbine, GA.)

298) 1055 Franklin Ave, #306, Garden City, NY 11530 — GM: Jane Bartsch. PD: D. Anthony. CE: Bill Schleinitz. AS. (Lic. to Hempstead, NY.)

299) 1468 W. 9th St, #805, Cleveland, OH 44113 — GM: Jim Meltzer. PD: Bobby Hatfield. CE: Dave Szucs. WTAM: N/t. WKNR Sports.

300) 16414 San Pedro Ave, #460, San Antonio, TX 78232 — Pres: Sam Morris. Rlg. 1130 (Sat/Sun 1200)-0600. (Lic. to Alamo Heights, TX.)

301) 720 S. White Ave, Bay Minette, AL 36507 — GM: Walter Bowen. CE: Lou Stokes. C&W.

303) 3580 Wilshire Blvd, Los Angeles, CA 90010 — GM: Bob Moore. PD: M. Wagner. CE: D. Ping. OLD, R&B. (Lic. to Pasadena, CA.)

304) 311-112th Ave. NE, St. Petersburg, FL 33716 — GM: Mike H. Smith. PD: Dave Guerin. CE: Art Karmgard. Rlg.

305) 2175 Click Rd, Petoskey, MI 49770 — GM: Rick Stone. PD: George Taylor. CE: Del Reynolds. N/t.

306) 5010 Underwood Ave, Omaha, NE 68132 — GM: Donn Seidholz. PD: Gary Sadlemyer. CE: Butch Dulaney. Talk, info.

307) 1 Julian Price Place, Charlotte, NC 28208 — GM: Rick Jackson. PD: Andy Bickel. CE: Ted Bryan. N/t.

308) P.O. Box 5037, Bend, OR 97708. **E-mail:** kbnd@webl.zland.com — GM: Mike Cheney. PD: Bob King. CE: James Boyd. N/t.

309) 870 N. Military Hwy, #211, Norfolk, VA 23502 — GM: Cynthia Johnson. RB.

311) 2131 Crimmins Lane, Falls Church, VA 22043 — GM: Alan Pendleton. Op's Mgr: Nick Martinelli. CE: Ed Bukont. Sp, ethnic, rel. foreign b'casters. (Lic. to Washington, DC.)

312) Rt 6 Box 735, Macon, GA 31201 — GM: Ralph Meachum. CE:

Tom McAfee. UC. (Lic. to Gordon, GA.)
313) 1 Memorial Drive, St. Louis, MO 63102 — CE: Paul J. Grundhauser. N/t.
314) P.O. Box 1120, Eugene, OR 97440 — GM: Dave Woodward. PD: Jerry Allen. CE: T. Mulder. N/t.
315) P.O. Box 3880, Ogden, UT 84409 — GM: Jack French. PD: Chris Staley. CE: Tim Hunt. Cont. Christian. (Lic. to Roy, UT.)
316) 5050 Murphy Canyon Rd, San Diego, CA 92123 — GM: Kevin McCarthy. PD: Cliff Albert. CE: Jack H. Rabell. N/t.
319) 2930 E. Jefferson Ave, Detroit, MI 48207 — G.M: Peter Connolly. PD: Gregg Henson. CE: Ralph Hunt. Sports.
320) 7900 Xerxes Ave. S, #102, Bloomington, MN 55431. **E-mail:** kfanprog@kfan.com **Web:** http://www.kfan.com — GM: Mick Anselmo. PD. Doug Westerman. CE: Matt Connor. Sports. (Lic. to Minneapolis, MN.)
321) 499 Park Ave, New York, NY 10022. **Web:** http://www.bloomberg.com — News Dir: Bob Leverone. CE: Rene Tetro. Financial news.
322) P.O. Box 1614, Laredo, TX 78044 — Spanish rlg. (Lic. to Edna, TX.)
323) 759 N. 19th St, Milwaukee, WI 53201 — GM: Chuck DuCoty. PD: Gary Jensen. CE: R.P. Johnson. N/t, sports.
324) 490 S. Farrell Dr, #C-210, Palm Springs, CA 92262. **E-mail:** kcmjamfm@worldnet.att.net — GM: Bruce Johnson. PD: Gary DeMaroney. CE: Jason Houts. AS.
325) 5244 Madison Ave, Sacramento, CA 95841 — GM: Doug McGuire. CE: Bruce Hirsh. Talk, sports.
327) 5601 Cassia St, Boise, ID 83705 — GM: Ken Koch. PD: Ty McLean. CE: Mac McGaha. AS.
330) 2600 S. Spring Ave, Sioux Falls, SD 57105 — GM: Roger Currier. CE: Vince Fuhs. N/t, sports.
331) 2926-D Huntsville Hwy, Fayetteville, TN 37334 — GM: Carla Brady Gonzales. CE: Don Roden. Rlg. (Lic. to Hazel Green, AL.)
332) 200 N. 22nd St, Richmond, VA 23223 — GM: Cral McNeill. PD: Tim Farley. CE: J. Francioni. N/t.
333) P.O. Box 352, Greybull, WY 82426 — GM: Jeff Keith. CE: W.R. Burckhard. CWM.
333a) 830 Oilfield Ave, Shelby, MT 59474 — ADC.
334) 130 M.L.K. Dr, Atlanta, GA 30309 — GM: R.J. Carter. C&W.
335) 4949 W. Belmont Ave, Chicago, IL 60641 — Sports.
336) P.O. Box 159, Skowhegan, ME 04976 — GM: Tim Gatz. PD: Denny Morin. CE: Andy Armstrong. Sports.
337) 1086 Teaneck Rd, Teaneck, NJ 07666 — GM: Ron Lustberg. AS. 1100-0500. (Lic. to Oakland, NJ.)
338) P.O. Box 279, Columbus, NC 28722 — GM: John Owen. PD: Diana Johnson. Rlg. (Lic. to Tryon, NC.)
339) P.O. Box 1020, Indiana, PA 15701 — GM: Mark A. Bertig. PD: Jack Benedict. CE: Larry Campbell. ADC. (Lic. to Homer City, PA.)
340) 1617 Lebanon Rd, Nashville, TN 37210 — Mgr: William O. Barry. PD: Kenneth R. Bramming. CE: Gary M. Brown. WAMB: BB. Spanish D0000-0100. WQDQ: F.Pl. (WAMB lic. to Donelson, TN. WQDQ lic. to Lebanon, TN.)
341) 5400 Fredricksburg Rd, San Antonio, TX 78229 — GM: Bill Hill. N/t.
342) 55 N. 300 West, Salt Lake City, UT 84180 — GM: Richard Mecham. PD: Rod Arguette. CE: John Dewhurst. N/t, C&W.
343) P.O. Box 210723, Montgomery, AL 36121 — GM: Brent Markwell. PD. Don Markwell. CE: Tom Jones. N/t.
344) 2905 S. King Rd, San Jose, CA 95122 — GM: Sue Bell. PD: Guillermo Prince. CE: John Burger. Spanish.
345) 5745 Kearny Villa Rd, Suite M, San Diego, CA 92123 — GM: Steve Jacobs. CE: Bill Lipis. Talk.
347) 4590 E. 29th, Tulsa, OK 74114 — GM: Mike DeMarco. PD: Mike Adams. CE: Joe Hancock. C&W.
348) 145 Branham View Rd, Lexington, SC 29072 — GM: Blanche Goodson. Gospel.
349) 2340 East Sunset Dr, Bellingham, WA 98226 — GM: Michael Pollock. PD: Mark Edwards. CE: Todd Merely. News, sports.
350) 1015 Main St, Wheeling, WV 26003 — GM: Larry Anderson. PD: Tom Miller. CE: Roy Humphrey. Talk, rlg, C&W.
351) 138 W. Rt. 66, Williams, AZ 86046 — N/t.
352) 110 S. Montclair St, Suite 205, Bakersfield, CA 93309. **E-mail:** keri@lightspeed.net **Web:** http://www.keri.com — GM: Don Bevilacqua. PD: Terri Blankenship. CE: Terry Gaiser. Rlg. talk, Sp. (Lic. to Wasco, CA.)
353) Marathon, FL. For details refer to **R. Martí.**
354) P.O. Box 1248, Jackson, MS 39215. **E-mail:** wjnt@teclink.net **Web:** http://www.wjnt.com— GM, PD: Thena Gunn. PD: Bob Rall. CE: Stan Carter. N/t.
355) P.O. Box 608, Kalispell, MT 59903 — GM: Jerry Adams. PD:

Dave Shannon. CE: Tony Mulligan. ADC. OLD.
356) 1001 Farnam-on-the-Mall, Omaha, NE 68102 — GM: Marty Riemenschneider. PD: Leon Turnthirsty. CE: Allen Sherrill. AS. (Lic. to Bellevue, NE.)
357) 207 Midtown Plaza, Rochester, NY 14604 — GM: Ken Spitzer. SM: Jeff Howlett. CE: Craig Kingcaid. N/t, C&W.
358) 482 Dove Haven Ln. SE, Winnabow, NC 28479 — GM: Chuck Langley. Moran. PD: Cheryl Sparks. CE: Paul Knight. Rlg. (Lic. to Carolina Beach, NC.)
359) 802 S. Central St, Knoxville, TN 37902 — Pres: Harry J. Morgan. PD: Tom Burchfield. CE: Scott Wyrick. Rlg.
360) 5821 SW Freeway, Ste 600, Houston, TX 77057 — GM: Roger Medvin. PD: David Darling. CE: Steve Halatyn. Rlg, ethnic (Vietnamese, Chinese, Hindi). (Lic. to Humble, TX.)
361) 1190 E. Ball Road, Anaheim, CA 92805 — GM: Milton Sexton. Op's Mgr: Lou Salatino. CE: Mark Moceri. Ethnic, variety.
362) P.O. Box 52128, Atlanta, GA 30355 — GM: Brenda Brown. Classical, Sun also ethnic.
363) 2915 Maples Road, Ft. Wayne, IN 46816 — GM: Tony Richards. PD: Gary Noe. CE: Ed Didier. N/t.
364) P.O. Box 631, Annapolis, MD 21404 — Ethnic, rel. WUST-1120.
365) 6320 Zuni Rd. SE, Albuquerque, NM 87108 — GM: José Molina. SM: Alex Bozzo. Spanish "La Super X".
366) 3 Park Ave, New York, NY 10016 — GM: Janie Washington. PD: Bob Frederick. CE: David Antoine. Black, Caribbean.
367) 4949 SW Macadam Ave, Portland, OR 97201 — GM: Dave Milner. PD: Duane Link. CE: Byron Swanson. Full service ADC, OLD, talk.
368) 1080 Ballpark Way, Arlington, TX 76011 — GM: Jerry Bobo. CE: Tracy Barnes. OLD. (Lic. to Dallas, TX.)
369) 900 Bluefield Ave, Bluefield, VA 24701 — GM: Sandy Frazier. CE: Jay Belt. Sports.
372) Soquel, CA.
373) 2824 Palm Beach Blvd, Ft. Myers, FL 33916 — Talk. (Lic. to Pine Island Center, FL.)
374) 625 N. Michigan Ave, 3rd Fl, Chicago, IL 60611 — GM: Jim Pagliai. PD: Mary Moreira. CE: John McGuinness. Spanish "La Equis".
375) 100 Mt. Wayte Ave, Framingham, MA 01701 — GM: Scott Gibbons. CE: Paul Andrews. N/t, Spanish.
377) 32790 Henry Ruff Rd, Inkster, MI 48141 — GM: Terry Arnold. Ag. CE: Tony Simpson. Talk. (Lic. to Taylor, MI.)
378) P.O. Box 2307, Newburgh, NY 12550 — GM: Joerg Klebe. PD: Pat McKay. CE: Shawn C. McGrath. OLD.
379) P.O. Box 900, Graham, NC 27253 — GM: Ted J. Gray Jr. PD: Larry Ingold. CE: Jim Davis. Rlg.
380) P.O. Box 6008, Fargo, ND 58108. **E-mail:** kfnw@rrnet.com — GM: Gary Herr. CE: Gary L. Ellingson. Rlg. (Lic. to West Fargo, ND.)
381) P.O. Box 3638, Williamsport, PA 17701 — GM: Jim Dabney. N/t. (Lic. to Hughesville, PA.)
382) P.O. Box 2088, Dothan, AL 36302 — GM: Tony Scott. CE: Neil Riddle. News. (Lic. to Ozark, AL.)
383) P.O. Box 1511, Tucson, AZ 85702 — GM: Raul B. Gamez. PD: Berta Gallego. CE: Peter Trowbridge. Spanish "R. Mundo". (Lic. to Sahuarita, AZ.)
385) 9255 Towne Centre Dr, #535, San Diego, CA 92121. **E-mail:** kprz@aol.com **Web:** http://www.kprz.com — GM. & PD: Mark Larson. Op's Mgr: Monica Murray. Rlg. (Sp. Mon-Fri 1000-1200, Sat/Sun 0800-1500). (Lic. to San Marcos, CA.)
386) 8400 NW 52nd St, #101, Miami, FL 33166 — Spanish. (Lic. to Miami Springs, FL.)
387) P.O. Box 1210, Dahlonega, GA 30533 — GM: Doug Roy. PD: Lee Ann Roy. CE: Dan Davis. C&W, Gospel.
388) 5555 Hilton Ave, #500, Baton Rouge, LA 70808 — Sports.
389) Frankenmuth, MI.
390) P.O. Box 130, Guymon, OK 73942 — GM: Ed Smith. PD: Bill Weldon. CE: Jim Stanford. CWM, farm.
391) 10 Monument Rd, Philadelphia, PA 19004 — GM: Chtis Claus. CE: Sam A. Virgillo. Talk.
392) 6080 Mt. Moriah Rd, Memphis, TN 38115 — GM: Fred Flinn. CE: Rob Herrin. Spanish. (Lic. to Bartlett, TN.)
393) P.O. Box 252, McAllen, TX 78502 — GM: Eloy Bernal. Spanish rlg. (Lic. to San Juan, TX.)
394) 720 S. River Rd, St. George, UT 84790 — GM: Carl Lamar. PD: David Combs. CE: Jed Wilkinson. OLD. (Lic. to Washington, UT.)
395) 1730 Minor Ave, #2000, Seattle, WA 98101 — GM: Steve Oshin. PD: Jay Kelly. CE: Clay Freinwald. OLD. (Lic. to Auburn, WA.)
396) P.O. Box 149, Sunnyside, WA 98944 — GM: Bob Powers. PD: Rene Taylor. CE: Dave Hebert. AS. 1400-0600 incl. Sp. MF0200-

0600, Sun 2100-2300.
397) P.O. Box 848, Laramie, WY 82070 — GM: Russ Jenks. OLD, talk.
400) P.O. Box 8182, Topeka, KS 66608 — Rlg.
401) 5668 143rd Ave, Holland, MI 49423 — Rlg. (Lic. to Zeeland, MI.)
401a) 1422 Euclid Ave, #604, Cleveland, OH 44115 — Rlg.
401b) 1396 Broad St, Clifton, NJ 07013 — Spanish Nt. (Lic. to New York, NY.)
402) 57 W. South Temple, #700, Salt Lake City, UT 84101 — GM: Stephen C. Johnson. MOR, BB.
403) P.O. Box 96, Neenah, WI 54947 – NOS.
403a) 2406 South Congress Ave, West Palm Beach, FL 33406 — Talk.
404) P.O. Box 1090, Jacksonville, OR 97530 — GM: Chris Thompson. Rlg. 1300-0800. (Lic. to Phoenix, OR.)
405) 1300 WWCR Ave, Nashville, TN 37218. **E-mail:** wwcr@aol.com **Web:** http://www.wwcr.com/wnqm.htm — GM: George McClintock. PD: John McClintock. CE: Watt Hairston. Rlg. Spanish: MF0100-0600.
407) 384 Embarcadero W, 3rd Flr, Oakland, CA 94607 — GM: Jim Smith. Childrens prgrs.
408) P.O. Box 465, Derry, NH 03038 — GM: Gail Rehse. Rlg. 1100-0300.
409) 10200 E. Girard Ave, #B-131, Denver, CO 80231 — NOS. (Lic. to Aurora, CO.)
410) 4155 Harrison Blvd, #206, Ogden, UT 84403 — GM: John Webb. NOS.
411) Salinas, CA.
412) 4660 S. Decatur Blvd, Las Vegas, NV 89103 — GM: Tony Bonnici. Sports.
413) 550 Market Ave. S, Canton, OH 44702 — GM: Ray Hexamer. NOS, talk, sports.
414) 501 Wooster Ave, San Jose, CA 95116 — GM: Bob Stewart. CE: David Williams. Asian, rlg.
415) 3400 Idaho Ave NW, Washington, DC 20016 — GM: Steve Swenson. PD: H. Cooke. CE: D. Garner. News.
416) 29200 Vassar Dr, #650, Livonia, MI 48152 — GM: John Yinger. PD: Brian Patrick. CE: Paul Goodpastor. Rlg. (Lic. to Detroit, MI.)
417) 2792 Maplewood Dr, Maplewood, MN 55109 — GM: Ginny Morris. CE: Bob Gagne. N/t. talk. (Lic. to St. Paul, MN.)
418) 4800 N. Central Ave, Phoenix, AZ 85012. **E-mail:** kfnnradio@jund.com — GM & PD: Ronald Cohen. CE: David Dixon. Business n/t. (Lic. to Mesa, AZ.)
419) 4750 N. Blackstone Ave, Fresno, CA 93726 — GM. & PD: Gary Tompkins. Rlg.
420) 9485 Haven Ave, #101, Rancho Cucamonga, CA 91730 — GM: Tom Henderson. CE: John Artal. Sports. (Lic. to Ontario, CA.)
421) 2559 Welton St, Denver, CO 80205 — GM: James Walker. PD: Danny Harris. CE: Daren McMullin. Urban cont. (Lic. to Littleton, CO.)
422) P.O. Box 1031, New London, CT 06320 — GM: Jim Reed. PD: Andy Russell. CE: Stephen Keefe. AS. 1000-0500.
423) 500 W. Cummings Park, #2500, Woburn, MA 01801 — GM: Mike Wheeler. CE: Michael J. Klein. Sports. (Lic. to Boston, MA.)
424) 10841 E. 28th St, Independence, MO 64052 — GM: Gary Acker. Sports.
424a) P.O. Box 95, Syracuse, NY 13201 — UC.
425) 10 Music Circle E, Nashville, TN 37203 — GM: John King. PD: Kelly Carls. CE: Watt Hairston. N/t, rlg, C&W.
426) 406 W. South Jordan Pkwy, #130, South Jordan 84095 — Gospel. (Lic. to West Jordan, UT.)
427) P.O. Box 30013, Spokane, WA 99223 — GM: Steve Cody. CE: Jim Bender. Rlg.
428) 1801 Coral Dr, Waukesha, WI 53186 — GM: Ed Walters. PD: Joe Till. CE: Don Hunjadi. Sports.
429) 2284 S. Victoria, #2M, Ventura, CA 93003 — GM: Harold A. Frank. CE: Martin Walker. Talk. (Lic. to Port Hueneme, CA.)
430) 202 Galbert Rd, Lafayette, LA 70506 — GM: Charles H. Wood. Childrens prgrs.
431) 267 Main St, Greenfield, MA 01301 — GM: Ed Skutnik. CE: Jim Hemingway. OLD.
432) 592 Pontaluna, Muskegon, MI 49443 — GM: Frank Landingham. CE: Walt Love. Rlg. (Lic. to Muskegon Heights, MI.)
433) 1220 4th Ave. SW, Rochester, MN 55902 — GM: Dick Radke. PD: Jim McCann. CE: Jim Casey. OLD.
434) 695 Delaware Ave, Buffalo, NY 14209 — GM: Terry Rodda. CE: Tom Atkins. Sports, C&W.
435) 820 SW 4th St, Moore, OK 73160 — GM: Vance Harrison Jr.. PD: Kent Jones. CE: Ray Klotz. OLD. (Lic. to Oklahoma City, OK.)

436) 4614 SW Kelly Ave, Portland, OR 97201 — GM: Tom Baker. PD: Steve Arena. CE: Larry Holtz. Sports. (Lic. to Oregon City, OR.)
437) P.O. Box 909, Moreno Valley, CA 92388 — CE: John Patterson. Ethnic, var.
439) 5900 Picketville Rd, Jacksonville, FL 32254 — Gospel.
440) P.O. Box 216, Dalton, GA 30722 — GM: Devona Poe. CE: Ed Allen. Rlg.
441) P.O. Box 352, Poplarville, MS 39470 — GM: Mike Porter. CE: Dominic Mitchum. Rlg.
442) 3013 Guess Rd, Durham, NC 27705 — PD: Mark Parker. CE: Jim Davis. Rlg. (Lic. to Chapel Hill, NC.)
444) 1707 N. Mays St, Round Rock, TX 78664 — Spanish. (Lic. to Creedmore, TX.)
445) 200 S. 10th St, #600, McAllen, TX 78501 — GM: Rogelio Botello. CE: N. Lindsay. Spanish "R. Panamericana". (Lic. to Harlingen, TX.)
446) 1445 W. Baseline Rd, Phoenix, AZ 85041 — GM: Moisés Herrera. CE: Jim Clark. E/S rlg "R. Casa".
447) 333 N. Glenoaks Rd, #690, Burbank, CA 91502 — Sports. (Lic. to Los Angeles, CA.)
448) Capitola, CA. Silent, licence cancelled.
449) P.O. Box 1540, Waterloo, IA 50704 — GM: Tim Mathews. CE: Leonard Tompkins. N/t, rlg, C&W.
450) 4243 Albany St, Albany, NY 12212 — GM: Nevin Larson. CE: Paul H. Thurst. Rlg.
451) P.O. Box 16408, Charlotte, NC 28297 — GM: Wayne Hammond. CE: David Anthony. Rlg.
452) 200 Monument Rd, #6, Bala-Cynwyd, PA 19004 — Corp. GM: Alan Pendleton. SM: Sam Speiser. Multicultural, foreign languages. (Licensed to Philadelphia, PA.)
453) 3931 White Horse Rd, Greenville, SC 29611 — GM: Charles Creager Sr. CE: Ted Caldwell. Rlg. (Licensed to Pickens, SC.)
454) 5307 E. Mockingbird Lane, #500, Dallas, TX 75206 — CE: Nathan Lindsey. Spanish. (Licensed to Fort Worth, TX.)
455) 6001 Wilkinson Rd, Richmond, VA 23227 — GM: Larry Jones, Sr. PD: Kevin Kofax. CE: Ray Mills. Rlg. (Licensed to Richmond, VA.)
456) 2523 Bronco Circle, Huntsville, AL 35816 — GM: Marcus Taylor. PD: Ed Gaines. CE: John Hain. Black Gospel.
457) Univ. of Arizona, Tucson, AZ 85721 — GM: Edward Kupperstein. PD: Ed Kesterson. CE: Tom Boone. News, jazz. (Sp. Sun 2000-0230).
458) 500 Washington St, 4th Fl, San Francisco, CA 94111 — C&W.
459) 730 West Hampden Ave, #300, Englewood, CO 80110 — Childrens prgrs. (Licensed to Arvada, CO.)
460) 2700 W. Martin Luther King Bövd, Tampa, FL 33607 — GM: Jeff Liss. PD: Manuel E. Semprit. CE: Arthur Karmgard. Spanish. 1100-0500.
461) 2460 N. Atlanta Rd, Smyrna, GA 30080 — GM/PD: Javier Macias. CE: John York. Spanish "R. Exitos".
462) 1300 Grimmett Dr, Shreveport, LA 71107 — Rlg.
463) 143 Rumford Ave, Newton, MA 02166 — SM: Rob Rudnick. PD: John Frassica. CE: Leo V. Sullivan, Jr. English, Chinese, Greek.
464) 2501 13th Ave. SW, Ste. 201, Fargo, ND 58103 — GM: Nancy Odney. CE: Aaron White. NOS, talk. (Lic. to West Fargo, ND.)
465) P.O. Box 1275, Bennettsville, SC 29512 — GM: Dwight M. Johnson. Talk. 1100-0500.
466) 600 Baltimore Dr, Wilkes-Barre, PA 18702 — GM: Buz Boback. PD: Greg Foster. CE: Ron Schact. Sports. (Licensed to Pittston, PA.)
467) P.O. Box 352, Salt Lake City, UT 84110 — Rlg. (Licensed to West Valley City, UT.)
468) 2043 10th St. NE, Roanoke, VA 24012 — Pres: Dave H. Moran. PD: Sharon M. Moran. CE: Ben Girardan. Rlg. (Licensed to Vinton, VA.)
469) P.O. Box 308, Ferndale, WA 98248 — GM: Glenda Hamilton. PD: George Hamilton. CE: Ron Cowell. Rlg. 1400-0800.
470) 901 E. 2nd, #110, Spokane, WA 99202. (Lic. to Opportunity, WA.)
471) P.O. Box 4507, Vancouver, WA 98662 — GM: David Granger. N/t, ADC.
472) 3561 Pegasus, #107, Bakersfield, CA 93308 — GM: Randy Warwick. PD: Rob Lang. CE: Greg Garcia. N/t.
473) 330 SW 27th Ave, #207, Miami, FL 33135 — GM: Jorge A. Rodríguez. PD: Emilio Milian, Sr. CE: "Eddy" Rodríguez. Spanish "Cadena Azul". (Lic. to Coral Gables, FL.)
474) P.O. Box 2397, Paducah, KY 42002 — GM: Gary Morse. PD: Jamie Richards. CE: Earl Abanathy. AS.
476) 122 5th Ave, New York, NY 10011. **Web:** http://www.wqxr.com — GM: Warren G. Bodow. SM: Stan Martin. CE: Herb Squire. AS.
477) P.O. Box 28, Lancaster, SC 29721 — GM: B. Len Phillips, Jr.

CWM.
478) P.O. Box 1770, Aberdeen, SD 57402 — GM: Deb O'Donnell. PD. & CE: Mike Johnson. CWM, talk. 1100-0700.
479) P.O. Box 3174, Tempe, AZ 85280 — GM: Michael C. Owens. PD: Larry Daniels. CE: Robert A. Van Buhler. C&W.
480) 1700 N. Alvarado, Los Angeles, CA 90026 — GM: Ron Thompson. CE: Ron Russ. Korean. (Lic. to Santa Monica, CA.)
481) P.O. Box 5668, Colorado Springs, CO 80931 — GM: Rick Martin. PD: John Boles. CE: Mel Rauh. Rlg.
482) P.O. Box 36, Chattahoochee, FL 32324 — Talk, R&B.
483) 3000 SW 60th Ave, Fort Lauderdale, FL 33314 — GM: Deborah Shane. Talk.
484) 1498 Trade Center Dr, Lexington, KY 40505 — GM: Keith Yarber. RB. (Lic. to Georgetown, KY.)
485) 6301 Ivy Lane, #800, Greenbelt, MD 20770 — GM: Benjamin Hill. CE: T. McGinley. Talk, gospel. (Lic. to Morningside, MD.)
486) 8009 Marble Ave. NE, Albuquerque, NM 87110 — GM: Mark Crump. Childrens prgrs.
487) Woodside Ave, Patchogue, NY 11772 — NOS. 1030-0300.
488) P.O. Box 40, Hubert, NC 28539 — Rlg. (Lic. to Camp Lejeune, NC.)
489) 41-30 58th St, Woodside, NY 11377 — Soul. (Lic. to New York, NY.)
490) P.O. Box 1166, Atlanta, TX 75551.
491) 805 North Main St, Atmore, AL 36502 — GM: John Mathis. Rlg.
492) 1230 High St, #120, Auburn, CA 95603.
493) P.O. Box 699, Blackfoot, ID 83221.
494) 2010 South Michigan St, South Bend, IN 46613.
496) 212 West Apache, Farmington, NM 87401.
497) P.O. Box 1210, Minot, ND 58702.
498) 1116 Ocala St, Myrtle Beach, SC 29577.
499) 2700 East Bypass, #5000, College Station, TX 77845.
501) 2600 South Jackson St, Seattle, WA 98114 — PD: Frank P. Barrow. Urban ADC. (Lic. to Renton, WA.)
502) 1071 West Shaw, Fresno, CA 93711. (Lic. to Clovis, CA.)
503) 1480 Eisenhower Dr, Augusta, GA 30904.
504) P.O. Box 2118, Iowa City, IA 52244.
505) 7901 Carpenter Freeway, Fort Worth, TX 75247.
506) 110 East 17th St, # 205, Cheyenne, WY 82001. (Lic. to Fox Farm, WY.)
507) 3267 Sonoma Blvd, Vallejo, CA 94590 — PD: Eric Brown. RB.
508) P.O. Box 1678, Mount Airy, NC 27030.
509) P.O. Box 952, Enid, OK 73702.
510) 820 North River St, #100, Portland, OR 97227 — CE: Cris Alexander. Rlg. (Lic. to Oswego, OR.)
512) 10348 South Redwood Rd, South Jordan, UT 84095. (Lic to Sandy, UT.)
513) West 223 North 3251 Shady Lane, Pewaukee, WI 53072 — Rlg. 1200-0402 (Sun 0230). (Lic. to Sussex, WI.)
514) 423 Garrison Ave, Fort Smith, AR 72901.
515) 1500 Cotner Ave, Los Angeles, CA 90025 — CE: Tom White. Traffic information.
516) 3131 South Vaughn Way, #601, Aurora, CO 80014. (Lic. to Denver, CO.)
517) 1201 Peachtree St. NE, Atlanta, GA 30361.
518) 721 Shirley St, Cedar Falls, IA 50613.
519) 5426 North Mesa, El Paso, TX 79912.
520) 2202 Jolliff Rd, Chesapeake, VA 23321. (Lic. to Portsmouth, VA.)
521) 514 East Bellevue Rd, Atwater, CA 95301. (Lic. to Merced, CA.)
522) 599 South Collier Blvd, #203, Marco, FL 33937. (Lic. to Marco Island, FL.)
524) 4200 West Main Street, Kalamazoo, MI 49006.
525) 9 Caldwell Place, Elizabeth, NJ 07201 — GM: John Quinn. CE: Don Neumuller.
526) 520 Hwy 29 North, Concord, NC 28027. (Lic. to Charlotte, NC.)
528) 1018 North Valley Mills, Waco, TX 76710.
529) 4455 South 5500 West, Hooper, UT 84315. (Lic. to Brigham City, UT.)
530) 4352-C Caterpillar Rd, Redding, CA 96003.
531) 7080 Industrial Hwy, Macon, GA 31206 — Sports.
532) P.O. Box U, Salisbury, MD 21802.
533) P.O. Box 2058, Madison, WI 53701 — Talk.
534) 139 West Olive Ave, Fresno, CA 93728.
535) P.O. Box 4808, Monroe, LA 71201.
536) 875 East Summit Ave, Muskegon, MI 49444.
537) 619 Alexander Rd, 3rd Fl, Princeron, NJ 08540.
539) 5301 Madison Ave, #402, Sacramento, CA 95841. (Lic. to Roseville, CA.)
540) 1200 West 4th St, Adel, GA 31620.

541) 1822 North Court St, Marion, IL 62959. (Lic. to Johnston City, IL.)
542) P.O. Box 600, Lexington Park, MD 20653 — Talk.
543) 2609 Jordan Lane NW, Huntsville, AL 35806.
543a) 1001 Ponce de Leon Blvd, Coral Gables, FL 33134 — GM: Claudia Puig. CE: Ralph Chambers. Spanish talk. (Lic. to Miami Springs, FL.)
544) 5161 Maple Dr, Des Moines, IA 50317.
545) 113 Rochester Hill Rd, Rochester, NH 03866.
546) 1006 West Marine Dr, Astoria, OR 97103.
547) 1050 McIntosh, Brownsville, TX 78521.
548) 4800 Shannon Rd, Denison, TX 75092.

UNITED STATES ARMED FORCES RADIO AND TELEVISION SERVICE

✉ 1363 Z Street, Bldg 2730, March ARB, CA 92518. 🖷 +1 (909) 413-2234
Web: http://www.dodmedia.osd.mil/afrts_bc/ahome.htm
L.P: Dir Programming: Robert W. Matheson. Chief Radio Div. Mgr: Emanuel J. Levy. Dir. Eng. & Op's: Bruce V. Ziemienski.
In service since 1942, the mission of the Armed Forces Radio and Television Service (AFRTS) is to inform and entertain US military personnel and their families stationed in more than 120 countries. Television and radio programming consisting of news, sports, public affairs and other timely information is relayed from Los Angeles, CA to overseas locations by satellite. Programming is derived from the US commercial radio and television networks, and other programming sources. Radio entertainment programs, which include a wide range of popular music formats, are transcribed or otherwise duplicated for shipment to stations. Television entertainment and news programs are videotaped for distribution to locations.
STATIONS: st's with free-to-air AM and FM sces are listed under the countries where they are located. Closed circuits sces are not listed.

INTERNATIONAL BROADCASTING

Government-operated, private and religious stations are listed in the separate International Broadcasting section.

CENTRAL AMERICA & THE CARIBBEAN

ANGUILLA (British West Indies)

L.T: UTC - 4h — **Pop:** 9.000 — **Radios:** 2.000 — **Pr.L:** English — **E.C:** 50Hz, 230V — **ITU:** AIA.

RADIO ANGUILLA (Gov.)

✉ The Valley, Anguilla, BWI. ☎ +1 (264) 497 2218 or 3620. 🖷 +1 (264) 497 5432 — **L.P:** Dir: N.A.Hodge. N. Editor: W.A. Richardson. Engineer: L.Richardson — **MW:** 1505kHz 1kW.
D.PRGR: 0930(Sun 1600)-0210. **N:** BBC: 1100W, 1200, 1300W, 1600, 2300, 0200. **VOA:** 0100. **Local:** 1115W, 1615Sun, 2315.
ANN: "1505 Radio" — **V.** by letter.

THE CARIBBEAN BEACON (Rlg.)

✉ Box 690, Anguilla, BWI. ☎ +1 (264) 497 4340. 🖷 +1 (264) 497 4311. **L.P:** Owner: Dr. Gene Scott. CE: Kevin Mooney.
MW: 690kHz 100kW day/50kW night, 1610kHz 200kW.
FM: 100.1MHz 35kW (ERP) — **D.PRGR:** 24h — **V.** by QSL-card.
SW: 1 x 100kW (see International Broadcasting section)

ZJF-FM, Sachasses, Anguilla, BWI. ☎ +1 (264) 497 3918: 105.3MHz.
Gem Radio Network: 107.9MHz (rel. Trinidad)

ANTIGUA & BARBUDA

L.T: UTC - 4h — **Pop:** 66.000 — **Radios:** 50.000 (est.) — **Pr.L:** English — **E.C:** 50/60Hz, 110/220V — **ITU:** ATG.

ANTIGUA & BARBUDA BROADCASTING SERVICE (Gov. Comm.)

✉ P. O. Box 590, St. John's. ☎ +1 (268) 462 0010. 🖷 +1 (268) 462 4442.
L.P: Ag. DG: Hollis Henry. SM (Radio): Norman Gus Thomas. CE: Denis Leandro — **MW:** 620kHz 10kW — **D.PRGR:** 0925-0500.
BBC N: 1100, 2300, 0400. **Local N:** 1600, 2310.

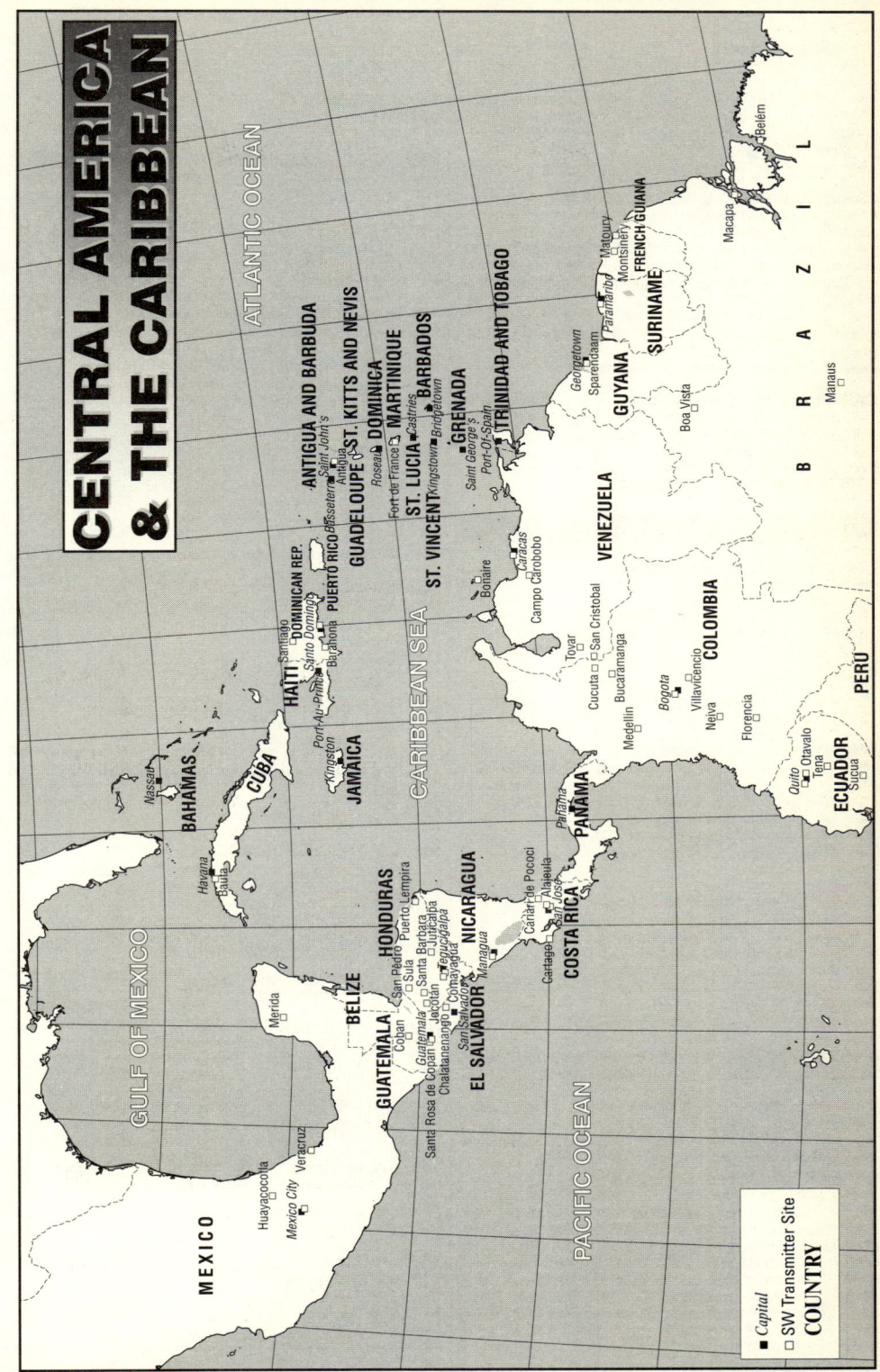

CENTRAL AMERICA & THE CARIBBEAN

ATLANTIC OCEAN

BRAZIL

Belém

Macapa

Manaus

Boa Vista

FRENCH GUIANA
Montsinéry
Matoury

SURINAME
Paramaribo
Sparendaam

GUYANA
Georgetown

VENEZUELA
Caracas
Campo Cárobobo

ANTIGUA AND BARBUDA
Saint John's
Antigua

ST. KITTS AND NEVIS
Basseterre

GUADELOUPE
PUERTO RICO

DOMINICA
Roseau

MARTINIQUE
Fort-de-France

ST. LUCIA
Castries

BARBADOS
Bridgetown

ST. VINCENT
Kingstown

GRENADA
Saint George's

TRINIDAD AND TOBAGO
Port-Of-Spain

Bonaire

COLOMBIA
Tovar
San Cristobal
Cucuta
Bucaramanga
Bogota
Villavicencio
Medellin
Neiva
Florencia

DOMINICAN REP.
Santiago
Santo Domingo
Bajahona

HAITI
Port-Au-Prince

JAMAICA
Kingston

CUBA
Havana
Bauta

BAHAMAS
Nassau

CARIBBEAN SEA

GULF OF MEXICO

MEXICO
Mexico City
Veracruz
Merida
Huayacocotla

BELIZE

GUATEMALA
Coban
Guatemala
Santa Rosa de Copán
Chalatenango
Jocotán
Quetzaltenango

EL SALVADOR
San Salvador
San Miguel

HONDURAS
San Pedro
Sula
Santa Barbara
Puerto Lempira
Juticalpa
Tegucigalpa
Comayagua

NICARAGUA
Managua

COSTA RICA
Cartago
San José
Alajuela
Cañar de Pococi

PANAMA
Panama

PACIFIC OCEAN

PERU

ECUADOR
Quito
Otavalo
Tena
Sucúa

Legend:
■ Capital
□ SW Transmitter Site
COUNTRY

ANN: "You are listening to 620 - ABS" — **V.** by letter.

RADIO ZDK (Comm.)

✉ Grenville Radio Ltd, P. O. Box 1100, St. John's. ☎ +1 (268) 462 1100. **Cable:** Grenville Radio Limited - Antigua.
L.P: Chief Exec: Mrs.Lydia Bird. PD: Ivor Bird. GM (Admin): Clayton Hinds. CE: E.Philip — **MW:** 1100kHz 10kW — **FM:** 99.0MHz.
D.PRGR: MF 0904-0504, Sat 0956-0504, Sun 1426-0204.
N: 1000W, 1400W, 1600, 1800.
ANN: "Magic Radio" — **V.** by QSL-card.

CARIBBEAN RADIO LIGHTHOUSE (Rlg.)

✉ Box 1057, St.John's. ☎ +1 (268) 462 1454.
E-mail: cradiolight@candw.ag
L.P: Dir: Curtis L.Waite. Asst. Dir. & Tech. Dir: Jerry Baker.
MW: 1165kHz 10kW — **FM:** 90.0MHz 10W.
D.PRGR: 0955-0130. **BBC N:** 1100, 1200, 1300, 1600, 2000, 2300.
V. by QSL-card.

Sun FM: 100.1MHz.
Observer Radio: 91.1MHz 12h daily. **F.PI:** 24h.
Gem Radio Network: 93.9MHz (rel. Trinidad)

CARIBBEAN RELAY CO. LTD. (Deutsche Welle/BBC)

✉ P. O. Box 1203, St. John's. ☎ +1 (268) 462 0994. 🖷 +1 (268) 462 0436. **E-mail:** acm_crc@candw.ag or cm-crc@candw.ag
FM: 100.1MHz (rel. BBCWS & Caribbean Report).
SW (G.C: 61.48W/17.06N): 4 x 250kW tr's (only 2 used by BBC)
D.PRGR: see International Broadcasting section.

ARUBA

L.T: UTC-4h — **Pop:** 67.269 — **Radios:** 40.000 (est.) — **Pr.L:** Papiamento, English, Dutch — **E.C:** 50+60Hz, 127/220V — **ITU:** ABW

	kHz	kW	MHz	kW	Station
1)	960	10	93.1	1	R. Victoria, Oranjestad (Rlg.)
2)	1270	1.5	107.5		R. 1270, San Nicolaas
3)	1320	1	89.9	10	Voice of Aruba, Oranjestad
4)	1440	1	106.7	6	R. Kelkboom, Oranjestad
5)			97.9		R. Caruso Booy, Oranjestad
6)			103.5		R. Carina FM, Oranjestad

Addresses and other information
1) Pos Chiquito, P.O. Box 410, Oranjestad. ☎ +297 847090. 🖷 +297 849052 — SM: E. Parsons. CE: W. Gandre. **MW:** MF 1000-2300, Sat 1100-2300, Sun 1130-0100. **N. in English:** 1130W, 1630W, 2230W. **FM:** 1200-0300W (music). **V.** by letter. Rp.
2) Bernardstraat 138, P.O. Box 28, San Nicolaas. ☎ +297 845602. 🖷 +297 827753 — SM: J.A.C. Alders. Dir: F.A. Leauer.
3) Van Leeuwenhoekstraat 26, P.O. Box 219, Oranjestad. ☎ +297 824134 (admin.), 821601 (studios) — Dir: A.M. Arends Jr. **MW:** 1000-0500 (Sat 0700). **FM** ("**Channel 90**"): 1000 (Sun 1200)-0500 — **V.** by QSL-card or letter.
4) Bloemond 14, P.O. Box 146, Oranjestad. ☎ +297 821899. 🖷 +297 834825 — GM: C.A. Kelkboom. CE & PD: E.A.M. Kelkboom. 1000 (Sun 1200)-0400. **Dutch:** 1030-1100, 1230-1300, 2315-2400.
5) De Bruynewijk 49, Savaneta. ☎ +297 847752. 🖷 +297 843351 — Owner: Ms. S. Booy.
6) Emanstraat 49A, Oranjestad. ☎ +297 826433/821450.

BAHAMAS

L.T: UTC - 5h — **Pop:** 266.000 — **Radios:** 80.000 — **Pr.L:** English — **E.C:** 60Hz, 120/220V — **ITU:** BAH.

BROADCASTING CORPORATION OF THE BAHAMAS (Comm.)

✉ P.O. Box N-1347, Nassau. ☎ +1 (242) 322 4623. 🖷 +1 (242) 322 392.
L.P: GM: Sandra Knowles. Asst. GM (Radio): Carl Bethel. Asst. GM (Eng): Michael Thompson. Asst. GM (Northern Sce): Edwin Lightbourn.
MW: Nassau ZNS1 1540kHz 50kW (d), ZNS2 1240kHz 1kW — Freeport ZNS3 810kHz 1kW — **FM:** ZNS-FM 104.5MHz.
D.PRGR: 24h. **N.** (ZNS1): 0800, 1300, 1830, 0000, 0300.
ANN: ZNS1: "This is Radio Bahamas". ZNS2: "Inspiration 1240". ZNS3: This is Northern Service, Radio Bahamas". ZNS-FM: "Power 104.5" — **V.** by letter.

Jamz, P.O. Box N-3207, Nassau. ☎ +1 (242) 328 4771. 🖷 +1 (242) 356 5343 — Nassau 100MHz, Freeport (**F.PI:** extend to national coverage).
Love-97 FM, P.O. Box N 3909, Nassau. ☎ +1 (242) 356 4960. 🖷 +1 (242) 356 7256 — Nassau 97MHz
More FM: Nassau
Cool 96 FM: Freeport 96MHz.

BELIZE

L.T: UTC - 6h — **Pop:** 230.000 — **Radios:** 29.620 — **Pr.L:** English, Spanish — **E.C:** 60Hz, 110/220V — **ITU:** BLZ.

BROADCASTING CORPORATION OF BELIZE (Gov. Semi-Comm.)

✉ Albert Cattouse Bldg, Regent Str, Belize City. ☎ +501 (2) 72468. 🖷 + 501 (2) 75040, 75174. **Cable:** Radio Belize. **E-mail:** rbgold@btl.net
L.P: GM: Mrs. Ruth Staine-Dawson. PD: Brian Mossiah.
FM: Ladyville 91.1MHz, Belize C. 92.MHz, Corozal 101.1MHz.
R. Belize Gold: 24h (0500-1200 rel. BBCWS). **N:** English: 1300, 1500, 1700, 1830, 2100, 2300, 0100, 0300. **Spanish:** 1845, 0115.
Friends FM on 88.9/91.3/94.7MHz: 1600-1100.
V. by QSL-card. Rec.acc.

RADIO KREM Ltd. (Comm.)

✉ 3304 Partridge Str, c/o P.O. Box 15, Belize C. ☎ +501 (2) 75929. 🖷 +501 (2) 74079 — **L.P:** SM: Eva S.Hyde, Jr. CE: J.C.Arzu.
FM: 96.5MHz 0.075kW, 89.9MHz (repeater).
D.PRGR: 1100-0500.

MY REFUGE CHRISTIAN RADIO (Rlg.)

✉ P. O. Box 275, Belmopan. ☎ +501 (8) 21080. 🖷 +501 (8)22601.
E-Mail: myrefuge@btl.net
Web: http://www.nauticom.net/www/jhorst/mrindex.htm
L.P: Dir: Richard Smith.
FM: Belmopan 97.3MHz 0.5kW.
D.PRGR: 24h. **N.** (rel. R. Belize): 1300MF, 1830. (rel. Love FM): 1430W, 1630MF, 2030MF, 0000W, 0230MF.

Love FM: Orange Walk 107.1MHz. **N:** 0100.
Estereo Mar: Belize C. 97.9MHz (Spanish)
Voice of the West: FM (no further details).

VOICE OF AMERICA RELAY STATION BELIZE

MW: 1530/1580kHz 100kW.
See International Broadcasting section.

BRITISH FORCES BROADC. SCE. BELIZE

✉ BFBS Belize, Airport Camp, BFPO 12. ☎ +501 25 2333. 🖷 +501 25 2334 — **L.P:** SM: Chris Pratt — **FM:** 99.1MHz 3.5kW.

CAYMAN ISLANDS

L.T: UTC - 5h — **Pop:** 37.660 — **Radios:** 46.000 — **Pr.L:** English — **E.C:** 60Hz, 110V — **ITU:** CYM.

RADIO CAYMAN (Gov. Comm.)

✉ Box 1110, George Town, Grand Cayman, Cayman Islands, British West Indies. ☎ +1 (345) 949 7799. 🖷 +1 (345) 949 6536. **Cable:** Broadcast Grandcayman.
L.P: Dir. of Inf. & Broadc: Loxley Banks.
MW: George Town 1205kHz 1kW.
FM: 105.3MHz 5kW, 89.9MHz 3kW — Brac 91.9/93.9MHz 0.25kW.
R. Cayman One on 1205kHz: 89.9/93.9MHz: 1100-0400. **N:** 1100W*, 1200W, 1300W, 1300Sun*, 1500W, 1700, 1900MF, 2000SS, 2100MF, 2300MF*, 2300SS, 0000W, 0300* — *) BBC.
R. Cayman Two on 91.9/105.3MHz: music and news.
V. by letter. Rec.acc.

ZFZZ (Comm.)

✉ Box 30110 SMB, George Town, Grand Cayman, Cayman Islands, British West Indies. ☎ +1 (345) 945 1166. 🖷 +1 (345) 945 1006.
L.P: Pres. & GM: Randy Merren. Op's Mgr: Tim Byrd. Music Dir: Peter Miles.
FM: George Town 99.9MHz 15kW — **D.PRGR:** 24h.
ANN: "This is ZFZZ, Grand Cayman", "Z-99", "Z-99-FM".

V. by letter or prepared card.

ICCI-FM (Educ.)

✉ International College of the Cayman Islands, Newlands,Savannah Post Office, Grand Cayman, Cayman Islands, British West Indies. **College** ☎ +1 (345) 947 1100. 🗎 +1 (809) 947 1210. **Station** ☎ +1 (345) 947 1212 (request line).
L.P: College Pres. & GM: Elsa M. Cummings, Ph.D.
FM: 101.1MHz 0.5kW.
D.PRGR: variable acc. to availability of student volunteers. Locally produced prgrs in English & Spanish for Grand Cayman residents.
ANN: "This is ICCI-FM, Newlands", "FM 101.1 ICCI".
V. by prepared card.

COSTA RICA

L.T: UTC -6h (Su: UTC -5h) – **Pop:** 3.350.000 – **Radios:** 780.000 – **Pr.L:** Spanish – **E.C:** 60Hz, 120/220V – **ITU:** CTR.

CONTROL NACIONAL DE RADIO (CNR)

✉ Ministerio de Gobernación y Policia, Ap.10006, 1000 San José.
☎ +506 2210992, 2219910. ① RadcontrolSJoséCostaRica.

CAMARA NACIONAL DE RADIO (CANARA)

✉ Ap.1583, 1002 San José. ☎ +506 2331845. 🗎 +506 2554483.
E-mail: canara@sol.racsa.co.cr **Web:** http://207.201.156.149/canara/
L.P: Exec. Dir: Luzmilda Vargas Gonzáles.

Mediumwaves: Call TI-,
Þ = also on shortwave, * = inactive, (r) = repeater, v = varying fq.

	Call	kHz	kW	Name and h. of tr.
1)	CAL	v530	18	R. Rumbo, Cartago: 1100-0600
2)	SCL	550	20/2	R. Santa Clara, Cd. Quesada: 1100-0300
3)	ELR	570	6.5	R. 570 "es la radio", Guadalupe: 1100-0500
4)	RN	*590	40	R. Nacional, San José: 1030 (Sun 1300)-0600
6)	RSU	610	4	R. Sistema Universal, Guadalupe: 1100-0700
5)	AD	640	20	R. Rica, San José: 1100-0400
7)	TNT	670	5	R. Monumental, San José: 1100-0400
12)	JC	700	10	R. Sonora, San José: 1100-0400
8)	HB	Þ730	20	R. Reloj, San José: 24h
9)	LX	760	20	Em. Columbia, San José: 24h
10)	RA	780	3	R. América, San José: 1000-0500 (Sun 0400)
43)	W	800	5	R. Tigre, San José: 24h
11)	GC	820	2.5	R. Centro, Tibás: 1130-0600
52)	JLS	850	20	R. Viva, Pavas: 1200-0500
13)	UCR	Þ870	10	R. Universidad de Costa Rica "R. U", San José: 1300-0600
7)	HOT	*890	10	R. Fabulosa, San José
14)	QM	910	1	R. Metrópolis Norte, Cd. Quesada: 1200-0600
14)	QM	910	1	R. Metrópolis Norte, Tibás (r: 910)
53)	RCR	930	5	R. Costa Rica, San José: 1200-0600
54)	CS	960	5	Continente Estéreo, Sto Domingo de Heredia: 1100-0600
16)	RI	980	20	R. Cordillera, Esparza: 1030-0600
17)	MIL	1000	10/1	R. Mil, San José: 24h
29)	TIC	v1020	2	R. Mil Veinte "LV de la Liberación", San José: 1130-0500
18)	AC	1040	5	R. Fides, San José: 1000-0500
9)	LX	1060	1	Em. Columbia, San Isidro del General (r: 760)
9)	LX	1060	1	Em. Columbia, Liberia (r: 760)
19)	FC	Þ1080	19	Faro del Caribe, San José: 24h
19)	FC	1080	1	Faro del Caribe, Chomes (r: 1080)
19)	FC	1080	1	Faro del Caribe, Liberia (r: 1080)
19)	FC	1080	1	Faro del Caribe, La Marina (r: 1080)
20)	SCR	1100	5	R. Chorotega, Santa Cruz: 0900-0600
15)	VAL	1100	1.5	R. Guápiles, Guápiles
8)	SHB	1120	3	R. Reloj, Liberia (r: 730)
7)	TNT	1140	1	R. Monumental, San Isidro del General (r: 670)
7)	TNT	1140	1	R. Monumental, San Carlos (r: 670)
9)	LX	1160	3	Em. Columbia, Puntarenas (r: 760)
22)	PJ	1180	20	R. Victoria, Heredia: 1200-0500
23)	AM	1200	5	R. Cucú, San José: 1000-0600
21)	Q	Þ1220	1	R. Casino, Limón: 1030-0600
9)	LX	1240	1	Em. Columbia, Nicoya (r: 760)
24)	HN	1260	6	R. Emaus, San Vito de Coto Brus
25)	HT	1280	2	R. Alajuela, Alajuela: 1000-0600
26)	LC	1300	7.5	R. La Fuente Musical, Cartago: 1030-0500
9)	LX	1320	1	Em. Columbia, San Carlos (r: 760)
27)	HR	1340	1	R. Sideral, San Ramón: 1000-0400
7)	TNT	1360	1	R. Monumental, Limón (r: 670)
7)	TNT	1360	1	R. Monumental, Puntarenas (r: 670)
28)	MS	1380	1	R. Guanacaste, Liberia: 1000-0500
28)	MS	1380	1	R. Guanacaste, San José (r: 1380)
24)	CJ	1400	12	R. Sinaí, San Isidro del General: 1000-0400
33)	RP	1420	5	R. Pampa, Nicoya: 1000-0300
30)	RSC	1440	12	R. San Carlos, Cd. Quesada: 1100-0400
9)	LX	1460	2	Em. Columbia, Limón (r: 760)
49)	AW	1480	1	R. Sol, Puntarenas: 1200-0400
58)	RC	1500	5	R. Cima, Cd. Quesada: 1100-0300
43)	SH	1520	1	R. Tigre, Turrialba (r: 800)
32)	RN	1560	6	R. Nicoya, Nicoya: 1000-0300
55)		1580	10	R. Mi País, Siquirres
34)	CLS	1580	0.25	R. Cultural Los Santos, San Marcos
35)	RCC	1580	0.25	R. Cultural de Corredores, Cd. Neily
36)	RSCLC	1580	0.25	R. Sist. Cultural, La Cruz
37)	RSCM	1580	0.25	R. Sist. Cultural Maleku, Tonjibe
38)	RSCN	1600	0.25	R. Sist. Cultural Nicoyano, Nicoya
39)	RCT	1600	0.25	R. Cultural, Turrialba
40)	RCBA	1600	0.25	R. Cultural, Buenos Aires
41)	RCP	v1600	0.25	R. Cultural, Pital
42)	RCU	1600	0.25	R. Cultural, Upala
44)	JV	Þ1600	1.5	R. 88 Estéreo, Pérez Zeledón: 1000 (Sun 1100)-0500
45)	LGJ	1600	1.5	R. 16, Grecia: 1100-0400
46)	MMCC	1600	2/1	R. Golfito, Pto Golfito: 24h
47)	MQ	1600	1.5	R. Pococí, Guápiles: 1100-0400

Shortwaves:

60)		3040		R. Puntarenas (r.) rel. 91.9 FM
59)		3210		R. Exterior de España, Cariari de Pococí: 1100-0400
8)	HB	v4832	3	R. Reloj, San José: 0100-1300
31)	AWR	5030	20	AWR Latin America, Cahuita: 1000-0600
19)	FC	5055	5	Faro del Caribe, San José: 2000-1200
21)	Q	v5955	1	R. Casino, Limón: 1030-2400
8)	HB	v6006	3	R. Reloj, San José: 1300-0100
44)	JV	6075	1.5	R. 88 Estéreo, Pérez Zeledón (**F.PI.**)
13)	UCR	6105	1	R. Universidad de Costa Rica "R. U", San José: 1300-0600
31)	AWR	6150	50	AWR Latin America, Cahuita: 1000-0600
19)	FC	6175	2.5	Faro del Caribe, San José: 1000-0600
31)	AWR	7375	50	AWR Latin America, Cahuita: 0300-0600, 0900-1400
48)		7385	30	R. For Peace Int., Cd. Colón: 2200-1700
48)		7585U	5	R. For Peace Int., Cd. Colón: 0000-1200
19)	FC	v9645	5	Faro del Caribe, San José: 1000-0600
31)	AWR	9725	50	AWR Latin America, Cahuita: 1100-1700, 2300-0600
31)	AWR	11870	50	AWR Latin America, Cahuita: 1400-1700
31)	AWR	13750	20	AWR Latin America, Cahuita: 1100-1700, 2300-0500
48)		15050	10	R. For Peace Int., Cd. Colón: 1700-0500
31)	AWR	15460	50	AWR Latin America, Cahuita: 2300-0300

For further R. Exterior de España relay freqs, see International Broadcasting section.

Addresses and other information:
1) Ap. 140, 7050 Cartago – Dir: Carlos Lafuente Fernández.
2) Ap. 221, 4400 Cd. Quesada -DG: Pbro Marco Solís V.
3) Más X Menos Guadalupe 25m este, P3, Tienda ADOC, 2100

Guadalupe.
4) Ap. 7-1980, 1000 San José – Dir: Lic. Iris Zamora Z.
5) Ap. 3835, 1000 San José – Dir: Lic. Armando Acuña Delgado.
6) Ap. 4666, 1000 San José – Dir: Luis Vargas McC.
7) Ap. 800, 1000 San José – DG: Guillermo Murillo Campos.
8) Ap. 341, 1000 San José – DG: Dr. Hernán Barquero M. de O.
9) Ap. 708, 1000 San José – Dir: C. Arnoldo Alfaro Chavarra.
10) Ap. 1951, 2100 Guadalupe – Dir: Iary Gómez.
11) Ap. 6133, 1000 San José – Own: Roberto Hernández R.
12) Ap. 708, 1000 San José – Dir: Javier Rojas Gonzales.
13) Ap. 2060, San Pedro – DG: Carlos Morales.
14) Ap. 25, 4400 Cd. Quesada – Dir: Limberth Quesada.
15) Ap. 266, 7210 Guápiles – Dir: Pbro Walter Marchena.
16) Ap. 497, 2050 San Pedro de Monte de Oca (or: Ap. 80, Esparza) – Own: Parmenio Medina.
17) Ap. 10. 001, 1000 San José – Dir: R. Giralt C.
18) Ap. 5079, 1000 San José – DG: Pbro William Lizano A.
19) Ap. 2710, 1000 San José – Mgr: Juan J. Ochoa F. **English: Shortwave** 0300-0400.
20) Ap. 92, 5175 Santa Cruz – Gte: Pbro E. Montes de Oca C.
21) Ap. 287, 7300 Puerto Limón – Dir: Luis Grau V. **FM:** 98.3.
22) Ap. 298, 3000 Heredia – Gte: Emilio Piedra G.
23) Ap. 1128, 1000 San José – Own: José Rodolfo Traube.
24) Ap. 262, San Isidro del General – Dir: Pbro Alvaro Coto O.
25) Ap. 122, 4050 Alajuela – DG: Flora Mª Herrera T.
26) Ap. 596, 7050 Cartago – Gte: Ramón Marrero R.
27) Ap. 73, 4250 San Ramón – Mgr: Roger Retana C.
28) Ap. 27, 5600 Liberia, (✉ Ap. 6462, 1000 San José) – DG: Mario Salgado S.
29) Ap. 8130, 1002 San José – Dir: Pastor Moises Cressente.
30) Ap. 25, 4400 Cd. Quesada – DG: José Angel Moya R.
31) Ap. 1177, 4050 Alajuela – DG: Victor Shepherd. Owned and operated by Adventist World Radio, Latin America. **Ann:** "Radio Mundial Adventista, La Voz de la Esperanza".
32) Ap. 50, 5200 Nicoya – DG: Juan Cheng Gutierrez.
33) Ap. 66, 5200 Nicoya – Dir: Carlos Ramirez Arrieta.
34-42) St`s are affiliated to Instituto Costarricense de Enseñanza Radiofónica, Ap. 132, 2050 San Pedro de Montes de Oca – Coordin: Marco Antonio Gómez V.
43) Ap. 695, 2100 Guadalupe – DG: Fernando Hernández.
44) Ap. 827, Pérez Zeledón – Dir: Juan Vega.
45) Ap. 16, 4100 Grecia – Own: Luis G. Jiménez R.
46) Ap. 11, Golfito, (✉ Ap. 5362, 1000 San José) – Dir: Marco A. Muñoz C.
47) Ap. 160, 7210 Guápiles – Gte: Jorge Quesada B.
48) Ap. 88, Santa Ana (✉ P.O. Box 29728, Portland, OR 97220, USA) – DG: Debra Latham. Coordin: Joe Bernard.
49) Cevichito 25 al oeste, Puntarenas – DG: Humberto Ramírez.
50) Ap. 400, 2150 Moravia.
51) Ap. 10.270, 1000 San José – DG: Roxie Blenviuda de Sotela.
52) Ap. 636, 1200 Pavas – Dir: Federico Sosto Peralta.
53) Ap. 6462, 1000 San José – Dir: Mario Alberto Salgado.
54) Ap. 130, 3100 Sto Domingo de Heredia – Dir: Eugenio Broune Ugarte.
55) Siquirres.
57) Ap. 21, 2010 Zapote, San José – DG: Arnoldo Alfaro Chavarra.
58) Ap. 300, 4400 Cd. Quesada – Gte: José Angel Moya Rodriguez.
59) Calle 31 y 33, Av. 14, San José. See International Broadcasting section.
60) Ap. 708, San José.

FM in San José and vicinities: 44) 88.5 – 31) 88.7 R. Lira – 3) 89.1 89 Ya – 49.5 R. Sendas FM – 50) 89.9 Hit90 – 6) 90.7 – 91.1 R. Juvenil – 3) 91.5 – 60) 91.9 – 9) 92.7 – 18) 93.1 – 7) 93.5 – 50) 93.9 Sonido Latino – 8) 94.3 – 14) 94.7 – 7) 95.1 "95.1" – 17) 95.5 Power95 – 54) 95.9 – 11) 96.3 – 13) 96.7 "R. U" – 19) 97.1 – 5) 97.5 – 51) 97.9 Titania – 25) 98.3 – 9) 98.7 – 57) 99.1 R. Sabrosa – 28) 99.5 R. Dos – 99.9 R. Azul – 23) 100.3 R. Cristal – 17) 100.7 – 4) 101.5 – 13) 101.9 "R. U" – 50) 102.3 Super R – 9) 102.7 Uno – 3) 103.1 R.103 – 8) 103.5 24) 103.9 R. Sinai – 51) 104.3 R. Sensación – 104.7 R. Emperador – 105.1 R. Omega – 105.5 R. Omega – 105.9 Puntarenas – 106.3 Alfa Estéreo – 107.1 R. Atlántida – 107.5 R. Paladín.

CUBA

LT: UTC-5h (Su: UTC-4h) – **Pop:** 11.057.000 – **Radios:** 3.608.000 – **PR. L:** Spanish – **E.C:** 60Hz, 115/120V – **ITU:** CUB.

MINISTERIO DE COMUNICACIONES (MC)

Dirección General de Telecomunicaciones
✉ Plaza de la Revolución, Ciudad de la Habana.
LP: Dir. Tec: Carlos Martínez Alburne.

INSTITUTO CUBANO DE RADIO Y TELEVISION (ICRT)
✉ Edif.Radiocentro, Av. 23 Nᵖ 258, Vedado, Habana 4. ☎ +53 (32) 1568. 🖷 +53 (31) 1723.
LP: Pres: Enrique Román. Dir. Int.Rel: Susana Sardiñas.

Mediumwaves: Call CM-, v = varying fq.

NATIONAL NETWORKS: Radio Enciclopedia, Radio Musical Nacional, Radio Progreso, Radio Rebelde, Radio Reloj, Radio Taíno.

Call	kHz	kW	Location
HV	540	1	Sancti Spíritus
DN	550	10	Guantánamo
	550	30	Pinar del Río
	550	1	Manzanillo
DC	570	10	Pilón
DF	580	5	Baracoa
AM	580	10	Mantua
HI	590	30	Santa Clara
KV	600	150	Urbano Noris
AN	610	1	Bahía Honda
HI	610	1	Trinidad
KF	620	1	Moa
GN	620	30	Colón
BC	640	50	Guanabacoa
DQ	640	10	Las Tunas
DD	640	5	Las Mercedes
KU	650	1	Stgo de Cuba
DC	650	1	Media Luna
HG	660	30	Santa Clara
Q	670	50	Arroyo Arenas
HN	680	1	Cienfuegos
JV	680	10	Ciego de Avila
DB	680	1	Stgo de Cuba
	690	20	Jovellanos
DU	700	1	Guantánamo
GA	700	1	Sancti Spíritus
W	710	150	La Julia
KJ	710	10	Holguín
HQ	710	50	Santa Clara
JN	710	30	Camagüey
HC	720	1	Cienfuegos
BB	730	10	La Fé
JL	740	10	Camagüey
KP	740	20	Sagua de Tánamo
HV	750	1	Trinidad
CD	760	5	Guanabacoa
DE	780	20	Las Mercedes
AQ	v790	30	Pinar del Río
DT	v800	1	Manzanillo
DW	810	10	Guantánamo
JT	820	10	Morón
DE	820	10	Contramaestre
DQ	840	1	Las Tunas
BQ	840	1	La Fé
HL	850	1	Trinidad
DB	850	5	Baracoa
BL	860	10	Arroyo Arenas
HL	870	10	Sancti Spíritus
DT	870	1	Pilón
AF	880	30	Pinar del Río
HD	890	1	Santa Clara
DZ	890	10	Stgo de Cuba
KP	900	50	Cacocum
GL	910	10	Bolondrón
GL	920	1	Unión de Reyes
GB	930	1	Cárdenas
JS	930	10	Ciego de Avila
KN	930	1	Stgo de Cuba
DP	930	10	Las Tunas
GU	940	1	Colón
KD	940	10	Holguín
HI	950	1	Sancti Spíritus
GF	960	1	Matanzas
DJ	960	10	Guantánamo
HJ	960	1	Cienfuegos

	kHz	kW	Location
JD	960	0.25	Ciego de Avila
DE	980	1	Bayamo
HB	1000	1	Sancti Spíritus
JB	1000	1	Camagüey
KM	1010	1	Holguín
	1100		Varadero
HA	1130	1	Santa Clara
CG	1140	5	Loma de la Cruz
BA	1180	50	Villa María
DB	1180	1	Mayarí Arriba
BM	v1180	1	Nueva Gerona
	1220		Cienfuegos
BF	1260	5	Arroyo Arenas
JG	1270	10	Camagüey
GF	1280	1	Varadero

Call	kHz	kW	Location
CS	1290	5	La Pastora
BQ	1290	1	La Habana
DA	1300	1	Las Tunas
DA	1320	1	Stgo de Cuba
HA	1320	1	Sancti Spíritus
GX	1400	1	Matanzas
AL	1410	1	Pinar del Río
DA	1500	1	Holguín
DA	1510	1	Moa
	1560	1	Ciego de Avila

PROVINCIAL NETWORKS AND LOCAL STATIONS

	Call	kHz	kW	Name and location
L01)	CA	820	10	R. Ciudad de la Habana "La Emisora Joven", Santa Catalina, CH
P01)	HW	840	10	Doblevé, Santa Clara, VC
P02)	KC	840	1	R. Revolución, Stgo de Cuba, SC
P03)	FA	910	10	R. Cadena Agramonte, Camagüey, CM
L02)	BL	950	1	R. Metropolitana, Arroyo Arenas, CH
L03)	CK	980	1	COCO-El Periodico del Aire, Sapo, CH
L04)	KR	980	1	La Voz del Níquel, Moa, HO
P04)	AC	1000	10	R. Guamá, Los Palacios, PR
P04)	AP	1010	10	R. Guamá, Guane, PR
P04)	AP	1020	1	R. Guamá, Bahía Honda, PR
P04)	AX	1030	1	R. Guamá, La Palma, PR
P05)	KT	1040	1	R. Victoria, Puerto Padre, LT
P05)	KT	1050	10	R. Victoria, Las Tunas, LT
P05)	KT	1060	1	R. Victoria, Amancio Rodríguez, LT
P06)	DX	1060	5	CMKS-Trinchera Antiimperialista, Baracoa, GU
P04)	AS	1070	1	R. Guamá, Pinar del Río, PR
P06)	KS	1070	10	CMKS-Trinchera Antiimperialista, Guantánamo, GU
P07)	CH	1080	10	R. Cadena Habana, Güines, HA
P04)	AP	1090	1	R. Guamá, Santa Lucia, PR
P07)	CH	1090	1.5	R. Cadena Habana, La Salud, HA
P08)	KO	1090	1	R. Angúlo, Moa, HO
P08)	KO	1100	1	R. Angúlo, Manes, HO
P08)	KO	1110	10	R. Angúlo, Holguín, HO
P07)	CH	1120	5	R. Cadena Habana, Artemisa, HA
P08)	KO	1120	1	R. Angúlo, Mayarí, HO
P08)	KO	1130	5	R. Angúlo, Sagua de Tánamo, HO
P09)	KX	1140	1	R. Bayamo, Media Luna, GR
P09)	KX	1150	10	R. Bayamo, Entronque Bueycito, GR
P09)	KX	1160	1	R. Bayamo, Pilón, GR
P06)	KS	1170	10	CMKS-Trinchera Antiimperialista, Maisí, GU
P10)	HT	1190	1	R. Sancti Spíritus, Yaguajay, SS
P11)		1190	1	R. 26, La Caridad, MA
L05)	BS	1200	0.25	R. Ariguanabo, San Antonio de los Baños, HA
P10)	HT	1200	1	R. Sancti Spíritus, Trinidad, SS
P02)	KC	1200	10	R. Revolución, Palma Soriano, SC
P10)	HT	1210	10	R. Sancti Spíritus, Sancti Spíritus, SS
P02)	KC	1210	1	R. Revolución, Chivirico, SC
P02)	KC	1210	1	R. Revolución, Mayarí Arriba, SC
P11)	GY	1220	10	R. 26, Central España, MA
P11)	GJ	1230	3	R. 26, Unión de Reyes, MA
P11)	GW	1240	10	R. 26, Bolondrón, MA
P01)	HW	1250	1	Doblevé, Caibarién, VC
P09)	KS	1250	0.25	R. Bayamo, Imías, GU
P11)	GH	1260	1	R. Victoria de Girón, Torriente

	Call	kHz	kW	Name and location
L06)	KW	1280	1	R. Mambí, Stgo de Cuba, SC
L07)	JC	1280	1	R. Rectángulo, Guáimaro, CM
P01)	HW	v1290	1	Doblevé, Rancho Veloz, VC
P01)	HR	1310	1	Doblevé, Sagua La Grande, VC
P12)	BN	1310	10	R. Caribe, Nueva Gerona, IJ
L13)	AD	1330	1	R. Jaruco, Artemisa, HA
L09)	DO	1340	1	R. Banes, Banes, HO
P13)	HU	1340	10	R. Ciudad del Mar, Palmira, CI
P13)	HU	1350	1	R. Ciudad del Mar, Aguada de Pasajeros, CI
L10)	KY	1350	10	R. Libertad, Puerto Padre, LT
L11)		1350	1	R. Portada de la Libertad, Holguín, HO
P03)	FA	1360	1	R. Cadena Agramonte, Rodolfo Ramírez Esquível, CM
P03)	FA	1370	1	R. Cadena Agramonte, Nuevitas, CM
L12)	DV	1370	1	R. Siboney, Stgo de Cuba, SC

	Call	kHz	kW	Name and location
P03)	FA	1380	10	R. Cadena Agramonte, Central Brasil, CM
L13)	BT	1390	1	R. Jaruco, Jaruco, HA
P03)	FA	1400	1	R. Cadena Agramonte, Guaimaro, CM
L14)	JW	1410	1	R. Santa Cruz, Santa Cruz del Sur, CM
L16)	HS	1420	1	R. Caibarién, Caibarién, VC
L15)		1420	0.25	R. Llanuras, Colón, MA
P14)	JY	1430	10	R. Surco, Morón, CA
L17)	BU	1440	0.25	R. Güines, Güines, HA
P14)	JP	1440	10	R. Surco, Ciego de Avila, CA
L18)	JF	1450	1	R. Maboa, Amancio Rodríguez, LT
P03)		1460	1	R. Cadena Agramonte, Solas
L19)	GE	1470	1	R. Ciudad Bandera, Cárdenas, MA
L20)	JI	1480	0.25	R. Florida, Florida, CM
L21)	BW	1490	0.25	R. Camoa, San José de las Lajas, HA
L22)	DH	1490	1	R. Mayarí, Mayarí, HO
L23)	KQ	1500	0.25	R. Majaguabo, San Luis, SC
L24)	KZ	1520	1	R. Baragua, Palma Soriano, SC
L25)	IX	1530	1	R. Morón, Morón, CA
L26)	ES	v1540	1	R. Sagua, Sagua La Grande, VC
L27)		1540	1	R. Juvenil, Holguín, HO
L28)	JQ	1550	1	R. Nuevitas, Nuevitas, CM
P03)	FA	1580	1	R. Cadena Agramonte,Santa Cruz del Sur, CM
L29)	DF	1590	0.5	R. Granma, Manzanillo, GR

Shortwaves:

		v5025	10	R. Rebelde, Bauta

R. Habana Cuba: see International Broadcasting section.

FM in La Habana: 87.8 COCO – 90.3 R. Progreso – 92.1 R. Taíno – 93.7 R. Enciclopedia – 95.0 R. Musical Nacional – 95.9 R. Reloj – 97.9 COCO – 100.7 R. Cadena Habana – 102.9 R. Metropolitana.

Provinces: CA = Ciego de Avila, CH = Ciudad Habana, CI = Cienfuegos, CM = Camagüey, GR = Granma, GU =Guantánamo, HA = Habana, HO = Holguín, IJ = Isla de laJuventud, LT = Las Tunas, MA = Matanzas, PR = Pinar del Río,SC = Santiago de Cuba, SS = Sancti Spíritus, VC = Villa Clara.

NATIONAL NETWORKS
R. Enciclopedia, Calle N.Nb 266, Vedado, La Habana 4 – **D.Prgr:** Light instrumental music, information about culturalaffairs, timechecks and encyclopedic "facts".
R. Musical Nacional (CMBF), Av. 23 Nb 258, Vedado, La Habana 4 – **D.Prgr:** Classical music and programmingof a cultural nature.
R. Progreso "La Onda de la Alegría", Áp. 30-42 (or: Infanta 105), La Habana 3.
R. Rebelde, Ap. 6277, La Habana 6 (or: Edif.Radiocentro, Av. 23 Nb 258, Vedado, La Habana 4) – Head of Public Rel. Dept: Jorge Luís Mas Zabala.
R. Reloj (CMCD), P y 23, Vedado, La Habana 4 – **D.Prgr:** Continuous news and commentary with background of clock-likepulses marking the seconds.
R. Taíno, Av. 23 Nb 258, Vedado, La Habana 4 – Bilingual ID's. Music, features and "commercials" for tourists and visitors to Cuba.

PROVINCIAL NETWORKS
P01) Ap. 376, (or: Parque Vidal el Martha Abreu y Pao Chao), Santa Clara 50100.
P02) Ap. 232, Santiago de Cuba.
P03) Ap. 140, Camagüey 70100.
P04) Ap. 14 (or: Colón entre Adela Azuay y Juan Gualberto Gómez),

Pinar del Río 20100.
P05) Ap. 21, Las Tunas 72510 – CE: Carlos López M.
P06) Donato Mármol 409, Guantánamo 95100.
P07) Av. 51 el 128 y 130, La Habana.
P08) Ap. 14, Holguín 80100.
P09) Ap. 74, Bayamo 85100.
P10) Circunvalación Olivos 1, Sancti Spíritus 60100.
P11) Ap. 51 (or: Contreras 69), Matanzas 40100.
P12) Nueva Gerona, Isla de la Juventud.
P13) Ap. 290, (or: Av. 58 Nⷠ 3311), Cienfuegos 55100.
P14) Ap. 183, Ciego de Avila.

LOCAL STATIONS
L01) Calle N.Nⷠ 266, Vedado, La Habana 4.
L02) Edif. Focsa, Calle 17 esq.N.Vedado, 10400 La Habana.
L03) Edif. Focsa, Calle 17 esq.N.Vedado, 10400 La Habana.
L04) Moa, HO.
L05) Av. 41 Nⷠ 5614, San Antonio de los Baños 32500.
L06) Stgo de Cuba, SC
L07) Guáimaro, CM.
L09) Banes, HO.
L10) Av. de la Libertad 95, Puerto Padre, LT.
L11) Niquero, Holguín 80100.
L12) Santiago de Cuba. **D.Prgr:** Instrumental musicwith timechecks by female announcer.
L13) Av. 25 Nⷠ 1810, Jaruco 32800.
L14) Calle F.Nⷠ 31, Santa Cruz del Sur 73200.
L15) Ap. 97, Colón, MA 42400.
L16) Caibarién, VC.
L17) Calle 76 Nⷠ 7707, Güines 33900.
L18) Amancio Rodríguez, LT.
L19) Calle 13 y 5a Av. Cárdenas, MA.
L20) Florida, CM.
L21) Av. 47 Nⷠ 10202, San José de las Lajas 32700.
L22) Mayarí, HO.
L23) San Luis, SC.
L24) Palma Soriano, SC.
L25) Ap. 157, Morón 67210.
L26) Sagua La Grande, VC.
L27) Holguín 80100.
L28) Ap. 46, Nuevitas 72510.
L29) Ap. 220, Manzanillo 87510.

AFRTS (US Navy)
✉ FPO New York, NY 09595, USA.
MW: Guantánamo Bay: 1340kHz 0.25kW
FM: 102.1 MHz 0.5kW (stereo), 103.1 MHz 0.5kW – **D.Prgr:** 24h on 1340kHz/102.1 MHz. Rel AFRTS satellite sce on 103.1 MHz.

L.T: UTC - 4h — **Pop:** 83.600 — **Radios:** 65.000 — **Pr.L:** English, Patois — **E.C:** 50Hz, 240V — **ITU:** DMA.

DOMINICA BROADCASTING CORP. (Gov. Comm.)
✉ P.O. Box 1, Victoria Str, Roseau. ☎ +1 (767) 448 3282/3. 🖷 +1 (767) 448 2918. **Cable:** RADOM Dominica. **E-mail:** dbs@cybersevices.com
Web: http://www.dbsradio.com/
L.P: GM: Dennis Joseph. PD: Shermaine Green-Brown. CE: Fred White.
MW: Hillsborough 595kHz 10kW.
FM: 88.1/88.6/89.5/103.2/103.6MHz.
D.PRGR: 0900-0300. **N:** 1000, 1030, 1100(BBC), 1400, 1715, 2000, 2100, 2200. **Patois:** 1800-1930MF — **ANN:** "DBS Radio".
V. by QSL-card — **F.PL:** 20kW.

GOSPEL BROADCASTING CORPORATION (Rlg.)
✉ P.O. Box 205, Roseau. ☎ +1 (767) 44 84391. (✉ in USA: P.O. Box 464, Orland Park, IL 60462-0464).
L.P: GM: W. K. DeBoer. CE: F. R. Guiste.
MW: ZGBC 740kHz 10kW — **FM:** 90.6/96.6/102.1MHz.
D.PRGR: 0900-0100 on MW, 24h on FM. **BBC N:** 1100W, 1200W, 2200W. **Local N:** 1355. **Patois:** Sun 1730-1800, Wed 1215-1230.
ANN: "Caribbean Voice of Life from Dominica" — **V.** by letter.

VOICE OF THE ISLANDS (Rlg.)
✉ P.O. Box 2402, Roseau. ☎ +1 (767) 44 84042. 🖷 +1 (767) 44 80938.
L.P: GM: Edward A. Alexander.
MW: 860kHz 10kW — **FM:** 96.1MHz 1kW (stereo).

D.PRGR: 1015-0300. **BBC N:** 1100, 1200.
ANN: "Voice of the Islands - VOI" — **F.PL:** additional st's.

Radio Caribbean International: FM 98.1MHz 24h (see St. Lucia).
Kairi FM: no further details received.

L.T: UTC-4h — **Pop:** 7.835.000 — **Radios:** 1.180.000 — **Pr.L:** Spanish — **E.C:** 60Hz, 110V — **ITU:** DOM.

DIRECION GENERAL DE TELECOMUNICACIONES
✉ Calle Isabel la Católica 73, Santo Domingo. ☎ +1 (809) 688-8955.
L.P: Rubén Montás.

ASOCIACION DOMINICANA DE RADIODIFUSORAS (ADORA)
✉ Calle Paul Harris 3, Centro de los Héroes, Santo Domingo. ☎ +1 (809) 535-4057 — **L.P:** Pres.Mrs.Ivelise de Torres.

Mediumwaves: Call HI-,
Þ = also on shortwave, * = inactive, v = varying fq.

	Call	kHz	kW	Name and h. of tr.
1)	CM	540	5	R. ABC, Sto Domingo: 0900-0400
2)	AA	560	5	R. Ritmos, Santiago: 1000-0400
52)	MS	570	10/5	R. Cristal, Sto Domingo
8)	AF	580	3	R. Monte Cristi: 1000-0400
4)	DV	590	10/5	R. Santa María, La Vega: 0900-0400
6)	JR	610	5/1	R. Acción, Santiago: 24h
7)	SD	620	10	R. Televisión Dominicana, Sto Domingo: 0900-0500
7)	SD	640	10	R. Televisión Dominicana, Santiago (r: 620)
9)	AT	650	15/10	R. Universal, Sto Domingo: 24h
10)	AM	660	3	Super Regional, Santiago: 1000-0400
59)	BS	670	5	R. Dial, San Pedro de Macorís: 24h
11)	JX	680	3	R. Zamba, San Ignacio de Sabaneta: 0830-0300
12)	AW	690	10	R. Guarachita, Sto Domingo: 0900-0400
13)	DC	700	1.5	R. Mao, Valverde: 1000-0400
14)	AQ	720	5	R. Norte, Santiago: 0930-0400
15)	Z	730	10	R. HIZ, Broadcasting Nacional, Sto Domingo: 0930-0500
16)	DB	*750	5	R. Alegre, Santiago
17)	CO	760	5	R. Cordillera, Santiago
18)	MD	770	10	R. Popular, Santiago: 1000-0400
19)	BO	780	0.5	R. Constanza: 1100-0200
20)	L	790	5	LV del Trópico, Sto Domingo: 0930-0400
24)	AV	810	1	R. Baní: 1100-0300
21)	AZ	820	5/1	R. Santiago: 1000-0400
22)	JB	830	10	R. HIJB, Sto Domingo: 1000-0500
23)	AB	840	1	R. Isabel de Torres, Puerto Plata: 0930-0330
72)	GA	850	5	R. Guarocuya, Barahona: 1000-0400
5)	UA	850	5	R. Clarín, Santiago (r: 860)
5)	UA	860	10	R. Clarín, Sto Domingo: 0900-0500
25)	VG	870	5	R. La Vega: 0930-0400
26)	OR	880	1	AM-88, Mao: 1000-0400
27)	PJ	890	4/5	R. Continental, Sto Domingo: 0930-0600
28)	BN	900	5/1	R. Puerto Plata: 0900-0400
29)	LB	910	1	R. 91, Bonao: 0930-0300
9)	BA	920	10	R. 9-20 AM-Stereo, Sto Domingo: 24h
31)	CK	930	10	Ondas del Yaque, Santiago: 0945-0400
32)	G	950	10	R. Popular, Sto Domingo: 0900-0400
33)	FF	960	5/1	LV del Atlántico, Puerto Plata: 1000-0500
25)	VP	970	5	R. Olímpica, La Vega: 0930-0400
35)	FA	*980	5	LV de las Fuerzas Armadas, Sto Domingo
36)	SA	990	5	R. Cibao, Santiago: 1000-0400
37)	HB	1000	5/1	R. Beller, Dajabón: 1000-0300
38)	JA	1010	10	R. Comercial, Sto Domingo: 0900-0400
38)	JA	1010	1	R. Comercial, Salcedo (r: 1010)
38)	JA	1010	1	R. Comercial, San Juan (r: 1010)
30)	TS	1020	10	R. Enriquillo, Tamayo: 0900-0400
39)	DL	1030	10	R. Novedades, La Vega: 1000-0400
40)	ON	1040	10	R. Cadena de Noticias, Sto Domingo: 24h
14)	CB	1050	10	R. Hispaniola, Santiago: 0930-0400

	Call	kHz	kW	Name and h. of tr.
42)	KF	1060	1	R. Azua: 1000-0200
43)	RV	1060	1	R. Mar, San Pedro de Macorís: 0900-0300
44)	BI	1070	5/1	HIBI R. 1070, San Francisco de Macorís: 0900-0400
45)	MC	1080	1	R. Ambar, Sto Domingo
46)	JM	1090	3/1	R. Amistad, Santiago: 0930-0400
50A)	RB	1090	1/0.5	R. Jimaní: 0900-0400
47)	HD	1100	1	R. Oriente, San Pedro de Macorís: 0900-0400
48)	MP	1100	1	R. Ocoa, San José de Ocoa: 1200-0200
49)	PS	1100	1	R. Nagua: 0900-0200
51)	TC	1110	2.5	R. Jarabacoa: 1000-0400
95)	OS	1110	1	R. Marién, Dajabón: 0930-0200
52)	CN	1120	10	R. Metro 1120 AM Stereo, Sto Domingo: 1000-0400
40)	RL	1130	10/1	R. Cadena de Noticias, Santiago: 24h
	Call	**kHz**	**kW**	**Name and h. of tr.**
54)	RA	1140	5	R. Anacaona, San Juan de la Maguana: 1100-0400
55)	AS	Þ1150	5	Onda Musical, Sto Domingo: 1000-0300
56)	BG	1160	5/1	Radiolandia, Santiago: 0900-0400
57)	BE	1180	10	R. Mil, Sto Domingo: 0930-0500
58)	AG	1190	10	R. Azul, Santiago: 1100-0400
50B)	MR	1200	1	R. Caracol, Azua: 1000-0400
61)	CJ	1210	5	R. Merengue, San Francisco de Macorís: 18h
62)	N	*1220	10	R. HIN, Sto Domingo
63)	NG	1230	1	R. Ideal, Moca: 1000-0300
64)	AU	1240	1	R. Revelación, Puerto Plata: 0900-0400
50C)	CV	Þ1240	5/0.5	R. Barahona: 0900-0400
66)	BC	1250	5	LV del Progreso, San Francisco de Macorís: 1000-0400
67)	RS	1250	5	El Sonido del Este, La Romana: 0930-0430
38)	T	1260	1	R. Recuerdos, Sto Domingo: 0900-0400
14)	DA	1270	5	R. Hit, Santiago: 0930-0400
69)	TA	1270	1	R. Ambiente, Baní: 1000-0400
71)	HZ	1280	1	R. Clave, Monte Plata: 1000-0300
70)	VM	1290	1	R. Bonao: 1000-0400
74)	KQ	1300	1	R. Radio, Sto Domingo: 0900-0600
75)	MH	1310	1	R. Real, La Vega: 1100-0400
7)	SD	1310	1	R. Televisión Dominicana, El Seibo (r: 620)
68)	VR	1330	1	R. Villa-La Sencilla, Sto Domingo
35)	FA	*1330	3	LV de la Fuerzas Armadas, Moca
76)	BZ	1340	1	R. Centro, San Juan de la Maguana: 1000-0300
77)	PM	1350	1	R. Rutas Musical, La Romana: 1000-0400
7)	SD	1360	1	R. Televisión Dominicana, Monte Cristi (r: 620)
7)	SD	1360	1	R. Televisión Dominicana, La Vega (r: 620)
7)	SD	1370	1	R. Televisión Dominicana, Barahona (r: 620)
35)	FA	*1370	5	LV de las Fuerzas Armadas, Elías Piña
79)	RP	1370	5	R. Seybo, El Seybo: 1000-0300
80)	SC	1380	5/1	R. Nacional, Santiago: 1000-0400
7)	SD	1380	1	R. Televisión Dominicana, San Juan (r: 620)
81)	AR	1390	1	R. San Cristóbal: 1100-0300
82)	AC	1400	1	Ondas del Valle, La Vega: 1100-0200
7)	SD	1400	1	R. Televisión Dominicana, San Juan (r: 620)
7)	SD	1400	1	R. Televisión Dominicana, Pedernales: (r: 620)
83)	AE	1410	1	R. Revelación en América, Sto Domingo: 24h
84)	RM	1410	3	R. Sol, Higüey: 1000-0300
85)	GG	1410	1/0.5	R. Grí-Grí, Río San Juan: 1000-0300
50D)	CH	1410	3/0.5	R. 14-10, Barahona: 0900-0400
86)	FD	1420	15	R. Oro, Cotuí: 18h
34)	SC	1430	3	R. Emanuel, Santiago: 1000-0500
88)	FS	v1440	1	R. Bahía, Nagua: 1000-0500
89)	AD	1440	5	R. San Juan: 1000-0400
90)	AK	1440	5	R. Cristocéntrico, Sto Domingo: 1000-0400
87)	LF	1440	1	R. Cayacoa, Higüey: 0900-0400
91)	AC	1450	10	R. Util, Salcedo: 0900-0400
92)	AN	1460	0.5	R. Magua, Hato Mayor
93)	DE	1470	1	R. Super 56, San Francisco de Macorís:

				1000-0400
98)	AH	1480	5	R. VEN, Sto Domingo: 1000-0400
7)	SD	1490	1	R. Televisión Dominicana, Puerto Plata (r: 620)
96)	AP	1490	1	La 54 AM, Moca: 1000-0400
97)	PA	1500	0.5	R. Color, Higüey: 0900-0400
98)	BL	1510	10/3	R. Pueblo, Sto Domingo: 1000-0300
99)	WJ	1520	1	R. Samaná "R. 15-20"; 0930-0400
38)	FB	1540	1	R. Criolla Comercial, Sto Domingo
41)	BU	v1540	1	LV de La Romana: 0930-0400
50E)	PZ	1560	1/0.5	R. Pedernales: 0900-0400
60)	AJ	Þv1570	10	R. Amanecer Internacional, Sto Domingo: 1000-0400
50F)	PK	1580	1	R. Neiba: 0900-0400
		1590		**F.PI:** Santiago
65)	EG	1600	5	R. Revelación en América, Sto Domingo: 24h

Shortwaves:

55)	AS	v4780	1	Onda Musical, Sto Domingo: 1000-0300 (r. 4766)
3)	MI	v5012	1	R. Cristal Int., Sto Domingo: 2100-0300
50C)	CV	5160	1	R. Barahona: irr.
60)	IJ	v6025	1	R. Amanecer Internacional, Sto Domingo: 1000-0400
73)	AM	v6235	0.7	R. Quisqueya, Sto Domingo: -0430

Addresses and other information

1) Ap. 517, Sto Domingo – DG: Freddy Mariano Cruz.
2) Ap. 581, Santiago – Own: Johannes W. Streese.
3) Ap. 894 (or: Calle Pepillo Salcedo 18, Altos) Sto Domingo – DG: Dario Badía.
4) Ap. 55, La Vega – Dir: Antonio Lluberes. **FM:** 97.9.
5) Ap. 205-2, Sto Domingo – Pres: José Penzo.
6) Ap. 556, Santiago – Dir: Jaime Nelson Rodríguez.
7) Ap. 869, Sto Domingo – Dir: George Rodríguez Dabas. CE: Andrés Cruz – **F.PI:** SW.
8) Ap. 52, Monte Cristi – DG: Felipe Lemolne.
9) Ap. 2000, Sto Domingo – DG: Héctor José Torres.
10) Ap. 26, Santiago – Dir: Marcos Tulio Capeda.
11) Ap. 2, San Ignacio Sabaneta – Dir: Manuel de Js. Thomas B. **FM:** 104.5.
12) Calle Palo Hincado 302, Sto Domingo – DG: R. A. Aracena.
13) Ap. 20, Valverde-Mao (✉ Ap. 789, Santiago) – Pres: José Raposo P.
14) Ap. 454, Santiago – DG: Ing. Juan Heriberto Medrano. **FM:** 107.3.
15) Ap. 68, Sto Domingo – Dir: Oscar Ureña.
16) Ap. 112, Santiago – Dir: A. Vanderhorst.
17) Calle Emilio A. Morel esq. Luis Pérez, Ensanche La Fé, Sto Domingo – Own: J. Jiménez Maxwell.
18) Ap. 1636 (or: Calle El Sol 51, 3a Planta, Edif. Lamarche Alvarez), Santiago – Dir: Rafael Corporán Jr.
19) Calle Duarte 17, Constanza – Dir: R. M. Tactuk.
20) Ap. 335, Sto Domingo – Own: J. Custals.
21) Ap. 282, Santiago – DG: Héctor José Torres. **FM:** 99.1.
22) Edif. Teleantillas, Carr. Duarte km 7½, Sto Domingo – DG: Hector Olivo.
23) Ap. 146, Puerto Plata – Own: Antonio Damián Brisso B.
24) Calle Pres. Billini esq. Duarte, Baní – DG: Manuel E. Bello.
25) Ap. 203, La Vega – DG: Ing. Juan Heriberto Medrano. **FM:** 100.7 + 104.9.
26) Ap. 80, Mao, Valverde – DG: Sergio García. **FM:** 106.7.
27) Ap. 156, Sto Domingo – Dir: José Lluberes.
28) Calle Beller 35, Puerto Plata – Dir: Waldo R. Musa.
29) Calle Mella 50, Boano – Dir: Salvador Chestaro.
30) Ap. 99, Tamayo – Dir: Pablo G. Schildermaus.
31) Ap. 225, Santiago.
32) Ap. 928, Sto Domingo – Dir: R. Corporán de los Santos.
33) Duarte 65, altos, Puerto Plata – Dir: Antonio Baduí.
34) Ap. 42, Santiago – DG: Manuel Martínez. **FM:** 89.1.
35) Ap. 1350, Sto Domingo – Dir: Osvaldo Capeda C.
36) Ap. 141, Santiago – Own: July Mermúdez.
37) Hno. Martin Juffermans, Dajabón – Dir: Nelson Ricards O.
38) Ap. 1322 (or: E. A. Morel 27), Sto Domingo.
39) Las Arboledas 5, La Vega – Dir: M. C. Fernández.
40) Unicentro Plaza. Av. 27 de Febrero, Esq. A. Lincoln, Sto Domingo – Pres. Adm: Ing. Irvin Pérez P.
41) Ap. 213, La Romana – Pres: N. Mariñez.
42) Calle Emilio Prud'homme 17A, Azua – Dir: J. A. Félix A.
43) Ap. 476, San Pedro de Macorís – Dir: D. Agulió Hidalgo.
44) Ap. 201, San Francisco de Macorís – Own: Julio A. Gonzales. **FM:**

102.3.
45) Edif. Jaar, Calle El Conde esq. Espaillat, Sto Domingo – Own: V. M. Furment U.
46) Ap. 561, Santiago – Dir: José Enrique McDougal. **FM:** 101.9.
47) Ap. 64, San Pedro de Macorís – Dir: Agustín Santana.
48) Calle Canada, San José de Ocoa – Dir: Felipe A. Isa P.
49) Calle Colón 66, Nagua – Pres: P. Riggio.
50A-F) Empresas Radiofónicas SA, Ap. 20339, Sto Domingo – Mgr: Rodolfo Lama Jaar. CE: Roberto Lama Sajour.
50A) 27 de Febrero 1, Jimaní; 50B Félix del Rosario 1, Azua; 50C-D) Ap. 201, Barahona; 50E) Duarte 1, Pedernales; 50F) Cambronal 8, Neiba.
51) Ap. 10, Jarabacoa – Dir: R. Angeles S. **FM:** 98.7.
52) Av. 27 de Febrero 514, Sto Domingo – DG: Marino Vasquez.
54) Ap. 37, San Juan de la Maguana – Dir: Marcos A. Rojas F.
55) Ap. 860, Sto Domingo – DG: Ramon Pacheco Saiz.
56) Ap. 187, Santiago – Dir: Lic. Américo Rodríguez A. **FM:** 93.1.
57) Ap. 1372, Sto Domingo – Own: Manuel Mª Pimentel.
58) Ap. 79, Santiago – Dir: Héctor José Torres.
59) Ap. 142, San Pedro de Macorís – Dir: Calazán Omar Cepeda. **FM:** 90.7 Sultana + 98.7 Estéreo 98.
60) Ap. 4680, Sto Domingo. **E-mail:** amanecer@tricom.net **Web:** http://www.tricom.net/amanecer/ – Dir: Joel Calderón. **F.PI:** 10 kW SW.
61) Ap. 57, San Francisco de Macorís – Dir: Ernesto A. Grullón. **FM:** 94.7.
62) Av. 27 de Febrero 375, Sto Domingo – DG: Dr. Julio Hazim.
63) Corazón de Jesús 61, Moca – Dir: P. Mirelis.
64) Av. Circunvalación Norte, Puerto Plata – Dir: Rafael Martínez Martínez.
65) Av. 25 de Febrero 144, Ensanche Las Américas, P3 Hotel Hostal Puerto Rico, Sto Domingo – DG: Julio Soto Medina.
66) Ap. 264 (or: Calle San Francisco 50), San Francisco de Macorís – Dir: Fabio Raposo Tabar.
67) Ap. 151, La Romana – DG: Francisco A. Micheli. **FM:** 107.5.
68) Ap. 804, Sto Domingo – Own: Agr. Roberto Vargas.
69) Sánchez esq. Mella, Baní – Dir: Pedro C. Guerrero.
70) Calle Libertad 97, Bonao – Dir: José A Aquino. **FM:** 88 7 Latina 88.
71) Miguel A. Monclús, Monte Plata – Dir: Ricardo Acosta.
72) Padre Billini esq. Jaime Mota, Barahona – Dir: Alci Pimentel.
73) Ap. 363, Puerto Plata – DG: Gregory & Bonny Castellanos.
74) Conde esq. 19 Marzo, Edif. El Palacio, Sto Domingo – Dir: R. A. Martínez G.
75) Juan Rodríguez 76-A, La Vega – Dir: Cuto F. Holguín.
76) Ap. 65, San Juan de la Maguana – Dir: Iván A. Ramírez. **FM:** 100.1.
77) Ap. 207, La Romana – Dir: Jorge Garib M. – **FM:** 94 5.
79) Ap. 266, El Seybo – DG: P. Ercilio Ant. Brito. **FM:** 93.7.
80) Av. Las Carreras, Santiago – Dir: Nombrito Wehbe. **FM:** 106.1.
81) Calle Padre Borbón 16, San Cristóbal – Dir: Marcos A. Bello.
82) Restauración 64, La Vega – Dir: Dionisio Moya.
83) Calle 18 Nᵖ 158, Ensanche La Fé, Sto Domingo.
84) Carr. Mella Km 1.5, Higüey – Dir: Juan Julio Campos.
85) Ap. 003, Río San Juan – Dir: Miguel A. Abdul G. FM: 98. 9.
86) Mª Trinidad Sánchez 75, Cotuí – Dir: Dagoberto López. **FM:** 97.3.
87) Calle General Santana 65, Higüey – Dir: Modesto Amado Mercedes.
88) Calle Duarte 84, Nagua – Dir: Lic. Rolando Santos. **FM:** 99.3 R. Trebol.
89) Ap. 88, San Juan de la Maguana – Dir: L. P. González.
90) Prol. Av. Bolívar 49, Sto Domingo – Dir: Lic. Braulio Portes.
91) Ap. 2, Salcedo – Dir: Eusebio H. Bencosme.
92) Calle Felipe de Castro, Hato Mayor – Dir: Domingo A. Ubiera.
93) Ap. 123, San Francisco de Macorís – Own: Pedro R. Fernández.
95) Pres. Henriquez 53, Dajabón – DG: P. José Lanz, S. J. **FM:** 105.1.
96) Ap. 213, Moca – Dir: Vda. Nuñez.
97) Calle 16 de Agosto, Higüey – Dir: Ramón Ant. Lescaille.
98) Ap. 30011 (or: Calle Pepillo Salcedo 18, La Fé), Sto Domingo – Dir: Dario Badia.
99) Av. Malecón, Samaná – Dir: Miguel Castillo.

FM in Sto Domingo: 88.5 La Brava – 15) 89.1 R. WAO – 89.5 Super 89 – 89.7 Renuevo – 12) 90.1 – 52) 90.5 Estrella 90 – 91.1 La 91 – 38) 91.7 La Rocha – 55) 92.1 Hits 92 – 40) 92.5 Mágica – 5) 92.9 – 93.5 R. Alfa Omega – 62) Viva – 94.5 KQ-94 – 38) 95.1 – 22) 95.7 La Nota Diferente – 52) 96.5 – 32) 97.1 Galaxia – 97.5 Clásica – 9) 98.1 – 20) 98.5 R. Prisma – 99.1 Sonido Suave – 99.5 R. Listín – 68) 100.5 Cima – 100.9 Super Q – 101.3 Z-101 – 101.7 R. Supra – 9) 102.1 R. 102 – 57) 103.1 – 9) 103.7 Power 103 – 17) 104.1 Amor – 104.5 La Super Potente – 1) 105.1 – 106.1 Disco 106 – 38) 106.5 La Rocha – 62) 107.9 Romántica.

EL SALVADOR

L.T: UTC -6h — Pop: 5.750.000 — **Radios:** 2.080.000 — **Pr.L:** Spanish — **E.C:** 60Hz, 110/220V — **ITU:** SLV.

ADMINISTRACION NACIONAL DE TELECOMUNICACIONES (ANTEL)
⊡ Centro de Gobierno, San Salvador. 🖃 +503 226849 - **L.P:** Pres: Lic.Saúl Suster, Head Div. Radioeléctrica: Ing.Carlos Pineda R.

ASOCIACION SALVADORENA DE EMPRESARIOS DE RADIODIFUSION (ASDER)
⊡ Ap.210, San Salvador - Pres: Manuel A.Flores B.

Mediumwaves: Call YS-,
* = inactive, (r) = repeater, v = varying fq. H.of tr: 1100-0400 exc.where indicated.

	Call	kHz	kW	Name and h.of tr.
1)	HV	540	5	R.Restauración, San Salvador: 1100-0600
2)	FG	550	2	R.Variedades, Sonsonate
3)	KT	570	10	R.Cadena Central, San Salvador: 1000-0800
3)	DR	570	5	R.Cadena Central, Sta Ana (r:570)
3)	KR	570	5	R.Cadena Central, San Miguel (r:570)
4)	NK	600	3	La Preferida, San Salvador
6)	LN	630	10	La Monumental, San Salvador: 1130-0400
5)	SS	655	10	R.Nacional, San Salvador: W 1100-0400, Sun 1200-0300
7)	UU	700	5	R.Cadena YSU, San Salvador
7)	UU	700	1	R.Cadena YSU, Ahuachapán (r:700)
7)	UU	700	1	R.Cadena YSU, Usulután (r:700)
7)	UU	700	1	R.Cadena YSU, San Miguel (r:700)
7)	UU	700	1	R.Cadena YSU, Sta Ana: (r:700)
7)	UU	700	1	R.Cadena YSU, Sonsonate (r:700)
9)	RA	720	1	Circuito YSR, San Salvador: 1100-0500
9)	RA	720	1	Circuito YSR, Sta Ana (r:720)
9)	RA	720	50	Circuito YSR, San Miguel (r:720)
10)	KL	770	10	R.Cadena YSKL "La Poderosa", San Salvador: 1030-0530
10)	KL	780	5	R.Cadena YSKL, San Miguel (r:770)
10)	KL	780	1	R.Cadena YSKL, Sonsonate (r:770)
10)	KL	780	1	R.Cadena YSKL, Usulután (r:770)
10)	KL	780	1	R.Cadena YSKL, Sta Ana (r:770)
11)	AX	800	1	LV Panamericana, San Salvador: 1045 (Sun 1200)-0300
12)	FA	810	2	R.Lorenzana, San Vicente
44)	DA	810	1	R.Imperial, Sonsonate: 1100-0300
13)	PX	830	5	R.Pax, San Miguel
14)	FB	840	10	R.Vanguardia, San Salvador: 1000-0430
15)	RC	860	1	R.Tecana, Sta Ana
16)	AR	870	1	R.Cadena de Oro, San Salvador: 1400-0900
8)	CD	880	1	R.Ritmo, Stgo de María
17)	LA	890	3	R.Musical, Sta Ana: 1000-0500
18)	QJ	900	2	R.El Tiempo, San Salvador: 1200-0600
19)	TG	930	5	R.Cadena Sonora, San Salvador: 1100-0500
19)	TG	930	1	R.Cadena Sonora, Sonsonate (r:930)
19)	TG	930	1	R.Cadena Sonora, San Miguel (r:930)
19)	TG	930	1	R.Cadena Sonora, Sta Ana (r:930)
19)	TG	930	1	R.Cadena Sonora, Ahuachapán (r:930)
19)	TG	930	1	R.Cadena Sonora, Usulután (r:930)
20)	HG	950	1	R.Mundo, San Miguel
21)	TW	960	0.5	R.Centro, Sonsonate
47)	MS	970	5	UTEC-R.Universidad Tecnológica, San Salvador: 1200-0400
23)	AT	990	1	R.Upa "La radio de los niños", San Salvador: 1200-0200
24)	HH	1000	1	Estación H, Sta Ana
25)	CA	v1020	10	R.Internacional, San Salvador: 1000-0700
27)	RM	1030	1	R.Frontera, Ahuachapán: 1200-0400
7)	QR	1050	2	Auto R., San Salvador
26)	AN	1070	1	LV de los Ausoles, Ahuachapán
3)	ME	1080	5	R.1080, San Salvador: 24h
3)	ME	1080	1	R.1080, San Miguel (r:1080)

28)	MG	v1090	3	R.1090, Atiquizaya
29)	RF	1100	3	R.Ranchera, San Salvador
30)	CL	1110	2.5	R.Horizonte, San Miguel (r.1160)
30)	CL	1110	1	R.Horizonte, Sta Ana (r:1160)
10)	LR	1120	3	La Romántica "LV del Amor", San Salvador: 1100-0500
20)	LG	1130	1	R.Chaparrastique, San Miguel
31)	AJ	1130	1	R.Moderna, Sta Ana: 1200-0400
23)	TS	1140	10	R.El Mundo, San Salvador: 1130-0600
32)	CF	1150	1	Ondas Orientales, San Miguel: 1930-0200
48)	RG	1160	1	R.Corporación, Sta Ana
30)	CL	1160	3	R.Horizonte, San Salvador
2)	CB	1175	0.5	LV del Pacifico, Sonsonate (n.1170)
33)	VE	1180	5	R.VEA, San Salvador: 1200-0300
29)	MM	1200	10	R.Fiesta, San Salvador
29)	MM	1200	1	R.Fiesta, Sta Ana (r:1200)
34)	KJ	1200	1	R.Sirama, San Miguel
22)	CG	1210	1	R.La Paz, Zacatecoluca
49)	MT	1240	0.5	R.Metapan, Sta Ana
50)	QR	1240	1	R.Norteña, San Miguel
35)	AA	1260	3	R.América, San Salvador
34)	QZ	1270	1	R.Mía, San Miguel
36)	QV	1280	1	R.Galaxia, Sta Ana
37)	MA	1290	1	R.Chalatenango, Chalatenango
38)	LV	1300	5	R.Cabal, San Salvador
51)	RV	1310	5	R.Veritas, Usulután
52)	AH	1320	1	R.Emanuel, La Unión
39)	HQ	1330	5	R.Progreso, San Salvador: 1100-0500
40)	XW	1340	1	R.Novedades, Usulután
53)	KO	1370	1	R.LV de la Amistad, San Miguel: 1100-0300
41)	JI	1400	1	LV del Litoral, Usulután
42)	UCA	1420	1	R.Universitaria, San Salvador
46)	FM	*1450	5	Super R., San Salvador
54)	KR	1450	1	R.Restauración, San Miguel: 1000-0400
43)	CS	1500	1	R.Fides, Usulután
45)	CZ	1550	5	Cadena Cuscatlán, San Salvador: 24h
45)	CZ	1570	5	Cadena Cuscatlán, Cuscatlán (r:1550)

Addresses and other information:
1) Ap.2854 (or: Calle al Matazano N° 1, Final Col.Sta Lucía, Ilopango), San Salvador - DG: Sergio Daniel Solórzano.
2) Av.Sta Monica 9, Sonsonate - DG: Larry A.Zedán B.
3) 25 Calle Poniente N° 113, San Salvador - DG: Dr. Enrique Restrepo.
4) Edif.TV2, Alameda Dr.Manuel E.Araujo, San Salvador.
5) Av.El Cocal N° 1509, Barrio San Jacinto, San Salvador - DG: Lic.Jaime Huete.
6) 4a Calle Oriente 224, Edif.Comercial, San Salvador.
7) Av.Olimpica y Alameda Dr.Manuel E.Araujo, Col.Escalón, San Salvador - DG: Julio Suvillaga Zaldívar.
8) 2a Av.Norte 24, Stgo de María, Usulután.
9) Ap.720, San Salvador - DG: Andrés Rovira.
10) Ap.1329 (or: 4a Calle Oriente 528), San Salvador - Pres: Manuel Antonio Flores.
11) Ap.1835 (or: 1a Calle Poniente 3412), San Salvador - Dir: Pbro.Roberto A.Torruella.
12) Carr.Amapulapa km 1, San Vicente.
13) 10 Calle Oriente 102 Bis, San Miguel.
14) Pasaje Vilanova 114, Entre 8 y 10 Av.Norte, San Salvador.
15) Altos del Cine Tecana, Sta Ana.
16) 27 Calle Poniente 544, San Salvador.
17) 4a Av.Sur, Entre 7a y 9a Calle Poniente, Edif.Plaza de Vidrio, Sta Ana - DG: Mª Emma de Peñate.
18) Ap.2156, San Salvador - DG: Sra.Rossy Castillo de Maynes.
19) Urb.La Esperanza. Diagonal Principal 1322, San Salvador - DG: Roberto Castañeda.
20) 4a Av.Sur 304, San Miguel - DG: Joaquín Aparicio Borja.
21) 5a Calle Oriente 44, Sonsonate - Dir: J.L.Gudiel G.
22) 2a Calle Poniente 22, Zacatecoluca - DG: Federico Alberto Pineda.
23) Ap.06-210, San Salvador - DG: Betty Suarez de Teaque.
24) 9a Calle Poniente 25, Sta Ana - Dir: M.Montes M.
25) 29a Calle Oriente 218, San Salvador - DG: Gustavo Bustillo.
26) Av.Morazán km 101, Ahuachapán.
27) Av.2 de Abril y 8a Calle Poniente, Ahuachapán - DG: Rafael Armando Ramos P.
28) Esq.2a Av.Sur y 1a Calle Oriente 3-93, Atiquizaya.
29) Calle y Col.Roma, Edif.3-B, San Salvador - DG: Ilse Marlene Valle Sillos.

30) 4a Calle Poniente, Entre 43 y 45 Av.Sur, Col.Flor Blanca, San Salvador - Dir: Julio René Vargas.
31) 8a Calle Poniente 11A, Sta Ana - DG: Julio A.Arriola.
32) Ap.19, San Miguel - DG: Carlos Alb.Morales T.
33) Carr.Panamericana km 18.5, Cantón La Palma, San Martín - DG: Enrique Ramírez Castillo.
34) Carr.Litoral km 134, Cantón Jalacatal, San Miguel.
35) Col.San Benito, Pasaje Las Palmas 182, San Salvador.
36) 2a Calle Poniente 43B, Col.Sta Lucía, San Salvador.
37) 4a Calle Poniente 11, Chalatenango - DG: Mario A.Espinoza.
38) 17 Calle Oriente 143, Barrio San Miguelito, San Salvador - DG: Margarita Herrera.
39) Ap.855, San Salvador - Dir: José M.Cárcamo.
40) 1a Oeste 18, Usulután - DG: Oscar Aparicio Borjas.
41) 4a Av.Norte 4, Usulután - DG: José Infantozzi. **FM:** 90.1.
42) UCA, Universitaria, Autopista Sur, San Salvador - Dir: Mario Torres C.
43) 4a Calle Oriente N° 2, Usulután - Dir: René R.Rivera A.
44) Ap.56, Sonsonate - Dir: Daniel M.Cardona.
45) Ap.2147 (or: Carr.Sta Tecla km 5.5, Contiguo Al Circulo Militar), San Salvador - DG: Lic.Patricia Botto de Melendez.
46) 12 Av.Norte 1712, San Salvador - Dir: Fco.A.Melara L.
47) Universidad Tecnológica, 17 Av.Norte 130, San Salvador - DG: Lic.Elsy Larios.
48) Sta Ana.
49) Sta Ana.
50) San Miguel.
51) Usulután.
52) La Unión.
53) Ap.507, San Miguel - DG: Daniel M.Cardona.
54) Ap.210, San Miguel - DG: Sergio Daniel Solórzano. **FM:** 98.1.

FM in San Salvador: 9) 88.9 - 89.7 R.Vida - 90.5 Familia - 23) 90.4 - 91.3 Club - 92.1 "92" - 92.9 R.Láser - 23) 93.7 - 4) 94.5 La Preferida - 95.3 Doble L - 96.1 Scan - 18) 96.5 - 96.9 R.Nacional - 45) 98.5 - 29) 99.3 - 100.1 ABC - 100.5 Venceremos - 100.9 Amor - 3) 101.7 - 102.1 Farabundo Martí - 29) 102.5 Femenina - 23) 103.3 Clásica - 3) 103.7 - 10) 104.1 - 19) 104.5 - 29) 104.9 Fiesta - 105.7 Doble S - 16) 106.5 - 7) 107.3 - 107.9 Cadena La Exitosa.

GRENADA

L.T: UTC - 4h — **Pop:** 95.767 — **Radios:** 53.000 — **Pr.L:** English — **E.C:** 50Hz, 230/400V — **ITU:** GRD.

GRENADA BROADCASTING CORPORATION - G.B.C. Radio (Statutory Body, Comm.)
✉ Observatory Road, P.O. Box 535, St. George's. ☎ +1 (473) 440 1252 or 1253. 🖷 +1 473 440 4180. **Cable:** Broadcast Grenada W.I.
L.P: Chmn: Joseph Charter. MD: Calvin Haywood. PD: Troy Garvey. Sen. Eng: John Phillip.
MW: 535kHz 2 x 10 kW — **FM:** 105.5MHz (+ 9 st's planned).
D.PRGR: 0925-0400(Sun 0200). **N:** BBC: 1200, 1600, 0200. **VOA:** 1000, 2400 — **ANN:** "This is GBC Radio." — **V.** by letter. Rp.

THE HARBOUR LIGHT OF THE WINDWARDS (Rlg.)
✉ Carriacou. ☎ +1 (473) 443 7628 — **L.P:** SM: Randy Cornelius.
MW: 1400kHz 5kW.
D.PRGR: 0600-2130 (Sat 2145, Sun 2200) — **N:** rel. BBC & VOA.
ANN: "This is the Harbour Light of the Windwards broadcasting from beautiful and friendly Carriacou" — **V.** by QSL-card. Rp.

GUADELOUPE

L.T: UTC - 4h — **Pop:** 408.000 — **Radios:** 85.000 — **Pr.L:** French, Créole Patois — **E.C:** 50Hz, 220V — **ITU:** GDL.

RADIODIFFUSION FRANÇAISE D'OUTRE MER (RFO)
✉ B.P. 402, F-97163 Point-à-Pitre Cédex. ☎ +590 939696. 🖷 +590 939682. ☼ 919064
L.P: Dir: R.Surjus. Editor-in-Chief: Philippe Goudé. PD: L.Francil. Head Communications Dept: Sonia Gémieux.
MW: Point-à-Pitre 640kHz 40kW.
FM: Point-à-Pitre 88.9 MHz, Marie Galante 89.1 MHz, Deshaies 96.8MHz, Basse-Terre 97.0MHz, Pointe-Noire 97.4 MHz.
D.PRGR: 24h. **N:** 1100, 1700, 2230, plus relays of France-Inter.

ANN: "Ici Point-à-Pitre, RFO Guadeloupe" or "RFO".
IS: "Biguin" (guitar) — **V.** by QSL-card. Rp.

RCI - RADIO CARAÏBES INTERNATIONAL GUADELOUPE (Comm.)

✉ RCI Guadeloupe, B.P. 1309, F-97187 Point-à-Pitre Cédex.
☎ +590 839696. 📠 +590 839697 — **FM:** 95.1/98.6/106.MHz.
D.PRGR: 24h. **N:** on the h. (rel. Europe 1).

RCI 2 (Comm.)

FM: 96.3/100.6/102.6MHz (rel. RCI 2 St. Martin) — **D.PRGR:** 24h.

Europe 2: Basse Terre 96.6MHz, Point-à-Pitre 103.4MHz.
R. France Internationale: via R. Basse Terre 98.2MHz and Ile FM 103.0MHz.
Other stations: over 30 FM st's are operating.

SAINT MARTIN & SAINT BARTH

RADIODIFFUSION FRANÇAISE D'OUTRE MER

FM: St. Martin 88.9MHz, St. Barth 88.6MHz (see main entry above)

RCI - RADIO CARAÏBES INTERNATIONAL GUADELOUPE (Comm.)

FM: St. Martin 105.0MHz (see main entry above)

RCI2 (Comm.)

✉ RCI2 Saint Martin, B.P. 173, Marigot, F-97150 Saint Martin.
☎ +590 875406. 📠 +590 878887.
FM: 102.1MHz + relays in Guadeloupe & Martinique.

RADIO SAINT MARTIN (Comm.)

✉ Port de Marigot, F-97150 Saint Martin — Mgr: H. Cocks.
FM: 95.3MHz — **D.PRGR:** 1000-0500(Sun 0400) in French & English exc. **Spanish:** 2000-2100W.

RADIO VOIX CHRETIENNES DE ST. MARTIN (Rlg.)

✉ B.P. 103, Marigot, F-97150 Saint Martin. ☎ +590 873159.
L.P: Mgr: Father Cornelius Charles — **FM:** 106MHz 0.25kW.
D.PRGR: 0845-0530 in English & French.

R. France Internationale: via R. St. Bath 100.7MHz.

GUATEMALA

L.T: UTC -6h (Su: UTC -5) — **Pop:** 9,978,000 — **Radios:** 570,000 — **Pr.L:** Spanish — **E.C:** 60Hz, 110/120/127/220V — **ITU:** GTM.

MINISTERIO DE TRANSPORTES Y OBRAS PUBLICAS

Dirección General de Radiodifusión y Televisión Nacional
✉ 18 Calle N° 6-72, Z-1, Edif. Tipografía Nacional, Nivel 3, Guatemala.
☎ +502 (2) 532539.
L.P: DG: Ricardo Gómez Flor, Dep. Dir: Enrique Hernán.

Mediumwaves: Call TG-,
° = also on shortwave, * = inactive, (r) = repeater, v = varying fq.

	Call	kHz	kW	Name and h. of tr.
25)		540		R. Cobán, Cobán
1)	RV	550	10	R. 5-60, Guatemala: 1200-0500
84)		560	1	R. Quetzal, Malacatán
2)	PA	570	1	R. Palmeras, Escuintla
3)	Y	580	3	R. Progreso, Guatemala: 1100-0700
4)	RQ	590	5	R. Quiché, Sta Cruz: 1200-0400
5)	RC	600	1	R. Campesina, Escuintla: 1100 (Sun 1200)-0500
5)		600		R. Campesina, Mazatenango (r:600)
6)	GA	610	0. 5	R. Alianza, Guatemala: 1200-0300
7)	PQ	620	5	R. 6-20, San Cristóbal: 1200-0400
63)	EL	630		R. El Porvenir, Sta Elena: 1000-0400
8)	W	°640	10	LV de Guatemala, Guatemala: 1100-0600
9)	Q	660	3	R. Nal. "LV de Quetzaltenago", Quetzaltenango: 1100-0400
10)	RT	670	10	Emisoras Unidas AM, Guatemala: 1000-0500
11)	VP	680	10	R. Norte, Cobán: 1100-0400
12)	VB	690	1	R. Tamazulapa, Jutiapa
28)	HU	690	1	R. Escuintla, Escuintla
13)	HR	700	15	R. Mundial, Guatemala: 1000-0600
14)	XL	710	1	R. Tecún Umán, Quetzaltenango: 1030-0400
15)	RO	720	1	R. Corona, Morales: 24h
16)	N	°730	10	R. Cultural, Guatemala: 24h
17)	HF	740	1	R. Tacaná, San Marcos: 1100-0500
18)	AJ	750	1	R. Tropicana, Escuintla
19)	HB	760	5	Nueva R. Super, Guatemala: 1000-0500
20)	BX	770	1	R. Fraternidad, Quetzaltenango: 1000-0600
85)	CK	780	1	R. Sultana del Oriente, Zacapa
21)	O	790	3	R. Festival, Guatemala: 1100-0400
79)	YZ	800	1	R. Rosa, Chiquimulilla: 24h
38)		810		R. Moapán, Sta Elena
22)	TO	820	10	R. Internacional, Guatemala: 1000-0600
23)	AV	830	5	R. Satélite, Mazatenango: 1100-0400
24)	SM	840	0. 35	LV de San Marcos, San Marcos: 1300-0400
49)	X	850	10	R. Ranchera, Guatemala: 24h
26)	FP	*860	1	R. Nal. Tikal, Flores Petén
59)	L	v870	0. 5	R. Victoria, Mazatenango
27)	J	880	10	R. Nuevo Mundo, Guatemala: 1030-0500
29)	MA	900	1	R. Amatique, Puerto Barrios
13)	KL	910	10	R. 910 "La Em. de los Exitos", Guatemala: 1100-0500
30)	RS	920	0. 2	R. Sur, Escuintla
31)	JL	v930	5	R. Imperial, San Pedro Carchá: 1000-0400
32)	TL	940	5	LV del Hogar. Guatemala: 1200-0500
33)	AF	950	1	R. Indiana, Mazatenango
34)	RU	960	1	R. Utatlán, Sta Cruz del Quiché: 1100-0300
35)	AX	970	5	R. Continental, Guatemala: 1200-0430
36)	MQ	980	1	R. Retama, San Marcos: 1200-0500
37)	AL	990	1	R. Perla de Oriente, Chiquimula: 24h
39)	RX	1010	1	R. Miramundo, Zacapa
61)	XI	1010	1	R. Ixil, Nebaj: 1100-0200
40)	CM	1020	5	R. Frontera, Pajapita: 1100-0400
41)	UX	1030	10	R. Panamericana, Guatemala: 1200-0500
42)	JP	1040	1	R. Oriental, Jalapa
43)	SL	1050	5/1	LV de los Cuchumatanes, Huehuetenango: 1100-0600
44)	T	1060	10	R. Sonora, Guatemala: 1100-0600
45)	D	1070	3	LV de Occidente, Quetzaltenango: 1130-0500
53)	LU	1080	1	R. Novedad, Zacapa
10)	Z	1090	10	Emisoras Unidas AM, Guatemala
101)	SR	1100	1	R. Superior, Coatepeque
46)	MK	1110	1	R. Verapaz, Cobán
47)	C	1120	0. 5	R. Uno 120 AM, Guatemala
48)	VR	1130	1	LV de la Costa Sur, Retalhuleu
49)	RR	1150	1	R. Fiesta, Guatemala: 24h
50)	RI	1160	1	R. Izabal, Morales: 1100-0130
51)	RL	1170	5	R. Cadena Landívar, Quetzaltenango: 0900-0300
44)	T	1180	10	R. Sonora, Guatemala: 1100-0600 (r:1060)
54)	RJ	1200	1	R. Jutiapa, Jutiapa
55)	MX	1210	10/5	R. Rumbos/Coco Radio, Guatemala: 24h
56)	MT	1220	1	R. Amiga, Antigua: 1100-0300
57)	AT	1230	1	R. Atlántida, Puerto Barrios: 1100-0400
52)		1230		R. América, Cuyotenango
58)	K	1240	5	R. Sensación "R. Luz", Guatemala
87)	PY	1250	1	R. Payakí, Esquipulas: 1100-0300
60)	PA	1260	1	R. Monumental, Antigua: 1100-0300
102)	CQ	1270	2. 5	R. Exclusica, Guatemala: 1100-0400
77)	VY	1280	2. 5	R. Zamaneb, Salamá: 1100-0200
62)	TU	1290	0. 5	R. Nal. de Totonicapán, Totonicapán: 0000-0400

97)		1300		R. Miramundo, Zacapa
64)	AN	1310	1	R. LV de los Altos, Quetzaltenango: 1100-0500
65)	ME	1320	1	R. Quesada, Jutiapa
66)	MU	°1330	5. 5	Unión Radio "LV de la Esperanza", Guatemala: 1100-0300
67)	CO	1340	10	LV del Trópico, Coatepeque; 24h
68)	MC	1350	1	R. Monja Blanca, Cobán
41)	LK	1360	10	R. Tic Tac, Guatemala: 1100-0500
69)	AC	1370	5	LV de Colomba, Colomba: 1100-0300
70)	EB	1380	1	R. Momostenango Educativa, Momostenango: 1100-1900
71)	YC	1390	5	R. Estrella, Guatemala: 1100-0330
72)	RB	1400	1	R. Porteña, Puerto Barrios: 24h
73)	GH	1410	5	R. Xelajú, Quetzaltenango: 1200-0600
74)	RP	1420	1	R. Viva, Guatemala: 1100-0400
75)	AG	1430	1. 2	LV de Huehuetenango: 1100-0400
76)	MS	1410	0. 5	R. Nal. "LV de Mazatenango", Mazatenango: 0000-0400
13)	LG	1450	1	R. Epoca, Guatemala: 24h
78)	RN	1460	5	R. Petén, Flores Petén
80)	HB	1480	5	R. Horizontes, Guatemala: 24h
81)	RE	1490	1	R. Modelo, Retalhuleu: 0900-0300
90)	DS	1490	1	R. LV de Atitlán, Santiago Atitlán
82)	DX	1510	5	R. Centroamericana "Nueva RCA", Guatemala: 1130-0500
98)		1520		R. Taysal, Sta Elena de la Cruz
83)	RS	1520	1	R. Superior, Coatepeque
86)	VE	1570	10	R. Voz Evengélica de América, "R. VEA", Guatemala: 1030-0600
88)	XC	1590	1	R. Triunfadora, Chimaltenango
8)	WC	1600	10	LV de Guatemala, Guatemala (r:640)

Shortwaves:

89)	BA	2360	0. 5	R. Maya, Barillas: 1030-1400, 2230-0330
90)	DS	v2390	1	R. LV de Atitlán, Santiago Atitlán: W 2200-01115, Sun 1800-2400
16)	NC	3300	10	R. Cultural, Guatemala: 0930-0630
89)	BA	v3325	1	R. Maya, Barillas: 1030-1400, 2230-0330
91)	VN	3360	1	R. LV de Nahualá, Nahualá: 1100-1400, 2100-0300
92)	TZ	3370	1	R. Sistema Cultural Tezulutlán, Cobán: 1100-1500, 2100-0300
93)	CH	3380	1	R. Chortís, Jocotán: 1100-1300, 2100-0330
99)	LT	v4780	1	R. Cultural Coatán, San Sebastián Coatán
95)	MI	v4800	1	R. Buenas Nuevas, San Sebastián, Huehuetenango: 1100-1430, 2200-0230
94)	MN	4825	0. 5	R. Mam, Cabricán: 1300-1700, 2000-2400
92)	TZ	v4835	5	R. Sistema Cultural Tezulutlán, Cobán: 1100-1500, 2100-0300
96)	VC	4845	1. 25	R. K'ekchi, Fray Bartolomé de las Casas: 1230-1700, 2100-0300
16)	NA	v5955	0. 5	R. Cultural, Guatemala: 0930-0630
66)	MUA	v5980	7. 5	Unión Radio "LV de la Esperanza", Guatemala: 1100-0200
8)	WB	6180	10	LV de Guatemala, Guatemala: 1100-0600

Addresses and other information:
1) 14 Calle 4-73, Z-1, Guatemala – Dir: Edna Castillo Obregón.
2) 15 Calle 2-48, Z-3, Escuintla – Dir: F. Tres G.
3) 9 Av. 0-32, Z-2, Guatemala – Dir: G. Humberto Andrino.
4) 7 Calle 3-67, Z-5, Sta Cruz del Quiché – Dir: Edgar A. López – **D. Prgr:** Vernaculars only.
5) Col. 15 de Junio, Z-3, Tiquisate, Escuintla – Dir: José M. Alvarez C. **FM:** 99.3.
6) 34 Av. "A" 7-60, Z-7, 01007 Guatemala – DG: Haroldo Castillo O.
7) Barrio La Cienaga, San Cristóbal, Totonicapán – Dir: José F. Herrera Nelson.
8) 18 Calle 6-70, Z-1, Guatemala – DG: Lic. Aida Arvizu López.
9) Ap. 113 (or: 13 Av. 8-19, Z-1), Quetzaltenango – Dir: Victor Manuel Villagrán Sajquim.
10) 4 Calle 6-84, Z-13, 01013 Guatemala – Dir: Rolando Archila

Dehesa.
11) 2 Calle 5-57, Z-3, Cobán, Alta Verapaz – DG: Guillermo Casa F.
12) Calle 15 Septiembre, Sta Cruz, Jutiapa – DG: Silvia L. de Azurdia.
13) 6 Av. 2-80, Z-1, Guatemala – Dir: Federico Azurdia.
14) 6 Av. 6-41, Z-1, Quetzaltenango – Dir: M. Sosa de Cifuentes.
15) Calle Principal, Morales, Izabal – Dir: Ana de Liu.
16) Ap. 601, 01901 Guatemala – Dir: Esteban Sywulka B. **English:** 0300-0430, **Kekchi:** 1045-1100.
17) 8 Calle 8-01, Z-2, San Marcos – DG: Sandra E. Rubio.
18) 4 Av. 6-26, Z-1, Escuintla – DG: Carlos Echeverría
19) 30 Av. 3-86, Z-11, Utatlán II, 01001 Guatemala – Dir: Dr. Ernesto Ponce Bedoya.
20) Ap. 90, Quetzaltenango – DG: Byron E. Rivera D.
21) 11 Calle 2-43, Z-1, Guatemala – DG: Pablo F. Nuiza S.
22) 37 Av. 1-15, Z-7, 01007 Guatemala – Dir: Claudia Flores.
23) Mazatenango – Dir: Herberto Díaz G.
24) Palacio Maya, San Marcos.
25) 5 Calle 1-06, Z-3, Cobán, Alta Verapaz.
26) Flores Petén.
27) 6 Av. 10-45, Z-1, Guatemala – Dir: Alfredo González G. Various st's relay news from this station under slogan: "Sistema de Emisoras Nuevo Mundo".
28) 4 Av. 11-38, Z-1, Escuintla – Dir: Freddy Azurdia.
29) Ruta Atlántico km 291, Puerto Barrios – Gte: J. Aquino R.
30) 6 Av. 0-18, Z-4, Escuintla – Dir: Violeta Gonzales G.
31) 5 Calle 7-53, Z-1, San Pedro Carchá, Alta Verapaz – DG: Rolando Archila Marroquín.
32) Ap. 31, Guatemala – Dir: Carlos Martínez.
33) 6 Av. 10-54, Z-1, Mazatenango – Dir: Guillermo Alcázar.
34) 5 Av. 4-22, Z-2, Sta Cruz del Quiché – Dir: C. I. de Archila.
35) 15 Calle 3-45, Z-1, Guatemala – Dir: Roberto Vizcaíno R.
36) 5 Calle 8-21, Z-1, San Pedro, San Marcos – Dir: Heberto Díaz M.
37) 7 Calle Av. 4-00, Z-1, Chiquimula (or: 6 Av. 0-60, Z-4, Torre Prof. II, Of. 904, 01004 Guatemala) – Gte: Eduardo Alfonso Liu.
38) Sta Elena de la Cruz, Petén.
39) Palacio Nal. , Ministerio de la Defensa Nacional, Guatemala.
40) Pajapita, San Marcos – Dir: Carlos Alcázar F.
41) 1 Av. 35-48, Z-7, Guatemala – Dir: Jaime J. Paniagua.
42) Av. Chipilapa "A" 1-03, Z-2, Jalapa – Dir: A. Morales H. 43) 2 Calle 4-42, Z-1, Huehuetenango – Dir: Héctor L. Alvarez G. 44) 2 Calle 18-07, Vista Hermosa I, Z-15, Guatemala – DG: Lic. Eduardo Mendoza.
45) 7 Av. 0-26, Z-2, Quetzaltenango – Dir: María Cristina R. de Ríos.
46) 2 Calle 5-57, Z-3, Cobán, Alta Verapaz.
47) 4 Av. 8-72, Z-1, Edif. Horizontal nivel 6, Guatemala.
48) Ap. 84, Retalhuleu – Dir: José E. Archila M.
49) Ap. 1526, (or: Grupo Radial El Tajín, 2 Av. "A" 13-45, Z-1), 01001 Guatemala – Pres: Eleonora Girón de Rubens.
50) Barrio El Carrizal, Morales, Izabal – Dir: Carlos Enrique Molina.
51) 14 Av. "A" 0-78, Z-1, Quetzaltenango – Dir: Herberto Díaz M.
52) 13 Av. 23-60, Z-12, Coyotenango, Such.
53) 4 Calle 10-34, Z-1, Zacapa – DG: Eduardo Liu.
54) Carr. Interamericana km 117, Jutiapa – Dir: C. Archila.
55) 4 Av. 1-14, Z-1, Guatemala – Dir: Roberto Bocaletti de León.
56) Av. del Desengaño 20, Antigua – Dir: C. Echeverría.
57) Ap. 425, Puerto Barrios – Dir: Juan Archila.
58) Ap. 281, Guatemala – Dir: Humberto Gonzales G.
59) La Libertad 9-91, Z-1, Mazatenango – Dir: Alfonso Medina.
60) 7 Av. Norte N° 8, Antigua – Dir: E. Cifuentes.
61) 5 Av. 1-14, Canton Batzbaca, 14013 Nebaj, Quiché – Dir: Fausto Cabeira Samayda.
62) Palacio Municipal, Totonicapán.
63) Sta Elena de la Cruz, Petén.
64) Ap. 107, Quetzaltenango – Dir: J. César García P.
65) Quezada, Jutiapa.
66) Ap. 51-C, 01015 Guatemala – Dir: Daniel Rolando García P. Owned and operated by Adventist World Radio, Panamérica.
67) 2 Av. 2-02, Z-3, Barrio San Francisco, Coatepeque – Dir: Gilberto Calderón V.
68) Edif. Municipalidad, 5a Calle 1-06, Cobán, Alta Verapaz – DG: Mauro R. Acchun.
69) Colomba, Quetzaltenango – Coord: José Molina.
70) Momostenango, Totonicapán – DG: José F. Guinea Y.
71) 24 Av. 23-39, Z-12, Centro San Pablo, 01012 Guatemala – DG: Byron Valdizón Catalán.
72) 8 Av. 15 y 16 Calle, Puerto Barrios – Dir: Mario A. Miranda M.
73) 4 Calle 15A-62, Z-1, Quetzaltenango – Dir: Rolando Oliva V.
74) 11 Calle 32-15, Z-7, Guatemala – Dir: Eduardo Soria S.
75) Ap. 13, Huehuetenango – Dir: Mauro Guzmán M.

76) Calle 30 de Junio 1a y 2a, Z-5, Mazatenango.
77) Inst. de Educación Básica, Barrio Abajo San Jerónimo, Salamá, Baja Verapaz – DG: Elías Santiago García Reyes. Prgrs. in Sp. , Achi and Q'eqchí.
78) Isla Sta Bárbara, 17001 Flores Petén – Dir: Mariano Colmenares.
79) Edif. Municipal, Chiquimulilla, Sta Rosa – Dir: José L. Castro.
80) 17 Av. N° 20-35, Res. Granati & Townson Ill, Z-11, Guatemala – DG: Juan Carlos Coronado.
1) 7 Av. 6-72, Retalhuleu (or: Ap. 183-A, Guatemala) – Dir: G. Andreu C.
82) 3 Av. 6-32, Z-1, Guatemala – Dir: Carlos Peynado.
83) 3 Calle 3-38, Z-1, Coatepeque, Retalhuleu – Dir: M. Plinio Q.
84) 5 Calle 3-58, Z-1, Malacatán, San Marcos
85) 4 Calle 12-54, Z-1, Zacapa.
86) Ap. 1213, 01901 Guatemala – Dir: José Adonias Corado.
87) 5 Av. 9-02, Z-1, 20007 Esquipulas – Dir: Luis Felipe Paz R. **FM:** 91.5.
88) 2 Calle 3-33, Z-3, Chimaltenango.
89) 4 Av. 0-14, Z-1, 13026 Barillas, Huehuetenango – DG: Esteban Sywulka B.
90) Cantón Xechivoy, Santiago Atitlán, 07019 Sololá – Mgr: José Miguel Pop Tzina. FM: 103. 5.
91) Asociación Pro-Desarrollo y Educación Popular, Nahualá, Depto Sololá – Dir: Manuel Sac. Chovón. **FM:** 93.1.
92) Ap. 19, 16001 Cobán, Alta Verapaz – Dir: Mons. Gerardo Flores Reyes. Prgrs in Sp., Achí, Q'eqchí and Pocomchí. **FM:** 103.5.
93) Centro Social, 20004 Jocotán – Mgr: P. Juan María Boxus.
94) Acu'Mam, Cabricán, Dept. Quetzaltenango – Dir: José B. Escalante Ramos.
95) 13020 San Sebastián, Huehuetenango – DG: Israel G. Rodas M.
96) 3 Calle 7-15, 2-1, Fray Bartolomé de las Casas, 16015 Alta Verapaz (or: Ap. 25, 53140 Bulevares, Edo Mex. , México) – DG: Anzelmo Cuc Chub. Prgrs in Sp. and Q'eqchí.
97) Zona Militar N° 7, Zacapa.
98) Sta Elena de la Cruz, Petén.
99) San Sebastián Coatán – Dir: Domingo Hernández.
101) 3 Calle 3-38, Z-1, Coatepeque – Dir: Marie P. Quintana.
102) Calz. Roosevelt 34-13, Z-11, Guatemala – Dir: Jaime J. Paniagua.

FM in Guatemala City: St. names: Estéreo . . . exc where indicated. 85.5 Galaxia – 10) 88.8 – 10) 89.7 – 10) 90.3 Sideral – 49) 91.9 R. Exitos – 92.3 R. Universidad – 92.7 Cristal – 44) 93.5 – 93.9 Jazz 94 – 55) 94.5 – 37) 94.9 "FM95" – 49) 95.9 Ranchera - 44) 96.9 – 97.3 Alfa Sigma – 98.1 Doble S – 37) 98.9 Globo – 1) 99.7 Conga – 16) 100.5 – 1) 101.0 Fresca – 102.0 "102" – 102.3 Eco la Cariñosa – 102.9 Metroestéreo – 49) 103.7 R. Fiesta – 66) 105.7 – 49) 106.1 Máxima – 106.5 Ejecutiva – 22) 106.9 – 8) 107.3 – 37) 107.5 Fama.

HAITI

L.T: UTC — 5h (Su: UTC -4) — **Pop:** 6.764.000 — **Radios:** 270.000 —**Pr.L:** Créole, French — **E.C:** 50+60Hz, 110/220V — **ITU:** HTI.

CONSEIL NATIONAL DES TELECOMUNICATIONS (CONATEL)
✉ B.P.2002 (or: Cité de l'Exposition 16), Port-au-Prince. ☎ +509 22-0300 - 🖷 +509 22-0579 - Chief Service Gestion des Frequences: Ing.Alfredo Estriplet.

Mediumwaves: Call 4V-, P-au-P = Port-au-Prince.

	Call	kHz	kW	Name and h.of tr.
1)	JS	610	0.5	R.L'Eternel est Grand. P-au-P: 1000-0200
2)	I	660	5/1	R.Lumière, P-au-P: 1000-0100
2)	IA	720	1	R.Lumière, Petite Riv: 1000-0100
2)	IE	740	1	R.Lumière, Pignon-le-Jeune: 1000-0100
34)	-	740	1	R.Reveille, Cap Haïtien
2)	U	760	5	R.Lumière, Les Cayes: 1000-0100
2)	o	780	0.5	R.Lumière, Jérémie: 1000-0100
3)	EF	840	10	R.4VEH, Cap Haïtien: 1000-0300
4)	MK	860	3	R.Men Kontre, Les Cayes: 1100-0100
5)	JV	870	1	R.Tele Express Continental, Jacmel: 1030-0300
6)	PM	880	0.8	R.Indépendence Nouvelle, Gonaïves: 1000-0300
7)	VB	895	0.5	R.Trans-Artibonite, Gonaïves: 1000-0300
35)		910		R.Kyskeya, P-au-P
8)	KB	920	1	R.Cap-Haïtien, Petite Anse (r:930)
8)	KB	930	10/3	R.Cap-Haïtien, Cap-Haïtien: 24h
9)	LF	940	0.5	R.Saint Marc, Saint-Marc: 1200-0200
10)	FF	940	0.2	R.Diff.Jacmélienne, Jacmel: 1000-0400
11)	CD	960	0.3	R.Carillon, P-au-P: 1100 (Sun 1200)-0300
12)	CPS	990	0.2	R.Cacique, P-au-P: 1100-0300
13)	JH	1020	3	R.Pétion-Ville, Pétion-Ville: 1000-0100
14)	DN	1080	20	R.Nationale, P-au-P: 1000-0500
36)		1120		R.Magic, P-au-P
15)	AB	1150	5	R.Caraïbes, P-au-P: 1000-0400 (rep. 1146)
16)	RS	1170	10	R.Soleil, P-au-P: 1000-0200
17)	AF	1190	0.2	La Voix de la Grande-Anse, Jérémie
30)	LS	1210	1	R.Plus, P-au-P: 24h
18)	AV	1230	1	R.La Voix de l'Avé-Maria, Cap Haïtien: 1000-0400
19)	SJ	1240	10	R.Antilles Internationales, P-au-P: 24h
31)	S	1250	1	La Voix du Plateau Central, Hinche
20)	AM	1280	10	R.Métropole, P-au-P: 1100-0500
21)	JLD	1330	10	R.Haïti-Inter, P-au-P: 1030-0400
32)	RL	1360	5	R.Liberté, P-au-P: 24h
22)	EE	1370	1	R.Citadelle, Cap Haïtien: 1000-0400
23)	MM	1370	1	R.Diffusion Cayenne, Les Cayes
24)	SS	1380	1	R.Port-au-Prince, P-au-P: 1030-0400
25)	TS	1410	1	La Voix du Nord Ouest, Port-de-Paix
26)	GM	1430	10	MBC, P-au-P: 1030 (Sun 1330)-0400
27)	EA	1460	0.2	La Voix du Nord, Cap Haïtien: 1000-0400
28)	AA	1470	1	R.Arc-en-Ciel (R.Lakansyel), P-au-P: 0900-0500
33)	OC	1500	0.5	R.Select, P-au-P
29)	VE	1560	10	La Voix de L'Espérance (R.Adventiste), P-au-P: 0900-1700, 2100-0200

Addresses and other information: P-au-P = Port au Prince.
1) B.P.1164, P-au-P - Dir: Simon Jean-Baptiste.
2) B.P.1050, P-au-P - Dir: Robinson Joseph.
3) B.P.1, Cap Haïtien (or: B.P.1739, P-au-P; or:OMS, Box A, Greenwood, IN 46142, USA) - Dir: David Shaferley. C.E: Louis Destine.
4) B.P.43, Les Cayes - DG: Albert Gouin. FM: 95.5.
5) 31 rue Stenio, Jacmel - DG: Jean-Pierre Jacques - **FM:** 88.9.
6) 116 rue Egalité, Gonaïves - Dir: Paul Mitton.
7) 32 rue Anténor Firmin, Gonaïves - Dir: Volny Bastien.
8) B.P.64, Cap-Haïtien - Dir: Kallil Bitar.
9) rue Armand Thoby 18, Saint Marc - Dir: Leon S.Fleury.
10) 32 rue D'Orleáns, Jacmel - DG: Francois Frenel. **FM:** 101.5.
11) Bourdon, Ave. John Brown 356, P-au-P - Dir: Mme. C.Desmangles.
12) B.P.1480, P-au-P - Dir: Mme. Jean-Claude Carrié.
13) rue Rigaud 71, Pétion-Ville - Mgr: Mme. Carlo Hubert.
14) B.P.1143, P-au-P - Dir: Pierre Raymond Dumas.
15) Ruelle Chavannes 23, P-au-P - Dir: Jacques G. Simeon.
16) B.P.1362, P-au-P - Dir: Pére Arnoux Chéry.
17) 52 rue Dr. Hyppolite, Jérémie - Dir: Héritos Felix.
18) B.P.22, Cap-Haïtien - Dir: Rev.Marcel Bussels.
19) B.P.81, P-au-P - Prés: J. Sampeur.
20) B.P.62, P-au-P - DG: Richard Widmaier. **FM:** 100.1.
21) B.P.737, P-au-P - Dir: Jean L.Dominique.
22) rue 16-A, Cap-Haïtien - Dir: Emmanuel C.Eugène.
23) B.P.54, Les Cayes - Dir: Pierre Yvon Chéry.
24) B.P.863, P-au-P - Dir: Georges L.Hérard.
25) 84 rue B.Sylvain, Port-de-Paix - Dir: St-Aubin Saintil.
26) B.P. 367, P-au-P - DG: Franck C.Magloire.
27) rue 20-A-B, Cap-Haïtien - Dir: Mme.Malherbe.
28) Route de Delmas, P-au-P - Dir: Alex St-Surin.
29) B.P.1339, P-au-P - Dir: Wébert Lahens.
30) B.P.1174, P-au-P - Dir: Lionel Benjamin.
31) rue Bon Coeur 10, Hinche - Dir: Jude Simon.
32) B.P.1485, P-au-P - Dir: Serge Beaulieu.
33) P-au-P - Adm: William Bonhomme.
34) Cap-Haïtien.
35) 17 rue Pavee, P-au-P.
36) 346 Route, P-au-P - Dir: Fritz Joassin.

FM in Port-au-Prince: 2) 88.1 - 35) 88.5 - 89.3 RFI, 29) 89.7 - 90.1 Phare - 90.5 Signal - 91.3 Tropic - 2) 92.1 - 92.9 Guinen - 32) 94.1 - 26)

94.9 - 96.7 Delta - 19) 96.9 - 2) 97.9 - 20) 100.1 - 36) 100.9 Magic - 101.3 Universo - 14) 102.1 - 102.9 Super - 20) 103.7 - 104.1 Sodec - 104.5 Galaxie - 14) 105.1 - 21) 106.1 - 106.9 Kadans.

RADIO FRANCE INTERNATIONALE
✉ B.P. 1126, Port-au-Frec. ☎ +509 22-4724 - 🖫 +509 22-9140 - **E-mail**: ablanc@acn2.net - Dir: Marie-Christine Mourral Bussenius. FM Port-au-Prince 89.3.

HONDURAS

L.T: UTC -6h (Su: UTC -5h) — **Pop:** 5,600,000 — **Radios:** 1,910,000 — **Pr.L:** Spanish — **E.C:** 60Hz, 110V — **ITU:** HND.

EMPRESA HONDURENA DE TELECOMUNICACIONES (HONDUTEL)
✉ Ap. 1794, Tegucigalpa. ☎ +504 22 2101. **L.P:** Dir. Ing. Camilo A. Pon Z. Jefe Depto. de Frec: Ing. Emilio A. Montesi P.

ASOCIACION NACIONAL DE RADIODIFUSORES DE HONDURAS (ANARH)
✉ Ap. 4039, Tegucigalpa. ☎ +504 39 1992.

Mediumwaves: Call HR-,
° = also on shortwave, * = inactive, (r) = repeater, v = varying fq. H. of tr. 1100-0500 exc. where indicated.

Call	kHz	kW	Name and h. of tr.
154) OW	540	1	R. Atlántida, La Ceiba
155) XT	550	1	R. X, Tegucigalpa: 0945-0445
2) XD	550	0.5	R. Manantial, Santa Rosa de Copán: 1115-0300
76) OY	*560	1	R. Jupiter, Comayagua
3) RZ	560	5	R. Juticalpa, Juticalpa: 1100-0400
4) PX	560	1	R. Tropical, San Pedro Sula
156) OS	560	1	R. Castilla, Tocoa
160) OT	560	1	R. Montserrat, Danlí
157) OX	570	1	R. El Triunfo, Choluteca
6) ZQ	580	3	R. Tegucigalpa, Tegucigalpa
158) OU	580	1	R. Unión, Gracias
5) LP3	590	1	R. América, San Pedro Sula (r:610)
159) OV	590	1	LV de Lepaguare, Juticalpa
5) LP	610	10	R. América, Tegucigalpa: 1045-0515
5) LP4	610	10	R. América, Santa Rosa de Copán (r:610)
5) LP5	620	1	R. América, Siguatepeque (r:610)
5) LP9	620	1	R. América, Juticalpa (r:610)
28) LP17	620	1	R. Continental, San Pedro Sula
5) LP3	630	1	R. América, Choluteca (r:610)
5) LP7	630	1	R. América, La Ceiba (r:610)
8) NN4	640	1	Exitos, Tegucigalpa: 24h.
7) VW	650	25	LV de Centroamérica, San Pedro Sula
5) LP6	650	15	R. América, Danlí (r:610)
5) LP8	650	1	R. América, Olanchito (r:610)
8) NN18	660	3	LV de Honduras, La Ceiba (r:670)
8) N	670	10	LV de Honduras, Tegucigalpa: 1045-0500
8) NN20	670	1	LV de Honduras, Santa Rosa de Copán (r:670)
8) NN8	680	10	LV de Honduras, San Pedro Sula (r:670)
8) NN11	680	10	LV de Honduras, Tocoa (r:670)
8) NN2	680	10	LV de Honduras, Siguatepeque (r:670)
8) NN7	680	1	LV de Honduras, Danlí (r:670)
8) NN10	680	1	LV de Honduras, Juticalpa (r:670)
8) NN3	690	1	LV de Honduras, Choluteca (r:670)
8) NN9	690	1	LV de Honduras, Tela (r:670)
151) GP	700	5	Hacer Radio, Tegucigalpa
9) RH	710	3	LV de Occidente, Santa Rosa de Copán
14) UP	710	1	Estéreo Rey, San Pedro Sula: 1100-0600
10) LK	710	2	R. Comayagua/LV Católica, Comayagua: 1200-0300
11) KN	710	3	LV de Olancho, Catacamas: 1000-0300
8) NN13	710	1	LV de Honduras, Olanchito (r:670)
79) NN3	720	1	R. Caribe, La Ceiba
161) NG	720	1	Super Stereo Costa Sur, Choluteca
8) TG	730	1	R. Televisión, Tegucigalpa
162) XG	730	0.25	R. Cadena Dial, Santa Bárbara
12) QQ	740	1	R. Intibucá, La Esperanza

13) IH	740	1	7-40 La Super, Juticalpa: 1200-0400
14) NN23	740	1	R. Eco, San Pedro Sula: 24h.
85) TU	750	1	R. Trujillo, Trujillo
16) XK	°750	1	LV de la Mosquitia, Puerto Lempira
18) XW	760	2.5	R. Comayagüela, Comayagüela: 1200-0600
137) CG	760	1	R. Copán Galel, La Entrada
14) NN21	v770	10	R. Norte, San Pedro Sula
19) MV	770	0.5	R. Aguán, Olanchito
135) RO	770	1	R. Majestad "LV del Guayape", Juticalpa: 1100-0300
163) SE	780	1	Estéreo Sol 2000, Choluteca
20) IR	790	1	R. Estéreo Relámpago, Santa Bárbara
8) TG2	790	3	R. Satélite, Tegucigalpa: 24h
17) MA	800	3	R. Mundial, San Pedro Sula
21) DL	v800	1	R. Corporación, Comayagua 1100-0400
22) LP26	800	1	R. Sonora, Danlí
90) VC	810	1	LV Evangélica, La Ceiba (r:1390)
25) LP24	810	3	R. Valle, Choluteca: 1000-0400
5) LP16	820	5	R. Moderna, Tegucigalpa
84) KW	820	7/3	R. Sultana, Santa Rosa de Copán: 1100-0400
24) RU	830	1	R. Uno, San Pedro Sula
26) JB	830	1	Cadena Radial Impacto, Comayagua
27) VQ	830	1	R. Excelsior, Juticalpa: 1100-0300
18) CR	840	1	Dif. Cristiana de Radio "DCR", Choluteca
8) UP	850	10	R. Centro, Tegucigalpa: 1100-0400
165) IF	850	0.5	R. Inspiración, La Entrada
28) BS	860	10	R. San Pedro, San Pedro Sula
110) LS	860	0.5	R. Dinorama, La Paz: 1200-0300
1) H9	870	5	R. Honduras, La Ceiba (r:880)
1) H10	870	5	R. Honduras, Puerto Lempira (r:880)
1) H4	870	3	R. Honduras, Nacaome (r:880)
1) H	880	10	R. Honduras, Tegucigalpa
1) H5	880	10	R. Honduras, Santa Rosa de Copán (r:880)
23) MD	880	5	R. Yoro, Yoro
1) H3	890	10	R. Honduras, San Pedro Sula (r:880)
1) H7	890	10	R. Honduras, Juticalpa (r:880)
1) H9	890	10	R. Honduras, Siguatepeque (r:880)
1) H2	890	1	R. Honduras, Comayagua (r:880)
1) H6	890	1	R. Honduras, El Paraíso (r:880)
1) H8	890	1	R. Honduras, Olanchito (r:880)
8) UP5	890	1	R. Centro, Danlí (r:850)
8) UP6	900	1	R. Centro, La Ceiba (r:850)
8) UP7	900	1	R. Centro, Choluteca (r:850)
29) VS	910	10	LV de Suyapa "R. Católica", Tegucigalpa: 1030-0400
166) VH	910	0.5	R. Corona, La Entrada
21) RM	920	1	R. Sistema, Comayagua: 1300-0300
31) ZV	920	1	R. Fabulosa, San Pedro Sula: 1200-0500
32) SK	920	5	R. Catacamas, Catacamas: 1200-0400
1) H11	920	1	R. Honduras, Danlí (r:880)
18) CR	940	1	R. Dif. Cristiana de Radio "DCR", Tegucigalpa: 1200-0400
15) BO	940	1	R. Cadena Occidental, Santa Rosa de Copán
34) QL	950	1	Centro Radial Hondureño, Siguatepeque: 1100-0300
35) ZE	950	1.2	R. Cortés, Puerto Cortés: 1200-0400
138) PS	950	0.5	R. Sistema Popular, Danlí
36) YF	960	1	R. Fergusón, Choluteca
37) RD3	960	1	R. Sangrelaya, Sangrelaya
38) TL	970	2	Tic Tac, Tegucigalpa: 24h.
139) AS	970	0.25	R. Señorial, Ocotepeque
41) YG	980	1	R. Tocoa, Tocoa
39) ZC	980	2	R. Monumental, San Pedro Sula: 1200-0600
42) RD2	980	1	R. Emperador, Campamento
140) VO	990	3.5	R. Paz, Choluteca: 1000-0400
44) XZ	1000	1	Unión R. , Tegucigalpa: 1000-0600
43) MH	1000	0.5	LV del Junco, Santa Bárbara: 1200-0400
89) CD	1010	1	R. Constelación. Juticalpa: 1200-0400
46) LP23	1010	1	R. Moderna, El Progreso
48) UW	1020	1	R. Michelle, La Ceiba
167) MP	1020	1	R. Moropocai, Nacaome

152)	YF	1030	1	R. Ticante, Ocotepeque: 1200-0400
8)	UP3	1030	1	Estéreo Mil, Tegucigalpa
14)	NN22	1040	3	Exitos, San Pedro Sula
11)	FX	1040	1	R. Musical, Catacamas: 1200-0200
33)	MJ	1040	1	R. Renovación, Comayagua: 1200-0400
49)	ZX	1040	10/5	La Perimerísima, Olanchito
178)		1050	1	Estéreo Ceiba, La Ceiba
7)	VW	1060	2	LV de Centroamérica, Tegucigalpa (r:650)
50)	FA	1060	0.5	R. Peña Blanca, Santa Barbara
52)	GR	1070	3	Cadena Guaymuras, El Paraíso: 1100-0400
53)	LE	1070	1	R. 1050, San Pedro Sula
54)	LP26	1070	2.5	R. Siguatepeque, Siguatepeque 1055-
55)	XM	1070	1	R. Meridiano, Choluteca 1200-
56)	ID	1080	1	R. Miramar, Téla: 1200-0400
8)	NN27	*1090	1	Exitos, Santa Rosa de Copán (r:640)
57)	WC	1090	1	R. Aeropuerto Internacional, Tegucigalpa: 24h.
58)	ND	1100	1	R. Esperanza, La Esperanza: 1100-0400
59)	VA	1100	1	R. Tiempo, San Pedro Sula
60)	VL	1100	1	R. Lux, Olanchito: 1000-0400
29)	VS	1100	1	LV de Suyapa "R. Católica", Juticalpa: 1030-0400
5)	LP25	1110	1	R. Sur, Choluteca
87)	ME	1110	0.5	R. El Patio, La Ceiba
38)	YL	1120	2	R. Fiesta, Tegucigalpa
62)	DG	1120	1	R. Oriental "RCO", Danlí: 1100-0600
66)	AV	1120	1	Ondas del Ulúa, Santa Barbara (r:1150)
61)	PL	1130	5	R. Progreso, El Progreso: 1000-0300
63)	BT	1130	1	R. San Francisco, San Francisco de la Paz: 1100-0200
99)	HP	1130	1	R. Pinares, Siguatepeque 1155-
64)	AP	1140	1	R. Mercurio, Choluteca
65)	UN	1140	1	R. Palmeras, La Ceiba
66)	AV	1150	5	Ondas del Ulúa, Santa Bárbara: 1000-0400
5)	LP12	1150	1	R. Universal, Tegucigalpa
67)	QN	1160	5	LV del Atlántico, Puerto Cortés: 1000-0400
68)	GF	v1160	0.5	R. El Paraíso, El Paraíso
34)	QL2	1160	1	R. Sensación, Siguatepeque
47)	AZ3	1170	1	R. Hits, La Ceiba: 24h.
45)	AF	1170	2	R. Atenea "La Internacional", Choluteca: 1000-0400
149)	CY	1180	1	R. Estéreo Congolon, Gracias: 1100-0400
57)	AZ	1180	1	La Exitosa, Tegucigalpa
70)	PO	1190	1	R. Santa María de la Luz, Gualaco
6)	ZQ	1190	1	R. Tegucigalpa, San Pedro Sula: 1030-0500 (r:580)
134)	GK	1190	1	R. Brassavola, Minas de Oro: 1000-0300
169)	FS	1190	0.5	R. Familiar, Morazan
71)	DS	1200	1	R. Nacaome, Nacaome
72)	SI	1210	1	R. Impacto, Téla: 1200-0400
73)	RO	1210	1	R. Capital, Comayagüela
114)	HO	1210	1	R. Maya, Santa Rosa de Copán
74)	QO	°1220	1	R. Internacional, San Pedro Sula: 1100-0600
75)	YS	1220	1	R. Suari, Marcala
148)	JM	1220	0.5	R. Sava, Sava Colón
170)	GW	1220	10/1	R. Patria, Catacamas: 1000-0400
56)	QW	1230	10	R. Téla, Téla: 1200-0200
171)	SM	1230	0.25	R. Samaritano, San Marcos de Colón
133)	ZC	1240	1	R. Monumental, Tegucigalpa
172)	VN	1240	1	R. Venus, Santa Bárbara
40)	AT	1250	1	Super R. , San Pedro Sula
51)	CC	1250	1	R. Cadena Continental, Comayagua
115)	YN	v1250	1	R. Latina, Danlí
141)	QV	1250	0.5	R. Subirana, Yoro
77)	YF2	1260	1	R. San Marcos, San Marcos de Colón
116)	ZR	1260	1	R. 1260, La Ceiba
5)	NQ	1270	1	R. Sonora, Tegucigalpa
5)	NQ	1270	1	R. Sonora, Danlí (r:1270)
117)	OF	1270	1	Ecos del Celaque, Lempira
78)	AM	1280	1	R. Olanchito, Olanchito
136)	BU	1280	1	R. Digital, San Pedro Sula
107)	BN	1280	1	R. San Miguel, Marcala: 1000-0400
81)	NN26	1290	1	R. Choluteca, Choluteca: 1050-0400
118)	GS	1290	1	R. HRGS, Utila
82)	LR	1300	5	R. Santa Rosa, Santa Rosa de Copán
83)	LH	1300	1	LV de la Amistad, Tegucigalpa
90)	VC	1310	2.5	LV Evangélica, San Pedro Sula (r:1390)
103)	RL	1310	1	R. Libertad, Marcala, La Paz: 1200-0500
142)	JH	1310	0.5	R. Colón, Tocoa
119)	MG	1320	1	R. Bahía "La Super Grande", La Ceiba
126)	GM	1320	1	R. Ilusión, Choluteca
86)	SW	1330	1	R. Evangélica, Tegucigalpa
173)	FL	1330	1	R. Florida, La Entrada
153)	HH	1340	10	R. El Mundo, San Pedro Sula
120)	JC	1340	1	R. Colonial, Comayagua
174)	ED	1340	1	R. Red, Olanchito
143)	JV	1350	1	LV de San Lorenzo, San Lorenzo
28)	BS	1360	1	R. San Pedro, Tegucigalpa (r:860)
66)	BH	1360	1	R. Santa Bárbara, Santa Bárbara
175)	SZ	1370	1	R. Santa Bárbara, Siguatepeque
69)	TR	v1370	1.5	R. Danlí, Danlí
88)	ST	*1370	1	R. Fraternidad, San Pedro Sula
127)	AH	1380	0.5	R. Jutiapa, Jutiapa: 1200-0600
176)	EJ	1380	1	R. Voz Evangélica, Choluteca
90)	VC	°1390	10/5	LV Evangélica, Tegucigalpa: 24h
90)	VC	1390	1	LV Evangélica, Santa Rosa de Copán (r:1390)
91)	JJ	1400	1	R. Estéreo Punto, Comayagua
80)	YT	1400	1	R. Estrella de Oro, San Pedro Sula: 1100-0200
177)	AU	1400	1	R. Alegre, Sava Colón
92)	DD	1410	1	LV de Atlántida, La Ceiba
93)	SY	1410	1	LV del Pacífico, San Lorenzo
94)	SL	1420	1	R. Actualidad, Santa Bárbara
168)		1420	1	LV de las Fuerzas Armadas, Comayagüela
95)	IC	1430	1	La R. del 70, Puerto Cortés
96)	SJ	1430	1	R. Mundial, Tocoa: 1100-0400
97)	VM	1430	1	R. Maranatha, La Paz
164)	TP	1430	1	R. Recuerdos, Juticalpa
98)	RD	1440	5	Dimensión R, La Ceiba
121)	RY	1440	0.5	R. Mía, San Marcos de Colón
44)	XZ2	1450	1	R. Titania, Tegucigalpa: 1200-0600
144)	BR	1450	1	R. Cultural, La Entrada
74)	GC	1460	2.5	R. Conga, San Pedro Sula: 24h
100)	QX	1460	1	Radiolandia, Comayagua
122)	CX	1460	0.5	LV de Patuca, Catacamas: 1000-0400
113)	OC	1460	0.5	R. Ranchera, Olanchito
123)	SA	1470	1	R. Luz y Verdad, La Ceiba
145)	WP	1480	1	R. Soberanía, San Marcos, Ocotepeque: 1100-0300
102)	MI	°1480	1	LV de Misiones "R. MI", Comayagüela: 1100-0300
35)	ER	1490	1	R. Porteña, Puerto Cortés: 1200-0300
104)	OM	1490	1	R. Omega "Sonido Internacional", La Esperanza: 1100-0400
129)	RA	1490	1	R. Juventud, Sonaguera: 1230-0200
105)	TX	1500	1	R. Victoria, Choluteca
124)	YK	1510	1	R. Gualcho, Tegucigalpa
106)	EM	1510	1	R. Emanuel, Ocotepeque
130)	RG	1520	5	R. Providencia, Danlí
108)	CR	1520	1	Dif. Cristiana de Radio "DCR", San Pedro Sula: 1200-0400
131)	HJ	1520	1	R. Santiago, Yoro
132)	JX	1550	1	R. Nueva Vida, San Pedro Sula: 1000-0200
146)	KR	1550	1	R. Kristell, Juticalpa
101)	JO	1550	1	R. Campeona, Comayagua
57)	RF	v1570	2.5	R. Cadena Nacional de Noticias "RCN", Tegucigalpa: 24h.
18)	CR	1580	1	Dif. Cristiana de Radio "DCR", La Esperanza
111)	PC	°1600	1	R. Luz y Vida, San Luís: 2200-0400
125)	IK	1600	1	R. San Antonio, Tegucigalpa

Shortwaves:

111)	PC	v3250	1	R. Luz y Vida, San Luís: 1115-1800, 2200-0400
90)	VC	v4820	5	LV Evangélica, Tegucigalpa: 1100-0600

147)	LW	4830	0. 5	R. Litoral, La Ceiba: 1300-2300
16)	XK	v4910	0. 75	LV de la Mosquitia, Puerto Lempira: 2300-0300
74)	QO2	v4930	1	R. Internacional, San Pedro Sula: 1100-0600
83)	JA	4940	1	R. Copán Internacional, Tegucigalpa
128)	ET	v4960	1	R. HRET, Puerto Lempira: W:1300-1500. Tues-Sun: 0100-0300
102)	MI	5890	0. 2	LV de Misiones Int. "R. MI", Comayagüela
43)	MH3	6075	1	LV del Junco, Santa Bárbara (r: R. Galaxia)
112)	RI	v6300	1	Sani R. , Puerto Lempira: irr.
83)	JA	7460	1	R. Copán Internacional, Tegucigalpa
83)	JA	15675	1	R. Copán Internacional, Tegucigalpa

Addresses and other information:
1) Ap. 403, Tegucigalpa – Dir: Miguel Rafael Zavala.
2) 1a Av. 439, Barrio El Calvario, Santa Rosa de Copán – Dir: José R. Bueso P.
3) Ap. 3, Juticalpa – Dir: Victor Rubío Zapala. **FM:** 98.9.
4) Ap. 24, San Pedro Sula.
5) Audio Video, Ap. 259, Tegucigalpa – DG: Amilcar Zelaya R.
6) Cadena Corp. de Radiodifusión, Edif. Landa Blanca, Calle la Fuente, Tegucigalpa – DG: Antonio Masariego V.
7) Corpocentro, Ap. 120, San Pedro Sula – DG: Nohemy Sikaffy. **FM:** 89.5.
8) Emisoras Unidas, Blvd. Suyapa (or: Ap. 642), Tegucigalpa – DG: Nahum Valladares.
9) Ap. 206, Santa Rosa de Copán – DG: Dr. Arturo Rendon P. **FM:** 92.1.
10) Ap. 347, Comayagua – Dir: P. Alfonso Esteban F. **FM:** 90.3.
11) 3a Calle, Barrio El Centro 46, Catacamas, Olancho – DG: Dr. Raúl Zaldívar G. **FM:** 91.9.
12) 1 Cda. Abajo del Santa Cecilia, La Esperanza, Intibucá – Gte: Ismael Martínez A.
13) Ap. 9, Juticalpa – DG: José Guitarro.
14) Emisoras Unidas, Ap. 163, San Pedro Sula.
15) El Triángulo, Santa Rosa de Copán – Dir: B. Rivera D.
16) Barrio El Centro, Puerto Lempira (or: Global Outreach, Box 1, Tupelo, MS 38802, USA) – Dir: L. Wilkinson.
17) Av. New Orleans 20C, San Pedro Sula.
18) Ap. 3448, Tegucigalpa) – DG: Humberto Andino N.
19) Coyoles Central, Olanchito, Yoro – Gte: Gregorio Irías.
20) Ap. 26, Santa Bárbara – DG: Isidro Rodríguez – **FM:** 90.9.
21) Barrio San Francisco, Fte Parque, Comayagua – Gte: René Martínez V.
22) Danlí, El Paraíso – Dir: H. Mendoza.
23) Yoro, Yoro – Dir: Isaias Martínez.
24) 3 Calle 7 y 8, San Pedro Sula.
25) Ap. 29, Choluteca – Dir: Sra. Lety Stercke de Madragón.
26) Ap. 33, Comayagua – Dir: Juan Bosco Campos.
27) Ap. 28, Juticalpa – DG: J. Orlando Aguirre.
28) Ap. 364, San Pedro Sula.
29) Av. Paz Barahona (or: Ap. 480), Tegucigalpa – Dir: Manuel Cerrato.
31) Ap. 2918, San Pedro Sula – Dir: Mike Handal. **FM:** 102.1.
32) Ap. 50, Catacamas – DG: María E. Mendéz.
33) Casa N° 220, Av. Ppal, Comayagua – Dir: Jesús Martínez V.
34) Av. 2-3 Calle 269, Siguatepeque – DG: José A. Baires.
35) Barrio Copán, 3 Av. 7 y 8 Calle 772, Puerto Cortés – DG: Mario E. Prieto A. **FM:** 89.1+102.7.
36) Calle Vicente Williams, Edif. Fergusón, Choluteca – DG: Arturo Fergusón Luna.
37) Municipio de Iriona, Sangrelaya, Colón.
38) Ap. 771, Tegucigalpa – DC: María Antonieta Mendoza.
39) Ap. 996, San Pedro Sula – DG: Sergio Canahuati. **FM:** 98.5.
40) Barrio Las Acacias, 10 Calle 2 Av. , San Pedro Sula.
41) Tocoa, Colón – Dir: José F. Mejía H.
42) Campamento Olancho – Dir: Héctor Rubilio Romero P.
43) Ap. 6, Santa Bárbara
44) Ap. 614, Tegucigalpa – DG: Antonio Lardizabal.
45) Ap. 78, Choluteca – DG: Marcía A. de Aguilera. **FM:** 97.3.
46) Col. Brisas del Ulúa, El Progreso, Yoro.
47) Barrio La Isla, 4 Calle, La Ceiba – DG: Amilcar Zelaya.
48) Calle 14 entre Av. Fco. Morazan y Av. Cabañas, La Ceiba – DG: José Armando Iriasy. **FM:** 96.9.
49) Av. Francisco J. Mejía, Barrio Sofoco, Olanchito – DG: Santiago Ruíz.
50) Peña Blanca, 10 km al norte de Las Vegas, Santa Bárbara – Dir: Gaspar A. Pineda.
51) Barrio Costado Norte Cine Valladolid, Comayagua – Dir: Oscar I.

Murillo B.
52) Barrio Santa Clara, El Paraíso – DG: M. A. Ordónez G.
53) 9 Av. 4 Calle, Edif. Las Fuentes, San Pedro Sula.
54) Barrio Fatima, Edif. Audiovideo, Siguatepeque.
55) Calle San Marcos, Choluteca – DG: Mario Antonio Flores H.
56) Av. Panamá, Edif. Canales N° 861, Téla – DG: Ela Corina e Canales. **FM:** 96.3.
57) Ap. 2250, Tegucigalpa – Dir: Frances C. Bodden.
58) Ap. 25, La Esperanza – Dir: Natanael del Cid.
59) Ap. 906, San Pedro Sula – DG: Victor Manuel Rodríguez A. **FM:** 97.9.
60) Calle El Calvario Frente Al Parque, Edif. Plaza, Olanchito – DG: Erlin Evelio Rubi J. FM: 88.7.
61) Ap. 20, El Progreso, Yoro – Gte: José Vicente Owens. FM: 103.3 Stereo Alegría.
62) Ap. 21, Danlí, El Paraíso – Dir: Hernan Mendoza M.
63) San Francisco de la Paz, Olancho – DG: Nolberta Caridad Mejía.
64) Contiguo Cine Rey, Choluteca – Dir: Luis A. Pavón G.
65) Barrio La Isla, La Ceiba – Dir: Oscar L. Irías.
66) Ap. 004, Santa Bárbara – DG: Benjamín E. Handal.
67) 12 Calle 2a Av. N° 206, Puerto Cortés – DG: A. Griffin C.
68) Barrio San Isidro, El Paraíso – Gte: Hernan Mendoza M.
69) Barrio El Centro, Danlí – DG: Carlos Castillo Valle.
70) Iglesia Católica, Gualaco, Olancho.
71) Barrio El Centro, Nacaome, Valle – DG: Enrique Rodríguez.
72) Calle José Trinidad Cabañas, Edif. Hotel Presidente, Téla – DG: Ricardo R. Patiño. **FM:** 89.9.
73) Col. El Prado 1C-107A, Comayagüela – DG: Maria D. Sagastume.
74) Ap. 1473, San Pedro Sula – DG: Víctor Antonio Handal H. **FM:** 91.9+93.7.
75) Calle Principal, Marcala, La Paz – DG: Mauro Suazo.
76) Barrio Abajo, Comayagua – DG: Juan Estebán Ortiz M.
77) San Marcos de Colón, Choluteca – Dir: Arturo Fergusón Luna.
78) Olanchito, Yoro – Dir: Carlos A. Muñoz.
79) Emisoras Unidas, Solares Nuevos, Av. República, La Ceiba.
80) Ap. 303, San Pedro Sula – DG: Merton Rundell III. **FM:** 97.3.
81) Barrio Campo Luna, Choluteca – Dir: M. Villeda T.
82) Ap. 203, Santa Rosa de Copán – Gte: P. Iván de Jesus A.
83) Ap. 955, Tegucigalpa – DG: J. A. Padilla H.
84) Ap. 204, Santa Rosa de Copán – DG: Noé L. Cruz. **FM:** 90.3. Rosa de Copán.
85) Barrio El Centro, Trujillo – Own: José Ham.
86) Ap. 3405, Tegucigalpa – Dir: Roberto Durón C.
87) Av. San Isidro, La Ceiba – Dir: Gregorio Irías.
88) Colonia Fesitran, San Pedro Sula.
89) Juticalpa – Dir: Roberto J. Torres Z.
90) Ap. 3252, Tegucigalpa – Dir: Venancio Mejía Cartagena. (Owned and operated by Conservative Baptist Home Mission Society, Box 828, Wheaton, IL 60187, USA).
91) 1a Av. N° 189, Camayagua – Dir: Oscar Ovidío Bueso.
92) Ap. 17, La Ceiba – Dir: Miguel R. Moncada R.
93) San Lorenzo, Valle – Dir: E. Pineda Hernández.
94) 2a Av. N° 38, Santa Bárbara.
95) 12 Calle 2a Ave 206, Barrio La Curva, Puerto Cortés – DG: Andrés Griffin Cubas.
96) Barrio El Centro, Tocoa, Colón – DG: Erlin Evelio Jiménez.
97) Santiago de la Paz, La Paz – Dir: Eva A. García V.
98) Av. San Isidro, Entre Calles 9 y 10, La Ceiba - DG: José Luis Asenjo.
99) Barrio Abajo, Siguatepeque – Dir: Helda Olinda Pinel O.
100) Calle Boulevar, Comayagua – Dir: Rolando Barahona.
101) Barrio Abajo 229, Comayagua – Dir: Juan Ortíz.
102) Ap. 20583, Comayagüela (or: IMF World Misiones, PO Box 6321, San Bernardino, CA 92412, USA) – Dir: Wayne Downs.
103) Barrio San Miguel, Calle Principal, Marcala, La Paz – Dir: Luis Enrique Gusmán.
104) Av. España, La Esperanza, Intibucá – Dir: M. A. Orellana.
105) Barrio La Cruz, Calle Chorotega, Choluteca – DG: Mario Gotto. **FM:** 96.2.
106) Barrio San Andrés, Ocotepeque – DG: José Ruben Martínez O.
107) Barrio Concepción, Marcala, La Paz – Dir: P. Lucio Nuñez C.
108) Ap. 2017, San Pedro Sula.
110) Parque Central, La Paz
111) San Luis, Santa Bárbara (or: Ap. 303, San Pedro Sula) – DG: Merton Rundell III.
112) Puerto Lempira – Gte: Edard A. Pfister.
113) Olanchito, Yoro – DG: Carlos A. Muños O.
114) Col. Progreso 2 y 3, Av. S. O. , Santa Rosa de Copán.
115) Barrio El Centro, Danlí, El Paraíso – Dir: Hernán Mendoza M.
116) 4a Calle N° 1185, La Ceiba – DG: Amilcar Zelaya Rodrígues.

117) 2a Av. 9C N° 9, Gracias, Lempira – DG: Oscar Felix Reyes S.
118) Col. de Jerico, Utila – Dir: Glenn R. Solomon.
119) 4c Av. Juan Ramón M, La Ceiba – DG: Miguel Velasquez V.
120) Calle del Comercio 12A, Comayagua.
121) San Marcos de Colón, Choluteca – DG: Migdonía Yanez.
122) Barrio La Mora, Catacamas – DG: José A. Euceda. **FM:** 99.1.
123) Barrio Loma Jackson N° 3, La Ceiba.
124) Col. 21 de Octubre, Sector 3, Bl. 1, Casa 4, Tegucigalpa – DG: Omar Rodríguez A.
125) Col. Rio Grande, Tegucigalpa – Dir: Ivonne B. Pagoaga.
126) Barrio Cafetal, San Marcos de Colón, Choluteca – Dir: G. Holner. **FM:** 91.3.
127) 1 Av. Calle Principal, Jutiapa, Atlántida – DG: José Fernando Cruz. **FM:** 95.7.
128) Misión La Mosquitia, Pto Lempira – Dir: Mateo McCollum.
129) Calle Central, Sonaguera, Colón – DG: José Tomas Vastillo F. **FM:** 93.5.
130) Danlí, El Paraíso – Dir: Roberto Gamero.
131) Yoro – Dir: Jamil N. Hawit Castro.
132) Ap. 2424, San Pedro Sula.
133) Ap. 914, Tegucigalpa – Dir: Fernando H. Gómez H.
134) Barrio La Manzana, Minas de Oro, Comayagua – Dir: Gladys de Kesler.
135) Ap. 15, 16101 Juticalpa, Olancho – DG: Héctor Robilio Romero P. – **FM:** 106.3. Prgrs in Sp. and E.
136) Col. Río Piedras, 5 Calle 26 Av. , San Pedro Sula.
137) La Entrada – Own: Armando Calidonio Alvarado.
138) Fte. Supermercado Demar, Danlí, El Paraíso.
139) Barrio El Centro, Fte. Coop, Ocotepeque – DG: Alejandro Sanchez.
140) Ap. 40, Choluteca – Dir: Carmen Galeas. **FM:** 95. 5.
141) Barrio El Centro, Yoro – Own: Amílcar Zelaya R.
142) Tocoa, Colón – DG: Jesús Mejía H.
143) San Lorenzo – Own: Julio César Villatoro.
144) La Entrada. DG: Julio Cesar Lopez.
145) Barrio San Sebastián 2 Calle, San Marcos, Ocotepeque - DG: Wilfredo Paz.
146) Barrio Jesus, Juticalpa – Own: Marco Tulio Torres Rodríguez.
147) Centro Comercial San José, La Ceiba – Dir: José E. Mejía.
148) Av. Principal, Sava Colón.
149) Gracias, Lempira – Dir: Marco Augusto Hernández E. **FM:** 95.1.
151) Ap. 5812, Tegucigalpa – DG: Lic. Marco Antonio S.
152) Media Cuadra Al Norte del Parque, B:o El Centro, Ocotepeque DG: Sandra Maria Rendon Medina.
153) Ap. 210, San Pedro Sula – DG: Joel Lopez Pineda. **FM:** 90.7.
154) Av. 19 de Julio, La Ceiba – DG: Jorge A. Tovar L.
155) Col. Miraflores, Tegucigalpa – DG: René Cepeda.
156) Calle Principal, Tocoa – DG: Jorge A. Tovar L.
157) Calle Vicente Williams 345, Choluteca.
158) Barrio El Centro, Gracias – DG: Jorge A. Tovar L.
159) Barrio El Centro, Juticalpa.
160) Calle del Comercio, Danlí, El Paraíso.
161) Barrio Sanpile, Carr. a Guasaule, Choluteca – DG: Hector O. Martínez.
162) 2 Av. Calle 38, Trinidad, Santa Bárbara – DG: Jorge G. Mathis.
163) Barrio La Esperanza 4A N° 142, Choluteca – DG: Arturo G. Espinal.
164) Barrio Jesus, Juticalpa.
165) Barrio Monte Sinai, La Entrada – DG: Santos A. Martínez.
166) Barrio El Progreso, La Entrada – DG: Exel. René Gamez C.
167) Barrio Santa Rosario, Nacaome, Valle – DG: Manuel R. Pino R.
168) Las Torres, Comayagüela.
169) Barrio San José, Morazán, Yoro.
170) Calle del Estadio, Barrio El Campo, Catacamas – DG: Marco A. Ramírez M.
171) San Marcos de Colón, Choluteca.
172) Av. Independencia, Santa Bárbara.
173) Barrio Miraflores, La Entrada – Dir: Camilo Vazques.
174) Olanchito, Yoro – Dir: Jaime R. Espinoza.
175) Barrio El Centro, Siguatepeque – DG: Benjamín Handal.
176) Barrio Guadalupe, Choluteca – DG: Pablo Medina.
177) Barrio El Coyol, Sava Colón – DG: Holber Velasquez.
178) La Ceiba.

FM in Tegucigalpa: 8) 88.1 R. Satélite – 83) 89.3 Estéreo Amistad – 89.9 Saturno FM – 8) 90.5 Exitos – 18) 91.1 Comayagüela – 38) 91.7 R. Fiesta – 8) 92.3 Mil – 8) 92.9 6i 93.5 – 8) 94.1 FM 94 – 18) 95.9 R. Panamericana – 38) 97.1 Tic Tac – 57) 97.7 Azul – 98.3 Concierto FM – 57) 98.9 La Exitosa – 18) 99.5 Suprema – 100.1 Super 100 – 1) 101.3 – 8) 101.9 "102" – 8) 102.5 R. Centro – 18) 103.1 DCR – 90) 103.7 Luz – 104.3 Momentos FM – 104.9 Amor – 151) 105.5 – 57) 106.1 – 7) 106.7

– 107.3 W107.

JAMAICA

JAMAICA

LT: UTC - 5h — **Pop:** 2.635.000 — **Radios:** 1.859.000 — **Pr.L:** English — **E.C:** 50Hz, 110/220V — **ITU:** JMC.

JAMAICA BROADCASTING CORP. (Gov. Comm.)
✉ Box 100, Kingston 10. ☎ +1 (876) 926 5620/9. 🖷 +1 (876) 929 1029. **Cable:** JARAD Jamaica. ➀ 2218 BROADCORP JA.
L.P: DG: Claude Robinson. Mgr. Eng. Sces (Radio): Lloyd Bolageer. Mgr. Pub. & Int. Affairs: Lois Gayle.

MW (JBC1)	kHz	kW	MW (JBC1)	kHz	kW
Kingston	560	5	Port Maria	750	10
Mandeville	620	5	Savanna La Mar	850	5
Montego Bay	700	10	Morant Bay	1090	1

FM: see below — **D.PRGR:** 24h — **V.** by letter. Rp.

RADIO JAMAICA LIMITED (Comm.)
✉ P.O. Box 23, Kingston 5. ☎ +1 (876) 926 1100. 🖷 +1 (876) 929 7467. **Cable:** Broadco. **E-mail:** rjrnews@toj.com
Web: http://www.rjr.com.jm/
L.P: Chmn. & MD: J.A.Lester Spaulding. PD: D. Topping. Prgr. Mgr. (FAME FM): Norma Brown-Bell. CE: Carroll Lawrence. Mktg. Mgr: Michael Johnston. N. & Current Affairs Editor: Jennifer Grant.

MW (RJR)	kHz	kW	MW (RJR)	kHz	kW
Montego Bay	550	10	Kingston	720	10
Galina	580	10	Spur Tree	770	10

FM: see below — **D.PRGR:** 24h — **V.** by QSL-card.

FM transmitters of JBC & Radio Jamaica:

Location (MHz)	JBC1	JBC-FM	RJR	FAME
Coopers Hill	97.1	91.1	94.5	92.7
Flower Hill	97.3			
Half Way Tree	99.7	105.7		
Kingston			104.5	95.7
Montego Bay			92.9	95.3
Oracabessa	103.9	100.3	101.3	91.5
Port Antonio	98.7			
Spur Tree	93.3		90.5	98.1

Island Broadcasting Services, 19 Caledonia Rd, Mandeville: KLAS-FM 24h on 89.3/89.5/89.9MHz — **Irie FM:** 105.1/105.5/107.1/107.7MHz — **Power FM:** 106.1/106.5MHz — **Love FM:** 101.1/101.7MHz — **Waves FM:** 102.MHz.

MARTINIQUE (French)

LT: UTC - 4h — **Pop:** 380.000 — **Radios:** 71.000 — **Pr.L:** French, Créole Patois — **E.C:** 50Hz, 220V — **ITU:** MRT.

RADIO-TÉLÉVISION FRANÇAISE D'OUTRE MER (RFO)
✉ B.P. 662, F-97263 Fort de France. ☎ +596 595200. ➀ 029659 —
L.P: Dir: Jean Claude Arrivé. CE: Jean Claude Arrivé.
FM: Fort-de-France 92.0/94.5 MHz, Morne-Rouge 94.3 MHz, Marin 93.2 MHz, Trinité 94.0 MHz, Macouba 92.0 MHz, St-Pierre 94.0 MHz.
D.PRGR: 0800-0400. **N:** 1000, 1030, 2000, 2300 + rel. France-Inter.
IS: Piano — **V.** by QSL-card.

RCI - RADIO CARAÏBES INTERNATIONAL MARTINIQUE (Comm.)
✉ 2 Boulevard de la Marne, F-97200 Fort de France. ☎ +596 639870. 🖷 +596 632659. **Web:** http://www.fwinet.com/rci.htm
L.P: Dir: Yann Duval. Editor-in-Chief: Jean Philippe Ludon (**E-mail:** 100444.2371@compuserve.com). CE: Daniel Toussaint (**E-mail:** 100430.2743@compuserve.com).
FM: 91.2/98.7/103.0/104.6MHz.
D.PRGR: 24h. **N:** on the h. (rel. Europe 1) — **V.** by letter. Rp.

R. France Internationale: rel. via R. Intertropical 99.9MHz, R. 105 Canal Antilles 105.0MHz & R. AS 106.2MHz.

Other stations: approx. 40 FM st's are operating.

MONTSERRAT

LT: UTC - 4h — **Pop:** 5.000 — **Radios:** 10.000 — **Pr.L:** English — **E.C:** 60Hz, 220V — **ITU:** MSR.

RADIO MONTSERRAT (Gov. Comm.)

◻ Montserrat.
L.P: SM: Ms. Rose Willock, OBE. Sen. Technician: Lowell Mason.
MW: 885kHz 10kW (+ 1kW reserve).
FM: 92.5MHz 1kW, 92.3MHz 0.1kW.
D.PRGR: W 1000-0300, Sun 0900-1600 & 2000-0100.
ANN: "ZJB Radio, the Voice of Montserrat".
NB: Due to the Volcano emergency on the island, information may have changed since editorial deadline.

NETHERLANDS ANTILLES

L.T: UTC-4h — **Pop:** 210.045 — **Radios:** 206.000 — **Pr.L:** Papiamento (Leeward Antilles), English (Windward Antilles), Spanish — **E.C:** 50 + 60Hz, 127/220V — **ITU:** ATN.

LANDSRADIO (Telecommunication Administration)

◻ Schouwburgweg 22, P.O. Box 103, Curaçao. ☎ +599 (9) 631111.
⟳ 1075 LRDIR NA.

Call	kHz	kW	MHz	kW	Name and location
1) PJB	800	500/50			TWR, Bonaire
2) PJZ-86	860	10	95.7	4	R. Curom, Willemstad
3) PJC-7	1010	3	101.9	5	R. Hoyer 1, Willemstad
4) PJL-3	1100	0.25			R. Caribe, Willemstad
5) PJE-3	1120	1			R. Statia, St. Eustatius
6) PJD-2	1300	1	102.7	3.5	PJD-2 Radio, Philipsburg
7) PJF-1	1410	5	93.9	??	Voice of Saba, Saba
3) PJC-9	1500	3	105.1	5	R. Hoyer 2, Willemstad
8)			92.3	5	R. Merkadeo, Willemstad
9)			93.9	20	R. Korsou FM, Willemstad
10)			94.7	1	Voz di Bonaire, Bonaire
11)			97.1	5	Ritmo FM, Bonaire
18)			97.9	0.5	Easy 97.9 FM, Willemstad
12)			98.7	2.5	R. Semiya,Willemstad
15)			100.3		R. Super Jumbo,Willemstad
8)			101.1	5	R. Korsou FM/Laser 101, Willemstad
10)			101.1	0.4	Voz di Bonaire, Bonaire
13)			103.1		R. Paradise, Willemstad
14)			107.9	1	R. Exito, Willemstad
16)			107.9	1	Gem Radio Network, Philipsburg
17)					R. Tropical, Willemstad

Addresses and other information

All st's comm. exc.1) (see separate listing under Trans World R.)
2) R. Curom, Roodeweg 64, P.O. Box 2169, Willemstad, Curaçao. ☎ +599 (9) 626586. 🖳 +599 (9) 625796 — Dir. & GM: Orlando Cuales. CE: C. Siegenthaler. 1000-0400 in Papiamento. Separate music prgrs. ("Z-FM") on 95.7MHz — **V.** by letter. Rp.
3) Plasa Hoyer 21, Willemstad, Curaçao. ☎ +599 (9) 611678. 🖳 +599 (9) 616528. **E-mail:** hoyer@cura.net **Web:** http://www.cura.net/radio-hoyer/ — MD: Ms. Helen Hoyer.W 0930-0430, Sun 1230-0330. **R. Hoyer 1** in Papiamento on 1010 kHz + 101.9 MHz, **R. Hoyer 2** in Dutch on 1500 kHz + 105.1 MHz. Sports and parliamentary coverage as necessary on AM, regular prgrs continue on FM only. **V.** by letter. **N.B.** FM-tr's are fully powered by solar energy.
4) Ledaweg, Brievengat, Willemstad, Curaçao. ☎ +599 (9) 369555. 🖳 +599 (9) 369569. **Cable:** R. Caribe — DG: C.R. Heilegger. TD: G.A. Heilegger. 1000 (Sun 1200)-0400. **Papiamento:** W 1000-1700, 1900-2200, 0100-0400. Sun 1400-1700, 1830-2200, 2300-0400. N: 1615W, 1700W. **Spanish:** W 1700-1900, 2200-0100. Sun 1700-1830. N: 1700W, 1800W (VOA), 1815Sun 2200W, 0015W. **English** (rlg. prgr's): Sun 1200-1400, 2200-2300. **V.** by letter. Rec. acc.
5) St. Eustatius Broadc. N.V., Korthals Weg, St. Eustatius. ☎ +599 (3) 82262.
6) Plaza 21 Shopping Centre, P.O. Box 366, Philipsburg, St. Maarten. ☎ +599 (5) 22580, 22764. 🖳 +599 (5) 22300 — GM/Dir: Donald R. Hughes. **MW:** 0930-0600 (Mon 0400). **N:** 1030, 1700, 0000. **FM:** 0930-0400. **N:** 1200, 1800, 0000.
7) P.O. Box 1, The Bottom, Saba. (Studios in St. Maarten). ☎ +599 (5) 63213 — Owner: Max Nicholson. **On 1410 kHz:** 1000-2330 (Sun 1700). **N:** on the h. also 1030. **On 93.9 MHz:** rel. ZGM-FM 94 Montserrat with local commercials inserted. **V.** by letter.
8) Generaalsweg 50, Willemstad, Curaçao. ☎ +599 (9) 376115. 🖳 +599 (9) 374514 — Dir: E. Leito. 1000-0400 in Papiamento.

9) Bataljonweg 7, P.O. Box 3250, Willemstad, Curaçao. ☎ +599 (9) 373377. 🖳 +599 (9) 372888 — Dir: J.P.C. Oosterhof. PD: Alan H. Evertsz. 24h. Separate prgrs ("Laser 101") on 101.1 MHz. 24h. **N. in Dutch:** 1000, 1600, 2100. **N. in Papiamento:** 2200. **English:** Tues 0000. **Portuguese:** Wed 2330. **Sranan Tongo:** Fri 0000.
10) P. O. Box 325, Kralendijk, Bonaire. ☎ +599 (8) 5971. 🖳 +599 (8) 5000 — Dir: Edsel Jesurun Jr. & Irwin E. Halley. 24h Fully automated music st. with N. breaks.
11) Kaya Gob. N. Debrot 2, Kralendijk, Bonaire. ☎ +599 (7) 7220 (offices), 8273 (studios). 🖳 +599 (7) 8220 — Dir: F. Piloto. 1000-0400 in Papiamento and Dutch. **N:** 2200. **N: in Dutch:** 1900, 2000.
12) P. O. Box 4709, Willemstad, Curaçao. ☎ +599 (9) 628488. 🖳 +599 (9) 648390 — Dir: Ferris Thode. 1000-0400 in Papiamento.
13) ITC Building, P.O. Box 6103, Willemstad, Curaçao. ☎ +599 (9) 636103. 🖳 +599 (9) 636404. **E-mail:** Radiop@IBM.net — MD: Jacques Visser. 1000-0400. **Dutch:** 1000-1300. **N. and tourist information: Dutch:** 1430, 1630, 1830, 2030, 0030. **English:** hourly 1300-2300. **Spanish:** 1530. **Portuguese:** 1330.
14) Wolkstraat 15, Willemstad, Curaçao. ☎ +599 (9) 658884. 🖳 +599 (9) 658886 — Dir: Donny Hernandez. 24h in Papiamento.
15) Willemstad. ☎ +599 (9) 628811. 🖳 +599 (9) 628868 — Dir: Ernest Willems.
16) Relays prgrs from Trinidad.
17) Willemstad. ☎ +599 (9) 652467. 🖳 +599 (9) 652470 — Dir: Dwight Rudolphina.
18) Arikokweg 19A, Willemstad. ☎ +599 (9) 462 3162. 🖳 +599 (9) 462 8712.**Info line:** +599 (9) 462 3611. **E-mail:** radio@easyfm.com . **Web:** http://easyfm.com— Dir: Kevin Carthy. 24h.

TRANS WORLD RADIO (Cult. Educ. Rlg.)

MW (G.C: 68.28W/12.11N): 800kHz 500/50kW.
Full details in International Broadcasting section.

RADIO NEDERLAND RELAY STATION

see International Broadcasting section.

NICARAGUA

L.T: UTC -6h — **Pop:** 4,130,000 — **Radios:** 925,000 — **Pr.L:** Spanish — **E.C:** 60Hz, 120V — **ITU:** NCG.

DIRECCION DE TELECOMUNICACIONES

◻ Ap. 232, Managua. ☎ +505 (2) 632171, 632181. **L.P:** Ing. Adolfo López Gutiérrez.

ASOCIACION NICARAGUENSE DE RADIODIFUSION (ANIR)

◻ c/o R. Ya, Ap. 1787, Managua. ☎ +505 (2) 785600. **L.P:** Pres. Carlos J. Guadamuz.

CAMARA NICARAGUENSE DE RADIODIFUSION

◻ c/o R. Corporación, Ap. 2442, Managua. ☎ +505 (2) 443824 – **L.P:** Pres. Fabio Gadea Mantilla.

Mediumwaves: Call YN-, * = inactive, (r) = repeater, v = varying fq.

	Call	kHz	kW	Name and h. of tr.
1)	OW	540	20	R. Corporación, Managua 0900-0400
2)	CH	550	10	R. 19 de Julio "la 19", Chinandega: 1000-0200
60)		560	10	Managua (F.PI.)
50)		570	5	R. 5-70, Chinandega: 1030-0250
47)	EA	580	1	R. 5-80, Managua 1000-
3)	LD	600	10	R. Ya, Managua: 1000-0600, Weekends 24h.
4)	N	620	5	R. Nicaragua, Managua: 0955-0500
61)	LN	640	10	R. Ranchera, Managua: 1000-
5)	RI	650	12	R. Septentrión, Matagalpa: 1100-0100
6)	RD	650	10/8	R. Diriangén "La Super D", Granada: 0950-2300
52)		670		R. Caribe, Puerto Cabezas
7)	AM	680	10/2	R. La Primerísima, Managua: 24h
53)		690	5/2	R. Hermanos, Matagalpa
8)	MM	700	10	R. Istmo, Managua: 1000-2400
9)	RC	720	10	R. Católica, Managua: 1100-0355
10)	NS	730	10	R. Segovia, Ocotal: 1100-0200
11)	RS	740	50	R. Sandino "La S Grande", Managua:

			1025-0400
54)		760	10	Ultravisión de Nicaragua, Managua
17)	AD	780	1	R. Deportes, Managua
51)		800	1	R. 800, Managua
12)	RR	810	5	R. Rumbos, Rivas
13)	OL	820	20	R. Ondas de Luz, Managua: 1000-0400
14)	RZ	v830	10	R. Zinica "La Voz Costeña", Bluefields
15)	RN	840	5	R. Noticias, Managua: 1030-0200
54)		860	5	Ultravisión de Nicaragua, Managua
16)	CC	870	10	R. Centro, Juigalpa
17)	AT	880	5	R. El Pensamiento, Managua: 1000-0420
18)	RT	900	5	R. Tiempo, Managua: 1050-0400
55)		910	5	R. Jinotega, Jinotega
19)	W	920	10	R. Mundial, Managua: 1100-0400
17)	AD	940	1	R. Deportes, Managua
56)		960	2.5	LV del Trópico Húmedo, San Carlos
20)	FF	1000	1	R. Mil, Managua: 24h
21)	HG	1010	5	R. LV del Pinar, Ocotal: 1100-0400
22)	LL	1030	2	R. Masaya, Masaya
23)	VJ	1040	2	LV de Jinotega, Jinotega
24)	JJ	1060	1	R. Juvenil, Managua: 24h
25)		1060	1	LV del Atlántico, Bluefields
57)		1090	5	R. Alma Latina, Estelí: 1100-0400
26)	MT	1110	1	R. Momotombo, La Paz Centro
27)	CP	1120	5	R. CEPAD "El Arco Iris Del Amor", Managua: 1100-0100
29)	UW	1150	5	R. Darío, León: 1000-0300
30)	HM	1160	1	R. Satélite, Estelí
58)		1170	5	R. Máxima, Masaya
17)	AD	1200	1	R. Democracia, Managua: 1000-0400
31)	NG	1230	5	R. Manantial, Nueva Guinea: 1000-0300
59)		1240	1	R. Restauración, Managua
32)		1250	5	Radial Samaritano, Condega
33)	RA	*1270	3	R. Amistad, Matagalpa
34)	R	1300	1	Canal 130 AM, Managua: 1200-2330
35)	SC	1310	10/1	R. San Cristóbal, Chinandega: 1000-0200
36)	OS	1340	1	R. Ondas Sonora, Managua
36)	GF	1350	1	R. Ondas del Sur, Jinotepe
37)	GA	*1370	1	R. Matagalpa, Matagalpa
38)	RE	1370	1	R. Somoto, Somoto
39)	RV	1410	3/1	La Estación de la Amistad, León: 1000-0200
40)	LE	1430	5	R. Liberación "La Tayacana", Estelí: 1100-0300
41)	RM	1440	25	R. Maranatha, Managua: 1000-0600
42)	RY	1470	1	R. Yarrince, Boaco
43)	PT	1500	1	R. Minuto, Managua: 1000-0300
44)	RF	*1520	1	R. Flash, Managua
28)		1530	0.5	R. LV de Tereza, Sta Tereza
45)	CN	1560	5	R. América, Managua: 1600-0200

Shortwaves:

49)	PM	5770	1	R. Miskut, Puerto Cabezas

Addresses and other information:

1) Ap. 2442 (or: Cd. Jardín Q-20), Managua – Gte: Fabio Gadea Mantilla – Dir: José Castillo Osejo.
2) Ap. 12 (or: Frente Iglesia Guadalupe), Chinandega – Dir: Alberto Jarquín Sáenz.
3) Ap. 1787 (or: Pista de la Resistencia), Managua – DG: Carlos José Guadamuz P.
4) Ap. 4665 (or: Costado Sur TELCOR, Villa Fontana), Managua – Dir: Franklin Sequeira Lopez.
5) Frente a la Catedral, Matagalpa – Dir: Celso Martínez.
6) Piedra Bocona. ½ cuadra abajo, Granada – Dir: Henry Ruiz Valdo A.
7) Ap. 4003 (or: Bolonia, Teatro Cabrera 2, 2 cuadras abajo, 3 cuadras al sur), Managua – Gte: William Grigsby V.
8) Ap. 700 (or: Del Zumen, 1 cuadra abajo, 2 cuadras al sur), Managua – Dir: Jesús Miguel Blandón
9) Ap. 2183 (or: Altamira D'Este N° 621, Etapa III), Managua – Mons. Bismarck Carballo M.
10) Detras de la Iglesia la Asunción, Ocotal – DG: Roger Solis Correa. **FM:** 97.3.
11) Ap. 4776 (or: Paseo Tiscapa este, Contiguo al Restaurante Mirador), Managua – Dir: Conrado Pineda Aguilar.
12) Carr. a San Jorge, Rivas – Dir: José María Cuadra.
13) Ap. 607 (or: Costado sur del Hospital Bautista), Managua – DG:

Eduardo Gutierrez N.
14) Ap. 25 (or: Av. del Cementerio 20), Bluefields – Dir: Arturo J. Valdez R.
15) Ap. A-150 (or: Col. Robles, IV Etapa 92), Managua – Dir: Agustín Fuentes.
16) Caracoles ½ cuadra al sur, Juigalpa – Dir: Jaime Zamora E.
17) Altamira D'Este N° 73, Managua – Dir: Allan Téfel Alba.
18) Ap. 2776 (or: Col. Los Robles, Repto. Pancasán N° 217, Etapa VII), Managua – Dir: Nelson Reyes
19) Ap. 3170 (or: Repto. Loma Verde, 36 Av. Oeste), Managua – Dir: Manuel Arana Valle.
20) Col. Los Robles, IV Etapa 70, Managua – DG: Dulce María Rivera Cruz.
21) De la Parroquita 1 cuadra al norte, Ocotal – Dir: Heriberto Gadea Mantilla. FM: 100.9.
22) Teatro Masaya 1½ cuadras al oeste N° 135, Masaya - Dir: Jorge Correa.
23) Banades, 1 cuadra al oeste, Jinotega – Dir: Medardo Contreras P.
24) Col. Los Robles del gimnasio Atlas, 1 cuadra al sur, Managua – DG: Miguel Blandón Montenegro. **FM:** 101.5.
25) Frente al Palacio Municipal, Condega
26) La Paz, Centro Depto, León – Dir: Rafael Quintana Ortiz.
27) Ap. 3091 (or: Del Portón Cementerio Occidental, 2 cuadras al lago), Managua – Dir: Nidia Aguirre. (Owned and operated by CEPAD – Consejo Evangélicas Pro Alianza Denominacional).
28) Entrada II Calle, ½ cuadra abajo, Sta Tereza – Dir: Raúl Cruz Guadamuz.
29) Residencial Posada del Sol, Casa N° 93, León – Dir: Juan Toruño L.
30) Costado oeste Inst. Mª Zeledón, ½ cuadra al oeste, Estelí – Dir: Santiago Hudiel E.
31) Calle Principal, Nueva Guinea – Dir: Marcos Antonio Urbina L.
32) Instituto Bíblico Samaria, Condega
33) Detrás de la Iglesia San José, Matagalpa – Dir: Mario Mairena M.
34) Ap. E-2 (or: Carr. a Masaya km 12¼, 450 m al este), Managua – Dir: Reinerio E. Montiel B.
35) Ap. 59 (or: Club Eden, 250 m vrs al sur), Chinandega - Dir: Lic. Wilfredo Romero B.
36) Barrio Cristo del Rosario, Cine Blanco 5 cuadras al norte, ½ cuadra al este, Managua – Dir: Diego Manuel Rodríguez C.
37) Sector 7, Del Parque Central 1½ cuadra al sur, Matagalpa – Dir: Germán Alfaro.
38) Somoto – Dir: Eli Ramón Alfaro.
39) UNAM 1½ cuadra al norte, León – DG: Edmundo Icaza N.
40) Calle Conocida, Estelí – Dir: Wilfredo Rodríguez R.
41) Ap. 2434 (or: Semáforos de Metrocentro, 1 cuadra al sur, ½ cuadra abajo), Managua – Dir: Rev. José Luis Soto.
42) Boaco.
43) Ap. 2442 (or: Cd. Jardín Q-20), Managua – Dir: Carlos Gadea Mantilla.
44) Cd. Jardín S-24, Managua – Dir: Freddy Rostrán A.
45) Col. Don Bosco, Foto Castillo 1 cuadra al sur, ½ cuadra abajo, Managua – Dir: Francisco Talavera U.
46) BANIC, 2½ cuadras al oeste, Jinotepe – Dir: Julio Hernández M.
47) Col. del Periodista, Casa N° 128, Managua.
48) Ap. 88 (or: Suc. 14 de Septiembre), Managua – Dir: Digna Bendaña.
49) Barrio Pancasan, Puerto Cabezas – Dir: Evaristo Mercado Pérez. **FM:** 104.0.
50) Frente a la Iglesia de Guadalupe, Chinandega.
51) Plaza El Sol, 2 cuadras al sur, 5 cuadras al este N° 35, Managua – DG: José E. Quesada G.
52) Puerto Cabezas.
53) Banco Mercantil, ½ cuadra al este, Matagalpa.
54) Iglesia El Carmen, 1 cuadra al norte, Managua.
55) Del Silais 75 vrs al este, Jinotega – DG: Medardo Contreras.
56) Contigua a Oficina de ENEL, San Carlos – Dir: Carlos Corea.
57) Del Quiabú, 1 cuadra al oeste, ½ cuadra al sur, Estelí – DG: Oscar Rodríguez Moreno.
58) Carr. a Managua km 24½, Masaya.
59) Montoya 5 cuadras al lago, ½ cuadra abajo, Managua - DG: José Ovidio Valladares.
60) Zumen 3 cuadras al sur, contigo al Cuerpo de Bomberos, Managua.
61) Ap. 4665 (or: Costado Sur TELCOR, Villa Fontana), Managua.

FM in Managua: 89.1 Exitos – 89.5 Canal 21 – 3) 90.5 – 91.7 Estéreo Amante – 92.1 Continental – 92.7 La Bonita – 93.1 Estéreo Linda – 93.5 Eco – 13) 94.3 – 94.7 Mujer – 95.5 Horizonte – 95.9 Ritmo – 54) 96.3 – 44) 96.7 – 1) 97.1 – 98.3 Estéreo Variedades – 17) 99.1 Estéreo La Grande – 99.5 Universidad – 99.9 Pirata – 100.7 Omega – 101.1

Güegüense – 101.5 Juvenil – 102.3 Universidad – 103.1 Bautista – 41) 103.5 – 103.9 Titania Estéreo – 43) 106.7 Minuto – 11) 107.5 – 107.9 Restauración.

PANAMA

L.T: UTC -5h — **Pop:** 2,635,000 — **Radios:** 527,000 — **Pr.L:** Spanish — **E.C:** 60Hz, 110/115/120/126V — **ITU:** PNR

DIRECCION NACIONAL DE MEDIOS DE COMUNICACION SOCIAL
✉ Ap. 1628, Panamá 1. ☎ +507 227300 — **L.P:** Dir: Edwin Cabrera. Asesor: Alfredo de Souza.

ASOCIACION PANAMENA DE RADIODIFUSION
✉ Ap. 55-1326, Panamá – Pres: Fernando Eleta Casanovas.

Mediumwaves: Call HO-, * = inactive, (r) = repeater, v = varying fq.

	Call	kHz	kW	Name and h. of tr.
1)	U23	540	5	R. Mía de Chiriquí, David (r:650) 24h
2)	PU	540	5	R. Líder, Panamá: 1000-0400
3)	H2	560	1	RPC Radio, Colón (r:610)
4)	S	570	1	R. Soberana Civilista, Panamá 1000-0300
3)	H4	580	10	RPC Radio, David (r:610)
3)	H3	590	10	RPC Radio, Chitré (r:610)
3)	HM	610	10	RPC Radio, Panamá: 24h
5)	J35	630	1	R. Provincias, Chitré
6)	K22	640	1	CPR, Colón: 1000-0300
1)	S22	650	10	R. Mía "Cadena Nacional", Panamá: 24h
3)	F33	660	1	RPC Radio. Bocas del Toro (r:610) 24h
7)	LY	670	5	R. Hogar, Panamá: 0955-0300
8)		680	5	Em. Voz Sin Fronteras, Metetí: 1300-2200
22)	F32	680	5	Super Z Estéreo, David
9)	R43	690	5	R. Veraguas, Santiago: W:1000-0300, Sun: 1100-1700
10)	Q51	710	5	KW Continente, Panamá: 24h
11)	B52	710	3.5	Ondas del Caribe, Bocas del Toro: 1000-0400
12)	B50	720	5	R. República, Chitré: 1100-0300
13)	N26	740	10	R. Cristal, David
14)	R44	740	10	CMQ, La Exitosa, Panamá: 24h
15)		750	5	R. Inolvidable, Chitré: 1000-2300
16)	XO	760	5	LV del Istmo, Panamá: 1145-0300
17A)	L83	770	1.5	LV de Herrera, Chitré: 1100-0100
18)	B55	780	10	R. Chiriquí, David: 1100-0300
20)	L60	800	15	La Exitosa Provincias Centrales, Chitré: 0900-0300, Weekends 24h
21)	G	810	10	R. Mundial, Panamá
22)	F28	820	10	R. Ritmo, David: 1100-0400
14)	T61	850	10	La Exitosa de Chiriquí, David
14)		850	3	CMQ, Colón (r:740)
24)	L55	860	5	R. Reforma, Chitré 1030-0400
25)	HO	870	5	R. Libre, Panamá: 24h
23)	R56	870	1	R. Península, Macaracas
26)	B51	880	5	R. Hit, Colón
12)	Q62	890	10	R. Ritmo Stereo, Chitré
27)	HA	900	1	LV del Pueblo, Panamá 1100-0500
17B)	L81	*910	1	R. Nal Guaymíe, David
17C)	L85	910	1	R. Nal Cristóbal, Colón
1)	S56	920	5	R. Mía de Los Santos, Los Santos (r:650): 24h
14)	R46	930	3	CMQ, La Chorrera (r:740)
29)	K85	930	2.5	Mi Preferida Estéreo, Pto Armuelles: 1000 (Sat/Sun 1200)-0400
17D)	L84	950	1	R. Nal Victoriano Lorenzo, Penonomé
60)	M33	950	0.5	R. Cadena Azul "RCA", La Concepción: 1100-0300, Sun 1200-1800
31)	K71	960	10	Onda Popular, Panamá: 1000-0500
32)	S97	970	1	Ondas Centrales, Santiago: W 1000-0300, Sun 1300-2400
1)		980	1	R. Mía de Colón, Colón (r. 650) 24h
33)	U44	*990	3	R. Monumental, San Miguelito: 24h
34)	K36	1000	10/5	R. Poderosa "La Reina del Espacio", Aguadulce: 1000-0400
17E)	L86	1015	3	LV del Teribe, Bocas del Toro
35)		1020	10	R. Ancón, Panamá: 1000-0400
36)	J2	1040	2.5	Ondas del Canajagua Stereo, Las Tablas: 1000-0400
43)	J60	1060	3	LV de Panamá, Panamá: 1100-0300
38)	J24	*1080	1	R. Mil, Panamá
17F)	L82	1090	5	R. Urracá Civilista, Santiago W 1000-0100, Sun 1100-1800
39)		1100	2.5	Stereo Suave, Panamá: 24h
28)	M21	1120	5	R. Sonora, Panamá: 1000-0400
40)	U80	1130	1	R. Sensación, Aguadulce
41)	B49	1140	1	R. Panamericana, Panamá: 1000-0500
1)	S25	1150	1	R. Mía Deportes, Panamá
42)	C20	1160	10	Ondas Chiricanas, David
43)	WK	1160	3	R. Metrópolis, Panamá
44)	U84	1180	10	R. Belén, Santiago: 1100-0400
21)	E91	1210	4	R. Diez, Panamá
47)	M56	1240	1	Faro de David: W 1100-0400, Sun 1200-1800
48)		1240	5	R. BB, Panamá: 24h
7)	LY	1250	5	R. Hogar, Penonomé: 1000-0300
14)	J22	1270	0.5	R. CMQ, Panamá
43)	S23	1290	10	R. Guadalupe, Panamá
49)	I417	1300	1	R. Baha'í, Boca del Monte: 1045-2300
10)	M92	1330	1	R. Sabrosa, Panamá
50)	Z38	1350	10	R. Cadena Millonaria, Panamá: 24h
61)	B64	1370	1	R. Sitrachilco, Pto Armuelles: 1000-0100, Sun 1200-2300
35)		1380	10	R. América, Panamá: 1130-0400
51)	L	1390	1	R. Super Sol, Colón
52)	T40	1400	5	R. Luz, La Chorrera
53)	H779	1410	5	R. Mensabé, Las Tablas: 1100-0500
30)		1450		R. 1450 AM Stereo, Panamá
56)	D42	1460	0,5	LV de Almirante, Bocas del Toro: 1400-0400
14)	R45	1490	3	CMQ, La Palma (r:740)
57)	A95	v1510	8	Hosanna R, Panamá: 24h
59)	E35	*1570	2.5	LV del Trópico, "La Grande", Colón

Addresses and other information:
1) Ap. 5117, Panamá 5 – DG: José Luis Gil Alvarez.
2) Panamá 9 – DG: Dr. Carlos Raúl Moeno.
3) Ap. 1795, Panamá 1 – DG: Mirabel Arias M.
4) Ap. 6-2323, El Dorado, Panamá.
5) Ap. 423, Chitré – Gte: Gaspar Reyes.
6) Ap. 33, Colón – DG: Jacobo L. Salas.
7) Ap. 102, Panamá 9-A – Gte: Fernando Guardia Jaén,S. J.
8) Calle Principal de Metetí, Provincia de Darién (or: Ap. 87-0871 Panamá 7) – DG: P. Vicente Sidera Plana.
9) Ap. 48, Santiago – Dir: Régulo José Franco.
10) Ap. 87-1324, Panamá 7 (or:Vía Argentina, Edif. Carillón, Panamá) – DG: Rodrigo Correa.
11) Finca 13, Empalme, Changuinola (Bocas del Toro) – Dir: F. Artola.
12) Ap. 191, Chitré – Dir: Elías S. Solís C.
14) Calle 46 Este, Panamá 5 – Dir: A. de Icaza.
15) Ap. 375, Chitré – DG: Rogelio Herrera.
16) Ap. 6-1192, El Dorado, Panamá.
17A-F) SER – Sistema Estatal de Radiodifusión Panameño: 17A) Av. Pérez, Chitré; 17B) Av. Primera Oeste, David; 17C) Av. Bolívar Calle 9, Colón; 17D) Villa Inter- americana, Penonomé; 17E) Av. Central, Bocas de Toro; 17F) Barriada Urracá, MIDA Central, Santiago de Veraguas. Gte: Damarís H. de Urriola.
18) Ap. 375, David – Dir: Ramón Manuel Guerra. **FM:** 107.5.
20) Ap. 38, Chitré – Dir: Osvaldo Ramos.
21) Calle J. La Gloria, Panamá – Dir: Carlos I. Zúñiga.
22) Calle D. Norte, David – Dir: Denis Arce.
23) Macaracas, Los Santos – Gte: Eida Monroe.
24) Ap. 194, Chitré – Dir: Pedro Solís Villalaz. **FM:** 98.5.
25) Panamá 4 – DG: N. de Icaza.
26) Calle 1 Paseo Washington, Colón – Dir: D. Caballero. **FM:** 105.3.
27) Ap. 6-3045, El Dorado, Panamá – Dir: Modesto Lombardo Vega.
28) Ap. 87-1165 (or: Calle 65, Urb. Industrial, Los Angeles), Panamá 7 – DG: Lic. Humberto Gonzáles V.
29) Ap. 44, Puerto Armuelles – Dir: José A. Mora.
30) Vía Fdez de Córdoba, Panamá.
31) El Dorado, Panamá – DG: Héctor Gonzalez A.
32) Ap. 131, Santiago – Dir: Héctor A. Santacoloma.
33) Ap. 7471, Panamá 5 – Dir: Victor R. Vásquez.
34) Ap. 90, Aguadulce, Coclé – Dir: Italo E. Rojas. **FM:** 99.9.

35) El Dorado, Panamá – Dir: René A. Hernández.
36) Ap. 10, Las Tablas – DG: Jorge R. Villareal C.
38) Ap. 3115, Panamá 3 – DG: Anabella A. de Nahem.
39) Ap. 7729 (or: Calle 74, San Francisco, Edif. El Golf), Panamá 9 – Dir: Delsa L. Fábrega.
40) Av. Arosemena, Aguadulce – Dir: N. González.
41) Ap. 6956, Panamá 5 – Dir: R. de Icaza.
42) Ap. 172, David – Dir: Sra. Abigail C. de Calvo.
43) Vía España y Calle 45, Edif. El Conquistador PB, Panamá – Dir: F. Nuñez Fábrega.
44) Calle Decima, Santiago – Dir: Roberto Chen B.
47) Entrega General, David – Dir: Hilario Pinzón.
48) Ap. 5316, Panamá 5 – Gte: Andrés Vega C.
49) Ap. 1187, David – Gte: Manuel Flores.
50) Ap. 6-8868, El Dorado, Panamá – Dir: Nino Macías.
51) Ap. 5011, Colón – Dir: Estéban Lam.
52) Ap. 473, La Chorrera – DG: F & E Brands.
53) Ap. 20, Las Tablas – Dir: Rigoberto Amaya M.
56) Calle 6 y Av. Almirante, Bocas del Toro – Dir Nicolás O. Dosman.
57) Ap. 6-8229 (or: Calle Erick del Valle y Vía Argentina, Edif. Vicky 2), El Dorado, Panamá – DG: Marco A. Gómez L.
59) Ap. 506, Colón – DG: David A. Simons. **FM:** 102.3.
60) Entrega Gral, La Concepción, Chiriquí.
61) Ap. 03, Puerto Armuelles – Dir: Isidro Cedeño.

FM in Panamá City. St. names: Estéreo . . . exc. where indicated. 21) 88.1 "10" – 16) 88.5 – 88.9 Tropical Moon – 89.3 "RM" – 89.9 89Digital – 14) 90.5 Super Q – 3) 90.9 – 2) 91.9 – 92.1 Festival – 48) 92.5 – 43) 92.9 "F" – 43) 93.5 Super – 1) 93.7 Mía – 93.9 Universal – 94.1 Bahía – 7) 94.3 – 94.9 Tricolor – 14) 95.3 Alegre – 10) 95.9 – 96.3 Selecta – 1) 96.7 Magic – 3) 97.3 – 97.7 Omega – 97.9 Mix – 98.5 Oeste – 98.9 Ultra 99.3 FM99 – 100.1 "21" – 101.1 Azul – 38) 101.7 – 102.6 Rey – 10) 103.1 FM Latina – 38) 103.5 Continental – 38) 103.9 – 39) 104.7 – 16) 105.1 Vida – 105.7 Bahía – 27) 106.7 Panamá – 16) 107.3 Universidad.

AFRTS – Southern Command Network
+ Drawer 919, APO Miami, FL 34004, USA

Mediumwaves:

ACA20	790	10kW	Pacific: 24h
ACB20	1410	10kW	Atlantic: 24h

FM:

ACA20	91.5	1kW	Pacific: 24h
ACB20	98.3	1kW	Atlantic: 24h

PUERTO RICO

L.T: UTC -4h — **Pop:** 3,653,000 — **Radios:** 2,480,000 — **Pr.L:** Spanish, English — **E.C:** 60Hz, 120V — **ITU:** PTR.

BROADCASTERS ASSOCIATION OF PUERTO RICO
Suite 212, Cobians Plaza, Ave. Ponce de León 1607, Santurce 00912.

Mediumwaves:
Most stations broadcast in Spanish only. °=English or mainly English, d=directional antenna. Locations: Hato Rey, Río Piedras and Santurce are districts in San Juan.

	Call	kHz	kW	Name and h. of tr.
1)	WPAB	550	5	La RedAlerta de Puerto Rico, Ponce: 0900-0400
2)	WKAQ	580	5d	R. Reloj, San Juan: 24h
3)	WAEL	600	1d	WAEL, Mayagüez: 0830-0400
4)	WEXS	610	0.25/1	WEXS, Patillas
5)	WSKN	630	5d	Super Kadena Noticiosa, San Juan: 24h
6)	WAPA	680	10	Guapa R, San Juan: 24h
7)	WKJB	710	10/0.75	KJB, Mayagüez: 0915-0400
8)	WIAC	740	10d	WIAC, San Juan: 0859-0400 (Weekends-0600)
9)	WORA	760	5d	Super Kadena Noticiosa, Mayagüez: 24h
10)	WKVM	810	50d	AM-81, San Juan: 0900-0500
11)	WXEW	840	5d	R. Victoria, Yabucoa: 0830-0300
12)	WABA	850	5/1	WABA/La Grande, Aguadilla: 24h
13)	WQBS	870	5d	R. Voz, San Juan: 0800-0300
14)	WYKO	880	1/0.5	La Poderosa 880, Sabana Grande: 1000-0400
15)	WFAB	890	0.25	WFAB, Ceiba
16)	WPRP	910	5	Super Kadena Noticiosa, Ponce: 24h Ponce: 24h
17)	WEKO	930	2.5	NotiUno, Cabo Rojo: 0840-0600
18)	WIPR	940	10d	La Emisora del Puebo, San Juan: 0930-0230 (Sun 1130-0400)
19)	WKVN	960	1/0.5	WKVN, Quebradillas
20)	WPRA	990	1	R. Mil, Mayagüez: 0900-0400
21)	WPJC	1020	1	R. Gigante, Adjuntas: 1000-0200
22)	WOSO	°1030	10d	WOSO/El Oso, San Juan: 24h
23)	WZNA	1040	0.25	WZNA, Moca
24)	WCGB	1060	5/0.5	Iniciativa Mil 60, Juana Díaz: 0930-0200
25)	WMIA	1070	2.5	R. Arecibo, Arecibo: 0925-0400 (Sun -0200)
26)	WLEY	1080	0.25	R. Ley/Motivos 1080, Cayey: 24h
27)	WSOL	1090	0.75	R. Sol, San Germán: 0930 (Sun 1000)-0400
28)	WVJP	1110	2.5/0.5	R. Caguas, Caguas: 24h
29)	WMSW	1120	2.6/5	La Gran W, Arecibo: 1000-0200
30)	WOIZ	1130	0.2/0.7	R. Antillas, Guayanilla: 0900-0200
31)	WQII	1140	10d	Once Q, San Juan: 24h
32)	WBQN	1160	5/2.5d	La Super B, Barceloneta: 1030-0200
33)	WZUR	1170	0.25	WZUR, Ponce: 24h
34)	WBMJ	1190	10/5d	WBMJ, San Juan: 0900 (Sat/Sun 1100)-0400
35)	WGDL	1200	1	R. Grito, Lares
36)	WHOY	1210	5d	R. Hoy, Salinas: 0830-0230
37)	WNIK	1230	1	WNIK/R. Unica, Arecibo
38)	WALO	1240	1/5	Walo/R. Oriental, Humacao: 0900-0400
39)	WJIT	1250	1/0.25d	WJIT, Sabana: (C. P.)
40)	WISO	1260	1	R. Wiso, Ponce: 1000-0300
41)	WCMN	1280	5/1	R. Centro/La Grande de Arecibo, Arecibo: 0900-0400
42)	WTIL	1300	1	R. Util, Mayagüez: 1000-0400
43)	WUNO	1320	5/2.3	NotiUno, San Juan: 24h
44)	WENA	1330	1/0.5	La Buena del Sur, Peñuelas: daytime
45)	WNOZ	1340	1	R. Nosotros, Aguadilla: 0900-0300
46)	WEGA	1350	2.5d	WEGA/R. Las Vegas, Vega Baja: 1300-0100
47)	WCHQ	1360	5/1d	C-H-Q, Camuy (F.Pl. 1660 kHz)
48)	WIVV	°1370	5/1	WIVV Missionary R. St., Viequez Isl: 0930 (Sun 1000)-0230
49)	WOLA	1380	1	R. Prócer, Barranquitas: 0900-0200
50)	WISA	1390	1	R. Noroeste, Isabela: 0930-0300
51)	WIDA	1400	1	R. Vida, Carolina: 24h
52)	WRSS	1410	1	R. Progreso, San Sebastián: 0900-0300
53)	WEUC	1420	1	R. Universidad Católica, Ponce: 0900-0200
54)	WNEL	1430	5	R. Tiempo, Caguas: 24h
55)	WCPR	1450	1	R. Coamo, Coamo: 1000-0200
56)	WRRE	1460	0.5d	La Fabulosa, Juncos
57)	WLRP	1460	2.5/0.5	R. Raíces, San Sebastián: 0900-0400
58)	WKCK	1470	1/2.5d	R. Cumbre, Orocovis: 0900-0200
59)	WMDD	1480	5	Sonido 14-80/R. El Conquistador, Fajardo: 1000-0400 (Sun 0200)
60)	WLEO	1490	5/1	WLEO, Ponce: 24h
61)	WMNT	1500	1/0.25	R. Atenas, Manatí: 1000-0200
62)	WAVB	1510	1	Super B, Lajas
63)	WVOZ	1520	10d	R. Aeropuerto Int, San Juan: 24h
64)	WUPR	1530	1/0.25	Exitos 15-30, Utuado: 1000-0300
65)	WIBS	1540	1	R. Caribe, Guayama: daytime
66)	WKFE	1550	0.25	R. Café, Yauco: 24h
67)	WRSJ	1560	5/0.75	R. San Juan, Bayamon: 24h
68)	WPPC	1570	1	R. Felicidad, Peñuelas
69)	WMTI	1580	5/2.5	R. Jefe, Morovis: 1000-0200
		1580	0.2/0.5	Manatí (synchr. WMTI)
		1580	0.2/0.5	Arecibo (synchr. WMTI)
70)	WXRF	1590	1	R. 15-90, Guayama: 24h
71)	WLUZ	1600	5d	R. Luz, Bayamon: 1000-0400

NotiUno
Box 363222, San Juan 00936. News 0930–1300, 1900–2100 on st's 6), 11), 17), 26), 32), 41), 43), 45), 46), 54), 57), 65) and 66).

R. Reloj

☞ Box 364668, San Juan 00936. Prgr 0930–1300, 1900–2100 on st's 2), 7), 12), 29) and 40).

Super Kadena Noticiosa
☞ Calle Eleanor Roosevelt 117, Hato Rey 00918. st's 4), 9), 15), 16), 25), 44), 47).

Addresses and other information:
1) Box 7243, Ponce 00732 – GM: Alfonso Jiménez P. FM: WOQI 93.3, WIOC 105. 1, WOYE 94. 1 Mayagüez.
2) Box 364668, San Juan 00936-4668 – GM: Huberto E. Biaggi. **FM:** 104.7.
3) Box 1370, Mayagüez 00681 – Mgr: Luis Pirallo. **FM:** WAEL-FM 96.1 Maricao.
4) Box 640, Patillas 00723 – GM: Enrique García.
5) Calle Eleanor Roosevelt 117, Hato Rey 00918 – GM: Reynoldo Royo.
6) 134 Domenech Ave, Hato Rey 00918 – GM: W. Blanco P.
7) Box 1293, Mayagüez 00681 – GM: Mabel Santo. **FM:** WKJB-FM 99.1.
8) Box 9023916, San Juan 00902 – GM: Luis Mejia. **FM:** 102.5.
9) Box 3822 Marina Station, Mayagüez 00682. GM: Reynoldo Royo.
10) Calle Carbonell 415, Hato Rey 00918 – GM: Ana Meléndez. **FM:** WORO 92.5 Corozal.
11) Box 840, Yabucoa 00767 – GM: Victor M. Calderón.
12) Box 188, Aguadilla 00605 – GM: Rosa M. Pellotzea.
13) Ave de Diego 129, San Francisco, San Juan 00927 – DG: Angel Román.
14) 63 Comercio St. , Yauco 00683 – GM: Juan Galiano R.
15) Box 1231, Yguncos 00777 – GM: Daniel Rosario.
16) Box 7302, Ponce 00732 – GM: Reynoldo Royo.
17) Box 681, Cabo Rojo 00623 – GM: María Ortiz. **FM:** WMIO 102.3.
18) Box 190909, Hato Rey 00919 – GM: Edgardo Gierbolini. **FM:** 91.3.
19) Box 7, Moca 00716. GM: Aureo Matos.
20) Box 1293, Mayagüez 00681 – GM: Mabel Santo.
21) Box 982, Adjuntas 00601 – GM: Luis Francisco Ojada.
22) Box 9023940, San Juan 00902 – GM: Sherman Wildman.
23) Box 7, Moca 00716 – GM: Aureo Matos.
24) Box 248, Juana Díaz 00795 – GM: José A. Rodríguez.
25) Box 1055, Arecibo 00613 – GM: Epifanio Rodríguez-Vélez.
26) Box 371300, Cayey 00737 – GM: Julio H. Conessa.
27) Box 6000442, San Germán 00683 – GM: Manuel Cruz.
28) Box 207, Caguas 00726 – GM: Berta Perrera. **FM:** 103.3.
29) Box 1652, Arecibo 00613 – GM: Zaida Santos.
30) Box 561130, Guayanilla 00656-1130 – GM: Luis A. Rodriguez.
31) Box 193779, Hato Rey 00919 – GM: N. Gonzales-Abreu.
32) Box 993, Manatí 00674 – GM: Luis R. Rivera.
33) Box 7213, Ponce 00732 – GM: Julio H. Conessa. FM: WZAR 101.9.
34) Box 367000-7000, San Juan 00936 – GM: Janet L. Luttrell. Often in parallel with 48).
35) Box 872, Lares 00669 – GM: Pedro Hernández.
36) Box 1148, Salinas 00751-1148 – GM: Martín Colón Jr.
37) Box 556, Arecibo 00613 – GM: Raul Santiago. **FM:** 106.5.
38) Box 1240, Humacao 00792 – GM: Mercy Padilla.
39) WJIT Broadcasting Corp, Sabana.
40) 134 Domenech Ave, Hato Rey 00918 – GM: W. Blanco P.
41) Box 436, Arecibo 00613 – GM: Byron Mitchell. **FM:** 107.3.
42) Box 1360, Mayagüez 00681 – GM: Lydia Basora.
43) Box 363222, San Juan 00936 – GM: Joe Pagan Jr. **FM:** WFID 97.7 Río Piedras.
44) Box 1338, Yauco 00698 – GM: Nephtali Rodríguez.
45) Box 1, Aguadilla 00605 – GM: Fernando Gallardo.
46) Box 1488, Vega Baja 00694 – GM: Carmello A. Santiago.
47) Box 629, Camuy 00627 – GM: José Cordero. **FM:** WCHQ-FM 102.9.
48) Box 338, Viequez Island 00765 – LP: As 34).
49) Box 669-A, Barranquitas 00794 – GM: Ramon Delgado.
50) Box 750, Isabela 00662 – GM: Luis Mejía. FM: WKSA 101.5.
51) Box 188, Carolina 00985 – GM: William Lebrón. FM: 90.5.
52) Box 1410, San Sebastián 00685 – GM: Ivan Feliu.
53) Box 529, Stn 6, Catholic University of Puerto Rico, Ponce 00732 – GM: Nestor Figueroa. **FM:** 88.9.
54) Box 487, Caguas 00724 – GM: Carmen Pagan. **FM:** WPRM 98.5 San Juan.
55) Box 316, Coamo 00769 – GM: José D. Soler.
56) Box 1460, Juncos 00771 – GM: Edgar Román.
57) Box 1670, San Sebastián 00685 – GM: Alfredo Peréz.
58) Box 1210, Orocovis 00720 – GM: Arnaldo García R.
59) Box 948, Fajardo 00738 – GM: Rita Friedman. **FM:** WDOY 96.5.
60) Box 7213, Ponce 00732 – GM: Julio H. Conessa.

61) Box 6, Manatí 00674-0006 – GM: José A. Ribas Dominicci.
62) Box 593, Lajas 00667 – GM: Liccie Negron.
63) Penthouse, Darlington Bldg. , Río Piedras 00925 – GM: Pedro Roman Collazo. **FM:** 107.7 Carolina.
64) Box 868, Utuado 00641 – GM & PD: José A. Martínez.
65) Box 1540, Guayama 00655 – GM: Wigberto Baez.
66) Box 324, Yauco 00698 – GM: Jaime L. Bermúdez.
67) Box 4039, Carolina 00984 – GM: Richard Rosado.
68) Box 9064, Ponce 00732 – GM: Roberto Peréz.
69) 155 San Antonio St., Floral Park, Hato Rey 00917 – GM: W. Blanco P.
70) Box 14590, Guayama 00785 – GM: Pedro Collazo.
71) Box 9394, San Juan 00908 – GM: Julia M. Acosta.

AFRTS (US Air Force)
☞ FPO AE 09503, USA (or: FPO AA 34051, USA).
Mediumwaves:

AFCN	1200	0.05kW	Roosevelt Roads.
FM:			
AFCN	92.9	0.3kW	Roosevelt Roads.

ST. KITTS & NEVIS

L.T: UTC - 4h — **Pop:** 42.045 — **Radios:** 25.000 — **Pr.L:** English — **E.C:** 60Hz, 220V — **ITU:** SCN.

ZIZ RADIO (Gov. Comm.)
☞ P.O. Box 331, Springfield, Basseterre, St. Kitts. ☎ +1 (869) 465 776 or 7081. 🖷 +1 (869) 465 5624. **Cable:** ZIZ St. Kitts.
L.P: Dir. of Broadc: Clement Juni Liburd. Sen. Admin. Officer: Lovina Maynard. Prgr. Mgr: Claudette Manchester. Sen. Eng: Bertill Browne.
MW: ZIZ 555kHz 10kW — **FM:** 96.0MHz — **D.PRGR:** 1000-0400. **BBC N:** 1100, 2300, 0200. **Local N:** 1115, 1615W, 1755, 1855, 1955, 2055, 2155, 2315 — **ANN:** "The Voice of the Federation of St. Kitts & Nevis, Radio.ZIZ" — **V.** by letter.

RADIO PARADISE (Rlg.)
☞ P.O. Box 423, Charlestown, Nevis (**Addr in USA:** P.O. Box A, Santa Ana, CA 92711. ☎ +1 (714) 832 2950. 🖷 +1 (714) 730 0661) — **L.P:** Dr. Paul F.Crouch. PD: Barry Phaeler. CE: Ben Miller. Head of PR: Rod Henke — **MW:** 830kHz 50kW.
D.PRGR: 0900-0400(Sat 0500) — **ANN:** "This is Radio Paradise, 830 on your AM dial, broadcasting from St. Kitts" — **V.** by QSL-card.

VOICE OF NEVIS (Comm.)
☞ P.O. Box 195, Charlestown, Nevis. ☎ +1 869 469 1616 or 1700. 🖷 +1 869 469 5329 — **L.P:** SM: Evered Herbert.
MW: 895kHz 10kW — **ANN:** "This is VON Radio on 895 AM".

ST. LUCIA

L.T: UTC - 4h — **Pop:** 162.894 — **Radios:** 100.000 — **Pr.L:** English, Creole — **E.C:** 50Hz, 220V — **ITU:** LCA.

ST. LUCIA BROADCASTING CORP. (Gov. Comm.)
☞ P.O. Box 660, Castries. ☎ +1 (758) 452 2337/9. 🖷 +1 (758) 456 2749. ⓣ 6285 SUNRADIO LC.
L.P: GM: Felix St. Hill. Prgr. Mgr: Mrs. Barbara Jacobs-Small. CE: Eliseus Louis.
MW: 660kHz 25kW — **FM:** 97.7/99.5/103.3/107.3MHz.
D.PRGR: 0845-0300 (Fri/Sat 0400). **N:** 1030, 1200, 1400, 1500, 1700, 2000, 2100. **N. in Creole:** 1800. **BBC N:** 1100, 1600, 0200.
ANN: "You are listening to Radio St. Lucia" — **V.** by QSL-card.

RADIO CARIBBEAN INTERNATIONAL (Comm.)
☞ P.O. Box 121, Castries. ☎ +1 (758) 45 22636. 🖷 +1 (758) 45 22637. ⓣ 6240 RADOCAR. **E-mail:** rci@candw.lc
Web: http://www.candw.lc/homepage/rci.htm
L.P: Pres: H.Coquerelle. PD: C.Bonne — **MW:** 840kHz 20kW.
FM: 101.1MHz + 3 relays (1 in Dominica) — **D.PRGR:** 24h on FM, 0500-1802 on 840kHz. **N:** 0600, 0700, 1230 — **ANN:** "This is Radio Caribbean International, the Caribbean Sound" — **V.** by letter.

Catholic TV Broadcasting Service, Micoud Str, Castries. ☎ +1 (758) 45 27050 — **FM:** 87.8MHz (rel. TV sound of EWTN, USA).
Gem Radio Network: 93.7 (So.) & 94.5 (No.)MHz (rel. Trinidad).

ST. VINCENT & THE GRENADINES

L.T: UTC - 4h — **Pop:** 119.647 — **Radios:** 65.000 — **Pr.L:** English — **E.C:** 50Hz, 230V — **ITU:** VCT.

ST. VINCENT & THE GRENADINES NATIONAL BROADCASTING CORPORATION (Gov. Comm.)

🖃 P.O. Box 705, Kingstown, St. Vincent, W.I. ☎ +1 (784) 457 1111. 🖷 +1 (784) 456 2749. **Cable:** Broadcast, St. Vincent.
L.P: GM: vacant. CE: Leslie McKie — **MW:** 705kHz 10kW.
D.PRGR: W 0930-0300 (Fri/Sat 0400), Sun 1000-0200. **N:** 1630, 2230, 2345W — **ANN:** "NBC 705 Radio" — **V.** by QSL-card.

PRIVATE STATIONS
Hitz-FM: 99.9MHz – **Nice-FM & We-FM** (freqs unknown).

TRINIDAD & TOBAGO

L.T: UTC - 4h — **Pop:** 1.275.740 — **Radios:** 550.000 — **Pr.L:** English — **E.C:** 60Hz, 115V — **ITU:** TRD.

INTERNATIONAL COMMUNICATIONS NETWORKS Ltd. (Gov. Comm.)

🖃 11a Maraval Rd, Port of Spain. ☎ +1 (868) 622 4141. 🖷 +1 (868) 628 6733. **Cable:** Voice Trinidad. **E-mail:** ttt@trinidad.net *or* yes98@trinidad.net **Web:** http://www.icn.co.tt/
L.P: Ag. Chief Exec. Officer: Ms. Ingrid Isaac. Prgr. Mgr: Mrs. Brenda de Silva — **MW:** 610kHz 50kW.
FM: Trinidad 98.9/100.0MHz, Tobago 91.1MHz 20kW.
D.PRGR: 24h on MW + 91.1MHz. **N:** on the h. (BBC 2000, 0200). **Wrp:** on the half h — **FM 100:** 24h on 100.0MHz.
Yes 98.9FM: 24h on 98.9MHz.
ANN: "This is NBS Radio, 610 AM and 91.1MHz on the FM band serving Tobago" — **IS:** s/on: "Yellow Bird, Chocoun Destiny" (guitar & xylophone) — **V.** by letter. Rp.

TRINIDAD BROADCASTING COMPANY (Comm.)

🖃 P.O. Box 716, 11b Maraval Rd, Port of Spain. ☎ +1 (868) 622 1151-7. 🖷 +1 (868) 622 2380. **Web:** http://www.ttol.co.tt/caribbeantempo/
L.P: MD: Grenfell Kissoon. Mktg. Mgr: Brandon Khan. CE: Rawlins Rambaran. Dir. of N. & Current Affairs: Andy Johnson. Promotions Mgr: Norvan Fullerton.
MW: 730kHz 20kW — **FM:** 95.1MHz 25kW, 105.1MHz 2.5kW.
R. Trinidad: 24h on 730kHz. **N:** 0200*, 1030, 1100*, 1110, 1200, 1300*, 1500, 1600, 1630, 1700, 1800, 2000, 2200 — *) BBC.
Rhythm R: 24h on 95.1MHz — **Caribbean Tempo:** 24h on 105.1MHz.
V. by letter. Rec. acc.

PRIME RADIO LTD. (Comm.)

🖃 35-37 Independence Square, Port of Spain. ☎ +1 (868) 627 7463. 🖷 +1 (868) 627 2721 — **L.P:** GM: Julian Rogers.
Classic R. 106FM: 24h on 106.1/106.5MHz. **BBC N:** 1100, 2300.

Music Radio 97 FM Stereo, Level 4, Long Circular Mall, Long Circular Road, St. James — 0900-0500 on 97.0MHz.
103FM (st. of Indian Community): 0845-0500 on 103.0MHz. **Main N:** 1500. **N. Summaries:** 1400, 17000, 1900. **BBC N:** 1100, 2000.
WEFM: 24h on 96.1MHz.
Gem Radio Network: 93.1MHz).
Power 102 FM: Chief Exec: Louis Lee Sing — 102.0MHz.

TURKS & CAICOS ISLANDS (British)

L.T: UTC - 5h (Su: UTC - 4h) — **Pop:** 14.973 — **Radios:** 2.022 — **Pr.L:** English — **E.C:** 60Hz, 110/220/440V — **ITU:** TCA.

RADIO TURKS & CAICOS (Gov. Comm.)

🖃 P.O. Box 69, Grand Turk. ☎ +1 (809) 946 2007. 🖷 +1 (809) 946 1705. ☏ 946 8212 GOVTCI TQ.
L.P: Ag. Mgr: Mrs. Lynette Smith — **MW:** 1460kHz 2.5kW (inactive).

D.PRGR: 1100-0400 (Sun 0315) — **ANN:** "This is Radio Turks & Caicos on Grand Turk, Turks & Caicos Islands" — **V.** by letter.

RADIO VISION CRISTIANA INTERNACIONAL (Rlg.)

🖃 North End, So. Caicos. ☎ +1 (809) 946 3311 or 6601. 🖷 +1 (809) 946 6600 — **L.P:** Mgr: Efrain Rivera. CE: Bob Janney.
MW: So. Caicos 535kHz 100kW.
D.PRGR. in Spanish: rel. WWRV 1330, NY, USA.
ANN: "R. Visión Cristiana Internacional, transmitiendo para todo el área del Caribe, Sudamérica y la parte sur de los Estados Unidos."

CARIBBEAN CHRISTIAN RADIO (Rlg.)

Operating under license from West Indies Broadcasting
🖃 P.O. Box 200, Grand Turk. ☎🖷 +1 (809) 946 1095. (🖃 **in USA:** Box 3, DeLand, FL 32721).
L.P: GM: Reo Stubbs. SM: Buddy Tucker. CE: Jerry Kiefer.
MW: Grand Turk 1020kHz 20kW (directional).
ANN: "Super Power 1020" — **V.** by QSL-card.

WIV FM Radio (Comm.), Box 108, Providenciales: 24h on 89.3/ 89.9/90.5/92.5MHz (4 separate pgrs)
WPRT Radio (Comm.), Box 262, Providenciales: 88.7MHz 0.25kW.
VIC Victory in Christ (Rlg.), Box 32, Providenciales: 96.7MHz

VIRGIN ISLANDS (American)

L.T: UTC - 4h — **Pop:** 105.000 — **Radios:** 100.000 — **Pr.L:** English, Spanish, Creole — **E.C:** 60Hz, 110V — **ITU:** VIR.

	MW	kHz	kW		MW	kHz	kW
1)	WSTX	970	5/1	3)	WRRA	1290	0.5
2	WVWI	1000	5/1*	4)	WSTA	1340	1
6)	WGOD	1090	0.25				
	FM	**MHz**	**kW**		**FM**	**MHz**	**kW**
14)	"Gem."	90.9		12)	WTBN	102.1	50*
11)	WAVI	93.5	15	5)	WIUJ	102.9	0.15
8)	WJKC	95.1	50	10)	WIYC	104.3	44.8*
6)	WGOD	97.9	50	9)	WVGN	105.3	50*
7)	WVIQ	99.5	10.5	13)	WVIS	106.1	50*
1)	WSTX-FM	100.3	50*				

*) silent at time of editing.

Addresses and other information
1) Family B'casting Inc., Box 3279, Christiansted, St. Croix 00822.
2) Box 5678, St. Thomas 00803 (silent).
3) Box 277, Frederiksted, St. Croix 00841-0277.
4) Box 1340, St. Thomas 00804-1340. **Web:** http://www.wsta.vi/ – 24h. **ANN:** "Lucky 13"
5) Virgin Isl. Youth Development Radio, GPO, Govt. of Virgin Isl., St. Thomas 00801
6) Box 5012, St. Thomas 00803-5012 (Rlg.)
8) Suite V-2, Caravelle Arcade, Christiansted, St. Croix 00820
9) P.O. Box 10772, St. Thomas 00801 (silent)
10) Box 5234, St. Thomas 00801 (silent)
11) Box 25016, Christiansted, St. Croix 00824
13) Box 487, Frederiksted, St. Croix 00840 (silent)
14) Gem Radio Network (rel. Trinidad).

VIRGIN ISLANDS (British)

L.T: UTC - 4h — **Pop:** 13.529 — **Radios:** 34.000 — **Pr.L:** English — **E.C:** 60Hz, 110V — **ITU:** VRG.

VIRGIN ISLANDS BROADCASTING LTD. (Comm.)

🖃 P.O. Box 78, Road Town, Tortola, BVI. ☎ +1 (284) 494 2250. 🖷 +1 (284) 494 1139. **E-mail:** zbvi@caribsurf.com **Web:** http://www.zbvi.vi
L.P: GM: Harvey Herbert. Op's Mgr: Mrs. Sandra Potter-Warrican. PD: Iris Jones. N. Dir: Angela Burns Piper. CE: Olin Hester.
MW: ZBVI 780kHz 10kW.
D.PRGR: W 1000-0200, Sun 1100-0100. **N:** W on the h. exc. 1000, 1700, 1900, 0100. Sun 1600, 2000. **Local N:** 1100, 2230.
ANN: "This is ZBVI Radio from Tortola" — **V.** by letter.

CARIBBEAN BROADCASTING SYSTEM (Comm.)

🖃 Box 3049, Roadtown, Tortola, BVI. ☎ +1 (284) 494 4990.
L.P: GM: Alvin Korngold — **FM:** 91.7/94.3/97.3MHz 10kW.
D.PRGR: 1000-0400 (three different formats).

GEM RADIO NETWORK (Comm.)
FM: Tortola 90.9MHz 1kW (rel. Trinidad).

SOUTH AMERICA

ARGENTINA

L.T: UTC-3h, -4h in Ma, SJ, SL, LR, Ca. and SE (Oct.-Mar.)
Pop: 33.750.000 — **Radios:** 21.500.000 — **Pr.L:** Spanish —
E.C: 50Hz, 220V — **ITU:** ARG.

PRESIDENCIA DE LA NACION
SECRETARIA DE COMUNICACIONES
✉ Sarmiento 151, piso 4, 1000 Buenos Aires. ☎ +54 (1) 3115030, 3115049 (int. 3700). 🖷 +54 (1) 3189432
LP: Dr. Germán Luis Kammerath (Secretario de Comunicaciones).

COMISION NACIONAL DE COMUNICACIONES (C.N.C.)
✉ Perú 103, 1067 Buenos Aires. ☎ +54 (1) 3479870, 3479873. 🖷 +54 (1) 3479244.
LP: Presidente: Roberto C. Catalán. Vicepresidente 1°: Roberto E. Uanini. Vicepresidente 2°: Hugo J. Zothner.

COMITE FEDERAL DE RADIODIFUSION (COMFER)
✉ Suipacha 765, 1008 Buenos Aires. ☎ +54 (1) 3204900. 🖷 +54 (1) 3946866.
LP: Pres: José Carmelo Aiello.
COMFER controls certain technical aspects of broadcasting, and also controls the prgrs. transmitted over all kinds of broadcasting stations

SECRETARIA DE PRENSA Y DIFUSION
Subject directly to the Presidencia de la Nación. Controls and administers the media.

SERVICIO OFICIAL DE RADIODIFUSION (S.O.R.)
✉ Maipú 555, 1006 Buenos Aires. ☎ +54 (1) 9259100. 🖷 +54 (1) 3259433 — **LP:** Dir: Patricia Parral.
All LRA st's belong to S.O.R. (incl. LRA36 in Antarctica). Common prgrs (originated from LRA1) in network called "Cadena Celeste y Blanca de Emisoras Argentinas".

Mediumwaves: Þ = also on shortwaves, v = varying freq.

	Call	kHz	kW	Name, location and h. of tr.
1)	LRA14	540	25/8	R. Nal. Santa Fé: 0700-0300
2)	LRA25	540	5	R. Nal. Tartagal: 1000-0400
25)	LU17	540	25/5	R. Golfo Nuevo, Pto. Madryn: 24h
3)	LRA9	560	15/7.5	R. Nal. Esquel: 1000-0400
4)	LRA13	560	25/5	R. Nal. Bahia Blanca: 0900-0400
5)	LT15	560	10/3	R. del Litoral, Concordia: 0900-0330
6)	LV1	560	25/5	R. Colón, San Juan: 0900-0600
7)	LRA16	560	25/15	R. Nal. La Quiaca: 0855-0400
239)		570		R. del Centro, Lomas de Mirador
8)	LU20	580	10	R. Chubut "La 20," Trelew:0800-0500
9)	LW1	580	25/5	R. Univ. Nal. de Córdoba, Córdoba: 24h
69)	LU5	580	20/10	R. Neuquén: 0900-0500
10)	LS4	590	50/25	R. Continental, Buenos Aires: 24h
11)	LRA30	590	25	R. Nal. Bariloche, San Carlos de Bariloche: 0855-0400
12)	LV12	590	25/10	R. Independencia, San Miguel de Tucumán: 24h
196)	LRH356	610	1	R. Solidaridad, Añatuya: Mon-Sat: 1100-2300, Sun 1200-0100
94)	LV4	620	10	R. San Rafael: 1000-0400
13)	LRA18	620	25/7	R. Nal. Río Turbio: 0845-0400
14)	LT17	620	25/5	R. Provincia de Misiones, Posadas: 0900-0500
85)	LRA26	620	10/5	R. Nal. Resistencia: 0830-0300
15)	LRA28	620	25/5	R. Nal. La Rioja: 0900-0400
16)	LS5	630	25/5	R. Rivadavia, Buenos Aires: 24h
52)	LW8	630	25/5	R. San Salvador de Jujuy: 0900-0400
65)	LU4	630	25/5	R. Dif. Patagonia Argentina, Comodoro Rivadavia: 0900-0500
17)	LRA24	640	25/5	R. Nal. Río Grande: 24h
18)	LU18	640	10/1	R. El Valle, "640 AM", General Roca: 0900-0300
19)	LV15	640	10/5	R. Villa Mercedes: 1000-0400
122)	LT41	660	5/1	R. LV del Sur Entrerriano, Gualeguaychú: 0900-0300
205)		660		R. Melody, Las Piedras
20)	LRA11	670	25/5	R. Nal. Comodoro Rivadavia: 24h
21)	LRA52	670	5	R. Nal. Chos Malal: 0900-0300
22)	LT4	670	25/8	R. Dif. Misiones, Posadas: 0800-0200
23)	LRI209	670	25/10	R. Mar del Plata, Mar del Plata: 24h
110)	LT3	680	10	R. Cerealista "AM 680", Rosario: 24h
24)	LU12	680	10/6	R. Río Gallegos: 1000-0500
26)	LV6	680	25/5	R. Nihuil, Mendoza: 1000-0700
27)	LRA4	690	25/5	R. Nal. Salta: 0900-0400
28)	LU19	690	10/5	LV de Comahue, Cipoletti: 0900-0500
99)	LV3	700	25/1	R. Córdoba: 24h
29)	LRA17	710	25/1	R. Nal. Zapala: 24h
29x)	LRA19	710	25/5	R. Nal. Pto. Iguazú: 0900-0300
168)	LRL202	710	10	R. Diez, Buenos Aires
31)	LRA59	720	1	R. Nal. Gobernador Gregores: 1100-2300
32)	LV10	720	25/1	R. de Cuyo, Mendoza: 0930-0700
33)	LRA3	730	5	R. Nal. Santa Rosa: 0900-0300
34)	LU23	730	7	Em. Lago Argentino, El Calafate: 1000-0300
35)	LRA27	730	25/5	R. Nal. Catamarca: 0900-0400
36)	LRA55	740	1	R. Nal. Alto Río Senguerr: 1100-0300
36x)	LRI200	740	1	R. Puerto Deseado: 24h
75)		740		R. Bonaerense, Llavallol
37)	LRA7	750	100/10	R. Nal, Córdoba: 24h
206)		760		R. Malvinas, Monte Grande
38)	LRA10	780	5/1	R. Nal. Ushuaía e Islas Malvinas, Ushuaía: 24h
39)	LRA12	780	5	R. Nal. Santo Tomé: 0900-0300
40)	LV8	780	8	R. Libertador, Mendoza: 0930-0600
83)	LV19	Þ790	5	R. Malargüe: 1100-0400
41)	LR6	790	25/5	R. Mitre "AM 80," Buenos Aires: 24h
42)	LRA22	790	25/5	R. Nal. San Salvador de Jujuy: 0945-0500
44)	LU15	800	10/7	R. Viedma: 0900-0300
45)	LV23	800	1/0.25	R. Rio Atuel, General Alvear: 1000-0400
50)	LT43	800	5/1	R. Mocoví, "la emisora regional", Charata: 0900-0300
46)	LRA8	820	25/5	R. Nal. Formosa: 0855-0400
48)	LU24	820	5/3	R. Tres Arroyos: 0900-0400
165)		v820		R. Federal, Lanús
49)	LT8	830	10/1	R. Rosario: 24h
114)	LV18	830	0.25	R. Municipal San Rafael: 1030-0230
118)	LT21	830	0.5	R. Municipal Alvear: 0900-0100
51)	LU14	830	25/20	R. Provincia de Santa Cruz, Río Gallegos: 0900-0500
54)	LT12	840	10/5	R. General Madariaga, Paso de los Libres: 0900-0300
56)	LU2	840	25	R. Bahía Blanca, "El Sonido de la Vida": 0900-0400
91)	LV9	840	25/5	R. Salta: 1000-0500
234)		840		R. General Belgrano, Buenos Aires
57)	LRA56	860	1	R. Nal. Perito Moreno: 0955-0300
53)	LRJ392	860	5/1	R. Municipal, Chilecito: 1000-0400
58)	LRA1	Þ870	100	R. Nal. Buenos Aires: 24h
51)	LU14	880	10	R. Provincia de Santa Cruz (r: LU14 830), Santa Cruz: 1000-0300
59)	LU33	890	20/1	Em. Pampeana, Santa Rosa: 24h
60)	LV11	890	25/5	Em. Santiago del Estero: 1000-0630
61)	LT7	900	25/10	R. Provincia de Corrientes: 0900-0300
62)	LRA23	910	50/25	R. Nal. San Juan: 1000-0500
63)	LR5	910	60/5	La Red, Buenos Aires: 24h
126)	LV28	930	1/0.25	R. Villa María, Villa María: 0900-0300
113x)	LRJ241	940	20/5	R. Dimensión, San Luís: 0900-0400
144)	LRH200	940	3/0.5	R. Chajarí: 0900-0300
55)	LT13	950	1	R. Oberá, Oberá: 0800-0200
64)	LR3	950	50/15	R. Libertad, Buenos Aires: 24h
66)	LRA6	Þ960	25/15	R. Nal. Mendoza: 1000-0300
68)	LU13	960	10/1	R. Necochea: 0900-0400
70)	LV2	970	25/7	R. General Paz, Córdoba: 24h
146)	LT25	970	1	R. Guaraní, Curuzú Cuatiá: 0900-0300

PANAMA
Panama

Caracas
Campo Carobobo

Tovar
Cúcuta
San Cristóbal
Bucaramanga

VENEZUELA

Sparendaam
Georgetown
GUYANA
Paramaribo
Montsinery
SURINAME
Matoury
FRENCH GUIANA

Medellín
Bogotá
Villavicencio
Neiva
COLOMBIA
Florencia

Boa Vista

Macapá

NORTH
ATLANTIC
OCEAN

Quito
Otavalo
Tena
ECUADOR
Sucua
Santa Rosa de Quijos

Manaus

Belém

Chachapoyas
Chiclayo
Celendín
Cajamarca

Humaitá

B R A Z I L

Recife

PERU
Tarma
Lima

Puerto Maldonado
Santa Ana de Ycuma
Reyes
Trinidad
San Borja
La Paz
Cochabamba
Montero
Oruro
Huanuni
Santa Cruz
Sucre
BOLIVIA
Animas
Tarija
Yacuiba

Brasília
Goiânia

Salvador

Belo Horizonte

Campos
São Paulo
Rio de Janeiro
Curitiba

PARAGUAY
Asunción

Encarnación

Florianópolis

SOUTH
PACIFIC
OCEAN

CHILE

Mendoza
Santiago
Malargüe

Artigas
Rivera
Porto Alegre
URUGUAY

San Fernando
Buenos Aires
General Pacheco
Montevideo

SOUTH
ATLANTIC
OCEAN

ARGENTINA

Temuco

Coyhaique

■ Capital
□ SW Transmitter Site
COUNTRY

SOUTH AMERICA

	Call	kHz	kW	Name and h. of tr.
109)	LU37	980	10/1	R. Gral. Pico: 0930-0300
136)	LT39	980	3/0.5	Em. Victoria: 0900-0300
63)	LR4	990	25/5	R. Splendid, Buenos Aires: 24h
70x)	LRJ201	990	1	R. Calingasta, Tamberias: 1000-0500
71)	LU16	1000	25/1	R. Río Negro, Villa Regina: 0900-0300
137)	LT42	1000	1	R. del Iberá, Medcedes: 0900-0300
72)	LV16	1010	20/5	R. Rio Cuarto: 1000-0500
130)	LW2	1010	1	Radioem. Tartagal: 1000-0300
73)	LT10	1020	10/5	"AM Universidad", Santa Fé: 0800-0500
15)	LRA28	1020	1	R. Nal. La Rioja, Chilecito
74)	LRA58	1020	1	R. Nal. Río Mayo: 1100-2300
76)	LS10	1030	25/5	R. del Plata, Buenos Aires: 24h
77)	LR1	1070	25/5	R. El Mundo, Buenos Aires: 24h
78)	LU3	1080	25/10	R. del Sur, "la emisora de Bahía Blanca": 0900-0400
79)		1080	10	R. Departamento Minas,Andacollo: 1000 (SS 1200)-2300
145)	LW4	1080	25	R. Orán: 1000-0600
30)	LS1	1110	25/5	R. Municipal, Buenos Aires (F.PI.) (still on 710)
79x)	LR2	1110	15/5	R. Argentina, Buenos Aires: 24h
80)	LV7	1110	10/1	R. Tucumán, San Miguel de Tucumán: 24h
81)	LU6	1120	25/10	R. Atlántica, Mar del Plata: 24h
82)	LV5	1120	25/5	R. Sarmiento, San Juan: 1000-0500
84)	LRA21	1130	25/5	R. Nal. Santiago del Estero: 0900-0400
207)		*1130		AM 1130, Remedio de Escalada
84x)	LRA2	Þ1150	10	R. Nal. Viedma: 0900-0400
86)	LRA51	1150	1	R. Nal. Jáchal: 1030-0300
87)	LT9	1150	25/1	R. Brigadier López, Santa Fé: 0700-0300
178)		1150	5/1	R. Sagrada Familia, San Justo: 24h
88)	LRA57	1160	1	R. Nal. El Bolsón: 0900-0300
89)	LRH253	1160	1.8	R. Cataratas del Iguazú, Pto. Iguazú: 1000-1600
90)	LU32	1160	10/5	R. Coronel Olavarría, Olavarría: 0900-0300
91)	LV9	1160	25/5	R. Salta: 1000-0500
92)	LRA29	1170	25/3	R. Nal. San Luis: 1000-0500
93)	LT16	1170	25/3	R. Esmeralda, Presidencia Roque Sáenz Peña: 0900-0300
242)		1170		R. Mi País, Hurlingham
95)	LRA15	1190	50	R. Nal. San Miguel de Tucumán: 0900-0400
96)	LR9	1190	50/10	R. América, Buenos Aires: 24h
66)		1200	1	R. Nal. Mendoza (r. LRA6 960), Valle de Uspallata: 1000-0500
97)	LT6	1200	10	R. Goya: 0900-0300
98)	LT18	1210	10/5	R. Eldorado: 0930-2130
215)		1220		LV del Aire, Buenos Aires: 24h
100)	LT2	1230	25/5	R. Gral. San Martín, "R. 2", Rosario: 24h
101)	LW5	1230	5/1	R. Libertador General San Martín, Ledesma: 1000-0300
47)		1230		R. Ciudad de Banfield
208)		v1240		R. Amanacer Argentino, Numancia
164)		1250	3/1	AM 1250 R. Estirpe Nal., Rafael Castillo: 24h
102)	LT14	1260	10/5	R. General Urquiza, Paraná: 0900-0500
103)	LRA20	1270	5	R. Nal. Las Lomitas: 0900-0300
104)	LS11	1270	50/10	R. Provincia de Buenos Aires, La Plata: 24h
240)		1270		R. Ciudad de Avellaneda, Wilde
163)		*1290	5/1	Em. del Lago, Junín
193)		1290		R. Amanacer, Reconquista
210)		1290		R. Cristal, Lanús: 24h
176)		1290		R. Itatí, Ituzaingo
168)		1300	0.15	R. Ecos, Llavallol: 1200-0300
105)	LU22	1300	5/1.5	R. dif. Tandil: 0900-0300
106)	LRA5	1300	10/5	R. Nal. Rosario: 0853-0303
235)		1300		R. Malvinas Argentinas, Caseros
106x)	LRA42	1310	1	R. Nal. Gualeguaychú (r. 1309.5)
106z)		1310	1	R. Dr. Gregorio Alvarez, "LV de la Comunidad", Piedra del Aguila: 24h
211)		1310		AM Panamericana, GregorioLaferrere

	Call	kHz	kW	Name and h. of tr.
107)	LRH251	1320	25/5	R. Chaco, Resistencia: 0800-0300
108)	LU10	1320	5/1	R. Azul: 0900-0300
212)		1320		R. Independencia, Lanús
213)		*v1325		R. Tango, Caseros
214)		1330		R. Mar, San Antonio de Padua
111)	LS6	1350	50	R. Buenos Aires, "RBA": 24h
112)	LT46	1360	1/0.4	R. Provincia Bernardo de Irigoyen: 0900-0300
113)	LRA54	1370	1	R. Nal. Ingeniero Jacobacci: 0900-0300
241)		1370		R. Cosmos, Isidro Casanova: 24h
236)		1380		LV del Futuro, Merlo
66)		1390	1	R. Nal. Mendoza (r. LRA 960), Valle de Uspallata: 1000-0500
115)	LR11	1390	10	R. Universidad Nacional de La Plata: 0800-0300
116)	LU11	1400	7.5/1	R. Em del Oeste, Trenque Lauquén: 0930-0300
187)		*1400		R. Fantasía, Tapiales
188)		*1400		R. 2001, Quilmes Oeste
216)		1400		R. Siglo XXI, Burzaco
217)		1400		R. Comunitaria Cristiana, San Fernando
197)	LRJ360	1410		R. Obispado de San Luís, Villa Mercedes
182)		1410		R. Folklorismo, José Léon Suárez: 1100-0100
166)		1410	0.1	R. Fortín Federación, Junín: 1000-0400
117)	LRK221	1420	1	R. Ciudad Perico, Perico: 1000-0400
194)	LRH362	1420	1	R. Tupá Mbaé, Posadas: 0800-0300
198)	LRH375	1420		R. Arzobispado de Corrientes, Itati
199)	LRJ359	1420		R. Opispado de San Luís, San Luís
200)	LRJ361	1420		R. Arzobispado de Córdoba, Villa Carlos Paz
218)		1420		R. Mágica, Lanús
237)		1420		R. Mailin, Libertad: irr.
119)	LT14	1430	5/1	R. San Nicolás: 24h
120)	LV26	1430	1/0.25	R. Río Tercero: 0930-0330
121)	LRI235	1430	1	R. Balcarce, Balcarce: 0900-0300
189)		1430		R. Libertad, Libertad
220)		1430		R. José de San Martín, "la Pionera", El Jagüel
221)		1430		R. Trinidad, Presidente Derqui
43)	LRA53	1440	1	R. Nal. San Martín de los Andes: 1000-0400
123)	LU36	1440	0.25	R. Coronel Suárez: 1000-0300
124)	LU40	1440	1	R. Laboulaye: 1000-0200
125)	LV27	1440	1	R. San Francisco: 0900-0300
177)		1440		R. Chascomús "RCH," Chascomús
190)		*1440		"RSO" R. Sudoeste, Marcos Paz
201)		1440		R. Impacto, Tapiales
183)		1450	0.5	R. Ciudad, Lanús: 1000-0400
67)	LT44	1450	1/0.25	R. Fortín Yunka, Formosa: 0900-0400
243)		1450		R. Ambar, Caseros
127)	LT29	1460	1/0.25	R. Venado Tuerto: 0930-0330
128)	LU30	1460	0.25	R. Maipú: 0900-2400
129)	LU34	1460	0.25	R. Pigüé: 1000-0300
130)	LW2	1460	0.25	R. Emis. Tartagal: 1000-0300
222)		1460		R. Estilo, Longchamps: 24h
131)	LT20	v1470	1/0.25	R. Junín: 0900-0300
132)	LT26	1470	1/0.25	R. Nuevo Mundo, Colón: 0900-0300
133)	LT28	1470	1	R. Rafaela: 0900-0300
134)	LU26	1470	1.5/0.6	Emis. Coronel Dorrego: 1030-0300 (r. 1475)
135)		1470	1	R. Municipal, Luis Beltrán: 1200-0100
181)		1470		R. Contacto, San Antonio de Padua: 24h
209)		1470		R. Ciudadana, Luis Guillón
223)		1470		R. Conurbana del Sur, Wilde
192)	LU27	1480	0.25	R. Dolores: 1100-0300
238)		1480		R. Temperley, Temperley
138)	LV22	1490	1	R. Huinca Renancó: 1000-0300
191)	LRI202	1490		R. Obispado de San Juan de los Arroy, Pergamino
224)		1490		La Nueva R. , Isidro Casanova
138x)		1500	1	R. Municipal, Gral. Conesa: 1000-2400

Call	kHz	kW	Name and h. of tr.
139) LT22	1500	0.25	R. Nueva Era, Pehuajó: 1000-0300
169) LRI208	1500	0.25	R. El Sol AM 1500, Lanús: 24h
140) LT34	v1500	0.25	R. Nuclear, Zárate: 0900-0000
141) LT45	1500	1/0.25	R. San Javier: MF: 1100-0500; SS 1100-2100
142) LV25	1500	5/1	R. Unión, Bell Ville: 0900-0400
170)	1510		LV del Oeste, Libertad: 24h
143) LV21	1510	3/0.25	R. Champanal, Villa Dolores: 1000-0100
179)	1520		R. Güemes, Quilmes Oeste
147) LT37	1520	0.25	Em. Rufino: 1000-0400
148) LT38	1520	0.25	R. Gualeguay: 0900-0300
149) LV24	1520	1.5	R. Manantiales, Tunuyán: 1000-0600
150) LRJ200	1530	1	R. Centro Morteros, Morteros: 0900-0300
151) LU25	1530	0.1	R. Carhué: 0900-0330
167)	1530		R. Popular, González Catán
171)	1530		RG R. Libertador José de San Martin, Glew
225)	1530		R. Esperanza, Bella Vista
152) LT35	1540	1	R. Mon, Pergamino: 0930-0400
153) LU28	1540	0.25	R. Tuyú, General Madariaga: 1000-0300
184)	*1540		R. Digital, Ituzaingó
226)	1540		R. Clube, Bánfield
227)	*1540		R. Universal, Pablo Podestá
154) LT23	1550	1	R. San Jenaro Norte: 0900-0300
155) LT32	1550	0.25	R. Chivilcoy: 1000-0400
156) LT40	1550	0.25	R. LV de la Paz, La Paz: 0900-0100
158) LT33	1550	1	R. Nueve de Julio, Buenos Aires: 0900-0300
172)	1550	0.4	R. La Cruz del Sur, Guernica: 0900-2300 (r. 1557)
180)	1550		R. Metropolitana, Ciudadela
157) LT11	1560	2.5/1.5	R. Gral. Francisco Ramírez, Concepción del Uruguay: 0900-0300
185)	1560		R. Castañares "RCI," Ituzaingó
228)	1560		R. Inolvidable, Villa Domínico
159) LRI229	1570	1	R. Las Flores: 0900-0300
186)	*1570		Carisma AM, San Justo
195)	1570		R. Líder, Central
202)	1570		R. Interactiva, Ciudad Madero
230)	1570		R. AM Rocha, La Plata
173)	*1580		"RCL" R. Cd. de Libertad, Libertad
160) LT27	1580	0.25	R. LV del Montiel, Villaguay: 0900-0300
161) LT36	1580	0.25	R. Chacabuco: 1000-0400
174)	*1580		R. General Martín Miguel de Güemes, Guernica
232)	1580		AM Ciudad de San Justo, San Justo
233)	1580		R. Urbana, Lanús
229)	1593		R. Emociones, Luis Guillón
162) LR14	1600	1	R. Ciudad Aluminé, Aluminé
175)	1600	1.2	R. Armonia, Caseros: 24h
231)	1610		R. Cultura, Lanús
203)	1620	5/1	R. Universidad de Buenos Aires (CP)
244)	1620		R. Universo, Glew
204)	1680	1/0.5	R. Universidad Tecnolólica Avallenada (CP)

Shortwaves

Call	kHz	kW	Name, h. of tr.
58) LRA31	6060	30	R. Nal. Buenos Aires: 2100-1500
83) LV19	6160.7	1	R. Malargüe: 1100-0100v
66) LRA34	6180.3	7.5	R. Nal. Mendoza: irr.
58) LRA31	9690	25	R. Argentina al Exterior: Tue/Sat 2300-0100
58) LRA35	11710	100	R. Argentina al Exterior: 1000-1400, SS 1100-0400
58) LRA31	15345.2	100	R. Argentina al Exterior:Ext. Sce: 0900-1200, 1500-1600, 1800-0200 r. of LRA1 Weekends 1100-2300

ASOCIACION DE RADIODIFUSORAS PRIVADAS ARGENTINA (ARPA)

Pte. Perón 1561, Piso 8, 1037 Buenos Aires. ☎ +54 (1) 3824412, 3824483 — **L.P:** Presidente: Domingo F. L. Elías.
ARPA is an association of privately owned commercial st's.

STATE ABBREVIATIONS *(Provincias):* BA = Provincia deBuenos Aires, C = Catamarca, Ca = Córdoba, Cs = Corrientes, Ch = Chaco, Ct = Chubut, ER = Entre Ríos, F = Formosa,J = Jujuy, LP = La Pampa, LR = La Rioja, Ms = Misiones, Ma = Mendoza, N = Neuquén, RN = Río Negro, S = Salta, SC = Santa Cruz, SE = Santiago del Estero, SF = Sante Fé, SJ = San Juan, SL = San Luis, T = Tucumán, (Territorio): TFAIA = Territorio Nacional de la Tierra del Fuego, Antártida e Islas del Atlántico Sur.

NB: These abbreviations should not be used for postal purposes.

Addresses and other information

1) Mendoza 2430, 3000 Santa Fé, SF. ☎ +54 (42) 533340. ▤ +54 (42) 528640 – DG: Alberto Fraga. **FM:** 94.9.
2) Ruta Nal. 34,km. 1433, 4560 Tartagal, S – DG: Eduardo Fortunato Esper.
3) Av. Alvear 1180, 9200 Esquel, Ct – DG: Raúl H. Pasarín.
4) Moreno 30, 8000 Bahía Blanca, BA – Dir: Martin Allica.
5) San Martín 371, 3200 Concordia, ER – Dir: Darío Gómez. **FM:** 89.3 "Encuentro".
6) Mendoza Sur 169, 5400 San Juan, SJ – DG: Jorge Ricardo Barassi. **FM:** 98.5.
7) Av. España 700 Sur, 4650 La Quiaca, J. ☎ +54 (885) 22356. ▤ +54 (885) 23488 – DG: Jesús Alberto Guerrero.
8) Hipólito Yrigoyen 1735, 9100 Trelew, Ct. ☎ +54 (965) 21289. ▤ +54 (965) 25457 – Mgr: Carlos Pérez Luces. **FM:** 95.7 "Galaxia".
9) R. Universidad Nal. de Córdoba, Miguel de Mojica 1600, 5000 Córdoba, Ca – Dir: Carlos López Araoz. **FM:** Power 102.3.
10) Rivadavia 835, 1002 Buenos Aires. ☎ +54 (1) 3425425. ▤ +54 (1) 3456812 – Dir: Norberto Mantiñán.
11) Av. 12 de Octubre 2421, 8400 San Carlos de Bariloche, RN – Dir: Sebastián A. Tetelboin. **FM:** 95.5.
12) Rivadavia 118-120, 4000 San Miguel de Tucumán, T. ☎ +54 (81) 312012 – DG: Ing. Eduardo J. Bader. **FM:** 98.9.
13) Comodoro Py 342, Planta Baja, 9407 Río Turbio, SC. ☎ +54 (902) 21131. ▤ +54 (902) 21129 – DG: Ramón Lozano. **FM:** 90.3.
14) Cas. 372, 3300 Posadas, Ms – Dir: Pedro D. Cabrera.
15) Yrigoyen 324, 5300 La Rioja, LR – DG: Ricardo R. Quiroga.
16) Arenales 2467, 1124 Buenos Aires. ☎ +54 (1) 823 2323 – DG: Luis María Cetrá.
17) Monseñor Fagnano y Leonardo Rosales, 9420 Río Grande, TF – Dir: Miguel Bersier.
18) Tucumán 1074, 8332 General Roca, RN – DG: H.X. de Cozzi. **FM:** 99.3 "Color".
19) Lavalle 291, P.lanta Alta, 5730 Villa Mercedes, SL. ☎▤ +54 (657) 21406, 24400 – DG: Lic. Alfredo Mario. Noroña. **FM:** 95.5 "Unica".
20) 25 de Mayo 453, 9000 Comodoro Rivadavia, Ct – DG: J. Horat.
21) Gral. Paz 536, 8353 Chos Malal, N – DG: Jorge Alberto Cotaro.
22) Bolívar 1867, 3300 Posadas, Ms. ☎ +54 (752) 30500, 33500. ▤ +54 (752) 30664 – Pte: Pedro Warenycia. DG: Sergio E. Lang. **FM:** 104.5.
23) La Rioja 2371, 7600 Mar del Plata, BA. ☎ +54 (23) 941039, 941040, 941041 – DG: Carlos Federico Infante. **FM:** 103.3 "Universo".
24) Zapiola 25, 9400 Río Gallegos, SC. ☎ +54 (966) 20095. ▤ +54 (966) 22608 – Dir: Oscar Trucco. **FM:** 92.9 "Laser".
25) Estivariz 226, 9120 Pto. Madryn, Ct – DG: H. Castro. **FM:** 88.3.
26) Echeverría 144, 5500 Mendoza, Ma – Dir: Daniel E. Vila. **FM:** 93.7 "Montecristo".
27) Deán Funes 140, 4400 Salta, S – DG: Gloria Beatriz Franco. **FM:** 102.9.
28) Cas. 14, 8324 Cipoletti, RN – Dir: Miguel Romay.
29) San Martín y Chanetón, 8340 Zapala, N – DG: Mario Jorge Centeno.
29x) Av. Victoria Aguirre 809, 3370 Puerto Iguazú, Ms. ☎▤ +54 (757) 20999 – DG: Alejandro Segundo Guerrero.
30) Sarmiento 1551, Piso 8, 1042 Buenos Aires. ☎ +54 (1) 3721085. ▤ +54 (1) 3745535 – Dir: Ruben Machado.
31) Av. San Martín 114, 9311 Governador Gregores, SC – DG: O. Muñoz.
32) Rioja 1093, 5500 Mendoza, Ma. ☎ +54 (61) 205100 – DG: Juan José Guaranda. **FM:** 100.9 "Estación del Sol".
33) Rivadavia 202, 6300 Santa Rosa, LP. ☎▤ +54 (954) 25102 – DG: Carlos Antonio Cuco.
34) Ruta Complementaria "0" s/n, 9405 El Calafate, SC – DG: Adrian Martrinez. **FM:** 88.1 "Glaciar". **F.PI:** 10kW.
35) Chacabuco 762, 4700 San Fernando del Valle de Catamarca, C – Dir: Raúl J. Acuña.
36) Av. Comandante Fontana y Dr. Mariano Moreno, 9033 Alto Río Senguerr, Ct – Dir: Roxana B. de Jensen.
36x) 12 de Octubre 633, 9050 Puerto Deseado, SC – DG: Lucio Ibiricu.
37) Santa Rosa 241, 5000 Córdoba, Ca – Dir: Marcos Daniel Sastre.
38) Av. San Martín 351, 9410 Ushuaía, TFAIA – Dir: Luis Donato Casielli.
39) Chacra 46, La Tablada, 3340 Santo Tomé, Cs – DG: Mariano Lovera.

40) Rioja 1484, 5500 Mendoza, Ma – DG: Norma S. Fontemachi.

41) Mansilla 2668, 1425 Buenos Aires. **Web:** http://www.clarin.com/chats/html/mitre.html – Dir: Dr. Jorge H. Santos.

42) Olavarría y Río Bermejo, 4600 San Salvador de Jujuy, J. – Mgr: José Ramón Casas. C.E: Roberto Pemberton.

43) Villegas 251, 8370 San Martín de los Andes, N – DG: Norberto Hilzerman.

44) Av. Alvaro Barros 1148, 8500 Viedma, RN – DG: Dr. Tomás A. Rébora. Italian: Sat 1500-1600. **FM:** 94.3 "FM Río".

45) Patricias Mendocines 72-78, 5620 General Alvear, Ma – Dir: José M. Ferraro.

46) Junín 665, 3600 Formosa, F. ☎ +54 (717) 26917. 🖷 +54 (717) 28570 – DG: Eduardo Cuevas. **FM:** 90.9.

47) Benito Perez Galdós 688, 1828 Banfield, BA. ☎ +54 (1) 2863148, 2860692.

48) Belgrano 457, 7500 Tres Arroyos, BA. ☎ +54 (983) 31620, 31630 – DG: Leonel Elías. **FM:** 95.3 "Ilusiones".

49) Córdoba 1843, 2000 Rosario, SF. ☎ +54 (41) 494000, 495000. 🖷 +54 (41) 495050, 496000 – DG: Manuel A. Fernández. **FM:** 99.5 "Estación del Siglo."

50) Cas. 83, 3730 Charata, Ch – Dir: Adolfo Zinser. **FM:** 95.7 "FM Líder".

51) Av. Roca 823, 9400 Río Gallegos, SC – DG: Justo G. Lerena.

52) Dr. H. Guzmán 496, 4600 San Salvador de Jujuy, J. ☎ +54 (882) 30035 – DG: Guillermo Jenefes.

53) José Hernández 46, 5360 Chilecihto, LR – Dir: Julio César Pedroza.

54) Av. Lib. Gral. San Martín 2031, 3230 Paso de los Libres, Cs – Dir: E. Nicholas.

55) Larrea 886, 3360 Oberá, Ms – Dir: Hugo W.R. Amable. **FM:** 99.7.

56) Sarmiento 64, 8000 Bahía Blanca, BA. ☎ +54 (91) 590002. 🖷 +54 (91) 555556 – Dir: José O. Trillini. **FM:** 105.7 "La Ciudad".

57) Saavedra 1318, 9040 Perito Moreno, SC – DG: Ing. Hugo Santos Rojas.

58) Maipú 555, 1006 Buenos Aires. ☎ +54 (1) 3254590, 3254313. 🖷 +54 (1) 3259433 – DG: Patricia Barral. PD: Elbio Petrocelli. CE: Roberto Core. **International Sce:** see International Broadcasting section.

59) Torre 474, 6300 Santa Rosa, LP. ☎ +54 (954) 33505, 33104 – DG: Carlos Luis Grotto. **FM:** 103.7 "Power".

60) 9 de Julio 390, 4200 Santiago del Estero, SE – Pres: José Maria Cantos. CE: Ricardo Mauro Rivero. **FM:** 89.5.

61) La Rioja 743, 3400 Corrientes, Cs. ☎ +54 (783) 23554. 🖷 +54 (783) 23149 – Dir: Jorge F. Gómez. **FM:** 95.3 "Capital".

62) Av. Ignacio de la Roza 293 Este, 5400 San Juan, SJ – Dir: Francisco Herrada.

63) Arenales 1925, 1124 Buenos Aires – Dir: Carlos D. Bautista.

64) Rivadavia 825, 1002 Buenos Aires. ☎🖷 +54 (1) 345 0495/6 – DG: Dr. Julio Rossi.

65) Cas. 229, 9000 Comodoro Rivadavia, Ct – DG: N. Dames. **FM:** 101.7 "Alta FM".

66) Emilio Civit 460, 5500 Mendoza, Ma. ☎ +54 (61) 380596 – Dir: Lic. J. Parvanoff. **FM:** 97.3.

67) España 376, 3600 Formosa, F – Dir: M.F. Bedoya.

68) Calle 64 No. 2946, 7630 Necochea, BA. ☎ +54 (262) 23462, 25630 – DG: Arturo Coupau. **FM:** 88.1 "Oceanica".

69) Cas. 204, 8300 Neuquén, N – Dir: Juan D. Kohon. **FM:** 94.7..

70) 27 de Abril 979, 5000 Córdoba, Ca. ☎ +54 (51) 224220. 🖷 +54 (51) 228164 – DG: Gustavo Viramonte. **FM:** 99.7.

70x) Juan Ramón Diaz s/n, 5401 Tamberias, SJ. ☎ +54 (648) 92050. 🖷 +54 (648) 92075 – DG: Carlos Ernesto Herrera.

71) Reconquista 135, 8336 Villa Regina, RN. ☎ +54 (941) 61102. 🖷 +54 (941) 62620 – DG: R. A. Musso. **FM:** 92.7 "Rio Negro".

72) Constitución 399, 5800 Río Cuarto, Ca. ☎ +54 (58) 638255. 🖷 +54 (58) 634800 – DG: Julio Cesar Castellina. **FM:** 93.9.

73) R. Univ. Nal. del Litoral, 9 de Julio 3560, 3000 Santa Fé, SF – DG: C.P.N. Guillermo R. Alvarez. **FM:** 107.3.

74) Ruta 40 s/n, 9030 Río Mayo, Ch – DG: Hugo R. Font.

75) Doyenhard 318, 1836 Llavallol, BA. ☎ +54 (1) 2313225.

76) Honduras 5673, 1414 Buenos Aires. ☎ +54 (1) 778 6767, 6160 – DG: Dr. Alberto G. Veiga. (AM Stereo)

77) Pte. Perón 646, 1038 Buenos Aires – Dir: Estéban Ferrari. **FM:** 94.3 "Horizonte".

78) Departamento de Relaciones Publicas, Av. Lamadrid 116, 8000 Bahía Blanca, BA. **FM:** 94.1.

79) Ramón Elías Troitiño 624, 8353 Andacollo, N – Dir: David L. Bilsky – **F.PI:** 5kW, 24h.

79x) Tacuarí 2035/37, 1139 Buenos Aires – DG: Carlos Fioroni.

80) Mendoza 273, 4000 San Miguel de Tucumán, T – DG: Jorge Antonio Tejada. **F.PI:** Freq change to 710kHz.

81) Córdoba 1865, 7600 Mar del Plata, BA. ☎ +54 (23) 913350. 🖷 +54

(23) 912355. **Web:** http://www.argenet.com.ar/atlantica/ – Dir: Florencio Aldrey Iglesias. **FM:** 88.1 "Payunia".

82) Mendoza 452 Sur, 5400 San Juan, SJ – Dir: Emilio Ventura.

83) Esquivel Aldao 350, 5613 Malargüe, Ma – DG: D. Porras.

84) Urquiza 332, 4200 Santiago del Estero, SE – DG: Antonio Medel.

84x) Cas. 424, 8500 Viedma, RN – Dir: Carlos E. Reyes.

85) Av. 9 de Julio 1855, 3500 Resistencia, Ch – DG: Arturo Allende – Occ. prgr. in Toba and Guaraní.

86) Ruta Nal. 150, Km. 3,5460 San José del Valle de Jáchal, SJ – DG: C. Paez.

87) 4 de Enero 2153, 3000 Santa Fé, SF – DG: Federico Caputto. **FM:** 92.5 "FM Laser".

88) Cas. 176, 8430 El Bolsón, RN – DG: Carlos Pogliano. **FM:** 92.5.

89) Cas. 55, 3370 Puerto Iguazú, Ms – DG: Alberto Gollan.

90) Alsina 3377, 7400 Olavarría, BA. ☎🖷 +54 (284) 30911 – Dir: Daniel Panarace. **FM:** 98.7 "Cristal".

91) Cas. 113, 4400 Salta, S – DG: José Luis Cambetta. **FM:** 96.9 "Genesis".

92) Av. Lafinur 488, 5700 San Luís, SL – DG: Carlos A. Moreno.

93) Justo José de Urquiza 585, 3700 Presidencia Roque Sáenz Peña, Ch – Dir: Reinaldo Rojas. **FM:** 93.3.

94) Hipólito Yrigoyen y Godoy Cruz, 5600 San Rafael, Ma. ☎ +54 (627) 30055. 🖷 +54 (627) 30065 – DG: Raul Oscar Odoriz.

95) San Martín 251, Piso 4, 4000 San Miguel de Tucumán, T. ☎ +54 (81) 226831, 310131. 🖷 +54 (81) 302409 – DG: Adrián Lomello. **FM:** 93.3.

96) Honduras 5663, 1414 Buenos Aires, BA. 🖷 +54 (1) 777 1234 – DG: Paul Burzoco. (All news).

97) Mariano I. Lloza 231, 3450 Goya, Cs. ☎ +54 (777) 31990, 32333 – DG: Luis Moises Cavalieri. **FM:** 98.3 "FM 6 Splendid".

98) Cas. 218, 3384 Eldorado, Ms – DG: Máximo Gunther Kreis. **FM:** 96.1 "Dorada".

99) Alvear 139, 5000 Córdoba, Ca. ☎ +54 (51) 241080. 🖷 +54 (51) 227227. **E-mail:** mailto:lv3@lv3.com.ar. **Web:** http://200.26.95.98/cor/ – DG: (Cr.) Carlos M. Molina. **FM:** 100.5MHz.

100) Av. Pres. Juan D. Perón 8101, 2000 Rosario, SF – Dir: Alberto C. Gollán. **FM:** 97.9 "Vida FM".

101) Jujuy 470, 4512 Lib. Gral. San Martín, J. ☎ +54 (41) 400017, 400018. 402490. 🖷 +54 (41) 218149 – Dir: Omar Gabriel Sapienza. **FM:** 104.5.

102) Rivadavia 126, 3100 Paraná, ER – DG: R. Galanti.

103) Intersección Rutas Nacional 81 y Provincial 32, 3630 Las Lomitas, F – Mgr: María Dolores Ramírez.

104) Av. 53 No. 810, 1900 La Plata, BA – Dir: V. Tomaselli.

105) Cas. 102, 7000 Tandil, BA – Dir: Juan V. Martínez.

106) Córdoba 1331, 2000 Rosario, SF – Dir: Omar Gabriel Sapienza. **FM** 104.5.

106x) Urquiza al Oeste, 2820 Gualeguaychú, ER.

106z) Calle Las Rosas 16, Barrio Jardín, 8315 Piedra del Aguila, N. ☎🖷 +54 (942) 93216 – Dir: Oscar Isaac Lillo. **FM:** 99.5.

107) Córdoba 710, 3500 Resistencia, Ch – DG: Edmundo Ordenavia.

108) Av. Mitre 819, 7300 Azul, BA – DG: Donato A. Santomauro. **FM:** 89.5 "Celestial".

109) Calle 29 y 40, 6360 General Pico, LP – DG: J. Diván. **FM:** 88.9 "Melodias".

110) Balcarce 840, 2000 Rosario, SF – DG: Teodoro Gluck. **FM:** 102.7.

111) Av. Entre Ríos 1931, 1133 Buenos Aires, BA. ☎ +54 (1) 3072201 – DG: Jorge A. Civit – DP: Eduardo Haida.

112) Av. Independencia s/n, 3366 Bernardo de Irigoyen, Ms.

113) Martín Coronado y José Hernández, 8418 Ingeniero Jacobacci, RN – Dir: José L. Hernández.

113x) Av. Pte. Illia 128, 5700 San Luís – DG: Enrique Finocchietti. **FM:** 102.5 "Maxima FM".

114) Cdte. Salas 287, 5600 San Rafael, Ma – DG: Juan C. Rivas.

115) Plaza Rocha 133, 1900 La Plata, BA. ☎ +54 (21) 220330 🖷 +54 (21) 224165 – Dir: Silvina Fernández Cortés. **FM:** 107.5 "Universa".

116) Cas. 36, 6400 Trenque Lauquen, BA – DG: Roberto Aguilera. **FM:** 88.5 "Proyección".

117) Villafañe y Calilegua, 4608 Perico, J – DG: M. Viotti.

118) Isaco Abitbol y Gral Paz, 3344 Alvear, Cs – Mgr: Jorge L. Simón.

119) Av. Moreno 124, 2900 San Nicolás, BA. ☎🖷 +54 (461) 25222, 24479. **E-mail:** lt24@cablenet.com.ar. **Web:** http://www.cablenet.com.ar/lt24/ – DG: Gustavo Fabian Ramini. **FM:** 88.3.

120) Libertad 455, 5850 Río Tercero, Ca – DG: Carlos Estévez. **FM:** 94.5 "Libra".

121) Cas. 77, 7620 Balcarce, BA – Dir: Juan H. Orofino. **FM:** 89.7MHz.

122) San Martín 814, 2820 Gualeguaychú, ER – DG: José Bignolo. **FM** 90.3.

123) Avellaneda 1144, 7540 Coronel Suárez, BA – Dir: Raúl Soguer.

124) Tucumán 159, 6120 Laboulaye, Ca. ☎🖷 +54 (385) 26259 – DG: Jorge F. López. **FM:** 89.9 "Líder".

125) Edif. Regio II, Centro Cívico, 2400 San Francisco, Ca – DG: Dionisio E. Peretti. **FM:** 88.7 "Galaxia".
126) Santa Fé 1490, 5900 Villa María, Ca – Dir: Roberto Kfuri.
127) Cas. 340, 2600 Venado Tuerto, SF – Dir: Hilmar H. Long. **FM:** 88.9 "Imaginación".
128) Lavalle Sud 312, 7160 Maipú, BA – DG: Sra. Griselda Olmo.
129) Cas. 101, 8170 Pigüe, BA – Dir: Isidro A. Bros.
130) Gorriti 524, 4560 Tartagal, S – Dir: Olimpia Perez del B de Oller. **FM:** 96.1 "Geminis".
131) Hipólito Yrigoyen 86, 6000 Junín, BA – DG: Miguel Chiarantano & Albelardo Scorsetti. **FM:** 89.1 "Nova".
132) Pte. Perón 117, 3280 Colón, ER – DG: Maria A. Benitez de Rouiller. **FM:** 93.7 "Palmares".
133) Bulevar Lehman 245, 2300 Rafaela, SF. ☎ +54 (492) 22150. ▤ +54 (492) 31855. **E-mail:** radioraf@santafe.com.ar – DG: Arquímedes R. Robert. PD:Roberto Robert. CE: José Pfisterer. **FM:** 96.5 "Fantasía".
134) San Martín y Uslenghí, 8150 Coronel Dorrego, BA – DG: Salvador E. Lavios.
135) 8361 Fray Luis Beltrán, RN – Dir: D. Natali.
136) Sarmiento 474, 3153 Victoria, ER – DG: Miguel Rodera. **FM:** 90.3.
137) Av. Atanacio Aguirre Km 2 s/n, 3470 Mercedes, Cs. ☎ +54 (773) 20097, 22260, 22259 – DG: Dr. L. Horacio Gutnisky.
138) Santa Fé y La Pampa, 6270 Huinca Renancó, Ca – DG: O. Berlaffa.
138x) Roca 570, 8503 General Conesa, RN – Dir: Jorge O. Monteoliva.
139) Perito Moreno 700, 6450 Pehuajó, BA – Dir: Raúl Negreira.
140) Independencia 501, 2800 Zárate, BA – Dir: Jorge E. Lynch.
141) Ordóñez 737, 3357 San Javier, Ms – Dir: H. Andersson.
142) Rivadavia 92, 2550 Bell Ville, Cs – DG: Constantino Tela. **FM:** 105.5 "Armonía".
143) Belgrano 33, 5870 Villa Dolores, Ca – Dir: Héctor O. Escudero. **FM:** 98.9.
144) Pablo Stampa 2430, 3228 Chajarí, ER. ☎ +54 (456) 20002 – DG: José Alberto Ponzoni. **FM** 88.7.
145) 9 de Julio 163, 4530 Orán, S – DG:David Taranto. **FM:** 90.9.
146) San Martín 1380, 3460 Curuzú Cuatiá, Cs. ☎ +54 (774) 22540. ▤ +54 (774) 23624 – DG: Ing. Luis Maria Mestres.
147) Chacabuco 123, 6100 Rufino, SF – Dir: Clara R. Vda. de Valeri.
148) Chacabuco 38, 2840 Gualeguay, ER – DG: G.J. Carbone. **FM:** 104.3 "Gamma".
149) Alem y Juan B. Justo, 5560 Tunuyán, Ma – Dir: F. Rossi. **FM:** 88.6.
150) 25 de Mayo 133, 2421 Morteros, Ca – Dir: José L. Brunone. **FM:** 90.3.
151) Av. Colón y Rivadavia, 6430 Carhué, BA – DG: Mario V. Fernández.
152) Dr. Alem 340, 2700 Pergamino, BA. ☎▤ +54 (477) 24022, 23848 – DG: Parios Eernesto Trincavelli. **FM:** 90.3 "Magica".
153) Cas. 123, 7163 General Madariaga, BA – DG: Hipólito Maceira.
154) Calle Juan Chavarri 458, 2147 San Jenaro Norte, SF – Dir: Rubén Cavallera.
155) Cas. 179, 6620 Chivilcoy, BA – DG: Alberto Mónaco. **FM:** 101.1 "Sonica".
156) Sáenz Peña 1082, 3190 La Paz, ER. ☎▤ +54 (437) 21568 – DG: Carlos A. Malvasio. **FM:** 89.1.
157) Onésimo Leguizamón 269, 3260 Concepción del Uruguay, ER – DG: Cont. Gladys Lindstrom. **FM:** 92.9 "FM Arenas".
158) Libertad 777, 6500 9 de Julio, BA – DG: Gustavo Tinetti. **FM:** 89.9.
159) Avellaneda 773, 7200 Las Flores, BA – DG: César Dantas. **FM:** 95.5 "Condor".
160) Av. Vélez Sarsfield 1111, 3240 Villaguay, ER. ☎ +54 (455) 21717 – DG: Alicia Faust. **FM:** 88.7 "FM 27".
161) Almirante Brown 135, 6740 Chacabuco, BA. ☎ +54 (352) 26346, 26347 – DG: Dr. Marcelo A. Loprete. **FM:** 91.7 "Universal".
162) 8345 Aluminé, N.
163) Av. Rep. Argentina 737, 6000 Junín, BA – Dir: Jorge Andreollo.
164) Chavarria 1675, 1755 Rafael Castillo, BA. ☎ +54 (1) 6254345, 4660024 – DG: Hector Franco.
165) Av. General Arias 1372, 1824 Lanús, BA. ☎ +54 (1) 2478237.
166) Além 469, 6000 Junín, BA.
167) 1759 González Catán, BA. ☎ +54 (202) 30111. **FM:** 89.7.
168) Ascasubi 212, 1836 Llavallol, BA. ☎ +54 (1) 2981441 – Dir: Omar E. Carusso. **FM:** 104.3.
169) Sarmiento 1436, 1824 Lanús, BA. ☎ +54 (1) 2419547, 2494545 – DG: Carlos Ranieri & Aldo Linares.
170) Alberto Echagüe y España, 1716 Libertad, BA.
171) Orden 65, 1856 Glew, BA. ☎ +54 (224) 21574
172) Calle 26 No. 28, 1862 Guernica, BA. ☎ +54 (224) 71532 – DG: Alfredo Camihort. **FM:** 89.3.
173) Cangallo 255, 1716 Libertad, BA.
174) Calle 32 entre 1 y 2, 1862 Guernica, BA.
175) Wenceslau Paunero 2921 (Partido 3 de Febrero), 1678 Caseros,

BA. ☎ +54 (1) 7347100. ▤ +54 (1) 7501744 – DG: Carlos H. Lavoro.
176) 1714 Ituzaingó, BA.
177) San Martín 178, 7130 Chascomús, BA. ☎ +54 (241) 25367, 31299 – **FM:** 90.9.
178) Pichincha 2849, 1754 San Justo, BA ☎ +54 (1) 4418196, 4843345 – DG: Hector Franco. **FM:** 104.5 "Sintonía".
179) Estanislao del Campo esq. Corrientes, 1879 Quilmes Oeste, BA. ☎ +54 (1) 2006766.
180) Julio A. Roca 3414, 1702 Ciudadela, BA.
181) Olegario Andrade 925, 1718 San Antonio de Padua, BA.
182) Calle Lacroze 7279, 1655 José León Suárez, BA. ☎ +54 (1) 7202688.
183) Fray Mamerto Esquiú 2855, 1826 Lanús, BA. ☎ +54 (1) 2403544. 2253823 – DG: Oscar N. Suarez.
184) Gral. Franco Rivera 3541, 1714 Ituzaingó, BA.
185) Calle Treinta y Tres No. 1033, Villa Ariza, 1714 Ituzaingó, BA.
186) 1754 San Justo, BA. ☎ +54 (1) 4418201.
187) Pastor Lacasa 485, 1770 Tapiales, BA. **FM:** 107.9 "Fantasma".
188) Calle 803 No. 1430, 1879 Quilmes Oeste, BA – **FM:** 106.7.
189) Zapiola 1520, 5ᵇ piso, 1716 Libertad, BA. ☎ +54 (220) 44214.
190) Bartolomé Mitre y Lavalle, 1727 Marcos Paz, BA.
191) Obispado de San Nicolás de los Arroy, 2700 Pergamino, BA.
192) Gral. Pico 424, 7100 Dolores, BA – Dir: N.G. Peñoñori.
193) 3560 Reconquista, SF.
194) Calle Felix de Azara 1648, 3300 Posadas, Ms. ☎ +54 (752) 20203 – Dir: Presbítero Jorge Luís Lagazio. **Guaraní:** Sun 0900-1500. **FM:** 102.7 & 105.9.
195) 7114 Central, BA.
196) Av. 25 de Mayo 69, 3760 Añatuya, SE. ☎▤ +54 (844) 21661 – DG: Mons. Antonio Baseotto.
197) Obispado de San Luís, 5730 Villa Mercedes, SL.
198) Arzobispado de Corrientes, 3414 Itati, Cs.
199) Obispado de San Luís, 5700 San Luís, SL.
200) Arzobispado de Córdoba, 5152 Villa Carlos Paz, Ca.
201) Juncal 12, 1ᵇ piso, Of. 3, 1770 Tapiales, BA. ☎ +54 (1) 4426333.
202) Mariquita Sánchez de Thompson 1850, 1768 Ciudad Madero, BA. ☎ +54 (1) 6222572, 4429055.
203) Universidad de Buenos Aires, Buenos Aires.
204) Universidad Tecnológica Nacional, Facultad Regional de Avallenada, 1870 Avallenada, BA.
205) Centenario Uruguayo 2222, 1825 Monte Chingolo, BA.
206) Berasain 659, 1842 Monte Grande, BA. ☎ +54 (1) 2965988, 2813740. **FM:** 91.7.
207) 1826 Remedios de Eascalada, BA. ☎ +54 (1) 2492374.
208) Máximo Baño y Jujuy, 1858 Numancia, BA. ☎ +54 (224) 73315.
209) Robertson 1247, 1838 Luis Gillón, BA. ☎ +54 (1) 2810901.
210) Fray Mamerto Esquiú 1161, 1824 Lanús, BA. ☎ +54 (1) 2251129.
211) Zárate 5942, 1757 Gregorio de Laferrere, BA. ☎ +54 (1) 6161620.
212) Pringles 2576, 1824 Lanús. ☎ +54 (1) 2479155.
213) 1678 Caseros, BA.
214) Echeverría 1435, 1718 San Antonio de Padua, BA. ☎ +54 (20) 51966.
215) Av. Rivadavia 10561, 3ᵇ piso, 1408 Buenos Aires. ☎ +54 (1) 9999903, 9999090 – **FM:** 90.3 "Eco".
216) Eetchegoyen 579, 1852 Burzaco, BA. ☎ +54 (1) 2999556.
217) Gral Pinto 1932, 1646 San Fernando, BA. ☎ +54 (1) 774 7228.
218) Las Piedras 1544, 1824 Lanús, BA. ☎ +54 (1) 2473890.
220) E. Santamarina 2642, 1842 El Jagüel, BA. ☎ +54 (1) 2320704.
221) Saavedra Lamas y Trinidad, 1635 Presidente Derque (Pdo. de Pilar), BA. ☎ +54 (320) 51665.
222) Florencio Sánchez 119 esq. Payró, Barrio Los Alamos, 1854 Longchamps, BA. ☎ +54 (1) 2331323.
223) 1875 Wilde, BA. ☎ +54 (1) 2073309.
224) Ibarrola 6026, Barrio San Alberto, 1765 isidro Casanova, BA. ☎ +54 (1) 6942169, 6940299.
225) Mármol 1590, 1661 Bella Vista, BA. ☎ +54 (1) 6663623.
226) Cabrera 352, 1828 Bánfield, BA. ☎ +54 (1) 2880759.
227) 1657 Pablo Modestá, BA. ☎ +54 91) 7394603.
228) San Nicolás 4356, 1874 Villa Domínico, BA. ☎ +54 (1) 2308826.
229) Barrio Siglo XX, 1838 Luis Guillón, BA.
230) Calle 37 No. 221, 1990 La Plata, BA. ☎ +54 (21) 823014.
231) Domingo Purita 2247, 1824 Lanús. ☎ +54 (1) 2400435. **FM:** 88.1 RL88 R. Lanús FM.
232) 1754 San Justo, BA.
233) 1824 Lanús, BA. ☎ +54 (1) 2286189.
234) Traful 3836, 1437 Buenos Aires. ☎ +54 (1) 9120497.
235) 1678 Caseros, BA. ☎ +54 (1) 7502613.
236) Gral. Paz 346, Barrio Los Vascos, 1722 Merlo, BA.
237) 1716 Libertad (Pdo. de Merlo), BA.
238) Barrio San José, 1834 Temperley, BA. ☎ +54 (1) 2647587.
239) Av. Mosconi 634, 1752 Lomas de Mirador, BA. ☎ +54 (1) 4532097. 4532098.

240) 1875 Wilde, BA. ☎ +54 (1) 2172208.
241) Saráchaga 5929, 1765 Isidro Casanova. ☎ +54 (1) 6251925, 4856499. **FM:** 92.1.
242) 1686 Hurlingham, BA. ☎ +54 (1) 4525052.
243) Paunero 2919, 1678 Caseros, BA. ☎ +54 (1) 7347100, 7347101, 7347102. **FM:** 90.5.
244) 1856 Glew, BA.

FM in Buenos Aires: 30) 92.7 Frec. Municipal – 77) 94.3 Horizonte – 76) 95.1 City – 63) 95.9 Rock & Pop – 58) 96.7 FM Folklore y Tango Nacional – 79x) 97.5 Clásica – 98.3 FM News – 58) 99.1 R. Nacional – 41) 99.9 FM Cien – 64) 100.7 FM Feeling – 63) 101.3 "FM Top" – 96) 102.3 Aspen 102 – 16) 103.1 R. Uno – 105.1 FM del Redentor – 10) 105.5 FM Hit – 111) 106.3 La Rocka.
In the City area there are over 150 unlicensed LP st's, about 900 in the rest of the country.
FM in Mar del Plata: 88.7 Cristiana – 89.5 Laser – 90.1 Stereo Rey – 91.7 Radioactiva – 92.3 Victoria – 92.7 Lider – 81) 93.3 – 94.1 FM del Mar – 94.5 Latina – 94.9 En Compañia – 95.1 Paraíso – 95.9 Uno – 96.5 Presidencia – 97.1 Hit – 97.5 Inolvidable – 98.1 Arena – 98.5 Brisas – 98.9 Rock'n Pop – 99.5 Stop – 99.9 Jovén Imaginación – 100.7 Del Sol – 101.1 Master L.C.5 – 101.5 Cosmos – 102.1 Bristol – 102.3 Oceano – 23) 103.3 Universo – 103.9 FM 103.9 (Bahía) – 104.5 FM 104.5 – 104.9 LV Amiga – 105.3 Encuentro – 106.3 Panamericana – 106.7 Sur.

BOLIVIA

L.T: UTC-4h – **Pop:** 7.680.000 – **Radios:** 4.250.000 – **Pr.L:** Spanish, Quechua, Aymará – **E.C:** 50Hz, La Paz 110/220V, Santa Cruz 220/380V – **ITU:** BOL.

DIRECCION GENERAL DE TELECOMUNICACIONES
✉ Av. Mariscal Santa Cruz, Edif. Palacio de Comunicaciones, P. 14, (Cas. 4475), La Paz. ☎ +591 (2) 36878. **Cable:** Gentel. ① 2595 Gentel BV – **L.P:** DG: Dr. Héctor Guzmán Hinojosa. Dir. de R. y TV: Dr. Ramón Oliden Ortuño.

CAMARA NACIONAL DE MEDIOS DE COMUNICACION
✉ Casilla 2431, La Paz.

Mediumwaves: Þ = also on shortwaves, * = inactive,.v = varying freq.

Call	kHz	kW	Name, location and h. of tr.
4) CP-	550	15	R. El Mundo, La Paz
1) CP91	Þ580	25	R. Panamericana, La Paz: 0930-0400 (Sun 1100-2400)
2) CP190	600	8	R. ACLO, Sucre: 0930-1230 (Sat 1430), 1800-0200, Sun 1000-1500
3) CP63	Þ620	20	R. San Gabriel, La Paz: 0900-1800, 2000-0200
5) CP204	640	10	R. Tarija, Tarija: 0900-1345, 1600-0140, Sun 1100-1600
17) CP263	Þ650	1	R. Dif. Integración, El Alto: 0900-0300, Sun 1100-0200
50) CP-	Þ660	1	R. ABC, Santa Cruz: 0915-0500
114) CP274	680	5	Radiodifusora Cristal, La Paz: 1100-0300 (Sun 1200-0100)
139) CP50	710	10	R. Pío XII, Siglo Veinte: 0900-0200 (Sun 1030-2300)
8) CP148	720	2.5	R. Yungas, Chulumani: 0900-1700, 2000-0100
9) CP165	*730	3	R. Mensaje, Montero
10) CP27	Þ730	10	R. La Cruz del Sur, La Paz: 0930-0230, (Sun 1400)
11) CP29	Þ760	18	R. Fides, La Paz: 1045 (Sun 1100)-0300
12) CP116	Þ770	10	R. Cosmos, Cochabamba: 0900-0200
13) CP265	Þ790	5	R. Libertad, La Paz: 1000-0200, (Sun 2400)
14) CP157	800	0.25	R. Santa Clara, Sorata: 0900-2400
16) CP35	Þ820	10	Radiodifusoras Altiplano, La Paz: 1000-0400 (Sun 1100-0300)
18) CP126	Þ840	3	R. Em. Juan XXIII, San Ignacio de Velasco: 0930-2400
19) CP160	850	1	R. 21 de Diciembre, Mina Catavi: 1000-0100
20) CP210	850	5	R. María Auxiliadora, Montero: 0900-0300 (Sun 1200-2400)

Call	kHz	kW	Name, location and h. of tr.
21) CP8	Þ860	10	R. Nueva América, La Paz: 1000-0300 (Sun 0130)
22) CP185	*Þ860		R. Paitití, Guayaramerín: 1030-0300
187) CP-	875		R. Eucaliptos, Eucaliptos
23) CP20	890	5	R. El Cóndor, La Paz: 0900-2400 (Sun 0930-2300) (n. 900)
24) CP79	900	0.25	R. Em. LV Nacional, Tarija: 1100-2300
49) CP83	Þ900	1.5	R. Norte, Montero: 0900-0200 (n. 990)
188) CP-	900	1	R. Central Misionera, Cochabamba 0900-0100
25) CP145	Þ930	10	R. Carlos Palenque, La Paz: 0900-0500 (Sat 0600, Sun 0030)
26) CP-	940.1	0.8	R. San Lorenzo, Colcapirhua: 0950-0200 (n. 960)
7) CP-	950	3	R. Yurac Molino, Chimboata
27) CP93	960	1	R. Kollasuyo, Potosí: 0930-2400 (Sun 2200)
28) CP30	Þ970	10	R. Santa Cruz, Santa Cruz: 0900-0100
29) CP118	980	2.5	R. Mar AM, La Paz: 1000-0100
30) CP192	990	2.5/5	R. Esperanza, Aiquile: 0930-0130
31) CP119	Þ1000	3	R. Dif. Trópico, Trinidad: 1300-0030/0100
32) CP28	1000	2.5/ 0.1	R. Cochabamba, "CBA": 1000-2200 (Sat. 2400)
189) CP	1000	1	R. Pirari, Santa Cruz
34) CP220	1000	5	R. Bahá'í de Bolivia, Caracollo: 0800-0100
35) CP4	Þ1020	10	R. Illimani, "Em. del Estado Boliviano", La Paz:1000 (Sun 1100)-0400
37) CP113	1040	1	R. Villazón, Villazón: 1100-0200(Sun 2200)
116) CP208	1045.2	1	R. Sipe Sipe, Quillacollo: 0900-0200 (n. 1045)
38) CP233	Þ1050	5	R. El Mundo, Santa Cruz: 1000-0200
15) CP-	1050	1.5	R. Noticias, Oruro: 1100-0600
190) CP-	1050	1	R. La Cumbre, Tiquipaya
40) CP181	1060	0.5	R. LV de la Frontera, Pto. Suárez: 0900-0300 (Sat 0230, Sun 2300)
191) CP57	1060	1	R. Eco 2000, La Paz: 1100-0100 (Sun 2400)
36) CP173	1075	0.5	R. Agricultura, Portachuelo: 1000-0400 (n. 1030)
43) CP291	1080	1	R. Dif. Colosal, Sucre: 0900-0300 (n. 1070)
45) CP45	1090	3	R. Cultura, Cochabamba: 1100-0100 (Sat 2200, Sun 1800)
48) CP55	1100	1/ 0.75	R. Universidad de Oruro: 1100(Sun 1200)-2300
6) CP137	1100	4	R. Mundial, La Paz: 0900-0330
46) CP184	1120	1	R. Estación El Dorado, Trinidad: 1000-0100
33) CP-	1125	0.3	R. Em. Cooperativa Poopó, Poopó
44) CP-	1125	0.5	R. Cruceña, Cotoca: 1130-2200 (n. 1530)
192) CP-	1125		R. Porvenir, Tiquipaya
193) CP-	1125		R. Unión, Vinto
51) CP-	1135	1	Kanata Radiodifusión, Cochabamba: 1000-0200 (n. 1125)
194) CP-	1140	1	R. Líder 1140, Tiquipaya
184) CP-	Þ1143	1	R. Colonia, Yapacani: 0900-0200
52) CP19	1145	1	R. Chuquiago Musical, La Paz: 0800-0200 (n. 1140)
195)	1150		R. 24 de Noviembre, Eucaliptos
47) CP194	1150	0.2	R. Chaco, "LV del Campesino", Yacuíba: 1000-2400, Sun 1100-1600 (n. 1100)
53) CP71	1150	0.5	R. El Cóndor, Oruro: 1100-2030
55) CP25	Þ1160	5	R. Centenario, "La Nueva", Santa Cruz: 0930-0230
56) CP78	1160	3/1	R. RTC, "La Deportiva", Cochabamba: 1030-2400
57) CP98	1170	1	R. Nuevo Mundo, Sucre: 1030-0300 (Sun 1100-0200)
39) CP-	1180	5	R. Central, Oruro: 0900-2400
58) CP235	1180	1	R. Emisora Ingavi, Viacha: 1000-0200 (Sun 1100-2400)
196) CP-	1180	1	Radioemisora 20 de Septiembre, Arbieto: 1100-0200

Call	kHz	kW	Name, location and h. of tr.
59) CP108	1195	1	R. Independencia, Quillacollo: 1100-0200
60) CP32	1200	5	R. Oriental, Santa Cruz: 0930-0200
62) CP171	1200	0.25	R. 24 de Noviembre, Arani: 1030-0400
63) CP67	1200	1	R. Splendid, La Paz: 1000-0200 (n. 1220)
64) CP162	Þ1220	1	R. Batallón Topáter, Oruro: 1000-0200 (Sat 1000-2300, Sun 1100-2300)
65) CP-	r1220		R. El Cóndor, Arque
197) CP-	1220		R. F 1220, La Paz
66) CP53	1235	1	R. Indoamérica, Potosí: 1030-2400 (n. 1250)
67) CP180	1240	1	R. San Miguel, Arani: 1000-0300
68) CP54	Þ1250	2.5	R. La Plata, "LV de la Capital", Sucre: 1000-0230 (Sun 2200)
69) CP17	Þ1250	0.5	R. Sararenda, Camiri: 1000-0400
71) CP26	1250	1	R. Amboró, Santa Cruz: 0845-0200 (Sun 1000-2400)
72) CP69	Þ1250	5	R. Nacional, Cochabamba: 0900-0100 (Sun 1100-2300)
73) CP14	1250	2	Emisoras Unidas, La Paz: 1000-1930 (SS 2200)
74) CP65	1250	0.4	R. Oruro, Oruro: 1300-0400
76) CP47	Þ1250	0.1	R. Frontera, Cobija: 1000-1800
77) CP-	1265	0.4	R. Uncía, Uncía: 1100-0300
78) CP134	1270	1	R. Vanguardia, Colquiri: 1000-0300
70) CP16	Þ1270	2	R. Los Andes, Tarija: 1000-2200
79) CP187	1275	0.5	R. Chané, Mineros: 1000-0400 (SS 1100-2200)
81) CP212	Þ1290	1	Radiodifusoras Minería, Oruro: 1000-2300
82) CP51	Þ1300	2.5	R. Loyola, Sucre: 1000-0200 (Sun 1100-2230)
83) CP127	*1300	0.15	R. Juan XXIII, Uyuni
84) CP82	1300	1	Coop. Radial Electra, Potosí: 1000-0100 (Sun 1100-0500)
85) CP168	1300	0.3	R. Chichas, Siete Suyos: 1100-0400
86) CP-	1300	1	R. Coronel Eduardo Avaroa, Santa Cruz: 0930-0230
87) CP68	1310	10	R. San Rafael, Cochabamba: 0900 (Sun 1000)-0300
54) CP176	1330	3/3.5	R. América, Oruro: 0930-2400
80) CP112	Þ1330	1	R. Frontera, Yacuíba: 0930-0330
97) CP-	1330	15/6	R. Sol, "Poder de Diós", La Paz: 1000-0200
90) CP24	1340v	1	R. Grigotá, Santa Cruz: 1000 (Sun 1100)-0100v
177) CP-	Þ1340	0.5	R. 11 de Octubre, Cobija: 1200-2200 (SS 1800)
91) CP146	1340	0.35	R. San Francisco, Apolo
198) CP-	1340		Radiodifusora Copacabana, Copacabana
92) CP214	1350	5	R. Ichilo, Yapacaní, San Carlos: 0930-0135
93) CP-	1350	1	América Radiodifusión, Sucre: 1000-2300, 0230-0600 (Sun 1100-2200)
94) CP154	1355v	0.25	R. Armonía, Cliza: 0900-0400v (n. 1350)
95) CP143	1360	1	R. Libertad, Villazón: 1100-0300v (Sun 2200)
41) CP270	1360	5	Difusoras Jiménez, El Alto: 0900-2400 (Sun 1100-1900)
199) CP-	1360		R. Stentor, La Paz
96) CP158	1370	0.5	R. LV de Minero, Siglo XX: 0900-1700, 2200-2300
75) CP288	1370	5/3	R. Intergración, Oruro: 1000-0300 (Sun 1100-1900)
98) CP133	1370	1	R. Agricultura, Achacachi: 1000-2400
99) CP186	1370	0.15	R. Libertad, Cliza: 0930-0300
100) CP221	1380	0.25	R. 16 de Noviembre, Sacaba: 1000-2300
101) CP227	Þ1380	0.5	R. Luis de Fuentes, Tarija: 0930-0400
102) CP3	Þ1390	5	R. Nacional de Bolivia, La Paz: 1000-0400 (Sun 1100-2300)
103) CP169	1390	0.25	R. LV Minera del Sud, Mina Telamayu: 1100-0300
104) CP174	*1400	1	R. Libertador, Santa Cruz
200) CP-	1400		R. Comunidad, Patacamaya
201) CP-	*1400		R. Nobel, La Paz
105) CP124	1410	0.25	R. Atlántida, Oruro: 1100-2400
106) CP-	1410	0.25	R. Roboré, Roboré: 1000 (Sun 1100)-0400
89) CP254	1420	1.5	R. Guadalquivir, Tarija: W 0930-2300, Sun 1100-1700
121) CP-	1420	1	R. Chaka, Pucarani: 0900-1330, 2030-0130
108) CP49	1420	1	R. Centro, Cochabamba: 1000 (Sun 1100)-0400
109) CP141	1430	0.25	R. Nuestro Señor de Burgos, Mizque: 1000-0200
110) CP193	*1430	0.25	R. 23 de Marzo, Tupiza
178) CP-	1430	0.15	R. Centinela, Tupiza
111) CP61	Þ1440	1	R. Batallón Colorados, La Paz: 1100-2400
112) CP107	1440	5/3	R. Yaguarí, Vallegrande: 1000-0300
186) CP-	1440	0.5	R. Oriente, Camiri
113) CP-	1445	1	Super Broadcasting Alborada "SBA", Santa Cruz: 1000-0200 (n.1475)
88) CP62	Þ1450	1	R. Em. Bolivia, Oruro: 0900-0400 (Sun 1000-0300)
202) CP-	1450		R. Amanacer, Huari
203) CP-	1450		R. Amazonia, Cobija
204) CP-	1450		Radiodifusora Capinota, Capinota
51)	1455	2	R. Estebán Arce, Colcapirhua: 0900-0200
118) CP215	1470	0.25	R. CORDECH, Alcalá: 1100-1600, 1900-0200
119) CP-	r1475		R. Tiraque, Tiraque
120) CP262	1475	1	Emisoras Verde y Blanco, Santa Cruz: 1000-0700 (n. 1450)
115) CP-	1480	0.1	Patrimonio R. , Potosí: 1000-0100
122) CP135	1485	1	R. LV del Valle, Punata: 1000-0300 (n. 1500)
123) CP172	Þ1490	1	R. San José, San José, Oruro: 1100-2400 (Sun 2300)
124) CP196	1490	0.25	R. Moxos, San Ignacio de Moxos
125) CP198	1490	0.35	R. Pedro Domingo Murillo, Quime: 1000-1400
126) CP-	1490	0.25	R. Mairana, Mairana: 1100-0300
205) CP-	1490		R. Domingo Savio, Villa Independencia
117) CP152	1495	2.5	El Mundo Radiodifusión, Sacaba: 1000-0300
127) CP238	1500	1	R. Sagrado Corazón, Mineros: 0930-0100
129) CP102	Þ1510	0.25	R. 27 de Diciembre, Villamontes: 1000-0200 (Sun 1100-0100)
130) CP1	1510	5/1	R. Chuquisaca, El Alto: 1100-0200
131) CP179	1520	1	R. Petrolera, Santa Cruz: W 1200-0330, SS 1400-2300
132) CP207	1520	0.25	R. LV del Cobre, Corocoro: 1000-0400
42) CP-	1520.2		R. Melodía, Oruro (n. 1520)
133) CP111	Þ1530	1	R. Em. Ballivián, San Borja: 1100-0200
134) CP237	1530	1	R. Don Bosco, Kami: 0900-1330, 2100-0200
135) CP200	1530	0.25	R. Litoral, Llica: 1400-0300
185) CP-	1540	1	R. Chiwalaki, Vacas
136) CP191	1545	0.35	R. Mejillones, Tarata: 1900-0400
158) CP-	Þ1545		R. Emissoras Villamontes, Villamontes
137) CP115	1550	10	R. Caranavi, Caranavi: 0930-1800, 2200-0200
138) CP205	1550	1	R. Tamengo, Pto. Quijarro (n. 1495)
169) CP255	1560	1	R. Occidental, Oruro: 0930-2400
140) CP-	1560	0.5	1Þ de Octubre "la radio," Capinota: 1030-0100
141) CP132	1568.9	10	R. Continental, La Paz: 1100-0300 (n. 1570)
142) CP-	1570	0.5	R. 1Þ de Mayo, 1Þ de Mayo: 1000-0200 (Sun 1800)
143) CP-	1570	0.3	R. Urkupiña, Quillacollo
173) CP-	1580	1	R. Andrés Ibáñez, Santa Cruz: 1000-0300
107) CP155	1590	3	R. Bermejo, Bermejo: 0900-0030
128) CP-	1590	0.5	R. Producciones Pusisuyu, Oruro: W 1000-1300, 0000-0400, Sun 1200-1300
206) CP-	1590		R. Globo, La Guardia
145) CP153	1600	0.5	R. Continental, Punata: 1000-0200

Shortwaves: * = inactive, v = varying freq.

Call	kHz	kW	Name, location and h. of tr.
146) CP-	*3200.4	0.5	R. 9 de Abril, Pulacayo
129) CP103	3350.2	0.1	R. 27 de Diciembre, Villamontes: 2230-0200v (n. 3350)
150) CP209	*3370.5	1.2	R. Florida, Samaipata
151) CP167	*3380.3	1	R. Cumbre, Tazna
152) CP175	3390.4	1	R. Emisoras Camargo, Camargo: 2300-0200
207) CP-	3420		R. Melodía, Bermejo
153) CP-	v3475	0.5	R. Padilla, Padilla: 2200-0200 (v. up to 3479)
177) CP-	v4183.2	0.25	R. 11 de Octubre, Cobija: 0900-1200, 2200-0200
208) CP-	4297		Radioemisora Orion, Huaca
155) CP-	4409.3	0.5	R. Eco, Reyes: 2230-0200
156) CP-	*4420.7	1	R. Santa Rosa, Santa Rosa del Yacuma
209) CP-	4422		Radioemisora Reyes, Reyes
76) CP59	v4449.8	0.25	R. Estación Frontera, Cobija: 2100-0100v (n. 4730)
157) CP142	v4472	1	R. Movima, Santa Ana del Yacuma: 1030-1600, 2100-0200v (n. 4835) (v4471-4472.9)
162) CP-	*4508.7		Radioemisora San Joaquín, San Joaquín
160) CP-	4530.3	0.12	R.Hitachi, Guayaramerín: 1100-1600, 2200-2400v (n. 4530)
31) CP120	4549.4	3	Radiodifusoras Trópico, Trinidad: 1030-1200, 1600-1830, 2230-0100v
31) CP120	4552.3	1	Radiodifusoras Trópico, Trinidad (alt. freq)
158) CP-	4599.3	1	Radio Emisoras Villamontes: 0900-1730, 1900-2400 (Sun 0900-1700)
159) CP-	4600	0.2	R. Perla del Acre, Cobija: W 0900-1100, 2130-0300
165) CP-	*4632.4	1	R. 11 de Octubre, Cobija
61) CP89	4649.1	1	R. Santa Ana, Santa Ana del Yacuma: 1055-1700, 1930-0330 (n. 4805)
22) CP185	v4681.8	1/0.2	R. Paitití (Paitití Radiodifusión), Guayaramerín: 1030-1900, 2100-0200 (alt. freq. 4682.3, n. 4865)
154) CP-	v4702.3		R. Eco, San Borja: W 1100-0300 (n. 4700)
161) CP136	v4719.9	0.5	R. Abaroa, Riberalta: 1000-0400 (Sun 0100) (n. 4760)
183) CP-	*4732.2	0.7	R. La Palabra, Santa Ana del Yacuma
88) CP-	*4756		R. Em. Bolivia, Oruro
70) CP84	*4775.2	3	R. Los Andes, Tarija
210) CP-	4777.7		R. A.N.D.E.S, Uyuni: 0930-0005
133) CP152	4785	1	R. Em. Ballivián, San Borja: 1100-0200
21) CP73	*4795.9	10	R. Nueva América, La Paz
102) CP144	*4815	1	R. Nacional de Bolivia, La Paz
90) CP70	*4830	1	R. Grigotá, Santa Cruz: 1700-1900
11) CP72	4845.0	5	R. Fides, La Paz: 1000 (Sun 1100)-1805, 2115-0200 (Sun 0600)
55) CP66	4855	5	R. Centenario "La Nueva",Santa Cruz: 0930-0230
167) CP-	v4864.5	1	R. Em. 16 de Marzo, Mina Bolívar: 1000-0200 (SS 2300)
10) CP75	4875	10	R. La Cruz del Sur, La Paz: 0930-1300, 2130-0030 (Sun 0930-1400)
69) CP77	4885.6	1	R. Sararenda, Camiri: W 1000-1200, 1800-2100
168) CP-	4900.7	0.75	R. Em. San Ignacio, San Ignacio de Moxos: 1050-1200, 1600-2000
147) CP114	v4925	5	R. San Miguel, Riberalta: 1000-0300 (Sun 0200)
49) CP110	4939.2	1.5	R. Norte, Montero: 0900-1300, 1600-1800, 2200-0100 (n.4935)
35) CP7	4945	10	R. Illimani, "Em. del Estado Boliviano", La Paz: 1000 (Sun 1100)-0400
18) CP90	4965	3	R. Em. Juan XXIII, San Ignacio de Velasco: W 0930-2400, Sun 1100-1700
64) CP-	4980	1	R. Batallón Topater. 1000-0200

Call	kHz	kW	Name, location and h. of tr.
170) CP163	4990.9	1	R. Animas, Animas: 1000-1800, (Sun 1100-1300)
13) CP265	5005	5	R. Libertad, La Paz: irr.
16) CP-	*5044.8	5	Radiodifusoras Altiplano, La Paz
172) CP218	5153.4	0.2	R. Galaxia, Guayaramerín: 1100-0200 (SS 1800) (n.5160)
211) CP-	5500		R. Luz del Oriente, Santa Cruz
175) CP109	5503.7	0.2	R. Em. 2 de Febrero, Rurrenabaque: 1100-0300 (n. 5020)
176) CP-	5580.2	0.25	R. San José, San José de Chiquitos: 1100-1700, 2100-0200v
81) CP213	5927.1	1	Radiodifusoras Minería, Oruro: 1000-0030 (n. 5925)
163) CP-	*5929.5		Radiodifusoras Amazonia "LV del Trópico", Las Romas de San Lorenzo San Antonio
139) CP60	5948	5	R. Pío XII, Siglo Veinte: 1000-
139) CP60	v5952.3	5	R. Pío XII, Siglo Veinte: 1000(Sun 1100)-1400, 2100-0200 (Sun 0100) (n. 5955)
179) CP177	5964.8	1	R. Nacional de Huanuni, Huanuni: 0900/1000-0200 (Sun 1100-2300)
72) CP200	*5974.2	1	R. Nacional, Cochabamba
123) CP222	*5985	2.5	R. San José, San José, Oruro
82) CP41	5996.4	1	R. Loyola, Sucre: 1000-0200 (n. 5955)
38) CP234	6014.8	10	R. El Mundo, Santa Cruz: 1000-0200 (n. 6015)
35) CP5	6025	10	R. Illimani, "Em. del Estado Boliviano", La Paz: 1000 (Sun 1100)-0400
50) CP-	*6030	1	R. ABC, Santa Cruz
13) CP266	*6045	1	R. Libertad, La Paz
3) CP229	6085	5	R. San Gabriel, La Paz: 0850-0200 (n. 6080)
12) CP216	*6095	10	R. Cosmos, Cochabamba
1) CP92	6105.3v	10	R. Panamericana, La Paz: 1000-0400
17) CP282	*6119.3	5	R. Dif. Integración, El Alto (n.6120)
28) CP32	6135	10	R. Santa Cruz, Santa Cruz: 0900-0100
101) CP81	*6140.7	1	R. Luis de Fuentes, Tarija
171) CP-	6142	1	R. Mauro Nuñez, Villa Serrano: 1000-0230 (n.6065)
11) CP12	6155	10	R. Fides, La Paz: W 1045-0300, Sun 1100-0600
111) CP-	*6185	3	R. Batallón Colorados, La Paz
25) CP	6195	5	R. Carlos Palenque, La Paz: 0900-2100
212) CP-	6325		R. Velmar, Dalence
184) CP-	v6557	0.3	R. Colonia, Yapacani: 2100-0200 (n. 6555)
11) CP	9625	15	R. Fides, La Paz
68) CP21	9717.1	1	R. La Plata "LV de la Capital," Sucre: 1400-1900 (Sun 2200) (n. 9715)

Addresses and other information

ERBOL (Educación Radiofónica de Bolivia), Calle Ballivián 1323, 4ᵖ piso (⌧ Cas. 5946), La Paz. ☎ +591(2) 354142, 324768. 🖷 +591 (2) 391985 — Pte: Jorge Trias S.J

Affiliated st's: 1), 2), 3), 5), 6), 8), 11), 14), 16), 18),20), 21), 25), 28), 29), 30), 35), 52), 92), 107), 111), 121), 134), 139), 141), 147), 185).

UNESBO (Unión de Emisoras Sindicales de Bolivia), Yanacocha 689, La Paz. ☎ +591 (1) 341881 — Pte: Jorge Bustillo Burgos.

Affiliated st's: 4), 19), 23), 53), 60), 65), 74), 78), 85), 88), 103), 123), 132), 141), 151), 165), 166), 167), 170), 179).

1) Av. 16 de Junio 1566, Of. 902 (⌧ Cas. 503), La Paz. ☎ +591 (2) 324606. 🖷 +591 (2) 325239 – DG: Daniel Sánchez R. PD: Ivan José Hidalgo. CE: Johnny Dueri. **N:** "El Panamericano" relayed by many st's.
2) Acción Cultural Loyola, Cas. 538, Sucre – DG: P.J. Javier Velazco. Dir: Prof. René Santillán. Prgr in Quechua.
3) Gral. Lanza 2001, Cas. 4792, La Paz. ☎ +591 (2) 414371. 🖷 +591 (2) 411174 – DG: Hno. Jaime Calderón M. Prgrs in **Aymara** exc. Sp. & Quechua 1400-1430, Sat 2100-2130.
4) 16 de Julio 1295, La Paz – DG: Carlos Olea A.
5) Obispado de Tarija, Cas. 1003, Tarija – Dir: Fernando Arandia. **FM:** 98.7.
6) Av. Sánchez Lima 2554, P. 2, La Paz – Dir: Jorge Gonzáles.
7) Chimboata (Cochabamba) – DG: Gualberto Villarroel.
8) Cas. 4535, La Paz – Dir: R. P. José Geldens. **Aymara:** 0930-1130, 2230-0030. **FM:** 104.0.
9) Igl. Metodista de Bolivia, Cas. 434, Santa Cruz – Dir: G. de Molina.

10) Nicaragua 1759, Cas. 1759, La Paz – DG: Pastor Rodolfo Moya Jiménez. **Aymara:** 0900-1110, 2200-0030, otherwise in Sp.

11) Jenaro Sanjinés 799, Cas. 9143, La Paz. ▤ +591 (2) 379030 – Dir: Eduardo Pérez Iribarne. **N:** "La hora del país", relayed by many st's, at 1100, 1630, 2230, 0130.

12) Av. de las Heroinas 467 (▢ Cas. 2925), Cochabamba. ☎ +591 (4) 250422, 250423. ▤ +591 (4) 251173 – Gte: Laureano Rojas. **Quechua:** 0930-1030, 0000-0200. **FM:** 95.1 "Fides".

13) Cas. 5324 or: Av. Sánchez Lima 2278, 3Þ piso (entre Fernando Guachalla y Rosendo Guttierez), La Paz. ☎ +591 (2) 361591. ▤ +591 (2) 363069 – Dir: Lic. Teresa Sanjinés Lora. **Aymara:** 0945-1015.

14) Cas. 2329, La Paz – Dir: Natalio Lucano.

15) Ayacucho 785 (Altos) esq. Presidente Montes (▢ Cas. 670), Oruro. ☎▤ +591 (52) 53534 – Dir: Jorge Lazzo Q. **Aymara & Quechua:** 1030, 2130, 0200.

16) Galería Heriba, Evaristo Valle 140, or: Cas. 1081, La Paz – Dir: Eduardo Ibáñez.

17) Cas. 1722, La Paz – Dir: Benjamin J.C. Blanco Q.

18) Vicariato Apostólico, San Ignacio de Velasco (Santa Cruz). DG: Pbro. Elías Cortezón R. **FM:** 100MHz. **F.PI:** MW to be suspended.

19) Plaza 6 de Agosto, Camamento Mina Catavi (Potosí) – Dir: José Carlos Guzmán Pacheco. **FM:** 105.7.

20) Potosí entre Rafael Terrazas y G. Busch, Barrio Floresta, Montero or Cas. 507, Santa Cruz – DG: Lorenzo Camporese SDB. **FM:** 105.7 "Concierto".

21) Calle Abdón Saavedra 1990 (or Cas. 8780), La Paz - DG: Yerco Mereán.

22) Cas. 172, Guayaramerín (Beni) – Own: Ancir Vaca Cuéllar. **FM:** 100.1.

23) Yanacocha 689, La Paz – Dir: Jorge Bustillo Burgos.

24) Cas. 404, Tarija – Dir: J. Jaime Fernández Mogro.

25) Juan de la Riva 1527 (▢ Cas. 8704), La Paz. ☎ +591 (2) 363745. ▤ +591 (2) 376785 – DG: Adolfo Paco. **Aymara:** 0900-1000.

26) Cajón 21, Cochabamba – Dir: J. Valverde Leiza.

27) Cobija 19, Potosí – Gte: Angel Moscoso M.

28) Calle Mario Flores esquina Güenda or Cas. 672, Santa Cruz. ☎ +591 (3) 531817. ▤ +591 (3) 532257 – DG: P. Francisco Flores S.I. **Guaraní:** 1830-1900. **FM:** 92.1 "Stereo 92".

29) Calle Jenaro Sanjinés 799, La Paz - Dir: Juan León Cornejo.

30) Loa s/n, Aiquile, or: Cas. 5736, Cochabamba – DG: P. Floriano Weiss. Prgr. in Quechua exc. Spanish: 1500-2200. **FM:** 100.1MHz.

31) Cas. 60, Trinidad (Beni) – Dir: Eduardo Avila Alberdi.

32) 25 de Mayo 214 esq. Bolívar (▢ Cas. 5500), Cochabamba. ☎ +591 (42) 51504. ▤ +591 4(2) 51561 – Dir: Marcial Moreira. **FM:** 104.1 "Gaviota".

33) Cooperativa Minera Poopó, F. Fontanilla y Oblitas, Oruro. ☎ +591 (511) 2113 – DG: Carlos Zenteno. **Quechua & Aymara:** 0900-1100.

34) Cas. 1019, Oruro. ☎ +591 (511) 2259 – Dir: Augusto E. Costas M. **Aymara & Quechua:** 0830-1400, 2200-0030.

35) Av. Camacho 1465, P. 6, (▢ Cas. 1042), La Paz – DG: Benedicto Hurtado C. Dir: Rubén D. Choque V.

36) Calle Warnes s/n, Portacuelo, Provincia Sarah (Santa Cruz) - Dir: Yavier Yabeta J.

37) Cas. 58, Villazón (Potosí) – Dir: G. Cocca A.

38) Cas. 1984, Santa Cruz – DG: David Terceros Banzer. Dir: Lic. José Luiz Vélez Ocampo.

39) Montesinos 436 entre 6 de Octubre y Potosí, Oruro - **Quechua & Aymara:** 0900-1230.**FM:** 98.7.

40) Cas. 18, Pto. Suárez (Sta. Cruz) – Dir: Leonardo Arteaga.

41) Av. Panamericana 93, Cd. Satélite, El Alto (La Paz) (or Cas. 6412, La Paz) – Dir: Arturo Jiménez Rocha.

42) Oruro.

43) Cas. 335, Sucre – Dir: Prof. Moisés Torres R. **FM:** 90.7.

44) Victoriano Gutiérrez 200, Cotoca (Santa Cruz).

45) Santiváñez 176 entre Junín y Ayacucho (▢ Cas. 719), Cochabamba – DG: Jorge Serrano. **Aymara & Quechua:** 1130-1300.

46) 18 de Noviembre 628 (▢ Cas. 720), Trinidad – Dir: Elizabeth Durán de Arias.

47) Inst. Politécnico Campesino, Cas. 42, Yacuíba (Tarija) – Dir: Jorge Arias Soto.

48) Calle Cochabamba esquina 6 de Octubre (▢ Cas. 49), Oruro. ☎ +591 (52) 50004. ▤ +591 (52) 42215 – DG: Omar Barro Velasquez. PD: Juan Mamani. CE: Tec. Loenardo Choque T. **F.PI:** FM stereo.

49) Warnes 195, Altos Cine Escorpio, Montero (Santa Cruz). ☎▤ +591 (92) 20970 – DG: Leonardo Arteaga Ríos. **FM:** 99.1.

50) Warnes 334 (▢ Cas. 629), Santa Cruz. ☎ +591 (3) 363990. ▤ +591 93) 369087- DG: María Eugenia Landivar. **FM:** 92.7.

51) Colcapirhua (Cbba.) or Cas. 3475 Cochabamba - Own: Silvestre Vallejos Gutierrez.

52) Calle Nueva York 140 Pasaje (or Cas. 8084), La Paz – Dir: Hugo A. Aspiazu

53) Junín 508, P.2, Oruro – Dir: Florentino Rocabado C.

54) Calle Cochabamba 998 y Camacho (▢ Cas. 41), Oruro. ☎ +591 (52) 54255, 54254. ▤ +591 (52) 40035 – DG: Eduardo Veneros Suarez. PD: Jorge Barrientos. CE: leonardo Choque Tapia. **F.PI:** FM.

55) Cas. 818, Santa Cruz. ☎ +591 (3) 529625. ▤ +591 (3) 524747 – DG: Julio Acosta Campos. Quechua & Aymara: 0930-1000, 0200-0230. **FM:** 90.7 "R. Super Color".

56) Lanza esq. Ecuador 261 (▢ Cas. 846), Cochabamba. ☎ +591 (42) 57289. ▤ +591 (42) 41414 – Dir: Carlos Dalence Loayza.

57) Cas. 25, Sucre – Dir: Víctor Hugo Hevia R.

58) Murillo 340, Viacha, Provincia Ingavi (La Paz) – Dir: Mario Mamani.

59) Cochabamba esq. Heroes del Chaco, Quillacollo (Cbba.) or: Cas. 108, Cochabamba. – Dir: René Rojas B.

60) Independencia 372 (▢ Cas. 186), Santa Cruz. ☎ +591 (3) 337194. ▤ +591 (3) 335778 – DG: Arturo Mendivil. **FM:** 96.3.

61) Calle Sucre 250, Santa Ana de Yacuma – Dir: Delicia Guardia Lima.

62) Arani (Cochabamba) – Dir: Oscar J. López A.

63) Cas. 1539, La Paz – Dir: Guido Velasco A. Prgr in Aymara.

64) Calle Junín y 6 de Agosto, Oruro. ☎ +591 (52) 60200, 60462 – DG: Tcnl. Castelo Giácoman P. **Aymara & Quechua:** 1000-1100. **FM:** 98.3.

65) Arque (Cochabamba)

66) Cas. 572, Potosí – Dir: Agustín Vera Duarte.

67) Plaza Progreso 20, Arani (Cbba.) – Dir: Wilfredo Rojas.

68) Abaroa 422, Cas. 276, Sucre. ☎ +591 (64) 23231 – Dir: Gregorio Donoso Daez. **FM:** 92.1.

69) Cas. 7, Camiri (Sta. Cruz) – Dir: Luís Domingo Eyzaguirre.

70) Cas. 344, Tarija – Dir: Jaime Rollano Monje. **FM:** 103.0.

71) Cas. 697, Santa Cruz – Dir: Hugo R. Tarabini.

72) Av. San Martín 342 (▢ Cas. 4274), Cochabamba – Dir: Mario Godoy.

73) Calle Tumusla 765, La Paz – Gte: Carlos M. Guzmán D.

74) Ayacucho y La Plata, Oruro – Dir: A. Quezada.

75) Av. 6 de Octubre 1042 y Montecinos (▢ Cas. 845), Oruro. ☎ +591 (52) 54143. ▤ +591 (52) 30645 – DG: Dr. Alfredo Luján M. **FM:** 97.1.

76) Cas. 179, Cobija (Pando) – Dir: L. Miahuchi von Ancken.

77) Cas. 15, Uncía (Potosí) – Dir: Roberto Velásquez.

78) Cas. 154, Colquiri (La Paz) – Dir: Alfredo Murillo M.

79) Mineros (Santa Cruz) – Dir: Manuel Emilio Peña.

80) Cas. 24, Yacuíba (Tarija) – Dir: Juan Castillo R.

81) Cas. 247, OrCo – Dir: Dr. José C. Gómez Espinoza. **Quechua & Aymara:** 1000-1100, 2200-2300. **FM:** 107.7.

82) Calle Ayacucho 161, Sucre. ☎ +591 (64) 53677, 54570. ▤ +591 (64) 42555 – DG: Lic. J.Weimar León G. **FM:** 98.3 "Onda Joven"

83) Cas. 28, Uyuni (Potosí) – Dir: Juan López Claros.

84) Cas 328, Potosí – Dir: J. Antonio González B.

85) Campamento Minero, Siete Suyos (Potosí) – Dir: C. Panama.

86) Av. Charcas 1051 lado octava División del Ejército, Santa Cruz. ☎ +591 (3) 360447. ▤ 591 (3) 372242 – DG: Tcnl. Mario Vasquez Zenteno. **FM:** 98.1.

87) Cas. 546, Cochabamba – DG: Salim Sauma. **Quechua/Aymara:** 0900-1100, 2300-0200. **FM:** 92.0.

88) Av. Velasco Galvarro 651, Oruro – DG: Juan de la Cruz. **Aymara:** 1000-1230. **FM:** 105.1.

89) Daniel Campos 824, Tarija – Dir: Roberto Vargas A.

90) Abaroa 223, Cas. 203, Santa Cruz. ☎ +591 (33) 22142 – Dir: Daniel Arteaga Farrell. **FM:** 90.1.

91) Apolo, Pcia. Franz Tamayo (La Paz).

92) Calle Calama s/n, Yapacani, Provincia Ichilo (Santa Cruz) - Dir: Oscar Enrique. **Quechua:** 0930-1100. **FM:** 101.1.

93) Eduardo Berdecio 568, Sucre – Dir: Jorge Poppé Avilés.

94) Cliza (Cochabamba) – Dir: Francisco Quiroz J.

95) Cas. 40, Villazón (Potosí) – Dir: Jaime Néstor Lima.

96) Campamento Minero Siglo XX (Potosí) – Dir: Luís Reyes Mercado.

97) Plaza del Estudiante 1905, La Paz – Dir: Dr. José Luis Paredes.

98) Achacani, Pcia. Omasuyos (La Paz) - Gte/Own: Castro Ordóñoz.

99) Calle Santa Cruz s/n, Cliza (Cochabamba) – Dir: Crescencio Escobar A.

100) Cas. 2522, Cochabamba – Dir: Juan José Camacho.

101) Cas. 125, Tarija – Own: Jimmy Borda. Dir: Carlos Pomarino Lora. **FM:** 93.1.

102) Isaac Tamayo 640, or: Cas. 2532, La Paz – C.E: Ing. Carlos Belmonte. Prgr. mainly in Aymara.

103) Campamento Minero, Telamuy (Potosí) – Dir: Raúl Chorocoro.

104) Arenales 630, Cas. 1333, Santa Cruz – Dir: Juan Flores Ortiz. **FM:** 107.7.

105) Linares 1160, Oruro – Dir: Ruben Garcia Zapata.

106) Roboré (Santa Cruz) – Gte: Ciro Rivero C.

107) Av. Barrientos esq. Ameller, Bermejo (Tarija). ☎▤ +591 (69) 61584 – Dir: P. José Rodriguez M.

108) Cas. 839, Cochabamba – Dir: V. Oropeza de los Llanos. **FM:** 96.0.

109) Cas. 893, Cochabamba – Dir: P. Mario Comi.

110) Cochabamba s/n, Tupiza (Potosí) – Dir: Mario Loza F.

111) Av. Saavedra esq. My. Zubieta Estado Mayor (Miraflores), La Paz.

☎ +591 (2) 372065– DG: Tcnl. Javier Cespedes A. **Aymara:** 1100-1200.
112) Florida esq. Montesclaros 143, Vallegrande (Sta. Cruz). ☎ +591 (942) 2033, 2034 – Dir: Prof. Edgar Bonilla M.
113) Mutualista y Las Garzas 284 (✉ Cas. 2024), Santa Cruz – Dir: Bismarck Gutiérrez.
114) Av. 16 de Julio 1490, or: Cas. 5303, La Paz – Dir: Mario Castro.
115) Victor Flores 410, Potosi – Dir: Luís Velasquez Pareja.
116) Plaza de Granos 44, Quillacollo (Cbba.) – Dir: Rodolfo Yucra F.
117) Plaza 6 de Agosto 44, Sacaba; Cas. 3230, Cochabamba. ☎ +591 (42) 86150 – Dir: Víctor Gutiérrez Iriarte. **English:** 1900-2230.
118) Cas. 156, Sucre – Dir: Jorge Crespo Avilés.
119) Calle Junín s/n, Tiraque (Cochabamba) – Dir: José Protasio Montaño.
120) Av. Virgen de Cotoca 131, Santa Cruz – Dir: Mario Ychasco A.
121) Casilla 204, Colegio Don Bosco, La Paz – Dir: Padre Esteban Bartolusso. Prgrs mainly in **Aymara**, but also in Spanish.
122) Cas. 1361, Cochabamba – Dir: Roberto Mendoza.
123) Caro 235 entre Pagador y Av. Velasco Galvarro, Oruro - Dir: Herán Soliz. Valdez.
124) San Ignacio de Moxos (Beni) – Dir: Mina Bauer de Suárez.
125) Quime, Inquisivi (La Paz) – Dir: Oscar Terrazas B.
126) Correo Central, Mairana, Provincia de Florida (Santa Cruz).
127) Cas. 507, Santa Cruz – DG: Luís Maistrello **FM:** 89.5. Prgrs also in Quechua.
128) Cas. 812, Oruro - Dir: Prof. Domingo Choque Quispe. Prgm. in **Aymara & Quecua.**
130) Cas. 3123, La Paz – Dir: Mercedes Camacho de Kuncar.
131) Barrio Petrolero 21 de Diciembre fte. Plaza Tarija, Santa Cruz – Dir: Nelson Bustos Q.
132) Correo Central, Corocoro, Provincia Pacajes (La Paz) – Dir: Tomas Tola.
133) Oruro s/n, San Borja (Beni). ☎▣ +591 (848) 3020 – Dir: Ignacio Viruez.
134) Congregación Salesiana, Kami (Cochabamba) or: Cas. 1151, Cochabamba. ☎▣ +591 (811) 9295 – DG: Roberto Ledezma. Prgrs in Sp, **Aymara & Quechua.**
135) Llica, Pcia. Daniel Campos (Potosí) – Dir: Jorge Nuayta C.
136) Tarata (Cochabamba) – Dir: Filiberto Rodríguez.
137) Liga de Oración en Misión Mundial, Cas. 266, La Paz or: Correo Central, Caranavi – Dir: Pablo Mikaelsen.
138) Puerto Quijarro, Provincia Angel Sandoval (Santa Cruz) – Dir: German Mendoza Arias.
139) Cas. 434, Oruro – Dir: Pbro. Roberto Durette OMI. ☎ +591 (52) 53163.
140) Calle Bolívar 13, Capinota (Cochabamba) – Dir: Roberto Gutierrez Merida.
141) Av. República 872, La Paz – Dir: Eduardo Calderón.
142) Calle 5 lado este, Villa 1Þ de Mayo (Santa Cruz) – Dir: Erlan Justiniano Fernandez.
143) Av. Blanco Galindo 800, Quillacollo (Cbba.) – Dir: J. Salazar.
145) Ayacucho 138, Punata (Cbba.) – Own: Orlando Grajeda.
146) Planta Industrial de Pulacayo, Pcia. Quijarro (Potosí) – Dir: Antonio Lafuente Azurduy.
147) Calle Fray Bernardino Ochoa 58 (✉ Cas. 102), Riberalta (Beni) – Dir: Gerin Pardo Molina. **FM:** 92.5 "Centenario".
150) Cas. 5901, Santa Cruz de la Sierra – Dir: Felipe Aponte. ▣ +591 (33) 60649.
151) Campamento Minero Tazna, Pcia. Nor Chichas (Potosí) – Dir's: R. Davila & J. Ricardi.
152) Cas. 09, Camargo, Pcia. Nor Cinti (Chuquisaca) – Dir: Pablo García Benito. **FM:** 100MHz.
153) Alcaldía Municipal, Padilla (Chuquisaca) – Dir: A. Paredes.
154) Av. Selim Majuli (✉ Correo Central), San Borja, Pcia. Ballivián (Beni) – Dir: Gonzalo Espinosa Cortez.
155) Reyes, Pcia. Ballivián (Beni) – DG: Gonzalo Espinoza Cortez.
156) C.C., Sta. Rosa del Yacuma (Beni) – Gte: Carlos Baeny.
157) Calle Baptista 24, Santa Ana del Yacuma (Beni) – Dir: Javier Roca Diaz.
158) Av. Méndez Arcos 157, Villamontes, Pcia. Gran Chaco (Tarija) –Dir: Gerardo Rocabado G.
159) Av. Prof. Alguira Gutierrez (✉ Cas. 209), Cobija (Pando) – Dir: Guadalupe M. de Vidal.
160) Sucre 320, Guayaramerín – Gte: Jéber Hitachi Banegas.
161) Calle Nicanor Gonzalo Salvatierra 249, Riberalta (Beni) – Dir: René Arias P. **FM:** 98.
162) La Paz y Av. Felipe Bezerra, San Joaquín, Mamoré (Beni). ☎ +591 (811) 9282 – DG: Ingeniero Asunta Figueroa.
163) Cas. 2494, Cochabamba – Dir: Oscar Ustáriz Aranda. **FM:** 100.1.
165) Cas. 200, Cobija (Pando).
167) Cas. 821, Zona Postal No. 5, Oruro – Dir: Víctor E. Flores Salas. **FM:** 102.0.

168) Ballivián s/n, San Ignacio de Moxos (Beni) – Dir: Carlos Salvatierra R.
169) Av. Bakovic 1027 entre Caro y Montecinos, or Cas. 705, Oruro – Dir: Olga Rojas de García. **Aymara & Quechua:** 0930-1030. **FM:** 93.1.
170) Dtto. Minero de Animas (Potosí) – Dir: Juan Aguanta.
171) CEDEC, Cas. 196, Sucre. ☎ +591 (64) 55008. ▣ +591 (64) 62628 - DG: Vladimir Gutierrez P. **FM:** 103.1.
172) Cas. 395, Guayaramerín (Beni) – Dir: Jeber Hitachi.
173) Av. del Micro UB-47, Plan 3000, Santa Cruz – Own: Wilson Caballero.
175) Vaca Diez 400, Rurrenabaque (Beni). ☎ +591 (832) 9999. ▣ +591 (832) 2205 – DG: Roger Arze Salmón.
176) Cas. 15, San José de Chiquitos (Santa Cruz) – Dir: Fabian Eugez Franco.
177) Cas. 200, Cobija (Pando) – Own: Carlos Arze C. **FM:** 101.0.
178) Cas. 180, Tupiza (Potosí) – Dir: Israel Jurado G.
179) Calle Sucre, Huanuni (✉ Cas. 681), Oruro – Dir: Rafael Linneo Morales.
183) Plaza Fr. Martín Baltasar de Espinosa, Santa Ana del Yacuma (Beni) – Dir: Yosú Arketa.
184) Calle Libertad 1 (✉ Correo Central), Yapacani, Provincia Ichilo (Santa Cruz). ☎▣ +591 (933) 6164 – Dir: Yrey Fausto Montano U. Prgrs in Sp, **Quechua, Aymara & Guaraní. FM:** 107.0.
185) Cas. 80, Cochabamba ✉ Vacas, Pcia. de Arani (Cochabamba). ☎▣ +591 (42) 55390 – Dir: Victor van Oeyen. Prgrs mainly in **Quechua**.
186) Cas. 30, Camiri (Santa Cruz) – DG: Ramón Severiche Gutiérrez
187) Eucaliptos, Provincia Tomas Baron (Oruro) – Dir: Wilfredo Cardenas Aguilar.
188) Cochabamba.
189) Cas. 1766, Santa Cruz – Dir: Walter Arzabe.
190) Calle Aroma, Tiquipaya, Provincia Quillacollo (Cochabamba) – Dir: Guillermo Barriga Chacón.
191) Plaza Alonso de Mendoza, 5Þ piso del Edificio Santa Anita (or Cas. 4973), La Paz.☎ +591 (2) 378904 - Dir: Walter Mur Bardales.
192) Tiquipaya, Provincia Quillacollo (Cochabamba).
193) Vinto (Cochabamba).
194) Villa Tiquipaya, Provincia Quillacollo (Cochabamba) – Dir: Mario Oropesa Rodriguez.
195) Eucaliptos, Provincia Tomas Baron (Oruro) – Dir: Gregorio Tellez.
196) Calle Sucre s/n, Arbieto, Provincia de Estaban Arce (Cochabamba) – Dir: Fidel Amurrio Rocha.
197) Licenciado Palacios 56, Zona 12 Octubre, El Alto, La Paz. ☎ +591 (2) 822717.
198) Calle Professor Pedro F. Mejía 6, Esquina Plaza Central de la Basílica Nacional Nuestra Senora de Copacabana, Copacabana, Provincia Manco Capac (La Paz).
199) Manco Capac 434, La Paz. ☎ +591 (2) 816602 – Dir: C. Chiappe.
200) Patacamaya (La Paz).
201) La Paz.
202) Huari (Oruro).
203) calle 9 de Octubre 2, Barrio Conavi, Cobija (Pando) – Dir: Juan Carlos Valdivia.
204) Calle Brasil 676, Capinota (Cochabamba) – Dir: Gerardo Cuevas Morales.
205) Villa Independencia, Provincia Ayopaya (Cochabamba) – Dir: Pacifico Peletti Gonella.
206) El Toro s/n, La Guardia, Provincia Andrés Ibañez (Santa Cruz) – Dir: Wilfredo Salvatierra Mercado.
207) Av. Bolívar 608, Bermejo (Tarija).
208) Huaca, Huaca Huamelies, Region Andrés Avelino Cáceres.
209) Reyes, Provincia de Ballivián (Beni) – Dir: Carmen C. de Vasquez.
210) Calle Final Uruguay s/n (✉ Cas. 16), Uyuni, Provincia Antonio Quijarro (Potosí). ☎ +591 ()693) 2145.
211) Mutualista y Las Garzas 284 (✉ Cas. 854), Santa Cruz – Dir: Bismark Guttierez.
212) Campamento Minero Santa Marta, Dalence (Oruro). **FM:** 103.1. **F.PI:** MW.

NB: Whenever listed, Casilla addresses should preferably be used for mailing purposes.

FM in La Paz: 25) 88.5 Doble 8 Latina – 89.1 FM 89 Punto Uno – 89.5 Salesiana – 16) 90.1 FM 90 – 90.5 Dimensión FM Estereo – 97) 91.1 Espirito Santo FM – 91.7 Compañera – 92.1 Estudio 92 – 92.7 Estelar – 141) 93.1 FM 93 – 35) 94.1 – 94.5 Perfección FM – 10) 95.1 – 95.5 Gigante – 52) 96.0 – 96.7 La Paz – 1) 97.1 – 97.6 Digital FM – 1) 98.1 Láser – 98.7 PCM La Super FM – 25) 99.3 Melodía –99.7 Maranata – 41) 100.1 FM Cien – 100.5 Cadena 100 - 101.0 Color – 11) 101.3 – 102.1 Infinita – 102.5 Reloj – 114) 102.9 – 103.3 Cumbre – 130) 103.7 Contemporánea – 141) 104.0 - 104.7 RCN – 105.3 Nuevo Amanecer – 105.7 Excelsior – 17) 106.5 – 107.5 Cosmos – 107.9 Central.

BRAZIL

L.T: Eastern (Coastal): UTC - 3h. Western: UTC - 4h. Acre: UTC - 5h (Su: UTC - 2h, 3h and 4h respectively) —**Pop:** 161.654.000 — **Radios:** 55.000.000 — **Pr.L:** Portuguese — **E.C:** 50/60Hz, 110-127/220V — **ITU:** B.

AGENCIA NACIONAL DE COMUNICAÇÕES (ANATEL)

SAS Quadra 06 Bloco 4, 2Þ andar, 70313-900 Brasília, DF. — **L.P:** Dir. Gen. Dr. Rubens Bussacos. Dir. of Radio: Roberto Blois Montes de Souza. Dir. Dept. of Authorizations: Domingo Poty Chabalgoity.

ASSOCIAÇÃO BRASILEIRA DE EMISSORAS DE RADIO E TELEVISÃO (ABERT)

◪ Mezanino do Hotel Nacional, Salas 5 a 8, 70322-900 Brasília, DF (C.P. 08780, 70312-970). ☎ +55 (61) 224 4600. 🖹 +55 (61) 321 7583. ◪ (061) 2001.
L.P: Joaquim Mendonça. Exec. Dir: Antônio Abelin.
N.B: all st's carry "A Voz do Brasil" MF 2200-2300 (official prgr.)

Mediumwaves: Call ZY----, Þ = also on shortwaves, * = inactive, v = varying freq.
The letters preceding the stn. number indicate the state or territory. Addresses are listed by state in alphabetical order.
H. of tr. usually 0800-0300. Larger st's may operate 24h.

Call	kHz	kW	Name and location
BA01) H481	540	1/0.25	R. Regional, Irecê
CE01) H610	540	1/0.25	R. Jornal, Canindé
GO01) H755	540	10/1	R. Riviera, Goiânia
MA01) H894	540	1/0.25	R. Guajajara, Barra do Corda
MG151) L331	540	1/0.5	R. Ipanema, Ipanema
MS37) I...	540	10	R. Nacional, Corumbá (CP)
PI31) I914	540	1/0.25	R. Primeiro de Julho, Agua Branca
PR110) J322	540	1/0.25	R. Nova Era, Borrazópolis
RJ01) J450	540	10/2.5	R. AM Fluminense, Niterói
RS01) K226	540	5/0.5	R. Visão, Canoas
RS02) K322	540	10/1	R. Sepe Tiaraju, Santo Angelo
SC01) J778	540	10/1	R. Mirador, Rio do Sul
SE01) J924	540	10/2.5	R. Jornal de Sergipe, Aracaju
SP01) K697	540	1/0.25	R. Uirapuru, Biriguí
SP02) K734	540	1/0.25	R. Nova Sumaré, Sumaré
CE59) H644	550	1/0.25	R. Em. Acopiara, Acopiara
MG01) L225	550	5/0.5	R. Cataguases, Cataguases
MG02) L263	550	20/5	R. Soc. Norte de Minas, Montes Claros
MT29) I429	550	1	R. Gaspar, Sinop
PE01) I796	550	5/1	R. Meridional, Garanhuns
PI01) I902	550	1/0.25	R. Serra da Capivara, São Raimundo Nonato
PI22) I907	550	10/0.5	R. Igaraçu, Parnaíba
PR139) J331	550	5/0.5	R. Banda B, Curitiba
RS03) K287	550	2.5/0.25	R. Sta. Cruz do Sul, Sta. Cruz do Sul
SP03) K578	550	5/0.5	R. Mantiqueira, Cruzeiro
SP04) K696	550	5/0.5	R. Educação e Cultura, Sertãozinho
BA02) H456	560	5/1	R. Jornal, Itabuna
CE41) H604	560	1/0.25	R. Educadora Jaguaribana, Limoeiro do Norte
GO25) H769	560	5/0.25	R. Emissora Sul Goiana, Quirinópolis
MA02) H887	560	25/5	R. Educadora do Maranhão, São Luís
MG05) L277	560	10/5	R. Difusora, Patrocínio
MT01) I395	560	10/2.5	R. Em. Aruanã, Barra do Garças
MT24) I419	560	10/1	R. Pioneira, Tangará da Serra
PB16) I695	560	1/0.25	R. Potiguara, Mamanguape
PR01) J214	560	1/0.5	R. Londrina, Londrina
PR02) J281	560	2.5/0.25	R. Cultura, Guarapuava
RJ02) J496	560	5/0.25	R. AM-560, Araruama
RS04) K231	560	5/1	R. São Francisco AM, Caxias do Sul
SP213) K761	560	25/10	R. Paulista AM, São Paulo (retransmits Sistema LBV Mundial 24h)

Call	kHz	kW	Name and location
AL01) H244	570	5/1	R. Novo Nordeste, Arapiraca
CE02) H613	570	5/0.25	R. Vale do Cariri, Juazeiro do Norte
CE03) H614	570	1/0.5	R. Uirapuru, Itapipoca
GO02) H750	570	1/0.25	R. Cultura, Catalão
MA06) H890	570	10/1	R. Imperatriz, Imperatriz
MA19) H...	570	10/0.25	Sist. Timon de Radiodifusão, Timon
MG03) L261	570	25/5	R. Capital de Minas, Belo Horizonte
MT30) N407	570	1/0.25	R. Jornal, São José dos Quatro Marcos
PR146) J349	570	1/0.5	R. Continental, Palotina
RS05) K267	570	1/0.5	R. Passo Fundo, Passo Fundo
SC02) J735	570	5/0.5	R. Eldorado AM, Criciúma
SC99) J794	570	1/0.25	R. Fronteira Oeste, Dionísio Cerqueira
SP05) K595	570	1/0.25	R. Clube, Itapeva
SP06) K672	Þ570	5/1	R. Difusora, Taubaté
SP195) K698	570	1/0.25	R. Jornal, Nhandeara
SP150) K717	570	1/0.25	Bariri R. Clube, Bariri
AM01) H290	580	10	R. Nacional, Tefé
BA03) H477	580	1/0.25	R. Difusora, Teixeira de Freitas
GO44) H799	580	1/0.25	R. Serra Azul, Caiapônia
MG04) L328	580	5/1	R. América, Uberlândia
MS01) I387	Þ580	10/1	R. Educação Rural, Campo Grande
PE02) I776	580	20/10	CBN AM, Recife
PI12) I905	580	5/1	R. Itamaraty, Piripiri (**F.PI.**) (still on 920)
P148)	580		CBN Meio Norte, Teresina
PR03) J330	580	1/0.25	R. Grande Lago, Santa Helena
RJ03) J465	Þ580	20/5	R. Relógio Federal, Rio de Janeiro
RS06) K299	580	2/0.5	R. São Gabriel, São Gabriel
RS07) K318	580	10/5	R. Fátima, Vacaria
SP07) K540	580	1/0.25	R. Clube, Americana
SP08) K724	580	1/0.25	R. Regional, Palmital
TO01) H785	580	10/1	R. Tocantins AM, Porto Nacional
BA04) H445	590	10/5	R. Cruzeiro da Bahia, Salvador
CE04) H627	590	5/0.25	R. Vale do Rio Poty, Crateús
ES01) I213	590	5/0.5	R. Tribuna, Vitória
GO03) H798	590	10/1	R. Manchester, Anápolis
MT03) I420	590	10/5	R. Gazeta, Cuiabá
PB25) I692	590	2/0.25	R. Serrana, Araruna
PR04) J234	590	10/5	R. Difusora, Curitiba
RR01) 0700	Þ590	10	R. Difusora de Roraima, Boa Vista
RS08) K210	590	5/0.5	R. Alegrete, Alegrete
SC03) J809	590	2/1	R. Progresso, Descanso
SP09) K534	590	10/1	R. Atlântica, Santos
SP10) K612	590	1/0.25	R. Clube, Mirandópolis (**F.PI.**) (still on 1420)
SP11) K643	Þ590	5/1	R. Ribeirão Preto, Ribeirão Preto
AM02) H287	Þ600	10	R. Nacional, São Gabriel da Cachoeira
BA05) H486	600	10/1	R. Vale do Rio Grande, Barreiras
BA64) H...	600	1/0.25	R. Difusora, Rio Real (CP)
CE38) H627	600	1/0.25	R. Cultura, Aracati
MA38) H...	600	10/1	R. Litoral Maranhense, São Luís (CP)
PE03) I789	600	1/0.25	R. Cardeal Arcoverde, Arcoverde
RS09) K278	Þ600	100	R. Gaúcha, Porto Alegre
AL10) H249	610	10/2	R. Cidade Imperial, Marechal Deodoro
AM18) H321	610	10	R. Sucesso Manaus AM, Iranduba
GO10) H786	610	25/2	R. Jornal, Luziânia
MG06) L268	Þ610	100/25	R. Itatiaia, Belo Horizonte
MT04) I425	610	10/5	R. Celeste, Sinop
PA01) I544	610	10/1	R. Oriente de Redenção, Redenção (**F.PI.**) (still on 1510)
PB01) I678	610	1/0.25	R. Progresso, Sousa
PI02) I899	610	1/0.25	R. Poty, Teresina
SC04) J746	610	5/0.5	Radiodifusão Indio Condá, Chapecó
SP12) K532	610	1/0.25	R. Chamonix, Mogi Mirim
SP13) K577	610	1/0.25	R. Em. A Voz de Catanduva, Catanduva
SP14) K589	610	1/0.25	R. Piratininga, Guaratinguetá
SP15) K726	610	1/0.25	R. Paranapanema, Piraju
CE05) H590	620	10	R. Assunção Cearense, Fortaleza

Call	kHz	kW	Name and location
GO04) H777	620	1/0.25	R. Cristã Educativa, Pires do Rio
MG125) L320	620	10/0.25	R. Educadora, Porteirinha
MG147) L348	620	1/0.25	R. Ibiá, Ibiá
MG145) L357	620	1/0.25	R. Catuaí, Manhuaçu
PR05) J332	620	2.5/0.25	R. Cidade Jandaia, Jandaia do Sul
RS10) K270	620	10/1	R. Pelotense, Pelotas
RS11) K315	620	1.5/0.5	R. Municipal, Tenente Portela
SC05) J779	620	5/0.25	R. Difusora Alto Vale, Rio do Sul
SP16) K521	620	50/20	R. Jovem Pan, São Paulo
AP01) H422	Þ630	20/10	R. Difusora, Macapá
CE58) H636	630	1/0.5	R. Cidade, Campos Sales
			(F.PI.) (still on 1480)
MG07) L299	630	1/0.5	R. Difusora, Uberaba
MS32) N603	Þ630	10/1	R. IPB AM, Campo Grande
MT05) I384	630	1/0.5	R. Difusora Bom Jesús, Cuiabá
PI03) I904	630	1/0.5	R. Difusora, Barras
PR06) J284	630	10/0.5	R. Educativa AM, Curitiba
PR07) J300	630	5/0.25	R. Educadora Marechal, Marechal Cândido Rondón
RJ04) J466	630	25/10	R. Roquette Pinto, Rio de Janeiro
RS12) K259	630	1/0.5	R. Cacique, Lagoa Vermelha
RS13) K289	630	5/1	R. Santamariense, Santa Maria
SC06) J800	630	1/0.25	R. Doze de Maio, São Lourenço d'Oeste
SE02) J920	630	10/5	R. Aperipe, Aracaju
SP17) K613	630	1/0.5	R. Difusora, Mirassol
SP18) K635	630	5/0.25	R. Cidade Globo, Presidente Prudente
BA12) H458	640	10/0.5	R. Difusora Sul da Bahia, Itabuna
ES02) I204	640	10/0.5	R. Vitória, Vitória
GO05) H757	640	25/5	R. Difusora, Goiânia
MG08) L308	640	1	R. Santa Cruz, Pará de Minas
MT06) I406	640	10/5	R. Progresso, Alta Floresta
MT18) I424	640	10/1	R. Tangará, Tangará da Serra
PI28) I924	640	1/0.25	R. Cruzeiro, Pedro II
PR08) J262	640	5/1	R. Tupi, Londrina
RJ05) J489	640	5/1	R. Agulhas Negras, Resende
RN01) J590	640	20/5	R. Cabugi, Natal
RS14) K277	640	50/10	R. Bandeirantes, Porto Alegre
SP19) K547	640	5/1	R. A Morada do Sol, Araraquara
BA06) H462	650	5/0.5	R. Clube, Valença
GO06) H790	650	1/0.25	R. Cultural do Araguaia, Jussara
MA16) H924	650	1/0.25	R. Macaru, Viana
MG09) L200	650	10/0.5	R. Princesa, Lagoa Formosa
MG85) L309	650	5/0.5	R. Veredas, Unaí
MG142) L372	650	10/1	R. Itatiaia AM Vale do Aço, Timóteo
MT19) I414	650	5	R. Educadora, Colider
PA20) I540	650	10/1	R. Tropical, Santarém
PB02) I672	650	5/0.5	R. Alto Piranhas, Cajazeiras
PI26) I925	650	1/0.25	R. Tapuio, Miguel Alves
PR09) J202	650	1/0.25	R. Cultura, Cambará
PR91) J250	650	5/1	R. Colméia, Cascavel
RS15) K238	650	5/0.5	Radiodifusão Sul Riograndense, Erechim
SP20) K508	650	1/0.25	R. Andradina, Andradina
SP21) K524	650	5/0.25	R. Difusora, Piracicaba
SP112) K656	Þ650	25/10	Super R. Tupi, São Paulo
BA07) H465	660	5/0.25	R. Jornal, Itapetinga
BA36) H480	660	10/0.25	R. Bom Jesus AM, Bom Jesus da Lapa
BA66) H510	660	1/0.25	R. Tribuna do Vale do São Francisco, Xique-Xique
BA65) H...	660	1/0.25	R. Planalto, Euclides da Cunha (CP)
CE06) H619	660	1/0.25	R. Rio das Garças, Itarema (Acaraú)
GO07) H778	660	5/0.25	R. Primavera, Itapuranga
MG11) L206	660	10/0.25	R. Clube, Curvelo
MT07) I401	660	10/0.5	R. Amorim, Rondonópolis
PA21) I552	660	1/0.25	R. Xinguara, Xinguara
PE04) I787	660	1	R. Jornal, Limoeiro
PE05) I795	660	1/0.25	R. da Grande Serra, Araripina
PI34) I925	660	1/0.25	R. Tacarijus, São Miguel do Tapuio
RJ06) J472	660	1	R. Sociedade de Friburgo, Nova Friburgo
R001) J673	660	10/5	R. Eldorado, Porto Velho
RS16) K286	660	1/0.25	R. Marajá, Rosário do Sul
RS17) K319	660	10/0.25	R. Esmeralda, Vacaria
SP22) K518	660	5/0.5	R. CBN, Santos
SP23) K639	Þ660	10/0.5	R. Clube, Ribeirão Preto

Call	kHz	kW	Name and location
AM04) H288	Þ670	10	R. Cabocla, Tabatinga
AM03) H297	Þ670	1/0.25	R. Vale do Rio Madeira, Humaitá
AP02) H420	Þ670	10/1	R. Equatorial, Macapá (F.PI.) (still on 1350)
CE07) H606	670	1/0.25	R. Cultura, Várzea Alegre
GO08) H747	670	10/1	R. São Francisco, Anápolis
MG12) L310	670	10/2.5	R. Educadora, Montes Claros
MG126) L347	670	1/0.25	R. Visão, Ponte Nova
MG123) L361	670	5/0.25	R. Uberaba, Uberaba
MG135) L3..	670	1/0.25	R. Cidade, Bambuí
MS23) I408	670	1/0.25	R. Patriarca, Cassilândia (F.PI.) (still on 1570)
MS28) N600	670	10	Super R. Fronteira, Ponta Porã
MT16) I422	670	5/1	Emissoras Reunidas, Poconé
PA02) I537	670	1/0.25	R. Rural, Altamira
PA22) I539	670	1/0.25	R. Atalaia, Obidos
PA27) I546	670	1/0.25	R. Tropical AM, Paragominas
PI32) I927	670	1/0.25	R. Livramento, José de Freitas
PR10) J231	670	2.5/0.25	R. Soc. Nova Esperança, Nova Esperança
PR11) J248	670	10/2	R. Cidade, Curitiba
RS18) K296	670	2.5/0.25	R. Cultura Jaguarão, Santa Vitória do Palmar
RS19) K370	670	10/0.5	R. Gazeta, Carazinho
SE03) J921	670	10/5	R. Cultura de Sergipe, Aracaju
SP24) K574	670	1/0.5	R. Oceânica, Caraguatatuba
SP25) K585	670	1/0.25	R. Centro Oeste AM, Garça
SP26) K598	670	1/0.5	R. Emissora Convenção, Itu
BA08) H471	680	10/0.5	R. Clube, Sto. Antônio de Jesús
GO09) H765	Þ680	1/0.25	R. Difusora, Jataí
MA03) H885	Þ680	10/5	R. Difusora do Maranhão, São Luís
MG173) L270	680	2/0.25	R. Difusora, Ouro Fino
MG13) L326	680	10/0.25	R. União, João Pinheiro
MS02) I389	680	5/1	R. Cultura, Campo Grande
PB26) I683	680	2.5/0.25	R. Integração do Brejo, Bananeiras
PE06) I793	680	10	R. do Grande Rio, Petrolina
PR155) J362	680	5/0.25	R. Poema, Pitanga
RJ07) J452	Þ680	20/5	R. Copacabana, Rio de Janeiro
RS69) K275	680	50	R. Farroupilha, Porto Alegre
SP27) K576	680	1/0.25	R. Difusora, Catanduva
SP28) K628	680	2/0.25	R. Piratininga, Piraju
BA09) H453	690	10/1	R. Cultura, Ilhéus
CE08) H587	690	25/10	R. Dragão do Mar, Fortaleza
ES10) I201	690	10/1	R. América, Vitória
GO48) H780	690	10/1	R. Sociedade, Ceres
MG14) L228	690	50/5	R. Mineira, Belo Horizonte
MS03) I402	690	10/5	R. Cultura, Navirai
MT31) I451	690	5/1	R. Parecis, Diamantino
PA03) I532	Þ690	20/5	R. Clube do Pará, Belém
PR13) J229	Þ690	5/0.25	R. Difusora, Londrina
PR14) J252	690	1/0.25	R. Difusora, Ponta Grossa
PR143) J360	690	1/0.25	R. A Voz do Sudoeste, Coronel Vivida
RS21) K252	690	5/0.5	R. Progresso, Ijuí
SC07) J772	690	5/1	R. Clube, Lages
SP29) K561	690	1/0.25	R. Bebedouro, Bebedouro
SP30) K588	690	1/0.25	R. Clube, Guaratinguetá
SP31) K625	690	1/0.25	R. Cidade Pereira Barreto, Pereira Barreto
SP220) K646	690	1/0.25	R. Brasil, Santa Bárbara d'Oeste
BA10) H500	700	25/1	R. Cultura AM, Feira de Santana
MT21) I428	700	20/1	R. Sorriso, Sorriso
PI04) I890	700	10/5	R. Clube, Teresina
RJ56) J507	700	0.25	R. Nova Italva, Italva
RS22) K356	700	1/0.25	R. Em. Batovi, São Gabriel
SP32) K686	700	50	R. Nova Eldorado AM, São Paulo
AL02) H240	710	5/1	R. Novo Tempo AM 710, Maceió
BA46) H490	710	10/0.25	R. Jacarandá, Eunápolis
CE09) H628	710	1/0.25	R. Difusora Asa Branca, Boa Viagem
DF07) H710	710	10/2.5	R. Nova Aliança, Brasília
MA12) H891	710	1	R. Verdes Campos, Pinheiro
MA27) H910	710	1/0.25	R. Verdes Vales, Grajaú
MG79) L219	710	5/0.5	R. Cancella, Ituiutaba (F.PI.) (still on 1300)
MG15) L258	710	5/0.25	R. Manhuaçu, Manhuaçu
MG80) L319	710	1/0.25	R. Planeta, Carmo do Paranaíba
MG16) L333	710	2/0.25	R. Difusora, Pouso Alegre
MT08) I386	Þ710	5/0.5	R. Cultura, Cuiabá

Call	kHz	kW	Name and location
MT23) I436	710	5/1	R. Nova Xavantina, Nova Xavantina
PA04) I534	Þ710	10/5	R. Rural, Santarém
PB03) I685	710	5/0.25	R. Educadora, Conceição
PI19) I901	710	1/0.25	R. Alvorada do Sertão, São João do Piauí
PI23) I933	710	1/0.25	R. Clube, Barras
PR141) J328	710	1/0.25	R. Cultura, Cândido de Abreu (F.PI.) (still on 1410)
RJ09) J451	710	10	R. Difusora Carioca, Rio de Janeiro
SC08) J793	710	1/0.25	R. Fraiburgo, Fraiburgo
SP33) K559	710	10/0.25	R. 710, Bauru
AC01) H202	Þ720	10	R. Integração, Cruzeiro do Sul
AM05) H281	720	1	R. Difusora, Itacoatiara
MG28) L330	720	2.5/0.5	R. Divinópolis AM, Divinópolis
MS04) I390	Þ720	5/1	R. Clube, Dourados
MT20) I411	720	5/1	R. Difusora, Barra do Garças
PE07) I770	720	100	R. Clube de Pernambuco, Recife
RS23) K276	Þ720	100	R. Guaíba, Porto Alegre
SP34) K575	720	1	R. Difusora, Casa Branca
SP35) K701	720	1/0.25	R. Sentinela, Ourinhos
SP36) K718	720	1/0.25	R. Cruzeiro, Cruzeiro
SP37) K722	720	1/0.25	R. Menina, Olímpia
CE45) H640	730	1/0.5	R. Sinal, Aracati
ES16) I217	730	10/0.5	R. Novo Tempo, Vitória
G031) H759	730	25/5	R. Clube, Goiânia
MA04) H896	730	1/0.25	R. Eldorado, Codó
MG17) L287	730	5/1	R. Soc. Triângulo Mineiro, Uberaba
MG18) L297	730	10/1	R. Manchester AM, Juiz de Fora
MS38) I452	730	1/0.25	R. Princesa do Vale, Camapuã
MT09) I410	730	10/2.5	R. Jornal, Cáceres
PE08) I780	Þ730	10	Em. Rural, A Voz do São Francisco, Petrolina
PR15) J208	730	10/0.25	R. Marumby, Curitiba
PR16) J323	730	1/0.25	R. Humaitá, Campo Mourão
PR147) J353	730	5/0.25	R. Integração Oeste, Corbélia (F.PI.) (still on 1440)
RS24) K268	730	5/1	R. Planalto, Passo Fundo
SC09) J787	730	2/0.25	R. Super Tubá, Tubarão
SP38) K523	730	5/0.25	R. Cidade, Jundiaí
SP39) K610	730	10/1	R. Cidade, Marília
AC02) H206	Þ740	20/10	Super R. Alvorada, Rio Branco
BA11) H446	740	100	R. Soc. da Bahia, Salvador
MT25) N403	740	1/0.5	R. Cidade, Alto Araguaia
PR17) J259	740	1/0.25	R. Goio-Erê, Goio-Erê
PR135) J354	740	1/0.25	R. Placar, Ortigueira
RS25) K265	740	2.5/0.25	R. Palmeira, Palmeira das Missões
RS26) K283	740	5/0.25	R. Nativa, Rio Grande
SC10) J753	740	5/1	CBN - Diário, Florianópolis
SP93) K519	740	1/0.25	R. Assunção, Jales
SP40) K553	740	1/0.25	R. Cultura, Bariri
SP41) K650	740	10/1	R. Trianon, Santo André
DF01) H709	750	50/25	R. CBN, Brasília
MG19) L213	750	50/5	R. América, Belo Horizonte
PA28) I...	750	1/0.25	R. Ximango, Alenquer (CP)
PB04) I682	750	1/0.25	R. Panati, Patos
PI05) I897	750	1/0.25	R. Heróis do Jenipapo, Campo Maior
RS27) K264	750	5/0.25	R. Osório, Osório
SC11) J815	750	5/0.25	R. Aliança, Concórdia
SE04) J927	750	10/0.25	R. Progresso, Lagarto
SP42) K516	750	1/0.5	R. Clube, Osvaldo Cruz
SP43) K642	750	3/0.5	R. Renascença, "CMN Canal 750", Ribeirão Preto
SP44) K661	750	2.5/0.25	R. Piratininga, São José dos Campos
T003) H792	750	1/0.25	R. Tocantins, Tocantinópolis
AL12) H...	760	1/0.25	R. Pioneira, Delmiro Gouveia (CP)
AP03) H...	760	20/10	Rede Amapaense de Radiodifusão, Macapá (CP)
CE10) H588	760	10	R. Record, Fortaleza
G043) H775	760	5/0.5	R. Rio Claro, Iporá
G011) H783	760	10/0.25	R. Pousada do Rio Quente, Caldas Novas
MG83) L257	760	1/0.25	R. Difusora, Machado
MG137) L360	760	10/1	R. Terra, Monte Claros
MT32) N408	760	10/5	R. Central, Chapada dos Guimarães
PR12) J343	760	2/0.25	R. Cacique, Guarapuava
RJ11) J478	760	25/1	R. Manchete AM, Niterói
RS28) K222	760	1/0.25	R. Jornal do Comércio, Candelária
RS29) K351	760	2.5/0.25	R. Ametista, Planalto
SC12) J742	760	10/0.5	R. Nereu Ramos, Blumenau
SP149) K541	760	1	R. Urubupungá, Andradina
SP45) K560	760	10/0.5	R. Auri Verde, Bauru
BA51) H491	770	1/0.25	R. Rio Corrente, Santa Maria da Vitória
CE11) H609	770	10/0.25	R. Vale do Salgado, Lavras da Mangabeira
ES03) I211	770	5/1	R. Nova Difusora, Cachoeiro de Itapemirim
G012) H745	Þ770	1/0.25	R. Carajá, Anápolis
MA28) H922	770	1/0.25	R. Jornal ("R. Guanabara"), Coelho Neto
MG20) L209	770	1/0.25	R. Cultura d'Oeste, Lavras
MG21) L302	770	10/0.5	R. Clube de Patos, Patos de Minas
MG108) L315	770	5/0.5	R. Iturama, Iturama
MG22) L337	770	1/0.25	R. Itabira AM, Itabira
MS11) I412	770	10/5	R. Caiuás, Dourados
MT28) N404	770	1	R. Cidade de Matupá, Matupá
PA29) I...	770	10/0.25	SNC—Sistema Norte de Comunicações ("R. Eldorado AM"), Marabá (CP)
PR131) J344	770	1/0.25	R. Metropolitana, Cambé
SE05) J922	770	10/5	R. Atalaia de Sergipe, Aracaju
SP46) K506	770	5/0.5	R. Jornal do Povo, Limeira
AM22) H291	780	10	R. Nacional, Eirunepê
CE55) H657	780	5/0.5	R. Difusora, Nova Russas
G013) H789	780	10/1	R. Soc. Vera Cruz, Goianésia
MA24) H919	780	10/5	R. Alvorada, Zé Doca
MG23) L246	780	1	R. Educadora, Uberlândia
MG103) L259	780	2.5/1	R. Manhumirim, Manhumirim (F.PI.) (still on 1510)
PE09) I771	780	20/10	R. Jornal do Comércio, Recife
PR18) J247	780	1/0.25	R. Porta Voz, Cianorte
PR19) J305	780	5/0.25	R. Chopinzinho, Chopinzinho
RS30) K229	780	5/2	R. Carazinho, Carazinho
RS31) K279	780	2.5	R. Princesa, Porto Alegre
SC54) J751	780	0.5	R. Brasil Novo, Jaraguá do Sul
SC13) J788	780	1	R. Marconi, Urussanga
SP161) K619	780	1/0.25	R. Difusora, Monte Aprazível
SP47) K695	Þ780	50/10	R. CBN (Globo), São Paulo
BA13) H484	790	10/1	R. Barreiras, Barreiras
BA14) H505	790	1/0.25	R. Regional, Serrinha
CE12) H629	790	1/0.25	R. Jornal Centro Sul, Iguatu
G028) H761	790	1/0.5	R. Xavantes, Ipameri
G014) H771	790	1/0.5	R. Eldorado, Mineiros
MA29) H904	790	1/0.25	R. Pérola do Turi, ("R. Rio Turiaçu"), Santa Helena
MA20) H915	790	1/0.25	R. Cultura, Açailândia
MA30) H...	790	1/0.25	Sistema Clube de Comunicação, Tuntum (CP)
MG24) L279	790	5/0.25	R. Soc. Ponte Nova, Ponte Nova
MG25) L311	790	1/0.25	R. Treze de Junho, Mantena (nf 1190)
MG26) L314	790	5/1	R. Tropical, Lagoa da Prata
MT22) I456	790	1	R. Difusora, Nortelândia
PB05) I679	790	2.5/1	R. Cultura, Guarabira
PI36) I931	790	1/0.25	R. Mafrense, Simplício Mendes
PR130) J316	790	2.5/0.25	R. Clube, Faxinal
PR20) J337	790	5/0.25	R. Panorama, Mandirituba
RS32) K285	790	1/0.25	R. Rio Pardo, Rio Pardo
SC14) J789	790	1/0.25	R. Videira, Videira
SP48) K538	790	1/0.25	R. Brasil, Adamantina (nf 1270)
SP49) K549	Þ790	5/0.25	R. Cultura, Araraquara
BA55) H...	Þ800	25/5	R. Educadora da Bahia, Salvador (CP)
DF02) H705	800	10/1	R. MEC, Brasília
PI08) I921	800	10	R. Antares, Teresina
RJ12) J457	Þ800	100	R. MEC, Rio de Janeiro
RS33) K292	800	10	R. Universidade, Santa Maria
BA54) H528	810	1	R. Nossa Senhora de Guadalupe AM, Riacho de Santana
CE13) H589	810	50/5	R. Verdes Mares, Fortaleza
G015) H767	810	5/0.5	R. Alvorada, Rialma
MG27) L202	810	1	R. Aimorés, Aimorés
MG76) L346	810	1/0.25	R. Clube, Nepomuceno
MG138) L354	810	1/0.25	R. Capinópolis, Capinópolis
MG156) L366	810	1/0.25	R. Capital do Triângulo, Patrocínio

Call	kHz	kW	Name and location
MG160) L...	810	1/0.25	R. Princesa do Vale, Itaobim (CP)
MT33) N402	810	1/0.25	Gaspar Radiodifusão, São José do Rio Claro
MT34) N406	810	1/0.25	Magalhães Barros Radiodifusão, Alta Floresta
PR111) J336	810	5/0.5	R. Esperança, Prudentópolis
SP50) K604	810	5/0.25	R. Difusora Jundiaiense, Jundiaí
SP89) K655	810	1/0.5	R. Universal, Santos
SP51) K732	810	5/0.5	R. Centro-América, São José do Rio Preto
AC03) H...	Þ820	1/0.25	R. Difusora 6 de Agosto, Xapuri
AM06) H294	820	1/0.25	R. Princesa do Solimões, Manacapuru
CE14) H624	820	1/0.25	R. União, Camocim
CE60) H655	820	1/0.25	R. Sul Cearense, Brejo Santo
ES04) I212	820	10/2.5	R. Gazeta AM, Vitória
GO16) H752	820	10/5	R. Jornal de Goiás, Goiânia
MG29) L255	820	5/0.25	R. Barbacena, Barbacena
MG30) L291	820	5/1	R. Difusora Paraisense, São Sebastião do Paraíso
MT10) I400Þ	Þ820	10/1	R. Difusora, Cáceres
PA05) I543	820	5/1	R. Regional do Araguaia, Conceição do Araguaia
PE10) I775	820	5/1	R. Universitária, Recife
PI06) I912	820	5/0.25	R. Cacique Bruenque, Regeneração
PR21) J238	820	10/5	R. Cultura, Foz do Iguaçu
RJ13) J477	820	5/0.25	R. Jornal, Macaé
RS34) K241	820	5/1	R. Alto Taquari, Estrela
SC15) J738	820	10/5	CBN, Blumenau
SP52) K542	Þ820	10/0.5	R. Aparecida, Aparecida
SP53) K602	820	1/0.25	R. Jauense, Jaú
SP54) K622	820	0.5/0.25	R. Clube, Ourinhos
SP55) K624	820	1/0.25	R. Difusora, Penápolis
BA67) H506	830	1/0.25	R. Extremo Sul da Bahia, Itamaraju
CE65) H659	830	1/0.25	R. Pioneira, Forquilha
GO50) H805	830	1/0.25	R. Cidade, Goiatuba
MA26) H905	830	10/1	R. Mirante do Maranhão, Imperatriz
MA21) H925	830	1/0.25	R. Boas Novas, Esperantinópolis
MG86) L207	830	5/0.25	R. Imbiara, Araxá (ff 900)
MG31) L244	830	50/5	Cultura R. Sucesso, Belo Horizonte
MG167) L273	830	1/0.25	R. Bom Sucesso, Minas Novas
MT11) I430	830	5/0.25	R. Xavantes, Jaciara
MT26) N401	830	1/0.25	R. Educadora, Juina
PA24) I556	830	1/0.25	R. Guaraní de Marajó, Soure
PI07) I906	830	1/0.25	R. Primeira Capital, Oeiras
PI37) I934	830	1/0.25	R. União, União
PR22) J224	830	5/0.5	R. Iguaçu, Araucária
PR24) J266	830	10/0.5	R. Tabajara, Londrina
PR23) J311	830	1/0.25	R. Progresso, Clevelândia
RJ41) J494	830	1/0.25	R. Stereo Sul, Volta Redonda (ff 920)
RN02) J595	830	1/0.5	Emiss. de Educação Rural, Caicó
RS35) K332	830	5/0.25	R. Independente, Cruz Alta
SC16) J773	830	1/0.25	R. Cruz da Malta, Lauro Müller
SE06) J926	830	20/1	R. Princesa da Serra, Itabaiana
SP56) K681	830	10/1	R. Clube, Votuporanga
SP227) K746	830	1/0.25	R. Liberal, Nova Odessa
AL13) H253	840	10/1	R. Quilombo, União dos Palmares (**F.PI.**) (still on 1460)
AM07) H298	840	1/0.25	R. Rio Madeira, Manicoré
BA16) H447	840	25/10	R. Excelsior da Bahia, Salvador
CE51) H648	840	1/0.25	R. Campo Maior, Quixeramobim
PI38) I930	840	1/0.25	R. Ribeirão, Demerval Lobão
PI39) I...	840	1/0.5	R. Vitória, Batalha
RO13) J679	*840	50	R. Nacional, Porto Velho
RS36) K248	840	10	R. Capital, Porto Alegre
SC17) J750	840	10/1	R. Rural, Concórdia
SP57) K687	Þ840	100/50	R. Bandeirantes, São Paulo
BA17) H474	850	5/0.25	R. Caraiba, Senhor do Bonfim
CE15) H599	850	1	R. Iracema de Juazeiro, Juazeiro do Norte
GO17) H776	850	10/5	R. Tropical, Porangatu
MA31) H923	850	1/0.25	R. Vitória, Vitória do Mearim
MG32) L233	850	1/0.25	R. Difusora Formiguense, Formiga
MG33) L254	Þ850	10/0.5	R. Por Um Mundo Melhor, Governador Valadares
MG34) L295	850	1	R. Tupaciguara, Tupaciguara
MS30) I438	850	1/0.25	R. Difusora Nor'estado, SãoGabriel

Call	kHz	kW	Name and location
MT02) I416	850	10/2	R. Cultura, Poxoréo
PA17) I538	850	10/1	R. Itacaíunas, Marabá
PA06) I555	850	1/0.25	R. Tocantins, Cametá
PA18) I557	850	5/1	R. Itaituba, Itaituba
PB22) I693	850	5/1	R. Rural, Guarabira
PI30) I909	850	1/0.25	R. Grande Picos, Picos
PR50) J254	850	5/0.25	R. Dif. Colméia, Campo Mourão
PR86) J291	850	1/0.5	R. Alvorada do Sul, Rebouças
RO03) J675	850	5/1	R. Ariquemes, Ariquemes
SC20) J808	850	1/0.25	R. Cidade, Brusque
SP59) K563	850	1/0.5	R. Clube, Biriguí
SP58) K644	850	1/0.25	R. Jornal, Rio Claro
CE16) H592	860	25/10	R. Cidade, Maracanaú
RJ14) J459	860	100	R. CBN, Rio de Janeiro
RS37) K288	860	10/1	R. Guarathan, Santa Maria
AL03) H245	870	5/1	R. Educadora Sampaio, Palmeira dos Indios
AM19) H322	870	1/0.25	R. Jutanópolis, Manacapuru
BA18) H457	870	10/0.25	R. Clube, Itabuna
BA84) H499	870	5/1	R. da Cidade AM, Juazeiro
CE17) H591	870	1/0.25	R. Iracema, Iguatu
CE66) H658	870	1/0.25	R. Tabajara, São Benedito
GO18) H749	870	5/0.5	R. Educadora, Uruaçu
GO32) H754	870	20/0.5	R. Universitária, Goiânia
MA05) H903	870	10/0.5	R. Mirante, Codó
MG78) L304	870	10/0.5	R. Juriti, Paracatu
MG38) L318	870	1/0.5	R. Cultura, Diamantina
MG66) L324	870	5/0.25	R. Sacramento, Sacramento
MG128) L349	870	5/0.25	R. Princesa da Mata, Muriaé
MG127) L350	870	5/0.25	R. A Voz do São Francisco, Januária
MT35) I...	870	1/0.25	R. Garça Branca, Guiratinga (CP)
PA11) I547	870	10/0.25	R. Marajó, Breves (**F.PI.**) (still on 1380)
PR25) J243	870	5/0.25	R. Nova Ingá, Maringá
RS170) K376	870	5/1	R. Educadora, São João da Urtiga (**F.PI.**) (still on 1510)
SC96) J784	870	5/0.25	R. São Francisco , São Francisco do Sul
SP60) K620	870	1/0.25	R. Novo Horizonte, Novo Horizonte
SP61) K705	870	5/1	R. Central, Campinas
TO04) H762	Þ870	1/0.25	R. Anhanguera, Araguaína
MG35) L275	Þ880	100	R. Inconfidência, Belo Horizonte
PB18) I680	880	1/0.25	R. Maringá, Pombal
RS38) K249	880	10/2.5	R. Itaí, Porto Alegre
RS87) K317	880	2.5/1	R. São Miguel, Uruguaiana
RS20) K363	880	1/0.25	R. Seberi, Seberi
CE46) H642	890	1/0.25	R. Itatiaia, Santa Quitéria
DF03) H706	890	10	R. Planalto, Brasília
MG36) L250	890	10/1	R. Santa Cruz, Jequitinhonha
MG154) L370	890	5/0.25	R. Clube, Inhapim
MS33) I453	890	10/0.5	R. Guaicurus, Fátima do Sul
PA13) I536	890	5/1	R. Ponta Negra, Santarém
PE11) I772	890	20/10	R. Tamandaré, Recife
PR117) J287	890	1/0.25	R. Ubá, Ivaiporã
PR26) J338	890	5/1	Super R. Itapuã, Pato Branco
RO16) J685	890	1	R. Planalto, Vilhena
RS39) K215	890	5/0.25	R. Viva AM, Bento Gonçalves
RS40) K295	890	1/0.5	R. Noroeste, Santa Rosa
SC52) J745	890	1/0.25	R. Clube, Canoinhas (**F.PI.**) still on 1350)
SC18) J755	890	1/0.25	R. Santa Catarina, Florianópolis
SP62) K690	Þ890	50/10	R. Gazeta, São Paulo
SP127) K703	890	2.5/0.25	R. Noticias, Matão (**F.PI.**) (still on 1230)
BA19) H488	900	5/1	R. Sisal, Conceição do Coité
GO41) H768	900	10/1	R. Sudoeste AM, Rio Verde
MG86) L207	900	5/0.25	R. Imbiara, Araxá (**F.PI.**) (still on 830)
MG124) L338	900	1/0.25	R. Vinícola, Andradas
MT36) I455	900	10/2.5	R. Difusora Arco-Iris, Araputanga
PR27) J272	900	3/0.25	R. Emiss. Sant'Ana, Ponta Grossa
PR28) J295	900	5/0.25	R. União, Toledo
RJ15) J454	900	50/10	R. Tamoio, Rio de Janeiro
RN03) J591	900	10	R. Nordeste, Natal
RO04) J672	900	1	R. Alvorada de Rondônia, Ji-Paraná
RS41) K211	900	2.5/0.5	R. Aratiba, Aratiba
RS42) K263	900	5/0.5	R. Progresso, Nôvo Hamburgo

Call	kHz	kW	Name and location
RS164) K303	900	1/0.25	R. Mun. Sampedrense, São Pedro do Sul
SP63) K511	Þ900	0.5/0.25	R. Difusora, Presidente Prudente
SP64) K664	900	10/0.25	R. CBN, São José do Rio Preto
SP65) K742	900	1/0.25	R. Clube, Itapetininga
CE61) H645	910	1/0.25	R. Assunção Cearense, Sobral
GO20) H763	910	5/0.25	R. Paranaíba, Itumbiara
GO23) H804	910	1/0.25	R. Cidade, Jaraguá
MG37) L292	910	1/0.25	R. Teófilo Otoni, Teófilo Otoni (nf 990)
MG132) L346	910	1/0.25	R. Difusora Industrial, Nova Serrana
MG149) L...	910	5/1	R. Juiz de Fora, Juiz de Fora
PE12) I785	910	5/1	R. Liberdade, Caruaru
PI41) I935	910	10/1	R. Tropical, Teresina
PR29) J207	910	1/0.25	R. Difusora, Apucarana
RS43) K320	910	5/0.5	R. Venâncio Aires, Venâncio Aires
SC19) J811	910	1/0.25	R. Difusora, Içara
SC90) J824	910	1/0.25	R. Rainha das Quedas, Abelardo Luz
SP66) K536	910	5/0.25	R. Alvorada, Piracicaba
SP228) K763	910	1/0.25	R. Princesa, Monte Azul Paulista
BA42) H476	920	5/0.25	R. Educadora Santana de Caetité, Caetité (**F.PI.**) (still on 1310)
BA57) H519	920	10/2	R. Novo Tempo, Salvador
ES05) I207	920	5/0.5	R. Cultura, Linhares
GO33) H788	920	5/1	R. Vale da Serra, São Luís de Montes Belos
MG39) L271	920	5/0.5	R. Cultura, Visconde do Rio Branco
PB31) I...	920	5/0.5	R. Cidade Verde, João Pessoa (CP)
PI11) I893	920	10/0.5	R. Educadora, Parnaíba (**F.PI.**) (still on 1210)
PI09) I895	920	1/0.25	R. Difusora, Picos
PI12) I905	920	10/0.5	R. Itamaraty, Piripiri (ff 580)
RJ41) J494	920	5/0.5	R. Stereo Vale, Volta Redonda (**F.PI.**) (still on 830)
RN04) J600	920	1/0.25	R. Currais Novos, Currais Novos
RS44) K348	920	20/2	R. Tramandaí, Tramandaí
SP67) K584	920	1/0.25	R. Franca do Imperador, Franca
SP222) K769	920	1/0.25	R. Icatu, Penápolis (**F.PI.**) (still on 1490)
SP221) K775	920	25/1	R. Nacional Gospel, Cotia
AM16) H296	930	10/5	R. Boas Novas, Manaus
CE18) H605	930	1/0.25	R. Salamanca, Barbalha
CE52) H646	930	7/0.25	R. Metropolitana, Fortaleza
GO45) H802	930	10/1	R. Caraiba AM, Aparecida de Goiânia
MG41) L220	930	1/0.25	R. Clube, Campo Belo
MG42) L229	930	5	R. Araguari, Araguari
MG87) L237	930	30/1	R. Ibituruna, Governador Valadares
MS05) I454	930	10/0.25	R. Capital, Campo Grande
MT12) I423	Þ930	5/0.25	R. Clube, Rondonópolis
MT37) N400	930	10/0.25	R. Jornal, Pontes e Lacerda
PA07) I600	930	5/1	R. Liberal, Castanhal
PR30) J227	930	1/0.25	R. Cultura, Rolândia
PR31) J232	930	10/1	R. Cultura, Curitiba
PR69) J235	930	10/1	R. Princesa, Francisco Beltrão
RS45) K230	930	20/2.5	R. Caxias, Caxias do Sul
RS46) K298	930	10/0.25	Super R. Sto. Ângelo, Santo Ângelo
SE07) J923	930	20/5	R. Liberdade de Sergipe, Aracaju
SP71) K500	930	1/0.25	R. Dinâmica de Santa Fé, Santa Fé do Sul
SP68) K503	930	1/0.25	R. Clube, Itapira
SP69) K652	930	10/1	Super R. Cultura, Santos
SP70) K713	930	5/1	R. Universal, Agudos
SP214) K747	930	1/0.25	R. Jóia, Adamantina
AC04) H204	Þ940	5/1	R. Verdes Florestas, Cruzeiro do Sul
PI25) I911	940	10/0.25	R. Sete Cidades, Piracuruca
RJ16) J453	940	100	R. Sistema LBV Mundial, Rio de Janeiro
BA50) H489	950	1/0.25	R. Bahia Noroeste, Paulo Afonso
CE19) H593	950	5/1	R. Educadora do Nordeste, Sobral
MA17) H916	950	10/0.25	R. Capital, João Lisboa
MG43) L212	950	25/10	R. Atalaia, Belo Horizonte
MG44) L281	950	5/1	R. Independência, Bueno Brandão
MT17) I439	950	5/1	R. Tucunare, Juara
PB17) I681	950	1/0.25	R. Jornal, Sousa

Call	kHz	kW	Name and location
PE13) I782	950	10/5	R. Temurinha, Carpina
PI20) I915	950	10/0.25	R. São José dos Altos, Altos
PI42) I923	950	1/0.25	R. Boa Esperança, Padre Marcos
PR114) J239	950	5/0.25	R. Difusora Cultural, Irati
RS47) K260	950	5/0.25	R. Independente, Lajeado
SC21) J736	950	1/0.25	R. Vale Tijucas, Tijucas
SP72) K510	950	5/0.25	R. 950, Marília
AL04) H241	960	10	R. Difusora de Alagôas, Maceió
CE37) H618	960	1/0.25	R. Cultura dos Inhamuns, Tauá
ES15) I216	960	10/0.25	R. Diocesana, Cachoeiro de Itapemirim
GO21) H764	960	5/0.25	R. Difusora, Itumbiara
PA26) I551	960	5/1	R. Clube, Itaituba
PR109) J217	960	1/0.25	R. Legendária, Lapa
PR32) J257	960	1/0.25	R. Difusora, Maringá
RS48) K291	960	10/1	R. Imembuí, Santa Maria
SC22) J733	960	5/0.25	R. Guarujá, Orleãns
SC23) J813	960	1/0.25	R. Difusora, Xanxerê
SP73) K689	960	50/10	R. São Paulo, São Paulo
BA20) H451	Þ970	10/5	R. Sociedade, Feira de Santana
CE20) H612	970	1/0.25	R. Monólitos, Quixadá
MG45) L243	970	5/0.25	R. Sociedade Caratinga, Caratinga
MG46) L285	970	2.5/0.25	R. São João Del Rei, São João Del Rei
MG47) L321	970	1/0.25	R. Central, Monte Alegre de Minas
MS18) I399	970	5/0.5	R. Vale do Taquari, Coxim
PB06) I684	970	1/0.25	R. Princesa Isabel, Princesa Isabel
PI43) I910	970	10/0.5	R. Vale do Parnaíba, Luzilândia
PR33) J260	Þ970	5/1	R. Alvorada, Londrina
PR34) J277	970	5/0.25	R. Difusora do Paraná, Marechal Cândido Rondón
RS49) K201	970	50/10	R. Caiçara, Porto Alegre
RS50) K349	970	10/5	R. do Povo, Humaitá
SC24) J730	970	1/0.25	R. Araguaia, Brusque
SP74) K505	970	5/0.25	R. Difusora 2000, Itapetininga
SP75) K529	970	1/0.25	R. Piratininga, São João da Boa Vista
SP76) K684	970	1	R. Noticia, Franca
SP215) K744	970	5/0.25	R. Alvorada, Estrela d'Oeste
DF04) H707	Þ980	50/600	R. Nacional, Brasília
AM23) H299	990	1/0.25	Rede Amazonense de Comunicação, Maués
BA21) H483	990	1/0.25	R. Alvorada de Teixeira de Freitas, Caravelas
PI44) I922	990	1/0.25	R. Vale do Canindé, Oeiras
PR121) J321	990	5/0.25	R. Regional, Cianorte (**F.PI.**) (still on 1530)
RJ53) J461	990	100/10	R. Record, Rio de Janeiro
RN05) J596	990	10/1	R. Rural, Mossoró
RS51) K314	990	1/0.25	R. Tupanciretã, Tupanciretã
RS52) K335	990	1/0.25	R. Sananduva, Sananduva
RS154) K360	990	1/0.25	R. Clube, Pedro Osório
SC25) J763	990	1/0.25	R. Peperi, Itapiranga
SC56) J821	990	1/0.25	R. Cidade, Itaiópolis (**F.PI.**) (still on 1380)
SP239) K579	990	10/0.25	R. Cultura Regional, Dois Córregos
PE14) I791	1000	1/0.25	R. Princesa Serrana, Timbaúba
SP77) K522	Þ1000	200	R. Record, São Paulo
BA22) H448	1010	25/5	R. Bahia, Salvador
CE21) H625	1010	50	CBN, Fortaleza
GO39) H772	1010	10/0.5	R. Santelenense, Santa Helena de Goiás (**F.PI.**) (still on 1520)
MG50) L230	1010	10/1	R. Educadora, Coronel Fabriciano
MG48) L264	1010	5/0.5	R. Solar AM, Juiz de Fora
MG49) L325	1010	2.5/0.25	R. Estância, Jacutinga
MT13) I421	1010	5/1	R. Difusora, Mirassol d'Oeste
PR35) J263	1010	25/5	R. Celinauta, Pato Branco
RS53) K232	1010	1/0.5	R. 1010 - R. Sucesso, Caxias do Sul
RS54) K344	1010	2.5/1.5	R. Missionário Sete Povos, São Luíz Gonzaga
SC71) J764	1010	10/0.5	R. Jaraguá, Jaraguá do Sul
SC26) J807	1010	1/0.25	R. Verde Vale, Braço do Norte
SP78) K556	1010	1/0.5	R. Independente, Barretos
SP79) K611	1010	1/0.5	R. Tuiuti, Martinópolis
AL05) H247	1020	25/1	R. CBN AM 1020, Maceió
CE22) H600	Þ1020	5/1	R. Educadora Cariri, Crato
CE79) H664	1020	1/0.25	R. Macambira, Ipueiras
ES06) I205	1020	20/10	R. Difusora, Colatina
GO52) H781	1020	1/0.25	R. Maranata, Firminópolis

Call		kHz	kW	Name and location
MG55)	L224	Þ1020	10/1	R. Congonhas, Congonhas
MG51)	L260	1020	10/0.25	R. Cultura, Uberlândia
MS06)	I381	1020	10/0.25	R. Independente, Aquidauana
PB19)	I686	1020	1/0.25	R. Cenecista, Picuí
PR36)	J244	1020	10/0.25	R. Colombo do Paraná, Curitiba
PR37)	J307	1020	10/0.25	R. Independência, Medianeira
RJ42)	J484	1020	5/0.25	R. A Noticia AM, Campos
RO02)	J680	1020	5/1	R. Educadora, Rolim de Moura
RS49)	K202	1020	10/5	R. Pampa, Porto Alegre
SC27)	J805	1020	1/0.25	R. Continental, Coronel Freitas
SP80)	K513	Þ1020	10/0.25	R. Canção Nova, Cachoeira Paulista
SP81)	K515	1020	5/1	R. Cultura, Assis
SP82)	K531	Þ1020	2.5/0.5	R. Educadora, Limeira
SP83)	K600	1020	5/0.25	R. Cultura, Jales
BA40)	H475	1030	10/1	R. Bahiana, Itaberaba
GO22)	H746	1030	10/1	R. Imprensa, Anápolis
MA07)	H892	1030	10/1	R. Jaínara, Bacabal
MS07)	I396	1030	2.5/0.25	R. Cidade Maracaju, Maracaju
PE15)	I777	1030	20/5	R. Olinda, Olinda
PR38)	J240	1030	5/0.25	R. Difusora, Cruzeiro do Oeste
PR39)	J271	1030	5/0.25	R. Atalaia, Londrina
PR120)	J312	1030	1/0.25	R. Clube, Realeza
PR40)	J329	1030	1/0.25	R. Difusora do Xisto, São Mateus do Sul
RJ18)	J467	Þ1030	100/25	R. Capital, Rio de Janeiro
RO10)	J683	1030	5/1	R. Colina de Machadinho, Ariquemes
RS129)	K224	1030	1/0.25	R. Cultura, Canguçu (F.PI.) (still on 1430)
RS55)	K253	1030	10/0.5	R. Repórter, Ijuí
SC28)	J771	1030	2.5/0.25	R. Princesa, Lages
SP84)	K525	1030	1/0.25	R. Difusora de Franca, Franca
SP85)	K554	1030	1/0.25	R. Emissora da Barra, Barra Bonita
SP86)	K606	1030	5/0.25	Lins Rádio Clube, Lins
TO05)	H791	1030	1	R. Colinas, Colinas do Tocantins
SP87)	K537	1040	200/100	R. Capital, São Paulo
BA47)	H494	1050	10/0.5	R. Metropolitana da Bahia, Camaçari
CE54)	H647	1050	10/0.5	R. Primeira Capital, Aquiraz
ES07)	I203	Þ1050	100/10	R. Capixaba, Vitória
MG52)	L236	1050	10/0.5	R. Rural, Tupaciguara
MS20)	I391	1050	10/0.5	R. Difusora Paranaibense, Paranaíba
PB07)	I676	1050	5/1	R. Caturité, Campina Grande
PI45)	I...	1050	1/0.25	Jet Radiodifusão, Teresina (CP)
PR66)	J226	1050	1/0.25	R. Dif. Platinense, Santo Antônio da Platina
PR99)	J286	1050	5/0.25	R. Club de Palmas, Palmas (F.PI.) (still on 1460)
RJ19)	J497	1050	10/0.5	R. Angra, Angra dos Reis
SP160)	K601	1050	10/0.5	R. Super Tupi, Jardinópolis
BA23)	H460	1060	5/1	R. Clube de Conquista, Vitória da Conquista
BA68)	H520	1060	2.5/0.25	R. Clube, Itapicuru
MG53)	L278	1060	50/5	R. Grande Belo Horizonte ("R. Grande Be Aga"), Belo Horizonte
MG54)	L306	1060	1/0.25	R. Itajubá, Itajubá
MS39)	N604	1060	5/1	R. Tupinambás, Dourados
PR42)	J246	1060	10/0.5	R. Paraná, Curitiba
PR43)	J298	1060	5/0.25	R. Colorado, Colorado
PR44)	J306	1060	10/0.25	R. Educadora, Francisco Beltrão
RJ20)	J495	1060	1/0.25	R. Tropical, Miguel Pereira
RN06)	J597	1060	5	R. Tapuyo, Mossoró
RS56)	K220	1060	5/0.25	R. Camaqüense, Camaquã
RS57)	K302	1060	2/0.25	R. São Luís, São Luís Gonzaga
RS81)	K307	1060	1/0.5	R. Cristal, Soledade
SP88)	K533	1060	5/0.25	R. Educadora, Piracicaba
SP229)	K765	1060	5/0.25	R. Universitária, Garça
BA48)	H492	1070	5/0.5	R. Rural ("R. Tropical"), Ipiaú
CE47)	H643	1070	1/0.25	R. Cultura, Paracuru (ff 1150)
MG56)	L316	1070	1/0.25	R. Rural, Muzambinho
MG150)	L355	1070	10/0.25	Super R. Patos, Patos de Minas
MT14)	I427	1070	10/2.5	R. Industrial ("Antena 1"), Várzea Grande
PB08)	I673	1070	20/2.5	Difusora R. Cajazeiras, Cajazeiras
PR45)	J203	1070	5/0.25	R. Difusora União, União da Vitória
PR46)	J319	1070	1/0.25	R. Guaraniaçu, Guaraniaçu
RJ21)	J483	1070	10/0.25	R. Record, Campos

Call		kHz	kW	Name and location
RS58)	K218	1070	2/0.25	R. Caçapava, Caçapava do Sul
RS59)	K343	1070	1/0.25	R. Metrópole, Crissiumal
RS60)	K357	1070	1/0.25	R. Bento Gonçalves, Bento Gonçalves
SC91)	J747	1070	1/0.25	R. Gralha Azul, Urubici
SP91)	K603	1070	1/0.25	R. Piratininga, Jaú
SP145)	K615	1070	10/0.25	R. Metropolitana, Mogi das Cruzes
SP92)	K633	1070	5/1	R. Presidente Prudente, Presidente Prudente
SP212)	K758	1070	5/0.25	R. Jornal, Barretos
BA24)	H470	1080	20/0.5	R. Subaé AM, Feira de Santana
BA25)	H485	1080	1/0.25	R. Fascinação, Itapetinga
CE82)	H670	1080	2.5/0.25	R. Cultura, Quixadá
DF05)	H708	1080	25/5	R. Capital, Brasília
MG109)	L232	1080	2.5/0.25	R. Cultura, Dores do Indaiá (F.PI.) (still on 1530)
MG57)	L251	1080	10/0.5	R. Capital, Juiz de Fora
MT27)	I437	1080	1/0.25	R. Gaspar, Itiquira
PA32)	I540	1080	10/5	R. Novo Tempo, Belém
PE16)	I784	1080	5/1	R. Difusora, Caruaru
PE33)	I824	1080	1/0.25	R. Voluntários da Pátria, Ouricuri (F.PI.) (still on 1540)
PR47)	J201	1080	2.5/0.5	R. Clube Pontagrossense, Ponta Grossa
PR48)	J245	1080	1/0.25	R. Cultura do Norte, Paranavaí
PR49)	J261	1080	1/0.25	R. Educadora, Cornélio Procópio
RS61)	K254	1080	1/0.25	R. Marabá, Iraí
RS62)	K280	1080	10	R. da Univ. Federal do Rio Grande do Sul, Porto Alegre
SC29)	J759	1080	1/0.25	R. Clube, Indaial
SP94)	K557	1080	5/0.25	R. Difusora, Batatais
SP95)	K607	1080	1/0.25	R. Alvorada, Lins
SP96)	K669	1080	5/0.5	CBN Sorocaba, Sorocaba
SP97)	K704	1080	1/0.25	R. Monumental, Aparecida
SP190)	K710	1080	1/0.25	R. Alvorada, Cardoso
AL15)	H...	1090	5/0.5	Radiodifusão Eldorado, Pão de Açucar (CP)
BA26)	H455	1090	1/0.25	R. Santa Cruz, Ilhéus
GO24)	H758	1090	10/1	R. Aliança, Goiânia
MA08)	H893	1090	10	R. Rio Balsas AM, Balsas
PR51)	J283	1090	1/0.25	R. Vicente Pallotti, Coronel Vivida
RJ22)	J468	1090	25/5	R. Metropolitana, Rio de Janeiro
RN07)	J592	1090	10/5	R. Rural de Natal, Natal
RS63)	K216	1090	5/1	R. Cachoeira, Cachoeira doSul
RS64)	K262	1090	1/0.25	R. Salette, Marcelino Ramos
RS65)	K341	1090	1/0.25	R. Giruá, Giruá
SC30)	J732	1090	1/0.25	R. Colón, Joinville
SC31)	J786	1090	5/0.5	R. Tabajara, Tubarão
SP98)	K609	Þ1090	1/0.5	R. Clube, Marília
SP99)	K618	1090	1/0.25	R. Cultura, Monte Alto
SP233)	K768	1090	1/0.25	R. Em. do Planalto, Paulínia
CE67)	H638	1100	1/0.25	R. Difusora dos Inhamuns, Tauá
CE73)	H668	1100	1/0.25	R. Difusora do Vale Acaraú, Acaraú
RN18)	J607	1100	1/0.25	R. A Voz do Seridó, Caicó
SP100)	K694	Þ1100	150	R. Globo, São Paulo
CE32)	H620	1110	5/0.25	R. Litoral, Cascavel
GO46)	H782	1110	25/2	R. Redentor AM, Santo Antônio do Descoberto
MG58)	L205	1110	1/0.25	R. Planalto, Araguari
MG59)	L267	1110	1	R. Aurilândia, Nova Lima
MS08)	I392	1110	1	R. Ponta Porã, Ponta Porã
PB09)	I678	1110	20/10	R. Tabajara, João Pessoa
PR52)	J241	1110	10/1	R. Paiquerê, Londrina
PR53)	J253	1110	5/0.5	R. Eldorado, São José dos Pinhais
PR151)	J356	1110	1/0.5	R. Clube, Ubiratã
RJ23)	J471	Þ1110	50/5	R. Cultura, Campos
RS66)	K257	1110	2.5/0.25	R. Cultura, Jaguarão
RS67)	K306	1110	1/0.25	R. Sobradinho, Sobradinho
RS68)	K325	1110	2.5/0.25	R. Cruzeiro do Sul, Itaqui
RS152)	K364	1110	1/0.25	R. Solaris, Antônio Prado
SC74)	J743	1110	1/0.25	R. Caçanjure, Caçador
SC32)	J752	1110	1/0.5	R. Eldorado, Florianópolis
SC33)	J812	1110	1/0.25	R. São Carlos, São Carlos
SP101)	K544	1110	1/0.25	R. Luz, Araçatuba
SP102)	K592	Þ1110	1/0.25	R. Ibitinga, Ibitinga
SP103)	K617	1110	1	R. Cultura, Mogi Mirim
BA72)	H658	1120	25/1	R. Belo Campo, Belo Campo (F.PI.) (still on 1450)
CE23)	H598	1120	1	R. Tupinambá, Sobral

Call	kHz	kW	Name and location
ES14) I215	1120	1	R. Cricaré-AM, São Mateus
MG10) L272	1120	2.5/0.5	R. Ouro Preto, Ouro Preto
MG60) L301	1120	5/1	R. Sete Colinas, Uberaba
MG61) L332	1120	1/0.25	R. Serra da Boa Esperança, Boa Esperança
MS40) I...	1120	25/1	R. Campograndense, Campo Grande
PB10) I687	1120	1/0.25	R. Independência, Catolé do Rocha
PE17) I778	1120	5/1	R. Paulista, Paulista
PR85) J285	1120	5/0.5	R. Educadora, Laranjeiras do Sul (F.PI.) (still on 1400)
RS69) K274	1120	50	R. CBN, Porto Alegre
SP104) K631	1120	1/0.25	R. Em. Portofelicense, Porto Feliz
SP105) K660	1120	10/1	R. Clube AM, São José dos Campos
SP106) K671	1120	5/1	R. Clube Imperial, Taquaritinga
PA08) I531	Þ1130	1/0.25	R. Marajoara, Belém
PE18) I783	1130	5/1	R. Cultura do Nordeste, Caruaru
PR55) J220	1130	1/0.25	R. Castro, Castro
PR54) J333	1130	5/0.25	R. Ingamar, Marialva
RJ17) J460	1130	100/50	R. Nacional, Rio de Janeiro
RO06) J677	1130	5/0.25	R. Ji-Paraná, Ji-Paraná
RS70) K290	1130	5/1	R. Medianeira, Santa Maria
SC34) J790	1130	5/1	R. Princesa d'Oeste, Xanxerê
SP107) K676	1130	1/0.25	R. Tupã, Tupã
BA81) H449	1140	10	R. Cultura da Bahia, Salvador
CE24) H607	1140	1/0.25	R. Progresso, Russas
GO26) H751	1140	1/0.25	R. Tapirapés, Formosa
GO47) H801	1140	5/0.25	R. Pouso Alto, Piracanjuba
MG62) L204	1140	1/0.25	R. Minas, Divinópolis
MG63) L248	1140	1/0.25	R. Difusora, Campanha
MG64) L253	1140	1/0.25	R. Muriaé, Muriaé
MG129) L362	1140	5/0.5	R. Clube, Bocaiuva
MS22) I398	1140	10/0.5	R. Uberlim, Fátima do Sul
MS36) I418	1140	5/0.25	R. Cidade, Aparecida do Taboado (F.PI.) (still on 1570)
PR144) J352	1140	1/0.25	R. San Thiago Dantas, Chopinzinho
RS71) K228	1140	2/0.25	R. Cruz Alta, Cruz Alta
RS72) K316	1140	5/0.5	R. Charrua, Uruguaiana
RS73) K330	1140	1/0.25	R. Sobral, Butiá
SC35) J748	1140	10/0.5	R. Coroado AM, Curitibanos
SP108) K550	1140	1/0.25	R. Difusora, Assis
SP109) K555	1140	2/0.5	R. Barretos, Barretos
SP110) K645	1140	1/0.25	R. Educação e Cultura, Rio Claro
SP169) K708	1140	1/0.25	R. Regional AM, Registro
SP111) K709	1140	5/0.25	R. Costa Azul, Ubatuba
AL11) H250	1150	20/1	R. Cultura, Arapiraca
CE47) H643	1150	1/0.25	R. Cultura, Paracuru (F.PI.) (still on 1070)
MG65) L283	1150	20/5	R. CBN, Belo Horizonte
PI10) I891	Þ1150	10/5	R. Pioneira, Teresina
RJ24) J456	1150	10/0.5	R. Três Rios, Três Rios
RN25) J617	1150	1/0.25	R. Cabugi do Seridó, Jardim do Seridó
SP232) K777	Þ1150	100/50	R. Mundial, São Paulo
AM24) H...	1160	1/0.25	R. Sociedade TV Manauara, Boca do Acre (CP)
CE62) H652	1160	1/0.25	R. Vale do Coreaú, Granja
CE80) H660	1160	1/0.25	R. Araripe, Cedro (F.PI.) (still on 1590)
ES08) I202	1160	20/10	R. Espírito Santo, Vitória
GO42) H784	1160	1/0.25	R. Itaberaí, Itaberaí (F.PI.) (still on 1570)
MT15) I385	1160	10/5	R. A. Voz d'Oeste, Cuiabá
PA30) I558	1160	1/0.25	R. Guamá AM, São Miguel do Guamá
PB11) I674	1160	1	R. Soc. Campina Grande, Campina Grande
PR56) J258	1160	1/0.25	R. Norte, Londrina
RS74) K242	1160	5/0.5	R. Miriam, Farroupilha
RS75) K245	1160	5/1	R. Luz e Alegria, Frederico Westphalen
RS76) K256	1160	1/0.5	R. Jaguari, Jaguari
RS77) K273	1160	1	R. Universidade Católica, Pelotas
SC36) J741	1160	1	R. Globo, Blumenau
SC37) J767	1160	1/0.25	R. Laguna, Laguna
SC38) J776	1160	2.5/0.25	R. Difusora Colméia, Porto União
SP113) K502	1160	1/0.25	R. Presidente Venceslau, Presidente Venceslau
SP114) K517	Þ1160	5/0.5	R. Cacique, Sorocaba
SP124) K558	1160	2.5/1	R. Bandeirantes, Bauru
SP115) K582	1160	5/0.25	R. Difusora, Fernandópolis
SP237) K673	1160	1/0.25	R. Cacique, Taubaté
SP116) K685	1160	1/0.25	R. Clube, Mococa
AC08) H...	Þ1170	1/0.25	R. Difusora, Brasiléia
AM08) H284	1170	5/2.5	R. Guaranópolis, Maués
AM25) H...	1170	1/0.25	R. Difusora, Anori (CP)
AM21) H...	1170	1/0.25	Juruá Comunicações, Carauari (CP)
BA27) H473	1170	5/0.25	R. Jornal, Eunápolis
MG152) L234	1170	1/0.25	R. Fronteira, Fronteira
MG104) L269	1170	5/0.25	R. Oliveira, Oliveira (F.PI.) (still on 1510)
MG75) L327	1170	10/0.25	R. Vanguarda do Vale do Aço, Ipatinga
MG67) L336	1170	5/0.25	R. Cidade, Araxá
PR57) J273	1170	20/10	R. Atalaia, Curitiba
PR154) J363	1170	1/0.25	R. Colméia, Mandaguaçu (F.PI.) (still on 1550)
RN08) J598	1170	10/1	R. Difusora, Mossoró
RS78) K207	1170	1/0.25	R. Itapuí, Santo Antônio da Patrulha
RS79) K213	1170	5/1	R. Difusora A Voz de Bagé, Bagé
RS80) K359	1170	5/0.25	R. Uirapuru, Passo Fundo
RS155) K380	1170	2/1	R. Pitangueira, Itaqui
SP117) K569	1170	10/5	R. Educadora, Campinas
AL06) H248	1180	1/0.25	R. Correio do Sertão, Santana do Ipanema
AM20) H280	Þ1180	10/2.5	R. Difusora do Amazonas, Manaus
MA09) H889	Þ1180	10/5	R. Ribamar, São Luís
MG118) L203	1180	10/0.25	Nova R. Cultura, Alfenas (F.PI.) (still on 1570)
MS34) I602	1180	10	R. Guanandi AM, Campo Grande
MT38) N405	1180	5/0.25	Sociedade R. Educadora Nova Xavantina, Guarantã do Norte
PB20) I690	1180	1/0.25	R. Bonsucesso, Pombal
PE19) I797	1180	1/0.25	Radio Cultural de Vitória, Vitória de Santo Antão
PR126) J223	1180	2.5/0.5	R. Atalaia, Guarapuava
PR58) J237	1180	5/0.25	R. Guaçu, Toledo
PR81) J314	1180	1/0.25	R. Educadora, São João do Ivaí
RJ25) J463	1180	50/10	R. CBN (Globo), Rio de Janeiro
RS128) K340	1180	10/0.5	R. Gazeta, Santa Cruz do Sul
SC39) J737	1180	5/0.25	R. Integração d'Oeste, São José do Cedro
SC72) J770	1180	1/0.5	R. Difusora, Lages
SP260) K567	1180	1/0.25	R. Brotense, Brotas (F.PI.) (still on 1560)
SP179) K647	1180	1/0.5	R. Difusora, Santa Cruz do Rio Pardo
SP217) K749	1180	5/0.25	R. Nova, Bebedouro
BA28) H459	1190	10/1	R. Juazeiro, Juazeiro
CE68) H663	1190	1/0.25	R. Guaraciaba, Guaraciaba do Norte
GO35) H800	1190	10/0.25	R. Rio Vermelho, Silvânia
MG68) L221	1190	50/5	R. Guarani, Belo Horizonte
MG40) L276	1190	10/1	R. Mineira do Sul, Passa Quatro
PR59) J309	1190	1/0.25	R. Pontal, Nova Londrina
RN09) J594	1190	10/1	CBN, Natal
RS82) K234	1190	5/0.5	R. Cerro Azul, Cerro Largo
RS102) K301	1190	2.5/0.25	R. São Lourenço, São Lourenço do Sul
RS83) K354	1190	1	R. Rosário, Serafina Corrêa
SC40) J783	1190	1/0.25	R. Clube, São João Batista
SC87) J817	1190	1/0.25	R. Planalto, Major Vieira
SC41) J820	1190	1/0.25	R. Clube, São Domingos
SP199) K512	1190	2.5/0.25	R. Clube Marconi, Paraguaçu Paulista
SP118) K700	1190	5/0.25	R. Cidade AM, Votuporanga
SP119) K729	1190	10/0.25	R. 31 de Março, Santa Crus das Palmeiras
SP120) K741	1190	1/0.25	R. Regional, Taquarituba
AL14) H251	1200	20/1	R. Jornal, Maceió
BA29) H482	1200	1/0.25	R. Clube Rio do Ouro, Jacobina
CE26) H585	Þ1200	10	Ceará R. Clube, Fortaleza
RO07) J...	1200	5/1	R. Tucumã, Ji-Paraná (CP)
RS84) K239	1200	1/0.5	R. Erechim, Erechim
RS85) K342	1200	1/0.5	R. Cotrisel, São Sepé
SP121) K520	Þ1200	50/20	R. Cultura, São Paulo

Call	kHz	kW	Name and location
BA30) H452	1210	10/1	R. Carioca, Feira de Santana
BA58) H498	1210	10/0.25	R. Bandeirantes, Vitória da Conquista
CE50) H637	1210	5/0.25	R. Principe Imperial, Crateús (**F.PI.**) (still on 1430)
CE48) H641	1210	5/0.25	R. Boa Esperança, Barro
DF08) H...	1210	50/2.5	R. Brasília, Brasília
ES09) I200	1210	25/1	R. Cachoeiro, Cachoeiro de Itapemirim
MG69) L238	Þ1210	10/0.5	R. Clube, Varginha
MG70) L305	1210	10/1	R. Difusora, Uberlândia
PE20) I786	1210	5/1	R. Jornal, Garanhuns
PI11) I893	1210	10/1	R. Educadora, Parnaíba (ff 920)
PR60) J219	Þ1210	10/5	R. Universo, Curitiba
PR140) J325	1210	1/0.5	R. Brotense, Porecatu
RN29) J620	1210	5/0.5	R. Potengi, São Paulo do Potangi (**F.PI.**) (still on 1500)
RS86) K240	1210	10/5	R. Record AM, Porto Alegre
RS88) K353	1210	1/0.5	R. Blau Nunes, Santa Bárbara do Sul
SC42) J785	1210	1/0.25	R. Super Santa, Tubarão
SP122) K509	1210	1/0.25	R. Vida Nova, Jaboticabal
SP123) K545	1210	1/0.25	R. Difusora, Araçatuba
SP125) K668	1210	5/0.25	R. Vanguarda, Sorocaba
RJ25) J458	Þ1220	150	R. Globo, Rio de Janeiro
BA59) H...	1230		R. Jornal, Ubatã
GO27) H756	Þ1230	10/2.5	R. CBN Anhanguera, Goiânia
MA23) H...	1230	1/0.25	R. Alecrim, Caxias
MG105) L208	1230	5/0.25	R. Correio da Serra, Barbacena
MG102) L216	1230	2.5/0.25	R. Passos AM, Passos
MG71) L296	1230	5/0.25	R. Educadora Novo Tempo, Governador Valadares
PB12) I670	1230	10/1	CBN, João Pessoa
RS146) K297	1230	1/0.25	R. Santiago, Santiago
RS89) K326	1230	1/0.25	R. Nonoai, Nonoai
RS90) K333	1230	1/0.25	R. Prata, Nova Prata
RS91) K352	1230	1/0.25	R. Encruzilhadense, Encruzilhada do Sul
SC88) J816	1230	10/1	R. Guararema, São José
SP126) K573	1230	1/0.25	R. Cidade, Capão Bonito
SP258) K637	1230	10/0.25	R. Difusora, Rancharia (**F.PI.**) (still on 1520)
SP127) K703	1230	2.5/0.25	R. Noticias, Matão (ff 890)
SP128) K716	1230	5/0.5	R. Jequitibá AM, Campinas
BA31) H463	1240	10/0.25	R. Emiss. de Alagoinhas, Alagoinhas
CE49) H654	1240	1/0.25	R. São Francisco, Canindé
MG97) L294	1240	5/0.25	R. Três Pontas, Três Pontas
MG84) L298	1240	1/0.25	Soc. R. Ubaense, Ubá
MG72) L303	1240	1/0.25	R. Platina, Ituiutaba
MG116) L317	1240	5/0.25	R. Pirapora, Pirapora
MS09) I388	1240	5/1	CBN, Campo Grande
PE21) I774	1240	5	R. Capibaribe, Recife
PR112) J215	1240	5/0.25	R. Arapongas, Arapongas
PR61) J280	1240	2/0.25	R. Matelândia, Matelândia
PR94) J301	1240	1/0.25	R. Difusora Ubiratanense, Ubiratã (**F.PI.**) (still on 1450)
RO08) J...	1240	1	R. Verde, Jaru (CP)
RS92) K200	1240	1/0.25	R. Aparados da Serra, Bom Jesus
RS93) K251	1240	5/0.25	R. Ibirubá, Ibirubá
RS94) K355	1240	1/0.25	R. São Jerônimo AM, São Jerônimo
SC43) J774	1240	5/0.5	R. São José AM, Mafra
SC44) J810	1240	1/0.25	R. Iracema, Cunha Porã
SP129) K565	1240	10/0.25	R. Municipalista, Botucatu
SP130) K621	1240	5/0.25	Orlândia R. Clube, Orlândia
SP131) K653	1240	10/2.5	R. Clube, Santos
SP132) K711	1240	5	R. Vale do Rio Tietê, José Bonifácio
AM09) H289	Þ1250	1/0.25	R. Educação Rural, Coari
CE27) H594	1250	1	R. Educadora, Crateús
CE69) H...	1250	1/0.25	R. Liberdade, Itarema (CP)
ES18) I218	1250	10/1	R. CBN, Vitória
GO29) H748	1250	5/1	R. Difusora São Patricio, Ceres
MG73) L282	Þ1250	5	R. Difusora, Poços de Caldas
MG153) L367	1250	20/0.25	R. Metropolitana, Vespasiano
MS10) I394	1250	1/0.25	R. Difusora, Três Lagoas
PB27) I...	1250	1/0.25	R. Sociedade de Soledade, Soledade (CP)
PI47) I932	1250	1/0.25	R. João de Paiva, Altos
PR62) J211	1250	5/0.5	R. Difusora, Guarapuava
PR63) J233	1250	1/0.25	R. Paranavaí, Paranavaí
PR64) J313	1250	1/0.25	R. Danúbio Azul, Santa Isabel do Oeste
RJ50) J500	1250	10/0.5	R. Litoral, Casimiro de Abreu
RS95) K233	1250	10/0.5	R. Difusora Caxiense, Caxias do Sul
RS96) K272	1250	1	R. Tupanci, Pelotas
RS142) K361	1250	1/0.25	R. Aguas Claras, Catuípe
SC45) J766	1250	5/0.25	R. Cultura, Joinville
SE08) J925	1250	10/1	R. Esperança, Estância
SP133) K702	1250	5/0.5	R. Difusora do Grande Vale, Caçapava
AL07) H242	1260	50/5	R. Gazeta de Alagoas, Maceió
CE28) H596	1260	1/0.25	R. Vale do Jaguaribe, Limoeiro do Norte
MG77) L273	1260	25/5	R. Record, Uberlândia
RO09) J670	Þ1260	5	R. Educadora, Guajará Mirim
RS97) K204	1260	1/0.25	R. Cultura, São Borja
RS98) K327	1260	5/0.25	R. Em. Fandango, Cachoeira do Sul
RS99) K345	1260	1/0.25	R. Gaurama, Gaurama
SC46) J740	1260	5/0.5	R. Blumenau, Blumenau
SP257) K629	1260	1/0.25	Pirajuí R. Clube, Pirajuí
SP134) K688	1260	100/25	R. Morada do Sol, São Paulo
AM10) H282	Þ1270	2	R. Educação Rural, Tefé
GO30) H753	Þ1270	50/10	R. Brasil Central, Goiânia
MG74) L227	1270	5/1	R. Carijós, Conselheiro Lafaiete
MG107) L300	1270	2.5/0.5	R. Estância, São Lourenço
MG155) L...	1270	5/1	R. Libertas do Vale do Aço, Ipatinga (CP)
PA09) I530	1270	10/2.5	R. TransPaz, Belém
PB28) I696	1270	5/0.25	R. Cidade, Sumé
PR65) J222	1270	5/0.5	R. Atual Guairacá, Mandaguari
PR67) J236	1270	5/0.5	R. Capital, Curitiba
PR68) J289	1270	5/0.5	R. Cidade Cariri, Cascavel
RJ26) J474	1270	10/0.5	R. Continental, Campos
RN10) J593	Þ1270	0.5	R. Poti, Natal
RS131) K206	1270	5/0.5	R. América do Rio Grande do Sul, Montenegro
RS101) K250	1270	5/0.5	R. Vera Cruz, Horizontina
SC47) J765	1270	5/0.25	R. Catarinense, Joaçaba
SC48) J768	1270	1/0.25	R. Garibaldi, Laguna
SP135) K640	1270	10/0.5	R. Thathi, Ribeirão Preto
SP136) K678	Þ1270	5/0.5	R. Brasil 5000, Campinas
PB21) I688	1280	10/1	R. Sinhauá, Bayeux
RJ27) J455	1280	100	Super R. Tupi, Rio de Janeiro
AM11) H286	Þ1290	10/2.5	R. Rio Mar, Manaus
BA32) H450	1290	10/1	R. Clube, Salvador
MA10) H888	Þ1290	50/5	R. Timbira do Maranhão, São Luís
MG164) L345	1290	1/0.25	R. Cidade, Arcos
PR73) J310	1290	10/0.5	R. Brasil Sul, Londrina
RN26) J619	1290	1/0.25	Sistema Potiguar de Informação, Caicó
RS103) K331	1290	5/2	R. Planetário, Espumoso
SC81) J734	1290	1	R. Araranguá, Araranguá (**F.PI.**) (still on 1350)
SC49) J804	1290	5/1	R. Camboriú, Balneário Camboriú
SP240) K662	1290	1	R. Difusora, São José do Rio Pardo
SP137) K663	1290	5/1	R. Independência, São José do Rio Preto
SP216) K745	1290	1/0.5	R. Vale, São José dos Campos
CE29) H586	1300	10	R. CBN, Fortaleza
ES11) I210	1300	1	R. Novo Tempo, Afonso Cláudio
MG79) L210	1300	1/0.25	R. Cancella, Ituiutaba (ff 710)
MG143) L339	1300	5/1	R. Eldorado, Sete Lagoas
PE22) I799	1300	1/0.25	R. Guarany AM Stereo, Camarajibe
PR71) J278	1300	1/0.25	R. Pitangui AM, Ponta Grossa
PR127) J288	1300	5/0.25	R. Educadora, Dois Vizinhos
RS104) K203	1300	50/10	Super R. Tupi, Porto Alegre
RS105) K337	1300	1/0.25	R. Regional, Santo Cristo
RS106) K347	1300	5/0.5	R. Marathan, Santana do Livramento
SC89) J819	1300	1/0.25	R. Alvorada, Santa Cecília
SP138) K535	1300	50/1	R. Universo, São Bernardo do Campo
SP252) K649	1300	5/0.25	R. Cultura, Santo Anastácio
SP226) K762	1300	1/0.25	R. Realidade AM, São Carlos (**F.PI.**) (still on 1590)

Call	kHz	kW	Name and location
BA33) H454	1310	1/0.25	R. Novo Tempo, Ilhéus
BA42) H476	1310	5/1	R. Educadora Santana de Caetité, Caetité (ff 920)
BA63) H501	1310	5/0.25	R. Canto da Sereia, Jacobina
CE30) H602	1310	1	R. Progresso de Juazeiro, Juazeiro do Norte
CE63) H656	1310	1/0.25	R. Liberdade, Boa Viagem
MG168) L351	1310	10/0.25	R. Difusora, Salinas (F.PI.) (still on 1540)
MG144) L359	1310	1/0.25	R. Montanhesa, Vazante
MS31) I426	1310	5/1	R. Pindorama, Sidrolândia
PB23) I691	1310	10/0.5	R. Cidade Esperança, Esperança
PR70) J274	1310	10/0.5	R. Atalaia, Maringá
RJ28) J504	1310	1/0.25	R. Difusora Coroados, São Fidélis
RO17) J684	1310	10/5	R. Tropical AM, Porto Velho
RS107) K305	1310	5/1	R. Sarandi, Sarandi
RS124) K329	1310	2.5/0.25	R. Integração, Restinga Seca
RS160) K371	1310	2.5/0.5	R. Horizonte, Capão da Canoa
SC85) J801	1310	5/0.25	R. Sintonia 1310 AM, Ituporanga
SP141) K566	1310	2.5/0.25	R. Bragança, Bragança Paulista
SP139) K596	1310	2/1	R. Difusora, Itápolis
AL08) H243	1320	10/0.25	R. Progresso de Alagoas, Maceió
BA69) H503	1320	5/1	R. Regional, Cícero Dantas
CE31) H597	1320	1	R. Regional, Sobral
CE70) H672	1320	1/0.5	R. Vento Leste AM, Aracati
MG136) L322	1320	1/0.25	R. Mucuri, Teófilo Otoni
PE31) I823	1320	1/0.25	R. Cultura, São José do Egito
PR72) J255	1320	8/0.5	R. Porto Alegre AM, Curitiba
PR145) J351	1320	5/0.5	R. Foz, Foz do Iguaçu
RJ29) J475	1320	25/5	R. Difusora Boas Novas, Rio de Janeiro
RS108) K223	1320	1/0.25	R. Clube de Canela, Canela
RS110) K271	1320	5/1	R. Cultura, Pelotas
SP140) K630	1320	1/0.25	R. Difusora, Pirassununga
SP241) K675	1320	5/1	R. Clube, Tupã
BA34) H468	1330	5/1	R. Difusora, Serrinha
PA10) I533	Þ1330	25/5	R. Liberal, Belém (vf)
PR74) J264	1330	10/0.5	R. Jaguariaíva, Jaguariaíva
PR75) J320	1330	5/0.25	R. Educadora Inconfidência, Umuarama
RN27) J621	1330	10/0.5	R. Eldorado, Natal
RS111) K236	1330	1/0.25	R. Upacaraí, Dom Pedrito
RS112) K323	1330	1/0.5	R. Diplomata, São Marcos
SC50) J739	1330	5/0.5	R. Clube, Blumenau
SC51) J749	1330	5/1	R. Chapecó, Chapecó
SP142) K638	1330	5/0.25	R. Paulista, Regente Feijó
SP143) K641	1330	1	R. Cultura, Ribeirão Preto
SP187) K736	Þ1330	50/5	R. Nova Iguatemi, Osasco
CE71) H661	1340	2.5/0.25	R. Planalto, Maracanaú
MA11) H886	1340	10/2	R. Club São Luís, São Luís
MG81) L241	1340	50/5	R. Cultura, Itabirito
MG139) L352	1340	1/0.5	R. Minas Liberdade, Passos
MG156) L366	1340	1/0.5	R. Capital do Triângulo, Patrocínio (ff 810)
MS12) I380	Þ1340	1	R. Difusora, Aquidauana
PB13) I671	1340	5/1	R. Arapuan, João Pessoa
PR76) J205	1340	2.5/0.25	R. Difusora, Rio Negro
PR77) J249	1340	1/0.25	R. Cultura, Arapongas
PR41) J368	1340	5/0.25	R. Cascavel, Cascavel
RJ40) J490	1340	5/0.5	R. Novo Tempo, Rio Bonito
RS113) K227	1340	10/1	R. Educadora, Porto Alegre
SP144) K543	1340	5/1	R. Cultura, Araçatuba
SP203) K571	1340	5/0.25	R. Em. Campos do Jordão, Campos do Jordão (F.PI.) (still on 1560)
SP146) K738	1340	1/0.25	R. Canoa Grande, Igaraçu do Tietê
AC05) H201	1350	10/5	R. Capital, Rio Branco
AP02) H420	Þ1350	5	R. Equatorial, Macapá (ff 670)
BA70) H520	1350	50/10	R. Cristal, Salvador
CE56) H662	1350	1/0.25	R. Liberal, Morada Nova
MG82) L214	1350	10/5	R. Cultura, Poços de Caldas
PB14) I675	Þ1350	0.5	Super R. Borborema, Campina Grande
RS114) K205	1350	1/0.25	R. Aurora, Guaporé
RS109) K266	1350	1/0.25	R. Panambi, Panambi
RS115) K313	1350	5/1	R. Difusora, Três Passos
RS116) K336	1350	1/0.25	R. Agudo, Agudo
SC81) J734	1350	1/0.25	R. Araranguá, Araranguá (ff 1290)
SC52) J745	1350	1/0.25	R. Clube, Canoinhas (ff 890)
SC53) J760	1350	1/0.25	R. Clube, Itajaí
BA35) H469	1360	10/1	R. Cultura, Paulo Afonso
CE57) H650	1360	5/0.25	R. Iracema, Ipu
MS14) I383	1360	2	R. Difusora Matogrossense, Corumbá
PR165) J265	1360	10/0.25	R. Cidade, Pato Branco
PR78) J268	1360	1/0.25	R. Pepita de Ouro, Assaí
PR79) J302	1360	5/0.25	R. Cultura, Iporã
RJ30) J464	1360	50/10	R. Guanabara, Rio de Janeiro
RN11) J605	1360	1/0.25	R. Ouro Branco, Currais Novos
RS117) K261	1360	2/0.5	R. Alvorada, Marau
SP147) K581	1360	5/1	R. Aguas Quentes, Fernandópolis
SP148) K739	1360	5/0.25	R. Regional, Dracena
SP235) K759	1360	1/0.5	R. Luzes da Ribalta, Santa Bárbara d'Oeste
BA82) H555	1370	0.25	R. Piquaraca, Monte Santo
CE81) H...	1370	1/0.25	R. Vanguarda, Caridade (CP)
PE34) I800	1370	1/0.25	R. Vale do Capibaribe, Santa Cruz do Capibaribe
PI13) I892	1370	10/1	R. Difusora, Teresina
PR80) J267	1370	50/7	R. Independência, Curitiba
RN28) J618	1370	2.5/0.25	Sistema Potiguar de Informação,São Miguel
RS118) K243	1370	25/0.5	R. Veneto, Flores da Cunha
RS119) K334	1370	1/0.25	R. Gazeta, Alegrete
RS173) K377	1370	2/0.25	R. Jornal da Manhã, Ijuí (F.PI.) (still on 1580)
SC55) J782	1370	5/0.25	R. Peperí, São Miguel do Oeste
SE09) J929	1370	5/0.5	R. Capital do Agreste, Itabaiana
SP223) K766	1370	100/50	R. Atual AM, São Paulo
AM12) H283	Þ1380	5/1	R. Alvorada, Parintins
BA83) H495	1380	5/0.25	R. União, Gandu
MG172) L218	1380	1/0.25	R. Difusora, Brasópolis (F.PI.) (still on 1450)
MG120) L284	1380	5/0.25	R. Paranaíba, Rio Paranaíba
MG130) L323	1380	1/0.25	R. Gorutubana AM, Janaúba
PA11) I547	1380	1/0.25	R. Marajó, Breves (ff 870)
PA31) I...	1380	1/0.25	R. Tocantins, Tucuruí (CP)
PE23) I773	1380	10/5	R. Continental, Recife
PR122) J276	1380	1/0.25	R. Bom Jesus, Siqueira Campos
PR152) J367	1380	1/0.25	Rede Integração de Comunicação, Toledo
RJ31) J470	1380	10/1	Campos R. Difusora, Campos
RS120) K293	1380	1/0.25	R. Cultura, Santana do Livramento
RS134) K311	1380	1/0.5	R. Maristela, Torres
RS121) K350	1380	1/0.25	R. Gazeta, Tapera
RS165) K372	1380	1/0.25	R. Chiru Comunicações, Palmitinho
SC56) J821	1380	1/0.25	R. Cidade, Itaiópolis (ff 990)
SC93) J827	1380	1/0.25	R. Barriga Verde, Capinzal
SP151) K507	1380	1/0.25	R. Difusora, Lençóis Paulista
SP152) K616	1380	1/0.5	R. Difusora, Mogi Guaçu
SP224) K751	1380	5/0.25	R. Diário, Presidente Prudente
SP234) K772	1380	1/0.25	R. República, Morro Agudo
CE72) H667	1390	10/0.25	R. Paty, Senador Pompeu
ES13) I209	1390	5/0.25	R. Educadora, Afonso Cláudio (F.PI.) (still on 1580)
MA22) H907	1390	10/0.5	R. Cultura do Rio Jordão, Coroatá
MG157) L358	1390	2.5/0.25	R. Ouro Verde, São Sebastião do Paraíso
PA12) I535	Þ1390	10/1	R. Educadora, Bragança
PE24) I788	1390	1	R. Jornal O Povo, Pesqueira
PR82) J242	1390	10/1	R. Cultura, Maringá
PR83) J335	1390	1/0.25	R. Independência, Salto do Lontra
RJ32) J473	1390	5/0.5	R. Sul Fluminense, Barra Mansa
RO18) J687	1390	5/1	R. Itapirema, Ji-Piraná
RR02) O701	1390	10/5	R. Roraima, Caracaraí
RS122) K209	1390	10	R. Esperança, Porto Alegre
RS166) K368	1390	1/0.25	R. Atlântica, Constantina
SC57) J769	1390	1/0.25	R. Diário da Manha, Lages
SP153) K570	1390	5/0.25	CBN, Campinas
SP90) K594	1390	2.5/0.25	R. Anchieta, Itanhaém
SP154) K636	1390	1/0.25	R. Cultura, Promissão
AC06) H200	Þ1400	10/1	R. Difusora Acreana, Rio Branco
BA71) H549	1400	1/0.25	R. Vale do Vasa-Barris, Jeremoabo
PB15) I677	1400	5/1	R. Espinharas, Patos
PI27) I926	1400	1/0.25	R. Cantagalo, Jaicós
PR84) J256	1400	1/0.25	R. CBN Londrina, Londrina
PR85) J285	1400	5/0.25	R. Educadora, Laranjeiras do Sul (ff 1120)
PR87) J299	1400	1/0.25	R. Fronteira d'Oeste, Terra Roxa
RJ33) J462	1400	50/5	R. Rio de Janeiro ("R. Rio AM"), Ri

Call	kHz	kW	Name and location
RS123) K247	1400	1/0.25	R. Sideral, Getúlio Vargas
SC58) J775	1400	1/0.25	R. Entre Rios, Palmitos
SP155) K527	1400	1/0.25	R. Difusora, Lucélia
SP156) K658	1400	5/0.25	R. Progresso, São Carlos
SP157) K682	1400	5/0.25	R. Metrópole AM, São José do Rio Preto
TO08) N660	1400	5/1	Radiodifusão Guaraí, Guaraí
BA37) H467	1410	10/0.5	R. São Gonçalo, São Gonçalo dos Campos
CE25) H639	1410	10/1	R. Boa Nova, Pacajus
GO19) H803	1410	10/1	R. Fraternidade Universal, Santo Antônio do Descoberto
MS15) I382	1410	5/1	R. Clube, Corumbá
PR141) J328	1410	1/0.25	R. Cultura, Cândido de Abreu (ff 710)
RJ34) J486	1410	10/0.5	R. Itaperuna, Itaperuna
RS137) K246	1410	5/1	R. Garibaldi, Garibaldi
RS125) K284	1410	1/0.25	R. Minuano, Rio Grande
RS126) K294	1410	5	R. Santa Rosa, Santa Rosa
SC100) J799	1410	0.5	R. Namba, Ponte Serrada
			(F.PI.) (still on 1590)
SP158) K691	1410	50/10	R. América, São Paulo
BA60) H504	1420	1/0.25	R. Difusora, Irecê
MG88) L286	1420	1	Multisom R. São João Nepomuceno, São João Nepomuceno
MG89) L288	1420	5/1	R. Cultura, Sete Lagoas
MG90) L313	1420	1/0.25	R. Montanhês Botelhos, Botelhos
MS16) I397	1420	1/0.25	R. Difusora Cacique, Nova Andradina
PR88) J269	1420	5/0.25	R. Cultura, Umuarama
PR89) J282	1420	1	R. Educadora, Jacarezinho
PR90) J334	1420	1	R. Entre Rios, Santo Antônio do Sudoeste
RN12) J609	1420	1/0.25	R. Tapuyo, Alexandria
RS149) K258	1420	10/0.25	R. 14 de Julho, Júlio de Castilhos
SC60) J744	1420	5/0.25	R. Cultura, Campos Novos
SC59) J754	Þ1420	5/2.5	R. Guarujá, Florianópolis
SP159) K597	1420	1/0.25	CRN, Itatiba
SP10) K612	1420	1/0.25	R. Clube, Mirandópolis (ff 590)
SP162) K674	1420	1/0.25	R. Cultura, Taubaté
SP163) K733	1420	1/0.25	R. Nova São Manuel, São Manuel
CE50) H637	1430	1	R. Principe Imperial, Crateús (ff 1210)
MG91) L239	1430	1/0.25	R. Clube, Guaxupé
MG92) L252	1430	1/0.25	R. Educadora, Ubá
MG158) L371	1430	1/0.25	R. Planalto, Perdizes
PE32) I826	1430	1/0.25	R. Independência, Goiana
PR134) J200	Þ1430	50/10	R. Clube Paranaense, Curitiba
RN13) J604	1430	10/0.5	R. Libertadora Mossoroense, Mossoró
RO11) J671	1430	1	R. Caiari, Porto Velho
RS129) K224	1430	1/0.25	R. Cultura, Canguçu (ff 1030)
RS167) K366	1430	1/0.25	R. Guarita, Coronel Bicaco
SP164) K666	1430	1/0.25	R. Serra Negra, Serra Negra
SP171) K707	1430	25/0.25	R. Universal, São Roque
			(F.PI.) (still on 1480)
AM13) H285	Þ1440	10	R. Baré, Manaus
BA38) H466	1440	10/0.5	R. Independência, Santo Amaro
CE33) H603	1440	10/1	R. Cidade, Crato
MG159) L365	1440	1/0.25	Empreendimentos Guimarães e Franzão ("R. Som 200"), Santa Vitória
MS41) I407	1440	1/0.25	R. Bela Vista, Bela Vista
PR147) J353	1440	1/0.25	R. Integração Oeste, Corbélia (ff 730)
RJ35) J469	1440	20/5	R. AM 1440 ("R. AM O Día"), Rio de Janeiro
RS168) K221	1440	1/0.5	R. Ceres, Naõ Me Toque
RS130) K328	1440	1/0.25	R. Excelsior, Gramado
RS153) K362	1440	1/0.25	R. Caibaté, Caibaté
SC61) J792	1440	1/0.25	R. Difusora, Maravilha
SE11) J930	1440	5/0.25	R. Educadora, Frei Paulo
SP253) K568	1440	1/0.25	R. Cultura, Cajuru
			(F.PI.) (still on 1460)
SP165) K634	1440	5/0.25	R. Comercial, Presidente Prudente
SP218) K752	1440	1/0.25	R. Azul Celeste, Americana
BA73) H531	1450	1/0.25	R. Ipirá, Ipirá
BA72) H658	1450	10/1	R. Belo Campo, Belo Campo (ff 1120)
CE34) H601	1450	1/0.25	R. Difusora Cristal, Quixeramobim
CE35) H623	1450	1/0.25	R. Pinto Martins, Camocim
ES12) I208	1450	1/0.5	R. Gaeta, Guarapari
GO51) H794	1450	1/0.25	R. Alvorada, Quirinópolis
MA32) H900	1450	1/0.25	R. Boa Esperança, São João dos Patos
MA13) H901	1450	1/0.25	R. Cultura, Pedreiras
MG172) L218	1450	1/0.25	R. Cidade, Brasópolis (ff 1380)
MG93) L249	1450	5/0.25	R. Cultura, João Monlevade
MG94) L312	1450	10/0.25	R. Diamante, Coromandel
MS13) I417	1450	1/0.25	R. Difusora, Rio Brilhante
PA33) I559	1450	1/0.25	R. Juruá, São Felix do Xingu
PB29) I699	1450	1/0.25	R. Itatiunga, Patos
PE25) I794	1450	1/0.25	R. Cultura, Palmares
PI21) I908	1450	1/0.25	R. Cultura do Gurguéia, Bom Jesus
PI35) I917	1450	1/0.25	R. Confederação Valenciana ("R. Valenciana"), Valença do Piauí
PR92) J225	1450	1/0.25	R. Soc. Monte Alegre, Telêmaco Borba
PR93) J279	1450	1/0.25	R. Cabiúna, Bandeirantes
PR128) J293	1450	1/0.25	R. Najuá, Irati
PR94) J301	1450	1/0.25	R. Difusora Ubiratanense, Ubiratã (ff 1240)
PR95) J317	1450	1/0.25	R. Rainha d'Oeste, Altônia
PR148) J346	1450	1/0.25	R. Jornal São Miguel, São Miguel do Iguaçu
RJ36) J480	1450	5/0.25	R. do Comércio, Barra Mansa
RJ37) J503	1450	1/0.25	R. Feliz, Santo Antônio de Pádua
RO12) J674	1450	1	R. Vilhena AM, Vilhena
RS177) K338	1450	1/0.25	R. Cultura, Arvorezinha
RS132) K346	1450	1/0.25	R. Cassino, Rio Grande
SC62) J802	1450	1/0.25	R. São Bento, São Bento do Sul
SC63) J822	1450	5/0.25	R. Soc. Hulha Negra, Criciúma
SC97) J828	1450	1/0.25	R. Belos Montes, Seara
SE12) J...	1450	1/0.25	R. Abais, Estância (CP)
SP238) K526	1450	1/0.25	R. Cultura, Ituverava
SP166) K587	1450	1/0.25	R. Difusora, Guararapes
SP167) K591	1450	10/0.5	R. Difusora Boa Nova, Guarulhos
SP168) K657	Þ1450	1/0.25	R. São Carlos, São Carlos
AL13) H253	1460	1	R. Quilombo, União dos Palmares (ff 840)
AM17) H300	1460	5	R. Clube, Parintins
BA39) H472	1460	1	R. Baiana, Jequié
BA74) H...	1460	1/0.25	R. Alvorada, Cruz das Almas (CP)
BA85) H...	1460	1/0.25	R. Ferro Doido, Morro do Chapéu (CP)
CE36) H616	1460	1/0.25	R. Uirapuru, Morada Nova
CE53) H...	1460	1/0.25	R. Dragão do Norte, Massapé
GO34) H766	1460	1/0.25	R. Morrinhos, Morrinhos
MA33) H917	1460	1/0.25	R. Vanguarda, Santa Luzia
MG95) L201	1460	1	R. Cultura de Porto Novo, Além Paraíba
MG161) L356	1460	1/0.25	R. Buritis, Buritis
MG131) L363	1460	1/0.25	R. Entre Rios, Raul Soares
PB30) I...	1460	1/0.25	R. Oeste da Paraíba, Cajazeiras (CP)
PI14) I903	1460	1/0.25	R. Cultura, Amarante
PR96) J204	1460	1/0.25	R. Difusora, Paranaguá
PR97) J228	1460	1/0.25	R. Central do Paraná, Ponta Grossa
PR98) J251	1460	1/0.25	R. Cultura, Apucarana
PR99) J286	1460	2.5/0.25	R. Club de Palmas, Palmas (ff 1050)
PR100) J297	1460	1/0.25	R. Guaíra, Guaíra
PR101) J308	1460	1/0.25	R. Ampere, Ampere
PR102) J318	1460	1/0.25	R. Educadora, Loanda
RN20) J615	1460	2.5/0.25	R. Agreste, Santo Antônio
RS133) K214	1460	1/0.25	R. Cultura, Bagé
RS135) K312	1460	1/0.25	R. Colonial, Três de Maio
RS175) K373	1460	0.25	R. Campinas do Sul, Campinas do Sul
RS176) K...	1460	0.25	R. Mostardas, Mostardas (CP)
SC64) J756	1460	1	R. Sentinela do Vale, Gaspar
SP242) K548	1460	1	R. Clube Ararense, Araras
SP253) K568	1460	0.25	R. Cultura, Cajuru (ff 1440)
SP170) K608	1460	1/0.25	R. Cultura, Lorena
TO02) H774	1460	1/0.25	R. Independência do Tocantins, Paraíso do Tocantins
BA86) H...	1470	0.25	R. Morro Verde, Mairi (CP)
CE64) H665	1470	1/0.25	R. Guanacés de Itapajé, Itapajé
ES17) I214	1470	1/0.25	R. São Francisco, Barra de São Fra

Call	kHz	kW	Name and location
GO36) H773	1470	1/0.25	R. Dif. Serra dos Cristais, Cristalina
GO37) H779	1470	1/0.25	R. Cidade, Goiás
MA34) H...	1470	1/0.25	R. Paranoá, Presidente Dutra
MG96) L247	1470	1	R. Difusora, Ituiutaba
MS29) I413	1470	1/0.25	R. Alvorada, Dourados
PA25) I548	1470	5/1.5	R. Moreno Braga, Vigia
PE35) I822	1470	1/0.25	R. Educadora de Belém, Belém de São Francisco
PE37) I827	1470	1/0.25	R. Bom Conselho, Bom Conselho
PI15) I900	1470	1/0.25	R. Difusora Vale do Uruçuí, Uruçuí
PI24) I913	1470	1/0.25	R. Ingazeira, Paulistana
PI33) I928	1470	1/0.25	R. AM Cidade, Castelo do Piauí
PR103) J294	1470	1/0.25	R. Novo Horizonte, Ibaiti
PR104) J304	1470	2/0.25	R. Jornal, Assis Chateaubriand
PR105) J327	1470	1/0.25	R. Auri Verde, Pitanga
RJ08) J476	1470	1/0.25	R. Jornal Fluminense, Campos
RJ38) J481	1470	1/0.25	R. Vale do Paraíba, Barra do Piraí
RN23) J616	1470	1/0.25	R. Rural, Parelhas
RO05) J676	1470	1/0.25	R. Soc. Rondônia, Cacoal
RS169) K219	1470	1/0.25	R. Cultura Cacequiense, Cacequi
RS136) K324	1470	2.5/0.25	R. Cinderela, Campo Bom
SC65) J781	1470	6/1	R. Gazeta, São José
SC66) J798	1470	1/0.25	R. Líder do Vale, Herval d'Oeste
SP172) K586	1470	1/0.25	R. Cultura, Guaíra
SP173) K599	1470	1/0.25	R. Clube, Jacareí
SP174) K632	1470	1/0.25	R. Primavera, Porto Ferreira
SP175) K712	1470	1/0.25	R. Jornal, Indaiatuba
SP243) K771	1470	1/0.25	R. Bastos AM, Bastos
BA61) H508	1480	5/1	R. Alvorada, Guanambi
BA75) H...	1480	1/0.25	R. Santana, Santana (CP)
CE58) H636	1480	1/0.25	R. Cidade, Campos Sales (ff 630)
CE74) H...	1480	1/0.25	R. Universal, Morrinhos (CP)
MA14) H897	1480	1/0.25	R. Itapecuru, Colinas
MG98) L235	1480	1/0.25	R. Nova Frutal AM, Frutal
MG99) L265	1480	5/0.25	R. Difusora, Nanuque
MG100) L307	1480	2.5/0.25	R. Emboabas, Tiradentes
MG148) L341	1480	5/1	R. Carangola, Carangola
MS17) I393	1480	1/0.25	R. Caçula, Três Lagoas
PE26) I790	1480	1/0.25	R. A Voz do Sertão, Serra Talhada
PE36) I825	1480	1/0.25	R. Vale do Ipojuca, Gravatá
PI29) I929	1480	1/0.25	R. Vale do Coroatá, Elesbão Veloso
PR153) J221	1480	1/0.25	R. Piraí do Sul, Piraí do Sul
PR106) J230	1480	1/0.25	R. Astorga, Astorga
PR107) J270	1480	1/0.25	R. Educadora, União da Vitória
PR142) J359	1480	1/0.25	Radiodifusão Campo Aberto, Laranjeiras do Sul
RJ39) J488	1480	1/0.25	R. Mauá Solimões, Nova Iguaçu
RN14) J601	1480	1/0.25	R. Princesa do Vale, Açu
RO15) J681	1480	1/0.25	R. Soc Rondônia, Pimenta Bueno
RS100) K244	1480	2.5/0.25	R. São Roque, Faxinal do Soturno
RS127) K321	1480	1/0.25	R. Veranense, Veranópolis
SC67) J731	1480	1/0.25	R. Difusora, Joinville
SC68) J762	1480	1/0.25	R. Difusora 26 de Abril, Imaruí
SC94) J826	1480	1/0.25	R. Caibi, Caibi
SE10) J928	1480	1/0.25	R. Atalaia, Simão Dias
SP176) K539	1480	1/0.25	R. Clube, Altinópolis
SP177) K551	1480	1/0.25	R. Técnica, Atibaia
SP178) K562	1480	1	R. Clube, Bilac
SP171) K707	1480	1/0.25	R. Universal, São Roque (ff 1430)
TO06) H795	1480	1/0.25	R. Anhanguera, Miracema do Tocantins
AL09) H246	1490	5/1	Em. Rio São Francisco, Penedo
BA41) H478	1490	1/0.25	R. Educadora, Ipiaú
BA77) H...	1490	1/0.25	Rio São Francisco, Bom Jesus da Lapa (CP)
BA87) H...	1490	0.25	R. Planalto d'Oeste, Correntina (CP)
BA88) H...	1490	2.5/0.5	R. Antena Um, Ribeira do Pombal (CP)
MG162) L231	1490	0.25	R. Cacique, Araguari
MG163) L274	1490	1/0.25	R. Paraisópolis, Paraisópolis
MG165) L353	1490	1/0.25	R. Pirapetinga, Pirapetinga
MG166) L...	1490	1/0.25	R. Matinada, Paraguaçu (CP)
MS19) I404	1490	1/0.25	R. Paiaguás, Glória de Dourados
PI40) I...	1490	1/0.25	R. Lagoa, Buriti dos Lopes (CP)
PR108) J210	1490	1/0.25	R. Cornélio Procópio, Cornélio Procópio
PR149) J347	1490	1/0.25	R. Difusora, São Jorge d'Oeste
RS138) K208	1490	1/0.25	Rádiodifusão Assissense, São Francisco de Assis
RS139) K225	1490	1/0.25	R. Liberdade, Canguçu
RS140) K309	1490	1/0.25	R. Taquara, Taquara
SC69) J757	1490	1/0.25	R. Sentinela de Alto Vale, Ibirama
SC70) J791	1490	1/0.25	R. Cultura, Xaxim
SP180) K530	1490	1/0.25	R. Difusora, Olímpia
SP181) K580	1490	1/0.25	R. Nova Dracena, Dracena
SP182) K583	1490	1/0.25	R. Educadora Santa Rita, Fernandópolis
SP244) K680	1490	1/0.25	R. Cultura, Vargem Grande do Sul
SP183) K764	1490	5/0.25	R. Mauá, Mauá
SP255) K767	1490	0.25	R. Boituva AM, Boituva
SP222) K769	1490	1/0.25	R. Icatu, Penápolis (ff 920)
BA49) H487	1500	0.25	R. Jacuípe, Riachão do Jacuípe
BA52) H493	1500	5/0.5	R. Difusora do Descobrimento, Porto Seguro
CE39) H615	1500	2.5/0.25	R. Macico, Baturité
MG101) L215	1500	5/0.25	R. Montanhesa, Viçosa
MG140) L340	1500	1/0.25	R. Aparecida do Sul, Ilicínea
PA19) I542	1500	1/0.25	R. Floresta, Tucuruí
PE27) I779	1500	1/0.25	R. Pajeu, Afogados da Ingazeira
PI46) I919	1500	1/0.25	R. Voz do Longa, Esperantina
PR163) J366	1500	1/0.25	R. Margueirinha, Margueirinha
RN29) J620	1500	2.5/0.25	R. Potengi, São Paulo do Potengi (ff 1210)
RS171) K308	1500	0.5	R. Tapense, Tapes (F.PI.) (still on 1520)
RS161) K365	1500	3/0.25	R. Fraternidade, Chapada
SP184) K549	1500	1/0.25	R. Fraternidade, Araras
SP185) K626	1500	1/0.25	R. Difusora, Pindamonhangaba
SP186) K706	1500	1/0.25	R. Vale do Rio Grande, Miguelópolis
SP211) K773	1500	1/0.25	R. Cumbica AM, Guarulhos
SP236) K776	1500	1/0.25	R. Cidade, Apiaí
CE40) H608	1510	1/0.25	R. Planalto da Ibiapaba, São Benedito
GO38) H770	1510	1/0.25	R. Goiatuba, Goiatuba
MG103) L259	1510	2.5/0.25	R. Manhumirim, Manhumirim (ff 780)
PA01) I544	1510	1/0.25	R. Oriente de Redenção, Redenção (ff 610)
PI16) I894	1510	1/0.25	R. Difusora, Floriano
PI17) I896	1510	1/0.25	R. Progresso, Corrente
PR113) J216	1510	1/0.25	R. Educadora, Venceslau Brás
PR115) J326	1510	1/0.25	R. União, Céu Azul
RJ43) J492	1510	1/0.25	R. Teresópolis, Teresópolis
RN15) J602	1510	1/0.25	R. Centenário, Caraúbas
RS170) K376	1510	1/0.25	R. Educadora, São João da Urtiga (ff 870)
SC73) J795	1510	1/0.25	R. Centro Oeste, Pinhalzinho
SC82) J797	1510	2.5/0.25	R. Educadora, Taió
SP188) K654	1510	10/1	R. Cacique, Santos
SP189) K665	1510	0.25	R. Clube, São Manuel
SP256) K770	1510	0.5/0.25	R. Vale do Tietê, Salto
SP230) K...	1510	1/0.25	R. Rural, Rinópolis
BA78) H530	1520	5/1	R. Bela Vista, Poções
CE83) H653	1520	1/0.25	R. Difusora, Solonópole
CE75) H...	1520	1/0.25	R. Araripe, Ipu (CP)
GO39) H772	1520	1/0.25	R. Santelenense, Santa Helena de Goiás (ff 1010)
GO53) H...	1520	1/0.25	R. Campos Belos, Campos Belos
MA15) H899	1520	1/0.25	R. Ribamar, Pindaré Mirim
MA35) H928	1520	1/0.25	R. Educadora Vale do Munin, Chapadinha
MG174) L223	1520	0.25	R. Cultura, Cássia
MG106) L245	1520	1/0.25	R. Clube, Itaúna
MS21) I405	1520	1/0.25	R. Jornal, Amambaí
MS42) N605	1520	1/0.25	R. Campo Alegre, Rio Verde de Mato Grosso
PE28) I801	1520	1/0.25	R. Educadora, Surubim
PR116) J218	1520	1/0.25	R. Antoniense, Antonina
PR118) J292	1520	1/0.25	R. Cultura Palotinense, Palotina
PR132) J340	1520	1/0.25	R. Internacional, Quedas do Iguaçu
PR156) J358	1520	1/0.25	R. Guairacá, Terra Rica
RJ44) J491	1520	10/0.5	R. Continental, Rio de Janeiro
RJ10) J499	1520	1/0.25	R. Musical; Cantagalo
RN22) J610	1520	1/0.25	R. Salinas; Macau
RS141) K217	1520	1/0.25	R. 1520, Cachoeira do Sul
RS171) K308	1520	1/0.25	R. Tapense, Tapes (ff 1500)
SC75) J758	1520	1/0.25	R. Difusora, Imbituba
SC76) J806	1520	1/0.25	R. Cultura, Timbó
SE13) J931	1520	1/0.5	R. Imperatriz dos Campos, Tobias B

Call	kHz	kW	Name and location
SP191) K614	1520	10/1	R. Diário, Mogi das Cruzes
SP192) K627	1520	1/0.25	Pinhal R. Clube, Espírito Santo do Pinhal
SP258) K637	1520	1/0.25	R. Difusora, Rancharia (ff 1230)
SP225) K760	1520	1/0.25	R. Manchester, Votorantim
T007) H797	1520	1/0.25	R. Cristal, Cristalândia
BA43) H479	1530	10/0.5	R. Cultura, Guanambi
CE76) H666	1530	1/0.25	R. Tres Fronteiras, Campos Sales
MG175) L262	1530	0.25	R. Progresso, Monte Santo de Minas
MG110) L280	1530	1/0.25	R. Clube, Pouso Alegre
PE29) I781	1530	1/0.25	R. Bitury, Belo Jardim
PR121) J321	1530	1/0.25	R. Regional, Cianorte (ff 990)
PR119) J339	1530	2.5/0.25	R. Difusora, Balsa Nova
PR157) J348	1530	1/0.25	R. Vale do Iguaçu, Verê
RJ57) J482	1530	1/0.25	R. Cabo Frio, Cabo Frio
RJ45) J502	1530	1/0.25	R. Princesinha do Norte, Miracema
RN16) J603	1530	1/0.25	R. Curimataú, Nova Cruz
RS143) K235	1530	1/0.25	R. Sulina, Dom Pedrito
RS144) K300	1530	5/0.25	R. Líder, São Leopoldo
RS145) K304	1530	1/0.25	R. Tapejara, Tapejara
SC77) J761	1530	1/0.25	R. Difusora, Itajaí
SC78) J780	1530	1/0.25	R. Difusora, São Joaquim
SC79) J796	1530	1/0.25	R. Porto Feliz, Mondaí
SP193) K677	1530	1/0.25	R. Brasil, Tupi Paulista
SP194) K699	1530	1/0.25	R. Noticias, Tatuí
SP231) K755	1530	1/0.25	R. Universal, Teodoro Sampaio
BA79) H...	1540	0.25	R. Jornal, Souto Soares (CP)
BA89) H...	1540	0.25	R. Sociedade, Itiruçu (CP)
CE42) H611	1540	1/0.5	R. Sant'Ana, Tianguá
CE77) H631	1540	1/0.25	R. Sertões, Mombaça
CE78) H...	1540	1/0.25	R. Aratanha, Pacatuba (CP)
ES19) I206	1540	0.25	R. Agricultura, Santa Teresa
GO40) H760	1540	1/0.25	R. Jornal, Inhumas
MA36) H921	1540	1/0.25	R. Santa Maura, Lago da Pedra
MG111) L217	1540	1/0.25	R. Bomdespachense, Bom Despacho
MG112) L226	1540	1/0.25	R. Clube de Minas Gerais, Conselheiro Lafaiete
MG113) L293	1540	1/0.25	R. Tropical, Três Corações
MG168) L351	1540	1/0.25	R. Difusora, Salinas (ff 1310)
MS35) N601	1540	1/0.5	R. Regional Piravevê, Ivinhema
PA14) I545	1540	1/0.25	R. Boa Vista, São Sebastião da Boa Vista
PB24) I694	1540	1/0.5	R. Santa Maria, Monteiro
PE33) I824	1540	1/0.25	R. Voluntários da Pátria, Ouricuri (ff 1080)
PR150) J357	1540	1/0.25	R. Princesa, Roncador
RJ54) J...	1540	1/0.25	R. Clube, Paraíba do Sul (CP)
RN30) J611	1540	1	R. Baixa Verde, João Câmara
RS157) K282	1540	1/0.25	R. Quaraí, Quaraí
RS156) K367	1540	1/0.25	R. Querência, Santo Augusto
SC80) J803	1540	1/0.25	R. Capinzal, Capinzal
SP245) K514	1540	1/0.25	R. Cultura, Leme
SP196) K564	1540	2/1	R. Em. de Botucatu, Botucatu
SP197) K723	1540	25/0.5	R. Difusora Oeste, Osasco
SP246) K737	1540	0.5/0.25	R. Central, Pompéia
BA44) H461	1550	1/0.25	R. Regional de Conquista, Vitória da Conquista
BA80) H518	1550	5/0.25	R. Independência do São Francisco, Juazeiro
BA15) H...	1550	1/0.25	R. Cultura, Utinga
MA37) H...	1550	1/0.25	Sistema Janaina de Radiodifusão, Vargem Grande (CP)
MG114) L211	1550	1/0.25	R. Cultura, Monte Carmelo
MG169) L222	1550	1/0.25	R. Difusora, Carmo do Rio Claro
MG115) L289	1550	1/0.25	R. Difusora Santarritense, Santa Rita do Sapucaí
PA34) I550	1550	1/0.25	R. Cabano, Maracanã
PB32) I700	1550	1/0.25	R. Jardim, Areia
PR133) J213	1550	1/0.25	R. Ipiranga, Palmeira
PR123) J303	1550	1/0.25	R. Pioneira de Formosa, Formosa do Oeste
PR124) J315	1550	1/0.25	R. Cristal, Marmeleiro
PR136) J355	1550	1/0.25	R. Cidade, Palmital
PR154) J363	1550	1/0.25	R. Colméia, Mandaguaçu (ff 1170)
RJ46) J479	1550	1/0.25	R. Imperial, Petrópolis
RN19) J606	1550	1/0.25	R. Gazeta d'Oeste, Areia Branca
RS173) K377	1550	1/0.25	R. Jornal da Manhã, Ijuí (ff 1370)
RS162) K...	1550	1/0.25	R. Soc. Zona Sul, Capão do Leão
SC92) J814	1550	1/0.25	R. Imigrantes, Turvo
SP198) K501	1550	1/0.25	R. Clube, Itararé
SP259) K528	1550	1/0.25	R. Tambaú, Tambaú
SP200) K572	1550	1/0.25	R. Cacique, Capivari
SP201) K590	1550	5/1	R. Guarujá Paulista, Guarujá
SP247) K623	1550	1	R. Cultura, Pederneiras
SP202) K659	1550	1/0.25	R. São Joaquim, São Joaquim da Barra
SP219) K740	1550	1/0.25	R. Auriflama, Auriflama
BA90) H...	1560	0.25	R. Transportadora, Ribeira do Pombal (CP)
CE43) H622	1560	1/0.25	R. Difusora Vale de Curu, Pentecoste
MA18) H902	1560	1/0.25	R. Agua Branca, Vitorino Freire
MG117) L256	1560	1/0.25	R. Jornal, Leopoldina
PR125) J275	1560	1/0.25	R. Capanema, Capanema
PR161) J361	1560	0.25	R. Cultura Serpin, Ribeirão do Pinhal
PR162) J364	1560	1/0.25	R. Clube, Mallet
RJ47) J501	1560	5/1	R. Comunicadora Grande Rio, Itaguaí
RN17) J608	1560	1/0.25	R. Cultura do Oeste, Pau dos Ferros
RS158) K310	1560	2.5/0.25	R. Açoriana, Taquari
RS172) K369	1560	0.25	R. Ourense ("R. Poata"), São José do Ouro
SC95) J825	1560	1	R. Cidade, São Miguel d'Oeste
SP260) K567	1560	0.25	R. Brotense, Brotas (ff 1180)
SP203) K571	1560	1/0.25	R. Em. Campos do Jordão, Campos do Jordão (ff 1340)
SP261) K593	1560	1/0.25	R. Igarapavense, Igarapava
SP248) K679	1560	1/0.25	R. Valparaíso, Valparaíso
SP249) K725	1560	0.25	R. Regional AM, Pedreira
SP250) K...	1560	1/0.25	R. Vale do Rio Paraná, Presidente Epitácio
BA56) H496	1570	1/0.25	R. Vale Aprazível, Jaguaquara
BA91) H533	1570	1/0.5	R. Líder, Central
CE44) H621	1570	1/0.25	R. Sertão Central, Senador Pompeu
GO42) H784	1570	1/0.25	R. Itaberaí, Itaberaí (ff 1160)
GO49) H787	1570	1/0.25	R. Mantiqueira, Niquelândia
MG118) L203	1570	1/0.25	Nova R. Cultura, Alfenas (ff 1180)
MG146) L242	1570	1/0.25	R. Universitária, Itajubá
MG141) L344	1570	1/0.25	R. Cidade, Corinto
MG170) L364	1570	10/0.25	R. Difusora, Piranga
MS23) I408	1570	1/0.25	R. Patriarca, Cassilândia (ff 670)
MS24) I409	1570	1/0.25	R. Caarapó, Caarapó
MS36) I418	1570	1/0.25	R. Cidade, Aparecida do Taboado (ff 1140)
PE30) I798	1570	5/1	R. Asa Branca, Salgueiro
PR158) J324	1570	0.25	R. Brasileira, Bela Vista do Paraíso
PR137) J341	1570	1/0.25	R. Club, Nova Aurora
PR159) J365	1570	1/0.25	R. Arapoti, Arapoti
RJ48) J493	1570	1/0.25	R. Cultura, Valença
RJ49) J498	1570	1/0.25	R. Bom Jesus, Bom Jesus de Itabapoana
RS147) K358	1570	1/0.25	R. Vale do Gravataí, Gravataí
SC83) J777	1570	1/0.25	R. Rio Negrinho, Rio Negrinho
SC98) J829	1570	1/0.25	R. Modelo, Modelo
SP204) K552	1570	1/0.25	R. Avaré, Avaré
SP205) K605	1570	1/0.25	R. Junqueirópolis, Junqueirópolis
SP262) K648	1570	1/0.25	R. Zequinha de Abreu, Santa Rita do Passa Quatro
SP206) K651	1570	2.5/0.5	R. Emissora ABC, Santo André
SP207) K667	1570	1/0.25	R. Socorro, Socorro
SP208) K670	1570	1/0.25	R. Clube, Tanabi
BA45) H464	1580	1/0.25	R. Emissora Radiovox, Muritiba
BA53) H497	1580	1/0.25	R. Barra do Mendes, Barra do Mendes
BA62) H502	1580	1/0.25	R. Atalaia, Canavieiras
BA92) H...	1580	0.25	R. União, Teolândia (CP)
ES13) I209	1580	1/0.25	R. Educadora, Afonso Cláudio (ff 1390)
MG119) L210	1580	1/0.25	R. Liberdade AM, Itapecirica
MG121) L290	1580	1/0.25	R. Cultura, Santos Dumont
MG122) L329	1580	1/0.25	R. Educadora, Espinosa
MG133) L335	1580	1/0.25	R. Rural Nova Guaranésia, Guaranésia
MS25) I415	1580	1/0.25	R. Laguna, Jardim
MS43) I...	1580	1/0.25	R. Difusora, Ivinhema (CP)

Call	kHz	kW	Name and location
PI18) I898	1580	1/0.25	R. Irapuá, Floriano
PR164) J342	1580	1/0.25	R. São João do Sudoeste, São João
RJ51) J487	1580	1/0.25	R. Popular Fluminense, Conceição de Macabu
RJ58) J505	1580	0.25	R. Geração 2000, Teresópolis
RN24) J613	1580	1/0.25	R. Novos Tempos, Ceará Mirim
RS148) K237	1580	1/0.25	R. Encantado, Encantado
RS150) K339	1580	1/0.25	R. Difusora Fronteira, Arroio Grande
RS159) K...	1580	1/0.25	R. Difusora das Missões, Palmeira das Missões (CP)
SC84) J818	1580	1/0.25	R. Pomerode, Pomerode
SP251) K504	1580	1/0.25	R. Difusora, Amparo
SP209) K743	1580	0.25	R. Trans Universal, Itaporanga
BA76) H...	1590	0.25	R. Vale do Jiquiriçá, Jiquiriçá (CP)
CE80) H660	1590	1/0.25	R. Araripe, Cedro (ff 1160)
MG171) L368	1590	1/0.25	Rede Dinâmica de Comunicação, Ubá
MG134) L369	1590	10/1	R. Guaicuí, Várzea da Palma
MS26) I403	1590	1/0.25	R. Independência, Eldorado
PR129) J290	1590	1/0.25	R. Cultura, Andirá
PR160) J296	1590	1/0.25	R. Havaí, Capitão Leônidas Marques
RJ55) J485	1590	0.25	R. Duque de Caxias, Duque de Caxias
RJ52) J506	1590	1/0.25	R. Resende AM, Resende
RN21) J614	1590	1/0.5	R. Santa Cruz AM, Santa Cruz
RS174) K212	1590	0.25	R. Clube, Bagé
RS151) K281	1590	1/0.25	R. Navegantes, Porto Lucena
SC10) J799	1590	0.25	R. Namba, Ponte Serrada (ff 1410)
SC101) J823	1590	1	R. Floresta Negra AM, Joinville
SP226) K762	1590	1/0.25	R. Jovem Pam, São Carlos (ff 1300)
SP254) K774	1590	1/0.5	R. Japi, Cabreúva
SP263) K...	Þ1600	100/20	R. Nove de Julho, São Paulo (CP)

Shortwaves: *= inactive.

Call	kHz	kW	Name and Location, h. of tr.
SP82) ZYG852	2380	0.25	R. Educadora, Limeira: 24h
AC07) ZYF202	*2410	5	R. Transamazônica, Senador Guiomard: (ff 3255)
SP168) ZYG862	2420	0.5	R. São Carlos, São Carlos: 0800-0300
AC02) ZYF204	2460	1	R. Alvorada, Rio Branco: 0900-1100, 2200-0200
SP114) ZYG851	*2470	0.25	R. Cacique, Sorocaba
SP210) ZYG866	2490.1	0.25	R. Oito de Setembro, Descalvado: 0900-0100
AM03) ZYF279	3205	5	R. Vale do Rio Madeira, Humaitá
SP11) ZYG861	3205	1	R. Ribeirão Preto: 0700-0400
SP98) ZYG860	3235	0.5	R. Clube, Marília: 0800-0400
MG69) ZYG203	3245	1	R. Clube, Varginha: 0830-0210
AC03) ZYF...	3255	1	R. Difusora 6 de Agosto, Xapuri: 1000-0230
AC07) ZYF202	*3255	5	R. Transamazônica, Senador Guiomard (F.PI.) (still on 2410)
CE22) ZYF533	3255	1	R. Educadora Cariri, Crato: 0900-1400, 1900-0200
SP112) ZYG867	3325	2.5	Super R. Tupi, São Paulo
SP49) ZYG855	3365	1	R. Cultura, Araraquara: 0700-0300
AM02) ZYF276	v3375	5	R. Nacional, São Gabriel da Cachoeira: 0800-0200
MS04) ZYF907	3375.3	5	R. Clube, Dourados: 0800-0400
AP02) ZYF361	*3375	5	R. Equatorial, Macapá
RO09) ZYG792	3375	5	R. Educadora, Guajará Mirim: 0930-1200 (Sun 1600), 2030-0200
AM10) ZYF271	3385	5	R. Educação Rural, Tefé: 0900-0100
AC08) ZYF...	v3560.4	1.5	R. Difusora, Brasiléia: W 1000-0400, Sun 1100-0300
AC09) ZYF...	*4116	0.25	R. Difusora, Sena Madureira: may operate occasionally
MA03) ZYF810	4753.8	5	R. Difusora do Maranhão, São Luís: 24h
MS01) ZYF904	4755	10	R. Educação Rural, Campo Grande: 0645 (Sun 0745)-0400
AC01) ZYF200	v4765	10	R. Integração, Cruzeiro do Sul: 0900-0300
PA04) ZYG363	4765	10	R. Rural, Santarém: 0800-0400
MG55) ZYG207	4775	1	R. Congonhas, Congonhas: 0800-0100
MT39) ZYF902	4775	1	R. Portal da Amazônia, Cuiabá: 0700-0300
PA10) ZYG430	4775	5	R. Liberal, Belém
RO14) ZYG794	4775	5	R. Amarela, Rolim do Moura (CP)
MA09) ZYF812	*4785	5	R. Ribamar, São Luís
RO11) ZYG790	4785	10	R. Caiari, Porto Velho: 0800-0400
SP136) ZYG857	4785	1	R. Brasil 5000, Campinas: 0800-0200
MS12) ZYF900	4795	1	R. Difusora, Aquidauana: 0755-0400
AM20) ZYF273	4805	10	R. Difusora do Amazonas, Manaus: 0900-0300
MG06) ZYG209	4805	0.5	R. Itatiaia, Belo Horizonte: 0800-1900
AM04) ZYF277	*4814.9	10	R. Cabocla, Tabatinga
PR13) ZYG640	4814.9	10	R. Difusora, Londrina: 0755 (Sun 0900)-0355
SP80) ZYG868	4824.9	10	R. Canção Nova, Cachoeira Paulista: 24h
PA12) ZYG364	4825	5	R. Educadora, Bragança: 0830-0200
MS27) ZYF908	*4835	5	R. Atalaia, Corumbá
PA23) ZYG366	4865	5	R. Sentinela da Amazônia, Obidos: 24h
SP102) ZYG869	4845	1	R. Meteorologia Paulista, Ibitinga: 0700-2300
AM14) ZYF278	4845.2	250	R. Cultura Ondas Tropicais, Manaus: 0800-0300
MG33) ZYG202	*4855	1	R. Por Um Mundo Melhor, Governador Valadares
MT40) ZYF905	4855	2.5	R. Tropical da Barra, Barra do Garças: 0900-0130
AC04) ZYF203	4865	5	R. Verdes Florestas, Cruzeiro do Sul: 1000-0300
BA20) ZYF390	*4865	1	R. Sociedade, Feira de Santana
PA23) ZYG366	4865	5	R. Missões da Amazônia, Obidos: 24h
PR33) ZYG641	4865	5	R. Alvorada, Londrina: 0700-0300
RR01) ZYG810	4875	10	R. Difusora de Roraima, Boa Vista: 0900-0300
GO12) ZYF692	4884.9	1	R. Carajá, Anápolis: 0800-1200, 2000-0300
AC06) ZYF201	4885	5	R. Difusora Acreana, Rio Branco: 0855-0200 (Sat 0400, Sun 0300)
PA03) ZYG362	4885	5	R. Clube do Pará, Belém: 0800-0500 (irr.)
AM13) ZYF270	4895	5	R. Baré, Manaus: 0800-0300
MS32) ZYR200	4895	5	R. IPB AM, Campo Grande: 24h
RJ03) ZYG683	4905	5	R. Relógio Federal, Rio de Janeiro: 0730-0330
TO04) ZYF693	4905	1	R. Anhanguera, Araguaína: 0800-0100
AP01) ZYF360	4915	10	R. Difusora, Macapá: 0800-0300
GO27) ZYF691	4915	10	R. CBN Anhanguera, Goiânia: 0900-0300
SP06) ZYG684	4924.9	0.5	R. Difusora, Taubaté: 24h
ES07) ZYF641	4935	1	R. Capixaba, Vitória: 24h
GO09) ZYF694	4935	2.5	R. Difusora, Jataí: 0900-1300, 1900-2200
MG73) ZYG201	4945	1	R. Difusora, Poços de Caldas: 0800-0400
PE08) ZYG525	4945	1	Emiss. Rural A Voz do São Francisco, Petrolina: 0800-0100
RO13) ZYG791	4945	7.5	R. Progresso, Porto Velho: 0900-0200
MT12) ZYF906	4955	2.5	R. Clube, Rondonópolis: W 0800-0200, Sun 0900-0300
PA08) ZYG361	*4955	10	R. Marajoara, Belém
RJ23) ZYG682	4955	1	R. Cultura, Campos: 0700-0300
AM12) ZYF275	4965	5	R. Alvorada, Parintins: 0900-0200
RN10) ZYG761	*4965	1	R. Poti, Natal
MA10) ZYF813	4975	5	R. Timbira do Maranhão, São Luís: 0800-0300
SP232) ZYG865	4975	1	R. Mundial, São Paulo: 24h
SP187) ZYG685	*4975	1	R. Iguatemi, Osasco

Call		kHz	kW	Name and Location, h. of tr.
GO30)	ZYF690	4985	10	R. Brasil Central, Goiânia: 0700-0300
PI10)	ZYG595	5014.6	1	R. Pioneira, Teresina: 0800-0300
MT08)	ZYF903	5015	1	R. Brasil Tropical, Cuiabá: 24h (at night rel. R. Cultura 710kHz)
RJ07)	ZYG685	5015	1	R. Copacabana, Rio de Janeiro: 24h
PA15)	ZYG365	5025	5	R. Jornal da Transamazônica, Altamira: 0900-0200
PB14)	ZYG481	*5025	1	Super R. Borborema, Campina Grande
RO06)	ZYG793	*5025	5	R. Morimoto, Ji-Paraná
SP52)	ZYG853	5034.9	10	R. Aparecida, Aparecida: 0700-0300
AM09)	ZYF272	5035	5	R. Educação Rural, Coari: 0900-0400
PA16)	ZYG360	5045	10	R. Cultura do Pará, Belém: 0800-0300
SP63)	ZYG850	*5045	1	R. Difusora, Presidente Prudente
AM15)	ZYF274	5055	5	R. Jornal A Crítica, Manaus
MT10)	ZYF901	5055	1	R. Difusora, Cáceres: 0800-0400
SP62)	ZYG965	5955.3	10	R. Gazeta, São Paulo: 0900-0300
RS163)	ZYE857	v5965	7.5	R. Nova Visão, Santa Maria: W 1100-0200, SS 1100-2300 (often rel. R. Transmundial)
MG06)	ZYE523	5970	10	R. Itatiaia, Belo Horizonte: 0800-1900
SC59)	ZYE891	5980	10	R. Guarujá, Florianópolis: 24h
RJ12)	ZYE766	*5990	7.5	R. MEC, Rio de Janeiro
MA25)	ZYE481	6000	5	R. Boas Nova, São Luís (CP)
RS23)	ZYE852	6000	10	R. Guaíba, Porto Alegre: 0800-0400
MG35)	ZYE521	6010	5	R. Inconfidência, Belo Horizonte: 1200-1600, 1800-2000
BA55)	ZYE299	6020	10	R. Educadora da Bahia, Salvador: 0900-0100
RS09)	ZYE850	6020	10	R. Gaúcha, Porto Alegre: 0900-0400
RJ25)	ZYE770	6030	10	R. Globo, Rio de Janeiro: 0900-0330
PR134)	ZYE725	6040.3	7.5	R. Clube Paranaense, Curitiba: 0900-2200 (Sun 2300)
MG68)	ZYE520	6050	10	R. Guarani, Belo Horizonte
PR60)	ZYE726	6060	10	R. Universo, Curitiba: 24h
RJ18)	ZYE765	6070	7.5	R. Capital, Rio de Janeiro: 0800-0300
GO27)	ZYE441	6080	5	R. CBN Anhanguera, Goiânia: 0900-0300
PR15)	ZYE726	6080	10	R. Novas de Paz, Curitiba: 1000-0300
SP57)	ZYE956	6090	10	R. Bandeirantes, São Paulo: 24h
PR138)	ZYE728	6104.8	5	R. Cultura, Foz do Iguaçu: 0800-0300. Rel. Rede Transamérica FM
SP80)	ZYE971	6105	10	R. Canção Nova, Cachoeira Paulista: 24h
SP100)	ZYE968	6120	7.5	R. Globo, São Paulo: 0900-0200
SP52)	ZYE954	6135	25	R. Aparecida, Aparecida: 0800-0300
SP77)	ZYE950	6150	10	R. Record, São Paulo: 0900-2400
AM11)	ZYE245	6160	10	R. Rio Mar, Manaus: 1900-2400
RS49)	ZYE854	6160	10	R. Rio Grande do Sul, Porto Alegre (Sistema LBV Mundial hires airtime)
SP121)	ZYE959	6170	7.5	R. Cultura, São Paulo: 0800-0400
DF06)	ZYE365	*v6180	250	R. Nacional da Amazônia, Brasília
SP77)	ZYE951	9505	7.5	R. Record, São Paulo: 0900-2400
PR15)	ZYE726	9515	10	R. Novas de Paz, Curitiba: 1000-0300
RS163)	ZYE859	9530	10	R. Nova Visão, Santa Maria: 1000-0200
BA55)	ZYE300	9540	10	R. Educadora da Bahia, Salvador: 0900-0100
RS49)	ZYE855	*9550	5	R. Rio Grande do Sul, Porto Alegre
PR60)	ZYE727	9565	15	R. Universo, Curitiba: 24h
SP48)	ZYE969	9584.9	10	R. CBN (Globo), São Paulo: 0900-1500
RJ12)	ZYE772	9600	7.5	R. MEC, Rio de Janeiro
SP121)	ZYE960	9615	7.5	R. Cultura, São Paulo: 0800-0400
SP52)	ZYE954	9630	10	R. Aparecida, Aparecida:

Call		kHz	kW	Name and Location, h. of tr.
SP57)	ZYE957	9645	7.5	R. Bandeirantes, São Paulo: 24h
SC86)	ZYE890	9665	10	R. Marumby, Florianópolis: 0900-0100
SP80)	ZYE971	9674.9	10	R. Canção Nova, Cachoeira Paulista: 24h
SP62)	ZYE963	9684.7	10	R. Gazeta, São Paulo: 0900-0300
AM11)	ZYE245	9695	7.5	R. Rio Mar, Manaus: 0900-1900
RJ17)	ZYE768	9705	7.5	R. Nacional, Rio de Janeiro: 0800-0300
PR134)	ZYE725	9725	7.5	R. Clube Paranaense, Curitiba: 0900-2200 (Sun 2300)
DF06)	ZYE365	9745	250	Radiobras, Brasília
SP263)	ZYE...	*9820	10	R. Nove de Julho, São Paulo (CP)
RS163)	ZYE858	11704.3	10	R. Nova Visão, Santa Maria: 1100-0200 (SS 2300) (often rel. R. Transmundial) (ff 11735)
PR15)	ZYE726	11725	10	R. Novas de Paz, Curitiba: 1000-0300
RS163)	ZYE858	*11735	50	R. Nova Visão, Santa Maria (F.PI.) (still on 11704.3)
PR60)	ZYE726	11765	10	R. Universo, Curitiba: 0900-0300
DF06)	ZYE365	11780	250	R. Nacional da Amazônia, Brasília: 0900-0200
RS23)	ZYE853	v11785	7.5	R. Guaíba, Porto Alegre: 0800-0400
RJ25)	ZYE775	11805.4	10	R. Globo, Rio de Janeiro: 0900-0330
GO30)	ZYE440	v11815	7.5	R. Brasil Central, Goiânia: 0700-0300
GO27)	ZYE441	11830	10	R. CBN Anhanguera, Goiânia: 0900-0300
SP52)	ZYE954	11854.8	1	R. Aparecida, Aparecida: 0800-0300
RS49)	ZYE856	11895	1	R. Rio Grande do Sul, Porto Alegre (Sistema LBV Mundial hires airtime)
RS09)	ZYE850	11915	10	R. Gaucha, Porto Alegre: 0800-0400
SP57)	ZYE958	11925	10	R. Bandeirantes, São Paulo: 24h (R. Juratel hires airtime)
PR134)	ZYE725	11935	7.5	R. Clube Paranaense, Curitiba: 0900-2200 (Sun 2300)
RJ12)	ZYE773	11950	7.5	R. MEC, Rio de Janeiro: irr.
SP77)	ZYE952	11964.5	7.5	R. Record, São Paulo: 0900-2400
SP77)	ZYE953	15135	7.5	R. Record, São Paulo
MG35)	ZYE522	15190	5	R. Inconfidência, Belo Horizonte: 0800-0300
DF06)	ZYE365	15200	250	R. Nacional da Amazônia, Brasília
MA10)	ZYE480	*15215	2.5	R. Timbira, São Luís
DF06)	ZYE365	15265	250	Radiobras, Brasília
SP100)	ZYE970	*15265	1	R. Globo, São Paulo
SP62)	ZYE964	15325	10	R. Gazeta, São Paulo: 0900-0300
SP23)	ZYE955	15415	1	R. Clube, Ribeirão Preto
DF06)	ZYE365	15445	250	Radiobras, Brasília
DF06)	ZYE365	17750	250	Radiobras, Brasília
SP121)	ZYE961	17815	1	R. Cultura, São Paulo: 0800-0300
RJ12)	ZYE774	17875	7.5	R. MEC, Rio de Janeiro: 0900-1100

Addresses and other information

AC00) ACRE
AC01) Rua Alagoas 270, 69980-000 Cruzeiro do Sul – Gte: Oscar Alves Bandeira
AC02) Av. Ceará 2150, 69900-460 Rio Branco – Dir: Edivaldo Souza.
AC03) Rua Pio Nazario 31, 69930-000 Xapuri – DG: Francisco Evangelista de Abreu.
AC04) C.P. 53, 69981-970 Cruzeiro do Sul – DG: Rui Matos Said.
AC05) Rua Epaminondas Jacome 236, 69908-420 Rio Branco.
AC06) Rua Benjamin Constant 161, 69900-160 Rio Branco.
AC07) Av. Castelo Branco 329, 69925-000 Senador Guiomard.
AC08) Rua Genni Assis s/n, 69932-000 Brasiléia – DG: Carlos Henrique S. de Oliveira.
AC09) Rua Avelino Chaves 707, 69940-000 Sena Madureira.

AL00) ALAGOAS
AL01) Av. Coronel Wilson Santa Cruz 6, 57314-000 Arapiraca.
AL02) Via Expressa 4360, Serraria, 57080-000 Maceió. ☎ +55 (82) 328 3066. 🖳 +55 (82) 328 3771 – DG: Eliezer de M. Fontes Jr.
AL03) Rua José Maria Passos 25, 57600-030 Palmeiras dos Indios.
AL04) Rua Barão José Miguel 400, Farol, 57055-160 Maceió.
AL05) Rua Miguel Palmeira 1513, 7Þ andar, Farol, 57055-330 Maceió.

AL06) Praça Senador Eneas Araújo 61, 57500-000 Santana do Ipanema.
AL07) Rua Aristeu de Andrade 355, Farol, 57021-900 Maceió – Dir. Matias Junior.
AL08) Rua Barão de Penedo 258, 57020-340 Maceió.
AL09) C.P. 6, 57201-970 Penedo – Dir: Raul Rodrigues de Lima.
AL10) Quadra A lote 04, 57160-000 Marechal Deodoro.
AL11) Rua Porcos s/n, 57300-000 Arapiraca
AL12) Parque Industrial Delmiro Gouveia ao lado da Rodovia BR-423, 57480-000 Delmiro Gouveia.
AL13) BR-104 Km. 36, Bairro Roberto Correia de Arajuó, 57800-000 União dos Palmares – DG: Sílvio Sarmento Neto.
AL14) Rua Cônego Machado 917, Farol, 57021-160 Maceió – Dir: Carlos Galdino.
AL15) Av. Braulio Cavalcante s/n, 57400-000 Pão de Açucar.

AM00) AMAZONAS
AM01) Rua Brasília s/n, 69470-000 Tefé – DG: F. H. de Santos.
AM02) Av. Alvaro Maia s/n, 69750-000 São Gabriel da Cachoeira – DG: Eugenio M.S. Vaz.
AM03) Av. Rio Madeira 1940, 69800-000 Humaitá.
AM04) Gleba Tocantins, lote 15, 69640-000 Tabatinga – Gte: Jaime A. Silva.
AM05) Rua Eduardo Ribeiro esq. Solimões, 69100-000 Itacoatiara.
AM06) Rua Joana D'Angelo 2255, 69400-000 Manacapuru.
AM07) Av. Major Santana 2502, 69280-000 Manicoré – DG: Hamilton Maia C.
AM08) Estrada dos Morães 1455, 69190-000 Maués.
AM09) Praça São Sebastião 137, 69460-000 Coari – Gte: José C. Martins Cabral. Dir: Lino Rodrigues Pessoa Neto.
AM10) Praça Santa Teresa 283, 69470-000 Tefé – DG: Mario Clemente Neto.
AM11) Rua José Clemente 500, 69010-070 Manaus – DG: Martin J. Laumann.
AM12) Travessa Governador Leopoldo Neves 503, 69151-460 Parintins. ☎ +55 (92) 533 2002– Dir: Raimunda Ribeiro da Silva. **FM:** 100.1.
AM13) Av. Santa Cruz Machado 170A, 69078-000 Manaus. – Dir: Clodoaldo J. Guerra
AM14) Av. Barcelos s/n, Praça 14, 69020-200 Manaus.
AM15) Av. Andre Araujo s/n, Praça 14, 69060-001 Manaus – DG: Rui Alencar. **FM:** 93.1.
AM16) Rua Osvaldo Góes 18, Bairro Santo Antônio , 69029-900 Manaus.
AM17) Av. Amazonas 1958, 69151-000 Parintins.
AM18) Rodovia Manoel Urbano, km. 02, 69405-000 Iranduba.
AM19) Av. Ribeiro Junior 381, 69400-000 Manacapuru.
AM20) C.P. 311, 69011-970 Manaus – Dir: José J. Marinho.
AM21) Estrada do Gavião, km. 05, 69500-000 Carauari.
AM22) 69880-000 Eirunepê.
AM23) Estrada dos Morães, km. 05, 69190-000 Maués.
AM24) Av. Leopoldo Neves 360, 69850-000 Bôca do Acre.
AM25) Rua Anori/Anama 327, 69440-000 Anori.

AP00) AMAPA
AP01) Rua Cândido Mendes 525, 68900-100 Macapá – DG: Benedito Rostan Costa Martins.
AP02) Rua Major Eliezer Levy 684, 68900-140 Macapá.
AP03) Av. Nações Unidas 256, 68906-100 Macapá.

BA00) BAHIA
BA01) Rua M. Dourado Sobrinho 78, 44900-000 Irecê.
BA02) Praça José Bastos 2, 45600-000 Itabuna – DG: Waldeny Andrade.
BA03) Praça da Independência 244, 45995-000 Teixeira de Freitas.
BA04) Rua Gabriel Soares 23, Ladeira dos Aflitos, 40060-040 Salvador – DG: Paulo Schettini Motta.
BA05) Rua Luis Augusto Fernandes Borges 306, 47800-000 Barreiras – DG: Otacilio Junior.
BA06) Av. Nilo Peçanha 572, 45400-000 Valença.
BA07) Praça Duque Caxias 3, 45700-000 Itapetinga.
BA08) Rua dos Sururu s/n, 44570-000 Santo Antônio de Jesus.
BA09) Rua Joana Angélica 125, 45660-000 Ilhéus – Gte: Robério Meneses. Dir: José Lucas.
BA10) Rua Germiniano Costa 47, 44025-070 Feira de Santana.
BA11) Rua Ferreira Santos 5, Federação, 40230-040 Salvador – Dir: Sílvio Santos.
BA12) C.P. 19, 45601-970 Itabuna – Gte: P. Nunes E.
BA13) Rua 24 de Outubro 167, 47800-000 Barreiras. ☎ +55 (73) 811 4188 – DG: Jorge A, Gentil da Silva.
BA14) Praça Luiz Nogueira 385, 48700-000 Serrinha.
BA15) Rodovia BA-052, km. 48, 46810-000 Utinga.

BA16) Rua Martin Afonso de Souza 270, Bairro Garcia, 40100-050 Salvador – DG: Pe. Aderbal Galvão.
BA17) Av. Visconde do Rio Branco 68, 48970-000 Senhor do Bonfim.
BA18) Travessa da Catedral s/n, 45600-000 Itabuna – DG: Geraldo Pedrassoli.
BA19) C.P. 15, 48730-000 Conceição do Coité – Dir: Edvaldo de Santiago.
BA20) Rua Frei Hermenegildo 300, Capuchinos, 44052-250 Feira de Santana – DG: Frei Rutivalder Brito. **FM:** 96.9.
BA21) Gleba Fazenda Ouro Verde, 45900-000 Caravelas.
BA22) Rua Gabriel Soares 23, 40060-040 Salvador – DG: Gilmar Teixeira.
BA23) Praça Barão do Rio Branco 42, 45100-000 Vitória da Conquista – Gte: A. Ribeiro N.
BA24) Av. Maria Quitéria 223, 44062-630 Feira de Santana – DG: Moacir Mansur. **FM:** 95.3 "Nordeste FM".
BA25) Rua Santos Dumont 22, 45700-000 Itapetinga.
BA26) Rua Marquês de Paranaguá 259, 45660-000 Ilhéus. ☎ +55 (73) 231 3612 – DG: Hamilton Fontes.
BA27) C.P. 29, 45825-000 Eunápolis – DG: José Silvano F.
BA28) Rua Aprigo Duarte 4, 48900-000 Juazeiro.
BA29) Morro do Peru Pelado, 44700-000 Jacobina.
BA30) Rua Monte Castelo 45, Sobradinha, 44018-210 Feira de Santana – DG: José Roberto Pazzi.
BA31) Rua Dom Pedro II 98, 48100-000 Alagoinhas.
BA32) Rua Lima e Silva 216, Liberdade, 40375-010 Salvador.
BA33) Av. 2 de Julho 140, 45660-000 Ilhéus – DG: Tony Neto
BA34) Praça Luiz Nogueira 385, 48700-000 Serrinha.
BA35) Rua São Francisco 159, 48600-000 Paulo Afonso.
BA36) Praça da Bandeira s/n, Centro, 47600-000 Bom Jesus da Lapa. ☎ +55 (77) 481 4329 – DG: Antônio Rosivaldo Motta.
BA37) Praça Ruy Barbosa s/n, 44330-000 São Gonçalo dos Campos – DG: Margarida Alves B.
BA38) Rua Marechal Deodoro 16, 44200-000 Santo Amaro.
BA39) Rua 2 de Julho 20, 45200-000 Jequié – DG: Renato Ferreira.
BA40) Rua Sítio Escurinha s/n, BR-242, km. 90, 46880-000 Itaberaba – DG: João Leão Carneiro.
BA41) Rua Castro Alves 207, 45570-000 Ipiaú.
BA42) Av. Contorno s/n, 46400-000 Caetité – Dir: F. H. Negretros.
BA43) Rua Otavio Mangabeira 1026, Fazenda Piranha, 46430-000 Guanambi.
BA44) Av. Santa Maria 422, Sumaré, 45100-000 Vitória da Conquista – DG: Raimundo Martins.
BA45) Rua Ramiro Costa 41, 44340-000 Muritiba.
BA46) Av. Porto Seguro 718, 1Þ andar, Ed. Pisolux, 45820-000 Eunápolis. ☎ +55 (73) 281 2208 – DG: Paulo Henrique Santos Tavares.
BA47) Rua da Bandeira 27, 42800-000 Camaçari.
BA48) Travessa do Contorno 26, 45570-000 Ipiaú.
BA49) Pe. A. Guimarães 32, 44640-000 Riachão do Jacuípe.
BA50) Av. Getúlio Vargas s/n, 48600-000 Paulo Afonso.
BA51) Rua Rio Corrente s/n, 47640-000 Santa Maria da Vitória.
BA52) Travessa Luis Viana Filho s/n 45820-000 Porto Seguro.
BA53) Rua Alvaro Campos 83, 44990-000 Barra do Mendes – DG: Sansão Sodré Bastos.
BA54) Av. do Agricultor s/n, 46470-000 Riacho de Santana – DG: Joaquim O. Nogueira.
BA55) Rua Pedro Gama 413/E, Alto Sobradinho, Federação, 40230-290 Salvador – DG: Nívea Almeida.
BA56) Praça Guilherme Silva 85, 1Þ andar, 45345-000 Jaguaquara – DG: Italo Rabêlo de Amaral.
BA57) Rua Professor Severo Pessoa 170, Federação, 40210-170 Salvador – DG: Lázaro P. de Brito.
BA58) Centro Industrial Imbares, BR-116 Km 837, 45100-000 Vitória da Conquista.
BA59) Rua Gonçalo Martins 19, 45550-000 Ubatã.
BA60) Rua Reggio Emilia 32, 44900-000 Irecê – DG: Eduardo Lopes.
BA61) C.P. 45, 46430-000 Guanambi – Dir: Ivana Teixeira Bastos.
BA62) Rua General Pederneiras 62, 45860-000 Canavieiras.
BA63) Rua Margem do Rio do Ouro 115, 44700-000 Jacobina.
BA64) Rua Farias Goes 164, 48330-000 Rio Real.
BA65) Rua Otavio Mangabeira 13, 48500-000 Euclides da Cunha.
BA66) Rua Rui Barbosa 119, 47400-000 Xique-Xique.
BA67) Praça Castelo Branco 294 – Centro, 45830-000 Itamaraju. ☎ +55 (73) 294 1106 – DG: Gustavo Liberato Paixão.
BA68) Av. José Candido dos Santos 20, 48475-000 Itapicuru.
BA69) Rua Frei Apolônio de Todi 10, 48410-000 Cícero Dantas – DG: Celso José da Anunciação.
BA70) Terreiro de Jesus 13, Centro Histórico, 40025-010 Salvador.
BA71) Rua Vicente Paula Costa 16, 48540-000 Jeremoabo – DG: Francisco J. de Oliveira.

BA72) Rua 2 de Julho s/n, 45160-000 Belo Campo. ☎ 🖃 +55 (77) 4372233 — DG: Amasilio Ramara. **F.PI:** 25kW, move to 1120.
BA73) Praça São José 279 44600-000 Ipirá. ☎ +55 (75) 2541450 – DG: Jurandi Oliveira.
BA74) Rua Rubens de Oliveira s/n, 44380-000 Cruz das Almas.
BA75) Rua Teixeira de Freitas s/n, 47700-000 Santana.
BA76) Rua Coronel Vicente s/n, 45470-000 Jiquiraçá.
BA77) Rua Barão do Rio Branco s/n, 47600-000 Bom Jesus da Lapa.
BA78) Rua 5 No. 6, Alto da Bela Vista, 45260-000 Poções. ☎ +55 (73) 4311135 – DG: José Roberto Pazzi.
BA79) Rua Idalina Pinto s/n, 46990-000 Souto Soares.
BA80) Rua Coronel João Evangelista 5, 48700-000 Juazeiro.
BA81) Rua Machado de Assis 16, Brotas, 40285-280 Salvador. 🖃 +55 (71) 2550195 – Superintendente: Luiz Mocellin.
BA82) Rua Desembargador Salvio Martins 321, 48800-000 Monte Santo.
BA83) Parque Emilia Costa s/n, 45450-000 Gandu.
BA84) Praça da Bandeira 47, 3Þ andar, Centro, 48900-000 Juazeiro – DG: Caio Coelho.
BA85) Rua Coronel Dias Coelho 249, 44850-000 Morro do Chapéu.
BA86) 44630-000 Mairi.
BA87) Rua Dr. Guerra 91, 47650-000 Correntina.
BA88) Rua Espírito Santo s/n, 48400-000 Ribeira do Pombal.
BA89) Rua João Brandão 233, Centro, 45350-000 Itiruçu.
BA90) Praça Getúlio Vargas 211, 48400-000 Ribeira do Pombal.
BA91) Rua do Comércio 31, 44940-000 Central – DG: Celso Rangel.
BA92) 45465-000 Teolândia.

CE00) CEARA

CE01) Rua Romeu Martins, Ed. 29 de Julho, 63700-000 Canindé – DG: Darlano Coelho.
CE02) Rua São Pedro 918, 63010-010 Juazeiro do Norte.
CE03) Av. Monsenhor Tabosa 2514, 62500-000 Itapipoca. ☎ +55 (85) 6312173. 🖃 +55 (85) 6310469 – DG: José I. Magalhães.
CE04) Rua Carlos Rolim/Praça da Matriz, 63700-000 Crateús.
CE05) Rua Marcondes Pereira 684, 60130-060 Fortaleza.
CE06) Praça da Matriz s/n, 62590-000 Itarema – DG: Romulo Cesar Junior.
CE07) Rua Major Joaquim Alves 221, 63540-000 Várzea Alegre.
CE08) C.P. 2651, 60121-970 Fortaleza.
CE09) Rua Agronomando Rangel 475, 63870-000 Boa Viagem – DG: Antonio R. Barbosa. **English:** 1000-1400.
CE10) Rua Marcondes Pereira 426, 60130-060 Fortaleza. ☎ +55 (85) 272 3733. 🖃 +55 (85) 272 3749 – DG: Cláudio Rodrigues da Silva.
CE11) Rua Hilda Augusto 201, 63300-000 Lavras da Mangabeira – DG: Miguel Vasaconcelos Arrud.
CE12) Av. Agenor Araujo 1194, 63500-000 Iguatu.
CE13) C.P. 851, 60001-970 Fortaleza – Dir: E. Gomes de Souza.
CE14) Praça Vicente Aguilar s/n, 62400-000 Camocim.
CE15) Rua São Luís 68, 62040-450 Juazeiro do Norte.
CE16) C.P. 2683, 60121-970 Fortaleza – DG: Miguel D. de Souza.
CE17) Rua Floriano Peixoto 358, 63500-000 Iguatu.
CE18) Rua Totonho Figueiras 244, 63180-000 Barbalha.
CE19) Av. Dom José 947, 62030-630 Sobral – DG: E. Rodrigues A.
CE20) C.P. 87, 63901-970 Quixadá – DG: Heilton Dantas.
CE21) Av. Aguanambi 282, 60055-402 Fortaleza. ☎ +55 (85) 2314233.
CE22) Rua Coronel Antônio Luiz 1068, Bairro do Pimenta, 63100-000 Crato – Dir: Pe. Gonçalo Farias Filho. DG: D. Vicente de Paulo Araujo Matos.
CE23) Rua Conselheiro José Júlio 126, 62010-820 Sobral.
CE24) Rua Raul Vieira 562, 62900-000 Russas.
CE25) Rua Conego Eduardo Araripe 1692, 62870-000 Pacajus. ☎ +55 (85) 3480005 – DG: Zilmar Junior.
CE26) C.P. 222, 60001-970 Fortaleza – Dir: M. Pinheiro C.
CE27) Rua Coronel Zezé 1151, 63700-000 Crateús.
CE28) Rua Coronel Malveira 1122, 62930-000 Limoeiro do Norte.
CE29) Rua Torres Câmara 597, Aldeota, 60150-060 Fortaleza.
CE30) C.P. 34, 63011-970 Juazeiro do Norte – DG: José de Menezes Barbosa.
CE31) Praça Duque de Caxias 552, 62011-300 Sobral.
CE32) Av. Dr. Pedro de Queiroz Ferreira 2129, 62850-000 Cascavel.
CE33) Rua São Francisco 139, 63100-000 Crato.
CE34) Rua Monsenhor Salviano Pinto 71, 63800-000 Quixeramobim.
CE35) Praça Pinto Martins 260, 62400-000 Camocim.
CE36) Av. Manoel Castro 815, 62940-000 Morada Nova.
CE37) Rua Moacir Pereira Gondim s/n, 63660-000 Tauá.
CE38) Rua Coronel Alexanzito 835, 62800-000 Aracati – DG: Ernesto Gurgel Valente.
CE39) Rua Hildo Furtado s/n, 62760-000 Baturité.
CE40) Rua Italiano Júlio Filizola 512, 62370-000 São Benedito – DG:

Tomaz Antônio Brandão.
CE41) Rua Coronel Antônio Joaquim 2148, 62930-000 Limoeiro do Norte.
CE42) Rua Maestro Quincas Bezerril s/n, 62320-000 Tianguá – DG: Benedito Braz.
CE43) Rua Poeta A. Martins, 62640-000 Pentecoste.
CE44) Av. Santos Dumont 414, 63600-000 Senador Pompeu. ☎ +55 (85) 829 0206 – DG: Francisco Fonseca Coelho.
CE45) Praça Adolfo Caminha 247, 62800-000 Aracati.
CE46) Maria de Lourdes 545, 62280-000 Santa Quitéria.
CE47) Rua Coronel Meireles 35, 62680-000 Paracuru.
CE48) Rua Firmino Tavares 246, 63380-000 Barro.
CE49) Travessa Mercedes Santos 762, 62700-000 Canindé.
CE50) Rua Coronel Lucio 489, 63700-000 Crateús. ☎ +55 (85) 811 0060 – DG: José Maria B. Bonfim.
CE51) Rua Dr. Monteiro Filho 10, 63800-000 Quixeramobim.
CE52) Av. Barão de Studart 2820, Aldeota, 60120-002 Fortaleza. ☎ +55 (85) 2722922, 2727635 – DG: José Edilson Alves.
CE53) Rodovia Massapê/Sobral (Mumbaba), 62140-000 Massapê.
CE54) Rua Tibúrcio Targino 155, 61700-000 Aquiraz – DG: Iva Lioba.
CE55) Rua Dr. Almir Farias 446, 62200-000 Nova Russas.
CE56) Rua Raimundo Nonato 81 – Centro, 62940-000 Morada Nova.
CE57) Rua Coronel Felix 1237 altos, 62250-000 Ipu.
CE58) Travessa 10 de Novembro 198, 63150-000 Campos Sales.
CE59) Rua Cazuzinha Marques 87, 63560-000 Acopiara.
CE60) Rua Manoel Inacio de Lucena 249, 63260-000 Brejo Santo.
CE61) BR-222 Km 220, 62100-000 Sobral.
CE62) Rua Conrado Porto s/n, 62430-000 Granja.
CE63) Rua Antônio Queiroz 343, 63870-000 Boa Viagem.
CE64) Rua Major Barreto 3000, 62600-000 Itapajé – DG: José Ezaclyr Montenegro.
CE65) Loteamento Pioneiro, Entrada Sobral Santa Quitéria, 62115-000 Forquilha.
CE66) Rua Paulo Marques 374 – Centro, 62370-000 São Benedito.
CE67) Rua Monsenhor Joviniano Barreto s/n, 63660-000 Tauá.
CE68) Rua Monsenhor Eurico s/n, 62380-000 Guaraciaba do Norte.
CE69) Av. Rios 92, 62590-000 Itarema.
CE70) Av. Coronel Alexanzito 369, 62800-000 Aracati – DG: Neto Nascimento.
CE71) Av. José de Borba Vasconcelos 3589, 61900-000 Maracanaú.
CE72) Rodovia BR-226 km. 20, Distrito de Bonfim, 63600-000 Senador Pompeu.
CE73) Rua Dom Pedro II 141, 62580-000 Acaraú.
CE74) Rua Padre Antônio Tomaz s/n, 62550-000 Morrinhos.
CE75) Rua Major Liberalino s/n, 62250-000 Ipu.
CE76) Rua Joaquim Távora 333, 63150-000 Campos Sales. ☎ +55 (85) 533 1530 – DG: Fernando de Assis da Silva.
CE77) Rua Manoel Alencar 35, 63610-000 Mombaça.
CE78) Rua Caio Prado 406 – Centro, 61800-000 Pacatuba.
CE79) Rua Raul Catunda Fontenele 61, 62230-000 Ipueiras – Dir: Francisco Airton de Pinho.
CE80) Margem Direita da Ferrovia Cedro - Fortaleza, 63400-000 Cedro.
CE81) Rua Placido Pinho 294, 62730-000 Caridade.
CE82) Rua Ovidio Maia 213, Centro, 63900-000 Quixadá – DG: José Nison Ferreira Gomes Filho.
CE83) Av. Cachoeira do Riacho do Sangue s/n, 63620-000 Solonópole.

DF00) DISTRITO FEDERAL

DF01) SRTN, Qda. 701, Bloco P, Salas 3119/3121, 70719-900 Brasília – DG: Jorge Guilherme M. Pontes.
DF02) Setor de Rádio e TV Sul, Palácio do Rádio, Bloco 1 6Þ andar, 70340-901 Brasília – Dir: Antônio Moraes e Souza.
DF03) Av. W-3, Q-16, lotes 26/28-Gr. 314, 70340-000 Brasília.
DF04) Av. W2 Sul Q-701, bloco A SRTV, 70340-000 Brasília – DG: Eduardo da Veiga R. Fajardo.
DF05) SRTV/Sul, Q-701, bloco E, Térreo, 70340-000 Brasília.
DF06) **Radiobras.** C.P. 070.747, 70359-970 Brasília. ☎ +55 (61) 321 3949. ① 061-1682. **Web:** http://www.cdsid.com.br/radiobras/ – Pres: Marcelo Netto. Dir. Radio: Iolando Lourenço. Dir. News: Miriam Moura. Dir. Adm: Januário Procópio. Dir. Tec: Toshihiro Kanagae. Dir. SW: Renato Geraldo de Lima.
DF07) SRTV-Sul, Q. 701, Conj. "E", Bloco 2 e 4, sala 316, 70340-902 Brasília. ☎ +55 (61) 226 0265/322 2963 – DG: Adécio Sartori.
DF08) SCS-Quadra 05, Bl. B, Lotes 47 a 57, Ns. 39/40, 70340-000 Brasília.

ES00) ESPIRITO SANTO

ES01) Av. Alberto Torres 345, 29040-700 Vitória.
ES02) C.P. 700, 29001-970 Vitória – Dir: Leo Amorim.
ES03) Rua Coronel Guardia 11, 29300-070 Cachoeiro de Itapemirim –

DG: José Afonso Coelho.
ES04) Rua Chafic Murad 902, Bento Ferreira, 29050-901 Vitória. ☎ +55 (27) 3218333. ▤ +55 (27) 2231525 – DG: Aurelice A. Lindenberg. **Italian:** Sat/Sun 1200-1300. **FM:** 90.1 "R. Antena 1", 92.5 "Jovem Pan", 102.3 "Litoral".
ES05) C.P. 125, 29900-971 Linhares – DG: Paulo Sérgio Gava.
ES06) C.P. 178, 29700-971 Colatina – DG: G. Pereira.
ES07) C.P. 509, 29001-970 Vitória – Dir: Jairo Gouvea Maia.
ES08) C.P. 809, 29001-970 Vitória – DG: Edivaldo Euzébio dos Anjos.
ES09) Rua 7 de Setembro 2/4, Centro, 29300-901 Cachoeiro de Itapemirim.
ES10) Rua Alberto de Oliveira Santos 42, 19º andar, salas 1916-1920, Centro, 29010-901 Vitória. ☎ +55 (27) 322 0082 – DG: Yolanda Therezinha Bruzamolin. **FM:** 101.5 "Cidade".
ES11) Rua José Cupertino 120, 29600-000 Afonso Cláudio – DG: Antonio Valderi Bernardes.
ES12) Rua da Matriz 85, 29200-000 Guarapari.
ES13) Av. Presidente Vargas 449, 29600-000 Afonso Cláudio – DG: Altair A. Ferreira.
ES14) C.P. 132, 29930-000 São Mateus – DG: Paulo Sérgio Gava.
ES15) Rua Costa Pereira 37, Centro, 29300-090 Cachoeiro de Itapemirim. ☎ +55 (27) 5227056 – DG: Pe. Joselito Ramalho. **Italian:** Sun 1300-1600.
ES16) Rua Graciano Neves 250, 29156-050 Cariacica.
ES17)Av. Perfeito Manoel Vila 660, 29800-000 Barra de São Francisco.
ES18) Rua Chafic Murad 901, Bento Ferreira, 29050-901 Vitória – DG: Aurelice Lindenberg.
ES19) 29650-000 Santa Teresa.

GO00) GOIAS
GO01) Av. Goiás 636, 74010-010 Goiânia – DG: M. Barbosa L.
GO02) C.P. 41, 75701-970 Catalão. ☎ +55 (62) 4412500, 4412700. ▤ +55 (62) 4412867 – DG: Verônica F. Margon. **F.PI:** 10kW.
GO03) Rua Coronel Barbosa 420, 75070-110 Anápolis.
GO04) Av. Egídio Francisco Rodrigues 54, Centro, 75200-000 Pires do Rio – DG: Ulysses B. de Oliveira. **FM:** 102.3 "Rio FM".
GO05) Av. 24 de Outubro 1854, Campinas, 74505-011 Goiânia. ☎ +55 (62) 8334000. ▤ +55 (62) 8334019 – DG: Pe. Jesus Flores.
GO06) Av. Marechal Rondón, Q. 18 Lote 9, 76270-000 Jussara – DG: Itamar Caiado. Dir: Luciano Henrique.
GO07) Rua 48 esq. 47A no. 498, Praça Castelo Branco, 76680-000 Itapuranga – Pres: Joaquim S. P. da Silva.
GO08) C.P. 670, 75001-970 Anápolis – DG: Mária Lúcio Gomes Santos. **FM:** 96.3.
GO09) C.P. 33, 75800-000 Jataí – DG: Zacarias Faleiros.
GO10) Rua Evangelino Meireles 26, 72800-000 Luziânia.
GO11) Rua Coronel Gonzaga 400, 56900-000 Caldas Novas – DG: Renato Cunha.
GO12) C.P. 520, 75001-970 Anápolis – DG: Nelson Silva Rosa.
GO13) Av. Brasil 272, 76380-000 Goianésia.
GO14) Rua 03 No. 30, Praça Coronel Carrijo, Centro, 78530-000 Mineiros – DG: Elias de Oliveira.
GO15) Av. Bernardo Sayao 371, 76310-000 Rialma – DG: José R. Rego.
GO16) Av. Anhanguera 3511, 74610-010 Goiânia.
GO17) Av. Floriano Peixoto 17, 76550-000 Porangatu – DG: Agenor H. Manzi.
GO18) Av. Tocantins 2011, 76400-000 Uruaçu.
GO19) C.P. 17, 72900-000 Santo Antônio do Descoberto (or Av. São João Batista 23, 77223-000 Cidade Eclética). ☎ +55 (61) 626 1391. ▤ +55 (61) 502 8777 – DG: Jovino Alves de Azevedo.
GO20) C.P. 149, 75503-970 Itumbiara – DG: Anair G. Machado.
GO21) Praça da Bandeira 94, 75500-000 Itumbiara.
GO22) C.P. 501, 75001-970 Anápolis – DG: José Cunha Gonçalves.
GO23) Rua do Contorno 702, 76330-000 Jaraguá.
GO24) Rua 148 No. 326, 74170-110 Goiânia – Gte: A. Cozar.
GO25) Av. Lazaro Xavier 18, 75860-000 Quirinópolis. ▤ +55 (62) 6511538 – DG: Nerivaldo Costa.
GO26) Rua Herculano Lobo 80, 73800-000 Formosa.
GO27) C.P. 4104, 74823-870 Goiânia. ☎ +55 (62) 250 1000. ▤ +55 (62) 2501243 – DG: Marcos Vilas Boas. **FM:** 92.7 "Executiva FM" and 97.1 "Araguaia".
GO28) C.P. 34, 75781-970 Ipameri – DG: Ir. Maria Inêz de Oliveira.
GO29) Rua 15 No. 83, 76300-000 Ceres.
GO30) Av. Presidente Costa e Silva, Jardim Bela Vista, 74001-970 Goiânia – Pres: Altair Pereira. **FM:** 90.1.
GO31) Av. Goiás 174, Quadra 4, Lote 24, 16º andar, 74010-970 Goiânia – Dir: Luis G. de Araújo.
GO32) Alameda das Rosas 2200, Setor Oeste, 74126-010 Goiânia – DG: Wilmar Antônio Alves.
GO33) Av. Amazonas 367, 76100-000 São Luís de Montes Belos – DG: João P. de Souza.

GO34) Rua Barão do Rio Branco 989, 75650-000 Morrinhos.
GO35) Av. Mario Ferreira 58, Sobreloja, 75180-000 Silvânia.
GO36) Rua Kisleu Maciel 113, 73850-000 Cristalina – DG: Waldenir Figueiredo.
GO37) C.P. 60, 76601-970 Goiás – DG: Wilians de Souza P.
GO38) C.P. 70, 75601-970 Goiatuba.
GO39) Rua Sebastião F. Souza 137, 75920-000 Santa Helena de Goiás – DG: Renato Pereira de Araújo.
GO40) C.P. 117, 75401-970 Inhumas – DG: Lúsio de Freitas B.
GO41) C.P. 131, 75901-970 Rio Verde – DG: Iris C. de Freitas. **FM:** 95.3.
GO42) Praça Alves Castro s/n, 76630-000 Itaberaí.
GO43) Rua Catalão 182, 76200-000 Iporá – Dir: S. Alves C.
GO44) Av. Lindolfo Alves Dias 51, Centro, 75850-000 Caiapônia.
GO45) Rua São Domingos 376, 75300-000 Aparecida de Goiânia.
GO46) Quadra 33, Lotes 23/24, 72900-000 Santo Antônio do Descoberto (C.P. 06-799, 71701-970 Brasília, DF). ☎ +55 (61) 626 1266. ▤ +55 (61) 626 1052 – DG: Elpídio Ferreira dos Santos.
GO47) Rua Percival Rebelo 375, 75640-000 Piracanjuba.
GO48) C.P. 185, 76301-970 Ceres – Dir: Marcela J. Silva.
GO49) Praça Silva Junior 184, 76420-000 Niquelândia – DG: Sebastião da Silva Rocha Vidal.
GO50) Rua Xingu 625, 75600-000 Goiatuba – DG: Leonardo Rogério Costa.
GO51) Av. Brasil 100, 75860-000 Quirinópolis.
GO52) Av. Joaquim David Ferreira 1390, 76105-000 Firminópolis.
GO53) Rua Carlos Antônio Cardoso, Qd. C1 lotes 5 e 6, 73840-000 Campos Belos.

MA00) MARANHÃO
MA01) Av. Eliézer Moreira s/n, Incra, 65950-000 Barra do Corda – DG: Raimundo Nonato Cruz.
MA02) Praça Dom Pedro II s/n, 65030-000 São Luís.
MA03) Av. Camboa 120, Bairro Camboa, 65020-260 São Luís – DG: Edilson Lobão Filho. **FM:** 94.3.
MA04) Rua Henrique de Figueiredo 485, 65400-000 Codó.
MA05) Av. São Benedito 1075, Bairro São Benedito, 65400-000 Codó – DG: Raimundo Leonel M. Araújo.
MA06) Rua Dorgival Pinheiro de Sousa s/n, 65900-110 Imperatriz.
MA07) Rua Manoel Alves de Abreu 373, 65700-000 Bacabal.
MA08) Av. Coronel Fonseca 200, 65800-000 Balsas – DG: João Leitão Neto.
MA09) Parque do Bom Menino s/n, 65025-180 São Luís.
MA10) Rua do Correio s/n, Bairro Fátima, 65030-340 São Luís.
MA11) Av. Presidente Médici 77, 65031-410 São Luís. ▤ +55 (98) 2214141 – DG: Paulo Falcão. **FM:** 102.5.
MA12) Rua Albino Paiva 655, 65200-000 Pinheiro – DG: Manoel Maria Soares Paiva.
MA13) Rua Manoel Trindade 410, 65725-000 Pedreiras.
MA14) Av. Keened, 65690-000 Colinas – DG: Raimundo Felix.
MA15) BR-316 km 115, 65370-000 Pindaré Mirim.
MA16) Fazenda São João, BR-14 Km 37, 65215-000 Viana.
MA17) Rua Guarani 3, Parque das Laranjeiras, 65922-000 João Lisboa.
MA18) Rua A. Bandeira 831, 65320-000 Vitorino Freire.
MA19) Av. Belo Horizonte s/n, 65634-000 Timon.
MA20) Rua Piauí 895, 65926-000 Açailândia.
MA21) C. Carneiro 177, 65750-000 Esperantinópolis.
MA22) Travessa Tiradentes 338, Centro, 65415-000 Coroatá.
MA23) Rua Aarão Reis s/n, 65604-060 Caxias.
MA24) Av. Coronel Stanley Fortes Batista 454, 65365-000 Zé Doca.
MA25) Av. Marechal Castelo Branco 605, sala 204, 65075-000 São Luís.
MA26) Rua Alagoas 497, 65900-490 Imperatriz.
MA27) Av. Amaral Raposo s/n, 65940-000 Grajaú.
MA28) Rua Rui Barbosa s/n, 65620-000 Coelho Neto.
MA29) Rua Dr. Paulo Ramos 495, 65208-000 Santa Helena.
MA30) Rua Frederico Coelho esquina com Av. Frei Aniceto, 65763-000 Tuntum.
MA31) Rua do Puraqueu s/n, 65350-000 Vitória do Mearim.
MA32) Parque da Bandeira 222, Edificio Ariana, Centro, 65665-000 São João dos Patos – DG: Maria da Guia Lima Sá.
MA33) Praça do Guarim s/n, 65390-000 Santa Luzia – Dir: Herberth Herland.
MA34) Rua Terra esquina com Rua Jupiter s/n, 65760-0000 Presidente Dutra.
MA35) Rua Coronel Luis Vieira 52, 65500-000 Chapadinha.
MA36) Rua Cel. Pedro Bogéa 227 - Centro, 65715-000 Lago da Pedra – DG: Marcelo Rodrigues.
MA37) Rua Hemeterio Leitão 103, 65430-00 Vargem Grande.
MA38) Av. Ana Jansen 200, 65000-000 São Luís.

MG00) MINAS GERAIS

MG01) C.P. 123, 36771-970 Cataguases – DG: Mauro Rios Valentim. **FM:** 89.5.

MG02) Rua General Carneiro 10, Edificio Milinardo, s 200 à 305, 39400-095 Montes Claros.

MG03) Av. Sinfronio Brochado 805, 30640-000 Belo Horizonte.

MG04) Praça Nossa Senhora Aparecida 134, 38406-063 Uberlândia – DG: Pe. Edvaldo P. Sousa. **FM:** 98.7.

MG05) Av. Padre Matias 1089, Bairro Marciano Brandão, 38740-000 Patrocínio. ☎ +55 (34) 8311546, 8311596. 🖷 +55 (34) 8311896 – DG: Marciano Luiz R. Alves. **FM:** 98.9.

MG06) Rua Itatiaia 117, Bairro Bonfim, 31210-170 Belo Horizonte. ☎ +55 (31) 4213588, 4214288. 🖷 +55 (31) 4462900. **Web:** http://www.itatiaia.com.br – DG: Cláudio Emanuel Carneiro.

MG07) C.P. 24, 38001-970 Uberaba.

MG08) Av. Presidente Vargas 372, Centro, 35661-000 Pará de Minas – DG: Pe. Geraldo Gabriel de Bessa.

MG09) Rua Euripides Ribeiro 739, 38720-000 Lagoa Formosa – DG: Silvio Ribeiro.

MG10) Rua Padre Antônio Gabriel Carvalho 17, 35400-000 Ouro Preto – DG: Januário Carneiro.

MG11) C.P. 30, 35791-970 Curvelo – DG: Nelson Rocha F.

MG12) Rua Prof. Monteiro Fonseca 119, 39400-149 Montes Claros.

MG13) Praça Coronel Hermógenes 292, 38770-000 João Pinheiro.

MG14) Rus Entre Rios 33, Bairro Carlos Prates, 30710-080 Belo Horizonte – DG: Gilberto Gil Costa. **F.PI:** 100kW.

MG15) Praça 5 de Novembro 399, 36900-000 Manhuaçu.

MG16) C.P. 37, 37551-970 Pouso Alegre – DG: Milton L. de Paula.

MG17) Travessa Coronel José Ferreira 45, 38010-320 Uberaba – DG: João Batista Rodrigues.

MG18) Rua Dr. João Penido Filho 269, 36021-600 Juiz de Fora.

MG19) Av. Itaú 515, Bairro Dom Cabral, 30730-280 Belo Horizonte. ☎ +55 (31) 464 8355. 🖷 +55 (31) 462 8598 – DG: F. Albergaria P. **F.PI:** st. on "62m" 5kW, AM 100kW.

MG20) Rua Dr. Getúlio Vargas 63, 37200-000 Lavras.

MG21) C.P. 69, 38701-970 Patos de Minas.

MG22) C.P. 10, 35901-970 Itabitra – DG: Luiz Paulo. **FM:** 93.3.

MG23) C.P. 401, 38409-970 Uberlândia – DG: A. de Castro.

MG24) Praça Getúlio Vargas 10, 35430-002 Ponte Nova.

MG25) C.P. 153, 35291-970 Mantena – DG: U.A. Moreira.

MG26) Rua Luiz 235, Bairro Américo Silva, 35590-000 Lagoa da Prata. ☎ +55 (37) 261 1999. 🖷 +55 (37) 261 2549 – DG: Samuel Martins de Almeida.

MG27) Rua Bias Fortes 140, 35200-000 Almorés – DG: Dênis Souza Diniz. **F.PI:** 10kW.

MG28) Rua Minas Gerais 655, 35500-007 Divinópolis.

MG29) Praça Dom Pedro Teixeira 49, 36200-000 Barbacena.

MG30) Travessa Padre Benatti 1200, 37950-000 São Sebastião do Paraíso. ☎ +55 (35) 5311351. 🖷 +55 (35) 5315154 – DG: Mons. Hilário Pardini.

MG31) Rua Itatiaia 117, Bairro Bonfim, 31210-170 Belo Horizonte – DG: Emanuel Soares Carneiro.

MG32) Rua Barão de Piunhi 31, 37290-000 Formiga.

MG33) Av. Brasil 2770, 35020-070 Governador Valadares – DG: Odilon Lagares de Aguilar. **FM:** 97.7.

MG34) Rua Coronel Joaquim Mendes 19. Ed. Cine T. Helena, 38430-000 Tupaciguara. ☎ +55 (34) 2812034, 2812146 – DG: A.M. Netto. **FM:** 91.9.

MG35) C.P. 1027, 30161-970 Belo Horizonte – DG: Eugenio Silva.

MG36) Rua Dr. Olinto Martins 207, 39960-000 Jequitinhonha. ☎ +55 (33) 741 1521. 🖷 +55 (33) 741 1332 – DG: Pe. Frei Eliseu Tijdink.

MG37) Rua Benedito Valadares 139, 39800-000 Teófilo Otoni.

MG38) Praça Dom João 122, Centro, 39100-000 Diamantina – DG: Ragosino Ferreira de Araujo.

MG39) Praça 28 de Setembro 95, 36520-000 Visconde do Rio Branco – DG: Adilson Cunha Honório.

MG40) Rua Tenente Viotti 131, 37460-000 Passa Quatro. ☎ +55 (35) 371 1260/1129 – DG: Letícia Pinho Scarpa.

MG41) C.P. 62, 37271-970 Campo Belo – DG: Eugênio Cardoso Furtado.

MG42) Av. Teodolino Pereira de Araújo 731, 38440-000 Araguari.

MG43) C.P. 890, 30161-970 Belo Horizonte – DG: Valeria Carmo Ferreira.

MG44) Av. Bom Jesus 464, 37578-000 Bueno Brandão. ☎🖷 +55 (35) 4631160 – DG: Elena S. Asbahr. Dir: Jair Asbahr.

MG45) Rua dos Viajantes 81, 35300-000 Caratinga.

MG46) Av. Tirandentes 209, Centro, 36300-000 São João Del Rei – Dir: Etel Nogueira da Silva.

MG47) Rua Antenor Airosa Machado 64, 38420-000 Monte Alegre de Minas – DG: Carlos Nogueira M.

MG48) C.P. 353, 36001-970 Juiz de Fora.

MG49) C.P. 88, 37591-970 Jacutinga. ☎ +55 (35) 443 2121 – DG: João Alves.

MG50) Rua Dr. Querubino 303, 35170-001 Coronel Fabriciano. ☎ +55 (31) 8421112, 8421492. 🖷 +55 (31) 8421400 – DG: Gaspar de Almeida Pinto.

MG51) C.P. 557, 38409-970 Uberlândia.

MG52) Rua Bueno Brandão 26, 38430-000 Tupaciguara.

MG53) Rua Tamoios 200, 21º andar, 30120-050 Belo Horizonte – DG: Manoel Diamantino da Costa.

MG54) Rua Olegário Maciel 200, Bairro Vila Podis, 37500-000 Itajubá – DG: José Luiz M. Sannini. **F.PI:** 5kW.

MG55) C.P. 5, 36404-970 Congonhas. (+55 (31) 7311538, 7311342. – DG: Pe. Benedito P. Rocha. **F.PI:** 25kW.

MG56) Av. Dr. Américo Luz 153, 37890-000 Muzambinho – DG: Carlos A. Poscidonio.

MG57) C.P. 262, 36001-970 Juiz de Fora – DG: G. Mendes.

MG58) C.P. 85, 38441-970 Araguari – Gte: W. Gebrim.

MG59) Rua Areião do Matadouro s/n, Parque da California, 34000-000 Nova Lima.

MG60) C.P. 253, 38001-970 Uberaba. 🖷 +55 (34) 313-3400. **E-mail:** sete@ldc.com.br **Web:** http://www.ldc.com.br/sete/ – DG: Fuad Miguel Hueb Filho. **FM:** 98.1.

MG61) Av. Juscelino Kubstchek 740, 37170-000 Boa Esperança. ☎ +55 (35) 8311000. 🖷 +55 (35) 8511475 – DG: José Maria Oliveira.

MG62) C.P. 10, 35500-970 Divinópolis – DG: Mayrinck Pinto de Aguiar. **FM:** 94.5.

MG63) Rua João Luiz Alves 122, 37400-000 Campanha. ☎ +55 (35) 2611985 – DG: Odair José da Cunha.

MG64) C.P. 41, 36881-970 Muriaé – DG: Paulo R. Barros. **FM:** 96.3.

MG65) Rua Pernambuco 1077, 30130-151 Belo Horizonte. **F.PI:** Change of name to R. Globo Rural.

MG66) Rua do Rádio 60, 38190-000 Sacramento. ☎ +55 (34) 3511735. 🖷 +55 (34) 3511432 – DG: Pe. Levi F. Marques.

MG67) Rua Cassiano Lemos 87, 38180-000 Araxá.

MG68) Av. Assis Chateaubriand 499, Floresta, 30150-101 Belo Horizonte – Superintendente: Cássia B.S. Cinque.

MG69) C.P. 102, 37002-970 Varginha – Dir: S. Sampaio M. DG: Carlos Ferreira da Costa Filho.

MG70) Av. Brasil 4460, 38405-378 Uberlândia – Dir: E. Costa.

MG71) Av. Brasil 3049, 35020-070 Governador Valadares.

MG72) C.P. 110, 38301-970 Ituiutaba.

MG73) C.P. 937, 37701-970 Poços de Caldas – DG: Marcelo de Oliveira.

MG74) Rua Capitão Henrique Albuquerque 55, 36400-000 Conselheiro Lafaiete. ☎ +55 (31) 751 1752 – Dir. Pres: Dr. Agostinho Campos Neto. **FM:** 89.9.

MG75) Rua Itajubá 80, 35160-035 Ipatinga.

MG76) Rua Francisco Ribeiro 59, 37250-000 Nepomuceno.

MG77) Rua Duque de Caxias 450, 16º andar, Edificio Chams, 38400-066 Uberlândia. ☎ +55 (34) 2366320. 🖷 +55 (34) 2362679 – DG: Aildo Rodrigues Ferreira.

MG78) Travessa Juriti 59, 38600-000 Paracatu.

MG79) Av. 13 No. 658, 6º andar, Edifício Ituiutaba, 38300-000 Ituiutaba – DG: Daniel Paulo Nascimento. **FM:** 93.7.

MG80) Av. Costa Junior 400, 38840-000 Carmo do Paranaíba – DG: Everton José Alves.

MG81) Rua Dr. Guilherme 286, 35450-000 Itabirito.

MG82) C.P. 143, 37701-970 Poços de Caldas. ☎ +55 (35) 7221687. 🖷 +55 (35) 7222687 – Dir: F. de Araújo.

MG83) Rua Prof. Maria Justiana 26, 37750-000 Machado.

MG84) Praça Guido Marliere 30, 36500-000 Ubá.

MG85) Rua Calixto Martins de Melo 391, 38610-000 Unaí.

MG86) Rua Calimeiro Guimarães 308, 38180-000 Araxá.

MG87) Rua Antônio Dias Adorno 1290, 35045-040 Governador Valadares – DG: Edson Gualberto de Souza. **FM:** 100.1 "Imparsom". **F.PI:** SW 31mb 10kW.

MG88) Rua Dr. Péricles de Mendonça 91, 36680-000 São João Nepomuceno. ☎ +55 (32) 2611344, 2612596 – DG: Aristides Santos Dias.

MG89) Rua Niquel 457, 35701-107 Sete Lagoas – DG: Geraldo Padrão. **FM:** 92.1 "Musirama".

MG90) Av. Major Antônio Alberto Fernandes 445, 37720-000 Botelhos – Dir: Marcionil Moreira da Silva.

MG91) Av. Conde Ribeirão do Vale 661, 37800-000 Guaxupé.

MG92) Rua XV de Novembro 62, 36500-000 Ubá.

MG93) Praça Aires Quaresma s/n, 35930-000 João Monlevade.

MG94) Rua Gerson Coutinho da Silva 1001, 38550-000 Coroman-del – Dir: Carlos Amado de Oliveira.

MG95) Rua Juliano Marques Duarte 110, 36660-000 Além Paraíba. ☎ +55 (32) 4621413 – DG: Reinaldo Tavares. **FM:** 95.5 "Juventude".

MG96) Av. 17 No. 1045, 38300-000 Ituiutaba. ☎ +55 (34) 2611417, 2617118. 🖷 +55 (34) 2611829 – DG: Jeuid Abdulmassih.

MG97) Praça Conego Victor 73, 37190-000 Três Pontas.
MG98) Rua Coronel Domiciano Ferreira 314, 38200-000 Frutal – Dir: Odair de Moura e Silva.
MG99) Av. Belo Horizonte 108, 39860-000 Nanuque.
MG100) Av. 7 de Setembro 109, 36325-000 Tiradentes.
MG101) Rua Floriano Peixoto 31, 36570-000 Viçosa. ☎ +55 (31) 8911242. ▤ +55 (31) 8913421 – DG: João Bosco Torres. **FM:** 97.9.
MG102) C.P. 28, 37901-970 Passos – Dir: M.A. Soares M.
MG103) C.P. 61, 36971-970 Manhumirim – DG: Pe. Antônio Otaviano.
MG104) C.P. 044, 35541-970 Oliveira – DG: Wander Assis Ribeiro.
MG105) Rua 13 de Maio 425, 36200-000 Barbacena.
MG106) C.P. 54, 35680-970 Itaúna – DG: Afonso H.S. Lima.
MG107) C.P. 189, 37471-970 São Lourenço – DG: José Alberto Jardim. **FM:** 94.3.
MG108) Rua Ribeirao São Domingos 689, 38280-000 Iturama.
MG109) Av. Magalhães Pinto 829, 35610-000 Dores do Indaiá. ☎ +55 (37) 5511622 – DG: Adélia de Souza
MG110) Rua Adalberto Ferraz 50, 2Þ andar, 37550-000 Pouso Alegre – DG: Murilio Maia.
MG111) Rua Dr. José Gonçalves 17, 35600-000 Bom Despacho.
MG112) Praça Getúlio Vargas 81, 36400-000 Conselheiro Lafaiete.
MG113) Rua Casemiro Avelar Filho 143, Centro, 37410-000 Três Corações. ☎ +55 (35) 231 1000/1540/1941 – DG: Victor Cunha. **FM:** 95.5
MG114) Praça Nossa Senhora do Carmo 224, 38500-000 Monte Carmelo.
MG115) C.P. 20, 37541-970 Santa Rita do Sapucaí.
MG116) Praça Coronel Ramos 3, 39270-000 Pirapora.
MG117) C.P. 35, 36701-970 Leopoldina – DG: José A. Barcellos.
MG118) Rua Bias Fortes 366 A, 37130-000 Alfenas – Dir: Albany Gandya.
MG119) Av. Ribeiro Pena 35, 35560-000 Itapecerica – DG: Wantuil Rodrigues Nascimento.
MG120) Rua Anastácio José Gonçalves 139, 38810-000 Rio Paranaíba – DG: Cleber Jonas Ribeiro.
MG121) Alto da Boa Vista s/n, 36240-000 Santos Dumont.
MG122) Av. Minas Gerais 584, 39510-000 Espinosa – DG: Antônio P. Tércio.
MG123) Praça Nossa Senhora da Abadia 24, 38025-430 Uberaba. ☎ +55 (34) 332 6200. ▤ +55 (34) 332 6430 – DG: José Antônio Silva Filho.
MG124) Av. Hermenegildo Donatti 199, 37795-000 Andradas – DG: Luiz Antônio Barbon. **FM:** 94.9.
MG125) Praça Coronel Odilon Coelho s/n, 39520-000 Porteirinha.
MG126) Rua Ordalino Rodrigues 351, 35430-148 Ponte Nova.
MG127) Travessa Dona Santinha 20, 39480-000 Januária – DG: Ottorino dal Moro Neto.
MG128) Rua Benedito Valadares 423, 36880-000 Muriaé.
MG129) Rua Padre Pedro 53, 39391-000 Bocaiúva.
MG130) Rua Rui Barbosa 74, 39440-000 Janaúba – Gte: Odilon Coelho.
MG131) Av. Getúlio Vargas 25, 35350-000 Raul Soares.
MG132) Praça José Batista de Frieitas 78, 3Þ andar, 35519-000 Nova Serrana – DG: Nilton Moreira.
MG133) Praça Getúlio Vargas 143, 37810-000 Guaranésia.
MG134) BR-496 Km 103, 39260-000 Várzea da Palma.
MG135) Rua dos Quarteis s/n, 38900-000 Bambuí.
MG136) Rua Alto da Catedral s/n, 39800-000 Teófilo Otoni. ☎ +55 (33) 5212938 – DG: Pe. Joel Ferriera Filho.
MG137) Rua Major Honor Sarmento 393, Alto São João, 39400-103 Montes Claros. ☎ +55 (38) 221 3030/3355. ▤ +55 (38) 2213271 – DG: José Nardel.
MG138) Rua 102 No. 498, 38360-000 Capinópolis.
MG139) Praça Monsenhor Messias Bragança 80, salas 201 a 207, 37900-000 Passos. ☎ +55 (35) 5216767, 5216196 – DG: José Neif Jabur Filho.
MG140) Rua Padre João Lourenço Leite 100, 37175-000 Ilicínea.
MG141) Rua Astor Goulart de Moura 51, 39200-000 Corinto – DG: Afonso Victor V. Andrade.
MG142) BR-381 Km. 195, Bairro Cachoeira do Vale, 35180-000 Timóteo – DG: Cláudio Emanuel Carneiro. **F.PI:** 50kW.
MG143) Rua Teófilo Otoni 422, 35700-007 Sete Lagoas – DG: João Carlos S. Oliveira.
MG144) Rua Alves Rosa 255, 38780-000 Vazante – DG: Valmir Borges.
MG145) Rua Leandro Gonçalves 88, 3Þ andar, 36900-000 Manhuaçu – DG: João Magalhães Bifano.
MG146) Rua Coronel Rennó 7, 37500-000 Itajubá – DG: Eng. Ismael Noronha.
MG147) Praça São Pedro 49, 38950-000 Ibiá.
MG148) Praça Getúlio Vargas 108, 36800-000 Carangola – DG: Dr. Nelson Rocha. **FM:** 102.7 "Caparaó".
MG149) Rua Oscar Vidal 416, 36016-290 Juiz de Fora – DG: Josino

Aragão.
MG150) Rua Ceará esquina com Rua Alagoas, 38700-000 Patos de Minas.
MG151) Av. 7 de Setembro 55-A, 36950-000 Ipanema – DG: Euler Pães Dias.
MG152) Rua Julio Cosi 5, 38230-000 Fronteira – Dir: Aramis Passuelo.
MG153) Alameda José Maria Alkimim 1700, Bairro Serra Verde, 30320-210 Belo Horizonte. ☎ +55 (31) 455 2066/1133. ▤ +55 (31) 455 2070 – DG: Luiz Carlos Gomes.
MG154) Praça Lindolfo Barbosa Vieira 40, 35330-000 Inhapim.
MG155) Rua Itajubá 62, 35160-035 Ipatinga – DG: Ronaldo de Souza.
MG156) Av. Rui Barbosa 621, 5Þ andar, Conj. 501, 38740-000 Patrocínio – DG: Elias José A. Neto.
MG157) Praça Comendador José Honorio 100, 37950-000 São Sebastião do Paraiso.
MG158) Rua Pref. Terêncio Pereira Vale 10, 38170-000 Perdizes – DG: Antonio José Machado. **FM:** 96.1.
MG159) Av. Joaquim Ribeiro de Gouveia 1651, 38320-000 Santa Vitória.
MG160) Rua Belo Horizonte 461, 39625-000 Itaobim.
MG161) Rua das Acacias s/n, 38660-000 Buritis.
MG162) Rua João Peixoto 100, 38440-000 Araguari.
MG163) Travessa Cônego Benedito Profício 95, 37660-000 Paraisópolis – DG: Pe. Braz Tenório Rocha.
MG164) Av. Progresso 177, 37292-000 Arcos. ☎ +55 (37) 351 2100 – DG: Roulien Ribeiro Lima.
MG165) Rua Antônio Ribeiro da Costa Junior 16, 36730-000 Pirapetinga – DG: Cesar Romero. **F.PI:** 5kW.
MG166) 37120-000 Paraguaçu.
MG167) Praça Dr. Badaró 112, 39650-000 Minas Novas. ☎ +55 (33) 7641181, 7641185 – DG: Dr. Murilo P. Badaró.
MG168) Rua G 226, 39560-000 Salinas.
MG169) Av. Rondón Pacheco 450, 37150-000 Carmo do Rio Claro – DG: Mário Pio de Faria.
MG170) Rua Vereador Maria Anselmo 33, 36480-000 Piranga.
MG171) Rua Coronel Carlos Brandão 175, 36500-000 Ubá.
MG172) Av. Dr. Pedro Rosas s/n, 37530-000 Brasópolis. ☎ +55 (35) 6411194 – DG: Antonio Vasconcelos.
MG173) Rua Silviano Brandão 795, 37570-000 Ouro Fino. ☎ +55 (35) 441 1433. ▤ +55 (35) 441 1800 – DG: Milton Lucca de Paula.
MG174) Praça Vital Brasil 56, 37980-000 Cássia – DG: Fredy Machado.
MG175) Praça Coronel Silverio de Melo 172, 37958-000 Monte Santo de Minas.

MS00) MATO GROSSO DO SUL
MS01) Av. Mato Grosso 530, Centro, 79002-233 Campo Grande – DG: Ailton Guerra.
MS02) Av. Senador Felinto Müller 59, 79080-190 Campo Grande.
MS03) C.P. 104, 79951-970 Naviraí.
MS04) C.P. 68, 79804-970 Dourados.
MS05) Rua Anchieta 871, 79081-180 Campo Grande.
MS06) Rua 15 de Agosto 98, 79200-000 Aquidauana.
MS07) Rua Melanio Garcia Barbosa 749, 79150-000 Maracaju – DG: Mirian Sirlei da Veiga.
MS08) Rua Joaquim Pereira Teixeira 135, 79900-000 Ponta Porã.
MS09) Av. Eduardo Elias Zahran 1100, 79050-010 Campo Grande.
MS10) C.P. 37, 79601-970 Três Lagoas – DG: Delcina Rosa S. de Carvalho.
MS11) Av. Marcelino Pires 1404, 79801-002 Dourados.
MS12) C.P. 18, 79201-970 Aquidauana. ☎ +55 (67) 241 3956 – DG: Elidio Telles de Oliveira. Dir: Aldo Bertoni.
MS13) Rua Antônio Lino Barbosa 1130, 79130-000 Rio Brilhante.
MS14) C.P. 138, 79301-970 Corumbá – DG: Uriel Raghiant.
MS15) C.P. 217, 79301-970 Corumbá – DG: Fause Anache.
MS16) Rua Walther Hubachi 643, 79750-000 Nova Andradina.
MS17) Av. Aldair Rosa de Oliveira 1045, 79640-100 Três Lagoas. ☎ +55 (767) 521 2305 – DG: Romeu de Campos Junior.
MS18) Rua Ferreira 69 B, Piracema, 79400-000 Coxim – DG: Moacir Moura.
MS19) Av. Pres. Vargas 1893, 79730-000 Glória de Dourados.
MS20) Rua Visconde de Taunay 895, 79500-000 Paranaíba.
MS21) C.P. 126, 79991-970 Amambaí – DG: Gilberto Pereira Guedes.
MS22) Rua Tenente Antônio João 1592, 79700-000 Fátima do Sul.
MS23) C.P. 200, 79541-970 Cassilândia – DG: João Lourenço Girotto.
MS24) Av. Presidente Vargas 669, 79940-000 Caarapó.
MS25) Rua Dr. A. Coelho de Oliveira 549, 79240-000 Jardim.
MS26) Rua Rui Barbosa 753, 79970-000 Eldorado – DG: Valério de Medeiros.
MS27) C.P. 129, 79301-970 Corumbá.
MS28) C.P. 199, 79901-970 Ponta Porã – DG: Elite José Sandri.

MS29) C.P. 68, 79804-970 Dourados.
MS30) Rua São Paulo 1359, 79490-000 São Gabriel d'Oeste.
MS31) Rua Marquês de Tamandaré 349, Bairro São Bento, 79170-000 Sidrolândia. ☎ +55 (67) 2721514, 2721543. 🖳 +55 (67) 2721868 – DG: Irineu Bellani dall'Agnol. Dir: Helena Fábio Feitosa.
MS32) Rua Itajaí 433, 79041-270 Campo Grande.
MS33) Rua Severino de Araujo Ferreira 1375, 79700-000 Fátima do Sul.
MS34) Av. Calogeras 1932, 79012-003 Campo Grande.
MS35) Av. Costa Rica 654, 79740-000 Ivinhema – DG: Dr. Elias Vieira.
MS36) Av. João Pedro Pedrossian 4058, 79570-000 Aparecida do Taboado. ☎ +55 (67) 5651075 – DG: Paulo Rondon de Souza. **English & Spanish:** 1100-1400.
MS37) 79300-000 Corumbá.
MS38) Rua Candido Severino 462, 79420-000 Camapuã.
MS39) Rua 01 No. 1550, 79815-190 Dourados. 🖳 +55 (67) 421 0276 – DG: Luiz Fernando Cersosimo. **F.PI:** 10kW.
MS40) Rua José Antônio Pereira 1488, sala 23, 79010-190 Campo Grande.
MS41) Rua Antônio Maria Coelho 289, 79260-000 Bela Vista.
MS42) Rua Porfirio Gonçalves 1240, 79480-000 Rio Verde de Mato Grosso.
MS43) Rua Atilio Reginato 355, 79740-000 Ivinhema.

MT00) MATO GROSSO
MT01) Rua Boróros 45, 78600-000 Barra do Garças. ☎ +55(65)8611345.
MT02) Av. Brasil 27, 78600-000 Poxoréo – Gte: Edisio Rodrigues Rocha.
MT03) Professora tereza Lobo 30, 78048-700 Cuiabá. ☎ +55 (65) 321 4144, 624 6699 - DG: Onofre Ribeiro.
MT04) Av. Gov. Júlio Campos 300, 78550-000 Sinop – DG: Ascânio Baptista de Carvalho. **F.PI:** 20kW.
MT05) Praça do Seminário 239, 78015-140 Cuiabá. ☎ +55 (65) 623 7000 – DG: João Luiz de Oliveira.
MT06) Av. Leste s/n, 78580-000 Alta Floresta. ☎ +55 (65) 521 2999.
MT07) C.P. 401, 78700-970 Rondonópolis ☎ +55 (65) 423 2806 – DG: Edson A. Pires.
MT08) Rua Joaquim Murtinho 1456, Palacio do Rádio, 78020-830 Cuiabá. **Web:** http://www.solunet.com.br/rcultura/☎ +55 (65) 321 6226, 321 6882 – Gte: Eduard Rueda Saraiva Filho.
MT09) Rua São Pedro 806, 78200-000 Cáceres. ☎ +55 (65) 223 1885. 🖳 +55 (65) 223 1663.
MT10) C.P. 297, 78201-970 Cáceres – DG: Ivo Vignardi. **FM:** 102.3.
MT11) C.P. 227, 78821-970 Jaciara – DG: Francisco A. de Moura.
MT12) Av. Cuiabá 829, Edifício Mikerinos, 12Þ andar, 78700-090 Rondonópolis. ☎ +55 (65) 421 3666/3451– DG: Antônio Carlos Rocha Filho.
MT13) Rua 28 de Outubro 1445, 78280-000 Mirassol d'Oeste. ☎ +55 (65) 2411288, 2411770 – DG: Geni Fernandu C. Leitte.
MT14) Rua Benedito Monteiro 68, 78110-390 Várzea Grande.
MT15) Rua Zulmira Canavarros 285, 78005-390 Cuiabá. ☎ +55 (65) 321 1223 – Gte: Romulo Brunini.
MT16) Rua 2 No. 32, 78175-000 Poconé.
MT17) Chacara 70/71, 78575-000 Juara.
MT18) Av. Brasil 780, 78300-000 Tangará da Serra. ☎ +55 (65) 7262080 – DG: Carlos Antonio B. Longhi.
MT19) Av. T. Neves 1682, 78500-000 Colider. ☎ +55 (65) 541 1233.
MT20) Rua Waldir Rabelo 789, 78600-000 Barra do Garças. ☎ +55 (65) 861 2361. 🖳 +55 (65) 861 1760 – DG: João Bosco de Áquino Araujo.
MT21) Av. Natalino João Prescancim 155, 78890-000 Sorriso.☎ +55 (65) 544 1299 – DG: Celito Barbieri.
MT22) Praça Brigadeiro Eduardo Gomes 28, 78430-000 Nortelândia.
MT23) Av. Mato Grosso 133, 78690-000 Nova Xavantina. ☎ +55 (65) 438 1218.
MT24) Rua 6 No. 498, 78300-000 Tangará da Serra. ☎ +55 (65) 726 1084.
MT25) Rua Benjamim Constant s/n, 78780-000 Alto Araguaia.
MT26) Av. 9 de Maio 65, 78320-000 Juina.
MT27) Av. Mario Correa 350, 78790-000 Itiquira.
MT28) Av. Agricolo Pães de Barros 924, 78525-000 Matupá.
MT29) Rua das Primaveras 3574, 78550-000 Sinop – DG: Juarez Alves de Costa.
MT30) Av. Amazonas s/n, 78285-000 São José dos Quatro Marcos. ☎ +55 (65) 251 1317.
MT31) Rua 6 s/n, 78400-000 Diamantino. ☎ +55 (65) 736 1316.
MT32) Rua Perimetral s/n, Bairro Bom Clima, 78195-000 Chapada dos Guimarães – DG: Fernando Luis de Almeida.
MT33) Santa Catarina 1284, 78435-000 São José do Rio Claro. ☎ +55 (65) 786 1363.
MT34) Rua 02 – Canteiro Central, 78580-000 Alta Floresta.
MT35) Av. Bahia 430, 78760-000 Guiratinga.
MT36) Rua Joaquim Nabuco 450, 78260-000 Araputanga – DG: José Coelho Pinto.
MT37) Rua São Paulo 1440, 78250-000 Pontes e Lacerda. ☎ +55 (65) 266 2598.
MT38) Margem Esquerda BR-163 km 723, Quadra 05, Lote 01, 78520-000 Guarantã do Norte.
MT39) Rua Dom Antônio Malan 674, 78015-600 Cuiabá – Gte: Celso Castilho.
MT40) Rua Carajás 69, 76800-000 Barra do Garças – Dir: Walter Francisco Dourado.

PA00) PARA
PA01) Av. Araguaia 247, 68551-000 Redenção.
PA02) C.P. 119, 68371-970 Altamira – Gte: Ivanese Braga.
PA03) C.P. 533, 66017-970 Belém – DG: Edyr Paiva Proença.
PA04) Av. São Sebastião 622-A, 68005-090 Santarém. ☎ +55 (91) 522 2439 – Dir: Silvana Maria Santos.
PA05) Av. Marechal Rondón 786, 68540-000 Conceição do Araguaia.
PA06) Praça dos Notáveis 1006, 68400-000 Cametá – DG: Gerson dos S. Peres.
PA07) Rodovia BR-316 Km 58, 68741-740 Castanhal.
PA08) Travessa Campos Sales 370, 66019-050 Belém. ☎ +55 (91) 241 2103 – DG: Jones Lara Tavares. **FM:** 100.9.
PA09) Travessa Vileta 2193, 66093-380 Belém.
PA10) C.P. 498, 66017-970 Belém.
PA11) C.P. 038, 68801-970 Breves – DG: Adilson Almeida.
PA12) Rua Barão do Rio Branco 1151, 68600-000 Bragança.
PA13) Av. Mendonça Furtado 1481, 68005-100 Santarém.
PA14) Av. Coronel Monfredo 42, 68820-000 São Sebastião da Boa Vista.
PA15) C.P. 226, 68371-970 Altamira.
PA16) Av. Almirante Barroso 735, 66090-000 Belém – DG: Afonso Klautau. Dir: Ronald Pastor. **FM:** 93.7.
PA17) Margem da Rodovia PA-130 Km 8/9, 68500-000 Marabá.
PA18) Rodovia Transamazonica Km. 01, 68180-010 Itaituba.
PA19) Rua Lauro Sodré 722, 68456-000 Tucuruí – DG: José A. Costa
PA20) Av. Rui Barbosa 825, 68005-000 Santarém.
PA21) Av. Xingu s/n, 68555-010 Xinguara.
PA22) Travessa E. Simões 230, 68250-000 Obidos.
PA23) C.P. 8, 68251-970 Obidos – Dir: Max Hamoy.
PA24) Travessa 18 No. 1863, entre 4 e 5 ruas, 68870-000 Soure.
PA25) Av. Visconde de Sousa Franco 116, Centro, 68780-000 Vigia – DG: Edmilsom Campos.
PA26) Rodovia BR-230 (Transamazônica) Km 01, 68180-010 Itaituba.
PA27) Rodovia PA-125 Km 15, 68625-620 Paragominas.
PA28) Rua 2 de Outubro s/n, 68200-000 Alenquer.
PA29) Rodovia Transamazônica Km 04, 68502-290 Marabá.
PA30) Rodovia BR-010 Km. 09, Bairro Industrial, 68660-000 São Miguel do Guamá – DG: Leonardo Rodrigues. **F.PI:** 10kW.
PA31) 68460-000 Tucuruí.
PA32) Travessa Mauriti 1006, Bairro da Pedreira, 66080-650 Belém.
PA33) Av. Beira Rio s/n, 68380-000 São Felix do Xingu.
PA34) Av. Augusto Montenegro s/n, 68710-000 Maracanã.

PB00) PARAÍBA
PB01) Rua Pres. João Pessoa 25, 58800-010 Sousa.
PB02) C.P. 26, 58900-970 Cajazeiras – DG: Francisco A. de Albuquerque.
PB03) Rua Padre Manoel Otaviano 340, 58970-000 Conceição – DG: George Luís S. Leite.
PB04) Rua Presidente Epitácio Pessoa 242, 58700-020 Patos.
PB05) C.P. 40, 58200-970 Guarabira – DG: H. Camilo.
PB06) Praça Pres. Epitácio Pessoa 167, 58755-000 Princesa Isabel.
PB07) C.P. 134, 58100-970 Campina Grande.
PB08) Rua Coronel Juvêncio Carneiro 160, 58900-000 Cajazeiras. ☎ +55 (83) 5311497, 5311237 – DG: Francisca Zélia Ribeiro.
PB09) C.P. 1089, 58001-970 João Pessoa – DG: Deodato Borges.
PB10) Rua Manoel Pedro s/n, 58884-000 Catolé do Rocha.
PB11) Rua Venâncio Neiva 287, 58100-060 Campina Grande – DG: Luiz Aguacar.
PB12) Av. Pedro II 657, 58013-420 João Pessoa.
PB13) Rua das Trincheiras 198, Centro, 58011-000 João Pessoa – DG: Alberto P. de Castro.
PB14) C.P. 160, 58100-970 Campina Grande – DG: L. Barbosa A.
PB15) C.P. 57, 58700-970 Patos. ☎ +55 (83) 421 3791/3792. 🖳 +55 (83) 221 3795 – DG: Gerardo Andrade Ponte.
PB16) Rua Antônio Martins s/n, 58290-000 Mamanguape.
PB17) Rua Dr. Carlos Pires 17, 58804-200 Sousa.
PB18) Rua Monsenhor Valeriano s/n, 58840-000 Pombal.
PB19) Rua Antônio Firmino 344, 58187-000 Picuí – DG: A. Ferreira D.
PB20) Rua Cândido de Assis 421, 58840-000 Pombal.
PB21) Rua Osvaldo Cruz 161, 58309-490 Bayeux.

PB22) Rua Epitácio Pessoa 8, 58200-000 Guarabira – DG: Maria José L. Aguiar. **F.PI:** 10kW.
PB23) Rua Monsenhor Palmeira s/n, 58135-000 Esperança.
PB24) Rua Getúlio Vargas 129, 58500-000 Monteiro – DG: Simorion Matos.
PB25) Rua Coronel Pedro Targino s/n, 58233-000 Araruna.
PB26) Rua Castro Pinto 234, 58220-000 Bananeiras – DG: Pe. Pedro Alexandre.
PB27) Rua Prefeito Inacio Claudino 121, 58155-000 Soledade.
PB28) Rua João Sabiá 56, 58540-000 Sumé – DG: Alberto Jorge B. Chaves.
PB29) Praça Frei Martinho s/n, 1Þ andar, Centro, 58700-100 Patos. ☎ +55 (83) 4214531. 🖳 +55 (83) 4213704 – DG: Nabor Wanderley. **FM:** 102.9.
PB30) Rua Epifanio Sobreira 14, 58900-000 Cajazeiras.
PB31) Av. Almirante Barroso 918, 58040-220 João Pessoa.
PB32) Rua São José s/n, 58397-000 Areia.

PE00) PERNAMBUCO
PE01) Av. Santo Antônio 324, 55290-000 Garanhuns – DG: Mario Barbosa Filho.
PE02) Rua Floriano Peixoto 780, 50020-060 Recife – DG: Fernando Espinoza. **F.PI:** Change of name to R. Globo Rural.
PE03) Av. Joaquim Nabuco 322, 56500-000 Arcoverde – DG: Aureo Bradley Netto.
PE04) Praça da Bandeira s/n, 55700-000 Limoeiro – DG: Valério Batista.
PE05) Rodovia Araraipina – Picos Km 3, 56280-000 Araripina.
PE06) Travessa 7 de Setembro s/n, Km. 02, 56300-000 Petrolina – DG: Geraldo de Souza Coelho. PD/CE: Francisco Fernandes.
PE07) Rua do Veiga 590, Santo Amaro, 50040-908 Recife – Gte: Marcos Leite.
PE08) Praça Maria Auxiliadora 205, 56300-000 Petrolina – DG: Dom Frei Paulo Cardoso.
PE09) Rua Capitão Lima 250, Santo Amaro, 50040-080 Recife. ☎ +55 (81) 421 1588. 🖳 +55 (81) 421 3868 – DG: Paulo Fernandes.
PE10) Cidade Universtitária, 50670-901 Recife.
PE11) Av. Mascarenhas de Morais 3404, 51170-002 Recife.
PE12) Rua da Conceição 16/22, 2Þ andar, Centro, 55000-000 Caruaru. **E-mail:** radio@liberdade.com.br **Web:** http://marte.netstage.com.br/liberdade/ – DG: Ivan de Barros Feitosa. **FM:** 94.7.
PE13) Av. Padre Rocha s/n, 55810-000 Carpina – DG: Paulo Abou Hana.
PE14) Av. Maria Emília Cavalcanti 570, 55870-000 Timbaúba – DG: Sandra Dias de Moura. **F.PI:** 5kW.
PE15) Estrada do Passarinho 1415, 53170-110 Olinda. **E-mail:** olinda@elogica.com.br **Web:** http://bbs.elogica.com.br/R_Olinda/ – Dir: R. Neto. Gte: Euclides Campelo.
PE16) C.P. 88, 55001-970 Caruaru – DG: José Severino do Carmo.
PE17) Rua Floriano Peixoto 780, 1Þ andar, 50020-060 Recife.
PE18) Rua do Expedicionário 30, 55000-000 Caruaru.
PE19) Rua dos Ferreiros s/n, Granja Fazenda Nova, 55600-000 Vitória de Santo Antão – DG: Eduard Glaser Querálvares.
PE20) C.P. 13, 55301-970 Garanhuns – DG: Aluízio Alves de Melo.
PE21) Rua Coronel Urbano Ribeiro de Sena 956, 52221-000 Recife – DG: Humberto de Mota Pinto.
PE22) Rua Lindaura Marinho Dias 95, 54750-000 Camarajibe.
PE23) Rua Pajussara 225, Bairro Tigipió, 50920-121 Recife – Dir: M.A. da Cunha.
PE24) Av. F. Pessoa de Queiróz s/n, 55200-000 Pesqueira – Gte: Geraldo Santana.
PE25) Bairro São Manoel a margem da BR-101, 55540-000 Palmares.
PE26) Rua Inocencio Gomes de Andrade 619, 56900-000 Serra Talhada.
PE27) Rua 03 de Maio 5, 56800-000 Afogados da Ingazeira – Dir" Rogério Luís de Oliveira.
PE28) Rua Benjamim Constant 16, 55750-000 Surubim.
PE29) Rua José Lopes da Silva s/n, 55150-000 Belo Jardim – DG: E.M.N. de Moura Bezerra.
PE30) Rua Antônio Figueira Soares s/n, 56000-000 Salgueiro.
PE31) Av. C. Colombo 16, 56700-000 São José do Egito.
PE32) Praça Duque de Caxias 818, 55900-000 Goiana – Gte: José Daniel da Silva.
PE33) Av. Fernando Bezerra 1123, 56200-000 Ouricuri – DG: Edvaldo C. Torres.
PE34) Rua Manoel Balbino 184, 55190-000 Santa Cruz do Capibaribe.
PE35) Av. Coronel Trapia s/n, 56440-000 Belém de São Francisco – Dir: Rosimary G. Torres.
PE36) Rodovia BR-232, Km. 86, 56440-000 Gravatá.
PE37) Rua José do Amaral 12, 55330-000 Bom Conselho.

PI00) PIAUI
PI01) Praça Júlio Paixão s/n, 64770-000 São Raimundo Nonato.
PI02) Rua Alvaro Mendes 972, 64000-060 Teresina – DG: Dr. João

Freitas Filho. **FM:** 94.1.
PI03) Rua Taumaturgo de Azevedo 995, 64100-000 Barras – DG: Francisco das Chagas.
PI04) Av. Valter Alencar 2120, Monte Castelo, 64017-500 Teresina – DG: Segisnando Alencar. **FM:** 99.1.
PI05) Av. Heróis do Jenipapo 37, 64280-000 Campo Maior.
PI06) Praça Pres. Kennedy 233, 64490-000 Regeneração.
PI07) Praça do Comércio 400, 1Þ andar, 64500-000 Oeiras – DG: Mauro Tapety.
PI08) Av. Prof. Valter Alencar 2021, 64017-500 Teresina – DG: José D. da Silva Lima.
PI09) Rua Joaquim Baldoíno 40, 64600-000 Picos - DG: José Nunes de Barros.
PI10) Rua 24 de Janeiro 150, 64001-230 Teresina – Dir: Tony Batista.
PI11) Av. Presidente Getúlio Vargas 266, 64200-000 Parnaíba.
PI12) Rua 18 de Setembro 678, 64260-000 Piripiri.
PI13) Av. Miguel Rosa 3775 Sul, 64001-490 Teresina – DG: V. Machado.
PI14) Av. Prefeito J. de Carvalho, 64400-000 Amarante – Gte: José A. Nunes N.
PI15) Av. Rio Branco 314, 64860-000 Uruçuí.
PI16) Rua Clementino Ribeiro 56, 2Þ andar, 64800-000 Floriano – Dir: Isaac Antão do Vale Reis.
PI17) Praça Emilio Cavalcante 29, 64980-000 Corrente.
PI18) Rua Antônio Neto 1065, 64800-000 Floriano.
PI19) Rua Sabino Paulo 696, 64760-000 São João do Piauí.
PI20) Av. João de Paiva 94, 64290-000 Altos – DG: Elvira Raulino.
PI21) Rua Arsénio Santos 555, 64900-000 Bom Jesus – DG: Adelmar M. Benvindo.
PI22) Rua Riachuelo 770, 64200-280 Parnaíba.
PI23) Rua General Taumaturgo de Azevedo 800, 64100-000 Barras.
PI24) Praça Presidente Castelo Branco s/n, 64750-000 Paulistana.
PI25) Rua Fernando Bacelar 480, 64240-000 Piracuruca.
PI26) Av. José de Deus Lacerda 584, 64130-000 Miguel Alves.
PI27) Av. Governador Chagas Rodrigues s/n, 64575-000 Jaicós.
PI28) Rua Corrinto de Andrade s/n, 64255-000 Pedro II.
PI29) Praça da Independência 69, 64325-000 Elesbão Veloso.
PI30) Rua Joaquim Baldoíno 40, 64600-000 Picos.
PI31) Av. João Ferreira 199, 64460-000 Agua Branca.
PI32) Rua Hugo Napoleão 940, 64110-000 José de Freitas.
PI33) Rua Pedro II 695, 64340-000 Castelo do Piauí – Gte: Alcides L. Martins.
PI34) Rua Pedro II s/n, 64330-000 São Miguel do Tapuio.
PI35) Rua Coronel Anibal Martins 481, 64300-000 Valença do Piauí. ☎ +55 (86) 4651244 – DG: Eunice Nunes de Sousa.
PI36) Rua Matias Gomes 510, 64700-000 Simplício Mendes.
PI37) Rua Coronel Narciso 728, 64120-000 União.
PI38) Rua Padre Joaquim Nonato s/n, 64390-000 Demerval Lobão.
PI39) Rua Coronel Messeas Melo s/n, 64190-000 Batalha – DG: J. Messias do Freitag. **F.PI:** 10kW.
PI40) Estrada Barra do Longa, Periferia de Cidade, 64230-000 Buriti dos Lopes.
PI41) Av. Antonino Freire 1356, 64001-040 Teresina.
PI42) Praça Waldemar Leal 42, 64680-000 Padre Marcos.
PI43) Rua Sete de Setembro 471, 64160-000 Luzilândia.
PI44) Praça da Bandeira 93, 64500-000 Oeiras.
PI45) Rua Professor Alceu Brandão 2397, Bairro Monte Castelo, 64016-150 Teresina.
PI46) Rua Coronel José Fortes 549, 64180-000 Esperantina – DG: Elias Medeiros Barros.
PI47) Rodovia BR-343 s/n, 64290-000 Altos.
PI148) Av. Professor Alceu Brandão 2750, 64016-150 Teresina. **Web:** http://www.servo.com.br/gmn/iradios.htm – All news. **FM:** 91.1.

PR00) PARANA
PR01) Rua Quintino Bocaiuva 41, 86020-100 Londrina.
PR02) C.P. 157, 85010-970 Guarapuava – DG: Luis V. Sanquetta. **FM:** 93.7.
PR03) C.P. 10, 85892-970 Santa Helena. ☎ +55 (45) 2681112, 2681212, 2681425. 🖳 +55 (45) 2681135 – DG: Genaro M. Prates.
PR04) C.P. 202, 80001-970 Curitiba – Dir: João L. Seiter B.
PR05) Av. Anunciato Sonni 1673, 86900-000 Jandaia do Sul.
PR06) Rua Cruz Machado 66, 80410-170 Curitiba – Dir-Presidente: Toefilo Bacha Filho.
PR07) Rua 7 de Setembro 520, 85960-000 Marechal Cândido Rondón – DG: Werner Wanderer.
PR08) C.P. 337, 86001-970 Londrina.
PR09) C.P. 78, 86390-970 Cambará – DG: Julio Vial Marques.
PR10) C.P. 218, 87600-970 Nova Esperança.
PR11) Av. Marechal Humberto de Alencar Castelo Branco 69, 82530-020 Curitiba – Dir: A. R. de Assis.
PR12) Rua Marechal Floriano Peixoto 1670, Centro, 85010-250 Guarapuava – DG: Gilberto Baroni.

PR13) Rua Sergipe 843, sala 05, 86010-360 Londrina – Gte: Walter Roberto Manganoti.
PR14) Rua 15 de Novembro 433, Centro, 84010-905 Ponta Grossa.
PR15) Av. Paraná 1896, Bairro Boa Vista, 82510-000 Curitiba – Dir: Matheus Jensen.
PR16) C.P. 528, 87300-970 Campo Mourão – DG: Valdete R. Almeida.
PR17) Av. Daniel Portela 614, 87360-000 Goio-Erê.
PR18) C.P. 209, 87201-970 Cianorte – DG: S. Garcia P.
PR19) Rua Diego A. Feijo s/n, 85560-000 Chopinzinho.
PR20) Av. Brasil 178, 83800-000 Mandirituba.
PR21) Rua Marechal Floriano Peixoto 1123, 85851-020 Foz do Iguaçu. ☎ +55 (45) 5231133 – Own: Milton Rodrigues Filho. DG: Wilson Batista.
PR22) Rodovia do Xisto Km. 20 No. 2018, 83705-740 Araucária – DG: José Alberto.
PR23) Rua Coronel Manoel Ferreira Bello 64, 85530-000 Clevelândia.
PR24) C.P. 255, 86001-970 Londrina – DG: Amarildo Lopes.
PR25) Av. Getúlio Vargas 266, 87013-130 Maringá.
PR26) Rua Iguaçu 808, Centro, 85501-270 Pato Branco. ☎ +55 (46) 225 1087. 🖳 +55 (46) 224 1319 – DG: Valdomiro Cantim.
PR27) Praça Marechal Floriano Peixoto 581, 84010-910 Ponta Grossa – DG: Pe. Francisco Carlos Bach.
PR28) Rua Guarani 1497, 85900-190 Toledo.
PR29) Rua Gastão Vidigal 777, 86800-050 Apucarana.
PR30) C.P. 178, 86600-970 Rolândia – DG: A. Pinceli.
PR31) Rua João Negrão 558, 80010-200 Curitiba.
PR32) Rua Joubert de Carvalho 623, 87013-200 Maringá.
PR33) Edificio Júlio Fuganti, Rua Senador Souza Naves 9, 9Þ andar, salas 903 á 911, 86010-170 Londrina. ☎ +55 (43) 336 0606. 🖳 +55 (43) 321 4745 – DG: Paulo Lauro C. de Oliveira.
PR34) Rua Santa Catarina 970, 85960-000 Marechal Cândido Rondón – DG: Rui Pires. **FM:** 95.1.
PR35) C.P. 591, 85501-140 Pato Branco – DG: Nelson Rabelo. **German:** Sun 2200-2300.
PR36) Praça Generoso Marques 90, 80020-230 Curitiba.
PR37) C.P. 219, 85884-970 Medianeira – DG: Irineu Pelissari.
PR38) Rua Paraná 650, Centro, 874-000 Cruzeiro do Oeste – DG: Osvaldo Farinazzo Medeiros.
PR39) Km. 6 da Estrada de Cambé, 86010-040 Londrina.
PR40) Rua Ulisses Faria 1077, 83900-000 São Mateus do Sul. ☎ +55 (42) 5321644. 🖳 +55 (42) 5321777 – DG: Gerson A. Amaral Ferreira.
PR41) Rua Carlos Gomes 1600, 85807-700 Cascavel.
PR42) Rua Rockefeller 1311, 80230-130 Curitiba.
PR43) Rua Rio de Janeiro 965, 86690-000 Colorado.
PR44) Rua Porto Alegre 21, 1Þ andar, 85601-480 Francisco Beltrão.
PR45) Rua Dario Antônio Bordin 313, Centro, 84600-000 União da Vitória.
PR46) C.P. 09, 85400-970 Guaraniaçu – DG: Caetano B. Filho.
PR47) Rua 15 de Novembro 344, Centro, 84010-020 Ponta Grossa – DG: Osires Batista Nadal.
PR48) Rua Getúlio Vargas 1050, 87772-000 Paranavaí.
PR49) C.P. 101, 86300-000 Cornélio Procópio – DG: Santiago Reghin.
PR50) Av. Capitão Indio Bandeira 1115, salas 509/512, 87300-000 Campo Mourão – DG: Irineu Luiz F. Lima.
PR51) Rua das Américas 255, 85550-000 Coronel Vivida – Dir: Ivanil Pimentel Vieira. DG: Mateus Cassol.
PR52) Av. Higienópolis 1200, 86015-905 Londrina. ☎ +55 (43) 3361110, 3235500. 🖳 +55 (43) 3391175 – DG: J.B. Faria. **FM:** 98.9.
PR53) C.P. 107, 83005-970 São José dos Pinhais.
PR54) Av. Cristóvão Colombo 1055, 86990-000 Marialva.
PR55) Praça Manoel Ribas 112, 84165-000 Castro – DG: Ronie Cardoso Filho.
PR56) C.P. 56, 86001-970, Londrina – DG: Toshiro Okada.
PR57) Rua Presidente Beaurepaire Rohan 44, 80050-340 Curitiba – DG: João B. Ramos da Silva.
PR58) C.P. 258, 85900-970 Toledo. ☎ +55 (45) 2523161, 2522953, 2521248 – DG: Milton Martins.
PR59) Av. Londrina 500, 87970-000 Nova Londrina. ☎ +55 (44) 432 1540 – DG: Humberto Pereira.
PR60) C.P. 7133, 80011-970 Curitiba – Dir: Luiz Andreu Rúbio.
PR61) Av. Paraná 596, 85887-000 Matelândia.
PR62) Rua Dr. Vicente Machado 1384, 85010-260 Guarapuava.
PR63) Av. Getúlio Vargas 500, 87702-000 Paranavaí.
PR64) Av. dos Pinheiros 1970, 85650-000 Santa Isabel do Oeste.
PR65) C.P. 239, 86975-970 Mandaguari. ☎ +55 (442) 331180 – DG: Cyllêneo Pessoa Pereira Jr.
PR66) C.P. 91, 86430-970 Santo Antônio da Platina.
PR67) Rua Pedro Eloy de Souza 51, 82820-130 Curitiba.
PR68) Rua Rio Grande do Sul 1110, 85806-010 Cascavel.
PR69) C.P. 71, 85601-600 Francisco Beltrão – DG: Agustinho Seleski.
PR70) C.P. 1300, 87001-970 Maringá.

PR71) Rua Lamenta Lins 36, Uvaranas, 84025-901 Ponta Grossa – Dir: Helio Porto.
PR72) Rua Desembargador Clotário Portugal 222, 80410-220 Curitiba.
PR73) Av. Paraná 343, 10Þ andar, Conj. 1008, 86010-370 Londrina – DG: Waldimir José Coutinho Mendes.
PR74) C.P. 26, 84201-970 Jaguariaíva – DG: A. da Silva C.
PR75) C.P. 13, 87502-970 Umuarama.
PR76) Rua Bom Jesus 511, 83880-000 Rio Negro – DG: Helena S. Aquino.
PR77) Rua Flamingos 357, 86701-390 Arapongas – DG: Cibele Pugliesi.
PR78) Av. Paul Harris 50, 86220-000 Assaí.
PR79) Rua Pedro Alvares Cabral 1609, 87560-000 Iporã.
PR80) Rua Amauri Lange Silvério 606, 82120-000 Curitiba.
PR81) Rua Meron Heuko 501, 86930-000 São João do Ivaí.
PR82) C.P. 76, 87001-970 Maringá. ☎ +55 (44) 2223413. 🖳 +55 (44) 2224969 – DG: Carlos Piovezan Filho. **FM:** 102.5.
PR83) Av. Nicolau Inácio 845, sala 2, 85670-000 Salto do Lontra – DG: Nery Maria.
PR84) Rua Minas Gerais 297, 19Þ andar, sala 192, 86010-180 Londrina.
PR85) Av. Dep. Ivan Ferreira do Amaral Filho 150, 85303-000 Laranjeiras do Sul – DG: Angelo Manoel da Cunha.
PR86) Rua Simão Domingues 26, 84550-000 Rebouças – DG: Mariangela S.P. Reichelt.
PR87) Rua Vereador Vicente Balan 12, 85990-000 Terra Roxa.
PR88) Av. Tiradentes 2113, 87505-090 Umuarama. ☎ +55 (44) 6221014, 6221286. 🖳 +55 (44) 6221091 – DG: Dr. Fumyo Sakabe . Dir: Maria Apda. Mori.
PR89) Rua Antônio Lemos 807, 86400-000 Jacarezinho. ☎ +55 (43) 722 0773. 🖳 +55 (43) 722 0393 – DG: João Crisóstomo M. Freire.
PR90) Av. Brasil 702, 85710-000 Santo Antônio do Sudoeste. ☎ +55 (465) 63 1541 – DG: Zelirio Ferrara.
PR91) C.P. 66, 85801-970 Cascavel – DG: Fernando Fontana.
PR92) Av. Horácio Klabin 383, 2Þ andar, 84261-000 Telêmaco Borba. ☎ +55 (422) 72 1515/1381 – DG: Jair Neves. **FM:** 92.9.
PR93) Av. Bandeirantes 958, 86360-000 Bandeirantes.
PR94) Rua Herculino Otaviano 817, 1Þ andar, 85440-000 Ubiratã. ☎ +55 (44) 5431317, 5431717 – DG: Adilson Gonçalves Netto.
PR95) C.P. 171, 87550-970 Altônia – DG: Antônio Duma.
PR96) Rua Prof. Cleto 281, Centro, 83221-320 Paranaguá – DG: Ludovico Mikosz.
PR97) Rua XV de Novembro 522, 84010-908 Ponta Grossa. ☎ +55 (42) 224 5677 – DG: Rogerio Bocchi Serman.
PR98) C.P. 777, 86800-970 Apucarana – DG: Cynyra Woiski Ignácio.
PR99) Rua Jesuino Alves da Rocha Loures 1764, 85555-000 Palmas. ☎ 🖳 +55 (46) 262 1918 – DG: Pedro Fistarol Araujo. **FM:** 96.5 "Horizonte".
PR100) C.P. 217, 85980-970 Guaíra – DG: Antonio Roque Ferreira.
PR101) Rua Londrina 410, 85640-000 Ampére – DG: Hélio Manoel Alves.
PR102) C.P. 171, 87900-970 Loanda – DG: Theodoro H. Neto.
PR103) C.P. 16, 86590-970 Ibati.
PR104) C.P. 54, 85935-970 Assis Chateaubriand. ☎ +55 (449) 281576. 🖳 +55 (449) 281335 – DG: Moacir Micheletto.
PR105) Rua Ebano Pereira 157, 85200-000 Pitanga.
PR106) C.P. 45, 86730-970 Astorga – DG: O. Bassan.
PR107) Rua Ipiranga 91, 84600-000 União da Vitória. ☎ +55 (425) 221098, 221340 – DG: João Samoel Jensen. **Ukraine & German:** Sun 1230-1400, 1500-1600. **F.PI:** 5kW.
PR108) C.P. 230, 86300-000 Cornélio Procópio.
PR109) Rua 7 de Setembro 42, 83750-000 Lapa. ☎ +55 (41) 822 2326 – DG: Sergio Frederico Baggio. **German:** Sun 1010-1110.
PR110) C.P. 195, 86925-970 Borrazópolis.
PR111) C.P. 121, 84400-970 Prudentópolis.
PR112) C.P. 304, 86700-970 Arapongas.
PR113) C.P. 72, 86500-970 Venceslau Brás.
PR114) Rua Coronel Emilio Gomes 281, Terreo, 84500-000 Irati.
PR115) Rua Curitiba 1534, 85840-000 Céu Azul.
PR116) Travessa Vale Porto 240, 83370-000 Antonina.
PR117) C.P. 267, 86870-970 Ivaiporã – DG: Sandra Arruda. **F.PI:** 5kW.
PR118) Rua 5 de Julho 1065, 85950-000 Palotina. ☎ +55 (44) 6495266 – DG: Antônio José Piccoli.
PR119) C.P. 10, 83650-970 Balsa Nova.
PR120) Rua Mauá 2518, 85770-000 Realeza – DG: Aldo A. Fachinello.
PR121) C.P. 358, 87200-970 Cianorte – DG: A. Prado Ruiz.
PR122) Rua Marechal Deodoro 414, 84940-000 Siqueira Campos. ☎ +55 (43) 772 1125 – DG: Moacir A. Nasato.
PR123) C.P. 11, 85830-970 Formosa do Oeste.
PR124) C.P. 10, 85615-970 Marmeleiro – DG: Agustinho Seleski.
PR125) Av. Brasil 502, 85760-000 Capanema.
PR126) Rua Senador Pinheiro Machado 1536, Centro, 85010-100 Guarapuava – DG: Luis A. Ferreira.

PR127) C.P. 134, 85660-970 Dois Vizinhos. ☎ +55 (46) 536 1930 –DG: Valdir Luiz Pagnoncelli. **FM:** 100.7 "Vizinhança".
PR128) C.P. 199, 84500-970 Irati – DG: Nagib Harmuch.
PR129) Rua São Paulo 180, 86380-000 Andirá – Dir: V. de Resende.
PR130) Rua São Paulo 489, 86840-000 Faxinal – DG: Carlos Augusto Pinto.
PR131) Rua Noruega 98, 86182-000 Cambé – DG: Antônio E. Casaroto.
PR132) Praça São Pedro 1210, 85460-000 Quedas do Iguaçu.
PR133) C.P. 46, 84130-970 Palmeira. ☎ +55 (42) 2523669, 2523939 – DG: Euclides A. de Paula F.
PR134) Rua Rockefeller 1311, Prado Velho, 80230-130 Curitiba – DG: Euclydes Cardoso de Almeida. Superintendente: Vicente Mickoz.
PR135) Av. Brasil 746, 84350-000 Ortigueira.
PR136) Av. Maximiliano Vicentin 240, 85270-000 Palmital – DG: Edmilson Cecura.
PR137) C.P. 101, 85410-970 Nova Aurora – DG: Waldemar W. dal Molin.
PR138) Rua Dom Pedro II 196, 85852-520 Foz do Iguaçu. ☎ +55 (45) 5743030 – Own: Antonio Cirilo. DG: Francisco dos Santos.
PR139) Rua Oyapock 649, 80050-450 Curitiba. ☎ +55 (41) 264 9945 – DG: Eli João Thomaz Aquino. **F.PI:** Change of name to R. Globo Rural.
PR140) Rua Urbano Lunardelli 875, 86160-000 Porecatu – DG: Dr. Glauco M. Ferrigno. **English & Italian:** 2330-0030. **F.PI:** 5kW.
PR141) Av. Paraná 126, 84470-000 Cândido de Abreu.
PR142) Av. Santos Dumont 971, 85302-080 Laranjeiras do Sul.
PR143) Av. Genoroso Marques 599, 2Þ andar, 85550-000 Coronel Vivida – DG: Cicélia G. Lang.
PR144) Rua 7 de Setembro 540, 85560-000 Chopinzinho – DG: Odílio Budine. **F.PI:** 5kW.
PR145) Av. Paraná 201, 85852-000 Foz do Iguaçu – DG: Nilton Vitorassi.
PR146) Av. Presidente Kennedy 170, Norte, 85950-000 Palotina. ☎ +55 (44) 6495253. ▤ +55 (44) 6496084 – DG: Nilo Barbieri. **FM:** SW.
PR147) Rua Amor Perfeito 1827, 85420-000 Corbélia. ☎ +55 (45) 2421799, 2421999 – Gte: Sergio A. Marafon.
PR148) Rua Souza Naves 102, 85877-000 São Miguel do Iguaçu – DG: Noeli T. Bissolotti.
PR149) Av. Iguaçu 288, Centro, 85575-000 São Jorge d'Oeste.
PR150) Av. Santo Antônio 826, 87320-000 Roncador.
PR151) Av. Yolanda Loureiro de Carvalho 1021, 87350-000 Ubiratã.
PR152) Rua XV de Novembro 1260, 85900-000 Toledo.
PR153) Praça Barão do Rio Branco 55, 84240-000 Piraí do Sul.
PR154) Rua Bernardino Bogo 68, 87160-000 Mandaguaçu. ☎ +55(442) 45 1776 – DG: Marcos Antônio Rocco.
PR155) Rua Rosalvo Petrechem, Lote 7G, Quadra 109, 85200-000 Pitanga.
PR156) Av. Euclides da Cunha s/n, Prolongamento, 87890-000 Terra Rica. ☎ +55 (44) 4411816 – DG: Arquimedes Ferreira Filho.
PR157) Av. Iguaçu s/n, Ed. Fabiane, 85585-000 Verê. ☎▤ +55 (46) 535 1305 – DG: Ólivo E. Pastro. Dir: Sérgio Fabiane. **Italian:** Sat mornings. **F.PI:** 5kW.
PR158) Rua Independência s/n, Prox. Escola de Aplica, 86130-000 Bela Vista do Paraíso.
PR159) Rua Luiz Pinheiro 1446, 84990-000 Arapoti – DG: E. Pascoal Nunes. **F.PI:** 10kW, change of freq. to below 1000kHz.
PR160) Av. Iguaçu 434, 85808-000 Capitão Leônidas Marques.
PR161) Av. Silveira Pinto 500, 86490-000 Ribeirão do Pinhal.
PR162) Rua Vicente Machado 385, 84570-000 Mallet.
PR163) Rua Marechal Deodoro 22, 85540-000 Mangueirinha.
PR164) Rua São Miguel 922, 85570-000 São João.
PR165) Av. Tupy 2239, 85501-000 Pato Branco.

RJ00) RIO DE JANEIRO

RJ01) Rua Visconde de Itaborai 184, 24030-091 Niterói – Gte: Alberto F. Torres.
RJ02) Rua Costa Rica 151, 28970-000 Araruama – DG: Manoel Diamantino da Costa.
RJ03) Av. Presidente Vargas 409, Grupo 2201, 20071-003 Rio de Janeiro – DG:Roberto Leal.
RJ04) Praça Mauá 7, 21Þ andar, 20083-900 Rio de Janeiro.
RJ05) Rodovia Presidente Dutra Km 303, Paraíso, 27536-000 Resende – Abelardo Galvão. **FM:** 93.9.
RJ06) Praça Demerval Barbosa Moreira 28, 28610-160 Nova Friburgo.
RJ07) Rua Visconde de Inhaúma 37, 20091-000 Rio de Janeiro – Dir: Marilea Sales.
RJ08) Av. 24 de Outubro 201, 28100-000 Campos.
RJ09) Rua México 111, Sobreloja, 20031-145 Rio de Janeiro.
RJ10) Av. Djalma Beda Coube 719, 28500-000 Cantagalo. ☎ +55 (24) 5554455 – DG: Antonio Brum Andrade.
RJ11) Rua do Russel 804, 22210-010 Rio de Janeiro.
RJ12) Praça da República 141-A, 3Þ andar, sala 306, 20211-350 Rio de Janeiro. ☎ +55 (21) 221 7447/252 8413/242 6328. ▤ +55 (21) 232 1931

– DG: Regina Amaral de Salles.
RJ13) Av. Rui Barbosa 749, 3Þ andar, 27910-260 Macaé.
RJ14) Rua do Russel 434, Glória, 22210-010 Rio de Janeiro.
RJ15) Av. Portugal 96, Urca, 22291-050 Rio de Janeiro. ☎ +55 (21) 2958770, 2950332. ▤ +55 (21) 2953920 – DG: Rômulo Siqueira.
RJ16) Av. Meriti 2584, Vila da Penha, 21250-000 Rio de Janeiro.
RJ17) Praça Mauá 7, 21Þ andar, 20081-240 Rio de Janeiro – DG: Osmar de A. Frazão.
RJ18) Rua do Bispo, 20261-060 Rio de Janeiro – DG: W. Luiz.
RJ19) Rua Julio Maria 10, Centro, 23900-900 Angra dos Reis – Dir: Maria de Fátima.
RJ20) Rua Machado Bittencourt 300, 26900-000 Miguel Pereira.
RJ21) Av. Sete de Setembro 380, 28013-000 Campos.
RJ22) Estrada Velha da Pavuna 3517, 20765-170 Rio de Janeiro.
RJ23) C.P. 79, 28100-970 Campos – DG: Salvador Macedo.
RJ24) Rua Duque de Caxias 627, 25802-120 Três Rios. ☎▤ +55 (242) 520720, 521797 – DG: Edson Elias B. Jorge. **FM:** 89.7. **F.PI:** SW.
RJ25) Rua do Russel 434, Glória, 22210-010 Rio de Janeiro – DG: Marcos Libretti. CE: Guilherme Viterbo.
RJ26) Rua Carlos Lacerda 75, 28013-030 Campos – DG: Diva Arreu B.
RJ27) Rua do Livramento 189, 8Þ andar, 20221-191 Rio de Janeiro – DG: Alfredo R. Filho.
RJ28) Rua Alberto Torres 410, 28400-000 São Fidelis – DG: Amary de Almeida.
RJ29) Av. 28 de Setembro 258, 20551-031 Rio de Janeiro.
RJ30) Av. Treze de Maio 23, 6Þ andar, 20031-000 Rio de Janeiro – DG: Laércio Alves.
RJ31) Rua Carlos Lacerda 52, 28013-030 Campos.
RJ32) C.P. 85919, 27301-970 Barra Mansa – Dir: O. Marcui N.
RJ33) Estrada do Dende 659, Ilha do Governador, 21920-000 Rio de Janeiro. ☎ +55 (21) 396-6969 – Pres: Joel de Mattos Alvarenga. **Esperanto:** Wed 2330, Sun 1100.
RJ34) Av. Cardoso Moreira 422, 28300-000 Itaperuna. ☎ +55 (248) 221405, 220084 – DG: João A. Naufel.
RJ35) Rua Miguel de Frias 67, 20211-190 Rio de Janeiro.
RJ36) Av. Joaquim Leite 279, 2Þ andar, Centro, 27330-042 Barra Mansa – DG: Ildeu Nodge A. da Silva.
RJ37) Rua Dr. Temistocles de Almeida 97, 28470-000 Santo Antônio de Pádua.
RJ38) Praça Nilo Peçanha 42, 27123-020 Barra do Piraí.
RJ39) Praça Procópio Ferreira 22, 26000-000 Nova Iguaçu.
RJ40) Rua Eugênio Cordeiro 28, 28800-000 Rio Bonito.
RJ41) Rua 100 No. 01, Bairro Laranjal, 27255-000 Volta Redonda – DG: Silvia R. M. Reis .
RJ42) Rua Barão de Cotegipe 90, 25215-390 Duque de Caxias.
RJ43) Rua José Augusto da Costa 14, 25953-160 Teresópolis – DG: Hélio Carracena.
RJ44) Rua Comandante Vergueiro da Cruz 151, Olaria, 21021-020 Rio de Janeiro.
RJ45) Rua Paulino Padilha 80, 28460-000 Miracema – DG: Marcus Felipe M. Linhares.
RJ46) Rua Alencar Lima 26, 25620-050 Petrópolis – DG: Roberto Santos Ribeiro. **FM:** 88.5.
RJ47) Rua Vereador Darcy Teixeira Fontes 556, 23815-270 Itaguaí – DG: Ivan Borghi. Dir-Prop: Ronald G. Levinsohn. **F.PI:** 10kW.
RJ48) Rua Coronel João Rufino 105, 27600-000 Valença – DG: Rodrigo da Silva Tjäder.
RJ49) Praça Governador Portela 18, 28360-000 Bom Jesus do Itabapoana – Dir: F. R. Nevei Xavier.
RJ50) Av. Amaral Peixoto 366, 28860-000 Casimiro de Abreu.
RJ51) Rua Frei Valerio 58, 28740-00 Conceição de Macabu.
RJ52) Rua Nilo Peçanha 320, 27542-210 Resende – Dir: Luiz Oliveira Fontes.
RJ53) Rua Visconde de Inhaúma 37, 12Þ andar, 20091-000 Rio de Janeiro.
RJ54) Praça Carmelo Dutra 155, 28540-000 Paraíba do Sul.
RJ55) Av. Presidente Kennedy 1763, 25010-001 Duque de Caxias.
RJ56) Rua Figueiras de Barros 100, 28210-000 Italva.
RJ57) Praça Porto Rocha 56, Grupos 102/103, 28905-250 Cabo Frio.
RJ58) Rua Coronel Santiago 250, 25950-000 Teresópolis.

RN00) RIO GRANDE DO NORTE

RN01) Av. Duque de Caxias 106, Ribeira, 59010-200 Natal – DG: Ricardo Luiz V. Alves.
RN02) Av. Otávio Lamartine 603, 59300-000 Caicó.
RN03) Rua Dos Transmissores 10, Bairro Nordeste, 59042-070 Natal. ▤ +55 (84) 2133315. **Web:** http://www2.eol.com.br/clientes/iea-dem.central.htm – Dir: F. Rodrigues.
RN04) Rua João Pessoa 22, 1Þ andar, 59380-000 Currais Novos – DG: Paulo Roberto S. Dutra.

RN05) C.P. 112, 59600-970 Mossoró – DG: Emery J. Costa.
RN06) C.P. 60, 59600-970 Mossoró – Dir. Edilson Souza Lemos.
RN07) Rua Açú 335, Tirol, 59020-110 Natal – DG: Bremen Sales Quixadá.
RN08) C.P. 101, 59600-970 Mossoró – DG: Angelo A. Fernandes.
RN09) Rua Romualdo Galvão 973, Tirol, 59056-100 Natal – Dir: Janio Maria Carlos Vidal.
RN10) C.P. 145, 59001-970 Natal – Gte: Cid Lobo.
RN11) Praça Tomaz Salustino 42, 59380-000 Currais Novos.
RN12) Rua Francisca Delfina 30, 59965-000 Alexandria – DG: Carlos Antônio de Oliveira.
RN13) Praça Bento Praxades 104, 59600-000 Mossoró – Gte: José G. Leite M.
RN14) Rua Otávio Amorim 643, 59650-000 Açu.
RN15) Rua Coronel José Bezerra 41, 59780-000 Caraúbas.
RN16) Rua Frei Alberto Cabral s/n, 59215-000 Nova Cruz.
RN17) Rua Getúlio Vargas 1296, 59900-000 Pau dos Ferros – DG: Raimundo Nonato O.
RN18) Rua Coronel Martiniano 784, 59300-000 Caicó.
RN19) Av. Rio Branco 173, 59655-000 Areia Branca – Gte: Jaltion Rodrigues.
RN20) Rua Ana de Pontes s/n, 59255-000 Santo Antônio.
RN21) Rua Odorico Ferreira de Souza s/n, 59200-000 Santa Cruz. ☎ +55 (84) 291 2300 – DG: Odorico Ferreira de S. Neto.
RN22) Rua Esperidião Coimbra 22, Centro Macau, 59500-000 Macao - Gte: Luiz Antonio Gomes.
RN23) Rua Cícero Tomaz de Acevedo s/n, 59360-000 Parelhas – Gte: Getúlio da Costa.
RN24) Rua São Francisco s/n, 59570-000 Ceará Mirim.
RN25) Rua Sebastião Guilherme Caldas s/n, 59343-000 Jardim do Seridó.
RN26) Rua Coronel Martiniano 1077, 59300-000 Caicó.
RN27) Av. Nascimento de Castro 1597, 59054-180 Natal.
RN28) Rua Padre Cosme 48, 59920-000 São Miguel.
RN29) Rua Coronel Freire 242, Centro, 59460-000 São Paulo do Potengi.
RN30) Av. 21 de Abril 460, BR-460, 59550-000 João Câmara – DG: Walter Miranda.

RO00) RONDÔNIA
RO01) Rua Joaquim Nabuco 1573, 79815-350 Porto Velho.
RO02) Av. Rondônio s/n, 78987-000 Rolim de Moura.
RO03) Av. Jamari 4218, 78932-000 Ariquemes.
RO04) Rua Capitão Silvio 526, Bairro Urupá, 78960-000 Ji-Paraná – DG: Luiz A. Vital Neto. **F.PI:** 5kW.
RO05) Rua Rui Barbosa 3375, 78975-000 Cacoal – Dir: Herivelton J. Bernard.
RO06) Rua Costa e Silva 1113, Vila Jotão, 78958-000 Ji-Paraná – Dir: Joel Nogueira.
RO07) Rua Dom Augusto 681, Centro, 78958-000 Ji-Paraná.
RO08) Rua Ricardo Catanhede esquina com a Rua Goiás s/n, 78941-000 Jaru.
RO09) Praça Mario Correa 90, 78957-970 Guajará Mirim – Dir: Wanderlei Marques da Silva **FM:** 93.7.
RO10) Rua Dourados 4, Setor de Areas Especiais, 78932-000 Ariquemes.
RO11) Av. Carlos Gomes 932, 78900-030 Porto Velho – DG: Carlos A. Diniz Martins.
RO12) Rua Princesa Isabel 128, 78995-000 Vilhena – DG: Ruth Morimoto.
RO13) C.P. 005, 78900-970 Porto Velho – DG: Angela Xavier.
RO14) Loteamento Monte Alegre, Quadras 35, 36, 38, 39, 41 e 42, 78987-000 Rolim de Moura.
RO15) Rua Carlos Doneje esquina com Monteiro Lobato, 78984-000 Pimenta Bueno.
RO16) Setor Prefeitura de Characas - Ara de Faculdade, 78995-000 Vilhena.
RO17) Rua Miguel Chakian 1300, Bairro Embratel, 78906-300 Porto Velho. ☎ +55 (69) 225 3053. 📠 +55 (69) 225 3002 – DG: Euzébio Lopes. **F.PI:** 50kW.
RO18) Rua 06 de Maio 211, Bairro Casa Preta, 78960-000 Ji-Paraná – DG: Joel Armando. **English:** 0000-0200.

RR00) RORAIMA
RR01) Av. Capitão Ene Garcez 830, 69304-000 Boa Vista – DG: Glavao Soares.
RR02) Rua Sebastião Diniz 363, 69360-000 Caracaraí – DG: Henrique Rodrigues.

RS00) RIO GRANDE DO SUL
RS01) Av. Victor Barreto 3056, Conj. 207, 92010-000 Canoas – DG: Amabilio J. Castro.

RS02) Av. Antunes Ribas 1535, 98801-630 Santo Angelo.
RS03) Rua Marechal Deodoro 1157, 96810-110 Santa Cruz do Sul.
RS04) Rua General Sampaio 161, Bairro Rio Branco, 95097-000 Caxias do Sul – Dir-Presidente: Clementino Dotti. Gte: Pedro D. Fattori. **Veneto dialect:** Sun 1300-1500. **FM:** 98.5.
RS05) Rua Coronel Chicuta 453, 99010-051 Passo Fundo.
RS06) Rua Mascarenhas de Morães 586,, 97300-000 São Gabriel – DG: Luiz Fares R. Semhan.
RS07) C.P. 67, 95201-970 Vacaria. 📠 +55 (54) 231 2555 – DG: Pedro Salame. **FM:** 101.5.
RS08) Praça Oswaldo Aranha 39, 97540-000 Alegrete.
RS09) Av. Ipiranga 1075, 90160-093 Porto Alegre. **E-mail:** gaucha@rdgaucha.com.br **Web:** http://www.rdgaucha.com.br – DG: Armindo Antônio Ranzolin. Satellite signal downlinked via 60 st's in southern Brazil forming Rede Gaúcha Sat.
RS10) C.P. 284, 96001-970 Pelotas – Dir: Paulo Luiz Góz.
RS11) Rua Suécia 255, 98500-000 Tenente Portela. ☎ +55 (55) 5511395, 5511588 – DG: Loir José Eckert.
RS12) Rua 14 de Julho 588, 95300-000 Lagoa Vermelha.
RS13) Praça Saldanha Marinho s/n, 97010-540 Santa Maria.
RS14) C.P. 1474, 90001-970 Porto Alegre. **E-mail:** rbradio@netalpha.com.br .**Web:** http://www.netalpha.com.br/bandeirantes/– DG: Jorge Seadi.
RS15) C.P. 195, 99701-970 Erechim – DG: Idyllio S. Badalotti.
RS16) C.P. 80, 97590-970 Rosário do Sul – Gte: Moacir Guazina.
RS17) Rua Farroupilha 110, 95200-000 Vacaria.
RS18) Rua 7 de Setembro 1713, 96230-000 Santa Vitória do Palmar.
RS19) C.P. 55, 99500-970 Carazinho – DG: I.B. Albuquerque.
RS20) Travessa 4 de Junho 84, 98380-000 Seberi.
RS21) Rua XV de Novembro 275, 9Þ andar, 98700-000 Ijuí. ☎ +55 (55) 3321414. 📠 +55 (55) 3322000 – DG: José Luís Bonamigo. **Italian:** 1300-1400.
RS22) Rua Marechal Mascarenhas de Morães 298, 97300-000 São Gabriel. ☎ +55 (232) 2244/2646 – DG: Victor Hugo Ries.
RS23) Rua Caldas Jr. 219, 90010-260 Porto Alegre – DG: Carlos A. Bastos Ribeiro.
RS24) Av. Scarpelini Ghezzi 353, 99072-000 Passo Fundo – DG: Daltro José Wesp. **FM:** 105.9.
RS25) C.P. 75, 96200-970 Palmeira das Missões.
RS26) C.P. 147, 96500-970 Rio Grande – DG: P. Hahuys C.
RS27) Av. Marechal Floriano 920, Sala 301, 95520-000 Osório. ☎ +55 (51) 6633435, 6633436. 📠 +55 (51) 6633344 – DG: Pedro Edir D. Farias.
RS28) C.P. 69, 96930-970 Candelária.
RS29) Rua Duque de Caxias 736, 98470-000 Planalto – DG: Moacir Zilio.
RS30) Rua Santiago Matiotti 670, 99500-000 Carazinho.
RS31) Av. João Pessoa 1282, 90040-001 Porto Alegre.
RS32) Rua São João 485, 96640-000 Rio Pardo – DG: Jandir Junior Rocha Guedes.
RS33) Campus de Universidade Federal de Santa Maria, 10Þ andar, Reitoria, 97119-900 Santa Maria – DG: Áurea dos Santos Fonseca.
RS34) Rua Fernando Abott 427, 2Þ andar, 95880-000 Estrela – DG: Luiz Carlos Freitag.
RS35) Rua Voluntários da Pátria 805, 98025-770 Cruz Alta.
RS36) Rua Casemiro de Abreu 1233, 90420-001 Porto Alegre.
RS37) Rua Dr. Bozano 1336, Galeria do Comercio, 97010-902 Santa Maria – DG: Luiz Carlos Cozer.
RS38) Rua Coronel Correa Lima 1950, 90850-250 Porto Alegre.
RS39) Rua Marechal Deodoro 101, 95700-000 Bento Gonçalves – DG: Carlos Domingos Piccoli. **FM:** 92.5 "Serrana".
RS40) C.P. 1021, 98900-970 Santa Rosa. ☎📠 +55 (55) 5125757, 5123838 – DG: Sergio Ambros M. **FM:** 97.7.
RS41) Rua 15 de Novembro 336, 99770-000 Aratiba.
RS42) Rua Joaquim Nabuco 957, 93310-002 Nôvo Hamburgo.
RS43) Rua 7 de Setembro 1441, 95800-000 Venâncio Aires – DG: Gilmar João Uhry.
RS44) Rua Saihydi Abrahão 315, 95590-000 Tramandaí.
RS45) Rua Garibaldi 789, 95084-900 Caxias do Sul – DG: Jaime Luiz Walker.
RS46) Av. Brasil 523, 98801-590 Santo Ângelo. **E-mail:** rd.sto.angelo@missoes.com.br **Web:** http://home.missoes.com.br/radio_sto_angelo/ – DG: Jairo Almedida and Laercio Pilau.
RS47) C.P. 49, 95900-970 Lajeado – DG: Lauro Mathias Müller.
RS48) Rua Venâncio Aires 1980, 97010-004 Santa Maria – DG: Rudy Seibel.
RS49) Rede Pampa de Comunicaçao, Rua Orfanatrófio 711, 90840-440 Porto Alegre. **E-mail:** pampa@pampa.com.br **Web:** http://www.pampa.com.br – Gte: Roger dos Reis. Addr to Sistema LBV Mundial: Legião da Boa Vontade, Rua Sergio Tomas 740, 01131-010 Bom Retiro
RS50) Av. Getúlio Vargas 410, 98670-000 Humaitá – DG: Sady

Manjabosco Sandri.
RS51) Rua Otacílio Tupanciretã de Azevedo 2, 98170-000 Tupanciretã. ☎ +55 (55) 2721753. ☎ 🖹 +55 (55)3521865 – DG: Miguel Puretz Filho. **FM:** 92.5.
RS52) Av. Fiorentino Bachi 791, 99840-000 Sananduva – DG: Itamar Jacob Belin. **FM:** 97.7.
RS53) Rua Garibaldi 789, 21Þ andar, 95084-900 Caxias do Sul – DG: Jaime Luiz Walker.
RS54) Rua Julio de Castilhos 2236, 97800-000 São Luís Gonzaga. ☎ +55 (55) 3523131. 🖹 +55 (55) 3521865 – DG: Alfredo L.L. Cavilli. Dir: Milton Rockembach.
RS55) C.P. 155, 98700-970 Ijuí. ☎ 🖹 +55 (55) 3328000 – DG: Mathilde Salete Mânica. **FM:** 101.5 "Iguatemi".
RS56) Rua Acindino Inácio Dias 122, 96180-000 Camaquã – DG: Luis Renato Barbosa.
RS57) C.P. 172, 97800-970 São Luís Gonzaga.
RS58) Rua 15 de Novembro 236, 96570-000 Caçapava do Sul.
RS59) Rua Tucunduva 758, 98640-000 Crissiumal.
RS60) Rua Marechal Deodoro 101, 7Þ andar, 95700-000 Bento Gonçalves – DG: Carlos Domingos Piccoli.
RS61) Rua João Carlos Machado 645, 98460-000 Iraí.
RS62) Rua Sarmento Leite 426, 90046-900 Porto Alegre – DG: Sergio Stosch.
RS63) Rua Ramiro Barcelos 2092, 96508-070 Cachoeira do Sul. ☎ +55 (51) 7224022, 7224313 – DG: Jacir José Somavilla. **German & Italian:** Sat 1600-1830, Sun 1100-1200.
RS64) C.P. 71, 99800-970 Marcelino Ramos – DG: Hedwig Ilse S. Basso.
RS65) Av. Bento Gonçalves 733, 98870-000 Giruá. ☎ +55 (55) 3612020 – DG: Darci Uhry.
RS66) Rua Dr. João Azevedo 220, 96300-000 Jaguarão.
RS67) C.P. 11, 96901-970 Sobradinho. ☎ +55 (51) 7421089, 7421090 – DG: João V.Z. de Castro. **FM:** 97.3.
RS68) C.P. 80, 97650-970 Itaqui – Gte: José I. Dornelles.
RS69) Av. Ipiranga 1075, 90163-093 Porto Alegre. 🖹 +55 (51) 2186680 – DG: Ricardo Ferro Gentelini.
RS70) C.P. 1277, 97001-970 Santa Maria – DG: P. Antônio Bonini.
RS71) Rua Pinheiro Machado 628, 98005-000 Cruz Alta. ☎ +55 (55) 322 2639 – DG: Loy Newton Pilau. **FM:** 105.1.
RS72) Rua Domingos de Almeida 2194, 97500-004 Uruguaiana – DG: João R. Cobelli.
RS73) C.P. 50, 96750-970 Butiá.
RS74) C.P. 200, 95180-970 Farroupilha. ☎ +55 (54) 2611204, 2612121 – DG: Agenor Moretto.
RS75) Rua Tenente Lira 950, 98400-000 Frederico Westphalen. ☎ +55 (55) 7443500 – DG: Euclides Argenta. **Italian & Polish:** Sun 0930 & 1600.
RS76) Rua General Osório 1160, 97760-000 Jaguari – DG: Silvia B. Bertoncheli.
RS77) C.P. 100, 96001-970 Pelotas – DG: José M. Cunha.
RS78) Av. Coronel Victor Villa Verde 491, 95500-000 Santo Antônio da Patrulha.
RS79) C.P. 157, 96400-970 Bagé – DG: João Henrique B. Gallo.
RS80) Rua Sete de Setembro 366, 99010-121 Passo Fundo.
RS81) Av. Maurício Cardoso 697, 99300-000 Soledade. ☎ +55 (54) 381 1144 – DG: Jatir Scortegagna. **FM:** 99.1.
RS82) Rua da Anunciação s/n, Morro do Convento, 97900-000 Cerro Largo – DG: João Pio Flach. Dir: Liane Flach. **FM:** 105.9 "Shamballa".
RS83) Rua João Batista Scalabrini 346, 99250-000 Serafina Corrêa. ☎ +55 (54) 2441212, 2441089 – DG: Pe. Ivo Pretto. **Dialeto Veneto:** Sun 1030-1230.
RS84) Rua Torres Gonçalves 33, 99700-000 Erechim.
RS85) Próximo ao Km 233 da BR-392, 97340-000 São Sepé – DG: Enilton J. Bolzan.
RS86) Travessa Francisco de Leonardo Truda 40, 90010-050 Porto Alegre.
RS87) Rua Domingos de Almeida 1722, 97500-001 Uruguaiana – DG: Miguel Augusto S. Ramos.
RS88) Rua Victor Dumonal 1756, 98240-000 Santa Bárbara do Sul.
RS89) C.P. 89, 99600-970 Nonoai – DG: José Reck.
RS90) Av. Fernando Luzatto 42, 95320-000 Nova Prata.
RS91) Av. Rio Branco 224, 96610-000 Encruzilhada do Sul.
RS92) Rua Júlio de Castilhos 605, 95290-000 Bom Jesus.
RS93) Rua General Osório 1134, 98200-000 Ibirubá – DG: Saulo A. Stefanello. **FM:** 96.6.
RS94) C.P. 82, 96700-970 São Jerônimo – DG: Gilton Paschoal.
RS95) C.P. 707, 95001-970 Caxias do Sul – DG: Raul Bento Alves.
RS96) Rua 15 de Novembro 717, 96015-000 Pelotas.
RS97) Rua Riachuelo 928, 97670-000 São Borja.
RS98) Rua Moron 1520, 96508-030 Cachoeira do Sul.
RS99) Rua José Sponchiado 418, 99830-000 Gaurama. ☎ 🖹 +55 (54)

3911134 – DG: Dolcimar Gonçalves. **Italian & Polish:** 1300-1400.
RS100) Rua Júlio de Castilho 1038, 97220-000 Faxinal do Soturno.
RS101) C.P. 112, 98920-970 Horizontina. ☎ +55 (55) 537 1414 – DG: Luiz C. Pilau. **German:** Sun 1600-1700.
RS102) Rua Coronel Alfredo Born 447, 96170-000 São Lourenço do Sul.
RS103) Av. Angelo Macalós 246, 99400-000 Espumoso. ☎ +55 (54) 383 1082. 🖹 +55 (54) 3831552 – DG: Verdi Ubiratan Moura. **FM:** 95.3.
RS104) Av. Farrapos 146, 90220-000 Porto Alegre – Gte: G. Gianuca.
RS105) Rua 25 de Julho 39, 98960-000 Santo Cristo.
RS106) Rua Thomaz Albornoz 1108, 97574-330 Santana do Livramento.
RS107) Av. Duque de Caxias 1503, 99560-000 Sarandi. ☎ +55 (54) 361 1455 – Dir: Francisco A. Rosin. **German & Italian:** Sun 1100-1150, 1700-1800.
RS108) Av. Júlio de Castilhos 232, 95680-000 Canela.
RS109) Rua General Osório 1276, 98280-000 Panambi – DG: Rene Beck. **Italian:** 1300-1330. **German:** Sun 1230-1255.
RS110) Rua Sete de Setembro 353, 96015-300 Pelotas.
RS111) Av. Rio Branco 401, 96450-000 Dom Pedrito – Dir: O. Silva.
RS112) C.P. 22, 95190-970 São Marcos – DG: Nelson Tomiello.
RS113) Av. Bastian 286, 90130-020 Porto Alegre – DG: Erika Coester Kramer.
RS114) Av. Scalabrini 777, 99200-000 Guaporé. ☎ +55 (54) 4431624, 4431212 – DG: Avelino Magagnin. **Veneto dialect:** Sat evenings.
RS115) Rua Daltro Filho 417, 98600-000 Três Passos.
RS116) Av. Concordia 1013, Conj. 102, 96540-000 Agudo – DG: Leo Pachaly.
RS117) C.P. 10, 99150-970 Marau – DG: Velton da Silva Cezar. **FM:** 94.7 "Kosmos".
RS118) Rua Frei Eugenio 657, 95270-000 Flores da Cunha.
RS119) Rua Gaspar Martins 55, 97542-000 Alegrete. ☎ +55 (55) 422 1829 – DG: Marcelo V. da Silva. Dir: Hélio R. dos Santos.
RS120) Rua Conde de Porto Alegre 135, 97573-581 Santana do Livramento.
RS121) Rua Tiradentes 16, 99490-000 Tapera. ☎ +55 (54) 3851166. 🖹 +55 (54) 3851213 – DG: Fábio David Crestani.
RS122) C.P. 844, 90001-970 Porto Alegre – DG: Antônio F. Marin.
RS123) Rua Pedro Toniolo 529, Centro, 99900-000 Getúlio Vargas. ☎ +55 (54) 341 1555 – DG: Jussara Enrigone.
RS124) C.P. 19, 97200-970 Restinga Seca. ☎ +55 (51) 664 1110 – DG: Pe. Sergio Felix.
RS125) Rua Marechal Floriano 176, 96205-190 Rio Grande.
RS126) C.P. 79, 98900-970 Santa Rosa – Dir: Roberto Antônio Donadel.
RS127) Rua 24 de Maio 181, 95330-000 Veranópolis.
RS128) Rua Ramiro Barcelos 1206, 96901-900 Santa Cruz do Sul – DG: Norberto Frantz. **FM:** 101.7.
RS129) Rua General Osorio 1080, 96600-000 Canguçu – DG: Gilmar Oriente Mussi.
RS130) Av. Coronel Diniz 78, 95670-000 Gramado – DG: Sergio Lima.
RS131) C.P. 120, 95780-970 Montenegro – DG: Amabilio J. Cabral.
RS132) Rua General Osório 531, 96200-400 Rio Grande – Dir: Izabelino Tavares.
RS133) Av. Sete de Setembro 672, 96400-000 Bagé. ☎ +55 (532) 42 1211/1471 – DG: Vacionir Silva Lopes.
RS134) Rua Borges de Medeiros 401, 95560-000 Torres – DG: Pe. Sergio F. Leonadelli. **F.PI:** 5kW.
RS135) Rua Dr. Bruno Dockhorn 18, 98910-000 Três de Maio – DG: Bruno Rogério Dockhorn.
RS136) C.P. 102, 93700-970 Campo Bom – DG: Henrique Schneider.
RS137) Rua Julio de Castilhos 325, 95720-000 Garibaldi – DG: Inaro Cruz.
RS138) Rua João Moreira 1733, 97610-000 São Francisco de Assis.
RS139) Rua General Osorio 943, 96600-000 Canguçu.
RS140) Rua Rio Branco 1006, 95600-000 Taquara.
RS141) Rua Sete de Setembro 960, 96508-010 Cachoeira do Sul.
RS142) C.P. 706, 98770-970 Catuípe.
RS143) C.P. 144, 96450-970 Dom Pedrito – DG: Pascoal G. Brandi.
RS144) Rua Independência 197, 93010-001 São Leopoldo.
RS145) Rua Amâncio Cardoso 596, 99950-000 Tapejara. ☎ +55 (54) 3441185, 3441600 – DG: Silvino B. Lamb. **Italian & German:** Sat 1230-1400, Sun 1300-1400.
RS146) Rua Benjamin Constant 126, 97700-000 Santiago.
RS147) Rua José Loureiro da Silva 1727, 94010-001 Gravataí.
RS148) C.P. 20, 95960-970 Encantado.
RS149) Rua Assis Brasil 83, 98130-000 Júlio de Castilhos.
RS150) Rua José Bonifácio 41, 96330-000 Arroio Grande. ☎ +55 (532) 621008 – DG: João Carlos Saraiva.
RS151) Rua Paraguai 42, 98980-000 Porto Lucena. ☎ +55 (55) 5651200. 🖹 +55 (55) 5651221.
RS152) Av. Valdomiro Bocchese 872, 95250-000 Antônio Prado. ☎ +55 (54) 2931110 – DG: Laureano Antônio Fortuna.

RS153) Av. Padre Reus 324, 97930-000 Caibaté.
RS154) Rua Arroio Grande, Galeria Martins, salas 01 e 04, 96360-000 Pedro Osório.
RS155) Rua Borges de Medeiros 1462, 97650-000 Itaqui.
RS156) C.P. 241, 98590-970 Santo Augusto. ☎ +55 (55) 7811255, 7811675, 7811439 – DG: Roberto Antônio Sperotto.
RS157) Rua Baltazar Brum 343, 97560-000 Quaraí.
RS158) Rua Leandro Ribeiro 28, 95860-000 Taquari.
RS159) Rua Rui Barbosa s/n, 98300-000 Palmeira das Missões.
RS160) Rua Peri 625, Conj. 201, 95555-000 Capão da Canoa. ☎ +55 (51) 665 2161 – DG: Ricardo Alexandre Wagner.
RS161) Rua 7 de Setembro 601, 99530-000 Chapada – DG: José Paulo Ledur.
RS162) Rua Edilio Vitória 833, 96160-000 Capão do Leão.
RS163) Ao lado do Açude Pozobom, Distrito de Camobi, 97110-150 Santa Maria. **NB:** re. to R. Transmundial, C.P. 18300, 04699-970 São Paulo, SP.
RS164) Rua Coronel Scherer 195, 97400-000 São Pedro do Sul.
RS165) Rua Duque de Caxias 255, 98430-000 Palmitinho. ☎ +55 (55) 791 1244. ▤ +55 (55) 791 1175 – DG: Neodi Antônio Brandão.
RS166) Rua João Maffesoni 10, 99680-000 Constantina. ☎ +55 (54) 3631330. ▤ +55 (54) 3631122 – DG: João Carlos Martini.
RS167) Av. Presidente Vargas 1443, 98580-000 Coronel Bicaco.
RS168) Av. Alto Jacuí 435, Terreo, 99470-000 Não Me Toque. ☎ +55 (54) 332 1488 – DG: Dante Riczaneck.
RS169) Rua Tiradentes 35, Próximo a Praça Getúlio Vargas, 97450-000 Cacequi.
RS170) Rua Sanaduva 178, 99855-000 São João da Urtiga. ☎ +55 (54) 532 1247 – DG: Jocelei Flôres. **Polish:** Sat 1630-1700.
RS171) Av. Borges de Medeiros 407, 96760-000 Tapes.
RS172) Rua Elias Mendes de Araujo 147, 99870-000 São José do Ouro.
RS173) 98700-000 Ijuí.
RS174) Rua General Sampaio 18, Edificio Consorcio, Bl. Central, Conj. 09, 96400-370 Bagé – DG: Aristides Kucera.
RS175) Rua dos Andradas 560, 99660-000 Campinas do Sul.
RS176) 96270-000 Mostardas.
RS177) Av. Barão do Triunfo 584, 95995-000 Arvorezinha – DG: Aniceto Pastório Paganin. **FM:** 92.3.

SC00) SANTA CATARINA
SC01) C.P. 61, 89161-970 Rio do Sul. ☎ +55 (478) 22 2111 – DG: Iolanda Weise Gonçalves. **FM:** 93.3.
SC02) C.P. D2, 88801-970 Criciúma – DG: Pedro V. Barcellos.
SC03) Av. Martin Piaseski 25, 89910-000 Descanso. ☎▤ +55 (498) 230307 – DG: Veldir Basso.
SC04) Rua João Goulart 1020, 89806-050 Chapecó.
SC05) C.P. 313, 89161-970 Rio do Sul – Dir: Edison Andrade.
SC06) C.P. 25, 89990-970 São Lourenço d'Oeste – DG: Cirio Hippler.
SC07) Rua Carlos Jofre do Amaral 67, 88501-010 Lages. ☎ +55 (49) 2213000. ▤ +55 (49) 2213030 – DG: Osni Provença Jr.
SC08) Av. Sete de Setembro 155, 89580-000 Fraiburgo. ☎ +55 (492) 462775, 462754, 463294. ▤ +55 (492) 462877 – DG: Evanilda S. Frey.
SC09) C.P. 71, 88701-970 Tubarão – DG: Ademar Paulo de Faveri.
SC10) Rua Alan Kardec 12, 88025-100, Florianópolis – DG: Gilberto Benati.
SC11) Rua Leonel Mosele 275, 89700-000 Concórdia. ☎▤ +55 (49) 4421366, 4421217 – DG: Maria Luisa Stodieck. **Italian & German:** Sun 1130, 1600.
SC12) Rua Angelo Dias 251, 89010-020 Blumenau – DG: Evelasio Vieira.
SC13) Rua da Criança s/n, 88840-000 Urussanga – Gte: Osmar Nunes.
SC14) Rua Pedro Andreazza 49, 89560-000 Videira.
SC15) C.P. 209, 89010-971 Blumenau – DG: Walter U. Mick.
SC16) Rua Itagiba 215, 88880-000 Lauro Müller. ☎▤ +55 (48) 4643004 – DG: Gildo Rotta Filho. **Italian:** Sat 1600.
SC17) C.P. 71, 89700-970 Concórdia. ☎ +55 (494) 42 2611 – DG: César Luiz Pichetti. **FM:** 96.3.
SC18) Rua Jaú Guedes da Fonseca 17, 88080-080 Florianópolis – Dir: Arnoldo Cruz Lima. **FM:** 101.7 "Transamérica".
SC19) Margens da Rodovia SC-444, Linha Tres Ribeirões, 88820-000 Içara.
SC20) Rua Rodrigues Alves 111, 88350-000 Brusque. ☎ +55 (473) 55 3311/3257 – DG: Sueli Jandt Benevutti.
SC21) Loteamento Jardim Portobello, Rua A, 88200-000 Tijucas – Dir: Walmor Silva Telles.
SC22) Rua Barão de Rio Branco 229, 88870-000 Orleãns.
SC23) Rua José de Miranda Ramos 100, sala 105, 89820-000 Xanxerê.
SC24) C.P. 96, 88350-970 Brusque. ☎ +55 (47) 3511744 – DG: Ciro Marcial Roza. **FM:** 104.1. **F.PI:** 20kW on AM, change of FM to 107.7MHz.
SC25) C.P. 63, 89896-970 Itapiranga – DG: A. Baldissera.

SC26) C.P. 67, 88750-970 Braço do Norte – DG: Mario Schlickmann.
SC27) Av. Santa Catarina 25, 89840-000 Coronel Freitas.
SC28) Rua Otacilio Vieira da Costa 40, 88501-050 Lages. ☎ +55 (492) 223040, 223011, 224784 –DG: Jorge Luiz Maciel. **FM:** 95.7 "Amizade".
SC29) C.P. 34, 89130-970 Indaial – DG: Aroldo Laemmel.
SC30) C.P. 25, 89201-970 Joinville.
SC31) Rua Princesa Isabel 300, 88702-200 Tubarão. ☎ +55 (486) 22 0993/4400 – DG: Carlos Kürten. **FM:** 98.9.
SC32) C.P. 1477, 88010-970 Florianópolis.
SC33) Av. Santa Catarina 828, 2Þ andar, 89885-000 São Carlos. ☎ +55 (497) 254355. ▤ +55 (497) 254183 – DG: Mario Luiz Rauber.
SC34) Rua João Winkler 15, 89820-000 Xanxerê. ☎ +55 (494) 331110, 331115. ▤ +55 (494) 330682 – DG: Rita Bertoluzzi. **FM:** 101.3.
SC35) C.P. 291, 89520-970 Curitibanos. ☎ +55 (492) 451244 – DG: Vigilio Tambosi. **FM:** 98.9.
SC36) Rua 15 de Novembro 600, sala 401, 89010-000 Blumenau.
SC37) Rua Oswaldo Cabral 146, Ap. 7, 88790-000 Laguna.
SC38) Rua Siqueira Campos 33, 89400-000 Porto União.
SC39) C.P. D4, 89930-970 São José do Cedro.
SC40) Rua Joaquim Geraldo 169, 88240-000 São João Batista.
SC41) C.P. 59, 89835-970 São Domingos – DG: Valentim Bigolin.
SC42) C.P. 312, 88701-970 Tubarão. ☎ +55 (48) 6280658, 6280020. ▤ +55 (48) 6280638 – DG: Arilton B. de Souza.
SC43) Rua Tenente Ary Rauen 1361, 89300-000 Mafra – Dir: Tito Lamaré Schultz. **FM:** 104.5 "Nova Era".
SC44) Rua Rui Barbosa 136, 89890-000 Cunha Porã – DG: Edsom Zanetti.
SC45) C.P. 335, 89201-970 Joinville – DG: R. da Silva.
SC46) Rua São Paulo 1120, 89012-000 Blumenau.
SC47) C.P. 30, 89600-970 Joaçaba. ☎ +55 (495) 22 0433 – DG: Nelson Paulo dos Santos. **Italian:** Sun 1215-1500. **FM:** 97.3.
SC48) Rua Osvaldo Cabral 68, Centro, 88790-000 Laguna – DG: Jacopo Teixeira Tasso.
SC49) Av. Alvim Bauer 585, 88330-000 Balneário Camboriú. ☎ +55 (473) 671044, 671045, 672733. ▤ +55 (473) 674949 – DG: Maria Fernanda Rosa.
SC50) Rua Buenos Aires 131, 89051-050 Blumenau – DG: Mario E. Binder.
SC51) C.P. 44, 89801-970 Chapecó. ☎ +55 (49) 7220688, 7220658. ▤ +55 (49) 7220429 – Dir: José Francisco Müller Bohner. **FM:** 107.1.
SC52) Rua Princesa Isabel 311, 89460-000 Canoinhas – DG: Cirlene F. de Oliveira.
SC53) Av. Gov. Adolfo Konder 1500, Bairro São Vicente, 88308-000 Itajaí.
SC54) Rua Reinoldo Rau 400, 89251-600 Jaraguá do Sul.
SC55) Rua Duque de Caxias 1302, 2Þ andar, 89900-000 São Miguel d'Oeste. ☎ +55 (498) 22 1877. ▤ +55 (498) 22 1270 – DG: Ademar P. Baldissera. **FM:** 104.9.
SC56) Rua José Gonçalves 333, 89340-000 Itaiópolis – DG: Darcy Zanghelini. **Polish:** Sat 1030-1100.
SC57) Rua Carlos Jofre do Amaral 67, 88501-010 Lages. ☎ +55 (49) 2213000. ▤ +55 (49) 2213030 – DG: Osni Pereira Jr.
SC58) Rua Visconde do Rio Branco 1028, 89887-000 Palmitos – DG: Pedro Peiter.
SC59) Rua Nunes Machado 14, 10Þ andar, 88010-460 Florianópolis – DG: Mario Silva. Dir: Acy Cabral Teive.
SC60) Praça Lauro Müller 255, 89620-000 Campos Novos.
SC61) Rua 7 de Setembro 341, 89874-000 Maravilha – DG: Julmir Viccari.
SC62) C.P. 307, 89290-970 São Bento do Sul.
SC63) C.P. D 2, 88801-970 Criciúma.
SC64) C.P. 63, 89110-970 Gaspar. ☎ +55 (473) 32 0783 – DG: Alcides Morastôni and Léopoldo Miglióli. **F.PI:** 5kW.
SC65) Rua Ademar da Silva 1209, Bairro Kobrasol, 88101-091 São José – DG: Daniel Espírito Santo.
SC66) C.P. 91, 89610-970 Herval d'Oeste. ☎ +55 (495) 22 1579 – DG: Lamemir Floriano Colombo. **Italian:** Sat 1530-1730.
SC67) Rua Dr. João Colin 572, 89204-000 Joinville – DG: Ramiro Gregório da Silva.
SC68) Rua Izau Luiz de Bittencourt s/n, 88770-000 Imaruí.
SC69) Rua Getúlio Vargas 183, 89140-000 Ibirama.
SC70) Av. Plínio Arlindo de Nes 476, 89825-000 Xaxim. ☎ +55 (49) 753 2425 – DG: Alexandre S.C. de Amorim. **F.PI:** 5kW.
SC71) C.P. 86, 89251-970 Jaraguá do Sul – DG: Waldenir Luiz Freiberger.
SC72) C.P. 104, 88502-970 Lages – DG; J.P. Bággio.
SC73) C.P. 74, 89870-970 Pinhalzinho – DG: Ernani Edison Horvath.
SC74) Rua Altamiro Guimarães 230, 89500-000 Caçador.
SC75) Av. Dr. João Rimse 797, 88780-000 Imbituba.
SC76) C.P. 234, 89120-970 Timbó – DG: J. Reinart.

SC77) C.P. 122, 88301-970 Itajaí.
SC78) Rua Lauro Müller 109, 88600-000 São Joaquim.
SC79) Av. Porto Feliz 151, 89893-000 Mondaí. ☎ +55 (498) 740122. ▤ +55 (498) 740023 – DG: Érica Knorr. **German & Spanish:** Sun 1300-1400. **F.PI:** 5/2.5kW.
SC80) C.P. 160, 89665-970 Capinzal.
SC81) Av. Getúlio Vargas 429, 88900-000 Araranguá.
SC82) C.P. 3, 89190-970 Taió – DG: Célio Kellermann.
SC83) C.P. 103, 89295-970 Rio Negrinho – DG: Goldwin Meier.
SC84) Rua 15 de Novembro 39, 89107-000 Pomerode – Dir: René Eskelsen.
SC85) Rua João Steffens 260, 88400-000 Ituporanga – DG: Luiz Carlos Brering.
SC86) Rua Angelo Laporta 155, 88020-600 Florianópolis – Dir: Davy Campos.
SC87) Rua João Florentino de Souza s/n, 89480-000 Major Vieira.
SC88) C.P. 1477, 88103-970 São José – DG: Ismael Fabião.
SC89) Rua Sargento Juvenil Pereira de Souza 476, 89540-000 Santa Cecília – DG: Dr. Roberto R. Amaral.
SC90) Av. Getúlio Vargas 974, 89830-000 Abelardo Luz.
SC91) Av. Antônio Francisco Guizoni s/n, 88650-000 Urubici.
SC92) Rua Rui Barbosa 1321, 88930-000 Turvo.
SC93) Rua Professor João Sobotka 222, Bairro São Cristovão, 89665-000 Capinzal. ☎▤ +55 (495) 551799, 551369 – DG: Ailton Viel. **Italian:** Sun 1500-1600.
SC94) Av. Progresso s/n, 89888-000 Caibi – DG: Domingos Turcato.
SC95) C.P. 001, 89900-970 São Miguel d'Oeste. ☎ +55 (498) 22 1877 – DG: Elsa Rosa Cesca Baldissera.
SC96) Rua Rafael Pardinho 249, 89240-000 São Francisco do Sul. ☎ +55 (47) 4440235, 4440555, 4440733. ▤ +55 (47) 4441900 – DG: Antônio José Bueri. **F.PI:** 10kW.
SC97) Rua Vicente Rech 399, 89770-000 Seara.
SC98) Rua do Comércio 214, 89872-000 Modelo.
SC99) Rua 7 de Setembro 495, 89950-000 Dionísio Cerqueira.
SC100) Rua José Bertolucci 201, 89683-000 Ponte Serrada.
SC101) Av. Dr. Albano Schultz 925, 2Þ andar, 89201-220 Joinville – DG: Antônio S. Peixer. **FM:** 103.1.

SE00) SERGIPE
SE01) Rua Claudio Batista 334, Bairro Santo Antônio, 49060-100 Aracaju – DG: Ronaldo Moreira de Melo.
SE02) Rua Propria 124, 49010-020 Aracaju – Dir: Pedro P. Valverde.
SE03) Rua Simão Dias 643, 49010-430 Aracaju – DG: Luiz Carlos Q. de Alencer.
SE04) Praça Dr. Filomeno Mora 4, 49400-000 Lagarto.
SE05) C.P. 409, 49001-970 Aracaju – DG: E. Ferreira S.
SE06) Rua João Pessoa 85, 49500-000 Itabaiana.
SE07) Rua Itabaianinha 162, 49010-190 Aracaju.
SE08) Praça Coronel Gonçalo Prado s/n, 49200-000 Estância – DG: Jorge Prado L.
SE09) Av. Dr. Luíz Magalhães 346, 49500-000 Itabaiana.
SE10) Rua Presidente Vargas 280, 49480-000 Simão Dias.
SE11) Av. José da Cunha 6, 49510-000 Frei Paulo.
SE12) Rua Barão do Rio Branco 262, 49200-000 Estância.
SE13) Travessa Santa Luzia 69, 49300-000 Tobias Barreto – DG: Walter Monteiro.

SP00) SÃO PAULO
SP01) Rua Padre Geraldo Goseling 798, 16200-000 Biriguí – DG: Leonardo Sabioni.
SP02) Av. 7 de Setembro 100, 13171-440 Sumaré.
SP03) Av. Nesralla Rubez 353, 12700-000 Cruzeiro. ☎ +55 (125) 441364, 441439 – DG: Lamartine Fiorentini Jr. **FM:** 100.7.
SP04) Rua José Bonini 1415, 14160-000 Sertãozinho.
SP05) Rua Cafélândia 161, Vila Nova, 18400-000 Itapeva – DG: Paulo Roberto Silva. **FM:** 93.5 "Cristal".
SP06) Rua Dr. Sousa Alves 960, 12020-030 Taubaté.
SP07) Rua Rui Barbosa 474, 13465-280 Americana. ☎ +55 (19) 4621076, 4617031. ▤ +55 (19) 4617081 – DG: Edilberto de P. Ribeiro. **FM:** 88.9 "Notícia".
SP08) Av. Rotary 85, 19970-000 Palmital – DG: Eduardo Negrão.
SP09) Rua General Câmara 5, 11010-121 Santos.
SP10) Rua das Nações Unidas 127, 16800-000 Mirandópolis.
SP11) C.P. 814, 14001-970 Ribeirão Preto – DG: Octavio S. Silveira.
SP12) Av. Luíz Gonzaga de Amoedo Campos 28, 13800-000 Mogi Mirim – DG: Genésio D. Teodoro.
SP13) Rua Pará 155, Centro, 15800-000 Catanduva. ☎ +55 (175) 22 7311 – DG: Dr. Rodrigo Gabas. **FM:** 94.9.
SP14) Rua Comendador João Galvão 58, 12500-000 Guaratinguetá.
SP15) C.P. 150, 18800-970 Piraju. ▤ +55 (143) 511066 – DG: Carlos Alberto Leal.

SP16) Av. Paulista 807, 24Þ andar, 01311-915 São Paulo. **E-mail:** info@jovempan.com.br **Web:** http://www.jovempan.com.br – DG: Antônio A.A. Carvalho. **F.PI:** national sce.via satellite.
SP17) Rua Capitão Neve 18/40, 15130-000 Mirassol.
SP18) Av. Marcondes Filho 1130, 19013-160 Presidente Prudente.
SP19) Praça José Palamone Lepre 99, 14804-010 Araraquara. ☎ +55 (16) 236 3622. ▤ +55 (16) 236 0114. **E-mail:** radiomorada@ sun-rise.com.br **Web:** http://200.246.21.2/morada/ – DG: Maria Amalia Montoro. **FM:** 98.1.
SP20) Rua Homero Rodrigues Silva 1090, 16900-000 Andradina – DG: Nivaldo Bueno F. Rocha. **FM:** 97.9 "Cidade Andradina".
SP21) Praça José Bonifácio 815, 13400-340 Piracicaba – Dir: José R. Soave.
SP22) Rua João Pessoa 129, 8Þ andar, 11013-900 Santos – DG: Roberto Clemente Santini. **FM:** 105.5.
SP23) C.P. 1252, 14001-970 Ribeirão Preto – DG: José I. Gennari Pizani. **FM:** 100.5.
SP24) Rua Teotonio Tibiriçá Pimenta 380, 11660-230 Caraguatatuba. ☎ +55 (12) 4224633. ▤ +55 (12) 4225000 – DG: Rosely Aparecida Bueno.
SP25) Rua Heitor Penteado 173, 17400-000 Garça.
SP26) Praça Regente Feijo 167, 13300-000 Itu.
SP27) Rua 13 de Maio 720, 15800-000 Catanduva.
SP28) Av. Dr. Alvaro Schmidt Gallo 317, 18800-000 Piraju – DG: Nelson Meira.
SP29) Rua Francisco Inácio 257, 14700-000 Bebedouro.
SP30) Praça Conselheiro Rodrigues Alves 182, 12500-000 Guaratinguetá – DG: José Luíz Sannini.
SP31) Rua Humberto Liedikie 1936, 15370-000 Pereira Barreto – DG: Nivaldo Bueno Franca da Rocha.
SP32) Rua Pires da Mota 820/830, 01529-000 São Paulo. ☎ +55 (11) 277 5500. ▤ +55 (11) 277 3988 – Dir: João Lara Mesquita.
SP33) Rua 1 de Agosto 927, 17010-011 Bauru – DG: Tobias Ferreira.
SP34) Rua dos Pelegrini s/n, Bairro do Desterro, 13700-000 Casa Branca. ☎ +55 (19) 6712101, 6711143. ▤ +55 (19) 6711048 – DG: José Carlos Bini. **F.PI:** 5kW.
SP35) C.P. 355, 19900-970 Ourinhos – DG: Ruy S. Karpstein.
SP36) C.P. 25, 12700-970 Cruzeiro – DG: Carlos C. Fernandes.
SP37) C.P. 361, 15400-970 Olímpia – DG: Silvio R. Mathas N.
SP38) Rua Barão de Jundiaí 954, 13200-002 Jundiaí.
SP39) C.P. 324, 17500-970 Marília – DG: Edson J. de Souza.
SP40) C.P. 88, 17250-970 Bariri – DG: Osvaldo Orides Gimenez.
SP41) Av. Paulista 900, 01310-100 São Paulo. ☎ +55 (11) 289 3765/3592. ▤ +55 (11) 289 3768 – DG: Fernando Luiz Vieira de Mello. **F.PI:** 25kW.
SP42) C.P. 171, 17700-970 Osvaldo Cruz. ☎ +55 (18) 5612326, 5611788, 5611088, 5611089 – DG: Alvaro Luís Borini. **FM:** 97.3.
SP43) Rua Visconde do Rio Branco 401, 14015-000 Ribeirão Preto – Dir: Carlos Orlandi.
SP44) Rua Euclides Miragaia 394, 12245-550 São José dos Campos.
SP45) C.P. 621, 17001-970 Bauru – DG: Dr. Antônio Carlos de Jesus.
SP46) Rua Piauí 493, 13480-255 Limeira – DG: Orlando J. Zovico. **FM:** 100.7.
SP47) Rua das Palmeiras 315, 01288-900 São Paulo – DG: Paulo Novis.
SP48) Av. Armando de Sales Oliveira 575, 17800-000 Adamantina.
SP49) Av. Feijó 583, Centro, 14801-140 Araraquara. ☎ +55 (16) 2321177, 2323790. ▤ +55 (16) 2323475. **E-mail:** cultura@techs.com.br **Web:** http://200.246.30.1/cultura/ – DG: Ricardo Lupo. **FM:** 97.3.
SP50) C.P. 30, 13200-970 Jundiaí – DG: Hélio L. Lorencini.
SP51) Rua Benjamim Constant 3327, 15015-600 São José do Rio Preto.
SP52) Av. Getúlio Vargas 185, 12570-000 Aparecida. ☎ +55 (12) 5651133. ▤ +55 (12) 5651138 – DG: Antonio César Moreira Miguel. **FM:** 90.9.
SP53) Rua Tenente Lopes 191, 17201-460 Jaú – DG: Humberto Fabris Neto. **FM:** 101.1.
SP54) Rua José Galvão 359, 19900-000 Ourinhos.
SP55) Rua Dr. Mário Sabino 131, 16300-000 Penápolis.
SP56) C.P. 1001, 15500-970 Votuporanga. ☎ +55 (17) 4217622 – DG: Dimas L. de Camargo. **FM:** 92.1.
SP57) Rua Radiantes 13,.Morumbi, 05699-900 São Paulo. ☎ +55 (11) 8457211 – DG: Samir Razuk.
SP58) Av. Visconde do Rio Claro 2128, 13500-580 Rio Claro.
SP59) C.P. 154, 16200-970 Biriguí – Dir: Libério Telies.
SP60) Rua Prudente de Morães 418, 14960-000 Novo Horizonte.
SP61) Rua Orlando Carpino 237, Centro, 13073-430 Campinas – DG: Alfredo Orlando Enz. **FM:** 103.7 "Nova".
SP62) Av. Paulista 900, 01310-100 São Paulo – DG: Alberto Helena Jr.
SP63) C.P. 5, 19001-970 Presidente Prudente.
SP64) Rua Siqueira Campos 3223, 15010-210 São José do Rio Preto. ☎ +55 (172) 32 0101. ▤ +55 (172) 32 0089 – DG: Jô Silva. **FM:** 102.1 "R. Onda Nova FM".

SP65) C.P. 281, 18200-970 Itapetininga. ▤ +55 (15) 2724202, 2718207 – DG: Francisco Carlos Silva Janez.
SP66) Rua Alferes José Caetano 1039, 13400-120 Piracicaba – DG: Orlando J. Zovico.
SP67) Rua Major Claudiano 1393, 14400-690 Franca – Gte: Maria Abadia de Almeida.
SP68) C.P. 52, 13970-970 Itapira. ☎ +55 (192) 630138, 632128 – DG: Luís Antônio R. Nucci. **FM:** 91.1 "Clube Líder FM".
SP69) Av. Senador Salgado Filho 537, Jockey Club S.V., 11360-200 São Vicente.
SP70) Rua Rio Branco 719, 17015-310 Bauru – Gte: Carlos R. Morgado.
SP71) C.P. 150, 15775-970 Santa Fé do Sul. ☎ +55 (17) 6311521. ▤ +55 (17) 6312410 – DG: Helena Coelho Rubinho. **FM:** 104.7.
SP72) Av. Sampaio Vidal 185, 17501-040 Marília – Dir: Wilson N. Matos.
SP73) C.P. 57, 01059-970 São Paulo – DG: Paulo Coev.
SP74) C.P. 56, 18200-970 Itapetininga – DG: José Abrão.
SP75) Praça Professor Roque Fiori 18-B, 13870-000 São João da Boa Vista – DG: Oscar M. Castellan.
SP76) C.P. 34, 14400-970 Franca – DG: Sidnei Franco da Rocha. **FM:** 96.5 "R. 10".
SP77) Rua São Carlos do Pinhal 696, 9º andar, 01333-000 São Paulo. ☎ +55 (11) 25 31566. ▤ +55 (11) 28 47710 – DG: Mário Luiz Catto.
SP78) Praça Joel Waldo Dal Moro 1, 14781-574 Barretos.
SP79) Rua José Maria Sanches 539, 19500-000 Martinópolis. ☎ +55 (182) 521333 – DG: Sebastião Rodrigues N. **F.PI:** 10kW.
SP80) C.P. 57, 12630-970 Cachoeira Paulista – DG: Wellington Silva Jardim.
SP81) Rua Capitão Francisco Rodrigues Garcia 187, 19800-000 Assis – DG: Antônio José de Camargo. **FM:** 100.1.
SP82) C.P. 105, 13480-970 Limeira – DG: Vitorio Bortolan Filho.
SP83) C.P. 16, 15700-970 Jales – DG: Alexandrina C. Garcia.
SP84) C.P. 284, 14400-970 Franca. **Web:** http://www.difusora.com.br/ – DG: Ricardo Pucci Pieri. **F.PI:** 10kW.
SP85) Rua Antônio Franco Pompeu 261, 17340-000 Barra Bonita. ☎ +55 (14) 641 0131 – DG: Herivelto Ottoboni. **FM:** 97.7.
SP86) Av. Floriano Peixoto 1840, 16400-000 Lins – DG: Nivaldo Bueno Franco da Rocha. **FM:** 103.1.
SP87) Av. 9 de Julho 3939, 01407-900 São Paulo. ☎ +55 (11) 887 4436/9963/6760 – DG: Fernando Asprino.
SP88) Rua Boa Morte 1122, 13400-140 Piracicaba. ☎▤ +55 (19) 4349444 – DG: Anna Maria M. Mattos. **FM:** 103.1.
SP89) Av. Ana Costa 90, 11060-000 Santos.
SP90) C.P. 31, 11740-970 Itanhaém – DG: Dr. Paulo Alves Corrêa.
SP91) Rua Marechal Bitencourt 346, 17201-430 Jaú – DG: Maria C. Lucio Bauab.
SP92) Rua Tenente Nicolau Maffei 357, 19010-010 Presidente Prudente. ☎ +55 (18) 221 3030 – DG: Nenrod Adiel Antonagi Pereira. **FM:** 101.1.
SP93) Rua Vinte 3011, 15700-000 Jales – DG: Ramon Lombardi. **FM:** 103.5 "Regional FM".
SP94) C.P. 85, 14300-970 Batatais – DG: João César Cangussú.
SP95) Rua Olavo Bilac 693, 16400-000 Lins – DG: Cilmar Machado dos Santos.
SP96) C.P. 565, 18001-970 Sorocaba – Dir: Gastão Lima N.
SP97) Av. Monumental Papa João Paulo II 221, 12570-000 Aparecida – DG: Liliane Freitas.
SP98) C.P. 326, 17500-970 Marília – Gte: José Marques Beato.
SP99) Rua Jeremias de Paulo Eduardo 916, 15910-000 Monte Alto.
SP100) Rua das Palmeiras 315, 01226-901 São Paulo – DG: Cesar Augusto Garcia.
SP101) Rua Nossa Senhora do Rosario 233, Jardim Nova Iorque, 16065-420 Araçatuba – Dir: Oscar L. Piconez.
SP102) C.P. 91, 14940-970 Ibitinga – DG: Roque de Rosa. **FM:** 99.3 "Ternura FM".
SP103) Av. Luíz Gonzaga de Amoêdo Campos 28, 13800-000 Mogi Mirim – DG: Marilda de Cássia Pelincer Donatti. **FM:** 93.9.
SP104) C.P. 95, 18540-970 Porto Feliz. ☎ +55 (152) 621219 – DG: Francisco Maria Grillo.
SP105) Av. Dr. Mário Galvão 463, Jardim Bela Vista, 12209-400 São José dos Campos – DG: Antônio de V. Jordão. **FM:** 97.5 "Band FM".
SP106) Rua Duque de Caxias 260, 15900-000 Taquaritinga. ☎ +55 (16) 352200, 3522999 – DG: Paulo Delgado Jr.
SP107) C.P. 258, 17600-970 Tupã – Gte: J.A. Neto.
SP108) Rua Gonçalves Dias 208, 19800-000 Assis.
SP109) C.P. 115, 14780-970 Barretos.
SP110) C.P. 139, 13500-970 Rio Claro – DG: Orlando José Zovico.
SP111) C.P. 153, 11680-970 Ubatuba – Gte: Eduardo de Souza Neto.
SP112) Rua Nadir Dias de Figueiredo 1329, 02110-000 São Paulo.
SP113) Rua Almirante Barroso 456, Sobre Loja, 19400-000 Presidente

Venceslau – DG: Antônio Carlos More. **FM:** 95.1 "R. Jovem Som".
SP114) C.P. 486, 18001-970 Sorocaba.
SP115) Rua São Paulo 731, 15600-000 Fernandópolis.
SP116) Rua Barão de Monte Santo 1211, 3º andar, 13730-000 Mococa. ☎ +55 (196) 550223. ▤ +55 (196) 551598 – DG: Jeferson Luiz Freitas.
SP117) Rua General Osório 1031, 21º andar, 13010-111 Campinas.
SP118) C.P. 380, 15500-970 Votuporanga – DG: Oliveira S. Prates.
SP119) C.P. 75, 13650-970 Santa Cruz das Palmeiras. ☎ +55 (196) 721659 – DG: Irany Baston.
SP120) Rodovia SP-255 Km 384, 18740-000 Taquarituba – DG: Francisco Rodrigues.
SP121) Rua Cenno Sbrighi 378, 05099-900 São Paulo – DG: Renato Alves Bittencourt.
SP122) Rua Rui Barbosa 546, 4º andar, 14870-000 Jaboticabal. ☎ +55 (163) 220266. ▤ +55 (163) 220866 – DG: Edson Pedro Peloso.
SP123) Rua Tupinambás 115, Bairro São João, 16025-180 Araçatuba – DG: José C.S. Hernandes.
SP124) C.P. 209, 17001-970 Bauru – DG; Maximo Duarte.
SP125) C.P. 250, 18001-970 Sorocaba – DG: Maria Aparecida F. Pavlovsky. **FM:** 94.9.
SP126) Rua Floriano Peixoto 375, 18300-000 Capão Bonito.
SP127) Rua João Pessoa 1990, 15990-000 Matão – Dir: Benedito J. Fernandes.
SP128) Rua Dr. Miguel Penteado 585, Jardim Chapadão, 13073-180 Santos.
SP129) Praça Comendador Emilio Pedutti 28, 18600-410 Botucatu.
SP130) Rua 8 No. 472, 14620-000 Orlândia. ☎ +55 (16) 7262230, 7262060, 7265000 – DG: Chéster A. Martins.
SP131) Rua José Caballero 60, 11055-300 Santos – DG: José Fornos Rodrigues.
SP132) Av. Nove de Julho 265, 15200-000 José Bonifácio.
SP133) Av. da Saudade 200, 12280-000 Caçapava.
SP134) Rua Granja Julieta 205, Granja Julieta, 04721-060 São Paulo.
SP135) Av. Treze de Maio 650, Jardim Paulista, 14090-260 Ribeirão Preto.
SP136) Av. Benjamin Constant 1214, 5º andar, Centro, 13010-141 Campinas – DG: Dr. Synésio Pedroso Jr. Gte: Wilson R. C. Vianna.
SP137) Rua Bernardino de Campos 3180, 15015-300 São José do Rio Preto – DG: José Luiz Spotti.
SP138) Rua Carina 05, 09732-060 São Bernardo do Campo.
SP139) C.P. 66, 14900-970 Itápolis – DG: Acácio Batista.
SP140) Rua Duque de Caxias 154, 13630-000 Pirassununga.
SP141) C.P. 75, 12900-970 Bragança Paulista – Dir: J.F. Ferreira.
SP142) Rua Brigadeiro Tobias 911, 19570-000 Regente Feijó.
SP143) Av. Costabile Romano 2201, Ribeirania, 14096-380 Ribeirão Preto – DG: Cecilio Ap. Matuanan.
SP144) C.P. 246, 16001-970 Araçatuba. ☎ +55 (18) 6238466, 6232052. ▤ +55 (18) 6226024 – DG: Gladys May Fares de Campos. **FM:** 95.5.
SP145) Rua Barão de Jaceguai 468, 08710-905 Mogi das Cruzes.
SP146) C.P. 20, 17350-970 Igaraçu do Tietê – DG: Francisco Perico.
SP147) C.P. 173, 15600-970 Fernandópolis. ☎ +55 (174) 42 1811 – DG: Jorge S. Ribeiro. **FM:** 90.5. **F.PI:** Move of AM to R. Aguas Quentes.
SP148) Av. José Bonifácio 1229, 17900-000 Dracena – DG: Antônio Carlos V. Borini.
SP149) Rua Cuiabá 2790, 16900-000 Andradina – DG: Adeliz R. Fernandes Rocha.
SP150) Av. Antônio J. de Carvalho 1671, 17250-000 Bariri – DG: Carlos E. Marcos Galizia.
SP151) Rua 15 de Novembro 843, 18680-030 Lençóis Paulista.
SP152) Rua Guanabara 144, 13840-000 Mogi Guaçu – Gte: Cleuzio Fonseca.
SP153) Rua Benjamin Constant 1214, 3º andar, 13010-141 Campinas. **E-mail:** cbncultura@supernet.com.br. **Web:** http://www.supernet.com.br/cbncultura/ – DG: Paulo Roberto Russo Pedroso. **FM:** 99.1.
SP154) Rua Dr. Erico de Abreu Sodré 542, 16370-000 Promissão.
SP155) Av. Brasil 1119, 17780-000 Lucélia.
SP156) C.P. 96, 13560-970 São Carlos – DG: Dr. Afranio R. Zambel. **FM:** 104.7.
SP157) C.P. 755, 15001-970 São José do Rio Preto.
SP158) Rua Doutor Pinto Ferraz 183, Vila Mariana, 04117-900 São Paulo. ☎ +55 (11) 2526272 – DG: José Dias Goulart.
SP159) C.P. 29, 13250-970 Itatiba – DG: Paulo Abreu.
SP160) Rua Cerqueira Cesar 481, 14010-130 Ribeirão Preto.
SP161) Rua Oswaldo Aranha 1040, 15150-000 Monte Aprazível.
SP162) Rua Dr. Souza Alves 960, 12020-030 Taubaté.
SP163) Rua Coronel Joaquim Floriano 287, 18650-000 São Manuel – Gte: José A. di Santis.
SP164) Praça Lourenço Franco de Oliveira 81, 13930-000 Serra Negra.
SP165) Av. Manoel Goulart 291, 1º andar, Centro, 19010-270 Presidente Prudente. ☎ +55 (18) 2212900, 2218200 – DG: Mauricio

Mescoloti. **F.PI:** 10kW.

SP166) Praça Nossa Senhora da Conceição 434, 16700-000 Guararapes – DG: Antônio Ap. Simões. **F.PI:** Change of freq.

SP167) Av. André Luís 723, 07082-050 Guarulhos.

SP168) C.P. 115, 13560-970 São Carlos. ☎ +55 (162) 711248. 🖷 +55 (162) 713498 – DG: Sylvia Yvonne Keppe Rossi.

SP169) Av. Clara Gianotti de Souza 1124, 11900-000 Registro – DG: Mucio A.P. Mattos.

SP170) C.P. 71, 12600-970 Lorena – DG: Jose A. dos Santos.

SP171) Rua Enrico dell'Aqua 332, 18130-000 São Roque.

SP172) Av. 15 No. 225, 14790-000 Guaíra.

SP173) Av. Malek Assad 535, Jardim Santa Maria, 12300-000 Jacareí – Dir: Pe. Edinei Batista.

SP174) C.P. 40, 13660-970 Porto Ferreira – Dir: José Wanderlei Klein.

SP175) C.P. 297, 13330-970 Indaiatuba – DG: Antônio R. Geiss.

SP176) Rua Renato Jardim 511, 14350-000 Altinópolis – DG: Mauro Matsuo.

SP177) C.P. 161, 12940-970 Atibaia – DG: J. Adibabi C.

SP178) Praça Oswaldo Martins 218, 16210-000 Bilac. ☎ +55 (18) 681 1339 – DG: Pe. Luiz Zavariz.

SP179) Rua Conselheiro Antônio Prado 518, 18900-000 Santa Cruz do Rio Pardo – Gte: Mauricio Azevedo F.

SP180) Av. Governador Dr. Ademar Pereira de Barros 134, 15400-000 Olímpia.

SP181) Rua Monte Castelo 941, 17900-000 Dracena.

SP182) Rua Brasil 1712, Centro, 15600-000 Fernandópolis. ☎🖷+55 (17) 4422623, 4423225 – DG: Teotonio Ap. de Oliveira.

SP183) Rua Sorocaba 150, Matriz, 09370-150 Mauá.

SP184) Praça Barão de Araras 201, 13600-000 Araras – **FM:** 97.9.

SP185) Av. Coronel Fernando Prestes 28, 12400-000 Pindamonhangaba.

SP186) Av. Leopoldo Carlos de Oliveira 1038, 14530-000 Miguelópolis.

SP187) C.P. 66, 06001-970 Osasco – DG: J. Gutier Navarro. Dir: Toni Gomide.

SP188) Rua Euclides da Cunha 5, sala 702/705, 11065-100 Santos – DG: Inivio da Silva Borda.

SP189) C.P. 13, 18650-970 São Manuel – DG: João C. Saad.

SP190) Av. Romeu Viana Romaneli 1510, 15570-000 Cardoso. ☎ +55 (17) 4531376, 4561330 – DG: Teotonio L. Tridapali.

SP191) Rua Dr. Ricardo Vilela 568, 08710-150 Mogi das Cruzes. ☎ +55 (11) 4708222. 🖷 +55 (11) 4616005 – Presidente: Tirreno de San Biagio.

Japanese: 0800-0900.

SP192) C.P. 66, 13990-970 Espírito Santo do Pinhal – DG: José V. Sales. **FM:** 102.7.

SP193) C.P. 125, 17930-970 Tupi Paulista – Gte: J. Matias.

SP194) Rua Capitão Lisboa 1080, 18270-000 Tatuí – DG: Luíz G. Vieira de Camargo. **FM:** 93.9 "Ternura".

SP195) Rua Benedito Carlos dos Reis 700, 15190-000 Nhandeara.

SP196) Rua Marechal Deodoro 320, 18600-320 Botucatu. ☎ +55 (14) 8221332, 8221227 – DG: E. Placio Paganni. **FM:** 93.1.

SP197) C.P. 403, 06001-970 Osasco – DG: Francisco Rossi.

SP198) Rua Dom José Carlos Aguirre 567, 18460-000 Itararé.

SP199) Rua Pedro de Toledo 205, 19700-000 Paraguaçu Paulista.

SP200) Rua Regente Feijó 121, 13360-000 Capivari – Dir: Antônio Mattar Jr.

SP201) Rua José Vaz Porto 175, Vila Santa Rosa, 11431-190 Guarujá – Gte: Jovanir Batista Rampazo.

SP202) C.P. 135, 14600-970 São Joaquim da Barra.

SP203) Av. Frei Orestes Girardi 1285, 12460-000 Campos do Jordão. **FM:** 94.9.

SP204) C.P. 199, 18700-970 Avaré – DG: Marilia P. de Almeida.

SP205) C.P. 221, 17890-970 Junqueirópolis – DG: J. Sinicato.

SP206) Rua Tatuí 321, Bairro Casa Branca, 09015-620 Santo André – DG: A.C. Netto.

SP207) Rua Campos Sales 94, 13960-000 Socorro – Dir: Assir Bujato.

SP208) C.P. 68, 15170-970 Tanabi – DG: Adirley Lustri.

SP209) Rua Dr. Felipe Vita 1616, 18480-000 Itaporanga – DG: Arnoldo Krubniki.

SP210) C.P. 8, 13690-970 Descalvado – Dir: Pedro Gaspar.

SP211) Rua Dr. Timóteo Penteado 2207, 07094-000 Guarulhos.

SP212) Av. 17 No. 560, 14780-000 Barretos.

SP213) Av. Paulista 2202, 8Þ andar, Conj. 81/82, 01310-300 São Paulo.

SP214) Av. Capitão José Antônio de Oliveira 544, 17800-000 Adamantina. ☎ +55 (18) 521 3547 – DG: Jonas Bonassa. **FM:** 93.7 "Antena 1".

SP215) Av. São Paulo 788, 15650-000 Estrela d'Oeste.

SP216) Av. João Guilhermino 429, 12210-131 São José dos Campos – Dir: Mauro Brandão.

SP217) Rua Brandão Veras 1274, 14700-000 Bebedouro – DG: João Pedro Negrini.

SP218) Rua Antônio Lobo 237, 13465-000 Americana.

SP219) Estrada Serrinha Km 200, 15350-000 Auriflama.

SP220) Av. Monte Castelo 225, Centro, 13450-185 Santa Bárbara d'Oeste ☎ +55 (194) 635255, 635068 – Own: Natale Giacomini.

SP221) Rua Almerim 435, Gramado, 06700-000 Cotia (24h Gospel prgrs).

SP222) Rua Dr. Mário Sabino 131, 16300-000 Penápolis.

SP223) Rua Jacofer 615, 02712-070 São Paulo. ☎ +55 (11) 2650667, 2656208, 2526544 – Dir: José Jantalia.

SP224) Rua Kametaro Morishita 95, 3Þ andar, 19050-700 Presidente Prudente – DG: Carlos A. Manfrim. **FM:** 106.7.

SP225) Rua Paula Ney 79, 18110-000 Votorantim.

SP226) Rua Bento Carlos 61, 13560-660 São Carlos – DG: Edgar Araujo. **FM:** 90.7.

SP227) Rua Rio Branco 76, 13460-000 Nova Odessa. ☎ +55 (194) 66 2198 – Dir: Urias Chagas.

SP228) Rua Américo Vespúcio 20, 14730-000 Monte Azul Paulista – DG: Paulo Alves Meita.

SP229) Av. Dr. Labiano da Costa Machado 1735, 17400-000 Garça.

SP230) Rua Professor Sud Menucci 464, 17740-000 Rinópolis.

SP231) Alameda Juscelino Kubitschek 1914, 19280-000 Teodoro Sampaio – DG: Alexandre Németh. **F.PI:** SW.

SP232) Rua da Consolação 2608, 1Þ andar, Conj. 11 e 12, 01416-000 São Paulo. ☎ +55 (11) 258 0152. 🖷 +55 (11) 258 5838 – DG: Luci Rothschild Abreu.

SP233) Av. 9 de Julho 304, 13140-000 Paulínia.

SP234) Rua José Jorge Junqueira 999, 14640-000 Morro Agudo.

SP235) Rua General Câmara 733, 13450-000 Santa Bárbara d'Oeste. ☎ +55 (19) 4635499. 🖷 +55 (19) 4633490 – DG: José Flavio Scavassa.

SP236) Rua Joaquim Eliziário de Campos 126, 18320-000 Apiaí – DG: Valter Luiz de Araújo.

SP237) Rua 5 No. 170, Bairro Cristo Redentor, 12100-000 Taubaté.

SP238) Rua Ademar de Barros 275, 14500-000 Ituverava.

SP239) Rua Frederico Ozanan 554, 17300-000 Dois Córregos.

SP240) Rua 13 de Maio 379, 13720-000 São João do Rio Pardo – DG: Vara Lucia G. Torres. **FM:** 88.7 "Cidade Livre".

SP241) Rua Bororos 344, 17600-020 Tupã.

SP242) Av. Washington Luís 214, 13600-000 Araras. ☎ +55 (19) 5411265, 5413714. 🖷 +55 (19) 5410477. **Web:** http://www.radio-clube.com.br/ – DG: João Franchozza.

SP243) Av. 18 de Junho 367, 17690-000 Bastos.

SP244) Rua Santana 440, Centro, 13880-000 Vargem Grande do Sul.

SP245) Rua Rafael de Barros 126, 13640-000 Lemé.

SP246) Rua Francisco Geraldino 71, 17580-000 Pompéia. ☎ +55 (144) 52 1142 – DG: Marco A.F. Castro.

SP247) Rua 7 de Setembro S-73, 17280-000 Pederneiras – DG: Celso Carlos Al-Haj. **FM:** 88.3.

SP248) Rua Tenente Adolfo Padilha 157, 16880-000 Valparaíso.

SP249) Rua 15 de Novembro 52, 13920-000 Pedreira. 🖷 +55 (19) 8932520 – Dir: Oswaldo Dale Junior.

SP250) Rua Vitória 162, 19470-000 Presidente Epitácio.

SP251) Av. Dr. Carlos Burgos 1680, 13900-000 Amparo.

SP252) Praça da Matriz 290, 19360-000 Santo Anastácio.

SP253) Rua 7 de Setembro 911, 14240-000 Cajuru.

SP254) Av. São Paulo 1220, 13310-000 Cabreúva – DG: Sonia Maria Ferres.

SP255) Rua Manoel dos Santos Freire 629, 18550-000 Boituva.

SP256) Rua José Revel 477, Centro, 13320-020 Salto. ☎ +55 (11) 483 2015 – DG: José Carlos Tonin.

SP257) Rua 9 de Julho 666, 16600-000 Pirajui. ☎ +55 (14) 5721352. 🖷 +55 (14) 5721941 – DG: Veldecir de S. Godi.

SP258) Rua dos Operários 1441, 19600-000 Rancharia – DG: Hallem Colombo Pereira.

SP259) Rua Coronel João de Carvalho 39, 1Þ andar, 13710-000 Tambaú – DG: Pe. José Mário Ribeiro. **F.PI:** 1kW.

SP260) Av. Rodolfo Guimarães 635, 1Þ andar, 17380-000 Brotas.

SP261) Rua Benjamim Constant 543, 14540-000 Igarapava.

SP262) Rua Inácio Ribeiro 592, 13670-000 Santa Rita do Passa Quatro.

SP263) Av. Higienópolis 890, 01238-000 São Paulo.

TO00) TOCANTINS

TO01) Av. Joaquim Aires 2393, 77500-000 Porto Nacional. ☎ +55 (63) 863 1608 – DG: Jon. Oscar Oliveira.

TO02) Praça José Tôrres 03, Centro, 77600-000 Paraíso do Tocantins. 🖷 +55 (63) 861 1135 – DG: José Antônio A. Cavalcente.

TO03) Av. Nossa Senhora de Fátima 894, 77900-000 Tocantinópolis.

TO04) BR-157 Km. 1103, Zona Rural, 77804-970 Araguaína – DG: Fábio de C. Roriz. **FM:** 99.7 "Araguaia".

TO05) Rua Raul do Espírito Santo 1334, 77760-000 Colinas do Tocantins – DG: Edna Lúcia.

TO06) Av. Tocantins 422, 77650-000 Miracema do Tocantins – DG:

Rodolfo Leão.
TO07) Almeda João Pires Querido 07, 77490-000 Cristalândia.
TO08) Av. Bernardo Sayão 2201, 77700-000 Guaraí – DG: Laercio Regino Saboia.

FM Stations in principal cities

Belo Horizonte: 88.7 Scala FM – 90.7 Cidade – 91.7 Horizontes de Minas – 94.9 Alvorada – MG06) 95.7 – MG68) 96.5 – MG53) 97.3 Altaneira FM – 98.3 98 (Del Rey) – 99.9 Terra – MG35) 100.9 – 102.1 BH FM – MG06) 103.9 – 105.1 Antena Um – MG65) 106.1 – 107.5 FM 107.

Brasília: 89.9 Brasília Super FM – 91.7 Brasília Comunicação – 93.7 Atlântida FM – DF01) 95.3 – DF04) 96.1 – 97.7 Manchete FM – 99.3 Antena 9 – 100.1 Transamérica – DF02) 100.9 Cultura FM – 101.7 R Jornal de Brasília – 105.5 FM 105 (Planalto) – 106.3 Sigma Radiodifusão – DF03) 107.1 Atividade.

Curitiba: PR15) 88.5 – PR139) 90.1 – 91.3 Gospel FM – 92.3 Scala FM – 93.9 Capital – 95.1 Exclusiva FM – 96.3 Studio 96 – PR06) 97.1 – 97.9 Melodia – 98.7 FM 98 – 100.3 Transamérica – PR134) 101.5 – 102.3 Caioba – 103.9 Jovem Pan – 105.5 Ouro Verde – 106.5 R Novo Tempo.

Fortaleza: 88.9 FM Jangadeiro – 89.9 Capital – 93.1 Tropical – 93.9 FM 93 – 94.7 Suceso – CE21) 95.5 FM do Povo – 99.1 Cidade FM – CE08) 99.9 Dragão FM – 100.9 Pajeu – 101.7 FM Casablanca – 103.9 FM O Tempo – 105.7 Atlântico Sul FM – 106.7 Hoje – 107.9 Universitária.

Porto Alegre: 89.3 Antena Um – 91.3 Metropolitana – 92.1 Cidade – 94.1 Atlântida – 94.9 Ipanema – 95.9 Liberdade FM – 96.7 Eldorado – 97.5 Jovem Pan FMI – 98.3 Continental – 99.3 Bandeirantes – 100.5 Capital – RS23) 101.3 – RS09) 102.3 – 104.1 FM 104 (Rede Pampa) – 106.3 Aliança – 107.7 Cultura.

Recife: 88.7 89 FM (Maranguape FM) – 90.3 J.C. FM – 91.9 Antena Um – 92.7 Transamérica – 94.3 Manchete – 95.9 Cidade – 97.5 Recife FM Stereo – 99.1 Caetes – PE10) 99.9 – 100.7 Evangélica do Brasil – 103.9 Maranata FM – 107.9 JMB Empreendimentos.

Rio de Janeiro: 88.5 Tribuna – RJ11) 89.3 – 90.3 FM O Día – 91.1 Diário – RJ25) 92.5 – 93.3 El Shaddai – RJ04) 94.1 – 94.9 Jovem Pan FM – 95.7 Alvorada FM – RJ27) 96.5 – 97.3 Melodia FM – RJ25) 98.1 FM 98 – RJ12) 98.9 – RJ16) 99.7 – RJ35) 100.5 – 101.3 Transamérica FM – 102.1 Imprensa FM – 102.9 Cidade – 103.7 Antena Um – 104.5 Tropical – RJ16) 105.1 105 FM – 106.3 Universidade – 106.7 Catedral – 107.1 107 – 107.9 Universidade.

Salvador: BA57) 90.1 Globo FM – 91.3 Itaparica – 92.3 Salvador FM – 94.3 Piata – 95.9 FM 96 (Aratu) – 97.5 Itapuã – 99.1 Bandeirantes – 100.1 Transamérica – 101.3 Cidade – BA04) 102.3 – 103.9 104 FM (R FM a Tarde) – 104.7 Manchete – BA55) 107.5.

São Paulo: SP62) 88.1 – 89.1 FM 89 – SP77) Nova FM 89.7 – SP47) 90.5 CBN) – 91.3 Manchete FM – SP32) 92.9 – 93.7 R. USP – 94.7 Antena Um FM – 95.3 Nativa FM – 95.5 Rede CBS – 95.7 Scala FM – SP57) 96.1 Band FM – 96.9 Cidade FM – SP232) 97.3 – 97.7 FM 97 – 98.5 Metropolitana FM – 99.3 Scala FM – 100.1 Transamérica – SP16) 100.9 – 101.7 Alpha FM – 102.5 Imprensa – SP121) 103.3 – 104.1 Líder FM – 104.7 Transcontinental – 105.1 105 FM – 105.7 Musical FM – SP41) 106.3 Mix FM – 106.9 Nova Omega FM – 107.3 Brasil 2000 – 107.5 Antena 1 – 107.9 Tropical FM.

CHILE

L.T: UTC - 4h (Su: UTC - 3h) — **Pop:** 13.900.000 — **Radios:** 4.400.000 — **Pr.L:** Spanish — **E.C:** 50Hz, 220/380V — **ITU:** CHL.

SUBSECRETARIA DE TELECOMUNICACIONES

Offices: Amunátegui 139, Santiago. ✉ Clasificador 120, Correo 21, Santiago. ☎ + 56 (2) 672 6503. 🖷 +56 (2) 699 5138. ① 341 156.
L.P: Subsecr of Telecommunocations: Roberto Pliscoff Vásquez.

ASOCIACION DE RADIODIFUSORES DE CHILE

✉ Cas. 10476, Santiago de Chile. ☎ +56 (2) 6398755. 🖷 +56 (2) 6394205. **Cable:** ARCHI.
L.P: Pres: César Molfino Mendoza. Dir: Alfredo Matte L.

STATIONS: SW: Call letters CE are used for all zones.
MW: Call letters CA, CB, CC and CD indicate: A=No. Zone, B=Central Zone, C=So. Zone and D=Antarctic Zone. The figures indicate the freq. in kHz minus one cipher, f. inst. CB82 = Central Zone 820kHz.

Mediumwaves: Þ = also on shortwaves, * = inactive, v=varying freq.

	Call	kHz	kW	Name, location and h. of tr.
1)	CB54	540	1	R. Ignacio Serrano, Melipilla: 1100-0400
2)	CD54	540	1	R. Calle Calle Saval, Valdivia: 1000-0400
2x)	CD55	550	1	R. LV. de la Tierra, Angol
3)	CB57	Þ570	50	R. Agricultura, Santiago: 1000-0400
4)	CA59	590	1	R. Minería, Antofagasta: 1000-0430
5)	CC59	590	1	R. Caracol, Concepción: 24h
6)	CD59	590	10	R. Pres. Ibáñez, Punta Arenas: 1000-0430
6x)	CB60	600	10	R. Monumental, Santiago: 24h
65)	CD60	600	10	R. Tricolor, Osorno: 1000-0500
24)	CD61	610	5	R. CD61, Puerto Aysén
7)	CA62	620	1	R. Norte Verde, Ovalle: 1100-0400
8)	CC62	620	10	R. Bío-Bío, Concepción: 24h
9)	CB63	630	10	R. Fundación, Valparaíso: 1100-0500
10)	CD64	640	10	R. Temuco Cooperativa AM, Temuco: 1000-0430
11)	CB66	660	50	R. Chilena, Santiago: 24h
12)	CA68	680	10	R. Chilena AM, Chuquicamata: 1000-0400
13)	CC68	680	10	R. Octava, Concepción
14)	CB69	690	10	R. Santiago, Santiago: 1000-0600
15)	CD69	690	10	R. Estrella del Mar, Ancud: 1100-0330
15x)	CA70	700	1	R. Amancay, Copiapó: 1030-0300
28)	CD70	700	1	R. Valdivia, Valdivia: 1030-0500
63)	CD70	700	5	R. Magallanes, Punta Arenas
81)	cd72	720	1	R. Alerce, Castro
16)	CB73	730	10	R. Valparaíso Cooperativa AM, Valparaíso: 1000-0430
17)	CD73	730	1	R. Angelina, Los Angeles
18)	CD73B	730	1	R. Aysén, Pto. Aysén: 1100-0400
16)	CB76	760	75	R. Cooperativa, Santiago: 1000-0430
19)	CD78	780	10	R. Sago AM, Osorno: 1030-0400
20)	CB80	800	5/1	R. Minería, Viña del Mar: 1000-0400
21)	CA82	820	10/1	R. Gabriela Mistral, La Serena: 1000-0400
127)	CB82	820	10	Radioemisora Carabineros de Chile, Santiago: 1000-0400
22)	CC82	820	1	R. Almirante Latorre, Talcahuano: 1000-0430
23)	CD82	820	1	R. Concordia, La Unión: 1100-0400
25)	CA82	820	0.25	R. Pampa, Pedro de Valdivia
26)	CB84	840	10	R. Portales, Valparaíso: 1000-0500
27)	CD84	Þ840	10	R. Santa María, Coyhaique: 1000-0400
95)	CC86	860	10	R. Nueva Inés de Suárez, Concepción: 24h
29)	CB88	880	10	R. Colo Colo, Santiago: 24h
30)	CA89	890	10	R. Chilena AM, Pozo Almonte: 1100-0200
31)	CC89	890	5	R. Interamericana, Concepción: 1030-0400
32)	CD89	890	20	R. Nal., Punta Arenas: 1000-0400
33)	CA90	900	5	R. Manantial, Copiapó: 1100-0400
9)	CB90	900	1	R. Amapola, Valparaíso: 1100-0500
34)	CC90	900	1	R. Ñuble, Chillán: 1100-0400
35)	CD90	900	1	R. LV de la Costa, Osorno: 1050-0400
35x)	CD92	920	1	R. 920, Temuco
36)	CA93	*930	10	R. El Cobre, Antofagasta
37)	CB93	930	10	R. Nuevo Mundo, Santiago: 1000-0530
38)	CD93	930	10	R. Reloncaví, Puerto Montt: 1000-0500
48)	CB94	940	1	R. Valentín Letelier, Valparaíso
152)	CA94	*940	1	R. 9-40, Copiapó
39)	CB96	960	10	R. Carrera, Santiago: 1100-0400
40)	CD96	960	10	R. Polar, Punta Arenas: 24h
41)	CA97	970	1	R. Calama, Calama: 1000-0400
42)	CC97	970	1	R. Lautaro, Talca: 1000-0500
43)	CD97B	Þ970	5	R. Patagonia Chilena, Coyhaique: 1000-0400
44)	CD97A	970	1	R. Austral, Valdivia: 1030-0400
45)	CA98	*980	1	R. Univ. de Tarapacá, Arica
3)	CB98	980	10	R. Agricultura, Valparaíso
145)	CC99	990	1	R. El Roble, Parral (F. PI.) (still on 1590)
46)	CB100	1000	10	R. Cien AM, Santiago: 1000-0500
47)	CD101	1010	10	R. Chilena-Niebol, Temuco: 1000-0430
91)	CC102	1020	5	R. Amiga, Talca
49)	CC103	1030	10	R. Chilena, Concepción: 1100-0430
139)	CB103	1030	1	R. Progreso, Talagante
50)	CD103A	1030	1	R. Chiloé, Castro: 1100-0330

	Call	kHz	kW	Name, location and h. of tr.
51)	CD103	1030	1	R. Payne AM, Puerto Natales
52)	CD104	1040	1	R. Raíces, Curacautín: 1100-0100
107)	CD105	1050	1	R. Bío Bío, Osorno: 1030-0630
53)	CB106	1060	50	R. Minería, Santiago: 1000-0500
54)	CA108	1080	1	R. Río Elqui, Vicuña
55)	CD108	1080	1	R. Los Confines, Angol: 1100-0400
56)	CC109	1090	5/1	R. Chilena, Talca: 1000-0400
169)	CC109	1090	5	Pentagrama AM, Constitución
58)	CA110	*1100	1	R. La Portada, Antofagasta
59)	CB110	1100	10	R. Recreo, Viña del Mar: 1100-0500
60)	CD111	1110	10	R. La Frontera, Temuco: 1000-0400
61)	CB114	1140	40/75	R. Nal., Santiago: 1000-0600
62)	CC116	1160	1	R. Ancoa, Linares: 1000-0600
99)	CD116	1160	1/0.25	R. Baha'i, Temuco: 1030-0230
98)	CD117	1170	5	R. Natales, Puerto Natales
173)	CB118	1180	50	R. Portales, "la primera de Chile", Santiago: 1030-0430
64)	CA120	*1200	1	R. Almirante Blanco Encalada, Tocopilla
3)	CD120	1200	10/5	R. Agricultura, Los Angeles: 1000-0400
66)	CB121	1210	5	R. Chilena AM, Valparaíso: 0930-0600
67)	CC121	1210	1	R. Universidad de Talca, Talca: 1100-0400
170)	CC121	1210	5	R. Armonía, Chillán
161)	CD121	1210	10	R. Nueva Galicia, Puerto Montt
162)	CD122	1220	10	R. Minería, Temuco
68)	CB124	1240	25	R. Universidad de Santiago, Santiago: 1100-0400
163)	CA124	1240	0.25	R. Principal Chuquicamata, Calama: 0950-0500
69)	CA124	1240	1	R. del Salitre, Iquique
70)	CA125	1250	1	R. Minería, La Serena
71)	CC125	1250	10	R. Universidad Austral de Chile, Valdivia: 1100-0400
72)	CA126	*1260	10	R. Nal., Arica
73)	CC126	1260	1	R. Condell, Curicó: 1100-0500
57)	CD126	1260	10	R. Minería, Punta Arenas: 1000-0400
74)	CB127	v1270	10	R. Festival, Viña del Mar: 1000-0700
75)	CD127	1270	10	R. Agricultura, Temuco: 1000-0400
151)	CC128	1280	1	R. Arturo Prat Chacón AM, San Carlos: 1050-0405
150)	CA129	1290	0.25	R. Coya, María Elena
164)	CD128	1280	10	R. del Sur, Osorno
165)	CD129	1290	1	R. Mulchen, Mulchen
76)	CB130	1300	5	R. Tierra, Santiago: 1200-0300
77)	CD130	1300	10	R. Pilmaiquén, Valdivia: 1000-0430
78)	CD130	1300	1	R. Cabo de Hornos, Pto. Williams: 1100-0500
79)	CA132	1320	1	R. Estrella del Norte, Vallenar
80)	CD132	1320	1	R. Lincoyan, Mulchén: 1100-0300
46)	CB133	v1330	3	R. Metropolitana, Santiago
83)	CD133	1330	3/1.5	R. Vicente Pérez Rosales, Puerto Montt: 1055-0401
84)	CA134	*1340	0.25	R. El Toco, Tocopilla
85)	CB134	1340	10	R. Caracola, Valparaíso: 0930-0300
86)	CC134	v1340	1	R. Doña Isabel Riquelme, Chillán: 1000-0430
87)	CD134	1340	1	R. Panguipulli, Panguipulli: 1200-0100
89)	CD135	1350	1	R. San Carlos, Ancud: 1130-0400
90)	CA135	1350	1	R. Riquelme, Coquimbo: 1030-0430
92)	CC136	1360	5	R. Universidad del Bío Bío, "UBB", Concepción: 1040-0300
153)	CD137	1370	1	R. Armonía, Temuco: 1100-0400 (Sat 0800)
93)	CB138	1380	50	R. Corporación, Santiago: 24h
154)	CD140	1400	5	R. La Amistad, Los Angeles: 1030-0400
94)	CA140	v1400	1	R. 140, "La Nueva Tarapacá", Iquique: 1045-0600
166)	CD140	1400	5	R. Belén, Puerto Montt
96)	CB141	1410	2.5	R. Color, "Canal 141", Valparaíso
97)	CD141	1410	1	R. Loncoche, Loncoche: 1100-0330
82)	CB142	v1420	1	R. Panamericana, Santiago: 1100-0400
100)	CC142	1420	1	R. Nueva Maule, Cauquenes: 1050-0430
101)	CC143	1430	1	R. Chilena AM, Rancagua: 1030-0500
102)	CA144	1440	1	R. Minería, Arica: 1000-0430
103)	CA144	1440	1	R. Agricultura, La Serena: 1000-0400
104)	CC144	1440	1	R. El Sembrador, Chillán: 1000-0430
105)	CB145	1450	1	R. Universidad Técnica "Federico Santa María", Valparaíso: 1100-0300
106)	CC145	1450	5	R. Libertad, Curicó: 1000-0400
108)	CD145	v1450	1	R. Minería, Puerto Varas: 24h
109)	CA146	1460	10	R. Antofagasta, Antofagasta: 1130-0300
110)	CC146	1460	1	R. Yungay, Santiago: 1100-0400
111)	CC146	1460	1	Nueva R. Talcahuano, Talcahuano: 18h
112)	CB147	1470	1	R. Sargento Aldea, San Antonio: 1000-0500
113)	CA148	1480	1/0.25	R. Amanecer, Ovalle: 1100-0400
114)	CC148	1480	1	R. La Amistad AM, Tomé: 1100-0230
115)	CD148	1480	1	R. Portales, Valdivia
116)	CA149	1490	5	R. Alicanto, Salvador: 1000-0400
117)	CB149	1490	1	R. El Canelo de Nos AM, San Bernardo: 1100-0400
88)	CD149	1490	5	R. Malleco, Victoria: 0300
118)	CA150	1500	1	R. Minería, Iquique: 1100-0500
119)	CC150	1500	1	R. Centenario, San Javier: 1100-0300
120)	CD150	1500	1	R. Tierra del Fuego, Puerto Porvenir: 1100-0400
136)	CB150	1500	1	R. Trasandina, Los Andes: 1100-0400
149)	CA151	1510	1/0.5	R. Luís Alvarez Sierra. Illapel: 1100-0400
121)	CC151	1510	1	R. Rancagua, Rancagua: 1100-0400
122)	CD151	1510	1	R. Teniente Merino, Lebu: 1130-0300
123)	CC152	1520	1	R. Nueva Soberanía, Linares: 1030-0430
124)	CD152	1520	0.1	R. Aníbal Pinto, Lautaro
155)	CB152	v1520	1	R. Integración, San Antonio: 0900-0500
125)	CA153	*1530	1	R. Juan Godoy, Copiapó
156)	CB153	1530	1	R. Nexo, Quillota: 1100-0600
126)	CC153	1530	1	R. Corporación, Lota
167)	CD153	1530	1	R. Calbuco, Calbuco
128)	CB154	1540	1	R. Sudamérica, Santiago: 1200-0030
157)	CC154	1540	1	R. Central, Chillán
129)	CC154	1540	1	R. San José de Alcudia, Río Bueno: 0955-0300
130)	CB155	1550	1	R. Putaendo, Putaendo
131)	CC155	v1550	1	R. Manuel Rodríguez, San Fernando (r 1555): 1100-0400
132)	CD155	1550	0.25	R. Regional, Traiguén
133)	CA156	1560	5/3	R. Parinacota, Putre: 0955-0430 (r. 1565)
134)	CB156	1560	1	R. Manantial, Talagante: 1100-0400
135)	CD156	1560	1	R. Parque Nacional, Villarrica: 1100-0400
137)	CC157	v1570	1	R. Nueva O'Higgins, Rancagua: 24h
159)	CC157	1570	1	R. Campanario, Talca: 24h.
138)	CD157	1570	0.25	R. Acuarela, Nueva Imperial: 1155-1810, 2200-0200
140)	CC158	v1580	1	R. Colchagua, Santa Cruz
141)	CD158A	1580	0.25	R. Millaray, Cañete
142)	CD158B	1580	1	R. Nueva Continental, Collipulli: 1100-0430
143)	CB159	v1590	1	R. Aconcagua, San Felipe: 1100-0430
144)	CC159	v1590	0.25	R. Rengo, Rengo: 1000-0400
145)	CC159	1590	0.1	R. El Roble, Parral (ff 990)
168)	CD159	1590	5	Em. Tepual, Llanquihue
146)	CB160A	1600	0.25	R. Armonía, Puente Alto
147)	CB160B	1600	0.25	R. Reloj, Viña del Mar
148)	CC160A	v1600	0.25	R. Regional, Concepción: 1000-0430
171)	CD160	1600	0.25	Millalebu-La Regalona AM, Temuco

Shortwaves: * = inactive

	Call	kHz	kW	Name, location and h. of tr.
160)		5825	0.05	R. Triunfal Evangélica, Talagante: 2100-2400 exc. Thurs & Sun
27)	CE603	6029.7	10	R. Santa María, Coyhaique: 0900 (Sun 1000)-0300 (n. 6030)
172)	CE607	6070	100	Voz Cristiana, Santiago: 0900-1200, 2000-0400
43)	CE608	6080	1	R. Patagonia Chilena,Coyhaique: 0930 (Sun 1000)-2400
158)	CE609	6090	10	R. Esperanza, Temuco: 24h
172)	CE936	9360	100	Voz Cristiana, Santiago: 2100-0600
3)	CE963	*9630	10	R. Agricultura, Santiago:
172)	CE963	9635	100	Voz Cristiana, Santiago: 1200-2000
172)	CE1169	11690	100	Voz Cristiana, Santiago: 0100-0600
172)	CE1189	11890	100	Voz Cristiana, Santiago: 0800-1100
172)	CE1537	15375	100	Voz Cristiana, Santiago: 1100-1400
172)	CE1768	17680	100	Voz Cristiana, Santiago: 1400-2100

Call	kHz	kW	Name, location and h. of tr.
172) CE215021500		100	Voz Cristiana, Santiago: 1100-2100
172) CE215521550		100	Voz Cristiana, Santiago: 1300-0100

Addresses and other information

E-mail & Web: http://www.chilnet.cl/rubros/RADIOE01.HTM.
A complete list of Chilean radio stations is maintained by Chilnet as part of the Chilean yellow pages, with links to the Web pages of individual stations and a *free* Internet Fax service to each station.

1) Cas. 110, Melipilla — DG: José Luís Massoud Pozo. **FM:** 104.5 "Caricia FM".

2) Chacabuco 210, piso 3, Validivia. ☎ +56 (63) 225087. 🖷 +56 (63) 213549 — DG: Waldemar Ritter Astudillo.

2x) O'Higgins 297, (🖃 Cas. 268), Angol — ☎ / 🖷 +56 (45) 712331, 714706 — Dir: Luis Humberto Robles.

3) CB57: Av. Manuel Rodriguez 15, Santiago. ☎ +56 (2) 6722749 — Gte: Ricardo Hoffmann L.; CD120: Cas. 327, Los Angeles — Gte: Jorge Rodríguez Vergaray; CB98: Cas. 90, Valparaíso — Gte: Carlos Prieto.

4) Washington 2562, Depto. 204 (🖃 Cas. 1060), Antofagasta — Gte: Rodolfo Pizarro N.

5) Cas. 486, Concepción — DG: Hugo Díaz Uribe.

6) Cas. 97-D, Punta Arenas — DG: P. Jorge Murillo.

6x) Av. Condell 910, Santiago. ☎ +56 (2) 2224500. 🖷 +56 (2) 2223093 — Gte: Claudio Jaque.

7) Cas. 355, Ovalle — Mgr: Francisco Morales C.

8) O'Higgins 680, Concepción. ☎ +56 (41) 225660. 🖷 +56 (41) 236742. **E-mail:** rbiobio@reuna.cl **Web:** http://www.reuna.cl/rbiobio/ — Gte: Nibaldo Mosciatti M.

9) Cas. 3218, Valparaíso. ☎ +56 (32) 251041. 🖷 +56 (32) 257557 — Gte: Luís Garcia Tapia.

10) Portales 775, Temuco. ☎ +56 (45) 210404. 🖷 +56 (45) 235005 — Gte: Hernán Espina Rojas. **FM:** 93.5 "Temuco Rock & Pop FM".

11) Phillips 40, 2þ piso, (🖃 Cas. 10277), Santiago. ☎ +56 (2) 6324202. 🖷 +56 (2) 6336052 — Gte: Juan Carrasco H.

12) Calle Domeyko esquina Los Pimientos, Chuquicamata — DG: Hector Basualto Bustamante. **FM:** 104.7 "Aurora".

13) Paicavi 119, 2þ piso, (Paza Peru) (or Cas. 2337), Concepción. ☎ +56 (41) 223207, 242563. 🖷 +56 (41) 234697 — DG: Olga Alicia Sanhueza.

14) Cas. 10195, Santiago — Mgr: Waldo Mora Longa.

15) Ramírez 207, or Cas. 260, Ancud. ☎ +56 (65) 622905. 🖷 +56 (65) 622722 — DG: Carlos González S.

15x) Vallejos 650, Copiapó — ☎ +56 (52) 214133 — DG: Miguel Vejar Olivares.

16) CB76: Antonio Bellet 223, Providencia, Santiago. ☎🖷 +56 (2) 2360066 — DG: Sergio Parra G; CB73: Lira 543, Valparaíso. ☎🖷 +56 (32) 213342. 🖷 +56 (32) 255027 — Mgr: Carlos Spahie Nasser. **FM:** 93.1 "Viña del Mar Rock & Pop FM".

17) Cas. 51, Los Angeles. ☎ +56 (43) 311015 — Dir: Julio Cerda C.

18) Sargento Aldea 586, Puerto Aysén — Dir: Sonia Ruiz Fernandez.

19) Mackenna 904 entrepisos (or Cas. 35-0), Osorno. ☎ +56 (64) 232160. 🖷 +56 (64) 233881 — DG: Hernan Landmann Commentz.

20) 5 Norte 168, Viña del Mar. ☎ +56 (32) 971201 — DG: Fernando Damke Marín.

21) Los Carrera 521, 3º piso, La Serena — ☎🖷 +56 (51) 211987 — DG: Anita Maria Ketterer Vorphal. **FM:** 98.5 "Intima".

22) Anibal Pinto 250 (🖃 Cas. 268), Talcahuano — Dir: Tamara Deij S.

23) Cas. 312, La Unión. ☎ +56 (63) 322322 — Mgr: César Figueroa F.

24) Puerto Aysén.

25) Of. SOQUIMICH, Pedro de Valdivia.

26) Cas. 89-V, Valparaíso — Mgr: Leandro Pons P. **FM:** 98.9 "Carolina".

27) Bilbao 681, Coyhaique. ☎ +56 (67) 232398. 🖷 +56 (67) 231306 — DG: Bruno Fredonzani.

28) Caupolican 597, of. 31, Valdivia — Dir: Rubén Rivas A.

29) Philips 56, 6þ piso (or Cas. 56042), Santiago. ☎ +56 (2) 6396825, 6394774 🖷 +56 (2) 6396826 — DG: Patricio Malatesta Garcia.

30) Ap. 6, Pozo Almonte, Iquique — Mgr: Jean Marie Tremblay.

31) Calle Barros Arana 871, 5º piso, (🖃 Cas. 3115), Concepción. ☎ +56 (41) 224704, 245705 — Dir: Manuel J. Fonseca A.

32) I. José Nogueira 1188, Punta Arenas — DG: Sofia Mansilla Alvarado.

33) Cas. 240, Copiapó. ☎ +56 (52) 212650. 🖷 +56 (52) 211245 — DG: Hugo Latorre P. **FM:** 96.5.

34) 5 de Abril 655 (🖃 Cas. 267), Chillán. ☎ +56 (42) 215530 — DG: Maria Eugenia Vaccaro C.

35) Cochrane 746 (🖃 Cas. 5-0), Osorno. ☎ +56 (64) 233366 — DG: W. van den Berg. **Mapuche:** 20h per week. **F.PI:** 10kW.

35x) Gral. Cruz 551 (or Cas. 1499), Temuco. ☎ +56 (45) 212707 — DG: Ricardo Rondón Ortega.

36) Cas. 1298, Antofagasta — Dir: German Aguirre T.

37) Estado 115, Santiago — Gte: Jorge Carrasco T.

38) Illapel 60 (🖃 Cas. 67), Puerto Montt — Gte: René Salinas P.

39) Arzobispo Vicuña 15 Dto. 111, Santiago. ☎ +56 (65) 252234. 🖷 +56

---

65) 256523 — Gte: Luis Fernando Pizarro R.

40) Bories 871, Punta Arenas. ☎ +56 (61) 248893 — Gte: René Venegas O. **FM:** 105.7 "Finísima".

41) Granaderos 1690, Calama — DG: Juan Luis Mauras N. **FM:** 104.7 "Aurora FM".

42) Cas. 214, Talca. ☎ +56 (71) 220061. 🖷 +56 (71) 231344 — DG: Pedro V. Letelier Donoso.

43) Simón Bolívar 26, Coyhaique. ☎ 56 (67) 232240. 🖷 +56 (67) 233287 — DG: Luis Arturo Ojeda. **FM:** 99.3.

44) Arauco 363 3º piso, Valdivia — Dir: Augusto E. Olave Pávez.

45) Cas. 6-D, Arica. ☎ +56 (58) 222380. 🖷 +56 (58) 222278 — DG: Ricardo Pizarro Cortes. **FM:** 95.9.

46) Av. Libertador Bernardo O'Higgins 2248, 3º piso, Santiago — Gte: Uros Domic Besic.

47) Arturo Prat 215 (🖃 Cas. 640), Temuco. ☎🖷 +56 (45) 212039 — Dir: Alfonso Ruiz Fernandez. **FM:** 97.9 "Aurora".

48) Av. Errazuriz 2120, Valparaíso — DG: Jorge Gonzalez Mancilla. **FM:** 97.3.

49) Aníbal Pinto 215, of 801, Concepción — Dir: Héctor Alarcón M. **FM:** 90.1 "Galaxia", 106.5 "Aurora".

50) O' Higgins 486 (🖃 Cas. 106), Castro. ☎ +56 (65) 532260 — DG: Arcadio Velásquez Carasco. **FM:** 90.1 "Martin Ruiz de Gamboa".

51) Calle H. Eberhardt 229, Puerto Natales — Mgr: Jorge Antonio Silva.

52) Cas. 136, Curacautín, Malleco — Gte: Marina E. Sánchez T.

53) Miguel Claro 161, Providencia, Santiago. ☎ +56 (2) 2358202 — DG: Jaime Herrera R.

54) Cas. 34, Vicuña — DG: Guillermo Olivares.

55) Lautaro 124 (🖃 Cas. 211), Angol — DG: Sergio Jimenez Rojas.

56) Uno Poniente 1239 (🖃 Cas. 516), Talca — Dir: Manuel Yáñez G. **FM:** 98.9 "Aurora".

57) Faguano 548 A, Punta Arenas — Gte: Pablo Cruz Nocetti.

58) Cas. 410, Antofagasta — Gte: Carlos Brown P.

59) Uno Oriente 938, (🖃 Cas. 596), Viña del Mar — Gte: Uros Domic B.

60) Claro Solar 536, Temuco — Gte: Eduardo Díaz. **FM:** 95.9 "La Araucana".

61) Cas. 244-V, Santiago — Dir: Eduardo Avila Lizana.

62) Cas. 7-D, Linares — Gte: Herbert Krebs Rosenberg.

63) Errazuriz 675, 2º piso, Punta Arenas. ☎ +56 (61) 223210 — DG: Hugo Enrique Guala H.

64) Cas. 2001, Tocopilla — Gte: Mario V. Reyes R.

65) Cas. 923, Osorno — Gte: G. Kittsteiner R.

66) Av. España 2000, Valparaíso — Gte: María Cecilia Venegas.

67) 2 Norte 685, Talca — Dir: Sergio Eugenio Torres P. **FM:** 102.1.

68) Cas. 442, Correo 2, Santiago — Dir: Erico Carrasco.

69) Cas. 290, Iquique. ☎ +56 (57) 411801 — Gte: Elsa Díaz Benavides. **FM:** 97.7.

70) Matta 591, La Serena. — Gte: Alejandro Pino Uribe.

71) General Lagos 2067 (or Cas. 567), Valdivia. ☎ +56 (63) 221843 — DG: Marcelo Capel. **FM:** 90.1.

71x) J. Nogueira 1277, Punta Arenas

72) Baquedano 575, (🖃 Cas. 49-D), Arica — DG: Federico Soto Figari.

73) Cas. 492, Curicó — Gte: Juana Cervera T. **FM:** 89.9 "Futura".

74) Cas. 337, Viña del Mar — Gte: Santiago Chiesa H.

75) Lynch 646, Temuco. ☎ +56 (45) 213854 — Dir: Sergio Jiménez R. **FM:** 105.7 "San Cristóbal".

76) Purísima 251, Recoleta, Santiago. ☎ +56 (2) 7779803.🖷 +56 (2) 7773840 — DG: Maria Eugenia Meza.

77) Cas. 597, Valdivia — Mgr: Luis Papic D.

78) Calle Cabo de Hornos, Puerto Williams. ☎🖷 +56 (61) 621122 — DG: Juan Ternicien Novoa. **FM:** 98.5.

79) Cas. 13, Vallenar — Gte: Gretta Pizarro B.

80) Gana 360, Mulchén — Mgr: Pedro Amigo A.

81) Thompson 255 (🖃 Cas. 174), Castro — DG: José Sergio Barria Perez.

82) Gran Av. José Miguel Carrera 5848, 4º piso, Santiago. 🖷 +56 (2) 5216302 — DG: Julio Vizcarra Sanchez.

83) Cas. 166, Puerto Montt — Gte: Augusto van der Stelt.

84) Cas. 2098, Tocopilla.

85) Plaza de la Justicia 45, Of. 702, Valparaíso — DG: Omar Garate Gamboa.

86) 18 de Septiembre 721, Chillán. ☎ +56 (42) 215518. 🖷 +56 (42) 213578 — Gte: Daniel Sepúlveda. **FM:** 94.7.

87) Bernard O'Higgins 572, Panguipulli. — Gte: J. Figueroa.

88) Cas. 257, Victoria — Gte: Roberto Muñoz Barra.

89) Cas. 283, Ancud, Isla de Chiloé — Gte: Ricardo Wagner.

90) Aldunate 1619, Coquimbo. ☎ +56 (2) 312948, 321051. 🖷 +56 (2) 327367 — Dir: Jorge Videla Aarancibia.

92) Cas. 5-C, Concepción — Dir: Alejandro Salazar.

93) Av. Ejército 136 (or Cas. 13962), Santiago. ☎ +56 (2) 6994668, 6954809. 🖷 +56 (2) 6980664 — DG: Juan Chaparro Soto.

94) Cas. 614, Iquique. ☎ +56 (57) 426606 — DG: Hector Vivero A.

95) Barros Arana 340, 2°- piso, Concepción. ☎ +56 (41) 234975. 🖹 +56 (41) 237920 — Dir: Hugo Diaz Uribe.
96) Pasaje Ross 149 Dpto. 1201, Valparaíso — DG: Leopoldo Moreno R.
97) Ignacio Serrano 264 (🖃 Cas. 61), Loncoche — DG: Osvaldo Núñez H. **FM:** 105.9.
98) Baquedaño 230, Puerto Natales — Own: Patricio E. Pavlovic Morrison. Gte: Ivan Lara P.
99) Cas. 56-D, Temuco. ☎ +56 (45) 253258. 🖹 +56 (45) 323657 — DG: Roberto Jara B. **Mapuche:** 7h daily.
100) Errazuriz 221 (or Cas. 196), Cauquenes. ☎ +56 (73) 511987 — DG: Pedro Amigo Anriquez. FM: 101.9 "Dinastia".
101) Cáceres 62-A, Rancagua. ☎ +56 (72) 236602. 🖹 +56 (72) 230796 — Dir: Patricio Tapia Urrutia.
102) General Lagos 678 (🖃 Cas. 225), Arica — DG: Juan Carlos Hernandez.
103) Cas. 536, La Serena — Gte: Rodrigo Jara.
104) Cas. 336, Chillán — Gte: José Fco. Arrau U.
105) Av. España 1680, Valparaíso — Gte: David Dahma B. **FM:** 99.7.
106) Yungay 737, Curicó — Gte: Nelson Tarud A. **FM:** 101.1MHz (rel. AM), 92.7 "Opus".
107) Manuel Rodríguez 741, Osorno — DG: Gian Piero Mosciatti P.
108) San Francisco 248, Of. 2, Pto. Varas — DG: Raul Palma Vera.
109) Cas. 248, Antofagasta — DG: Miguel Vejar Olivares.
110) José Domingo Cañas 1061, Santiago — Gte: Abel Velázquez A.
111) Aníbal Pinto 399, Talcahuano — Gte: Rubén Sáez S. (Evangelical).
112) Av. Barros Luco 1678 (🖃 Cas. 68, Correo 2), San Antonio. ☎🖹 +56 (35) 211321 — Gte: G. González Vera. **FM:** 90.9 "Cristalina".
113) Libertad 786 (🖃 Cas. 34), Ovalle — Gte: Sergio Peralta P.
114) Cas. 148, Tomé. ☎ +56 (41) 650657 — DG: Humberto Mena Pinto.
115) Cas. 758, Valdivia — Mgr: Carlos Cockbaine Monsalve.
116) Av. El Tofo 535, El Salvador — DG: Carlos Silva Jimenez. **English:** 2100-2300.
117) Av. Portales 3020 (🖃 Cas. 380), San Bernardo. ☎ +56 (2) 8571943. 🖹 +56 (2) 8571160 — DG: Luís Gallegos.
118) Esmeralda 594 (🖃 Cas. 290), Iquique. ☎ +56 (57) 422693 — Dir: Juan Carlos Hernández.
119) Cas. 18-D, San Javier — Gte: Mario Gaete L.
120) Bulnes 449,Puerto Porvenir — DG: Juan Nesbet Garcia.
121) Cas. 8-D, Rancagua — Gte: Jorge Romero.
122) Cas. 76, Lebu — Gte: Abel Peña Peña.
123) Cas. 11-D, Linares — Gte: Alfonso Astete B.
124) O'Higgins 828, 2° piso, Of. 5 (🖃 Cas. 15), Lautaro — DG: Elena Bravo Sepulveda.
125) Colipí 371, Copiapó. ☎ +56 (52) 212031 — Gte: Orlando Rojas.
126) Cas. 66, Lota — Mgr: Nibaldo Mosciatti M.
127) Santiago — DG: Colonel Nelson Flores.
128) Cas. 1346, Santiago — Gte: Manuel Casabianca L.
129) Pedro Lagos 295, Río Bueno — DG: Ana Maria Cañas Sandoval.
130) Cas. 75, Putaendo — Gte: Francisco Arancibia Cadiz.
131) Chacabuco Esq. España, San Fernando. ☎🖹 +56 (72) 714267 — Mgr: Flor Quiroz S.
132) General Lagos 662 (🖃 Cas. 186), Traiguén — Gte: Rosa Valenzuela V. **FM:** 101.3 "Granero".
133) Calle José Miguel Carrera s/n, Putre (🖃 Cas 82, Arica). ☎ +56 (58) 222521 — DG: Tomislav E. Simunovic G.
134) Cas. 223, Talagante — Gte: Francisco Leyton F.
135) Pedro Montt 478 (🖃 Cas. 110), Villarrica — DG: Max Wenger M.
136) Cas. 307, Los Andes. ☎🖹 +56 (34) 514343 — Gte: Carlos Grez T.
137) Cas. 309, Rancagua — Gte: Sergio Díaz C. **FM:** 103.5 "Bonita".
138) Cas. 18, Nueva Imperial — Gte: Luis E. Rodríguez C.
139) Enrique Alcalde 1081, Talagante. ☎ +56 (2) 8151666 - DG: Moises Leyán Carcur. **FM:** 103.9 "Contacto".
140) Cas. 170, Santa Cruz — Gte: Jorge Villalón G. **FM:** 105.5 "Ensueño".
141) Arturo Prat 399, Cañete — Gte: Lilian Vallejos D.
142) Alcázar 1138, 2° piso, Collipulli. ☎ +56 (45) 811623 — DG: Jorge Vazquez Inostroza.
143) Cas. 100, San Felipe — Gte: Carlos Grez Tellez. **FM:** 91.7 "Colunquén FM".
144) Cas. 147, Rengo—Gte: Arturo Vergara.
145) Cas. 37, Parral — Gte: Juan Rogazy C.
146) Arturo Prat 53, Puente Alto (Santiago) — Gte: Hernán Claro.
147) Cas. 972, Viña del Mar — Gte: Salvador Cortés P.
148) Cas. 2311, Concepción — DG: Juan Antonio Ríos C.
149) Independencia 175, Illapel. 🖹 +56 (53) 522726 — DG: Sergio A. Guzmán Leiva. **FM:** 100.9.
150) Of. SOQUIMICH, María Elena.
151) Cas. 265, San Carlos — DG: Jorge Vaccaro Collao.
152) Maipú 370, Copiapó — Gte: Carlos Porchille.

153) Rudecindo Ortega 691 (or Cas. 12), Temuco. ☎ +56 (45) 220408 — Dir: Jorge Figueroa M.
154) Cas. 541, Los Angeles — Dir: Jaime Sala S.
155) Cas. 33, Llolleo, San Antonio — Gte: Emilio Alcaíno.
156) Cas. 529, Quillota — Gte: Luis Pardo S.
157) Av. Libertad 247 (🖃 Cas. 33), Chillán — Gte: Orlando Villamán.
158) Cas. 830, Temuco. ☎🖹 +56 (45) 213790 — Gte: Eleazar J. Salazar. **English:** 0800-0830. **German:** Sun 1230-1300. **FM:** 106.9.
159) 1 Sur 865, Talca. ☎🖹 +56 (71) 232143 — DG: Maria Elisa Gonzalez C.
160) Villa Monseñor Larraín, Talagante — Dir: Fernando González S.
161) Cas. 808, Puerto Montt — Dir: Franklin Molina Fuenzalida.
162) Antonio Varas 920, Temuco — DG: Jorge Collao Collao.
163) Cas. 127, Calama — Dir: Iris Contreras M.
164) Patricio Lynch 1848, Osorno. ☎ +56 (64) 237440 — DG: Juan Isidro Carrasco.**FM:** 101.5 "La Palabra".
165) Aníbal Pinto 720, Mulchen (Bío Bío).
166) Benavente 385, Puerto Montt. ☎ +56 (65) 258909. 🖹 +56 (65) 258048 — DG: P. Nelson Gonzalez Andrade. **FM:** 92.3 "Aurora".
167) Camino Pte. Ibáñez, Calbuco (Llanquihue).
168) E. Ramirez y Pte. Ibáñez, Llanquihue (Puerto Montt).
169) Montt 1320 (or Cas. 210), Constitución. ☎ / 🖹 +56 (71) 671579 — DG: Alberto E. Garcia Valenzuela.
170) Chillán.
171) Ziem 2676, Temuco. ☎ +56 (45) 220271 — DG: Hugo Gomez.
172) P O Box 2889, Miami, Fl 33144, USA - CE: Andrew Flynn.
173) Av. Ejército 136 (or Cas. 13962), Santiago. ☎ +56 (2) 6968278, 6963502: 🖹 +56 (2) 6957159 - DG: Fernando Ocaranza Y.

FM in Santiago. **Power** 1-10 kW. **Slogans:** Name + "FM".
11) 88.1 Aurora – 61) 88.9 Futuro – 11) 89.7 Duna – 90.5 Pudahuel – 91.3 El Conquistador – 3) 92.1 San Cristóbal – 92.9 La Clave – 93.3 La Cooperativa – 93.7 Universo – 16) 94.1 Rock and Pop – 68) 94.5 – 95.3 Caracol – 76) 95.9 Tiempo – 96.5 Beethoven – 97.1 Clásica – 97.7 Zero – 98.5 Andrés Bello – 29) 99.3 Carolina – 76) 100.1 Infinita – 100.9 La Ciudad – 101.3 Corazon – 101.7 Concierto – 102.5 Univ. de Chile – 103.3 Horizonte – 104.1 Romantica – 104.9 Niña – 105.7 Manquehue – 46) 106.9 Sintonia.

COLOMBIA

LT: UTC-5h – **Pop:** 36.800.000 – **Radios:** 5.400.000 – **Pr.L:** Spanish – **E.C:** 60Hz, 110/120V – **ITU:** CLM.

MINISTERIO DE COMUNICACIONES
Dirección General de Telecomunicaciónes y Servicios Postales

🖃 Edificio Murillo Toro, Santafé de Bogotá, DC. ☎ +57 (1) 286 6911.
LP: Dir.Gral de Telecomunicaciónes: Dr. Vicente Alonso S.

Mediumwaves: Call HJ--, Þ = also on shortwave, * = inactive, (r) = repeater, v = varying fq, SF de Bogotá=Santafé de Bogotá.

The letters preceding the st. number indicate the departamento. Addresses are listed by departamento in alphabetical order.

	Call	kHz	kW	Name and h. of tr.
DC01)	KA	540	30	R. Auténtica, SF de Bogotá: 24h
DC02)	HF	550	50	R. Dif. Nal. , Marinilla (r: 570)
DC02)	GS	560	10	R. Dif. Nal., Tunja (r: 570)
GU01)	PF	560	25	LV de la Pampa "R. 5-60", Maicao: 0900-0300
DC02)	ND	Þ570	100	R. Dif. Nacional de Colombia, SF de Bogotá: 0900-0500
DC02)	HP	*580	50	R. Dif. Nal., Cali (r: 570)
AN01)	CR	590	100	R. Net, Medellín: 24h
AT01)	HJ	600	50	R. Libertad, Barranquilla: 24h
DC03)	KL	610	30	6-10 Bolero RCN, SF de Bogotá: 24h
BO01)	VP	620	10	Colmundo, Cartagena: 24h
VA01)	EL	620	50/20	Colmundo, Cali: 24h
CL01)	FD	630	20	R. Manizales, Manizales: 24h
GN01)	WC	630	10	LV del Guainía, Puerto Inírida
MA01)	BJ	640	10	RCN, Santa Marta: 24h
DC03)	KH	650	30	RCN Antena 2, SF de Bogotá: 24h
NS01)	QS	660	25	Colmundo, Cúcuta: 24h
VA02)	JN	660	10	R. Auténtica, Cali: 24h
AN02)	PL	670	25/10	RCN Antena 2, Medellín: 1100-0500
DC02)	BU	v680	100	R. Dif. Nal., Zambrano (r: 570)

Call		kHz	kW	Name and h. of tr.
DC04)	CZ	690	50	R. Recuerdos, SF de Bogotá: 24h
VA03)	CX	700	120	R. Net, Cali: 24h
AN03)	NX	710	50	R. Super, Medellín: 24h
AT01)	AN	720	30	Emisoras Unidas, Barranquilla
QU01)	VO	720	25	Transm. del Quindío, Armenia
DC05)	CU	Þ730	30	Cadena Melodía, SF de Bogotá: 24h
CE01)	NS	Þ740	50	R. Guatapurí, Valledupar: 0945-0500
NA01)	HB	740	10	Ecos de Pasto, Pasto: 1000-0300
AN01)	DK	750	50	Caracol Colombia, Medellín: 24h
AT02)	AJ	760	30	RCN, Barranquilla: 24h
DC03)	JX	770	50	RCN, SF de Bogotá: 24h
GU02)	ZW	780	30	R. Almirante, Riohacha: 1000-0500
VA04)	ZG	780	30	LV del Valle, Cali: 24h
AN01)	DC	790	10	R. Recuerdos, Medellín: 24h
SS01)	BW	800	10	RCN Santander/La Emisora de Atalaya, Bucaramanga
DC04)	CY	Þ810	250	Caracol Colombia, SF de Bogotá: 24h
BO02)	AD	v820	25	R. Vigía, Cartagena: 1000-0500
VA03)	ED	820	50	Caracol Colombia, Cali: 24h
AN01)	DM	830	10	R. Deportes, Medellín: 24h
HU01)	KK	840	30	H J Doble K, Neiva: 24h
MA02)	BI	840	10	Ondas del Caribe, Santa Marta: 1100-0500
DC23)	KC	850	50	R. Net, SF de Bogotá: 24h
CE02)	NJ	860	50	LV del Cañaguate, Valledupar
VA05)	FP	860	10	Voces de Occidente, Buga: 24h
AT03)	SB	870	25	R. Mar Caribe Internacional, Barranquilla: 0900-0400
TO01)	LA	870	10	LV del Tolima, Ibagué: 24h
SS02)	GE	880	20	Caracol Bucaramanga, Bucaramanga: 24h
AT13)		890		R. Ecos de Soledad, Soledad
DC06)	CE	v890		R. Continental, SF de Bogotá: 24h
MA03)	PM	v890	30	R. Galeón, Santa Marta: 24h
NS02)	DD	900	15/5	R. Super, Cúcuta: 1000-0400
VA04)	EY	v900	30	LV de Cali, Cali: 24h
AN04)	DO	v910	25	LV del Río Grande, Medellín: 24h
IS01)	MY	910	30	RCN San Andrés, San Andrés: 24h
BO03)	AA	v920	10	Emisoras Fuentes, Cartagena: 1000-0420
NA02)	JN	v920	25	Ondas del Mayo, Pasto: 1000-0200
TO02)	SJ	920	15	Colmundo, Ibagué: 24h
DC07)	CS	930	10	LV de Bogotá, SF de Bogotá: 24h
NS03)	TL	940	25	RCN, Cúcuta: 24h
VA04)	GB	v940	10	R. Calima, Cali: 24h
RI01)	FN	950	15	Caracol Colombia, Pereira: 24h
BO08)	HN	960	120	Caracol Colombia, Magangué: 24h
CA01)	VK	Þ970	30	Armonías del Caquetá, Florencia: 1000-0300
DC08)	CI	970	10	R. Super, SF de Bogotá: 24h
GU03)	ME	970	50	RCN Guajira, Maicao: 24h
NS04)	JV	980	30	R. Punto, Cúcuta
VA06)	ES	980	100	RCN, Cali: 24h
AN02)	DB	990	10	RCN, Medellín: 24h
BO04)	AQ	1000	15	RCN, Cartagena: 24h
CC01)		1000	0.8	R. Panamericana, Cajibío: 1300-2300
DC02)	JG	v1000	10	R. Dif. Nal., Manizales (r: 570)
NA03)	BN	1000	10/5	LV del Galeras, Pasto: 24h
AT04)	OP	1010	10	Caracol Barranquilla, Barranquilla: 24h
CO01)	ZD	v1010	15	R. Panzenú, Montería
DC09)	CC	1010	10	R. Deportes, SF de Bogotá
HU02)	JR	v1010	15	Caracol Colombia, Neiva: 24h
SS03)	IX	1010	10	R. Yarima, Barrancabermeja
AN04)	DQ	1020	15	Emisora Claridad, Medellín: 24h
ME01)	KS	Þ1020	30	LV del Llano, Villavicencio: 24h
RI02)	FQ	1020	10	RCN, Pereira: 24h
SS04)	DZ	v1020	15	R. Primavera, Bucaramanga: 0900-0400
TO03)	FT	Þ1020	10	R. Super, Ibagué: 24h
BY01)	DJ	1030	10	LV de los Libertadores, Duitama: 24h
CE03)	RF	1030	15	Ondas del Cesar, Aguachica
CO02)	GX	1030	10	R. Progreso de Córdoba, Lorica: 1000-0400
VA06)	ER	1030	30	RCN Antena 2, Cali: 24h
VP01)		1030	5	Ondas del Vaupés, Mitú
AT05)	AI	1040	15	R. Net, Barranquilla: 24h
CC02)	SY	1040	10	R. Mil 40, Popayán: 0900-0400
DC10)	CJ	Þ1040	40	Colmundo, SF de Bogotá: 24h
NA04)	UB	1040	15	Colmundo, Pasto

Call		kHz	kW	Name and h. of tr.
NS05)	BF	1040	15	LV del Norte, Cúcuta
QU02)	FM	1040	15	LV de Armenia, Armenia: 24h
AN04)	DR	1050	15	R. Sonorama, Medellín: 24h
AR01)	LZ	Þ1050	10	LV del Cinaruco/Caracol, Arauca: 24h
CE04)	BB	1050	15	R. Valledupar: 0900-0500
CO03)	TJ	1050	15	R. Uno, Montería: 1000-0100
ME02)	IO	1050	5	LV de la Conquista, Granada: 24h
TO04)	FZ	1050	10	RCN Antena 2, Espinal: 24h
VA07)	NG	1050	5	R. Palmira, Palmira: 1000-0500
AN05)	MG	1060	1	Caracol Colombia, Turbo
BY02)	MV	1060	10	R. Furatena, Chiquinquirá: 0900-0600
CL02)	FJ	1060	15	RCN Caldas, Manizales: 24h
GU04)	LY	1060	10	R. Delfín, Riohacha: 0900-0300
HU03)	OV	1060	15	R. Cultural Surcolombiana, Neiva: 0955-0530
SS05)	GU	1060	10	R. Bucarica, Bucaramanga
SU01)	YX	1060	1	R. Caracol Colombia, Sincelejo: 24h
AT06)	AH	v1070	20/10	Em. Atlántico, Barranquilla: 1000-0455
CC03)	VR	1070	15	R. Super, Popayán: 24h
DC11)	CG	Þ1070	30	R. Santa Fé, SF de Bogotá: 24h
AN01)	AX	v1080	10	Caracol Medellín, Medellín: 24h
CL03)	JS	1080	15	R. Pontoná, La Dorada: 1000-0500
CO04)	AW	v1080	10	LV de Montería, Montería: 1000-0300
ME03)	KT	Þ1080	15	R. Macarena, Villavicencio: 24h
SS06)	MH	1080	10	Melodía AM, Bucaramanga
VA04)	JF	1080	10	R. Eco, Cali: 24h
BO05)	OM	1090	5	R. Bucanero, Cartagena
BY03)	IH	1090	10	Caracol Colombia, Sogamoso: 24h
CA02)	IG	Þ1090	10	Caracol Colombia, Florencia
CL01)	IA	1090	10	Ondas del Nevado, Manizales: 24h
NS06)	BC	1090	15	R. Net, Cúcuta: 24h
TO05)	JY	1090	1	LV de los Pijaos, Guamo: 1000-0300
AN06)	GQ	1100	10	Transmisora Surandes, Andes: 1100-0500
AT04)	AT	v1100	15	Caracol Colombia, Barranquilla: 24h
CO05)	MK	1100	5	Emisora Ideal, Planeta Rica
DC04)	CN	1100	10	Caracol Bogotá, SF de Bogotá: 24h
HU04)	YZ	1100	15	R. Super, Neiva
SS07)	GI	1100	5/1	Em. José António Galán/LV de Colombia, El Socorro: 0930-0300
VI01)	EF	*1100	2	LV del Vichada, Puerto Carreño
AN07)	DI	1110	8/10	R. Bolivariana, Medellín: 24h
AR02)	GP	Þ1110	5	LV del Río Arauca, Arauca
IS02)	PA	1110	5/1	LV de las Islas, San Andrés: 24h
ME04)	JP	1110	10	RCN, Villavicencio: 24h
SU02)	ZE	1110	15	R. Piragua, Sincelejo
TO03)	NC	1110	1	Ecos del Combeima, Ibagué: 1000-0400
VA03)	EW	1110	10	Caracol Cali, Cali: 24h
BY04)	KQ	1120	10	Caracol Colombia, Tunja
NS01)	TI	1120	10	Colmundo R. 24, Cúcuta
RI03)	JC	1120	5	R. Matecaña, Pereira: 24h
SS02)	GH	1120	15	R. Net, Bucaramanga: 24h
AT07)	AC	1130	10	Em. Riomar, Barranquilla: 0930-0500
BO06)	NN	1130	1	Ondas del Río, Magangué
DC12)	VA	1130	15	R. "K", SF de Bogotá: 24h
NA05)	QQ	1130	15	Caracol Pasto, Pasto
AN02)	DL	1140	15/5	R. Paisa, Medellín: 24h
BO07)	KO	1140	10	LV de la Victoria, Cartagena
CC12)		1140		R. Piendamo, Piendamo
CL04)	FH	1140	10	R. Regional Independiente, Anserma: 0900-0300
CU01)	CL	1140	10	R. Girardot, Girardot
ME05)	RW	Þ1140	10	LV de Centauros/Caracol, Villavicencio: 0900-0500
SS08)	RN	1140	10	RCN/LV de la Provincia, Barbosa: 24h
BY05)	GJ	1150	1	R. Net, Duitama: 24h
CH01)	TE	1150	1	LV del Chocó, Quibdó: 24h
HU05)	FP	1150	10	R. Neiva, Neiva: 24h
NS07)	BT	1150	10	R. Catatumbo, Ocaña
QU03)	FI	1150	15	Caracol Colombia, Armenia: 24h
AT01)	BL	1160	10	R. Aeropuerto, Barranquilla
CA03)	AU	Þ1160	10	Ondas del Orteguaza, Florencia: 1030-0300
CO06)	AZ	*1160	5	LV de Sinú, Montería
DC13)	OC	1160	15	Ecos de Colombia, SF de Bogotá: 24h
NA06)	ZV	1160	5	R. Las Lajas, Ipiales: 1100-0200
NS08)	EC	1160	10	R. San José de Cúcuta, Cúcuta:

	Call	kHz	kW	Name and h. of tr.
RI04)		v1160		Ondas del Puerto, La Virginia
SS09)	E75	1160	5	Voz de la Paz, Barrancabermeja
VA04)	EV	1160	10	La Caleñísima, Cali: 24h
AN04)	KW	1170	10	R. Nutibara, Medellín: 24h
BO08)	NW	1170	10	Caracol Colombia, Cartagena: 24h
BY04)	GA	1170	5	R. Recuerdos, Tunja
CE06)	PB	1170	10	Ondas de Macondo, Valledupar
GV01)	WA	Þ1170	5	LV del Guaviare, San José del Guaviare 1130-0200
ME01)	BX	Þ1170	30	Ondas del Meta, Villavicencio
VA08)	JE	1170	1	RCN, Tuluá: 24h
AN08)		1180		Em. Coorpurabá, Apartadó
CL05)	FX	1180	15	Caracol Colombia, Manizales: 24h
SS10)	GK	1180	20	R. Uno, Bucaramanga: 24h
TO06)	JT	1180	10/5	RCN, Ibagué: 24h
AT05)	CT	1190		LV de la Costa, Barranquilla: 1030-0200
CO07)	KI	1190	1	R. Barají, Sahagún: 1000-0500
DC07)	CV	1190	10	R. Cordillera, SF de Bogotá: 24h
NA07)	KG	Þ1190	10	R. Mira, Tumaco: 1100-0400
VA09)	EO	1190	15	Ondas del Valle, Cartago: 24h
GU05)		v1195		Ondas del Ranchería, Barrancas
AN09)	IJ	1200	15	R. 1200 "LV de la Raza", Medellín: 24h
BO09)	BV	1200	10	R. Net, Cartagena: 24h
BY06)	LR	1200	10	RCN Antena 2, Sogamoso: 1100-0500
CE07)		1200	0.5	R. Tamalemeque, Tamalemeque
CU02)	CD	1200	10	Em. Nueva Época, Fusagasugá: 0900-0400
GU06)	BZ	1200	10	Ondas del Riohacha, Riohacha
VA10)	NF	v1200	10	R. Super, Cali: 24h
HU02)	FR	1210	10	R. Net, Neiva: 24h
NS03)	BE	1210	10	RCN Antena 2, Cúcuta: 24h
RI02)	BQ	1210	10	LV Amiga, Pereira: 24h
AT05)	FF	1220	15	Em. 12-20, Barranquilla: 1030-0200
CO08)	AV	1220	10	RCN, Montería: 24h
DC04)	KR	1220	10	La Vallenata 12-20, SF de Bogotá: 24h
NA08)	NM	v1220	10	Em. Cultural Bolívar, Ipales: 1100-0200
SS11)	MT	1220	10	RCN, San Gil: 24h
AN10)	IL	1230	10	Minuto de Dios, Medellín: 24h
BY04)	BR	1230	10	Caracol Tunja, Tunja: 1100-0500
CU03)	TP	1230	1	R. Colina, Girardot: 18h
GU03)	MJ	1230	5	RCN Antena 2, Maicao
NA15)	FV	1230	1.5	R. Viva, Pasto: 1100-0500
SS12)	EH	1230	15	Colmundo, Bucaramanga
VA06)	LK	1230	10	R. Calidad, Cali: 24h
AN11)	DA	1240	15	R. Auténtica, Medellín: 24h
AR03)	GO	1240	1	R. Caribabare, Saravena
QU04)	FG	1240	10	RCN, Armenia: 24h
SS13)	GN	1240	5	R. Barrancabermeja: 1000-0300
VA11)	JA	Þ1240	3	R. Buenaventura: 1030-0500
AT07)	OK	1250	10	Em. ABC, Barranquilla: 24h
DC14)	CA	1250	10	R. Capital, SF de Bogotá: 24h
NS06)	HS	1250	15	Caracol Cúcuta, Cúcuta: 24h
SU03)	EM	1250	1	LV de Corozal, Corozal: 1100-0500
AM01)	OU	1260	2	Ondas del Amazonas, Leticia: 1000-0500
BY05)	NO	1260	5	R. Recuerdos, Duitama: 1000-0300
CE08)	OH	1260	10/5	RCN Cesar, Valledupar: 24h
IS03)	HU	1260	1	Caracol Colombia, San Andrés
ME06)	LX	1260	5	Eco Llanero Radio, Villavicencio: 24h
NS10)	TM	1260	15/10	R. Sonar, Ocaña: 1030-0300
TO07)	DV	1260	5	Caracol Colombia, Ibagué: 24h
VA03)	ET	1260	5	R. Deportes, Cali: 24h
BO04)	AR	1270	1	RCN Antena 2, Cartagena: 24h
CE05)	KJ	1270	1.5	LV de Curumaní, Curumaní: 1000-0100
CU04)	XQ	1270	1	R. Melodía, Ubaté
PU01)	SV	1270	1	LV de Orito, Orito: 1100-2300
RI05)	IM	v1270	1	Colmundo, Pereira: 0930-0515
SS02)	TX	1270	5	R. Net, Bucaramanga: 24h
AN12)	MB	1280	5	R. Suroeste, Concordia
AT01)	SO	*1280	5	R. Playa Mendoza, Barranquilla
DC07)	KN	1280	5	R. Musical, SF de Bogotá: 24h
GU07)	HO	1280	5	Impacto Popular, San Juan del Cesar
NA05)	LR	1280	5	Caracol Colombia, Pasto: 24h
NS11)	RP	1280	5	Ecos de Tibú, Tibú: 1000-2400
SS14)	NQ	1280	3	LV del Río Suárez, Barbosa: 0730-2330
VA12)	TK	1280	1	R. Super, Caicedonia
AN13)	TH	1290	5	LV de las Estrellas, Medellín: 24h
AR04)	SZ	1290	5	Ecos del Sarare, Saravena
BY07)	HI	1290	5	RCN/LV de Garagoa, Garagoa: 0930-0330
CU05)	KY	v1290	5	RCN, Girardot: 24h
HU06)	CM	1290	10	R. Sur, Pitalito: 0900-0400
MA04)	EB	1290	5	LV del Turismo, Santa Marta
ME07)	NE	1290	15/1	LV del Ariari, Granada: 24h
SU04)	OI	1290	5	R. Chacurí, Sampués: 1100-0300
VA13)	MC	1290	5	Acuario AM, Cali: 24h
BO10)	AP	1300	15	LV de las Antillas, Cartagena: 24h
BY08)	RB	1300	5	R. Super, Tunja
CC04)	TN	1300	5	R. Eucha, Belalcázar: 1000-2400
PU02)	UA	1300	5	R. Sindamanoy, Mocoa: 0900-0300
RI01)	LD	1300	5	Caracol Pereira, Pereira: 24h
SS02)	NB	1300	5	Caracol Bucaramanga, Bucaramanga: 24h
TO08)	EA	1300	5	R. Lumbí, Mariquita
AN14)	LM	1310	5	R. Santa Bárbara: 1000-0500
AN15)	IR	1310	5	RCN Urabá, Apartadó: 24h
AT08)	AK	1310	5	LV de la Patria, Barranquilla: 0900-0500
CO09)	DG	1310	5	Caracol Colombia, Monteria
DC09)	JZ	1310	10	Colorín Color R, SF de Bogotá: 1100-0200
HU07)	WD	1310	5	Micrófono Cívico, Palermo: 0900-0300
NS12)	TQ	1310	5	R. Tasajero, Cúcuta: 1000-0500
AN16)	TA	1320	1	NVI Radio, Medellín: 24h
BY09)	HT	1320	5	R. Guateque, Guateque: 1100-0300
CU06)	NO	1320	1	R. Uno, Girardot: 24h
IS04)	QI	1320	5	R. Leda Internacional, San Andrés: 1100-0500
MA05)	LV	1320	5	La Esquina del Progreso, Fundación
SS15)	MS	1320	5	R. El Sol, Barrancabermeja: 0900-0400
VA14)	NK	1320	1	R. Luna, Palmira
AN17)	RD	1330	1	R. Coopeñol, El Peñol
CE09)	MP	1330	1	LV de Aguachica, Aguachica
CC05)	LS	1330	5	Caracol Colombia, Popayán
RI02)	FE	v1330	10	RCN Antena 2, Pereira: 24h
SS16)	NR	1330	5	Caracol San Gil, San Gil: 0900-0300
AN18)	NP	1340	1	R. Comunal, Nariño: 1100-0100
AT03)	FA	1340	5	R. Alegre, Barranquilla: 24h
DC03)	FB	1340	5	R. Uno, SF de Bogotá: 24h
HU05)	KD	1340	5	RCN, Neiva: 24h
NA10)	HA	1340	5	RCN Nariño, Pasto: 24h
NS04)	PY	1340	5	R. Mil, Cúcuta
NS13)	VL	1340	0.5	Brisas del Catatumbo, Tibú: 1030-1700, 2100-0200
SS05)	NY	1340	1	R. Sonorama, Bucaramanga
SU05)	HY	1340	5	RCN Sucre, Sincelejo: 24h
VA15)	IS	1340	1	R. El Sol, Buenaventura: 24h
AN19)	DS	1350	1	Ondas de la Montaña, Medellín
AN20)	LO	1350	5	R. El Sol, Caucasia: 24h
BY10)	HW	1350	5/1	Em. Ecos del Río, Puerto Boyacá: 0900-0400
CE10)	MU	1350	1	R. Perijá, Codazzi
CE11)		1350		Fascinación AM, Becerril
CE12)		1350	1	R. Cultural 2001, Pailitas
CS01)	LH	Þ1350	2	LV de Yopal, Yopal: 1000-0500
MA01)	OC	1350	5	RCN Antena 2, Santa Marta: 24h
TO09)	HL	v1350	5	R. Festival, Ibagué
VA16)	EN	1350	5	R. Melodía, Cali: 24h
AN21)	PK	1360	10/5	Voz de Abejorral: 0800-0500
AN22)		1360	0.5	R. Segovia, Segovia: 1100-0300
BO08)	UT	1360	5	Caracol Cartagena, Cartagena
BY21)	RX	1360		LV de la Camarca, Miraflores
ME08)	SD	1360	1	R. Morichal, San Martín
RI06)	RA	1360	10	Ecos del Risaralda, Pereira: 24h
SS17)	KV	1360	1	R. Láser, Zapatoca
TO18)	MI	1360	15	R. Autentica, Melgar: 0900-0500
AN23)	NU	1370	1	RCN, Rionegro
AT09)	BO	1370	10	Minuto de Dios, Barranquilla: 24h
CC06)	EQ	1370	5	RCN Cauca, Popayán: 24h
DC01)	KX	1370	5	R. Mundial, SF de Bogotá: 24h
NS15)	BD	1370	5/2	La Nueva R. Guaimaral, Cúcuta: 1000-0500
SU06)	NI	1370	1	Caracol Colombia, Sincelejo: 24h
VA17)	JQ	1370	2.5	RCN Antena 2, Zarzal: 24h
AN24)	JD	1380	2.5	Colmundo Medellín: 24h
BY11)	EE	1380	5	RCN, Tunja: 24h
CE13)	MM	1380	1	R. Net, Valledupar: 24h

	Call	kHz	kW	Name and h. of tr.
CL06)	LG	1380	5	LV de La Dorada, La Dorada: 1000-0500
HU08)	ID	1380	1	Em. Villa del Páez, La Plata: 1000-0100
VA18)	EJ	1380	1	Armonías del Palmar, Palmira: 1130-0300
AN25)		1390	0.1	R. Ciudad de Antioquia, Santa Fé de Antioquia: 1000-2300
CL07)	FO	1390	5/3.5	LV de los Andes, Manizales: 24h
CU07)	YW	1390	15	R. Suramerica, Pacho: 24h
SS18)	ZY	v1390	1	Lemas de Colombia, Bucaramanga: 24h
TO10)	FY	1390	5	R. Avenida, Espinal: 1000-0400
AN26)	CC	1400	5	RCN Antena 2, Santa Bárbara: 24h
AT02)	AS	1400	5	RCN Antena 2, Barranquilla: 24h
CC07)	WY	1400	1	LV de los Samanes: Quilichao: 1100-2400
CC13)		1400	0.45	R. Cañaveral, Morales: 1130-1700, 1900-2200
CH02)	IT	Þ1400	1	Ecos del Atrato, Quibdó 1000-0400
CO10)		1400	0.25	Brisas del Sinú, Tierralta
CO11)	DF	1400	5	LV de Niquel, Montelíbano
DC16)	KM	1400	5	Em. Mariana de Bogotá, SF de Bogotá: 1100-0130
NA11)	JJ	1400	5	R. Ipiales, Ipiales
NA12)		1400	1.5	LV de Samaniego, Samaniego
NS16)	BK	1400	1	LV de la Gran Colombia, Cúcuta: 24h
QU05)	HM	1400	1	RCN Antena 2, Calarcá: 24h
SS19)	D31	1400	1	LV de Cimitarra, Cimitarra: 0900-0300
AN27)	DU	1410	5	Em. Cultural Universidad de Antioquia, Medellín: 1100-0300
SS20)	TY	v1410	1	Caracol Colombia, Vélez
TO11)	FS	1410	5	RCN Antena 2, Honda: 24h
VA19)	EI	1410	10	R. Guadalajara, Buga: 1055-0300
AN28)	D23	1420	1	Ecos de Frontino, Frontino
BO02)	OG	1420	5	LV de las Antillas, Cartagena: 24h
CL05)	HK	1420	5	Caracol Manizales, Manizales: 24h
MA06)	BH	1420	5	Caracol Colombia, Santa Marta: 24h
SS21)	SN	1420	1	R. Lenguerque, Zapatoca
TO06)	LE	1420	10/5	RCN Antena 2, Ibagué
AN29)	CK	1430	1	R. Sensación, Yarumal
AN30)	MF	1430	1	R. Venus, Puerto Berrío
AT10)	PW	1430	5	Colmundo, Barranquilla: 24h
CC08)	EG	1430	5/1.5	LV de Belalcázar, Popayán: 0930-0400
CL08)	IU	1430	1	Armonías del Ingrumá, Riosucio: 1100-0300
DC17)	KU	1430	12	Em. Kennedy, SF de Bogotá: 1055-0300
NS17)	BP	1430	1	R. Cariongo, Pamplona: 0900-0400
PU03)		1430	0.5	R. Preferencial, Sibundoy: 1300-2300
RI07)	RL	1430	15/2	Antena de los Andes, Santa Rosa de Cabal: 24h
SU07)	QX	1430	5	R. Majagual, Sincelejo
AN24)	NZ	1440	5	R. 24, Medellín: 1100-0600
BY06)	GM	1440	5	RCN, Sogamoso
CA04)	IB	1440	1	RCN Caquetá, Florencia: 24h
TO12)	BM	1440	5	R. Internacional, Honda: 1000-0300
VA20)	EK	1440	5	Caracol Tuluá, Tuluá: 1000-0500
CU08)		1445	0.5	Em. R. Unión, La Palma
AN31)	E20	1450	1	R. Ciudad de Urrao, Urrao: 0900-0300
AN32)		1450	0.2	R. LV del Nordeste, Remedios: 1100-2400
BO11)	MX	1450	1	R. Mancomoján, Carmen de Bolívar: 1030-0400
BY12)	TT	1450	10/5	Ondas del Porvenir, Samacá: 0900-0300
CC09)		1450	0.5	LV del Cauca, El Bordo
CL02)	NL	1450	5	RCN Antena 2, Manizales: 1000-0500
SS22)	HH	1450	5	R. Metropolitana, Bucaramanga: 24h
TO13)	VY	1450	5	R. Ciudad de Flandes, Flandes
AN33)	TF	Þ1460	1	Ondas del Darién, Turbo: 1000-0400
AN34)	MN	1460	2	LV de Amalfi, Amalfi: 1000-0300
AN45)	E26	1460	1	R. Capiro, La Ceja
AT02)	VH	1460	5	R. Uno, Barranquilla
DC18)	JW	1460	10	Em. Nuevo Continente, SF de Bogotá: 24h
HU09)	FL	1460	3	Atalaya Agustiniana, San Agustín: W 1100-2400, Sat/Sun 1100-2300
NA10)	ZU	1460	5	RCN Antena 2, Pasto: 24h

	Call	kHz	kW	Name and h. of tr.
NS18)	IW	1460	3	R. Monumental, Cúcuta: 1000-0400
QU06)	JH	1460	10	R. Ciudad Milagro, Armenia: 1000-0500
SU08)	AL	1460	1	R. Sincelejo, Sincelejo: 1000-0400
AN04)	II	1470	10	R. Musical, Medellín: 24h
BO12)	PX	1470	5	Colmundo, Cartagena: 24h
BY13)	B63	1470	1	R. Uno, Iza: 24h
CU09)	HQ	1470	5	R. Futurama, Pacho: 0930-0400
PU04)	IF	1470	.1	R. Tres Fronteras, Puerto Asís: 1100-0200
TO14)	TB	1470	5	Ondas de Ibagué, Ibagué: 1000-0400
VA01)	NT	1470	10/5	R. Total, Cali: 1000-0500
AN35)	TC	1480	1	R. Sonsón, Sonsón (n. 1490)
MA07)	OD	1480	5	R. Rodadero, Santa Marta: 1000-0400
NS14)		1480	0.25	LV del Samán, Bochalema: 1100-0500
RI03)	FC	1480	5	R. Sonorama, Pereira: 1000-0500
SS10)	TZ	1480	5	Antena 2 RCN, Bucaramanga: 24h
TO15)	VB	1480	1	R. Guayabal, Guayabal
AT11)	AY	1490	5	Onda Nueva, Barranquilla: 1100-0500
BY14)	YD	1490	15	R. La Paz Caracol, Paipa: 1000-0400
DC19)	BS	1490	5	Em. Punto Cinco, SF de Bogotá: 1100-2300
HU10)	AG	1490	1	R. Garzón, Garzón: 24h
SU09)	JO	1490	1	LV de San Marcos, San Marcos
VA21)	ZB	1490	5	LV de los Robles, Tuluá: 1100-0500
AN09)	ZH	1500	5	R. 1500 La Vallenata, Medellín: 24h
BY15)	SH	1500	5	Caracol Moniquirá, Moniquirá: 1030-0300
CL09)	UW	1500	5	LV de la Feria, Manizales: 1015-0500
CU10)	TW	1500	5	R. Sumapaz, Fusagasugá: 0900-0300
NA13)		1500	2	R. Cultural "LV del Telembí", Baracoas: 1300-2300
VA22)	LJ	1500	5	R. Mil 500, Cali: 24h
AN37)	D24	1510	5	LV de la Unión, La Unión: 1000-0200
BY16)	A22	1510	1	LV de San Luis, San Luis Gaceno 0900-0300
MA08)	UH	1510	5	LV de Ciénaga Espectacular, Santa Marta
QU07)	ZA	1510	5	R. Estrella, Armenia
SS23)	HX	1510	1	LV Panamericana, Bucaramanga: 1000-0500
TO16)		1510	0.5	LV de los Cedros, Libanó: 1000-0300
AN38)		1520	0.3	Brisas del Palmar, Caucasia: 1130-0400
AN39)	MA	1520	1	LV de Suroeste, Jericó: 1000-0500
AT03)	LQ	v1520	5	R. Minuto, Barranquilla: 24h
CC11)		1520	0.5	R. Cristalares Timbío, Timbío: 1200-2400
CL10)		1520		Sonoradio 1520 AM, Viterbo
DC20)	LI	1520	5	Ecos del Palmar, SF de Bogotá: 1100-0500
NS19)		v1520	1	Em. Una Voz de la Frontera, Puerto Santander: 1000-2200
RI08)	RL	1520	1	R. Calina, Santa Rosa de Cabal
SU10)	MZ	1520	1	Ecos de la Sierra Flor, Sincelejo: 1030-0430
TO17)	AM	1520	1	R. Altamizal, Dolores: 0945-0300
AN24)	DN	1530	5	R. Total, Medellín: 24h
BY17)	GD	1530	5	Em. Reina de Colombia, Chiquinquirá: 0930-0300
CC10)	R73	1530		Ecos del Pacífico, Guapí: 1300-2300
CC14)		1530		R. Integración, Morales
CE14)	OZ	1530	5	LV de la Prov. de Padilla, San Juan del Cesar: 1000-0400
VA23)	JB	1530	1	R. Sevilla, Sevilla
AN40)		1540	0.25	LV Dorada, Segovia
AN41)	B89	1540	1	Brisas del Río Chico, Belmira
CC15)		1540	0.3	R. Valle de Pubenza, Popayán
CL11)	ZF	1540	1	R. Cóndor "Em. Universitaria", Manizales: 1000-0400
NA09)	RQ	1540	2	R. Austral, Túquerres: 0900-0400 (cfr 1300)
SS24)		1540	0.25	R. El Sur, San Vicente de Chucurí
SS25)	HD	1540	1	LV del Petróleo, Barrancabermeja: 1000-0400
TO18)	D89	1540		LV del Nevado Cumbal, Melgar
AN36)		v1550	0.5	Ondas del Nechí, Campamento: 1100-2300
AT02)	CB	1550	5	R. El Sol, Barranquilla: 24h

Call	kHz	kW	Name and h. of tr.
BY18) UN	1550	1	LV del Río Arma, Aguadas: 1000-0300
DC21) ZI	1550	10	MCI Radio 15-50, SF de Bogotá: 24h
QU03) QD	1550	5	Caracol Armenia, Armenia: 24h
VA01) LT	1550	1	R. 24, Cali: 1040-0500
AN42)	1555	0.5	R. Parroquial, El Santuario: 1230-1500, 2200-2400
AN24) XZ	1560	5	Colmundo, Medellín
CE15) PZ	1560	1	R. Codazzi, Codazzi
CU11) CP	1560	5	RCN Antena 2, Arbeláez: 24h
SS26) HE	1560	5	Voces Rovirenses, Málaga: 1000-0100
VA08) LP	1560	5	RCN Antena 2, Tuluá: 1000-0300
AN43) C22	1570	1	R. Ciudad Dabeiba, Dabeiba: 0900-0300
BY19) UJ	1570	1	Armonías Boyacenses, Motavita
CL12) ZT	v1570	1	R. Sensación, Manizales
CU12) TG	1570	1	R. Camino, Machetá
NS09)	1570		LV de Fomeque, Fomeque: 1000-1700, 2200-0100
AT12) QZ	1580	5	FSI Radio "LV de las Estrellas", Barranquilla: 0930-0130
DC22) QT	1580	5	R. María "LV Católica de su hogar", SF de Bogotá: 24h
MA09) LC	1580	1	LV del Banco, El Banco: 1000-0400
NS20) KB	1580	1	R. Zulima, Villa del Rosario
SU11) RM	1580	1	R. Costanera, Sincelejo: 1000-0430
TO19) DE	1580	1	R. Miraflores, Rovira
VA24) NA	1580	5	R. Robledo, Cartago: 0900-0500
AN44) IP	1590	5	Em. Nuevo Continente, Envigado: 24h
CC16)	1590		R. Espacial, Andalucía
CL13) QM	1590	1	Ecos de la Miel, Samaná
CU14)	1590		Ondas del Rioseco, Rioseco
SS27) WB	1590	5	R. Sra del Socorro: 0930-0400
BY20)	1600		R. Fortaleza, Sogamoso
BY23)	1600		R. Bello Horizonte, Pesca (r. 1620-1630)
CL14)	v1600	0.25	LV de Aranzazu: 1100-1300, 1700-1900, 2100-2300
CL15)	v1600	2.5	Armonías de Risaralda, Risaralda: 1030-0300
CL16)	1600		LV de Colina, Risaralda
CU13) HB	1600	5	Armonías Zipaquireñas, Zipaquirá
BY22)	1640		Toquilla Estéreo 1640, Aquitania: Sat 2300-0100, Sun 1000-1300, 2130-0300

Shortwaves:

Call	kHz	kW	Name and h. of tr.
TO03)LW	v4785	5	R. Super, Ibagué: 1000-1400, 2200-0100
CE01) SG	*4815	10	R. Guatapurí, Valledupar
VA11) AM	v4835	1	R. Buenaventura: irr.
AR01) LZ	4865	5	LV del Cinaruco/Caracol, Arauca: 0900-0030
ME01) IE	v4885	5	Ondas del Meta, Villavicencio: 1100-0500
AR02)	4895	10	LV del Río Arauca, Arauca: irr.
VI02)	4905		Ecos del Orinoco, Puerto Carreño
CA01) RI	v4915	3	Armonías del Caquetá, Florencia: 1000-0300
DC02) CQ	4955	20	R. Dif. Nal. de Colombia, SF de Bogotá: 0900-0500
DC11) AE	*4965	2.5	R. Santa Fé, SF de Bogotá
CA03) QA	4975	1	Ondas del Orteguaza, Florencia: 1700-2300
CH02) CP	v5020	1	Ecos del Atrato, Quibdó: 1000-0400
CS01) PV	5040	5	LV de Yopal, Yopal: 1000-0500
DC04) GC	5075	50	Caracol Colombia, SF de Bogotá: 24h
SS28)	v5570	0.1	R. Nueva Vida, Bucaramanga: irr.
ME05) KW	5955	5	LV de los Centauros/Caracol, Villavicencio: 0900-0500
ME03) HZ	5975	5	R. Macarena, Villavicencio: 24h
NA07) OW	6015	1	R. Mira, Tumaco
GV01) OY	6035	5	LV del Guaviare, San José del Guaviare: 1130-0200
DC05) PT	6045	5	Cadena Melodía, SF de Bogotá: irr.
DC10) CR	6065	1	Colmundo, SF de Bogotá: 24h
AN33) TF	*6085	1	Ondas del Darién, Turbo
ME01) IQ	v6115	10	LV del Llano, Villavicencio: 0900-0400
DC04) OR	6150	10	Caracol Colombia, SF de Bogotá: irr.

Call	kHz	kW	Name and h. of tr.
CA02) KF	v6170	2.5	Caracol Colombia, Florencia
DC02) ZO	9655	20	R. Dif. Nal. de Colombia, SF de Bogotá: 1755-0300
DC02) ZO	11785	20	R. Dif. Nal. de Colombia (alt. fq.)

CARACOL (Primera Cadena Radial Colombiana)

A.A. 9291 (Cra. 39A Nº 15-81), Santafé de Bogotá, DC. ☎ +57 (1) 337 8866. 🖷 +57 (1) 337 7126. **Web:** http://www.caracol.com.co/
L.P: Pres: Ricardo Alarcón Gaviria. Vice-Pres: María del Socorro Valencia. CE: Luís Hernán Moreno.
Stations: AM01, AN01, AN05, AN34, AR01, AT04, AT05, B008, B009, BY02, BY03, BY05, BY15, CA02, CC05, CE04, CE13, CH02, CL05, CL06, CO09, CO11, DC04, DC09, DC15, DC23, GU01, HU02, HU08, HU09, IS03, MA06, ME05, NA05, NS06, NS10, QU03, R001, RI07, SS02, SS16, SS20, SU06, TO07, TO09, TO13, TO17, VA03, VA07, VA11, VA19, VA20.

RADIONET (All-news 24h.)

Diagonal 22A Nº 43-77, Santafé de Bogotá, DC. ☎ +57 (1) 268 6700, 269 4123. 🖷 +57 (1) 269 4770.
L.P: Mgr. Yamid Amat. Dir. Tec: Jaime Ruiz.
Stations: AN01, AT05, B009, BY05, DC23, HU02, NS06, SS02, VA03. Radionet belongs to Caracol.

RCN (Radio Cadena Nacional)

A.A. 4984 (Calle 37 Nº 13A-19), Santafé de Bogotá, DC. ☎ +57 (1) 288 2288. 🖷 +57 (1) 288 6130.
L P: Pte: Ricardo Londoño L. Gte Nal: António Pardo García. Dir. Tec: Hernán Medina.
Stations: AN02, AN15, AN20, AN23, AN26, AT02, B004, BY01, BY06, BY11, BY12, BY13, CA04, CC06, CE08, CL02, CL03, CO03, CO08, CU05, CU06, CU11, DC03, GU03, GV01, HU05, HU06, IS01, MA01, ME04, NA10, NS03, NS07, QU04, QU05, RI02, SS01, SS08, SS10, SS11, SS15, SU05, TO04, TO06, TO11, VA06, VA08, VA15, VA17, VA24.

TODELAR (Circuito Todelar de Colombia)

A.A. 27344 (Av. 13 Nº 84-42), Santafé de Bogotá, DC. ☎ +57 (1) 616 1011. 🖷 +57 (1) 616 0056 — **L.P:** Bernardo Tobón de la Roche. Dir. Prgrs: Luís Guillermo Troya. Dir. Tec: Guillermo Escobar B.
Stations: AN04, AN06, AN21, AN44, AT07, B002, BY12, BY16, VA03, CC08, CE02, CE06, CE09, CE10, CE14, CE15, CH01, CL01, CL08, CO06, CU03, DC06, DC07, DC14, HU01, HU03, HU08, MA07, ME02, NA03, NA08, NS12, NS15, NS17, QU02, RI03, SS05, SS21, SS25, SU08, TO08, TO10, TO14, TO16, VA04, VA05, VA14, VA23.

SUPER RADIO

AM Network: A.A. 23316 (Calle 39A Nº 18-12), Santafé de Bogotá, DC. ☎ +57 (1) 287 7777. 🖷 +57 (1) 287 8678.
FM Network: Cra. 16A Nº 86A-78, Santafé de Bogotá, DC.
L.P: Pres: Juan Carlos Pava.
Stations: AN03, B006, BY08, CC03, CL12, CO01, CU01, DC08, GU04, HU04, MA04, ME01, NA15, NS02, QU06, RI04, SS04, SS22, SU10, TO03, VA10, VA12.

COLMUNDO

A.A. 36750 (Diag. 58 Nº 26A-29), Santafé de Bogotá, DC. ☎ +57 (1) 217 8911. 🖷 +57 (1) 217 9358
L P: Pte: Néstor Chamorro. Dir: Fabio Restrepo G.
Stations: AN24, AT10, B001, CO04, DC10, NA04, NS01, QU07, RI05, SS12, TO02, VA01.

CADENA RADIAL AUTENTICA DE COLOMBIA (Rlg.)

A.A. 18350, (Calle 32 Nº 16-12), Santafé de Bogotá, DC. ☎ +57 (1) 285 3360. 🖷 +57 (1) 285 2505 — **L.P:** DG: Jorge E. Herrera H.
Stations: AN11, B010, CU07, DC01, ME03, TO18, VA02.

WV RADIO

Calle 56 Nº 37-34, Santafé de Bogotá, DC. ☎ +57 (1) 222 0001. 🖷 +57 (1) 222 4188 — L. P: Pres: William Vinasco Ch.
Stations: AN09, AT08, B005, DC13, DC22, SS23, VA13.

CREAR (Mil.)

Brigada de Apoyo Logístico, Depto E-5, Cra 50 Nº 18-62, Santafé de Bogotá, DC — **L.P:** My Héctor Beltrán.

State abbreviations: (Departamentos) AM = Amazonas, AN = Antioquia, AR = Arauca, AT = Atlántico, BO = Bolívar, BY = Boyacá, CA

= Caquetá, CC = Cauca, CE = Cesar, CH = Chocó, CL = Caldas, CO =
Córdoba, CS = Casanare, CU = Cundinamarca, DC = Distrito Capital,
GN = Guainía, GU = Guajira, GV = Guaviare, HU = Huila, IS = Islas San
Andrés y Providencia, MA = Magdalena, ME = Meta, NA = Nariño, NS
= Norte de Santander, PU = Putumayo, QU = Quindó, RI = Risaralda,
SS = Santander del Sur, SU = Sucre, TO = Todelar, VA = Valle del Cauca,
VI = Vichada, VP = Vaupés.

N.B: These abbreviations are not officially recognized by the
Colombian Post Office. Letters should therefor carry full name.

Addresses and other information: A.A. = Apartado Aéreo.

AM00) AMAZONAS
AM01) A.A. 236, Leticia – DG: Luis M. Usoga.

AN00) ANTIOQUIA
AN01) Cra. 79A Nº 39-45, Medellín – Gte: Dr. Luis Salazar Gómez.
AN02) A.A. 1244, Medellín – DG: Francisco Plata R.
AN03) A.A. 3399, Medellín – Dep. Mgr: Henry Pava C.
AN04) A.A. 1431, Medellín – DG: Carlos Arturo Hoyos B.
AN05) Cra. 19 Nº 20-66, Turbo – DG: Fabio Patiño G.
AN06) A.A. 1431, Andes – DG: Carlos Arturo Hoyos B.
AN07) A.A. 56006, Medellín – Gte: Humberto Mesa R. **FM:** 92.4.
AN08) Apartadó
AN09) A.A. 4552, Medellín.
AN10) Calle 56 Nº 41-57, Medellín – Gte: Eduardo Villalba.
AN11) Cra. 74 Nº 49-97, Medellín.
AN12) Cra. 3 Calles 2 y 3, Concordia.
AN13) A.A. 4300, Medellín – Gte: Marco F. Eusse.
AN14) A.A. 3854, Medellín – Gte: Ignazio Ramírez B.
AN15) Calle 94 Nº 99-51, Apartadó – Dir: Carlos Vélez.
AN16) A.A. 65103, Medellín – Gte: Victor Castaño Y.
AN17) Centro Cooperativo, El Peñol – Dir: Edgar A. Garcia.
AN18) Plazuela Antonio Nariño, Nariño – Dir: Jorge H. Zuluaga Z.
AN19) Medellín – Gte: Nora Roldán.
AN20) Cra. 2 Nº 21-54, Caucasia – Dir: Hdo. Garcia
AN21) Cra. 49 Nº 49-11, Abejorral – DG: Fernando Gonzales C.
AN22) Segovia – Gte: Miguel Angel Mendoza V.
AN23) Cra. 51 Nº 49-09, Rionegro – Gte: A. Arenas B.
AN24) Calle 48B Nº 79-38, Medellín – Gte: Rogelito Tangarife.
AN25) Casa de la Cultura, Santa Fé de Antioquia – Dir: Nelson M.
Duque.
AN26) Cra. Bolívar, Calle López, Santa Bárbara – Dir: H. Gómez C.
AN27) A.A. 1226, Medellín – DG: Alba Lucía Esnaola. FM: 101. 9
AN28) Frontino – Dir: Hernando Martínez Borja.
AN29) Cra. 20 Nº 20-21, Yarumal – Gte: Alba Cecilia Buitrago.
AN30) Calle 6 Nº 1-23, Puerto Berrio.
AN31) Urrao – Gte: Raul Escobar.
AN32) Remedios.
AN33) A.A. 1289, Medellín – Gte: Aura Luz García.
AN34) Cra. 19 Nº 18-33, Amalfi – DG: Maria Helena Gonzales M.
AN35) Calle 7 Nº 7-54, Sonsón – Gte: J. Cardona H.
AN36) Casa Cural, Campamento.
AN37) Plaza Principal Nº 9-37, La Unión (or: A.A. 4897, Medellín) – Gte:
Nelson López Buitrago.
AN38) Batallón de Infantería Nº 29 "Rifles", Barrio El Palmar, Caucasia.
AN39) Calle 7, Cras. 3 y 4, Jericó – DG: P. Sánchez P.
AN40) Batallón Bomboná, Segovia.
AN41) Parroquia N. S. del Rosari, Belmira.
AN42) Parroquia N. S. de Chiquinquirá, El Santuario.
AN43) Carr. Uribe Nº 12-08, Dabeiba – Own: Manuel T. Bustamente T.
AN44) A.A. 81095, Enviago – DG: Antonio Neira V.
AN45) Alcaldia Municipal, La Ceja.

AR00) ARAUCA
AR01) Calle 19 Nº 19-62, Arauca – Gte: Efrahim Varela N.
AR02) Cra. 22 Nº 18-21. Arauca (or: A.A. 16555, SF de Bogotá) – Dep.
Mgr: Gloria Jeanne Silvia R.
AR03) A.A. 6558, Saravena.
AR04) Grupo de Caballería Rebéiz Pizarro, Saravena.

AT00) ATLANTICO
AT01) A.A. 3143, Barranquilla – Gte: Roberto Esper R. **FM:** 96.9.
AT02) A.A. 1883, Barranquilla – Gte: Rubén Dario García Zuluaga.
AT03) Calle 72 Nº 41C-64, Barranquilla – Dir: Efraín Tejada.
AT04) A.A. 1688, Barranquilla – Gte: Alberto Suárez L.
AT05) Cra. 53 Nº 82-132, Barranquilla – Gte: Janeth Flórez.
AT06) A.A. 51266, Barranquilla – Gte: Miguel Char Abdala.
AT07) A.A. 2010, Barranquilla – DG: Wenceslao Herrera.

AT08) A.A. 231, Barranquilla.
AT09) Calle 53 Nº 50-11, P2, Barranquilla – DG: José Martínez.
AT10) Cra. 44 Nº 70-61, Barranquilla – Dir: Margarita Bermúdez.
AT11) A.A. 2647, Barranquilla – Gte: Jaime Jiménez V.
AT12) A.A. 50906, Barranquilla – Dir: Gabriel Forero Sanmiguel.
AT13) Alcaldia Municipal de Soledad, Soledad.

BO00) BOLIVAR
BO01) Av. Venezuela, Edif. Banco Internacional, La Matuna 8B-05,
Cartagena – Dir: Valerio Mejira.
BO02) Calle 35 Nº 8B-10, Edif. Banco Internacional, Cartagena.
BO03) A.A. 1771, Cartagena.
BO04) A.A. 246, Cartagena – Gte: Luis Lopera.
BO05) Av. Venezuela Nº 9-111, Cartagena.
BO06) A.A. 180, Magangué.
BO07) A.A. 2456, Cartagena – Gte: J. Franco E.
BO08) A.A. 97, Cartagena – Dir: Humberto González Kerquelen.
BO09) Av. 3 Nº 21-62, La Manga, Cartagena.
BO10) Calle Real del Pie de La Popa Nº 20-217, Cartagena – Gte: Juan
Gallo.
BO11) Calle 25 Nº 49-33, Carmen de Bolívar – Dor: E. Sanabria T.
BO12) Av. Venezuela, Edif. Suramericana, Of. 801, Cartagena – Gte:
María F. Gutiérrez R.

BY00) BOYACA
BY01) A.A. 1030, Duitama – Gte: Gloria Guatibonza.
BY02) Cra. 10 Nº 16-36, Chiquinquirá – Adm: José Elías Ballesteros.
BY03) A.A. 282, Sogamoso – Dir: Carlos Lamprea R. **FM:** 88.5 + 107.3.
BY04) Calle 20 Nº 8-54, Tunja – Gte: Miguel Hernando Lozano A.
BY05) Cra. 15 Nº 14-47, Duitama.
BY06) A.A. 019, Sogamoso – Dir: E. Cubillos C.
BY07) Cra. 9 Nº 8-65, Garagoa (or: A.A. 13729, SF de Bogotá) – DG:
Hernando Gámez R.
BY08) Calle 20 Nº 10-64, Tunja – DG: Victor Umaña.
BY09) Cra. 7 Nº 9-57, Guatepeque (or: A.A. 17387, SF de Bogotá) – Gte:
Tito E. Piñeros.
BY10) Cra. 5a Alto Buenos Aires, Puerto Boyacá – Gte: José A.
Velasquez.
BY11) A.A. 1186, Tunja – Dir: Alfonso Escobar F.
BY12) Calle 4, P2, Parque Santander, Samacá – DG: Gonzalo Parra
Pamplona.
BY13) Iza (or: Cra. 7 Nº 17-51, Of. 610, SF de Bogotá).
BY14) Cra. 6 Nº 6-93, Paipa – Dir: Abraham Castelblanco.
BY15) Calle 7 Nº 3-69, Moniquirá – Adm: G. Pardo A.
BY16) Calle 6 Nº 5-42, San Luis Gaceno – DG: Rosalbina Gil de R.
BY17) Calle 18 Nº 12-81, Chiquinquirá – DG: Luís Sastoque. **FM:** 92.6.
BY18) Cra. 3 Nº 7-25, Aguadas – Dir: J. F. Gómez E.
BY19) Plaza Princ. Izquierda C. Cural, Motavita.
BY20) Cra. 10 Nº 1495, Sogamoso.
BY21) Calle 4 Nº 8-66, Miraflores – Own: Silvio Nel Huertas R.
BY22) Calle 10 Nº 11-47, Sogamoso – Own: Pedro Julio Rosas.
BY23) Pesca – Dir: Juan de Jesús Rivera.

CA00) CAQUETA
CA01) Florencia – Dir: Padre Alvaro Serna A.
CA02) A.A. 465, Florencia – Dir: Alonso Orozco.
CA03) A.A. 209, Florencia – Gte: J. D. Santos C.
CA04) A.A. 150, Florencia – Gte: Carlos Moya.

CC00) CAUCA
CC01) Barrio El Porvenir, Cajibío (or: A.A. 945, Popayán) – Dir: Orlando
Castro M.
CC02) A.A. 1321, Popayán – Gte: Marco A. Segura.
CC03) A.A. 1700, Popayán – Dir: Eliécer León Martínez.
CC04) Casa Cural, Parque Principal, Belalcázar-Páez (✉ A.A. 987, SF
de Bogotá) – DG: Jorge Luis Rodriguez.
CC05) Calle 5A Nº 11-25, Popayán – Gte: Carlos Alberto Lenis.
CC06) A.A. 535, Popayán – Dir: Javier Perdomo F.
CC07) Cra. 13 Nº 9-20, Santander de Quilichao – DG: Olmes Llera
Vargas.
CC08) A.A. 759, Popayán – DG: Victoria Eugenia Valencia A.
CC09) Batallón José Hilario López, Bordo.
CC10) Fundación Civica Atarraya, Guapí.
CC11) Calle 15 Cra. 17 Esq. Casa de la Cultura, Timbío (or: Calle 12B
Nº 13B-22, Popayán) – DG: Faber Hernan Astaiza.
CC12) Cra. 4 Nº 9-42, Piendamo – Dir: Angel Ledezma Muñoz.
CC13) Barrio Sagrada Familia, Cra. 3 esq., Morales – DG: Celimo
Calambas Rivera.
CC14) Casa de la Cultura, Morales – Dir: José Luis Hurtado Quintero.
CC15) Cra. 8 Nº 4-55, Popayán – DG: Carlos Tosse.

CC16) Calle 11 Nº 4A-63, Andalucía.

CE00) CESAR
CE01) Calle 17 Nº 15-67, Valledupar – DG: Liliana Baron Amaris.
CE02) Cra. 5 Nº 13-52, Valledupar.
CE03) Calle 7 Nº 16-39, Aguachica – Gte: Alberto Cabrales.
CE04) A.A. 22, Valledupar – Gte: Eduardo Dangond.
CE05) Calle 6 Nº 19-66, Curumaní – DG: Antonio Sanchez Rodríguez.
FM: 95. 7
CE06) Calle 16B Nº 13-74, Valledupar – Gte: Alfonso Campos S.
CE07) Casa de la Cultura, Tamalemeque – DG: Alberto Aguilera P.
CE08) A.A. 250, Valledupar – Gte: Santiago Marmolejo M.
CE09) Cra. 10a Nº 4-24, Aguachica – Gte: Carlos Navarro.
CE10) Cra. 16 Nº 11-102, Codazzi.
CE11) Becerril.
CE12) Pailitas – DG: Antonio Acevedo.
CE13) Cra. 9 Nº 5-02, Valledupar.
CE14) Calle 1 Nº 5-63, San Juan del Cesar.
CE15) Calle 14 Nº 13A-04, Codazzi.

CH00) CHOCO
CH01) A.A. 482, Quibdó – DG: Gildardo F. Rentería
CH02) A.A. 196, Quibdó – Adm: Absalón Palacios Aqualimpia.

CL00) CALDAS
CL01) A.A. 67, Manizales – Gte: Carlos A. Mejía.
CL02) A.A. 244, Manizales – Gte: Alvaro Fonseca C.
CL03) Cra. 2 Nº 13-31, P3, La Dorada – DG: Fabian Ordoñez Medina.
CL04) Cra. 4 Nº 8-58, P3, Anserma – Gte: Jorge Ivan Giraldo Ramirez.
CL05) A.A. 2000, Manizales – DG: Ricardo Osorio G.
CL06) A.A. 34, La Dorada – Gte: Javier Comas Peñaranda.
CL07) Cra. 23 Nº 24-27, Manizales – DG: Amparo Arango de Mejía.
CL08) Cra. 5 Nº 11-102, Av. Los Fundadores, Riosucio – Gte: Omar Cuesta Salazar.
CL09) A.A. 990, Manizales – Gte: Myriam Uribe B.
CL10) Viterbo.
CL11) A.A. 441, Manizales – DG: Lelia María de Toro.
CL12) Cra. 23 Calle 24, Edif. Cuéllar, Of. 505, Manizales.
CL13) C. A. M, Samaná – Dir: Roberto Govez Arenas.
CL14) Parroquia N. Sra. del Rosario, Aranzazu.
CL15) Cra. 2a Nº 13-49, Risaralda – Gte: Uriel Román C.
CL16) Av. Joaquín 1-09, Salida a San José, Risaralda.

CO00) CORDOBA
CO01) Cra. 3 Nº 30-12, Montería – DG: José V. Muskus.
CO02) Calle 1 Nº 16A-03, Lorica – DG: Lucy Alvarez Osorio.
CO03) Calle 23 Nº 1-53, Montería.
CO04) A.A. 497, Montería – Own: Magola Gómez G.
CO05) Cra. 8 Nº 17-56, Planeta Rica – DG: A. Bula C.
CO06) A.A. 148, Montería – DG: A. Gómez D.
CO07) Calle de Comercio, Sahagún – DG: B. E. Nader
CO08) Calle 27 Nº 8-25, Montería – DG: F. Rangel R.
CO09) A.A. 364, Montería – DG: Francisco T. Zubiría R.
CO10) Brigada Nº 11, Tierralta.
CO11) Cra. 5 Nº 14-35, Montelibano – DG: Alberto Nader.

CS00) CASANARE
CS01) Edif. Cine Casanare, P2, Yopal – Gte: Pedro Antonio Socha P.

CU00) CUNDINAMARCA
CU01) Cra. 10 Nº 13-36, Girardot – DG: José C. Cifuentes G.
CU02) Calle 8 Nº 5-08, Fusagasugá – Own: F. Carrera Otero.
CU03) Calle 14 Nº 9-100, Of. 302, Girardot – DG: Germán Rodríguez P.
CU04) Cra. 6 Nº 6-38, Ubaté.
CU05) A.A. 416, Girardot – DG: Carlos A. Barón A.
CU06) Girardot – DG: Carlos A. Barón A.
CU07) Calle 7 Nº 14-66, Pacho – Gte: Dilier Eduardo Valencia.
CU08) La Palma – DG: Luis Eduardo Moreno Salazar.
CU09) Calle 7a Nº 16-44, Pacho – Gte: Luis A. Forero.
CU10) Av. 17 Nº 10-45, Fasagasugá – Gte: Santiago Munévar S.
CU11) Cra. 3 Nº 2-36, Arbeláez (or: Av. 37 Nº 75-84, SF de Bogotá).
CU12) Cra. 7 Nº 8-06, Machetá – Gte: José Sánchez.
CU13) Calle 3 Nº 7-56, Zipaquirá – Gte: Pedro Rozo S.

DC00) DISTRITO CAPITAL
DC01) A.A. 18350, SF de Bogotá – Gte: Jorge E. Herrera H.
DC02) Edif. Inravisión, CAN, Av. El Dorado, SF de Bogotá – Dir: Jimmy García Camargo.
DC03) A.A. 4984, SF de Bogotá. **Web:** http://www.rcn.com.co/ – Dir:

Antonio Pardo García.
DC04) A.A. 9291, SF de Bogotá – DG: R. Deportes: Hernan Peláez. Caracol Colombia: Dario Arizmendi. R. Recuerdos: Fernando Sarmiento V.
DC05) A.A. 19823, SF de Bogotá – DG: Gerardo Paez Mejia.
DC06) Calle 48 Nº 18-77, SF de Bogotá – DG: Germán Tobón M.
DC07) Av. 13 Nº 84-42, SF de Bogotá – Dir: Carlos A. Tobón.
DC08) A.A. 23316, SF de Bogotá – Gte: Henry Pava Camelo.
DC09) A.A. 10556, SF de Bogotá – Gte: Diana Díaz.
DC10) Diagonal 58 Nº 26A-29, SF de Bogotá – DG: María Teresa Gutierrez.
DC11) A.A. 9339, SF de Bogotá – DG: María Luisa Bernal M.
DC12) A.A. 55438, SF de Bogotá – Pres: Hernando Herrera L.
DC13) Calle 56 Nº 37-34, SF de Bogotá – Dir: William Vinasco Ch.
DC14) Cra. 30 Nº 91-84, SF de Bogotá – GD: Germán Tobón Martinez.
DC15) A.A. 9291, SF de Bogotá.
DC16) A.A. 3201, SF de Bogotá – Gte: Luis Alberto Monroy B.
DC17) A.A. Cd. Kennedy 72825, SF de Bogotá – DG: José Vicente Echeverri.
DC18) A.A. 10900, SF de Bogotá – Gte: Antonio Neira V.
DC19) Av. 15 Nº 123-61, Of. 408, SF de Bogotá – DG: Jorge Baron Ortíz.
DC20) A.A. 90883, SF de Bogotá – Dir: José Paipilla P.
DC21) Calle 22C Nº 31-01, SF de Bogotá – Dir: Orlando Castañeda Serrano.
DC22) Cra. 34 Nº 159-78, Barrio Villa Magdala, SF de Bogotá – Mgr: Magdalena Clopastofsky.
DC23) Diagonal 22A Nº 43-77, SF de Bogotá – Mgr: Yamid Amat.

FM in Santafé de Bogotá: 88. 3 Ecos de Centro Oriente – DC08) 88. 9 – 89. 9 HJCK El Mundo en Bogotá – 90. 4 Universidad Distrital – DC03) 90. 9 La Mega – 91. 9 Javeriana – 92. 5 Policía – 92. 9 La Z – 93. 4 CREAR – DC03) 93. 9 Amor – DC03) 94. 9 Rumba – DC02) 95. 9 Nacional – 96. 3 El Mundo Diners – DC05) 96. 9 Melodía – DC04) 97. 9 Tropicana – 98. 5 Universidad Nacional – DC02) 99. 1 Nacional – DC04) 99. 9 Caracol Stereo – DC04) 100. 9 Caracol Colombia (rel. 810 kHz) – DC13) 101. 9 Candela – DC04) 102. 9 Radioactiva – DC07) 103. 9 La X – 104. 9 Acuario – 105. 9 Olímpica – 106. 9 Universidad Jorge Tados Lozano – 107. 5 Big Up – 107. 9 Minuto de Dios.

GN00) GUAINIA
GN01) Calle 6 con Cra. 3, Puerto Inírida – Gte: Luis Fernando Román.

GU00) GUAJIRA
GU01) Cra. 9 Nº 12-29, Maicao – Gte: Carlos Serrano.
GU02) Cra. 8 Nº 3-27, Riohacha – Gte: Roberto Gutiérrez.
GU03) A.A. 125 & 256, Maicao – Gte: William Gómez.
GU04) Carr. a Maicao, Riohacha.
GU05) Barrancas.
GU06) A.A. 3, Riohacha – Gte: José I. Vives.
GU07) Cra. 6 Nº 55-37, San Juan del Cesar.

GV00) GUAVIARE
GV01) Cra. 3 Calle 2, San José del Guaviare – Dir: Luis Fernando Roman R.

HU00) HUILA
HU01) A.A. 727, Neiva – DG: Luís F. Escogar M.
HU02) A.A. 150, Neiva – DG: Josué Gómez Aristizábal.
HU03) A.A. 496, Neiva – DG: Samuel Sánchez C.
HU04) Cra. 6 Nº 6-51, P2, Neiva.
HU05) A.A. 192, Neiva – Gte: Ernesto Cruz D.
HU06) Calle 6 Nº 5-36, P4, Pitalito – DG: Manuel Castro Tovar.
HU07) Cra. 7 Nº 18A-21, Neiva – DG: Alvaro Falla Ramirez.
HU08) Calle 4a Nº 5-59, La Plata – DG: Humberto Orozco P.
HU09) Cra. 14 Nº 2-47, San Agustín – DG: Wilson Martin Ramirez C.
HU10) Cra. 7 Nº 7-05, Garzón – Gte: A. Losada L.

IS00) ISLAS SAN ANDRES Y PROVIDENCIA
IS01) A.A. 354, San Andrés Isla – DG: Humberto Paffen G.
IS02) A.A. 1034, San Andrés Isla – DG: Mauricio Gallardo A.
IS03) Edif. Bermuda, P2, Av. de las Américas, San Andrés Isla.
IS04) A.A. 665, San Andrés Isla – Own: León Cybul.

MA00) MAGDALENA
MA01) A.A. 540, Santa Marta – DG: Javier Orozoco R.
MA02) A.A. 757, Santa Marta – Gte: Roberto Esper R.
MA03) A.A. 103, Santa Marta – Own: Rodrigo Ahumada L.
MA04) A.A. 1529, Santa Marta – Gte: P. de Andreís A.
MA05) Cra. 5 Nº 6-12, Fundación – Gte: R. Arango Galvis.
MA06) A.A. 1240, Santa Marta – Gte: Gustavo Morales.
MA07) A.A. 052, Santa Marta – Gte: Edgard Polo F.

MA08) Calle 18 NÞ 5-58, P6, Santa Marta.
MA09) A.A. 45, El Banco – DG: J. Vanegas T.

ME00) META
ME01) Calle 38 NÞ 30A-106, Villavicencio – Gte: Alcides Antonio Jauregui Bautista.
ME02) Calle 16 NÞ 15-90, Granada – DG: Javier Ruiz A.
ME03) Calle 38 NÞ 32-41, P7, Edif. Parque Santander, Villavicencio – Gte: Carlos Alberto Pimienta.
ME04) Cra. 30 NÞ 36-14, P4, Villavicencio – Dir: J. A. Martínez.
ME05) A.A. 2472, Villavicencio – DG: Guillermo Torres B.
ME06) A.A. 2284, Villavicencio – Gte: Omar López R.
ME07) Calle 13 # 28-05 *(✉ A.A. 001), Granada – DG: Luis A. Rojas.
ME08) Vía Villavio km 3, San Martín – Gte: Hernando Sánchez.

NA00) NARINO
NA01) A.A. 375, Pasto – Gte: E. del Castillo C.
NA02) A.A. 635, Pasto – DG: Alejandro Ortega P.
NA03) A.A. 454, Pasto – DG: Felipe Chaves.
NA04) Calle 20 NÞ 24-37, P6, Pasto – DG: Ovidio Palacio C.
NA05) Cra. 27 NÞ 19-30, Pasto – DG: Gonzalo Ardila.
NA06) A.A. 1005, Ipiales – DG: Fidencio A. Erazo R.
NA07) A.A. 165, Tumaco – DG: Gabriel Osorio.
NA08) Cra. 8 NÞ 4-48, Ipiales – DG: Alvaro A. Enriquez.
NA09) Calle 20 NÞ 15-02, Túquerres – DG: J. Aguirre Benavides.
NA10) A.A. 516, Pasto – Dir: Carlos E. Pérez L.
NA11) Cra. 6 NÞ 10-60, Ipiales – Gte: Roberto Ramírez.
NA12) Cra. 5 NÞ 3-15, Samaniego – DG: William Montenegro Chamorro.
NA13) Fundación Tomás Cipriano de Mosquera, Calle Polo, Barbacoas – Gte: Amado Waldemar Ortíz.
NA14) Nevado Cumbal.
NA15) Cra. 24 NÞ 17-75, Pasto – DG: Manuel Eraso.

NS00) NORTE DE SANTANDER
NS01) Calle 5 NÞ 3-26, Cúcuta – Gte: Ernestina Nuñez B.
NS02) Centro Comercial Bolívar, Local E4, Cúcuta – DG: Claudia L. Montañez.
NS03) A.A. 400, Cúcuta – Gte: Fernando Fonseca R.
NS04) Calle 5 NÞ 0-45, Cúcuta – Dir: Luis E. Mantilla.
NS05) A.A. 624, Cúcuta – Gte: Gustavo Sánchez Chacón.
NS06) A.A. 519, Cúcuta – Dir: Carlos Pérez A.
NS07) Calle 9 NÞ 13-10, Ocaña – Gte: Carlos Navarro.
NS08) A.A. 2284, Cúcuta – DG: José Urbina A.
NS09) Cra. 4 Calle 5 junto Almacén Fotorubio, Fomeque.
NS10) A.A. 19, Ocaña – DG: Jorge Cabrales Romero.
NS11) Cra. 5a Calle 7a Esq., Tibú – DG: William Vanegas Moncada.
NS12) A.A. 473, Cúcuta – Gte: Carlos Ernesto Díaz M.
NS13) Base Militar "San Jorge", Tibú – DG: Oscar Hinestroza M.
NS14) Av. 2 NÞ 4-11, Bochalema – DG: Oscar Alberto Arguello.
NS15) A.A. 2582, Cúcuta – DG: Alberto Buitrago.
NS16) A.A. 1303, Cúcuta – Gte: José Neira Rey.
NS17) A.A. 1074, Pamplona – DG: Sergio Luis Peña G.
NS18) Av. 6a NÞ 14-95, Cúcuta – DG: Pablo Emilio Cortés P.
NS19) Cra. 2 NÞ 1-10, Puerto Santander – Dir: Noifer Camperos P.
NS20) A.A. 151, Villa del Rosario.

PU00) PUTUMAYO
PU01) Calle Principal, Orito – Dir: Jeppy Navarro.
PU02) A.A. 011, Mocoa – Own: Jaime Herrera A.
PU03) Sibundoy – Dir: Luis Ernesto Jojoa.
PU04) A.A. 9, Puerto Assis – DG: Carlos A. Herrera R.

QU00) QUINDO
QU01) Cra 16 NÞ 19-23, Armenia – DG: Anibal Mejía.
QU02) A.A. 2361, Armenia – DG: Mrs. Iliana O. de Ramírez.
QU03) A.A. 2481, Armenia – Gte: Carmen E. Valencia.
QU04) A.A. 556, Armenia – Gte: Jorge Orozco.
QU05) Cra. 25 NÞ 41-38, Calarcá – Dir: Jorge Orozco
QU06) Cra. 14 NÞ 21-26, P2, Armenia – DG: Alvaro Jaramillo Ospina.
FM: 104. 7 Robles FM Stereo.
QU07) A.A. 617, Armenia – Gte: Leonardo Marín I.

RI00) RISARALDA
RI01) A.A. 354, Pereira – Gte: Arney Ocampo.
RI02) A.A. 045, Pereira – Gte: Fernando Sánchez.
RI03) A.A. 221, Pereira – DG: Hernán Ospina G.
RI04) La Virginia.
RI05) Cra. 7a NÞ 18-21, Of. 1103, Pereira – DG: Luis Alberto Delgado.

RI06) A.A. 1262, Pereira – DG: Patricia Liliana Salazar S.
RI07) Cra. 15 NÞ 11-80, Santa Rosa de Cabal (or: Calle 19 NÞ 8-74, Pereira) – DG: Gildardo Cardona R.
RI08) Santa Rosa de Cabal.

SS00) SANTANDER DEL SUR
SS01) A.A. 915, Bucaramanga – Own: Gustavo Serrano F.
SS02) A.A. 223, Bucaramanga – Gte: Germán Gómez Vahos.
SS03) Calle 10 NÞ 17-73, Barrancabermeja.
SS04) Cra. 27 NÞ 45-80, Bucaramanga.
SS05) A.A. 007, Bucaramanga.
SS06) Calle 35 NÞ 18-13, Bucaramanga.
SS07) Calle 16 NÞ 15-01, El Socorro – Gte: Helio G. Zembrano.
SS08) Transv. 6 NÞ 9-56, Barbosa – DG: Henry Cortés T.
SS09) Batallón de Artillería de Defensa Aerea NÞ 2 "Nueva Granada" (✉ A.A. 036), Barrancabermeja. DG: Jairo A. Lombana Trujillo.
SS10) A.A. 1100, Bucaramanga – Gte: Fernando Ardila.
SS11) Calle 10 NÞ 9-71, San Gil – Dir: T. Rueda R.
SS12) Cra. 19 NÞ 33-52, Bucaramanga – Gte: Carlos Soto.
SS13) A.A. 23, Barrancabermeja – DG: Cecilia Carvajal de D.
SS14) Calle 7 NÞ 17-44, Barbosa – DG: Dora I. Fonseca.
SS15) A.A. 578, Barrancabermeja – Gte: Enrique Fuentes S.
SS16) Calle 12 NÞ 10-30, San Gil – Gte: Mario Guillermo Rojas V.
SS17) Calle 16 NÞ 4-47, Zapatoca – DG: Angel J. Jaimes.
SS18) A.A. 3104, Bucaramanga – Gte: Saúl Gómez Calderón.
SS19) Cra. 6 NÞ 4-116, P2, Cimitarra – Gte: Jesús Martín Contreras.
SS20) Cra. 3 NÞ 3-42, Vélez – Gte: Rodrigo Correa F.
SS21) Calle 20 NÞ 7-18, Zapatoca – Gte: Alvaro Grass.
SS22) Cra. 27 NÞ 34-44, Bucaramanga.
SS23) Calle 41 NÞ 19-87, Bucaramanga.
SS24) Batallón Luciano D'Ahuyar, San Vicente de Chucurí.
SS25) A.A. 250, Barrancabermeja – Gte: J. García R.
SS26) Pasaje Carrillo Casas, P3, Málaga – DG: Constantino Páez Albarracin.
SS27) Socorro.
SS28) A.A. 3068, Bucaramanga – Dir: Marco Antonio Caicedo.

SU00) SUCRE
SU01) A.A. 750, Sincelejo – DG: J. L. Feris.
SU02) A.A. 448, Sincelejo – Gte: Juan Guerra T.
SU03) A.A. 100, Corozal – Gte: Nelson Martelo O.
SU04) A.A. 191, Sincelejo – DG: Marino Diconza.
SU05) Calle 20 NÞ 24-93, Av. las Penitas, Sincelejo.
SU06) Cra. 18 NÞ 18-31, Sincelejo – Gte: N. Miserque.
SU07) A.A. 542, Sincelejo – Gte: Antonio Turbay.
SU08) A.A. 307, Sincelejo – Gte: Carlos Severich.
SU09) Cra. 28 Calle 18, San Marcos.
SU10) Calle 25A NÞ 18, Sincelejo – Gte: Roberto Angulo M.
SU11) A.A. 167, Sincelejo – Gte: C. Cohén.

TO00) TOLIMA
TO01) Calle 12 NÞ 1-17, Ibagué – Gte: Carmen Laserna.
TO02) Calle 14 NÞ 2A-14, P2, Ibagué – Gte: Alberto Viancha.
TO03) A.A. 1187, Ibagué – DG: German Acosta Ramos.
TO04) Cra 7 con Calle 10, Espinal – Dir: Carlos Barón A.
TO05) Calle 11 NÞ 10-22, Guamo.
TO06) A.A. 2419, Ibagué – DG: Alonso Botero Palacio.
TO07) A.A. 1094, Ibagué – Gte: Fabio Zúñiga.
TO08) Calle 3 NÞ 3-62, Mariquita – Gte: Gustavo García Y.
TO09) Calle 9 NÞ 1-124, P3, Ibagué.
TO10) A.A. 64, Espinal – Gte: Antonio Forero B.
TO11) A.A. 536, Honda – Dir: Rafael Pabón B.
TO12) A.A. 509, Honda – DG: Eduardo Cañas E.
TO13) Cra. 2 NÞ 11-27, Flandes – DG: Antonio Forero B.
TO14) A.A. 589, Ibagué – Gte: Jorge Alvarez Agudelo.
TO15) Guayabal.
TO16) Cra. 13 NÞ 5-61, Libanó – Gte: José Antonio Polanía C.
TO17) Cra. 7a NÞ 6-25, Dolores – Dir: Héctor Hernández N.
TO18) Calle 7 NÞ 20-70, Melgar – Gte: Hernando Sierra M.
TO19) Cra. 2 NÞ 3-74, Rovira – Gte: Luis F. Tovar.

VA00) VALLE DEL CAUCA
VA01) Cra. 26 NÞ 5C-25, San Fernando, Cali – DG: Gustavo Adolfo Kuri G.
VA02) Cra. 24a NÞ 4-35, Cali – Gte: Homero Vásquez.
VA03) A.A. 1941, Cali – DG: Antonio José Uribe.
VA04) A.A. 4666, Cali – DG: Bernardo Tobón de la Roche.
VA05) A.A. 96, Buga – Gte: Sonia Govea.
VA06) A.A. 2180, Cali – Gte: Jairo Dominguez A.
VA07) A.A. 280, Palmira – DG: Otto Burckhardt C.

VA08) A.A. 126, Tuluá – Gte: Luis F. Ossa P.
VA09) A.A. 145, Cartago – Gte: Cristóbal de los Rios R.
VA10) Calle 15 NⱣ 8-75, P7, Cali – Dir: H. Pava C.
VA11) A.A. 383, Buenaventura – DG: Padre Antonio Bayter Abud.
VA12) Cra. 16 NⱣ 6-22, Calcedonia – Gte: Carlos Alberto Orozco.
VA13) Calle 24N NⱣ 2N-24, Of. 901, Cali – Adm: Vicky Albán.
VA14) Cra. 20 NⱣ 29-09, Palmira – Gte: Ma. Consuelo Valdez.
VA15) Cra. 6 NⱣ 2-16, Buenaventura – Gte: C. Astorquiza.
VA16) Av. 3 Bis. NⱣ 23CN-71, Cali – DG: Efraín Páez E.
VA17) Cra. 11 NⱣ 11-43, P2, Zarzal – DG: Anibal Jurado Carmona.
VA18) A.A. 201, Palmira – Gte: Jaime Bejarano D.
VA19) Cra. 14 NⱣ 5-77, Buga – Gte: Zoraida Sanchez de G.
VA20) Cra. 26 NⱣ 28-72, Tuluá – Gte: Godofredo Sánchez V.
VA21) Calle 27 NⱣ 33-55, Tuluá – DG: Francisco Ujueta.
VA22) Calle 4D NⱣ 34-23, Cali – Dir: Victor Sánchez.
VA23) Cra. 51 NⱣ 49-21, Sevilla – Gte: Oscar Salazar J.
VA24) A.A. 127, Cartago – Dir: Edgar Arroyave Londoño.

VI00) VICHADA

VI01) Av. Orinoco, Puerto Carreño.
VI02) Puerto Carreño – **FM:** 100.3.

VP00) VAUPES
VP01) Mitú.

ECUADOR

L.T: UTC -5h — **Pop:** 12,000,000 — **Radios:** 3,240,000 — **Pr.L:** Spanish, Quechua — **E.C:** 60Hz, 110/220 V — **ITU:** EQA.

INSTITUTO ECUATORIANO DE TELECOMUNICACIONES (ANDINATEL)
+ Av. 6 de Diciembre y Colón, Quito.
L.P: Dir. Nal. de Frecuencias: Ing. Mario Cuesta Bastidas. Jefe Administration Tec: Osvaldo Benavides.

ASOCIACION ECUATORIANA DE RADIODIFUSION (AER)
+ Av. Inglaterra 31-82, Quito – **L.P:** Lenin Andrade.

Mediumwaves: Call HC-, ° = also on shortwave, * = inactive, (r) = repeater,
v = varying fq.

The letters preceding the st. number indicate the Province.
Addresses are listed by Province in alphabetical order.

Call		kHz	kW	Name and h. of tr.
PI01)	DC1	530	1	R. Iris, Quito: 1000-0500
GU01)	FA2	540	25	R. Tropicana "Canal 540", Guayaquil: 1100-0600
PI02)	GM1	550	50	R. Reloj "5-50", Quito: 1100-0400
GU01)	AJ2	560	25	C. R. E. Satelital, Guayaquil: 24h
PI03)	CE1	570	10	R. El Sol, Quito: 24h
GU02)	PC2	580	10	R. Uno, Guayaquil: 24h
PI04)	SP1	590	10	R. Carrousel, Quito: 1100-0200
GU03)	XY2	600	10	R. Nal. del Ecuador, Quito: 1100-0400
PI05)	MJ1	610	10	Caravana Satélite, Quito: 1000-0400
LO01)	XY3	620	50	R. Nal. del Ecuador, Loja: 1100-0400
LR01)	HA2	630	10	Ondas Quevedeñas, Quevedo: 2300-0300
GU04)		640		R. Morena, Guayaquil
PI06)	XY1	640	50	R. Nal. del Ecuador, Quito: 1100-0400
MA01)	FD4	650	5	R. Visión de Manta, Manta: 0900-0500
GU05)	LG2	660	30	R. Carrousel, Guayaquil: 24h
PI07)	FF1	°670	12. 5	R. Jesús del Gran Poder, Quito: 0945-0100
GU06)	VP2	680	25/12	Sistema de Emis. Atalaya, Guayaquil: W 0900-0500, Sun 1000-0300
MA02)	FA4	690	5	Sucre Portoviejo, Portoviejo
PI08)	JB1	°690	50d	LV de los Andes, Quito: 24h
GU07)	RS2	700	50	Sucre Guayaquil, Guayaquil: 24h
CR01)	ER5	°710	7	Escuelas Radiofónicas Populares, Riobamba: 0900-0300
EO01)	UE3	720	10	Sucre Machala, Santa Rosa
LO02)	MO3	720	5	R. Matovelle "HCM-3", Loja: 1000-0200

Call		kHz	kW	Name and h. of tr.
MA03)	GB4	v720	10	LV de Portoviejo, Portoviejo: 1000-0400
PI09)	IC1	°720	5	R. Municipal, Quito: 1030-0300
GU08)	MG2	730	10	R. Guayaquil, Guayaquil: 24h
MA04)	SE4	v740	10	R. Libertad, Chone: 1100-0600
PI02)	GC1	740	10	R. Melodía "Canal 7-40", Quito: 1100-0400
GU09)	RC2	750	30	R. Caravana, Guayaquil: 1000-0430
PI10)	QR1	°760	25	R. Quito "LV de la Capital", Quito: 24h
GU10)	MF2	770	25/12	R. El Telégrafo, Guayaquil: 1000-0500
MA05)	RG4	780	1. 5	R. Mía, Manta: 1000-0200
PI11)	AA1	780	1	R. Noticia, Quito: 1000-0300
PI12)		790		R. Paraíso, Maldonado
GU05)	ML2	800	25	R. Super K, Guayaquil: 1200-0500
PI13)	FB1	800	5	R. Sensación 800, Quito: 1000-0300
GU11)	VT2	810	5	R. Atalaya, El Milagro: 2300-0300
TU01)		810		Sucre Ambato, Ambato
CA01)	VI5	820	5	R. LV de Ingapirca, Cañar: 0900-0330
IM01)	CR1	820	3	Super R. , Otavalo
MA06)	RF4	820	1	Canal Manabita, Portoviejo
PI14)	FB1	820	1	R. Monumental, Sto Domingo de los Colorados
PI15)		820		Radiaxsión, Quito
CR02)	RP5	830	4. 5	R. Promoción, Riobamba: 0900-1400, 2200-0200
GU12)	RM2	830	25	R. Huancavilca, Guayaquil: 24h
MA07)	EM4	840	1	R. Costa Azul, Portoviejo: 1100-0500
PI16)	PN1	840	50	R. Vigía "LV de la Policía Nal. ", Quito: 1100-0500
GU13)	VS2	850	15	R. San Francisco, Guayaquil: 0930-0300
PA01)	GB7	850	0. 5	R. Nal. Espejo, El Puyo
PI17)	PC1	v860	10	R. Visión, Quito: 1015-0400
GU14)	NY2	870	20	R. Cristal "RCQ", Guayaquil: 1000-0500
TU02)	GS6	870	1	R. Píllaro, Píllaro: 1100-0400
PI18)	RP1	°880	50/40	R. Católica Nacional, Quito: 1000 (Sun 1100)-0200
CR03)	TL5	890	1	Ondas del Chimborazo, Riobamba: 1100-0500
EO02)	RS3	890	25/20	R. Superior, Machala: 0900-0500
AZ01)	RR5	900	1	R. Carrousel, Cuenca: 1100-0200
MA08)	OF4	v900	5	R. Chone, Chone: 1100-0400
PI19)	VA1	900	10	Sucre Quito, Quito: 1100-0400
CR04)	GE5	910	2. 5/2	R. Mundial, Riobamba: 1000-0400
GU15)	BO2	910	2	R. Colón, Guayaquil
EO03)	RU3	920	10	C. R. O. , Machala: 0930-0430
GU16)		920		R. Peripa, El Empalme
PI20)	CM1	920	20	R. Colón, Quito: 1100-0500
GU12)	VI2	930	5	Ondas Antillanas, Guayaquil: 24h
TU03)	BA6	*930	5	R. Ambato, Ambato
PI21)	BZ1	*940	5	R. Dif. de la Casa de la Cultura Ecuatoriana, Quito
CR05)	UE5	950	3	R. Colta "LV de la Asociación", Colta: 0900-0200
GU17)	DE2	950	10	R. Latina, Guayaquil: 1100-0700
IM02)		950		R. AM Norte, Ibarra
AZ02)	SA5	960	1	Sononda Internacional, Cuenca: W 0925-0430, Sun 1200-0400
PI22)	NC1	960	1	R. Cosmopolita, Quito: 1100-0200
TU04)	JX6	960	1	LV del Santuario, Baños: 1000-0300
PI23)	OT1	*965	10	R. Católica, Sto Domingo de los Colorados
GU18)	AW2	970	20	R. Católica Nal. del Ecuador, Guayaquil: 1000-0500
IM03)	MB1	970	1	R. Imperio, Ibarra: 1030-0300
CR06)	JI5	980	1	R. El Prado, Riobamba: 1200-0400
AZ03)	OL5	990	2	R. América, Cuenca: 1000-0300
PI24)	GH1	990	25	R. Tarqui, Quito: 1015-0400
GU19)	EW2	990	15	Frecuencia Mil, Guayaquil: 24h
LO03)		1000		Dinamita Mil, Catamayo
PI25)	CR1	1000	1	R. Alegría, Sto Domingo de los Colorados: 1000-0500
AZ04)	RV5	*1010	2. 5	R. Visión, Cuenca: 1130-0400
GU20)	RZ2	1010	3	R. Amiga, Guayaquil: 24h
MA09)	RC4	1010	3	R. Cenit, Manta: 1200-0500
TU05)	NR6	1010	10	T. S. B. R. Lider, Ambato: 1000-0500
BO01)	CR6	1020	5/3	R. Surcos, Guaranda: 1030-0100
EO04)	GO3	1020	3	Canal Estelar, Santa Rosa
PI26)	HR1	1020	5	R. Quitumbe, Quito: 24h
GU21)	RF2	1030	5	R. Punto 1030, Guayaquil: 1100-0500

Call		kHz	kW	Name and h. of tr.
MA10)	KL4	v1030	5	R. Unión, Chone: 1100-0500
NA01)	FO7	1030	3	R. Francisco de Orellana, Coca
AZ05)	EV3	v1040	10/5	R. Splendid, Cuenca: 24h
PI27)	CW1	1040	3	LV del Valle, Machachi: 1100-1700, 2130-0300
TU06)	GB6	1040	3	R. Colosal, Ambato: 0930-0500
GU01)	RQ2	1050	5	Super R. , Guayaquil: 1030-0400
IM04)	IM1	1050	5	R. Municipal LV de Imbabura, Ibarra: 0900-1900
CP01)	MG6	v1060	5	R. Ecos del Pueblo, Saquisilí: 1045-0330 (Sun -0200)
LO04)		1060		R. Simbolo, Pindal
AZ06)	CJ5	1070	5	R. LV de Tomebamba, Cuenca: 1000-0500
LR02)		1070		R. Richi, El Empalme
PI28)	VP1	1070	1	R. Libertad, Quito: 1100-0300
PI29)	RS1	1070	1	R. Lubacán, Santo Domingo de los Colorados: 0950-0200 (Sun -2300)
CP02)	BH6	1080	10	R. Latacunga, Latacunga: 0900-0230
GU22)	KD2	1080	10	R. Tigre, Guayaquil: 1100-0500
MA11)	AB4	v1080	5	R. Contacto, Manta: 1000-0400
PI30)	VI1	v1090	5	Fé y Alegría, Quito
CP03)	GR6	1100	5/2	R. Novedades, Latacunga: 1000-0500
GU23)	FW2	v1100	10	R. Alegría, Guayaquil: 24h
NA02)	LE7	°1100	1. 5	R. Oriental, Tena: 0955-0300
AZ07)	JC5	1110	5	R. Ondas Azuayas, Cuenca: 1030-0500
PI31)	JR1	1110	10	Hoy La Radio, Quito: 24h
TU07)	RP6	v1110	5	R. Pelileo, Pelileo: 1100-0400
CC01)	EB1	1120	2	Ecos de Montúfar, San Gabriel: 1200-0200
GU24)	FV2	1120	5	Estación Intercontinental, Guayaquil: 1100-0500 (rep. on 1220 kHz)
LO05)	NT3	1120	5	R. Once Catamayo, La Toma: 1100-0300
PA02)	AS7	1120	3	R. Variedades del Puyo, El Puyo
PI32)	LE1	1120	10	R. Dif. Marañon, Sto Domingo de los Colorados: 1300-0200
IM05)	RD1	1130	5/3	R. Punto, Ibarra: 1000-0400
LR03)		1130		R. Sibimbe, Ventanas
TU08)	PV6	1130	5	R. Centro, Ambato: 24h
AZ08)	AZ5	1140	1	R. Alfa Musical, Cuenca: 1100-0600
GU25)	FB2	1140	1. 5	R. Cóndor, Guayaquil: 1130-0500
MA12)	MF4	1140	4	R. Momento, Portoviejo: 24h
PI33)	IR1	1140	5	Colorin Colorado, Quito: 1130-2400
BO02)		1150		R. E. 2, Guaranda
CR07)	GB5	v1150	10	LV de Riobamba "Antena 1", Riobamba: 24h
LO06)	AV3	°1150	10	R. Luz y Vida, Loja: 1000-0330, (Sat -0400, Sun -0700)
SU01)	BC7	1150	1	R. El Cisne, Nueva Loja: 1000-0200
CA02)		1160		LV del Pueblo, Azogues
CP04)	UR6	1160	1	R. Runatacuyaj "LV de la Asociación". Latacunga: W 1000-0200 (Sat -2000, Sun 1300-2100)
EO05)	VR3	1160	2	R. Vía, Machala
MA13)	WD4	v1160	1	R. Cenit, Portoviejo: 1200-0500
PI34)	CP1	v1160	2	R. Presidente, Quito: 24h
CR08)	JV5	1170	5	R. Central, Riobamba: 0900-0500
EO06)		1170		R. Trébol, Zaruma: 1030-0200
ES01)	JM4	1170	10	R. Antena Libre, Esmeraldas: 1100-0300
GU26)	RV2	1170	10	R. Filadelfia, Guayaquil: 1100-0400
AZ09)	DP5	1180	4	R. Cuenca "LV de los Cuatro Ríos", Cuenca: 1200-0300
CC02)	RV1	1180	1. 2	R. Familiar, Julio Andrade: 1100-1800, 2200-0200
PI35)	LR1	1180	12. 5	Nueva Em. Central, Quito: 1100-0400
CP05)	RF6	1190	1	R. El Sol, Pujilí: 1100-0200
GU22)	DE2	1190	2	Estudio 11-90, Guayaquil: 24h
AZ10)	RM5	1200	5	R. El Mercurio, Cuenca: 0900-0500
EO07)		1200		Sucre Machala, Machala
LR04)	RE2	1200	5	LV del Trópico, Quevedo: 1000-0400
MA14)	MP4	1200	1	R. La Grande, Bahía de Caráquez
PI36)	CS1	v1200	5	R. Super K, Sangolquí: 1030-0100
GU27)	BJ2	1210	20	R. El Mundo, Guayaquil: 1200-0300, (Sat -0100, Sun -0400)
LO07)	VC3	°1210	10	R. Centinela del Sur "CDS", Loja: 1100-0400
TU09)	JM6	1210	3	R. Sira, Ambato: 1000-0700
BO03)		1220		Ecos de Bolívar, Guaranda
PI32)	AP1	1220	10	R. Dif. Marañon, Quito: 0900-0430
AZ11)	MV5	1230	3	R. Popular, Cuenca: 1045-0500
ES02)	FG4	1230	3	Sucre Esmeraldas, Esmeraldas: 1000-0500, Weekends 24h
GU48)	GT2	1230	15	R. Galáctica, Guayaquil: 1000-0400
IM06)	RI1	1230	3	Centro Radiofónico de Imbabura "CRI", Ibarra: 1100-0300
CP06)	RL6	1235	1	LV de Saquisilí y Libertador, Saquisilí: 1045-0300
CR09)	LA5	v1240		R. Sonorama, Riobamba
EO08)	RF3	1240	5	R. Nacional Fenix, Zaruma: 1000-0100
PI37)	PA1	1240	1	R. Metropolitana, Quito: 1200-0300
CC03)	EM1	1250	10	Ondas Carchenses, Tulcán: 1000-0400
GU28)	HB2	1250	2	R. Linda, Guayaquil: 1000-0400
PI38)	MY1	°1250	3	LV del Triunfo, Sto Domingo de los Colorados: 1000-0500
AZ12)	PB5	1260	2	XG Radio, Cuenca: 1100-0300
EO09)	RB3	1260	1	R. Benemérita, Machala
PI39)	MO1	1260	10	LV del Santuario del Quinche "HCM-1", Quito: 1000-0300
TU10)	RO6	1260	3	R. Calidad, Ambato: 0930-0600
CP07)		1270		Casa de la Cultura, Latacunga
GU22)	UM2	1270	15	R. Universal, Guayaquil: 0900-0500
MA15)	LD4	1270	3	R. Junín, Junín: 1100-0500
CR10)	NW5	1280	1	R. Canal Tropical, Riobamba: 1100-0100
EO10)	RP3	1280	2	R. Continental, Arenillas
MA16)	IN4	1280	1	LV del Sur de Manabí, Jipijapa: 1000-0500
PI40)	AB1	1280	1	R. Democracia "La Genial", Quito: 1000-0400
AZ13)	JA5	°1290	3	LV del Río Tarqui, Cuenca: 0900-0200
GU29)	OF2	1290		Canal Milagreño, El Milagro
IM07)	NS1	1290	1	R. Popular, Atuntaqui: 1100-0300
CP08)	VM6	1295	0. 5	R. Once de Noviembre, Latacunga: 1200-0400
BO04)		1300		R. La Paz, Guaranda
GU30)	DC2	1300	5	R. Cenit, Guayaquil: 1200-0400
MS01)		1300		Ecuadoradio, Macas
PI41)	RU1	1300	10	R. Festival, Sto Domingo de los Colorados: 1000-0200
SU02)	RS7	1300	2/1	R. Sucumbíos, Lago Agrio: 1100-2400
EO11)	CP3	v1310	1	LV de El Oro, Pasaje
PI02)	GB1	°1310	20	R. Nal. Espejo, Quito: 24h
CA03)	CI5	1320	3	T. V. O. "El Poder Mágico de la Fé", Biblián
LR05)	FR2	1320	3	R. Guayaquil, Babahoyo: 1030-0300
MS02)	OB7	°1320	0. 5	R. Nacional Limón, Limón Indanza: 2100-0300
TU11)	JD6	1320	10	R. Continental, Ambato: 0930-0400
AZ14)	LW5	1330	2	R. Scala, Cuenca: 0930-0100
CC04)	OV1	1330	3	Ondas del Volante, El Angel: 1200-2400
EO12)	RV3	1330	5	R. Nal. El Oro, Machala: 1000-0600
GU31)		1330		R. Hispaña, Guayaquil
PI42)	JA1	v1330	3	R. Sideral, Quito: 24h
CR11)	VP5	1340	2. 5	R. San Francisco, Penipe: 0900-0300
ES03)		1340		LV de su Amigo "Estéreo Musical", Esmeraldas
LO08)		1340	1	R. Regional, Loja: 1100-0300
LR06)	SF2	1340	1	R. Estéreo Fluminense, Babahoyo: 1100-0500
MA17)	DR4	*1340	1	R. Cenit, Bahía de Caráquez:
TU12)	RT6	1340	5	R. Paz y Bien, Ambato: 0930-0130
AZ15)	SF5	1350	2/1. 5	R. LV de San Fernando, San Fernando: 0900-0100
CC05)	PZ1	1350	5	R. Rumichaca, Tulcán: 1045-0415
GU47)	VP2	1350	3	R. Reloj, Guayaquil: 24h
PI43)	PU1	1350	1	LV de Sto Domingo, Sto Domingo de los Colorados
CR12)	RJ5	1360	1	R. América, Riobamba: 1100-0300
EO13)	HG3	1360	5	R. El Sol, Machala: 1045-0500
MA18)	EG4	1360	3	LV del Carmen, El Carmen
PI44)	MT1	v1360	3	Oyambaro AM, Tumbaco: 1000-0300
CA04)		1370		R. El Rocío, Biblián
GU32)	VO2	1370	5	LV del Milagro, El Milagro
IM08)	JS1	v1370	2	Ecos Andinos, Pimampiro
LO09)	ER3	°1370	5	R. Progreso, Loja: 1000-0315

	Call	kHz	kW	Name and h. of tr.
PA03)	RP7	1370	2	R. Pastaza, El Puyo: 1100-0100 (Sun 1200-2300)
EO14)	OA3	1380	5	Impacto Piñas, Piñas: 1100-0500
MS03)	WV7	*1380	3	R. Morona, Macas
PI45)	CV1	1380	5	R. Cristal "RCQ", Quito: 24h
TU13)	JR6	v1380	5	R. Mera, Ambato: 24h
AZ16)	EA5	1390	4	R. Tropicana, Cuenca: 1200-0300
CR13)	DN5	1390	3	R. Atenas, Riobamba: 0900-0500
ES04)	HE4	1390	1	LV de Esmeraldas, Esmeraldas: 1000-0500
IM09)	IE1	1390	1.5	R. Uno, Urcuquí: 0800-0200
LR07)	WA2	1390	5	Estación Radial, Quevedo
CC06)	VL7	1400	7	R. Multicolor, Tulcán
CP09)		1400		Corporativa del Cotopaxi, Latacunga
GU02)	FL2	1400	10	R. Z Uno, Guayaquil: 24h
ZC01)	VZ7	1400	5	R. LV de Zamora, Zamora: 1000-0430
AZ17)	GC5	1410	1	R. Centro Gualaceo, Gualaceo: 1200-0300 (Sun 1100-2300)
CR14)		1410		Ondas Cisnerias, Riobamba
ES05)	FR4	1410	1	LV de Quinindé, Quinindé
GU33)	CQ2	1410	1	R. Presidente "LV del Pueblo", El Milagro
NA03)	MS7	1410	1	LV del Río Coca, El Coca
PI05)	EC1	1410	1	R. El Tiempo, Quito: W 1130-1900, at other times (r: HCAP1 1220 kHz)
CP10)	MA6	v1420	3	R. Nuevos Exitos, Salcedo: 1000-0400
EO15)	NR3	1420	1	LV de Huaquillas, Huaquillas
IM10)	RN1	°1420	3	R. Bahá'í, Otavalo: 0900-1500, 1930-2300
BO05)	JC6	1430	5	R. Guaranda, Guaranda: 1100-0300
GU34)	MB2	1430	10	Caribe Ritmo, Guayaquil: 24h
LO10)	CV3	1430	5	Ondas del Zamora, Loja: 1130-0330
PI46)	GF1	1430	1	R. Continental, Quito: 1300-2300
BO06)		1440		R. Antología, Caluma
CA05)	OV5	1440	3	Ondas del Volante, Azogues
CP11)		1440		Fenix AM, Latacunga
ES06)	DY4	1440	2.5	R. Iris, Esmeraldas: 1000-0400
IM11)	DF1	1440	5	R. Panorama, Ibarra: 1030-0400
PA04)	MD7	*1440	5	R. Puyo, El Puyo
CR15)	SC5	1450	10	R. Calidad, Riobamba: 0800-0400
GU35)	DR2	v1450	1	R. Minutera, Guayaquil
GU36)	SE2	1450	1	R. Santa Elena, Santa Elena: 2200-0200
LR08)	HW2	1450	3	R. Rey, Quevedo: 1000-0500
NA04)	RI7	*1450	1	R. Interoceánia, Santa Rosa de Quijos
PI47)	SC1	1450	1	R. Sensación, Cayambe
CP12)	IC6	1460	5	R. Nuevos Horizontes, Latacunga: 1000-0200
LO11)	CL3	1460	5	R. Cariamanga, Cariamanga: 1100-0300
MS04)		1460		LV de Gualaquiza, Gualaquiza
GU37)	LD2	1470	1	R. Ecos de Naranjito, Naranjito: 1300-0300
MA19)	RD4	1470	1	R. Richi, Calceda
PI48)	JC1	1470	5	Ecos de Cayambe, Cayambe: 1000-0330
CR16)	WP5	1480	3	R. Atlántida, Alausí: 1000-0400
CP13)	CY6	1480	5	R. Popular de la Maña, La Maña
EO16)	BS3	v1480	3	Ondas del Jubones, Pasaje: 1000-0500
IM12)	MC1	1480	1	R. Municipal, Cotacachi: 1000-0100
LO12)		1480	0.5	R. Luciérnaga del Condór, Yanzatza
MA20)	JB4	1480	3	R. LV de Jipijapa, Jipijapa: 1100-0400
CA06)	SM5	1490	5	R. Santa María, Azogues: 1000-0100
ES07)	AE4	1490	2.5	R. Unión, Esmeraldas: 1000-0300
GU38)	VY2	1490	1	La R. Dinámica, Guayaquil
PI55)		1490		R. Alfaro, Quito
TU14)	AI6	1490	3	R. Moderna, Píllaro: 1300-0200
CR17)	WN5	1500	2	R. Puruhá, Riobamba: 0945-0500
IM13)	RO1	1500	1	R. Otavalo, Otavalo: 1200-0300
LR09)	HG2	1500	5	LV del Río Vinces, Vinces: 1100-0500
MA21)	AD4	1500	5	R. Satélite, El Carmen: 1000-0500
ZC02)	OY7	1505	1.5	Ondas del Río Yacuambí, Yacuambí: 1000-1400, 2000-2400
BO07)	RY6	1510	1	R. Runacunapac Yachana "R. El Saber del Hombre", Guaranda
CA07)	RC5	1510	0.5	LV de la Juventud, Cañar
GU39)	HD2	°1510	1	Inst. Oceanográfico de la Armada, Guayaquil: time signals 24h
LO13)	UC3	1510	10	R. Unión Calvense, Cariamanga
PI56)		1510	5	R. Monumental, Quito: 1300-0400
SU03)	LV7	1510	3	R. Ecos del Oriente, Lago Agrio: 1000-0300
CR18)	RI5	1520	2.5	LV de Guamote, Guamote
GU40)	RN2	1520	1	LV de Naranjal, Naranjal
IM14)	TI1	1520	1	R. Ibarra, Ibarra: 1000-0400
MA22)	EB4	1520	5	R. Manta, Manta: 1100-0500
CA08)	CC5	1530	5	Ondas Cañaris, Azogues: 1030-0400
CR19)	VP5	1530	3	R. LV de Pallatanga, Pallatanga: 1100-0300
ES08)	JY4	1530	5	R. Uno, La Concordia: 1000-2400
GU41)	MP2	v1530	3.5	LV de la Península, La Libertad: 1130-0330
TU15)	MZ6	1530	1	LV del Dorado, Pelileo: 1130-0230
CC07)	PV1	1540	1	R. Mira. Mira
CP14)	MH6	v1540	0.5	Organisación Radiofónico Cotopaxi, "ORC", Latacunga: 1000-0400
LR10)	FM2	1540	3	R. Cristal, Ventanas
MS05)	VB7	°1540	0.25	R. Dif. Católica Cultural "Voz del Upano", Macas: 1045-0300
PI49)	DP1	1540	1	R. Caracol, Quito: 1000-0400
AZ18)	AD5	1550	5	R. Chaguarurco, Santa Isabel: 1000-0300
GU42)	AD2	1550	2	LV del Triunfo, El Triunfo: 1100-0400
NA05)	RA7	1550	1	R. Amazonas, Archidona
TU16)	EI6	1550	2	R. Montalvo, Ambato: 1130-0300
EO17)	TR3	1560	2	LV del Guabo, El Guabo: 1100-1300, 2300-0400
GU43)	CS2	1560	2	R. Sideral, Daule: 1300-0500
IM15)	ZD1	1560	1.5	Ecos Culturales de Urcuquí, Urcuquí
PI50)	NG1	1560	1	LV de El Dorado, Puembo
CR20)	AI5	1570	0.5	Eco de los Andes, Cumandá: 1000-0200
MA23)		1570		R. Jet, Manta
PI51)	PG1	1570	10	R. Nucanchic, Maldonado
AZ19)	TP5	1580	3	Ecos del Portete, Girón: 1200-0330
BO08)		1580	1.5	LV de Libertador, Chimbo
ES09)	VA4	1580	5	Estación de la Alegría, Esmeraldas: 24h
GU44)	CP2	1580	0.5	Canal del Pueblo, Sanborondón
LO14)	AB3	1580	0.25	Ondas de Paltas, Catacocha
PI52)	LF1	1580	1	Ecos de Orellana, Machachi: 1030-0230
CP15)	CS6	1590	1	LV de La Maná, La Maná
GU45)	AS2	1590	0.25	R. Record, La Libertad
PI53)	RZ1	1590	1	R. Mensaje, Tabacundo: 1000-1400, 2130-0230
TU17)	QT6	1590	1	R. Panamericana, Quero: 1000-0200, (Sun -2400)
AZ20)	PB5	1600	3	R. Intiñán, Girón: 1100-0300
BO09)		1600		Ondas de Caluma, Caluma
GU46)	JP2	v1600	1	R. Consular, Playas: 1200-0100
LR11)		1600		R. Consular, Buena Fe
PI54)	UP1	1600	25	R. Unión, Quito: 1100-0100

Shortwaves:

	Call	kHz	kW	Name and h. of tr.
PI08)	JB1	3220	10	LV de los Andes, Quito: 0830-1330, 2100-0300
PI38)	MY1	v3250		LV del Triunfo, Sto Domingo de los Colorados: 1000-0500
MA24)	OS4	3260	0.5	R. Estéreo Carrizal, Calceta: 1000-0300
NA06)	VN7	v3280	2.5	LV del Napo, Tena: 1000-1215, 2200-0300
AZ13)	JA5	v3285	4	LV del Río Tarqui, Cuenca: 0800-1500, 1900-0500
TU08)	PV6	3290	1	R. Centro, Ambato
MS05)	VB7	3360		R. Dif. Católica Cultural "Voz del Upano", Macas
MS02)	OB7	3370	0.25	R. Nacional Limón, Limón Indanza: 2100-0300
ZC03)		v3569		R. Dif. Luciernaga del Condor, Yanzatza
PI09)	IC1	*4750		R. Municipal, Quito: 1030-0300
LO07)	VC3	v4770	5	R. Centinela del Sur "CDS Internacional", Loja: 1100-1400, 1645-1800, 2300-0030
NA02)	LE7	v4780	3	R. Oriental, Tena: 0955-1500, 2100-0300

	Call	kHz	kW	Name and h. of tr.
MS06)SK7		v4785	10	R. Federación Shuar, Sucúa: irr.
MA25)	AS4	v4795	3	LV de los Caras, Bahía de Caráquez: 1200-0400
LO15)	AX3	4815	1	R. El Buen Pastor, Saraguro -0305
LO06)	AV3	4850	3	R. Luz y Vida, Loja: 1100-1230, 2300-0300 (Sat -0400, Sun -0700)
MS06)	SK7	4860	10	R. Federación Shuar, Sucúa: 2100-0300
PI02)	GB1	v4880	1	R. Nal. Espejo, Quito: 1100-0400
PI10)	QR1	v4920	12	R. Quito "LV de la Capital", Quito: 0900-1430, 1645-1900, 2245-0500
IM10)	RN1	v4950	1	R. Bahá'í, Otavalo: 0900-1500, 1930-2300
MS06)	SE7	v4960	5	R. Federación Shuar, Sucúa: 1100-1300, 2100-0300
CR01)	ER5	v5010	0. 8	Escuelas Radiofónicas Populares, Riobamba: 0900-1400, 2200-0300
PI18)	RP1	5030	10	R. Católica Nacional, Quito: 1000 (Sun 1100)-0200
MS05)	VB7	5040	10	R. Dif. Católica Cultural "Voz del Upano", Macas: 1100-0300
PI07)	FF1	v5050	5	R. Jesús del Gran Poder, Quito: 0930-1500, 2300-0100
LO09)	ER3	v5060	5	R. Progreso, Loja: 1000-0315
PI08)	JB1	5860	250	LV de los Andes, Quito
MS06)	SK7	5980	10	R. Federación Shuar, Sucúa: irr.

For further LV de los Andes freqs: see International Broadcasting section.

Province abbreviations: AZ=Azuay, BO=Bólivar, CA=Cañar, CC=Carchi, CP=Cotopaxi, CR=Chimborazo, EO=El Oro, ES=Esmeraldas, GU=Guayas, IM=Imbabura, LO=Loja, LR=Los Ríos, MA=Manabí, MS=Morona Santiago, NA=Napo, PA=Pastaza, PI=Pichincha, SU+Sucumbios, TU=TUngurahua, ZC=Zamora Chinchipe. Galapagos, see Pacific.

N.B.: These abbreviations are not recognized by the Ecuadorian Post Office. Letters should therefore carry the full name.

Addresses and other information:
AZ00) AZUAY
AZ01) Bolívar 368, Cuenca – Dir: Jorge Antonio Chamoun.
AZ02) Larga 1043, Cuenca – Dir: Rolando Fáres Lucero.
AZ03) Cas. 01-01-0820, Cuenca – Gte: Fabián Molineros N.
AZ04) Cas. 198, Cuenca – Dir: Paúl Rodas Moscoso.
AZ05) Cas. 01-01-1352, Cuenca – Dir: Edmundo Valencia V. **FM:** 90.5+92.5.
AZ06) Cas. 01-01-0493, Cuenca – Dir: Jorge Piedra Ledesma. **FM:** 102.1.
AZ07) Cas. 01-01-4980, Cuenca – DG: Ing. Gustavo Cardoso.
AZ08) Edif. Hermano Miguel, Of. 301, Hermano Miguel, Entre Mariscal Lamar y Gran Colombia, Quito – Dir: Oswaldo Flores de la T.
AZ09) Cas. 1598, Cuenca – Dir: Daniel Pinos Guaricela.
AZ10) Cas. 60, Cuenca – Dir: Germán Morales Campoverde.
AZ11) Av. Loja 2408, Cuenca – Dir: Manena E. de Villavicencio.
AZ12) J. Dávila y C. Merchán, Cuenca.
AZ13) Manuel Vega 653 y Presidente Córdova, Cuenca – DG: Manuel Pulla C.
AZ14) Av. Ordóñez Lazo 2-247, Cuenca – DG: Dr. Kaupolikan Ochoa.
AZ15) Av. José María Quito, San Fernando – Dir: Humberto Romero V.
AZ16) Cas. 830 (or: Pumapungo E-25), Cuenca – Dir: Fernando Pinos L.
AZ17) Gran Colombia y 9 de Octubre 3102, Frente al Parque Central, Gualaceo – DG: Marco Sarmiento O.
AZ18) Convento Parroquial, Santa Isabel (or: Cas. 01-01-46, Cuenca) – Dir: Francisco Aperador.
AZ19) Antonio Flor 6-57, Girón – Dir: Thelmo Pinos Guaricela. AZ20) Girón.

BO00) BOLIVAR
BO01) Cas. 21, Guaranda – DG: Fausto F. Gaibor García. **FM:** 97.3.
BO02) A. Chavez y Davila, Guaranda.
BO03) Av. 10 de Agosto 612, Guaranda.
BO04) G. Moreno y 7 de Mayo, Guaranda.
BO05) Cas. 86, Guaranda – Dir: Jorge W. Caravajal G.
BO06) Jesús del Gran Poder, Caluma – Dir: Pomerio G. Garófalo M.

BO07) Simiátug, Guaranda – Dir: Mariano Poaquiza.
BO08) Guayas 194, San José de Chimbo – Dir: Oswaldo Jiménez.

CA00) CANAR
CA01) Av. Ingapirca, Cdla. El Vergel, Cañar (or: Cas. 01-01-0447, Cuenca) – Dir: Martín Samaniego C. Quechua: 0900-1300. **FM:** 94.3.
CA02) Veintimilla 110, Azogues.
CA03) Tomás Sacoto 161 y Sucre, Biblián (or: Cas. 729, Azogues) – Gte: Telmo V. Ortega.
CA04) Sucre 195, Biblián.
CA05) Cas. 720, Azogues – Dir: Lic. Wilson G. León.
CA06) Cas. 03-01-730, Azogues – Dir: Fray Raúl Pinto E.
CA07) Av. San Antonio y Panam, Cañar – Dir: Manuel Córdova E.
CA08) Cas. 731, Azogues – Dir: Humberto Vicuña N.

CC00) CARCHI
CC01) Colón 619, San Gabriel – Dir: Yolando Pozo P.
CC02) 13 de Abril, Julio Andrade – DG: Mons. José Avelino Fuertes.
CC03) Ap. 30, Tulcán – Dir: Fausto Almeida C.
CC04) Calle Salinas, El Angel – Dir: Ignacio Mena.
CC05) Cas. 42, Tulcán – Dir: Efraín Cabezas M.
CC06) Cas. 60, Tulcán – Dir: Ruth Muñoz L.
CC07) G. Suárez 824, Mira – Dir: Leopoldo V. Padilla.

CP00) COTOPAXI
CP01) Cas. 251, Saquisilí – Dir: Galo Morocho T.
CP02) Cas. 05-01-0392, Latacunga – DG: Victoriano Naranjo. **FM:**97.3+102.1.
CP03) Cas. 504, Latacunga – DG: Oscar Fabricio Erazo.
CP04) Cas. 208, Latacunga.
CP05) B. Quevedo 555, Pujilí – Dir: Jorge Barreno A.
CP06) Cas. 669, Saquisilí – Dir: Arturo Mena Herrera.
CP07) M. Benavides 7411 y Pastaza, Latacunga.
CP08) Cas. 286, Latacunga – Dir: Hilda Nieto de Muñoz.
CP09) Maldonado y 2 de Mayo, Latacunga.
CP10) Cas. 05-17-0294, Salcedo – Dir: Wilson R. Abril G.
CP11) Belisario Quevedo y Guayaquil, Latacunga.
CP12) Cas. 252, Latacunga – Dir: Isaías Carrera N.
CP13) Calle Calavi, La Maná – Dir: José María Lema Yugcha.
CP14) Cas. 474, Latacunga – Gte: Carlos Morán M.
CP15) B. León y La Prensa, La Maná – Dir: José Gonzalo S.

CR00) CHIMBORAZO
CR01) Cas. 06-01-0693, Riobamba – Dir: Juan Pérez Sarmiento. Quechua: 0900-1100, 2300-0300 – **FM:**91.7.
CR02) Cas. 06-01-0242, Riobamba – Dir: Ciro G. Cazar N.
CR03) Pichincha 1363, Riobamba – Dir: Luis A. Tasambay.
CR04) Cas. 06-01-0572, Riobamba – DG: Lcda. Mª. Dolores Encalada.
CR05) Majipamba, Colta – Prgrs. in Quechua only.
CR06) Cas. 20, Riobamba – Dir: Marcelo Vizcaíno M.
CR07) Cas. 348, Riobamba – Dir: Lic. Patricio Pumagualle.
CR08) 10 de Agosto 1742, Riobamba – Dir: J. Segundo León V.
CR09) M. E. Flor 4009, Riobamba – Dir: Luis F. Aliaga Dávalos.
CR10) Cas. 06-01-0471, Riobamba – DG: Luz. V. Luzuriaga N.
CR11) Av. Amazonas y Chávez, Penipe – Dir: Pedro V. Padilla.
CR12) Cas. 82, Riobamba – DG: Lic. Raúl Lomas B. **FM:**100.1.
CR13) Cas. 287, Riobamba – Dir: Galo Silva.
CR14) México y La Paz, Riobamba.
CR15) Cas. 06-01-0376, Riobamba – DG: Doria N. Guevara.
CR16) Cas. 06-03-0805, Alausí – DG: Antonio Ordóñez A.
CR17) Olmedo 3128, Riobamba – Dir: N. Fausto M.
CR18) Maldonado y Vela, Guamote – Dir: Miguel Alcócer.
CR19) Panamericana y Eloy Alfaro, Pallatanga – DG: Hemilton Novilos Toledo.
CR20) Abdon Calderón y Simon Bolívar, Cumanda.

EO00) EL ORO
EO01) Franco y Colón, Santa Rosa – Dir: Boanerges Ugarte V.
EO02) Cas. 221, Machala – Dir: Rodrigo Pineda I.
EO03) Cas. 473, Machala – Dir: Raúl Cedillo S.
EO04) Cdla. Marlene Nieto, Calle Eloy Alfaro, Santa Rosa – Dir: A. Acosta.
EO05) Cas. 07-01-0086, Machala (or: 9 de Octubre y Páez), Machala – Dir: Héctor Bravo.
EO06) Av. H. Márquez 105, Zaruma – Dir: José R. Valdiviezo L.
EO07) 9 de Mayo y Pasaje, P2, Machala.
EO08) Cas. 23, Zaruma – DG: Prof. Luis A. Berrezueta A.
EO09) Chávez Franco y Guayas, Edif. Sind. de Choferes, Machala – Dir: Joffre Unda Ch.

EO10) Cap. Chiriboga y J. J. Olmedo s/n, Arenillas –
Dir: Franco Ordoñez.
EO11) Ochoa León 505, Pasaje – Dir: César Prado A.
EO12) Cas. 895, Machala – Dir: Vicente Serrano Zambrano.
EO13) Cas. 474, Machala – Dir: Silvio Pinos G.
EO14) Sucre 1105, Piñas – Dir: G. Olmedo Torres.
EO15) Cdla. Jaime Roldós, Calle Martha Bucaram, Huaquillas – Gte:
Carlos Naranjo M.
EO16) Rocafuerte 619, Pasaje – Dir: Alfredo H. Sarmiento.
EO17) Av. del Ejército, El Guabo – Dir: Rosa R. Vda. de Ramón.

ES00) ESMERALDAS
ES01) Cas. 08-01-0065, Esmeraldas – Dir: Luis E. Velasco León.
ES02) Mejía 325 y Olmedo, Esmeraldas – Dir: Rónald Murillo C.
ES03) Manuela Cañizares y Olmedo, Esmeraldas.
ES04) Cas. 108, Esmeraldas – Dir: Héctor Endara Endara. **FM:** 92.1.
ES05) Plaza Torres, Quinindé – Dir: Fulton Villagómez.
ES06) Cas. 116, Esmeraldas – Dir: Homero López S.
ES07) Cas. 155, Esmeraldas – Dir: Walter Delgado Clavijo.
ES08) Edif. Benegas, La Concordia – Dir: Jesús Yánez B.
ES09) Cas. 379, Esmeraldas – Own: Jaime Recalde B.

GU00) GUAYAS
GU01) Cas. 4144, Guayaquil – Dir: Rafael Guerrero V. **FM:** 90.5.
GU02) Cas. 2119, Guayaquil (or: Amazonas 743 y Veintemilla, P8,
Quito) – Dir: Raúl Marcos Salcedo C.
GU03) V. M. Rendón y Garaycoa, Edif. Huancavelica, Guayaquil.
GU04) Av. Quito 933 y Velez, Guayaquil.
GU05) Cas 9974 (or: Av. de Las Américas junto Canal 10), Guayaquil
– Dir: Kléber Chica.
GU06) Cas. 204, Guayaquil – Gte: A. Mendoza Paladines.
GU07) Cas. 11714, Guayaquil – Dir: Vicente Arroba D.
FM: 95.3.
GU08) Cas. 2440 (or: Escobedo 1504 y Aguirre, P9), Guayaquil.
GU09) Cas. 716, (or: P. Icaza 437 y Cordova), Guayaquil – DG: Ab.
Mario Canessa O. **FM:**88.9.
GU10) Cas. 09-01-4203 (or: Colón 548 y Boyacá, P7), Guayaquil –
DG: Hugo Miranda.
GU11) Juan Montalvo 1042, El Milagro – Gte: Ecuador Martínez C.
GU12) Cas. 856 (or: Edif. Gran Pasaje. Of. 906/908), Guayaquil -
Dir: A. Feraud M.
GU13) Cas. 09-01-5762, Guayaquil – Dir: P. Alfredo Velasco R.
GU14) Cas. 5062 (or: Laque 1407 y Antepara), Guayaquil – Dir:
Armando Romero Rodas.
GU15) Machala 2605 y Gomez, Guayaquil – Dir: C. Córdova V.
GU16) Pista Papín, El Empalme – Dir: Oswaldo Vélez B.
GU17 Cas. 11548 (or: 9 de Octubre y Los Ríos), Guayaquil – Dir:
Wilson Chávez.
GU18) Ballén 501, Guayaquil – Dir: Mons. Elías Dávila.
GU19) Cas. 10863 (or: 1° de Mayo), Guayaquil – Dir: Guillermo
Rendón.
GU20) Cas. 9406, Guayaquil.
GU21) Cas. 09-01-4719 (or: García Moreno 913), Guayaquil – DG:
Julio Coll Suárez.
GU22) Cas. 5177, Guayaquil – Dir: Gabriel Arroba D.
GU23) Cas. 856, Guayaquil – Dir: Fausto Feraud Manssur.
GU24) Cas. 4839, Guayaquil – Dir: Luis Albán.
GU25) Cas. 1054, Guayaquil – Dir: V. Villacís G.
GU26) Cas. 8529, Guayaquil – Dir: P. Juan Villamar.
GU27) Cas. 3311, Guayaquil – Dir: Gabriel Vergara J.
GU28) L. de Garaycoa 2615, Guayaquil – Dir: H. Alvarado P.
GU29) Laurel y Guayacanes, El Milagro – Dir: Francisco Ortiz G.
GU30) Cas. 4065, Guayaquil – Dir: W. Delgado C.
GU31) Av. 9 de Octubre 1703, Guayaquil.
GU32) Av. 17 de Septiembre, El Milagro – Dir: Carlos H. Vega V.
GU33) Av. Cuba y Ecuador, El Milagro – Dir: Stelvio H. Centenario
Q.
GU34) Pte. Portete, Guayaquil.
GU35) Quito 1520 entre Sucre y Colón, Guayaquil.
GU36) Guayaquil y 9 de Octubre, Santa Elena – Dir: Walter Gellibert
L.
GU37) Av. 5 de Octubre 150, Naranjito – Dir: Leoncio Delgado H.
GU38) Coll. 7 Lagos, Guayaquil.
GU39) Cas. 5940, Guayaquil – Cf. listing under Standard Frequency
and Time Signal St's.
GU40) Pastaza y Guayaquil, Naranjal – Dir: José D. Toapanta Ch.
GU41) 4a Av. 619 y Robles, La Libertad – Own: M. Pinoargote P.
GU42) Av. 8 de Abril y Roldós Aguilera, El Triunfo – Gte: A. Torres.
GU43) Sitio El Belén, Daule – Dir: Francisco Dávila de la R.
GU44) El Oro y Los Ríos, Sanborondón – Dir: W. Gellibert L.

GU45) 12 de Octubre 1032, La Libertad – Dir: A. N. Suárez B.
GU46) Gral. Villamil, Playas (or: Cas. 10983, Guayaquil) – Dir: Jorge
Pazos S.
GU47) Urdesa Central, Guayaquil.
GU48) Velez y Pedro Moncayo, Guayaquil – Dir: Marlene de
Sanchez.

IM00) IMBABURA
IM01) San Pablo del Lago, Otavalo – Dir: F. Beltrán Proaño.
IM02) Flores 1115, Ibarra – Dir: Luis Alfredo Mejía M.
IM03) Cas. 413, Ibarra – Dir: Luis Viteri A.
IM04) Cas. 10-01-0179, Ibarra – DG: Ricardo M. Ruiz.
IM05) Cas. 10-01-0099, Ibarra – Dir: Augusto Báez R. **FM:** 98.5.
IM06) Cas. 10-01-0181, Ibarra – Dir: Silvio Morán Madera.
IM07) Cas. 3, Atuntaqui – Dir: Carlos A. Calderón
IM08) G. Suarez s/n y Montufar, Pimapiro – Dir: José Valle N.
IM09) Antonio Ante 100, Urcuquí – Dir: G. Ruiz S.
IM10) Cas. 10-02-1464, Otavalo – DG: William Rodriguez B. Quicha:
0900-1200, 1930-2300, Sp: 1200-1500.
IM11) Cas. 694, Ibarra – Dir: Cristóbal Bonifaz A. **FM:** 93.7.
IM12) Palacio Municipal, Cotacachi – Dir: Ulpiano Saltos T.
IM13) Cas. 1474, Otavalo – Dir: Marco A. Chicaiza T.
IM14) Av. Pérez Guerrero 812, Ibarra – Gte: Carlos Arciniega.
IM15) Antonio Ante s/n, Urcuquí – Dir: Fausto Dueñas C.

LO00) LOJA
LO01) Cas. 521, Loja – Dir: A. Maldonado.
LO02) Cas. 474, Loja – DG: P. Alvaro J. Chamorro. **FM:** 100.3.
LO03) 24 de Mayo s/n, Catamayo.
LO04) Celia y El Comercio, Pindal.
LO05) 24 de Mayo de Espejo, La Toma – Dir: Norberto A. Torres C.
LO06) Cas. 11-01-222, Loja – DG: Hna. Ana Maza Reyes – **FM:** 88.3.
LO07) Cas. 196, Loja – Dir: José A. Coronel V.
LO08) Av. Lauro Guerrero 14-25 y Venezuela, Loja – Gte: Lucía E.
Vda. de García.
LO09) Cas. Letra V, Loja – Dir: Efraín Herrera G.
LO10) Valdivieso 859, Loja – Dir: Victor M. Sánchez.
LO11) 18 de Noviembre s/n, Cariamanga – Dir: Jorge Montero R.
LO12) Yanzatza.
LO13) R. Carrasco y Paz, Cariamanga – Dir: Luis P. Cuenca Valle.
LO14) Coop. Ahorro y Crédito 3 de Diciembre, Isidro Ayora 235,
Catacocha – Dir: Artemio Ríos G.
LO15) Asociación Cristiana de Indigenas Saraguros, Saraguro – DG:
Segundo Poma – Quicha 1200-1400, 2100-2300, Sp: 1400-1600,
2300-0100. ID in Quechua: "R. Alli Michic".

LR00) LOS RIOS
LR01) 12 Calle N° 207, Quevedo – Dir: Humberto Alvarado P.
LR02) Manabí y J. L. Mera, El Empalme.
LR03) Bolívar y Velasco, Ventanas.
LR04) Cas. 104, Quevedo – Dir: Rosendo V. Escobar C.
LR05) Cas. 77, Babahoyo – Dir: Fernando E. Ronquillo F.
LR06) Cas. 31, Babahoyo – Gte: Gustavo von Lippke. **FM:** 100.9.
LR07) Av. 7 de Octubre 917, Quevedo – Dir: Julio César Jalón F.
LR08) Cas. 84, Quevedo – Dir: Benito Wong Y.
LR09) Olmedo 109, Vinces – Dir: Selvi Elizalde C.
LR10) 28 de Mayo y 6 de Octubre, Ventanas – Dir: Arnaldo Rivas.
LR11) G. Rivera 11 de Octubre, Buena Fe.

MA00) MANABI
MA01) Cas. 4817, Manta – Dir: César Dávalos H.
MA02) Olmedo 426 y 10 de Agosto, Portoviejo – Dir: Jorge
Gutiérrez.
MA03) Cas. 89, Portoviejo – Dir: Guido Bonilla.
MA04) Cas. 6, Chone – Dir: Stalin Espinel A.
MA05) Cas. 4810, Manta – Dir: Enrique Quevedo A.
MA06) Cas. 69, Portoviejo – Dir: Lauro Fernández.
MA07) Colón 180, Portoviejo – Dir: Sergio Enrique Morlas A.
MA08) Cas. 8, Chone – Dir: Argemiro Andrade.
MA09) Calle 13 y Av. 14, Manta – Dir: W. López Mero.
MA10) Cas. 5, Chone – Dir: Alberto E. Espinel A.
MA11) 9 y Malecón, Edif. "Jacob Vera", P1 Of. 7, Manta -
DG: Lcdo. Frank Zambrano M.
MA12) C. Central, Portoviejo – Dir: Ab. Joffre Cedeño Molina.
MA13) Cas. 145, Portoviejo – Dir: Dúmar Iglesias M.
MA14) Bolívar 1219, Bahía de Caráquez – Dir: Trajano Velástegui.
MA15) 10 de Agosto 180, Junín – Dir: Ricardo Villaviconcion.
MA16) Cas. 13-04-0705, Jipijapa – Own: José R. Falconí Y.
MA17) Bolívar 920, Bahía de Caráquez – Dir: W. Delgado C.
MA18) Chone y Carlos Aray, El Carmen – Dir: Enrique García E.

MA19) Av. Principal, Tosagua, Calceda – Dir: Luis Grijalva Delgado.
MA20) Noboa y Colón, Jipijapa – Dir: Ulbio Delgado H.
MA21) 4 de Diciembre y Alfaro, El Carmen – Dir: Enrique García E.
MA22) Cas. 13-05-4869, Manta – DG: Lic. Cínerman Miranda.
MA23) Av. 4 entre Calles 12 y 13, Manta – Dir: José G. España T.
MA24) Av. Estudiantil, Quinta Velásquez, Calceta – Dir: Ovidio Velásquez A.
MA25) Cas. 13-01-629, Bahía de Caráquez – Dir: Marcelo Nevares Faggioni. **FM:** 95.3.

MS00) MORONA SANTIAGO

MS01) Centro de Macas, Macas.
MS02) Quito s/n, Limón Indanza – Dir: Antonio Ochoa.
MS03) Tarquí 6-34 y 24 de Mayo, Macas (or: 10 de Agosto 4106 y Rumipampa, Quito) – Dir: Eduardo Romero.
MS04) G. Pesántez 375, Gualaquiza – Dir: J. V. Hernández G.
MS05) Misión Selesiana, Calle 10 de Agosto s/n, Macas. Dir: P. Domingo Barrueco. Shuar: 1200-1230, 2230-2300 on SW.
MS06) Federación de Centros Shuar, Domingo Comín 17-38, Sucúa (or: Cas. 17-01-4122, Quito) – DG: Ruben Yurank. Educational prgr. in Shuar only. ID in Shuar: "Shuar Achuara Tuntuiri".

NA00) NAPO

NA01) Fco. de Orellana, El Coca – Dir: Leonor Cargua de Jiménez.
NA02) Cas. 260, Tena – Dir: Luis Enrique Espín Espinoza.
NA03) El Coca – Dir: Gonzalo Mansilla Santos.
NA04) Santa Rosa de Quijos (or: Cas. 11294, Quito) – Dir: Walther Manzo B.
NA05) Elchaco, Archidona – Dir: Aurelio Rosero Castillo.
NA06) Misión Josefina, Tena – Dir: Ramiro Cabrera.

PA00) PASTAZA

PA01) Cas. 744, El Puyo – Dir: Julio Zamora.
PA02) Vía Macas km 1. 5, El Puyo – Dir: Alberto Salvador S.
PA03) Cas. 728, El Puyo – Dir: Walter Moyano.
PA04) Cas. 06-01-0777, El Puyo – DG: Monseñor Frumencio Escudero A. **FM:** 88.9.

PI00) PICHINCHA

PI01) Cas. 1018, Quito – Dir: Daniel Alvarez T.
PI02) Cas. 352, Quito – Dir: Marco Caicedo P.
PI03) Cas. 8270 (or: 9 de Octubre 215 y 18 de Septiembre), Quito – Dir: Jaime Bowen.
PI04) Juan Severino 391, Quito.
PI05) Cas. 17-01-2246, Quito – Dir: Nancy Cevallos Castro.
PI06) Cas. 60 (or: Mariano Echeverría y Brasil), Quito – Dir: Gustavo Cevallos.
PI07) Cas. 17-01-0133, Quito – Dir: R. P. Jorge Enríquez Silva.
PI08) Cas. 17-17-0691, Quito. See Int. Broadcasting section.
PI09) García Moreno 887 y Espejo, Quito – Dir: Lic. Sammy de la Torre.
PI10) Cas. 17-21-1971 (or: Av. Amazonas 1491 y Colón, Edif. España, P,4) Quito – DG: Dr. Xavier Almeida M.
PI11) Cas. 2938, Quito – Dir: Arcesio Arcentales R.
PI12) Principal s/n, Maldonado.
PI13) Av. Amazonas 1538 y La Niña, Quito – DG: Dr. Polibio Córdova C.
PI14) Cas. 038, Sto Domingo de los Colorados.
PI15) Paris N 8 y Av. Los Granados, Quito.
PI16) Ramírez Dávalos 612 y 10 de Agosto, Quito – Dir: My. Mario Valenzuela Alvarado.
PI17) Cas. 251-A (or: Checoeslovaquia 323), Quito – Dir: Lic. Lulio Pazmiño.
PI18) Cas. 17-03-540, Quito – DG: Dr. John Siguenza.
PI19) Palacio 303 y Av. La Gasca, Quito – Dir: L. Pumagualle.
PI20) Av. América y Villalengua esq. P3, Quito – Dir: Margarita Molina.
PI21) Cas. 17-01-0067, Quito – DG: Francisco Velasco A.
PI22) Morales 1224, Quito – Dir: Numa P. Castro.
PI23) Cas. 17-24-0567, Sto Domingo de los Colorados – Dir: Padre Gualberto Pérez Paredes.
PI24) Cas. 2558, Quito – Own: Diego Chriboga.
PI25) Cas. 045, Sto Domingo de los Colorados – Dir: C. Reinoso M.
PI26) J. L. Mera 565 y Carrión, Quito.
PI27) García Moreno 486, Machachi – Gte: S. B. Cueva S.
PI28) Cas. 17-10-7006 (or: Tarquí 785 y Estrada), Quito – DG: Fernando Espinoza N.
PI29) Guayaquil 124, Sto Domingo de los Colorados – Dir: M. Zavala V.
PI30) Cas. 17-11-6014, CCI, Quito – Dir: Jorge Aguilar V.

PI31) San Ignacio y Orellana, Quito- DG: Rodrigo Coronel.
PI32) Cas. 17-01-2263, Quito – Dir: Ligia Wilches de Garcés. In Sto Domingo de los Colorados studios at Av. Quevedo 405.
PI33) Cas. 17-01-0638, Quito – Dir: Dr. Raúl Izurieta M.
PI34) Cas. 2493 (or: Marquesa Dosantos y Montúfar), Quito – DG: Manuel Cabezas.
PI35) Cas. 6110, Quito – Dir: Carlos E. Machado.
PI36) Cas. 47, Sangolquí – Dir: Abdón Calderón González.
PI37) 12 de Octubre 227, Quito – Dir: Eduardo Loza C.
PI38) Cas. 17-24-0043, Sto Domingo de los Colorados – Dir: M. Yánez O.
PI39) Cas. 17-01-3386, Quito – Dir: Padre Jesús Palomino I.
PI40) Edif. Doral Mariscal, Of. 86, Páez y Mercadillo, Quito – DG: Gonzalo Rosero.
PI41) San Miguel y 29 de Mayo, Sto Domingo de los Colorados – Dir: Rómmel Velástegui I. **FM:** 89.7.
PI42) Cas. 17-12-0861 (or: Madrid 1234 y Andalucía), Quito – Dir: Ing. Enrique Grímbal B.
PI43) Edif. Vidal Gómez, P3, Sto Domingo de los Colorados – Dir: Joseph Sánchez.
PI44) Interoceánica y Panecillo, Tumbaco – DG: Lcdo. Eduardo Barros Proaño. **FM:** 104. 1.
PI45) Av. de la Prensa N60-22, Entre Angel Ludeña y Av. del Maestro, Quito – Dir: Ricardo Romero Rivas.
PI46) Olmedo 122 y Ríos, Quito.
PI47) Bolívar y Rocafuerte esq. , Cayambe – Own: Ramiro Almeida.
PI48) Terán 5-27, Entre Calderón y 10 de Agosto, Cayambe – DG: Juan M. Córdoba Solá.
PI49) Venezuela 701 y Espejo, Quito – Dir: J. A. Buenaño.
PI50) Av. 24 de Mayo s/n, Puembo – Dir: G. Hernández N.
PI51) Pedro Vicente, Maldonado (or: Concejo Provincial de Pichincha, Manuel Larrea y Antonio Ante, Cas. 298, Quito).
PI52) Luis Cordero 226, Machachi – Dir: José J. Quinga.
PI53) Plaza Gutiérrez 255, Tabacundo (or: Cas. 7179, Suc. 10, Quito) – Dir: Isaías Barriga N.
PI54) Cas. 5184 (or: Unión Nacional de Periodistas, Iñaquito y N. N. U. U.), Quito – DG: Hector Espin.
PI55) Av. Marín, Dorado, Quito.
PI56) Av. 6 de Diciembre y Julio Moreno, Edif. Jockey Club, Torre Granderví Mezzanine 31, Quito. Own: Coronel Edgar Salinas.

FM in Quito: 88.1 Latina – 88.5 Metro – PI08) 89.3 – 89.7 Majestal – 90.1 Tropicalida – 90.5 Concierto – 90.9 Platino – 91.3 Planeta – 93.7 Galaxia – PI18) 94.1 – 94.5 Rumba – 94.9 Radio Q – 95.3 Teleonda Musical – 95.7 Cumbre – 96.1 Joya – PI30) 96.5 – PI31) 97.3 – 97.7 Centro – 98.1 "98.1" – 98.5 Alpha – OI20) 98.9 – 99.3 La Luna – 99.7 La Rumbera – PI23) 100.5 – 101.7 "101.7" – 102.1 "102.1" – PI07) 102.5 – 102.9 Armonía – 103.3 Sonora – 104.5 América – 104.9 Ecuashyris – 105.3 Sintonía – 105.7 CRE – 106.1 Pichincha – 106.5 Bonita – 107.3 La Bruja.

SU00) SUCUMBIOS

SU01) Cas. 21-01-45, Nueva Loja – Gte: Víctor J. Campoverde – **FM:** 101.7.
SU02) Cas. 21-01-14, Lago Agrio – DG: Noemi Torres A. – **FM:** 105.3.
SU03) Cas. 40 (or: Mariscal Sucre y 12 de Febrero), Lagos Agrio – Dir: Fausto Velástegui G.

TU00) TUNGURAHUA

TU01) Cevallos 345, Ambato.
TU02) Bolívar 917, Píllaro (or: Cas. 18-01-244, Ambato) – Dir: Angel G. Salazar.
TU03) Cas. 18-01-0181, Ambato – Dir: Mario R. Barona Andrade. **FM:** 96.7 R. Amor.
TU04) Cas. 18-02-1921, Baños – DG: P. Neptali Acosta M.
TU05) Mariano Equez 367 Ambato – Dir: Alberto Salazar.
TU06) Chayán y Daquilema, Ciudadela Presidencial, Ambato – Dir: Ramón Salazar C.
TU07) Av. 22 de Julio y P. Chacón 447, Pelileo – DG: Lic. José M. Rodríguez.
TU08) Cas. 18-01-0574, Ambato – Dir: Luis A. Gamboa T. **FM:** 93.7.
TU09) Cas. 818, Ambato – Dir: Angel Nuñez.
TU10) Cas. 18-01-0198, Ambato – Dir: Luis Ramiro Vela G.
TU11) Cas. 368, Ambato – Dir: Gerardo Velástegui G.
TU12) Cas. 18-01-0094, Ambato – Dir: Padre Carlos Avedaño. **FM:** 92.9+104.5+106.9.
TU13) Cas. 618, Ambato – Dir: Isabel Yolanda Andaluz O.
TU14) Bolívar y Carlos Tamayo, Píllaro – Dir: Hugo Ibarra S.
TU15) Cas. 146, Pelileo – Dir: Luis Jordán M.

TU16) Cas. 173, Ambato – Dir: Pedro Calvache.
TU17) Montalvo 106, Quero – Dir: Fauri Llerena.

ZC00) ZAMORA-CHINCHIPE
ZC01) Sevilla de Oro 312 y Pasaje San Francisco, Zamora – DG:
Juan Cartagena Alvarado – **FM:**102.7.
ZC02) Parroquia 28 de Mayo, Cantón Yacuambí.
ZC03) Yanzatza.

FALKLAND ISLANDS (British)

L.T: UTC - 3h — **Pop:** 2.564 (excl. military personnel) — **Radios:** 1.000
— **Pr.L:** English — **E.C:** a/c 50, 220V — **ITU:** FLK.

FALKLAND ISLANDS BROADC. STATION (Gov.)
✉ John Str, Stanley. ☎ +500 27277/8. 🖹 +500 27279.
L.P: Broadc. Officer: Patrick Watts, MBE. Secr: Steff Hanlon.

BRITISH FORCES BROADCASTING SERVICE
✉ Rockhopper Road, RAF Mount Pleasant. BFPO 655. ☎ +500 32179.
🖹 +500 32193 — **L.P:** SM: Colin McDonald.

Transmitters shared by FIBS and BFBS:
MW: Bush Rincon 550kHz 10kW.
FM: Pt. Stanley 88.3MHz 0.015kW, March Ridge 90MHz 2kW, Bryon
Heights 90.0MHz 0.015kW, Mt. Alice 92.5MHz 0.3kW, Sapper Hill
96.5MHz 0.03kW, MPA 98.5MHz 0.01kW, Mt. Maria 102.0 MHz 2kW,
Mt. Byron 104.7MHz 0.3kW.

D.PRGR: 24h (FIBS: approx 1000-1215 and 1715-2200, other times
BFBS). **N:** 0900, 1000, 1100, 1200, 1500, 1600, 1700, 1900, 2000 (relays
of BBCWS or BBC News fed from BFBS London by satellite) — **V.** by
QSL-card. Rp.

GUIANA (French)

L.T: UTC - 3h — **Pop:** 120.000 — **Radios:** 71.000 — **Pr.L:** French —
E.C: 50Hz, 127/220V — **ITU:** GUF.

R. F.O. GUYANE (RADIODIFFUSION FRANÇAISE D'OUTRE-MER)
✉ B.P. 7013, F-97305 Cayenne. ☎ +594 301500. 🖹 +594 302649.
L.P: Dir: Henri Neron. Head of N: Anastasie Bourquin. Dir. Tec: Daniel
Beugin. PD: Jean-Pierre Karam — **MW:** Matoury 1070kHz 10kW, St.
Laurent du Maroni 1060kHz 0.05kW.
SW: Matoury (G.C: 52.20W/04.54N): 5055kHz 10kW.
FM: Cayenne 91.5/95.2MHz 0.05kW, 94.0MHz 0.3kW + 5 relays.
D.PRGR: 24h on 1060/5055kHz + FM. **N:** 1030, 1630, 2230.
Rel. France-Inter: 24h on 1070kHz + 91.5MHz.
ANN: "Ici Cayenne, R. F.O. Guyane".
IS: "Nos richesses" on guitar — **V.** by QSL-folder. Rec. acc.

Local Radio: 6 private FM-st's are operating.

RADIO FRANCE INTERNATIONALE RELAY STATION
✉ TDF Montsinery, B.P. 97307, Cayenne Cedex.
FM: Cayenne 98.7MHz.
SW: see International Broadcasting section.

SWISS RADIO INTERNATIONAL RELAY STATION
See International Broadcasting section.

GUYANA (Co-operative Republic of)

L.T: UTC - 4h — **Pop:** 700.000 — **Radios:** 350.000 — **Pr.L:** English,
Hindi, Amerindian dialects — **E.C:** 50Hz, 240V — **ITU:** GUY.

GUYANA BROADCASTING CORP. (Gov. Comm.)
✉ Broadc. House, P.O. Box 10760, Georgetown. ☎ +592 (2) 58734. 🖹
+592 (2) 58756 — **L.P:** Ag. Chmn: Earl Bousquet. GM: Fazil Azeez. Prgr.
Mgr: Margaret Lawrence. CE: Shiroxley Goodman.
MW: Georgetown 560/760kHz 10kW, Linden 700kHz 1kW

SW: Sparendaam (G.C: 58.10W/06.49N): 3290/5950kHz 10kW
FM: Georgetown 100/102MHz, Linden 104MHz.
R. Roraima: 0800-0200 on 760kHz + 100MHz. **N:** 0900, 1000, 1100,
1330, 1500, 1900, 2100, 2230 (Sun), 2300 (W), 0100.
Voice of Guyana: 24h on 560/700kHz + 102/104MHz, 0900-2200 on
5950kHz, 2200-0900 on 3290kHz. **N:** as R. Roraima.
V. by letter.

PARAGUAY

L.T: UTC - 4h (SU: UTC - 3h) — **Pop:** 5.225.000 — **Radios:** 700.000
— **Pr.L:** Spanish, Guaraní — **E.C:** 50Hz, 220V — **ITU:** PRG.

ADMINISTRACION NACIONAL DE TELECOMUNICACIONES
Offices: Administración General, Edificio Gral. Bernardino Caballero,
Alberdi y Oliva 4⁰ piso (✉ Cas. de Correo 84), Asunción.☎ +595 (21)
44001. **Cable:** Antelco.
L.P: Pres: Ing. Miguel Horacio Gini. Dir. Admin: F.F.Duarte. Dir. Tech.
Dept: Ing. Osmar G. López R. Dir. Radio: Ing. A.Barboza.

Mediumwaves: Þ) also on shortwave, *) inactive, v) varying freq.
All st's comm. exc. R. Nacional.

kHz		kW		Name, location and h. of tr. (Call)
1)	ZP16	550	20/12	R. Parque, Cd. del Este: 0900-0200
2)	ZP15	vÞ570	1	R. LV del Amambay, Pedro Juan Caballero: 0930 (Sun 1000)-0100
3)	ZP32	590	5	R. Ycuamandyú, Villa de San Pedro: 0900-0200
4)	ZP30	610	10/1	R. ZP30, LV del Chaco Paraguayo, Filadelfia: 0930-0200
5)	ZP40	v620	1	R. Nasindy, San Estanislao (r. 619-616.6v)
6)	ZP19	640	8	R. Caaguazú, Coronel Oviedo: 0830-0300
44)		650	15	R. Uno, Asunción: 0900 (Sun 0930)-0400 (SS 0300)
7)	ZP29	650	5	R. Vallemi, Vallemi: 0800-0200
8)	ZP26	660	6/10	R. Itapirú, Cd. del Este: 0800-0100
29)	ZP11	680	10/1	R. Cáritas, Asunción: 0800 (Sun 0900)-0400
9)	ZP41	680	2	R. Ypacaraí, Ypacaraí: 0920-0200
10)	ZP12	v700	10	R. Carlos Antonio López, "LV del Neembucú", Pilar: 0900-0200 (r. 699.6)
45)		720	25	R. Pa'i Puki, 25 Leguas
11)	ZP7	730	50/5	R. Cardinal AM Stéreo, Asunción: 24h.
12)	ZP5	760	10	R. Encarnación: 0800-0400
13)	ZP70	780	25	R. Primero de Marzo, Asunción: 0830-0400
14)	ZP27	800	5/3	R. Mbaracayú, Saltos del Guairá: 0900-0300
15)	ZP6	vÞ840	5	R. Guairá, Villarrica: 0900-0200
16)	ZP28	v860	25	LV de la Cordillera, Caacupé: 0900 (Sat/Sun 1000)-0400 (r. 860.4)
17)	ZP33	v885	5/10	R. Tres de Febrero, Itá: 0900-0300 (n. 890)
18)	ZP1	Þ920	100	R. Nal. del Paraguay, Asunción: 0745 (Sun 0845)-0400
19)	ZP9	970	10	R. Monte Carlo AM, Asunción: 0900-0400
20)	ZP31	980	5	R. Mburucuyá, Pedro Juan Caballero: 0900-0400
21)	ZP36	1000	5/1	R. Ypoá 1000, Paraguari: 0900-0100
22)	ZP14	1020	25/10	R. Nandutí, Asunción: 24h.
23)	ZP43	1040	5	R. Arapizandú, San Ignacio(Misiones): 0900-0200
24)	ZP13	v1060	3	R. Boquerón, Alberdi: 1000-0300 (r. 1060.5)
25)	ZP25	1080	10	Radiodif. Nanawa, Luque: 0800 (Sun 0900)-0300
26)	ZP71	1100	5	R. Ñu Verá, Capitán Badó: 0900-0200
27)	ZP24	1120	10	R. Nuevo Mundo, San Lorenzo: 0900-0200
28)	ZP22	1140	5/3	R. Panambí Verá, Villarrica: 0900-0200
46)	ZP--	1160	10	R. Antena 3, Asunción
43)	ZP52	1180	3	R. Coronel Oviedo, Coronel Oviedo
34)		1200	10	R. Libre, Itauguá
30)	ZP21	1230	3	R. Oriental, Caaguazú: 0800-0300
31)	ZP3	1250	10/15	R Asunción, "LV del Paraguay", "La R. Capital": 0900-0600

	Call	kHz	kW	Name, location and h. of tr.
42)	ZP34	1260		R. Cultura, Villarrica
32)	ZP10	*1300	5	Em. Paraguay, Asunción
41)	ZP53	1310		LV del Este, Cd. del Este
33)	ZP4	*1330	10	R. Chaco Boreal, Asunción
36)	ZP37	1360	1	R. Yby Yau, Carapeguá: 0900-0300
35)	ZP8	1380.7	1	R. Concepción, "LV del Norte": 0830-0030
37)	ZP35	1410	2	R. Dif. Mangore, San Juan Bautista Misiones: 1000-0130 (n. 1430)
38)	ZP42	1417	5	R. Güyrá Campana, Horqueta (n. 1420)
39)	ZP23	1480	1	R. Mariscal Francisco Solano López, Bella Vista Norte: 1000-0100
40)	ZP20	v1480	1	R. America, Ñemby: 0800-0300

Shortwaves: *) inactive, v) varying freq.

	Call	kHz	kW	Name and location, h. of tr.
15)	A6	5976.1v	1	R. Guairá, Villarrica:(n. 5975)
2)	A15	*5995	2	R. LV del Amambay, Pedro Juan Caballero
32)	A10	*6014.4v	0.3	Em. Paraguay, Asunción:
18)	A1	*6025	0.7	R. Nal. del Paraguay, Asunción
18)	A1	9736.2v	100	R. Nal. del Paraguay, Asunción: 0800-0400 (n. 9735)
12)	A5	11939.5	1	R. Encarnación: 0800-0400 (n. 11940)

Addresses and other information

1) km 10, Ruta Internacional, Cd. del Este. ☎ +595 (61) 570262 — DG: Eduardo Maria Barreto. **Guaraní:** 0900-1130, 2200-2300. **FM:** 101.1.
2) 14 de Mayo 475, Pedro Juan Caballero — DG: L.C. Rolón P.
3) Ruta 11 Juana M. de Lara, Villa de San Pedro — Dir: Ing. Juan B. Ibáñez.
4) Av. Trebol 137E, Filadelfia, Chaco (✉ Casilla 984, Asunción). ☎ +595 (91) 2330. 🖷 +595 (91) 2501 — Dir: Erwin Wiens. Prgrs in Guaraní & ethnic, German, E. and Sp. operated by Mennonite Mission. **F.PI:** SW.
5) 13 de Noviembre y Boquerón, San Estanislao — Dir: Demetrio Sánchez Candia.
6) Ruta Mcal. Estigarribia Km 131, Coronel Oviedo. ☎🖷 +595 (21) 202251 — DG: Anibal Espinola.
7) Fulgencio R. Moreno 1060, Asunción.
8) Av. San Bkas Km 3,800 Ruta Internacional, Cd. del Este — Dir: Robero Simbron. **FM:** 96.1.
9) Calle Paraguay, Ruta 2, km 36, Ypacaraí.
10) Alberdi 998, Pilar — Gte: Miguel Angel Rodríguez Hermosa.
11) Rio Paraguay y Guaranies, Lambaré (✉ Cas. 247, Asunción) — DG: Rodolfo Schaerer Peralta.
12) Gral. Artigas 724 y Capitán Caballero, Encarnación — Dir: Ing. Pedro Gomez Falcón. **Guaraní:** 0800-1000, 1730-1830MF, 0245-0400W. **FM:** 95.7.
13) Av. General Perón y Concepción Prieto Yegros (✉ Cas. 1456), Asunción — DG: Angel Guerreros A. **FM:** Asunción 97.1 + Chaco 96.9 "R. Marechal Estigarribia". **F.PI** 50kW.
14) Defensa Nacional y Av. Paraguay, Saltos del Guairá — Dir: M. Godoy.
15) Pres. Franco y Alejo Garcia, Villarrica. ☎ +595 (541) 2130/2385 — DG: Lídice Rodriguez. CE: Enrique Traversi. **FM:** 103.5. **Guaraní:** 0900-1100, 2100-2300.
16) Dr. Venancio Pino y 3ra. Proyectada, Caacupé.
17) Av. Enrique Doldán Ibieta y Presidente Franco, Itá. ☎ +595 (24) 543 — DG: Carlos Ruben Gutierrez. **Guaraní:** 0900-1000, 1330-1430, 1800-1845.
18) Niamad Blas Garay 241 c/Iturbe, Asunción — Dir: Efraín Martinez
19) Estados Unidos 2019, Asunción - Dir: P. Sandaval D.
20) Villa María Victoria Fracción San Jorge, Pedro Juan Caballero. **Web:** http://www.infonet.com.py/holding/mburu/mburu.htm
21) Derlis Caceres y Uruguay, Paraguarí. ☎ +595 (21) 204284 — DG: Carmelo Daniel Ruggilo Castro. **Guaraní:** 0900-1030, 1730-1830.
22) Choferes del Chaco 1194, Asunción. ☎ +595 (21) 604308. 🖷 +595 (21) 606074. **Web:** http://www.infonet.com.py/holding/nanduam/ — Dir: Leonardo Rubín.
23) Mariscal López y Capitan del Puerto, San Ignacio (Misiones).
24) Mariscal López 379, Alberdi — DG: J. Isidro Gamarra
25) Av. General Aquino km. 12, Luque — Own: Juan Carlos Bernabé.
26) Estrella esquina 4 de Enero, Capitán Badó — Dir: Víctor Castro Monte.
27) Coronel Romero y Florida, San Lorenzo (Asunción: General Díaz 488, Ofic. 34) — Dir: Julio César Pereira. **Guaraní:** 0900-1030, 1730-1830, 0000-0200.
28) General Caballero 748, Villarrica — Dir: Alba G. de González.
29) Azara y Kubitschek (✉ Cas. 1313), Asunción. 🖷 +595 (21) 204161 — Dir: Pbro. Cristóbal López.

30) Av. Delfín Chamorro 1631, Caaguazú — Dir: Robustiano Cabrera J.
31) Capitán Lombardo 174 y Av. Ártigas, Asunción — DG: Miguel G. Fernández.
32) Av. Gral. Santos 2525 y 18 de Julio, Asunción — Dir: Roque A. Fleytas.
33) Av. Mcal. López 2948, Asunción — Own: Humberto Domínguez D. Dir: Maria Contrera.
34) Itauguá. ☎ +595 (500) 9876 — **F.PI:** move to Fernando de la Mora.
35) Casilla de Correo 78, Concepción. ☎ +595 (21) 312318. 🖷 +595 (21) 312254 — DG: S.E. Dacak. Prgrs. in **Spanish & Guaraní**.
36) Carapeguá — Dir: Alfonzo Romero.
37) Coronel Alfredo A. Ramos esq. San Juan, San Juan Bautista, Misiones — Dir: Johnny Sánchez Martínez. **F.PI:** 5kW.
38) Horqueta — Dir: Dr. Salvador R. Paredes.
39) Calle Iturbe 146 y Riachuelo, Bella Vista Norte — Dir: Blanca Cardoso.
40) Cas. 2220, Ñemby.
41) Km. 213 Ruta 7a, Caaguazú.
42) Angostura y Olimpio, Bo. Ybarotu, Villarrica.
43) Mariscal Estigarribia y C. A. López, Coronel Oviedo.
44) Benjamin Constant 973, Edificio Arasá II, piso 9, Asunción. ☎ +595 (21) 497983, 497984, 497957 — Dir: Oscar Acosta.
45) 25 Leguas, Chaco.
46) Estados Unidos 2019, Asunción.

FM in Asunción: 89.1 Conquistador — 18) 90.0 – 90.7 Ysapy – 91.5 R. 91.5 – 11) 92.3 – 93.1 R. Florida – 93.9 Universal – 22) 95.5 Rock & Pop – 13) 97.1 "FM 97" – 97.9 Santa Mónica – 98.5 Yacyretá – 99.1 Nova FM/99.1FM – 100.1 R. Canal 100 – 100.9 Monte Carlo – 102.1 Obedira FM – 33) 103.1"FM Fem" – 105.1 R. Venus – 32) 106.1 – 22) 107.7.

L.T: UTC - 5h (UTC - 4h. Jan. 15-Apr. 15) — **Pop:** 25.055.000 — **Radios:** 5.300.000 — **Pr.L:** Spanish, Quechua, Aymara — **E.C:** 60Hz, 220V — **ITU:** PRU.

MINISTERIO DE TRANSPORTES, COMUNICACIONES, VIVIENDA Y CONSTRUCCION

Dirección General de Telecomunicaciones.
✉ Av. 28 de Julio 800, Lima 1. ☎ +51 (14) 4337800, 4331212, 4330570. **Cable:** Digecom.
L.P: Dir. de Telecomunicaciones: Ing. Carlos A. Romero Sanjinés. Dir. Freq. Div: Ing. José Villa Gamboa.

ASOCIACION DE RADIO Y TELEVISION DEL PERU (AR&TV)

✉ Av. Roma 140, San Isidro, Lima 27. ☎ +51 (14) 470 3734.
L.P: Pres: Humberto Maldonado Balbín. Dir: Daniel Linares Bazan.

INSTITUTO NACIONAL DE COMUNICACION SOCIAL

✉ Jr. de la Unión 264, Lima — **L.P:** Hernán Valdizán C. Dir. Bc: Sra. Clarisa P. de Olivera.

UNION DE RADIOEMISORAS DE PROVINCIAS DEL PERU (UNRAP)

✉ Mariano Carranza 754, Santa Beatriz, Lima 1.
Mediumwaves: Þ = also on shortwave, * = inactive, v = varying freq.

	Call	kHz	kW	Name and h. of tr.
400)	OBX4E	540	12	R. Inca del Perú, Lima: 24h
260)	OCX2D	540	1	R. San Antonio, Trujillo: 0900 (Sun 1100)-0500
621)	OAX6R	540	1	R. El Tiempo, Arequipa: 1000-0300
401)	OBZ4L	v560	2	R. Oriente, Lima: 1000-0500 (Sun 1100-2300) (r. 559.8-564)
333)	OAX3X	570	1	R. Progreso, Chimbote: 1100-0500
719)	OAX7G	Þ570	1	R. Cusco, Cusco: 1000-0300
200)	OAX2E	Þ580	10	R. Marañón, Jaén: 1000 (Sun 1200)-0200
314)	OCY2L	580	1	R. El Sol, La Esperanza
434)	OAX4S	580	15	R. Maria, Lima: 1100-0400
820)	OAZ8C	590	1	R. Sideral, Pucallpa
626)	OCX6A	590	1	R. Frecuencia Catedral, Arequipa
690)	OAX6N	*600	0.3	R. Cultural, Toquepala
403)	OBZ4W	Þ600	10	R. Cora del Perú, Lima: 24h
262)	OAX2M	620	1	R. Chepén, Chepén
404)	OCY4A	620	10	R. Estrella del Sur, Cañete: 24h
453)		620	10	R. Ovación, Lima

Call	kHz	kW	Name and h. of tr.
750) OBX7B	Þ640	10	Onda Azul, Puno: 0900-0400
405) OAZ4K	Þ640	10	R. Del Pacífico, Lima: 1030-0430
540) OAX5R	650	1	R. Regionalista, Ica: 1030-0530
263) OAX2N	650	2	R. Noticias del Perú, Trujillo: 1100-0500
101) OCX1U	660	3	R. JHC, Chiclayo: 1000-0500
172) OCX1V	660	1	R. Tumbes, Tumbes: 1000-0300
406) OCX4R	660	10	R. Omega, Lima: 1100-0700
618) OBX6O	660	1	R. Internacional, Arequipa
704) OAZ7P	660	1	R. Prensa al Día, Cusco
751) OAX7H	670	10	R. Nal. del Perú, Puno
407) OBX4A	680	5	RBC, la estación: Lima: 24h
264) OBX2L	680	0.5	R. Amauta, Chócope: 1100-0500
541) OAX5E	680	10	R. Independencia, Ica: 1100-0600
127) OCX1W	690	1	R. Ondas Populares, Chiclayo
354) OBX3F	690	2	R. Chimbote, Chimbote
600) OBX6Q	690	0.5	Radiodif. La Joya, La Joya
408) OBZ4H	700	10	R. 700, "la grande," Lima: 1000-0500
131) OBX1U	700	10	R. Cutivalú "LV del Desierto", Castilla: 0900-2300
543) OBX5Q	710	5	(r. RPP 730), Ica
291) OCY2N	700	1	R. Sausal Superior, Sausal
628) OCX6G	710	1	R. ABC, Cerro Colorado
741) OCX7T	710	10	R. Nal. del Perú, Puerto Maldonado
255) OCY2Y	720	50	R. San Luís (r. RPP 730), Jaén
460) OAU4E	720	10	R. Sideral, La Oroya
265) OAX2J	*720	10	R. Nal. del Perú, Trujillo
409) OAX4G	730	10	R. Programas del Perú, "RPP", Lima: 24h
132) OAX1D	740	10	R. Progreso, Piura: 1100-0500
266) OCX2X	*740	1	R. El Puerto, Pacasmayo
601) OAX6C	Þ740	10	R. Continental, Arequipa: 0900-0600
335) OAX3T	750	1	R. Ondas del Norte, Chimbote: 1100-0500
490) OCX4B	Þ750	2.5	R. Altura, Cerro de Pasco:1000-0400
410) OBZ4X	760	10	Radiomar AM, Lima: 24h
122) OCX1T	Þ768.9	3	R. Horizonte, Chiclayo: 0800-0500 (n. 770)
559) OAU5B	770	1	Radiodifusora Litoral, Ica
629) OAX6H	770	5	Radiomar (r. RPP 730), Arequipa
771) OAU7D	770	2.5	R. OAU7D, Puno
801) OAX8M	Þ770	5	LV de la Selva, Iquitos: 1000-0300 (Sat 1000-0300, Sun 1100-1700)
171) OAX1K	780	10	R. Nal. del Perú, Tumbes
411) OAX4X	Þ780	10	R. Victoria, Lima: 24h
767) OAZ7S	Þ780	5/7	Radiodifusoras LV de la Esperanza, Juliaca: 0800-1400, 2100-0200 (n. 730)
268) OAX2I	790	10	R. CRU, (r: RPP 730), Trujillo: 24h
706) OAZ7H	790	5	R. Armonía, Cusco
150) OCX1P	800	1	Telecomunicaciones del Norte, Piura
412) OAZ4U	800	0.5	R. Sol, Huaral
544) OBX5B	800	0.5	R. Sur Medio, Ica
670) OBX6A	800	1	R. La Porteña, Ilo
602) OBX6A	800	0.75	R. Texao, Arequipa: 0900-0430
102) OCX1F	809.8	1	R. Continente, Chiclayo: 1100-0400 (n. 810)
330) OAZ3M	810	1	R. Marinos, Santa
752) OAX7V	810	10	Radial Andina (r. RPP 730), Juliaca: 24h
413) OAX4O	820	10	R. Libertad, Lima: 24h
630) OCW6I	820	1	R. Paraíso, Camaná
290) OCX2Y	830	1	R. Antena 1, Trujillo: 1000-0200
461) OAU4C	830	1	R. Antena 1, Piura, Huancayo
360) OAX3Y	830	1	R. La Selva, Tingo María: (n. 800)
701) OCX7H	830	1	R. Inti, Santiago
691) OAX6D	830	10	R. Nal. del Perú, Tacna: 1000-0500
731) OAZ7N	830	3	R. OAZ7N, Cusco
151) OCX1N	840	1	R. Antena 1, Huancayo: 1100-0500
603) OBX6Y	840	1	R. Azul, Arequipa: 24h
336) OAX3S	840	1	R. Casma, Casma: 1100-0300
414) OAX4A	Þ850	20	R. Nal. del Perú, Lima: 1100-0500
297) OCX2F	850	1	R. Selecta, Virú
152) OCX1M	860	3	R. Nuevo Norte, Sullana
203) OCY2A	860	1	R. Antena 1, Cjamarca
111) OBX1F	870	10	R. CRU, (r: RPP 730), Chiclayo
631) OCX6F	870	1	R. OCX6F, Paucarpata
542) OBX5R	870	1	R. Sport, Nazca
702) OCX7R	Þ870	1	R. Mundo, Cusco: 0900-0100
415) OBZ4N	Þ880	50	R. Unión, Lima: 24h
303) OAX2P	880	3	R. Sintonía, Trujillo: 0900-0300
258) OCY2N	890	1	R. Libertad, Baños del Inca
753) OBX7S	890	1	R. Bahá'í, Puno: 0845-0200
259) OCY2T	900	1	R. Nor Oriental del Marañón, Jaén
416) OBX4X	Þ900	10	R. El Sol, Lima: 0900-0500
361) OAX3E	900	1	R. Ribereña, Aucayacu: 1000-0500
604) OBX6K	900	3/1	Nevada R. , Arequipa: 24h
269) OAX2H	Þ910	1	R. Huamachuco: 1000-0300
772) OAU7G	910	1	R. Visión del Altiplano, Juliaca
103) OCX1D	910.1	1	R. El Tigre, Ferreñafe: 1000-0300
270) OBX2S	920	1	R. Ollantay, Virú
133) OBX1J	920	10	(r: RPP 730), Piura
478) OAU4D	920	1	R. Super Red, Santa Rosa de Sacco: 1000-0400, Sun 1100-0200
560) OCX5C	920	1	R. Stelar, Chincha
695) OAU6G	920	1	R. La Heróica, Tacna
727) OCX7M	*920	1	R. Andina, Cusco
925) OAX9V	Þ920	1	R. Marginal, Tocache
299) OCX2V	930	1	R. Inti, Chepén: 1000-0300
417) OAX4E	930	10	R. Moderna "R. Papá", Lima: 24h
480) OCY4K	930	1	R. Satipo, Satipo
605) OBX6T	930	2.5	R. Yaraví, Arequipa: 0900-0400
703) OBX7L	940	5/3	R. Willkamayu, Cuzco: 24h
294) OCY2F	940	1	R. Estudio, Huanchaco: 0900-0200, SS 1000-2300
134) OBX1C	940	1	R. Presidente, Paita
337) OBX3S	950	1	(r: RPP 730), Chimbote
201) OBX2G	Þ950	1	R. Cutervo, Cutervo: 1100-0200 (Sun 2300)
773) OAZ7T	950	10	R. OAZ7T, Puno
941) OBX9L	950	1	R. Estación Láser, Rioja
237) OCY2K	Þ958.1	1	R. Líder del Norte, Cajamarca: 1000-0400 (Sun 0300) (n. 970)
104) OBX1Y	960	3	R. Amor, Chiclayo: 1100-0400
300)	960	0.5	R. Industrial. Trujillo
418) OAX4D	960	10	R. Panamericana AM, Lima: 24h
462) OCY4V	960	2/1	R. Constelación, Huancayo: 1000-0200
606) OBX6S	Þ960	3/1	R. Hispana, Arequipa: 1000-0500
809) OBX8H	Þ960	1	R. Diez, Iquitos
546) OBX5A	970	1	Latinoamericana R. , Ica: 1100-0500
135) OBX1V	970	1.5	R. La Capullana, Sullana: 1100-0700
732) OAX7D	970	1	R. OAX7D, Cusco
607) OAX6I	980	1	R. Universidad, Arequipa: 0900-0300
271) OAX2F	980	1	R. La Hora, Trujillo: 1100-0455
511) OAX5T	Þ980	5	R. LV de Huamanga, Ayacucho: 0900-2330 (Sun. 1900)
911) OBX9G	980	1	R. Cosmos, Bagua
419) OBX4J	990	10	R. Latina, Lima: 0900-0500
153) OCX1J	990	0.1	R. San Marcos, Piura
338) OBX3L	*990	1.2	R. Onda Nueva, Chimbote
463) OBZ4R	990	1	R. LV del Valle de Mantaro, Jauja
692) OAX6K	990	10	R. Bicolor, Tacna: 0900-0500
810) OBX8E	990	1	Radial Andina, Iquitos
202) OBX2M	v990	0.5	R. Contumazá: 1100-0400
608) OBX6R	1000	1	R. Endesa, Arequipa
298)	1005	2	R. Cristal, Trujillo
175) OBZ1C	1010	1	R. Sonora, Tumbes
267) OBX2K	1010	1/0.8	R. Andino, Otuzco: 0900-0300
420) OAX4U	*Þ1010	10	R. América, Lima
928) OBX9D	1010	1	R. Moyobamba, Moyobamba: 1100-0300
112) OBX1G	1020	1	R. Heroica, Chiclayo: 1000-0500
261) OBX2N	*1020	0.5	R. Esperanza 2000, La Esperanza
353) OBX3U	1020	1	R. Nal del Perú, Chimbote
754) OAX7N	1030	5	R. LV del Altiplano, Puno: 1000-0400
296) OAX2U	Þ1030	3	R. Los Andes (r. RPP 730), Huamachuco: 1000-0200
612) OCX6L	1030	1	R. Andina, Arequipa
705) OCX7O	1030	1	Super R. , Cusco
162) OAZ1O	1040	1	R. Vecinal, Piura
242) OBX2O	1040	1	R. Nor Oriente, Jaén
301)	1040	1	R. Fraternidad, El Porvenir
421) OBX4O	*1040	10	R. Red, Lima
609) OBX6B	Þ1050	1	R. Arequipa: 1000-0200
119) OBX1H	1050	10	Radiomar, Chiclayo
274) OCX2B	1050	1	R. San Sebastián, Chepén: 0900-0500
339) OAX3A	1050	1	R. Alpamayo, Huaraz
464) OBZ4J	1050	5	R. Bolognesi, Huancayo:

Call	kHz	kW	Name and h. of tr.
545) OBX5U	1050	1	R. Andina del Pacífico, Ica
762) OAZ7Q	1050	1	R. San Agustín, Juliaca
811) OBX8F	1050	1	Antena 1, Iquitos
331) OAZ3N	1060	1	R. Sudamericana, Chimbote
446) OCY4D	1060	1	R. Exito, Lima: 1000-0600
500) OBX5D	*Þ1060	2	R. Andahuaylas
226) OCY2O	Þ1060.1	1.5	R. Sudamérica, Cutervo: 1130-0300, Sun 1200-2300 (n. 1060)
488) OBX4G	1070	1	R. Visión, San Ramón
547) OAX5A	1070	0.25	R. Centinela, Pto. de San Juan
632) OCX6C	1070	3	R. Ondas del Sur, Camaná
768) OAZ7L	1070	3	R. OAZ7L, Puno
921) OBX9J	1070	3	R. Andes, Tarapoto
123)	Þ1072.5	1	R. Latina, Chiclayo: 0900-1400, 2100-0500 (n. 1070)
136) OBX1D	v1080	5	R. San Miguel, Piura: 0900-0600
195) OCY2B	1080	1	R. Andes, Cajamarca
275)	1080		R. Inca, Trujillo
676) OCX6X	1080	1	Radiodifusora Futura, Ilo
707) OAX7S	1080	1	R. Salcantay, Cusco: 0900-0400 (Sun 1000-0200)
802) OAX8N	1080	1	R. Amazonas, Iquitos: 1130-0500
491) OCX4I	Þ1080	1	R. Huayllay, Huayllay: 1030-0200 (Sun 0100)
204) OBX2A	1090	1.2/3	R. Cajabamba, Cajabamba: 0900-0500
332) OAZ3I	1090	1	R. AST, Chimbote
362) OAX3I	Þ1090	1	R. Televisión Tingo María (n. 1100)
610) OBX6X	1090	1	R. Amistad, Arequipa
424) OAZ4W	1100	1	R. Musical (r: RPP 730), Barranca
425) OCX4S	1100	0.2	Imperial, "LV de la Provincia", Cañete: 0900-0300
548) OAX5Z	1100	1	R. Oasis, Ica: 1100-0200
465) OCY4G	1100	1	Sonorama R. , Huancayo: 1000-0200
113) OBX1L	v1100	1	R. Star, Chiclayo: 24h
755) OBX7Z	Þ1100	1/0.5	R. LTC/RCC (R. Leoncio Torres Collao/R. Comercial Collao), Juliaca: 0900-0200
922) OAX9J	1100	0.5	R. Lamas, Lamas
426) OAU4J	1109.5	1	R. Antarqui, Lima: 1000-0400 (n. 1110)
154) OCX1R	1110		R. Centro Popular, La Unión: 1000-0500
340) OBX3B	1110	1	R. Heroica, Huaraz
219) OCX2U	Þ1110	1	R. Jaén, "LV de la Frontera", Jaén
708) OCX7T	Þ1110	5	R. Líder, Cusco: 0900-0200, Sun 1000-1700
677) OCX6P	1110	1	R. Austral, Ilo
276) OBX2I	1120	3	R. Dinámica, Trujillo: 0900-2400
633) OCX6U	1120	1	R. Municipal, Cerro Colorado
427) OAX4N	1130	3/1	R. Selecta, Lima: 1000-0500
466) OAZ4S	1130	1	R. Chanchamayo, La Merced: 1000-0300
924) OAX9Q	*Þ1130	1	R. San Martín, Tarapoto: 1000-0200
205) OAX2V	1130	1	R. Los Andes (r: RPP 730), Cajamarca: 24h
209) OCX2D	Þ1130	1/0.6	R. LV de Cutervo, Cutervo: 1300-0200
143) OBX1W	1140	3	R. Piura: 1000-0330
341) OAX3R	Þ1140	2	R. Bahía, Chimbote
611) OAX6L	Þ1140	12	R. Concordia, Arequipa: 0900-0500
467) OCY4C	1140	10	R. Centro (r: RPP 730), Huancayo: 24h
558) OAX5W	1140	0.5	R. Chinchaysuyo, Chincha
221)	1150Þ	0.65	R. Bambamarca, "Frecuencia Líder", Bambamarca: 1400-0300 (n. 980)
236)	1150	1	R. Naylamp "Estación Pucará, Pucará: 0900-0100
709) OCX7Q	Þ1150	10	R. Universal, Cusco: 1000-2400
803) OAX8F	1150	1.5	R. Loreto, Iquitos: 1000-0500
238) OCY2E	1150	0.5	R. Chasqui Llacta, San Marcos: 2300-0400
907)	Þ1150		R. San Juan, Lonya Grande
105)	1160	0.5	Estación Sasape, Cruce: 1300-2300
428) OAX4C	1160	10	R. 11-60, Lima: 1100-0600
671) OBX6G	1160	1	R. Nal. del Perú, Moquegua
933) OAX9A	Þ1160	5	R. Imagen, Tarapoto
271) OAX2C	1160	5	R. Libertad, Trujillo: 1100-0400
350) OAX3B	1160	0.5	R. Huaraz, Huaraz: 1000-0500 (Sun 1100-0300)
516) OBX5O	Þ1160	1/0.5	R. Huanta 2000, Huanta: 1030-0130
742) OCX7Z	1160	1	R. OCX7Z, Puerto Maldonado
155) OCX1B	1170	1	R. San Juan, Talara
359) OAZ3K	1170	1	R. Nor Peruana, Chimbote
468) OCX4Y	Þ1170	1	R. COSAT, Satipo: 1100-0200
640) OBX6L	1170	10	(r. RPP 730), Arequipa
760) OCX7Y	1170	1	R. Constelación, Puno: 0900-0300
163) OAZ1C	1180	1	R. Superior, Chulucanas
469) OCY4Z	Þ1180	1	R. Libertad "RLJ," Junín: 24h
277) OCX2A	1180	1.5	R. América, Quiruvilca
912) OBX9I	1180	3	R. Kuelap, Chachapoyas
114) OAX1E	1190	1	R. Chiclayo, Chiclayo: 1100-0600
549) OBX5M	1190	1	R. Lurén, Ica: 1000-0500
641) OCX6G	1190		R. Frecuencia 11-90, Arequipa
343) OBX3D	Þ1190	3	R. Ancash, Huaraz: 24h
248)	1190		R. Membrillar, Cascas
946)	1200		R. Las Palmas, Nuevo Bambamarca: -0200
756) OCX7S	1200	1	R. San Ramón, Juliaca
926) OBX9C	Þ1200	1	R. Estación "C", Moyobamba
672) OAX6N	1200	1	R. El Faro, Ilo: 1000-0500
196) OAU2A	1200	1	Frecuencia Pedagógica, Cajamarca
367)	1200	1	R. Satélite (r: RPP730), Tingo María
429) OAX4B	1200	4	R. Cadena, Lima: 1000-0500
725) OAX7B	Þ1200	5	R. Tawantinsuyo, Cusco: 1000-0300
247)	1200		R. San Mateo, Contumazá
485) OAU4G	1200	3	R. Andes, Huancayo
506) OBX5X	1200	1	R. Prensa al Día, Abancay
206) OCX2T	1200.2	1	R. Notícias, Cajamarca: 1100-0300 (n. 1190)
430) OBZ4A	*1210	1	R. Paramonga, Barranca
470) OCY4T	1210	1	R. Galaxia, Satipo
278) OAX2Q	1210	1	R. Universo, Trujillo (rel. Radiomar, Lima 760): 24h
710) OAX7M	Þ1210	1	R. Quillabamba: 1000-0300
106) OCX1X	1219.9	3	R. Cadena Nor-Oriental, Chiclayo: 1100-0600 (Sun 0300) (n. 1200)
357) OAZ3J	1220	1	Radiodifusora Sánchez, Chimbote
550) OAX5S	1220	1	R. San Pedro, Leoncio Prado
613) OAX6X	Þ1220	10/5	R. Melodía, Arequipa: 24h
733) OCX7N	1220	1	R. Inti, Cusco
900) OAX9T	1220	3	R. Bagua, Bagua: 1000-0300
279) OAX2T	1230	1	R. Albújar, Guadalupe: 1200-2400
471) OBZ4Y	1230	1	R. Selecciones, Tarma: 1100-0200 (Sun 1800)
740) OBX7J	Þ1230	1	R. Madre de Dios, Puerto Maldonado: 1030-0200
156) OCX1C	1240	1	R. Sechura. Sechura
280) OCX2K	1240	1	R. Indoamérica, Trujillo: 0900-0400
431) OBZ4Q	1240	5	R. Oro, Huacho
489) OAU4V	1240	1	R. Nuestra Señora del Valle, Huancayo
643) OAU6D	1240	1	R. Líder, Arequipa
711) OBX7M	1240	1	R. Túpac Amaru, Sicuani: 1000-0100
675) OCX6M	1240	1	R. Minería, Moquegua
769)	1240		R. Campesino, Puno: 0900-0300
164) OBZ1B	1250	1	R. OBZ1B, Talara Alta
214) OCY2J	1250	1	R. OCY2J, Cajamarca
821) OAX8P	Þ1250	2	R. Pucallpa, Pucallpa: 1030-0500
432) OAX4L	1250	8	R. Miraflores, Lima: 1100-0500
927) OAX9K	*Þ1250	1	Estación Tarapoto,Tarapoto: 0900-0400
109) OCX1O	1260	1	R. Nor Peruana, Chiclayo
110) OAZ1A	1260	1	R. Ferreñafe, Ferreñafe
281) OBX2C	1260	1	R. Otuzco, Otuzco
510) OBX5S	*1260	10	R. Ayacucho, Ayacucho
759) OAZ7K	1260	5	R. OAZ7K, Juliaca
614) OBX6D	1260	1	R. Red, Arequipa
304) OCX2Z	1270	1	R. Estación Latina, Chepén: 1000-0400
433) OAZ4H	1270	0.4	R. Huacho, Huacho
551) OAX5B	1270	1	R. Ica, Ica
804) OAX8T	Þ1270	1	R. Eco, Iquitos: 0900-0400
615) OBX6H	1270	0.35	R. Mollendo, Mollendo
712) OBX7A	1270	0.7	R. Solar, Cusco: 1000-0500
757) OBX7G	Þ1270	0.5	R. Ondas del Titicaca, Puno: 1000-0200
901) OAX9F	Þ1270	1	R. Municipal Nor Peruana, Chachapoyas: 1200-0515

Call	kHz	kW	Name and h. of tr.
472)OBX4T	Þ1270.2	0.4	R. La Merced, La Merced: 1100-1900, 2200-0200 (Sun 1200-1900) (n. 1270)
138) OAX1H	1270	1.5/3	R. Grau, Piura: 24h
334) OBX3C	1280	1	R. El Puerto, Chimbote
422) OCX4Q	1280	1	Radiodifusora Comercial RAGA, Cañete
616) OBX6P	1280	1	R. Fénix, Camaná
208) OBX2F	1280.4	1	R. Moderna, Cajamarca: 1000-0400 (Sun 0300) (n. 1280)
174) OCX1Q	1290	1	Radial Andina (r. RPP 730), Tumbes: 24h
282) OBX2Z	1290	1	R. Casa Grande, Chócope: 1000-0200
634) OCX6B	1290	1	R. Telecomunicaciones del Norte, Arequipa
908) OAX9N	1290	10	R. Dif. Nor Oriental, Bagua
115) OAX1I	1300	1	R. Lambayeque, Chiclayo: 1000-0400
210) OCY2C	1300	1	R. Paraíso, Cajabamba
344) OAX3O	1300	0.5	R. Huascarán, Huaraz: 1100-0400
822) OAX8B	Þ1300	3	R. Nuevo Mundo, Pucallpa: 24h
402) OAX4M	Þ1300	5	R. Comas, Comas: 1000-0500
693) OAX6P	1300	2	R. Latina (r. RPP 730), Tacna
758) OAX7X	Þ1300	1.5	R. Juliaca, Juliaca: 0900-0300 (Sun 2300)
713) OAX7P	Þ1300	10	R. Onda Imperial, Cusco: 1200-2130
473) OAZ4B	Þ1300.2	1.5	R. Andina, Huancayo: 0900-0400, Sun 1000-0300v (n. 1300)
552) OBX5K	1310	1	R. San Martín, Pisco
231) OBX2D	Þ1310	1	R. Chota, Chota: 1130-0300
442) OAX4I	Þ1320	10	R. La Crónica, Lima: 1000-0500
307) OBX2Q	1320	1	R. Frecuencia Popular, Chepén: 0900-0500
617) OBX6W	1320	0.8	R. Premier, Camaná
345) OAX3U	1320	1	R. Miramar, Chimbote
492) OAZ4I	Þ1320	1	R. LV de Oxapampa: 24h
139) OAX1F	1320	0.35	R. Talara, Talara
509) OAU5C	1320	1	R. Perú, Abancay
947) OCZ9K	1320	0.25	R. Pajatén, Juanjuí
100) OAU1A	1330	1	R. Dos Mil, José Leonardo Ortiz
283) OAX2P	1330	0.5	R. Paiján, Paiján: 1100-0400
310) OCX2W	1330	1	R. Regional, Trujillo
635) OCX6E	1330	1	R. Ondas del Misti, Mariano Melgar
714) OCX7K	Þ1330	1	R. San Miguel, Cusco: 0900-0300 (Sun 1100-0200)
929) OAX9C	Þ1340	3	R. Tropical, Tarapoto: 0900-0500 (Sun. 1300-1500)
474) OAZ4Q	1340	1	R. Jauja, Jauja: 1000-0300
553) OAX5D	1340	0.5	R. Chincha, Chincha Alta
140) OBX1K	1340	1.5	R. San Francisco, Piura: 24h
436) OAZ4M	1340	0.2	R. Huaral, Huaral
211) OAX2D	Þ1346.4	1	R. Cajamarca, Cajamarca: 0800-0300 (SS 0400) (n. 1370)
823) OAX8D	1350	1.5	R. Super AM-FM, Pucallpa: 1000-0500
673) OBX6F	1350	1	R. Ilo, Ilo: 24h
363) OAX3N	Þ1351.2	2	R. Ondas del Huallaga, Huánuco: 0930 (Sun 1100)-0300 (n. 1350)
107)	1356.8		R. Regional, Chiclayo
245) OCY2T	1360	2	R. Santa Mónica, Chota: 1200-0100 (Sun 1100-1800)
284) OBX2N	1360	1	R. Super Uno, Cartavio
407) OAX4Y	*1360	10	R. Excelsior, Lima
459) OAU4O	1360	1	Radiodifusión Hecaburt, Tarma
561) OBZ5Z	1360	1	R. Cruz del Sur, Palpa
636) OCX6T	1360	1	Radiodifusora Luza, Paucarpata
715) OAX7R	Þ1362	1	R. Sicuani, Sicuani: 0930-0300 (n. 1365)
486)	1362.5	0.5	R. Sudamericana, Tarma: 1000-0400 (n. 1360)
346) OAX3H	1370	1	R. Interamericana, 'LV del Pacífico', Chimbote
674) OAX6T	Þ1370	1/0.5	R. Moquegua, Moquegua: 0930-0500
728) OAZ7J	1370	1	R. Santa Mónica, Cusco: 0900-0300
438) OAZ4O	1370	1	R. Cosmos, Huacho
932) OBX9A	1370	1	R. Palmera, Bellavista: 1100-0500
212) OAX2W	Þ1379.8	0.3	R. Atahualpa, Cajamarca: 1100-0500 (n. 1380)
108)	1380	0.3	Estación Fanupe. Fanupe: 0900-0100
554) OBX5C	1380	1	R. El Pueblo, Ica: 18 h
124) OBX1V	1380	1	Sistema Radial, Ferreñafe
165) OBZ1D	1380	1	R. Bellavista, Sullana
439) OCY4U	1380	1	R. ABC, Chosica: 1000-0300
619) OAX6O	1380	3	R. San Martín, Arequipa: 0930-0300
364) OBX3I	1381.9	1	R. Pilcomozo, Huánuco (n. 1380)
302) OCX2M	1390	0.5	R. Espacial, Otuzco
910)	*Þ1390	0.5	R. San Nicolás, Rodriguez de Mendoza
766) OAX7F	*Þ1395	1	R. Ayaviri, "LV de Melgar", Ayaviri
157) OCX1A	1400	1	Radiodifusora El Milagro, El Alto
250)	Þ1400		R. La Merced, Tongod
230) OCY2D	Þ1400	1	R. Internacional del Perú, San Pablo: 1000-1700, 2100-0300
285) OAX2O	1400	1	R. Perú, Chepén
476) OAZ4P	1400	1	R. San Juan, Tarma: 1030-0230 (Sun 1100-2100)
931)	Þ1400	0.2	R. Juanjuí, Juanjuí
948) OBX9E	1400	1	R. El Dorado, Villa Picota
440) OBX4W	1400	2.5	Callao Super R. , Callao: 1100-0500
505)	Þ1400		R. LV de Andahuaylas, Andahuaylas
515)	Þ1400	1	Estación Wari, Ayacucho: 1000-0200
562) OCX5B	1400	1	R. San Clemente, Pisco
620) OAX6J	Þ1400	3	R. Super Landa, Arequipa
716) OAX7I	Þ1400	3	R. La Hora, Cusco: 1000-0200 (Sun 1100-1700)
729)	Þ1400		Ondas del Suroriente, Quillabamba
347) OBZ3F	1400	3	R. Willaq, Huaraz: 1000-0300
116) OAX1N	1409.9	1	R. Ideal, Lambayeque: 1000-0500 (n. 1410)
286) OAX2Y	1410	5	R. Heróica, Trujillo: 0900-0600
441) OBZ4V	1410	1	R. Universal, Huacho: 1000-0600
824) OBX8I	1410	1	R. Concierto, Pucallpa
487) OAU4A	1410.1	1	R. Láser, Santa Bosa Sacco: 0930-0400 (n. 1480)
501) OBX5H	Þ1417.5	0.5	R. Apurímac, Abancay: 1000-0200 (SS 1100-2300) (n. 1420)
761) OCX7C	Þ1420	1	R. Qollasuyo, Juliaca: 1000-0200 (Sun 0100)
805) OAX8Z	Þ*1420	1	R. Oriente, Yurimaguas
447) OBZ4G	1420	1	R. San Isidro, Lima: 1000-0600 (Sun 0400)
158) OCX1H	1420	1	R. San José, La Unión
222)	*Þ1420	1	R. Nor Andina, Celendín
223)	Þ1420		R. Nor Agricultura, Baños del Inca: 1100-0400 (Sun 1200-0300)
477) OAZ4V	1429.9	1	R. Universal, Huancayo: 1100-0500 (n. 1430)
125)	Þ1430		R. Uno, Chiclayo
287) OBX2T	1430	1	R. Santa Bárbara, Ascope
352) OBX3E	1430	1	Telesistema Peruano, Huarmey
355) OAZ3H	1430	1	R. Chavín, Santa
423) OAU4H	1430	1	R. OAU4H, Chancay
734) OAZ7M	1430	1	Antena Uno, Cusco
913) OBX9H	1430	1	R. Utcubamba, Utcubamba
117) OBX1T	v1440	2	R. Cooperativa Tumán, Chiclayo: 1100-0300 (r. 1440.3)
349) OAZ3O	Þ1440	1	R. LV de Pomabamba, Pomabamba
443) OAX4K	1441.2	5	R. Imperial 2, Lima: 1000-0500 (n. 1440)
272) OCX2J	Þ1450	1.5	R. San Juan, Trujillo
507)	1450	1	R. Panorama, Andahuaylas: 1100-0300
718) OCX7W	1450	1	R. Santa Rosa, Quillabamba
923) OAX9D	Þ1450	1.5/2	R. Rioja, Rioja: 0800-0500
622) OBX6C	1460	1	R. Bahía, Mollendo
141) OAX1V	1460	1	R. Sullana "LV de Chira", Sullana
475) OAZ4F	Þ1460	1	R. La Oroya, La Oroya: 0930-0500
935) OBX9B	Þ1460	1	R. Frec. Popular, Rioja: 1000-0500
763) OAX7W	Þ1460	5	R. El Sol de los Andes, Juliaca: 0900-0200
555) OAX5K	1460	2.5	R. Internacional, Pisco
288) OBX2Y	1460	2	R. Estelar, Guadalupe: 1000-0600
348) OBX3A	1460	1	R. Chimú (r. RPP 730), Chimbote: 24h
213) OBX2U	Þ1470	1	R. Ilucan, Cutervo; 1100 (Sun 1030)-0300
315) OCY2G	1470	1	R. Occidental, Quiruvilca
452) OAX4B	1470	25	CPN Radio (Cadena Peruana de Noticias), Lima
479) OCX4D	Þ1470	1	R. Huancayo, Huancayo: 24h

	Call	kHz	kW	Name and h. of tr.
644)	OAU6E	1470	1	Radiodifusora del Sur, Arequipa
694)	OAX6M	Þ1470	1.5	R. Tacna, Tacna: 0900-0500 (Sun 0300)
142)	OCX1L	1480	1	R. Frecuencia Popular, Piura: 1100-0500
244)		Þ1480		
254)		1480	0.25	R. Nueva Imagen, Hualgayoc: 2300-0300
289)	OCX2C	1480	0.6	Radiodifusora Comercial San Pedro, Virú
435)	OCX4V	1480	1	R. K'ler, Barranca
717)	OAZ7G	1480	1	R. Espinar, Yauri
118)	OAX1L	1490	1	R. Imperio, Chiclayo: 1100-0300
311)	OCX2S	1490	1	R. San Pedro de Lloc, San Pedro de Lloc
358)	OAZ3G	1490	1	R. San Jacinto, Santa
556)	OAX5N	1490	0.5	R. Nasca, Nasca
623)	OAX6Q	1490	1.2	R. Fidelidad, Arequipa: 1100-0500
806)	OAX8J	Þ1490	1	R. Atlántida, Iquitos
942)	OBX9N	Þ1490	1	R. Ondas del Río Mayo, Nueva Cajamarca: 0900-0200 (Sun 1100-2400)
159)	OCX1E	1500	1	R. Jupiter, Sechura
366)	OBX3I	Þ1500	1	R. Luz y Sonido, Huánuco: 1530-2130
764)	OCX7P	1500	2	R. Em. Frontera, Puno: 0900-0500
305)	OBX2X	Þ1500	1	R. JCL, Florencia de Mora: 0845-0300
316)		Þ1500		R. LV de Bolívar, Bolívar
445)	OBX4I	Þ1500	10/5	R. Santa Rosa, Lima: 24h
624)	OBX6V	1500		Nueva R. 2000, Orcopampa
696)	OAU6B	1500		Radiodifusora Bulevar, Tacna
700)	OCX7G	1500	1.5	Radiodifusora Canchis, Sicuani
437)	OAU4M	1510	1	R. Organización Flores del Campo, Barranca
481)	OCX4J	Þ1510	1.5/3	R. Tarma, Tarma: 1000-0500 (Sun 1100-0200)
557)	OAX5F	1510	1	R. LV de Nasca, Nasca: 24h
637)	OCX6Q	1510	1	R. Alegría, Arequipa
720)	OBX7P	1510	1	R. Red, Huánchac
504)	OCX5A	1512.1		Inti R. , Abancay: 1000-2300 (Sun 1000-1500) (n. 1580)
218)	OCX2Q	1517	2	R. LV del Inca, Baños del Inca: 0800-0300 (n. 1510)
934)	OAX9X	1520	0.25	R. Visión, Juanjuí (r. 1545)
273)		1520	1	R. Virú, Virú
292)	OCX2G	1520	1.2	R. Santiago, Santiago de Chuco: 0900-0500
482)	OAZ4D	1520.4	2	R. Minería, "la R. Estrella". La Oroya: 1100-0300 (n. 1520)
166)	OCX1Y	1530	1	Radiodifusora Leomar, Sullana
483)	OBZ4S	1530	1	R. 15-50, Huancayo: 1000-0600
356)	OBX3H	1530	1	Antena 1, Chimbote
216)	OBX2R	1530	3	R. Oriental, Jaén
454)		Þ1530	10	CollectivoRadialFeminista,Lima(F.PL)
721)	OAZ7F	1530	0.5	R. Confraternidad, Yauri
293)	OBZ2B	1540	1	R. Star, Trujillo: 24h
448)	OBZ4U	1540	1	R. Barranca, Barranca: 1100-0500
173)	OBX1B	1540	1	LV de la Frontera, Tumbes
722)	OCX7V	Þ1540	1	R. Chasqui, Cusco: 1000-0200
449)	OBX4P	1550	1	R. Independencia, Lima: 1000-0600 (Sun 2400)
765)	OCX7A	*1550	1	R. Cultura, Puno
120)	OAX1O	1550	1	R. Superior, Monsefú: 1100-0400
914)	OAX9Y	1550	0.5	R. Amazonas, Bagua
723)		1560		R. Municipal de Cusco, Cusco
217)	OBX2J	1560	1/ 0.85	R. Nuevo Continente, Cajamarca: 1000-0300
167)	OCX1Z	1570	1	R. Nueva Esperanza, Tambo Grande
351)	OBX3N	1570	1	Radiodifusora Chasqui, Yungay
368)	OBX3M	1570	1	R. San Martín, Huánuco
638)	OCX6I	1570	1	R. Willy, Uraca
215)	OAU2E	1577.5	1	R. Frecuencia San Ignacio, San Ignacio: 1200-2300
525)	OAX5M	Þ1580	0.5	R. Virgen del Carmen, Huancavelica: 1000-0300
295)	OBX2H	Þ1580	1	R. Tayabamba, Tayabamba: 1000-0500 (Sun 1100-1500)
920)	OAX9M	Þ1580	1.5	R. Central, Bellavista: 1100-0300
121)	OBX1M	Þ1580.4	1	R. Naylamp, Lambayeque: 1000-0300 (n. 1580)
724)	OBX7Q	1584	3/1	RPP (r: RPP 730), Cusco: 24h
508)	OAU5A	1587.1	1	R. Armonía, Abancay: 1030-0300 (n. 1130)
312)	OCX2L	1590	1	R. JJ, Chócope
450)	OAZ4Z	1590	5	R. Agricultura, "La Peruanísima", Lima: 1000-0400
639)	OCX6S	1590	1	R. Mundo, Cerro Colorado
246)		Þ1598	1	R. Jerusalén, Coop. Atahualpa-Jerusalén: 0900 (Sun 1000)-1300, 2300-0300
625)		1600		R. Camaná, Camaná

Shortwaves: * = inactive, v = varying freq.

	Call	kHz	kW	Name and h. of tr.
246)		3194.6	1	R. Jerusalén: cf. 1598kHz
763)	OBX7D	3230.2	1	R. El Sol de los Andes, Juliaca: 0930-1430, 2130-0200
366)	OAW3A	3234.8	1	R. Luz y Sonido, Huánuco: 1100-0200 (n. 3235)
761)	OCX7C	3250.1	1	R. Qollasuyo, Juliaca: 1000-0200 (Sun 0100)
402)		3250.8	1	R. Comas, Comas: 1000-0430 (n. 3250)
492)	OAZ4I	3260	1	R. LV de Oxapampa, Oxapampa: 2300-0400
213)	OAX2G	*3280	1	R. Ilucán, Cutervo (F.PI.)
515)	OAZ5D	3283.4	1	Estación Wari, Ayacucho (n. 3280)
363)	OAX3Q	v3329.5	5	R. Ondas del Huallaga, Huánuco: 0930 (Sun 1100)-0300
490)	OBZ4B	3339.9	5	R. Altura, Cerro de Pasco: 1000-0430
224)	OAZ2H	*3350	0.5	R. Visión 2000, Bambamarca (F.PI.)
805)		3350	1	R. Oriente, Yurimaguas: (F.PI)
230)	OAZ2G	v3397.4		R. Internacional del Perú, San Pablo: 1000 (Sun 1100)-0400 (n. 3395)
306)		v3502.1		R. Chillia, Chillia: 2300-0200
451)		*3800.3	1	R. Oyón, Oyón
938)		3818.2	0.25	Estación Laser, Rioja: 1000-0230v
235)		3869.3	0.7	R. Adventista Mundial "LV de la Esperanza," Celendín: 2330-0300 (n. 3870)
462)	OAZ4R	4000	1	R. Constelación, Huancayo: 1000-0200
272)		v4000.5	0.5	R. San Juan, Trujillo: 0100-0500
925)		4039.1	1	R. Marginal, Tocache: 1100-0300v (Sun 0200)
232)		v4182.3		R. LV de Sayapullo, Sayapullo: 1100-
915)		4190.4		R. San Juan, Aramaango
313)		4190.5		R. Super San Ignacio, San Ignacio
943)		v4191.7		R. Selva Superior, Moyobamba (n. 4200) (v. 4187.9-4194)
211)		4279.4		R. Cajamarca, Cajamarca: 1000-0200v (tr. also used by st. 223)
197)		4300		R. Hualgayoc, Hualgayoc
221)		4300		Estación Norte, Bambamarca (F.PI.)
904)		4319.3		R. Condorcanqui, Santa Maria de Nieva: 2300-0300 (n. 4320)
126)		v4394.9		R. Nueva Sensación, Chiclayo (v. 4387-4394.9)
221)		4419.9	0.85	R. Bambamarca "Frecuencia Líder," Bambamarca: 2300-0400v
222)		4460.9	1	R. Nor Andina, Celendín: 2300-0400
239)		4485	0.5	R. Frecuencia VH, Celendín: 2300-0300
122)		4505.1	0.5	R. Horizonte, Chiclayo: 0800-1400v, 2300-0500
493)		v4510.1	0.25	R. Paucartambo "RP," Paucartambo: 0800-1500, 2100-0300, SS 1000/1100-0200
125)		v4567		R. Uno, Chiclayo: 1100-v0520, (v. 4548-4573)
123)		v4574.5		R. Latina, Chiclayo: 0900-1400, 2100-0500 (n. 4580)
807)		4593.6		R. Estación Yurimaguas, Yurimaguas
766)	OBX7F	4606.5	1	R. Ayaviri "LV de Melgar", Ayaviri: 1000-1500, 2100-0300 (n. 5035)

Call	kHz	kW	Name and h. of tr.
234)	4627.3		R. Cosmos, Celendín
309)	v4631		R. Soledad, Parcoy: 1000v-0300v (v. 4549-4632.5)
516) OAZ5B	v4747.4	0.3	R. Huanta 2000, Huanta: 1000-1300, 2200-0100 (alt. freq. 4756.8)
148)	4750.1		R. San Francisco Solano, Sondor: -0250v
362) OAX3P	4759	0.5	R. Televisión Tingo María: 0900-1200, 2300-0400
503) OAX5A	4764	1	R. Chincheros, Chincheros: 2200-0330 (n. 4765)
481) OCX4E	v4774.9	0.5	R. Tarma, Tarma: 1000-0500
341)	4780.1	0.8	R. Bahía, Chimbote
233) OAX2L	v4781	0.3	R. Satélite, Santa Cruz: 2300-0300v (alt. freq. to 6725.6)
468) OCX4Z	4784.9	1	R. COSAT, Satipo: 1000-0200
806) OAX8F	4790.1	3	R. Atlántida, Iquitos: 0900 (Sun 1000)-0500
750) OBX7C	4801	1	Onda Azul, Puno: 0930-0400v (n 4800)
924) OAX9R	*4810.2	3	R. San Martín, Tarapoto
212) OAZ2F	*4820.8	1	R. Atahualpa, Cajamarca
801) OAX8S	4824.4	10	LV de la Selva, Iquitos: 0950 (Sat 1000)-0300, Sun 1100-1700
674) OAX6Z	4825	0.5	R. Moquegua, Moquegua: 0930-0500
715) OAX7T	v4826.3	1.5	R. Sicuani, "LV de Canchis", Sicuani: 1000-0300 (n. 4835)
528) OBX5G	4830.6	1	R. Lircay, Lircay
200) OCX2E	4834.8	1	R. Marañón, Jaén: 1000-0300, Sun 1100-0200
500) OBX5E	4840	2	R. Andahuaylas: 1030-0200v
708) OAX7-	4845	1	R. Líder, Cusco: 0900-0200, Sun 1100-1700
716) OAZ7A	4855.4	1	R. La Hora, Cusco: 1000-1500, 1700-0100, Sun 1100-1700
458) OCX4T	4870	1	R. Majestad, Huancayo
822) OAZ8H	4881.5	1	R. Nuevo Mundo, Pucallpa: 24h (n. 4880)
525) OAX5X	4886.5	0.8	R. Virgen del Carmen, Huancavelica: 1000-0300 (n. 4885)
517) OAZ5C	4888.9	0.5	R. Dif. Huanta, Huanta: 0900-0300, Sun 1000-0100
231) OAZ2B	v4890.5	1	R. Chota: 1100-0300v (n. 4890)
466) OAZ4T	4895	0.4	R. Chanchamayo, La Merced: 1000-0300
475) OAZ4G	v4904.8	1	R. La Oroya, La Oroya: 1000-0500 (n. 4905)
725) OBX7V	*4908	5	R. Tawantinsuyo, Cusco
403) OAZ4N	4914.5	10	R. Cora del Perú, Lima: 1030-1500, 2300-0500 (n. 4915)
920) OAZ9A	*4920	0.5	R. Central "LV del Huallaga Central", Bellavista
757) OBX7H	*4922	1	R. Ondas del Titicaca, Puno (n. 4920)
929) OAX9E	4935	3	R. Tropical, Tarapoto: 1100-0200
740) OBX7I	4950.1	5	R. Madre de Dios, Puerto Maldonado: 1000 (Sun 1100)-0200 (alt.freq. 4953)
513) OAX5S	4955.4	5	R. Cultural Amauta, Huanta: 1000-0100 (Sun 1000-1500, 2100-2400)
472) OCX4T	v4964.7	1	R. La Merced, La Merced: 1030-1900, 2200-0100, Sun 1200-1900 (n. 4965)
714) OAZ7B	4965.8	1	R. San Miguel, Cusco: 0900v (Sun 1000)-0100
933) OAX9M	4970.9	1	R. Imagen, Tarapoto: 0900 (Sun 1000)-0100v (n. 4970) (alt. freq. 5144)
405) OCX4H	4974.8	5	R. del Pacífico, Lima: 1030-1400, 2300-0430 (n. 4975)
343) OAZ3B	v4990.9	5	R. Ancash, Huaraz: 24h (n. 4990)
473) OAZ4G	4995.6	2	R. Andina, Huancayo: 2100-0400 (Sun 0300) (n. 4995)
219) OAX2S	5005.6	0.25	R. Jaén, "LV de la Frontera", Jaén: 1130-0400
927) OAZ9B	*5014.8	0.7	Estación Tarapoto, Tarapoto
606) OAX6Z	*5015	0.1	R. Hispana, Arequipa
758) OAX7Z	v5015.6	0.5	R. Juliaca, Juliaca: 0900-0300 (Sun 2300)
902) OBX9K	5018.6	1.5	R. Horizonte, Chachapoyas: 1100-0200 (n. 5020)
710) OAX7Q	5025	5	R. Quillabamba, Quillabamba: 1000-0300
767)	*5030		Radiodifusora LV de la Esperanza, Juliaca (F.PI.)
296) OAZ2A	5030.6	5	R. Los Andes, Huamachuco
469) OCY4Y	5039.3	1	R. Libertad "RLJ", Junín: 0945 (Sun 1100)-1500, 1900-0300 (Sun 0100)
923) OAX9Z	5045	1	R. Rioja, Rioja: 0800-0500 (n. 5015)
504)	5046.3		R. Integración, Abancay: 1000-0500 (Sun 1500)
532)	5049.6	0.1	R. Tayacaja, Pampas: 1105- (n. 4670)
342) OAZ3P	*5050	1	R. Amistad, Chiquian
484)	5053.6		R. Acobamba, Acobamba: 1100-0300
713)	5056.2	0.15	R. Onda Imperial, Cusco: 1200-2130
729)	5067.1	1	R. Ondas del Suroriente, Quillabamba: 1000-0100
905) OAZ9C	*5081	0.5	R. Huayabamba, Rodríguez de Mendoza
702) OAZ7D	5083.7	0.5	R. Mundo, Cusco: 0900-1500, 2000-2400 (n. 5045)
243)	5085	0.1	R. Celendín: 2300-0500
804) OAX8V	5097.3	1	R. Eco, Iquitos: 0900/1000-0500v (n. 5010)
809)	5116.4		R. Diez, Iquitos
224)	5131.4	0.5	R. Visión 2000, Bambamarca: 2300-0300v
251)	5139.8		R. Amauta, San Pablo
933) OAX9M	5144	1	R. Imagen, Tarapoto: 0900 (Sun 1000)-0100 (n. 4970) (alt. freq. to 4969.5)
937)	v5175.4		R. Máster, Moyobamba
225)	*5179		R. Oriental, San Ignacio
501)	5236.3		R. Apurimac, Abancay: 1000-0200 (Sun 1100-2300)
906)	v5264.8		R. LV de Chiriaco, Chiriaco: 1100-1700, 0000-0400
909)	5271.8		R. Nororiental, Rodríguez de Mendoza: irr. 1100-0200
944)	5278.8		R. Estación Soritor, Moyobamba
228)	5305		R. La Inmaculada, Santa Cruz: 2000-0230
526)	v5323.7		R. LV de Anta, Anta
529)	v5323.8	0.5	R. Origen, Huancavelica: 1100-0200 (Sun 0100) (n. 5322)
237)	*5339.1	0.8	R. Líder del Norte, Cajamarca
349) OAZ3Q	*5344	1	R. Pomabamba, Pomabamba (n. 3215)
295)	5397		R. Tayabamba: 1000-, 2300-0500 (n. 4500, also r. 4505v, 4713v)
316)	5460.4		R. LV de Bolívar, Bolívar: 0930-
910) OBX1M	5470.6	0.5	R. San Nicolás, Rodríguez de Mendoza: 1100-0200 (n. 5470)
903)	5486.6	0.06	La Reina de la Selva, Chachapoyas
198)	v5500.6		R. Regional, Celendín: -0200
226)	v5522.5	0.25	R. Sudamérica, Cutervo: 1130-0300 (Sun 1200-2300) (v. 5521.4-5522.7) (n. 5525)
241)	v5547.4		R. Paccha, Paccha (alt. freq. 5277)
244)	5558.2		R. Lajas, Lajas: 23315-0400
256)	5560.7		R. El Sol, Pucará: 1100-2400 (n. 5560)
146)	5602.7		R. LV del Campesino, San Miguel de El Faique: -0030 (n. 5600)
227)	v5604.4		R. LV de Marañon, Cajabamba: -0100v (n. 5610) (alt. freq. 5636.8)
257)	2637.4		R. Perú, San Ignatio: 1000-1400, 2100-0200 (n. 5635)
227)	5636.8		R. LV de Marañon, Cajabamba (alt. freq. to v5604.4)
249)	v5645.8	0.25	R. Cultural Amauta, Bambamarca: 1130-, 2300-0400v (n. 5645)
209)	5660.5	0.6	R. LV de Cutervo, Cutervo: 1100-1300, 2200-0200 (Sun 1100-0200)

Call	kHz	kW	Name and h. of tr.
213) OAX2G	5678	1	R. Ilucán, Cutervo: 1100-0300
215)	5699.8	0.15	R. Frecuencia San Ignacio: 1200-, 2330-0400v (n. 5700)
121)	5728.5	0.5	R. Naylamp, Lambayeque: 0900-0400 (n. 5730)
916)	5730		R. Santiago, Puerto Galilea: 1100-2300
930)	5774.9		R. Power, Moyobamba
907)	5775.3		R. LV de San Juan, Lonya Grande: 0315
730) OAW7A	5894.7	1	R. Paucartambo, Paucartambo (n. 3200)
609) OAX6A	5951.3	1	R. Arequipa, Arequipa: 1000-0200
252)	5955.1		R. Estación Uno, Pucará: -0300 (n. 5960)
479) OAX4P	*5955.3	1	R. Huancayo, Huancayo
642)	5956.8		R. Selecciones, Chuquibamba
416) OBX4Q	5969.7	10	R. El Sol, Lima: irr.
755)	5980	1	R. LTC/RCC (R. Leoncio Torres Collao/R. Comercial Collao), Juliaca: 1100-1300, 0000-0200, SS 1100-1900
722)	5981.5		R. Chasqui, Cusco: 1100-1300, 0000-0200
613) OBX6L	5995.2	1	R. Melodía, Arequipa: 24h
420) OAX4W	6010.7	1	R. América, Lima
411) OAX4Q	6019.8	5	R. Victoria, Lima: 1100-1400, 2300-0200 (n. 6020)
620) OAX6B	6033.1	0.25	R. Super Landa, Arequipa (n. 6035)
445) OCY4H	v6045.2	5	R. Santa Rosa, Lima: 1000-0500
601) OAX6D	6055.3	2	R. Continental, Arequipa: 24h
365) OAX3L	6060.5	3	R. JSV, Huánuco: 1000-0000
709) OAZ7C	*6090.8	2	R. Universal, Cusco
414) OAX4H	6095	10	R. Nal. del Perú, Lima: 1100-1500, 2200-0500
770)	6096.4		R. Interoceánica, Azangaro
415) OBZ4O	6115	10	R. Unión, Lima: 24h
491) OBZ4P	6140.1		R. Huayllay, Huayllay: irr.
611) OBX6B	v6141	1	R. Concordia, Arequipa: 0845-2300v (v. up to 6141.5) (n. 6150)
767)	*6150		Radiodifusoras LV de la Esperanza, Juliaca (F.PI.)
821) OAX8Q	*6155	0.5	R. Pucallpa, Pucallpa
454)	*6160	10	Colectivo Radial Feminista, Lima (F.PI.)
725) OAX7C	6173.8	5	R. Tawantinsuyo, Cusco: 1000-0300 (n. 6175)
805) OAX8I	v6188.2	5	R. Oriente, Yurimaguas: 1000-0200 (n. 6190)
511) OAX5U	v6201.2	2	R. LV de Huamanga, Ayacucho (n. 6070)
719) OAX7A	v6203.8	1	R. Cusco, Cusco: 1000-0300 (Sun 1100-0200) (n. 6195)
808)	v6238		R. Estación Yurimaguas, Yurimaguas: 1000-0300
726)	v6242.1	0.12	R. Calca, Calca: 1000-1500, -0500
505)	v6250		R. LV de Andahuaylas, Andahuaylas: -0200
945)	6261.1		R. J-V-L, Centro Comercial de Consuelo
137)	v6281.1	1.5	R. Huancabamba, Huancabamba: 1000-0200v (n. 6290)
926)	6329	0.8	R. Estación C, Moyobamba: 1130-2400
229)	6339.5		R. San Miguel Arcangel, San Miguel: -0300v
149)	v6405.5		R. Huarmaca, Huarmaca, -0415
207)	6420.3		R. Mi Frontera, Chirinos
147)	6479.8		R. Altura, Huarmaca: -0200
145)	6535.8		Radiodifusoras Huancabamba, Huancabamba: -0200
308)	6670.2		R. Santa Mónica, Santiago de Chuco
917)	6675.4		R. Ondas del Río Marañón, Aramango: 1130-0100 (n. 6660)
269)	v6676.3		R. Huamachuco, Huamachuco
201)	*6690	1	R. Cutervo, Cutervo
233) OAX2L	v6725.6	0.3	R. Satélite, Santa Cruz: 2300-0300 (alt. freq. 4781)

Call	kHz	kW	Name and h. of tr.
250)	6754.7		R. La Merced, Tongod: irr. 2300-0200v (n. 6760)
220)	6761		R. San Andrés de Llapa, Llapa: 2300-0200
942)	6797.7		R. Ondas del Río Mayo, Nueva Cajamarca: 0900-0200 (Sun 1100-2400)
627)	*6851.2		R. Cultural, Cailloma
160)	6895.3	0.25	R. San Miguel, San Miguel de El Faique: 1100-2300
168)	7003.3		R. LV de Huarinjas, Huancabamba: 2100-0200
240)	7040.5		R. San Ignacio, San Ignacio: 1200-2400 (n. 7015)
253)	v7048.4		R. LV de Santa Cruz, Santa Cruz: 2345-0430 (v. 7049.1-7050.9)
144)	v7141.8	0.25	R. Ayabaca, Ayabaca: 1000-0300
161)	7205.6		R. Dif. Paratón, Huarmaca: 1030-2400
940)	7446		R. Cristal, San Hilarión: 2330v-0300
694) OAX6H	9504.7	0.2	R. Tacna: W 0900-0100 (n. 9505)
442) OAX4J	*9520	5	R. La Crónica, Lima
611) OBZ6A	*9580	1	R. Concordia, Arequipa
901) OAX9G	9660	5	R. Municipal Nor Peruana, Chachapoyas
405) OAZ4L	9674.8	5	R. del Pacífico, Lima: 1400-2300
411)	9721.6		R. Victoria, Lima (n. 9720)

Addresses and other information

(Names of **departamentos** listed under each region (cf. call-sign numbers) should be added to address info, main capitals excluded).

REGION 1
Lambayeque
100) Nicolas de Pierola 335, José Leonardo Ortiz
101) Juan Guglíván 984, Chiclayo.
102) Calle Paraguay 1000, Urbanazación Barsallo, Chiclayo.
103) Jr. Victoria 289, Ferreñafe.
104) Juan Guelman de Carneros s/n, Chiclayo.
105) Caserio Cruce, Hacienda Vieja, Túcume, Provincia de Lambayeque – Own: Juan José Grandez Vargas.
106) C.C. Gran Plaza Edificio Grau, Av. Miguel Grau 350, Oficina 222, Chiclayo.
107) Chiclayo.
108) Caserio Fanupe, Barrio Nuevo Distrito Mórrope, Provincia de Lambayeque – Own: Juan José Grandez Vargas. PD: Luis Chaponan.
109) Calle Las Violetas s/n, Chiclayo.
110) Calle Monseñor Francisco Gonzales Burga 717, Ferreñafe.
111) Urb. Patazca M-41, Chiclayo – **FM:** 102.9.
112) Ap. 490, Chiclayo – Dir: H. Hubert Limo Mendoza.
113) Cas. 448, Chiclayo – Dir: Jorge E. Alfaro R. **FM:** 98.3
114) Cas. 210, Chiclayo – Pte: Carlos Montjoy S.
115) Av. Balta 415, Chiclayo.
116) 2 de Mayo 371, 2Þ piso, Lambayeque – DG: César Limo Quiñonez.
117) Ap. 77, Chiclayo – Dir: Víctor Sandoval G.
118) Jr. Arica 972, Chiclayo – Gte: Mesías Sánchez F.
119) Km. 4 Carretera Pimental, Chiclayo - **FM:** 105.1
120) Av. Mcal. Castilla 839, Monsefú, Chiclayo – Gte: Víctor Yaipen U.
121) Av. Huamachuco 1080, 1Þ piso, Lambayeque. ☎ +51 (74) 283353. 🖷 +51 (74) 283087 – DG: Dr. Juan José Grández Vargas. CE: Jorge Noriega. **FM:** 105.9.
122) Jr. Incanato 387, Altos, Distrito José Leonardo Ortiz, Chiclayo – Own: Enrique Becerra Rojas.
123) Av. Saenz Peña 1558, Chiclayo – DG: Carlos Tipara Gonzalez.
124) Bolívar 219, Ferreñafe.
125) Calle Primero de Mayo 278, Urbanización Urrunaga, Distrito José Leonardo Ortiz, Chiclayo. ☎ +51 (74) 256090.
126) Panamerica Norte 361, Urb. Cultural Ricardo Palma, Chiclayo.
127) Zona Santa Isabel, Chiclayo.

Piura
131) Cas. 165, Piura. ☎ +51 (74) 342802. 🖷 +51 (74) 342965 – Dir: Rodolfo Aquino Ruiz. **FM:** 100.5
132) Cas. 254, Piura – Dir: Raquel la Madrid J. **FM:** 96.1.
133) Cayetano Heredia 498, Piura.
134) Cas. 55, Paita – Dir: César Minaya V.
135) San Martin 1045, Sullana – DG: Ena Ognio de Silva. **FM:** 95.7.
136) Calle Cusco 1312, Piura – Dir: Geraldo Garrido P.
137) Unión 610, Huancabamba – Dir: César Colunche B.
138) Cuzco 842-850, Piura – DG: Walter Ruiz Zelada.
139) Cas. 74, Talara – Gte: Max H. Cornejo P.

140) Moquegua 458, Piura – Dir: Cecilia Cruzado Solari.
141) Jr. Sucre 556, Sullana – Dir: Jorge Geller G.
142) Jr. Blas Atienza, Galería Anita, Piura – Gte: Segundo N. Rojas E.
143) Jr. Callao 318, P. 2, Piura – DG: Hernán Vanegas F.
144) Jr. Montero 437, Ayabaca – Own: José Heli Requejo Aldean.
145) Unión 409, Huancabamba – DG: Fredy Alberca. Adm: Edwin Arrieta.
146) Av. Piura 1015, Pampa Alegre, Distrito de San Miguel de El Faique, Provincia de Huancabamba - Dir: Alberto Soto Santos
147) Jr. Antonio Raymondi 312, Huarmaca, Provincia de Huancabamba.
148) Calle San Miguel 207, Distrito de Sondor, Región Grau, Provincia de Huancabamba.
149) Av. Grau 454, Huarmaca, Provincia de Huancabamba – DG: Prof. Gregorio Zavaleta Perez.
150) Santa Maria C., Piura.
151) Jr. Cusco 750, Piura. **FM:** 98.7
152) Ugarteche 490, Sullana.
153) Calle Arequipa 970, 3Þ piso, Piura.
154) Calle Unión 817, La Unión.
155) Urbanización Aproviser A2-1, Talara.
156) Calle San Martín 554, Sechura.
157) Las Azucenas J-16, Urbanización El Milagro, El Alto, Provincia de Talara.
158) Calle Tumbes 641, La Unión.
159) Jr. Bolívar 317, 2Þ piso, Sechura.
160) Av. Piura 1200, Pampa Alegre, , Distrito de El Faique, Provincia de Huancabamba.
161) Jr. Alfonso Ugarte 109, esq. del Parque Leoncio Pardo, Huarmaca, Provincia de Huancabamba – DG: Prof. Hernando Huancas Huancas.
162) Calle Cusco 670, Piura.
163) Jr. Buenos Aires 405, Chulucanas.
164) Calle 8 s/n, Talara Alta.
165) Calle Madre de Diós 258, Bellavista, Sullana.
166) Calle Cajamarca 485, Bellavista, Sullana.
167) Av. Grau cuadra 5, Cruceta San Lorenzo, Tambo Grande.
168) Barrio El Altillo, Huarinjas, Huancabamba - Own: Alfonso García Silva. Re. to Correo Central, Huancabamba.

Tumbes

171) Pza. Alipio Rosales s/n, Tumbes – Admor: C. Fernández P.
172) Jr. Bolívar 117, Tumbes. ☎ +51 (74) 523003 – DG: María del Pilar Medina Aponte.
173) Piura 1010, Tumbes.
174) Panamericana Norte Km. 1321, Tumbes – **FM:** 100.5.
175) Calle Tarapaca 163, Tumbes.

REGION 2
Cajamarca

195) Av. Arequipa 101, Cajamarca.
196) Av. El Maestro 290, Cajamarca.
197) Jr. San Martín, Hualgayoc. - Own: Máximo Zamora Medina.
198) Jr. Grau s/n, frente al Colegio Nuestra Señora del Carmen, Celendín.
199) Ap. 50, Jaén (via Chiclayo) – DG: Luis Távara Martín. ▨ +51 (74) 731245.
200) Parque José Galvez Egusquiza s/n, Cutervo – Adm: Emperatriz Arteaga Ramirez.
202) David León 601, Contumazá.
203) Jr. Amazonas 725, Cajamarca – **FM:** 99.3.
204) Ap. 61, Cajabamba – Own: Salustiano Asmad Vergara
205) Prolongación La Mar 530, Cajamarca.
206) Jr. Etén 243, Cajamarca – DG: Prof. Hugo Vásquez Camacho.
207) Calle San Ignacio 520, Distrito de Chirinos, Provincia de San Ignacio, Región Nororiental de Marañón.
208) Jr. Apurímac 694, 2Þ piso, Oficina 5, Cajamarca – Dir: Luciano Perez Sangay.
209) Jr. Maria Elena Medina 644-650, Cutervo.
210) Jr. Silva 673, Cajabamba.
211) Jr. La Mar 675, Cajamarca – Dir: Porfirio Cuzco Potosí.
212) Juan XXIII s/n (Plaza Bolognesi), Cajamarca – DG: José Suarez. **FM:** 89.9.
213) Jr. Lima 290, Cutervo. ☎ +51 (74) 737010, 737022. ▨ +51 (74) 737231 – DG: José Gálvez Salazar. **FM:** 96.1.
214) Jr. Lima s/n, Cajamarca.
215) Jr. Villanueva Pinillos 330, San Ignacio. ☎ +51 (74) 716100 – Dir: Franklin R. Hoyos Condor.
216) Av. Mesones Muro 157, Jaén.
217) Jr. Cinco Esquinas 53, Cajamarca – Dir: Eduardo Cabrera U.
218) Cas. 418, Baños del Inca – Dir: J. Gonzalo Urteaga V.
219) Calle Mariscal Castilla 439, Jaén. **FM:** 105.1.
220) Jr. El Castillo 850, Llapa, Provincia de San Miguel – DG: Carlos

A. Cabanillas.
221) Jr. Jorge Chávez 416, Bambamarca. ☎ +51 (74) 713249 – Own: Valentín Peralta Diaz.
222) Jr. José Gálvez 1271, Celendín – Dir: Misael Alcántara G.
223) Av. Manco Kapac 686, Baños del Inca – Adm: Marisol Cuzco Salazar.
224) Jr. Prolongación Mariscal Sucre s/n, Bambamarca – Own: Victor Marino Tello C.
225) San Ignacio.
226) Jr. Ramón Castilla 704, tercer nivel, Plaza de Armas, Cutervo. ☎ +51 (74) 736090, 737443 –Own: Adameo Mario Muñoz Cuivar. **FM:** 95.3.
227) Jr. Bolognesi 1300, Barrio La Alameda, Sector Los Delfines, Cajabamba, Provincia de Cajabamba, Región Nororiental de Marañón – Dir: Eduardo Diaz Coronado.
228) Parroquia de la Inmaculada Concepción, Frente al Parque Central, Santa Cruz: DG: R.P. Jorge Carrasco
229) Jr. Bolívar 356, a media cuadra de la Plaza de Armas, San Miguel de Pallaques – Own: Miguel Lozano T.
230) Av. Bolognesi 532, San Pablo – Own: Agustin Alva H. **FM:** 100.7.
231) Jr. Anaximandro Vega 690 (or Ap. 14), Chota. ☎ +51 (74) 711240 – Dir: Mauro Vásquez Q.
232) Sayapullo, Provincia de Cayabamba.
233) Jr. Cutervo 570, Santa Cruz – Dir: Sabino Llamo Ch.
234) Jr. San Martín 484, Celendín.
235) Jr. 2 de Mayo 218, Celendín – DG: Francisco Guicoche Ortiz.
236) Pucará, Jaén, Provincia de Jaén.
237) Jr. Huánuco 2367, Cajamarca. ☎ +51 (44) 823363. ▨ +51 (44) 828637 – Own: Elías Machuca Alcarde. **FM:** 90.3.
238) Jr. Leoncio Prado 324, San Marcos.
239) Jr. José Gálvez 730, Celendín – Dir: Fernando Vásquez Castro. **FM:** 104.1.
240) Av. Libertador 751, San Ignacio, Región Nororiental de Marañón – DG: Pedro Alfonso Morales y Saenz.
241) Jr. Ramón Castilla 52, Paccha, Provincia de Chota.
242) Jr. Bolívar 1020, Jaén.
243) Jr. 2 de Mayo 501, Celendín – Gte: Jaime Silva Díaz.
244) Jr. Recinto Manuel 589, Lajas, Provincia de Lajas.
245) Calle 6 de Agosto 460, Chota – Dir: Prof. Jacinto Anaya Morán.
246) Pje. Las Américas, or: Cas. 367, Cajamarca – Mgr: Máximo Terán.
247) Av. Benjamín Gálvez 118, Contumazá.
248) Jr. San Martín 610, Cascas, Contumazá.
249) Jr. Jaime de Martinez, Bambamarca.
250) Distrito de Tongod, Santa Cruz – Dir: Roberto Ramos Chanas.
251) Jr. Manuel Iglesias s/n, a pocos pasos de Plazuela San Juan, San Pablo, Provincia de San Pablo, Región Nororiental de Marañón.
252) Barrios Altos, Pucará, Provincia de Jaén, Región Nororiental de Marañón.
253) Av. Zurumilla 190, Santa Cruz.
254) Hualgayoc, Provincia de Bambamarca – Own: Oscar Lino Peralta.
255) Jaén.
256) Pucará, Provincia de Jaén.
257) Jr. Atahualpa 191, San Ignacio **FM**: 97.7.
258) Caserio Ventanilla de Otuzco, Baños del Inca.
259) Calle Zurumilla 1328, Jaén. FM: 94.9.

La Libertad

260) Av. Gonzáles Prada 695, piso 12, Trujillo.
261) Condorcanqui 1757, La Esperanza.
262) Calle Lima 599, Chepén – Dir: José A. Tejada T.
263) Independencia 543, Trujillo. ☎ +51 (44) 251792, 252974.
264) Casagrande 128, Chócope, Provincia de Ascope – Dir: J. Daniel Cabellos R. **FM:** 102.5.
265) Francisco Pizarro 532, Of. 205, Trujillo – Dir: Carlos Martínez P.
266) Jr. Ayacucho 65, Pacasmayo. ☎ +51 (44) 523042 – DG: Enrique Becerra Rojas.
267) Jr. Trujillo 597, Otuzco – DG: Julio Acevedo Castañeda.
268) Panamericana Norte 546, Trujillo. ☎ +51 (44) 242501.
269) Bolívar 937, Huamachuco – Dir: Manuel D. Gil G.
270) Alfonso Ugarte s/n, Virú.
271) Zepita 452, Trujillo. ☎ +51 (44) 249326, 233644. ▨ +51 (44) 252970 – DG: Carlos Burmester.
272) Pasaje San Martín 300, Urb. Las Mochicas, Trujillo – Gte: Santiago Lopez Valderrama.
273) Virú.
274) Jr. Atahualpa 795, Chepén – Dir: Jorge Acevedo Rojas.
275) Av. Hiamán 530, Vista Alegre, or: Ap. 1215, Trujillo – Pte: Carmela Castro de Manrique.
276) Av. Perú 374, Of. 5, Trujillo. ☎ +51 (44) 244315 – DG: Jorge L. Mantilla O.

277) Jr. Progreso 121, Quiruvilca – Dir: Walter Díaz Ramos.
278) Bolívar 780 (✉ Cas. 1029), Trujillo. ☎ +51 (44) 233981 – Pte: Antonio Cassinelli R.
279) Victoria 229, Guadalupe, Provincia de Pacasmayo.
280) Av. América Sur 478, Urb. Palermo, Trujillo. ☎ +51 (44) 222675 – DG: Pedro Avalos Graos.
281) San Antonio 880, Otuzco.
282) Urbanización Casa Grande, Chócope, Provincia de Ascope.
283) Jr. Chocope 684, Paiján.
284) Camal 13, Hacienda Cartavio, Santiago de Cao, Provincia de Ascope.
285) Lima 392, Chepén.
286) Jr. Alfonso Ugarte 614, Of. 502, Trujillo. ☎ +51 (44) 246211 – Dir: Carlos Casanova B.
287) Ascope.
288) Calle Victoria 310, Guadalupe, Pcia. Pacasmayo.
289) Jr. Grau s/n, Virú, Provincia de Trujillo.
290) Jr. Francisco Pizarro 208, Of. 302,.Trujillo. ☎ +51 (44) 242666. **FM:** 107.5.
291) Calle Junín 23, Sausal, Provincia de Ascope.
292) Jr. Bolognesi 1247, Santiago de Chuco – Dir: Saturnino Asmad Vergara.
293) Pasaje San Luís 149, Trujillo. ☎ +51 (44) 246066.
294) Pasaje Educación 220, Huanchaco, Provincia de Trujillo.
295) Bolívar 565, Tayabamba – DP: Israel Alva L.
296) Damián Nicolau 108-110, Huamachuco – Dir: Mons. Sebastián Ramis Torrens.
297) Av. Grau s/n, Virú, Provincia de Trujillo.
298) Jr. Pancho Fierro, L-1:17, Urb. Santo Dominguito, Trujillo.
299) Calle Trujillo 699-A, Chepén, Provincia de Pacasmayo.
300) Trujillo.
301) Mateo Pumacahua 1661, El Porvenir, Trujillo.
302) Jr. Bolognesi 937, Otuzco.
303) Plaza Diaz 184, Trujillo.
304) Jr. Progreso 759, Chepén, Provincia de Pacasmayo.
305) Florencia de Mora.
306) Jr. Miraflores y San Cristóbal, Chillia, prov. Pataz.
307) Urb. El Recodo L-17, 2b piso, Chepén. ☎ +51 (44) 562522 – DG: Carlos E. Rojas Esquerre. **FM:** 100.3.
308) Mcal. Cáceres 453, Santiago de Chuco – Gte: Faustino L. Rodríguez Rebaza.
309) Centro Minero de Retama, Distrito de Parcoy, Provincia de Pataz.
310) Av. España 2239, Oficina 402, Trujillo.
311) Jr. A. Razuri 606, San Pedro de Lloc, Provincia de Pacasmayo.
312) Jr. Nemecio Caceda 77, Chócope, Provincia de Ascope.
313) Av. Victor Larco 104, San Ignacio, Distrito de Sinsicap, Provincia de Otuzco.
314) Jr. Benito Juarez 1753, La Esperanza.
315) Jr. Trujillo 281, Quiruvilca.
316) Bolívar.

FM in Trujillo: 88.5 Nor Peru – 89.1 Armonía – 90.3 R. Studio 92 – 90.9 Canal 90-90 – 91.5 R. Renovación Cristiana – 92.1 Stereo Diplomat R. – 92.7 R. Universal – 93.5 Stereo Norte – 410) 93.9 – 94.3 R. Estación 94 – 94.7 R. Tropicana – 94.9 Stereo Latina – 410) 95.3 – 96.1 R. 96.1 FM – 96.5 R. Melody – 96.9 R. Mundial – 268) 97.5 – 284) 97.9 R. Super 1 – 293) 98.3 R. Star – 276) 98.9 – 99.1 R. La Grande – 99.7 R. Club – 100.5 R. Hit – 418) 101.1 – 101.9 R. Frec. 100 – 264) 102.5 – 428) 102.7 – 103.6 Pentagrama – 104.1 R. Maravillosa FM – 104.5 Melodía FM – 105.1 R. Nova Stereo – 105.7 R. Sonorama – 410) 106.5 – 107.1 R. Sintonía FM – 107.3 R. King Stereo – 290) 107.5 Antena 1 R. (Lima st's are satellite relays).

REGION 3
Ancash
330) Jr. Pacasmayo 530, Coishco, Santa.
331) Manzana E-2, Lote 19, Urbanización Los Heroes, Chimbote – **FM:** 94.9.
332) Pasaje 3 de Octubre, Chimbote.
333) Cas. 117, Chimbote – Dir: Darío C. Moreno.
334) Panam. Nte. Km 440 (or: Leoncio Prado 660), Chimbote.
335) Prolongación Parco s/n (or Ap. 294), Chimbote – Gte: Humberto Luna L.
336) Av. Nepeña Mza. 8-C, Lote 3, Casma. ☎ +51 (44) 711266 – DG: Roberto Achú P. **FM:** 92.5.
337) Aguirre y Pizarro, Chimbote. **FM:** 95.5.
338) Av. José Gálvez 337, 3° piso, Of. 4, (✉ Ap. 72), Chimbote. ☎ +51 (44) 331073 – DG: Fausto León Rivas. **FM:** 97.5.
339) Ap. 240, Huaraz – Dir: Antonio Camones Blas.

340) Av. Mariscal Luzuriaga 1239, Huaraz.
341) Av. Alfonso Ugarte 309, Chimbote – Dir: F. Cruz M.
342) Plaza de Armas s/n, Chiquian, Provincia de Bolognesi.
343) Av. Raymondi y Jr. Cajamarca, Huaraz – DG: Roque Moreno Neglo. **FM:** 101.1.
344) Jr. San Martin 655, Huaraz. ☎ +51 (44) 721002. 🖷 +51 (44) 721124 – Dir: Francisco Sotelo López. PD: Francisco Sotelo Alvarado. CE: César Echandia.
345) Jr. Alfonso Ugarte 433, Of. 101, Chimbote – DG: Guillermo Arias Saavedra.
346) Ap. 300, Chimbote – DG: Juana Arias O.
347) Jr. Pallasca 348 (or Ap. 150), Huaraz. ☎ +51 (44) 721051 - DG: Lorgio Sanchez. Dir: Antonio Lirio. Prgm. in **Sp. & Quechua**. **FM:** 91.7.
348) Prolongación Pardo, Barrio 1 de Mayo, Chimbote.
349) Jr. Huamachuco 207, Pomabamba, Región Chavín.
350) Jr. José de Sucre 1124, Huaraz – Own: Enrique Alzamora Arévalo.
351) Barrio Luchapamba, Yungay.
352) Av. Cabo Alberto Reyes 281, Huarmey.
353) Jr. Leoncio Prado 660, Oficina 416, Chimbote – **FM:** 105.1.
354) Elias Aguirre 346, Chimbote.
355) Urbanización San Juan Zona 5, Santa.
356) Av. Aviación 298, Chimbote.
357) Manzana-U, Lote 27, Pasaje 3 de Octubre, Chimbote.
358) Calle Solidex s/n, Santa.
359) Av. José Pardo 120, Chimbote.

Huánuco
360) Av. Raimondi 432, Tingo María – DG: Carlos Quiroz Flores.
361) Malecón Huallaga 1038, Aucayacu, José Crespo y Castillo.
362) Ap. 25, Tingo María – DG: Ricardo Abad Vásquez.
363) Leoncio Prado 373 (✉ Cas. 343), Huánuco. ☎ +51 (64) 511525. 🖷 +51 (64) 512428 – Gte: Flaviano Llanos M. Prgr. in Sp. & Quechua. **FM:** 99.3.
364) Ruben Dario 128, Zona Cerro Paucarbamvilla, Distrito Amarilis, Huánuco – DG: Constantino Leandro Crispin.
365) Jr. Aguilar 742, Huánuco. **FM:** 100.5.
366) Jr. 2 de Mayo 1286, Of. 205, Huánuco. ☎🖷 +51 (64) 518500. 🖷 +51 (64) 511985 – DG: Carlos Ortega Obregón. **Quechua:** 1000-1200, 2300-0200. **FM:** 105.7.
367) Tingo María.
368) Jr. Aguilar 744-746, Huánuco – **FM:** 100.1.

REGION 4
Lima
400) Bernardo Alcedo 375, Lince, Lima 14. ☎ +51 (14) 723882 – Dir: Augusto S. Irei S.
401) Jr. Camaná 615, Of. 605, Lima 1 – Dir: Henry Delgado.
402) Av. Estados Unidos 327, Urb. Huaquillay, Km. 10 de la Av. Tupacamaros, Distrito de Comas, Lima 7. ☎ +51 (1) 5250859. 🖷 +51 (1) 5250094 – Dir: Julio Saldaña G. **FM:** 101.7.
403) Paseo de la República 144, Centro Cívico, Of. 5, Lima 1 – Dir: Juan Ramírez Lazo. **F.PI:** change from 60m to 31m band.
404) Cas. 55, San Vicente de Cañete – DG: A. Alvarez Calderón R. **FM:** 91.5.
405) Ap. 4236, Lima 1 – Gte: Pedro Ferreyra G.
406) Av. Javier Prado Este 309, Lima 27 – Gte: Alberto Carreño B.
407) Av. Manco Cápac 333, La Victoria, Lima 13 – Gte: Gonzalo Belmont C. Prgr. in Quechua 4h./day.
408) Av. de la Marina 3099, San Miguel, Lima 32 – Gte: Martha de Láinez.
409) Jr. Alejandro Tirado 217, Santa Beatriz, Lima 1. ☎ +51 (14) 246042. **Web:** http://www.rpp.com.pe/ – Gte: Rolando Estremadoyro R. **F.PI:** 50kW.
410) Justo Pastor Dávila 197, Chorillos, Lima 9. ☎ +51 (1) 2515151. 🖷 +51 (1) 4675557 – DG: Abraham Zavala Falcón. Signal downlinked to 76 repeaters.
411) Victor Reynel 320, Mirones Bajo, Lima 1 – Gte: Santiago Salcedo G.
412) Ap. 8, Huaral – Gte: Santiago Kong M.
413) Av. Salaverry 1082, Jesús Maria, Lima 11 – DG: Humberto Panfichi.
414) (Studios) Av. Petit Thouars 447, (Adm.) Av. José Gálvez 1040, Santa Beatriz, Lima – Dir: Henry Aragón Ibarra.
415) Av. Central 717, 12° piso, San Isidro, Lima 27 or Ap. 833, Lima 1 – DG: Carlos A. Gonzales.
416) Av. Uruguay 355, 7 piso, Lima 1 – Gte: Emilio García Lara.
417) Av. Nicolás Dueña 813, Lima 1 – Dir: Francisco Sambuceti P.
418) Paseo Parodi 340, San Isidro, Lima 27. ☎ +51 (1) 4226787 – DG: Fernando Torres D. Satellite signal downlinked by 60 FM repeaters.
419) Ignacio Merino 230, Lima 18 – DG: Delia Ll. A. de Caballero.
420) León Valverde 1140, Lince, Lima 14 – Dir: Karen Crousillat Carreño.

Relays FM 94.1.
421) Av. Petit Thouars 4328, Miraflores, Lima – DG: Pedro Tello B.
422) Jr. Jorge Chávez 144, San Vicente de Cañete.
423) Urbanización San Francisco, Manzana H, Lote 5, Chancay.
424) Barranca.
425) 2 de Mayo 573, San Vicente de Cañete – Own: Roberto Pachas.
426) Calle Gerardo Unger 6347, Los Olivos, Lima 39.
427) Av. Petit Thouars 4830, Miraflores. ☎ +51 (14) 2423522 – Gte: Oscar Artacho S.
428) Jr. General Ortegoso 140, Breña, Lima.
429) Los Angeles 129, Miraflores, Lima – DG: Cesar Lecca Arrieta.
430) Av. Francisco Vidal 921, Paramonga, Prov. de Chancay – Gte: Francisco Niquín Quezada.
431) Ap. 156, Huacho – DG: Jorge H. Bisso J.
432) Av. Manco Cápac 495, Lima 18 – DG: Ing. Ricardo Palma Michelsen.
433) Jr. Echenique 140, Huacho.
434) Calle El Alhelí, Manzana 'N', Lote 3, Urbanización Los Sauces de Higuereta, Distrito Surquillo, Lima 34. ☎ +51 (14) 2711084. ▤ +51 (14) 2711699.
435) Av. Recreo 317, Altos, Barranca.
436) Cas. 66, Huaral.
437) MZ-17, Urbanización El Olivar, Barranca.
438) Av. Grau 538, Huacho – Gte: M. Arakaki O.
439) Cas. 40, Chosica, Lima 15 – Gte: Luis Bordo Rosell.
440) Av. Alte. Miguel Grau 487, Callao 1 – DG: Bruno Espósito M.
441) Jr. Ausejo Salas 153, Huacho. ☎ +51 (34) 321976 – DG: César Minaya Velasquez. **FM:** 101.7.
442) Av. José Gálvez 1004, Santa Beatriz, Lima.
443) Av. Separadora Industrial s/n, Villa El Salvador, Lima – Gte: Aníbal Torres A.
445) Jr. Camaná 170 (▧ Cas 4451), Lima 1. ☎ +51 (14) 277488 – Gte: R. P. Sokolich A. **Quechua:** 1300. **English:** 0130.
446) Av. José Prado 313, Lince, Lima.
447) Av. Petit Thouars 1806, Lince, Lima.
448) Plaza de Armas 132, Barranca – DG: Wesley S. Portnoy. **FM:** 99.1.
449) Jr. Yahuar Huaca 108 y Tahuantinsuyo, Independencia, Lima 28 – Dir: Claudio Gabiño.
450) Av. Alfonso Ugarte 1428, Of. 202, Lima 5 (▧ Cas. 11-0625, Lima 11). ☎ +51 (14) 244302. ☎ +51 (14) 246677 – DG: Bartolomé Dextre.
451) Av. Huánuco 144, Oyón – Dir: Aurelio Liberato A.
452) Paseo de la República 3662, San Isidro, Lima 27 – DG: Antonio Saavedra L.
453) Lima.
454) Colectivo Radial Feminista, Lima.
FM in Lima: 88.3 Telestereo 88 – 416) 88.9 Sol y Armonía – 409) 89.7 – 452) 90.5 – 91.1 R. San Borja – 407) 91.9 – 92.5 Studio 92 – 410) 93.1 R. Ritmo – 93.7 Carolina – 420) 94.3 – 94.9 R. "A" – 95.5 Z "Solo rock & Pop" – 432) 96.1 – 97.1 R. Omega – 428) 98.1 – 99.1 Doble Nueve – 100.1 Estéreo 100 – 418) 101.1 – 402) 101.7 – 405) 102.1 Ke Buena – 415) 102.7 Unión FM – 414) 103.9 – 407) 104.7 Excelsior – 105.5 R. San Isidro – 410) 106.3 Radiomar Plus – 400) 107.1 Inca – 107.7 Star.

Junín
458) Calle Real 1033, Of. 302, Huancayo.
459) Av. Castilla 100, Tarma.
460) Tayacaja 324, Of. 202, La Oroya.
461) Calle Ancash 543, Of. 208, Huancayo – **FM:** 103.1.
462) Calle 28 de Julio 264, Chilca, Huancayo. ☎ +51 (64) 212554 – DG: Ramón Chang Pelayo. **Quechua & English:** 1h daily.
463) Jr. Junín 1008, Jauja – Gte: Jonás Barreto P.
464) Calle Real 692, Of. 402, El Tambo, Huancayo – **FM:** 92.9.
465) Calle Real 270, El Tambo, Huancayo. ☎ +51 (64) 245396 – DG: Guido Barreto Rivera. PD: Werne Porta Morales. CE: Angel Gomez Gonzeales. **FM:** 96.7 "R. Futura".
466) Jr. Tarma 551, La Merced – Gte: Augusto Espejo P. **FM:** 97.6.
467) Calle Calixto 548, Huancayo – **FM:** 89.7.
468) Jr. Manuel Prado 459, Satipo – Dir: Fermín Zúñiga T.
469) Jr. Cerro de Pasco 582, or Ap. 2, Junín. ☎ +51 (64) 344026 – Dir: Mauro Chaccha Guere. **FM:** 97.7.
470) Av. Manuel Prado 239, Satipo.
471) Jr. Moquegua 648, Tarma – Own: César Anaya Chagua.
472) Jr. Junín 171, La Merced – Dir: Teodoro Fidehl F.
473) Jr. Real 175, Chilca, Huancayo – Dir: Fernando Cruz M. Adm: Teodosio Muños Inga.
474) Jr. Bolognesi 450, P. 3, Jauja. ☎ +51 (64) 362428 – Dir: Godofredo Castellares.
475) Calle Lima 190, 3º piso, Of. 3, La Oroya – Dir: Jacinto Manuel

Figueroa. **FM:** 100.1. **F.PI:** AM & FM in Junín as R. Junín, AM, OCT & FM in Huancayo as R. Wanka.
476) Jr. Moquegua 461, 2þ piso, Tarma – DG: Augusto Mayor Quito.
477) Av. José Carlos Mariátegui 699, El Tambo, Huancayo. ☎ +51 (64) 233681. ▤ Jesus Pomalaza Baldeón. **FM:** 102.5.
478) Av. Las Palmeras 285, Santa Rosa sacco, Provincia de Yauli – DG: Javier Chaccha.
479) Jr. Real 517, Huancayo – Gte: Vicente Crisóstomo M. **FM:** 104.3.
480) Zona Aledaña a Satipo a 4 km, Satipo – **FM:** 98.9.
481) Cas. 167, Tarma. ☎ +51 (64) 321167, 321510 – DG: Mario Monteverde P. **FM:** 99.3 & 103.3 "R. Tropicana" in La Merced.
482) Dario León 425, La Oroya – DG: Horacio Gago.
483) Huancavelica 430, 2þ piso (or Ap. 230), Huancayo – Dir: Horacio Gago Priale. **FM:** 89.1.
484) Acobamba.
485) Av. Ayacucho 300, Huancayo.
486) Jr. Jauja 494, Tarma – DG: Hector Abel Hurtado Pomja. **FM:** 102.4.
487) Av. Arevalo 484, Santa Rosa Sacco, Provincia de Yauli – DG: Raul Espinoza Machacuay.
488) Calle Mercado 194, San Ramón.
489) Jr. Puno 430, Huancayo.

Pasco
490) Plazuela Gamaniel Blanco 127, Altos, Cerro de Pasco – Own: O. de la Cruz Vásquez. **FM:** 98.9.
491) Cooperativa Comunal de Huayllay, Prolongación Lima s/n, Huayllay – Dir: Alejandro Baracco.
492) Jr. Mullenbruck 469, Oxapampa. ☎▤ +51 (64) 762249 – Own: Pascual Villafranca G. **German:** 0300-0400.
493) Jr. Conde de La Laguna, 2þ piso, Paucartambo – Dir: Irwin Junio Berrios P. **FM:** 99.8.

REGION 5
Apurímac.
500) Tresierra 188, Andahuaylas – Gte: Daniel Andreu C.
501) Jr. Cuzco 206, Abancay – DG: Antero Quispe Allcaca. **FM:** 104.5.
503) Jr. Apurímac s/n, Chincheros.
504) Av. Seoane 200, Abancay. **FM:** 97.7 "Sudamericana".
505) Jr. Cesar Vallejo, Comercial Kelly, Andahuaylas.
506) Av. Muñoz 401, Abancay.
507) Frente de la Plaza de Armas, Jr. Ayacucho 1ra cuadra, Andahuaylas – Own: Romulo Ripa.
508) Jr. Cusco 319, Abancay – Own: Manuel Mendivil Ramírez. **FM:** 90.5.
509) Jr. Cusco 206, Abancay.

Ayacucho
510) Jr. Piura s/n, Ayacucho.
511) Calle Nazareño 200. Pasaje No. 163-A, Ayacucho – DG: Fernando Cruz Rondinel.
513) Cas. 24, Huanta – DG: Santos Angel Buendía Torres. **FM:** 99.9.
515) Calle Nazareno 108, Ayacucho – Dir: Walter Muñoz Inga. **FM:** 95.3.
516) Jr. Gervasio Santillana 455, Huanta – Dir: Humberto Sapaico Salazar. **FM:** 92.7.
517) Cas. 43, Huanta – Dir: Clemente Ramos Q. **FM:** 92.5.

Huancavelica
525) Virrey Toledo 468 (or: Ap. 92), Huancavelica – Dir: Fernando Cruz M. Adm: Elías Ccanto Pacheco. Mgr: Rvdo. Samuel Moran Cardenas. **FM:** 105.3.
526) ANta, Acobamba.
528) Jr. Maravilla, Barrio de Bellavista, Lircay, Angaraes – v/s: Gilmar Zorrilla Ll.
529) Av. Augusto B. Leguía 196, Huancavelica – Own/Dir: Oscar Alvarez Alvarado Yalico. **FM:** 93.1.
532) Jr. Manco Capac s/n. Pampas, Provincia de Tayacaja – DG: Ing. Larry Flores Lezama. Prgrs in **Sp.** and **Quechua**. **FM:** 100.0.

Ica
540) Jr. Salaverry 355, Ica – Dir: Juana F. de Nieri.
541) Fray Ramón Rojas 120, Urb. Luren, Ica – DG: Luis A. Umbert.
542) Pasaje los Girasoles 103, Nazca – **FM:** 99.9.
543) Km 308, Panamericana Sur, Los Aquijes, Ica.
544) Av. Arenales 1050, Ica – DG: Oscar F. Drago V. **FM:** 90.7.
545) Av. Arenales 1370, Ica - **FM:** 98.7.
546) Jr. Cajamarca 195, Ica – DG: Américo Aquije P.
547) Av. Los Incas 130, Puerto de San Juan.
548) Calle Independencia 159, Ica – DG: Lucio H. Fox M.

549) Jr. Zurumilla 360, Ica – DG: Fernando Cruz Mendoza.
550) Av. Progreso 400, Grocio Prado, Chincha.
551) Jr. San Martín 251, Ica.
552) Barrionuevo 142, Pisco.
553) Av. Progreso s/n, Chincha Alta.
554) Ap. 91, Ica – DG: Rosa Ch. de Flores.
555) Cas. 24, Pisco – DG: Wilfredo Villa M.
556) Av. Los Incas 117, Nasca.
557) Ap. 57, Nasca – DG: Fernando Cruz R.
558) Ap. 54, Chincha Alta – DG: Augusto Córdova B.
559) Santa Rosa de San Joaquin 892, Ica - **FM:** 90.1.
560) Av. San Martín 305, 2° piso, Chincha. - **FM:** 89.7.
561) Portal de Escribanos s/n, Palpa.
562) Calle Perez Figuerola 353, Pisco.

REGION 6
Arequipa
600) Lote 64-5-C, La Joya.
601) Av. Independencia 56, Arequipa – DG: J. Antonio Umbert D. **FM:** 93.5.
602) Av. Mariscal Castilla 683, Arequipa.
603) Francia 120, Urb. Satélite Chico-Paucarpata, Arequipa. ☎ +51 (54) 424237, 236086, 457806 – DG: Víctor Falcón C. **Quechua:** Sun 0830-1030. **FM:** 103.5.
604) Av. Victor A. Belaúnde C-8, Umacollo, Arequipa. ☎ +51 (54) 255888. 🖷 +51 (54) 251822 – DG: Leopoldo Rondón Fudinaga. **English:** 1900. **FM:** 97.1.
605) Junín s/n (🖃 Ap. 1706), Arequipa. ☎ +51 (54) 263324. 🖷 +51 (54) 263079 – DG: Eloy Arribas Lazaro. **Quechua:** 2h per day. **FM:** 106.3.
606) Bolivar 118, Arequipa. ☎ +51 (54) 219928 – Dir./CE: David Rivera R. PD: Fernando Rivera. **FM:** 98.5 in Majes, 102.1 in Tacna.
607) Av. Independencia s/n, Ciudad Universitaria (or Cas. 23), Arequipa. ☎🖷 +51 (54) 287771 – DG: Alejandra Cornejo.
608) Av. Jorge Chávez 201, Arequipa.
609) Palacio Viejo 401 (or Cas. 657), Arequipa – Dir: Pedro Rufino Zúñiga G.
610) Av. Bolívar 317, Arequipa.
611) Av. La Paz 512-A, Arequipa – Dir: Victor R. Fernández B. **FM:** 95.9.
612) Calle Santa Martha 304, Arequipa – DG: Victor Raúl Ponce de León Roca. **FM:** 107.7.
613) San Camilo 501, Arequipa – Dir: Elba Alvarez de Delgado.
614) Calle Pierola 209, Of. 205, Arequipa.
615) Puno 801, Mollendo.
616) Av. Lima 481, Camaná.
617) Av. Lima 301, Camaná.
618) Av. Parra 239, Arequipa.
619) Deán Valdivia 221 (or Cas. 56), Arequipa – DG: Fr.Lucio Tapia Flores, O.P. **Quechua:** Sun 1000. **FM:** 97.7.
620) Sucre 409, Arequipa – Dir: Máximo Landa D.
621) Santa Maria 310, Arequipa. ☎ +51 (54) 244421 – Dir: Tirso Borja Vizcarra.
622) Cas. 128, Mollendo – Gte: J.A. Bernedo Ll.
623) Ap. 2330, Arequipa – DG: Gustavo Quintanilla P. **FM:** 99.9.
624) Urb. Buena Ventura AN-7, Orcopampa, Castilla.
625) Camaná.
626) Calle San Juan de Dios 210, Arequipa.
627) Comunidad Minera de Cailloma, Cailloma.
628) Calle Deán Valdivia 203, Cerro Colorado, Provincia de Arequipa.
629) Justo Pastor Davila 197, Chorillos, Arequipa. – **FM:** 91.7.
630) Calle Samuel Pastor s/n, Camaná. **FM:** 100.1.
631) Arequipa 109, Urbanización Apima, Paucarpata, Provincia de Arequipa.
632) Av. Lima 172, Camaná.
633) Calle Mariano Melgar 500, Cerro Colorado, Provincia de Arequipa.
634) Parque Azángaro 150, Distrito Miraflores, Arequipa.
635) Av. República de Chile 123, Mariano Melgar, Provincia de Arequipa.
636) Zona Rural Huayracpampa, Paucarpata, Provincia de Arequipa.
637) Calle Nueva 313, Arequipa.
638) Av. Progreso 58, Corire, Uraca, Provincia de Castilla.
639) Jr. 28 de Julio 111, Alto Libertad, Cerro Colorado, Provincia de Arequipa.
640) Km. 5 Variante Uchumayo,Arequipa.
641) Calle Letitia 218, Arequipa.
642) Chuquibamba, Provincia de Condesuyos.
643) Pasaje San Vicente, Lote B-S, Vítor, Arequipa.
644) Av. Independencia 935, Urb. Municipal, Arequipa.

Moquegua

670) Zepita 743, Ilo – Dir: Guillermo Fernández P.
671) Jr. Tarapaca 260, Moquegua.
672) Ap. 31, Ilo – Dir: Javier Valencia Alponte. (Ann. OAX70).
673) Moquegua 123, 5ᵇ piso, Ilo. ☎ +51 (54) 781909. 🖷 +51 (54) 781867 – DG: Rolando Rodríguez M. **FM:** 105.5.
674) Jr. Ayacucho 639 (🖃 Ap. 22), Moquegua – DG: Guillermo Kuong Flores. **FM:** 105.3.
675) Calle Ancash 273, Moquegua.
676) Alto Ilo, Sector Arenal G-6, Ilo.
677) Mz. E lote 48 P, Joven J.F. Kennedy, Ilo.

Tacna
690) Cas. 303, Tacna – DG: Jaime Cáceres Chavez. **FM:** 89.9.
691) Cas. 113, Tacna – Adm: Royce Zeballos R. **FM:** 99.9.
692) Cas. 281, Tacna.
693) Ap. 115, Tacna – Dir: Rosa L. R. de Liendo. **FM:** 89.5.
694) Aniceto Ibarra 436 (🖃 Cas. 370), Tacna. ☎ +51 (54) 714871. 🖷 +51 (54) 723745. **E-mail:** radiotac@principal.unjbg.edu.pe **Web:** http://principal.unjbg.edu.pe/radio/radta.htm – DG: Yolanda Vda. de Cáceres. Dir. Tec: Alfonso Cáceres Contreras. **FM:** 104.3.
695) Av. Varela 705, Tacna.
696) Av. San Martín de Porres 2096, Tacna.

REGION 7
Cusco
700) Jr. Pumacahua 219, Sicuani. ☎ +51 (84) 351323 – DG: Victor Curie Deza. **Quechua:** 0930-1130. **FM:** 100.3.
701) Matto Turner 340, Santiago.
702) Calle Tecte 214, Cusco. ☎ +51 (84) 227071. – DG: Valentín Olivera P.
703) Av. Infancia 527, Cusco – DG: Julio C. Tello A.
704) Portal de Carnes 260, Cusco.
705) Belén 886, Cusco.
706) Av. de la Cultura 731, Of. 402, Huánchac, Cusco – **FM:** 92.1.
707) Calle Triunfo 379, Cusco – DG: Carmela Pinto.
708) Portal Belén 115, 2° piso, Plaza de Armas, Cusco – Own: Mauro Calvo Acurio. **FM:** 100.1.
709) Jr. José Santos Chocano G-11, Urbanización Santa Monica, Huánchac, Cusco – Dir: Luis Villasante C. **FM:** 103.3.
710) Ap. 76, Quillabamba – Dir: P. Francisco Panera. CE: Luis Verde I. **Quechua:** 1300-1430, 2100-0100.
711) Av. Manuel Callo Zevallos 111, Sicuani – Dir: Paul R. Infantas C.
712) Av. El Sol 346, 4° piso, Of. 402, Cusco - Gte: Gabino Pérez C.
713) Calle Sacsayhuaman K-10, Urb. Manuel Prado, Cusco – Dir: Alberto Barrionuevo S. **FM:** 104.1.
714) Av. Garcilazo 604, Huánchac, Cusco – Gte: Sra. Catalina Perez de Alencastre.
715) Jr. 2 de Mayo 206, Sicuani. 🖷 +51 (84) 223308 – Dir: Mario Ochoa Vargas. **Quechua:** 0930-1100, 2300-0300. **FM:** 91.1.
716) Av. Garcilazo 411, Distrito de Wanchác, Cusco – Dir: Carlos Gamarra M.
717) Av. El Sol 150, Yauri, Provincia de Espinar – Dir: Marcelino Taco Quispe.
718) Jr. Independencia 4308, Quillabamba.
719) Cas. 251, Cusco – DG: Raúl Siu Almonte.
720) Huayna Capac 154, Huánchac.
721) Av. Cusco s/n, Yauri, Provincia de Espinar.
722) Calle Triunfo 124, 2° piso, Cusco.
723) Calle Heladeros 220, Wanchác.
724) Urbanización Quinta Jardin C-12, Cusco. ☎ +51 (84) 233101. 🖷 +51 (84) 231881 – CE: Angel Reyes P. **FM:** 93.3.
725) Cas. 39, Cusco – Dir: Raúl Montesinos Espejo. **FM:** 94.6.
726) Edif. Municipal, Calca – DG: José Victor del Castillo Farfán &. J. Serapio Surco Ramírez.
727) Calle "C" 13, Urb. Jardin, Cusco.
728) Urb. Marcavalle, P-20, Huánchac, Cusco – DG: Walter Farian Flower. **FM:** 93.9.
729) Jr. Ricardo Palma 510, Quillabamba. **FM:** 96.5.
730) Plaza de Armas, Paucartambo. **FM:** 104.5.
731) Calle Heladeros 220, Huánchac, Cusco.
732) Plaza Túpac Amaru L5, Huánchac, Cusco.
733) Calle Matto de Turner 340, Santiago, Cusco.
734) Jr. Matara 526, Cusco.

Madre de Dios
740) Calle Daniel Alcides Carrión 385 (🖃 Ap. 37), Puerto Maldonado. ☎ +51 (84) 571050 – DG: Rufino Lobos Alonso. Prgrs also in **Quechua** and **native languages. FM:** 92.5.
741) Jr. Guillermo Billingurst 406, Puerto Maldonado – **FM:** 101.3.
742) Plaza de Armas 200, Puerto Maldonado – **FM:** 100.5.

Puno

750) Cas. 210, Puno. ☎ +51 (54) 351562. 🖷 +51 (54) 352233 – DG: Rvdo. José Loits M. **Quechua & Aymara:** 5h daily. **FM:** 95.7

751) Jr. Arequipa 385, Puno.

752) Parque Industrial C-16, Juliaca – **FM:** 89.5.

753) Asociación Cultural Nueva Era, Carretera Panamericana Sur Km. 17.5 Chucuito - Puno (or. Ap. 299), Puno. ☎ +51 (54) 622129 – DG: Nancy G. Cutipa Laqui. **Aymara:** 6h. **Quechua:** 6h. **Spanish:** 5h.

754) Ap. 130, Puno – Dir: G. Humpiri C.

755) Jr. Unión 214, Juliaca. ☎ +51 (54) 321453. 🖷 +51 (54) 322570 – DG: Leoncio Z. Torres. Gte: Mario Leonidas Torres. **FM:** 102.7. **Quechua & Aymara:** 1000-1100, 1800-1900.

756) Jr. Union 229, 3º piso, Juliaca.

757) Jr. Arequipa 833 (or Ap. 3), Puno. ☎ +51 (54) 352522. 🖷 +51 (54) 351217 – Dir: Alberto Lescano R. **Quechua & Aymara:** 1000, Sat 2300-0200. **FM:** 100.1 "Sudamericana".

758) Jr. Ramón Castilla 949, Juliaca – Dir: Robert Harry Therán Escobedo. **FM:** 90.1.

759) Jr. C. Arestegui 734, Juliaca.

760) Jr. Piura 167, Puno. ☎ +51 (54) 353680 – DG: Victor H. Avila Macedo. **Aymara & Quechua:** 0900-1100.

761) Hipolito Uname 240, La Rinconada, Juliaca – Own: Carlos Chanvilla Farfán.

762) Mariano Pandía 166, 2Þ piso, Juliaca.

763) 2 de Mayo 257, Juliaca – DG: Enrique Cáceres Escalante. **Aymara & Quechua:** 0900-1100, 1900-2100. **FM:** 104.5 "El Sol de los Andes FM".

764) Av. Titicaca 160-C, Puno – Dir: Atalo Gutiérrez V.

765) Cas. 340, Puno – Gte: Jorge Aramayo G.

766) Ap. 8, Ayaviri – Gte: J. Solórzano M. **Quechua:** 0900-1100, 1900-2300. **FM:** 101.5 "Radio A". **F.PI:** Tests on 1395kHz.

767) Jr. Salvador Allende 109-B (or Cas. 344) Juliaca. ☎ +51 (54) 620750, 322338. 🖷 +51 (54) 352784 – DG: Alfredo Padilla Chavez. Quechua & Aymara: 0900-1100.

768) Jr. Ancash 239, Puno.

769) Puno.

770) Azangaro, Provincia de Azangaro.

771) Plaza 28 de julio s/n, Macusani, Puno – **FM:** 90.5.

772) Jr. Altiplano 206, Urb. La Pampilla, Juliaca. – **FM:** 107.9.

773) Jr. Deustua 901, Puno. **FM:** 101.7.

REGION 8

Loreto

801) Ap 207, Iquitos. ☎ +51 (94) 265244 🖷 +51 (94) 264531 – Gte: Julia Jáuregui R. **FM:** 93.9.

802) Jr. Lima 960, Iquitos – Dir: Teddy R. Bendayan D.

803) Arica 228, Iquitos. ☎ +54 (94) 23302 – DG: Celia Cruzado Solari. **FM:** 103.5.

804) Cas. 174, or: Jr. Próspero 645, piso 2, Iquitos – Dir: Marcos Panduro R. **FM:** 105.9.

805) Calle Progreso 112, Yurimaguas – Dir: J.A. López-Manzanares. C.E: Pedro Capó Moragues. **FM:** 100.1.

806) Ap. 786, (Reports to: Pablo Rojas Bordales, Jr. Bermúdez 445), Iquitos – Dir: Julio Reátegui B.

807) Plaza de Armas 106, Yurimaguas, Provincia de Alto Amazonas. – **FM:** 97.1.

808) Calle Comercio 808, Yurimaguas – Dir: Prof. Ramirez Vecez. **FM:** 103.3.

809) Jr. Elias Agurrie 857, Iquitos. **FM:** 104.5.

810) Jr. Raymondi 137, 5Þ piso, Iquitos. **FM:** 98.9.

811) Jr. Lima 821, Iquitos. **FM:** 98.1.

Ucayali

820) Av. Cnel. Portillo 252, Pucallpa.

821) Cas. 263, Pucallpa – Dir: Julio R. Zevallos M.

822) Av. 9 de Diciembre 646, Pucallpa – Dir: Luis Salas B. **FM:** 102.5.

823) Jr. Coronel. Portillo 448, Pucallpa. ☎ +51 (64) 573876. 🖷 +51 (64) 571540 – DG: Néstor Rodríguez R. **FM:** 103.3.

824) Zona San Fernando, Callería, Pucallpa.

REGION 9

Amazonas

900) Av. Principal 794-796, Bagua – Gte: Julio Carzado G.

901) Jr. Octavio Ortiz Arrieta 588, Chachapoyas. ☎ +51 (74) 757002. 🖷 +51 (74) 757454 – DG: Jorge Luís Zubiate Mas. **FM:** 99.1.

902) Jr. Amazonas 1177 (🖂 Ap. 69), Chachapoyas. ☎ +51 (74) 757793. 🖷 +51 (74) 757004 – DG: Soledad Sánchez C. Adm: Ing. Maria Dolores Gutierrez Atienza. **FM:** 99.9.

903) Jr. Ayacucho 944, Chachapoyas – DG: Jorge Oscar Reina Noriega.

904) Santa Maria de Nieva, Provincia de Condorcanqui – Dir: Eduardo Weepiu Daekat.

905) Jr. Huayabamba 513, Santa Rosa de Huayabamba, Rodríguez de Mendoza – Dir: José M. Calderón S.

906) Jr. Ricardo Palma s/n, Chiriaco, Imaza, Provincia de Bagua – DG: Santos Castañeda Cubas. Dir: Hildebrando López Pintado.

907) 28 de Julio 420, Lonya Grande, Provincia de Utcubamba.

908) Simón Bolívar 433, Bagua Grande.

909) Jr. Matiaza Rimache 201, Rodriguez de Mendoza – DG: Luis Elber Julca Ll.

910) Jr. Amazonas 114, Rodriguez de Mendoza – DG: Juan J. Grandez S. PD: Salomon Grandez. CE: Jorge Noriega. **FM:** 98.5.

911) Av. Circunvalación 1249, Bagua.

912) Jr. Ayacucho 711, Chachapoyas.

913) Jr. F. Villareal 400, Utcubamba.

914) Av. Principal 1348, Bagua.

915) Aramango, Bagua.

916) Puerto Galilea, Provincia de Condorcanqui. – Dir: Guillermo Gómez García

917)Jr. Amazonas 3-5, Aramango, Bagua.

San Martín

920) Jr. Progreso 389, Bellavista. ☎ +51 (94) 544179 – Dir: Francisco González Meléndez.

921)Av. Compagñón 410, Tarapoto..

922) San Martín 317, Lamas.

923) Jr. Faustino Maldonado 930-932, Rioja. ☎ +51 (94) 558040. 🖷 +51 (94) 558454 – DG: Jhony Mori del Aguila. **FM:** 101.1.

924) Jr. Progreso 225, Tarapoto – DG: Fernando Tofur Arevalo. **FM:** 97.5.

925) Jr. San Martín 257, Tocache – Dir: Nélson Chavez S.

926) Ap. 210, Moyobamba – Dir: Porfirio Centurión Tapia. **FM:** 96.9.

927) Jr. Federico Sánchez 702, Tarapoto. ☎ +51 (94) 522709 – DG: Luis H. Hidalgo Sánchez. **FM:** 101.3 & 103.3.

928) Jr. San Filomeno 495, Moyobamba – Own: Prof. César Augusto Arévalo Seijas & Susana López del Aguila. **FM:** 99.9.

929) Cas. 31, Tarapoto – Gte: Luis F. Mori R.

930) Moyobamba.

931) Juanjuí, Provincia de Mariscal Cáceres. Re. to John Wiley Villanueva, Jr. Cerro de Pasco 461, Urb. "Aranjuez", Trujillo, Cajamarca.

932) Jr. Bolognesi 312, Bellavista. ☎ +51 (94) 544170. 🖷 +51 (94) 544135 – Dir: Severo Ruiz R. **FM:** 95.1.

933) Jr. San Martín 328-330, Tarapoto – Pres: Víctor Daniel Coral P. **FM:** 102.3.

934) Peña Meza 467, Juanjuí – Admor: Heber Arévalo Ríos.

935) Prolongación Santo Toribio s/n, Complejo Turístico Venecia, Rioja. ☎ +51 (94) 560212-3071 – Dir: Tito Pilco Mori. **FM:** 92.1.

937)Jr.20 de Abril 308, Moyobamba. – Own: Américo Vásquez Hurtado

938) Jr. Santor Toribio 270, Rioja – Dir: Juan Bolaños C.

940) Jr. Ucayali s/n, Distrito de San Hilarión, Provincia de Picota. – Own: Lucho Garcia Gaona & Marina Gaona.

941) Jr. Santo Toribio 1252, Rioja.

942) Jr. Huallaga 348,Nueva Cajamarca, Rioja. ☎ +51 (94) 556006 – Gte: Edilberto Peralta Lozada. **FM:** 100.7, though assigned to 94.5.

943) Jr. Reyes Guerra 579, Moyobamba – **FM:** 93.7.

944) Moyobamba.

945) Jr. Tupac Amaru 105, Centro Poblado Consuelo, Distrito de San Pablo, Provincia de Bellavista – Own: John Villanueva Lara.

946) Nuevo Bambamarca, Tocache.

947) Jr. Progreso s/n, Bo. La Victoria, Juanjuí.

948) Jr. Túpac Amaru s/n, Villa Picota, Picota.

L.T: UTC - 3h — **Pop:** 420.000 — **Radios:** 262.000 — **Pr.L:** Dutch, Sranan Tongo, Sarnami Hindi — **E.C:** 60Hz, 110/115/127/220V — **ITU:** SUR.

TELESUR N.V. (Gov. Administration)

🖃 P.O. Box 1839, Paramaribo. ☎ +597 473944. **Cable:** Latel.

STICHTING RADIO-OMROEP SURINAME (SRS) (Gov. Comm.)

🖃 P.O. Box 271, Paramaribo. ☎ +597 498115. 🖷 +597 498116. **MW:** 725kHz 5kW (inactive) — **FM:** Paramaribo 96.3MHz 1kW (stereo), 93.1MHz 0.1kW — Nieuw Nickerie 95.7MHz 0.01kW. **D.PRGR:** 0730-0400 (Sat 0430, Sun 0300). **Sarnami Hindi:** 0730-

0830 on all freqs, W 1900-2030 & 2315-2400 on 600kHz + 93.1MHz. **Javanese:** 0830-0930. Other times in Dutch. **N. in Dutch:** W 1000, 1100, 1200, 1730, 2000, 2345, 0100; Sun 1600, 1900, 0000.— **ANN:** "Dit is de Stichting Radio Omroep Suriname, de SRS in Paramaribo" — **V.** by QSL-folder or letter. Rec. acc.

PRIVATE COMMERCIAL STATIONS

	kHz	kW	Name and Location
2)	600	10	R. Paramaribo, Paramaribo
3)	820	1	R. Apintie, Paramaribo
4)	914	3	R. Nickerie, Nieuw Nickerie

SW: 3) 4990.9 (alt 5005) kHz 0.35kW.
FM: 2) 89.7MHz 1.5kW, Nieuw Nickerie 103.5MHz 0.01kW — 3) 97.1MHz 1kW, Nieuw Nickerie 92.1MHz 0.01kW, Moengo 92.1MHz 0.01kW— 4) 99.9MHz — 5) 98.3MHz 1kW — 6) 99.3MHz 1.5kW — 7) 103.1MHz 1.5kW, Nieuw Nickerie 101.1MHz 0.01kW.

Addresses and other information

2) P.O. Box 975, Paramaribo. ☎ +597 499995 — Dir: R. Pierkhan. Mgr: F. Pierkhan. 24h. **Sarnami Hindi:** W 0845-0930, 1500-1600; Mon-Thurs 2100-0100; Fri 2200-0200; Sat 2200-0300; Sun 0845-1330, 1930-2100. **Javanese:** W 0800-0845, 1900-2100. Other times in Dutch. **N. in Dutch:** 1830 — **ANN:** "Dit is R.P. Internationaal, the hot one", or "dit is k. Paramaribo, RAPAR N.V. op de 500 meter AM Band en 89.7 FM".
3) P.O. Box 595, Paramaribo. ☎ +597 400500 📠 +597 400684. **Cable:** Apintie. **E-mail:** apintie@sr.net — Mgr: Ch. Vervuurt. 0730-0400 (Sun 0300). **Sarnami Hindi:** W 1900-2030, Sun 1800-1900. **Javanese:** W 2030-2100, Sun 1900-1930. Other times in Dutch. **N. in Dutch:** 1730, 2200 — **ANN:** "U luistert naar R. Apintie op AM-FM stereo en special voor het binnenland op de kortegolf in de zestig meter band". **IS:** The beat of the Apintie drum — **V.** by letter.
4) R. RANI N.V, Waterloostraat 3, Nieuw Nickerie. ☎ +597 231462 — Dir: Dj.Lalta. 0900-0030. **Javanese:** 2100-2230. Other times in Sarnami Hindi — **IS:** "Surinam hamara pyara desh" sung by G. Kallasing.
5) RADIKA, Indira Ghandiweg 165, Paramaribo. ☎ +597 481175 or 482910 — Mgr: Mrs. Radhakishun. 0800-0300 in Sarnami Hindi plus some Dutch.
6) R. Sangeet Mala, Indira Gandhiweg 40, Paramaribo ☎ +597 423902 — Dir: Radjen Soekhradj. 0800-0400 (Sat 0500) in Sarnami Hindi. **N. in Dutch:** 0830, 1830. **N. in Hindi:** 0915, 2300.
7) Kara's Broadc. Corp, Verlengde Keizerstraat 5-7, Paramaribo. ☎ +597 475032 — Mgr: Orlando Karamat Ali. 0800-0400 (Fri/Sat 0600). **ANN:** "This is the spirit of the nation".

URUGUAY

L.T: UTC - 3h (Su: UTC - 2h) — **Pop:** 3.177.000 — **Radios:** 1.850.000 — **Pr.L:** Spanish — **E.C:** 50Hz, 220V — **ITU:** URG.

MINISTERIO DE DEFENSA NACIONAL

✉ Av. 8 de Octubre 2628, Montevideo.

DIRECCION NACIONAL DE COMUNICACIONES

✉ Bvar. Artigas 1520, 11000 Montevideo. ☎ +598 (2) 960215. **Cable:** 23213 Dinaco Uy — **LP:** Dir: Dr. Ernesto Dehl.

ASOCIACION NACIONAL DE BROADCASTERS URUGUAYOS (ANDEBU)

✉ Calle Yi 1264, 11100 Montevideo. ☎ + 598 (2) 900053. **Cable:** Andebu.
All st's affiliated exc. 4), 22), 32), 80); All affiliated st's carry "Cadena Andebu" prgr. D 1455-1500.
Mediumwaves: Þ = also on shortwave, * = inactive, v = varying freq.

	Call	kHz	kW	Name, location & h. of tr.
1)	CW1	550	25	R. Colonia, Colonia: 24h
2)	CX58	580	5	R. Clarín, Montevideo: 24h
3)	CX4	610	50/25	R. Capital, Montevideo: 0800-0600 (Sat/Sun 0300)
4)	CX6	Þ650	50/25	S.O.D.R.E.. Montevideo: 1000-0300 (classical)
71)	CW68	680	0.25	R. Young, Young: 0900-0300
6)	CX8	690	20/10	R. Sarandí, Montevideo: 24h
7)	CX10	730	5/2.5	R. Continente, Montevideo: 24h
5)	CW27	740	5/1	R. Tabaré, Salto: 0900 (Sun 1100)-0300

	Call	kHz	kW	Name, location & h. of tr.
8)	CX12	Þ770	100	R. Oriental, Montevideo: 0930-0300
9)	CX14	Þ810	50/25	R. El Espectador, Montevideo: 0830-0600
10)	CW23	820	1/0.5	R. Cultural, Salto: 0800-0300
11)	CX16	850	50	R. Carve, Montevideo: 0800-0500
49)	CX18	Þ890	50/10	R. Sarandi Sport, Montevideo: 24h
12)	CW17	900	2.5	R. Frontera, Artigas: 0900-0300
8)	CX20	Þ930	50	R. Monte Carlo, "la Super R.", Montevideo: 24h
13)	CW96	960	1	Radio Yi, Durazno: 1000-0200
14)	CX22	970	10	R. Universal, Montevideo: 24h
11)	CX24	1010	20	R. Nuevo Tiempo, Montevideo: 0900-0400
86)	CW102	1020	0.1	R. Libertadores, Salto: 0900-0300
4)	CX26	Þ1050	25	S.O.D.R.E.. Montevideo: 1000-0300
15)	CX28	1090	15/7	R. Imparcial, Montevideo: 24h
78)	CX111	1110	0.1	R. Paso de los Toros, Paso de los Toros: 1100-0200
16)	CW31	1120	5/10	R. Salto, Salto: 1000 (Sun 1100)-0300
17)	CX30	Þ1130	20	R. Nacional, Montevideo: 24h, Sun 0800-0500
18)	CW116	1160	0.5	R. Agraria del Uruguay, Cerro Chato: 0830-0130
87)	CV116	1160	1/0.25	R. Impacto, Mercedes: 0900-0400
19)	CX32	Þ1170	10	Radiomundo, Montevideo: 1100-0300
43)	CX118	Þ1180	10	LV de Artigas, Artigas: 0930-0300
20)	CW33	1200	2	La Nueva Radio, Florida: 24h
22)	CX121	v1210	1	Difusora Soriano, Mercedes: 24h
23)	CV121	1210	0.25	Em. RBC, Piriápolis: 0900-0400 (su), 1000-0200 (wi)
24)	CW121	1210	0.25	Em. El Libertador, Villa Vergara: 0830-0300
21)	CX122	1220	1	R. Reconquista, Rivera: 0945-0300
25)	CW35	1240	2.5/5	R. Paysandú, "la emisora sin fronteras", Paysandú: 0900-0400
26)	CX36	1250	10	R. 36, Montevideo: 24h
27)	CW125	1250	5	R. Bella Unión, Bella Unión: 0930-0200
82)	CW37	1260	4.5/1	Dif. Rochense, Rocha: 0900-0300
28)	CV127	1270	3/1	R. Cuareim, Artigas: 0900-0300
81)	CX128	1280	1	R. "Noticias"Tacuarembó, Tacuarembó: 0900-0300
4)	CX38	Þ1290	10	S.O.D.R.E.. Montevideo: 0900-0330
30)	CW39	1320	1/0.5	R. LV de Paysandú, Paysandú: 0858-0400
31)	CW132	1320	0.3	R. Fortaleza, Rocha: 0900-0300
32)	CX40	1330	5	R. Fenix, Montevideo: 1000-0600
33)	CW53	1340	10	LV de Melo, Melo: 0800-0300
34)	CW136	1360	1/0.5	R. Rio Branco, Rio Branco: 0930-0230
35)	CW41	1360	2.5/0.5	Broadcasting San José, San José: 0900-0300
36)	CX42	Þ1370	10/2.5	Emis. Ciudad de Montevideo: 1100-0300
37)	CV137A	1370	0.5/0.25	Emis. Real, Minas de Corrales: 0930-0130
88)	CW137	1370	0.25/0.1	R. San Javier, San Javier: 1000-2400
39)	CW45	v1390	2.5/1	Dif. Treinta y Tres, Treinta y Tres: 0800 (Sun 0900)-0300
40)	CX140	v1400	25	R. Zorrilla de San Martin, Tacuarembó: 0900-0300
42)	CX44	1410	5/3.5	Emisora del Siglo, Montevideo: 0800-0400
84)	CW141	1410	1/0.5	R. Turística, Termas del Arapey
43)	CW142	Þ1420	1/0.25	LV de Artigas, Artigas: 0900-0400
44)	CW43	1420	5	R. Lavalleja, Minas: 0830-0300
45)	CX142	1420.1	0.1	R. Felicidad, Paysandú: 1000-0300
46)	CW25	1430	5/1	R. Durazno, Durazno: 0900-0300
47)	CV144	v1440	3	R. Chuy, Chuy: 24h
48)	CX144	1440	3/0.5	R. Rivera, Rivera: 0830 (Sun 1000)-0300
38)	CX46	1450	5	R. América, Montevideo: 0900-0630
83)	CW145	1450	1/0.25	R. Arapey, Salto: 24h
50)	CW146	1460	1	R. Carmelo, Carmelo: 0900-0300
29)	CV146	1460	0.25	R. José Batlle y Ordoñez, José Batlle y Ordóñez: 1100-0300
51)	CX147	1470	5	R. Cristal del Uruguay, Las Piedras: 1000-0400
52)	CW147	1470	1/0.5	R. Tacuarí, Melo: 0830-0330

Call	kHz	kW	Name, location & h. of tr.
53) CW148	1480	3/1	R. Universo, Castillos: 1000-0300
54) CW43B	1480	3/1.5	R. Internacional, Rivera: 0800-0300
85) CX148	1480	1/0.7	Difusora Rio Negro, Young
41) CV149	1490	0.1	R. del Centro, Baltasar Brum
55) CX149	1490	5/4	R. del Oeste, Nueva Helvecia: 0930-0300
56) CX151	1510	0.25	R. Rincón, Fray Bentos: 1000-0300
57) CW57	1510	1/0.75	R. San Carlos, San Carlos: 0830-0300
58) CW151	v1510	0.25	R. Ibirapitá, San Gregorio de Polanco: 1000-0200
59) CX152	v1520	2	R. Cerro Largo, Melo: 0900-0300
60) CV152	1520	0.1	R. Paz, "la nueva R. ", Guichón: 1000-2300
61) CX50	v1530	1.2/0.25	R. Independencia, Montevideo: 1000-0400 (r. 1510)
62) CW153	1530	0.25	Emisora Cono Sur, Nueva Palmira: 0900-0300
63) CW154	1540	0.1	R. Charrúa, Paysandú: 1000-0300
64) CX154	1540	0.25	R. Patria, Treinta y Tres: 0800 (Sun 0900)-0300
65) CV154	1540	0.1	R. Centro, Cardona: 0900-0200
66) CV155	1550	1/0.25	R. Agraciada, Mercedes: 0800-0300
67) CW155	1550	0.25	R. Sarandí del Yí, Sarandí del Yí: 1000-0200
68) CW51	1560	2	R. Maldonado, Maldonado: 24h
69) CX156	1560	0.25	Dif. Americana, Trinidad: 0930-0130
70) CV156	1560	1/0.25	R. Vichadero: 1000-0200
72) CX157	1570	1	R. Canelones: 1055-0200
89) CV157A	1570	0.25/0.1	R. Celeste, Tomás Gomensoro: 0900-0300
73) CW54	v1580	0.5	Emisoras del Este, Minas: 0800-0200
74) CW158	1580	0.25	R. San Salvador, Dolores: 24h
75) CW159	1590	0.25	R. Regional, Lascano: 0900-0300
76) CV159	1590	0.25	R. Regional, Constitución: 1000-0200
77) CX159	v1590.5	2	R. Real, Colonia: 1000-0300
79) CV160	1600	5	Emisora Continental, Pando: 0900-0300
80) CX160	1600	1	R. Litoral, Fray Bentos: 1000-0300

Shortwaves: Call CX--, * = inactive, v = varying freq.

Call	kHz	kW	Name, location & h. of tr.
19) A2	*6000	2	Radiomundo, Montevideo
4) A5	*6010L	0.15	S.O.D.R.E., Montevideo: (r: CX26 1050)
17) A30	*6035	1	R. Nacional, Montevideo: 1300-0500
49) A61	6045.1	1	R. Sarandí Sport, Rivera: 1300-1400, 0100-0300
43) A3	*6074	5	LV de Artigas, Artigas: (n. 6075)
4) A4	6125	0.3	S.O.D.R.E.: Montevideo: 1000-0330 (r: CX38 1290)
8) A20	6139.9	2.5	R. Monte Carlo, Montevideo: 1030-1630 (n. 6140)
8) A72	9595	2.5	R. Monte Carlo, Montevideo: 2330-0300
4) A6	9620	0.25	S.O.D.R.E., Montevideo: 1000-0300 (r: CX6 650)
36) A42	9650	10	Em. Ciudad de Montevideo: 1300-2100
8) A7	11735	2.5	R. Oriental, Montevideo: 1630-2330
9) A19	*11835	1	R. El Espectador, Montevideo
36) A142	*15230	125	Em. Ciudad de Montevideo
4) A14)	*15275	1	S.O.D.R.E. Montevideo: (r: CX26 1050)

Addresses and other information

1) Rivadavia 383, 70000 Colonia or: Av. Sta. Fé 1970 Piso 14 "A," Buenos Aires, Argentina – DG: Mario R. Kaminsky.
2) Av. 18 de Julio 1516, 11200 Montevideo – DG: F.D. Torrellies.
3) Colonia 2212, 11100 Montevideo – Dir: Cr. Pedro Burunsusian.
4) Sarandí 430, 11000 Montevideo. ☎ +598 (2) 957865. 📠 +598 (2) 961933 – DG: Dr. Jorge T. Mascheroni. CE: José A. Cuello. CX6: classical music; CX26 music & spoken word. **DX Prgr:** SS1400-1500, "Radioactividades" on 1050kHz. Re. to Cas, 7011, 11000 Montevideo.
5) Av. Uruguay 1416, 50000 Salto. ☎📠 +598 (73) 33222 – Dir: Carlos Gelpi.
6) Enriqueta Compte y Rique 1250, 11800 Montevideo. ☎ +598 (2) 282612. 📠 +598(2) 236906. **Web:** http://www.netgate.com.uy/sarandi/ – Dir: Dr. Ramiro Rodriguez Villamil. News and talk.
7) Plaza Independencia 846, 11100 Montevideo. ☎ +598 (2) 925641–

Own: Heber Pinto.
8) Av. 18 de Julio 1224, 11100 Montevideo. ☎ +598 (2) 914433. **E-mail:** cx12@netgate.cominter.com.uy or cx20@netgate.comintur.com.uy **Web:** http://netgate.com.uy/cx12/ orhttp://www.netgate.com.uy/cx20/ – DG: Daniel Romay S. 930kHz is key st. for nationwide N. Net "CORI" comprising 42 AM & FM st's. **Shortwave relays:** The st's relay alternately CX12 770 and CX20 930kHz.
9) Río Branco 1481, 11100 Montevideo. **E-mail:** espectad@zfm.com **Web:** http://www.zfm.com/espectador/ – Dir: Luis De María.
FM: 106.5 "Em. del Exodo."
11) Mercedes 973, 11100 Montevideo – Own: Pablo Fontaina.
12) Av. Lecueder 803, 55000 Artigas – Dir: Basilio Borgato.
13) Zorrilla de San Martín 875, 97000 Durazno – Dir: T.B. Rodri-guez C.
14) Av. 18 de Julio, 1220, 3er piso, 11100 Montevideo – DG: R. Baleato – Dir: Oscar Imperio.
15) Av. del Libertador Brg. Gral. Lavalleja 1708, Edificio Carioca, 11800 Montevideo – DG: José Félix Iglesias.
16) Brasil 715, 50000 Salto – DG: Dr. Enrique Pera Erro. **FM:** 88.3 "Emisora del Lago".
17) Pza. Independencia 846, 11100 Montevideo – DG: Juan Carlos Bugallo. CE: Jorge Iglesias.
18) Juan Muñoz s/n, 30204 Cerro Chato.
19) Convención 1343, 9º piso, Montevideo. ☎ +598 (2) 921440. 📠 +598 (2) 906560 – Dir: Pablo Lecueder.
20) Antonio Ma. Fernández 800, 94000 Florida – DG: W. Pastorini C.
21) Figueroa 887, 40000 Rivera. ☎📠 +598 (622) 3807 – Own: Everildo Viera.
22) Castro y Careaga 568, 75000 Mercedes. ☎ +598 (53) 23430 – DG: Miguel M. Racioppi. **FM:** 89.3 "Em. del Hum".
23) Chacabuco y Moreno, 20200 Piriápolis – Dir: C. Repetto F.
24) Basilio Araujo 273, Villa Vergara, 33000 Treinta y Tres – DG: Geraldo Alzugaray.
25) 18 de Julio 1044, 60000 Paysandú. ☎ +598 (722) 23980. 📠 +598 (722) 22687 – DG: Aldo Rubens Bayce.
26) 18 de Julio 1357, 11200 Montevideo – Dir: José Bermat.
27) Enrique Ferreira 1550, 55100 Bella Unión – DG: Luis Murillo Muraro. **FM:** 105.5 "Stereo Norte FM".
28) Av. Lecueder 167, 55000 Artigas – ☎ +598 (642) 2867. 📠 +598 (642) 3867 – Dir: José H. Meirelles.
29) Ruta N 7 s/n, 30200 José Batlle y Ordoñez – Dir: Elbio Erazu P.
30) 18 de Julio 614, 60000 Paysandú. ☎ +598 (722) 22267. 📠 +598 (722) 24970 – Dir: Eduardo A. Estefanell.
31) Zorrilla de S. Martin 200, 27000 Rocha – Dir: Homero Casals O.
32) Canelones 1969, 11200 Montevideo – Dir: Miguel M. Racioppi.
33) Remigio Castellanos 721, 37000 Melo – Gte: Luis A. Bengochea.
34) Virrey Arredondo 986, 37100 Rio Branco – Dir: Raquel Balero González.
35) Treinta y Tres 860, 80000 San José – Dir: Marcos Marin.
36) Canelones 2061, 11200 Montevideo. ☎ +598 (2) 420142. 📠 +598 (2) 420700 – Dir: Aramazd Yizmeyián.
37) Dr. Dávison s/n, 40002 Minas de Corrales. ☎ +598 (622) 161173 – Dir: Rabindranath Eduardo Andina Lisboa.
38) Emilio Frugoni 1312, Montevideo – DG: Martin Zalcberg.
39) Pablo Zufriátegui 1076, 33000 Treinta y Tres – Dir: A. Pinho.
40) 18 de Julio 302, 45000 Tacuarembó. ☎ +598 (632) 2538 – Dir: Luis O. Dini. **FM:** 88.9"Em. de la Música".
41) Batlle y Ordóñez s/n, Baltasar Brum, Artigas – Own: Julio C. Da Rosa.
42) Dr. Aquiles Lanza 1314, 11100 Montevideo – Own: Nelson Marroco. CE: Daniel Hernández.
43) Av. Lecueder 483, 55000 Artigas. ☎ +598 (772) 2447 – DG: Luis Murillo Muraro. **FM:** 90.7 "Amatista FM".
44) José E. Rodó 530, 30000 Minas – Dir: Juan José Volante.
45) Calle 18 de Julio 817, 60000 Paysandú – Gte: Nery R. Cabillon.
46) Artigas 487, 97000 Durazno. ☎ +598 (362) 2015. 📠 +598 (362) 2058 – Dir: Sandy S. de Lerena.
47) Laguna Negra 174, 27100 Chuy – Gte: Gerardo Jara.
48) Dr. Anolles 441, 40000 Rivera – Dir: C.A. Redezert.
49) Enriqueta Compte y Rique 1250, 11800 Montevideo – Own: Santiago Rodriguez Villamil.
50) 19 de Abril 444, 70100 Carmelo – Dir: A. Lobecio P.
51) Av. Artigas 781, 90200 Las Piedras, Canelones – Gte: Juan Angel Maciel.
52) Treinta y Tres 949, 37000 Melo – Own: Myrian Suárez G.
53) Ferrer 1404, 27000 Castillos ☎ +598 (47) 24269. 📠 +598(47) 23611 – Own: Juan Brañas Brañas.
54) Av. Sarandí 792, 40000 Rivera. ☎ +598 (62) 24210. **E-mail:** internac@asinet.com.uy 📠 +598 (62) 23422. **Web:** http://www.asinet.com.uy/logos_de_radios/clientes/internacional/ – DG: Claudio Yanuzzo. Dir: Patricia Iglesias.

55) Calle Berna 1375, 70201 Nueva Helvecia. ☎ +598 (552) 4409 – Dir: Rafael M. Ravazzani. **FM:** 90.7 "Reflejos".
56) 25 de mayo 3164, 65000 Fray Bentos. ☎ +598 (562) 2653, 2022 – Dir: Omar Donato.
57) Sarandí 775, 20400 San Carlos – DG: Enrique Lamaison. Gte: Hugo H. Lafuente.
58) Gral. Artigas 193, 42500 San Gregorio de Polanco, Tacuarembó. ☎ +598 (639) 4017 – Own: Luis A. Gancio.
59) Ramigio Castellanos 437, 37000 Melo – Dir: Sra. Teresa Cortazzo.
60) Calle Fraternidad 382, 60008 Guichón. ☎ ▤ +598 (725) 2055 – Dir: Pe. Juan Bendinelli.
61) Paysandú 1186, 11100 Montevideo – Gte: Antonio Rupenian.
62) Chile 1162 y Gral. Artigas, 70101 Nueva Palmira. ☎ +598 (544) 6068. ▤ +598 (544) 6053 – Dir: Lorenzo Pagés.
63) Leandro Gómez 1089, Local 18, 60000 Paysandú – Dir: Lucio F. Francolino.
64) Lavalleja 1530, 33000 Treinta y Tres – Dir: José Eduardo Rodríguez.
65) Boulevard y Rivera, 75200 Cardona. ☎ ▤ +598 (536) 9315 – Dir: A. Martínez D.
66) José E. Rodó 781, 75000 Mercedes – Dir: Nestor Grana. **FM:** 100.3 "Galicia".
67) Sarandí 328, 97100 Sarandí del Yí. ☎ ▤ +598 (363) 9155 – Dir: Arnol Franco.
68) Zelmar Michelini 819, 20000 Maldonado. ☎ +598 (42) 24800. ▤ +598 (42) 22555 – Dir: Enrique Lamaison. **N:** every ½h. **FM:** 103.5 "FM Punta del Este".
69) 25 de Agosto 724, 85000 Trinidad – Dir: R. Estrada B.
70) Bld. Artigas s/n, 40003 Vichadero – Dir: José E. Mariño.
71) Rincón 1689, 65100 Young – Dir: Gerardo Arévalo.
72) J.T. González 434, 90000 Canelones – Dir: Walther Espiga. **FM:** 101.1.
73) Treinta y Tres 632, 30000 Minas. ☎ +598 (442) 3092 – DG: Carlos Francisco Falco.
74) Av. Asencio, 1695, 75100 Dolores. ☎ +598 (534) 2110. ▤ +598 (534) 2691 – DG: Nelly Sosa de A. **FM:** 89.7 "Skorpio".
75) Ituzaingó 1149, 27300 Lascano – Dir: Mireya Brun de Punales.
76) Av. Gral. Artigas y Av. Domingo Pérez, 50002 Constitución – Dir: Dr. Carlos A. Gelpi.
77) Av. Gral. Flores 468, 70000 Colonia. ☎ ▤ +598 (52) 22030 – DG: Jorge Newton Urrutia.
78) 18 de Julio 743, 45100 Paso de los Toros – Dir: Nelson Gómez.
79) Av. Artigas 932, Galería Solari, 91000 Pando. ☎ +598 (392) 2512. ▤ +598 (392) 4440 – DG: Julio Olivar Cabrera & Dr. Alfredo Fernández Vicente.
80) 18 de Julio y 25 de Agosto, 65000 Fray Bentos. ☎ +598 (535) 3100. ▤ +598 (535) 3528 – DG: Miguel M. Racioppi.
81) Ituzaingó 246, 45000 Tacuarembó – Dir: Lauro Siqueira.
82) 25 de Agosto 108, 27000 Rocha – Dir: Edgardo R. Ramírez G.
83) Amorín 35, 50000 Salto – Dir-Own: Juan Manuel Gutiérrez & Luis Felipe Gutiérrez.
84) Estación Afe, Termas del Arapey – Dir: Juan Carlos Arredondo.
85) Rincón 1811, 65100 Young – Dir: Eduardo A. Machado Machado. **FM:** 89.1 "Imagen FM".
86) Uruguay 1416, 50000 Salto. ☎ +598 (73) 22672 – Dir: Inocencio di Giacomo Rizzo.
87) 18 de Julio 291, 75000 Mercedes. ☎ +598 (532) 24840. ▤ +598 (532) 23340 – Dir-Own: Danilo González Ultra & Graciela Romero Cejas. Gte: Walter Martínez Ortega.
88) 27 de Julio casi Basilio Lubkov, San Javier. ☎ +598 (569) 2089 – Dir-Own: Eduardo Luis Añasco Federovsky.
89) Lavalleja 25 casi 25 de Agosto, Tomás Gomensoro – Dir-Own: Mario Franco Keel.

FM in Montevideo: All st's 10-100 kW. St. name Emisora (del)...
91.1 Gardel FM – 6) 91.9 – 19) 93.9 Océano – 4) 94.7 Concierto – 95.5 Plata – 96.3 Alfa – 4) 97.1 – 98.7 Diamante – 99.5 Sol – 100.3 FM X – 101.9 Azul – 4) 103.9 – 105.9 Galaxia – 106.7 Cabildo.
In the rest of the country there are well over 90 FM outlets.

VENEZUELA

L.T: UTC -4h - **Pop:** 21.675.000 - **Radios:** 8.300.000 - **Pr.L:** Spanish - **E.C:** 50+60Hz, 120/240V - **ITU:** VEN.

MINISTERIO DE TRANSPORTES Y COMUNICACIONES

▤ Dir. General Sectorial de Comunicaciónes, Torre Este, piso 35, Parque Central, Caracas. ▤ : +58 (2) 574 0753.
LP: Dir. General Sectorial de Comunicaciónes: Luis E. García. Head International Affairs: Nidya Crocker.

CAMARA VENEZOLANA DE LA INDUSTRIA DE RADIODIFUSION
▤ Ap.3955, Caracas. ☎ +58 (2) 261 1651, ▤ +58 (2) 261 4783.
LP: Pres.Marietta Hernández de Gómez.
Mediumwaves: Call YV-,
° = also on shortwave, * = inactive, v = varying fq.

Call	kHz	kW	Name and h.of tr.
2) OY	540	50/25	R.Perijá, Villa del Rosario: 0900-0400
3) KE	550	50	YVKE Mundial, Caracas: 24h
11) RH	560	50	R.Nal., Cd. Guayana (r:630)
114) PJ	560	20/10	R.Exitos "Latina 5-60", Rubio: 1000-0400
5) LX	570	100	R.Rumbos, Villa de Cura: 24h
6) MJ	580	50/10	LV de la Fé, Maracaibo: 24h
7) KL	590	20	R.Continente, Caracas: 0900-0500
8) QB	600	10	Mundial Sucre, Cumaná: 24h
149) SW	600	15	R.Alto Llano, Sta Bárbara de Barinas: 0900-0500
9) SE	610	10	R.Cristal, Barquisimeto
4) XY	610	10	R.Centro 6-10, Cantaura
10) NO	620	10	R.Libertad, Cabimas: 0900-0400
11) KA	630	50	R.Nal."Canal Ligero", Caracas: 24h
12) QO	640	30	Porteñas 6-40 "Unión R", Puerto La Cruz: 24h
74) MU	640	10/5	R.Carora, Carora: 1000-0400
13) LH	650	50/20	Aragueña 650, Maracay: 0900-0500
14) NA	660	10	Ondas de los Médanos, Coro: 0900-0400
134) QZ	660	5	R.Anaco, Anaco: 0900-0400
5) LL	°670	100	R.Rumbos, Caracas: 1000-0500, Weekends: 24h
15) QR	680	10	R.Continente, Cumaná: 1000-0500
69) MR	690	50/20	R.Barquisimeto, Barquisimeto
16) MH	700	10	R.Popular, Maracaibo: 1000-0400
160) PQ	700	5/2	R.Sur, Puerto Ordaz: 24h
17) KY	710	50	R.Capital, Caracas: 1000-0600
18) QE	720	50	R.Oriente, Porlamar: 24h
30) XE	720	10	R.Elorza, Elorza
28) OO	730	10	R.Frontera, San Antonio del Táchira
9) MT	730	10	R.Universo, Barquisimeto
19) NC	740	10	R.Maracaibo, Maracaibo: 0900-0400
55) NQ	740	50	R.Caroní, Puerto Ordaz
20) KS	750	100	R.Caracas Radio "RCR", Caracas: 0900-0400
21) QQ	760	10	Doble Q, Puerto La Cruz: 0955-0300
118) SO	760	10	R.Popular, Trujillo
11) KK	770	50/20	R.Nal., Valencia (r:630)
22) OD	°780	50/20	Ecos del Torbes, San Cristóbal: 0900-0400, Weekends -0600
87) MN	780	10	R.Coro, Coro: 24h
11)	790		R.Nal., Cd.Bolívar (r:630)
157) KC	790	10	R.Dif.Venezuela, "AM 7-90", Caracas
107) XM	790	50	R.Minuto "La Barquisimetana", Barquisimeto: 24h
71) ZC	800	50/10	R.Fé y Alegría, Guasdualito: 1100-0200
25) LP	810	10	Super R.810, Valencia: 0900-0500
169) XG	820	25/10	R.Guadalupana, Coro: 0900-0400
26) SH	820	50	R.Guayana, Upatá: 0900-0500
142) KU	820	10	R.Altura, La Grita: 1000-0400
27) LT	830	25	R.Sensación, Caracas: 0900-0500
9) MY	840	10	R.Juventud, Barquisimeto
166) KZ	840	10/5	Guarapiche 8-40, Maturín: 24h
175) RV	850	10	RV-850, Valencia
108) ZC	850	10	R.Fé y Alegría, Maracaibo: 0900-0500
173) YE	860	20/10	R.Enlace 8-60, Valle de la Pascua
29) OL	860	10	Mundial 8-60, San Cristóbal: 0900-0500
163) RU	870	10	R.Pueblo, Puerto La Cruz
9) MP	870	10	R.Lara, Barquisimeto: 0958-0400
135) YM	880	20/10	Mundial, Puerto Ordaz
23) KV	880	10	R.Deportiva 8-80, Caracas
174)	880	10	R.Paraguaná, Punto Fijo: 24h.
31) LW	890	10	R.América, Valencia: 0900-0400
45) VO	890	1	R.Oriente, El Tigre
32) MD	900	25	R.Mara "Ritmo 900", Maracaibo
33) RQ	910	50/20	RQ 910, Caracas
34) QX	920	10	R.Nueva Esparta, Porlamar: 1000-0400

Call	kHz	kW	Name, location & h. of tr.
116) QU	920	10/5	R.San Carlos, San Carlos: 0955-0400
35) LJ	930	10	R.Maracay, Maracay: 1000-0600
36) NN	940	10	R.Punto Fijo, Punto Fijo: 0900-0500
110) ZR	°940	15	R.Continental, Barinas
153) LU	940	10	R.Fé y Alegría, Campo Mata: 0900-0300
37) KG	950	50	R.Informativa 9-50, Caracas: 24h
38) RB	960	50/20	R.Monagas, Maturín
176)	960	25	Mundial Llanera, Acarigua
39) SS	960	10	R.San Sebastián, San Cristóbal: 1000-0500
40) LR	970	10	R.Continente 970, Maracay: 0900-0400
41) SD	970	15	Mundial Turismo, Valera: 0855-0355
177)	970	10	Mundial Oriental, Barcelona: 24h
42) QM	980	10	Mundial El Tigre, El Tigre: 1000-0300
43) RT	990	20	R.Tropical "99-0", Caracas
44) TA	990	10	Mundial Tricolor, Barquisimeto: 24h
22) OA	°1000	10	R.Táchira, San Cristóbal: 1000-0400
46) NM	1000	10	Mundial Mil "La Caribeña", Morón: 0900-0400
47) PC	1010	10	R.Aragua, Cagua: 0900-0400
48) QF	1010	10	Mundial Bolívar, Cd.Bolívar: 0900-0400
49) MX	1020	50/10	R.Calendario, Maracaibo: 24h
50) RS	1020	10	R.Margarita, La Asunción: 0955-0500
124) TW	1020	25	R.Alegría 1020, Chivacoa
51) TD	1030	25/1	R.Valles del Tuy, Ocumare del Tuy: 0930-0400
52) QY	1030	20	R.Onda, Guanare: 0900-0600
53) ON	°1040	20/10	R.Los Andes, Mérida
54) LB	1040	20	LV de Carabobo, Valencia: 0900-0400
11) KZ	1050	20/10	R.Nal. "Canal Clásico", Caracas: 1000-0400
11) PO	1050	20	R.Nal. "Canal Clásico", Cabudare: 1000-0400 (r:1050)
56) LN	1060	10	R.Guárico, San Juan de los Morros: 1030-0330
57) OE	1060	10	R.San Cristóbal "RSC", San Cristóbal: 1000-0600
81) PX	1070	5	R.El Sol, La Fría
59) MA	1070	10	Mundial Zulia, Maracaibo: 24h
181)	1070	25	Contacto 1070, Ispino: 1000-0400
185)	1070	10	R.10-70 AM, Maiquetía: 1000-0400
60) NR	1080	10	Mundial 1080, Maracay: 24h
61) QJ	1080	10	R.Barcelona, Barcelona
62) PB	1090	10	R.Yaracay 1090 AM, San Felipe
63) SZ	1090	20	Unión R.1090, Caracas: 24h
64) TG	1090	3	Melódica 1090, Machiques: 0900-0500
65) OP	°1100	10	R.Occidente, Tovar: 0900-0300
66) SV	1100	10	R.Angostura, Cd.Bolívar: 0900-0430
67) QT	1110	10	R.Carúpano, Carúpano: 0900-0400
182) RX	1110	10	Unión R., Valencia
155) MF	1120	10	Ondas del Lago "Super Ondas", Maracaibo: 24h
171) SK	*1120	20/10	R.Dif.del Sur, San Fernando de Apure
70) XZ	1120	5	R.República "La Estación Feliz", Maturín
184)	1120		R.Fé y Alegría, Tovar
44) KQ	1130	10	R.Popular, Barquisimeto: 0900-0400
72) RL	1130	20/10	R.Ideal, Maiquetía
73) PY	°1130	10	R.Amazonas, Puerto Ayacucho: 0900-0400

Call	kHz	kW	Name, location & h. of tr.
183)	1140		R.Porlamar "LV del Caribe", Porlamar: 1000-0400
76) MV	1150	10	Mundial Caribe, Punto Fijo: 24h
77) QD	1150	10	Ecos del Orinoco, Cd.Bolívar: 24h
78) OK	1160	1	R.Universidad, Mérida
79) RR	1160	20/10	R.Industrial "Doble Uno 6-0", Guarenas
80) QV	1170	20/10	R.Acarigua, Acarigua
58) KW	1170	10	Señal 11-70, Maiquetía: 1000-0300
82) OR	1180	10	R.Maturín, Maturín: 0900-0400
94) LQ	1180	10	Super Suave 11-80 "LV de la Victoria", La Victoria
83) NJ	1180	10	R.Petrolera, Cd.Ojeda: 0900-0700
84) RE	1190	20/10	R.Barinas 1190 AM Estéreo, Barinas
85) ZD	1190	10	R.Dif.Cultural del Táchira, San Cristóbal: 24h
105) PF	1190	20/10	R.Canaima, San Félix: 0900-0300
95) SF	1200	10	R.Dimensión, Caripito: 1000-0300
93) OZ	1200	10	R.Tiempo, Caracas: 24h
86) NH	1200	1	Ondas del Escalante, Sta Bárbara del Zulia: 1000-0300
88) ZT	1210	10	R.Anzoátegui, Barcelona: 24h
89) RD	1220	10	LV de Apure, San Fernando de Apure
178) VM	v1220	10/5	Mundial Valencia, Valencia
90) ZO	1220	20/10	R.Aeropuerto 1220, Maracaibo: 24h
91) NT	1230	10	R.Barlovento, Caucagua
92) OH	°1230	10	R.Valera, Valera: 0900-0400
11) NV	1240	50	R.Nal.de Venezuela, Punta Tumatey (program Antena Populares)
96) PZ	1250	20/10	R.Puerto Ordaz: 0900-0400
97) ML	1250	1	R.Cabimas, Cabimas: 1000-0300
1) RM	1260	10	R.Super Suave, Caracas
99) RY	1260	10	R.Horizonte, Nirgua: 1000-0200
100) OU	1270	10	Ondas Panamericanas, El Vigía
101) TR	1270	5	R.Tucupita, Tucupita: 0900-0500
102) OF	1280	10	R.Trujillo, Trujillo
103) QS	1280	10/5	R.Zaraza, El Tamarindo: 1000-0300
104) LF	1290	10	R.Puerto Cabello, Puerto Cabello: 0900-0400
106) KH	1300	10/8	R.1300, Caracas: 1000-0400
16) NS	1300	10	R.Reloj, Maracaibo: 24h.
109) TS	1310	5	R.Andina "Sonido 13-10", Isnotú: 0900-0500
11) SM	1310	10	R.Nal., Barcelona (r:630)
11) SL	1310	1	R.Nal., Guri (r:630)
11)	1310	1	R.Nal., Elena de Uairén (r:630)
111) WP	1320	10/5	R.Apolo, Turmero: 24h
112) SG	1320	10	R.Colonial, El Tocuyo
113) OY	1330	5	R.Los Llanos, Calabozo: 0900-0300
115) TU	1330	10	Mundial Regional, Ciudad Ojeda: 0900-0500
11)	1330	1	R.Nal., La Paragua (r:630)
117) NE	1340	10	R.Uno "AM 1340", Caracas: 24h
119) ZZ	1350	5	R.Guanipa "R.13-50", El Tigrito: 24h
121) TJ	1350	5	R.Falcón, Puerto Cumarebo
122) TI	1360	10	R.Internacional, Maracaibo: 24h
123) TZ	1360	5	R.Armonia, Charallave: 1000-0300
125) OQ	1370	5	R.La Pascua, Valle de la Pascua: 1000-0400
126) LG	1370	5	R.Cumbre, Ejido
127) SV	1370	5	R.Portuguesa, Araure: 1000-0400
128) ME	1380	5	R.Fantasía, Soledad
129) NG	1380	10	Ondas del Mar, Puerto Cabello: 0900-0400

Call		kHz	kW	Name, location & h. of tr.
120)	TL	1380	10	R.Mundial, Caja Seca: 0900-0400
130)	TT	1390	10	R.Terepaima, Cabudare: 24h
131)	ZA	1390	10	R.Fé y Alegría, Caracas: 1000-0300
165)		1390		R.Lumen 2000, Maracaibo: 1030-0500
132)	NF	1400	1	R.Sabana, El Sombrero: 1000-0200
133)	ZJ	1400	10	R.1400, Barinas: 0900-0500
136)	SP	1410	10	R.Simpatía, Valera
137)	ST	1410	5	R.Turén, Turén: 0900-0400
139)	NZ	1420	5	R.Marabina, Maracaibo: 0900-0500
140)	RW	1420	10/5	R.Cardenal, Carora: 1000-0400
180)		1420	5	R.Sintonía, San Antonio de los Altos
141)	NB	1430	10	R.Satélite, Guacara: 0930-0400
143)	TM	1430	10/5	R.Caicara, Caicara del Orinoco
138)	TP	1430	25	R.Bahía, Puerto La Cruz: 24h
144)	RF	1440	5	R.Orituco, Altagracia del Orituco
145)	TY	1440	1	R.Sucesos, Táriba: 0950-0400
146)	ZI	1440	5	R.Estelar, Guanare: 1000-0300
24)	XC	1450	10/5	R.Icabarú, Puerto Ordaz
147)	ZQ	1450	10	R.Informativa, Puerto de Altagracia: 0900-0500
148)	KJ	1450	10	Sonera 14-50, Caracas
150)	RJ	1460	5	R.Jardín, Boconó
151)	JW	1470	10	R.Latina, Valencia: 24h
152)	SY	1470	10	R.Vibración, Carúpano
154)	SQ	1490	1	R.Mérida 14-90, Mérida: 24h
155)	RP	1490	10/5	R.El Sol, Maracaibo
158)	XD	1490	10	Dinámica 14-90, Caracas
156)	RZ	1500	10/5	R.Dos Mil, Cumaná: -0400
179)		1510	20	Informativa "LV del Centro", Güigüe
159)	IC	1520	25	R.Bonita Guapa, Guatire: 0930-0400
161)	NP	1530	10	R.San Felipe, San Felipe
170)	XO	1550	10/5	R.Ojeda, Cd. Ojeda
162)	MW	1550	10	R.Metropolitana, Los Teques: 0900-0500
164)	LZ	1560	10/5	R.Dif.Andina, Mérida
167)	TK	1580	10/5	Manzanares 15-80, Cumaná: 1000-0600
168)	YV	1580	10/5	R.Venezolana, Calabozo: 1000-0200
172)	YO	1580	10	R.Occidental, Sta Rita
98)		1590	10	R.Deporte 15-90, Caracas: -0400

Shortwaves:

Call		kHz	kW	Name, location & h. of tr.
22)	OB	4830	10	R.Táchira, San Cristóbal: 1000-1400, 1930-0400
92)	OI	v4840	1	R.Valera, Valera: irr.
73)	PA	4940	1	R.Amazonas, Puerto Ayacucho: 0900-0400
5)	LK	4970	5	R.Rumbos, Villa de Cura
22)	OC	4980	10	Ecos del Torbes, San Cristóbal: 0900-1400, 1800-0400
53)	SB	v6010	1	R.Mundial Los Andes, Mérida
11)	NV	v9540	25	R.Nal.de Venezuela, Caracas
22)	SC	9640	10	Ecos del Torbes, San Cristóbal: 0900-1400, 1800-0400
5)	LM	v9660	20	R.Rumbos, Caracas
65)	OS	v9750	1	R.Occidente, Tovar: 0900-0300
11)	NV	*11695	1	R.Nal.de Venezuela, Caracas

State abbreviations (Estados): Ama = Amazonas; Amro = Amacuro; Anz = Anzoátegui; Apu = Apure; Ara = Aragua; Bar = Barinas; Bol = Bolívar; Car = Carabobo; Coj = Cojedes; Fal = Falcón; Guá = Guárico; Lar = Lara; Mér = Mérida; Mir = Miranda; Mon = Monagas; Nes = Nueva Esparta; Por = Portuguesa; Suc = Sucre; Tách = Táchira; Tru = Trujillo; Yar = Yaracuy; Zul = Zulia.**N.B:** These abbreviations are not officially recognized by the Venezuelan Post Office. Letters should therefor carry full name.

Addresses and other information:

1) Av.Los Mangos, Qta.Marisabel, La Florida, Caracas 1050 - Pte: Samuel Robinson.

2) Calle Central, Villa del Rosario 4047, Zul (or: Edif. La Palmera, P1, Las Delicias, Maracaibo 4001, Zul) - Dir: Ali Rachid L.

3) Edif.Tequendama, Av.Tamanaco, El Rosal, Caracas 1060 - Dir: Antonio Mejía Jaimes.

4) Cantaura 6007, Anz - Dir: Pedro Fermín V.

5) Ap.2618, Caracas 1010A (or: Multicentro Empresarial del Este, Edif.Libertador, P7, Núcleo A.Chacao, Caracas 1060) - Pte: Andrés Serrano Trías. Dir: Felipe Serrano García. ☎ +58 (43) 335179, 333734. **E-mail:** rumbos@tycom.com.ve. **Web:** http://www.tycom.com.ve/rumbos/.

6) Ap.459, Maracaibo 4002, Zul - Dir: Padre Angel Rodríguez del Palacio.

7) Edif.Cosmos, PH, Calle La Joya, Chacao, Caracas 1010 - DG: Pedro Francisco Grespan M.

8) Ap.26, Cumaná 6101A. Suc - Dir: Jaime Muñoz.

9) Av.Venezuela, Entre Calles 13 y 14, Barquisimeto 1301, Lar - DG: Ligia Villanueva A.

10) Ap.99, Cabimas 4013, Zul - Dir: Denis Contreras

11) Final Calle las Marias, El Pedregal de Chapellin, Caracas 1050 - DG: Maritza Esparragoza. GM Eng: Ing. A Miranda B. (630 kHz+relays: Light music; 1050 kHz Classical music).

12) Ap.4123, Puerto La Cruz 6023, Anz - DG: Jaime Ross.

13) Hotel Maracay, Maracay 2101, Ara - DG: Rafael E.Mena.

14) Edif.Don Beto, Ca.Bolívar, Coro 4101, Fal - Dir: José Jatar Diaz.

15) Edif.La Copeta, Torre 2, P3, Av.Fernández de Zerpa, Cumaná 6101, Suc.

16) Ap.347, Maracaibo 4002, Zul - Dir: José Higuera M.

17) Centro Comercial Los Ruices, Av.Francisco de Miranda, Caracas 1071 - DG: Radamés Lebrón.

18) Hotel Bella Vista, Calle Velásquez, Porlamar 6301, Nes - Dir: Antonio Antepaz.

19) Calle Paraíso con Calle 67 N° 24-88, Maracaibo 4002, Zul - Dir: Rafael E.Mena.

20) Av.José A.Paez, Edif.RCR, El Paraiso, Caracas 1021 - DG: Ing.Julian Isaac.

21) Ap.4082, Puerto La Cruz 6023, Anz - Dir: Sergio Ramos.

22) Ap.152, San Cristóbal 5001, Tách - Own: Albertina Zerpa de Gonzales.

23) Edif.Mundial, P2, Av.Tamanaco, El Rosal, Caracas 1060 - Gte: Oscar Montenegro.

24) Edif.Arosa, Puerto Ordaz 8015, Bol - Dir: Luis A.Pallares.

25) C.C.Big Low Center, Nave H, Valencia 2001, Car - Dir: Celso Pacheco.

26) Av.Raúl Leoni, Edif.Antonelly, Upatá 8026, Bol - Dir: Edgar H. Rodríguez.

27) Av.Stgo de Chile, Qta.Radio Sensación, Los Caobos, Caracas 1050 - Dir: Hosé Gregorio Guillot.

28) Centro Cívico, PH Of.7-1, San Antonio del Táchira 5007, Tách - Dir: M.Marchena

29) Av.Libertador, Edif.Lotería del Táchira, Planta Baja, San Cristóbal 5001, Tach - Dir: Nelson V.Chacín.

30) Av.Aeropuerto esq. Calle 9, Municipio Rómulo Gallegos, Elorza 7007, Apu - Dir: Luis H.Borjas.

31) Ap.117, Valencia 2001, Car - Mgr: Manolo Fachín V.

32) Ap.1969, Maracaibo 4001A, Zul - Dir: Gustavo E.Vecino.

33) Edif.Mundial, P3, Av.Tamanaco, El Rosal, Caracas 1060 - Dir: Wilfredo Arévalo.

34) Av.Miranda, Edif.Porlamar, Planta Alta, Porlamar 6301, Nes - Dir: Luis J.Arreaza A.

35) Av.Boyacá, Edif.Centro, Of.01, Maracay 2101, Ara - Dir: G. Mewa M.

36) Ap.25, Punto Fijo 4102, Fal - Dir: Román Oswaldo Aguillar.

37) Edif.Mundial, P2, Av.Tamanaco, El Rosal, Caracas 1060 - Dir: Saúl Martínez.

38) Edif.R.Monagas, P1, Maturín 6201, Mon - Juan Carlos García Salcedo.

39) Ap.338, San Crostóbal 5001, Tách - Mgr: Jesús Contreras. Te: +58 (76) 26444, 26445.

40) Ap.4545, Maracay 2101, Ara - Dir: Freddy Contreras 41) Ap.12, Valera 3101, Tru - Own: Pedro José Fajardo.

42) Ap.430, El Tigre 6034, Anz - Dir: Francisco González P.

43) Torre Oeste, P1, Puente Nuevo a Puerto Escondido, Caracas 1010 - Dir: María Victoria de Grasso.

44) Final Av.Vargas, Edif.Radiorama, Barquisimeto 3001, Lar - Dir: Bernardo A.Morales.

45) Calle Miguel Otero Silva. Edif.Ingrid, El Tigre 6034, Anz - Dir: Calazans Guzmán.

46) Edif.Cumboto, Calle Comercio, Puerto Cabello 2024, Car.

47) Calle Boyacá Nte 9, Edif.R.Aragua, Oficina 01, Cagua 2122, Ara - Dir: Jesus Mª Labrador U.

48) Ap.123, Cd. Bolívar 8001, Bol - Dir: Luvén Rossi V.

49) Ap.789, Maracaibo 4001, Zul - Mgr: León Magno Montiel.

50) Ap.145, La Asunción 6311, Nes - Mgr: Bernardo Pirela.

51) Calle Urdaneta, Edif.Radio, Ocumare del Tuy 1209, Mir - Dir: Eleázar Ramos B.

52) Calle 15 esq.Cra.9, Guanare 3310, Por - Pte: Rafael Vicente Camacho.

53) Calle 4 N° 3-57, Av.Urdaneta, Mérida 5101 - Dir: Celso Pacheo. Te: +58 (74) 639286, 636015.

54) Torre Trébol, Urb.Lomas del Este, Valencia 2001, Car - Dir: Guillermo Degwitz.

55) Altavista Calle Caura, Edif.Los Bancos, Puerto Ordaz 8015, Bol.

56) Ap.25, San Juan de los Morros 2301, Guá - DG: J.Ghersy.

57) Ap.374, San Cristóbal 5001, Tách - Mgr: Désiree González Zerpa.

58) Av.Soublette, Edif.Américas B, P16, Maiquetía 1160.

59) Av.26, N° 22-23, Maracaibo 4001, Zul - DG: Pedro Carreño.

60) Calle 96, Nte 6, Calicanto, Maracay 2101, Ara - Dir: Abraham Granadillo.

61) Av.Miranda, Edif.R.Barcelona, Barcelona 6001, Anz - Dir: Ruffo José Vivas.

62) 5a Av.con Calle 15, San Felipe 3201, Yar - Dir: Orión Oviedo Pinto.

63) Edif.Esplendor, N° 42, Av.Mohedano, La Castellana, Caracas 1060 - Pte: Sergio Gómez Letón.

64) Av.Gral Trias, Machiques 4021, Zul - Gte: António Perozo.

65) Cra.4 N° 6-46, Tovar 5143, Mér - DG: Pedro R.Silguero.

66) Ap.31, Cd.Bolívar 8001, Bol - Dir: Tomás León.

67) Calle Juncal, Edif.Siglo XX, Mezzanina, Carúpano 6124 - Dir: Pedro Luis Regnault Villalba.

68) Edif.Ramos, Calle Páez N° 97-53, Valencia 2001, Car.

69) Calle 29, Entre Cra.18 y 19, Barquisimeto 3001, Lar.

70) Edif.Isnotú, Calle Monagas esq.Mariño, Maturín 6201, Mon.

71) Carr.Nal., Vía Elorza, La Arenosa, Guasdualito 5063, Apu - Dir: Ignacio Ibañez.

72) Calle El Trébol, Qta.Anna, Maiquetía 1161 - Dir: Luciano Landaeta Lovera.

73) Av.Bolívar 4, Puerto Ayacucho 7101, Ama - DG: Lic.Hugo Ali Bina.

74) La Casita, Sucre e/Cras. 7 y 8, Carora 3040, Lar - DG: Marconi González C.

75) Ap.23, San Cristóbal 5001, Tách - DG: Modesto Marchena.

76) Calle Ecuador, Punto Fijo 4102, Fal - DG: Bartolo Cárdenas.

77) Ap.47, Cd.Bolívar 8001, Bol - Dir: Luis Galeano Martínez.

78) Av.2 N° 21-22, Mérida 5101, Mér - DG: Lourdes E.Dubuc.

79) Calle Régulo Franquiz, Edif.La Electricidad, Guarenas 1220, Mir.

80) Av.35 con Calle 29, Acarigua 3301, Por - Gte: Oswaldo A.Ramírez.

81) Calle 4, Edif.Illinois, La Fría 5020, Tách - Dir: Luis Orlando Soto.

82) Calle 15 Antigua Sucre N° 73, Maturín 6201, Mon - Dir: Rafael D. Sifontes N.

83) Ap.201, Cd.Ojeda 4019, Zul - Dir: H.Zabala P.

84) Final Av.Sucre, Barinas 5201, Bar - DG: Rubén Darío Rodríguez.

85) Ap.150, San Cristóbal 5001, Tách - Dir: Ello R.Ríos R.

86) Av.5 N° 2-21, Sta Bárbara del Zulia 5148, Zul - Dir: J. Alvarado S.

87) Ap.7421, Coro 4101A, Fal - Dir: José G.Polanco.

88) C.C.Tricentenaria, Planta Alta, Of.9 y 4, Barcelona 6001, Anz - Gte: Dally V.Arreaza G.

89) Av.14, San Fernando de Apure 7001, Apu - Dir: J.Barbella.

90) Edif.Plaza, Locales 2 y 3, Calle 78 Doctor Portillo, esq. con Av.3H, Maracaibo 4002, Zul - Dir: José Bonacia.

91) Calle Pantoja, Caucagua 1246, Mir - Dir: Rómulo Raymondo.

92) Av.10 entre calles 9 y 10, Valera 3101, Tru - Own: Carlos Rumbos. Te: +58 (51) 53744, 53978.

93) Av.Los Mangos N° 49, La Florida, Caracas 1050 - Pte-Dir: Marieta Hernández de Gómez.

94) Edif.Belis, Blvd.Aldao, La Victoria 5064, Ara - Dir: R.Vásquez.

95) Av.Jerusalem al lado de CANTV, Caripito 6211, Mon.

96) Centro Comercial San Vicente, Carr.El Tocuyo, Pto Ordaz 8015, Bol - Dir: Angel de Jesús Coa.

97) Av.Andrés Bello, Cabimas 4013, Zul - DG: Ivan Arzola.

98) Centro Profesional Sta Paula, Torre A, P5 Of.51, Av. Curcunvalación del Sol, Sta Paula, Caracas 1061 - Pte: Asdrúbal Fuenmayor.

99) Urb.Las Tunitas, Nirgua 3205, Yar - Dir: Iván Arzola.

100) Av.Bolívar 10-87, El Vigía 5145, Mér - Dir: Dr.Serafín González Alvarez.

101) Calle Petión esq.La Paz, Tucupita 6401, Amro.

102) Calle Independencia 10-14, 11-10, Trujillo 3102, Tru - DG: L.Torres. ☎ +58 (72) 33080, 33081.

103) Calle Concordia 35, Qta.Pto Arturo, Zaraza 2332, Guá - DG: Luís Solórzano F.

104) Av.La Marina, Edif.Diproca, Local 3, Puerto Cabello 2024, Car - DG: Henry R.Oviedo.

105) Av.Antonio de Berrios, San Félix 8024, Bol - Dir: L.J. Pastrano.

106) Edif.Perú, P1-E, Centro Res.El Bosque, Av.Libertador, El Bosque, Caracas - Pres: Juan Manuel Chinea.

107) C.C.Venrol, Av.Pedro León Torres, Barquisimeto 3001, Lar - Own: Freddy Andrade.

108) Av.3E N° 63-50 (or Ap.10247), Maracaibo 4002, Zul - DG: Lic.Gerardo Lombardi.

109) Ap.63, Valera 3101, Tru (or: Calle Iglesia J.G. Hernández, Isnotú 3109, Tru) - Dir: Leonardo Salas.

110) Ap.202, Barinas 5201, Bar - Dir: Angel M.Pérez. M. Pérez.

111) Calle Bermúdez, Edif.Torre Apolo, Turmero 2115, Ara - DG: Mercedes de Salazar.

112) Av.Lisandro Alvarado, El Tocuyo 3018, Lar - DG: Norton Ramírez P.

113) Edif.Páez, Carr.12, Calabozo 2312, Guá - Dir: Eleázar Ramos B.

114) Av.9 N° 13-61, Rubio 5030, Tách - Dir: Belmira de Rojas.

115) Calle Vargas 141, Cd.Ojeda 4019, Zul - Dir: Héctor Toyo.

116) Edif.Gral. Manrique, Av.Sucre, San Carlos 2201, Coj - Dir: Peter Taffin M.

117) Edif.Mundial, Av.Tamanaco, El Rosal, Caracas 1060.

118) Ap.4, Trujillo 3102, Tru - Dir: Brixio Oliveros.

119) Av.Fernández Padilla 26-80, San José de Guanipa, El Tigrito 6035, Anz - Mgr: Fernando Zaurín.

120) C.C.Nuevo Mundo, Loc.2, Caja Seca 3156, Zul - Dir: A.Granadillo.

121) Urb.Alta Vista, Puerto Cumarebo 4167, Fal - DG: L.Hildago G.

122) Calle 27 con Av.12 N° 12-10, Edif.Camsa, Maracaibo 4002, Zul - DG: Gerardo Pozo.

123) Calle Zamora 8, Charallave 1210, Mir - Dir: A.Sarmiento C.

124) Av 7 N° 61-1, Chivacoa 3202, Yar - Dir: Victor Moreno.

125) Calle 5 de Julio 20, Valle de la Pascua 2307, Guá.

126) Calle Industria, Ejido 5111, Mér - Dir: Ender Ramírez.

127) Av.26, Centro Radial, Araure 3303, Por - DG: Armando Roque López.

128) Ap.240, Cd.Bolívar 8001, Bol - Dir: Marcos Pitter.

129) Av.Bolívar, Edif.Sabatinos, Puerto Cabello 2024, Car - Own: Carlos Rodríguez.

130) Av.Libertador, Altos Farmacia San Rafael, Cabudare 3023, Lar - Dir: Cruz Pantoja C.

131) Final Av.Intercomunal El Valle, Urb.La Rinconada, Colegio Fé y Alegría, P2, Caracas 1091 - DG: Carlos Correa.

132) Calle Alegría, Qta.Gallia, El Sombrero 2319, Guá.

133) Qta.Clari, Av.Sucre, Barinas 5201, Bar - Dir: Ing. Rafael González V.

134) Calle Cajigal 39, Anaco 6003, Anz - DG: Otman H.Quiñones.

135) C.C.Venezuela, Local 15, P1, Urb.Villa Colombia, Puerto Ordaz 8015, Bol - Dir: Luiz Alexander Pallares.

136) Av.11 N° 12-59, Valera 3101, Tru - Dir: Eladio Pacheco.

137) Edif.Los Andes, Av.Peñalver, Turén 3308, Por - DG: N. Pérez L.

138) Av.Municipal, Torre Pelicano, P8 N° 8-4, Puerto La Cruz 6023, Ang - Dir: Fernando Zaurín.

139) Calle 85 Falcón con Av.8, Santa Rita, Edif.Nerilú, Planta Baja, Local 2 N° 8-81, Maracaibo 4002, Zul - Dir: Agustín Arteaga.

140) Calle 3 N° 6-100, Carora 3040, Lar - Dir: Jorge D.González L.

141) Av.Jacinto Lara, Edif.R.S., Guacara 2015, Car - Dir: S.Hidalgo F.

142) Av.Fco. de Cáceres 9-88, Qta.Delia Mercedes, La Grita 5022, Tách - DG: Prof.Luis A.Aguilar Ch.

143) Caicara del Orinoco 7107, Bol - DG: Horacio Romero.

144) Calle Andrés Eloy Blanco, Altagracia de Orituco 2320, Guá - DG: Enrique Morón A.

145) Carrera 4 N° 1-35, Tariba 5017, Tách - DG: Eliseo A.Suárez U.

146) Av.Circunvacación, Edif.Gómez Galerío, Guanare 3310, Por.

147) Sector La Salina, Puertos de Altagracia 4036, Zul - Dir: Luis Enrique Torres.

148) Edif.Cavendes, P13 Of.1301, Av.Fco de Miranda, Los Palos Grandes, Caracas 1062 - Dir: Alcides Delgado.

149) Cra.3 N° 7-39, Santa Bárbara de Barinas 5210, Bar - Dir: R.Carvajal.

150) Av.Bolívar, Plaza la Alameda, Boconó 3103, Tru - DG: Fermín Mejías.

151) Ap.1624, Valencia 2001, Car - Dir: Nelson Villalba.

152) Av.Independencia 141, Edif.Plaza, Carúpano 6124, Suc - Dir: A.Hernández.

153) Campo Mata, Dtto.Freites, Cantaura 6007, Anz - DG: Gerardo Monreal.

154) Av.2 N° 36-18, Mérida 5101, Mér - Gte: Gustavo Arévalo P.

155) Calle 74, Av.3D y 3E, Edif.Super Ondas, Sector Lalago, Maracaibo 4002, Zul - Dir: Angel Orizza.

156) Av.Santa Rosa 18, Cumaná 6101, Suc - Dir: J.Alfonzo.

157) Edif.Oeste 10-1, P2, Esq.de Cipreses, Av.Lecuna, Caracas 1010 - Dir: Ricardo L.Siblesz.

158) Edif.Corporación Felman, PH, N° 2, Miracielas a Cipreses, Av.Lecuna, Caracas 1010 - Dir: Manuel Felipe García.

159) Calle Soledad con Av.Concepción 16, Guatire 1221, Mir - DG: G.Obelmejias.

160) Puerto Ordaz 8015, Bol - DG: Jesús R.Sifontes.

161) Av.Cartagena, Entre 19 y 20, San Felipe 3201, Yar.

162) Calle Ribas, Edif.Centro Empresarial, P7, Los Teques 1201, Mir - DG: Omar A.García.

163) Torre Porteña, Of.2-4, Av.Municipal, Puerto La Cruz 6023, Ang - Dir: Luis Alberto Paruta.

164) Ap.265, Mérida 5101, Mér - DG: Gildardo Moreno G.

165) Arquidiocesis de Maracaibo, Maracaibo, Zul - Dir: Sra.Dunia Mavares.

166) Carrera 5 N° 33, Maturín 6201, Mon - Dir: Bernardo Pirela.

167) Av.Miranda, Edif.de la Radio, Cumaná 6101, Suc - Dir: Luis Martinéz.

168) Carrera 12 N° 3-57, Calabozo 2312, Guá - DG: G.Helimenas Rojo P.

169) Ap.7342, Coro 4101A, Fal - Dir: Pbro.David Gutiérrez.

170) Crtra N.Av.51, Zona Industrial, Cd.Ojeda 4019, Zul - Dir: Lic.Tina Tomasi.

171) Gobernación del Estado Apure, San Fernando 7001, Apu - Dir: Francisco Javier Loreto.

172) Av.Pedro Lucas Uribarri, Santa Rita 4020, Zul.

173) Av.Rómulo Gallegos, Res.Flor de Pascua, Loc 2, Valle de la Pascua 2307, Guá - Mgr: Ramiro Seijas.

174) 75 Calle Falcón, Urb.Los Caciques, Qta.Paraguaná, Punto Fijo 4102, Fal - DG: Wagib Latuff Vargas.

175) Av.Mérida 107, Edif.Felpo, Valencia 2001, Car - Dir: Armando Moure G.

176) Barrio Reja, Acarigua 3301, Por. Own: José Ignacio Casal.

177) Av.Intercomunal, C.C.Géminis, P3, Barcelona 6001, Anz - Dir: Douglas Ramos Ossio.

178) Urb.Lomas del Este, Av.Rotaria, Edif.El Parque, PB2, Valencia 2001, Car - Dir: Alfonso Nava M.

179) Av.Bolívar, Edif.Padre Cecilio Avila, PH, Güigüe 2010, Car - Dir: Antonio Saqueiro.

180) C.C.El Pichacho, P8, San Antonio de los Altos 1204, Mir - Gte: Luis A.Alarcón.

181) Intersección Autopista José António Paéz, Troncal 5, Ospino 3319, Por - Own: Dr.Elias D'onghia.

182) Valencia 2001, Car

183) Av.4 de Mayo, Centro Comercial Real, P2, Porlamar 6301, Nes. Dir: Antonio Antepaz. Te: +58 (95) 636806. Fax: +58 (95) 638455.

184) Továr 5143, Mér.

185) Maiquetía 1161.

FM in Caracas: 88.1 Imagen - 37] 88.9 Nueve FM - 89.7 X FM - 11) 91.1 - 91.9 Avila - 20] 92.9 - 37] 94.1 Hot 94 - 95.5 Jazz 95 - 37] 96.0 Estrella - 96.9 X FM - 97.7 Em.Cultural - 99.1 Mágica - 63] 99.9 Hits 100 - 101.5 Kys - 117] 102.3 CNB - 3] 103.3 Radiorama - 116] 104.5 Operadora - 37] 106.5 Fiesta 106 - 148] 106.9 Playa 107 - 63] 107.3 Megaestación - 107.9 Playa.

ANTARCTICA

Pop: 4.120 (Su), 1.066 (Wi) — **ITU:** ATA.

RADIO NACIONAL ARCANGEL SAN GABRIEL

✉ LRA36 Radio Nacional Arcangel San Gabriel, Base de Ejercito Esperanza, 9411-Antartida Argentina, Argentina.

SW: (G.C: 63.24S/56.59W): LRA36 15476kHz 1kW.

Spanish: MF 1800-2000v — **V.** by QSL-card & letter. Re. to Comando de Comunicaciones, Comando en Jefe del Ejército.

AMERICAN FORCES ANTARCTIC NETWORK (AFAN McMurdo)

ADDR: AFAN McMurdo, US Naval Support Force Antarctica, 651 Lyons Str, Port Hueneme, CA 93043-4345, USA.

FM: 93.9/104.5MHz 0.03/0.05kW.

D.PRGR: 24h on both freqs. Rel AFRTS exc. some local prgrs on 104.5MHz.

NATIONAL TELEVISION
EUROPE

<table>
<tr><td>St. Pölten</td><td>38</td><td>150</td><td>21</td><td>600</td><td>H</td></tr>
<tr><td>Poysdorf</td><td>51</td><td>10</td><td>43</td><td>10</td><td>H</td></tr>
<tr><td>Lienz</td><td>7</td><td>0.3</td><td>41</td><td>15</td><td>H</td></tr>
<tr><td>Gmunden</td><td>11</td><td>0.2</td><td>59</td><td>10</td><td>H</td></tr>
</table>

+ 867 low power tr's — *) also carries local prgr.
N.B: ORF 1&2 partly stereo.
D.Prgr: Prgr. 1: 24h. **Prgr. 2:** 24h.

ALBANIA

TV-sets: 300.000 — **Systems:** B & G — **Colour:** PAL.

RADIOTELEVISIONE SHQIPTAR (Gov.)
✉ Rr. "Ismail Qemali" 11, Tirana. ☎ +355 (42) 27512. 🖷 +355 (42) 27512. **L.P:** DG: Albert Minga. Dir. TVSH: Eduard Mazi.

Stations:	ch	kW/Pol	Stations:	ch	kW/Pol
Tirana	N-	60H	Butrint	12	5H
Elbasan	6	0.5H	Kukës	12	100H
Gjirokaster	7	10H	Vlora	12	10H
Peshkopi	8	2H	Peskopi	32	10H
Berat	9	10H	Letaj	39	25H
Tirana	11	2H	Tirana	57	800H
Pogradec	11	100H			

D.Prgr: 0600-2300. **F.Pl:** 2nd Channel.

ANDORRA

ANDORRA TELEVISIÓ
(Telivisió Pública)
✉ Baixada del Molí, nþ24, Andorra la Vella. ☎ +376 873777. 🖷 +376 864999. **L.P:** DG: Enric Castellet. Ed. in Chief: Montserrat Talarn. Prod: Raimon Cartró. Chief Tech: Josep Mª Samper. Chief of Prog: Lluis Quintana. Chief of Info: Jordi Pifarré.

Stations:	ch	Stations:	ch
Pic de Maiá	37	Arans	50
Bosc de Pal	44	Espiolets	52
Coll de Gomá	46	Roc de l'àguila	53
Sant Juliá	46	Bony d. l. Neres	59
La Comella	47	Erts	61

AUSTRIA

TV-sets: 3.036.000 — **Colour:** PAL — **Systems:** B & G.

ÖSTERREICHISCHER RUNDFUNK
✉ ORF-Zentrum Wien, A-1136 Wien, Würzburggasse 30. ☎ 43 (1) 87 8780. **Web:** http://www.orf.at
L.P: DG: Gerhard Weis. Dir. of TV: Dr. Rudolf Napilles, Kathrin Zechner. Tech.Dir: Dr. Wolfgang Pasewald. Head of PR: Thomas Prantner.
Stations: System B ch1-12, System G ch21-68.

Location	Prgr 1	kW	Prgr 2	kW	Pol
St. Polten	2A	60	21	500	H
Innsbruck	4	80	23	800	H
Neumarkt	4	1	47	10	H
Bad Ischl	5	2	38	5	H
Bregenz	5	100	24	350	H
Kufstein	5	3	24	50	H
Wien	5	100	24	500	H
			34*	500	H
Linz	6	100	43	500	H
Rechnitz	6	3			V
			43	55	H
Graz	7	100	23	800	H
Weitra	7	5	58	90	H/V
Salzburg	8	100	32	800	H
			36	300	H
Wofsberg	8	2	28	20	H
Bruck/Mur	9	20	41	200	H
Bludenz	9	2	33	30	H
Klagenfurt	10	150	24	1250	H
Semmering	10	10	36	65	H/V
Schladming	11	10	40	80	H
Spittal/Drau	12	2	54	20	H
Mattersburg	38	30	52	30	V

AZORES (Portuguese)

TV-sets: 3380—**Colour:** PAL—**System:** B.

RADIOTELEVISÃO PORTUGUESA (RTP)
✉ Ponta Delgada, S. Miguel.
Stations: Pico da Barrosa chE7 150kW H, Santa Barbara chE9 100kW H, Lages (Isle of Terceira) chE4 1 kW + 8 repeaters (only relay of RTP's 1st. prgr.).
D. Prgrs: 2000-2400.

AFRTS (US Air Force)
✉ Detachment 3, Air Force European Broadc. Squadron, APO New York, NY. 09406-5000, USA
Station: (System M): Lajes Field chA8 1kW H.
D. Prgr: 0900-0200 (Fri/Sat 0400).

BELARUS

TV-sets: 3.600.000 — **Colour:** SECAM — **Systems:** D & K

BELARUSKAE TELEBACHANNE
(National State TV and Radio Company)
✉ Makaenka 9, Miensk, 220807. ☎ +375 (172) 649286, 🖷 +375 (172) 648182, ✆ 252267 TV SU.
L.P: President: A.R. Stljarou.

Stations	ch	Pol	Stations	ch	Pol
Miensk	1	H	Bragin	27	H
Mjadzel	8	H	Grodna	3	H
Brest	7	H	Slonim	10	H
Pinsk	4	H	Heraneny	7	V
Viciebsk	2	H	Smargon	36	H
Ushachy	9	H	Magileu	4	H
Homel	10	H	Babrujsk	12	V
Smjatanichy	5	H	Kastjukovichy	9	V
Zlobin	22	H			

+ 3 low power repeaters.
D.Prgr: 0600-2200

TELEVISION BROADCASTING NETWORK (TBN)
✉ 15a F. Skariny Street, Miensk 22072. ☎🖷 +375 (172) 394171, 394536. **E-mail:** mmc@glas.apc.org
TBN unites 12 private TV companies in the biggest cities of Belarus:

City	TV Company	Ch.
Mogilev	2nd Channel	2
Piensk	Varyag	7
Miensk	NTRC Bel TV	8
Bobruysk	Tele-Vesta	9
Baranovici	Intex	23
Soligorsk	Soltec	34
Kobrin	Inform TV	25
Zhlobin	Nuans	29
Orsha	Skif	34
Gomel	Nireya	35
Svetlogorsk	Ranak	36
Vitebsk	Delta TV	48

BELGIUM

TV sets: 4.200.000 — **Colour:** PAL — **Systems:** B & H.

VLAAMSE RADIO EN TELEVISIE (VRT)
⌨ 1043 Brussels. **Cable:** BRT-TV. ☎ +32 (2) 741 3111. 🗎 +32 (2) 734 9351.
L.P: DG: J. Ceuleers. Dir. Prgr. Planning: J. Bauwens.
Stations: ch2-11 System B, ch21-68 system H. NICAM stereo audio.

Location	Prgr 1	kW	Prgr 2	kW	Pol
Antwerpen	2			0.1	V
Waver/Overijse	10	100	25	10	H
Egem	43	1000	46	1000	H
Genk	44	200	47	200	H
Schoten			62	200	H

D.Prgr: BRT1: 1400-2200. **BRT2:** 1800 (Sun 1300)-2230.

RADIO TELEVISION BELGE DE LA COMMUNAUTE CULTURELLE FRANCAISE (RTBF)
⌨ 1044 Brussels. ☎ +32 (2) 737 2111. 🗎 +32 (2) 737 4357. **Cable:** RTBF-TV — **L.P:** PD: G. Lovebites.
Stations: ch2-11 System B, ch21-68 system H. NICAM stereo audio.

Location	Prgr 1	kW	Télé 21	kW	Pol
Liège	3	100	42	1000	H
Wavre	8	100	28	500	H
Couvin			30		H
Léglise	57	10	60	200	H
Bruxelles			45	0.5	H
Profondeville	52	200	49	200	H
Tournai	57	20	63	20	V

+ 8 low power relay st's.
D.Prgr: RTBF1: 24h. **Télé 21:** 1800 (Sat 1600, Sun 1300)-2200.

VTM (Vlaamse Televisie Maatschappij)
⌨ 1818 Vilvoorde. ☎ +32 (2) 254 5611. 🗎 +32 (2) 252 5016
L.P: MD: J. Merckx. Commercial TV Service in Dutch on cable only.

CANAL PLUS (Comm.)
⌨ Chaussee de Lauvain 656, 1050 Brussels. ☎ +32 (2) 7300 211. 🗎 +32 (2) 732 1848

Location	ch	kW	Pol
Liège	39	200	H
Wavre	50	500	H
Anderlues	58	200	H
Leglise	63	?	H

RTL-TVi (Comm.)
⌨ 1051 Brussels. ☎ +32 (2) 640 51 50. 🗎 +32 (2) 640 9307. ☺ 64430.
Station: ch27 (tr located in Dudelange, Luxembourg).
D.Prgr in French: 12h daily.

TV-5 - Europe
Station: Bruxelles ch56 1kW H.
Rebroadcasts the TV-5 satellite sce.

AMERICAN FORCES NETWORK - SHAPE
Colour: NTSC — **System:** M.
Station: chE33 1kW V, ch 34 4.5kW (rel. AFN Germany).

BOSNIA/HERCEGOVINA

TV-sets: 1.012.094 — **Colour:** PAL — **Systems:** B&H.

RADIO TELEVIZIJA BOSNE I HERCEGOVINE
⌨ VI Proleterske brigade 4, 71000 Sarajevo. ☎ +38 (71) 522333.
Stations: System B=ch2-12, System H=ch21-68. **ChE.** audio powers 1/10 of vision powers indicated.

TV Sarajevo 1st Prgr.	ch	kW (ERP)		ch	kW (ERP)
Majevica	5	10	Hum	8	6
Bjelasnica	5	6	Trovrh	9	10
Kozara	6	18	Leotar	10	6
Tusnica	6	10	Pljesvica-B	10	2
Velez	7	10	Vlasić	11	100

+ 152 low power sts.
D.Prgr: Mon-Fri 0500-1130, 1345-0030. Sat 1100-0015, Sun 0700-0015.

TV Sarajevo 2nd Prgr.

	ch	kW		ch	kW
Trovrh	21	100	Tusnica	43	7
Velez	26	1000	Majevica	46	750
Kozara	27	1000	Bjelasnica	47	60
Vlasić	29	1000	Pljesevica	47	125
Leotar	37	100	Hum	52	22

+ 140 repeaters.
D.Prgr: Mon-Fri 1355-1600, 1730-2230. Sat 0900-2230, Sun 0635-2200.

TV Sarajevo 3rd Prgr.

	ch	kW
Hum	37	100

D.Prgr: 1700-2400.

BULGARIA

TV-sets: 3.127.000 — **Colour:** SECAM — **Systems:** D & K.

BALGARSKA TELEVIZIJA (Gov.)
⌨ UI. San Stefano 29, 1504 Sofia. ☎ +359 (2) 446329. 🗎 +359 (2) 662388. **Cable:** BT SOF BG. ☺ 22581 bt sof bg.
L.P: DG: Ivan Granitsky.
Stations: (System D ch1-12, System K ch21-60):

Prgr. 1	ch	kW		ch	kW
Shumen	R5	100	Kjustendil	R10	50
Smoljan	R6	5	Botev Vrâh	R11	250
Burgas	R7	100	Sliven	R12	20
Sofija	R7	10	Belogradčik	R12	50
Silistra	R8	1	Dobrich (ex-		
Zelena Glava	R9	50	Tolbuhin tr)	R12	5
Kârdjali	R9	10	Ruse	R9	20
Varna	R9	5			

D.Prgr: approx 70h per week.

Prgr. 2	ch	kW		ch	kW
Botev Vrâh	24	1000	Sliven	31	200
Gotze Delchev	25	?	Varna	33	100
Burgas	26	1000	Kârdjali	34	100
Mihajlovgrad	26	1000	Kjustendil	34	100
Ruse	27	250	Silistra	35	100
Dobrich (ex-			Stara Zaora	37	?
Tolbuhin tr)	28	100	Smoljan	38	100
Plovdiv	28	?	Shumen	39	1000
Sofija	29	?	Belogradčik	46*	1000
Belogradcik	30	?			

+ 250 low power tr's (Prgr. 1 + 2)
D.Prgr: approx 70h per week.

NOVA TV (Comm.)
⌨ 16 Sveta Nedelja Sq, 1000 Sofia. ☎ +359 (2) 805025. 🗎 +359 (2) 870298 — **L.P:** Exec. Mgr: Rumen Kovachev.
Station: Kopitoto ch48 1kW (covers greater Sofia area).
D.Prgr: 80h per week.

RUSSIAN TELEVISION RELAY
Station: Sofia ch36 2kW, Ruse ch32 100kW.
D.Prgr: Mon-Fri 1230-2300; Sat/Sun 0630-2330. Relays of OK-1 via satellite from Moscow.

TV 5 EUROPE
Station: Plovdiv chR6, Sofija ch41 0.2kW, Burgas ch32.

CROATIA

TV-sets: 950.000 — **Colour:** PAL — **System:** B&H

HRVATSKA TELEVIZIJA (HTV)
⌨ Prisavlje 3, Zagreb, Croatia. ☎ + 385 (1) 616 3366. 🗎 + 385 (1) 616 3392. ☺ 21477 HTV RH. **Web:** http://www.hrt.hr
L.P: GM: Ivan Parac. Head of Prgrs: Hloverka Srzic-Novak. Head of Int. Rel. Dept: Marija Nemcic.

Stations (main st's in bold)

Transmitter	HTV1	HTV2	HTV3
Babino Polje	11	38	-
Bakar	7	45	41
Biokovo	9	41	45
Blato na Korculi	38	55	-
Bolfan	29	21	52
Brac	11	29	53
Brezje	58	34	62
Brezovica	44	47	54
Brinje	35	38	30
Brodski Stupnik	51	28	-
Buje	21	24	56
Cabar	10	33	36
Cavtat	56	59	-
Celavac	8	31	25
Cres	8	23	37
Crni Lug	23	47	5
Cucerje	30	36	42
Delnice	37	49	53
Dinjiska	24	27	-
Doljani	-	-	-
Donji Lapac	6	32	35
Drenovci	39	51	27
Dvor na Uni	22	42	60
Fara	44	51	-
Fuzine	8	60	49
Fuzine-Jezero	10	-	-
Gerovo	8	47	50
Gospic	6	-	-
Govedjari	11	36	-
Grobnik	5	47	53
Gruda	38	41	-
Gunjavci	40	43	52
HRT Bldg.	49	52	59
Hvar	6	39	31
Imotski	12	27	35
Ist	26	41	44
Ivanec	27	-	-
Ivanscica	36	40	63
Jablanac	37	45	50
Jelsa	27	32	57
Kalnik	5	43	-
Kasina	31	38	44
Klis	55	58	-
Knin	6	44	53
Komiza	7	55	59
Komolac	12	36	46
Koprivnica	33	37	48
Koromacno	7	36	22
Korcula	11	21	33
Kostajnica	38	43	31
Krapinske Toplice	21	55	45
Kriz	49	53	56
Kuna	11	43	-
Kupjacki Vrh	40	55	58
Kutjevo	26	29	-
Labinstica	4	23	34
Lastovo	53	59	-
Lepoglava	51	58	47
Lic	10	31	24
Licka Pljesivica	5	53	57
Lokve	6	21	32
Lopud	10	47	43
Majkovi	7	55	-
Mali Losinj	7	53	36
Mandicevac	31	40	-
Maranovici	36	43	-
Markusevec	55	44	41
Martinscica	6	52	-
Metkovic	47	53	31
Milna	31	43	39
Mirkovica	7	43	46
Mokosica	10	33	39
Molunat	11	21	-
Moslavacka Gora	67	21	34
Mrkopalj	8	36	33
Murter	25	28	43
Nova Gradiska	31	42	48
Novalja	41	49	23

Transmitter	HTV1	HTV2	HTV3
Novigrad	42	45	53
Novigrad (Zadar)	6	25	28
Obrovac	10	34	-
Omis	7	37	40
Oriovac	36	45	-
Osijek	6	23	33
Osijek Donji Grad	27	29	49
Ostra	56	58	50
Otes	30	38	-
Otok	34	59	-
Pag	7	35	32
Pakrac	44	47	31
Papuk	53	56	21
Peljesac	5	38	58
Planina	41	48	60
Plocice	43	46	-
Plomin	55	58	46
Podvinje	54	60	-
Porec	26	31	59
Prezid	9	23	26
Primosten	29	21	-
Promina	38	59	25
Psunj	4	50	58
Pucisca	36	39	28
Pula	35	26	48
Rabac	7	22	53
Razromir	9	55	58
Rasa	8	47	50
Resetari	21	38	25
Rovinj	43	49	31
Ruda	21	24	37
Sibenik-Martinska	5	52	58
Sibenik-Subicevac	11	49	55
Sibinj	21	24	32
Sinj	5	29	26
Skradin	51	54	-
Slano	49	46	-
Slatina	29	44	47
Slavonski Brod	9	42	48
Srb	10	-	-
Sljeme	9	28	25
Srinjine	51	59	-
Srdj	6	28	22
Starigrad Paklenica	5	41	-
Stipanov Gric	12	24	27
Ston	11	53	56
Straza	12	30	60
Strigova	44	48	30
Suvaja	-	-	-
Sv.Gera	50	58	63
Sv.Martin	45	47	-
Sv.Nedjelja	12	39	54
Svilno	26	42	34
Trstenik	8	40	-
Trsce	52	59	49
Ugljan	51	57	37
Uljenje	12	35	48
Umag	25	32	52
Unije	24	27	32
Ucka	11	29	39
Vela Luka	7	29	43
Veleb. Pljesivica	34	44	51
Velika	52	38	-
Velika Petka	8	45	48
Vinkovci	12	44	36
Virovitica	42	45	37
Vis	51	55	59
Visovac	36	50	53
Vrgorac-Gomila	21	31	10
Vrgorac-Polje	57	60	52
Vrlika	11	26	29
Vrsar	23	36	44
Vrucica	6	12	-
Vucinici	52	22	32
Zadar	-	-	-
Zaton	36	40	-
Zlarin	32	45	-
Zlatarevac	25	28	34
Zupa	50	54	40
Zupanja	49	58	65

D.Prgr: HTV1: 0700-2300; HTV2: 1000-2300; HTV3: 0800-2400.

OTV (Open TV)

📧 Teslina 7, Zagreb, 10000. ☎ +385 (1) 424 124. 🖹 +385 (1) 455 1386.

REGIONAL STATIONS
SLAVONSKA TELEVIZIJA OSIJEK

📧 Hrvatske Republike 20, Osijek 31000. ☎ +385 (31) 124 666. 🖹 +385 31 124 111.

TV MARJAN

📧 Savska bb, Split 21000. ☎ +385 (21) 364 525. 🖹 +385 (21) 523 455.

VINKOVACKA TELEVIZIJA

📧 Genschera 2, Vinkovci, 32000 Croatia. ☎ +385 (32) 331 990. 🖹 +385 (32) 331 985.

ZADARSKA TELEVIZIJA

📧 Molotska bb, Zadar, 23000 Croatia. ☎ +385 (23) 311 791. 🖹 +385 (23) 314 749.

CZECH REPUBLIC

TV-sets: 3.800.000 (est.) — **Colour:** SECAM — **System:** D & K

CZECH TELEVISION (Public Sce.)

📧 Kavcíhory, 140 70 Praha 4. ☎ +42 (2) 61131111. 🖹 +42 (2) 6927202.
Web: http://www.czech-tv.cz
L.P: DG: Ivo Mathé. PD: Jirí Pittermann. TD: Jan Horský. Head of PR: Jiri Moc.

Location	CT1	kW	CT2	kW
Brno	29	20	46	20
Brno-mesto	35	2	52	2
Ceske Budejovice	39	20	49	20
Domazlice	41	2	12	0.2
Frydek-Mistek	37	17		
Hodonin			33	1
Hradec Kralove	22	20	57	20
Cheb	36	3.5	53	2
Chomutov	52	6	35	5
Jachymov	38	5		
Jesenik	36	20	50	20
Jihlava	25	10	42	10
Klatovy	22	5	58	0.1
Liberec	31	5	43	0.01
Mikulov	26	10		
Novy Jicin	34	5		
Olomouc	33	2		
Ostrava	31	20	51	2
Pacov	36	5		
Plzen	31	20	48	20
Plzen-mesto	34	5	27	1.5
Praha	26	50	53	1
			29	1
Praha-mesto	51	10	41	10
Rychnov nad Kneznou	28	5		
Susice	35	5	52	0.08
Svitavy	24	10	58	0.1
Tabor	27	1		
Trutnov	23	16	40	20
T	28	10	45	10
Uhersky Brod			47	20
Usti nad Labem	33	20	50	20
Valasske Klobouky	24	2	42	2
Vimperk	32	5	47	0.08
Votice	30	5	56	5
Zilina	22	5	51	0.2
Z	32	5	49	0.1

D.Prgr: ET1: 0630-2330 (approx.). **ET2:** 0600-0030 (approx.)

PRIVATE STATIONS
NOVA (Comm.)

📧 Vladislavova 20, 11313 Praha 1. ☎ +42 (2) 2110 0111. 🖹 +42 (2)

2110 0565 — **L.P:** Gen Dir: Vladimir Zelezny. PD: Jan Vit. Head of PR: Karel Soukup.

Stations	ch	kW	Pol.		ch	kW	Pol.
Domazlice	12	1.6	H	Liberec	31	100	H/V
Klatovy	22	100	H	Vimperk	32	100	H
Hradec Kralove	22	600	H	Usti n. Labem	33	600	H
Trutnov	23	1000	H	Plzen (city)	34	100	H
Praha (mesto)	24	60	H	Chomutov-Jedl	35	100	H
Svitavy	24	100	H	Brno (city)	35	20	H
Jihlava	25	1	H	Cheb	36	100	H
V. Klobouky	25	10	H	Pacov	36	1.6	H
Praha	26	1000	H	Jesenik	36	600	H
Tabor	27	7.8	H	Bardejov	37	10-0	H
Trebec	28	100	H	Jáchymov	38	300	H
Rychnov	28	100	H	Jihlava-Javorice	42	300	H
Brno	29	600	H	Brno-Kojal	46	600	H
Blatna	29	600	H	Plzen-Krasow	48	600	H
Ostrava	31	600	H	Jesenik-Praded	50	300	H
Plzen	31	600	H	Hradec Kralove	57	600	H

DENMARK

TV-sets: 2.700.200 — **Colour:** PAL — **System:** B.

TELECOM A/S

📧 Telegade 2, DK-2630 Taastrup, Denmark. ☎ +45 (42) 529111. 🖹 +45 (42) 529331. **L.P:** Man. Dir: Jens Kiil.

DANMARKS RADIO (Gov.)

📧 Danmarks Radio, TV-Byen, DK-2860 Søborg. ☎ +45 (35) 203040. 🖹 +45 (35) 202644. ☺ 22695.
Web: http://www.dr.dk
L.P: Dir TV: Finn Rowold; Dir of Prgrs: Thomas Dahlberg; TV Fakta: Bo Lynnerup; TV Int: Mogens Vemmer.
Stations: Pol H exc. Ølgod: V — Stereo: NICAM digital system, only on Copenhagen-Hove ch 31.

Location	ch	kW	Location	ch	kW
Sdr. Højrup (Fyn)	3	10	Søndenjylland	7	60
Copenhagen-	4	50/5	Århus	8	60/6
Bornholm	5	10	Vestjylland	10	60
Sydvestjylland	5	5	Copenhagen-	31	600/60
Sydsjælland	6	60	Tolne, Vendsyssel	57	40

+ 26 low-powered repeaters not mentioned.
D.PRGR: MF 1400-2230, SS 1300-2300.

TV2 (Comm.)

📧 Rugaardsvej 25, DK-5100 Odense C. ☎ +45 (65) 91 12 44. 🖹 +45 (65) 91 33 22.
L.P: Dir. Gen: Tøger Seidenfaden. Head of News: Svenning Dalgaard. Head of Prgr's: Jørgen Steen Nielsen. Head of Facts Dept: Lally Hoffmann. Head of Fiction Dept: Lone Bastholm. Head of Tech. Dept: Lars Esben Hansen.
Stations: Pol H, exc. Thisted: V — Stereo: NICAM digital system.

Location	ch	kW	Location	ch	kW
Odense	22	500	Ringkøbing	40	600
(Tommerup)			(Videbæk)		
Hadsten	26	600	Jyderup	48	600
Åbenrå	27	600	Nakskov	52	100
Thisted	28	250	Copenhagen-	53	600
Vejle (Hedensted)	30	500	Hove		
Svendborg	32	150	Bornholm (Rø)	56	600
Varde	33	500	Viborg	56	500
Nibe	35	600	Vordingborg	58	600
Tolne	37	50			

+25 low-power translators.
D.PRGR: 1600-2300 MF, 1300-2300 SS (all colour) — **N:** 1800-1830 & 2100-2115 MF. **Comm:** between prgr's.
Regional Prgr's: 1830-1900 (exc. Sat.)
Regional Addr's:
TV2/Bornholm, Ravnsgade 5, 3720 Aakirkeby. ☎ +45 (53) 97 57 10. 🖹 +45 (53) 97 45 85. **Dir:** Bent Nørby Bonde. **D.Prgr.** on ch. 56.
TV/Fyn, Kongensgade 68 — I, 5000 Odense C. ☎ +45 (66) 14 18 18. 🖹 +45 (66) 12 44 24. **Dir:** Ebbe Larsen. **D.Prgr.** on ch. 22 & 32.
TV2/Lorry, Allegade 7-9, 2000 Frederiksberg. ☎ +45 (38) 88 00 11. 🖹 +45 (38) 88 31 11. **Dir:** Dan Tschernia. **D.Prgr.** on ch. 53.

TV/Midt-Vest, Søvej 2, P.O. Box 1460, 7500 Holstebro. ☎ +45 (97) 40 33 00. 📠 +45 (97) 40 14 44. **Dir:** Ivar BrÆndgaard. **D.Prgr.** on 28, 40 & 56.
TV/Nord, Søparken 4, 9440 Åbybro. ☎ +45 (98) 24 46 00. 📠 +45 (98) 24 46 98. **Dir:** Bent Bjørn. **D.Prgr.** on ch. 35 & 37.
TV/Syd, Laurids Skausgade 12, 6100 Haderslev. ☎ +45 (74) 53 05 11. 📠 +45 (74) 53 23 16. **Dir:** Helge Lorenzen. **D.Prgr.** on ch. 27, 30, 33 & 42.
TV/Øst, Kildemarksvej 7, 4760 Vordingborg. ☎ +45 (55) 34 02 00. 📠 +45 (55) 34 01 34. **Dir:** Ole Dalgaard. **D.Prgr.** on ch. 48, 52 & 58.
TV2/Østjylland, Niels Brocks Gade 16, 8900 Randers. ☎ +45 (86) 40 40 00. 📠 +45 (86) 40 56 86. **Dir:** Erling Bundgaard. **D.Prgr.** on ch. 26.
ALF: DKK 195 of the Radio Denmark license income, plus commercial income.

TV3 (Comm.)
📧 Overgaden oven Vandet 10, DK-1415 København K. — ☎ +45 (31) 57 30 80. 📠 +45 (31) 57 80 81.
L.P: Dir: Klaus Fog. Head of prgr's: Susanne Teilmann.
Station: see satellite section.
D.PRGR: 1600-2400. Commercials interrupt prgr's.
Local TV Stations: Approx. 35 organizations are on the air on UHF with 0.2-3 kW ERP.

TV-sets: 615.000 — **Colour:** PAL/SECAM — **Systems:** B/D&K

TALLINN EESTI TELEVISIOON (ETV)
📧 12 Faehlmanni Street, 0100 Tallinn. ☎ +372 (2) 434102. 📠 +372 (2) 434155. ① 173869 ETV EE. **E-mail:** etv@etv.ee **Web:** http://www.etv.ee
L.P: DG: Hagi Shein. Dep. DG & Editor in Chief: Raul Rebane.

EVTV (Comm.)
📧 Peterburi 81, 0014 Tallinn. ☎ +372 (6) 328228. 📠 +372 (6) 323650 — **L.P:** Chmn: Victor Siilats. MD: Enn Eesmaa. Mktg. Mgr: Antero Laanela. PR: Kristina Haiba.

KANAL KAKS (Channel 2) (Comm.)
📧 Harju 9, 0001 Tallinn. ☎ +372 (2) 442356. 📠 +372 (2) 446862 — **L.P:** Chmn: Ilmar Taska. Sen. VP: Eva Banhidi. PD: Liina Kirt. Int. Rel: Vanessa Vogel.

Reklaamitelevisioon (Comm.)
📧 Endla 3, 0106 Tallinn. ☎ +372 (2) 666743. 📠 +372 (6) 311077 — **L.P:** DG: Tomas Lepp.

TIPP TV (Comm.)
📧 Regati Pst-1-6, 0019 Tallinn. ☎ +372 (2) 238535. 📠 +372 (2) 238555 — **L.P:** Pres: Juri Makarov.

Transmitting stations (composite listing for all networks):

Site	ch.	prgr.
Kohtla-Nômme	R1	EVT/RTV
Kunda	R1	ORT/BFD/ORS
Pôltsamaa	R1	ORT
Tallinn Kl.	R2	ETV
Narva-Keskus	R3	ETV
Vôru-Keskus	R3	?
Rakvere-Tabani	R3	ETV
Tallinn Kl.	R3	KK
Haapsula-Uuemôisa	R3	ORT/BFD
Kallaste-Torila	R3	ORT
Sillamäe	R3	?
Pärnu-Tammiste	R4	ETV
Tartu-Soinaste	R4	EVT/RTV/TAT
Narva-Keskus	R4	ORT/BFD
Kohtla-Nômme	R5	KK
Jôgeva-Eristvere	R5	ORT/BFD
Ruhnu-Majakas	R5	ETV
Valgjärve	R6	ETV
Paide-Keskus	R8	ETV
Sôrve-Lôopôllu	R8	ETV
Pärnu-Tammiste	R9	ORT/BSD
Järvakandi	R10	ORT/BFD/ORS
Kohtla-Nômme	R11	ETV
Orissaare	R11	ETV
Vôhma	R11	ORT
Tallinn-Kl.	R12	ORT/BFD/ORS
Tartu-Puiestee	R12	ETV
Viljandi-Viiratsi	22	ORT
Orissaare	22	ORT/BFD
Tallinn-Ülemiste	22	ALO
Narva-Keskus	23	KK
Rakvere-Tabani	27	ORT/BFD
Kihelkonna	27	ORT
Koeru	27	?
Tartu-Tamme	28	ALO/RTV
Tallinn-Kl.	28	EVT/RTV
Kohtla-Nômme	29	ORT/BFD
Valgjärve	30	ORT/BFD
Kuressaare-Keskus	31	ORT/BFD
Kärdla-Rehemäe	34	ORT/BFD
Paide-Keskus	39	ORT/BFD
Koeru	39	?
Sôrve-Lôopôllu	41	ORT
Tallinn-Kl.	42	KK
Tallinn-Kl.	45	SEE

ETV=Eesti Televisioon; EVT=EVTV; ORT=Ostankino TV; RTV=Reklaamitelevisioon; ALO=Alo TV Tartu (comm.); KK=Channel 2 (comm.); BFD=BFD Reklaamiklubi (comm.); ORS=TV Orsent (loc. comm.)

TV-sets: 14.000 — **System:** B&G — **Colour:** PAL.

SJÓNVARP FØROYA
📧 M.A. Winthersgøta 2, Postboks 21, 110 Tórshavn. ☎ +298 317780. . 📠 +298 311345
L.P: GM: Tróndur Djurhus. Dir. of Adm: F. Lómstein. Head of Technical Dept: R.C. Joensen. Head of programmes: Ivan Niclasen.
Stations: Tórshavn ch6 145 kW H, Hesturin ch9 6 kW H, Stoedlafjell ch 10 600 W V + 44 low power repeaters.
D.Prgrs: Tues 1845-0030; Wed. 1800-2400, Thurs 1900-0030; Fri 1900-0230; Sat 1400-0200; Sun 1400-2400.

Sat. relays: ch22/28 MTV Europe, ch30/37 BBC Prime, ch46/53 BBC World, ch34/44 TV3, ch 41/48 Euro sport.

TV-sets: 2.081.000 — **Colour:** PAL — **Systems:** B & G.

YLEISRADIO OY (Non-comm, Public Broadc. Sce.)
📧 TV-1, Box 96, FIN-00024, YLEISRADIO, Finland. ☎ +358 (0) 14801. **Cable:** YLEtv. ① 121270. 📠 +358 (0) 1480 5148 **FST (Swedish Language TV)**, Box 83, FIN-00024 YLEISRADIO. ☎ +358 (0) 14801. 📠 +358 (0) 1481256. **TV-2**, Box 196, FIN-33101 Tampere. ☎ +358 (31) 345 6111.① 22749. 📠 +358 (31) 345 6892.
L.P: Dir. TV: H. Lehmusto; Dir. Prgr TV-1: A. Gartz; Dir Prgr TV-2: A. Hoffren; Dir. Swedish lge Radio & TV: A. Sandelrin; Dir Prgrs: J. Harms.

MTV Oy (Comm.)
📧 00033 MTV3, Finland ☎ +358 (0) 15001. 📠 +358 (0) 1500707. **Cable:** Comtele. ① 124319 COMTV SF.
L.P: Pres: E. Pilkama. Exec. Vice Pres: J. Paavela. Prgr. Dir: T. Äijälä. News Dir. (Editor-in-chief): P. Hyvârinen. Tec. Dir: H. Marsalo. Marketing Dir: Eero Aalto. Sales Dir: Heikki Rotko. Vice Pres (communications): J. Mietinen. Comm. Mgr: M. Paaso. Proj. Mgr: M. Rainbird (FinnImage)
Prgrs: Own production & acquisition programmes on 3rd network (MTV3).

CHANNEL THREE FINLAND/OY KOLMOSTELEVISIO AB (Comm., subsidiary of MTV Oy)
📧 00033 MTV3, Finland. ☎ +358 (0) 15001. ① 126068 CHANT SF. 📠 +358 (0) 150 0677.
L.P: Man. Dir: J-P. Louhelainen. Sales Dir: H. Vahala. V.P (aquisitions): J. Sairanen. Hd. of Sports: T. Lehmuskallio. Prgr Mgr: S. Kievari. Producer: R. Haavisto.
Prgrs: Sport and other subcontracted programming on 3rd network (MTV3 channel).

Prgr. I	ch	kW	Pol		ch	kW	Pol
Tervola	3	20	H	Vuotso	8	0.25	H
Vuokatti	4	40	H	Haapavesi	9	60	H
Kuttanen	5	3	H	Kerimäki	9	10	H
Lapua	5	80	H	Lahti	9	80	H
Pyhätunturi	5	60	H	Joutseno	10	10	H
Anjalankoski	5	30	H	Pihtipudas	10	60	H
Ahvenanmaa	5	10	H	Posio	10	10	H
Espoo	6	180	H	Tammela	10	10	H
Iisalmi	6	1	H	Inari	10	30	H
Mikkeli	6	10	H	Eurajoki	11	20	H
Pyhävuori	6	20	H	Jyväskylä	11	10	H
Taivalkoski	6	60	H	Vaasa	11	1	H
Utsjoki	6	1	H	Ylläs	11	60	H
Karigasniemi	7	1	H	Ähtäri	26	100	H
Kruunupyy	7	10	H	Fiskars	31	100	H
Kuopio	7	80	H	Kiihtelysvaara	35	50	H
Oulu	7	60	H	Ruka	36	300	V
Turku	7	60	H	Pernaja	39	100	H
Koli	8	20	H	Rovaniemi	53	100	H
Tampere	8	80	H				

+ 142 low power repeaters.
YLE-prgrs: approx. 82h weekly.

Prgr. II	ch	kW	Pol		ch	kW	Pol
Espoo	8	60	H	Pyhävuori	10	20	H
Turku	9	30	H	Oulu	11	20	H
Vaasa	9	1	H	Iisalmi	22	20	H
Tervola	22	1000	H	Pernaja	36	100	H
Taivalkoski	23	600	H	Tammela	37	600	H
Karigasniemi	23	20	H	Mikkeli	38	600	H
Lapua	24	1000	H	Ähtäri	39	100	H
Jyväskylä	25	600	H	Lahti	40	600	H
Haapavesi	28	1000	H	Inari	40	500	H
Ruka	28	300	H	Fiskars	41	100	H
Ahvenanmaa	28	300	H	Kiihtelysvaara	41	50	H
Kruunupyy	30	200	H	Anjalankoski	49	600	H
Posio	30	500	H	Koli	51	600	H
Joutseno	32	600	H	Tampere	53	1000	H
Pihtipudas	32	1000	H	Ylläs	55	900	H
Pyhätunturi	32	650	H	Kerimäki	55	600	H
Eurajoki	33	600	H	Rovaniemi	56	100	H
Utsjoki	34	20	H	Vuokatti	56	1000	H
Vuotso	35	1	H	Kuttanen	60	20	H
Kuopio	36	600	H				

+ 142 low power repeaters.
YLE-prgrs: approx. 65h weekly.
Relay st. FINLAND'S TV in Stockholm: see under "Sweden".

Prgr. III	ch	kW	Pol		ch	kW	Pol
Pyhävuori	12	20	H	Lapua	40	1000	H
Espoo	24	600	H	Pihtipudas	43	1000	H
Pyhätunturi	24	650	H	Kuopio	49	600	H
Tervola	25	1000	H	Rovaniemi	49	100	H
Posio	27	500	H	Kuttanen	50	20	H
Vuotoso	28	1	H	Vaasa	51	20	H
Mikkeli	28	600	H	Ylläs	52	900	H
Ähtäri	29	100	H	Lahti	51	600	H
Iisalmi	29	20	H	Ruka	51	300	H
Tammela	30	600	H	Inari	54	500	H
Haapavesi	31	1000	H	Turku	54	1000	H
Utsjoki	31	20	H	Pernaja	55	100	H
Oulu	33	600	H	Anjalankoski	56	600	H
Kruunupyy	33	200	H	Koli	57	600	H
Jyväskylä	35	600	H	Kerimäki	58	600	H
Taivalkoski	35	600	H	Fiskars	58	100	H
Eurajoki	36	600	H	Vuokatti	59	1000	H
Karigasniemi	37	20	H	Tampere	59	1000	H
Kiihtelysvaara	38	50	H	Joutseno	60	600	H

+ 141 low power repeaters.
MTV-3 prgrs: approx. 105h weekly.

TV-4	ch	kW	Pol	TV-4	ch	kW	Pol
Fiskars	23	100	H	Espoo	35	600	H
Pernaja	26	100	H	Turku	57	1000	H

+6 low power repeaters.
Prgrs: (produced by Swedish television, SVT) approx. 60h weekly.

FRANCE

TV-sets: 29.300.000 — **Colour:** SECAM — **System:** L.

TÉLÉVISION FRANÇAISE 1 (TF1) (Priv, Comm)
⊡ Société TF1, 1 quai du Point du Jour 92100 Boulogne-Billancourt.
☎ +33 (1) 4141 1234.
L.P: Pres. & Dir. Gen: P. Le Lay. Dir. Tec: F. Hericourt.
N.B: Using NICAM stereo on main transmitters

FRANCE 2 (Public Television)
⊡ Société Nationale Antenne 2, 22 Ave Montaigne, F-75008 Paris. ☎ +33 (1) 4421 4242. ▤ +33 (1) 4421 5145. ① 204 068. **L.P:** Pres: Xavier Gouyou-Beauchamp.
N.B: Using NICAM stereo on main transmitters

FRANCE 3 (FR3) (Public Television)
⊡ 116 Avenue du President Kennedy, 75016 Paris. ☎ +33 (1) 4230 1313. ▤ +33 (1) 4289 0327.
L.P: Pres: Xavier Gouyou-Beauchamp.

CANAL PLUS (Private)
⊡ 85/89 Quai Andre Citroën, 75015 Paris. ☎ +33 (1) 4425 1000. ▤ +33 (1) 4425 1234. ① CPLUS 201 141 F.
L.P: Man. Dir: Pierre Lescure, Tec. Dir: Lucien Banton, Film Buyer: Rene-Bonnell.
Canal Plus is a subscription network. The signals are mostly coded and subscribers need a decoder. Times for uncoded signals: W 0600-0630, 1130-1230, 1715-1930.
Stations (Pol=H, exc. where indicated)
*) Lower power applies to Canal Plus only.

Location	TF1	F2	F3	Canal* Plus	kW(ERP)
Abbeville	63	57	60	—	250
Ajaccio	31	21	24	4	500/8
Albertville	45	39	42	7	5/0.1
Alençon	48	51	54	—	100
Ales	27	21	24	65	100/3
Amiens	41	47	44	10	500/0.01
Angers	47	44	41	10	20/0.04
Annemasse	—	—	66	—	1.75
Argenton	46	40	43	—	80
Aurillac	59	65	62	9V	500/4
Aurillac	—	—	54	—	350
Autun	48	51	54	10	500/0.1
Auxerre	37	31	34	6	300/0.03
Avignon	42	45	39a	—	125
Avignon	—	—	33	—	55
Bar-le-Duc	54	51	48	6	200/0.05
Bastia	41	47	—	2V	500/17
Bayonne	64	58	61	7V	500/2.5
Bergerac	37	34	31	—	250
Besançon	47	41	44	3V	500/60
Besançon	29	23	26	10	250/0.02
Bordeaux	63	57	60	8	1000/50
Boulogne	29	34	37	10	100/0.5
Bourges	23	26	29	8	1000/211
	—	—	43	—	34
Brest	27	21	24	10	1000/230
Brive	23	29	26	6	150/0.5
Caen	22	25	28	9	1000/198
Carcassonne	64	58	66/61	3V	1000/100
Chambery	29	23	26	8	100/1
Chambery	—	—	49	—	4
Chamonix	25b	28	22	7	50/0.02
Champagnole	58	61	64	—	80
Chartres	55	50	53	9	250/0.18
Chartres	—	—	34	—	30
Chaumont	52&62	49	55	—	80
Cherbourg	65	59	62	6	100/8
Clermont-Fd.	22	28	25	5	1000/200
Cluses	56	50	53	6	8/0.1
Corte	51	61	54	7	100/1.5
Dijon	—	—	56	—	30
Dijon	59	62	65	9	1000/30
Dunkerque	42	39	45	10	200
Epinal	65	60	63	10	100/0.1
Forbach	47	22	25	28	20/2

Location	TF1	F2	F3	Canal* Plus	kW(ERP)
Gap	27	21	24	9	8/0.3
Gex	27	21	24	5V	1000/30
Grenoble	56	50	53	6	50/0.6
Gueret	64	58	61	9	80/0.3
Hirson	54	48c	51	32	200/0.25
Hyères	65	59	62	6	35/2
Laval	63	57d	60	8	80/0.02
Le Creusot	35	33	30	67	20/10
Le Havre	46	43	40	5	100/2
Le Havre	—	—	35	—	2
Le Mans	24	27	21	5V	1000/200
LePuy	63	57	60	6	100/0.13
Lesparre	39	45	42	—	10
Le Vilhain	—	—	36	—	3.6
Lille	27	21	24	5	1000/200
Limoges	—	—	32	—	450
Limoges	56	50	53	10	1000/260
Longwy	52	47	44	8	100/0.1
Lyon (Mt. Pilat)	46	40	43	10	1000/400
Lyon	61	58	64	66	10/10
Macon	57	55	49	—	5
Mantes	64	58	61	9	80/0.02
Marseille	29	23	26	5	1000/200
Marseille	40	46	43	57	22/4.5
Maubeuge	39	42	45	29	5/2.5
Mende	37	31	34	68	80/1
Menton	62	50	56	68	50/17
Metz	37	34	31	5V	1000/45
Mezières	29	23	26	36	500/?
Millau	47	44	41	10	20/0.15
Montmelian	64	58	61	9	3/?
Montpellier	56	50	53	9	1000/2.5
Mortain	50	52	55	—	10
Morteau	48	54	51	10	20/0.01
Mulhouse	27	21	24	5	1000/300
Nancy	23	29	26	8	500/30
Nantes	23	29	26	9V	1000/300
Neufchatel	51	48	54	65	80/8
Nice	64	58	61	66	10/6
Niort	28	22	25&58e	6V	1000/405
Orleans	42	39	45	—	100
Paris	25	22	28	6	700/104
Paris Est	43	46	40	53	5/5
Paris Nord	45	39	56	59	10/4
Paris Sud	49	52	62	65	3/3
Parthenay	52	49	55	—	80
Perpignan	22	25	28	7	10/3
Pignans	46	43	40	56	2/3
Porto-Vecchio	40	37	34	6	25/?
Privas	64	58	61	6	50/0.02
Reims	43	46	40	9	1000/80
Rennes	39	45	42	7	1000/500
Rouen	23	33	26	7	500/65
Sarrebourg	50	40	53	—	250
Sens	57	63	50/60	5	100/1.3
Serres	50	53	56	4	2/1
St. Etienne	35	30	33	38	10/12
St. Flour	52	49	55	7	6/0.4
St. Martin	48	51f	54	8	5/0.03
St. Raphael	25	28	22	10V	1000/70
Strasbourg	62	56	43	10V	1000/20
Tarascon	52	55	49	8	5/0.5
Toulon	51	48	54	9	100/1.6
Toulouse	27g	21	24	5	500/100
Toulouse	—	—	47	—	600
Toulouse (town)	45	39	42	7	10/1
Tours	65	59	62	10	200/0.1
Troyes	27	24	38/21	7	1000/100
Ussel	42	45	39	—	50
Utelle	47	44	41	—	10
Vannes	50	56	53	5	500/16
Verdun	65	59	62	8	500/0.08
Villers-Cotterets	65	59	62	—	4
Vittel	30	35	32	—	80
Wissembourg	54	48	51	—	50

+ low power st's not mentioned under 1kW.
a=300kW; b=50kW; c=500kW; d=100kW; e=310kW; f=3kW; g=1000kW.
D.Prgrs: TF1: 24h — **F2:** 24h — **FR3:** 0500-0000 — **CanalPlus:** 0600-0200, Sat/Sun 24h.

LA CINQUIEME (Educational, Public Television)
⌕ 10 rue Horace Vernet, 92130 Issy les Moulineaux. ☎ +33 (1) 4146 5555. 🖷 +33 (1) 4108 0222
L.P: Jean-Marie Cavada
D.Prgrs: 0515-1800 (using ARTE txs)

ARTE (Cultural, Public Television)
⌕ 2A rue de la Fonderie, 67080 Strasbourg Cedex. ☎ +33 (3) 8814 2222. 🖷 +33 (3) 8814 2200 — **L.P:** Jérôme Clément.
D.Prgrs: 1800-0100.

M6 Metropole TV (Priv. Comm.)
⌕ 16 Cours Albert 1er, 75008 Paris. ☎ +33 (1) 4421 6666. 🖷 +33 (1) 4563 7852.
L.P: Dir. Gen: Jean Drucker.
D.Prgrs: 24h.

Stations:

	ARTE	kW (ERP)	M6	kW (ERP)
Abbeville	45	80	63	80
Ajaccio	41	110	44	110
Alencon	42	12	45	12
Amiens	49	5	52	(5)
Angers	50	20	53	(15)
Angouleme	31	5	34	(5)
Annemasse	54	3.8	51	3.8
Argenton	38	18	51	18
Aurillac	57	80	67	80
Autun	45	300	42	(300)
Auxerre	55	50	49	(100)
Avignon	47	40	36	40
Bar-Le-Duc	—	—	38	70
Bastia	49	5	35	5
Bayeux	52	1.6	49	1.6
Bayonne	56	40	45	(40)
Beauvais	49	100	52	(100)
Bergerac	66	90	58	90
Bordeaux	65	130	43	10
Boulogne	59	3.2	62	(3.2)
Bourges	21	130	56	(130)
Brest	34	100	60	(100)
Brive	21	2	48	(2)
Caen	64	100	61	100
Carcassonne	46	56	43	56
Chalon-Sur-Saone	44V	2	41V	(2)
Chambery	55	4	52	(4)
Charleville	44	5	41	(5)
Chartres	47	50	44	(50)
Chaumont	57	24	39	20
Cherbourg	35	10		
Cluses	32	1.6	63	1.6
Clermont-Ferrand	30	100	33	(100)
Dijon	46V	1		
Dunkerque	59	1	62	1.8
Gueret	30	10	33	8
Haguenau	—	—	36	1
Hyeres	67	34	38	10
La Rochelle	48	1	51	1
Laval	33	20	30	(20)
Le Creusot	38	2.5	60	(2.5)
Le Havre	53	20	56	(20)
Le Mans	32	150	35	(150)
Lens	51	9.5	54	9.5
Les Sables D'Olonne	61	1	63	1
Le Touquet	55	2.1	32	2.1
Lille	65	6	53	(6)
Limoges	38	2	35	2
Lorient	62	1	65	1
Lyon	28	10	22	(10)
Lyon	59	250	62	(250)
Mantes	55	10	53	(10)
Marseille	32	150	38	(100)
Marseille	54	22	49	(22)
Maubeuge	32	3	37	10
Mende	50	1.5	53	(1.5)
Metz	39	200		
Montpellier	48	250	40	100
Nancy	55	7	43	(5)
Nantes	21	50	65	(50)
Neufchatel	34	10	31	10
Nice	51	2	45	2
Niort	38	200	64	(200)
Orleans	52	20	36	(20)

Stations:	kW ARTE	(ERP)	M6	kW (ERP)Paris
30	100	33	(100)	
Paris-Est	48	5	58	(5)
Paris-Nord	65	3	62	(3)
Paris-Sud	59	3	42	(3)
Parthenay	60	30	67	(30)
Pau	60	2	63	(2)
Perpignan	38	2	35	(2)
Pignans	58	3	61	3
Poitiers	41	1	44	1
Reims	53	25	56	(25)
Rennes	34	75	31	(75)
Rouen	59	100	62	(100)
Royan	—	—	33	2.1
Saint-Etienne	65	10	55	10
Saint-Flour	—	—	58	1
Saint-Nazaire	55	4	52	(4)
Saint-Raphael	36	80	39	(4)
Selestat	67	2.1		
Sens	47	40	44	(40)
Toulon	57	50	60	(50)
Toulouse	32	5	34	(5)
Toulouse	29	80	38	(80)
Tours	57	80	54	(80)
Troyes	29	250	56	(250)
Ussel	63	50	60	50
Valence	56	1.5	53	1.5
Valenciennes	49	6	34	(6)
Vannes	58	75	48	(75)
VillersCotterets	39V	25	57V	(25)

+1345 stations under 1kW.

PRIVATE STATIONS

TELE MONTE CARLO (relays in France)

16 boulevard Princesse Charlotte, Monte Carlo 98090, France. ☎ +377 9315 1617. 🖷 +377 9325 0109.
Stations: Toulonch 33 25kW, Marseille ch35 150kW, Marseille (town) ch51 4.5kW all using L/SECAM.
D.Prgr: 1030-2400.

TELE TOULOUSE (Private, Comm.)

3 Place Alphonse Jourdain, 31069 Toulouse Cedex. ☎ +33 (5) 6123 6565. 🖷 +33 (5) 612471.
L.P: Etienne Mallet.
Station: Toulouse ch37 2kW ERP, Muret ch63 40W.
D.Prgr: 0630-0000.

TELE BLEUE (Private, Comm.)

Rue Brousan,30128 Garons.☎ +33 (66) 700123. 🖷 +33 (4) 66700701
— **Station:** Nimes ch61 400W ERP, Ch66 500W.
D.Prgr: MF 1700-2230, Sat 0900-1130/1800-2200, Sun 1800-2230.

TELE LYON METROPOLE (Priv., Comm.)

15 bd Yves Farge, 69007 Lyon. ☎ +33 (4) 7271 1090. 🖷 +33 (4) 7271 1095 — **L.P:** Jerome Bellay.
Station: Lyon ch25 2kW ERP, Lyon (south) ch56 1kW.
D.Prgr: 1700-2305.

8 Mont Blanc (Priv., Comm.)

Route Pontets, 74320 Sevrier. ☎ +33 (4) 5052 6969. 🖷 +33 (4) 5052 4991 — **L.P:** André Campana.
Stations: Cluses ch47 1.8kW, Annemasse ch57 3.8kW +29 low power stations under 1kW — **D.Prgr:** 1800-2300.

AQUI TV (Priv., Comm.)

19 rue Campniac, 24000 Perigueux. ☎ +33 (5) 5335 3000.
L.P: François Carrier.
Stations: Bergerac ch25 3.5kW + 10 stations under 1 kW
D. Prgr: 1130-2200

GERMANY (Federal Republic)

TV-sets: 30.500.000 — **Colour:** PAL — **System:** B.

ARD (PROGRAMMDIREKTION DEUTSCHES FERNSEHEN)

Arnulfstrasse 42, 80335 München. ☎ +49 (89) 59 00 01 🖷 +49 (89) 5900 3249
L.P: PD: Dr. Günter Struve. TD. Chrmn: Ingo Dahrendorf. Film Buyer: Klaus Lackschewitz
NB: The ARD is an umbrella organisation representing regionalized German public radio- and tv broadcasters. The ARD is responsible for the first public tv network (ARD Eins) and third public tv programs.
ARD members:
-**Bayerischer Rundfunk Fernsehen**, Rundfunkplatz 1, 80335 München. ☎ +49 (89) 59 00 01 🖷 +49 (89) 5900 2375
-**Hessischer Rundfunk Fernsehen**, Bertramstrasse 8, 60320 Frankfurt. ☎ + 49 (69) 1551 🖷 +49 (69) 1552 900
-**MDR Fernsehen**, Kantstrasse 71-73, 04275 Leipzig. ☎ +49 (341) 22760 🖷 +49 (341) 5663 544
-**NDR Fernsehen**, Rothenbaumchaussee 132, 20149 Hamburg. ☎ +49 (40) 4131 🖷 +49 (40) 4476 02
-**ORB Fernsehen**, August-Bebel-Strasse 25-53, 14482 Potsdam-Babelsberg. ☎ +49 (331) 72 36 00 🖷 +49 (331) 77395
-**Radio Bremen Fernsehen**, Hans-Bredow-Strasse 10, 28307 Bremen. ☎ +49 (421) 2460 🖷 +49 (421) 246 2010/1010
-**Saarländischer Rundfunk Fernsehen**, Funkhaus Halberg, 66100 Saarbrücken. ☎ +49 (681) 6020 🖷 +49 (681) 6023 874
-**SDR Fernsehen**, Neckarstrasse 230, 70190 Stuttgart. ☎ +49 (711) 929 1 🖷 +49 (711) 929 2600
-**SFB Fernsehen/B 1**, Masurenallee 8-14, W-14057 Berlin. ☎ +49 (30) 3031 0 🖷 +49 (30) 301 50 62
-**SWF Fernsehen**, Hans-Bredow-Strasse, 76530 Baden-Baden. P.O. Box 820, 76485 Baden-Baden. ☎ +49 (7221) 92 0 🖷 +49 (7221) 92 20 13
-**WDR Fernsehen**, Appellhoffplatz 1, D-50667 Köln. ☎ +49 (221) 22 01 🖷 +49 (221) 2204 800
D.Prgr: ARD Eins, Germany's first public tv network, broadcasts nationwide 24h (except for 2 ½ hrs. in the evening when regional tv from 13 tv centers is relayed). In 1992 ARD together with Germany's second public tv network ZDF, started broadcasting a common breakfast tv program called "Morgenmagazin" (Mon-Fri 0500-0800). Germany's third program comprises a combination of seven separate channels: NDR/RB in "N 3"; MDR in "MDR 3"; ORB in "ORB 3"; WDR in "West 3"; HR in "HR 3"; SWF/SDR/Saarländischer Rundfunk in "Südwest 3"; Bayerischer Rundfunk in "BR 3".

ZWEITES DEUTSCHES FERNSEHEN (ZDF)

P.O. Box 4040, 55030 Mainz. ☎ +49 (6131) 70 1 🖷 +49 (6131) 7021 57
L.P: Gen.Dir: Prof.Dr. Dieter Stolte; TD: Dr. Albert Ziemer; Film Buyer: Dr. Hans-Jürgen Steimer
D.Prgr: The ZDF is Germany's second public tv network, broadcasting nationwide 24h.

COMMERCIAL TV STATIONS

SAT EINS

Martin Luther Strasse 1, 10777 Berlin, Germany. ☎ +49 (30) 21241 0 🖷 +49 (30) 21241 140
L.P: Man Dir: Hans Grimm; Hd. of News: Heinz Klaus Mertes; Film Buyer: Akim Andorfer
D.Prgr: This commercial satellite tv station, set up by a number of German publishing houses, broadcasts 24h. Reg. prgr. from Hannover, Dortmund, Mainz, Stuttgart, München at 1630-1700 UTC.

RTL FERNSEHEN

Aachenerstrasse 1036, 50858 Cologne, Germany. ☎ +49 (221) 456 0 🖷 +49 (221) 456 4290
D.Prgr: Originally Luxembourg's German language tv channel, but due to German legal rules currently operating from Cologne. RTL Fernsehen broadcasts 24 hrs. with loc. prgrs. (Hamburg, Essen, Frankfurt, Mannheim, München, Berlin) from 1700-1730.
N.B: Also relayed via ECS II f1 (13Þ East)

RTL-2

Bavariafilmplatz 7, 82031 Grünwald. ☎ +49 (89) 641850. 🖷 +49 (89) 64185999.
Stations: Aschaffenburg ch21; Augsburg ch58; Bayreuth ch46; Deggendorf ch52; München ch27; Nürnberg ch53; Regensburg ch48; Rosenheim ch50; Weilheim ch47; Würzburg ch34/56 (ASTRA 19.2Þ East)

SUPER RTL
✉ Richard-Byrd-Strasse 6, D-50829, Cologne. ☎ +49 (221) 9155 0
Stations: Hamburg ch34 (local prgr "Hamburg 1").

DEUTSCHES SPORTFERNSEHEN (DSF)
✉ Bahnhofstrasse 27, 85774 Unterföhring. ☎ +49 (89) 95002 0 ▤
+49 (89) 9500 2392
L.P: Man.Dir: Dr. Dieter Hahn; Prgr. Dir: Rudolf Brückner; Marketing
Dir: Kai Blasberg
D.Prgr: 24h. Sports & Leisure.

VOX
✉ Richard-Byrd-Strasse 6, D-50829, Cologne. ☎ +49 (221) 9534 0 ▤
+49 (221) 9534 440
Stations: Bielefeld ch36 (1kW); Bochum ch50 (100W); Bonn ch34
(1.5W); Bonn ch51 (200W); Bremen ch49 (63kW); Bremerhaven ch57
(63kW); Burscheid ch41 (4kW); Chemnitz ch47; Dortmund ch43
(500kW); Düren ch22 (400W); Düsseldorf/Witzhelden 39 (100kW);
Gevelsberg ch5 (2.5W); Herdecke ch22 (20W); Iserlohn ch7 (2W); Köln
ch27 (200W); Köln/Bonn ch40 (3000W); Lippstadt ch24 (30W);
Mülheim ch27 (250W); Münster ch34 (200W); Paderborn ch22
(100W); Rheine ch57 (4kW); Saarbrücken/ Winterberg 56 (1kW);
Siegburg/Troisdorf ch51 (200W); Siegen ch36 (30W); Stolberg ch29
(1500W); Unna ch51 (200W); Wesel/ Büderich ch59 (200kW);
Wuppertal ch8 (1kW).

PRO SIEBEN
✉ Bahnhofstrasse 27, 85774 Unterföhring. ☎ +49 (89) 9507 1000.

N-TV
✉ Taubenstr. 1, 10117 Berlin. ☎ +49 (30) 201900.

PRINCIPAL TRANSMITTING STATIONS:

1) BAYERISCHER RUNDFUNK FERNSEHEN
BR. third program = "Bayerisches Fernsehen", at some moments split
up in an "Altbayern" and "Frankenland" version.

Area	ARD	ZDF	3	SAT1	RTL	DSF	kW
Amberg	—	37	43	50	52	—	350/320
Ansbach	—	—	—	—	—	49	
Augsburg	—	23	44	38	30	58	310/430
Bamberg	52	24	56	45	48	54	50/85/90
Bayreuth	—	30	54	39	46	—	98/96
Brotjacklr./D.	7	33	40	22	35	52	100/380/270
Büttelberg	55	—	—	—	—	—	400
Coburg	—	22	41	—	—	—	190/41
Dillb./Nürnb.	6	34	59	40	36	53	100/400/492
Grünten-Allgäu	2	28	46	—	—	—	100/470/470
H.Linie/Rgnsb.	53	21	42	38	34	46	75/370/400
H.Peissenberg	25	22	53	42	33	—	10/350/350
Högl	—	42	50	—	—	—	40
Hoher Bogen	55	28	59	—	—	—	200
Hühnrb./Hsslb.	60	32	47	—	—	—	400/140/200
Kreuzberg/Röhn	3	29	49	44	—	21	100/350/270
Landshut	—	39	58	—	—	—	27/55
Ochsenk./Hof	4V	23	57	—	—	—	100/500/500
Passau	—	30	60	—	—	—	41/34
Pfaffenberg/S.	59	35	51	—	—	—	100/230/250
Pfaffenhofen	—	31	41	—	—	—	480/420
Pfarrkirchen	—	27	57	—	—	—	250
Regensburg	—	—	—	—	—	38	?
Schnaitsee	—	26	54	—	—	—	360/280
Wendelstein/	10	35	56	59	24	37	100/200/200
München	—	—	—	—	—	38	
Würzburg	10	25	45	21/	57/	—	5/410/360
				38	60	—	

2) HESSISCHER RUNDFUNK FERNSEHEN
HR. third program = "Hessen 3".

Area	ARD	ZDF	3	SAT1	RTL	DSF	kW
Biedenkopf/A.	2	24	52	—	—	—	100/500/500
Fulda/Rhön	47	29	37	60	41	—	10
Gr. Feldberg	8	34	54	—	—	—	100/500/500
Habichtswald	56	28	41	—	—	—	50/100/100
Hardb./Krehb.	5	33	43	—	—	—	10/100/100
Hoher Meissner	7	32	55	—	—	—	100/390/470
Hohes Lohr	—	32	45	—	—	—	63/63
Kassel	—	—	—	42	35	58/60	
Rimberg	57	25	39	—	—	—	400/400/332
Würzberg	56	—	—	—	—	—	100/350/350

3) MITTELDEUTSCHER RUNDFUNK FERNSEHEN
MDR. third program = "Mitteldeutsches Fernsehen"

Area	ARD	ZDF	3	SAT1	RTL	DSF	kW
Brocken	6H	49	34H	—	—	—	100/500/1000
Chemnitz	8H	49	32	—	—	—	100
Dequede	12V	—	31H	—	—	—	2/500
Dresden	10V	46	29H	—	—	—	100/570/1000
Görlitz	6	—	—	—	—	—	1
Halle	—	—	46	57	—	—	
Inselsberg	5H	—	31H	—	—	—	100/500
Kapaunberg	—	40	37	—	—	—	?
Keula	—	—	45	—	—	—	25
Kulpenberg	8	—	57	—	—	—	0.12/1
Leipzig	9V	42	22H	—	—	—	100/500/600
Leipzig-Zeitz.	35	—	52	—	—	—	85/100
Löbau	27H	56	39H	—	—	—	100/500/200
Naumburg	35	—	52	—	—	—	85/200
Remda	21	50	27	—	—	—	200/500/200
Ronneburg	59	—	25	—	—	—	20/2
Saalfeld-Remda	21	50	27	—	—	—	200/500/200
Sieglitzberg	7	—	29	—	—	—	0.8/5
Schöneck	28	—	—	—	—	—	5
Sonneberg	12H	—	33H	—	—	—	30/500
Weida	59	—	25	—	—	—	20/50
Wittenberg	30	38	55	—	—	—	150/5/210

D.Prgr: Mitteldeutsches Fernsehen: Mon-Fri 1300-2330 approx.
(includes relays of local "Fenster" (Window) programmes. Sat/Sun:
1200-0100 approx.)

4) NDR FERNSEHEN/B1/RB FERNSEHEN
NDR/RB. third program = "N 3". NDR1 Hamburg on ch E56 (500kW),
NDR1 Cuxhaven on ch E51 (250kW).
NB: On October 1st 1992 Sender Freies Berlin Fernsehen started up
a new regional prgr. under the name **"B 1"**. This prgr. replaces "N 3"
in the Berlin area.

Area	ARD	ZDF	3	SAT1	RTL	DSF	kW
Aurich	53	33	43	—	—	—	230/410/400
Bannenb./Höhb.	43	21	45	—	—	—	250/500/440
Berlin	7	33	39	25	22	9V	100/430/200
Bremen	22	32	42	29	46/	—	100/500/500
	—	—	—	—	49	—	
Bremerh./Cuxh.	45	24	48	5	8/11	—	30/330/330
Bungsb./Eutin	50	21	47	31	44	—	260/235/280
Flensburg	4	39	57	28	24	—	50/160/150
Garz (Rügen)	40	21	29	—	—	—	250/250/250
Hamburg	9/26	30	40	48	46	34	20/100/460/500
Hannover	8	24	44	40	36	—	5/500/500
Harz/Göttingen	4	39	59	39	29	—	100/160/250
Heide/Eiderst.	10V	31	44	—	—	—	25/500/380
Helpterberg	37	52	22	—	—	—	500/200/1000
Kiel	5	35	55	53	24	—	2/250/200
Lingen	41	24	59	30V	—	—	400/500/400
Lübeck	—	23	33	42	36	—	240
Marlow	8	43V	24V	—	—	—	100/100
Minden	—	26	54	—	—	—	500/20
Neumünster/S.	28	26	45	—	—	—	500/100/100
Niebüll	29	34	60	—	—	—	100/200/150
Osnabrück	50	39	56	44V	36	—	160/250/138
Schwerin	11	—	29	—	—	—	100
Stadthagen	47	—	—	—	—	—	100
Steinkimmen/C.	55	37	40	27V	35V	—	400/47/47
Torfhaus	—	23	53	—	—	—	500
Ülzen	—	27	58	—	—	—	500/400
Visslh./Verden	7V	25	60	—	—	—	20/66/50
Züssow	36	51	23	—	—	—	70/190/190

5) OSTDEUTSCHER RUNDFUNK BRANDENBURG FERNSEHEN
ORB. third program = "Fernsehen Brandenburg"

Area	ARD	ZDF	3	SAT1	RTL	DSF	kW
Belzig	—	—	54	—	58	—	50/5
Berlin	7	33	27	—	—	—	100/700
Casekow	—	—	54	—	—	—	100
Cottbus	53	57	23	—	—	—	355/100/577
Frankfurt/Oder	11	50	43	—	—	—	0.8/83
Höhbeck	51	21	35	—	—	—	200
Treplin	—	50	43	—	—	—	?
Woldegk	—	—	34	—	—	—	2.9

6) SWF FERNSEHEN/SDR FERNSEHEN/SR FERNSEHEN

SWF/SDR/SR. third program = "Südwest 3" (further regional split-ups as indicated under a-d).

a) SWF-Mainz:

Area	ARD	ZDF	3	SAT1	RTL	DSF	kW
Ahrweiler	—	33	56	—	—	—	170/210
Bad Marienberg	47	21	44	—	—	—	50/190/165
Boppard	—	28	41	—	—	—	160/120
Donnersberg	10	37	60	—	—	—	100/250/250
Eifel	23	30	40	—	—	—	20/195/180
Haardtkopf	25	35	55	—	—	—	400
Hohe Derst	23	44	46	—	—	—	1
Kaiserslautern	3	22	44	50	33	—	25/25
Koblenz	6	31	51	57	36	—	50/85/83
Mainz	11	—	—	36	—	44	1
Saarbr./Tflsk.	29	45	53	26	—	—	20/50/50
Trier	5	22	48	56	—	—	0.6/41/22
Weinbiet	6	—	—	—	—	—	30

b) SR-Saarbrücken

Area	ARD	ZDF	3	SAT1	RTL	DSF	kW
Göttelborner Höhe/ Saarbrücken	2	45	42	26	—	—	100/470/500
Nelb./Piesb.	28	36	44	56	—	—	?

c) SWF-Baden-Baden

Area	ARD	ZDF	3	SAT1	RTL	DSF	kW
Brandenkopf	48	28	45	—	—	—	2.5/25/50
Eggb./Hochrhn.	22	39	52	—	—	—	0.8/225/295

Area	ARD	ZDF	3	SAT1	RTL	DSF	kW
Feldberg/ Donausch.	8	22	57	50/ 60	36/ 38	—	100/100/100
Freib.im Brsg.	7	33	58	—	38	—	0.9/500/500
Grünten	43	—	—	—	—	—	500
Hornisg./B-B	9	31	49	49	—	55	80/220/290
Raichenbach/R.	4	35	55	—	—	—	100/260/330
Ravensburg	—	37	40	27	—	30V	270/290
Waldshut	21	—	—	—	—	—	1
Wannenberg	30	—	—	—	—	—	5

d) SDR-Stuttgart

Area	ARD	ZDF	3	SAT1	RTL	DSF	kW
Aalen/Heubach	8	29	52	—	—	—	50/209/135
Bd Mergentheim	48	—	—	—	—	—	10
Heidelberg	7/50	27	53	—	—	—	100/500/440/ 492
Heilbron/Eberb.	49	30	58	—	—	—	10/215/195
Langenbrand/P.	21	34	59	23	—	—	100/170/250
Stuttgart	11	26	39	—	—	—	100/404/234
Neu-Ulm	—	33	54	36	—	48/ 51	370/330/2
Waldenburg/L.	9	28	42	—	—	—	100/165/185

7) WESTDEUTSCHER RUNDFUNK FERNSEHEN

WDR. third program = "West 3".

Area	ARD	ZDF	3	SAT1	RTL	DSF	kW
Aachen	24	37	58	27	26	26	200/320/400
Bonn	43	26	49	—	5/36	—	100/81/100
Dortmund	—	25	53/	47	58	58	500/470
Ederk./Hochs.	50	27	40	—	—	—	50/250/250
Eggebirge	—	31	48	—	—	—	250
Kleve/ Wesel	46	35	48/ 59	—	52	52	100/500/475/ 80
Köln	11	—	—	—	36	—	10
Langenberg/ Düsseldorf	9	29	39/ 55	44	36	36	100/500/100/ 450
Minden	—	26	57	—	—	—	500/200
Münster	32	21	45	51	38	—	500/170/250
Nordh./Lüdens.	30	37	60	—	—	—	240/200/250
T.Wald/Bielef.	11	33	46	38	59	59	100/310/450
Wuppertal	6	22	42	—	—	—	0.6/70/85

OTHER TV STATIONS

AMERICAN FORCES TELEVISION GERMANY

✉ Headquarters AFN Europe, Bertramstrasse 6, 60320 Frankfurt a/M.(or APO New York, NY 09757, USA). ☎ +49 (611) 1516 101
System: NTSC (M)
Stations:

Site	ch	kW	Pol		ch	kW	Pol
Wiesbaden AB	22	2	H	Karlsruhe	39	0.1	H

Berlin	29	0.5	H	Bitburg	51	5	V
Kaiserslautern	30	3	H	Mannheim	56	0.15	V
H. Oldendorf	38	0.1	V	Kitzingen LB	58	0.2	V

+ many lps less than 0.1kW
D.Prgr: 24h.

SSVC TELEVISION (U.K. Mil.)

✉ BFPO 40
System: PAL (I)
Stations:

Site	ch	kW	Pol	Site	ch	kW	Pol
Münster	23	0.8	H	Osnabrück	48	0.9	H
Lemgo	37	0.15	V	Paderborn	49	0.2	H
Geilenkirchen	40	0.4	V	Wassenberg	49	0.4	H
Hannover	40	0.4	V	Mülheim	54	0.1	H
Bielefeld	41	0.1	V	Celle	51	0.2	V
Hildesheim	41	0.2	V	Herford	52	0.1	V
Berlin Westend	41	0.2	V	Niederkrüchten	52	1	H
Minden	41	0.4	H	Hamm	55	0.1	V
Gütersloh	44	0.4	V	Wolfenbüttel	56	0.14	V
Hameln	44	0.1	V	Düsseldorf	57	0.12	H

+ many lps less than 100W
D.Prgr in English: 1100-2200. Special service intended solely for H.M. Forces.

BELGIAN FORCES TELEVISION

Stations: Bensberg ch39 (600W); N.-Hinsbeck ch47H (200W); Bensberg & N.-Hinsbeck ch51H (800/200W); Titz ch39H (600W); Arolsen ch9 (500W)

TV-sets: 7.500 — **Colour:** PAL — **System:** B & G.

GBC TELEVISION (partly comm.)

✉ Broadcasting House, 18 So. Barrack Rd, Gibraltar.
☎ +350 79760. **Cable:** Broadcasts. 🖷 +350 78673. ☉ 2229 GBCEE GK.
L.P: GM: George Valarino. Senior Eng: John Tewkesbury.
Station: chE6 0.2/0.4kW H (+ low power repeaters ch12, 53, 56).
D.Prgr: rel BBC World + local prgrs.

TV-sets: 2.300.000 — **Colour:** PAL — **Systems:** B & G.

ELLINIKI TILEORASSI-1 (ET-1)

✉ Leophoros Messogeion 432, GR-153 42 Aghia Paraskevi Attikis. ☎ +30 (1) 63 95 970. 🖷 +30 (1) 63 92 263. ☉ 216066 — **L.P:** DG: George Stamatelppoulos.

Stations	ch	kW	Pol		ch	kW	Pol
Akarnanika	3	1.5/ 0.3	H	Rhodes Isl.	9	10/2	V
				Mytilini	9	10/2	H
Thessaloniki	5	30/6	H	Tripolis	10	30/6	H
Pilion (Volos)	6	30/6	H	Heraklion	10	10/2	H
Alexandroupolis	6	30/6	V	Ioannina	10	10/2	H
Kalamata	6	10/2	H	Parnis (Athens)	11	30/6	V
Chania	7	10/2	H	Hymettos			
Kavala	7	30/6	H	(Athens)	21	15/3	H
Kastoria	7	30/6	V	Tholopotamos	22	30/6	H
Kefalinia Isl.	8	30/6	V	Alexandroupolis	33	15/3	V
Thira	8	30/6	H	Thassos Isl.	39	550/55	H
Gerania	9	10/2	H	Euros	50	18/7	H
Corfu Isl.	9	10/2	V				

+ 700 low power repeaters.
D.Prgr: 0800-2400 (approx).

ELLINIKI TILEORASSI-2 (ET-2)

✉ Leophoros Messogeion 136, GR-115 25 Athens. ☎ +30 (1) 7701911. 🖷 +30 (1) 77 97 776. ☉ 210886.
L.P: DG: Panos Panayotu.

Stations	ch	kW	Pol		ch	kW	Pol
Hymettos				Pilion (Athens)	41	750	H
(Athens)	5	30/7.5	V	Rhodes Isl.	42	500	H
Alexandroupolis	21	170	H	Tripolis	42	500	H
Thassos Isl.	23	1000	H	Akarnanika	43	1000	H
Ioannina	25	250	H	Mytilini	48	200	H

Thira Isl.	29	500	H	Vitsi (Kastoria)	49	250	H
Thessaloniki	30	1000/	H	Corfu Isl.	50	530	H
		200	H	Kefalinia Isl.	57	430	H
Parnis (Athens)	34	450	H	Kavala	59	650	H

+300 low power repeaters
D.Prgr: 0800-2400 (approx).

ET-3 (regional channel for Macedonia)

✉ Aggelaki 2, GR-546 21 Thessaloniki. ☎ +30 (31) 27 87 84. 🖷 +30 (3) 23 64 66 — **L.P:** DG: Hichalis Alexandridis.
Station: Thessaloniki ch23/27 H (local prgrs), Pilion ch 44H, Pag-gaion ch 35H, Parnis ch 52H, Hymettos ch 31H, Thassos Isl. ch 26H, Florina ch 39H, Polygiros ch 21H.
D.Prgr: 0800-2400 (approx).

PRIVATE STATIONS

Antenna TV, 10-12 Kifissias Ave, Maroussi, 15125 Athens. ☎ +30 (1) 6842220. 🖷 +30 (1) 3890304.
Argo TV, Metamorphosseos 9, 55132 Kalamaria, Thessaloniki. ☎ +30 (31) 351733. 🖷 +30 (31) 351739.
Channel Seven-X, 64 Leoforos Kiffissoas, Athens. ☎ +30 (1) 68976042. 🖷 +30 (1) 6897608.
City Channel, 14 Leoforos Kastoni, 41223 Larissa. ☎ +30 (1) 232839. 🖷 +30 (1) 232013.
Tele City, 58 Praxitelous, 17674 Athens. ☎ +30 (1) 9429222. 🖷 +30 (1) 9413589.
Jeronimo Groovy TV, Ag konstantin 40, 15124 Athens. ☎ +30 (1) 6896360. 🖷 +30 (1) 6896950.
Mega Channel, 10 Alamanas Str, 15125 Amarousion, Athens. ☎ +30 (1) 689900014. 🖷 +30 (1) 6899016.
Neo Kanali SA, 9-11 Pireos, 10552 Athens. ☎ +30 (1) 5238230. 🖷 +30 (1) 5247325.
Serres TV, Nigritis 27, 62124 Serres.
Skai TV, 2 Phalereos & Ethnarchou, Macaroiu, N. Phaliro.
Star Channel, 37 Dimitras, 1178 Tayros, Athens. ☎ +30 (1) 3450626. 🖷 +30 (1) 3452190.
Teletora, 17 Lycabetous, 10672 Athens. ☎ +30 (1) 3617285. 🖷 +30 (1) 3638712.
Traki TV, Central Square, 67100 Xanthi. ☎ +30 (541) 20670. 🖷 +30 (541) 27368.
TRT, 5 Zachou Str, 38333 Volos. ☎ +30 (421) 30500. 🖷 +30 (421) 36888.
TV Macedonia, 222 Nea Egnatia, 54642 Thessaloniki. ☎ +30 (31) 850512. 🖷 +30 (31) 850513.
TV Plus, Syngrou Ave. 97, 11745 Athens. ☎ +30 (1) 9028707. 🖷 +30 (1) 9028310.
TV-100 (Thessaloniki municipal st.), 16 Aggelaki Str, 54621 Thessaloniki. ☎ +30 (31) 265828. 🖷 +30 (31) 267532.

AFN TV (U.S. Mil.)

Station: chA2 Iraklion, A6.

HUNGARY

TV-sets: 4.261.600 — **Colour:** PAL — **Systems:** D & K.

MAGYAR TELEVÍZÍO (MTV)

✉ Szabadság ter 17, 1810 Budapest, 5. ☎ +36 1111 4059. 🖷 +36 1115 74979. ① 325558.
L.P: Pres: Gyula Berecky.
Stations: Prgr. 1 (System D), Prgr. 2 (System K)

Loc.	Prgr 1	(kW)	Pol	Prgr 2	(kW)	Pol
Aggtelek	R28	200	H	R45	200	V
Budapest	R1	150/50	H	R24	600	H
Budapest	R41	600	H	—	—	—
Csávoly	R28	80/8	H	R7	9/0.9	V
Csengod	25	200	H	42	200	H
Fehergyarmat	R24	600	V	R41	600	H
Györ	R8	10/1	V	35	105/10	H
Kabhegy	R12	150/15	V	22	760/76	H
Kékes	R8	30/3	H	36	880/88	H
Komádi	R7	50	H	R32	200/20	H
Nagykanizsa	R1	50/5	V	R31	380	V
Ozd	35	200	H	—	—	—
Pécs	R2	60/6	V	32	400/40	H

Sopron	R9	5/1	V	R32	200	V
Szeged	26	200	V	31	?	H
Szentes	R10	200/20	V	R23	480	H
Tokaj	R43	80/20	H	26	420/42	H
Tokaj	R4	80	H	—	—	—
Vasvar	R33	600	V	46	600	H

low power st's not mentioned.
D.Prgr: Prgr .1: Tue/Thu 0655-1045, 1440-2245; Mon/Wed 0800-1045, 1440-2245; Fri 0800-1045, 1440-2345; Sat 0530-1140, 1415-2300; Sun 0700-1140, 1250-2210. **Prgr. 2:** 1600-2215; Sat 1300-2320; Sun 1300-2300.

PRIVATE STATIONS

NAP TV

✉ Angol utca 13, 1149 Budapest. ☎ +36 (1) 251 0490. 🖷 +36 (1) 251 3372 — **L.P:** Chmn: Tamas Gyarfas.
Nap TV provides morning news and inf. sce. on Prgr. 1 of MTV.

A3 (Pest-Buda TV) (local st.)

✉ Rona utca 140, 1147 Budapest. ☎ +36 (1) 251 4749.
L.P: Gen. Dir: Adam Namenyi.

TV3 BUDAPEST (local st.)

✉ Budakeszi utca 51, 1021 Budapest. ☎ +36 (1) 275 1800. 🖷 +36 (1) 275 1801 — **L.P:** Dir: Peter Kolin.

SIO TELEVISIO (local st.)

✉ Fo ter 2, 8600 Siofok. ☎ +36 (84) 317111. 🖷 +36 (84) 310887 (provides prgrs for visitors to Lake Balaton area).

ICELAND

TV-sets: 93.595 — **Colour:** PAL — **Systems:** B&G.

RÍKISÚTVARPID — SJÓNVARP
ICELANDIC NATIONAL TELEVISION

✉ Laugavegur 176, 150 Reykjavik. ☎ +354 515 3900. 🖷 +354 515 3008. ① 2035 VISION IS. **E-mail:** istv@ruv.is
L.P: MD: Petur Gudfinnsson. CE: E. Valdimarsson.
Stations: Pol H; Audio power 10% of visual power indicated.

Location	chE	kW		Location	chE	kW
Stykkishólmur	3	90	Háfell		7	1100
Skálafell	4	300	Girdisholt		7	2.5
Gagnheidi	4	80	Hegranes		8	12
Vestmannaeyjar	5	39	Hnjúkar		9	20
Heidarfjall	5	10	Vatnsendi		10	20
Vadlaheidi	6	490	Vidarfjall		11	8.4

+ 156 low power relay st's.
D.Prgr: Mon-Thurs 1800-2300; Fri 1800-2400; Sat 1600-0100; Sun 1600-2330.

OMEGA (Rlg.)

✉ Grensasvegur 8, 108 Reykjavik. ☎ +354 568 3131. 🖷 +354 568 3741 — **L.P:** General Dir: Erik Eriksson.

STOD 2 (Channel 2) & STOD 3 (Channel 3) (Subscription sces.)

✉ Krokhals 6, 112 Reykjavik. ☎ +354 515 6000 (Stod 2), +354 533 5600 (Stod 3). 🖷 +354 515 6810.

ICE TV Channel 3

✉ c/o Laufey Gudjonsdottir, Kringlan 7, Reykjavik 103. ☎ +354 533 5633. 🖷 +354 533 5699.

IRELAND

TV-sets: 1.056.000 — **Colour:** PAL — **System:** I.

RADIO TELEFIS EIREANN (Statutory Corporation)

✉ Donnybrook, Dublin 4. ☎ +353 (1) 208 3111. 🖷 +353 (1) 208 3080.
Web: http://www.rte.ie/tv/
L.P: Contr. of TV. Prod: Liam Miller.

Stations	I	II	kW	Pol	Stations	I	II	kW	Pol
Maghera	B		100	H	Mt.Leinster	F	I	100	V

	E	H	100	V	Kippure	E	H	100	H
Mullaghanish	D	G	100	V	Truskmore	I	G	100	H
N.E.Donegal	23	26	20	H	Longford	40	43	800	H
Dublin	29	33	25	H	Co.Louth	52	56	250	V

+36 low power transposer st's.

D.Prgr: RTE One: Mon-Fri 1100-2345 (Fri 0100 appr.), Sat 0900-0100, Sun 1005-2345. **RTE Two:** Mon-Fri 1430-2330, Sat 1230-0030, Sun 1015-2345.

ITALY

TV-sets: 17m — **Colour:** PAL — **Systems:** B & G.

RADIOTELEVISIONE ITALIANA
✉ Direzione Centrale TV, Viale Mazzini 14, 00195 Roma.
☎ +39 (6) 38781. 🖷 +39 (6) 3226070. ✆ 614432
L.P: Chmn: Enzo Siciliano. GM: Franco Iseppi. Dir RAIUNO: Giovanni Tantillo. Dir RAIDUE: Carlo Freccero. Dir RAITRE: Giovanni Minoli. Dir TG1: Rodolfo Brancoli. Dir TG2: Clemente Mimun. Dir TG3: Lucia Annunziata. Dir. TGR: Nino Rizzo Nervo. Dir. TGS: Marino Bartoletti. Dir. Televideo: Marcello Del Bosca. Dir. RAI-International: Roberto Morrione. PR: Carlo Sartori.

Stations: System: B/PAL in VHF band — G/PAL in UHF band. Pol. H except where indicated differently.

Location	RAIUNO	RAIDUE	RAITRE	kW
Agricento Giache (AG)	D	—	—	2
Alcamo M. Bonifato (TP)	Ev	25	44	18/12
Bari (BA)	Fv	—	—	10
Benevento (BE)	—	33	51	12
Bertinoro (FO)	Fv	30	40	40/8

Location	RAIUNO	RAIDUE	RAITRE	kW
Bologna C. Barbiano (BO)	31	28	48	200/100
Bra (CN)	—	22v	—	20
Brescia Vedetta (BS)	—	—	33	20
Cagliari Capoterra	—	28v	—	10
Camerino (MC)	F	—	—	3
Campo dei Fiori (VA)	59	28	22	20
Canepina (VT)	—	—	68v	12
Capo Milazzo (ME)	47	25	22	15
Capo Spartivento (RC)	H	23v	39v	30/6
Casteluono (PA)	F	22v	62v	3/2
Catanzaro M. Tiriolo (CZ)	Fv	30	55	20/2
Cima Penegal (TN)	—	27	69	20/10
Col Visentin (BL)	H	34	49	50/4
Crotone (CZ)	Bv	27	40	20/8
Crotone (CZ)	58	—	—	20
Firenze M. Morello (FI)	—	29	—	15
Fiuggi (FR)	—	25	67	40
Gambarie (RC)	D	26v	44v	400/20
Gambarie (RC)	—	—	51v	400
Genova Portofino (GE)	H	29	45	1000/160
Golfo di Palicastro (SA)	—	33v	50v	12
Golfo di Salerno (SA)	En	30v	69V	20
L'Aquila M. Luco (AQ)	—	24	—	15
Martina Franca (TA)	D	32	49	1000/200
Messina M. Piselli (ME)	—	29	42	40/20
Milano (MI)	G	26	33	40/20
Mione (TN)	69v	—	51	10
Monte Argentario (GR)	E	24v	46v	500/25
Monte Argentario (GR)	—	—	68v	300
Monte Ascensione (AP)	H2	56	—	50/5
Monte Beigua (SV)	—	32	42	120/60
Monte Caccia (BA)	A	25	44	1000/200
Monte Caccia (BA)	54	—	—	1000
Monte Cammarata (AG)	A	34	48	400/30
Monte Canata (PR)	42v	31v	37v	30/10
Monte Cavo (RM)	H2	35	39	230/40
Monte Cimarani (AQ)	—	22	—	10
Monte Conero (AN)	E	26	42	400/100
Monte Creo (BG)	—	27	41	10
Monte D'Elio (FG)	Bv	24	40	20/2
Monte D'Elio (FG)	63	—	—	20
Monte Faito (NA)	B	23	39	200/44
Monte Faito (NA)	—	23v	39v	1200
Monte Favone (FR)	H	29	37	40/3
Monte Lauro (SR)	F	24	41	400/200
Monte Limbara (SS)	H	32	52	40/3

Location	RAIUNO	RAIDUE	RAITRE	kW
Monte Luco (SI)	H1	23	39	40/3
Monte Luco (SI)	H1v	—	—	3
Monte Maddalena (BS)	F	43	D	13/2
Monte Maddalena (BS)	—	43v	—	10
Monte Nerone (PS)	A	33	54	400/34
Monte Peglia (TR)	H	31	49	400/30
Monte Peglia (TR)	—	—	37	400
Monte Penice (PV)	B	23	35	2000/100
Monte Penice (PV)	—	—	36	250
Monte Pierfaone (PZ)	—	21	42	20
Monte Sambucco (FG)	H	27	47	400/35
Monte Sambucco (FG)	—	—	50	400
Monte Scuro (CS)	G	28	46	60/8
Monte Serpeddi (CA)	G	30	49	500/30
Monte Serpeddi (CA)	67	—	—	10
Monte Serra (PI)	D	27	43	400/200
Monte Soro (ME)	E	32	67	50/20
Monte Subasio (PG)	51	35	47	10
Monte Subasio (PG)	51v	35v	47v	10
Monte Turu (TO)	H1v	28v	51v	10/2
Monte Venda (PD)	D	25	32	1000/100
Monte Vergine (AV)	D	31	43	200/20
Paganella (TN)	—	21	47	30/15
Palermo M. Pallegrino (PA)	F	30v	46v	80/40
Palermo M. Pallegrino (PA)	—	27V	40V	80/40
Pantelleria C. Glindo (TP)	Gv	—	—	25
Pescara S. Silvestro (PE)	F	30v	46v	60/30
Pescara S. Silvestro (PE)	60v	—	—	110
Piane di Mocogno (MO)	—	30v	—	13
Pigazzano (PC)	—	—	37	30
Punto Badde Urbana (OR)	D	27	47	400/160
Roma M. Mario (RM)	G	28	43	300/36
Roseto Capo Spulico (CS)	Fv	—	68	40/12

Location	RAIUNO	RAIDUE	RAITRE	kW
Saint Vincent (AO)	—	31	41	20/10
Salento Turrisi (LE)	—	34	47	60
San Cerbone (FI)	G	—	—	50
Sassari M. Oro (SS)	—	30v	—	20
Sezze (LT)	F	31v	68v	5/4
Stazzona (CO)	—	—	53	10
Terracina (LT)	—	28	—	10
Torino Eremo (TO)	C	30	40	400/15
Torino Eremo (TO)	5	—	—	400
Trapani Erice (TP)	Hv	31	43	50/20
Trapani Erica (TP)	—	31v	43v	8/4
Trieste M. Belvedere (TS)	G	31	44	50/3
Trieste Muggia (TS)	35v	28v	48v	20
Udine (UD)	F	22	47	300/160
Val Venosta (BZ)	A	22	36	3/2
Velletri (RM)	—	26	50	400/200

+ over 5200 st's below 2kW (VHF)/10kW (UHF) not mentioned

RUNDFUNKANSTALT SÜDTIROL (RAS)
(Public Statutory Body of the Autonomous Province of Southern Tyrol)
✉ Europaalee 164A, I-39100 Bozen. ☎ +39 (471) 3317 4258.
L.P: Pres: Helmuth Hendirch. MD: Klaus Gruber.
Stations: RAS1 (relay ORF-FS1); RAS2 (relay ZDF); RAS3 (relay ORF-FS2+SRG).

Location	ch	kW	Netw.	Location	ch	kW	Netw.
Perdonig	23	2.5	III	Penegal	55	10	II
Penegal	48	10	III	Perdonig	59	7.5	I
Vinschgau	50	3	III	Vinschgau	63	3	II
Penegal	52	10	I	Perdonig	65	7.5	II
Vinschgau	53	3	I				

+ 122 low power st's not mentioned.

PRIVATE TV STATION NETWORKS
There are about 900 privately operated tv sts in Italy, mostly on local service. Due to space limitations, only those sts with nationwide networks are mentioned. Readers requiring more information should write (with Rp pls) to: D. Monferini, Via Davanzati 8, I- 20158 Milano, Italy.
Stations:

1) Amica 8 & Amica 9	24h	(key st. Firenze)
2) Canale 5	24h	(key st. Milano)
2) Italia 1	24h	(key st. Milano)
2) Rete Quattro	24h	(key st. Milano)
3) Circuito 5 Stelle	24h	(key st. Milano)
4) Junior TV	24h	(key st. Firenze)
5) Cons.Italia 9 Netw.	24h	(key st. Firenze)

6) Video Music/TMC2	24h	(Key st. Castelv. Pasceli)
7) Italia 7	24h	(key st. Milano)
8) Odeon	24h	(key st. Rho-Milano)
9) Rete A	24h	(key st. Lucca)
10) Rete Mia	24h	(key st. Lucca)
11) Tivultalia	24h	(Key st. Rho-Milano)
12) Rete MTV	24h	(key st. Milano)
13) Super Six	24h	(key st. Milano)
14) Tele Montecarlo	0700-0130	(key st. Roma)
15) Tele Più	24h	(key st. Milano)

Addresses and other information

1) Sation Group DAPS, Via Vecchi 15, I 200-94 Assago (MI). ☎+39 (2) 48843984. ▤ +39 (2) 488 43990. 20 st's affiliated.
2) Reti Televisive Italiane, Via Europa 48, I-20093 Cologno Monzese (MI). ☎ +39 (2) 25125. ▤ +39 (2) 2138019. 20 st's affiliated.
3) Società Circuito Cinque Stelle, Via Plinio 44, I-00193 Roma. ☎ +39 (6) 688001. ▤ +39 (6) 68800400. 17 st's affiliated.
4) Junior TV, Corso Garibaldi 35, I-20121 Milano. ☎ +39 (2) 801376. ▤ +39 (2) 72001156. 20 st's affiliated
5) Società Consorzio Italia 9 Network Television, Via Settala 57, I-20124 Milano. ☎ +39 (2) 29524964. ▤ +39 (2) 29404892. 20 -st's affiliated.
6) TMC2, Piazza delle Balduina 49, I-00136 Roma. ☎ +39 (6) 35584271. ▤ +39 (6) 35584227. 18 st's affiliated.
7) Italia 7, Via Einstein 21, I-20094 Assago (MI). ☎ +39 (2) 45701747. ▤ +39 (2) 45701724. 19 st's affiliated.
8) Odeon, Via Tavecchia 43/45, I-20017 Rho (MI). ☎ +39 (2) 935151. ▤ +39 (2) 93504423. 16 st's affiliated.
9) Rete A, Viale E Marelli 165, I-20099 Sesto San Giovanni (MI). ☎ +39 (2) 22477241. ▤ +39 (2) 2401630. 7 st's affiliated.
10) Rete Mia, Vallan Italiana Promomarket s.r.l., Via Tempaegnano 40, I-55100 Lucca. ☎ +39 (583) 490555. ▤ +39 (583) 490459; 20 st's affiliated.
11) Tivuitalia, Via Tavecchia 43/45, I-20017 Rho (MI). ☎ +39 (2) 935151. ▤ +39 (2) 93504423; 20 st's affiliated
12) MTV, Corso Europa 7, I-20122 Milano. ☎ +39 (2) 7621171. ▤ +39 (2) 7621227. 9 st's affiliated.
13) Super Six, GGS International s.r.l., Via Bolzano 29, I-20127 Milano. ☎ +39 (2) 26144822. ▤ +39 (2) 26144832; 19 st's affiliated.
14) Tele Montecarlo, Piazza della Balduina 49, I-00136 Roma. ☎ +39 (6) 35584271. ▤ +39 (6) 35584257. 18 st's affiliated.
15) Tele Più, Via Piranesi 44/a, I-20137 Milano. ☎ +39 (2) 700271. ▤ +39 (2) 70027201. 20 st's affiliated.

TV-sets: 1.230.000 — **Colour:** SECAM/PAL — **System:** D&K

LATVIJAS TELEVIZIJA (LTV) (Gov.)
✉ Zakusalas krastmala 3, Riga LV-1509 ☎ +371 (2) 200314. ▤ +371 (2) 200025 ➀ 161188 video lv — **L.P:** Dir. of TV: Imants Rakins. Dir. of Prgs: Daina Ostrovska. Int. Rel: Inese Vitkus.
Stations (pol = H)

	Netw 1	Netw 2	Netw 3	
	LTV-1	LTV-2	LNT	L.
Location	ch/kW	ch/kW	ch/kW	ch/kW
Cesvaine	R8/30	R5/20	41/603	36/0.6
Dagda	36/0.06	24/0.06	—	
Daugavpils	R7/30	R10/0.65	40/603	21/0.6
Dundaga	25/47	30/47	R8/0.3	
Engure	—	—	—	R2/0/015
Jekabpils	—	—	29/0.6	—
Kandava	R12/0.0015	—	—	
Kraslava	—	R4/0.02	—	
Kuldiga	R6/30	R1/10	40/0.6	—
Liepaja	33/203	21/203	35/81	R5/0.165
Malpils	—	—	—	R2/0.015
Preili	—	R4/0.015	R2/0.015	—
Rekava *				
Rezekne	39/603	27/603	6/0.475	R10/0.155
Riga	R3/150	R10/158	28/380	R7/2.3
	—	—	—	31/95
Roja	—	—	—	R8/0/015
Sabile	R8/0.015	R5/0.015	R11/0.015	—
Valmiera	33/603	21/540	R11/0.55	R9/0.2
Ventspils	R5/0.165	R9/0.38	R12/0.38	R7/0.055
				38/0.9
				43/1
				45/1

Viesite *

Vilani	—	—	R11/0.015	R2/0.015

L = State transmitters, leased by private broadcasters
* = under construction

D.Prgr: Netw. I (LTV national ch): 1600 (SS 0700)-2200.
Netw. 2 (LTV multilingual ch.): 1420 (SS 1000)-2200 in Latvian & Russian incl. rel. of private, regional and int. broadcasters.
Netw. 3: rel. of Lavijas Neatkariga Televizija (LNT) in Russian.

PRIVATE, MUNICIPAL AND REGIONAL STATIONS

Location	ch	kW (ERP)	Station
Dagda	R4Þ	0.04	TV Ezerzeme
Dobele	R5	0.05	Dobeles TV
Dundaga	R11	0.01	Dundagas TV
Gulbene	R11	0.1	Gulbenes TV 11. kanals
Jelgava	33	0.04	Jelgavas TV
Kraslava	R2	0.01	Kraslavas TV
Kuldiga	R8	0.07	Kuldigas TV
Ledurga	R12	0.04	Ledurgas TV
Liepaja	R23Þ	0.1	Kurzemes TV
Livani	R12	0.02	Livanu TV
Malpils	R12	0.02	Malpils TV "Spektrs"
Ogre	R2	0.015	Ogres TV
Riga	33	0.05	TV Miraza
	43Þ	1	TV Riga
Rujiena	R7	0.02	Rujienas TV
Salaogriva	R8	0.04	Salaogrivas TV
Skrunda	R5	0.01	Skrundas TV
Smiltene	R7	0.02	Smiltenes TV
Talsi	R4	0.02	Talsu TV
Valmiera	R7	0.02	Valmieras TV 7. kanals
Viesite	R6	1	Viesites TV

Þ = PAL colour, others SECAM.

TV-sets: 1.570.000 — **Colour:** PAL B&G/SECAM D&K.

LIETUVOS TELEVIZIJA (Gov.)
✉ see radio. ☎ +370 (2) 637961. ➁ 261151 litvr. lt. ▤ +370 (2) 263282.
L.P: Dir: Saulius Sondeckis.
Stations: pol = H, exc. where stated.

Location	ch R-	kW	Location	ch R-	kW
Druskininkai	1	0.316	Klaipéda	8	50
Vilnius	2V	177	Juragiai	9	2500
Jurbarkas	3	0.15	Pazagieniai	10	0.144
Mazeikiai	4	0.15	Bubiai	11	70
Ignalina	5	0.098	Viesintos	12	58.5
Naujoji Akmené	5	0.15	Neringa	22	0.355
Plunksniai	6	0.15	Visaginas	30	448
Utena	6	0.019	Tauragé	39	600

D.Prgr: MF 0545-0645 & 1600-2130, Sat 0700-2300, Sun 0700-2130 (closing times vary). **Russian:** MF 1650-1700, Sat 60 min. around 0900. **Polish:** Tue 1705-1735. **Byelorussian:** Mon 1705-1715. **Ukrainian:** fortnightly 10 min. within the Sat Russian broadcast.

AIDAS (Priv., Comm.)
✉ Birutes 42, LT-4050 Trakai. ☎ +370 (38) 52480.
L.P: Dir: Ceslovas Rulevicius.
Stations: Vievis ch R8 V 60W(ERP); Kaisaidorys ch R6 H 60W (ERP)
D.Prgr: Mainly relays German Pro-7 prgrs. Own prgrs are broadcast irreg.

BALTIJOS TELEVIZIJA (Priv., Comm.)
✉ Laisvés pr. 60, LT-2044 Vilnius. ☎ +370 (2) 417134. ▤ +370 (2) 428907. **L.P:** Dir. Gen: Gintaras Songaila.
Stations: Alytus ch 33 10kW; Vilnius ch 38 H 500kW; Taurage ch 44 10kW; Juragiai ch 45 10kW; Klaidepa ch 46 10kW; Vilnius ch 48 10kW; Utena ch 49 10kW; Bubiai ch51 10kW
D.Prgr: 24h. on all channels except ch38. Relays CNN Int (Mon-Fri 2200-0500, SS 2200-0630), Deutsche Welle TV (Mon-Fir 0630-1615, Sat 1100-1615, Sun 1015-1615). relays en in PAL, own prgrs in SECAM. ch 38: Mon-Fri 0500-2400, SS 0630-2400. Relays TV Polonia 2000-2030, Mon-Fri 1400-1615, Sat 1100-1615, Sun 1015-1615, Tue/Sun 2030-2400, Sat 2215-2400.

KAUNAS PLIUS (Priv., Comm.)

UAB "Style TV", P.D. 2040, LT-3000 Kaunas. ☎ +370 (7) 220 650/224 429. 🗎 +370 (7) 220 640.
L.P: Pres: Henrikas Zukauskas.
Stations: Juragiai chR7 1.5kW H; Vilnius chR11 1kW H
D.Prgr: Mon-Fri 1630-1230, Sat 0700-2200, Sun 0700-2100 (closing times vary)
F.PI: Stations in several major cities.

KEDAINIU KRASTO TV (Municipal)
Kedainiu TV, Valdybos Culturos Centras, Radvilu 5, LT-5030 Kedainiai. ☎ +370 (57) 53947. 🗎 +370 (57) 65894. **L.P:** Dir: Vytoldas Burneika.
Stations: Kedainiai ch35 300W, Juragiai chR9 2500kW
D.Prgr: Tue, Thu, Sat 1800-2130 on ch35, Sat 0615-0655 on chR9. Relays TV Polonia (Poland) Wed 1700-2200 on ch35. Also prepares prgrs for TELE-3 which a re broadcast Mon 1630-1700

KLAIPÉDOS TV (Municipal)
Klaipédos municipaliné televizijos studija, Vilties 12, LT-5800 Klaipéda. ☎ +370 (61) 11701. **L.P:** Dir: Vytautas Peckus.
Station: Klaipéda ch R-12 H 0.2kW (ERP).
D.Prgr: own prgrs are broadcast irreg., mainly during late evening hours. Other times relays **Baltijos Televizija** and satellite prgrs.

LNK (LITPOLIINTER TV) (Priv., Comm.)
Laisvas Nepriklausomas Kanalas, Mazosios Lapes, Kauno raj. ☎ +370 (7) 531 269. 🗎 +370 (7) 712 444. **L.P:** Dir: Audrius Girzadas
Stations: Pol=H, exc, where stated.

Location	ch R-	kW	Location	ch R-	kW
Pazagieniai	1	0.15	Ignalina	10	0.098
Vilnius	4V	177	Alytus	11	0.15
Raseiniai	2	1.15	Plunksniai	12	0.15
Neringa	5	0.15	Tauragé	27	600
Mazeikiai	7	0.4	Juragiai	28	800
Druskininkai	8	0.32	Klaipéda	29	700
Location	**ch R-**	**kW**	**Location**	**ch R-**	**kW**
Viesintos	32	382	Visaginas	35	112
Bubiai	34	500			

D.Prgr: Mon-Fri 0455-0800 & 1455-2300, Sat/Sun 0650-2400

MARIJAMPOLÉS TELEVIZIJA (Municipal)
Kauno 13, LT-4520 Mirijampolé. ☎ +370 (43) 51687. 🗎 +370 (43) 50606. **L.P:** Dir: Stasys Baublys.
Station: Marijampolé ch R-6 H 0.1kW.
D.Prgr: Mon-Wed 1600-2100, Thurs 1600-2200, Fri 0700-0900 & 1600-2400, Sat 1200-2300, Sun 0700-2130 (closing times vary). **Rel. SAT.1** and occ. other satellite TV prgrs. Own prgrs, which comprise about 10% of the total programming, are broadcast irreg.

PAN-TV (Priv., Comm.)
Laisves a. 26, LT-5300 Panevezys. ☎ +370 (54) 62121. 🗎 +370 (54) 64980. **L.P:** Dir: Saulius Bukelis.
Station: Pazagieniai ch R-8 H 0.2kW(ERP).
D.Prgr: 24h. Own prgrs. Mon-Fri 15 minutes in the evening (time varies). Other times relays Baltijos TV and (when Baltijos TV is not on the air) Deutsche Welle TV or other satellite TV prgrs.

SIAULIU TV (Priv.)
Ausros al. 48, LT-5400 Siauliai. ☎ +370 (14) 25211. 🗎 +370 (14) 25211. **L.P:** Dir: Stasys Susinskas.
Station: Bubiai ch R-2 H 0.33kW.
D.Prgr: Mon-Sat 1600-2200. Relays Baltijos TV and satellite TV prgrs. Own prgrs are broadcast irregularly.

TAURAGÉS TV-3 (Priv., Comm.)
Vytauto 16, LT-5900 Tauragé. ☎ +370 (46) 54503. **L.P:** Dir: Liudas Urbonas.
Station: Tauragé ch R-3 H 10W(ERP).
D.Prgr: 24h exc. W 0100-0300. Mostly relays satellite prgrs.

TELE-3 (Priv., Comm.)
Kalvariju 35, LT-2000 Vilnius. ☎ +370 (2) 752533. 🗎 +370 (2) 355032. **L.P:** Acting Dir: Gintautas Gaidamavicius.
Station: Pol=H.

Location	ch R-	kW	Location	ch R-	kW

Taurage	1	0.15	Viesintos	26	1.53
Pazagieniai	5	0.15	Naujoji Akmené	27	0.2
Alytus	6	0.19	Utena	29	0.158
Klaipéda	10	0.15	Vilnius	31	800
Juragiai	21	800	Visaginas	33	112
Bubiai	22	500	Neringa	40	0.32
Raseiniai	25	0.2	Mazeikiai	41	0.92

D.Prgr: MF 0530-1000 & 1155-2230, Sat/Sun 0700-2230 (closing times vary). Mostly relays Rossijskoe TV, CNN Int'l, Discovery Channel, MTV, etc. Own prgrs, which comprise 20% of the total programming, are broadcast MF 1630-1900 (regional N. in Lithuanian), Sun 1700-1725 (national N. in English), and sometimes at other times irregularly.

VILNIAUS TV (Priv., Comm.)
Vivulskio 25, LT-2600 Vilnius. ☎ & 🗎 +370 (2) 661560. **L.P:** Dir: Vytautas Kvietkauskas.
Station: Vilnius ch26 1.12kW H.
D.Prgr: Mon-Fri 0515-0800 & 1650-2100, Sat 0500-0930 & 1700-2130, Sun 0500-0930 & 1600-2100 (closing times vary). Own prgrs Mon-Fri 1650-1700. Other times relay TV-6 Moskva from Russia.

5 KANALAS (Priv., Comm.)
Vanagupes 20, LT-5720 Palanga. ☎ +370 (36) 51131. 🗎 +370 (36) 52878. **L.P:** Mrs. Janina Norviliene.
Stations: same as TELE-3, plus Klaipeda chR12 200W H.
D.Prgr: On TELE-3 network: Tue, Thu 2130-2330 (own prgrs start after s/off of TELE-3);. On chR12 in Klaipeda: Mon-Fri 1500-1830, Sat 1500-2200

ORBITA (Priv., Comm)
Biliuno 6, LT-5720 Palanga
Station: Palanga ch R6 H 20kW.
D.Prgr: 24h., mostly relays satellite TV

TV-sets: 100.500 — **Colour:** PAL & SECAM — **Systems:** B, L&G.

CLT Multi Media
45, blvd Pierre Frieden, L-1543 Luxembourg-Kirchberg. ☎ +352 421421. 🗎 +352 42142-2760
Web: http://www.cltmulti.com — **L.P:** MD: Rémy Sautter.
CLT Multi Media operates various terrestrial and satellite-delivered TV services serving Belgium, France, German, Italy, Luxembourg, The Netherlands and Poland. It is also a major shareholder in Channel 5 (UK).
Terrestrial tr's: Dudelange ch7 (System B) 100/25kW H, Dudelange ch21 (System L) 1000/100kW, Dudelange ch27 (System G) 1000/100kW (2 studios), ch24 100kW.
RTL 9 in French: W 1045-1210, 1600-2230 (Sat 1400-2240); Sun 1100-2200 on ch21 (SECAM) for French viewers and ch27 (PAL) for Belgium & Luxembourg.
RTL-4 in Dutch & Multilingual: Luxembourg ch. 41 (30kW).
Club RTL (French for Belgium): ch27.

RTL Télé Lëtzebuerg (local sce. of CLT)
177, rue de Luxembourg, L-8077 Bertrange. ☎ +352 252 72 51. 🗎 +352 252 725431. **Web:** www.rtl.lu
D. Prgr: in Luxembourgish: MF 1700-1900, Sat 1730-1900, Sun 1800-1930.

TV-sets: N/A — **Colour:** PAL — **Systems:** B&H

TELEVIZIJA MAKEDONIJE
Dolno Nerezi bb, 91000 Skopje. ☎ +38 (91) 258 230.
Stations:
TV Makedonije 1st Prgr.

Pelister	4	30	Mali Vlaj	9	10

Golak	5	10
Cnri Vrv	6	100
Stogovo	6	10
Belasica	7	6.6
Boskija	8	10

Vodno	9	10
Popova Sapka	10	10
Turtel	11	50
Vodno	12	10

+58 low power sts.

D.Prgr: Mon-Thu 1110-2200, Fri 1445-2200, Sat 0820-2145, Sun 0830-2100.

TV Skopje 2nd Prgr.

Ohrid	21	4
Turtel	22	1000
Pelister	29	600
Crn Vrv	30	600
Stogovo	31	10
Belasica	35	13

Oteseva	37	10
Golak	38	200
Popova Sapka	38	10
Mali Vlaj	44	200
Boskija	57	20

+56 low power sts.

D.Prgr: Mon-Fri 1355-2230, Sat 0900-2230, Sun 0650-2200.

MALTA

TV-sets: 180.000 — **Colour:** PAL — **System:** B.

PUBLIC BROADCASTING SERVICES LTD.

✉ P.O. Box 70, Msida 01. ☎ +356 225051. 🖹 +356 244601.
L.P: Chmn: Chief Justice Prof. J.J. Cremona. Chief Exec: Albert Marshall. Head of TV: Joe Galea. TV Prgr. Mgr: Sylvana Cristina.
Station: chE10 10/2.5kW H.
D. Prgr: 0700 - midnight.

SUPER ONE TV

(Operated by Maltese Labour Party)
Station: Ch 29 10kW
D.Prgr: 0500-2230.

MOLDOVA

TV-sets: N/A — **Colour:** SECAM — **System:** D & K.

TV MOLDOVA

Location	TV Moldova	Românâ-1	ORT 1
Balti	2*	41*	8**
Cahul	8*	1**	31*
Camenca	10	5	8
Causeni	40*	—	28*
Ciadîr-Lunga	—	—	12
Cimislia	33**	—	2**
Comrat	—	—	10
Dnestrovsk	6	—	10
Edinet	7	27	31*
Hîncesti	—	7	10
Leovo	4	—	39
Rezina	12	25**	27*
Soroca	9	4	28
Stefan-Voda	9	—	—
Straseni	3*	11*	30*
Tighina	—	1	8
Ungheni	32*	7	29*

*) high power — **) medium power — others low power
N.B: Very low power repeaters not mentioned

VTV (priv-comm)

✉ str. Hinncesti 61, Chisinau 277028
L.P: Dir: F.D. Bulan
Station: Chisinau ch23 600W
D.Prgr: 1700-2100 rel TV6 Moscow and own ads.

MONACO

TV-sets: 20.000 — **Colour:** SECAM & PAL — **Systems:** L & G.

TELE MONTE CARLO (Comm.)

✉ 16 Blvd. Princess Charlotte, MC 98090 Monaco-Cedex.

Cable: Tele-Carlo. ☎ +3393 505940. 🖹 +3393 250109. ① 469823 Tele Carlo. **Web:** http://www.cecchigori.com/tv/
L.P: Pres: Jean-Louis Medecin. Dir. Prgrs: George Giaveret.
Stations: chF8 (L/SECAM) 50kW(ERP), ch30(L/SECAM) 500kW, ch33(G/PAL) 0.05kW, ch35(G/PAL) 40kW, ch39(L/SECAM) 0.02kW.
D.Prgr: French 1800-2400 on ch 8, 30, 39. (See also listing of relay st's in France). **Italian:** TMC1 & TMC 2: 1000-2400 on ch 33, 35.

NETHERLANDS

TV-Sets: 6.5m — **Colour:** PAL — **System:** B & G.

NEDERLANDSE OMROEPPROGRAMMA STICHTING (NOS)

✉ Sumatralaan 45, 1217 GP Hilversum, P.O. Box 26600, 1202 JT Hilversum. ☎ +31 (35) 6779222. 🖹 +31 (35) 6772649 ① 43287.
L.P: Dir. Radio and TV: Ruurd Bierman; Hd. Comm: F. de Vries
Board Members: drs. A. Grewel; mr A. Herstel (NCRV); ds. A. van der Veer (EO); S.H. Piersma (VPRO); K. van Doodewaard (TROS); mr H.J.E. Bruins Slot (NPS); G.H. Veringa (SEO); dr. D.Th. Kuiper; prof mr A.J.C.M. Geers; mr B. Staal.

NEDERLANDSE PROGRAMMA STICHTING (NPS)

✉ P.O. Box 29000, 1202 MA Hilversum. ☎ +31 (35) 6779333. 🖹 +31 (35) 6774517
L.P: Dir: W.J.M. van Beusekom

NB: The Dutch prgrs. are provided by the NOS, NPS and seven broadcasting organizations:
-Algemene Omroepvereniging AVRO, 's Gravelandseweg 52, 1217 ET Hilversum, Postbus 2 1200 JA Hilversum. ☎ +31 (35) 6717911 🖹 +31 (35) 6717439
-Vereniging Evangelische Omroep EO, Oude Amersfoortseweg 79a, 1213 AC Hilversum, Postbus 21000, 1202 BB Hilversum. ☎ +31 (35) 6474747. 🖹 +31 (35) 6474727
-Katholieke Radio Omroep KRO, Emmastraat 52, 1213 AL Hilversum, Postbus 23000, 1202 EA Hilversum. ☎ +31 (35) 6713911 🖹 +31 (35) 6237345
-Nederlandse Christelijke Radio Vereniging NCRV, Bergweg 30, 1217 SC Hilversum, Postbus 25000, 1202 HB Hilversum. ☎ +31 (35) 6719911 🖹 +31 (35) 6719285
-TROS, Lage Naarderweg 45-47, 1217 GN Hilversum, Postbus 28450, 1202 LL Hilversum. ☎ +31 (35) 6715715 🖹 +31 (35) 6715236
-Omroepvereniging VARA, Heuvellaan 50, 1217 JN Hilversum, Postbus 175, 1200 AD Hilversum. ☎ +31 (35) 6711911 🖹 +31 (35) 6711333
-Omroepvereniging VPRO, 's Gravelandseweg 63-73, 1217 EH Hilversum, Postbus 11, 1200 JC Hilversum. ☎ +31 (35) 6712911 🖹 +31 (35) 6712254

Stations:

Loc.	Ned. 1	Ned. 2	Ned. 3	kW
Arnhem	50	53	43	30
Eys	51	54	48	1
Goes	29	32	35	250
Hulsberg	57	60	43	0.1
Lopik	4	27	30	100/2 x 1000
Losser	—	26	34	3
Maastricht	53	56	59	1
Markelo	7	54	51	30/2 x 300
Noorbeek	46	49	52	0.01
Roermond	5	31	34	50/2 x 250
Slenaken	29	35	32	0.02
Smilde	6	47	44	40/2 x 1000
St.Pietersberg	26	33	23	0.25
Wieringermeer	39	45	42	300
Wijk aan Zee	33	49	21	0.004

+ 5 low power repeaters.
D.Prgr: Ned 1: AVRO/KRO/NCRV; Ned 2: EO/NOS/TROS; Ned 3: NPS/VARA/VPRO.

NORWAY

TV-sets: 2.000.000 — **Colour:** PAL — **Systems:** B & G.

NORSK RIKSKRINGKASTING

⌧ N-0340 Oslo 3. ☎ +47 (22) 459050. 🖷 +47 (22) 457440. ☉
International Relations 76820. Eurovision: 71794. TV News: 18530.
Prgr. Purchases: 19676. **Web:** http://www.nrk.no
L.P: Dir. Gen (Radio & TV): Einar Førde. Dep. TV Dir: Anne Torjusson
Diesen. PD: Kent Nilssen. Tech Dir: Geir Sundal. Dir of Inf. and PR:
Hanne Løchstøer. Head. of Int. Rel: Kjell Lokvam.

TV-2 (Priv. Comm.)

⌧ Postboks 2, 5002 Bergen. ☎ +47 (55) 908070. 🖷 +47 (55) 908090.
L.P: MD: Arne Jensen. Dir. of Prgrs: Finn H. Andreassen.

Stations: all sts. NICAM stereo.

Location	NRK	TV-2	kW
Alta	7*	—	5
Andenes	6*/35*	49	4/14/15.5
Arendal	—	25	1.95
Bagn	3	—	55
Ballstad	—	26	5.5
Bangsberget	35	29	3.3/10.5
Bergen	9**	12	80
Biri	—	45	1.1
Bjerkreim	6+	27	15/42
Bokn	8+	44	100/270
Bransøy	—	30	1.1
Bremanger	4**	—	80
Dokka	—	26	1.05
Fannrem	—	25	1.45
Fennefossknipa	—	28	1.423
Fister	11+	—	1.5
Flisa	32	22	7/13.5
Frekhaug	—	23	1.1
Fyresdal	—	26	1.323
Førde	5**	—	1.2
Gamlemsveten	3	24	60/49
Gausta	8	—	45
Gol	—	26	5.445
Gran	—	52	6
Greipstad	2	—	60
Grisvågøy	—	25	1.07
Grong	5	24	100/9.45
Location	NRK	TV-2	kW
Grøtevær	42*	—	1.6
Gulen	2**	—	48
Hadsel	4*	—	60
Halden	11	32	100/600
Hammerfest	9*	—	90
Harstad	—	47	4.8
Hasvik	—	26	1.25
Hemnes	3*	39	70/1
Hermansverk	—	27	1.25
Hestmannen	6*	54	3.3/1.115
Hol	—	28	1.296
Hommelfjell	—	39	14.6
Hovdefjell	7	—	62.5
Hvitingen	—	47	4.8
Ibestad	—	52	1.5
Jetta	8	—	95
Kappfjell	8*	—	2
Karasjok	5*	—	10
Kautokeino	3*	—	6
Kistefjell	8*	23	87.5/3.87
Kongsberg	4	43/28	100/175/1.1
Kongsvinger	9	28	60/250
Kopparen	11	23	45/60
Kristiansund	—	40	2.35
Kveøy	—	31	1.8
Kvisvik	—	32	2.15
Lifjell	52	56	1/1.3
Lifjell, Stavanger	—	37	24
Lillehammer	—	53	4.8
Lyngen	5*	—	3
Lyngdal	9	33	80/1.3
Lønahorgi	11**	—	25
Melhus	2	30	100/10.5
Mistberget	44	54	1.15/5.752
Mofjellet	—	37	1.4
Mosvik	7	37	34/360
Narvik	10*	24	130/100
Nittedal	—	25	1.3
Nordbykollen	—	22	3.5

Location	NRK	TV-2	kW
Nordfjordeid	10**	27	3/1.45
Norhue	5	27	40/28
Nordkapp	6*	—	70
Norefjell	—	28	3.3
Oppstad	—	42	2.38
Oslo	6	12	100/90
Raufoss	—	33	1.519
Reinsfjell	6	29	85/28.4
Ringerike	41	23	2/17.447
Rubbestadfjell	—	29	6.667
Salten	7*	30	55/44.55
Skien	10	24	110/270
Skotterud	—	33	1.288
Snertingdal	37	46	1.05/1.1
Sogndal	7**	—	20
Sokna	—	26	1.512
Sollihøgda	—	39	1.676
Steigen	2*	—	80
Stord	5**	47	60/188.17
Spåkenes	52*	45	3.3/2.4
Stadlandet	—	47	2.6
Storberget	—	25	1.488
Store Jækkir	11*	22	4/1.2
Storhaugen, Målselv	—	45	11.756
Stryn	—	23	5.356
Sundalsøra	—	47	1.65
Tana	8*	—	60
Torsvarde	—	25	1.32
Toåsen	—	52	1.07
Tresfjord	—	40	3.736
Trolltind	11*	—	110
Tromsø	—	33	2.566
Tron	9	23	67.5/5.77
Trysilfjell	40	54	6.9/13.9
Ullandhaug	35+	—	1.1
Varanger	2*	—	90
Vardheia	—	53	1.088
Vega	9*	22	50/105
Veggen	—	41	1.75
Vestre Slidre	—	25	1.25
Location	NRK	TV-2	kW
Viktjernåsen	—	48	1.2
Vinstra	—	24	1.2
Volda	—	46	2.8
Øksnes	—	25	2.9
Øvre Eiker	—	37	1.95
Åkersten	—	40	3.2
Ål	—	52	1.819
Ålesund	—	45	4.7
Ålmenberget	—	24	1.3
Åndalsnes	—	22	1.45
Ånsmarka	—	49	1.396

*) carriers regional prgrs from NRK TV Nord
**) carries NRK Vestlandsrevyen
+) carries NRK Rogeland
D.Prgr: NRK: basically 1600-2300, school programming 0800-1600.
Regional TV: Vestlandsrevyen: Mon, Tue, Thu, Fri 1745-1800, Wed.
1730-1800. TV Rogeland: Wed 1730-1800. TV Nord: (see
Vestlandsrevyen). **TV-2:** 1300-2300

PHILIPS PETROLEUM 66

TV1: Ekofisk ch55. V. relays NRK.
TV2: Ekofisk ch52. V. relays Film Net.

POLAND

TV-sets: 12m — **Colour:** PAL — **Systems:** D & K.

TELEWIZJA POLSKA S.A. (Gov.)

⌧ Ul. Woronicza 17, 00-999 Warszawa. ☎ +48 (22) 433361/445432.
🖷 +48 (22) 447419/435779. **Cable:** Telewar. ☉ 825331. **Web:**
http://www.tvp.com.pl
L.P: Dir TVP1:vacant. Dir TVP2: Maciej DomanskTech. Dir: Rajmund
Gruszka. Dir. Int. Rel: Jerzy Romanski.
Power = ERP Pol H exc. where indicated.

Prgr I	chR	kW/Pol		chR	kW/Pol
Bydgoszcz	1	120	Szczecin	12	100
Pila	2	50	Wroclaw	12	100

Prgr I	chR	kW/Pol		chR	kW/Pol
Warszawa	2	60	Konin	22	1
Kielce	3	100	Przemysl	24	100
Zielona Góra	3	200	Warszawa	27	20
Suwalki	5	100	Kalisz	28	1
Lódź	7	80	Plock	29	1000
Bialystok	8	100	Jelenia Góra	30	200
Katowice	8	200	Rabka	31	2
Koszalin	8	60V	?	32	?
Lublin	9	80	Poznan	33	1
Poznań	9	150	Bydgoszcz	36	650
Olsztyn	9	100V	Skierniewice	37	1
Gdańsk	10	100	Opole	40	700
Kraków	10	200	Bydgoszcz	41	650
Zamość	10	50V	Siedlce	52	300
Luban	11	1	Klodzko	52	100
Rzeszów	12	100V	Lezajsk	58	1

+124 low power st's

Prgr II	chR	kW/Pol		chR	kW/Pol
Wroclaw	2	1	Kielce	28	1000
Lublin	2	0.4V	Zielona Gora	29	1000
Gdynia	3	2	Zakopane	34	30
Kielce	5	2	Lobez	35	40
Rzeszów	7	0.5V	Zamosc	36	300
Lódź	10	1V	Luban	37	1
Warszawa	11	50	Klodzko	38	300
Bialystok	11	1V	Lomza	38	7
Gizycko	11	1	Czluchów	39	13
Koszalin	11	0.5V	Plock	39	1000
Bydgoscz	12	0.6V	Wroclaw	25	1000
Katowice	21	450	Olsztyn	26	270
Elblag	21	20	Poznań	27	250
Tarnów	22	20	Kielce	28	1000
Lublin	23	500	Zielona Góra	29	1000
Opole	23	600	Rzeszów	29	700
Pila	24	150	Szczecin	30	600
Lebork	25	40	Kalisz	31	1
Wroclaw	25	1000	Krakow	33	300
Lezajsk	26	1	Swinoujscie	33	10
Olsztyn	26	130	Konin	34	2
Poznan	27	300	Jelenia Góra	35	200
Lobez	35	25	Plock	39	1000
Bydgoszcz	36	650	Czluchow	39	30
Suwalki	36	1000	Szczawnica	39	10
Rabka	36	10	Slupsk	40	15
Zamość	36	250	Przemysl	41	100
Gdansk	37	600	Wisla	41	30
Siedlce	37	300	Skierniewice	57	10

+ 87 low power st's.

Prgr. III (local on Progr. II transmitters): Katowice ch 21. **D. Prgr:** W 0800-1000, 1500-1530.
Satellite TV Relay: Katowice ch R6 20kW — **Echo (priv):** Wroclaw ch 28 1kW — **RAI Uno Relay:** Krakow ch 50 50kW.

TELE-9
Kielce ch R5, Tarnow ch R22, Kielce ch R28, Rzeszow ch R29, Kraków ch R33, Przemy'sl ch R41.

TV Polsat
Marchakovska 83, P-00517, Warsaw 84. ☎ +48-22-295684.
Stations:

Loc.	ch/Pol	Loc.	ch/Pol
Biala Potlaska	32H	Olsztyn	60H
Chelm	21H	Ostroleka	21H
Choragewica	53H	Palac Kultury	35H
Chrzelice	57H	Pila	57H
Chiechanow	52H	Piotrków Trybun.	34H
Czestochawa	34H	Plock	21H
Elblag	23H	Poznan-Piatkowo	50H
Gorzow Wielk.	26H	Przemysl	56H
Góra Siéza	59H	Radom	53H
Góra Skrzyczne	58H	Rzeszów	48H
Jezów Sudecki	57H	Siedlce	57H
Kalisz	56H	Sieradz	57H
Katowice-Bytków	47H	Skierniewice	24H
Kielce	22H	Suwalki	41H
Klodzko	R8H	Tarnobrzeg	41H
Kolowo	48H	Tarnów	60H
Konin	58H	Trebice	53H
Koszalin	60H	Urzud Miasta	60H
Krosno	60H	Walbrzych	49H
Lebork	57H	Waly Pistwskie	30H
Leszno	57H	Wloclawek	22H
Lomza	57H	Zakopane	51?
Lódz	49H	Zamosc	53H
Lublin	35H	Zielona Góra	R10H
Okskywie	57H		

PORTUGAL

TV-sets: 3m — **Colour:** PAL — **System:** B&G.

RADIOTELEVISÃO PORTUGUESA (RTP) (Comm.)
Head Office: Av. 5 de Outubro 197, 1094 Lisboa. ☎ +351 (1) 7931 774. **Cable:** Televisão. ① 14527-RTP RE P. 🖷 +351 (1) 7931 758.
Mailing ✉ Apartado 2934, Lisboa.
L.P: SM: José Eduardo Moniz. TD: Ismael Augusto. Film Buyer: José Eduardo Moniz.
Pol H unless stated otherwise.

RTP Channel I

Stations	Ch	kW	Stations	Ch	kW
Muro	2	10/67V	Valença	7	1/7
Lousa	3	12/60	Foia	8	2/20
Marofa	5	2/16	Leiranco	8	2/4
Mendro	5	5/30	Gardunha	8	1/3
Montejunto	6	5/22	Porto	9	10/100
S. Miguel	6	2/20	Palmela	22	2/115
Marao	6	5/40	Mosteiro	24	1/9
Bornes	7	5/12	S. Macario	47	2/67V
Lisboa	7	10/100			

+42 repeaters.
D.Prgr: Mon-Fri 0900-0030, Sat 0900-0230, Sun 0900-0030.

RTP Channel II

Stations	Ch	kW	Stations	Ch	kW
Mosteiro	21	1/9	Lousa	26	20/540
Lisboa	25	10/405	Mendro	27	20/560
Bornes	25	10/200	Muro	27	10/500
S. Miguel	31	10/250	Montejunto	46	10/200
Palmela	32	2/128	Valença	46	2/70
Leiranco	34	5/40	Foia	47	10/550
Gardunha	34	2/18	Marofa	48	10/300
Marao	35	10/300	S. Macario	50	2/67V
Porto	41	10/100			

+20 repeaters.
D.Prgr: Mon-Fri 1200-0050/0130, Sat 0900-0130, Sun 0900-0100.

PRIVATE STATIONS

SIC-Sociedade Independente de Comunicacao, SA
✉ Estrada da Outorela 119, Carnaxide 2795 Linda-a-vehla, Lisboa. ☎ +351 (1) 417 3138. 🖷 +351 (1) 418 7156.
L.P: Chmn: Francisco Pinto Balsemao. Dir. of Prgrs: Dr. Emilio Rangel. Dir. of Mktg: Dr. Hugo Correia Pires. Tech. Dir: Eng. triga de Sousa.

TVI, Televisao Independente, SA
✉ Rua 3, Matinha, Edifício Altejo, 6°, 1900 Lisboa. ☎ +351 (1) 4347500. 🖷 +351 (1) 4355075
E-mail: tvi-sa@individual.eunet.pt **Web:** http://www.tvi.pt
L.P: Chmn: Roberto Carmeiro. PD: Diogo Gaspar Ferreira. Tech. Dir: Joao Penha Lopes.

ROMANIA

TV-sets: 7.277.000 — **Colour:** PAL — **System:** D & G.

RADIOTELEVIZIUNEA ROMÂNÂ (Gov.)
✉ 191 Calea Dorobanti, Bucharest. ☎ +40 (1) 212 0290. 🖷 +40 (1) 312 0381. ① 10182 TVR. **Cable:** Radioteleviziunea.
L.P: DG: Dumitru Popa. PD: Mamase Radnev.
Stations: Pol. H (except where indicated)
1st. National Program

Location	ch.	kW	Location	ch.	kW
Balota-Turnu S.	2	15	Rarau-Cimpulung	8	15H
Baneasa-Dobrogea	3	15	Cerbu-Novaci	8	150
Bistrita-Heniu	3	15	Iasi	9	25

Oradea	3	120	Magura-Odobesti	9	10V
Semenic-Resita	3	15	Timisoara	9	50
Bucaresti	4	50	Brasov	10	10
Suceava	4	100	Topolog-Constanta	10	50
Harghita-Gheorghieni	5	50	Mogosa-Bia Mare	10	15
Birlad	5	50	Bacau	10	10
Bucegi-Ploiesti	6	150	Paring-Petrosani	10	10
Mahmudia-Delta	6	50V	Mangalia	11	1
Zalau	6	15	Feleac-Cluj	11	25
Piatra Neamt	6	5	Botosani-Hirlau	11	15
Vacareni-Galati	7	15V	Cozia-Rimnicu V.	12	10
Varatec	7	50	Tulcea	12	2
Sibiu-Paltinis	7	50	Tirgu Mures	12	10
Vaslui	7	5.0	Comanesti	12	20
Litoral-Constanta	8	10V	Magura Boiului-Deva	12	20
Bihor-Vascau	8	2.5V	Siria-Arad	12	20
+ 400 lps.					

2nd National Prgram

Location	ch.	kW	Location	ch.	kW
Bucaresti	2	150	Timisoara	21	1000
Nucet II	5	0.05	Craiova	21	1
Iasi	6	1.3	Vaslui	28	0.1
Cluj	8	1.3	Brasov	34	50
Pitesti	9	0.5	Resita	36	1.5
Piatra Neamt	11	0.1	Constanta	42	0.05

D.Prgr: Prgr 1: Mon-Fri 0800-2200, Sat/Sun 0700-2300; Prgr 2: 1500-2200

PRIVATE STATIONS

More than 80 local st's are operating incl. 8 st's in Bucharest are as follows:

Antena 1, Poesti 25-27, 70000 Bucharest. ☎ +40 (1) 212 0619. 🖃 +40 (1) 212 0188 — DG: F. Bratescu.

Canal 31, 155 Piata Victorei Di, 7th Floor, 70411 Bucharest. ☎ +40 (1) 210 6628 — DG: Adrian Sirlon.

Canalul de Stiri, Calea Victorei 133-5, s 1, Bucharest. ☎ +40 (1) 312 4348. 🖃 +40 (1) 312 0349.

Pro TV, Bd Carol 1 109, etaj s2, Bucharest. ☎ +40 (1) 312 4218. 🖃 +40 (1) 312 4228.

Tele 7 ABC, Christo Botev 8, s3, Bucharest. ☎ +40 (1) 312 1695. 🖃 +40 (1) 611 6576 — DG: Mihai Tatulia.

Tele America, Blvd. Armata Poporului 1-3, s6, Bucharest. ☎ +40 (1) 311 0419. 🖃 +40 (1) 311 0417.

Tele Europa Nova, Dr. Lister 6, s5, Bucharest. ☎ +40 (1) 623 6661. 🖃 40 (1) 312 1324.

TV Sigma, Armata Poporului Blvd. 1, complex LeuFacultatea de Electronic Crp A, Et 8, Bucharest. ☎ +40 (1) 631 4734 — MD: Constantin Crbu.

RUSSIA

TV-sets: 50m — **Colour:** SECAM — **Systems:** D & K.

NB: Due to incomplete information, details of channels used by the various broadcasters cannot be given. These details will be added in future editions as information become available.

ORT (formerly Ostankino) (Public sce.)
🖃 12 Academika Korolyova Str, Moscow 127000. ☎ +7 (095) 217 7898. 🖃 +7 (095) 288 9542 — **L.P:** DG: Sergei Blagovolin.

MOSCOW TV
🖃 12 Akademika Korolyova Str, Moscow 127000. ☎ +7 (095) 217 5158. 🖃 +7 (095) 216 5401.
L.P: Gen. Dir: Algar Misan. MD: Natalya Smirnova.

TV Channel 2x2
🖃 12 Akademika Korolyova Str, Moscow 127000. ☎ +7 (095) 217 7094. 🖃 +7 (095) 215 2063.
L.P: GM: Vladimir Troepolski. PD: Victor K. Litenko.

Telexpress - Channel 31
🖃 15 Akademika Koroleva Str, Moscow 127000 ☎ +7 (095) 282 4260. 🖃 +7 (095) 276 8892.

NTV
🖃 Novy Arbat Str 36, Moscow 121102. ☎ +7 (095) 290 7077.

🖃 +7 (095) 290 9757.

MOSCOW TV 6
🖃 Ilyinka 15, Moscow. ☎ +7 (095) 206 8423. 🖃 +7 (095) 206 0886 —
L.P: Exec. Mgr: Tatiana Voronovich.

Independent Broadcasting System (MVS)
🖃 Suite 303, 8A, Suvorovsky Blvd, Moscow 121019. ☎ +7 (095) 291 1787. 🖃 +7 (095) 291 2174.

RTR Network
🖃 5 Yamskogo Polya 19/21, Moscow 125124. ☎ +7 (095) 250 0511. 🖃 +7 095 250 0105.
L.P: Chmn: Oleg Poptsov. Dep. Chmn: Segei Skvortsov. DG: Anatoly Lysenko. Int. Dir: Sergei Erofeev. Tech. Dir: Stanislav Bunevic.

Russian University Channel
🖃 5 Yamskogo Polya 19/21, Moscow 125124. ☎ +7 (095) 213 1754. 🖃 +7 (095) 213 1436.

Regional Television
🖃 Akademika Pavlova Str 3, St Petersburg 197022. ☎ +7 (812) 238 6073. 🖃 +7 (812) 238 5807.

6TR Petersburg - Channel 5
🖃 6 Chapygina Str, St Petersburg 197022. ☎ +7 (812) 234 3763. 🖃 +7 (812) 234 1416 — **L.P:** Chmn: Oleg Rudny.

Russkoye Video - Channel 11
🖃 Malaya Nevka 4, St Petersburg 197022. ☎ +7 (812) 234 4207. 🖃 +7 (812) 234 0088.

As Baikal TV
🖃 6 Shelehov, Shelehov 666020. ☎ +7 (395) 109 3303. 🖃 +7 (395) 103 1917.

Kaliningrad State TV
🖃 19 Klinicheskaja Str, Kaliningrad 236016. ☎ +7 (0112) 452 700. 🖃 +7 (0112) 452 233

State National TV and Radio Company of Sakha Republic
🖃 48 Ordjonikidze Str, Yakutsk 677892. ☎ +7 (411) 225 3169. 🖃 +7 (411) 225 3176.

TV Comp Fourth Channel
🖃 P.O. Box 751, Yekaterinburg 620069. ☎ +7 (3432) 232 041. 🖃 +7 (3432) 236 033.

SAN MARINO

Colour: PAL — **Systems:** B&G

SAN MARINO RTV spa
🖃 V. le J. F. Kennedy, 13 - Città. ☎ +39 549 882000.
L.P: DG: Raviele Gianni.

SLOVAKIA

TV-sets: 1.800.000 — **Colour:** PAL System B; SECAM — **Systems** D&K

SLOVAK TELEVISION
🖃 Mlynska Dolina 28, Bratislava 845 45. ☎ +42 (7) 727 448. 🖃 +42 (7) 729 440.
L.P: DG: Jozef Darmo. PD: Milan Polak. Int. Rel: Mikulas Gavala.

ST1: (PAL)

Stations	ch	kW	Pol.		ch	kW	Pol.
Bratislava	2	150	H	Stúrovo	9	1.5	V
Námestovo	4	0.8	H	Trencin	10	1.6	V
Bardejov	4	1.2	H	Zilina	11	100	V
Poprad	5	80	V	M. Kamen	12	15/6	V
Kosice	6	100	V	Bratislava	31	2	H
B. Bystrica	7	100	H	B. Stiavnica	40	9	H
Ruzemberok	9	1	H	Zilina	41	10	H

Sucise   9   1.6   H
D.Prgr: Mon-Fri 0900-1200, 1600-2330, Sat 0830-0045, Sun 0825-2400

ST2 (SECAM)

Stations	ch	kW	Pol.		ch	kW	Pol.
Modrý Kamen	21	100	H	Poprad	30	600	H
Kosice (city)	21	2	H	Sturovo	31	100	H
Zilina	22	100	H	B. Brystica	32	600	H
Roznava	22	100	H	S. N. Sazavou	32	100	H
Trencin	23	300	H	Lucenec	33	100	H
Snina	23	15	H	Olomouc	33	100	H
Kosice	25	600	H	Novi Jisin	34	100	H
Mikulov	26	300	H	Zilina	35	1000	H
B. Stiavnica	26	300	H	Susice	35	100	H
Ruzomberok	27	18	H	Frýdec-Mistec	37	300	H
Bratislava	27	1000	H	B. Mikulás	37	100	H
Stara Lubovna	27	100	H	C. Budejovice	39	600	H
Námestovo	29	100	H	N. Mesto	39	600	H

D.Prgr: Mon-Fri 1530-2330, Sat/Sun 1000-1230, 1530-2330
F.PI: ST2 will be privatized during 1997. 9 bids have been received.

Markyza (Comm.)

✉ Palisady 39, Bratislava 81106. ☎ +42 (7) 531 6610. 🖷 +42 (7) 531 4061 — **L.P:** CEO: Pavol Rusko.

TV-sets: 600.000 — **Colour:** PAL — **Systems:** B & H

TELEVIZIJA SLOVENIJA

✉ Kolodvorska 24, 61000 Ljubljana. ☎ +38 (61) 131 1333. 🖷 +38 (61) 131 9171. ① 32283. **Web:** http://www.rtvs.si
L.P: DG: Zarko Petan. PD (TV): Janez Lombergar. Int. Rel: Boris Bergant.
Stations: System B=ch2-12, System H=ch21-68. **ChE.** audio powers 1/10 of vision powers indicated.

Main stations	Ch. 1	kW	Ch.2	kW	Pol
Beli Kriz	33	20	46	20	H
Krim	31	10	44	10	H

Main stations	Ch. 1	kW	Ch.2	kW	Pol
Krvavec	5	50	21	27	H
Kuk	40	12.5	26	12.5	V
Kum	3	50			H
	38	20	32	20	H
Nanos	6	35	400	41	H
Plesivec	6	35	49	400	H
Pohorje	11	30	56	1000	H
Trdinov Vrh	35	25	48	25	H
Tinjan	38	10	49	100	H
Tsrtelj			60	10	H

+194 low power sts (ch. 1), 180 low power st's (ch. 2)
Prgr. 1: Mon-Fri 0750-2215 (Thu 2300, Fri 2330), Sat 0700-1115, 1330-2345, Sun 0725-2230.
Prgr. 2: Mon-Fri 1430-2300, Sat 1530-0030, Sun 0900-2300.

TV Koper-Capodistria

Location	ch	kW	Pol
Nanos	27	400	H
Beli Kriz	58	20	H

+30 low power sts.
D.Prgr: Mon-Fri 1245-2400, Sat 0900-2400, Sun 0930-2400.

PRIVATE STATIONS

KANAL A

✉ Tivolska 50 pp 44, 61101 Ljubljana. ☎ +386 (61) 133 4133. 🖷 +386 (61) 133 4222 — **L.P:** GM: Frank Dovecar. Mktg. Dir: Vlasta Bostjancic. Tech. Dir: Mladen Uhlik.

MMTV

✉ Zorgova 70, 61231 Ljubljana. ☎ +386 (61) 161 2525. 🖷 +386 (61) 374 554.
L.P: Pres: Marjan Meglic. PD: Andrej Meglic. CE: Tomislav Kalan.

PRO PLUS/POP TV

✉ Kranjceva 26, 61113 Ljubljana. ☎ +386 (61) 189 3200. 🖷 +386 (61) 189 3204 — **L.P:** Gen. Dir: Marjan Jurenec.

TV-sets: 17.000.000 — **Colour:** PAL — **Systems:** B&G.

RADIOTELEVISION ESPAÑOLA (RTVE) (Gov.)

✉ Prado del Rey, 280023 Madrid. Torrespaña C/ O'Donnell 77, 28007 Madrid. ☎ +34 (1) 346 8754. 🖷 +34 (1) 581 7125. ① 22053. **Web:** http://www.rtve.es/rtve/
L.P: Dir. Gen: Jordi Garcia Candau. Ex.Dir.TVE, S.A: Ramón Colom Esmatges. Dir. of Int. Coop: Alfonso Callego. TD: Francisco Baquedano. Film Buyer: Fernando Moreno.
Stations: (System B&G): Powers = ERP. Audio powers 10% of vision powers indicated.
N.B: TVE's 1st. and 2nd. prgr. are also relayed via satellite. The 1st. prgr is relayed under the name TVE Internacional. All Pol H. unless otherwise indicated.

TVE 1st Prgr.

Station	ch	(kW)		ch	(kW)
Pechina	41	100	Lierganes	40	100
Jerez de la	8	6	Villadiego	8	13
Frontera	26	32	Villadiego	47	40
Jerez de la			Castropodame	7	22V
Frontera			Matadeon	8	12
Cordoba	11	10	El Cabaco	9	63
Parapanda	9	24	Soria	39	16
Sierra de Lujar	7	150	Zamora	31	160
Huelva	39	250	Chinchilla	11	40
Sierra Almaden	5	40	La Mancha	9	60
Mijas	65	200	Isleta	6	4
Guadalcanal	4	120	Montaña Mina	32	6
Valencina	63	79	Pozo de las	10	10
Arguis	64	36	Nieves		
Camarena de la	9	53V	Fuencaliente	7	10
Sierra			Izaña	3	200
Inoges	5	24	Tibidabo	4	150
La Muela	3	35	Gerona	49	126
Gamoniteiro	3	50	Alpicat	5	15
Alfabia	6	50	Soriguera	39	40
Alfabia	54	119	La Musara	8	22

Station	ch	(kW)		ch	(kW)
Monte Caro	10	60	Domayo	10	10
Aitana	3	60	Madrid	4	15
Benicasim	7	10V	Navacerrada	2	250
Torrente	5	20	Navacerrada	34	604
Montanchez	11	45	Torrespaña	49	117
Ares	5	50	Murcia	59	160
Santiago de	4	112	Monreal	29	158
Compostela	Sollube		4	60	
Paramo	9	9	Jaizquibel	54	158
Parada del Sil	5	40V	Zaldiaran	45	15

TVE 2nd Prgr.

Station	ch	(kW)		ch	(kW)
Pechina	47	100	Zamora	37	160
Jerez de la	23	32	Chinchilla	43	158
Frontera			La Mancha	58	200
Cordoba	21	200	Isleta	28	20
Parapanda	23	160	Montana Mina	35	6
Sierra de Lujar	57	250	Pozo de las Nieves	59	48
Huelva	45	250	Temejereque	52	25
Sierra Almaden	39	260	Fuencaliente	27	148
Mijas	59	200	Izaña	45	87
Guadalcanal	40	32	Tibidabo	41	28
Valencina	52	79	Tibidabo	31	144
Alpicat	55	112	Gerona	55	126
Arguis	58	36	Alpicat	49	160
Camarena de la	41	200	Soriguera	45	40
Sierra			La Musara	57	300
Monte Caro	43	60	Monte Caro	26	225
Inoges	57	200	Aitana	32	200
La Muela	46	158	Benicasim	59	200
La Muela	33	158	Monte Caro	23	232
Gamoniteiro	39	158	Torrente	22	160
Alfabia	48	119	Montanchez	23	158
Lierganes	46	100	Ares	22	200
Villadiego	41	200	Santiago de	45	316
Castropodame	21	200	Compostela		
Matadeon	33	28	Santiago de	2	40
El Cabaco	39	315	Compostela		
Soria	45	16	Paramo	57	158

Parada del Sil	47	400	Murcia	65	160
Domayo	39	141	Monreal	23	158
Madrid	21	100	Archanda	22	50
Navacerrada	24	870	Jaizquibel	48	158
Navacerrada	27	604	Zaldiaran	39	15
Torrespaña	55	117			

D.Prgr: TVE 1 0800-0230 (24 H. at weekends), TVE 2 0800-0200.

TV Networks in autonomous areas
TELEVISIO DE CATALUNYA (Aut)

Televisió de Catalunya, S.A. (TV3 and Canal 33) Jacint Verdaguer, s/n 08970- Sant Joan Despí, Catalunya. ☎ +34 (9) 3 499 9333. 🖹 +34 (9) 3 473 1964 ① 53280 TVDC E. **L.P:** Dir: Jaume Ferrús i Estopa. T.D: Pere Vila. Film Buyer: Jaume Santacana.

Location	TV3	C33	Power
Alt Camp			
Pont d'Armentera	34	42	0.6W
Alt Empordà			
Cadaqués	42	40	5W
Maçenet de Cabrenys	62	43	200W
Portbou	41	47	20W
Alt Urgell			
Arsequel	22	26	20W
Coll de Nargó	38	40	1W
Oliana	34	27	5W
Organyà	47	51	200W
Os de Civis	30	34	1W
Valls de Valira	22	40	20W
Vilanova de Banat	21	23	0.6W
Alte Ribagorça			
Campament de Tor	23	27	1W
El Pont de Suert	31	51	100W
Senet	52	49	1W
Taüll	45	37	1W
Vall de Boí	52	55	200W
Vilaller	22	26	5W
Anoia			
Calaf	59	63	20W
Castellfollit de Riubregós	33	43	1W

Location	TV3	C33	Power
Anoia			
Igualada	62	60	20W
Pobla de Clar/Capellades	37	33	5W
Bages			
Boixadors	36	38	20W
Cardona	62	64	5W
Ministrol	65	61	5W
Ministrol de Calders	50	56	1W
Montserrat	28	40	200W
Mura	43	35	0.6W
Sallent	57	38	5W
Baix Camp			
Escornalbou	33	30	5W
Pratdip	52	31	0.6
Riudecols	22	26	1W
Vandellós	33	22	5W
Centre Emissor la Mussara	63	25	10/5kW
Baix Erbre			
El Perelló	45	47	5W
Centre Emissor Montcare	29	36	5/1kW
Baix Empordà			
Calonge	61	63	100W
L'Estartit	60	26	20W
Palafrugell	23	27	20W
Sant Feliu de Guixols	24	28	20W
Baix Llobregat			
Baix Llobregat	49	21	20W
Begues	22	25	1W
Molins de Rei	64	60	20W
La Palma	24V	36V	1W
Vallirana	51	54	5W
Baix Penedès			
El Vendrell	17	21	100W
Barcelonès			
Centre Emissor Collserola	44	23	20/10kW
Berguedà			
Guardiola I	60	62	100W
Guardiola II	24	28	1W
Puig-Reig	36	57	20W
Saldes	24	28	1W
La Cerdanya			

	TV3	C33	Power
Aràncer	43	47	1W
Alp	51	—	0.6W
La Cerdanya	62	41	200W
La Molina	24	28	1W
Martinet	44	50	1W
Prullans	51	54	1W
Conca del Barberà			
Montblanc	43	23	100W
S. Coloma Queralt	36V	38V	5W
Garraf			
Garraf	52	56	5W
S. Pere de Ribes	42	32	100W
Garrotxa			
La Vall de Bianya	61	65	—
Les Lloses	52	45	—
Olot	21	40	40W
Montagut	36	43	20W
Gironès			
Sarrià de Ter	40	57	20W
Centre Emissor Rocacorba	52	45	10/5kW
Maresme			
Tiana	57	61	—
Argentona	25	37	1W
Cabrils/Mataró	54	52	100W
Calella	53	59	100W
Calella/Blanes	29	37	10W
Canet	52	45	1W
Llavaneres	29	37	1W
Teià/Alella	29	37	5W
Montsià			
Alcanar	27	25	5W
S. Carles de la Ràpita	24	30	20W
Ulldecona	42	32	100W
La Noguera			
Ager	35	54	20W
Artesa de Segre	46	34	20W
Camarasa	33	25	1W
Ponts	35	25	5W
Vila Nova de Meià	51	33	5W

Location	TV3	C33	Power
Osona			
Bellmunt d'Osona	46	43	100W
Collsuspina	26	60	100W
Olost	52	45	1W
Viladrau	40	44	1W
Pollars Jusà			
Boixols	60	63	1W
Central de Cabdella	55	59	0.6W
Comiols	61	54	20W
Erinyà	65	62	1W
Llimiana	21	25	5W
Mur	28	32	100W
Pobla de Segur	23	25	5W
Senterada	53	50	5W
Torre de Cabdella	47	37	5W
Xerallo	22	24	1W
Pollars Sobirà			
Alins	36	34	5W
Alt Aneu	22	26	10W
Aneu	51	53	1W
Boldis	30	32	0.5W
Escaló	23	21	2W
Esterri de cardos	48	61	1W
Gerri de la Sal	47	49	5W
Llavorsí	31	35	1W
Centre Pic de l'Orri	42	56	5/2.5kW
Rialp	48	51	1W
Tavascan	23	27	0.6W
Unarre	60	62	5W
Tornafort-Sort	24	26	5W
El Priorat			
El Priorat	34	38	100W
Ulldemolins	25	31	5W
Ribera d'Ebre			
Flix I	61	54	20W
Torre de lEspanyol	53V	60V	5W
Riba-Roja	44	46	1W
Flix II	34	38	0.6W
Ripollès			
Camprodon	22	26	20W

Espinabell	21	25	1W
Gombrèn	33	29	1W
Núria	22	—	0.6W
Ribes de Freser	54	44	20W
Ripoll I	39	63	100W
Ripoll II	21	23	1W
Setcases	57	61	1W
S. Joan de les Abadesses	29	—	0.6W
Taga	25	23	5W
Segarra			
Torà	37	64	1W
Segrià			
Almenar	35	25	20w
Mangraners	25	33	20W
Centre Emissor Alpicat	52	38	10/8kW
La Selva			
Lloret	50	56	5W
Arbúcies	42	36	20W
Osor	28	24	1W
S. Hilari Sacalm	26	22	5W
Tossa	25	37	5W
Solsonès			
La Coma	35	29	5W
S. Llorenç de Morunys	26	32	5W
Solsona	36	29	5W
Tarragonès			
Tarragona	21	27	0.5W
Terra Alta			
Gandesa	61	54	—
Val d'Aran			
Bagergue	41	43	5W
Bausen	54	52	1W
Bossost	43	37	20W
Cap de Vaqueira	30	28	20W
Sa Nela	61 53	5W	
Vallès Occidental			
Gallifa	39	37	1W
Montcada	65	21	100W
S. Llorenç Savall	37	39	1W
Location	**TV3**	**C33**	**Power**
Vallès Oriental			
Aiguafreda	42	38	5W
Bigues	59	61	5W
El Figaró	37	39	1W
Montseny	22V	36V	2W
Sant Celoni	61V	65V	5W
S. Fost de Campsentelles	37	39	1W
Sant Quirze Safaja	21	25	0.6W
Vallromanes-Monternés	50	52	5W
Andorra			
Andorra la Vella			
Les Escaldes/ Sant Julià/			
S. Coloma/El Serrat	33-26	63-47	—
Pas de ka Casa/Grau Roig	40	37	—
Encamp i Conillo	65	58	—
Soldeu i Vall d'Incles	49	52	—
La Massana i Ordino	40	46	—
Pal	38	44	—
Arinsal	59	53	—
Arans	56	50	—
Sant Julià i Fontaneda	40	46	—
Mallorca			
Palma/Inca/Manacor/			
Pla de Palma i Soller	46	51	—
Alcúdia	29	35	—
Cap de Pera	21	33	—
Pollensa	27	21	—
Calvià i Palma Nova	24	—	—
Son Cervera	28	—	—
Sant Salvador	33	—	—
Andraitx	30	—	—
Cala Sant Vicenç	27	—	—
Sa Racó	49	—	—
Sant Telm	27	—	—
Bunyola	21	—	—
Llunch	24	21	—
Lloseta	27	—	—
Mancor de la Vall	26	23	—
Menorca			
Maó i sud de l'illa	27	31	—

Ciutadella i nord de l'illa	46	51	—
Pitiüses			
Eivissa/S. Antoni/S. Eulàlia/			
S. Gertrudis i Formentera	25	38	—
S. Josep i part de S. Antoni	56	52	—
Eivissa Ciutat	43	50	—
Cala Sant Vicenç	56	63	—
Castelló			
Castelló General	34	—	—
Vilafranca de Maestrat	50	—	—
Benicarló/Baix Maestrat	29	—	—
Vall d'Uxo-Plana Baixa	45	—	—
Morella Els Ports	22	—	—
Cinctorres Ports	38	—	—
València			
València, General	37	—	—
Sueca/Sollana/Carlet/Benifaió/			
part Ribera Alta i Baixa	37	34	—
Vilamarxant (Camp de Túria)	41	—	—
Algemesí (Ribera Alta)	30	—	—
Montroi, Montserrat d'Alcalà/			
Reial de Montroi	24	—	—
Cullera (Ribera Baixa)	56	—	—
Tavernes de Valldigna/Simat/			
Benefaió (Safor)	31	—	—
Gandia/Font d'en Carrós/			
Oliva Bellreguard (Safor)	48	—	—
Ontinyent (Val d'Albaida)	39	—	—
Sagunt (Camp de Morvedre)	34	—	—
Alacant			
Alacant/Xixona/S. Vicent del Raspeig/			
S.Joan d'Alacant i Mutxamel	41	—	—
Alcoi	45	—	—
Callosa/Toravella/Almoradi	41	—	—
Crevillent (Baix Vilanopó)	48	—	—
Santa Pola (Baix Vilanopó)	39	—	—
Dènia i La Marina Alta	44	—	—
Benidorm/			
Alfas del Pi, VilaJoiosa	47	—	—

D.Prgr: TV3 1115-0030, Canal 33 1900-2400.

TV DE GALICIA (TVG)

Apt. 707, San Marcos (Santiago de Compostela). ☎ +34 (81) 565141. ☒ +34 (81) 562886. ☼ 97012.
L.P: Dir: Xerardo R. Rodríguez.
Station: ch 42. In Galician: 1200 (Sun 1030)-2300 (approx).

TELEVISIO VALENCIANA (Gov.)
CANAL 9

☒ 46100 Burjassot, Valencia. ☎ +34 (6) 364 1100. ☒ +34 (6) 363 9516.
L.P: GM: Amaden Fabregat-Manes. MD: Rafael Cano-Baron.

TV VASCA-EUSKAL TELEBISTA (Gov.)

☒ Barria Lurreta, 48200 Durango (Vizcaya). ☎ +34 (94) 6816600. ☒ +34 (94) 6816526. ☼ 34441.
L.P: Dir: Koldo Anasagasti.
Stations: ch 42 (ETB-1) & ch 49 (ETB-2) +160 low power st's.
D.Prgr: ETB-1 1400-2000 — ETB-2 1430-2100.

COMMERCIAL STATIONS

ANTENA TELEVISION (Antena 3)

☒ (See Satellite Section)
L.P: Pres: Javier de Godó; G.M: Manuel Martín Ferrand; Film Buyer: Condorcet da Silva Costa.

Projected Autonomous Networks:
Radiotelevision Madrid, Radiotelevision Navarra, TV Andalucia, TV Cantabria & TV Valenciana.

Stations	Tele-5	Antena 3	C. Plus	kW
Albacete-Chinchilla	50	53	56	50
Alicante-Aitana	50	53	60	100
Almería-Pechina	58	61	64	100
Avila-San Mateo	59	62	65	1
Barcelona-Tibidado	27	34	47	100
Bilbao-Archanda	59	62	65	50
Burgos	30	33	36	1
Cádiz-San Cristobal	49	53	55	30
Castellón-Desierto	49	52	55	50
Ciudad Real-Atalaya	23	26	29	1

Córdoba-Lagar	58	61	64	100
Cuenca-San Cristóbal	37	60	63	1
Gerona-Rocacorba	32	35	38	50
Granada-Parapanda	50	53	56	100
Guadalajara	31	37	51	1
Huelva-Punta Umbría	32	35	56	100
Huesca-Arguis	21	24	27	30
Jaen-Sierra Almadén	32	35	49	100
La Coruña-Ares	35	62	65	100
Las Palmas-La Isleta	32	35	38	20
León-El Portillo	49	52	55	1
Lérida-Alpicat	59	62	65	50
Logroño-Moncalvillo	40	46	48	20
Lugo-Páramo	23	41	44	100
Madrid-Torrespaña	59	62	65	100
Málaga-Mijas	39	42	45	100
Mallorca-Alfabia	58	61	64	100
Montánchez	59	62	65	100
Murcia-Carrascoy	38	42	44	50
Orense-Barbadanes	26	40	43	1
Oviedo-Gamoniteiro	28	32	35	100
Palencia-Villamuriel	48	51	54	1
Pamplona-S. Cristóbal	49	52	55	1
Pontevedra-Tomba	31	35	41	1
Salamanca-Teso	29	60	63	1
San Roque-Carbonaras	21	24	27	20
San Sebastián-Ulia	31	41	44	1
Santander-Peña Cabarga	29	60	63	50
Santiago-Pedroso	38	56	59	100
Segovia	48	51	54	1
Sevilla-Valencia	38	41	44	50
Soria-Santa Ana	21	24	27	1
Tarragana-La Musara	37	50	53	20
Tenerife-Izaña	23	26	29	100
Teruel-Santa Bárbara	26	30	33	1
Toledo-Los Palos	30	53	56	1
Valencia-Torrente	40	43	46	100
Valladolid-Contienda	50	53	56	1
Vigo-Domayo	54	61	64	100
Vitoria-Zaldiarán	29	32	35	15
Zamora-El Viso	58	61	64	100
Zaragoza-La Muela	22	30	54	100

TELEVISION MURCIANA

Plateraa 44, 230001 Murcia
☎ +34 (68) 212 224. 🖷 +34 (68) 214 673.

AFRTS (US Air Force Europe)

Det. 1 AFEBS, Zaragoza AB, c/o APO New York 09286-5000, USA.
Station: Zaragoza chA4 (Power: cable distribution), Torrejon chA4 2kW.
D.Prgrs: W 12h; Sat/Sun 15h.
NBS TV: Rota chA2 2kW.

SWEDEN

TV-sets: 3.750.000 — **Colour:** PAL — **Systems:** B & G.

TERACOM SVENSK RUNDRADIO AB

This company has the responsibility for the distribution of the prgrs produced by Sveriges Television AB and TV4 Nordisk TV.
HQ: Medborgarplatsen 3, Stockholm (🖃 Box 17666, S-118 92 Stockholm). ☎ +46 (8) 671200. 🖷 +46 (8) 6712001.
L.P: Pres. & CEO: Valdemar Persson.

SVERIGES TELEVISION AB (Non-comm.)

Oxenstiernsgatan 26-34, Stockholm.
Postal 🖃 S-105 10 Stockholm. ☎ +46 (8) 7840000 & 7847400. 🖷 +46 (8) 7841500. ① 10000. **Cable:** Broadcast. **Web:** http://www.svt.se
L.P: Chmn. Board of Governors: Anna-Greta Leijon. Man. Dir: S.Nilsson. Dir. Prgr. (SVT1): I. Bengtsson. Dir. Prgr: (SVT2): Hans Bonnevier. Dir. Eng: S.O. Ekholm. Dir. Staff: Peter Fellmarck. Contr. News and Curr. Aff. (SVT1). Stig Fredrikson. Head News. and Curr. Aff. (SVT2): Jan Axelsson. Contr. SVT International: Åke Källqvist. Hd. Legal Dpt/Adm. Britt-Marie Blanck. Hd International Rel: Cecilia Hallgren Järeborg. Press and Information:Jan-Olof Gurinder.

TV4 NORDISK TELEVISION Co. (Comm.)

🖃 S-115 79 Stockholm. ☎ +46 (8) 644 4400 🖷 +46 (8) 644 4440 ①

14124. **Web:** http://www.tv4.se
LP: Chmn: Erik Belfrage; Man Dir: Lars Weiss; Hd. of Inf. & Pub. Rel: Helga Baag¢e; Dept Dir of Prgrs: Thomas Nilsson; Hd. of News: Kerstin Persdotter; Hd. of Eng: Olle Mossberg.

Stations:

Location	Kanal 1 Sverige		TV 2 Sverige		Nordisk tv 4	
	Ch	kW	Ch	kW	Ch	kW
Arvidsjaur	5	60	21	1000	24	1000
Älvsbyn	4	60	36	1000	52	1000
Ånge	8	15	42	250	52	250
Bäckefors	8	60	29	1000	49	1000
Bollnäss	6	60	29	1000	49	1000
Borås	6	3	42	1000	55	1000
Borlänge	10	60	47	1000	60	1000
Emmaboda	8	60	31	1000	47	1000
Filipstad	9	5	33	1000	23	1000
Finnveden	41	1000	48	1000	58	1000
Gällivare	9	60	33	1000	43	1000
Gävle	9	60	27	1000	30	1000
Göteborg	9	60	30	1000	46	1000
Hagfors	—	—	—	—	38	?
Halmstad	7	60	24	1000	45	1000
Helsingborg	9	1	30	10	41	10
Hörby	43	1000	33	1000	50	1000
Hudiksvall	11	4	31	1000	44	1000
Jönköping	8	1	28	15	31	15
Kalix	8	60	35	1000	29	1000
Karlshamn	11	30	26	1000	49	1000
Karlskrona	9V	3	34	20	41	20
Karlstad	5	1	43	20	46	20
Kiruna	6	60	29	1000	42	1000
Kisa	11	30	49	1000	56	1000
Köpmannholmen	—	—	—	—	52	?
Loffstrand	—	—	—	—	44	?
Lycksele	8	15	45	1000	48	1000
Malmö	10	3	27	150	47	150
Mora	8	20	22	1000	25	1000
Motala	7	10	52	1000	39	1000
Nässjö	10	60	22	1000	25	1000
Norrköping	5	60	31	1000	54	1000

Stations:

Location	Kanal 1 Sverige		TV 2 Sverige		Nordisk tv 4	
	Ch	kW	Ch	kW	Ch	kW
Örebro	2	60	48	1000	58	1000
Örnsköldsvik	6	7	39	400	42	400
Östersund	4	100	27	1000	45	1000
Östhammer	11	30	26	1000	48	1000
Överkalix	10	5	45	1000	48	1000
Pajala	7	60	34	1000	47	1000
Skellefteå	6	35	46	1000	49	1000
Skövde	3	60	37	1000	47	1000
Sollefteå	7	60	46	1000	49	1000
Stockholm	4	60	23	1000	42	1000
Storuman	10	60	33	1000	43	1000
Sundsvall	5	60	47	1000	59	1000
Sunne	7	60	50	1000	53	1000
Sveg	2	60	21	1000	24	1000
Tåsjö	9	60	37	1000	40	1000
Trollhättan	7	1	51	20	41	20
Uddevalla	33	1000	23	1000	43	1000
Uppsala	6	10	49	200	52	200
Varberg	10	1	49	200	43	200
Väddö	8	200				
Vännäs	2	60	47	1000	50	1000
Västerås	8	10	31	1000	51	1000
Västervik	6	60	26	1000	43	1000
Virserum	—	—			42	?
Visby	9	60	41	1000	44	1000
Vislanda	39	1000	32	1000	56	1000

N.B: All sts. H unless indicated otherwise.

FINNISH TELEVISION RELAY

Stockholm (Nacka) ch39 1000kW(ERP). Pol H.
Relay of Finnish Prgr II: Mon-Fri 1530-2115, Sat 1200-2200, Sun 1200-2115.
+ 2 lp st.

SWITZERLAND

TV-sets: 2.602.023 — **Colour:** PAL — **Systems:** B&G.

SBC—SWISS BROADCASTING CORPORATION
The SBC's Television programme services are an integral part of the Swiss Broadcasting Corporation.

SBC, Giacomettistrasse 3, CH-3000 Berne 15. **Cable:** Ra-dif. ☎ +41 (31) 3509111. ① 911590 SSR ch. 🖷 +41 (31) 3509256.
L.P: Pres. SRG: Eric Lehmann. DG: Antonio Riva. Secr. Gen & Dir. Legal Dept: Beat Durrer. Dir. Finance: François Landgraf. Dir. Eng: Daniel Kramer. Dir. Human Resources: Raymond Zumsteg. Television Affairs: Tiziana Mona. Radio Affairs: Félix Bollmann. Dir. Communication & Marketing: Roy Oppenheim. Press Officer: Dr. Oswald Sigg.
Prgr. Sce. in German: TV Dir: Peter Schellenberg, Schweizer Fernsehen DRS, Fernsehstrasse 1-4, 8052 Zürich. ☎ +41 (1) 305 6611. ① 823823 TVZ. 🖷 +41 (1) 305 5660.
Web: http://www.srg-ssr.ch/srg/
Prgr. Sce. in French: TV Dir: Guillaume Chenevière, Télévision suisse romande, TSR, 20 Quai Ernest Ansermet. B.P. 234, 1211 Geneva 8. ☎ +41 (22) 708 9911. 🖷 +41 (22) 7811908. TV Prgr. Dir:Raimond Vouillamoz. **Web:** www.tsr.ch
Prgr. Sce. in Italian: RTSI, Radiotelevisione Svizzera di lingua italiana, Via Canevascini, P.O. Box 6903, Lugano.Reg. Dir: M. Blaser, TV Prgr Dir: D. Balestra. ☎ +41 (91) 58 5111. 🖷 +41 (91) 589150. Studio Televisione, Casella postale, CH-6949 Comano. ☎ +41 (91) 585111. 🖷 +41 (91) 585355.
Web: http://www.rtsi.ch/

SCHWEIZ 4/SUISSE 4/SVIZZERA 4 (Pub.)
Giacomettistrasse 1. CH-3000 Bern 15, Switzerland. ☎ +41 (31) 350 9444. 🖷 +41 (31) 350 9725. ① 911 590 ssr ch.
L.P: Dir: Dario Rabbiani
D.Prgr: National multi-lingual TV channel in cooperation with private program producers.

Prgrs as follows:

Schweizer Fernsehen DRS:	DRS1, TSR2, TSI2
Télévision Suisse romande:	TSR1, DRS2, TSI1
Televisione Svizzera italiana:	TSI1, DRS3, TSR3
Schweiz 4:	DRS4, TSR4, TSI4

Stations (Systems B&G)

Loc.	Progr.	ch	Power
Ausserberg	DRS2	37	1kW
Ausserberg	DRS3	40	1kW
Bantiger	DRS1	2	47kW
Bantiger	DRS2	50	120kW
Bantiger	DRS3	40	123kW
Bantiger	DRS4	43	92kW
Bantiger	TSR1	10	26kW
Cardada	TSI2	21	3.5kW
Cardada	TSI3	31	3.6kW
Cardada	TSI4	68	4.5kW
Castel San Pietro	TSI1	56	7.7kW
Castel San Pietro	TSI2	39	5.4kW
Castel San Pietro	TSI3	42	6.1kW
Celerina	DRS1	9	3.7kW
Celerina	DRS2	49	5.2kW
Celerina	DRS3	33	4.3kW
Celerina	TSI1	7	3.6kW
Chamossaire	TSR2	57	2kW
Chamossaire	TSR3	60	2kW
Chasseral	DRS1	62	17.5kW
Chasseral	TSR1	22	12.1kW
Chasseral	TSR2	59	15.1kW
Chasseral	TSR3	56	15.8kW
Chasseral	TSR4	25	12.4kW
Chaux-de-Fonds, La	TSR1	9	8.5kW
Chaux-de-Fonds, La	TSR2	32	9.1kW
Chaux-de-Fonds, La	TSR3	35	9.8kW
Dôle, La	TSR1	4	107.1kW
Dôle, La	TSR2	31	192.7kW
Dôle, La	TSR3	34	186.2kW
Dôle, La	TSR4	69	251.1kW
Feldis	DRS2	24	1.7kW
Feldis	DRS3	21	1.7kW
Gebidem	DRS1	11	5.1kW
Gebidem	DRS2	52	15.4kW
Gebidem	DRS3	55	15.1kW
Gorduno	TSI4	63	1.0kW
Haute-Nendaz	TRS1	7	2.3kW
Haute-Nendaz	TSR2	43	10.7kW
Haute-Nendaz	TSR3	46	11.2kW
Klewenalp	DRS3	55	2kW
Klewenalp	DRS3	58	2kW
Mont-Pèlerin	TSR1	11	2.6kW
Mont-Pèlerin	TSR2	44	2.4kW
Mont-Pèlerin	TSR3	47	2.4kW
Mont-Pèlerin	TSR4	52	2.7kW
Monte Ceneri-Passo 1	TSI1	5	4.9kW
Monte Ceneri-Passo 1	TSI2	46	34kW
Monte Ceneri-Passo 1	TSI3	49	34kW
Monte Ceneri-Passo 1	TSI4	55	37.9kW
Monte Morello	TSI1	68	6.7kW
Monte Morello	TSI2	62	4.2kW
Monte Morello	TSI3	66	6.7kW
Monte San Salvatore	TSI1	10	16.1kW
Monte San Salvatore	TSI2	54	106.7kW
Monte San Salvatore	TSI3	57	113.5kW
Monte San Salvatore	TSI4	60	114.8kW
Niederhorn	DRS1	12	0.9kW
Niederhorn	DRS1	53	19.4kW
Niederhorn	DRS2	27	8.4kW
Niederhorn	DRS3	30	9.2kW
Niederhorn	DRS4	65	16kW
Olten	DRS1	42	13.8kW
Olten	DRS2	45	13.8kW
Olten	DRS3	63	13.8kW
Ordens, Les	TSR1	7	3.6kW
Ordens, Les	TSR2	31	6 3kW
Ordens, Les	TSR3	34	6.3kW
Pfänder	DRS2	49	13.4kW
Pfänder	DRS3	62	14.2kW
Piz Lagalb	TSI1	55	1.1kW
Piz Lagalb	TSI2	37	1.1kW
Piz Lagalb	TSI3	40	1.1kW
Pizzo Matro	TSI2	29	7kW
Pizzo Matro	TSI3	32	7.3kW
Pizzo Matro	TSI4	22	5.9kW
Ravoire	TSR1	9	3.9kW
Ravoire	TSR2	51	5.6kW
Loc.	**Progr.**	**ch**	**Power**
Ravoire	TSR3	54	5.6kW
Rigi	DRS1	6	38.5kW
Rigi	DRS2	32	70.8kW
Rigi	DRS3	29	70.8kW
Säntis	DRS1	7	36.6kW
Säntis	DRS2	31	59.5kW
Säntis	DRS3	34	62.3kW
Schaffhausen	DRS1	47	5.3kW
Schaffhausen	DRS2	50	5.3kW
Schaffhausen	DRS3	60	3.6kW
Schüpfheim	DRS2	51	1kW
Schüpfheim	DRS3	54	1kW
Sedrun	DRS2	55	1kW
Sedrun	DRS3	52	1kW
St. Chrischona	DRS1	11	37.6kW
St. Chrischona	DRS2	46	104.7kW
St. Chrischona	DRS3	49	104.7kW
St. Niclaus/VS	DRS1	64	1.2kW
Uetliberg	DRS1	3	74.1kW
Uetliberg	DRS2	23	105.9kW
Uetliberg	DRS3	26	105.9kW
Vallée de Joux	TSR2	46	1kW
Vallée de Jouz	TSR3	49	1.1kW
Valzeina	DRS1	10	0.7kW
Valzeina	DRS2	56	11.6kW
Valzeina	DRS3	53	11.6kW
Wattenwil	DRS2	57	6kW
Wattenwil	DRS3	60	6kW
Ziegelbrücke	DRS2	51	1.3
Ziegelbrücke	DRS2	54	1.3

UKRAINE

TV-Sets: 1000 — **Colour:** SECAM — **Systems:** D&K.

UKRAJINSKA TELEBACENNJA (Gov.)
✉ vul. Chrescatik 26, 252001 Kyjiv. ☎ +380 (44) 2290638. 🖷 +380 (44) 2296945.

UT-1
(*Not ERP)

Stations	ch	kW*		ch	kW*
L'viv	1	5	Kirovograd	6	5
Sovetskij	2	5	Simferopol'	6	5
Izmajil	3	0.1	Ivano-Frankivs'k	7	5
Komis Zorja	3	5	Kam'jans'ke	7	5
Nikopol'	3	2.5	L'viv	8	25
Kyjiv	4	50	Kryvyj Rig	9	5
Kovel'	5	5	Krasnogorivka	10	25
Sevastopol'	5	5	Mykolajiv	10	5
Ternopil'	5	25			

+various low power translators.

NB: The above list of main transmitters is not complete.
D.Prgr: 0600-2200.

UT-2
(*Not ERP)

Stations	ch	kW*		ch	kW*
Kramators'k	1	0.1	Krasnoperekops'k	10	0.1
Ivano-Frankivs'k	2	0.1	Pervomejs'k	12	0.1
Simferopol'	3	5	Pryluky	12	0.1
Kirovograd	5	0.1	Kirovograd	21	5
L'viv	6	5	Mariupol'	25	5
Sevastopol'	8	0.1	Krasnogorivka	28	—
Kyjiv	9	50	Komis Zorja	28	0.1
Odesa	9	5	Kam'jans'ke	29	—

*various low power translators.

NB: The above list of main transmitters is not complete.
D.Prgr: 0500-2200. Includes relays of Rossijskoje televidenije (RTV) from Moscow.

Relays of ORT (Russia)
(*Not ERP)

Stations	ch	kW*		ch	kW*
L'viv	1	40	Kryvyj Rig	1	5
Simferopol'	1	25	Kyjiv	2	50
Stations	**ch**	**kW***		**ch**	**kW***
Mykolajiv	2	5	Komis Zorja	8	5
Rivne	3	5	Krasnogorivka	8	25
Kam'jans'ke	4	5	Starobil's'k	8	5
Mariupol'	5	5	Ivano-Frankivs'k	9	5
Odesa	5	5	Izjum	11	25
Kovel'	6	25	Kirovograd	11	5
Kotovs'k	6	5	Melitopol'	11	5
Pervomejs'k	7	5	Sevastopol'	11	5
Sovetskij	7	5			

+various low power translators.

NB: The above list of main transmitters is not complete.
D.Prgr: Relays of ORT from Moscow. Further details see under Russia.

UNITED KINGDOM

TV-sets: approx 20m — **Colour:** PAL — **System:** I.

BRITISH BROADCASTING CORPORATION
✉ Television Centre, London W12. ☎ +44 (181) 743 8000. ① 265781. 🖷 +44 (181) 749 7520.
L.P: MD: Will Wyatt. Asst. MD: Jane Drabble. Contr. BBC-1: J. Powell. Contr. BBC-2: Alan Yentob. CE: Peter Marchant. Publ: Keith Samuel.
Stations: See below.
D.Prgr: BBC1: Mon-Fri 0600-2400; Sat/Sun 0700-2400 (closing times vary). **BBC2:** 1500-2400 (Sat/Sun 0100) approx.

S4C, Welsh Fourth Channel Authority
✉ Clos Sophia, Cardiff, CF1 9XY. ☎ +44 (1222) 747444. 🖷 +44 (1222)

754444. **Web:** http://www.s4c.co.uk

INDEPENDENT TELEVISION COMMISSION
✉ 33 Foley Street, London W1P 7LB. ☎ 0171 255 3000. 🖷 0171 306 7800. **E-mail:** 100731.3515@compuserve.com
L.P: Chief Exec: Peter Rogers. Dep. Chief Exec: Clare Mulholland. Dir. of Finance: Sheila Cassells. Dir. of Public Affairs: Paul Smee. Dir. of Prgrs and Cable: Sarah Thane.Dir. of Eng: Gary Tonge. Dir. of Advertising and Sponsorship: Frank Willis.Contr. of Admin: Don Horn. Secr. to the Commission: Michael Redley. Dep. Dir. of Prgrs: Robin Duval. Dep. Dir. of Cable:Anthony Hewitt.
Function: The ITC is the public body responsible for licensing and regulating commercially funded television services provided in and from the UK. These include Channel 3 (ITV), Channel 4, Channel 5, public teletext and a range of cable, local delivery and satellite services. They do not include services provided by the BBC or by S4C, the fourth channel in Wales.

Program contractors licensed by the ITC:
Anglia Television (1), Anglia House, Norwich NR1 3JG. ☎ +44 (1603) 615151. 🖷 +44 (1603) 631032. **Web:** http://www.anglia.co.uk/
Border Television (3), The Television Centre, Carlisle, CA1 3NT. ☎ +44 (1228) 25101. 🖷 +44 (1228) 41384. **E-mail:** ian@border-tv.com **Web:** http://www.border-tv.com
Carlton Television Ltd. (8), 101 St. Martin's Lane, London WC2N 4AZ. ☎ +44 (171) 240 4000. 🖷 +44 (171) 240 4171. **Web:** http://www.carl-tontv.co.uk
Central Independent Television (2), Central House, Broad Street, Birmingham, B1 2JP. ☎ +44 (121) 643 9898. 🖷 +44 (121) 616 4766.
Channel Four Television Corporation, 124 Horseferry Road, London SW1P 2TX. ☎ +44 (171) 306 8333. **Web:** http://www.channel4.com
Channel Television (4), The Television Centre, St. Helier, Jersey, Channel Islands. ☎ +44 (1534) 68999. 🖷 +44 (1534) 59446/24770.
Data Broadcasting International Ltd, Allen House, Station Road, Egham, Surrey, TW20 9NT. ☎ +44 (1784) 471515.
GMTV Ltd, The London Television Centre, London SE1 9LT. ☎ +44 (171) 827 7000. 🖷 +44 (171) 827 7001.
Grampian Television (5), Queen's Cross, Aberdeen, AB9 2XJ. ☎ +44 (1224) 646464. 🖷 +44 (1224) 635127.
Granada Television (6), Granada Television Centre, Manchester, M60 9EA. ☎ +44 (161) 832 7211. 🖷 +44 (161) 839 0454. **Web:** http://www.granadatv.co.uk
HTV Wales (7), The Television Centre, Culverhouse Cross, Car-diff, CF5 6XJ. ☎ +44 (1222) 590590. 🖷 +44 (1222) 597183.
HTV West (7a), The Television Centre, Bath Road, Bristol, BS4 3HG. ☎ +44 (1272) 778366. 🖷 +44 (1272) 722400.
London Weekend Television (LWT) (8), South Bank Television Centre, London, SE1 9LT. ☎ +44 (171) 620 1620. 🖷 +44 (171) 928 6948.
Meridian Broadcasting Ltd (10), Television Centre, Southampton, SO14 0PZ. ☎ +44 (1703) 222 555. 🖷 +44 (1703) 335050. **Web:** http://www.meridian.tv.co.uk
S4C, Welsh Fourth Channel Authority, Clos Sophia, Cardiff, CF1 9XY. ☎ +44 (1222) 747 444. 🖷 +44 (1222) 75 4444. **Web:** http://www.s4c.co.uk
Scottish Television (9), Cowcaddens, Glasgow, G2 3PR. ☎ +44 (141) 300 3000. 🖷 +44 (141) 300 3030. **Web:** http://www.scotnet.co.uk/stv/
Teletext UK Ltd, 101 Farm Lane, London SW6 1QJ. ☎ +44 (171) 386 5000. 🖷 +44 (171) 386 5002.
Tyne Tees Television (11), The Television Centre, City Road, Newcastle upon Tyne, NE1 2AL. ☎ +44 (191) 261 0181. 🖷 +44 (191) 261 2302.
Ulster Television (12), Havelock House, Ormeau Road, Belfast, BT7 1EB. ☎ +44 (1232) 328122. 🖷 +44 (1232) 246695.
Westcountry Television Ltd (13), Western Wood Way, Lan-guage Science Park, Plymouth, PL7 5BG. ☎ +44 (1752) 333333. 🖷 +44 (1752) 333 444.
Yorkshire Television (14), The Television Centre, Leeds, LS3 1JS. ☎ +44 (532) 438283. 🖷 +44 (532) 445107.

INDEPENDENT TELEVISION NEWS (ITN)
✉ 200 Gray's Inn Road, London WC1X 8XZ. ☎ +44 (171) 833 3000.

INDEPENDENT TELEVISION ASSOCIATION (ITVA)
✉ Knighton House, 56 Mortimer Street, London W1N 8AN. ☎ +44 (171) 612 8000. 🖷 +44 (171) 580 7892.

STATIONS:	BBC1	BBC2	ITV	Ch4	kW
(Pol=H)					(ERP)
So. Ea. England (+ 66 relays)					

	BBC1	BBC2	ITV	Ch4	kW
Bluebell Hill (10)	40	46	43	65	30
Crystal Palace (8)	26	33	23	30	1000
Dover (10)	50	56	66	53	100
Heathfield (10)	49	52	64	67	100
Oxford (2)	57	63	60	53	500

So. We. England & Channel Isl. (+ 81 relays)

Beacon Hill (13)	57	63	60	53	100
Caradon Hill (13)	22	28	25	32	500
Huntshaw Cross (13)	55	62	59	65	100
Redruth (13)	51	44	41	47	100
Stockland Hill (13)	33	26	23	29	250
Fremont Point (4)	51	44	41	47	20

So. England (+ 38 relays)

Hannington (10)	39	45	42	66	250
Midhurst (10)	61	55	58	68	100
Rowridge (10)	31	24	27	21	500

We. England (+ 51 relays)

Mendip (7a)	58	64	61	54	500

Ea. England (+ 18 relays)

Sandy Heath (1)	31	27	24	21	1000
Sudbury (1)	51	44	41	47	250
Talcolneston (1)	62	55	59	65	250

Ce. England (+ 60 relays)

Ridge Hill (2)	22	28	25	32	100
Sutton Coldfield (2)	46	40	43	50	1000
The Wrekin (2)	26	33	23	29	100
Waltham (2)	58	64	61	54	250

No. England (+ 52 relays)

Belmont (14)	22	28	25	32	500
Emley Moor (14)	44	51	47	41	870

No. We. England (+ 66 relays)

Winter Hill (6)	55	62	59	65	500

No. Ea. England (+ 64 relays)

Bilsdale (11)	33	26	29	23	500
Caldbeck (3)	30	34	28	32	500
Chatton (11)	39	45	49	42	100
Pontop Pike (11)	58	64	61	54	500

Scotland (+ 188 relays)

Angus (5)	57	63	60	53	100
Black Hill (9)	40	46	43	50	500
Sandale	22	—	—	—	500
Craigkelly (9)	31	27	24	21	100
Darvel (9)	33	26	23	29	100
Durris (5)	22	28	25	32	500
Eitshal (5)	33	26	23	29	100
Keelylang Hill (5)	40	46	43	50	100
Knock More (5)	33	26	23	29	100
Rosemarkie (5)	39	45	49	42	100

STATIONS:	BBC1	BBC2	ITV	Ch4	kW
Rumster Forest (5)	31	27	24	21	100
Selkirk (5)	55	62	59	65	50

No. Ireland (+ 41 relays)

Brougher Mt. (12)	22	28	25	32	100
Divis (12)	31	27	24	21	500
Limavady (12)	55	62	59	65	100

Wales (+ 162 relays)	BBC1	BBC2	ITV	S4C	kW (ERP)
Blaen-plwyf (7)	31	27	24	21	100
Carmel (7)	57	63	60	53	100
Llanddona (7)	57	63	60	53	100
Moel-y-Parc (7)	52	45	49	42	100
Presely (7)	46	40	43	50	100
Wenvoe (7)	44	51	41	47	500

NB: The number in brackets indicates the ITC company responsible for prgrs on ITV (3rd column).

Channel 5 Broadcasting Limited

22 Long Acre, London WC2E 9LY. +44 (171) 550 5555. +44 (171) 550 5554. **Web:** http://www.channel5.co.uk

L.P: Chief Exec. Officer: David Elstein. Chief Operating Officer: Ian Ritchie. Dir. of Programming: Dawn Airey. Dir. of Sales: Nick Milligan. Dir. of Marketing & Communications: David Brook. Dir. of Finance: Damian Harte.

Location	ch	kW (ERP)	Pol.
Plympton	30	2	V
Londonderry	31	10	V
Tay Bridge	34	4	V
Nottingham	34	2	V
Fawley	34	1	H
Fenton	35	F.Pl.	V
Hannington	35	F.Pl.	H
Ridge Hill	35	F.Pl.	H
Sudbury	35	F.Pl.	H
The Wrekin	35	F.Pl.	H
Waltham	35	F.Pl.	H
Emley Moor	37	870	H
Black Hill	37	500	H
Croydon	37	250	H
Mendip	37	126	H
Lichfield	37	100	H
Presely	37	100	H
Black Mountain	37	50	H
Cambret Hill	37	20	H
Redruth	37	3	H
Sandy Heath	39	10	H
Storeton	39	2.8	V
Craigkelly	48	4	H
Winter Hill	48	12.5	H
Churchdown Hill	48	1	H
Oxford	49	40	H
Selkirk	52	50	H
Tacolneston	52	4	H
Perth	55	2	V
Belmont	56	50	H
Caldbeck	56	10	H
Blaen Plwyf	56	4	H
Fenham	56	2	V
Chelmsford	63	1	H
Durris	67	100	H
Mounteagle	67	100	H
Sheffield	67	2.5	V
Huntshaw Cross	67	2	H
Burnhope	68	50	H

YUGOSLAVIA (Federal Republic of)

TV-sets: 1.642.522 — **Colour:** PAL—**System:** B&G.

UDRUZENJE JUGOSLOVENSKIH RADIOTELEVIZIJA d.O.O. (JRT)

JRT-Permanent Services — Television Department: Hartvi-gova 70/I, 11000 Beograd. +381 (11) 434-910. 11469 yu jurate — **Televizija Srbije-Televizija Beograd:** Takovska 10, 11000 Beograd. +381 (11) 342-001. 11884 — **Televi-zija Srbije-Televizija Novi Sad:** Kamenicki put 45, 21000 Novi Sad. +381 (21) 56-855. **Televizija Srbije-Televizija Pristina:** Beogradska 66, 38000 Pristina. +381 (38) 31 211 — **Televizija Crne Gore:** Cetinski put bb, 81000 Podgorica. +381 (81) 41 529. **Stations** (System B) (Ch 2-12), System G (Ch 21-69). **Ch E.** Audio powers 1/10 of vision powers indicated. St's below 3kW not mentioned.

MONTENEGRO

TV Crne Gore	I	II	III	IV	kW
Bjelasica	6	12	37	62/43	100/100/1/500/1
Sudjina Glava	12	6	24	39	5/5/15/15
Sjenica	6	33	23	25/29	1.5/15/15/15
Lustica	4	26	33	39/42	1/10/10/10
Lov´cen	8	31	10	35/67	100/1000/20/1000/500
Volujica	6	12	24	38	1/1/10/10
Muzura	12	23	33	43/53	1/10/10/10/10

+321 low power st's.

Local/private TV sts: Montena TV, 81000 Podgorica (ch 53); Montenegro TV, 81000 Podgorica (ch 59)

SERBIA

TV Srbije-TV Beograd	I	II	III	kW
Kopaonik	3	41		50/1000
Jastrebac	5	27	33	100/1000/1000
Avala	6	22	28	100/1000/30
Deli Jovan	6	23	43	10/50/100
Tornik	7	53	59	1/50/50
Besna Kobila	8	49	59	10/300/300
Ovcar	8	42	56	10/400/10
Maljen	9	26	32	5/2.5/250
Crveni Cot	10	24	30	100/1000/1000
Tupiznica	10	25	31	35/500/500
Crni Vrh (Svetozarevo)	11	35	38	35/500/500

Cer	7	37	34	3/300/300

+288 low power st's.

Local/private TV sts in Beograd:
TV Pink, Bul. Lenjina 2, 11070 Beograd (ch 59+5 relays)
BK Telecom, Omlandinskih Brigada 1, 11070 Beograd (ch 12 + 4 relays)
TV Politika, Makedonska 29, 11000 Beograd (ch 43 + 4 relays)
NTV Studio "B", Masarikova 5, 11000 Beograd (ch 53 + 4 relays)
TV Art, Vl. Kovacevica 6, 11000 Beograd (ch 38 + 2 relays)
TV Palma. Banijska 2, 11080 Beograd (ch 34 + 1 relay)

TV Srbije-

TV Novi Sad	I	II	III	kW
Subotica	5	43		35/1000
Venac	41		48	1000/1000
Vrsac	39	56		1000/1000

+13 low power st's.

TV Srbije-

TV Pristina	I	II	III	kW
Goles	7	44		35/600
Cviljen	9	21		10/400

+63 low power st's.

Other stations: there are numerous local/private TV st's in Serbia and Vojvodina.

AFRICA

TV-sets: 2m — **Colour:** PAL — **System:** B.

ENTREPRISE NATIONALE DE TÉLÉVISION (E.N.T.V.) (Gov.)
✉ 21 Blvd. des Martyrs, Algiers 16000. ☎ +213 (2) 780310. ① 66101 or 65282. 🖷 +213 (2) 601922.
L.P: DG: Zemzoum Zoubir. Dir. of Inf: M. Ibrahim. Dir. Tec: M. El Ksouri. Dir. Ext. Rel: M. Bey.
Stations: Pol: H.

	ch	kW (ERP)		ch	kW (ERP)
M. Cid	5	150/30	Akfadou	6	100/20
Ain-n-Sour	5	100/20	Nador	6	150/30
Reggane	5	25/2.5	Ouargla	6	5/0.5
Ghardaia	5	11/1.1	Constantine	7	100/20
	ch	kW (ERP)		ch	kW (ERP)
Aouilef	7	19/1.9	Metlili	10	150/30
Adrar	7	17/1.7	Tessala	10	150/30
Mecheria	8	30/6	Algiers	11	20/4
Chrea	9	120/24	Aflou	11	150/30
Touggourt	9	12/1.2	In Amenas	11	10/1

Low power st's not mentioned.
D.Prgr: 1500-2300 (Thurs/Sat/Sun 1300-2400).

TV-sets: 50.500 — **Colour:** PAL — **System:** I.

TELEVISÃO POPULAR DE ANGOLA (Gov.)
✉ Avenida Ho-Chi-Min, P.O. Box 2604, Luanda. ☎ +244 (2) 320025. 🖷 +244 (2) 391091. ① 3238, 4153, 4157 TPA-AN. **Cable:** TPA.
L.P: MD: Carlos Chunha. Head of Prgrs: António Pedreira.
Station: ch9 13kW (ERP) + st's at Benguela, Huambo, Lubango, Namibe, Cabinda, Bié
D.Prgrs: 1730-2300 (Mon-Fri), 1400-2300 (Sat), 0900-2300 (Sun).

TV-sets: 30.000 — **Colour:** SECAM. — **System:** K.

OFFICE DE RADIODIF. ET TV DE BENIN (ORTB) (Gov.)
✉ P.O. Box 366, Cotonou. ☎ +229 3010628.
L.P: Dir. Gen: Nicolas Benon. Dir. TV: Michèle Badarou. Chief of Sce.

(TV): Marcellin Illougbade. Head of Prgrs: Didier Falde.
Station: ch4 20kW.
D.Prgr: Mon-Fri 1800-2100, Sat/Sun 1700-2200.

TV-sets: 13.800 — **Colour:** SECAM — **System:** K.

GABORONE TELEVISION CORP.
✉ Private Bag 0060, Gaborone. ☎ +267 352541. 🖷 +267 357 138.
L.P: Dep. Dir. of Prgrs: Mrs. B. Tafa; Dep. Dir. of Prod: Mr. S. Moribame.
Station: ch. not known.
D.Prgr. in English: 1700-2000.

TV-sets: 45.500 — **Colour:** SECAM — **System:** K.

TÉLÉVISION NATIONALE BURKINA (Gov.)
✉ 1 T.N.B. 995 Boulevard de la Révolution: B.P. 2530, Ouagadougou 01. ☎ + 226 354773/306621. 2 + 32 48 09 BURKINA FASO① 5327.
E-mail: aline.koala@mcc.gov.bf **L.P:** DG: Mme Aline Koala.
Station: Ouagadougou & Bobodioulasso ch6 50/10W V.
D.Prgr: Tues-Fri 1833-2220; Sat 1508-2000; Sun 1205-1300, 1833-2200 — **Projected:** Ouagadougou 10kW H.

TV sets: 4500 — **Colour:** SECAM — **System:** K.

TÉLÉVISION NATIONALE DU BURUNDI (Gov.)
✉ B.P. 1900, Bujumbura. ☎ +257 2247 60. ① 5119.
L.P: DG: Louis-Marie Nindorera; Hd. of Prgr: Leonidas; Director TV: Clément Kirahagazwi.
Station: ch 25 0.5kW — **D.Prgr:** 1600 (Sat/Sun 1400)-2200.

TV-sets: 15.000 — **Colour:** PAL — **System:** B

CAMEROON RADIO AND TELEVISION (CRTV) (Gov.)
✉ P.O. Box 1634, Yaoundé. ☎ +237 20 4366/21 4088. 🖷 +237 204340. ① 8888 KN. Douala ☎ +237 42 6060/7211/9440.
L.P: Dir. Gen: Pr. Gervais Mendoze.
Stations: ch 5 10kW

TV-sets: 240.000 — **Colour:** PAL — **Systems:** B & G.

TELEVISION ESPAÑOLA EN CANARIAS
✉ 69 Calle Buenos Aires, Santa Cruz de Tenerife. ☎ +34 (22) 216200.
Stations: TV1: Santa Cruz de Tenerife (Izana) chE3 350/35kW H, Fuencaliente chE7 4.7/0.47 H, Poso de las Nieves ch10 10/1H, Arrecife ch 32 6/0.6 kW H + 38 relays.
TV2: Fuencaliente ch27 170/17kW H, La Isleta ch28 18.6/1.86kW H, Arrecife ch38 6/0.6kW H, Izana ch45 87/8.7kW H, Puerto Rosario ch52 23.8/2.38kW H, Los Christianos ch 57 6/0.6kW H, Pozo de las Nieves ch59 48/4.8kW H + 13 relays.
D.Prgr. TV1: W 0615-0020 (Sun 0745-0020); **TV2:** W 1745-2300; Sun 0845-2325 (closing time may vary).

COMMERCIAL STATIONS

CANAL BUENAS NUEVAS
✉ Calle Sao Paulo 45, Santa Cruz de Tenerife. ☎ +34 (22) 279442 —
Station: Cebadal ch21.

CANARYVISION
✉ Calle Arequipa 10, Santa Cruz de Tenerife. ☎ +34 (22) 470366 —
Station: Cebadal ch25.

TELE GRAN CANARIA

☞ Calle Sao Paulo 46, Santa Cruz de Tenerife. ☎ +34 (22) 464722 — **Station:** ch40.

ONDA TELEVISION MASPALOMAS (OTM 6)
☞ Calle Galdar 48, San Agustin, Playa del Ingles. ☎ +34 (22) 772445, 773737 — **Stations:** Guia ch42, Cumbre ch46.

LIBERTAD TELEVISION
☞ Avda. Escaleritas 112, Escaleritas. ☎ +34 (22) 251440.
Station: Escaleritas ch50.

ANTENA 3 TELEVISION
☞ Eduardo Benot 3, Santa Cruz de Tenerife. ☎ +34 (22) 275242 — **Stations:** Cumbre ch36, Isleta ch38,

CENTRAL AFRICAN REPUBLIC

TV-sets: 7.500 — **Colour:** SECAM — **System:** K.

RADIODIFFUSION-TÉLÉVISION CENTRAFRIQUE
☞ P.O. Box 940, Bangui. ☎ +236 613242.
LP: MD: Paul Service; Hd. of Prgrs: Henri-Gustav Hytayu.
Stations: N/A.

CHAD

TV-sets: 50.000 — **Colour:** SECAM — **System:** D.

TÉLÉTCHAD (Gov.)
☞ B.P. 74, N'Djamena. ☎ +235 51 29 23. ① 5307.
LP: Dir: Hourmadji Houssa Doumgor. Adj. Dir: Houssa Dago.
Station: N'Djamena ch7 (offset) 100W.
D.Prgr. in French/Arabic: 1800-2100 4 days per week.

CONGO (People's Rep.)

TV-sets: 8.500 — **Colour:** SECAM — **System:** K.

RADIODIFFUSION TÉLÉVISION CONGOLAISE
☞ 2241, Brazzaville. ☎ +242814574/814273/814030.
LP: D.G: J.F. Sylvestre SOUKA.
Station: ch7 10/20kW H.
D.Prgr: 1730-2300

CONGO (Dem. Rep. of)

TV-sets: 22.000 — **Colour:** SECAM — **System:** K.

OZRT (Gov.)
☞ B.P. 3171, Kinshasa, Gombe 3164. ☎ +243 23171.
① 21583.Vozakin ZR.
LP: Dir. Gen: B. Dongo. Dir. Tec: S. Lepamabla
Stations: a) regular; b) regular but not fulltime; c) irregular.

Location	ch	kW	Location	ch	kW
Kinshasa	5a	27	Uvira	6c	24
Kananga	4b	2	Kambove	7a	2
Kamina	4b	2	Goma	7a	2
Gbadolite*	4a	2	Gemena	8c	24
Kolwezi	5a	1	Kisangani*	8a	2
Mbuji Mayi	6b	2	Bandundu	8b	2
Mbandaka	6b	2	Lubumbashi*	9a	24
Kalemie	6a	2	Bukavu	9a	2
Kindu	6b	2	Isiro	9c	2

*) own studio facilities.
D.Prgr: Mon/Tues/Thurs/Fri 1130-1330 & 1630-2300, Wed 1130-2300, Sat/Sun 0900-2300. Relayed on Intelsat 66ÞE, C-band. Tr's from inland towns are dependent on power supplies (i.e. fuel availability).

PRIVATE/COMMERCIAL STATIONS

ANTENNE A (Private/Comm.)

☞ Av. du Port 4, Building Forescom 2e floor, Kinshasa/Gombe. P.O. Box 2681 Kinshasa 1. ☎ +243 21736/24818/25308
LP: P.D.G: A Pinhas; Dir Gen: Igal Avivi Neirson; Tech Dir: Ranny Ranny Shoket.
D.Prgr: Mon-Fri: 1430-0030, Sat1430-0130, Sun 1000-2330.
Station: ?

CANAL Z (Comm.)
☞ 6, av. du Port, Kinshasa/Gombe, P.O. Box 614 Kinshasa I. ☎ +243 20239.
LP: Dir Gen: Frederic Flasse
D.Prgr: 1500-2400
Station: Kinshasa UHF ch23

COTE D'IVOIRE

TV-sets: 810.000 — **Colour:** SECAM — **System:** K.

TÉLÉVISION IVOIRIENNE (Gov.)
☞ 08 B.P. 883, Abidjan 08. ☎ +225 439039. 🖷 +225 222297. ① 26110 ditele.
LP: Dir. Gen: Mamadou Berté.

Stations	chK	kW	Pol		chK	kW	Pol
Koun	4	10/2.5	V	Touba	6	1/1	H
Tiémé	4	10/2.5	V	Man	7	10/2.5	H
Bouake	4	0.1	V	Dabakala	7	2	H
Abobo*	4	0.1	H	Abidjan	8	10/2.5	H
Séguéla	5	10/2.5	H	Niangbo	8	10/2.5	H
Digo	5	2/1	H	Niangué	8	10/2.5	H
Dimbroko	6	2/1	H	Bouaflé	9	10/2.5	V

*) 2nd prgr for Abidjan only.
D.Prgrs: 1st Prgr: Mon-Wed 1200-1330 & 1900-2300, Thurs 1200-1300 & 1600-2300, Sat 1200-0130, Sun 1030-2330. **2nd Prgr:** Mon-Fri 2030-2300, Sat 1600-2030 (Sun no transmissions). **Bouaké Regional Prgrs:** Thurs 1200-1300, Fri 1700-1830.

DJIBOUTI (Rep.)

TV-sets: 17.000 — **Colour:** SECAM — **System:** K.

RADIO TÉLÉVISION DE DJIBOUTI (Gov.)
☞ P.O. Box 97, Djibouti. ☎ +253 352294. ① 5863 DJ.
LP: DG: Mohamed Tara Moussa.
Station: Djibouti chK'6 0.05kW H.
D.Prgr: 1500-2000. **F.PI:** 5 relay st's.

EGYPT

TV-sets: 5m — **Colour:** PAL — **System:** B&G.

EGYPTIAN RADIO AND TV UNION (Gov.)
☞ TV Bldg, Cornish El-Nil, Maspero Cairo. ☎ +20 (2) 757155. **Cable:** Cibrotev. ① 92152 Karadun.
LP: Chmn: Amin Bassounia. Head of TV: Sohair El Atreby.

Stations	I	II	kW	Pol	Stations	I	II	kW	Pol
Asswan	5	9	67	H	Baharia Baris	7	5	10	V
Cairo	5	9	200	H	El Dakhla	8	6	22.4	H
Hurghada	5	7	89	V	El Minya	8	5	165	V
Nag Hamadi	5	8	17	H	Idfu	81	11	65	H
Port Said	5	7	10	V	Mahalla	8	10	1600	H
Sidi Barani	5	7	11.2	H	Abu Simbil	9	11	0.05	H
El Farafra	5	7	10	V	Ras Gharib	9	11	66	H
Alexandria	6	11	110	H	Salum	9	11	6.9	H
Beni Ali	6	9	5	V	Qena	9	6	80	H
El Arish	6	10	182	V	Assiut	10	6	67.2	H
Isna	6	9	17	H	El Kharga	10	8	40	V
Siwa	6	8	10	V	Kom Ombo	10	7	40	H
Dahab	6	8	9.33	V	Matruh	10	8	39.2	H
El Daba	7	5	2.95	V	Bawiti	10	8	21.9	H
Quseir	7	5	50	H	Beni Suef	11	7	110	V
Sohag	7	11	52	H	Ismailia	11	9	260	V
Suez	7	5	20	H	Luxor	11	7	19	H
Managem	7	5	2.5	H	Safaga	11	9	50	V

Stations	I	II	kW	Pol	Stations	I	II	kW	Pol
Negala	22	25	5	H	Taba	32	37	25.7	H
Nuweiba	26	35	25.7	V	Barnis	24	29	830	H
Sh. El-Sheikh	27	33	8.91	H	El Hamam	48	51	69.2	V

D.Prgr: Prgr. I: 0400-0100.
Prgr. II: 0500-2400.
Prgr. III: Cairo ch7 100/10kW H, ch21 100/10kW H.
D.Prgr: 1100-2330.
Prgr. IV: Suez ch54 20kW H, Ismailia ch31 260kW V, Port Said ch42 10kW V, Cairo ch40 8.9kW.
D.Prgr: 1400-2330 (Mon: 1200-2330).
Prgr. V: Alexandria ch36 670kW H, Cairo ch46 9.5kW
D.Prgr: 1400-2400 (Thu 1400-0500), (Fri 0800-2400)
Prgr VI: Cairo ch43 9kW H, Mahalla ch49 32kW H
D.Prgr: 1500-2400 (Fri 1400-2400)
Prgr VII: Beni Ali ch22 1.32kW V, El Minya ch39 5.6kW V, Cairo ch42 6.5kW V, Assiut ch48 11.7kW H, Beni Suef ch51 4.3kW V, Fayoum ch55 10.7kW V
D.Prgr: 1400-2200
Esat Prgr: Shalatin ch96 10kW H, Cathrine ch12 1kW H, Marsa Alam ch21, Halayeb ch25 1kW, Natron ch12 1kW, Hasana ch 34, Nekhel ch34, Ras Sedr ch27 0.4kW, Abu Zonema ch33 o.4kW
D.Prgr: 24h.

EQUATORIAL GUINEA

TV-sets: 2.500 — **Colour:** SECAM — **System:** B.

TELEVISION NACIONAL (Gov.)
Malabo Bioko Norte.
L.P: Dir: Antonio Nkulu Oye.
Station: chE2.
No further information available.

ETHIOPIA

TV Sets: 150.000 — **Colour:** PAL — **System:** B (Pol=H).

ETHIOPIAN TELEVISION (Gov.)
P.O. Box 5544, Addis Ababa. ☎ +251 (1) 516977. ▤ +251 (1) 512686.
L.P: St. Mgr: Teshome Asrat. TD: Taye Zewde. Film Buyer: Mrs. Meaza Zewde.

City	ch	kW	City	ch	kW
Bahrdar	5	0.5	Gambella	8	1
Diredawa	5	0.5	Arbaminch	9	1
Jimma	5	0.5	Assaita	9	1
Shashemene	5	1	Axum	9	1
Debrebirhan	6	1	Debrezeit	9	0.2
Debre Markos	6	1	Dessie	9	1
Goba	6	1	Godie	9	1
Addis Ababa	7	5	Nekemptie	9	0.5
Gondar	7	0.5	Assosa	11	1
Harar	7	1	Dilla	11	0.25
Mekele	7	1	Jijiga	11	0.2
Metu	7	0.5	Nazereth	11	1
Yirgalem	7	0.2			

D.Prgr: Mon-Fri 1900-2230, Sat/Sun 1800-2400 (Sun 2300).

GABON

TV-sets: 40.000 — **Colour:** SECAM — **System:** K.

RADIODIFFUSION-TÉLÉVISION GABONAISE (Gov.)
B.P. 10150, Libreville. ☎ +241 732152. ▤ +241 732153. **Cable:** Radiotelegabon.
L.P: DG (TV): Jules César Lekogho.
Stations: Libreville chK4 & K8 2kW H, Port Gentil chK10 0.1kW H, Moanda 1kW (relay) + 5 low power relay sts.
D.Prgr: 1800-2200; ch10: 1800-2100 (relay Libreville).

GAMBIA

TV-Sets: 6.000 — **Colour:** PAL — **System:** B.

RADIO GAMBIA
Mile 7, Banjul. ☎ +220 495101/495921. ▤ +220 495102/495923.
L.P: Senior Op's Mgr (Broadcasting): Momodou Cham. Contr. TV Prgrs: Lasana Jobarteh.

GHANA

TV-sets: 800.000 — **Colour:** PAL — **System:** B.

GHANA BROADCASTING CORPORATION
P.O. Box 1633, Accra. ☎ +233 (21) 221161. ▤ +233 (21) 773240 .
Ⓣ 2114 GBC GH.
L.P: DG: Dr Kofi Frimpong. Dir. of TV: Prof. Mark Duodu. Dep. Dir. of TV (News): Berfi Apenteng. Dep. Dir. of TV (Prgrs): H. Torto-Gilbertson.
Stations: 1) Adjangote Greater Accra-10kW. 2) Jamasi Ashati-10kW. 3) Kissi Central-10kW. 4) Han Upper West-5kW. 5) Bolgatanga Upper East-5kW. 6) Tamale Northern-10kW. 7) Sunyaani Brong Ahafo-10kW. 8) Amedzofe Volta-5kW. 9) Akatsi Volta-5kW. 10) Tarkwa Western-5kW. 11) Sefwi Wiawso Western-5kW.
D.Prgr: 1630 (SS 0930)-0100. Incl. relays of CNN International.

GUINEA (Republic)

TV sets: 65.000 — **Colour:** PAL — **System:** K.

RADIODIFFUSION TÉLÉVISION GUINÉENNE (Gov.)
B.P. 391, Conakry. ☎ +224 442205. Ⓣ 22341 rtg ge conakry - Guinée.
L.P: Dir. Gen: B. Camara. Dir. of Prgrs: B. Kaba.
Stations: Conakry ch 5 1kW + Kindia ch4 0.2 kW, Faranah ch5 0.5kW, Labé ch7 8kW, Mamou/Mali ch9 0.2kW, Kankan ch9 1kW.
D.Prgr: 1700-2000 (approx).

KENYA

TV-sets: 500.000 — **Colour:** PAL — **System:** B.

KBC TV (Gov, Comm.)
Box 30456, Nairobi. ☎ +254 (2) 334567. ▤ +254 (2) 220675 Ⓣ 25361 KBC KE.
L.P: MD: Philip. Okundi. Contr. TV Prgrs: Joseph Murema. Asst. Mgr. Tech. Sces (TV): Ben Muinde.
Stations: ch4 (Limuru) 10/1kW, Timboroa ch2 10/1 kW, Mombasa ch6 10/1kW, Mazeras ch6 5/1kW, Nyeri ch10 10/1kW, Nyambene ch11 10/1kW
D.Prgrs: 1400 (Sat/Sun 1100)-2100.

COMMERCIAL STATIONS

KENYA TELEVISION NETWORK (KTN-TV)
P.O. Box 56985, Nairobi. ☎ +254 (2) 227122. ▤ +254 (2) 214467.
Web: http://www.kenyaweb.com/ktn/ktn.html
L.P: Chmn: Mwakio Sio. MD: Sam Shollei. CEO: Steve Crozier. Mktg. Mgr: Patrick Ndeda. Tech. Mgr: Francis Kimore.
Station: UHF ch 62 (Nairobi), ch59 (Ngong).
D.Prgr: 24hrs with own programming, also incl. relays of CNN.
F.Pl: extend coverage to coast and we. Kenya by the end of 1997.

LESOTHO

TV-sets: 250.000 — **Colour:** PAL — **System:** I.

LESOTHO NATIONAL BROADCASTING SERVICES
P.O. Box 552, Maseru 100. ☎ +266 323561. ▤ +266 310003. Ⓣ 4340 LENA 10
L.P: Ag. Cont. of TV: Litebo Tshola.
D.Prgr: No information available.

LIBERIA

TV-sets: 45.000 — **Colour:** PAL — **System:** B.

LIBERIAN BROADCASTING CORPORATION (Gov. Comm.)
P.O. Box 10-594, Monrovia. ☎ +231 271250. ① 44249 BROADCAST.
L.P: DG: Jesse B. Karnley.
Station: ELTV chE6 1/0.1kW + 4 low power repeaters.
D.Prgr: W 1815 (Sat 1615)-2300; Sun 1415-2230.

LIBYA

TV-sets: 550.000 — **Colour:** PAL — **System:** B.

PEOPLES REVOLUTION BROADCASTING TV (Gov.)
P.O. Box 333, Tripoli — **L.P:** Dir: Youssif Debri.
Stations:

	chE	kW (ERP)	Pol		chE	kW (ERP)	Pol
Benghazi	5	10/2	V	Khoms	8	5/1	V
Tripoli	6	20/4	H	Tobruk	8	5/1	H
Derna	6	5/1	H	Yefren	9	20/4	H
Elmarj	7	5/1	V	El Beida	9	5/1	H
Houn	7	1/0.2	H	Misurata	10	5/1	H
Sirte	7	1/0.2	H	Egdabia	11	15/3	H

+ 1 low power repeater.
D.Prgr's: 1700-2230.

MADAGASCAR

TV-sets: 130.000 — **Colour:** SECAM — **System:** K.

RADIO TELEVISION MALAGASY (Gov.)
P.O. Box 442, Antananarivo. ☎ +261 (2) 21784. ① 22506.
L.P: Dir: M. Rabesahala.
Stations: Antananarivo ch5 1kW H + 36 low power st's.
D.Prgr: 1600-1930.

MADEIRA

TV-sets: 80.700 — **Colour:** PAL — **System:** B.

RADIO TELEVISÃO PORTUGUESA, E.P. (Comm.)
Rua das Maravilhas 42, 9000 Funchal. ☎ +351 45197/8. 🖷 +351 48859. ① 72478.
L.P: Gen. Mgr: A. Abreu.
Stations: Funchal (Pico do Silva) chE5 20/4kW H, with repeaters on ch5 100kW H, 6 20kW H, 8 100kW H (+4 low power st's not mentioned.)
N.B: only RTP's 1st. prgm. is broadcast.
D.Prgr: 1800 (Sat/Sun 1400)-2300.

MALI

TV-sets: 10.000 — **Colour:** SECAM — **System:** K.

RADIODIFFUSION TÉLÉVISION DU MALI
B.P. 171, Bamako. ☎ +223 (22) 2019 2243 08.
L.P: DG: Abdoulaye Sidibe. DG Adj: Sidki Konate. CE: Mahamadou Sow.
Stations: Bamako ch5 10kW + 2 repeaters.
D.Prgr: 1854 (Sat/Sun 1454)-2300.

MAURITANIA (Islamic Rep.)

TV-sets: 1100 — **Colour:** SECAM — **System:** B.

TÉLÉVISION NATIONALE (Gov.)
TVM (Television du Mauritanie)
B.P. 5522, Nouakchott. ☎ +222 53303. ① 5817.
Station: Nouakchott chE5 (2x1kW).
D.Prgr: 2000-2245.

MAURITIUS

TV-sets: 156.850 — **Colour:** SECAM — **System:** B.

MAURITIUS BROADCASTING CORP. (Comm.)
1 Louis Pasteur Str, Forest Side. ☎ +230 675 5001/2. 🖷 +230 675 7332. ① 4320 MAUBROAD
L.P: Chairm: Mr Denis Rivet; DG: Mr Bijaye Madhou; Dep Chief Eng: Mr. Pather Amoordalingum
Stations:
MBC 1: Malherbes ch. 10 kW (H) ERP + 12 relay st's 0.6kW ERP.
MBC 2: Malherbes ch. 15kW (H) ERP + 14 relay st's 0.6kW ERP.
MBC 3: Malherbes ch. 23kW (H) ERP + 12 relay st's 0.6kW ERP.
MBC 4: *Malherbes ch. 23kW (H) ERP + 3 relay st's 0.6kW ERP.
MBC 5: *Malherbes ch. 23kW (H) ERP + 3 relay st's 0.6kW ERP.
Rodrigues TV: Pointe Venus ch. 0.3kW (H) ERP + 4 relay st's 1kW ERP.
(*= pay TV)
D.Prgr: 1200 (Sat 0530/Sun 0600)-1900.

MAYOTTE

TV-sets: 3.500 — **Colour:** SECAM — **System:** K.

R.F.O.—MAYOTTE
B.P. 103, F-97610 Pamandzi, Ile de Mayotte. ☎ +269 601017. 🖷 +269 601852.
L.P: St. Dir: Robert Xavier. Dir. Tec: Serge Sulpice-Timothee.
Stations: Lavigie ch 9H 100W, Mamadzou ch 7H 50W, Lima Combanich 4H 200W.
D.Prgr: Mon-Fri 1100-2100, Sat/Sun 0800-2100.

MOROCCO

TV-sets: 1.210.000 — **Colour:** SECAM — **System:** B.

RADIODIFFUSION TÉLÉVISION MAROCAINE (Gov.)
1, rue Al Brihi, Rabat. ☎ +212 (7) 704963. 🖷 +212 (7) 722047. ① 36577.
L.P: DG: Mohamed Tricha. Dir. of TV: Mohamed Issari. Head of Ext. Rel: A. Bekkali Abdellatif.
Stations: Pol: H.

	ch	kW(ERP)		ch	kW(ERP)
Zerhoun	4	120/12	Tan Tan	8	11/1.1
Zaio	4	9/1.8	Safi	8	20/2
Laayoune	E4	316/31.6	Tazerkount	8	90/18
Oujda	5	267/26.7	Tazekka	8	4/0.8
Boukhouali	5	150/15	Touzarine	9	9/1.8
Tanger	5	20/4	Tiguelmamine	9	9/1.8
Sidi Bounouara	5	11/1.1	S. Bounouara	9	11/1
Oukaimeden	6	18/3.6	Biougra	9	4/0.8
Azougar	6	9/1.8	Casablanca	10	180/18
Dakhla	6	11/1.1	Hafa Safa	10	9/1.8
Figuig	6	9/1.8	Ourzazate	10	267/26.7
Rabat	7	180/18	Bouarfa	10	267/26.7
Izeft	7	14/2.8	Essaouira	11	20/2

+ 35 low power relay sts.
N.B: All ch's M except where otherwise indicated.
D.Prgr. in Arabic/French: Mon-Thurs 1215-1415 & 1700-0100v, Fri/Sat/Sun 1215-0100v.

2M INTERNATIONAL (Comm.)
KM7, 300 route de Rabat, Casablanca. ☎ +212 (2) 354444. 🖷 +212 (2) 354071.
L.P: MD: Tawfik Bennani-Smires. Tech. Dir: Driss Anouar. Comm. Dir: Mouhaddab Khadija.
D.Prgr: 15h daily, partly subscription.
F.PI: Pan-African satellite sce.

MOZAMBIQUE

TV sets: 35.000 — **Colour:** PAL — **System:** B.

TELEVISÃO DE MOÇAMBIQUE TVM (Gov.)
C.P. 2675, Maputo. ☎ +258 744788 or 741395.

① 6-346 TEVEMP MO.
L.P: Dir: Botelho Moniz. Dt.Tec: Jaime Ferreira. Film Buyer: Arlando Tembe.
Station: Maputo ch33 1kW.
D.Prgr: 24h weekly.

NAMIBIA

TV-sets: 38.500 — **Colour:** PAL — **System:** I.

NAMIBIAN BROADCASTING CORPORATION
✉ P.O. Box 321, Windhoek 9000. ☎ +264 (61) 215811. 🖷 +264 (61) 2912291. ① 50908 - 622/708. **Cable:** Broadcast.
L.P: DG: Dan Tjongarero. Sen. Contr. Prgrs: Gabriel Haindaka. CE: Martin Venter. Head of PR: Cyril Lowe.

Stations	ch	Stations	ch
Rundu	4	Windhoek	6
Keetmanshoop	4	Erongo	7
Paresis	5		

+ 3 low power st's and approx. 20 private low power relay st's.
D.Prgrs: Mon-Sat: 1400-2100, Sun: 0600-1100, 1400-2100 in English.

NIGER

TV-sets: 25.000 — **Colour:** SECAM — **System:** K.

TÉLÉ-SAHEL (Gov.)
✉ B.P. 309, Niamey. ☎ +227 723155. ① 5229 NI.
L.P: Dir. Gen: DG: Abdou Souley. TD: Zoudi Issouf.

Stations	ch	kW	Pol		ch	kW	Pol
Agadez	4	10	H	Dogondoutchi	7	1	H
Dosso	4	10	H	Gaya	8	1	V
Zinder	5	10	H	Niamey	9	10	H
Arlit	6	1	H	Konni	9	10	H
Maradi	7	10	V	Diffa	9	10	H

+ 7 low power relay st's.
D.Prgr: 0630-1130, 1430-1730.

NIGERIA

TV-sets: 6.100.000 — **Colour:** PAL — **Systems:** B.

NIGERIAN TELEVISION AUTHORITY (Gov.)
✉ Television House, PMB 120005, Victoria Island, Lagos. ☎ +234 (1) 614966/615154/612529. 🖷 +234 (1) 2610289. ① 21245 NTA HQ.
L.P: D G: Mohammed Ibrahim. Dir. Tec: Isaac Wakombo. Dir. of Prgrs: Prince Bayo Sanda.
National TV Production Centre.
TV House, Victoria Island, PMB 12005, Lagos.
L.P: Man. Dir: A. Micika.

REUNION

TV-sets: 90.500 — **Colour:** SECAM — **System:** K.

SOCIÉTÉ NATIONALE DE RADIO-TÉLÉVISION FRANÇAISE D'OUTRE-MER (RFO)
✉ 1, rue Jean-Chatel, F-97716 Saint-Denis. ☎ +262 406767. 🖷 +262 406771.
L.P: Regional Dir: Jean-Philippe Roussy.
Stations: P. Textor chK'9 0.5kW + 18 low power repeaters.
D.Prgr: Prgr.I: 1300 (Sat/Sun 1100)-1930; Prgr.II: no times available.

PRIVATE STATIONS

ANTENNE REUNION (Priv., Comm.)
✉ 33 Chemin Vavangues, 97400 Saint Denis. ☎ +262 48 2828. 🖷 +26248 2829
Stations: Saint Leu ch36 1kW, Sainte Suzanne ch42 2kW, Le Port ch57 9kW, Saint Pierre ch61 2kW, Saint Joseph ch55 1,7kW, Saint Benoit

ch26 2,4kW, Saint Denis ch33 2kW, Piton Textor ch56 2kW.
D.Prgr: 0830-1830

CANAL REUNION (Priv., Comm.)
✉ 35 Chemin Vavangues, 97400 Saint Denis. ☎ +262 29 0202. 🖷 +262 29 1709
Stations: Saint Joseph ch52 1.7kW, Sainte Suzanne ch39 2kW, Le Port ch54 9kW, Saint Denis ch25 2kW, Piton Textor ch53 2kW, Saint Pierre ch26 2kW.
D. Prgr: 0245-2200, Sun 24h

TV SUD (Priv., Comm.)
✉ 10 rue Aristide Briand, 97430 Le Tampon. ☎ +262 57 4242
Stations: Saint Pierre ch58 2kW, Les Avirons ch60 0.72kW
D.Prgr: 1400-1800

TV-4 (Priv., Comm.)
✉ 8 chemin Fontbrune, 97400 Saint Denis. ☎ +262 52 7373
Stations: Saint Leu ch49 1kW, Saint Denis ch52 2kW, Sainte Suzanne ch31 2kW, Le Port ch65 9kW, Piton Textor ch63 2kW
D.Prgr: 0230-2130

SÃO TOMÉ E PRINCÍPE

TV-Sets: 21.000 — **Colour:** PAL — **System:** B & G.

TELEVISÃO DE SÃO TOMÉ E PRINCÍPE
✉ P.O. Box 393, S. Tomé, Republica de S. Tomé e Princípe, Africa. ☎ +239 (12) 21041/22970 🖷 +239 (12) 21942.
L.P: Dir: Carlos Teixeira d'Alva
Stations: One 2kW transmitter in S. Tomé and one 10Watts transmitter in Principe covering 80% of the area. Channels: 11, 7, 5.
D.Prgr: RTP Internaçional is also relayed.

SENEGAL

TV-sets: 61.000 — **Colour:** SECAM — **System:** K

Radiodiffusion Télévision Sénégalaise (RTS) (Gov.)
✉ B.P. 1765, Dakar. ☎ +221 21780. 🖷 +221 223490.
L.P: DG: Guila Thiam. Dir. of TV Sces: Babacar Diagne. Dir. of Tech. Sces: Seydou Diallo. Head of Ext. Affairs: Ka Aissatou.
Station: Dakar ch7 10kW
D.Prgr: 1900-2300 (Sat-Sun 1200-2330)

CANAL HORIZONS Sénégal (Pay TV Sce.)
✉ 31 ave. Albert Sarraut (B.P. 1390), Dakar. ☎ +221 232525. 🖷 221 233030.
L.P: Pres: Fara N'Diaye. DG: Jacques Barbier de Crozes. Dep. DG: Anne Marie Senghor. Tech. Coordinator: Issa Laye Diop.

SEYCHELLES

TV-sets: 14.000 — **Colour:** PAL — **System:** B.

SEYCHELLES BROADCASTING CORPORATION
✉ P.O. Box 321, Hermitage, Mahe. ☎ +248 22416. 🖷 +248 22564. ① 2315 INFO TV SEZ.
L.P: MD: Ibrahim Afif. Dir. Admin. and Personnel: Fauzia Rose. Sen. Eng: (TV): Joyvani Chetty. Chief Editor (TV): Ms. Marie-Claire Elizabeth. Prgr. Mgr (TV): Jean-Claude Matombe.
Stations: La Misère chE2 1kW, St. Louis chE7 6kW + 9 low power repeaters.
D.Prgr: Mon-Fri 1345-1830, Sat 1200-1830, Sun 1000-1830

SIERRA LEONE

TV-sets: 25.000 — **Colour:** PAL — **System:** B.

SIERRA LEONE TELEVISION (Gov. Comm.)

✉ Private Mailbag, Freetown. ☎ + 232 (22) 40403/40906. ① RADTEX 3334 SL.
LP: Ag. Head of Prgrs: G. Felix-George. Sen Eng. (TV): W.A.I. George.
Station: chE2 1kW H, chE7 126kW H.
D.Prgr: 1745-2330.

SOMALIA (Rep.)

TV sets: N/A — **Colour:** PAL — **System:** B.

MINISTRY OF INFORMATION
✉ P.O. Box 1748. ① 999621 Mogadishu.
LP: Dir. of TV project: A. Ali Askar. Dir. Tec: A. Hassan.
Station: ch6 1kW.
D.Prgrs: 1700-2000 (Fri 2100).

SOUTH AFRICA

TV-sets: 3.485.000 — **Colour:** PAL — **System:** I.

SENTECH (PTY) LTD.
✉ Private Bag X06, Honeydew 2040. ☎ +27 (11) 475 1596. 🖷 +27 (11) 475 5112. **Web:** http://www.sentech.co.za
Sentech is responsible for the signal distribution of all radio and TV services in South Africa.

SOUTH AFRICAN BROADCASTING CORPORATION (SABC)
✉ (Head Office): Broadcasting Centre, Auckland Park, Johannesburg 2092/Private Bag XI, Auckland Park 2006. ☎ +27 (11) 714 9111. 🖷 +27 (11) 714 3106. **Cable:** Broadcast.
Web: www.sabc.co.za
LP: Chmn (Board): Dr. Ivy Matsepe-Casaburri. Group Chief Exec: Zwelakhe Sisulu. Chief Exec. (Signal Distribution): Neël Smuts. Chief Exec. (Operations): Gert Claassen. Chief Exec. (Human Resources): Ntombi Langa. Ag. Group GM (Finance): Talib Sadik. Sen. GM Strategic Planning: Solly Mokoetle. Group Sces. Co-ordinator: Leslie Xinwa.

CCIR System 1 (PAL colour) used on bands III/IV/V using ch4-13 (174-254MHz) and 21-68 (470-845MHz), chs 12/38 not used. Sound/Vision spacing is + 6MHz.

SABC-1:
Stations: 100kW (ERP) exc. *) 1-10kW

Location	ch	Pol		ch	Pol
Durban	4	H	Ladismith (C.P.)*	22	H
Kimberley	4	H	Bedford*	23	H
Potgietersrus	4	H	Boesmanskop*	23	H
De Aar	5	H	Nelspruit	24	H
George*	5	V	Petrus Steyn*	24	H
Pretoria	5	H	Kareedouw*	25	H
Theunissen	5	H	Glencoe	27	H
Donnybrook*	6	H	Schweizer Reneke	33	H
Graaff-Reinet*	6	V	Eshowe	36	H
Pomfret*	6	H	Mooi River*	37	H
Prieska*	6	V	Springfontein*	37	H
Springbok*	6	V	Hoedspruit*	39	H
Volkrust*	6	V	Matjiesfontein*	39	H
Walvis Bay*	6	V	Vryheid*	39	H
Port Elizabeth	7	H	Carnarvon*	40	H
Queenstown	7	H	Matatiele*	40	H
Welverdiend	7	H	Cradock*	40	H
Villiersdorp	7	H	Middelburg (Tvl)	41	H
Cape Town	8	V	Senekal*	42	H
Garies*	8	H	Williston*	42	V
Grahamstown	8	H	Klerksdorp	45	H
Kuruman	8	H	Blouberg*	45	V
Louis Trichardt*	8	V	Ubombo*	45	H
Port Shepstone	8	V	Carolina*	50	H
Bloemfontein	9	H	Piet Plessis*	50	H
East London	9	H	Zeerust*	52	H
Johannesburg	9	H	Dullstroom*	53	H
Napier*	9	V	Greytown*	53	H
Piketberg	9	H	Noupoort*	54	H
Oudshoorn	9	H	Bethlehem	55	H
Thabazimbi	9	H	King William'sTown*	56	H
Victoria West	9	H	Ladybrand*	56	H
Vanrhynsdorp	10	H	Rustenburg*	56	H
Beaufort West*	10	H	Douglas*	57	H
Pofadder	10	V	Kroonstad	57	H
Upington	10	H	Willowmore*	57	H
Piet Retief*	11	H	Suurberg*	59	H
Riversdale	13	H	Aliwal North	61	H
Davel	22	H	Christiana*	62	H
Calvinia*	22	H	Tzaneen	64	H
Enzelsberg	22	H			

+ 70 gap fillers and estimated 400 privately owned low power tr's.
D.Prgr: 0400-0615 (Mon-Fri), 1300 (Sat 0400, Sun 1130)-2200 (Sat 2230).

SABC-2

Location	ch	Pol		ch	Pol
Hoedspruit	43	H	Kimberley	24	H
Nelspruit	36	H	Klerksdorp	37	H
Potgietersrus	13	H	Durban	13	H
Pretoria	11	V	Glencoe	23	H
Tzaneen	56	H	Cape Town	34	V
Johannesburg	13	H	East London	4	H
Welverdiend	10	H	Port Elizabeth	13	H
Bloemfontein	44	H			

+27 gap fillers

SABC-3

Location	ch	Pol		ch	Pol
Aliwal North	53	H	Middelburg	45	H
Bethlehem	63	H	Nelspruit	32	H
Bloemfontein	13	H	Nongoma	58	H
Boesmanskop	27	H	Oudtshoorn	6	H
Cape Town	5	V	Piet Retief	5	H
Christiana	58	H	Piketberg	6	H
Cradock	48	H	Port Elizabeth	4	H
Davel	30	H	Port Shepstone	5	V
Donnybrook	9	H	Potgietersrus	7	H
Durban	7	H	Pretoria	8	V
Dullstroom	61	H	Queenstown	4	H
East London	13	H	Riversdale	8	H
Eshowe	28	H	Rustenburg	64	H
George	11	V	Schweizer-Reneke	25	H
Glencoe	31	H	Senekal	38	H
Grahamstown	5	H	Thabazimbi	9	H
Greytown	61	H	Theunissen	8	H
Hoedspruit	47	H	Tzaneen	60	H
Johannesburg	6	H	Ubombo	37	H
Kimberley	7	H	Upington	25	H
King William's Town	60	H	Villiersdorp	10	H
Klerksdorp	41	H	Volksrust	9	V
Kroonstad	61	H	Vryheid	47	H
Kuruman	11	H	Welverdiend	4	H
Ladybrand	60	H	Zeerust	44	H
Louis Trichardt	11	V			

+ 47 Gap fillers.

D.Prgrs: broadcasts in various languages.

M-NET TELEVISION (Pay channel, Comm, Priv.)
✉ P.O. Box 4950, Randburg 2125. ☎ +27 (11)329 5156. 🖷 +27 (11) 329 5166 — **LP:** PD: Sheryl Raine. Mktg. Dir: Etienne Heyns. Head of PR: John Badenhorst.
Stations: Bloemfontein ch6 (H), Alverstone & Pt. Elizabeth ch10 (H), Constantiaberg ch11 (V), Pretoria ch21 (H), Johannesburg ch39 (H) + 16 repeaters, Durban ch10, Cape Town ch11, George ch7, Newcastle ch62.
D.Prgr: Mon-Fri 0830-1030 & 1300-2300; Sat/Sun 0500-2300. **Indian Prgr:** Sun 0530-0830. **Portuguese Prgr:** Sun 0830-1130.

SUDAN

TV-sets: 250.000 — **Colour:** PAL — **System:** B.

SUDAN TELEVISION (Gov, Comm.)
✉ P.O. Box 1094, Omdurman. ☎ +249 (11) 55022. ① 28002, 28053.
Cable: Sudan TV.
L.P: Head of Directorate: Hadid al-Sira.
Stations: Omdurman chE5 5kW H, Gezira chE7 10kW, Atbara chE9 0.5kW.
D.Prgr: 1500-2200.

SWAZILAND

TV sets: 12.500 — **Colour:** PAL — **Systems:** B & G.

SWAZILAND TELEVISION AUTHORITY
✉ Swazi TV, P.O. Box A146, Swazi Plaza, Mbabane. ☎ +268 43036/7.
🖹 +268 42093. ① 2138 WD.
L.P: MD: Dan S. Lamini.
Stations: Bulembu ch5 1.5kW H, Ntondozi ch15 15kW H + 7 relay st's.
D.Prgr. in English: 1600-2100.

TANZANIA

TV-sets: 80.000 — **Colour:** PAL — **System:** B.

TELEVISION ZANZIBAR TVZ (Gov.)
✉ P.O. Box 314, Zanzibar. ☎ +255 (54) 32816/7. ① 57200 TVZ TZ.
L.P: MD: Jama A. Simba. TD: George H. Majaliwa. Film Buyer: Jaffar S. Kassingo; Prod. Mgr: Abdulhamid H. Dau.
Stations: Unguja chE21 40kW, Pemba chE9 40kW (ERP).
D.Prgr 1645-1900 (Sat/Sun/National Holidays 1645-2000).

INDEPENDENT TELEVISION (ITV)
Stations: Dar es Salaam UHF ch24
D. Prgr: 1400-1505/1700-1800 (weekdays), local prgr. Sundays and Tuesdays from 1600-1630

TOGO

TV sets: 150.000 — **Colour:** SECAM — **System:** K.

TÉLÉVISION TOGOLAISE (Gov.)
✉ B.P. 3286, Lomé. ☎ +228 215357. 🖹 +228 215786. ① 5320 ATOP.
L.P: Dir. of TV: Yao Martin Ahiavee. Tech. Dir: Vokou Raphaël Soumsa. Film Buyer: Ayi Léopold Mamavi.
Stations: Mt. Agou ch6 10KW V, Lomé ch8 1kW H, Aldjo-Kadara ch8 10kW H + 2 relay st's.
D.Prgr: Mon-Fri 1830-2230, Sat/Sun 1230-2400.

TUNISIA

TV-sets: 650.000 — **Colour:** SECAM — **System:** B&G.

ENTREPRISE DE LA RADIODIFFUSION-TÉLÉVISION TUNISIENNE E.R.T.T. (GOV.)
✉ 71, Ave de la Liberté, 1002 Tunis Belvedere. ☎ +216 (1) 287300, 782700 . 🖹 +216 (1) 781058. ① 14960.
L.P: Dir: Abdeh Afidh Hardudm.
Stations: Network I (Arabic)

	ch	kW	Pol		ch	kW	Pol
Remada	4	32/6.4	H	Gafsa	10	400/80	V
Kasserine	6	340/68	H	Zaghouan	11	280/56	H
Boukornine	7	12.5/1.25	V	Ain Draham	12	32/6.4	H
Sfax	8	365/73	H	Zarzis	12	180/36	H
Goraa	9	32/6.4	H				

Network II: (French) Pol=H.

	ch	kW		ch	kW
Zarzis	22	1500	Ain Drahan	35	87
Biadha	23	780	Chambi	40	1410
Boukornine	26	44	Ghraba	51	1800
Zaghouan	33	1425	Bizerte	53	40
Remada	33	89	Goroa	55	89

low power st's not mentioned.
D.Prgr: Netw. I & II: 1600-2300.

UGANDA

TV-sets: 115.000 — **Colour:** PAL — **System:** B (Pol=H).

UGANDA TELEVISION (Gov.)
✉ P.O. Box 7142, Kampala. ☎ +256 (41) 254461. ① KNOLLEDGE 61084.
L.P: Ag. Dir: J.M.A. Obo.
Stations: Kampala chE5, Lira chE7, Masaka chE8, Mbale chE8, Mbarara chE10, Soroti chE10. Gulu chE9, Jinja chE11 + low power repeater at Kabale.
D.Prgr: Mon-Fri 1500-2100; Sat/Sun 1200-2100.

ZAMBIA

TV-sets: 200.000 — **Colour:** PAL — **System:** B.

TELEVISION-ZAMBIA (Gov.)
✉ Broadc. House, P.O. 50015, Lusaka 229648. ☎ +260 (1) 220864-74; P.O. Box 20748, Kitwe. ☎ +260 (2) 223555. ① 41221 ZA. **Cable:** Broadcasts TV.
L.P: Act. Controller: Miss Emelda Yumbe.

Stations:	chE	kW	Pol		chE	kW	Pol
Solwezi	3	0.075	H	Kitwe	9	200	H
Kapiri Mposhi	6	200	H	Lusaka	10	200	H
Pemba	8	200	V	Senkobo	10	200	V
Kasama	8	16	H	Chipata	11	16	H
Mubwa	8	0.1	H				

D.Prgrs: Mon-Thur. 1700-2230; Fri-Sat 1500-2400; Sun 1500-2230.

ZIMBABWE

TV-sets: 375.000 — **Colour:** PAL — **System:** B.

ZIMBABWE BROADCASTING CORPORATION (Independent Statutory Body, Comm.)
✉ P.O. Box HG 444, Highlands, Harare. ☎ +263 (4) 498659, 498670, 498630, 498620. 🖹 +263 (4) 498613. ① 24175 ZBCHOVZW.
L.P: DG: Edward Moyo. Dir. Eng. and Tech. Sces: Elliot Muchimb ri. Contr. TV Prgrs: Mr Nyasha Masiwa.

Stations	chE	kW	Pol	Stations	chE	kW	Pol
Gweru	2	17.6	H	Kadoma	6	4	H
Bulawayo	3	3	H	Bulawayo	6	6	V
Harare	4	20	H	Mutare	7	12	V
Harare	5	30	H	Masvingo	7	6	H
Gwendingwe	5	7	H	V. Falls	7	1	H
Rukotso	5	6	V	Gwanda	9	6	H
Kariba	5	1	H	Gweru	11	25	H
Kamativi	5	16	V	Mutoroshanga	12	8	H

D.Prgr: English/Shona/Ndebele 0700-2200.

NEAR & MIDDLE EAST

AFGHANISTAN

TV-sets: 100.000 — **Colour:** PAL — **System:** B.

RADIO-TELEVISION OF AFGHANISTAN (RTA) (Gov.)
✉ P.O. Box 544, Kabul. ☎ +93 25460/25373. ① 24288 (AFGRTV AF).
L.P: Gen Pres: Shamsul Haq Arayanfar. Tech Adv: Eng. Faizuddin Ferogh
Station: Kabul ch5 10/1kW H.
D.Prgr: 1330-1830.
Reg. st's are operating in 9 different provinces at 1330-1530 or 1430-1630 (approx.).

ARMENIA

TV-sets: N/A — **Colour:** SECAM — **Systems:** D&K.

ARMENIAN TELEVISION (Gov.)
✉ Alek Manukyan 5, 375025 Yerevan. ☎ +374 (2) 552502. 🖹 +374 (2) 551513.

Stations (Prgr. 1)	ch	Stations (Prgr. 2)	ch
Yerevan	1	Yerevan	11

D.Prgr: 1730-2100.   **D.Prgr:** 1730-2030.
NB: Prgr. 1 is carried by various main and relay transmitters throughout the country. Further details not known.

ORT

Stations	ch
Yerevan	8

D.Prgr: Relays of ORT from Moscow. Further details see under Russia.

ROSSIJSKOJE TELEVIDENIJE (RTV)

Stations	ch
Yerevan	5

D.Prgr: Relays of RTV from Moscow. Further details see under Russia.
NB: ORT and RTV are carried by various main and relay transmitters throughout the country. Further details not known.

AZERBAIJAN

TV-sets: N/A — **Colour:** SECAM — **Systems:** D&K

AZERBAYCAN RESPUBLIKASI RABITA NAZIRLIYI (Gov.)

✉ Azerbaycan pr. 33, 370139, Baki. ☎ +994 (12) 93 0004. 🖷 +994 (12) 98 3325. ℂ 142105 INAM SU
LP: Minister: Sirus Abasbeyli

TELERADIO (Gov.)

✉ Azerbaycan pr. 33, 370139, Baki. ☎ +994 (12) 98 8066. 🖷 +994 (12) 98 3325. ℂ 142472 DALGA

AZERBAYCAN RADIOTELIVIZIYA SIRKETL (Gov.)

✉ Mehdi Huseyin küçäsi 1, 370011 Baki. ☎ +994 (12) 39 8585. 🖷 +994 (12) 39 5452. ℂ 142214 TEMBR

AZERBAYCAN MILLI TELEVIZIYASI (Nat. TV, Gov.)

D.Prgr: Progamming in Azeri and Russian. 14 hrs. a day.
Stations: 17 powerful (5kW) tx, 48 100 W tx, 83 (1-30 W) tx.

BM-TV

Private TV program, 500 W. transmitter relay in Baki
D.Prgr: In Azeri and Russian 5 hrs.a day

TRT-1

Turkish Gov. TV program, 100 W transmitter relay prgr in Baki
D.Prgr: In Turkish

Stations:	National TV	Moscow TV-1	Russian TV
Baku	3 (5kW)	7 (5kW)	12 (5kW)
Gyandzha	11 (5kW)	8 (5kW)	31 (5kW)
Shusha	2 (5kW)	12 (5kW)	30 (1kW)
Duzdag	3 (5kW)	9 (5kW)	5 (5kW)
Ali-Bairamly	4 (0.1kW)	10 (5kW)	28 (1kW)
Lerik	6 (5kW)	2 (5kW)	? (5kW)
Ordubad	11 (5)	8 (5)	2 (0.1)
Dzhalilabad	1 (5kW)	—	7 (0.1kW)
Danachi	3 (5kW)	7 (5kW)	5 (0.1kW)
Lenkorañ	8 (5kW)	3 (0.1kW)	10 (0.1kW)
Dzhebrail	6 (5kW)	3 (0.1kW)	12 (0.1kW)
Imishly	11 (5kW)	7 (0.1kW)	21 (1kW)
Kuba	6 (5kW)	12 (0.1kW)	8 (0.1kW)
Shakuk	2 (5kW)	7 (0.1kW)	—
Astaka	9 (20kW)	5 (5kW)	2 (0.1kW)
Akstafa	7 (5kW)	—	—
Yvanovka	5 (5kW)	33 (2kW)	—
Kelbadzhari	3 (5kW)	6 (0.1kW)	—
Geokchay	3 (0.1kW)	8 (5kW)	25 (0.1kW)
Dashkesan	7 (0.1kW)	9 (0.1kW)	2 (0.1kW)
Sabirabad	3 (0.1kW)	2 (0.1kW)	12 (0.1kW)
Sheki	5 (0.1kW)	12 (0.1kW)	10 (0.1kW)
Mingechaur	1 (0.1kW)	9 (0.1kW)	4 (0.1kW)
Agdam	7 (0.1kW)	3 (0.1kW)	10 (0.1kW)
Kedabek	1 (0.1kW)	8 (0.1kW)	4 (0.1kW)
Zangelan	11 (0.1kW)	3 (0.1kW)	1 (0.1kW)
Shemakha	9 (0.1kW)	11 (0.1kW)	6 (0.1kW)
Divichi	12 (0.1kW)	5 (0.1kW)	26 (0.1kW)
Yardymly	5 (0.1kW)	8 (0.1kW)	26 (0.1kW)

Stations:	National TV	Moscow TV-1	Russian TV
Oguz	3 (0.1kW)	9 (0.1kW)	7 (0.1kW)
Kubatly	11 (0.1kW)	3 (0.1kW)	—
Lachin	9 (0.1kW)	12 (0.1kW)	—
Kürdamir	3 (0.1kW)	6 (0.1kW)	—
Gabala	1 (0.1kW)	4 (0.1kW)	—
Bakda	1 (0.1kW)	—	—
Salyany	—	8 (0.1kW)	3 (0.1kW)
Almaly	2 (0.1kW)	—	6 (0.1kW)
Dzhafarabad	6 (0.1kW)	—	1 (0.1kW)
Ger-Ger	12 (0.1kW)	10 (0.1kW)	—
Tauz	12 (0.1kW)	—	—
Kemekly	3 (0.1kW)	—	—
Shikhly	12 (0.1kW)	—	—
Sadhrak	—	5 (0.1kW)	—
Shakur	—	7 (0.1kW)	—
Neftechala	—	4 (0.1kW)	—
Jergündzh	—	10 (0.1kW)	—

\+ 81 low power st's not mentioned.

BAHRAIN

TV-sets: 270.000 — **Colour:** PAL — **Systems:** B&G.

BAHRAIN TELEVISION (Gov, Comm.)

✉ P.O. Box 1075, Bahrain. ☎ +973 781888/686000. 🖷 +973 681544. ℂ 8311.
LP: Dir: Dr. H. Al-Umran. Head of Prgrs: Fowzia Zainal. Head of Mktg: Maria Khoury. Dir. Pub. Rel: Ahmed Al Sherooqi.
Station: chE4, chE44, chE55, ch57.
D.Prgr: ch4: Main Arabic ch. loc. & int. news, cultural and variety prgrs, Arabic & English feature films.
ch44: Satellite broadcasts, loc. & int. sports. Also rel. Egypt satellite ch.
ch55: Main English ch. Also rel. CNN.
ch57: Transmits BBC World Sce. TV, 24h.

CYPRUS

TV-sets: 103.000 — **Colour:** PAL — **Systems:** B/G.

CYPRUS BROADCASTING CORPORATION

✉ P.O. Box 4824, 1397 Nicosia. **Cable:** Broadcasts. ☎ +357 (2) 422231. 🖷 +357 (2) 335010.
LP: DG: George Potamites. Dir. of Tech. Sces: Andreas Michaelides. Dir. of Prgrs: Panos Ioannides. Head of Public and Int. Rel: Nayia Roussou. Head of News & Current Affairs: Evangelos Louca. Head of TV prgrs: A. Papayrinnis. Head of TV and Radio Studios: Chrysanthos Hadjioannou.
Stations: Mt. Olympus chE6 40/4kW V + 35 low power st's.
Ch.1 in Greek: 1300 (Sun 0800)-2300.
Ch. 2 in Greek, Turkish, English: 1400 (Sun 1000)-2300.

LOGOS TV (Comm.)

✉ 20 St. Avgoustinou, Archangelos, Nicosia. ☎ +357 (2) 355595. 🖷 +357 (2) 355737.
LP: Chmn: Michalis Colocasides. DG: Frixos Klenatous.

LUMIERE TELEVISION (Comm. Pay-TV sce.)

✉ Papabisiliou Bldg, 70 Kennedy Ave, Nicosia. ☎ +357 (2) 311080. 🖷 +357 (2) 415767.
LP: Pres: Chris Economides. Dir: Joe Avraamides.

BFBS Akrotiri (SSVC)

✉ BFPO 57, Dhekelia Mil 381. ☎ +357 (474) 8518
Stations: ch 69 and 66 in ESBA, ch 60 and 68 in the WSBA.
D.Prgrs: relays of English prgrs. + live prgrs. from satellite.

NORTHERN CYPRUS

TV-sets: 75.000 — **Colour:** PAL — **Systems:** B & G.

BAYRAK RADIO & TELEVISION CORP.

✉ Yeni Organize Sanayi Bolgesi, Lefkosa via Mersin 10, Turkey. ☎ +90 392 228 5555. 🖷 57264 brtk. **E-mail:** brt@cc.emu.edu.tr
Web: http://www.cc.emu.edu.tr/press/brt/brt.htm
LP: DG: Ismet Kotak. Head of Tr's: A.Ziya Dïncer. Head of Admin:

Süleyman Türem. Head of Prgrs: Hüseyin Cobanoglu. Head of Sales: Mehmet Kircailiar. Head of Int. N: Huriye Dimililer.
Stations: Sinan dagi ch8 100kW (Prgr. 2) — Selvilitepe ch40 450kW (Prgr. 3), ch44 450kW (Prgr. 1) + 4 relay st's.
D.Prgrs: no details received.

IRAN

TV-sets: 7.000.000 — **Colour:** SECAM — **System:** B.

ISLAMIC REPUBLIC OF IRAN TELEVISION (Gov.)
✉ P.O. Box 19395 3333, Tehran; P.O. Box 15875-4344, Tehran ☎ +98 (21) 298053, 290079, 96715150. 🖷 +98 (21) 295056/ 291051. **Cable:** "IRIB IR". ① 213910/212431/212797/213253.
L.P: Pres: H.E. Dr. Ali Larijani. Gen Dir. Int. Affairs: Dr. A. Ghasemzadeh.
D.Prgr: Network I: 0545-1200 & 1600-2400 (on holidays 1400-2400). **Network II:** 0800-1300 & 1400-2400. **Network III:** 0830-1100 & 1730-2300
Local st's (28) Abadan, Ahwaz, Alamdeh, Ardebil, Baharlo, Bakhtaran, Bandar-Abbas, Booshehr, Esfahan, Hamedan, Kerman, Khoramabad, Kohe Genon, Kohe Noor, Kosangar, Mashhad, Oromieh, Rasht, Sanandaj, Sari, Shahrkord, Shiraz, Tabriz, Tehran, Yazd, Zahedan, Zanjan, Zibakenar + 450 low power repeaters.

IRAQ

TV-sets: 1 m — **Colour:** SECAM — **System:** B.

IRAQI BROADCASTING AND TELEVISION ESTABLISHMENT (IBTE)
✉ Salhiya Baghdad. ☎ +964 (1) 884 4412, 884 4413. 🖷 +964 (1) 541 0480. ① 212246 IDAH.
L.P: DG: Dr. Sabah Yaseen; Dir PR: Mr. Abdul-Ilah Al-Musfir.
Stations: 1=1st Prgr; 2=2nd Prgr; R=Regional Prgr.

Location	chE	kW/Pol	Location	chE	kW/Pol
Misan (1)	5	144H	Baghdad (1)	9	360H
Muthanna (1)	6	144V	Basrah (1)	9	360V
Al-Taamin (R)	6	144H	Nenavah (1)	9	180V
Baghdad (2)	7	288H	Al-Taamin (2)	10	144H
Nenavah (2)	7	144V	Misan (2)	11	180H
Al-Taamin (1)	8	180H	Basrah (2)	12	144V
Muthanna (2)	8	180V			

D.Prgrs: Prgr 1: 1400 (Fri 0700)-1930 (Fri 2100); Prgr 2: 1400-2100.

ISRAEL

TV-sets: 1.500.000 — **Colour:** PAL — **Systems:** B&G.

ISRAEL TELEVISION (operated by the IBA)
✉ P.O. Box 7139, Jerusalem 91071. ☎ +972 (2) 291888. 🖷 +972 (2) 292944. ① 25301 — **L.P:** DG: Ayre Mekel.

Stations	chE	kW	Stations	chE	kW
M. Ramon	5	15/1.5	Jerusalem	27	1/0.2
Acco	7	20/2	Tel Aviv	28	60/6
Eilat	7	0.15/0.015	Zefat	34	22/2.2
Eitanim	8	200/20	Haifa	36	40
Manara	8	1.5/0.15	Karmiel	39	1.5/0.15
Arad	9	10/1	Jerusalem	40	3/0.3
Grofit	10	15/1.5	Manara	44	20/2
Haifa	10	100/10	Haifa	46	200/20
Bar Jehuda	10	1.5/0.15	Jerusalem	56	1.5/0.15
Beersheba	11	100/10	K. Hayarden	60	20/2
Acco	24	50/5			

+31 low power repeaters.
Ch. 1 in Hebrew/Arabic: 0400-0600, 1430-2300v
Ch. 3: broadcasts only via satellite.

ISRAEL EDUCATIONAL TELEVISION
✉ 14 Klausner Str, Tel Aviv. ☎ +972 (3) 5434343. ① 342325 ITCIL.
L.P: Gen. Mgr: Y. Lorberbaum. Dir. of Eng: A. Kaplan. Dir. of Prgrs: Mrs. Y. Prener.
Stations: uses same tr's as the IBA (see above).
D.Prgr. Educational: 0600-1530 (Fri 1430). (Sat: no trs).

CHANNEL 2
✉ 97 Jaffa St, Jerusalem 94340. ☎ +972 (2) 242776/242750. ① 25678 ITRA. 🖷 +972 (2) 242720.
L.P: Man. Dir: Oren Tokatly.

Stations: Eitanim chE22 50kW (ERP), Acco ch E27 2kW, Beersheba chE35 6kW — **D.Prgr:** 1100 (Sat 0800)-2200v.

PALESTINIAN BROADCASTING CORP. TV
✉ Ramallah- Um Al Sharyet. ☎ +972 (2) 656 4017/9. 🖷 +972 (2) 6564029.
L.P: Dir: Radwan Abu Hayash. Tech. Coordinator: Hisham Makki.
Station: UHF ch23, VHF ch4.
D.Prgr: 1600-2000 incl two news bulletins

JORDAN

TV-sets: 250.000 — **Colour:** PAL — **Systems:** B.

JORDAN RADIO & TELEVISION CORP.
✉ P.O. Box 1041 or 2333, Amman. ☎ +962 (6) 773 111/9, 777 151/5, 779 111, 638 760, 638 766/7. 🖷 +962 (6) 751 503, 788 115. **Cable:** Television. ① 24213 RTVENG JO.
L.P: DG: Mr. Ihsam Ramzi Shikim; Dir Tel: Mr. Nasser Judeh; Dir Int. Rel: Mrs Fatima Masri
Stations: 1st Prgr (in stereo): Suweilih chE3 104kW H, Aqaba ch9 5kW, Ras Munif-Ajlun ch9 500kW H, Deir Alla ch26 6kW H + low power repeaters. **2nd Prgr:** Aqaba ch5 5kW H, Suweilih ch6 108kW H, Ras Munif-Ajlun ch11 500kW H, Deir Alla ch29 6kW H + low power repeaters. **1st Prgr in Arabic (in stereo):** 1330 (Fri 0800)-2200. **2nd Prgr:** 1600-2200. **French:** 1600-1730. **N:** 1700. **Hebrew:** 1730-1800. **Arabic:** 1800-1830 (rel. 1st Prgr). **English:** 1830-2200. **N:** 2000.+ 43 low power sts.

KUWAIT

TV-sets: 800,000 — **Colour:** PAL — **System:** B & G

KUWAIT TELEVISION (Gov.)
✉ (Administration) Ministry of Information, P.O. Box 193 Safat, 13002 Safat, Kuwait. ☎ +965 (24) 150301. 🖷 +965 (24) 34511. **Cable:** ALIR-SHAD. (Television) **Kuwait Television,** P.O. Box 621 Safat, 13007 Safat, Kuwait. ☎ +965 242 3774. 🖷 +965 245 6660/243 9667. ① (496) MI 46285 KT, (496) KTV 22169 KT
L.P: Minister: Sheikh Saud Nasir Al-Saud Al-Sabah; Asst. Under Secr. for TV Affairs: Rida Yousef Al-Feeli; Dir. of Eng. (TV): Abdulazeez Al-Baghli.
Stations: 1st Prgr. in Arabic: ch8, ch9, ch24, ch26, ch38, ch45. 0500-2100.
2nd Prgr. in English: ch10, ch11, ch39, ch47: 1100-2100.
3rd Prgr (sports): ch11 & ch47 (occasional)
4th Prgr (Arabic and English entertainment): 2100-0500 on tr's of 1st and 2nd Prgr.

OTHER STATIONS

MBC (Middle East Broadcasting Centre), Arabic relay ch12, Sat-Thu. 1100-0030, Fri 0900-0030, Sun-Wed. 1200-2400
ESC (Egyptian Satellite Channel), Arabic relay ch5, Sat-Fri. 24h.
Kuwait Space Channel: see satellite section ARABSAT 1C 31Þ East.

LEBANON

TV-sets: 1.100.000 — **Colour:** SECAM — **Systems:** B & G.

TÉLÉ-LIBAN (Gov.)
✉ B.P. 115054, Hazmieh, Beirut. ☎ +961 (1) 405100. 🖷 +961 (1) 457253 — **L.P:** DG: Jean-Claude Boulos.

Stations	ch	kW)	Prgr		ch	kW)	Prgr
Fih	2	1/0.2	1	Fih	8	10/1	2
Jounieh	2	1/0.2	2	Zahle	8	10/1	2
Beit Mery	2	1/0.2	3	Beyrouth	9	5/0.5	3
M. el Chouf	4	60/6	2	Fih	10	10/1	1
Beyrouth	5	50/5	1	Zahle	10	10/1	1
Fih	6	10/1	3	Beit Mery	11	10/1	2
Beyrouth	7	5/0.5	2				

+ 4 low power repeaters
D.Prgrs: 1st and 2nd Prgr (Arabic/French/English): 1000-2200; **N. in English:** 1600 (Prgr. 1). **N. in French:** 1615 (Prgr. 2). **3rd Prgr.** (French/English) 1630-2200.

MIDDLE EAST TELEVISION (Comm.)

✉ P.O. Box 5689, Nicosia, Cyprus (Studios in Marjayoun).
Addr. in USA: 977 Centreville Turnpike, Virginia Beach, VA 23463-0001, USA. ☎ +1 (804) 579 3419. 🖷 +1 (804) 579 3417.
L.P: GM: Wes Hylton. Prgr. Mgr: Tom Foley.
Station: chE12 100kW (Pol:V) — chE5 10W.
D.Prgrs: 1130-2200 (approx) in English & Arabic.
N. in English: 1700.

LEBANESE BROADC. CORP. INT. (Comm.)

✉ P.O. Box 16-5853 Beirut. ☎ +961 (9) 938938. 🖷 +961 (9) 937916.
E-mail: lbci@lbci.com.lb **Web:** http://www.lbci.com.lb
L.P: SM: Pierre Al Daher. TD: Nasim Boustany. Film Buyer: Selim El-Azar.
Stations: ch12H 60kW; ch33H 325kW; ch10H 35kW; ch9H 35kW; ch5H 35kW.
D.Prgr: 0445-2200.

FUTURE TELEVISION (Comm.)

✉ White House, Rue Spears, Sanayeh, Beirut, Lebanon. ☎ +961 (1) 347121/4/5/8, 340303, 341303. 🖷 +961 (1) 602310.
E-mail: future@future.com.lb **Web:** http://www.future.com.lb
Stations: ch28, ch37, ch46, ch52.
D.Prgr: 1100-2200.

MURR TELEVISION (MTV) (Comm.)

✉ Fouad Chehab Ave (P.O. Box 166000) - Fassouh -MTV Bldg. ☎ +961 (1) 217000. 🖷 +961 (1) 423121. **E-mail:** mtv@dm.net.lb
Web: http://www.dm.net.lb/mtv/
L.P: Pres: Michel El Murr.
Stations: ch28, ch38, ch48, ch68.
D.Prgr: 0700-2200.

OMAN (Sultanate of)

TV-sets: 1.500.000 — **Colour:** PAL — **Systems:** B & G.

SULTANATE OF OMAN TELEVISION (Gov.)

✉ P.O. Box 600, 113 Muscat, Oman. ☎ +968 603222. 🖷 +968 602381. ☻ 5454/5151.
L.P: DG: Ali Bin Abdallah Al Mujeni. DG (Tech. Affairs): H.Y.Al-Kindy.

Location	ch	kW	Pol	Location	ch	kW	Pol
Qurum	6	120/24	H	Shinas	5	7	
Sur	7	6/1.2		Salalah	10	300/60	V
Buraimi	8	160/32		Saham	10	25/5	
Thumrait	8	140/28	V	Nizwa	10	200/20	
Barka	51	70		Bahla	5	4	
Bilad Bani				Fine Peak	56	10	
Bu Ali	9	5		Ibri	55	20	

+ 25 low power relay st's.
D.Prgr: 0400-2100.

QATAR

TV-sets: 250.500 — **Colour:** PAL — **System:** B

QATAR TELEVISION SERVICE (Gov.)

✉ Min. of Information and Culture, QTV. P.O. Box 1944, Doha. ☎ +974 89 4444. 🖷 +974 86 4511. **Cable:** TELEVISION DOHA. ☻ 4040 TEEVEE DH
L.P: Asst. Under Secretary for Radio & TV: Mr. Abdul Rahman Saif Al-Madhadi. Dir. of TV: Mr. Saad Al-Rumehi. Dir of Eng: Mr. Hussain A. Jaffar. Dir of News: Mr. Abdullah Al Haj
Stations: Doha ch9 200kW, Jamiliyah ch11 (repeater) 600kW, Doha ch37 695kW, ch49 (repeater), Jamiliyah ch 52 (repeater)
D.Prgr: 1200-2200 UTC (Fri. 0600-2200), 0600-2200 UTC (during Summer)

SAUDI ARABIA

TV-sets: 4.700.000 — **Colour:** SECAM & PAL — **System:** B & G.

SAUDI ARABIAN TELEVISION (Gov.)

✉ P.O. Box 57137, Riyadh 11574. ☎ +966 (1) 4014440. 🖷 +966 (1) 4044192/4054176. ☻ 401030 SJ.

L.P: Asst. Dep. Minister for TV Affairs: Dr. Ali M. Al Najai.

1st Prgr:	ch	kW	Pol		ch	kW	Pol
Abu Qumais	5	200	H	Ranyah	10	7	H
Riyadh	5	45	H	Al Ardiyah	10	7	H
Madinah	5	170	H	Bani Malik	10	2	H
Abha	5	170	V	Ar Rass	10	5	H
Hail	5	42	H	Wadi Al Freah	10	9	H
Zilfi	5	9	V	Jebel Garah	10	3	H
Arar	5	14	H	Al Summan	10	29	H
Tayma	5	5	H	Tathlith	10	9.5	H
Bijadiah	5	9	V	Al Howaidh	10	4.7	H
Harradh	5	5	V	Al Kharj	11	5	V
Halat Ammar	5	8	H	Al Majmaah	11	160	H
Al Hariq	5	2	H	Al Sulaiyal	11	5	H
Linah	5	5		Bani Saad	12	6	H
Dammam	6	800	H	Owegliyah	12	26	H
Taif	6	47	H	Al Harjah	21	2	H
Yanbu	6	46	H	Al Ahmer	22	16	H
Baqa	6	5	V	Shaqra	22	100	H
Sharourah	6	42	V	Durma	22	8	H
Al Quwaiyah	6	5	V	Khybar	22	7	H
Hanakiyah	6	5	H	Al Qunfudah	22	7	H
Athnen	6	6	H	Al Nuariyah	22	8	H
Mahd Ad Dahab	6	5	H	Al Sudah	22	19	H
Khamasin	6	6	H	Artawiyah	23	9	H
Makkah	7	52	H	Turaif	24	3	H
Al Baha	7	94	H	Hufuf	25	25	H
Jizan	7	160	H	Al Musayjid	25	11	H
King Khaled City	7	2	H	Al Khurma	26	7	H
Al Qurayyat	7	1	H	Dammam	27	44	H
Al Ula	7	5	H	Layla	27	7	H
Buraydah	7	98	H	Haql	27	7	H
Tabarjal	8	12	V	Rafha	28	7	H
Hafr Al Batin	9	16	V	Al Nimas	30	6	H
Al Hassi	9	8	H	Al Alaya	31	10	H
Uqlat Assuqr	9	9	H	Bishah	39	6	H
Al Muwayh	9	9	V	Turabah	40	7	H
Nabq Abu Qasr	9	2	H	Umm Lajj	41	7	H
Skaka	9	5	H	Rabigh	42	8	H
Badr Hunayn	9	5	H	Al Hawtah	45	7	H
Sajir	9	13	V	Qaryat Al Ulya	46	17	H
Jebel Khasheb	9	10	H	Tuwal	50	100	H
Al Bad	9	1	H	Al Khafji	51	40	H
Jeddah	10	27	H	Al Wajh	53	7	H
Tabuk	10	15	V				

+ 25 Stations not mentioned.

2nd Prgr:	ch	kW	Pol		ch	kW	Pol
Makkah	5	52	H	Sharourah	8	42	H
Skaka	5	5	H	Khamasin	8	6	H
King Khaled City	5	2	H	Abha	9	170	V
Buraydah	6	98	H	Al Baha	11	106	H
Riyadh	7	45	H	Hafr Al Batin	11	17	V
Madinah	7	170	H	Jeddah	12	27	H
Hail	7	42	H	Al Howaidh	12	4.7	H
Zilfi	7	9	V	Shaqra	25	100	H
Arar	7	14	H	Hufuf	28	25	H
Taif	8	47	H	Dammam	29	890	H
Tabuk	8	15	V	Tuwal	53	100	H
Yanbu	8	51	H	Al Khafji	53	40	H

+ 6 Stations not mentioned.

D.Prgr: 1st Prgr: (Arabic): Sat-Wed 0700-0930, 1400-2130; Thurs/Fri: 0700-2130. **2nd Prgr:** (English): 1400-2130 (Thurs/Fri also 0600-0900).

CHANNEL 3 TV (Non-Comm, Private Co.)

✉ Bldg. 3030 LIP, Dhahran.
☎ 875-4634. **Cable:** Aramco, Dhamedia.
Station: chE3 Dhahran 5kW.
D.Prgr: 0600-2030 (or 2100). **Colour:** PAL.

SYRIAN ARAB REP.

TV-sets: 700.000 — **Colour:** PAL — **System:** B.

SYRIAN ARAB TELEVISION (Gov.)

Ommayyad Square, Damascus. ☎ +963 (11) 720700. 🖷 +963 (11) 720700. ✆ 411223.
L.P: DG: Khudr Omran. Dir. Eng: M. Bara. Dir. PR: Mrs. Awafet Haffar.
Stations: Pol = H. +) Prgr 2, others Prgr 1.

Location	chE	kW		chE	kW
Holms	2	?+	Aein-Al Arab	7	100/10
Abou-Kmal	3	200/20	Tabqua	8	100/10
Nabi-Saleh	3	100/10	Kaldoun	8	10/1
Hassakeh	4	200/20	Slenfeh	9	200/20
Aleppo	5	200/20	Salhieh	11	30/3
Damascus	5	100/10	Afrien	11	10/1
Deir-Al-Zoor	6	100/10	Palmyra	11	10/1
Soueida	7	350	Al-Malkieh+	12	200/20
Homs	7	200/20	Lattakia	26	60

+ 36 low power & repeater st's.
Prgr 1: 1100 (Fri 0700)-2130.
Prgr 2 in Arabic/English/French: 1700-2130.

TV-sets: 10.530.000 — **Colour:** PAL — **System:** B.

TURKISH RADIO TELEVISION CORPORATION

TRT-TV Department, TRT Sitesi Katib Ablok Oran 06450, Ankara.
☎ +90 (312) 4904 983/986. 🖷 +90 (313) 4904 985.
L.P: Head of TV Dept: Çetin Izbul; Dir Ankara TV: Sabahattin Alpdogan; Tech Dir: Sükrü Sipka.

TV-1	chE	kW	TV-1	chE	kW
Ankara	5	100	Hatay	8	10
Istanbul	5	100	Canakkale	8	30
Antalya	5	30	Usak	8	30
Agri	5	30	Ersincan	8	30
Giresun	5	30	Nigde	8	30
Cizre	5	30	Afyon	9	30
Adana	6	30	Amasya	9	100
Erzurum	6	100	Edirne	9	30
Bursa	6	100	Diyarbakir	9	30
Kastamonu	6	30	Trabzon	9	30
Aksehir	6	30	Kars	9	100
Malatya	7	30	Zonguldak	9	30
Eskisehir	7	100	Izmir	10	100
Kirsehir	7	30	Gaziantep	10	30
Samsun	7	30	Konya	10	100
Silifke	7	30	Adapazari	10	30
Van	7	30	Bingöl	10	30
Aydin	7	30	Tokat	10	30

+ 476 low power repeaters not mentioned.

TV-2*	chE	TV-2*	chE
Elbistan	5	Trabzon	28
Adapazari	21	Agri	32
Cizre	21	Elbistan	34
Kars	21	Ankara	37
Adana	22	Isparta	40
Canakkale	22	Ersincan	42
Eskisemir	22	Kastamonu	48
Van	22	Mus	48
Zonguldak	23	Istanbul	51
Erzurum	23	Samsun	52
Izmir	23	Bursa	55
Kayseri	26	Diyarbakir	56
Gaziantep	27	Giresun	57
Denizli	28	Mugla	55
Edirne	28		

*) all tr's 450kW (ERP).
D.Prgr: TV-1: 24 hrs. **TV-2:** Mon-Fri 1700-0100, Sat/Sun 0930-0300(Sun 0100)
TV-3, TV-4, TV-5
Stations: (Istanbul area) TV-3 ch48, TV-4 ch54, TV-5 ch24.
D.Prgr: TV-3 Wed/Thu/Sat/Sun1340-2400, Mon/Tue/Fri 1700-0100; TV-4 1330-2400.

UNIVERSITE TECHNIQUE

Stations: Istanbul chE4, 0.5kW.
NB: Turkish Private Comm. TV Stations see Satellite section.

TV-sets: 170.000 — **Colour:** PAL — **Systems:** B & G.

UNITED ARAB EMIRATES TELEVISION SERVICE (Gov.)

P.O. Box 637, Abu Dhabi. ☎ +971 (2) 452000. 🖷 +971 (2) 461823. ✆ 22557 teevee em.
Web: http://www.ecssr.ac.ae/05uae.6television.html
L.P: Hd. of Eng: Mustafa Hamouda Ishag. Dir. Gen: Ali Obaid.

Stations	Prgr. 1	Prgr. 2	kW
Mohawi	7	48	40/5
Abu Dhabi	11	—	2
Al-Ain	5	35	40/5
Umm Al Quwain	6	43	40
Ras Al Khaimah	11	—	10
Fujairah	9	36	10
Habshan	3	—	40
Dibba	21	23	0.5
Jabel Al Dhanna	31	—	40
Khorfakkan	28	—	0.02
F.PI:	**Prgr. 1**	**Prgr. 2**	**kW**
Al Khan	62	—	?
Liwa	25	28	?
Al-Wagen	29	32	?
Habshan	—	59	?
Jabel Dhanna	—	61	?

D.Prgr: Prgr. 1 (Arabic main prgr), Prgr.2I (Arabic/English/French): 0600-1005, 1200-2030 (Fri & holidays 0600-2030).

U.A.E. RADIO AND TELEVISION-DUBAI (Gov.)

P.O. Box 1695, Dubai. ☎ +971 (4) 470255. ✆ 45605 DRCTV EM.
Web: http://www.ecssr.ac.ae/05uae.6television.html
L.P: Dir. Gen: Abdul G. Al Sayeed Ibrahim. Contr. of Prgrs: Nasib Bitar. Contr. of Eng: Ahmed Najeeb.
Stations (Systems B & G): Audio powers are 1/10 of the vision powers indicated.

Location	ch	kW (ERP)	Pol	Prgr
Trade Centre	2	150	H	1
Zabeel	10	455	H	1
Zabeel	33	1700	H	2
Jebel Hatta	41	1600	H	1

D.Prgrs: Arabic (prgr. 1) on chs 2, 10 & 41: 1200 (Fri 1000)-2130 (or 2030). **English** (prgr. 2) on ch33: 1300 (Fri 1200)-2000 (approx).

SHARJAH TV (Gov.)

P.O. Box 111, Sharjah. ☎ +971 (6) 547755.
Stations: ch22, 28, 54, 57.
D.Prgr: 1300-2000 (in Arabic).

TV-sets: 100.000 — **Colour:** PAL/NTSC — **System:** B.

YEMEN RADIO & TV CORP. (Gov.)

P.O. Box 2182, Sana'a. ☎ +967 (1) 230654. 🖷 +967 (1) 230761. ✆ 2645 YARTV.
L.P: Dir's: Ali Caleh Algamrah, Mohammed Abdul Gawi.
Stations: chE5 (0.2kW), chE6 (2 st's, 4 & 1kW), chE7 (2kW), chE8 (4kW), chE10 (4kW), chE11 (1kW), chE12 (2kW) + low power repeaters.
D.Prgr: 1300-2100.

ASIA

TV-sets: 600.000 — **Colour:** PAL — **System:** B/G.

NATIONAL BROADCASTING AUTHORITY BANGLADESH TELEVISION (Gov, Comm.)

Television Bhaban, P.O. Box 456, Dhaka-1219. ☎ +880 (2) 933036/7, 9330131-39. 🖷 +880 (2) 832927. ✆ 675624 BTV BJ.
L.P: DG: Syed Salahuddin Zaki. Dep. DG (Prgrs): Mustafa Kamal Syed. Dep. DG (News): Faruoq Alamgir. Chief Eng: Anisur Rahman.

Stations	chE	kW		chE	kW
Chittagong	5	15	Dhaka	9	60
Rangpur	6	360	Cox's Bazar	10	7
Sylhet	7	136	Khulna	11	91
Satkhira	7	250	Mymensingh	12	390

Rangamati	8	55	Noakhali	12	390
Natore	8	90	Bramangaria	5	—
Jenaidah	5	—	Txakurgaon	10	—
Patuakhali	7	—			

D.Prgr: 8½hrs except Fri 13½, Sat 11½hrs. Also carries prgrs of BBC, CNN and the Open University — **F.PI:** 2nd channel.

Private TV: The govt. has decided to allow private TV st's to operate in the future.

BRITISH INDIAN OCEAN TERRITORY

TV-sets: 650 — **Colour:** NTSC — **System:** M.

AF DIEGO GARCIA TELEVISION (AFRTS)
US Navy Broadcasting Service, Detachment, Box 14, US Navy Support Facility, F.P.O. San Francisco, Calif. 96685, USA.
L.P: St. Mgr: J.A. Riccio. Dir. Tec: R.L. Newton.
Station: chA8/chA10 0.2kW H.
D.Prgr: ch 8: 0600 (Sat/Sun 0300)-2200; ch 10 24h.

BRUNEI DARUSSALAM

TV-sets: 95.000 — **Color:** PAL — **System:** B

RADIO TELEVISION BRUNEI (RTB) (Gov.)
Jabatan Perdana Menteri, Bandar Seri Begawan 2042, Negara Brunei Darussalam. ☎ +673 (2) 243 111. 📠 +673 (2) 241882. **Cable:** Broadcast
L.P: Dir: Pg Dato Haji Badaruddin. Bin Pg Hj Ghani. Dep. Dir: Haji Md Yusof Hj Abd Rahman. Head of Eng: Kim Sam Lee. Hd of Prgrs: Mrs. Pg Hjh Normah Pg Hj Daud. Prgr. Prod. Mgr (TV): Haji Mohd Hussain Abdul Rahman.
Stations: Bt. Subok ch5 10kW H; Bt. Andulau ch8 20kW H.
D.Prgr: 0800 (Fri/Sun 0030) -1600.

CAMBODIA

TV-sets: 70.000 — **Colour:** PAL — **System:** B/G.

CAMBODIAN TELEVISION (Gov.)
19, Street 242, Chaktomuk, Daun Penh. Phnom Penh.
☎ +855 (2983) 22349/(4449) 24149
L.P: St. Mgr: Tan Yan. TD: Uy Thuon.
Stations: ch7 1kW.
D.Prgr: 0715 PM-0745 PM.

INTERNATIONAL BROADCASTING CORP. Ltd. (IBC)
Borei Keila Street No. 169, Sangkat Vealvong, Phnom Penh City. ☎ +855 (23) 66061, 66064. 📠 +855 (23) 66063

CHINA (People's Republic of)

TV sets: 300m — **Colour:** PAL — **System:** D.

CHINA CENTRAL TELEVISION (CCTV Gov.)
11 Fuxing Lu, Haidian, Beijing 100859, China. ☎ +86 (10) 850 0000. 📠 +86 (10) 851 3025. ☏ 222299 CCTVE CN.
Web: http://www.wtdb.com/CCTV/about.htm
L.P: Pres: Yang Weigwang. VP: Yu Guanghgua. GM TV Prgr Agency: Xu Xiongxiong. Dir. Int. Rel: Zhao Yuhui.
1st Prgr (news & information network): 2250-1700 on ch2 (Beijing)/Teletext Prgr: 0200-1600.
2nd Prgr (economic and social education prgrs): 0030-1700 on ch8 (Beijing).
3rd Prgr (entertainment prgrs): 0030-1600 on ch15 (Beijing).
4th Prgr: news and entertainment prgrs. Channel 4 is also broadcast outside of China to dozens of television stations in Asia, Australia, Africa, and in the former Soviet Union republics. 2020-2220 on ch29 (Beijing/cable).

Location	1st Prgr	2nd Prgr	3rd Prgr
Beijing	2	8	15

Tianjin	5	36	—
Shijiazhuang	1	13	—
Taiyuan	7	14	—
Hohhot	2	26	—
Shenyang	5	33	—
Changchun	2	13	—
Harbin	8	14	—
Shanghai	5	—	

Location	1st Prgr	2nd Prgr	3rd Prgr
Nanjing	38	15	—
Hangzhou	6	—	—
Hefei	11	2	—
Fuzhou	10	—	—
Nanchang	9	—	—
Jinan	8	25	—
Zhengzhou	2	7	—
Wuhan	8	25	—
Changsha	21	—	—
Guangzhou	8	—	—
Nanning	12	20	—
Haikou	7	—	—
Chengdu	8	10	—
Guiyang	2	—	—
Kunming	4	15	23
Lhasa	4	12	—
Xi'an	8	16	—
Lanzhou	4	10	—
Xining	11	15	—
Yinchuan	4	24	—
Urumqi	12	15	—

Other TV Stations: There are more than 31 Provincial TV Stations and nearly 3,000 City-TV Stations.

Location	Provincial TV chs				City TV chs	
Beijing	6	21	27			
Tianjin	12	29	17	23		
Shijiazhuang	10	32			4	
Taiyuan	9	14			12	
Hohhot	10	20			4	
Shenyang	10	12	21		2	27
Changchun	7	19			9	39
Harbin	1				6	20
Shanghai	8	14	26			
	(Shanghai Eastern TV) 20					
Nanjing	1+10	4	32		12	28
Hangzhou	4	22			11	41
Hefei	4				9	
Fuzhou	2	7			4	
Nanchang	7	13			1	
Jinan	2	15			6	
Zhengzhou	9				12	
Wuhan	4	2			19	
Changsha	9	27			4	40
Guangzhou	2	14	21		34	
Nanning	4	14			10	
Haikou	2	12			32	
Chengdu	2	27	21		15	
Guiyang	4				9	
Kunming	2+9	21	27		11	30
Lhasa	8	6			10	
Xi'an	4				10	
Lanzhou	2	8			18	
Xining	4	8	27		21	
Yinchuan	8				18	
Urumqi	4	6	8	21	27	33
Zhengzhou	9				12	
Wuhan	4	2			19	
Changsha	9	27			4	40
Guangzhou	2	14	21		34	
Nanning	4	14			10	
Haikou	2	12			32	
Chengdu	2	27	21		15	
Guiyang	4				9	
Kunming	2+9	21	27		11	30
Lhasa	8	6			10	
Xi'an	4				10	
Lanzhou	2	8			18	
Xining	4	8	27		21	

Yinchuan	8			18	
Urumqi	4	6	8	21	27	33

BEIJING TELEVISION (BTV)
⊡ Bejing — **Station:** ch33 — **D.Prgr:** 0030-1600.

ORIENTAL TV
⊡ Shanghai. ☎ +86 (21) 322 3007. 🖹 +86 (21) 320 7368.

CHINA (Republic of) TAIWAN

TV-sets: 7m — **Colour:** NTSC — **System:** M.

CHINA TELEVISION COMPANY (Comm.)
⊡ No. 120 Chung-Yang Road, Nankang District, Taipei. ☎ +886 (2) 783 8308. 🖹 +886 (2) 782 6007. ① 25080 CHINA TV.
L.P: Pres: Hu Ping Chung. CE: Mr. Chen.
Stations: BEF21 chA9 180/40kW (No. Taiwan), BEF22 chA10 150/17kW (Ce. Taiwan), BEF23 chA9 180/80/20kW (So. Taiwan).
D.Prgr: W 0400-0500, 0930-1515; Sat 0450-1615; Sun 0250-1515.

CHINESE TELEVISION SERVICE (Comm.)
⊡ 100 Kuang Fu South Road, Taipei. ☎ +886 (2) 751 0321. 🖹 +886 (2) 751 6019.
L.P: Chmn: Chien-Chiu Yee. Pres: Shih-shung Wu. CE: Shiao Ho Whu.
Stations: BET31 chA11 200kW (No.We.), BET32 chA8 190kW (Ce.), BET33 chA11 131kW (So.We.), BET34 chA8 24.5kW (No.Ea.), BET35 chA11 18kW (Ce.Ea. Taiwan), BET36 chA8 9.5kW (So.Ea. Taiwan), BET37 ch35 1702kW (No.We.), BET39 ch36 1383kW (Ce.), BET40 ch34 1383kW (So.We.), BET41 ch33 316kW (No.Ea.), BET42 ch35 316kW (Ce.Ea.), BET43 ch33 316 kW (So.Ea.), BET38 ch33 283kW (We.) + 2 relay st's.
D.Prgr: W 2320-0120, 0340-1600; Sun 2320-1600.

EDUCATIONAL TV
BET37 chA35 1717kW (No.We. Taiwan), BET38 chA33 285kW (Ce.No. Taiwan), BET39 chA36 1500kW (Ce. Taiwan), BET40 chA34 1500kW (So.We. Taiwan), BET41 chA33 354kW (No.Ea. Taiwan), BET42 chA35 343kW (Ce.Ea. Taiwan), BET43 chA33 334kW (So.Ea. Taiwan).
D.Prgr: W 1030-1435, Sun 1130-1505.

TAIWAN TELEVISION ENTERPRISE Ltd. (Comm.)
⊡ No. 10, Pa Te Rd, Section 3, Taipei 10560. ☎ +886 (2) 7711515. 🖹 +886 (2) 7413626. **Cable:** Television. ① 25714 TV TAIWAN.
L.P: Chmn: Ching-Teh Hsu. Pres: Walter C.H. Wang. Executive vice president: Wei-Yung Lee. Vice president: Hsiang-chuan Hsiung. Vice president: Shen-Wen Lee. Manager, Engineering Department: To-Hui Yang. Manager, Business Department: Wen-Lung Liu. Film buyer: Nancy Hu. Manager, News Department: Victor S.T. Chang. Manager, Sports Department: Jason Liao. Manager, Program Department: Sheng Chu-yu. Dir. Inf. Center: Ko-Jan Hwang.
Stations: BET21 chA7 282/95kW (No. Taiwan), BET22 chA12 162/136/6.5kW (Ce. Taiwan), BET23 chA7 136/162/5.5kW (So. Taiwan), BET24 chA10 4.0/0.4kW (I-Lan), BET25 chA7 12/36kW (Hua Lien), BET 26 chA10 1.9kW (Taitung) + 7 Transposer stations.
D.Prgr: Mon-Fri 2159-0030, 0311-0810, 0839-1610; Sat 0244-1740; Sun 2339-1610.

GEORGIA

TV-sets: N/A — **Colour:** SECAM — **Systems:** D & K.

GEORGIAN TELEVISION (Gov.)
⊡ M. Kostavas 68, 380071 Tbilisi. ☎ +7 (8832) 362294. 🖹 +7 (8832) 362319 (manual).
L.P: Chief-editor, Information Sector: Areshidze Mamuka.

Stations (Prgr. 1)

	ch		ch
Tbilisj	4	Kutaisi	10
Suchumi	5	Gori	12
Batumi	8		

+ various translators.
D.Prgr: Mon-Fri 1000-1100, 1200-2100. Sat/Sun 0900-2100.

Stations (Prgr. 2) **ch**
Tbilisi 28
D.Prgr: Current schedule not available.

ORT Relays

Stations	ch		ch
Batumi	3	Tbilisi	9
Kutaisi	7	Gori	31
Suchumi	9		

+ various translators.
D.Prgr: Relays of ORT-1 from Moscow. Further details see under Russia.

ROSSIJSKOJE TELEVIDENIJE (RTV)
Stations ch
Tbilisi 6
+ various translators.
D.Prgr: Relays of RTV from Moscow. Further details see under Russia.

HONG KONG

TV-sets: 1.749.000 — **Colour:** PAL — **System:** I.

TELEVISION BROADCASTS LIMITED
⊡ TV City, Clear Water Bay Road, Kowloon. ☎ +852 2719 4828. 🖹+852 2358 1337.
L.P: MD: Louise Page. GM (Corporate Affairs): Alex Ying Ki Luen. GM (International Operations): Ken Lam Kon Leung. Film Buyer: Sophia Chan (Jade Channel), Musetta Wu (Pearl Channel).
Stations: chE21 & chE25, 10kW.
D.Prgr: Jade network on ch21: 2245-2030 (approx). Pearl network on ch25: 0000 (Sat 0100, Sun 0030)-1730 (approx).

ASIA TELEVISION LIMITED
⊡ 81 Broadcast Drive, Kowloon. ☎ +852 2992 8888. 🖹 +852 2338 0438. ① HX44680. **Cable:** ASIATV. **E-mail:** atv@hkatv.com **Web:** http://www.hkatv.com
L.P: CEO: Mr. Mark Lee; Dep CEO: Mr. Clarence Chang; Controller: Mr. Jermyn Lynn
Stations: chE23 & chE27, 10kW.
D.Prgr: Home Channel in Chinese on ch23: 2230-1930. World Channel in English on ch27: 24h.

INDIA

TV sets: 50m (est.) — **Colour:** PAL — **System:** B

DOORDARSHAN INDIA (Gov., semi-comm.)
⊡ Directorate General of Doordarshan, Mandi House, Copernicus Marg, New Delhi-110 001. ☎ +91 (11) 382094-99. 🖹 +91 (11) 3386507. ① 81-31-65290/66413 dgdd in. **Cable:** tvgeneral.
L.P: DG: K.S. Sarma. CE: K.C.C. Raja.
Programming/News:
Doordarshan Kendra, Akashvani Bhavan, 1 Sansad Marg, New Delhi-110 001. ☎ +91 (11) 3382021/3715411(pabx). Add. DG (News): D.C. Bhaumik. PD: M.B. Pahari.
Programming:
Central Production Centre, Doordarshan, Asiad Village Complex, Siri Fort, New Delhi-110 016. ☎ +91 (11) 6462539/6462446.

Sations: Only high power tr's (1 kW and above) are listed below.

Location	ch	kW	Relay
Mumbai	4	10	
Calcutta	4	10	
Delhi	4	10	
Hyderabad	4	10	
Lucknow	4	10	
Chenai	4	10	
Srinagar	4	10	
Pathankot	4	10	Delhi
Ahmedabad	5	10	
Bangalore	5	10	
Bhopal	5	10	
Jaipur	5	10	
Kanpur	5	10	Lucknow
Pune	5	6	Mumbai
Raipur	5	1	Delhi
Silchar	5	10	Delhi
Mumbai 2	6	1	Delhi 2
Calcutta 2	6	10	Delhi 2
Imphal	6	1	Delhi

Location	ch	kW	Relay
Kasauli	6	10	Delhi
Chennai 2	6	10	Delhi 2
Muzaffarpur	6	1	Delhi
Rajkot	6	10	Ahmedabad
Allahabad	7	10	Lucknow
Amritsar	7	10	Jaladhar
Asansol	7	10	Calcutta
Delhi 2	7	10	
Gulbarga	7	1	Bangalore
Location	**ch**	**kW**	**Relay**
Kodaikanal	7	10	Chennai
Nagpur	7	10	Mumbai
Panaji	7	10	Mumbai
Sambalpur	7	1	Cuttak
Pij	7	1	Ahmedabad
Vijayawada	7	10	Hyderabad
Visakapatnam	7	10	Hyderabad
Agartala	8	10	Delhi
Cuttack	8	10	
Kohima	8	1	Delhi
Kurseong	8	10	Calcutta
Varanasi	8	10	Lucknow
Aizawl	9	1	Delhi
Agra	9	10	Lucknow
Gorakhpur	9	10	Lucknow
Guwahati	9	10	
Indore	9	10	Bhopal
Jalandhar	9	10	
Thiru'puram	9	10	
Murshidabad	10	10	Calcutta
Mussoorie	10	10	Lucknow
Ranchi	10	10	Patna
Tura	10	10	Delhi
Dibrugarh	11	10	Guwahati
Patna	11	10	
Poonch	11	10	Delhi
Bathinda	12	10	Jalandhar
Kochi	12	10	Thiru'puram
Dwarka	12	10	Ahmedabad
Jammu	12	10	Delhi
Shillong	12	1	

+ 23 additional tr's 10kW or 1kW (channels unknown).
Total transmitters = 562.
D Prgr: National: 1500-1800v (extended telecasts during live coverage). **Network:** 0130-0315 (Sun 0830), 0830-1043, 1420-1430. **General/Regional:** W 1130-1510, Sun 0315-0330/1400-1415. **Countrywide Classroom:** 0730-0830, 1030-1130 (W). Open University: 0100-0130 (Mon/Wed/Fri) Educational Television: 0500-0530 (W), summer vacations till 0730. The **Metro** (Entertainment) channels are networked from Delhi to Ch. 2 of Mumbai, Calcutta & Chennai, and are also available on Insat-2B. Regional Channels: available on Insat-2C.

INDONESIA

TV sets: 11.000.000 — **Colour:** PAL — **System:** B.

TELEVISI REPUBLIK INDONESIA TVRI (Gov.)
✉ Jalan Jerbang Pemuda, Senayan, Jakarta. ☎ +62 (21) 5733135/2279. 🖷 +62 (21) 5732408. **Cable:** telviri Jakarta. ☺ 073-46154 tvri jkt.
L.P: MD: Azis Husein. TD: Djoko Widayat. Film Buyer: Adi Kasno.
Key Stations

Loc.	chE	kW(ERP)	Loc.	chE	kW(ERP)
Ujung Padang	4	1	Pontianak	7	10
Medan	5	10	Jakarta	8	10
Banda Aceh	5	10	Yogyakarta	8	10
Padang	5	5	Denpasar	8	5
Banjarmasin	5	10	Kupang	8	5
Jakarta	6	5	Palembang	9	5
Bandung	6	10	Balikpapan	9	1
Semarang	6	5	Surabaya	9	10
Ambon	7	5	Manado	9	1

+ 10 relay st's.
D.Prgr: W 0930-1630 (Sat 1700), Sun & Holidays 0100-0630, 0930-1630. **English:** 1130-1200. **N:** 1130.

COMMERCIAL STATIONS

RCTI (PT Rajawali Citra Televisi Indonesia)
✉ Jl. Raya Perjuangan No. 3, kb. Jeruk, Jakarta 11000.
☎ +62 (21) 530 3540/3550/3564. 🖷 +62 (21) 549 3852/3846.
L.P: Pres Dir: Mr. Muchamad Ralie Siregar; TM: Mr Doopy Irwan
Stations: Ambon ch11, Balikpapan ch4, Batam ch43, Denpasar ch35, Jakarta ch43, Malang ch40, Manado ch30, Semarang ch30, Ujung Pandang ch33
D.Prgr: 90 hrs per week.

SCTV (PT Surya Citra Televisi)
✉ JL Raya Darmo Permai III, Surabaya 60189. ☎ +62 (31) 714 567/714 033. 🖷 +62 (31) 717 273
Stations: Balikpapan ch11, Bandung ch52, Banjarmasin ch11, Batam ch47, Dili ch11, Malang ch46, Manado ch34, Mataram ch11, Medan ch35, Solo ch44, Surabaya ch43, Ujung Pandang ch35, Yogjakarta ch34

TPI (PT Cipta Televisi Pendidikan Indonesia)
✉ Jalan Pintu II - Taman Mini Indonesia Indah, Pondok Gede, Jakarta Timur 13810. ☎ +62 (21) 841 2473 to 83 (HQ). 🖷 +62 (21) 841 2470/1
L.P: Sa'Dullah Sulchan; GM: Syamsudin C. Haesy
Stations: Jakarta ch34 & 37
D.Prgr: Mon-Fri: 2230-1800, Sat: 2230-1800

ANTEVE (PT Cakrawala Andalas Televisi)
✉ Mulia Center Building, 19th Floor, Jl. HR Rasuna Said Kav. X-6 No.8, Jakarta 12940. ☎ +62 (21) 522 2084 to 86, 522 9175. 🖷 +62 (21) 522 2087, 522 9174.
L.P: GM: Mr. Dennis M. Cabalfin
Stations: unknown

IVM (PT. Indosiar Visual Mandiri)
✉ Jl. Damai No 11, Daan Mogot, Jakarta 11510. ☎ +62 (21) 567 2222, 568 8888. 🖷 +62 (21) 565 2221
Stations: unknown

JAPAN

TV-sets: 100m — **Colour:** NTSC — **System:** M.

NIPPON HOSO KYOKAI
(Japan Broadc. Corp.) (non Gov., non Comm.)
✉ 2-2-1, Jinnan, Shibuya-ku, Tokyo 150-01. ☎ +81 (3) 3485-6517, 3481-1362. 🖷 +81 (3) 3481-1576.
Web: http://www.nhk.or.jp
L.P: Pres: Mikio Kawaguchi; Exec. Dir: Michio Futami; Eng: Hiroaki Ohtsuka.
1)= General Sce; 2)= Educational Sce. Call: JO(call) TV.

Main st's	Call	ch	kW	Prgr
Tokyo	AK	1	50	1
Kofu	KG	1	3	1
Fukuyama	DP	1	1	1
Yamaguchi	UC	1	1	2
Nagasaki	AC	1	1	2
Asahikawa	CC	2	1	2
Kushiro	PC	2	1	2
Muroran	IZ	2	1	2
Akita	UB	2	5	2
Fukushima	FD	2	3	2
Nagano	NK	2	1	1
Shizuoka	PB	2	1	1
Osaka	BK	2	10	1
Matsuyama	ZB	2	5	2
Kumamoto	GB	2	1	2
Okinawa	AP	2	5	1
Sapporo	IK	3	10	1
Kitami	KP	3	1	1
Sendai	HK	3	10	1
Tsuruoka	-	3	1	1
Aomori	TG	3	5	1
Tokyo	AB	3	50	2
Kofu	KC	3	3	2
Nagoya	CK	3	10	1
Fukui	FC	3	3	2
Toyama	IG	3	3	1
Hiroshima	FK	3	10	1
Okayama	KB	3	10	2
Tottori	LG	3	1	1
Tokushima	XK	3	1	1

Fukuoka	LK	3	10	1
Nagasaki	AG	3	1	1
Kagoshima	HG	3	5	1
Oita	IP	3	3	1
Hakodate	VK	4	1	1
Obihiro	OG	4	1	1
Yamagata	JC	4	3	2
Morioka	QG	4	3	1
Kanazawa	JK	4	3	1
Main st's	**Call**	**ch**	**kW**	**Prgr**
Hamamatsu	DG	4	1	1
Tottori	LC	4	1	1
Kochi	RK	4	1	1
Sendai	HB	5	10	2
Aomori	TC	5	5	2
Okayama	KK	5	10	1
Kagoshima	HC	5	5	2
Tsuruoka	-	6	1	2
Matsue	TK	6	1	1
Matsuyama	ZK	6	5	1
Kochi	RB	6	1	2
Fukuoka	LB	6	10	2
Kitakyushu	SK	6	1	1
Hiroshima	FB	7	10	2
Fukuyama	DO	7	1	2
Yamagata	JG	8	3	1
Morioka	QC	8	3	2
Niigata	QK	8	5	1
Kanazawa	JB	8	3	2
Hamamatsu	DC	8	1	1
Miyazaki	MG	8	1	1
Asahikawa	CG	9	1	1
Kushiro	PG	9	1	1
Muroran	IQ	9	1	1
Akita	UK	9	5	1
Fukushima	FP	9	3	1
Nagano	NB	9	1	2
Nagoya	CB	9	10	2
Shizuoka	PK	9	1	1
Fukui	FG	9	3	1
Yamaguchi	UG	9	1	1
Kumamoto	GK	9	1	1
Hakodate	VB	10	1	2
Toyama	IC	10	3	2
Sapporo	IB	12	10	2
Obihiro	OC	12	1	2
Kitami	KD	12	1	2
Niigata	QB	12	5	2
Osaka	BB	12	10	2
Matsue	TB	12	1	2
Kitakyushu	SB	12	1	2
Miyazaki	MC	12	1	2
Oita	ID	12	3	2
Okinawa	AD	12	5	2
Kobe	PP	28	10	1
Otsu	QP	28	1	1
Tsu	NP	31	5	1
Kyoto	OK	32	10	1
Wakayama	RP	32	1	1
Takamatsu	HP	37	10	1
Tokushima	XB	38	10	2
Saga	SP	38	5	1
Gifu	CP	39	5	1
Takamatsu	HD	39	10	2
Saga	SD	40	5	2
Nara	UP	51	1	1

+ approx. 6900 st's.
D.Prgr: 1)Gen. TV: 2055-1507. 2)Educ. TV: 2100-1502

THE NATIONAL ASSOCIATION OF COMMERCIAL BROADCASTERS IN JAPAN

✉ 3-23, Kioi-cho, Chiyodaku, Tokyo 102.
☎ +81 (3) 5213-7700. 🖷 +81 (3) 5213-7701. **Cable:** Mimporen Tokyo.
☏ 2325163 NABTYO J.

TV NETWORKS & KEY STATIONS

Japan News Network (28 st's)
Tokyo Broadc. System, Inc. (TBS)

✉ 3-6, Akasaka 5-chome, Minato-ku, Tokyo 107-06.
☎ +81 (3) 3746 1111. 🖷 +81 (3) 3588-6378. **LP:** Pres: Hirozono Isozaki.
Station: JOKR-TV ch6 50kW(+ 89 relay st's). D.Prgr: 24h

Nippon News Network (30 st's)
Nippon Television Network Corp. (NTV)
✉ 14, Niban-cho, Chiyoda-ku, Tokyo 102-40.
☎ +81 (3) 5275-1111. 🖷 +81 (3) 5275-4008. **LP:** Pres: Seiichiro Ujiie.
Station: JOAX-TV ch4 50kW (+ 89 relay st's). D.Prgr: Approx 2000-1800

Fuji News Network (26 st's)
Fuji Television Network, Inc. (CX)
✉ 3-1, Kawada-cho, Shinjuku-ku, Tokyo 162
☎ +81 (3) 3353-1111. 🖷 +81 (3) 3358-1747. **LP:** Pres: Hisashi Hieda.
Station: JOCX-TV ch8 50kW(+ 89 relay st's). D.Prgr: Approx 2010-1900.

All-Nippon News Network (23 st's)
Asahi National Broadc. Co., Ltd. (TV Asahi) (ANB)
✉ 1-1, Roppongi 1-chome, Minato-ku, Tokyo 106.
☎ +81 (3) 3587-5111. 🖷 +81 (3) 3505-3539. **LP:** Pres: Kunio Ito.
Station: JOEX-TV ch10 50kW (+ 88 relay st's). D.Prgr: Approx 2030-1800.

TXN Network (6 st's)
Television Tokyo Channel 12 Ltd. (TX)
✉ 3-12, Toranomon 4-chome, Minato-ku, Tokyo 105-12.
☎ +81 (3) 3432-1212. 🖷 +81 (3) 5473-3447. **LP:** Pres: Naomichi Sugino.
Station: JOTX-TV ch12 50kW (+ 86 relay st's). D.Prgr: Approx 2055-1800.

ARMED FORCES RADIO & TV SERVICE (U.S. Mil.)

✉ (Misawa) OLAA, AFPBS, APO San Francisco 96519 - (Okinawa) Det 2, AFPBS, APO San Francisco 96239
Stations: Misawa ch66 1kW, Iwakuni ch66 0.4kW — Okinawa ch8 40kW + additional st's at Iwakuni & Sasebo on cable only.

KAZAKHSTAN

TV-sets: N/A — **Colour:** SECAM — **Systems:** D&K.

KAZAKH TELEVISION (Gov.)

✉ Jeltoksan kösesi 175, 480013 Almati. ☎ +7 (3272) 695188. 🖷 +7 (3272) 631207.

Stations	ch		ch
Karagandi	1	Akmola	10
Semey	4	Öskemen	10
Aktöbe	8	Oral	11
Almati	10	Petropavlovsk	12

+ further main transmitters and translators. Complete details not known.
D.Prgr: Mon-Fri 0255-0800, 1200-1830. Sat/Sun 0300-1830.

PRIVATE STATIONS (details Incomplete)

Rakhat: ch26
TAN: channel unknown
TRK 32: ch31
TRK Shakhar [Kazakhcity] Ltd: ch46.

ORT-1 RELAYS

Stations	ch		ch
Aktöbe	1	Akmola	7
Öskemen	1	Semey	8
Oral	2	Karagandi	12
Almati	3		

+ further main transmitters and translators. Complete details not known.
D.Prgr: Relays of ORT-1 from Moscow.

OTHER STATIONS

Location	ch	Location	ch
Almati	7	Almati	40

D.Prgr: The above transmitters carry relays of private programme producers, Kyrgyz and Turkish TV.

KOREA (Dem. People's Rep.)

TV-sets: 2.000.000 — **Colour:** PAL/NTSC — **System:** D & K/M

THE RADIO AND TELEVISION BROADCASTING COMMITTEE OF THE DEMOCRATIC PEOPLES REPUBLIC OF KOREA (KRT) (Gov.)

✉ Chonsung-dong, Moranbong District, Pyongyang. ☎ +850 (2) 816 035. 📠 +850 (2) 812 100
L.P: Chairm: Chong, Ha-Chol; Dir: Chun, Li-Ji; Head of Tech: Chol, Li-Yong.
Stations (ERP over 10kW):

Location	ch	kW	Location	ch	kW
Sangmasan	1	10	Kanggye	8	70
Chayubong	2	30	Jaedoksan	9	30
Suryongsan	2	30	Unjubong	9	70
Pegebong	3	30	Wangjesan	9	30
Hamhung	3	70	Sepo	9	70
Wonsan	4	10	Sinyang	9	30
Songjinsan	4	30	Wonsan	10	70
Jajiryong	5	30	Haeju	11	70
Paekam	5	10	Sambongsan	11	10
Sambongsan	5	10	Jonchon	11	10
Kangryong	5	30	Songsan	12	10
Kumgangsan	5	30	Jajiryong	12	30
Chongjin	6	30	Chongjin	12	70
Hyangsan	6	10	Haksongsang	12	20
Sepo	6	70	Misan	12	70
Sinuiju	6	70	Pyongyang	12	700
Sariwon	7	30	Rimbong	12	10
Chayubong	8	30	Sobaeksan	12	30
Haksongsan	8	20	Tokusan	12	20

D.Prgr: W 0900-1400; Sun: 0000-0300, 0600-1400
NB: In early 1997, transmissions were irr. due to power shortages.

MANSUDAE TELEVISION

✉ Mansudae, Pyongyang.
Station: ch5 350kW(ERP).
D.Prgr: Sun: 0100-0400, 0700-1300.

KAESONG TELEVISION

✉ Kaesong.
Stations: Kaesong ch8 30kW(ERP), Pyongyang ch9 140kW(ERP).
D.Prgr: W 0900-1400; Sun: 0000-0300, 0600-1400

KOREA (Rep.)

TV-sets 10.430.000 — **Colour:** NTSC — **System:** M

KOREAN BROADCASTING SYSTEM (KBS) (Public Corporation)

✉ 18 Yoido-dong Youngdungpo-gu, Seoul 150-790. ☎ +82 (2) 781 2001/2, 781 1460/1, 781 5108. 📠 +82 (2) 781 2099, 781 1496-7, 781 5199. **Web:** http://www.kbs.co.kr
L.P: Pres: Hong, Too-Pyo; Dir Int. Rel: Ms Cha, Myonh-hee; Tech Coord: Ahn, Dong-Su
Stations: 1=Prgr. 1; 2=Prgr. 2. Call: HL—TV.

Location	ch	kW	Prgr.	Location	ch	kW	Prgr.
Chinju	3	1	1	Ch'unch'on	12	1	1
Yosu	4	10	1	Chonju	13	2	2
Ch'ongju	4	1	1	Cheju	13	1	2
Seoul	5	10	1	P'ohang	13	1	1
Ulsan	5	2	1	P'ohang	20	10	2
Namwon	5	1	1	Ch'unch'on	22	30	2
Ch'angwon	6	10	1	Andong	23	10	2
Taejon	6	10	1	Yosu	24	30	2
Kangnung	6	10	2	Ch'ongju	24	5	2
Ch'unch'on	6	1	2	Kwangju	25	30	2
Seoul	7	50	2	Seoul	26	10	2
Chonju	7	10	1	Ulsan	27	10	2
Pusan	7	10	2	Chinju	27	5	2
T'aebaek	7	1	1	Mokp'o	27	0.5	1
Taegu	8	10	1	Kangnung	28	5	2
Ch'unch'on	8	10	1	Ch'ungju	30	10	2
Seoul	9	50	1	Wonju	31	10	2
Pusan	9	10	1	Seoul	32	10	1
Kangnung	9	10	1	Taejon	35	5	2
Cheju	9	1	1	Seoul	37	10	2

Location	ch	kW	Location	ch	kW
Cheju	10	5	2	T'aebaek	37	5	2
Ch'ongju	10	1	1	Teagu	38	30	2
Wonju	10	1	1	Taejon	42	10	2
Chinju	10	1	1	Chinju	42	5	2
Kwangju	11	10	1	Ch'angwon	45	30	2
Wonju	11	1	1	Taejon	47	10	1
Andong	11	1	1	Ch'ongju	48	0.5	2
Cheju	12	5	1	Taejon	51	10	2
Kangnung	12	1	1	Wonju	52	5	2
Taejon	12	1	1	Andong	53	5	2
Ch'ungju	12	1	1	Namwon	57	10	2
Andong	12	1	1				

+ approx 500 relay st's less than 1kW.
D.Prgr: 1st prgr: W 2100-0100, 0830-1500; Sat/Sun: 2100-1500; 2nd prgr: W 2100-0100, 0830-1500; Sat/Sun: 2100-1530.

EDUCATION BROADCASTING SYSTEM

✉ 92-6., Umyeon-dong, Seocho-gu. Seoul 137-791. ☎ +82 (2) 521 1586/1988/1989/1357. 📠 +82 (2) 521 0241/522 8043
L.P: Pres: Dr. Chung, Yun Choon; Dir: Ms Chung, Hyo-soon; GD: Park, Myung-ha
Stations (operated by KBS): Call: HLQK-TV.

Location	ch	kW	Location	ch	kW
Seoul	13	10	Taejon	29	5
Kwangju	19	30	Yosu	30	30
Kangnung	19	10	Kangnung	34	5
Cheju	20	10	Chinju	36	5
Ch'ungju	21	30	Ch'angwon	39	30
Ulsan	21	10	Taejon	39	30
Namwon	22	10	Wonju	40	10
Pusan	23	30	Seoul	43	1
T'aebaek	25	10	Taegu	44	30
P'ohang	26	10	Chonju	45	30
Cheju	26	1	Andong	47	5
Ch'unch'on	28	30	Taejon	53	10
Andong	29	10	Ch'ongju	54	5

+ approx 200 relay st's less than 1kW.
D.Prgr: W 0730-0010; Sun: 2200-1500.

MUNHWA BROADC. CORP. (Comm.)

✉ 31 Yoido-dong Youngdungpo-gu, Seoul 150-728. ☎ +82 (2) 789 2851/3521. 📠 +82 (2) 782 3094/0294.
Web: http://www.mbc.co.kr
L.P: Pres: Mr. Kang, Sung-Koo; Dir: Mr. Song, Iljun; Dir TV Eng: Mr Jung, Jai-Soon
Call: HL—TV.

Location	ch	kW	Location	ch	kW
Andong	5	1	Yosu	28	2
P'ohang	6	1	Kangnung	31	10
Cheju	7	5	Ulsan	33	10
Mokp'o	7	0.5	Ch'ongju	33	5
Taejon	8	5	Wonju	34	10
Chinju	8	1	Ch'ungju	36	10
Kwangju	9	10	Seoul	38	10
Taegu	10	5	Taejon	40	5
Chonju	10	5	Seoul	41	10
Ch'unch'on	10	1	Samch'ok	43	5
Seoul	11	50	Taejon	45	10
Pusan	11	10	Masan	48	5
Cheju	11	5	Wonju	49	5
Masan	13	5	Kwangju	51	10
Samch'ok	22	5	Andong	59	5
Ch'unch'on	24	30			

+ approx 150 relay st's less than 1kW.
D.Prgr: W 2100-0100, 0830-1500; Sat/Sun: 2100-1500.

SEOUL BROADCASTING SYSTEM (Comm.)

Addr. 10-2 Yoido-dong Youngdungpo-gu, Seoul 150-010. ☎ +82 (2) 786 0792, 780 0006. 📠 +82 (2) 785 6171
L.P: Pres: Mr Yoon, Hyuck-ki; Mr. Park, Jin
Station: HLSQ-TV ch6. **D.Prgr:** W 2100-0100, 0830-1500; Sat/Sun: 2100-1500.

AMERICAN FORCES KOREA NETWORK (US Mil.)

✉ Unit #15324, APO AP 96205-0097, USA. ☎ +82 (2) 7914 6495 —
L.P: Commanding Officer: LTC Cad C. Starr.
Stations: chA2: P'algonsan, Tongduch'on, Pusan, Chinhae, Hoi-dok, Wonju, Kunsan — chA6: Munsan — chA12: Ch'unch'on, Taegu, Taejon — chA13: Kwangju — chA34: Seoul — chA49: Osan, Susong, Taegu — chA70: P'yongt'aek, Tongduch'on — chA75: Tongduch'on.
D.Prgr: 24h.

KYRGYZSTAN

TV-sets: N/A — **Colour:** SECAM — **Systems:** D & K.

KYRGYZ TELEVISION (Gov.)
☑ pr. Molodoj Gvardii 63, 720885 Biskek. ☎ +7 (3312) 253404. 📠 +7 (3312) 257930 — **L.P:** Gen. Dir: Tugelbay Kazakov.
Stations: current details not known.
D.Prgr: 1300-1600.

TV RELAYS
Various main and relay transmitters throughout the country relay prgrs from Moscow, as well as Uzbek and Kazakh TV, and prgrs of TRT from Turkey.

LAOS (Peoples Dem. Rep.)

TV-sets: 80.000 — **Colour:** PAL — **System:** B

LAO NATIONAL TELEVISION (TVNL) (Gov.)
☑ P.O. Box 310 Vientiane Lao PDR. ☎ +856 4475/4523/4425
L.P: DG: Dr. Khekkeo Soisaya; Dir. Tech: Mr. D. Sisombath
Stations: Vientiane stations: ch9 (5kW), ch23 (0.1kW), ch12 (1kW); Savannakhet station: ch12 (1kW)
D.Prgr: ch9, 2330-0030 (UTC) (Sat-Sun 2330-0230 UTC), evening prgr.:1130-1600 (UTC); ch23, 2330-0400 (UTC); ch12, 2330-0030 (UTC), 1000-1600 (UTC); Savannakhet station: 1130-1600 (UTC

MACAU

TV-sets: 70.300 — **Colour:** PAL — **System:** I.

TELEDIFUSÃO DE MACAU (TDM SARL)
☑ P.O. Box 446, Macau. ☎ +853 520204/6. 📠 +853 520 208. ① 88309 RADIO OM
L.P: Chairman: Stanley Ho; Exec. Vice Chairman: Maria do Carmo Figueiredo.
Stations: Portuguese Ch. — Ch30 0.2kW+1 repeater of 10W. Chinese Channel — Ch32 0.2kW+1 repeater of 10W.
D.Prgr: Portuguese Channel: appr. 1858-2400; Chinese Channel 0730-0900, 1815-2400.

MALAYSIA (Federation of)

TV-sets: 9.4m* — **Colour:** PAL — **System:** B.

Peninsular Malaysia

RADIO TELEVISION MALAYSIA (Gov.)
☑ Dept. of Broadc, Angkasapuri, Kuala Lumpur 50614.
☎ +60 (3) 282 5333/3140. 📠 +60 (3) 282 4735. **Cable:** Tivimalsia, Kuala Lumpur. ① MA 31383 Kuala Lumpur.
Web: http://www.asiaconnect.com.my
L.P: DG: Dato' Jaafar Kamin; Int Rel: Ms Nawiyah Che'Lah; Dir Eng: Mr. Lal Singh

Location	TV1 ch	kW(ERP)	TV2 ch	kW(ERP)
Genting Sempah	2	112	10	112
Johor Baru/Gunong Pulai	3	38	10	142.5
Tangkak	4	154	7	77.4
Kuala Lumpur	5	100	8	100
Alor Setar	5	100	8	100
Kuantan	5	127	8	127
Kuala Pilah	5	7.95	8	7.95
Kuala Lipis	5	20	8	20
Mersing	6	100	9	100
Ipoh	6	190	9	190
Melaka	6	185	9	185
Kota Bharu	6	250	9	250

+ low power relay st's.
D.Prgr: TV1: Mon-Wed 0900-1600, Thurs/Sat 0700-1600; Fri/Sun 0200-1600. **TV2:** 0900 (Fri/Sun 0700)-1600.

SYSTEM TV MALAYSIA BERHAD (TV3)
☑ Sri Pentas (Ground Floor, South Wing) No. 3, Persiaran Banjar Utama, 47800 Petaling Jaya Selangor Darul Ehsan. ☎ +60 (3) 716

6333. 📠 +60 (3) 716 133. ① 33014 STMB MA.
L.P: MD: Khalid Hj Ahmad. Eng. Mgr: Rahmad A Kadir.
Stations: Kuala Lumpur ch12 124/25kW H.
Repeater st's of high power in Kuantan (ch 11), Ipoh (ch 11), Gunung Ledang (ch 12), Johore Bahru (ch 26), Kedah (ch 29), Kota Bharu (ch 27), Kuala Terengganu (ch 11), Dungun (ch 27), Tampin (ch 23), Taiping (ch 41), Kuching (ch 12), Sibu (ch 11), Kota Kinabalu (ch 29), Ulu Kali (ch 29) & Machang (ch 11).
D. Prgrs: Mon-Wed 0830-1615, Thurs-Fri 0630-1615, Sat/Sun 0100-1615.

METROVISION
☑ 33 Jln Delima 1/3 Subang, Hi-Tec Industrial Park 40000, Shahalam.
☎ +60 (3) 732 8000. 📠 +60 (3) 732 8932.
L.P: GM (Prgrs): Tunku Yahaya. GM (Mktg.): Lim Eng Kien.

East Malaysia

TV MALAYSIA SABAH AND SARAWAK (Gov.)
☑ P.O. Box 1016, 88614 Kota Kinabalu. ☎ +60 (88)52711. **Cable:** Broadcasts, Sabah. ① MA 80061 Kota Kinabalu.
L.P: Prgr. Contr: M.A. Mahmood.
Stations: Pol = H.

Location	TV1 chE	TV2 chE	kW (ERP)
Sipitang	11	3	13.6
Lahad Datu	7	10	100
Sandakan	8	5	100
Tawau	9	6	33
Tambunan/Keningau	9	12	36
Kudat	9	6	100
K.K. (Lawa Mandau)	10	7	100
Limbang	2	4	13
Sibu	8	5	100
Bintulu	9	6	25
Simangang (Sri Aman)	9	6	120
Miri	10	12	55
Kuching	10	7	220
Kapit	4	10	18

D.Prgrs: as for Peninsular Malaysia.

MALDIVES (Rep. of)

TV sets: 4.750 — **Colour:** PAL — **System:** B.

TELEVISION MALDIVES (Gov.)
☑ Buruzu Magu, 20-04 Male'. ☎ +960 323 105, 324 105. 📠 +960 325 083. ① 66183 TVM MF.
L.P: DG: Mr. Hussain Mohamed
Station: ch7 1kW H.
D.Prgr: 0300-0500 (Fri), 1200-1620.

PAKISTAN

TV-sets: 2.080.000 — **Colour:** PAL — **System:** B.

PAKISTAN TELEVISION CORPORATION LTD.
☑ Federal Television Complex, P.O. Box 1221, Islamabad. ☎ +92 (51) 828723. 📠 +92 (51) 823406/812202. **Cable:** Pakteevee. ① 5833 PPV RP. (Prgr. Centres at Lahore, Karachi, Peshawar, Quetta and Islamabad).
L.P: MD: Mrs Raana Shaikh. TD: Mohammad Kamil.

Stations	ch	kW		ch	kW
Karachi	4	60	Mangora	7	10
Sakessar	4	114	Muree	8	180
Lak Pass	4	8	Shikarpur	8	213
Kala Shah Kaku	5	400	Quetta	8	1.25
Nurpur	5	170	Shujabad	8	178
Ghazaband	5	10	Thana Bola Khan	9	205
Quetta	6	30	Cherat	10	170
Thandiani	6	5	Sahiwal	10	277
Faisalabad	6	20	J. din Wali	10	200
Sibi	6	6	Pasrur	10	6.8
Sakessar	7	168	Tando Allahyar	11	400

+7 low power st's.
D.Prgr: 1130-1930. **N:** 1400, 1600.

SHALIMAR TELEVISION NETWORK (STN)
☑ P.O. Box 1246, Islamabad. ☎ +92 (51) 856 171. 📠 +92 (51) 261 225.
Cable: SUPERSOUND Islamabad
L.P: MD: M Arshad Choudhry. CE: Agha Nasir.

PHILIPPINES (Rep. of the)

TV-sets: 7m — **Colour:** NTSC — **System:** M.

NATIONAL TELECOMMUNICATIONS COMMISSION
(Department of Transportation and Communications)

✉ 855 Vibal Bldg, Esda Corner Times Str, Quezon City.
LP: Commissioner: Josefina Lichauco. Dep. Commissioners: Aloysius R. Santos, Florentino L. Ampil. Chief, Broadcast Sce. Dept: Carlos D. Saliuan Jr.
Stations C=City.

Call	chA	kW	Location
1) DWWX-TV	2	35	Quezon C.
2) DXRV-TV	2	1	Ihgan C.
3) D-3-Z0-TV	3	100	Baguio, C.
12) —	3	1	Metro Manila
1) DYCB-TV	3	10	Cebu C.
11) DYLL-TV	3	500	Zamboanga C.
3) DWGT-TV	4	25	Metro Manila
1) DYXL-TV	4	2	Bacolod C.
1) DXAS-TV	4	10	Davao C.
2) DYXX-TV	6	1	Iloilo C.
2) DZBB-TV	7	25	Quezon C.
2) DYSS-TV	7	5	Cebu C.
5) DXSS-TV	7	1	Davao C.
2) DWAI-TV	7	1	Naga C.
4) DYKB	8	5	Bacolod C.
4) DZKB-TV	9	25	Quezon C.
4) DYKC-TV	9	5	Cebu C.
6) DXLA-TV	9	2.5	Zamboanga C.
2) DYAF-TV	10	5	Bacolod C.
2) DWMJ-TV	10	1	Baguio C.
2) DWMT-TV	10	25	Benguet
2) DWLA-TV	12	10	Legaspi C.
2) D-12-ZB	12	1	Batangas
4) DZBS-TV	12	5	Baguio C.
8) PR-TV	12	?	Tacloban C.
9) DXNS-TV	12	1	Cotabato C.
3) —	13	20	Metro Manila
10) DZTV-TV	13	6.25	Quezon C.
7) DXGL-TV	13	2	Butuan C.
10) DXTV-TV	13	5	Davao C.
10) DYTV-TV	13	12.5	Cebu C.

low power st's not mentioned.
Addresses & other information:
1) ABS-CBN Broadc. Corp, Mother Ignacia Ave, 1100 Quezon C. — Mon-Fri 2300-1530 (Sat/Sun 0000-1730).
2) Republic Broadc. System Inc. EDSA, Diliman, Quezon City, Metro Manila. Mon-Fri 0800-0130, Sat 0730-0130, Sun 0730-0200.
3) PTV Channel 4, Media Center, Bohol Ave, Quezon City — 0425 (Sat/Sun 0300)-1730.
4) RPN Channel 9, Broadcast City, Capitol Hills, Quezon City — 0230-1600.
5) Southern Broadc. Netw, 3881 E. Vallejo Str, Santol Sta. Mesa, Metro-Manila.
6) First United Broadc. Corpo, Lozenzo Bldg, 787 Vito Cruz Str, Metro-Manila. ☎ +63 583082. 🖷 +63 583082.
7) Philippine Electronic & Communications Institute, Montilla Blvd, Butuan C — Mon-Fri 1000-1500, 0500-2200.
8) East Visayas Broadc, 2647 Donada Str, Malate, Metro-Manila.
9) Cotabato TV Corpo, Regional Complex, Catabato City.
10) Intercontinental Broadc. Corp., Quezon C.
11) Zamboanga TV Corp., Zomboanga C.
12) RT Broadcasting Spec. Rm.9, RC Poblete Bld., 17 Sen. Gil Y. Puyat Av. Makati, Metro Manila.

PEOPLE'S TELEVISION NETWORK, INC (Comm.)

✉ Broadcast Complex, Visayas, Avenue, Quezson City 1100. ☎ +63 (2) 921 2344/2451. 🖷 +63 (2) 921 1777/7310/886, 922 9112/6064
LP: Chairm: Ms Lourdes I Ilustre; GM: Mr Ramon S. Diez; CE: Mr Antonio M Leduna

ARMED FORCES RADIO & TV SERVICE (U.S. Mil.)

✉ Det 1, AFPBS, APO San Francisco 96274, USA.
Stations: Olongapo ch14 0.25kW, Angeles City ch17 1kW, San Miguel ch40 0.03kW, Baguio ch14 0.3kW, San Fernando ch17 0.25kW, Capas ch17 25W.

SINGAPORE

TV-sets: 1.060.000 — **Colour:** PAL— **System:** B.

TELEVISION CORPORATION OF SINGAPORE (TCS)

✉ Caldecott Broadcast Centre, Andrew Rd. Singapore, 299939. ☎ +65 2560401. 🖷 +65 2538808. ➀ RS 39265 SBCGEN.
Web: http://rock.tcs.com.sg
LP: Chief Exec. Officer: Leo Cheok Yew. PR Mgr: Julie Lee. VP (Prgrs): Daniel Yun. VP Eng: Tay Joo Thong.
Stations: Singapore chE5/chE8 120kW.
D.Prgr: chE5: Sun-Thurs 0600-0300, Fri & Sat 0600-0400; chE8: Mon-Fri 0800-1230, Sat 0800-0230, Sun 0800-1230

TELEVISION TWELVE (TV-12)

✉ 12 Prince Edward Road, #05-00 Bestway Building, Singapore 0207. ☎ +65 225 8133. 🖷 +65 220 3881, 225 0966.
LP Chief Exec. Officer: Sandra Buenaventura. VP Programming: Amy Chua. Transmission Mgr: Mr. H.T. Lau
Stations: chE12 120kW & E24
D.Prgr: Mon-Fri 0600-1230, Sat 0200-1230, Sun 0900-1230

SRI LANKA

TV-sets: 1.500.000 — **Colour:** PAL — **System:** B.

INDEPENDENT TELEVISION NETWORK (ITN)

✉ Wickramasinghepura, Battaramulla. ☎ +94 (1) 864591. 🖷 +94 (1) 864591. ➀ 22445.
LP: Chairm: Mr G.B Rajapakse; GM: Mr Bertie Galahitiyawa; DE: Mr W.S.E. Fermando
Station: Wickramasinghepura ch12 10kW (ERP), Yatiyantota ch9 10kW, Deniyaya ch5 1kW.
D.Prgr: 1800-2300 (approx).

SRI LANKA RUPAVAHINI CORPORATION (SLRC)

✉ P.O. Box 2204, Colombo 7. ☎ +94 (1) 580136. 🖷 +94 (1) 580929.
Cable: Rupavahini. ➀ 22148 SLTV CE.
LP: DG: Mr. W.D Jayasinghe. Dep. DG: A. Senadheera. Dep. DG (Prgrs): Lucien Bulathsinghala. Dep. DG (Eng): Upali Arambewela.
Stations: Kokavil ch8 20 kW, ch11 1 kW Sooriyakanda +2 low power repeaters.
D.Prgrs: 1130-1700. Sun 1000-1700.

TELSHAN NETWORK (PVT) Ltd. (TNL TV)

✉ Innagale Estate Dampe-Piliyandala. ☎ +94 (1) 575436 430 859 🖷 +94 (1) 575436,574 962
LP: Chairm MD: Mr. Shantilal Nilkant Wickremesinghe
Stations: Colombo ch 21 22kW; Piliyandala ch 26 22kW, ch 3 20kW; Nuweraeliya ch 4 40kW; Polgahawela ch 3 1kW; Ratnapura ch 26 1kW; Hantana (Kandy) ch21 22kW.

EAP NETWORK (PVT) LTD.

✉ 676 Galle Rd, Colombo 3. ☎ +94 (1) 503819 (9 lines). 🖷 +94 (1) 503788. **E-mail:** eapnet@slt.lk
LP: Chairperson: Mrs. Soma Edirisinghe. MD: Jeevaka Edirisinghe. Dir/GM: Rosmand Senaratne.

Location	ETV1	ETV2	kW
Colombo	37	35	1
Deniyaya	31	35	1
Hantana	37	33	1
Matale	31	35	1
Nuwara Eliya	37	33	1

D.Prgr: 24h. ETV1 rel. Sky News. ETV2 rel. Star TV.

MTV CHANNEL (PVT) LTD.

✉ 109 Collets Bldg., Rt. Hon. D. S. Senanayake Mawatha, Colombo 8. ☎ +94 (1) 689324-6. 🖷 +94 (1) 689328.
Stations: Depanama ch23 1kW, Nuwaraeliya ch25 1kW.
D.Prgr: 24h. Rel. BBC World + local prgrs 1130-1700.

TAJIKISTAN

TV-sets: N/A — **Colour:** SECAM — **Systems:** D & K.

TAJIK TELEVISION (Gov.)
✉ Behzod küça 7, 734013 Dushanbe. ☎ +7 (3772) 224357.
L.P: Head of TV: Mirbobo Mirrahimov.
Stations and **D.Prgr:** current details not available.

TV RELAYS
Various main and relay transmitters throughout the country relay programmes of ORT-1 from Moscow, as well as TRT from Turkey and IRIB from Iran. Further details not known.

THAILAND

TV-sets: 3.300.140 — **Colour:** PAL — **Systems:** B&M.

TELEVISION OF THAILAND (Gov.)
✉ Public Rel. Dept., 26th Floor Fortune Town Bldg, Ratchadaphisek Road, Huay Khwang, Bangkok 10310. ☎ +66 (2) 248 1601/8088 to 94. 🖹 +66 (2) 248 1601/1655/2155. ① 72243 PRDTHAI TH.
Stations: Bangkok ch11 200kW (ERP) + 27 relay st's.
D.Prgrs: 0930-1400.

BANGKOK ENTERTAINMENT CO. Ltd.
(Licensed through Mass Communications Organisation of Thailand).
🖂 1126/1 New Petchbury Road, Bangkok 10400.
☎ +66 (2) 2539970-3. 🖹 +66 (2) 2539978. ① 82616 BECOM TH. **Cable:** BANGERTAIN.
L.P: Prgr. Dir: Pravit Maleenont. TD: Manoontham Thachai.
Stations: Bangkok ch3 650 kW + 32 relay st's not mentioned.
D.Prgr: Mon-Fri 1400-2400, SS 0800-2400.

MASS COMMUNICATIONS ORG. OF THAILAND
🖂 222 Thanon Ysok Asok-Dindaeng, Bangkok 10210.
☎ +66 (2) 2450700. ① 84577 MOT BKK TH.
Stations: Bangkok ch9 20/4kW + 32 relay st's not mentioned.
D.Prgrs: 0850 (Sat/Sun 0025)-1700.

THE ARMY TELEVISION HSA-TV (Gov, Comm.)
🖂 Phaholyothin Rd, Sanampao. Bangkok 10400. ☎ +66 (2) 2710069.
🖹 +66 (2) 2712510. ① 81080 ATV TH.
L.P: DG: Maj. Gen. Vijit Junapart.
Stations: Bangkok ch5 20/4kW + 18 relay st's not mentioned.
D.Prgr: 0900 (Sat/Sun 0100)-1700.

BANGKOK BROADCASTING & TV CO. LTD. (Comm.)
🖂 P.O. Box 4-56, Bangkok 10900. ☎ +66 (2) 2781255. 🖹 +66 (2) 2701976. ① 82730 BBTV TH.
L.P: St. Man: Chatchur Karnasuta. TD: Supoch Sangsayan.
Stations: Bangkok ch7 20kW + 22 relay st's not mentioned.
D.Prgr: 0900 (Sat/Sun 0100)-1700.

TURKMENISTAN

TV-sets: N/A — **Colour:** SECAM — **Systems:** D&K.

TURKMEN TELEVISION (Gov.)
🖂 Machtumkuli 89, 744000 Asgabat. ☎ +7 (3632) 251515. 🖹 +7 (3632) 251421 (manual).
Stations: Chardzhev chR10 (TMT-1), R8 (TMT-2), R3 (ORT), all 5kW.
D.Prgr: TMT-1: MF 0150-0540, 655-1020, SS 0150-2030; TMT-2: MF 0200-0300, 1300-1930, SS 0200-0830, 1200-1930. rel: ORT, Moscow (24hrs) & TRT-1 (Turkey): 1200-1600.
TMT-1/2, ORT on nationwide networks, TRT-1 in Asgabat.

UZBEKISTAN

TV-sets: N/A — **Colour:** SECAM — **Systems:** D&K.

UZBEK TELEVISION (Gov.)
🖂 Navoii küçasi 69, 700011 Toskent. ☎ +7 (3712) 495214.
L.P: Chmn: Eerkin K. Haitboev.

Stations: (Prgr. 1)	**ch**	**Stations:** (Prgr. 2)	**ch**
Toskent	5	Toskent	1

D.Prgr: Current schedule not available.

ORT
Stations	**ch**
Toskent	3

D.Prgr: Relays of ORT from Moscow. Further details see under Russia.

ROSSIJSKOJE TELEVIDENIJE (RTV)
Stations	**ch**
Toskent	9

D.Prgr: Relays of RTV from Moscow. Further details see under Russia.

TV RELAYS
Stations	**ch**
Toskent	40

D.Prgr: Relays of Kazakh, Kyrgyz and Tadzhik TV, as well as TRT from Turkey.

VIETNAM

TV-sets: 2.500.000 — **System:** M — **Colour:** NTSC/SECAM

TELEVISION VIETNAM (Gov.)
🖂 59 Giang Vo Street, Hanoi. ☎ +84 (43) 43188/ 55933. 🖹 +84 (43) 55332. ① 412279 THVN/VT.
L.P: DG: Ho Anh Dung. Dir of loc. TV: Nguyen Van Nhuong.
Stations: Hanoi ch2, ch6 (SECAM); Ho Chi Minh City ch9 240kW; Hue, Cantho & Quinhon ch7 and DaNang ch13 (NTSC). Details of st's in Vinh & Nha Trang not available. 13 additional relay tr's.
D.Prgr: Central Prgr. from Hanoi: Mon-Fri 90 min's a day, Sat/Sun 3h (prgrs in black & white only).
Ch 6: Hanoi City Sce. **Ch. 2:** Provincial Sce. for Red River delta. **Other st's:** 25% in colour (no further details available).

PACIFIC

AUSTRALIA

TV-sets: 8m — **Colour:** PAL. **System:** B.

AUSTRALIAN BROADCASTING CORPORATION (ABC)
The 'National Television Service' is controlled by the Australian Broadcasting Corporation which is responsible to Parliament through the Minister for Transport and Communications.

In addition to the terrestrial transmitters listed below, ABC also broadcasts to remote areas using five transponders of the Australian domestic satellite system to cover the five main geographic regions and time zones. This service is known as the 'Homestead and Community Broadcasting Satellite Service' (HACBSS) and is carried in the frequency band 12 250 to 12 750 MHz on satellites which are owned and operated by Optus Communicatons Pty Ltd. Except in the States of Victoria and Tasmania, most transmitters outside the State capital cities are fed by satellite. The satellite signal is 'B-MAC' encoded and carries three sound-only programs, one of which is stereophonic, in addition to the television signal and its associated sound channels.

🖂 GPO Box 9994, Sydney NSW 2001. ☎ +61 (2) 437 8000. 🖹 +61 (2) 9950 3055. **Web:** http://www.abc.net.au
L.P: Ag. Dir. TV: Penny Chapman.
Stations: Audio powers are 1/10 of vision powers when a single sound channel is used. Stations below 10 kW ERP are not listed. Pattern: 'O': omnidirectional, 'D': directional. Channels 3, 4, 5 and 5A are progressively being phased out to make way for FM radio and other services — the expected new channel numbers and other information covering changes arising from the first stages of the clearance of stations from 'Band II' are given in parentheses in the table.

Call	ch	ERP(kW)	POL	Pattern	Location
Australian Capital Territory					
ABC	3	100	V	O	Canberra
ABN	2	100	H	O	Sydney
New South Wales					
ABMN	0	100	H	O	Wagga Wagga
ABCN	1	100	V	O	Orange
ABTN	6	100	V	D	Taree

ABDN	2	100	H	D	Grafton/Kempsey
ABQN	11	200	V	O(D)	Dubbo
ABHN	5A	100	H	O	Newcastle
ABWN	56	600	H	D	Illawarra
ABRN	6	100	H	O	Lismore
ABGN	7	100	H	O	Griffith
ABUN	7	100	H	O	Tamworth
ABSN	8	100	V	D	Bega/Cooma

Victoria

ABAV	1	100	H	D	Albury
ABEV	1	100	V	D	Bendigo
ABV	2	100	H	O	Melbourne
ABSV	2	100	V	O	Swan Hill
ABGV	40	100	V	O	Shepparton
ABRV	11	100	H	D	Ballarat
ABLV	40	100(1000)	H	D	Traralgon

Call	ch	ERP(kW)	POL	Pattern	Location

Victoria

ABMV	6	100	H	D	Mildura
ABWV	5A	66	H	D	Western Victoria
ABWV	45	100	H	D	Horsham/Dimboola

Queensland

ABSQ	1	100	H	D	Warwick
ABQ	2	100	H	O	Brisbane
ABDQ	3(32)	100(1000)	H	O(D)	Toowoomba
ABRQ	9	100	H	D	Rockhampton
ABTQ	3	100	H	D	Townsville
ABMQ	8	180	H	D	MacKay
ABWQ	6	150	V	D	Maryborough
ABNQ	9	100	H	D	Cairns
ABQ	49	50	H	D	Gold Coast

South Australia

ABGS	1	100	H	D	Mount Gambier
ABNS	1	100	V	D	Port Pirie
ABS	2	100	H	D	Adelaide
ABRS	3	150	V	D	Loxton
ABS	30	10	H	D	Wudinna

Western Australia

ABW	2	100	H	O	Perth
ABAW	2	100	V	O	Albany
ABSW	5	100	H	O	Bunbury
ABCW	5A	100	H	O	Central Agricultural
ABGW	6	10	H	D	Geraldton
ABW	8	60	H	D	Wagin
ABCMW	8	10	H	D	Morawa
ABW	46	10	H	D	Dalwallinu
ABW	60	60	H	D	Moora

Tasmania

ABT	2	100	H	O	Hobart
ABNT	32	100(300)	H	D	Launceston

Northern Territory

ABD	6	30	H	D	Darwin

+ 343 low power services with ERPs below 10kW, made up of 110 translators and 233 satellite-fed transmitters.
+ 46 (approx.) low power (up to 150 watts ERP), privately owned transmitters/translators which are licensed under a 'self-help' scheme to carry ABC television programs.
Daily Prgrs: Mon-Fri 2100-1400; Sat/Sun 24h.

SPECIAL BROADCASTING SERVICE (SBS)

✉ Locked Bag 028, Crows Nest, NSW 2065. ☎ +61 (2) 9430 2828. 📠 +61 (2) 9430 3700 — P.O. Box 294, So. Melbourne, VIC 3205. ☎ +61 (3) 9685 2828. 📠 +61(3) 9686 7501.
Web: http://www.sbs.com.au
LP: MD: Malcolm Long.
Stations: All stations horizontal polarity unless specified.

	Ch.	kW(ERP)		Ch.	kW(ERP)
Sydney	28	300	Hobart	28	225
Melbourne	28	300	Gold Coast	61	50
Brisbane	28	300	Newcastle	45	600
Adelaide	28	300	Illawarra	53	600
Perth	28	300	Ulladulla	30(V)	400
Bendigo	29	100	Ballarat	30	300
Canberra	28	200	Traralgon	34	800

+ estimated 35 low power transmitters with kw(ERP) below 10 kw. not mentioned.

IMPARJA TELEVISION PTY LTD.

✉ PO Box 2924, Alice Springs, NT 0871. ☎ +61 (89) 523744, 📠 +61 (89) 531014. ③ AA81166.
This st. is owned by the Central Australian Aboriginal Media Association (CAAMA).
Station: ch9 250kW.

COMMERCIAL TV STATIONS

FEDERATION OF AUSTRALIAN COMMERCIAL TELEVISION STATIONS

✉ 44A Avenue Rd, Mossman, NSW 2088. ☎ +61 (2) 960 2622. 📠 + 61 (2) 969 3520. ③ 121542.

Main Networks

THE SEVEN NETWORK

✉ Television Centre, Mobbs Lane, Epping, NSW 2121. ☎ +61 (2) 858 7777. 📠 +61 (2) 858 7888. ③ AA20250.
Web: http://www.seven.com.au
(5 owned and 9 affiliated st's).

THE NINE NETWORK

✉ P.O. Box 27, Willoughby, NSW 2068. ☎ +61 (2) 9906 9999. 📠 +61 (2) 9958 2279. **Web:** ninemsn.com.au

NETWORK 10 AUSTRALIA

✉ P.O. Box 10, Lane Cove, NSW 2066. ☎ +61 (2) 887 0222.

Stations: The first two letters are an abbreviation of the name of the licence, the third indicates the state and the numeral signifies the channel

Call	on air ID	affiliation	tx's
1) RTQ	WIN Television Northern Queensland	9 Network	70
2) STQ	Sunshine Television	7 Network	67
3) TNQ	TEN North Queensland	10 Network	64
4) BTQ	Seven Network	7 Network	4
5) QTQ	Nine Network	9 Network	5
6) TVQ	Ten Network	10 Network	3
7) NBN	NBN Television	9 Network	38
8) NEN	Prime Television	7 Network	36
9) NRN	TEN Northern NSW	10 Network	37
10) CBN	Prime Television	7 Network	39
11) CTC	TEN Capital	10 Network	44
12) WIN	WIN Television	9 Network	46
13) ATN	Seven Network	7 Network	6
14) TCN	Nine Network	9 Network	6
15) TEN	Ten Network	10 Network	6
16) AMV	Prime Television	7 Network	80
17) BCV	TEN Victoria	10 Network	25
17) GLV	TEN Victoria	10 Network	41
18) VTV	WIN Television Victoria	9 Network	71
19) ATV	Ten Network	10 Network	9
20) GTV	Nine Network	9 Network	9
21) HSV	Seven Network	7 Network	9
22) TNT	Southern Cross Network	7/10 Netw.	30
23) TVT	WIN Television Tasmania	9 Network	28
24) ADS	Ten Network	10 Network	9
25) NWS	Nine Network	9 Network	9
26) SAS	Seven Network	7 Network	9
27) NEW	Ten Network	10 Network	4
28) STW	Nine Network	9 Network	4
29) TVW	Seven Network	7 Network	4
30) BKN7	Broken Hill TV	—	1
31) MTM	—	—	3
32) STV8	Sunrayia TV	—	1
33) GTS	—	—	8
34) RTS 5a	—	1	
35) SES	—	—	5
36) GTW	—	—	4
37) SSW	Golden West Network	—	10
37) VEW	Golden West Network	—	8
38) NTD	—	—	2
39) ITQ 8	—	—	1
--) QQQ	—	—	75
40) IMP	Imparja Television	—	68
--) WAW	—	—	100

Addresses:
1) WIN Television, P.O. Box 568, Rockhampton QLD 4700
2) Sunshine Television, P.O. Box 30, Maryborough QLD 4650
3) TEN North Queensland, P.O. Box 1016, Townsville QLD 4810
4) Seven Network Limited, P.O. Box 604, Brisbane QLD 4001
5) Queensland Television Limited, P.O. Box 72, Brisbane QLD 4001

6) Ten Network, P.O. Box 751, Brisbane QLD 4001
7) NBN Television, P.O. Box 750L, Newcastle NSW 2300
8) Prime Television (Northern) Pty Ltd, P.O. Box 317, Tamworth NSW 2340.
9) TEN Northern NSW, P.O. Box 920, Coffs Harbour NSW 2450
10) Prime Television (Southern) Pty Ltd, P.O. Box 465, Orange NSW 2800
11) Capital Television, Private Mailbag 10, Dickson ACT 2602
12) WIN Television, Television Ave, Mt St Thomas NSW 2500
13) Seven Network Ltd, Mobbs Lane, Epping NSW 2121
14) Nine Network, P.O. Box 27, Willoughby NSW 2068
15) Ten Network, 44 Bay St, Ultimo NSW 2007
16) Prime Television (Victoria) Pty Ltd, Union Road, Lavington NSW 2641
17) TEN Victoria, P.O. Box 888, Bendigo VIC 3550
18) WIN Television, P.O. Box 464, Ballarat VIC 3350
19) Ten Network, Level 3-6, Como Centre, TRB West, 620 Chapel Street, South Varra VIC 3141
20) Nine Network, P.O. Box 100, Richmond VIC 3121
21) Seven Network Ltd, P.O. Box 215D, Melbourne VIC 3001
22) Southern Cross Television, 37 Watchorn Street, Launceston TAS 7250
23) WIN Television, P.O. Box 1209M, Hobart TAS 7001
24) Ten Network, 125 Strangeways Tce, North Adelaide SA 5006
25) Southern Television Corp. Pty Ltd, P.O. Box 9, North Adelaide SA 5006
26) Seven Network Ltd, 45-49 Park Tce, Gilberton SA 5081
27) Ten Network, P.O. Box 1010, Mirrabooka WA 6061
28) Sunraysia Television Ltd, P.O. Box 99, Tuart Hill WA 6060
29) Seven Network, P.O. Box 77, Tuart Hill WA 6060
30) Broken Hill Television Ltd, P.O. Box 472, Broken Hill NSW 2880
31) Lochfield Consultants Pty Ltd, P.O. Box 493, Griffith NSW 2680
32) Sunrayia Television Ltd, P.O Box 1157, Mildura VIC 3500
33) Spencer Gulf Telecasters Ltd, P.O. Box 305, Port Pirie SA 5540
34) Riverland Television Ltd, P.O. Box 471, Loxton SA 5333
35) South East Telecasters Ltd, P.O. Box 821, Mt Gambier SA 5290
36) Geraldton Telecasters Pty Ltd, Cnr Fifth & Howard street, Geraldton WA 6530
37) Golden West Network Pty, Ltd, P.O. Box 112, Bunbury WA 6230
38) Territory Television Pty Ltd, P.O. Box 1764, Darwin NT 0800
39) Mt Isa Television Pty Ltd, P.O. Box 1557, Mt Isa QLD 4825
40) Imparja Television Pty Ltd, 14 Leichhardt Tce, Alice Springs NT 0871

COOK ISLANDS

TV-sets: 3500 — **Colour:** PAL — **System:** B.

COOK ISLANDS BROADCASTING CORPORATION (Gov.)
✉ P.O. Box 126, Avarua, Rarotonga. ☎ +682 29460. 🖷 +682 21907 Cook Islands Television (CITV).
Stations: VHF ch1 & 6 — **D.Prgr:** 6h a day, 5 days a week.

EASTER ISLAND

TV-sets: N/A — **Colour:** PAL — **System:** B.

TV RAPANUI
✉ Hanga Roa, Isla de Pascua. ☎ +5639 223291
L.P: Dir. Gen: J. Edmund Paoa. Head Tec. Sce's: J. Pont Chavez.
Station: ch (unknown). **D.Prgr:** 0000-0600.

FIJI

TV-sets: N/A — **Colour:** NTSC — **System:** M.

Fiji Television
✉ GPO Box 2442, Suva. ☎ +679 305100. 🖷 +679 305077. **E-mail:** fijitv@is.com.fj — **L.P:** Chief Exec: Peter Wilson.

Pay-TV: the govt. is considering the award of a franchise to operate a Pay-TV sce.

GALAPAGOS ISLANDS

TV-sets: 4.000 — **Colour:** NTSC — **System:** M.

TELEGALAPAGOS (Cult)
✉ Misión Franciscana, Puerto Baquerizo Moreno, Isla San Cristobal, Galapagos, Ecuador.
LP: Dir. Gen: Mons. Manuel Valarezo. Dir. Tec: Germán Chiriboga. Film Buyer: Remigio Andrade.
Station: chA13 — **D.Prgr:** 2000-0400.

GUAM (US Terr.)

TV-sets: 75.000 — **Colour:** NTSC — **System:** M.

KUAM TELEVISION (Comm.)
✉ Pacific Telestations, P.O. Box 368, Agana 96910. ☎ +671 6375826, 6376397. 🖷 +671 6379865, 6379870. **Cable:** Kuam.
L.P: Pres: L.S. Berger. Gen. Mgr: Greg Perez. Dir. Tec: K. Tydingco.
Station: chA8 21.9/3.8kW
D.Prgr: Mon-Fri 2000-1400 (also rel. CBS/NBC prgrs).

KGTF TELEVISION (Educ.)
✉ Guam Educational Telecommunications Corporation, P.O. Box 21449, Guam, Marianas Is. 96921. ☎ +671 7342207. 🖷 +671 7345483. ⌖ 6467 KGTF TV GM.
L.P: GM: Joseph E. Tighe. TD: Edmond Cheung. Film Buyer: Doris Gallo.
Station: chA12 27.6/5.5kW. **F.Pl.** ch 14 (Saipan & Tinian); ch 16 (Rota).
D.Prgr: 1100-1300.

KTGM-TV
✉ 692 Marine Dr, Tamuning 96911. ☎ +671 6498814. 🖷 +671 6490371 — Station: chA14.

HAWAII (US State)

TV-sets: 552.500 — **Colour:** NTSC — **System:** M.

All st's comm. exc. 7) and 9) Educational, 18) religious.

Call	ch	kW (ERP)		Call	ch	kW (ERP)
1) KHON-TV	2	100/20		7) KMEB-TV	10	5/0.5
1a) KHBC-TV	2	2.29/0.458		9) KHET	11	150/30
2) KGMV	3	14.1/2.82		9a) KHAW-TV	11	2.09/0.27
3) KITV	4	100/20		10) KMAU-TV	12	30/4.36
16) KFVE	5	100/20		11) KHNL-TV	13	316/63.2
14) KVHF	6	52.6/6.7		12) KHVO	13	4.68/0.59
4) KAII-TV	7	29.8/5.9		13) KHAI-TV	20	537/63
5) KGMB-TV	9	209/41.8		14) KMGT	26	100/10
6) KGMD	9	9.68/1.71		15) KBFD	32	146/14.6

1) Burnham Broadc. Co., 1170 Auahi Str, Honolulu 96814-4975.
1a) Hilo Broadc. Corp, Box 4250 Hilo, Big Island 96720-0520.
2) Wailuku (Relay of st. 5).
3) TAK Communications, 1290 Ala Moana Blvd, Honolulu 96814-4299.
4) Wailuku (Relay of st. 1).
5) Lee Enterprises, 1534 Kapiolani Blvd, Honolulu 96814-3799. Mon-Fri 1600-1130; Sat/Sun 24h.
6) Hilo (Relay of st. 5).
7) Wailuku (Relay of st. 9).
9) Hawaii Public Broadc. Authority, 2350 Dole Str, Honolulu 96822-2495.
9a) Hilo (Relay of st. 1).
10) Wailuku (Relay of st. 3).
11) King Broadc. Co, 150-B Puuhale Rd, Honolulu, HI 96819.
12) Hilo (Relay of st. 3).
13) Honolulu Family TV Ltd, 735 Sheridan Str, Honolulu 96814-3095.
14) Mauna Kea Broadc. Co, 970 N. Kalaheo Ave, Honolulu 96734-1892.
15) The Allen Broadc. Corp, 1188 Bishop Str, Honolulu 96813-3314.
16) 315 Sand Island, Access Rd, Honolulu 96819-2245.

KIRIBATI

TV-sets: 685 — **Colour:** PAL — **System:** B.

RADIO KIRIBATI
✉ Broadcasting House, P.O. Box 78, Bairiki, Taiwara, Rep. of Kiribati, Central Pacific. ☎ (686) 21187. 🖷 (686) 21096
Station & D.Prgr: details unknown

MARSHALL ISLANDS

TV-sets: N/A — **Colour:** NTSC — **System:** M.

MBC-TV
☑ Marshall Islands Broadcasting Company, Majuro 96960. ☎ +692 6253413.

AFRTS TELEVISION (Department of Defense)
☑ Box 23, APO San Francisco, CA 96555, USA.
LP: Netw. Mgr: Larry Malinowski.
Stations: ch9, ch13 0.25kW (24 h to Kwajalein Island & Roi-Namur Island).

MICRONESIA (Federated States of)

TV-sets: 7.000 — **Colour:** NTSC — **System:** M

TV STATION POHNPEI (Comm.)
☑ KPON-TV, Central Micronesia Communications, P.O. Box 460, Kolonia, Pohnpei, FSM 96941.
LP: Pres: Bernard Hegenberger. Dir. Tec: David Cliffe.
Station: Pohnpei chA7 1kW + cable TV on ch's 4,5,9.

TV-STATION TRUK (Comm.)
☑ Truk State, FSM 96942.
Station: TTTK chA7 0.1kW.
D.Prgr: 0400-1200 (approx).

TV-STATION YAP (Gov.)
☑ WAAB-TV, Yap State, FSM 96943.
Station: chA7 1kW.
D.Prgr: 0400-1200 (approx).

NAURU (Republic of)

TV-sets: ? — **Colour:** SECAM — **System:** K

Nauru Television
☑ Rep. of Nauru, Ce. Pacific. ☎ +674 4443190

NEW CALEDONIA

TV-sets: 35.500 — **Colour:** SECAM — **System:** K.

RFO-TV (Gov.)
☑ Radio Télévision Française d'Outre Mer (RFO), BP G3 Mont Coffin, F-98848 Nouméa cedex. ☎ +687 274327. 🖷 +687 281252 — **LP:** Dir: Alain Le Garrec.
Stations: Mont Do chK4 2 x 1kW H, Noumea chK8 0.4kW H, Lifou chK7 0.4kW H (+ 25 low power repeaters).
D.Prgr: 2 channels, 8h a day.

PRIVATE STATIONS

CANAL CALEDONIE (Priv, Comm.)
☑ 8 rue de Verneilh, Noumea
Stations: Noumea Mt Coffyn ch43 11kW, Noumea (Town) ch33 0.06kW, Noumea Mt Koghi ch25 9kW
N.B: Canal Caledonie is a subscription sce and the signal is encrypted except: 1945-2020, 0125-0225, 0715-0910.
D. Prgr: W1945-1400 Sun 24h.

NEW ZEALAND

TV-sets: 1.100.000 — **Colour:** PAL — **System:** B.

TELEVISION NEW ZEALAND Ltd.
☑ P.O. Box 3819, Auckland. ☎ +64 (9) 377 0630. 🖷 +64 (9) 375 0918. ⓣ TVNZACQ 60056.
LP: Dep Group Exec: Darryl Dorrington. Contr of Eng:Neville Lane.
Stations: Channel One: Wellington ch1 100/20 kW H, Auckland ch2 100/20 kW H, Christchurch ch3 100/20 kW H, Dunedin ch2 50/10 kW

H + relay st's at Hamilton ch1 100/20 kW V, Palmerston North ch2 100/20 kW V, Invercargill ch1 1/0.1 kW H + 18 medium and 412 low powered relay st's.
D.Prgr: 2100 (Sat 1900, Sun 2000)-1200 (Fri 1300, Sat 1400).
Channel Two: Auckland ch4 300/30 kW H, Wellington ch5 300/30 kW H, Christchurch ch3 300/30 kW H, Dunedin ch4 300/25 kW H + repeaters at Hamilton ch3 100/20 kW H, Palmerston North ch4 300/30 kW V, Invercargill ch3 1/0.1 kW H + 17 medium and 240 low powered repeaters.
D.Prgr: 1830-1200. Weekends 24hrs.

ACTION TV (Trackside)
☑ P.O. Box 388-99 Wellington. ☎ +64 (4) 576 6999. 🖷 +64 (4) 576 6942
Stations: Auckland ch55 150/15kW, ch53 100/20kW, ch58 50/5kW; Waikato ch56 100/20kW, ch50 50/5kW, ch47 50/5kW; Palmerston North ch?; Wellington ch56 150/15kW, ch58 100/20kW; Christchurch ch48 65/7kW.
D.Prgr: Fri 2200-0930, Mon 2330-0600, Tues 2345-0610, Thur 2250-1000.

TV3 (Comm.)
☑ P.O. Box 5185, Auckland. ☎ +64 (9) 779 730. 🖷 +64 (9) 366 7029.
LP: Head of Netw. Programming: Kel Geddes.
Stations: Auckland ch7 100/20kW; Wellington ch11 100/20kW; Christchurch ch6 100/20kW; Dunedin ch10 100/20kW; Hamilton ch9 100/20kW; Palmerston North ch7 100/20kW; Invercargill ch7 1/0.1kW. (+ 11 medium powered relay sts.)
D.Prgr: Auckland ch7 H, Christchurch ch6 H, Hamilton ch9 V, Invercargill ch7 H, Wellington ch11 H, Dunedin ch10 H, Palmerston North ch7 V.

CANTERBURY TELEVISION (CTV)
☑ 196 Gloucester Street, Christchurch.
D.Prgr: 0500-1200 UTC, Fri 2300-Sat 1200 — **Station:** ch48 65kW.

SKY NETWORK TELEVISION (pay-tv)
☑ P.O. Box 9059, Auckland. ☎ +64 (9) 525 5555. 🖷 +64 (9) 525 5725
Stations: Auckland ch27/29/30V (movies), ch31/33/52V (sports), ch43/45/54 (news); Waikato ch28/30V/31V (movies), ch32/34V/27V (sports), ch44/46V/51V (news); Palmerston North ch30 (movies), ch34 (sports), ch46 (news; Wellington ch28/30V/30/49 (movies), ch32/34V/34/53 (sports), ch44/54V/54/47 (news); Christchurchch30 (movies), ch34 (sports), ch46 (news).

NIUE ISLAND

TV-sets: ? — **Colour:** SECAM — **System:** K

NIUE TV
☑ Broadcasting Corp. of Niue, P.O. Box 23, Alofi. ☎ +683 4026. 🖷 +683 4217 — **LP:** GM: Hima Douglas

NORFOLK ISLAND

TV-sets: 900 — **Colour:** PAL — **System:** B.

NORFOLK ISLAND TELEVISION SCE. (Gov.)
☑ New Cascade Rd, Norfolk Island 2899, Australia. ☎ +672 (3) 22137. 🖷 +672 (3) 23298. ⓣ NV 30003
Station: Mt. Pitt ch7 0.02kW V.
D.Prgr: rel. ABC & SBS Australia from Optus Satellites.

N. MARIANA ISLANDS (US Commonw.)

TV-sets: 4.100 — **Colour:** NTSC — **System:** M.

MICRONESIA BROADC. CORPO. (Comm.)
☑ c/o KUAM, Box 368, Agana, Guam 96920.
LP: Pres: H. Scott Killgore. Gen. Mgr: T. Dickey. Asst. Gen. Mgr: A. Ocambo. Technician: M. Madaing.
Stations: WSZE-TV ch10 0.5kW (Saipan)
D.Prgr: 0600-1400.

PALAU (Rep.)

TV-sets: 1.600 — **Colour:** NTSC — **System:** M.

STV-TV KOROR (Comm.)
✉ Koror, Palau 96940.
L.P: Mgr: David Nolan. Technician: Ray Omelen.
Station: Ngermit, Koror ch7 0.1kW.
D.Prgr: 0400-1400.

PAPUA NEW GUINEA

TV-sets: 100.000 — **Colour:** PAL — **Systems:** B & G.

EMTV (Comm.)
✉ Media Niugini Pty. Ltd, P.O. Box 443, Boroko NCD. ☎ +675 3257322.
🖷 +675 3254450.
L.P: Chief Exec: John Taylor. CE: Geoff Kong.
Stations: Burns Peak ch9 1.1kW, Air Niugini Hill ch31 0.17kW, Garden City ch68 0.02kW (all Port Moresby area)
F.PI: st's at Mt. Hagen, Goroka, Lae, Rabaul.
D.Prgr: 12 h daily in English and Tok Pisin, 7 days a week.

POLYNESIA (French)

TV-sets: 26.500 — **Colour:** SECAM — **System:** K.

TELE TAHITI
✉ Radio Télévision Française d'OutreMer (RFO), B.P. 125, Papeete, F-98 713 Polynésie Française. ☎ +689 430551. 🖷 +689 413155. **E-mail:** rfopolyfr@mail.pf
Web: http://www.tahiti-explorer.com/rfo.html
L.P: Dir: Claude Ruben. Chief Editor: Patrick Durand Gaillard. Dir. of Prgrs: Jean-Raymond Bodin.
Stations: Papeete chK4 0.1kW H, Mont Marau chK8 0.5kW H, Vaitape chK7 0.2kW V, Taravao chK4 0.1kW H (+ 8 low power repeaters).
D.Prgr: 0400 (Sat/Sun 0200)-0830.

COMMERCIAL STATIONS

CANAL POLYNESIE (Priv., Comm.)
✉ Colline de Putiaoro, Papeete
Stations: Taravao ch26 2kW, Mont Marau ch43 55 kW, Punaauia ch55 1.2kW + 9 low power repeaters.
N.B: Canal Polynesie is a subscription sce and the signal is encrypted except: 1630-1720, 2225-2325, 0415-0610.
D.Prgr: W 1630-1100 Sun 24h.

SAMOA (American)

TV-sets: 8.000 — **Colour:** NTSC — **System:** M.

KVZK-TV (Gov.)
✉ Office of Public Information, PO Box 3511, Pago Pago 96799
☎ + 684 6334191. 🖷 +684 6331044
L.P: Dir: Vaoita Sava. Dir. Tec: Robert Blauvelt.
Stations: chA2/A4/A5 (72kW).
D.Prgr: ch2 & 5: 1830-1000; ch4: 1830-1100.

SAMOA (Western)

TV-sets: 5.000 — **Colour:** PAL — **System:** B.

TELEVISE SAMOA
✉ P.O. Box 1868, Apia. ☎ + 685 26641. 🖷 +685 24789.
L.P: GM: Tupai Kuka Brown.
Stations: Apia ch11 10W; Mount Vaea ch8 50W; Faleasiu ch10 10W; Mount Aflau ch4 50W, Mount Fiamoe ch6 10W; Api Park ch5 5W
D.Prgr: 0400-1030.

TONGA

TV-sets: 2500 — **Colour:** NTSC — **System:** M.

ASTL-TV3 (Comm.)
✉ P.O. Box 66, Nuku'alofa. ☎ +676 22325. 🖷 +676 22811
L.P: Pres: Latu Tupouniua
Station: Nuku'alofa ch3 50W
D.Prgr: Mo-Sat 1800-2000 & 0300-1000; Sun 0400-0900.

OCEANIA BROADCASTING NETWORK LTD
✉ P.O. Box 91, Nuku'alofa. ☎ +676 23314. 🖷 +676 23658

VANUATU

TV BLONG VANUATU
✉ Vanuatu Broadcasting and Television Corp, PMB 927, Port Vila. ☎ +678 25412. 🖷 +678 22026 — **L.P:** Mgr: Gaile Dantec..

WALLIS & FUTUNA

TV-sets: N/A — **Colour:** SECAM — **System:** K.

RADIODIFFUSION FRANÇAISE D'OUTRE-MER (RFO)
✉ B.P. 102, Mata Utu, 98600 Uvea, Iles de Wallis-et-Futuna, Pacifique sud (par Nouméa, Nouvelle-Calédonie). ☎ +681 722020. 🖷 +681 722346 — B.P. 20, Sigave, F-98620 Futuna. ☎ +681 723531. 🖷 +681 723534 — **L.P:** SM: Joseph Blasco.
Stations: ch6 & ch9 **D.Prgr:** 5-6h each evening.

NORTH AMERICA

ALASKA (US State)

TV-sets: 2000 — **Colour:** NTSC — **System:** M.

TV-STATIONS: All comm. exc. 2, 4, 5, 7 (Educational).

	Call	ch	kW(ERP)		Call	ch	kW(ERP)
1)	KTUU-TV	2	100/10	6)	KJUD-TV	8	0.24/0.05
1a)	KATN-TV	2	28.2/5.5	7)	KUAC-TV	9	46.8/8.33
2)	KTOO-TV	3	2.45/0.49	8)	KTVA-TV	11	50.7/5.07
3)	KJNP-TV	4	19.1/3.31	9)	KTVF-TV	11	27/5.5
4)	KYUK-TV	4	4.67/1.16	10)	KIMO-TV	13	39/7.8
4a)	KTBY-TV	4	42.5/8.5	11)	KTNL-TV	13	0.2/0.03
5)	KAKM-TV	7	163/16.3				

1) Box 102880, Anchorage, AK 99510 — Gen. Mgr: Al Bramstedt Jr. 2230 (Sat/Sun 0000)-0900.
1a) P.O. Box 74730, Fairbanks, AK 99701.
2) Capital Community Broadc. Inc, 224 Fourth Str, Juneau 99801 — Pres. & Gen. Mgr: D. Rinker. Dir. Tec: J.W. Foster. 1930 (Sat/Sun 1530)-0800.
3) Evangelistic Alaska Missionary Fellowship, Box "O", North Pole 99705 — Pres. & Dir: D.L. Nelson. Dir. Tec: E. Nichols.
4) Bethel Broadc. Inc, Box 468, Bethel 99559 — Dir. Gen:J. Brigham.
4a) KTBY Inc, 1840 S. Bragnaw Str, Anchorage 99508 — Gen. Mgr: R.V. Bradley. Dir. Tec: E. Gjernes.
5) Alaska Public Television Inc, 2677 Providence Dr, Anchorage 99508 — Gen. Mgr: E. Sackett. Dir. Tec: F. Mengel.
6) 1107 West Eighth St., Suite 2, Juneau, AK 99801.
7) University of Alaska, Fairbanks 99701 — St. Mgr: Kathryn Jensen. Dir. Tec: David L. Walstad. 1800-0900.
8) Northern TV Inc, Box 102200, Anchorage 99510 — Dir. Tec: D. Milsap. 1700-1100.
9) Northern TV Inc, Box 950, Fairbanks 99707 — 1700-1100.
10) Alaska 13 Corp, 2700 Tudor Rd., Anchorage 99507 — Pres: D.L. Triplett.
11) Sitka Broadc. Co. Inc, Box 2668, Sitka 99835 — Gen. Mgr: D. Etulain. Mon-Fri 1900-0900, Sat 1700-0900, Sun 1400-0800.

ARMED FORCES RADIO & TV SCE. (US Mil.)
✉ Navsta Box 14, FPO Seattle, WA 98791, USA
Station: Adak ch 8 & 10 (cable only).

BERMUDA

TV-sets: 30.000 — **Colour:** NTSC — **System:** M.

BERMUDA BROADCASTING COMPANY Ltd.
P.O. Box HM 452, Hamilton. ☎ +1 (809) 295-2828. 🖶 +1 (809) 295-4282. ① 3702 ZBMBA.

LOCAL TV STATIONS
There are three local commercial television stations in Bermuda. Reception is island-wide and no special cabling or antennas are required.
Stations: ZFB-TV, ch7, operated by the Bermuda Broadcasting Company Ltd. (US ABC affiliate); **ZBM-TV**, ch9, operated by the Bermuda Broadcasting Company Ltd (US CBS afilate); **VSB-TV**, ch11, operated by DeFontes Broadcasting Television Ltd. (US NBC Affiliate).

CANADA

TV-sets: 19.400.000 — **Colour:** NTSC — **System:** M.

CANADIAN BROADCASTING CORPORATION
SOCIÉTÉ RADIO-CANADA (Publicly owned)
1500 Bronson Avenue. P.O. Box 8478, Ottawa, Ontario K1G 3J5. ☎ +1 (613) 724 1200. 🖶 +1 (613) 738 6887. Cable: Broadcasts: ① 053-4260. **Web:** http://www.tv.cbc.ca
L.P: Pres. and CEO: Perrin Beatty. Chmn, Board of Directors: Guylaine Saucier. Sen. VP, Resources: Louise Tremblay. Sen. VP, Media (Vacant). VP/Sen. Advisor, Office of the Pres. and CEO: Michael McEwen. VP, Internal Audit: Robert Hertzog. VP, General Counsel and Corporate Secr: Gerald Flaherty, Q.C. VP, Human Resources: George C. B. Smith. Exec. Dir, Media Accountability: Donna Logan. Sen. Dir. of Corporate Communications and Public Affairs: Charlotte O'Dea.
English Networks: P.O. Box 500, Station "A", Toronto, ON M5W 1E6. **Cable:** Broadcast. ① 062-17796. ☎ +1 (416) 975-3311.
L.P: VP English Television Networks: Jim Byrd. Exec. Dir, News, Current Affairs & Newsworld, Television: Bob Culbert. Head, CBC Newsworld: Slawko Klymkiw. Exec. Dir, Media Operations: Michael Harris. Sen. Dir, Media and Public Relations: Tom Curzon. Sen. Dir, Broadcast Communications: Diane Kenyon.
French Networks: P.O. Box 6000, Montreal, PQ H3C 3A8. **Cable:** Radcan. ① 05-267417. ☎ +1 (514) 285-3211.
L.P: VP, French Television: Michèle Fortin. DG, Communications: Raymond Guay. Dir. of Public Rel: Micheline Savoie. Exec. Dir, RDI, Renaud Gilbert. GM TV5 (Consortium Québec-Canada): Guy Gougeon.
CBC Engineering: 7925 Côte St-Luc Road, Montreal, PQ H4W 1R5. ① 055-66437. ☎ +1 (514) 485-1301.
L.P: Sen. Dir, Eng: Brian D. Baldry.

Affiliated Stations: CKVR Barrie, Ont; CKX Brandon, MB; CJDC Dawson Creek, BC; CFJC Kamloops, BC; CHBC Kelowna, BC; CKWS Kingston, Ont; CKSA Lloydminister, AB; CHAT Medicine Hat, AB; CHNB North Bay, Ont; CHEX Peterborough, Ont; CKBI Prince Albert SK; CKPG Prince George BC; CDMI Quebec, PQ; CKRD Red Deer, AB; CHSJ Saint John, NB; CJIC Sault St. Marie, Ont; CKNC Sudbury, Ont; CJFB Swift Current, SK; CFTK Terrace, BC:CKPR Thunder Bay, Ont; CFCL Timmins, Ont.
Stations: Pol: H.

English Network	Call	chA	kW
St. John's, Nfld.	CBNT	8	196
Charlottetown, P.E.Q.	CBCT	13	178
Halifax, N.S.	CBHT	3	56
Montreal, Que	CBMT	6	100
Ottawa, Ont.	CBOT	4	100
Toronto, Ont.	CBLT	5	84
Winnipeg, Man.	CBWT	6	100
Regina, Sask.	CBKT	9	140
Edmonton, Alta.	CBXT	5	318
Vancouver, B.C.	CBUT	2	50
Yellowknife, N.W.T.	CFYK	8	2.4

CBC English TV is also distributed through 200 rebroadcasters

Addresses:
P.O. Box 12010 Stn. "A", St. Johns, Nfld. A1B 3T8.
P.O. Box 2230, Charlottetown, P.E.I. C1A 8B9.
P.O. Box 3000, Halifax, N.S. B3J 3E9.
P.O. Box 6000, Montreal, Que. H3C 3A8.
P.O. Box 3220, Stn. "C", Ottawa, Ont. K1Y 1E4.
P.O. Box 500, Stn. "A", Toronto, Ont. M5W 1E6.

P.O. Box 160, Winnipeg, Man. R3C 2H1.
2440 Broad Str, Regina, Sask. S4P 4AI.
P.O. Box 555, Edmonton, Alta. T5J 2P4.
P.O. Box 4600, Vancouver, B.C. V6B 4A2.
P.O. Box 160, Yellowknife, N.W.T. X1A 2N2.

French Netw.	Call	chA	kW
Moncton, N.B.	CBAFT	11	163
Montreal, Que.	CBFT	2	100
Quebec, Que.	CBVT	11	252
Ottawa, Ont.	CBOFT	9	128
Winnipeg, Man.	CBWFT	3	59.0
Regina, Sask	CBKFT	13	140
Edmonton, Alta.	CBXFT	11	90
Vancouver, B.C.	CBUFT	26	105

CBC French TV is also distributed through 200 rebroadcasters

Addresses:
P.O. Box 950, Moncton, N.B. E1C 8N8.
P.O. Box 10400, St-Foy, Que G1V 2X2.
Others: See English Netw.

Private TV Networks

ATLANTIC TELEVISION LTD. (ATV)
P.O. Box 1653, Halifax, Nova Scotia B3J 2Z4. ☎ +1 (902)453-4000. 🖶 +1 (902)454-3302. **E-Mail:** ASN@asn.ca **Web:** http://www.atv.ca

CITY-TV
299 Queen Str. W. Toronto, Ontario, M5V 2Z5. ☎ +1 (416) 591-5757. 🖶 +1 (416) 591-7791. ① 06218283.
Web: http://www.bravo.ca/citytv.html
L.P: GM: Dennis Fitzgerald.
Stations: Toronto ch57 310kW (ERP), Woodstock ch31 929kW (ERP). Ottawa ch65.
D.Prgr: 24h.

CANWEST GLOBAL SYSTEM
81 Barber Greene Rd, Don Mills, ON, M3C 2A2. ☎ +1 (416) 446-5311. 🖶 +1 (416) 446-5371/5490.
Stations: CKVU Vancouver, BC; CKKX Calgary, AB; CISA Lethbridge, AB; CFRE Regina, SK; CFSK Saskatoon, SK; CKND Winnipeg, MB; CIII Toronto, Ont; MITV Saint John, NB; MITV Halifax/Dartmouth, NS.

CTV TELEVISION NETWORK LIMITED
Head Office: 42 Charles Str. E., Toronto, Ontario M4Y 1T5. ☎ +1 (416) 928-6000. ① 06-22080. 🖶 +1 (416) 928-0907.
L.P: Pres. & C.E.O: Murray H. Chercover; Chief Financial Officer and Treasurer: Duncan Morrison; VP Network Relations: Marge Anthony; VP Gov't Relations & Corporate Planning: John T. Coleman; VP Operations: Joseph A. Colson; VP Sports: Johnny Esaw; VP News, Features & Information Programming: Tim Kotcheff; VP Finance: Peter O'Neill; VP Sales: Peter Sisam; VP Programming: Philip Wedge; VP Entertainment Programming: Arthur Weinthal.
Affiliated Stations: NTV St. John's, Nfld; CJCB Sydney, N.S; CJCH Halifax, N.S.; CKCW Moncton, N.B.; CKNY North Bay, Ont.; CITO Timmins, Ont.; CJOH Ottawa, Ont.; CFTO Toronto, Ont.; MCTV Sudbury, Ont.; CKCO Kitchener, Ont.; CFCF Montreal, Que.; CKY Winnipeg, Man.; CKTV Regina, Sask.; CFQC Saskatoon, Sask.; CFRN Edmonton; Alta.; CFCN Calgary, Alta.; BCTV Vancouver, B.C.; CHEK Victoria, B.C.; CIPA Prince Albert, Sask.; CHBX Sault Ste. Marie, Ont.; CHFD Thunder Bay, Ont.; CICC Yorkton, Sask.; CITL Lloydminster, Alta.; CJBN Kenora, Ont.; CHRO Pembroke, Ont.

LE RESEAU DE TELEVISION (TVA)
2600 boul. de Maisonneuve, Montreal, PQ, H2L 4P2. ☎ +1 (514) 526-9251.
Stations (ch no in brackets): CFCM Quebec (4), CJPM Chicoutimi (6), CHLT Sherbrooke (7), CHEM Trois-Rivières (8), CIMT Rivière d. Loup (9), CFTM Montreal (10), CFER Rimouski (11), CFEM Rouyn-Noranda (13), CHOT Hull/Ottawa (40), CHAU-TV Carleton (5).

TÉLÉVISION QUATRE SAISONS
405 Oglivy Ave., Montreal, PQH3N 1M4. ☎ +1 (514) 495-6884. 🖶 +1 (514) 495-6231.
Stations: CFJP Montreal (35); CFGS Hull (49); CFRS Jonguière (4); CFAP Quebec (2); CFPC Rimouski (18); CFTF Rivière-du-Loup (29); CFKS Sherbrooke (30); CFKM Trois Rivières (16); CFVS Val d'Or (20/25).

Public TV Networks

SOCIETE DE RADIO-TELEVISION DU QUEBEC
✉ 800 rue Fullum, Montreal, Quebec H2K 3L7. ☎ +1 (514) 521-2424. ✆ 05-25808.
L.P: Dir. Gen: Ms. Françoise Bertrand. Head of Prgrs: Pierre Roy.
Affiliated Stations: CKRS Jonguière; CKRN Rouyn-Noranda; CKRT Rivière-du-Loup; CKSM Sherbrooke; CKTM Troit-Rivières.

TVO (English) & TFO (French)
Ontario Educational Communications Authority.
✉ Box 200, Station Q, Toronto, Ont. M4T 2T1.
☎ +1 (416) 484-2600. 🖷 +1 (416) 484-7771. ✆ 06-23547. **E-mail:** online@tvo.org. **Web:** http://www.tvo.org
L.P: Chmn. & Chief Exec. Officer: B. Ostry. Chief Operating Officer: P. Bowers. Man. Dir. French Prgr. Sces: Jacques Bensimon. Man. Dir. English Prgr. Sces: Don Duprey. MD. Ext. Rel: Bill Roberts/Judith Tobin.
Station: CICA-TV chA19 1080/108kW + low power relays.

TV-sets: 21.000 — **Colour:** PAL — **System:** B.

KNR TV (Gov.)
✉ Kalaallit Nunaata Radioa TV, P.O. Box 1007, DK-3900 Nuuk. ☎ +299 325333. 🖷 +299 325042.
Web: http://www.knr.gl
L.P: MD: Peter Frederik Rosing. Head of Prgr's: Niels Pavia Lynge. Head of engineering: Nathan Biilmann.
Stations; The KNR TV prgr's are aired locally using band 3 channels with 1 to 100 watts tr. power.
D.Prgr: 1330-0030 MF, 1000-0100 SS.
N: Greenlandic: 2130-2155. **Danish:** 1930-2000 and 2345-0045: rebroadcasts from the Danish TV.
Prgrs mainly originate from DR TV, Denmark, plus a few hours a week from the Danish TV2 + local commercials.
Text-TV: 24 hrs.**PRIVATE STATIONS (local TV):TV Aasiaat,** Box 20, 3950 Aasiaat: ch. 12 (10 W)**Arctic TV,** Box 420, 3952 Ilulissat: ch. 12 (10 W)
Arfivik-TV, Box 138, Qeqertarsuaq: ch. 8 (2 W)
Nuuk TV, Box 1016, 3900 Nuuk: ch. 5/6/11/12 (2/20/10/10 W)
Qaqortoq TV: Box 55, 3920 Qaqortoq: ch. 8 (10 W)
Qeqertaq TV: B-84, Qeqertaq, 3952 Ilulissat: ch. 12 (3 W)
Sisimiut TV: Box 1004, 3911 Sisimiut: ch. 7 (10 W)

TV-sets: 2.200 — **Colour:** SECAM — **System:** K.

SOCIETE NATIONALE DE RADIO TÉLÉVISION FRANÇAISE D'OUTRE MER (RFO)
✉ BP 4227, F-97500 St. Pierre et Miquelon. ☎ +508 411111. 🖷 +508 412219 — **L.P:** Dir: Joseph Eden.
Stations (Pol H)
1st Prgr: St. Pierre chK4 0.1kW, chK8 0.5kW, chK39 5W — Miquelon chK6 0.1kW.
2nd Prgr: St. Pierre chK31 0.5kW, chK55 0.05kW — Miquelon chK56 0.2kW.
D.Prgr: 1st Prgr: 1800 (SS 1500)-0400. **2nd Prgr:** 2115-0330.

TV-sets: 215m — **Colour:** NTSC — **System:** M.

Networks providing programming to local st's nationwide:

ABC TELEVISION DIVISION (Comm.)
✉ 77 West 66th Street, New York NY 10023. ☎ +1 (212) 456 6400. 🖷 +1 (212) 456 2795. ✆ 422003.
E-mail: abcaudr@ccabc.com (Audience Information Section)
Web: http://www.abctelevision.com
Owned stations: WABC-TV/New York, WLS-TV/Chicago, KGO-TV/San Francisco, KABC-TV/Los Angeles, KTRK-TV/Houston, KFSN-TV/Fresno, CA, WPVI-TV/Philadelphia, WTVD-TV/Durham, North Carolina. **Affiliates:** approx. 220.

America One Television (Comm.)
✉ 100 E. Royal Ln, Irving, TX 75039. ☎ +1 (214) 868 1000. 🖷 +1 (214) 868 1662. **E-mail:** a1tv@airmail.net
Web: http://www.americaone.com

CBS, Inc. (Comm.)
✉ 51 West 52nd Str, New York, NY 10019. ☎ +1 (212) 975-4321. 🖷 +1 (212) 975 7452. **E-mail:** marketing@cbs.com
Web: http://www.cbs.com
Owned Stations: WCBS-TV, New York; WBBM-TV, Chicago; KCBS-TV, Los Angeles; WCIX-TV Miami; WCCO TV, Minneapolis; WBAY TV, Green Bay. **Affiliated Stations:** 200.

FOX TELEVISION NETWORK (Comm.)
✉ P.O. Box 900, Beverly Hills CA 90213. ☎ +1 (310) 277 2211. **E-mail:** foxnet@delphi.com
Web: http://www.foxnetwork.com

NATIONAL BROADCASTING COMPANY (Comm.)
✉ 30 Rockefeller Plaza, New York, NY 10112. ☎ +1 (212) 664-2074. 🖷 +1 (212) 664 7541. ✆ 662131.
Web: http://www.nbc.com
NBC Television Stations: WNBC-TV, New York; WRC-TV, Washington; WMAQ-TV, Chicago; WKYC-TV, Cleveland; KNBC-TV, Los Angeles; WCAU-TV, Philadelphia; KCNC-TV, Denver; WTVJ-TV, Miami. **Affiliated stations:** 208.

PUBLIC BROADCASTING SERVICE (Non-comm., Educ.)
✉ 1320 Braddock Place, Alexandria, VA 22314-1698. ☎ +1 (703) 739-5000. 🖷 +1 (703) 739-0775. **E-mail:** viewer@pbs.org (viewer mail).
Web: http://www.pbs.org
Member Stations: 314.

United Paramount Network (UPN) (Comm.)
✉ 5555 Melrose Avenue, Marathon 1200, Los Angeles, CA 90038.
Web: http://www.upn.com

Univision (Spanish Language Network) (Comm.)
✉ 605 Third Ave, New York, NY 10158-0180. ☎ +1 (212) 455 5200.
Web: http://www.univision.net

Warner Brothers Television Network (WB)
✉ 4000 Warner Blvd, Bldg. #34R, Burbank, CA 91522. ☎ +1 (818) 954 6479. **E-mail:** wbnetwork@aol.com
Web: http://www.tv.warnerbros.com

Local stations: There over 1,500 TV stations across the U.S. The major cities have st's affiliated to each of the above-mentioned networks, and there may be additional local st's which generally broadcast movies, and re-runs of older programs. For details of satellite-delivered services, refer to the satellite section.

Armed Forces Radio & TV Service Broadcast Center (AFRTS-BC) (Mil.)
✉ 1363 Z Street, Bldg 2730, March ARB, CA 92518. 🖷 +1 (909) 413-2234.
Web: http://www.dodmedia.osd.mil/afrts_bc/ahome.htm
L.P: Dir. Programming: Robert W. Matheson. Dir. Eng. & Op's: Bruce V. Ziemienski.
The AFRTS Broadcast Center delivers programming to the AFRTS overseas audience through land-based outlets or Navy ships at sea. AFRTS has more than 450 land-based outlets located in over 140 countries and U.S. territories (including remote areas of Alaska).
TV stations with over-the-air transmissions may be found at the end of listings for the respective countries.

TV Marti (Gov.)
TV Marti is the television broadcasting service of the United States Information Agency, Office of Cuban Broadcasting.
✉ Washington, D.C. 20547, USA. ☎ +1 (202) 501-7210. 🖷 +1 (202) 208-7808.
L.P: Ag. Dir. Office of Cuba Broadcasting: Dr. Rolando Bonachea. Dir. of Tech. Op's (OCB): Michael Pallone.
Stations: VHF ch13 (0830-1100).

CENTRAL AMERICA & THE CARIBBEAN

ANTIGUA & BARBUDA

TV-sets: 28.000 — **Colour:** NTSC — **System:** M.

ANTIGUA & BARBUDA BROADC. SERVICE (Gov.)
✉ P. O. Box 590, St. John's. ☎ +1 (268) 4620010. 🖷 +1 (268) 4621622.
L.P: Prgr. Mgr. (TV): James Tanny Rose. CE: Denis Leandro. Film Buyer: J.T. Rose.
Stations: ABS-TV chA10 50/20kW H.
Relay: Montserrat ch13 10kW.
D.Prgr: 1900-0700.

CTV ENTERTAINMENT SYSTEMS (Comm.)
✉ 25 Long Str, St. Johns. ☎ + 1 (809) 4620346. 🖷 + 1 (809) 4624211 — **L.P:** Prgr. Dir: J. Cox.

ARUBA

TV-sets: 19.000 — **Colour:** NTSC — **System:** M.

TELE ARUBA (Comm.)
✉ P.O. Box 392, Oranjestad. ☎ + 297 (8) 47302.
🖷 + 297 (8) 41683. ① 5195 TELAR.
L.P: Gen. Mgr: Mrs. Jane Lampkin. CE: Miguel Roga.
Station: chA13 3/0.6kW H.
D.Prgr: 2030-0400.

BAHAMAS

TV-sets: 50.000 — **Colour:** NTSC — **System:** M.

BAHAMAS TELEVISION (owned and operated by the Broadc. Corp. of the Bahamas)
✉ P.O. Box N-1347, Nassau. ☎ +1 (242) 322 4623. 🖷 +1 (242) 322 3924 — **L.P:** Ag. Prgr. Dir (TV): Ms. R. Simmons.
Station: ZNS ch13 50kW (ERP).
D.Prgr: 2200 (Sat 2100, Sun 2000)-0400.

BARBADOS

TV-sets: 65.000 — **Colour:** NTSC — **System:** M.

CARIBBEAN BROADCASTING CORP. (Gov, Comm.)
✉ P.O. Box 900, Bridgetown. ☎ +1 (246) 429 2041. 🖷 +1 (246) 429 4795.
L.P: Prgr. Mgr (TV): Mrs. O. Cumberbatch.
Station: CBC-TV ch8 60/30kW, ch 9, ch14, ch18, ch22, ch26.
D.Prgr: Mon-Fri 0940-1400 and 2000-0330, Sat 1200-0430, Sun 1200-0330.
Cable TV: A Gov. Sce, STV, provides two additional subscription ch's.
N.B: Wednesday and Friday closing may be after 0400 as movies are shown on these days. During close times classified commercials and coming attractions are broadcast continuously.

BELIZE

TV-sets: 23.457 — **Colour:** NTSC — **System:** M.

TROPICAL VISION (Comm.)
✉ 48 Albert Cattouse Bldg, Regent Str, P.O. Box 89, Belize City. ☎ +501 77246/7/8 — 🖷 +501 (2) 75040 — **Station:** ch 7.

BAYMEN BROADCASTING NETWORK (Comm.)
✉ 27 Baymen Ave., Belize City. ☎ + 501 (2) 44400. 🖷 +501 (2) 31242 — **Station:** ch 9.

COSTA RICA

TV-sets: 340.000 — **Colour:** NTSC — **System:** M.

Stations	chA	kW	Stations	chA	kW
2)	4	10	2)	9	10
1)	6	300	4)	13	2.5
3)	7	3605)	15	0.1	

Addresses & other information:
1) Corporación Costaricense de Televisión (Comm.), P.O. Box 2860, 1000 San José. ☎ +506 312222 — Dir. Gen: M. Sotela B. — 1530-0700.
2) Multivisión, Apt 4666, 1000 San José. ☎ +506 334444.
① 3043 Televi. 🖷 +506 211734 — Dir: Arnaldo Vargas V. — 1600-0600.
3) Televisora de Costa Rica, Apt 3876, San José. ☎ +506 322222. ① 2220 Teletica — 1730-0600 (+ 9 repeaters).
4) Rede Nacional de Televisión, Apt 7-1980, San José. ☎ +506 200071 — Dir. Gen: Dr. Ch. Zelaya Goodman — 2200-0400 (+ 2 repeaters).
5) Universidad de Costa Rica, San Pedro, Montes de Doa, San José — Dir. Gen: Dr. Sergio Guevara Fallas — 1600-2200. ☎ +506 340463. 🖷 +506 256950.

CUBA

TV-sets: 2.500.000 — **Colour:** NTSC — **System:** M.

INSTITUTO CUBANO DE RADIODIFUSION (Gov.)
✉ Television Nacional, Calle M No. 313, Vedado, La Habana. Estudios en Pinar del Río, Ciudad de la Habana, Santa Clara, Nueva Gerona, Camagüey, Holgüí, Santiago de Cuba & Guantánamo.

TELE REBELDE
✉ Mazón No. 52, Vedado, La Habana. ☎ +537 (32) 3369. ① 511661.
L.P: VP: Gary Gonzalez.
Studios in Santiago de Cuba, Holguín & La Habana.
Stations: Pol H (no calls used).

ch	Location	kW	ch	Location	kW
2	Babiney, P.Río	8	11	Cienfuegos	8
2	La Habana	132	11	Salón, P.Río	32
4	Camagüey	53	12	Prov. Granmá	54
5	La Palma	8	12	Chivirico	56
5	Stgo. de Cuba	77.5	12	Ciego de Avila	56
5	Santa Clara	60	13	Pinar del Río	63
8	Holgüín	27.5	13	Guantánamo	79
9	Baracoa	12	13	Prov. Matanzas	120
10	I. Juventud	100			

+ 12 st's below 4kW.
D. Prgr: 2257-0500.

CUBAVISION (Gov.)
✉ Calle M No. 313, Vedado, La Habana.

ch	Location	kW	ch	Location	kW
2	Stgo. de Cuba	32	7	Salón, P. Río	32
3	La Palma	8	8	Ciego de Avila	51
3	Holguín	27.5	8	I. Juventud	100
3	Santa Clara	60	9	Pinar del Río	63
6	Babiney, P.Río	6	9	Prov. Matanzas	120
6	Camagüey	56	10	Prov. Granma	54
6	La Habana	129	10	Chivirico	56
7	Baracoa	12	11	Guantánamo	79
7	Cienfuegos	8			

+ 12 st's below 4kW.
D.Prgr: 2227-0500.

AFRTS (US Navy)
✉ US Naval Base, P.O. Box 22, FPO New York, NY 09406.
Station: Guantanamo Bay chA8 0.35kW.

DOMINICA (Commonwealth of)

TV-sets: 5.200 — **Colour:** NTSC — **System:** M.

MARPIN-TV (Comm.)
✉ P.O. Box 382, Roseau. ☎ +1 (767) 4484107. 🖷 +1 (767) 4482965 — **L.P:** Prgr. Mgr: Ron Abraham.

DOMINICAN REPUBLIC

TV-sets: 728.000 — **Colour:** NTSC — **System:** M.

VHF Stations		chA	Stations		chA
1)	Sto. Domingo	2	7)	Sto. Domingo	11
2)	Cibao/Costa Norte	2	5)	Cibao/Costa Norte	11
3)	Sto. Domingo	4	3)	Barahona	12
3)	Cibao/Costa Norte	4	3)	Dajabon	12
3)	Alto Bandera	5	3)	Descubierta	12
4)	Sto. Domingo	6	3)	El Cercado	12
4)	Cibao/Costa Norte	6	3)	Enriquillo	12
5)	Sto. Domingo	7	3)	La Romana	12
6)	Santiago	7	3)	Pedernales	12
2)	Sto. Domingo	9	3)	Puerto Plata	12
2)	La Romana	9	8)	Sto. Domingo	13
7)	Cibao/Costa Norte	9	1)	Cibao/Costa Norte	13

UHF Stations		chA
5)	La Naviza	70

N.B: Because of contradictory information, no transmitter power figures are given.

Addresses and other information

1) Teleantillas, Autopista Duarte Km.7, Sto. Domingo. ☎ +1 (809) 567-7751.
2) Color Vision, Corporación Dominicana de Radio & TV, Emilio Morel esq. Lulu Perez, Ens. La Fé, Sto. Domingo. ☎ +1 (809) 556-5876.
3) RadioTelevisión Dominicana (Gov.), Dr. Tejeda Florentino 8, Sto. Domingo. ☎ +1 (809) 689-2120 — Dir: George Rodriguez.
4) Canal 6, Circuito Independencia C-A, Mariano Cestero esq. Enrique Henriquez, Sto. Domingo. ☎ +1 (809) 689-8151.
5) Rahintel, Av. Indepencia, Centro de los Heroes, Sto. Domingo. ☎ +1 (809) 532-2531.
6) Canal 7 Cibao, Edificio Banco Universal, Calle del Sol, Santiago. ☎ +1 (809) 583-0421.
7) Telesistema, 27 de Febrero, Sto. Domingo. ☎ +1 (809) 567-1251.
8) TV 13, Av. Pasteur esq., Santiago, Sto. Domingo. ☎ +1 (809) 687-9161.

EL SALVADOR

TV-sets: 500.700 — **Colour:** NTSC — **System:** M.

	chA	kW (ERP)		chA	kW (ERP)		chA	kW (ERP)
1)	2	100/25	3)	6	150/30	4)	10	109/5
2)	4	75/37.5	4)	8	109/5			

Addresses & other information:

1) Canal Dos SA(Comm.), Ap. Postal 720, San Salvador. ☎ +503 236744 — 1730-0530.
2) Canal Cuatro(Comm.), Ap. Postal 720, San Salvador. ☎ +503 244633 — 2100(Sat/Sun 1500)-0600.
3) Canal Seis(Comm.), Km. 6, Carretera Panamericana a Santa Tecla, San Salvador. ☎ +503 235122 — 2300 (Sat/Sun 1700)-0600.
4) Television Cultural Educativa, Ap. Postal 4, Santa Tecla. ☎ +503 280499 — Dir. Gen: Maura Echaverria — 1300-0500.

GRENADA

TV-sets: 15.000 — **Colour:** NTSC — **System:** M.

GRENADA BROADCASTING CORPORATION (GBC).

✉ Morne Jaloux, P.O. Box 535, St. George's, Grenada. ☎ +1 (473) 444 5521/22 (PBX). 📠 +1 (473) 444 5054.
LP: MD: Cecil Benjamin. Sen. Eng: John Phillip
Station: ch7 (4kW), ch11 (5kW).

GUADELOUPE

TV-sets: 150.000 — **Colour:** SECAM — **System:** K.

RFO-GUADELOUPE

✉ B.P. 402, F-97163 Point-à-Pitre Cedex. ☎ +590 939696. 📠 +590 939682. ☯ 919064
LP: Dir: R.Surjus. Editor-in-Chief: Philippe Goudé. PD: L.Francil. Head Communications Dept: Sonia Gémieux.
Stations: Basse Terre chK5 2kW + low power repeaters.
D.Prgr: 1900-0230.

SAINT MARTIN
An RFO st. is operating on chK7 0.1kW relaying RFO Guadeloupe.

COMMERCIAL STATIONS

ARCHIPEL 4 (Priv., Comm.)
✉ Résidence Les Palmiers, Gabarre 2, 97110 Pointe a Pitre. ☎ +590 8363 50
Stations: Morne a Louis ch53 1.3kW

CANAL ANTILLES (Priv., Comm.)
✉ 2 lot. Les Jardins de Houelbourg, 97122 Baie Mahault. ☎ +590 26 8179
Stations: Morne a Louis ch 58 1.3kW, Basse-Terre ch42 60kW + 5 low power repeaters under 1kW.
N.B: Canal Antilles is a subscription sce and the signal is encrypted except: 1120-1130, 1630-1730, 1820-0010.
D. Prgr: W 1100-0500 Sun 24h

TCI GUADELOUPE (Priv., Comm.)
✉ Montauban, 97190 Gosier.
Station: Basse-Terre ch32 60kW

GUATEMALA

TV-sets: 475.000 — **Colour:** NTSC — **System:** M.

Stations: 2)Gov, others comm.

Call	chA	kW/Pol	Call	chA	kW/Pol
1) TGV-TV	3	240H	1) TGV-TV	10	25H
2) TGCE-TV	5	?	4) TGMO-TV	11	316/63
3) TGVG-TV	7	108/22H	5) TGSS-TV	13	25/5

Addresses & other information:

1) Radio-Television Guatemala, Apt. 1367, Guatemala. ☎ +502 (2) 922491. **Cable:** Teletenango — Pres: Lic. M. Kestler F. Dir. Tec: E. Sandoval — 1200-0600 (+ 2 repeaters).
2) Television Cultural Educativa, 4a Calle 18-38, Zona 1, Guatemala. ☎ +502 (2) 531913 — 2200-0500.
3) Televisiete, Apt. Postal 1242, Guatemala. ☎ +502 (2) 62216. **Cable:** TV siete — 1800-0600(+ 3 repeaters).
4) Teleonce, Ca. 20, 5-02, Zona 10, Guatemala. ☎ +502 (2) 682165 — 1800-0600 (+ 2 repeaters).
5) Trecevision, 3a Calle 10-70, Zona 10, Guatemala. ☎ +502 (2) 63266. ☯ 6070 Trece GU — 1800-0600 (+ 12 repeaters).

HAITI

TV-sets: 25.000 — **Colour:** NTSC — **System:** M.

TÉLÉVISION NATIONALE D'HAITI (Gov., Cult.)
✉ P.O. Box 13400 Delmas 33, Port-au-Prince. ☎ +509 (1) 63324/64049/62202.
LP: Dir: Mme. Jacqueline André.
Stations: Port-au-Prince chA8 0.3kW, chA10 5kW.
D.Prgr: Mon-Fri 2100-0400 in French & Creole.
F.PI: chA12 Cap. Haïtien.

TÉLÉ HAITI S.A. (Comm.)
✉ B.P. 1126, Port-au-Prince. ☎ +509 (1) 23000.
LP: Dir: Walter Bussenius.
Operates cable TV 24h in Port-au-Prince area.

HONDURAS (Rep.)

TV-sets: 160.000 — **Colour:** NTSC — **System:** M.

Stations (all comm.)

Call	chA	kW	Call	chA	kW
1) HRJS-TV	2	25	2) HRTS-TV	7	10/2
2) HRCV-TV	3	2/0.4	4) HRLP-TV	7	2/0.4
4) HRLP-TV	4	2/0.4	1) HRJS-TV	9	25
3) HRTG-TV	5	10/2	3) HRTG-TV	9	5/1
3) HRTG-TV	5	5/1	5) HRNQ-TV	13	2/0.4
6) HRGJ-TV	6	10/2			

Addresses & other information:

1) Corp. Centroamericana de Comunicaciones, S.A. de C.V, Apt. Postal

120, San Pedro Sula — Pres: J.J. Sikaffy. Exec. Vice Pres: F.J. Sikaffy. Dir. Tec: R. Beurket & A. Pinto — 1530-0400 (+ 2 relays).
2) Telesistema Hondureño, Apt. Postal 642, Tegucigalpa — 1530-0400 (+ 6 relays).
3) Compañia Televisora Hondureña, Apt. Postal 734, Tegucigalpa — 1230-0400 (+ 9 relays). ☎ +504 (32) 7835. 🖷 +504 (32) 0097.
4) Compañia Centroamericana de TV, Apt. Postal 68, Tegucigalpa — 1430-0400.
5) Cruceña de Televísion, Casilla 3424, Tegucigalpa — Pres: Lic. Ivo Kuljis F. Dir. Gen: Lic. Walter Gasser Diaz C. — 1700-0800.
6) Compañia Broadcasting, Apt. Postal 882, Barrio Rio Piedras — 1130-0500.

JAMAICA

TV-sets: 500.000 — **Colour:** NTSC — **System:** M.

JAMAICA BROADCASTING CORPORATION (Comm.)
🖂 Box 100, Kingston 10. ☎ +1 (876) 926 5620/9. 🖷 +1 (876) 929 1029.
Cable: JARAD Jamaica. ⓓ 2218 BROADCORP JA.
LP: DG: Claude Robinson. Dir. of Television: Keith Campbell. Mgr. Eng. Sces (TV): Norman Mighty.
Station: HWT ch 7, 8, 9, 10, 11, 12, 13.
D.Prgr: 2200(Sat 1930, Sun 1900)-0500.

MARTINIQUE

TV-sets: 65.000 — **Colour:** SECAM — **System:** K.

SOCIÉTÉ NATIONAL DE RADIO-TÉLÉVISION D'OUTRE MER (RFO)
🖂 B.P. 662, F-97263 Fort de France Cedex. ☎ +595 595200.
LP: Dir: Fred Jouhoud. CE: Jean Claude Arrivé.
Stations: Fort de France chK4 1kW (+ 9 relay st's). Pol: V.
D. Prgr: 2200 (Sun 2000)-0300.

COMMERCIAL STATIONS

ATV ANTILLES TELEVISION (Priv., Comm.)
🖂 28 rue Arawaks, 97200 Fort de France. ☎ +596 75 4444. 🖷 +596 75 5565
Stations: La Trinite ch39 7kW, Fort de France ch44 8 kW, La Morne Rouge ch52 1.4 kW, Riviere Pilote ch34 0.19kW

CANAL ANTILLES (Priv., Comm.)
🖂 Centre Commercial La Galléria, 97232 Le Lamentin. ☎ +596 50 5787
Stations: La Trinite ch25 7 kW, Le Morne Rouge ch46 1.4kW, Port de France ch29 8kW, Saint Pierre ch34 0.6kW, Riviere Pilote ch50 0.19 kW.

TCI MARTINIQUE-TELE CARAIBES INTERNATIONAL (Priv., Comm.)
🖂 Immeuble RCI/TCI-Zone industrielle-97232 Le Lamentin.
☎ +596 510606. 🖷 +596 518562.
D. Prgr: 11:00 a.m. - NOON-(rel. T.F.I. & Euronews)

MEXICO

TV-sets: 15m — **Colour:** NTSC. — **System:** M.

INSTITUTO MEXICANO DE TELEVISIÓN (IMEVISIóN) (Gov. Agency)
🖂 Ave. Periférico Sur 4121, Colonia Fuentes del Pedregal, 14141 México, DF. ☎ +52 (5) 5685684, 5681313.
LP: Dir.Gral: Lic. Jose Antono Alvarez Lima.
STATIONS: México, D.F: IMT(7), DF(13), IMT(22) — Chihuahua: CH(2) — Monterrey: FN(8).
Red Nacional 7: Aguascalientes: LGA(10), Cd. Obregón: BK(10), Culiacán: BL(13), Guadalajara: SFJ(11), Hermosillo: TH(10), Irapuato: CCG(7), Jalapa: IC(13), Matamoros: OR(14), Mérida: DH(11), México DF: IMT(7), Nogales: FA(2), Oaxaca: PSO(10), Tampico: WT(12) + 51 others.
Red Nacional 13: Aguascalientes: JCM(4), Cd. Obregón: CSO (6), Cuernavaca: CUR(13), Guadalajara: JAL(13), Hermosillo: HSS(4), Irapuato: MAS(12), Mérida: MEY(7), Mexicali: AQ(5), México DF: DF(13), Monterrey: WX(4), Oaxaca: IG(12), Puebla: TEM(12), Saltillo:

SAO(11), S.L. Potosi: CLP(6), Tuxtla Gutiérrez: AO(4) + 39 others. ch7 & 13 are national, 11 & 22 regional.
N.B: Early 1993, the Mexican Radio and Television Corporation ceased to exist and **Azteca Television** was established.

TELEVISA, S.A. (Priv, Comm.)
🖂 Av. Chapultepec 28, 06724 México, D.F — ☎ +52 (5) 709-3333. ⓓ XEWTM 77-3154 — 🖷 +52 (5) 709-3021 — Chmn. of the Adm. Council: R. O'Farrill Jr. Pres. TV: Emilio Azcárraga.
STATIONS: XEW-TV(2), XHTV(4), XHGC(5), XEQ(9).
Cadena Canal 2: Acapulco: AP(2), Cd. Juárez: EPM(2), Cd. Obregón: BS(4), Córdoba/Orizaba: AH(8), Culiacán: BT(7), Chihuahua: FI(5), Guadalajara: EWO(2), Hermosillo: XEWH(6), Jalapa: AH(8), León: L(10), Los Mochis: BS(4), Mazatlán: OW(12), Matamoros: AB(7), Mérida: TP(9), México DF: XEW(2), Mexicali: BM(14), Monterrey: XHX(10), Morelia: KW(10), Nuevo Laredo: BR(11), Oaxaca: BN(7), Saltillo: AE(5), Tampico: XHD(4), Torreón: O(11), Tuxtla Gutiérrez: TX(8), Villahermosa: LL(13) + 28 others.
Cadena Canal 4: Córdoba/Orizaba: AI(10), Jalapa: AI(10), México DF: TV(4), Puebla: XEX-TV(7) + 2 others.
Cadena Canal 5: Aguascalientes: AG(13), Guadalajara: GA(9), Hermosillo: TH(10), Jalapa: AJ(5), México DF: GC(5), Monterrey: XET-TV(6), Saltillo: AD(7) + 3 others.
Cadena Canal 9: Mexico City.
Ch2 & 5 are national, 4 & 9 regional.
Part-time affiliates (after 2200) noted below.

STATIONS originating local programming.
Call: XH— (unless otherwise stated).
*) see Televisa above — **)see Imevision above.

No.	Call	ch	kW	No.	Call	ch	kW
1)	IA	2	30	6)	I	2	86.5
2)	CH	2	6	7)	RIO	2	100
*)	XEW-TV	2	100	8)	FE	2	60
3)	AP	2	1.2	9)	FM	2	14.2
4)	KG	2	0.5	10)	BC	3	100
5)	FB	2	100	11)	PN	3	5.9

No.	Call	ch	kW	No.	Call	ch	kW
12)	JMA	3	13	29)	TVL	9	250
13)	P	3	30	30)	FW	9	50
14)	Q	3	100	*)	Q	9	100
15)	Y	3	100	31)	K	10	8
16)	LN	4	4	32)	A	10	0.85
*)	TV	4	100	33)	L	10	150
17)	AL	4	2	34)	IPN	11	300
18)	G	4	88	35)	WT	12	66
19)	CC	5	3.5	36)	MH	12	1
20)	XEJ-TV	5	67.4	37)	ND	12	1
*)	GC	5	100	38)	AW	12	100
21)	ETV	6	100	39)	CG	12	7
22)	EDK	6	84	40)	AK	12	5
23)	EWH	6	1.6	41)	DE	13	50
**)	IMT	7	?	42)	UT	13	2
24)	GEM	7	1	43)	ST	13	54
25)	MZ	7	72	**)	IMT	22	?
26)	GO	7	50	44)	S	23	18
27)	FN	8	17.5	45)	IJ	44	330
28)	US	8	1.6				

MONTSERRAT

TV-sets: 5.000 — **Colour:** NTSC — **System:** M.

ANTILLES TV LIMITED (Comm.)
🖂 P.O. Box 342, Plymouth, Montserrat. ☎ +1 (664) 491 2226. 🖷 +1 (664) 491 4511.
LP: Gen. Mgr: K. Osborne. Dir. Tec: Z.A. Joseph.
Station: Chance Pic chA7 (48kW towards Dominica, 3kW towards Antigua & St. Kitts).
D.Prgrs: 1000-1110, 2000-0400 (Fri/Sat 0700).

NETHERLANDS ANTILLES

TV-sets: 35.000 — **Colour:** NTSC — **System:** M.

TELE CURAÇAO (Gov, Comm.)
🖂 P.O. Box 415, Curaçao. ☎ +599 (9) 61288. 🖷 +599 (9) 614138.
LP: Gen. Mgr: Norbert Hendrikse. Prgr. Mgr: H. van der Beist. Dir. Tec:

J. Rufina.
Station: chA8 20/5kW H/A6 — **D.Prgr:** 2000-0345.
Cable TV: Telecuraçao also provides a cable service relaying various U.S. satellite networks and two Venezuelan channels.

LEEWARD BROADCASTING CORPORATION — TELEVISION
✉ P.O. Box 375, Philipsburg, St. Maarten.
Station: chA7 5kW.
D.Prgr: 2030 (Sun 1800)-0300 (approx); Sat also 1200-1300.

A Gov. st. on chA6 is reported operating from Saba.

TV-sets: 210.000 — **Colour:** NTSC — **System:** M.

SISTEMA SANDINISTA DE TELEVISIÓN (Gov.)
✉ Km 31/2 Carretera Sur, Contig o Shell, Las Palmas, Managua. ☎ +505 (2) 660028/660879. 🖷 +505 (2) 662411. ① 1226 Sandino.
L.P: Hd. of Sales: Miguel Chivel.
Stations: chA2 25kW, chA6 25kW H, Chg 1 kW.
+ 7 low power repeaters.
D.Prgr: 2030-0600 (approx.)

TV-sets: 204.539 — **Colour:** NTSC — **System:** M.

Location	chA	kW		Location	chA	kW
1) Cerro Azul	2	650	6)	Fort Davis	10	5
2) Cerro C. Chilibre	4	32	4)	Cerro Oscuro	11	110
3) Cerro Azul	5	20	2)	El Valle	12	5
6) Cerro Ancón	8	5	5)	Cerro Oscuro	13	10
1) El Valle	9	130				

Addresses and other information
1) Televisora Nacional, Apt. 6-3092, El Dorado, Panamá — Dir. Gen: Lic. Alejandro Ayala V. — 1700-0500 (+6 relays).
2) RPC Television, Apt. 1795, Panamá 1 — 1330-0600 (+8 relays).
3) Panavision, Apt. 6-2605, El Dorado, Panamá (+ 1 relay).
4) Sistema de TV Educativa, Estafeta Universitaria, Universidad de Panamá — Dir. Gen: I. Velasquez de Cortes — Mon-Fri 2200-0400, Sat 1500-2200. **F.PI:** 1 relay st at Ninguna in 1992.
5) Telemetro, P.O. Box 8-116, Panamá 8 — Dir. Gen: B. Marques — 1730-0530 (+ 1 relay).
6) Armed Forces Television (US Mil), Drawer 919, Bldg. 209, Fort Clayton — 24h.

TV-sets: 830.000 — **Colour:** NTSC — **System:** M.

Call	ch	kW		Call	ch	kW
1) WKAQ-TV	2	53.7/10.5	9a)	WPRV-TV	13	170/17
2) WIPM-TV	3	72.4/7.24	9b)	WSJU-TV	18	759
3) WAPA-TV	4	53.7/8.1	9c)	WKPV	20	100/10
4) WORA-TV	5	95.5/19.1	8b)	WJNX	2	?
5) WIPR-TV	6	53.7/5.37	9d)	WSJN-TV	24	4384/490
6) WLUZ-TV	7	166/22.4	12)	WMTY	40	—
7) WSUR-TV	9	57.5/7.24	10)	WVEO-TV	44	993/198
8) WLII	11	200/39.8	11)	WATX-TV	54	11.78/6.35
9) WOLE-TV	12	316/31.6	11a)	WUJA	58	55/5.5

1) Telemundo of Puerto Rico Inc., 383 Roosevelt Av., Hato Rey, PR 00918 — GM: Jose Ramos. CE: Jose Medina. 1210-0500 (+ 8 relay sts).
2) Dept. of Education, Box 909, Hato Rey 00919 — 1800-0500 (Sat/Sun 1330-0400).
3) SFN Communications, Inc, GPO Box 2060, San Juan 00938.
4) Telecinco Inc, Box 43, Mayagüez 00708 — St. Mgr. & Film Buyer: E. Bado. Dir. Tec: G.A. Bonet. 1200-0400.
5) Avda. Hostos Hato Rey. Esquina Tous Urbanization Baldrich, 00919 Puerto Rico.
6) Ponce TV Corpo, Isabel Esq Montaner, Ponce 00731 — Pres: L.T. Muniz.
7) La Ramble, Ponce (belongs to 8).
8) American Colonial Broadc. Corp, Box S-4189, San Juan 00905. 1230-0430 — 8b) rel. 8.
9) Western Broadc. Corpo. of Puerto Rico, Box 1200, Mayagüez 00709 — 1245-0500 (partly rel. st. 3).
9a) No. 10 Simon Madeira Rio Piedras, 00929 — GM: Nacha Rivera.

9b) Three Star Telecast Inc, Box 18, Carolina 00628 — Dir. Gen.: Barakat Saleh. 1030-0400.
9c) Multi Media TV, Box 2556, San Juan 00936.
9d) belongs to 9c) 24h (mostly rel. st's 1 and 4). All News.
10) Siglares Iglesia Catolica Inc, Buzon C-339, Quebradillas 00742.
11) Arecibo Video Corp, Arecibo 00612. Pres: F. Velasquez.
11a) Community TV of Cagua, Box 6556, Caguas 00626.
12) Ana G. Mendez Foundation, Box 21345, Rio Piedras 00928 — GM: Gloria Hernandez. CE: Ariel Diaz.

AFRTS (U.S. Mil.)
✉ U.S. Naval Station, Box 3029, FPO Miami 34051.
Stations: Roosevelt Roads chA40 1kW, Ft. Allen ch40 0.05kW, Viques Isl. ch56 0.05kW.

TV-sets: 9500 — **Colour:** NTSC — **System:** M.

ZIZ TELEVISION (Gov, Comm.)
✉ P.O. Box 331, Basseterre, St. Kitts. ☎ +1 (809) 465 2621. 🖷 +1 (809) 465-5202.
L.P: GM: Mrs. Claudette Manchester. Ag. Producer TV: Barry Thomas.
Stations: Basseterre chA5 20/5kW H. + 3 repeaters.
D.Prgr: 2000-0430.

TV-sets: 25.000 — **Colour:** NTSC — **System:** M.

HELEN TV (Comm.)
✉ P.O. Box 621, Le Morne Castries. ☎ +1 (758) 4522 693. 🖷 +1 (758) 454 1737. ① 6254 HTSTV.
L.P: MD: Linford Fevrier. CE: Stephenson Anius.
Station: ch's 4 & 11.

CABLEVISION
✉ George Gordon Bld., Bridge Str. Castries, St. Lucia. P.O. Box 111, Castries. ☎ (758) 452 3301. 🖷 +1 (758) 453 2544. ① 6362 — **Station:** ch's 16 on CATV.

TV-sets: 20.000 — **Colour:** NTSC — **System:** M.

ST. VINCENT & THE GRENADINES BROADCASTING CORPORATION Ltd.
✉ P.O. Box 617, Kingstown. ☎ +1 (809)-4561078. 🖷 +1 (809) 4561015.
L.P: Chief Eng: R.P. MacLeish.
Station: ch9. Relay sts. ch's 7, 11, 13 varying between 5-30 W. 1500-0900.
Coverage area: St. Vincent, Grenadines, St. Lucia & Grenada.

TV-sets: 250.000 — **Colour:** NTSC — **System:** M.

TRINIDAD & TOBAGO TELEVISION CO. LTD.
✉ Television House. 11A Maraval Rd., P.O. Box 665, Port of Spain. ☎ +1 (809) 622-4141-4. 🖷 +1 (809) 622-0344. **Cable:** Television Trinidad.
L.P: GM: Grenfell Kissoon. Chief Eng: D. Dhani.
Stations: chA2 12/6kW, chA13 2/0.2kW (ch's 9 & 14 since 1983).
D.Prgr: 0940-0400.

TV-sets: 31.500 — **Colour:** NTSC — **System:** M.

WBNB-TV
✉ Box 1947, Charlotte Amalie, St. Thomas 00801.
☎ +1 (809) 774-0300. 🖷 +1 (809) 776-3511.
L.P: Sen. Vice Pres: J. Potter. St. Mgr: P. Stull.
Station: WBNB-TV chA10 113/76kW.

CARIBBEAN COMM. CORP.

✉ 1 Beltjen Place, St. Thomas, VI 00802. ☎ +809 77621 50. 📠 +809 774 5029.
L.P: Hd. of Sales: Randolph H. Knight; MD: Andrea L. Martin.

VIRGIN ISLANDS PUBLIC TV-SYSTEM
✉ Box 7879, Charlotte Amalie, St. Thomas 00801. ☎ +1 (809) 774-6255. 📠 +1 (809) 774 7092.
L.P: TD: Leslie Hayes; Film Buyer: Lori Elskoe.
Station: WTJX-TV chA12 31.6/6.32kW.

VIRGIN ISLANDS (British)

TV-sets: 3.000 — **Colour:** NTSC — **System:** M.

TELEVISION WEST INDIES LTD. (Comm.)
✉ P.O. Box 34, Broadcast Peak, Chawell, Tortola, BVI.
☎ +1 (809) 409 43332
Station: ZBTV (Tortola) chA5 30/3kW.

BVI CABLE TV
✉ P.O. Box 694, Road Town, Tortola. ☎ (809) 495 3205.
L.P: MD: Todd Klindworth.

SOUTH AMERICA

ARGENTINA

TV-sets: 9.800.000 — **Colour:** PAL — **System:** N.

ASOCIACIÓN DE TELERADIODIFUSORAS ARGENTINAS (ATA)
✉ Av. Córdoba 323, 6to., 1054 Buenos Aires. ☎ +54 (1) 312-4208/4219/4533. 📠 +54 (1) 312-4208. ⚙ 17253 ATA AR. **Cable:** Teleradio Baires — **L.P:** Pres: Alejandro Enrique Massot. CE: Enrique Parodi.

	Call	chA	kW (ERP)	City of location
1)	LS86	2	100/10	La Plata
2)	LT83	3	90/20	Rosario
3)	LU89	3	30/3	Santa Rosa
4)	LT84	5	36/6	Rosario
6)	LV84	6	1.5/0.15	San Rafael
5)	LU93	6	1.5/0.15	San Carlos de Bariloche
7)	LS82	7	212/20	Buenos Aires
8)	LU81	7	28/3	Bahía Blanca
9)	LU84	7	24/3	Neuquén
10)	LU90	7	30/3	Rawson
11)	LV89	7	45/5	Mendoza
12)	LW80	7	10/1	San Salvador de Jujuy
13)	LW81	7	28/3	Santiago del Estero
14)	LRI486	8	75/7.5	Mar del Plata
15)	LV82	8	10/1	San Juan
16)	LV85	8	18/1.8	Córdoba
17)	LRK458	8	5/0.5	San Miguel de Tucumán
18)	LS83	9	62/6	Buenos Aires
19)	LU80	9	43/4	Bahía Blanca
20)	LU83	9	6/0.6	Comodoro Rivadavia
21)	LV83	9	52/5	Mendoza
22)	LU85	9	1.5/0.15	Río Gallegos
23)	LT81	9	1.5/0.15	Resistencia
24)	LV91	9	5/0.5	La Rioja
25)	LU82	10	295/30	Mar del Plata
26)	LV80	10	80/8	Córdoba
27)	LW83	10	6/0.6	San Miguel de Tucumán
28)	LU92	10	50/5	General Roca
29)	LRH450	10	22/2.2	Junín
30)	LS84	11	180/20	Buenos Aires
31)	LT88	11	230/20	Formosa
32)	LU87	11	15/1.5	Ushuaia
33)	LW82	11	11/1	Salta
34)	LV81	12	170/20	Córdoba
35)	LT85	12	20/2	Posadas
36)	LU91	12	4/0.4	Trenque Lauquén
37)	LS85	13	116/20	Buenos Aires
38)	LT80	13	65/6	Corrientes
39)	LT82	13	80/8	Santa Fé
40)	LU88	13	15/1.5	Río Grande
41)	LV86	13	5/0.5	Río Cuarto
42)	LV90	13	12/1.2	San Luis

Associated to A.T.A.: 2), 4), 5), 8), 9), 12), 13), 14), 15), 16), 17), 19), 20), 21), 23), 25), 33), 36), 38), 39), 41).
Operated by State Administration: 3), 7), 10), 22), 24), 28), 29), 30), 31), 32), 34), 35), 37), 40), 42).
Addresses and other information:
1) Radiodifusora El Cármen S.A., Calle 27 e/530 y 531, 1900 La Plata. Pcia. Buenos Aires. (+5 relays).
2) Televisión Litoral S.A., Av. Godoy 8101., 2000 Rosario. Pcia. Santa Fé. (+ 5 relays).
3) Ruta 35 Km.322, Casilla Correo 139, 6300 Santa Rosa. Pcia. de La Pampa. (+7 relays).
4) Rader S.A., Av. Belgrano 1055, 2000 Rosario. Pcia. de Santa Fé. (+5 relays).
5) TV Río Diamante, Luzuriaga 360, 5600 San Rafael. Pcia. de Mendoza. (+3 relays).
6) Bariloche Televisión SRL., Elflein 251, 8400 San Carlos de Bariloche. Pcia. de Río Negro. (+5 relays).
7) Argentina Televisora Color S.A., Figueroa Alcorta 2977, 1425 Buenos Aires. Capital Federal. (+163 relays).
8) Telba S.A., Blandengues 225, 8000 Bahía Blanca. Pcia. de Buenos Aires. (+4 relays).
9) Neuquén TV S.A., Av. Argentina 1700, 8300 Neuquén. Pcia. de Neuquén. (+30 relays).
10) Av. Fontana 50, 9103 Rawson. Pcia. de Chubut. (+9 relays).
11) Garibaldi 7 Piso 5, 5500 Mendoza. Pcia. de Mendoza. (+8 relays).
12) R. Visión Jujuy S.A., Av. 19 de Abril 749, 4600 San Salvador de Jujuy. Pcia. de Jujuy. (+8 relays).
13) Pellegrini 345, 4200 Santiago del Estero. Pcia. de Santiago del Estero. (+7 relays).
14) Emisora Arenales de Radiodifusión S.A., Av. Luro 2907, 7600 Mar del Plata. Pcia. de Buenos Aires. (+ 8 relays).
15) Mitre 59 Oeste, 5400 San Juan. Pcia. de San Juan. (+5 relays).
16) DICOR S.A., Vélez Sarsfield 4300, 5000 Córdoba. Pcia. de Córdoba. (+16 relays)
17) Televisora Tucumana Color S.A., Av. Salta y Delfín Gallo. 4000, San Miguel de Tucumán. Pcia. de Tucumán.
18) Telearte S.A., México 990, 1097 Buenos Aires. Capital Federal.
19) Telenueva S.A., Sarmiento 64, 8000 Bahía Blanca. Pcia. de Buenos Aires. (+13 relays).
20) Comodoro Rivadavia TV, Rawson 1459, 9000 Comodoro Rivadavia. Pcia. de Chubut. (+7 relays).
21) Cuyo Televisión S.A., San Martín 1027, Galería Piazza, Local 10. 5500 Mendoza. Pcia. de Mendoza. (+5 relays)
22) H. Irigoyen 250, 9400 Río Gallegos. Pcia. de Santa Cruz. (+15 relays).
23) TV Resistencia S.A., Av. Alvear 50, 3500 Resistencia. Pcia. de Chaco. (+9 relays).
24) Av. Ortiz de Ocampo 1700, 5300 La Rioja. Pcia. de La Rioja. (+6 relays).
25) TV Mar del Plata S.A., Independencia 1163, 7600 Mar del Plata. Pcia. de Buenos Aires. (+4 relays).
26) SRTV S.A., Rivera Indarte 170, 5000 Córdoba. Pcia. de Córdoba. (+24 relays)
27) Televisora de Tucumán S.A., Av. Buenos Aires 296, 4000 San Miguel de Tucumán. Pcia. de Tucumán. (+2 relays).
28) Mitre y Sarmiento Piso 1, 8332 General Roca. Pcia. de Río Negro. (+10 relays).
29) Junín TV S.A., Belgrano 84, 6000 Junín. Pcia. de Buenos Aires. (+1 relay).
30) Telefé S.A., Pavón 2444, 1248 Buenos Aires. Capital Federal.
31) Tucumán 56, 3600 Formosa. Pcia. de Formosa. (+2 relays).
32) Magallanes 1310, 9410 Ushuaia. Tierra del Fuego. (+1 relay).
33) CorTe S.A., España 475, 4400 Salta. Pcia. de Salta. (+4 relays).
34) Telecor S.A., Av. Fader 3469, Cerro de las Rosas. 5000 Córdoba. Pcia. de Córdoba. (+21 relays).
35) Rioja 161, 3300 Posadas. Pcia. de Misiones. (+11 relays).
36) Av. Leandro N. Alem 351, 6400 Trenque Lauquén. Pcia. de Buenos Aires.
37) ArTeAr S.A., Cochabamba 1153, 1147 Buenos Aires. Capital Federal.
38) Río Paraná TV SRL., Calle 13 s/n., 3400 Corrientes. Pcia. de Corrientes. (+3 relays).
39) Televisora SantaFesiona S.A., Blvd. Gálvez 840, 3000 Santa Fé. Pcia. de Santa Fé. (+6 relays).
40) Alberti 739, 9420 Río Grande. Tierra del Fuego.
41) Imperio Televisión S.A., Alberdi 823, 5800 Río Cuarto. Pcia. de Córdoba. (+6 relays).

42) Colón 925, 5700 San Luis. Pcia. de San Luis. (+6 relays). Federal.
38) Río Paraná TV SRL., Calle 13 s/n., 3400 Corrientes. Pcia. de Corrientes. (+3 relays).
39) Televisora SantaFesiona S.A., Blvd. Gálvez 840, 3000 Santa Fé. Pcia. de Santa Fé. (+6 relays).
40) Alberti 739, 9420 Río Grande. Tierra del Fuego.
41) Imperio Televisión S.A., Alberdi 823, 5800 Río Cuarto. Pcia. de Córdoba. (+6 relays).
42) Colón 925, 5700 San Luis. Pcia. de San Luis. (+6 relays).

BOLIVIA

TV-sets: 750.000 — **Colour:** NTSC — **System:** M&N.

Location	chA	kW	Location	chA	kW
1) La Paz	2		14) Trinidad	8	
2) Cochabamba	2		20) Tarija	8	1
3) Santa Cruz	2		15) La Paz	9	
4) Trinidad	2		13) Cobija	9	
4a) Oruro	3		16) Cochabamba	9	
5) La Paz	4	5	17) Santa Cruz	9	
5a) Cochabamba	4		18) Sucre	9	
6) Santa Cruz	4	8	20) Potosi	9	0.1
13) Trinidad	4		1) Oruro	10	
7) Santa Cruz	5	2	19) La Paz	11	
8) Oruro	5		13) Tarija	11	
7) Trinidad	5	2	20) Cochabamba	11	1
10) La Paz	6		20) Santa Cruz	11	10
11) Cochabamba	6		13) Potosi	11	
10) Oruro	6		20) Trinidad	11	0.1
1) Cochabamba	7		20) Sucre	12	0.1
1) Sucre	7		1) Trinidad	12	
12) La Paz	7		20) La Paz	13	10
1) Potosi	7		7) Cochabamba	13	5
1) Tarija	7		22) Santa Cruz	13	2
13) Oruro	8		20) Oruro	13	2
Location	**chA**	**kW**	**Location**	**chA**	**kW**
14) Trinidad	13		24a) La Paz	15	
23) La Paz	15		25) Santa Cruz	15	1
24) Cochabamba	15		26) La Paz	20	

Addresses & other information
1) Television Boliviano, Cas. 4837, La Paz.
2) CCA-TV, Av. Heroinas 467, Cochabamba. 1500-0430.
3) Cristal de TV, Cas. 4399, Santa Cruz.
4) Benivision, Cas. 54, Trinidad.
4a) TV O Ltda, Cas 631, Oruro.
5) TV-Popular, Calle Juan de la Riva 1527, Casilla #8704, La Paz — 1200-0600 (+76 relays).
5a) Casilla 4573, Cochabamba.
6) Galavision, Cas. 495, Santa Cruz — 1000-0600.
7) TV-Universo, Av. Circunvalación, Santa Cruz — 24h.
8) Orureña TV, Cas. 14, Oruro.
10) America TV, Cas. 10076, La Paz. 24h.
11) TV Cochabamba, Cas. 1009, Cochabamba.
12) Empresa Nacional de TV, Cas. 900, La Paz. 24h.
13) Illimani de Comunicaciones, Oruro.
14) Trinivision, Cas. 333, Trinidad.
15) Rede ATB, Av. 6 de Agosto 2972, La Paz.
16) Cochabamba TV, Cas. 4545, Cochabamba.
17) Teleoriente, Cordillera 550, Santa Cruz.
18) Chuquisaqueña TV, Cas. 187, Sucre.
19) Teleandina, Cas. 1665, La Paz.
20) TV Universitaria, Cas. 21982, La Paz. 1900-0300.
22) Cruceña TV, Cas. 3424, Santa Cruz.
23) Sonomac/Tricolor, Av. 16 de Julio 1810, La Paz.
24) TV Integral, Cochabamba.
24a) SONOMAC, Cas. 21375, La Paz.
25) Grigota TV, Santa Cruz.
26) Tecnitron, Cas. 4410, La Paz.

BRAZIL

TV-sets: 32.600.000 — **Colour:** PAL — **System:** M.
All st's comm. exc. where indicated.

ASSOCIACÃO BRASILEIRA DE EMISSORAS DE RADIO E TELEVISÃO (ABERT).
✉ Hotel Nacional, s/5 a 8, C.P. 040-280, 70322-900 Brasilia, DF. ☎ +55 (61) 224 4600. 🖷 +55 (61) 321 7583.

SEARA — Serviços Associados de Rádio Ltda, Rua do Livramento 189, 20021 Rio de Janeiro, RJ. ☎ +55 (61) 243 2225.

STATIONS (relay st's omitted):

ch	kW	Name and location
1)	2	13.17	TV Educativa, Manaus
2)	2	70/7	TV Guaiba, Porto Alegre
3)	2	100/10	TV Anhanguera, Goiânia
5)	2	78.25/8	TV Jornal do Comercio, Recife
6a)	2	6	TV Globo, Bauru
7)	2	57	TV Educativa, Rio de Janeiro.
8)	2	282/60	TV Cultura, São Paulo
8)	2	282	TV Geradora
9)	2	10	RBS TV, Erexim
9a)	2	120	TV Manchete, Fortaleza
10)	2	42	TV Educativa, São Luiz
11)	2	37.8	TV Educativa da Bahia, Salvador
12)	2	10.6	TV Educativa Espirito Santo, Vitória
63)	2	1	TV Liberal Marabé
12a)	2	8.5	TV Curitiba, Curitiba
13)	3	30.4	TV Nacional de Brasilia
6j)	3	10	TV Coroados, Londrina
16)	3	50	RBS TV, Blumenau
18)	3	1.6	TV Studios Rio de Janeiro, Nova Friburgo
108)	3	10	RBS TV, Cruz Alta
108)	4	5	Santa Angelo
19)	4	7.2	TV À Critica, Manaus
20)	4	100/10*	TV Aratu, Salvador
21)	4	10.31	TV Gazeta, Vitória
22)	4	10/1*	TV Record, Franca
6)	4	100	TV Globo, Rio de Janeiro
23)	4	14.77	TV Goyá, Goiânia

ch	kW	Name and location
106)	4	3.16	TV Paranaiba-Prata
24)	4	1.8	TV Montes Claros, Montes Claros
25)	4	12.3	TV Difusora do Maranhão, São Luis
27)	4	120	TV Manchete, Belo Horizonte
29)	4	15.6	TV Guajará, Belém
30)	4	10	TV Iguaçu, Curitiba
31)	4	3.1	TV Tapajós, Santarém
32)	4	6.7	TV Radio Clube Teresina, Teresina
33)	4	15	RBS TV, Pelotas
34)	4	30/6	TV Rio Grandense, Porto Alegre
38a)	4	116*	TVS, São Paulo
39)	4	0.5	TV Roraima
39)	4	4	TV Rondônia
39)	4	0.5	TV Acre
92a)	4	100	TV Santa Cruz
85)	4	1	TV O Estado Florianopolis
88)	4	—	TV Bandeirantes, Brasilia
4)	5	100	TV Alterosa, Belo Horizonte
39)	5	13	TV Amazonas, Manaus
40)	5	2/0.5	TV Universitaria, Natal
40a)	5	5	TV Educativa, Teresina
41)	5	100/10*	TV Itapoan, Salvador
42)	5	10	RBS TV, Joinville
43)	5	55	TV Educativa, Fortaleza
6f)	5	2	TV Globo, Juiz de Fora
46)	5	38/1	TV Alagoas, Maceio
6b)	5	75	TV Globo, São Paulo
38b)	5	100*	TVS, Porto Alegre
38c)	5	100*	Sistema B de TV, Belem
92a)	5	100	TV Sudoeste
39)	6	1	TV Amapá
47)	6	24	TV Brasilia, Brasilia
48)	6	8	TV Morena, Campo Grande
49)	6	12/2	TV Parana, Curitiba
17)	6	30/3*	TV Cruz Alta, Cruz Alta
50)	6	70*	TV Tarobá, Cascavel
96)	6	20	RBS TV, Santa Cruz (F.Pl.)
51)	6	8.5	TV Cultura, Florianópolis
52)	6	400	TV Manchete, Rio de Janeiro
52a)	6	180	TV Manchete, Recife
53)	6	10	RBS TV, Bagé

	ch	kW	Name and location
54)	6	14	TV Ribamar, São Luiz
54a)	6	27.6	TV Vitoria, Vitoria
55)	7	189/18.9*	TV Bandeirantes, Salvador
56)	7	14/1.2	TV Tropical, Londrina
57)	7	20	TV Esplanada, Ponta Grossa
58)	7	31.6*	TV Sudoeste do Parana, Pato Branco
59)	7	316	TV Record, São Paulo
88)	7	316/31.6	TV Bandeirantes, Belo Horizonte
6h)	7	5	TV Ribeirão Preto, Ribeirão Preto
62)	7	12	TV Gazeta de Alagoas, Maceió
63)	7	60	TV Liberal, Belém
64)	7	57.48	TV Bandeirantes, Rio de Janeiro
65)	7	100	TV Educativa, Porto Alegre
66)	7	40	RBS TV, Passo Fundo
66a)	7	7.7	TV Uberaba, Uberaba
99)	7	?	TV Independência, Curitiba
67)	8	15.8	TV Capital, Brasilia
68)	8	3.16/0.36*	TV Sul Fluminense, Barra Mansa
70)	8	9.6	TV Brasil Oeste, Cuiabá
71)	8	5	TV Campo Grande
72)	8	31.6	RBS TV, Caxias do Sul
73)	8	20/4	TV Uberlândia, Uberlândia
74)	8	80 /8*	TV Record, São José do Rio Preto
75)	8	9.2	TV Ajuricaba, Manaus
76)	8	87.9	TV Atalaia, Aracaju
77)	8	8.91	TV de Fortaleza, Fortaleza
6k)	8	5	TV Cultura, Maringá
108)	9	5	Santa Rosa
79)	9	2/0.5	TV Borborema, Campina Grande
80)	9	31.6	RBS TV, Rio Grande
81)	9	21.6	TV Educativa, Belo Horizonte
82)	9	76	TV Morado do Sol, Araraquara
83)	9	1.32	TV Eldorado, Criciuma
106)	9	316	TV Paranaiba—Frutal
83a)	9	320	TV Manchete, São Paulo, SP
83b)	9	16.2	TV Record, Rio de Janeiro
84)	10	138	TV Verde Mares, Fortaleza
85)	10	2	TV O Estado Chapeco
ch		**kW**	**Name and location**
85)	10	36	TV Planalto, Lages
106)	10	3.16	TV Paranaiba—Araxa
106)	10	31.6	TV Paranaiba—Uberlândia
6c)	10	47	TV Globo Capital, Brasilia
87)	10	30/6	TV Paranaiba, Uberlândia
87a)	10	?	TV Carima, Cascavel
88)	10	126	TV Bandeirantes, Porto Alegre
88)	10	8.8	TV Bandeirantes, Pres. Prudente
88b)	11	70	TVS, Rio de Janeiro
63)	11	2	TV Liberal—Castanhal
105)	11	10	TVU
90)	11	5.6	TV Tibagi, Apucarana
91)	11	5/1.5	TV Universitaria, Recife
92)	11	10	TV Gazeta, São Paulo
92a)	11	316	TV Bahia, Salvador
108)	11	1	São Luiz Gonzaga
6d)	12	35	TV Globo, Belo Horizonte
6i)	12	25	TV Paranaense, Curitiba
95)	12	5	TV Norte Fluminense, Campos
96)	12	520	RBS TV, Porto Alegre
97)	12	31.6	RBS TV, Santa Maria
6g)	12	1	TV Campinas, Campinas
99)	12	13	TV Independência, Cornelio Procopio
99)	12	?	TV Independência, Maringá
100)	12	31.6	TV Cultura, Chapeco
101)	12	316	RBS TV, Florianópolis
38e)	12	10	TVS, Belem
105)	12	7.5	Televisão Imembui
102)	13	200	TV Bandeirantes, São Paulo
6e)	13	330	TV Globo, Recife
103)	13	80	TV Brasil Central, Goiânia
104)	13	31.6	RBS TV, Uruguaiana

*) ERP.

Addresses and other information:

1) Rua Major Gabriel s/n, 69000 Manaus, AM — 2200-0400.
2) Rua Caldas Jr. 219, 90000 Porto Alegre, RS — 1200-0400.
3) Rua Thomaz Edson, Qd.7, 74000 Goiânia, GO—0900-0600.
4) Av. A. Chateaubriand 499, 30150 Belo Horizonte, MG — 1100-0600.
5) Rua do Lima 250, Recife, PE — 1200-0430.
6) Rua Lopez Quintas 303 (Studio: Rua Von Martins), Jardim Botanico, 22463 Rio de Janeiro, RJ. Dir. Tec: A. Pontes Malta. 0900-0600.

6a) Rua Padre Anchieta 941, 17100 Bauru, SP.
6b) Alameda Santos 1893, 01419 São Paulo, SP — 0930-0300.
6c) SCS Q2, Bl. B. no 81, Edif. Bradesco, 70300 Brasilia, DF.
6d) Rua Rio de Janeiro 1279, 30000 Belo Horizonte. 1200-0300.
6e) Av. Dantes Barreto 1186, 50000 Recife, PE.
6f) Rua Ewbank da Camara 46, 36100 Juiz de Fora, MG — 0930-0530.
6g,h) Rua Javari 3099, 14110 Ribeirão Preto, SP.— Dir. Gen: A.C. Coutino Nogueira. Dir. Tec: A. João Filho. 0900-0600.
6i) Av. Batel 1393, 80000 Curitiba, PR — 1430-0500.
6j) Av. Tirandentes 1370, 86100 Londrina, PR — 2000-0300.
6k) Rua Sta. Joaquina de Vedruna 625, 87100 Maringá, PR.
7) Av. Gomes Freire 474-B, Rio de Janeiro, RJ.— 0930-0345.
8) Rua Cenno Sbrighi 378, Agua Branca, 05099 São Paulo, SP — St. Mgr: R. Muylaert. Dir. Tec: J. Munhoz — 1100-0300.
9) Rua Soledade 277, Erexim, RGS.
9a) Av. Antonio Sales Esq, 60000 Fortaleza — 1100-0500.
10) Av. Kennedy s/n, São Luis, MA — 1000-0300.
11) Rua Pedro Gama 413/E, 40000 Salvador, BA — 1330-0300.
12) Rua Pedro Palacio 99, 29000 Vitoria, ES.
12a) Rua Francisco Caron 29, Curitiba, PR.
13) Praça 31 de Marco s/n, 70000 Brasilia — 1400-0300.
16) Rua Getulio Vargas 32, 89100 Blumenau, SC.
17) Rua Jango Vidal 427, 98100 Cruz Alta, RS — 1000-0500.
18) Rua E. Brasilio 30, 28600 Nova Friburgo.
19) Estrada do Aleixo Km 3, 69060 Manaus, AM — 0900-0400.
20) Rua Pedro Gama 31, 40230 Salvador, BA — 0900-0500.
21) CP. 1070, 29050 Vitória, ES — 0900-0600.
22) Rua José Maria Medeiros 5120, 14400 Franca, SP — 1100-0500.
☎ +55 (16) 727 0400.
23) Av. Goiás 187, 74000 Goiânia, GO.
24) Praça dos Morrinhos S/N, Morrinhos. 0930-0530 (Sat/Sun 24h).
25) Av. Camboa do Maio 120, 65000 São Luis, Maranhão — 1000-0600.
29) Av. Governador José Malcher 1332, Belém, PR — 1400-0400.
30) Rua João Tscharnell 800, Jardim Merces, 80000 Curitiba, PR.
31) Av. Ismael Aranju 266, 68100 Santarem, PA.
32) CP. 209, 64000 Teresina, PI—Pres: S.F. de Alénear. Dir Tec: H.P. de Carvalho. 0900-0600.
33) Rua 15 de Novembro 612, Pelotas, RGS. 1200-0600.
34) Rua Orfanotrôfio 711, Alto Teresópolis, 90000 Porto Alegre, RS — 1000-0500.
35) Rua Alto do Morro, 49000 Aracaju, SE.
38a-e) Rua Dona Santa Velozo 535, 02050 São Paulo — 1000-0500 (on all st's).
39) Ave. Carvalho Leal 1270, Cachoeirinha, Manaus, AM — 0930-0500.
40) Rua Princesa Isabel 758, 59020 Natal, RN — Dir: C.A. Martins. Dir. Tec: R. de Andrade Martins — 1000-0300.
40a) Ave. Prof. V. Alencar s/n, 64065 Teresina, PI — Dir. Tec: F.J. de Paiva Ribeiro — 1000-0400.
41) Rua Ferreira Santos 5, 40000 Salvador, BA — Dir: A. Moraes. 0930-0430.
42) Rua Saguaçu s/n, Joinville, SC.
43) Rua Oswaldo Cruz, 1985, 60000 Fortaleza, CE — 1000-0200.
46) Rua Cel. Paranhos 305, Jacintinho, 57000 Maceió, AL.
47) Av. W-3, Setor de Radio e TV, Brasilia — 1300-0500.
48) Av. Eduardo E. Zahran s/n, Campo Grande, MS — 2030-0300.
49) C.P. 7061, 80000 Curitiba, PR — 1000-0500.
50) C.P. 1169, 85800 Cascavel, PR — 1030-0500.
51) Alto do Morro Antão, Florianópolis, SC.
52) Rua do Russel 766-804, 20000 Rio de Janeiro — Dir. Gen: R. Furtado. Dir. Tec: F. Cavalcanti — 1100-0300.
52a) 1100-0500.
53) Rua do Acampamento 2550, 96400 Bage, RGS.
54) Praça Bom Menino s/n, São Luiz, Maranhão.
54a) Av. Presidente Florentino Avidos, 350 7º anolar Vitoria ES.
55) Largo do Candomblé 19A, Salvador, B.A — Dir. Gen: F.H. Chagas. Dir. Tec: R. Blum. 0900-0500.
56) Rodovia Celso Garcia, Londrina, PR — 1300-0500.
57) Rua João França Silva 2885, Ponta Grossa, PR. 1030-0300.
58) C.P. 591, 85500 Pato Branco, PR — Dir: V. Hillesheim. Dir. Tec: L. Schmitz — 1000-0300.
59) Av. Miruna 713, Aeroporto, 01000 São Paulo, SP — 1100-0500. ☎ +55 (11) 542 9000.
59a) Campo de São Cristovão 105, Rio de Janeiro, RJ.
60) Rua Tomé de Souza 1251, 30000 Belo Horizonte, MG — Dir. Gen: M. Pereira Leite. Dir. Tec: O. Dominco Dalip. Film Buyer: R. Hachich Maluf. 1200-0600.
62) Av. Aristeu de Andrade s/n, Maceió, AL — 1830-0330.
63) Av. Nazaré, 350 Belem Pará, PA — 1200-0400.

64) Rua Alvaro Ramos 492, Bota fogo, Rio de Janeiro, RJ — 1230-0300.
65) Rua Correa Lima 2118, Morro Santa Tereza, 90640 Porto Alegre, RS — St. Mgr: A.C. Fedrizzi — 1145-0330.
66a) Rua Princesa Isabel, 99100 Passo Fundo, RGS.
66a) Rua General Osorio 755, 38100 Uberaba, MG — 2000-0400.
67) Torre de Televisão de Brasília, Box 1, Brasilia, DF — 1440-0200.
68) C.P. 85919, 27400 Barra Mansa, RJ—Dir: O.M. Nora. Dir. Tec: C. Pina. 1000-0500.
70) Rua Giboia, s/nÞ, Bairro Concil, 78000 Cuiaba, MT—1030-0600.
71) Av. Calogeras 315, 79300 Campo Grande, MT — 0600-0200.
72) Av. Rio Grande do Sul, Caxias do Sul, RS — 1530-1730, 1900-0500.
73) Rua R.G. do Norte 1069, 38400 Uberlândia, MG.
74) Via Washington Luiz, km 436, 15100 São José do Rio Preto, SP — Dir: P.M. de Carvalho F. Dir. Tec: C. Victor Donato. 1100-0500.
75) Rua O.G. 18 Santo Antonio, 69000 Manaus.
76) Rua Claudio Batista 122 (ex Cláudio Batista S/N), 49045 Aracaju, SE.
77) Av. Desembargador Moreira 2565, Fortaleza, CE. 1300-0400.
79) Rua Venâncio Neiva 287, 2Þ andar, 58100 Campina Grande, PB — 1355-0400.
80) Rua Duque de Caxias 63, 1Þ e 2Þ andares, Rio Grande, RGS.
81) Av. Assis Chateaubriand 167, 30150 Belo Horizonte, MG — 0900-0500.
82) Praça José Palmores Lepre 99, 14800 Araraquara, SP — 24h.
83) Rua Silva Jardim 216, Cricuma, SC.
83a) Rua Bruxelas 193 Sumare, 01000 São Paulo, S.P.
83b) Rua G. Padilha 144, 20000 Rio de Janeiro.
84) Av. Desembargador Moreira 2430, Fortaleza, CE. 1405-0220.
85) Rua Carlos Jofre do Amaral 67, 88500 Lages, SC.
87) C.P. 210, 38400 Uberlândia, MG — St. Mgr: A. de Castro jun. Dir. Tec: M.J. Rodrigues dos Reis — 0900-0500.
87a) Av. Barao do Rio Branco, 1960, Sao Paulo.
88) Rua Delfino Riet 183, C.P. 1474, Porto Alegre, RS — 1300-0400.
88) Rua Radiantes 13, Sao Paulo, Brazil 05699.
90) Av. Santos Dumont 11, 86800 Apucarana, PR — 1200-0400.
91) Av. Norte 68, Santo Amaro, Recife, PE. 1900-0300.
92) Av. Paulista 900, São Paulo, SP. 1500 (Sat/Sun 1200)-0300.
92a) Estrada de São Lázaro 540, Federação, Salvador-BA, Brazil.
95) Av. 24 de Outubro 201, 28100 Campos, RJ — 0945-0530.
96) Rua Radio y TV Gaucha 189, 90650 Porto Alegre, RS — 0900-0500.
97) Av. 2 de Novembro s/n, 97100 Santa Maria, RS.
99) SSC — Sistema Sul de Comunicacão, R. Amaury Lange Silvério, 450 Pilarzinho, 82000 Curitiba, PR.
100) Estrada de Seara Km 3, 89800 Chapeco, SC.
101) Morro da Cruz, 88000 Florianópolis, SC.
102) Rua Radiantes 13, 05699 São Paulo, SP — 1300-0400.
103) Rua 201 no. 430, Vila Nova, Goiânia, GO.
104) Rua Domingos de Almeida 1722, 97500 Uruguaiana, RGS. 24h.
105) Avenida Norte, 68 Sante Amaro, Recife, Pernambuco.
106) Av. Prof. José Ignácio de Souza, 2710, 38400, Uberlandia.
107) Av. Mauricio Sirotsky Sobrinho, 25.
108) Caixa Postal 324, 98100, Cruz Alta.

TV-sets: 2.371.520 — **Colour:** NTSC — **System:** M.

TELEVISION NACIONAL DE CHILE (Gov.)
▣ Bellavista 0990, Providencia, Santiago. ☎ +56 (2) 7077660. ▤ +56 (2) 7077761. ① 241375 TVNCH CL. **Cable:** TV Chile. **Web:** http://www.tvn.cl
L.P: DG: Jorge Navarrete Martinez. GM: Bartolome Dezerega Salgado. CE: Jaime Sancho Martinez.

Stations		kW			kW
Location	ch	(ERP)	Location	ch	(ERP)
El Roble	2	175	Copiapo	7	4
Valdivia	3	20	Pto. Williams	8	1
Puerto Montt	4	20	Ancud	8	2
Coquimbo	4	20	Iquique	10	30
San Fernando	5	25	Castro	10	50
Antofagasta	6	20	Salvador	10	4
Osorno	6	20	Talca	10	20
Punta Arenas	6	20	Ovalle	10	3
Cayumanqui	6	175	Vallenar	12	5
Santiago	7	340	Valparaíso	12	100
Temuco	7	115	San Antonio	12	10

+ 96 low power repeaters less than 1 kW.
D.Prgr: 1200 (Sat/Sun 1400)-0500.

CHILEVISION S.A. (Comm.)
▣ Ines Matte Urrejola, 0825, Santiago. ☎ +56 (2) 7372227. ▤ 56 (2) 7377923.① 340 492 TVCH. **E-mail:** chilevis.ionsa001@chilnet.cl **Web:** http://www.cisneros.com/companies/broadcast/chilevision.htm
Stations: chA11 60/30kW (+ relay st. at Valparaíso ch10).
D.Prgr: 2145-0430.

TELETRECE
▣ Inés Matte Urrejola 0848, Santiago. ☎ +56 (2) 514000. ① 440182 TRECE CZ. ▤ +56 (2) 377044.
Web: http://www.reuna.cl/teletrece/

Location	ch	kW	Location	ch	kW
Arica	8	21.7	Talca	8	27
Iquique	8	28.8	Chillán	13	245
María Elena	11	20	Concepción	5	56
Chuquicamata	12	20	Angol	10	1.2
Antofagasta	13	41.6	Victoria	10	1.2
Copiapó	11	9.5	Temuco	4	45
La Serena	13	21.6	Villarrica	9	2
Ovalle	5	2	Valdivia	12	45
San Felipe	7	21	Osorno	9	47
Valparaíso	8	137	Puerto Montt	13	66.4
Santiago	13	220	Ancud	5	7
San Antonio	10	21.7	Castro	12	73.8
San Fernando	5	37			

+ 25 low power tr's less than 1 kW.

RED DE RADIOTELEVISION DE LA UNIVERSIDAD DEL NORTE (TELENORTE)
▣ Carrera 1625, Antofagasta. ☎ +56 222496 — **L.P:** GM: Juan Carlos Salas Floras.
Stations: ch3 Antofagasta, ch11 Arica, ch12 Iquique +7 low power sts.
D. Prgr: 1735-0435.

CORPORACION DE TELEVISION DE LA UNIVERSIDAD CATOLICA DE VALPARAÍSO
▣ Agua Santo Alto 2455 (Casilla 4059), Viña del Mar. ☎ +56 (32) 610140. ▤ +56 (32) 610505.
E-mail: ucvtelev.ision@chilnet.cl
Stations: chA4 (Valparaíso), chA5 (Santiago), ch7 (Puerto Montt), chA8 (La Serena).

RADIO COOPERATIVA TELEVISION S.A.
▣ Antonio Bellet 223, Santiago. ☎ +56 (2) 2360066. ▤ +56 (2) 2352320. **E-mail:** canalroc.kpop002@chilnet.cl
Station: chE2.

MEGAVISION S.A.
▣ Av. Vicuña Mackenna 1348, Santiago. ☎ +56 (2) 5555400. ▤ +56 (2) 5518916. **E-mail:** megavisi.onsa001@chilnet.cl
Station: chE9.

TV-sets: 7.029.000— **Colour:** NTSC — **System:** M.

INSTITUTO NACIONAL DE RADIO Y TELEVISION (INRAVISION)
▣ Centro Administrativo Nacional, Via Eldorado, Bogotá.
☎ +57 (1) 2220700. ▤ +57 (1) 222 0800. ① 43311 INRACO — **L.P:** Exec. Dir: Jose Jorgo Dangorich Castro.
Inravision leases airtime to 26 comm. companies. The three largest are:

Caracol
▣ AA 9291, Santafé de Bogot. ☎ +57 (1) 337 8866. ▤ +57 (1) 337 7126. **Web:** http://latina.latina.net.co/empresa/caracol/
Key st: Manjui ch7.

Punch
▣ Carrera 28, 49-98 Bogotá. ☎ +57 (1) 2174750.
Key st: Manjui ch9.

RTI
▣ Calle 19 N 4-56 Piso 2, Bogotá. ☎ +57 (1) 282 7700. ▤ +57 (1) 284 9012. ① 43294.
L.P: Pres: Patricio Wills. Head of Prgrs: Patricio Wills.
Key st: Manjui ch11.

Netw I   Netw II

Station/Area	ch	kW (ERP)	ch	kW (ERP)
Manjui (1)	7	668	9	668
La Rusia (2)	10	54	8	54
C. Oriente (3)	8	336	3	336
Tasajero (4)	5	150	11	100
Jurisdicciones (5)	6	468	4	372
Saboya (6)	13	27	11	20
Chigorodo (7)	7	15	11	15
El Ruiz (8)	12	54	10	54
Padre Amaya (9)	3	152	13	252
Galeras (10)	12	393	9	393
Planadas (11)	4	16	2	125
Munchique (12)	10	54	3	54
Paramo de Dominguez (13)	8	426	13	426
La Popa (14)	7	10	9	10
La Pita (15)	10	?	2	?
Monteria (16)	7	27	9	25
Cerro Kennedy (17)	13	700	11	700
El Alguacil (18)	7	270	12	270
Arauca (19)	11	1.2		
Gabinete (20)	6	10	4	10
Buenavista (21)	9	5	7	5
Leticia (22)	8	54	13	54
Nieva (23)	12	54	10	54
Alto del Tigre (24)	12	16	2	10
San Andres (25)	11	27	9	15

+ 7 low power stations not mentioned.

Areas (Departments) served: 1) Bogotá (Sabana); Tolima, Huila; 2) Boyaca, Santander (South); 3) Santander (North), Arauca; 4) Santander (North & Central); 5) Santander, Magdalena (South); 6) Boyaco; 7) Uraba; 8) Manizales, Caldas; 9) Medellin, Antioquia (Central); 10) Nariño (Central); 11) Quindio; 12) Popayan; 13) Cauca & Cauca Valley; 14) Cartagena, Bolivar (North); 15) Bolivar, Sucre; 16) Cordoba; 17) Atlantico, Magdalena (North); 18) Guajira; 19) Arauca; 20) Caqueta, Huila; 21) Huila (South); 22) Leticia, Amazonas; 23) Huila (Central & North); 24) Meta; 25) San Andres.
Network III: Manjui (1) ch A 11 668kW.
D. Prgrs: 1630-1830 (Comm), 1830-2130 (Educ.), 2100-0500 (Comm).

ECUADOR

TV-sets: 900.000 — **Colour:** NTSC — **System:** M.

Stations	chA	kW	Stations	chA	kW
1) Guayaquil	2	6/0.6	9) Portoviejo	7	5
2) Quito	2	1/0.1	7) Quito	8	10/1
3) Cuenca	2	8	11) Quito	8	10
4) Quito	4	5/0.5	8) Portoviejo	9	3
5) Guayaquil	4	3/0.3	9) Guayaquil	10	10/1
) Quito	5	?	9) Quito	10	10
6) Esmeraldas	6	0.5/0.05	10) Quito	13	1/0.1

Addresses and other information:
1) Corporación Ecuatoriana de Televisión, Casilla 1239, Guayaquil. ☎ +593 (4) 300150. ▤ +593 (4) 303677. ① 3409 TVDOSG. **Cable:** Teledos — 1200-0600.
2) Canal 2 Quito, Murgeon 732, Quito. ☎ +593 (2) 540877 — 1930-0730 (Sun 1330-0700).
3) Telecuenca, Canal Universitaria Catolica, Casilla 400, Cuenca. ☎ +593 827862. ① 4775 TELCUE ED — 1600-0400.
4) Teleamazonas, Av. Diguja 529 y Brazil, Quito. ☎ +593 (2) 430313.
5) Canal 4 Guayaquil S.A, 9 de Octubre 1200, Guayaquil. ☎ +593 (4) 308194.
6) Canal 6 Esmeraldas, Cas. 108, Esmeraldas. ☎ +593 (2) 710090. **Cable:** TECEM — 2300-0400.
7) Canal 8 Quito, Cas. 3888, Quito. ☎ +593 (2) 244888 — 1730-0530.
8) Manavision S.A, Apt. 50, Portoviejo — 1930-0530.
9) Canal 10, Guayaquil, Casilla 673, Guayaquil. ☎ +593 (4) 391555 — 1130-1430, 1630-0500 (+14 relays).
10) Canal 13 Quito, Rumipampa 1039, Quito. ☎ +593 (2) 242758.
11) Televisora Nacional, Bosmediano 447 y José Carbo, P.O. Box 6615, Quito.

FALKLAND ISLANDS

TV-sets: N/A — **Colour:** PAL — **System:** I.

BRITISH FORCES BROADCASTING SERVICE (BFBS)

▤ BFBS Falkland Islands, Mount Pleasant, BFPO 655. ☎ 32179. ▤ 32193.
L.P: St.Mgr: Steve Brearton. Gen. Eng: S. Brown. **Station:** Mount Pleasant ch 24 UHF/100 W. Port Stanley ch 30 UHF/15 W. Rebros ch40UHF.
D.Prgr: 4 h. of taped broadcasts from BBC and ITV London.

GUIANA (French)

TV-sets: 6.500 — **Colour:** SECAM — **System:** K.

RFO-GUYANE

▤ BP 7013, Cayenne Cedex. ☎ +594 299900. ▤ +590 302649 — **L.P:** Dir: Henri Neron. T.D: Daniel Beugin.
Stations: Cayenne chK4 0.1kW, +8 low power repeaters.
D.Prgr: 2100-0130.

ANTENNE CREOLE (Priv., Comm.)

▤ 31 avenue Louis Pasteur, 97300 Cayenne. ☎ +594 31 2020
Stations: Kourou ch44 1kW, Cayenne ch39 3kW

GUYANA

TV-sets: 15.000 — **Colour:** NTSC — **System:** M.

GUYANA TELEVISION (Gov.)

▤ 68 Hadfield St., Lodge Georgetown. ☎ +592 (2) 69231-4/62691-4/58584.
L.P: GM: A. Brewster; CE: S. Goodman.
Station: Georgetown ch10 0.04kW.
D.Prgr: Sun 1500-1600, repeated 2100-2200.
Two private TV stations relay U.S. satellite sces.

PARAGUAY

TV-sets: 350.000 — **Colour:** PAL — **System:** N.

Stations	chA	kW	Stations	chA	kW
1) Encarnación	7	60/12	4) Asunción	9	60/12
5) Illar	7	10	4) P.J. Caballero	9	5/1
5) Encarnacion	7	10	5) Misiones	10	10
5) C. del Este	8	10	5) Villarrica	12	10
2) Pto. Stroessner	8	25/53)	4) Asunción	13	30/6

Addresses:
1) Television Itapua, Avda Irrazabal y 25 de Mayo, Encarnación — Gen. Mgr: J. Mateo G. — 1600-0400.
2) Televisora del Este, Area 5, Cd. Puerto Stroessner — Gen. Mgr: A. Villalba V. — 2000-0300 (Sat/Sun 1100-0330).
3) Canal 9 TV Cerro Cora SA. ▤ Avenida Carlos Antonio Lopez 572, Asuncion. ☎ +595 (84) 22226. ▤ +595 (84) 498911. **L.P:** Hd. of Sales: Hugo Montgomery.
4) Teledifusora Paraguaya S.A., 8 Proy Lambare, Asunción — 1600-0400. ☎ +595 (21) 443093.
5) Sistema Nacional de Television

PERU

TV-sets: 2m — **Colour:** NTSC — **System:** M.

Stations: *) in stereo.

	chA	kW (ERP)		chA	kW (ERP)
1)	2	22.54/2.25	5)	8	5
5)	3	20	5)	9	315
2)	4	25/12.5	6)	11	30/6
5)	4	20	1)	13	?
3)	5	290/29	7)	27*	7
5)	6	2	8)	33*	10
4)	7	10/5			

Addresses & other information:

1) Compañia Latinoamericana de Radiodifusion S.A. Av. San Felipe 968, Jesús Mariá, Lima 11. ☎ +51 (14) 707272. 🖺 +51 (14) 712688.
2) Compañia Peruana de Radiodifusion, Cas. 1192, Lima. Dir. Gen: M. M. Arbulu B. Man. Dir: N. Gonzalez U. Dir. Tec: D. Capella. Operates st's in Piura(ch2 2 kW), Chiclayo(ch 4 5 kW), Trujillo(ch 6 2 kW), Tacna(ch 9 2 kW), Huancayo(ch 4 2 kW) + 59 repeaters — 1000-0600.
3) Panamericana de Television, Av. Arequipa 1110, Lima. 1500-0600 (+ 14 relay st's).
4) Empresa de Cine, Radio y Television Peruana, José Galvez 1040, Lima. Pres: Carlos Guillen B. 1900-0400 (+ 39 repeaters).
5) Andina de Television, Arequipa 3570, San Isidro, Apartado 270077, Lima. 1900-0400.
6) RBC Television, Juan de la Fuente 453, Miraflores, Lima.
7) Difusora Universal de Television, Paseo de la República 6099, San Antonio, Miraflores, Lima — Exec. Pres: J.L. Banchero H. 2200-0400.
8) Empresa Radiodifusora 1160 TV, Apt. Postal 2355, Lima. 800-0500.

SURINAME

TV sets: 43.000 — Colour: NTSC — System: M.

SURINAAMSE TELEVISIE STICHTING (STVS) (Gov./Comm.)
🖃 P.O. Box 535, Paramaribo. ☎ +597 473100. 🖺 +597 477216. ① 271 STVS. Cable: Surteve.
Stations: Paramaribo chA8 6kW + 5 relay st's.
D.Prgr: 2130-0300 (+ Sat 1500-2130, Sun 1400-1700).

ALTERNATIEVE TELEVISIE VERZORGING (ATV Telesur)
🖃 Adrianusstraat 1, Paramaribo. ☎ +597 410027, +597 470425. 🖺 +597 479260. ① 488 ATV TLS SN.
L.P: Mgr: Roy Doorson.
Stations: Paramaribo chA2 + chA12 + 2 relay st's.
D.Prgr: 1115-0320.

URUGUAY

TV-sets: 600.000 — Colour: PAL — System: N.

St's 1), 3), 8), 9), 13), 15), 16), 17), 18), 20), 23), 24) & 26) are affiliated to ANDEBU (Asociación Nacional de Broadcasters Uruguayos) — All st's are comm.

Location	ch	kW	Location	ch	kW
1) Artigas	3	0.5	14) P. de los Toros	9	0.1
2) Colonia	3	1/0.5	15) Maldonado	9	12
3) Paysandú	3	1/0.5	16) Bella Unión	10	0.1
4) Rio Branco	3	0.1	17) Montevideo	10	600
5) Montevideo	4	300	18) Rivera	10	1/0.1
6) Montevideo	5	96/9	19) Punta del Este	11	5/0.5
7) Rivera	5	7/0.7	20) Treinta y Tres	11	1/0.5
8) Rocha	7	16	21) Durazno	11	0.1
9) Tacuarembó	7	75	22) Fray Bentos	12	30/3
10) M. de Corrales	7	0.1	23) Melo	12	1/0.5
11) Melo	8	0.5/0.05	24) Montevideo	12	600/180
12) Rosario	8	75/7.5	25) Chuy	12	0.1
13) Salto	8	2/1	26) Minas	13	1/0.5

+10 st's 0.1kW (all rel. st 6).

Addresses and other information:
1) Tele-Artigas, Lecueder 291, 55000 Artigas — Dir. Gen: Carlos F. Falco. 1800-0300.
2) Televisora Colonia, W. Barbot 172, 70000 Colonia — Dir. Gen. G. de Gonzáles.
3) Río de los Pájaros TV, Av. España 1629, 60000 Paysandú — Dir Gen: A. Davison. Dir. Tec: H. Caporale.
4) 37100 Río Branco, Cerro Largo.
5) Monte Carlo TV Color, Paraguay 2253, 11800 Montevideo — Dir. Gen: H. Romay. Film Buyer: M. Fonticiella. Dir. Tec: J. Spinella — 1830/0400 (Sat/Sun 1300-0400).
6) S.O.D.R.E., Bul Artigas 2552, 11600 Montevideo.
7) Canal 5 S.O.D.R.E, Bulevar Artigas 2552, Montevideo. Gen. Man: Julio Frade. Dir. Tec: Pedro Narancio.
8) Tele-Rocha, Av. O. de los Santos 105, 27000 Rocha — Dir. Gen: M. Scherchener. Dir. Tec: J. Regalo. Film Buyer: L. Castillo.
9) Radiotelevisión "Zorrilla de San Martín", 18 de Julio 302, 45000 Tucuarembó — Dir. Gen: D. Dini. S. St. Mgr: G. Valdés G. 2030 -0330.

Dir. Tec: G. Acosta. Film Buyer: Jose Abbondanza.
10) 40002 Minas de Corrales, Rivera.
11) Canal 8 TV Melo, 18 de Julio 572, 37000 Melo, Cerro Largo — Dir: Raul Figueredo. Dir. Tec: Eduardo Baptista.
12) Canal 8 Rosario TV Color, Ruta 2 Km. 136.500, 70200 Rosario, Colonia — Dir: H. Fripp.
13) Televisora Salto Grande, Av. Viera 1280, 50000 Salto — Dir. Gen: Dr. Carlos A. Gelpi. Dir. Tec: K. Muguerza.
14) 45100 Paso de los Toros, Tacuarembó.
15) Canal 9 del Este TV Color, Av. Artigas 879, 20000 Maldonado — Dir. Gen: M. Scherschener. Dir. Tec: Fernando Bareño. Film Buyer: J. López — 2000-0400.
16) Telediez, General Rivera s/n, 55100 Bella Unión, Artigas — Dir. Gen: C. Gelpi.
17) SAETA TV Canal 10, Dr. Lorenzo Carnelli 1234, 11200 Montevideo — Dir. Gen: J. de Feo. Dir. Tec: Oscar Inchausti. Film Buyer: H. Villar — Mon-Fri 1930.
18) Tevediez, Sarandí 705, 40000 Rivera — Dir. Gen: A. Pereira. 2000-0300.
19) Canal 11 Punta del Este, Cantegril Country Club, 20100 Punta del Este, Maldonado — Dir. Gen: D. Romay — 1830-0400.
20) Televisora Treinta y Tres, Pablo Zufriategui 226, Treinta y Tres — Dir. Gen: A. Pinho. St. Mgr: A. Lagos. Dir. Tec: D. Ponce.
21) 97000 Durazno.
22) Canal 12 Río Uruguay, Cno. San Salvador s/n, 65000 Fray Bentos, Río Negro — Dir. Gen: D. Romay.
23) Melo TV, Castellanos 723, 37000 Melo, Cerro Largo — Dir. Gen: R. Lucas. Dir. Tec: C. Britos. Film Buyer: J. Lucas.
24) Teledoce Televisora Color, Enriqueta Compte y Riqué 1276, 11800 Montevideo — Dir. Gen: H. Scheck. Dir. Tec: M. Donnangelo. Film Buyer: C. Restano — 1810 (Sat 1600, Sun 1330)-0400.
25) 27100 Chuy, Rocha.
26) TV Cerro del Verdún, Treinta y Tres 632, 30000 Minas, Lavalleja — Dir. Gen: C. Falco. Dir. Tec: J. Rodriguez.

VENEZUELA

TV-sets: 3.855.480 — Colour: NTSC — System: M.

CAMARA VENEZOLANA DE LA TELEVISION
(Organization for private TV stations).
🖃 Ap. 60423, Chacao, Caracas 1050. ☎ +58 (2) 7814608. Cable: CAVETEL. ① 21144 — L.P: Pres. H. Ponsdomenech.

TELEVISORA NACIONAL TVN (Gov.)
🖃 Ap. 3979, Caracas 1010-A. ☎ +58 (2) 239 9811
D.Prgr: 1800-0400 (actually relaying VTV).
Stations: Anzoategui ch13 64kW; Carobobo ch6 67kW; Bolivar ch5 57kW; D. Federal ch5 279kW; Falcon ch5 50kW; Lara ch13 64kW; Merida ch13 2kW; Tachira ch13 72kW; Tachira ch2 5kW; Zulia ch6 61kW.

VENEZOLANA DE TELEVISION "5" (Gov.)
🖃 Ap. 2739, Caracas 1010-A. ☎ +58 (2) 2399811. 🖺 +58 (2) 35734. ① 25401.
Stations: Caracas (Central St): chA5 210/105kW.

Repeaters	chA	kW(ERP)		chA	kW(ERP)
Margarita	5	99.6/49.8	El Tocuyo	9	2/1
La Grita	6	75.9/37.9	Terepalma	13	72/36
Los Olivos	6	40.9/20.9	Pto. La Cruz	13	72/36
Maracaibo	6	108/54	Táchira	13	72/36
Aleton	6	67.2/38.6	Litoral Central	13	60.25/30.12
Güigüe	6	67.2/38.6	Escuque	13	26.25/13.12
Cd. Bolivar	6	12.5/6.25	Caricuao	13	0.631

D.Prgr: 2200-0400 (approx.)

VENEZOLANA DE TELEVISION "8" (Gov.)
🖃 Ap. 2739, Caracas 1010-A. ☎ +58 (2) 349571. Cable: VTV. ① 25401-25412.
Stations: Caracas (Central st): chA8 190/95kW.

Repeaters	chA	kW(ERP)		chA	kW(ERP)
Mérida	5	5/2.5	Valencia	11	151/75.5
Barquisimeto	7	107.5/53.5	Litoral	11	79/49.5
Maracaibo	8	157/78.5	Pto. La Cruz	11	23/11.5
Boconó	8	2/1	Cd. Bolivar	11	23/11.5
Cd. Piar	8	0.15	Vidoño	11	23/11.5
Anaco	9	12.5/6.25	Pto. Cabello	12	20/10

D.Prgr: 1600-0600 (approx.)

COMMERCIAL STATIONS

AMAVISION (Cult & Rlgs)

✉ Calle Selesiano, Colegio Pio XI, Puerto Ayacucho, Amazonas. ☎ +58 (2) 987 6190.
D.Prgr: 2200-0200
Stations: Puerto Ayacucho ch7 6kW

CANAL 10

✉ Av. Francisco de Miranda, con Principal de los Ruices, Centro Empresarial Miranda PHD, Caracas. ☎ +58 (2) 239 8679. ▤ +58 (2) 239 7757.
Station: Caracas ch10.

CANAL METROPOLITANO DE TELEVISION (Comm.)

✉ Av. Circumvalacion El Sol, Centro Professional Santa Paula, Torre B, Piso 4, Santa Paula, Caracas. ☎ +58 (2) 987 6190. ▤ +58 (2) 985 4856
D. Prgr: 2100-0400
Stations: Caracas ch51

NCTV (Comm.)

✉ Urv. La Paz, Avenida 57 y Maracaibo, Maracaibo. ☎ +58 (61) 512662. ▤ +58 (61) 512729.
L.P: Dir: Gustavo Ocando Yamarte.
Station: Maracaibo chA11 108/54kW (est).
D.Prgr: 1600-0400.

OMNIVISION (Comm.)

✉ Calle Milan, Edif. Omnivision, Los Ruices Sur, Caracas. ☎ +58 (2) 256 3586/256 5011. ▤ +58 (2) 256 4482
D. Prgr: 24 hrs.
Stations: Anzoategui ch25 10kW; Aragua ch26 10kW; Bolivar ch26 52kW; Carabobo ch24 10kW; D. Federal ch12 605kW, ch13 10kW; Lara ch21 10kW; Tachira ch24 10kW; Zulia ch23 10kW.

RADIO CARACAS TELEVISION RCTV(Comm.)

✉ Ap. 2057, Caracas. ☎ +58 (2) 256 3696. ▤ +58 (2) 256 1812. ☉ 21527.
D.Prgrs: 24 hrs.
Stations: Anzoategui ch3 210kW, ch2 20kW; Carabobo ch7 264kW; Bolivar ch2 80kW, ch3 2kW; Carabobo ch10 22kW; D. Federal ch10 22kW, ch2 132kW, ch7 22kW; Falcon ch10 330kW; Lara ch3 300kW; Monagas ch10 50kW; Portuguesa ch2 12kW; Tachira ch10 14kW, ch7 75kW; Trujillo ch7 5kW; Zulia ch2 400kW.

TELEVISORA ANDINA DE MERIDA (Cult & Rlgs)

✉ Av. Bolivar, Calle 23 entre Av. 4-5, Merida 5101. ☎ +58 (74) 525 785. ▤ +58 (74) 520 098
D.Prgr: 1400-0200
Stations: Merida ch6 20kW; Tachira ch3 33kW.

TELE BOCONO (Cult.)

✉ Calle 3, Qta. Caleuche, El Saman. Bocono. ☎ +58 (72) 521 27. ▤ +58 (72) 524 85
D. Prgr: 2100-0300
Stations: Trujillo ch13 4kW.

TELECARIBE (Comm.)

✉ Centro Banaven (Cubo Negro), Torre C, Piso 1, of C-12, Chuao, Caracas. ☎ +58 (2) 911 964/913 089.
D. Prgr: 1000-0400
Stations: Anzoategui ch9 50kW; Nueva Esparta ch12 30kW.

TELECENTRO (Comm.)

✉ Avenide Pedro León Torres, esquina de la calle 47, Edificio Telecentro, Barquisimeto, (3001) Lara. ☎ +58 (51) 460 917/4525 27. ▤ same as telepone.
L.P: Dir: Jorge Felix
D.Prgr: 1030-0400
Station: Lara ch11 100kW.

TV GUAYANA (Comm.)

✉ Puerto Ordaz, Bolivar. ☎ +58 (86) 2299 08
D.Prgr: 2100-0300
Stations: Bolivar ch12 125kW, ch13 80kW

TELESOL (Comm.)

✉ Calle Sucre no 15, Cumana, Sucre. ☎ +58 (93) 6620 59. ▤ +58 (93) 6627 75
D.Prgr: 1800-0200
Station: Sucre ch7 12kW

TELEVEN (Comm.)

✉ C.C. Los Chaguaramos, Caracas. ☎ +58 (2) 6617 511. ▤ +58 (2) 6625 300
D.Prgr: 24 hrs.
Stations: Anzoategui ch6 94kW; Carabobo ch13 360kW; Bolivar ch10 125kW, ch9 125kW; D. Federal ch10 139kW, ch21 188kW, ch13 10kW, ch6 23kW; Falcon ch3 111kW; Lara ch9 420kW; Miranda ch13 61kW; Nueva Esparta ch10 375kW; Tachira ch3 120kW; Zulia ch13 182kW.

TELEVISORA REGIONAL DEL TACHIRA (Comm.)

✉ Av. Libertador, edif. Servicios Unidos, Piso 3, San Cristobal, Tachira. ☎ +58 (76) 4473 66. ▤ +58 (76) 4652 77
D. Prgr: 2000-0200
Stations: Tachira ch6 144kW.

TELEVISORA DE ORIENTE, TVO (Comm.)

✉ Puerto la Cruz, Anzoategui. ☎ +58 (82) 6621 63
D.Prgr: 1800-0300
Station: Anzoategui ch5 50kW.

VENEVISION (Comm.)

✉ Av. La Salle, Edif, Venevision,Colinas de Los Caobos, Caracas. ☎ +58 (2) 782 0111/4444/4356/4267. **Cable:** Ventel.
Web: http://www.venevision.com
L.P: Pres: Carlos Bardasano. GM: Manuel Fraiz Grijalba. Eng. Mgr: German Landaeta. PR: Mariela Castio.
Stations: Caracas (Central st): chA4 132/66kW.
Repeaters: Trujillo ch3(6kW), Valencia ch4(20), Puerto Cabello ch4(20), Maracaibo ch4(336), Puerto Ordaz ch4(20), El Tigre ch4(12), Barquisimeto ch6(198), Barcelona ch7(25), Ci. Bolivar ch7(10), Maturin ch7(25), La Guaira ch9(21), Maracay ch9(476), Mérida ch9(16), San Cristobal ch9(40), Valera ch9(12), Anyarito ch9(10), Caricuao ch9(11), Curimagua ch12(40), Guaramacal ch12(13kW).
D.Prgr: 24 hrs.

TV-sets: N/A — **Colour:** NTSC — **System:** M

AMERICAN FORCES ANTARCTIC NETWORK (AFAN McMurdo)

✉ "Operation Deep Freeze", Fleet Post Office, San Francisco, California 96692, USA.
D.Prgr: The US Navy Antarctic support group operates six cable TV channels. incl. occ. local prgrs on ch13.

CHARACTERISTICS OF TELEVISION SYSTEMS
(as indicated in CCIR Report 624-3, XVIth Plenary Assembly, Dubrovnik, 1986)

System	Number of lines	Channel width MHz.	Vision band-width MHz.	Vision/Sound separation MHz.	Vestigial side-band MHz.	Vision mod.	Sound mod.
B	625	7	5	+5.5	0.75	Neg.	FM
D	625	8	6	+6.5	0.75	Neg.	FM
G	625	8	5	+5.5	0.75	Neg.	FM
H	625	8	5	+5.5	1.25	Neg.	FM
I	625	8	5.5	+5.996	1.25	Neg.	FM
K	625	8	6	+6.5	0.75	Neg.	FM
L	625	8	6	+6.5	1.25	Pos.	AM
M	525	6	4.2	+4.5	0.75	Neg.	FM
N	625	6	4.2	+4.5	0.75	Neg.	FM

N.B: Channels L2, L3, L4, Vision/Sound separation is -6.5 MHz. (France)

CHANNEL INFORMATION
(frequencies in MHz.)

VHF Channels:

West European "E" Channels
2 = 48.25	5 = 175.25	9 = 203.25
2A = 49.75	6 = 182.25	10 = 210.25
3 = 55.25	7 = 189.25	11 = 217.25
4 = 62.25	8 = 196.25	12 = 224.25

Italy
A = 53.75	D = 175.25	G = 201.25
B = 59.75	E = 183.75	H = 210.25
C = 82.75	F = 192.25	H1 = 217.25

Ireland
A = 45.75	E = 183.75	H = 207.25
B = 53.75	F = 191.25	I = 215.25
C = 61.75	G = 199.25	J = 223.25
D = 175.25		

France
2 = 55.75	5 = 176	8 = 200
3 = 60.50	6 = 184	9 = 208
4 = 63.75	7 = 192	10 = 216

East European "R" Channels
1 = 49.75	5 = 93.25	9 = 199.25
2 = 59.25	6 = 175.25	10 = 207.25
3 = 77.25	7 = 183.25	11 = 215.25
4 = 85.25	8 = 191.25	12 = 223.25

North/South America
2 = 55.25	6 = 83.25	10 = 193.25
3 = 61.75	7 = 175.25	11 = 199.25
4 = 67.25	8 = 181.25	12 = 204.25
5 = 77.25	9 = 187.25	13 = 211.25

Japan
1 = 91.25	5 = 177.25	9 = 199.25
2 = 97.25	6 = 183.25	10 = 205.25
3 = 103.25	7 = 189.25	11 = 211.25
4 = 171.25	8 = 193.25	12 = 217.25

Australia
0 = 46.25	5 = 102.25	8 = 189.25
1 = 57.25	5A = 138.25	9 = 196.25
2 = 64.25	6 = 175.25	10 = 209.25
3 = 86.25	7 = 182.25	11 = 216.25
4 = 95.25		

New Zealand
1 = 45.25	4 = 175.25	7 = 196.25
2 = 55.25	5 = 182.25	8 = 203.25
3 = 62.25	6 = 189.25	9 = 210.25

China (P.R.)
1 = 49.75	5 = 85.25	9 = 192.25
2 = 57.75	6 = 168.25	10 = 200.25
3 = 65.75	7 = 176.25	11 = 208.25
4 = 77.25	8 = 184.25	12 = 216.25

South Africa
4 = 175.25	7 = 199.25	10 = 223.25
5 = 183.25	8 = 207.25	11 = 231.25
6 = 191.25	9 = 215.25	13 = 247.43

Morocco
4 = 163.25	7 = 187.25	9 = 203.25
5 = 171.25	8 = 195.25	10 = 211.25
6 = 179.25		

French Overseas Territories
4 = 175.25	6 = 191.25	8 = 207.25
5 = 183.25	7 = 199.25	9 = 215.25

UHF Channels:
North/South America
14 = 471.25	33 = 585.25	52 = 699.25
15 = 477.25	34 = 591.25	53 = 705.25
16 = 483.25	35 = 597.25	54 = 711.25
17 = 489.25	36 = 603.25	55 = 717.25
18 = 495.25	37 = 609.25	56 = 723.25
19 = 501.25	38 = 615.25	57 = 729.25
20 = 507.25	39 = 621.25	58 = 735.25
21 = 513.25	40 = 627.25	59 = 741.25
22 = 519.25	41 = 633.25	60 = 747.25
23 = 525.25	42 = 639.25	61 = 753.25
24 = 531.25	43 = 645.25	62 = 759.25
25 = 537.25	44 = 651.25	63 = 765.25
26 = 543.25	45 = 657.25	64 = 771.25
27 = 549.25	46 = 663.25	65 = 777.25
28 = 555.25	47 = 669.25	66 = 783.25
29 = 561.25	48 = 678.25	67 = 789.25
30 = 567.25	49 = 681.25	68 = 795.25
31 = 573.25	50 = 687.25	69 = 801.25
32 = 579.25	51 = 693.25	

UHF TV channels 70-83 were discontinued. (to be used by radio)

Europe/Africa
21 = 471.25	38 = 607.25	55 = 743.25
22 = 479.25	39 = 615.25	56 = 751.25
23 = 487.25	40 = 623.25	57 = 759.25
24 = 495.25	41 = 631.25	58 = 767.25
25 = 503.25	42 = 639.25	59 = 775.25
26 = 511.25	43 = 647.25	60 = 783.25
27 = 519.25	44 = 655.25	61 = 791.25
28 = 527.25	45 = 663.25	62 = 799.25
29 = 535.25	46 = 671.25	63 = 807.25
30 = 543.25	47 = 679.25	64 = 815.25
31 = 551.25	48 = 687.25	65 = 823.25
32 = 559.25	49 = 695.25	66 = 831.25
33 = 567.25	50 = 703.25	67 = 839.25
34 = 575.25	51 = 711.25	68 = 847.25
35 = 583.25	52 = 719.25	69 = 855.25
36 = 591.25	53 = 727.25	
37 = 599.25	54 = 735.25	

Australia
28 = 527.25	42 = 625.25	56 = 723.25
29 = 534.25	43 = 632.25	57 = 730.25
30 = 541.25	44 = 639.25	58 = 737.25
31 = 548.25	45 = 646.25	59 = 744.25
32 = 555.25	46 = 653.25	60 = 751.25
33 = 562.25	47 = 660.25	61 = 758.25
34 = 569.25	48 = 667.25	62 = 765.25
35 = 576.25	49 = 674.25	63 = 772.25
36 = 583.25	50 = 681.25	64 = 779.25
37 = 590.25	51 = 688.25	65 = 786.25
38 = 597.25	52 = 695.25	66 = 793.25
39 = 604.25	53 = 702.25	67 = 800.25
40 = 611.25	54 = 709.25	68 = 807.25
41 = 618.25	55 = 716.25	69 = 814.25

China (P.R.)
13 = 471.25	21 = 534.25	29 = 637.25
14 = 479.25	22 = 543.25	30 = 645.25
15 = 487.25	23 = 551.25	31 = 653.25
16 = 495.25	24 = 559.25	32 = 661.25
17 = 503.25	25 = 605.25	33 = 669.25
18 = 511.25	26 = 613.25	34 = 677.25
19 = 519.25	27 = 621.25	36 = 693.25
20 = 527.25	28 = 629.25	

N.B: Japan Channel 13-62 = No./So. America Channel 14-63

LONG & MEDIUMWAVE STATIONS

Compiled by Bengt A. Ericson

EUROPE, AFRICA, NEAR & MIDDLE EAST

Abbreviations: AFN = American Forces Network — AFRTS = US Armed Forces Radio & TV Service — AWR = Adventist World Radio — CI = Cadena Ibérica — COPE = Cadena de Ondas Populares Españolas — DLF = Deutschlandfunk — DR = Deutschlandradio — E1 = Europe No. 1 — EI = R. Euskadi/Euskadi Irratia — HR = Hessischer Rundfunk — IND = Emissores Independentes — NDR = Norddeutscher Rundfunk — NRK = Norsk Rikskringkasting — OCR = Onda Cero Radio — RA = Radio Altitude — RFI = Radio France International — RFE = Radio Free Europe — RMC = Radio Monte Carlo — VoR = Voice of Russia relay station — RNE = Radio Nacional de España — RRE = Rádio Renascença — SDR = Süddeutscher Rundfunk — SER = Sociedad Española de Radiodifusion — SWF = Südwestfunk — TWR = Trans World Radio — VOA = Voice of America — WDR = Westdeutscher Rundfunk.

---

kHz	Country	kW	Station	kHz	Country	kW	Station
153	ALG	1000	Bechar	252	ARM	150	Yerevan
153	D	500	Donebach	252	IRL	500/100	Clarkestown
153	ROU	1200	Bod	252	RUS	150	Kazan
153	RUS	300	Ufa	261	BUL	60	Sofia
153	RUS	300	Taldom	261	D	80	Burg
162	F	2000	Allouis	261	RUS	2500	Taldom
162	TUR	1000	Agri	270	CZE	750	Topolná
171	BLR	1000	Sasnovy	279	BLR	500	Sasnovy
171	MRC	2000	Nador	520	D	0.2	Würzburg/Hof(BR)
171	RUS	1250	Elektrostal	531	ALG	600/300	Ain-El-Beida
171	RUS	150	Murmansk	531	BOT	50	Maun
171	RUS	1200	Tbilisskaya	531	E	25	Pontevedra (RNE5)
171	RUS	150	Syktyvkar	531	E	20	Oviedo (RNE5)
171	RUS	1200	Bolshakovo	531	E	10	Pamplona (RNE5)
171	UKR	1000	Lviv	531	E	10	Cordoba (RNE5)
177	D	500	Oranienburg (DRB)	531	FRI	100(200)	Akraberg
180	TUR	1200	Polatli	531	IRN	20	Iranshahr
183	D	2000	Saarlouis	531	IRQ	-	Unk. location
189	GEO	500	Tbilisi	531	ISR	100	Jerusalem
189	I	10	Caltanissetta	531	MDR	10	Porto Santo
189	ISL	300	Gufuskálar	531	NIG	50	Akure
198	ALG	1000	Ouargla	531	ROU	14	Petrosani
198	G	500	Droitwich	531	RUS	30	Cheboksary
198	G	50	Burghead	531	SUI	500	Beromünster
198	G	50	Westerglen	531	TZA	10	Dar es Salaam
198	RUS	150	Ufa	531	YUG	10	Uzice
198	RUS	150	Olgino	531	YUG	1	Vranje
198	RUS	150	Noginsk	540	AFS	100/50	Ga-Rankuwa
198	TUR	120	Etimesgut	540	BEL	150	Wavre-Overijse
207	D	500	Aholming	540	E	50	Barcelona (OCR)
207	ISL	100	Eidar	540	HNG	1000	Solt
207	JOR	600	Al Karanah	540	IRL	2	Conamara
207	MRC	800	Azilal	540	IRN	200	Mashhad
207	UKR	500	Kyyiv	540	KEN	100	Voi
216	F	1400	Roumoules (RMC)	540	KWT	600	Sulaibiyah
218	AZE	500	Gäncä	540	MLI	100	Bamako
225	POL	600	Konstantynow 1	540	MRC	600	Sidi Bennour
225	TUR	600	Van	540	MWI	10	Mangochi
234	ARM	500	Kamo	540	NIG	50	Sokoto
234	LUX	2000	Junglinster	540	RUS	50	Orenburg
234	RUS	2000	Syzran	540	SDN	50	Nyala
234	RUS	1200	Krasnyy Bor	549	ALG	600/300	Les Trembles
234	RUS	500	Arkhangelsk	549	ARS	2000	Qurayyat
243	DNK	300	Kalundborg	549	ARS	20	Rafha
243	TUR	200	Erzurum	549	ARS	1	Gizan
252	ALG	1500/750	Tipaza	549	AZE	50	Gäncä

kHz	Country	kW	Station	kHz	Country	kW	Station
549	BLR	1000	Sasnovy	576	D	200	Mühlacker(SWR)
549	D	100	Thurnau	576	E	100	Barcelona (RNE5)
549	D	100	Nordkirchen	576	IRN	750	Abadan
549	GAB	20	Oyem	576	ISR	200	Tel Aviv
549	IRL	4	"C,Monaghan"	576	NIG	50	Moniya
549	IRN	400	Sirjan	576	OMA	100	Haima
549	NIG	50	Tukun Tawa	576	POR	10	Braga (RRE)
549	RUS	1200	Krasnyy Bor	576	RUS	50	Astrakhan + 1 st
549	RUS	1000/500	Kurovskaya	576	RUS	25	Nalchik
549	RUS	50	Kalingrad	576	SDN	100	Khartoum
549	RUS	50	Novocherkassk	576	SDN	7	Omdurman
549	RUS	150	Syktyvkar	576	UGA	100	Mawagga
549	SVN	20	Beli Kriz	582	IRN	100	Tehran
549	UKR	500	Mykolayiv-Luch	585	ARS	1200	Riyadh
549	UKR	150	Kyyiv + 7 st's	585	E	600	Madrid (RNE1)
549	YUG	100	Pristina	585	F	10	Paris
549	ZMB	100	Kitwe	585	G	2	Dumfries
558	AFS	100	Umtata	585	IRQ	-	Unk. location
558	ARS	1	Jeddah	585	LBN	100	Ehden
558	CYP	10	Paphos	585	NIG	50	Abakaliki + 1 st
558	CZE	1	Broumov + 2 st's	585	RUS	150	Perm
558	E	50	Valencia (RNE5)	585	TUN	350	Gafsa
558	E	25	La Coruna (RNE5)	585	TZA	10	Chumbum
558	E	10	San Sebastian (RNE5)	594	ARS	2000	Duba
558	EGY	40	Cairo	594	BUL	250	Pleven
558	FIN	50	Espoo	594	BUL	5	Dobrich
558	G	0.8	London	594	BUL	10	Burgas
558	IRN	1000	Gheslagh	594	D	300	Frankfurt (HR)
558	IRQ	300	Rutba	594	D	200	Hoher Meissner (HR)
558	KEN	20	Kisumu	594	HRV	10	Osijek
558	LBR	10	Gbarnga	594	IRN	400	Shiraz
558	MWI	10	Karonga	594	MRC	100	Oujda
558	POR	10	Faro	594	MWI	25	Lilongwe
558	ROU	400	Tirgu Jiu	594	NIG	100	Jaji
558	SUI	300	Monte Ceneri-Cima	594	NMB	100	Tsumeb
558	SVN	20	Maribor	594	POR	100	Muge (RRE)
558	TUR	600	Denizli	594	RUS	40	Izhevsk
567	AFS	25	Cape Town	594	RUS	25	Vladikavkaz
567	ARS	15	Afif	594	SVN	1	Cerkno
567	ARS	5	Abha + 1 st	594	TUR	600	Malatya
567	D	50	Berlin (B2)	594	YUG	10	Niksic
567	E	50	Murcia (RNE5)	603	AFS	100	Umzimkulu
567	E	10	Socuéllamos (RNE5)	603	CYP	100	Nicosia
567	E	5	Marbella (RNE5)	603	E	50	Sevilla (RNE5)
567	I	20	Bologna + 4 st's	603	E	5	Palencia (RNE5)
567	IRL	500	Tullamore	603	EGY	100	Barnis
567	KEN	50	Garissa	603	F	300	Lyon-Tramoyes
567	MDN	10	Strumica	603	G	0.1	Littlebourne
567	NIG	50	Alaho	603	G	0.1	Cheltenham
567	ROU	50	Bod/Satu Mare	603	G	2	Newcastle
567	RUS	1000	Volgograd	603	GUI	25	Kankan
567	SVK	10	R. Sobota	603	IRN	100	Zahedan
567	SVK	7	Zilina	603	IRQ	300	Nineva
567	SYR	1000	Damas-Adra	603	MDR	10	Areeiro
567	YUG	10	Pljevlja	603	NIG	30	Ibese
567	ZMB	10	Kasama	603	ROU	50	Bucuresti/Botosani
576	AFS	50	Meyerton	603	ROU	14	Oradea + 1 st
576	ALG	400	Bechar	603	TUN	10	Monastir
576	BUL	500	Vidin	603	TZA	100/10	Dodoma
576	CNR	25	Las Palmas	603	ZMB	10	Mongu

kHz	Country	kW	Station	kHz	Country	kW	Station
612	ARS	15	Khamasin	639	MKD	1	Shtip
612	ARS	5	Hail	639	NIG	25	Kaduna
612	AZE	50	Baki	639	NIG	5	Warri
612	BHR	100	Manama	639	SDN	200	Juba+ 1 st
612	BIH	600	Sarajevo	639	UGA	50	Kampala
612	D	10	Kiel	639	YUG	1	Soko Banja
612	E	10	Lleida (RNE1)	648	ALB	50	Kukes
612	E	10	Vitoria (RNE1)	648	ARS	2000	Jeddah
612	IRL	100	Athlone	648	AZR	1	Lajes
612	JOR	200	Amman	648	BOT	50	Mopipi
612	KEN	100	Nairobi	648	BUL	30	Plovdiv
612	LTU	40	Vilnius	648	E	10	Badajoz (RNE1)
612	LTU	20	Klaipéda	648	G	500	Orfordness
612	MRC	300	Sebaa-Aioun	648	GMB	50	Bonto
612	NIG	50	Ilorin	648	LBY	300	Tobruk
612	RUS	150	Petrozavodsk	648	SVN	10	Murska Sobota
612	RUS	40	Kurkino	648	TZA	100/10	Nachingwea
612	UKR	5	Kyyiv (BBC)	648	UKR	150	Oktyabrskoye
612	UKR	10	Kharkiv	648	UKR	150	Kharkiv
621	ALB	25	Korce	648	YUG	1	Lazarevac
621	BEL	300	Wavre-Overijse	657	AFS	50	Bloemendal
621	BOT	100	Selebi-Phikwe	657	ARS	20	Rafha
621	CNR	600	"Santa Cruz, Tenerife"	657	E	50	Madrid (RNE5)
621	E	10	Jaén (RNE1)	657	G	2	Wrexham
621	E	10	Avila (RNE1)	657	G	0.5	Bodmin
621	E	50	Palma de Mallorca (RNE1)	657	I	120	Napoli
621	EGY	1000	Batrah	657	I	100	Firenze
621	IRN	20	Birjand	657	I	50	Torino + 2 st's
621	NIG	20	Akwa + 1 st	657	ISR	200	Tel Aviv
621	RUS	50	Syktyvkar	657	NIG	100	Ibadan
621	RUS	50	Makhachkala	657	RUS	150	Murmansk
621	RUS	5	Kochubey	657	RUS	50	Groznyy
621	SVK	7	Orava	657	TZA	2*50	Dar-es-Salaam
621	TZA	50/10	Mbeya	657	UAE	2*50	Sadiyat
630	ARS	20	Gizan	657	UKR	25	Chernivtsi
630	ARS	10	Najran	666	ALG	5	Tindouf
630	G	2	Redruth	666	D	150	Messkirch Rohrdorf
630	G	0.2	Luton	666	E	50	Barcelona (SER)
630	IRQ	20	Kirkuk	666	G	0.34	Exeter
630	KWT	10	Kuwait city	666	G	0.5	Fulford
630	MDG	150/75	Antananarivo	666	GRC	15	Athens
630	NIG	50	Maiduguri	666	LTU	500	Sitkunai
630	NIG	50	Pt. Harcourt	666	MAU	1	Rodrigues
630	NOR	100	Vigra	666	NIG	50	Benin
630	POR	50	Montemor + 1 st	666	POR	10	Lisboa + 5 st's
630	ROU	400	Timisoara + 1 st	666	REU	20	St. Pierre
630	RUS	42	Saratov	666	RUS	25	Sochi
630	TUN	600	Tunis-Djedeida	666	SDN	5	Kassala
630	TUR	300	Tarsus	666	SYR	600	Damas-Sabboura
630	ZMB	500	Lusaka	666	YUG	10	Sombor
639	CYP	500	Ladies' Mile	666	ZMB	10	Chipata
639	CZE	1500	Praha-Liblice + 1 st	675	ARS	20	Afif
639	E	300	La Coruna (RNE1)	675	ARS	5	Abha
639	E	20	Almeria (RNE1)	675	HOL	120	Lopik
639	E	50	Bilbao (RNE1)	675	ISR	100	Ramallah city
639	E	50	Zaragoza (RNE1)	675	KEN	50	Marsabit
639	E	10	Albacete (RNE1)	675	LBY	100	Benghazi
639	IRN	400	Bonab	675	MWI	50	Mzuzu
639	KEN	50	Garissa	675	NIG	25	Obomasho
639	MDN	1	Stip	675	NOR	10	Bodö

kHz	Country	kW	Station	kHz	Country	kW	Station
675	QAT	100	Al- Khaisah + 1 st	711	LBY	50	Sebha
675	RUS	5	Ukhta	711	MRC	600	Laayoune
675	UKR	25	Uzhgorod + 1 st	711	ROU	50	Sighet
675	YUG	10/5	Bosilegrad	711	RUS	7	Naryan-Mar
675	YUG	10/5	Dimitrovgrad	711	TZA	100/10	Kigoma
680	LBN	5	Marjayoun	711	UKR	50	Dokuchayevsk
684	ARS	50	Jeddah	711	YEM	200	Mocha
684	ARS	10	Riyadh	711	YUG	100/20	Nis
684	E	600	Sevilla (RNE1)	720	CNR	20	"Santa Cruz, Tenerife"
684	ETH	100	Metu	720	CYP	500	Ladies' Mile
684	IRN	100	Mashhad	720	D	200	Langenberg (WDR)
684	IRQ	-	Tanaf	720	G	10	Lisnagarvey
684	LBN	-	Baalbekk	720	G	0.75	London
684	MAU	10	Curepipe	720	G	0.75	Londonderry
684	NIG	20	Damaturu	720	IRN	400	Tayebad
684	RUS	10	Olgino	720	MKD	1	Struga/Veles
684	YUG	2000	Beograd	720	NIG	50	Owerri
693	ALB	50	Shoder	720	OMA	750	Masirah Island (BBC)
693	ALG	5	Ain-el-Hamam + 1 st	720	POR	100	Porto + 8 st's
693	AZR	10	Santa Barbara + 4 st's	720	ROU	14/7	Sighèt + 2 st's
693	CYP	10	Nicosia	720	TCH	0.1	Sarh
693	D	5	Berlin (VoR)	720	TUN	100	Sfax-Sidi Mansour
693	E	10	Toledo (RNE1)	720	TZA	50/10	Mwanza
693	E	10	Santa Barbara (RNE1)	722	NIG	20	Bali
693	G	150	Droitwich	729	AFS	20	Cape Town
693	G	25/10/1	10 st's R5	729	E	100	Oviedo (RNE1)
693	GEO	-	Batumi	729	E	20	Logrono (RNE1)
693	I	100	Milano + 1 st	729	E	20	Malaga (RNE1)
693	IRQ	-	Basrah	729	E	10	Alicante (RNE1)
693	MDG	2*75	Antananarivo	729	E	10	Cuenca (RNE1)
693	NIG	10	Ayanga	729	E	20	Valladolid (RNE1)
693	RUS	30	Ufa	729	G	0.2	Manningtree
693	RUS	10	Moskva (DW)	729	GRC	150	Athens
693	RUS	0.2	Astrakhan	729	IRL	10	Cork
693	YUG	10/5	Negotin	729	NIG	25	Jogana
695	NIG	10	Suleja	729	REU	20	St. Andre
702	AFS	100	Ga-Rankuwa	729	UAE	2*750	Sadivat
702	AGL	10	Mulenvos	729	UGA	100	Butebo
702	ARM	150	Yerevan	729	ZMB	10	Livingstone
702	ARS	40	Duba	738	ALG	5	Ain-Amenas
702	BUL	40	Pirin	738	E	600	Barcelona (RNE1)
702	D	7.5	Flensburg (NDR)	738	F	5	Paris (RFI)
702	EGY	10	El Kharga + 1 st	738	G	0.04	Worcester
702	KEN	100	Meru	738	IRN	10	Tabas
702	MCO	40	Monte Carlo	738	ISR	1200	Tel-Aviv
702	MRC	140	Sebaa-Aioun	738	MDN	1	Tetovo
702	NIG	25	Wukari	738	MKD	1	Tetovo
702	NOR	20	Vadsö	738	MOZ	50	Maputo
702	SVK	200	Presov	738	RUS	5	Moskva
702	SVK	200	Banska Bystrica	738	RUS	-	Izhevsk
702	TUR	150	Catalca	738	YUG	10	Krusevac
702	YUG	10	Sabac	738	YUG	5	Bar
711	BUL	1	Silistra	747	ARS	1	Najran
711	D	5	Heidelberg + 3 st(SWR)	747	BFA	100	Ougadougou
711	E	25	Murcia (COPE)	747	BUL	300	Petrich + 2 st's
711	EGY	100	Tanta	747	CNR	20	Las Palmas
711	F	300	Rennes	747	E	10	Cádiz (RNE5) F.Pl. 936
711	INT.W	-	Arutz Sheva	747	GMB	10	Basse
711	LBY	50	Ghadames	747	HOL	400	Flevoland
711	LBY	50	Jefren	747	IRN	400	Bandar-e-Torkaman

kHz	Country	kW	Station	kHz	Country	kW	Station
747	IRN	150	Sirjan	783	MOZ	5	Lichinga(782)
747	KEN	100	Nairobi	783	NIG	50	Ochaja
747	SYR	100	Sarakeb	783	POR	10	Porto + 1 st
756	BIH	10	Mostar	783	SDN	5	Atbara
756	D	200	Braunschweig (DLF)	783	SYR	600	Tartus
756	D	100	Ravensburg (DLF)	783	UKR	150	Kyyiv
756	E	25	Bilbao (EI)	783	YUG	0.2	Stara Pazova
756	EGY	10	Qena	792	ARS	50	Jeddah
756	G	2	Redruth	792	CZE	40	Praha
756	G	0.63	Powys	792	D	5	Lingen (NDR)
756	G	1	Carlisle	792	E	50	Sevilla (SER)
756	IRQ	-	Baghdad	792	F	300	Limoges
756	LBN	-	Marjayoun	792	G	0.2	Bedford
756	MWI	25	Blantyre	792	GRC	500	Kavala
756	NIG	100	Ibadan	792	LBY	20	Sirte
756	NIG	50	Minna	792	NIG	10	Aha
756	NIG	25	Damagun	792	RUS	50	Astrakhan
756	POR	1	Lamego	792	RUS	5	Moskva
756	ROU	400	Lugoj	792	SVK	5	Bratislava
756	RUS	10	Ufa	792	YUG	1	Arandjelovac
756	YEM	-	Mukalla	801	AZE	150	Pirsagat + 1 st's
765	ARS	20	Khamasin	801	BHR	100	Manama
765	ARS	10	Qurayyat	801	D	100	München-Ismaning (BR)
765	G	0.5	Chelmsford	801	D	50	Dillberg (BR)
765	GRC	10	Ioannina	801	E	20	Ciudad Real (RNE1)
765	IRN	600	Chahbahar	801	E	20	Lugo (RNE1)
765	MOZ	5	Nampula	801	E	10	3 st's (RNE1)
765	RUS	150	Petrozavodsk	801	G	2	Barnstaple
765	SDN	50	Omdurman	801	JOR	2000	Ajlun
765	SEN	300/200	Dakar	801	NIG	20	Damaturu
765	SUI	600	Sottens	801	NIG	1	Zuru
765	TUR	600	Gaziantep	801	RUS	600	Krasnyy Bor
765	UKR	50	Petrivka	810	E	50	Madrid (SER)
774	AGL	5	Luanda	810	G	100	Burghead
774	BIH	2	Bihac	810	G	100	Westerglen
774	BUL	50	Sofia	810	G	5	Redmoss
774	BUL	5	Varna	810	GEO	200	Tbilisi
774	D	5	Bonn	810	HNG	15	Siófok
774	E	100	Valencia (RNE1)	810	MDN	-	Skopje
774	E	50	Cáceres (RNE1)	810	MDN	-	Skopje
774	E	25	San Sebastián (RNE1)	810	MKD	1000	Skopje
774	E	25/10	5 st's (RNE1)	810	MWI	10	Bangula
774	EGY	500	Abis	810	RUS	500	Volgograd
774	G	1	Enniskillen + 1 st	810	RUS	-	Moskva
774	G	0.7	Littlebourne	810	SEN	1	Podor
774	G	0.5	Farnley	810	UAE	50	Maqtaa
774	G	0.14	Gloucester	810	UGA	100	Bobi
774	HRV	50	Hvar	810	UKR	25	Tokmak + 2 st's
774	IRN	100	Arak	819	E	10	San Sebastian(EI)
774	MOZ	5	Xai-Xai	819	EGY	1000	Batra
774	MRC	50	Agadir	819	F	20	Toulouse
774	NIG	10	Bali	819	I	20	Trieste
774	NIG	10	Wukari	819	IRN	20	Sari
774	RUS	30	Somovo	819	MAU	10	Forest Side
783	ALG	5	Djanet	819	MRC	25	Rabat
783	ARS	100	Dammam	819	SDN	5	Dongola
783	D	100	Leipzig(MDR)	819	UKR	8	Novodnistrovsk
783	E	50	Barcelona (COPE)	819	ZMB	200	Lusaka (818)
783	HRV	10	Buje	828	ARS	20	Medinah
783	IRN	10	Saravan	828	AZR	1	Mt.das Cruzes + 1 st

kHz	Country	kW	Station	kHz	Country	kW	Station
828	BUL	500	Shumen	855	ETH	100	Harar
828	BUL	50	Sofia	855	G	2	Postwick
828	D	50/10	Hannover (NDR)	855	G	1	Plymouth
828	D	10	Freiburg (SWR)	855	G	1.5	Preston
828	G	0.27	Bournemouth	855	G	0.15	Ludlow
828	G	0.12	Leeds	855	JOR	10	Amman
828	G	0.2	Dunstable	855	NIG	10	Port Harcourt(854)
828	G	0.8	Cookstown	855	ROU	1500	Bucuresti
828	G	0.2	Birmingham	855	RUS	50	Kamenka
828	HOL	20	Heijenoord	855	ZMB	10	Chipata
828	IRL	1	Ballydavid	864	ALB	30	Sarande
828	LBY	300	Sebha	864	ARM	1000	Kamo(F.S)
828	MRC	100	Oujda	864	ARS	100	Dammam
828	NIG	25	Enugu	864	BIH	1	Sokolac
828	POR	1	Coimbra + 4 st's	864	BUL	150	Blagoevgrad
828	RUS	10	Sankt Peterburg	864	BUL	10	Samuil
828	RUS	50	Nizhiny Novgorod	864	CZE	7	Strakonice
828	RUS	-	Unknown E.	864	E	5	Socuellamos (RNE1)
828	SYR	100	Deir-Ez-Zoor	864	EGY	500	Santah
828	UAE	1	Al Ain	864	F	300	Paris
828	ZMB	10	Mongu	864	MRC	15	Errachidia
837	ALG	5	Ben-Abbas	864	SVK	1	Stakein
837	AZR	10	Pico da Barossa + 1 st	873	ALG	5	Ghardaia
837	CNR	10	Las Palmas	873	ARS	10	Qassim
837	E	50	Sevilla (COPE)	873	BLR	150	Minsk
837	E	10	Burgos (COPE)	873	BOT	-	"Gantsi(F.Pl,)"
837	E	2	El Ferrol (COPE)	873	BUL	25	Stara Zagora
837	E	2	Ibiza	873	D	150	Frankfurt (AFN)
837	ERI	10	Asmara	873	E	10	Sant. de Compostela (SER)
837	F	200	Nancy	873	ETH	100	Addis Ababa
837	G	1.5	Barrow in Furness	873	G	0.3	West Lynn
837	G	0.45	Freemen´s Common	873	G	1	Enniskillen
837	G	0.4	Leicester	873	HNG	20	Lakihegy
837	IRN	200	Esfahan	873	HNG	20	Pecs
837	LBN	100	Amchit	873	LBN	80	Beirut
837	MOZ	5	Chimoio	873	MOZ	50	Beira (872)
837	TZA	1	Dar-es-Salaam	873	RUS	2000	Novosemeykino
837	UKR	150	Kharkiv + 1 st	873	RUS	1200	Elektrostal
837	YEM	30	Sana´a	873	RUS	150	Olgino
837	YUG	50	Sombor	873	RUS	50	Kaliningrad
840	TCD	20	N`djamena	873	RUS	1	2 st's
846	AFS	100	Komga	873	SDN	5	Medani
846	CZE	7	Ceske Budejovice	873	SYR	10	Kharabo (877)
846	CZE	2	Ostrava	873	UKR	50	Kyyiv + 4 st's
846	I	1200	Roma	873	UKR	10	Dnipropetrovsk
846	IRQ	.	Nasiriya	882	ARS	100	Dammam
846	ISR	5	Safad	882	CNR	20	La Laguna
846	KEN	100	Kisumu	882	D	20	Wachenbrunn (MDR)
846	NIG	10	Azare	882	E	20	Barcelona
846	NIG	10	Katsina Ala	882	E	5	Málaga (COPE)
846	RUS	150	Noginsk	882	E	5	Valladolid (COPE)
846	RUS	42	Elista	882	E	2	Gijón (COPE)
846	RUS	-	Perm	882	E	2	Alicante (COPE)
846	UAE	20	Umm al Qiwain	882	EGY	10	Matruh
850	TCD	1	Sarh	882	G	100/1	Washford + 3 st's
855	D	5	Berlin(irr.)	882	IRN	10	Mahabad
855	E	300	Murcia (RNE1)	882	ISR	1	Beit Hillel
855	E	50	Santander (RNE1)	882	KEN	50	Kitale
855	E	20	Pontevedra (RNE1)	882	NIG	25	Kaduna
855	E	10/5	7 st's (RNE1)	882	RUS	30	Stavropol

kHz	Country	kW	Station	kHz	Country	kW	Station
882	TUN	1	Remada	927	ISR	50	Haifa
882	YUG	600/300	Podgorica + 5 st's	927	ISR	1	Eilat
882	ZMB	10	Kasama	927	KEN	100	Malindi
891	ALG	600/300	Alger	927	POR	1	Evora(RRE)
891	AZE	30	Baki	927	SVK	50	Nitra
891	HOL	5	Hulsberg	927	SVK	10	Kosice
891	IRN	20	Yasuj	927	TUR	200	Izmir
891	LSO	50	Maseru	927	ZMB	10	Livingstone
891	POR	10	Vilamoura (RRE)	930	LBN	-	Saida
891	RUS	50	Tyumen	936	ARS	5	Riyadh
891	TUR	600	Antalya	936	D	100	Bremen (RB)
891	UKR	150	Uzhgorod	936	E	20	Valladolid (RNE5)
900	ARS	1000	Qurayyat	936	E	20	Zaragoza (RNE5)
900	CZE	25	Brno	936	E	10	Alicante (RNE5)
900	E	25	Bilbao (COPE)	936	EGY	10	Salum
900	E	5	Vigo (COPE)	936	G	1	Hawes
900	E	5	Granada (COPE)	936	G	0.18	Naish Hill
900	E	5	Cacéres (COPE)	936	I	20	Venezia
900	I	600	Milano	936	I	10	Trapani/Genova
900	IRN	50	Tehran	936	IRN	50	Urumiyeh + 1 st
900	KEN	100	Meru	936	MOZ	50	Quelimane
900	NIG	25	Abeoukuta	936	MRC	600	Agadir
900	RUS	25	Sovietskiy	936	NIG	10	Ikorodu
900	SVK	7	Poprad (Tatry)	936	RUS	5	Matveyevka
900	YUG	2	Beograd(Studio B)	936	UKR	1000	Lviv
902	LBN	-	Aley	936	UKR	3	Starobilsk
909	ALB	50	Gjirokaster	936	YUG	10	Djakovica + 1 st
909	ALG	5	Tamanrasset	945	AGL	25/1	Cazenga (944)
909	AZR	10	Angra do Heroismo	945	ARS	5	Hail
909	BOT	500	Selebi-Phikwe (VOA)	945	ERI	50	Asmara
909	E	10	Palma de Mallorca (RNE5)	945	ERI	50	Asmara
909	G	200	Moorside Edge + 10 st's R5	945	F	300	Toulouse
909	GMB	2.5	Banjul	945	G	0.7	Bexhill
909	IRQ		Baghdad	945	G	0.2	Derby
909	LBY	20	Giaghboub	945	GRC	5	Larissa
909	LBY	10	Kufra	945	IRN	20	Sanandaj
909	MRC	5	Safi C	945	LBN	50	Marjayoun
909	NIG	50	Fwagwa Lada	945	LVA	150	Riga-Ulbroka
909	ROU	200	Cluj + 1 st	945	MKD	2	Kumanovo
909	RUS	-	"Unk,So.E."	945	NIG	10	Birnin Kebbi
909	UGA	20	Kampala	945	ROU	14	Miercurea Ciuc
909	ZMB	10	Solwezi	945	RUS	150	Rostov-na-Donu
918	CYP	10	Paphos	945	YUG	1	Smederevo
918	E	50	Madrid (IND)	954	CZE	200	Brno-Prostejov
918	EGY	10	Bawiti	954	CZE	20	Karlovy Vary
918	EGY	10	Hurghada	954	CZE	40	Plzen
918	IRN	20	Jirof	954	CZE	30	C. Budejovice
918	MOZ	50	Maputo	954	CZE	3	Liberec
918	NIG	50	Makurdi	954	E	50	Madrid (IND)
918	NIG	50	Ikeja	954	G	0.32	Torquay
918	RUS	150	Arkhangelsk	954	G	0.16	Hereford
918	RUS	50	Makhachkala	954	GRC	10	Heraklion
918	RUS	75	Balashikha	954	ISR	100	Jerusalem
918	SVN	600	Domzale	954	KEN	100	Kisumu
918	SYR	200	Al-Hassake	954	NIG	10	Enugu
918	ZMB	10	Mansa	954	QAT	1500	Al-Arish
927	ALG	5	Timimoun	954	SWZ	50/30	Sidvokodvo
927	BEL	100	Wolvertem	954	SYR	60	Deir El Zawr
927	GRC	50	Zakynthos	954	TUR	300	Trabzon
927	IRN	10	Dorud	954	YUG	10	Gnjilane

kHz	Country	kW	Station	kHz	Country	kW	Station
963	BEN	16/10	Parakou	990	G	0.1	Wolverhampton
963	BUL	25	Shumen	990	GAB	20	Epila
963	BUL	40	Sofia	990	I	10	Potenza
963	BUL	20	Pirin	990	IRN	400	Shiraz (v.f.)
963	BUL	5	Malko Tarnovo	990	LBN	10	Amchit
963	CYP	100	Nicosia	990	LBR	5	Enoanyi
963	E	10	Vitoria (EI)	990	NIG	50	Bauchi
963	FIN	600	Pori	990	NIG	10	Ikeja
963	G	1	London	990	RUS	-	Moskva
963	G	1	London	990	TZA	100/10	Songea
963	G	1	Southall	990	YUG	1	Pozarevac
963	G	0.8	Blackburn	999	ARS	20	Tabuk
963	IRL	10	Derrybeg	999	E	50	Madrid (COPE)
963	IRN	20	Birjand	999	G	1	Fareham
963	IRQ	-	Unk. location	999	G	0.8	Preston
963	KWT	20	Kuwait city	999	G	0.25	Nottingham
963	LBN	100	Zuk Musbih	999	I	20	Torino + 3 st's
963	POR	10	Seixal (RRE)	999	MDA	500	Maiac
963	RUS	15	Moskva	999	MLT	5	Bizbizja
963	SDN	100	Soba	999	MRC	10	Ad-Dakhla
963	SEN	1	Matam	999	QAT	50	Al -Khaisah
963	SOM	150	Mogadishu (962)	999	UGA	100	Kabale
963	TUN	200	Tunis-Djedeida	999	YUG	10/5	Kladovo + 1 st
972	BIH	2	Vukovar	1000	LBN	-	-
972	BOT	50	Sebele	1008	AGL	1	Huambo (1010)
972	D	100	Hamburg (NDR)	1008	BFA	10	B.Dioulasso
972	E	4	Cabra (RNE1)	1008	BLR	50	Slonim + 4 st's
972	E	2	Monf. de Lemos (RNE1)	1008	CNR	10	Las Palmas
972	G	1	London	1008	E	5	Badajoz (SER)
972	HRV	1	Knin	1008	E	5	Girona (SER)
972	IRN	10	Ilam	1008	E	5	Alicante (SER)
972	MRC	5	Melilla	1008	EGY	100	El Arish + 1 st
972	NIG	50	Katsina	1008	GRC	50	Kerkyra
972	NIG	10	Otite	1008	HOL	400	Flevoland
972	UAE	2*50	Fujairah	1008	IRN	20	Semnan
972	UKR	500	Mykolayiv-Luch	1008	MOZ	50	Maputo
981	ALG	600/300	Alger	1008	NIG	10	Kontagora
981	ARS	20	Madinah	1008	NIG	10	Ibadan
981	BUL	150	Varna	1008	NIG	10	Osogbo
981	CZE	7	Jihlava	1008	RUS	25	Tuapse
981	EGY	10	Assiut + 2 st's	1008	YEM	600	Sana'a
981	GRC	200	Megara	1008	YUG	200	Beograd
981	I	10	Trieste	1008	YUG	200	Aleksinac
981	KEN	100	Voi	1014	LBN	-	Tripoli
981	NIG	10	Port Harcourt	1017	BIH	1	Cazin(R. Mir)
981	NIG	10	New Bussa	1017	BUL	50	Kardzhali
981	POR	10	Coimbra + 3 st's	1017	BUL	40	Vidin
981	YUG	10	Bor	1017	D	200	Rheinsender (SWF)
981	YUG	10	Cacak	1017	E	10	Granada (RNE5)
990	ALB	15	Kukes (996)	1017	E	10	Burgos (RNE5)
990	BIH	1	Foca	1017	G	2	Shrewsbury
990	D	100	Berlin (DRB)	1017	G	0.6	Shropshire
990	E	10	Bilbao (SER)	1017	KEN	20	Nyeri
990	E	5	Cádiz (SER)	1017	NIG	10	Yola
990	ETH	1	Addis Ababa	1017	RUS	500	Lesnoy
990	G	1	Exeter	1017	SVK	50	R. Sobota
990	G	1	Tywyn	1017	SVK	5	Bratislava
990	G	1	Redmoss	1017	TUR	1200	Mundanya
990	G	1	Aberdeen	1026	ALG	5	Hassi Messaoud
990	G	0.25	Doncaster	1026	BLR	50	Mahiliou + 5 st's

kHz	Country	kW	Station	kHz	Country	kW	Station
1026	E	5	Oviedo (SER)	1062	IRN	20	Kerman
1026	E	2	Salamanca (SER)	1062	NIG	10	Onitsha
1026	E	2	Jaén (SER)	1062	RUS	25	Saransk
1026	E	2	Jerez (SER)	1062	TUR	300	Diyarbakir
1026	E	10	Reus (SER)	1062	YUG	10	Jagodina
1026	E	5	Vigo (SER)	1062	YUG	1	Novi Pazar
1026	G	1	Jersey	1071	BIH	40	Mostar
1026	G	1.7	Belfast	1071	BOT	25	Jwaneng
1026	G	0.5	Cambridge	1071	CZE	25	Moravske Budejovice
1026	IRN	100	Tabriz	1071	CZE	10	Hradec Králové
1026	ISR	50	Tel Aviv	1071	E	50	Bilbao (EI)
1026	MDN	5	Skopje	1071	EGY	100	Cairo
1026	MOZ	5	Chimoio	1071	G	1	3 st's
1026	MRC	1	Rabat	1071	GNB	100	Bissau
1026	NIG	25	Dutse	1071	NIG	-	Nsukka
1026	RUS	7	Nyandoma + 4 st's	1071	RUS	10	Moskva
1026	YUG	10	Kragujevac	1071	SVK	40	Presov
1026	YUG	1	Bar	1071	SYR	60	Tartus
1035	AFS	100	Tshisahulu	1071	UKR	50	Dnipropetrovsk+ 1 st
1035	G	1	Sheffield	1071	ZMB	100	Kitwe
1035	G	1	London	1080	E	10	Granada (SER)
1035	G	0.32	Ayr	1080	E	5	Palma de M. (SER)
1035	G	0.78	Aberdeen	1080	E	5	La Coruna (SER)
1035	GNB	5	Bissau (1034)	1080	E	5	Huesca (SER)
1035	HOL	2	Echt(F.Pl)	1080	E	5	Toledo (OCR)
1035	I	50	Milano + 7 st's	1080	EGY	10	El Minya + 1 st
1035	IRN	20	Yazd	1080	GRC	10	Orestiada
1035	OMA	100	Salalah	1080	IRN	750	Abadan
1035	POR	100	Lisboa (inactive)	1080	ISR	5	Jerusalem
1035	SVK	7	Banská Bystrica	1080	LBN	50	Saida
1035	TZA	10	Dar es Salaam	1080	LBY	40	Ajedabia + 1 st
1035	YUG	10	Mitrovica	1080	MRC	1	Casablanca
1044	CYP	10	Limassol	1080	NIG	50	Minna
1044	D	20	Dresden(MDR)	1080	RUS	30	Murmansk
1044	E	10	San Sebastian (SER)	1080	YEM	30	Taiz
1044	E	5	Valladolid (SER)	1089	AGL	25/100	Mulenvos (1088)
1044	GEO	200	Tbilisi	1089	ALB	150	Durres
1044	GRC	150	Thessaloniki	1089	ALG	5	Adrar
1044	IRQ	-	Unk. location	1089	COM	40	Haboho
1044	KEN	100	Malindi	1089	G	100	Brookmans Park + 7 st's
1044	MRC	300	Sebaa-Aioun	1089	NIG	50	Yola
1044	RUS	7	Kotlas	1089	NIG	20	Sogunle
1044	RUS	20	Kurkino	1089	RUS	1200	Tbilisskoye
1044	YUG	1	Temerin	1089	RUS	20	Sankt Peterburg
1050	LBN	-	Beirut	1089	YUG	10	Subotica
1053	E	10	Zaragoza (COPE)	1098	AFS	100	Ga-Rankuwa
1053	E	5	Castellón (COPE)	1098	ALG	5	Ouargla
1053	G	100/1	Droitwich + 13 st's	1098	ARS	10	Qurayyat + 1 st
1053	IRN	20	Khorramabad	1098	CYP	100	Yeni Iskele
1053	LBY	50	Tripoli	1098	E	20	Lugo (RNE5)
1053	MRC	600	Tanger	1098	E	25	Almeria (RNE5)
1053	ROU	1000	Iasi	1098	E	10	Avila (RNE5)
1053	RUS	10	Sankt-Peterburg	1098	E	10	Huelva (RNE5)
1062	CZE	20/1	Praha	1098	IRN	200	Zabol
1062	DNK	250	Kalundborg	1098	RUS	7	Vologda + 4 st's
1062	I	25	Cagliari	1098	SVK	750	Nitra
1062	I	25	Squinzano	1107	AFG	-	Kabul
1062	I	10	Ancona	1107	AGL	1	Sumbe (1110)
1062	I	2	4 st's	1107	CME	5	Douala (1106)

kHz	Country	kW	Station	kHz	Country	kW	Station
1107	D	10	Grafenwöhr (AFN Bavaria)	1134	IRN	20	Bojnurd
1107	D	10	Kaiserslautern (AFN)	1134	KEN	50	Kitale
1107	E	20	Caceres (RNE5)	1134	KWT	10	Sulaibiyah
1107	E	25	Logrono (RNE5)	1134	LBN	-	Beirut
1107	E	20	Santander (RNE5)	1134	NIG	10	Ogoja
1107	E	10	Ponferrada (RNE5)	1134	RUS	75	Murmansk
1107	E	10	Teruel (RNE5)	1134	RUS	30	Saransk + 5 st's
1107	E	25	Toledo (RNE5)	1134	UKR	5	Luhansk
1107	EGY	600	Batrah	1143	BIH	0.25	Tuzla (AFN Bosnia)
1107	G	1.5	Inverness	1143	BUL	50	Varna
1107	G	2/1	Lydd + 4 st's	1143	D	10	Stuttgart (AFN)
1107	I	6	Roma	1143	D	1	Heidelberg (AFN)
1107	KEN	100	Maralal	1143	D	1	Mönchen-Gladbach (AFN)
1107	LTU	150	Sitkunai + 3 st's	1143	D	0.3	5 low power st's (AFN)
1107	MWI	1	Nkhotakota	1143	E	2	Réus (COPE)
1107	NIG	25	Jaji	1143	E	5	"Oviedo (COPE,inactive)"
1107	RUS	150	Samara	1143	E	2	Orense (COPE)
1107	RUS	50	Arkhangelsk	1143	E	5	Jaen (COPE)
1107	RUS	25	Nalchik + 1 st	1143	EGY	10	Sohag
1107	UAE	10	Dubai	1143	I	10	Sassari + 1 st
1107	YUG	150	Novi Sad	1143	INT.W	-	Arutz Sheva
1116	AGL	10	Sumbe (1115)	1143	IRN	20	Yasuj
1116	E	5	Pontevedra (SER)	1143	ISR	-	Arutz Sheva
1116	E	5	Albacete (SER)	1143	NIG	10	Bida
1116	E	5	Tarrasa (SER)	1143	RUS	150	Bolshakovo
1116	G	1.2	Derby	1143	RUS	150	Mekhzavod
1116	G	0.5	Guernsey	1148	AGL	10	Menongue
1116	G	0.5	Welsh valleys	1152	AGL	10	M'Banza Kongo
1116	HOL	0.5	Bloemendaal	1152	CME	300/1	Bamenda
1116	I	150	Bari	1152	E	10	Lerida (RNE5)
1116	I	60	Bologna	1152	E	10	Zamora (RNE5)
1116	I	25	Pisa	1152	E	10	Cartagena (RNE5)
1116	I	20	Cuneo	1152	E	10	Albacete (RNE5)
1116	I	10	Palermo + 1 st	1152	E	20	Malaga (RNE5)
1116	IRN	125	Bandar Lengeh	1152	G	23.5	London
1116	IRQ	-	Rutba	1152	G	1.8	Newcastle
1116	RUS	30	Kaliningrad	1152	G	3.6	Glasgow
1116	RUS	30	Sochi	1152	G	3.0	Birmingham
1116	RUS	10	Moskva	1152	G	1.5	Manchester
1116	RUS	30	Perm	1152	G	0.8	Norwich
1125	BLR	150	Minsk + 1 st	1152	G	0.3	Plymouth
1125	E	10	Castellon (RNE5)	1152	HRV	2	Dubrovnik
1125	E	10	Soria (RNE5)	1152	IRN	100	Tabriz
1125	E	10	Vitoria (RNE5)	1152	KEN	50	Wajir
1125	E	10	Toledo (RNE5)	1152	ROU	950	Cluj
1125	E	10	Badajoz (RNE5)	1152	RUS	150	Kupavna
1125	G	1	Llandrindod Wells	1152	UAE	50	Ras al Khaimah
1125	HRV	100	Deanovac	1160	AGL	10/1	Huambo
1125	LBY	500	El Beida	1161	ALG	5	Ain-Salah
1125	MDR	1	Ponta do Pargo	1161	BUL	300	Stara Zagora + 3 st's
1125	NGR	20	Niamey	1161	E	50	San Sebastian (EI)
1125	RUS	150	Olgino	1161	EGY	100	Tanta
1125	SYR	200	Al-Hassake	1161	G	1	Bexhill
1125	YEM	-	Hudaydah	1161	G	1.4	Dundee
1125	ZMB	10	Solwezi	1161	G	0.16	Swindon
1125	ZMB	1	Kabwe(1124)	1161	G	0.35	Hull
1134	E	5	Puertollano (COPE)	1161	G	0.1	Bedford
1134	E	2	5 st's (COPE)	1161	IRN	10	Abadan(1160)
1134	HRV	600	Zadar	1161	MOZ	5	Tete (1160)

kHz	Country	kW	Station	kHz	Country	kW	Station
1161	RUS	75	Volgograd	1206	F	300	Bordeaux
1161	YUG	1	Ulcinj	1206	GAB	1	Oyem
1161	YUG	0.2	Kovin	1206	ISR	50	Haifa
1161	ZAI	50	Matadi (1160)	1206	MAU	10	Rodrigues
1161	ZMB	1	Kabwe(1157)	1206	MOZ	5	Inhambane
1170	BLR	1000	Sasnovy	1206	NIG	10	Yola
1170	BOT	-	Loc.unknown	1206	SRL	50/10?	Goderich
1170	DJI	20	Djibouti	1206	YUG	1	Majdanpek
1170	G	0.32	Stockton	1215	ALB	500	Fllake
1170	G	0.58	Swansea	1215	BOT	-	Loc.unknown
1170	G	0.28	Ipswich	1215	BOT	50	Mahalapye
1170	G	0.12	Portsmouth	1215	E	5	Santander (COPE)
1170	G	0.2	Stoke-on-Trent	1215	E	5	Leon (COPE)
1170	G	0.25	High Wycombe	1215	E	2	Córdoba (COPE)
1170	IRN	2	Damghan	1215	E	2	Lorca (COPE)
1170	NIG	25	Abeoukuta	1215	G	100/0.004	Moorside Edge + 13 st's
1170	POR	10	Vila Real + 3 st's	1215	NGR	0.1	Tahoua
1170	RUS	1200	Tbilisskoye	1215	RUS	1200	Bolshakovo (inactive)
1170	SVN	300	Beli Kriz	1215	TZA	50/10	Arusha
1170	SWZ	100	Mpangela Range	1215	UKR	7	Dnieperopetrovsk
1170	YUG	2	V. Banja	1215	YUG	1	Mladenovac
1170	YUG	0.2	B. Topola	1224	BUL	300	Vidin
1175	UAE	10	Ras al Khaimah	1224	E	10	San Sebastián
1179	CNR	20	Santa Cruz	1224	E	2	6 st's (COPE)
1179	E	50	Valencia (SER)	1224	G	0.5	Manningtree
1179	E	2	Logrono (SER)	1224	HOL	50	Ijsselmer
1179	E	2	Logrono	1224	IRQ	-	Nasiriya
1179	EGY	10	Quena	1224	ISR	20	Beersheba
1179	GRC	50	Thessaloniki	1224	MOZ	5	Pemba
1179	ROU	200	Bacau	1224	NIG	50	Jos
1179	ROU	7	Vascau	1224	RUS	-	Elista
1179	S	600	Sölvesborg	1224	SEN	20/1	Ziguinchor
1179	YUG	1	Kacanik	1224	YUG	1	B.Petrovac
1188	AGL	10/1	Malanje	1233	AGL	10	Lubango (1232)
1188	BEL	5	Kuurne	1233	BEL	2.5	Liège
1188	D	3	Reichenbach (MDR)	1233	CYP	600	Cape Greco
1188	EGY	10	Ras Gharib	1233	CZE	400	Melnik
1188	HNG	135/5	Szolnok	1233	CZE	30	Ostrava
1188	HNG	25	Szombathely	1233	CZE	25	Brno-Komarov
1188	I	6	San Remo	1233	CZE	20	Brno-Dobrochov
1188	IRN	100	Tehran	1233	CZE	7	Jihlava
1188	LBN	-	Saida	1233	G	0.2	Reading / Swindon
1188	MRC	1	Casablanca	1233	KEN	50	Marsabit
1188	RUS	10	Sankt-Peterburg	1233	MDG	0.25	Maintirano
1197	BIH	2	Bjeljina + 1 st	1233	MRC	200	Tanger
1197	BIH	1	Bosanski Nova	1233	QAT	100	Al-Khaisah
1197	BLR	50	Minsk + 3 st's	1233	RUS	25	Velsk
1197	D	300	München (VOA)	1233	RUS	150	Chkalovskaya
1197	E	10	Vitoria (EI)	1242	AGL	5	Saurimo (1241)
1197	EGY	25	Alexandria	1242	BIH	50/10	Pale
1197	G	1/0.2	Hoo + 9 st's	1242	CTI	1	Bondoukou (1240)
1197	HRV	2	Pula	1242	F	150	Marseille
1197	IRN	20	Chalus	1242	G	1	Dundee + 2 st's
1197	IRQ	-	Nineva	1242	G	0.5	Isle Of Wight
1197	LSO	100	Lancer's Gap (BBC)	1242	G	0.32	Maidstone
1197	MRC	20	Agadir	1242	IRN	10	Kerman
1197	ROU	14	Bod	1242	OMA	200	Muscat
1200	AGL	1	Benguela	1242	UKR	50	Uzhhorod + 7 st's
1200	TUR	-	Izmir	1242	UKR	50	Oktyabrskoye + 6 st's

kHz	Country	kW	Station	kHz	Country	kW	Station
1242	YUG	10	Kraljevo	1287	E	2	Burgos (SER)
1251	G	0.76	Bury St. Edmunds	1287	ISR	100	Tel Aviv
1251	HNG	135	Szeged	1287	MDG	1	Diego Suarez
1251	HOL	10	Hulsberg	1287	MKD	1	Kochani
1251	LBY	500	Tripoli	1287	POR	2	Portalegre
1251	POR	10	Porto	1287	RUS	50	Yazykovo + 3 st's
1251	POR	10	Viseu	1287	SEN	4	Kaolack
1251	POR	10	Chaves + 1 st	1287	SVK	400	Velké Kostolany (RFE)
1251	RUS	7	Izhevsk + 4 st's	1287	SVK	14	Poprad (RFE)
1251	RUS	1	Neftekumsk	1296	AZE	500	Pirsagat
1251	UAE	600	Dubai	1296	BUL	150	Kardzali
1259	AGL	10/1	N'dalatando	1296	BUL	30	Pleven
1260	ALB	50	Korce	1296	E	20	Valencia (COPE)
1260	AZR	1	Espalamaca	1296	G	500	Orfordness
1260	E	5	Murcia (SER)	1296	G	6.4	Birmingham
1260	E	5	Algeciras (SER)	1296	GUI	1	Conakry (1295)
1260	G	1.6	Bristol	1296	I	2	La Spezia + 1 st
1260	G	0.64	Wrexham	1296	NIG	20	Maiduguri
1260	G	0.29	Leicester	1296	SDN	1500	Reba
1260	G	0.5	Guildford	1296	YUG	10	Loznica
1260	G	0.5	Scarborough	1296	YUG	10	Vranje
1260	GRC	500	Rhodos	1305	ALB	15	Gjirokaster
1260	MDG	10	4 st's	1305	ALG	20	Constantine
1260	RUS	10	Moskva	1305	ARS	1	Taif
1260	RUS	10	Sankt-Peterburg	1305	BIH	1	B. Krupa
1260	UKR	50	Kharkiv	1305	E	25	Orense (RNE5)
1269	CNR	20	Las Palmas	1305	E	20	Ciudad Real (RNE5)
1269	D	300	Neumunster	1305	E	10	Bilbao (RNE5)
1269	E	10	Ciudad Real (COPE)	1305	E	10	León (RNE5)
1269	E	10	Figueras (COPE)	1305	EGY	10	Assiut
1269	E	5	Zamora (COPE)	1305	G	0.5	Enfield
1269	E	5	Badajoz (COPE)	1305	G	0.5	Ewell
1269	KEN	20	Nyeri (1268)	1305	G	0.2	Newport
1269	KWT	100	Sulaibiyah	1305	G	0.15	Barnsley
1269	NIG	10	Jalingo	1305	I	12	Genova+ 1 st
1269	YUG	750	Novi Sad	1305	IRN	20	Bushehr
1269	ZAI	2	Mbandaka(1268)	1305	ISR	20	Haifa
1278	AFG	100	Kabul	1305	ISR	5	Eilat
1278	BLR	10	Brest	1305	KEN	50	Wajir
1278	EGY	10	Asswan	1305	POL	60	Bialystok
1278	F	300	Strasbourg	1305	RUS	300	Kupavna
1278	G	0.4	Bradford	1305	SEN	20	Dakar
1278	GRC	10	Florina	1305	YUG	1	Herceg Novi
1278	IRL	10	Dublin	1314	AGL	1	Lubango (1313)
1278	IRL	10	Cork	1314	AGL	1	Namibe (1313)
1278	IRN	200	Kermanshah	1314	ARM	1000	Kamo
1278	MDG	1	Low power st's	1314	E	10	Cuenca (RNE5)
1278	MWI	1	Chitipa	1314	E	10	Salamanca (RNE5)
1278	RUS	50	Dubovka + 7 st's	1314	E	20	Reus (RNE5)
1278	UKR	150	Petrivka + 1 st	1314	EGY	10	Hurghada + 2 st's
1287	AFS	2	Welgedacht	1314	GRC	10	Tripoli
1287	ALG	5	El Golea	1314	I	10	Ancona + 3 st's
1287	ARS	5	Makkah	1314	IRN	20	Ardebil
1287	BUL	1	Smolyan	1314	LBN	10	Ehden (1310)
1287	CZE	150	Litomysl	1314	MDN	100	Skopje
1287	CZE	30	C.Budejovice	1314	NIG	10	Jos
1287	CZE	20	Plzen + 2 st's	1314	NOR	1200	Kvitsoy
1287	E	10	Lerida-Lleida (SER)	1314	ROU	30	Timisoara
1287	E	2	Lugo	1314	ROU	14	Constanta

kHz	Country	kW	Station	kHz	Country	kW	Station
1314	ROU	7	Craiova	1359	G	0.3	Coventry
1314	SYR	10	Aleppo	1359	G	0.3	Chelmsford
1314	UAE	2*1000	Dabiya	1359	NIG	10	Iwo
1323	ALB	15	Shkoder	1359	RUS	40	Perm
1323	CYP	100	Zyyi (BBC)	1359	RUS	150	Chkalovskaya
1323	D	1000/150	Wachenbrunn (VoR)	1359	UKR	50	Dokuchayevsk + 1 st
1323	G	0.63	Taunton	1359	YUG	2	Vrbas
1323	G	0.5	Brighton	1367	AGL	100	Mulenvos
1323	G	0.001	Leeds	1368	EGY	10	El Kharga + 1 st
1323	GEO	30	Tshinvali	1368	G	20	Douglas
1323	IRN	20	Jolfa	1368	G	0.5	Duxhurst
1323	MDG	1	Tulear	1368	G	0.1	Swindon
1323	MRC	5	Safi C (1325)	1368	G	2	Lincoln
1323	ROU	7	Targu Mures	1368	I	20	Venezia
1323	RUS	10	Sankt-Peterburg	1368	I	12	Milano
1323	SEN	1	Linguére	1368	I	10	Napoli
1323	YUG	1	Sid	1368	I	2/1	8 st's
1331	NGR	20	Meninsoroua	1368	IRN	20	Sari
1332	AFS	10	Gauteng	1368	ISR	50	Rosh-pina + 1 st(IDF)
1332	CZE	50	Moravské Budéjovice	1368	KEN	20	Nyeri
1332	G	1	Bow(London)	1368	LBN	-	Antélias
1332	G	0.6	Peterborough	1368	NIG	50	Calabar
1332	G	0.3	Lacock	1368	SEN	20	St Louis
1332	HOL	2	Utrecht (F.PI)	1368	SEY	50	Victoria
1332	I	600	Rome	1368	YUG	10	Valjevo
1332	I	50	Bari	1377	ARM	-	Sisian
1332	I	25	Pescara	1377	F	300	Lille
1332	I	10	Palermo	1377	G	1	Rochdale
1332	IRN	100	Tehran	1377	HRV	5	Zadar
1332	MDR	1	Funchal	1377	LBN	-	Murabitun
1332	POR	1	Elvas	1377	SWZ	50	Sandlane
1332	ROU	50	Galati	1377	UKR	50	Chernivtsi + 5 st's
1332	UKR	7	Odesa	1377	YUG	10	Prizren
1341	BFA	1	Ouagadougou	1386	BIH	1	Glamoc
1341	BIH	1	Prnjavor	1386	EGY	10	Luxor + 1 st
1341	E	2	Leon (SER)	1386	G	0.001	Oldham
1341	E	2	Almeria (OCR)	1386	G	0.5	Devizes
1341	E	2	Ciudad Real (OCR)	1386	GRC	50	Athens
1341	EGY	100	Cairo + 3 st's	1386	GUI	50	Labe
1341	G	100	Lisnagarvey	1386	IRN	400	Ahwaz
1341	IRN	1	Bam	1386	KEN	50	Maralal
1341	KWT	10	Magwa	1386	MKD	1	Kratovo
1341	YUG	10	Zajecar	1386	RUS	1200	Bolshakovo
1350	BOT	50	Tsabong	1386	RUS	5	Obninsk + 1 st
1350	EGY	10	Quseir	1390	NIG	10	Egbe
1350	GEO	50	Sokhumi	1395	ALB	1000	Fllake
1350	GRC	4	Pyrgos	1395	ARM	-	Yerevan
1350	HNG	10	Pecs	1395	HOL	120	Lopik
1350	IRN	10	Bandar-Abbas	1395	ISR	10	Mitspe Ramon (IDF)
1350	LVA	50	Cesvaine/Kuldiga	1395	MDG	4	Antananarivo (1394)
1350	MTN	20	Nouakchott (1349)	1395	NIG	10	Abak
1350	RUS	10	Ufa	1395	RUS	50	Petrozavodsk + 3 st's
1350	UKR	8	Novodnistrovsk + 3 st's	1395	RUS	50	Ufa
1350	YUG	10	Beograd	1395	TGO	20/1	Lomé (1394)
1359	ALB	50	Tirana	1395	UKR	50	Oktyabrskoye
1359	E	600	Madrid (RNE-FS)	1395	YUG	0.2	Kovacica
1359	EGY	450	Batra	1397	NIG	10	Warri
1359	G	0.85	Bournemouth	1403	AGL	1	Lobito
1359	G	0.2	Cardiff	1404	AFS	2	Welgedacht

kHz	Country	kW	Station	kHz	Country	kW	Station
1404	AGL	1	Bie	1449	I	50	Squinzano
1404	F	20/5	4 st's	1449	I	6	Catania
1404	GRC	50	Komotini	1449	I	2/0.5	19 low power st's
1404	GUI	200	Conakry (inactive)	1449	IRN	800	Bandar-e-Torkaman
1404	IRN	800	Kiashahr	1449	LBY	20	Misurata
1404	ISR	10	Jerusalem	1449	MDA	30	Chisinau
1404	LBY	20	Tripoli	1449	RUS	42	Monchegorsk + 8 st's
1404	NIG	10	Gombe	1449	UKR	50	Chernivtsi
1404	ROU	50	Sighet + 1 st	1449	YUG	2	Srbobran
1404	UKR	50	Lviv + 8 st's	1449	ZAI	2	Kinshasa (1448)
1413	E	10	Jaen (RNE5)	1458	ALB	500	Fllake
1413	E	5	Gerona (RNE5)	1458	ARM	30	Kamo
1413	E	5	Vigo (RNE5)	1458	BHR	1	Manama
1413	G	0.5	Heathrow	1458	G	50/0.5	London + 5 st's
1413	G	0.5	Dartford	1458	G	5	Manchester
1413	G	0.1	Skipton	1458	GIB	0.5/2	Wellington Front
1413	G	0.001	Leeds	1458	ISR	10	Jerusalem
1413	G	0.001	Manchester	1458	ISR	10	Eilat
1413	G	0.001	Newcastle	1458	MYT	100	Pamandzi
1413	G	0.001	Blackburn	1458	NIG	1	Sogunle
1413	G	0.5	Bourton-on-the-Water	1458	ROU	50	Constanta
1413	OMA	750	Zakaki (BBC)	1458	RUS	7	Kudymkar + 3 st's
1413	RUS	20	Dubovka	1458	UKR	3	Mariupol
1413	RUS	30	Stavropol	1458	YUG	1	Kladovo
1413	RUS	7	Volgograd	1458	YUG	1	Niksic
1413	YUG	100	Pristina	1467	BIH	1	Zvornik
1416	NIG	25	Kaduna	1467	F	1000	Romoules (RMC/TWR)
1422	ALG	50	Alger	1467	MDA	150	Maiac
1422	ALG	-	Constantine	1467	UKR	0.1	Yalta
1422	ARS	20	Riyadh	1467	YUG	2	Zrenjanin
1422	D	600	Heussweiler(DLF)	1476	AUT	60	Wien-Bissamberg
1422	EGY	10	Salum + 1 st	1476	AZE	150	Gäncä + 2 st's
1422	GEO	30	Batumi	1476	BEN	50/20	Cotonou
1422	IRN	100	Kermanshah	1476	COG	20	Brazzaville
1422	LVA	25	Rezekne/Valmiera	1476	EGY	10	El Minya + 1 st
1422	ROU	7	Olanesti	1476	G	1	Guildford
1422	RUS	300	Samara	1476	IRN	10	Marivan
1431	G	0.48	Rayleigh	1476	NIG	10	Mokwa
1431	G	0.14	Reading	1476	RUS	1	Aleksin
1431	I	2	Foggia	1476	UAE	1500	Dubai
1431	I	2	Pesaro	1476	UKR	30	Lviv
1431	I	2	Taranto	1476	UKR	7	Sevastopol
1431	IRN	200	Isfahan	1485	AFS	1	Johannesburg (R. Today)
1431	IRQ	-	Missan	1485	ARS	1	Jeddah
1431	MDN	1	Probistip	1485	ASC	0.5	English Bay
1431	MKD	1	Probishtip	1485	BIH	1	26 st's
1431	UKR	500	Mykolayiv-Luch	1485	BUL	1	Haskovo
1431	YUG	0.2	Beocin	1485	D	0.3	Hohenfels (AFN Bavaria)
1440	AGL	1/10	Luachimo	1485	D	0.3	3 st's (AFN)
1440	ARS	1600	Dammam	1485	D	1	Baden-Baden (SWR)
1440	CAF	20/50	Bangui	1485	E	5/2	"4 st's (SER, OCR)"
1440	LUX	1200	Marnach	1485	EGY	1	El Tur
1440	NIG	58	Yola	1485	ETH	1	3 st's
1440	RUS	10	Moskva (RFI)	1485	G	2/1	Wallasey + 4 st's
1440	RUS	10	Sankt-Peterburg (RFI)	1485	G	1	Basingstoke (F.Pl.)
1440	TZA	10	Dar-es-Salaam	1485	GRC	1	Volos
1440	YUG	20/10	Jagodina	1485	GRC	1	Orestiada
1449	G	2	Redmoss	1485	GRC	1	Patra
1449	G	0.15	Gunthorpe	1485	HNG	1	Mohács

kHz	Country	kW	Station	kHz	Country	kW	Station
1485	HNG	1	Debrecen	1512	BEL	300	Wolvertem
1485	HOL	1	Tilburg (F.Pl.)	1512	GRC	50	Chania
1485	HOL	1	Haag (F.Pl.)	1512	I	5	Bari
1485	HRV	1	Dubrovnik	1512	RUS	7	Ibresi
1485	I	2/0.15	10 low power st's	1512	RUS	7	Sochi + 2 st's
1485	IRN	10	Abadeh + 2 st's	1512	UKR	10	Kyyiv
1485	ISL	0.02	Raufarhöfn	1512	YUG	10	Pristina
1485	JOR	5	Aqaba	1521	ARS	2000	Duba
1485	LBN	5	Marjayoun	1521	BHR	1	Manama
1485	LBY	1	Brak	1521	E	2	Castellón (SER)
1485	MDR	1	Funchal	1521	G	0.64	Reigate
1485	MKD	1	Kavadarci	1521	G	-	Craigavon
1485	MRC	1	Casablanca	1521	GAB	1	Port Gentil (1520)
1485	MRC	4	Melilla	1521	IRN	100	Kiashahr
1485	NGR	0.1	Difa (1484)	1521	RUS	20	Kazan + 1 st
1485	NOR	1	Longyearbyen (NRK)	1521	SVK	10	Bratislava
1485	RUS	1	Oktyabrski	1521	SVK	14	Banská Bystrica
1485	RUS	20	Kurkino	1530	ARS	-	El Afif (?)
1485	RUS	-	Saratov	1530	AZE	7	Baki
1485	SUI	1	Savièse	1530	BIH	1	Donji Vakuf
1485	SVN	1	Ptuj	1530	CVA	300/600	Vatican City
1485	SVN	1	Radlje	1530	G	0.74	Huddersfield
1485	SYR	10	Homs	1530	G	0.52	Worcester
1485	UGA	10	Arua	1530	G	0.15	Southend-on-Sea
1485	UKR	7	Kharkiv + 2 st's	1530	IRN	20	Yazd
1485	YUG	2	Novi Sad	1530	ISL	0.25	Keflavik (AFRTS)
1485	YUG	1	Crna Trava + 6 st's	1530	MDR	1	Funchal
1485	YUG	1	Priboj	1530	ROU	14	Mahmudia + 1 st
1485	YUG	1	Cetinje + 3 st's	1530	STP	600	Pinheira (VOA)
1494	CTI	20	Abidjan	1530	UKR	30	Vinnytsya + 1 st
1494	CYP	10	Yeni Iskele	1539	D	700	Mainflingen
1494	F	20	Clermont-Ferrand	1539	DJI	20	Djibouti
1494	F	4	Bayonne	1539	E	2	Manresa (SER)
1494	F	5	Besancon	1539	E	2	Elche-Elx (SER)
1494	F	20	Bastia	1539	IRN	-	Unid
1494	GRC	50	Rhodos	1539	IRQ	150	Arbat
1494	JOR	1000	Al Karanah	1539	LVA	7	Daugavpils/Liepaja
1494	MDA	30	Cahul	1539	RUS	5	Sochi
1494	MDA	30	Edinet	1539	RUS	150	Lesnoy
1494	MOZ	5	Pemba (1493)	1539	SEN	10	Dakar
1494	RUS	1200	Krasnyy Bor	1539	UAE	2*50	Sadiyat
1494	YUG	0.2	Apatin	1539	UKR	7	Kharkiv
1503	AGL	10	Benguela (1502)	1539	YUG	10	Pec
1503	AZR	0.05	Lajes	1539	YUG	0.2	Odzaci
1503	BIH	1	Zavidovici	1544	MRC	-	L.V.D. de Sahara
1503	E	2	Monforte de Lemos (RNE5)	1548	G	97.5	London
1503	EGY	25	El Arish	1548	G	4.4	Liverpool
1503	G	1	Stoke-on-Trent	1548	G	2.2	Edinburgh
1503	G	0.001	Northants	1548	G	0.74	Sheffield
1503	IRN	100	Bushehr	1548	G	5	Bristol
1503	MDG	4	Antananarivo (1502)	1548	I	1	Vicenza(R. Star)
1503	POL	300	Szczecin	1548	IRN	20	Sanandaj
1503	RUS	7	Belgograd	1548	KWT	600	Kuwait (VOA)
1503	SEN	20	Tambacounda	1548	MKD	1	Negotino
1503	TGO	10	Kara (1502)	1548	MRC	-	LVD Sahara
1503	YUG	10	"Beograd""202"""	1548	SHN	1/0.1	Jamestown
1503	YUG	10	Ulcinj	1548	YUG	1	Bar
1512	ARS	1000	Jeddah	1548	YUG	1	Stara Pazova

kHz	Country	kW	Station	kHz	Country	kW	Station
1554	GAB	20	Melen	1584	I	2.5	Bologna(private)
1557	F	300	Nice	1584	I	2	Momigno (private)
1557	G	0.8	Northampton	1584	MKD	1	Prilep/Radovish
1557	G	0.5	Southampton	1584	MRC	5	Ceuta
1557	G	0.25	Oxcliffe	1584	MRC	5	Ceuta
1557	G	0.125	Clacton	1584	POR	1	Guarda (R.A.)
1557	HOL	2	Amsterdam (F.Pl.)	1584	RUS	7	Khunzakh + 2 st's
1557	HRV	20	Osijek	1584	SVN	1	Brezice
1557	IRN	50	Ardabil	1584	SVN	1	Kovor
1557	LTU	20	Klaipeda + 1 st	1584	UKR	1	Simferopol + 3 st's
1557	MDG	1	Diego-Suarez (1554)	1584	UKR	0.1	Sevastopol
1557	RUS	-	Stavropol	1584	UKR	0.1	Chernihiv
1557	UKR	8	Chernivtsi	1584	YUG	1	Pancevo
1566	ARM	5	Yerevan	1584	YUG	1	Prijepolje
1566	AZR	10	Vila De Porto	1584	YUG	1	Urosevac
1566	BLR	10	Miadziel + 4 st's	1584	YUG	1	Urosevac
1566	G	0.001	RSL-stations	1593	D	150	Holzkirchen
1566	IRN	100	Bandar Abbas	1593	EGY	10	Matruh
1566	MDG	1	Tamatave (1562)	1593	MDA	5	Chisinau
1566	TUN	1200	Sfax	1593	MRC	1	Marrakech
1566	UKR	7	Odesa + 1 st	1593	ROU	14	Constanta + 3 st's
1566	YUG	1	Smed. Palanka	1593	UKR	1	Polonne
1570	AGL	10	Cabinda	1593	YEM	1	Beihan
1575	E	5	Cordoba (SER)	1593	YUG	0.2	Indjija
1575	E	5	Pamplona (SER)	1602	ASC	1	ZD8VR(AFRTS)
1575	EGY	10	Queseir	1602	E	10	Vitoria (EI)
1575	G	0.001	Wolverhampton	1602	E	5	Linares (SER)
1575	GAB	20	Tchibanga	1602	E	2	Cartagena (SER)
1575	I	50	Genova	1602	E	2	Onteniente (SER)
1575	I	20	Perugia	1602	E	2	Segovia (SER)
1575	I	2/0.15	13 st's	1602	G	0.25	Rusthall
1575	ISR	5	Haifa	1602	G	0.001	Kilmarnock
1575	MAU	1	Forest Side	1602	G	0.001	Liverpool
1575	NGR	1	Niamey	1602	G	0.001	Norwich
1575	POR	1	Braga + 2 st's	1602	G	0.001	Donnington
1575	UAE	50	Sharjah	1602	G	0.001	Fife
1575	UKR	0.1	Botanichnoye	1602	G	0.001	Silverstone
1575	YUG	1	Bac + 1 st	1602	G	0.001	RSL-stations
1580	TUR	0.01	Adana(AFRS)	1602	GRC	1	Kavala/Kozani/Samos
1584	AGL	10	Luena (1586)	1602	HNG	1	Kaposvar(Somogyi R.)
1584	BHR	1	Manama	1602	HNG	1	Zalaegerszeg
1584	BIH	1	Mostar	1602	HNG	1	Komádi
1584	CLA	-	Voice of Dicle(PKK)	1602	HOL	1	Leeuwarden (F.Pl.)
1584	E	2	Gandía (SER)	1602	I	2	7 st's
1584	E	2	Orense (SER)	1602	IRN	2	Zanjan
1584	E	2	Zamora (SER)	1602	MDG	1	Diego-Suarez
1584	EGY	10	Idfu + 2 st's	1602	QAT	1	Al-Khaisah
1584	G	1	Clipstone	1602	RUS	5	Gayny
1584	G	0.2	Perth	1602	SVN	1	Ormoz
1584	G	0.04	Kettering	1602	UKR	7	Sevastopol + 2 st's
1584	G	0.3	Wofferton	1602	UKR	1	Uzhhorod (BBC)
1584	G	0.2	London (LTR)	1602	UKR	1	Luhansk
1584	G	0.001	Wigan	1602	YUG	1	Vrsac
1584	G	0.5	Shrewsbury	1602	YUG	1	Negotin
1584	G	0.01	Corby	1602	YUG	1	Leskovac
1584	GRC	1	Serres	1602	YUG	1	Sjenica
1584	HOL	1	Utrecht (F.Pl.)	1602	YUG	1	Budva + 4 st's
1584	I	2.5	Bologna	1611	CVA	100	Vatican City

EUROPE, AFRICA, NEAR & MIDDLE EAST

BY COUNTRY

Country	kHz	kW	Location	Country	kHz	kW	Location
					1367	100	Mulenvos
Afghanistan	1107	-	Kabul		1403	1	Lobito
	1278	100	Kabul		1404	1	Bie
Albania	1089	150	Durres		1440	1/10	Luachimo
	1215	500	Fllake		1503	10	Benguela (1502)
	1260	50	Korce		1570	10	Cabinda
	1305	15	Gjirokaster		1584	10	Luena (1586)
	1323	15	Shkoder		702	10	Mulenvos
	1359	50	Tirana		774	5	Luanda
	1395	1000	Fllake		945	25/1	Cazenga (944)
	1458	500	Fllake	Armenia	1314	1000	Kamo
	621	25	Korce		1377	-	Sisian
	648	50	Kukes		1395	-	Yerevan
	693	50	Shoder		1458	30	Kamo
	864	30	Sarande		1566	5	Yerevan
	909	50	Gjirokaster		234	500	Kamo
	990	15	Kukes (996)		252	150	Yerevan
Algeria	1026	5	Hassi Messaoud		702	150	Yerevan
	1089	5	Adrar		864	1000	Kamo(F.S)
	1098	5	Ouargla	Ascension	1485	0.5	English Bay
	1161	5	Ain-Salah		1602	1	ZD8VR(AFRTS)
	1287	5	El Golea	Austria	1476	60	Wien-Bissamberg
	1305	20	Constantine	Azerbaijan	1296	500	Pirsagat
	1422	50	Alger		1476	150	Gäncä + 2 st's
	1422	-	Constantine		1530	7	Baki
	153	1000	Bechar		218	500	Gäncä
	198	1000	Ouargla		549	50	Gäncä
	252	1500/750	Tipaza		612	50	Baki
	531	600/300	Ain-El-Beida		801	150	Pirsagat + 1 st's
	549	600/300	Les Trembles		891	30	Baki
	576	400	Bechar	Azores	1260	1	Espalamaca
	666	5	Tindouf		1503	0.05	Lajes
	693	5	Ain-el-Hamam + 1 st		1566	10	Vila De Porto
	738	5	Ain-Amenas		648	1	Lajes
	783	5	Djanet		693	10	Santa Barbara + 4 st's
	837	5	Ben-Abbas		828	1	Mt.das Cruzes + 1 st
	873	5	Ghardaia		837	10	Pico da Barossa + 1 st
	891	600/300	Alger		909	10	Angra do Heroismo
	909	5	Tamanrasset	Bahrain	1458	1	Manama
	927	5	Timimoun		1521	1	Manama
	981	600/300	Alger		1584	1	Manama
Angola	1008	1	Huambo (1010)		612	100	Manama
	1089	25/100	Mulenvos (1088)		801	100	Manama
	1107	1	Sumbe (1110)	Belarus	1008	50	Slonim + 4 st's
	1116	10	Sumbe (1115)		1026	50	Mahiliou + 5 st's
	1148	10	Menongue		1125	150	Minsk + 1 st
	1152	10	M'Banza Kongo		1170	1000	Sasnovy
	1160	10/1	Huambo		1197	50	Minsk + 3 st's
	1188	10/1	Malanje		1278	10	Brest
	1200	1	Benguela		1566	10	Miadziel + 4 st's
	1233	10	Lubango (1232)		171	1000	Sasnovy
	1242	5	Saurimo (1241)		279	500	Sasnovy
	1259	10/1	N'dalatando		549	1000	Sasnovy
	1314	1	Lubango (1313)		873	150	Minsk
	1314	1	Namibe (1313)	Belgium	1188	5	Kuurne

Country	kHz	kW	Location	Country	kHz	kW	Location
	1233	2.5	Liège		828	50	Sofia
	1512	300	Wolvertem		864	150	Blagoevgrad
	540	150	Wavre-Overijse		864	10	Samuil
	621	300	Wavre-Overijse		873	25	Stara Zagora
	927	100	Wolvertem		963	25	Shumen
Benin	1476	50/20	Cotonou		963	40	Sofia
	963	16/10	Parakou		963	20	Pirin
Bosnia Herc.	1017	1	Cazin(R. Mir)		963	5	Malko Tarnovo
	1071	40	Mostar		981	150	Varna
	1143	0.25	Tuzla (AFN Bosnia)	Burkino Faso	1008	10	B.Dioulasso
	1197	2	Bjeljina + 1 st		1341	1	Ouagadougou
	1197	1	Bosanski Nova		747	100	Ougadougou
	1242	50/10	Pale	Clandestine	1584	-	Voice of Dicle(PKK)
	1305	1	B. Krupa	Cameroon	1107	5	Douala (1106)
	1341	1	Prnjavor		1152	300/1	Bamenda
	1386	1	Glamoc	Cen. African Rep.	1440	20/50	Bangui
	1467	1	Zvornik	Christmas Island	1008	10	Las Palmas
	1485	1	26 st's		1179	20	Santa Cruz
	1503	1	Zavidovici		1269	20	Las Palmas
	1530	1	Donji Vakuf		576	25	Las Palmas
	1584	1	Mostar		621	600	"Santa Cruz, Tenerife"
	612	600	Sarajevo		720	20	"Santa Cruz, Tenerife"
	756	10	Mostar		747	20	Las Palmas
	774	2	Bihac		837	10	Las Palmas
	864	1	Sokolac		882	20	La Laguna
	972	2	Vukovar	Comoros	1089	40	Haboho
	990	1	Foca	Congo	1161	50	Matadi (1160)
Botswana	1071	25	Jwaneng		1269	2	Mbandaka(1268)
	1170	-	Loc.unknown		1449	2	Kinshasa (1448)
	1215	-	Loc.unknown		1476	20	Brazzaville
	1215	50	Mahalapye	Cote d'Ivoir	1242	1	Bondoukou (1240)
	1350	50	Tsabong		1494	20	Abidjan
	531	50	Maun	Croatia	1125	100	Deanovac
	621	100	Selebi-Phikwe		1134	600	Zadar
	648	50	Mopipi		1152	2	Dubrovnik
	873	-	"Gantsi(F.Pl,)"		1197	2	Pula
	909	500	Selebi-Phikwe (VOA)		1377	5	Zadar
	972	50	Sebele		1485	1	Dubrovnik
Bulgaria	1017	50	Kardzhali		1557	20	Osijek
	1017	40	Vidin		594	10	Osijek
	1143	50	Varna		774	50	Hvar
	1161	300	Stara Zagora + 3 st's		783	10	Buje
	1224	300	Vidin		972	1	Knin
	1287	1	Smolyan	Cyprus	1044	10	Limassol
	1296	150	Kardzali		1098	100	Yeni Iskele
	1296	30	Pleven		1233	600	Cape Greco
	1485	1	Haskovo		1323	100	Zyyi (BBC)
	261	60	Sofia		1494	10	Yeni Iskele
	576	500	Vidin		558	10	Paphos
	594	250	Pleven		603	100	Nicosia
	594	5	Dobrich		639	500	Ladies' Mile
	594	10	Burgas		693	10	Nicosia
	648	30	Plovdiv		720	500	Ladies' Mile
	702	40	Pirin		918	10	Paphos
	711	1	Silistra		963	100	Nicosia
	747	300	Petrich + 2 st's	Czech Rep.	1062	20/1	Praha
	774	50	Sofia		1071	25	Moravske Budejovice
	774	5	Varna		1071	10	Hradec Králové
	828	500	Shumen		1233	400	Melnik

Country	kHz	kW	Location
	1233	30	Ostrava
	1233	25	Brno-Komarov
	1233	20	Brno-Dobrochov
	1233	7	Jihlava
	1287	150	Litomysl
	1287	30	C.Budejovice
	1287	20	Plzen + 2 st's
	1332	50	Moravské Budéjovice
	270	750	Topolná
	558	1	Broumov + 2 st's
	639	1500	Praha-Liblice + 1 st
	792	40	Praha
	846	7	Ceske Budejovice
	846	2	Ostrava
	864	7	Strakonice
	900	25	Brno
	954	200	Brno-Prostejov
	954	20	Karlovy Vary
	954	40	Plzen
	954	30	C. Budejovice
	954	3	Liberec
	981	7	Jihlava
Denmark	1062	250	Kalundborg
	243	300	Kalundborg
Djibouti	1170	20	Djibouti
	1539	20	Djibouti
Egypt	1008	100	El Arish + 1 st
	1071	100	Cairo
	1080	10	El Minya + 1 st
	1107	600	Batrah
	1143	10	Sohag
	1161	100	Tanta
	1179	10	Quena
	1188	10	Ras Gharib
	1197	25	Alexandria
	1278	10	Asswan
	1305	10	Assiut
	1314	10	Hurghada + 2 st's
	1341	100	Cairo + 3 st's
	1350	10	Quseir
	1359	450	Batra
	1368	10	El Kharga + 1 st
	1386	10	Luxor + 1 st
	1422	10	Salum + 1 st
	1476	10	El Minya + 1 st
	1485	1	El Tur
	1503	25	El Arish
	1575	10	Queseir
	1584	10	Idfu + 2 st's
	1593	10	Matruh
	558	40	Cairo
	603	100	Barnis
	621	1000	Batrah
	702	10	El Kharga + 1 st
	711	100	Tanta
	756	10	Qena
	774	500	Abis
	819	1000	Batra
	864	500	Santah

Country	kHz	kW	Location
	882	10	Matruh
	918	10	Bawiti
	918	10	Hurghada
	936	10	Salum
	981	10	Assiut + 2 st's
Eritrea	837	10	Asmara
	945	50	Asmara
	945	50	Asmara
Ethiopa	1485	1	3 st's
	684	100	Metu
	855	100	Harar
	873	100	Addis Ababa
	990	1	Addis Ababa
Finland	558	50	Espoo
	963	600	Pori
France	1206	300	Bordeaux
	1242	150	Marseille
	1278	300	Strasbourg
	1377	300	Lille
	1404	20/5	4 st's
	1467	1000	Romoules (RMC/TWR)
	1494	20	Clermont-Ferrand
	1494	4	Bayonne
	1494	5	Besancon
	1494	20	Bastia
	1557	300	Nice
	162	2000	Allouis
	216	1400	Roumoules (RMC)
	585	10	Paris
	603	300	Lyon-Tramoyes
	711	300	Rennes
	738	5	Paris (RFI)
	792	300	Limoges
	819	20	Toulouse
	837	200	Nancy
	864	300	Paris
	945	300	Toulouse
Faroe	531	100(200)	Akraberg
Gabon	1206	1	Oyem
	1521	1	Port Gentil (1520)
	1554	20	Melen
	1575	20	Tchibanga
	549	20	Oyem
	990	20	Epila
Gambia	648	50	Bonto
	747	10	Basse
	909	2.5	Banjul
Georgia	1044	200	Tbilisi
	1323	30	Tshinvali
	1350	50	Sokhumi
	1422	30	Batumi
	189	500	Tbilisi
	693	-	Batumi
	810	200	Tbilisi
Germany	1017	200	Rheinsender (SWF)
	1044	20	Dresden(MDR)
	1107	10	Grafenwöhr (AFNBavaria)
	1107	10	Kaiserslautern (AFN)
	1143	10	Stuttgart (AFN)

Country	kHz	kW	Location
	1143	1	Heidelberg (AFN)
	1143	1	Mönchen-Gladbach (AFN)
	1143	0.3	5 low power st's (AFN)
	1188	3	Reichenbach (MDR)
	1197	300	München (VOA)
	1269	300	Neumunster
	1323	1000/150	Wachenbrunn (VoR)
	1422	600	Heussweiler(DLF)
	1485	0.3	Hohenfels (AFN Bavaria)
	1485	0.3	3 st's (AFN)
	1485	1	Baden-Baden (SWR)
	153	500	Doneback
	1539	700	Mainflingen
	1593	150	Holzkirchen
	177	500	Oranienburg (DRB)
	183	2000	Saarlouis
	207	500	Aholming
	261	80	Burg
	520	0.2	Würzburg/Hof(BR)
	549	100	Thurnau
	549	100	Nordkirchen
	567	50	Berlin (B2)
	576	200	Mühlacker(SWR)
	594	300	Frankfurt (HR)
	594	200	Hoher Meissner (HR)
	612	10	Kiel
	666	150	Messkirch Rohrdorf
	693	5	Berlin (VoR)
	702	7.5	Flensburg (NDR)
	711	5	Heidelberg + 3 st(SWR)
	720	200	Langenberg (WDR)
	756	200	Braunschweig (DLF)
	756	100	Ravensburg (DLF)
	774	5	Bonn
	783	100	Leipzig(MDR)
	792	5	Lingen (NDR)
	801	100	München-Ismaning (BR)
	801	50	Dillberg (BR)
	828	50/10	Hannover (NDR)
	828	10	Freiburg (SWR)
	855	5	Berlin(irr.)
	873	150	Frankfurt (AFN)
	882	20	Wachenbrunn (MDR)
	936	100	Bremen (RB)
	972	100	Hamburg (NDR)
	990	100	Berlin (DRB)
Gibraltar	1458	0.5/2	Wellington Front
Greece	1008	50	Kerkyra
	1044	150	Thessaloniki
	1080	10	Orestiada
	1179	50	Thessaloniki
	1260	500	Rhodos
	1278	10	Florina
	1314	10	Tripoli
	1350	4	Pyrgos
	1386	50	Athens
	1404	50	Komotini
	1485	1	Volos
	1485	1	Orestiada

Country	kHz	kW	Location
	1485	1	Patra
	1494	50	Rhodos
	1512	50	Chania
	1584	1	Serres
	1602	1	Kavala/Kozani/Samos
	666	15	Athens
	729	150	Athens
	765	10	Ioannina
	792	500	Kavala
	927	50	Zakynthos
	945	5	Larissa
	954	10	Heraklion
	981	200	Megara
Guinea	1296	1	Conakry (1295)
	1386	50	Labe
	1404	200	Conakry (inactive)
	603	25	Kankan
	1188	135/5	Szolnok
Guinea Bissau	1035	5	Bissau (1034)
	1071	100	Bissau
Hungary	1188	25	Szombathely
	1251	135	Szeged
	1350	10	Pecs
	1485	1	Mohács
	1485	1	Debrecen
	1602	1	Kaposvar(Somogyi R.)
	1602	1	Zalaegerszeg
	1602	1	Komádi
	540	1000	Solt
	810	15	Siófok
	873	20	Lakihegy
	873	20	Pecs
Iceland	1485	0.02	Raufarhöfn
	1530	0.25	Keflavik (AFRTS)
	189	300	Gufuskálar
	207	100	Eidar
Internat. Waters	1143	-	Arutz Sheva
Internat. Waters	711	-	Arutz Sheva
Ireland	1278	10	Dublin
	1278	10	Cork
	252	500/100	Clarkestown
	540	2	Conamara
	549	4	"C,Monaghan"
	567	500	Tullamore
	612	100	Athlone
	729	10	Cork
	828	1	Ballydavid
	963	10	Derrybeg
Iran	1008	20	Semnan
	1026	100	Tabriz
	1035	20	Yazd
	1053	20	Khorramabad
	1062	20	Kerman
	1080	750	Abadan
	1098	200	Zabol
	1116	125	Bandar Lengeh
	1134	20	Bojnurd
	1143	20	Yasuj
	1152	100	Tabriz

Country	kHz	kW	Location
	1161	10	Abadan(1160)
	1170	2	Damghan
	1188	100	Tehran
	1197	20	Chalus
	1242	10	Kerman
	1278	200	Kermanshah
	1305	20	Bushehr
	1314	20	Ardebil
	1323	20	Jolfa
	1332	100	Tehran
	1341	1	Bam
	1350	10	Bandar-Abbas
	1368	20	Sari
	1386	400	Ahwaz
	1404	800	Kiashahr
	1422	100	Kermanshah
	1431	200	Isfahan
	1449	800	Bandar-e-Torkaman
	1476	10	Marivan
	1485	10	Abadeh + 2 st's
	1503	100	Bushehr
	1521	100	Kiashahr
	1530	20	Yazd
	1539	-	Unid
	1548	20	Sanandaj
	1557	50	Ardabil
	1566	100	Bandar Abbas
	1602	2	Zanjan
	531	20	Iranshahr
	540	200	Mashhad
	549	400	Sirjan
	558	1000	Gheslagh
	576	750	Abadan
	582	100	Tehran
	594	400	Shiraz
	603	100	Zahedan
	621	20	Birjand
	639	400	Bonab
	684	100	Mashhad
	720	400	Tayebad
	738	10	Tabas
	747	400	Bandar-e-Torkaman
	747	150	Sirjan
	765	600	Chahbahar
	774	100	Arak
	783	10	Saravan
	819	20	Sari
	837	200	Esfahan
	882	10	Mahabad
	891	20	Yasuj
	900	50	Tehran
	918	20	Jirof
	927	10	Dorud
	936	50	Urumiyeh + 1 st
	945	20	Sanandaj
	963	20	Birjand
	972	10	Ilam
	990	400	Shiraz (v.f.)
Iraq	1044	-	Unk. location

Country	kHz	kW	Location
	1116	-	Rutba
	1197	-	Nineva
	1224	-	Nasiriya
	1431	-	Missan
	1539	150	Arbat
	531	-	Unk. location
	558	300	Rutba
	585	-	Unk. location
	603	300	Nineva
	630	20	Kirkuk
	684	-	Tanaf
	693	-	Basrah
	756	-	Baghdad
	846	.	Nasiriya
	909	.	Baghdad
	963	-	Unk. location
Israel	1026	50	Tel Aviv
	1080	5	Jerusalem
	1143	-	Arutz Sheva
	1206	50	Haifa
	1224	20	Beersheba
	1287	100	Tel Aviv
	1305	20	Haifa
	1305	5	Eilat
	1368	50	Rosh-pina + 1 st(IDF)
	1395	10	Mitspe Ramon (IDF)
	1404	10	Jerusalem
	1458	10	Jerusalem
	1458	10	Eilat
	1575	5	Haifa
	531	100	Jerusalem
	576	200	Tel Aviv
	657	200	Tel Aviv
	675	100	Ramallah city
	738	1200	Tel-Aviv
	846	5	Safad
	882	1	Beit Hillel
	927	50	Haifa
	927	1	Eilat
	954	100	Jerusalem
Italy	1035	50	Milano + 7 st's
	1062	25	Cagliari
	1062	25	Squinzano
	1062	10	Ancona
	1062	2	4 st's
	1107	6	Roma
	1116	150	Bari
	1116	60	Bologna
	1116	25	Pisa
	1116	20	Cuneo
	1116	10	Palermo + 1 st
	1143	10	Sassari + 1 st
	1188	6	San Remo
	1296	2	La Spezia + 1 st
	1305	12	Genova+ 1 st
	1314	10	Ancona + 3 st's
	1332	600	Rome
	1332	50	Bari
	1332	25	Pescara

Country	kHz	kW	Location	Country	kHz	kW	Location
	1332	10	Palermo		846	100	Kisumu
	1368	20	Venezia		882	50	Kitale
	1368	12	Milano		900	100	Meru
	1368	10	Napoli		927	100	Malindi
	1368	2/1	8 st's		954	100	Kisumu
	1431	2	Foggia		981	100	Voi
	1431	2	Pesaro	Kuwait	1134	10	Sulaibiyah
	1431	2	Taranto		1269	100	Sulaibiyah
	1449	50	Squinzano		1341	10	Magwa
	1449	6	Catania		1548	600	Kuwait (VOA)
	1449	2/0.5	19 low power st's		540	600	Sulaibiyah
	1485	2/0.15	10 low power st's		630	10	Kuwait city
	1512	5	Bari		963	20	Kuwait city
	1548	1	Vicenza(R. Star)	Latvia	1350	50	Cesvaine/Kuldiga
	1575	50	Genova		1422	25	Rezekne/Valmiera
	1575	20	Perugia		1539	7	Daugavpils/Liepaja
	1575	2/0.15	13 st's		945	150	Riga-Ulbroka
	1584	2.5	Bologna	Lebanon	1000	-	-
	1584	2.5	Bologna(private)		1014	-	Tripoli
	1584	2	Momigno (private)		1050	-	Beirut
	1602	2	7 st's		1080	50	Saida
	189	10	Caltanissetta		1134	-	Beirut
	567	20	Bologna + 4 st's		1188	-	Saida
	657	120	Napoli		1314	10	Ehden (1310)
	657	100	Firenze		1368	-	Antélias
	657	50	Torino + 2 st's		1377	-	Murabitun
	693	100	Milano + 1 st		1485	5	Marjayoun
	819	20	Trieste		585	100	Ehden
	846	1200	Roma		680	5	Marjayoun
	900	600	Milano		684	-	Baalbekk
	936	20	Venezia		756	-	Marjayoun
	936	10	Trapani/Genova		837	100	Amchit
	981	10	Trieste		873	80	Beirut
	990	10	Potenza		902	-	Aley
	999	20	Torino + 3 st's		930	-	Saida
Jordan	1485	5	Aqaba		945	50	Marjayoun
	1494	1000	Al Karanah		963	100	Zuk Musbih
	207	600	Al Karanah		990	10	Amchit
	612	200	Amman	Lesotho	1197	100	Lancer's Gap (BBC)
	801	2000	Ajlun		891	50	Maseru
	855	10	Amman	Liberia	558	10	Gbarnga
Kenya	1017	20	Nyeri		990	5	Enoanyi
	1044	100	Malindi	Libya	1053	50	Tripoli
	1107	100	Maralal		1080	40	Ajedabia + 1 st
	1134	50	Kitale		1125	500	El Beida
	1152	50	Wajir		1251	500	Tripoli
	1233	50	Marsabit		1404	20	Tripoli
	1269	20	Nyeri (1268)		1449	20	Misurata
	1305	50	Wajir		1485	1	Brak
	1368	20	Nyeri		648	300	Tobruk
	1386	50	Maralal		675	100	Benghazi
	540	100	Voi		711	50	Ghadames
	558	20	Kisumu		711	50	Jefren
	567	50	Garissa		711	50	Sebha
	612	100	Nairobi		792	20	Sirte
	639	50	Garissa		828	300	Sebha
	675	50	Marsabit		909	20	Giaghboub
	702	100	Meru		909	10	Kufra
	747	100	Nairobi	Lithuania	1107	150	Sitkunai + 3 st's

Country	kHz	kW	Location
	1557	20	Klaipeda + 1 st
	612	40	Vilnius
	612	20	Klaipéda
	666	500	Sitkunai
Luxembourg	1440	1200	Marnach
	234	2000	Junglinster
Macedonia	1026	5	Skopje
	1314	100	Skopje
	1431	1	Probistip
	567	10	Strumica
	639	1	Stip
	738	1	Tetovo
	810	-	Skopje
	810	-	Skopje
	1287	1	Kochani
	1386	1	Kratovo
	1431	1	Probishtip
	1485	1	Kavadarci
	1548	1	Negotino
	1584	1	Prilep/Radovish
	639	1	Shtip
	720	1	Struga/Veles
	738	1	Tetovo
	810	1000	Skopje
	945	2	Kumanovo
Madagascar	1233	0.25	Maintirano
	1260	10	4 st's
	1278	1	Low power st's
	1287	1	Diego Suarez
	1323	1	Tulear
	1395	4	Antananarivo (1394)
	1503	4	Antananarivo (1502)
	1557	1	Diego-Suarez (1554)
	1566	1	Tamatave (1562)
	1602	1	Diego-Suarez
	630	150/75	Antananarivo
	693	2*75	Antananarivo
Madeira	1125	1	Ponta do Pargo
	1332	1	Funchal
	1485	1	Funchal
	1530	1	Funchal
	531	10	Porto Santo
	603	10	Areeiro
Malawi	1107	1	Nkhotakota
	1278	1	Chitipa
	540	10	Mangochi
	558	10	Karonga
	594	25	Lilongwe
	675	50	Mzuzu
	756	25	Blantyre
	810	10	Bangula
Mali	540	100	Bamako
Malta	999	5	Bizbizja
Mauritania	1350	20	Nouakchott (1349)
Mauritius	1206	10	Rodrigues
	1575	1	Forest Side
	666	1	Rodrigues
	684	10	Curepipe
	819	10	Forest Side

Country	kHz	kW	Location
Mayotte	1458	100	Pamandzi
Moldova	1449	30	Chisinau
	1467	150	Maiac
	1494	30	Cahul
	1494	30	Edinet
	1593	5	Chisinau
	999	500	Maiac
Monaco	702	40	Monte Carlo
Morocco	1026	1	Rabat
	1044	300	Sebaa-Aioun
	1053	600	Tanger
	1080	1	Casablanca
	1188	1	Casablanca
	1197	20	Agadir
	1233	200	Tanger
	1323	5	Safi C (1325)
	1485	1	Casablanca
	1485	4	Melilla
	1544	-	L.V.D. de Sahara
	1548	-	LVD Sahara
	1584	5	Ceuta
	1584	5	Ceuta
	1593	1	Marrakech
	171	2000	Nador
	207	800	Azilal
	540	600	Sidi Bennour
	594	100	Oujda
	612	300	Sebaa-Aioun
	702	140	Sebaa-Aioun
	711	600	Laayoune
	774	50	Agadir
	819	25	Rabat
	828	100	Oujda
	864	15	Errachidia
	909	5	Safi C
	936	600	Agadir
	972	5	Melilla
	999	10	Ad-Dakhla
Mozambique	1008	50	Maputo
	1026	5	Chimoio
	1161	5	Tete (1160)
	1206	5	Inhambane
	1224	5	Pemba
	1494	5	Pemba (1493)
	738	50	Maputo
	765	5	Nampula
	774	5	Xai-Xai
	783	5	Lichinga(782)
	837	5	Chimoio
	873	50	Beira (872)
	918	50	Maputo
	936	50	Quelimane
Namibia	594	100	Tsumeb
Netherlands	1008	400	Flevoland
	1035	2	Echt(F.Pl)
	1116	0.5	Bloemendaal
	1224	50	Ijsselmer
	1251	10	Hulsberg
	1332	2	Utrecht (F.Pl)

Country	kHz	kW	Location	Country	kHz	kW	Location
	1395	120	Lopik		666	50	Benin
	1485	1	Tilburg (F.Pl.)		675	25	Obomasho
	1485	1	Haag (F.Pl.)		684	20	Damaturu
	1557	2	Amsterdam (F.Pl.)		693	10	Ayanga
	1584	1	Utrecht (F.Pl.)		695	10	Suleja
	1602	1	Leeuwarden (F.Pl.)		702	25	Wukari
	675	120	Lopik		720	50	Owerri
	747	400	Flevoland		722	20	Bali
	828	20	Heijenoord		729	25	Jogana
	891	5	Hulsberg		756	100	Ibadan
Niger	1125	20	Niamey		756	50	Minna
	1215	0.1	Tahoua		756	25	Damagun
	1331	20	Meninsoroua		774	10	Bali
	1485	0.1	Difa (1484)		774	10	Wukari
	1575	1	Niamey		783	50	Ochaja
Nigeria	1008	10	Kontagora		792	10	Aha
	1008	10	Ibadan		801	20	Damaturu
	1008	10	Osogbo		801	1	Zuru
	1017	10	Yola		828	25	Enugu
	1026	25	Dutse		846	10	Azare
	1062	10	Onitsha		846	10	Katsina Ala
	1071	-	Nsukka		855	10	Port Harcourt(854)
	1080	50	Minna		882	25	Kaduna
	1089	50	Yola		900	25	Abeoukuta
	1089	20	Sogunle		909	50	Fwagwa Lada
	1107	25	Jaji		918	50	Makurdi
	1134	10	Ogoja		918	50	Ikeja
	1143	10	Bida		936	10	Ikorodu
	1170	25	Abeoukuta		945	10	Birnin Kebbi
	1206	10	Yola		954	10	Enugu
	1224	50	Jos		972	50	Katsina
	1269	10	Jalingo		972	10	Otite
	1296	20	Maiduguri		981	10	Port Harcourt
	1314	10	Jos		981	10	New Bussa
	1359	10	Iwo		990	50	Bauchi
	1368	50	Calabar		990	10	Ikeja
	1390	10	Egbe	**Norway**	1314	1200	Kvitsoy
	1395	10	Abak		1485	1	Longyearbyen (NRK)
	1397	10	Warri		630	100	Vigra
	1404	10	Gombe		675	10	Bodö
	1416	25	Kaduna		702	20	Vadsö
	1440	58	Yola	**Oman**	1035	100	Salalah
	1458	1	Sogunle		1242	200	Muscat
	1476	10	Mokwa		1413	750	Zakaki (BBC)
	531	50	Akure		576	100	Haima
	540	50	Sokoto		720	750	Masirah Island (BBC)
	549	50	Tukun Tawa	**Poland**	1305	60	Bialystok
	567	50	Alaho		1503	300	Sz cin
	576	50	Moniya		225	600	Konstantynow 1
	585	50	Abakaliki + 1 st	**Portugal**	1035	100	Lisboa (inactive)
	594	100	Jaji		1170	10	Vila Real + 3 st's
	603	30	Ibese		1251	10	Porto
	612	50	Ilorin		1251	10	Viseu
	621	20	Akwa + 1 st		1251	10	Chaves + 1 st
	630	50	Maiduguri		1287	2	Portalegre
	630	50	Pt. Harcourt		1332	1	Elvas
	639	25	Kaduna		1575	1	Braga + 2 st's
	639	5	Warri		1584	1	Guarda (R.A.)
	657	100	Ibadan		558	10	Faro

Country	kHz	kW	Location		Country	kHz	kW	Location
	576	10	Braga (RRE)			1107	150	Samara
	594	100	Muge (RRE)			1107	50	Arkhangelsk
	630	50	Montemor + 1 st			1107	25	Nalchik + 1 st
	666	10	Lisboa + 5 st's			1116	30	Kaliningrad
	720	100	Porto + 8 st's			1116	30	Sochi
	756	1	Lamego			1116	10	Moskva
	783	10	Porto + 1 st			1116	30	Perm
	828	1	Coimbra + 4 st's			1125	150	Olgino
	891	10	Vilamoura (RRE)			1134	75	Murmansk
	927	1	Evora(RRE)			1134	30	Saransk + 5 st's
	963	10	Seixal (RRE)			1143	150	Bolshakovo
	981	10	Coimbra + 3 st's			1143	150	Mekhzavod
Qatar	1233	100	Al-Khaisah			1152	150	Kupavna
	1602	1	Al-Khaisah			1161	75	Volgograd
	675	100	Al- Khaisah + 1 st			1170	1200	Tbilisskoye
	954	1500	Al-Arish			1188	10	Sankt-Peterburg
	999	50	Al -Khaisah			1215	1200	Bolshakovo (inactive)
Reunion	666	20	St. Pierre			1224	-	Elista
	729	20	St. Andre			1233	25	Velsk
Romania	1053	1000	Iasi			1233	150	Chkalovskaya
	1152	950	Cluj			1251	7	Izhevsk + 4 st's
	1179	200	Bacau			1251	1	Neftekumsk
	1179	7	Vascau			1260	10	Moskva
	1197	14	Bod			1260	10	Sankt-Peterburg
	1314	30	Timisoara			1278	50	Dubovka + 7 st's
	1314	14	Constanta			1287	50	Yazykovo + 3 st's
	1314	7	Craiova			1305	300	Kupavna
	1323	7	Targu Mures			1323	10	Sankt-Peterburg
	1332	50	Galati			1350	10	Ufa
	1404	50	Sighet + 1 st			1359	40	Perm
	1422	7	Olanesti			1359	150	Chkalovskaya
	1458	50	Constanta			1386	1200	Bolshakovo
	153	1200	Bod			1386	5	Obninsk + 1 st
	1530	14	Mahmudia + 1 st			1395	50	Petrozavodsk + 3 st's
	1593	14	Constanta + 3 st's			1395	50	Ufa
	531	14	Petrosani			1413	20	Dubovka
	558	400	Tirgu Jiu			1413	30	Stavropol
	567	50	Bod/Satu Mare			1413	7	Volgograd
	603	50	Bucuresti/Botosani			1422	300	Samara
	603	14	Oradea + 1 st			1440	10	Moskva (RFI)
	630	400	Timisoara + 1 st			1440	10	Sankt-Peterburg (RFI)
	711	50	Sighet			1449	42	Monchegorsk + 8 st's
	720	14/7	Sighèt + 2 st's			1458	7	Kudymkar + 3 st's
	756	400	Lugoj			1476	1	Aleksin
	855	1500	Bucuresti			1485	1	Oktyabrski
	909	200	Cluj + 1 st			1485	20	Kurkino
	945	14	Miercurea Ciuc			1485	-	Saratov
Russia	1008	25	Tuapse			1494	1200	Krasnyy Bor
	1017	500	Lesnoy			1503	7	Belgograd
	1026	7	Nyandoma + 4 st's			1512	7	Ibresi
	1044	7	Kotlas			1512	7	Sochi + 2 st's
	1044	20	Kurkino			1521	20	Kazan + 1 st
	1053	10	Sankt-Peterburg			153	300	Ufa
	1062	25	Saransk			153	300	Taldom
	1071	10	Moskva			1539	5	Sochi
	1080	30	Murmansk			1539	150	Lesnoy
	1089	1200	Tbilisskoye			1557	-	Stavropol
	1089	20	Sankt Peterburg			1584	7	Khunzakh + 2 st's
	1098	7	Vologda + 4 st's			1602	5	Gayny

Country	kHz	kW	Location	Country	kHz	kW	Location
	171	1250	Elektrostal		873	1200	Elektrostal
	171	150	Murmansk		873	150	Olgino
	171	1200	Tbilisskaya		873	50	Kaliningrad
	171	150	Syktyvkar		873	1	2 st's
	171	1200	Bolshakovo		882	30	Stavropol
	198	150	Ufa		891	50	Tyumen
	198	150	Olgino		900	25	Sovietskiy
	198	150	Noginsk		909	-	"Unk,So.E."
	234	2000	Syzran		918	150	Arkhangelsk
	234	1200	Krasnyy Bor		918	50	Makhachkala
	234	500	Arkhangelsk		918	75	Balashikha
	252	150	Kazan		936	5	Matveyevka
	261	2500	Taldom		945	150	Rostov-na-Donu
	531	30	Cheboksary		963	15	Moskva
	540	50	Orenburg		990	-	Moskva
	549	1200	Krasnyy Bor	**Saint Helena**	1548	1/0.1	Jamestown
	549	1000/500	Kurovskaya	**S.T. e Principe**	1530	600	Pinheira (VOA)
	549	50	Kalingrad	**Saudi Arabia**	1098	10	Qurayyat + 1 st
	549	50	Novocherkassk		1287	5	Makkah
	549	150	Syktyvkar		1305	1	Taif
	567	1000	Volgograd		1422	20	Riyadh
	576	50	Astrakhan + 1 st		1440	1600	Dammam
	576	25	Nalchik		1485	1	Jeddah
	585	150	Perm		1512	1000	Jeddah
	594	40	Izhevsk		1521	2000	Duba
	594	25	Vladikavkaz		1530	-	El Afif (?)
	612	150	Petrozavodsk		549	2000	Qurayyat
	612	40	Kurkino		549	20	Rafha
	621	50	Syktyvkar		549	1	Gizan
	621	50	Makhachkala		558	1	Jeddah
	621	5	Kochubey		567	15	Afif
	630	42	Saratov		567	5	Abha + 1 st
	657	150	Murmansk		585	1200	Riyadh
	657	50	Groznyy		594	2000	Duba
	666	25	Sochi		612	15	Khamasin
	675	5	Ukhta		612	5	Hail
	684	10	Olgino		630	20	Gizan
	693	30	Ufa		630	10	Najran
	693	10	Moskva (DW)		648	2000	Jeddah
	693	0.2	Astrakhan		657	20	Rafha
	711	7	Naryan-Mar		675	20	Afif
	738	5	Moskva		675	5	Abha
	738	-	Izhevsk		684	50	Jeddah
	756	10	Ufa		684	10	Riyadh
	765	150	Petrozavodsk		702	40	Duba
	774	30	Somovo		747	1	Najran
	792	50	Astrakhan		765	20	Khamasin
	792	5	Moskva		765	10	Qurayyat
	801	600	Krasnyy Bor		783	100	Dammam
	810	500	Volgograd		792	50	Jeddah
	810	-	Moskva		828	20	Medinah
	828	10	Sankt Peterburg		864	100	Dammam
	828	50	Nizhniy Novgorod		873	10	Qassim
	828	-	Unknown E.		882	100	Dammam
	846	150	Noginsk		900	1000	Qurayyat
	846	42	Elista		936	5	Riyadh
	846	-	Perm		945	5	Hail
	855	50	Kamenka		981	20	Madinah
	873	2000	Novosemeykino		999	20	Tabuk

Country	kHz	kW	Location	Country	kHz	kW	Location
Senegal	1224	20/1	Ziguinchor		1008	5	Girona (SER)
	1287	4	Kaolack		1008	5	Alicante (SER)
	1305	20	Dakar		1017	10	Granada (RNE5)
	1323	1	Linguére		1017	10	Burgos (RNE5)
	1368	20	St Louis		1026	5	Oviedo (SER)
	1503	20	Tambacounda		1026	2	Salamanca (SER)
	1539	10	Dakar		1026	2	Jaén (SER)
	765	300/200	Dakar		1026	2	Jerez (SER)
	810	1	Podor		1026	10	Reus (SER)
	963	1	Matam		1026	5	Vigo (SER)
Seychelles	1368	50	Victoria		1044	10	San Sebastian (SER)
Sierra Leone	1206	50/10?	Goderich		1044	5	Valladolid (SER)
Slovakia	1017	50	R. Sobota		1053	10	Zaragoza (COPE)
	1017	5	Bratislava		1053	5	Castellón (COPE)
	1035	7	Banská Bystrica		1071	50	Bilbao (EI)
	1071	40	Presov		1080	10	Granada (SER)
	1098	750	Nitra		1080	5	Palma de M. (SER)
	1287	400	Velké Kostolany (RFE)		1080	5	La Coruna (SER)
	1287	14	Poprad (RFE)		1080	5	Huesca (SER)
	1521	10	Bratislava		1080	5	Toledo (OCR)
	1521	14	Banská Bystrica		1098	20	Lugo (RNE5)
	567	10	R. Sobota		1098	25	Almeria (RNE5)
	567	7	Zilina		1098	10	Avila (RNE5)
	621	7	Orava		1098	10	Huelva (RNE5)
	702	200	Presov		1107	20	Caceres (RNE5)
	702	200	Banska Bystrica		1107	25	Logrono (RNE5)
	792	5	Bratislava		1107	20	Santander (RNE5)
	864	1	Stakein		1107	10	Ponferrada (RNE5)
	900	7	Poprad (Tatry)		1107	10	Teruel (RNE5)
	927	50	Nitra		1107	25	Toledo (RNE5)
	927	10	Kosice		1116	5	Pontevedra (SER)
Slovenia	1170	300	Beli Kriz		1116	5	Albacete (SER)
	1485	1	Ptuj		1116	5	Tarrasa (SER)
	1485	1	Radlje		1125	10	Castellon (RNE5)
	1584	1	Brezice		1125	10	Soria (RNE5)
	1584	1	Kovor		1125	10	Vitoria (RNE5)
	1602	1	Ormoz		1125	10	Toledo (RNE5)
	549	20	Beli Kriz		1125	10	Badajoz (RNE5)
	558	20	Maribor		1134	5	Puertollano (COPE)
	594	1	Cerkno		1134	2	5 st's (COPE)
	648	10	Murska Sobota		1143	2	Réus (COPE)
	918	600	Domzale		1143	5	"Oviedo (COPE,inactive)"
Somalia	963	150	Mogadishu (962)		1143	2	Orense (COPE)
South Africa	1035	100	Tshisahulu		1143	5	Jaen (COPE)
	1098	100	Ga-Rankuwa		1152	10	Lerida (RNE5)
	1287	2	Welgedacht		1152	10	Zamora (RNE5)
	1332	10	Gauteng		1152	10	Cartagena (RNE5)
	1404	2	Welgedacht		1152	10	Albacete (RNE5)
	1485	1	Johannesburg (R. Today)		1152	20	Malaga (RNE5)
	540	100/50	Ga-Rankuwa		1161	50	San Sebastian (EI)
	558	100	Umtata		1179	50	Valencia (SER)
	567	25	Cape Town		1179	2	Logrono (SER)
	576	50	Meyerton		1179	2	Logrono
	603	100	Umzimkulu		1197	10	Vitoria (EI)
	657	50	Bloemendal		1215	5	Santander (COPE)
	702	100	Ga-Rankuwa		1215	5	Leon (COPE)
	729	20	Cape Town		1215	2	Córdoba (COPE)
	846	100	Komga		1215	2	Lorca (COPE)
Spain	1008	5	Badajoz (SER)		1224	10	San Sebastián

Country	kHz	kW	Location	Country	kHz	kW	Location
	1224	2	6 st's (COPE)		621	50	Palma de Mallorca (RNE1)
	1260	5	Murcia (SER)		639	300	La Coruna (RNE1)
	1260	5	Algeciras (SER)		639	20	Almeria (RNE1)
	1269	10	Ciudad Real (COPE)		639	50	Bilbao (RNE1)
	1269	10	Figueras (COPE)		639	50	Zaragoza (RNE1)
	1269	5	Zamora (COPE)		639	10	Albacete (RNE1)
	1269	5	Badajoz (COPE)		648	10	Badajoz (RNE1)
	1287	10	Lerida-Lleida (SER)		657	50	Madrid (RNE5)
	1287	2	Lugo		666	50	Barcelona (SER)
	1287	2	Burgos (SER)		684	600	Sevilla (RNE1)
	1296	20	Valencia (COPE)		693	10	Toledo (RNE1)
	1305	25	Orense (RNE5)		693	10	Santa Barbara (RNE1)
	1305	20	Ciudad Real (RNE5)		711	25	Murcia (COPE)
	1305	10	Bilbao (RNE5)		729	100	Oviedo (RNE1)
	1305	10	León (RNE5)		729	20	Logrono (RNE1)
	1314	10	Cuenca (RNE5)		729	20	Malaga (RNE1)
	1314	10	Salamanca (RNE5)		729	10	Alicante (RNE1)
	1314	20	Reus (RNE5)		729	10	Cuenca (RNE1)
	1341	2	Leon (SER)		729	20	Valladolid (RNE1)
	1341	2	Almeria (OCR)		738	600	Barcelona (RNE1)
	1341	2	Ciudad Real (OCR)		747	10	Cádiz (RNE5) F.Pl. 936
	1359	600	Madrid (RNE-FS)		756	25	Bilbao (EI)
	1413	10	Jaen (RNE5)		774	100	Valencia (RNE1)
	1413	5	Gerona (RNE5)		774	50	Cáceres (RNE1)
	1413	5	Vigo (RNE5)		774	25	San Sebastián (RNE1)
	1485	5/2	"4 st's (SER, OCR)"		774	25/10	5 st's (RNE1)
	1503	2	Monforte de Lemos (RNE5)		783	50	Barcelona (COPE)
	1521	2	Castellón (SER)		792	50	Sevilla (SER)
	1539	2	Manresa (SER)		801	20	Ciudad Real (RNE1)
	1539	2	Elche-Elx (SER)		801	20	Lugo (RNE1)
	1575	5	Cordoba (SER)		801	10	3 st's (RNE1)
	1575	5	Pamplona (SER)		810	50	Madrid (SER)
	1584	2	Gandía (SER)		819	10	San Sebastian(EI)
	1584	2	Orense (SER)		837	50	Sevilla (COPE)
	1584	2	Zamora (SER)		837	10	Burgos (COPE)
	1602	10	Vitoria (EI)		837	2	El Ferrol (COPE)
	1602	5	Linares (SER)		837	2	Ibiza
	1602	2	Cartagena (SER)		855	300	Murcia (RNE1)
	1602	2	Onteniente (SER)		855	50	Santander (RNE1)
	1602	2	Segovia (SER)		855	20	Pontevedra (RNE1)
	531	25	Pontevedra (RNE5)		855	10/5	7 st's (RNE1)
	531	20	Oviedo (RNE5)		864	5	Socuellamos (RNE1)
	531	10	Pamplona (RNE5)		873	10	Sant. de Compostela (SER)
	531	10	Cordoba (RNE5)		882	20	Barcelona
	540	50	Barcelona (OCR)		882	5	Málaga (COPE)
	558	50	Valencia (RNE5)		882	5	Valladolid (COPE)
	558	25	La Coruna (RNE5)		882	2	Gijón (COPE)
	558	10	San Sebastian (RNE5)		882	2	Alicante (COPE)
	567	50	Murcia (RNE5)		900	25	Bilbao (COPE)
	567	10	Socuéllamos (RNE5)		900	5	Vigo (COPE)
	567	5	Marbella (RNE5)		900	5	Granada (COPE)
	576	100	Barcelona (RNE5)		900	5	Cacéres (COPE)
	585	600	Madrid (RNE1)		909	10	Palma de Mallorca (RNE5)
	603	50	Sevilla (RNE5)		918	50	Madrid (IND)
	603	5	Palencia (RNE5)		936	20	Valladolid (RNE5)
	612	10	Lleida (RNE1)		936	20	Zaragoza (RNE5)
	612	10	Vitoria (RNE1)		936	10	Alicante (RNE5)
	621	10	Jaén (RNE1)		954	50	Madrid (IND)
	621	10	Avila (RNE1)		963	10	Vitoria (EI)

Country	kHz	kW	Location	Country	kHz	kW	Location
	972	4	Cabra (RNE1)		882	1	Remada
	972	2	Monf. de Lemos (RNE1)		963	200	Tunis-Djedeida
	990	10	Bilbao (SER)	Turkey	1017	1200	Mundanya
	990	5	Cádiz (SER)		1062	300	Diyarbakir
	999	50	Madrid (COPE)		1200	-	Izmir
Sudan	1296	1500	Reba		1580	0.01	Adana(AFRS)
	540	50	Nyala		162	1000	Agri
	576	100	Khartoum		180	1200	Polatli
	576	7	Omdurman		198	120	Etimesgut
	639	200	Juba+ 1 st		225	600	Van
	666	5	Kassala		243	200	Erzurum
	765	50	Omdurman		558	600	Denizli
	783	5	Atbara		594	600	Malatya
	819	5	Dongola		630	300	Tarsus
	873	5	Medani		702	150	Catalca
	963	100	Soba		765	600	Gaziantep
Sweden	1179	600	Sölvesborg		891	600	Antalya
Switzerland	1485	1	Savièse		927	200	Izmir
	531	500	Beromünster		954	300	Trabzon
	558	300	Monte Ceneri-Cima	Uganda	1485	10	Arua
	765	600	Sottens		576	100	Mawagga
Swaziland	1170	100	Mpangela Range		639	50	Kampala
	1377	50	Sandlane		729	100	Butebo
	954	50/30	Sidvokodvo		810	100	Bobi
Syria	1071	60	Tartus		909	20	Kampala
	1125	200	Al-Hassake		999	100	Kabale
	1314	10	Aleppo	Ukraine	1071	50	Dnipropetrovsk+ 1 st
	1485	10	Homs		1134	5	Luhansk
	567	1000	Damas-Adra		1215	7	Dniperopetrovsk
	666	600	Damas-Sabboura		1242	50	Uzhhorod + 7 st's
	747	100	Sarakeb		1242	50	Oktyabrskoye + 6 st's
	783	600	Tartus		1260	50	Kharkiv
	828	100	Deir-Ez-Zoor		1278	150	Petrivka + 1 st
	873	10	Kharabo (877)		1332	7	Odesa
	918	200	Al-Hassake		1350	8	Novodnistrovsk + 3 st's
	954	60	Deir El Zawr		1359	50	Dokuchayevsk + 1 st
Tanzania	1035	10	Dar es Salaam		1377	50	Chernivtsi + 5 st's
	1215	50/10	Arusha		1395	50	Oktyabrskoye
	1440	10	Dar-es-Salaam		1404	50	Lviv + 8 st's
	531	10	Dar es Salaam		1431	500	Mykolayiv-Luch
	585	10	Chumbum		1449	50	Chernivtsi
	603	100/10	Dodoma		1458	3	Mariupol
	621	50/10	Mbeya		1467	0.1	Yalta
	648	100/10	Nachingwea		1476	30	Lviv
	657	2*50	Dar-es-Salaam		1476	7	Sevastopol
	711	100/10	Kigoma		1485	7	Kharkiv + 2 st's
	720	50/10	Mwanza		1512	10	Kyyiv
	837	1	Dar-es-Salaam		1530	30	Vinnytsya + 1 st
	990	100/10	Songea		1539	7	Kharkiv
Tchad	840	20	N`djamena		1557	8	Chernivtsi
	850	1	Sarh		1566	7	Odesa + 1 st
	720	0.1	Sarh		1575	0.1	Botanichnoye
Togo	1395	20/1	Lomé (1394)		1584	1	Simferopol + 3 st's
	1503	10	Kara (1502)		1584	0.1	Sevastopol
Tunisia	1566	1200	Sfax		1584	0.1	Chernihiv
	585	350	Gafsa		1593	1	Polonne
	603	10	Monastir		1602	7	Sevastopol + 2 st's
	630	600	Tunis-Djedeida		1602	1	Uzhhorod (BBC)
	720	100	Sfax-Sidi Mansour		1602	1	Luhansk

Country	kHz	kW	Location	Country	kHz	kW	Location
	171	1000	Lviv		1152	0.8	Norwich
	207	500	Kyyiv		1152	0.3	Plymouth
	549	500	Mykolayiv-Luch		1161	1	Bexhill
	549	150	Kyyiv + 7 st's		1161	1.4	Dundee
	612	5	Kyyiv (BBC)		1161	0.16	Swindon
	612	10	Kharkiv		1161	0.35	Hull
	648	150	Oktyabrskoye		1161	0.1	Bedford
	648	150	Kharkiv		1170	0.32	Stockton
	657	25	Chernivtsi		1170	0.58	Swansea
	675	25	Uzhgorod + 1 st		1170	0.28	Ipswich
	711	50	Dokuchayevsk		1170	0.12	Portsmouth
	765	50	Petrivka		1170	0.2	Stoke-on-Trent
	783	150	Kyyiv		1170	0.25	High Wycombe
	810	25	Tokmak + 2 st's		1197	1/0.2	Hoo + 9 st's
	819	8	Novodnistrovsk		1215	100/0.004	Moorside Edge + 13 st's
	837	150	Kharkiv + 1 st		1224	0.5	Manningtree
	873	50	Kyyiv + 4 st's		1233	0.2	Reading / Swindon
	873	10	Dnipropetrovsk		1242	1	Dundee + 2 st's
	891	150	Uzhgorod		1242	0.5	Isle Of Wight
	936	1000	Lviv		1242	0.32	Maidstone
	936	3	Starobilsk		1251	0.76	Bury St. Edmunds
	972	500	Mykolayiv-Luch		1260	1.6	Bristol
UAE	1107	10	Dubai		1260	0.64	Wrexham
	1152	50	Ras al Khaimah		1260	0.29	Leicester
	1175	10	Ras al Khaimah		1260	0.5	Guildford
	1251	600	Dubai		1260	0.5	Scarborough
	1314	2*1000	Dabiya		1278	0.4	Bradford
	1476	1500	Dubai		1296	500	Orfordness
	1539	2*50	Sadiyat		1296	6.4	Birmingham
	1575	50	Sharjah		1305	0.5	Enfield
	657	2*50	Sadiyat		1305	0.5	Ewell
	729	2*750	Sadivat		1305	0.2	Newport
	810	50	Maqtaa		1305	0.15	Barnsley
	828	1	Al Ain		1323	0.63	Taunton
	846	20	Umm al Qiwain		1323	0.5	Brighton
	972	2*50	Fujairah		1323	0.001	Leeds
United Kingdom	1017	2	Shrewsbury		1332	1	Bow(London)
	1017	0.6	Shropshire		1332	0.6	Peterborough
	1026	1	Jersey		1332	0.3	Lacock
	1026	1.7	Belfast		1341	100	Lisnagarvey
	1026	0.5	Cambridge		1359	0.85	Bournemouth
	1035	1	Sheffield		1359	0.2	Cardiff
	1035	1	London		1359	0.3	Coventry
	1035	0.32	Ayr		1359	0.3	Chelmsford
	1035	0.78	Aberdeen		1368	20	Douglas
	1053	100/1	Droitwich + 13 st's		1368	0.5	Duxhurst
	1071	1	3 st's		1368	0.1	Swindon
	1089	100	Brookmans Park + 7 st's		1368	2	Lincoln
	1107	1.5	Inverness		1377	1	Rochdale
	1107	2/1	Lydd + 4 st's		1386	0.001	Oldham
	1116	1.2	Derby		1386	0.5	Devizes
	1116	0.5	Guernsey		1413	0.5	Heathrow
	1116	0.5	Welsh valleys		1413	0.5	Dartford
	1125	1	Llandrindod Wells		1413	0.1	Skipton
	1152	23.5	London		1413	0.001	Leeds
	1152	1.8	Newcastle		1413	0.001	Manchester
	1152	3.6	Glasgow		1413	0.001	Newcastle
	1152	3.0	Birmingham		1413	0.001	Blackburn
	1152	1.5	Manchester		1413	0.5	Bourton-on-the-Water

Country	kHz	kW	Location
	1431	0.48	Rayleigh
	1431	0.14	Reading
	1449	2	Redmoss
	1449	0.15	Gunthorpe
	1458	50/0.5	London + 5 st's
	1458	5	Manchester
	1476	1	Guildford
	1485	2/1	Wallasey + 4 st's
	1485	1	Basingstoke (F.Pl.)
	1503	1	Stoke-on-Trent
	1503	0.001	Northants
	1521	0.64	Reigate
	1521	-	Craigavon
	1530	0.74	Huddersfield
	1530	0.52	Worcester
	1530	0.15	Southend-on-Sea
	1548	97.5	London
	1548	4.4	Liverpool
	1548	2.2	Edinburgh
	1548	0.74	Sheffield
	1548	5	Bristol
	1557	0.8	Northampton
	1557	0.5	Southampton
	1557	0.25	Oxcliffe
	1557	0.125	Clacton
	1566	0.001	RSL-stations
	1575	0.001	Wolverhampton
	1584	1	Clipstone
	1584	0.2	Perth
	1584	0.04	Kettering
	1584	0.3	Wofferton
	1584	0.2	London (LTR)
	1584	0.001	Wigan
	1584	0.5	Shrewsbury
	1584	0.01	Corby
	1602	0.25	Rusthall
	1602	0.001	Kilmarnock
	1602	0.001	Liverpool
	1602	0.001	Norwich
	1602	0.001	Donnington
	1602	0.001	Fife
	1602	0.001	Silverstone
	1602	0.001	RSL-stations
	198	500	Droitwich
	198	50	Burghead
	198	50	Westerglen
	558	0.8	London
	585	2	Dumfries
	603	0.1	Littlebourne
	603	0.1	Cheltenham
	603	2	Newcastle
	630	2	Redruth
	630	0.2	Luton
	648	500	Orfordness
	657	2	Wrexham
	657	0.5	Bodmin
	666	0.34	Exeter
	666	0.5	Fulford

Country	kHz	kW	Location
	693	150	Droitwich
	693	25/10/1	10 st's R5
	720	10	Lisnagarvey
	720	0.75	London
	720	0.75	Londonderry
	729	0.2	Manningtree
	738	0.04	Worcester
	756	2	Redruth
	756	0.63	Powys
	756	1	Carlisle
	765	0.5	Chelmsford
	774	1	Enniskillen + 1 st
	774	0.7	Littlebourne
	774	0.5	Farnley
	774	0.14	Gloucester
	792	0.2	Bedford
	801	2	Barnstaple
	810	100	Burghead
	810	100	Westerglen
	810	5	Redmoss
	828	0.27	Bournemouth
	828	0.12	Leeds
	828	0.2	Dunstable
	828	0.8	Cookstown
	828	0.2	Birmingham
	837	1.5	Barrow in Furness
	837	0.45	Freemen's Common
	837	0.4	Leicester
	855	2	Postwick
	855	1	Plymouth
	855	1.5	Preston
	855	0.15	Ludlow
	873	0.3	West Lynn
	873	1	Enniskillen
	882	100/1	Washford + 3 st's
	909	200	Moorside Edge + 10 st's R5
	936	1	Hawes
	936	0.18	Naish Hill
	945	0.7	Bexhill
	945	0.2	Derby
	954	0.32	Torquay
	954	0.16	Hereford
	963	1	London
	963	1	London
	963	1	Southall
	963	0.8	Blackburn
	972	1	London
	990	1	Exeter
	990	1	Tywyn
	990	1	Redmoss
	990	1	Aberdeen
	990	0.25	Doncaster
	990	0.1	Wolverhampton
	999	1	Fareham
	999	0.8	Preston
	999	0.25	Nottingham
Vatican City	1530	300/600	Vatican City
	1611	100	Vatican City

Country	kHz	kW	Location	Country	kHz	kW	Location
Yemen	1008	600	Sana'a		1584	1	Pancevo
	1080	30	Taiz		1584	1	Prijepolje
	1125	-	Hudaydah		1584	1	Urosevac
	1593	1	Beihan		1584	1	Urosevac
	711	200	Mocha		1593	0.2	Indjija
	756	-	Mukalla		1602	1	Vrsac
	837	30	Sana'a		1602	1	Negotin
Yugoslavia	1008	200	Beograd		1602	1	Leskovac
	1008	200	Aleksinac		1602	1	Sjenica
	1026	10	Kragujevac		1602	1	Budva + 4 st's
	1026	1	Bar		531	10	Uzice
	1035	10	Mitrovica		531	1	Vranje
	1044	1	Temerin		549	100	Pristina
	1062	10	Jagodina		567	10	Pljevlja
	1062	1	Novi Pazar		594	10	Niksic
	1089	10	Subotica		639	1	Soko Banja
	1107	150	Novi Sad		648	1	Lazarevac
	1161	1	Ulcinj		666	10	Sombor
	1161	0.2	Kovin		675	10/5	Bosilegrad
	1170	2	V. Banja		675	10/5	Dimitrovgrad
	1170	0.2	B. Topola		684	2000	Beograd
	1179	1	Kacanik		693	10/5	Negotin
	1206	1	Majdanpek		702	10	Sabac
	1215	1	Mladenovac		711	100/20	Nis
	1224	1	B.Petrovac		738	10	Krusevac
	1242	10	Kraljevo		738	5	Bar
	1269	750	Novi Sad		783	0.2	Stara Pazova
	1296	10	Loznica		792	1	Arandjelovac
	1296	10	Vranje		837	50	Sombor
	1305	1	Herceg Novi		882	600/300	Podgorica + 5 st's
	1323	1	Sid		900	2	Beograd(Studio B)
	1341	10	Zajecar		936	10	Djakovica + 1 st
	1350	10	Beograd		945	1	Smederevo
	1359	2	Vrbas		954	10	Gnjilane
	1368	10	Valjevo		981	10	Bor
	1377	10	Prizren		981	10	Cacak
	1395	0.2	Kovacica		990	1	Pozarevac
	1413	100	Pristina		999	10/5	Kladovo + 1 st
	1431	0.2	Beocin	Zambia	1071	100	Kitwe
	1440	20/10	Jagodina		1125	10	Solwezi
	1449	2	Srbobran		1125	1	Kabwe(1124)
	1458	1	Kladovo		1161	1	Kabwe(1157)
	1458	1	Niksic		549	100	Kitwe
	1467	2	Zrenjanin		567	10	Kasama
	1485	2	Novi Sad		603	10	Mongu
	1485	1	Crna Trava + 6 st's		630	500	Lusaka
	1485	1	Priboj		666	10	Chipata
	1485	1	Cetinje + 3 st's		729	10	Livingstone
	1494	0.2	Apatin		819	200	Lusaka (818)
	1503	10	"Beograd""202"""		828	10	Mongu
	1503	10	Ulcinj		855	10	Chipata
	1512	10	Pristina		882	10	Kasama
	1539	10	Pec		909	10	Solwezi
	1539	0.2	Odzaci		918	10	Mansa
	1548	1	Bar		927	10	Livingstone
	1548	1	Stara Pazova				
	1566	1	Smed. Palanka				
	1575	1	Bac + 1 st				

EAST ASIA & PACIFIC

Abbreviations peculiar to the E.Asia/Pacific section of MW freq. lists: AF = allocated freq. C. = City. PO = Present operation on. Proj. = Projected station. Rptr. = repeater. Trtr = translator. Unk. = Unknown location.

Australia: The numeral preceding the call letters indicates the state: 2 = New South Wales. 3 = Victoria. 4 = Queensland. 5 = South Australia. 6 = Western Australia. 7 = Tasmania. 8 = Northern Territory. ACT = Australian Capital Territory.

China, P.R: If several locations are listed for one frequency, the power listed applies to the first entry. For full details see country section. A 'v' in brackets under location: variable frequency.

Indonesia: Only RRI stns included. For details of other stns see country section.

Philippines: Province Abbreviations: Ak = Aklan; Ant = Antique; Boh = Bohol; Bat = Batangas; Buk = Bukidnon; Bul = Bulacan; Cag = Cagayan; Cam Nte = Camarines Norte; Cam Sur = Camarines Sur; Dvo Nte = Davao Norte; Dvo Sur = Davao Sur; Isa = Isabela; IN = Ilocos Norte; IS = Ilocos Sur; LU = La Union; Mas = Masbate; MM = Metropolitan Manila; M Octal = Mindoro Occidental; M Ortal = Mindoro Oriental; Pang = Pangasinan; Que = Quezon; Riz = Rizal; S Sur = Surigao del Sur; Sor = Sorsogon; Tar = Tarlac; Z Nte = Zamboanga del Norte; Z Sur = Zamboanga del Sur.

Russia: Regions in the Asian parts of Russia: Sib. = Siberia. FE = Far East.

Country	kW	Call	Location	Country	kW	Call	Location
153 kHz				**531 kHz**			
Russia	1200		Komsomolsk, FE	Australia	5	2MC	W. Kempsey
Turkmenistan	500		Asgabat		5	3GG	Warragul
162 kHz					5	4KZ	Innisfail
Mongolia	1000		Ulaanbaatar		0.5	5UV	Adelaide (Univ.)
Uzbekistan	150		Toskent		10	6DL	Dalwallinu (RR)
Russia	150		Norilsk, Sib.	China, PR	10		ZJ
171 kHz				India	100		Jodhpur A
Russia	250		Oyash, Sib.	Japan	10	JOQG	Morioka (NHK1)
	150		Yakutsk, Sib.	New Zealand	2		Alexandra (rel 4XO)
180 kHz							Auckland
Kazakhstan	250		Almaty	Philippines	5	DZBR	Lipa C., Bat.
	150		Aktöbe		5	DZAH	Davao C.
	50		Türkistan	Sri Lanka	40		Ambewela
Russia	150		Chita, Sib.	Thailand	50		Maha Sarkham
	150		Petropavlovsk, FE	**540 kHz**			
189 kHz				Australia	10	4QL	Longreach (RR)
Russia	1200		Belogorsk, Sib.		5	7SD	Scottsdale
198 kHz				China, PR			(C1) Synchro
Kyrgyzstan	150		Biskek	India	20		Aizawl
Russia	250		Angarsk, Sib.	Indonesia	10		Bandung (RRI)
207 kHz				Japan	5	JOJG	Yamagata (NHK1)
Mongolia	60		Ulgii, Bayan-Ulgii		5	JOMG	Miyazaki (NHK1)
	150		Dalanzadgad, So. Gobi		1	JOSK	Kitakyushu (NHK1)
	150		Choibalsan, Ea. Aimak	Korea, Rep.	10	HLCZ	Hongsong (KBS)
Russia	150		Tynda, Sib.		1	HLSC	Chomchòn (KBS)
216 kHz					1	HLSM	Changhung (KBS)
Russia	150		Krasnoyarsk, Sib.		1	HLSN	Changsu (KBS)
	30		Birobidzhan, FE	New Zealand	5	1XC	Tauranga (R. Rhema)
225 kHz					4	2XV	New Plymouth (R. Rhema)
Russia	1000		Surgut, Sib.	Pakistan	300		Peshawar 1
227 kHz				Philippines	10	DYCU	Cebu C.
Mongolia	150		Altai, Gobi		10	DZWT	Baguio C., Benguet
234 kHz				Samoa, W.	10	2AP	Apia
Russia	1000/300		Angarsk, Sib.	Thailand	10		Bangkok
	1000		Arman, FE	**549 kHz**			
243 kHz				Australia	50	2CR	Orange (RR)
Kazakhstan	1000		Almaty	China, PR	100		FJ (T1);NM;HEN
	1000		Karagandy	China (Taiwan)			various (Fuhsing)
Russia	1200		Razdolnoye, FE	India	100		Ranchi
252 kHz				Japan	10	JOAP	Okinawa (NHK1)
Tajikistan	150		Yangi-Yul	Kazakhstan	1000		Almaty
261 kHz				Korea, Rep.	5		Seoul (AFKN)
Russia	1200		Chita (Atamanovska), Sib.	Malaysia	20		Kuching, Sarawak
270 kHz				New Zealand		F.PI.	Kaitaia (R. Rhema)
Russia	150		Novosibirsk, Sib.	Philippines	10	DWRP	Naga C. Cam. Sur
	50		Orenburg, Sib.		5	DXHM	
	150		Khabarovsk, FE		1	DYAF	
Tajikistan			Orzu		1	DXIM	Makati, MM
279 kHz				Russia	50-500		4 st's (sync.)
Russia	150		Selenginsk, Sib.	Tajikistan	40		Dushanbe
	50		Gorno-Altaysk, Sib.	Thailand	10		Krabi
	150		Yekaterinburg, Sib.		100		Lampang
	1000		Yuzhno-Sakhalinsk, FE		10		Mukdahan
Turkmenistan	150		Asgabat	Uzbekistan	5		Nukus

Country	kW	Call	Location
Vietnam			Mê Tri
550 kHz			
Hawaii	5	KMVI	Wailuku, Maui
Vietnam			Lam Dong
558 kHz			
Australia	5	4AM	Atherton (Mareeba)
	5	4GY	Gympie
	50	6WA	Wagin (RR)
	2	7BU	Burnie
Bangladesh	100		Khulna
China, PR			FJ;XJ;NM
China (Taiwan)	1	BEV58	Chingshui (Cheng Sheng)
	1	BEH7	Taipei (Fu Hsing)
Fiji	20	VRH	Suva (Naulu)
India	100		Mumbai B
Japan	10		Kobe (AM Kobe)
Korea, Rep.	250	HLQH	Taegu (KBS)
Philippines	40	DZXL	Makati, MM
Russia	25		Ulan-Ude, Sib.
Sri Lanka	20		Diyagama (Colombo)
Thailand	10		Chaiyaphum
	50		Songkhla
Vietnam			Quan Tre
567 kHz			
Australia	0.5	2BH	Broken Hill
	10	4JK	Julia Creek (RR)
	0.1	6...	4 st's W.A. (RR)
China, PR			(C1);TJ
Guam	5	KGUM	Agana
Hong Kong	20		Golden Hill (RTHK3)
India	300		Dibrugarh
Japan	100	JOIK	Sapporo (NHK1)
Korea, Rep.	100	HLKF	Chonju (KBS)
Malaysia	10		Tenom, Sabah
New Zealand	50	2YA	Wellington
Pakistan	300		Khuzdar
Philippines	10	DYCA	Agoo, LU
Russia	150		Kyzyl, Sib.
Sri Lanka	10		Senkadagala (Kandy)
570 kHz			
Hawaii	1	KQNG	Lihue, Kauai (F.PI: 3 kW)
Vietnam			Ha Giang BS
			Mê Tri
572 kHz			
Vietnam			Dong Dé
576 kHz			
Australia	50	2RN	Sydney (RN)
China (Taiwan)	10	BEM2	Taipei (ICRT)
China, PR			YN;ZJ(v);JL;HEN
India	100		Alappuzha
Japan	10	JOHG	Kagoshima (NHK1)
Korea, Rep.	1	HLKZ	Sunch'on (KBS)
	5		Munsan (AFKN)
Kyrgyzstan	40	Os	
Malaysia	100		Tronoh
	50		Johore Bharu
	20		Miri, Sarawak
Nepal	100		Surkhet
Philippines	10	DYMR	Cebu C
	1	DZYZ	Laoag C, IN
	5	DZMQ	
	5	DXCH	
	1	DXMF	
Russia	100-1000 5 st's. (sync)		
Thailand	20		Bangkok
Turkmenistan	150		Asgabat
Uzbekistan	50		Toskent
	7		Vobkent
577 kHz			
Myanmar	200		Yangon
580 kHz			
Easter Isl.	0.4		Mataveri Airport

Country	kW	Call	Location
Laos	150		Vientiane
Pakistan	1000		Lawrencepur (AF585)
Vietnam	50		Nhatrang
			Yên Bai
			Khanh Hoa/Tay Ninh
585 kHz			
Australia	5	6PB	Perth (PNN)
	10	7RN	Hobart (RN)
	2	2WEB	Bourke (Educ)
China, PR	10		JS;JX;FJ
China (Taiwan)	1200		Fangliao (VO Asia)
India	100		Nagpur A
Indonesia	50		Surabaya (RRI)
Japan	10	JOPG	Kushiro (NHK1)
Korea, Rep.	5		Taegu (AFKN)
New Zealand	2	2XR	Ruatoria
		F.PI.	Blenheim (R. Rhema)
Pakistan	1000		Lawrencepur (r. 580)
Philippines	5	DXCP	Makati, MM
	10	DYCI	Iloilo C
Papua N. Guinea	5		Port Moresby
Russia	1200		Belogorsk, Sib. (FS)
Sri Lanka	20		Kantale
Thailand	10		Chumporn
	20		Prachinburi
	10		Phitsanulok
	100		Phrae
590 kHz			
Hawaii	7.5	KSSK	Honolulu
594 kHz			
Australia	50	3WV	Horsham (RR)
Brunei	200		Tutong
China, PR	300		XZ;FJ(T2);SD
China (Taiwan)	10	BEH2	Taipei + 2 st's (Fu Hsing)
India	1000		Chinsurah
Japan	300	JOAK	Tokyo (NHK1)
Korea, Rep.	10	HLAG	Yongju (KBS)
New Zealand	2	3XL	Timaru (R. Rhema)
			Wanagnui (R. Rhema)
Niue	0.4	ZK2ZN	Alofi
Philippines	20	DZBB	Quezon C, MM
	10	DYWR	
	5	DXOB	
Russia	1000		Surgut, Sib.
	150		Krasnoyarsk, Sib.
Sri Lanka	20		Weeraketiya
Thailand	10		Khonkaen
600 kHz			
China, PR			SC
Vietnam			Quang Nam/Da Nangh
603 kHz			
Australia	10	2RN	Nowra (RN)
	10	4CH	Charleville (RR)
	2	6PH	Port Hedland (RR)
China, PR			GD + 10 st's
China (Taiwan)	1000		Lukang (CBS)
India	200		Ajmer
Japan	5	JOKK	Okayama (NHK1)
	5	JOOG	Obihiro (NHK1)
Korea, Rep.	500	HLSA	Namyang (Seoul) (KBS)
Malaysia	10		Batu Pahat
	10		Kota Kinabalu, Sabah
New Zealand	5	1XO	Auckland
Philippines	10	DWLV	Naga C.
	10	DXPR	Pagadian C, Zamb. Sur
	5	DZVV	Vigan, I.S.
Russia	50		Salekhard, Sib.
	30		Belogorsk/Skovorodino
Sri Lanka	10		Ratnapura
Thailand	20		Bangkok
610 kHz			
Vietnam	20		Ho Chi Minh C.

Country	kW	Call	Location
612 kHz			
Australia	50	4QR	Brisbane (MS)
	0.2	6NM	Northam (ABC3)
	10	6RN	Dalwallinu (RR)
China, PR	100		FJ;LN;SC
China (Taiwan)	6		various (Fuhsing)
Guam	10	KUAM	Agana
India	200		Bangalore A
Japan	100	JOLK	Fukuoka (NHK1)
Kyrgyzstan	150		Bishkek
Malaysia	10		Kangar
New Zealand	2	3XG	Christchurch (R. Rhema)
Philippines	5	DWHB	Baguio C., Benguet
	10	DYHP	Cebu C. (r 615)
	5	DZTG	Teguegarao, Cag.
Russia	25		Norilsk, Sib.
Thailand	20		Lopburi
620 kHz			
Hawaii	10	KIPA	Hilo
	10	KIPA-1	Kona/Keaan (rptr KIPA)
621 kHz			
Australia	50	3RN	Melbourne (RN)
China, PR	100		HL;QH
Hong Kong	20		Golden Hill (RTHK)
India	100		Patna A
Japan	3	JOCG	Asahikawa (NHK1)
	1	JOOK	Kyoto (NHK1)
Korea, DPR	500		Ch'ongjin (PBS+FS)
Korea, Rep.	10	HLCF	Sogwipò (KBS)
	10	HLSJ	T'aebaek (KBS)
	1	HLAY	Yongdong (KBS)
Malaysia	20		Sibu, Sarawak
	100		Segamat
New Zealand	2	4XG	Dunedin (R.Rhema)
Philippines	10	DYHP	Davao C.
	5	DWBH	
	1	DWHB	Legaspi, MM
Russia	50		Khabarovsk, FE
	25		Yakutsk, Sib.
Sri Lanka	20		Diyagama (Colombo)
Thailand	100		Khon Kaen
Tuvalu	10	T2U2	Funafuti
630 kHz			
Australia	10	2PB	Sydney (PNN)
	50	4QN	Townsville (RR)
	5	6AL	Albany (RR)
	0.4	7QN	Queenstown (RR)
Bangladesh	100		Dhaka B
China, PR			(C2) synchro;NX
China (Taiwan)	10	BEG51	Ilan (BCC3)
	1		Sungling (TBC)
Cook Is.	5	ZK1ZC	Rarotonga (Black Rock)
India	100		Trichur
Indonesia	50		Ujung Pandang (RRI)
Korea, Rep.	10	HLCY	Yosu (KBS)
	5	HLSE	Inje (KBS)
New Zealand	20	2YZ	Napier (Opapa)
Pakistan	100		Lahore I
Philippines	50	DZMM	Quezon C., MM
Russia	500		Komsomolsk, FE (FS)
Thailand	10		Bangkok
Vietnam			Playcu, Gia Lai prov.
			Dong Hoi
639 kHz			
Australia	5	2CS	Coff's Harbour
	1	4MS	Mossman (RR)
	10	5CK	Port Pirie (RR)
	2	8RN	Katherine (RN)
China, PR			(C1)
China (Taiwan)	1		Tahsi (TBC)
Fiji	10		Lautoka (Drasa)
India	50		Kohima
Japan	10	JOPB	Shizuoka (NHK2)
	5/1	JOWN	Hakodate (STV)
Korea, Rep.	50		Kaebong (Seoul) (EBS)
New Zealand	2	4YW	Alexandra
Pakistan	10		Karachi II
Philippines	5	DYAG	Cadiz, Neg. Occ.
	5	DXKR	Marbel, Cot Sur
	1	DZRL	Batac, IN
Russia	150		Omsk, Sib.
Sri Lanka	50		Maho
Thailand	20		N. Si Thammarat
	10		Phuket
640 kHz			
Laos	10		Vientiane
648 kHz			
Australia	10	2NU	Tamworth (RR)
	2	6GF	Kalgoorlie (RR)
China, PR	150		GD;SH;AH
India	100		Indore A
Indonesia	1/5		Dili (RRI)
Japan	10		Okinawa C. (FEN)
	5	JOIG	Toyama (NHK1)
Korea, Rep.	1	HLSL	Posong (KBS)
Malaysia	20		Kuala Lipis
	20		Limbang, Sarawak
Nepal	100		Dhankuta
New Zealand		2XC	Napier (R. Rhema)
Philippines	5	DXMB	Malaybalay, Buk.
	2.5	DXRH	Tacloban C, Leyte
	5	DWPS	Puerto Princesa, Palawan
	10	DWRH	
	5	DYRC	
Russia	1000		Ussuriysk, FE (FS)
Samoa, Am.	10	WVUV	Pago Pago (Vailoa)
Sri Lanka	20		Ambewela
Tajikistan	1000		Orzu
Thailand	55		Khon Kaen
Uzbekistan			Samarkand
Vietnam			An Nhon
650 kHz			
Hawaii	10	KHNR	Honolulu
Vietnam			Ha Bac
655kHz			
Vietnam			Ho Chi Minh C.
657 kHz			
Australia	10	2BY	Byrock (RR)
	2	8RN	Darwin (RN)
China, PR	300		HEN;JL;ZJ
China (Taiwan)	20	BED34	Taipei (BCC1)
	10	BEV59	Taichung (Cheng Sheng)
India	100		Calcutta A
Korea, DPR	1500		Kangnam (KCBS)
Korea, Rep.	50	HLKM	Chùnch'on (KBS)
Malaysia	20		Grik
New Zealand	60	2YC	Wellington
Philippines	5	DXDD	Ozamis C., Mis. Occ.
	5	DWRN	Naga C., Cam. Sur
	1	DYFL	Borongan, E. Samar
	1	DZXC	San Fernando, LU
	1	DZLU	
	1	DYFL	
Thailand	10		Samut Sakhon
	1		N. Raichasima
Vietnam			Quan Tre
666 kHz			
Australia	5	2CN	Canberra, ACT (MS)
	2	4LM	Mt. Isa
	2	4CC	Biloela (trtr)
	1	6LN	Carnarvon
China, PR	200		QH+ 7 st's
China (Taiwan)			various (Fuhsing)
India	100		Delhi B

Country	kW	Call	Location
Japan	100	JOBK	Osaka (NHK1)
Malaysia	10		Penang (r 665)
New Caledonia	20		Noumea (RFO)
Philippines	10	DXRP	Davao C.
	25	DZRH	Navotas, MM
Russia	7		Aldan, Sib.
	10		Yekaterinburg
	150		Komsomolsk, FE
Thailand	10		Tak
Uzbekistan	30		Toskent
	1		Samarkand
Vietnam			Dong De

670 kHz

Country	kW	Call	Location
Hawaii	10	KPUA	Hilo

675 kHz

Country	kW	Call	Location
Australia	10	2CO	Corowa (RR)
	5	6BE	Broome (RR)
Brunei	200		Serasa
China, PR	150		NM;XJ
Hong Kong	10		Peng Chau (RTHK)
India	100		Itanagar
	20		Chhattarpur
	1		Bangalore B
Japan	5	JOUG	Yamaguchi (NHK1)
	5	JOVK	Hakodate (NHK1)
Korea, Rep.	10	HLAS	Kunsan (KBS)
Malaysia	10		Lahad Datu, Sabah
New Zealand	20	3YA	Christchurch
Papua N. Guinea	2		Lae/Wewak
Philippines	5	DYKC	Mandaue (Cebu)
	5	DWLW	Laoag C., IN
	1	DXCD	Bongao, Sulu
Russia	250		Oyash, Sib.
Sri Lanka	20		Weeraketiya
Thailand	20		Chieng Mai
Turkmenistan	150		Asgabat
Uzbekistan			Zarafson
	20		Koson
	5		Küngirod
Vietnam			Hanoi

684 kHz

Country	kW	Call	Location
Australia	10	2KP	Kempsey (RR)
	4	6BS	Busselton (RR)
	1	8RN	Tennant Creek (RN)
China, PR			FJ + 7 st's
China (Taiwan)	10	BEC22	Taipei (Han Sheng)
Fiji	10		Labasa
India	10		Kozhikode A
	20		Port Blair
Japan	5	JOAG	Nagasaki (NHK1)
	5	JOUG	Yamaguchi (NHK1)
	5	JODF	Morioka (IBC)
Kazakhstan	30		Kostanay
Korea, DPR	250		Samgo (PBS)
Nepal	100		Pokhara
Philippines	5	DWGW	Legaspi C., Albay
	1	DWAU	Cabanatuan, Nva. Ecija
	10	DYEZ	Bacolod, Neg. Occ.
	5	DZCV	Tuguegarao, Cag.
Thailand	20		Bangkok

690 kHz

Country	kW	Call	Location
Hawaii	10	KQMQ	Honolulu

693 kHz

Country	kW	Call	Location
Australia	5	3EE	Melbourne
	5	4KQ	Brisbane (Newstead)
	0.5	4KZ	Tully (trtr)
	0.5	4LM	Cloncurry (trtr)
	2	5SY	Streaky Bay (RR)
	2	6TZ	Bunbury
	5	6WR	Kununurra
Bangladesh	1000		Dhaka A
China, PR	100		SN

Country	kW	Call	Location
China (Taiwan)	10	BEC25	Taoyuan (Han Sheng)
	10	BEC32	Tainan (Han Sheng)
	10	BEH39	Taichung (Fu Hsing)
	1	BEH82	Hualein (Fu Hsing)
Indonesia			Sungai Liat (RRI)
Japan	500	JOAB	Tokyo (NHK2)
Malaysia	10		Kota Kinabalu, Sabah (r690)
Philippines	5	DXBC	Butuan, Agusan Nte
	1	DXDX	Gen. Santos C., Cot Sur
	10	DYPH	
	1	DXDN	
Russia	25		Anadyr, FE
Sri Lanka	20		Amparai
Thailand	20		Saraburi
Vietnam			Dac Lac
			An Hai (PO 694)

700 kHz

Country	kW	Call	Location
China, PR			LN
Vietnam			Quang Ninh

702 kHz

Country	kW	Call	Location
Australia	50	2BL	Sydney (MS)
	10	6KP	Karratha (RR)
China, PR	150		JS + 4 st's
China (Taiwan)	10	BEP24	Taichung (Ching Cha)
India	300		Jalandhar B
Japan	10	JOFB	Hiroshima (NHK2)
	10	JOKD	Kitami (NHK2)
Korea, DPR	50		Ch'ongjin (KCBS)
Malaysia	20		Kota Bharu
New Zealand	5	1XP	Auckland
Philippines	40	DZAS	Valenzuela, MM
Russia	7		Bratsk, Sib.
Sri Lanka	25		Diyagama (Colombo)
Tajikistan	150		Orzu
Thailand	10		Lopburi
Vietnam			An Hai

705 kHz

Country	kW	Call	Location
Laos			Luang Prabang

710 kHz

Country	kW	Call	Location
Vietnam			Dong Thap

711 kHz

Country	kW	Call	Location
Australia	10	4QW	St. George (RR)
	10	7NT	Launceston (RR)
Brunei	20		S. Hanching
China, PR			7 st's
China (Taiwan)	250	BEC72	Hsinfeng (Kung Chun)
India	200		Siliguri
Korea, Rep.	500	HLKA	Seoul (Sorae) (KBS)
New Zealand		2XP	Wellington
Philippines	5	DXIC	Iligan C., Lanao Nte
	5	DXRD	Davao C.
	5	DZYI	Tacloban C, Leyte
	5	DYBR	
Russia	7		Okhotsk, FE
Thailand	20		Bangkok
	20		Chieng Mai
	20		U. Ratchathani
Uzbekistan			Buhoro
			Urgonç

720 kHz

Country	kW	Call	Location
Australia	0.4	2AN	Armidale (RN)
	0.4	2ML	Murwillumbah (RR)
	2	3MT	Omeo (RR)
	4	4AT	Atherton (RR)
	50	6WF	Perth (MF)
China, PR			(C2);SC
China (Taiwan)	10	BED58	Taichung (BCC1)
Hawaii	5	KUAI	Eleele, Kauai
India	200		Madras A
Indonesia	10		Ambon (RRI)
Japan	1	JOIL	Kitakyushu (KBC)

Country	kW	Call	Location	Country	kW	Call	Location
Korea, DPR	500		Wiwon (KCBS)		50		Uralsk
New Zealand	20	4YZ	Invercargill	Vietnam			Quan Tre
Philippines	10	DYOK	Iloilo C	**750 kHz**			
	5	DZSO	San Fernando, LU	China, PR			LN;GX;SC
	5	DZJO		**756 kHz**			
Russia	1000		Yuzhno-Sakhalinsk, FE	Australia	2	2TR	Taree (RR)
Turkmenistan	1		Ek-Arça/Gyzyletrek		10	3RN	Wangaratta (RN) (rptr)
Vietnam			Thua Thien Hué/Dong Nai		0.1	6KW	Kununurra (RR)
729 kHz							(C1)
Australia	50	5RN	Adelaide (RN)	China, PR			
	2	7PB	Hobart (PNN)	China (Taiwan)	1		Makung (Shengli)
China, PR	150		JX;HEN	India	20		Jagdalpur
China (Taiwan)	3	BEE43	Taipei (Shi Hsin)	Indonesia	2/10		Purwokerto (RRI)
India	100		Gowahati A	Japan	10	JOGK	Kumamoto (NHK1)
Japan	50	JOCK	Nagoya (NHK1)	Korea, Rep.	100		Suwon (KBS)
Korea, DPR	50		Sep'o (PBS)	New Zealand	20	1YA	Auckland
Malaysia	20		Kuching, Sarawak	Pakistan	150		Quetta 1
New Zealand	0.9	4XX	Ranfurly	Philippines	1	DWHL	Olongapo C, Zamb.
			Tokoroa		10	DWGC	
Philippines	10	DWPE	Tuguegarao, Cag		10	DZGC	Tayug, Pang. (inactive?)
	5	DXOR	Cagayan de Oro, Mis. Or.		5	DWNW	Naga C., Cam. Sur.
	5	DZGB	Legaspi C., Albay		1	DXJM	Butuan C., Agusan Nte
	10	DZPE		Thailand	50		Narathiwat
	5	DXAR		Uzbekistan	50		Piskent
Sri Lanka	10		Ratnapura		5		Samarkand
Thailand	50	HSKT	N. Raichasima	**760 kHz**			
735 kHz				Hawaii	10	KGU	Honolulu
Macau	10		Em. Vila Verde	Philippines	5	DYCB	Cebu C. (AF 765)
738 kHz				Vietnam			Tanan. (Long An BS)
Australia	50	2NR	Grafton (RR)	**765 kHz**			
	5	6MJ	Manjimup (RR)	Australia	3.5	2EC	Bega
China, PR	150		HN; JL; XJ;ZJ		0.5	4GC	Hughenden (trtr)
China (Taiwan)	100	BEL2	Penghu (Taiwan Yuyeh)		5	5CC	Port Lincoln
India	50		Hyderabad A		0.1		Tom Price/Parab. (trtr)
Indonesia	1		Jermber (RRI)	China (Taiwan)			various (Fu Hsing)
Japan	5	JOLR	Toyama (KNB)	China, PR			C1(v);FJ (T1);GZ; NM; GD
	1		Takaoka (KNB)	India	200		Dharwar
	5	JORR	Naha, Okinawa (RBC)	Indonesia	0.2		Banjarmasin (RRI)
Korea, Rep.	100	HLKG	Taegu (KBS)	Japan	5	JOJF	Kofu (YBS)
Philippines	10	DZFM	Quezon C. MM		5	JOPF	Tokuyama (KRY)
Polynesia (Fr.)	20		Papeete, Tahiti	Korea, DPR	50		Hyesan (KCBS)
Russia	40		Chelyabinsk, Sib.	Korea, Rep.	10	HLCQ	Taejon (MBC)
	25		Palana, FE	Malaysia	10		Kuala Trengganu
Thailand	20		Bangkok	New Zealand		2XT	Napier
	10		Chieng Mai	Philippines	1	DXGS	Gen. Santos C., S.Cotab.
Tonga	5	A3R	Tangatapy (UCB)		5	DYCB	Cebu C. (r 760)
Vietnam			An Nhon		5	DYFR	Puerto Princesa, Pal.
740 kHz					5	DZYT	Tuguegarao, Cag.
Cambodia	120		Phnom Penh	Thailand	20		Lampang
747 kHz				Turkmenistan	30		Çärýev
Australia	10	4QS	Toowoomba (RR)	Vietnam			Binh Thuan
	5	6SE	Esperance	**774 kHz**			
	0.25	8JB	Jabiru (RR)	Australia	50	3LO	Melbourne (MS)
China, PR			9 st's		5	4TO	Townsville
China (Taiwan)	250		Minhsiung (CBS)	China, PR	100		HB
India	300		Lucknow A	China (Taiwan)	10	BEV88	Taoyuan (Hsien Sheng)
Indonesia	5		Bengkulu (RRI)		10	BEV94	Taichung (TBC)
Japan	500	JOIB	Sapporo (NHK2)	Fiji	20		Suva (Naulu)
Kazakhstan	20		Karagandy	India	100		Shimla
	50		Oral	Japan	500	JOUB	Akita (NHK2)
			Köksetau	Korea, Rep.	10	HLAJ	Cheju (MBC)
Korea, Rep.	100	HLKH	Kwangju (KBS)		10	HLAN	Chùnch'on (MBC)
Malaysia	10		Tawau, Sabah	Malaysia	10		Sungai Petani
Philippines	5	DXND	Kidapawan, Cotab. Nte		10		Tenom, Sabah
	10	DYHB	Bacolod, Neg. Occ.	Philippines	1	DXSO	Marawi C., Lanao del Norte
	10	DZJC	Laoag C, IN		10	DWAT	Quezon C., Manila
Samoa, W.	10	2AP	Apia (Afiamalu)	Sri Lanka	10		Anuradhapura
Sri Lanka	20		Kantalai	Thailand	10		Udon Thani
Thailand	10		Songkhla	Vietnam			Quang Ngai
	10		Surin	**780 kHz**			
	10		Udon Thani	Vietnam			Can Tho (Hua Giang B.S.)
	1		3 st's	**783 kHz**			
				Australia	10	4VL	Charleville (**F.PI.**)

Country	kW	Call	Location
	2	6VA	Albany
	2	8AL	Alice Springs (RR)
China PR	100		HEB
China (Taiwan)	10	BED92	Tainan (BCC2)
Hong Kong	20		Golden Hill (RTHK5)
India	20		Madras C
Japan	10	JOXR	Naha, Ok (ROK)
Kazakhstan	150		Türkistan
Korea, Rep.	10	HLCV	Yongwol (KBS)
	5		Pusan (AFKN))
Malaysia	10		Sandakan, Sabah
New Zealand	20	2YB	Wellington
Philippines	10	DXRA	Davao C.
	5	DYME	Masbate, Mas
	1	DZNL	San Fernando, LU
Russia	500		Tavrichanka, FE
Thailand	20		Lopburi
Vietnam			Nghê An

790 kHz

Country	kW	Call	Location
Hawaii	5	KKON	Kealakekua

792 kHz

Country	kW	Call	Location
Australia	20	4RN	Brisbane (RN)
China, PR	5	100	GX;FJ(T1) + 5 st's
China (Taiwan)		BEC33	Hualien (Han Sheng)
	1	BEV79	Keelung BS
India	100		Pune A
Korea, Rep.	50	HLSQ	Koyang (Seoul) (SBS)
Nepal	100		Kathmandu
New Zealand	0.4	1XSR	Hamilton
Pakistan	150		Muzaffarabad, Kashmir
Philippines	10	DXBN	Butuan
	5	DYRR	Ormoc C.
	1	DZAP	Angeles C, Pampanga
Russia	25		Abakan, Sib.
	50		Aleksandrovsk, FE
Thailand	20		Bangkok
	20		Kalasin

800 kHz

Country	kW	Call	Location
Laos	1		Oudomxay

801 kHz

Country	kW	Call	Location
Australia	5	2RF	Gosford
	2	4QY	Cairns (RR)
	2	5RM	Berri
China, PR	50		GD + 9 st's
China (Taiwan)	250		Kuanyin (Kuanghua)
	1		Hsinhua (Chien Kuo)
	250		Kuanyin (CBS)
Guam	10	KTWG	Agana
India	200		Jabalpur
Indonesia	10		Semarang (RRI)
	1		Medan (RRI)
Kazakhstan	30		Arkalyk
	30		Öskemen
Korea, DPR	500		Hwadae
Malaysia	10		Kudat, Sabah
New Zealand	1	2XL	Nelson
Philippines	1	DWFA	Sorsogon, Sor.
	1	DXBL	Bislag, Surigao S.
	5	DXES	
	5	DXMZ	Gen. Santos C., S.Cotab.
	5	DYKA	San José, Ant.
	5	DYWC	Dumaguete, Neg. Or.
	10	DZNC	Cauayan, Isa.
Russia	1200		Chita (Atamanovka), Sib. (FS)
Sri Lanka	20		Maho
Tajikistan			Orzu
Thailand	10		Chieng Mai
	5-1		4 st's.

810 kHz

Country	kW	Call	Location
Australia	10	2BA	Bega (RR)
	10	6RN	Perth (RN)
China, PR	200		ZJ + 5 st's

Country	kW	Call	Location
China (Taiwan)	1		Kuanhsi (TBC)
	10	BEV54	Changhua (Kuo Sheng)
Fiji	10		Labasa
India	300		Rajkot A
Indonesia	7.5		Merauke (RRI)
Japan	50		Tokyo (FEN)
Korea, DPR	50		Kaesong
Korea, Rep.	20	HLCT	Taegu (MBC)
Malaysia	10		Kuantan
Nepal	10		Dandeldhura
New Zealand	20	4YA	Dunedin (Highcliff)
	2	1XU	Auckland
Papua N. Guinea	2		Rabaul
Philippines	10	DZRJ	Quezon C, MM
Russia	50		Yekaterinburg, Sib.
	150		Razdolnoye, FE
Thailand	7		Khon Kaen
	20		Nong Kai
	10		Trang

819 kHz

Country	kW	Call	Location
Australia	10	2GL	Glen Innes (RR)
China, PR	200		SX; XJ
China (Taiwan)	10	BED28	Taitung (BCC1)
	5	BEP28	Kaohsiung (Ching Cha)
	5	BEV35	Taipei (Cheng Sheng)
India	200		Delhi A
Japan	5	JONK	Nagano (NHK1)
	5	JOIP	Oita (NHK1)
	1	JOAQ	Sasebo (NHK1)
Korea, DPR	500		Pyongyang (KCBS)
Korea, Rep.	20	HLCN	Kwangju (MBC)
Malaysia	20		Miri, Sarawak
New Zealand	10	1YZ	Rotorua (Paengaroa)
Philippines	5	DXMC	Davao C
	10	DXUM	
	5	DWRI	
	1	DXSC	Zamboanga C, Z Sur
	5	DYVL	Tacloban, Leyte
Sri Lanka	20		Senkadagala (Kandy)
Thailand	10		Bangkok
	10		Satoon

820 kHz

Country	kW	Call	Location
Easter Isl.	0.25	CB82	R. Manukena
Vietnam			Son La/Trien Giang

828 kHz

Country	kW	Call	Location
Australia	10	3GI	Sale (RR)
	1	4GC	Charters Towers
	10	6GN	Geraldton (RR)
China, PR	50		BJ; GD; HEN (2 st's); HB(v)
India	10		Silchar
Japan	300	JOBB	Osaka (NHK2)
Malaysia	10		Lahad Datu, Sabah
New Zealand	2	2XS	Palmerston No.
Pakistan	100		Karachi I
Philippines	5	DWZR	Legaspi C., Albay
	10	DXCC	Cagayan de Oro, Mis. Or.
	1	DZTC	Tarlac, Tar.
Russia	150		Kyzyl, Sib.
Singapore	50		Singapore
Thailand	20		Royang
	10		Khon Kaen
	10		Sukhothai

830 kHz

Country	kW	Call	Location
Hawaii	10	KIKI	Honolulu

837 kHz

Country	kW	Call	Location
Australia	10	4RK	Rockhampton (RR)
	1	6ED	Esperance (RR)
	0.5	7XS	Queenstown
China, PR			(C1); H;XJ
China (Taiwan)	1	BEV56	Tainan (Shengli Chih S.)
	10	BED27	Taichung (BCC2)
India	100		Vijayawada A

Country	kW	Call	Location
Japan	10	JOQK	Niigata (NHK1)
Korea, Rep.	50	HLKY	Seoul (CBS)
New Zealand	2	1YX	Kaitaia
	2	1YX	Whangarei
Pakistan	10		Gilgit (Kashmir)
Philippines	5	DXRE	Gen. Santos C., S.Cotab.
	5	DYFM	Iloilo C.
	5	DZXE	Vigan, IS
Thailand	50		Sakon Nakhon
846 kHz			
Australia	10	2RN	Canberra, ACT (RN)
	5	4CA	Cairns
	0.2	6CA	Carnarvon (RR)
Bangladesh	100		Rajshahi
China, PR	20		SX + 12 st's
China (Taiwan)	250		Kuanyin (Kuang Hua)
	10	BEC38	Penghu (Han Sheng)
	10	BEH56	Kaohsiung (Fu Hsing)
India	200		Ahmedabad A
Kazakhstan	50		Jambyl
Kiribati	10	T3K1	Naranakei, Betio I. (v)
Korea, Rep.	10	HLAU	Ulsan (MBC)
	5	HLSY	Yanggu (KBS)
Malaysia	10		Jerantut
	20		Kuching, Sarawak
New Zealand	2	2ZD	Masterton
Philippines	50	DZNN	Quezon C., MM
Thailand	10		Bangkok
Vietnam			Quang Binh
850 kHz			
China (PR)			GD
Hawaii	5	KHLO	Hilo
Indonesia	1		RRI Bogor
Laos			Luang Prabang
Vietnam			Than Hoa
855 kHz			
Australia	1	3CR	Melbourne
	10	4QB	Maryborough (RR)
	10	4QO	Eidsvold (RR)
China, PR			(C2) Synchro
China (Taiwan)	1	BEV24	Taipei (Min Pen)
	5	BEV72	Chia-i (Cheng Sheng)
	10	BED27	Hualien (BCC1)
Indonesia	50		Medan (RRI)
	2/10		Mataram (RRI)
Korea, DPR	500		Sangwon (PBS)
Korea, Rep.	10	HLCX	Chonju (MBC)
New Zealand	2	1XH	Hamilton (rel.3KG)
Pakistan	10		Quetta II
Philippines	1	DXWG	Ilagan C. Lanao Nte.
	5	DXLA	Zamboanga C., Z Sur
	10	DZGE	Naga C., Cam. Sur
	10	DZPE	
	5	DXDH	
	5	DXZH	
	1	DXCT	
Russia	20		Petropavlovsk, FE
Sri Lanka	20		Amparai
860 kHz			
Pakistan	10		Quetta (r)
861 kHz			
Vietnam			Lai Cau
864 kHz			
Australia	2	4GR	Toowoomba
	2	6AM	Northam
	5	7RPH	Hobart
China, PR	50		AH; ZJ
China (Taiwan)	10	BED25	Kaohsiung (BCC1)
	10	BEV52	Taichung (Chung Sheng)
Hong Kong	10		Peng Chau (Comm. R.)
India	100		Shillong
Indonesia	2/10		Cirebon (RRI)

Country	kW	Call	Location
Japan	5	JOPR	Fukui (FBC)
	3	JOHE	Asahikawa (HBC)
	3	JOQF	Muroran (HBC)
	1	JOSO	Matsumoto (SBC)
	1	JOXN	Nasu (CRT)
Korea, DPR	250		Shinuiju (KCBS)
Korea, Rep.	100	HLKR	Kangnung (KBS)
Malaysia	10		Penang
New Zealand	10	4ZA	Invercargill (Dacre)
Papua N. Guinea	2		Madang
Philippines	1	DWSF	San Fernando, Pampanga
	5	DYRE	Mandaue C, Cebu
	1	DZSP	San Pablo C., Laguna
	5	DWSI	
	5	DYJS	
Russia	25		Blagoveshchensk, Sib.
Sri Lanka	20		Mannar
Thailand	10		Bangkok
	10		Tak
870 kHz			
Hawaii	50	KAIM	Honolulu
873 kHz			
Australia	5	2GB	Sydney
	2	6DB	Derby (RR)
Bangladesh	10		Chittagong
China, PR	200		HL;GS + 5 st's
India	300		Jalandhar A
Japan	500	JOGB	Kumamoto (NHK2)
Malaysia	20		Limbang, Sarawak
New Zealand	1	3ZE	Ashburton (rel. 3ZB)
Philippines	5	DXRB	Butuan C., Agusan Nte
	5	DXRT	Tawitawi, Sulu
	5	DZPA	Bangued, Abra
	5	DZRC	Legaspi C., Albay
	10	DXJS	
Sri Lanka	20		Diyagama (Colombo)
Russia			Unk., FE
Thailand	10		Phattalung
	10		Phetchabun
882 kHz			
Australia	2	3YB	Warrnambool
	5	4BH	Brisbane
	2	6PR	Perth
China, PR	100		FJ + 9 st's
China (Taiwan)	10	BEG77	Hsinchu (BCC1)
	1		Penghu (Feng Ming)
India	50		Imphal
Indonesia			Kendari (RRI)
Japan	10	JOPK	Shizuoka (NHK1)
	3	JOWS	Kushiro (STV)
	1		Esashi (STV)
Korea, Rep.	20	HLKI	Taejon (KBS)
Kyrgyzstan	500		Biskek
Mongolia	150		Murun, Hövsgöl
	75		Choibalsan, Ea. Aimak
New Zealand	10	1YC	Auckland
Philippines	40	DWIZ	Navotas, MM
	10	DXMS	Cotabato C., Cot. Nte
	1	DXRG	Gingoog, Misamis Or.
	10	DYJR	Calbayog, W. Samar
Russia	10		Aleksandrovsk, FE
Sri Lanka	400		Puttalam
886kHz			
Indonesia	0.5		Serui (RRI)
891 kHz			
Australia	5	4RR	Townsville (Ayr)
	50	5AN	Adelaide (MS)
China, PR	100		NX; LN
China (Taiwan)	10	BED24	Tainan (BCC)
	3.5	BEG78	Miaoli (BCC3)
Fiji	10		Lautoka (Drasa)

Country	kW	Call	Location
India	10		Rampur
	1		Tiruchirapalli B
Indonesia	10		Malang (RRI)
Japan	20	JOHK	Sendai (NHK1)
Korea, Rep.	250	HLKB	Pusan (KBS)
New Zealand	5	2XW	Wellington
Philippines	10	DYSR	Dumaguete C., Neg. Or.
	5	DWMM	
	5	DWHQ	
Thailand	1000		Saraburi
899 kHz			
Vietnam			Vin Phu
900 kHz			
Australia	5	2LM	Lismore
	5	2LT	Lithgow
	2	6BY	Bridgetown
	2	7AD	Devonport
	2	8HA	Alice Springs
China, PR	10		HL + 15 st's
China (Taiwan)	300	BED96	Tanshui (CBS)
Hawaii	5	KNUI	Kahului, Maui
India	100		Cuddapah
Japan	5	JOHF	Yonago (BSS)
	5	JOZR	Kochi (RKC)
	5	JOHO	Hakodate (HBC)
Kazakhstan	150		Karaturuk
Korea, Rep.	50	HLKV	Seoul (MBC)
Macau	1		Teledif. de M.
New Zealand	10	4YC	Dunedin (Highcliff)
		F.PI.	Whangarei (R. Rhema)
Papua N. Guinea	2		Goroka
Philippines	10	DXSS	Davao C.
	5	DXRZ	Zamboanga C.
	5	DYOW	
	5	DWNE	
Vietnam			Can Tho/Ha Tinh
909 kHz			
China, PR	100		FJ;TJ + 4 st's
China (Taiwan)	10	BED79	Kaohsiung (BCC2)
	10	BEG38	Taichung (BCC)
	10	BEH3	Taipei (Fu Hsing)
	1	BEH79	Hualien (Fu Hsing)
India	100		Gorakhpur
Indonesia			Sorong (RRI)
Japan	10	JOCB	Nagoya (NHK2)
	5	JOVX	Abashiri (STV)
Korea, Rep.	10	HLQY	Kumi (KBS)
Malaysia	20		Sibu, Sarawak
New Zealand	0.8		Napier
Philippines	5	DYLA	Cebu C.
	5	DZEA	
Russia	5		Dalnerechensk, FE
	10		Yekaterinburg
Thailand	50	HSKS	Surin
Vietnam			Lang Son
918 kHz			
Australia	2	2XL	Cooma (F.PI. 5kW)
	2	4VL	Charleville (F.PI. 918kHz)
	2	6NA	Narrogin
China, PR	10		SD; GX
India	300		Suratgarh
Japan	5	JOEF	Yamagata (YBC)
	1		Yonezawa (YBC)
	1	JOPM	Shimonoseki (KRY)
	1	JOPN	Iwakuni (KRY)
Korea, Rep.	50		Yonch'on (KBS)
New Zealand			Timaru
Philippines	50	DZRB	Quezon C., MM
Russia	5		Makushino/Shumikha, S.
	7		Shadrinsk, Sib.
Thailand	10		Chieng Mai
	1		Chanthaburi
Vietnam			Cao Bang
927 kHz			
Australia	5	3UZ	Melbourne
	5	4CC	Gladstone
	2	6NR	Perth
China, PR	50		BJ;GZ;FJ + 7 st's
China (Taiwan)	1200		Chanhchih (CBS)
Fiji	2.5		Sigatoka
India	100		Visakapatnam
Indonesia	25		Pekanbaru (RRI)
Japan	5	JOFG	Fukui (NHK1)
	5	JOKG	Kofu (NHK1)
Korea, DPR	50		Hwangju (KCBS)
Korea, Rep.	10	HLQA	Puyo (KBS)
	1	HLQD	Hongch'on (KBS)
	1	HLSU	Hadong (KBS)
Malaysia	10		Tawau, Sabah
New Zealand	2	2ZA	Palmerston No.
Pakistan	100		Khairpur
Philippines	5	DXMM	Sulu
	5	DXDA	
	5	DWRS	Vigan, IS
Russia	25		Komsomolsk, FE
Thailand	10		Bangkok
Turkmenistan	50		Çärÿev
	5		Krasnovodsk
	1		Bekdas
930 kHz			
Vietnam			Ben Tre
936 kHz			
Australia	5	4BK	Brisbane (PO 1296)
	10	4PB	Brisbane
	10	7ZR	Hobart (ABC)
China, PR	200		AH;NM
China (Taiwan)	10	BEC24	Taipei (Kuo Kuang)
India	100		Tiruchirapalli A
Japan	5	JONF	Miyazaki (MRT)
	5	JOTR	Akita (ABS)
	1		4 st's (MRT)
Kazakhstan	1000		Karagandy
Korea, Rep.	10	HLKD	Ch'angwon (KBS)
N. Mariana Is.	10	KSAI	Saipan
Philippines	5	DXON	Davao C.
	10	DXIM	Cagayan de Oro C., Mis. Or.
	1	DYKW	Binalgaban, Neg. Occ.
	1	DYCC	Calbayog C, W. Samar
	1	DZXT	Tarlac, Tar.
Russia	5		Krasnoyarsk, Sib.
	5		Dalnegorsk, FE
Thailand	10		Chieng Mai
	10		N. Raichasima
	50		N. Sawan
	10		N. Si Thammarat
	1.5		Pattani
940 kHz			
Hawaii	10	KDEO	Waipahu, Oahu
945 kHz			
Australia	1	4HI	Dysart (trtr)
China, PR			(C2) synchro
India	20		Sambalpur
Japan	1	JOQP	Hikone (NHK1)
	3	JOIQ	Muroran (NHK1)
	5	JOXK	Tokushima (NHK1)
Kazakhstan	50		Kaçiry
Korea, Rep.	10	HLQW	Poun (KBS)
Malaysia	100		Tronoh
	50		Johore Bharu
New Zealand	2	2ZG	Gisborne
Philippines	5	DXRO	Cotabato C., Cot. Nte
	5	DYRO	Roxas C., Capiz
Solomon Is.	10	H4B	Gizo
Turkmenistan	20		Gusgy

Country	kW	Call	Location
Vietnam			Hanoi
950 kHz			
Vietnam			Vinh Long
954 kHz			
Australia	5	2UE	Sydney
	0.5	4CA	Gordonvale (trtr)
China, PR	30		HAN + 6 st's
China (Taiwan)	20	BED55	Taipei (BCC2)
	1		Wuho (Yen Sheng)
	10	BEV85	Hsinying (Chien Kuo)
			Tunglo (Tien Sheng)
India	100		Najibabad
Indonesia	10		Kendari (RRI)
Japan	100	JOKR	Tokyo (TBS)
Kazakhstan			Kyzylorda
Malaysia	20		Kuching, Sarawak (r 960)
New Zealand	2	1XW	Hamilton
Philippines	10	DZEM	Quezon City, MM
	1	DXRI	Iligan C, Misamis Or
	10	DWFB	
	10	DWSB	
	5	DZAL	
	2.5	DYMM	
Russia			Unk., FE (FEBC)
Thailand	5		7 st's.
960 kHz			
Malaysia	20		Kuching, Sarawak (AF954)
963 kHz			
Australia	5	2RG	Griffith
	5	4WK	Warwick
	5	5SE	Southern Cross Br. c.
Bangladesh	20		Sylhet
China, PR	10		LN; ZJ; XJ
China (Taiwan)	10		Matsu (Kuang Hua)
	10	BEV84	Chunghsing (TBC)
India	20		Jalgaon
Indonesia	10		Jember (RRI)
Japan	5	JOTG	Aomori (NHK1)
	5	JOZK	Matsuyama (NHK1)
	1	JOSP	Saga (NHK1)
Korea, Rep.	10	HLKS	Cheju (KBS)
	10	HLCR	Andong (KBS)
Malaysia	10		Kuala Trengganu
New Zealand	10	3YC	Christchurch
Philippines	10	DYMF	Cebu C.
	5	DXYZ	Zamboanga C.
		DZNS	Vigan
Russia	20		Zakamensk, Sib.
	1		Guzino-Ozersk, Sib.
Thailand	50		Krabi
	10		N. Ratchasima
970 kHz			
Hawaii	5	KJRI	Waimea (CP)
Vietnam			Cao Bang/Song Be
			Rach Gia (Kiên Giang BS)
972 kHz			
Australia	0.3	2DU	Cobar (trtr)
	5	2MW	Murwillumbah
Brunei	1		Kuala Belait
China, PR	150		HEN + 2 st's
India	100		Cuttack A
Indonesia	50		Surakarta (RRI)
Kazakhstan	30		Akmola
Korea, Rep.	1500	HLCA	Tangjin (KBS)
New Zealand	5	2XG	Wellington (R. Rhema)
Philippines	5	DWTI	Lucena C, Que
	5	DXKH	
	5	DXMO	Cagayan de Oro, Mis.Or.
	1	DYSM	Catarman, N. Samar
Russia	30		Yuzhno-Sakhalinsk, FE
Sri Lanka	20		Amparai
Tajikistan	1000		Orzu
Thailand	50		Phetchaboon
981 kHz			
Australia	5	2NM	Muswellbrook
	2	3HA	Hamilton
	2	6KG	Kalgoorlie
			(C1) synchro
China, PR			
China (Taiwan)	250	BEC70	Hsinfeng (Kung Chun)
	1	BEV68	Kaohsiung (Feng Ming)
India	100		Raipur
Kazakhstan			Aktöbe
New Zealand	2	1YE	Kaikohe
Philippines	10	DXOW	Davao C.
	5	DXDR	Dipolog, Zamb. Nte
	20	DYBQ	Iloilo C.
	5	DZRD	Dagupan C., Pang.
	10	DXBR	
	5	DWTH	
	5	DWRS	
Russia	5		Krasnoyarsk, Sib.
Thailand	50	HSKZ	Yala
	10		Bangkok
	10		Nakhon Phanom
	5	HSKH	Mae Hong Son
Vietnam			Kon Tum
990 kHz			
Australia	0.25	3RN	Albury/Wodonga (RN)
	5	4RO	Rockhampton
	2	6RPH	Victoria Park
	0.5	8GO	Gove (RR)
China, PR	50		SH; YN + 2 st's
China (Taiwan)	10	BEP34	Hualien (Ching Cha)
	1	BEP38	Ilan (Ching Cha)
	10	BEV58	Taichung (Cheng Sheng)
Fiji	5		Lautoka (Vomo)
Hawaii	10	KIKI	Honolulu
India	300		Jammu A (Kashmir)
Japan	10	JORK	Kochi (NHK1)
Korea, Rep.	10	HLAP	Masan (MBC)
Mongolia	150		Altai
	500		Ulaanbaatar
New Zealand	1		Auckland
	1		Nelson
Philippines	15	DWRT	Makati, MM
	10	DY..	
	5	DZDH	
	5	DYTH	
Russia	1		Yuryuzan, Sib.
Thailand	10		N. Ratchasima
999 kHz			
Australia	2	2NB	Broken Hill (RR)
	5	2ST	Nowra
Bangladesh	10		Thakurgaon
China, PR.	50		LN + 3 st's
India	10		Coimbatore
	1		Almora
Indonesia	1/150		Jakarta (RRI)
Japan	1	JODG	Hamamatsu (NHK1)
	1	JODP	Fukuyama (NHK1)
	1	JOTQ	Hachinohe (NHK1)
Korea, DPR	250		Hamhung
Korea, Rep.	10	HLCL	Kwangju (CBS)
Philippines	1	DXHP	Bislig, Surigao S.
	1	DXPT	
	10	DYSS	Cebu C.
	1	DZEQ	Baguio C., Benguet
Russia	7.5		Birobidzhan, FE
Vietnam			Quang Tri
1000 kHz			
China, PR			SC
Laos			Houa Phan
India	100		Itanagar

Country	kW	Call	Location
1001kHz			
Indonesia	0.3		Bukittinggi (RRI)
1008 kHz			
Australia	0.3	2XX	Canberra, ACT
	5	4TAB	Brisbane (Ipswich)
	5	7EX	Launceston
China, PR			YN (C1) + 5 st's
China (Taiwan)	600		Lukang (CBS)
	10	BED88	Taitung (BCC2)
	5	BEV60	Kaohsiung (Cheng Sheng)
	1	BEV92	Taipei (Tien Nan)
India	100		Calcutta B
Indonesia	10		Madiun (RRI)
Japan	50	JONR	Osaka (ABC)
Kazakhstan	50		Jilaydy
Korea, Rep.	50	HLCS	Sokchò (KBS)
New Zealand	10	1ZD	Tauranga (Paengaroa)
Pakistan	120		Hyderabad I
Philippines	25/10	DXXX	Zamboanga C.
	5	DWBS	
Russia	50		Petropavlovsk, FE
	25		Khabarovsk, FE
Thailand	10		Bangkok
1010 kHz			
Vietnam	100		Hanoi (FS)
1017 kHz			
Australia	5	2KY	Sydney
	0.1	6WH	Wyndham (RR)
China, PR			(C1); JL (FS); GD
China (Taiwan)	10	BED53	Hsinchu (BCC2)
India	20		Delhi D
	1		Mysore
Japan	50	JOLB	Fukuoka (NHK2)
Korea, Rep.	10	HLAW	Andong (MBC)
Malaysia	100		Segamat
New Zealand		F.PI.	Christchurch
Philippines	10	DWDW	Dagupan C., Pang.
	10	DWLC	Lucena C., Que.
	10	DXRR	
	5	DXGE	Davao C.
	5	DXSN	Surigao, S. Nte
	5	DYRP	Iloilo C. (v)
Tonga	10	A3Z	Nuku'alofa
1023 kHz			
China, PR			HN (v)
1026 kHz			
Australia	5	3PB	Melbourne (PNN)
	2	6NW	Port Hedland
China, PR	50		GZ;BJ + 3 st's
China (Taiwan)			Yuanli (Tien Sheng)
	1	BEV51	Sanchung (Chung Hua)
India	20		Allahabad
Indonesia	5		Serui (RRI)
Korea, Rep.	1	HLKW	Koch'ang (KBS)
	1	HLCG	Hwach'on (KBS)
New Zealand	2	1ZK	Kaitaia (rel. 1ZN)
	2	1ZN	Whanggarei
	1	4XSR	Invercargill
Philippines	25	DZAM	Quezon C, MM
	10	DWXX	
	1	DXMI	Iligan C., Lanao Nte
Russia	1000		Oyash, Sib.
Sri Lanka	10		Galle
Thailand	50		Phitsanulok
	10		Yala
1030 kHz			
Laos	10		Vientiane (FS)
1035 kHz			
China, PR			(C1)
China (Taiwan)	10	BED26	Chia-i (BCC2)
India	10		Gowahati B
Indonesia			Palu (RRI)

Country	kW	Call	Location
	1/5		Tanjungkarang (RRI)
Japan	1	JOIC	Toyama (NHK2)
	1	JOHD	Takamatsu (NHK2)
Korea, Rep.	10	HLCP	Pohang (KBS)
Malaysia	10		Penang
New Zealand	20	2ZB	Wellington
Pakistan	120		Multan
Philippines	2.5	DXCP	Gen. Santos C., S. Cotabato
	5	DXWB	
	5	DZWX	Baguio C., Benguet
Solomon Is.	10	H4B1	Honiara
Thailand	20		Bangkok
1040 kHz			
Hawaii	5	KLHT	Honolulu
1044 kHz			
Australia	1	2UH	Muswellbrook (RR)
	0.5	4WP	Weipa (RR)
	2	5CS	Port Pirie
	1	6BR	Bridgetown (RR)
China, PR	300		JS (FS) +2 st's
China (Taiwan)	1	BEV98	Kaohsiung (Cheng Kung)
	5	BEV64	Hualien (Yen Sheng)
Hongkong	10		Peng Chau (Metro Broadc.)
India	100		Mumbai A
Indonesia	10		Biak (RRI)
Korea, Rep.	10	HLCI	Samch'ok (KBS)
	10	HLCD	Chech'on (KBS)
	1		Chùnch'on (AFKN)
Malaysia	20		Sri Aman, Sarawak
New Zealand	10	4ZB	Dunedin (Highcliff)
Philippines	1	DXML	Digos, Dvo. Sur
	5	DXCO	Cayagan de Oro, Mis. Or.
	5	DXLL	Zamboanga C.
	3	DYMS	Catbalogan C., W. Samar
	10	DZDR	Naga C., Cam. Sur
Russia	10		Yuzhno-Sakhalinsk, FE
Thailand	10		Khon Kaen
1050 kHz			
China, PR			HEB;ZJ
1053 kHz			
Australia	5	2CA	Canberra, ACT
	0.5	4EB	Brisbane (West End)
Bangladesh	20		Rangpur
China P.R.	10		SD + 4 st's
China (Taiean)	100		Kuanyin (Kuang Hua)
India	200		Tuticorin
	10		Leh (Kashmir)
Indonesia	10		Jayapura (RRI)
	1		Surakarta (RRI)
Japan	50	JOAR	Nagoya (CBC)
Malaysia	20		Mersing (r 1055)
N. Mariana Is.	1	KCNM	Navy Hill, Saipan
New Zealand	2	2ZP	New Plymouth
Philippines	5	DYSA	
	1	DXKD	Dipolog, Zamb. Nte
	5	DZEL	Lucena, Que. (v)
Russia	25		Belogorsk, Sib.
			Krasnoyarsk, Sib.
Thailand	10		N. Si Thammarat
	10		Nakhon Sawan
	10		Lampang
Vietnam			Tra Vinh
1060 kHz			
Hawaii	1	KAHU	Hilo (CP)
1062 kHz			
Australia	2	4TI	Thursday Isl (RR)
China, PR	150		GD; HL
China (Taiwan)	10	BED85	Taipei (BCC3)
	10	BED23	Taiching (BCC3)
	5	BEV82	Ilan (Cheng Sheng)
	5	BEV74	Pingtung (Min Li)
Korea, Rep.	50	HLKQ	Ch'ongju (KBS)

Country	kW	Call	Location
Malaysia	20		Sibu, Sarawak
Philippines	5	DXKI	Marbel, S. Cotabato
	10	DZEC	Quezon C., MM
Russia	1		Ust-Kamchatsk, FE
Thailand	20		Bangkok
Uzbekistan	50		Andijon
			Toskent/Nukus
1068 kHz			
Vietnam			unk. (r)
1071 kHz			
Australia	5	3CV	Maryborough
	2	4SB	Kingaroy
	2	6WB	Katanning
China, PR	50		TJ + 7 st's
China (Taiwan)	1	BEV96	Tainan (Tien Sheng)
India	1000		Rajkot
Japan	20	JOFK	Hiroshima (NHK1)
	5	JOWM	Obihiro (STV)
Kazakhstan	100		Öskemen
New Zealand			Masterton
Philippines	1	DXKT	Davao C.
	5	DYXT	Tagbilaran C., Bohol
	5	DZAL	Iriga C., Cam Sur
Russia	7		Zeya, Sib.
1080 kHz			
Australia	2	2MO	Gunnedah
	0.2	4MI	Mt. Isa (RR)
	2	6GL	Perth (**F.PI.** 5kW)
	5	7TAB	Hobart
Bangladesh	10		Rajshahi
China, PR	200		YN (FS) + 6 st's
Hawaii	5	KWAI	Honolulu
Indonesia	2/10		Singaraja (RRI)
	1		Gorontalo (RRI)
Korea, DPR	1500		Haeju (KCBS)
Korea, Rep.	10	HLAT	Yosu (MBC)
Malaysia	10		Sandakan, Sabah
New Zealand	10	1ZB	Auckland
Pakistan	50		Lahore II
Philippines	2.5	DWAM	Batangas C
	5	DWIN	Dagupan C., Pang.
	5	DWRL	Albay
	5	DXRH	
	1	DWCL	
	1	DXKS	Surigao, S. Nte
	1	DYSJ	Kalibo, Ak
Russia	1000		Angarsk, Sib. (FS)
Thailand	10		N. Sawan
	10		Yala
Turkmenistan	7		Krasnovodsk
	5		Gusgy
1089 kHz			
Australia	5	2GZ	Orange
	5	3WM	Horsham
China, PR	200		LN; FJ; HN
China (Taiwan)	1	BED56	Jiyuetan (BCC)
	10	BEH80	Hualien (Fu Hsing)
	5	BEH5	Taipei (Fu Hsing)
	10	BEC31	Yunlin (Han Sheng)
	10	BEG28	Kaohsiung (City Council)
Fiji	1		Suva/mobile
India	20		Udipi
	1		Jammu B (Kashmir)
Japan	10	JOHB	Sendai (NHK2)
Korea, Rep.	10	HLCH	Chùngju (KBS)
Malaysia	20		Grik
Philippines	5	DXCM	Cotabato C., Cot. Nte
	1	DYHR	Calbayog C., W. Samar
Russia	5		Tilichiki, FE
Thailand	10		Udon Thani
Vietnam			Dac Lac
1090 kHz			

Country	kW	Call	Location
Wallis & Futuna	1		Alu, Futuna (RFO)
1098 kHz			
Australia	0.2	2RN	Goulburn (RN)
	2	4LG	Longreach
	2	6MD	Merredin
	5	7LA	Launceston
China (Taiwan)	300		Tanshui (CBS)
China, PR			GD + 6 st's
Fiji	1		Suva (mob./spec.prgr.)
Indonesia	10		Jambi(RRI)
	0.5/10		Sumenep (RRI)
Japan	5	JOGF	Oita (OBS)
	1	JOMF	Sasebo (NBC)
	5	JOSR	Nagano (SBC)
	1	JOSW	Iida (SBC)
	5	JOWO	Koriyama (RFC)
Kazakhstan	150		Almaty + 7 st's
Korea, Rep.	20	HLCJ	Chinju (KBS)
Marshall Is.	5	WSZO	Majuro
New Zealand	10	3ZB	Christchurch
Pakistan	10		Hyderabad II
Philippines	10	DWAD	Mandaluyong, MM
	5	DXCL	Cagayan de Oro, Mis. Or.
Russia			Vladivostok, FE
Thailand	10		Songkhla
	10		Tak
1100 kHz			
China, PR			HEB(v);SD(v)
Vietnam			Binh Thuan
1107 kHz			
Australia	5	2EA	Sydney
China P.R.	20		XJ + 6 st's
China (Taiwan)	10		Kinmen (Kuang Hua)
India	10		Gulbarga
Indonesia	1/10		Yogyakarta (RRI)
	1/5		Kupang (RRI)
Japan	5	JOMR	Kanazawa (MRO)
	1		Nanao (MRO)
	1		3 st's (MBC)
	20	JOCF	Kagoshima (MBC)
Korea, Rep.	10	HLAV	Po'hang (MBC)
New Zealand	0.1		Waihi
Papua N.G.	5		Kavieng (Proj.)
Philippines	10	DWDY	Cagayan Isabela
	10	DXMF	Davao C.
	5	DYRI	Iloilo C.
	5	D...	
Thailand	20		Chon Buri
1110 kHz			
Hawaii	5	KAOI	Kihei (Pukalani), Maui
1116 kHz			
Australia	5	3BM	Richmond
	5	4BC	Brisbane
	2	6MM	Mandurah
China P.R.			HL(C2); FJ(T1); SC + 2 st's
China (Taiwan)	10	BEC22	Taipei (Han Sheng)
	10	BEC30	Ilan (Han Sheng)
	10	BEP25	Kaohsiung (Chingcha)
	10	BEH34	Taiching (Fu Hsing)
	5	BEP26	Hsinchu (Chingcha)
	3.5	BED72	Yuli (BCC1)
India	20		Srinagar A (Kashmir)
Indonesia	0.3		Pekanbaru (RRI)
Japan	5	JOAF	Matsuyama (RNB)
	5	JODR	Niigata (BSN)
	1	JOAL	Niihama (RNB)
	1	JOAM	Uwajima (RNB)
New Zealand	2	2YX	Nelson
Philippines	5	DXAS	Zamboanga C.
	1	DYTR	Tagbilaran C, Bohol
	5	DZLB	Los Banos, Laguna
Russia	50		Chita, S

Country	kW	Call	Location
	25		Birobidzhan, FE
Thailand	20		Samut Sakhon
1120 kHz			
Vietnam			Minh Hai
1125 kHz			
Australia	2	1RPH	Canberra, ACT
	5	5MU	Murray Bridge
China, PR			HEB; HB
China (Taiwan)	5	BEV36	Yunlin (Cheng Sheng)
	1	BEP40	Taitung (Ching Cha)
India	10		Udaipur
Indonesia			Pontianak (RRI)
	1		Palu (RRI)
Japan	10	JOAD	Naha, Okinawa (NHK2)
	1	JOIZ	Muroran (NHK2)
	1	JOLC	Tottori (NHK2)
	1	JOOC	Obihiro (NHK2)
New Zealand	0.2		Dunedin
			Napier
Philippines	5	DWAS	Daraga, Albay
	10	DXGL	
	1	DYRM	
	10	DZWN	Dagupan C., Pang.
Thailand	50	HSKI	Chanthaburi
Turkmenistan	1000		Asgabat
Vanuatu	10	YJB2	Port Vila (Malapoa)
1130 kHz			
Hawaii	10	KLEI	Kailua (inact.)
1134 kHz			
Australia	2	2AD	Armidale (**F.Pl.** 5kW)
	5	3CS	Colac
	2	6TZ	(trtr)
China, PR	10		GD + 4 st's
China (Taiwan)	10	BEG26	Taipei (City Council)
India	1000		Calcutta (Mogra)
Indonesia	25/1		Banjarmasin (RRI)
Japan	100	JOQR	Tokyo (NCB)
Korea, Rep.	500		Hwasong (KBS)
New Zealand	2	4YQ	Queenstown
Philippines	1	DZPT	
	10	DZAF	Camp Aguinaldo, MM
	5	DXMV	Valencia, Buk.
	1	DYRM	Dumaguete, Neg. Or.
Thailand	10		Kanchanaburi
	10		Lampang
	10		Phang Nga
1140 kHz			
Malaysia	10		Kuantan (AF 1143)
1143 kHz			
Australia	2	2HD	Newcastle
	5	4HI	Emerald
China, PR			ZJ (v) + 13 st's
China (Taiwan)	100	BEL3	Penghu (Taiwan Yuyeh)
Fiji	2.5		Labasa ? (r)
India	20		Ratnagiri/Rohtak
Japan	20	JOBR	Kyoto (KBS)
Malaysia	10		Kuantan (r 1140)
Nepal	10		Dhalkebar
New Zealand	2	1YW	Hamilton
Philippines	1000	DWVA	Poro Pt, LU (VOA)
		DYAF	Bacolod
Russia	7		Tayshet, Sib.
Tajikistan	150		Yangi-Yul
Thailand	25		Bangkok
1147 kHz			
China, PR			JS
1152 kHz			
Australia	2	2WG	Wagga Wagga
China, PR	7.5		NM (2 st's) + 2 st's
China (Taiwan)	1	BED68	Puli (BCC2)
	5	BEV70	Taipei 1 (Hua Sheng)
India	1		Ranchi B

Country	kW	Call	Location
Japan	10	JOPC	Kushiro (NHK2)
	10	JORB	Kochi (NHK2)
Korea, Rep.	10	HLCW	Wonju (KBS)
New Zealand	2	3ZC	Timaru
Pakistan	10		Rawalpindi
Philippines	5	DXSR	
	5	DYCM	
Russia	20		Khanty-Mansiysk, Sib.
	50		Komsomolsk, FE
Thailand	20		Chieng Mai
	20		Khon Kaen
	1		Songkhla
1161 kHz			
Australia	2	4MB	Maryborough
	10	5PA	Naracoort (RR)
	1	7FG	Fingal (RR)
Bangladesh	10		Rangamati
China, PR			(C1) + 3 st's
China (Taiwan)	10	BED86	Ilan (BCC2)
	10	BED89	Miaoli (BCC2)
	1	BEV67	Kaohsiung (Feng Ming)
India	10		Thiruananthapuram
Korea, Rep.	20	HLKU	Pusan (MBC)
	0.25		Uijongbu (AFKN)
Malaysia	20		Sri Aman, Sarawak
	10		Melaka
New Zealand	5	2XM	Wellington
Philippines	5	DYRD	Tagbilaran C, Bohol
	5	DWCM	
	1	DXDS	Digos, Dvo. Nte
	2.5	DZMD	Daet, Cam Nte
	1	DYKR	
Thailand	20		Nakhon Phanom
	10		Rayong
Tajikistan	40		Orzu
1170 kHz			
Australia	5	2CH	Sydney
China, PR			3 st's
Hawaii	5	KOHO	Honolulu
India	1		Hyderabad C
Indonesia	50		Semarang (RRI)
Korea, Rep.	500	HLSR	Kimje (KBS)
New Zealand	0.2	1ZW	Te Kuiti (rel. ZH-FM)
Philippines	10	DXMR	Zamboanga C.
	10	DZCA	Quezon C, MM
	50	DZRP	
Russia	250		Angarsk, S.
Thailand	20		Bangkok
1179 kHz			
Australia	5	3RPH	Melbourne
China, PR	10		NM + 3 st's
China (Taiwan)	2.5		Erhlin (Kuo Sheng)
India	20		Rewa
Indonesia	0.3/10		Padang (RRI)
Japan	50	JOOR	Osaka (MBS)
New Zealand	0.1	4XE	Wanaka (rel. 4XC)
Philippines	10	DXMT	Marawi C., Lanao Sur
	5	DYCX	
Thailand	20		Chanthaburi
	10		Chiangrai
1180 kHz			
Indonesia	1/10		Tanjung Pinang (RRI)
Vietnam			Ha Tay
1188 kHz			
Australia	2	2NZ	Inverell
	2	6XM	Exmouth (RR)
China, PR	300		YN (FS) + 3 st's
China (Taiwan)	10	BED32	Hualien (BCC2)
	10	BEV46	Taipei (TBC2)
		BEV57	Tainan (Shengli Chih Sheng)
India	50		Mumbai C
Indonesia	0.3		Pekanbaru (RRI)

Country	kW	Call	Location
Japan	10	JOKP	Kitami (NHK1)
			4 st's
Kazakhstan			
Korea, Rep.	100	HLKX	Inchon (FEBC)
New Zealand	0.4	1YR	Rotorua
Philippines	5	DXIF	Cagayan de Oro
	1	DYRV	Catbalogan, W. Samar
	5	DZLT	Lucena C, Que.
	5	DZXO	Cabanatuan, Nva. Ecija
Thailand	10		Sakon Nakhon
	10		Phitsanulok
	10		Prachin Buri
Wallis & Futuna	2		Mata Utu, Wallis (RFO)

1190 kHz

Country	kW	Call	Location
Laos			Vientiane

1194 kHz

Country	kW	Call	Location
Indonesia	0.5		Mataram (RRI)

1195 kHz

Country	kW	Call	Location
Indonesia	5		Bogor (RRI)
Vietnam			Hai Hung

1197 kHz

Country	kW	Call	Location
Australia	2	5RPH	Adelaide
			Gold Coast
China, PR	10		HL; SH
India	10		Tirunelveli
	1		Jodhpur B
	1		Shillong
Indonesia	1/5		Palangkaraya (RRI)
Japan	5	JOBF	Kumamoto (RKK)
	1		3 st's. (RKK)
	5	JOYF	Mito (IBS)
	1	JOFO	Kitakyushu (RKB)
	1	JOWL	Asahikawa (STV)
	1	JOWN	Hakodate (STV)
	1		2 st's (STV)
	1		Nakamura (RKC)
Kazakhstan			Akmola/Almaty, Unk.
Korea, Rep.	1		Tongduchòn (AFKN)
Malaysia	10		Kudat, Sabah
New Zealand	2	2ZW	Wanganui
Philippines	5	DXFE	Davao C.
	5	DYRH	Bacolod, Neg. Occ.
	5	DWBA	
Russia	5		Berozovo, Sib.
Thailand	20		Bangkok

1200 kHz

Country	kW	Call	Location
Laos	1		Xieng Khouang
Vietnam			Soc Trang

1206 kHz

Country	kW	Call	Location
Australia	5	2CC	Canberra, ACT
	5	2GF	Grafton
	2	6TAB	Perth
China, PR	300		YN (FS); JL; SD (v); GD
China (Taiwan)	100		Minhsiung (CBS)
	1	BEV62	Hsinchu (TBC)
Fiji	2.5		Sigatoka
India	200		Bhawanipatnam
Indonesia	10/0.5		Denpasar (RRI)
	1		Surabaya (RRI)
	1		Manado (RRI)
Korea, Rep.	1	HLSW	Chongson (KBS)
	1	HLQR	Ch'ongsong (KBS)
Malaysia	20		Miri, Sarawak
New Zealand	2	4XO	Dunedin
	0.5	1XHC	Hamilton (Access CR.)
Philippines	10	DWAN	Quezon C, MM
	1	DXRS	Surigao C, S. Nte

1210 kHz

Country	kW	Call	Location
Hawaii	1	KZOO	Honolulu

1215 kHz

Country	kW	Call	Location
Australia	0.35	2ST	Bowral (trtr)
	0.1	4HI	Moranbah (trtr)
	0.4	6NM	Northam (RR)

Country	kW	Call	Location
China, PR			(C2) synchro; GD (C7); HB; XJ
India	20		Delhi (Kingsway)
	1		Pudducheri
Indonesia	10/0.5		Samarinda (RRI)
			Jakarta (RRI)
Japan	1	JOBO	Maizuru (KBS)
	1	JOBW	Hikone (KBS)
Korea, Rep.	10	HLAK	Chinju (MBC)
New Zealand	2	1ZE	Kaikohe (rel. 1ZN)
Philippines	10	DYRF	Cebu C.
Thailand	50	HSKV	Surat Thani
	10		Phrae
	10		U. Ratchathani

1224 kHz

Country	kW	Call	Location
Australia	5	2RPH	Sydney
	5	3EA	Melbourne (SBS)
China, PR			FJ + 3 st's
China (Taiwan)	10	BED52	Kaohsiung (BCC)
	1	BEV71	Taipei (Hua Sheng 2)
India	1		Srinagar D (Kashmir)
	10		Calcutta D
Japan	10	JOJK	Kanazawa (NHK1)
Kazakhstan			3 st's
Korea, Rep.	20	HLAA	Kwangju (KBS)
Malaysia	50		Johore Bharu
	100		Tronoh
Marshall Is.	1		Kwajalein (AFRS)
New Zealand	2	4XF	Invercargill
Philippines	10	DXED	Davao C.
	1	DXSM	Jolo, Sulu
	5	DZLH	
	5	DWBF	
	2.5	DZAG	
Russia	7		Ust-Kut, Sib.
Thailand	10		Bangkok
	10		Chieng Rai

1233 kHz

Country	kW	Call	Location
Australia	10	2NC	Newcastle (MS)
China, PR			HN
China (Taiwan)	1		Juifang (Chung Hua)
India	1		Bhopal B
	20		Tura
Indonesia	0.2/1/5		Pontianak (RRI)
Japan	5	JOGR	Aomori (RAB)
	5	JOUR	Nagasaki (NBC)
Korea, Rep.	1	HLCC	P'yongch'ang (KBS)
	1	HLQG	Yongyang (KBS)
Mongolia	75		Sainshand
New Zealand	0.6		Wellington (Southern Star)
Philippines	5	DYVS	Bacolod, Neg. Occ.
	5	DWRV	
Thailand	10		Udon Thani
	10		Prachuap Khiri Khan
	1		Uttaradit
Turkmenistan	40		Syrtagta

1235 kHz

Country	kW	Call	Location
Indonesia	0.5		Bogor (RRI)

1240 kHz

Country	kW	Call	Location
Vietnam			Hanoi (FS)

1242 kHz

Country	kW	Call	Location
Australia	5	3TR	Sale
	2	4AK	Toowoomba (**F.Pl.**: 5kW)
	2	5AU	Port Augusta
	2	8TAB	Darwin
China, PR			LN; HB;YN
China (Taiwan)	10	BED 77	Taichung + 7 st's (BCC)
India	100		Varanasi
Indonesia	1		Bengkulu (RRI)
Japan	100	JOLF	Tokyo (NBS)
Korea, Rep.	10	HLSB	Wonju (MBC)
New Zealand	2	1XX	Whakatane

Country	kW	Call	Location
	0.1	1XX	Galatea (rptr)
Philippines	10	DWBL	Pasig, MM
	5	DXZB	Zamboanga C
	5	DXSY	Ozamis C, Mis. Occ.
Russia	10		Angarsk, Sib.
Thailand	10		Roi Et
	10		Lampang
	10		Phetchabun
	1		Phayao
	50		Surat Thani
1250 kHz			
China, PR			ZJ; AH
Indonesia	1		Bogor (RRI)
Philippines	1	DYRG	Kalibo, Ak
Vietnam			Thai Binh
1251 kHz			
Australia	2	2DU	Dubbo
China, PR	100		QH + 7 st's
China (Taiwan)	100		Kuanyin (Kuang Hua)
	5	BEC29	Kaohsiung (Han Sheng)
Hong Kong	10		Tam Mei (BFBS)
India	20		Sangli
Indonesia	10		Banda Aceh (RRI)
	1		Mataram (RRI)
	1		Singaraja (RRI)
Korea, Rep.	10	HLKT	Taegu (CBS)
New Zealand	5	1XG	Auckland (R. Rhema)
Philippines	2.5	DZMS	Sorsogon, Sor
	1	DYRG	
Russia	600		Ussuriysk, FE (+ FS)
Thailand	20		Bangkok
1254 kHz			
China, PR			ZJ (v)
1260 kHz			
Australia	2	3SR	Shepparton
	1	6KA	Karratha
China, PR			LN; XZ
China (Taiwan)	10	BEP22	Taipei (Ching Cha)
	1		Putzu (Cheng Sheng)
India	20		Ambikapur
Japan	20	JOIR	Sendai (TBC)
Korea, Rep.	10	HLKL	Namwon (KBS)
Malaysia	5		Dungun
New Zealand	2	3XA	Christchurch
Philippines	5	DWMC	Rosales, Pang. (r 1265)
	10	DYDS	Cebu C.
Russia	10		Yekaterinburg, Sib.
Thailand	50		Chieng Rai
Uzbekistan	100		Nukus
1265 kHz			
China, PR			HEB (v)
1269 kHz			
Australia	5	2SM	Sydney
China, PR	200		FJ (H1); YN (FS) + 2 st's
China (Taiwan)	1	BEV37	Taitung (Cheng Sheng)
	10	BEC44	Penghu (Han Sheng)
India	20		Agartala
	1		Jaipur B
	10		Madurai
Japan	5	JOHW	Obihiro (HBC)
	5	JOJR	Tokushima (JRT)
	1	JOFM	Esashi (HBC)
	1		Okeda (JRT)
Korea, Rep.		HLSI	Kurye (KBS)
	10		Yangju (KBS)
New Zealand	0.4	2ZT	Takata (rptr 2ZN)
Philippines	5	DZVX	Daet, Cam. Nte
	10	DYWB	Bacolod, Neg. Occ
Thailand	20		Songkhla
Uzbekistan	50		Zarafson
	5		Namangan
1270 kHz			

Country	kW	Call	Location
Hawaii	5	KNDI	Honolulu
1278 kHz			
Australia	5	3AW	Melbourne
China, PR	100		HEB + 5 st's
China (Taiwan)	10	BEC40	Taipei (Fu Hsing Kang)
India	10		Lucknow C
Japan	50	JOFR	Fukuoka (RKB)
Korea, Rep.	1	HLQV	Hapch'on (KBS)
Kyrgyzstan	150		Biskek
New Zealand	2	2ZC	Napier
Philippines	10	DZRM	Quezon C, MM
	10	DXAM	
Russia	20		Barguzin, Sib.
	5		Bagdarin, Sib.
	7		Severobaykalsk, Sib.
Thailand	25		Bangkok
1280 kHz			
Vietnam			Nam Ha
1287 kHz			
Australia	2	2TM	Tamworth
China, PR			(C1) + 4 st's
China (Taiwan)	5	BEC27	Taichung (Han Sheng)
	1		Fangliao (Min Li)
India	100		Panaji A
Indonesia	0.5/50		Palembang (RRI)
Japan	50	JOHR	Sapporo (HBC)
Korea, Rep.	10	HLAF	Kangnung (MBC)
	10	HLAX	Ch'ongju (MBC)
New Zealand	2	3ZW	Westport (rel. 3ZA)
Philippines	1	DZRS	Sorsogon
	5	DZZH	
	1	DYJJ	Roxas C.
Russia	75		Novosibirsk, Sib.
	7		Shagonar/Chadan, Sib.
	5		Kyakhta, Sib.
Thailand	10		Chieng Rai
	10		U. Ratchathani
	10		Uttaradit
	10		N. Si Thammarat
Turkmenistan			Bekdas
1296 kHz			
Australia	5	4RPH	Brisbane
China, PR	300		YN (FS) + 5 st's
China (Taiwan)	10	BED47	Tainan (BCC1)
	1	BEV23	Taipei (Min Pen 1)
India	10		Darbhanga
Indonesia			Surabaya (RRI)
Japan	10	JOTK	Matsue (NHK1)
New Zealand		1ZH	Hamilton
Philippines	5	DWPR	Dagupan C, Pang.
	5	DWLQ	Lucena C., Que. (v;r 1293)
Thailand	10		Pattani
1305 kHz			
Australia	2	5MW	Renmark (RR)
China, PR			(C2)
India	10		Parbhani
Korea, Rep.	10	HLSV	Uljin (KBS)
New Zealand	2	4XD	Dunedin
Philippines	10	DYFX	Cebu C. (v)
Russia	10		Irkutsk, Sib.
	7		Serov, Sib.
Sri Lanka	1		Welikada
Thailand	10		Bangkok
1310 kHz			
Hawaii	0.5	KKAI	Honolulu (CP)
Malaysia	20		Mersing (AF 1314)
1314 kHz			
Australia	5	2ZW	Wollongong (trtr)
	5	3BA	Ballarat
China, PR	100		NM (FS) + 4 st's
China (Taiwan)	10	BEV76	Chunan (Tien Sheng)
	5	BEP32	Hsinying (Chingcha)

Country	kW	Call	Location
India	10		Bhuj
	1		Cuttack B
Japan	50	JOUF	Osaka (OBC)
Korea, Rep.	10	HLCM	Iri (CBS)
Malaysia	20		Mersing (r1310)
New Zealand	2	2YW	Gisborne
Philippines	10	DWXI	Paranaque, MM
Thailand	20		Khon Kaen
1320 kHz			
Galapagos Is.	5	HCVG8	Puerto Baqueriz
1323 kHz			
Australia	2	5AD	Adelaide
China, PR	100		XZ; JL; SN (all FS) + 4 st's
China (Taiwan)	1	BEV45	Taipei (TBC1)
Fiji	2.5		Rakiraki
India	20		Calcutta C
Japan	1	JOFP	Fukushima (NHK1)
Kazakhstan	20		Atyrau
Korea, Rep.	1	HLQJ	Yong-gwang (KBS)
	1	HLCU	Ullung (KBS)
Kyrgyzstan	30		Biskek
Nauru I.	0.2	C2AM	Nauru
New Zealand			Christchurch
Philippines	10	DYSI	Iloilo C.
	1	DZRK	Kalinga Apayao, Beng.
	5	D...	
Tajikistan			Dushanbe
Thailand	10		Chieng Mai
	10		Songkhla
	1		N. Pathom
	10		N. Surat Thani
Uzbekistan	5		Dangara/Koson
1332 kHz			
Australia	2	3SH	Swan Hill
	5	4BU	Bundaberg
China, PR	10		HEN; JL; FJ; GS
China (Taiwan)	10	BEC36	Tsaoying (Han Sheng)
	1		Puli (TBC)
India	10		Tezu
	1		Mangalore
Indonesia	10		Jakarta (RRI)
	1		Wamena (RRI)
Japan	50	JOSF	Nagoya (Tokai R.)
Korea, Rep.	10	HLAO	Chùngju (MBC)
New Zealand	5	1XI	Auckland
Philippines	5	DWAY	Cabanatuan, Nva. Ecija
	5	DYBB	Baybay, Leyk
	1	DZKI	Iriga C., Cam. Sur
	1	DWKI	
Russia			Tyumen, Sib.
Thailand	10		Bangkok
Uzbekistan			Muborak
1341 kHz			
Australia	5	2HH	Newcastle (trtr)
China, PR	100		GD (FS); HL + 5 st's
China (Taiwan)	20	BEC48	Matsu (Kuang Hua)
India	1		Kohima
Indonesia	1/5		Tanjung Pinang
Kazakhstan	25		Aktau
	30		Almaty
Korea, Rep.	10		Kimpò (KBS)
New Zealand	2	2ZN	Nelson
Pakistan	10		Bahawalpur
Philippines	5	DYRL	Bacolod, Neg. Occ.
	5	DZYS	Santiago, Isa.
Thailand	50		Ubon Ratchathani
	10		Loei
	10		Phang Nga
1350 kHz			
Australia	5	2LF	Young

Country	kW	Call	Location
China, P.R.	50		YN; NM; + 2 st's
China (Taiwan)	10	BED63	Chia-i (BCC1)
	2.5	BEV50	Sanchung (Chung Hua 1)
India	1		Dharwar B./Jalandhar C
Japan	20	JOER	Hiroshima (RCC)
Korea, Rep.	10	HLAQ	Samch'ok (MBC)
Laos	10		Pakse
Micronesia		V6A	Moen. Chuuk State
Mongolia	150		Dalanzadgad
Philippines	10	DZXQ	Makati, MM
	5	DZER	Boac, Marinduque
Russia	5		Onguday/Ust-Kan/Ust-Ulagan, Sib.
	1		Choya/Shebalino, Sib.
Thailand	10		Lampang
	10		Trang
1359 kHz			
Australia	0.25	4WK	Toowoomba (trtr)
China, PR			(C1) Synchro
China (Taiwan)	5	BEC40	Hualien (Han Sheng)
India	20		Bhadravathi
Indonesia	1		Ujung Padang (RRI)
Korea, Rep.	1		Songt'an (AFKN)
New Zealand	1	4XC	Queenstown (Resort R.)
Philippines	1	DYSL	Sogod, So. Leyte
	5	DZYR	S. Fernando, LU
Russia	7		Ust-Ilimsk, Sib.
Thailand	10		Bangkok
1360 kHz			
Cambodia	20		Phnom Penh (FS)
1368 kHz			
Australia	2	2GN	Goulburn
China, PR			FJ (v) + 2 st's
India	10		Delhi C
Japan	5	JOHP	Takamatsu (NHK1)
	1	JOLG	Tottori (NHK1)
	1	JOJS	Wakkanai (HBC)
Korea, PDR			Pyongyang (FS)
Korea, Rep.	1	HLKO	Muju (KBS)
New Zealand	1	1XT	Tauranga (Village R.)
Philippines	1	DZRA	Virac, Catanduanes
	5	DXKO	Cagayan de Oro, Mis. Or.
	5	DZBS	Baguio C., Benguet
	1	D...	
Thailand	50		Nan
	1		Chumporn
1377 kHz			
Australia	5	3MP	Mornington
China, PR	200		FJ (H1) + 2 st's
China (Taiwan)	1		Chihsan (Cheng Sheng)
	1	BEH9	Taipei (Fu Hsing)
India	10		Hyderabad B
Japan	5	JOUC	Yamaguchi (NHK2)
	1	JOAC	Nagasaki (NHK2)
New Zealand	2	2XX	Levin
Pakistan	10		Peshawar II
Philippines	1	DXKP	Pagadian C., Zamb. Sur
	5	DYVR	Roxas C., Capiz
Russia	75		Tavrichanka, FE
	5		Okha, FE
	50		Yekaterinburg, Sib.
Thailand	10		Phitsanulok
1380 kHz			
Hawaii	5	KIFO	Pearl City
1386 kHz			
Australia	5	2EA	Sydney (SBS)
China, PR	20		TJ + 3 st's
China (Taiwan)	3.5	BED87	Hsinchu (BCC3)
	3.5	BED74	Yuli (BCC2)
	1		Peikang (Cheng Sheng)

Country	kW	Call	Location
India	10		Gwalior
Japan	10	JOQC	Morioka (NHK2)
	10	JOJB	Kanazawa (NHK2)
	10	JOHC	Kagoshima (NHK2)
	5	JOKB	Okayama (NHK2)
Korea, Rep.	10	HLAM	Mokpò (MBC)
New Zealand	5/1		Auckland
			Napier (Sports Roundup)
Philippines	5	DXCR	Malaybalay, Buk.
Solomon Is.	5		Lata, Sta. Cruz I.
Thailand	20		Pathum Thani
1395 kHz			
Australia	0.2	2LG	Lithgow (RR)
	5	5AA	Adelaide
China, PR	10		AH;NM
India	10		Bikaner
	10		Madras B
Japan	1		Toyooka (AMK)
	1	JOWE	Wakamatsu (RFC)
Kazakhstan			Panfilov
Korea, Rep.	10	HLCO	Ch'olwon (KBS)
New Zealand	2	4ZW	Oamaru
Philippines	1	DWMG	Solano, Nva. Visc.
	10	DYCH	
	5	DZVT	
Russia	5		Buguruslan, Sib.
Thailand	25		Mae Chau Chiangrai
1398 kHz			
Indonesia	50		RRI Bandung
1404 kHz			
Australia	2	2PK	Parkes
China, PR	10		LN + 2 st's
China (Taiwan)	10	BED65	Ilan (BCC1)
	5	BEV78	Keelung (Yi Shih)
Cocos Is.	0.1	VKW	Keeling
India	20		Gangtok
Japan	10	JOVR	Shizuoka (SBS)
	1		Hamamatsu (SBS)
	5	JOQL	Kushiro (HBC)
Korea, Rep.	10	HLKP	Pusan (CBS)
Kyrgyzstan			5 st's
New Zealand	2.5	4XL	Invercargill (R. Rhema)
Pakistan	10		Dera Ismail Khan
Philippines	5	DYKB	Bacolod, Neg. Occ.
	2	DZOR	Olongapo, Zamb.
Samoa, W	2.5	2AP	Unk.
Thailand	10		Yaso Thon
	10		Phichit
	50		Songkhla
1410 kHz			
Galapagos Is.	1	HCSC8	Puerto Ayora
1413 kHz			
Bangladesh	10		Comilla
China, PR	100		XJ + 4 st's
China (Taiwan)	10	BED54	Miaoli (BCC1)
	1	BED67	Puli (BCC1)
	3.5	BED80	Taitung (BCC3)
Indonesia			Sibolga (RRI)
Japan	50	JOIF	Fukuoka (KBC)
New Zealand	2	1ZO	Tokoroa
	0.1	3XP	Christchurch
Philippines	10		DYDW
	5		DYXW
Russia			Sangar, Sib.
			Vladivostok, FE
Thailand	10		Bangkok
1420 kHz			
Hawaii	5	KCCN	Honolulu
1422 kHz			
Australia	1	4AM	Port Douglas (trtr)
	5	3XY	Melbourne
China, PR	20		SH + 2 st's
China (Taiwan)	1		Kuanyin (Chien Kuo)
	50		Minhsiung (BCC1)
Christmas I.	0.25	VLU2	Phosphate Hill
India	1		Rajkot B
Japan	50	JORF	Yokohama (R. Nippon)
Kazakhstan			Unk.
New Zealand	0.1	4XK	Kingston
Philippines	15	DWBC	Quezon C, MM
	5	DXMU	Musuan, Buk.
Russia	5		Kuanda, Sib.
Thailand	10		Phitsanulok
1430 kHz			
Laos			Savannakhet
1431 kHz			
Australia	2	2RN	Wollongong (RR)
China P.R.	200		NM + 4 st's
China (Taiwan)	10	BEC73	Chingshui (Kung Chun)
India	1		Kozhikode B/Chandigarh
Japan	5	JOVF	Wakayama (WBS)
	5	JOZF	Gifu (GBS)
	1	JOWW	Iwaki (RFC)
	1		Fukue (NBC)
	1	JOHL	Tottori (BSS)
	1		Izumo (BSS)
Kyrgyzstan			4 st's
New Zealand	2	2XKC	Hastings
Philippines	1	DYRS	San Carlos, Neg. Occ.
Russia	7		Neryungri, Sib.
			Nogliki, FE
Thailand	20		Songkhla
1440 kHz			
Australia	2	2PB	Canberra (PNN)
China, PR	50		NM + 4 st's
China (Taiwan)	10	BEC60	Taipei (Kung Chun)
India	1		Ahmedabad B
	1		Pt. Blair (standby tr.)
Japan	50	JOWF	Sapporo (STV)
	3		Muroran (STV)
	1		Tomakomai (STV)
Kazakhstan			4 st's
Korea, Rep.	1-0.25		4 st's (AFKN)
Myanmar			Mayawadi R. Stn.
New Zealand	0.2	1XK	Tauranga
	0.1		Lawrence
			Nelson
Philippines	10	DWDH	Dagupan C., Pang.
	1	DZOM	Calapan, M Ortal
Russia	5		Ust-Koksha/Kosh-Agach, Sib.
Thailand	10		Samut Sakhon
	10		N. Phanom
			Nan
1449 kHz			
Australia	5	2MG	Mudgee
China, PR	20		JX + 3 st's
China (Taiwan)	10	BED48	Kaohsiung (BCC3)
India	1		Kanpur
Japan	5	JOKF	Takamatsu (RNC)
	5	JOQM	Abashiri (HBC)
	5	JOKF	Marugame (RNC)
Korea, Rep.	10	HLQB	Ulsan (KBS)
Micronesia	10	V6AH	Kolonia, Pohnpei
New Zealand	2	2YM	Palmerston No.
Philippines	5	DXRM	Marawi C, Lanao Sur
	5	DYAC	Baybay, Leyte
Russia	5		Yakutsk, Sib.
Thailand	5		Chumporn
1453 kHz			

Country	kW	Call	Location
Cambodia			Unk., Batambang Province
1458 kHz			
Australia	2	2PB	Newcastle (PNN)
China, PR	150		NM
China (Taiwan)	10	BED57	Taipei (BCC Ladys/Dream Channel)
India	10		Bhagalpur
	20		Barmer
Japan	1	JOUO	Saga (NBC)
	1	JOYL	Tsuchiura (IBS)
	1	JOWR	Fukushima (RFC)
	1		Shobara (RCC)
			Sekijo (IBS)
Kazakhstan			Arkalyk/Buran
Korea, Rep.	1	HLSH/SD	Hamyang/Ponghwa (KBS)
Maldives	5		Malé
New Zealand	2	3YW	Westport
Philippines	10	DWRF	Iba, Zamb.
	10	DYMT	
	10	DZOE	
Russia	5		Tara, Sib.
Thailand	20		Phuket
	10		Sri Sa Ket
1460 kHz			
Hawaii	5	KULA	Kailua
1467 kHz			
Australia	2	3MA	Mildura
China, PR	50		FJ (v) + 3 st's
China (Taiwan)	10	BED81	Hualien (BCC3)
	10	BED82	Chia-i (BCC3)
	50		Kinmen (Kuang Hua)
Fiji	2.5		Rakiraki
India	20		Jeypore
Japan	1		4 st's (NHK2)
Korea, Rep.	50	HLKN	Mokpo (KBS)
Kyrgyzstan	30		Biskek
	15		Os/Batken/Kara-Kul
Philippines	10	DZTP	
	5	DXVP	
Thailand	100		Bangkok
1476 kHz			
Australia			Penrith (not yet on air)
	2	4ZR	Roma
China, PR	200		HL (C2) + 8 st's
China (Taiwan)	10	BEH57	Kaohsiung (Fu Hsing)
India	1		Jaipur A
Malaysia	600		Kota Kinabalu, Sabah (r 1475)
New Zealand	5	1XD	Auckland Airport
Philippines	1	DZYA	Angeles C, Pamp.
	1	DWRB	Lipa C, Bat
Russia	1000		Tavrichanka, FE
	20		Onguday, Sib.
Thailand	100		Chiengmai (tr: Lamphun)
1485 kHz (Low-power ch.)			
Australia	0.1-0.2		4 st's
Br.Ind.Oc.Terr.	0.25		Diego Garcia (AFRTS)
China, PR			19 st's
India	1		3 st's
Indonesia	0.3		Padang (RRI)
Japan	1	JOGO	Hachinohe (RAB)
	1	JOPL	Hagi (KRY)
Korea, Rep.	1	HLQS	Kongju (KBS)
	1	HLQU	Kohung (KBS)
New Zealand	1		Gisborne
Pakistan	10		Faisalabad
Philippines	1	DYKR	Kalibo, Aklan
	1	DYDH	Santiago, Isa.
	1	DY..	
Russia	1		Ust-Maya/Yakutsk, Sib.
	1		Kamenskoye, FE
	5		Kartaly, Sib.
	1		Tazovskiy/Krasnoselkup, Sib.
	1		Gaz Sale/Cherskiy, Sib.
	1		Chakalovskoye/Plastun, Sib.
	1		Irkutsk/Omsk, Sib.
			Tomtor, Sib.
Sri Lanka	1		Mahiyangana
Thailand			5 st's
Uzbekistan			Jizza/Nurobod/Munyok
1494 kHz			
Australia	2	2AY	Albury
China, PR	120		XJ + 2 st's
China (Taiwan)	10	BEE32	Taipei (Chiao Yu)
	5	BEE34	Changhua (Chiao Yu)
India	1		Thiruananthapuram B
Japan	1	JOTL	Nayoro (HBC)
	10	JOYR	Okayama (RSK)
	1		4 st's (RSK)
Micronesia	10	V6A	Colonia, Yap
New Zealand	2.5	1YT	Taupo
Papua N.Guinea	10		Wabag
Philippines	1	DXOC	Ozamis C., Mis.Occ.
	10	DWLR	Quezon C, MM
Thailand	25		Bangkok
1500 kHz			
Hawaii	50	KUMU	Honolulu
Vietnam			Ba Ria Vung tau
1503 kHz			
Australia	5	2BS	Bathurst
	20	3AK	Melbourne
China, PR			3 st's
India	1		Srinagar D/Vijayawada B
Japan	10	JOUK	Akita (NHK1)
Korea, Rep.	1	HLSK	Kimchòn (KBS)
New Zealand			Christchurch
			Wellington
Russia	5		Salekhard, Sib.
Tajikistan	7		Dushanbe
Thailand	10		Surat Thani
	10		Sakon Nakhon
1512 kHz			
Australia	10	2RN	Newcastle (RN)
China, PR	10		GS; NM
China (Taiwan)	10	BEP27	Taipei (Ching Cha)
	5	BEH36	Taichung (Fu Hsing)
		BEH84	Hualien (Fu Hsing)
Japan	5	JOZB	Matsuyama (NHK2)
Korea, Rep.	0.25-0.1		4 st's (AFKN)
New Zealand	1	1ZU	Taumarunui
Pakistan	10		Muzaffarabad
	10		Gilgit, Kashmir (r 1510)
Philippines	10	DYAB	
	5	DYCR	
Thailand	10		Suphan Buri
	10		Pha Yao
	10		Surin
1521 kHz			
Australia	5	2QN	Deniliquin
China, PR	500		XJ (FS) + 7 st's
China (Taiwan)	1200		Changchih (CBS)
India	1		Aurangabad
	10		Tawang
Japan	1		4 st's (NHK2)
Russia	5		Boguchany, Sib.
Thailand	25		Bangkok
1530 kHz			
Australia	2	2VM	Moree

Country	kW	Call	Location
China, PR	10		SX; ZJ; + 3 st's
India	10		Agra
Japan	1	JODO	Joetsu (BSN)
	5	JOXF	Utsunomiya (CRT)
	1	JOEO	Fukuyama (RCC)
	1		Mihara (RC)
Kyrgyzstan	5		Tas-Kumyr
New Zealand	2	2YP	New Plymouth
Philippines	10	DZME	Quezon C., MM
	5	DXDV	
Russia	5		Krasnyy Ckikoy, Sib.
			Arginskoye, Sib.
Thailand	10		Uttaradit
	10		Chanthaburi
1539 kHz			
Australia		2RF	Sydney
	5	5TAB	Adelaide
China (Taiwan)	10	BED78	Tainan (BCC3)
India	20		Panaji B
Kazakhstan			Kostanay
Korea, Rep.	1	HLQC	Kosan (KBS)
New Zealand	1	2ZE	Blenheim
Philippines	1	DZYM	San José, Mind. Occ.
Thailand	10		Udon Thani
	10		Ratchaburi
Uzbekistan			Samarkand
1540 kHz			
Hawaii	5	KISA	Honolulu
1548 kHz			
Australia	50	4QD	Emerald (RR)
China, PR	100		SD; HN
Kazakhstan			Tekeli/Jansügirov/Lepsi/Andreyevka
New Zealand	1	1XN	Rotorua
Philippines	5	DYDM	
Russia	10		Aleksandrovsk-Sakhalinsk, FE
Sri Lanka	600		Trincomalee (DW relay)
Thailand	10		Bangkok
1551 kHz			
Indonesia	0.1		Manado (RRI)
1557 kHz			
Australia	2	2RE	Taree
China, PR			HEB
Kazakhstan			Uç-Aral/Kapal/Antonovka
New Zealand	2	2ZH	Hawera (rel. 2ZP)
Pakistan	10		Skardu (Kashmir)
Philippines	2.5	DWGO	Olongapo C., Zamb.
Russia	2		Irkutsk, Sib.
	10		Vladivostok,. FE
Thailand	10		Lom Sak, Phetchaboon
1566 kHz			
Australia	5	3NE	Wangaratta
	0.2	4GM	Gympie (RR)
China, PR			HEB (2 st's) + 4 st's
India	1000		Nagpur (Buttibori)
Korea, Rep.	250/100	HLAZ	Cheju (FEBC)
Norfolk Is.	0.05	VL2NI	Kingston
Philippines	10	DXID	Pasay C, MM
	10	DZHH	
Russia			Barnaul, Sib.
			Omsk, Sib.
1570 kHz			
China (Taiwan)	10		Kaohsiung (ICRT)
Hawaii	1	KUAU	Haiku, Maui
1575 kHz			
Australia	0.25	2RF	Wollongong
China, PR	7.5		ZJ
Japan	1		Iwakuni (FEN)

Country	kW	Call	Location
	0.3		Misawa/Sasebo (FEN)
Kazakhstan	10		Semipalatinsk, Kaz.
New Zealand	2.5	4XS	Dunedin
Russia	5		Yuzhno-Sakhalinsk, FE
Thailand	1000		Bangkok (Ayutthaya)
1584 kHz (Low-power ch.)			
Australia	0.1-0.5		7 st's
China, PR	10		HL + 12 st's
Hong Kong	0.1		Chung Hom Kok (RTHK3)
India	1		6 st's
Korea, Rep.	1	HLDK	Tanyang (KBS)
	1	HLQZ	Kumsan (KBS)
	1		Sanch'ong (KBS)
Micronesia	1	V6AJ	Tofol, Kosrae
New Zealand	0.4	2ZF	Picton (rel. 2ZE)
Pakistan	0.25		Turbat
	0.25		Chitral
Palau	3	WSZB	Koror (Malakai Is.)
Philippines	10	DYSG	
	1	D...	
Russia	1		Klyuchi/Tigil, FE
	5		Aykhal, Sib.
			Ust-Nera/Yakutsk, Sib.
			Taksimo, Sib.
Thailand	1		5 st's
Uzbekistan	1		Muynak
1593 kHz			
Australia	5	3RG	Melbourne
China, PR	100		XJ + 3 st's
China (Taiwan)	1		Kaohsiung (Pai Yun)
	1		Ilan (Taiwan Yuyeh)
India	10		Bhopal A
Japan	10	JOQB	Niigata (NHK2)
	10	JOTB	Matsue (NHK2)
Micronesia	5	V6AK	Weno, Chuuk
New Zealand	2.5		Auckland (Southern Star)
Papua N.Guinea	10		Vanimo
Philippines	1	DZUP	Quezon C, MM
Russia	50		Irkutsk, Sib.
			Neftegorsk, FE
Thailand	10		Ranong
	1		Ratchaburi
Uzbekistan			Jar-Kürgon
1600 kHz			
Hawaii	5	KHNL	Honolulu (CP)
1602 kHz (Low-power ch.)			
Australia	0.05-0.5		3 st's (RR)
China, PR	1		JS; HB
India	1		7 st's
Japan	1		4 st's (NHK2)
Korea, Rep.	1	HLQE	Sabuk (KBS)
New Zealand	1	2XA	Levin
Philippines	10	DZUP	
Russia			Novo Ilinsk/Ust-Barguzin, Sib.
	1		Kurilsk/Severo-Kurilsk, FE
Sri Lanka	1		Mahiyangana
Thailand	1		Buriram
	1		N. Si Thammarat
1610 kHz			
Australia	5		Caringbah (not yet on air)
1629 kHz			
Australia	0.01		Newcastle (Hospital RN)
1656 kHz			
Australia			Gold Coast (TAFE R.)
1720 kHz			
Australia	0.2		Kensington (trtr VL2UV)

NORTH AMERICA

Compiled by Olle Alm
of the
Arctic Radio Club, Sweden

Explanation of listing:—
FIRST COLUMN — Call letters. These are required identification for US sts, usually on top of the hour.
SECOND COLUMN — Location. The official Post Office abbreviations are used for US states and Canadian provinces.
THIRD COLUMN — Transmitter powers. A daytime station listed with two powers uses the second one during critical hours (cf. D and U6 below). For stations using a directional antenna the ERP in most cases varies greatly with the direction from the station.
FOURTH COLUMN — Contains authorized operating hours and directional antenna information. The symbols mean as follows: **D** is daytime operation (between local sunrise and local sunset), **D1** without directional antenna, **D3** with directional antenna, **D4** with directional antenna during critical hours only, **D5** with directional antenna except during critical hours, **D6** with directional antenna, different patterns during critical and non-critical hours. **L** is limited time, and means a st. West of the dominant st. can operate from as early as sunrise at the dominant sts location; A st. East of the dominant st. can operate as late as the dominant sts sunset. Number indicates directional pattern as under "U" below. **SH** equals "Specified Hours" and usually means a st. that would normally operate unlimited time files a schedule of reduced hours with the FCC because of economic factors. **ST** means "Shares Time". This usually finds a college, university, or church st. on the air two or three hours a day, during which time the commercial st. signs off. **U** means the st. is authorized to operate any time up to 24 hours a day. **U1** means without directional antenna; **U2** means with directional antenna at night; **U3** means directional antenna, same pattern, day and night; **U4** means directional antenna, different pattern day and night; **U5** means directional antenna daytime, non-directional at night; **U6** means directional antenna at night and during critical hours (until two hours after sunrise and from two hours before sunset). **U7** means different directional patterns for day, critical hours and night. **U8** as U7 but non-directional day. **U9** means directional day and night (different patterns), but non-directional during critical hours (usually on reduced power).
FIFTH COLUMN — Gives an indication of actual operating hours as follows: **NH** normal hours, usually from 18 to 20h a day. **AN** all night operation, i.e. 24h a day most days. **AN7** continuous operation. **AN6** 24h six days a week, off a few hours during the early morning local time on the seventh day for maintenance, usually Sun or Mon morning. **AN5** off both Sun and Mon morning, etc. **D** daytime, i.e. local sunrise to sunset only. **D*** 0600 local time to local sunset, but with low power (from a few watts to a maximum of 500 watts) before local sunrise if sunrise is later than 0600. A similiar extension of evening hours also exists. **SH** specified hours as fourth column; or special hours, i.e. less than NH, often with breaks during the day.
SCOPE — Because of the vast number of stations in operation in North America it has been found necessary to limit the scope of this list. All US and Canadian st's on the "local" channels (1230, 1240, 1340, 1400, 1450, 1490kHz), except those having more than 1kW power, have been omitted. For the remaining frequencies a lower power limit of 5kW day and/or 1kW night has been set for US st's and 1kW day for Canadian st's. For each frequency the approximate number of omitted low power (**Ip**) stations is mentioned. Full details of these stations may be obtained from specialized publications, cf. next paragraph.
ADDRESSES — In most cases the US Post Office requires a full address to deliver a letter. Complete addresses for all US MW stations may be obtained from the National Radio Club AM Radio Log (see club list for address) and from trade publications (The M Street Radio Directory and others). Zip codes for stations not mentioned in the respective country sections are no longer listed below.

530kHz

Call	Location	Power	Col4	Col5
CKHL	High Level, AB	1/0.25	U1	AN
CIAO	Brampton, ON	1/0.25	U1	AN

540kHz

Call	Location	Power	Col4	Col5
KUHB	St. Paul, AK	2.5	U1	F.Pl.
KNOB	Costa Mesa, CA	25/0.36	U7	NH
KIEZ	Carmel Valley, CA	10/0.5	U4	AN7
WQTM	Pine Hills, FL	50	U2	AN6
WDAK	Columbus, GA	5/0.5	U4	AN6
KWMT	Fort Dodge, IA	5/0.2	U3	AN7
KNOE	Monroe, LA	5/1	U4	AN7
KNMX	Las Vegas, NM	5	D3	D
WETC	Wendell, NC	8/0.5	U4	NH
WWCS	Canonsburg, PA	5/0.5	U4	AN
CBT	Grand Falls, NF	10	U1	AN
CBEF	Windsor, ON	2.5/5	U3	AN7
CBGA1	New Carlisle, PQ	10	U4	NH
CBK	Watrous, SK	50	U1	AN

+Ip: USA 12

550kHz

Call	Location	Power	Col4	Col5
WASG	Atmore, AL	25/0.1	U1	NH
KTZN	Anchorage, AK	5	U1	AN7
KOY	Phoenix, AZ	5/1	U1	AN7
KUZZ	Bakersfield, CA	5	U4	AN7
KRAI	Craig, CO	5/0.5	U2	AN7
WAYR	Orange Park, FL	5/0.26	U3	AN
WDUN	Gainesville, GA	10/2.5	U2	NH
KFRM	Salina, KS	5/0.1	U3	NH
KTRS	St. Louis, MO	5	U2	AN7
KBOW	Butte, MT	5/1	U2	AN6
WGR	Buffalo, NY	5	U2	AN7
KFYR	Bismarck, ND	5	U2	AN6
WKRC	Cincinnati, OH	5/1	U4	AN7
KOAC	Corvallis, OR	5	U4	NH
WJMW	Bloomsburg, PA	1	U4	AN7

Call	Location	Power	Col4	Col5
WLKW	Pawtucket, RI	4.6/3.4	U4	AN7
KCRS	Midland, TX	5/1	U4	AN
KTSA	San Antonio, TX	5	U2	AN7
WDEV	Waterbury, VT	5/1	U4	NH
WSVA	Harrisonburg, VA	5/1	U2	AN
KARI	Blaine, WA	5/2.5	U4	AN7
WSAU	Wausau, WI	5	U4	AN7
CFJC	Kamloops, BC	25/5	U4	AN
CKPG	Prince George, BC	10	U2	AN7
CHNO	Sudbury, ON	50/10	U4	AN
CHLN	Trois-Rivières, PQ	10/5	U4	AN7

+Ip: USA 5

560kHz

Call	Location	Power	Col4	Col5
WOOF	Dothan, AL	5	D1	D*
KVOK	Kodiak, AK	1	U1	NH
KBLU	Yuma, AZ	1	U2	AN
KSFO	San Franciso, CA	5	U2	AN7
KLZ	Denver, CO	5	U3	AN7
WQAM	Miami, FL	5/1	U1	AN7
WIND	Chicago, IL	5	U4	AN6
WGAN	Portland, ME	4.8	U3	AN7
WFRB	Frostburg, MD	5	D1	D
WHYN	Springfield, MA	5/1	U3	AN7
WEBC	Duluth, MN	5	U4	AN
KWTO	Springfield, MO	5/4	U2	AN6
KMON	Great Falls, MT	5	U2	AN6
WFIL	Philadelphia, PA	5	U4	AN7
WVOC	Columbia, SC	5	U2	AN
WHBQ	Memphis, TN	5/1	U4	AN7
KLVI	Beaumont, TX	5	U2	AN
KPQ	Wenatchee, WA	5	U2	AN7
WJLS	Beckley, WV	4.5/0.5	U2	AN7
CKNL	Ft. St. John, BC	1	U2	AN7
CHTK	Prince Rupert, BC	1/0.25	U1	AN7
CHVO	Carbonear, NF	5	U2	AN7

Call	Location	Power	Type	Notes
CJKL	Kirkland Lake, ON	5	U2	AN
CFOS	Owen Sound, ON	7.5/1	U4	NH
CKCN	Sept-Iles, PQ	10/5	U4	AN
+lp USA 4/Can 2				

570kHz

Call	Location	Power	Type	Notes
WAAX	Gadsden, AL	5/0.5	U2	AN
KKFJ	Alturas, CA 5		U4	
KLAC	Los Angeles, CA	5	U2	AN7
WHNZ	Pinellas Park, FL	5	U4	AN7
WACL	Waycross, GA	5/1	U2	NH
WWRC	Bethesda, MD	5/3	U4	AN6
WVMI	Biloxi, MS 5/1		U4	AN7
KGRT	Las Cruces, NM	5	D1	D*
WMCA	New York, NY	50/30	U4	AN7
WSYR	Syracuse, NY	5	U4	AN7
WWNC	Asheville, NC	5	U2	AN7
WKBN	Youngstown, OH	5	U2	AN7
WNAX	Yankton, SD	5	U2	AN6
KLIF	Dallas, TX 5		U4	AN7
KNRS	Salt Lake City, UT	5	U3	AN7
KVI	Seattle, WA	5	U1	AN7
CKEK	Cranbrook, BC	10/1	U3	AN6
CKWL	Williams Lake, BC	1	U2	AN6
CFCB	Corner Brook, NF	1	U1	AN7
CKGL	Kitchener, ON	10	U3	AN7
CKSW	Swift Current, SK	10	U4	AN
CFWH	Whitehorse, YT	5/1	U1	AN7
	Nuuk (Godthåb), Greenland	25	U1	
SH				
+lp: USA 4, Can 3				

580kHz

Call	Location	Power	Type	Notes
KRSA	Petersburg, AK	5	U1	AN
KSAZ	Marana, AZ	5/0.8	U2	
KMJ	Fresno, CA 5		U1	AN7
KUBC	Montrose, CO	5/1	U2	NH
WDBO	Orlando, FL 5		U2	AN7
KFXD	Nampa, ID 5		U2	AN6
WILL	Urbana, IL	5/0.1	U4	NH
KKSU	Manhattan, KS	5/0.5	ST1	SH
WIBW	Topeka, KS 5		ST2	AN
KLBG	Alexandria, LA	5/1	U2	NH
WTAG	Worcester, MA	5	U4	AN6
WTCM	Traverse City, MI	15/0.8	U4	AN
KCMX	Ashland, OR	1	U4	AN6
WHP	Harrisburg, PA	5	U2	AN7
WCHS	Charleston, WV	5	U2	AN
WKTY	LaCrosse, WI	5/0.74	U4	AN
CKUA	Edmonton, AB	10	U4	NH
CKXR	Salmon Arm, BC	10/1	U2	AN7
CKY	Winnipeg, MB	50	U4	AN7
CJFX	Antigonish, NS	25	U3	AN7
CKAP	Kapuskasing, ON	10/1	U2	AN7
CFRA	Ottawa, ON	50/10	U4	AN7
CKPR	Thunder Bay, ON	5/1	U1	AN
CHLC	Baie-Comeau, PQ	10/2.5	U4	AN2
+lp: USA 8/Can 1				

590kHz

Call	Location	Power	Type	Notes
KHAR	Anchorage, AK	5	U1	AN
KBHS	Hot Springs, AR	5	D	D*
KSZZ	San Bernardino, CA	1	U4	AN7
KTHO	South Lake Tahoe, CA	2.5/0.5	U2	AN7
KCSJ	Pueblo, CO 1		U4	AN6
WDIZ	Panama City, FL	1.7/2.5	U2	AN
WDWD	Atlanta, GA	5/4.5	U4	AN7
KID	Idaho Falls, ID	5/1	U2	NH
KFNS	Wood River, IL	1	U4	AN7
WVLK	Lexington, KY	5/1	U4	AN6
WEZE	Boston, MA	5	U3	AN7
WJMS	Ironwood, MI	5/1	U2	NH
WKZO	Kalamazoo, MI	5	U2	NH

Call	Location	Power	Type	Notes
WOW	Omaha, NE 5		U1	AN7
WROW	Albany, NY 5/1		U4	AN
WGTM	Wilson, NC 5		U4	AN7
KUGN	Eugene, OR5		U2	AN6
WARM	Scranton, PA	5	U4	AN7
WMBS	Uniontown, PA	1	U2	AN
KLBJ	Austin, TX 5/1		U2	AN7
KSUB	Cedar City, UT	5/1	U2	AN
WLVA	Lynchburg, VA	5/1	U4	AN6
KAQQ	Spokane, WA	5	U1	AN7
CFTK	Terrace, BC1		U3	AN7
CFAR	Flin Flon, MB	10/1	U5	NH
CJCW	Sussex, NB 1/0.25		U4	AN6
VOCM	St. John's, NF	20/10	U2	AN7
CJCL	Toronto, ON	50	U4	AN7
CKRS	Jonquière, PQ	25/10	U4	AN
+lp: USA 4/Canada 1				

600kHz

Call	Location	Power	Type	Notes
KVNA	Flagstaff, AZ	5/0.5	U2	
KNRO	Indio, CA 1		U3	AN
KOGO	San Diego, CA	5	U3	AN7
KIIX	Fort Collins, CO	5/0.5	U4	AN7
WBWL	Jacksonville, FL	5	U2	AN7
WMT	Cedar Rapids, IA	5	U2	AN7
WKLW	Paintsville, KY	5	D1	D
WFST	Caribou, ME	5	D1	D*
WCAO	Baltimore, MD	5	U3	AN7
KGEZ	Kalispell, MT	5/1	U4	NH
WSJS	Winston-Salem, NC	5	U4	AN7
KSJB	Jamestown, ND	5	U3	AN7
WREC	Memphis, TN	5	U4	AN7
KROD	El Paso, TX 5		U2	AN7
KTBB	Tyler, TX 5/2.5		U4	AN
CKBD	Vancouver, BC	10	U3	AN7
CBNA	St. Anthony, NF	10	U4	NH
CKCL	Truro, NS	10	U3	AN
CKAT	North Bay, ON	10/5	U4	AN
CIQC	Montreal, PQ	10/5	U4	AN7
CJWW	Saskatoon, SK	25/10	U4	AN7
+lp: USA 14/Canada 3				

610kHz

Call	Location	Power	Type	Notes
WEZN	Birmingham, AL	5/1	U2	AN7
KAVL	Lancaster, CA	4.9/4	U4	NH
KFRC	San Francisco, CA	5	U1	AN7
KSKE	Vail, CO 5		D1	D
WIOD	Miami, FL 10		U4	AN7
KDAL	Duluth, MN5		U2	AN5
WDAF	Kansas City, MO	5	U1	AN7
KOJM	Havre, MT 1		U4	NH
WGIR	Manchester, NH	5/1	U4	NH
KZSS	Albuquerque, NM	5	U2	AN7
WFNZ	Charlotte, NC	5/1	U4	AN7
WTVN	Columbus, OH	5	U2	AN7
KRTA	Medford, OR	2.5/5	U4	AN2
WIP	Philadelphia, PA	5	U3	AN7
KILT	Houston, TX	5	U4	AN7
KVNU	Logan, UT 5/1		U2	AN6
WSLC	Roanoke, VA	5/1	U4	AN6
KONA	Tri-Cities, WA	5	U4	AN7
CKYL	Peace River, AB	10	U2	AN7
CHNL	Kamloops, BC	25/5	U4	AN7
CJAT	Trail, BC	10/1	U5	AN
CHTM	Thompson, MB	1	U1	AN7
CKTB	St. Catherines, ON	10/5	U3	AN6
CFLO	Mont-Laurier, PQ	1	U2	AN
CHNC	New Carlisle, PQ	10/5	U3	AN7
CKRW	Whitehorse, YT	1	U1	AN7
+lp: USA 7/Canada 2				

620kHz

Call	Location	Power	Type	Notes
WKNI	Lexington, AL	5	D1	D

Call	Location	Power	Ant	Code
KGTL	Homer, AK 5	U1	NH	
KTAR	Phoenix, AZ	5	U2	AN7
KIGS	Hanford, CA	1	U2	AN
KKGM	Grand Junction, CO	5/0.85	U1	D*
WSUN	St. Petersburg, FL	5	U2	AN7
KWAL	Wallace, ID1	U2	NH	
KMNS	Sioux City, IA	1	U4	AN
WZON	Bangor, ME5	U2	AN7	
WJDS	Jackson, MS	5/1	U2	AN7
WJWR	Jersey City, NJ	3/7.6	U4	AN7
WHEN	Syracuse, NY	5/1	U2	AN7
WDNC	Durham, NC	5/1	U4	AN7
KEWS	Portland, OR	5	U2	AN7
WRJZ	Knoxville, TN	5	U2	AN7
KAAM	Plano, TX	5/4.5	U2	NH
WVMT	Burlington, VT	5	U4	AN6
WTMJ	Milwaukee, WI	50/10	U4	AN7
CJCI	Prince George, BC	10	U2	AN7
CKCM	Grand Falls, NF	10	U3	AN7
CKOY	Timmins, ON	10/5	U4	NH
CFRP	Forestville, PQ	1	U3	AN
CKCK	Regina, SK 10	U4	AN7	
+lp: USA 6				
630kHz				
KJNO	Juneau, AK5/1	U1	NH	
KIAM	Nenana, AK	10/3.1	U1	NH
KIDD	Monterey, CA	1	U4	AN6
KHOW	Denver, CO 5	U4	AN7	
WMAL	Washington, DC	5	U4	AN7
WBMQ	Savannah, GA	5	U2	AN
KIDO	Boise, ID 5	U4	AN7	
WLAP	Lexington, KY	5/1	U4	AN6
KJSL	St. Louis, MO	5	U4	AN67
KPTT	Reno, NV 5/1	U2	AN6	
WMFD	Wilmington, NC	1	U4	AN7
KWRO	Coquille, OR	5	D1	D
WPRO	East Providence, RI	5	U2	AN7
KSLR	San Antonio, TX	5	U4	AN7
KCIS	Seattle, WA	3.9/2.5	U2	AN
CHED	Edmonton, AB	50	U4	AN7
CKOV	Kelowna, BC	5/1	U1	AN7
CFCO	Chatham, ON	10/1	U4	AN7
CJET	Smiths Falls, ON	10	U4	AN7
CFCY	Charlottetown, PEI	10	U4	AN7
CHLT	Sherbrooke, PQ	10/5	U4	NH
+lp: Can 4/USA 9				
640kHz				
KYUK	Bethel, AK 10	U1	NH	
KFI	Los Angeles, CA	50	U1	AN7
WLVJ	West Palm Beach, FL	7.5/0.4	U2	NH
WGST	Atlanta, GA50/1	U4	AN7	
WOI	Ames, IA 5/1	U2	AN7	
KTIB	Thibodeaux, LA	5/1	U4	
WNNZ	Westfield, MA	50/15	U4	AN
KGVW	Belgrade, MT	10/1	U2	NH
WWJZ	Mt. Holly, NJ	50/0.95	U3	
WFNC	Fayetteville, NC	10/1	U1	AN7
WHLO	Akron, OH 5/0.5	U4	AN7	
WWLS	Moore, OK 1	U2	NH	
WGOC	Blountsville, TN	5/0.81	U4	AN7
WCRV	Collierville, TN	50/0.48	U2	AN7
CBN	St. John's, NF	10	U1	AN
CFOB	Ft. Frances, ON	1	U2	AN
CHOG	Toronto, ON	50	U4	AN7
+lp: Can 1/USA 2				
650kHz				
KENI	Anchorage, AK	50	U1	AN7
KSTE	Rancho Cordova, CA	25/10	U4	NH
WNMT	Nashwauk, MN	10/1	U2	AN
WSM	Nashville, TN	50	U1	AN7
KMTI	Manti, UT 10/0.9	U4	AN	
KGAB	Orchard Valley, WY	8.6/0.5	U2	
CISL	Richmond, BC	10/9	U4	AN7
CKGA	Gander, NF 5	U3	AN7	
CKOM	Saskatoon, SK	10	U4	AN7
	Qeqertarsuaq (Godhavn),			
	Greenland 5	U1	SH	
+lp: Can 1/USA 2				
660kHz				
WDLT	Fairhope, AL	10/0.85	U2	
KFAR	Fairbanks, AK	10	U1	AN
KTNN	Window Rock, AZ	50	U2	NSP
KGDP	Orcutt, CA 5/7	U4	AN7	
WORL	Altamonte Springs, FL1	U3	F.Pl.	
WBHR	Sauk Rapids, MN	10/0.25	U4	
WFAN	New York, NY	50	U1	AN7
KEYZ	Williston, ND	5	U4	AN
KZTU	Junction City, OR	10/0.07	U1	
WESC	Greenville, SC	50/10	D4	D*
KSKY	Balch Springs, TX	10/0.66	U2	AN
KAPS	Mt. Vernon, WA	10/1	U1	AN
CFFR	Calgary, AB50	U4	AN7	
+lp: USA 2				
670kHz				
WYLS	York, AL 5	D1	D	
KDLG	Dillingham, AK	10	U1	NH
KWXI	Glenwood, AR	5	D1	D*
KVCA	Simi Valley, CA	5/3	U4	AN
KLTT	Commerce City, CO	50/1.4	U4	
WWFE	Miami, FL 50/1	U4		
KBOI	Boise, ID 50	U2	AN7	
WMAQ	Chicago, IL 50	U1	AN7	
WIEZ	Lewiston, PA	5.4	D1	D
WVNS	Claremont, VA	20/5	D3	D
CKXB	Musgravetown, NF	10	U4	AN7
+lp: USA 1				
680kHz				
KBRW	Barrow, AK 10	U1	NH	
KNBR	San Francisco, CA	50	U1	AN7
WCNN	Atlanta, GA50/10	U4	AN	
WCBM	Baltimore, MD	10/5	U4	AN7
WRKO	Boston, MA	50	U4	AN7
WNZK	Dearborn Heights, MI	2.5	N2	N
WDBC	Escanaba, MI	10/1	U4	NH
KFEQ	St. Joseph, MO	5	U4	AN7
KKGR	East Helena, MT	5	D1	D
WINR	Binghamton, NY	5/0.5	U4	
WPTF	Raleigh, NC	50	U2	AN7
WJCE	Memphis, TN	10/5	U2	AN7
KKYX	San Antonio, TX	50/10	U2	AN7
KOMW	Omak, WA 5	D1	D	
WCAW	Charlestown, WV	10/0.25	U4	AN
CHFA	Edmonton, AB	10	U4	AN7
CJOB	Winnipeg, MB	50	U4	AN7
CKXG	Grand Falls, NF	10	U4	AN7
CFTR	Toronto, ON	50	U4	AN7
+lp: USA 10/Can 1				
690kHz				
WJOX	Birmingham, AL	50/0.5	U2	AN
WOKV	Jacksonville, FL	50/10	U2	AN7
KGGF	Coffeyville, KS	10/5	U4	NH
WTIX	New Orleans, LA	10/5	U4	AN7
KHEY	El Paso, TX 10	U4	AN7	
WZAP	Bristol, VA 10/0.02	U1	AN	
CBU	Vancouver, BC	50	U3	AN
CBKF-1	Gravelbourg, SK	5	U2	AN
+lp: USA 17/Can 6				
700kHz				
KBYR	Anchorage, AK	10	U1	AN7
WWTL	Walkersville, MD	5	D3	F.Pl.

Call	Location	Power		
WLW	Cincinnati, OH	50	U1	AN7
KGRV	Winston, OR	25/0.5	U1	NH
KSEV	Tomball, TX	15/1	U4	NH
KWLW	Salt Lake City, UT	50/1	U4	AN7
KJMY	Newport, WA	10/1	U2	NH
CKRD	Red Deer, AB	50/25	U2	AN7

+lp: USA 3

710kHz

Call	Location	Power		
WNTM	Mobile, AL	1	U2	AN6
KUET	Black Canyon City, AZ	50/4.2	U3	F.PI.
KFIA	Carmichael, CA	25/1	U4	AN
KDIS	Los Angeles, CA	50/10	U2	AN7
KNUS	Denver, CO	25/2.7	U4	AN6
WAQI	Miami, FL	50	U4	AN6
KEEL	Shreveport, LA	50/5	U4	AN7
WREM	Monticello, ME	5	D1	D
KCMO	Kansas City, MO	10/5	U4	AN7
WOR	New York, NY	50	U3	AN7
KXMR	Bismarck, ND	10/4	U7	
KGNC	Amarillo, TX	10	U4	AN6
WFNR	Christiansburg, VA	10	D3	D
KIRO	Seattle, WA	50	U2	AN7
WDSM	Superior, WI	10/5	U2	AN6
CKVO	Clarenville, NF	10	U3	AN7
CJRN	Niagara Falls, ON	10/5	U4	AN6
CIPC	Port-Cartier, PQ	1	U3	AN
CKVM	Ville-Marie, PQ	10/1	U2	AN

+lp: USA 14/Can 3

720kHz

Call	Location	Power		
KOTZ	Kotzebue, AK	10	U1	NH
WRZN	Hernando, FL	10/0.25	U2	
WMXY	Hogansville, GA	8	D3	D
WGN	Chicago, IL	50	U1	AN7
WWDF	Richland, MS	5	D1	D
KDWN	Las Vegas, NV	50	U2	AN6
WQTH	Hanover, NH	50/0.5	U4	F.PI.
WGCR	Pisgah Forest, NC	10	D1	D
KSAH	Universal City, TX	10/0.89	U4	AN7
CHTN	Charlottetown, PEI	10/7.5	U2	AN7
	Simiutaq, Greenland	20	U1	SH

+lp: USA 1

730kHz

Call	Location	Power		
WSNI	Thomasville, GA	5	D1	D*
KBSU	Boise, ID	15/0.5	U4	NH
WMTC	Vancleve, KY	5	D3	D
WACE	Chicopee, MA	5	D1	D*
KURL	Billings, MT	5/0.23	U1	NH
WPIT	Pittsburgh, PA	5/0.02	U1	NH
WVPA	Alexandria, VA	5/0.38	U1	AN
CKLG	Vancouver, BC	50	U4	AN7
CKDM	Dauphin, MB	10/5	U2	AN
CKAC	Montreal, ON	50	U3	AN7

+lp: USA 25

740kHz

Call	Location	Power		
WMSP	Montgomery, AL	10/0.34	U4	
KBRT	Avalon, CA	10	D3	D
KCBS	San Francisco, CA	50	U4	AN6
KTWK	Colorado Springs, CO	3.3/1.5	U4	AN7
WWNZ	Orlando, FL	5/1	U2	AN7
WGSM	Huntington, NY	25/0.04	U3	AN
WPAQ	Mt. Airy, NC	10/1	D1	D
KRMG	Tulsa, OK	50/25	U4	AN7
WVCH	Chester, PA	50/0.05	U3	F.PI.
KTRH	Houston, TX	50	U4	AN7
KCMC	Texarkana, TX	1	U3	AN7
CBX	Edmonton, AB	50	U4	AN
CHCM	Marystown, NF	10	U2	AN7

+lp: USA 16/Can 15

750kHz

Call	Location	Power		
KFQD	Anchorage, AK	50	U1	AN6
WSB	Atlanta, GA	50	U1	AN7
KBNN	Lebanon, MO	5	D1	D
KERR	Polson, MT	50/1	U2	AN
KMMJ	Grand Island, NE	10	L3	D*
KXL	Portland, OR	50/20	U4	AN7
KAMA	El Paso, TX	10/1	U4	AN
KOAL	Price, UT	10/6.8	U2	AN7
CBGY	Bonavista, NF	10	U4	NH
CKGB	Timmins, ON	10/5	U2	AN7
CJVR	Melfort, SA	25	U2	AN

+lp: USA 7

760kHz

Call	Location	Power		
KMTL	Sherwood, AR	10	D1	D
KFMB	San Diego, CA	50/5	U2	AN7
KTLK	Thornton, CO	50/1	U1	AN7
WBDN	Brandon, FL	5/1	U4	F.PI.
KCCV	Overland Park, KS	6	D3	D
WVNE	Leicester, MA	25/8.6	D1	D
WJR	Detroit, MI	50	U1	AN7
WCHP	Champlain, NY	35	D3	D
KTKR	San Antonio, TX	50/1	U4	AN7
CFLD	Burns Lake, BC	1/0.5	U1	AN7
CKQR	Castlegar, BC	20/5	U2	AN7

+lp: USA 3

770kHz

Call	Location	Power		
WVNN	Athens, AL	10/0.25	U2	AN
KCHU	Valdez, AK	9.7	U1	NH
KCBC	Riverbank, CA	50/1	U4	AN6
WWCN	North Fort Myers, FL	10/1	U4	AN7
WFBN	Lynn Haven, FL	5/0.5	U4	F.PI.
KUOM	Minneapolis, MN	25/2.2	U2	D
KATL	Miles City, MT	10/1	U2	AN7
KKOB	Albuquerque, NM	50	U2	AN7
WABC	New York, NY	50	U1	AN7
WTOR	Youngstown, NY	5	D3	D
WLWL	Rockingham, NC	5	D1	D
KPBC	Garland, TX	10/1	U4	AN7
WYRV	Cedar Bluff, VA	5	D1	D
KNWX	Seattle, WA	50/5	U4	AN7
CHQR	Calgary, AB	50	U2	AN7

+lp: USA 5

780kHz

Call	Location	Power		
WZZX	Lineville, AL	5	D1	D
KNOM	Nome, AK	25/14	U1	NH
KAZM	Sedona, AZ	5/0.25	U2	NH
WBBM	Chicago, IL	50	U1	AN7
WIIN	Ridgeland, MS	5	D1	D
WJAG	Norfolk, NE	1	L1	D*
WWOL	Forest City, NC	10	D1	D
KKOH	Reno, NV	50	U2	AN7
WABS	Arlington, VA	5	D1	D
CFDR	Dartmouth, NS	50/10	U4	AN7

+lp: USA 4

790kHz

Call	Location	Power		
WTSK	Tuscaloosa, AL	5	D1	D
KCAM	Glenallen, AK	5	U1	AN
KNST	Tucson, AZ	5/0.5	U3	AN7
KURM	Rogers, AR	5/0.5	U2	
KOOR	Clovis, CA	5/2.5	U4	
KWSW	Eureka, CA	5	D1	D
KABC	Los Angeles, CA	5	U2	AN7
WLBE	Leesburg, FL	5/1	U2	AN
WAXY	South Miami, FL	5	U4	AN7
WQXI	Atlanta, GA	28/1	U2	AN7
KBRV	Soda Springs, ID	5	D1	D*
KXXX	Colby, KS	5/0.02	U1	AN
WWKY	Louisville, KY	5/1	U4	AN7
WSGW	Saginaw, MI	5/1	U4	AN6
KGHL	Billings, MT	5	U2	AN7
WTNY	Watertown, NY	1	U2	AN7

KFGO	Fargo, ND 5	U2	AN7	
KWIL	Albany, OR 1	U4	AN6	
WAEB	Allentown, PA	3.6/1.5	U4	AN7
WSKO	Providence, RI	5	U2	AN7
WETB	Johnson City, TN	5	D1	D*
WMC	Memphis, TN	5	U2	AN7
KBME	Houston, TX	5	U4	AN7
KFYO	Lubbock, TX	5/1	U4	AN
WNIS	Norfolk, VA 5	U3	AN7	
KGMI	Bellingham, WA	5/1	U2	AN7
KJRB	Spokane, WA	5/3.8	U4	AN7
WAYY	Eau Claire, WI	5	U2	AN6
CFCW	Camrose, AB	50	U4	AN7
CFAN	Newcastle, NB	5/1	U4	AN
CFNW	Port au Choix, NF	1	U3	AN7
CIGM	Sudbury, ON	50	U4	AN6
+lp: USA 17				
800kHz				
KINY	Juneau, AK 10/7.8	U1	AN	
WTMR	Camden, NJ	5/0.5	U4	
WSVS	Crewe, VA 5/0.3	U1	AN7	
WKEE	Huntington, WV	5	D1	D*
WDUX	Waupaca, WI	5/0.5	U4	
CKOR	Penticton, BC	10/0.5	U1	AN7
VOWR	St. John's NF	10/2.5	SH1	SH
CJBQ	Belleville, ON	10	U4	AN6
CKDR	Dryden, ON 1/0.7	U1	AN	
CKLW	Windsor, ON	50	U4	AN7
CJAD	Montreal, PQ	50/10	U4	AN7
CHRC	Quebec, PQ 50	U3	AN7	
CHAB	Moose Jaw, SK	10	U2	AN7
+lp: USA 21				
810kHz				
WNSI	Jacksonville, AL	50/0.5	U4	
KGO	San Francisco, CA	50	U3	AN7
WEKG	Jackson, KY	5	D1	D*
WHB	Kansas City, MO	50/5	U2	AN7
WSJC	Magee, MS	50/0.5	U4	NH
KSWV	Santa Fe, NM	5	D1	D*
WGY	Schenectady, NY	50	U1	AN7
WQIZ	St. George, SC	5	D1	D*
KBHB	Sturgis, SD 25	D1	D*	
WAPB	Murfreesboro, TN	5/6W	U1	
KTBI	Ephrata, WA	50/23	D3	
WJJQ	Tomahawk, WI	10	D3	D
CKJS	Winnipeg, MB	10	U3	AN
CJVA	Caraquet, NB	10/5	U2	AN
	Upernavik, Greenland 5	U1	SH	
+lp: USA 15				
820kHz				
KCBF	Fairbanks, AK	10	U1	AN6
WZTM	Largo, FL 50/1	U2		
WYPA	Chicago, IL 5/1.2	U2	D	
WNYC	New York, NY	10/1	U4	NH
WOSU	Columbus, OH	5/0.8	U2	NH
WBAP	Forth Worth, TX	50	U1	AN7
WGGM	Chester, VA	10/1	U4	AN
KGNW	Seattle, WA	50/5	U4	AN7
CHAM	Hamilton, ON	50/10	U4	AN7
+lp: USA 4				
830kHz				
KSDP	Sand Point, AK	1	U1	F.Pl.
KFLT	Tucson, AZ 50/1	U2	AN	
KNCO	Grass Valley, CA	5/1	U2	NH
KNCO	Grass Valley, CA	25/10	U2	F.Pl.
KPLS	Orange, CA 2.5/1	U2	AN	
KPLS	Orange, CA 50/20	U2	F.Pl.	
WXTO	Hialeah, FL 1	U4		
WFNO	Norco, LA 5/0.75	U4	AN	
WCRN	Worcester, MA	7/3	U4	F.Pl.

WCCO	Minneapolis, MN	50	U1	AN7
KOTC	Kennett, MO	10	D1	D
WXII	Kernersville, NC	50/10	U2	
KUYO	Evansville, WY	10	D1	D
CKKY	Wainwright, AB	10/3.5	U2	AN7
CFJR	Brockville, ON	5/1	U2	AN7
+lp: USA 3/Can 1				
840kHz				
WBHY	Mobile, AL 10	D1	D	
KABN	Long Island, AK	8	U1	F.Pl.
WHGH	Thomasville, GA	10	D1	D
WHAS	Louisville, KY	50	U1	AN7
KWDF	Ball, LA 8	D1	D	
KTIC	West Point, NE	5	D1	D
KXNT	North Las Vegas, NV	50/25	U3	AN
WCTG	Columbia, SC	50	D3	F.Pl.
KVJY	Pharr, TX 5/1	U3		
WKTR	Earlysville, VA	8.2	D3	F.Pl.
KMAX	Colfax, WA 10/0.28	U1		
CJXX	Grande Prairie, AB	25/10	U4	AN7
CKBX	100 Mile House, BC	1/0.5	U1	AN
CHUR	North Bay, ON	10	U2	AN
+lp: USA 9				
850kHz				
WYDE	Birmingham, AL	9/1	U2	AN
KICY	Nome, AK 50	U2	NH	
KOA	Denver, CO 50	U1	AN6	
WRUF	Gainesville, FL	5	U2	AN7
WDJA	West Palm Beach, FL	5/1	U4	AN7
WPTB	Statesboro, GA	1	U2	NH
WEEI	Boston, MA	50	U4	AN6
WKBZ	Muskegon, MI	1	U3	NH
WWJC	Duluth, MN 10	D1	D	
WQST	Forest, MS 10	D3	D	
KFUO	Clayton, MO	5	L1	D
WRBZ	Raleigh, NC	10/5	U2	AN6
WRMR	Cleveland, OH	10/5	U4	AN7
WJAC	Johnstown, PA	10	U3	AN
WEEU	Reading, PA	1	U2	AN6
WJBZ	Knoxville, TN	50	D3	D
KEYH	Houston, TX	10/0.185	U4	
WTAR	Norfolk, VA 50/25	U4	AN7	
KHHO	Tacoma, WA	10/1	U4	AN7
CKBA	Athabasca, AB	1	U2	AN7
CKMA	Abbotsford, BC	10	U4	AN6
CKVL	Montreal, PQ	50/10	U4	AN7
+lp: USA 10				
860kHz				
KMVP	Phoenix, AZ	1	U2	
KTRB	Modesto, CA	50/10	U2	AN7
WGUL	Dunedin, FL 2/1.5	U4	NH	
WAEC	Atlanta, GA 5/0.5	U2		
WDMG	Douglas, GA	5	U2	NH
KKOW	Pittsburg, KS	10/5	U2	AN7
KPAM	Troutdale, OR	50/5	U2	
WTEL	Philadelphia, PA	10/0.03	U3	NH
KONO	San Antonio, TX	5/0.9	U2	AN7
KCNR	Salt Lake City, UT	10/0.2	U1	AN
WOAY	Oak Hill, WV	10/0.01	U1	NH
CFPR	Prince Rupert, BC	10/2.5	U3	AN
CHAK	Inuvik, NWT	1	U1	AN
CJBC	Toronto, ON	50	U1	AN7
CBKF-2	Saskatoon, SK	10	U4	AN
+lp: USA 26/Can 26				
870kHz				
WQRX	Valley Head, AL	10/4.7	D1	D
KSKO	McGrath, AK	10	U1	NH
KIEV	Glendale, CA	20/3	U4	AN7
WWL	New Orleans, LA	50	U3	AN7
WLAM	Gorham, ME	10/1	U4	AN

Call	Location	Power	Class	Notes
WKAR	East Lansing, MI	10	D3	D
KPRM	Park Rapids, MN	25/1	U2	NH
KLSQ	Laughlin, NV	5/0.3	U2	AN
WHCU	Ithaca, NY 5/1		U2	AN
WPWT	Colonial Heights, TN	10	D3	D
KFLD	Pasco, WA 10/0.25		U1	AN
CKIR	Invermere, BC	1/0.25	U3	AN7
CFBV	Smithers, BC	1/0.5	U3	AN7

+lp: USA 5/Can 1

880kHz

Call	Location	Power	Class	Notes
KGHT	Sheridan, AR	50/0.2	U2	
KKMC	Gonzales, CA	10/1	U4	AN7
W...	Oviedo, FL 10/1		U2	F.Pl.
WBKZ	Jefferson, GA	5	D3	D
KJJR	Whitefish, MT	10/0.5	U2	AN6
KRVN	Lexington, NE	50	U4	AN
KHAC	Tse Bonito, NM	10/0.44	U1	NH
WCBS	New York, NY	50	U1	AN7
WRFD	Worthington, OH	23/6.1	D3	D
KWIP	Dallas, OR 5/1		U1	NH
KTMT	Phoenix, OR	1	U1	AN
KJOJ	Conroe, TX 10/1		U4	AN7
KEIO	Hamby, TX 2.5/1		U4	F.Pl.
KIXI	Mercer Island, WA	50/10	U4	AN7
WMEQ	Menomonie, WI	10/0.21	U2	AN
CHQT	Edmonton, AB	50	U2	AN7
CKKC	Nelson, BC 1/0.7		U3	AN
CKLQ	Brandon, MB	10	U2	AN

+lp: USA 2/Can 1

890kHz

Call	Location	Power	Class	Notes
KBBI	Homer, AK 10		U1	NH
WLS	Chicago, IL 50		U1	AN7
WBPS	Dedham, MA	25/5	U4	
WQIS	Laurel, MS 10		D1	D
WBAJ	Blythewood, SC	10.4	D1	F.Pl.
KVOZ	Laredo, TX 10/1		U2	AN
KDXU	St. George, UT	10	U2	AN
WKNV	Fairlawn, VA	10	D3	F.Pl.
CJDC	Dawson Creek, BC	10	U2	AN

+lp: USA 5

900kHz

Call	Location	Power	Class	Notes
KZPA	Fort Yukon, AK	5	U1	NH
KRRA	West Covina, CA	1	U4	NH
WJWL	Georgetown, DE	10/1	U4	AN7
WLSI	Pikeville, KY	5/0.16	D1	D*
KTIS	Minneapolis, MN	25/0.3	U4	AN7
CJVI	Victoria, BC10		U3	AN7
CKDH	Amherst, NS	1	U4	AN7
CHML	Hamilton, ON	50	U4	AN7
CHYC	Sudbury, ON	10/1	U4	AN7
CJBR	Rimouski, PQ	10	U2	NH
CJER	Saint-Jerome, PQ	1	U3	AN
CKTS	Sherbrooke, PQ	10	U4	AN7
CKBI	Prince Albert, SK	10	U2	AN7
	Uummannaq, Greenland	5	U1	

SH

+lp: USA 51/Can 6

910kHz

Call	Location	Power	Class	Notes
KIYU	Galena, AK 5		U1	NH
KFYI	Phoenix, AZ	5	U2	AN7
KLCN	Blytheville, AR	5	D1	D*
KECR	El Cajon, CA	5	U4	AN7
KNEW	Oakland, CA	5	U2	AN7
KOXR	Oxnard, CA 5/1		U4	NH
KPOF	Denver, CO 5/1		U1	AN7
WNEZ	New Britain, CT	5	U2	AN7
WFNS	Plant City, FL	5	U3	NH
WFVR	Valdosta, GA	5	U2	AN7
WSUI	Iowa City, IA	5	U4	AN7
WNDC	Baton Rouge, LA	1	U3	AN6

Call	Location	Power	Class	Notes
WABI	Bangor, ME5		U2	NH
WFDF	Flint, MI 5/1		U3	AN
WALT	Meridian, MS	5/1	U1	AN
KBIM	Roswell, NM	5/0.5	U2	NH
WSTK	Jacksonville, NC	5	U2	AN7
KCJB	Minot, ND 5/1		U4	AN6
WYLI	Marietta, OH	5	D3	D*
KVIS	Miami, OK 1		U3	NH
WAVL	Apollo, PA 5		D3	D*
WSBA	York, PA 5/1		U4	AN7
WJCW	Johnson City, TN	5/1	U4	AN6
WEPG	South Pittsburg, TN	5	D1	D
KRIO	McAllen, TX	5	U4	AN6
KXEB	Sherman, TX	1	U3	AN7
KALL	Salt Lake City, UT	5/1	U4	AN6
WRNL	Richmond, VA	5/1.5	U2	AN7
KFXX	Vancouver, WA	5	U4	AN7
WHSM	Hayward, WI	5	D1	D*
CKDQ	Drumheller, AB	50	U4	AN7
CHRL	Roberval, PQ	10	U2	AN7

+lp: USA 12

920kHz

Call	Location	Power	Class	Notes
WKYD	Andalusia, AL	5/0.04	U1	NH
KSRM	Soldotna, AK	5	U1	AN
KARN	Little Rock, AR	5	U2	AN6
KLOC	Modesto, CA	0.5/2.5	U4	
KPSI	Palm Springs, CA	5/1	U4	AN6
KLMR	Lamar, CO 5/0.5		U2	NH
WMEL	Melbourne, FL	5/1	U4	
WAFS	Atlanta, GA5/1		U1	AN7
WBAA	West Lafayette, IN	5/1	U2	NH
KYFR	Shenandoah, IA	5/2.5	U2	AN7
WTCW	Whitesburg, KY	5	D1	D*
WBOX	Bogalusa, LA	1	U2	AN
KDHL	Faribault, MN	5	U4	NH
KWAD	Wadena, MN	1	U2	NH
KBAD	Las Vegas, NV	5/0.5	U4	AN7
KQLO	Reno, NV 4.6/0.85		U2	AN7
WCHR	Trenton, NJ	1	U3	AN7
WGHQ	Kingston, NY	5/0.5	U4	
WIRD	Lake Placid, NY	5/0.09	U1	AN7
WPCM	Burlington, NC	5/0.06	U1	AN7
KSHO	Lebanon, OR	1	U3	AN6
WHJJ	Providence, RI	5	U4	AN7
WYMB	Manning, SC	5/1	U2	
KKLS	Rapid City, SD	5	D3	D*
KYST	Texas City, TX	5/1	U4	AN7
KVEL	Vernal, UT 5/1		U2	
KGHO	Olympia, WA	5/0.5	U4	AN7
KXLY	Spokane, WA	5	U1	AN6
WMMN	Fairmont, WV	5/0.2	U1	NH
WOKY	Milwaukee, WI	5/1	U4	AN7
CKCQ	Quesnel, BC	10/1	U2	AN6
CFRY	Portage la Prairie, MB	25/10	U4	AN6
CJCJ	Woodstock, NB	10/1	U4	AN7
CJCH	Halifax, NS 25		U2	AN7
CKNX	Wingham, ON	10/1	U4	AN

+lp: USA 19/Can 4

930kHz

Call	Location	Power	Class	Notes
WYNI	Monroeville, AL	5	D1	D
WJBY	Rainbow City, AL	5/0.5	U2	
KTKN	Ketchikan, AK	5/1	U1	NH
KNSA	Unalakleet, AK	2.5	U1	
KAFF	Flagstaff, AZ	5	D1	D*
KKHJ	Los Angeles, CA	5	U2	AN7
KIUP	Durango, CO	5/0.1	U1	AN6
WNZS	Jacksonville, FL	5	U2	AN7
WKXY	Sarasota, FL	5/2.5	U4	AN7
WMGR	Bainbridge, GA	5/0.5	U2	NH
KSEI	Pocatello, ID	5	U2	AN6

Call	Location	Power	Type	Code
WTAD	Quincy, IL 5/1		U2	NH
WAUR	Sandwich, IL	2.5/4.2	U4	AN7
WKCT	Bowling Green, KY	5/0.5	U2	NH
WFMD	Frederick, MD	5/2.5	U4	AN
WBCK	Battle Creek, MI	5/1	U4	AN
WSLI	Jackson, MS	5/3.2	U2	NH
KWOC	Poplar Bluff, MO	5/0.5	U2	NH
KLCY	East Missoula, MT	5/1	U2	
KOGA	Ogallala, NE	5/0.5	U4	NH
WZNN	Rochester, NH	5	U2	NH
WPAT	Paterson, NJ	5	U4	AN7
WBEN	Buffalo, NY5		U2	AN7
WYFQ	Charlotte, NC	5/1	U2	AN6
WDLX	Washington, NC	5/1	U2	AN7
WEOL	Elyria, OH 1		U4	AN6
WKY	Oklahoma City, OK	5	U2	AN7
KAGI	Grants Pass, OR	5/1	U2	AN7
KSDN	Aberdeen, SD	5/1	U4	AN
WSEV	Sevierville, TN	5	D1	D
KLUP	San Antonio, TX	5/1	U2	AN7
WLLL	Lynchburg, VA	5	D1	D*
WRVC	Huntington, WV	5/1	U2	AN7
WLBL	Auburndale, WI	5	D1	D
KROE	Sheridan, WY	5	D1	D*
CJCA	Edmonton, AB	50	U4	AN7
CFBC	St. John, NB	50	U4	AN7
CJYQ	St. John's, NF	50	U4	AN7

+lp: USA 14

940kHz

Call	Location	Power	Type	Code
KCEE	Tucson, AZ 5/1		U4	AN7
KFRE	Fresno, CA 50		U4	AN7
WINZ	Miami, FL 50/10		U2	AN7
WMAC	Macon, GA 50/10		U2	AN
WMIX	Mount Vernon, IL	5/1.5	U4	AN7
KXTK	Des Moines, IA	10/5	U4	AN7
WYLD	New Orleans, LA	10/0.5	U4	AN6
WMKC	St. Ignace, MI	5	D1	D*
WCPC	Houston, MS	50/0.25	U4	NH
KVSH	Valentine, NE	5	D1	D*
WKYK	Burnsville, NC	5/0.25	U2	NH
KXUX	Bend, OR 10		U4	NH
WECO	Wartburg, TN	5	D1	D*
KIXZ	Amarillo, TX	5/1	U4	AN7
KBRE	Cedar City, UT	10	D1	D*
WNRG	Grundy, VA 5		D1	D*
WKGM	Smithfield, VA	10/3.1	U2	NH
CJIB	Vernon, BC 10		U2	AN7
C...	Montreal, PQ	50	U3	
CJGX	Yorkton, SK10		U2	AN6

+lp: USA 14/Can 3

950kHz

Call	Location	Power	Type	Code
KSWD	Seward, AK	1	U1	AN
KXJK	Forrest City, AR	5	D1	D*
KFSA	Fort Smith, AR	1/0.5	U4	NH
KAHI	Auburn, CA5		U4	AN
KKFN	Denver, CO 5		U3	AN7
WTLN	Orlando, FL 5		U2	AN7
WGTA	Summerville, GA	5	D1	D*
WGOV	Valdosta, GA	5/0.06	U1	
KKIC	Boise, ID 5		D1	D*
KOZE	Lewiston, ID	5/1	U4	AN
WIDB	Chicago, IL 1/5		U2	AN7
WXLW	Indianapolis, IN	5/0.11	U3	NH
KOEL	Oelwein, IA5/0.5		U4	AN6
WROL	Boston, MA 5		D1	D*
WWJ	Detroit, MI	12/50	U4	AN7
KSGS	St. Louis Park, MN	1	U4	AN7
WBKH	Hattiesburg, MS	5	D1	D*
KLIK	Jefferson City, MO	5/0.5	U2	NH
KMTX	Helena, MT5		U2	NH

Call	Location	Power	Type	Code
KNFT	Bayard, NM	5/0.22	U1	AN7
WEZO	Rochester, NY	1	U4	AN7
WIBX	Utica, NY 5		U3	AN
WPEN	Philadelphia, PA	5	U2	AN7
WSPA	Spartanburg, SC	5	U2	AN7
KWAT	Watertown, SD	1	U2	NH
KPRC	Houston, TX	5	U2	AN7
KXTQ	Lubbock, TX	5/0.5	U4	AN6
KJR	Seattle, WA	50	U2	AN7
WQBE	Charleston, WV	5/1	U2	AN7
KMER	Kemmerer, WY	5	D1	D*
CFAM	Altona, MB 10		U4	AN7
CKNB	Campbellton, NB	10/1	U4	AN6
CHER	Sydney, NS10		U2	AN6

+lp: USA 14

960kHz

Call	Location	Power	Type	Code
WERC	Birmingham, AL	5	U2	AN6
WLPR	Pritchard, AL	5/1	U2	
KPXQ	Phoenix, AZ	5	U2	AN7
KCGS	Marshall, AR	5	D1	D
KIXW	Apple Valley, CA	5	D1	D*
KABL	San Francisco, CA	5	U3	AN7
WELI	New Haven, CT	5	U2	AN6
WGRO	Lake City, FL	0.5/1	U2	NH
WJCM	Sebring, FL 5/1		U3	NH
WJYZ	Albany, GA 5		D3	D*
WRFC	Athens, GA 5/2.5		U2	AN6
WSBT	South Bend, IN	5	U4	AN
KMA	Shenandoah, IA	5	U2	NH
WPRT	Prestonsburg, KY	5	D1	D*
WTGM	Salisbury, MD	5	U4	AN7
WFGL	Fitchburg, MA	1	U4	
WHAK	Rogers City, MI	5	D1	D*
KLTF	Little Falls, MN	5	D1	D*
KZIM	Cape Girardeau, MO	5/0.5	U2	NH
KFLN	Baker, MT 5		D1	D*
KNDN	Farmington, NM	5	D1	D*
WEAV	Plattsburgh, NY	5	U4	NH
WRNS	Kinston, NC	5/1	U2	AN7
KGWA	Enid, OK 1		U3	NH
KLAD	Klamath Falls, OR	5	U2	NH
WHYL	Carlisle, PA5		D3	D*
WATS	Sayre, PA 5		D	D*
KGKL	San Angelo, TX	5/1	U2	AN
KOVO	Provo, UT 5/1		U2	AN7
WFIR	Roanoke, VA	5	U2	AN7
KALE	Richland, WA	5/1	U2	AN6
WTCH	Shawano, WI	1	U2	NH
CFAC	Calgary, AB50		U2	AN7
CHNS	Halifax, NS 10		U2	AN7
CFFX	Kingston, ON	10/5	U4	AN7

+lp: USA 15

970kHz

Call	Location	Power	Type	Code
WERH	Hamilton, AL	5	D1	D*
WTBF	Troy, AL 5/0.5		U2	NH
KIAK	Fairbanks, AK	5	U1	AN7
KVWM	Show Low, AZ	5	D1	D*
KAFY	Bakersfield, CA	1/5	U4	AN7
KCLB	Coachella, CA	5/1	U4	AN7
KANM	Modesto, CA	1	U4	AN
WFLA	Tampa, FL 25/11		U4	AN7
WNIV	Atlanta, GA5/0.04		U3	NH
WVOP	Vidalia, GA 5		D1	D*
WFSR	Harlan, KY 5		D1	D
WLKY	Louisville, KY	5	U4	AN6
KSYL	Alexandria, LA	1	U2	AN
WESO	Southbridge, MA	5	U4	
WZAN	Portland, ME	5	U2	AN
WZAM	Ishpeming, MI	5	D1	D*
WKHM	Jackson, MI	1	U4	AN7

Call	Location	Power	Class	Notes
KNFX	Austin, MN	5/0.5	U4	AN
KBUL	Billings, MT	5	U2	AN7
KJLT	North Platte, NE	5	D1	D*
KNUU	Paradise, NV	5/0.5	U4	
WWDJ	Hackensack, NJ	5	U4	AN7
WNED	Buffalo, NY	5	U3	AN
WWIT	Canton, NC	5	D1	D*
WDAY	Fargo, ND	5	U2	AN7
WFUN	Ashtabula, OH	5/1	U4	NH
KCFO	Tulsa, OK	2.5/1	U4	AN6
KUPL	Portland, OR	5	U2	AN7
WWSW	Pittsburgh, PA	5	U4	AN6
WJMX	Florence, SC	10/3	U2	NH
KIXL	Del Valle, TX	1	U4	
WVAO	Waynesboro, VA	5/1	U4	NH
KTRW	Spokane, WA	5/1	U2	AN7
WHA	Madison, WI	5	D1	D*
CJYR	Edson, AB	10	U3	AN

+lp: USA 21/Can 1

980kHz

Call	Location	Power	Class	Notes
KZXX	Kenai, AK	1	U1	AN7
KCAB	Dardanelle, AR	5	D1	
KINS	Eureka, CA	5/0.5	U2	AN7
KFWB	Los Angeles, CA	50	U3	AN7
KCTY	Salinas, CA	10/0.25	U4	NH
WTEM	Washington, DC	50/5	U4	AN7
WLUS	Gainesville, FL	5	D1	D*
WRNE	Gulf Breeze, FL	2.5/1	U4	AN
WHSR	Pompano Beach, FL	5/1	U5	AN7
WPGA	Perry, GA	5	D1	D*
KUPI	Ammon, ID	5/1	U4	AN
WITY	Danville, IL	1	U3	NH
KOKA	Shreveport, LA	5	D1	D*
WCAP	Lowell, MA	5	U4	AN
KKMS	Richfield, MN	5	U3	AN7
WAPF	McComb, MS	5	D1	D*
KMBZ	Kansas City, MO	5	U2	AN7
KVLV	Fallon, NV	5	D1	D*
KICA	Clovis, NM	1	U2	NH
KMIN	Grants, NM	1	U1	AN
WTRY	Troy, NY	5	U2	AN7
WAAV	Wilmington, NC	5	U2	AN7
WONE	Dayton, OH	5	U2	AN7
WILK	Wilkes-Barre, PA	5/1	U2	AN
KDSJ	Deadwood, SD	5/1	U2	NH
WYFN	Nashville, TN	5	U2	AN7
KRTX	Rosenberg, TX	4.6/4	U4	
KSVC	Richfield, UT	5/1	U2	
WXBQ	Bristol, VA	5/1	U2	AN7
KJOX	Selah, WA	5/0.5	U4	AN
WCUB	Two Rivers, WI	5	U4	AN
CKNW	New Westminster, BC	50	U3	AN7
CFPL	London, ON	10/5	U4	AN7
CKRU	Peterborough, ON	10/5	U4	AN7
CKRM	Regina, SA	10/5	U4	AN7

+lp: USA 27

990kHz

Call	Location	Power	Class	Notes
KTKT	Tucson, AZ	10/1	U4	AN7
KATD	Pittsburg, CA	5	U4	AN7
KQSB	Santa Barbara, CA	5/0.5	U4	NH
KRKS	Denver, CO	5/0.4	U2	AN7
WFBA	Miami, FL	5	U4	
WHOO	Orlando, FL	50/5	U4	AN7
WWCM	Ypsilanti, MI	9.2/0.25	U4	AN7
WDCZ	Rochester, NY	5/2.5	U4	AN
WEEB	Southern Pines, NC	10/0.5	U1	AN5
WZZD	Philadelphia, PA	50/10	U4	AN7
WVSC	Somerset, PA	10/0.25	U3	NH
WALE	Greenville, RI	50/5	U4	NH
WNOX	Knoxville, TN	10	U2	AN7
KWAM	Memphis, TN	10/0.45	U4	AN
KZZB	Beaumont, TX	1	U3	AN
KWFT	Wichita Falls, TX	10/1	U4	AN6
WNRV	Narrows, VA	5	D1	D
CBW	Winnipeg, MB	50/46	U1	AN
CBY	Corner Brook, NF	10	U3	AN
CKGM	Montreal, PQ	50	U4	AN7

+lp: USA 25/Can 21

1000kHz

Call	Location	Power	Class	Notes
WDJL	Huntsville, AL	5	D1	D
KFLG	Bullhead City, AZ	5	D1	D
WMVP	Chicago, IL	50	U4	AN7
WXTN	Lexington, MS	5	D1	D
KKIM	Albuquerque, NM	10/0.04	U1	D*
WLNL	Horseheads, NY	5	D1	D
KTOK	Oklahoma City, OK	5	U4	AN7
KXRB	Sioux Falls, SD	10	D3	D*
WMUF	Paris, TN	5/2.5	D3	D
KOMO	Seattle, WA	50	U2	AN7
CKBW	Bridgewater, NS	10	U2	AN
CFLP	Rimouski, PQ	10	U4	AN7

+lp: USA 21

1010kHz

Call	Location	Power	Class	Notes
WPYK	Dora, AL	5	D1	D
KXEG	Tucson, AZ	7.5/0.25	U1	AN7
KCHJ	Delano, CA	5/1	U4	NH
KIQI	San Francisco, CA	10/1.5	U4	
KSIR	Brush, CO	25/0.28	U3	
WIOJ	Jacksonville Beach, FL	10/0.14	U4	
WQYK	Seffner, FL	50/5	U4	AN7
WGUN	Atlanta, GA	50/0.3	U1	AN7
WMOX	Meridian, MS	10/1	U4	NH
KXEN	Festus, MO	50/0.5	U4	D
WINS	New York, NY	50	U3	AN7
WFGW	Black Mountain, NC	50/0.5	U7	
WHIN	Gallatin, TN	5	D1	D
KTNZ	Amarillo, TX	5/0.5	U4	AN7
KLAT	Houston, TX	5/3.6	U4	AN7
KBBW	Waco, TX	10/2.5	U4	NH
KTUR	Tooele, UT	50/0.013	U1	AN
WPMH	Portsmouth, VA	5	D3	D
CBR	Calgary, AB	50	U4	NH
CKXD	Gander, NF	1	U2	AN7
CFRB	Toronto, ON	50	U4	AN7

+lp: USA 20/Can 7

1020kHz

Call	Location	Power	Class	Notes
KAXX	Eagle River, AK	10	U2	AN
KTNQ	Los Angeles, CA	50	U4	AN7
WJEP	Ochlocknee, GA	10	D1	D
KJJK	Fergus Falls, MN	2/1	U3	
WNTK	Newport, NH	10	D1	D
KCKN	Roswell, NM	50	U4	NH
KDKA	Pittsburgh, PA	50	U1	AN7
WRIX	Anderson, SC	10/3	D1	D
KYXE	Selah, WA	5/0.5	U5	AN7
KWIQ	Moses Lake, WA	2.5/0.5	U5	AN
CKVH	High Prairie, AB	1/0.4	U1	AN7

+lp: USA 2

1030kHz

Call	Location	Power	Class	Notes
WAUL	Brantley, AL	5/0.4	U2	F.Pl.
KFAY	Farmington, AR	10/1	U2	AN
KEVT	Cortaro, AZ	10/1	U4	F.Pl.
KIOQ	Folsom, CA	50/1	U4	
KJDJ	San Luis Obispo, CA	2.5/0.8	U4	AN7
WONQ	Oviedo, FL	10/1.7	U4	
KBUF	Holcomb, KS	1	U1	AN
WBZ	Boston, MA	50	U3	AN7
WWGB	Indian Head, MD	50	D3	D
WUFL	Sterling Heights, MI	5	D3	D

WCTS	Maplewood, MN	50/1	U4	NH
WNOW	Mint Hill, NC	10	D3	D
WFTK	Wake Forest, NC	50/2.5	D3	D
KLLU	Reedsport, OR	10/0.63	U1	AN7
WSFZ	Memphis, TN	50/1	U2	AN7
KCTA	Corpus Christi, TX	50	L1	D
KMAS	Shelton, WA	10/1	U1	AN
WBGS	Point Pleasant, WV	10/2.9	D5	D
KTWO	Casper, WY	50	U2	AN6

+lp: USA 8

1040kHz

KCBR	Monument, CO	5/2	D1	D
WJNO	Boynton Beach, FL	25/1.1	U4	
WWBA	Pinellas Park, FL	5/0.5	U2	
WHO	Des Moines, IA	50	U1	AN7
WJHR	Flemington, NJ	4.7/1	U4	
WSGH	Lewisville, NC	10/0.3/	U3	
WJTB	North Ridgeville, OH	5/2.5	D1	D
WSKE	Everett, PA 10/4	D1	D	
WQBB	Powell, TN 10/3	D3	D	
CKST	Langley, BC10	U4	AN7	

+lp: USA 11

1050kHz

KTBA	Tuba City, AZ	5	D1	D
KTRJ	Frazier Park, CA	10	D3	D
KTCT	San Mateo, CA	50/10	U4	AN
WROS	Jacksonville, FL	5	D3	D
WFAM	Augusta, GA	5	D1	D*
WTKA	Ann Arbor, MI	10/0.5	U3	AN
KLOH	Pipestone, MN	9/0.4	U4	
KMTA	Miles City, MT	10/0.13	U1	
WEVD	New York, NY	50	U3	AN7
KORE	Springfield, OR	5	D1	D
WJXY	Conway, SC	5	D1	D*
WCMS	Norfolk, VA5/0.36	U3	AN	
KEYF	Dishman, WA	5/0.34	U1	AN
KBLE	Seattle, WA	5/0.44	U1	NH
WADC	Parkersburg, WV	5	D1	D*
CICF	Vernon, BC 10/4	U2	AN7	
CKSB	Winnipeg, MB	10	U2	AN7
CHUM	Toronto, ON	50	U4	AN7
CJNB	North Battleford, SK	10	U2	AN

+lp: USA 44

1060kHz

KDUS	Tempe, AZ 5/0.5	U2	AN7	
KLMO	Longmont, CO	10/0.5	U1	NH
WAMT	Titusville, FL	10/5	U4	AN6
WKNG	Tallapoosa, GA	5	D1	D*
KBGN	Caldwell, ID	10	D1	D*
WLNO	New Orleans, LA	50/5	U4	AN7
WHFB	Benton Harbor, MI	5/2.5	D1	D*
KKVV	Las Vegas, NV	5	D1	D
WRCW	Canton, OH5	D3	D*	
KYW	Philadelphia, PA	50	U3	AN7
KGFX	Pierre, SD 10/1	U4	AN	
KFNA	El Paso, TX 10	D1	D*	
KIJN	Farwell, TX 10	D3	D	
KBNB	Gilmer, TX 10	D1	D*	
KKDS	Salt Lake City, UT	10/1	U4	AN
CKMX	Calgary, AB50	U2	AN7	

+lp: USA 20

1070kHz

WAPI	Birmingham, AL	50/5	U2	AN
KNX	Los Angeles, CA	50	U1	AN7
WFRF	Tallahassee, FL	10	D1	D
WIBC	Indianapolis, IN	50/10	U4	AN7
KFDI	Wichita, KS	10/1	U2	AN7
KHMO	Hannibal, MO	5/1	U4	NH
KATQ	Plentywood, MT	5	D1	D
WDOT	Plattsburgh, NY	5	D1	D

WNCT	Greenville, NC	10	U4	AN7
WKOK	Sunbury, PA	10/1	U2	NH
WHYZ	Greenville, SC	50/1.5	U4	AN
WFLI	Lookout Mountain, TN50/2.5	U4	AN7	
WDIA	Memphis, TN	50/5	U4	AN6
KOPY	Alice, TX	1	U2	AN
KENR	Houston, TX	10/5	U4	AN7
WINA	Charlottesville, VA	5	U2	NH
WIWS	Beckley, WV	10	D1	D*
WTSO	Madison, WI	10/5	U4	AN7
CFAX	Victoria, BC10	U3	AN7	
CBA	Moncton, NB	50	U1	NH
CHOK	Sarnia, ON 10	U4	AN7	

+lp: USA 6/Can 1

1080kHz

WKAC	Athens, AL 5/2.5	D1	D	
KASH	Anchorage, AK	10	U1	AN7
KSCO	Santa Cruz, CA	10/5	U2	NH
WTIC	Hartford, CT	50	U2	AN7
WVCG	Coral Gables, FL	50/10	U4	AN7
WFIV	Kissimmee, FL	10	D3	D
WFTD	Marietta, GA	10	D3	D
KVNI	Coeur d'Alene, ID	10/1	U2	AN7
WKJK	Louisville, KY	10/1	U4	AN7
WSSI	Carthage, MS	5	D3	D
WKGX	Lenoir, NC 5/2.5	D1	D*	
WKKE	St. Pauls, NC	50/25	D3	D
KOTK	Portland, OR	50/10	U3	AN6
WPGR	Pittsburgh, PA	50/25	D3	D*
KRLD	Dallas, TX 50	U2	AN7	
KRPX	Price, UT 10/5	D1	D	
CKSA	Lloydminster, AB	50/10	U2	AN6

+lp: USA 17/Can 1

1090kHz

KAAY	Little Rock, AR	50	U2	AN7
KAJK	Fortuna, CA10	D1	D	
KMXA	Aurora, CO 50/0.5	U4	AN7	
WNVY	Cantonment, FL	5	D1	D
WSLG	Gonzales, LA	10	D3	D
WBAL	Baltimore, MD	50	U2	AN7
WILD	Boston, MA	5/1	D1	D
KBOZ	Bozeman, MT	5	U2	AN7
WKCV	Kingsport, TN	10/1.8	D1	
KKYN	Plainview, TX	5/0.25	U4	AN
KRPM	Seattle, WA	50	U4	AN7
WAQE	Rice Lake, WI	5	D1	D
CKKW	Kitchener, ON	10	U4	AN7

+lp: USA 23/Can 4

1100kHz

KCCF	Cave Creek, AZ	50/1	U4	F.PI.
KZPM	Bakersfield, CA	4.2/0.8	U4	F.PI.
KFAX	San Francisco, CA	50	U3	AN7
KNZZ	Grand Junction, CO	50/10	U2	NH
WWWE	Carrollton, GA	5/3.8	D1	D
WCGA	Woodbine, GA	10	D1	
KKLL	Webb City, MO	5	D1	D
WHLI	Hempstead, NY	10	D3	D
WTAM	Cleveland, OH	50	U3	AN6
KDRY	Alamo Heights, TX	11/1	U2	NH
WYKR	Wells River, VT	5/2	D1	D

+lp: USA 1

1110kHz

WBCA	Bay Minette, AL	10/2.5	D1	D
KGFL	Clinton, AR 5/1	D1	D	
KRLA	Pasadena, CA	50/20	U4	AN7
KRCX	Roseville, CA	5/0.5	U4	NH
WTIS	Tampa, FL 10	D3	D*	
WUHN	Pittsfield, MA	5	D3	D
WJML	Petoskey, MI	10	D3	D
KFAB	Omaha, NE 50	U2	AN6	

WNNW	Salem, NH 5	D3	D	
KYKK	Hobbs, NM 5	D1	D*	
WBT	Charlotte, NC	50	U2	AN7
KEOR	Atoka, OK 5	D3	D	
KBND	Bend, OR 10/5	U2	AN7	
WPMZ	East Providence, RI	5	D3	D
WCKO	Norfolk, VA 50	D3	D	
CKTY	Sarnia, ON 10/1	U4	AN7	

+lp: USA 22/Can 2

1120kHz

KZSJ	San Martin, CA	5/0.15	U1	F.Pl.
WUST	Washington, DC	20/3	D1	D
WBNM	Gordon, GA 10/2.5	D1	D*	
WADN	Concord, MA	5/1	U4	
WTWZ	Clinton, MS	5/2.5	D1	D
KMOX	St. Louis, MO	50	U1	AN7
KPNW	Eugene, OR 50	U3	AN7	
KANN	Roy, UT 10/1	U2	AN7	

+lp: USA 13

1130kHz

KRDU	Dinuba, CA 5/6	U4	AN7	
KSDO	San Diego, CA	50/10	U4	AN6
KHTH	Dillon, CO 5	D1	D	
WWBF	Bartow, FL 2.5/0.5	U2	NH	
WSJP	Murray, KY 2.5/0.5	U4	AN	
KWKH	Shreveport, LA	50	U2	AN7
WDFN	Detroit, MI 50/10	U4	AN7	
KFAN	Minneapolis, MN	50/30	U4	AN7
KCIB	Milan, NM 5/2.5	D1	D	
WBBR	New York, NY	50	U2	AN7
KBMR	Bismarck, ND	10	D1	D
WASP	Brownsville, PA	5/1	D5	D*
KTMR	Edna, TX 10	D3	D	
WISN	Milwaukee, WI	50/10	U4	AN7
CKWX	Vancouver, BC	50	U4	AN7

+lp: USA 21/Can 1

1140kHz

WBXR	Hazel Green, AL	15	D3	D
KSLD	Soldotna, AK	10	U1	AN
KCMJ	Palm Springs, CA	10/2.5	U4	AN
KHTK	Sacramento, CA	50	U4	AN7
WQBA	Miami, FL 50/10	U4	AN6	
KGEM	Boise, ID 10	U2	AN6	
WVEL	Pekin, IL 5/3.2	D1	D*	
WKWM	Kentwood, MI	5	D3	D
WSAO	Senatobia, MS	5	D1	D
KSFN	Las Vegas, NV	10/2.5	U2	AN6
WCJW	Warsaw, NY	5/2.3	D3	D
KSOO	Sioux Falls, SD	10/5	U2	AN
KCHC	Conroe, TX 5	D1	D	
WRVA	Richmond, VA	50	U3	AN7
KZMQ	Greybull, WY	10	D1	D
CHRB	High River, AB	50/46	U4	AN7
CBI	Sydney, NS 10	U2	AN	

+lp: USA 14/Can 4

1150kHz

WSPZ	Tuscaloosa, AL	5/1	U2	AN7
KCKY	Coolidge, AZ	5/1	U4	AN7
KLRG	North Little Rock, AR	5/1	U2	AN7
KXTA	Los Angeles, CA	50/44	U2	AN7
KMXN	Santa Rosa, CA	5	U4	AN6
KCUV	Englewood, CO	5/1	U4	AN7
WDEL	Wilmington, DE	5	U4	AN7
WNDB	Daytona Beach, FL	1	U2	AN7
WTMP	Egypt Lake, FL	5/0.5	U1	AN7
WJEM	Valdosta, GA	5	D3	D*
WGGH	Marion, IL 5	D3	D*	
KWKY	Des Moines, IA	1	U4	NH
KSAL	Salina, KS 5	U2	AN7	
WJBO	Baton Rouge, LA	5	U3	AN

WNFT	Boston, MA	5	U4	AN7
KSEN	Shelby, MT 10/5	U4	NH	
KDEF	Albuquerque, NM	5/0.5	U2	AN7
WRVB	Utica, NY 5/1	U4	AN	
WGBR	Goldsboro, NC	5	U1	NH
WCUE	Cuyahoga Falls, OH	5/0.5	U4	
WIMA	Lima, OH 1	U2	AN7	
KNED	McAlester, OK	1	U2	AN
KAGO	Klamath Falls, OR	5/1	U2	AN
KKGT	Portland, OR	5	D3	D*
WHUN	Huntingdon, PA	5/0.04	U1	AN
KIMM	Rapid City, SD	5/0.5	U2	
WGOW	Chattanooga, TN	5/1	U2	AN7
WCRK	Morristown, TN	5/0.5	U2	NH
KSVE	El Paso, TX 5/0.38	U1	24h	
KEZX	Seattle, WA	5	U2	AN6
WELC	Welch, WV 5	D1	D*	
WEAQ	Chippewa Falls, WI	5	D1	D*
WHBY	Kimberly, WI	5	U3	AN7
CKBL	Kelowna, BC	10	U2	AN7
CKX	Brandon, MB	50/10	U4	AN
CKOC	Hamilton, ON	50	U4	AN7
CJRC	Ottawa, ON	50/5	U4	AN
CHGM	Gaspé, PQ 5	U2	AN	

+lp: USA 31/Can 7

1160kHz

WKWA	Mobile, AL 10/0.8	U2	F.Pl.	
WELX	Callahan, FL	5/0.25	U5	F.Pl.
WKGE	East Point, GA	10/0.4	U4	F.Pl.
WSCR	Chicago, IL 50/5	U4	AN7	
WBOB	Florence, KY	5/1	U4	AN
WKCM	Hawesville, KY	2.5/1	U2	AN
WSKW	Skowhegan, ME	10/0.73	U1	AN
WWON	Fenton, MI 1	U1		
WOBM	Lakewood, NJ	20/8.9	U4	
WVNJ	Oakland, NJ	10/2.5	U4	
WMVI	Mechanicville, NY	5/0.47	U1	NH
WPIE	Trumansburg, NY	5/0.31	U4	
WYRU	Red Springs, NC	5/0.25	U1	NH
WJFJ	Tryon, NC 10/0.5	U2		
WCCS	Homer City, PA	10/1	U4	AN7
WYNS	Lehighton, PA	4/1	U4	AN2
WAMB	Donelson, TN	50/1	U2	AN
KENS	San Antonio, TX	10/1	U4	AN7
KSL	Salt Lake City, UT	50	U1	AN7
WODY	Fieldale, VA	5/0.25	U4	
VSB3	Hamilton, Bermuda	1	U1	SH

+lp: USA 3/Can 1

1170kHz

WACV	Montgomery, AL	10/1	U4	AN6
KJNP	North Pole, AK	50/21	U2	NH
KCBQ	San Diego, CA	50/1.5	U4	AN7
KLOK	San Jose, CA	50/5	U4	AN7
WAVS	Davie, FL 5/0.25	U2	NH	
WHNE	Cumming, GA	5/2.5	D1	D
WLBH	Mattoon, IL 5	D3	D*	
KJOC	Davenport, IA	1	U4	AN7
WCXN	Claremont, NC	7.7/1	D1	D
WCLN	Clinton, NC 5/1	D1	D	
KVOO	Tulsa, OK 50	U2	AN7	
WLGO	Lexington, SC	10/2.5	D1	D
KPUG	Bellingham, WA	10/5	U2	AN6
WWVA	Wheeling, WV	50	U2	AN7
CKGY	Red Deer, AB	50/14	U2	AN7

+lp: USA 15/Can 3

1180kHz

KYET	Williams, AZ	10/0.25	U1	NH
KERI	Wasco, CA	50/10	U4	AN
"VOA"	R Marti, Marathon Key, FL	50	U3	
AN				

Call	Location	Power	Class	Notes
WSAF	Trion, GA	5	D1	D
WJNT	Pearl, MS	50/0.5	U2	AN
KOFI	Kalispell, MT	50/10	U2	AN6
KOIL	Bellevue, NE	5/1	U4	
WHAM	Rochester, NY	50	U1	AN6
WMYT	Carolina Beach, NC	10	D3	D
WHJM	Knoxville, TN	10/2.6	D1	D
KGOL	Humble, TX	8.5/4.7	U4	AN7
KLAY	Lakewood, WA	5/1	U2	

+lp: USA 7

1190kHz

Call	Location	Power	Class	Notes
KMYL	Tolleson, AZ	5/0.25	U4	AN7
KORG	Anaheim, CA	10/1.3	U4	AN7
KVCU	Boulder, CO	5/0.11	U1	AN7
WAJL	Pine Castle, FL	5	D1	D
WPSP	Royal Palm Beach, FL	0.89/1	U2	
WGKA	Atlanta, GA	10/2.5	D1	D
WOWO	Fort Wayne, IN	50/9.8	U2	AN7
WANN	Annapolis, MD	10	D3	D
KKOJ	Jackson, MN	5	D3	D
WBSL	Bay St. Louis, MS	5	D1	D
KHAD	DeSoto, MO	5	D3	D*
KPHN	Kansas City, MO	5/0.25	U2	AN7
KXKS	Albuquerque, NM	10/0.024	U1	
WLIB	New York, NY	10/30	U4	AN
KEX	Portland, OR	50	U2	AN7
WSDQ	Dunlap, TN	5/1	D1	D
KOOO	Dallas, TX	50/5	U4	AN7
WBDY	Bluefield, VA	10	D3	D
CFSL	Weyburn, SK	10/5	U2	AN

+lp: USA 18

1200kHz

Call	Location	Power	Class	Notes
KTCD	Eureka, CA	10/1	U4	F.Pl.
KRDE	Pismo Beach, CA	5/1	U4	F.Pl.
KOQI	Soquel, CA	25/1	U2	F.Pl
WTLQ	Pine Island Center, FL	10/1	U4	AN7
WLXX	Chicago, IL	10/2.5	U4	AN
WNSW	Brewer, ME	10	U3	AN
WKOX	Framingham, MA	50	U4	AN
WCHB	Taylor, MI	50/2.1	U4	
WGNY	Newburgh, NY	10/2.5	U4	
WTLA	North Syracuse, NY	1	U2	
WSML	Graham, NC	10/1	U2	
KFNW	West Fargo, ND	10/0.7	U2	NH
WBZY	New Castle, PA	5/1	U2	AN
WRKK	Hughesville, PA	10/0.25	U4	D
WMIR	Atlantic Beach, SC	6.5/0.01	U1	F.Pl.
WQDQ	Lebanon, TN	10/0.5	U2	AN
WOAI	San Antonio, TX	50	U1	AN7
WAGE	Leesburg, VA	5/1	U2	AN6
CKXM	Victoria, BC	50	U4	AN7
CJBZ	Ottawa, ON	50	U4	AN7

+lp: USA 2

1210kHz

Call	Location	Power	Class	Notes
WQLS	Ozark, AL	10/0.004	U1	
KQTL	Sahuarita, AZ	10/1	U1	AN
KEBR	Rocklin, CA	5/0.5	U5	AN7
KPRZ	San Marcos, CA	20/5	U4	AN7
WNMA	Miami Springs, FL	25/2.5	U4	
WDGR	Dahlonega, GA	10/2.5	D1	D
WSKR	Denham Springs, LA	10/1	U2	AN
WJZZ	FRankenmuth, MI	50/2.5	D1	
KGYN	Guymon, OK	10	U2	AN7
WPHT	Philadelphia, PA	50	U1	AN7
KOKK	Huron, SD	5/0.9	U1	
WGSF	Bartlett, TN	10/0.25	U2	AN
KUBR	San Juan, TX	10/5	U4	
KONY	Washington, UT	10/0.25	U1	AN
KBSG	Auburn, WA	27.5/10	U1	AN7
KREW	Sunnyside, WA	10/1	U1	NH

Call	Location	Power	Class	Notes
KRSV	Afton, WY	5/0.25	U1	AN
KLDI	Laramie, WY	10/1	U2	
CKWA	Slave Lake, AB	1	U1	AN7
VOAR	St. John's, NF	10	U4	NH
CFYM	Kindersley, SK	1	U1	AN

+lp: USA 3/Can 1

1220kHz

Call	Location	Power	Class	Notes
KBPA	Palo Alto, CA	5/0.14	U1	NH
WSLM	Salem, IN	5	D3	D*
WEZU	Stillwater, MN	5	D1	
WENC	Whiteville, NC	5	D1	D*
WKNR	Cleveland, OH	50	U3	AN7
WFAX	Falls Church, VA	5	D1	D*
CJOC	Lethbridge, AB	10/5	U2	AN7
CJRB	Boissevain, MB	10	U3	AN7
CKCW	Moncton, NB	25	U2	AN6
CJSS	Cornwall, ON	1	U4	AN
CJRL	Kenora, ON	5/1	U1	AN
CHSC	St. Catharines, ON	10	U4	AN7
CFVM	Amqui, PQ	10/5	U2	AN7
CKSM	Shawinigan, PQ	10/2.5	U3	AN

+lp: USA 42

1230kHz

USA 172 st's mostly 1kW, Canada 19 st's 40W to 1kW

Call	Location	Power	Class	Notes
CFFB	Iqaluit, NWT	4/1	U1	NH
CHVD	Dolbeau, PQ	10/1	U1	AN7
ZFB	Hamilton, Bermuda	1	U1	NH

1240kHz

USA 156 st's mostly 1kW, Canada 30 st's 40W to 1kW

1250kHz

Call	Location	Power	Class	Notes
WZOB	Fort Payne, AL	5	D1	D
WAPZ	Wetumpka, AL	5	D1	D*
KHIL	Willcox, AZ	5	D1	D
KLIH	Little Rock, AR	2/1.2	U4	AN7
KEYT	Santa Barbara, CA	2.5/1	U4	AN7
KLLK	Willits, CA	5/2.5	U4	
WDAE	Tampa, FL	5	U3	AN7
WGL	Fort Wayne, IN	2.3/1.5	U4	NH
KFKU	Lawrence, KS	5	ST2	
WREN	Kansas City, KS	25/5	U4	AN7
WARE	Ware, MA	5/2.5	U4	NH
KBRF	Fergus Falls, MN	5/2.2	U2	AN
WHNY	McComb, MS	5/1	U2	NH
KIKC	Forsyth, MT	5	D1	D*
WKBR	Manchester, NH	5	U4	AN
WMTR	Morristown, NJ	5	U4	AN
WGHB	Farmville, NC	5/2.5	U4	NH
WBRM	Marion, NC	5	D1	D
WTAE	Pittsburgh, PA	5	U2	AN7
WTMA	Charleston, SC	5/1	U2	AN7
KALO	Port Arthur, TX	5/1	U2	AN
KXDC	San Antonio, TX	1	U2	AN
KNEU	Roosevelt, UT	5	D1	D*
WDVA	Danville, VA	5	U2	AN7
WPRZ	Warrenton, VA	5	D3	D
KWSU	Pullman, WA	5/2.5	U1	NH
KKDZ	Seattle, WA	5	U2	D
WYKM	Rupert, WV	5	D1	D*
WEMP	Milwaukee, WI	5	U4	AN7
CHSM	Steinbach, MB	10/5	U3	AN7
CHWO	Oakville, ON	10	U4	AN7
CBGA	Matane, PQ	10/5	U2	AN

+lp: USA 33

1260kHz

Call	Location	Power	Class	Notes
WDJC	Birmingham, AL	5/1	U2	
KGIL	Beverly Hills, CA	5	U4	NH
KOIT	San Francisco, CA	5/1	U1	AN7
WWDC	Washington, DC	5	U4	AN7
WSUA	Miami, FL	5	U4	AN6
WUFE	Baxley, GA	5	D1	D*

Call	Location	Power		
WTJH	East Point, GA	5/0.04	U1	AN7
KICN	Idaho Falls, ID	5	D1	D*
WIBV	Belleville, IL	5	U4	AN7
WNDE	Indianapolis, IN	5	U2	AN6
KFGQ	Boone, IA 5		D3	D
KBRH	Baton Rouge, LA	5/0.127	U1	AN
WPZE	Boston, MA	5	U2	AN6
WWJQ	Zeeland, MI	10/1	U4	
KTTS	Springfield, MO	5	U2	AN7
WBUD	Trenton, NJ	5/1	U4	AN7
KVSF	Santa Fe, NM	5/1	U1	F.PI.
WNSS	Syracuse, NY	5	U2	AN7
WKXR	Asheboro, NC	5/0.5	U4	NH
WMIH	Cleveland, OH	10/5	U4	AN7
WNXT	Portsmouth, OH	5/1	U4	NH
KWSH	Wewoka, OK	1	U2	NH
KLYC	McMinnville, OR	1	U2	NH
WRIE	Erie, PA	5	U4	AN7
WPHB	Philipsburg, PA	5	D1	D*
WMUU	Greenville, SC	5	D1	D*
WHYM	Lake City, SC	5	D1	D
KWYR	Winner, SD 5		D1	D*
WNOO	Chattanooga, TN	5	D1	D
WDKN	Dickson, TN	5	D1	D*
WCHV	Charlottesville, VA	5/2.5	U4	AN6
WXCE	Amery, WI 5		U4	NH
KPOW	Powell, WY5/1		U2	NH
CFRN	Edmonton, AB	50	U2	AN7
CIHI	Fredericton, NB	10	U2	AN7
+Ip: USA 34				
1270kHz				
WKSJ	Prichard, AL	5	D1	D*
KDJI	Holbrook, AZ	5	D1	D*
KPBA	Pine Bluff, AR	5	D1	
KXPS	Thousand Palms, CA	5/0.75	U4	
KJUG	Tulare, CA 5/1		U2	AN
WRLZ	Eatonville, FL	5	U2	AN
WNOG	Naples, Fl 5/1.9		U4	NH
WNLS	Tallahassee, FL	5	U2	NH
WTMQ	Columbus, GA	5	D1	D*
WJJC	Commerce, GA	5	D1	D
KTFI	Twin Falls, ID	5/1	U1	AN
WKBF	Rock Island, IL	5	U2	AN7
WFRN	Elkhart, IN 5/1		U4	AN
WWCA	East Chicago, IN	2.5	U3	AN7
WCBC	Cumberland, MD	5/1	U4	AN7
WLAW	North Dartmouth, MA 5		U4	F.PI.
WSPR	Springfield, MA	5/1	U4	AN7
WMKT	Charlevoix, MI	5	U2	AN
WXYT	Detroit, MI 5		U2	AN7
WWWY	Baxter, MN5		U4	
KWEB	Rochester, MN	5/1	U4	AN6
KPLY	Sparks, NV 5		U2	
WTSN	Dover, NH 5		U4	AN5
WHLD	Niagara Falls, NY	5/0.14	U3	NH
WDLA	Walton, NY5		D1	D*
WCGC	Belmont, NC	5/0.5	U4	NH
WMPM	Smithfield, NC	5	D1	D*
KTRT	Claremore, OK	1	U4	
KAJO	Grants Pass, OR	5	D1	D*
WLBR	Lebanon, PA	5/1	U4	NH
KNWC	Sioux Falls, SD	2.5	U2	NH
WLIK	Newport, TN	5/0.5	U2	NH
KFCC	Bay City, TX	1	U2	NH
KESS	Fort Worth, TX	5	U3	AN7
WTJZ	Newport News, VA	1	U2	AN
WHEO	Stuart, VA 5		D1	D*
KBAM	Longview, WA	5	D1	D*
KIML	Gillette, WY	5/1	U2	AN7
CHAT	Medicine Hat, AB	10	U2	AN7

Call	Location	Power		
CHWK	Chilliwack, BC	10	U3	AN6
CJCB	Sydney, NS10		U2	AN7
CJTN	Trenton, ON	1	U4	AN
CFGT	Alma, PQ 10/5		U2	AN7
+Ip: USA 30/Can 1				
1280kHz				
WWPG	Tuscaloosa, AL	5/0.5	U2	AN
KKAL	Arroyo Grande, CA	5/2.5	U4	AN
KFRN	Long Beach, CA	1	U5	AN7
KJAX	Stockton, CA	1	U2	AN7
KRRF	Denver, CO 5		U4	AN7
WSVE	Jacksonville, FL	5	D1	D*
WLCG	Macon, GA 5		D1	D*
WBIG	Aurora, IL 2.5/0.5		U4	AN6
WGBF	Evansville, IN	5/1	U2	AN7
WODT	New Orleans, LA	5	U3	AN7
WFAU	Gardiner, ME	5	U2	NH
WEIM	Fitchburg, MA	5/1	U4	NH
WWTC	Minneapolis, MN	5	U2	AN6
KVOX	Moorehead, MN	5/1	U4	AN7
KDOX	Henderson, NV	5/0.05	U1	
KRZE	Farmington, NM	5	D1	
WADO	New York, NY	50/5	U4	AN6
WHTK	Rochester, NY	5	U2	AN6
WSAT	Salisbury, NC	1	U2	NH
WYAL	Scotland Neck, NC	5	D1	D*
KRVM	Eugene, OR5/1.5		U2	AN7
WHVR	Hanover, PA	5/0.5	U4	NH
WKST	New Castle, PA	5/1	U2	NH
WANS	Anderson, SC	5/1	U2	AN7
WMCP	Columbia, TN	5/0.5	U4	
KDYL	Salt Lake City, UT	10/0.6	U4	AN7
KUDY	Spokane, WA	5	D3	D*
KIT	Yakima, WA	5/1	U1	AN7
WNAM	Neenah, WI	20/5	U4	AN7
CHQB	Powell River, BC	1	U3	NH
CFMB	Montreal, PQ	50	U4	
CJSL	Estevan, SK	10	U4	AN6
VSB2	Hamilton, Bermuda	1	U1	AN
+Ip: USA 32/Can 1				
1290kHz				
WOPP	Opp, AL 2.5/0.5		U4	NH
KCUB	Tucson, AZ 1		U1	AN7
KDMS	El Dorado, AR	5	D1	D*
KUOA	Siloam Springs, AR	5	D1	D*
KPAY	Chico, CA 5		U2	AN
KAZA	San Jose, CA	5	D3	D*
KMRZ	San Bernardino, CA	5	U4	AN7
WTMC	Ocala, FL 5/1		U2	AN
WBZT	West Palm Beach, FL	10/4.9	U4	AN
WCHK	Canton, GA5/0.5		U2	NH
WCHY	Savannah, GA	5	U2	AN7
WIRL	Peoria, IL 5		U4	AN7
WDGS	New Albany, IN	0.5/1	U4	
KWLS	Pratt, KS 5/0.5		U4	AN
WCBL	Benton, KY 5		D1	D*
WKLB	Manchester, KY	5	D1	D
WHGR	Houghton Lake, MI	4.9/4.6	U2	NH
KGVO	Missoula, MT	5	U2	AN7
KKAR	Omaha, NE 5		U2	AN6
WKNE	Keene, NH 5		U3	AN
WNBF	Binghamton, NY	5	U2	AN7
WHKY	Hickory, NC5/1		U2	AN
WHIO	Dayton, OH5		U2	AN7
KUMA	Pendleton, OR	5	U2	NH
KKSL	Lake Oswego, OR	5	U3	AN7
WFBG	Altoona, PA	5/1	U2	AN
WRNI	Providence, RI	5	U4	NH
WQMC	Sumter, SC 1		U2	AN
WATO	Oak Ridge, TN	5/0.5	U4	NH

Call	Location	Power		
KRGE	Weslaco, TX	5	U2	NH
KWFS	Wichita Falls, TX	5/0.07	U5	AN6
WZOD	Colonial Heights, VA	25/0.41	U1	D
WVOW	Logan, WV 5/1	U2	AN	
WMCS	Greenfield, WI	5	U4	AN
WCOW	Sparta, WI 5	D1	D*	
KOWB	Laramie, WY	5/1	U2	NH
CIFX	Winnipeg, MB	10	U4	AN7
CJBK	London, ON10	U3	AN7	
CHRM	Matane, PQ	10	U4	AN7

+lp: USA 29

1300kHz

Call	Location	Power		
WKXM	Winfield, AL	5	D1	D*
KWCK	Searcy, AR 5	D1	D*	
KYNO	Fresno, CA 5/1	U2	AN7	
KPMO	Mendocino, CA	5	D3	D*
KAZN	Pasadena, CA	5/1	U4	AN7
KVOR	Colorado Springs, CO 5/1	U1	AN7	
WAVZ	New Haven, CT	1	U2	AN6
WXXU	Cocoa, FL 5/1	U4	NH	
WFFG	Marathon, FL	2.5	U3	NH
WQBN	Tampa, FL 5/1	U4	NH	
WMTM	Moultrie, GA	5	D1	D*
KLER	Orofino, ID 5/1	U2	NH	
WTAQ	La Grange, IL	5/4.5	U4	AN2
KGLO	Mason City, IA	5	U2	AN7
WLXG	Lexington, KY	2.5/1	U2	AN7
WIBR	Baton Rouge, LA	5/1	U4	AN7
KFLO	Shreveport, LA	5	D1	D
WJFK	Baltimore, MD	5	U4	AN7
WOOD	Grand Rapids, MI	5	U2	AN
WOAD	Jackson, MS	5/1	U1	AN
KBRL	McCook, NE	5	D3	D*
KPTL	Carson City, NV	5/0.5	U2	AN
WPNH	Plymouth, NH	5	D1	D*
WIMG	Ewing, NJ 3.2/1.3	U4		
WXRL	Lancaster, NY	2.5	U4	AN
WTMM	Rensselaer, NY	5	U4	AN
WSYD	Mount Airy, NC	5/1	U2	NH
WERE	Cleveland, OH	5	U3	AN7
KAKC	Tulsa, OK 5/1	U4	AN7	
KAPL	Phoenix, OR	20/5	U2	
WILP	West Hazelton, PA	5/0.5	U4	
KOLY	Mobridge, SD	5	D1	D*
WMTN	Morristown, TN	5	D1	D*
WNQM	Nashville, TN	50/5	U2	AN7
KVET	Austin, TX 5/1	U4	AN7	
WKCY	Harrisonburg, VA	5	D1	D*
KKOL	Seattle, WA	35/16	U4	AN7
CJME	Regina, SK 5	U4	AN7	

+lp: USA 30

1310kHz

Call	Location	Power		
WAJO	Marion, AL 5	D1	D*	
KXAM	Mesa, AZ 5/0.5	U2	AN7	
KIQQ	Barstow, CA	5/0.5	U4	
KMKY	Oakland, CA	20	U3	AN7
KFKA	Greeley, CO	5/1	U2	NH
WICH	Norwich, CT	5	U4	AN
WYND	DeLand, FL 5	D1	D*	
WAUC	Wauchula, FL	5/0.5	U3	
KLIX	Twin Falls, ID	5/2.5	U2	AN7
WTLC	Indianapolis, IN	5/1	U2	AN7
WDOC	Prestonsburg, KY	5	D1	D
KMBS	West Monroe, LA	5	D1	D*
WLOB	Portland, ME	5	U4	NH
WORC	Worcester, MA	5/1	U4	NH
WYUR	Dearborn, MI	5	U4	AN7
WCCW	Traverse City, MI	5	D1	D*

Call	Location	Power		
KOCR	Joplin, MO 5/1	U4	AN7	
KEIN	Great Falls, MT	5/1	U1	AN7
WADB	Asbury Park, NJ	2.5/1	U4	AN7
KIVA	Corrales, NM	5/0.5	U2	AN7
WCGR	Canandaigua, NY	1	U4	
WVIP	Mount Kisco, NY	5	D3	D*
WTLB	Utica, NY 5/0.5	U4	AN6	
WISE	Asheville, NC	5/1	U2	AN6
WTIK	Durham, NC	5/1	U4	NH
KNOX	Grand Forks, ND	5	U2	AN6
KNPT	Newport, OR	5/1	U2	NH
WNAE	Warren, PA5	D1	D*	
WDKD	Kingstree, SC	5	D1	D*
WDOD	Chattanooga, TN	5	U2	AN7
WDXI	Jackson, TN	5/1	U2	NH
KTCK	Dallas, TX 5	U4	AN7	
KPOZ	San Antonio, TX	5/0.28	U3	
WDCT	Fairfax, VA 5/0.5	U4	NH	
WGH	Newport News, VA	5	U2	AN7
KZXR	Prosser, WA	5	D1	D*
WSLW	White Sulphur Springs, WV	5	D1	D*
WIBA	Madison, WI	5	U2	AN7
CHLW	St. Paul, AB	10	U2	AN
CIWW	Ottawa, ON	50	U4	AN7

+lp: USA 26

1320kHz

Call	Location	Power		
WAGG	Birmingham, AL	5	D1	D*
WAGF	Dothan, AL 1	U2	AN	
KWHN	Fort Smith, AR	5	U2	NH
KCTC	Sacramento, CA	5	U4	AN
WATR	Waterbury, CT	5/1	U4	NH
WLQY	Hollywood, FL	5	U4	AN7
WJGR	Jacksonville, FL	5	U2	AN7
WAMR	Venice, FL 5/1	U2	AN7	
WHIE	Griffin, GA 5/0.08	U1	NH	
KNCB	Vivian, LA 5	D1	D*	
WJYT	Attleboro, MA	5	U4	NH
WILS	Lansing, MI5/1	U4	AN7	
WDMJ	Marquette, MI	5/1	U2	NH
KOZY	Grand Rapids, MN	5	U2	NH
WRJW	Picayune, MS	5	D1	D*
KSIV	St. Louis, MO	5/0.27	U2	AN7
KOLT	Scottsbluff, NE	5/1	U2	AN
WDER	Derry, NH 10/1	U4	NH	
WHHO	Hornell, NY5	D1	D*	
WTCK	Greensboro, NC	5	U4	AN7
WKRK	Murphy, NC	5	D1	D*
WOBL	Oberlin, OH1	U4	AN	
WTKZ	Allentown, PA	5/1	U4	AN7
WJAS	Pittsburgh, PA	5	U2	AN7
WISW	Columbia, SC	5/2.5	U2	AN
KELO	Sioux Falls, SD	5	U2	AN
WKIN	Kingsport, TN	5/0.5	U2	AN
WMSR	Manchester, TN	5	D1	D*
KXYZ	Houston, TX	5.4	U2	AN7
KFNZ	Salt Lake City, UT	5	U3	AN7
WLEE	Richmond, VA	5	D3	D*
KXRO	Aberdeen, WA	5/1	U2	NH
WFHR	Wisconsin Rapids, WI5/0.5	U2	NH	
CHMB	Vancouver, BC	50	U4	AN7
CKEC	New Glasgow, NS	25	U2	AN7
CJMR	Mississauga, ON	20	U4	AN

+lp: USA 31/Can 1

1330kHz

Call	Location	Power		
WPRN	Butler, AL 5	D	D	
WZCT	Scottsboro, AL	5/0.5	U2	
KMRR	Tucson, AZ 2/5	U2	AN	

Call	City	Power		
KWKW	Los Angeles, CA	5	U2	AN7
KLBS	Los Banos, CA	0.5/5	U2	
WJNX	Fort Pierce, FL	5/1	U2	AN
WEBY	Milton, FL 5	D1		
WCVC	Tallahassee, FL	5	D1	D*
WMLT	Dublin, GA 5/0.5	U2	NH	
WKTA	Evanston, IL	5/0.5	U4	NH
WVHI	Evansville, IN	5/1	U2	AN
KWLO	Waterloo, IA	5	U4	AN7
KFH	Wichita, KS	5	U2	AN7
WKDP	Corbin, KY 5	D3	D*	
KVOL	Lafayette, LA	5/1	U2	AN7
WASA	Havre de Grace, MD	5/0.5	U2	D
WRCA	Waltham, MA	5	U4	AN7
WTRX	Flint, MI 5/1	U4		
WMNN	Minneapolis, MN	9.7/5.1	U4	AN6
WFTO	Fulton, MS 5	D1	D*	
KGAK	Gallup, NM5/1	U2	AN	
WWRV	New York, NY	5	U3	AN7
WEBO	Owego, NY5	D1	D*	
WSPQ	Springville, NY	1	U4	NH
WASN	Youngstown, OH	0.5/1	U4	AN7
KKPZ	Portland, OR	5	U3	AN7
WFLP	Erie, PA 5	U4	AN7	
WYSN	Somerset, PA	5	D3	D*
WPJS	Conway, SC	5/0.5	U2	NH
WYRD	Greenville, SC	5	U2	AN7
KLBO	Monahans, TX	5/1	U2	NH
WBTM	Danville, VA	5/1	U2	NH
WOLD	Marion, VA5	D1	D	
WESR	Onley, VA 5	D1	D*	
KMBI	Spokane, WA	5	D1	D*
WHBL	Sheboygan, WI	5/1	U4	NH
KOVE	Lander, WY5/1	U2	NH	
CKLD	Thetford Mines, PQ	10	U3	AN7
CJYM	Rosetown, SK	10	U3	AN7

+lp: USA 30

1340kHz

USA 174 st's, mostly with 1kW, Canada 27 st's 20W to 1kW.

Call	City	Power		
CFYK	Yellowknife, NWT	2.5	U1	AN
CJLS	Yarmouth, NS	5/4	U4	AN
ZBM	Hamilton, Bermuda	1	U1	NH

1350kHz

Call	City	Power		
WGAD	Gadsden, AL	5/1	U2	NH
KCKC	San Bernardino, CA	5/0.6	U4	AN6
KSRO	Santa Rosa, CA	5	U2	NSP
KGHF	Pueblo, CO	4.7/0.75	U2	AN7
WNLK	Norwalk, CT	2.5/0.5	U4	AN
WINY	Putnam, CT5	D1	D*	
WMMV	Cocoa, FL 1	U2	AN7	
WGIA	Blackshear, GA	5	D1	D*
WCOP	Warner Robins, GA	5/0.5	U2	
KRLC	Lewiston, ID	5/1	U2	AN7
WOAM	Peoria, IL 1	U4	AN7	
WIOU	Kokomo, IN5/1	U4	AN7	
KRNT	Des Moines, IA	5	U2	AN7
WSMB	New Orleans, LA	5	U2	AN7
WEZS	Laconia, NH	5	D1	D*
WHWH	Princeton, NJ	5	U4	AN6
KABQ	Albuquerque, NM	5/0.5	U2	NH
WTOU	Akron, OH 5	U3	AN7	
WOYK	York, PA 5/1	U2	AN7	
KTXJ	Jasper, TX 5/0.04	U1	NH	
KCOR	San Antonio, TX	5	U2	AN7
WNVA	Norton, VA5	D1	D*	
WGPL	Norfolk, VA5	U4	AN7	
CKCI	Parksville, BC	10	U4	AN7
CKAD	Middleton, NS	1	U3	AN

Call	City	Power		
CKDO	Oshawa, ON	10/5	U4	AN7

+lp: USA 37/Can 5

1360kHz

Call	City	Power		
WMOB	Mobile, AL 5	D3	D*	
KGME	Glendale, AZ	5/1	U2	AN7
KFFA	Helena, AR 1	U2	NH	
KFIV	Modesto, CA	4/1	U4	AN7
KPOP	San Diego, CA	5/1	U1	AN7
WDRC	Hartford, CT	5	U2	AN7
WHNR	Cypress Gardens, FL	5/2.5	U4	NH
WCGL	Jacksonville, FL	5	D1	D
WKAT	Miami, FL 5/1	U1	AN7	
KSCJ	Sioux City, IA	5	U2	AN
WWLG	Baltimore, MD	5/1.5	U2	AN7
WKYO	Caro, MI 1	U4	NH	
WKMI	Kalamazoo, MI	5/1	U2	AN6
KKBJ	Bemidji, MN	5/2.5	U2	AN6
WNJC	Washington Township, NJ	5/0.8	U4	AN7
KBUY	Ruidoso, NM	5/0.2	U1	NH
WKOP	Binghamton, NY	5/0.5	U4	AN7
WCHL	Chapel Hill, NC	5/1	U2	AN7
WCKY	Cincinnati, OH	5	U2	AN7
KOHU	Hermiston, OR	4.3/1	U2	NH
KUIK	Hillsboro, OR	5	U2	NH
WPTT	McKeesport, PA	5/1	U2	AN6
WPPA	Pottsville, PA	5/0.5	U4	NH
WELP	Easley, SC 5/0.036	U1		
KWWJ	Baytown, TX	5/1	U4	AN7
KRYS	Corpus Christi, TX	1	U1	AN7
KAHZ	Fort Worth, TX	5/1	U2	AN7
WWWJ	Galax, VA 5	D1	D*	
WHBG	Harrisonburg, VA	5	D1	D*
KKMO	Tacoma, WA	5	U1	AN6
WGEE	Green Bay, WI	5	U2	AN7
KRKK	Rock Springs, WY	5/1	U2	AN7
CKBC	Bathurst, NB	10	U2	AN
CJVL	Ste-Marie-de-Beauce, PQ	10/5	U4	AN

+lp: USA 41

1370kHz

Call	City	Power		
KWRM	Corona, CA 5/2.5	U4	AN6	
KPCO	Quincy, CA 5/0.5	U4		
KZSF	San Jose, CA	5	U4	AN7
KTMG	Deer Trail, CO	5	D3	D
WOCA	Ocala, FL 5	D1	D*	
WCOA	Pensacola, FL	5	U2	AN6
WLOP	Jesup, GA 5	D1	D*	
WGCL	Bloomington, IN	5/1	U4	NH
KDTH	Dubuque, IA	5	U2	AN7
KGNO	Dodge City, KS	5/0.23	U1	AN
WGOH	Grayson, KY	5	D1	D
WDEA	Ellsworth, ME	5	U4	NH
WKJF	Cadillac, MI	5/1	U4	NH
KSUM	Fairmont, MN	1	U4	AN6
KXTL	Butte, MT 5	U1	AN7	
WFEA	Manchester, NH	5	U4	AN7
WELV	Ellenville, NY	5	D1	D*
WXXI	Rochester, NY	5	U2	AN7
WLTC	Gastonia, NC	5	D1	D*
WLLN	Lillington, NC	5	D3	D*
WTAB	Tabor City, NC	5	D1	D*
WSPD	Toledo, OH 5	U2	AN7	
KAST	Astoria, OR1	U2	NH	
WKMC	Roaring Spring, PA	5	D3	D*
WDEF	Chattanooga, TN	5	U2	AN7
KFRO	Longview, TX	1	U2	AN7
KSOP	Salt Lake City, UT	5/0.5	U4	AN7

WHEE	Martinsville, VA	5	D1	D*		WRSC	State College, PA	2/1	U2	AN
WVMR	Frost, WV	5	D1	D*		WXTC	Charleston, SC	5	U2	AN7
WMJT	Moundsville, WV	5	D1	D*		WTJS	Jackson, TN	5/1	U2	NH
WCCN	Neillsville, WI	5/0.04	U1	AN7		KLGN	Logan, UT 5/0.5	U2	AN	
CFOK	Westlock, AB	10	U2	AN		WKDR	Burlington, VT	5	U2	AN

+lp: USA 37/Can 2

1375kHz

RFO	St. Pierre & Miquelon 20		U1	NH		WZHF	Arlington, VA	5	U4	AN

1380kHz

WVSA	Vernon, AL 5	D1	D*			KBBO	Yakima, WA	5/0.5	U2	NH
KRNN	Little Rock, AR	5/2.5	U4			WRIG	Schofield, WI	5	U4	NH
KTKZ	Sacramento, CA	5	U4	AN7		CJCY	Medicine Hat, AB	10	U2	AN7

+lp: USA 26

1400kHz

USA 176 st's, mostly 1kW; Canada 14 st's, 40W-1kW.

KTOM	Salinas, CA 5	U4	AN			CBG	Gander, NF 4	U1	AN	
WFNW	Naugatuck, CT	5/0.5	U4	NH		CJFP	Rivière-du-Loup, PQ	10/5	U2	AN7

1410kHz

WELE	Ormond Beach, FL	5/2.5	U4	AN7		WIQR	Prattville, AL	5/1	U4	
WRBQ	St. Petersburg, FL	5	U2	AN7		WLVV	Mobile, AL 3.9	U3	AN6	
WAOK	Atlanta, GA 5/4.2	U2	AN7			KERN	Bakersfield, CA	1	U1	AN7
WSIZ	Ocilla, GA 5	D1	D*			KRML	Carmel, CA 2.5/2	U4	NH	
WBEL	South Beloit, IL	5	U2	NH		KMYC	Marysville, CA	5/1	U4	NH
WHWD	Fort Wayne, IN	5	U4	AN7		KCAL	Redlands, CA	5/4	U4	NH
KCIM	Carroll, IA 1	U4	NH			KCOL	Fort Collins, CO	5/1	U4	AN7
WYNK	Baton Rouge, LA	5	D3	D*		WPOP	Hartford, CT	5	U4	AN7
WPHM	Port Huron, MI	5	U4	NH		WDOV	Dover, DE 5	U4	AN7	
KLIZ	Brainerd, MN	5	U2	AN6		WMYR	Fort Myers, FL	5	U2	AN
KKWK	St. Louis, MO	5/1	U4	AN		WQBQ	Leesburg, FL	5	D1	D*
WTMN	Portsmouth, NH	2.2/5	U4	AN		WHBT	Tallahassee, FL	5/0.18	U1	
WKDM	New York, NY	5	U3	AN		WLAQ	Rome, GA 1	U2	AN	
WKJV	Asheville, NC	5/1	U2	NH		KKLO	Leavenworth, KS	5/0.5	U4	AN
WTOB	Winston-Salem, NC	5/2.5	U4	NH		KMYR	Wichita, KS	5/1	U4	AN7
KXCA	Lawton, OK 1	U4	NH			WHLN	Harlan, KY 5	D1	D*	
KSRV	Ontario, OR 5/1	U2	NH			KRWB	Roseau, MN	1	U2	NH
KOTA	Rapid City, SD	5	U2	NH		KOOQ	North Platte, NE	5/0.5	U2	AN
WOOM	Millington, TN	2.5/1	U4	NH		WELM	Elmira, NY 5/1	U4	AN7	
KTSM	El Paso, TX 5/0.5	U1	AN7			WUZZ	Watertown, NY	5/1	U2	AN6
WSYB	Rutland, VT 5/1	U2	AN7			WSRC	Durham, NC	5/0.29	U4	AN7
WTVR	Richmond, VA	5	U4	AN7		WING	Dayton, OH 5	U2	AN7	
KRKO	Everett, WA	5	U2	AN6		KBNP	Portland, OR	5	D1	D
WFCL	Clintonville, WI	3.9/1.8	U3			WLSH	Lansford, PA	5/0.06	U3	AN
CKPC	Brantford, ON	25	U4	AN7		KQV	Pittsburgh, PA	5	U4	AN7
CKLC	Kingston, ON	10	U4	AN7		KLEV	Cleveland, TX	1	U4	NH
CFDA	Victoriaville, PQ	10	U4	AN7		KRIL	Odessa, TX 1	U2	AN7	

+lp: USA 36/Can 4

1390kHz

						WRIS	Roanoke, VA	5	D1	D*
WHMA	Anniston, AL	5/1	U2	NH		WSCW	South Charleston, WV5	D1	D*	
KKDV	Fields Landing, CA	5	D1	F.PI.		WIZM	La Crosse, WI	5	U2	AN7
KLTX	Long Beach, CA	5	U4	AN6		KWYO	Sheridan, WY	5/0.5	U1	NH
KVIN	Turlock, CA 5	U4	AN6			CFUN	Vancouver, BC	50	U4	AN7
KJME	Denver, CO 5/0.14	U1	AN7			CIGO	Port Hawkesbury, NS 10	U4	AN7	
WAJD	Gainesville, FL	5	D1	D*		CKSL	London, ON10	U4	AN7	
WGCI	Chicago, IL 5	U4	AN7			CFMB	Montreal, PQ	10	U3	AN7

+lp: USA 38

1420kHz

WKIC	Hazard, KY 5	D1	D*			WACT	Tuscaloosa, AL	5	D1	D*
WEGP	Presque Isle, ME	5	U2			KXOW	Hot Springs, AR	5	D1	D*
WPLM	Plymouth, MA	5	U4	NH		KSTN	Stockton, CA	5	U4	AN7
WLCM	Charlotte, MI	5	D3	D*		WLIS	Old Saybrook, CT	5/0.5	U2	NH
KXSS	Waite Park, MN	2.5/1	U4	AN		WDBF	Delray Beach, FL	5/0.46	U4	NH
WROA	Gulfport, MS	5	U4	AN7		WBRD	Palmetto, FL	2.5/1	U4	
WMER	Meridian, MS	5	D1	D*		WRCG	Columbus, GA	5	U2	AN
KENN	Farmington, NM	5/1.3	U2	AN		WLET	Toccoa, GA 5	D1	D*	
KUCU	Hobbs, NM 5/0.5	U2	AN7			WIMS	Michigan City, IN	5	U4	NH
WEOK	Poughkeepsie, NY	5	D3	D		WOC	Davenport, IA	5	U4	AN6
WDCW	Syracuse, NY	5	U4	AN6		WVJS	Owensboro, KY	5/1	U4	AN7
WEED	Rocky Mount, NC	5/2.5	U2	NH		WBSM	New Bedford, MA	5/1	U3	AN7
KRRZ	Minot, ND 5/1	U1	AN6			WBEC	Pittsfield, MA	1	U2	AN
WMPO	Middleport, OH	5/0.12	U1	AN		KTOE	Mankato, MN	5	U4	AN7
WRTK	Youngstown, OH	9.5/4.2	U2	AN7		WIGG	Wiggins, MS	5	D1	D*
KCRC	Enid, OK 1	U3	NH			WASR	Wolfeboro, NH	5	D1	D*
KSLM	Salem, OR 5/0.69	U1	NH							
WLAN	Lancaster, PA	5/1	U4	AN7						

Call	Location	Power	Class	
WACK	Newark, NY	5/0.73	U4	NH
WLNA	Peekskill, NY	5/1	U4	AN
WHK	Cleveland, OH	5	U2	AN7
WCOJ	Coatesville, PA	5	U2	NH
WCED	DuBois, PA 5/0.5		U2	NH
WEMB	Erwin, TN 5		D1	D*
WKSR	Pulaski, TN 1		U2	NH
WKCW	Warrenton, VA	5	D1	D*
KITI	Chehalis, WA	5	U4	AN6
KUJ	Walla Walla, WA	5	U2	AN7
WTCR	Kenova, WV	5/0.5	U2	AN
CKDY	Digby, NS 1		U3	AN
CKPT	Peterborough, ON	10/5	U4	AN7
CKTL	Plessisville, PQ	1/0.5	U2	AN7

+lp: USA 44/Can 1

1430kHz

Call	Location	Power	Class	
WFHK	Pell City, AL	5	D1	D
KFIG	Fresno, CA 5		U4	AN7
KALI	San Gabriel, CA	5	U4	AN7
KVVN	Santa Clara, CA	1/2.5	U4	AN6
KEZW	Aurora, CO 10/5		U2	AN7
WOIR	Homestead, FL	5/0.5	U2	AN
WLKF	Lakeland, FL	5/1	U1	AN
WLTG	Panama City, FL	5	U4	AN7
WGFS	Covington, GA	5	D1	D*
WWGS	Tifton, GA 5/1		U2	NH
WEEF	Highland Park, IL	1	U4	
WMYS	Indianapolis, IN	5	U2	AN7
WYMC	Mayfield, KY	1	U2	AN
WNAV	Annapolis, MD	5/1	U2	NH
WTTT	Amherst, MA	5	D3	D*
WXKS	Everett, MA	5/1	U2	NH
WION	Ionia, MI 5		D3	D*
WLAU	Laurel, MS 5		D1	D*
WRTH	St. Louis, MO	5	U4	AN7
KRGI	Grand Island, NE	5/1	U2	NH
WNJR	Newark, NJ	5	U2	AN7
KCRX	Roswell, NM	5/1	U2	NH
WENE	Endicott, NY	5	U2	AN7
WDEX	Monroe, NC	2.5	U4	AN7
WFOB	Fostoria, OH	1	U4	NH
KQLL	Tulsa, OK 5		U2	AN7
KYKN	Keizer, OR 5		U2	AN
WVAM	Altoona, PA	5/1	U2	AN6
WBLR	Batesburg, SC	5	D1	D*
WOWW	Germantown, TN	2.8/2.5	U4	
WMAK	Madison, TN	5/1	U2	
KEES	Gladewater, TX	5/1	U2	NH
KCOH	Houston, TX	5/0.33	U4	AN7
KLO	Ogden, UT 10/5		U4	AN7
WHAN	Ashland, VA	5	D3	D*
WDIC	Clintwood, VA	5	D1	D*
KBRC	Mount Vernon, WA	5/1	U2	NH
KCLK	Asotin, WA 5/1		U4	
WEIR	Weirton, WV	1	U4	AN
WBEV	Beaver Dam, WI	1	U2	NH
CHTK	Toronto, ON	50	U4	AN7

+lp: USA 32

1440kHz

Call	Location	Power	Class	
WHHY	Montgomery, AL	5/1	U2	AN7
KSLX	Scottsdale, AZ	5	D1	D*
KITA	Little Rock, AR	5/0.24	U2	
KVON	Napa, CA 5/1		U4	AN
KDIF	Riverside, CA	1	U1	AN
KUHL	Santa Maria, CA	5/1	U2	AN6
KRDZ	Wray, CO 5		D1	D
WWCL	Lehigh Acres, FL	5/1	U4	AN
WPRD	Winter Park, FL	5/1	U2	AN6

Call	Location	Power	Class	
WGIG	Brunswick, GA	5/1	U2	AN
WGEM	Quincy, IL 5/1		U4	NH
WROK	Rockford, IL 5/0.27		U5	AN7
KMAJ	Topeka, KS 5/1		U3	AN7
WCDS	Glasgow, KY	5	D3	
KMLB	Monroe, LA 5/1		U2	AN
WJAE	Portland, ME	5	U2	AN
WWTM	Worcester, MA	5	U4	AN7
WMAX	Bay City, MI	5/2.5	U4	AN6
WMKM	Inkster, MI 1		U4	AN7
KDIZ	Golden Valley, MN	5/0.5	U2	AN6
WRBE	Lucedale, MS	5	D1	D
WJJL	Niagara Falls, NY	5/0.75	U4	
WBLA	Elizabethtown, NC	5	D1	D*
WLXN	Lexington, NC	5/1	U2	AN7
WRRO	Warren, OH	5	U4	AN6
KMED	Medford, OR	5/1	U1	AN6
KODL	The Dalles, OR	5/1	U2	NH
WCDL	Carbondale, PA	5	D1	D*
WGVL	Greenville, SC	5	U2	AN7
WZYX	Cowan, TN 5		D1	D*
KPUR	Amarillo, TX	5/1	U2	AN7
KEYS	Corpus Christi, TX	1	U2	AN
KTNO	Denton, TX 5/0.5		U2	NH
KETX	Livingston, TX	5	D1	D*
WKLV	Blackstone, VA	5	D1	D*
KJUN	Puyallup, WA	5/1.7	U4	F.Pl.
WHIS	Bluefield, WV	5/0.5	U1	AN6
WAJR	Morgantown, WV	5/0.5	U4	AN6
WNFL	Green Bay, WI	5/0.5	U4	AN7
CKJR	Wetaskiwin, AB	10	U2	AN
CFCP	Courtenay, BC	1	U2	AN7

1450kHz

USA 178 st's mostly with 1kW, Canada 21 st's 40W-1kW

Call	Location	Power	Class	
CHUC	Cobourg, ON	8/1	U4	AN
CHEF	Granby, PQ 10/5		U4	AN7
VSB1	Hamilton, Bermuda	1	U1	AN

1460kHz

Call	Location	Power	Class	
WFMH	Cullman, AL	5/0.5	U2	AN
KTYM	Inglewood, CA	5/0.5	U4	AN7
KDON	Salinas, CA 10		U3	
KKCS	Colorado Springs, CO	5/0.5	U2	AN
WZEP	DeFuniak Springs, FL	5/0.19	U1	AN7
WZNZ	Jacksonville, FL	5	U2	AN7
WXEM	Buford, GA 5		D1	D*
KDMI	Des Moines, IA	5	U2	AN7
WBPA	Elkhorn City, KY	5	D1	D
WXOK	Baton Rouge, LA	5/1	U2	AN6
WBET	Brockton, MA	5/1	U2	NH
WBRN	Big Rapids, MI	5/2.5	U2	NH
KDMA	Montevideo, MN	1	U2	NH
KIRL	St. Charles, MO	5/0.5	U4	AN
KKPR	Kearney, NE	5	D1	D*
KENO	Las Vegas, NV	10/0.65	U4	AN7
WIFI	Florence, NJ	5/0.5	U3	AN
WGNA	Albany, NY 5		U2	NH
WWWG	Rochester, NY	5	U2	AN7
WCRY	Fuquay-Varina, NC	5	D3	D*
WEWO	Laurinburg, NC	5	U4	AN7
WHBK	Marshall, NC	5/0.14	U1	
KLTC	Dickinson, ND	5	U2	AN
WBNS	Columbus, OH	5/1	U2	AN6
WWKL	Harrisburg, PA	5	U2	AN7
WEMR	Tunkhannock, PA	5/1	U4	NH
WBCU	Union, SC 1		U2	NH
WKTK	Waco, TX 1		U2	AN7
WKDV	Manassas, VA	5	U4	NH
WRAD	Radford, VA	5/0.5	U2	NH

Call	City	Power		
KARR	Kirkland, WA	5/2.5	U4	AN
KMWX	Yakima, WA	5/3.7	U2	AN7
WBUC	Buckhannon, WV	5	D1	D*
CJOY	Guelph, ON	50/10	U3	AN7
CKRB	St-George-de-Beauce, PQ	10/5	U2	AN7

+lp: USA 36/Can 2

1470kHz

Call	City	Power		
KUTY	Palmdale, CA	5	U4	AN7
KQPT	Sacramento, CA	5/1	U4	AN7
WMMW	Meriden, CT	2.5	U4	AN
WRTB	Dunedin, FL	5	D1	D*
WWNN	Pompano Beach, FL	50/2.5	U4	AN
WRGA	Rome, GA	5	U2	AN7
WCFJ	Chicago Heights, IL	1	U4	AN
WMBD	Peoria, IL	5	U4	AN7
KWSL	Sioux City, IA	5	U4	AN6
KAIR	Atchison, KS	1	U3	NH
KLCL	Lake Charles, LA	5/0.5	U1	AN7
WZOU	Lewiston, ME	5	U3	AN7
WJDY	Salisbury, MD	5	D3	D*
WTTR	Westminster, MD	1	U2	NH
WSRO	Marlboro, MA	5	U2	AN
WFNT	Flint, MI	5/1	U4	AN6
WQSN	Kalamazoo, MI	0.8/1	U2	
WLOL	Brooklyn Park, MN	5	U4	AN7
WTKO	Ithaca, NY	5/1	U2	AN6
WWBG	Greensboro, NC	3.5/5	U2	F.Pl.
WPNC	Plymouth, NC	5	D1	D*
WTOE	Spruce Pine, NC	5	D1	D*
WLQR	Toledo, OH	1	U4	AN7
WKAP	Allentown, PA	5	U2	AN7
WQXL	Columbia, SC	5	D1	D*
WVOL	Berry Hill, TN	5/1	U4	NH
KBBA	Abilene, TX	5/1	U2	AN7
KWRD	Henderson, TX	5	D1	D
WBTX	Broadway, VA	5	D1	D*
WTZE	Tazewell, VA	5/0.75	U1	
KELA	Centralia, WA	5/1	U1	NH
KBSN	Moses Lake, WA	5/1	U4	NH
WHRD	Huntington, WV	5	D1	D*
WBKV	West Bend, WI	2.5	U4	NH
CJVB	Vancouver, BC	50	U4	AN6
CHOW	Welland, ON	10	U4	AN7

+lp: USA 39/Can 1

1480kHz

Call	City	Power		
WLPH	Irondale, AL	5	D1	D*
WABB	Mobile, AL	5/4.4	U2	AN7
KPHX	Phoenix, AZ	5/0.5	U4	AN
KTHS	Berryville, AR	5	D1	D*
KRHT	Concord, CA	0.5/5	U4	
KGOE	Eureka, CA	5/0.68	U1	AN
KYOS	Merced, CA	5	U2	AN6
KVNR	Santa Ana, CA	5	U4	AN7
WODX	Marco Island, FL	1	U4	NH
WYZE	Atlanta, GA	5/0.04	U1	AN7
WRDW	Augusta, GA	5	U2	AN7
WTHI	Terre Haute, IN	5/1	U4	AN7
KQAM	Wichita, KS	5/1	U4	AN6
WEZC	Neon, KY	5	D1	D*
WSAR	Fall River, MA	5	U3	NH
WGVU	Kentwood, MI	2.5/5	U2	AN
WSDS	Ypsilanti, MI	0.5/3.8	U4	NH
KAUS	Austin, MN	1	U4	AN6
KKCQ	Fosston, MN	5/0.09	U1	NH
KLMS	Lincoln, NE	5/1	U4	AN7
KKEL	Hobbs, NM	5/1	U2	AN
WZRC	New York, NY	5	U4	AN7

Call	City	Power		
WADR	Remsen, NY	5/0.02	U1	AN
WTLT	Charlotte, NC	5	U4	AN7
WPFJ	Franklin, NC	5/0.01	U1	
WHBC	Canton, OH	15/5	U4	AN6
WCNS	Latrobe, PA	0.5/1	U2	NH
WDAS	Philadelphia, PA	5/1	U4	AN7
WISL	Shamokin, PA	1	U2	
WBBP	Memphis, TN	5	D1	D*
KDXX	Dallas, TX	5/1.9	U4	AN7
WCFR	Springfield, VT	5	D1	D*
WTOY	Salem, VA	5	D1	D*
KBMS	Vancouver, WA	1/2.5	U2	AN7
WTDA	Madison, WI	5	U4	AN7

+lp: USA 39/Can 2

1490kHz

USA 177st's, mostly 1kW, Canada 14 st's 40W-1kW

1500kHz

Call	City	Power		
KRCK	Los Angeles, CA	50/14	U4	AN
KSJX	San Jose, CA	10/5	U4	AN7
WFIF	Milford, CT	5	D3	D
WTOP	Washington, DC	50	U4	AN7
WBRI	Indianapolis, IN	5	D3	D
WLQV	Detroit, MI	50/5	U4	AN7
KSTP	St. Paul, MN	50	U2	AN7
WSMX	Winston-Salem, NC	10/1	D6	D
CKAY	Duncan, BC	10/1	U3	AN

+lp: USA 30

1510kHz

Call	City	Power		
KFNN	Mesa, AZ	22/0.1	U4	D*
KIRV	Fresno, CA	10	D3	D*
KMSL	Ontario, CA	10/1	U4	AN6
KNOB	San Rafael, CA	8	D3	D
KDKO	Littleton, CO	10/1.3	U4	AN7
WWJY	New London, CT	10	D3	D
WNRB	Boston, MA	50	U4	AN7
WJKN	Jackson, MI	5	D3	D*
KCTE	Independence, MO	10	D3	D*
WAHT	Annville, PA	5	D3	
KMSD	Milbank, SD	5/1	D1	D*
WLAC	Nashville, TN	50	U2	AN7
KQHN	Nederland, TX	5	D3	D
KLLB	West Jordan, UT	10	D1	D
KGA	Spokane, WA	50	U2	AN7
WAUK	Waukesha, WI	10	D3	D*
CKOT	Tillsonburg, ON	10	D3	D

+lp: USA 36/Can 1

1520kHz

Call	City	Power		
WTLM	Opelika, AL	5	D3	D
KMPG	Hollister, CA	5/3.5	D3	D
KTRO	Oxnard, CA	10/1	U4	AN6
WHIM	Apopka, FL	5/0.35	U4	D
WHOW	Clinton, IL	5/1	D1	D
WLGC	Greenup, KY	5/0.3	D1	D
KDYS	Lafayette, LA	10/0.5	U2	AN
WGAM	Greenfield, MA	10	D3	D
WQWQ	Muskegon, MI	10/5	U4	
WMLM	St. Louis, MI	1	U4	NH
KOLM	Rochester, MN	10/0.8	U6	
KRHW	Sikeston, MO	5/2	U7	NH
WWKB	Buffalo, NY	50	U3	AN7
WDSL	Mocksville, NC	5/1	D1	D
WARR	Warrenton, NC	5	D1	
WDMN	Toledo, OH	1	U4	AN
KOMA	Oklahoma City, OK	50	U2	AN7
KKSN	Oregon City, OR	50/15	U4	AN7

+lp: USA 32

1530kHz

Call	City	Power		
KFBK	Sacramento, CA	50	U3	AN7

Call	Location	Power		
KHPY	Moreno Valley, CA	10/1.3	D3	D
WDJZ	Bridgeport, CT	5	D3	D
WOBS	Jacksonville, FL	50	D3	D
WTTI	Dalton, GA 10		D3	D
WLMR	Lapeer, MI 5		D3	D
KKCM	Shakopee, MN	8.6/0.01	U3	
WRPM	Poplarville, MS	10/1	D1	D*
WRTP	Durham, NC	10	D3	D
WSAI	Cincinnati, OH	50	U2	AN7
KXTD	Wagoner, OK	5	D3	
KQQA	Creedmore, TX	10/1.1	D3	
KGBT	Harlingen, TX	50/10	U4	AN
KCLR	Ralls, TX 5/1		D1	D*
KSHY	Fox Farm, WY	10/1	U4	F.PI.

+lp: USA 31/Can 2

1540kHz

Call	Location	Power		
KASA	Phoenix, AZ	10	D3	D
KCTD	Los Angeles, CA	50/10	U4	AN7
KXEL	Waterloo, IA	50	U2	AN7
WACA	Wheaton, MD	5	D1	D
WMYF	Exeter, NH 5/2.5		D1	D
WDCD	Albany, NY 50		U3	AN7
WNWR	Philadelphia, PA	50	D3	D
WECZ	Punxsutawney, PA	5/1	D1	D
WTBI	Pickens, SC	10/1	D1	D*
KZMP	University Park, TX	35/0.89	U4	NH
KEDA	San Antonio, TX	5/1	U4	AN7
WREJ	Richmond, VA	10	D3	D*
KXPA	Bellevue, WA	5	U2	AN7
CHIN	Toronto, ON	50/30	U4	AN7

+lp: USA 37/Can 2

1550kHz

Call	Location	Power		
WLOR	Huntsville, AL	50/0.5	U4	AN7
KUAT	Tucson, AZ 50		D1	D
KWRN	Apple Valley, CA	5/0.5	U2	
KXEX	Fresno, CA 5/2.5		U4	AN
KYCY	San Francisco, CA	50/10	U4	AN7
KQXI	Arvada, CO 10/0.475		U2	D*
WDZK	Bloomfield, CT	5/2.5	U4	NH
WAMA	Tampa, FL 10		D1	D*
WTHB	Augusta, GA	5	D1	D
WAZX	Smyrna, GA	50/0.5	U4	AN7
WPFC	Baton Rouge, LA	5	D1	D
WNTN	Newton, MA	10	D1	D
KAPE	Cape Girardeau, MO	5/0.05	U3	AN7
KLFJ	Springfield, MO	5/0.03	U1	AN7
KSFT	St. Joseph, MO	5	U2	AN7
KQWB	West Fargo, ND	10/5	U2	AN
WKQV	Pittston, PA10/0.5		U4	NH
WBSC	Bennettsville, SC	10/5	U2	NH
WBCV	Bristol, TN 5		D1	D*
WCTZ	Clarksville, TN	2.5/0.25	U4	NH
KMRI	West Valley City, UT	10/0.5	U2	
WKBA	Vinton, VA 10		D3	D
WVAB	Virginia Beach, VA	5/0.01	U1	AN
KNTR	Ferndale, WA	50/10	U2	NH
KSVY	Spokane, WA	10/2.5	U2	NH
KVAN	Vancouver, WA	10	U2	AN
WXVA	Charles Town, WV	5	D1	D*
WHIT	Madison, WI	5	D3	D*
CBE	Windsor, ON	10	U3	NH

+lp: USA 43/Can 2

1560kHz

Call	Location	Power		
WTKN	Daleville, AL	5	D1	D
KNZR	Bakersfield, CA	25/10	U2	AN7
WRHC	Coral Gables, FL	10/4.4	U4	AN
WINV	Inverness, FL	5	D1	D
WTMS	Melbourne, FL	5	D1	D*
KCJJ	Iowa City, IA	1	U4	AN7
WPAD	Paducah, KY	10/5	U4	AN
KQYX	Joplin, MO	10	D3	D*
WQEW	New York, NY	50	U4	AN6
WCNW	Fairfield, OH	5/1	L3	D*
WTOD	Toledo, OH 5		D3	D*
WAGL	Lancaster, SC	50	D6	D*
KKAA	Aberdeen, SD	10/5	U4	AN2
WWRO	Nashville, TN	10	D3	D*
KTXZ	West Lake Hills, TX	2.5	U4	AN7
KZIZ	Sumner, WA	5/1	U2	D*

+lp: USA 36/Can 1

1570kHz

Call	Location	Power		
KCVR	Lodi, CA 5/0.5		U4	AN
KPRO	Riverside, CA	5/0.2	U1	AN7
KTGE	Salinas, CA5/0.5		U4	AN
WTWB	Auburndale, FL	5	D1	D*
WYHI	Fernandina Beach, FL 5		D1	
WSSA	Morrow, GA	5/0.05	U1	
WFRL	Freeport, IL 5/0.5		U3	AN7
WKDB	Baltimore, MD	5	D1	D*
WFLR	Dundee, NY	5/0.44	U1	NH
WBUX	Doylestown, PA	5/0.95	U3	AN
WNKX	Centerville, TN	5	D1	D*
WCLE	Cleveland, TN	5/0.08	U1	NH
WTRB	Ripley, TN 28/0.5		U4	D
WMQA	Minocqua, WI	5	D1	D
CKTA	Taber, AB 10/5		U2	AN7
CKEG	Nanaimo, BC	10	U4	AN7
CKMW	Winkler, MB	10	U4	AN7

+lp: USA 63

1580kHz

Call	Location	Power		
KCWW	Tempe, AZ 50		U2	AN7
KBLA	Santa Monica, CA	50	U4	AN6
KWYD	Colorado Springs, CO	10	D1	D
WTCL	Chattahoochee, FL	10	D1	D
WSRF	Fort Lauderdale, FL	10/5	U4	AN7
WEAM	Columbus, GA	2.3/1	U2	AN6
WTKT	Georgetown, KY	10	D3	D
KXZZ	Lake Charles, LA	1	U2	AN7
WPGC	Morningside, MD	50/0.27	U4	AN7
WZZJ	Pascagoula, MS	5	D3	D
KDZZ	Albuquerque, NM	5/0.06	U1	AN
WLIM	Patchogue, NY	10/0.5	U7	NH
WCTJ	Camp LeJeune, NC	10	D1	D
KGAL	Lebanon, OR	1	U3	AN
WDAB	Travelers Rest, SC	5/0.01	U1	AN
WAHI	Knoxville, TN	5/1	D1	D
WLIJ	Shelbyville, TN	5	D1	D
WPUV	Pulaski, VA 5/1		D1	D
CBJ	Chicoutimi, PQ	50	U3	AN

+lp: USA 50

1590kHz

Call	Location	Power		
WVNA	Tuscumbia, AL	5/1	U2	AN7
KLIV	San Jose, CA	5	U2	AN7
KXFS	Ventura, CA	5	U4	AN7
WPSL	Port St. Lucie, FL	5	D1	D
WRXB	St. Petersburg Beach, FL	5/1	U4	AN
WALG	Albany, GA 5/1		U4	AN7
WQCH	LaFayette, GA	5	D1	D*
WONX	Evanston, IL	3.5/1	U2	NH
WAIK	Galesburg, IL	5/0.55	U5	
WNTS	Beach Grove, IN	5/0.5	U4	NH
KVGB	Great Bend, KS	5	U2	NH
WASY	Gorham, ME	5/2.5	U4	
WJRO	Glen Burnie, MD	1	U4	AN7

Call	Location	Power		
WTVB	Coldwater, MI	5/1	U2	AN
KCNN	East Grand Forks, MN	5/1	U2	NH
WZRX	Jackson, MS	5/1	U2	AN7
KIHM	Sun Valley, NV	5	D1	D*
WSMN	Nashua, NH	5	U3	NH
WAUB	Auburn, NY	0.5/1	U4	AN
WASB	Brockport, NY	1	U4	NH
WGGO	Salamanca, NY	5	D1	D*
WHPY	Clayton, NC	5	D3	D*
WAKR	Akron, OH	5	U2	AN7
KMBD	Tillamook, OR	5/1	U2	NH
WCBG	Chambersburg, PA	5/1	U2	AN
WPWA	Chester, PA	2.5/1	U2	NH
WARV	Warwick, RI	5	U4	
WKTP	Jonesborough, TN	1.6/5	U4	AN
KELP	El Paso, TX	5/0.8	U4	NH
KYOK	Houston, TX	5	U2	AN7
KDAV	Lubbock, TX	1	U4	AN7
WFTH	Richmond, VA	5	D1	D*
KLFE	Seattle, WA	5	U2	AN7
WIXK	New Richmond, WI	5	D1	D*

+lp: USA 50

1600kHz

Call	Location	Power		
WEUP	Huntsville, AL	5/0.5	U2	AN7
WXVI	Montgomery, AL	5/1	U4	AN
KXEW	S. Tucson, AZ	2.5/1	U4	AN7
KNWA	Bellefonte, AR	5	D1	D
KGST	Fresno, CA	5	U2	AN7
KMNY	Pomona, CA	5	U2	NH
KUBA	Yuba City, CA	5/2.5	U2	AN
KYGO	Lakewood, CO	5	U2	AN7
WQVL	Dover, DE	5/1	U4	NH
WQOP	Atlantic Beach, FL	5	D1	D*
WPOM	West Palm Beach, FL	5/4.7	U3	AN7
WOKB	Winter Garden, FL	5	U4	AN6
WAOS	Austell, GA	5	D1	D
KCRG	Cedar Rapids, IA	5	U2	AN7
WINX	Rockville, MD	1/0.5	U2	NH
WUNR	Boston, MA	5	U3	AN7
WMRE	East Longmeadow, MA	2.5	U4	AN7
WAAM	Ann Arbor, MI	5	U4	AN
WMHG	Muskegon, MI	5	U2	AN
KWOM	Watertown, MN	5	U3	AN
KATZ	St. Louis, MO	5	U2	AN7
WWRL	New York, NY	25/5	U4	AN7
WGIV	Charlotte, NC	1	U2	AN7
WTZQ	Hendersonville, NC	5	D3	D*
KEED	Eugene, OR	5/1	U2	AN7
WBFD	Bedford, PA	5	D1	D*
KBOR	Brownsville, TX	1	U4	NH
KRVA	Cockrell Hill, TX	5/1	U4	AN
KOGT	Orange, TX	1	U4	AN6
KOKE	Pflugerville, TX	5/0.7	U4	
KCPX	Centerville, UT	5/1	U2	AN
WCPK	Chesapeake, VA	5/0.03	U1	AN7
WXMY	Saltville, VA	5	D1	D
WZZW	Milton, WV	5	D1	D*
WOHZ	Wheeling, WV	5	D1	D*
WCWC	Ripon, WI	5	U4	NH

+lp: USA 37

1610kHz

Call	Location	Power		
	Atlanta, TX	10/1	U1	F.Pl.

1620kHz

Call	Location	Power		
WPHG	Atmore, AL	10/1	U1	AN
KSMH	Auburn, CA	10/1	U1	F.Pl.

Call	Location	Power		
KBLI	Blackfoot, ID	10/1	U1	F.Pl.
WJVA	South Bend, IN	10/1	U1	F.Pl.
KAZP	Bellevue, NE	10/1	U1	F.Pl.
KNNT	Farmington, NM	10/1	U1	F.Pl.
	Minot, ND	10/1	U1	F.Pl.
WAZG	Myrtle Beach, SC	10/1	U1	F.Pl.
KAZW	College Station, TX	10/1	U1	F.Pl.
KYIZ	Renton, WA	10/1	U1	AN7

1630kHz

Call	Location	Power		
KBEG	Clovis, CA	10/1	U1	F.Pl.
WAWX	Augusta, GA	10/1	U1	F.Pl.
KCJK	Iowa City, IA	10/1	U1	F.Pl.
KBCM	Fort Worth, TX	10/1	U1	F.Pl.
KKWY	Fox Farm, WY	10/1	U1	F.Pl.

1640kHz

Call	Location	Power		
KDIA	Vallejo, CA	10/1	U1	AN7
WLHJ	Mount Airy, NC	10/1	U1	F.Pl.
KBFQ	Enid, OK	10/1	U1	F.Pl.
KKJY	Oswego, OR	10/1	U1	AN
	Sandy, UT	10/1	U1	F.Pl.
WKSH	Sussex, WI	10/1	U1	NH
KHFS	Fort Smith, AR	10/1	U1	F.Pl.

1650kHz

Call	Location	Power		
KKTR	Costa Mesa, CA	10/0.93	U1	AN7
	Denver, CO	10/1	U1	F.Pl.
WAZJ	Atlanta, GA	10/1	U1	F.Pl.
	Cedar Falls, IA	10/1	U1	F.Pl.
	El Paso, TX	8.5/0.85	U1	F.Pl.
WHKT	Portsmouth, VA	10/1	U1	F.Pl.

1660kHz

Call	Location	Power		
KAXW	Merced, CA	10/1	U1	F.Pl.
WMIB	Marco Island, FL	10/1	U1	F.Pl.
	Kansas City, KS	10/1	U1	F.Pl.
	Kalamazoo, MI	10/1	U1	F.Pl.
WBAH	Elizabeth, NJ	10/1	U1	AN7
	Charlotte, NC	10/1	U1	AN7
KQJD	West Fargo, ND	10/1	U1	F.Pl.
KAXY	Waco, TX	10/1	U1	F.Pl.
KXOL	Brigham City, UT	10/1	U1	AN7

1670kHz

Call	Location	Power		
KAZT	Redding, CA	10/1	U1	F.Pl.
WNML	Warner Robins, GA	10/1	U1	AN7
WAWR	Salisbury, MD	10/1	U1	F.Pl.
WTDY	Madison, WI	10/1	U1	AN7

1680kHz

Call	Location	Power		
KAVT	Fresno, CA	10/1	U1	F.Pl.
	Monroe, LA	10/1	U1	F.Pl.
	Ada, MI	10/0.68	U1	F.Pl.
WTTM	Princeton, NJ	10/1	U1	F.Pl.
KAZJ	Seattle, WA	10/1	U1	F.Pl.
KSXX	Roseville, CA	10/1	U1	F.Pl.

1690kHz

Call	Location	Power		
KAYK	Arvada, CO	10/1	U1	AN
WAZD	Adel, GA	10/1	U1	F.Pl.
WHTE	Johnston City, IL	10/1	U1	F.Pl.
WMDM	Lexington Park, MD	10/1	U1	AN7

1700kHz

Call	Location	Power		
	Huntsville, AL	10/1	U1	F.Pl.
WCMQ	Miami Springs, FL	10/1	U1	AN7
KBGG	Des Moines, IA	10/1	U1	AN
WAYU	Rochester, NH	10/1	U1	F.Pl.
KCHT	Astoria, OR	10/1	U1	F.Pl.
KQXX	Brownsville, TX	10/1	U1	F.Pl
KTBK	Denison-Sherman, TX	10/1	U1	F.Pl.

CENTRAL AMERICA AND THE CARIBBEAN
(incl. México)

Country Codes used in this section: AIA=Anguilla, ATG=Antigua & Barbuda, ABW=Aruba, ATN=Netherlands Antilles, BAH=Bahamas, BRB=Barbados, BLZ=Belize, CYM=Cayman Islands, CTR=Costa Rica, CUB=Cuba, DMA=Dominica, DOM=Dominican Rep., GDI-Guadeloupe, GRD=Grenada, GTM= Guatemala, HND=Honduras, HTI=Haiti, JMC=Jamaica, LCA=St. Lucia, MEX=México, MRT=Martinique, MSR=Montserrat, NCG=Nicaragua, PNR=Panama, PTR=Puerto Rico, SLV=El Salvador, TCA=Turks & Caicos, TRD=Trinidad & Tobago, VCT=St. Vincent & The Grenadines, VIR=Virgin Islands (American), VRG=Virgin Islands (British).-
Other abbreviations: *=inactive, n=nominal freq., (r)=reported, (r:)=relays station indicated, LV=La Voz, vf=variable frequency.
Others: see respective country listings.

530 kHz			
18	TICAL	CTR	R. Rumbo, Cartago (vf)
535			
	20/10	GRD	GBC Radio, Grenada
	100	TCA	R. Visión Cristiana Internacional, So. Caicos
540 kHz			
1	CMHV	CUB	Sancti Spíritus
5	CMCM	DOM	R. ABC, Sto Domingo
	TG-	GTM	R. Cobán, Cobán
1	HROW	HND	R. Atlántida, La Ceiba
150	XEWA	MEX	Super Estelar, Monterrey
10	XETIN		X-Bach AM, Tijuana
20	YNOW	NCG	R. Corporación, Managua
5	HOPU	PNR	R. Líder, Panamá
5	HOU23		R. Mía de Chiriquí, David (r:650)
5	YSHV	SLV	R. Restauración, San Salvador
550 kHz			
20/2	TISCL	CTR	R. Santa Clara, Cd. Quesada
30	CM--	CUB	Pinar del Río
10	CMDN		Guantánamo
1	CM--		Manzanillo
10	TGRV	GTM	R. 5-60, Guatemala
1	HRXT	HND	R. X, Tegucigalpa
0.5	HRXD		R. Manantial, Santa Rosa de Copán
10		JMC	R. Jamaica, Montego Bay
5	XEPL	MEX	La Super Estación, Cd. Cuauhtémoc
2/0.35	XEQW		QW La Poderosa, Mérida
1/0.25	XEKL		Primerísima 550, Jalapa
1/0.25	XEUC		R. Felicidad, Tehuantepec *
1	XEACD		R. Lobo, Acapulco
1	XEHLL		R. Mar, Salina Cruz
10	YNCH	NCG	R. 19 de Julio "la 19", Chinandega
5	WPAB	PTR	La RedAlerta de Puerto Rico, Ponce
2	YSFG	SLV	R. Variedades, Sonsonate
555 kHz			
10	ZIZ	SCN	ZIZ Radio, Basseterre, St. Kitts
560 kHz			
5	CMAA	DOM	R. Ritmos, Santiago
1	TG-	GTM	R. Quetzal, Malacatán
5	HRRZ	HND	R. Juticalpa, Juticalpa
1	HROS		R. Castilla, Tocoa
1	HROY		R. Jupiter, Comayagua *
1	HROT		R. Montserrat, Danlí
1	HRPX		R. Tropical, San Pedro Sula
5		JMC	JBC1, Kingston, Jamaica
5d	XESRD	MEX	La Ke Suave, Santiago Papasquiaro
5/0.75	XEOC		R. Chapultepec, México
5	XEMZA		La Buena Onda del Pacífico, Manzanillo
5	XEXZ		R. Cañon, Zacatecas
1	XEYO		R. Lobo/R. 5-60, Huatabampo
10	YN-	NCG	Managua (F.P.I.)
1	HOH2	PNR	RPC Radio, Colón (r:610)
570 kHz			
6.5	TIELR	CTR	R. 570 "es la radio", Guadalupe
10	CMDC	CUB	Pilón
10/5	CMMS	DOM	R. Cristal, Sto Domingo
1	TGPA	GTM	R. Palmeras, Escuintla
1	HROX	HND	R. El Triunfo, Choluteca
5	XEKZX	MEX	KZX Tejano 57, Cd. Guzmán
5	XEBJB		La Estación del Barrilito, Monterrey
5	XEOA		R. Oaxaca, Oaxaca
2.5	XEVX		R. Fórmula/La Grande de Tabasco, Villahermosa
1/0.25	XETJ		La Tropical de Oro, Torreón
1	XEVJP		Espacio 570, Xicotepec de Juárez
1	XELQ		R. Imagen, Morelia
1	XEME		R. Valladolid, Valladolid
1	XEUK		UK AM Stereo, Caborca
5	YN-	NCG	R. 5-70, Chinandega
1	HOS	PNR	R. Soberana Civilista, Panamá
5	YSKR	SLV	R. Cadena Central, San Miguel (r:570)
5	YSDR	V	R. Cadena Central, Sta Ana (r:570)
10	YSKT		R. Cadena Central, San Salvador
580 kHz			
5	CMDF	CUB	Baracoa
10	CMAM		Mantua
3	CMAF	DOM	R. Monte Cristi
3	TGY	GTM	R. Progreso, Guatemala
3	HRZQ	HND	R. Tegucigalpa, Tegucigalpa
1	HROU		R. Unión, Gracias
10		JMC	R. Jamaica, Galina
5/2.5	XEMU	MEX	La Rancherita del Aire, Piedras Negras
5/0.25	XEFI		R. Mexicana, Chihuahua
10/1	XEAV		Canal 58/R. Guadalajara, Guadalajara
1/0.25	XEYI		R. Festival, Cancún
1	XEDZ		Canal 58, Córdoba
1	XEUE		LV de Chiapas, Tuxtla Gutiérrez
1	XEHO		R. Centro, Cd.Obregón
1	XEHP		R. Exitos, Cd.Victoria
1	YNEA	NCG	R. 5-80, Managua 1000-
10	HOH4	PNR	RPC Radio, David (r:610)
10d	WKAQ	PTR	R. Reloj, San Juan
590 kHz			
40	TIRN	CTR	R. Nacional, San José *
30	CMHI	CUB	Santa Clara
10/5	CMDV	DOM	R. Santa María, La Vega
5	TGRQ	GTM	R. Quiché, Sta Cruz
1	HROV	HND	LV de Lepaguare, Juticalpa
1	HRLP3		R. América, San Pedro Sula (r:610)
5	XEFD	MEX	La Super F-D, Reynosa
1d	XEHQ		Globo 590, Hermosillo
10/5	XEPH		La Sabrosita, México
1	XEOM		R. ACIR, Coatzacoalcos
1	XEE	X	R. Festival, Durango
10	HOH3	PNR	RPC Radio, Chitré (r:610)
595 kHz			
10		DMA	Dominica Broadc. Corp, Hillsbrough
600 kHz			
150	CMKV	CUB	Urbano Noris
1	TGRC	GTM	R. Campesina, Escuintla
	TG-		R. Campesina, Mazatenango (r:600)
5/1	XECV	MEX	La Gran Compañía, Cd.Valles
5	XEHW		Digital 600, Rosario
5	XEBB		La Comadre, Acapulco
5	XEZ		R. 600, Mérida
10	XEGTO		R. Cañon, León
1	XEDN		La Mexicana, Torreón
1	XEMN		La Regiomontana, Monterrey
0.5	XETA		R. Sensación, Zitácuaro
10	YNLD	NCG	R. Ya, Managua
5d	WAEL	PTR	WAEL, Mayagüez
3	YSNK	SLV	La Preferida, San Salvador
610 kHz			
4	TIRSU	CTR	R. Sistema Universal, Guadalupe
1	CMAN	CUB	Bahía Honda
1	CMHI		Trinidad
5/1	CMJR	DOM	R. Acción, Santiago
0.5	TGGA	GTM	R. Alianza, Guatemala
10	HRLP4	HND	R. América, Santa Rosa de Copán (r:610)
10	HRLP		R. América, Tegucigalpa
0.5	4VJS	HTI	R. L'Eternel est Grand. Port-au-Prince
5/2.5	XEEL	MEX	Super Canal 610, Fresnillo
5/0.5	XEBX		La Primera Estación, Sabinas
5	XEUF		R. Variedades, Uruapan

1	XEKZ		LV del Istmo, Tehuantepec
1	XEJA		R. Fiesta, Jalapa
1	XEGS		R. Guasave, Guasave
10	HOHM	PNR	RPC Radio, Panamá
0.25/1	WEXS	PTR	WEXS, Patillas
50		TRD	NBS Radio, Port of Spain
620 kHz			
10		ATG	Antigua & Barbuda Broadc. Sce.
30	CMGN	CUB	Colón
1	CMKF		Moa
10	CMSD	DOM	R. Televisión Dominicana, Sto Domingo
5	TGPQ	GTM	R. 6-20, San Cristóbal
1	HRLP9	HND	R. América, Juticalpa (r:610)
1	HRLP5		R. América, Siguatepeque (r:610)
1	HRLP17		R. Continental, San Pedro Sula
5		JMC	JBC1, Mandeville, Jamaica
5/1	XEBU	MEX	La Norteñita, Chihuahua
5/1	XEOO		R. Triunfadora, Tepic
2.5	XEWZ		R. Novedades, San Luis Potosí
10/5	XENK		R. 6-20, México
1/0.5	XECK		Digital 620, Durango
1	XEHGR		Globo 620, Villahermosa
50	YNN	NCG	R. Nicaragua, Managua
630 kHz			
	TGEL	GTM	R. El Porvenir, Sta Elena
1	HRLP3	HND	R. América, Choluteca (r:610)
1	HRLP7		R. América, La Ceiba (r:610)
5/0.25	XETK	MEX	R. Hits, Mazatlán
5	XEACA		R. ACIR, Acapulco
5	XEJB		R. Gobierno, Guadalajara
5	XETS		R. Tapachula, Tapachula
20	XEFB		Radar FB, Monterrey
10/0.75	XEFU		LV Amiga de la Cuenca, Cosamaloapan
1	XEFX		La Profesionales de la Radio, Guaymas
1	XECCQ		R. Sensación, Cancún
1	XEERO		R. Tamaulipas, Altamira
1	HOJ35	PNR	R. Provincias, Chitré
5d	WSKN	PTR	Super Kadena Noticiosa, San Juan
10	YSLN	SLV	La Monumental, San Salvador
640 kHz			
20	TIAD	CTR	R. Rica, San José
50	CMBC	CUB	Guanabacoa
5	CMDD		Las Mercedes
10	CMDQ		Las Tunas
10	CMSD	DOM	R. Televisión Dominicana, Santiago (r:620)
40		GDL	RFO, Guadeloupe
10	TGW	GTM	LV de Guatemala, Guatemala
1	HRNN4	HND	Exitos, Tegucigalpa
5	XEJUA	MEX	Super Estelar, Cd.Juárez
20/2.5	XENQ		NQ La Vista/LV de la Provincia de la Provincia, Tulancingo
1d	XEWM		R. 6-40, San Cristóbal de las Casas
10	XEYQ		La Poderosa, Fresnillo
1	XETAM		Fiesta Mexicana, Cd.Victoria
1	XEHHI		R. Uno, Hidalgo del Parral
10	YNLN	NCG	R. Ranchera, Managua
1	HOK22	PNR	CPR, Colón
650 kHz			
1	CMDC	CUB	Media Luna
1	CMKU		Stgo de Cuba
15/10	CMAT	DOM	R. Universal, Sto Domingo
25	HRVW	HND	LV de Centroamérica, San Pedro Sula
15	HRLP6		R. América, Danlí (r:610)
1	HRLP8		R. América, Olanchito (r:610)
5/2	XEPX	MEX	LV del Angel, Puerto Angel
5/1	XEZM		La Expresión de Zamora, Zamora
2.5	XEVSS		Notiflash, Hermosillo
2.5	XETNT		R. 65/La Ley, Los Mochis
10	XEEJ	X	La Grande, Puerto Vallarta
1	XEVG		R. Fórmula, Mérida
	XECHH		Portavoz de la Amistad, Chilpancingo
	XEVLL		R. Felicidad, Villahermosa
12	YNRI	NCG	R. Septentrión, Matagalpa
10/8	YNRD		R. Diriangén "La Super D", Granada
10	HOS22	PNR	R. Mía "Cadena Nacional", Panamá
655 kHz			
10	YSSS	SLV	R. Nacional, San Salvador
660 kHz			
30	CMHG	CUB	Santa Clara
3	CMAM	DOM	Super Regional, Santiago
3	TGQ	GTM	R. Nal. "LV de Quetzaltenango", Quetzaltenango
3	HRNN18	HND	LV de Honduras, La Ceiba (r:670)
5/1	4VI	HTI	R. Lumière, Port-au-Prince
25		LCA	R. St. Lucia, Castries
50	XEDTL	MEX	La Candela 660, México
5/3	XEACB		ACB AM Numero Uno, Cd.Delicias
5	XEAR		La Mexicana, Tampico
20d	XEEY		La Consentida, Jalpa
10	XEFZ		FZ Metrópoli, Monterrey
1/0.5	XEWX		R. Mexicana, Durango
1	XECPR		R. Chan Santa Cruz-LV de los Mayas, Felipe Carillo Puerto
1	XEYG	MEX	R. Consentida, Matías Romero
1	HOF33	PNR	RPC Radio. Bocas del Toro (r:610)
670 kHz			
5	TITNT	CTR	R. Monumental, San José
50	CMQ	CUB	Arroyo Arenas
5	CMBS	DOM	R. Dial, San Pedro de Macorís
10	TGRT	GTM	Emisoras Unidas AM, Guatemala
10	HRN	HND	LV de Honduras, Tegucigalpa
1	HRNN20		LV de Honduras, Santa Rosa de Copán (r:670)
5d	XEIS	MEX	La Rancherita Consentida, Cd.Guzmán
5	XEOB		La Máquina Musical, Pichucalco
1	XETOR		R. Ranchito, Torreón
0.5	XESIC		R. Festival, Córdoba
YN-		NCG	R. Caribe, Puerto Cabezas
5	HOLY	PNR	R. Hogar, Panamá
680 kHz			
10	CMJV	CUB	Ciego de Avila
1	CMHN		Cienfuegos
1	CMDB		Stgo de Cuba
3	CMJX	DOM	R. Zamba, San Ignacio de Sabaneta
10	TGVP	GTM	R. Norte, Cobán
10	HRNN8	HND	LV de Honduras, San Pedro Sula (r:670)
10	HRNN2		LV de Honduras, Siguatepeque (r:670)
10	HRNN11		LV de Honduras, Tocoa (r:670)
1	HRNN7		LV de Honduras, Danlí (r:670)
1	HRNN10		LV de Honduras, Juticalpa (r:670)
5/0.5	XEKQ	MEX	R. Soconusco, Tapachula
1d	XEFO		Fiesta Norteña, Chihuahua
10	XELG		LG La Grande, León
1	XEORO		La Tremenda, Guasave
1	XESON		R. ACIR, Hermosillo
1	XECHG		R. Fiesta, Chilpancingo
1	XEFJ		R. Teziutlán, Teziutlán
10/2	YNAM	NCG	R. La Primerísima, Managua
5	HO	PNR	Em.Voz Sin Fronteras, Metetí
5	HOF32		Super Z Estéreo, David
10	WAPA	PTR	Guapa R, San Juan
690 kHz			
100/50		AIA	The Caribbean Beacon, Anguilla
20	CM--	CUB	Jovellanos
10	CMAW	DOM	R. Guarachita, Sto Domingo
1	TGHU	GTM	R. Escuintla, Escuintla
1	TGVB		R. Tamazulapa, Jutiapa
1	HRNN3	HND	LV de Honduras, Choluteca (r:670)
1	HRNN9		LV de Honduras, Tela (r:670)
77.5	XETRA	MEX	69 Extra Gold/Sport 6-90, Tijuana
5	XERG		La Deportiva 6-90, Monterrey
5	XECS		La Grande, Manzanillo
20/5	XEN		Ondas del Lago, México
2.5	XEUY		R. Sensación, Coatzacoalcos
10	XEMA		La Onda Musical del Altiplano, Fresnillo
10	XEST		R. Sensación, Mazatlán
5/2	YN-	NCG	R. Hermanos, Matagalpa
5	HOR43	PNR	R. Veraguas, Santiago
700 kHz			
10	TIJC	CTR	R. Sonora, San José
1	CMDU	CUB	Guantánamo
1	CMGA		Sancti Spíritus
1.5	CMDC	DOM	R. Mao, Valverde
15	TGHR	GTM	R. Mundial, Guatemala
5	HRGP	HND	Hacer Radio, Tegucigalpa
10		JMC	JBC1, Montego Bay, Jamaica
2.5	XEVC	MEX	Canal 70, Córdoba
1	XEPUJ		LV del Corazón de la Montaña, X'pujil
1	XEGD		La Poderosa, Hidalgo del Parral

kW	Call	Ctry	Station
1	XEDKR		R. Red, Guadalajara
10	YNMM	NCG	R. Istmo, Managua
5	YSUU	SLV	R. Cadena YSU, San Salvador
1	YSUU		R. Cadena YSU, Ahuachapán (r:700)
1	YSUU		R. Cadena YSU, San Miguel (r:700)
1	YSUU		R. Cadena YSU, Sonsonate (r:700)
1	YSUU		R. Cadena YSU, Sta Ana
1	YSUU		R. Cadena YSU, Usulután (r:700)

705 kHz

kW	Call	Ctry	Station
10		VCT	R. St. Vincent, Kingstown

710 kHz

kW	Call	Ctry	Station
50	CMHQ	CUB	Santa Clara
30	CMJN		Camagüey
150	CMW		La Julia
10	CMKJ		Holguín
1	TGXL	GTM	R. Tecún Umán, Quetzaltenango
3	HRRH	HND	LV de Occidente, Santa Rosa de Copán
3	HRKN		LV de Olancho, Catacamas
2	HRLK		R. Comayagua/LV Católica, Comayagua
1	HRUP		Estéreo Rey, San Pedro Sula
1	HRNN13		LV de Honduras, Olanchito (r:670)
5/0.25	XEYK	MEX	R. Mágica, Mérida
5	XEBL		Fiesta Mexicana, Culiacán
5	XERPO		R. Variedades, Oaxaca
3	XEON		R. Mexicana, Tuxtla Gutiérrez
10/1	XEDP		La Ranchera de Cuauhtémoc, Cd.Cuauhtémoc
10	XEMP		Alma Musical de México, Mexico
1/0.25	XERK		R. Korita, Tepic
1/0.1	XEPQ		La Sa Sa Sabrosita, Cd.Muzquiz
1	XEOLA		La Reina Grupera, Tampico
1	XERL		La Super R-L, Colima
1	XEPS		R. Amistad, Empalme
1	XELZ		R. Consentida, Torreón
1	XEMAR		R.-Felicidad, Acapulco
1	XESMR		R. Fórmula, San Luis Potosí
1	XEKV		R. Sensación, Villahermosa
5	HOQ51	PNR	KW Continente, Panamá
3.5	HOB52		Ondas del Caribe, Bocas del Toro
10/0.75	WKJB	PTR	KJB, Mayagüez

720 kHz

kW	Call	Ctry	Station
1	CMHC	CUB	Cienfuegos
5	CMAQ	DOM	R. Norte, Santiago
1	TGRO	GTM	R. Corona, Morales
1	HRNN3	HND	R. Caribe, La Ceiba
1	HRNG		Super Stereo Costa Sur, Choluteca
1	4VIA	HTI	R. Lumière, Carrefour Paye
10		JMC	R. Jamaica, Kingston
5/0.25	XEDE	MEX	Audio Digital 720, Saltillo
5	XEAVR		R. Fórmula, Veracruz
1d	XEQZ		Ritmo 7-20, San Juan de los Lagos
1	XEVU		720 Super Radio, Mazatlán
10	YNRC	NCG	R. Católica, Managua
5	HOB50	PNR	R. República, Chitré
50	YSRA	SLV	Circuito YSR, San Miguel (r:720)
1	YSRA		Circuito YSR, San Salvador
1	YSRA		Circuito YSR, Sta Ana (r:720)

730 kHz

kW	Call	Ctry	Station
20	TIHB	CTR	R. Reloj, San José
10	CMBB	CUB	La Fé
10	CMZ	DOM	R. HIZ, Broadcasting Nacional, Sto Domingo
10	TGN	GTM	R. Cultural, Guatemala
1	HRTG	HND	R. Televisión, Tegucigalpa
0.25	HRXG		R. Cadena Dial, Santa Bárbara
100	XEX	MEX	La X de México, México
10/1	XEVF		R. Villaflores/Stereo 730, Villaflores
0.5	XEGDL		La Tremenda 730, Zapotlanejo
	XEEBC		R. ACIR, Ensenada
10	YNNS	NCG	R. Segovia, Ocotal
20		TRD	R, Trinidad, Port of Spain

740 kHz

kW	Call	Ctry	Station
20	CMKP	CUB	Sagua de Tánamo
10	CMJL		Camagüey
10	ZGBC	DMA	Gospel Broadc. Corp, Dominica
1	TGHF	GTM	R. Tacaná, San Marcos
1	HRIH	HND	7-40 La Super, Juticalpa
1	HRNN23		R. Eco, San Pedro Sula
1	HRQQ		R. Intibucá, La Esperanza
1	4VIE	HTI	R. Lumière, Pignon-le-Jeune

kW	Call	Ctry	Station
1	4V–		R. Reveille, Cap Haïtien
5	XEHS	MEX	Audio A, Los Mochis
5	XEQN		R. Capullo 7-40, Torreón
10	XEPET		LV de los Mayas, Peto
1/0.5	XEVAY		La Mexicana, Puerto Vallarta
1	XELTZ		Super Fiesta, Aguascalientes
50	YNRS	NCG	R. Sandino "La S Grande", Managua
10	HOR44	PNR	CMQ, La Exitosa, Panamá
10	HON26		R. Cristal, David
10d	WIAC	PTR	WIAC, San Juan

750 kHz

kW	Call	Ctry	Station
1	CMHV	CUB	Trinidad
5	CMDB	DOM	R. Alegre, Santiago *
1	TGAJ	GTM	R. Tropicana, Escuintla
1	HRXK	HND	LV de la Mosquitia, Puerto Lempira
1	HRTU	HND	R. Trujillo, Trujillo
10		JMC	JBC1, Port Maria, Jamaica
5	XEJMN	MEX	LV de los Cuatro Pueblos, Jesús María
5	XEKOK		R. Sensación, Acapulco
1d	XEOH		Chiquitita, Camargo
1	XETI		La Fabulosa, Tempoal
1	XECSI		R. Mexicana, Culiacán
5	HO	PNR	R. Inolvidable, Chitré

760 kHz

kW	Call	Ctry	Station
20	TILX	CTR	Em. Columbia, San José
5	CMCD	CUB	Guanabacoa
5	CMCO	DOM	R. Cordillera, Sto Domingo
5	TGHB	GTM	Nueva R. Super, Guatemala
2.5	HRXW	HND	R. Comayagüela, Comayagüela
1	HRCG		R. Copán Galel, La Entrada
5	4VU	HTI	R. Lumière, Les Cayes
50/5	XEABC	MEX	La Estación de la palabra, México
5/0.1	XENY		R. Geny, Nogales
5	XEDGO		La Super Grupera, Durango
2.5/0.5	XEYW		La Tropiranchera, Mérida
10/0.25	XEZZ		R. Gallito, Guadalajara
1	XEEB		La Rancherita, Cd.Obregón
1	XERA		R. Chiapas, San Cristóbal las Casas
0.25d	XEEQ		R. Fiesta, San Luis Potosí
10	YN–	NCG	Ultravisión de Nicaragua, Managua
5	HOXO	PNR	LV del Istmo, Panamá
5d	WORA	PTR	Super Kadena Noticiosa, Mayagüez

770 kHz

kW	Call	Ctry	Station
1	CMMD	DOM	R. Popular, Santiago
1	TGBX	GTM	R. Fraternidad, Quetzaltenango
10	HRNN21	HND	R. Norte, San Pedro Sula (vf)
1	HRRO		R. Majestad "LV del Guayape", Juticalpa
0.5	HRMV		R. Aguán, Olanchito
10		JMC	R. Jamaica, Spur Tree
5	XEFTA	MEX	Fiesta 77, Los Mochis
5	XEQRV		La Costeñita, Veracruz
5	XEACH		R. Centro, Monterrey
1d	XEHB		R. Jilguerita, San Francisco del Oro
1d	XEML		R. Tropical, Apatzingán
10/5	XEIH		La Nueva Dimensión en Radio, Fresnillo
1.5	XESUR		R. Chilapa, Chilapa
1.5	HOL83	PNR	LV de Herrera, Chitré
10	YSKL	SLV	R. Cadena YSKL "La Poderosa", San Salvador

780 kHz

kW	Call	Ctry	Station
3	TIRA	CTR	R. América, San José
20	CMDE	CUB	Las Mercedes
0.5	CMBO	DOM	R. Constanza
1	TGCK	GTM	R. Sultana del Oriente, Zacapa
1	HRSE	HND	Estéreo Sol 2000, Choluteca
0.5	4V-	HTI	R. Lumière, Jérémie
5d	XELD	MEX	R. Costa, Autlán
5/1	XESFT		La Bronca Norteña, San Fernando
5	XEGLO		LV de la Sierra Guelatao de Juárez
1d	XEXY		LV del Balsas, Cd.Altamirano
1	XEMTS		R. Fórmula, Tampico
0.5d	XEZN		La Tremenda, Celaya
1	YNAD	NCG	R. Deportes, Managua
10	HOB55	PNR	R. Chiriquí, David
5	YSKL	SLV	R. Cadena YSKL, San Miguel (r:770)
1	YSKL		R. Cadena YSKL, Sonsonate (r:770)
1	YSKL		R. Cadena YSKL, Sta Ana (r:770)
1	YSKL		R. Cadena YSKL, Usulután (r:770)
10	ZBVI	VRG	ZBVI Radio, Tortola, B.V.I.

790 kHz

30	CMAQ	CUB	Pinar del Río (vf)
5	CML	DOM	LV del Trópico, Sto Domingo
3	TGO	GTM	R. Festival, Guatemala
3	HRTG2	HND	R. Satélite, Tegucigalpa
1	HRIR		R. Estéreo Relámpago, Santa Bárbara
50	XERC	MEX	El Fonógrafo, México
5/0.75	XENT		R. La Paz, La Paz
5/0.25	XERPC		R. Ranchito, Chihuahua
5	XEBI		R. B-I, Aguascalientes
1d	XEGZ		La Pantera, Torreón
10/0.25	XEVA		R. Hogar, Villahermosa
1/0.5	XEFE		R. Fiesta, Nuevo Laredo
1	XEUP		La Bonita de Oriente, Tizimín
1	XESU		R. 7-90, Mexicali
1	XEGAJ		R. 790, Guadalajara
1	XECOV		R. Lobo, Poza Rica
10	ACA20	PNR	AFRTS Southern Command Netw.

800 kHz

500/50	PJB	ATN	Trans World Radio, Bonaire
5	TIW	CTR	R. Tigre, San José
1	CMDT	CUB	Manzanillo (vf)
1	TGYZ	GTM	R. Rosa, Chiquimulilla
3	HRMA	HND	R. Mundial, San Pedro Sula
1	HRDL		R. Corporación, Comayagua (vf)
1	HRLP26		R. Sonora, Danlí
5	XEQT	MEX	La Poderosa, Veracruz
3	XEZV		LV de la Montaña, Tlapa de Comonfort
1d	XEAN		R. Alegría/R. Ocotlán, Ocotlán
150	XEROK		R. Cañon, Cd.Juárez
1	XEGX		R. Cañon, San Luis de la Paz
0.5	XEMMM		R. 80/Onda Ranchera, Tijuana
1	YN-	NCG	R. 800, Managua
15	HOL60	PNR	La Exitosa Provincias Centrales, Chitré
10	YSAX	SLV	LV Panamericana, San Salvador

810 kHz

1	ZNS3	BAH	R. Bahamas Northern Sce, Freeport
10	CMDW	CUB	Guantánamo
1	CMAV	DOM	R. Baní
	TG-	GTM	R. Moapán, Sta Elena
3	HRLP24	HND	R. Valle, Choluteca
1	HRVC		LV Evangélica, La Ceiba (r:1390)
5d	XERSV	MEX	R. Alegría, Cd.Obregón
50/35	XEFW		R. Estrella, Tampico
5/1	XERB		Sol Stereo, Cozumel
3/0.25	XEMAX		Radiomax, Armería
2.5	XEHT		R. Huamantla, Huamantla
2	XEMQ		R. Fórmula Mérida, Mérida
10/0.25	XEUX		La Legendaria, Tuxpán
1/0.25	XEZC		R. Felicidad, Río Grande
1/0.15	XEAGR		R. Fórmula, Acapulco
1	XEIN		LV del Valle, Cintalapa
1	XEOE		R. Amistad, Tapachula
1	XEIM		R. Capital, Saltillo
1	XEEMM		R. Salmantina, Salamanca
1	XERI		Realmente Música, Reynosa
5	YNRR	NCG	R. Rumbos, Rivas
10	HOG	PNR	R. Mundial, Panamá
50d	WKVM	PTR	AM-81, San Juan
2	YSFA	SLV	R. Lorenzana, San Vicente
1	YSDA		R. Imperial, Sonsonate

820 kHz

2.5	TIGC	CTR	R. Centro, Tibás
10	CMDE	CUB	Contramaestre
10	CMJT		Morón
10	CMCA		R. Ciudad de la Habana "La Emisora Joven", Santa Catalina, CH
5/1	CMAZ	DOM	R. Santiago
10	TGTO	GTM	R. Internacional, Guatemala
7/3	HRKW	HND	R. Sultana, Santa Rosa de Copán
5	HRLP16		R. Moderna, Tegucigalpa
3/0.1	XEZQ	MEX	R. Futurama, Huimanguillo
3.5d	XEYX		R. Hablado YX, Mexicali
2.5	XEKG		Golden Hits, Córdoba
1d	XESB		R. Mexicana, Santa Bárbara
10/1	XEBA		La Consentida, Guadalajara
1/0.5	XEDRD		La Poderosa, Durango
1	XEUDO		R. Universidad de Occidente, Los Mochis
1	XEYN		Sonovida, Oaxaca

0.75d	XEESC		R. Escárcega, Escárcega
20	YNOL	NCG	R. Ondas de Luz, Managua
10	HOF28	PNR	R. Ritmo, David

830 kHz

10	CMJB	DOM	R. HIJB, Sto Domingo
5	TGAV	GTM	R. Satélite, Mazatenango
1	HRJB	HND	Cadena Radial Impacto, Comayagua
1	HRVQ		R. Excelsior, Juticalpa
1	HRRU		R. Uno, San Pedro Sula
5	XEVQ	MEX	La Grande de Sinaloa, Culiacán
3	XELN		R. Linares/La Llegadora, Linares
10/0.5	XELK		R. Sensación, Zacatecas
10	XELA		Buena Música desde México, México
1	XEPUR		LV de los Purépechas, Cheran
10	YNRZ	NCG	R. Zinica "LV Costeña", Bluefields (vf)
50		SCN	R. Paradise, Charlestown, Nevis
5	YSPX	SLV	R. Pax, San Miguel

840 kHz

10	CMHW	CUB	Doblevé, Santa Clara, VC
1	CMBQ		La Fé
1	CMDQ		Las Tunas
1	CMKC		R. Revolución, Stgo de Cuba, SC
1	CMAB	DOM	R. Isabel de Torres, Puerto Plata
0.35	TGSM	GTM	LV de San Marcos, San Marcos
1	HRCR	HND	Dif.Cristiana de Radio "DCR", Choluteca
10	4VEF	HTI	R. 4VEH, Cap Haïtien
20		LCA	R. Caribbean International, Castries
5d	XEFG	MEX	R. Juventud, Celaya
5	XEXXX		Fiesta Mexicana/Fiesta Digital, Tamazula
5	XEOJN		LV de la Chinantla, San Lúcas Ojitlán
2.5	XEPV		La Fiera Grupera, Papantla
10d	XEIO		R. Capital, Tuxtla Gutiérrez
1	XEACC		LV del Puerto, Puerto Escondido
1	XETEY		R. Sensación, Tepic
0.5	XECUC		Casa de la Cultura de Campeche, Campeche
0.5	XEMY		La Cañerita, Cd.Mante
5	YNRN	NCG	R. Noticias, Managua
5d	WXEW	PTR	R. Victoria, Yabucoa
10	YSFB	SLV	R. Vanguardia, San Salvador

850 kHz

20	TIJLS	CTR	R. Viva, Pavas
5	CMDB	CUB	Baracoa
1	CMHL		Trinidad
5	CMUA	DOM	R. Clarín, Santiago (r:860)
5	CMGA		R. Guarocuya, Barahona
10	TGX	GTM	R. Ranchera, Guatemala
10	HRUP	HND	R. Centro, Tegucigalpa
0.5	HRIF		R. Inspiración, La Entrada
5		JMC	JBC1, Savanna La Mar, Jamaica
5/0.5	XEM	MEX	R. Exitos, Chihuahua
1d	XEMIA		R. ACIR, Guadalajara
1d	XEUS		R. Universidad de Sonora, Hermosillo
10	XETQ		La Q Orizabeña, Orizaba
1/0.25	XEZR		La Formula 85, Zaragoza
1	XEJAQ		R. Felicidad, Jalpan
0.5d	XEZF		La Rancherita Consentida, Mexicali
3	HO	PNR	CMQ, Colón (r:740)
10	HOT61		La Exitosa de Chiriquí, David
5/1	WABA	PTR	WABA/La Grande, Aguadilla

860 kHz

10	PJZ-86	ATN	R. Curom, Willemstad
10	CMBL	CUB	Arroyo Arenas
10		DMA	Voice of the Islands, Dominica
10	CMUA	DOM	R. Clarín, Sto Domingo
1	TGFP	GTM	R. Nal.Tikal, Flores Petén *
10	HRBS	HND	R. San Pedro, San Pedro Sula
0.5	HRLS		R. Dinorama, La Paz
3	4VMK	HTI	R. Men Kontre, Les Cayes
50	XEUN	MEX	R. UNAM-Universidad Nacional Autónoma de México, México
5	XEDU		D-U la que le gusta, Durango
5	XEDB		DB La Máquina Musical, Tonalá
5	XEMO		La Poderosa AM Estéreo, Tijuana
5	XECCN		R. Caribe, Cancún
5	XECTL		R. Chetumal, Chetumal
5	XEAL		R. Mundo, Manzanillo
5	XENL		R. Recuerdo, Monterrey
10	XETW		R. Fiesta/La Sabrosona, Tampico

Power	Call	Country	Station
1/0.15	XEZX		Voz del Usumacinta, Tenosique
1	XENW		La Radio de la Ciudad, Culiacán (vf)
1	XEPLA		R. Mexicana, Aguascalientes
1	XEZOL		R. Noticias 860, Cd.Juárez
5	YN-	NCG	Ultravisión de Nicaragua, Managua
5	HOL55	PNR	R. Reforma, Chitré
1	YSRC	SLV	R. Tecana, Sta Ana

870 kHz

Power	Call	Country	Station
10	TIUCR	CTR	R. Universidad de Costa Rica "R. U", San José
10	CMHL	CUB	Sancti Spíritus
1	CMDT		Pilón
5	CMVG	DOM	R. La Vega
0.5	TGL	GTM	R. Victoria, Mazatenango (vf)
5	HRH9	HND	R. Honduras, La Ceiba (r:880)
5	HRH10		R. Honduras, Puerto Lempira (r:880)
3	HRH4		R. Honduras, Nacaome (r:880)
1	4VJV	HTI	R. Tele Express Continental, Jacmel
1d	XELY	MEX	R. Moderna, Morelia
10	XETAR		LV de la Sierra Tarahumara, Guachochi
1/0.5	XEAMO		R. Amor, Irapuato
1	XEGRO		R. Guerrero, Acapulco
1	XEFIL		R. Maz, Mazatlán
1	XENG		R. Nueva Generación, Huauchinango
10	YNCC	NCG	R. Centro, Juigalpa
5	HOHO	PNR	R. Libre, Panamá
1	HOR56		R. Península, Macaracas
5d	WQBS	PTR	R. Voz, San Juan
1	YSAR	SLV	R. Cadena de Oro, San Salvador

880 kHz

Power	Call	Country	Station
30	CMAF	CUB	Pinar del Río
1	CMOR	DOM	AM-88, Mao
10	TGJ	GTM	R. Nuevo Mundo, Guatemala
5	HRMD	HND	R. Yoro, Yoro
10	HRH5		R. Honduras, Santa Rosa de Copán (r:880)
10	HRH		R. Honduras, Tegucigalpa
0.8	4VPM	HTI	R. Indépendence Nouvelle, Gonaïves
5d	XEPNK	MEX	Super Canal 88, Los Mochis
5	XEV		R. Fórmula, Chihuahua
5	XEYV		R. Variedades, Huatusco
3	XEQQQ		Super Q, Villahermosa
20/1	XEAAA		La Triple A/Noticias 880, Guadalajara
1d	XEEM		R. Alegría, Río Verde
10	XETC		R. Mayrán, Torreón
1	XEIG		RCN La Grande de Iguala, Iguala
5	YNAT	NCG	R. El Pensamiento, Managua
1	HOB51	PNR	R. Hit, Colón
1/0.5	WYKO	PTR	La Poderosa 880, Yauco
1	YSCD	SLV	R. Ritmo, Stgo de María

885 kHz

Power	Call	Country	Station
10/1		MSR	R. Montserrat

890 kHz

Power	Call	Country	Station
10	TIHOT	CTR	R. Fabulosa, San José *
10	CMDZ	CUB	Stgo de Cuba
1	CMHD		Santa Clara
4/5	CMPJ	DOM	R. Continental, Sto Domingo
10	HRH7	HND	R. Honduras, Juticalpa (r:880)
10	HRH3		R. Honduras, San Pedro Sula (r:880)
10	HRH9		R. Honduras, Siguatepeque (r:880)
1	HRUP5		R. Centro, Danlí (r:850)
1	HRH2		R. Honduras, Comayagua (r:880)
1	HRH6		R. Honduras, El Paraíso (r:880)
1	HRH8		R. Honduras, Olanchito (r:880)
5/1	XEPC	MEX	Sonido Estrella, Zacatecas
5	XEPOR		Por... Tavos de la Amistad, Putla de Guerrero
20	XEFRT		R. Frontera, Comitán
10/0.5	XENZ		Fiesta Mexicana, Culiacán
10	HOQ62	PNR	R. Ritmo Stereo, Chitré
0.25	WFAB	PTR	WFAB, Yguncos
3	YSLA	SLV	R. Musical, Sta Ana

895 kHz

Power	Call	Country	Station
0.5	4VVB	HTI	R. Trans-Artibonite, Gonaïves
10		SCN	Voice of Nevis, Charlestown, Nevis

900 kHz

Power	Call	Country	Station
50	CMKP	CUB	Cacocum
5/1	CMBN	DOM	R. Puerto Plata
1	TGMA	GTM	R. Amatique, Puerto Barrios
1	HRUP7	HND	R. Centro, Choluteca (r:850)
1	HRUP6		R. Centro, La Ceiba (r:850)
50	XEWB	MEX	LV de la América Latina/La W, Veracruz (r:900)
250	XEW	X	LV de la América Latina/La W, México
1	XETAK		Super K-90, Tapachula
5	YNRT	NCG	R. Tiempo, Managua
1	HOHA	PNR	LV del Pueblo, Panamá
2	YSQJ	SLV	R. El Tiempo, San Salvador

910 kHz

Power	Call	Country	Station
1	TIQM	CTR	R. Metrópolis Norte, Cd. Quesada
1	TIQM		R. Metrópolis Norte, Tibás (r:910)
10	CMGL	CUB	Bolondrón
10	CMFA		R. Cadena Agramonte, Camagüey, CM
1	CMLB	DOM	R. 91, Bonao
10	TGKL	GTM	R. 910 "La Em.de los Exitos", Guatemala
10	HRVS	HND	LV de Suyapa "R. Católica", Tegucigalpa
0.5	HRVH		R. Corona, La Entrada
5	XEACN	MEX	R. Metropoli, León
10	XENAY		La Buena Onda del Pacífico, Tepic
1	XEAO		R. Mexicana, Mexicali
5	YN-	NCG	R. Jinotega, Jinotega
1	HOL85	PNR	R. Nal Cristóbal, Colón
1	HOL81		R. Nal Guaymíe, David *
5	WPRP	PTR	Super Kadena Noticiosa, Ponce

920 kHz

Power	Call	Country	Station
1	CMGL	CUB	Unión de Reyes
10	CMBA	DOM	R. 9-20 AM-Stereo, Sto Domingo
0.2	TGRS	GTM	R. Sur, Escuintla
5	HRSK	HND	R. Catacamas, Catacamas
1	HRZV		R. Fabulosa, San Pedro Sula
1	HRH11		R. Honduras, Danlí (r:880)
1	HRRM		R. Sistema, Comayagua
1	4VKB	HTI	R. Cap-Haïtien, Petite Anse (r:930)
5/1	XEBH	MEX	Los Correcaminos, Hermosillo
1	XECQ		La Rancherita de Culiacán, Culiacán
5	XEBM		La Rancherita, San Luis Potosí
5	XERE		R. Cañon, Celaya
5	XETAA		R. Exitos, Torreón
2.5	XESDD		Onda 92/R. Mil, Ensenada
10/0.25	XELT		R. Escucha, Guadalajara
10	XEVV		R. Diversión, Tuxtla Gutiérrez
1	XEQD		La Divertida, Chihuahua
1	XEMJ		La Nueva MJ del 92, Piedras Negras
1	XEOK		R. ACIR, Monterrey
1	XEPNX		R. Costa, Pinotepa Nacional
1	XEOP		R. Estelar 920, Monclova
1	XEZAR		R. Fiesta, Puebla
1	XELCM		XELCM, Cd.Lázaro Cárdenas
10	YNW	NCG	R. Mundial, Managua
5	HOS56	PNR	R. Mía de Los Santos, Los Santos (r:650)

930 kHz

Power	Call	Country	Station
5	TIRCR	CTR	R. Costa Rica, San José
10	CMJS	CUB	Ciego de Avila
10	CMDP		Las Tunas
1	CMGB		Cárdenas
1	CMKN		Stgo de Cuba
10	CMCK	DOM	Ondas del Yaque, Santiago
5	TGJL	GTM	R. Imperial, San Pedro Carchá (vf)
10/3	4VKB	HTI	R. Cap-Haïtien, Cap-Haïtien
1	XEU	MEX	La U de Veracruz, Veracruz
5	XEMK		R. Mexicana, Huixtla
2.5	XEUL		La Barracuda/Tu Nueva Onda, Progreso
10/5	XEQS		Romance en Radio, Fresnillo
1	XETTT		Caliente AM, Colima
0.7	XETLA		LV de la Mixteca, Tlaxiaco
3	HOR46	PNR	CMQ, La Chorrera (r:740)
2.5	HOK85		Mi Preferida Estéreo, Pto Armuelles
2.5	WEKO	PTR	NotiUno, Cabo Rojo
5	YSTG	SLV	R. Cadena Sonora, San Salvador
1	YSTG		R. Cadena Sonora, Ahuachapán (r:930)
1	YSTG		R. Cadena Sonora, San Miguel (r:930)
1	YSTG		R. Cadena Sonora, Sonsonate (r:930)
1	YSTG		R. Cadena Sonora, Sta Ana (r:930)
1	YSTG		R. Cadena Sonora, Usulután (r:930)

940 kHz

Power	Call	Country	Station
10	CMKD	CUB	Holguín
1	CMGU		Colón
5	TGTL	GTM	LV del Hogar, Guatemala
1	HRBO	HND	R. Cadena Occidental, Santa Rosa de Copán

1	HRCR		R. Dif.Cristiana de Radio "DCR", Tegucigalpa
0.5	4VLF	HTI	R. Saint Marc, Saint-Marc
0.2	4VFF		R. Diff. Jacmélienne, Jacmel
50	XEQ	MEX	La Q 940 AM Stereo/Fonoteca 9-40, México
1d	XERKS		R. Cañon, Reynosa
1	XEWV		Fiesta Mexicana, Mexicali
1	XEREC		R. 9-40, Reforma
1	YNAD	NCG	R. Deportes, Managua
10d	WIPR	PTR	La Emisora del Puebo, San Juan

950 kHz

1	CMBL	CUB	R. Metropolitana, Arroyo Arenas, CH
1	CMHI		Sancti Spíritus
10	CMG	DOM	R. Popular, Sto Domingo
1	TGAF	GTM	R. Indiana, Mazatenango
1.2	HRZE	HND	R. Cortés, Puerto Cortés
1	HRQL		Centro Radial Hondureño, Siguatepeque
0.5	HRPS		R. Sistema Popular, Danlí
5	XEMEX	MEX	950 La Mexicana, Cd.Guzmán
5	XECEL		R. Lobo, Celaya
2.5	XEZE		La Cotorra de Tu Radio, Santiago Ixcuintla
1d	XERN		R. Naranjera, Monterey
10/5	XEKAM		R. California, Tijuana
10	XEYJ		R. 95, Nueva Rosita
10	XEPB		R. Amor, Hermosillo
1	XEMAB		Canal Internacional, Cd.del Carmen
1	XECAA		La Doble A, Aguascalientes
1	XEORF		R. Exitos, Los Mochis
1	XEFA		R. Rama/La Poderosa, Chihuahua
1	XETUG		Sonovida, Tuxtla Gutiérrez
1	XETO		Tu Recuerdo, Tampico
1	HOL84	PNR	R. Nal Victoriano Lorenzo, Penonomé
0.5	HOM33		R. Cadena Azul "RCA", La Concepción
1	YSHG	SLV	R. Mundo, San Miguel

960 kHz

10		ABW	R. Victoria, Oranjestad, Aruba
5	TICS	CTR	Continente Estéreo, Sto Domingo de Heredia
10	CMDJ	CUB	Guantánamo
1	CMHJ		Cienfuegos
1	CMGF		Matanzas
0.25	CMJD		Ciego de Avila
5/1	CMFF	DOM	LV del Atlántico, Puerto Plata
1	TGRU	GTM	R. Utatlán, Sta Cruz del Quiché
1	HRYF	HND	R. Fergusón, Choluteca
1	HRRD3		R. Sangrelaya, Sangrelaya
0.3	4VCD	HTI	R. Carillon, Port-au-Prince
5/1	XEK	MEX	La Estación Grande de Nuevo, Laredo
5	XEROO		La Pirata del Caribe, Chetumal
5	XETAP		La Tropical de Oro, Tapachula
5	XECZ		Perfil 960, San Luis Potosí
1/0.5	XEIQ		R. ACIR, Cd.Obregón
1/0.5	XEUQ		R. Variedades, Zihuatanejo
1/0.25	XEHK		LV de Guadalajara, Guadalajara
1	XEMM		R. ACIR, Morelia
1	XEKS		R. Exitos, Saltillo
1	XEOZ		R. Festival, Jalapa
1	XEGB		R. Fiesta, Coatzacoalcos
0.25	XECC		XECC, Cd.Camargo
2.5	YN-	NCG	LV del Trópico Húmedo, San Carlos
10	HOK71	PNR	Onda Popular, Panamá
0.5/1	WKVN	PTR	WKVN, Isabela
0.5	YSTW	SLV	R. Centro, Sonsonate

970 kHz

5	CMVP	DOM	R. Olímpica, La Vega
5	TGAX	GTM	R. Continental, Guatemala
2	HRTL	HND	Tic Tac, Tegucigalpa
0.25	HRAS		R. Señorial, Ocotepeque
50	XEDF	MEX	R. Fórmula/R. Noticias, México
5/2.5	XEVT		Stereorey, Villahermosa
5/0.25	XEVOX		Fiesta Mexicana/Fiesta Digital, Mazatlán
5	XEMH		La Grande, Mérida
5	XEEZ		La Super Z, Caborca
10/5	XEJ		R. Mexicana, Cd.Juárez
10	XEZAZ		De Mil Amores 9-70, Zacatecas
1	XEMF		La Tremenda, Monclova
1	XEBJ		R. 9-70, Cd.Victoria

1	XECJ		R. Apatzingán, Apatzingán
1	XEO		R. Gallito, Matamoros
0.5	XEUG		R. Universidad de Guanajuato, Guanajuato
1	HOS97	PNR	Ondas Centrales, Santiago
5	YSMS	SLV	UTEC-R. Universidad Tecnológica, San Salvador
5/1	WSTX	VIR	WSTX, Christiansted, St. Croix

980 kHz

20	TIRI	CTR	R. Cordillera, Esparza
1	CMDE	CUB	Bayamo
1	CMCK		COCO-El Periodico del Aire, Sapo, CH
1	CMKR		LV del Níquel, Moa, HO
5	CMFA	DOM	LV de las Fuerzas Armadas, Sto Domingo *
1	TGMQ	GTM	R. Retama, San Marcos
2	HRZC	HND	R. Monumental, San Pedro Sula
1	HRRD2		R. Emperador, Campamento
1	HRYG		R. Tocoa, Tocoa
5/0.5	XENR	MEX	La N-R, Nueva Rosita
5	XELC		Dual Stereo, La Piedad
5	XEQO		R. Romance, Cosamaloapan
10	XETU		R. Tampico, Tampico
1	XEQG		Canal 98, Querétaro
1	XEKE		KE-98 Acción Digital, Navojoa
1	XEFQ		LV de la Ciudad del Cobre, Cananea
1	XEXT		R. Fiesta, Tepic
1	HO	PNR	R. Mía de Colón, Colón (r.650)

990 kHz

5	CMSA	DOM	R. Cibao, Santiago
1	TGAL	GTM	R. Perla de Oriente, Chiquimula
2.5	HRPR	HND	R. Paz "LV Católica", Choluteca
0.2	4VCPS	HTI	R. Cacique, Port-au-Prince
5	XECL	MEX	La L de Mexicali, Mexicali
5	XEER		R. Fórmula, Cd.Cuauhtémoc
20	XETG		L Grande del Sureste, Tuxtla Gutiérrez
2.5	XEIU		Estéreo Crystal, Oaxaca
150	XET		La T Grande de Monterrey, Monterrey
10	XEFP		R. Alegría, Jalpa
1/0.25	XEHZ		R. 9-90, La Paz
1	XEATM		A Toda Máquina, Morelia
1	XEBC		La Divertida, Cd.Guzmán
1	XEUM		R. ACIR, Valladolid
1	XEOL		R. Impacto, Teziutlán
3	HOU44	PNR	R. Monumental, San Miguelito *
1	WPRA	PTR	R. Mil, Mayagüez
1	YSAT	SLV	R. Upa "La radio de los niños", San Salvador

1000 kHz

10/1	TIMIL	CTR	R. Mil, San José
10	CMAC	CUB	R. Guamá, Los Palacios, PR
1	CMJB		Camagüey
1	CMHB		Sancti Spíritus
5/1	CMHB	DOM	R. Beller, Dajabón
1	HRXZ	HND	Unión R. , Tegucigalpa
0.5	HRMH		LV del Junco, Santa Bárbara
50	XEOY	MEX	R. Mil "La Estación de la Ciudad", México
5/0.25	XEMYL		Super MYL, Mérida
1d	XETAC		Audio Mil de Chiapas, Tapachula
1	XEMIL		La Comadre, Los Mochis
1	XEFV		La Rancherita, Cd.Juárez
1	XECSV		La Tremenda, Coatzacoales
1	XENLT		Laredo Radio, Nuevo Laredo
1	XEMMS		R. Ranchito, Mazatlán
1	YNFF	NCG	R. Mil, Managua
10/5	HOK36	PNR	R. Poderosa "La Reina del Espacio", Aguadulce
1	YSHH	SLV	Estación H, Sta Ana
5/1	WVWI	VIR	WVWI, St. Thomas *

1010 kHz

3	PJC-7	ATN	R. Hoyer 1, Willemstad
10	CMAP	CUB	R. Guamá, Guane, PR
1	CMKM		Holguín
10	CMJA	DOM	R. Comercial, Sto Domingo
1	CMJA		R. Comercial, Salcedo (r:1010)
1	CMJA		R. Comercial, San Juan (r:1010)
1	TGXI	GTM	R. Ixil, Nebaj
1	TGRX		R. Miramundo, Zacapa
1	HRCD	HND	R. Constelación. Juticalpa

1	HRLP23		R. Moderna, El Progreso
5d	XEWS	MEX	R. Capital, Culiacán
50/5	XEHL		La Poderosa, Guadalajara
5	XEVK		Sonovida, Torreón
2	XEDX		R. Variedades, Ensenada
1/0.25	XETX		Voz de la Sierra Madre Occidental, Nuevo Casas Grandes
1	XEFM		La Máquina Tropical, Veracruz
1	XELO		R. Lobo, Chihuahua
0.5	XEKD		K de Oro, Cd.Acuña
0.5	XEXN		R. Ures, Ures
5	YNHG	NCG	R. LV del Pinar, Ocotal

1015 kHz

3	HOL86	PNR	LV del Teribe, Bocas del Toro

1020 kHz

2	TITIC	CTR	R. Mil Veinte "LV de la Liberación", San José (vf)
1	CMAP	CUB	R. Guamá, Bahía Honda, PR
10	CMTS	DOM	R. Enriquillo, Tamayo
5	TGCM	GTM	R. Frontera, Pajapita
1	HRUW	HND	R. Michelle, La Ceiba
1	HRMP		R. Moropocai, Nacaome
3	4VJH	HTI	R. Pétion-Ville, Pétion-Ville
2	XEGF	MEX	R. Fiesta, Gutiérrez Zamora
1d	XEWO		R. Chetumal, Chetumal
1d	XEXL		R. Láser, Pátzcuaro
1	XEVE		La Comadre, Colima
1	XEKH		R. Centro, Querétaro
10	HO	PNR	R. Ancón, Panamá
1	WPJC	PTR	R. Gigante, Adjuntas
10	YSCA	SLV	R. Internacional, San Salvador (vf)
20		TCA	Caribbean Christian Radio, Grand Turk

1030 kHz

1	CMAX	CUB	R. Guamá, La Palma, PR
10	CMDL	DOM	R. Novedades, La Vega
10	TGUX	GTM	R. Panamericana, Guatemala
1	HRUP3	HND	Estéreo Mil, Tegucigalpa
1	HRYF		R. Ticante, Ocotepeque
50	XEQR	MEX	R. Centro, México
5/1	XELJ		R. Central, Lagos de Moreno
5	XEIE		R. Alegría, Matehuala
4	XEVFS		LV de la Frontera Sur, Las Margaritas
10	XEMPM		R. Fama, Los Mochis
1	XEZON		LV de la Sierra Zongolica, Zongolica
1	XEBCC		La Gaviota Musical del Golfo, Cd. del Carmen
1	XEVP		R. Fiesta, Acapulco
2	YNLL	NCG	R. Masaya, Masaya
10d	WOSO	PTR	WOSO/El Oso, San Juan
1	YSRM	SLV	R. Frontera, Ahuachapán

1040 kHz

5	TIAC	CTR	R. Fides, San José
1	CMKT	CUB	R. Victoria, Puerto Padre, LT
10	CMON	DOM	R. Cadena de Noticias, Sto Domingo
1	TGJP	GTM	R. Oriental, Jalapa
3	HRNN22	HND	Exitos, San Pedro Sula
10/5	HRZX		La Perimerísima, Olanchito
1	HRFX		R. Musical, Catacamas
1	HRMJ		R. Renovación, Comayagua
1d	XEGR	MEX	R. Favorita, Jalapa
10/1	XEBBB		R. Mujer, Guadalajara
1	XEHES		La Poderosa, Chihuahua
1	XEJAQ		R. Felicidad, Jalpan
1	XETRE		R. Palenque 10-40, Palenque *
1	XETLX		R. Tlaxiaco, Tlaxiaco
2	YNVJ	NCG	LV de Jinotega, Jinotega
2.5	HOJ2	PNR	Ondas del Canajagua Stereo, Las Tablas
0.25	WZNA	PTR	WZNA, Moca

1050 kHz

10	CMKT	CUB	R. Victoria, Las Tunas, LT
10	CMCB	DOM	R. Hispaniola, Santiago
5/1	TGSL	GTM	LV de los Cuchumatanes, Huehuetenango
1	HR	HND	Estéreo Ceiba, La Ceiba
5	XEJF	MEX	R. Variedades, Tierra Blanca
35/2.5	XEQOO		R. Pirata, Cancún
100	XEG		La Gigante del Norte, Monterrey
10/1	XEBCS		R. Cultural Surcalifornia, La Paz
10	XED		La Gran D, Mexicali
1	XEURM		R. Mexicana, Uruapán

	XEZUM		R. Poder, Chilpancingo
	XEDC		XEDC, Aguascalientes
3	YSQR	SLV	Auto R. , San Salvador

1060 kHz

1	TILX	CTR	Em. Columbia, Liberia (r:760)
1	TILX		Em. Columbia, San Isidro del General (r:760)
5	CMDX	CUB	CMKS-Trinchera Antiimperialista, Baracoa, GU
1	CMKT		R. Victoria, Amancio Rodríguez, LT
1	CMKF	DOM	R. Azua
1	CMRV		R. Mar, San Pedro de Macorís
10	TGT	GTM	R. Sonora, Guatemala
2	HRVW	HND	LV de Centroamérica, Tegucigalpa (r:650)
0.5	HRFA		R. Peña Blanca, Santa Barbara
100/20	XEEP	MEX	R. Educación, México
1	YN-	NCG	LV del Atlántico, Bluefields
1	YNJJ		R. Juvenil, Managua
3	HOJ60	PNR	LV de Panamá, Panamá
5/0.5	WCGB	PTR	Iniciativa Mil 60, Juana Díaz

1070 kHz

10	CMKS	CUB	CMKS-Trinchera Antiimperialista, Guantánamo, GU
1	CMAS		R. Guamá, Pinar del Río, PR
5/1	CMBI	DOM	HIBI R. 1070, San Francisco de Macorís
3/2	TGD	GTM	LV de Occidente, Quetzaltenango
3	HRGR	HND	Cadena Guaymuras, El Paraíso
2.5	HRLP26		R. Siguatepeque, Siguatepeque 1055-
1	HRLE		R. 1050, San Pedro Sula
1	HRXM		R. Meridiano, Choluteca 1200-
5/1	XESP	MEX	R. Juventud Unica, Guadalajara
5	XEANT		LV de las Huastecas, Tancanhuitz de Santos
1/0.25	XEGY		Tecno Radio, Tehuacán
1	XEMI		La M Grande, Minatitlán
1	XEOBS		R. Mexicana, Cd.Obregón
1	XEEI		R. Mexicana, San Luis Potosí
1	XEIT		Un Amor Frente Al Mar, Cd. del Carmen
	XEAGS		XEAGS, Acapulco
2/2.5	WMIA	PTR	R. Arecibo, Arecibo
1	YSAN	SLV	LV de los Ausoles, Ahuachapán

1080 kHz

19	TIFC	CTR	Faro del Caribe, San José
1	TIFC		Faro del Caribe, Chomes (r:1080)
1	TIFC		Faro del Caribe, La Marina (r:1080)
1	TIFC		Faro del Caribe, Liberia (r:1080)
10	CMCH	CUB	R. Cadena Habana, Güines, HA
1	CMMC	DOM	R. Ambar, Sto Domingo
1	TGLU	GTM	R. Novedad, Zacapa
1	HRID	HND	R. Miramar, Téla
20	4VDN	HTI	R. Nationale, Port-au-Prince
50	XECAQ	MEX	R. Fórmula, Cancún
5	XEDY		R. Gallo, San Luis Río Colorado
5	XEGM		Sistema XEGEM "La Radio de la Gente", Buena Vista (r:1600)
2	XESAC		La Tremenda Tropical, Saltillo
10	XEXK		1080 R. Mundo, Poza Rica
1	XEDT		La Divertida, Cd.Cuahtémoc
1	XECN		Lasser Hits, Irapuato
1	XEUU		R. Variedades, Colima
1	HOJ24	PNR	R. Mil, Panamá *
0.25	WLEY	PTR	R. Ley/Motivos 1080, Cayey
5	YSME	SLV	R. 1080, San Salvador
1	YSME		R. 1080, San Miguel (r:1080)

1090 kHz

1.5	CMCH	CUB	R. Cadena Habana, La Salud, HA
1	CMKO		R. Angúlo, Moa, HO
1	CMAP		R. Guamá, Santa Lucia, PR
3/1	CMJM	DOM	R. Amistad, Santiago
1/0.5	CMRB		R. Jimaní
10	TGZ	GTM	Emisoras Unidas AM, Guatemala
1	HRNN27*	HND	Exitos, Santa Rosa de Copán (r:640)
1	HRWC		R. Aeropuerto Internacional, Tegucigalpa
1		JMC	JBC1, Morant Bay, Jamaica
5d	XEMCA	MEX	La Emisora de las Huastecas, Pánuco
50	XEPRS		R. Express, Tijuana
5	XELB		R. La Barca, La Barca
2.5	XEAU		R. AU, Monterrey

10/0.25	XEFC		Estéreo Romantica, Mérida
1	XEIL		La Comadre, Veracruz
1	XEWL		Nostalgia, Nuevo Laredo
1	XEHR		R. ACIR, Puebla
5	YN-	NCG	R. Alma Latina, Estelí
5	HOL82	PNR	R. Urracá Civilista, Santiago
0.25	WSOL	PTR	R. Sol, San Germán
3	YSMG	SLV	R. 1090, Atiquizaya (vf)
0.25	WGOD	VIR	WGOD, St. Thomas

1100 kHz

10		ATG	R. ZDK, St. John's, Antigua
0.25	PJL-3	ATN	R. Caribe, Willemstad
5	TISCR	CTR	R. Chorotega, Santa Cruz
1.5	TIVAL		R. Guápiles, Guápiles
1	CMKO	CUB	R. Angúlo, Manes, HO
	CM--		Varadero
1	CMPS	DOM	R. Nagua
1	CMMP		R. Ocoa, San José de Ocoa
1	CMHD		R. Oriente, San Pedro de Macorís
1	TGSR	GTM	R. Superior, Coatepeque
1	HRVS	HND	LV de Suyapa "R. Católica", Juticalpa
1	HRND		R. Esperanza, La Esperanza
1	HRVL		R. Lux, Olanchito
1	HRVA		R. Tiempo, San Pedro Sula
5/0.5	XETGO	MEX	R. Alegría Digital, Tlaltenango
5	XEBV		R. Alegría, Moroleón
1	XEPO	X	R. ACIR, San Luis Potosí
2.5	HO	PNR	Stereo Suave, Panamá
3	YSRF	SLV	R. Ranchera, San Salvador

1110 kHz

10	CMKO	CUB	R. Angúlo, Holguín, HO
2.5	CMTC	DOM	R. Jarabacoa
1	CMOS		R. Marién, Dajabón
1	TGMK	GTM	R. Verapaz, Cobán
1	HRLP25	HND	R. Sur, Choluteca
0.5	HRME		R. El Patio, La Ceiba
5d	XELEO	MEX	La Rancherita, León
50/25	XERED		R. Red, México
1d	XEWR		11-10 Stereomania, Cd.Juárez
1d	XEOQ		La Pantera Cumbiambera, Reynosa
1d	XEACM		R. Exitos, Cárdenas
1d	XEVS		R. Festival, Hermosillo
1/0.5	XEES		R. Alegría, Chihuahua
1	XEPVJ		R. Vallarta, Puerto Vallarta
1	YNMT	NCG	R. Momotombo, La Paz Centro
2.5/0.5	WVJP	PTR	R. Caguas, Caguas
2.5	YSCL	SLV	R. Horizonte, San Miguel (r.1160)
1	YSCL		R. Horizonte, Sta Ana (r:1160)

1120 kHz

1	PJE-3	ATN	R. Statia, St. Eustasius
3	TISHB	CTR	R. Reloj, Liberia (r:730)
5	CMCH	CUB	R. Cadena Habana, Artemisa, HA
1	CMKO		R. Angúlo, Mayarí, HO
10	CMCN	DOM	R. Metro 1120 AM Stereo, Sto Domingo
0.5	TGC	GTM	R. Uno 120 AM, Guatemala
2	HRYL	HND	R. Fiesta, Tegucigalpa
1	HRAV		Ondas del Ulúa, Santa Barbara (r:1150)
1	HRDG		R. Oriental "RCO", Danlí
1d	XETR	MEX	R. Panorámica, Cd.Valles
1/0.5	XEUNO		R. Centro, Guadalajara
1/0.25	XERUY		R. Universidad, Mérida
1/0.1	XEZB		La Tremenda, Oaxaca
1	XEMX		Sonido de Mexicali, Mexicali
0.5d	XEGV		R. ACIR, Villa del Pueblito
5	YNCP	NCG	R. CEPAD "El Arco Iris Del Amor", Managua
5	HOM21	PNR	R. Sonora, Panamá
2.6/5	WMSW	PTR	La Gran W, Arecibo
3	YSLR	SLV	La Romántica "LV del Amor", San Salvador

1130 kHz

5	CMKO	CUB	R. Angúlo, Sagua de Tánamo, HO
1	CMHA		Santa Clara
10/1	CMRL	DOM	R. Cadena de Noticias, Santiago
1	TGVR	GTM	LV de la Costa Sur, Retalhuleu
5	HRPL	HND	R. Progreso, El Progreso
1	HRHP		R. Pinares, Siguatepeque 1155-
1	HRBT		R. San Francisco, San Francisco de la Paz
1d	XEHN	MEX	R. Fiesta, Nogales

1d	XEFN		R. Moderna, Uruapan
10	XEZL		La Tropical, Jalapa
10	XETOL		R. Lobo, Toluca
1	XELUP		R. Lupita, Compostela
1	HOU80	PNR	R. Sensación, Aguadulce
0.2/0.7	WOIZ	PTR	R. Antillas, Guayanilla
1	YSLG	SLV	R. Chaparrastique, San Miguel
1	YSAJ		R. Moderna, Sta Ana

1140 kHz

1	TITNT	CTR	R. Monumental, San Carlos (r:670)
1	TITNT		R. Monumental, San Isidro del General (r:670)
5	CMCG	CUB	Loma de la Cruz
1	CMKX		R. Bayamo, Media Luna, GR
5	CMRA	DOM	R. Anacaona, San Juan de la Maguana
1	HRAP	HND	R. Mercurio, Choluteca
1	HRUN		R. Palmeras, La Ceiba
5d	XEXF	MEX	R. ACIR, León
50	XEMR		La Quebradita, Monterrey
1d	XETE		R. Manantial, Tehuacán
1	XELIA		La Tremenda, Morelia
1	HOB49	PNR	R. Panamericana, Panamá
10d	WQII	PTR	Once Q, San Juan
10	YSTS	SLV	R. El Mundo, San Salvador

1150 kHz

10	CMKX	CUB	R. Bayamo, Entronque Bueycito, GR
5	CMAS	DOM	Onda Musical, Sto Domingo
1	TGRR	GTM	R. Fiesta, Guatemala
5	HRAV	HND	Ondas del Ulúa, Santa Bárbara
1	HRLP12		R. Universal, Tegucigalpa
5	4VAB	HTI	R. Caraïbes, Port-au-Prince (r. 1146)
5d	XEXP	MEX	XEXP, Tuxtepec
5/0.5	XERRF		R. Juvenil, Mérida
5	XESO		La Poderosa, Cd.Obregón
20	XECMQ		Formato 21, México
20	XEAD		R. Metrópoli, Guadalajara
1/0.5	XERTM		R. Variedades, Macuspana
1.5	XETVR		Azul 11-50, Tuxpam
1	XERM		Baja R. Mexicali, Mexicali
1	XEJS		JS Digital, Hidalgo del Parral
1	XEBF		R. San Pedro, San Pedro
5	YNUW	NCG	R. Darío, León
1	HOS25	PNR	R. Mía Deportes, Panamá
1	YSCF	SLV	Ondas Orientales, San Miguel

1160 kHz

3	TILX	CTR	Em. Columbia, Puntarenas (r:760)
1	CMKX	CUB	R. Bayamo, Pilón, GR
5/1	CMBG	DOM	Radiolandia, Santiago
1	TGRI	GTM	R. Izabal, Morales
5	HRQN	HND	LV del Atlántico, Puerto Cortés
1	HRQL2		R. Sensación, Siguatepeque
0.5	HRGF		R. El Paraíso, El Paraíso (vf)
5/0.25	XEBE	MEX	R. Perote, Perote
2.5	XEVW		R. Sensación, Acámbaro
1d	XEIW		Canal Juvenil, Uruapan
1	XEGI		Reyna de las Huastecas, Tamazunchale
1	YNHM	NCG	R. Satélite, Estelí
3	HOWK	PNR	R. Metrópolis, Panamá
10	HOC20		Ondas Chiricanas, David
5/2.5d	WBQN	PTR	La Super B, Manatí
3	YSCL	SLV	R. Horizonte, San Salvador
1	YSRG		R. Corporación, Sta Ana

1165 kHz

10		ATG	Caribbean Radio Lighthouse, Antigua

1170 kHz

10	CMKS	CUB	CMKS-Trinchera Antiimperialista, Maisí, GU
5	TGRL	GTM	R. Cadena Landívar, Quetzaltenango
2	HRAF	HND	R. Atenea "La Internacional", Choluteca
1	HRAZ3		R. Hits, La Ceiba
10	4VRS	HTI	R. Soleil, Port-au-Prince
5d	XERT	MEX	La Rancherita, Reunosa
5	XEUVA		Uva Digital, Aguascalientes
2	XECD		R. 11-70 AM, Puebla
1d	XEJTF		Prisma Musical, Zacoalco de Torres
1/0.5	XEJE		R. XEJE, Dolores Hidalgo
1	XEYH		R. Mil, Hermosillo
1	XEIB		R. Sensación, Caborca
1	XERLK		Super Stereo Miled, Atlacomulco
0.5	XEZS		R. Hit 11-70, Coatzacoalcos

5	YN-	NCG	R. Máxima, Masaya
0.25	WZUR	PTR	WZUR, Ponce
1175 kHz			
0.5	YSCB	SLV	LV del Pacífico, Sonsonate (n.1170)
1180 kHz			
20	TIPJ	CTR	R. Victoria, Heredia
50	CMBA	CUB	Villa María
1	CMDB		Mayarí Arriba
1	CMBM		Nueva Gerona (vf)
10	CMBE	DOM	R. Mil, Sto Domingo
10	TGT	GTM	R. Sonora, Guatemala
1	HRAZ	HND	La Exitosa, Tegucigalpa
1	HRCY		R. Estéreo Congolon, Gracias
10	XEFR	MEX	Oxido 11-80, México
10	XEUBS		R. Universidad Autonoma de Baja California Sur, La Paz
10	HOU84	PNR	R. Belén, Santiago
5	YSVE	SLV	R. VEA, San Salvador
1190 kHz			
1	CM	CUB	R. 26, La Caridad, MA
1	CMHT		R. Sancti Spíritus, Yaguajay, SS
10	CMAG	DOM	R. Azul, Santiago
1	HRGK	HND	R. Brassavola, Minas de Oro
1	HRPO		R. Santa María de la Luz, Gualaco
1	HRZQ		R. Tegucigalpa, San Pedro Sula (r:580)
0.5	HRFS		R. Familiar, Morazan
0.2	4VAF	HTI	La Voix de la Grande-Anse, Jérémie
50	XEWK	MEX	La W de Guadalajara, Guadalajara
5	XETOT		La Poderosa, Tampico
5	XEPP		Super Mil Con El Poder de la Música, Orizaba
2d	XEDO		R. Lobo, Cuernavaca
1d	XECT		Nucléo Deportes 11-90, Monterrey
1d	XEPZ		R. Norteña, Cd.Juárez
1d	XESOL		R. Sol, Cd.Hidalgo
10	XERV		R. Villa, Villahermosa
0.5	XEMBC		La Picosita, Mexicali
10/5d	WBMJ	PTR	WBMJ, San Juan
1200 kHz			
5	TIAM	CTR	R. Cucú, San José
10	CMKC	CUB	R. Revolución, Palma Soriano, SC
1	CMHT		R. Sancti Spíritus, Trinidad, SS
0.25	CMBS		R. Ariguanabo, San Antonio de los Baños, HA
1	CMMR	DOM	R. Caracol, Azua
1	TGRJ	GTM	R. Jutiapa, Jutiapa
1	HRDS	HND	R. Nacaome, Nacaome
1/0.3	XEPW	MEX	La Tremenda, Poza Rica
1	XEZI		Canal Festivo 120, Zacapu
1	XEWT		La Nueva WT, Culiacán
1	XEQY		La Tremenda, Toluca
1	XEAGA		R. Felicidad, Aguascalientes
	XEYF		R. Fórmula, Hermosillo
1	YNAD	NCG	R. Democracia, Managua
1	WGDL	PTR	R. Grito, Caguas
0.05	AFCN		AFRTS, Roosevelt Roads, Puerto Rico
10	YSMM	SLV	R. Fiesta, San Salvador
1	YSMM		R. Fiesta, Sta Ana (r:1200)
1	YSKJ		R. Sirama, San Miguel
1205 kHz			
1		CYM	R. Cayman, George Town, Cayman Isl.
1210 kHz			
10	CMHT	CUB	R. Sancti Spíritus, Sancti Spíritus, SS
1	CMKC		R. Revolución, Chivirico, SC
1	CMKC		R. Revolución, Mayarí Arriba, SC
5	CMCJ	DOM	R. Merengue, San Francisco de Macorís
10/5	TGMX	GTM	R. Rumbos/Coco Radio, Guatemala
1	HRRO	HND	R. Capital, Comayagüela
1	HRSI		R. Impacto, Téla
1	HRHO		R. Maya, Santa Rosa de Copán
1	4VLS	HTI	R. Plus, Port-au-Prince
50/10	XEBCO	MEX	R. Occidente/La Poderosa Voz de Colima, Colima
10	XEBD		R. Centro/R. Punto, Jalapa
4	HOE91	PNR	R. Diez, Panamá
5d	WHOY	PTR	R. Hoy, Salinas
1	YSCG	SLV	R. La Paz, Zacatecoluca
1220 kHz			
1	TIQ	CTR	R. Casino, Limón
10	CMGY	CUB	R. 26, Central España, MA

	CM--		Cienfuegos
10	CMN	DOM	R. HIN, Sto Domingo *
1	TGMT	GTM	R. Amiga, Antigua
10/1	HRGW	HND	R. Patria, Catacamas
1	HRQO		R. Internacional, San Pedro Sula
1	HRYS		R. Suari, Marcala
0.5	HRJM		R. Sava, Sava Colón
2.5d	XEZAJ	MEX	Sono Ritmo 12-20, Guadalajara
100	XEB		La B Grande, México
1230 kHz			
3	CMGJ	CUB	R. 26, Unión de Reyes, MA
1	CMNG	DOM	R. Ideal, Moca
1	TGAT	GTM	R. Atlántida, Puerto Barrios
	TG-		R. América, Cuyotenango
10	HRQW	HND	R. Téla, Téla
0.25	HRSM		R. Samaritano, San Marcos de Colón
1	4VAV	HTI	R. La Voix de l'Avé-Maria, Cap Haïtien
2.5	XEID	MEX	R. Alamo, Alamo
1/0.5	XETBH		R. Tabasco, Villahermosa
1	XEEX		R. ACIR, Culiacán
0.25	XELP		R. Pia, La Piedad
5	YNNG	NCG	R. Manantial, Nueva Guinea
1	WNIK	PTR	WNIK/R. Unica, Arecibo
1240 kHz			
1	ZNS2	BAH	Inspiration 1240, Nassau, Bahamas
1	TILX	CTR	Em. Columbia, Nicoya (r:760)
10	CMGW	CUB	R. 26, Bolondrón, MA
5/0.5	CMCV	DOM	R. Barahona
1	CMAU		R. Revelación, Puerto Plata
5	TGK	GTM	R. Sensación "R. Luz", Guatemala
1	HRZC	HND	R. Monumental, Tegucigalpa
1	HRVN		R. Venus, Santa Bárbara
10	4VSJ	HTI	R. Antilles Internationales, Port-au-Prince
2.5	XEOV	MEX	La Picosa, Orizaba
2.5	XECE		R. Hit 12-40, Oaxaca
1/0.25	XEBQ		Estelar 12-40/R. Mexicana, Guaymas
1/0.25	XEBN		Radiola, Cd.Delicias
1	XES		Energía 12-40, Tampico
1	XERZ		Heavy Radio, León
1	XEIZ		La Potranquita, Monterrey
1	XEWG		Mágico 12-40, Cd.Juárez
1	XESI		R. Afirmación, Santiago Ixcuintla
1	XEPK		R. Fiesta, Pachuca
1	XECG		R. Mexicana, Nogales
1	XERPA		R. Ranchito, Morelias
1	XEVM		R. Sensación, Piedras Negras
1	XELM		R. Sensación, Tuxtla Gutiérrez
1	YN-	NCG	R. Restauración, Managua
5	HO	PNR	R. BB, Panamá
1	HOM56		Faro de David
1/5	WALO	PTR	Walo/R. Oriental, Humacao
1	YSQR	SLV	R. Norteña, San Miguel
0.5	YSMT		R. Metapan, Sta Ana
1250 kHz			
1	CMHW	CUB	Doblevé, Caibarién, VC
0.25	CMKS		R. Bayamo, Imías, GU
5	CMRS	DOM	El Sonido del Este, La Romana
5	CMBC		LV del Progreso, San Francisco de Macorís
1	TGPY	GTM	R. Payakí, Esquipulas
1	HRCC	HND	R. Cadena Continental, Comayagua
1	HRYN		R. Latina, Danlí (vf)
1	HRAT		Super R. , San Pedro Sula
0.5	HRQV		R. Subirana, Yoro
1	4VS	HTI	La Voix du Plateau Central, Hinche
5	XEDL	MEX	R. Ambiente, Hermosillo
5	XESJ		R. Saltillo, Saltillo
5	XEDK		Radiorama DK, Guadalajara
2.5/0.5	XEPI		La Consentida, Chilpancingo
1/0.25	XEMG		La Poderosa, Arriaga
1	XEJX		Antena 12-50, Querétaro
1	XEZT		Fórmula Stereo Z, Puebla
1	XETF		La Jarocha, Veracruz
1	XESC		R. 1250, Sabinas
1	XEAT		R. Onda, Hidalgo del Parral
0.5	XETEJ		Sistema XEGEM "La Radio de la Gente", Tejupilco
5	YN-	NCG	Radial Samaritano, Condega
5	HOLY	PNR	R. Hogar, Penonomé

1/0.25d	WJIT	PTR	WJIT, Sabana
1260 kHz			
6	TIHN	CTR	R. Emaus, San Vito de Coto Brus
5	CMBF	CUB	Arroyo Arenas
1	CMGH		R. Victoria de Girón, Torriente
1	CMT	DOM	R. Recuerdos, Sto Domingo
1	TGPA	GTM	R. Monumental, Antigua
1	HRZR	HND	R. 1260, La Ceiba
1	HRYF2		R. San Marcos, San Marcos de Colón
5/1	XEXR	MEX	R. Mensajera, Cd.Valles
5	XEJAM		LV de la Costa Chica, Santiago Jamiltepec
5	XESA		R. Hits, Culiacán
10	XEL		R. ACIR, México
10	XEJY		R. Sistema del Suroeste/LV del Valle, Autlán
1	XEWGR		Estéreo Vida, Monclova
1	XEQL		La Catedral de la Música, Zamora
1	XER		R. Linares, Linares
1	XEOG		R. Ranchito, Ojinaga
1	XEMW		R. San Luis, San Luis
0.5d	XEZH		La Estación que se oye, Salamanca
0.25	XEMTV		R. Lobo, Minatitlán
1	WISO	PTR	R. Wiso, Ponce
3	YSAA	SLV	R. América, San Salvador
1270 kHz			
1.5		ABW	R. 1270, San Nicolaas, Aruba
10	CMJG	CUB	Camagüey
5	CMDA	DOM	R. Hit, Santiago
1	CMTA		R. Ambiente, Baní
2.5	TGCQ	GTM	R. Exclusica, Guatemala
1	HROF	HND	Ecos del Celaque, Lempira
1	HRNQ		R. Sonora, Danlí (r:1270)
1	HRNQ		R. Sonora, Tegucigalpa
5/0.5	XEAX	MEX	Express 12.7, Oaxaca
5	XEAZ		R. Zeta 13, Tijuana
2	XERRT		R. Ranchito, Cd.Madero
10	XERPL		La Poderosa, León
1/0.5	XERRR		R. Ritmo, Poza Rica
1	XEGL		La Pionera del Radio, Navojoa
1	XEQH		R. Sinfonia, Ixmiquilpán
0.5d	XEHD		Universidad Juárez del Estado de Durango, Durango
0.5	XEZU		R. Fiesta, Zacapu
0.5	XEWN		R. Variedades, Gómez Palacio
3	YNRA	NCG	R. Amistad, Matagalpa *
0.5	HOJ22	PNR	R. CMQ, Panamá
1	YSQZ	SLV	R. Mía, San Miguel
1280 kHz			
2	TIHT	CTR	R. Alajuela, Alajuela
1	CMKW	CUB	R. Mambí, Stgo de Cuba, SC
1	CMJC		R. Rectángulo, Guáimaro, CM
1	CMGF		Varadero
1	CMHZ	DOM	R. Clave, Monte Plata
2.5	TGVY	GTM	R. Zamaneb, Salamá
1	HRBU	HND	R. Digital, San Pedro Sula
1	HRAM		R. Olanchito, Olanchito
1	HRBN		R. San Miguel, Marcala
10	4VAM	HTI	R. Métropole, Port-au-Prince
2.5	XECAM	MEX	LV de las Murallas, Campeche
2	XEAG		R. Capital, Córdoba
10	XEAW		La Macroestación, Monterrey
1/0.5	XEBW		Canal 12-80, Chihuahua
1	XEKY		Dimensión 12-80, Huixtla
1	XEEG		La Tribuna/La Gigante, Puebla
1	XETUT		R. Tamaulipas, Tula
1	XEQP		R. Triunfadora, Guadalajara
0.5	XESQ		R. San Miguel. San Miguel de Allende
5/1	WCMN	PTR	R. Centro/La Grande de Arecibo, Arecibo
1	YSQV	SLV	R. Galaxia, Sta Ana
1290 kHz			
5	CMCS	CUB	La Pastora
1	CMHW		Doblevé, Rancho Veloz, VC (vf)
1	CMBQ		La Habana
1	CMVM	DOM	R. Bonao
0.5	TGTU	GTM	R. Nal.de Totonicapán, Totonicapán
1	HRNN26	HND	R. Choluteca, Choluteca
1	HRGS		R. HRGS, Utila
5	XEFAC	MEX	La Mera Mera, Salvatierra
2.5	XEQIN		LV del Valle, San Quintín
1d	XEIY		R. Juventud, Río Verde
10	XENX		La Poderosa NX, Mazatlán
10	XEDA		R. 13, México
1	XEIX		La Pantera, Sahuayo
1	XEAP		R. Sensacion, Cd.Obregón (vf)
0.25	XETH		R. Palizada, Palizada
10	HOS23	PNR	R. Guadalupe, Panamá
1	YSMA	SLV	R. Chalatenango, Chalatenango
0.5	WRRA	VIR	WRRA, Frederiksted, St. Croix
1300 kHz			
1	PJD-2	ATN	PJD-2 Radio, Philipsburg
7.5	TILC	CTR	R. La Fuente Musical, Cartago
1	CMDA	CUB	Las Tunas
1	CMKQ	DOM	R. Radio, Sto Domingo
	TG-	GTM	R. Miramundo, Zacapa
10	HRLR	HND	R. Santa Rosa, Santa Rosa de Copán
1	HRLH		LV de la Amistad, Tegucigalpa
1d	XESW	MEX	R. Madera, Cd.Madera
10	XEXV		Stereo R. 13, León
1	XEP		R. 13, Cd.Juárez
1	XELE		R. 13, Tampico
1	XEKW		R. Festival, Morelia
1	XEJL		R. Guamuchil, Guamuchil
1	XEXW		R. Norteña, Nogales
1	XEHU	X	R. Tropical, Martinez de la Torre
1	YNR	NCG	Canal 130 AM, Managua
1	HOI417	PNR	R. Baha'í, Boca del Monte
1	WTIL	PTR	R. Util, Mayagüez
5	YSLV	SLV	R. Cabal, San Salvador
1310 kHz			
10	CMBN	CUB	R. Caribe, Nueva Gerona, IJ
1	CMHR		Doblevé, Sagua La Grande, VC
1	CMMH	DOM	R. Real, La Vega
1	CMSD		R. Televisión Dominicana, El Seibo (r:620)
1	TGAN	GTM	R. LV de los Altos, Quetzaltenango
2.5	HRVC	HND	LV Evangélica, San Pedro Sula (r:1390)
1	HRRL		R. Libertad, Marcala, La Paz
0.5	HRJH		R. Colón, Tocoa
5	XEAM	MEX	La M Grande, Matamoros
5	XEHIT		La Tremenda, Puebla
5	XEVB		R. Alegría, Monterrey
10	XETIA		R. Contacto, Guadalajara
1/0.25	XERU		R. Universidad, Chihuahua
1	XEC		R. Enciso, Tijuana
1	XETRC		R. Gazeta de la Tierra Norte, Tepetzintla
1	XEHY		R. Mexicana, Querétaro
1	XEFH		R. Plan de Agua Prieta, Agua Prieta
1	XEHV		R. Trópico, Veracruz
0.5	XEPO		R. Voz, San Luis Potosí *
0.25d	XEHJ		R. Petatlán, Petatlán
10/1	YNSC	NCG	R. San Cristóbal, Chinandega
5	YSRV	SLV	R. Veritas, Usulután
1320 kHz			
1		ABW	Voice of Aruba, Oranjestad, Aruba
1	TILX	CTR	Em. Columbia, San Carlos (r:760)
1	CMHA	CUB	Sancti Spíritus
1	CMDA		Stgo de Cuba
1	TGME	GTM	R. Quesada, Jutiapa
1	HRMG	HND	R. Bahía "La Super Grande", La Ceiba
1	HRGM		R. Ilusión, Choluteca
5	XERJ	MEX	R. Mazatlán, Mazatlán
1d	XEJZ		R. Variedades, Cd.Jimenez
10	XECPN		R. 1320, Monclova
10	XEJP		R. Variedades, México
1/0.25	XECY		R. Diversión, Huejutla
1	XEUI		R. Comitán, Comitán
1	XENI		R. Festival, Uruapán
1	XEUH		R. U-H, Tuxtepec
0.5	XESR		R. Cachanía, Santa Rosalia
5/1	WUNO	PTR	NotiUno, San Juan
1	YSAH	SLV	R. Emanuel, La Unión
1330 kHz			
1	CMAD	CUB	R. Jaruco, Artemisa, HA
3	CMFA	DOM	LV de la Fuerzas Armadas, Moca *
1	CMVR		R. Villa-La Sencilla, Sto Domingo
5.5	TGMU	GTM	Unión Radio "LV de la Esperanza", Guatemala
1	HRSW	HND	R. Evangélica, Tegucigalpa

1	HRFL		R. Florida, La Entrada
10	4VJLD	HTI	R. Haïti-Inter, Port-au-Prince
5/1	XEUAS	MEX	R. Universidad, Culiacán
5	XEBO		R. Variedades, Irapuato
1	XEWQ		La Superestación, Monclova
1	XERP		R. Principal, Tampico
1	XEUZ		R. Veracruzana, Martínez de la Torre
0.5d	XEEV		Alma Musical, Icúcar de Matamoros
0.5	XEAJ		La Revolución del Radio, Saltillo
0.5	XEAH		R. Hit, Juchitán
	XEMAC		La Turquesita, Manzanillo
1	HOM92	PNR	R. Sabrosa, Panamá
1/0.5	WENA	PTR	La Buena del Sur, Peñuelas
5	YSHQ	SLV	R. Progreso, San Salvador
1340 kHz			
6	TIHR	CTR	R. Sideral, San Ramón
10	CMHU	CUB	R. Ciudad del Mar, Palmira, CI
1	CMDO		R. Banes, Banes, HO
1	CMBZ	DOM	R. Centro, San Juan de la Maguana
10	TGCO	GTM	LV del Trópico, Coatepeque; 24h
10	HRHH	HND	R. El Mundo, San Pedro Sula
1	HRJC		R. Colonial, Comayagua
1	HRED		R. Red, Olanchito
5	XELU	MEX	R. Esmeralda, Cd.Serdán
2	XESL		R. Centro, San Luis Potosí
1d	XEQE		La Perla Camaronera, Escuinapa *
1/0.2	XECW		Sonido Zeta, Los Mochis
1	XEQB		La Divertida, Tulancingo
1	XECI		La Nueva 13-40, Acapulco
1	XENV		La Sabrosita, Monterrey
1	XEDH		La Tremenda Tropical, Cd.Acuña
1	XEMT		La Tremenda Tropical, Matamoros
1	XEASM		Q'Bonita, Cuernavaca
1	XEYR		R. 13, Teapa
1	XEAA		R. Ambiente, Mexicali
1	XEJK		R. Amistad, Cd.Delicias
1	XERPV		R. Festival, Cd.Victoria
1	XEDKT		R. Ranchito, Guadalajara
1	XEAPM		R. Romántica, Apatzingán
1	XEBY		R. Tuxpam, Tuxpam
1	XEOS		R. Variedades, Cd.Obregón
1	XECR		R. Variedades, Morelia
1	XEBK		Super Grupera, Nuevo Laredo
0.25	XERCH		R. Exitos, Ojinaga
1	YNOS	NCG	R. Ondas Sonora, Managua
1	WNOZ	PTR	R. Nosotros, Aguadilla
1	YSXW	SLV	R. Novedades, Usulután
1	WSTA	VIR	WSTA, St. Thomas
1350 kHz			
10	CMKY	CUB	R. Libertad, Puerto Padre, LT
1	CMHU		R. Ciudad del Mar, Aguada de Pasajeros, CI
1	CM		R. Portada de la Libertad, Holguín, HO
1	CMPM	DOM	R. Rutas Musical, La Romana
1	TGMC	GTM	R. Monja Blanca, Cobán
1	HRJV	HND	LV de San Lorenzo, San Lorenzo
5/0.5	XETB	MEX	R. Laguna, Torreón
5	XEQK		La Hora Exacta, México
5	XECAH		R. Chiapas/LV Soconusco, Cacahoatán
2.5	XECTZ		LV de la Sierra Norte, Cuetzalán del Progreso
1d	XETM		El Heraldo de la Frontera, Naco
0.5d	XELBL		R. Centro, San Luis
0.25	XEZD		La Doña de la Frontera, Camargo
1	YNGF	NCG	R. Ondas del Sur, Jinotepe
10	HOZ38	PNR	R. Cadena Millonaria, Panamá
2.5d	WEGA	PTR	WEGA/R. Las Vegas, Vega Baja
1360 kHz			
1	TITNT	CTR	R. Monumental, Limón (r:670)
1	TITNT		R. Monumental, Puntarenas (r:670)
1	CMFA	CUB	R. Cadena Agramonte, Rodolfo Ramírez Esquível, CM
1	CMSD	DOM	R. Televisión Dominicana, La Vega (r:620)
1	CMSD		R. Televisión Dominicana, Monte Cristi (r:620)
10	TGLK	GTM	R. Tic Tac, Guatemala
1	HRBS	HND	R. San Pedro, Tegucigalpa (r:860)
1	HRBH		R. Santa Bárbara, Santa Bárbara
5	4VRL	HTI	R. Liberté, Port-au-Prince
5	XEUD	MEX	La U de Tuxtla, Tuxtla Gutiérrez
1/0.25	XEY		R. Celaya, Celaya
1/0.1	XEXM		R. Jerez, Jerez de García Salinas
1	XEIK		La Super Invasora, Piedras Negras
1	XEDQ		R. Alegría, San Andrés Tuxtla
1	XEKF		R. Independiente, Iguala
1	XEDI		R. Sensación, Chihuahua
5/1d	WCHQ	PTR	C-H-Q, Camuy (F.PI. 1660 kHz)
1370 kHz			
1	CMFA	CUB	R. Cadena Agramonte, Nuevitas, CM
1	CMDV		R. Siboney, Stgo de Cuba, SC
5	CMFA	DOM	LV de las Fuerzas Armadas, Elías Piña *
5	CMRP		R. Seybo, El Seybo
1	CMSD		R. Televisión Dominicana, Barahona (r:620)
1	TGAC	GTM	LV de Colomba, Colomba
1.5	HRTR	HND	R. Danlí, Danlí (vf)
1	HRST		R. Fraternidad, San Pedro Sula *
1	HRSZ		R. Santa Bárbara, Siguatepeque
1	4VEE	HTI	R. Citadelle, Cap-Haïtien
1	4VMM	I	R. Diffusion Cayenne, Les Cayes
5	XEHF	MEX	R. Arizona, Nogales
5	XEPJ		R. Capital, Guadalajara
10	XEPA		La Super Fiera, Puebla
10	XEMON		R. Fórmula, Monterrey
1	XEA		La Grande A, Campeche
1	XERPU		Sonidoz, Durango
0.5d	XESV		R. Universidad Michoacana de San Nicolas de Hidalgo, Morelia
0.5	XEHG		R. Norteña, Mexicali
0.25d	XEGNK		R. Mexicana, Nuevo Laredo
1	YNGA	NCG	R. Matagalpa, Matagalpa *
1	YNRE		R. Somoto, Somoto
1	HOB64	PNR	R. Sitrachilco, Pto Armuelles
5/1	WIVV	PTR	WIVV Missionary R. St., Viequez Isl
1	YSKO	SLV	R. LV de la Amistad, San Miguel
1380 kHz			
1	TIMS	CTR	R. Guanacaste, Liberia
1	TIMS		R. Guanacaste, San José (r:1380)
10	CMFA	CUB	R. Cadena Agramonte, Central Brasil, CM
5/1	CMSC	DOM	R. Nacional, Santiago
1	CMSD		R. Televisión Dominicana, San Juan (r:620)
1	TGEB	GTM	R. Momostenango Educativa, Momostenango
1	HREJ	HND	R. Voz Evangélica, Choluteca
0.5	HRAH		R. Jutiapa, Jutiapa
1	4VSS	HTI	R. Port-au-Prince, Port-au-Prince
5	XETP	MEX	R. Sensación. Jalapa
5	XEGW		Super Tropical, Cd.Victoria
2.5	XEAK		R. Consentida, Acámbaro
10	XECO		Romántica 13-80, México
1/0.1	XEVD		R. Allende, Allende
1	XERS		R. Sinfonía, Torreón
1	XEKT		R. Variedades, Tecate
10	HO	PNR	R. América, Panamá
1	WOLA	PTR	R. Prócer, Barranquitas
1390 kHz			
1	CMBT	CUB	R. Jaruco, Jaruco, HA
1	CMAR	DOM	R. San Cristóbal
5	TGYC	GTM	R. Estrella, Guatemala
10/5	HRVC	HND	LV Evangélica, Tegucigalpa
1	HRVC		LV Evangélica, Santa Rosa de Copán (r:1390)
5/1	XETL	MEX	LV de la Huasteca, Tuxpam
5/1	XEXO		La Super Buena, Cd.Mante
5/1	XETY		R. Tecomán, Tecomán
5	XECTA		R. Cuantla, Cuantla Morelos
1d	XEQC		La Reina del Mar, Puerto Peñasco
10	XERW		Sonido 13-90, León
1/0.25	XERUY		R. Universidad, Mérida
1	XEOR		R. Gallito, Reynosa
0.25	XEZG		R. Mezquital y Huasteca Hidalguense, Ixmiquilpan
1	HOL	PNR	R. Super Sol, Colón
1	WISA	PTR	R. Noroeste, Isabela
1400 kHz			
12	TICJ	CTR	R. Sinaí, San Isidro del General
1	CMGX	CUB	Matanzas

1	CMFA		R. Cadena Agramonte, Guaimaro, CM
1	CMAC	DOM	Ondas del Valle, La Vega
1	CMSD		R. Televisión Dominicana, Pedernales
1	CMSD		R. Televisión Dominicana, San Juan (r:620)
5		GRD	Harbour Light of the Windwards, Grenada
1	TGRB	GTM	R. Porteña, Puerto Barrios
1	HRAU	HND	R. Alegre, Sava Colón
1	HRYT		R. Estrella de Oro, San Pedro Sula
1	HRJJ		R. Estéreo Punto, Comayagua
5	XEAC	MEX	La A Grande, Aguascalientes
1d	XELH		La Gardenia Musical, Acaponeta
1/0.25	XEKJ		La Tremenda, Acapulco
1	XEXI		La I de Ixtapan, Ixtapan de la Sal
1	XEPF		La Ranherita, Ensenada
1	XEOJ		R. Horizonte, Cd.Lázaro Cárdenas
1	XEFS		R. Matamoros, Izúcar de Matamoros
1	XEI		R. Morelia, Morelia
1	XEAB		R. Santana, Santa Ana
0.5	XEVI		Energía 14, San Juan del Río
0.5	XEUBJ		R. Universidad Benito Juárez, Oaxaca
0.25	XEWU		XEWU, Matehuala
5	HOT40	PNR	R. Luz, La Chorrera
1	WIDA	PTR	R. Vida, Carolina
1	YSJI	SLV	LV del Litoral, Usulután

1410 kHz

5	PJF-1	ATN	Voice of Saba, Saba
1	CMAL	CUB	Pinar del Río
1	CMJW		R. Santa Cruz, Santa Cruz del Sur, CM
3/0.5	CMCH	DOM	R. 14-10, Barahona
3	CMRM		R. Sol, Higüey
1/0.5	CMGG		R. Grí-Grí, Río San Juan
1	CMAE		R. Revelación en América, Sto Domingo
5	TGGH	GTM	R. Xelajú, Quetzaltenango
0.5	TGMS		R. Nal. "LV de Mazatenango", Mazatenango
1	HRDD	HND	LV de Atlántida, La Ceiba
1	HRSY		LV del Pacífico, San Lorenzo
1	4VTS	HTI	La Voix du Nord Ouest, Port-de-Paix
5/0.5	XETAB	MEX	R. Juvenil, Villahermosa
5	XEIR		Estelar 14-10, Cd.Valles
5	XESH		R. Sabinas, Cd.Sabinas
25/10	XEKB		Canal 14-10, Guadalajara
10	XEBS		R. Sinfonola, México
1/0.5	XECF		La Mexicana, Los Mochis
1/0.1	XEYD		R. Madero, Torreón
1	XEAS		La Tamaulipeca, Nuevo Laredo
3/1	YNRV	NCG	La Estación de la Amistad, León
5	HOH779	PNR	R. Mensabé, Las Tablas
10	ACB20		AFRTS Southern Command Netw.
2.5	WRSS	PTR	R. Progreso, San Sebastián

1420 kHz

5	TIRP	CTR	R. Pampa, Nicoya
1	CMHS	CUB	R. Caibarién, Caibarién, VC
0.25	CM		R. Llanuras, Colón, MA
15	CMFD	DOM	R. Oro, Cotuí
1	TGRP	GTM	R. Viva, Guatemala
1	HR	HND	LV de las Fuerzas Armadas, Comayagüela
1	HRSL		R. Actualidad, Santa Bárbara
5	XEF	MEX	Diamante Stereo, Cd.Juárez
5	XEH		La Tremenda Tropical, Monterrey
2	XEXX		Super XX, Tijuana
1d	XEKMX		La Super X, Sayula
10/1	XEWE		La Estación Familiar, Irapuato
1/0.25	XEJV		La Super Joven, Minatitlan
1/0.25	XEWJ		R. Popular, Tehuacán
1	XEWF		Fiesta Mexicana, Cuernavaca
1	XEEW		LV del Bajo Bravo, Matamoros
1	XERD		R. Lobo, Pachua
1	WEUC	PTR	R. Universidad Católica, Ponce
1	YSUCA	SLV	R. Universitaria, San Salvador

1430 kHz

10	CMJY	CUB	R. Surco, Morón, CA
3	CMSC	DOM	R. Emanuel, Santiago
1.2	TGAG	GTM	LV de Huehuetenango
1	HRIC	HND	La R. del 70, Puerto Cortés
1	HRVM		R. Maranatha, La Paz
1	HRSJ		R. Mundial, Tocoa

1	HRTP		R. Recuerdos, Juticalpa
10	4VGM	HTI	MBC, Port-au-Prince
5	XEOX	MEX	La Super Estación, Cd.Obregón
2	XEWD		La Grande, Cd.Miguel Alemán
10	XETT		R. Tlaxcala, Tlaxcala
1	XERAC		La Juvenil, Campeche
1	XECOC		La Querida, Colima
1	XECA		R. Exitos, Ixtepec
1	XELL		R. Onda, Veracruz
1	XESHT		R. Mexicana, Saltillo
5	YNLE	NCG	R. Liberación "La Tayacana", Estelí
5	WNEL	PTR	R. Tiempo, Caguas

1440 kHz

1		ABW	R. Kelkboom, Oranjestad, Aruba
12	TIRSC	CTR	R. San Carlos, Cd. Quesada
10	CMJP	CUB	R. Surco, Ciego de Avila, CA
0.25	CMBU		R. Güines, Güines, HA
5	CMAK	DOM	R. Cristocéntrico, Sto Domingo
5	CMAD		R. San Juan
1	CMFS		R. Bahía, Nagua (vf)
1	CMLF		R. Cayacoa, Higüey
5	HRRD	HND	Dimensión R, La Ceiba
0.5	HRRY		R. Mía, San Marcos de Colón
10/1	XECCC	MEX	Fiesta Mexicana, Guadalajara
10	XEEST		Sonido Cristal/Oro 14-40, México
1	XEVSD		La Señal del Progreso, Cd. Constitución
0.5	XENAC		LV de los Chontales, Nacajuca
25	YNRM	NCG	R. Maranatha, Managua

1450 kHz

1	CMJF	CUB	R. Maboa, Amancio Rodríguez, LT
10	CMAC	DOM	R. Util, Salcedo
1	TGLG	GTM	R. Epoca, Guatemala
1	HRBR	HND	R. Cultural, La Entrada
1	HRXZ2		R. Titania, Tegucigalpa
10	XEBP	MEX	R. Sensación, Torreón
1/0.5	XENA		R. Alegría, Querétaro
1/0.5	XEKM		R. Mina/Onda 14-50, Minatitlán
1/0.25	XEARE		R. Pegüis, Ojinaga
1/0.1	XERY		La Poderosa Voz del Sur, Arcelia
1	XEJM		CNN Headline News, Monterrey
1	XETD		Canal de la Juventud, Tecuala
1	XEPUE		Ella AM, Puebla
1	XEDJ		R. Clave, Magdalena
1	XECM		R. Festival, Cd.Mante
1	XESS		R. Fiesta, Ensenada
1	XEVH		R. Galáctica, Matamoros
1	XEGC		R. Impacto, Sahuayo
1	XECB		R. Ranchito, San Luis
1	XEJD		R. Tropicana, Poza Rica
1	XEYZ		R. Variedades, Aguascalientes
0.5	XEPY		La Tropical Ardiente, Mérida
	HO	PNR	R. 1450 AM Stereo, Panamá
1	WCPR	PTR	R. Coamo, Coamo
5	YSFM	SLV	Super R. , San Salvador *
1	YSKR		R. Restauración, San Miguel

1460 kHz

2	TILX	CTR	Em. Columbia, Limón (r:760)
1	CM	CUB	R. Cadena Agramonte, Solas
0.5	CMAN	DOM	R. Magua, Hato Mayor
5	TGRN	GTM	R. Petén, Flores Petén
2.5	HRGC	HND	R. Conga, San Pedro Sula
1	HRQX		Radiolandia, Comayagua
0.5	HRCX		LV de Patuca, Catacamas
0.5	HROC		R. Ranchera, Olanchito
0.2	4VEA	HTI	La Voix du Nord, Cap Haïtien
5/1	XEKC	MEX	Stereo Exitos, Oaxaca
1d	XEHE		LV de Jalisco, Atotonilco el Alto
10	XEIP		La Poderosa, Uruapán
10	XELX		R. Mexicana, Zitácuaro
1	XEGRA		14-60 La Estación Familiar, Acapulco
1	XEHX		La Consentida, Cd.Obregón
1	XECPQ		La Estrella Maya, Felipe Carillo Puerto
1	XEJH		R. ACIR, Jalapa
1	XEYC		R. Sensación, Cd.Juárez
0.25	XEXQ		R. Universidad de San Luis Potosí, San Luis Potosí
0.5	HOD42	PNR	LV de Almirante, Bocas del Toro
5d	WRRE	PTR	La Fabulosa, Juncos
0.5	WLRP		R. Raíces, San Sebastián
2.5		TCA	R. Turks & Caicos *

1470 kHz

1	CMGE	CUB	R. Ciudad Bandera, Cárdenas, MA
1	CMDE	DOM	R. Super 56, San Francisco de Macorís
0.5	HRSA	HND	R. Luz y Verdad, La Ceiba
1	4VAA	HTI	R. Arc-en-Ciel (R. Lakansyel), Port-au-Prince
6	XEQF	MEX	R. Loma, Loma Bonita
5	XERCN		RCN-R. Cadena Nacional, Tijuana
3/0.25	XEHI		El Heraldo Internacional, Cd.Miguel Alemán
20	XEAI		Radio A-I, México
2.5	XEBAL		R. Voz Maya de México, Bécal
1/0.25	XEACE		Stereo Exitos/La Gran Máquina Musical, Mazatlán
1	XECAV		Estéreo Imagen, Durango
1	XEIND		LV Sierra Hidalguense, Tlanchinol
1	XEYA		La Picosa, Irapuato
1	XECU		La Rancherita, Los Mochis
1	YNRY	NCG	R. Yarrince, Boaco
1/2.5d	WKCK	PTR	R. Cumbre, Orocovis

1480 kHz

1	TIAW	CTR	R. Sol, Puntarenas
0.25	CMJI	CUB	R. Florida, Florida, CM
5	CMAH	DOM	R. VEN, Sto Domingo
5	TGHB	GTM	R. Horizontes, Guatemala
1	HRMI	HND	LV de Misiones "R. MI", Comayagüela
1	HRWP		R. Soberanía, San Marcos, Ocotepeque
5	XEPR	MEX	R. Imagen, Poza Rica
5	XEVIC		R. Tamaulipas, Cd.Victoria
1d	XEHM		HM Radio, Cd.Delicias
1	XETKR		La TKR/Rancherita y Regional, Monterrey
1	XENS		R. Ambiente, Navojoa
1	XEXU		R. Variedades, Monclova
1	XEXC		Super Mil de Guerrero, Taxco
1	XEZJ		XEZJ El 1480 AM, Guadalajara
0.5d	XEOU		R. Joya, Huajuapan de León
5	WMDD	PTR	Sonido 14-80/R. El Conquistador, Fajardo

1490 kHz

1	CMDH	CUB	R. Mayarí, Mayarí, HO
0.25	CMBW		R. Camoa, San José de las Lajas, HA
1	CMAP	DOM	La 54 AM, Moca
1	CMSD		R. Televisión Dominicana, Puerto Plata (r:620)
1	TGDS	GTM	R. LV de Atitlán, Santiago Atitlán
1	TGRE		R. Modelo, Retalhuleu
1	HRRA	HND	R. Juventud, Sonaguera
1	HROM		R. Omega "Sonido Internacional", La Esperanza
1	HRER		R. Porteña, Puerto Cortés
2.5	XEKN	MEX	R. Variedades, Huetamo
1/0.25	XEED		ED Contigo, Ameca
1/0.25	XEDR		La Grande de Guaymas, Guaymas
1/0.2	XEXE		R. Hit 14-90, Querétaro
1	XEFF		Canal 14-90, Matehuala
1	XEPOP		La Comadre, Puebla
1	XERO		La Inolvidable, Aguascalientes
1	XECJC		La Pantera, Cd.Juárez
1	XECH		R. ACIR, Toluca
1	XEAQ		R. Internacional, Agua Prieta
1	XEMS		R. Mexicana, Matamoros
1	XEVZ		R. Sensación, Acayacan
1	XEYT		R. Teocelo, Teocelo
1	XEGT		Stelaris 14-90 AM, Zamora
0.25	XESK		La Super K, Cd.Ruiz
3	HOR45	PNR	CMQ, La Palma (r:740)
5/1	WLEO	PTR	WLEO, Ponce

1500 kHz

3	PJC-9	ATN	R. Hoyer 2, Willemstad
5	TIRC	CTR	R. Cima, Cd. Quesada
1	CMDA	CUB	Holguín
0.25	CMKQ		R. Majaguabo, San Luis, SC
0.5	CMPA	DOM	R. Color, Higüey
1	HRTX	HND	R. Victoria, Choluteca
0.5	4VOC	HTI	R. Select, Port-au-Prince
50	XESM	MEX	Vida 1500, México
1	XEGN		R. Gigante, Piedras Negras
0.4	XEJQ		R. Parras, Parras
0.25d	XEFL		R. Santa Fé de Guanajuato, Guanajuato

1	YNPT	NCG	R. Minuto, Managua
1/0.25	WMNT	PTR	R. Atenas, Manatí
1	YSCS	SLV	R. Fides, Usulután

1505 kHz

1		AIA	R. Anguilla, The Valley

1510 kHz

1	CMDA	CUB	Moa
10/3	CMBL	DOM	R. Pueblo, Sto Domingo
5	TGDX	GTM	R. Centroamericana "Nueva RCA", Guatemala
1	HREM	HND	R. Emanuel, Ocotepeque
1	HRYK		R. Gualcho, Tegucigalpa
5d	XEOF	MEX	Stereo Carnaval, Cortázar
0.5	XEJPM		R. Veracruz, La Antigua Veracruz
	XEQI		Sistema R. Nuevo Leon, Monterrey
8	HOA95	PNR	Hosanna R, Panamá (vf)
1	WAVB	PTR	Super B, Lajas

1520 kHz

1	TISH	CTR	R. Tigre, Turrialba (r:800)
1	CMKZ	CUB	R. Baragua, Palma Soriano, SC
1	CMWJ	DOM	R. Samaná "R. 15-20"; 0930-0400
1	TGRS	GTM	R. Superior, Coatepeque
	TG-		R. Taysal, Sta Elena de la Cruz
5	HRRG	HND	R. Providencia, Danlí
1	HRCR		Dif.Cristiana de Radio "DCR", San Pedro Sula
1	HRHJ		R. Santiago, Yoro
2	XEART	MEX	La Señal de las Estrellas, Jojutla
1d	XEVUC		La Estación de los Super Grupos, Zaragoza
1d	XEYP		La Juvenil, El Limón
1d	XEEH		R. Exitos, San Luis
1d	XEVO		R. San Rafael, San Rafael
1	XEJCC		R. Paso del Norte, Cd.Juárez
1	XEATL		Sistema XEGEM "La Radio de la Gente", Atlacomulco
0.25d	XEUAA		Universidad Autonomica de Aguascalientes, Aguascalientes
1	YNRF	NCG	R. Flash, Managua *
10d	WVOZ	PTR	R. Aeropuerto Int, San Juan

1530 kHz

100		BLZ	Voice of America, Belize
1	CMIX	CUB	R. Morón, Morón, CA
20	XEUR	MEX	La Poderosa/R. Onda, México
1d	XEGQ		R. Variedades, Los Reyes
1	XESD		R. Actualidades, Silao
0.5	YN-	NCG	R. LV de Tereza, Sta Tereza
1/0.25	WUPR	PTR	Exitos 15-30, Utuado

1540 kHz

50	ZNS1	BAH	R. Bahamas, Nassau
1	CM	CUB	R. Juvenil, Holguín, HO
1	CMES		R. Sagua, Sagua La Grande, VC (vf)
1	CMBU	DOM	LV de La Romana (vf)
1	CMFB		R. Criolla Comercial, Sto Domingo
5	XEHOS	MEX	La Poderosa, Hermosillo
5	XESTN		R. Red, Monterrey
1/0.25	XENC		R. Voz Musical, Celaya
1	XERTP		R. Impacto, San Martín Texmelucán
1	WIBS	PTR	R. Caribe, Guayama

1550 kHz

1	CMJQ	CUB	R. Nuevitas, Nuevitas, CM
1	HRJO	HND	R. Campeona, Comayagua
1	HRKR		R. Kristell, Juticalpa
1	HRJX		R. Nueva Vida, San Pedro Sula
5d	XENU	MEX	La Poderosa, Nuevo Laredo
10	XERUV		R. Universidad Veracruzana, Jalapa
5	XEXOO		La O de Oro, El Oro *
1	XEBG		R. 15-50/La B-G, Tijuana
0.5	XERE		R. Michoacán, Morelia *
0.25	WKFE	PTR	R. Café, Yauco
5	YSCZ	SLV	Cadena Cuscatlán, San Salvador

1560 kHz

6	TIRN	CTR	R. Nicoya, Nicoya
1	CM--	CUB	Ciego de Avila
1/0.5	CMPZ	DOM	R. Pedernales
10	4VVE	HTI	La Voix de L'Espérance (R. Adventiste), Port-au-Prince
5d	XESE	MEX	LV de Campeche, Champotón
5d	XERIO		R. Triunfadora, Ixtlán del Río
5	XELAC		R. Azul, Cd.Lázaro Cárdenas

20/10	XEFAJ		R. Consentida, México
1d	XEJPV		Mágica Digital, Cd.Juárez
1d	XEZW		R. Diversión, Cerritos
1	XEMAS		La Estación de los Exitos, Salamanca
0.5	XEDD		D-D R. Amistad, Montemorelos
5	YNCN	NCG	R. América, Managua
5/0.75	WRSJ	PTR	R. San Juan, Guaynabo

1570 kHz

10	CMAJ	DOM	R. Amanecer Internacional, Sto Domingo (vf)
10	TGVE	GTM	R. Voz Evangélica de América, "R. VEA", Guatemala
2.5	HRRF	HND	R. Cadena Nacional de Noticias "RCN", Tegucigalpa (vf)
50	XERF	MEX	La Poderosa, Cd.Acuña
2.5	HOE35	PNR	LV del Trópico, "La Grande", Colón *
1	WPPC	PTR	R. Felicidad, Peñuelas
5	YSCZ	SLV	Cadena Cuscatlán, Cuscatlán (r:1550)

1580 kHz

100		BLZ	Voice of America, Belize
10	TI	CTR	R. Mi País, Siquirres
0.25	TICLS		R. Cultural Los Santos, San Marcos
0.25	TIRCC		R. Cultural de Corredores, Cd. Neily
0.25	TIRSCM		R. Sist. Cultural Maleku, Tonjibe
0.25	TIRSCLC		R. Sist. Cultural, La Cruz
1	CMFA	CUB	R. Cadena Agramonte, Santa Cruz del Sur, CM
1	CMPK	DOM	R. Neiba
1	HRCR	HND	Dif.Cristiana de Radio "DCR", La Esperanza
50	XEDM	MEX	La Grande de Sonora, Hermosillo
1/0.25	XEAF		R. Luz, Celaya
1	XELI		LV del Sur, Chilpancingo
0.25	XEVAB		Super Stereo Miled, Valle del Bravo

5/2.5	WMTI	PTR	R. Jefe, Morovis
0.2/0.5			Arecibo (synchr.WMTI)
0.2/0.5			Manatí (synchr.WMTI)

1590 kHz

0.5	CMDF	CUB	R. Granma, Manzanillo, GR
		DOM	F. PI
1	TGXC	GTM	R. Triunfadora, Chimaltenango
5/0.25	XEFD	MEX	La Super F-D, Reynosa
20	XEVOZ		Bonita AM, México
1	XEBZ		R. ACIR, Cd.Delicias
1	XEHC		R. Bahía, Ensenada
1	XEPT		R. Misantla, Misantla
1	WXRF	PTR	R. 15-90, Guayama

1600 kHz

2/1	TIMMCC	CTR	R. Golfito, Pto Golfito
1.5	TILGJ		R. 16, Grecia
1.5	TIJV		R. 88 Estéreo, Pérez Zeledón
0.25	TIRCBA		R. Cultural, Buenos Aires
0.25	TIRCP		R. Cultural, Pital (vf)
0.25	TIRCT		R. Cultural, Turrialba
0.25	TIRCU		R. Cultural, Upala
0.25	TIRSCN		R. Sist. Cultural Nicoyano, Nicoya
5	CMEG	DOM	R. Revelación en América, Sto Domingo
10	TGWC	GTM	LV de Guatemala, Guatemala (r:640)
1	HRPC	HND	R. Luz y Vida, San Luís
1	HRIK		R. San Antonio, Tegucigalpa
5	XEGEM	MEX	Sistema XEGEM "La Radio de la Gente", Mectepec
1/0.25	XEZK		Sonido 16, Tepatitlán de Morelos
1	XEAE		La Super Juvenil, Cd.Acuña
1	XEKTT		La Tremenda, Tecate
5d	WLUZ	PTR	R. Luz, Santurce

1610 kHz

200		AIA	The Caribbean Beacon, Anguilla

SOUTH AMERICA (excl. Brazil)

Country Codes used in this section: ARG=Argentina, BOL=Bolivia, CHL=Chile, CLM=Colombia, EQA=Ecuador, FLK=Falkland Islands, GUF=French Guiana, GUY=Guyana (Rep.), PRG=Paraguay, PRU=Peru, SUR=Suriname, URG=Uruguay, VEN=Venezuela.
Other abbreviations: *=inactive, n=nominal freq., (r)=reported, (r:)=relays station indicated, LV=La Voz, RPP=Radio Programas del Peru. vf=variable frequency.
Others: see respective country listings.
Brazil: all st's are listed in country section.

kW	Call	Land	Station & Location
540 kHz			
5	LRA25	ARG	R. Nal. Tartagal
25/8	LRA14		R. Nal. Santa Fé
25/5	LU17		R. Golfo Nuevo, Pto. Madryn
1	CB54	CHL	R. Ignacio Serrano, Melipilla
1	CD54		R. Calle Calle Saval, Valdivia
25	HCFA2	EQA	R. Tropicana "Canal 540", Guayaquil
12	OBX4E	PRU	R. Inca del Perú, Lima
1	OCX4D		R. San Antonio, Trujillo
1	OAX6R		R. El Tiempo, Arequipa
550 kHz			
15	CP-	BOL	R. El Mundo, La Paz
1	CC55	CHL	R. LV. de la Tierra, Angol
50	HJHF	CLM	R. Dif. Nal., Marinilla (r:570)
50	HCGM1	EQA	R. Reloj, Quito
10		FLK	FIBS/BFBS, Bush Rincon, Falkland Isl.
20/12	ZP16	PRG	R. Parque, Cd. del Este
10	CW1	URG	R. Colonia, Colonia
50	YVKE	VEN	YVKE Mundial, Caracas
560 kHz			
25/5	LRA13	ARG	R. Nal. Bahia Blanca
25/5	LV1		R. Colón, San Juan
25/15	LRA16		R. Nal. La Quiaca
15/7.5	LRA9		R. Nal. Esquel
10/3	LT15		R. del Litoral, Concordia
25	HJPF	CLM	LV de la Pampa "R. 5-60", Maicao
10	HJGS		R. Dif. Nal., Tunja (r:570)
25	HCAJ2	EQA	C.R. E.Satelital, Guayaquil

kW	Call	Land	Station & Location
10		GUY	V. of Guyana, Georgetown
2	OBZ4L	PRU	R. Oriente, Lima (r. 559.8-564) (vf)
50	YVRH	VEN	R. Nal., Cd. Guayana (r: 630)
20/10	YVPJ		R. Exitos "Latina 5-60", Rubio
570 kHz			
		ARG	R. del Centro, Lomas de Mirador
50	CB57	CHL	R. Agricultura, Santiago
100	HJND	CLM	R. Dif. Nacional de Colombia, SF de Bogotá
10	HCCE1	EQA	R. El Sol, Quito
1	ZP15	PRG	R. LV del Amambay, Pedro Juan Caballero (vf)
1	OAX3X	PRU	R. Progreso, Chimbote
1	OAX7G		R. Cusco, Cusco
100	YVLX	VEN	R. Rumbos, Villa de Cura
580 kHz			
25/5	LW1	ARG	R. "Universidad 580", Córdoba
20/10	LU5		R. Neuquén
10	LU20		R. Chubut "La 20," Trelew
25	CP91	BOL	R. Panamericana, La Paz
50	HJHP	CLM	R. Dif. Nal., Cali (r:570) *
10	HCPC2	EQA	R. Uno, Guayaquil
15	OAX4S	PRU	R. Maria, Lima
10	OAX2E		R. Marañón, Jaén
5	CX58	URG	R. Clarín, Montevideo
50/10	YVMJ	VEN	LV de la Fé, Maracaibo
590 kHz			
50/25	LS4	ARG	R. Continental, Buenos Aires

kW	Call	Land	Station & Location
25/10	LV12		R.Independencia, San Miguel de Tucumán
25	LRA30		R. Nal. Bariloche, San Carlos de Bariloche
10	CD59		R. Pres. Ibáñez, Punta Arenas
1	CA59		R. Minería, Antofagasta
1	CC59		R. Caracol, Concepción
100	HJCR	CLM	R. Net, Medellín
10	HCSP1	EQA	R. Carousel, Quito
5	ZP32	PRG	R. Ycuámandyyú, Villa de San Pedro
1	OAZ8C	PRU	R. Sideral, Pucallpa
1	OCX6A		R. Frecuencia Catedral, Arequipa
20	YVKL	VEN	R. Continente, Caracas

600 kHz

kW	Call	Land	Station & Location
8	CP190	BOL	R. ACLO, Sucre
10	CD60	CHL	R. Tricolor, Osorno
10	CB60		R. Monumental, Santiago
50	HJHJ	CLM	R. Libertad, Barranquilla
50	HCXY2	EQA	R. Nal.del Ecuador, Guayaquil
10	OBZ4W	PRU	R. Cora del Perú, Lima
0.3	OAX6X		R. Cultural, Toquepala *
10		SUR	R. Paramaribo, Paramaribo
15	YVSW	VEN	R. Alto Llano, Sta Bárbara de Barinas
10	YVQB		Mundial Sucre, Cumaná

610 kHz

kW	Call	Land	Station & Location
1	LRH356	ARG	R. Solidaridad, Añatuya
5	CD61	CHL	R. CD61, Puerto Aysén
30	HJKL	CLM	6-10 Bolero RCN, SF de Bogotá
10	HCMJ1	EQA	Em.Gran Colombia, Quito
10/1	ZP30	PRG	R. ZP30, LV del Chaco Paraguayo, Filadelfia
50/25	CX4	URG	R. Capital, Montevideo
10	YVSE	VEN	R. Cristal, Barquisimeto
10	YVXY		R. Centro 6-10, Cantaura

620 kHz

kW	Call	Land	Station & Location
25/7	LRA18	ARG	R. Nal. Río Turbio
25/5	LT17	ARG	R. Provincia de Misiones, Posadas
25/5	LRA28		R. Nal. La Rioja
10/5	LRA26		R. Nal. Chaco
10	LV4		R. San Rafael
20	CP63	BOL	R. San Gabriel, La Paz
10	CC62	CHL	R. Bío-Bío, Concepción
1	CA62		R. Norte Verde, Ovalle
50/20	HJEL	CLM	Colmundo, Cali
10	HJVP		Colmundo, Cartagena
50	HCXY3	EQA	R. Nal.del Ecuador, Loja
1	ZP40	PRG	R. Ñasindy, San Estanislao (r. 619-616.6v) (vf)
10	OCY4A	PRU	R. Estrella del Sur, Cañete
1	OAX2M		R. Chepén, Chepén
10	YVNO	VEN	R. Libertad, Cabimas

630 kHz

kW	Call	Land	Station & Location
25/5	LS5	ARG	R. Rivadavia, Buenos Aires
25/5	LU4		R. Dif. Patagonia Argentina, Comodoro Rivadavia
10	CB63	CHL	R. Fundación, Valparaíso
20	HJFD	CLM	R. Manizales, Manizales
10	HJWC		LV del Guainía, Puerto Inírida
10	HCHA2	EQA	Ondas Quevedeñas, Quevedo
50	YVKA	VEN	R. Nal. "Canal Ligero", Caracas

640 kHz

kW	Call	Land	Station & Location
25/5	LRA24	ARG	R. Nal. Río Grande
10/5	LV15		R. Villa Mercedes
10/1	LU18		R. El Valle, "640 AM", General Roca

kW	Call	Land	Station & Location
10	CP204	BOL	R. Tarija, Tarija
10	CD64	CHL	R. Temuco Cooperativa AM, Temuco
10	HJBJ	CLM	RCN, Santa Marta
50	HCXY1	EQA	R. Nal.del Ecuador, Quito
8	ZP19	PRG	R. Caaguazú, Coronel Oviedo
10	OAZ4K		R. Del Pacífico, Lima
10	OBX7B		Onda Azul, Puno
30	YVQO	VEN	Porteñas 6-40 "Unión R", Puerto La Cruz
10/5	YVMU		R. Carora, Carora

650 kHz

kW	Call	Land	Station & Location
1	CP263	BOL	R. Dif. Integración, El Alto
30	HJKH	CLM	RCN Antena 2, SF de Bogotá
5	HCFD4	EQA	R. Visión de Manta, Manta
5	ZP29	PRG	R. Vallemi, Vallemi
15	ZP-		R. Uno, Asunción
2	OAX2N	PRU	R. Noticias del Perú, Trujillo
1	OAX5R		R. Regionalista, Ica
40	CX6	URG	S.O.D.R.E. Montevideo
50/20	YVLH	VEN	Aragu'eña 650, Maracay

660 kHz

kW	Call	Land	Station & Location
5/1	LT41	ARG	R. LV del Sur Entrerriano, Gualeguaychú
1	CP-	BOL	R. ABC, Santa Cruz
50	CB66	CHL	R. Chilena, Santiago
25	HJQS	CLM	Colmundo, Cúcuta
10	HJJN		R. Auténtica, Cali
30	HCLG2	EQA	R. Carrousel, Guayaquil
6/10	ZP26	PRG	R. Itapirú, Cd. del Este
3	OCX1U	PRU	R. JHC, Chiclayo
10	OCX4R		R. Omega, Lima
1	OCX1V		R. Tumbes, Tumbes
1	OAZ7P		R. Prensa al Día, Cusco
1	OBX60		R. Internacional, Arequipa
5	YVQZ	VEN	R. Anaco, Anaco
10	YVNA		Ondas de los Médanos, Coro

670 kHz

kW	Call	Land	Station & Location
5	LRA52	ARG	R. Nal. Chos Malal
25/8	LT4		R. Dif. Misiones, Posadas
25/5	LRA11		R. Nal. Comodoro Rivadavia
25/10	LRI209		R. Mar del Plata, Mar del Plata
25/10	HJPL	CLM	RCN Antena 2, Medellín
12.5	HCFF1	EQA	R. Jesús del Gran Poder, Quito
10	OAX7H	PRU	R. Nal. del Perú, Puno
100	YVLL	VEN	R. Rumbos, Caracas

680 kHz

kW	Call	Land	Station & Location
25/5	LV6	ARG	R. Nihuil, Mendoza
10/6	LU12		R. Río Gallegos
10	LT3		R. Cerealista, Rosario
5	CP-	BOL	Radiodifusora Cristal, La Paz
10	CC68	CHL	R. Octava, Concepción
10	CA68		R. Chilena AM, Chuquicamata
100	HJBU	CLM	R. Dif. Nal., Zambrano (r:570) (vf)
25/12	HCVP2	EQA	Sistema de Emis.Atalaya, Guayaquil
2	ZP41	PRG	R. Ypacaraí, Ypacaraí
10/1	ZP11		R. Cáritas, Asunción
5	OBX4A	PRU	RBC, la estación
10	OAX5E		R. Independencia, Ica
0.5	OBX2L		R. Amauta, Chócope
0.25	CW68	URG	R. Young, Young
10	YVQR	VEN	R. Continente, Cumaná

690 kHz

kW	Call	Land	Station & Location
25/5	LRA4	ARG	R. Nal. Salta
10/5	LU19		La Voz de Comahue, Cipoletti

kW	Call	Land	Station & Location
10	CB69	CHL	R. Santiago, Santiago
10	CD69		R. Estrella del Mar, Ancud
50	HJCZ	CLM	R. Recuerdos, SF de Bogotá
50d	HCJB1	EQA	LV de los Andes, Quito
5	HCFA4		Sucre Portoviejo, Portoviejo
	HC--		Sucre Machala, Machala
	HC--		Sucre Esmeraldas, Esmeraldas
2	OBX3F	PRU	R. Chimbote, Chimbote
0.5	OBX6Q		Radiodif. La Joya, La Joya
20/10	CX8	URG	R. Sarandí, Montevideo
50/20	YVMR	VEN	R. Barquisimeto, Barquisimeto
700 kHz			
25/1	LV3	ARG	R. Córdoba
5	CD70	CHL	R. Magallanes, Punta Arenas
1	CD70		R. Valdivia, Valdivia
1	CA70		R. Amancay, Copiapó
120	HJCX	CLM	R. Net, Cali
50	HCRS2	EQA	Sucre Guayaquil, Guayaquil
1		GUY	V. of Guyana, Linden
10	ZP12	PRG	R. Carlos Antonio López, "LV del Neembucú", Pilar (r. 699.6) (vf)
10	OBX1U	PRU	R. Cutivalú "LV del Desierto", Castilla
1	OCY2H		R. Sausal Superior, Sausal
1	OBZ4H		R. 700, "la grande," Lima
5/2	YVPQ	VEN	R. Sur, Puerto Ordaz
10	YVMH		R. Popular, Maracaibo
710 kHz			
25/5	LRA19	ARG	R. Nal. Pto. Iguazú
25/5	LS1		R. Municipal, Buenos Aires
25/1	LRA17		R. Nal. Zapala
10	CP50	BOL	R. Pío XII, Siglo Veinte
50	HJNX	CLM	R. Super, Medellín
7	HCER5	EQA	Escuelas Radiofónicas Populares, Riobamba
5	OBX5Q	PRU	(r. RPP 730), Ica
10	OCX7T		R. Nal. del Perú, Puerto Maldonado
1	OCX6G		R. ABC, Cerro Colorado
50/20	YVKY	VEN	R. Capital, Caracas
720 kHz			
25/1	LV10	ARG	R. de Cuyo, Mendoza
1	LRA59		R. Nal. Gobernador Gregores
2.5	CP148	BOL	R. Yungas, Chulumani
30	HJAN	CLM	Emisoras Unidas, Barranquilla
25	HJVO		Transm. del Quindío, Armenia
5	HCIC1	EQA	R. Municipal, Quito
5	HCMO3		R. Matovelle "HCM-3", Loja
10	HCUE3		R. Sucre-Machala, Santa Rosa
10	HCGB4		LV de Portoviejo, Portoviejo (vf)
25	ZP-	PRG	R. Pa'i Puki, 25 Leguas
50		PRU	R. San Luís (r: RPP 730), Jaén
10	OAU4E		R. Sideral, La Oroya
10	OAX2J		R. Nal. del Perú, Trujillo *
50	YVQE	VEN	R. Oriente, Porlamar
10	YVXE		R. Elorza, Elorza
730 kHz			
7	LU23	ARG	Em. Lago Argentino, El Calafate
5	LRA3		R. Nal. Santa Rosa
25/5	LRA27		R. Nal. Catamarca
3	CP165	BOL	R. Mensaje, Montero *
10	CP27		R. La Cruz del Sur, La Paz
10	CB73	CHL	R. Valparaíso Cooperativa AM, Valparaíso
1	CD73B		R. Aysén, Pto. Aysén
1	CD73		R. Angelina, Los Angeles

kW	Call	Land	Station & Location
30	HJCU	CLM	Cadena Melodía, SF de Bogotá
10	HCMG2	EQA	R. Guayaquil, Guayaquil
50/5	ZP7	PRG	R. Cardinal AM Stéreo, Asunción
10	OAX4G	PRU	R. Programas del Perú, "RPP", Lima
5/2.5	CX10	URG	R. Continente, Montevideo
10	YVMT	VEN	R. Universo, Barquisimeto
10	YVOO		R. Frontera, San Antonio del Táchira
740 kHz			
1	LRI200	ARG	R. Puerto Deseado
1	LRA55		R. Nal. Alto Río Senguerr
			R. República, Monte Chingolo
			R. Bonaerense, Llavallol
50	HJNS	CLM	R. Guatapurí, Valledupar
10	HJHB		Ecos de Pasto, Pasto
10	HCGC1	EQA	R. Melodía, Quito
10	HCSE4		R. Libertad, Chone (vf)
10	OAX1D	PRU	R. Progreso, Piura
10	OAX6C		R. Continental, Arequipa
1	OCX2X		R. El Puerto, Pacasmayo *
5/1	CW27	URG	R. Tabaré, Salto
50	YVNQ	VEN	R. Caroní, Puerto Ordaz
10	YVNC		R. Maracaibo, Maracaibo
750 kHz			
100/10	LRA7	ARG	R. Nal, Córdoba
50	HJDK	CLM	Caracol Colombia, Medellín
30	HCRC2	EQA	R. Caravana, Guayaquil
2.5	OCX4B	PRU	R. Altura, Cerro de Pasco
1	OAX3T		R. Ondas del Norte, Chimbote
100	YVKS	VEN	R. Caracas Radio "RCR", Caracas
760 kHz			
		ARG	R. Malvinas, Monte Grande
18	CP29	BOL	R. Fides, La Paz
75	CB76	CHL	R. Cooperativa, Santiago
30	HJAJ	CLM	RCN, Barranquilla
15	HCQR1	EQA	R. Quito "LV de la Capital", Quito
10		GUY	R. Roraima, Georgetown
10	ZP5	PRG	R. Encarnación
10	OBZ4X	PRU	Radiomar AM, Lima
10	YVSO	VEN	R. Popular, Trujillo
10	YVQQ		Doble Q, Puerto La Cruz
768.9 kHz			
3	OCX1T	PRU	R. Horizonte, Chiclayo (n. 770)
770			
10	CP116	BOL	R. Cosmos, Cochabamba
1	CD77	CHL	R. Alerce, Castro
50	HJJX	CLM	RCN, SF de Bogotá
25/12	HCMF2	EQA	R. El Telégrafo, Guayaquil
5	OAX6H	PRU	Radiomar (r: RPP 730), Arequipa
5	OAX8M		LV de la Selva, Iquitos
100	CX12	URG	R. Oriental, Montevideo
50/20	YVKK	VEN	R. Nal., Valencia (r: 630)
780 kHz			
8	LV8	ARG	R. Libertador, Mendoza
5/1	LRA10		R. Nal. Ushuaía e Islas Malvinas, Ushuaía
5	LRA12		R. Nal. Santo Tomé
10	CD78	CHL	R. Sago AM, Osorno
30	HJZW	CLM	R. Almirante, Riohacha
30	HJZG		LV del Valle, Cali
1.5	HCRG4	EQA	R. Mía, Manta
1	HCAA1		R. Noticia, Quito
25	ZP70	PRG	R. Primero de Marzo, Asunción
5/7	OAZ7S		Radiodifusoras LV de la Esperanza, Juliaca (n. 730)
10	OAX4X		R. Victoria, Lima

kW	Call	Land	Station & Location
10	OAX1K		R. Nal. del Perú, Tumbes
50/20	YVOD	VEN	Ecos del Torbes, San Cristóbal
10	YVMN		R. Coro, Coro
790 kHz			
5	LV19	ARG	R. Malargüe
25/5	LRA22		R. Nal. San Salvador de Jujuy
25/5	LR6		R. Mitre "AM 80," Buenos Aires
5	CP265	BOL	R. Libertad, La Paz
10	HJDC	CLM	R. Recuerdos, Medellín
5	OAZ7H	PRU	R. Armonía, Cusco
50	YVXM	VEN	R. Minuto "La Barquisimetana", Barquisimeto
10	YVKC		R. Dif.Venezuela, "AM 7-90", Caracas
	YV--		R. Nal., Cd.Bolívar (r: 630)
800 kHz			
10/7	LU15	ARG	R. Viedma
1/0.25	LV23		R. Rio Atuel, General Alvear
0.25	CP157	BOL	R. Santa Clara, Sorata
5/1	CB80	CHL	R. Minería, Viña del Mar
10	HJBW	CLM	RCN Santander/La Emisora de Atalaya, Bucaramanga
5	HCFB1	EQA	R. Sensación 800, Quito
25	HCML2		R. Super K, Guayaquil
5/3	ZP27	PRG	R. Mbaracayú, Saltos del Guairá
1	OCX1P	PRU	Telecomunicaciones del Norte, Piura
1	OBX6A		R. La Porteña, Ilo
0.75	OBX6A		R. Texao, Arequipa
0.5	OBX5B		R. Sur Medio, Ica
0.5	OAZ4U		R. Sol, Huaral
50/10	YVZC	VEN	R. Fé y Alegría, Guasdualito
809.8 kHz			
1	OCX1F	PRU	R. Continente, Chiclayo (n. 810)
810 kHz			
250	HJCY	CLM	Caracol Colombia, SF de Bogotá
10	OAX7V	PRU	Radial Andina (r: RPP 730), Juliaca
1	OAZ3M		R. Marinos, Santa
50/25	CX14	URG	R. El Espectador, Montevideo
50	YVLP	VEN	Super R. 810, Valencia
815 kHz			
5	HCVT2	EQA	R. Atalaya, El Milagro
820 kHz			
5/3	LU24	ARG	R. Tres Arroyos
25/5	LRA8		R. Nal. Formosa
			R. Federal, Lanús (vf)
10	CP35	BOL	R. Altiplano, La Paz
10/1	CA82	CHL	R. Gabriela Mistral, La Serena
10	CB82		Radioemisora Carabineros de Chile, Santiago
1	CD82		R. Concordia, La Unión
1	CC82		R. Almirante Latorre, Talcahuano
0.25	CA82		R. Pampa, Pedro de Valdivia
50	HJED	CLM	Caracol Colombia, Cali
25	HJAD		R. Vigía, Cartagena (vf)
5	HCVI5	EQA	R. LV de Ingapirca, Cañar
3	HCCR1		Super R., San Pablo del Lago
1	HCFB1		R. Monumental, Sto Domingo de los Colorados
1	HCRF4		Canal Manabita, Portoviejo
10	OAX4O	PRU	R. Libertad, Lima
1	OCW6I		R. Paraiso, Camaná
1		SUR	R. Apintie, Paramaribo
1/0.5	CW23	URG	R. Cultural, Salto
50	YVSH	VEN	R. Guayana, Upata
25/10	YVXG		R. Guadalupana, Coro

kW	Call	Land	Station & Location
10	YVKU		R. Altura, La Grita
830 kHz			
5/1	LT43	ARG	R. Mocoví, "la emisora regional", Charata
25/5	LW8		R. San Salvador de Jujuy
25/20	LU14		R. Provincia de Santa Cruz, Río Gallegos
10/5	LT8		R. Rosario
0.5	LT21		R. Municipal Alvear
0.25	LV18		R. Municipal San Rafael
10	HJDM	CLM	R. Deportes, Medellín
4.5	HCRP5	EQA	R. Promoción, Riobamba
25	HCRM2		R. Huancavilca, Guayaquil
3	OAZ7N	PRU	R. OAZ7N, Cusco
10	OAX6D		R. Nal. del Perú, Tacna
1	OAX3Y		R. La Selva, Tingo María
1	OCX7H		R. Inti, Santiago
1	OCX2Y		R. Antena 1, Trujillo
1	OAU4C		R. Antena 1, Piura, Huancayo
25	YVLT	VEN	R. Sensación, Caracas
840 kHz			
25	LU2	ARG	R. Bahía Blanca, "El Sonido de la Vida"
10/5	LT12		R. General Madariaga, Paso de los Libres
			R. General Belgrano, Buenos Aires
3	CP126	BOL	R. Em. Juan XXIII, San Ignacio de Velasco
10	CD84	CHL	R. Santa María, Coyhaique
10	CB84		R. Portales, Valparaíso
30	HJKK	CLM	H J Doble K, Neiva
10	HJBI		Ondas del Caribe, Santa Marta
50	HCPN1	EQA	R. Vigía "LV de la Policía Nal.", Quito
1	HCEM4		R. Costa Azul, Portoviejo
5	ZP6	PRG	R. Guairá, Villarrica (vf)
1	OAX3S	PRU	R. Casma, Casma
1	OBX6Y		R. Azul, Arequipa
	OCX1N		R. Antena 1, Piura
10/5	YVKZ	VEN	Mundial 8-40, Maturín
10	YVMY		R. Juventud, Barquisimeto
850 kHz			
5	CP210	BOL	R. María Auxiliadora, Montero
1	CP160		R. 21 de Diciembre, Mina Catavi
50	HJKC	CLM	R. Net, SF de Bogotá
15	HCVS2	EQA	R. San Francisco, Guayaquil
20	OAX4A	PRU	R. Nal. del Perú, Lima
1	OCX2F		R. Selecta, Virú
50	CX16	URG	R. Carve, Montevideo
10	YVRV	VEN	RV-850, Valencia
10	YVZC		R. Fé y Alegría, Maracaibo
855 kHz			
0.5	HCGB7	EQA	R. Nal.Espejo, El Puyo *
860 kHz			
5/1	LRJ392	ARG	R. Municipal, Chilecito
1	LRA56		R. Nal. Perito Moreno
10	CP8	BOL	R. Nueva América, La Paz
	CP185		R. Paitití, Guayaramerín *
10	CC86	CHL	R. Nueva Inés de Suárez, Concepción
50	HJNJ	CLM	LV del Cañaguate, Valledupar
10	HJFP	CLM	Voces de Occidente, Buga
10	HCPC1	EQA	R. Visión, Quito (vf)
25	ZP28	PRG	LV de la Cordillera, Caacupé (r. 860.4) (vf)
3	OCX1M	PRU	R. Nuevo Norte, Sullana

kW	Call	Land	Station & Location
1	OCY2A		R. Antena 1, Cjamarca
20/10	YVYE	VEN	R. Enlace 8-60, Valle de la Pascua
10	YVOL		Mundial 8-60, San Cristóbal
870 kHz			
100	LRA1	ARG	R. Nal. Buenos Aires
25	HJSB	CLM	R. Mar Caribe Internacional, Barranquilla
10	HJLA		LV del Tolima, Ibagué
20	HCNY2	EQA	R. Cristal "RCQ", Guayaquil
1	HCGS6		R. Píllaro, Píllaro
10	OBX1F	PRU	R. CRU, (r: RPP 730), Chiclayo
1	OBX5R		R. Sport, Nazca
1	OCX6F		R. OCX6F, Paucarpata
1	OCX7R		R. Mundo, Cusco
10	YVRU	VEN	R. Pueblo, Puerto La Cruz
10	YVMP		R. Lara, Barquisimeto
875 kHz			
	CP-	BOL	R. Eucaliptos, Eucaliptos
880 kHz			
10	LU14	ARG	R. Provincia de Santa Cruz
10	CB88	CHL	R. Colo Colo, Santiago
20	HJGE	CLM	Caracol Bucaramanga, Bucaramanga
50/40	HCRP1	EQA	R. Católica Nacional, Quito
50	OBZ4N	PRU	R. Unión, Lima
3	OAX2P		R. Sintonía, Trujillo
20/10	YVYM	VEN	Mundial 8-80, Puerto Ordaz
10	YV--		R. Paraguaná, Punto Fijo
10	YVKV		R. Deportiva 8-80, Caracas
885 kHz			
5/10	ZP33	PRG	R. Tres de Febrero, Itá (n. 890) (vf)
890 kHz			
25/5	LV11	ARG	Em. Santiago del Estero
20/1	LU33		Em. Pampeana, Santa Rosa
5	CC89	CHL	R. Interamericana, Concepción
20	CD89		R. Nal., Punta Arenas
10	CA89		R. León Trece, Pozo Almonte
30	HJPM	CLM	R. Galeón, Santa Marta (vf)
10	HJCE		R. Continental, SF de Bogotá
	HJ--		R. Ecos de Soledad, Soledad
25/20	HCRS3	EQA	R. Superior de El Oro, Machala
1	HCTL5		Ondas del Chimborazo, Riobamba
1	OBX7S	PRU	R. Bahá'í, Puno
50/10	CX18	URG	R. Sarandi Sport, Montevideo
10	YVLW	VEN	R. América, Valencia
1	YVVO		R. Oriente, El Tigre
900 kHz			
25/2.5	LT7	ARG	R. Provincia de Corrientes
1.5	CP83	BOL	R. Norte, Montero
1	CP-		R. 24 Exitos, Cochabamba
0.25	CP79		R. Em. La Voz Nacional, Tarija
0.18	CP20		R. El Cóndor, La Paz (vf)
5	CA90	CHL	R. Regional, Copiapó
1	CC90		R. Ñuble, Chillán
1	CD90		R. La Voz de la Costa, Osorno
1	CB90		R. Amapola, Valparaíso
30	HJEY	CLM	LV de Cali, Cali (vf)
15/5	HJDD		R. Super, Cúcuta
5	HCOF4	EQA	R. Chone, Chone (vf)
10	HCVA1		Sucre Quito, Quito
1	HCRR5		R. Carrousel, Cuenca
3/1	OBX6K	PRU	Nevada R., Arequipa
10	OBX4X		R. El Sol, Lima
1	OAX3E		R. Ribereña, Aucayacu
2.5	CW17	URG	R. Frontera, Artigas
25	YVMD	VEN	R. Mara "Ritmo 900", Maracaibo
910 kHz			
50/5	LR5	ARG	La Red, Buenos Aires
50/25	LRA23		R. Nal. San Juan
30	HJMY	CLM	RCN San Andrés, San Andrés
25	HJDO		LV del Rio Grande, Medellín (vf)
2.5/2	HCGE5	EQA	R. Mundial, Riobamba
2	HCBO2		R. Espectáculo, Guayaquil *
1	OAX2H	PRU	R. Huamachuco
50/20	YVRQ	VEN	RQ 910, Caracas
910.1 kHz			
1	OCX1D	PRU	R. El Tigre, Ferrañafe (n. 900)
914 kHz			
3		SUR	R. Nickerie, Nieuw Nickerie
920 kHz			
1	CD92	CHL	R. 920, Temuco
25	HJJN	CLM	Ondas del Mayo, Pasto (vf)
15	HJSJ		Colmundo, Ibagué
10	HJAA		Emisoras Fuentes, Cartagena (vf)
20	HCCM1	EQA	R. Colón, Quito
10	HCRU3		C.R.O., Machala
	HC--		R. Peripa, El Empalme
100	ZP1	PRG	R. Nal. del Paraguay, Asunción
10	OBX1J	PRU	(r: RPP 730), Piura
1	OAU4D		R. Super AM, Santa Rosa de Sacco
1	OBX2S		R. Ollantay, Virú
1	OAX9V		R. Marginal, Tocache
1	OCX7M		R. Andina, Cusco *
20	YVQX	VEN	R. Nueva Esparta, Porlamar
10/5	YVQU		R. San Carlos, San Carlos
930 kHz			
1/0.25	LV28	ARG	R. Villa María, Villa María
6/1	CP145	BOL	R. Carlos Palenque, La Paz
10	CD93	CHL	R. Reloncaví, Puerto Montt
10	CB93		R. Nuevo Mundo, Santiago
10	CA93		R. El Cobre, Antofagasta
10	HJCS	CLM	LV de Bogotá, SF de Bogotá
5	HCBA6	EQA	R. Ambato, Ambato *
5	HCVI2		Ondas Antillanas, Guayaquil
2.5	OBX6T	PRU	R. Yaraví, Arequipa
10	OAX4E		R. Moderna "R. Papá", Lima
1	OCY4K		R. Satipo, Satipo
1	OCX2V		R. Inti, Chepén
50	CX20	URG	R. Monte Carlo, "la Super R.", Montevideo
10	YVLJ	VEN	R. Maracay, Maracay
940 kHz			
3/0.5	LRH200	ARG	R. Chajarí
20/5	LRJ241		R. Dimensión, San Luís
0.8	CP-	BOL	R. San Lorenzo, Colcapirhua (n. 960)
1	CB94	CHL	R. Valentín Letelier, Valparaíso
1	CA94		R. 9-40, Copiapó
25	HJTL	CLM	RCN, Cúcuta
10	HJGB		R. Calima, Cali (vf)
5	HCBZ1	EQA	R. Dif.de la Casa de la Cultura Ecuatoriana, Quito *
5/3	OBX7L	PRU	R. Willkamayu, Cuzco
1	OBX1C		R. Presidente, Paita
1	OCY2F		R. Estudio, Huanchaco
15	YVZR	VEN	R. Continental, Barinas
10	YVNN		R. Punto Fijo, Punto Fijo
10	YVLU		R. Fé y Alegría, Campo Mata
950 kHz			
50/15	LR3	ARG	R. Libertad, Buenos Aires

kW	Call	Land	Station & Location	kW	Call	Land	Station & Location
1	LT13		R. Oberá, Oberá	100	HJES		RCN, Cali
3	CP-	BOL	R. Yurac Molino, Chimboata	1	HCJI5	EQA	R. El Prado, Riobamba
15	HJFN	CLM	Caracol Colombia, Pereira	5	ZP31	PRG	R. Mburucuyá, Pedro Juan Caballero
3	HCUE5	EQA	R. Colta "LV de la Asociación", Colta	5	OAX5T	PRU	R. LV de Huamanga, Ayacucho
10	HCDE2		R. Latina, Guayaquil	1	OAX6I		R. Universidad, Arequipa
	HC--		R. AM Norte, Ibarra	1	OAX2F		R. La Hora, Trujillo
1	OBX9L	PRU	R. OBX9L, Rioja	1	OBX9G		R. Cosmos, Bagua
1	OBX2G		R. Cutervo, Cutervo	10	YVQM	VEN	Mundial El Tigre, El Tigre
1	OBX3S		(r: RPP 730), Chimbote	**985 kHz**			
50	YVKG	VEN	R. Informativa 9-50, Caracas	10	OBX4J	PRU	R. Latina, Lima
958.1 kHz				**990 kHz**			
4/5	OCY2K	PRU	R. Líder del Norte, Cjamarca (n. 970)	25/5	LR4	ARG	R. Splendid, Buenos Aires
960 kHz				1	LRJ201		R. Calingasta, Tamberias
25/15	LRA6	ARG	R. Nal. Mendoza	2.5/5	CP192	BOL	R. Esperanza, Aiquile
10/1	LU13		R. Necochea	1	CC99	CHL	R. El Roble, Parral
1	CP93	BOL	R. Kollasuyo, Potosí	100	HJDB	CLM	RCN, Medellín
10	CD96	CHL	R. Polar, Punta Arenas	25	HCGH1	EQA	R. Tarqui, Quito (vf)
10	CB96		R. Carrera, Santiago	2	HCOL5		R. América, Cuenca
120	HJHN	CLM	Caracol Colombia, Magangué	15	HCEW2		Frecuencia Mil, Guayaquil
1	HCSA5	EQA	Sono Onda Internacional, Cuenca	10	OAX6K	PRU	R. Bicolor, Tacna
1	HCNC1		R. Cosmopolita, Quito	1.2	OBX3L		R. Onda Nueva, Chimbote *
1	HCJX6		LV del Santuario, Baños	1	OBX4E		Radial Andina, Iquitos
3/1	OBX6S	PRU	R. Hispana, Arequipa	1	OBZ4R		R. LV del Valle de Mantaro, Jauja
3	OBX1Y		R. Amor, Chiclayo	0.5	OBX2M		R. Contumazá (vf)
2/1	OCY4V		R. Constelación, Huancayo	0.1	OCX1J		R. San Marcos, Piura
10	OAX4D		R. Panamericana AM, Lima	20	YVRT	VEN	R. Tropical "99-0", Caracas
1	OBX8H		R. Diez, Iquitos	10	YVTA		Mundial Tricolor, Barquisimeto
0.5			R. Industrial. Trujillo	**1000 kHz**			
1	CW96	URG	Radio Yi, Durazno	25/1	LU16	ARG	R. Río Negro, Villa Regina
50/20	YVRB	VEN	R. Monagas, Maturín	1	LT42		R. del Iberá, Medcedes
10	YV960		Mundial 9-60, Acarigua	3	CP119	BOL	R. Dif. Trópico, Trinidad
10	YVSS		R. San Sebastián, San Cristóbal	2.5/	CP28		R. Cochabamba, "CBA"
965 kHz				1	CP		R. Pirari, Santa Cruz
10	HCOT1	EQA	R. Católica, Sto Domingo de los Colorados *	10	CB100	CHL	R. Cien AM, Santiago
				15	HJAQ	CLM	RCN, Cartagena
970 kHz				10/5	HJBN		LV del Galeras, Pasto
25/7	LV2	ARG	R. General Paz, Córdoba	10	HJJG		R. Dif. Nal., Manizales (r:570) (vf)
1	LT25		R. Guaraní, Curuzú Cuatiá	0.8	HJ--		R. Panamericana, Cajibío
10	CP30	BOL	R. Santa Cruz, Santa Cruz	1	HCCR1	EQA	R. Alegría, Sto Domingo de los Colorados
5	CD97B	CHL	R. Patagonia Chilena, Coyhaique	5/1	ZP36	PRG	R. Ypoá 1000, Paraguari
1	CC97		R. Lautaro, Talca	1	OBX6R	PRU	R. Endesa, Arequipa
1	CA97		R. Calama, Calama	1			R. El Sol, Saposoa
1	CD97A		R. Austral, Valdivia	10	YVOA	VEN	R. Táchira, San Cristóbal
50	HJME	CLM	RCN Guajira, Maicao	10	YVNM		Mundial Mil, "La Caribena", Morón
30	HJVK		Armonias del Caquetá, Florencia	**1005 kHz**			
10	HJCI		R. Super, SF de Bogotá	1		PRU	R. Cristal, Trujillo
20	HCAW2	EQA	R. Católica Nal.del Ecuador, Guayaquil	**1010 kHz**			
1	HCMB1		R. Imperio, Ibarra	20/5	LV16	ARG	R. Rio Cuarto
10	ZP9	PRG	R. Monte Carlo AM, Asunción	1	LW2		Radioem. Tartagal
1.5	OBX1V	PRU	R. La Capullana, Sullana	5	CP220	BOL	R. Bahá'í de Bolivia, Caracollo (n. 1000)
1	OAX7D		R. OAX7D, Cusco	10	CD101	CHL	R. Chilena-Ñielol, Temuco
1	OBX5A		Latinoamericana R., Ica	15	HJZD	CLM	R. Panzenú, Montería (vf)
1	CX22	URG	R. Universal, Montevideo	15	HJJR		Caracol Colombia, Neiva (vf)
15	YVSD	VEN	Mundial Turismo, Valera	10	HJIX		R. Yarima, Barrancabermeja
10	YVLR		R. Continente 970, Maracay	10	HJCC		R. Deportes, SF de Bogotá
10	YV--		Mundial Oriental, Barcelona	10	HJOP		Caracol Barranquilla, Barranquilla
980 kHz				3	HCRZ2	EQA	R. Amiga, Guayaquil
3/0.5	LT39	ARG	Em. Victoria	2.5	HCRV5		R. Visión, Cuenca
10/1	LU37		R. Gral. Pico	10	HCNR6		R. Variedades "Onda Mil", Ambato
2.5	CP118	BOL	R. Mar AM 980, El Alto	1	HCRC4		R. Cenit, Manta
10	CB98	CHL	R. Agricultura, Valparaíso	10	OAX4U	PRU	R. América, Lima *
1	CA98		R. Univ. de Tarapacá, Arica	1/0.8	OBX2K		R. Andino, Otuzco
30	HJJV	CLM	R. Punto, Cúcuta				

kW	Call	Land	Station & Location
1	OBX9D		R. Moyobamba, Moyobamba
20	CX24	URG	R. Nuevo Tiempo, Montevideo
10	YVPC	VEN	R. Aragua, Cagua
10	YVQF		Mundial Bolívar, Cd.Bolívar
1020 kHz			
10/5	LT10	ARG	AM Universidad
1	LRA58		R. Nal. Río Mayo
1	LRA28		R. Nal. La Rioja, Chilecito
10	CP4	BOL	R. Illimani, "Em. del Estado Boliviano", La Paz
5	CC102	CHL	R. Amiga, Talca
30	HJKS	CLM	LV del Llano, Villavicencio
15	HJDZ		R. Primavera, Bucaramanga (vf)
15	HJDQ		Emisora Claridad, Medellín
10	HJFQ		RCN, Pereira
10	HJFT		R. Super, Ibagué
5/3	HCCR6	EQA	R. Surcos, Guaranda
5	HCHR1		R. Hit, Quito
3	HCGO3		R. Estelar, Santa Rosa
25/10	ZP14	PRG	R. Ñandutí, Asunción
1	OBX3U	PRU	R. Nal del Perú, Chimbote
1	OBX1G		R. Heroica, Chiclayo
0.5	OBX2N		R. Esperanza 2000, La Esperanza *
0.1	CW102	URG	R. Libertadores, Salto
50/10	YVMX	VEN	R. Calendario, Maracaibo
25	YVTW		R. Alegría 1020, Chivacoa
10	YVRS		Mundial Margarita, La Asunción
1030 kHz			
25/5	LS10	ARG	R. del Plata, Buenos Aires
10	CC103	CHL	R. Chilena, Concepción
1	CB103		R. Progreso, Talagante
1	CD103		R. Paine, Puerto Natales
1	CD103A		R. Chiloé, Castro
5	HJ--	CLM	Ondas del Vaupés, Mitú
30	HJER		RCN Antena 2, Cali
15	HJRF		Ondas del Cesar, Aguachica
10	HJGX		R. Progreso de Córdoba, Lorica
10	HJDJ		LV de los Libertadores, Duitama
5	HCKL4	EQA	R. Unión, Chone (vf)
5	HCRF2		R. Punto 1030, Guayaquil
3	HCFO7		R. Francisco de Orellana, Coca
5	OAX7N	PRU	R. LV del Altiplano, Puno
3	OAX2U		R. Los Andes (r: RPP 730), Huamachuco
1	OCX7O		Super R., Cusco
1	OCX6L		R. Andina, Arequipa
25/1	YVTD	VEN	R. Valles del Tuy, Ocumare del Tuy
20	YVQY		R. Onda, Guanare
1040 kHz			
1	CP113	BOL	R. Villazón, Villazón
1	CD104	CHL	R. Raíces, Curacautín
40	HJCJ	CLM	Colmundo, SF de Bogotá
15	HJAI		R. Net, Barranquilla
15	HJBF		LV del Norte, Cúcuta
15	HJFM		LV de Armenia, Armenia
15	HJUB		Colmundo, Pasto
10	HJSY		R. Mil 40, Popayán
3	HCGB6	EQA	R. Colosal, Ambato
3	HCCW1		LV del Valle, Machachi
10/5	HCEV3		R. Splendid, Cuenca (vf)
5	ZP43	PRG	R. Arapizandú, San Ignacio (Misiones)
10	OBX4O	PRU	R. Red, Lima *
1	OBX2O		R. Nor Oriente, Jaén
1			R. Fraternidad, El Porvenir

kW	Call	Land	Station & Location
20/10	YVON	VEN	R. Mundial Los Andes, Mérida
20	YVLB		LV de Carabobo, Valencia
1045 kHz			
1	CP208	BOL	R. Sipe Sipe, Quillacollo (vf)
1050 kHz			
5	CP233		R. El Mundo, Santa Cruz
1.5	CP-		R. Noticias, Oruro
1	CP-		R. La Cumbre, Tiquipaya
1	CD105	CHL	R. Bío Bío, Osorno
5	HJNG	CLM	R. Palmira, Palmira
5	HJIO		LV de la Conquista, Granada
15	HJBB		R. Valledupar
15	HJTJ		R. Uno, Montería
15	HJDR		R. Sonorama, Medellín
10	HJFZ		RCN Antena 2, Espinal
10	HJLZ		LV del Cinaruco/Caracol, Arauca
5	HCRQ2	EQA	Super R., Guayaquil
5	HCIM1		R. Municipal LV de Imbabura, Ibarra
5	OBZ4J	PRU	R. Bolognesi, Huancayo
1	OCX2B		R. San Sebastián, Chepén
1	OAZ7Q		R. San Agustín, Juliaca
1	OBX6B		R. Arequipa
1	OBX5U		R. Andina del Pacífico, Ica
1	OAX3A		R. Alpamayo, Huaraz
1	OBX8F		Antena 1, Iquitos
20	CX26	URG	S.O.D.R.E. Montevideo
20/10	YVKZ	VEN	R. Nal. "Canal Clásico", Caracas
20	YVPO		R. Nal. "Canal Clásico", Cabudare (r: 1050)
1060 kHz			
0.5	CP181	BOL	R. La Voz de la Frontera, Pto. Suárez
0.25	CP-		R. Splendid, Vinto *
	CP-		R. Eco 2000, La Paz
50	CB106	CHL	R. Minería, Santiago
15	HJFJ	CLM	RCN Caldas, Manizales
15	HJOV		R. Cultural Surcolombiana, Neiva
10	HJMV		R. Furatena, Chiquinquirá
10	HJLY		R. Delfín, Riohacha
10	HJGU		R. Bucarica, Bucaramanga
1	HJYX		R. Caracol Colombia, Sincelejo
1	HJMG		Caracol Colombia, Turbo
5	HCMG6	EQA	R. Ecos del Pueblo, Saquisilí (vf)
0.05		GUF	RFO, St. Laurent du Maroni, French Guiana
3	ZP13	PRG	R. Boquerón, Alberdi (r. 1060.5) (vf)
2	OBX5D	PRU	R. Andahuaylas *
1	OAZ3N		R. Sudamericana, Chimbote
1	OCY4D		R. Exito, Lima
10	YVOE	VEN	R. San Cristóbal "RSC", San Cristóbal
10	YVLN		R. Guárico, San Juan de los Morros
1060.1 kHz			
1.5		PRU	R. Sudamérica, Cutervo (n. 1060)
1070 kHz			
25/5	LR1	ARG	R. El Mundo, Buenos Aires
0.5	CP-	BOL	R. Quillacollo, Quillacollo
30	HJCG	CLM	R. Santa Fé, SF de Bogotá
20/10	HJAH		Em. Atlántico, Barranquilla (vf)
15	HJVR		R. Super, Popayán
10	HCKD2	EQA	R. Tigre, Guayaquil
1	HCRS1		R. Lubacán, Santo Domingo de los Colorados
1	HCVP1		R. Libertad, Quito
10		GUF	RFO, Matoury, French Guiana
3	OCX6C	PRU	R. Ondas del Sur, Camaná

kW	Call	Land	Station & Location
3	0AZ7L		R. OAZ7L, Puno
0.25	0AX5A		R. Centinela, Pto. de San Juan
5	YVPX	VEN	R. El Sol, La Fría
25	YV--		R. 10-70 AM, Maiquetía
10	YVMA		Mundial Zulia, Maracaibo
1072.5 kHz			
1		PRU	R. Latina, Chiclayo (n. 1070)
1075 kHz			
0.5	CP-	BOL	R. Agricultura, Portachuelo
1080 kHz			
25/10	LU3	ARG	R. del Sur, "la emisora de Bahía Blanca"
25	LW4		R. Orán
10			R. Departamento Minas, Andacollo
1	CP291	BOL	R. Dif. Colosal, Sucre (n. 1070)
1	CA108	CHL	R. Río Elqui, Vicuña
1	CD108		R. Los Confines, Angol
15	HJJS	CLM	R. Pontoná, La Dorada
15	HJKT		R. Macarena, Villavicencio
10	HJJF		R. Eco, Cali
10	HJMH		Melodía AM, Bucaramanga
10	HJAW		LV de Montería, Montería (vf)
10	HJAX		Caracol Medellín, Medellín (vf)
5	HCAB4	EQA	R. Contacto, Manta (vf)
10	HCBH6		R. Latacunga, Latacunga
10	ZP25	PRG	Radiodif. Nanawa, Luque
5	OBX1D	PRU	R. San Miguel, Piura (vf)
1	0AX7S		R. Salcantay, Cusco
1	OCX4I		R. Huayllay, Huayllay
1	0AX8N		R. Amazonas, Iquitos
			R. Inca, Trujillo
10	YVQJ	VEN	R. Barcelona, Barcelona
10	YVNR		Mundial 1080, Maracay
1090 kHz			
3	CP45	BOL	R. Cultura, Cochabamba
5/1	CC109	CHL	R. Chilena, Talca
5	HJOM	CLM	R. Bucanero, Cartagena
15	HJBC		R. Net, Cúcuta
10	HJIA		Ondas del Nevado, Manizales
10	HJIH		Caracol Colombia, Sogamoso
10	HJIG		Caracol Colombia, Florencia
1	HJJY		LV de los Pijaos, Guamo
5	HCVI1	EQA	Fé y Alegría, Quito
1.2/3	OBX2A	PRU	R. Cajabamba, Cajabamba
1	0AX3I		R. Televisión Tingo María (n. 1100)
1	OBX6X		R. Amistad, Arequipa
1	0AZ3I		R. AST, Chimbote
15/7	CX28	URG	R. Imparcial, Montevideo
3	YVTG	VEN	Melódica 1090, Machiques
20	YVSZ		Unión R. 1090, Caracas
10	YVPB		R. Yaracay 1090 AM, San Felipe
1100 kHz			
4	CP137	BOL	R. Mundial, La Paz
1/0.75	CP55		R. Universidad de Oruro
10	CB110	CHL	R. Recreo, Viña del Mar
1	CA110		R. La Portada, Antofagasta
5/1	HJGI	CLM	Em. José António Galán/LV de Colombia, El Socorro
5	HJMK		Emisora Ideal, Planeta Rica
15	HJYZ		R. Super, Neiva
15	HJAT		Caracol Colombia, Barranquilla (vf)
10	HJGQ		Transmisora Surandes, Andes
10	HJCN		Caracol Bogotá, SF de Bogotá

kW	Call	Land	Station & Location
5/2	HCGR6	EQA	R. Novedades, Latacunga
10	HCFW2		R. Alegría, Guayaquil (vf)
1.5	HCLE7		R. Oriental, Tena
5	ZP71	PRG	R. Ñu Verá, Capitán Badó
1/0.5	OBX7Z	PRU	R. LTC/RCC, Juliaca
1	OCY4G		Sonorama R., Huancayo
1	OBX1L		R. Star, Chiclayo (vf)
1	0AX5Z		R. Oasis, Ica
1	0AZ4W		R. Musical (r
0.5	0AX9J		R. Lamas, Lamas
0.2	OCX4S		Imperial, "LV de la Provincia", Cañete
10	YVOP	VEN	R. Occidente, Tovar
10	YVSV		R. Angostura, Cd.Bolívar
1109.5 kHz			
1	0AU4J	PRU	R. Antarqui, Lima (n. 1110)
1110 kHz			
25/5	LS1	ARG	R. Municipal, Buenos Aires
			(F.PI.) (still on 710)
15/5	LR2		R. Argentina, Buenos Aires
10/1	LV7		R. Tucumán, San Miguel de Tucumán
10	CD111	CHL	R. La Frontera, Temuco
8/10	HJDI	CLM	R. Bolivariana, Medellín
5/1	HJPA		LV de las Islas, San Andrés
5	HJGP		LV del Río Arauca, Arauca
15	HJZE		R. Piragua, Sincelejo
10	HJJP		RCN, Villavicencio
10	HJEW		Caracol Cali, Cali
1	HJNC		Ecos del Combeima, Ibagué
5	HCRP6	EQA	R. Pelileo, Pelileo (vf)
5	HCJC5		R. Ondas Azuayas, Cuenca
10	HCJR1		Hoy La Radio, Quito
5	OCX7T	PRU	R. Líder, Cusco
1	OCX2U		R. Jaén, "LV de la Frontera", Jaén
1	OBX3B		R. Heroica, Huaraz
	OCX1R		R. Centro Popular, La Unión
10	YVRX	VEN	Unión R. , Valencia
10	YVQT		R. Carúpano, Carúpano
1120 kHz			
25/5	LV5	ARG	R. Sarmiento, San Juan
25/10	LU6		R. Atlántica, Mar del Plata
1	CP184	BOL	R. Estación El Dorado, Trinidad
5	HJJC	CLM	R. Matecaña, Pereira
15	HJGH		R. Net, Bucaramanga
10	HJTI		Colmundo R. 24, Cúcuta
10	HJKQ		Caracol Colombia, Tunja
5	HCNT3	EQA	R. Once Catamayo, La Toma
5	HCLA2		R. América, Guayaquil
3	HCAS7		R. Variedades del Puyo, El Puyo
2	HCEB1		Ecos de Montúfar, San Gabriel
10	HCLE1		R. Dif.Marañon, Sto Domingo de los Colorados
10	ZP24	PRG	R. Nuevo Mundo, San Lorenzo
3	OBX2I	PRU	R. Dinámica, Trujillo
1	OCX6U		R. Municipal, Cerro Colorado
5/10	CW31	URG	R. Salto, Salto
5	YVXZ	VEN	R. República "La Estación Feliz", Maturín
20/10	YVSK		R. Dif.del Sur, San Fernando de Apure *
10	YVMF		Ondas del Lago "Super Ondas", Maracaibo
	YV--		R. Fé y Alegría, Tovar
1125 kHz			

kW	Call	Land	Station & Location
0.5	CP-	BOL	R. Cruceña, Cotoca (n. 1530)
0.3	CP-		R. Em. Cooperativa Poopó, Poopó
	CP-		R. Unión, Vinto
	CP-		R. Porvenir, Tiquipaya
1130 kHz			
25/5	LRA21	ARG	R. Nal. Santiago del Estero AM 1130, Remedio de Escalada *
15	HJVA	CLM	R. "K", SF de Bogotá
15	HJQQ		Caracol Pasto, Pasto
10	HJAC		Em. Riomar, Barranquilla
1	HJNN		Ondas del Río, Magangué
5/3	HCRD1	EQA	R. Punto, Ibarra
5	HCPV6		R. Centro "Estación 112", Ambato
3/1	OAX4N	PRU	R. Selecta, Lima
1/0.6	OCX2D		R. LV del Cutervo, Cutervo
1	OAX9Q		R. San Martín, Tarapoto
1	OAX2V		R. Los Andes (r: RPP 730), Cajamarca
1	OAZ4S		R. Chanchamayo, La Merced
20	CX30	URG	R. Nacional, Montevideo
20/10	YVRL	VEN	R. Ideal, Maiquetía
10	YVKQ		R. Popular, Barquisimeto
10	YVPY		R. Amazonas, Puerto Ayacucho
1135 kHz			
1	CP-	BOL	Kanata Radiodifusión, Cochabamba (r. 1125)
1140 kHz			
	CP-		R. Latinoamericana, Tiquipaya
40/75	CB114	CHL	R. Nal., Santiago
15/5	HJDL	CLM	R. Paisa, Medellín
10	HJRN		RCN/LV de la Provincia, Barbosa
10	HJFH		R. Regional Independiente, Anserma
10	HJCL		R. Girardot, Girardot
10	HJKO		LV de la Victoria, Cartagena
10	HJRW		LV de Centauros/Caracol, Villavicencio
	HJ--		R. Piendamo, Piendamo
5	HCIR1	EQA	Colorin Colorado, Quito
4	HCMF4		R. Momento, Portoviejo
1.5	HCFB2		R. Cóndor, Guayaquil
1	HCAZ5		R. Alfa Musical, Cuenca
5/3	ZP22	PRG	R. Panambí Verá, Villarrica
3		PRU	R. Piura
2	OAX3R		R. Bahía, Chimbote
12	OAX6L		R. Concordia, Arequipa
10	OCY4C		R. Centro (r. RPP 730), Huancayo
0.5	OAX5W	PRU	R. Chinchaysuyo, Chincha
	YV--	VEN	R. La Margariteña 11-40 AM, Porlamar
1143 kHz			
1	CP-	BOL	R. Colonia, Yapacani
1150 kHz			
5/1		ARG	R. Sagrada Familia, San Justo
25/1	LT9		R. Brigadier López, Santa Fé
10	LRA2		R. Nal. Viedma *
1	LRA51		R. Nal. Jáchal
1	CP19	BOL	R. Chuquiago Musical La Paz (n. 1140)
0.5	CP71		R. El Cóndor, Oruro
0.2	CP194		R. Chaco, "LV del Campesino", Yacuíba (n. 1100)
			R. 24 de Noviembre, Eucaliptos
15	HJFI	CLM	Caracol Colombia, Armenia
10	HJFP		R. Neiva, Neiva
10	HJBT		R. Catatumbo, Ocaña
1	HJGJ		R. Net, Duitama

kW	Call	Land	Station & Location
1	HJTE		LV del Chocó, Quibdó
10	HCAV3	EQA	R. Luz y Vida, Loja
10	HCGB5		LV de Riobamba "Antena 1", Riobamba (vf)
1	HCBC7		R. El Cisne, Nueva Loja
10	OCX7Q	PRU	R. Universal, Cusco
1.5	OAX8F		R. Loreto, Iquitos
1			R. Naylamp "Estación Pucará", Pucará
0.65			R. Bambamarca, "Frecuencia Líder", Bambamarca (n. 980)
0.5	OCY2E		R. Chasqui Llacta, San Marcos
			R. San Juan, Lonya Grande
10	YVMV	VEN	Mundial Caribe, Punto Fijo
10	YVQD		Ecos del Orinoco, Cd.Bolívar
1160 kHz			
25/5	LV9	ARG	R. Salta
10/5	LU32		R. Coronel Olavarría, Olavarría
1.8	LRH253		R. Cataratas del Iguazú, Pto. Iguazú
1	LRA57		R. Nal. El Bolsón
5	CP25	BOL	R. Centenario, "La Nueva", Santa Cruz
3/1	CP78		RTC, La Deportiva, Cochabamba
1/0.25	CD116	CHL	R. Baha'i, Temuco
1	CC116		R. Ancoa, Linares
5	HJE75	CLM	Voz de la Paz, Barrancabermeja
5	HJZV		R. Las Lajas, Ipiales
5	HJAZ		LV de Sinú, Montería *
15	HJOC		Ecos de Colombia, SF de Bogotá
10	HJEC		R. San José de Cúcuta, Cúcuta
10	HJBL		R. Aeropuerto, Barranquilla
10	HJAU		Ondas del Orteguaza, Florencia
10	HJEV		La Caleñísima, Cali
	HJ--		Ondas del Puerto, La Virginia (vf)
2	HCVR3	EQA	R. Vía, Machala
2	HCCP1		R. Presidente, Quito (vf)
1	HCUR6		R. Runatacuyaj "LV de la Asociación", Latacunga
1	HCWD4		R. Cenit, Portoviejo (vf)
10	ZP--	PRG	R. Antena 3, Asunción
5	OAX2C	PRU	R. Libertad, Trujillo
5	OAX9A		R. Imagen, Tarapoto
5			Estación Sasape, Cruce
10	OAX4C		R. 11-60, Lima
1/0.5	OBX50		R. Huanta 2000, Huanta
1	OCX7Z		R. OCX7Z, Puerto Maldonado
1	OBX6G		R. Nal. del Perú, Moquegua
1	OAX3B		R. Huaraz, Huaraz
1/0.25	CV116	URG	R. Impacto, Mercedes
0.5	CW116		R. Agraria del Uruguay, Cerro Chato
20/10	YVRR	VEN	R. Industrial "Doble Uno 6-0", Guarenas
1	YVOK		R. Universidad, Mérida
1170 kHz			
25/3	LRA29	ARG	R. Nal. San Luis
25/3	LT16		R. Esmeralda, Presidencia Roque Sáenz Peña
			R. Mi País, Hurlingham
1	CP98	BOL	R. Nuevo Mundo, Sucre
5	CD117	CHL	R. Natales, Puerto Natales
5	HJGA	CLM	R. Recuerdos, Tunja
5	HJWA		LV del Guaviare, San José del Guaviare
30	HJBX		Ondas del Meta, Villavicencio
10	HJKW		R. Nutibara, Medellín

kW	Call	Land	Station & Location
10	HJPB		Ondas de Macondo, Valledupar
10	HJNW		Caracol Colombia, Cartagena
1	HJJE		RCN, Tuluá
5	HCJV5	EQA	R. Central, Riobamba
10	HCJM4		R. Antena Libre, Esmeraldas
10		PRU	(r. RPP 730), Arequipa
1	OCX1B		R. San Juan, Talara
1	OCX7Y		R. Constelación, Puno
1	OCX4Y		R. COSAT, Satipo
10	CX32	URG	Radiomundo, Montevideo
20/10	YVQV	VEN	R. Acarigua, Acarigua
10	YVKW		Señal 11-70, Maiquetía

1180 kHz

kW	Call	Land	Station & Location
1	CP235	BOL	Radiodifusora Ingavi, Viacha
1	CP-		R. 20 de Septiembre, Arbieto
50	CB118	CHL	R. Portales, "la primera de Chile", Santiago
20	HJGK	CLM	R. Uno, Bucaramanga
15	HJFX		Caracol Colombia, Manizales
10/5	HJJT		RCN, Ibagué
	HJ--		Em. Coorpurabá, Apartadó
4	HCDP5	EQA	R. Cuenca "LV de los Cuatro Ríos", Cuenca
12.5	HCLR1		Nueva Em.Central, Quito
10/5	HCAH3		R. Trébol, Zaruma
1.2	HCRV1		R. Familiar, Julio Andrade
3	ZP52	PRG	R. Coronel Oviedo, Coronel Oviedo
3	OBX9I	PRU	R. Kuelap, Chachapoyas
1.5	OCX2A		R. América, Quiruvilca
1	OCY4Z		R. Libertad "RLJ," Junín
10	CX118	URG	LV de Artigas, Artigas
10	YVLQ	VEN	Super Suave 11-80 "LV de la Victoria", La Victoria
10	YVNJ		R. Petrolera, Cd.Ojeda
10	YVOR		R. Maturín, Maturín

1190 kHz

kW	Call	Land	Station & Location
50/10	LR9	ARG	R. América, Buenos Aires
50	LRA15		R. Nal. San Miguel de Tucumán
15	HJEO	CLM	Ondas del Valle, Cartago
10	HJKG		R. Mira, Tumaco
10	HJCV		R. Cordillera, SF de Bogotá
10	HJCT		LV de la Costa, Barranquilla
1	HJKI		R. Barají, Sahagún
2	HCDE2	EQA	R. Excelsior, Guayaquil
1	HCRF6		R. El Sol, Pujilí
3	OBX3D	PRU	R. Ancash, Huaraz
1	OBX5M		R.Lurén, Ica
1	OAX1E		R. Chiclayo, Chiclayo
			R. Membrillar, Cascas
			R. Frecuencia 11-90, Arequipa
20/10	YVPF	VEN	R. Canaima, San Félix
20/10	YVRE		R. Barinas 1190 AM Estéreo, Barinas
10	YVZD		R. Dif.Cultural del Táchira, San Cristóbal

1195 kHz

kW	Call	Land	Station & Location
1	CP108	BOL	R. Independencia, Quillacollo
	HJ--	CLM	Ondas del Ranchería, Barrancas (vf)

1200 kHz

kW	Call	Land	Station & Location
10	LT6	ARG	R. Goya
1			R. Nal. Mendoza (r. LRA6 960),
5	CP32	BOL	R. Oriental, Santa Cruz
0.25	CP171		R. 24 de Noviembre, Arani
0.15	CP37		R. Santa Ana, Santa Ana de Yacuma *

kW	Call	Land	Station & Location
10/5	CD120	CHL	R. Agricultura, Los Angeles
1	CA120		R. Almirante Blanco Encalada, Tocopilla *
15	HJIJ	CLM	R. 1200 "LV de la Raza", Medellín
10	HJLR		RCN Antena 2, Sogamoso
10	HJNF		R. Super, Cali (vf)
10	HJBV		R. Net, Cartagena
10	HJBZ		Ondas del Riohacha, Riohacha
10	HJCD		Em. Nueva Epoca, Fusagasugá
0.5	HJ--		R. Tamalemeque, Tamalemeque
5	HCCS1	EQA	R. Super K, Sangolquí (vf)
5	HCRM5		R. El Mercurio, Cuenca
5	HCRE2		LV del Trópico, Quevedo
1	HCMP4		R. La Grande, Bahía de Caráquez
10	ZP-	PRG	R. Libre, Itauguá
5	OAX7B	PRU	R. Tawantinsuyo, Cusco
4	OAX4B		R. Cadena Nal., Lima
3	OAU4G		R. Andes, Huancayo
1			R. Satélite (r: RPP 730), Tingo María
1			R. Santa, Santa
1	OCX7S		R. San Ramón, Juliaca
1	OBX5X		R. Prenda al Día, Abancay
1	OBX9C		R. Estación "C", Moyobamba
1	OAX6N		R. El Faro, Ilo
			R. San Mateo, Contumazá
			R. Las Palmas, Nuevo Bambamarca
2	CW33	URG	La Nueva Radio, Florida
10	YVOZ	VEN	R. Tiempo, Caracas
10	YVSF		R. Dimensión, Caripito
1	YVNH		Ondas del Escalante, Sta Bárbara del Zulia

1200.2 kHz

kW	Call	Land	Station & Location
1	OCX2T	PRU	R. Noticias, Cajamarca

1210 kHz

kW	Call	Land	Station & Location
10/5	LT18	ARG	R. Eldorado
5	CB121	CHL	R. Chilena AM, Valparaíso
10	CD121		R. Nueva Galicia, Puerto Montt
1	CC121		R. Universidad de Talca, Talca
10	HJBE	CLM	RCN Antena 2, Cúcuta
10	HJFR		R. Net, Neiva
10	HJBQ		LV Amiga, Pereira
20	HCBJ2	EQA	R. El Mundo, Guayaquil
10	HCVC3		R. Centinela del Sur "CDS", Loja
1	OAX2Q	PRU	R. Universo, Trujillo (rel. Radiomar, Lima 760)
1	OAX7M		R. Quillabamba
1	OBZ4A		R. Paramonga, Barranca *
1	OCY4T		R. Galaxia, Satipo
1	CX121	URG	Difusora Soriano, Mercedes (vf)
0.25	CV121		Em. RBC, Piriápolis
0.25	CW121		Em. El Libertador, Villa Vergara
10	YVZT	VEN	R. Anzoátegui, Barcelona

1219.9 kHz

kW	Call	Land	Station & Location
3	OCX1X	PRU	R. Cadena Nor-Oriental, Chiclayo

1220 kHz

kW	Call	Land	Station & Location
		ARG	LV del Aire, Buenos Aires
1	CP67	BOL	R. Splendid, La Paz
1	CP162		R. Batallón Topáter, Oruro
	CP-		R. F 1220, La Paz
	CP-		R. El Cóndor, Arque (r.)
10	CD122	CHL	R. Minería, Temuco
15	HJFF	CLM	Em. 12-20, Barranquilla
10	HJMT		RCN, San Gil
10	HJAV		RCN, Montería
10	HJKR		La Vallenata 12-20, SF de Bogotá

kW	Call	Land	Station & Location
10	HJNM		Em. Cultural Bolívar, Ipales (vf)
10	HCAP1	EQA	R. Dif.Marañon, Quito
3	OAX9T	PRU	R. Bagua, Bagua
10/5	OAX6X		R. Melodía, Arequipa
1	OAZ3J		Radiodifusora Sánchez, Chimbote
1	OAX5G		R. San Pedro, Leoncio Prado
1	OCX7N		R. Inti, Cusco
1	CX120	URG	R. Reconquista, Rivera
20/10	YVZO	VEN	R. Aeropuerto 1220, Maracaibo
10/5	YVVM		Mundial Valencia, Valencia (vf)
10	YVRD		LV de Apure, San Fernando de Apure
1225 kHz			
3	HCJM6	EQA	R. Sira, Ambato
1230 kHz			
5/1	LW5	ARG	R. Libertador General San Martín, Ledesma
25/5	LT2		R. Gral. San Martín, "R.2", Rosario
			R. Ciudad de Banfield
5	HJMJ	CLM	RCN Antena 2, Maicao
15	HJEH		Colmundo, Bucaramanga
10	HJLK		R. Calidad, Cali
10	HJIL		Minuto de Dios, Medellín
10	HJBR		Caracol Tunja, Tunja
1.5	HJFV		R. Viva, Pasto
1	HJTP		R. Colina, Girardot
5	HCFG4	EQA	R. Sucre, Esmeraldas
3	HCMV5		R. Popular Independiente, Cuenca
3	HCRI1		Centro Radiofónico de Imbabura "CRI", Ibarra
15	HCGT2		R. Galáctica, Guayaquil
3	ZP21	PRG	R. Oriental, Caaguazú
1	OBZ4Y	PRU	R. Selecciones, Tarma
1	OBX7J		R. Madre de Dios, Puerto Maldonado
1	OAX2T		R. Albújar, Guadalupe
10	YVOH	VEN	R. Valera, Valera
10	YVNT		R. Barlovento, Caucagua
1235	1	CP53 BOL	R. Indoamérica, Potosí (n. 1250)
1	HCRL6	EQA	LV de Saquisilí y Libertador, Saquisilí
1240 kHz			
		ARG	R. Amanacer Argentino, Numancia (vf)
1	CP180	BOL	R. San Miguel, Arani
25	CB124	CHL	R. Universidad de Santiago, Santiago
1	CA124		R. del Salitre, Iquique
0.25	CA124		R. Principal Chuquicamata, Calama
5	HJGN	CLM	R. Barrancabermeja
3	HJJA		R. Buenaventura
15	HJDA		R. Auténtica, Medellín
10	HJFG		RCN, Armenia
1	HJGO		R. Caribabare, Saravena
5	HCRF3	EQA	R. Nacional Fenix, Zaruma
1	HCPA1		R. Metropolitana, Quito
	HCLA5		R. Sonorama, Riobamba (vf)
5	OBZ4Q	PRU	R. Oro, Huacho
1	OBX7M		R. Túpac Amaru, Sicuani
1	OCX1C		R. Sechura. Sechura
1	OCX6M		R. Minería, Moquegua
1	OCX2K		R. Indoamérica, Trujillo
			R. Campesino, Puno
2.5/5	CW35	URG	R. Paysandú, "la emisora sin fronteras", Paysandú
50	YVNV	VEN	R. Nal.de Venezuela, Punta Tumatey
1250 kHz			
3/1		ARG	AM 1250 R. Estirpe Nal., Rafael Castillo
5	CP69	BOL	R. Nacional, Cochabamba
2.5	CP54		R. La Plata, "LV de la Capital", Sucre
2	CP14		Emisoras Unidas, La Paz
1	CP26		R. Amboró, Santa Cruz
0.5	CP17		R. Sararenda, Camiri
0.4	CP65		R. Oruro, Oruro
0.1	CP47		R. Frontera, Cobija
10	CD125	CHL	R. Universidad Austral de Chile, Valdivia
1	CA125		R. Minería, La Serena
15	HJHS	CLM	Caracol Cúcuta, Cúcuta
10	HJCA		R. Capital, SF de Bogotá
10	HJOK		Em. ABC, Barranquilla
1	HJEM		LV de Corozal, Corozal
3	HCMY1	EQA	LV del Triunfo, Sto Domingo de los Colorados
2	HCHB2		R. Linda, Guayaquil
10	HCEM1		Ondas Carchenses, Tulcán
10/15	ZP3	PRG	R Asunción, "LV del Paraguay", "La R. Capital"
8	OAX4L	PRU	R. Miraflores, Lima
2	OAX8P		R. Pucallpa, Pucallpa
1	OCY2J		R. OCY2J, Cajamarca
1	OAX9K		Estación Tarapoto,Tarapoto
5	CW125	URG	R. Bella Unión, Bella Unión
10	CX36		R. 36, Montevideo
20/10	YVPZ	VEN	R. Puerto Ordaz
1	YVML		R. Cabimas, Cabimas
1260 kHz			
10/5	LT14	ARG	R. General Urquiza, Paraná
10	CA126	CHL	R. Nal., Arica
10	CD126		R. Minería, Punta Arenas
1	CC126		R. Condell, Curicó
5	HJNO	CLM	R. Recuerdos, Duitama
5	HJET		R. Deportes, Cali
5	HJLX		Eco Llanero Radio, Villavicencio
5	HJDV		Caracol Colombia, Ibagué
2	HJOU		Ondas del Amazonas, Leticia
15/10	HJTM		R. Sonar, Ocaña
10/5	HJOH		RCN Cesar, Valledupar
1	HJHU		Caracol Colombia, San Andrés
3	HCRO6	EQA	R. Calidad, Ambato
2	HCPB5		XG Radio, Cuenca
10	HCMO1		LV del Santuario del Quinche "HCM-1", Quito
1	HCRB3		R. Benemérita, Machala
	ZP34	PRG	R. Cultura, Villarrica
5	OAZ7K	PRU	R. OAZ7K, Juliaca
10	OBX5S		R. Ayacucho, Ayacucho *
1	OBX6D		R. Red, Arequipa
1	OBX2C		R. Otuzco, Otuzco
4.5/1	CW37	URG	Dif. Rochense, Rocha
10	YVRM	VEN	R. Super Suave, Caracas
10	YVRY		R. Horizonte, Nirgua
1265 kHz			
0.4	CP-	BOL	R. Uncía, Uncía
1270 kHz			
50/10	LS11	ARG	R. Provincia de Buenos Aires, La Plata
5	LRA20		R. Nal. Las Lomitas
			R. Ciudad de Avellaneda, Wilde
2	CP16	BOL	R. Los Andes, Tarija
1	CP134		R. Vanguardia, Colquiri

kW	Call	Land	Station & Location
10	CB127	CHL	R. Festival, Viña del Mar (vf)
10	CD127		R. Agricultura, Temuco
5	HJTX	CLM	R. Net, Bucaramanga
1.5	HJKJ		LV de Curumaní, Curumaní
1	HJAR		RCN Antena 2, Cartagena
1	HJXQ		R. Melodía, Ubaté
1	HJSV		LV de Orito, Orito
1	HJIM		Colmundo, Pereira (vf)
3	HCLD4	EQA	R. Junín, Junín
15	HCUM2		R. Universal, Guayaquil
1.5/3	OAX1H	PRU	R. Grau, Piura
1	OAX9F		R. Municipal Nor Peruana, Chachapoyas
1	OAX5B		R. Ica, Ica
1	OCX2Z		R. Estación Latina, Chepén
1	OAX8T		R. Eco, Iquitos
0.7	OBX7A		R. Solar, Cusco
0.5	OBX7G		R. Ondas del Titicaca, Puno
0.4	OAZ4H		R. Huacho, Huacho
0.35	OBX6H		R. Mollendo, Mollendo
3/1	CV127	URG	R. Cuareim, Artigas
5	YVTR	VEN	R. Tucupita, Tucupita
10	YVOU		Ondas Panamericanas, El Vigía

1270.2 kHz

kW	Call	Land	Station & Location
0.4	OBX4T	PRU	R. La Merced, La Merced

1275 kHz

kW	Call	Land	Station & Location
0.5	CP187	BOL	R. Chané, Mineros
10	CD128	CHL	R. del Sur, Osorno
1	CC128		R. Arturo Prat, San Carlos
5	HJMB	CLM	R. Suroeste, Concordia
5	HJSO		R. Playa Mendoza, Barranquilla *
5	HJKN		R. Musical, SF de Bogotá
5	HJHO		Impacto Popular, San Juan del Cesar
5	HJRP		Ecos de Tibú, Tibú
5	HJLR		Caracol Colombia, Pasto
3	HJNQ		LV del Río Suárez, Barbosa
1	HJTK		R. Super, Caicedonia
2	HCRP3	EQA	R. Continental, Arenillas
1	HCAB1		R. Democracia, Quito
1	HCNW5		R. Canal Tropical, Riobamba
1	HCIN4		LV del Sur de Manabí, Jipijapa
1	OCX4Q	PRU	Radiodifusora Comercial RAGA, Cañete
1	OBX6P		R. Fénix, Camaná
1	OBX3C		R. El Puerto, Chimbote
0.75/0.5	CW64	URG	R. Tacuarembó, Tacuarembó
10/5	YVQS	VEN	R. Zaraza, El Tamarindo
10	YVOF		R. Trujillo, Trujillo

1280.4 kHz

kW	Call	Land	Station & Location
1	OBX2F	PRU	R. Moderna, Cajamarca (n. 1280)

1290 kHz

kW	Call	Land	Station & Location
5/1		ARG	Em. del Lago, Junín *
			R. Cristal, Lanús
			R. Amanacer, Reconquista
			R. Itatí, Ituzaingó
1	CP212	BOL	Radiodifusoras Minería, Oruro
1	CD129	CHL	R. Mulchen, Mulchen
0.25	CA129		R. Coya, María Elena
5	HJKY	CLM	RCN, Girardot (vf)
5	HJHI		RCN/LV de Garagoa, Garagoa
5	HJOI		R. Chacurí, Sampués
5	HJEB		LV del Turismo, Santa Marta
5	HJTH		LV de las Estrellas, Medellín
5	HJSZ		Ecos del Sarare, Saravena

kW	Call	Land	Station & Location
5	HJMC		Acuario AM, Cali
15/1	HJNE		LV del Ariari, Granada
10	HJCM		R. Sur, Pitalito
3	HCJA5	EQA	LV del Río Tarqui, Cuenca
1	HCNS1		R. Popular, Atuntaqui
1	HCOF2		Canal Milagreño, El Milagro
10	OAX9N	PRU	R. Dif. Nor Oriental, Bagua
1	OCX1Q		Radial Andina (r. RPP 730), Tumbes
1	OCX6B		R. Telecomunicaciones del Norte, Arequipa
1	OBX2Z		R. Casa Grande, Chócope
10	CX38	URG	S.O.D.R.E. Montevideo
10	YVLF	VEN	R. Puerto Cabello, Puerto Cabello

1295 kHz

kW	Call	Land	Station & Location
0.5	HCVM6	EQA	R. Once de Noviembre, Latacunga

1300 kHz

kW	Call	Land	Station & Location
5/1.5	LU22	ARG	R. dif. Tandil
10/5	LRA5		R. Nal. Rosario
0.15			R. Ecos, Llavallol
			R. Malvinas Argentinas, Caseros
2.5	CP51	BOL	R. Loyola, Sucre
15/6	CP-		R. Sol, Poder de Diós, La Paz
1	CP-		R. Coronel Eduardo Avaroa, Santa Cruz
1	CP82		Coop. Radial Electra, Potosí
0.3	CP168		R. Chichas, Siete Suyos
0.15	CP127		R. Juan XXIII, Uyuni *
5	CB130	CHL	R. Tierra, Santiago
10	CD130		R. Pilmaiquén, Valdivia
1	CD130		R. Cabo de Hornos, Pto. Williams
5	HJUA	CLM	R. Sindamanoy, Mocoa
5	HJEA		R. Lumbí, Mariquita
5	HJTN		R. Eucha, Belalcázar
5	HJLD		Caracol Pereira, Pereira
5	HJNB		Caracol Bucaramanga, Bucaramanga
15	HJAP		LV de las Antillas, Cartagena
5	HJRB		R. Super, Tunja
5	HCDC2	EQA	R. Cenit, Guayaquil
2/1	HCRS7		R. Sucumbíos, Lago Agrio
10	HCRU1		R. Festival, Sto Domingo de los Colorados
5	ZP10	PRG	Em. Paraguay, Asunción *
5	OAX4M	PRU	R. Comas, Comas
3	OAX8B		R. Nuevo Mundo, Pucallpa
2	OAX6P		R. Latina (r: RPP 730), Tacna
10	OAX7P		R. Onda Imperial, Cusco
1.5	OAX7X		R. Juliaca, Juliaca
1	OCY2C		R. Paraiso, Cjabamba
1	OAX1I		R. Lambayeque, Chiclayo
0.5	OAX3O		R. Huascarán, Huaraz
10/8	YVKH	VEN	R. 1300, Caracas
10	YVNS		R. Reloj, Maracaibo

1300.2 kHz

kW	Call	Land	Station & Location
1.5	OAZ4B	PRU	R. Andina, Huancayo (n. 1300)

1310 kHz

kW	Call	Land	Station & Location
1	LRA42	ARG	R. Nal. Gualeguaychú (r. 1309.5)
1			R. Dr. Gregorio Alvarez, "LV de la Comunidad", Piedra del Aguila
			AM Panamericana, Gregorio de Laferrere
10	CP68	BOL	R. San Rafael, Cochabamba
5	HJIR	CLM	RCN Urabá, Apartadó
5	HJTQ		R. Tasajero, Cúcuta
5	HJLM		R. Santa Bárbara

kW	Call	Land	Station & Location
5	HJWD		Micrófono Cívico, Palermo
5	HJAK		LV de la Patria, Barranquilla
5	HJDG		Caracol Colombia, Monteria
10	HJJZ		Colorín Color R, SF de Bogotá
20	HCGB1	EQA	R. Nal.Espejo, Quito
1	HCCP3		LV de El Oro, Pasaje (vf)
	ZP53	PRG	LV del Este, Cd. del Este
1	OBX5K	PRU	R. San Martín, Pisco
1	OBX2D		R. Chota, Chota
5	YVTS	VEN	R. Andina "Sonido 13-10", Isnotú
10	YVSM		R. Nal., Barcelona (r: 630)
1	YVSL		R. Nal., Guri (r: 630)
1	YV--		R. Nal., Elena de Uairén (r: 630)
1320 kHz			
5/1	LU10	ARG	R. Azul
25/5	LRH251		R. Chaco, Resistencia
			R. Independencia, Lanús
1	CD132	CHL	R. Lincoyan, Mulchén
1	CA132		R. Estrella del Norte, Vallenar
5	HJQI	CLM	R. Leda Internacional, San Andrés
5	HJHT		R. Guateque, Guateque
5	HJMS		R. El Sol, Barrancabarmeja
5	HJLV		La Esquina del Progreso, Fundación
1	HJNO		R. Uno, Girardot
1	HJNK		R. Luna, Palmira
1	HJTA		NVI Radio, Medellín
3	HCCI5	EQA	T.V.O., Biblián
3	HCFR2		R. Guayaquil, Babahoyo
10	HCJD6		R. Continental, Ambato
0.5	HCOB7		R. Nacional Limón, Limón Indanza
10	OAX4I	PRU	R. La Crónica, Lima
1	OAX3U		R. Miramar, Chimbote
1	OAZ4I		R. LV de Oxapampa
1	OBX2Q		R. Frecuencia Popular, Chepén
0.8	OBX6W		R. Premier, Camaná
0.35	OAX1F		R. Talara, Talara
0.25	OCZ9K		R. Pajatén, Juanjuí
1/0.5	CW39	URG	R. LV de Paysandú, Paysandú
0.3	CW132		R. Fortaleza, Rocha
10/5	YVWP	VEN	R. Apolo, Turmero
10	YVSG		R. Colonial, El Tocuyo
1324 kHz			
		PRU	R. Continental, Caraz (r.)
1325 kHz			
		ARG	R. Tango, Caseros * (vf)
1330 kHz			
			R. Mar, San Antonio de Padua
3/3.5	CP176	BOL	R. América, Oruro
1	CP112		R. Frontera, Yacuíba
3/1.5	CD133	CHL	R. Vicente Pérez Rosales, Puerto Montt
3	CB133		R. Metropolitana, Santiago (vf)
5	HJNR	CLM	Caracol San Gil, San Gil
5	HJLS		Caracol Colombia, Popayán
10	HJFE		RCN Antena 2, Pereira (vf)
1	HJRD		R. Coopeñol, El Peñol
1	HJMP		LV de Aguachica, Aguachica
5	HCRV3	EQA	R. Nal.El Oro, Machala
3	HCJA1		R. Sideral, Quito (vf)
3	HCOV1		Ondas del Volante, El Angel
2	HCLW5		R. Scala, Cuenca
10	ZP4	PRG	R. Chaco Boreal, Asunción *
1	OCX7K	PRU	R. San Miguel, Cusco
1	OCX2W		R. Regional, Trujillo

kW	Call	Land	Station & Location
1	OCX6E		R. Ondas del Misti, Mariano Melgar
0.5	OAX2P		R. Paiján, Paiján
5	CX40	URG	R. Fenix, Montevideo
5	YVNX	VEN	R. Los Llanos, Calabozo
10	YVTU		Mundial Regional, Ciudad Ojeda
1	YV--		R. Nal., La Paragua (r: 630)
1340 kHz			
1	CP24	BOL	R. Grigotá, Santa Cruz (vf)
0.5	CP-		R. 11 de Octubre, Cobija
0.35	CP146		R. San Francisco, Apolo
	CP-		Radiodifusora Copacabana, Copacabana
10	CB134	CHL	R. Caracola, Valparaíso
1	CD134		R. Panguipulli, Panguipulli
1	CC134		R. Doña Isabel Riquelme, Chillán (vf)
0.25	CA134		R. El Toco, Tocopilla *
5	HJKD	CLM	RCN, Neiva
5	HJHY		RCN Sucre, Sincelejo
5	HJHA		RCN Nariño, Pasto
5	HJFB		R. Uno, SF de Bogotá
5	HJPY		R. Mil, Cúcuta
5	HJFA		R. Alegre, Barranquilla
1	HJNY		R. Sonorama, Bucaramanga
1	HJIS		R. El Sol, Buenaventura
1	HJNP		R. Comunal, Nariño
0.5	HJVL		Brisas del Catatumbo, Tibú
5	HCRT6	EQA	R. Paz y Bien, Ambato
2.5	HCVP5		R. San Francisco, Penipe
1	HC--		R. Regional, Loja
1	HCSF2		R. Estéreo Fluminense, Babahoyo
1	HCDR4		R. Cenit, Bahía de Caráquez *
	HC--		LV de su Amiga "Estéreo Musical", Esmeraldas
3	OAX9C	PRU	R. Tropical, Tarapoto
1.5	OBX1K		R. San Francisco, Piura
1	OAZ4Q		R. Jauja, Jauja
0.5	OAX5D		R. Chincha, Chincha Alta
0.2	OAZ4M		R. Huaral, Huaral
10	CW53	URG	LV de Melo, Melo
10	YVNE	VEN	R. Uno "AM 1340", Caracas
1346.4 kHz			
1.5	OAX2D	PRU	R. Cajamarca, Cajamarca (n. 1370)
1350 kHz			
50	LS6	ARG	R. Buenos Aires, "RBA"
1	CP214	BOL	R. Ichilo, Yapacaní, San Carlos
1	CP-		América Radiodifusión, Sucre
1	CD135	CHL	R. San Carlos, Ancud
1	CA135		R. Riquelme, Coquimbo
5/1	HJHW	CLM	Em. Ecos del Río, Puerto Boyacá
5	HJOC		RCN Antena 2, Santa Marta
5	HJEN		R. Melodía, Cali
5	HJHL		R. Festival, Ibagué (vf)
5	HJLO		R. El Sol, Caucasia
2	HJLH		LV de Yopal, Yopal
1	HJMU		R. Perijá, Codazzi
1	HJ--		R. Cultural 2001, Pailitas
1	HJDS		Ondas de la Montaña, Medellín
	HJ--		Fascinación AM, Becerril
5	HCPZ1	EQA	R. Rumichaca, Tulcán
3	HCVP2		R. Reloj, Guayaquil
2/1.5	HCSF5		R. LV de San Fernando, San Fernando
1	HCPU1		LV de Sto Domingo, Sto Domingo de los Colorados
1.5	OAX8D	PRU	R. Super AM-FM, Pucallpa

kW	Call	Land	Station & Location
1	OBX6F		R. Ilo, Ilo
5	YVZZ	VEN	R. Guanipa "R. 13-50", El Tigrito
5	YVTJ		R. Falcón, Puerto Cumarebo
1351.2 kHz			
2	OAX3N	PRU	R. Ondas del Huallaga, Huánuco (n. 1350)
1355 kHz			
0.25	CP154	BOL	R. Armonía, Cliza (n. 1350) (vf)
1356.8 kHz			
		PRU	R. Regional, Chiclayo
1360 kHz			
1/0.4	LT46	ARG	R. Provincia Bernardo de Irigoyen
1	CP143	BOL	R. Libertad, Villazón
1	CP-		R. Dif. Jiménez, El Alto
	CP-		R. Stentor, La Paz
5	CC136	CHL	R. Universidad del Bío Bío, "UBB", Concepción
5	HJUT	CLM	Caracol Cartagena, Cartagena
15	HJMI		R. Autentica, Melgar
10/5	HJPK		Voz de Abejorral
10	HJRA		Ecos del Risaralda, Pereira
1	HJSD		R. Morichal, San Martín
1	HJKV		R. Láser, Zapatoca
0.5	HJ--		R. Segovia, Segovia
	HJRX		LV de la Camarca, Miraflores
5	HCHG3	EQA	R. El Sol, Machala
3	HCMT1		Oyambaro AM, Tumbaco (vf)
3	HCEG4		LV del Carmen, El Carmen
1	HCRJ5		R. América, Riobamba
1	ZP37	PRG	R. Yby Yau, Carapeguá
2	OCY2T	PRU	R. Santa Mónica, Chota
10	OAX4Y		R. Excelsior, Lima *
1	OCX6T		Radiodifusora Luza, Paucarpata
1	OBX2N		R. Super Uno, Cartavio
2.5/0.5	CW41	URG	Broadcasting San José, San José
1/0.5	CW136		R. Rio Branco, Rio Branco
5	YVTZ	VEN	R. Armonia, Charallave
10	YVTI		R. Internacional, Maracaibo
1362 kHz			
1	OAX7R	PRU	R. Sicuani, Sicuani (n. 1365)
1362.5 kHz			
0.5			R. Sudamerica, Tarma (n. 1360)
1370 kHz			
1	LRA54	ARG	R. Nal. Ingeniero Jacobacci
			R. Cosmos, Isidro Casanova
5/3	CP249	BOL	Radiodifusoras Coral, Oruro
1	CP33		R. Agricultura, Achacachi
0.5	CP158		R. La Voz de Minero, Siglo XX
0.15	CP186		R. Libertad, Cliza
1	CD137	CHL	R. Conun Huenu - La Popular, Temuco
5/2	HJBD	CLM	La Nueva R. Guaimaral, Cúcuta
5	HJEQ		RCN Cauca, Popayán
5	HJKX		R. Mundial, SF de Bogotá
2.5	HJJQ		RCN Antena 2, Zarzal
10	HJBO		Minuto de Dios, Barranquilla
1	HJNU		RCN, Rionegro
1	HJNI		Caracol Colombia, Sincelejo
5	HCER3	EQA	R. Progreso, Loja
5	HCVO2		LV del Milagro, El Milagro
2	HCRP7		R. Pastaza, El Puyo
2	HCJS1		Ecos Andinos, Pimampiro (vf)
	HC--		R. El Rocio, Biblián
1/0.5	OAX6T	PRU	R. Moquegua, Moquegua
1	OAZ7J		R. Santa Mónica, Cusco
1	OBX9A		R. Palmera, Bellavista
1	OAX3H		R. Interamericana, "LV del Pacífico", Chimbote
1	OAZ40		R. Cosmos, Huacho
			R. Tocache, Tocache
10/2.5	CX42	URG	Emis. Ciudad de Montevideo
0.5/0.25			Emis. Real, Minas de Corrales
0.25/0.1	CW137		R. San Javier, San Javier
5	YVSV	VEN	R. Portuguesa, Araure
5	YVOQ		R. La Pascua, Valle de la Pascua
5	YVLG		R. Cumbre, Ejido
1379.8 kHz			
0.3	OAX2W	PRU	R. Atahualpa, Cajamarca (n. 1380)
1380 kHz			
		ARG	LV del Futuro, Merlo
0.5	CP227	BOL	R. Luis de Fuentes, Tarija
0.25	CP221		R. 16 de Noviembre, Sacaba
50	CB138	CHL	R. Corporación, Santiago
5	HJEE	CLM	RCN, Tunja
5	HJLG		LV de La Dorada, La Dorada
2.5	HJJD		Colmundo Medellín
1	HJMM		R. Net, Valledupar
1	HJID		Em. Villa del Páez, La Plata
1	HJEJ		Armonías del Palmar, Palmira
5	HCJR6	EQA	R. Mera, Ambato (vf)
5	HCOA3		R. Impacto, Piñas
5	HCCV1		R. Cristal "RCQ", Quito
3	HCWV7		R. Morona, Macas *
3	OAX6O	PRU	R. San Martín, Arequipa
1	OBX1V		Sistema Radial, Ferreñafe
1	OBX5C		R. El Pueblo, Ica
1	OCY4U		R. ABC, Chosica
0.5			Estación Fanupe. Fanupe
5	YVME	VEN	R. Fantasía, Soledad
10	YVTL		R. Mundial, Caja Seca
10	YVNG		Ondas del Mar, Puerto Cabello
1380.7 kHz			
1	ZP8	PRG	R. Concepción, "LV del Norte"
1381.9 kHz			
1		PRU	R. Pilcomozo, Huánuco
1390 kHz			
10	LR11	ARG	R. Universidad Nacional de La Plata
1			R. Nal. Mendoza (r. LRA 960), Valle de Uspallata
5	CP3	BOL	R. Nacional de Bolivia, La Paz
0.25	CP169		R. La Voz Minera del Sud, Mina Telamayu
5/3.5	HJFO	CLM	LV de los Andes, Manizales
5	HJFY		R. Avenida, Espinal
15	HJYW		R. Suramerica, Pacho
1	HJZY		Lemas de Colombia, Bucaramanga (vf)
0.1	HJ--		R. Ciudad de Antioquia, Santa Fé de Antioquia
5	HCEA5	EQA	R. Tropicana, Cuenca
5	HCAA2		R. Sucre, Quevedo
3	HCDN5		R. Atenas, Riobamba
1.5	HCIE1		R. Uno, Urcuquí
1	HCHE4		LV de Esmeraldas, Esmeraldas
0.5		PRU	R. San Nicolás, Rodriguez de Mendoza
0.5	OCX2M		R. Espacial, Otuzco
2.5/1	CW45	URG	Dif. Treinta y Tres, Treinta y Tres (vf)
10	YVTT	VEN	R. Terepaima, Cabudare
10	YVZA		R. Fé y Alegría, Caracas

kW	Call	Land	Station & Location
	YV--		R. Lumen 2000, Maracaibo
1395 kHz			
1	OAX7F	PRU	R. Ayaviri, "LV de Melgar", Ayaviri *
1400 kHz			
7.5/1	LU11	ARG	R. Em del Oeste, Trenque Lauquén
			R. Siglo XXI, Burzaco
			R. Fantasía, Tapiales *
			R. Comunitaria Cristiana, San Fernando
			R. 2001, Quilmes Oeste *
1	CP174	BOL	R. Libertador, Santa Cruz *
	CP-		R. Nobel, La Paz *
	CP-		R. Comunidad, Patacmaya
5	CD140	CHL	R. La Amistad, Los Angeles
5	CD140		R. Belén, Puerto Montt
1	CA140		R. 140, "La Nueva Tarapacá", Iquique (vf)
5	HJCC	CLM	RCN Antena 2, Santa Bárbara
5	HJAS		RCN Antena 2, Barranquilla
5	HJJJ		R. Ipiales, Ipiales
5	HJDF		LV de Niquel, Montelíbano
5	HJKM		Em. Mariana de Bogotá, SF de Bogotá
1.5	HJ--		LV de Samaniego, Samaniego
1	HJHM		RCN Antena 2, Calarcá
1	HJWY		LV de los Samanes
1	HJBK		LV de la Gran Colombia, Cúcuta
1	HJD31		LV de Cimitarra, Cimitarra
1	HJIT		Ecos del Atrato, Quibdó
0.45	HJ--		R. Cañaveral, Morales
0.25	HJ--		Brisas del Sinú, Tierralta
7	HCVL7	EQA	R. Multicolor, Tulcán
5	HCVZ7		R. LV de Zamora, Zamora
10	HCFL2		R. Z-1, Guayaquil
3	OAX6J	PRU	R. Super Landa, Arequipa
3	OAX7I		R. La Hora, Cusco
2.5	OBX4W		Callao Super R., Callao
1	OCX1A		Radiodifusora El Milagro, El Alto
1	OBX3G		R. Willaq, Huaraz
1	OAZ4P		R. San Juan, Tarma
1	OAX2O		R. Perú, Chepén
1	OCY2D		R. Internacional del Perú, San Pablo
1	OBX9E		R. El Dorado, Villa Picota
1			Estación Wari, Ayacucho
0.2			R. Juanjuí, Juanjuí
			R. La Merced, Tongod
			Ondas del Suroriente, Quillabamba
25	CX140	URG	R. Zorrilla de San Martín, Tacuarembó (vf)
10	YVZJ	VEN	R. 1400, Barinas
1	YVNF		R. Sabana, El Sombrero
1409.9 kHz			
1	OAX1N	PRU	R. Ideal, Lambayeque (n. 1410)
1410 kHz			
0.1		ARG	R. Fortín Federación, Junín
	LRJ360		R. Obispado de San Luis,
			R. Folklorismo, José Léon Suárez
0.25	CP-	BOL	R. Roboré, Roboré
0.25	CP124	L	R. Atlántida, Oruro
2.5	CB141	CHL	R. Color, "Canal 141", Valparaíso
1	CD141		R. Loncoche, Loncoche
5	HJFS	CLM	RCN Antena 2, Honda
5	HJDU		Em. Cultural Universidad de Antioquia, Medellín
10	HJEI		R. Guadalajara, Buga

kW	Call	Land	Station & Location
1	HJTY		Caracol Colombia, Vélez (vf)
1	HCCQ2	EQA	R. Presidente "LV del Pueblo", El Milagro
1	HCEC1		R. El Timepo, Quito
1	HCGC5		R. Centro Gualaceo, Gualaceo
1	HCMS7		LV del Río Coca, El Coca
1	HCFR4		LV de Quinindé, Quinindé
2	ZP35	PRG	R. Dif. Mangore, San Juan Bautista Misiones (n. 1430)
5	OAX2Y	PRU	R. Heróica, Trujillo
1	OBZ4V		R. Universal, Huacho
1	OBX8I		R. Concierto, Pucallpa
5/3.5	CX44	URG	Emisora del Siglo, Montevideo
1/0.5	CW141		R. Turística, Termas del Arapey
5	YVST	VEN	R. Turén, Turén
10	YVSP		R. Simpatía, Valera
1410.1 kHz			
1	OAU4A	PRU	R. Láser, Santa Bosa Sacco (n. 1480)
1417 kHz			
5	ZP42	PRG	R. Güyrá Campana, Horqueta (n. 1420)
1417.5 kHz			
0.5	OBX5H	PRU	R. Apurímac, Abancay (n. 1420)
1420 kHz			
1	LRH362	ARG	R. Tupá Mbaé, Posadas
1	LRK221		R. Ciudad Perico, Perico
	LRJ359		R. Opispado de San Luís, San Luís
			R. Mágica, Lanús
			R. Mailin, Libertad
	LRJ361		R. Arzobispado de Córdoba,
	LRH375		R. Arzobispado de Corrientes, Itati
1.5	CP254	BOL	R. Guadalquivir, Tarija
1	CP-		R. Chaka, Pucarani
1	CP49		R. Centro, Cochabamba
1	CB142	CHL	R. Panamericana, Santiago (vf)
1	CC142		R. Nueva Maule, Cauquenes
5	HJOG	CLM	LV de las Antillas, Cartagena
5	HJHK		Caracol Manizales, Manizales
5	HJBH		Caracol Colombia, Santa Marta
10/5	HJLE		RCN Antena 2, Ibagué
1	HJSN		R. Lenguerque, Zapatoca
1	HJD23		Ecos de Frontino, Frontino
3	HCMA6	EQA	R. Nuevos Exitos, Salcedo (vf)
3	HCRN1		R. Bahá'í, Otavalo
1	HCNR3		LV de Huaquillas, Huaquillas
1	OCX1H	PRU	R. San José, La Unión
1	OBZ4G		R. San Isidro, Lima
1	OCX7C		R. Qollasuyo, Juliaca
1	OAX8Z		R. Oriente, Yurimaguas *
1			R. Nor Andina, Celendín *
			R. Nor Andina, Baños del Inca
5	CW43	URG	R. Lavalleja, Minas
1/0.25	CW142		LV de Artigas, Artigas
5	YV--	VEN	R. Sintonía, San Antonio de los Altos
5	YVNZ		R. Marabina, Maracaibo
10/5	YVRW		R. Cardenal, Carora
1420.1 kHz			
0.1	CX142	URG	R. Felicidad, Paysandú
1429.9 kHz			
1	OAZ4V	PRU	R. Universal, Huancayo (n. 1430)
1430 kHz			
5/1	LT14	ARG	R. San Nicolás
1/0.25	LV26		R. Río Tercero
1	LRI235		R. Balcarce, Balcarce

kW	Call	Land	Station & Location
			R.Libertad, Libertad
			R. Trinidad, Presidente Derqui
			R. José de San Martín, "la Pionera", El Jagüel
0.25	CP141	BOL	R. Nuestro Señor de Burgos, Mizque
0.25	CP193		R. 23 de Marzo, Tupiza
1	CC143	CHL	R. Chilena AM, Rancagua
5/1.5	HJEG	CLM	LV de Belalcázar, Popayán
5	HJQX		R. Majagual, Sincelejo
5	HJPW		Colmundo, Barranquilla
15/2	HJRL		Antena de los Andes, Santa Rosa de Cabal
12	HJKU		Em. Kennedy, SF de Bogotá
1	HJMF		R. Venus, Puerto Berrío
1	HJCK		R. Sensación, Yarumal
1	HJBP		R. Cariongo, Pamplona
1	HJIU		Armonías del Ingrumá, Riosucio
0.5	HJ--		R. Preferencial, Sibundoy
5	HCCV3	EQA	R. Musical "Punto 14-30", Loja
5	HCJC6		R. Guaranda, Guaranda
10	HCMB2		R. Caribe, Guayaquil
1	HCGF1		R. Continental, Quito
1	OBX9H	PRU	R. Utcubamba, Utcubamba
1	OBX2T		R. Santa Bárbara, Ascope
1	OAU4H		R. OAU4H, Chancay
1	OAZ3H		R. Chavín, Santa
1	OAZ7M		Antena Uno, Cusco
			R. Gotas de Oro, Chiclayo
5/1	CW25	URG	R. Durazno, Durazno
25	YVTP	VEN	R. Bahía, Puerto La Cruz
10/5	YVTM		R. Caicara, Caicara del Orinoco
10	YVNB		R. Satélite, Guacara

1440 kHz

kW	Call	Land	Station & Location
1	LV27	ARG	R. San Francisco
1	LRA53		R. Nal. San Martín de los Andes
1	LV20		R. Laboulaye
0.25	LU36		R. Coronel Suárez
			R. Impacto, Tapiales
			R. Chascomús "RCH," Chascomús
			"RSO" R. Sudoeste, Marcos Paz *
5/3	CP107	BOL	R. Yaguarí, Vallegrande
1	CP61		R. Batallón Colorados, La Paz
0.5	CP-		R. Oriente, Camiri
1	CA144	CHL	R. Minería, Arica
1	CC144		R. El Sembrador, Chillán
1	CA144		R. Agricultura, La Serena
5	HJGM	CLM	RCN, Sogamoso
5	HJBM		R. Internacional, Honda
5	HJNZ		R. 24, Medellín
5	HJEK		Caracol Tuluá, Tuluá
1	HJIB		RCN Caquetá, Florencia
5	HCMD7	EQA	R. Puyo, El Puyo *
5	HCDF1		R. Panorama, Ibarra
3	HCOV5		Ondas del Volante, Azogues
2.5	HCDY4		R. Iris, Esmeraldas
	HC-		R. Antología, Caluma
2	OBX1T	PRU	R. Cooperativa Tumán, Chiclayo (r. 1440.3) (vf)
1	OAZ30		R. LV de Pomabamba, Pomabamba
3/0.5	CX144	URG	R. Rivera, Rivera
3	CV144		R. Chuy, Chuy (vf)
5	YVRF	VEN	Super Suave Orituco, Altagracia del Orituco
5	YVZI		R. Estelar, Guanare

kW	Call	Land	Station & Location
1	YVTY		R. Sucesos, Táriba
5	OAX4K	PRU	R. Imperial 2, Lima (n. 1440)

1445 kHz

kW	Call	Land	Station & Location
1	CP-	BOL	Super Broadcasting Alborada "SBA", Santa Cruz (n.1475)
0.5	HJ--	CLM	Em. R. Unión, La Palma

1450 kHz

kW	Call	Land	Station & Location
1/0.25	LT44	ARG	R. Fortín Yunka, Formosa
0.5			R. Ciudad, Lanús
			R. Ambar, Caseros
1	CP61	BOL	R. Em. Bolivia, Oruro
	CP-		Radiodifusora Capinota, Capinota
	CP-		R. Amazonia, Cobija
	CP-		R. Amanacer, Huari
5	CC145	CHL	R. Libertad, Curicó
1	CB145		R. Universidad Técnica "Federico Santa María", Valparaíso
1	CD145		R. Minería, Puerto Varas (vf)
5	HJNL	CLM	RCN Antena 2, Manizales
5	HJHH		R. Metropolitana, Bucaramanga
5	HJVY		R. Ciudad de Flandes, Flandes
10/5	HJTT		Ondas del Porvenir, Samacá
1	HJMX		R. Mancomoján, Carmen de Bolívar
1	HJE20		R. Ciudad de Urrao, Urrao
0.2	HJ--		R. LV del Nordeste, Remedios
0.5	HJ--		LV del Cauca, El Bordo
3	HCHW2	EQA	R. Rey, Quevedo
10	HCSC5		R. Calidad, Riobamba
1	HCSE2		R. Sta Elena, Sta Elena
1	HCSC1		R. Sensación, Cayambe
1	HCDR2		R. Minutera, Pascuales (vf)
1	HCRI7		R. Interoceánia, Sta Rosa de Quijos
1.5/2	OAX9D	PRU	R. Rioja, Rioja
1.5	OCX2J		R. San Juan, Trujillo
1	OCX7W		R. Santa Rosa, Quillabamba
1			R. Panorama, Andahuaylas
5	CX46	URG	R. América, Montevideo
1/0.25	CW145		R. Arapey, Salto
10/5	YVXC	VEN	R. Icabarú, Puerto Ordaz
10	YVKJ		Sonera 14-50, Caracas
10	YVZQ		R. Informativa, Puerto de Altagracia

1455 kHz

kW	Call	Land	Station & Location
2		BOL	R. Estévan Arze, Tarata

1460 kHz

kW	Call	Land	Station & Location
1/0.25	LT29	ARG	R. Venado Tuerto
0.25	LU34	ARG	R. Pigüé
0.25	LU30		R. Maipú
0.25	LW2		R. Emis. Tartagal
			R. Estilo, Longchamps
10	CA146	CHL	R. Antofagasta, Antofagasta
1	CB146		R. Yungay, Santiago
1	CC146		Nueva R. Talcahuano, Talcahuano
5	HJZU	CLM	RCN Antena 2, Pasto
5	HJVH		R. Uno, Barranquilla
3	HJIW		R. Monumental, Cúcuta
3	HJFL		Atalaya Agustiniana, San Agustín
2	HJMN		LV de Amalfi, Amalfi
10	HJJH		R. Ciudad Milagro, Armenia
10	HJJW		Em. Nuevo Continente, SF de Bogotá
1	HJAL		R. Sincelejo, Sincelejo
1	HJE26		R. Capiro, La Ceja
1	HJTF		Ondas del Darién, Turbo
5	HCIC6	EQA	R. Nuevos Horizontes, Latacunga
5	HCCL3		R. Cariamanga, Cariamanga

kW	Call	Land	Station & Location
	HC--		LV de Gualaquiza, Gualaquiza
5	0AX7W	PRU	R. El Sol de los Andes, Juliaca
2.5	0AX5K		R. Internacional, Pisco
2	OBX2Y		R. Estelar, Guadalupe
1	0AX1V		R. Sullana "LV de Chira", Sullana
1	0AZ4F		R. La Oroya, La Oroya
1	OBX9B		R. Frec. Popular, Rioja
1	OBX3A		R. Chimú (r: RPP 730), Chimbote
1	OBX6C		R. Bahía, Mollendo
1	CX146	URG	R. Carmelo, Carmelo
0.25	CV146		R. José Batlle y Ordoñez, José Batlle y Ordoñez
5	YVRJ	VEN	R. Jardín, Boconó
1470 kHz			
1/0.25	LT26	ARG	R. Nuevo Mundo, Colón
1/0.25	LT20		R. Junín (vf)
1.5/0.6	LU26		Emis. Coronel Dorrego
1	LT28		R. Rafaela
1			R. Municipal, Luis Beltrán
			R. Conurbana del Sur, Wilde
			R. Contacto, San Antonio de Padua
			R. Ciudadana, Luis Guillón
0.25	CP215	BOL	R. CORDECH, Alcalá
1	CB147	CHL	R. Sargento Aldea, San Antonio
5	HJHQ	CLM	R. Futurama, Pacho
5	HJTB		Ondas de Ibagué, Ibagué
5	HJPX		Colmundo, Cartagena
10/5	HJNT		R. Total, Cali
10	HJII		R. Musical, Medellín
1	HJB63		R. Uno, Iza
1	HJIF		R. Tres Fronteras, Puerto Asís
5	HCJC1	EQA	Ecos de Cayambe, Cayambe
1	HCRD4		R. Richi, Tosagua
1	HCLD2		R. Ecos de Naranjito, Naranjito
25	0AX4B	PRU	CPN Radio (Cadena Peruana de Noticias), Lima
1.5	0AX6M		. Tacna, Tacna
1	OCX4D		R. Huancayo, Huancayo
5	CX147	URG	R. Cristal del Uruguay, Las Piedras
1/0.5	CW147		R. Tacuarí, Melo
10	YVSY	VEN	R. Vibración, Carúpano
10	YVJW		R. Latina, Valencia
1475 kHz			
1	CP262	BOL	Emisoras Verde y Blanco, Santa Cruz (n. 1450)
	CP-		R. Tiraque, Tiraque (r.)
1476.4 kHz			
1		PRU	R. Ilucán, Cutervo (n. 1470)
1480 kHz			
0.25	LU27	ARG	R. Dolores
			R. Temperley, Temperley
0.1	CP-	BOL	Patrimonio R., Potosí
1/0.25	CA148	CHL	R. Amanecer, Ovalle
1	CC148		R. Tomé, Tomé
1	CD148		R. General Baquedano, Valdivia
5	HJFC	CLM	R. Sonorama, Pereira
5	HJOD		R. Rodadero, Santa Marta
5	HJTZ		Antena 2 RCN, Bucaramanga
1	HJTC		R. Sonsón, Sonsón (n. 1490)
1	HJVB		R. Guayabal, Guayabal
0.25	HJ--		LV del Samán, Bochalema
5	HCCY6	EQA	R. Popular de la Maña, La Maña
3	HCJB4		R. LV de Jipijapa, Jipijapa
3	HCWP5		R. Atlántida, Alausí
3	HCBS3		Ondas del Jubones, Pasaje (vf)
1	HCMC1		R. Municipal, Cotacachi
0.5	HC--		R. Luciérnaga del Condór, Yanzatza

kW	Call	Land	Station & Location
	HC		R. Sucre, Machala
1	ZP23	PRG	R. Mariscal Francisco Solano López, Bella Vista Norte
1	ZP20		R. America, Ñemby (vf)
1	OCX4V	PRU	R. K'ler, Barranca
1			R. Frecuencia Popular, Piura
1	0AZ7G		R. Espinar, Yauri
0.6	OCX2C		Radiodifusora Comercial San Pedro, Virú
0.25			R. Nuevo Imagen, Hualgayoc
			R. Lajas, Chota
3/1.5	CW43B	URG	R. Internacional, Rivera
3/1	CW148		R. Universo Castillos
1/0.7	CX148		Difusora Rio Negro, Young
1485 kHz			
1	CP135	BOL	R. La Voz del Valle, Punata
1490 kHz			
1	LV22	ARG	R. Huinca Renancó
	LRI202		R. Obispado de San Juan de los Arroy, Pergamino
			La Nueva R., Isidro Casanova
1	CP172	BOL	R. San José, San José, Oruro
0.35	CP198		R. Pedro Domingo Murillo, Quime
0.25	CP196		R. Moxos, San Ignacio de Moxos
0.25	CP-		R. Mairana, Mairana
	CP-		R. Domingo Savio, Villa Independencia
5	CD149	CHL	R. Malleco, Victoria
5	CA149		R. Alicanto, Salvador
1	CB149		R. El Canelo AM, San Bernardo
5	HJAY	CLM	Onda Nueva, Barranquilla
5	HJZB		LV de los Robles, Tuluá
5	HJBS		Em. Punto Cinco, SF de Bogotá
15	HJYD		R. La Paz Caracol, Paipa
1	HJAG		R. Garzón, Garzón
1	HJJO		LV de San Marcos, San Marcos
5	HCSM5	EQA	R. Santa María, Azogues
3	HCAI6		R. Moderna, Píllaro
2.5	HCAE4		R. Unión, Esmeraldas
1	HCVY2		La R. Dinámica, Guayaquil
1.2	0AX6Q	PRU	R. Fidelidad, Arequipa
1	OCX2S		R. San Pedro de Lloc, San Pedro de Lloc
1	0AZ3G		R. San Jacinto, Santa
1	OBX9N		R. Ondas del Río Mayo, Nueva Cajamarca
1	0AX1L		R. Imperio, Chiclayo
1	0AX8J		R. Atlántida, Iquitos
0.5	0AX5N		R. Nasca, Nasca
5/4	CX149	URG	R. del Oeste, Nueva Helvecia
0.1	CV149		R. del Centro, Baltasar Brum
10/5	YVRP	VEN	R. El Sol, Maracaibo
10	YVXD		Dinámica 14-90, Caracas
1	YVSQ		R. Mérida 14-90, Mérida
1495 kHz			
2.5	CP152	BOL	El Mundo Radiodifusión, Sacaba
1500 kHz			
5/1	LV25	ARG	R. Unión, Bell Ville
1/0.25	LT45		R. San Javier
1			R. Municipal, Gral. Conesa
0.25	LT22		R. Nueva Era, Pehuajó
0.25	LT34		R. Nuclear, Zárate (vf)
0.25	LRI208		R. El Sol AM 1500, Lanús
1	CP238	BOL	R. Sagrado Corazón, Mineros
1	CB150	CHL	R. Trasandina, Los Andes
1	CD150		R. Tierra del Fuego, Porvenir
1	CA150		R. Minería, Iquique
1	CC150		R. Centenario, San Javier
5	HJTW	CLM	R. Sumapaz, Fusagasugá
5	HJLJ		R. Mil 500, Cali

kW	Call	Land	Station & Location
5	HJZH		R. 1500 La Vallenata, Medellín
5	HJUW		LV de la Feria, Manizales
5	HJSH		Caracol Moniquirá, Moniquirá
2	HJ--		R. Cultural "LV del Telembí", Baracoas
5	HCAD4	EQA	R. Satélite, El Carmen
5	HCEB4		R. Manta, Manta
5	HCHG2		LV del Río Vinces, Vinces
2	HCWN5		R. Puruhá, Riobamba
1	HCRO1		R. Otavalo, Otavalo
2	OCX7P	PRU	R. Em. Frontera, Puno
10/5	OBX4I		R. Santa Rosa, Lima
1.5	OCX7G		Radiodifusora Canchis, Sicuani
1	OBX3I		R. Luz y Sonido, Huánuco
1	OCX1E		R. Jupiter, Sechura
1	OBX2X		R. JCL, Florencia de Mora
1	OBX6V		Nueva R. 2000, Orcopampa
10/5	YVRZ	VEN	R. Dos Mil, Cumaná

1505 kHz

kW	Call	Land	Station & Location
1.5	HCOY7	EQA	Ondas del Río Yacuambí, Yacuambí

1510 kHz

kW	Call	Land	Station & Location
3/0.25	LV21	ARG	R. Champaqui, Villa Dolores
			La Voz del Oeste, Libertad
5/1	CP1	BOL	R. Chuquisaca, El Alto
0.25	CP80		R. Litoral, Cochabamba *
0.25	CP102		R. 27 de Diciembre, Villamontes
1/0.5	CA151	CHL	R. Luís Alvarez Sierra. Illapel
1	CD151		R. Teniente Merino, Lebu
1	CC151		R. Rancagua, Rancagua
5	HJZA	CLM	R. Estrella, Armenia
5	HJD24		LV de la Unión, La Unión
5	HJUH		LV de Ciénaga Espectacular, Santa Marta
1	HJA22		LV de San Luis, San Luis Gaceno
1	HJHX		LV Panamericana, Bucaramanga
0.5	HJ--		LV de los Cedros, Libanó
3	HCUC3	EQA	R. Unión Calvense, Cariamanga
3	HCLV7		R. Ecos del Oriente, Lago Agrio
1	HCRY6		R. Runacunapac Yachana "R. El Saber del Hombre", Simiátug
1	HCHD2		Inst.Oceanográfico de la Armada, Guayaquil
0.5	HCRC5		LV de la Juventud, Cañar
1.5/3	OCX4J	PRU	R. Tarma, Tarma
1	OBX7P		R. Red, Huánchac
1	OAU4M		R. Organización Flores del Campo, Barranca
1	OAX5F		R. LV de Nasca, Nasca
1	OCX6Q		R. Alegría, Arequipa
1/0.75	CW57	URG	R. San Carlos, San Carlos
0.25	CX151		R. Rincón, Fray Bentos
0.25	CW151		R. Ibirapitá, San Gregorio de Polanco (vf)
20	YV--	VEN	LV del Centro/Informativa 1510, Güigüe

1512.1 kHz

kW	Call	Land	Station & Location
		PRU	Inti R., Abancay

1517 kHz

kW	Call	Land	Station & Location
2	OCX2Q	PRU	R. LV del Inca, Baños del Inca (n. 1510)

1520 kHz

kW	Call	Land	Station & Location
1.5	LV24	ARG	R. Manantiales, Tunuyán
0.25	LT38		R. Gualeguay
0.25	LT37		Em. Rufino
			R. Güemes, Quilmes Oeste
1	CP179	BOL	R. Petrolera, Santa Cruz
0.25	CP207		La Voz del Cobre, Corocoro
1	CC152	CHL	R. Nueva Soberanía, Linares
1	CB152		R. Integración, San Antonio (vf)
0.1	CD152		R. Aníbal Pinto, Lautaro *

kW	Call	Land	Station & Location
5	HJLQ	CLM	R. Minuto, Barranquilla (vf)
5	HJLI		Ecos del Palmar, SF de Bogotá
1	HJRL		R. Calina, Santa Rosa de Cabal
1	HJAM		R. Altamizal, Dolores
1	HJMA		LV de Suroeste, Jericó
1	HJ--		Em. Una Voz de la Frontera, Puerto Santander (vf)
1	HJMZ		Ecos de la Sierra Flor, Sincelejo
0.5	HJ--		R. Cristalares Timbío, Timbío
0.3	HJ--		Brisas del Palmar, Caucasia
	HJ--		Sonoradio 1520 AM, Viterbo
3	HCJJ2	EQA	R. Punto Verde, Quevedo
2.5	HCRI5		LV de Guamote, Guamote
1	HCTI1		R. Ibarra, Ibarra
1	HCRN2		LV de Naranjal, Naranjal
1.2	OCX2G	PRU	R. Santiago, Santiago de Chuco
1			R. Virú, Virú
0.25	OAX9X		R. Visión, Junjuí (r. 1545)
2	CX152	URG	R. Cerro Largo, Melo (vf)
0.1	CV152		R. Paz, "la nueva R.", Guichón
25	YVIC	VEN	R. Bonita "La Guapa", Guatire

1520.4 kHz

kW	Call	Land	Station & Location
2	OAZ4D	PRU	R. Minería, "la R. Estrella. La Oroya (n. 1520)

1530 kHz

kW	Call	Land	Station & Location
1	LRJ200	ARG	R. Centro Morteros, Morteros
0.1	LU25		R. Carhué
			RG R. Libertador José de San Martin, Glew
			R. Popular, González Catán
			R. Esperanza, Bella Vista
1	CP111	BOL	R. Em. Ballivián, San Borja
1	CP237		R. Don Bosco, Kami
0.25	CP200		R. Litoral, Llica
1	CB153	CHL	R. Nexo, Quillota
1	CA153		R. Juan Godoy, Copiapó
1	CC153		R. El Carbón, Lota
1	CD153		R. Calbuco, Calbuco
5	HJDN	CLM	R. Total, Medellín
5	HJOZ		LV de la Prov. de Padilla, San Juan del Cesar
5	HJGD		Em. Reina de Colombia, Chiquinquirá
1	HJJB		R. Sevilla, Sevilla
	HJ--		R. Integración, Morales
	HJR73		Ecos del Pacífico, Guapí
5	HCJY1	EQA	R. Uno, La Concordia
5	HCCC5		Ondas Cañarias, Azogues
3.5	HCMP2		LV de la Península, La Libertad (vf)
3	HCVP5		R. LV de Pallatanga, Pallatanga
1	HCMZ6		LV del Dorado, Pelileo
3	OBX2R	PRU	R. Oriental, Jaén
1	OBZ4S		R. 15-50, Huancayo
1	OBX3H		Antena 1, Chimbote
0.5	OAZ7F		R. Confraternidad, Yauri
1.2/0.25	CX50	URG	R. Independencia, Montevideo (r. 1510) (vf)
0.25	CW153		Emisora Cono Sur, Nueva Palmira
10	YVNP	VEN	R. San Felipe, San Felipe

1540 kHz

kW	Call	Land	Station & Location
1	LT35	ARG	R. Mon, Pergamino
0.25	LU28		R. Tuyú, General Madariaga
			R. Universal, Pablo Podestá *
			R. Digital, Ituzaingó *
			R. Clube, Bánfield
1	CP-	BOL	R. Chiwalaki, Vacas
1	CB154	CHL	R. Sudamérica, Santiago
1	CD154		R. San José de Alcudia, Río Bueno
1	CC154		R. Central, Chillán
2	HJRQ	CLM	R. Austral, Túquerres (cfr 1300)
1	HJZF		R. Cóndor "Em. Universitaria",

kW	Call	Land	Station & Location
			Manizales
1	HJHD		LV del Petróleo, Barrancabermeja
1	HJB89		Brisas del Río Chico, Belmira
0.3	HJ--		R. Valle de Pubenza, Popayán
0.25	HJ--		R. El Sur, San Vicente de Chucurí
0.25	HJ--		LV Dorada, Segovia
	HJD89		LV del Nevado Cumbal, Melgar
3	HCFM2	EQA	R. Cristal, Ventanas
1	HCPV1		R. Mira. Mira
1	HCDP1		R. Caracol, Quito
0.5	HCMH6		O.R. C., Latacunga (vf)
0.25	HCVB7		R. Dif.Católica Cultural "Voz del Upano", Macas
1	OBX2B	PRU	R. Star, Trujillo
1	OCX7V		R. Chasqui, Cusco
1	OBZ4U		R. Barranca, Barranca
1	OBX1B		LV de la Frontera, Tumbes
0.25	CX154	URG	R. Patria, Treinta y Tres
0.1	CW154		R. Charrúa, Paysandú
0.1	CV154		R. Centro, Cardona

1545 kHz

kW	Call	Land	Station & Location
0.35	CP191	BOL	R. Mejillones, Tarata

1550 kHz

kW	Call	Land	Station & Location
1	LT23	ARG	R. San Jenaro Norte
1	LT33		R. Nueve de Julio, Buenos Aires
0.4			R. La Cruz del Sur, Guernica
0.25	LT40		R. La Voz de la Paz, La Paz
0.25	LT32		R. Chivilcoy
			R. Metropolitana, Ciudadela
5/10	CP115	BOL	R. Caranavi, Caranavi
1	CP205		R. Tamengo, Pto. Quijarro (n. 1495)
0.3	CP-		R. Unión, Quillacollo
1	CB155	CHL	R. Putaendo, Putaendo
1	CC155		R. Manuel Rodríguez, San Fernando (r 1555) (vf)
0.25	CD155		R. Regional, Traiguén
5	HJCB	CLM	R. El Sol, Barranquilla
5	HJQD		Caracol Armenia, Armenia
10	HJZI		MCI Radio 15-50, SF de Bogotá
1	HJLT		R. 24, Cali
1	HJUN		LV del Río Arma, Aguadas
0.5	HJ--		Ondas del Nechí, Campamento (vf)
5	HCAD5	EQA	R. Chaguarurco, Santa Isabel
2	HCEI6		R. Montalvo, Ambato
2	HCAD2		LV de El Triunfo, El Triunfo
1	HCRA7		R. Amazonas, Archidona
1	OAX1O	PRU	R. Superior, Monsefú
1	OBX4P		R. Independencia, Lima
1	OCX7A		R. Cultura, Puno *
0.5	OAX9Y		R. Amazonas, Bagua
1/0.25	CV155	URG	R. Agraciada, Mercedes
0.25	CW155		R. Sarandí del Yí, Sarandí del Yí
10/5	YVXO	VEN	R. Ojeda, Cd. Ojeda
10	YVMW		R. Metropolitana, Los Teques

1555 kHz

kW	Call	Land	Station & Location
0.5	HJ--	CLM	R. Parroquial, El Santuario

1560 kHz

kW	Call	Land	Station & Location
2.5/1.5	LT11	ARG	R. Gral. Francisco Ramírez, Concepción del Uruguay
			R. Inolvidable, Villa Dominico
			R. Castañares "RCI," Ituzaingó
1	CP255	BOL	R. Occidental, Oruro
0.5	CP-		1 de Octubre "la radio," Capinota
5/3	CA156	CHL	R. Parinacota, Putre (vf)
1	CD156		R. Parque Nacional, Villarrica
1	CB156		R. Manantial, Talagante
5	HJHE	CLM	Voces Rovirenses, Málaga
5	HJLP		RCN Antena 2, Tuluá
5	HJCP		RCN Antena 2, Arbeláez
5	HJXZ		Colmundo, Medellín

kW	Call	Land	Station & Location
1	HJPZ		R. Codazzi, Codazzi
2	HCCS2	EQA	R. Sideral, Daule
2	HCTR3		LV del Guabo, El Guabo
1.5	HCZD1		Ecos Culturales de Urcuquí, Urcuquí
1	HCHG1		LV de El Dorado, Puembo
1/0.85	OBX2J	PRU	R. Nuevo Continente, Cajamarca
			R. Municipal de Cusco, Cusco
2	CW51	URG	R. Maldonado, Maldonado
1/0.25	CV156		R. Vichadero
0.25	CX156		Dif. Americana, Trinidad
10/5	YVLZ	VEN	R. Dif.Andina, Mérida

1570 kHz

kW	Call	Land	Station & Location
1	LRI229	ARG	R. Las Flores
			R. Líder, Central
			R. Interactiva, Ciudad Madero
			R. AM Rocha, La Plata
			Carisma AM, San Justo *
10	CP132	BOL	R. Continental, El Alto
0.5	CP-		R. 1 de Mayo, 1 de Mayo
0.3	CP-		R. Urkipiña, Quillacollo
1	CC157	CHL	R. Nueva O'Higgins, Rancagua (vf)
1	CC157		R. Campanario, Talca
0.25	CD157		R. Acuarela, Nueva Imperial
1	HJZT	CLM	R. Sensación, Manizales (vf)
1	HJC22		R. Ciudad Dabeiba, Dabeiba
1	HJTG		R. Camino, Machetá
1	HJUJ		Armonías Boyacenses, Motavita
	HJ--		LV de Fomeque, Fomeque
10	HCPG1	EQA	R. Nucanchic, Maldonado *
0.5	HCAI5		Ecos de los Andes, Chunchi
	HC--		R. Jet, Manta
1	OCX6I	PRU	R. Willy, Uraca
1	CX157	URG	R. Canelones
0.25/0.1	CV157A		R. Celeste, Tomás Gomensoro

1577.5 kHz

kW	Call	Land	Station & Location
1	OAU2E	PRU	R. Frecuencia San Ignacio, San Ignacio

1580 kHz

kW	Call	Land	Station & Location
0.25	LT27	ARG	R. La Voz del Montiel, Villaguay
0.25	LT36		R. Chacabuco
			"RCL" R. Cd. de Libertad, Libertad *
			R. Urbana, Lanús
			R. General Martín Miguel de Güemes, Guernica *
			AM Ciudad de San Justo, San Justo
1	CP-	BOL	R. Andrés Ibáñez, Santa Cruz
1	CD158B	CHL	R. Nueva Continental, Collipulli
1	CC158		R. Colchagua, Santa Cruz (vf)
0.25	CD158A		R. Millaray, Cañete
5	HJNA	CLM	R. Robledo, Cartago
5	HJQT		R. María "LV Católica de su hogar", SF de Bogotá
5	HJRM		R. Costanera, Sincelejo
5	HJQZ		FSI Radio "LV de las Estrellas", Barranquilla
1	HJKB		R. Zulima, Villa del Rosario
1	HJDE		R. Miraflores, Rovira
1	HJLC		LV del Banco, El Banco
5	HCVA4	EQA	R. Bolívar, Esmeraldas
3	HCTP5		Ecos del Portete, Girón
1.5	HC--		LV de Libertador, Chimbo
1	HCLF1		R. Orellana, Machachi
0.5	HCCP2		Canal del Pueblo, Sanborondón
0.25	HCAB3		Ondas de Paltas, Catacocha
1.5	OAX9M	PRU	R. Central, Bellavista
1	OBX2H		R. Tayabamba, Tayabamba
0.5	OAX5M		R. Villa Rica, Huancavelica
0.5	CW54	URG	Emisoras del Este, Minas (vf)
0.25	CW158		R. San Salvador, Dolores
10/5	YVYV	VEN	R. Venezolana, Calabozo

kW	Call	Land	Station & Location
10/5	YVTK		Manzanares 15-80, Cumaná
10	YVYO		R. Occidental, Sta Rita
1580.4 kHz			
1/1.5	OBX1M	PRU	R. Naylamp, Lambayeque
1584 kHz			
3/1	OBX7Q		RPP (r: RPP 730), Cusco
1587.1 kHz			
1			R. Armonía, Abancay (n. 1570)
1590 kHz			
3	CP155	BOL	R. Bermejo, Bermejo
	CP-		R. Globo, La Guardia
5	CD159	CHL	Em. Tepual, Llanquihue
1	CB159		R. Aconcagua, San Felipe (vf)
0.25	CC159		R. Rengo, Rengo (vf)
0.1	CC159		R. El Roble, Parral
5	HJWB	CLM	R. Sra del Socorro
5	HJIP		Em. Nuevo Continente, Envigado
1	HJQM		Ecos de la Miel, Samaná
	HJ--		R. Espacial, Andalucía
	HJ--		Ondas del Rioseco, Rioseco
1	HCQT6	EQA	R. Panamericana, Quero
1	HCRZ1		R. Mensaje, Tabacundo
1	HCCS6		LV de La Maná, La Maná
0.25	HCAS2		R. Record, La Libertad
5	OAZ4Z	PRU	R. Agricultura, "La Peruanísima", Lima
1	OCX6S		R. Mundo, Cerro Colorado
1	OCX2L		R. JJ, Chócope
0.25	CW159	URG	R. Regional, Lascano
0.25	CV159		R. Regional, Constitución
0.1	CX159A		R. Paso de los Toros, Paso de los Toros
10	YV--	VEN	R. Deporte 15-90, Caracas
1590.5 kHz			
2	CX159	URG	R. Real, Colonia (vf)
1593 kHz			
		ARG	R. Emociones, Luis Guillón
1598 kHz			
1		PRU	R. Jerusalén, Coop. Atahualpa-Jerusalén
1600 kHz			
1.2		ARG	R. Armonia, Caseros
1	LR14		R. Ciudad Aluminé, Aluminé
0.5	CP153	BOL	R. Continental, Punata
0.25	CB160B	CHL	R. Reloj, Viña del Mar
0.25	CC160A		R. Regional, Concepción (vf)
0.25	CB160A	CHL	R. Armonía, Puente Alto
5	HJHB	CLM	Armonías Zipaquireñas, Zipaquirá
2.5	HJ--		Armonías de Risaralda, Risaralda (vf)
0.25	HJ--		LV de Aranzazu (vf)
	HJ--		R. Fortaleza, Sogamoso
	HJ--		R. Bello Horizonte, Pesca (r. 1620-1630)
	HJ--		LV de Colina, Risaralda
3	HCPB5	EQA	R. Intiñán, Girón
25	HCUP1		R. Unión, Quito
1	HCJP2		R. Consular, Playas (vf)
		PRU	R. Camaná, Camaná
5	CV160	URG	Emisora Continental, Pando
1	CX160		R. Litoral, Fray Bentos
1610 kHz			
		ARG	R. Cultura, Lanús
1620 kHz			
5/1			R. Universidad de Buenos Aires (CP)
			R. Universo, Glew
1640 kHz			
	HJ--	CLM	Toquilla Estéreo 1640, Aquitania
1680 kHz			
1/0.5		ARG	R. Universidad Tecnolólica Avallenada (CP)

WORLD RADIO TV HANDBOOK

WRTH

THE DIRECTORY OF INTERNATIONAL BROADCASTING

Published by:
**WRTH Publications Limited
PO Box 7373
Milton Keynes
MK12 5ZL**

Fax: +44 1908 321030
E-mail: editor@wrth.demon.co.uk

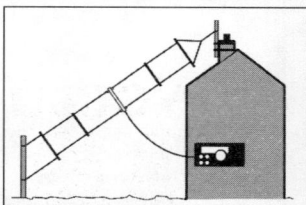
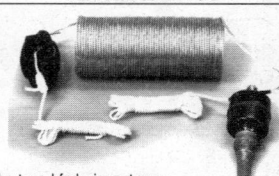

NOW LISTEN HERE!

Every month *Short Wave Magazine* has essential information on scanning, decode, maritime beacons, propagation & satellite TV, etc.

In fact it has all the information a short wave listener could possibly want...

SUBSCRIBE NOW!

And read Britain's best selling radio magazines every month

CALL US TODAY!

...and with features on new products, antennas, packet radio, h.f. bands, vintage radio & construction, etc.

Practical Wireless is the monthly magazine no radio amateur should be without.

MAIL ORDER HOTLINE

Tel: (+44) (0) 1202 659930

Fax: (+44) (0) 1202 659950

PW PUBLISHING
ARROWSMITH COURT
STATION APPROACH
BROADSTONE
DORSET BH18 8PW
United Kingdom

SUBSCRIPTION RATES – 1 YEAR

PW	SWM	JOINT SUBSCRIPTION
£25.00 (UK)	£30.00 (UK)	£50.00 (UK)
£30.00 (Europe)	£35.00 (Europe)	£59.00 (Europe)
£32.00 (Rest of World)	£38.00 (Rest of World)	£63.00 (Rest of World)
$50.00 (USA)	$60.00 (USA)	$100.00 (USA)

Cheques - dollars or sterling - payable to PW Publishing Ltd. (All credit cards accepted). If you have never seen a copy of *Practical Wireless* or *Short Wave Magazine*, send 3 IRC's to: Circulation Dept. for a recent sample copy.

BOTH MAGAZINES CARRY A COMPREHENSIVE SPECIALIST BOOK SERVICE

Mail-Order – Low cost – Large stock – Swift delivery – Friendly service

SHORTWAVE STATIONS OF THE WORLD

October 30th, 1998 World Copyright by WRTH Publications Ltd.

Country Codes: Are ITU country codes other than: CLA = clandestine & XXX = site unknown (see page 635). **Relay Stations** are listed under the code for the country where they are actually located. **Transmitter Sites and Powers:** Where a station has more than one transmitter site and/or power output, this information is given for each frequency wherever possible. However, due to schedule changes made during the currency of this edition, such information should not be regarded as definitive. **Abbreviations:** * = inactive, r = reported, vf = variable frequency, irr.= irregular, alt.= alternative frequency, **F.Pl.**= Future Plans, wi = winter frequency su = summer frequency.

kHz	kW	Land	Station
2310	50	AUS	VL8A, Alice Springs, Australia
2325	50	AUS	VL8T, Tennant Creek, Australia
2340		CHN	Fujian N
2350		KRE	KCBS, Sariwon, No. Korea
2360	0.5	GTM	R. Maya, Barillas
2380	0.25	B	R. Educadora, Limeira
2390	0.5	MEX	R. Huayacocotla, Huayacocotla
	1	GTM	R. LV de Atitlán, Santiago Atitlán (vf)
	100	USA	WWCR, Nashville, TN
2410	5	B	R. Transamazônica, Senador Guiomard *
	10/2	PNG	R. Enga, Papua New Guinea
2415	1	CHN	Wenzhou
2420	0.5	B	R. São Carlos, São Carlos
2445		CHN	Jiangxi 1
2460	1	B	R. Alvorada, Rio Branco
		CHN	Yunnan G
2470	0.25	B	R. Cacique, Sorocaba *
2475	20	CHN	Zhejiang
2485	50	AUS	VL8K, Katherine, NT, Australia
2490	2	INS	RRI Ujung Pandang (vf)
	0.25	B	R. Oito de Setembro, Descalvado
2500	1	UZB	RCH, Tashkent (Time Signals)
	1	AUS	VNG, Llandilo (Time Signals)
	5	HWA	WWVH, Hawaii (Time Signals)
2560	15	CHN	Xinjiang-Ch
2580		INS	RSPDT2 Timor Tengah Selatan, Soe
2695	0.5	INS	RPDT2 Ende
2850	120	KRE	KCBS, Pyongyang, No. Korea
2899		INS	RPDT2 Ngada, Bajawa
2960	0.3	INS	RPDT2 Manggarai, Ruteng
3040		CTR	R. Puntarenas, San José (r)
3115		INS	RSPDT2 Halmahera Tengah, Soasio (vf)
3195	1	PRU	R. Jerusalén
3200		CHN	Haixia 2 *
	50/25	SWZ	Trans World Radio Africa, Swaziland
	0.5	BOL	R. 9 de Abril, Pulacayo
3204	10	INS	RRI Bandung *
3205	1	B	R. Ribeirão Preto
	5	B	R. Vale do Rio Madeira, Humaitá
	10	PNG	R. West Sepik, Vanimo
3210	10	MOZ	Em. Nacional, Maputo
	100	CTR	Radio Exterior de España, Costa Rica
	100	USA	WWCR, Nashville, TN
3215	200	MDG	AWR via Madagascar
	10	INS	RRI Manado
	100	USA	WWCR, Nashville, TN
3220	50	CHN	CNR 1
		KRE	Hamhung Reg. St, No. Korea
	8/9	EQA	HCJB, LV de los Andes, Quito
	10	EQA	LV de los Andes, Quito
	10	PNG	R. Morobe, Lae, Papua N.G.
3222	10	TGO	Radio Kara, Togo (vf)
3223	50	IND	All India Radio, Shimla
3223	5	INS	RRI Mataram *
3225	10	INS	RRI Tanjung Pinang
3230	100	NPL	R. Nepal (wi only)
	100	AFS	Radio Oranje, So. Africa
	50	AUS	VL8A, Alice Springs (Alt.freq)
	1	PRU	R. El Sol de los Andes, Juliaca
3232	10	INS	RRI Bukittinggi
3235	1	PRU	R. Luz y Sonido, Huánuco
	0.5	B	R. Clube, Marília
	2	PNG	R. West New Britain, Kimbe
3240		BOL	R. Melodía, Bermejo
	25	SWZ	Trans World Radio Africa, Swaziland
3242	10	ISL	Ríkisútvarpid, Reykjavík
3245	50	IND	All India Radio, Lucknow
	1	B	R. Clube, Varginha
	10	PNG	R. Gulf, Kerema, Papua N.G
3250		EQA	LV del Triunfo, Sto Domingo de los Colorados (vf)
	120	KRE	Pyongyang BC St, No. Korea (vf)
	1	HND	R. Luz y Vida, San Luís (vf)
	200	KRE	R. Pyongyang, Korea (D.P.R.)
	20	INS	RRI Banjarmasin
	1	PRU	R. Qollasuyo, Juliaca
3255	500/250	AFS	BBC World Sce. via Meyerton
	1	B	R. Difusora 6 de Agosto, Xapuri
	1	B	R. Educadora Cariri, Crato
	5	B	R. Transamazônica, Senador Guiomard * (F.Pl.) (still on 2410)

kHz	kW	Land	Station
3260	10	CHN	Guizhou Sat.
	4	NGR	La Voix du Sahel, Niger
	0.5	EQA	R. Estéreo Carrizal, Calceta
	1	PRU	R. LV de Oxapampa, Oxapampa
	10	PNG	R. Madang, Papua New Guinea
3263	0.6	J	NHK Fukuoka 1 (SSB) (3262.75)
3265	10	INS	RRI Bengkulu
	10	INS	RRI Gorontalo (vf)
3270	100	NMB	Namibian Broadc. Corp. Ch.2
3275	1	EQA	R. Ecos del Oriente, Lago Agrio
	10	PNG	R. Southern Highlands, Mendi
3280	2.5	EQA	LV del Napo, Tena (vf)
	0.25	BOL	R. Em. Chaco, Yacuíba *
	1	PRU	R. Ilucán, Cutervo (F.Pl.)
		CHN	V.O.Pujiang
3282	100	MDG	Radio Television Malagasy (vf)
3283	1	PRU	Estación Wari, Ayacucho (n. 3280)
3285	4	EQA	LV del Río Tarqui, Cuenca (vf)
3290	100	NMB	Namibian Broadc. Corp. Ch.1
	10	PNG	R. Central, Boroko
		INS	RRI Samarinda
	5	GUY	Voice of Guyana, Georgetown
3295	10	ISL	Ríkisútvarpid, Reykjavík
3300	10	IND	All India Radio, Ranchi
	10	GTM	R. Cultural, Guatemala
	10	PNG	R. Western, Daru
3306	100/20	ZWE	ZBC R3/R4, Gweru, Zimbabwe
3315	50	IND	All India Radio, Bhopal
	2	PNG	R. Manus, Lorengau
	50	AUS	VL8T, Tennant Creek (Alt.freq)
3316	10	SRL	SLBS, Goderich, Sierra Leone
3320	120	KRE	Pyongyang BC St, No. Korea (vf)
	100	AFS	R. 2000, So. Africa
	100	AFS	Radio Sonder Grense, So. Africa
3325	1	GTM	R. Maya, Barillas (vf)
	10	PNG	R. North Solomons, Kieta
	10	INS	RRI Palangkaraya
	2.5	B	Super R. Tupi, São Paulo *
3326	50	NIG	R. Nigeria Ch.1, Lagos
3330	5	PRU	R. Ondas del Huallaga, Huánuco (vf)
	100	ZMB	Christian Voice, Zambia
	3	CAN	CHU (Time Signals), Canada
	0.5	BOL	R. Charcas, Monteagudo *
3335	100	TWN	CBS Network 1, Taipei, Taiwan
	10	PNG	R. East Sepik, Wewak
3340	5	PRU	R. Altura, Cerro de Pasco
	50	UGA	R. Uganda, Kampala
	1	BOL	R. Víloco, Víloco
3345	50	IND	All India Radio, Jaipur
	500/100	AFS	Channel Africa, Johannesburg
	1	PHL	DZB-2 Mindoro, Philippines
	2/10	PNG	R. Northern, Popondetta
	10	INS	RRI Ternate
3350		KRE	KCBS P'yongsong, No. Korea
	3	EQA	R. Cumandá, Coca (vf)
	5	PRU	R. Oriente, Yurimaguas (F.Pl)
	0.5	PRU	R. Visión 2000, Bambamarca (F.Pl.)
	0.1	BOL	R. 27 de Diciembre, Villamontes (n. 3350)
3355	50	IND	All India Radio, Kurseong
	10	AGL	R. Nacional., Cazenga, Angola
	10	PNG	R. Simbu, Kundiawa
	1	INS	RRI Jambi
	0.6	INS	RRI Sumenep
3356	50	BOT	R. Botswana, Gaborone
3359	10	MDG	Radio Television Malagasy
3360		EQA	R. Dif.Católica Cultural "Voz del Upano", Macas
	1	GTM	R. LV de Nahualá, Nahualá
3365	20	IND	All India Radio, Delhi
	1	B	R. Cultura, Araraquara
	10	PNG	R. Milne Bay, Alotau
3366	50	GHA	GBC Radio 2, Ghana
3370	10	MOZ	Beira, Mozambique (vf)
	0.25	EQA	R. Nacional Limón, Limón Indanza
	1	GTM	R. Sistema Cultural Tezulutlán, Cobán
	50	AUS	VL8K, Katherine (Alt.freq)
3371	1.2	BOL	R. Florida, Samaipata
3375		CHN	CNR 1 (Winter only)
	10	MOZ	Em. Nacional, Maputo (vf)
	10	AGL	Em. Nacional, Mulenvos, Angola (vf)
	5	B	R. Educadora, Guajará Mirim
	5	B	R. Equatorial, Macapá *

kHz	kW	Land	Station
	5	B	R. Nacional, São Gabriel da Cachoeira (vf)
	10	PNG	R. Western Highlands
	20	INS	RRI Medan *
	5	B	R. Clube, Dourados
3380	50	MWI	MBC, Blantyre, Malawi
	1	GTM	R. Chortís, Jocotán
	1	BOL	R. Cumbre, Tazna *
3381	0.3	J	NHK Osaka 2 (SSB) (3381.25)
3385	10	PNG	R. East New Britain, Rabaul
	5	B	R. Educação Rural, Tefé
	10	INS	RRI Kupang
	10	MLA	RTM Miri, Sarawak, Malaysia
3390	50	IND	All India Radio, Gangtok
	500/250	AFS	BBC World Sce. via Meyerton
	1	ZAI	Bunia, Congo (Dem. Rep.)
3391	1	BOL	R. Em. Camargo, Camargo (vf)
3395	2	PNG	R. Eastern Highlands, Goroka
	10	INS	RRI Tanjung Karang
3396	100/20	ZWE	ZBC R3, Gweru, Zimbabwe
3398		PRU	R. Internacional del Perú, San Pablo (n. 3395) (vf)
3400	10	LBR	Star R. Monrovia (F.Pl.)
3475	0.5	BOL	R. Padilla, Padilla (v. up to 3479) (vf)
3480		CLA	Voice of National Salvation
3502		PRU	R. Chillia, Chillia (vf)
3542		INS	RSPDT2 Sumba Timur, Waingapu
3550		CLA	Voice of the Mojahed (vf)
3560	200	KRE	R. Pyongyang, Korea (D.P.R.)
	1.5	B	R. Difusora, Brasiléia (vf)
3568		EQA	R. Dif.Luciernaga del Condor, Yanzatza
3578	0.4	INS	RSPDT2 Maluku *
3611	0.9	J	NHK Tokyo 1 (SSB) (3611.25)
3630		INS	RPDT2 Buol, Tolitoli (vf)
3665	1	PAK	Azad Kashmir R, Muzaffarabad (vf)
3760		CLA	Voice of Iraqi Kurdistan (vf)
3800	1	PRU	R. Oyón, Oyón *
3818	0.25	PRU	Estación Laser, Rioja
3850		CLA	Voice of the Mojahed (vf)
3869	0.7	PRU	R. Adventista Mundial
3875		CLA	Voice of the Communist Party of Iran (vf)
		CLA	Voice of the Iranian Revolution (vf)
3880		CLA	Radio Free Bougainville
		CLA	Radio United Bougainville
3900		CHN	Haixia 1
		CHN	Hulun Buir-Ch
3905	2	PNG	R. New Ireland, Kavieng
		CLA	Freedom Radio Station
	50	INS	RRI Banda Aceh
	10	INS	RRI Merauke
		CLA	Voice of the Communist Party of Iraqi Kurdistan (vf)
		CLA	Voice of the Iraqi People (vf)
3912		CLA	Voice of the People
3915	250/100	SNG	BBC Far Eastern Relay Station
3920		KRE	KCBS Hyesan, No. Korea
3925	0.5	I	R. 106 Comunitaria, Rubiera (vf)
	50/10	J	R. Tanpa 1, Tokyo / Sapporo
3930	10	KOR	KBS R.1 & Liberty Prgr, Hwasong
	50	CHN	Nei Menggu-Mo
3934	05-Oct	INS	RRI Semarang *
3935		IRQ	R. Baghdad, Iraq
	1	NZL	ZLXA, Levin
3940	10	CHN	Hubei 1
		CLA	Voice of Iranian Kordestan (vf)
3945	50	IND	All India Radio Ext. Sce, Gorakhpur
	10	J	R. Tanpa 2, Tokyo, Japan
	10	VUT	R. Vanuatu, Malapoa
		CLA	Voice of the Worker (vf)
3946	2.5	INS	RRI Tanjung Karang
3950		CHN	Qinghai 1
3955	500/250	G	BBC World Sce, United Kingdom
3960	12	MNG	Dalanzadgad, Mongolia (vf)
		KRE	KCBS Kanggye, No. Korea
	1	INS	RRI Padang
	10	INS	RRI Palu
	1	MNG	Uliosxai, Mongolia (vf)
	50	CHN	Xinjiang-Ch
3965	250	G	Merlin Communications, UK
3970	4	CME	CRTV, Buea, Cameroon (vf)
	100	NIG	Enugu National St., Nigeria
		KRE	KCBS Wonsan, No. Korea
	50	CHN	Nei Menggu-Ch
	0.6	J	NHK Sapporo 1 (DSB)
	250	G	R. Korea Int, Skelton, UK
3974	0.3	J	NHK Nagoya 1 (3973.75)
3975	250	HNG	R. Budapest, Jászberény
	100/20	HNG	R. Budapest, Székesfehérvár
3976		INS	RRI Pontianak
3980	10	I	IRRS, Milano, Italy
		KRE	KCBS Ch'ongjin, No. Korea
3985	250	SUI	China Radio International via Switerland
	250	SUI	China Radio International, Lenk
	10	I	IRRS, Milano, Italy
	250	G	Merlin Communications, UK
		CLA	Radio Echo of Hope

kHz	kW	Land	Station
3987	1	INS	RRI Manokwari *
3990	50	CHN	Xinjiang-Ug
	500/100	D	Deutsche Welle, Germany
	50	RUS	Khabarovsk, FE
3996		INS	RRI Kendari
3996		INS	RRI Pontianak
4000	20	CME	CRTV, Bafoussam, Cameroon *
	1	PRU	R. Constelación, Huancayo
	5	INS	RRI Kendari
		INS	RRI Padang (alt. to 4003)
		ERI	V. of the Broad Masses of Eritrea
		CLA	Voice of the Iraqi Communist Workers' Party
4001		PRU	R. San Juan, Trujillo *
4003	10	INS	RRI Padang
4005	500/100	CVA	Vatican Radio
4010	100	KGZ	Biskek 1, Kyrgyzstan
	200	RUS	Vladivostok, FE
4035		CHN	Xizang-Tb
4040	1	PRU	R. Marginal, Tocache
	5	RUS	Tura, S
4050	100	KGZ	Biskek 2, Kyrgyzstan
4075		CLA	Harim Radio (vf)
		CLA	Voice of Iraqi Kurdistan (vf)
4078	50	MNG	Ulaanbaatar, Mongolia (vf)
4116	0.25	B	R. Difusora, Sena Madureira *
		CLA	Voice of Independent Kashmir (vf)
4120		CLA	Voice of National Salvation
		CLA	Voice of the People of Kurdistan (vf)
4135		CLA	Radio of Islam
4182		PRU	R. LV de Sayapullo, Sayapullo (vf)
4183	0.25	BOL	R. 11 de Octubre, Cobija (vf)
4190		CHN	CNR Minor.
		CLA	Voice of the Worker (vf)
4191		PRU	R. Super San Ignacio, San Ignacio
4192		PRU	R. Selva Superior, Moyobamba (n. 4200) (v. 4187.9-4194)
4211		BOL	R. Perla del Oriente, Roboré
4215		VTN	Lai Cau, Vietnam (vf)
4220	15	CHN	Xinjiang-Mo
4233		HKG	VPS8 (Time Sigs), Hong Kong
4270	1	EQA	R. Gonzanamá, Loja
4279		PRU	R. Cajamarca, Cajamarca
4291	10	AFS	ZSC (Time Signals), So. Africa
4297		BOL	Radioemisora Orion, Huaca
4298		CHL	CBV (Time Signals), Chile
4300		PRU	Estación Norte, Bambamarca (F.Pl.)
4319		PRU	R. Condorcanqui, Santa Maria de Nieva (n.
4320)			
4330		CLA	Voice of the Struggle of Iranian Kordestan (vf)
		CHN	Xinjiang-Kz
4375		CLA	Voice of the Communist Party of Iran (vf)
4385		CLA	Voice of the Iranian Revolution (vf)
4395	15	RUS	Yakutsk, S
4400		CLA	Radio of Islam
		CLA	Voice of National Salvation
4405	200	KRE	R. Pyongyang, Korea (D.P.R.)
4409	0.5	BOL	R. Eco, Reyes
4420	0.85	PRU	R. Bambamarca
		PRU	R. Sam Juan, Aramango
4421	1	BOL	R. Santa Rosa, Santa Rosa del Yacuma *
4422		BOL	Radioemisora Reyes, Reyes
4441	0.3	BOL	R. Ecología Internacional, San Matías
4444		CLA	Voice of the Communist Party of Iraqi Kurdistan
4445		PHL	NPO (Time Sigs), Philippines
4450	0.25	BOL	R. Estación Frontera, Cobija (n. 4730) (vf)
		CLA	Voice of National Salvation
		CLA	Voice of the Mojahed (vf)
4460		CHN	CNR 1
4461	1	PRU	R. Nor Andina, Celendín
4472	1	BOL	R. Movima, Santa Ana del Yacuma (n. 4835) (v 4471-4472.9) (vf)
4485	0.5	PRU	R. Frecuencia VH, Celendín
	17	RUS	Yazykovo, E
4500		CHN	Xinjiang-Ch
4505	0.5	PRU	R. Horizonte, Chiclayo
4509		BOL	Radioemisora San Joaquín
4510		UZB	Fargona, Uzbekistan
	0.25	PRU	R. Paucartambo
4520	5	RUS	Khanty-Mansiysk, S
	2	RUS	Palana, FE
4530	0.12	BOL	Hitachi Radiodifusión, Guayaramerín (n.
4530)			
4545	20	KAZ	Almaty. Kazakhstan
4549	3/1	BOL	R./Dif./Trópico, Trinidad
4552	3/1	BOL	R./Dif./Trópico, Trinidad (alt. freq)
4557		CLA	Voice of National Salvation
4567		PRU	R. Gotas de Oro, Chiclayo (vf)
4575		PRU	R. Latina, Chiclayo (n. 4580) (vf)
4594		PRU	Estación X, Yurimaguas
4599	1	BOL	Radiodifusora Villamontes
4600	0.2	BOL	R. Perla del Acre, Cobija
4606	1	INS	RRI Serui

kHz	kW	Land	Station
4607	1	PRU	R. Ayaviri
4615		IRQ	R. Baghdad, Iraq
4624		PRU	R. Soledad, Parcoy
4626	0.2	BOL	R./Dif./Machupo, San Ramón (vf)
4627		PRU	R. Cosmos, Celendín
4631		BOL	R. Em. Entre Ríos, Entre Ríos (vf)
4632	1	BOL	R. 11 de Octubre, Cobija
4635	50	TJK	Yangi-Yul, Tadzhikistan
4649	1	BOL	R. Santa Ana, Santa Ana del Yacuma (n. 4805) (vf)
4650		CLA	Voice of the Mojahed (vf)
4675		VTN	Lam Dong, Vietnam (vf)
4677		VTN	Lam Dong, Vietnam
4682	1/0.2	BOL	R. Paitití (Paitití Radiodifusión), Guayaramerín (alt. freq.)
4690	1	LAO	Xam Nua, Laos (vf)
4697	2.5	INS	R. Khusus Informasi Pertanian (RKIP), Surabaya
4702		BOL	R. Eco, San Borja (n. 4700) (vf)
4719	50	INS	RRI Ujung Pandang (irr.)
	0.5	BOL	R. Abaroa, Riberalta (n. 4760) (vf)
4723		VTN	R. TV Gia-Lai, Vietnam (vf)
4725	50	BRM	R. Myanmar, Yangon
4730	100	TJK	Yangi-Yul, Tadzhikistan
4732	0.7	BOL	R. La Palabra, Santa Ana del Yacuma
4735		CHN	Xinjiang-Ug
4747	0.3	PRU	R. Huanta 2000, Huanta (alt. freq. 4752.5) (vf)
4750	20	CME	CRTV, Bertoua, Cameroon *
		CHN	Hulun Buir-Mo
		EQA	R. Municipal, Quito
	12	MNG	Ulgii, Mongolia (vf)
		CHN	Xizang-Ch
		PRU	R. San Francisco Solano, Sondor
4753	20	INS	RRI Ujung Pandang
4754	5	B	R. Difusora do Maranhão, São Luís
4755	10	B	R. Educação Rural, Campo Grande
4759	0.5	PRU	R. Televisión Tingo Maria
4760	50	IND	All India Radio, Leh
	20	IND	All India Radio, Port Blair
	25	SWZ	Trans World Radio Africa, Swaziland
		CHN	Yunnan G
4764	1	PRU	R. Chincheros, Chincheros (n. 4765)
4765	100	COG	R. Congo, Brazzaville
	10	B	R. Integração, Cruzeiro do Sul (vf)
	10	B	R. Rural, Santarém
	0.3	BOL	Radiodif. Guanay, Guanay
4766	20	INS	RRI Medan *
4770	5	EQA	R. Centinela del Sur "CDS Internacional", Loja (vf)
	50	NIG	R. Nigeria, Kaduna Ch.2
4775	50	IND	All India Radio, Imphal
	5	B	R. Amarela, Rolim de Moura (CP)
	1	B	R. Congonhas, Congonhas
	5	B	R. Liberal, Belém
	1	B	R. Portal da Amazônia, Cuiabá
	0.5	PRU	R. Tarma, Tarma (vf)
	50	SWZ	Trans World Radio Africa, Swaziland
	3	BOL	R. Los Andes, Tarija *
4777		GAB	Radiodiffusion Gabonaise, Moyabi
	7.5	INS	RRI Jakarta (Kebayoran)
4778		BOL	R. A.N.D.E.S, Uyuni
4780	1	DOM	Em. Onda Musical, Sto. Domingo (r. 4766)
	5	AGL	EP do Kuando Kubango, Angola * (vf)
	1	GTM	R. Cultural Coatán, San Sebastián Coatán (vf)
	3	EQA	R. Oriental, Tena (vf)
	100	PAK	R. Pakistan, Rawalpindi (vf)
	0.8	PRU	R. Bahía, Chimbote
4781	0.3	PRU	R. Satélite, Santa Cruz (alt. freq. to 6725.6) (vf)
4783	100	MLI	Radiodif.-TV Malienne, Bamako (vf)
4785	1	PRU	R. COSAT, Satipo
	50	AZE	Gäncä, Azerbaijan *
	1	B	R. Brasil 5000, Campinas
	10	B	R. Caiari, Porto Velho
	1	BOL	R. Em. Ballivián, San Borja
	10	EQA	R. Federación Shuar, Sucúa (vf) (irr.)
	5	B	R. Ribamar, São Luís *
	5	CLM	R. Super, Ibagué (vf)
	20	CHN	Zhejiang
4789	1	INS	RRI Fak-Fak
4790	100	IND	All India Radio Ext. Sce, Madras
	100	PAK	Azad Kashmir R, Rawalpindi
	3	PRU	R. Atlántida, Iquitos
4795	100	CME	CRTV, Douala, Cameroon *
	3	EQA	LV de los Caras, Bahía de Caráquez (vf)
	1	B	R. Difusora, Aquidauana
	50	RUS	Ulan-Ude, S
4796	10	BOL	R. Nueva América, La Paz *
4800	50	IND	All India Radio, Hyderabad
		CHN	CNR 2/Minor.
		VTN	Dac Lac, Vietnam (vf)
	1	GTM	R. Buenas Nuevas, San Sebastián H (vf)
	5	EQA	R. Popular Independiente, Cuenca *

kHz	kW	Land	Station
		VTN	R. Son La 1, Vietnam (vf)
	5	MEX	R. Transcontinental de América, México (nighttime)
		VTN	R. TV Kontum, Vietnam (vf)
4801	1	PRU	Onda Azul, Puno (n 4800)
4805	10	B	R. Difusora do Amazonas, Manaus
	0.5	B	R. Itatiaia, Belo Horizonte
4810	100	RUS	Vladivostok, FE
	50	ARM	Voice of Armenia
	3	PRU	R. San Martín, Tarapoto *
4815	10	B	R. Cabocla, Tabatinga *
	10	B	R. Difusora, Londrina
		CHN	China Radio International
	50	BFA	R. Burkina, Ouagadougou
	1	EQA	R. El Buen Pastor, Saraguro -0305
	10	CLM	R. Guatapurí, Valledupar *
	1	BOL	R. Nacional de Bolivia, La Paz *
	100	PAK	R. Pakistan, Islamabad (vf)
	50	IND	All India Radio, Calcutta
	3	AGL	EP da Huíla, Angola (vf)
	50	RUS	Khanty-Mansiysk, S
	5	HND	LV Evangélica, Tegucigalpa (vf)
	50	BOT	R. Botswana, Gaborone
		CHN	Xizang-Tb
4821	1	PRU	R. Atahualpa, Cajamarca *
4823	100	MTN	R. Mauritanie (alt. to 4845)
4824	10	PRU	LV de la Selva, Iquitos
4825	10	B	R. Canção Nova, Cachoeira Paulista
	5	B	R. Educadora, Bragança
	0.5	GTM	R. Mam, Cabricán
	0.5	PRU	R. Moquegua, Moquegua
	50	RUS	Yakutsk, S
4826	8	IND	VWC, Calcutta (Time Sigs)
	1.5	PRU	R. Sicuani,
4828	12	MNG	Altai, Mongolia
	100/20	ZWE	ZBC R3/R4, Gweru, Zimbabwe
4830		CHN	China Huayi BC
	1	BOL	R. Grigotá, Santa Cruz
	0.5	HND	R. Litoral, La Ceiba
	10	VEN	R. Táchira, San Cristóbal
	10	THA	R. Thailand, Pathumthani (vf)
4831	1	PRU	R. Lircay, Lircay
4832	3	CTR	R. Reloj, San José (vf)
4835	1	PRU	R. Marañón, Jaén
	5	B	R. Atalaia, Corumbá *
	1	CLM	R. Buenaventura (vf)
	5	GTM	R. Sistema Cultural Tezulutlán, Cobán (vf)
	100	MLI	Radiodiff.-TV Malienne, Bamako (vf)
	10	MLA	RTM Kuching, Sarawak, Malaysia
	50	AUS	VL8A, Alice Springs, Australia
4840	50	IND	All India Radio, Mumbai
		CHN	Heilongjiang
	2	PRU	R. Andahuaylas
	1	EQA	R. Interoceánica, Sta Rosa de Quijos *
	1	VEN	R. Valera, Valera (vf) (irr.)
4845	5	BOL	R. Fides, La Paz
	1.25	GTM	R. K'ekchi, Fray Bartolomé de las Casas
	1	PRU	R. Líder, Cusco
	100	MTN	R. Mauritanie, Nouakchott
	1	B	R. Meteorologia Paulista, Ibitinga
	10	INS	RRI Ambon
	100	MLA	RTM R6, Kuala Lumpur, Malaysia
	250	B	R. Cultura Ondas Tropicais, Manaus
4850	50	IND	All India Radio, Kohima
		CHN	CNR 1
	3	EQA	R. Luz y Vida, Loja
	50	MNG	Ulaanbaatar, Mongolia
	50	UZB	UZR2, Toskent, Uzbekistan
4853	20	YEM	Yemen Radio & TV Corp, Sana'a
4855	20	MOZ	Em. Nacional, Maputo (vf)
	10	MAU	Mauritius Broadcasting Corp. (F.P.I.)
	5	BOL	R. Centenario
	1	B	R. Por Um Mundo Melhor, Governador Valadares *
	2.5	B	R. Tropical da Barra, Barra do Garças
	1	PRU	R. La Hora, Cusco
4857	0.75	BOL	R. El Cóndor, Uyuni (n. 4860) *
4860	50	IND	All India Radio Ext. Sce, Delhi
	50	IND	All India Radio, Delhi
	80	RUS	Chita
	10	EQA	R. Federación Shuar, Sucúa
4865	1	BOL	R. Em. 16 de Marzo, Centro Polimetalúrgico Bolívar (vf)
		CHN	Gansu 1
	5	CLM	LV del Cinaruco/Caracol, Arauca
	5	B	R. Alvorada, Londrina
	5	B	R. Sentinela da Amazônia, Obidos (rel. R. Líder FM, São Paulo)
	1	B	R. Sociedade, Feira de Santana *
	5	B	R. Verdes Florestas, Cruzeiro do Sul
	12	MNG	Sainshand, Mongolia (vf)
	5	MNG	Ulangoom, Mongolia (vf)
4866	0.3	INS	RRI Wamena *
4870	30	BEN	ORTB, Cotonou, Benin
	1	PRU	R. Majestad, Huancayo

kHz	kW	Land	Station
	10	CLN	SLBC Sinhala Comm. Sce.
4871	0.05	INS	RRI Wamena (alt. to 4867) *
4875	100	GEO	Dusheti, Georgia
	10	B	R. Difusora de Roraima, Boa Vista
	10	BOL	R. La Cruz del Sur, La Paz
	10	INS	RRI Sorong
4880	50	IND	All India Radio, Lucknow
	100	BGD	R. Bangladesh Home Sce. (vf)
	1	EQA	R. Nal.Espejo, Quito (vf) (irr.)
		VTN	Than Hoa, Vietnam *
4882	1	PRU	R. Nuevo Mundo, Pucallpa (n. 4880)
4883		CHN	China Radio International
4885	1	B	R. Carajá, Anápolis
	5/100	KEN	KBC Eastern Sce, Nairobi
	5	CLM	Ondas del Meta, Villavicencio (vf)
	5	B	R. Clube do Pará, Belém (irr.)
	5	B	R. Difusora Acreana, Rio Branco
4886	1	BOL	R. Sararenda, Camiri
4887	0.8	PRU	R. Villa Rica, Huancevelica
4890	0.5	PRU	R. Dif. Huanta, Huanta
	100	PNG	NBC Port Moresby, Papua N.G.
	100	SEN	ORTS Nat.Sce, Dakar, Senegal *
	F		R. France Internationale
4891	1	PRU	R. Chota (n. 4890) (vf)
4895	20	IND	All India Radio, Kurseong
	10	CLM	LV del Río Arauca, Arauca
	12	MNG	Murun, Mongolia (vf)
	5	B	R. Baré, Manaus
	0.4	PRU	R. Chanchamayo, La Merced
	5	B	R. IPB AM, Campo Grande
	100	PAK	R. Pakistan, Islamabad (vf)
	10	MLA	RTM Kuching, Sarawak, Malaysia
	50	RUS	Tyumen, S
4900	18	GUI	Conakry, Guinea Rep. (vf)
		CHN	Haixia 2
	1	EQA	LV de Saquisilí y Libertador, Saquisilí (vf)
4902	10	CLN	SLBC Sinhala Nat. Sce.
4903	0.75	BOL	R. Em. San Ignacio, San Ignacio de Moxos
4905	100	TCD	Radiodiff. Nat. Tchadienne (vf)
	1	PRU	R. La Oroya, La Oroya (n. 4905) (vf)
	10	CHN	CNR 2
		CLM	Ecos del Orinoco, Puerto Carreño
	1	B	R. Anhanguera, Araguaína
	5	B	R. Relógio Federal, Rio de Janeiro
4907	50	CBG	National Radio of Cambodia
4908	5	PRU	R. Tawantinsuyo, Cusco *
4910	50	IND	All India Radio, Jaipur
	0.75	HND	LV de la Mosquitia, Puerto Lempira (vf)
	50	AUS	VL8T, Tennant Creek, Australia
	50	ZMB	ZNBC Radio One, Lusaka, Zambia (vf)
4911		INS	RRI Bukittinggi
4915	10	PRU	R. Cora del Perú, Lima (n. 4915)
	3	CLM	Armonías del Caquetá, Florencia (vf)
	50	GHA	GBC Radio 1, Ghana
		CHN	Guangxi Sat.
	100	KEN	KBC Central Sce, Nairobi
	10	B	R. CBN Anhanguera, Goiânia
	10	B	R. Difusora, Macapá
	100	PAK	R. Pakistan, Islamabad
4920	50	IND	All India Radio, Chennai
	0.5	PRU	R. Central
	12	EQA	R. Quito "LV de la Capital", Quito (vf)
	1	INS	RRI Tanjung Pinang
4922	1	PRU	R. Ondas del Titicaca, Puno (n. 4920) *
4925	0.5	B	R. Difusora, Taubaté
	7.5	MOZ	Em. Nacional, Maputo (vf)
	5	BOL	R. San Miguel, Riberalta (v.4922-4926) (vf)
	10	INS	RRI Jambi
4926	50	GNE	Bata, Equatorial Guinea (alt. to 5003.5) (vf)
4927	7.5	INS	RRI Jambi
4930	50	IND	All India Radio, Shimla
	50	TKM	Asgabat, Turkmenistan
		CHN	Honghe
	100	NMB	Namibian Broadc. Corp. Ch.2
	1	HND	R. Internacional, San Pedro Sula (vf)
4931.6	10	INS	RRI Surakarta
4935	5/100	KEN	KBC General Sce, Nairobi
	1	B	R. Capixaba, Vitória
	2.5	B	R. Difusora, Jataí
	1.5	BOL	R. Norte, Montero
	3	PRU	R. Tropical, Tarapoto
4940	50	IND	All India Radio, Guwahati
		CHN	Haixia 1
	1	VEN	R. Amazonas, Puerto Ayacucho
	1	HND	R. Copán Internacional, Tegucigalpa
	25	CTI	R. Côte d'Ivoire, Abidjan *
	10	CLN	SLBC English Comm. Sce
	100	TJK	Yangi-Yul, Tadzhikistan
4945	1	B	Emiss. Rural A Voz do São Francisco, Petrolina
	1	B	R. Difusora, Poços de Caldas
	10	BOL	R. Illimani,
	7.5	B	R. Progresso, Porto Velho
4950	50	IND	All India Radio, Srinagar
	10	AGL	Em. Nacional, Mulenvos, Angola (vf)
	1	EQA	R. Bahá'í, Otavalo (vf)
	10	MLA	RTM Kuching, Sarawak, Malaysia
		CHN	V.O.Pujiang
	100	STP	Voice of America, São Tomé
		CHN	Xilingol-Mo
4950.1	5	PRU	R. Madre de Dios, Puerto Maldonado (alt.freq. 4953)
4955	2.5	B	R. Clube, Rondonópolis
	1	B	R. Cultura, Campos"
	20	CLM	R. Dif. Nal. de Colombia, SF de Bogotá
	10	B	R. Marajoara, Belém *
4955.4	5	PRU	R. Cultural Amauta, Huanta
4957.5	15	AZE	Baki, Azerbaijan *
4960		CHN	China Radio International
		VTN	Hanoi 2, Vietnam
	5	EQA	R. Federación Shuar, Sucúa (vf)
	1	HND	R. HRET, Puerto Lempira (vf)
	100	STP	Voice of America, São Tomé
4964.5	1	PRU	R. La Merced, La Merced (n. 4965) (vf)
4965	100	ZMB	Christian Voice, Zambia
	5	B	R. Alvorada, Parintins
	3	BOL	R. Em. Juan XXIII, San Ignacio de Velasco
	1	B	R. Poti, Natal *
	2.5	CLM	R. Santa Fé, SF de Bogotá *
		VTN	R. Son La, Vietnam (vf)
4965.8	1	PRU	R. San Miguel, Cusco
4969.5	1	PRU	R. Imagen, Tarapoto (n. 4970) (alt. freq. 5144)
4970	50	IND	All India Radio, Shillong
	5	AGL	EP da Cabinda, Angola * (vf)
	5	VEN	R. Rumbos, Villa de Cura
	10	MLA	RTM Kota Kinabalu, Malaysia
	10	CLN	SLBC, Sri Lanka (irr.)
		CHN	Xinjiang-Kz
4974.8	5	PRU	R. del Pacífico, Lima (n. 4975)
4975		CHN	Fujian N
		CLM	Ondas del Orteguaza, Florencia
	1	B	R. Iguatemi, Osasco *
	1	B	R. Mundial, São Paulo
	5	B	R. Timbira do Maranhão, São Luís
	100	TJK	Yangi-Yul, Tadzhikistan
4980	5	VEN	Ecos del Torbes, San Cristóbal
	1	BOL	R. Batallón Topater, Oruro
		CHN	Xinjiang-Mo
4985	10	B	R. Brasil Central, Goiânia
4990	50	IND	All India Radio, Itanagar
	10	CHN	Hunan Sat.
	50	NIG	R. Nigeria Ch.1, Lagos
	100	ARM	Voice of Armenia
4990.9	1	BOL	R. Animas, Chocaya
4990.9	0.35	SUR	R. Apintie, Paramaribo, Surinam
4991.6	5	PRU	R. Ancash, Huaraz (n. 4990) (vf)
4995	12	MNG	Choibalsan, Mongolia (vf)
4995.6	2	PRU	R. Andina, Huancayo (n. 4995)
4996	5	RUS	RWM (Time Signals), Moscow
5000	8	IND	ATA, New Delhi (Time Sigs.)
		VTN	Binh Thuan, Vietnam (vf)
	2	TWN	BSF (Time Signals), Taiwan
		KOR	HLA (Time Signals), Korea
	2	J	JJY, Japan (Time Signals)
	2	ARG	LOL (Time Signals), Argentina
		VTN	R. Yên Bai 1, Vietnam (vf)
	1	UZB	RCH, Tashkent (Time Signals)
		ERI	V. of the Broad Masses of Eritrea
	10	AUS	VNG, Llandilo (Time Signals)
	10	HWA	WWVH, Hawaii (Time Signals)
5003.5	100	GNE	Bata, Equatorial Guinea (vf)
5004	1	RUS	RID (Time Signals), Irkutst
5005	5	BOL	R. Libertad, La Paz
	100	NPL	R. Nepal
		MLA	RTM Sibu, Sarawak, Malaysia
5005.6	0.25	PRU	R. Jaén,
5009.5	100	MDG	Radio Television Malagasy
5010	20	IND	All India Radio, Thiru'puram
	100	CME	CRTV, Garoua, Cameroon
	0.8	EQA	Escuelas Radiofónicas Populares, Riobamba (vf)
		CHN	Guangxi EBS
5012	1	DOM	R. Cristal Int., Sto Domingo (vf)
5014.6	1	B	R. Pioneira, Teresina
5014.8	0.7	PRU	Estación Tarapoto, Tarapoto *
5015	100	TKM	Asgabat, Turkmenistan
		AGL	EP do Namibe, Angola
	1	B	R. Brasil Tropical, Cuiabá (at night rel. R. Cultura 710kHz)
		B	R. Copacabana, Rio de Janeiro
	100	RUS	Tavrichanka, FE
5015.9	0.5	PRU	R. Juliaca, Juliaca (vf)
5019.8	5	PRU	R. Horizonte, Chachapoyas
5020	1	CLM	Ecos del Atrato, Quibdó (vf)
		CHN	Jiangxi 1
	30/100	NGR	La Voix du Sahel, Niger (vf)
	10	SLM	SIBC Honiara, Solomon Islands
	10	CLN	SLBC Tamil Comm. Sce.
		CHN	Xizang-Tb

kHz	kW	Land	Station
5025	10	BEN	ORTB, Parakou, Benin
	5	B	R. Jornal da Transamazônica, Altamira
	5	B	R. Morimoto, Ji-Paraná *
	10	PAK	R. Pakistan, Quetta (vf)
	5	PRU	R. Quillabamba, Quillabamba
	10	CUB	R. Rebelde, Bauta, Cuba
	1	B	Super R. Borborema, Campina Grande *
	50	AUS	VL8K, Katherine, NT, Australia
5026	20	UGA	R. Uganda, Kampala (vf)
5030	50/20	CTR	AWR Latin America, Cahuita
	50/10	BTN	BBS, Thimpu, Bhutan
		VTN	Hanoi Hmong Sce, Vietnam (vf)
	10	EQA	R. Católica Nacional, Quito
		PRU	Radiodifusora LV de la Esperanza, Juliaca (F.Pl.)
	10	MLA	RTM Kuching, Sarawak, Malaysia
	1	TON	Tonga Broadcasting Commission (F.Pl.)
5030.6	5	PRU	R. Los Andes, Huamachuco
5033.5	100	CAF	R. TV Centrafricaine, Bangui (vf)
5034.9	10	B	R. Aparecida, Aparecida
	5	B	R. Educação Rural, Coari
	100	UZB	R. Tashként Ext.Sce, Uzbekista
5039.3	1	PRU	R. Libertad
5040	50	GEO	Dusheti, Georgia
		CHN	Fujian N
	5	CLM	LV de Yopal, Yopal
	10	EQA	R. Dif.Católica Cultural "Voz del Upano", Macas
	100	UZB	R. Tashként Ext.Sce, Uzbekistan
		INS	RRI Pekanbaru
5043	1	AGL	EP de Benguela, Angola (vf)
5044.8	5	BOL	R. Altiplano, La Paz
5045	10	B	R. Cultura do Pará, Belém
	1	B	R. Difusora, Presidente Prudente *
	1	PRU	R. Rioja, Rioja (n. 5015)
5046	20	INS	RRI Yogyakarta *
5046.3		PRU	R. Integración, Abancay
5047	50/100	TGO	Radiodiffusion Togolaise, Lomé
5049.6	0.1	PRU	R. Tayacaja, Pampas(n. 4670)
5050	10	IND	All India Radio, Aizawl
		CHN	Guangxi FBS
		CHN	Haixia 1
	1	PRU	R. Amistad, Chiquian *
	5	EQA	R. Jesús del Gran Poder, Quito (vf)
5050.1	10	TZA	R. Tanzania National / Ext. Sce
5050.4	0.5	PRU	R. Municipal, Cangallo
5053.6		PRU	R. Acobamba, Acobamba
5055	5	CTR	Faro del Caribe, San José
	1	B	R. Difusora, Cáceres
	5	B	R. Jornal A Crítica, Manaus
	100	PAK	R. Pakistan, Rawalpindi (vf)
	10	GUF	RFO Cayenne, French Guiana
		INS	RRI Nabire
5056.2	0.15	PRU	R. Onda Imperial, Cusco
5059		INS	RRI Yogyakarta
5060	5	EQA	R. Progreso, Loja (vf)
	100	UZB	R. Tashként Ext.Sce, Uzbekistan
		CHN	Xinjiang-Mo
5066		CLA	La Voix du Peuple
5066.4	1	ZAI	Bunia, Congo (Dem. Rep.)
5068.7	1	PRU	R. Ondas del Suroriente, Quillabamba
5070	100	USA	WWCR, Nashville, TN
5075	50	CLM	Caracol Colombia, SF de Bogotá
5081	0.5	PRU	R. Huayabamba, Rodríguez de Mendoza *
5083.7	0.5	PRU	R. Mundo, Cusco (n. 5045)
5085	0.1	PRU	R. Celendín
	100	PAK	R. Pakistan, Islamabad
	50	USA	WGTG, McCaysville, GA
5090		CHN	CNR Chinese (Winter only)
5097.3	1	PRU	R. Eco, Iquitos (n. 5010)
5100	7.5	LBR	R. Liberia Int., Totota
5116.4		PRU	R. Diez, Iquitos
5123.7	0.2	BOL	R. Galaxia, Guayaramerín (n.5160) (vf)
5125		CHN	CNR Taiwan (Winter only)
5131.4	0.5	PRU	R. Visión 2000, Bambamarca
5139.8		PRU	R. Amauta, San Pablo
5144	1	PRU	R. Imagen, Tarapoto (alt. freq. to 4969.5) (n. 4970)
5145		CHN	China Radio International
5150		CLA	Voice of the Mojahed (vf)
5160	1	DOM	R. Barahona (irr.)
5163	50	CHN	CNR 2
5179		PRU	R. Oriental, San Ignacio
5200		VTN	Nghê An, Vietnam (vf)
5220		CHN	China Radio International
5236.3		PRU	R. Apurímac, Abancay
5240		CHN	Xizang-Ch
5250		CHN	China Radio International
5264.8		PRU	R. 5264 (Cinco mil doscientos sesenta y cuatro), Chiriaco (vf)
5271.8		PRU	R. Nororiental, Rodríguez de Mendoza
5275.5		PRU	Estación NC, Nueva Cajamarca
5286.2	5	TCD	R. Moundou, Chad *
5290	50	RUS	Krasnoyarsk, S
	5	RUS	Perm, E
5295		CHN	CNR 1
5300		CLA	Voice of Independent Kashmir
5320		CHN	CNR 1
5323.8	0.5	PRU	R. Origen, Huancavelica (n. 5322) (vf)
5339.1	0.8	PRU	R. Líder del Norte, Cajamarca *
5344	1	PRU	R. Pomabamba, Pomabamba (n. 3215) *
5385.5		PRU	R. Huarmaca, Huarmaca (vf)
5397		PRU	R. Tayabamba (n. 4500, also r. 4505v, 4713v)
5407		CLA	R. of the Prov. Govt. of Nat. Union of Cambodia (vf)
5420		CHN	CNR Minor.
5431.8	0.3	J	NHK Osaka 2 (SSB) (5431.75)
5470		CLA	Voice of the Mojahed (vf)
5470.6		PRU	R. San Nicolás, Rodriguez de Mendoza (n. 5470)
5486.6	0.06	PRU	La Reina de la Selva, Chachapoyas
5498.5		PRU	R. Lajas, Lajas (n. 5500)
5500		BOL	R. Luz del Oriente, Santa Cruz
		CLA	Voice of the Tigray Revolution
5503.7	0.2	BOL	R. Em. 2 de Febrero, Rurrenabaque (n. 5020)
5522.5	0.25	PRU	R. Sudamérica, Cutervo (v. 5521.4-5522.7) (n. 5525) (vf)
5547.4		PRU	R. Paccha, Paccha (alt. freq. 5277) (vf)
5556		PRU	R. Santa Fé, Santa Cruz
5560.7		PRU	R. El Sol, Pucará (n. 5560)
5570	0.1	CLM	R. Nueva Vida, Bucaramanga (vf)
5580.2	0.25	BOL	R. San José, San José de Chiquitos
5597		VTN	R. Lao Cai, Vietnam (vf_
5604.4		PRU	R. LV de Marañón, Cajabamba (n. 5610) (alt. freq. 5636.8) (vf)
5620	500/250	AFS	AWR via Meyerton
5629.9	1	PRU	R. Ilucán, Cutervo
5636.8		PRU	R. LV de Marañón, Cajabamba (alt. freq. to v5604.4)
		PRU	R. Perú, San Ignacio
5645.8	0.25	PRU	R. LV de San Antonio, Bambamarca (vf)
5660.5	0.6	PRU	R. LV de Cutervo, Cutervo
5678.1	0.15	PRU	R. Frecuencia San Ignacio (n. 5700)
5691.2	0.5	INS	RPDT2 Berau, Tanjungredeb
5728.5	0.5	PRU	R. Naylamp, Lambayeque (n. 5340)
5730		PRU	R. Santiago, Puerto Galilea
5745	250	USA	Voice of America, Greenville
	100	USA	WHRI, South Bend, IN
5750		CLA	Voice of Independent Kashmir
		CLA	Voice of the Mojahed (vf)
5767.1		PRU	R. Master, Moyobamba
5770	1	NCG	R. Miskut, Puerto Cabezas
5775.6		PRU	R. San Juan, Lonya Grande (vf)
5800		PRU	R. Nuevo Cajamarca,
	100	TJK	Yangi-Yul, Tadzhikistan
5810	100	USA	KAIJ, Denton, TX
	100/50	USA	R. Taipei International via WYFR
5824.6		MOZ	Em. Nacional, Maputo
5825	500/100	BUL	R. Bulgaria
	100	PAK	R. Pakistan, Islamabad (vf)
	0.05	CHL	R. Triunfal Evangélica, Talagante
	500	USA	WEWN, Vandiver, AL
5835	500	HOL	R. Netherlands, Flevo
	100	TCH	R. Prague, Czech Rep.
5840	100	D	AWR Europe via Jülich
5850		CHN	China Radio International
	100/10	HRV	Hrvatski Radio, Zagreb
	500/100	BUL	R. Bulgaria
	500	USA	WSHB, Cypress Creek, SC
	500	USA	WVHA, Greenbush, ME
	100	USA	WYFR-Family Radio, Okeechobee, FL
5855		XXX	R. Free Asia
5860	50	CHN	V.O.Jinling
5865	250	EQA	LV de los Andes, Quito
	500/100	BUL	R. Bulgaria
5875	250/100	CYP	BBC East Mediterranean Relay Station
	500/250	G	BBC World Sce, United Kingdom
5880		CHN	CNR 1
	500/100	CVA	Vatican Radio
5890	100/10	HRV	Hrvatski Radio, Zagreb
	0.2	HND	LV de Misiones Int. "R. MI", Comayagüela
	100	AUS	R. Australia, Shepparton
	500/100	BUL	R. Bulgaria
	250	USA	R. Martí, Delano
	10	LBR	Star R. Monrovia (F.Pl.)
5894.9		PRU	R. Paucartambo, Paucartambo
5895	250/200	RUS	Deutsche Welle, Petropavlovsk
	100/10	HRV	Hrvatski Radio, Zagreb
	250/100	GRC	Voice of Greece
5900	100	POL	Polish Radio, Warsaw
	1	TCD	R. Abéché, Chad
	500/100	BUL	R. Bulgaria
		F	R. France Internationale
		CHN	Sichuan GB
5905		UKR	R. Ukraine International
5910	200	BEL	R. Vlaanderen Internationaal
	200	RUS	Voice of America, Novosibirsk
		CHN	CNR 1
5915	500/20	ISR	Kol Israel, Arabic prgr

kHz	kW	Land	Station
	100	POL	Polish Radio, Warsaw
	100/20	HNG	R. Budapest, Székesfehérvár
		F	R. France Internationale
	100	SVK	R. Slovakia Int, Rimavská Sobota
5920		CHN	Guangxi 3
	100/10	HRV	Hrvatski Radio, Zagreb
		F	R. France Internationale
5925	250/200	RUS	Deutsche Welle, Samara
	0.15	BOL	R. Centinela,
		F	R. France Internationale
	100	EST	Tallinn (Laitse), Estonia
	500	IRN	Voice of the Islamic Republic of Iran
5927.1	1	BOL	Radiodifusoras Minería, Oruro (n. 5925)
5929.5		BOL	R. CVU
5930	50	RUS	Monchegorsk, E
		XXX	R. Free Asia
	200	MDG	R. Netherlands, Madagascar
	100	TCH	R. Prague, Czech Rep.
	100	SVK	R. Slovakia Int, Rimavská Sobota
5932	10	MOZ	Em. Nacional, Maputo
5935	250/200	RUS	Deutsche Welle, Samara
	500/100	BUL	R. Bulgaria
	100	LVA	Riga, Latvia
	100	USA	WWCR, Nashville, TN
	50	CHN	Xizang-Ch
5940	100	RUS	Arman, FE
	500/100	BUL	R. Bulgaria
		UKR	R. Ukraine International
		RUS	Voice of Vietnam via Russia
5945	250/200	RUS	Deutsche Welle, Samara
	500/100	AUT	R. Austria International
		F	R. France Internationale
	500/100	CVA	Vatican Radio
5948	5	BOL	R. Pío XII, Siglo Veinte
5950	50	CHN	Heilongjiang EBS
	100/50	USA	R. Taipei International via WYFR
	5	GUY	Voice of Guyana, Georgetown
	100	USA	WYFR-Family Radio, Okeechobee, FL
	50	CHN	Xizang-Tb
	300	YEM	Yemen Radio & TV Corp, Sana'a
5951.3	1	PRU	R. Arequipa, Arequipa
5952.3	5	BOL	R. Pío XII, Siglo Veinte (n. 5955) (vf)
5955	500/100	AFS	Channel Africa, Johannesburg
	50	CHN	CNR 1
	5	CLM	LV de los Centauros/Caracol, Villavicencio
	1	CTR	R. Casino, Limón (vf)
	0.5	GTM	R. Cultural, Guatemala (vf)
			R. Free Europe/R. Liberty
	500	HOL	R. Netherlands, Flevo
	100	UZB	R. Tashkent Ext.Sce, Uzbekistan
	250/50	ROU	Radio România, Bucharest
	100	I	RAI Ext. Sce, Rome, Italy
	250	PHL	Voice of America, Philippines
	500	TUR	Voice of Turkey, Emirler
5955.1		PRU	R. Estación Uno, Pucará (n. 5960)
5955.3	10	B	R. Gazeta, São Paulo
5955.3	1	PRU	R. Huancayo, Huancayo *
5956.8		PRU	R. Selecciones, Chquibamba
5960	500/250	AFS	AWR via Meyerton
	250	ATN	Deutsche Welle, Bonaire
	250	CAN	Deutsche Welle, Sackville
	250	CAN	R. Canada International, Sackville
	500	NOR	R. Norway International/R. Denmark
	100	D	R. Vlaanderen Internationaal via Jülich
	250	CAN	Radio Japan via Sackville
	250	G	Radio Japan via Skelton
	500	TUR	Voice of Turkey, Çakirlar
	50	CHN	Yunnan ED
5964.8	1	BOL	R. Nacional de Huanuni, Huanuni
5965	250	THA	BBC East Asian Relay Station
	100	OMA	BBC Eastern Relay Station
	500/250	G	BBC World Sce, United Kingdom
	250	CAN	BBC World Sce. via Sackville
		CHN	China Radio International
	500/200	RUS	Deutsche Welle, Krasnodar
	10-Feb	PNG	NBC Mount Hagen, Papua N.G.
	10	NIG	Plateau R.TV Corp, Jos
	100	HNG	R. Budapest, Diósd
	250	HNG	R. Budapest, Jászberény
	100/20	HNG	R. Budapest, Székesfehérvár
	50	CUB	R. Habana, Cuba
	500	NOR	R. Norway International/R. Denmark
	7.5	B	R. Nova Visão, Santa Maria (vf) (often rel. R. Transmundial)
	100	AFS	Radio Oranje, So. Africa
	250/50	ROU	Radio România, Bucharest
	100	MLA	RTM R1, Kuala Lumpur, Malaysia
	100	SWZ	Trans World Radio Africa, Swaziland
	250	GRC	Voice of America, Kavala
		CLA	Voice of Free Tajikistan (vf)
	500	TUR	Voice of Turkey, Emirler
5969.7	10	PRU	R. El Sol, Lima
5970	100	KAZ	Almaty, Kazakhstan
	250	ASC	BBC Atlantic Relay Station
	1	CHN	Gannan

kHz	kW	Land	Station
	250	EQA	HCJB, LV de los Andes, Quito
	100	POL	Polish Radio, Warsaw
		F	R. Free Europe
	10	B	R. Itatiaia, Belo Horizonte
	100	CTR	Radio Exterior de España, Costa Rica
	100	I	RAI Ext. Sce, Rome, Italy
	250/100		Voice of America (several locations)
	500	TUR	Voice of Turkey, Çakirlar
5973		BRM	Mayawadi R. St, Myanmar
5974.2	1	BOL	R. Nacional, Cochabamba *
5975	250	ATG	BBC Caribbean Relay Station
	250	THA	BBC East Asian Relay Station
	250/100	SNG	BBC Far Eastern Relay Station
	500/250	USA	BBC World Sce. via Greenville
	10	KOR	R. Korea Int, Hwasong
	5	CLM	R. Macarena, Villavicencio
	100	UZB	R. Tashkent Ext.Sce, Uzbekistan
	250	G	Radio Japan via Skelton
	100	I	RAI Ext. Sce, Rome, Italy
	250	G	SLBC Ext. Sce via Skelton
	100		Voice of America (several locations)
	100/20	ZWE	ZBC R3/R4, Gweru, Zimbabwe
5976.1	1	PRG	R. Guairá, Villarrica (n. 5975) (vf)
5980	500/100	D	Deutsche Welle, Germany
	0.5	MEX	LV Internacional de la Provincia Mexicana, Linares *
	10	BLR	Mahiliou, Belarus
	50	CHN	Nei Menggu-Ch
	10	B	R. Guarujá, Florianópolis
	1	PRU	R. LTC/RCC (R. Leoncio Torres Collao/R. Comercial Collao)
	10	MLA	RTM Kota Kinabalu, Sabah
	7.5	GTM	Unión Radio "LV de la Esperanza", Guatemala (vf)
	500/100	CVA	Vatican Radio
	250	TUR	Voice of Turkey, Çakirlar
5985	10	PNG	NBC Rabaul, Papua N.G.
	100	COG	R. Congo, Brazzaville
	250	FIN	R. Finland
			R. Liberty
	10	MEX	R. México Internacional, México
	2.5	BOL	R. San José, San José, Oruro *
	50	ALB	R. Tirana, Cërrik
	200	BEL	R. Vlaanderen Internationaal
	300/250		Voice of America (several locations)
	100	USA	WYFR-Family Radio, Okeechobee, FL
5986	50	BRM	R. Myanmar, Yangon (vf)
	100	TZA	R. Tanzania National Sce. *
5987.7	10	INS	RRI Manado
5990	250	IND	All India Radio Ext. Sce, Aligarh
	250	THA	BBC East Asian Relay Station
	500/250	G	BBC World Sce, United Kingdom
	500/100	D	Deutsche Welle, Germany
	250	CLN	Deutsche Welle, Trincomalee
		F	R. France Internationale
	7.5	B	R. MEC, Rio de Janeiro *
	100	CTR	Radio Exterior de España, Costa Rica
	250/50	ROU	Radio România, Bucharest
	100	I	RAI Ext. Sce, Rome, Italy
	250	THA	Voice of America, Udorn
	100	ETH	Voice of Ethiopia
	500	TUR	Voice of Turkey, Çakirlar
5995	250	USA	BBC World Sce. via Delano
	10	ZAI	Mbandaka, Congo (Dem. Rep.)
	20/100	MWI	MBC, Blantyre, Malawi
	100	POL	Polish Radio, Warsaw
	100	AUS	R. Australia, Shepparton
	300	G	R. Canada International via Skelton
	250	CAN	R. Canada International, Sackville
			R. Liberty
	2	PRG	R. LV del Amambay, Pedro Juan Caballero*
	50	MLI	Radiodiff.-TV Malienne, Bamako
	100	I	RAI Ext. Sce, Rome, Italy
	100	UZB	UZR1, Toskent, Uzbekistan
	250		Voice of America (several locations)
		CLA	Voice of Palestine (from Iran)
		CLA	Voice of Palestine, Voice of the Palestinian Islamic Revolution
	500	IRN	Voice of the Islamic Republic of Iran
	50	CHN	Xizang-Tb
5995.2		PRU	R. Melodía, Arequipa
5996.4	1	BOL	R. Loyola, Sucre (n. 5955)
5998.5		MLD	R. Maldives (irr.)
6000	50	IND	All India Radio, Leh
	500/100	AFS	Channel Africa, Johannesburg
	250	SNG	City Sounds, Singapore
	250/200	RUS	Deutsche Welle, Samara
	50	CHN	Haixia 2
	100	POL	Polish Radio, Warsaw
	5	B	R. Boas Nova, São Luís (CP)
	10	B	R. Guaíba, Porto Alegre
	250	CUB	R. Habana, Cuba
			R. Liberty
	250	SNG	R. Singapore International
	500	S	R. Sweden, Hörby

kHz	kW	Land	Station
	350	E	Radio Exterior de España
	100	AFS	Radio Sonder Grense, So. Africa
	2	URG	Radiomundo, Montevideo *
6005	250	ASC	BBC Atlantic Relay Station
	250	SEY	BBC Indian Ocean Relay Station
	0.5	CAN	CFCX Montreal (rel.CFCF)
	4	CME	CRTV, Buea, Cameroon
	100	D	Deutschlandradio Berlin, Germany
	0.6	J	NHK Sapporo 1 (DSB)
	100/20	HNG	R. Budapest, Székesfehérvár
	500	IRN	Voice of the Islamic Republic of Iran
6006	3	CTR	R. Reloj, San José (vf)
6008.8	0.3	J	NHK Nagoya 1 (SSB) (6008.75)
6010	250/100	CYP	BBC East Mediterranean Relay Station
	500/250	G	BBC World Sce, United Kingdom
		CHN	China Radio International
	500/100	D	Deutsche Welle, Germany
	60	BHR	R. Bahrain *
	100/20	HNG	R. Budapest, Székesfehérvár
	5	B	R. Inconfidência, Belo Horizonte
			R. Liberty
	1	MEX	R. Mil "La Estación de la Ciudad", México
	1	VEN	R. Mundial Los Andes, Mérida (vf)
		UKR	R. Ukraine International
		CLA	Radio United Bougainville
	100	I	RAI Ext. Sce, Rome, Italy
	0.15	URG	S.O.D.R.E. Montevideo (r: CX26 1050) (LSB)
		CLA	Voice of National Salvation
	500	TUR	Voice of Turkey, Emirler
6010.7	1	PRU	R. América, Lima
6012	5	I	Tele Radio Stereo, Roma, Italy
6014.4	0.3	PRG	Em. Paraguay, Asunción * (vf)
6014.6	50	TZA	R. Tanzania Zanzibar
6014.8	10	BOL	R. El Mundo, Santa Cruz (n. 6015)
6015	250/100	CYP	BBC East Mediterranean Relay Station
	500/250	G	BBC World Sce, United Kingdom
	50	CHN	CNR Taiwan
	500/100	D	Deutsche Welle, Germany
	250	RRW	Deutsche Welle, Kigali
	100	KOR	KBS Liberty 1 Prgr, Hwasong
	250	CAN	R. Austria International via Sackville
	500	CTI	R. Côte d'Ivoire, Abidjan
	1	CLM	R. Mira, Tumaco
	200	MDG	R. Netherlands, Madagascar
	250	SNG	R. Singapore International
	100	ALB	R. Tirana, Cërrik
	250	THA	Voice of America, Udorn
	500	IRN	Voice of the Islamic Republic of Iran
		CLA	Voice of the People of Kurdistan
	100	USA	WYFR-Family Radio, Okeechobee, FL
6018.2	5	PRU	R. Victoria, Lima
6020	50	IND	All India Radio, Shimla
	500/50	ARS	Broadc. Sce. of the Kingdom of Saudi Arabia
	0.25	MEX	La U de Veracruz, Veracruz *
	10	PNG	NBC Kieta, Papua N.G.
	10	B	R. Educadora de Bahia, Salvador
	10	B	R. Gaúcha, Porto Alegre
			R. Liberty
	250	ATN	R. Netherlands, Bonaire
	500	HOL	R. Netherlands, Flevo
	200	MDG	R. Netherlands, Madagascar
		UKR	R. Ukraine International
	500/100	CVA	Vatican Radio
	250	G	Voice of America, Woofferton
	500	IRN	Voice of the Islamic Republic of Iran
		CLA	Voice of the Mojahed (vf)
6025	15	CHN	Alxa-Ch
	10	MOZ	Beira, Mozambique (vf)
	10	NIG	Enugu National St., Nigeria
	1	DOM	R. Amanecer Internacional, Sto. Domingo (vf)
	250	HNG	R. Budapest, Jászberény
	300	G	R. Canada International via Skelton
	500	D	R. Canada International via Wertachtal
	10	BOL	R. Illimani,
			R. Liberty
	0.7	PRG	R. Nal. del Paraguay, Asunción *
		INS	RRI Banjarmasin
	100	MLA	RTM R5, Kuala Lumpur, Malaysia
	250/100		Voice of America (several locations)
		CLA	Voice of Palestine (from Iran)
		CLA	Voice of Palestine, Voice of the Palestinian Islamic Revolution
	500	IRN	Voice of the Islamic Republic of Iran
6029.7	10	CHL	R. Santa María, Coyhaique
6030	10	OMA	BBC Eastern Relay Station
	50/10	BTN	BBS, Thimpu, Bhutan
	0.1	CAN	CFVP Calgary (rel.CFCN)
		CHN	CNR 1
	1	BOL	R. ABC, Santa Cruz
	250	KOR	R. Canada International via Kimje
	10	B	R. Globo, Rio de Janeiro
	250	USA	R. Martí, Greenville
	250	ATN	R. Netherlands, Bonaire
	20	D	Süddeutscher Rundfunk
			Voice of America (several locations)
	500	IRN	Voice of the Islamic Republic of Iran
6031	10	INS	RRI Banjarmasin
6033.1		PRU	R. Super Landa, Arequipa (n. 6035)
6035	250	KOR	BBC World Sce. via Kimjae
	500/100	D	Deutsche Welle, Germany
	5	CLM	LV del Guaviare, San José del Guaviare
	100	POL	Polish Radio, Warsaw
	1	URG	R. Nacional, Montevideo *
	300/100	J	Radio Japan
	250		Radio Japan via Singapore
	12.5	CLN	Trans World Radio, Sri Lanka
			Voice of America (several locations)
	500	TUR	Voice of Turkey, Emirler
	50	CHN	Yunnan BS
6040	100	TWN	CBS Network 2, Taipei, Taiwan
	250	ATG	Deutsche Welle, Antigua
	50	CHN	Jiangxi 1
	10	PNG	NBC Alotau, Papua N.G.
	250	CAN	R. Canada International, Sackville
			R. Liberty
	500	THA	R. Thailand, Udon Thani
	100	SWZ	Trans World Radio Africa, Swaziland
	250	GRC	Voice of America, Kavala
	100	USA	WHRI, South Bend, IN
6040.3	7.5	B	R. Clube Paranaense, Curitiba
6045	250/50	IND	All India Radio Ext. Sce, Delhi
	5	CLM	Cadena Melodía, SF de Bogotá
	500/100	D	Deutsche Welle, Germany
	250	POR	Deutsche Welle, Sines
	100	D	Hrvatski Radio, Croatia via Jülich
	50	CHN	Nei Menggu-Ch
		F	R. France Internationale
	5	BOL	R. Libertad, La Paz *
	100	D	R. Netherlands, Jülich
	0.25	MEX	R. Universidad de San Luis Potosí, San Luis Potosí*
	350	E	Radio Exterior de España
	250		Voice of America (several locations)
	100/20	ZWE	ZBC R3/R4, Gweru, Zimbabwe
6045.1	1	URG	R. Sarandí Sport, Rivera
6045.2	5	PRU	R. Santa Rosa, Lima (vf)
6050	250/100	CYP	BBC East Mediterranean Relay Station
	500/250	G	BBC World Sce, United Kingdom
	30	EQA	HCJB, LV de los Andes, Quito
	50	NIG	Ibadan National St., Nigeria
	10	B	R. Guarani, Belo Horizonte
			R. Liberty
	10	MLA	RTM Sibu, Sarawak, Malaysia
	10	CLN	SLBC, Sri Lanka (irr.)
	250	G	Voice of America, Woofferton
	500	IRN	Voice of the Islamic Republic of Iran
6055	250	KOR	BBC World Sce. via Kimjae
	500/100	D	Deutsche Welle, Germany
	250	CLN	Deutsche Welle, Trincomalee
	500/250	KWT	R. Kuwait
	100	TCH	R. Prague, Czech Rep.
	250	RRW	R. Rwanda, Kigali
	100	SVK	R. Slovakia Int, Rimavská Sobota
	50	J	R. Tanpa 1, Tokyo, Japan
	350	E	Radio Exterior de España
	250	G	radio Japan via Skelton
6055.3	2	PRU	R. Continental, Arequipa
6060	250/100	SNG	BBC Far Eastern Relay Station
	100	NMB	Namibian Broadc. Corp. Ch.1
	30	ARG	R. Nal. Buenos Aires
	100	PAK	R. Pakistan, Islamabad
	10	B	R. Universo, Curitiba
	25	I	RAI Radio 1, Caltanissetta
	100	I	RAI Radio 1, Rome, Italy
	10	MLA	RTM Miri, Sarawak, Malaysia
	50	CHN	Sichuan 1
			Voice of America (several locations)
	500	TUR	Voice of Turkey, Emirler
6061	3	PRU	R. JSV, Huánuco *
6064		CLA	Radio of Jihad
6065	50	IND	All India Radio, Kohima
	250	THA	BBC East Asian Relay Station
	100	OMA	BBC Eastern Relay Station
	300	J	BBC World Sce. via Yamata
	100	ZMB	Christian Voice, Zambia
		CLM	Colmundo, SF de Bogotá
	500/100	D	Deutsche Welle, Germany
	1	BOL	R. Mauro Nuñez, Villa Serrano
	500	S	R. Sweden, Hörby
	50	PHL	Voice of America, Philippines
	100	USA	WYFR-Family Radio, Okeechobee, FL
6070	500/250	AFS	BBC World Sce. via Meyerton
	1	CAN	CFRX Toronto (rel.CFRB)
	500/100	D	Deutsche Welle, Germany
	7.5	B	R. Capital, Rio de Janeiro
	100/50	CUB	R. Habana, Cuba

kHz	kW	Land	Station
	100	NZL	R. New Zealand International (from March 29)
	250/50	PAK	R. Pakistan Ext. Sce. (vf)
	100	PAK	R. Pakistan, Islamabad
	200	KRE	R. Pyongyang, Korea (D.P.R.)
	10	THA	R. Thailand Prgr. 1
	50	ATN	R. Vlaanderen Internationaal via Bonaire
	500	GUF	Radio Japan via Montsinery
	20	INS	RRI Jayapura
6074	5	URG	LV de Artigas, Artigas (n. 6075) (vf)
6075	250	KEN	KBC National Sce, Nairobi
	1	HND	LV del Junco, Santa Bárbara (r: R. Galaxia)
	1.5	CTR	R. 88 Estéreo, Pérez Zeledón (F.Pl.)
	250	PHL	R. Veritas Asia, Philippines
	10	CLN	SLBC Sinhala Nat. Sce.
6080	250/100	SNG	BBC Far Eastern Relay Station
	10	BLR	Brest, Belarus
	0.01	CAN	CKFX Vancouver (rel.CKWX)
	8	EQA	HCJB, LV de los Andes, Quito
	10	PNG	NBC Daru, Papua N.G.
	100	AUS	R. Australia, Shepparton
	5	B	R. CBN Anhanguera, Goiânia
	100	GEO	R. Georgia, Tbilisi
	10	B	R. Novas de Paz, Curitiba
	1	CHL	R. Patagonia Chilena, Coyhaique
	100	STP	Voice of America, São Tomé
6085	50	IND	All India Radio, Gangtok
	100	D	Bayerischer Rundfunk, Germany
	100	TWN	CBS Network 2, Taipei, Taiwan
	250	CAN	Deutsche Welle, Sackville
	1	CLM	Ondas del Darién, Turbo *
	100	OMA	R. Oman, Thumrait
	100	RUS	Voice of America, Petropavlovsk
		CLA	Voice of Rebellious Iraq (vf)
	100	USA	WYFR-Family Radio, Okeechobee, FL
6085.3	5	BOL	R. San Gabriel, La Paz (n. 6080) (vf)
6087.9		PRU	R. Chasqui, Cusco
6090	500/250	G	BBC World Sce., United Kingdom
	1	MEX	Música Romántica, Cd.Mante *
	15	CBG	National Radio of Cambodia *
	10	B	R. Bandeirantes, São Paulo
	250	CAN	R. Canada International, Sackville
	10	CHL	R. Esperanza, Temuco
	250/50	NIG	R. Nigeria, Kaduna Ch.1 (vf)
		UKR	R. Ukraine International
	300/100	J	Radio Japan
	100	AIA	The Caribbean Beacon, Anguilla
	100	BOT	Voice of America, Botswana
	1000	RUS	Voice of America, Irkutsk
6090.8	2	PRU	R. Universal, Cusco *
6095	100	OMA	BBC Eastern Relay Station
	500/250	G	BBC World Sce., United Kingdom
	2.7	RUS	Khanty-Mansiysk, S
	100	POL	Polish Radio, Warsaw
	10	BOL	R. Cosmos, Cochabamba
			R. Free Europe/R. Liberty
	10	PRU	R. Nal. del Perú, Lima
	250	G	Radio Japan via Skelton
	250	G	Radio Japan via Woofferton
	500/100	CVA	Vatican Radio
	250	GRC	Voice of America, Kavala
	500	TUR	Voice of Turkey, Emirler
	500	USA	WSHB, Cypress Creek, SC
6096.4		PRU	R. Interoceánica, Azangaro
6100	500/250	AFS	AWR via Meyerton
	10	NIG	Borno R. Corp, Nigeria
	500/100	D	Deutsche Welle, Germany
	200	KRE	KCBS, Kanggye, No. Korea
	7.5	LBR	R. Liberia Int., Totota
	100	NZL	R. New Zealand International (from March 29)
	100	ALB	R. Tirana, Cërrik
	500/250	BIH	R. Yugoslavia, Bijeljina, Bosnia
	50	SWZ	Trans World Radio Africa, Swaziland
	250	THA	Voice of America, Udorn
	100	MLA	Voice of Malaysia, Kuala Lumpur
6104.8	5	B	R. Cultura, Foz do Iguaçu (rel. Rede Transamérica FM)
6105	10	B	R. Canção Nova, Cachoeira Paulista
	0.25	MEX	R. Fórmula Mérida, Mérida
	500	JOR	R. Jordan, Al Karanah
			R. Liberty
	100	NZL	R. New Zealand International
	30	TZA	R. Tanzania National Sce. *
	1	CTR	R. Universidad de América R. U, San José
	250/50	ROU	Radio România, Bucharest
	125	D	Voice of America, Holzkirchen
6105.3	7.5	BOL	R. Panamericana, La Paz
6109	5	BOL	R. Dif. Integración, El Alto (n.6120)
6110	50	IND	All India Radio, Srinagar
	250	ASC	BBC Atlantic Relay Station
	250	ATG	BBC Caribbean Relay Station
	250	THA	BBC East Asian Relay Station
	500/250	G	BBC World Sce., United Kingdom
	300	J	BBC World Sce. via Yamata

kHz	kW	Land	Station
		CHN	China Radio International
		CHN	CNR 1
	170	AZE	Gäncä, Azerbaijan
	100	EQA	HCJB, LV de los Andes, Quito
	250	G	Merlin Communications, UK
	250	CAN	Radio Japan via Sackville
	100	I	RAI Ext. Sce, Rome, Italy
	250	ASC	RAI Ext. Sce. via Ascension
	250	PHL	Voice of America, Philippines
6115	50	CHN	Haixia 1
	75	BLR	Kalodziscy, Belarus
	10	CLM	LV del Llano, Villavicencio (vf)
	100	COG	R. Congo, Brazzaville
			R. Free Europe
	50	J	R. Tanpa 2, Tokyo, Japan
	100	ALB	R. Tirana, Cërrik
	10	PRU	R. Unión, Lima
6120	250	CAN	Deutsche Welle, Sackville
	250	POR	Deutsche Welle, Sines
	250	HNG	R. Budapest, Jászberény
	250	EGY	R. Cairo, Abis
	250/100	FIN	R. Finland
		F	R. France Internationale
	7.5	B	R. Globo, São Paulo
	200	MDG	R. Netherlands, Madagascar
	100	OMA	R. Oman, Seeb
	250	SNG	R. Singapore International
	100	ALB	R. Tirana, Shijak
	100	D	R. Vilnius, Jülich relay
	250	CAN	Radio Japan via Sackville
	100	STP	Voice of America, São Tomé
	300	G	Voice of America, Woofferton
	250	TUR	Voice of Turkey, Çakirlar
6125	100	D	BBC World Sce. via Lampertheim
	50	CHN	CNR 1
	100	EQA	HCJB, LV de los Andes, Quito
	10	ZAI	Kananga, Congo (Dem. Rep.) *
	250	G	Merlin Communications, UK
	200	KRE	R. Pyongyang, Korea (D.P.R.)
		CLA	Radio Agatashya
	350	E	Radio Exterior de España
	100	I	RAI Ext. Sce, Rome, Italy
	0.3	URG	S.O.D.R.E, Montevideo (r: CX38 1290)
	250	G	Voice of America, Woofferton
6130	500/250	G	BBC World Sce., United Kingdom
	250	USA	BBC World Sce. via Delano
	0.5	CAN	CHNX Halifax (rel.CHNS)
	250	CLN	Deutsche Welle, Trincomalee
	50	GHA	GBC Radio 2, Ghana
	25	LAO	Lao National Radio, Vientiane
			R. Liberty
	350	E	Radio Exterior de España
	10	CLN	SLBC English Comm.Sce.
	100	SWZ	Trans World Radio Africa, Swaziland
	250		Voice of America (several locations)
6133.8	0.6	J	NHK Fukuoka 1 (SSB) (6133.75)
6134	1	INS	RRI Samarinda
6135	500/250	AFS	BBC World Sce. via Meyerton
	10	KOR	KBS Liberty 2 Prgr, Hwasong
	25	B	R. Aparecida, Aparecida
	250	HNG	R. Budapest, Jászberény
	10	KOR	R. Korea Int, Hwasong
			R. Liberty
	10	BOL	R. Santa Cruz, Santa Cruz
	250	SNG	R. Singapore International
	250	ASC	Radio Japan via Ascension Island
	30	MDG	Radio Television Malagasy
	100/50	SWZ	Trans World Radio Africa, Swaziland
	500	TUR	Voice of Turkey, Çakirlar
6139.9	2.5	URG	R. Monte Carlo. Montevideo (n. 6140)
6140	10	IND	All India Radio, Ranchi
	100	OMA	BBC Eastern Relay Station
		CHN	China Radio International
	10	PNG	NBC Wewak, Papua N.G.
	500/100	BUL	R. Bulgaria
		CHN	R. Canada International via X'ian
			R. Free Europe/R. Liberty
	100	ALB	R. Tirana, Cërrik
	250	G	Radio Japan via Skelton
	100	BDI	Radiodiff. Nat., Burundi (irr.)
6140.1		PRU	R. Huayllay, Huayllay
6140.7	1	BOL	R. Luis de Fuentes, Tarija *
6141	1	PRU	R. Concordia, Arequipa (v. up to 6141.5) (n. 6150) (vf)
6142		BOL	R. Mauro Nuñez, Villa Serrano
6145	500/100	D	Deutsche Welle, Germany
	250	G	R. Korea Int, Skelton, UK
	50	ALB	R. Tirana, Cërrik
	300/100	J	Radio Japan
	250	TUR	Voice of Turkey, Emirler
6149.8	10	THA	Or. Sor., Thailand
6150	10	IND	All India Radio, Itanagar
	50/20	CTR	AWR Latin America, Cahuita
	10	CLM	Caracol Colombia, SF de Bogotá
		F	China Radio International via France

kHz	kW	Land	Station
	1	AGL	EP de Benguela, Angola (vf)
	250	KEN	KBC National Sce, Nairobi
	100	ALS	KNLS, Anchor Point, AL
	7.5	CYP	R. Bayrak II, No.Cyprus (vf)
	250	KOR	R. Canada International via Kimje
	250	POR	R. Canada International via Sines
	300	G	R. Canada International via Skelton
		CHN	R. Canada International via X'ian
	300	J	R. Canada International via Yamata
			R. Free Europe/R. Liberty
	10	B	R. Record, São Paulo
	250	SNG	R. Singapore International (from March 29)
	300/100	J	Radio Japan
	250	G	Radio Japan via Skelton
	250/50	ROU	Radio România, Bucharest
		PRU	Radiodifusoras LV de la Esperanza, Juliaca (F.PI.)
	300/100	POR	RDP Internacional (R. Portugal)
	10	CLN	SLBC Sinhala Comm.Sce.
6153		INS	RRI Biak
6155	100	IND	All India Radio Ext. Sce, Delhi
	200	KAZ	Almaty. Kazakhstan
	250	ASC	BBC Atlantic Relay Station
	100	GUI	Conakry, Guinea Rep. (irr.)(vf)
	500	LBY	LJB, Tripoli/Sabrata. Libya
	500/100	AUT	R. Austria International
	250	CAN	R. Canada International, Sackville
	10	BOL	R. Fides, La Paz
	0.5	PRU	R. Pucallpa, Pucallpa (vf)
	250	SNG	R. Singapore International (to March 28)
	250	G	Radio Japan via Skelton
	20/50	TGO	Radio Kara, Togo (vf)
	250	SNG	Radio One, Singapore
	250/50	ROU	Radio România, Bucharest
	100	SWZ	Swazi Radio (alt.9704kHz)
6160	200	KAZ	Almaty. Kazakhstan
	40	RUS	Arkhangelsk, E
	0.3	CAN	CKZN St.John's (rel.CBN)
	0.5	CAN	CKZU Vancouver (rel.CBU)
	250	CLN	Deutsche Welle, Trincomalee
	10	B	R. Rio Mar, Manaus
	350	E	Radio Exterior de España
	500/100	CVA	Vatican Radio
	250		Voice of America (several locations)
6160.2	10	B	R. Rio Grande do Sul, Porto Alegre *
6160.7	1	ARG	R. Malargüe
6165	100	IND	All India Radio Ext. Sce, Delhi
	250	SUI	China Radio International via Switzerland
	250	ATN	R. Netherlands, Bonaire
	300/100	J	Radio Japan
	100	TCD	Radiodiff. Nat. Tchadienne
	250	SUI	Red Cross Broadc. Sce, Geneva
	100	D	Swiss R. International, via Jülich
	250	USA	Voice of America, Greenville
	500	IRN	Voice of the Islamic Republic of Iran
	50	ZMB	ZNBC Radio Two, Lusaka, Zambia
6170	2.5	CLM	Caracol Colombia, Florencia (vf)
	250	CLN	Deutsche Welle, Trincomalee
	7.5	B	R. Cultura, São Paulo
			R. Liberty
	500	S	R. Sweden, Hörby
	100	ALB	R. Tirana, Cërrik
	100	I	RAI Ext. Sce, Rome, Italy
	250		Voice of America (several locations)
6173.8	5	PRU	R. Tawantinsuyo, Cusco (n. 6175)
6175	250	USA	BBC World Sce. via Delano
	250	CAN	BBC World Sce. via Sackville
	2.5	CTR	Faro del Caribe, San José
	250	HNG	R. Budapest, Jászberény
	100/20	HNG	R. Budapest, Székesfehérvar
		F	R. France Internationale
	50	MLA	Voice of Malaysia, Kuala Lumpur
	500	IRN	Voice of the Islamic Republic of Iran
		CLA	Voice of the Mojahed
		CLA	Voice of the Mojahed (vf)
	100	D	West Coast R. Ireland via Jülich
	100	USA	WYFR-Family Radio, Okeechobee, FL
6176	15	CHN	Shaanxi Sat.
6178.8	0.9	J	NHK Tokyo (SSB) (6178.75)
6180	100	KAZ	Almaty. Kazakhstan
	250/100	CYP	BBC East Mediterranean Relay Station
	500/250	G	BBC World Sce, United Kingdom
	300	TWN	CBS Network 1/2, Taipei, Taiwan
	250	CYP	CyBC via BBC Relay St.
	10	GTM	LV de Guatemala, Guatemala
	250	FIN	R. Finland
	100	GEO	R. Georgia, Tbilisi
	100/50	CUB	R. Habana, Cuba
	250	B	R. Nacional da Amazônia, Brasília (vf) *
	500	UAE	R. of the UAE, Abu Dhabi
	250	G	Radio Japan via Skelton
6180.3	7.5	ARG	R. Nal. Mendoza
6185	250	USA	BBC World Sce. via Delano
		CHN	China Huayi BC
		CHN	China Radio International

kHz	kW	Land	Station
	250	ATG	Deutsche Welle, Antigua
	100	LBY	LJB, Benghazi, Libya
	250	G	Merlin Communications, UK
	3	BOL	R. Batallón Colorados, La Paz
	300	J	R. Canada International via Yamata
	5	MEX	R. Educación, México
	100	ALB	R. Tirana, Cërrik
	250	BIH	R. Yugoslavia, Bijeljina, Bosnia
	10	CLN	SLBC Tamil Comm.Sce.
	500/100	CVA	Vatican Radio
	250	PHL	Voice of America, Philippines
6188.2	5	PRU	R. Oriente, Yurimaguas (vf)
6189	0.75	INS	RRI Manokwari.
6190	50	IND	All India Radio, Delhi
	500/250	AFS	BBC World Sce. via Meyerton
	250	PHL	R. Veritas Asia, Philippines
	50	D	Radio Bremen
6194.5	5	BOL	R. Carlos Palenque, La Paz (n. 6195) (vf)
6195	50	IND	All India Radio, Shillong
	250	ATG	BBC Caribbean Relay Station
	250/100	CYP	BBC East Mediterranean Relay Station
	250/100	SNG	BBC Far Eastern Relay Station
	500/250	G	BBC World Sce, United Kingdom
	50	CHN	Nei Menggu-Mo
	500	BIH	R. Yugoslavia, Bijeljina, Bosnia
6200	100	TCH	R. Prague, Czech Rep.
	500	S	R. Sweden, Hörby
		CHN	Xizang-Tb
6201.2	2	PRU	R. LV de Huamanga, Ayacucho (n. 6070) (vf)
6203.8	1	PRU	R. Cusco, Cusco (n. 6195) (vf)
6205		CLA	R. Kudirat Nigeria
	500	AFS	R. Kudirat Nigeria via Meyerton
6210		CLA	Radio Fana
	500	IRN	Voice of the Islamic Republic of Iran
6230	100	KAZ	Almaty. Kazakhstan
	7	I	European Christian Radio, Italy
	250	EGY	R. Cairo, Abis
	100	UZB	R. Tashkent Ext.Sce, Uzbekistan
6231		CLA	La Voz de la Resistencia (vf)
6235	0.7	DOM	R. Quisqueya, Santo Domingo (vf)
6238		PRU	R. Estación Yurimaguas, Yurimaguas (n. 6190) (vf)
6242.1	0.12	PRU	R. Calca, Calca (vf)
6250	10	GNE	Malabo, Equatorial Guinea
		KRE	Pyongyang BC St, No. Korea (vf)
	10	PAK	R. Pakistan, Peshawar
6251		CLA	Radio Patria Libre
6260		PRU	R. LV de Andahuaylas, Andahuaylas (vf)
	250/100	GRC	Voice of Greece
6261.1		PRU	R. Juanjuí, Juanjuí
6261.1		PRU	R. J-V-L, Centro Comercial de Consuelo
6280	10	LBN	Voice of Hope, Lebanon
6281.1	1.5	PRU	R. Huancabamba, Huancabamba (n. 6290) (vf)
6295	0.5	I	Good Fun R., Treviso, Italy
6300		TWN	Family Radio via VOFC, Taiwan
	1	HND	Sani R., Puerto Lempira (vf) (irr.)
6305		CLA	Radio of Islam
6306		CLA	La Voz del CID - la Voz de Cuba, Independiente y Democrática
6325		BOL	R. Velmar, Dalence
6325.1		TUR	Istanbul Polis Radyosu, Turkey (r)
6329	0.8	PRU	R. Estación C, Moyobamba
6330		CLA	Estación Fariana (vf)
6339.5		PRU	R. San Miguel Arcangel, San Miguel
6348		CLA	Radio Echo of Hope
6355		CLA	Thabye Radio (vf)
6400		KRE	Pyongyang BC St, Kanggye (vf)
6405		CLA	Voice of Iranian Kordestan (vf)
		CLA	Voice of the Iranian Revolution (vf)
6420.3		PRU	R. Mi Frontera, Chirinos
6450		CLA	Voice of the Mojahed (vf)
6479.8		PRU	R. Altura, Huarmaca
6480	250/100	KOR	R. Korea Int, Kimje
6500	50	CHN	Qinghai-Tb
		CLA	Radio Awdal
6518		CLA	Voice of the People
6520	200	KRE	R. Pyongyang, Korea (D.P.R.)
6520.2		PRU	R. Ondas del Río Marañon, Aramango
6535.8		PRU	Radiodifusoras Huancabamba, Huancabamba
6544		CLA	Holy Koran Radio (vf)
6545		SOM	Holy Quran R, Somalia (vf)
6550		LBN	R. Voice of Lebanon
6557	0.3	BOL	R. Colonia, Yapacani (n. 6555) (vf)
6560		IRQ	R. Baghdad, Iraq
6570	10	BRM	Defence Forces BC, Myanmar
6575	200	KRE	R. Pyongyang, Korea (D.P.R.)
6590		CHN	China Radio International
6600		CLA	Voice of the People
6670.2		PRU	R. Santa Mónica, Santiago de Chuco
6676.3		PRU	R. Huamachuco, Huamachuco (vf)
6679		HKG	HK Aeradio (Time Signals)
6690	1	PRU	R. Cutervo, Cutervo *

kHz	kW	Land	Station
6702		VTN	R. Lao Cai, Viettnam (vf
6725.6	0.3	PRU	R. Satélite, Santa Cruz (alt. freq. 4781) (vf)
6732		SOM	R. Mogadishu, V. of the Somali Pacification (vf)
6750		CHN	CNR 1
6754.7		PRU	R. La Merced, Tongod (n. 6760)
6761		PRU	R. San Andrés de Llapa, Llapa
6790		CHN	CNR Chinese
6797.8		PRU	R. Ondas del Río Mayo, Nueva Cajamarca
6815.7	0.1	PRU	R. Universo, Saposoa *
6822		SOM	R. Mogadishu, V. of the Somali Rep. (vf)
6840		CHN	CNR 1
		E	EBC (Time Signals), Spain
6851.2		PRU	R. Cultural, Cailloma *
6890		CHN	CNR 2
		SOM	R. Mogadishu, V. of the Masses of the Somali Rep. (vf)
6895.3	0.25	PRU	R. San Miguel, San Miguel de El Faique
6900	5	TUR	Turkish State Met.Sce.
6920		CHN	China Radio International
6933		CHN	China Radio International
6937		CHN	Yunnan Minor.
6940		CLA	Radio Fana
6950		CHN	China Radio International
6955		CHN	China Radio International
6974	50	CHN	Nei Menggu-Mo
6975	1	LAO	Luang Prabang Regional R, Laos (vf)
7008		CLA	Freedom Radio Station
7008		CLA	Voice of the Iraqi People (vf)
7020		ERI	V. of the Broad Masses of Eritrea
7030		CLA	Voice of Independent Kurdistan (vf)
7040		CLA	R. Democracy (vf)
7040.4		PRU	R. San Ignacio, San Ignacio
7048.4		PRU	R. LV de Santa Cruz, Santa Cruz (v. 7049.1-7050.9) (vf)
7060		CLA	Radio Butembo, the Voice of the People
7070		CLA	Voice of Palestine (from Iran)
	500	IRN	Voice of the Islamic Republic of Iran
		CLA	Voice of the Mojahed (vf)
7082		CLA	Radio Afghanistan (vf)
7085		ERI	V. of the Broad Masses of Eritrea
7093		CLA	Voice of Southern Azerbaijan (vf)
7099	7.5	INS	RRI Yogyakarta
7100		CLA	A Voz da Resistencia de Galo Negro (VORGAN)
		ZAI	Kinshasa, Congo (Dem. Rep.)
		CLA	Voice of Free Tajikistan (vf)
	500	IRN	Voice of the Islamic Republic of Iran
7103.5		TUR	R. Izmir, Turkey (Oct-May only) (r)
7105	50	IND	All India Radio, Lucknow
	10	I	AWR Europe, Forli
	250	ASC	BBC Atlantic Relay Station
	250	THA	BBC East Asian Relay Station
	250/100	CYP	BBC East Mediterranean Relay Station
	500/250	G	BBC World Sce, United Kingdom
	300	TWN	CBS Network 1/2, Taipei, Taiwan
	300	CYP	CyBC via BBC Relay St.
	250	BLR	Kalodziscy, Belarus (wi)
	50	CHN	Nei Menggu-Ch
	100	UZB	R. Tashkent Ext.Sce, Uzbekistan
	100	ALB	R. Tirana, Cërrik
	250/50	ROU	Radio România, Bucharest
	250	ASC	Voice of America, Ascension Isl.
	250	GRC	Voice of America, Kavala
7110		IND	All India Radio, Delhi
	250/100	SNG	BBC Far Eastern Relay Station
		CHN	China Radio International
	100	PAK	R. Pakistan, Rawalpindi
	100	ALB	R. Tirana, Cërrik
	20-Oct	UGA	R. Uganda, Kampala
	300/100	POR	RDP Internacional (R. Portugal)
	250	THA	Voice of America, Udorn
	100	ETH	Voice of Ethiopia
	500/250	TUR	Voice of Turkey, Emirler
7110	50	CHN	Xizang-Tb
7115	20	IND	All India Radio, Port Blair
			R. Free Europe/R. Liberty
	500	S	R. Sweden, Hörby
	10	THA	R. Thailand Prgr. 1
		UKR	R. Ukraine International
	500	BIH	R. Yugoslavia, Bijeljina, Bosnia
	100	I	RAI Ext. Sce, Rome, Italy
	10	CLN	Voice of America, Colombo
	250	GRC	Voice of America, Kavala
	500	IRN	Voice of the Islamic Republic of Iran
		CLA	Voice of the Islamic Revolution in Iraq
7120	50	IND	All India Radio, Jaipur
	500/250	G	BBC World Sce, United Kingdom
		CHN	China Radio International
	250	RRW	Deutsche Welle, Kigali
	500	LBY	LJB, Tripoli/Sabrata. Libya
			R. Liberty
	100	I	RAI Ext. Sce, Rome, Italy
	100	PHL	Voice of America, Philippines
7125	100	IND	All India Radio Ext. Sce, Delhi

kHz	kW	Land	Station
7125	500/250	G	BBC World Sce, United Kingdom
	500/250	AFS	BBC World Sce. via Meyerton
		CHN	China Radio International
	10	I	IRRS, Milano, Italy
	100	PAK	R. Pakistan, Islamabad
	300/100	J	Radio Japan
	250	THA	Voice of America, Udorn
	500	IRN	Voice of the Islamic Republic of Iran
7125.4	100	GUI	Conakry, Guinea Rep. (irr.) (vf)
7130	50	IND	All India Radio, Shillong
	250/100	CYP	BBC East Mediterranean Relay Station
	500/100	D	Deutsche Welle, Germany
	250	POR	Deutsche Welle, Sines
	500	HOL	R. Netherlands, Flevo
	250/50	TWN	R. Taipei International, Taiwan
	10	MLA	RTM Kuching, Sarawak, Malaysia
	250	THA	Voice of America, Udorn
	300	G	Voice of America, Woofferton
	500	IRN	Voice of the Islamic Republic of Iran
7135	250	THA	BBC East Asian Relay Station
	250/100	SNG	BBC Far Eastern Relay Station
	100	HNG	R. Budapest, Diósd
	250	HNG	R. Budapest, Jászberény
		F	R. France Internationale
	250/50	ROU	Radio România, Bucharest
	100	STP	Voice of America, São Tomé
7140	100	IND	All India Radio Ext. Sce, Delhi
	50	IND	All India Radio, Hyderabad
	250/100	CYP	BBC East Mediterranean Relay Station
	5	BLR	Hrodna, Belarus
	100	KEN	KBC National Sce, Nairobi
		F	R. France Internationale
	15	I	R. Italia Int, Spoleto, Italy
	300/100	J	Radio Japan
	250	THA	Voice of America, Udorn
	50	RUS	Yakutsk, S
7145	500/250	G	BBC World Sce, United Kingdom
	500/100	D	Deutsche Welle, Germany
	10	LAO	Lao National Radio (Ext. Sce.) *
	10	BLR	Mahiliou, Belarus
	100	POL	Polish Radio, Warsaw
			R. Liberty
	250/50	PAK	R. Pakistan Ext. Sce. (vf)
	250/50	ROU	Radio România, Bucharest
	10	MLA	RTM Kuching, Sarawak, Malaysia
	100	ALG	RTV Algerienne, Bouchaoui
	100		Voice of America (several locations)
	100	STP	Voice of America, São Tomé
	500	IRN	Voice of the Islamic Republic of Iran
7147.7		PRU	R. Ayabaca, Ayabaca (vf)
7150	250	IND	All India Radio Ext. Sce, Delhi
	50	IND	All India Radio, Imphal
	500/250	G	BBC World Sce, United Kingdom
	1	ZAI	Bunia, Congo (Dem. Rep.)
	20	AFS	Capital Radio, So. Africa (vf)
		CHN	China Radio International
	100	CME	CRTV, Douala, Cameroon
	1	AGL	EP do Lobito, Angola (vf)
		IRQ	R. Baghdad, Iraq
		CHN	R. Canada International via X'ian
	500	S	R. Sweden, Hörby
	50	ALB	R. Tirana, Cërrik
		UKR	R. Ukraine International
	250	PHL	Voice of America, Philippines
	250	THA	Voice of America, Udorn
	500	TUR	Voice of Turkey, Emirler
7153		VTN	R. TV Bac Thai, Vietnam (vf)
7155	500/250	G	BBC World Sce, United Kingdom
	100	D	BBC World Sce. via Lampertheim
			R. Free Europe/R. Liberty
	500	JOR	R. Jordan, Al Karanah
	100	ALB	R. Tirana, Cërrik
	100	MDG	Radio Television Malagasy
	100	ALB	Trans World Radio, Cërrik
	250	MRC	Voice of America, Morocco
7155.5	20	NGR	La Voix du Sahel, Niger
7160	50	IND	All India Radio, Chennai
	250	ASC	BBC Atlantic Relay Station
	250	THA	BBC East Asian Relay Station
	250/100	SNG	BBC Far Eastern Relay Station
		F	R. France Internationale
	100	ALB	R. Tirana, Cërrik
	10	MLA	RTM Kuching, Sarawak, Malaysia
	500/100	MCO	Trans World Radio, Monte Carlo
	100	PHL	Voice of America, Philippines
	500	IRN	Voice of the Islamic Republic of Iran
7165	250/100	CYP	BBC East Mediterranean Relay Station
	100/10	HRV	Hrvatski Radio, Zagreb
	100	HNG	R. Budapest, Diósd
	250	HNG	R. Budapest, Jászberény
			R. Free Europe/R. Liberty
	100	NPL	R. Nepal (su only)
	10	TZA	R. Tanzania National Sce.
	500	BIH	R. Yugoslavia, Bijeljina, Bosnia
	250	G	Voice of America, Woofferton

kHz	kW	Land	Station
	100	ETH	Voice of Ethiopia
7170	500	USA	Merlin Comms, via Rampisham
	100	SNG	Olikkalanjiam, Singapore
	100	SEN	ORTS Nat.Sce, Dakar, Senegal
	100	OMA	R. Oman, Seeb
	10	PAK	R. Pakistan, Quetta
	100	ALB	R. Tirana, Cërrik
	100	ALB	Trans World Radio, Cërrik
	300	G	Voice of America, Wooferton
	500	IRN	Voice of the Islamic Republic of Iran
	50	CHN	Xizang-Ch
7173	0.5	INS	RRI Serui
7175	500/100	D	Deutsche Welle, Germany
	250	RRW	Deutsche Welle, Kigali
			R. Liberty
	250/50	ROU	Radio România, Bucharest
	100	I	RAI Ext. Sce, Rome, Italy
	5	I	RAI Radio 2, Caltanissetta
	100	SWZ	Trans World Radio Africa, Swaziland
	250	E	Voice of America, Playa de Plas
	500	IRN	Voice of the Islamic Republic of Iran
7180	50	IND	All India Radio, Bhopal
		CHN	China Radio International
	100	POL	Polish Radio, Warsaw
			R. Free Europe/R. Liberty
	500	NOR	R. Norway International/R. Denmark
		UKR	R. Ukraine International
	250	MRC	Voice of America, Morocco
	300	G	Voice of America, Wooferton
	500	IRN	Voice of the Islamic Republic of Iran
7185		CHN	China Radio International
	100	POL	Polish Radio, Warsaw
	250	BGD	R. Bangladesh Ext. Sce. (vf)
	250	HNG	R. Budapest, Jászberény
	50	BRM	R. Myanmar, Yangon
	100	AFS	Radio Sonder Grense, So. Africa
7190		CHN	China Radio International
	100	GNE	R. Africa, Malabo, Eq. Guinea
			R. Liberty
	500	S	R. Sweden, Hörby
	300/100	J	Radio Japan
	100	I	RAI Ext. Sce, Rome, Italy
	250	THA	Voice of America, Udorn
		CLA	Voice of Palestine, Voice of the Palestinian Islamic Revolution
	500	IRN	Voice of the Islamic Republic of Iran
	500	TUR	Voice of Turkey, Emirler
			R. Liberty
7195	20-Oct	UGA	R. Uganda, Kampala (vf)
	250/50	ROU	Radio România, Bucharest
	300	G	Voice of America, Wooferton
	50	CHN	Xinjiang-Ug
		CHN	Xizang-Tb
7200	100	SDN	Omdurman, Sudan
	100	AFG	R. Afghanistan, Yakatut
	500	D	R. Canada International via Wertachtal
	200	KRE	R. Pyongyang, Korea (D.P.R.)
	300/100	J	Radio Japan
	100	YUG	Radiotelevizija Srbije, Stubline, Serbia
	250/100		Voice of America (several locations)
	100	RUS	Yakutsk, S
	5	RUS	Yoshkar-Ola, E
7202	10	ZAI	Lubumbashi, Congo (Dem. Rep.) *
7205	250	THA	BBC East Asian Relay Station
	100	OMA	BBC Eastern Relay Station
	250/100	SNG	BBC Far Eastern Relay Station
	100	POL	Polish Radio, Warsaw
	250	GRC	Voice of America, Kavala
7210	50	IND	All India Radio, Calcutta
	250/100	CYP	BBC East Mediterranean Relay Station
	500/250	G	BBC World Sce, United Kingdom
		CHN	CNR 1
	75	BLR	Kalodziscy, Belarus
	100	RUS	Khabarovsk, FE
	250/100	QAT	Qatar Broadc. Sce, Doha
	250	G	Radio Japan via Skelton
	250	GRC	Voice of America, Kavala
	250	MRC	Voice of America, Morocco
	15	CHN	Yunnan G
7210.2	20	BEN	ORTB, Cotonou, Benin
7215	10	AGL	Antena 2, Cazenga, Angola
	100	POL	Polish Radio, Warsaw
	20	CTI	R. Côte d'Ivoire, Abidjan
	500	UAE	R. of the UAE, Abu Dhabi
	500/250	AFS	Trans World Radio Africa, Meyerton
	50	CHN	V.O.Jinling
	10	CLN	Voice of America, Colombo
	250/50	PHL	Voice of America, Philippines
	500	IRN	Voice of the Islamic Republic of Iran
7220			R. Liberty
	100	CAF	R. TV Centrafricaine, Bangui
	250	BIH	R. Yugoslavia, Bijeljina, Bosnia
	100	D	Voice of America, Lampertheim
	250	E	Voice of America, Playa de Plas
	50	ZMB	ZNBC Radio One, Lusaka, Zambia
7225	100	IND	All India Radio Ext. Sce, Delhi
	250	THA	BBC East Asian Relay Station
	500/100	D	Deutsche Welle, Germany
	250	CLN	Deutsche Welle, Trincomalee
	300/100	J	Radio Japan
	250/50	ROU	Radio România, Bucharest
	15	CHN	Sichuan 1
		CLA	Voice of the Mojahed (vf)
7230	50	IND	All India Radio, Kurseong
	100	D	AWR Europe via Jülich
	10	I	AWR Europe, Forli
	500/250	AFS	BBC World Sce. via Meyerton
	100	HNG	R. Budapest, Diósd
	50	BFA	R. Burkina, Ouagadougou
	50	OMA	R. Oman, Seeb
	100	PAK	R. Pakistan, Islamabad
	500	BIH	R. Yugoslavia, Bijeljina, Bosnia
	250	G	Radio Japan via Skelton
	500	IRN	Voice of the Islamic Republic of Iran
7231	0.5	INS	RRI Fak-Fak
7235	250/100	SVK	AWR Europe via Slovakia
	250	THA	BBC East Asian Relay Station
	250/100	CYP	BBC East Mediterranean Relay Station
		CHN	China Radio International
	250	CLN	Deutsche Welle, Trincomalee
	300	G	R. Canada International via Skelton
	300	J	R. Canada International via Yamata
	250	PHL	R. Veritas Asia, Philippines
	100	I	RAI Ext. Sce, Rome, Italy
		INS	RRI Palu
	250	THA	Voice of America, Udorn
7240	50	IND	All India Radio, Mumbai
	100	CME	CRTV, Garoua, Cameroon
	500/100	D	Deutsche Welle, Germany
	100	AUS	R. Australia, Shepparton
			R. Liberty
		UKR	R. Ukraine International
	100	I	RAI Ext. Sce, Rome, Italy
	500/250	AFS	Trans World Radio Africa, Meyerton
	500	TUR	Voice of Turkey, Emirler
7245	250/100	CYP	BBC East Mediterranean Relay Station
		CHN	China Radio International
	100	AGL	Em. Nacional, Mulenvos, Angola
			R. Free Europe/R. Liberty
	500	THA	R. Thailand, Udon Thani
	250/50	ROU	Radio România, Bucharest
	100	ALG	RTV Algerienne, Bouchaoui
	100	D	Voice of America, Lampertheim
	250	MRC	Voice of America, Morocco
	100	TJK	Yangi-Yul, Tadzhikistan
7245.2	100	MTN	R. Mauritanie, Nouakchott
7250	50	IND	All India Radio Ext. Sce, Gorakhpur
	250	IND	All India Radio Ext. Sce, Panaji
	100	TWN	CBS Network 1, Taipei, Taiwan
		F	China Radio International via France
	100	ZMB	Christian Voice, Zambia
	250	HNG	R. Budapest, Jászberény
	250	SNG	R. Singapore International
		CLA	Radio Free Somalia
	300/100	J	Radio Japan
	500/100	CVA	Vatican Radio
	250	SNG	Warna, Singapore
7255	100	KAZ	Almaty. Kazakhstan
	250/100	CYP	BBC East Mediterranean Relay Station
	500/100	D	Deutsche Welle, Germany
	50	BOT	R. Botswana, Gaborone
			R. Free Europe/R. Liberty
	250/50	PAK	R. Pakistan Ext. Sce. (vf)
	100	MCO	Trans World Radio, Monte Carlo
	250	GRC	Voice of America, Kavala
	250	THA	Voice of America, Udorn
	250	NIG	Voice of Nigeria, Lagos
7260	10	I	AWR Europe, Forli
	250/100	CYP	BBC East Mediterranean Relay Station
	100	D	BBC World Sce. via Lampertheim
		CHN	China Radio International
	100	HNG	R. Budapest, Diósd
	250	KOR	R. Canada International via Kimje
			R. Liberty
		RUS	R. Netherlands, Petropavlovsk
7265	20	VUT	R. Vanuatu, Malapoa
	300/100	J	Radio Japan
	250	PHL	Voice of America, Philippines
	500	IRN	Voice of the Islamic Republic of Iran
	100	IND	All India Radio Ext. Sce, Delhi
	100	PAK	Azad Kashmir R, Rawalpindi
		CHN	China Radio International
	5	BLR	Hrodna, Belarus
			R. Liberty
	100	TCH	R. Prague, Czech Rep.
	500	S	R. Sweden, Hörby
	250	PHL	R. Veritas Asia, Philippines
	50/100	TGO	Radiodiffusion Togolaise, Lomé (vf)
	20	D	Südwestfunk, Rohrdorf
	500/250	AFS	Trans World Radio Africa, Meyerton

kHz	kW	Land	Station
	100	STP	Voice of America, São Tomé
7270	500/250	G	BBC World Sce, United Kingdom
	100	D	BBC World Sce. via Lampertheim
	500/100	D	Deutsche Welle, Germany
	100/75	SEY	FEBA Radio, Seychelles
	50	CHN	Nei Menggu-Mo
	100	POL	Polish Radio, Warsaw
			R. Liberty
	100	PAK	R. Pakistan, Islamabad
	100/50	ALB	R. Tirana, Cërrik
	350	E	Radio Exterior de España
	100	AFS	Radio Oranje, So. Africa
	100	GAB	Radiodiffusion Gabonaise, Moyabi
	10	MLA	RTM Kuching, Sarawak, Malaysia
	100	D	Voice of America, Lampertheim
	250	TUR	Voice of Turkey, Emirler
		VTN	Voice of Vietnam, Hanoi
7275	500/250	G	BBC World Sce, United Kingdom
	7.5	CHN	Guizhou Sat.
	10	LBR	LBS, Monrovia, Liberia
	250	KOR	R. Korea Int, Kimje
			R. Liberty
	100	NIG	R. Nigeria, Kaduna
	350	E	Radio Exterior de España
	250	GRC	Voice of America, Kavala
7280	50	IND	All India Radio, Guwahati
	250	RRW	Deutsche Welle, Kigali
	250	RRW	Deutsche Welle, Kigali
	100	TUN	ERTT, Tunis
	50	CHN	Haixia 1
		F	R. France Internationale
			R. Liberty
	200	MDG	R. Netherlands, Madagascar
	500	S	R. Sweden, Hörby
	30	TZA	R. Tanzania National Sce. *
	250	PHL	Voice of America, Philippines
	100	TUR	Voice of Turkey, Emirler
7285	500/100	D	Deutsche Welle, Germany
	250	POR	Deutsche Welle, Sines
	100	POL	Polish Radio, Warsaw
	250	KOR	R. Korea Int, Kimje
	200	MDG	R. Netherlands, Madagascar
	50	NIG	R. Nigeria Ch.1, Lagos
	250/50	TWN	R. Taipei International, Taiwan
	100	UZB	R. Tashkent Ext.Sce, Uzbekistan
	50	MLI	Radiodiff.-TV Malienne, Bamako
	250	MRC	Voice of America, Morocco
	100	TWN	Voice of Asia, Taiwan
		VTN	Voice of Vietnam, Hanoi
	100/20	ZWE	ZBC R4, Gweru, Zimbabwe
7290	20	IND	All India Radio, Thiru'puram
		CHN	CNR 1
			R. Liberty
	250/50	PAK	R. Pakistan Ext. Sce. (vf)
	500	S	R. Sweden, Hörby
	100	I	RAI Ext. Sce, Rome, Italy
	100	STP	Voice of America, São Tomé
	250	THA	Voice of America, Udorn
	500	IRN	Voice of the Islamic Republic of Iran
		CHN	Xizang-Tb
	1	NZL	ZLXA, Levin. New Zealand
7294	0.5	I	R. Europe, Pioltello, Italy
7295	10	IND	All India Radio, Aizawl
	500/250	G	BBC World Sce, United Kingdom
	100	TWN	BCC News Network, Paochung
	10	ZAI	Mbujimayi, Congo (Dem. Rep.) *
	100	POL	Polish Radio, Warsaw
			R. Free Europe/R. Liberty
	100	I	RAI Ext. Sce, Rome, Italy
	100	MLA	RTM R4, Kuala Lumpur, Malaysia
	250	MRC	Voice of America, Morocco
	250	THA	Voice of America, Udorn
	500	TUR	Voice of Turkey, Emirler
7300		VTN	Ha Giang BS, Vietnam (vf)
	0.5	I	R. Europe (alt. freq)
	100	SVK	R. Slovakia Int, Velké Kostolany
	250	RUS	Voice of America, Petropavlovsk
	100	TUR	Voice of Turkey, Emirler
7305	999/500	RUS	Deutsche Welle, Novosibirsk
	250	CLN	Deutsche Welle, Trincomalee
		F	R. France Internationale
	500/100	CVA	Vatican Radio
	500/100	IRN	Voice of the Islamic Republic of Iran
7310	500/100	CVA	Vatican Radio
7315		CHN	China Radio International
		CLA	La Voz de Alpha 66
		F	R. France Internationale
	500	S	R. Sweden, Hörby
	50	SWZ	Trans World Radio Africa, Swaziland
	100	USA	WHRI, South Bend, IN
7320	100	RUS	Arman, FE
	250/100	CYP	BBC East Mediterranean Relay Station
	500/250	G	BBC World Sce, United Kingdom
	300	CYP	CyBC via BBC Relay St.
		UKR	R. Ukraine International

kHz	kW	Land	Station
7325	500/250	G	BBC World Sce, United Kingdom
	500/100	AUT	R. Austria International
	500	S	R. Sweden, Hörby
7330		RUS	BBC World Sce. via Chita
7335		CHN	China Radio International
	10	CAN	CHU (Time Signals), Canada
	500/100	CVA	Vatican Radio
7340	100	BOT	Voice of America, Botswana
	500	IRN	Voice of the Islamic Republic of Iran
7345	100	TCH	R. Prague, Czech Rep.
	100	SVK	R. Slovakia Int, Velké Kostolany
	500/100	CVA	Vatican Radio
	50	RUS	Yakutsk, S
7350		CHN	China Radio International
		AGL	EP da Huíla, Angola
	100/75	SEY	FEBA Radio, Seychelles
		CHN	Heilongjiang
		F	R. France Internationale
7355	999/500	RUS	Deutsche Welle, Novosibirsk
	100	ALS	KNLS, Anchor Point, AL
	100	ALB	Trans World Radio, Cërrik
	500	MCO	Trans World Radio, Monte Carlo
	50	USA	WRNO, New Orleans, LA
	100	USA	WYFR-Family Radio, Okeechobee, FL
7360		CHN	China Radio International
	500/100	CVA	Vatican Radio
7365	100	ALS	KNLS, Anchor Point, AL
	250	USA	R. Martí, Greenville
	500/100	CVA	Vatican Radio
7370	1	TUR	Turkish Police R, Ankara
	500/250	USA	Voice of America, Greenville/Delano
7375	50/20	CTR	AWR Latin America, Cahuita
		CHN	China Radio International
	500/100	BUL	R. Bulgaria
	200	MDG	R. Netherlands, Madagascar
7380	500/100	BUL	R. Bulgaria
		UKR	R. Ukraine International
7385		RUS	BBC World Sce. via Irkutsk
		CHN	China Radio International
	30-May	CTR	R. For Peace Int., Cd. Colón
	100	ALB	Trans World Radio, Cërrik
	50	CHN	Xinjiang-Ch
7390		RUS	BBC World Sce. via Moscow
	999/500	RUS	Deutsche Welle, Novosibirsk
	500/100	BUL	R. Bulgaria
		ERI	V. of the Broad Masses of Eritrea
7395	500/100	D	Deutsche Welle, Germany
	500/20	ISR	Kol Israel, Jerusalem
	50	USA	WRNO, New Orleans, LA
	500	USA	WSHB, Cypress Creek, SC
	100	USA	WYFR-Family Radio, Okeechobee, FL
7400	500/100	BUL	R. Bulgaria
		VTN	Voice of Vietnam, Hanoi
7405		CHN	China Radio International
	250	USA	R. Martí, Delano
	250	USA	Voice of America, Greenville
	500	USA	WVHA, Greenbush, ME
7410	250	IND	All India Radio Ext. Sce, Aligarh
	500	IND	All India Radio Ext. Sce, Bangalore
		XXX	R. Free Asia
7415	100	BOT	Voice of America, Botswana
7425	500/100	BUL	R. Bulgaria
	500	USA	WEWN, Vandiver, AL
7430	35	GRC	ERT, Thessaloniki, Greece
	250/100	GRC	Voice of Greece
7435	100	USA	WWCR, Nashville, TN
7440	10	CHN	CNR 2
	500	NOR	R. Norway International/R. Denmark
7445	250/50	TWN	R. Taipei International, Taiwan
	100	TWN	Voice of Asia, Taiwan
7450	250/100	GRC	Voice of Greece
7460	1	HND	R. Copán Internacional, Tegucigalpa
		XXX	R. Free Asia
7465	500/100	ISR	Kol Israel, Jerusalem
	500/100	BUL	R. Bulgaria
	500	NOR	R. Norway International/R. Denmark
	500	USA	WSHB, Cypress Creek, SC
	500	USA	WVHA, Greenbush, ME
7470		CHN	China Radio International
7475	100	TUN	ERTT, Tunis
	0.5	I	R. Strike, Palermo, Italy
7480		CHN	China Radio International
	500	NOR	R. Norway International/R. Denmark
	1000	ARM	Voice of Armenia
7485	500	NOR	R. Norway International/R. Denmark
	250/50	PAK	R. Pakistan Ext. Sce. (vf)
	50	USA	WJCR, Millerstown, KY
7490	500/20	ISR	Kol Israel, Jerusalem
7495	500/20	ISR	Kol Israel, Jerusalem
	500/100	BUL	R. Bulgaria
7500	500/100	BUL	R. Bulgaria
	120	ROU	R. Moldova International via Romania
7504		CHN	CNR 1
7510	100	USA	KTBN, Salt Lake City, UT
	500	USA	WSHB, Cypress Creek, SC
	100	TJK	Yangi-Yul, Tadzhikistan

kHz	kW	Land	Station
7516		CHN	CNR 2
7520	100	USA	KVOH, Rancho Simi, CA
		CLA	Radio Pridnestrovye
	100	USA	WYFR-Family Radio, Okeechobee, FL
7525	500/100	BUL	R. Bulgaria
7535	500/100	BUL	R. Bulgaria
	500	USA	WSHB, Cypress Creek, SC
7536	1	SOM	R. Hargeisa
7545	500/100	BUL	R. Bulgaria
	500	NOR	R. Norway International/R. Denmark
7550	100	KOR	R. Korea Int, Kimje
		CLA	Voice of the Mojahed (vf)
7555	50	USA	KJES, Vado, NM
7560	500	NOR	R. Norway International/R. Denmark
7565	500	NOR	R. Norway International/R. Denmark
7580	500	NOR	R. Norway International/R. Denmark
	200	KRE	R. Pyongyang, Korea (D.P.R.)
7590		CHN	China Radio International
7620		CHN	CNR Taiwan
7660		CHN	China Radio International
7670	15	BUL	Bulgarian National R., Stolnik
7695		CLA	Voice of the Communist Party of Iraqi Kurdistan (vf)
7770		CHN	CNR 2
7800		CHN	China Radio International
7820		CHN	China Radio International
7870	10	ISL	Ríkisútvarpid, Reykjavík
7935		CHN	CNR 1
8000	2	J	JJY, Japan (Time Signals)
		CLA	Voice of Sudan
8260		CHN	China Radio International
8300		CLA	New Star Broadcasting Station No. One
8461	10	AFS	ZSC (Time Signals), So. Africa
8473		CLN	4PB (Time Signals), Sri Lanka
8539		HKG	VPS35 (Time Sigs), Hong Kong
8566		CHN	CNR Minor.
8638	10	AUS	VNG, Llandilo (Time Signals)
8660		CHN	China Radio International
8677		CHL	CBV (Time Signals), Chile
8828		HKG	HK Aeradio (Time Signals)
9022	500	IRN	Voice of the Islamic Republic of Iran
9024	10	SDN	Omdurman, Sudan (vf)
9025		CLA	Voice of Sudan (vf)
9064		CHN	CNR 2
9080	50	CHN	CNR 1
9170		CHN	CNR Chinese
9184.8	0.3	J	NHK Osaka 2 (SSB) (9184.75)
9275	10	ISL	Ríkisútvarpid, Reykjavík
9280	100	TWN	BCC Popular Network, Paochung
	100	TWN	Family Radio via VOFC, Taiwan
	100	TWN	Voice of Asia, Taiwan
9282	10	ISL	Ríkisútvarpid, Reykjavík
9285		CLA	Republic of Iraq Radio/Voice of the Iraqi People (vf)
9290		CHN	CNR 1
9310	100	GEO	Voice of Hope, Dusheti, Georgia
9325	200	KRE	R. Pyongyang, Korea (D.P.R.)
9335	250/50	PAK	R. Pakistan Ext. Sce. (vf)
9340		CHN	CNR 1
9345	200	KRE	R. Pyongyang, Korea (D.P.R.)
9355	100	GUM	AWR Asia, Guam
	100	MRA	KHBI, Agingan Point, Saipan
	500	USA	WSHB, Cypress Creek, SC
	100	USA	WYFR-Family Radio, Okeechobee, FL
9365		CHN	China Radio International
	500/20	ISR	Kol Israel, Jerusalem
		XXX	R. Free Asia
9370	100	GUM	AWR Asia, Guam
	100	USA	WYFR-Family Radio, Okeechobee, FL
9375	250/100	GRC	Voice of Greece
9380		CHN	CNR Taiwan
	250/100	GRC	Voice of Greece
9385	100	MRA	KHBI, Agingan Point, Saipan
	500/100	BUL	R. Bulgaria
9390	500/20	ISR	Kol Israel, Jerusalem
9395		XXX	R. Free Asia
	250/100	GRC	Voice of Greece
9400	500/100	BUL	R. Bulgaria
	120	ROU	R. Moldova International via Romania
	50	USA	WGTG, McCaysville, GA
9405	100	PHL	FEBC Manila, Philipppines
9410	250/100	CYP	BBC East Mediterranean Relay Station
	500/250	G	BBC World Sce, United Kingdom
9415	100	AUS	R. Australia, Shepparton
	500/100	BUL	R. Bulgaria
	500	NOR	R. Norway International/R. Denmark
9420	250/100	SVK	AWR Europe via Slovakia
	100/75	SEY	FEBA Radio, Seychelles
	0.5	I	R. Europa Int., Brescia, Italy
		XXX	R. Free Asia
	100	TCH	R. Prague, Czech Rep.
	250/100	GRC	Voice of Greece
9430	100	MRA	KHBI, Agingan Point, Saipan
	100	GUM	KTWR-Trans World Radio, Guam

kHz	kW	Land	Station
	100	TCH	R. Prague, Czech Rep.
		CLA	Voice of the Mojahed (vf)
	500	USA	WSHB, Cypress Creek, SC
9435	500/20	ISR	Kol Israel, Jerusalem
	100	AUS	R. Australia, Shepparton
	250	GRC	Voice of America, Kavala
9440		CHN	China Radio International
	100	SVK	R. Slovakia Int, Rimavská Sobota
	100	SVK	R. Slovakia Int, Velké Kostolany
	500	S	R. Sweden, Hörby
9445	250/100	SVK	AWR Europe via Slovakia
		XXX	R. Free Asia
	500	TUR	Voice of Turkey, Çakirlar
9450	10	PAK	R. Pakistan, Islamabad
9455	130/75	SEY	FEBA Radio, Seychelles
	250	USA	Voice of America, Greenville
	500	USA	WEWN, Vandiver, AL
	500	USA	WSHB, Cypress Creek, SC
9457		CHN	China Radio International
9460	500	TUR	Voice of Turkey, Çakirlar
9465	250/100	SVK	AWR Europe via Slovakia
		TWN	Family Radio via VOFC, Taiwan
	100	MRA	KFBS, Marpi, Saipan
	50	USA	WMLK, Bethel, PA
9470	100	D	AWR Europe via Jülich
	100	D	Hrvatski Radio, Croatia via Jülich
9475	100	D	AWR Europe via Jülich
	50	PHL	FEBC Manila, Philippines
	100	GUM	KTWR-Trans World Radio, Guam
	250	EGY	R. Cairo, Abis
	100	USA	WWCR, Nashville, TN
9480		CHN	China Radio International
	500/250	USA	Voice of America, Greenville
9485	500/100	BUL	R. Bulgaria
		F	R. France Internationale
	250/50	PAK	R. Pakistan Ext. Sce. (vf)
	100	SVK	R. Slovakia Int, Velké Kostolany
	100	ALB	Trans World Radio, Cërrik
	100	MCO	Trans World Radio, Monte Carlo
9490		F	R. France Internationale
	100	S	R. Sweden, Hörby
	100	ALB	Trans World Radio, Cërrik
	100	MCO	Trans World Radio, Monte Carlo
9494		CLA	Radio Respublikij Abkhazie (vf)
9495	100	PHL	FEBC Manila, Philippines
	100	D	Hrvatski Radio, Croatia via Jülich
	100	MRA	KFBS, Marpi, Saipan
		CLA	La Voz de Alpha 66
	500/100	AUT	R. Austria International
		F	R. France Internationale
		GEO	Rep. of Abkhazia Radio, Sukhumi
	100	USA	WHRI, South Bend, IN
9500	100/75	SEY	FEBA Radio, Seychelles
	500/100	BUL	R. Bulgaria
	100	SWZ	Trans World Radio Africa, Swaziland
	500/100	CVA	Vatican Radio
9504.7	0.2	PRU	R. Tacna (n. 9505)
9505	100	KAZ	Almaty, Kazakhstan
	500/100	D	Deutsche Welle, Germany
	50	CHN	Haixia 2
	300	AUT	R. Canada International via Vienna
	500	D	R. Canada International via Wertachtal
			R. Free Europe/R. Liberty
	250	CUB	R. Habana, Cuba
	100	TCH	R. Prague, Czech Rep.
	7.5	B	R. Record, São Paulo"
	100	SVK	R. Slovakia Int, Velké Kostolany
	250	PHL	R. Veritas Asia, Philippines
	250	THA	Voice of America, Udorn
	250	G	Voice of America, Woofferton
	100	USA	WYFR-Family Radio, Okeechobee, FL
9510	100	OMA	BBC Eastern Relay Station
	250/100	SNG	BBC Far Eastern Relay Station
	100	OMA	R. Oman, Thumrait
	250/50	ROU	Radio România, Bucharest
	250	THA	Voice of America, Udorn
		CHN	Xizang-Ch
9515	100	GUM	AWR Asia, Guam
	250	USA	BBC World Sce. via Delano
	500/250	AFS	BBC World Sce. via Meyerton
	250	CAN	BBC World Sce. via Sackville
	50	BFA	R. Burkina, Ouagadougou
	100	KOR	R. Korea Int, Kimje
	10	B	R. Novas de Paz, Curitiba
	250/50	PAK	R. Pakistan Ext. Sce. (vf)
	300/100	J	Radio Japan
	5	I	RAI Radio 1, Caltanissetta
9520	250	USA	Voice of America, Greenville
	500/250	AFS	BBC World Sce. via Meyerton
	100	D	Hrvatski Radio, Croatia via Jülich
	5	PRU	R. La Crónica, Lima *
			R. Liberty
	250/50	PAK	R. Pakistan Ext. Sce. (vf)
	250	PHL	R. Veritas Asia, Philippines
	100	D	Voice of America, Lampertheim

kHz	kW	Land	Station
	250	E	Voice of America, Playa de Plas
9525	500/100	AFS	Channel Africa, Johannesburg
	100	POL	Polish Radio, Warsaw
	100	SWZ	Trans World Radio Africa, Swaziland
	250	INS	V. of Indonesia (Cimanggis)
	250	USA	Voice of America, Greenville
	100/50	PHL	Voice of America, Philippines
	250/100	GRC	Voice of Greece
	500	IRN	Voice of the Islamic Republic of Iran
	500	TUR	Voice of Turkey, Emirler
9530	100	RUS	Arman, FE
	500/250	G	BBC World Sce, United Kingdom
	100	D	BBC World Sce. via Biblis
	100	D	BBC World Sce. via Lampertheim
		CLA	Holy Medina Radio
	10	B	R. Nova Visão, Santa Maria
	500	GUF	Radio Japan via Montsinery
	250/50	ROU	Radio România, Bucharest
	300/250		Voice of America (several locations)
9535	100	AGL	Antena 2, Mulenvos, Angola
		CHN	China Radio International
	250	POR	Deutsche Welle, Sines
	0.6	J	NHK Sapporo 1 (DSB)
		CHN	R. Canada International via X'ian
	250	CAN	R. Canada International, Sackville
	100	SVK	R. Slovakia Int, Rimavská Sobota
	250	PHL	R. Veritas Asia, Philippines
	300/100	J	Radio Japan
	100	ALG	RTV Algérienne, Bouchaoui
		D	Swiss R. International, via Jülich
	250	THA	Voice of America, Udorn
9538.8	0.6	J	NHK Fukuoka 1 (SSB) (9538.75)
9540		POL	Polish Radio, Warsaw
	10	B	R. Educadora da Bahia, Salvador *
			R. Liberty
	25	VEN	R. Nal. de Venezuela (vf) (irr.)
	100	UZB	R. Tashkent Ext.Sce, Uzbekistan
	100	I	RAI Ext. Sce, Rome, Italy
	500	SNG	Swiss R. International, via Kranji
	100	UZB	UZR2/FS, Toskent, Uzbekistan
	100	STP	Voice of America, São Tomé
9545	250	IND	All India Radio Ext. Sce, Delhi
	100	GUM	KTWR-Trans World Radio, Guam
	0.25	MEX	La Jarocha, Veracruz *
	10	SLM	SIBC Honiara, Solomon Islands
	100	UZB	UZR2/FS, Toskent, Uzbekistan
	250	PHL	Voice of America, Philippines
9548	250	BGD	R. Bangladesh Ext. Sce. (vf)
9550	250	IND	All India Radio Ext. Sce, Aligarh
		CHN	China Radio International
		F	R. France Internationale
	100	CUB	R. Habana, Cuba
	5	B	R. Rio Grande do Sul, Porto Alegre *
		UKR	R. Ukraine International
	250/50	ROU	Radio România, Bucharest
	100	USA	WYFR-Family Radio, Okeechobee, FL
9552	7.5	INS	RRI Ujung Pandang
9553.8	0.6	J	NHK Tokyo (SSB) (9553.75)
9555	500/50	ARS	Broadc. Sce. of the Kingdom of Saudi Arabia
	250	CLN	Deutsche Welle, Trincomalee
	300	G	R. Canada International via Skelton
	250	G	R. Korea Int, Skelton
	250	PHL	R. Veritas Asia, Philippines
	250	THA	Voice of America, Udorn
	500	IRN	Voice of the Islamic Republic of Iran
9560	250	ASC	BBC Atlantic Relay Station
		CAN	China Radio International via Canada
	500	USA	Merlin Comms, via Rampisham
	500	FIN	R. Finland
		UKR	R. Ukraine International
		CLA	Radio Amahoro
	300/100	J	Radio Japan
		CLA	Radio Voice of Peace
	100	ETH	Voice of Ethiopia
	500	TUR	Voice of Turkey, Emirler
	50	CHN	Xinjiang-Ch
9562		CLA	Republic of Iraq Radio/Voice of the Iraqi People (vf)
9565	100	IND	All India Radio Ext. Sce, Delhi
	250	IND	All India Radio Ext. Sce, Panaji
	50	IND	All India Radio, Delhi
	500/250	G	BBC World Sce, United Kingdom
		CHN	China Radio International
	500/250	D	Deutsche Welle, Germany
	250	RRW	Deutsche Welle, Kigali
	250	CLN	Deutsche Welle, Trincomalee
			R. Liberty
	250	USA	R. Marti, Greenville
	15	B	R. Universo, Curitiba
	250	INS	RRI Jakarta (Bonto Sunggu)
	250	MRC	Voice of America, Morocco
	300	G	Voice of America, Woofferton
9568	250	BGD	R. Bangladesh Ext. Sce. (vf)
		CLA	Republic of Iraq Radio/Voice of the Iraqi
			People (vf)
9570	250/100	QAT	Qatar Broadc. Sce, Doha
	100	KOR	R. Korea Int, Kimje
	50	NIG	R. Nigeria, Kaduna Ch.2
	250/50	ROU	Radio România, Bucharest
	300/100	POR	RDP Internacional (R. Portugal)
	500	IRN	Voice of the Islamic Republic of Iran
9575	500/100	D	Deutsche Welle, Germany
	250	MRC	R. Mediterranee Int, Nador
	100	I	RAI Ext. Sce, Rome, Italy
	500	SNG	Swiss R. International, via Kranji
	500	USA	Voice of America, Greenville
	250	GRC	Voice of America, Kavala
	500	IRN	Voice of the Islamic Republic of Iran
9580	500	GAB	Africa No.1, Moyabi, Gabon
	250	THA	BBC East Asian Relay Station
	500/50	ARS	Broadc. Sce. of the Kingdom of Saudi Arabia
	100	AUS	R. Australia, Shepparton
	1	PRU	R. Concordia, Arequipa *
	250/50	ROU	Radio România, Bucharest
	250	USA	Voice of Greece via VOA
9584.9	10	B	R. CBN (Globo), São Paulo
9585	500/100	D	Deutsche Welle, Germany
			R. Liberty
	250/50	PAK	R. Pakistan Ext. Sce. (vf)
	500/100	CVA	Vatican Radio
	250	GRC	Voice of America, Kavala
	250	MRC	Voice of America, Morocco
	500	IRN	Voice of the Islamic Republic of Iran
9590	250	USA	BBC World Sce. via Delano
	250	CAN	BBC World Sce. via Sackville
		USA	BBC World Sce. via WYFR
	100	GUM	KTWR-Trans World Radio, Guam
	250	ATN	R. Netherlands, Bonaire
	200	MDG	R. Netherlands, Madagascar
	500	NOR	R. Norway International/R. Denmark
	250	USA	Voice of America, Greenville
	250	USA	Voice of Greece via VOA
	500	TUR	Voice of Turkey, Emirler
9595	250/100	IND	All India Radio Ext. Sce, Delhi
	250/100	SNG	BBC Far Eastern Relay Station
	100	D	Hrvatski Radio via Jülich
	100	D	Hrvatski Radio, Croatia via Jülich
			R. Liberty
	2.5	URG	R. Monte Carlo, Montevideo
	50	J	R. Tanpa 1, Tokyo, Japan
	250	GRC	Voice of America, Kavala
	250	MRC	Voice of America, Morocco
	50	CHN	Xinjiang-Ug
9600	100	RUS	Arman, FE
	100	OMA	BBC Eastern Relay Station
	250/100	SNG	BBC Far Eastern Relay Station
	500	LBY	LJB, Tripoli/Sabrata. Libya
	50	BOT	R. Botswana, Gaborone
			R. Free Europe
	7.5	B	R. MEC, Rio de Janeiro
	250/50	PAK	R. Pakistan Ext. Sce. (vf)
	200	KRE	R. Pyongyang, Korea (D.P.R.)
		UKR	R. Ukraine International
	1	MEX	R. UNAM-Universidad Nacional Autónoma de México
	250	G	Radio Japan via Skelton
	500/100	CVA	Vatican Radio
	250	MRC	Voice of America, Morocco
9605	250	THA	BBC East Asian Relay Station
	100	OMA	BBC Eastern Relay Station
	250/100	SNG	BBC Far Eastern Relay Station
	250	ATG	Deutsche Welle, Antigua
		F	R. France Internationale
			R. Free Europe
	500	UAE	R. of the UAE, Abu Dhabi
	100	I	RAI Ext. Sce, Rome, Italy
	500/100	CVA	Vatican Radio
	500	IRN	Voice of the Islamic Republic of Iran
	100	USA	WYFR-Family Radio, Okeechobee, FL
9610	250	ASC	BBC Atlantic Relay Station
	500/250	AFS	BBC World Sce. via Meyerton
	250	TWN	BCC News Network, Annan
		CHN	CNR 1
	100	COG	R. Congo, Brazzaville
	250/50	TWN	R. Taipei International, Taiwan
		CLA	Voice of the Islamic Revolution in Iraq
9615	50/10	BTN	BBS, Thimpu, Bhutan
	500/100	D	Deutsche Welle, Germany
	250	RRW	Deutsche Welle, Kigali
	250	POR	Deutsche Welle, Sines
	250	CLN	Deutsche Welle, Trincomalee
	500	D	R. Canada International via Wertachtal
	7.5	B	R. Cultura, São Paulo
	350	E	Radio Exterior de España
	250	GRC	Voice of America, Kavala
	500	IRN	Voice of the Islamic Republic of Iran
9619	120	MOZ	Em. Nacional, Maputo
9620	500/50	ARS	Broadc. Sce. of the Kingdom of Saudi

kHz	kW	Land	Station
			Arabia
		CHN	China Radio International
	500/100	D	Deutsche Welle, Germany
	250/100	EGY	R. Cairo, Abu Zaabal
	250	BIH	R. Yugoslavia, Bijeljina, Bosnia
	350	E	Radio Exterior de España
	120	CHN	Radio Exterior de España via Beijing
	0.25	URG	S.O.D.R.E, Montevideo (r: CX6 650)
	100	D	Swiss R. International, via Jülich
	500	SUI	Swiss Radio International
	250/50	PHL	Voice of America, Philippines
9625	100	CAN	CBC No. Quebec Sce, Canada
	15	BOL	R. Fides, La Paz
			R. Free Europe/R. Liberty
	250/50	ROU	Radio România, Bucharest
	100	D	Voice of America, Lampertheim
	100	USA	WYFR-Family Radio, Okeechobee, FL
9630	250	SEY	BBC Indian Ocean Relay Station
	300	TWN	CBS Network 2, Taipei, Taiwan
	10	CHL	R. Agricultura, Santiago *
	10	B	R. Aparecida, Aparecida
	500	JOR	R. Jordan, Al Karanah
	350	E	Radio Exterior de España
	100	CTR	Radio Exterior de España, Costa Rica
	250	INS	RRI Jakarta (Bonto Sunggu)
	500/250	TUR	Voice of Turkey, Emirler/Çakirlar
9635	250/100	CYP	BBC East Mediterranean Relay Station
	500/250	G	BBC World Sce, United Kingdom
	50	MLI	Radiodiff.-TV Malienne, Bamako
	250	MRC	Voice of America, Morocco
	250	THA	Voice of America, Udorn
	500	TUR	Voice of Turkey, Emirler
9637	100	MOZ	Beira, Mozambique
9640	100	D	AWR Europe via Jülich
	250	ATG	Deutsche Welle, Antigua
	500/100	D	Deutsche Welle, Germany
	250	POR	Deutsche Welle, Sines
	10	VEN	Ecos del Torbes. San Cristóbal
	100	EQA	HCJB, LV de los Andes, Quito
		CLA	La Voz de Sahara Libre
	250	CAN	R. Canada International, Sackville
	250/100	KOR	R. Korea Int, Kimje
	200	KRE	R. Pyongyang, Korea (D.P.R.)
		UKR	R. Ukraine International
	50	ALG	RTV Algerienne, Bouchaoui
9645	5	CTR	Faro del Caribe, San José (vf)
	7.5	B	R. Bandeirantes, São Paulo
	300	G	R. Canada International via Skelton
			R. Liberty
	100/10	PAK	R. Pakistan, Islamabad
	500	S	R. Sweden, Hörby
	500/100	CVA	Vatican Radio
			Voice of America (several locations)
9650	250	IND	All India Radio Ext. Sce, Aligarh
	100	GUI	Conakry, Guinea Rep. (irr.) (vf)
	500/100	D	Deutsche Welle, Germany
	10	URG	Em. Ciudad de Montevideo
	250	CAN	R. Korea Int, Sackville, Canada
	200	KRE	R. Pyongyang, Korea (D.P.R.)
	350	E	Radio Exterior de España
	250	GRC	Voice of America, Kavala
	500	IRN	Voice of the Islamic Republic of Iran
9655		CHN	China Radio International
	500/100	AUT	R. Austria International
	100/20	HNG	R. Budapest, Székesfehérvár
	20	CLM	R. Dif. Nal. de Colombia, SF de Bogotá
	250	FIN	R. Finland
		F	R. France Internationale
	5	PRU	R. Municipal Nor Peruana, Chachapoyas
	500	S	R. Sweden, Hörby
	100	THA	R. Thailand, Pathumthani
	500/250	TUR	Voice of Turkey, Emirler
9660	500/250	G	BBC World Sce, United Kingdom
	10	AUS	BBC World Sce. via Brandon
	100	EQA	HCJB, LV de los Andes, Quito
	10	AUS	R. Australia, Brandon
	250	HNG	R. Budapest, Jászberény
		F	R. France Internationale
			R. Liberty
	20	VEN	R. Rumbos, Caracas (vf)
	250	PHL	R. Veritas Asia, Philippines
	500	GUF	Radio Japan via Montsinery
	100	ALB	Trans World Radio, Cërrik
	500/100	CVA	Vatican Radio
	100	D	Voice of America, Lampertheim
9665	200	KRE	KCBS, Pyongyang, No. Korea
			R. Liberty
	10	B	R. Marumby, Florianópolis
	250/50	ROU	Radio România, Bucharest
	100	ALB	Trans World Radio, Cërrik
	250	MRC	Voice of America, Morocco
		CLA	Voice of Palestine, Voice of the Palestinian Islamic Revolution
	500	IRN	Voice of the Islamic Republic of Iran
9670	250/100	CYP	BBC East Mediterranean Relay Station

kHz	kW	Land	Station
	250	USA	BBC World Sce. via Delano
		CHN	China Radio International
	500/100	D	Deutsche Welle, Germany
	250	RRW	Deutsche Welle, Kigali
	250	CLN	Deutsche Welle, Trincomalee
	100	MRA	KFBS, Marpi, Saipan
	100	POL	Polish Radio, Warsaw
			R. Liberty
	250	PHL	R. Veritas Asia, Philippines
	100	I	RAI Ext. Sce, Rome, Italy
	100	ALB	Trans World Radio, Cërrik
			Voice of America (several locations)
		CLA	Voice of Palestine, Voice of the Palestinian Islamic Revolution
		CLA	Voice of the Islamic Revolution in Iraq
9674.8	5	PRU	R. del Pacífico, Lima
9674.9		CHN	China Radio International
9674.9	10	B	R. Canção Nova, Cachoeira Paulista
9675			R. Liberty
	100	I	RAI Ext. Sce, Rome, Italy
	100	STP	Voice of America, São Tomé
9680	250/100	SNG	BBC Far Eastern Relay Station
	100/20	HNG	R. Budapest, Székesfehérvár
			R. Free Europe/R. Liberty
	100/50	USA	R. Taipei International via WYFR
	250	PHL	R. Veritas Asia, Philippines
	100	INS	RRI Jakarta (Cimanggis)
	250		Voice of America (several locations)
	100	USA	WYFR-Family Radio, Okeechobee, FL
9684.7	10	B	R. Gazeta, São Paulo
9685	50	IND	All India Radio, Delhi
	500/250	G	BBC World Sce, United Kingdom
	100	EQA	HCJB, LV de los Andes, Quito
	300/100	J	Radio Japan
	500	IRN	Voice of the Islamic Republic of Iran
9690	2	TWN	CBS Network 2, Taipei, Taiwan
		E	China Radio International via Spain
	500/100	D	Deutsche Welle, Germany
	500	USA	Merlin Comms, via Rampisham
	25	ARG	R. Argentina al Exterior
			R. Free Europe
	100/50	USA	R. Taipei International via WYFR
	350	E	Radio Exterior de España
	250/50	ROU	Radio România, Bucharest
	10	MDG	Radio Television Malagasy
	100	I	RAI Ext. Sce, Rome, Italy
9695	250	MRC	Voice of America, Morocco
			R. Free Europe/R. Liberty
	120	UAE	R. of the UAE, Makta
	7.5	B	R. Rio Mar, Manaus
	300/100	J	Radio Japan
	250	MRC	Voice of America, Morocco
	250	E	Voice of America, Playa de Plas
9700		CLA	A Voz da Resistencia de Galo Negro (VOR GAN)
	250	IND	All India Radio Ext. Sce, Aligarh
		CHN	China Radio International
	250	ATG	Deutsche Welle, Antigua
	500/100	D	Deutsche Welle, Germany
	500/100	BUL	R. Bulgaria
	250/100	EGY	R. Cairo, Abu Zaabal
	100	NZL	R. New Zealand International
	500	S	R. Sweden, Hörby
	500	THA	R. Thailand, Udon Thani
	250	GRC	Voice of America, Kavala
	250	PHL	Voice of America, Philippines
9704	100	SWZ	Swazi Radio (alt.to 6155 kHz)
9705	250	IND	All India Radio Ext. Sce, Aligarh
	500/50	ARS	Broadc. Sce. of the Kingdom of Saudi Arabia
	100	NGR	La Voix du Sahel, Niger
	10	MEX	R. México Internacional, México
	7.5	B	R. Nacional, Rio de Janeiro
	500	S	R. Sweden, Hörby
	50	INS	RRI Pontianak
		CHN	V.O.Pujiang
	250	GRC	Voice of America, Kavala
	300	G	Voice of America, Woofferton
	100	ETH	Voice of Ethiopia
	50	USA	WYFR-Family Radio, Okeechobee, FL
9710	500/250	G	BBC World Sce, United Kingdom
		CHN	China Radio International
		MLI	China Radio International via Mali
	10	MAU	Mauritius Broadcasting Corp. (F.Pl.)
	100	AUS	R. Australia, Shepparton
	50	LTU	Sitkunai, Lithuania
9715	500/50	ARS	Broadc. Sce. of the Kingdom of Saudi Arabia
		CHN	China Radio International
		IRQ	R. Baghdad, Iraq
		F	R. France Internationale
	250	ATN	R. Netherlands, Bonaire
	500	S	R. Sweden, Hörby
	100	UZB	R. Tashkent Ext.Sce, Uzbekistan
	500	IRN	Voice of the Islamic Republic of Iran

kHz	kW	Land	Station
	50	USA	WYFR-Family Radio, Okeechobee, FL
9717.1	1	BOL	R. La Plata
9720	500/50	ARS	Broadc. Sce. of the Kingdom of Saudi Arabia
	100	AGL	Em. Nacional, Mulenvos, Angola
	250	ATN	R. Netherlands, Bonaire
	500	BIH	R. Yugoslavia, Bijeljina, Bosnia
	100	I	RAI Ext. Sce, Rome, Italy
	250	THA	Voice of America, Udorn
	250	MNG	Voice of Mongolia
	500	IRN	Voice of the Islamic Republic of Iran
9721.6		PRU	R. Victoria, Lima (n. 9720)
9725	50/20	CTR	AWR Latin America, Cahuita
		CHN	China Radio International
		CLA	New Star Broadcasting Station No. One
	250	CAN	R. Canada International, Sackville
	7.5	B	R. Clube Paranaense, Curitiba
		XXX	R. Free Asia
			R. Free Europe/R. Liberty
	250	MRC	Voice of America, Morocco
	50	PHL	Voice of America, Philippines
9730	250/100	SNG	BBC Far Eastern Relay Station
	500/50	ARS	Broadc. Sce. of the Kingdom of Saudi Arabia
		CHN	China Radio International
	50	BRM	R. Myanmar, Yangon
	100	CLN	SLBC Ext. Sce, Colombo
		VTN	Voice of Vietnam, Hanoi
9735	250	RRW	Deutsche Welle, Kigali
	300	G	R. Canada International via Skelton
	100	OMA	R. Oman, Thumrait
	500	IRN	Voice of the Islamic Republic of Iran
9736.2	100	PRG	R. Nal. del Paraguay, Asunci (n. 9735) (vf)
9740	250/100	CYP	BBC East Mediterranean Relay Station
	250/100	SNG	BBC Far Eastern Relay Station
	250	EGY	R. Cairo, Abis
	300	G	R. Canada International via Skelton
	100	ALB	R. Tirana, Cërrik
	250	GRC	Voice of America, Kavala
9743	10	INS	RRI Sorong
9745	500/250	AFS	AWR via Meyerton
9745	250/100	CYP	BBC East Mediterranean Relay Station
		CHN	CNR 1
	100	EQA	HCJB, LV de los Andes, Quito
	60	BHR	R. Bahrain *
	250	FIN	R. Finland
		F	R. France Internationale
			R. Liberty
	100	CTR	Radio Exterior de España, Costa Rica
	250	B	Radiobras, Brasília
	250	GRC	Voice of America, Kavala
	500	IRN	Voice of the Islamic Republic of Iran
9750	250/100	CYP	BBC East Mediterranean Relay Station
	100	D	BBC World Sce. via Lampertheim
	500/100	D	Deutsche Welle, Germany
			R. Liberty
	1	VEN	R. Occidente, Tovar (vf)
	300/100	J	Radio Japan
	250/50	ROU	Radio România, Bucharest
	100	D	Voice of America, Biblis
	100	MLA	Voice of Malaysia, Kuala Lumpur
9755	50	CHN	CNR 2
		IRQ	R. Baghdad, Iraq
	100/50	EGY	R. Cairo, Mokattam
	250	CAN	R. Canada International, Sackville
	250/50	ROU	Radio România, Bucharest
9760	250/100	CYP	BBC East Mediterranean Relay Station
		CHN	China Radio International
		MLI	China Radio International via Mali
	250	CAN	R. Canada International, Sackville
	50	J	R. Tanpa 2, Tokyo, Japan
			Voice of America (several locations)
9765	500/100	AFS	Channel Africa, Johannesburg
	500/100	D	Deutsche Welle, Germany
	100	EQA	HCJB, LV de los Andes, Quito
		RUS	V. of the Mediterranean, Malta via Russia
9770	250	SEY	BBC Indian Ocean Relay Station
		CHN	China Radio International via Mali
	500/100	D	Deutsche Welle, Germany
	250	CLN	Deutsche Welle, Trincomalee
	100	AUS	R. Australia, Shepparton
	250/100	EGY	R. Cairo, Abu Zaabal
			R. Liberty
	500	UAE	R. of the UAE, Abu Dhabi
	250	THA	Voice of America, Udorn
9775		CHN	CNR 2/Minor.
	100	EQA	HCJB, LV de los Andes, Quito
	500/100	BUL	R. Bulgaria
	250/100		Voice of America (several locations)
	250	USA	Voice of Greece via VOA
9779.4	50	YEM	Yemen Radio & TV Corp, Sana'a
9780	50	PHL	FEBC Manila, Philipppines
		CHN	Qinghai 1
	250	CAN	R. Canada International, Sackville
	300/100	POR	RDP Internacional (R. Portugal)

kHz	kW	Land	Station
	100	STP	Voice of America, São Tomé
9785		CHN	China Radio International
	100/75	SEY	FEBA Radio, Seychelles
	250/50	PAK	R. Pakistan Ext. Sce. (vf)
9790		F	R. France Internationale
	500	UAE	R. of the UAE, Abu Dhabi
		CLA	Radio Amahoro
	250	PHL	Voice of America, Philippines
9795	100/75	SEY	FEBA Radio, Seychelles
	100	PHL	FEBC Manila, Philipppines
			R. Liberty
	100	NZL	R. New Zealand International (from March 29)
	500/100	MCO	Trans World Radio, Monte Carlo
	250	PHL	Voice of America, Philippines
9800		CHN	CNR 1
	250/100	EGY	R. Cairo, Abu Zaabal
		F	R. France Internationale
9805	500/100	D	Deutsche Welle, Germany
	250	POR	Deutsche Welle, Sines
	300	G	R. Canada International via Skelton
	250	CAN	R. Canada International, Sackville
		F	R. France Internationale
9810	100/75	SEY	FEBA Radio, Seychelles
	1/0.25	KIR	R. Kiribati, Tarawa
	100	NZL	R. New Zealand International (to March 29)
		RUS	V. of the Mediterranean, Malta via Russia
9815	500/250	G	BBC World Sce, United Kingdom
	250	POR	Deutsche Welle, Sines
	50	USA	KAIJ, Denton, TX
	500/20	ISR	Kol Israel, Arabic prgr
	100	GUM	KTWR-Trans World Radio, Guam
	300/100	POR	RDP Internacional (R. Portugal)
	100	BOT	Voice of America, Botswana
9820		CHN	China Radio International
	100/75	SEY	FEBA Radio, Seychelles
	100	GUM	KTWR-Trans World Radio, Guam
	100	CUB	R. Habana, Cuba
	250	ATN	R. Netherlands, Bonaire
	10	B	R. Nove de Julho, São Paulo (CP) *
	100	STP	Voice of America, São Tomé
9825	250	ASC	BBC Atlantic Relay Station
	250/100	CYP	BBC East Mediterranean Relay Station
	500/250	G	BBC World Sce, United Kingdom
	250	USA	R. Martí, Greenville
	300/100	J	Radio Japan
	500/100	CVA	Vatican Radio
9830	100/10	HRV	Hrvatski Radio, Zagreb
		F	R. France Internationale
	30	CUB	R. Habana, Cuba (USB)
9835	100	IND	All India Radio Ext. Sce, Delhi
	250/100	CYP	BBC East Mediterranean Relay Station
	500	USA	Merlin Comms, via Rampisham
	250	HNG	R. Budapest, Jászberény
			R. Liberty
	300/100	J	Radio Japan
	500/100	CVA	Vatican Radio
	500	USA	WSHB, Cypress Creek, SC
9840	100	D	BBC World Sce. via Biblis
	250	HNG	R. Budapest, Jászberény
		VTN	Voice of Vietnam, Hanoi
	500	USA	WSHB, Cypress Creek, SC
9845		CHN	CNR 1
	100	MRA	KHBI, Agingan Point, Saipan
		F	R. France Internationale
	250	ATN	R. Netherlands, Bonaire
	100	I	RAI Ext. Sce, Rome, Italy
	250	GRC	Voice of America, Kavala
	250	PHL	Voice of America, Philippines
9850	250	EGY	R. Cairo, Abis
	100/50	USA	R. Taipei International via WYFR
	500/250	AFS	Trans World Radio Africa, Meyerton
	500/100	CVA	Vatican Radio
	250	GRC	Voice of America, Kavala
	250	THA	Voice of America, Udorn
9852		CLA	Republic of Iraq Radio/Voice of the Iraqi People (vf)
9855		CHN	China Radio International
	500	FIN	R. Finland
	500/250	KWT	R. Kuwait
	500	D	R. Netherlands, Wertachtal
	300/100	J	Radio Japan
			Voice of America (several locations)
9860	500/200	RUS	Deutsche Welle, Krasnodar
			R. Liberty
	200	MDG	R. Netherlands, Madagascar
	300/100	J	Radio Japan
9865	500/100	G	BBC World Sce, United Kingdom
	100	GUM	KTWR-Trans World Radio, Guam
	500	S	R. Sweden, Hörby
	250	RUS	R. Vlaanderen Internationaal via Petropavlovsk
9870	250/100	CYP	BBC East Mediterranean Relay Station
	500/50	ARS	Broadc. Sce. of the Kingdom of Saudi Arabia

kHz	kW	Land	Station
	100	GUM	KTWR-Trans World Radio, Guam
	500/100	AUT	R. Austria International
	100	HNG	R. Budapest, Diósd
	250/100	KOR	R. Korea Int, Kimje
		UKR	R. Ukraine International
9875	100	MCO	Trans World Radio, Monte Carlo
	250/100	CYP	BBC East Mediterranean Relay Station
	100	PHL	FEBC Manila, Philipppines
	250	G	R. Korea Int, Skelton , UK
	500/100	CVA	Vatican Radio
	100	D	West Coast R. Ireland via Jülich
9880		CHN	China Radio International
	500/250	KWT	R. Kuwait
9885	100	D	Swiss R. International, via Jülich
	500	SUI	Swiss Radio International
	500	GUF	Swiss Radio International, Montsinery
9890	250/100	PHL	Voice of America, Philippines
9895	500/250	G	BBC World Sce, United Kingdom
	500	USA	Merlin Comms, via Rampisham
	250	ATN	R. Netherlands, Bonaire
	500	HOL	R. Netherlands, Flevo
	200	MDG	R. Netherlands, Madagascar
9900		CHN	China Radio International
	250	EGY	R. Cairo, Abis
9903	250/50	PAK	R. Pakistan Ext. Sce. (vf)
9905		XXX	R. Free Asia
		UKR	R. Ukraine International
	500	GUF	Swiss Radio International, Montsinery
	100	TJK	Yangi-Yul, Tadzhikistan
9910	250	IND	All India Radio Ext. Sce, Aligarh
		XXX	R. Free Asia
	500	NOR	R. Norway International/R. Denmark
9915	250/100	CYP	BBC East Mediterranean Relay Station
	500/250	G	BBC World Sce, United Kingdom
	250	G	Merlin Communications, UK
	250/100	GRC	Voice of Greece
9920		CHN	China Radio International
		CHN	CNR Minor.
9925	500	NOR	R. Norway International/R. Denmark
	200/100	BEL	R. Vlaanderen Internationaal
9930	100	HWI	KWHR, Naalehu, Hawaii
		XXX	R. Free Asia
		CLA	Voice of Oromo Liberation
	500	USA	WVHA, Greenbush, ME
9935	35	GRC	ERT, Thessaloniki, Greece
	500	NOR	R. Norway International/R. Denmark
9940		XXX	R. Free Asia
	500	NOR	R. Norway International/R. Denmark
	100	BEL	R. Vlaanderen Internationaal
	500/100	CVA	Vatican Radio
9941		CLA	La Voz del CID - la Voz de Cuba, Independiente y Democrática
9945		CHN	China Radio International
	500	NOR	R. Norway International/R. Denmark
9950	250/50	IND	All India Radio Ext. Sce, Delhi
	250	EGY	R. Cairo, Abis
		CLA	Voice of Iraq (vf)
9955		TWN	Family Radio via VOFC, Taiwan
		CLA	Foro Militar Cubano
	50	PAL	KHBN, Koror, Palau
		CLA	La Voz del Puente de Jovenes Profesionales Cubanos en el Exili
	250/50	TWN	R. Taipei International, Taiwan
	50	USA	WRMI, Miami, FL
9960	25	LBN	Voice of Hope, Lebanon
9965		CHN	China Radio International
	50	PAL	KHBN, Koror, Palau
	1000	ARM	R. AR-Intercontinental
	1000	ARM	Voice of Armenia
	25	LBN	Voice of Hope, Lebanon
9975	50	USA	KVOH, Rancho Simi, CA
		XXX	R. Free Asia
	200	KRE	R. Pyongyang, Korea (D.P.R.)
9980		XXX	R. Free Asia
9985	50	PAL	KHBN, Koror, Palau
	100/50	USA	R. Taipei International via WYFR
	100	USA	WYFR-Family Radio, Okeechobee, FL
9990	250	EGY	R. Cairo, Abis
9995	500	SYR	R. Damascus, Syria
9996	5	RUS	RWM (Time Signals), Moscow
10000	8	IND	ATA, New Delhi (Time Sigs.)
	2	J	JJY, Japan (Time Signals)
	2	ARG	LOL (Time Signals), Argentina
	1	UZB	RCH, Tashkent (Time Signals)
	5	RUS	RTA (Time Signals), Russia
	10	HWA	WWVH, Hawaii (Time Signals)
10004	1	RUS	RID (Time Signals), Irkutsk
10010		VTN	Voice of Vietnam, Hanoi
10060		VTN	Hanoi 1, Vietnam
10260		CHN	CNR 2/Minor.
10330	50	IND	All India Radio, Delhi
10440.5		PHL	NPO (Time Sigs), Philippines
11000		CHN	CNR Chinese
11040		CHN	CNR 2
11092.5	1	SHL	R. St. Helena (SSB) occ. tests

kHz	kW	Land	Station
11100		CHN	CNR Taiwan
11292		IRQ	R. Baghdad, Iraq (vf)
11335	200	KRE	R. Pyongyang, Korea (D.P.R.)
11375		CHN	CNR Minor.
11402	10	ISL	Ríkisútvarpid, Reykjavík
11430		CLA	New Star Broadcasting Station No. One
11445		CHN	China Radio International
11500		CHN	China Radio International
11515		CHN	China Radio International
11530		XXX	R. Free Asia
11540		XXX	R. Free Asia
11550		TWN	Family Radio via VOFC, Taiwan
	100	MRA	KHBI, Agingan Point, Saipan
	500	USA	WSHB, Cypress Creek, SC
11555	500/100	BUL	R. Bulgaria
11560	250	EGY	R. Cairo, Abis
11565	250/50	PAK	R. Pakistan Ext. Sce. (vf)
11570	250/50	PAK	R. Pakistan Ext. Sce. (vf)
		CLA	Voice of Tibet
11575		CHN	China Radio International
		XXX	R. Free Asia
11580		XXX	R. Free Asia
	500	USA	WVHA, Greenbush, ME
	100	USA	WYFR-Family Radio, Okeechobee, FL
11585	250	IND	All India Radio Ext. Sce, Aligarh
	500	IND	All India Radio Ext. Sce, Bangalore
	500/20	ISR	Kol Israel, Jerusalem
11590		CHN	Haixia 1
	500/20	ISR	Kol Israel, Jerusalem
		XXX	R. Free Asia
11595	35	GRC	ERT, Thessaloniki, Greece
	250/100	GRC	Voice of Greece
11600		CHN	China Radio International
	100/75	SEY	FEBA Radio, Seychelles
		F	R. France Internationale
		XXX	R. Free Asia
11605	500/20	ISR	Kol Israel, Jerusalem
	250	TUR	Voice of Turkey, Çakirlar
11610	250/100	SVK	AWR Europe via Slovakia
		CHN	CNR 2
	100	PHL	FEBC Manila, Philipppines
11615	500/100	D	Deutsche Welle, Germany
		F	R. France Internationale
		XXX	R. Free Asia
11620	500	IND	All India Radio Ext. Sce, Bangalore
	250	IND	All India Radio Ext. Sce, Delhi
	250	IND	All India Radio, Khampur
11625	100	GUM	AWR Asia, Guam
	500/100	CVA	Vatican Radio
	250	TUR	Voice of Turkey, Emirler
11630		CHN	CNR 2/Minor.
11635	50	PHL	FEBC Manila, Philipppines
	100/10	HRV	Hrvatski Radio, Zagreb
	100	ALB	Trans World Radio, Cërrik
	100	ALB	Trans World Radio, Shijak
11640	100/75	SEY	FEBA Radio, Seychelles
	500/100	CVA	Vatican Radio
11645	250/100	GRC	Voice of Greece
11650		CHN	China Radio International
	50	PHL	FEBC Manila, Philipppines
	100	MRA	KFBS, Marpi, Saipan
	100	GUM	KTWR-Trans World Radio, Guam
		F	R. France Internationale
	500	S	R. Sweden, Hörby
11655	250	ATN	R. Netherlands, Bonaire
	200	MDG	R. Netherlands, Madagascar
11660	100	GUM	AWR Asia, Guam
	250/100	CYP	BBC East Mediterranean Relay Station
	500/250	G	BBC World Sce, United Kingdom
		CHN	China Radio International
	100	AUS	R. Australia, Shepparton
	500/100	BUL	R. Bulgaria
	500	GUF	Swiss Radio International, Montsinery
11665	250	THA	BBC East Asian Relay Station
	250/100	CYP	BBC East Mediterranean Relay Station
	100	GUM	KTWR-Trans World Radio, Guam
	250/100	EGY	R. Cairo, Abu Zaabal
	300/100	J	Radio Japan
	250	AFS	Radio Netherlands via Meyerton
11670	100	GUM	KTWR-Trans World Radio, Guam
		F	R. France Internationale
	250	GRC	Voice of America, Kavala
11675		CHN	China Radio International
	100/75	SEY	FEBA Radio, Seychelles
	500/250	KWT	R. Kuwait
	200	BEL	R. Vlaanderen Internationaal
11680	250/100	CYP	BBC East Mediterranean Relay Station
	500/250	G	BBC World Sce, United Kingdom
	100	D	BBC World Sce. via Biblis
	200	KRE	KCBS, Kanggye, No. Korea
11685	250	THA	BBC East Asian Relay Station
	250/100	CYP	BBC East Mediterranean Relay Station
	250/100	SNG	BBC Far Eastern Relay Station
		CHN	China Radio International
		F	R. France Internationale

kHz	kW	Land	Station
	300/100	J	Radio Japan
11690	100	GUM	AWR Asia, Guam
	10	BLR	Brest, Belarus
		CHN	China Radio International
	100/75	SEY	FEBA Radio, Seychelles
	100	GUM	KTWR-Trans World Radio, Guam
	500	JOR	R. Jordan, Al Karanah
		RUS	R. Netherlands, Petropavlovsk
	300/100	J	Radio Japan
	100	ALB	Trans World Radio, Cërrik
11695	500	IND	All India Radio Ext. Sce, Bangalore
	250/100	CYP	BBC East Mediterranean Relay Station
	500/250	G	BBC World Sce, United Kingdom
		RUS	BBC World Sce. via Vladivostok
		CHN	China Radio International
	100/75	SEY	FEBA Radio, Seychelles
	500/20	ISR	Kol Israel, Jerusalem
	100	AUS	R. Australia, Shepparton
	1	VEN	R. Nal. de Venezuela (vf) (irr.)
	500	HOL	R. Netherlands, Flevo
	250	USA	Voice of America, Greenville
11700	10	I	AWR Europe, Forli
		CHN	China Radio International
	500/100	BUL	R. Bulgaria
		F	R. France Internationale
	200	KRE	R. Pyongyang, Korea (D.P.R.)
11704.3	10	B	R. Nova Visão, Santa Maria (often rel. R. Transmundial)
11705	100/75	SEY	FEBA Radio, Seychelles
	300	J	R. Canada International via Yamata
		F	R. France Internationale
			R. Liberty
		UKR	R. Ukraine International
	250	CAN	Radio Japan via Sackville
	35	CLN	Voice of America, Colombo
	250	PHL	Voice of America, Philippines
	500	IRN	Voice of the Islamic Republic of Iran
11710	100	IND	All India Radio Ext. Sce, Delhi
	250	IND	All India Radio, Aligarh
	500/50	ARS	Broadc. Sce. of the Kingdom of Saudi Arabia
	100	ARG	R. Argentina al Exterior
		F	R. France Internationale
	250/50	PAK	R. Pakistan Ext. Sce. (vf)
	250	G	Radio Japan via Skelton
	250/100	GRC	Voice of Greece
11713		CLA	Republic of Iraq Radio/Voice of the Iraqi People (vf)
11715	250	IND	All India Radio Ext. Sce, Aligarh
	500	IND	All India Radio Ext. Sce, Bangalore
		IND	All India Radio Ext. Sce.
	250	THA	BBC East Asian Relay Station
	50	MLI	China Radio International via Mali
	50	USA	KJES, Vado, NM
	250	EGY	R. Cairo, Abis
	300	AUT	R. Canada International via Vienna
	250	CAN	R. Korea Int, Sackville, Canada
	200	KRE	R. Pyongyang, Korea (D.P.R.)
	350	E	Radio Exterior de España
	300/100	J	Radio Japan
	100	I	RAI Ext. Sce, Rome, Italy
	50	ALG	RTV Algerienne, Bouchaoui (vf)
	500/100	CVA	Vatican Radio
	250	GRC	Voice of America, Kavala
	50	PHL	Voice of America, Philippines
11720	500/100	BUL	R. Bulgaria
	100/50	USA	R. Taipei International via WYFR
	50	PHL	Voice of America, Philippines
	250	THA	Voice of America, Udorn
	500	IRN	Voice of the Islamic Republic of Iran
11725	100	TWN	BCC News Network, Paochung
	250/100	KOR	R. Korea Int, Kimje
			R. Liberty
	10	B	R. Novas de Paz, Curitiba
	250	GRC	Voice of America, Kavala
	100	D	Voice of America, Lampertheim
	100	USA	WYFR-Family Radio, Okeechobee, FL
11730	250	IND	All India Radio Ext. Sce, Aligarh
	250	SEY	BBC Indian Ocean Relay Station
	500/50	ARS	Broadc. Sce. of the Kingdom of Saudi Arabia
	100	TUN	ERTT, Tunis
	300	J	R. Canada International via Yamata
			R. Liberty
	250	ATN	R. Netherlands, Bonaire
	500/250	AFS	Trans World Radio Africa, Meyerton
	250	THA	Voice of America, Udorn
11735	500	IND	All India Radio Ext. Sce, Bangalore
	100	NZL	R. New Zealand International
	2.5	URG	R. Oriental, Montevideo
	200	KRE	R. Pyongyang, Korea (D.P.R.)
11740	500	IND	All India Radio Ext. Sce, Bangalore
	250/100	CYP	BBC East Mediterranean Relay Station
	250/100	SNG	BBC Far Eastern Relay Station
	50	CHN	CNR 2
			R. Free Europe
	100/50	USA	R. Taipei International via WYFR
	250	SNG	Radio Japan via Singapore
	250/50	ROU	Radio România, Bucharest
	500/100	CVA	Vatican Radio
	100	USA	WYFR-Family Radio, Okeechobee, FL
11745	100/75	SEY	FEBA Radio, Seychelles
	250/50	TWN	R. Taipei International, Taiwan
		CLA	Voice of Palestine (from Iran)
		CLA	Voice of Palestine, Voice of the Palestinian Islamic Revolution
	500	IRN	Voice of the Islamic Republic of Iran
11750	100	OMA	BBC Eastern Relay Station
	250/100	QAT	Qatar Broadc. Sce, Doha
	200	ATN	R. Vlaanderen Internationaal via Bonaire
	250	INS	RRI Jakarta (Bonto Sunggu)
	250	MRC	Voice of America, Morocco
	500	IRN	Voice of the Islamic Republic of Iran
11755		CHN	China Radio International
	250	CAN	Merlin Comms, via Sackville
	500	FIN	R. Finland
	500	BIH	R. Yugoslavia, Bijeljina, Bosnia
	100	I	RAI Ext. Sce, Rome, Italy
	250	ASC	RAI Ext. Sce. via Ascension
11760	100	OMA	BBC Eastern Relay Station
	500/250	G	BBC World Sce, United Kingdom
	100	GEO	R. Georgia, Tbilisi
	100	CUB	R. Habana, Cuba
	250/50	PAK	R. Pakistan Ext. Sce. (vf)
	300/100	J	Radio Japan
	250	PHL	Voice of America, Philippines
11765	250	ASC	BBC Atlantic Relay Station
	250/100	SNG	BBC Far Eastern Relay Station
		CHN	China Radio International
		XXX	R. Free Asia
	10	B	R. Universo, Curitiba
	250	ASC	RAI Ext. Sce. via Ascension
11770	500/250	G	BBC World Sce, United Kingdom
	500	LBY	LJB, Tripoli/Sabrata. Libya
			R. Liberty
	10	MEX	R. México Internacional, México *
	250	GRC	Voice of America, Kavala
	100	USA	WYFR-Family Radio, Okeechobee, FL
11775	100	GUM	AWR Asia, Guam
	300	TWN	CBS Network 1/2, Taipei, Taiwan
	500/100	D	Deutsche Welle, Germany
	250/50	PAK	R. Pakistan Ext. Sce. (vf)
	350	E	Radio Exterior de España
	250/50	ROU	Radio România, Bucharest
	100	AIA	The Caribbean Beacon, Anguilla
	100	STP	Voice of America, São Tomé
11780	250/100	CYP	BBC East Mediterranean Relay Station
	500/50	ARS	Broadc. Sce. of the Kingdom of Saudi Arabia
		CHN	China Radio International
	250	B	R. Nacional da Amazônia, Brasília
	300/100	POR	RDP Internacional (R. Portugal)
11785	500/50	ARS	Broadc. Sce. of the Kingdom of Saudi Arabia
	500/100	D	Deutsche Welle, Germany
	250	RRW	Deutsche Welle, Kigali
	250	CLN	Deutsche Welle, Trincomalee
		CLA	Holy Medina Radio
		IRQ	R. Baghdad, Iraq (vf)
	250/100	EGY	R. Cairo, Abu Zaabal
	20	CLM	R. Dif. Nal. de Colombia (alt. fq.)
		XXX	R. Free Asia
	7.5	B	R. Guaíba, Porto Alegre (vf)
			R. Liberty
	250	INS	RRI Jakarta (Cimanggis)
	250	THA	Voice of America, Udorn
11790	500	IND	All India Radio Ext. Sce, Bangalore
	50	CHN	China Radio International via Mali
	300/100	J	Radio Japan
	250/50	ROU	Radio România, Bucharest
	500	IRN	Voice of the Islamic Republic of Iran
	500	TUR	Voice of Turkey, Emirler
11795	100	GUM	AWR Asia, Guam
		CHN	China Radio International
	250	ATG	Deutsche Welle, Antigua
	500/100	D	Deutsche Welle, Germany
	250	RRW	Deutsche Welle, Kigali
			R. Liberty
	500/300	UAE	UAE Radio & TV, Dubai
	100	PHL	Voice of America, Philippines
	250	TUR	Voice of Turkey, Çakirlar
11800		CHN	CNR 1
	500/100	BUL	R. Bulgaria
	500	BIH	R. Yugoslavia, Bijeljina, Bosnia
		CLA	Radio Voice of Peace
	100	I	RAI Ext. Sce, Rome, Italy
	100	CLN	SLBC Ext. Sce, Colombo
	100	ETH	Voice of Ethiopia
11805	250/100	CYP	BBC East Mediterranean Relay Station

kHz	kW	Land	Station
			R. Liberty
	500	THA	R. Thailand, Udon Thani
	250	PHL	R. Veritas Asia, Philippines
	100	D	R. Vlaanderen Internationaal via Jülich
	250	GRC	Voice of America, Kavala
	250	PHL	Voice of America, Philippines
11805.4	10	B	R. Globo, Rio de Janeiro
11810	250	RRW	Deutsche Welle, Kigali
	250	POR	Deutsche Welle, Sines
	250	CLN	Deutsche Welle, Trincomalee
	100	GUM	KTWR-Trans World Radio, Guam
	500	JOR	R. Jordan, Al Karanah
	250/100	KOR	R. Korea Int, Kimje
	250/100	ROU	Radio România, Bucharest
	500/100	CVA	Vatican Radio
11815		CHN	China Radio International
		AGL	EP do Lobito (vf)
	100	HWI	KWHR, Naalehu, Hawaii
	100	LBY	LJB, Sebha, Libya
	100	POL	Polish Radio, Warsaw
	7.5	B	R. Brasil Central, Goiânia (vf)
	250	USA	R. Martí, Delano
	100	CTR	Radio Exterior de España, Costa Rica
	300/100	J	Radio Japan
11818	120	MOZ	Em. Nacional, Maputo
11820	250/100	CYP	BBC East Mediterranean Relay Station
	500/50	ARS	Broadc. Sce. of the Kingdom of Saudi Arabia
	500/100	D	Deutsche Welle, Germany
	250	PHL	R. Veritas Asia, Philippines
	100	I	RAI Ext. Sce, Rome, Italy
11825		CHN	China Radio International
	250	CLN	Deutsche Welle, Trincomalee
	100	POL	Polish Radio, Warsaw
	500/100	BUL	R. Bulgaria
	250/50	TWN	R. Taipei International, Taiwan
	250	GRC	Voice of America, Kavala
	300	G	Voice of America, Woofferton
11830		CLA	A Voz da Resistencia de Galo Negro (VOR GAN)
	100	IND	All India Radio, Delhi
	10	B	R. CBN Anhanguera, Goiânia
		UKR	R. Ukraine International
	250/250	ROU	Radio România, Bucharest
	500/100	CVA	Vatican Radio
	500	IRN	Voice of the Islamic Republic of Iran
	100	USA	WYFR-Family Radio, Okeechobee, FL
11835	250/100	SNG	BBC Far Eastern Relay Station
	500/250	G	BBC World Sce, United Kingdom
	500/50	ARS	Broadc. Sce. of the Kingdom of Saudi Arabia
	1	URG	R. El Espectador, Montevideo *
	500	JOR	R. Jordan, Al Karanah
	500	BIH	R. Yugoslavia, Bijeljina, Bosnia
	35	CLN	SLBC Ext. Sce, Colombo
	250	GRC	Voice of America, Kavala
	250	MRC	Voice of America, Morocco
	250	TUR	Voice of Turkey, Çakirlar
11840	500	IND	All India Radio Ext. Sce, Bangalore
	250/100	CYP	BBC East Mediterranean Relay Station
		CHN	China Radio International
			R. Liberty
		UKR	R. Ukraine International
		J	Radio Japan
	250/50	ROU	Radio România, Bucharest
	100	I	RAI Ext. Sce, Rome, Italy
	15	RUS	Yuzhno-Sakhalinsk, FE
11845	250/100	CYP	BBC East Mediterranean Relay Station
	100	D	BBC World Sce. via Lampertheim
		F	R. France Internationale
	200	KRE	R. Pyongyang, Korea (D.P.R.)
	500	THA	R. Thailand, Udon Thani
11850	100	IND	All India Radio Ext. Sce, Delhi
	250/100	CYP	BBC East Mediterranean Relay Station
	250/100	SNG	BBC Far Eastern Relay Station
		CLA	Democratic Voice of Burma
	250	PHL	R. Veritas Asia, Philippines
	300/100	J	Radio Japan
	300/100	POR	RDP Internacional (R. Portugal)
	250	MRC	Voice of America, Morocco
	50	PHL	Voice of America, Philippines
11854.8	1	B	R. Aparecida, Aparecida
11855	250	IND	All India Radio Ext. Sce, Panaji
	250	ASC	BBC Atlantic Relay Station
	250	CAN	R. Canada International, Sackville
			R. Liberty
	250		Voice of America (several locations)
	500	IRN	Voice of the Islamic Republic of Iran
	100	USA	WYFR-Family Radio, Okeechobee, FL
11860	250	SEY	BBC Indian Ocean Relay Station
	100	MRA	KHBI, Agingan Point, Saipan
	250/50	TWN	R. Taipei International, Taiwan
	250	SNG	Radio Japan via Singapore
	100	D	Swiss R. International, via Jülich
11865	250	SEY	BBC Indian Ocean Relay Station
	500/100	D	Deutsche Welle, Germany
	250	CLN	Deutsche Welle, Trincomalee
	250	CAN	R. Canada International, Sackville
	250		Voice of America (several locations)
	100	USA	WYFR-Family Radio, Okeechobee, FL
11870	250	PHL	Voice of America, Philippines
11875	50	MLI	China Radio International via Mali
	100/20	HNG	R. Budapest, Székesfehérvar
	250	EGY	R. Cairo, Abis
			R. Liberty
	250	MRC	Voice of America, Morocco
	500	IRN	Voice of the Islamic Republic of Iran
		CLA	Voice of the Mojahedin of Iranian Baluchestan
	500	USA	WEWN, Vandiver, AL
11880	250	ASC	BBC Atlantic Relay Station
	100	AUS	R. Australia, Shepparton
	500	S	R. Sweden, Hörby
	100	CTR	Radio Exterior de España, Costa Rica
	300	CLN	Radio Japan via Sri Lanka
	100	I	RAI Ext. Sce, Rome, Italy
	100	STP	Voice of America, São Tomé
11885	100	TWN	BCC Popular Network, Paochung
			R. Free Europe/R. Liberty
	500	UAE	R. of the UAE, Abu Dhabi
	250	INS	RRI Jakarta (Bonto Sunggu)
	250	D	Voice of America, Holzkirchen
11890		F	R. France Internationale
	100	OMA	R. Oman, Thumrait
	350	E	Radio Exterior de España
	250	USA	Voice of America, Greenville
11895	500/250	G	BBC World Sce, United Kingdom
	250	HNG	R. Budapest, Jászberény
			R. Liberty
	500	NOR	R. Norway International/R. Denmark
	1	B	R. Rio Grande do Sul, Porto Alegre *
		CLA	Radio Egypt of Arabism
	500	GUF	Radio Japan via Montsinery
	100	D	Voice of America, Lampertheim
	250	PHL	Voice of America, Philippines
11900	500	IND	All India Radio Ext. Sce, Bangalore
	500/100	AFS	Channel Africa, Johannesburg
	500/100	BUL	R. Bulgaria
	500	FIN	R. Finland
	250/100	SNG	RAI via BBC Far Eastern Relay
	250	USA	Voice of America, Delano
11905	500/100	D	Deutsche Welle, Germany
	300	G	R. Canada International via Skelton
	100	NZL	R. New Zealand International (to March 29)
	50	THA	R. Thailand, Pathumthani
	100	CLN	SLBC Ext. Sce, Colombo
	250	GRC	Voice of America, Kavala
	250	MRC	Voice of America, Morocco
11910		CHN	China Radio International
		F	R. France Internationale
	100	GEO	R. Georgia, Tbilisi
	250/50	PAK	R. Pakistan Ext. Sce. (vf)
	120	CHN	Radio Exterior de España via Xi'an
	300/100	J	Radio Japan
	100	D	Swiss R. International, via Jülich
	500	IRN	Voice of the Islamic Republic of Iran
	250	TUR	Voice of Turkey, Çakirlar
	50	USA	WWBS, Macon, GA (F.Pl.)
11915		CHN	China Radio International
	250/200	RUS	Deutsche Welle, Petropavlovsk
	250	POR	R. Canada International via Sines
			R. Free Europe
	10	B	R. Gaucha, Porto Alegre
	250/50	TWN	R. Taipei International, Taiwan
	250	GRC	Voice of America, Kavala
11920	100	OMA	BBC Eastern Relay Station
	250/100	SNG	BBC Far Eastern Relay Station
	100	EQA	HCJB, LV de los Andes, Quito
	500	CTI	R. Côte d'Ivoire, Abidjan
	250	SNG	Radio Japan via Singapore
	250/100	SNG	RAI via BBC Far Eastern Relay
	50	MRC	RTV Marocaine, Tanger, Morocco
	250	THA	Voice of America, Udorn
11925	100	GUM	AWR Asia, Guam
	500/250	G	BBC World Sce, United Kingdom
	10	B	R. Bandeirantes, São Paulo (R. Juratel hires airtime)
	250	PHL	Voice of America, Philippines
	500	TUR	Voice of Turkey, Emirler
11930	250	USA	R. Martí, Greenville
	250/50	PHL	Voice of America, Philippines
	500	IRN	Voice of the Islamic Republic of Iran
11935	100	IND	All India Radio Ext. Sce. Mumbai
	500/50	ARS	Broadc. Sce. of the Kingdom of Saudi Arabia
		CHN	CNR Taiwan
	300	G	R. Canada International via Skelton
	7.5	B	R. Clube Paranaense, Curitiba
	500	JOR	R. Jordan, Al Karanah
	500	HOL	R. Netherlands, Flevo

kHz	kW	Land	Station
	250	USA	Voice of America, Greenville
		CLA	Voice of Southern Azerbaijan (vf)
11939.5	1	PRG	R. Encarnación (n. 11940)
11940	500/250	AFS	BBC World Sce. via Meyerton
	50	CBG	National Radio of Cambodia
	250/50	ROU	Radio România, Bucharest
		CLA	Voice of China
	250	TUR	Voice of Turkey, Çakirlar
11945	250/100	SNG	BBC Far Eastern Relay Station
	300	J	BBC World Sce. via Yamata
		CHN	China Radio International
	250	CAN	R. Canada International, Sackville
		XXX	R. Free Asia
	350	E	Radio Exterior de España
	500/300	UAE	UAE Radio & TV, Dubai
	250	E	Voice of America, Playa de Plas
11950	100	KAZ	Almaty. Kazakhstan
	500/50	ARS	Broadc. Sce. of the Kingdom of Saudi Arabia
	250	POR	Deutsche Welle, Sines
	7.5	B	R. MEC, Rio de Janeiro
	250	GRC	Voice of America, Kavala
	50	USA	WINB, Red Lion, PA
	15	CHN	Xizang-Tb
11955	250	THA	BBC East Asian Relay Station
	250/100	CYP	BBC East Mediterranean Relay Station
	250/100	SNG	BBC Far Eastern Relay Station
	500/100	D	Deutsche Welle, Germany
	100	AGL	Em. Nacional, Mulenvos, Angola
	100	EQA	HCJB, LV de los Andes, Quito
	300	J	R. Canada International via Yamata
		XXX	R. Free Asia
	250	PHL	Voice of America, Philippines
	250	TUR	Voice of Turkey, Çakirlar
11960	100	EQA	HCJB, LV de los Andes, Quito
	250	BLR	Kalodziscy, Belarus (su)
	50	MLI	Radiodiff.-TV Malienne, Bamako
	250	GRC	Voice of America, Kavala
	500	IRN	Voice of the Islamic Republic of Iran
11964.5	7.5	B	R. Record, São Paulo
11965	500/50	ARS	Broadc. Sce. of the Kingdom of Saudi Arabia
		CHN	China Radio International
		F	R. France Internationale
	250	PHL	Voice of America, Philippines
	250	THA	Voice of America, Udorn
11970	250	THA	BBC East Asian Relay Station
	250/100	CYP	BBC East Mediterranean Relay Station
	100	TWN	CBS Network 1, Taipei, Taiwan
	250	CUB	R. Habana, Cuba
			R. Liberty
	250/50	PAK	R. Pakistan Ext. Sce. (vf)
	250/50	ROU	Radio România, Bucharest
	100	D	Voice of America, Lampertheim
	100	USA	WYFR-Family Radio, Okeechobee, FL
11975	200	RUS	Deutsche Welle, Vladivostok
	250/100	EGY	R. Cairo, Abu Zaabal
		F	R. France Internationale
	100	STP	Voice of America, São Tomé
11980	100	GUM	AWR Asia, Guam
	500/250	G	BBC World Sce, United Kingdom
		CHN	China Radio International
	250	EGY	R. Cairo, Abis
	250/100	EGY	R. Cairo, Abu Zaabal
11985	500	USA	Merlin Comms, via Rampisham
	250	HNG	R. Budapest, Jászberény
11990	250	HNG	R. Budapest, Jászberény
	250	EGY	R. Cairo, Abis
	500/250	KWT	R. Kuwait
			R. Liberty
	100	TCH	R. Prague, Czech Rep.
	100	SVK	R. Slovakia Int, Rimavská Sobota
	100	SVK	R. Slovakia Int, Velké Kostolany
	250	GRC	Voice of America, Kavala
11995	100	PHL	FEBC Manila, Philipppines
	100	GUM	KTWR-Trans World Radio, Guam
		F	R. France Internationale
	500	IRN	Voice of the Islamic Republic of Iran
12000	100	TUN	ERTT, Tunis
	50	PHL	FEBC Manila, Philipppines
	500/100	BUL	R. Bulgaria
		F	R. France Internationale
	300/100	J	Radio Japan
	300/100	J	Radio Japan
12005	250/100	CYP	BBC East Mediterranean Relay Station
			R. Liberty
	100	D	Voice of America, Lampertheim
	250	PHL	Voice of America, Philippines
12008		E	EBC (Time Signals), Spain
		CLA	Voice of Sudan
12010	250	PHL	Voice of America, Philippines
12015		CHN	China Radio International
	500/100	D	Deutsche Welle, Germany
	250	POR	Deutsche Welle, Sines
	250/30	EQA	HCJB, LV de los Andes, Quito
		F	R. France Internationale
12020		VTN	Voice of Vietnam, Hanoi
12025	250	POR	Deutsche Welle, Sines
	250/100	EQA	HCJB, LV de los Andes, Quito
		F	R. France Internationale
	100	ARM	Voice of America, Erevan
	250	USA	Voice of America, Greenville
12030	250/100	CYP	BBC East Mediterranean Relay Station
	300/100	J	Radio Japan
	50	PHL	Voice of America, Philippines
12035		VTN	Hanoi 2, Vietnam
	350	E	Radio Exterior de España
		VTN	Voice of Vietnam, Hanoi
12040	250/100	CYP	BBC East Mediterranean Relay Station
	500/250	G	BBC World Sce, United Kingdom
		UKR	R. Ukraine International
	250	PHL	Voice of America, Philippines
12045	250/200	RUS	Deutsche Welle, Petropavlovsk
		UKR	R. Ukraine International
12050	250	EGY	R. Cairo, Abis
		UKR	R. Ukraine International
12055		CHN	China Radio International
	500/100	D	Deutsche Welle, Germany
12060		RUS	V. of the Mediterranean, Malta via Russia
		RUS	BBC World Sce. via Vladivostok
		RUS	R. Netherlands, Irkutsk
12075		F	R. France Internationale
12080	100	D	AWR Europe via Jülich
	10	AUS	BBC World Sce. via Brandon
	10	AUS	R. Australia, Brandon
	100	D	R. Vlaanderen Internationaal via Jülich
	100	ALB	Trans World Radio, Shijak
	100	BOT	Voice of America, Botswana
12085	500	SYR	R. Damascus, Syria
	250	MNG	Voice of Mongolia
12090	200	MDG	R. Netherlands, Madagascar
		RUS	R. Netherlands, Petropavlovsk
12095	250	ASC	BBC Atlantic Relay Station
	250/100	CYP	BBC East Mediterranean Relay Station
	500/250	G	BBC World Sce, United Kingdom
12105	250/100	GRC	Voice of Greece
12110		CHN	China Radio International
12120		CHN	CNR 1
12130	500/250	AFS	AWR via Meyerton
12160	100	USA	WWCR, Nashville, TN
12724	10	AFS	ZSC (Time Signals), So. Africa
12745	8	IND	VWC, Calcutta (Time Sigs)
12804		PHL	NPO (Time Sigs), Philippines
12984	10	AUS	VNG, Llandilo (Time Signals)
13020.4		HKG	VPS60 (Time Sigs), Hong Kong
13282		HKG	HK Aeradio (Time Signals)
13525	40	AUS	Australian Defence Forces Radio, Canberra
13580	100	TCH	R. Prague, Czech Rep.
13585	250/50	PAK	R. Pakistan Ext. Sce. (vf)
13590		UKR	R. Ukraine International
13595	50	USA	WJCR, Millerstown, KY
13600	500/100	BUL	R. Bulgaria
	100	STP	Voice of America, São Tomé
13605	100	AUS	R. Australia, Shepparton
	500	UAE	R. of the UAE, Abu Dhabi
		RUS	V. of the Mediterranean, Malta via Russia
	500	IRN	Voice of the Islamic Republic of Iran
13610	500	IRN	Voice of the Islamic Republic of Iran
13615	500/250	G	BBC World Sce, United Kingdom
	100	PHL	Voice of America, Philippines
	500	USA	WEWN, Vandiver, AL
13620	500	IND	All India Radio Ext. Sce, Bangalore
	500/250	KWT	R. Kuwait
13625	100	HWI	KWHR, Naalehu, Hawaii
		F	R. France Internationale
	500	S	R. Sweden, Hörby
	300/100	POR	RDP Internacional (R. Portugal)
13630	500	JOR	R. Jordan, Al Karanah
	300/100	J	Radio Japan
	250	G	Radio Japan via Skelton
13635	500	SUI	Swiss Radio International
13640		F	R. France Internationale
13645	250	G	Merlin Communications, UK
	999/500	FIN	R. Finland
13650	250	CAN	R. Canada International, Sackville
	200	KRE	R. Pyongyang, Korea (D.P.R.)
	100	D	R. Vlaanderen Internationaal via Jülich
13660	250/100	CYP	BBC East Mediterranean Relay Station
	500/250	G	BBC World Sce, United Kingdom
	250	G	Merlin Communications, UK
13665	250/50	PAK	R. Pakistan Ext. Sce. (vf)
13670	500/250	G	BBC World Sce, United Kingdom
	250/100	SVK	AWR Europe via Slovakia
	500/250	G	BBC World Sce, United Kingdom
	100	KOR	R. Korea Int, Kimje
	200	ATN	R. Vlaanderen Internationaal via Bonaire
13675	500/250	G	BBC World Sce, United Kingdom
	500/300	UAE	UAE Radio & TV, Dubai
	100	STP	Voice of America, São Tomé
13680	250	G	Merlin Communications, UK
	100	CUB	R. Habana, Cuba

kHz	kW	Land	Station
13685	500/250	G	BBC World Sce, United Kingdom
	250	HNG	R. Budapest, Jászberény
	500	SUI	Swiss Radio International
13690	500/100	D	Deutsche Welle, Germany
	250	CLN	Deutsche Welle, Trincomalee
	250	CAN	R. Canada International, Sackville
13695	100	USA	WYFR-Family Radio, Okeechobee, FL
13700	500	IND	All India Radio Ext. Sce. Bangalore
	500/100	BUL	R. Bulgaria
	500	HOL	R. Netherlands, Flevo
	500	GUF	Swiss Radio International, Montsinery
13710	500	IND	All India Radio Ext. Sce. Bangalore
	500/100	AUT	R. Austria International
	500/100	BUL	R. Bulgaria
	XXX		R. Free Asia
	100	BOT	Voice of America, Botswana
	250	MRC	Voice of America, Morocco
13715	500/100	BUL	R. Bulgaria
	100	CUB	R. Habana, Cuba
	100	SVK	R. Slovakia Int, Rimavská Sobota
13720	500	IND	All India Radio Ext. Sce. Bangalore
	100	GUM	AWR Asia, Guam
	250	G	Merlin Communications, UK
		UKR	R. Ukraine International
13725	20	CUB	R. Habana, Cuba (USB)
13730	500/100	AUT	R. Austria International
13740	50	USA	KAIJ, Denton, TX
	500/100	BUL	R. Bulgaria
	500	S	R. Sweden, Hörby
	250	USA	Voice of America, Greenville/Delano
13745	500/250	G	BBC World Sce, United Kingdom
	200	BEL	R. Vlaanderen Internationaal
13750	50/20	CTR	AWR Latin America, Cahuita
	500	USA	Voice of America, Greenville
	100	D	Voice of America, Lampertheim
	500	TUR	Voice of Turkey, Emirler
13755	500/20	ISR	Kol Israel, Jerusalem
	100	AUS	R. Australia, Shepparton
		UZB	R. Netherlands, Tashkent relay
13760	200	KRE	R. Pyongyang, Korea (D.P.R.)
	250	THA	Voice of America, Udorn
	100	USA	WHRI, South Bend, IN
13765	500/100	CVA	Vatican Radio
13770	500	IND	All India Radio Ext. Sce. Bangalore
	500/100	BUL	R. Bulgaria
			R. Liberty
	250	PHL	R. Pilipinas via VOA Tinang
	250	USA	Voice of America, Greenville
	500	TUR	Voice of Turkey, Emirler
	500	USA	WSHB, Cypress Creek, SC
13780	500	IND	All India Radio Ext. Sce. Bangalore
	500/100	D	Deutsche Welle, Germany
	200	KRE	R. Pyongyang, Korea (D.P.R.)
	50	USA	WINB, Red Lion, PA
13790	500/100	D	Deutsche Welle, Germany
	500/100	BUL	R. Bulgaria
	500	NOR	R. Norway International/R. Denmark
13795	500	IND	All India Radio Ext. Sce. Bangalore
13800	500/100	BUL	R. Bulgaria
	XXX		R. Free Asia
	500	NOR	R. Norway International/R. Denmark
13815	100	USA	KAIJ, Denton, TX
13820	250	USA	R. Martí, Greenville
13825	500	USA	WVHA, Greenbush, ME
13830	100/2.5	HRV	Hrvatski Radio, Zagreb
13835	10	ISL	Ríkisútvarpid, Reykjavík
13840	100	MRA	KHBI, Agingan Point, Saipan
13845	100	USA	WWCR, Nashville, TN
13855	10	ISL	Ríkisútvarpid, Reykjavík
13860	10	ISL	Ríkisútvarpid, Reykjavík
14670	3	CAN	CHU (Time Signals), Canada
14996	8	RUS	RWM (Time Signals), Moscow
15000	8	IND	ATA, New Delhi (Time Sigs.)
	2	TWN	BSF (Time Signals), Taiwan
	2	ARG	LOL (Time Signals), Argentina
	5	RUS	RTA (Time Signals), Russia
	10	HWA	WWVH, Hawaii (Time Signals)
15004	1	RUS	RID (Time Signals), Irkutsk
15010		VTN	Voice of Vietnam, Hanoi
15020	100	IND	All India Radio Ext. Sce. Delhi
15050	250	IND	All India Radio Ext. Sce. Aligarh
	250	IND	All India Radio Ext. Sce. Delhi
	10	CTR	R. For Peace Int., Cd. Colón
15060	500/50	ARS	Broadc. Sce. of the Kingdom of Saudi Arabia
15075	250	IND	All India Radio Ext. Sce. Aligarh
	250	IND	All India Radio Ext. Sce. Bangalore
	50	IND	All India Radio Ext. Sce. Delhi
15084	500	IRN	Voice of the Islamic Republic of Iran
15095	50	PHL	FEBC Manila, Philipppines
	500	SYR	R. Damascus, Syria
15100		CHN	China Radio International
15105	250	ASC	BBC Atlantic Relay Station
	500/250	G	BBC World Sce, United Kingdom
	250	ATG	Deutsche Welle, Antigua

kHz	kW	Land	Station
	100	GUM	KTWR-Trans World Radio, Guam
	250/50	ROU	Radio România, Bucharest
	500	TUR	Voice of Turkey, Emirler
15110	100	USA	WHRI, South Bend, IN
	50	MLI	China Radio International via Mali
	500/250	KWT	R. Kuwait
15115	500/250	G	BBC World Sce, United Kingdom
	100	EQA	HCJB, LV de los Andes, Quito
	250/100	EGY	R. Cairo, Abu Zaabal
	100	NZL	R. New Zealand International
	250	MRC	Voice of America, Morocco
15120	50	CHN	China Radio International via Mali
	999	FIN	R. Finland
	5	MEX	R. Transcontinental de América, México (daytime)
	300/100	J	Radio Japan
	250	ASC	Radio Japan via Ascension Island
	250	USA	Voice of America, Delano
15125	250	TWN	BCC Popular Network, Annan
	100	CTR	Radio Exterior de España, Costa Rica
	250	INS	RRI Jakarta (Cimanggis)
	250	MRC	Voice of America, Morocco
15130	50	CHN	China Radio International via Mali
	200	KRE	R. Pyongyang, Korea (D.P.R.)
	100/50	USA	WYFR-Family Radio, Okeechobee, FL
15135	50	IND	All India Radio, Delhi
		CHN	China Radio International
	250	RRW	Deutsche Welle, Kigali
		F	R. France Internationale
	7.5	B	R. Record, São Paulo
15140	500	IND	All India Radio Ext. Sce, Bangalore
	100	EQA	HCJB, LV de los Andes, Quito
	100	OMA	R. Oman, Thumrait
15145	500/100	D	Deutsche Welle, Germany
	250	MDG	Deutsche Welle, Madagascar
	250	CLN	Deutsche Welle, Trincomalee
	500	FIN	R. Finland
	50	PHL	Voice of America, Philippines
	100	USA	WYFR-Family Radio, Okeechobee, FL
15150	250	CAN	R. Canada International, Sackville
	200	BEL	R. Vlaanderen Internationaal
	250	INS	RRI Jakarta (Cimanggis)"
	250/100	GRC	Voice of Greece
15155	250/100	EGY	R. Cairo, Abu Zaabal
		F	R. France Internationale
15160	100	ALG	RTV Algerienne, Bouchaoui
	250	GRC	Voice of America, Kavala
15165	500/50	ARS	Broadc. Sce. of the Kingdom of Saudi Arabia
	300/100	J	Radio Japan
	100	UZB	UZR2, Toskent, Uzbekistan
	500	IRN	Voice of the Islamic Republic of Iran
	5	RUS	Yoshkar-Ola, E
15168	20	OCE	RFO, Tahiti, French Polynesia (vf)
15170		CLA	Democratic Voice of Burma
	XXX		R. Free Asia
	250	MNG	Voice of Mongolia
	100	USA	WYFR-Family Radio, Okeechobee, FL
15175	500	IND	All India Radio Ext. Sce, Bangalore
	5	BLR	Mahiliou, Belarus
	500/100	BUL	R. Bulgaria
	500	NOR	R. Norway International/R. Denmark
	250/50	PAK	R. Pakistan Ext. Sce. (vf)
	500	BIH	R. Yugoslavia, Bijeljina, Bosnia
	250/100	GRC	Voice of Greece
15180	250	IND	All India Radio Ext. Sce, Bangalore
	250/100	CYP	BBC East Mediterranean Relay Station
	500/250	G	BBC World Sce, United Kingdom
		CHN	China Radio International
	200	KRE	R. Pyongyang, Korea (D.P.R.)
	50	PHL	Voice of America, Philippines
15185	50	IND	All India Radio, Delhi
	500/100	D	Deutsche Welle, Germany
	250	CLN	Deutsche Welle, Trincomalee
	999	FIN	R. Finland
	500	SUI	Swiss Radio International
	50	PHL	Voice of America, Philippines
15185.7	100	GNE	R. Africa, Malabo, Eq. Guinea
15190	250	ASC	BBC Atlantic Relay Station
		CHN	CNR 1
	100	COG	R. Congo, Brazzaville
	5	B	R. Inconfidência, Belo Horizonte
	250/50	PAK	R. Pakistan Ext. Sce. (vf)
	250	PHL	R. Pilipinas via VOA Tinang
15195	100	GUM	KTWR-Trans World Radio, Guam
		F	R. France Internationale
	200	BEL	R. Vlaanderen Internationaal
15200	100	GUM	KTWR-Trans World Radio, Guam
		F	R. France Internationale
	250	B	R. Nacional da Amazônia, Brasília
	300/100	POR	RDP Internacional (R. Portugal)
	100	UZB	UZR2/FS, Toskent, Uzbekistan (su.)
	500	IRN	Voice of the Islamic Republic of Iran
15205	250	ATG	Deutsche Welle, Antigua
	500/100	D	Deutsche Welle, Germany

kHz	kW	Land	Station
			R. Liberty
	100	ALG	RTV Algerienne, Bouchaoui (vf)
	250/50		Voice of America (several locations)
15210	250	EGY	R. Cairo, Abis
	F		R. France Internationale
	250	PHL	Voice of America, Philippines
15215	500/100	AFS	Channel Africa, Johannesburg
		CLA	La Voz de Sahara Libre
			R. Free Europe/R. Liberty
	100/50	USA	R. Taipei International via WYFR
	2.5	B	R. Timbira, São Luís *
	100	ALG	RTV Algerienne, Bouchaoui (vf)
	250	E	Voice of America, Playa de Plas
	100	USA	WYFR-Family Radio, Okeechobee, FL
	100	USA	WYFR-Family Radio, Okeechobee, FL
15220	250	ATG	BBC Caribbean Relay Station
	250	CAN	BBC World Sce. via Sackville
	250	EGY	R. Cairo, Abis
	250/100	EGY	R. Cairo, Abu Zaabal
	250	ASC	Radio Japan via Ascension Island
	100	I	RAI Ext. Sce, Rome, Italy
	250	MRC	Voice of America, Morocco
15225	100	GUM	AWR Asia, Guam
	500/250	G	BBC World Sce, United Kingdom
	250	ASC	Voice of America, Ascension Isl.
15230	500/50	ARS	Broadc. Sce. of the Kingdom of Saudi Arabia
	125	URG	Em. Ciudad de Montevideo *
	200	KRE	R. Pyongyang, Korea (D.P.R.)
	300/100	J	Radio Japan
	100	I	RAI Ext. Sce, Rome, Italy
	500	IRN	Voice of the Islamic Republic of Iran
15235	100	OMA	BBC Eastern Relay Station
	500	LBY	LJB, Tripoli/Sabrata. Libya
	500	FIN	R. Finland
		CLA	Republic of Iraq Radio/Voice of the Iraqi People (vf)
15240	100	AUS	R. Australia, Shepparton
	500	S	R. Sweden, Hörby
15245	250/100	CYP	BBC East Mediterranean Relay Station
	100	ZAI	Kinshasa, Congo (Dem. Rep.)
	500	TUR	Voice of Turkey, Emirler
15250			R. Liberty
	200	ATN	R. Vlaanderen Internationaal via Bonaire
	250/50	ROU	Radio România, Bucharest
	100	I	RAI Ext. Sce, Rome, Italy
	35	CLN	Voice of America, Colombo
15255	250/100	EGY	R. Cairo, Abu Zaabal
	250	ATN	R. Netherlands, Bonaire
	50	PHL	Voice of America, Philippines
	100	USA	WYFR-Family Radio, Okeechobee, FL
15260	50	IND	All India Radio, Delhi
	250/100	CYP	BBC East Mediterranean Relay Station
		CHN	China Radio International
		CLA	Voice of Palestine (from Iran)
	500	IRN	Voice of the Islamic Republic of Iran
15265	1	B	R. Globo, São Paulo *
	500	UAE	R. of the UAE, Abu Dhabi
	250	B	Radiobras, Brasília
	250	USA	Voice of America, Greenville
15270	250	TWN	BCC News Network, Annan
	250/50	TWN	R. Taipei International, Taiwan
	1000	ARM	Voice of Armenia
15275	250	RRW	Deutsche Welle, Kigali
	250	CLN	Deutsche Welle, Trincomalee
	100	I	RAI Ext. Sce, Rome, Italy
	1	URG	S.O.D.R.E. Montevideo (r: CX26 1050) *
	500	SUI	Swiss Radio International
15280	250	THA	BBC East Asian Relay Station
	250/100	SNG	BBC Far Eastern Relay Station
	500/50	ARS	Broadc. Sce. of the Kingdom of Saudi Arabia
	100	I	RAI Ext. Sce, Rome, Italy
		CLA	Voice of China
15285	250/100	EGY	R. Cairo, Abu Zaabal
15290	250	PHL	Voice of America, Philippines
15295	100	EQA	HCJB, LV de los Andes, Quito
	500	MLA	Voice of Malaysia, Kuala Lumpur
	500	TUR	Voice of Turkey, Emirler
15300	F		R. France Internationale
	250	THA	Voice of America, Udorn
15305	250	GRC	Voice of America, Kavala
	35	PHL	Voice of America, Philippines
15310	250	THA	BBC East Asian Relay Station
	100	OMA	BBC Eastern Relay Station
	250/100	SNG	BBC Far Eastern Relay Station
15315	500/250	USA	BBC World Sce. via Greenville
	500/100	BUL	R. Bulgaria
	F		R. France Internationale
	120	ROU	R. Moldova International via Romania
	250	ATN	R. Netherlands, Bonaire
	500	HOL	R. Netherlands, Flevo
	500	UAE	R. of the UAE, Abu Dhabi
15320	100	TWN	CBS Network 1, Taipei, Taiwan
	250	ASC	RAI Ext. Sce, via Ascension
	500/300	UAE	UAE Radio & TV, Dubai
15325	500/250	G	BBC World Sce, United Kingdom
	250	POR	R. Canada International via Sines
	250	CAN	R. Canada International, Sackville
	10	B	R. Gazeta, São Paulo
	250/50	PAK	R. Pakistan Ext. Sce. (vf)
	500	IRN	Voice of the Islamic Republic of Iran
15330	100	GUM	KTWR-Trans World Radio, Guam
	250	USA	R. Martí, Delano
	250/50	PAK	R. Pakistan Ext. Sce. (vf)
	250	PHL	R. Pilipinas via VOA Tinang
	100	UZB	R. Tashkent Ext.Sce, Uzbekistan
	100	I	RAI Ext. Sce, Rome, Italy
	100	SWZ	Trans World Radio Africa, Swaziland
	100	UZB	UZR2/FS, Toskent, Uzbekistan (wi.)
	500	IRN	Voice of the Islamic Republic of Iran
15335	250/100	CYP	BBC East Mediterranean Relay Station
	500/50	ARS	Broadc. Sce. of the Kingdom of Saudi Arabia
	250	EGY	R. Cairo, Abis
	250	PHL	R. Veritas Asia, Philippines
	250/50	ROU	Radio România, Bucharest
	50	MRC	RTV Marocaine, Tanger, Morocco
15340	500	IND	All India Radio Ext. Sce, Bangalore
	250/100	CYP	BBC East Mediterranean Relay Station
	100	OMA	BBC Eastern Relay Station
	500/250	G	BBC World Sce, United Kingdom
		IRQ	R. Baghdad, Iraq
	100/50	CUB	R. Habana, Cuba
			R. Liberty
	200	KRE	R. Pyongyang, Korea (D.P.R.)
15345	500/50	ARS	Broadc. Sce. of the Kingdom of Saudi Arabia
	250/50	TWN	R. Taipei International, Taiwan
	100	MRC	RTV Marocaine, Tanger, Morocco
	500/300	UAE	UAE Radio & TV, Dubai
15345.2	100	ARG	R. Argentina al Exterior
15350	100	GUI	Conakry, Guinea Rep. (irr.) (vf)
	250	USA	Voice of America, Delano
	500	IRN	Voice of the Islamic Republic of Iran
	500	TUR	Voice of Turkey, Çakirlar
15355	500	JOR	R. Jordan, Al Karanah
	300/100	J	Radio Japan
	250	USA	Voice of America, Greenville
15360	250/100	SNG	BBC Far Eastern Relay Station
	500/250	G	BBC World Sce, United Kingdom
	500/100	D	Deutsche Welle, Germany
	100	BOT	Voice of America, Botswana
15365	F		R. France Internationale
	250/50	ROU	Radio România, Bucharest
	250	USA	Voice of America, Greenville
	500	IRN	Voice of the Islamic Republic of Iran
15370	250	RRW	Deutsche Welle, Kigali
			R. Free Europe/R. Liberty
	500	THA	R. Thailand, Udon Thani
	200	BEL	R. Vlaanderen Internationaal
	250/50	ROU	Radio România, Bucharest
	100	D	Voice of America, Lampertheim
15375	250/100	EGY	R. Cairo, Abu Zaabal
	350	E	Radio Exterior de España
	100	BOT	Voice of America, Botswana
	500	USA	WEWN, Vandiver, AL
15380	250	THA	BBC East Asian Relay Station
	500/50	ARS	Broadc. Sce. of the Kingdom of Saudi Arabia
	100	MRA	KFBS, Marpi, Saipan
	250/50	ROU	Radio România, Bucharest
15385	50	USA	KJES, Vado, NM
	250	USA	Voice of America, Greenville
	500	TUR	Voice of Turkey, Emirler
15388		CLA	New Star Broadcasting Station No. One
15390	250	ASC	BBC Atlantic Relay Station
	500/250	G	BBC World Sce, United Kingdom
	50	CHN	CNR 1
	250/50	ROU	Radio România, Bucharest
	250	ASC	Voice of America, Ascension Isl.
15395	250/100	QAT	Qatar Broadc. Sce, Doha
	500/300	UAE	UAE Radio & TV, Dubai
	35	CLN	Voice of America, Colombo
	250	PHL	Voice of America, Philippines
15400	250	ASC	BBC Atlantic Relay Station
	500/250	G	BBC World Sce, United Kingdom
		CHN	China Radio International
	500	FIN	R. Finland
	250	ASC	Radio Japan via Ascension Island
	500/300	UAE	UAE Radio & TV, Dubai
15405	250	THA	BBC East Asian Relay Station
	100	OMA	BBC Eastern Relay Station
	F		R. France Internationale
	250/50	ROU	Radio România, Bucharest
15410	500/100	D	Deutsche Welle, Germany
	250	RRW	Deutsche Welle, Kigali
	500/100	AUT	R. Austria International
	250	MRC	Voice of America, Morocco
	250	THA	Voice of America, Udorn

kHz	kW	Land	Station	kHz	kW	Land	Station
15415	500	LBY	LJB, Tripoli/Sabrata. Libya	15605		F	R. France Internationale
	100	AUS	R. Australia, Shepparton	15610	100	GUM	AWR Asia, Guam
	1	B	R. Clube, Ribeirão Preto	15615	500/20	ISR	Kol Israel, Jerusalem
	250/100	GRC	Voice of Greece		250	TUR	Voice of Turkey, Çakirlar
15420	250	ASC	BBC Atlantic Relay Station	15625	250/50	PAK	R. Pakistan Ext. Sce. (vf)
	250/100	CYP	BBC East Mediterranean Relay Station		100	D	West Coast R. Ireland via Jülich
	250	SEY	BBC Indian Ocean Relay Station	15630	250/100	GRC	Voice of Greece
	500/250	AFS	BBC World Sce. via Meyerton	15640	500/20	ISR	Kol Israel, Jerusalem
	250/100	EGY	R. Cairo, Abu Zaabal		100	TCH	R. Prague, Czech Rep.
	50	USA	WRNO, New Orleans, LA	15650	250/100	GRC	Voice of Greece
15425	250	CAN	R. Canada International, Sackville	15660		XXX	R. Free Asia
	35	CLN	SLBC Ext. Sce, Colombo	15665	100	MRA	KHBI, Agingan Point, Saipan
	50	PHL	Voice of America, Philippines		500	USA	WEWN, Vandiver, AL
15430	100/75	SEY	FEBA Radio, Seychelles		500	USA	WSHB, Cypress Creek, SC
	10	MEX	R. México Internacional, México *		100	USA	WYFR-Family Radio, Okeechobee, FL
15435		CHN	China Radio International	15670		CHN	CNR Minor.
	500	LBY	LJB, Tripoli/Sabrata. Libya	15675	500/100	BUL	R. Bulgaria
		F	R. France Internationale		1	HND	R. Copán Internacional, Tegucigalpa
	500	JOR	R. Jordan, Al Karanah	15685	100	USA	WWCR, Nashville, TN
	500/300	UAE	UAE Radio & TV, Dubai	15695	100	USA	WYFR-Family Radio, Okeechobee, FL
15440		CHN	China Radio International	15700	500/100	BUL	R. Bulgaria
	500	FIN	R. Finland	15705	500	NOR	R. Norway International/R. Denmark
		F	R. France Internationale	15710		CHN	CNR Taiwan
	100/50	USA	R. Taipei International via WYFR	15715	100	USA	WYFR-Family Radio, Okeechobee, FL
	100	USA	WYFR-Family Radio, Okeechobee, FL	15725	100	GUM	AWR Asia, Guam
15445	100/75	SEY	FEBA Radio, Seychelles		50	USA	KAIJ, Denton, TX
			R. Liberty		100	MRA	KHBI, Agingan Point, Saipan
	250	B	Radiobras, Brasília	15745	500	USA	WEWN, Vandiver, AL
	250/100		Voice of America (several locations)	15770	250	IND	All India Radio Ext. Sce, Aligarh
15450	100	TUN	ERTT, Tunis		10	ISL	Rikisútvarpid, Reykjavík
	100/50	PHL	FEBC Manila, Philipppines	15790	10	ISL	Rikisútvarpid, Reykjavík
15460		F	R. France Internationale	15880		CHN	CNR Chinese
	100	SVK	R. Slovakia Int, Rimavská Sobota	16000	5	AUS	VNG, Llandilo (Time Signals)
	250	SNG	Radio Japan via Singapore	17018	10	AFS	ZSC (Time Signals), So. Africa
	250	THA	Voice of America, Udorn	17096		HKG	VPS80 (Time Sigs), Hong Kong
15465	500/250	G	BBC World Sce, United Kingdom	17387	250	IND	All India Radio Ext. Sce, Aligarh
	50	PHL	FEBC Manila, Philipppines		250	IND	All India Radio Ext. Sce, Delhi
15475	500	GAB	Africa No.1, Moyabi, Gabon	17485	100	TCH	R. Prague, Czech Rep.
	300/100	J	Radio Japan	17500	100	TUN	ERTT, Tunis
15480		CHN	CNR 1		500/100	BUL	R. Bulgaria
	500/20	ISR	Kol Israel, Arabic prgr.	17510	10	HWI	KWHR, Naalehu, Hawaii
15485	500/250	G	BBC World Sce, United Kingdom		500	USA	WSHB, Cypress Creek, SC
	250/50	PAK	R. Pakistan Ext. Sce. (vf)	17535	500	USA	WVHA, Greenbush, ME
	250	USA	Voice of Greece via VOA	17540	500/20	ISR	Kol Israel, Jerusalem
15495	500/250	KWT	R. Kuwait		250/50	PAK	R. Pakistan Ext. Sce. (vf)
15500		CHN	CNR 2	17545	500/20	ISR	Kol Israel, Jerusalem
15505	500/250	KWT	R. Kuwait	17550		CHN	CNR 1
	100	BOT	Voice of America, Botswana		500/100	CVA	Vatican Radio
15510	100	AUS	R. Australia, Shepparton	17555	500/250	G	BBC World Sce, United Kingdom
	100	STP	Voice of America, São Tomé		250/50	PAK	R. Pakistan Ext. Sce. (vf)
15515		F	R. France Internationale		100	USA	WYFR-Family Radio, Okeechobee, FL
		XXX	R. Free Asia	17560		F	R. France Internationale
	250	THA	Voice of America, Udorn	17570		RUS	V. of the Mediterranean, Malta via Russia
15520	50	PHL	FEBC Manila, Philipppines	17575		F	R. France Internationale
	250/100	BGD	R. Bangladesh Home Sce. (vf)	17580	200	MDG	R. Netherlands, Madagascar
	250/50	PAK	R. Pakistan Ext. Sce. (vf)	17585	500/250	G	BBC World Sce, United Kingdom
		UKR	R. Ukraine International	17595	250	EGY	R. Cairo, Abis
15525	500/100	D	Deutsche Welle, Germany		50	MRC	RTV Marocaine, Tanger, Morocco
	250/200	RUS	Deutsche Welle, Samara	17605		CHN	CNR 1
		F	R. France Internationale		250	ATN	R. Netherlands, Bonaire
15530		F	R. France Internationale	17610	10	I	AWR Europe, Forli
	250	PHL	R. Veritas Asia, Philippines		500/250	G	BBC World Sce, United Kingdom
	500	IRN	Voice of the Islamic Republic of Iran	17615	500/100	AUT	R. Austria International
15540	100/75	SEY	FEBA Radio, Seychelles	17620		F	R. France Internationale
		F	R. France Internationale	17630	500	GAB	Africa No.1, Moyabi, Gabon
	500/100	CVA	Vatican Radio		500	USA	Merlin Comms, via Rampisham
	250	THA	Voice of America, Udorn			F	R. France Internationale
15545	500/100	AFS	Channel Africa, Johannesburg		250	THA	Voice of America, Udorn
15550		CHN	CNR 1	17640	250/100	CYP	BBC East Mediterranean Relay Station
	100	EQA	HCJB, LV de los Andes, Quito		500/250	G	BBC World Sce, United Kingdom
		UKR	R. Ukraine International		250	USA	Voice of America, Greenville
		RUS	V. of the Mediterranean, Malta via Russia	17645	100	GUM	AWR Asia, Guam
15555	250/100	CYP	BBC East Mediterranean Relay Station		500	UAE	R. of the UAE, Abu Dhabi
	100/75	SEY	FEBA Radio, Seychelles	17650	500/100	BUL	R. Bulgaria
	250/50	PAK	R. Pakistan Ext. Sce. (vf)			F	R. France Internationale
	300/100	POR	RDP Internacional (R. Portugal)		100	BOT	Voice of America, Botswana
15560	200	MDG	R. Netherlands, Madagascar	17670	250	HNG	R. Budapest, Jászberény
		UZB	R. Netherlands, Tashkent relay		250/100	EGY	R. Cairo, Abu Zaabal
15565	500/250	G	BBC World Sce, United Kingdom	17675		RUS	BBC World Sce. via Samara
15570	500/100	CVA	Vatican Radio			UZB	BBC World Sce. via Tashkent
15575	250/100	CYP	BBC East Mediterranean Relay Station	17680		CHN	China Radio International
	500/250	G	BBC World Sce, United Kingdom			UKR	R. Ukraine International
	250/100	KOR	R. Korea Int, Kimje		300/100	POR	RDP Internacional (R. Portugal)
15580	250	USA	Voice of America, Greenville	17685	300/100	J	Radio Japan
	250/100	CYP	BBC East Mediterranean Relay Station	17690	250	EGY	R. Cairo, Abis
	200	MDG	R. Netherlands, Madagascar		200	BEL	R. Vlaanderen Internationaal
	500/100	CVA	Vatican Radio		200	MDG	R. Vlaanderen Internationaal via Madagascar
15590	100	USA	KTBN, Salt Lake City, UT		100	UZB	R. Vlaanderen Internationaal via Tashkent
	300/100	J	Radio Japan	17695	250	ASC	BBC Atlantic Relay Station
15595	500/250	G	BBC World Sce, United Kingdom		250/100	CYP	BBC East Mediterranean Relay Station
	250/50	PAK	R. Pakistan Ext. Sce. (vf)		500/250	G	BBC World Sce, United Kingdom
	500/100	CVA	Vatican Radio		100	D	BBC World Sce. via Lampertheim
15600		CHN	China Radio International		500	USA	WEWN, Vandiver, AL
	100	BOT	Voice of America, Botswana	17700	50	CHN	CNR 2
	100	USA	WYFR-Family Radio, Okeechobee, FL				

kHz	kW	Land	Station
	500/100	CVA	Vatican Radio
17705	500	IND	All India Radio Ext. Sce, Bangalore
	500/250	G	BBC World Sce, United Kingdom
	250/50	PAK	R. Pakistan Ext. Sce. (vf)
	100	BOT	Voice of America, Botswana
	250	USA	Voice of Greece via VOA
17710	250	EGY	R. Cairo, Abis
17715	100	GUM	AWR Asia, Guam
	500/250	USA	BBC World Sce. via Greenville
	250	CAN	R. Canada International, Sackville
		UKR	R. Ukraine International
	350	E	Radio Exterior de España
	500	TUR	Voice of Turkey, Emirler
17720	250/50	ROU	Radio România, Bucharest
17725	300/100	POR	RDP Internacional (R. Portugal)
	250	USA	Voice of America, Greenville
	100	USA	WYFR-Family Radio, Okeechobee, FL
17730			R. Free Europe
	250	PHL	R. Pilipinas via VOA Tinang
17735	200	KRE	R. Pyongyang, Korea (D.P.R.)
	250	PHL	Voice of America, Philippines
17740	250	USA	Voice of America, Greenville
	250	PHL	Voice of America, Philippines
17745	250	EGY	R. Cairo, Abis
	250/50	ROU	Radio România, Bucharest
17750	100	ALG	RTV Algerienne, Bouchaoui
	100	AUS	R. Australia, Shepparton
			R. Liberty
	100/50	USA	R. Taipei International via WYFR
	250	B	Radiobras, Brasília
	100	BOT	Voice of America, Botswana
	100	USA	WYFR-Family Radio, Okeechobee, FL
17755	500/50	ARS	Broadc. Sce. of the Kingdom of Saudi Arabia
	350	E	Radio Exterior de España
	250	ASC	Voice of America, Ascension Isl.
	500	TUR	Voice of Turkey, Emirler
17760	250	THA	BBC East Asian Relay Station
	250/100	SNG	BBC Far Eastern Relay Station
	500	UAE	R. of the UAE, Abu Dhabi
	250	PHL	R. Pilipinas via VOA Tinang
17765	10	MEX	R. México Internacional, México *
	50	PHL	Voice of America, Philippines
17770	250	EGY	R. Cairo, Abis
	250	THA	Voice of America, Udorn
17775	500/50	ARS	Broadc. Sce. of the Kingdom of Saudi Arabia
	50	USA	KVOH, Rancho Simi, CA
	250/50	ROU	Radio România, Bucharest
17780	500/100	D	Deutsche Welle, Germany
	300/100	J	Radio Japan
	100	I	RAI Ext. Sce, Rome, Italy
	250	PHL	Voice of America, Philippines
17785	500	IND	All India Radio Ext. Sce, Bangalore
	100	OMA	BBC Eastern Relay Station
	500	S	R. Sweden, Hörby
	250	MRC	Voice of America, Morocco
17790	250	ASC	BBC Atlantic Relay Station
	250	THA	BBC East Asian Relay Station
	100	OMA	BBC Eastern Relay Station
	250/50	ROU	Radio România, Bucharest
17795	100	EQA	HCJB, LV de los Andes, Quito
	100	AUS	R. Australia, Shepparton
		F	R. France Internationale
17800	500/100	D	Deutsche Welle, Germany
	250	RRW	Deutsche Welle, Kigali
	250/100	EGY	R. Cairo, Abu Zaabal
		F	R. France Internationale
	250	USA	Voice of America, Greenville
17805	5	BLR	Hrodna, Belarus
		XXX	R. Free Asia
			R. Liberty
	350	E	Radio Exterior de España
	250/50	ROU	Radio România, Bucharest
	250	ASC	Voice of America, Ascension Isl.
	250	E	Voice of America, Playa de Plas
	100	USA	WYFR-Family Radio, Okeechobee, FL
17810	250	ASC	BBC Atlantic Relay Station
	300/100	J	Radio Japan
17815	1	B	R. Cultura, São Paulo
	250	ASC	Radio Japan via Ascension Island
	50	MRC	RTV Marocaine, Tanger, Morocco
	500	SUI	Swiss Radio International
	500	TUR	Voice of Turkey, Emirler
17820	250	ASC	BBC Atlantic Relay Station
	500/100	D	Deutsche Welle, Germany
	250	CLN	Deutsche Welle, Trincomalee
	250	CAN	R. Canada International, Sackville
	100	PHL	Voice of America, Philippines
17825	500/100	D	Deutsche Welle, Germany
	250	CLN	Deutsche Welle, Trincomalee
17830	500/300	UAE	UAE Radio & TV, Dubai
17835	250/50	PAK	R. Pakistan Ext. Sce. (vf)
17840	500	IND	All India Radio Ext. Sce, Bangalore
	250	ATG	BBC Caribbean Relay Station
	250	CAN	BBC World Sce. via Sackville
	250	EGY	R. Cairo, Abis

kHz	kW	Land	Station
	250	PHL	R. Pilipinas via VOA Tinang
17845	100/50	USA	R. Taipei International via WYFR
	350	E	Radio Exterior de España
	100	USA	WYFR-Family Radio, Okeechobee, FL
17850		F	R. France Internationale
	35	CLN	SLBC Ext. Sce, Colombo
17855	500	UAE	R. of the UAE, Abu Dhabi
17860	100	IND	All India Radio Ext. Sce, Delhi
	500/100	D	Deutsche Welle, Germany
		F	R. France Internationale
17865	250	IND	All India Radio, Aligarh
17870	500/100	AFS	Channel Africa, Johannesburg
	500/100	AUT	R. Austria International
	100	I	RAI Ext. Sce, Rome, Italy
	500/250	TUR	Voice of Turkey, Emirler
17875	250	CAN	R. Canada International, Sackville
	7.5	B	R. MEC, Rio de Janeiro
	250	ASC	Voice of America, Ascension Isl.
17880	500/250	G	BBC World Sce, United Kingdom
	500/50	ARS	Broadc. Sce. of the Kingdom of Saudi Arabia
17885	250	ASC	BBC Atlantic Relay Station
	250/100	CYP	BBC East Mediterranean Relay Station
	250	SEY	BBC Indian Ocean Relay Station
	250	KWT	R. Kuwait
17890	350	E	Radio Exterior de España
17895	250	IND	All India Radio Ext. Sce, Aligarh
	500/50	ARS	Broadc. Sce. of the Kingdom of Saudi Arabia
	500/100	AFS	Channel Africa, Johannesburg
	500/100	D	Deutsche Welle, Germany
	250	CLN	Deutsche Welle, Trincomalee
	250/100	QAT	Qatar Broadc. Sce, Doha
			R. Free Europe
	250	MRC	Voice of America, Morocco
18930	500	USA	WSHB, Cypress Creek, SC
18950	500	NOR	R. Norway International/R. Denmark
21455	0.5	EQA	HCJB, LV de los Andes, Quito (USB)
	250	PHL	R. Pilipinas via VOA Tinang
21470	250/100	CYP	BBC East Mediterranean Relay Station
	100	EQA	HCJB, LV de los Andes, Quito
21480	200	MDG	R. Netherlands, Madagascar
21485	500/300	UAE	UAE Radio & TV, Dubai
	250	USA	Voice of America, Greenville
21490	250	ASC	BBC Atlantic Relay Station
	250/100	CYP	BBC East Mediterranean Relay Station
	250	ASC	Radio Japan via Ascension Island
21500	500	TUR	Voice of Turkey, Emirler
21510	200	BEL	R. Vlaanderen Internationaal
21515	300/100	POR	RDP Internacional (R. Portugal)
21520	100	I	RAI Ext. Sce, Rome, Italy
21525	100	AUS	R. Australia, Shepparton
	100	USA	WYFR-Family Radio, Okeechobee, FL
21530	500/50	ARS	Broadc. Sce. of the Kingdom of Saudi Arabia
	500/100	AFS	Channel Africa, Johannesburg
21535	100	I	RAI Ext. Sce, Rome, Italy
21550	250	G	Merlin Communications, UK
21560	250	HNG	R. Budapest, Jászberény
	250	THA	Voice of America, Udorn
21580		F	R. France Internationale
	250	PHL	R. Pilipinas via VOA Tinang
21590	500/250	G	BBC World Sce, United Kingdom
	250	ATN	R. Netherlands, Bonaire
21600	500/100	D	Deutsche Welle, Germany
21605	500/300	UAE	UAE Radio & TV, Dubai
21620		F	R. France Internationale
21630	250	ASC	BBC Atlantic Relay Station
	500	UAE	R. of the UAE, Abu Dhabi
21640	250	ASC	BBC Atlantic Relay Station
	500/250	G	BBC World Sce, United Kingdom
21660	250	ASC	BBC Atlantic Relay Station
	250	THA	BBC East Asian Relay Station
	250/100	SNG	BBC Far Eastern Relay Station
21685		F	R. France Internationale
21695	500/100	D	Deutsche Welle, Germany
	250	CLN	Deutsche Welle, Trincomalee
21700	300/100	J	Radio Japan
	500/300	UAE	UAE Radio & TV, Dubai
21705	500/50	ARS	Broadc. Sce. of the Kingdom of Saudi Arabia
	100	SVK	R. Slovakia Int, Rimavská Sobota
	100	STP	Voice of America, São Tomé
21710	100	I	RAI Ext. Sce, Rome, Italy
21715	250/100	SNG	BBC Far Eastern Relay Station
	500	TUR	Voice of Turkey, Emirler
21720	300/100	POR	RDP Internacional (R. Portugal)
21725	500	NOR	R. Norway International/R. Denmark
21740	100	AUS	R. Australia, Shepparton
21765	500/100	AUT	R. Austria International
21770	100	GUM	AWR Asia, Guam
21850	500/100	CVA	Vatican Radio
22455	10	AFS	ZSC (Time Signals), So. Africa
22536		HKG	VPS22 (Time Sigs), Hong Kong
25700	100	HNG	R. Budapest, Diósd

INTERNATIONAL BROADCASTING SECTION

AFGHANISTAN

RADIO AFGHANISTAN
P.O. Box 544, Ansari Wat, Kabul. ☎ +93 25241. ① 24288 (AFGRTV AF) Kabul. **Cable:** RADIOTAF Kabul — **L.P:** Gen. Pres of Radio, TV & Afghan Film: Shamsul Haq Arayanfar. Pres. of Planning & Foreign Rel: Abdul Rahman. Tech. Advisor: Eng. Faizuddin Ferogh.
MW: Kabul 1278/1600kHz 100kW — **SW:** Yakatut 7200kHz 100kW.

English	Area	kHz
1645-1700	ME	1278, 1600, 7200
Urdu		
1630-1645	ME	1278, 1600, 7200

IS: A record on a national flute — **V.** by QSL-card. Rp.

ALASKA

KNLS – The New Life St. (Rlg.) Owned and operated by World Christian Broadc. Corp.
Box 473, Anchor Point, AK 99556, USA. ☎ +1 (904) 235-8262.
E-Mail: knls@aol.com **Web:** http://www.knls.org
Op's Center: P.O. Box 681706, Franklin, TN 37068. ☎ +1 (615) 371-8707. +1 (615) 371-8791. ① 6504393650.
L.P: Pres: Charles Caudill. SM & CE: Kevin K. Chambers. Exec. Producer: Dale Ward. Freq. Coordinator: F.M. Perry.
SW: Anchor Point (G.C: 151.44 W/59.45 N). 1 x 100kW tr.
kHz: *6150, 7355, 7365*

Chinese	Area	kHz	Russian	Area	kHz
1000-1100	CHN	7365	0900-1000	RUS	7365
1200-1300	CHN	7365	1100-1200	RUS	6150
1400-1700	CHN	7355	1700-1800	RUS	7355
English					
0800-0900	As	7365			
1300-1400	As	7365			

V. by QSL-card. Rec. acc.

ALBANIA

RADIO TIRANA
Rruga "Ismail Qemali" 11, Tirana. ☎ +355 (42) 23239. +355 (42) 23650 — **L.P:** Dir: Martin Leka.
MW: Flakkë 1215kHz 500kW, 1395kHz 1000kW, 1458kHz 500kW.
SW: Cërrik (G.C: 20.00E/41.00N): 4 x 100kW, 2 x 50kW — Shijak (G.C: 19.35E/41.21N): 1 x 100kW.
kHz: *5985, 6015, 6100, 6115, 6140, 6170, 6190, 7110, 7115, 7150, 7155, 7160, 7200, 7270, 7295, 9515, 9570, 9740*

Albanian	Area	kHz
2200-0500	Am	7270, 6190
0400-2300	Eu	6100 (rel. Home Sce. 1st Prgr.)
0630-0700	Eu	1215
0800-1100	Eu	7270, 1395
1000-1500	Eu	7150 (rel. Home Sce. 2nd Prgr.)
1500-1800	Eu	7270, 5985, 1215
1800-1830	Eu	1215
2000-2300	Eu	6170, 1458
English		
0245-0300	Am	7160, 6115
0330-0400	Am	7160, 6140
1945-2000	Eu	9570, 7270, 1458
2200-2230	Eu	9515, 7110, 1215
French		
1830-1900	Eu	9570, 7295, 1458
German		
1915-1945	Eu	9570, 7295, 1458
Greek		
1845-1900	Eu	7155, 5985, 1215
Italian		
1800-1830	Eu	7200, 6015, 1458
Serbian		
1730-1800	Eu	7110, 6015, 1458
Turkish	Area	kHz
1830-1845	Eu	7155, 5985, 1215

ANN: Albanian: "Ju flet Tirana". E: "This is R. Tirana".
V. by QSL-card.

TRANS WORLD RADIO (Rlg.)
B.P.349, M-98007 Monte Carlo. ☎ +377 165600. +377 165601.
Web: http://www.gospelcom.net/twr/twr_index.htm

L.P: Pres: Tom Lowell. MD: Werner Krömer. Mgr. Broadc. Dept: Beate Kiebel. Freq. Coordinator: Bernhard Schraut. Head of PR: Felix Widmer
MW/SW: as R. Tirana. (additional sces are listed under 'Monaco')
kHz: *7250, 7385, 9475, 9490, 9495, 9665, 9950, 9970, 11635, 11880, 12080, 12085*

Winter Schedule 1998/99

Armenian	Days	kHz	Kurdish	Days	kHz
1602-1617	67	7170	1544-1559	1234567	7155
1602-1617	67	9660	**Hungarian**		
1602-1632	12345	7170	0930-0945	23456	9490
1602-1632	12345	9660	0930-0945	23456	7385
Belorussian			**Polish**		
1915-1945	1	7355	0615-0630	123-567	7385
1915-1945	1	9485	1600-1630	1234567	7385
Farsi			0615-0630	1234567	9670
1700-1745	1234567	9665	1600-1630	1234567	9485
1700-1745	1234567	11690	**Russian**		
1745-1800	2345	9665	1900-1915	1234567	7355
1745-1800	2345	11690	1900-1915	1234567	9485
Kurdish			1915-1945	234567	7355
0600-0615	1234567	11635	1915-1945	234567	9485

VOICE OF AMERICA RELAY
MW: 1395 & 1458kHz (as R. Tirana).
1395kHz: 0430-0500 & 2300-2330 **Serbian**, 0530-0600 **Croatian**.
1458kHz: 0630-0700 & 2300-2330 **Serbian**.

ALGERIA

RADIO ALGIERS
21 Blvd. des Martyrs, Alger 16000. ☎ +213 (2) 590700. +213 (2) 605814. ① 65265 DZ — **L.P:** DG: Lamine Bechichi. Dir. Tec: Sadek Laskri. Dir. Coop: Abdelkader Lalmi.
LW: Tipaza 252kHz 750kW.
SW: Bouchaoui (G.C: 02.56E/36.44N): 100/50kW

HOME SERVICE RELAYS

kHz	kW	Ch.	Area	Times
7245	100	1/4	ME	1600-2300
9535	100	4	ME	1600-1900
9640	50	1/4	SAm	2200-0300
11715	50	3	CEu	1200-1700
15160	100	3	CEu	1000-1500, 1800-2000
15205	50	1/4	ME	1200-1600, 1700-2000
15215	100	1/4	SAm	2200-2400
17745	100	4	EAf	1500-1900

INTERNATIONAL SERVICE

English	Areas	kHz
1600-1700	Eu/ME/Af	15160, 11715
2000-2100	EuME/Af	15160, 11715, 252
Spanish		
1700-1800	Eu/ME/Af	15160, 11715
1900-2000	Eu/ME/Af	11715, 252

ANN: A: "Huna El Djazair, idha'atu-El Djoumhouriya El Djazairia". C: "al-Idha'atu al-Thaqafiyah". F: "Alger chaîne 3, Radiodiffusion Algerienne". E: "This is the Int. Sce. of R. Algiers" — **IS:** Oriental Lute (Ud) — **V.** by QSL-card.

ANGOLA

RADIO NACIONAL DE ANGOLA
C.P. 1329, Luanda, Angola. +244 (2) 391234.
MW: Cazenga 944kHz 1kW — **FM:** Luanda 104.1MHz.
SW: Cazenga 3355kHz 10kW, Mulenvos 9535kHz 100/50kW

English	Area	kHz
2000-2100	Af	3355, 9535, 944, 104.1MHz
French		
1900-2000	Af	3355, 9535, 944, 104.1MHz
Lingala		
2200-2300	Af	3355, 9535, 944, 104.1MHz
Portuguese		
2100-2200	Af	3355, 9535, 944, 104.1MHz

ANGUILLA

THE CARIBBEAN BEACON (Rlg.)
Box 690, Anguilla, BWI. ☎ +1 (264) 497 4340. +1 (264) 497 4311

Web: http://www.drgenescott.org/.
LP: Owner: Dr. Gene Scott. CE: Kevin Mooney.
MW: 690kHz 100kW day/50kW night, 1610kHz 200kW.
SW: 1 x 100kW.

kHz	Area	H. of tr.	kHz	Area	H. of tr.
6090	NAm	2300-1100	11775	NAm	1100-2300

FM: 100.1MHz 35kW (ERP) — **D.PRGR:** 24h — **V.** by QSL-card.

ANTIGUA

CARIBBEAN RELAY CO. LTD. (Deutsche Welle/BBC)
⌨ P.O. Box 1203, St. John's. ☎ +1 (268) 462 0994. 🖷 +1 (268) 462 0436.
E-mail: acm_crc@candw.ag or cm-crc@candw.ag
FM: 98.1MHz (rel. BBCWS and Caribbean Report).
SW: (G.C: 61.48W/17.06N): 4 x 250kW tr's.
DW kHz: 6040, 6185, 9640, 9700, 11810, 11945, 15105, 15205
BBC kHz: 5975, 6110, 6195, 15220, 17840
D.PRGR: see listings under United Kingdom and Germany.

ARGENTINA

RADIODIFUSION ARGENTINA AL EXTERIOR (RAE)
Operated by Servicio Oficial de Radiodifusión (SOR).
Studios: Maipú 555, 1006 Buenos Aires. ⌨ Casilla de Correo 555, 1000 Buenos Aires. ☎ +54 (1) 3259100. 🖷 +54 (1) 3256368 ☼ 21250 LRA AR.
E-mail: miglia1@gmtech.com
Web: http://www.primenet.com/~miglia/rae.htm
LP: Directora Nacional: Patricia Barral. Directora RAE: Perla Damuri.
SW: General Pacheco (G.C: 58.22W/34.36S): 1 x 100kW, 2 x 50kW.
kHz: 9690, 11710, 15345

English	Areas	kHz	Japanese	Areas	kHz
0200-0300	NAm	11710	1000-1200	FE	11710
1800-1900	Eu/Af	15345	**Portuguese**		
French			0000-0200	SAm	11710
0300-0400	NAm	11710	**Spanish**		
2000-2100	Eu/Af	15345	1200-1400	SAm	11710
German			2200-2400	Eu/Af	9690
2100-2200	Eu/Af	15345			15345
Italian					
1900-2000	Eu/Af	15345			

NB: Foreign Language Broadcasts from Mon to Fri only. On Sat & Sun relays LRA1 R. Nacional — **ANN:** English: "This is R.A.E., the International Sce. of the Argentine Radio" — **V.** by QSL-card.

ARMENIA

VOICE OF ARMENIA
⌨ 5 Alek Manoukyan, 3rd Floor, 375025 Yerevan. ☎ +374 (2) 552650. 🖷 +374 (2) 151947.
LP: DG: Dr. Levon V. Ananigyan. Dep. DG: (Armenian) Varujan Vosganyan. Dep. Head (European languages): Sayat Kuyumkuyan. Dep. Head (Asiatic languages): Penyamin Mindjian. Dep. Head (Slavic languages): Zara Ohanyan. Dir. Finance: Laura Baghdassarian. Head Eng: Dr. Vladimir Khazandjian.
MW: Kamo 864kHz 1000kW — **SW:** Yerevan (G.C: 44.40E/40.10N): 3 x 50kW, 2 x 100kW, Kamo (G.C: 45.10E/40.23N): 4 x 100, 3 x 1000kW.

Arabic	Area	kHz
1900-2000	ME	864, 4810, 4990, 7480
Armenian		
1630-1830 su	ME	864, 4810, 4990, 7480
1730-1830 wi	ME	864, 4810, 4990, 7480
2000-2200 wi	ME	864, 4810, 4990
2000-2100*	Eu/Am	7480, 9965
0200-0300	LAm	9965
Azerbaijani		
1500-1530* MF	ME	864, 4810, 7480
English		
0930-1000* Sun	Eu	15270
1845-1900 MF	ME	864, 4810, 4990, 7480
2130-2200*	Eu/Am	7480, 9965
French		
0900-1000* Sun	Eu	15270
2100-2130*	Eu/Am	7480, 9965
Georgian		
1530-1545*	ME	864, 4810, 7480
Kurdish		
1545-1630*	ME	864, 4810, 4990, 7480
Persian		
1630-1700*	ME	864, 4810, 4990, 7480
Russian		
1700-1730*	Eu	864, 4810, 4990, 7480

Spanish	Area	kHz
2200-2215*	Eu/Am	7480, 9965
0300-0315 Sun	LAm	9965
Turkish		
1830-1845 MF	ME	864, 4810, 4990, 7480
1830-1900 SS	ME	864, 4810, 4990, 7480

*) = one hour earlier during summer.
ANN: Armenian: "Yerevan e khosum" or "Yeterum Hayastani Dzaynn e". E: "This is Yerevan, the English Service of the Voice of Armenia". F: "Ici Erevan. Vous écoutez la Voix d'Armenie."
IS: "Spring" by Gomitas Vartabed, and "Dance of the Rose Maidens" from the Kayane ballet suite by Aram Khachaturian — **V.** by QSL-card

RELAY SERVICES
MW: Kamo 864kHz 1000kW, 1314 kHz 1000kW.
SW: Kamo: 4 x 100, 3 x 1000kW (cf. above).
Relays of V. of Russia, VOA, AWR. Cf. these stations for details.

RADIO AR-INTERCONTINENTAL
⌨ 28 Wartanantz, Apartment 34, Yerevan 375070.
SW: Kamo 9965kHz 1000kW.
D.PRGR: 1900-1915 (1930SS) religious prgr in **German**.

TRANS WORLD RADIO (Rlg.)
⌨ and other information: see 'Monaco'.
MW: Kamo 864kHz 1000kW.

English	Days	kHz	Kirmanji	Days	kHz
1810-1815	1234567	864	1846-1916	12345	864
Farsi			1846-1901	67	864
1815-1845	1234567	864	**Tajik**		
Kazak			0045-0100	12	864
0000-0015	12345	864	**Turkish**		
Kyrgysz			1901-1916	67	864
0015-0030	1234	864	**Uzbek**		
Korean			0045-0100	567	864
0030-0045	2345	864			

ASCENSION ISLAND

BBC ATLANTIC RELAY STATION
⌨ English Bay, Ascension Island, So. Atlantic.
SW (G.C: 14.23W/07.54S): 6 x 250kW tr's.
kHz: 5970, 6005, 6110, 6155, 7105, 7160, 9560, 9600, 9610, 9825, 11750, 11765, 11835, 11855, 11880, 15105, 15190, 15390, 15400, 17695, 17790, 17810, 17820, 17830, 17885, 17895, 21490, 21640, 21660
D.PRGR: see United Kingdom.

VOICE OF AMERICA RELAY STATION
SW: 2 x 250kW tr's operated by the BBC (see above).
D.PRGR: see USA.

AUSTRALIA

RADIO AUSTRALIA
Studios: 120 Southbank Boulevard, South Melbourne, Vic. 3205. ⌨ GPO Box 428G, Melbourne, Vic. 3001. ☎ +61 (3) 9626 1800 (management), +61 (3) 9626 1915 (Tech). 🖷 +61 (3) 9626 1899 (management), +61 (3) 9626 1917 (Tech). **Openline:** +61 (3) 9626 1825.
E-mail: raelp@radioaus.abc.net.au **Web:** http://www.abc.net.au/ra/
LP: GM: Jean-Gabriel Manguy. PD: Tony Hastings. CE: vacant at editorial deadline. Publicity: Caroline Bilney. Correspondence: Lorraine D'Cruz.
SW: Shepparton, Vic. (G.C: 145.25E/36.20S): 7 x 100kW — Brandon (G.C: 147.20E/19.31S): 2 x 10kW.
kHz: 5890, 5995, 6080, 7240, 9415, 9435, 9580, 9660, 9710, 9770, 11660, 11695, 11880, 12080, 13605, 13755, 15240, 15415, 15510, 17750, 17795, 21525, 21740

Chinese	Area	kHz
0400-0500MF	EAs	17750
0400-0500	EAs	15415
1100-1330	EAs	9435, 11880
2200-2300	EAs	15240
English		
0000-0600	Pac	17795
0000-0800	PNG	9660, 13605
0000-0400	As	17750
0000-0900	Pac	12080, 15510
0100-0400	As	15415
0100-0800	PNG/Pac	15240
0600-0830	Pac/NAm	11880
	As	15415, 17750, 21525
0800-0900	PNG	9710
0800-1400	Pac/NAm	9580
0830-1200	PNG	6080

English	Area	kHz
0900-1100	As	11880
1200-1800	PNG/Pac/NAm	5890
	As	6080
1200-2130	PNG/Pac/NAm	9415
1330-1700	As	11660
1430-2200	As	9435
1700-2130	Pac/NAm	11880
1800-2000	PNG	6080
	PNG/Pac	7240
2100-2200	PNG/Pac	7240
2100-0000	PNG	9660 (BBC 2200-2300)
	Pac	12080 (BBC 2200-2300)
	Pac	17795, 21740
2130-0000	Pac/NAm	13755
2200-0000	Pac	15510
Indonesian		
0830-1100	SEAs	9770
2130-0000	SEAs	15415
Khmer		
0500-0600	SEAs	15415, 17750MF
2300-0000	SEAs	15240
Tok Pisin		
0900-1200	PNG/Pac	5890, 5995, 9710, 12080
2000-2100	PNG/Pac	6080, 7240
Vietnamese		
0000-0100	SEAs	15415
1330-1430	SEAs	9435
2300-0000	SEAs	11695

ANN: "This is Radio Australia broadcasting from studios in Melbourne, Victoria." — **IS:** "Waltzing Matilda" prior to opening on all freqs. Foreign language broadcasts start with the laugh of the Kookaburra.
PUB: English & language prgr. guides available on request. All free.
V. by QSL-card.

RADIO AUSTRIA INTERNATIONAL

A-1136 Vienna. ☎ +43 (1) 87878 3636 or 2130. 🖷 +43 (1) 87878-4404.
Web: http://www.orf.at/roi **E-mail:** roi.service@orf.at (reception reports, letters, schedule requests), hfbc@orf.at (technical mail), roi.intermedia@orf.at (mail to Intermedia programme *only*)
L.P: Dir: Prof. Paul Lendvai. Dep. Dir: Dr. Edgar Sterbenz. Chief Eng: Ing. Klaus Hollndonner. Head of German Dept: Helmut Blechner. Head of English Dept: David Ward. Head of French Dept: Robert Denis. Head of Spanish Dept: Jacobo Naar-Carbonell. Internet Sce: Marianne Veit & Oswald Klotz. Listeners' Sce: Mag. Vera Bock and Christine Soucek
Satellite: Radio Austria International with all its language prgrs can be heard in Europe via Astra Digital Radio (ADR) 24h a day under the code ROI WIEN. The ROI Report from Austria is available via WRN on the following satellites:
Europe: Astra 1B (11.538GHz V, audio subcarrier 7.38MHz).
No. America: Galaxy 5 (3.820GHz V, audio subcarrier 6.8MHz).
Near East/Africa: Intelsat 707 (3.9115GHz RHC, SR 8.022, FEC 3/4).
Asia/Oceania: AsiaSat 2 (4.000GHz H, SR 28.125, FEC 3/4).
The ROI newscasts and some orgrs in German are transmitted by World Radio Network on the following satellites:
Europe: Astra 1C (10.964GHz H, subcarrier 7.56MHz.channel B).
Europe: Astra 1A (11.436GHz V, audio subcarrier 7.38MHz).
North America: Galaxy 5 (3.820GHz V, 6.2MHz).
MW: Wien/Bisamberg 1476kHz 60kW (2200-2300).
SW: Moosbrunn (G.C: 16.28E/48.00N): 500/300/100kW.
C=rel. via Sackville, Canada.
kHz: *5945, 6015, 6155, 7325, 9495, 9655, 9870, 13710, 13730, 15410, 15455, 17870*

Winter Schedule 1998/99

Arabic	Area	kHz
0505-0530Sun	Eu, ME	6155, 13730, 15410, 17870
1805-1830Sun	ME	9655, 13730
German		
0000-0030Thu/Sun	NAm/SAm	7325, 9495, 9870
0100-0130	NAm/SAm	7325, 9495, 9870
0200-0300	NAm/CAm/SAm	7325, 9495, 9870
0300-0330	CAm/SAm	9495, 9870
0400-0430Mon/Sat	Eu	6155, 13730
0430-0500	Eu	6155, 13730
0500-0530	NAm	6015C
0050-0530Mon/Sat	Eu/Pac	6155, 13730, 15410, 17870
0600-0630	Eu/NAm/ME	6015C, 6155, 13730, 15410, 17870
0630-0730	Eu/ME	6155, 13730, 15410, 17870

Arabic	Area	kHz
0800-0830	Eu/Pac/FE	6155, 13730, 17615, 21765
0900-0930	Eu/FE/Pac	6155, 13730, 17615, 21765
0930-1000	Eu	6155, 13730
1000-1100	Eu/FE/Pac	6155, 13730, 17615, 21765
1100-1230	Eu/NAm	6155, 13730
1300-1330	Eu/NAm	6155, 13730
1400-1430	Eu/WAf	6155, 13730
1500-1530Mon/Sat	Eu/Waf/ME/SAs/SEAs	6155, 9655, 13710, 13730
1530-1730	Eu/Waf/ME/SAs/SEAs	6155, 9655, 13710, 13730
1800-1830	Eu	5945, 6155
1800-1830Mon/Sat	ME	9655, 13730
1900-2130	Eu/SAf	5945, 6155, 13730
2200-2230	Eu/SAf	5945, 6155, 13730
2300-2330	Eu/SAm	6155, 5945, 9870
English		
0030-0100	NAm	7325
0130-0200	NAm/SAm	7325, 9495, 9870
0530-0600	Eu/ME/NAm	6015C, 6155, 13730, 15410, 17870
0630-0700	NAm	6015C
0830-0900	Eu/FE/Pac	6155, 13730, 17615, 21765
0930-1000	FE/Pac	17615, 21765
1330-1400	Eu/NAm	6155, 13730
1730-1800	Eu/ME/SAs/SEAs, Waf	6155, 9655, 13710, 13730
2230-2300	Eu/SAf	5945, 6155, 13730
Esperanto		
0405-0430Sun	Eu	6155, 13730
1505-1530Sun	Eu/Waf/ME/SAs/SEAs	6155, 9655, 13710, 13730
0005-0030Mon	NAm/SAm	7325, 9495, 9870
French		
0730-0800	Eu, ME	6155, 13730, 15410, 17870
1230-1300	Eu, NAm	6155, 13730
1830-1900	Eu, ME	5945, 6155, 9655, 13730
2330-2400	Eu	6155, 5945
Spanish		
0030-0100	SAm	9495, 9870
0330-0400	CAm/SAm	9495, 9870
1430-1500	Eu/WAf	6155, 13730
2130-2200	Eu/SAf	5945, 6155, 13730
2330-2400	SAm	9870

Via Satellite:
ROI can be heard in Europe via ASTRA 1G (DVB-Audio) 19°E, 12.692MHz, horizontal & Astra 1C, 19°E, 11.141MHz, horizontal, 8.10MHz on ASTRA DIGITAL RADIO (ADR) under the code ROI WIEN.

IS: First bars of the Blue Danube — **V.** by QSL-card. Rec. acc.

VOICE OF AZERBAIJAN

1 M. Hüseyin Street, 370011 Baku. ☎ +994 (12) 927851 (Editor-in-Chief), +994 (12) 392885 (Turkish sce.). 🖷 +994 (12) 398505.
L.P: Editor-in-Chief: Cahanqik Baqirov. Dep. Dir: Mehman Valiyev.
Mediumwave: Pirsaqat 1296kHz 125kW.
Shortwave: Gäncä (Gyandzha) 6110kHz 170kW.
Prgr's 1 h earlier during summer.

Arabic	Area	kHz
1700-1800	ME	1296, 6110
Azerbaijani		
0315-0400	ME	1296, 6110
1300-1500	ME	1296, 6110
English		
1800-1830	ME	1296, 6110
French		
1730-1800	ME	On AZR2 (see domestic section)
German		
1630-1700	ME	On AZR2 (see domestic section)
Persian		
1500-1600	ME	1296, 6110
Russian		
1830-1900	ME	1296, 6110

Turkish	Area	kHz
1200-1300	ME	1296, 6110

ANN: Azerbaijani: "Danişir Baki" — **V.** by QSL-card.

BANGLADESH

BANGLADESH BETAR (Radio Bangladesh)
✉ P.O. Box 2204, Dhaka. ☎ +880 (2) 503688.
LP: Dir: Mrs. Dilruba Begum.
SW: Khabirpur (G.C: 90.26E/23.43N): 2 x 250kW tr's.
kHz: *7185, 9548, 9568, 15520*

Arabic	Area	kHz	Hindi	Area	kHz
1600-1630	ME	9568, 7185	1515-1545	As	9548, 7185
Bengali			**Nepali**		
1630-1730	ME	9568, 7185	1315-1345	As	9548, 7185
1915-2000	Eu	9568, 7185	**Urdu**		
English			1400-1430	As	9548, 7185
1230-1300	As	9548, 7185	**Voice of Islam**		
1815-1900	Eu	15520, 9568, 7185	1745-1815	Eu	15520, 9568 7185

ANN: E: "This is the External Service of Bangladesh Betar".
IS: Local composition of violin and tanpura — **V.** by letter. Re. to Senior Eng. (Research Wing), NBA House, 121 Kazi Nazrul Islam Ave, Dhaka-1000.

BELARUS

RADIO BELARUS (RADIO MINSK)
✉ ul. Krasnaja 4, 220807 Minsk. ☎ +375 (172) 395875. 🖷 +375 (172) 2366643. **Web:** http://www.nestor.minsk.by/radiod/
L.P: Editor-in-Chief: Micjail Tondel.
MW: Sasnovy 1170kHz 1000kW.
SW: Kalodziscy (G.C: 53.58N/27.47E): 7210kHz 75kW ND, 7105kHz (wi) or 11960kHz (su) 250kW 252Þ + relays via Kiev (Kv) (100kW) in the Ukraine.

Belarusian	Days	Area	kHz
1900-1930	1-7	Eu	1170, 7105, 7210 + Kiev
1930-2000	1245	Eu	1170, 7105, 7210 + Kiev
2000-2030	1-7	Eu	1170, 7105, 7210
2030-2100	1245	Eu	1170, 7105, 7210
German			
1930-2000	367	Eu	1170, 7105, 7210 + Kiev
2030-2100	367	Eu	1170, 7105, 7210

ANN: Belarusian: "Havoryc Minsk" — **V.** by QSL-card.

VOICE OF RUSSIA RELAY
SW: Minsk (Kalodziscy, G.C: 53.57N/27.48E) 1 x 250kW. Transmissions included in V. of Russia schedule.

BELGIUM

RADIO VLAANDEREN INTERNATIONAAL
✉ P.O. Box 26, B-1000 Brussels. ☎ +32 (2) 741 3802. 🖷 +32 (2) 732 6295. **Cable:** BRT Brussels. ① 22486 BRTBIN B. **E-mail:** rvi@brtn.be
Web: http://www.brtn.be/rvi/
L.P: Dir: J. Vandersichel. Freq. Mgr: Willy Devos.
Satellite: Astra 1D transp. 63
MW: Wolvertem 1512 kHz 300kW (0600-0730/0830* & 1800-2330). 20kW (0870/0830*-1800). Rel. European sces. *) 0830 till Jan 31st, 0730 Feb 1st onwards.
SW: Wavre (G.C: 04.35E/50.45N): ORU1/ORU2: 250kW, ORU8: 100kW B=via Bonaire, Netherlands Antilles, J = via Jülich, Germany; M=via Madagascar, N = via Nauen, Germany, P=via Petropavlovsk-Kamchatskiy, T=via Tashkent
kHz: 5910, 5960, 5985, 6070, 9925, 9940, 9865, 11675, 11750, 17690, 11805, 12080, 13650, 13670, 13745, 15150, 15195, 15250, 15370, 17690, 21510

NB: All transmissions will be one hour *earlier* March 29th-Oct. 25th.

Winter Schedule 1998/99

Dutch	Area	kHz
0430-0456	WNAm	11750
0600-0656	CAf/SEEu	13745
0600-0656	CAf	15195
0700-0826	SEu	5985
0700-0826	SWEu	9925
1000-1026	SAf	17690
1030-1130Sun	CAf	17690
1030-1130Sun	CAf	21510
1200-1226	SAm	6070
1200-1226	SAm	15250
1200-1226	EAs/AUS	9865

Dutch	Area	kHz
1200-1226	SEAs/AUS	17690
1200-1256	SEu	5985
1200-1256	SWEu	13745
1300-1326	WAf	17690
1400-1700Sun	SEu	9925
1400-1700Sun	SWEu	13745
1400-1700Sun	CAf	21510
1700-1730	SEu	5910
1700-1730	SWEu	9925
1700-1730	EEu	12080
1800-1830	CAf	13745
1800-1830	CAf	15370
1900-1956	Eu	5960
1900-1956	CAf	9925
1900-1956	ME	11805
1900-1956	CAf	13745
1900-1956Sat	Eu	5960
1900-1956Sat	ME	11805
1900-1956Sat	SEu	5910
1900-1956Sat	SWEu	9925
1900-1956Sat	CAf	13745
2000-2056	SEu	5910
2000-2056	SWEu	9925
2000-2100Sat	SEu	5910
2000-2100Sat	SWEu	9925
2000-2100Sat	CAf	11675
2100-2200Sat	SEu	5910
2100-2200Sat	SWEu	9925
2100-2200Sat	CAf	11675
2300-2326	ENAm	13670
English		
0830-0856	EEu	9925
0830-0856	SWEu/AUS	9940
1130-1156	NEu	9925
1130-1156	SEEu	13745
1730-1756	SEu	5910
1730-1756	EEu	12080
1730-1756	ME	13650
1830-1856	CAf	13745
2230-2256	ENAm	13670
French		
0930-0956MF	NAf	15150
0930-0956MF	CAf	21510
2130-2156Fri	ENAm	9925
German		
0900-0926MF	SEu	5985
0900-0926MF	EEu	9925

ANN: "Dit is Radio Vlaanderen Internationaal" (in all languages). E: "Brussels Calling"; F: "Ici Bruxelles"; D: "Hier ist Brüssel".
IS: Starting: "Tussen Maas en Schelde".
V. by QSL-card — **PUB:** Detailed schedule (Free).

DEUTSCHE WELLE RELAY: 1500-1700 on 1512kHz.
RADIO NETHERLANDS RELAY: 2130-2330 on 1512kHz
See listings under Germany and Netherlands respectively.

BELIZE

VOICE OF AMERICA RELAY STATION BELIZE
✉ Unit 7407, APO AA 34025, USA. ☎ +501 (7) 22091 🖷 +501 (7) 22147 — **L.P:** SM: Glenn Britt (**E-mail:** glenn_britt@beng.voa.gov)
MW: 1530/1580kHz 100kW.
D.PRGR: 0100-0400 (SS 0500). See VOA schedule under "USA". Airtime available for lease at 0400-1200 UTC. Contact St. Mgr.

Winter Schedule 1998/99

English		
0200-0400Su/Mo	1530, 1580
0200-0500Tu-Sa	1530, 1580
Spanish		
2300-2400Su/Mo	1530, 1580
2300-2400Tu-Sa	1530, 1580

BRAZIL

RÁDIO NACIONAL DO BRASIL (RADIOBRÁS)
✉ C.P. 070.747, 70359-970 Brasília. ☎ +55 (61) 321 3949. 🖷 +55 (61) 321 7602. ① 061-1682. **E-mail:** maurilio@radiobras.gov.br
Web: http://www.radiobras.gov.br/
L.P: Pres: Maurílio Ferreira Lima. Dir. Radio: Iolando Lourenço. Dir. News: Miriam Moura. Dir. Adm: Januário Procópio. Dir. Tec: Toshihiro Kanagae. Dir. SW: Renato Geraldo de Lima.
SW: Brasilia (G.C: 47.56W/15.51S): 250kW.

kHz: *9745, 11765, 11780, 15265, 15445, 17750*

English	Area	kHz	Portuguese	Area	kHz
1200-1320	NAm	15445	0700-0800	Af	9745
1800-1920	Eu	15265	1630-1750	Eu	15265
German			1800-1920W	Af	17750
1930-2050	Eu	15265	**Spanish**		
Portuguese			1000-1120	SAm	9745
0115-0215	NAm	11780	1330-1450	SAm	15445
0415-0515	NAm	11765			

V. by QSL-card.

BULGARIA

RADIO BULGARIA
✉ 4, Dragan Tsankov Blvd., 1040 Sofia.
☎ +359 2 854 733, 🖷 +359 2 650560
LP: Dir. Atanas Tzenov. Editor-in-Chief: Spassianna Rousseva. Freq. Mgr:Ivo Ivanov.
MW: Petritch (G.C: 23.18E/41.42N): 747 500kW - Vidin (G.C: 22.40E/43.49N): 1224 500kW
SW: Plovdiv (G.C: 24.42EE/42.10N): 2x500 kW, 3x250 kW - Sofia (G.C: 23.13E/42.49N): 1x150kW, 2x100kW, 2x50kW, 1x15kW.

kHz: *5825, 5850, 5865, 5890, 5900, 5935, 5940, 6140, 7375, 7380, 7390, 7400, 7425, 7465, 7495, 7500, 7525, 7535, 7545, 9385, 9400, 9415, 9485, 9500, 9700, 9775, 11555, 11660, 11700, 11720, 11800, 11825, 11900, 12000, 13600, 13700, 13710, 13715, 13740, 13770, 13790, 13800, 15175, 15315, 15675, 15700, 17500, 17650.*

Winter Schedule 1998/99

Albanian	Area	kHz
0530-0600MF	Balkans	1224, 5940
0700-0800SS	Balkans	1224, 5940
1645-1730	Balkans	1224, 5865
2000-2045	Balkans	747, 1224
Bulgarian		
0100-0200	SAm	5865, 9415
0100-0200	NAm	7375, 9485
0400-0500	Balkans	1224
0400-0500	EEur	5890, 7390
0400-2100	WEu	9400, 9700, 11555
1300-1600	Balkans(H)	1224, 5865
1300-1600	WEur	12000
1600-1700	ME	7500
1600-1700	SAm	17500
1600-1700	EEur	5935, 7425
1730-1815	Balkans	747
1900-2000	Balkans	747, 1224
1900-2200	WEur	5865, 7545
1900-2200	ME	7500

(H) Home Sce. "Horizont" prgr.

English	Area	kHz
0000-0100	NAm	7375, 9485
0300-0400	NAm	7375, 9485
1200-1300	WEu	15700, 17500
2000-2100	WEu	5850, 7535
2200-2300	WEu	7535, 7545
French		
0200-0300	NAm	7375, 9485
0700-0800	WEu	7500, 9500
1800-1900	WEu	7535, 7545
2100-2200	WEu	5850, 7535
German		
0615-0700	WEu	7500, 9500
1100-1200	WEu	15700, 17500
1715-1800	WEu	7535, 7545
1900-2000	WEu	5850, 7535
Greek		
0500-0530MF	Balkans	1224, 5940
0500-0600SS	Balkans	1224, 5940
1730-1815	Balkans	1224, 5865
2045-2130	Balkans	747, 1224
Russian		
0300-0400	EaEUR	5890, 7390
1500-1600	EaEUR	5935, 7425
1500-1600	CeAS	9385
1900-2000	EaEUR	5935, 7425
Serbian		
0600-0700MF	Balkans	1224, 5940
0800-0900SS	Balkans	1224, 5940
1600-1645	Balkans	1224, 5865
2130-2215	Balkans	747, 1224
Spanish		
0000-0100	SoAM	5865, 9415
0200-0300	SoAM	5865, 9415

Spanish	Area	kHz
0200-0300	CeAM	7500
1700-1800	SoEUR	11900, 13800
2215-2315	SoEUR	5865, 7500
Turkish		
0515-0600MF	Balkans	ME	5900, 7390
0600-0700SS	Balkans	ME	1224 5900, 7390
1815-1900	Balkans	ME	747, 1224 5865, 7500

ANN: Albanian:"Ju flet Radio Bullgaria". Bulgarian:"Tuk e Radio Balgaria". English:"This is Radio Bulgaria". French:"Ici Radio Bulgarie". German:"Hier spricht Radio Bulgarien". Greek:"Akute to Radio Vulgaria". Russian:"V efire Radio Balgaria". Serbian:"Radio Bugarska". Spanish:"Esta es Radio Bulgaria". Turkish:"Burasi Bulgaristan Radiosu".

IS: Starting: The first music phrase from "Bulgarian Suite" for orchestra by Pancho Vladiguerov-12 sec. - **PUB:** Detailed schedule (free) **V.** by QSL-card Rp (1IRC) Rec acc.

CAMBODIA

NATIONAL VOICE OF CAMBODIA
✉ 106, Street Preah Kossamak, Phnomh Penh. ☎ +855 (2) 3369 or 2869. 🖷 +855 (23) 27319.
MW: Phnom Penh 1360kHz 1kW
SW: Phnom Penh (G.C: 104.51E/11.34N): 11940vkHz 50kW.

English	Area	kHz
0000-0015	As	11940v, 1360
1200-1215	As	11940v, 1360
French		
0015-0030	As	11940v, 1360
1215-1230	As	11940v, 1360
Laotian		
0045-0100	As	11940v, 1360
1245-1300	As	11940v, 1360
Thai		
0030-0045	As	11940v, 1360
1230-1245	As	11940v, 1360
Vietnamese		
0100-0115	As	11940v, 1360
1300-1315	As	11940v, 1360

V. by letter.

CANADA

RADIO CANADA INTERNATIONAL
(Canadian Broadcasting Corporation/Société Radio Canada) (Publicly owned)
✉ P.O.Box 6000, Montreal, Canada H3C 3A8. ☎ +1 (514) 597 7555. 🖷 +1 (514) 284 0891. ☉ 0526417. **E-mail:** rci@montreal.src.ca
Web: http://www.rcinet.ca
LP: Exec. Dir: Bob O'Reilly. Prgr. Dir: Allan Familiant. Dir. of Eng: Jacques Bouliane. Head of Pub. Rel: Ousseynou Diop. N .Editor: Gordon MacDougall.

STATION: Sackville, New Brunswick (G.C: 64.19W/45.53N): 6 x 250kW, 3 x 100kW tr's. **Relays:** K=Kimje (Korea), V=Vienna (Austria), S=Sines (Portugal), U=Skelton (UK), W=Wertachtal (Germany), X=X'ian (China), Y=Yamata (Japan).
kHz: *5925, 5960, 5995, 6025, 6030, 6040, 6050, 6140, 6150, 6155, 6185, 7150, 7200, 7210, 7235, 7260, 9505, 9535, 9555, 9615, 9640, 9645, 9650, 9735, 9740, 9755, 9760, 9780, 9805, 11705, 11715, 11730, 11855, 11865, 11905, 11915, 11935, 11945, 11955, 13650, 13690, 15150, 15325, 15425, 17820*

Winter Schedule 1998/99

Arabic	Area	kHz
0330-0359	ME	6025W, 6150S, 9505V, 9615W
0430-0459	ME	7200W, 9505W, 11715V
2146-2215	ME	1233C
Chinese		
1300-1325	CHN	6150Y, 7260K, 11955Y
1430-1455	CHN	6030K, 6150Y, 7235Y
English		
0000-0029	CAm/SAm/NAm	6040, 11855
(Tu-Sa)	CAm/SAm	9535
0000-0059	Carib/NAm	5960
(Su&Mo)	Carib/NAm	9755
0200-0259	CAm/SAm/NAm	6155, 9535, 9755, 9780, 11865
0300-0329	MEX/NAm	6155

English	Area	kHz
(TUE-SAT)	MEX/NAm	9755, 9780
0300-0359	Carib/NAm	6155
(Su&Mo)	Carib/NAm	9755, 9780
0400-0429	ME	6150S, 9505V, 9645U
0600-0629	Af	9740U
(Mo-Fr)	Af/Eu/ME	6090, 6150U, 9760, 11905U
1200-1229	As	6150Y, 11730Y
1300-1359	Carib/NAm	9640, 13650, 17715
1330-1357	EAs	6150X, 9535X
1400-1459	Carib/NAm	9640
(Mo-Fr)	Carib/NAm	13650, 17715
1400-1659:30	Carib/NAm	9640, 17715
(Su)	Carib/NAm	13650
1430-1459	Eu/Af/ME	9555U, 11915S, 15325S
1630-1657	IND	6140X, 7150X
2100-2159	Eu/Af	5995U, 7235U, 9725, 9805U, 11945, 13650, 13690, 15150, 17820
2200-2229	Eu/Af	5995, 7235U, 9735U, 9805, 11945,13690,15150
	SEAs	11705Y
2300-2329	CAm/SAm/NAm	5960, 6040, 9535, 9755, 11865
2330-2359	CAm/SAm/NAm	5960, 6040, 9755, 11865
(SS)	CAm/SAm	9535
2330-0059	Carib/NAm	5960
(Mo-Fr)	Carib/NAm	9755

French

0100-0159	ANT/Nam/SAm	9535&, 9755&, 11865
0300-0329	ME	6025U, 9505V
0330-0359	MEX/NAm	6155, 9780
(Tue-Sat)	MEX/NAm	9755
0630-0659MF	Eu/Af/ME	6090, 6150U, 9740U, 9760, 11906U
1230-1259	As	6150Y, 11730Y
1300-1359	ANT/NAm	11855, 15425
1400-1659:30		
(Sun)	ANT/NAm	17875
1500-1559	Eu/Af/ME	9555U, 11915S, 11935U, 17820
2000-2059	Eu/Af	5995U, 7235U, 9805U, 13650, 13690, 15150, 17820
2229:30-2259	Eu/Af	5995, 7235U, 9735U, 11945, 13690
	SEAs	11705Y
	ANT/NAm	5960, 9755

Spanish

0000-0059	CAm/LAm	6040, 11855
(Su&Mo)	CAm/LAm	9535
0030-0059	CAm/LAm	6040, 11865
(Tu-Sa)	CAm/LAm	9535
0130-0159	CAm/LAm	9535
(Tu-Sa)	CAm/LAm	11865
2330-2359	Carib/LAm	6040, 11865
(Mo-Fr)	Carib/LAm	9535, 11865

Russian

1600-1659	RUS	6150K, 6185Y
	CEu/EEu	9555U, 11935U, 17820#
1800-1859	CEu/Eeu	7235U, 11935U, 13650, 15325, 17820

Ukrainian

1700-1759	CEu/EEu	9555U, 11935U, 15325, 17820
1900-1959	CEu/EEu	7235U, 11935U, 11945, 13650, 15325, 17820

ANN: E: "This is Radio Canada International". F: "Ici Radio Canada International" — **IS:** first bar of Canadian National Anthem. **V:** by QSL-card. Re. handled by Canadian International DX Club. **PUB:** Radio Canada International programme schedule, published twice a year, is sent free upon request. Records (spoken-word or music), topical discs or tapes can be provided on request to broadcast organizations.

CHINA

CHINA RADIO INTERNATIONAL (CRI)
(Zhongguo Guoji Guangbo Diantai)

Jia 16, Shijingshan Lu, Shijingshan Qu, Beijing 100039. ☎ +86 (10) 6889 1001. CE's Office: +86 (10) 6889 1533. Audience Rel. +86 (10) 6889 1149. Foreign Affairs Office: +86 (10) 6889 1110.
English Sce: ☎ +86 (10) 6851 3135, 6609 2274, 6609 2760. 📠 +86 (10) 6851 3174. **E-mail:** crieng@public.bta.net.cn — Dir: Xia Jixuan.
L.P: Gen. Dir: Zhang Zhenhua. CE: Yu Jikai.
Mediumwave key freqs (kHz): 1017(JL), 1044(JS), 1080/1188/1206/

1269/1296 (YN), 1116(HL), 1314(NM), 1323(JL/XZ), 1341(GD), 1521(XJ).
Shortwave: Baoding, HEB (G.C: 115.44E/38.39N) 240/120kW – Beijing, BJ (G.C:116.27E/39.57N) 240/120/50kW – Jinhua, ZJ (G.C:119.39E/28.07N) 500kW – Kunming, YN 120/50kW – Shijiazhuang, HEB (G.C:114.28E/38.04N) 500kW – Urumqi, XJ (G.C: 87.30E/43.35N): 500/100kW – Xi'an, SN (G.C:108.54E/34.12N) 300/150kW. Further trs in border areas are used at times.
Overseas relay st's: CRI exchanges transmitting facilities with R. Canada Int, R. France Int, R. Exterior de Espana, Swiss Radio Int and V. of Russia, and also uses relay facilities in Mali.
Relays: C=Canada, E=Spain, F=France, G=Fr. Guiana, M=Mali, R=Russia, S=Switzerland.

Albanian	Area	kHz
1900-1927	Eu	9965, 7385
1930-1957	Eu	9965, 7405
2100-2127	Eu	6150F
Amoy		
1400-1457	SEAs	11685, 11650, 7335
Arabic		
1600-1657	ME	15540R, 9760, 9440
1830-1927	ME	12035R, 9770, 7480
2100-2157	ME	11515, 9440, 7250
Bengali		
1500-1527	SAs	11825, 8660, 7335, 1206
1530-1557	SAs	11825, 8660, 7335, 1206
Bulgarian		
1830-1857	Eu	9860, 7385, 7255, 6933
2030-2057	Eu	9860, 6150F
Burmese		
1130-1157	SEAs	11825, 9880, 8260, 1269, 1188
1300-1327	SEAs	11780, 9880, 8260, 1269
1330-1357	SEAs	11780, 9880, 8260, 1269
Cambodian		
1030-1127	SEAs	9870, 9440, 8260, 1080
1200-1257	SEAs	11825, 9870, 8260, 1080
1400-1457	SEAs	9440, 8260
Cantonese		
0100-0157	NAm/SAm	11695M, 9710M
1000-1057	SPac	11915, 11650
1100-1157	SEAs	12015, 11945, 11685
1200-1257	NAm	11965
1700-1757	SAs/SEAf	9900, 7180
1900-1957	Eu	11945, 7780
2330-2357	SEAs	15400, 12065, 12015, 11685, 9870, 9457, 9440, 8260, 7335, 7190, 6140, 5250
Chaozhou		
0030-0057	SEAs	15400, 12065, 12015, 11685, 9870, 9457, 9440, 8260, 7335, 7190, 6140
Chinese		
0200-0257	SAm/NAm	15435, 11695M, 9710M, 9690E
0300-0357	NAm	9730G
0400-0457	NAm	11695M, 9710M
0900-0957	NEAs/SPac	15180, 12015, 11945, 11840, 11685, 11650, 11500, 9945, 9480, 6590, 6010, 5250
1000-1057	NEAs	12015, 11945, 11840, 11685, 6590, 6010, 5250
1200-1357	NEAs/SPac	15260, 11945, 11685
1500-1557	SAs/EAf	11980, 9457, 8260, 7110
1730-1827	WAs/NAf/Eu	9820, 9670, 7800, 7335, 7110
2000-2027	Eu	6165S
2000-2057	WAs/NAf/Eu	11650, 11515, 9730, 9710, 7660, 7185
2230-2257	NAm	11790M, 9770M, 9535
2230-2327	NEAs/SEAs	15400, 12065, 12015, 11685, 9870, 9457, 9440, 8260, 7335, 7190, 7170U, 6140, 5250
Czech		
1900-1927	Eu	7375, 7255, 6933
1930-1957	Eu	7375, 7305F
English		
0000-0057	NAm	11695M, 9710M
0300-0357	NAm	11695M, 9710M, 9690E
0400-0457	NAm	9730G
0500-0557	NAm	9560C
0900-0957	SPac	15440, 11755
1000-1057	SPac	15440, 11755
1200-1257	SEAs	11660, 9715, 1341
	SPac	11795, 7385
1300-1357	SEAs	11660, 9715, 1341
	SPac	7385
1400-1457	WNAm	7405
	SAs	9785, 9535, 7405
1500-1557	WNAm	7405
	SAs	9785, 9535, 7405
1600-1657	EAf/SAf	15130M, 15110M

English	Area	kHz
1700-1757	EAf/SAf	9570, 7405, 7150
1900-1957	WAf/NAf	9440, 6955
2000-2057	EAf/SAf	15110M, 11715M, 9440
	WAf/NAf	9440, 6955
	Eu	9920, 6950
2100-2127	EAf/SAf	15110M, 11715M
2100-2157	Eu	9920, 6950
2200-2227	Eu	3985S
2200-2257	Eu	7170R

Esperanto

1100-1127	NEAs	11575, 9570, 7170
1300-1327	SEAs	11840, 11600
2000-2027	Eu	9965, 7405, 4960
2230-2257	SAm	9860, 6950

Filipino/English

1130-1157	EAs	15135, 11700, 8660, 1341
1200-1227	EAs	12110, 11700, 8660
1430-1457	EAs	11445, 9880, 8660, 1341

French

1830-1927	Af/Eu	9670, 7800, 7335, 7110, 5250
1930-2027	Af/Eu	7800, 7335, 7110, 5250
2030-2057	Eu	3985S
2030-2127	Af/Eu	9890R, 9820, 7800, 7335, 7110, 5250
2130-2227	Af/Eu	11790M, 9820, 9770M, 7800, 7335, 7175U, 7110, 5250

German

1800-1857	Eu	9920, 6950, 5220
1900-1957	Eu	9920, 9900, 6950, 5220
2000-2027	Eu	3985S

Hakka

0000-0027	SEAs	15400, 12065, 12015, 11685, 9870, 9457, 9440, 8260, 7335, 7190, 6140
1600-1657	SEAf	11825, 9900

Hausa

1730-1757	Af	11790M, 9770M, 7235
1800-1827	Af	11790M, 9770M, 7405, 7235, 4960

Hindi

1500-1557	SAs	11675, 9920, 7590, 1323
1600-1657	SAs	11675, 9920, 8660, 1269

Hungarian

1800-1827	Eu	9900, 9860, 7385, 7255, 6933
2030-2057	Eu	9365, 6933
2130-2157	Eu	6150F

Indonesian

0830-0927	SEAs	17680, 15600, 15135, 8660
1030-1127	SEAs	15135, 11445, 8660
1330-1427	SEAs	15135, 11445, 8660, 1341

Italian

1830-1857	Eu	9965, 9900, 7405
2030-2057	Eu	9965, 7385
2100-2127	Eu	9965, 9365
2130-2157	Eu	3985S

Japanese

0930-1527	NEAs	9855, 7190, 1044

Korean

1100-1257	NEAs	5965, 1017
1300-1457	NEAs	5965, 1017

Lao

1230-1327	SEAs	7350, 6590, 6010
1430-1527	SEAs	7350, 6590, 6010, 1080

Malay

0930-1027	SEAs	17680, 15135, 8660
1230-1327	SEAs	15135, 11445, 8660

Mongolian

1100-1157	CAs	5850, 5145
1200-1257	CAs	5850, 5145, 4883, 4815, 1314
1400-1457	CAs	5850, 5145, 4883, 4815

Nepalese

1500-1527	SAs	9535, 7150, 1269
1530-1557	SAs	9535, 7150, 6590, 1269

Pashto

1500-1527	ME	11515, 7480, 5220
1530-1557	ME	11515, 7480, 5220

Persian

1500-1527	ME	11500, 9440
1800-1827	ME	12065R, 11575, 11515, 9440
1830-1857	ME	11515, 9785, 9440

Polish

2000-2027	Eu	7375, 6933, 6150F
2030-2057	Eu	7405, 7375, 7255

Portuguese

0000-0027	SAm	11850G
0030-0057	SAm	7435 ,11500, 11850
1900-1927	Af	9535, 6920
1930-1957	Af/Eu	9785, 9670, 7385

Portuguese	Area	kHz
2130-2157	Eu	6165S
2200-2227	SAm	7435, 6950

Romanian

1900-1927	Eu	9860, 7305F
1930-1957	Eu	9860, 7255, 6933

Russian

0000-0057	Siberia	9725, 7110
0300-0357	CAs	15435, 11755, 11515
1000-1057	FE/Siberia	11980, 9725, 7820, 7110, 5145, 1521, 1323, 1116
1100-1157	FE/Siberia	9725, 7820, 7110, 1521, 1323, 1314, 1116
1200-1257	Siberia	1521, 1323, 1116, 963
1300-1357	E.Rus/CAs	7820, 5850, 5145, 4883, 4815, 1521, 1323, 1314, 1116, 963
1400-1457	Siberia	1521, 1323, 1116, 963
1500-1557	Eu/C.Rus	11650, 9730, 7375, 5850, 5145, 4883, 4815, 1521, 963
1600-1657	Eu/CAs	11650, 11500, 9965, 9860 9655, 9585 7375, 7255, 1521
1700-1757	Eu/CAs	11650, 11500, 9860, 9655, 9585, 7780, 7420, 7375, 7255, 6950, 1521, 1116
1800-1857	Eu	11685, 9655, 9535, 9365, 7420, 1521, 1116
1900-1957	Eu	11685, 9655, 9365, 7420, 1521, 1116
2000-2057	Eu	11915, 11765, 7420, 1116
2300-2357	Siberia	9725, 7110, 4883,s 4815

Serbian/Croatian

2000-2027	Eu	9860, 9365, 7385, 7255, 6933
2030-2057	Eu	6165S
2100-2127	Eu	9860, 7405, 4960

Sinhalese

1400-1427	SAs	11780, 7120
1430-1457	SAs	11780, 7120

Spanish

0000-0057	SAm	11650, 7265, 7160, 5250
0100-0157	SAm	11650, 9565
0200-0257	SAm	11650, 9730G
0300-0357	SAm	9560C
2100-2127	Eu	6165S
2100-2157	Eu	7360, 6933
2200-2257	Eu	7360, 6933
2300-2357	SAm	15120M, 11875M, 7160

Swahili

1600-1627	Af	9457, 7335, 5250
1630-1657	Af	9457, 7335, 5250
1700-1727	Af	11790M, 9770M, 9457, 7260, 5250

Tamil

1400-1427	SAs	11575, 9457
1430-1457	SAs	11575, 9457

Thai

1130-1227	SEAs	7350, 6590, 6010, 1080
1330-1427	SEAs	7350, 6590, 6010

Turkish

1400-1427	WAs	11515, 9480
1600-1627	WAs	9785, 7480
1900-1927	WAs	9785, 7405, 1170R

Urdu

1600-1627	SAs	9880, 7590, 7160, 1323
1630-1657	SAs	9880, 7590, 7160, 1323

Vietnamese

1100-1157	SEAs	9550, 6110, 1296
1200-1257	SEAs	9550, 6110, 1296
1300-1357	SEAs	9550, 6110, 1296
1400-1457	SEAs	9550, 6110, 1296
1500-1557	SEAs	9550, 6110
1600-1657	SEAs	7360, 6590, 6185, 6010

ANN: Arabic: "Idha'at as-Sin ad-Duwaliyah". Chinese: "Zhongguo guoji guangbo diantai." English: "This is China Radio International, broadcasting from Beijing". Esperanto: "C^ina Radio Internacia parolas en Pekino". German: "Hier ist Radio China International". Indonesian: "Inilah Radio CRI, China Radio Internasional". Japanese: "Kochirawa Pekin Hoso, Chugoku Kokusai Hosokyoku desu". Malay: "Inilah Radio Antarabangsa China, dalam bahasa Melayu". Mongolian: "Hyatadyn Olon Ulsyn Radio". Spanish: "Esta es Radio Internacional de China". Vietnamese: "Day la dai phat thanh quoc te Trung quoc".
IS: First 4 bars of "The East is Red"— **V.** by QSL card.
PUB: Free literature, schedules, calendars and English magazine "The Messenger" on request.

CHINA HUAYI BROADCASTING COMPANY (Priv.)

📧 P.O. Box 251, Fuzhou, Fujian 350001 — **LP:** GM: Cao Jinwang.
D.PRGR: on 666/4830/6185kHz + 99.6/107.1MHz for Hong Kong, Macao and Southeast Asia: 0255-0700(Tues 0600), 0855-1600.
V. by letter.

CHINA (Republic of) (Taiwan)

CBS - TAIPEI RADIO INTERNATIONAL

✉ 55, Pei'an Road, Tachih, Taipei 104, Taiwan.
L.P: Dir. Overseas Dept: Lee Hou-sheng.
MW: Fangliao: 585kHz 1200kW, Minhsiung: 1422kHz 50kW.
SW: Annan (G.C: 120.38E/23.11N: 4 x 250kW – Paochung (G.C: 120.18E/23.43N) 4 x 100kW, 1 x 50kW, 2 x 25kW – Minhsiung (G.C: 120.25E/23.33N): 1 x 50kW —*) relay via WYFR, USA.
NOTE: Due to the reorganization of External Services from Jan 1st 1998, the new schedule was not available at press time. The schedule given here is a composite listing of existing services from Voice of Free China and Voice of Asia, which have merged into one organization.
kHz: 5950*, 7130, 7285, 7445, 7520*, 9280, 9610, 9680*, 9955, 11740*, 11745, 11775*, 11805*, 11825, 11915, 15215*, 15600*, 15270, 15345, 15440*, 17805*, 17845*

Amoy	Area	kHz
0300-0400	SEAs	7130, 11915
0900-1000	CHN	1422
1000-1100	SEAs	7130, 15270, 15345
	FE	11745
	Pac	9610
1400-1500	FE	7130
	SEAs	11745+
Arabic		
2000-2100	Af/ME	9955, 15270
Cantonese		
0100-0200	NAm	15440*
0300-0400	CAm	11740*
0500-0600	SEAs	11825, 11915, 15270, 15345
	NAm	5950*, 9680*
0500-0700	CHN	7285
0800-0900	CHN	7445
1000-1100	SEAs	7285, 7445, 11915
1100-1200	SEAs	15270
1300-1400	SEAs	11915, 15345
Chinese		
0100-0200	SAm	15215*, 17845*
0100-0300Sun	CHN	15270
0400-0500	NAm	5950*, 9680*
	SEAs	7130, 11825, 15270,15345
0700-0800	SEAs	7130
0700-0900Sat	CHN	7285
0700-1100	CHN	9280
0900-1000	SEAs	7445, 11915, 15270, 15345
	FE	11745
	Pac	9610
1048-1100*	SEAs	585
1300-1445	SEAs	585, 7445
1200-1300	SEAs	15270
	FE	11745
1448-1500*	SEAs	585, 7445
1900-2000	Eu	15600*,17750*
	ME/NAf	9955

*) History of China Prgr.

English		
0200-0300	FE	7130, 15345
	SEAs	11825
	NAm	5950*, 9680*
	CAm	11740*
0300-0400	FE	11745, 15345
	SEAs	11825
	NAm	5950*, 9680*
0700-0800	NAm	5950*

English		
1100-1200	SEAs	585, 7445
1200-1300	FE	7130
	Pac	9610
2200-2300	Eu	5810*, 9985*
French		
0700-0800	Eu	7520*
2000-2100	Eu	9610, 15600*
German		
1900-2000	Eu	9610
2100-2200	Eu	15600*,17750*
Hakka		
0900-1000	CHN	7285
1100-1200	SEAs	11860, 15345
1400-1500	SEAs	11915, 15345
Indonesian		
0600-0700	SEAs	7130, 11825, 11915
1200-1300	SEAs	585, 7445, 11860, 15345
1600-1700	SEAs	585, 7445

Japanese	Area	kHz
0100-0200	FE/NAm	11745, 15345
1100-1200	FE/NAm	7130, 11745
1300-1400	FE/NAm	7130, 11745
Russian		
1705-1805	CIS	9955
Spanish		
0200-0300	SAm	15215*, 17845*
0400-0500	CAm	11740*
2100-2200	Eu	15270
2300-2400	SAm	15130*, 17805*
Thai		
0600-0700	SEAs	15270, 15345
1500-1600	FE/NAm	7130
	SEAs	585, 7445, 9610
2200-2300	SEAs	7445, 15345
2300-2400	SEAs	7445
Vietnamese		
1500-1600	SEAs	11915, 15345

V. by QSL-card. Rec.acc —**PUB:** Prgr. schedule free.

COSTA RICA

RADIO FOR PEACE INTERNATIONAL

✉Ap. 88, Santa Ana. (✉ in USA: P.O. Box 29728, Portland, OR 9722).
E-mail: rfpicr@sol.racsa.co.cr
Web: http://www.clark.net/pub/cwilkins/rfpi/
L.P: DG: Debra Latham. Prgr. Coordinator (English): Joe Bernard.
SW: Ciudad Colon.

English	Days	kHz
0100-1200	Daily	7385, 6205USB
1700-2200	Daily	15050
2200-2400	Daily	15050, 7385
French		
1500-1600	Thurs/Sun	15050, 7385
Spanish		
0000-0100	Daily	7385, 6205USB
1200-1500	Daily	7385
1500-1600	exc. Thurs/Sun	7385
1600-1700	Daily	7385

World of Radio with Glenn Hauser: Tues 1900; Wed 0300, 1000; Fri 2000; Sat 0430, 1130, 1800; Sun 0200, 0900. In the third week of each month, WoR is followed by Glenn Hauser's **Continent of Media**.

ADVENTIST WORLD RADIO-PANAMERICA (Rlg.)

✉ P.O. Box 1177, 4050 Alajuela. ☎ +506 483 0550. ▤ +506 483 0555
Web: http://www.awr.org/awr-panamerica/
L.P: Americas Region Dir: Gordon Retzer. St. Dir: Victor Shepherd. CE: Karl Thompson
SW:

Arabic	Days	kHz
0000-0100	67	5030, 6160, 9725, 15460
Dutch		
0200-0230	6	5030, 6150, 9725, 15460
English		
0500-0600	67	5030, 6150, 7375, 9725
1100-1300	1234567	5030, 6150, 7375, 9725
1600-1700	67	9725, 11870
2300-2400	1234567	5030, 6150, 9725, 15460
French		
0000-0100	12345	5030, 6150, 9725, 15460
Papiamento		
0230-0300	6	5030, 6150, 9725, 15460
Portuguese		
0100-0200	1234567	5030, 6150, 9725, 15460
Spanish		
0200-0300	123457	5030, 6150, 9725, 13750, 15460
0300-0400	1234567	5030, 6150, 7375, 9725, 13750
0400-0500	1234567	5030, 6150, 7375, 9725
0400-0500	67	13750
1000-1100	1234567	5030, 6150, 7375
1300-1400	1234567	5030, 6150, 7375
1300-1500	1234567	9725, 13750
1400-1600	67	11870
1500-1600	67	9725, 13750

RADIO EXTERIOR DE ESPAÑA RELAY STATION

✉ Radio Nacional de España,Delegación in Costa Rica, Ap. Postal 677-2010, Zapote, San José. ☎ +506 2329471, 2328476.
SW: Cariari de Pococí (G. C: 83.43W/10.25N). ☎ +506 7677308, 7677311: 3 x 100 kW tr`s.
kHz: 3210, 5970, 5990, 9630, 9745, 11815, 11880, 15125
D.PRGR: see Spain.

CROATIA

HRVATSKI RADIO (HR)

✉ Prisavlje 3, Zagreb. ☎ +385 (1) 6163355. 🖷 +385 (1) 616-3347. ℃
21154. **Web:** http://www.hrt.hr — **L.P:** Dir: Tomislav Dakaric.
SW: Deanovec (G.C: 16.29E/45.42N): 1 x 100kW, 2 x 10kW, 1 x 2.5kW
— J) rel. via Jülich, Germany.
kHz: *5850, 5890, 5895, 5920, 7165, 9595, 9830, 11635, 13830*

Winter Schedule 1998/99

Croatian*	Area	kHz
0000-0200	SAm	5890
0200-0400	ENAm	5890
0400-0600	WNAm	5895
0600-1230	Eu	9830, 7165, 5920
1230-1600	Eu	7165, 5920
1230-2100	ENAm	13830
1600-2300	Eu	7165, 5895
2100-2200	AUS/NZL	5850
	SAf	9595J
2100-2300	ENAm	11635

* **N. in English:** 0703W, 0803Sun, 0903W, 1303, 2203.
ANN: "Hrvatska Radio, Kratki Val" (Croatian Radio, Short Wave). E:
"This is Croatian Radio" — **IS:** Tune to Dubrovnik's poem "Lovely, Dear,
Sweet Liberty" played on celeste — **V.** by letter. Re. in SINPO code to
Zelimir Klasan, HRT, Prisavlje 3, Zagreb, Croatia. ☎ +385 (1) 6163355.
🖷 +385 (1) 6163347. **E-Mail:** Zelimir.Klasan@hrt.hr

CUBA

RADIO HABANA CUBA

✉ Apartado 6240, La Habana. ☎ +53 (7) 791053. 🖷 +53 (7) 795007. **E-mail:** rhc@radiohc.org *Engineering, technical, DX programs, QSL requests:* inforhc@mail.infocom.etecsa.cu. *Programming:* radiohc@mail.infocom.etecsa.cu. **Web:** http://www.radiohc.org
L.P: MD: Milagro Hernandez Cuba. Head of Prgrs: Ignacio Canel. CE: Luis
Pruna — **SW:** Bauta (G.C: 82.30W/23.00N): 50/75/100/250kW.
kHz: *5965, 6000, 6070, 6180, 9505, 9550, 9820, 9830, 11760, 11970,
13680, 13715, 13725, 15340*

Arabic	Area	kHz
2000-2030	ME/Af	11760
2030-2100	ME/Af	13680
Creole		
0000-0030	CAm	6180
0100-0130	CAm	6180
2130-2200	CAm	6180
English		
0100-0500	NAm	6000, 9820, 9830USB
0500-0700	NAm	9830USB, 9820
2030-2130	Eu/NAm/WAf	13715, 13725USB
Esperanto		
0700-0730 Sun	Am	9820
Esperanto	**Area**	**kHz**
1500-1530 Sun	Am	11760
1930-2030 Sun	Eu	13715
2200-2230 Sun	Eu	13715
2330-2400 Sun	Am	11760, 6070
French		
0030-0100	NAm	6180
0130-0200	NAm	6180
2000-2030	Eu	13715, 13725USB
2130-2200	Eu	13715, 13725USB
2200-2230	NAm	6180
2330-2400	NAm	6180
Guarani		
2230-2300	SAm	15340
2330-2400	SAm	15340
Portuguese		
2000-2030	Eu	13680
2030-2100	Eu	11760
2200-2230	SAm	15340
2300-2330	SAm	15340, 11970
2330-2400	SAm	11970
Quechua		
0000-0030	SAm	15340
Spanish		
0000-0100	ENAm	6000
0000-0500	LAm	13715, 11760, 9505, 6070, 5965
0000-0300	SAm	11970, 9550
0200-0500	ENAm	6180
1100-1500	CAm	11760, 6180
1100-1400	SAm	9550

Spanish	Area	kHz
1200-1400	SAm	15340
1200-1500	Mexico	6070
2100-2300	Eu	13680, 11760, 9830USB

V. by QSL-card. Prgr. schedule free on request.

CYPRUS

CYPRUS BROADCASTING CORPORATION

✉ P.O. Box 4824, 1397 Nicosia. ☎ +357 (2) 422231. 🖷 +357 (2) 314050.
℃ 2333 CYBC CY. **Cable:** Broadcasts.
E-mail: rik@cybc.com.cy **Web:** http://www.cybc.com.cy
L.P: DG: vacant. PD: Panos Ioannides. Dir. Tech Sces: Andreas
Michaelides. Head of Radio Prgrs: Kyriacos Charalambides. Head of Pub.
& Int. Rel: Nayia Roussou.
SW: via British East Mediterranean Relay St. (see below)

Greek	Days	Area	kHz
2215-2245	Fri/Sat/Sun	United Kingdom	6180, 7105, 9760

ANN: "Radiofonikon Idryma Kyprou"— **IS:** "Avkoritssa" (guitar).
V. by QSL-card. Rec.acc.

TRANS WORLD RADIO (Rlg.)

MW: Cape Greco 1233kHz 600kW
Arabic: 0410-0430 (Sat/Sun/Mon 0415), 2140-2230 (Sat).
Armenian: 0355-0410.
Persian: 0325-0355.
Other details: see Monaco.

BRITISH EAST MEDITERRANEAN RELAY STATION

✉ P.O. Box 4912, Limassol. ☎ +357 (4) 332511/12, 332341. 🖷 +357 (4)
332595. ℃ 2665 BEMRS CY — **L.P:** Sen. Eng: S. Welch.
MW: 639/720kHz 500kW, 1323kHz 100kW.
SW: Zygi (G.C: 33.19E/34.43N): 8 x 300kW, 2 x 250kW tr's.
kHz: *3990, 5875, 5965, 6010, 6015, 6050, 6050, 6125, 6180, 7105, 7115,
7125, 7130, 7140, 7155, 7320, 9410, 9585, 9605, 9635, 9740, 9750, 9765,
9815, 9825, 9825, 9870, 9915, 11665, 11680, 11740, 11805, 11820,
11840, 11845, 11845, 11850, 11970, 12005, 12030, 12095, 13660, 15180,
15225, 15245, 15325, 15340, 15360, 15420, 15555, 15565, 15575, 15585,
17640, 17695, 17705, 17770, 17885, 21470, 21490*

CZECH REPUBLIC

RADIO PRAGUE

✉ Vinohradská 12, Praha 120 99, Czech Rep. ☎ + 420 (2) 240 94 608.
🖷 +420 (2) 242 18239 or 24222236. **E-mail:** cr@radio.cz
Web: http://www.radio.cz *or* http://www.prague.org/
Direct lines to depts: Czech: +42 (2) 24222236. **English:** +42 (2)
24218349. **French:** +42 (2) 24222211. **German:** +42 (2) 24222235.
Spanish: +42 (2) 24222237. **Internet:** +42 (2© 24215456.
L.P: Head of Prgrs: Dusan Palka. Ce: Oldrich Cip.
SW: L=Litomysl (G.C: 16.10E/49.48N): 2 x 100kW.
kHz: *5930, 6055, 6200, 7265, 7345, 7350, 9420, 9430, 9505, 11600,
13580, 15640, 17485, 21705*
NB: At press time, the Czech government was considering applications
from organizations interested in taking over the running of the External
Service as of Jan 1st 1998. This schedule is valid until December 31st
1997. The situation after Jan 1st 1998 may change.

Winter Schedule 1998/99

Czech	Area	kHz
0930-0957	SWEu	15260
	EAf/ME	21745
1030-1057	WAf	17485
	MEast/Sas	21745
1200-1227	NEu	11640
	SAs/AUS	21745
1330-1357	CEu	6055
	WEu	7345
1430-1457	EAf/NAm	21745
1630-1657	WEu	5930
	Caf/WAf	17485
1830-1857	WEu	5930
	EEu/As/AUS	7315
2030-2057	WEu	5930
	SWEu/WAf	9430
2330-2357	NAm	9435
	SAm	11600
0130-0157	NAm	6200
	SAm	7345
0230-0257	NAm	7345, 9435
German		
0730-0757	Eu	5930, 7345

German	Area	kHz
1100-1127	Eu	7345, 11640
1300-1329	Eu	6055, 7345
1600-1627	Eu	5930

English

	Area	kHz
0800-0827	Eu	11600, 15260
1000-1030	WAf	17485
	ME/SAs	21745
1130-1157	NEu	11640
	EAf/ME	21745
1230-1257	CEu	6055
	SAs/AUS	21745
1400-1430	EAf/NAm	21745
1700-1727	NWEu	5930
	Caf/WAf	17485
1800-1827	NWEu	5930
	EEu/As/AUS	7315
2100-2127	NWEu/NAm	5930, 7345
2230-2257	NAm	7345, 9435
0000-0027	NAm	7345, 9465
0100-0127	NAm	6200, 7345
0300-0327	NAm	7345, 9435
0330-0357	ME/SWAs	9585, 11600
0400-0427	NAm	9955*

Spanish

	Area	kHz
0830-0857	SWEu	11600, 15260
1500-1527	SWEu	11600, 13580
1900-1927	SWEu	5930, 9430
2000-2027	SWEu	5930, 9430
2130-2157	SWEu/SAm	7345, 9435
2300-2327	SAm	9435, 11600
0030-0057	CAm	7345, 9465
0200-0227	CAm	6200
	SAm	7345

French

	Area	kHz
0700-0727	WEu	5930, 7345
0900-0927	WEu	11600
	SWEu	15260
1530-1557	WEu	11600
	SWEu/WAf	13580
1730-1757	WEu	5930
	CAf	17485
1930-1957	WEu	5930
	SWEu/WAf	9430
2200-2227	NAm	7345, 9435

*Relayed via WRMI Miami, Florida 80W22 25N54

IS: Fanfare from Dvorák's 9th Symphony ("From the New World") played on French horn — **V.** by QSL-card. Rec. acc.

RADIO FREE EUROPE/RADIO LIBERTY Inc.

This is a non-profit, privately managed network broadcasting news and information to the peoples of East Central and South Eastern Europe, the Baltic Countries and the successor states to the USSR.
✉ Vinohradská 1, 110 00 Praha 1. ☎ +42 (2) 2112 1111. 🖷 + 42 (2) 2421 1501 (Tech. Op's). **Web:** http://www.rferl.org
L.P: Pres: Thomas Dine. Dir: Luke Springer. Dir. N. & Current Affairs: Robert McMahon. Public Rel. Coordinator: Alena Fendrychova (**E-mail:** fendrychovaa@rferl.org).
Washington Office: 1201 Connecticut Ave NW, Washington DC 20036, USA. ☎ +1 (202) 457 6900. 🖷 +1 (202) 457 6992.
MW/FM: see details under respective target countries.
SW: Biblis, Germany (100kW), Holzkirchen, Germany (250kW), Lampertheim, Germany (100kW), Playa de Pals, Spain (250kW).
kHz: 5955, 5970, 5985, 5995, 6020, 6035, 6040, 6050, 6095, 6105, 6115, 6135, 6140, 6150, 6170, 7115, 7145, 7155, 7165, 7180, 7190, 7215, 7220, 7240, 7245, 7255, 7260, 7265, 7270, 7275, 7280, 7295, 9505, 9520, 9540, 9565, 9585, 9595, 9600, 9605, 9625, 9645, 9660, 9665, 9670, 9675, 9680, 9690, 9695, 9710, 9725, 9745, 9750, 9770, 9835, 9860, 11705, 11725, 11730, 11770, 11795, 11805, 11840, 11855, 11875, 11885, 11905, 11915, 11970, 11980, 11990, 15205, 15215, 15250, 15340, 15370, 15410, 15445, 17725, 17750, 17805

Armenian	Area	kHz
0500-0600	ARM	5955, 7180, 9680, 11730
1500-1600	ARM	9665, 11770, 15410
1700-1800	ARM	6020, 9540, 11875
Azerbaijani		
0400-0500	AZE	7180, 9595, 9680
1600-1700	AZE	7240, 9540, 11875
1900-2000	AZE	6140, 7180, 9670
Belorussian		
0400-0500	BLR	5995, 7295, 9750

Belorussian	Area	kHz
1600-1730	BLR	6170, 7155, 7295
2000-2130	BLR	6095, 6170, 7155, 9750
Bulgarian		
0700-0800	BUL	9625, 11855, 15215 + FM in Bulgaria
1100-1200	BUL	9725, 11915, 17725 + FM in Bulgaria
1500-1600	BUL	9625, 15370, 17725
1600-1700	BUL	9625, 11980, 15370
1700-1800	BUL	9625 11915 11980
1500-2100	BUL	FM in Bulgaria
2000-2100	BUL	6115, 7295, 11915
Estonian		
2000-2100	EST	FM in Estonia
2200-2300	EST	5970, 7155, 7255 + FM in Estonia
Georgian		
0400-0500	GEO	5985, 6040, 9725
1700-1800	GEO	6150, 7280, 9595
1900-2000	GEO	6095, 7190, 9585 11905
Kazakh		
0000-0100	KAZ	6135, 7145, 9625
0100-0200	KAZ	1341 + FM in Kazakhstan
0200-0300	KAZ	6135, 7145, 9625
1200-1300	KAZ	11795, 15340, 11750
1300-1400	KAZ	1341 + FM in Kazakhstan
1400-1500	KAZ	9695, 11840, 11970
Kyrgyz		
0100-0200	KYR	1323, 6050, 7295, 9835 + FM
1300-1400	KYR	1323, 7215, 9595, 11990
1600-1700	KYR	1323, 7270, 9660, 9695 + FM
Latvian		
1600-1700	LAT	1071, 1485, 1539 + FM (all Latvia)
1900-2000	LAT	5955, 7255, 9505
Lithuanian		
1800-1900	LIT	612, 666, 1557, 9710 + FM (Lithuania)
2100-2200	LIT	5970, 6150, 7255
Romanian		
0500-0600W	ROU	6095, 7165, 9725
0600-0700	ROU	7165, 9605, 9725 + FM in Romania
0700-0800Sun	ROU	7165, 9605, 9725 + FM in Romania
1600-2000	ROU	6115, 7165, 9725
1800-2000	ROU	FM in Romania
Russian		
0000-0100	RUS	5955, 5985, 7155, 7220, 7245, 9520
0000-0200	RUS	5955, 6035, 7155, 7220, 7245, 9520
0300-0400	RUS	5955, 6105, 7155, 7220, 7245, 9520
0400-0500	RUS	5955, 6105, 6140, 7220, 9520, 11885
0500-0600	RUS	6105, 6140, 7220, 9520, 11875, 11885
0600-0700	RUS	6105, 7190, 7220, 9520, 9680, 11875, 11885
0700-0800	RUS	6105, 7220, 9520, 9680, 11875, 11885, 15205, 15250
0800-0900	RUS	9520, 9625, 11885, 11970, 15205, 15250, 15370
1100-1200	RUS	9520, 11725, 11885, 15215, 15370, 17805
1200-1300	RUS	9520, 11725, 11885, 15215, 15370, 17805
1300-1400	RUS	7220, 9520, 11725, 11885, 15370, 17805
1500-1600	RUS	7190, 7220, 9520, 11885, 11970, 15445

Russian	Area	kHz
1600-1700	RUS	6105, 7220, 9520, 9860, 11805, 11885
1700-1800	RUS	5985, 6105, 7220, 9505, 9520, 9860, 11885
2000-2100	RUS	6105, 6140, 7220, 7265, 7270, 9520, 9660
2100-2200	RUS	6105, 6140, 7220, 7265, 7270, 9520, 9645
2200-2300	RUS	5955, 6105, 7220, 7245, 9520, 9645
2300-2400	RUS	5955, 5985, 7155, 7220, 7245, 9520

NB: In addition to SW transmissions, R. Liberty's Russian sce. is relayed in part by numerous MW and FM tr's in major Russian cities.

Serbo-Croat		
1700-1800	SEEu	1197, 1458, 7115, 9695 + FM
1800-2000	SEEu	1593
2000-2030	SEEu	1593, 7180, 9600, 9690
2100-2200	SEEu	1593, 5955, 7165, 9680 + FM
2230-2300	SEEu	1593
Tajik		
0200-0300	TJK	6050, 7295, 9585
1500-1600	TJK	7280, 9695, 11705
1600-1700	TJK	7190, 9675, 11705, 11730
Tatar-Bashkir		
0400-0500	CAs	7190, 7275, 9540
0600-0700	CAs	9540, 11770, 11855

Tatar-Bashkir	Area	kHz
1600-1700	CAs	7115, 9505, 9745
2000-2100	CAs	5955, 7245, 7280
Turkmen		
0200-0300	CAs	6170, 7255, 9680
1400-1500	CAs	9660, 11770, 11875
1600-1700	CAs	9565, 9770, 11770
1800-1900	CAs	7280, 9540, 9565
Ukrainian		
0400-0500	UKR	873, 1260, 1278, 1359, 1476, 6170, 7165, 7245
0600-0700	UKR	873, 1260, 1278, 1359, 1476, 6170, 7245, 9695
1800-2100	UKR	873, 1260, 1278, 1359, 1476, 5985, 7115, 9625
Uzbek		
0200-0400	UZB	5985, 7190, 9725
1300-1400	UZB	6170, 11770, 15215
1700-1800	UZB	7260, 9565, 9660

ANN: RFE, Sign on Sign off: Bulgarian: "Tuk e Radio Svobodna Evropa"; Polish: "Mowi Rozglosnia Polska Radio Wolna Evropa"; Romanian: "Aici e Radio Europa libera"; Estonian: "Siin Vaabaduseradio"; Latvian: "Te Runa Brivibas Raiditajs"; Lithuanian: "Kalba Lasives Radijas". RL, Azerbaijani: "Danyshyr Asadlyk Radiosu"; Armenian: "Chosume Azatutiun Rdiokayane"; Belarus:
"Haworyts Radyio Svoboda"; Georgian: "Laparakobs Radio Thavisupleba"; Russian: "Govorit Radio Svoboda"; Tatar-Bashkir: "Monda Azatlyk Radiostansyese Suyili"; Ukranian: "Hovoryit Radio Svoboda"; Uzbek: "Azadiq Radiosidan Qapiramiz"; Turkmen: "Qepleyaer Azatik Radiosi"; Tadzhik: "Inja Radioi Azadi"; Kazakh: "Azzatiq Radioyosinan Sövlep Turmiz"; Kyrgyz: "Azattiq Radiosunan Söylöbüz"; Dari: "Inja Radioi Afghanistan Azad Ast".
IS: RFE: Bulgarian: Part of an old patriotic Bulgarian folk song.; Polish: Polish patriotic song "Hail Glorious Dawn of May" (brass instruments); Romanian: "Romanian Rhapsody" by G. Enescu. RL: Russian sce: "Hymn to Freedom" by Gretchaninoff — **V:** by QSL-card .

DENMARK

RADIO DENMARK
✉ Radiohouse, DK-1999 Frederiksberg C. ☎ +45 35 205785. 🖷 +45 35 205781. **E-mail:** rdk.ek@login.dknet.dk.
Web: http://www.dr.dk/radioavi/radiodk.htm
L.P: Ag. Head of SW section: Flemming Heltø.
STATION: Radio Denmark is broadcasting via the facilities of Radio Norway International. The main broadcast being at 1430.

Winter Schedule 1998/99

Times	Areas	kHz
00.30-00.55	SEAs/AUS	9935
	Nam/CAm	9945
01.30-01.55	Nam	7465
	SAs	7545
	Nam/CAm	9945
02.30-02.55	NAm	7465
	SAs	7545
	Nam/CAm	9945
03.30-03.55	ME	7545
	Seu/EAf	7565
	NAm	9945
04.30-04.55	Eeu/SEEu/ME	7545
	NAm	7560
	SEEu/EAf	9940
05.30-05.55	EEu	7440
	Eu	7480
06.30-06.55	NEu/CEu	5965
	Eu	7180
	SWEu/NWAf	9590
	Af	13800
07.30-07.55	Weu	7180
	CEu	9590
	Af	18950
08.30-08.55	AUS/SAm	13800
	ME	15175
09.30-09.55	AUS/SAm	13800
	FE/Pac	15175
10.30-10.55	SEu	13800
	SAm	18950
11.30-11.55	Eu/SEu	13800
	SAm	18950
12.30-12.55	FE	9910
	ENAm/CAm	18950
	Eeu/SEAs/AUS	21725
13.30-13.55	Eur	9590
	FE	9910

Times	Areas	kHz
	ENAm/CNAm	18950
	Eeu/SEAs/AUS	21725
14.30-14.55	Eeu/ME/SAs	15705
	ENAm/CNAm	18950
15.30-15.55	WNAm	13790
	EEu/SEEu/ME/SAs	13800
	ME	15705
16.30-16.55	Eeu/SEEu/ME/Sas	9590
	SEEu/ME/EAf	13800
	WNAm	15705
17.30-17.55	Eu	7485
	Eeu/SEu	11895
	Eeu/EAfr	15705
	ENAm/CAm	18950
18.30-18.55	Eu	5960
	Eu	7485
	WAf/SAf	15705
	WNAm	18950
19.30-19.55	NEu	243/1062
	Eu	7485
	FE/Pac	9935
	Afr	13800
	WNAm	15705
20.30-20.55	Eu	7580
	WNAm	15705
	AUS	7560
21.30-21.55	Eu	7580
	AUS	9945
22.30-22.55	SAm	9415
	FE	9925
23.30-23.55	SAm	9415
	SEAs/AUS	9935
	Nam/CAm	9945

D.PRGR: All prgrs are in Danish. **Letterbox:** Saturdays in even weeks, hourly (24 times) from 1543. **Sport:** Mondays.
IS and ANN: All trs preceded by repetition of the first bars of a piece of Danish music by Carl Nielsen, followed by announcements in English and Danish.
Message to Listeners: For the Danish speaking audience Radio Denmark operates an automatic telephone recorder for important reports/messages on +45 35 205791. The current schedule is available in Danish on +45 35 205796 (Eu/Af), + 45 35 205797 (Eastern Globe) and + 45 35 205798 (Western Globe).
V: By QSL-card. Rp. (one IRC or US$1). Send 1 IRC for schedule.

ECUADOR

LA VOZ DE LOS ANDES (HCJB)
✉ Casilla 17-17-691, Quito. ☎ +593 (2) 466808. 🖷 +593 (2) 447263. ① 22734 HCJB ED. **E-mail:** info@hcjb.org (✉ **USA Office:** PO Box 39800, Colorado Springs, CO 80949-9800). **Web:** http://www.hcjb.org
L.P: MD: Glen Volkhardt. International PD John Beck. Dir. of Local Radio: Mark Irwin. Dir. of Eng: Daniel Enns. Head of PR: Harold Goerzen. Dir. Cooperating B'casters: Pat Talbot. Satellite Prgr. Distribution Coordinator: Horace Easterling. Freq. Mgr: David M. Lewis.
MW: Quito 690kHz 50kW (+ 25kW standby tr. at Pifo).
SW: Pifo (G.C: 78.20W/00.14S): 250/100/50/30/10/0.5kW.
kHz: 3220, 5860, 6050, 6080, 6110, 6125, 9415, 9445, 9745, 9765, 11615, 11960, 12005, 12025, 15140, 15295, 15550, 21455U

Winter Schedule 1998/99

Czech	kHz	Area	Days
0530-0600	9765	Europe	S-TWT-S
1830-1900	17795	Europe	S-TWT-S
1830-1900	21470	Europe	S-TWT-S
English			
0000-0130	9745	NAm	SMTWTFS
0130-0400	9745	NAm	SMTWTFS
0000-1600	21455	Eu/Pac	SMTWTFS
0000-0700	12015	NAm	SMTWTFS
0400-0700	9745	NAm	SMTWTFS
0700-0900	9775	Eur	SMTWTFS
0700-1100	9640	SPac	SMTWTFS
	kHz	**Area**	**Days**
1100-1600	12005	Carib	SMTWTFS
1100-1600	15115	NAm/SAm	SMTWTFS
1900-2200	15115	Eur	SMTWTFS
1900-2200	21455	Eur/Pac	SMTWTFS
French			
0030-0100	9660	CAN	SMTWTFS
0630-0700	9765	Eur	SMTWTFS

French	kHz	Area	Days
1900-1930	15550	Eur	SMTWTFS
1900-1930	17795	Eur	SMTWTFS
2000-2030	15550	Eur	SMTWTFS
2000-2030	17795	Eur	SMTWTFS

German

0430-0500	5970	Eur	SMTWTFS
0430-0500	9685	Eur	SMTWTFS
0600-0630	9765	Eur	SMTWTFS
0930-1000	6125	SAm	SMTWTFS
1930-2000	15550	Eur	SMTWTFS
1930-2000	17795	Eur	SMTWTFS
2100-2130	15550	Eur	SMTWTFS
2100-2130	17795	Eur	SMTWTFS
2300-0000	11955	SAm	SMTWTFS

Japanese

0430-0500	9765	NAm	SMTWTFS
1130-1200	9660	Japan	SMTWTFS
2030-2100	15550	Eur	SMTWTFS
2030-2100	17795	Eur	SMTWTFS
2200-2300	11920	SAm	SMTWTFS
2200-2300	21455	Eur/Pac	SMTWTFS

Kikongo

0500-0510	9765	WAf	-MTWTF-

Portuguese

0800-0930	9745	Brazil	SMTWTFS
1530-1800	15295	Brazil	SMTWTFS
2300-0100	11920	Brazil	SMTWTFS
0100-0230	11920	Brazil	SMTWTFS

Quichua

0830-1000	6110	SAm	SMTWTFS
0830-1030	690	Ecuador	SMTWTFS
0830-1400	6080	SAm	SMTWTFS
0830-1400	3220	SAm	SMTWTFS
2100-0300	6080	SAm	SMTWTFS
2100-0300	3220	SAm	SMTWTFS
2130-0000	9745	SAm	SMTWTFS

Russian

0200-0430	5970	WRuss	SMTWTFS
0515-0700	5970	WRuss	SMTWTFS
0515-0700	9685	WRuss	SMTWTFS
2230-2300	15295	NAm/SAm	SMTWTFS

Slovak

0530-0600	9765	EEur	-M——F-
1830-1900	17795	EEur	-M——F-
1830-1900	21470	EEur	-M——F-

Spanish

0100-0500	15140	Mexico	SMTWTFS
0700-0730	15550	Eur	SMTWTFS
0900-1100	9765	SAm	SMTWTFS
1030-1100	690	Ecuador	SMTWTFS
1030-1100	6050	SAm	SMTWTFS
1100-1300	11960	Cuba	SMTWTFS
1100-1500	15140	SAm	SMTWTFS
1300-1500	15295	Mexico	SMTWTFS
1500-1700	15140	SAm	SMTWTFS
1600-1900	21455	Eur/Pac	SMTWTFS
1700-1900	15140	N/AmSAm	SMTWTFS
1900-2100	15140	Mexico	SMTWTFS
2100-2300	15140	SAm	SMTWTFS
2130-2230	15550	Eur	SMTWTFS
2130-2230	12025	Eur	SMTWTFS
2300-0000	21455	Eur/Pac	SMTWTFS
2300-0100	15140	NAm/SAm	SMTWTFS

Ukranian

0500-0515	5970	WRuss	SMTWTFS
0500-0515	9685	WRuss	SMTWTFS

U = Upper Side Band mode.

ANN: E: "This is the Voice of the Andes, HCJB, in Quito, Ecuador, South America. The time at the tone will be ...Eastern Standard Time, the same as...Coordinated Universal Time. A program follows in...".

V. by QSL-card. Rp (1 IRC). Rec.acc. but tapes cannot be returned.

EGYPT (Arab Republic of)

EGYPTIAN RADIO & TV UNION (Gov.)
P.O. Box 566, Cairo 11511. ☎ +20 (2) 5787120, 5757155, 770355.
Cable: CIBROTEV. ☼ 92152 (KARADUN).
Web: http://www.sis.gov.eg/vidaudio/html/audiofm.htm
L.P: Pres. ERTU: Amin Bassiouni. Chmn. Eng. Sector: Ibrahim A. Ibrahim. Chmn. Sound Prgr. Sector: Farouk Susha. Chmn. of TV Prgr. Sector: Suhair El Attriby.

SW: Abu Zaabal (G.C: 31.22E/30.16N): 4 x 500kW, 1 x 250kW, 13 x 100kW — Abis (G.C: 30.05E/31.10N): 1 x 500kW, 8 x 250kW — Mokattam (G.C: 31.15E/30.03N): 1 x 100kW, 2 x 50kW.
kHz: 6120, 6230, 9475, 9620, 9700, 9740, 9755, 9770, 9800, 9850, 9900, 9950, 9990, 11560, 11665, 11715, 11785, 11875, 11975, 11980, 11990, 12050, 15115, 15160, 15210, 15220, 15255, 15285, 15335, 15375, 15420, 17595, 17670, 17690, 17710, 17745, 17770, 17800, 17840.

Relays of Home Services in Arabic
General Prgr.

kHz	Area	Times
9620	NAf/SEu	0350-0700
9770	NAf/SEu	0350-0700
9800	Gulf	0350-2400
9850	NAf/SEu	1100-2400
11785	NAf/SEu	0700-1500
12050	NAm/Eu	0200-2400
	EAf	0600-1500
15115	WAf	0700-1100
17670	NAf/SEu	1300-1900

Voice of the Arabs

9700	NAf	1800-0030
9850	NAf/SEu	0300-0600
11665	EAf/CAf	1900-0030
11980	NAf/SEu	0600-1400
	Gulf	0600-1100
15285	Gulf	0300-0030

Holy Quran Prgr.

9755M	NAf/ME	0200-2200

FOREIGN LANGUAGE SERVICES

	Area	kHz
Afar		
1530-1630	EAf/CAf	15155
Albanian		
1600-1800	Eu	9950
Amharic		
1730-1900	EAf/CAf	15155
Arabic		
0030-0430	ENAm	9900
1015-1215	ME/AFG	17745
1100-1130	CAf/SAf	17800
1300-1600	WAf	15220
2000-2200	AUS	11990
2345-0045	SAm	17770
	CAm	15220
Bambara		
1930-2030	WAf	15375
Bengali		
1330-1430	SAs	17595
English		
0200-0330	NAm	9900, 9475
1215-1330	SAs	17595
1630-1830	CAf/SAf	15255
2030-2200	WAf	15375
2115-2245	Eu	9900
Fulani		
1930-2030	WAf	17840
French		
2000-2115	Eu	9900
2030-2230	WAf	15335
German		
1900-2000	Eu	9900
Hausa		
1800-2100	WAf	15210
Hindi		
1500-1600	SAs	17690
Indonesian		
1320-1450	SEAs	17770
Italian		
1800-1900	Eu	9990
Lingala		
1830-1915	CAf/SAf	11875
Malay		
1215-1315	SEAs	17770

Oulof (Wolof)	Area	kHz
1830-1930	WAf	15375
Persian		
1300-1330	TJK	15160
1330-1530	ME	11560
Portuguese		
2215-2330	SAm	15420
Pushtu		
1500-1600	AFG	17710
Russian		
1800-1900	wRUS	6120
Shona		
1645-1730	CAf/SAf	11875
Ndebele		
1730-1815	CAf/SAf	11875
Somali		
1630-1730	EAf/CAf	15155
Spanish		
0045-0200	NAm	9475
	CAm	9740
	SAm	11715
Swahili		
1530-1730	EAf/CAf	11975
Thai		
1115-1215	SEAs	17770
Turkish		
1600-1800	TUR	6230
Urdu		
1600-1800	SAs	17690
Uzbek		
1530-1600	UZB	9620
Yoruba		
2100-2200	WAf	15210
Zulu		
1600-1645	CAf/SAf	11875

ANN: General Prgr: "Idha'atu jumhuriya misr al'arabbiya min al-qahira". Voice of the Arabs: "Saut al-'arab, min al-qahira". Quran prgr: "Idha'atu-l-Quran min al-qahira". English to Africa: "This is the Voice of Africa from Cairo" — **V.** by QSL-card. Re.to Box 1186, Cairo. Schedules available on request.

EQUATORIAL GUINEA

RADIO AFRICA & RADIO EAST AFRICA (Comm.)

✉ P.O. Box 851 Malabo (**USA** ✉ Pan American Broadcasting, 20410 Town Center Lane, Suite 200, Cupertino, CA 95014-3230. ☎ +1 (408) 996 2033. 🖷 +1 (408) 252 6855). **SW:**

kHz	kW	Station	Times	Days
7190	50	R. Africa	1700-2300	Daily
15185.7	50	R. East Africa	0700-1700	Sat/Sun
15185.7	50	R. Africa #2	0700-1100	Daily

Prgrs: religious prgrs & music in **English** — **V.** by QSL-card.

ESTONIA

RADIO ESTONIA

✉ Gonsiori 21, EE-0100 Tallinn. ☎ +372 (2) 2434115. (English Sce: +372 (2) 2434252). 🖷: +372 (2) 2434457. ✆ 173271 RADIO.
Web: http://www.er.ee/tallinn/
STATION: Tallinn (Laitse, G.C: 24.24/E59.12N): 1035kHz 150kW, 5925kHz 100kW ND — **FM:** Tallinn 103.5MHz.

English	Days	Area	kHz
1620-1630	12345	Eu	5935, 1035, 103.5MHz
2000-2030	14	Eu	5935, 1035, 103.5MHz
Esperanto			
2030-2045	2	Eu	5935, 1035, 103.5MHz
Finnish			
0800-0900	7	Eu	5935, 1035, 103.5MHz
1600-1610	12345	Eu	5935, 1035, 103.5MHz
2030-2100	12345	Eu	5935, 1035, 103.5MHz
German			
1610-1620	12345	Eu	5935, 1035, 103.5MHz
2000-2030	25	Eu	5935, 1035, 103.5MHz
Swedish			
0900-0930	7	Eu	5935, 1035, 103.5MHz
2100-2130	1235	Eu	5935, 1035, 103.5MHz
2100-2115	4	Eu	5935, 1035, 103.5MHz

V. by QSL-card. Rec. not acc. Rp. not req.

ETHIOPIA

RADIO ETHIOPIA (Gov.)

✉ P.O. Box 654, Addis Ababa: ☎ +251 (1) 711111.
L.P: Head of International Sce: Moges Taffese.
MW: Addis Ababa 990kHz 1kW.
SW: Gedja (G.C: 38.38E/08.47N): 2 x 100kW.
kHz: *9560, 11800*

Afar	Area	kHz
1300-1400	Af	9560
Arabic		
1400-1500	Af	9560
English		
1600-1700	Af	11800, 9560, 7165
French		
1700-1800	Af	9560
Somali		
1200-1300	Af	9560

ANN: Nat. Sce: "Yeh Ye-Ethiopia Radio Naw". Ext. Sce: "This is R. Ethiopia broadcasting in English".
IS: Electronic keyboard — **V.** by letter (IRC's acc.).

VOICE OF PEACE
(UN-funded Humanitarian Sce.)
D.PRGR: 1100-200 & 1900-2000 in Somali via R. Ethiopia on 9560 & 11800kHz.

RADIO AMAHORO

✉ Rue du Noyer, B-1040 Brussels, Belgium. ☎🖷 +32 (2) 735 3915 .
D.PRGR: SS 0430-0530 in Kinyarwanda via R. Ethiopia on 9560kHz (see also listing under Gabon).

FINLAND

RADIO FINLAND

✉ YLE Radio Finland, Box 78, 00024 Yleisradio, Finland. ☎ +358 (9) 1480 4320. 🖷 +358 (9) 148 1169 — ✉ **in USA:** Box 462, Windsor, CT 06095. ☎🖷 +1 203 688 5540. **Audience Sce.** ☎ 1-800-221-9539.
E-mail: rfinland@yle.fi **Web:** http://www.yle.fi/fbc/radiofin.html
L.P: Head of Ext. Broadc: Juhani Niinistö.
MW: Helsinki 558kHz, Pori 963kHz (see Finland listing in World Radio Europe section).
FM: Helsinki Capital FM 103.7MHz 3kW (rel. R. Finland + VOA, BBC, DW, RFI, NPR, R. Australia & C-SPAN).
Satellite: Eutelsat II f1, Astra 1B, Galaxy5, AsiaSat 2, Intelsat 702.
SW: Pori (G.C: 21.35E/61.28N): 3 x 500kW, 1 x 250kW, 1 x 100kW.
kHz: *5985, 6120, 6180, 9560, 9655, 9745, 9855, 11755, 11900, 13645, 15120, 15145, 15185, 15235, 15400, 15440*

English	Area	kHz
0900-0930*	EAs/Pac	13645
1230-1300	NAm	15400
1330-1400*	NAm	15400, 11900
2100-2130*	Eu/Af	6120, 9855

*) incl. weekly news in Latin (see below)

Finnish/Swedish		
0400-0430	EEu	6120

Finnish/ Swedish	Area	kHz
0500-0630	NEu	6120
	ME/EAf	11755
0500-0615	EEu	9655
0500-0630	ME	15440
0700-1700	NEu	6120
0700-1030	Eu	11755
0700-0800	Eu	9560
0700-0830	EEAs/FE	15335, 9760
0800-0930	Eu	9760
0800-0900	Eu	6180
1000-1100	EAs/FE	13645, 11805
1100-1130	SEAs/Pac	17685, 15245
	NAm	15400, 11785
1100-2000	Eu	11755
1300-1330	SEAs	13645, 15185
1400-1500	ME/EAf	13645, 15145
1415-1430	EEu	6180
1500-1530	EEu	6180
1500-1600	NAm	15400, 11900
1600-1700	Me/Eaf	11880, 9680
1700-2000	Eu	9855
1700-2230	Eu	6120
French		
0645-0700	Eu/WAf	6120, 9560

1015-1030	Eu	11755
German	**Area**	**kHz**
0630-0645	Eu	6120, 9560
1030-1100	Eu	11755
Latin (News)		
1353-1400SS	NAm	15400, 11900
2153-2200SS	Eu	6120
Russian		
0430-0500	RUS	6120
0930-1000	RUS	13645, 15235
1430-1500	RUS	6180

V. R. Finland does not verify direct, but QSL-cards can be obtained through their No. American bureau, Box 462, Windsor, CT 06095, USA. Att: QSL Manager.

FRANCE

RADIO FRANCE INTERNATIONALE
Société Nationale de Radiodiffusion sonore pour l'étranger (RFI).
✉ B.P. 9516, F-75016 Paris. (**Studios:** 116, Av. du Président Kennedy, F-75016 Paris). ☎ +33 (01) 42 30 22 22. 📠 +33 (01) 42 30 44 81. ⏰ RFI 614171F. **Web:** http://www.rfi.fr
LP: Pres:Jean-Paul Cluzel. DG: Christian Charpy. Dir. of Communication: Christine Berbudeau. Dir. of Inf: Anne Toulouse. Editor-in-Chief French Sces: Henri Perilhou. Editor-in-Chief RFI Afrique: Jean-Karim Fall. Dir. Foreign Language Sces: Nicolas Levkov. Dir. of Prgrs: Alex Taylor. Dir. Admin & Finance: Philippe Tarie. Dir. Human Resources: Catherine Dessein. Tech. Dir: Patrice Berestetsky.

HOME SERVICE
Polish on Lille 1377 kHz: MF 1800-1830.
✉ R.F.I. Emissions pour les travailleurs étrangers immigrés, 116 Av. Prés. Kennedy, 75016 Paris.

FOREIGN SERVICE
MW: Paris 738kHz 5kW (RFI Plus Afrique in French 0400-0800, foreign languages 0800-0400), Toulouse 945kHz 300kW, Strasbourg 1278kHz 300kW, Lille 1377kHz 300kW, Cyprus 1233kHz 600kW, Kunming (China) 1296kHz 1000/500kW.
FM: Paris 89.0MHz 10kW (rel. World Sce./RFI plus Afrique in French 24h).
SW: Allouis (G.C: 02.13E/47.10N): 4 x 100kW, 1 x 4kW – Issoudun (G.C: 01.54E/46.56N); 20 x 500kW, 9 x 100kW.
Relay stations: Moyabi, Gabon 500kW; Montsinéry, French Guiana 500kW; Yamata, Japan 300kW; Beijing & Xian, China 120kW.
A new relay st. is under construction in Thailand.
kHz: *4890, 5900, 5915, 5920, 5925, 5945, 5990, 6045, 6120, 6175, 7135, 7140, 7160, 7280, 7305, 7315, 7350, 9485, 9490, 9495, 9550, 9605, 9655, 9660, 9715, 9745, 9790, 9800, 9805, 9830, 9830, 9845, 9845, 9845, 9845, 9845, 9845, 11600, 11615, 11650, 11670, 11685, 11700, 11705, 11710, 11845, 11890, 11910, 11965, 11975, 11995, 12005, 12015, 12025, 12075, 13625, 13640, 15135, 15155, 15195, 15200, 15210, 15300, 15315, 15365, 15405, 15435, 15440, 15460, 15515, 15525, 15530, 15540, 15605, 17560, 17575, 17620, 17630, 17650, 17795, 17800, 17850, 17850, 17860, 21580, 21620, 21685*

Arabic	**Area**	**kHz**
1600-1700	NAf/ME	12025, 11650, 9655
1700-1800	NAf/ME	12025, 11650, 9845, 9655
Cambodian		
1200-1300	CBG	12025
Chinese		
0930-1030	CHN	12025, 9660
1200-1300	CHN	12075, 7140
Creole		
1330-1400	CAm	15515, 13640
English		
1200-1300	Eu/Af/As/Pac/Am	17575, 15540, 15195, 15155, 13625, 11600, 9805
1400-1500	ME/.As	17560, 15405, 11910
1600-1700	Eu/ME/Af	15530, 15460, 15210, 12015, 11700, 11615
1700-1730	ME/Af	15460, 15210
French (RFI Monde)		
0000-0100	Eu/As/Pac/Am	15440 (to 0030), 12025, 11670, 9805, 9800, 9790
0100-0130	As/Pac/Am	15440, 9790, 9715
0130-0200	As/Am	15440, 11995, 11670, 9800, 9790, 9715
0200-0300	Am	9800, 9790, 9715, 5920
0300-0400	Eu/ME/Af/Am	11700, 11685, 9805, 9800, 9790, 9550, 7315, 7280, 7135, 6045, 5990
0400-0430	Eu/ME/Af/Am	11995, 11700, 11685, 9805, 9800, 9790, 9745, 9550, 7280, 7135, 6175, 6045, 5990, 5925, 5920, 4890, 1233
0430-0500	Eu/ME/Af/Am	11685, 9805, 9800, 9745, 9550, 7280,

		6045, 5990 (to 0445,) 5925, 5920,
French		
(RFI Monde)	**Area**	**kHz**
		1233
0500-0530	Eu/M/Af	15605, 15155, 15135, 11995, 11975, 11700, 11685, 9805, 9790, 9745, 7305, 7280, 7135, 6045, 5945, 4890
0530-0600	Eu/ME/Af	15605, 11975, 11685, 9805, 9745, 7305, 7280
0600-0630	Eu/ME/Af	17800, 17620, 15605, 15300, 15155, 15135, 11975, 11700, 9845, 9805, 9790, 9745, 7305, 7280, 7135
0630-0700	Eu/ME/Af	15605, 11975, 9845, 9805, 9745, 7305, 7280
0700-0730	Eu/ME/Af	17850, 17800, 17620, 15605, 15315, 15300, 15135, 11975, 11700, 11670, 9845, 9805, 9790, 7280, 7135
0730-0800	Eu/ME/Af	15605, 11975, 11670, 9845, 9805, 7280
0800-0830	Eu/ME/As	17650, 15605, 15195, 15155, 11975, 11670, 9805
0830-0900	Eu/ME/Af/As	17850, 17800, 17620, 17620, 15605, 15315, 15300, 15155, 11975, 11845, 11670, 9805
0900-1000	Eu/ME/A/fAs	21620, 17850, 17795, 17650, 17620, 15605, 15315, 15300, 15195, 15155, 11845, 11670, 9805
1000-1030	Eu/ME/A/fAs	21620, 17850, 17795, 17650, 17620, 15605, 15315, 15300, 15195, 15155, 11845, 11670, 9805
1030-1100	Eu/ME/Af/As Pac/Am	21620, 17850, 17795, 17650, 17620 17575 15605 15315, 15300 15195, 15155, 13625, 11845, 11710, 11700, 11670, 9830, 9805
1100-1200	Eu/ME/Af/As/ Pac/Am	21620, 21580, 17850, 17795, 17650, 17620, 17575, 15435 (to 1130), 15365, 15315, 15300, 15195, 15155, 13625, 11890, 11845, 11710, 11700 (to 1130,) 11670, 9830, 9805, 6175
1200-1230	Eu/ME/Af/Am	21620, 21580, 17850, 17650, 17620, 15515, 15435, 15315, 15300, 13640, 11845, 11670, 9790
1230-1300	Eu/ME/Am	17650, 15515, 15435, 13640, 11670
1300-1330	Eu/ME/As/Am	17650, 15195, 13625, 11615, 9805, 684
1330-1400	Eu/ME/Af/As/ Pac/Am	21620, 21580, 1785, 17650, 17620, 17560, 15435, 15315, 15300, 15195, 13625, 11845, 11615, 9805, 9790, 684
1400-1500	Eu/ME/Am	17860 (from 1430), 17650, 15605, 15525, 15515=(from 1430), 15195, 15155, 11615
1500-1600	Eu/M/EAf/As/ Am	21620, 21580, 17860, 17850, 17795, 17620, 15605, 15515, 15405, 15315, 15300, 15195, 15155, 11845, 11615
1600-1630	Eu/ME/Af/As	21685, 17620, 15300, 12030, 11995, 11705, 1296
1630-1700	Eu/ME/Af/As/ Am	21685, 17860, 17620, 15525, 15515, 15300, 11995, 11705, 1296
1700-1730	Eu/ME/Af/Am	21685, 17620, 15525, 15300, 11995, 11705, 11670, 9805, 1278
1730-1800	Eu/ME/Af/Am	21685, 17620, 15525, 15460, 15300, 15210, 11995, 11705, 11670, 9805, 1278
1800-1830	Eu/ME/Af/Am	21685, 17630, 17620, 15460, 15300, 11995, 11705, 9790, 9495, 7160
1830-1900	Eu/ME/Am	17630, 11995, 9495, 7135
1900-2000	Eu/ME/Af/Am	17630, 15300, 11995, 11965, 11705, 11670, 979,0 9605, 9495, 9485, 7160
2000-2100	Eu/Am	17630, 9605, 9495, 7350, 5915
2100-2130	Eu/ME/Af/Am	17630, 15300, 9805, 9790, 9485, 7350, 7315, 7160, 6175, 5900
2130-2200	Eu/Am	17630, 9805, 7350, 5900
2200-2300	Am	11670, 9800, 9790, 9715
2300-2330	As/Am	15440, 12025, 12005, 11670, 9805, 9790, 9715
2330-2400	As/Pac/Am	17620, 15440, 15200, 12025, 12005, 11995, 11670, 9805, 980,0 9790, 9715
French (RFI Plus Afrique)		
0430-0500	Af/ME	11995, 11700, 9790, 7135, 6175, 4890
0530-0600	Af/ME	15155, 15135, 11995, 11700, 9790, 7135, 5945, 4890
0630-0700	Af/ME	17800, 17620, 15300, 15155, 15135, 11700, 9790, 7135
0730-0800	Af/ME	17850, 17800, 17620, 15315, 15300, 15135, 1170,0 9790, 7135
0800-0830	Af/ME	17850, 17800, 17620, 15315, 15300,

11845

French (RFI Plus Afrique)

	Area	kHz
1230-1330	Af/ME	21685, 21620, 21580, 17850, 17620, 15315, 15300, 11845, 9790
1400-1500	Af/ME	21620, 21580, 17850, 17795, 17620 15315, 15300, 11845
1830-1900	Af/ME	21685, 17620, 15460, 15300, 11705, 979, 7160
2000-2100	Af/ME	15300, 11995, 11965, 11705, 9790, 9485, 7315, 7160
2130-2200	Af/ME	15300, 9790, 9485, 7315, 7160, 6175

Lao

1100-1200	LAO	12025

Persian

1800-1900	ME	12025, 9845

Polish

0545-0600*	Eu	5990
1700-1800*	Eu	11670, 9805
1800-1830*	Eu	1377
2200-2300*	Eu	7135, 5915

Portuguese

0000-0100	Am	11995, 9490
1700-1800	Af	15530, 12015
2200-2300	Am	15200, 13625, 11995

Romanian

1600-1700*	Eu	11670, 9805
2100-2200*	Eu	9805, 7135

Russian

1400-1500*	RUS	15605, 15155, 11670
1900-2000*	RUS	11670, 9805

Serbian/Croatian

1500-1600*	Eu	11670, 9805
2000-2100*	Eu	9805, 7135

Spanish

0100-0130	Am	11995, 11670, 9800
1000-1030	Am	9790, 9715, 5920
1130-1200	Am	15435, 13640, 11670
1300-1330	Am	17560, 15515, 15435, 13640
1400-1430	Am	17860, 15515
1600-1630	Am	17860, 15525, 15515

Spanish	Area	kHz
2200-2230*	Eu	945
2300-2330	Am	17620, 15200, 11995, 11670, 9800

Vietnamese

1400-1500	VTN	6120
1500-1600	VTN	1296

NB: Transmissions to Europe marked * are 1h earlier in summer.
ANN: "Radio France Internationale" or "Ici Paris, Radio France Internationale" — **IS:** "La Marseillaise" — **V.** by QSL-card.

GABON

AFRICA No. 1 (Comm.)

B.P. 1, Libreville. ☎ +241 760001. 🖷 +241 742133. ① 5558 GO. |
in France: Eurafripub, 27 rue Guersant, F-75017 Paris or B.P. 944, F-75829 Paris Cédéx 17. ☎ +33 (1) 4574 8383. 🖷 +33 (1) 4574 1769).
E-mail: africagc@club-internet.fr　**Web:** http://www.sit.ca/africa/
L.P: Pres: Louis Barthélémy Mapangou. Dir Délegué: Michel Koumbangoye. Pierre Devoluy. Dir. Tec: Gaston Ombolo Ki-Obo. Dir. Prgrs. & Adv: Augustin Letamba. Dir. Inf: Jean Valère Mbina Mandza.
SW: Moyabi-Moanda (G.C: 13.31E/01.40S): 5 x 500kW tr's.

kHz	Language	Area	Times
9580	French	CAf	0500-2300
9790	Kinyarwanda+	Rwanda	1700-1900
15475	French	WAf`	1600-2100
17630	French	WAf	0700-1600

Main N: 0530, 0630, 0730, 1115W, 1215, 1700, 1830, 2200.
+) R. Amahoro, Rue de Noyer 332, B-1040 Brussels, Belgium.
Relays of other broadcasters: See R. France Internationale, R. Japan.
V. by letter (for own prgrs only).

GEORGIA

RADIO GEORGIA (Gov.)

68 M. Kostava Street, 380071 Tbilisi. ☎ +995 (32) 368885.
L.P: Editor-in-Chief: Shamil Cheabrishvili.
SW: Dusheti (G.C: 44.41E/42.03N): 100/50kW.
kHz: *6080, 6180, 11760, 11910*

English	Days	Area	kHz
0630-0700	1234567	NEu	11805
0830-0900	1234567	Eu	11910
0930-1000	1234567	ME	11910

English	Days	Area	kHz
1930-2000	1234567	Eu	11760

German			
0700-0730	1234567	NEu	11910
2000-2030	1234567	Eu	11760

Georgian			
0500-0600	24	ME	6080
1630-1730	67	ME	6180
1700-1730	12345	ME	6180

Romanian			
0600-0630	1234567	EEu	11910
2030-2100	1234567	EEu	11760

ANN: Georgian: "Laparakobs Tbilisi". E: "This is Georgia". R: "Govorit Radiostantsiya Gruziya" — **V.** by QSL-card.

VOICE OF HOPE (Rlg.)

ADDR: P.O. Box 109, Hereford HR4 9XR, United Kingdom.
E-mail: voh@broadcast.net **Web:** http://www.highadventure.org/
SW: Dusheti (G.C: 44.41E/42.03N): 100kW.

English	Area	kHz
1430-1515 MF	India	17525
1800-2000	Europe	9310

GERMANY

DEUTSCHE WELLE (DW)

P.O.Box 100 444, 50968 Köln. ☎ +49 (221) 3890 🖷 +49 (221) 389 3000
Cable: 888485 ① 221 386 DWK. **E-mail:** online@dwelle.de
Web: http://www.dwelle.de/
L.P: DG: Dieter Weirich. Head of Prgrs: Joseph M. Gerwald. CE: Dr. Hans-Dieter Godtmann. Press & P.R: Ralf Siepmann. Head of English Prgrs: Michael Behreus.
Satellite: Deutsche Welle is available on the following satellites:
Europe: Astra 1A - 19.2E (11.229 GHz V - subcarriers 7.38/7.56 MHz (stereo) and 7.74/7.92 MHz (mono) – Astra 1B (digital) 11.493 GHz subcarriers 6.66/6.84MHz.
Europe/Middle East: Eutelsat - 13.0E 11.163 GHz V - subcarriers 7.02/7.20 MHz (stereo), 7.74/7.92/8.28/8.46MHz (mono).
North America/South America: Intelsat-K – 338.5W 11.735 GHz H - subcarriers 7.38/7.56 MHz (stereo), 7.74MHz (mono).
North America/Central America/Hawaii: GE-1 - 257.0W 4.14GHz V - subcarriers 7.38/7.56 MHz (stereo), 7.74MHz (mono).
Europe/Africa/West Asia: Intelsat 707 -359.0E (3.911 GHz RHCP) - digital subcarriers
Asia/Pacific: Asiasat 2 - 100.5E 4.000GHz H - digital subcarriers
MW: Trincomalee, Sri Lanka 1548kHz 400kW – Moscow, Russia 693kHz 10kW – Krasnodar, Russia 1170kHz 600kW – St. Petersburg, Russia 1188kHz 1kW – Wolvertem, Belgium 1512kHz 25kW – Grigoriopol, Moldova 999kHz 500kW.
SW: Wertachtal (G.C: 10.41E/48.05N): 13 x 500kW – Nauen (G.C: 12.54E/52.38N): 4 x 500kW, 1 x 100kW – Jülich (G.C: 06.22E/50.57N): 1 x 100kW – Kigali, Rwanda (G.C: 30.07E/01.53S): 4 x 250kW – Sines, Portugal (G.C: 08.45W/37.57N): 3 x 250kW – Antigua (G.C: 61.48W/17.06N): 2 x 250kW – Trincomalee, Sri Lanka (G.C: 81.10E/08.44N): 3 x 250kW. Also via Sackville (Canada), Bonaire & Madagascar (R. Netherlands relay st's).
Relays: A=Antigua, B=Bonaire, C=Canada, G=Grigoriopol (Moldova), Ik=Irkutsk, K=Kigali, Kr=Krasnodar, Mo=Moscow, N=Novosibirsk, P=Petropavlovsk, S=Sines, Sa=Samara, So=Sofia, St-P=St. Petersburg, T=Trincomalee, Tl=Talata-Volondry (Madagascar), V=Vladivostok, W=Wolvertem (Belgium).
kHz: *3995, 5895, 5925, 5935, 5945, 5925, 5960, 5965, 5980, 6000, 6010, 6015, 6035, 6040, 6045, 6055, 6065, 6085, 6100, 6120, 6130, 6145, 6170, 6185, 7120, 7130, 7145, 7150, 7175, 7225, 7255, 7240, 7255, 7265, 7270, 7285, 7305, 7355, 7390, 7395, 9450, 9505, 9515, 9535, 9555, 9565, 9585, 9615, 9620, 9640, 9650, 9670, 9690, 9700, 9715, 9735, 9750, 9765, 9770, 9815, 9860, 9865, 9870, 11615, 11735, 11765, 11785, 11795, 11810, 11820, 11825, 11865, 11905, 11945, 11950, 11955, 12015, 12025, 12045, 13690, 13730, 13790, 15105, 15135, 15145, 15185, 15205, 15275, 15360, 15370, 15410, 15415, 15425, 15515, 15525, 17715, 17780, 17800, 17820, 17825, 17860, 17895, 21600, 21705*

Winter Schedule 1998/99

GERMAN LANGUAGE PROGRAMMES

Times	Area	kHz
0000-0200	EuAsAm	15410, 15275, 13780, 13750, 11785, 9765, 9750, 9730, 9545, 7130, 6100, 6075, 3995
0200-0400	EuAfAsAm	11785, 9545, 7235, 7130, 6145, 6100, 6075, 3995
0400-0600	EuAfAm	9735, 9545, 7235, 7195, 6145, 6075, 6045, 3995, 1188

Times	Area	kHz
0600-0800	EuAfAsAu	21640, 17845, 13780, 12000, 11865, 11795, 9735, 9690, 9545, 6140, 6075, 3995
0800-1000	EuAfAsAu	21640, 21560, 17845, 15135, 13780, 11865, 11795, 9735, 9690, 9545, 6140, 6075,1188
1000-1200	EuMEAfAs	21640, 17845, 15275, 15135, 13780, 12080, 11865, 9545, 9480, 7340, 7315, 6140, 6075
1200-1400	EuMEAfAsAm	17845, 17765, 17760, 17730, 15590, 15275, 15135, 13780, 12080, 11865, 9545, 9480, 7340, 7315, 6140, 6075
1400-1600	EuMEAfAsAm	17765, 17730, 15285, 15275, 15135, 13780, 11795, 9780, 9620, 9545, 7315, 6140, 6075
1600-1800	EuMEAfAsAm	17765, 17730, 15285, 15135, 13780, 11795, 9620, 9545, 7445, 7315, 6140, 6130
1800-2000	EuMEAfAsAuAm	17860, 11795, 9735, 9655, 9545, 7445, 6140, 6075, 3995
2000-2200	EuAfAsAuAm	17860, 17810, 11795, 9735, 9655, 9545, 6075, 3995
2200-0000	EuAsAm	17860, 15410, 15275, 13780, 11785, 9750, 9730, 9715, 9545, 9525, 7375, 7315, 7185, 6100, 6075, 3995, 1188

FOREIGN LANGUAGE PROGRAMS
International Service

Amharic	Area	kHz
1400-1450	Af	17800, 15410, 11810, 9735,

Arabic		
1300-1600	MEAf	17860, 15360, 13690, 12025, 9860

Bengali		
0100-0150	As	9615, 7355, 7285, 6055

Chinese		
1000-1050	As	17895, 17820, 15185, 7390, 5895
1330-1355	As	15205, 11955, 11825, 7305, 5895

Dari		
0900-0915	MEAs	21695, 17895
1355-1420	ME	17825, 15525

English		
0100-0150	Am	9640, 6145, 6085, 6040, 5960
0200-0250	As	9815, 9615, 9575, 7285, 7225, 6035
0300-0350	Am	9640, 9535, 6120, 6045
0400-0450	Af	9565, 7280, 7225, 6045, 6015
0500-0550	Am	6185, 6120, 6100, 5960
0600-0650	MEAfAs	21695, 17820, 11785, 9565, 7225, 6045
0900-0950	AfAsAu	21600, 17820, 17800, 15410, 15145, 12055, 11775, 9565, 6160
1100-1150	Af	17800, 17780, 15410, 15370
1600-1650	AfAs	21600, 17800, 17780, 15410, 15370, 15145, 11810, 9735, 9585, 7305, 7225, 7120, 6170
1900-1950	Af	15275, 15135, 13690, 11810, 11785, 9765, 9640
2100-2150	AfAsAuAm	15275, 11865, 11785, 9765, 9690, 9670, 9615
2300-2350	As	7235, 6130, 6045, 5990

French		
1200-1300	Af	17800, 17780, 15410, 15370, 12015
1700-1800	Af	17800, 11810, 9735, 9640, 7120

Hausa		
0700-0730	Af	15185, 13790, 11810, 9565
1300-1350	Af	17800, 17780, 15410, 15370
1800-1900	Af	15135, 11810, 11785, 9765

Hindi		
1515-1600	As	11865, 9585, 7305, 7225

Indonesian		
1100-1125	As	21695, 17825, 11775
1400-1425	As	15145, 13790, 11825
2200-2250	As	11865, 9805, 9505, 5990

Japanese		
1230-1300	As	11775, 9620, 9555, 7305

Pashto		
0915-0930	MEAs	21695, 17895
1330-1355	As	17825, 15525

Persian		
1000-1050	ME	21695, 17780, 15525
1800-1850	ME	11865, 9770, 7175, 5965, 5935

Portuguese		
0500-0515	Af	11785, 9615, 9565, 7280
1000-1040	Am	15205, 11795, 9605

Portuguese	Area	kHz
2000-2050	Af	13780, 11810, 11785, 9670, 7175
2230-2300	Am	15105, 11820, 9700, 9640, 6145

Sanskrit*		
1545-1600		6170

Spanish		
0200-0250	Am	11795, 9700, 9640, 6045
1100-1130	Am	15205, 11795
2300-0050	Am	15105, 11820, 9700, 9640, 6145

Swahili		
0300-0400	Af	9565, 7280, 7225, 6015
1000-1050	Af	21600, 17800, 15410, 11785, 9565
1500-1600	Af	17800, 15145, 11810, 9735, 7120

Turkish		
0600-0650	EuME	13690, 12015, 9750
1700-1750	EuME	11785, 7225

Urdu		
1430-1515	As	11865, 9585, 7305, 7225, 6170,

* = Biweekly transmission.

European Service

Albanian	Area	kHz
0650-0700	Eu	7270, 6010
1200-1230	Eu	9770, 7255
1600-1630	Eu	11950, 9805
1830-1845	Eu	9805, 7130

Bosnian		
0700-0715	Eu	7270, 6070
1300-1330	Eu	9770, 7175

Bulgarian		
1030-1150	Eu	11905, 11615, 9650,
1700-1800	Eu	11950, 9805, 7130

Croatian		
1330-1400	Eu	9770, 7175
1800-1830	Eu	9805, 7130

English		
2000-2050	Eu	7285

Hungarian		
1230-1300	Eu	7130, 6015

Macedonian		
1400-1430	Eu	9770, 7175

Polish		
1300-1330	Eu	9700, 6015
1730-1800	Eu	7240, 3995
2130-2200	Eu	9535, 7130

Romanian		
0930-1030	Eu	11905, 11615, 9650
1630-1700	Eu	11950, 9805, 7130
2000-2100	Eu	9615, 6000

Russian		
0000-0100	Eu	12045, 11975, 11915, 9565, 7285, 6035, 5925
0400-0500	Eu	7145, 6055,
1600-2200	Eu	7395-(2000-2200), 7145, 5980, 5945-(to 2000)

Serbian		
0715-0730	Eu	7270, 6010
0930-0945	Eu	9670, 7130
1030-1100	Eu	11795, 7255
1430-1530	Eu	9770, 7175

IS: Theme from "Fidelio" by Beethoven. "Es sucht der Bruder seine Brüder" played on celeste.
V. by QSL-card. **PUB:** Monthly prgr.schedule free.

Universelles Leben/UNIVERSAL LIFE (Rlg.)

Universelles Leben e.V., P.O.Box 5643, D-97006 Würzburg. ☎ +49 (931) 3903-0. 🖷 +49 (931) 3903 233. **E-mail:** info@universelles-leben.org
Web: http://universelles-leben.org
Universal Life - The Inner Religion, P.O.Box 3549, Woodbridge, CT 06525, USA. ☎ +1 (203) 281 7771. 🖷 +1 (203) 230 2703.

English	Day	Area	kHz
0800-0830	Sun	Japan/Pac	9930
1300-1330	Sun	Eu	9710a
		India	11700
		Af/Asia	9830, 9950
1400-1430	Sun	NAm	9955
1430-1500	Sun	WAf/Eu	15460
2330-2400	Sun	NAm	5745

French			
2130-2200	Fri	Eu/Af	11960, 9720, 9630, 7310, 1323

German			
1130-1200	Sat/Sun	Eu	7215
1300-1400	Sun	Eu	9710a
1900-1930	Tues	Eu	12010, 11980, 9720, 9450, 7310,

German	Day	Area	kHz
			1386, 1215
1930-2000	Sat	Eu	12010, 11980, 9720, 9450, 7310,
			1386, 1215
2100-2130	Fri	Eu	11960, 7210
2100-2130	Sat	Eu	12010, 11980, 9720, 9450, 7310,
			1386, 1215
Italian			
1100-1130	Sat	Eu	7125
2100-2130	Sun	Eu	3955
Russian			
1300-1330	Sun	Eu	9710

VOICE OF AMERICA/RADIO FREE EUROPE/ RADIO LIBERTY

MW: Munich 1197kHz 300kW, Holzkirchen 1593kHz 150kW.
SW: Wertachtal (G.C: 10.41E/48.05N): 500kW – Holzkirchen (G.C: 11.44E/47.52N): 250kW – Biblis (G.C: 08.30E/49.41N): 100kW – Lampertheim (G.C: 08.33E/49.30N): 100kW.
Schedules listed under Czech Republic (RFE/RL) and USA (VOA).

GREECE

THE VOICE OF GREECE

P.O. Box 60019, 153 10 Aghia Paraskevi, Athens. ☎+30 (1) 6066308. +30 (1) 6066309. **Eng. Dept:** ☎ +30 (1) 6397108, 6014700. +30 (1) 6009608. **E-mail:** skalai@leon.nrcps.ariadne-t.gr (technical), fonel@hol.gr (programming).
Web: http://alpha.servicenet.ariadne-t.gr/Docs/Era5_1.html
L.P: MD: Pantelis Trogadis. Head of Prgrs: Angeliki Mparka. Planning Engineer: Dionysios Angelogiannis. Technical Support: Efstratios Kalaitzis.
MW: Kavala 792kHz 500kW (operated by the VOA).
SW: Avlis (G.C: 23.36E/38.23N): 2 x 100kW – Kavala (G.C: 24.50E/40.52N): 1 x 250kW (operated by the VOA) — V) via VOA tr's in the USA.
kHz: 5895, 6260, 7430, 7450, 9375, 9380, 9395, 9420, 9425, 9580, 9590, 9775, 9915, 11595, 11645, 11710, 12105, 15150, 15175, 15415, 15485, 15630, 15650, 17705

Times	Area	kHz
0000-0350	NAm	5895, 6260, 7450, 9425
0400-0555	ME/Indian Oc.	9420, 11645, 15175, 15650
0600-0800	Eu/Australia	7430, 7450, 9425, 9775V, 11645
0900-0950	EAs/Australia	9590V, 15415, 15650
1000-1150	ME	9420, 9915
1200-1350	Eu/NAm	9580V, 12105, 15175, 15630
1200-1250	CAs	11645
1300-1350	Af	11645
1400-1450	ME/Indian Oc.	9420, 11645, 15630, 15650
1500-1700	Balkans	7450, 9425, 9375, 11645
1710-1750	CIS/Poland	7450, 9375, 9425, 11645
1800-1850	Af	11645, 15150
1800-2050	Eu	7450, 9420
1800-2200	CAN	15485V
	CAm/SAm	17705V
2050-2150	Eu	6260, 7450
1900-1950	Balkans	7430, 9380
2000-2050	Eu	7430, 9380
2100-2250	Australia	11645, 11595
2200-2300	SAm/CAm	9395
2300-2350	Aus/CAm	9395, 9425, 11595, 11710

N: Albanian: 1600, 1915. **Arabic:** 0525, 1230, 1430. **Bulgarian:** 1630, 1900. **English:** 0130, 0330, 0745, 1235 (Eu/NAm), 1240 (Af), 1840, 2000, 2240, 2335. **French:** 2010. **German:** 2020. **Greek:** 0000, 0200, 0400, 0400, 0600, 0900, 1000, 1200, 1300, 1400, 1500, 1710, 1800, 2100, 2200. **Italian:** 2030. **Polish:** 1740. **Portuguese:** 2305. **Romanian:** 1646, 1930. **Russian:** 1335, 1725. **Serbian/Croatian:** 1615, 1940. **Spanish:** 2320. **Swedish:** 2040. **Turkish:** 0745, 1135.
V. by QSL-card. Re. in SINPO code to Direction of Engineering & Development. No recordings.

VOICE OF AMERICA RADIO STATION KAVALA

MW: 792kHz 500kW — **SW:** (G.C: 24.50E/40.52N): 10 x 250 kW.
D.PRGR: see USA.

VOICE OF AMERICA RADIO STATION RHODES

MW: 1260kHz 500kW — **SW** (G.C: 28.10E/36.15N): 2 x 50kW.
D.PRGR: see USA.

GUAM

ADVENTIST WORLD RADIO - ASIA (Rlg.)

P.O. Box 8990, Agat, Guam 96928. ☎ +671 565 2000. +671 565

2983. **Cable:** Adventist. ① 721-6548 KSDA GM. **E-mail:** awr_asia@compuserve.com
Web: http://www.awr.org/awr-asia/guam/index.htm
L.P: GM: Gordon L. Retzer. PD: Greg Scott. CE: Elvin Vence. Freq. Consultant: George Jacobs.
SW: KSDA Agat (G.C: 144.39E/13.20N): 4 x 100kW tr's.
kHz: 5950, 7400, 7455, 9370, 9385, 9495, 9650, 11775, 11790, 11895, 11980, 12160, 13720, 13840, 15170, 15225, 15310, 15610, 17645

Arabic		
0500-0700	Af/ME	11610
0700-0800	Af	7260
0700-0900	Af	12080
0930-1000	Eu	7230
1000-1100	Af	17610
1230-1300	Eu	7230
1330-1430	Eu	11700
2100-2300	Af	5890
Armenian		
1800-1900	Eu	9475
Bangla		
1300-1330	SAs	15225
Bulgarian		
0400-0500	Eu	5840
1700-1800	Eu	5890
Burmese		
1400-1500	SEAs	11625
Cantonese		
1000-1030	CHN	11980
Chin		
0000-0030	SEAs	17645
Dyula		
2130-2200	Af	9640
English		
0500-0530	Af	5960
0500-0530	Af	6100
1000-1100	NEAs	11660
1030-1100	NEAs	11795
1230-1300	SEAs	15225
1330-1400	NEAs	11660
1430-1500	SAs	11980
1530-1600	SAs	11625
1600-1700	SAs	11980
1730-1800	SAs	9355
1730-1800	ME	9355
1730-1800	Af	12130
1800-1830	Af	5960, 6100
2030-2100	Af	9640, 9745
2130-2200	NEAs	15610
2300-2330	SEAs	11775
2330-0000MF	SEAs	11775
Farsi		
0330-0400	ME	9465
1630-1700	ME	9420
1700-1730	ME	9355
French		
0230-0300	Af	3215
0630-0700	Af	7260
0900-0930	Af	12080
1300-1330	Eu	11700
1700-1730	Af	3215
2000-2030	Af	9640, 9745
2300-2330	Af	5890
German		
0900-0930	Eu	7230
1100-1230	Eu	7230
Hakka		
1030-1100	CHN	11980
1000-1100	Af	17610
1800-2000	Af	9445
Hindi		
0230-0300	SAs	7235
1500-1530	SAs	11625
1530-1600	SAs	13670
Indonesian		
1100-1130	SEAs	13720
2200-2230	SEAs	15225
Italian		
0800-0900	Eu	7230
Japanese		
1300-1330	J	11660
2100-2130	J	15610
Javanese		
1130-1200	SEAs	13720
2230-2300	SEAs	15225

Kannada	Area	kHz
1530-1600	SAs	11980

Kazakh

1630-1700Su/We	WCAs	9475

Kiswahili

1700-1730	Af	12130

Korean

1200-1300	FE	11660
2000-2100	FE	9370, 9515

Kyrgyz

1630-1700Mo/Th	WCAs	9475

Malagasy

0300-0330	Af	3215
1530-1600	Af	3215

Malayalam

1500-1530	SAs	13720

Mandarin

0200-0400SS	CHN	21770
1000-1100	CHN	13720
1100-1200	CHN	11660
1100-1500	CHN	15725
1300-1400	CHN	11980
1400-1500	CHN	15225
2100-2200	CHN	9370
2200-2300	CHN	11785, 15610
2300-0000	CHN	17645
2300-0200	CHN	17715

Marathi

1530-1600	SAs	15225

Minnan

1100-1130	CHN	11980

Punjabi

0200-0230	SAs	7235
1500-1530	SAs	13670
1530-1600	SAs	13720

Romanian

1800-1900	Eu	5840

Russian

1000-1030	RUS	11795
1630-1700Sa	WCAs	9475

Shanghainese

1030-1100	CHN	11980

Sinhalese

1400-1430	SAs	11980

Slovakia

1600-1800	ME	11690
1800-2000	ME	9445

Somali

1630-1700	Af	5620

Spanish

0000-0400	CAm	9725
1000-1200SS	CAm	5030, 6150, 7375, 13750
1000-1300	CAm	9725
1300-1600SS	CAm	9725

Tagalog

1200-1230	SEAs	15225
1700-1800	ME	9535

Tamil

1500-1530	SAs	11980

Telugu

1500-1530	SAs	15225

Turkish(fromJan99)

0500-0530	Eu	5890
2100-2130	Eu	9470
2130-2200	Eu	7105

Urdu

0300-0330	SAs	9465
1600-1630	SAs	9420

Uzbek

1630-1700Tu/Fr	WCAs	9475

Vietnamese

1200-1300	SEAs	11980
1600-1700	SEAs	11625
2200-2300	SEAs	11775

a) Mon/Wed/Fri, b) Tues/Thurs/Sat/Sun, c) Thurs/Fri/Sat/Sun
IS: "Lift Up the Trumpet".
ANN: "This is Adventist World Radio, the Voice of Hope".
V. by QSL-card. Rp — **PUB:** "AWR Asiawaves", quarterly.

TRANS WORLD RADIO PACIFIC (Rlg.)
P.O. Box CC, Agana,.Guam 96910-8980. ☎ +1 (671) 477 9701.
+1 (671) 477 2838. **E-mail:** ktwr@twr.hafa.net.gu
Frequency Coordination: 1868 Halsey Drive, Asan, Guam 96922-1505.
☎ +1 (671) 828 8637. +1 (671) 828 8636.
E-mail: ktwrfreq@twr.hafa.net.gu

Web: http://www.guam.net/pub/twr/ktwrguam.htm
L.P: MD: Edward Stortro. PD: Gary Whitmore. CE: George Zensen. Head of PR: Wayne Frost.
SW: KTWR Merizo (G.C: 144.40E/13.17N): 5 x 100kW tr's.
kHz: *9430, 9475, 9545, 9590, 9815, 9820, 9865, 9870, 11650, 11665, 11670, 11690, 11810, 11995, 15105, 15195, 15200, 15330*

Burmese	Area	kHz
1415-1500	BRM	9430

Chinese dialects

1330-1500	CHN	9870
2200-2300	CHN	11650

English	Area	kHz
0730-0915	SEAs	15200
1000-1100	FE	9865
1500-1630	SAs	15105

Hakka

0815-0900	CHN	11665

Indonesian

0915-1100	INS	15200
2100-2200	INS	11810
2200-2245	INS	15195

Japanese

1200-1300	J	15330
2100-2200	J	11690

Korean

2030-2130	EAs	9475

Mandarin

0900-1130	CHN	9815
0900-1400	CHN	11670
1100-1330	CHN	9590
1130-1530	CHN	9820
1400-1530	CHN	9545
2130-2345	CHN	11995

South Asian languages

1330-1400	SAs	15200

Tamil

1400-1430	SAs	15195

Vietnamese

1130-1200	VTN	9870
1500-1530	VTN	9430

a) Wed-Sun, b) Mon/Tues.
ANN: "This is your station for inspiration, KTWR, Agana" or "This is the voice of Trans World Radio Pacific, KTWR, Agana" .
IS: "We've a Story to Tell the Nations" played on an organ.
V. by QSL-card. 3 IRC's for airmail reply, one for surface mail. Rec. acc.
PUB: Frequency schedule on request.

GUIANA (French)

RADIO FRANCE INTERNATIONALE RELAY STATION
TDF Montsinery, B.P. 97307, Cayenne Cedex.
FM: Cayenne 98.7MHz.
SW: Montsinery (G.C: 53.00W/05.00N): 5 x 500kW tr's.
kHz: *5920, 9715, 9790, 9800, 11670, 11995, 13625, 13640, 15190, 15200, 15300, 15435, 15515, 17560, 17575, 17620, 17630, 17860, 21645, 21685, 21765*
D.PRGR: See RFI listing under "France". St. also used by R. Japan & China R. Int. (see respective listings).
V. by QSL-card for direct re.

HAWAII (U.S. State)

KWHR – WORLD HARVEST RADIO (Rlg.)
Naalehu, Hawaii. ☎ +1 (219) 291 8200. +1 (219) 291 9043.
E-mail: kwhr@lesea.com **Web:** http://www.kwhr.com
L.P: GM: Peter Sumrall. Sales Mgr: Joe Hill. CE: Douglas W. Garlinger. Freq. Consultant: George Jacobs.
SW: Naalehu (G.C: 155.40W/19.02N): 4 x 100kW tr's.
kHz: *9930, 11815, 13625, 17510*

Times	Area	kHz	Times	Area	kHz
0000-0500	EAs	17510	1800-2000	Oce	13625
0500-1400	EAs	9930	2000-2200	EAs	11815
1400-1600	SAs/SEAs	9930	2200-2400	EAs	17510
1600-1800	SEAs	9930			

IS: "Onward Christian Soldiers" — **ANN:** "This is World Harvest Radio, the Int.Sce. of LeSea Broadc. Corp." — **V.** by QSL-card. Re. & monitors wanted — **PUB:** World Harvest Magazine.

HUNGARY

RADIO BUDAPEST
Bródy Sándor u. 5-7. H-1800 Budapest. ☎ +(36-1) 328-8320. ▤ +(36-1) 328-8517. **Answerline:** +36 (1) 138 8320. **E-mail:** angol1@kaf.radio.hu (English), nemet1@kaf.radio.hu (German Service), szulofold1@kaf.radio.hu (Hungarian), orosz1@kaf.radio.hu (Russian) **Web:** http://www.eunet.hu/radio/
L.P: Dir. Foreign Broadc: Antal Réger. Vice-Dir. Foreign Broadc: Zsuzsa Mészáros. Freq. Mgr: Ferenc Horváth.
SW: Jaszberény (G.C: 19.52E/47.35N: 2 x 250kW – Diósd (G.C: 18.57E/47.25N): 2 x 100kW – Szekesfehérvár (G.C: 18.24E/47.10N): 1 x 100kW, 1 x20kW.
kHz: 3975, 5905, 5945, 5965, 5995, 6005, 6010, 6025, 6030, 6165, 7135, 7155, 7180, 7185, 7190, 7225, 7255, 7260, 9530, 9535, 9805, 9710, 9795, 9805, 9835, 9840, 9870, 11660, 15395, 17790, 21560

Winter Schedule 1998/99

Croatian	Area	kHz
05.30-06.00Mo-Su	Eu	3975, 5965
16.00-16.30Mo-Su	Eu	3975, 5965

English		
02.00-02.30Mo-Su	NAm	6135, 9835
03.30-04.00Mo-Su	NAm	9835, 11990
20.00-20.30Mo-Su	Eu	3975, 6025
22.00-22.30Mo-Su	Eu	3975, 7250

German		
18.30-19.00Mo-Sa	Eu	3975, 6175
20.30-21.00Mo-Sa	Eu	3975, 7165
13.00-14.00Su Eu	7135	9660
18.00-19.00Su Eu	3975	6175

Hungarian		
00.00-01.00Mo	Sam	11985, 13685
01.00-02.00Mo-Su	NAm	6120, 9835
02.30-03.30Mo-Su	NAm	9835, 11990
11.00-12.00Mo-Su	AUS	17670, 21560, 25700
12.00-13.00Su AUS	17670	21560, 25700
12.00-13.00Su Eu	7135	9655
	Area	kHz
14.00-15.00Su Eu	7135, 9840	
19.00-20.00Mo-Su	Eu	3975, 7185
21.00-22.00Mo-Su	Eu	3975, 7230
21.00-22.00Mo-Su	NAm	11895
23.00-00.00Mo-Su	Sam	11985, 13685

Romanian		
05.00-05.30Mo-Su	Eu	3975, 7260
16.30-17.00Mo-Su	Eu	3975, 5965

Russian		
04.00-04.30Mo-Su	Eu	3975, 6005
15.30-16.00Mo-Su	Eu	9660, 11875

Serbian		
06.30-07.00Mo-Su	Eu	5915, 9870
18.00-18.30Mo-Sa	Eu	3975, 6175
20.30-21.00Su Eu	3975, 7165	

Slovak		
06.00-06.30Mo-Su	Eu	3975, 5915
17.30-18.00Mo-Su	Eu	3975, 6010

Ukrainian	Area	kHz
04.30-05.00Mo-Su	Eu	3975, 6005
17.00-17.30Mo-Su	Eu	3975, 9680

Satellite to Europe
Radio Budapest now available via the EUTELSAT Hotbird IV satellite at 13 degrees East: transponder 115 (Duna TV - 10,815.08MHz - horizontal polarisation), audio sub-carrier 7.56MHz.

Monday-Saturday		Saturday night - Sunday morning:	
22.00-23.00	Hungarian	22.00-23.00	Hungarian
23.00-23.30	English	23.00-00.00	Hungarian
23.30-00.00	German	00.00-00.30	English
00.00-00.30	English	00.30-01.30	German
00.30-01.00	Slovak	01.30-02.00	Slovak
01.00-01.30	Croatian	02.00-02.30	Croatian
01.30-02.00	Serbian	02.30-03.00	Serbian
02.00-02.30	Romanian	03.00-03.30	Romanian
02.30-03.00	Russian	03.30-04.00	Russian
03.00-03.30	Ukrainian	04.00-04.30	Ukrainian
04.30-05.00	English		

V. by QSL-card. Rec. acc.

ICELAND

RÍKISÚTVARPID
Efstaleiti 1, 150 Reykjavík. ☎ +354 515 3000. ▤ +354 515 3010. ① 2066 ISRADIO IS. **E-mail:** isradio@ruv.is **Web:** http://www.ruv.is
L.P: DG: Heimir Steinsson. Chmn. Prgr. Council: Guunlaugur Saeyar Gunnlaugsson. Dir. Admin: Hördur Vihjámsson. Dir. Radio: Markús Örn Antonsson. PD: Margaret Oddsdóttir (Ch. 1), Sigurdur Tomasson (Ch. 2). Dir. N: Karl Jónasson. Head of Int. Rel: Dóra Ingvadóttir.
SW: Reykjavík (G.C: 21.50W/64.05N): 2 x 10kW (USB mode).
kHz: 3242, 3295, 7870, 9275, 9282, 11402, 13835, 13855, 13860, 15770, 15790

Times	Area	kHz
1215-1300	Eu	11402, 13860
1410-1440	NAm	11402, 13860
1855-1930	Eu	7740, 9275
1935-2010	NAm	11402, 13860
2300-2335	NAm	9275, 11402

All prgrs in Icelandic.
NB: Frequencies change often, but will always be selected from the 11 registered freqs listed above.
ANN: "Utvarp Reykjavík" (P1), "Rás tvö" (P2) — **V.** by QSL-card.

RADIO ALPHA & OMEGA (Rlg.)
Omega Television, Grensasvegur 8, 108 Reykjavík. ☎ +354 568 3131 or 2777. ▤ +354 568 3741 — **L.P:** Dir: Erik Sigurbjoernsson.
English: Sun 1800-1900 on 6110kHz via Deutsche Welle,.Germany.
F.PI: Daily prgrs.

INDIA

ALL INDIA RADIO
External Services Division, All India Radio, Broadcasting House, 1 Sansad Marg, (P.O. Box 500), New Delhi-110 001. ☎ +91 (11) 3715411. **Cable:** Airforeign. **E-mail:** air@kode.net
Web: http://www.allindiaradio.org/
L.P: Dir: P.M. Iyer.
MW: See Home Sce
SW: A=Aligarh (78.06E/28.00N): 4 x 250kW – B=Bangalore (G.C: 77.38E/12.58N): 6 x 500kW – D=Delhi-Kingsway (G.C: 77.12E/26.45N): 2 x 100kW, 6 x 50kW & Delhi-Khampur: 2 x 250 kW, 2 x 100kW, 3 x 50kW, 2 x 20kW – G=Gorakhpur (G.C: 83.24E/26.45N): 1 x 50kW – Ma=Madras (G.C: 80.17E/13.04N): 2 x 100kW, 2 x 10kW – Mu=Mumbai (72.49E/19.11N): 1 x 100kW – P=Panaji (G.C: 73.52E/15.31N): 2 x 250kW.
F.PL: Pune 2 x 500kW.
kHz: 3945, 4790, 4860, 5990, 6045, 6150, 6155, 6165, 7110, 7125, 7140, 7150, 7225, 7250, 7265, 7410, 9545, 9550, 9565, 9595, 9650, 9700, 9705, 9835, 9910, 9950, 10330, 11585, 11620, 11695, 11710, 11715, 11730, 11735, 11740, 11790, 11830, 11840, 11850, 11855, 11870, 11900, 11935, 13620, 13700, 13710, 13720, 13770, 13780, 13795, 15020, 15050, 15075, 15135, 15140, 15175, 15180, 15185, 15260, 15340, 15770, 17387, 17705, 17785, 17840, 17860, 17895

Arabic	Area	kHz
0430-0530	ME	17785B, 15050A, 13620B, 11730A
1730-1945	ME	13620B, 9910A
Baluchi		
1500-1600	PAK	11585B, 6165D, 1071
Bamar		
0100-0130	BRM	11870D, 9950D
1215-1315	BRM	11710D, 11620D
Bengali		
0300-0430	BGD	594
0800-1100	BGD	594
1445-1515	BGD	1134
1600-1730	BGD	1134
Chinese		
1145-1315	CHN	17705B, 15050D, 11840B
Dari		
0300-0345	AFG	13620B, 11735B, 9910A, 7225D
1315-1415	AFG	9910A, 7140D
English		
1000-1100	NEAs	17840B, 15050A, 13700B, 11585A
	Pac	17387D, 15050A, 13700B, 11735B
	CLN	1053
1330-1500	SEAs	13710B, 11620B, 9545D
1745-1945	EAf	15075A, 11935Mu
	NAf/NWAf	13780B, 9650A
	Eu	11620B, 9950D, 7410B
2045-2230	Eu	11620B, 9950B, 9650A, 7410B
	Pac	11715A, 11620B, 9910A, 7150D
Farsi		
0400-0430	IRN	17785B, 15050A, 13620B, 11730A
1615-1730	IRN	11585B, 9910A, 7265D

French	Area	kHz
1945-2030	WAf/NWAf	13780B, 13620B, 9910A
Gujarati		
0415-0430	EAf	17387A, 15075B
1515-1600	EAf	15175B, 11620B
Hindi		
0315-0415	EAf	17387A, 15180D, 15075B
	ME	15075B, 11855P
1615-1730	EAf	15075A, 13720B, 9950D
	ME	13770B, 7410A
1945-2045	Eu	11620B, 9950D, 7410B
2300-2400	SEAs	13795B, 11740B, 9910A
Indonesian		
0845-0945	INS	17387D, 15050A
Nepali		
0130-0228	NPL	11715B, 9550A, 7250P, 6045D, 3945G, 594
0700-0800	NPL	11850D, 9595D, 7250G
1330-1430	NPL	11695B, 7410A, 6045D, 4860D, 3945G, 1134
Punjabi		
0800-0830	PAK	702
1230-1430	PAK	702
Pushtu		
0215-0300	AFG/PAK	13620B, 11735B, 9910A, 7225D
1415-1530	AFG/PAK	9910A, 7140D
Russian		
1615-1715	RUS	15140B, 11620B
Sindhi		
0100-0200	PAK	11790B, 7125D, 5990A, 1071
1230-1500	PAK	11585B, 6165D, 1071
Sinhala		
0045-0115	CLN	1053
1300-1500	CLN	15020D, 9700A, 1053
Swahili		
0430-0530	EAf	17387A, 15075B
1515-1615	EAf	15075A, 13720B, 9950D
Tamil		
0000-0045	SEAs	13795B, 11745B, 9910A
	SAs	9835D, 4790Ma, 1053
0115-0330	CLN	1053
1100-1300	CLN	1053
1115-1215	SEAs	15770A, 13710B
	SAs	17860D, 15075D
1500-1530	CLN	1053
Telugu		
1215-1245	SEAs	15770A, 13710B
Thai		
1115-1200	THA	17895A, 15340B
Tibetan		
0130-0200	CHN	13700B, 11900B, 9565P
1215-1330	CHN	11695B, 9565D, 7410A, 1134
Urdu		
0015-0100	PAK	1071
0015-0430	SAs	9595D, 6155D, 702
0100-0430	SAs	11620B
0200-0430	PAK	1071
0830-1130	SAs	11620D, 9595D, 1071, 702
1430-1930	SAs	6045D, 4860D, 702
1600-1930	PAK	1071

News in English: 2300, 1000, 1055, 1330, 1455, 1800, 1900, 2100, 2200.

ANN: Home Sce: Hindi:"Ye Akashvani... (name of regional st.) hai". English : "This is All India Radio ... (name of regional st.)". Tamil: "Idi Akashvani ... (name of regional st.), vanali nilayam".

Ext Sce: English: "This is the General Overseas Service of All India Radio". Hindi: "Yeh Akashvani ki videsh prasaran sewa hai". Tamil: "Idi Akashvani videsh sewai". Sinhala: "Me All India Radio videshiya sevayai". Nepali: "Yo All India Radio ho". Dari: "Injaw Delhi". Pushto: "Da All India Radio de." Farsi: "Inja Delhi". Indonesian: "Inilah All India Radio" — **IS:** All India Radio's signature tune is a melody of 8 seconds' duration repeated with intervals of 10 seconds in between. The instruments used are violin, cello and tanpura.

V: Re. to Director of External Services, All India Radio, Post Box No. 500, New Delhi-110001. Colorful QSL cards are issued after checking with station logs of resp stations. Some stations QSL direct via card/letter. Verification normally takes a few weeks.

THE VOICE OF INDONESIA
✉ P.O. Box 1157, Jakarta 10001.
SW: Jakarta (Cimanggis) 9525kHz 250kW.
FM: Jakarta 103.0MHz.
Arabic: 0300-0400. **Chinese:** 1000-1100. **English:** 0100-0200, 0800-

0900, 2000-2100. **French:** 1900-2000. **German:** 1800-1900. **Indonesian:** 0200-0300, 1200-1300. **Japanese:** 1130-1200. **Malay:** 0900-1000. **Spanish:** 0030-0100, 1730-1800. **Thai:** 1100-1130.
IS: Rayuan Pulau Kelapa (the song of the Coconut Islands) played on a Hammond Organ, Flute, Vibe and Piano.
V. by QSL-card or letter. Listeners reports are welcomed.
PUB: Weekly: RRI Radio dan Televisi (Home Prgr.)

THE VOICE OF THE ISLAMIC REPUBLIC OF IRAN
✉ P.O. Box 3333, Tehran. ☎ +98 (21) 2041051.
E-mail: irib@dci.iran.com
Web audio: http://netiran.com/PersianRadio.html
L.P: Dep. MD for Foreign Broadcasting: M. Sarafraz.
MW: See Home Sce. list.
SW: Ahwaz 250kW — Kamalabad (G.C: 51.27E/35.46N): 500/100kW — Mashhad (G.C: 59.33E/36.15N): 500kW — Sirjan (G.C: 55.41E/29.27N): 500kW — Zahedan (G.C: 60.53E/29.28N): 500kW
kHz: 5925, 5995, 6005, 6015, 6020, 6025, 6030, 6050, 6165, 6175, 6210, 7070, 7100, 7115, 7125, 7130, 7145, 7160, 7170, 7175, 7180, 7190, 7215, 7230, 7260, 7290, 7305, 7340, 9022, 9525, 9555, 9570, 9575, 9585, 9605, 9615, 9650, 9665, 9685, 9715, 9720, 9735, 9745, 11705, 11720, 11745, 11750, 11790, 11830, 11855, 11875, 11910, 11930, 11960, 11995, 13605, 13610, 15084, 15165, 15200, 15230, 15260, 15325, 15330, 15350, 15365, 15530

Albanian	Areas	kHz
1830-1900	Eu	9720, 9715, 9615, 7340, 7145, 9525
Arabic		
0030-0230	Am	7190, 6025
0230-0330	Am	15530, 9685, 9022, 7190, 6025
0330-0630	ME	15350, 13605, 11910, 7130, 5995, 1224, 576
0630-0930	ME	15350, 13605, 11910, 5995, 1224, 576
0930-1130	ME	15350, 13605, 11745, 1224, 576
1130-1330	ME	11745, 9650
1330-1530	ME	11995, 9650
1530-1700	ME	11995, 1224, 1080, 765
1700-1930	ME/Af	9665, 7190, 6025
2030-2130	ME/Af	9665, 7190, 6025
2130-2230	Eu/ME	9665, 9022, 7260, 6025
Armenian		
0330-0330	WAs	6210
0930-1000	WAs	11705, 9615
1630-1730	WAs	9615, 5995
Assyrian		
1930-2030	ME	936
Azeri		
1230-1400	WAs	9735, 702
1630-1830	WAs	1404, 1323, 1026
Baluchi		
1430-1530	EAs	594, 531
Bengali		
0030-0130	SAs	9570
1430-1530	EAs/ME	11790, 7215, 1224
Chinese		
1300-1330	CHN	15330, 15260, 6050
1330-1430	CHN	15200, 13605, 11855
2330-0030	CHN	9570, 7215, 6030
Dari		
0130-0230	AFG	720
1430-1530	AFG	720
English		
0030-0130	NAm/SAf	9685, 9022, 6050
1100-1230	ME/As	15260, 11875, 11830, 9585, 9555, 7180, 7160, 7115, 702
1530-1630	As/Pac	11790, 7215
1930-2030	Eu	9022, 7260, 1404
2130-2230	As/Pac	6175, 6165
French		
0630-0730	Eu/Af	15325, 15260, 11790
1830-1930	Eu	9022, 7260, 7160
2230-2330	NAm	9022, 7260, 7215, 7100, 6030
German		
1730-1830	Eu	9022, 7260
Hausa		
0545-0630	WAf	11720
1930-2030	WAf	15200, 11720, 9605
Italian		
0630-0700	Eu	15230, 15084, 13605, 9022
1200-1300	Eu	15350, 15084, 13605
1900-1930	Eu	9615, 7340, 7145
Kazakh		
1500-1530	As	11960, 9735, 9585

Kurdish	Area	kHz
1330-1830	ME	1422, 639

Malay

1300-1330	SEAs	15365, 15260, 11790

Persian

0000-0630+	Eu	15084
0700-1200+	Eu	15084
0130-1330+	ME	15365
1300-2400+	Eu	15084
1330-2000+	As	6020
1630-1730+	Eu	7230
1700-2130+	Af/ME	9665, 7190, 1224
1930-2030+	ME	7130
2030-2130	Am	15165, 11930, 6175, 6165

+) rel. Home Sce.

Pushtu

0230-0330	AFG	9575, 6005, 765
1230-1330	AFG/ME	11875, 11830, 7180, 7115, 765
1430-1530	AFG	11930, 11745, 1098

Russian

1400-1500	RUS	11960, 9735, 1449, 702
1530-1630	RUS	9735, 9585, 9575, 9022, 1404
1730-1830	RUS	7290, 7115
1930-2030	RUS	9745, 7305, 7290, 7175

Serbo-Croat

0530-0630	Eu	13610, 9615, 9525, 7340, 7100, 5925
1730-1830	Eu	9720, 9615, 7340, 7145, 5925
2130-2230	Bosnia	7170, 7145, 7125

Spanish

0030-0130	LAm	9650, 6175
0130-0230	LAm	9685, 9022, 6175, 6050, 6015
0230-0330	SAm	6050, 6015
0530-0630	Eu	15260, 11790
2030-2130	Eu	9022, 7260, 7100

Swahili

0400-0430	EAf	11750
1700-1730	EAf	9585, 7160

Tajik

1530-1600	As	7070, 720

Turkish

0430-0530	TUR	15260, 11790, 9525
1630-1730	TUR	9735, 9022, 7260
1800-1900	TUR	9720
2130-2230	TUR	6025, 1404, 639

Turkmen

1500-1930	NWAs	1449

Urdu

0130-0230	As	9575, 6005, 765
1330-1430	As	11830, 11790, 9585, 7115, 1224, 765
1430-1530	As	11830, 765
1530-1730	As	11830, 1098, 765

Uzbek

1600-1630	As	7070, 720

ANN: E: "This is Tehran the Voice of the Islamic Republic of Iran"; Persian: "Inja Tehran ast, seda-ye jomhuri-ye eslami-ye Iran"; F: "Ici Tehran la Voix de la République Islamique de l'Iran"; R: "Govorit Tegeran, Golos Islamskoy Respubliki Iran"; Turk: "Burasi Tehran, Iran Islam Gumhuriyetinin Sesi Radyosu"; Arabic: "Huna Tehran saut al-gumhuriy-ati-l-islamiyya fi l-Iran".
V. By QSL-card. Rec. Acc — **F.Pl:**Japanese & Georgian sces

IRAQ

RADIO IRAQ INTERNATIONAL
📧 P.O. Box 8145, Baghdad — **L.P:** Dir: Muzaffar Abd al-Al.
MW: Basrah 693kHz 600kW, Tanaf 1377kHz 1000kW.
SW: Abu Ghraib (G.C: 44.15E/33.19N), Babylon (G.C: 44.30E/ 32.30N), Salah el Deen (G.C: 44.10E/33.58N), Salman Pak (G.C: 44.35E/33.09N)

Arabic

	Area	kHz
0000-0300	NAm	11785v
1700-2000+	ME	9715, 1377, 693
2000-2030	Eu/ME/Af	11785v

English

0300-0400	NAm	11785v
2000-2100	Eu/ME/Af	11785v
2230-2300	Eu/ME/Af	11785v

French

1930-2000	Eu/ME/Af	11785v
2200-2230	Eu/ME/Af	11785v

+) Mother of Battles Radio.
ANN: A: "Huna Baghdad, Idha'at a;-'Iraq al-Duwaliyah". Mother of Battles R: "Idha'at umm al-ma'arik, idha'atu kul al-arab".

IRELAND

RADIO TELEFÍS ÉIREANN (Statutory Corporation)
📧 Donnybrook, Dublin 4. ☎ +353 (1) 208 3111. 📠 +353 (1) 208 3080. ⏱ 93700. **Cable:** Broadcasting. **Web:** http://www.rte.ie
L.P: Chmn: Farrel Corcoran. DG: Bob Collins. MD Organisation & Development: Liam Miller. MD Commercial: Conor Sexton. Dir. of Finance: Gerry O'Brien. Dir. of Radio: Helen Shaw. Dir. of News: Ed Mulhall. Dir. of Public Affairs: Kevin Healy. Dir. of Corporate Affairs: Tom Quinn.
SW: via the tr's of WWCR, USA (rebroadcast of Home Sce. N).

English	Days	Area	kHz
0900-0930	12345	Pac	5070
1000-1030	67	Pac	5070
1830-1900	12345	NAm/Eu/Af/ME	12160
1900-1930	6	NAm/Eu/Af/ME	12160
2000-2030	7	NAm/Eu/Af/ME	12160

WEST COAST RADIO IRELAND (Comm.)
📧 Murneen Post Office, Claremorris, Co Mayo. 📠 +353 (94) 81531.
E-Mail: wcri@mayo-ireland.ie
Web: http://www.mayo-ireland.ie/shortwave.htm
SW: Jülich, Germany (G.C: 06.37E/50.95N): 1 x 100kW tr (operated by Deutsche Welle).

English	Day	Area	kHz
0100-0200	Thurs	NAm	9875
1500-1600	Sat	Eu	6175
1900-2000	Thurs	Af	15625

ISRAEL

KOL ISRAEL – THE VOICE OF ISRAEL
Main Studios: Heleni Hamalka 21, P.O. Box 1082, Jerusalem 91010. ☎ +972 (2) 302222. 📠 +972 (2) 253282. ⏱ 25263. **Foreign Sce.** 📠 +972 (2) 302327. **News & Actuality Studios:** Torah M'Zion 15, Jerusalem. ☎ +972 (2) 383150 or +972 (2) 383173. **Cable:** Kolisrael **Web:** http://www.israel-mfa.gov.il/gov/brdcast.html. **Current Schedule:** gopher://israel-info.gov.il:70/00/cul/media/950900.med **Audio:** http://www.artificia.com/html/news.cgi (membership reqd.).
L.P: Dir. & PD: Amnon Nadav. CE: David Cohen. Dir. Israel Radio International: Shmuel Ben-Zvi. Dir. of Liaison & Coordination: Raphael Kochanowski. Head of Western Broadc: Sarah Gabbai.
SW: Tel Aviv (G.C: 34.47E/32.04N): 1 x 100kW, 1 x 50kW, 2 x 20kW, 1 x 7.5kW — Yavne (G.C: 34.45E/31.52N): 1 x 500kW, 4 x 300kW, 1 x 100kW.
kHz: *5915, 7395, 7465, 7495, 9365, 9390, 9435, 9815, 11585, 11590, 11605, 11695, 13755, 15480, 15615, 15640, 17540, 17545*

Relays of Home Sce. "Reshet Bet" in Hebrew

Times	Areas	kHz
0400-0655	NAm/WEu	9390
0700-1555	NAm/WEu	15615
0700-1455	NAm/WEu	17545
0800-0900 Sat	NAm/WEu	11585
0800-0900 Sat	NAm/WEu	13755
1000-1200 x Sat.	NAm/WEu	11585
1000-1200 x Sat	NAm/WEu	13755
Times	**Areas**	**kHz**
1000-1200	NAm/WEu	17540
1500-2200	NAm/WEu	9390
1500-1655	NAm/WEu	11590
1600-0655	NAm/WEu	7495
1700-1845	NAm/WEu	7395
1900-2100	NAm/WEu	11585
2030-2215	NAm/WEu	15640

Relays of Home Sce. "Reshet Gimel"

1600-1625 Sat	NAm/WEu	7465, 9435, 11605

Relays of Home Sce Prgr. D in Arabic

0400-2230	ME	5915
0400-0630	ME	9815
0630-1130	ME	15480
1300-2230	ME	9815, 15480

Bukharian

1300-1315	CAs	11605, 11695

English

0500-0515	Eu/NAm/Pac	7465, 9435, 17545
1130-1135	Eu/NAm	15640
1500-1530	Eu/NAm	12080
1645-1700	Eu/NAm	7465, 9435, 11605
2000-2025	Eu/NAm/Lam	7465, 9365, 9435, 15640

French	Areas	kHz
0600-0615	Eu/NAm	7465, 9435
1100-1130	Eu/NAm	15640
1630-1645	Eu/NAm	7465, 9435, 11605
2030-2050	Eu/NAm/Lam	7465, 9365, 15640
Georgian		
1315-1330	CAs	11605, 11695
Hungarian		
1745-1800	Eu/NAm	7465, 9365, 9435, 11585
Judeo-Tat		
1600-1625 Fri	Eu/NAm	9435, 11695
Ladino		
1745-1800	Eu	7395
Maghribi		
1635-1645	Eu	7395
Persian		
1500-1600	ME/Eu/NAm	7465, 9435, 11605
1600-1625 x Fri/Sat	ME/Eu/NAm	7465, 9435, 11605
Romanian		
1730-1745	Eu/NAm	7465, 9365, 9435, 11585
Russian		
1830-1955	RUS	7465, 9435
Spanish		
1725-1735	Eu/NAm	7395
2050-2100	Eu/NAm/Lam	7465, 9365, 9435, 15640
Yiddish		
1700-1730	Eu/NAm	7395, 7465, 9365, 9435, 11585
1800-1830	Eu/NAm	9365, 11585

ANN: Yiddish: "Hert zu der yisrael odizie im yiddish fyn yerushaalayion". Ladino: "Emission de yisrael en a langua Judeo-espagnol". Sp: "Esta es Kol Israel". E: "This is Kol Israel broadcasting from Jerusalem". F: "Ici Kol Israel, Radiodiffusion Israelienne".
IS: Foreign Sce: National Athem (Hatiqua).
V. by QSL-card. Re. to Israel Radio International, P.O. Box 1082, Jerusalem 91010. Correspondence concerning technical matters should be sent to BEZEQ, The Israel Telecommunications Corp.Ltd, P.O. Box 29555, Tel Aviv 61290 — **PUB:** Int. prgr. schedule free.

ITALY

RAI - INTERNATIONAL
✎ Largo Willy De Luca, 00188 Roma. ☎ +39 (6) 3317 4258.
📄 +39 (6) 3317 1885 **Web:** http://www.mix.it/raiinternational/
L.P: Dir: Roberto Morrione.
SW: Prato Smeraldo (G.C: 12.31E/41.48N): 9 x 100kW — A) rel. via BBC Ascension Island — S) relay via BBC Singapore.
kHz: 5975, 5990, 6010, 6015, 6030, 6035, 6110, 7105, 7110, 7200, 7215, 7235, 7240, 7270, 7275, 7290, 9565, 9575, 9670, 9685, 9710, 9755, 9855, 11765, 11800, 11840, 11850, 11880, 11905, 11925, 15230, 15240, 15245, 15250, 15310, 15320, 17780, 17870, 21520, 21535, 21710

	Winter Schedule 1998/99	
Albanian	**Area**	**kHz**
13.35-13.55	EEu	7240, 9690
Amharic		
04.15-04.35	EAf/ME	11840, 15220, 17780
Arabic		
05.35-05.55	EAf/ME	9690, 11840
14.30-14.55	SEU/Naf	7295, 9670, 11880
Arabic	**Area**	**kHz**
16.30-16.55	EAf/ME	9690, 11880, 15230
20.25-20.45	SEu/NAf	5955, 7290
21.35-21.55	SEu/NAf	5955, 7290
Bulgarian		
15.40-16.00	EEu	7240, 9720
Croatian		
14.35-14.55	EEu	7240, 9690
Czech		
18.15-18.30	EEu	6170, 7240
21.35-21.55	EEu	6125, 7115
Danish		
20.00-20.20SU/Tu/Th	Eu	5970, 7120
English		
00.50-01.10	NAm	6010, 9675, 11800
04.15-04.40	SEu/Naf	5975, 7235
19.35-19.55	Eu	5970, 7120

English	Areas	kHz
20.25-20.45	EAf/ME	7115, 9670, 11715
22.00-22.25	FE/As	5990, 9675,11900
Esperanto		
20.00-20.20Sa	Eu	5970, 7120
French		
01.10-01.25	NAm	6010, 9675, 11800
15.30-15.55	Eu	7295, 9670, 11880
16.30-16.55	SEu/NAf	9670, 11800
German		
14.15-14.35	EEu	7240, 9690
18.05-18.25	Eu	5990, 7120, 9540
Greek		
15.20-15.40	EEu	7240, 9720
Hungarian		
19.35-19.55	EEu	6110, 7240
Italian		
01.30-03.05	NAm	6010, 9675, 11800
01.30-02.30	CAm	11765
01.30-02.30	SAm	6110
01.30-02.30	SAm	6110
01.30-03.05	SAm	9575, 11755
03.05-03.25	SAm	9575, 11755
04.15-04.40	SEu/Naf	5975, 7235
04.35-05.10	EAf/ME	11840, 15220, 17780
05.00-23.00	Eu/NAf/NE	6060,7175, 9515
06.00-13.00	EEu	7240
06.00-13.00	EAf/ME	17780, 21520
07.00-13.00	EEu	9670
10.00-11.00	AUS	11920
13.45-17.00Sp	CEu	9670
13.45-17.00Sp	NAm	17780
13.45-17.00Sp	SAm	21535
13.45-17.00Sp	EAf	21520
13.45-17.00Sp	CAm/SAm	21710
14.00-14.25	NAm	17780, 21520
15.00-15.25	SEu/Naf	7295, 9670, 11880
17.00-18.00	NEu/NAf	9670, 11840
17.00-18.00	EAf/ME	15330
17.00-18.00	Af	15220, 15320, 17870
15.55-16.25	Eu	7295, 9670, 11880
18.30-19.05	NAm	11800, 15250, 17780
22.30-00.50	NAm	6010, 9675, 11800
22.30-00.50	SAm	9575, 11755
23.00-05.00	Eu/NAf/NE	6060
Lithuainian		
04.45-05.05	EEu	5975, 9670
Polish		
18.45-19.05	EEu	6170, 7240
22.10-22.25	EEu	6125, 7115
Portuguese		
01.10-01.25	SAm	9575, 11755
20.50-21.10	SEu/NAf	5955, 7290
20.50-21.10	EAf/ME	9670, 11715
Romanian		
05.10-05.30	EEu	5975, 7235
21.15-21.35	EEu	6125, 7115
Russian		
03.30-03.50	EEu/CIS	5975, 7235, 9690
05.35-05.55	EEu/CIS	9670, 11800, 15280
Russian		
16.05-16.25	EEu/CIS	9605, 11820, 15280
20.00-20.20	EEu/CIS	5995, 7190, 9845
Serbian		
19.10-19.30	EEu	6110, 7240
Slovak		
18.30-18.45	EEu	6170, 7240
21.5522.10	EEu	6125, 7115
Slovenian		
14.00-14.15	EEu	7240, 9690

Somali	Areas	kHz
05.10-05.30	EAf/ME	11840, 15220, 17780
19.10-19.30	EAf/ME	11840, 15275, 17780

Spanish		
03.05-03.25	NAm	6010, 9675, 11800
00.50-01.10	SAm	9575, 11755
21.10-21.30	SEu/NAf	5955, 7290

Swedish		
20.00-20.20MWF	Eu	5970, 7120

Turkish		
15.00-15.20	EEu	7240, 9720

Ukrainian		
03.50-04.10	EEu	5975, 7235, 9690

Sp = Sunday programmes are subject to change for sport and news coverage.

ANN: Home Sce: "Radiouno", "Radiodue", "Radiotre" as appropriate. Night Prgr: "RAI—Radiotelevisione Italiana stazioni a onda media di Roma kHz 846 pari a m.355, e di Milano kHz 900 pari a m.333, e stazione ad onda corta di kHz 6060 pari a m .49.50 RAI International "Notturno Italiano" — Foreign Sce: I. "RAI International programmi in lingua Italiana per..." E: "This is the Italian R. and Television Sce. Broadc: from Rome". F: "Ici la R. Italienne". G: "Hier ist der Italienische Auslandssendedienst in Rom".
IS: Bird chirping; opening signal: Bells (composition by A. Parelli).
V. by QSL-card — **F.PI:** new SW-st. with 500kW tr's.
PUB: Prgr schedule 3 times a year in E, F, G, I. (Foreign Service).

ADVENTIST WORLD RADIO EUROPE (Rlg.)

✉ AWR-Europe, Casella Postale 383, I-47100 Forli. ☎ +39 (543) 766655. 🖷 +39 (543) 768198. **Web:** http://www.awr.org/awr-europe/
L.P: Europe Region Dir: Bert Smit. Chief Eng./Freq Coordinator: Claudius Dedio. Dir. of Op's: Stefano Losio. Treasurer: Markus Czettl. Engineer, Italy: Christoph Gysin. Operator, Italy: Daryl Gungadoo. Broadcast Coordinator, Slovakia: Bronislav Soos. Listener Mail, England: Becky De Oliveira. Listener Mail, Italy: Erika Gysin. Monitor Sces, Europe: Simone Smit.
MW: leased tr. in Armenia: K=Kamo 1000kW.
SW: F=Forli (No. Italy) 2.5kW
Leased tr's in Germany: J=Jülich 100kW.
Leased tr's in Slovakia: R=Rimavská Sobota 250kW.
kHz: 5900, 5905, 5940, 6055, 6090, 7100, 7215, 7230, 7275, 7315, 9450, 9465, 9475, 11610, 11655, 13715, 15620

Arabic	Area	kHz
0500-0700	Af/ME	11610R, 13715R
0800-0900	Af	13580R
1000-1100	Af	11795F
1430-1530	Eu	7135F
1500-1600	ME	1314K
1800-2000	Af	11610R
2100-2200	Af	9455R
Bulgarian		
0400-0500	SEEu	5890J
1700-1800	SEEu	9835J
Czech		
0900-0930	CEu	5940R
0930-1000Sat	CEu	5940R
1400-1430SS	CEu	5940R
Czech		
1430-1500	CEu	5940R
Dari		
1300-1400Sun	SAs	13580R
Dyula		
2130-2200	Af	9830J
English		
0330-0400	SAs	11610R
0400-0430	Af	9465R
0600-0630	Af	11640R
0800-0900	Eu	9440R
0930-1000Sun	Eu	5940R
1000-1100	Eu	9450R
1630-1700	ME	13580R
1700-1730	Af	15620R
1630-1700	SAs	11600R
2030-2130	Af	9830J, 11610R
2200-2230	Eu	6055R
Farsi		
0400-0430	ME	13715R
1700-1730	ME	11600R

French	Area	kHz
0830-0900	Af	7270F, 11640R
0900-0930	Af	13580R
1400-1430	Eu	7135F
2000-2030	Af	11610R
2130-2200	Af	11610R
2200-2230	Af	9455R
German		
0800-0900SS	Eu	5940R
1000-1100	Eu	7230F
1300-1400	Eu	7230F
1500-1600	Eu	5940R
Hindi		
0200-0300	SAs	11610R
1530-1600	SAs	13580R
Hungarian		
0600-0630	Eu	5890J
2000-2030	Eu	9835J
Italian		
0900-1000	Eu	7230F
1200-1300	Eu	7230F
KiSwahili		
0330-0400	Af	9465R
1630-1700	Af	15620R
Malagasy		
0430-0500	Af	13580R
1730-1800	Af	15620R
Nepali		
0130-0200	SAs	9465R
1430-1500	SAs	13580R
Punjabi		
0200-0230	SAs	11610R
1500-1530	SAs	13580R
Romanian		
0500-0600	Eu	5890J
Romanian		
1900-2000	Eu	9835J
Somali		
0400-0430	Eu	9465R
1700-1730	Eu	15620R
Urdu		
0300-0330	SAs	11610R
1600-1630	PAK	13580R

ANN: E: "This is Adventist World Radio, the Voice of Hope". F: "Ici la Radio Mondiale Adventiste, la Voix d l'Esperance". G: "Sie hören Adventist World Radio, die Stimme der Hoffnung". I: "Questa è la Radio Mondiale Adentista, la Voce della Speranza".
IS: Different arrangements of the melody "Lift Up the Trumpet"
V. by letter or QSL-card — **PUB:** "AWR Current", bi-monthly.
F.PI: AWR has received permission to construct a major SW-st. near Argenta. The st. which will broadcast in more than 30 languages, is due for completion by 1999.

NEXUS – INTERNATIONAL BROADCASTING ASSOCIATION

✉ P.O. Box 10980, I-20110 Milano. ☎ +39 (2) 266 6971, +39 (337) 297788. 🖷 +39 (2) 706 38151. **E-mail:** info@nexus.org
Web: http://www.nexus.org/
L.P: Pres: Alfredo Cotroneo. Vice-Pres: Flavia F. Cotroneo. PR: Anna S. Boschetti — **FM:** Globe Radio Milan FM 88.8MHz.
SW: Milano (G.C: 09.11E/45.27N): IRRS: A3A 10kW USB with 6dB carrier reduction. **Target Area:** Europe/No. Africa/Middle East.

Times	kHz	Times	kHz
0600-0830	3985	1430-2100	3985
0830-1430	7125	2100-2300	3955

Languages: English, French, Spanish, Russian, German. Includes prgrs. from NEXUS-IBA members incl. European Gospel Radio, United Nations, UNESCO, BBC, SWR Switzerland and other smaller prgr. producers, as well as non-members: Universal Life.
IS: S/on: Triumphal Scene from Aida (Giuseppe Verdi). S/off: Prisoners' Chorus (Giuseppe Verdi) — **V.** by QSL-card. Rp.

JAPAN

FOREIGN SERVICE (NHK WORLD - R. JAPAN)

✉ 2-1, Jinnan 2-chome, Sibuya-ku, Tokyo 105-01. ☎ +81 (3) 3465 111. 🖷 +81(3) 3481 1350.) J34179 RADJAPAN. **E-mail:** info@intl.nhk.or.jp
Web: http://www.nhk.or.jp/rjnet/
L.P: D.G: K. Irisawa. Head of Prgrs: K. Miyamoto.
SW: Yamata (G.C: 139.50E/36.10N): 7 x 300kW and 4 x 100kW.
Relays: B = Skelton, UK 250kW — C = Sackville, Canada 250kW — F = Montsinery, French Guiana 500kW — G = Moyabi, Gabon 500kW — I = Kranji, Singapore 250kW — S = Ekala, Sri Lanka 300kW — A = Ascension Island 250kW.
kHz: 5960B, 5960C, 5975B, 6035, 6035I, 6055B, 6070F, 6090, 6110C,

6115B, 6120C, 6135A, 6145, 6150G, 6155B, 6165, 6180B, 6190, 7110, 7125, 7140, 7200, 7210B, 7225, 7230B, 7240, 7260, 9515G, 9530F, 9535, 9560, 9600B, 9660F, 9685, 9685F, 9695, 9750, 9825, 9835, 9855, 11665, 11690, 11705C, 11710B, 11715, 11730, 11740I, 11760, 11815, 11840, 11850, 11860I, 11880S, 11890S, 11895F, 11910, 11920I, 12030, 12030G, 12045S, 13630, 13650, 13700, 15120, 15120A, 15220A, 15230, 15230G, 15355G, 15400A, 15475, 15500, 15550, 15590, 17630G, 17685, 17770S, 17780G, 17810, 17815A, 17820S, 21490A, 21610

Freqs marked *) are 300kW

Arabic	Areas	kHz
0400-0430	ME/NAf	9515G
0600-0630	ME/NAf	17770S
Bengali		
0200-0230	SAs	11890S
1130-1200	SAs	12045S
Burmese		
1130-1200	SEAs	9695*
1345-1415	SEAs	7260*
2340-2400	SEAs	13630
Chinese		
0420-0520	EAs	11910*, 15475
0430-0530	SEAs	15120, 17685*
1130-1230	EAs	6190
1200-1300	SEAs	9695*
1500-1600	EAs/SEAs	6090, 9750*
2235-2320	SEAs	13650*
English		
0000-0030R	SEAs	13630, 13650*
0000-0100	Eu	6155B, 6180B
0100-0200	SEAs/IND	11860I, 17810
	EAs	15570*
	SAs	11890S
	NAm	5960C, 11790, 13630*
	Oceania	21610*
0300-0400	Oceania	17685
0500-0530R	Am/HWI/CAm	9835F, 13630*, 15230
0500-0600	EAs	11840*, 11910*
	Oceania	11920I
	NAm	6110C
	Eu	7230B
0600-0700	SEAs/IND	11740I, 17810*
	EAs	11840*, 11910*
	SAs	15550*
	Oceania	11920I
	Am/HWI	9505. 9635
	Eu	5975B, 7230B
0700-0800	SEAs/IND	11740I, 17810*
	EAs	11840*, 11910*
	Oceania	11850, 15570I
	Eu	7230B
	ME/NAf	15230G
	CAf	17815A
1100-1200	SEAs	11815*
	EAs	7125
	NAm	6120C
1200-1300	SEAs	11815*
	EAs	7125
1400-1500	SEAs	7200
1500-1600	SEAs	7200
	EAs	9750*
	SAs	7240, 11730*
	NAm	9535*
	SAf	15355G
1700-1800	SEAs	7200
English		
	EAs	6035*
	SAs	7225, 11730
	ME/NAf	11880S
	Am/HWI	6190*
	Eu	7110*
2100-2130R	EAs	6090
2100-2200	SEAs/IND	6035I
	NAm	13630*

R) Regional Sce, others Gen. Sce.

French		
0630-0700	ME/NAf	17770S
0930-1000	Eu	9600B
	Af	17815A
German		
0600-0630	Eu	11760*, 12030G
1100-1130	Eu	9600B, 11710B
Hindi		
0230-0300	SAs	11890S
1200-1230	SAs	12045S

Indonesian	Area	kHz
0915-1015	SEAs	11690
1030-1130	SEAs	9695*
2235-2320	SEAs	13630
Italian		
0545-0600	Eu	11760*, 12030G
1115-1130	Eu	17630G
Japanese		
0200-0300	SEAs/IND	11860I
	EAs	11840*, 15570*
	SAs	15590*
	NAm	5960C, 13630*
	Oceania	21610*
0300-0330	Af	13700* (N1)
0300-0400	SEAs	17810 (N1)
	EAs	11840* (N1)
	SAs	15590* (N1)
	NAm	13630* (N1)
	CAm	11895F (N1)
	SAm	9660F (N1)
	Am/HWI	15230
	Oceania	21610*
	Eu	11760*
Japanese		
	ME/NAf	9515G (N1)
0400-0430	SAm	9660F
	Am/HWI	15230
0400-0500	SEAs	17810*
	EAs	11840*
	NAm	6110C
	Eu	5960B, 7230B
	ME/NAf	17820S
	Oceania	11920I
	CAf	17815A
	WAf	15220A
0800-0930	SAm	6070F, 9685
0800-1000	SEAs/IND	11740I, 11815*
	EAs	7125
	Am/HWI	9835*
	Oceania	11850
	ME/NAf	15230G
0900-1500	SEAs	9855* (N1)
	EAs	9750* (N1)
1000-1100	SEAs	11815*
	EAs	7125
	NAm	6120C
	SAf	17780G
1300-1400	CAf	21490A (N1)
	WAf	15400A (N1)
1300-1500	SAs	12045S (N1)
	NAm	11705C (N1)
1600-1700	SEAs	7200
	EAs	6035*
	NAm	9535*
	Am/HWI	6190*
1800-1900	SEAs	7200
	Oceania	7140*
	ME/NAf	11880S
1800-2000	EAs	6035*
	NAm	9535*
	Eu	7110*
1900-2000	SEAs	7200*
	Oceania	7140
2000-2100	SEAs	7200* (N1)
	EAs	6090 (N1)
	NAm	13630* (N1)
Japanese		
	Oceania	6035I (N1)
	Eu	7110* (N1), 7210B (N1)
2000-2400	SEAs	9825 (N1)
	EAs	9560 (N1), 11910* (N1)
2100-2300	Oceania	11850* (N1)
2100-2400	SEAs	9825* (N1)
2200-2300	SAm	13680F (N1)
	Eu	6055B (N1), 6115B (N1)
2300-2400	Eu	6180B (N1)

N1) relay home sce. R. One.

Korean		
0320-0420	EAs	11910*
1030-1130	EAs	6190
1400-1500	EAs	6090
2130-2200	EAs	6090
Malay		
1400-1430	SEAs	9695*
2320-2340	SEAs	13650*

Persian	Area	kHz
0230-0300	ME/NAf	6150G, 11760*
Portuguese		
0930-1000	SAm	9530F, 9685
Russian		
0430-0530	Eu	11760*, 12030G
0500-0600	EAs	11715
0745-0845	FE	6145*, 6165*
1000-1100	Eu	11710B
1300-1400	EAs	6090
Spanish		
0430-0500	SAm (W)	9660F
	CAm	11895F, 9835*
	Am/HWI	15230
0630-0700	Eu	12030G
1000-1030	SAm	9530F, 9685*
Swahili		
0330-0400	Af	6135A, 13700*
Swedish		
0515-0545	Eu	11760*, 12030G
Thai		
1130-1200	SEAs	11740I
1300-1330	SEAs	9695*
2340-2400	SEAs	13650*
Urdu		
0300-0330	SAs	11890S
1230-1300	SAs	12045S
Vietnamese		
1200-1230	SEAs	11740I
1330-1400	SEAs	9695*
2320-2340	SEAs	13630

F.PI: Increasing the number of transmissions per day and decreasing the time period to a half per transmission in Chinese, Korean, Indonesian, Russian in the near future.
ANN: English: "This is Radio Japan, NHK World Network, Tokyo" – Japanese: "Kochirawa NHK Warudo, Rajio Nippon, NHK no kokusaihoso desu" – Chinese: "Zheli shi Riben Guoji Guangbo Diantai, NHK huanqiu guangbowang" – German: "Hier ist Radio Japan, NHK World" – Korean: "Yoginun Radio Ilbon, NHK Worldumnida"
IS: Opening music of Reg. Sce's "Sakura". Interval signal music "Kazoe Uta".
V. by QSL-card. —— **PUB:** In Japanese only: Weekly STERA (weekly), Hoso Bunka or Broadcast Culture (monthly), NHK Year Book (annual). The prgr. schedules for Overseas Service (R. Japan)
FI: In accordance with the Broadcasting Law, NHK's operations are financed by the revenue from receiving fee.

RELAY TRANSMISSIONS FOR INTERNATIONAL BROADCASTERS
SW: Yamata 300 kW/100kW (see R. Japan, above) — **Call:** JOD#, # is the number of frequency in MHz, ex. JOD6 for 6065 kHz.
D.PRGR: see R. Canada Int., R. France Int. and BBC.

JORDAN RADIO & TELEVISION CORP. (JRTV) (Gov.)
⬚ P. O. Box 1041 or 2333, Amman. ☎ +962 (6) 77311/9. 🖷 +962 (6) 751503 (Admin) or 788115 (Eng.). ① 23544 (JTVINT JO) International Relations or 24213 (RTVENG JO) Technical. **E-mail:** rj@jrtv.gov.jo (prgrs), eng@jrtv.gov.jo (techical), general@jrtv.gov.jo (general).
Web: http://www.jrtv.com
L.P: DG: Ihsan Ramzi Shikim. Dir. Radio: Hashim Khuraysat. Dir. Eng: Fawzi Saleh. Prgr. Dir's: J. Hajjat (Arabic), Jawad Zada (English). Dir. Admin: A. Zabi. Dir. Int. Rel: Mrs. Fatima Masri.
SW: Al Karanah (G.C: 36.26E/31.44N): 3 x 500kW tr's.
kHz: 6105, 7155, 9630, 9830, 11690, 11810, 11835, 11935, 13630, 15355, 15435

Arabic	Areas	kHz
0300-1400	ME/FE/Pac	11810
0300-0500	EAf/ME	9630
0300-0800	NAm/WEu	15435
0500-0800	EEu/RUS	11835
1000-1200	NAf/CAm	15355
1200-1530	EAf/ME	13630
1500-2100	EEu/RUS	7155
1500-1600	ME/FE/Pac	6105
1700-2100	WEu	9830
2100-0100	NAm/WEu	11935
	SAm	15435
English		
1100-1730	WEu/NAm	11690

ANN: Arabic: "Huna 'amman, Idha'atu-l-mamlaka al-urduniyya al-hashimiyya". E: "This is R. Jordan, broadc. from Amman".
V. by QSL-card. Rec. acc. (Re. to P.O. Box 909, Amman).

RADIO ALMATY
⬚ Zheltoksan Street 175 A, 480013 Almaty. ☎ +7 (3272) 627694.
L.P: Editor-in-Chief: Nadya Usayeva.
SW: Almaty (G.C: 77.00E/43.50N): 7 x 100kW, 1 x 200kW.

English	Days	Area	kHz
0100-0130	1234567	Asia	900, 6160, 6230
1000-1020	256	General	KR2 freq's, 9560K
1020-1040	3	General	KR2 freq's, 9560K
1400-1430	1234567	Asia	900, 6155, 6230
German			
1000-1020	256	General	KR2 freq's, 9560K
Kazakh			
0000-0030	1234567	Asia	900, 6160, 6230
0130-0200	1234567	Asia	900, 6160, 6230
1330-1400	1234567	Asia	900, 6155, 6230
1500-1530	1234567	Asia	900, 6155, 6230
Korean			
1620-1640	1234567	Asia	900, 6155, 6230
Russian			
0720-0800	1234567	General	KR2 freq's
1730-1800	1234567	General	KR2 freq's, 5940K
Tatar			
1600-1620	1234567	Asia	900, 6155, 6230
1640-1700	1234567	Asia	900, 6155, 6230
Turkish			
1040-1100	2	General	KR2 freq's, 9560K

Su: All prgr's 1h earlier. K = via Kiev, Ukraine.
ANN: Kazakh: "Almatydan söylep turmiz" — **V.** Rarely by letter.

RELAY SERVICES
SW: Almaty 7 x 100kW, 1 x 200kW; Karaturuk (G.C: 43.39N/77.56E, tentative location): 4 x 1000kW. Relays of V. of Russia, R. Netherlands, BBC, R. Liberty. Cf. these stations for details.

RADIO PYONGYANG
⬚ R. Pyongyang, Pyongyang.
MEDIUMWAVE: Chongjin 621 kHz 500kW - Pyongyang 1368kHz 2kW.
SHORTWAVES: Kanggye (G.C: 126.36E/40.58N): 5 x 200kW, Kujang (G.C: 125.05E/40.05N): 5 x 200kW, Pyongyang (G.C: 125.33E/39.05N): 10 x 200kW.
kHz: 3250, 3560+, 4405+, 6070, 6125, 6520, 6575, 7200, 7580, 9325, 9345, 9600, 9640, 9650, 9975, 11335, 11700, 11735, 11740, 11845, 13650, 13760, 13790, 15130, 15180, 15230, 15340, 17735
+)These are standby/emergency feeder tr's.

Arabic	Area	kHz
1500-1600	ME/Af	6520, 9600
1800-1900	ME/Af	6520, 9600, 9975
2000-2100	ME/Af	6520, 9600, 9975
Chinese		
0700-0800	CHN	4405, 6125, 7200, 9345
0900-1000	CHN	4405, 6125, 7200, 9345
1100-1200	CHN	4405, 6125, 7200, 9345
1300-1400	CHN	4405, 6125, 7200, 9345
English		
0000-0100	SEAs/LAm	11845, 13650, 15230
0500-0600	Eu	11740, 13790
1100-1200	ME/Af	3560, 9640, 9975
	SEAs/LAm	11335, 13650, 15230
1500-1600	ME/Af	3560, 9640, 9975
	SEAs/LAm	11335, 11735, 13650

English	Area	kHz
1800-1900	Eu	4405, 6575, 9345
	NAm	11700, 13760
1900-2000	ME/Af	6520, 9600, 9975
2100-2200	Eu	4405, 6575, 9345
	NAm	11700, 13760
2300-2400	NAm	11335, 11700, 13760, 15130
French		
0400-0500	Eu	11740, 13790
0500-0600	SEAs/LAm	13650, 15180
0500-0600	NAm	15340, 17735
1200-1300	ME/Af	3560, 9640, 9975
1200-1300	SEAs/LAm	11335, 13650, 15230
1500-1600	Eu	4405, 6575, 9345
2000-2100	Eu	4405, 6575, 9345
2000-2100	NAm	11700, 13760
2100-2200	ME/Af	6520, 9600, 9975
German		
1800-1900	Eu	3560, 9325, 13790
2000-2200	Eu	3560, 9325, 13790

Japanese	Area	kHz
0700-1000	J	621, 3250, 6070, 7580, 9650
1100-1400	J	621, 3250, 6070, 6520, 7580
2300-0100	J	621, 3250, 7580, 9650
Korean		
0400-0600	CHN	7200, 9345
0600-0650	Eu	11740, 13790
0600-0650	SEAs/LAm	13650, 15180
Korean		
0600-0650	ME/Af	15340, 17735
0800-0900	CHN	4405, 6125, 7200, 9345
1200-1300	CHN	4405, 6125, 7200, 9345
1400-1450	ME/Af	9640, 9975
1400-1450	SEAs/LAm	11335, 13650, 15230
1600-1650	ME/Af	3560, 6520, 9600, 9640, 9975
1600-1650	Eu	4405, 6575, 9345
1600-1650	SEAs/LAm	11335, 11735, 13650
2200-2250	Eu	4405, 6575, 9345, 11700, 13760
Korean/Russian		
0800-0900	URS	3560 ,6575, 9975, 11740, 13790
1700-1800	URS	6520, 9600, 9975
Russian		
0700-0800	URS	3560 ,6575, 9975, 11740, 13790
0900-1000	URS	3560 ,6575, 9975, 11740, 13790
1700-1800	Eu	3560, 9325, 13790
1900-2000	Eu	3560, 9325, 13790
Spanish		
0000-0100	NAm	11335, 11700, 13760, 15130
0400-0500	SEAs/LAm	13650, 15180, 15340, 17735
1300-1400	ME/Af	3560, 9640, 9975
1300-1400	SEAs/LAm	11335, 13650, 15230
1700-1800	Eu	4405, 6575, 9345
1700-1800	NAm	11700, 13760
1900-2000	Eu	4405, 6575, 9345
1900-2000	NAm	11700, 13760
2300-0000	SEAs/LAm	11845, 13650, 15230

Some transmissions are also on 1368kHz for Pyongyang area.
IS: Song of General Kim Il Sung. Opening & closing music for Korean sce: Nat. Anthem.
ANN: Korean: "Joson Jung-ang Pangsong-imnida" or "Pyongyang Pangsong-imnida". Arabic: "Idaat Jumhuriat Kuria Democratiat At Shabia min Pyongyang". Chinese: "Pingrang Guangbo Diantai". English: "This is Radio Pyongyang". French: "Ici Radio Pyongyang". German: "Hier ist Radio Pyongyang". Japanese: "Kochirawa Pyongyang Hoso desu". Russian: "Gavarit Penyan". Spanish: "Aqui Radio Pyongyang".
V. by QSL-card — **PUB:** "Radio Pyongyang" upon request.

KOREA (Republic of)

RADIO KOREA INTERNATIONAL (Radio Hankuk)
18, Yo-ui-do-dong, Yongdungp'o-gu, Seoul 150-790. ☎ +82 (2) 781-3710. +82(2)781-3798/3799 — **L.P:** DG: Kim, Sang-Soo.
MEDIUMWAVES: Kimje 1170kHz 500kW.
SHORTWAVES: Kimje (G.C: 126.50E/35.50N): 3 x 100kW, 3 x 250kW (+1 x 100kW reserve) – H) Hwasong (G.C: 126.47E/37.13N): 2 x 10kW, 2 x 100kW(+1 x 10kW reserve) – S) via Sackville, Canada – B) via Skelton, United Kingdom.
kHz: 3970B, 5975H, 6135H, 6145B, 6480, 7275, 7285, 7550, 9515, 9555B, 9570, 9640, 9650S, 9870, 9875B, 11715S, 11725, 11810, 13670, 15575 ND) Omnidirectional.

Arabic	Area	kHz
1600-1700*	Eu	6480
	ND	7275
	Eu	15575
1900-2000	Eu	6480, 7550, 15575
	ME/Af	9515, 9870
Chinese		
1130-1230	ND	9640
1300-1400	ND	1170, 5975, 6135
	CHN	7285
2200-2300*	ND	5975, 7275
	SEAs	9640
English		
0800-0900	AUS	9570
	Eu	13670
1030-1100	SAm	9650S
1200-1300	China	7285
1230-1300	SEAs	9570, 13670
	SEAs	13670
1600-1700	ND	5975
	ME/Af	9515, 9870
1900-2000*	Eu	5975, 7275
2100-2130	Eu	3970B
2100-2200	Eu	6480, 15575
0200-0300*	ND	7275

English	Area	kHz
	SAm	11725, 11810
	NAm	15575
French		
1300-1400	SEAs	9570
1700-1800*	ND	7275
	ME/Af	9515, 9870
1900-2000	Eu	6145B
German		
1800-1900*	Eu	6480
	ND	7275
2000-2100	Eu	6145B, 7550, 15575
Indonesian		
1130-1230	SEAs	9570, 13670
1400-1500	SEAs	13670
2300-0000*	ND	7275
	SEAs	9640
Japanese		
0800-0900*	ND	5975, 7275, 9640
1200-1300*	ND	1170, 5975, 6135, 7275
1400-1500	ND	5975
0000-0100	SEAs	9640
	SAm	11810
Korean		
0700-0800	Eu	7550, 9555B
0900-1000	SAm	7550
0900-1100*	ND	5975, 7275
	AUS	9570
	Eu	13670
1000-1100	ND	1170, 6135
1100-1130	NAm(E)	6145S. 9650S
	ND	9640
	SAm	11725
1300-1400	ND	9640
	SEAs	13670
1700-1900	ND	5975
	Eu	7550, 15575
2100-2200*	ND	5975, 7275
	SEAs	9640
2300-0100	ND	5975
	NAm	15575
0100-0200*	ND	7275
0300-0400*	ND	7275
	SAm	11725, 11810
	NAm	15575
Russian		
1100-1200*	ND	1170, 5975, 6135, 7275
1700-1800	Eu	6480
1800-1900	Eu	9875B
2000-2100	ND	5975
Spanish		
1000-1030	SAm	11715S
1000-1100	SAm	7550, 11725
1800-1900	ME/Af	9515. 9870
2000-2100*	Eu	6480
	ND	7275
	ME/Af	9870
2200-2300	Eu	6480
0100-0200	SAm	11725, 11810
	NAm	15575

IS: Korean children's song "Tar-a Tar-a Palgun Tar-a (Oh, bright moon)" played by a glockenspiel. Original music "Dawn" composed by Kim, Hee-jo with KBS symphony orchestra — **V.** by QSL-card.
ANN: Korean:"Yoginun Taehanminguk Sour-eso Ponaedurinun Hanguk Pasong Konsa, KBS-e Kukche Pangsong, Radio Hangug-imnida". English:"This is Radio Korea International of the KBS". Japanese: "Kochirawa Rajio Kankoku, KBS-no Kokusai Hoso desu".
PUB: Prgr. schedules, pennant, folder, RKI Newsletter (each lang.), KBS Handbook (English), KBS information Sce.

KUWAIT

RADIO OF THE STATE OF KUWAIT (Gov.)
P.O. Box 397 Safat, 13004 Safat. ☎ +965 2423774. +965 2456660, 2415498, 2415946. **Cable:** ALIRSHAD. ① (496) MI 46285 KT. **E-mail:** radiokuwait@radiokuwait.org
Web: http://www.radiokuwait.org/
L.P: Minister: Sheikh Saud Nasir Al-Saud Al-Sabah. Under-Secr: Faisal Mohamed Al-Hajji. Asst. Under Secretaries: Admin & Finance: Faisal Al-Malek. TV Affairs: Rida Yousef Al-Feeli. Broadc. Affairs: Dr. Abdulazeez Al-Mansour. Culture, Press & Documentation Affairs: Salman Al-Dawood. Eng. Affairs: Jawad A. Al-Mazeedi. Int. Media: Mrs. Amal Al-Hamad. News Affairs & Political Prgrs: Mohamed Hamed A. Al-Qahtani. Tech. Adviser: Yacoub Y. Dashty. Dir. Eng. (TV): Abdulazeez Al-Baghli. Dir. Eng. Communications: Bader F. Al-Mazeedi. Dir. Eng. (Radio): Maher N.

Al-Mutawa.
SW: Sulabiyah (G.C: 47.53E/29.16N): 500/250kW.

kHz	Times	Area	Prgr.
6055	0200-1305	Gulf	Main Arabic
9880	1315-2130	ME	Main Arabic
9855	1800-2400	Eu/NAm	Main Arabic
11675	0000-0530	WNAm	Main Arabic
11990	1615-1800	Eu/NAm	Main Arabic
	1800-2100	Eu/NAm	**English**
13620	0930-1605	Eu/NAm	Main Arabic
15110	0445-0930	SEAs	Main Arabic
	1315-1730	SEAs	Main Arabic
15495	0200-1305	ME	Main Arabic
	1800-2400	Af	Main Arabic
15505	0400-0805	WAs	Main Arabic
	0815-1740	Af	Main Arabic
	1745-2300	Eu/NAm	Main Arabic
15495	1800-2400	Af	Main Arabic
17885	0900-1505	FE	Main Arabic

ANN: A: "Huna al-Kuwait". E: "This is Radio Kuwait".
IS: old Kuwaiti tune on clarinet — **V.** by QSL-folder. Rec.acc.

VOICE OF AMERICA RELAY STATION
MW: 1548kHz 600kW — **D.PRGR:** see USA.

LAOS (People's Democratic Republic)

LAO NATIONAL RADIO AND TELEVISION (Gov.)
✉ B.P. 310, Vientiane. ☎ +856 4475. **Cable:** LAO NATIONAL RADIO
STATION VIENTIANE — **L.P:** DG: Dr. Khekkeo Soysaya. Dep. DG:
Keungkham Vilayasith. Tech. MD: D. Sisombath.
MW: Vientiane 1030kHz 10kW.
SW: Vientiane (G.C: 102.33E/17.58N): 7145vkHz 50kW (inactive)

Cambodian	Area	kHz
0000-0030	As	1030
1230-1300	As	1030
English		
0600-0630	As	1030
1330-1430	As	1030
French		
0530-0600	As	1030
1300-1330	As	1030
Thai		
0500-0530	As	1030
1130-1200	As	1030
Vietnamese		
1200-1230	As	1030
2330-2400	As	1030

ANN: National Sce: "Thini Withayu heng Sat, krachaisiang chak
Wianchan, nakhong-lwang khong sathanalat pasatipatai pasason Lao".
Vientiane Capital Sce: "Vitthayou Krachaistang Nakhonluang Vientiane".
Regional st's : "Thini sathani vithayu krachaisiang heng Luang Prabang",
"Thini vithayu krachaisiang heng sat Houa Phan", "Thini vithayu
krachaisiang heng sat Ke Thai otaa Pakse", "Thini vithayu krachaisiang
heng sat kueng Savannakhet", "Thini vithayu krachaisiang heng Sat
Oudomxay".
IS: (National Sce.): Music on Khéne (mouth organ) & Solo (bamboo instru-
ment) — **V.** by QSL-card. IRC's not acc.

LATVIA

RADIO LATVIA
✉ P.O. Box 266, LV-1098 Riga. ☎ +371 (7) 206735. ▤ +371 (7) 206709
Web: http://www.radio.org.lv.
L.P: Dir: Jogita Cinkus
SW: R¦ga (Ülbroka, G.C: 24.16E/56.57N) 5935kHz 100kW. **Beams:** 94Þ
(to Russia) + 250Þ (to UK). All transmissions on 5935kHz.

English	Area	kHz
2000-2030Sat	Eu	5935
2130-2135MF	Eu	5935
Latvian	**Area**	**kHz**
2100-2130MF	Eu	5935
Russian		
2030-2100Sat	Eu	5935
2050-2100MF	Eu	5935

ANN: Latvian: "Jus klausaties Latvijas Valsts Radio". E: "This is R. Latvia".
— **V.** by QSL-card — **PUB:** Free sched. on request.

LEBANON

VOICE OF HOPE (Rlg.)
✉ P.O. Box 3379, Limassol, Cyprus. ☎ + 972 (6) 959174 (English), +972

(6) 959889 (Arabic). ▤ + 972 (6) 997827 (English), +972 (6) 959889 (Arabic).
(**USA** ✉ P.O. Box 100, Simi Valley, CA 93062. ☎ +1 (805) 520-9460).
▤ +1 (805) 520-7823) **E-mail:** voh@broadcast.net
Web: http://www.highadventure.org/voh_midd.html
L.P: CEO: George Otis. Pres: John Tayloe. Asst. to the Pres: Paul Johnson.
GM: Gary Hull.
MW: 945kHz 50kW — **SW:** Hill of the Doves (G.C: 35.36E/33.22N):
6280kHz 10kW, 9960/9965kHz 25kW
NB: SW-tr's off air at editorial deadline due to fire at transmitter site.

Arabic	Area	kHz
2100-1530	ME	945, 6280
2100-1300	ME	9960
English		
1300-1630*	ME/Eu	9960
1600-2100	ME	945, 6280
Farsi		
1730-1800	ME	9965
French		
1530-1600	ME	945, 6280
Russian		
1830-2100	ME/Eu	9965
Turkish		
1630-1700	ME/Eu	9965
1800-1830	ME/Eu	9965

*) also incl. prgrs in Amharic, Armenian, Azerbaijani, Czech, Farsi,
Georgian, Romanian, Russian, Swedish and Ukrainian.
ANN: E: "High Adventure Ministries, the Voice of Hope for the whole of
the Middle East". A: "Saut al Amal le Kul Sharq al Awsat".
V. by QSL-card or letter.

LIBERIA

RADIO LIBERIA INTERNATIONAL
✉ P.O. Box 1103, 1000 Monrovia 10. ☎ +231 226963. ▤ +231 226003.
Web: http://www.afric-network.fr/afric/liberia/liberia.html
L.P: Dep. Mgr: James Kassoyen. Engineer-in-Charge: Isaac P. Davis.
SW: Totota (G.C: 10.00W/06.05N): 7.5kW.

English	kHz	English	kHz
1800-0300	5100	0600-1800	6100

ANN: "Radio Liberia International, the Voice of Peace"

LIBYA

LIBYAN JAMAHIRIYA BROADCASTING
✉ Box 4677, Souq al Jama, Tripoli. ☎ +218 (21) 603191/5. ☉ 20010.
Cable: Ida Shaabia Tripoli.
SW: Tripoli/Sabrata (G.C: 13.11E/32.45N): 500kW — *) Sebha (G.C:
14.50E/25.52N): 100kW — Þ) Benghazi (G.C: 20.04E/32.08N): 100kW

Arabic	Area	kHz
1045-1230	Eu/ME/Af	15435, 15415, 15235, 1251, 711
1645-1800	Eu/ME/Am	15435, 15235, 1251, 711
1800-0400	Eu/ME/Am	15435, 15415, 15235, 1251, 711
Foreign languages*		
1600-1645	Eu/ME	15435

*) **Bulgarian:** 10th, 11th, 25th, 26th of each month. **Czech/Slovak:**
12th, 13th, 27th, 28th of each month. **German:** 4th, 5th, 19th, 20th of
each month. **Hungarian:** 6th, 7th, 21st, 22nd of each month. **Polish:** 8th,
9th, 23rd, 24th of each month. **Romanian:** 14th, 29th of each month.
Russian: 1st, 2nd, 3rd, 16th, 17th, 18th, 29th of each month. **Serbo-
Croat:** 15th, 30th, 31st of each month.
ANN: H: "Idha'at al jamahiria al-'arrabiya al-libyya ash-sha'abiyya al-
sitirakkiya". V: "Idha at saout al-watan al'arabbiy al-Kabir".
IS: Prgrs open and close with National Anthem — **V.** by QSL-card.

LITHUANIA

RADIO VILNIUS
ADDR: Kornaskio 43, LT-2674 Vilnius. ☎ +370 (2) 233526, 236627.
▤ +370 (2) 233526. **E-mail:** ravil@rtv.lrtv.ot.lt
Web: http://www.lrtv.lt/lr.htm
L.P: MD: Mrs. Laima Grumadiene. PD: Mrs. Kazimiera Mazgeliene. TD:
Juozas Algirdas Vilciauskas.
SW: Kaunas-Sitkunai (G.C: 23.49E/55.02N): 1 x 50kW.
*) rel. via Jülich, Germany

English	Area	kHz
0930-0959	Eu/Atlantic	9710
2000-2029	Eu	612, 666, FM1, FM3
2230-2259	Eu	612, 666, FM1, FM3
0030-0059	NAm	6120*
Lithuanian		
0900-0929	Eu/Atlantic	9710
2200-2229	Eu	612, 666, FM1, FM3

Lithuanian | **Area** | **kHz**
0000-0029 NAm 6120*
ANN: E: "This is R. Vilnius". L: "Vilniaus radijas uzsieniui".
IS: L: song "Lietuviais esame mes gime" ("We Were Born Lithuanians").
E: melody played on kanklés (Lithuanian national instrument).
V. by QSL-card. Rec.acc. IRC's appreciated. Re. to respective language dept.

MADAGASCAR

RADIO NETHERLANDS RELAY STATION
(also used by Deutsche Welle)
✉ P.O. Box 404, Antananarivo.
SW: Talata/Volondry (G.C: 43.37E/18.43S): 2 x 300kW tr's.
R. Netherlands kHz: *7285, 9590, 9605, 9690, 9860, 9895, 11655, 12090, 15150, 15500, 15585, 17580, 21480*
Deutsche Welle kHz: *15145*
D.PRGR: see Netherlands, Germany.
V. by own QSL-card for direct re.

MALAYSIA

SUARA MALAYSIA (Voice of Malaysia)
✉ P.O. Box 11272, 50740 Kuala Lumpur. ☎ +60 (3) 2825333.
☼ MA 30283 (Radio).
Web: http://www.asiaconnect.com.my/rtm-net/
L.P: Contr: Stephen Sipaun.
MW: Kota Kinabalu (Sabah) 1476kHz 600kW
SW: Kajang (G.C: 101.46E/03.01N): 2 x 500kW, 9 x 100kW, 4 x 50kW (incl. transmitters used for Home Sce.)

Arabic	**Area**	**kHz**
1530-1700	As	15295
Bahasa Malaysia		
0600-0800	As	6100
0830-1025	As	15295
1700-1900	As	6175, 9750, 15295
Bamar (Burmese)		
1430-1530	As	6100
English		
0455-0825	As	6175, 9750, 15295
Tagalog		
1030-1300	As	1476
Indonesian		
0900-1600	As	6175, 9750
2200-2400	As	6100, 6175, 9750
Mandarin		
1030-1230	As	15295
Thai		
1300-1430	As	6100

N. in English: 0600, 0800 — **ANN:** "Inilah Suara Malaysia". E: "This is the Voice of Malaysia" — **IS:** First bar of Nat. Anthem "Negara Ku" (Chimes) — **V.** by QSL-card. Rec. acc.

MALTA

VOICE OF THE MEDITERRANEAN/R. MELITA
✉ P .O. Box 143, Valletta CMR 01. ☎ +356 248080, 240421. 🖷 +356 241501. **Web:** http://www.woden.com/~falcon/schedule.html
L.P: MD: Dr. Richard Vella Laurenti. Dep. MD: Said M. T. Shain. PD & Head of N: Ali Abdul Aziz El Kish.
NB: services to Eu/ME/Af are funded by the govts of Malta and Libya, but transmissions to other areas are financed by the Maltese govt.
SW (via tr's in Russia): *9765, 9810, 12060, 13605, 15550, 17570kHz*

Arabic	**Days**	**Area**	**kHz**
1900-2000	5	Eu/ME/Af	12060, 9765
2000-2100	12346	Eu/ME/Af	12060, 9765
English			
0100-0200	1	NAm	13605
0200-0330	7	As/Pac	17570, 15550
1900-2000	123467	Eu/ME/Af	12060, 9765
French			
0300-0330	1	NAm	13605
2000-2030	7	Eu/ME/Af	12060, 9765
2100-2130	7	NAm	9810
German			
0330-0400	1	NAm	13605
2030-2100	7	Eu/ME/Af	12060, 9765
2130-2200	7	NAm	9810
Japanese			
0400-0430	1	NAm	13605
0500-0530	7	EAs	17570
2200-2230	7	NAm	9810

Maltese	**Days**	**area**	**kHz**
0200-0300	1	NAm	13605
0330-0500	7	As/Pac	17570, 15550
2000-2100	7	NAm	9810

ANN: "V. of the Mediterranean" or "R. Melita" or "Valetta Calling".

MEXICO

RADIO MÉXICO INTERNATIONAL
✉ Ap. Postal 21300, 04021 Mexico, D.F.
E-mail: imer@mpsnet.com.mx — **LP:** Gte: Lic. Martín Rizo Gavira
SW (G.C: 99.03W/19.16N): 5985kHz 10kW, 9705kHz 5kW
(inactive tr's: 11770kHz 10kW, 15430kHz 50kW, 17765kHz 10kW)

English*	**Area**	**kHz**
1400-1430	Am	5985, 9705
1500-1530	Am	5985, 9705
1900-1930	Am	5985, 9705
2000-2030	Am	5985, 9705
0300-0330W	Am	9705
0400-0430MF	Am	9705
Spanish		
1200-1400	Am	5985, 9705
1800-1900	Am	5985, 9705
2100-2300	Am	5985, 9705
2300-0300	Am	9705
0300-0400Sun	Am	9705 (rel. La Hora Nacional)

*) United Nations R: Mon 0400-0430, Wed 2000-2030, Fri 1500-1530.

MOLDOVA

RADIO MOLDOVA INTERNATIONAL
✉ str. Miorița 1, MD-2028 Chișinau. ☎ +373 (2) 723379 or 723385. 🖷 +373 (2) 723329 or 723307. ☼ 163210 RTV SU.
L.P: Dir. of Dept: Valeriu Vasilica. Dep. Dir: Iurie Moraru.
MW: 1593kHz (cf. HS above). **FM:** 71.24/72.02MHz .
SW: Galbeni, Romania 2 x 120kW tr's.
kHz: *7500, 9400, 15315*

English	**Area**	**kHz**
0230-0255	MDA	MW+FM
0330-0355	NAm	7500
0430-0455	NAm	7500
2200-2225	Eu	7500
2300-2325	Eu	7500
French		
2000-2025	Eu	7500
2100-2125	Eu	7500
2330-2355	MDA	MW+FM
Romanian	**Area**	**kHz**
0300-0330	NAm	7500
1230-1300	NAm	9400
Russian		
0400-0425	RUS	7500
1930-1955	RUS	9400 or 7500
Spanish		
0230-0255	CAm	9400
0030-0055	MDA	MW+FM
1200-1225	CAm	15315
2030-2055	Eu	7500
2130-2155	Eu	7500

ANN: E: "This is RMI". F: "Vous êtes à l'écoute de RMI". Romanian: "Aici R. Moldova Internațional". Russian: "Govorit Mezhdunarodnoye Radio Moldovy", Sp: "Aquí RMI" — **V.** by letter and/or QSL card.
F.PI: SW relays via 1000kW tr's at Maiac, Pridnestrovye. Prgr's in German, Turkish, Ukrainian.

PRIDNESTROVYE

TRANS WORLD RADIO (Rlg.)
✉ and other information: see 'Monaco'.
MW: Maiac (Grigoriopol) 999kHz 1000kW.

Ukrainian	**Days**	**Area**	**kHz**
2030-2100	12345	Ukraine	999

MONACO

TRANS WORLD RADIO (Rlg.)
✉ B.P. 349, M-98007 Monte Carlo. ☎ +377 165600. 🖷 +377 165601.
Web: http://www.gospelcom.net/twr/
L.P: Pres: Tom Lowell. MD: Werner Krömer. Mgr. Broadc. Dept: Beate Kiebel. Freq. Coordinator: Bernhard Schraut. Head of PR: Felix Widmer.
MW: Romoules, France 1467kHz 1000kW.
Additional broadcasts via Albania, Armenia, Cyprus, Moldova, Poland:

see separate entries for each country.
SW: Mont Agel (G.C: 07.25E/43.46N): 1 x 500kW, 2 x 100kW.
kHz: 6230, 7115, 7160, 7255, 7355, 7385, 9490, 9925

Albanian	Days	kHz
2000-2015	346	1467
2015-2030	1267	1467

Arabic

1945-2015	1267	1467
2115-2200	7	1467
2130-2200	6	1467
2200-2300	12345	1467

Berber

2100-2130	12345	1467

Croatian

1330-1345	123567	7255
1345-1415	12345	7255
2015-2030	4	1467
2300-2320	1234567	1467
2320-2335	157	1467
2335-2350	17	1467
2350-2400	7	1467

Czech

1615-1700	1245	7355
1630-1700	3	7355

English

0745-0935	6	7115
0745-0920	7	7115
0755-0920	12345	7115
1230-1255	7	7115
1230-1300	6	7115
2300-2350	1	1467
2300-2320	2345	1467
2300-2315	6	1467
2300-2330	7	1467

French

0330-0400	1234567	216

German

0445-0515	1234567	1467
0930-0945	123456	7160
		6230
0830-1015	7	7160

Albanian

		6230
0945-1015	4	7160
		6230
1105-1120	123456	7160
		6230
1430-1500	1234567	7160
		6230
2030-2100	67	1467
2030-2130	12345	1467

Hungarian

1200-1245	7	9490
1700-1715	7	7355
1700-1730	123456	7355

Italian

2015-2030	3	1467

Kabyle

2130-2200	12345	1467

Macedonian

1345-1415	7	7255

Mandarin

2100-2115	7	1467

Romanian

1700-1730	1567	9925
1700-1715	234	9925

Serbian

1330-1345	4	7255

Slovak

1615-1630	367	7355

Slovene

1345-1415	6	7255

Spanish

2015-2030	5	1467

Tamazight

2115-2130	2	1467

Turkish

2030-2045	6	7385
2030-2100	12345	7385

ANN: This is Monte Carlo. The following programme of Trans World Radio is in the ... language".
IS: Musical Box — **V.** by QSL-card. Rp (2 IRC's). Rec. not wanted.

MONGOLIA

VOICE OF MONGOLIA (Gov.)
✉ C.P.O. Box 365, Ulaanbaatar 13. ☎ +976 (1) 327900 or 321624. 🖷 +976 (1) 323096. ① 223 RTV MH. **Cable:** Mongolian Radio.
E-mail: mrtv@magicnet.mn
Web: http://www.magicnet.mn/monradio/
L.P: Dir. Foreign Sce: B. Narantuya. Head of the Int. Rel. Dept: Neeneegin Tuvdennyam. Protocol Officer, Int. Rel. Dept: Mrs Jamyanjamts Narantsetseg
MW: Ulaanbaatar 990kHz 500kW.
SW: Khonkhor (G.C: 107.00E/47.55N): 1 x 500kW, 1 x 250kW, 1 x 100kW.
kHz: 9720, 12085, 15170

Chinese	Area	kHz
1000-1030	East Asia	990, 15170
1200-1230	Asia	990, 12085
1430-1500	Asia	9720, 12085

English

0900-0930	East Asia	15170
1230-1300	Australia	12085
1500-1530	South Asia	9720, 12085
1930-2000	Europe	9720, 12085

Japanese

1030-1100	East Asia	15170
1345-1415	East Asia	12085

Mongolian

0930-1000	East Asia	990, 15170
1130-1200	Asia	9720, 12085

Russian

0830-0900	Far East	15170
1300-1330	Siberia	12085
1900-1930	Europe	9720, 12085

ANN: Mongolian: "Ullaanbaataraas yaridz baina".
V. by QSL-card (9 different cards available).Re. to relevant language section. Rec. acc. Rp (2 IRC's or $1) appreciated.

MOROCCO

RADIODIFFUSION-TÉLÉVISION MAROCAINE (Gov.)
✉ 1, Rue El Brihi (or B.P. 1042), 10000 Rabat. ☎ +212 (7) 709613. 🖷 +212 (7) 703208. ① 0407 dt rtm 36687 m.
Web: http://www.maroc.net/rc/
Regional ✉ B.P. 459, Laayoune.
L.P: DG: Mohamed Tricha. Dir. Radio: Abderrahman Achour. Dir. TV: Mohamed Issari. Dir. Tec: Jamal Eddine Tanane. Dir. Ext. Rel: Ali M'Barek. Dir .Finance & Admin: Mehdi Bouzekri.
SW: Tanger (G.C:05.50W/35.45N): 100/50kW.

Arabic	Area	Times
11920	EEu/ME	0000-0500
15345	ME	0900-2200
15335	WEu	1100-1500
		2200-2400

English

17595	Eu/Af	1400-1500Sun

French

17595	Eu/Af	1400-1700W
		1500-1700Sun
17815	Eu/Af	1700-1900

ANN: Arabic: "Huna Ribat, idha'atu-l-mamlaka al Maghribiyya". French: "Ici Rabat, Radiodiffusion Télévision Marocaine".
V. by QSL-card. Rec.acc. Rp.

RADIO MEDITERRANÉE INTERNATIONALE (Gov. Comm.)
✉ 3 et 5, rue Emsallah (B.P. 2055), Tanger. ☎🖷 +212 (9) 936363. ① RADMED 33711M. (✉ **in France:** 78 av.Raymond Poincaré, 75116 Paris). **Web:** http://www.medi1.com/
LW/MW/FM: see domestic radio section.
SW: Nador (G.C: 02.55W/ 34.58N): 9575kHz 250kW

Arabic/French	Area	kHz
0500-0100	Eu/Af	9575

ANN: "Médi 1" — **V.** by letter.

VOICE OF AMERICA RELAY STATION
SW: Briech (G.C: 05.58W/35.34N): 10 x 500kW tr's.
D.PRGR: see VOA entry under "USA"

MOZAMBIQUE

RÁDIO MOÇAMBIQUE (Gov. Comm.)
✉ C.P. 2000, Maputo. ☎ +258 (1) 34041/5. 🖷 +258 (1) 421816. ① 6317 & 6340 RADIOMO. **Cable:** Moradio

L.P: Dir: Ext. Sce: Ian Christie (**E-mail:** christie@christie.uem.mz)
FM: 97.9MHz — **SW** (G.C: 32.28E/25.57S): 3262*/4855*/11818*kHz.

English	Area	kHz
1100-1135	Af	3262*, 11818*, 97.9MHz
1800-1900	Af	3262*, 4855*, 97.9MHz

N: 1102, 1802.
*) inactive at editorial deadline.
ANN: "This is the Ext. Sce. of R. Mozambique" or "This is R. Maputo, the Ext. Sce. of R. Mozambique" — **V.** by QSL-card.

NETHERLANDS

RADIO NEDERLAND WERELDOMROEP (RNW)

✉ P.O. Box 222, 1200 JG Hilversum. Office: Witte Kruislaan 55, 1217 AM Hilversum. ☎ +31 35 6724211. 🖹 +31 35 6724207.
E-mail: letters@rnw.nl **Web:** http://www.rnw.nl
Radio Netherlands maintains production & training bureaus in Bénin, Costa Rica & USA.
L.P: DG: Lodewijk Bouwens. Dir. of Prgrs: Jonathan Marks. Dir of Fin & Logistics: Jan Hoek. Gen Mgr Training Centre: Jaap Swart. Chief Editor: Corine Spoor. Head of English Language Service:Nick Meanwell. Head of Latin America Service: Jose Zepeda. Head of Indonesian Service: Indra Leihitu. Head of Television & Projects: Hans de Wildt. Head of Interactive Media: Diana Janssen. Ag. Head of Music: Hans Quant. Head of Programme Distribution: Jan Willem Drexhage. Ag. Head of Marketing & Research: Frans van Heteren
STATIONS: "Flevo": 4 x 500kW (+ 1 x 100kW reserve) (G.C: 05.27E/52.21N)."Bonaire RNW" (Neth. Antilles) 3 x 250kW (G.C: 68.19W/12.13N). "Talata/Volonondry" (Madagascar) 2 x 300kW. (G.C: 47.37E/18.45S). Additional relays via Germany (Jülich & Nauen), South Africa (Meyerton), Luxembourg, Belgium and CIS.
Satellite: Audio subcarriers via Astra in Europe, Intelsat & Asiasat 2 in Africa & Asia and Galaxy 4/5 in North America. RN English Programmes are also relayed domestically by the ABC Australia, CBC, Canada and SAFM in South Africa.
Re-broadcasters: RNW radio prgrs in English, Spanish, Papiamento, Portuguese, French and Dutch are relayed by local st's in many countries. The station is heavily involved with music and television co-productions too.
kHz: 1440, 1512, 5905, 5930, 5955, 5965, 5975, 6015, 6020, 6030, 6045, 6145, 6165, 6190, 7120, 7130, 7260, 7285, 7305, 7310, 7375, 9590, 9605, 9615, 9715, 9720, 9810, 9820, 9830, 9845, 9855, 9860, 9890, 9895, 9925, 11655, 11660, 11680, 11695, 11715, 11730, 11895, 12065, 12090, 13665, 13700, 13710, 13755, 13770, 15155, 15315, 15525, 15560, 15585, 15590, 17580, 17605, 17655, 21480, 21590

Schedule until 29th March 1999

Dutch

0130-0225	NAm	6165, 9895
0130-0225	SAm	15315
0130-0225	CAm/NAm	6020
0330-0425	EAf	9860
0330-0425	ME	15560
0530-0625	NAm	6165, 9715
0600-0657	CEu/SEEu	7130
0600-0700	SEu	6020
0600-0700Mo-Sa	CEu/WEu	1440
0600-0900	SEEu	9895
0600-1900	WEu	5955
0630-0730Su	CEu/WEu	1440
0700-0800	NZL/AUS	9720, 11655
0700-0900	SEu	11935
0700-1800	SWEu	9895
0930-1015Mo-Sa	Carib/SAm	6020
1030-1125	NZL/AUS	9720
1030-1125	SEAs	17580
1030-1125	EAs	21480
1100-1300	SEu	13700
1330-1425	SAs/AUS/ME	13700
1330-1425	SAs	15585, 17580
1330-1425	FE/EAs	5930
1330-1425	EAs/SEAs/AUS	7375
1600-1800	SEEu/ME	9895
1600-1800	Eu/NWAf	13700
1630-1725	SAf	6020
1630-1725	EAf	11655
1700-1800	WEu	1512
1830-1925	WAf	17605
1830-1925	CAf/WAf	21590
1830-1925	ME/EAf	11695
1830-1925	WAf	13700
2030-2125	WAf	15315
2030-2125	CAf/WAf	17605
2030-2125	Eu/NWAf	6020

Dutch	Area	kHz
2030-2125	Eu/NE	9895
2030-2125	WAf	9895
2030-2125	SAf	6015
2030-2125	EAf/Eu	11655
2030-2125	CEu/WEu	5835
2130-2225	SAm	6030
2130-2225	NAm	11730
2130-2225	SAm	15315
2130-2225	NAm	6020
2130-2225	Carib	9895
2330-0025	SEAs	7280
2330-0025	EAs	12090
2330-0025	FE	15560
2330-0025	SEAs	9590

English		
0430-0525	NAm	6165, 9590
0930-1125	NZL/AUS	9820
0930-1125	EAs	12065
0930-1125	FE/EAs	7260
1130-1325	CEu/WEu	6045
1130-1325	WEu	9855
1430-1625	SAs	12090, 15585, 13755
1730-2025	SAf	6020
1730-2025	EAf	9605
1830-2025	WAf	9895, 11655, 15315
2130-2330	CEu/WEu	1512
2330-0125	NAm	6165, 9845

Indonesian		
1130-1325	SEAs	11690, 17580, 21480
2130-2325	SEAs	7285, 9590
2230-2325	SEAs	6120

Spanish		
0230-0425	CAm	6165, 9590
1130-1155	SAm	6020, 9715, 15255
1200-1225	CAm	6020, 9715
2230-2325	SAm	9895, 15315
2230-2325	SAm	11730

Listener Answer Line: +31 35 6724222. This number is open 24 hrs a day for comments and questions concerning English language programmes. Fax line for English programmes: +31 35 6724239. The Spanish department also operates a 24 hr line for comments to their programmes: +31 35 6724444.
ANN: Dutch: "Dit is R. Nederland Wereldomroep in Hilversum". E: "This is Radio Netherlands, the Dutch International Service". Sp: "Transmite R. Nederland desde la ciudad de Hilversum en Holanda". Indonesian: "Inilah Radio Nederland di Hilversum dengan siaran dalam bahasa Indonesia".
IS: Traditional Dutch folk song "Merck toch hoe sterck" played on Breda Carillion three minutes before start of transmissions.
V: by QSL card. Programme feedback always reaches the producers. Reception reports appreciated if they follow the guidelines set down in "Writing Useful Reception Reports". Extensive series of free printed publications designed to assist newcomers to international radio listening. Also available on-line via RNW's extensive Internet pages. E-mail daily news service in Dutch, Indonesian & Spanish. Radio Netherlands issues two Newsletters per year in March and September. Available in Dutch "De Wereld", English "On Target", Spanish "En Sintonia" and Indonesian "Ranesi". Write or E-mail for details.
F.PL: The second series of tests for the Dutch language TV project "Zomer TV" went well. A permanent solution was being discussed at press time. This publicly funded station is actively expanding its co-operation ventures with commercial and public stations, including National Public Radio, CBC, SABC, ABC Australia, & over 950 stations in Latin America. More emphasis on interactive programming, including INTERNET projects. In early 1998, Radio Netherlands Bonaire relay station will conduct tests using digital AM .

NETHERLANDS ANTILLES

TRANS WORLD RADIO (Cult. Educ. Rlg.)

✉ Trans World Radio, P.O. Box 388, Bonaire, Netherlands Antilles.
☎ +599 (7) 8800. 🖹 +599 (7) 8808. **Cable:** VOTAN.
Web: http://www.gospelcom.net/twr/
L.P: MD: Thomas F. Corcoran. PD: McDaniel E. Phillips. CE: David J. Magnuson. Head of PR: A. Cort Tangeman.
MW (G.C: 68.28W/12.11N): 800kHz: 1 x 500kW, 1 x 50kW.
English: 1100-2300 (50kW). **N:** on the h.
Portuguese: 0700-0900 (exc. Sat 0845-0900 Baniua), 2300-2355.
Spanish: 2355-0145, 0900-1100.
ANN: "This is the international sound of the Caribbean, Trans World Radio, Bonaire".
IS: "Stand up, Stand up for Jesus" played on various instruments.
V. by QSL-card. Rp. preferred. IRC's requested. All correspondence replied by airmail. Re. in E, Sp or Portuguese, no tapes. Do not address

reports to individuals — **PUB:** Prgr. schedule free.

RADIO NEDERLAND RELAY STATION
✉ P.O. Box 45, Kralendijk.
SW: Bonaire (G.C: 68.19W/12.13N): 2 x 250kW tr's.
kHz: *5965, 5995, 6020, 6030, 6165, 9620, 9715, 9720, 9830, 11655, 11660, 11730, 11895, 15315, 15550, 17605, 21590*
D.PRGR: See Netherlands — **V.** by own QSL-card for direct re.

NEW ZEALAND

RADIO NEW ZEALAND INTERNATIONAL
✉ P.O. Box 123, Wellington, New Zealand. ☎ +64 (4) 474 1437.
🖷 +64 (4) 474 1433. **E-mail:** rnzi@actrix.gen.nz
Web: http://www.actrix.gen.nz/biz/rnzi/
L.P: Mgr: Linden Clark. CE: Adrian Sainsbury.
SW: Rangitaiki (C.C: 176.25E/38.50S): 100kW.
kHz: *6070, 6100, 6105, 9700, 9795, 9810, 11735, 11905, 15115*

Schedule 26 October 1997 - 29 March 1998

Time	Days	kHz
1650-1852	Mon-Fri	9810
1853-2052	Sun-Fri	11735
1959-2058	Sat	11735
2053-0458	Sun-Fri	15115
2059-0458	Sat	15115
0459-0815	Mon-Fri	11905
0459-0758	Sat & Sun	11905
0816-1206	Mon-Fri	9700
0758-1206	Sat & Sun	9700
1207-1650*	Occasional Use	6105

Schedule 29 March 1997 to 03 May 1997

Time	Days	kHz
1650-1852	Mon-Fri	6070
1853-2050	Sun-Thurs	11735
1850-2057	Fri	11735
1859-2057	Sat	11735
2051-0458	Sun- Thurs	15115
2058-0458	Fri & Sat	15115
0459-0758	Mon-Fri	9795
0459-0758	Sat & Sun	9795
0817-1206	Mon-Fri	6100
0758-1206	Sat & Sun	6100
1207-1650*	Occasional Use	6105

*) used for special events coverage outside normal hours, e.g. sport. Times vary as required.
D.PRGR: Mainly in English with news in various Pacific languages.
ANN: Maori: "Te Reo Irirangi O te Moana-nui-a-kiwa".
IS: Call of the NZ Bellbird — **V.** by QSL-card. Rp (3 IRC's). No rec.

NIGERIA

VOICE OF NIGERIA (Statutory Corp.)
✉ Broadc. House, PMB 40003, Falomo, Ikoyi, Lagos. (+234 (1) 2693245.
Listeners Answerline: +234 (1) 2693078. 🖷 +234 (1) 2691944 — **L.P:** DG: Y. Abubakar. PD: L.Y. Saulawa. Dir. Eng: G. Ugwu. Head of PR: Livy Iwok.
SW: Ikorodu (G.C: 03.56E/07.23N): 5 x 250kW tr's

Arabic	Area	kHz		Fulfulde	Area	kHz
1700-1800	WAf	7255		0900-1000	WAf	7255
English				1300-1400	WAf	7255
0500-0700	WAf	7255		2100-2200	WAf	7255
1000-1100	WAf	7255		**Hausa**		
1500-1700	WAf	7255		0800-0900	WAf	7255
1900-2100	WAf	7255		1200-1300	WAf	7255
French				2200-2300	WAf	7255
0700-0800	WAf	7255		**Swahili**		
1100-1200	WAf	7255		1400-1500	WAf	7255
1800-1900	WAf	7255				

N. in English: 0530, 0630, 1030, 1045 (Nigerian N.), 1530, 2000.
ANN: "This is the Voice of Nigeria".
IS: As Home Sce. Also bells playing the first bars of the National Anthem 15 mins. before the commencement of each block period.
V. by QSL-card.

NORTHERN MARIANA ISLANDS

FAR EAST BROADCASTING CO. (Rlg.)
✉ Box 209, Saipan, MP 96950-0209. ☎ +670 234 6520 (KSAI), +670 322 9088 (KFBS). 🖷 +670 322 3060. **E-mail:** febc@itecnmi.com
Web: http://www.febc.org

L.P: Dir: Chris Slabaugh. PD: Frank Gray. CE: Bob Springer. Head of Public Rel: Bob Stiles.
MW: Susupe (G.C: 145.50E/15.08N): KSAI 936kHz 10kW.
D.PRGR. in English: 2000-1200. **N.** on the h.
SW: KFBS Marpi (G.C: 145.48E/15.16N): 4 x 100kW tr's.
kHz: *9465. 9495, 9670, 11650, 15380*

Bugis	Area	Days	kHz
0930-1000	INS	1234567	15380
Burmese			
1430-1530	BRM	36	9670
English			
1830-1900	RUS	1234567	9465
German			
1530-1600	RUS/KAZ	7	9465
`830-1900	RUS/KAZ	1	9465
1845-1900	RUS/KAZ	6	9465
Gorontalo			
0830-0900	INS	1234567	15380
Indonesian			
1100-1230	INS	1234567	15380
Jarai			
1400-1430	VTN	3	9670
Javanese			
1030-1100	INS	1234567	15380
Kazakh			
1545-1600	KAZ	56	11665
Koho			
1400-1430	VTN	7	9670
Kyrgyz			
1530-1545	KGZ	56	9495
Makascar			
0900-0930	INS	1234567	15380
Malay			
1230-1300	MLA	1234567	15380
Mongolian			
1100-1130	CHN/MNG	1234567	11650
Osetin			
1545-1600	GEO	4	9495
Polish			
1915-1930	POL	2	9465
Romanian			
1830-1900	ROU	1	9465

Russian	Area	Days	kHz
0900-1100	RUS	1234567	11650
1130-1400	RUS	1234567	11650
1400-1530	RUS	1234567	9465
1530-1545	RUS	23	9465
1600-1800	RUS	1234567	9465
1830-1900	RUS	4	9465
Sundanese			
1000-1030	INS	1234567	15380
Uighur			
1600-1615	CHN	1234567	9495
Ukrainian			
1530-1600	UKR	1	9465
1800-1830	UKR	6	9465
1800-1845	UKR	5	9465
1845-1900	UKR	3	9465
Uzbek			
1530-1545	UZB	4	9495
Vietnamese			
1300-1430	VTN	12456	9670
1300-1400	VTN	37	9670

V. by QSL-card. Rp. (2 IRC's). Rec. acc.

CHRISTIAN SCIENCE PUBLISHING SOCIETY - SHORTWAVE BROADCASTS (Rlg.)
✉ KHBI Saipan, P.O. Box 1387, Saipan, Mariana Islands, CM 96950-1387, USA. ☎ +1 (670) 234 6515. 🖷 +1 (670) 234 5452.
Web: http://www.tfccs.com/GV/shortwave/shortwave_schedule.html
L.P: SM: Doming F. Villar. Contacts: Alexander U. Igisaiar (non-technical), Jess Emmanuel Domingo (technical). Rp. appreciated.
Other information: see listing under USA.
SW: KHBI, Agingan Point (G.C: 145.41E/15.07N): 2 x 100kW tr's.
kHz: *9355, 9385, 9430, 13840, 15665*

Times	Area	kHz	Area	kHz
0800-0900	Eu	15665	Oce/INS	13840
0900-1000	NEAs	9430	Oce/INS	13840
1200-1300	NEAs	9355	Oce/INS	9430
1300-1400	NEAs	9385	SEAs/IND	9355
1400-1500	NEAs	9355	Oce/INS	9385
1500-1600	NEAs	9355		
1600-2000	Eu	9355	Af	9385
2000-2200			Oce/INS	13840
2200-2300	NEAs	13840		

D.PRGR: see USA.

NORWAY

RADIO NORWAY INTERNATIONAL
N-0340 Oslo 3. ☎ +47 23 04 84 41. 🖷 +47 23 04 71 34 . **Schedules via Automatic Answer Lines: Eu/Af/Am** +47 23 04 80 08. **Rest of World** +47 23 04 80 09. **E-mail:** info@nrk.no
Web: http://www.nrk.no/radionorway/
MW: Kvitsøy 1314kHz 1200kW.
SW: Kvitsøy (G.C.: 05.15E/59.08N): 2 x 500kW — Sveio (G.C.: 05.19E/59.37N): 2 x 500kW.
kHz: *5960, 5965, 7180, 7295, 7405, 7420, 7465, 7485, 7520, 7540, 7545, 7560, 7565, 7570, 9590, 9905, 9960, 9975, 9985, 11625, 11840, 11850, 12055, 13800, 13805, 15175, 15220, 15605, 17840*

Winter Schedule 1998/99

Norwegian		
0000-0030	As/AUS/Am	9945, 9935
0100-0130	As/Am	9945, 7545, 7465
0200-0230	As/Am	9945, 7545, 7465
0300-0330	Eu/ME/Am	9945, 7565, 7545
0400-0430	Eu/ME/Af/Am	9940, 7560, 7545
0500-0530	Eu	7480, 7440
0600-0630	Eu/Af	13800, 9590, 7180, 5965
0700-0730	Eu/Af	18950, 9590, 7180
0800-0830	ME/As/AUS	15175, 13800
0900-0930	As/AUS	15175, 13800
1000-1130	Eu/Am	18950, 13800
1200-1230	As/AUS/Am	21725, 18950, 9910
1300-1330	Eu/As/AUS/Am	21725, 18950, 9910, 9590
1400-1430	ME/As/Am	18950, 15705
1500-1530	ME/As/Am	15705, 13800, 13790
1600-1630	Eu/Am	15705, 13800, 9590
1700-1730	Eu/Am	18950, 15705, 11895, 7485
1800-1830	Eu/Af/Am	18950, 15705, 7485, 5960
1900-1930	Eu/Af/AUS/Am	15705, 13800, 9935, 7485,, 1314
1930-2000	Eu	1314
2000-2030	Eu/Af/AUS/Am	15705, 7580, 7560
2100-2130	Eu/Af/AUS	9945, 7580
2200-2230	As/AUS/Am	9925, 9415
2300-2330	As/AUS/Am	9945, 9935, 9415

+) exc. Sun.
ANN: "Radio Norway International. Detter er Norge, utenlandssendigen NRP, program 1/2" — **IS:** Each SW transmission begins with theme which originates from an ancient folk tune from Hallingdal.
V. by QSL-card. No Rp — **PUB:** SW prgr/freq guide. Free on request

OMAN (Sultanate of)

RADIO SULTANATE OF OMAN (Gov.)
Ministry of Information, P.O. Box 600, 113 Muscat. ☎ +968 603222. 🖷 +968 603812. ① 5265 INFORM ON. **Cable:** Information.
Web: http://www.oman-tv.gov.om
L.P: DG: Ali Bin Abdallah Al Mujeni. DG (Tech. Affairs): H.Y.Al-Kindy. CE (Radio): Abood Al-Sawafi. Head of Maintenance: Rashid Haroon Al Jabry. PD: Hamdan S.Al-Humeidy. Dir. Foreign Sce: Fatiyah Al Hinai.
SW: S=Seeb (G.C.: 58.10E/23.40N), T=Thumrait (G.C.: 53.56E/ 17.38N): 50/100kW.

kHz	Area	Times
6085T	ME	0200-0400
6120S	ME	0200-0400+, 1800-2130
7170S	ME	1400-1700
7230S	ME	0200-0700, 1100-2130
9510T	ME	0700-1000
9735T	ME	1600-2130
11890T	ME	1000-1400;
15140T	ME	0800-1600

ANN: A: "Idha'atu Saltanat Oman min Muscat."
V. by QSL-folder.

BBC EASTERN RELAY STATION
P.O.Box 6898, 112 Ruwi, Sultanate of Oman.
MW: Masirah Island 720/1413kHz 750kW.
SW: Masirah Island (G.C.: 58.53E/20.36N): 4 x 100kW tr's.
kHz: *5965, 5975, 6030, 6065, 6095, 7200, 9510, 9605, 11750, 11760, 11845, 11920, 11955, 15310, 15340, 17790*
F.PI: This st. will be de-commissioned in 2001. A new tr. site is under construction at Al-Ashkharah. It will house 4 x 300kW SW and 2 x 600kW MW-tr's.

PAKISTAN

RADIO PAKISTAN
Broadcasting House, Constitution Avenue, Islamabad 4400. ☎ +92 (51) 214947. 🖷 +92 (51) 811861 or 223827. ① 5816 PBCNO PK.
MW: Islamabad 1260kHz 500kW — Quetta 756kHz 150kW.
SW: Islamabad (G.C.: 73.12E/33.27N): 2 x 250kW & 5 x 100kW tr's, Karachi (G.C.: 67.00E/24.55N): 2 x 50kW tr's.
kHz: *6070, 7145, 7255, 7290, 7485, 9335, 9485, 9515, 9520, 9585, 9600, 9785, 9903, 11565, 11570, 11710, 11760, 11775, 11910, 11970, 13585, 13665, 15175, 15190, 15325, 15530, 15485, 15520, 15555, 15595, 15625, 17540, 17555, 17705, 17835*

Arabic	Area	kHz
1800-1845	ME	7145, 9335
Bengali		
1200-1245	SAs	15625, 11970
English		
0230-0245	SAs	17705, 15485, 15190, 7290
1105-1120	Eu	17835, 15520 (Slow Speed News)
1600-1630	ME/EAf	11565, 9600, 9515, 7485, 6070
Farsi		
1300-1345	ME	11910, 9785
French		
1930-2000	Eu/Af	11570, 9585
Gujarati		
0330-0400	Af	15325
Hindi		
1100-1145	SAs	9520, 11775, 13665
1615-1700	SAs	9600, 7485, 6070
Indonesian		
0900-0930	SEAs	15625, 17540
Swahili		
0645-0745	EAf	15555
Tamil		
1000-1030	SAs	13665, 15330
Turkish		
1730-1800	ME	9515, 7255
Urdu		
0100-0215	SEAs	11760, 15485, 9515
0430-0630	ME	17555, 15175, 11710, 11570
0800-1103	Eu	15520, 17835
1330-1530	ME	15595, 13585, 11570, 9903, 9485
1700-1900	Eu	11570

ANN: "This is Radio Pakistan" — **V.** by QSL-card. Re. to Contr. Freq. Management at 1st addr. (Home Sces: Peshawar & Quetta by letter for direct rpts) — **PUB:** Ahang; Pakistan Calling (monthly).

PALAU

KHBN (Rlg.)
P.O. Box 66, Koror, Republic of Palau 96940. ☎ +680 488 2162. 🖷 +680 488 2163. **E-mail:** voh@broadcast.net
Web: http://www.highadventure.org/voh_china.html
L.P: Freq. Consultant: George Jacobs.
SW: Babeldaob (G.C.: 144.42E/13.28N): 2 x 100kW.
kHz: *9730, 9965, 9985*

English/Hindi/	Area	kHz
Tagalog		
1100-1800	SAs	9985
English		
2200-2400	SAs	9965
Mandarin		
0800-1100	CHN	9730
1200-1400	CHN	9730
2100-2300	CHN	11775
Korean		
1130-1200	KRE	9730
Urdu		
1700-1800W	SAs	9965
Vietnamese		
1100-1130	VTN	9730

ANN: "Voice of Hope", "Holy Spirit Radio Station".

PHILIPPINES

RADYO PILIPINAS
Philippine Broadc. Sce, 4th Flr. Media Center, Visayas Ave, Diliman, Quezon C., Metro Manila 1103. ☎ +63 (2) 924 2267. 🖷 +63 (2) 924 2745 — **L.P:** Dir's: Rafael Dante A. Cruz, Ben B. Tabisaura. Ag. Chief Eng. Div: Armando C. Remedo.
SW: Tinang (see Voice of America Philippines).

English	kHz
0230-0330	17760, 17840, 21580
Filipino	
1730-1930	11815, 11890, 15190

V. by QSL-card. Rp. (2 IRC's). Rec. acc.

RADIO VERITAS ASIA
Philippine Radio Education & Information Center
✉ P.O. Box 2642, Quezon C. 1166. ☎ +63 (2) 9390012. 🖷 +63 (2) 9381940. **Cable:** 64420 VERTAS PN. ☏ 632-900014.
E-mail: veritas@mnl.sequel.net
Web: http://www.catholic.org.tw/vntaiwan/veritas/
L.P: Chmn: H.E.J.Cardinal Sin. Mgr: Ms. Erlinda G.So. PD: Msgr. Pietro Nguyen Van Tai. Tech.Consultant: Rev. Fr. Hugo Delbaere, CICM. Prgr. Consultant: Fr.J.Desautels. Tech.Dir: Engr. Floremundo Kiguchi. Freq.Planning: Alfonso L. Macaranas. Head of PR: Cleote Labindao.
SW: Palauig, Zambales (G.C: 119.50E/15.28N): 3 x 250kW tr's.
kHz: 6075, 6190, 7235, 7265, 9505, 9520, 9535, 9555, 9660, 9670, 9680, 11805, 11820, 11850, 15335, 15530

Bengali	kHz	Filipino	kHz
0100-0125	15335	1500-1525a	9680
1400-1425	9555	1500-1555b	9680
Hindi		Russian	
0130-0155	15335	0230-0325	11805
1330-1355	9555	1430-1525	9660
Hmong		Sinhala	
1000-1025	9555	0000-0025	11820
Indonesian		1330-1355	9660
2300-2325	11820	Tamil	
2300-2325	9505	0030-0055	11820
1200-1225	9505	1400-1425	9520
Kachin		Telugu	
0030-0055	9535	0100-0125	15530
1230-1255	9670	1430-1455	9520
Karen		Urdu	
0000-0025	9535	0200-0225	15335
1200-1225	9670	1530-1555	9670
Mandarin		Vietnamese	
2100-2255	6190	2330-2355	9670
1000-1155	9520	0130-0225	15530
Burmese		1030-1125	11850
2330-2355	7235	1300-1325	7265
1130-1155	9670	Zomi-Chin	
Filipino		1530-1555c	9555
2230-2255	6075		

a) Mon/Tues/Thurs/Sat, b) Wed/Fri/Sun, c) Mon/Sat.
ANN: "This is R. Veritas Asia, Quezon C., Philippines".
IS: "O Via, Vita, Veritas" — **V.** by QSL-card.

FAR EAST BROADCASTING CO. (Rlg.)
✉ Box 1, 0560 Valenzuela, Metro-Manila. ☎ +63 (2) 361 1010. 🖷 +63 (2) 359490. **Cable:** FEBCOM MANILA. ☏ 40048 FEBCOM PM. **E-mail:** febcomphil@febc.jmf.org.ph **Web:** http://www.febc.org
L.P: MD: Efren M. Pallorina. Head of Prgrs: Peter McIntyre. CE: Romualdo M. Lintag. Head of PR: Priscilla R. Calica.
SW: Bocaue, Bulacan (G.C: 120.55E/14.48N): 1 x 100kW, 3 x 50kW — Iba, Zambales (G.C: 119.58E/15.20N): 2 x 100kW.
kHz: 9405, 9475, 9495, 9780, 9795, 9875, 11610, 11635, 11650, 11995, 12005, 15095, 15450, 15465, 15520

	Area	kHz	kW
Aceh			
1045-1115	Indonesia	15095	50
Akha			
2300-2315 SS	China/Laos/Thai	11650	50
Batak			
0945-1015	Indonesia	15095	50
Amoy			
1000-1030	China	9475	50
Ao Naga			
0145-0200 SS	India-NE	15465	50
Burmese			
1230-1345	Myanmar	15095	50
2330-0115	Myanmar	15465	50
Burmese Tribal			
1330-1345	Myanmar	9795	100
1145-1230	Myanmar	15095	50
1345-1430	Myanmar	15095	50
0115-0130	Myanmar	15465	50
Cantonese			
1100-1230	China	9475	50
English			
0100-0300	India/Asia SE	15450	50
0930-1100	China	11635	50
1400-1600	India/Asia-SE	11995	100
Hmong			
2300-2330	Laos	9875	100
1100-1130	Laos	11610	100

Indonesian	Area	kHz	kW
2230-2330	Indonesia	9780	50
0900-1000	Indonesia	12005	50
Malay			
1015-1045	Malaysia	15095	50
Javanese			
0830-0900	Indonesia	12005	50
Khmer			
2230-2315	Kampuchea	9495	100
1200-1300	Kampuchea	11610	100
1045-1100	Laos	11610	100
Kuki			
0130-0145 SS	India-NE	15465	50
Lahu			
2315-2330	China/Myanmar	11650	50
Lao			
2230-2300	Laos	9875	100
1130-1200	Laos	11610	100
Mandarin			
0500-0900	China	15450	100
0900-1600	China	9405	100
1230-1600	China	9475	50
2200-2400	China	9405	100
Meitei			
0130-0145 MF	India-NE	15465	50
Mien			
1030-1045	China/Vietnam	11610	100
1345-1400	China/Vietnam	9795	100
Minang			
1000-1030	Indonesia	12005	50
Mon			
1130-1145	Myanmar	15095	50
Ogan			
0915-0945	S. Sumatra	15095	50
Sasak			
0800-0830	Indonesia	12005	50
Shan			
2345-0030	Myanmar	15520	50
Swatow			
1030-1100	China	9475	50
Vietnamese Tribal			
1115-1200	Vietnam	9795	100
1300-1330	Vietnam	9795	100
Vietnamese			
1200-1300	Vietnam	9795	100
2315-2400	Vietnam	9495	100
Wa			
2330-2345	China/Myanmar	11650	50
Zoukam			
0200-0215 Sat	India-NE	15465	50

ANN: "Broadcasting from the Philippines, this is FEBC Radio International, The Sound Alternative"2200-2315
IS: "We have heard the Joyful Sound".
V. by QSL-card (B). Rp. preferred (3 IRCs for Airmail).
PUB: Prgr. schedule free on request. Signal (quarterly).

VOICE OF AMERICA RADIO STATION PHILIPPINES
MW: Poro 1143kHz 1000kW
SW: Poro (G.C: 120.17E/16.37N): 2 x 100kW, 3 x 50kW, 1 x 35kW tr's — Tinang (G.C: 120.37E/15.21N): 12 x 250kW, 3 x 50kW tr's.
D.PRGR: See USA.

POLAND

POLISH RADIO, WARSAW
✉ P.O. Box 46, 00-977 Warszawa. ☎ +48 (2) 645 9305. 🖷 +48 (22) 444123. **Listeners' Answerlines: English:** +48 (2) 645 9262. **German:** +48 (2) 645 9333. **E-mail:** piatka@radio.com.pl
Web: http://apollo.radio.com.pl/piatka/
L.P: Dir. of Foreign Sce. (Prgr. 5): Jerzy Marek Nowakowski.
SW: Warszawa (G.C: 20.53E/52.02N): 7 x 100kW.
kHz: 5900, 5915, 5970, 5995, 6000, 6035, 6095, 7145, 7180, 7185, 7205, 7215, 7270, 7285, 7295, 9525, 9540, 9670, 11815, 11825

English	kHz
1300-1355	11815, 9525, 7145, 6095, 5995
1800-1855	7270, 6095, 6000
2030-2125	7285, 6095, 6035
German	
1230-1255	9525, 7145, 6095
1500-1525	9525, 7285, 7145, 6095
1600-1625	7270, 7145, 6035, 6000
1730-1755	7285, 7270, 6095 ;6035, 6000
Esperanto	
1430-1455	9525, 7285, 7270, 7145, 6095, 6000
2130-2155	7285, 6095, 6035

Belorussian	kHz
1230-1255	7285, 5995, 5970
1430-1455	7270, 6095, 6000
Belorussian	
1630-1655	7270, 7205
1900-1925	7180, 6000, 5915
Czech	
1700-1725	7215, 6095
2000-2025	7215, 6035
Slovak	
1500-1525	7205, 6000 Sunday only
Lithuanian	
1300-1325	9525, 7270, 7145
1600-1625	9525, 6035
1830-1855	7180, 5915
Polish	
1200-1225	7285, 7270
1630-1725	7285, 7145, 6035, 6000
2200-2255	7285, 6095, 6035
Russian	
1200-1225	9525, 7145, 6095
1400-1425	9525, 7295, 7270, 7145
1530-1555	9670, 6095
1900-1925	7285, 7270, 6000
Ukrainian	
1530-1555	7285, 6035, 5995
1800-1825	7185, 7180, 5915
1930-1955	7270, 6095, 6035

ANN: Polish: "Tu Polskie Radio, Warszawa." E: "This is the Polish Radio, Warsaw"— **IS:** Revolutionary Etude (Chopin).
V. by QSL-card. Rec. Acc.

TRANS WORLD RADIO (Rlg.)
☞ and other information: see Monaco.
MW: Szczecin (Stargard) 1503kHz 300kW.

Finnish	Days	Area	kHz
2030-2100	12345	Finland	1503
Sweden			
2100-2130	12345	Sweden	1503

PORTUGAL

RDP INTERNACIONAL – RADIO PORTUGAL
☞ Apartado 1011, P-1001 Lisboa Codex. ☎ +351 (1) 347 5065.
📠 +351 (1) 347 4475. **Web:** http://www.rdp.pt/rdp/portugal
L.P: Dir: Jaime Marques de Almeida.
SW (G.C: 08.40W/38.45N): 1 x 300kW, 7 x 100kW.
S=via Sines (R. Trans Europe).
kHz: 6150, 7110, 9570, 9600, 9614S, 9630S, 9655, 9780, 9815, 11740, 11840, 15200, 17595, 17725, 17745, 21515, 21655, 21720

English	Area	kHz
0430-0500MF	NAm	9570, 6150
1330-1400MF	IND/ME	21515
2100-2130MF	Eu/Af	9815, 9780, 7110
French		
2030-2100MF	Eu/Af	9815, 9780, 7110
Tetum		
1205-1300MF	Timor	17595

Winter Schedule 1998/99

Portuguese	kHz
0500-0600	15555MF
0600-0645	9780MF, 15555MF
0645-0700	9780MF, 11780MF, 15555MF
0700-0800	9780, 11780MF, 15555, 17725SS
0800-0830	9780, 15555, 17725SS
0830-0900	9780, 11850SS, 15555, 17725SS
0900-1000	9780, 11850SS, 15555, 17725, 21720MF
1000-1055	9780, 15555, 17725, 21720MF
1055-1200	9780, 15555, 17725SS
1600-1700	11780MF, 13625MF, 15555MF, 17680MF
1700-2000	9780SS, 11780MF, 13625MF, 15200SS, 15555, 17680

ANN: P: "RDP Internacional - Rádio Portugal a emitir dos seus estúdios em Lisboa". E: "This is RDP - Radio Portugal International broadcasting from Lisbon". F: "Ici la Radiodiffusion Portugaise Internationale, qui vous parle de Lisbonne, dans son émission en langue française, émettant em ondes courtes". T: "RDP - Internacional hahu emissão ba Extrema Oriente, iha frequência rihun sanulo resin hitu, atus lima sianulo resin lima kHz, banda metro sanulo resin nen".
IS: Opens and closes with Nat.Anthem preceded by time gong.
V. by QSL-card. Rec. acc.

RÁDIO RENASCENÇA (Rlg. Comm.)
☞ Rua Ivens 14, 1294 Lisboa Codex. ☎ +351 (1) 347 5270. 📠 +351 (1) 342 2658.
L.P: MD: Oliveria Pires. PD: Pedro Tojal. CE: João Ramos.
SW: Muge (G.C: 08.41W/39.05N): 1 x 100kW (currently inactive).

RADIO TRANS EUROPE
SW: Sines (G.C: 08.46W/37.57N): 3 x 250kW tr's.
This st. is also used by various international broadcasters. Times and freqs are included in the International Broadcasting Section.

QATAR

QATAR BROADCASTING SERVICE (Gov.)
☞ QBS, P.O. Box 1414, Doha. ☎ +974 864111. 📠 +974 822888. **Cable:** Broadcasting. ☼ 4597 Q Radio DH.
L.P: Asst. Under-Secr. for Radio & TV: Abdul Rahman Saif Al-Madhadi (☎ +974 864823). Dir. of Broadc: Mubarak Jaham Al-Kawari (☎ +974 864805). Dir. Eng. Radio & TV: Hussain A.Jaffar. (☎ +974 864518).
SW: Al Khaisah (G.C: 51.25E/25.25N): 1 x 500kW, 1 x 250kW, 1 x 100kW.
kHz: 7210, 9570, 11750, 15395, 17895

Times	Area	kHz
0245-0706	ME	9570, 7210
0707-1304	ME	17895 (alt: 15395), 9570
1305-1704	ME	11750, 9570
1705-2130	ME	9570, 7210

ANN: A: "Idha'at at Qatar Min Al-Doha". E: "This is the Qatar Broadcasting Station, Doha".
IS: "Arabic melody on string instrument (23 notes) — **V.** by QSL-card.

ROMANIA

RADIO ROMÂNIA INTERNATIONAL
☞ P.O. Box 111,Bucharest. ☎📠 +40 (1) 312 3640 or 312 9262. **E-mail:** rri@radio.ror.ro **Web:** http://indis.ici.ro/romania/news/rri.html
L.P: MD: Eugen Preda. PD: Ion Ghitulescu. CE: Ilie Mihai. Dir. Int. Rel: Titus Voijeu. Head of PR: Dan Ursuleanu.
MW: Timisoara 630kHz, Lugoj 756kHz (see Home Sce.)
SW: Tiganesti (G.C: 26.05E/44.45N): 4 x 250kW — Galbeni (G.C: 26.41E/46.45N): 2 x 250kW, 2 x 120kW — Saftica (G.C: 26.05E/44.38N): 1 x 100kW, 1 x 50kW, 2 x 18kW.

kHz: 5955, 5965, 5990, 6105, 6150, 6155, 7105, 7135, 7145, 7175, 7195, 7225, 7245, 9510, 9530, 9550, 9570, 9580, 9625, 9665, 9690, 9750, 9755, 11740, 11775, 11790, 11810, 11830, 11840, 11940, 11970, 15105, 15250, 15335, 15365, 15380, 15390, 15405, 17720, 17745, 17775, 17790, 17805

Arabic	Area	kHz
1230-1256	ME/Af	15365, 15380, 15405, 17720, 17745, 17790
1400-1426	ME/Af	11740, 15365, 15380, 15405, 17745
1530-1626	ME/Af	11740, 11790, 15365, 15380, 15405
Aromanian		
1630-1659	Eu	756
1930-1959	Eu	756, 1458, 100.1MHz
English		
0530-0556	Af	11940, 15250, 15365, 17745, 17790
0631-0640	Eu	7105, 9510, 9570, 9665, 11775
0645-0745	Pac/FE	15250, 15405, 17720, 17805
1300-1356	Eu	11940, 15390, 17745
1430-1526	SAs	11740, 11810, 15335
1900-1956	Eu	6105, 7105, 7195, 9510
1730-1756	Af	9750, 11740, 11940
2100-2156	Eu	5955, 5990, 7105, 7195
0200-0256	NAm	5990, 6155, 9510, 9570, 11940
0400-0426	NAm	5990, 6155, 9510, 9570, 11940
French		
0613-0622	Eu	7105, 9510, 9570, 9665, 11940
1130-1156	Eu	15390, 17775
	Af	15380, 17790
	NAm	15250, 17745
1430-1456	Af	15365, 17745
1630-1656	Eu	9625, 9690, 11940
1900-1926	Eu	7225, 9665
2000-2956	Eu	5990, 7105, 7195, 9510
German		
0622-0632	Eu	7105,9510, 9570, 9665, 11775
1200-1226	Eu	9690, 11940, 15390
1630-1656	Eu	5990, 7195, 9690
1800-1856	Eu	6105, 7195, 9690
Greek		
1700-1729	Eu	756
1830-1859	Eu	756

Hungarian	Area	kHz
0500-0526	HNG	5955, 7245
1930-1959	HNG	630

Italian

1530-1556	Eu	9665, 11790
1800-1826	Eu	756
1900-1926	Eu	756, 6150
2000-2028	Eu	756, 6150

Persian

1500-1526	ME	15250, 17745
1700-1726	ME	9530, 11940

Portuguese

1800-1826	Af	9570, 11740
2130-2226	Eu	7145, 9665, 11790
0100-0130	SAm	5990, 6155, 9510, 9570, 11940

Romanian

0430-0600+	WEu	7225, 9570
0600-0614	Eu .	7105, 9510, 9570, 9665, 11775
0600-0900+	WEu	11970, 15105
0900-1500+	WEu	15105, 17720
1300-1326	AUS	15335, 17790
1500-2000+	WEu	9570, 11970
1630-1656	ISR	7105, 9665
1730-1756	Eu	5990, 6105, 7195
2000-2026	Eu	7175, 9690
2030-2058	CEu	756
2000-0430+	WEu	6105, 7225
2230-2256	LAm	9580, 9755, 11830
2300-2356	NAm/AUS	5990, 9550, 9570, 11740, 11940
0130-0156	NAm	5990, 6155, 9510, 9570, 11940

Russian

1330-1356	RUS	9570, 11775
1700-1726	RUS	5955, 6105, 7105, 7225
1830-1856	RUS	5995, 6105, 7105, 7175, 7225

Serbian

1700-1728	CEu	756
1830-1856	CEu	756

Spanish

1930-1956	Eu	7145, 9665, 11790
2200-2256	LAm/Eu	9510, 9570, 11940
0000-0056	LAm	5990, 6155, 9510, 9570, 11940
0300-0356	LAm	5990, 6155, 9510, 9570, 11940

Turkish

1400-1426	TUR	7105, 9510
1600-1626	TUR	7105, 9510

Ukrainian

0430-0456	UKR	5965, 7135
1500-1526	UKR	5955, 6105

+) rel. Home Sce. 1st Prgr. (România Actualitati).

VOICE OF AMERICA RELAY

MW: Lugoj 756kHz 400kW.
D.PRGR: 0430-0500 & 2200-2230 **Serbian**, 0530-0600 **Croatian**.

RUSSIA

VOICE OF RUSSIA

📧 ul. Pyatnitskaya 25, 113326 Moskva. ☎ +7 (095) 2337801. 📠 +7 (095) 2337648. ① 411137 SU. **E-mail:** letters@vor.ru
Web: http://www.vor.ru/ or http://www.vr.ru
LP: MD: Armen Oganesyan. Dir. Foreign Rel: Victor Kopyin. TD: Maria
MEDIUMWAVE: See Home Sce listing for locations and powers.
kHz: 585, 603, 612, 630, 648, 693, 720, 801, 972, 1080, 1089, 1125, 1143, 1170, 1215, 1251, 1269, 1323, 1386, 1494, 1548
SHORTWAVE: Official information regarding transmitting facilities is not available. This list is based on information collected by WRTH collaborators. Some transmitter sites may be inactive or operating below capacity. (H) = used for Home Sce relays, (F) = used for Foreign Sce.
Moscow area: Zelenogradskaya (Lesnoy, G.C: 37.58E/56.04N): 1 x 150kW, 15 x 250kW (F) operated as 9 units – Taldom (Severnyy, G.C: 37.38E/56.44N): 3 x 1000kW, 4 x 250kW, 12 x 100kW (H/F) – Chkalovskaya (Shchelkovo, G.C: 38.04E/55.54N): 3 x 100/200kW (H/F) – Kurovskaya (Mistsevo, G.C: 39.09E/55.34N): 2 x 80kW, 8 x 100kW, 1 x 150kW, 6 x 250kW (H/F) – Elektrostal (Noginsk, G.C: 38.22E/55.50N): At least 5 tr's of 50-120kW (H) – Kupavna (Svetlyy, G.C: 38.09E/55.44N): 100kW SSB (inactive) – Balashikha (G.C: 37.54E/55.47N): Several 20kW tr's (H) – Oktyabrskoye polye (Ce. Moscow, G.C: 37.30E/55.46N): 12 x 5kW on 4055kHz (inactive).
Other sites (west to east): Sovetsk (Bolshakovo, G.C: 21.43E/54.54N): 9 x 80kW (H/F) – Sankt-Peterburg (Popovka/Krasnyy Bor, G.C: 30.42E/59.39N): 18 x 200kW, operated as 10 units (H/F) – Kovylkino (Zelenaya Roshcha, G.C: 44.01E/54.09N): 4 x 60kW, 1 x 80kW (H) – Samara (G.C: 50.15E/53.17N): Approximately 9 x 100kW, 3 x 200kW, 6 x 250kW (H/F) – Krasnodar-Tbilisskaya (Oktyabrskiy, near Tbilisskaya, G.C: 40.07E/45.29N): 8 x 100kW, 1 x 250kW, 4 x 1000kW (H/F) –

Yekaterinburg (G.C: 60.36E/56.55N): Approximately 20 tr's, mostly 100kW, normally used in pairs (H/F) – Novosibirsk: Site A: 23 x 100kW (H/F); Site B appr. 4 x 60kW (H) – Novosibirsk-Oyash (Raduga, G.C: 83.45E/55.31N): 3 x 1000kW (F), 60kW (H) – Angarsk (G.C: 103.40E/52.25N): 2 x 100kW, 4 x 250kW, 2 x 2000kW (H/F) – Irkutsk: 2 x 50, 1 x 100kW (H) – Chita (Kruchina, S. of Atamanovka, G.C: 113.44E/51.51N): 250kW & 2 x 1000kW (F), 1 x 50kW (H) – Yakutsk: 5 x 15/100kW (H) – Khabarovsk: Many tr's of 60/200kW (H/F) – Komsomolsk: Site A (G.C: 136.55E/50.39N?): 1 x 15 + 4 x 100kW (H); Site B: 1 x 100, 1 x 200kW, 1 x 250kW (F) – Ussuriysk (Razdolnoye, G.C: 131.57E/43.32N) & Vladivostok (Tavrichanka, G.C: 131.54E/43.20N): Appro x imately 2 x 100kW, 1 x 120kW, 2 x 200kW 3 x 250kW, 2 x 1000kW (H/F) – Magadan (Arman, G.C: 150.13E/59.43N): 4 x 100kW (H/F) – Petropavlovsk-Kamchatskiy (Yelizovo?, G.C: 158.24E/53.13N?): 3 x 100kW, 1 x 200kW, 2 x 250kW, (H/F).
Foreign relays: Through transmitters leased from Armenia, Belarus, Germany, Kazakstan, Moldova, Tajikistan, Uzbekistan. Relays on an exchange basis via China (People's Rep.). Relay facilities in Russia are also leased to broadcasters from other countries.

Winter Schedule 1998/99

SW kHz: 4010, 4040, 4730, 4920, 4940, 4975, 5905, 5915, 5920, 5925, 5930, 5935, 5935, 5940, 5950, 5965, 5975, 6000, 6005, 6020, 6030, 6065, 6090, 6100, 6110, 6145, 6175, 6185, 6205, 7105, 7115, 7125, 7130, 7135, 7140, 7150, 7150, 7155, 7160, 7170, 7175, 7180, 7185, 7195, 7205, 7210, 7215, 7220, 7230, 7235, 7245, 7255, 7260, 7275, 7280, 7290, 7295, 7300, 7305, 7310, 7320, 7325, 7330, 7340, 7350, 7355, 7360, 7370, 7380, 7390, 7420, 7440, 7490, 9370, 9375, 9400, 9440, 9450, 9470, 9480, 9490, 9505, 9550, 9560, 9585, 9630, 9635, 9675, 9680, 9710, 9715, 9720, 9725, 9745, 9775, 9780, 9795, 9800, 9810, 9820, 9825, 9830, 9840, 9860, 9865, 9875, 9890, 9895, 9905, 9940, 9975, 11655, 11820, 11865, 11880, 11895, 11975, 12000, 12005, 12010, 12025, 12050, 12055, 12065, 13640, 13665, 13670, 15120, 15425, 15460, 15470, 15490, 15535, 15540, 15570, 17755, 17795, 17860, 21790

Albanian	Area	kHz
16.00-16.30	Eu	11630**, 9775**, 9450*, 7370, 7330*, 7300**, 5975*

Arabic

17.00-18.00	ME/Af	12055*, 12040**, 11695**, 9975, 9865, 9840, 7130, 6005, 5935, 5925, 1314, 1089
18.00-19.00	ME/Af	12040**, 9975, 9865, 9840*, 7130, 5925, 1314, 1170
19.00-20.00	ME/Af	12040**, 9975, 9865, 9840*, 7130, 6005, 5935, 5925, 1314, 1089

Bengali

15.30-16.00	Sas	17665**, 12055*, 11665**, 11655, 9860**, 9490*, 7400*

Bulgarian

18.00-19.00	Eu	15465**, 11630**, 9675, 9470**, 7420*, 7380*, 7330**, 7300**, 6000* 5975*

Chinese

11.00-12.00	FE/SEAs	15545**, 12025**, 11640**, 9895**, 9875**, 9810*, 9480**, 9450*, 7490, 7400*, 7355*, 7340**, 7135*, 6205*, 5940**, 5905, 4010, 1251, 801, 648,585
12.00-13.00	FE/SEAs	11765**, 9875**, 9480**, 9450*, 7400**, 7340**, 7205*, 6205*, 5905, 1251, 585
13.00-14.00	FE/SEAs	11765**, 9480**, 9450*, 7205*, 5905, 4010, 1251, 1080, 801, 585
14.00-15.00	FE/SEAs	12025**, 11765**, 9830**, 9480**, 9450*, 7205*, 7135*, 6145*, 5905, 4010, 1251, 1080, 801, 648, 585
11.00-12.00	FE/SEAs	1080

Czech

18.45-19.30	Eu	12030**, 7440**, 7280*, 7185*, 5975

Dari

14.00-15.00	As	15460, 9860**, 9490**, 4975, 4940, 4730, 972, 648
18.00-19.00	Af	12065**, 7305*
19.00-20.00	Af	15470**, 12065**, 9775*, 9745*, 9475**, 7305*
06.00-07.00	AUS/NZL	21790, 17665**, 17570, 15525, 15470*, 15460
07.00-08.00	AUS/NZL	21790, 17665**, 17570, 17495, 15525, 15470*, 15460
08.00-09.00	AUS/NZL	21790, 17665**, 17495, 15525, 15470*, 9905
09.00-10.00	AUS/NZL	17665**, 17495, 15525, 15470*, 9905
04.00-06.00	Eu	693
06.00-10.00	Eu	1323, 693
15.00-16.00	Eu	9710**, 9480**, 7440*, 6030**, 1323, 693

Dari	Area	kHz
18.00-19.00	Eu	12010**, 9890, 9820**, 9785**, 9480*, 7340, 5965*, 5940*, 1494, 1467
19.00-20.00	Eu	12010**, 9890, 9820**, 9785**, 9480*, 7340, 7310, 5965*, 5940*, 5920, 1494, 1467, 1143
20.00-21.00	Eu	12070**, 1143, 12010**, 9890, 9820**, 9785**, 9480*, 7340, 5965*, 5940*, 1494, 1467, 1386
20.00-21.00	Eu	12070**, 12010**, 9890, 9820**, 7360**, 7340, 7300*, 5965*, 5940*, 1494, 1323
16.00-17.00	ME	6005 1170 1089
16.00-17.00	ME	12065**, 9775**, 9475**, 7210, 4975, 4940, 4730
17.00-18.00	ME	12065**, 9830**, 9470**, 7305*
02.00-03.00	NAm	15595**, 12020, 9875, 9865*, 7180
03.00-04.00	NAm	15595**, 12020, 9875, 9865*, 7250**, 7180, 5940*
04.00-05.00	NAm	13640, 12040**, 12020, 12000, 9875*, 9850, 7180, 7125, 6065*, 5920*
04.00-04.30Su	Nam	12060*
05.00-06.00	NAm	13640, 12040**, 12020, 12000, 9875*, 9850, 7180, 7125, 6065*, 5920*
15.00-16.00	As	15550**, 11500, 9800*, 9775*, 9475**

Finnish

18.00-18.30Th-Fr	Eu	12010**, 9820**, 5965*, 5940*, 1494

French

17.00-19.00	Af	15470**, 9775*, 9745*, 9475**
19.00-21.00	Af	9840**, 7370*
17.00-18.00	Eu	12020**, 9810, 9710**, 9450**, 7360*, 6030*, 5905*
18.00-19.00	Eu	12020**, 9840**, 9710**, 9470*, 9450**, 7360*, 7290, 6030*, 5905*
19.00-20.00	Eu	12020**, 9840**, 9470**, 9450**, 7370, 7360*, 7290, 7205, 5905*
20.00-21.00	Eu	12020**, 9840**, 9470*, 9450**, 7370, 7360*, 7300*, 7290, 5905*,1323

German

10.00-11.00	Eu	17660, 15540, 12010, 1323, 693
11.00-13.00	Eu	1323, 693
16.00-17.00	Eu	11980**, 1386, 9810, 1323, 9795, 1215, 9710**, 693, 9480**, 7440*, 7390*, 7360**, 7300*, 6030*, 5950**, 5920*
17.00-18.00	Eu	12070**, 5920, 11980**, 1386, 9795, 1323, 9785**, 1215, 9480, 693, 7440*, 7390, 7360**, 7340, 7300*, 7215*, 5950
18.00-19.00	Eu	11980**, 9480**, 7390, 7360**, 7300*, 7215*, 5950, 1386, 1323, 1215, 693
19.00-20.00	Eu	9480**, 7390, 7360**, 7215*, 5950, 1386, 1323, 1215, 693

Hindi

13.00-14.00	SAs	17675**, 9490*, 17665**, 7400*, 17610**, 1269, 15550**, 972,15460, 12055*, 11665**, 11655, 11520, 9860, 9800*, 9780*
15.00-15.30	SAs	17665**, 15550**, 12055*, 11665**, 11655, 9860**, 9800*, 9775*, 9490*, 9475** 7400*, 972

Hungarian

18.00-18.45	Eu	12030**, 7440**, 7280*, 7185*, 5975

Italian

18.00-19.00	Eu	9810, 9775**, 9450*, 7370, 7320

Japanese

12.00-13.00	FE/SEas	11640**, 9895**, 7490, 7355*, 5940*, 720, 630
13.00-14.0	FE/SEas	12025**, 11640**, 9895**, 7355**, 7135*, 5940*, 720, 630

Korean

10.00-11.00	FE/SEas	15470**, 12055**, 12025**, 11640**, 9895**, 7490, 7355*, 7135*, 5940*, 4010, 1251, 648

Mongolian

12.00-13.00	FE/SEas	12025**, 7135*, 4010, 648
12.00-13.00	FE/SEas	1080, 801

Norwegian

18.30-19.00Mo-Sa	Eu	12010**, 9820**, 5965*, 5940*, 1494

Persian	Area	kHz
16.00-17.00	ME	9975, 9840, 5935, 5925

Pushtu

14.00-15.00	As	15460, 9860**, 9490*, 4975, 4940, 4730, 972, 648

Polish

18.00-19.00	Eu	7390**, 7215*, 5920, 1143

Portuguese

00.00-01.00	LAm	12070**, 12010**, 9965, 9810*, 9470, 9450, 7440, 7390*, 7330, 7310

Romanian

17.00-18.00	Eu	15465**, 9675, 9470**, 7420*, 7380*, 7330**, 6000*

Russian

13.00-15.00	As	15510**, 12015*, 9830*, 9470**, 7 315**, 7170*
02.00-03.00	CAm	15520**, 15470**, 15425**, 13665*, 13640, 12060*, 12040**, 12010*, 12000, 9850*, 7125
03.00-04.00	CAm	15520**, 15470**, 15425**, 13665*, 13640, 12060*, 12040**, 12010*, 12000, 9850*, 7125
13.00-15.00	Eu	1323, 693
18.00-19.00	Eu	9795, 7310
20.00-22.00	Eu	15465**, 9795, 9775**, 9450*, 7440**, 7380*, 7320, 7310, 7185*, 1323, 693

Serbo-Croat

16.30-18.00	Eu	11630**, 9775**, 9450*, 7370, 7350*, 7300**, 5975*

Spanish

01.00-02.00	LAm	12070**, 12010**, 9965, 9890, 9860**, 9810*, 9470, 9450, 7570, 7440, 7390*, 7350, 7330, 7310, 7260, 7180, 7125, 6185*
02.00-03.00	LAm	12070**, 12010**, 9965, 9890, 9860**, 9810*, 9470, 9450, 7570, 7440, 7390*, 7330, 7310, 6185*

Swedish

18.30-19.00Su	Eu	12010**, 9820**, 5965*, 5940*, 1494

Turkish

15.00-16.00	Eu/ME	12065**, 9830**, 9470**, 7210**, 1170

Urdu

14.00-15.00	SAs	17675**, 17665**, 15550**, 12055*, 11665**, 11655, 9800*, 9780*, 7400*

Vietnamese

12.00-13.00	SEas	17795**, 17470, 9470*, 603

* - from 25.10.98 to 7.03.99
** - from 7.03.99 to 28.03.99

VOICE OF AMERICA RELAYS
MW: Ussuriusk, FE 648kHz 1000kW (1300-1400 in **Korean**).
SW: Irkutsk 1000kW, Novosibirsk 100kW, Petropavlovsk-Kamchatskiy 250kW (Schedules listed in VOA entry under 'USA)'.

SÃO TOMÉ E PRINCÍPE

VOICE OF AMERICA RELAY STATION
MW: Pinheira 1530kHz 600kW.
SW: Pinheira (G.C: 06.42E/00.18N): 4 x 100kW tr´s.
D.PRGR: see USA.

SAUDI ARABIA

BROADCASTING SERVICE OF THE KINGDOM OF SAUDI ARABIA (Gov.)
P.O. Box 570, Riyadh-11161. ☎ +966 (1) 4425170. 🖷 +966 (1) 4041692. ① 401040 SJ.
SW: Jeddah (G.C: 39.10E/21.32N): 2 x 100kW, 6 x 50kW — Riyadh (G.C: 46.23E/24.30N): 6 x 500kW, 4 x 350kW.
kHz: *6020, 9555, 9580, 9620, 9705, 9715, 9720, 9730, 9870, 11710, 11730, 11780, 11785, 11820, 11835, 11935, 11950, 11965, 15060, 15165, 15230, 15280, 15335, 15345, 15380, 17755, 17775, 17880, 17895, 21530, 21705*

Main Prgr.	Area	kHz
0600-0900	NAf/SWAs	11820, 11710, 9720
0600-1200	ME	11820
0900-1500	NAf	15060
1200-1500	SWAs/WEu	15380, 15230
1700-1800	SEas/WEu/NAf	11965, 11950, 11780
1800-2300	NAf	11935, 9870, 9555
2nd Prgr.		
0300-1700	ME	9580

2nd Prgr.	Area	kHz
1700-2100	ME	6020

Holy Quran

	Area	kHz
0300-0600	NAf/ME/As	17895, 11820, 9720, 9620, 9555
0600-0900	NAf/ME/As	21530, 17880, 11785, 9715
0900-1200	SEAs	17880, 21530
1200-1600	NAf/ME	15280, 15165
1600-1800	NAf/CAf	11835, 11710, 9730
1800-2100	SWAs/NAf	11935, 9705

Call of Islam

	Area	kHz
1500-1700	NAf/SWAs/WEu	11965, 11950, 11780

Bambara	Area	kHz		Swahili	Area	kHz
1600-1700	CAf	17775		0500-0600	EAf	17755
Bengali				**Turkmen**		
1400-1500	SAs	15345		1500-1600	CAs	9730
Indonesian				**Turkish**		
1000-1200	SEAs	21705		0400-0600	TUR	15060
Persian				**Urdu**		
1400-1600	IRN	11730		1200-1400	SAs	15345
Somali						
0400-0500	EAf	15335				

ANN: Arabic: "Ithaa till Mamlakah till Arabiyah al-Saudiyah". E: "This is the Broadcasting Sce. of the Kingdom of Saudi Arabia".
IS: 'Ud' (oriental lute). Opens and closes with National Anthem.
V. by QSL-card. (Re. to Freq. Mgr.)

SEYCHELLES

FAR EAST BROADCASTING ASSOCIATION (FEBA) (Missionary)

Box 234, Mahe, Seychelles, Indian Ocean. ☎ +248 224449. ▤ +248 225171. **Cable:** FEBCOM. (**HQ** ▤ FEBA Radio, Ivy Arch Rd, Worthing, Sussex BN14 8BX, United Kingdom).
E-Mail: reception@febaradio.org.uk **Web:** http://www.feba.org.uk
L.P: Pres: Douglas Malton. Dir: Stewart Pepper. PD: Tony Ford. CE: Graham Kimber.
SW (G.C. 55.28E/04.36S): 3 x 100kW.
kHz: 7270, 7350, 9420, 9455, 9500, 9785, 9795, 9810, 9820, 11600, 11640, 11675, 11690, 11695, 11705, 11745, 15430, 15445, 15540, 15555

Days: 1=Mon, 2=Tues. 3=Wed, 4=Thurs, 5=Fri, 6=Sat, 7=Sun.

Amharic	Days	Area	kHz
1600-1630	567	ETH	11640
1630-1700	1234567	ETH	11640
Arabic			
0345-0415	12346	ME	11695
0345-0430	57	ME	11695
1100-1130	1234567	ME	15540
1900-1945	1234567	ME	9455
Azeri			
1832-1847	1567	AZE	9455
Badaga			
0830-0845	7	IND	15540
1500-1515	17	IND	7270
Baluchi			
0230-0245	67	AFG	9785
Bengali			
1412-1427	5	IND/BGD	15445
1412-1442	123467	IND/BGD	15445
Bhili			
1300-1315	56	IND	15445
Bhojpuri			
1315-1330	123567	IND	15445
Chattisgarhi			
1330-1345	123467	IND	15445
Creole			
1200-1205	1234567	MAU	11675
Dari			
0245-0300	1234567	AFG	11690
0430-0500	5	AFG	15555
1615-1630	1234567	AFG	11600
Dinka			
1545-1600	1567	MAU	11705
English			
0800-0900	5	SAs	15540
1500-1530	23456	SAs	11600
1500-1545	17	SAs	11600
1500-1558	123456	IND	9810
Farsi			
0300-0330	1234567	IRN	11690
1802-1832	1234567	IRN	9455
French			
0915-0930	1234567	EAf	15430
1205-1220	1234567	ZAI	11675

French	Days	Area	kHz
1833-1848	23	ZAI	9500
1833-1903	14567	ZAI	9500
Gujarati			
1330-1345	1234567	IND	11600
Guragena			
1615-1630	123	ETH	11640
Hazaragi			
1600-1615	1234567	IND	11600
Hindi			
0815-0900	4	IND	15540
1345-1400	1234567	IND	15445
1415-1458	1234567	IND	11600
Hindko			
0200-0215	1	PAK	9785
1443-1458	456	PAK	9810
Kannada			
0845-0900	6	IND	15540
1500-1530	6	IND	7270
1515-1530	123457	IND	7270
Konkani			
1500-1515	23	IND	7270
Kui			
1300-1315	27	IND	15445
Makondi			
1532-1547	3456	EAf	9795
Malagasy			
0930-0945	67	MDG	15430
Malayalam			
0800-0830	7	IND	15540
1300-1330	1234567	IND	11600
Marathi			
1345-1400	7	IND	11600
Marwari			
1330-1345	5	IND	15445
Mundari			
1245-1300	123567	IND	15445
Nagpuria			
1245-1300	4	IND	15445
Nepali			
1230-1245	1234567	IND/NPL	15445
Nuer			
1530-1545	1567	EAf	11705
Oriya			
0800-0815	4	IND	15540
1300-1315	13	IND	15445
1300-1330	4	IND	15445
Oromo			
1700-1730	7	ETH	11640
Portuguese			
0400-0430	123567	MOZ	9420
0400-0415	4	MOZ	9420
Punjabi			
0215-0230	12367	PAK	11690
1345-1400	123456	IND	11600
Pushtu			
0230-0245	1234567	PAK/AFG	11690
Sindhi			
0215-0230	134567	IND/PAK	9785
Sinhala			
0530-0600	5	ME	15555
1330-1400	1234567	CLN	9810
Siraiki			
0200-0215	567	PAK	9785
Somali			
1700-1730	1234567	SOM	9810, 11745
Swahili			
0315-0400	13467	EAf	9820
0315-0345	25	EAf	9820
1613-1658	123567	EAf	9810
1613-1643	4	EAf	9810
Tagalog			
0500-0558	7	ME	15555
1000-1058	5	ME	15540
Tamil			
0115-0200	1234567	IND/CLN	7350
0845-0900	7	IND/CLN	15540
1300-1330	1234567	IND/CLN	9810
1445-1500	123456	IND/CLN	7270
Telugu	**Days**	**Area**	**kHz**
0200-0230	1234567	IND	7350
0800-0845	6	IND	15540
Tibetan			
1200-1230	1234567	As	15445
Tsangla			
1145-1200	67	BHU	15445

Tulu	Days	Area	kHz
1500-1515	45	IND	7270

Urdu

	Days	Area	kHz
0200-0215	123467	PAK	11690
0200-0230	5	PAK	11690
1400-1415	1234567	IND	11600
1400-1443	123456	PAK	9810
1400-1458	7	PAK	9810

N. in English: 1530.
ANN: "This is FEBA Radio" — **IS:** "What a friend we have in Jesus" (xylophone) and spoken ID — **V.** by QSL-card. Rp (2 IRC's). Rec.Acc.
PUB: "Programme Guide" (half yearly).

BBC INDIAN OCEAN RELAY STATION

P.O. Box 448, Victoria, Seychelles. ☎ +248 78496, 78269. 🖷 +248 78500. ☼ 2273 BBC SZ.
SW (G.C: 55.28E/04.36S): 2 x 250kW tr's.
kHz: 6005, 6135, 9610, 9630, 11730, 11860, 15420, 17885
Full schedule included in BBC entry under United Kingdom.

SINGAPORE

RADIO SINGAPORE INTERNATIONAL (RSI)

Farrar Rd, P.O. Box 5300, Singapore 912899. ☎ +65 3597662. 🖷 +65 2591380. **E-Mail:** rsi@mediacity.com.sg
Web: http://www.rsi.com.sg
SW (G.C: 103.51E/01.24N): 6 x 250kW.

Times	Language	Area	kHz
1100-1400	English	SEAs	6015, 6155 (6150 from March 29)
1100-1400	Chinese	SEAs	6000, 6120
1100-1400	Malay	SEAs	6135, 7250

BBC FAR EASTERN RELAY STATION

P.O. Box 434, Singapore. ☎ +65 2601511. 🖷 +65 6690834.
FM: 88.9MHz 4kW (24h rel. of BBCWS in English).
SW (G.C: 103.44E/01.25N): 5 x 250kW, 4 x 100kW tr's.
kHz: 3915, 3955, 5975, 6060, 6065, 6080, 6140, 6195, 7110, 7135, 7160, 7205, 9510, 9600, 9605, 9680, 9725, 9740, 11685, 11750, 11765, 11850, 11920, 11955, 15280, 15340, 15360, 17760, 21660
V. by QSL-card for direct reports.
D.PRGR: see United Kingdom

SLOVAKIA

RADIO SLOVAKIA INTERNATIONAL

P.O. Box 55, 810 05 Bratislava. ☎ +42 (7) 496281 (**Int. Rel:** +42 (7) 496601). 🖷 +42 (7) 496282, 498247. **E-mail:** slrozv@ba-cvt.sanet.sk
Web: http://www.slovakradio.sk/rsi.html
L.P: Editor-in-Chief: Dr.Karol Palkovic.
SW: R=Rimavská Sobota (G.C: 20.08E/48.24N): 4 x 250kW, V=Velké Kostolany (G.C: 17.44E/48.31N): 2 x 100kW.
kHz: 5915, 5930, 5940, 6055, 7150, 7300, 7345, 9440, 9485, 9505, 11990, 13715, 17485, 21705

English

	Area	kHz
0100-0127	NAm	5930R, 7300V, 9440R
0830-0857	AUS	11990V, 17485R, 21705R
1630-1657	WEu	6055R, 7345V
1730-1757	WEu	5915R, 6055R, 7345V
1930-1957	WEu	5915R, 6055R, 7345V

French

0200-0227	NAm	5930R, 7300V, 9440R
1700-1730	Eu	5915R, 6055R, 7345V
2030-2057	WEu	5915R, 6055R, 7345V

German

0800-0830	WEu	6055R, 7300R
1430-1457	WEu	6055R, 7345V, 9505V
1900-1927	WEu	5915R, 6055R, 7345V

Russian

	Area	kHz
1400-1427	EEu/As	9440V, 11990R, 13715R
1830-1857	EEu	5915R, 7150R, 9485V

Slovak

0130-0157	NAm	5930R, 7300V, 9440R
0900-0927	AUS	11990V, 17485R, 21705R
1600-1700	SEu	5940R, 6055R, 7345V
1800-1827	WEu	5915R, 6055R, 7345V
2000-2027	WEu	5915R, 6055R, 7345V

IS: Motif from patriotic song "Who Burns for the Truth".

SOUTH AFRICA

CHANNEL AFRICA

P.O. Box 91313, Auckland Park 2006.
L.P: Exec. Editor: Hans-Dieter Winkens (☎ +27 (11) 714 2551. 🖷 27 (11) 714 2546. **E-mail:** winkensh@sabc.co.za).
Web: http://www.sabc.co.za/units/chanafr/
SW: Meyerton (G.C: 28.08E/26.35S): 3 x 500kW, 4 x 250kW, 10 x 100kW (incl tr's used for Home Sce and relays of other broadcasters).
kHz: 3345, 5955, 6000, 9525, 9440, 9675, 11900, 15240, 15330, 17675

Winter Schedule 1998/99

English

0300-0325	EAf/CAf	9525
0400-0430	SAf	5955
0500-0530	WAf	15215
0600-0630	WAf	15215
1300-1455SS	EAf/CAf	17895
1300-1455SS	WAf	21530
1500-1530	EAf/CAf	17870
1300-1455	SAf	11900
1600-1630	SAf	6000
1700-1730	WAf	17870
1800-1830	WAf	17870

French

0330-0355	MDG	9525
0430-0455	CAf	9525
1530-1555	MDG	17870
1630-1655	CAf	11900
1830-1855	WAf	17870

Portuguese

0430-0455	NMOZ	5955
0430-0455	SMOZ	3345
0530-0555	AGL	15215
0630-0655	AGL/WAf	15215
1630-1655	SMOZ	3345
1630-1655	NMOZ	6000
1730-1755	AGL/WAf	17870

Swahili

0300-0325	EAf	9765
1500-1525	EAf	15545
1600-1630	CAf/EAf	11900

V. by QSL-card. Rp. Re: to Data Section, Channel Africa, P.O. Box 91313, Auckland Park 2008, South Africa.

SPAIN

RADIO EXTERIOR DE ESPAÑA

Apartado 156.202, 28080 Madrid. ☎ +34 (1) 346 10 81. 🖷 +34 (1) 346 18 15. **E-mail:** radioexterior.espana@rtve.es
Web: http://www.rtve.es/rtve/_rne/_radios/_ree/radioe00.htm
L.P: Dir: Fermín Bocos Rodríguez. Asst. Dir: Nuria Alonso Veiga. Head of Inf. Sce: Diego Armario López. Head of Current Affairs Prgrs: Aurora Sánchez Párez. Foreign Language Prgr: José J. Amorena Zabalza. LF Dept: Manuel Fernández Muñoz. Tech. & Prgr. Secr: Josefina R. Peña Márquez. Head of Freqs: Gonzalo Díaz Palenzeula.
SW: Noblejas, Toledo (G.C: 03.27W/39.56N): 6 x 350kW. 2 x 250kW – C =Cariari, Costa Rica (G.C: 83.30W/10.00N): 3 x 100kW – B= Beijing, China (G.C:116.27E/39.57N): 120kW – X=Xi'an, China (G.C: 108.54E/34.12N): 120kW.
kHz: 3210, 5970, 5990, 6000, 6045, 6055, 6125, 6130, 6160, 7270, 7275, 9615, 9620, 9630, 9650, 9690, 9745, 11715, 11775, 11815, 11880, 11890, 11945, 12035, 15125, 15375, 17715, 17755, 17805, 17845, 17890

Arabic

	Area	kHz
0900-1400	ME/NAf	17890
1400-1900	ME/NAf	11715
1900-2300	NAf	7270

English

	Area	kHz
0000-0200	NAm	6055
0500-0600	NAm	6055
2000-2100MF	Eu/Af	6125, 11775
2200-2300SS	Eu/Af	6125, 11775

French

1800-1900MF	Eu/Af/ME	6125
1900-2000	Af/ME	6045
2300-2400	NAm	6055

German

1730-1800Mon	Eu	6125

Russian	Area	kHz
1700-1730MF	Eu	9620
Sefardi		
1830-1900Mon	ME/Af	6130
0115-0145Tues	SAm	11775
0415-0445Tues	NAm	9690
Spanish *		
0000-0100SS	NAm/LAm	9745C, 11815C, 11880C
0000-0200	NAm	11945
0000-0500	NAm	6125
0100-0400MF	NAm/LAm	5990C, 5970C, 3210C
0200-0500	NAm	6055
0500-0700	Eu/Af	6000, 11890
0500-0800	Eu	7275
0700-0900	Pac	9615, 9650
0700-1700	Eu	12035
0900-1700Sun	Eu	9620
0900-1900	SAm/Af	17715, 17755
1000-1200	Japan	9620B
1100-1400MF	CAm	5970C, 9630C, 11815C
1200-1400	Philippines	11910X
1200-1600W	Eu	9620
1200-1800	LAm	17805
1300-2400SS	NAm	11815C
1400-1800SS	CAm	11880C
1400-2400SS	NAm	9745C
1600-1700W	Af	15375
1700-2100Sun	Eu	6125
1700-2200Sat	Eu	6125
1700-2300	Eu	7275
1800-2100SS	LAm	17845
1800-2200	LAm	15125C
1800-2400MF	NAm	11815C
1800-2400Sat	NAm	9745C
1900-2300	NAm	9630
2200-2400	CAm	11880C
2300-2400	NAm/LAm	6125, 6160, 9620, 11945

*) incl. Catalan, Galican, Basque

ANN: S: "Radio Exterior de España". E: "Spanish Foreign Radio". F: "Vous êtes à l'écoute de Radio Exterieure d'Espagne". A: "Idha'atu al-Watania, Idhaa'atu Isbania al-Jariyía" — **V.** by QSL-card.

SRI LANKA

SRI LANKA BROADCASTING CORPORATION (Public Corporation)

✉ P.O. Box 574, Independence Square, Colombo 7. ☎ +94 (1) 695661. **SW:** Ekala (G.C: 79.54E/07.06N): 2 x 300kW, 2 x 100kW, 3 x 35kW, 1 x 12.5kW, 10 x 10kW (incl. tr's used by Home Sce, VOA, R. Japan and Trans World R.) – S) relay via BBC Skelton, United Kingdom
kHz: 5975S, 9730, 11800, 11835, 11905, 15425, 17850

English	Area	kHz
0030-0430	As	15425, 9730
1030-1130	As/AUS	11835, 17850
1230-1600	SAs	15425, 9730
1600-1700	SAs	9730 (SS), 15425
1900-2000Sat	Eu	5975S
Hindi		
0030-0430	SAs	11905
0630-1030	SAs	11800
1510-1700	SAs	11905
Malayalam		
1030-1130	SAs	11800
Sinhala		
1745-1900	ME	11800
Tamil		
1130-1330	SAs	11800
1905-1950	ME/Af	11800

V. by QSL-card. Rp.

TRANS WORLD RADIO (Rlg.)

✉ P.O. Box 364, 91 Wijerama Mawatha, Colombo 7. ☎ +94 (1) 685235/6/7. ✉ + 94 (1) 685245. **Cable:** Votan.
L.P: Dir. of Op's: Mark Blosser. CE: Darryl van Dyken.
Prgr. ✉ Trans World Radio-India, Box 4407, L-15, Green Park, New Delhi 110 016. ☎ +91 (11) 662058. ✉ +91 (11) 686 8049.
L.P: Regional Dir: Dr. N. Emil Jebasingh.
MW: Puttalam 882kHz 400kW.
SW: Ekala (G.C: 79.54E/07.06N): 6035kHz 12.5kW.

Banjara	Days	Area	kHz
0115-0130	17	SAs	882
1450-1505	17	SAs	882
Bengali			
2230-2300	1234567	SAs	882

Chatt	Days	Area	kHz
1605-1620	17	SAs	882
Chowdhary			
1535-1550	1	SAs	882
English			
2300-2330	23	SAs	882
Gujarati			
2300-2330	14567	SAs	882
1450-1505	23456	SAs	882
Kannada			
0045-0115	1234567	SAs	882
0115-0130	23456	SAs	882
Maithili			
1550-1605	1	SAs	882
Malayalam			
2330-2400	1234567	SAs	882
0030-0045	1234567	SAs	882
0830-0845	6	SAs	6035
Manipuri			
1550-1605	7	SAs	882
Marathi			
1505-1535	1234567	SAs	882
1535-1550	23456	SAs	882
Oriya			
1405-1420	1	SAs	882
1420-1435	23456	SAs	882
1435-1450	1234567	SAs	882
1550-1620	23456	SAs	882
Soura			
1420-1435	17	SAs	882
Tamil			
0000-0030	1234567	SAs	882
1620-1635	1234567	SAs	882
Telegu			
1335-1405	1234567	SAs	882
1405-1420	234567	SAs	882
Vasavi			
1535-1550	7	SAs	882

ANN: "Vishwa Vani" — **V.** by QSL-card — **IS:** Indian tune.

RADIO JAPAN RELAY STATION

SW: Ekala (G.C: 79.54E/07.06N): 1 x 300kW (operated by SLBC).
kHz:

DEUTSCHE WELLE RELAY STATION

✉ 92/2 Rt. Hon. D.S. Senanayake Mwta, Colombo 8.
MW: Trincomalee 1548kHz 400kW.
SW: Trincomalee (G.C: 81.10E/08.44N): 3 x 250kW.
kHz: 6055, 6130, 6170, 7225, 7235, 7265, 9555, 9565, 9585, 9615, 9670, 9770, 11735, 11785, 11810, 11825, 11865, 13690, 15185, 15415, 15425, 17820, 17825, 17895
V. by QSL-card for direct re.

VOICE OF AMERICA RELAY STATION

SW: Ekala (G.C: 79.54E/07.06N): 10/35kW.
D.PRGR: see USA.
F.PI: 3 x 500kW tr's under installation at Chilaw.

SUDAN

SUDAN NATIONAL BROADCASTING CORP.

✉ P.O. Box 572, Omdurman ☎ +249 (11) 53151. **Cable:** IZZA — **L.P:** DG: Salah al-Din al-Fadhil Usud. DG Eng. & Tech. Sces: Abbas Sidig.
MW: Omdurman 765kHz 50kW — **SW:** Omdurman (G.C: 32.28E/15.30N): 7200kHz 100kW, 9200kHz 10kW.

Amharic	Area	kHz
1600-1630	Af	9200, 765
Arabic		
1245-1500+	Af	9200, 765
English		
1500-1530+	Af	9200, 765
1700-1900	Af	7200, 765
1800-1900	Af	9200
French		
1600-1700	Af	7200, 765
1700-1800	Af	9200
Somali		
1530-1600	Af	9200, 765

+) National Unity R.

ANN: E "Sudan National Broadcasting Corporation, Omdurman".

SWAZILAND

TRANS WORLD RADIO AFRICA (Rlg.)
✉ P.O. Box 64, Manzini. ☎ +268 52781-5. 🖷 +268 55333.
Web: http://www.icon.co.za/~ttatlow/Welcome.htm
L.P: Pres: Thomas J. Lowell. Regional Dir (Africa): Rev. Stephen Boakye-Yiadom. St. Dir: Peter Prediger. Head of Prgrs: Rob Lincoln. CE: James Burnette. Freq. Mgr: Lars Gunnarsson.
MW: 1170kHz 50kW — **SW:** Mpangela Ranch (G.C: 31.59E/26.34S): 2 x 100kW, 1 x 50kW, 2 x 25kW — M) via Meyerton, South Africa.
kHz: 3200, 3240, 4775, 5965, 6040, 6100, 6115, 6040, 7175, 7215M, 7240M, 7265M, 7315, 9500, 9525, 9650, 9850M, 11730M, 15195

Winter Schedule 1998/99

Afrikaans	Days	Area	kHz
1700-1800	1234567	So Africa	1170
Chewa			
0400-0430	13467	Malawi	6040
0430-0500	1234567	-	6040
1600-1630	134567	Malawi	6135
1630-1658	1234567	-	6135
Chokwe			
1930-1945	67	-	6130
English			
0430-0600	1234567	SAf/Zimbabwe	3200
0430-0705	1234567	SAf/Zimbabwe	4775
0505-0705	1234567	CAf	9500
0605-0705	1234567	SAf/Zimbabwe	6100
1600-1830	1234567	EAF	9500
1730-1745	1234	-	3200
1745-2015	1234567	-	3200
1800-2045	1234567	So Africa	1170
French			
1440-1510	1234567	Madagascar	7175
1930-2000	1234567	-	9525
2000-2015	67	-	9525
German			
0400-0430	1234567	So Africa	3200
0400-0430	1234567	Namibia	4775
KiKongo			
1900-1915	345	-	6130
Kimbundu			
1945-2000	1234567	-	6130
Kimwani			
1540-1555	67	-	7315
Lingala			
1900-1930	1234567	Congo	9525
Lomwe			
0342-0358	1234567	Mozambique	4775
1525-1540	1234567	-	7315
1540-1555	1234	-	7315
Luchazi			
1915-1930	6	-	6130
Lunyaneka			
1900-1915	6	-	6130
Makua			
1510-1525	1234567	Mozambique	7315
Malagache			
1510-1555	1234567	-	7175
Ndau			
1900-1915	1234567	Mozambique	3240
Ndebele			
0300-0330	1234567	Zimbabwe	3200
1800-1830	1234567	-	3240
Pedi			
1730-1745	7	-	3200
Portuguese			
1435-1440	12345	Mozambique	7315
1440-1510	1234567	Mozambique	7315
1540-1555	5	Mozambique	7315
1600-1700	6	Mozambique	4760
1630-1645	1234	Mozambique	4760
1900-1915	12	Mozambique	6130
1900-1930	7	Mozambique	6130
1915-1945	12345	Mozambique	6130
2000-2005	12345	Mozambique	6130
Shangaan			
1600-1630	7	Mozambique	4760
1630-1645	57	-	4760
Shona			
0300-0345	1234567	Zimbabwe	3240
1830-1900	1234567	-	3240
SiSwati			
1530-1630	1234567	Swaziland	1170

Sotho	Days	Area	kHz
1730-1745	5	So Africa	3200
Swahili			
0300-0330	1234567	EAf	5965
0330-0345	6	-	5965
1700-1800	1234567	EAf	6135
1800-1830	1234567	-	6135
Tigrinya			
1540-1555	57	-	9500
Tshwa			
1600-1630	12345	Mozambique	4760
Tswana			
1730-1745	6	-	3200
Tumbuka			
0400-0430	25	-	6040
1600-1630	2	-	6135
Umbundu			
1845-1900	1234567	Angola	6130
Urdu			
1400-1415	1234567		15330
Zulu			
1630-1700	1234567	Swaziland	1170

ANN: "This is Trans World Radio - Swaziland" or "from the beautiful Kingdom of Swaziland, this is Trans World Radio".
IS: Last bar of "We've a story to tell the nations" on Hand Bells.
V. by QSL-folder. Rec. acc. preferably from target areas. IRC's appreciated (3 IRC's for airmail reply).
PUB: Programme guides available in English, free on request.

SWEDEN

RADIO SWEDEN
✉ S-105 10 Stockholm. ☎ + 46 (8) 7847288. 🖷 +46 (8) 6602990.
① 10 000 SRCENT S. **Listener contact numbers:** ☎ +46 (8) 7847287.
Polling/Swedish schedule by 🖷 +46 (8) 6673701.
E-Mail: info@rs.sr.se (**MediaScan:** wood@rs.sr.se)
Web: http://www.sr.se/rs/
L.P: MD: Finn Norgren. PD: Ulla Lindskog. Head of PR: Ch. Adler.
MW: Sölvesborg (G.C: 14.40E/55.59N): 1179kHz 600kW.
Satellite: Tele-X and Astra and Eutelsat.
SW: Hörby (G.C: 13.44E/55.49N): 3 x 500kW.

kHz: 6000, 6065, 6170, 6200, 7115, 7150, 7190, 7265, 7280, 7290, 7315, 7325, 9440, 9490, 9645, 9655, 9700, 9705, 9715, 9865, 11650, 11880, 13625, 13740, 15240, 17785

English	Areas	kHz
0130-0200	As/Pac	7265
0230-0300	NAm	7280
0330-0400	NAm	7115
1230-1300	NAm	15240, 11650 or 13740
1330-1400	As/Pac	9705, 13740
1430-1500	NAm	11650, 15240
	As/Pac	11880
1830-1900 W	Eu/Af/ME	1179, 6065, 9645
1830-1900 Sun	Af/ME	9645
2030-2100	Eu	1179, 6065
2130-2200 SS	Eu/Af/ME	1179, 6065, 9655
2230-2300	Eu/Af	1179, 6065, 7325
Estonian		
1530-1545	Eu	1179, 6000, 6065
1700-1715 Sun	Eu	1179, 6065
1800-1815 W	Eu	1179, 6065
German		
1830-1900 W	Eu	1179, 6065
1830-1930 Sun	Eu	1179, 6065
Latvian		
1715-1730	Eu	1179
1745-1800	Eu	6200 or 6170
1815-1830 W	Eu	6065, 1179
Russian		
1300-1330	As/Eu	9865
1400-1430	Eu	6000, 9865
1730-1800	Eu	1179, 6065
2000-2030 W	Eu	1179, 6065, 7190 (alt. 7150)
2000-2030 Sun	Eu	7190 (alt. 7150)
Swedish		
0000-0030	LAm	9440 (alt. 6065)
0100-0130	As/Pac	7265
	Lam	7290
0200-0230	NAm	7280
0300-0330	NAm	7115
0500-0715 MF	Eu/Af/ME	6065, 9490
0600-0715 MF	Eu/Af/ME	13625
0700-0900Sat	Eu/Af/ME	6065, 9490

Swedish	Area	kHz
0800-1000Sun	Eu/Af/ME	6065, 9490
1100-1130 SS	Eu	6065
	As/Pac	9865 (alt. 7150)
1130-1200	Eu	6065, 13740
	NAm	11650 (alt. 13740)
1200-1230	As/Pac	9865, 15240
	LAm	17785
1230-1245 MF	As	13740
1230-1330 SS	As	13740
1300-1315 MF	As	13740
1300-1330	NAm	15240
1500-1530	NAm	11650, 15240
1545-1600	Eu	1179, 6000, 6065
	NAm	15240
1600-1645	Eu	1179, 6065
1645-1715 W	Eu/Af/ME	1179, 6065, 9700, 13625
1645-1700 SS	Eu/Af/ME	1179, 6065, 9700, 13625
1800-1830 Sun	Eu	1179, 6065
1900-1930	Eu/Af/ME	1179, 6065, 9655, 11605
2100-2130 SS	Eu/Af	1179, 6065, 9655
2100-2200 MF	Eu/Af	1179, 6065, 9655
2200-2230	Eu/Af	1179, 6065, 7315
2300-2330	Eu	1179

ANN: E: "This is Radio Sweden". G: "Hier ist Radio Schweden, Stockholm". Sw: "Radio Sverige, Stockholm". R: "Govorit Stokgolm".
IS: "To the Wide, Wide World" (electronic music composed by Ralph Lundsten)— **V.** by QSL-card. Cassettes not wanted.
PUB: Schedule twice yearly (available by fax — see above).

SWITZERLAND

SWISS RADIO INTERNATIONAL
✉ Giacomettistrasse 1, P.O. Box, CH-3000 Berne 15. ☎ +41 (31) 350 9222. ▤ +41 (31) 350 9544. ① 911 358. **Cable:** Swissinter.
E-mail: *language*@sri.srg-ssr.ch **Web:** http://www.swissinfo.org
L.P: Dir: Mrs. Carla Ferrari. Vice-Dir. & Head of SRI-Enterprises: Nicolas D. Lombard. Head of Communication & Mktg. Sces: Walter Fankhauser. Head of Contr. & Admin: Bernhard Fischer. Head of Engineering: Dept: Pauld Badertscher.
SW: Schwarzenburg (G.C: 07.23E/46.49N): 3 x 150kW, 1 x 100kW — Lenk (G.C: 07.27E/46.27N): 2 x 250kW — Sottens (G.C: 06.44E/46.39N): 1 x 500kW. **Relays:** C=Beijing, China 2 x 120kW, M=Montsinery, French Guiana 1 x 500kW.
kHz: *5840, 5850, 6135, 6165, 7230, 7410, 7480, 9535, 9620, 9885, 9905, 11650, 11725, 11860, 12075, 13635, 13685*

Satellite Sce: SRI also broadcasts by satellite to Europe, the Americas, Australia and the Pacific.

Arabic	Areas	kHz
17.00-18.00	NE/EAf	6165,9620,15275
19.00-20.00	Af	9620,9885,11910
English		
01.00-01.30	NAm/CAm	9885,9905
04.00-05.00	NAm/CAm	9885,9905
05.00-05.30	SEEu	6165
06.30-07.00	SEEu	6165
07.30-08.00	Af	9885,11860,13635
08.30-09.00	AUS	9885,13685
11.00-11.30	SWEu	9535
11.00-12.00	FE/SEAs	9540,17815
13.00-13.30	SWEu	9535
14.00-15.00	CAs/SAs	9575,15185
16.00-16.15	CAs/SAs	9575,15185
20.00-20.30	NEu	6165
20.00-20.30	Af	9620,9885,11910,13700
French		
02.00-02.30	NAm/CAm	9885,9905
05.00-05.30	NAm/CAm	9885
05.30-06.00	SEEu	6165
06.00-07.00	Af	9885,11860,13635
06.15-06.30	SEEu	6165
10.00-10.30	AUS	9885,13685
12.00-12.30	SWEu	9535
12.30-13.00	FE/SEAs	9540,17815
15.30-16.00	CAs/SAs	9575,15185
18.00-18.15	NE/EAf	6165,9620,15275
19.30-20.00	NEu	6165
21.00-21.30	Af	9620,9885,11910,13700
22.00-22.30	SAm	9885,11660
German		
00.30-01.00	NAm/CAm	9885,9905
03.30-04.00	NAm/CAm	9885,9905

German	Area	kHz
06.00-06.15	SEEu	6165
08.00-08.15	Af	9885,11860,13635
09.30-10.00	AUS	9885,13685
11.30-12.00	SWEu	9535
12.00-12.30	FE/SEAs	9540,17815
15.00-15.30	CAs/SAs	9575,15185
18.30-19.00	NEu	6165
20.30-21.00	Af	9620,9885,11910,13700
22.30-23.00	SAm	9885,11660
Italian		
03.00-03.30	NAm/CAm	9885,9905
05.30-05.45	NAm/CAm	9885
07.00-07.30	SEEu	6165
07.00-07.30	Af	9885,11860,13635
09.00-09.30	AUS	9885,13685
12.30-13.00	SWEu	9535
13.00-13.30	FE/SEAs	9540,17815
16.30-17.00	NE/EAf	6165,9620,15275
18.30-19.00	Af	9620,9885,11910
19.00-19.30	NEu	6165
23.00-23.30	SAm	9885,11660
Spanish		
01.30-02.00	NAm/CAm	9885,9905
02.30-03.00	NAm/CAm	9885,9905
23.30-24.00	SAm	9885,11660

IS: Traditional Swiss-German folk tune.
V. by QSL-card on request if reports include program comments.
PUB: Program schedule twice a year.

ICRC RADIO
✉ Red Cross Broadcasting Service, 19 Ave de la Paix, 1202 Geneva. ☎ +41 (22) 7346001. ▤ +41 (22) 733 2057. ① 414 226 CCR CH.
L.P: Ag. Head of Communication Dept: Michèle Mercie. Head of Press Div: Carlos Bauverd. Editor-in-Charge: Patrick Piper.
ICRC Radio broadcasts monthly 5 minute prgrs in English, French, German and Spanish on the frequencies of Swiss Radio International. The 1998 schedule was not available at press time.
ANN: E: "This is Geneva—Red Cross Broadcasting Service". F: "Ici Genève—Service de Radiodiffusion du Comité International de la Croix-Rouge". Sp: "Aqui Ginebra — Servicio de Radiodifusión del Comité Internacional de la Cruz Roja". G: "Hier ist der Rundfunkdienst des Internationalen Komitees vom Roten Kreuz in Genf". P: "Aqui Genebra, Servico de Radiodifusão do Comite Internacional da Cruz Vermelha".
V. by QSL-card. IRC appreciated. (B)

SYRIAN ARAB REPUBLIC

ORGANISME DE LA RADIO-TÉLÉVISION ARAB SYRIENNE (Gov.)
✉ Ommayad Square, Damascus. ☎ +963 (11) 720700. ① SY 411138 GD — **L.P:** DG: Khudr Omran. Dir. Eng: M.Bara. Dir. Public Rel: Mrs. Awafet Haffar. Dir. Admin & Finance: Zuheir Breidi.
SW: Adra (G.C: 36.30E/33.27N): 4 x 500kW.
kHz: *9995, 12085, 13610*

Arabic	Area	kHz
0400-0530	IRQ	9995, 1125, 828
0600-1700*	ME	12085, 13610
2000-2200	IRQ	12085, 1125, 828
2215-2315	Eu/Am	12085, 13610
*) rel. Home Sce.		
English		
2005-2105	Eu	13610
2105-2205	Am/Pac	12085, 13610
French		
1905-2005	Eu/Am	13610
German		
1805-1905	Eu	12085, 13610
Hebrew		
1600-1830	ME	783
Portuguese		
1530-1600	ISR	12085
Russian		
1730-1800	ME	783
Spanish		
2315-0030	Eu/Am	12085, 13610
Turkish		
1700-1800	Eu	13610

ANN: A: "Idha'atu-l-gumhuriyati-l'arabiyya as-suriyya min dimashq". E: "This is the Syrian Arab Republic Broadcasting Service from Damascus". F: "Ici Damas". Hebrew: "Kol Damasek" — **IS:** Guitar.
V. by QSL-card.

TAJIKISTAN

TAJIK RADIO (Gov.)
✉ Chapayev Str. 31, 734025 Dushanbe. ☎ +7 (3772) 276569.
☼ 201392 TELE.
MW: Orzu 648kHz 1000kW, Yangi-Yul 1143kHz 150kW.
SW: Yangi-Yul (G.C: 68.48E/38.29N): 2 x 100kW.

Arabic	Area	kHz
0400-0415	ME	648, 1143, 7245
1700-1715	ME	648, 1143, 7245
Dari		
0100-0200	As	648, 1143, 7245
0300-0345	As	648, 1143, 7245
1400-1500	As	648, 1143, 7245
1600-1645	As	648, 1143, 7245
English		
0345-0400	As	648, 1143, 7245
1645-1700	As	648, 1143, 7245
Persian	**Area**	**kHz**
0200-0300	ME	648, 1143, 7245
0415-0500	ME	648, 1143, 7245
1500-1600	ME	1143, 7245
1715-1800	ME	1143, 7245

ANN: Tajik: "In jo Dushanbe" — **V.** by letter.

RELAY SERVICES
Mediumwave: Orzu 648kHz 1000kW.
Shortwaves: Orzu (G.C: 37.32N/68.42E): 2 x 1000kW. Yangi-Yul (G.C: 38.29N/68.48E): 3 x 100kW on 4730/4940/4975kHz + 2 x 100kW. Relays of V. of Russia, BBC, VOA, R. Netherlands. Cf. these stations for details

TANZANIA

RADIO TANZANIA - DAR ES SALAAM (Gov. Comm.)
✉ Nyerere Rd, P.O. Box 9191, Dar es Salaam. ☎ +255 (51) 860760-6.
▤ +255 (51) 865577. ☼ 41201 SAUTI.
LP: Dir. of Broadc: Abdul Ngarawa. Dep. Dir. of Broadc: Juma Seleka. Contr. of Prgrs: Mrs. Edda Sanga. Head of English Sce. & Int. Rel. Unit: Abisay Steven. CE: Taha Usi. Dep. CE: Emmanuel Mangula.
MW: Dar es Salaam 1035kHz 10kW, 837kHz 1kW (standby tr.)
SW: Dar es Salaam (G.C:39.14E/06.50S): 5050.1kHz 10kW.
FM: Dar es Salaam 89.9MHz.

English	Area	kHz
0330-0430	Af	5050.1, 1035 (alt. 837), 89.9MHz
0900-1000	Af	5050.1, 1035 (alt. 837), 89.9MHz
1530-1915	Af	5050.1, 1035 (alt. 837), 89.9MHz

N: 0400, 1000, 1300SS, 1600, 1800SS, 1900SS.
ANN: "This is the External Service of Radio Tanzania Dar es Salaam".
IS: Celeste — **V.** by QSL-card

THAILAND

RADIO THAILAND
✉ 236 Vibhavadi Rangsit Highway, Huaykhwang, Bangkok 10400.
E-mail: amporn@usa.net **Web:** http://www.radiothailand.com/
LP: DG, Govt. Pub. Rel. Dept: Bangern Musikapong. Dir, Radio Thailand: Somphong Visuttipat.
MW: Bangkok 918kHz 10kW.
SW: Ban Dung District, Udon Thani Province (G.C: 102.48E/17.42N): 500kW (operated by VOA).
kHz: 6040, 7145, 7150, 7165, 7285, 9645, 9690, 9830, 11805, 11850, 11905

Burmese	Area	kHz
1145-1200	BRM	918, 6030
Cambodian		
1115-1130	CBG	918, 7260
Chinese		
1315-1330	FE	918, 11850
English		
0000-0030	Eu/Af	918, 9690
0030-0100	ENAm	918, 15395
0300-0330	WNAm	918, 15370
0530-0600	Eu	918, 15115
1230-1300	As/Pac	918, 9885
1400-1430	As/Pac	918, 9830
1900-2000	Eu	918, 7165
2030-2045	Eu	918, 11805
French		
2015-2030	Eu	918, 11805

German	Area	kHz
2000-2015	Eu	918, 11805
Indonesian		
1215-1230	INS	918, 11805
Japanese		
1300-1315	J	918, 7145
Laotian		
1130-1145	LAO	918, 6030
Malay		
1200-1215	MLA	918, 11805
Thai		
0100-0200	ENAm	918, 15370
0330-0430	WNAm	918, 15370
Thai		
1330-1400	As	918, 7145
1800-1900	ME/Af	918, 7150
2045-2115	Eu	918, 11805
Vietnamese		
1100-1115	VTN	918, 15115

IS: Gong — **ANN:** E: "This the World Service of Radio Thailand".
V. by QSL-card.

VOICE OF FREE ASIA (Wor. Or. Sor.)
✉ P.O. Box 2-131, Rajdamnoen, Bangkok 10200.
LP: Dir: Anucha Osathanond.
STATION: Bangkok 1575kHz 1000kW (tr. operated by VOA).
D.PRGR: Mon-Fri: 1030-1130, 1500-1530 & 2230-2400. (1030-1100 Thai, 1100-1130 Malay, 1500-1530 **English**, 2230-2300 Vietnamese, 2300-2330 Cambodian, 2330-2400 Laotian).
IS: Thai National Anthem — **ANN:** "This is the Voice of Free Asia" —
V. by QSL-card and letter.

VOICE OF AMERICA RELAY STATION THAILAND
✉ Rangsit-Bangpoon Rd, Bangkok. ☎ +66 (2) 581 5191.
MW: Bangkok 1575kHz 1000kW.
SW: Ban Dung District, Udon Thani Province (G.C: 102.48E/17.42N): 7 x 500kW tr's.
D.PRGR: see USA.

BBC EAST ASIAN RELAY STATION
SW: Nakhon Sawan (G.C: 100.04E/15.03N): 4 x 250kW.
kHz: *5965, 5975, 5990, 6065, 7160, 7205, 7225, 7235, 7275, 9580, 9600, 9750, 11685, 11750, 11850, 11955, 15280, 15310, 15380, 15405, 17790, 21660*
D.PRGR: see BBC schedule under "United Kingdom".

RADIO FRANCE INTERNATIONALE RELAY STATION
F.PI: A new SW-st is planned. No further details at deadline time.

TURKEY

VOICE OF TURKEY
✉ P.O. Box 333, 06443 Ankara. ☎ 90 (312) 490 9800. ▤ +90 (312) 490 9845. ☼ 46727 DISY TR — **Transmitting St:** Çakirlar Kisa Dalga Verici Istasyonu, 06790 Etimesgut, Ankara.
LP: MD: Savas Kiratli. CE: Erdal San. FSD: Rafet Esit. Chief English Dept: Osman Erkan.
SW: Çakirlar (G.C: 32.40E/39.58N) 2 x 500kW, 3 x 250kW — Emirler (G.C: 32.51E/39.29N): 5 x 500kW.
kHz: *5960, 5970, 5980, 5990, 6035, 6110, 6120, 6135, 6145, 7110, 7150, 7200, 7240, 7245, 7255, 7265, 7300, 9445, 9460, 9505, 9560, 9565, 9630, 9655, 9670, 9675, 9685, 9710, 11760, 11795, 11800, 11835, 11900, 11910, 11955, 13710, 15130, 15145, 15200, 15290, 15350, 15385, 15415, 17705, 17775, 21715*

Albanian	Area	kHz
1230-1330	ALB	9670
Arabic		
1000-1200	ME/Am	13710, 15200
1500-1700	ME/NAf/Am	6120, 7265
Azerbaijani		
0830-0930	SWAs	11835, 15130
1500-1600	NWAs	6050
Bulgarian		
1230-1330	BUL	7110
1700-1800	BUL	5960
Chinese		
1200-1300	FE	11760 (13710 from Feb 1)
English		
0400-0500	Eu/As/Pac/NAm	7300, 9685, 17705
1330-1430	Eu/As/Pac	9630, 15290
1930-2030	Eu/Af	5960, 6110
2130-2230	Pac	7200
2300-2400	Eu/Af/NAm	6135, 9655

French	Area	kHz
2030-2130	Eu/Af	7245, 7255
German		
1430-1530	Eu	11800
1830-1930	Eu	5970, 6035
Greek		
1130-1230	SEEu/As	7150, 9445
1530-1630	SEEu	6145
Hungarian		
1630-1730	HNG	7240
Macedonian		
1730-1800	SEEu	5970
Persian		
0930-1030	IRN	11795, 13770
1400-1500	IRN	9710
Romanian		
1030-1130	ROU	9630
Russian		
1400-1500	NWAs	9675
1800-1900	NWAs	6135
Serbo-Croat		
1030-1130	SEEu	9505
1900-2000	BIH	5990
Turkish		
0500-0800	As	17775
0500-1000	As/AUS	9565, 15145, 21715
0500-1700	As/Af	11955
0800-2200	Eu/NAm	9460
1000-1300	As/AUS	15415
1000-1700	Eu	15350
1100-1600Fri	As/Af	11910
1300-2300	As/AUS	9560
1600-2300	Eu	5980
1700-2300	Af	7110
1700-0500	As/Af	6120
1700-1000	Eu	15385
2200-0800	Eu/NAm	9445, 9460
2300-0500	Eu/NAm	5980
Urdu		
1300-1400	As	11900

ANN: E: "This is the voice of Turkey". G: "Hier ist die Stimme der Turkei". Turkish: "Burasi Turkiyenin Sesi Radyosu" — **IS:** As for Home Sce — **V.** by QSL-card — **PUB:** "The Voice of Turkey Prgr sch" in Arabic, English, French, German; "Turkiyenin Sesi Radyosu Program Bülteni" (All languages, every 6 months, free). **F.PI:** Italian and Spanish

TURKMENISTAN

TURKMEN RADIO
✉ Mollanepes kucasi 3, 744000 Ashkhabad. ☎ +7 (3832) 251515. 🖷 +7 (3832) 251421.

English	Days	Area	kHz
1400-1410	2456	As	5015, 675, 279

UKRAINE

RADIO UKRAINE INTERNATIONAL
✉ vul. Khreshchatyk 26, 252001 Kyyiv. ☎ +380 (44) 2294586.
L.P: Chief Editor: Viktor I. Nabrusko.
MW: Cf. Home Sce — **SW:** Kharkiv (Ternivka, G.C: 36.07E/49.38N) 3 x 100kW, 1 x 5kW — Mykolayiv (Kopani G.C: 32.14E/46.49N): 1 x 100kW, 2 x 250kW, 4 x 1000kW — Kyyiv (Brovary G.C: 30.46E/50.31N): 3 x 100kW, 5 x 200kW, Lviv (Krasne G.C: 24.40E/49.51N) 2 x 1000kW. **FM:** Re. on 68.45MHz (Kyyiv-Brovary, 17kW).
kHz: 5905, 5940, 6010, 6020, 6090, 7115, 7150, 7180, 7240, 7320, 7380, 9550, 9560, 9600, 9640, 9870, 9905, 11705, 11830, 11840, 12040, 12045, 12050, 13590, 13720, 15520, 15550, 17680, 17715
Note: Su. all prgr's 1h earlier. Most prgr's also on 1512kHz in Kyyiv, 1350kHz (Mykolayiv?) and on 1197/1440/1494kHz in Kazakhstan.

English	Area	kHz
0100-0200	NAm/Eu	12040, 9905, 9560, 9550, 7240, 7180, 7150, 6090, 6020, 6010, 5905, 171
0400-0500	NAm/Eu	12040, 9550, 7150, 171
1300-1400	NAm/Eu/Pac	12050, 12045, 7180, 7150
2200-2300	Eu	13720, 13590, 12040, 9640, 9560, 9550, 7380, 7240, 7180, 6090, 6020, 6010, 5905, 1359, 171
German		
0000-0100	Eu	9560, 7240, 7180, 6090, 6020, 6010, 5905, 171
1800-1900	Eu	17680, 15520, 13590, 11830, 11705, 9870, 9640, 7320, 6020, 5905
2100-2200	Eu	15520, 13590, 11830, 9640, 9560, 7240, 7180, 7115, 6090, 6020, 6010, 5905, 171

Romanian	Area	kHz
1800-1830	Eu	657, 891, 1431
2030-2100	Eu	1431
2200-2230	Eu	891, 1431
Ukrainian		
0200-0400	World Sce	12040, 9560, 9550, 7180, 7150, 6090, 6020, 6010, 5905, 171
0400-0500	World Sce	6090, 6020, 6010
0500-0600	World Sce	11705, 9870, 7320, 7150, 6090, 6020, 6010
0600-0700	World Sce	11840, 11705, 9870, 7320, 7150, 6020
0700-0800	World Sce	13590, 11840, 11705, 9870, 9600, 7320, 7150, 6020
0800-0900	World Sce	13590, 11840, 11705, 9870, 9600, 7320, 7150
0900-1100	World Sce	13590, 11840, 11705, 9870, 9600, 7320, 7150, 6020
1100-1300	World Sce	13590, 12050, 12045, 11840, 11705, 9870, 9600, 7320, 7150, 6020
1300-1400	World Sce	13590, 11840. 11705, 9870, 9600, 7320, 6020
1400-1500	World Sce	15520, 13590, 12050, 11840, 11705, 9870, 9600, 7320, 7150, 6020
1500-1700	World Sce	17680, 15520, 13590, 12050, 11840, 11705, 9870, 9600, 7320, 7150, 6020
1700-1800	World Sce	17680, 15550, 15520, 13590, 12050, 11830, 11705, 9870, 9640, 7320, 6020
1800-1900	World Sce.	15550, 13720, 12050, 171
1900-2000	World Sce	17680, 15550, 15520, 13720, 13590, 12050, 11830, 11705, 9870, 9640, 6090, 6020, 5905, 171
2000-2100	World Sce	17715, 17680, 15520, 13720, 13590, 12050, 11830, 11705, 9640, 9560, 7240, 7180, 6090, 6020, 6010, 5940, 5905, 171
2100-2200	World Sce.	13720, 7380
2300-2400	World Sce	13720, 13590, 12040, 9640, 9560, 9550, 7240, 7180, 6090, 6020, 6010 5905, 936

ANN: Ukrainian: "Hovorit Kyyiv". "Vsesvitnya sluzhba Radio Ukrayiny" (World Sce.) — **V.** by QSL-card.

VOICE OF RUSSIA RELAYS
SW: Mykolayiv 2 x 250kW (operated as 1 x 500kW) & 1 x 1000kW of RUI facilities used for Golos Rossii FS in Russian, cf. Russia.

UNITED ARAB EMIRATES

RADIO OF THE UNITED ARAB EMIRATES (Gov.)
✉ P.O. Box 63, Abu Dhabi. ☎ +971 (2) 451000. 🖷 +971 (2) 451155.
L.P: DG: Abdul Wahab al Radhwan. Contr. of Prgrs: Abdul Hadi al Mobarak. Chief Eng (Tr's): Rushdi Hattab. Dir Foreign Prgrs: Ms. Aida Hamza.
SW: Dhabayya (G.C: 54.14E/24.11N): 4 x 500kW
kHz: 6180, 7215, 9605, 9695, 9770, 9790, 11885, 13605, 15265, 15315, 17645, 17760, 17855, 21630

Arabic	Area	kHz
0000-0200	NAm	9605, 9770, 11885
0200-0400	ME/FE/AUS	6180, 15315, 17760, 17855
0200-0600	SAs	9695
0400-0600	ME/FE	9770, 13605, 21630
0600-1200	Eu	17855
0600-0900	NAf/Eu	15315, 21630
0900-1100	NAf	17760
0900-1200	ME	15265
0900-1300	FE	21630
1100-1300	FE	17760
1200-1600	ME	13605
1200-1700	Eu	15315
1300-1600	AUS	15265, 17645
1600-1800	Eu	13605
1600-2000	Eu/SAs	9695+, 11885
1600-2200	ME	6180
1700-2000	Eu	9790
1800-2200	NAf	9770
2000-2200	ME	7215
English		
2200-2400	NAm	9605, 9770, 11885

ANN: A: "Idha'atu-l-imarat al-arabiyya al-muttahida min Abu Zabiy". **V.** by QSL-folder. R. to Dir, Foreign Prgrs.

U.A.E. RADIO & TELEVISION – DUBAI (Gov.)
✉ P.O. Box 1695, Dubai. ☎ +971 (4) 370255. 🖷 +971 (4) 379275. ⟳ 95845605 BRCAST EM — **L.P:** Dir. of Inf: Sheikh Hasher al Maktoum. Ag. DG: Ahmed Saeed al Gaoud. Contr. Radio: Hassan Ahmed. Contr. of

Eng: Abdul Rehman Al Ali.
SW (G.C: 55.16E/25.14N): 1 x 500kW, 3 x 300kW tr's.
kHz: *11795, 11945, 13675, 15320, 15345, 15395, 15400, 15435, 17830, 21485, 21605, 21700*

Arabic	Area	kHz
0230-0330	NAm	21485, 15400, 13675, 11945
0415-0530	FE	21700, 17830, 15435
0615-1030	Eu	21605, 15395, 15345, 13675
1055-1330	Eu	21605, 15395, 15345, 13675
1355-1600	Eu	21605, 15395, 15345, 13675
1640-1700	Eu	21605, 15395, 15345, 13675
1700-2100	Eu	15395, 15320, 13675, 11795

English		
0330-0400	NAm	21485, 15400, 13675, 11945
0530-0600	FE	21700, 17830, 15435
1030-1055	Eu	21605, 15395, 15345, 13675
1330-1355	Eu	21605, 15395, 15345, 13675
1600-1640	Eu	21605, 15395, 15345, 13675

ANN: A: "Idha'at al imarat al Arabiyyah al Mutahhida min Dubai". E: "This is United Arab Emirates Radio in Dubai" — **V.** by QSL-card.

UNITED KINGDOM

BBC WORLD SERVICE
P.O. Box 76, Bush House, The Strand, London WC2B 4PH.
☎ +44 (171) 240 3456. 🗏 +44 (171) 257 8258.
E-Mail: worldservice.letters@bbc.co.uk
Web: http://www.bbc.co.uk/worldservice
L.P: MD: Sam Younger. Dir. Strategy & Corporate Affairs: Caroline Thomson. Head of News Commissioning: Bob Jobbins. Head of English Programmes Commissioning: Penny Tuerk. Head of Resources Commissiong: Chris Gill. Financial & Commercial Dir: Andrew Hind. Dir. Monitoring: Andrew Hills. Chief Personnel Officer: Kate Poulton. Heads of Region: Benny Ammar (Africa & Middle East), Jerry Timmins (Americas), Elizabeth Wright (Asia/Pacific), Andrew Taussig (Europe), David Morton (former Soviet Union & SW Asia), Barry Langridge (So. Asia). Head of Press & PR: Richard McCarthy.
LW: Droitwich 198kHz 500kW
MW: Orfordness 648/1296kHz 500kW
SW: Rampisham (G.C: 02.38W/50.48N): 10 x 500kW — Skelton (G.C: 02.55W/54.44N): 6 x 300kW, 11 x 250kW — Woofferton (G.C: 02.43W/52.19N): 4 x 300kW, 6 x 250kW
Overseas Relay St's: A=Ascension Island, An=Antigua, B=Brandon (Australia), Bi = Biblis, C=Cyprus, D=Delano (USA), E=Masirah Isand (Oman), G=Greenville (USA), Gr=Grigoriopol (Moldova), H=Chita (Russia), I=Indian Ocean Relay (Seychelles), Ik=Irkutsk, K=Kimjae (Korea), L=Lesotho, La = Lampertheim, M=Meyerton (So. Africa), Mo=Moscow, N=Nakhan (Thailand), Oz=Orzu (Tajikistan), R=Rangitaiki (New Zealand) S=Singapore, Sa=Sackville (Canada), Sm = Samara, St=St. Petersburg, T=Tashkent, V=Vladivostok, W=WYFR, Okeechobee (USA), Y=Yamata (Japan), Ye=Yekaterinburg (Russia), Z=Almaty (Kazakstan)
Rebroadcasters: BBC World Sce.prgrs in English and other languages are relayed by local st's in many countries. Full details can be found in BBC On Air magazine, available on subscription (see end of entry).

Frequencies used by SW tr's in the UK
kHz: *3955, 3980, 3995, 5875, 5955, 5990, 6010, 6040, 6050, 6110, 6150, 6190, 6195, 7105, 7120, 7125, 7135, 7145, 7150, 7165, 7210, 7245, 7260, 7270, 7275, 7320, 7325, 9410, 9530, 9635, 9685, 9710, 9765, 9825, 9835, 9840, 9850, 9870, 9915, 11660, 11680, 11760, 11765, 11780, 11845, 11895, 12040, 12095, 13660, 13670, 13745, 15120, 15180, 15225, 15235, 15325, 15485, 15525, 15565, 15575, 17640, 17695, 17705, 17715, 17885, 21590, 21640*

Winter Schedule 1998/99

Albanian	Time	Area	kHz
	0630-0700	SEEu	7210
	0630-0700	SEEu	9915C, 11845C
	1415-1445MF	SEEu	11680C, 15115
	1800-1830	SEEu	6050C, 7325, 9685
	1830-1845M	SEEu	6050C, 7325, 9685
	2100-2115MF	SEEu	6050, 7295
Arabic	0330-0445	ME	9825, 9875C
	0330-0600	ME	11740C, 15235E
	0330-0715	ME	7140C
	0445-0600	NAf	6110, 7325, 12040C
	0445-0715	ME	9825
	0445-0715	ME	13660C
	0600-0715	NAf	7125
	0600-0715	ME	12040
	1200-1225	ME	7140C, 11820C, 15555C
	1200-1225	NAf	13660, 15180, 17585
	1225-1255Fr-Sa	NAf	13660, 15180, 17585
	1225-1255Fr-Sa	ME	15555C, 7140C, 11820C
	1255-1630	ME	11820C, 15555C

Arabic	Time	Area	kHz
	1255-1630	NAf	15180, 17585
	1255-1800	NAf	13660
	1255-2115	ME	7140C
	1630-1800	NAf	11680, 15180C
	1630-1800	ME	11820C
	1630-2115	ME	6030E
	1800-2115	NAf	6110, 9915C, 11680
	1800-2115	ME	9915C
Azeri	1800-1900	CAs	5875C, 7255C
	1800-1900	CAU	9750La
BBC English	1115-1130MF	Eu	9815
	1615-1630	Eu	9915
	0030-0045	SEAs	6080S, 7110S, 9600S
	0030-0100	FE	15280N, 17790N
	0230-0245	SAs	9825C, 15405E, 17790N
	0245-0300	CSAm	5995D
	0245-0300	SAm	6110An
	0245-0300	CAm	6110An, 7325
	0245-0300	EAf	7235C, 9610M, 11865I
	0245-0300	CSAm	9515D
	0300-0330	CHN	15280N, 17760N, 21715S
	0515-0530SS	SAf	6135M, 9515M, 11940M
	0530-0545	SEEu	6010, 7270, 9915
	0545-0600	NAf	7145, 9710
	0600-0615	WAf	6155A
	0630-0645SS	SEEu	6010, 9870C
	0645-0700	CEu	5875, 9635C
	0715-0730	AF	6005A, 15420A
	0730-0745	WEu	5875, 9635
	0915-0945	FE	6065Y, 9580N, 11955N, 15280N
	0915-0945	SEAs	11945S
	0915-0945MF	SEAs	6195S, 6195S, 9740S
	0915-0945MF	OC	11765S, 15360S
	0915-0945MF	SEAs	17760S, 21660S
	0945-1000Su-Fr	FE	6065Y, 9580N, 15280N
	0945-1000Su/Fr	FE	11955N
	0945-1000Su-Fr	SEAs	11945S
	1115-1130	RUS	11805C, 13685, 15325, 15340C, 17695C
	1115-1130MF	Eu	5875
	1115-1130Mo-Th	SEEu	9745C, 11970C
	1130-1200MF	SAs	15310N, 17785E, 17785E
	1200-1215	CHN	7135S, 9605S, 11920S
	1215-1230Su	Eu	5875
	1215-1230Su	SWEu	9635
	1225-1240Su-Th	ME	7140C
	1225-1240Su-Th	NAf	13660, 15180, 17585
	1225-1240Su-Th	ME	15555C
	1225-1240Su/Th	ME	11820C
	1230-1245	NAf	15105, 17555
	1230-1245	WCAf	21640A
	1230-1300	Eu	5875, 9745C, 11895, 15325
	1230-1300	NEu	9635, 13670, 15325, 17610
	1230-1300	SWEu	9635, 15325, 11680Bi
	1230-1300	RUS	15260C
	1240-1255Su-Th	ME	7140C, 15555C
	1240-1255Su-Th	NAf	13660, 15180, 17585
	1240-1255Su/Th	ME	11820C
	1330-1345	WAf	15105A, 17810A, 21640
	1400-1415MF	EAf	11860I, 15420I, 17880, 21490C
	1400-1430Su	EAf	11860I, 15420I, 17880
	1400-1430Su	AF	21490A
	1515-1530	WEu	11980
	1515-1530SS	Eu	5875
	1530-1545	SAs	9595S, 11685S
	1615-1630	WEu	6015
	1730-1745	Af	3390M, 6070M, 9520M
	1730-1800	WEu	5875
	1730-1800	WEu	9530
	1730-1800	UKR	7155La, 11770, 13685
	1800-1815	SAs	6065E, 9605S, 11665N
	1830-1845W	SEEu	6050C, 7325, 9685
	1900-1915	AF	11880A, 15105A, 17885A
	2130-2200	WEu	5875, 6050, 9635C
	2330-2345	SEAs	3915S, 5875
	2345-2400	SAm	5875, 6110A, 6110An
	2345-2400	CAm	6110An
	2345-2400	SAm	9825A, 11765A
Bengali	0030-0045	SAs	6065N
	0030-0045	SAs	11850S
	0800-0810	SAs	11685N, 15405N, 17675T

	Time	Area	kHz
Bengali	1330-1415	SAs	7225N, 11835S, 12065V
	1630-1700	SAs	5990N, 7205S, 9605S
Brazilian	2130-2215	SAm	6110An
	2130-2215	SAm	9560A, 9865, 11765A, 15390A
Bulgarian	0430-0500MF	SEEu	5875, 7210C
	1100-1115	SEEu	9745C, 11970C
	1300-1330SS	SEEu	9745C, 11680Bi
	1700-1730	SEEu	5875C, 9530Bi
	2000-2015	SEEu	6050C, 7105
	2015-2100SS	SEEu	6050C, 7105
	2200-2215MF	SEEu	6180C, 7105
Burmese	0000-0030Su-Th	SEAs	6065N, 9600E, 11850S
	0100-0215SS	SEAs	7135N, 11685C, 11850S
	1345-1430	SEAs	7135S, 9730S, 11740S, 11850C
Cantonese	1300-1330	CHN	7225N, 11945S
Croatian	0600-0615	SEEu	7210, 9915C
	0600-0615	SEEu	11845C
	1200-1230	SEEu	11895, 15325
	1800-1830	SEEu	6090, 7105, 9895
	2215-2230	SEEu	5965, 7165C, 9660
English	0000-0030	SEAs	3915S
	0000-0030	FE	17790N
	0000-0100	SEAs	6195S
	0000-0100	SAs	6195S
	0000-0200	SAs	5965E
	0000-0200	CSAm	9590D
	0000-0300	SAm	5970A, 9410C
	0000-0300	SAs	11955N, 15310S
	0000-0330	SEAs	15360S
	0100-0200	SAs	6195S
	0100-0300	FE	15280N
	0200-0300	SAs	9605E
	0200-0300	EAf	9770I
	0200-0500	CSAm	6185D
	0300-0330	SAs	17790E
	0300-0400	SAf	6005A
	0300-0400	RUS	9410C
	0300-0400	EAf	11730I
	0300-0430	NEu	6195
	0300-0500	SAs	11955C
	0300-0500	EAf	12095C
	0300-0530	FE	21660N
	0300-0600	SAf	3255M, 6190M
	0300-0700	WCAf	7160A
	0300-0700	SAf	11765A
	0300-0800	ME	11760E
	0300-0815	SAs	15310E
	0330-0500	FE	17760N
	0330-0530	FE	15280N
	0330-0530	SAs	17790N
	0330-0630	EAf	15420I
	0400-0430	NEu	3955
	0400-0430	Eu	6180C, 6195
	0400-0430	NEu	9410
	0400-0430	Eu	9410
	0400-0630	CNAm	5975An
	0400-0715	WAf	6005A
	0400-0730	WAs	15575C
	0430-0500Mo-Sa	NEu	3955, 6195, 9410
	0430-0500Mo-Sa	Eu	6180C, 6195, 9410
	0430-0500Su	NEu	3955, 6195, 9410
	0430-0500Su	Eu	6180C, 6195, 9410
	0500-0530	NEu	3955, 6195, 9410
	0500-0530	Eu	3955, 6180C, 6195, 9410
	0500-0530	RUS	12095C
	0500-0630	CSAm	6175D
	0500-0630	EAf	17885I
	0500-0700	EAf	17640C
	0500-0845	SEAs	15360S
	0500-0845	OC	15360S
	0500-0900	OC	11955S
	0500-0900	SEAs	17760S
	0500-0915	SEAs	9740S
	0530-0600MF	NEu	3955, 6195, 9410
	0530-0600MF	Eu	3955, 6180C, 6195, 9410
	0530-0600MF	RUS	12095C
	0530-0600SS	NEu	3955, 6195, 9410
	0530-0600SS	Eu	3955, 6180C, 6195, 9410
	0530-0600SS	RUS	12095C
	0530-0900	FE	21660N
	0600-0630	NEu	3955, 6195
	0600-0630	Eu	3955, 6195
	0600-0700	Eu	15565

English	Time	Area	kHz
	0600-0730	Eu	6180C
	0600-0800	NEu	9410
	0600-0800	SAs	17790E
	0600-0810	OC	7145
	0600-0900	Eu	7325, 9410
	0600-1100	Eu	9410
	0600-1700	SAf	6190M, 11940M
	0600-1800	NEu	12095
	0600-1800	Eu	12095
	0630-0700MF	NEu	3955, 6195
	0630-0700MF	Eu	3955, 6195
	0630-0700MF	EAf	15420C, 17885C
	0630-0700SS	NEu	3955
	0630-0700SS	Eu	3955
	0630-0700SS	NEu	6195
	0630-0700SS	Eu	6195
	0630-0700SS	EAf	15420I
	0630-0800Mo-Sa	CNAm	5975An
	0630-0800Mo-Sa	CSAm	6175D
	0630-0800Su	CNAm	5975An
	0630-0800Su	CSAm	6175D
	0630-0900SS	EAf	17885I
	0700-0800	Eu	6195
	0700-0800	WAf	11765A, 15400
	0700-1500	NEu	17640
	0700-1500	Eu	17640
	0700-1700	Eu	15565
	0700-1800	Eu	15485
	0700-1700	NEu	15565
	0730-0900SS	WAs	15575C
	0800-0810	SAs	17790N
	0800-1000	WAf	15400A
	0800-1000	SAf	15400A
	0815-0900	SAs	15310N
	0845-0915	SEAs	15360S
	0845-0915	OC	15360S
	0900-0915	FE	6065Y
	0900-0915	SEAs	6195S
	0900-0915	SAs	6195S
	0900-0915	FE	9580N
	0900-0915	OC	11765S
	0900-0915	SEAs	11945S
	0900-0915	FE	11955N
	0900-0915	SEAs	17760S, 21660S
	0900-1000	SAm	15190A
	0900-1100	SAs	15310E
	0900-1100	SAs	17790N
	0900-1200	WAs	15575C
	0900-1200	ME	15575C
	0900-1400	ME	11760E
	0900-1400	EAf	17885I
	0900-1515	Eu	17705
	0915-0945SS	SEAs	6195S, 9740S
	0915-0945SS	SAs	6195S
	0915-0945SS	OC	11765S, 15360S
	0915-0945SS	SEAs	17760S, 21660S
	0945-1000Sa	FE	6065Y, 9580N
	0945-1000Sa	SEAs	11945S
	0945-1000Sa	FE	11955N, 15280N
	0945-1030	OC	15360S
	0945-1030	SEAs	17760S, 21660S
	0945-1100	SEAs	9740S
	0945-1100	OC	11765S
	0945-1130	SEAs	6195S
	0945-1130	SAs	6195S
	1000-1100	CNAm	6195An
	1000-1100SS	WAf	15400A
	1000-1130SS	SAm	15190A
	1100-1130	FE	9580N, 11955N, 15280N
	1100-1130	SAs	15310N, 17785E
	1100-1130	WAf	15400A
	1100-1130	SAm	17790A
	1100-1130MF	Carib	6195An, 15220An
	1100-1130SS	CNAm	6195An, 15220An
	1100-1130SS	SAm	15220An
	1100-1130Su	SEAs	9740S
	1100-1130Su	OC	9740S
	1100-1400	NAm	5965Sa
	1100-1430Mo-Sa	SEAs	9740S
	1100-1430Mo-Sa	OC	9740S
	1100-1500	Eu	9410
	1100-1700	SAf	21660A
	1130-1200	SAm	15220An
	1130-1200	CNAm	15220An
	1130-1200SS	SAs	15310N, 17785E

English	Time	Area	kHz
	1130-1209	CNAm	6195An
	1130-1300Mo-Sa	SEAs	6195S
	1130-1300Mo-Sa	SAs	6195S
	1130-1300Mo-Sa	FE	9580N, 11955N, 15280N
	1130-1300Su	SEAs	6195S, 9740S
	1130-1300Su	FE	9580N, 11955N, 15280N
	1200-1209	SAm	15220An
	1200-1209	CNAm	15220An
	1200-1300	SAs	17785E
	1200-1400	SAs	15310N, 17785E
	1200-1500	WAs	15575C
	1200-1615	NAm	9515Sa
	1209-1215MF	Carib	6195An, 15220An
	1209-1215SS	CNAm	6195An, 15220An
	1209-1215SS	SAm	15220An
	1215-1400	CNAm	6195An, 15220An
	1215-1400	SAm	15220An
	1300-1400	EAf	15420I
	1300-1430	FE	5990N
	1300-1430	SEAs	6195S
	1300-1430	SAs	6195S
	1300-1430Su	SEAs	9740S
	1300-1430Su	OC	9740S
	1300-1455	SAs	11750E
	1400-1455	SAs	15310E
	1300-1600	NAm	9590W
	1400-1600	NAm	15220Sa
	1400-1700	SAm	17840An
	1400-1700	CNAm	17840An
	1400-1700	EAf	21470C
	1430-1600SS	FE	5990N
	1430-1600SS	SEAs	6195S, 9740S
	1430-1600SS	SAs	6195S
	1430-1600SS	OC	9740S
	1455-1600MF	FE	5990N
	1455-1600MF	SEAs	6195S, 9740S
	1455-1630Mo-Sa	SAs	11750E, 15310E
	1455-1630Su	SAs	11750E, 15310E
	1500-1529	EAf	11860I, 15420I
	1500-1529	CAf	21490A
	1500-1630Mo-Sa	SAs	5975N
	1500-1630Su	SAs	5975N
	1500-1830	Eu	9410
	1500-2100	Eu	9410
	1500-2100	WAf	15400A
	1529-1530Su-Fr	EAf	11860I, 15420I
	1529-1530Su-Fr	CAf	21490A
	1600-1615	FE	5990N
	1600-1700	SEAs	6195S
	1600-1700	Eu	6195
	1600-1700	SAs	6195S
	1600-1800	SEAs	3915S, 7160S
	1600-1800	SAs	9740S
	1615-1645SS	EAf	11860I
	1615-1700	EAf	15420I
	1615-1700Sa	NAm	9515Sa
	1630-1700	SAs	11750E, 15310E
	1630-1830	SAs	5975N
	1645-1700	EAf	11860I
	1700-1745	EAf	6005I, 9630I
	1700-1830	SAs	9510E
	1700-1830	NAm	17840Sa
	1700-1900	EAf	15420M
	1700-2000	ME	11970N
	1700-2100	NEu	6195
	1700-2100	Eu	6195
	1700-2100	SAf	15400A
	1700-2100	SAf	3255M, 6190M
	1700-2330	NEu	3955
	1730-1800Su	CAs	7390Mo, 9750C, 11660
	1800-1830	Eu	6180M
	1800-2000	Eu	12095
	1800-2100	Eu	6195
	1800-2200	OC	9740S
	1800-2330	Eu	3955
	1830-1900	Eu	6180C
	1830-1900Mo-Sa	Eu	9410
	1830-1900Mo-Sa	NAm	17840Sa
	1830-1900Su	RUS	6180C
	1830-1900Su	Eu	9410
	1830-2100	EAf	6005I, 9630I
	1900-2100	RUS	6180C
	1900-2100	Eu	6180C
	1900-2200	Eu	9410
	2000-0200	SAm	12095A

English	Time	Area	kHz
	2000-2200	OC	5975S
	2000-2200	Eu	7325
	2000-2300	WAf	11835
	2100-2115	CNAm	5975An
	2100-2200	SEAs	3915S
	2100-2200	EAf	6005I
	2100-2200	FE	6110Y
	2100-2200	NEu	6180
	2100-2200	Eu	6180
	2100-2300	WAf	15400A
	2100-2330	NAf	6195C
	2100-2330	Eu	6195C
	2100-2400	FE	5965N
	2100-2400	SEAs	6195S
	2100-2400	SAs	6195S
	2115-2130MF	Carib	5975An, 15390A, 17715G
	2115-2130SS	CNAm	5975An
	2130-0400	CNAm	5975An
	2130-2145Tu/Fr	SAs	11680
	2200-0030	SEAs	7110S
	2200-0300	SAm	9915
	2200-0500	NAm	6175Sa
	2200-2300	FE	7385Ik
	2200-2300	SEAs	9660B
	2200-2300	OC	12080B
	2200-2400	NAm	9590Sa
	2200-2400	SEAs	11955S
	2200-2400	OC	11955S
	2300-0030	FE	11945Y, 15280N
	2300-2330	SEAs	3915S
	2300-2400	FE	6035K
	2330-2400Sa	NEu	3955
	2330-2400Sa	Eu	3955, 6195C
	2330-2400Sa	NAf	6195C
	2345-2400MF	SEAs	3915S
French	0430-0500	WCAf	6155A
	0430-0500	WCAf	7105A
	0430-0500	EAf	17885I
	0600-0630	WCAf	7105A
	0600-0630	NAf	7275, 9610A, 9710
	0700-0730	WAf	15105A
	0700-0730	WCAf	17695A
	1200-1230	NAf	15105, 17555
	1200-1230	WCAf	21640A
	1800-1830	SAf	7230M
	1800-1830	WCAf	15105A, 21630A
	1800-1830	NAf	15180
	1800-1830	WAf	17885A
German	0600-0630MF	WEu	5875
	0600-0630MF	WEu	11665C
	0900-0915MF	WEu	5875, 9815
	1100-1115MF	WEu	5875, 9815
	1300-1315	WEu	5875, 11895
	1500-1515MF	WEu	5875, 11980
	1630-1700	WEu	6015, 9915
	1830-1900MF	WEu	6015, 9835C
Hausa	0530-0600	WAf	6155A, 7105A, 9610A
	1345-1415	WAf	15105A, 17810A, 21640
	1915-1945	WAf	11880A, 15105A, 17885A
Hindi	0045-0130	SAs	6065E, 7320C, 15380N
	0810-0820	SAs	15405N, 17675Sm, 17790N
	1400-1500	SAs	6140E, 7205N, 11920E, 15245C
	1700-1730	SAs	6060S, 7205E, 7235N, 9605S
Hungarian	0530-0545MF	CEu	6050, 9870C
	0630-0645	CEu	5875, 9635C
	1000-1130Su	CEu	11680, 13745
	1315-1400Su	CEu	9635, 11895
	1800-1830	CEu	6010, 9835C
	2200-2230	CEu	6010C, 9635C
Indonesian	1100-1130	SEAs	7135S
	1100-1130	SEAs	9605S, 11920S
	1100-1130	ME	21590
	1300-1345	SEAs	7135S, 9730S, 11740S, 15340E
	2200-2300	SEAs	5875, 3915S, 6080S, 9740S
Kazakh	1500-1530MF	CAs	11780C, 15575
Krwanda	1615-1645MF	EAf	11860I, 21490A
Mandarin	1000-1100	CHN	9580N, 11955N, 15280N
	1000-1100	FE	11695V
	1000-1300	FE	11945S
	1000-1530	CHN	6065Y

	Time	Area	kHz
Mandarin	1300-1530	CHN	6055K, 7330H, 9605N
	1300-1530	FE	15280S
	2200-2300	CHN	6035K, 6110N, 7160N, 7330H, 9580N, 11945Y
Nepali	1500-1530	SAs	9595S, 11685S
Pashto	0200-0230	WAs	5875, 6065E
	0200-0230	WAs	7165C
	0200-0230	SWAs	15380N
	0745-0945F	WAs	12030C, 15585C
	1530-1615	WEu	3955, 7105C, 15575
Persian	0230-0300	ME	5875, 7165C, 15380N
	0945-1230Th-Fri	ME	12030C, 15585C
	1445-1530	ME	7105C, 15595
	1615-1700	WEu	3955
	1615-1700	ME	6095E, 7105C, 15575
	1830-1900	ME	6095E, 7105N, 11715N
Portuguese	0430-0500	SAf	3390M
	0430-0500	SAf	6135M, 7125M
	0530-0600	SAf	6135M, 9515M, 11940M
	2030-2115	SAf	3390M, 6135M, 7125M
	2030-2115	WAf	7150, 11695, 11855A
Romanian	0400-0430MF	SEEu	6130, 7210C
	0600-0615	SEEu	6010, 9870C
	0615-0630SS	SEEu	6010, 9870C
	0900-0915	SEEu	11660C, 15465
	1200-1230	SEEu	9745C, 11680Bi
	1600-1630	SEEu	6050C, 9840Bi
	1900-1930	SEEu	6050C, 7155
Russian	0300-0330	CAs	6095, 7320C, 9825C
	0330-0430Mo-Sa	RUS	5875, 6095, 7130C, 7320, 9670C
	0700-0715	RUS	9635, 11680, 11845C, 13745, 15335C
	0900-0915	RUS	11680, 13745, 15325, 15340, 17695La
	1030-1100Su	RUS	11805C, 13685, 15325. 15340C, 17695C
	1100-1115	RUS	11805C, 13685, 15325, 15340C, 17695C
	1300-1400	RUS	11840C, 13670, 15260C, 17610
	1500-1600MF	RUS	9635C, 12005C, 15225, 17695
	1600-1800	RUS	13615, 15225
	1600-1900	RUS	9825, 11760, 11845C
	1600-2130	RUS	9635C
	1730-1800Mo-Th/Sa	CAs	7390Mo, 9750C, 11660
	1800-2130	RUS	7120, 9825
	1900-2130	RUS	5875C, 5990, 11845C
Serbian	0545-0600	SEEu	6010
	0545-0600	SEEu	7270, 9915C
	1000-1015MF	SEEu	11685, 13675, 15325
	1130-1145	SEEu	11680, 15325
	1700-1745	SEEu	6050C, 9915, 11925
	2200-2215	SEEu	5965, 7165C, 9660
Sinhala	1515-1545	SAs	6140E, 9510S
Somali	0900-0915	EAf	15420I, 21470C
	1415-1430MF	EAf	11860I, 15420I, 17880, 21490C
	1430-1500	EAf	11860I, 15420I, 17880, 21490C
	1800-1830	EAf	6005I, 9630I, 15360
Spanish	0000-0115	SAm	5875, 6110A, 9825A, 11765A
	0000-0115	CAm	5875, 6110An
	0115-0130Tu-Sa	SAm	5875, 6110A, 6110An, 9825A, 11765A
	0115-0130Tu-Sa	CAm	5875, 6110An
	0300-0345	CSAm	5995D, 9515D
	0300-0345	SAm	6110An
	0300-0345	CAm	6110An, 7325
	0345-0400Tu-Sa	CSAm	5995D, 9515D
	0345-0400Tu-Sa	SAm	6110An
	0345-0400Tu-Sa	CAm	6110An, 7325
	1100-1130MF	CSAm	5975G, 6130D, 9670D
	1100-1130MF	SAm	15190A, 17820A
	1300-1330MF	CSAm	6130D, 9670D, 15315G
Swahili	0300-0330	EAf	7235C, 9610M, 11865I
	0400-0430	EAf	7105A, 9610M, 11730I
	1529-1530Sa	EAf	11860I, 15420I, 21490A

	Time	Area	kHz
Swahili	1530-1615	EAf	11860I, 15420I, 21490A
	1745-1800	EAf	6005I, 7230M, 9630I
Tamil	1545-1615	SAs	6140E, 7205N, 9510S
Thai	1230-1300	SEAs	7135S, 9605S
	1230-1300	ME	21590
	2330-2400	SEAs	6080S, 7160S
Turkish	0500-0530MF	SEEu	6010
	0500-0530MF	SEEu	7270, 9915
	0900-1000Su	SEEu	6015C, 9740C
	1000-1015MF	SEEu	6015C, 9740C
	1600-1700	SEEu	5875C, 9530La, 15390
Ukrainian	0500-0600MF	UKR	5875, 7260La, 9760C
	1700-1730	UKR	7270La, 11770, 13685
	2000-2030	UKR	6125La, 7260C, 9565
	2030-2100SS	UKR	6125La, 7260C, 9565
Urdu	0130-0200	SWAs	6065E, 7320C
	0130-0200	SAs	15380N
	0820-0830	SAs	15405N, 17675Sm, 17790N
	1500-1545	SAs	7205N, 9680S
	1500-1545	SWAs	11920E, 15245C
	1730-1800	SAs	6060S, 7235N, 9605S
	1730-1800	SWAs	7205E
Uzbek	0200-0230MF	CAs	7245C, 9825C, 17790N
	1700-1730	CAs	7390Mo, 9750C, 11660
	1730-1800F	CAs	7390Mo, 9750C
Vietnamese	1130-1200	SEAs	7135S, 9605S, SEAs 11920S
	1430-1500	SEAs	7135S, 9730S, 11740S
	2300-2330	SEAs	6080S, 7160S, 9740S

IS: World Sce.in English: Bow Bells prior to transmission. Other Services: BBC in tonic scale.

PUB: BBC On Air is the monthly English-language magazine and programme guide of the BBC World Service Radio & Television. Available on subscription at £18/US$30 from P.O. Box 76S, Bush House, London WC2B 4PH. ☎ +44 (171) 257 2211. 🖷 +44 (171) 240 4899. **BBC English Magazine,** a monthly publication for language learners. Available on subscription - details from BBC English Magazine, 6 Bourne Entreprise Centre, Wrotham Road, Borough Green, Kent TN15 8DG. ☎ +44 (1732) 884023. 🖷 +44 (1732) 884034. **Focus on Africa Magazine,** a quarterly publication for listeners throughout Africa. Available on subscription - details from Focus on Africa Magazine, 6 Bourne Entreprise Centre, Wrotham Road, Borough Green, Kent TN15 8DG. ☎ +44 (1732) 884034. Many other publications, promotional items, videos and audio cassettes available from BBC World Service Information Centre & Shop, Bush House, Strand, London WC2B 4PH. ☎ +44 (171) 257 2576. 🖷 +44 (171) 240 4811.

MERLIN COMMUNICATIONS INTERNATIONAL

🖃 R20 Lincolns Inn Fields, London WC2A 3ES. ☎ +44 171 9690000. 🖷 +44 171 3966223. **WEB:** wwww.merlincommunications.co.uk E-mail: info@mercom.co.uk
L.P: Business Development Mgr: Caroline Soanes.
Merlin Communications International owns and operates the shortwave transmitter sites in the UK used by BBC World Service and other broadcasters. It also operates the British overseas relay stations under a management contract.UK

Winter Schedule 1998/99

Time	Area	kHz
0000-0300	NAm	9560
0300-0600	NAm	9895
0600-0800	Eu	6110
0600-0900	OC	13720
0700-1600	Eu	9915, 17630
0700-1900	Af	21550
0800-1200	Eu	13660
1200-1400	Eu	13645
1400-1600	Eu	13680
1600-1800	Eu	6185
1700-1900	Eu	3965
1800-2000	Eu	6125
1900-2000	Eu	9690
2000-2200	Eu	11755
2000-2200	NAm	11985
2200-2400	FE	7170
2200-2400	NAm	9835
2300-0600	Eu	3985

D.PRGR: English language music mix from various commercial broadcasters

UNITED STATES OF AMERICA

BROADCASTING BOARD OF GOVERNORS
The Broadcasting Board of Governors was established by the U.S. International Broadcasting Act of 1994 to "direct and supervise" and to ensure the editorial independence and effectiveness of the elements of U.S. government funded international broadcasting. These elements include the International Broadcasting Bureau of the Voice of America Agency (Voice of America, Radio-TV Marti, and WorldNet-TV), Radio Free Europe/Radio Liberty, Inc. and Radio Free Asia, Inc.
⌨ 330 Independence Ave. S.W., Washington, DC 20547. ☎ +1 (202) 401 3736. 🖷 +1 (202) 401 6605.
L.P: Chmn: David Burke. Chief of Staff: Kathleen Harrington

VOICE OF AMERICA (Gov.)
The Voice of America is the worldwide radio broadcasting service of the United States Information Agency, International Broadcasting Bureau.
⌨ Voice of America, 330 Independence Ave S.W., Washington, DC 20547. ☎ +1 (202) 619 2538. 🖷 +1 (202) 619 1241. ① 7607706 VOA USA.
E-mail: *General business:* pubaff@voa.gov *Listeners' correspondence:* letters@VOA.GOV *Reception reports:* qsl@VOA.GOV
Web: http://www.voa.gov
L.P: International Broadcasting Bureau: Dir: Geoffrey Cowan. Dir. of Eng: Robert Kamosa. Dir. of Ext. Affairs: Joseph O'Connell.
(The IBB is responsible for the operation of VOA, Radio & TV Marti, and WORLDNET-TV)
Voice of America: Dir: Mrs. E.S. Lieberman. Dep Dir: Alan Heil. Dir. of B'cast Op's: Edward DeFontaine.
Transmitters in the USA: Greenville, No.Carolina (G.C: 77.25W/35.35N): 11 x 500kW, 6 x 250kW and 6 x 50kW tr's — Delano, California (G.C: 119.17W/35.45N): 4 x 250kW tr's.
Overseas Relay Stations: Albania, Ascension Island, Belize, Botswana, Bulgaria, Germany, Greece, Kuwait, Morocco, Philippines, Romania, Russia, Sao Tome, Singapore, Sri Lanka, Thailand, United Arab Emirates, United Kingdom.
kHz: *4950, 4960, 5745, 5910, 5955, 5965, 5970, 5975, 5985, 6015, 6020, 6025, 6030, 6035, 6040, 6045, 6050, 6060, 6070, 6085, 6090, 6105, 6120, 6125, 6130, 6150, 6160, 6170, 6185, 7110, 7115, 7120, 7130, 7135, 7140, 7145, 7150, 7155, 7160, 7170, 7180, 7190, 7195, 7200, 7205, 7210, 7215, 7220, 7235, 7245, 7250, 7255, 7260, 7265, 7270, 7280, 7290, 7295, 7300, 7340, 7370, 7415, 9480, 9505, 9510, 9515, 9520, 9525, 9530, 9535, 9540, 9545, 9555, 9575, 9580, 9590, 9600, 9615, 9620, 9625, 9635, 9645, 9650, 9660, 9665, 9670, 9675, 9680, 9690, 9695, 9700, 9705, 9720, 9725, 9745, 9760, 9780, 9790, 9810, 9815, 9820, 9845, 9855, 9885, 9890, 11670, 11715, 11720, 11725, 11730, 11750, 11760, 11770, 11775, 11785, 11805, 11820, 11825, 11835, 11850, 11855, 11865, 11870, 11875, 11880, 11885, 11895, 11900, 11905, 11915, 11925, 11930, 11935, 11945, 11960, 11965, 11970, 12010, 12030, 12040, 12080, 13600, 13655, 13710, 13740, 13760, 15120, 15125, 15145, 15160, 15205, 15210, 15215, 15220, 15300, 15305, 15360, 15365, 15370, 15375, 15385, 15395, 15410, 15445, 15485, 15505, 15515, 15640, 17705, 17740, 17755, 17765, 17780, 17785, 17800, 17805, 17840, 21485*

Winter Schedule 1998/99

	Area	kHz
Afan Oromo		
1845-1900MF	Afr	7150, 9845, 15445
Albanian		
0600-0630	Eu	6030, 6125, 7295
1400-1415	Eu	7165, 15125, 17785
1700-1730	Eu	5970, 9705, 11865
1930-2000	Eu	7155, 9600, 9680
Amharic		
0300-0330	Af	7145, 7210, 9775
1800-1830	Af	7150, 9845, 15445
Arabic		
0400-0600	ME/Naf	1260, 1548, 5965, 7115, 9665, 11670, 11865, 15160,
0730-0830	ME/NAf	6045, 7170, 9565, 9705, 11805
1700-1800	ME/NAf	1260, 7105, 13750
1800-1900	ME/NAf	1260, 7180, 7205, 9530, 11825, 11905, 11960, 15305
1900-2000	ME/NAf	1260, 1548, 6040, 7195, 7205, 9530, 9615, 11825, 11905, 17740
2000-2100	ME/NAf	1260, 1548, 6040, 6060, 6160, 7195, 7205, 9530, 9650, 11825, 11895, 11905, 17740
Azerbaijani		
1830-1900	CIS	6060, 9595, 9695
Bangla		
0130-0200	SAs	7145, 11870, 15210, 17780
1600-1700	SAs	1575, 5955, 7280, 9700
Bosnian		
1600-1630MF	Eur	1197
2230-2300	Eur	792

	Area	kHz
Bulgarian		
0400-0430	Eur	792
Burmese		
1130-1200	EAs	1575, 6100, 9890, 11850, 15205
1200-1230	EAs	6100, 9890, 11850, 15205
2330-2400	EAs	6065, 6185, 7200, 7260
Cantonese		
1500-1600	EAs	1143, 6030
Chinese		
0000-0100	EAs	6045, 7190, 9545, 11925, 13615, 15395, 17765
0100-0200	EAs	7190, 9545, 11925, 13615, 15395, 17765
0200-0300	EAs	11925, 13615, 15395, 17765
0400-0500	EAs	12010, 15300, 15385, 15485, 17740, 17840
0700-0900	EAs	11855, 11965, 12010, 13760, 15515
0900-1100	EAs	9845, 11855, 11965, 12010, 13760, 15515
1100-1200	EAs	1143, 6160, 7120, 9530, 9680, 11785, 11965, 12040
1200-1300	EAs 12040	6160, 7120, 9530, 9680, 11785, 11965,
1300-1330	EAs	5910, 6160, 7120, 9680, 9790, 11785, 11965, 12040
1330-1400	EAs	1143, 5910, 6160, 7120, 9680, 9790,
Chinese		11785, 11965, 12040
1400-1500	EAs	5910, 6160, 7160, 9680, 9790, 11730, 11965
2200-2300	EAs	6025, 6045, 6090, 7140, 7200, 9545, 11925, 15395
Creole		
1230-1300MF	Carib	9525, 9670, 11935
1730-1800	Carib	13740, 15355, 15385
2200-2230	Carib	9670, 11900, 15120
Croatian		
0530-0600	Eu	756, 792, 1197, 1395, 6025, 7210, 9585
1830-1900	Eu	792, 1197
1930-2000	Eu	6050, 6170, 7245
Czech		
0430-0500	Eu	1197
2000-2030	Eu	1197
Dari		
0200-0215	ME	6025, 9505, 9670
1515-1600	ME	7295, 9850, 11990
English (special)		
0030-0100	As	1575, 7215, 9890, 11760, 15185, 15290, 17735, 17820
0130-0200Sa	SAm	7405, 9775, 13740
1500-1530	As	1575SS, 6110, 9760, 9845, 12040, 15460
1530-1600	As	1575, 6110, 9760, 9845, 12040, 15460
1600-1700	Af	13600, 15445, 17895
2100-2200	Eu/ME	1548, 9595, 9760
English		
0000-0030	Asia/Pac	1575, 7215, 9890, 11760, 15185, 15290, 17735p, 17820
0000-0030	ME/CAs	1260, 1548
0000-0100Tu-Sa	Carib/CAm	5995, 6130, 7405, 9455, 9775, 11695, 13740
0000-0400	Eu/NAf	1197
0030-0300	ME/CAs	1548
0100-0130Tu-Sa	Carib/CAm	5995, 6130, 7405, 9455, 9775, 13740
0100-0300	SAs	7115, 7200, 9740, 9850, 11705, 15250, 15300, 17740, 17820
0130-0200Tu-Sa	Carib/CAm	5995, 6130, 9455
0200-0300Tu-Sa	CAm	1530, 1580
0300-0330	Af	909, 1530, 4960MF, 6035, 6080, 7105, 7290, 7340, 7415, 9575
0300-0400	CAm	1530, 1580
0330-0400	Af	909, 1530, 6035, 6080, 7105, 7290, 7415, 9575
0400-0430	Af	909, 1530, 6035, 6080, 7290, 7415, 9575, 9775
0400-0430	Eu/NAf	1197, 7170

English

English	Area	kHz
0400-0500Su/Mo	CAm	1530, 1580
0430-0500	Eu/NAf	7170
0430-0500	Af	909, 6035, 6080, 7290, 7415, 9575, 9775
0500-0530	Eu/NAf	792, 1197, 7170, 9700, 11825
0500-0600	Af	909, 5970, 6035, 6080, 7295, 9775, 12080
0500-0600	ME/CAs	15205
0500-0700	SAs	15205
0530-0600	Eu/NAf	7170, 9700
0600-0630	Af	909, 1530, 5970, 6035, 6080, 7285, 11805, 12080, 15600
0600-0700	ME/CAs	1260, 15205
0600-0700	Eu/NAf	792, 1197, 5995, 7170, 11825, 11950
0630-0700SS	Af	909, 1530, 5970, 6035, 6080, 7285, 11805, 12080, 15600
0900-1000	Asia/Pac	9355, 11660, 15665
1000-1100	Asia/Pac	5985p, 9355, 11660, 11720p, 15425, 15665
1100-1130	Asia/Pac	1575SS, 5985p, 6110, 9645, 9760, 11705, 11720p, 15160, 15425
1130-1200	Asia/Pac	5985p, 6110, 9645, 9760, 11705, 11720p, 15425
1200-1230	Asia/Pac	6110, 9645, 9760, 11705, 11715p, 15425
1200-1400MF	Eu/NAf	1197
1230-1300	Asia/Pac	6110, 9645, 9760, 11705, 11715p, 15425
1300-1330	Asia/Pac	6110, 9355, 9645, 9760, 11705, 11715p, 15425
1330-1400	Asia/Pac	6110, 9355, 9645, 9760, 11705, 15425
1400-1500	Asia/Pac	1143, 6110, 7125, 7215, 9645, 9760, 11705, 15395, 15425
1400-1500	Eu/NAf	1197MF, 15255
1400-1500	ME/CAs	15205
1400-1500	SAs	7125, 7215, 9645, 15205, 15395
1500-1530	Asia/Pac	7125, 7215, 9645, 15395
1500-1530	Eu/NAf	1197
1500-1700	SAs	7125, 7215, 9575, 9645, 15205, 15395
1500-1800	ME/CAs	9575, 15205
1530-1600	Asia/Pac	7125, 7215, 9645, 15395
1600-1700	Asia/Pac	6110, 7125, 7215, 9645, 9760, 15395
1000-1100	Carib	6165, 7405, 9590
1600-1700	Af	909, 1530, 6035, 11920, 12040, 13710, 15225, 15410
1630-1700	Eu/NAf	1197
1700-1730	Af	909, 1530SS, 11920, 12040, 15410, 15445, 17895
1700-1800	SAs	1575MF, 7125, 7215, 9645, 9760, 15205, 15395
1700-1800	Asia/Pac	1143MF, 1575MF, 5990MF, 6045MFp, 6110, 7125, 7215, 9525MF, 9645, 9670MF, 9795MF, 11955MF, 12005MF, 15255MFp, 15395
1700-1900	Eu/NAf	6040
1730-1800	Af	909MF, 11920, 12040, 15410, 15445, 17895
1800-1830	Af	909, 6035, 7150MF, 9845MF, 11920, 11975, 13710, 15410, 15445MF, 15580
1800-1900	Asia/Pac	13820
1800-1900	SAs	9760, 13820
1800-2000	ME/CAs	9760
1830-1900	Af	909, 6035, 11920, 11975, 13710, 15410, 15580
1900-2000	Af	909, 4950Sa, 6035, 7415, 11920, 11975, 13710, 15410, 15580
1900-2000	Asia/Pac	9525, 11870, 15180
1900-2000	Eu/NAf	1197
1900-2000	SAs	9760
2000-2030	Af	909, 1530, 4950, 6035, 7415, 11855, 11975, 13710, 15410, 15580, 17725, 17755
2000-2100	ME/CAs	6095, 9760
2000-2200	SAs	6095, 9760
2030-2100	Af	909, 1530, 4950SS, 6035, 7415, 11975, 13710, 15410, 15580, 17725, 17755
2030-2100	Eu/NAf	1197
2100-2130	Af	909, 1530, 6035, 7415, 11975, 13710, 15410, 15580, 17725
2100-2200	ME/CAs	6095
2100-2200	Asia/Pac	11870, 15185, 17735
2130-2200Su-Fr	Af	909, 1530, 6035, 7415, 11975, 13710, 15410, 15580, 17725
2200-2230	Asia/Pac	7215, 9890, 9770, 11760, 15185, 15290, 15305, 17735p, 17820
2200-2230MF	Af	909, 1530, 6035, 7415, 11975, 12080, 13710
2230-2400	Asia/Pac	1575, 7215, 9760(until, 2330), 9770, 9890, 11760, 15185, 15290, 15305, 17735p, 17820

Farsi

English	Area	kHz
0300-0400	ME	1548, 6060, 7200, 9435
0400-0430	ME	6060, 7200, 9435
1700-1800	ME	648, 1548, 6160, 9680, 11835
1800-1900	ME	1548, 6160, 9680, 11835

French

English	Area	kHz
0530-0600MF	Af	1530, 4960, 5745, 6120, 7265, 7370, 9480, 9505, 11915, 13710, 15375
0600-0630MF	Af	4960, 5745, 6120, 7265, 7370, 9480, 9505, 11915, 13710, 15375
1100-1130MF	Af	13675, 15510, 17650, 17750, 21705
1830-1900	Af	1530, 4950, 9780, 9815, 11775, 12080, 15220, 15365, 17640, 17800
1900-2000	Af	1530, 9780, 9815, 11775, 12080, 15220, 15365, 17640, 17800
2000-2100	Af	9780, 9815, 11775, 12080, 15220, 15365, 17640
2100-2130MF	Af	4950, 5985, 9780, 9815, 11775, 12080, 15220, 15365, 17640, 17755

Georgian

English	Area	kHz
1430-1500	Eu	9745, 15125, 15215
1630-1700	Eu	7175, 9885, 12005

Hausa

English	Area	kHz
0500-0530	Af	1530, 4960, 6120, 7415, 9885
1500-1530	Af	7135, 9540, 11880
2030-2100MF	Af	4950, 9780, 9815, 11775, 12080, 15220, 15365, 17640

Hindi

English	Area	kHz
0030-0100	SAs	6170, 7275, 9505
1600-1700	SAs	6060, 9595, 11720

Indonesian

English	Area	kHz
1130-1230	EAs	7215, 7255, 9620, 9720, 11930
2200-2330	EAs	7130, 9535, 9620, 11805, 15205

Khmer

English	Area	kHz
1330-1500	EAs	1575, 5955, 9885, 11930
2200-2230	EAs	1575, 6060, 7260, 9760

Kinyarwanda/Kirunda

English	Area	kHz
0330-0400SS	Af	5975, 7145, 7340
0400-0430	Af	5975, 6170, 7145, 7340

Korean

English	Area	kHz
1300-1400	EAs	648, 6030, 7235, 11795, 11895
2130-2200	EAs	6160, 7110, 7300

Kurdish

English	Area	kHz
1600-1700	ME	6160, 7255, 11855

Lao

English	Area	kHz
1230-1300	EAs	1575, 6030, 7215, 11930
1300-1330	EAs	1575

Pashto

English	Area	kHz
0130-0145	ME	6025, 9505, 9585
1430-1515	ME	9850, 11725, 11990

Polish

English	Area	kHz
2100-2300	Eu	1197

Portuguese

English	Area	kHz
0430-0500	Af	1530, 5745, 5975, 6090, 6170, 7145, 7370, 9480, 9675
1700-1730MF	Af	1530, 6035, 13600, 15505
1730-1800	Af	909SS, 1530, 6035, 7290, 9815, 11775, 13600, 15505, 17785, 21485
1800-1830MF	Af	1530, 7290, 9815, 11775, 13600, 15505, 17785, 21485

Romanian

English	Area	kHz
0430-0500	Eu	6120, 7155, 9635

Russian

English	Area	kHz
0900-1100	CIS	9520, 9625, 9855, 11885, 11930E, 11970, 15205, 15370, 15410, 15445
1400-1500	CIS	7220, 9615, 11895, 11945, 15445, 17805
1800-2000	CIS	6105, 7220, 9520, 9660, 9670, 11770

Serbian

English	Area	kHz
0430-0500	Eu	756, 792, 1395, 6060, 6130, 7130
0630-0700	Eu	1458, 6035, 6125, 7295
1800-1830	Eu	792, 6060, 9695, 11750
2030-2100	Eu	792, 1593, 7180, 9600, 9690
2200-2230	Eu	756, 1593, 7195, 9530, 9725

Slovak

English	Area	kHz
0500-0515	Eu	1197
1800-1830	Eu	1197, 6020, 7245, 9645
2045-2100	Eu	1197

Spanish | **Area** | **kHz**
| 1200-1230 | CAm | 7??0C, 11890C, 12025, 13770, 15265, 15??0, 17875 |

Spanish	**Area**	**kHz**
1200-1230	CAm	7??0C, 11890C, 12025, 13770, 15265, 15??0, 17875
2300-2400	CAm	153??C, 1580C, 9480, 9515C, 12025, 1375??, 15350
Swahili		
1630-1700MF	Af	7290, 9820, 11765, 12080, 13670, 17705, 17785
1700-1730	, Af	7290, 9820, 11765, 12080, 13670, 17705, 17785
Tibetan		
0000-0100	EAs	6015, 7200, 7255, 9555
0500-0700	EAs	15540, 17630, 17770, 21560
1400-1500	EAs	6015, 7290, 9635, 12025, 12040
Tigrigna		
1830-1845MF	Af	7150, 9845, 15445
Turkish		
1800-1900	Eu/ME	792, 6160, 9750, 11855
Ukrainian		
0500-0600	CIS	6170, 7245, 9695
2100-2200	CIS	7180, 9565, 11875
Urdu		
0100-0130	SAs	6025, 7255, 9505
1330-1430	SAs	9510, 9850, 11715
Uzbek		
1500-1530	CIS	9745, 11835, 11850
1530-1545SS	CIS	9745, 11835, 11850
Vietnamese		
1230-1330	EAs	1143, 9890, 12030, 15145
1500-1600	EAs	5955, 6085, 7150, 9725
2230-2330	EAs	6060, 7260, 9760

Note: MF=Mon-Fri local time, SS=Sat/Sun local time.
ANN: E: At start and end of transmission period on each frequency: "This is the Voice of America, Washington DC, signing on/off". Before all foreign language prgrs: "This is the Voice of America. The following program is in ... (language)".
IS: Musical signature, "YANKEE DOODLE", is played at the beginning & conclusion of transmissions & within prgrs.
PUB: Prgr. schedule available upon request to persons outside the United States — **V.** by QSL-card.
VOA Express, a 24-hour English-language service of contemporary music, news, and features, is available on cable systems and FM stations in many cities of Europe and other continents. For more information, write to: VOA Express, Voice of America, Washington, D.C. 20547 USA.

RADIO MARTI (Gov.)
Radio Marti is the Cuban radio broadcasting service of the United States Information Agency, International Broadcasting Bureau (see Voice of America).
✉ Office of Cuba Broadcasting, 5325 N.W. 77th Avenue, Miami, Fla. 33166. ☎ +1(305) 994-1720. 🖷 +1(305) 597-4665. **E-Mail:** ocb@usia.gov. **Gopher:** gopher://gopher.voa.gov:70/11/marti
L.P: Ag. Dir. Office of Cuba Broadcasting & Dir. Radio Marti: Dr. Rolando Bonachea. Dir. of Tech. Op's (OCB): Michael Pallone. Dir Prgrs: Oscar Barcelo. Dir. N: Margarita Rojo. Ag. News Director: Ligia Guillen.
MW: Marathon, FL (G.C.: 81.06W/24.41N): 1180kHz 2 x 50kW (only one tr. normally used) — **SW:** see Voice of America
kHz: 5890, 6030, 7365, 7405, 9565, 9825, 11815, 11930, 13820, 15330

Spanish	**kHz**
0000-0300	6030, 7365
0300-0400	6030, 7365, 7405
0400-0500Tues-Sun	6030, 7405
0500-0900Tues-Sun	6030
0900-1000Tues-Sun	5890
1000-1200	5890, 6030
1200-1400	7405, 9565
1400-1500	11930, 13820
1500-1700	11815, 11930, 13820
1700-1800	9825, 11815, 11930, 13820
1800-2200	9825, 11930, 13820
2200-2300	11930, 13820, 15330
2300-2400	6030, 13820, 15330

RADIO FREE ASIA (Gov.)
✉ 2025 M Street NW, Washington, DC 20036. ☎ +1 (202) 530 4900. 🖷 (Tech. Op's Dept): +1 (202) 822 6234. **Web:** http://www.rfa.org
L.P: Pres: Richard Richter. VP Admin & Management: Craig Perry. VP

Programming/Exec. Editor: Dan Southerland. Chief Financial Officer: Patrick Taylor. Dir. Tech. Op's: David Baden.
SW: RFA hires airtime on tr's in various Asian and Pacific locations, and also uses tr's located within U.S. territory.

Burmese	**Area**	**kHz**
0030-0130	Myanmar	11580, 11600
1500-1600	Myanmar	11530, 11590
Korean		
1530-1630	Korea	9980, 15660
2200-2300	Korea	7460, 9395, 9455, 11785
Laotian		
1130-1230	Laos	15170, 17805
2200-2300	Laos	5930, 9365, 9725, 9940
Mandarin		
1500-1630	China	9445, 9905, 11540, 11615, 11945, 11955
1630-1700	China	9455, 9905, 11540, 11615, 11945, 11955
1700-1800	China	9455, 9905, 11540, 11945, 11955
2100-2200	China	9395, 9420, 9725, 11765, 15515
2300-2400	China	9395, 11785, 13800, 15515
Tibetan		
1300-1400	Tibet	11575, 11590
2300-2400	Tibet	7410
Vietnamese		
1400-1500	Vietnam	9455, 9910, 9930, 11540, 11590
2330-0030	Vietnam	9975, 11580, 13710

V. by QSL-card — **F.PI:** Khmer & Cantonese, expanded Chinese sce.

BBC RELAYS VIA VOA TRANSMITTERS
kHz (Delano): 5995, 6130, 6175, 9515, 9590, 9670, 17715
kHz (Greenville): 5975, 15315
D.PRGR: see United Kingdom.

KAIJ INTERNATIONAL (Rlg.)
✉ Two If By Sea Broadcasting Corp, 22720 S.E. 410th Str, Enumclaw, WA 98022. 🖷 +1 (818) 606 1254.
Tr. site: RR #3, P.O. Box 120, Frisco, TX 75034.
SW: KAIJ, Denton, TX (G.C: 96.52W/33.13N): 100/50kW.

Times	Area	kHz	kW
0000-1400	WNAm/FE	5810	100
0100-1400	NAm/CAm	9815	50
1400-2400	NAm/CAm/FE	13815	100
1400-2400	NAm/CAm	15725	50
2300-0100	NAm/CAm	13740	50

V. by QSL-card. Rp.

RADIO STATION KJES (Rlg.)
✉ The Lord's Ranch, Star Route Box 300, Mesquito, NM 88048. ☎ +1 (505) 233 2090. 🖷 +1 (505) 233 3019 — **L.P:** SM: Michael Reuter.
SW: 9.6km Ea. of Vado (G.C: 106.35W/32.08N): KJES 50kW

Times	Area	kHz	Beam
0200-0300	wCAN	7555	335
0300-0330	wCAN	7555	20
1400-1500	eCAN	11715	70
1500-1600	wCAN	11715	350
1600-1700	MEX	11715	150
1900-2000	AUS	15385	270
2100-2200	PTR	15385	100

V. by letter. Rp.

KTBN INTERNATIONAL RADIO (Rlg.)
✉ P.O. Box A, Santa Ana, CA 92711, USA. ☎ +1 (714) 832 2950. 🖷 +1 (714) 730 0661. **E-Mail:** tbntalk@aol.com
Web: http://www.tbn.org/ktbn.html
L.P: Dr. Paul F.Crouch. PD: Barry Phaeler. CE: Ben Miller. Head of PR: Rod Henke. Freq. Consultant: George Jacobs.
SW: KTBN, Salt Lake City, UT (G.C: 112.03W/40.39N): 100kW.

Times	Area	kHz	Times	Area	kHz
0000-1600	NAm	7510	1600-2400	NAm	15590

V. by QSL-card. Rp. Re. to Chris Diane Hiser, 2442 Michelle Drive,.Tustin, CA 92680 , USA.

KVOH (Rlg.)
✉ High Adventure Ministries Inc, Box 100, Simi Valley, CA 93062. ☎ +1 (805) 520 9460. 🖷 +1 (805) 520 7823. **E-mail:** kvoh@themall.net
Web: http://www.highadventure.org/voh_amer.html
L.P: Freq. Consultant: George Jacobs.
SW: KVOH Rancho Simi, CA (G.C: 118.38W/34.15N): 2 x 80kW.

Times	Area	kHz
0000-1600	NAm/Carib	9975

Times	Area	kHz
1600-2400	CAm/Carib	17775
1800-2400	Eu/NAf	7520

Spanish: 1400-1900, 2100-0200. Other times in English.
V. by QSL-card or letter.

WGTG (Rlg.)

✉ P.O. Box 1131, Copper Hill, TN 37517. ☎ +1 (315) 443 9237.
LP: Owner & CE: Dave Frantz — **SW:** 2 miles S of McCaysville, GA (G.C: 84.21W/34.57N): WGTG 50kW

Times	Area	kHz
1000-2200	NAm	9400
2200-0600	NAm	5085

V. by QSL-card.

RADIO MIAMI INTERNATIONAL (Comm.)

✉ P.O. Box 526852, Miami, FL 33152. ☎ +1 (305) 267 1728. 🖷 +1 (305) 267 9253. **E-mail:** wrmi@compuserve.com.
Web: http://home.nexus.org/WRMI/
LP: GM: Jeff White. CE: Indalecio Espinosa.
SW: WRMI (G.C: 80.21W/25.54N): 9955kHz 50kW (beam 160°).
D.PRGR: W 1045-0630& Sun 1200-0430 in **English & Spanish** exc.
Dutch: Thurs 1415-1430. **French:** Wed 1415-1430. **German:** Sun 0000-0030 (Mon UTC). **Portuguese:** Mon/Wed 2330-2400, Sun 2200-2230 —
V. by QSL-card. Prgr. schedules free on request.

WEWN - WORLDWIDE CATHOLIC RADIO (Rlg.)

✉ P.O. Box 100234, Birmingham, AL 35210. ☎ +1 (205) 672 7200.
🖷 +1 (672) 9988. **E-mail:** wewn@msn.com.
Web: http://www.ewtn.com/WEWN/radio1.htm
LP: SM: Richard Jones. Freq. Mgr: Joseph A. Dentici.
SW: Vandiver (G.C: 86.28W/33.30N): WEWN 4 x 500kW tr's.
kHz: 5825, 7425, 9455, 11875, 13615, 15375, 15665, 15745, 17695

English	Area	kHz
0000-0500	NAm	5825
0600-1400	NAm	5825
1200-1400	Eu	15745
1400-1600	NAm	9455, 11875
1400-1900	Eu	15665
1600-2200	NAm	11875, 13615
1900-2200	Eu	17695
2200-2400	NAm	5825, 13615
2200-1200	Eu	5825
Spanish		
0500-0600	NAm	5825
1100-1300	LAm	7425
1300-2300	LAm	15375
2300-0600	LAm	7425

V. by QSL-card. Rp (3 IRC's).

WHRI - WORLD HARVEST RADIO (Rlg. Comm.)

✉ P.O. Box 12, South Bend, IN 46624. ☎ +1 (219) 291 8200. 🖷 +1 (219) 291 9043. **E-mail:** whri@lesea.com **Web:** http://www.whri.com.
LP: GM: Peter Sumrall. Sales Mgr: Tom Lewis. CE: Douglas W. Garlinger.
Freq. Consultant: George Jacobs.
SW: WHRI, Noblesville, IN (G.C: 85.57W/40.01N): 2 x 100kW tr's —
KWHR, Hawaii (see listing under Hawaii).
kHz: 5745, 6040, 7315, 9495, 13760, 15105

Times	Area	kHz	Times	Area	kHz
0000-1000	CAm	15375	1500-2200	eCAN/Eu	13760
1000-1500	eCAN	6040	1800-2400	CAm	9495
1000-1100	CAm	9495	2200-1000	eCAN/Eu	5745
1300-1800	CAm	15105			

Russian: Sun 1815-1830 (Eu). **Spanish:** Sat 2300-2330.
IS: "Onward Christian Soldiers"— **ANN:** "This is World Harvest Radio, the International Sce. of LeSea Broadc. Corp — **V.** by QSL-card. Reports & Monitors wanted — **PUB:** World Harvest magazine.

WMLK (Rlg.)
Operated by the Assemblies of Yahweh

✉ P.O. Drawer C, Bethel, PA 19507. ☎ +1 (717) 933 4518.
Web: http://www.AssembliesofYahweh.com/log.htm
LP: Dir: Elder Jacob O. Meyer. CE: Gary McAvin.
SW (G.C: 76.17W/40.29N): 1 x 50kW tr.

Times	Area	KHz
0400-0900a	Eu/ME	9465
1700-2200a	Eu/ME	9465

a) exc. Sat — **V.** by QSL-card. Rp — **F.PL:** 100kW.

WORLD INTERNATIONAL BROADCASTERS INC. (Comm.)

✉ P.O. Box 88, Red Lion, PA 17356-0088. ☎ +1 (717) 246 1681. 🖷 +1 (717) 244 9316 — **LP:** Freq. Consultant: George Jacobs.
SW: WINB (G.C: 76.34W/39.54N): 2 x 50kW tr's.

Times	Area	kHz	Times	Area	kHz
2200-0400	CAm	11950	1700-2200	Eu/Af	13790

D.PRGR: Mostly English rlg. prgrs. and music
V. by QSL-card and sched.

WORLD WIDE CHRISTIAN RADIO (Comm.)

✉ 1300 WWCR Avenue, Nashville, TN 37218. ☎ +1 (615) 255 1300. 🖷 +1 (615) 255 1311. **Toll Free Number:** 1-800-238-5576.
E-mail: wwcr@aol.com **Web:** http://www.wwcr.com
LP: Pres: Fred P. Westenberger. GM: George McClintock. Op's Mgr: Adam W.Lock, Sr.
SW: WWCR (G.C: 86.53W/36.12N): 4 x 100kW
kHz: 2390, 3210, 3215, 5070, 5935, 7435, 9475, 12160, 13845, 15685

To Feb 28	From March 1	Area	kHz	Tr
0500-1100**	0500-1100**	Eu/Af	3210	1
1100-2100	1100-2200	Eu/Af	15685	1
To Feb 28	**From March 1**	**Area**	**kHz**	**Tr**
2100-2300	2200-2400	Eu/Af	9475	1
2300-0505**	0000-0500**	Eu/Af	3215	1
1400-2400	1300-0100	Af	13845	2
0000-1400	0100-1300	Af	5935	2
1400-2200	1400-2200	eCAN/Eu	12160	3
2200-1400	2200-1400	eCAN/Eu	5070	3
0200-1200	0300-1200	ENAm	2390	4
1200-1400	1200-1400	ENAm	7435	4
1400-2100	1400-2200	ENAm	9475	4
2100-0200	2200-0300	ENAm	7435	4

**) change from 3215 to 3210 at 0505 Mon-Fri, 0500 Sat/Sun.
D.PRGR: 24h (different prgrs.on each tr.).
V. by QSL-card. Rp preferred (SAE or one IRC). Rec.acc.

WRNO WORLDWIDE (Comm.)

✉ 4539 I-10 Service Rd. N, Metairie, LA 70002. (**Postal Addr:** P.O. Box 100, New Orleans, LA 70181). ☎ +1 (504) 889 2424. 🖷 +1 (504) 889 0602.
Web: http://www.wrnoworldwide.com
LP: CE: Jack Bruce. Head of PR: David Schneider. Freq. Consultant: George Jacobs.
SW: WRNO, Marrero, LA (G.C: 90.07W/29.50N): 1 x 100kW.

kHz	Area	Times	kHz	Area	Times
7355	ENAm/Eu	2300-0400	7395	ENAm/Eu	0400-0900
7355	CAN/Eu	0900-1600	15420	ENAm/Eu	1600-2300

V. by QSL-card. Rp.(2 IRC's).

CHRISTIAN SCIENCE PUBLISHING SOCIETY - SHORTWAVE BROADCASTS (Rlg.)

✉ P.O. Box 1524, Boston, MA 02117-1524. ☎ +1 (617) 450 2929. **Toll Free in US:** (800) 288 7090. With either number, extension 2060 to hear recorded frequency information, or 2929 for Shortwave Helpline to request printed schedules and information.
E-Mail: Letters and reception reports: letterbox@csps.com
Religious questions: Sentinel@csps.com
Web: http://www.tfccs.com/GV/shortwave/shortwave_schedule.html
LP: Dir. Int. Broadc, Herald Broadc. Syndicate: Catherine Aitken-Smith.
Freq. & Production Coordinator: Tina Hammers.
SW: WSHB, Cypress Creek, SC (G.C: 81.07W/32.41N): 3 x 500kW.
✉ **WSHB Cypress Creek**, Rt. 2, Box 107A, Pineland, SC 29934 (for technical feedback on WSHB transmissions only. Other correspondence to boston addr.). ☎ +1 (803) 625 5555 (general), +1 (803) 625 5551 (SM), +1 (803) 625 5554 (Eng.). 🖷 +1 (803) 625 5559.
LP: Senior SM: C. Ed Evans (**E-mail:** cee@csms.com). Chief Tr./Sce. Engineer: Antonio L. (Tony) Kobatake (**E-mail:** tony@csms.com). QSL Co-ordinator: Judy P. Cooke (**E-mail:** judy@csms.com).
kHz: 5850, 6095, 7395, 7535, 9430, 9455, 9840, 11550, 13770, 17510, 18930

Times	Area	kHz	Area	kHz
0000-0200	CAm/SAm	9430	NAm	7535
0200-0300	NAm	5850	Mexico	9430
0300-0400	WNAm	5850	Af	7535
0400-0500	Eu	7535	Af	9840
0500-0800	Eu	7535		
0800-0900	Eu	7535	Oce/INS	11550
0900-1000	Eu	7535	CAm/SAm	7395
1000-1200	NAm/Carib	6095	SAm	7395
1200-1300	NAm/Carib	6095	CAm/SAm	9455

Times	Area	kHz	Area	kHz
1300-1400	NAm	6095	Mexico	9455
1600-1800			Af	18930
1800-1900	Eu	11550	Af	18930
1900-2000	EEu	11550	Af	17510
2000-2200	Eu	7510	Eu	5850
2200-2300	Eu	7510	SAm	13770
2300-2400	Eu/Af	7510	SAm	13770

D.PRGR: Mon-Fri News programming in English. Sat/Sun religious programming in English, French, German, Portuguese, Russian and Spanish.
SW: KHBI, Saipan (see No. Mariana Islands).

WJCR WORLDWIDE (Rlg.)
World Wide Gospel Radio Inc., P.O. Box 91, Upton, KY 42784. ☎ +1 (502) 369 8614.
Web: http://www.mindspring.com/~brunner/wjcr.html
L.P: Pres: Pastor John Powell.
SW: WJCR, 3.2km SE of Millerstown, KY (G.C: 86.02W/37.26N): 2 x 50kW tr's — **D. PRGR:** 24h on 7490kHz (English) and 13595kHz (12h English and 12h Chinese).

WVHA (Rlg.)
P.O. Box 1844, Mt. Dora, FL 32751. ☎ +1 (904) 735 1844. 🖹 +1 (904) 735 4055 — **SW:** 2.75 miles SE of Greenbush, ME (G.C: 68.33W/45.08N): 1 x 500kW (r. using 250kW).
kHz: *5850, 7465, 9930, 11580, 13825, 15745*

Times	Area	kHz			
0200-0400	Eu/NAf	7465	1100-1800	Eu/NAf	15745
0400-0800	Eu/NAf	5850	1800-1900	Eu	11580
0800-1000	Eu/NAf	7465	1900-2200	Eu	9930
1000-1300	Eu	13825	2200-0200	Eu/NAf	5850

WYFR - FAMILY RADIO (Rlg.)
Family Stations Inc., 290 Hegenberger Rd, Oakland, CA 94621. ☎ +1 (510) 568 6200. 🖹 +1 (510) 633 7983. ① 703259 FAMILY RAD.UD. **E-mail:** shortwave@familyradio.com
Web: http://www.familyradio.com/wyfr.htm
L.P: Pres: Harold Camping. Dir. Tec: Wesley D. Becker. Prgr. Mgr: Thomas A. Schaff. Dir. Int. Rel: Richard H.Homeres.
SW: Okeechobee, FL (G.C: 80.56W/27.27N): 100/50kW.
+) via tr. in Taiwan.
kHz: *5810, 5825, 5850, 5850, 5950, 5985, 6065, 6085, 6105, 6175, 6300, 7355, 7435, 7520, 9280, 9355, 9355, 9455, 9505, 9555, 9575, 9605, 9630, 9680, 9690, 9705, 9715, 9955, 9985, 11550, 11580, 11665, 11720, 11725, 11740, 11825, 11830, 11855, 11885, 11970, 13695, 15130, 15170, 15215, 15355, 15400, 15440, 15565, 15695, 17555, 17760, 17845, 21525, 21725*

Arabic	Area	kHz
0500-0600	ME	5850, 9455
1600-1700	ME	15695
1700-1800	ME	17760

English		
0000-0100	NAm	6085, 9505*
0100-0200	IND	11550 +
0100-0400	NAm	6065*, 9505*
0400-0500	NAm	6085, 9505
0400-0500	Eu/Af	9985
0500-0700	NAm	5985
0500-0600	Eu/Af	9985, 11550
0600-0700	Eu/Af	7355, 9985
0700-0800	Eu/Af	7355, 9455, 9985
1000-1100	NAm	5950*
1100-1200	NAm	5950*, 7355
1200-1300	NAm	5950*, 11830*, 11970*
1300-1400	NAm	5950*, 11830*, 11970*, 13695
1400-1500	NAm	5950*, 11830*, 17760*
1500-1600	NAm	11830*, 17760*
1600-1700	NAm	11830*, 15215, 17760*
1600-1700	Eu/Af	15695, 17555, 21525
1610-1810	IND	11550+
1700-1900	Eu/Af	15695, 17555
1900-2000	Eu/Af	17555
2000-2100	Eu/Af	5810, 7355, 15566, 21525
2100-2200	Eu/Af	7355, 11580, 15565
2200-2300	Eu/Af	11580, 15565

*) Network programs Mon-Sat

French		
0600-0700	Eu/Af	5850, 9455, 11580
1000-1100	Carib	9680, 11740
1100-1300	CAN	13695

French	Area	kHz
1800-1900	Eu/Af	17760, 21720, 21525
2000-2100	Eu/Af	5825, 11665
2300-2400	CAN	6085
2300-0100	Carib	15400

German		
0300-0400	Eu	9985
0400-0500	Eu	7355, 11550
0500-0600	Eu	5850, 7355
1700-1800	Eu	15565
1900-2000	Eu	5825, 15565

Hindi		
0000-0100	IND	11550+
1510-1610	IND	11550+

Italian		
0600-0700	Eu	7520, 11550
0700-0800	Eu	5850
1800-1900	Eu	15565
2100-2200	Eu	5825, 11665

Mandarin		
2100-2200	CHN	6300+, 9280+
2200-2400	CHN	6300+, 9280+, 9630+
1102-1602	CHN	6300+, 9280+

Portuguese		
0100-0200	SAm	9690, 11885
0400-0500	Af	11580
0800-0900	SAm	6175, 9605
0900-1000	SAm	6175, 9605, 9575, 9680
1900-2000	Af	21525
2200-2300	SAm	15130
2300-0100	SAm	11885

Russian		
0300-0400	CIS	7435, 9355
0400-0500	CIS	5850
1505-1705	CIS	9955+
1700-1800	CIS	17760, 21725

Spanish		
0000-0500	CAm	9715, 11855
0100-0200	SAm	9605
0100-0300	Carib	5985
0300-0500	Carib	5985, 9985
0500-0600	CAm	9705, 11855
0800-1000	SAm	6105, 9555
1000-1100	Carib	6085
	SAm	6105, 9555, 9575
1100-1200	CAm	9605, 11725
	SAm	9575, 11830
1100-1300	Carib	6085, 11740
1200-1300	CAm	9575, 9605, 11725
1300-1400	Carib	6085
	CAm	11725, 15130, 15355
1400-1500	CAm	15130
2200-2300	Eu	7355
2200-0100	Carib	5985
2300-0100	SAm	15170, 15215, 17845

ANN: "This is your Family Radio, International Broadcast Station WYFR, Okeechobee, Florida, the United States of America". G: "Dies ist Ihr Familien Radio, die Internationale Radio-Station WYFR, in Okeechobee, Florida, der Vereinigten Staaten von Amerika". Sp: "Esta es Family Radio WYFR con el Sonido de la Nueva Vida transmitiendo desde Okeechobee, Florida, y con estudios de produccion ubicados en Oakland, California, Estados Unidos de Norte America".
IS: "To God be the Glory" — **V.** by QSL-card. No tapes.
PUB: International Program and Frequency Guide.

WWBS (Comm./Rlg.)
300 Polar Str, (✉ P.O. Box 18174), Macon, GA 31209. ☎ +1 (912) 745 1485.
L.P: Owner: Charles C. Josey (**E-mail:** charlesK4LBL@june.com)
SW: Macon, GA (G.C: 83.37W/32.49N): 11910kHz 50kW. Beam 30/330þ.
F.PI: 0000-1200 religious prgrs (scheduled to start early 1998).

UNITED NATIONS RADIO
✉ Room S850, New York, NY 10017. 🖹 +1 (212) 963 0765.
Web: http://www.internetbroadcast.com/un.htm (audio sce.)
United Nations radio prgrs. can be heard on radio st's throughout the world. Produced at UN HQ in New York, they comprise transcriptions as well as news an current affairs magazines which by special arrangement

are fed by radio/telephone circuits to a number of international broadcasting organizations, mostly on shortwave.

UZBEKISTAN

RADIO TASHKENT

📧 Khorezm Str. 49, 700047 Tashkent. ☎ +7 (3712) 441210. 📠 +7 (3712) 440021.
L.P: Dir: Gulnor Nadjim-kizi Bobojon.
Stations: Tashkent, exc. *) = Almaty, Kazakhstan, all 100kW.
kHz: *900*, 5035, 5060, 5955, 5975, 6025, 6155*, 6160*, 6230, 6230*, 7105, 7285, 9540, 9715, 11905, 15330*

Arabic	Area	kHz
1700-1730	ME	5975, 6025, 7285
1900-1930	ME	5975, 6025, 7105, 7285, 9540
Chinese		
2330-2400	As	5060
Dari		
0130-0200	ME	5955, 5975, 7285
1520-1550	ME	5975, 6025, 7285
English		
0100-0130	As	5955, 5975, 7285
1200-1230	SAs	5060, 5975, 7285, 9715
1330-1400	SAs	5060, 5975, 7285, 9715
Farsi		
1630-1700	ME	5975, 6025, 7285
1830-1900	ME	5975, 6025, 7105, 7285, 9540
German		
1935-2030	Eu	5035, 5060, 6230, 7105, 9540, 11905
Hindi		
1300-1300	SAs	5060, 5975, 6025, 9715
1430-1500	SAs	5060, 5975, 6025, 9715
Pushtu		
0200-0230	As	5955, 5975, 7285
Turkish		
0400-0430	ME	9540, 15330
1700-1730	ME	7105, 9540
Urdu		
1230-1300	SAs	5060, 5975, 7285, 9715
1400-1430	SAs	5060, 5975, 7285, 9715
Uigur		
1230-1500	As	900, 5060, 6155, 6230
1430-1500	As	900, 5060, 6155, 6230
1530-1600	As	900, 5060, 6155, 6230
0030-0100	As	5060
Uzbek		
0230-0330	ME	5955, 5975, 7105, 7285, 9540
1550-1630	ME	5975, 6025, 7105, 7285, 9540
1730-1830	ME	5975, 6025, 7105, 7285, 9540

Relays via Almaty tr's 1h earlier during summer.
ANN: Uzbek: "Toşkentdan gapiramiz".
V. by QSL-card.

VOICE OF RUSSIA RELAY STATION

SW: Tashkent (G.C: 69.09E/41.13N): 19 x 100kW used by Uzbek Radio, Radio Tashkent, Voice of Russia, Ostankino. V. of Russia transmissions: Cf. Russia — Voice of America, BBC, R. Netherlands relays: Cf. these stations.

VATICAN CITY STATE

RADIO VATICANA (Rlg.)

📧 Vatican Radio, I-00120 Vatican City. ☎ +39 (6) 6988 3551. (**Int. Rel:** +39 (6) 6988 3945). 📠 +39 (6) 6988 4565. ☼ 2023 VA. **Cable:** RADVAT.
E-mail: mc6778@mclink.it
Web: http://www.wrn.org/vatican-radio/
L.P: DG: Rev. Pasquale Borgomeo S.I. PD: Rev. Federico Lombardi S.I. TD: Rev. Eugenio Matis S.I. CE: Pier Vincenzo Giudici. Head of Int. Rel: Mrs. Solange de Maillardoz.
MW: 526kHz 5kW, 1530kHz 300/600kW, 1611kHz 100kW.
FM: 93.3/96.3/103.8/105.0MHz 10kW (stereo).
SW: S.M. Galeria, Italy (G.C: 12.19E/42.03N): 5 x 500kW, 4 x 100kW — Vatican City (G.C: 12.27E/41.54N): 1 x 80kW. 1 x 10kW.
kHz: *4005, 5880, 5945, 5980, 6020, 6095, 6160, 6185, 7250, 7305, 7310, 7335, 7345, 7360, 7365, 9500, 9585, 9600, 9605, 9645, 9660, 9825, 9835, 9850, 9875, 9940, 11625, 11640, 11715, 11740, 11810, 11830, 13765, 15540, 15570, 15585, 15595, 15605, 17550, 17700, 21850*

(Su: 1h earlier exc. transmissions marked * — H) Holy Days.
FM: All listed freqs are intended for local reception in Rome.

Albanian	Area	Days	kHz
0330-0345	Eu	1234567	1530, 4005, FM93.3
1730-1745	Eu	1234567	1530, 4005, 5880, 7250, FM93.3
Amharic/Tigre			
0405-0430	Af	1234567	7360, 9660, FM103.8
1635-1700	Af	1234567	13765, 15570, FM103.8
Arabic			
0500-0530	ME/Af	1234567	7305, 9645, FM96.3
0745-0800	Eu/ME/Af	123456	1530, 5880, 7250, 9645, 11740, 15595, FM93.3
1630-1700	ME/Af	1234567	9645, 11810, FM96.3
2145-2200	Eu/Af	1234567	1530, 4005, 5880, 7250, FM93.3
Armenian			
0310-0330	Eu	1234567	7345, 9545
1650-1710	Eu	1234567	1611, 9585, 11715
Bulgarian			
0540-0600	Eu	1234567	1611, 6185, 7345
1920-1940	Eu	1234567	1611, 6185, 7365, FM96.3
Belorussian			
0420-0440	Eu	1234567	6185, 7345
1800-1820	Eu	1234567	1611, 6185, 7365, 9585, FM96.3
Chinese	**Area**	**Days**	**kHz**
1230-1315*	As	123457	6020, 11625, 13765, 15585, FM103.8
2200-2245*	As	1234567	6160, 7305, 9600, 11830, FM103.8
Croatian			
0400-0415	Eu	1234567	1530, 4005, FM93.3
1800-1815	Eu	1234567	1530, 4005, 5880, 7250, FM93.3
Czech			
0415-0430	Eu	1234567	1530, 4005, 5880, FM93.3
1830-1845	Eu	1234567	1530, 4005, 5880,7250, FM93.3
English			
0140-0200*	As	1234567	5980, 7335, FM103.8
0250-0315*	NAm	1234567	6095, 7305, 9605, FM103.8
0310-0340*	Af	1234567	7360, FM103.8
0500-0530*	Af	1234567	7360, 9660, 11625, FM103.8
0600-0620	Eu	1234567	527, 1530, 4005, 5880, 7250, FM93.3, FM105.0
0630-0700*	Af	1234567	9660, 11625, 13765, FM103.8
1345-1405*	As/Pac	1234567	13765, 15540, FM103.8
1535-1550*	As	1234567	9875, 11640, FM103.8
1715-1730	Eu/As	1234567	527, 1530, 4005, 5880, 7250, 9645, 11810, FM93.3, FM105.0
1730-1800*	Af	1234567	11625, 13765, 15570, FM103.8
2000-2030*	Af	1234567	7365, 9660, 11625, FM103.8
2050-2110	Eu	1234567	527, 1530, 4005, 5880, 7250, FM93.3, FM105.0
2245-2305*	As/Pac	1234567	6160, 7305, 9600, 11830, FM103.8
Esperanto			
2020-2030	Eu	3	1611, 6185, FM96.3
2020-2030	Eu	7+H	527, 1530, 4005, 5880, FM93.3, FM105.0
French			
0230-0250*	Am	1234567	6095, 7305, 9605, FM103.8
0240-0320*	Af	1234567	7360, FM103.8
0430-0500*	Af	1234567	7360, 9660, 11625, FM103.8
0540-0600	Eu	1234567	527, 1530, 4005, 5880, 7250, FM93.3, FM105.0
0600-0630*	Af	1234567	9660, 11625, 13765, FM103.8
1130-1150	Eu	6	11740, FM93.3
1700-1715	Eu/As	1234567	527, 1530, 4005, 5880, 7250, 9645, 11810, FM93.3, FM105.0
1700-1730*	Af	1234567	13765, 15570, FM103.8
2030-2050	Eu	1234567	527, 1530, 4005, 5880, 7250, FM93.3, FM105.0
2030-2100*	Af	1234567	7365, 9660, 11625, FM103.8
German			
0520-0540	Eu	1234567	1530, 4005, 5880, 7250, FM93.3
1500-1515	Eu	1234567	1530, 5880, 7250, 9645, FM93.3
1920-1940	Eu	1234567	1530, 4005, 5880, 7250, FM93.3
Hindi			
0040-0100*	As	1234567	5980, 7335, FM103.8
1450-1505*	As	1234567	9875, 11640, FM103.8

Hungarian	Area	Days	kHz
0445-0500	Eu	1234567	1530, 4005, 5880, FM93.3
1815-1830	Eu	1234567	1530, 4005, 5880, 7250, FM93.3
Italian			
0620-0630	Eu	1234567	527, 1530,4005, 5880, 6185, 7250, FM93.3, FM105.0
1115-1130	Eu	7	527, 1530, 5880, FM93.3, FM105.0
1300-1330	Eu/Af/As	1234567	527, 1530, 5880, 9645, 11740, 15595, 17550, FM93.3, FM105.0
1530-1600	Eu	5	527, 1530, 5880, 7250, 9645, FM93.3, FM105.0
1630-1700	Eu	1234567	527, 1530, 5880, 7250, FM93.3, FM105.0
2000-2030		123456	527, 1530, 4005, 5880, FM93.3, FM105.0
2000-2020	Eu	7	527, 1530, 5880, 7250, FM93.3, FM105.0
2200-2310	Eu	1234567	527, 1530, 4005, 5880, FM93.3, FM105.0
Japanese			
1315-1345*	As	1234567	9875, 11625, FM103.8
2130-2200*	As	1234567	6160, 7310, 9600, FM103.8
Latvian			
0500-0520	Eu	1234567	6185, 7345
1840-1900	Eu	1234567	1611, 6185, 7365, 9585, FM96.3
Lithuanian			
0440-0500	Eu	1234567	6185, 7345
1820-1840	Eu	1234567	1611, 6185, 7365, 9585, FM96.3
Malayalam			
0120-0140*	As	1234567	5980, 7335, FM103.8
1520-1535*	As	1234567	9875, 11640, FM103.8
Polish			
0500-0520	Eu	1234567	1530, 4005, 5880, 7250, FM93.3
1515-1530	Eu	1234567	1530, 5880, 7250, 9645, FM93.3
1900-1920	Eu	1234567	1530, 4005, 5880, 7250, FM93.3
Portuguese			
0030-0100*	SAm	1234567	7305, 9605, FM96.3
0530-0600*	Af	1234567	7360, 9660, 11625, FM103.8
1000-1030*	SAm	1234567	17700, 21850, FM103.8
1415-1430	Eu	1234567	1530, 9645, 11740, FM93.3
1800-1830*	Af	1234567	11625, 13765, 15570, FM103.8
2130-2145	Eu	1234567	1530, 4005, 5880, 7250, FM93.3
Romanian			
0520-0540	Eu	1234567	1611, 6185, 7345
1900-1920	Eu	1234567	1611, 6185, 7365, FM96.3
Russian			
0330-0400	Eu	1234567	6185, 7345, 9545
1330-1400	Eu	1234567	11740, 15595, FM96.3
1710-1740	Eu	1234567	1611, 7365, 9585, 11715, FM96.3
2100-2120	Eu	1234567	6160, 7305, FM96.3
Scandinavian Sce.			
0600-0630	Eu	1234567	1611, 6185, 7345
2000-2020	Eu	1234567	1611, 6185, 7250, FM96.3
Slovak			
0430-0445	Eu	1234567	1530, 4005, 5880, FM93.3
1845-1900	Eu	1234567	1530, 4005, 5880, 7250, FM93.3
Slovene			
0345-0400	Eu	1234567	1530, 4005, FM93.3
1745-1800	Eu	1234567	1530, 4005, 5880, 7250, FM93.3
Somali			
1620-1635	Af	6	13376, 15570, FM103.8
Spanish			
0100-0145*	LAm	1234567	5945, 7305, 9605, FM96.3
0145-0230*	LAm	1234567	5945, 7305, 9605, FM96.3
0315-0400*	LAm	1234567	6095, 7305, 9605, FM96.3
1130-1200*	LAm	1234567	17700, 21850, FM103.8
1150-1210	Eu	6	11740, FM93.3
1400-1415	Eu	1234567	1530, 9645, 11740, FM93.3
1900-1930	Af	6	9660, 11625, FM103.8
2110-2130	Eu	1234567	527, 1530, 4005, 5880, 7250, FM93.3, FM105.0
Swahili			
0340-0405*	Af	1234567	7360, 9660, FM103.8
1610-1635*	Af	123457	13765, 15570, FM103.8
Tamil			
0100-0120*	As	1234567	5980, 7335, FM103.8
1505-1520*	As	1234567	9875, 11640, FM103.8

Ukrainian	Area	Days	kHz
0400-0420	Eu	1234567	6185, 7345
1740-1800	Eu	1234567	1611, 6185, 7365, 9585, FM96.3
Vietnamese			
1405-1450*	As	1234567	13765, 15540, FM103.8
2315-2400*	As	1234567	6160, 7305, 9600, FM103.8

OTHER BROADCASTS

News in Italian/French/English			
0700-0745	Eu/As	123456	527, 1530, 4005, 5880, 6185, 7250, 9645, 11740, 15595, FM93.3, FM105.0
1100-1200	Eu/As	123456	527, 1530, 5880, 9645, 11740, 15595, 17550, FM93.3, FM105.0
Music			
1530-1600	Eu	12346	527, 1530, 5880, 7250, 8645, FM93.3, FM105.0
2310-0140	Eu	123467	527, 1530, FM93.3, FM105.0
Papal Audience			
1000-1100	Eu	3	527, 5880, FM105.0

PRAYERS & LITURGY

Angelus			
1100-1115	Eu/As	7+H	527, 1530, 5880, 9645, 11740, 15595, 17550, FM93.3, FM105.0
Holy Rosary			
1940-2000	Eu/Af/As	1234567	527, 1530, 1611, 4005, 5880, 6185, 7250, 7365, 9660, 11625, FM93.3, FM96.3, FM 103.8, FM105.0
Mass in Arabic			
0930-1100	Eu	7+H	1530, 11740, 15595, 17550, FM93.3
Mass in Chinese			
1230-1315	As	6	6020, 11625, 13765, 15585, FM103.8
Mass in English			
1130-1200	Af/As	5	15595, 17550, FM103.8
1550-1620	Af/As	6	9875, 11640, FM93.8
Mass in Italian			
0830-0930	Eu	7+H	527, 1530, 7250, FM93.3, FM105.0
Mass in Latin			
0630-0700	Eu/Af/As	1234567	527, 1530, 4005, 5880, 6185, 7250, 7305, 9645, 11740, 15595, FM93.3, FM105.0
Mass in Ukrainian			
0715-0845	Eu	7+H	1611, 9825, 11740, FM96.3
Oriental Liturgy			
0930-1030	Eu	1	1530, 11740, 15595, 17550, FM93.3
Romanian Liturgy			
0710-0830	Eu	7+H	1530, 5880,7250, 9645, FM 93.3
Vespers			
1600-1630	Eu	12345	527, 1530, 5880, 7250, 9645, FM93.3, FM105.0

ANN: Before all transmissions: Latin: "Laudetur Jesus Christus" (Praised be Jesus Christ), repeated in the language of the broadcast, then station identification — **IS:** "Christus Vincit".
V. by QSL-card.

VENEZUELA

RADIO NACIONAL DE VENEZUELA
✉ P.O. Box 3979, Caracas 1010A.
L.P: DG: Maritza Esparragoza. Dir. Foreign Sce: Gregorio Montiel Cupello.
GM Eng: Ing. A. Miranda B.
SW: 9540vkHz 25kW.

Spanish	Area	kHz
0000-0100	Am	9540v
0300-0400	Am	9540v
1100-1200	Am	9540v
1400-1500	Am	9540v
1800-1900	Am	9540v
2100-2200	Am	9540v

VIETNAM

VOICE OF VIETNAM
✉ 45 Ba Trieu Str, Hanoi. ☎ +84 42 57870. 🖷 +84 42 55765.
① 412256. **Web:** http://www.ioit.ac.vn/tieng_noi_vn/tnvn.html
LP: Dir: Dao Dinh Tuan.
MW: Hanoi 1008kHz — **SW:** (G.C: 105.52E/20.59N):
kHz: 5940, 7270, 7285, 7400, 9730, 9840, 10010, 12020, 12035, 15010

Cambodian	Area	kHz
0000-0030	As	9730, 7285
1200-1230	As	9730, 7285
1330-1400	As	9730, 7415, 1008
Cantonese		
1230-1300	CHN	12035, 10010, 1008
1330-1400	CHN	12035, 10010, 1008
1430-1500	CHN	9730, 7285
1530-1600	CHN	15010 (16/10-15/2: 12020), 9840, 7400, 7270
2230-2300	CHN	15010 (16/10-15/2: 12020), 9840, 7400, 7270, 7400, 7270, 1008
Chinese		
0830-0930	CHN	15010 (16/10-15/2: 12020), 9840, 7400, 7270
1200-1230	CHN	12035, 10010
1300-1330	CHN	12035, 10010, 1008
1500-1530	CHN	9730, 7285
2200-2230	CHN	15010 (16/10-15/2: 12020), 9840, 7400, 7270, 7400, 7270 1008
English		
0200-0400	NAm	5940 (via Russian tr.)
1000-1030	As	15010 (16/10-15/2: 12020), 9840, 7400, 7270
1100-1130	As	9730, 7285, 1008
1230-1300	As	15010 (16/10-15/2: 12020), 9840, 7400, 7270
1330-1400	FE	15010 (16/10-15/2: 12020), 9840, 7400, 7270
1600-1630	Af	15010 (16/10-15/2: 12020), 9840, 7400, 7270
1800-1830	Eu	15010 (16/10-15/2: 12020), 9840, 7400, 7270
1900-1930	Eu	15010 (16/10-15/2: 12020), 9840, 7400, 7270
2030-2100	Eu	15010 (16/10-15/2: 12020), 9840, 7400, 7270
2330-2400	Eu/As	15010 (16/10-15/2: 12020), 9840, 7400, 7270
French		
1300-1330	As	15010 (16/10-15/2: 12020), 9840, 7400, 7270
1830-1900	Eu	15010 (16/10-15/2: 12020), 9840, 7400, 7270
1930-2000	Eu	15010 (16/10-15/2: 12020), 9840, 7400, 7270
2100-2130	Eu	15010 (16/10-15/2: 12020), 9840, 7400, 7270
Indonesian		
0930-1000	INS	15010 (16/10-15/2: 12020), 9840, 7400, 7270
1030-1100	INS	15010 (16/10-15/2: 12020), 9840, 7400, 7270
1430-1500	INS	15010 (16/10-15/2: 12020), 9840, 7400, 7270, 7400, 7270 1008
Japanese		
1200-1230	FE	15010 (16/10-15/2: 12020), 9840, 7400, 7270, 7400, 7270
1400-1430	FE	15010 (16/10-15/2: 12020), 9840, 7400, 7270, 7400, 7270
2130-2200	FE	15010 (16/10-15/2: 12020), 9840, 7400, 7270
Laotian		
0030-0100	LAO	9730, 7285,
1230-1300	LAO	9730, 7285
1400-1430	LAO	9730, 7285, 1008
Russian		
1130-1200	Eu	15010 (16/10-15/2:12010), 9840,
1300-1330	As	9730, 7285

Russian	Area	kHz
1430-1500	As	12035, 10010
1630-1700	Eu/As	15010 (16/10-15/2: 12020), 9840, 7400, 7270, 7400, 7270, 1008
Spanish		
1100-1130	LAm	15010 (16/10-15/2: 12020), 9840, 7400, 7270, 7400, 7270 1008
2000-2030	Eu	15010 (16/10-15/2: 12020), 9840, 7400, 7270
Thai		
1130-1200	THA	9730, 7285, 1008
1400-1430	THA	12035, 10010, 1008
1500-1530	THA	15010 (16/10-15/2: 12020), 9840, 7400, 7270, 7400, 7270 1008
2300-2330	THA	15010 (16/10-15/2: 12020), 9840, 7400, 7270
Vietnamese		
0000-0100	Am	15010 (16/10-15/2: 12020), 9840, 7400, 7270
1530-1630	As	1008
1700-1800	Eu/As	15010 (16/10-15/2: 12020), 9840, 7400, 7270

ANN: E: "This is the Voice of Vietnam, broadcasting from Hanoi".

YUGOSLAVIA

RADIO YUGOSLAVIA
✉ Hilandarska 2, 11000 Beograd. ☎ +381 (11) 346-884. 🖷 +381 (11) 332-014. (**Listener response:** ☎ +381 (11) 344-455. 🖷 +381 (11) 332014). ① 12432. **E-mail:** radioyu@beograd.com
Web: http://www.beograd.com/radioyu/
LP: DG: Nicola Ivanovic. Head of Prgrs:Dr. Zivorad Djordjevic. CE: Rodoljub Medan. Head of Public Rel: Aleksandar Popovic.
FM: Beograd 100.4MHz 0.5kW: 24h in Serbian.
MW: Pristina, Kosovo: 1413kHz 1000kW.
SW: Bijeljina, Bosnia (G.C: 19.10E/44.42N): 4 x 500kW
kHz: 6100, 6185, 6195, 7115, 7165, 7220, 7230, 9620, 9720, 11755, 11800, 11835, 15175

Albanian	Area	kHz
1800-1815MF	Eu	6100, 1413
Arabic		
1530-1600	ME/Af	11800, 15175
Bulgarian		
1815-1830MF	Eu	6100
English		
0100-0130W	Eu/NAm	7115, 6195
0200-0230	Eu/NAm	7230, 6100
1330-1400	AUS	11835
1930-2000	Eu/Af	6100, 9720
2200-2230	Eu	6100, 6185
French		
1700-1730	Eu/Af	6100, 11800
2130-2200	Eu	6100
German	**Area**	**kHz**
1730-1800	Eu	6100
2100-2130x	Eu	6100
Greek		
1645-1700MF	Eu	6100
Hungarian		
1630-1645MF	Eu	6100
Russian		
1600-1630	RUS	6100, 11755
1900-1930	RUS	6100, 7165
Serbian		
0030-0100W	Eu/NAm	7115, 6195
0030-0130Sun	Eu/NAm	7115, 6195
0130-0200+	NAm	7115, 6195
1400-1430+	AUS	11835
2000-2100Sat	Eu/AUS	6100, 7230
2030-2100x	Eu/AUS	6100, 7230
2100-2130+	AUS	7230
Spanish		
0000-0030	LAm	9720, 9620
2000-2030	Eu	7220

+) rel. Home Sce. 1st Prgr, x) exc. Sat — **V.** by QSL-card. Rec. acc.

INTERNATIONAL ORGANIZATIONS

ASBU

ARAB STATES BROADCASTING UNION

General Secretariat: 22a Taha Hussein Str, Zamalek, Cairo, Egypt. ☎ +20 (2) 805825. ① 347-ASBUUN. **Cable:** Asbunion, Cairo.

Technical Centre: P. O. Box 2231, Khartoum, Sudan — **Training Centre:** P. O. Box 6333, Damascus, Syria — **Audience Research Centre:** Iraqi Establishment for Radio & TV, Baghdad, Iraq.

Nature: Intergovernmental union within the framework of the League of Arab States comprising the Arab Radio & TV organizations (established February 1969).

Supreme Body: General Assembly, meeting in ordinary sessions once a year. The Presidency of the General Assembly is held in rotation among the full members according to the alphabetical order of the names of the member States. The term of office is one year. The Vice-Presidency is normally held by the member who shall assume the Presidency at the next session.

Admin. Council: Pres. & Vice-Pres. + one full member according to the alphabetical order of the names of the member States.

ABU

ASIA-PACIFIC BROADCASTING UNION

Headquarters: Secr. Gen's Office/Programme Dept./Tech. Dept: 2nd Floor, Bangunan IPTAR, 50614 Kuala Lumpur, Malaysia.

Postal ✉ P. O. Box 1164, 59700 Kuala Lumpur, Malaysia. ☎ +60 3 282-3592, 282-2480, 282-3108. 🖷 +60 3 282 5292. ① 32227 (ABU MA) Kuala Lumpur. **Cable:** ASBUN Kuala Lumpur.

Web: http://www.rthk.org.hk/rthk/abu/

L.P: Pres: Dato' Jaafar Kamin, DG, RTM-Malaysia. Vice-Presidents: Mikio Kawaguchi, Pres, NHK-Japan & Mehdi Tabeshian, Ag. Pres, IRIB-Iran. Secr-Gen: Hugh Leonard. Dir. Tech Dept: O. P. Khushu. Dir. Programme Dept: Naohiro Kato.

Administrative Council: Radio and Television of the People's Rep. of China, Television Broadcasts Ltd. (Hong Kong), Doordarshan India, Radio Republik Indonesia/Televisi Republik Indonesia, Islamic Rep. of Iran Broadcasting, NHK (Japan), KBS (Korea), Radio Television Malaysia, Television New Zealand Ltd., Shalimar Television Network (Pakistan), Republic Broadcasting System Inc. (Philippines), Sri Lanka Rupavahini Corporation, National Broadc. Sces. of Thailand.

Full Members (41): Radio-Television of Afghanistan, Australian Broadcasting Corporation, Special Broadcasting Service Corporation (Australia), National Broadcasting Authority of Bangladesh, Radio-Television Brunei, International Broadc. Corp. (Cambodia) Ltd., Radio & Television of the Peoples' Rep. of China, Egyptian Radio & Television Union, Fiji Broadcasting Commission, All India Radio, Doordarshan (India), Radio Republik Indonesia & Televisi Republik Indonesia, IRIB Iran, Iraqi Broadcasting & Television Establishment, NHK Japan, Tokyo Broadcasting System Inc. (Japan), Jordan Radio & Television, KBS Korea, MBC Korea, Lao National Radio & Television, RTM Malaysia, Dept. of Inf. & Broadcasting Maldives, Nepal Television Corporation, Radio Nepal, Television New Zealand Ltd, Pakistan Broadcasting Corporation, Pakistan Television Corporation Ltd, NBC Papua New Guinea, Peoples' Television Network Inc. (Philippines), Republic Broadcasting System Inc. (Philippines), Saudi Arabian Radio & Television, Television Corp. of Singapore, Television Twelve Pte. Ltd. (Singapore), SIBC (Solomon Islands), SLBC (Sri Lanka), Sri Lanka Rupavahini Corporation, National Broadcasting Sces. of Thailand, TV Pool of Thailand, TRT Turkey, The Voice of Vietnam, Vietnam Television.

Additional Full Members (28): Federation of Australian Commercial Television Stations, Asia Television Ltd. (Hong Kong), Hong Kong Commercial Broadcasting Co. Ltd, Metro Broadcast Corporation Ltd. (Hong Kong), Radio Television Hong Kong, Television Broadcasts Ltd. (Hong Kong), Pt.Cakrawala Andalas Televis (Indonesia), Pt. Rajawali Citra Televisi (Indonesia), Pt. Surya Citra Televisi (Indonesia), Pt. Cipta Televisi Pendidikan Indonesia,

Pt. Indosiar Visual Mandiri (Indonesia), Asahi National Broadcasting Co. Ltd. (Japan), Fuji Television Network, Inc. (Japan), Educational Broadc. System (Korea), Seoul Broadc. System (Korea), CBS Korea, Radio & Television Broadcasting Committee of the Democratic Peoples' Rep. of Korea, Teledifusão de Macau SARL, Sistem Televisiyen Malaysia Berhad, Mongolian Radio & Television, Norfolk Island Broadcasting Service, Shalimar Television Network (Pakistan), ABS-CBN Broadcasting Corporation (Philippines), Philippine Broadcasting Service, Radio Philippines Network Inc, Independent Television Network Ltd. (Sri Lanka), MTV Channel (Pte) Ltd. (Sri Lanka), Telshan Network (Pte) Ltd. (Sri Lanka).

Associate Members (28): Federation of Australian Radio Broadcasters Ltd, Nine Network Australia Ltd, CBC Canada, YLE Finland, OFRT France, ARD Germany, ZDF Germany, National Association of Commercial Broadcasters in Japan, Radio Kiribati, Mauritius Broadcasting Corporation, Federated States of Micronesia Broadcasting Service, Radio Netherlands, Radio New Zealand Ltd, NRK Norway, Islands Broadc. Corp. (Philippines), KBP (Philippines), Russian State TV and Radio Company, Radio Corp. of Singapore, Sveriges Radio AB (Sweden), Swiss Broadcasting Corporation, BBC (UK), ABC (USA), CBS (USA), CPB/PBS/NPR/APRN (USA), National Association of Broadcasters (USA), Turner Broadcasting System (USA), Voice of America.

Aims: To foster and coordinate the development of broadcasting in the Asia-Pacific region, to develop ways and means of establishing closer collaboration and cooperation among broadcasting organizations, and to serve the professional needs of the broadcasters in Asia and the Pacific.

PUB: ABU News & ABU Technical Review (both bi-monthly in English) — **Date of establishment:** 1.7.1964.

AIBD

ASIA-PACIFIC INSTITUTE FOR BROADCASTING DEVELOPMENT

✉ P. O. Box 1137, Pantai, 59700 Kuala Lumpur, Malaysia. ☎ +60 (3) 2821046, 2824618, 2823719. 🖷 +60 (3) 2822761. ① MA 30083 APBRO. **Cable:** UNESBROAD Kuala Lumpur. **Web:** http://aibd.org.my

L. P: Dir. & Secr. to Council: Mr Javad Mottaghi.

Nature: Intergovernmental institute for the development of broadcasting, operating since 1972 and established formally in 1977, serving countries of the United Nations Economic and Social Commission for Asia and the Pacific (UN ESCAP) region in the development and upgrading of their broadcasting personnel and systems.

Supreme Body: Governing Council of 11 voting member countries plus UNDP, UNESCO, ITU, ABU, CFTC and other non-voting representative private or public funding agencies invited to attend annual meeting.

Training Centre: at Malaysian Government National Broadcasting Headquarters and Training Centre (IPTAR), Angkasapuri, Kuala Lumpur, Malaysia.

Members: Afghanistan, Bangladesh, Brunei Darussalam, China, Fiji, France, India, Indonesia, Iran, Rep. of Korea, Laos, Malaysia, Maldives, Micronesia, Nepal, Pakistan, Papua New Guinea, Philippines, Samoa, Singapore, Sri Lanka, Thailand and Vietnam.

Aims: To foster and coordinate the development of broadcasting in the Asia-Pacific region, to develop ways and means of establishing close collaboration and cooperation among broadcasting organizations, and to serve the professional needs of the broadcasters in Asia and the Pacific.

PUB: Broadcaster (quarterly in English).

AIB

ASSOCIATION FOR INTERNATIONAL BROADCASTING

✉ P. O. Box 990, London SE3 9XL, UK. ☎ +44 (0) 1255 676996. 🖷 +44 (0) 1255 676996. **E-mail:** aib@mcmail.com **Web:**

http://www.aibcast.demon.co.uk
Nature: The Association is a co-operative forum for all organisations and individuals professionally connected with international broadcasting.
Aims: The AIB acts as the voice of the international broadcasting industry. It encourages high standards and co-operation within the industry and promotes international broadcasting to the public.
PUB: The AIB publishes a magazine "The Channel", and a twice-yearly listing of listening information the "Global Radio Guide".

CBU

CARIBBEAN BROADCASTING UNION
The CBU is an association of national broadcasting systems of the Commonwealth, Caribbean and other regional states formed for the purpose of promoting and facilitating interchange between members through the medium of broadcasting.
Secretariat: "Wilkins Lodge", Two Mile Hill, St. Michael, Barbados, W. I. ☎ +1 (809) 429 9146. ▤ +1 (809) 429 2171.
Web: http://www.caribunion.com/
Board of Directors: Pres: Vic Fernandes, Barbados Rediffusion Service Ltd. Vice-Pres (TV): Stewart Krohn, Great Belize Productions. Vice-Pres (Radio): James "Tanny" Rose, Antigua and Barbuda Broadcasting Sce. Vice-Pres. (Technology): Ronald Abraham, Marpin Television.
Board Members: Dennis Joseph (DBS, Dominica), Lester Spaulding (Radio Jamaica), Claude Robinson (JBC, Jamaica), Theodore Linscheer (ATV, Suriname).
Members: *Anguilla:* Radio Anguilla. *Antigua:* Antigua & Barbuda Broadcasting Service. *Aruba:* Tele Aruba. *Bahamas:* Broadcasting Corp. of the Bahamas. *Barbados:* Barbados Rediffusion Service Ltd, Caribbean Broadcasting Corp. *Belize:* Broadcasting Corp. of Belize, Great Belize Productions Ltd. *Bermuda:* Bermuda Broadcasting Co. Ltd. *Dominica:* Dominica Broadcasting Corp., Marpin TV Co. Ltd. *Grenada:* Grenada Broadcasting Corp. *Guyana:* Guyana Broadcasting Corp., Guyana Television Broadc. Co. Ltd., Vieira CommunicationsLtd., WRHM Television. *Jamaica:* Jamaica Broadcasting Corp., Radio Jamaica Ltd., CVM Television, Super Supreme Television. *Montserrat:* Radio Montserrat. *Netherlands Antilles:* Tele Curacão. *St. Kitts:* ZIZ Radio & TV. *St. Lucia:* St. Lucia Broadcasting Corp., Helen Television System, Cablevision. *St. Vincent:* National Broadcasting Corp. of St. Vincent & The Grenadines, St. Vincent and the Grenadines Broadcasting Corp. *Suriname:* Surinaamse Televisie Stichting, ATV Telesur. *Trinidad & Tobago:* Trinidad Broadcasting Co. Ltd, Caribbean Communications Network - TV6, International Communications Network. *Turks & Caicos:* Radio Turks & Caicos.
Associate Members: *Canada:* CBC, Radio Canada Int. *Germany:* Transtel GmbH. *Jamaica:* Caribbean Institute of Mass Communication, Creative & Production Training Centre Ltd, Radio Education Unit, Independent Radio Co. Ltd. *Trinidad:* Banyan Ltd. *United Kingdom:* BBC. *USA:* CNN, NAB, VOA. *British Virgin Islands:* Dept. of Inf. & Pub. Rel.

CBA

COMMONWEALTH BROADCASTING ASSOCIATION
▤ CBA Secretariat, 17 Fleet Street, London EC4Y 1AA.
☎ +44 (171) 583 5550. ▤ +44 (171) 583 5549. **E-mail:** cba@cba.org.uk
Web: http://www.oneworld.org/cba/
L.P: Secr. Gen: Mrs. Elizabeth Smith. Manager Training and Development: Colin Lloyd.
Members: Antigua & Barbuda Broadcasting Service, Australian Broadcasting Corp., Broadcasting Corp. of the Bahamas. Radio Bangladesh, Bangladesh Television, Caribbean Broadcasting Corp. (Barbados), Broadcasting Corp. of Belize, Radio Botswana, BBC, Radio Television Brunei, Cameroon Radio Television, CBC Canada, Radio Cayman, Cyprus Broadcasting Corp., Dominica Broadcasting Corp., Communications Fiji Ltd., Fiji TV Ltd., Island Networks Corp., Fiji, Marpin TV Co. Ltd., Dominica, Falkland Islands Broadcasting Station, Radio Gambia, Ghana Broadcasting Corp., Gibraltar

Broadcasting Corp., Grenada Broadc. Corp., Guyana Broadcasting Corp., All India Radio, Doordarshan India, Super Supreme TV Jamaica, RJR Jamaica, Kenya Broadcasting Corp., Radio Kiribati, Lesotho National Broadcasting Service, Malawi Broadcasting Corp., Radio Television Malaysia, Malta Broadcasting Authority, Mauritius Broadcasting Corp., Antilles Radio Corporation Limited, Radio Montserrat, Radio Moçambique, Televisão de Moçambique, Namibian Broadcasting Corp., Nauru Media Beaureau, Radio New Zealand, Television New Zealand, Pakistan Broadcasting Corp., Pakistan Television Corp., NBC Papua New Guinea, ZIZ Radio & Television (St. Kitts-Nevis), St. Lucia Broadcasting Corp., St. Vincent & The Grenadines National Broadcasting Corp., Samoa Broadcasting Service, Seychelles Broadcasting Corp., Sierra Leone Broadcasting Service, Radio Corporation of Singapore Pte. Ltd., Television Corporation of Singapore, South African Broadcasting Corporation, SLBC Sri Lanka, Sri Lanka Rupavahini Corp., Swaziland Broadcasting Service, Swaziland Television Authority, Radio Tanzania, ITV & Radio One Ltd., Tanzania, Tonga Broadcasting Commission, Trinidad & Tobago International Communications Network Ltd, Radio Turks & Caicos, Radio Uganda & Uganda Television, Vanuatu Broadcasting and TV Corp.,Western Samoa Broadcasting Service, Zambia National Broadcasting Corp., Zimbabwe Broadcasting Corp. Affiliates: Australian Broadcasting Authority. Broadcasting Standards Authority of New Zealand. The Commonwealth of Learning, Canada. Tanzanian Broadcasting Commission. The Television Corporation plc of the UK. UTV International, Canada. Orbicom (PTY) Ltd., South Africa. Sentech (PTY) Ltd., South Africa. IBA, South Africa. NAB, South Africa. BFBS, UK. Merlin Communications International Ltd., UK. national Transcommunications Ltd., UK. The Radio Authority, UK. The Thomson Foundation, UK. RTHK, Hong Kong. PCI, USA.
Nature: The CBA was established in London in February 1945 and is an association of broadcasting organizations in 54 Commonwealth countries pledged to work for the professional improvement of broadcasting in member organizations through collective study and mutual assistance. It carries out some 40-50 training courses in broadcasting skills each year.
General Conferences of the Association are held every two years to discuss and exchange ideas, information and experience on all aspects of radio and television broadcasting - in programming, administration and engineering.
In between General Conferences, regional meetings are held in Africa, the Caribbean, the Pacific and Asia as part of the Association's policy of encouraging more effective co-operation among members in its main regions.
The Secretariat publishes a quarterly magazine, "Commonwealth Broadcaster", which has a wide international circulation, and the CBA Directory.

EBU

EUROPEAN BROADCASTING UNION
UNION EUROPÉENNE DE RADIO-TÉLÉVISION
(EUROVISION/EURORADIO)
▤ Ancienne Route 17A, CH-1218 Grand-Saconnex (GE), Switzerland. ☎ +41 (22) 7172111. ▤ +41 (22) 7172481. **E-mail:** ebu@ebu.ch. **Website:** http://www.ebu.ch
Admin. Council: Pres: A. Scharf (ARDBR, Germany). Vice-Presidents: X. Goyou-Beauchamps (GRF, France),Will Wyatt (BBC/United Kingdom), Boris Bergant (RTVSLO/Slovenia).
Officials: Secr. Gen: Dr. J. B. Münch. Dir. Operations: H. Pérez. Dir. TV Dept: G. Stucchi. Dir. Radio Dept: T. Alexanderson. Dir. Legal Affairs Dept: Dr. W. Rumphorst. Dir. Tech. Dept: P. Laven. Head of Finance: J. Ekiert.
Purpose: The Geneva based EBU (founded 1950), is a professional association of broadcasting organizations whose activities include the Eurovision & Euroradio news and programme exchanges. Its objectives are to promote co-operation among its members in 49 countries and with broadcasting organizations all over the world, and to represent the interests of its members in the programme, legal, technical and other fields.
Organization: Supreme authority is vested in the General Assembly composed of all members. The Assembly elects a 19 member Administrative Council with executive powers. The activ-

ities of the EBU are channelled through its Committees - Television, Radio, Legal and Technical. The Committees, although independent of each other, work closely together on matters of common interest. In addition the EBU has a permanent staff of some 230 peoplelocated in the Geneva headquarters and the bureaux in Brussels, New York, Washington and Moscow, serving EBU members in the radio and television industry.

PUB: The EBU official magazine "Diffusion", published quarterly in English & French, The "EBU Technical Review": quarterly in English & French, technical documents & information sheets, technical monographs, reports of conferences & seminars.

EUTELSAT

EUROPEAN TELECOMMUNICATIONS SATELLITE ORGANIZATION

✉ Tour Maine-Montparnasse 33, Avenue de Maine, 75755 Paris Cedex 15, France. ☎ +33 (1) 45 38 4757. 🖷 +33 (1) 45 38 3700 **Web:** http://www.eutelsat.org

Signatories: Albania (Albanian Telecommunications Enterprise), Andorra (Servei de Télécommunications d'Andorra), Armenia (Ministry of Telecommunications), Austria (Austrian Administration of Posts and Telegraphs), Azerbaijan (Ministry of Communications), Belarus (Ministry of Posts, Telecommunications and Informatics), Belgium (Belgacom), Bosnia-Hercegovina (Public Enterprise of PTT Transport), Bulgaria (Bulgarian Telecommunications Company Ltd.), Croatia (Ministry of Maritime Affairs, Transport and Communications), Cyprus (Cyprus Telecommunications Authority), Czech Republic (Ceske Radiokomunikace a.s), Denmark (Tele Danmark Ltd), Finland (Telecom Finland Ltd, France (France Telecom), Georgia (Georgian Ministry of Communications), Germany (Deutsche Bundespost Telekom), Greece (OTE), Hungary (Hunsat), Iceland (Post and Telecommunications Administration of (Iceland), Ireland (Telecom Eireann), Italy (Telecom Italia), Latvia (Lattelekom), Liechtenstein (Government of the Principality of Liechtenstein), Lithuania (Lithuanian Ministry of Communications and (Informatics), Luxembourg (Administration luxembourgeoise des Postes et (Télécommunications), Malta (Government of the Republic of Malta), Moldova (Ministry of Communications and Informatics), Monaco (Gouvernement de la Principauté de Monaco, Netherlands (Royal PTT Nederland NV), Norway (Telenor Satellite Services AS), Poland (Polish Telecommunications SA), Portugal (Companhia Portuguesa Radio Marconi), Romania (Romanian Ministry of Communications), Russia (Ministry of Posts and Telecommunications), San Marino (Government of the Republic of San Marino), Slovakia (Slovak Telecommunications), Spain (Telefonica), Sweden (Telia AB), Switzerland (Direction générale de l'Entreprise des Postes, Télégraphes et Téléphones suisses), Turkey (Türk Telekomunikasyon AS), United Kingdom (BT), Ukraine (KRRT), Vatican (Government of the Holy See), Yugoslavia (Community of Yugoslav Posts, Telegraphs and Telephones).

IIC

INTERNATIONAL INSTITUTE OF COMMUNICATIONS

✉ Tavistock House South, Tavistock Square, London WC1H 9LF. ☎ +44 (171) 3880671. 🖷 +44 (171) 380 0623. **Web:** http://www.iicom.org

Exec. Committee Members: Pres: Henri Pigeat (France). Vice-Presidents: Moriyoshi Saito (Japan), Sheelagh Whittaker (Canada). Treasurer: Gaston Melo (Mexico). Chairman of Exec. Committee: Matti Packalen (Finland). Secretary: Gerry Jayasuriya.

Staff: Exec. Dir: Roger Hayes. Dep. Dir: Gerry Jayasuriya. Editor, InterMedia: Daniella Goldman. Consultant Editor: Rex Winsbury.

Mission Statement: The IIC, as an independent association of leading decision-makers and other professionals, aims to ensure that converging communications serve the values of free and civilized people everywhere. It does so through multi-disciplinary dialogue on the interactions between information technology and human societies worldwide.

ITU

INTERNATIONAL TELECOMMUNICATION UNION

✉ Place des Nations, CH-1211 Geneva 20, Switzerland. ☎ +41 (22) 730 5111. 🖷 +41 (22) 733 7256. ① 421 000 UIT CH. **Teletex:** 288 468 151 00 = UIT. **Web:** http://www.itu.ch

L. P: Secr. Gen: Dr. Pekka J. Tarjanne (Finland). Dep. Secr. Gen: Dr. Henry Chasia (Kenya).

The ITU was founded in 1865 and as such is the oldest intergovernmental organization. In 1947 it became a specialized agency of the United Nations and now has membership of over 184 Member States and 363 members (scientific and industrial companies, public and private operators, broadcasters, regional/international organizations) of the three Sectors. In essence, the Union's mission covers the following domains: a **technical domain:** to promote the development and efficient operation of telecommunication facilities, in order to improve the efficiency of telecommunication services, their usefulness, and their general availability to the public; a **development domain:** to promote and offer technical assistance to developing countries in the field of telecommunications, to promote the mobilization of the human and financial resources needed to develop telecommunications, and to promote the extension of the benefits of new telecommunications technologies to people everywhere; a **policy domain:** to promote, at the international level, the adoption of a broader approach to the issues of telecommunications in the global information economy and society.

Radio Regulations Board (RRB): consists of nine members who are elected by the Plenipotentiary Conference. They perform their duties on a part-time basis, meeting up to four times a year in Geneva.

Telecommunication Standardization Bureau (TSB) — L.P: Dir: Thedor Irmer.

Radiocommunication Bureau (BR) — L.P: Dir: Robert W. Jones. Study Group 10 (Broadcasting): Chmn: C. Terzani (Italy). Study Group 11 (Television): Chmn: M.I. Krivocheev (Russia).

Council: Region A (Americas): Argentina, Bahamas, Brazil, Canada, Chile, Cuba, Mexico and the USA. Region B (Western Europe): Denmark, France, Germany, Italy, Portugal, Spain, Switzerland and the United Kingdom. Region C (Eastern Europe): Bulgaria, Poland, Romania, Russia and the Ukraine. Region D (Africa): Algeria, Benin, Burkina Faso, Cameroon, Cape Verde, Egypt, Kenya, Mali, Morocco, Nigeria, Senegal, South Africa and Tanzania. Region E (Asia and Australasia): Australia, China (P.R.), India, Indonesia, Japan, Korea (Rep. of), Kuwait, Pakistan, Philippines, Saudi Arabia, Thailand and Vietnam.

Members: all UN members except Dominica, St. Kitts & Nevis, St. Lucia, Seychelles. Other members are Kiribati, Nauru, Switzerland, Tonga and Vatican City State.

INTELSAT

NTERNATIONAL TELECOMMUNICATIONS SATELLITE ORGANIZATION

✉ 3400 International Drive, N.W., Box 63, Washington D.C. 20008, USA. ☎ +1 (202) 944 6872. 🖷 +1 (202) 944 7925

Web: http://www.intelsat.com

Member countries: Afghanistan, Algeria, Angola, Argentina, Armenia, Australia, Austria, Azerbaijan, Bahamas, Bahrain, Bangladesh, Barbados, Belgium, Benin, Bhutan, Bolivia, Bosnia-Hercegovina, Botswana, Brazil, Brunei Darussalam, Burkina Faso, Cameroon, Canada, Cape Verde, Central African Rep., Chad, Chile, China (People's Rep. of), Colombia, Congo, Costa Rica, Côte d'Ivoire, Croatia, Cyprus, Czech Republic, Denmark, Dominican Rep., Ecuador, Egypt, El Salvador, Ethiopia, Fiji, Finland, France, Gabon, Germany, Ghana, Greece, Guatemala, Guinea, Haiti, Honduras, Hungary, Iceland, India, Indonesia, Iran, Iraq, Ireland, Israel, Italy, Jamaica, Japan, Jordan, Kazakhstan, Kenya, Korea (Rep. of), Kuwait, Kyrghyz Rep., Lebanon, Libya, Liechtenstein, Luxembourg, Madagascar, Malawi, Malaysia, Mali, Malta, Mauritania, Mauritius, Mexico, Micronesia, Monaco, Morocco, Mozambique, Namibia, Nepal, The Netherlands, New Zealand, Nicaragua, Niger, Nigeria, Norway, Oman, Pakistan, Panama,

Papua New Guinea, Paraguay, Peru, Philippines, Poland, Portugal, Qatar, Romania, Russian Federation, Rwanda, Saudi Arabia, Senegal, Singapore, Somalia, South Africa, Spain, Sri Lanka, Sudan, Swaziland, Sweden, Switzerland, Syria, Tajikistan, Tanzania, Thailand, Togo, Trinidad & Tobago, Tunisia, Turkey, Uganda, United Arab Emirates, United Kingdom, United States of America, Uruguay, Vatican State, Venezuela, Vietnam, Yemen, Yugoslavia, Zaire, Zambia, Zimbabwe.

ISBO

ISLAMIC STATES BROADCASTING SERVICE ORGANIZATION

📧 P. O. Box 6351, Jeddah 21442, Saudi Arabia. ☎ +966 (2) 66721121, 6722269. 🖹 +966 (2) 6722600. ① 601442 ISBO SJ.

Establishment: Founded as an affiliate of the Organization of the Islamic Conference (OIC) in 1975 in accordance with a resolution adopted by the Sixth Conference of Islamic Foreign Ministers. Membership: 45 Islamic States.

Objectives: a) Propagation of Islamic teaching — b) Protection of Islamic heritage — c) Strengthening fraternal relations among Muslim peoples — d) Promotion of Islamic solidarity — e) Defending Islamic causes — f) Furthering co-operative relations with regional and international information and cultural organizations.

Means: Production of radio & TV programmes — Exchange of programmes between broadcasting corp.orations of member states — Teaching Arabic language by audio-visual means — Training of personnel for broadcasting corp.orations in member states.

Governing Bodies

General Assembly: composed of all member states.

Executive Council: composed of 15 members including the host state, Palestine and representatives of the OIC as permanent members — **Chmn:** Dr. Foad Abdel-Salam Al-Farsi (Host State).

General Secretariat: responsible for implementation of resolutions adopted by the General Assembly and Executive Council and for running the affairs of the organization.

Secr. Gen: Hussain El-Askary.

International Relations: concluded co-operation agreements with UNESCO -and ALECSO. Attends Islamic Summits, Islamic Foreign & Information Ministers Conferences, meetings of Arab League and Arab Gulf Information Ministers and Committees.

NANBA

NORTH AMERICAN NATIONAL BROADCASTERS ASSOCIATION

📧 c/o CBC, P.O. Box 500, Station A, Toronto M5W 1E6, Canada. ☎ +1 (416) 205 8533. 🖹 +1 (416) 205 8601. **E-mail:** info@nanba.org

World Wide Web: http://nanba.org/broadcasting/

L.P: Secr. Gen: Bill Roberts.

Board of Directors: Pres: Larry M. Loeb (Capital Cities/ABC). Vice-Presidents: Jorge Kanahuati (Televisa s. a.), Tom Peddie (CTV Television Network Ltd.). Advisor to the Pres. & CEO: Michael McEwan (CBC). Other Directors: John A. Frazee (CBS), Ronald A. Gnidziejko (NBC), Eason Jordan (CNN).

Advisory Council: comprises the Board of Directors, Secr. Gen, Committee Chairs and representatives of NANBA member and associate member organizations.

Committees: Technical - Chmn: Brian Baldry (CBC), Legal - Chmn: Larry Loeb (Capital Cities/ABC Inc.), News & Operations - Chmn: Brian Knoblock (CBS).

Sports Group: Jim Thompson (TSN-Canada).

Members: Canada: CBC, CTV. Mexico: Televisa s. a. USA: Capital Cities/ABC Inc, CBS, NBC, Turner Broadcasting Systems Inc.

Associate Members: ESPN, HBO, NAB, TSN-Canada.

Meetings: The Advisory Council usually meets four times a year, including the annual meeting held in January. The Board of Directors may meet separately, The Committees meet as frequently as their work requires.

NANBA started as an ad hoc group in 1972. It became a formal organization in its present form in 1978. The Association is concerned with providing a framework for the identification, study and active solution of international questions affecting broadcasting and, working with other international Unions and Associations of broadcasters, to gain an effective voice in international forums on matters that affect broadcasting. Its objectives are not commercial in nature.

PUB: NANBA bulletin published at intervals during the year. SNG/ENG Handbook - a publication on satellite news gathering aimed at news operations staff and SNG producers. 3rd edition published June 1993.

PIBA

PACIFIC ISLANDS BROADCASTING ASSOCIATION

📧 P.O. Box 116, Port Vila, Vanuatu. ☎ +678 24250/1. 🖹 +679 24252.

L.P: Chmn: Hima Douglas. Projects officer: Bob Makin.

PIBA provides training for Pacific island broadcasters through its training project PACBROAD, and runs a news exchange based on the news output of the public broadcasters of the island countries.

SABA

SOUTHERN AFRICAN BROADCASTING ASSOCIATION

📧 P.O. Box 50752, Ridgeway, Lusaka, Zambia. ☎ +260 (1) 263595. 🖹 +260 (1) 263110. **E-mail:** pazasabc@zamnet.zm

L.P: Pres: Ted Makgekenene.

UNESCO

(United Nations Educational, Scientific & Cultural Organization)

📧 7 Place de Fontenoy, F-75700 Paris, France. ☎ +33 (1) 4568 1000.

Web: http://www.unesco.org/

L. P: Dir. Gen: Federico Mayor Zaragoza.

Office of Public Information (OPI). Dir: Hélène-Marie Gosselin.

Audio-Visual Division: Executive Producer: Claude van Engeland.

Communication, Information & Informatics Sector (CII): Asst. Dir. Gen: Henrikas Iouchkiavitchious.

Communication Division (COM): Dir: Alain Modoux.

International Programme for the Development of Communication (IPDC): Dir: Claude Ondobo.

UNION OF NATIONAL RADIO & TELEVISION ORGANIZATIONS OF AFRICA (URTNA)

General Secretariat: 101 rue Carnot, B. P. 3237, Dakar, Senegal. ☎ +221 211625, 215970. 🖹 +221 225113. ① 51650 URTNA SG — **L.P:** Secr. Gen: Efoé Adodo Mensah.

Technical Centre: B. P. 39, Bamako, Mali. ☎ +223 222593. 🖹 +223 224809. **Cable:** URTNA Bamako. ① 2505.

Programme Exchange Centre: P. O. Box 50518, Nairobi, Kenya. ☎/🖹 +254 (2) 241954. ① 22675 Nairobi.

Inter-African Centre for Rural Radio Studies (CIERRO): B. P. 385, Ouagadougou, Burkina Faso. ☎ +226 306686. 🖹 +226 312866. ① CIERRO 5349 BF.

Co-ordination Centre TV News & URTNA: 21 Bd. des Martyrs. B. P. 184 El Mouradia, Alger, Algeria. ☎ +213 (2) 602507/593414. 🖹 +213 (2) 590270. ① 66476/66101.

Aims: URTNA has no commercial aim. Its objectives, among others, are: a) to promote, maintain and develop co-operation amongst its members — b) to support in every domain the interests of radio and television organizations which have accepted its statutes and to establish relations with other such organizations — c) to promote and co-ordinate the study of all matters relating to radio and television and to ensure the exchange of information on all questions of general interest — d) to use its best endeavours to ensure that all its members honour the provisions of international and inter-African agreements in all matters relating to radio and tele-

vision — e) to co-ordinate and collaborate closely with member organizations in the coverage of national and international events.
Governing Bodies: President, 1st Vice-President, 2nd Vice-President, Legal Adviser. The Presidents and Vice-Presidents constitute the Bureau of the Union.
Administrative Council: Constituted by 15 active Member Organizations. The Administrative Council is the policy making body, responsible to the General Assembly.
Specialized Commissions: 1) Administrative & Legal, 2) Technical, 3) Programme Exchange and Cultural Affairs, 4) TV News Exchange. Each Commission is constituted by 10 active Member Organizations.
Active Members: Algeria (ENRS/ENTV/TDA), Angola (TPA/RNA), Benin (ORTB), Botswana (RB), Burkina Faso (RTBF), Burundi (RTNB), Cameroon (CRTV), Cape Verde (RTNCV), Chad (RTNT), Central African Rep. (RTC), Comoros (RC), Congo (RTC), Côte d'Ivoire (RTI), Djibouti (RTD), Egypt (ERTU), Ethiopia (ETS), Gabon (RTG), Gambia (RG), Ghana (GBC), Guinea (RTG), Guinea-Bissau (RTGB), Kenya (KBC), Lesotho (LBS), Liberia (LBS), Libya (LJB), Madagascar (RTM), Malawi (MBC), Mali (ORTM), Mauritania (OMRT), Mauritius (MaBC), Morocco (RTM), Mozambique (RM/TM), Namibia (NBC), Niger (ORTN), Nigeria (FRCN/NTA), Senegal (RTS), Seychelles (SBC), Sierra Leone (SLBS), Somalia (SBS), South Africa (SABC), Sudan (SNBC), Swaziland (STV), Tanzania (RTD), Togo (RTT), Tunisia (ERTT), Uganda (RU/UTV), Zaire (OZRT), Zambia (ZNBC), Zimbabwe (ZBC).
Supplementary Active Members: Africa No.1 (Gabon), Kenya Television Network (KTN).
Associate Members: Germany (ARD/ZDF), France (GRF), The Netherlands (RN), Palestine (RP), Portugal (RTP/RDP), Switzerland (SSR), United Kingdom (BBC), Vatican (RV).
Supplementary Associate Members: CNN (USA), Africa Center (Honolulu, Hawaii).
All Active and Associate Members constitute the General Assembly, the supreme body of the Union.
PUB: "URTNA Review" bi-annual in English/French, studies/reports etc.

UNITED NATIONS

⊡ New York, NY 10017, USA. ☎ +1 (212) 963 6982 / 1807/ 6939.
📄 +1 (212) 963 6869. **Web:** http://www.un.org
E-mail: audio-visual@un.org
L. P: Secr. Gen: Koffi Annan. Under Secr. Gen. for Communications & Public Inf: Kensaku Hogen. Chief, Audio-Visual Promotion & Distribution: Barbara Sue-Ting-Len.

URTNA

UNION OF NATIONAL RADIO AND TELEVISION ORGANIZATIONS OF AFRICA
International Secretariat: 101 rue Carnot, B.P. 3237, Dakar, Senegal. ☎ +(221) 821 16 25/821 59 70. 📄 +(221) 822 51 13. ⊙ 51650 URTNA SG. **E-mail:** urtnadkr@telecomplus.sn
LP: Secr. Gen: Abdelhamid Bouksani.
Technical Centre: B.P. 39, Bamako, Mali. ☎ +(223) 24 25 39. 📄 +(223) 24 19 54. ⊙ URTNA Bamako 2505
Programme Exchange Centre: P.O. Box 50518, Nairobi, Kenya. ☎ +(254-2) 24 29 39. 📄 +(254-2) 24 19 54. ⊙ 22675 Nairobi. **E-mail:** urtnapec@arcc.or.ke
Inter-African Centre for Rural Radio Studies (CIERRO): B.P. 385, Ouagadougou, Burkina Faso. ☎ +(226) 30 66 86. 📄 +(226) 31 28 66. **E-mail:** cierro@fasnet.bf
TV News Exchange Centre: 21 Bd. des Martyrs, B.P. 184 El Mouradia, Alger, Algeria. ☎ +(213-2) 60 25 07/59 34 14. 📄 +(213-2) 59 02 70. ⊙ 66476/66101.
Aims: URTNA has no commercial aim. Its objectives, among others, are: (a) to promote, maintain and develop co-operation amongst its members; (b) to support in every domain the interests of radio and television organizations which have accepted its statutes and to establish relations with other such organisations; (c) to promote and co-ordinate the study of all matters relating to

radio and television and to ensure the exchange of information on all questions of general interest; (d) to use its best endeavours to ensure that all its members honour the provisions of international and inter-African agreements in all matters relating to radio and television; (e) to co-ordinate and collaborate closely with member organizations in the coverage of national and international events.
Governing Bodies: President, 1st Vice-President, 2nd Vice-President, Secretary General, Legal Adviser. The President, Vice-Presidents and countries hosting URTNA permanent centres (Algeria, Burkina Faso, Kenya, Mali and Senegal) constitute the Executive Committee.
Administrative Council: Constituted by 15 active member organisations. The Administrative Council is the policy making body, responsible to the General Assembly.
Specialized Commissions: 1) Administrative, Legal and Financial; 2) Technical; 3) Programme; 4) TV News Exchange; 5) Training
Each commission is constituted by 11 active member organisations.
Active Members: Algeria (ENRS/ENTV/TDA), Angola (TPA/RNA), Benin (ORTB), Botswana (RB), Burkina Faso (RTBF), Burundi (RTNB), Cameroon (CRTV), Cape Verde (RTNCV), Chad (RTNT), Central African Republic (RTC), Comoros (RC), Congo (RTC), Cote d'Ivoire (RTI), Democratic Republic of Congo (RTNC), Djibouti (RTD), Egypt (ERTU), Ethiopia (ETS), Gabon (RTG), Gambia (RG), Ghana (GBC), Guinea (ORTG), Guinea-Bissau (RTGB), Kenya (KBC), Liberia (LBS), Libya (LJB), Madagascar (ORTM), Malawi (MBC), Mali (ORTM), Mauritania (OMRT), Mauritius (MABC), Morocco (RTM), Mozambique (RM/TM), Namibia (NBC), Niger (ORTN), Nigeria (FRCN/NTA), Senegal (RTS), Seychelles (SBC), Sierra Leone (SLBS), Somalia (SBS), South Africa (SABC), Sudan (SNBC), Swaziland (STV), Tanzania (RTD), Togo (RTT), Tunisia (ERTT), Uganda (RU/UTV), Zambia (ZNBC), Zimbabwe (ZBC).
Supplementary Active Members: Africa No.1 (Gabon), Kenya Television Network (KTN).
Associate Members: Germany (ARD/ZDF), France (GRF), The Netherlands (RN), Palestine (RP), Portugal (RTP/RDP), Switzerland (SSR), United Kingdom (BBC), Vatican (RV).
Supplementary Associate Members: CNN (USA), Africa Centre (Honolulu, Hawaii), World Space (USA).
All active and associate members constitute the General Assembly, the supreme body of the Union.
PUB: "URTNA Review" bi-annual in English/French, studies/reports, etc.

AMARC

WORLD ASSOCIATION OF COMMUNITY RADIO BROADCASTERS
International Secretariat: 3575 blvd. St.-Laurent, Suite 611, Montreal, Quebec H2X 2T7, Canada. ☎ +1 (514) 982 0351. 📄 +1 (514) 849 7129. **E-mail:** amarcho@amarc.org
Web: http://www.amarc.org
Latin American Regional Office: Av. Athualpa 333, Casilla 17-08-8489, Quito, Ecuador. ☎ +593 (2) 501180.
E-mail: latem@amarc.org
European Regional Office: 15 Paternoster Row, Sheffield S1 2BX, England. ☎ +44 (114) 221 0592. 📄 +44 (114) 279 8976.
E-mail: europe@amarc.org
AMARC Africa: c/o National Community Radio Forum, Cosatu House, Private Deck X42, Braamfontein 2017, South Africa. ☎ +27-11-403-751. 📄 +27-11-403-7514. **E-mail:** amarc@bridges.co.za
Board of Directors: Pres: Rafael Roncagliolo. Vice-Presidents: Lumko Mitimde (Austral. & Ea. Africa), Oumar Mariko (We. & Ce. Africa), Gjorgy Simo (Eastern Europe), Steve Buckley (Western Europe), Ernesto Lamas (Lat. America & Carrib.), Barbara Losier (N. America), Tiga Bayles (Oceania), Maria-Victoria Polanco (International Women's Network).
Secr. Gen: Sophie K. Ly.
PUB: "InteRadio" and various other books in English, French and Spanish.

CLUBS FOR DX-ERS AND INTERNATIONAL LISTENERS

This section lists non-commercial hobby clubs serving international radio and TV enthusiasts. Some are oriented to programme listening, and others to DX-ing, the reception of low power or distant stations. In most cases, bulletins are produced on a regular basis. Sample copies are generally available for return postage (3 or 4 IRCs). Unless otherwise stated, all clubs are international, publish bulletins in E and cover all aspects of the hobby. This list does not include clubs run by commercial publications or by individual broadcasters. Information is given according to information received at the WRTH editorial office, but WRTH cannot accept responsibility for changes not notified to us.

EUROPE

European DX Council, P.O. Box 214, 00101, Helsinki, Finland. Umbrella organization of DX Clubs in Europe. Send 2 IRC's for more info. *E-mail:* risto.vahakainu@helsinki.fi (Secr. Gen.), arto.mujunen@ ibb.fi (Asst. Secr. Gen.)

AUSTRIA: Assoziation Junger DX-er in Österreich, Postfach 1000, 1081 Wien (G)

BELGIUM: DX-Antwerp, Postbus 16, 2660 Hoboken (Dutch). *E-mail:* dxa@dxa.be *Web:* http://www.dxa.be – **Belgique Radio-Loisirs,** B.P.12, 7160 Chappelle-Lez-Herlaimont (F).

BULGARIA: Association of Balkan Crossband DXers, ABCDX, c/o Rumen Pankov, P.O.Box 199, 1000 Sofia-C – **Bulgarian DX Club plus Satellite,** c/o Ivan Penev, P.O. Box 47, Sofia 11, Bulgaria 1111. *E-mail:* kip@biscom.net

CZECH REPUBLIC: Czechoslovak DX Club, c/o Václav Dosoudil, Horní 9, 76821 Kvasice. (Czech/Slovak/English). *E-mail:* mail@csdxc.cz *Web:* http://www.csdxc.cz/

DENMARK: The Danish Shortwave Club International, Tavleager 31, DK-2670 Greve. *E-mail:* DSWCI@centrum.dk *Web:* http://www.nordicdx.com – **Dansk DX Lytter Klub,** P.O.Box 392, 8100 Aarhus C (Danish). *E-mail:* hartvig@wmr.dk *Web:* www.wmr.dk/ddxlk

FINLAND: Suomen DX Liitto, P.O.Box 454, 00101 Helsinki (Umbrella organization of Finnish-language DX Clubs) *E-mail:* sdxl@sci.fi *Web:* http://www.sci.fi/~sdxl/ – **Finlands Svenska DX-Förbund r.f,** PB 402, SF-20101 Åbo, Finland (Umbrella organization of various Swedish-language DX clubs in Finland). *BBS:* +358 68 831 6468.

FRANCE Amitié Radio, B.P.56, 94002 Creteil Cédéx (F). *Web:* http://www.cpod.com/monoweb/Amitie_Radio/ – **Association Union des Ecouteurs Français,** B.P. 31, 92242 Malakoff Cédéx (F, also publishes magazine on IBM compatible floppy disk in F/E) *E-mail:* tsfinfo@magic.fr *Web:* http://www.radioecouter.com – **Club Européen de DX Radio TV,** c/o Bernard Fontaine, 22, La Marjolaine, 13420 Gemenos (F) *E-mail:* Bernard-A.Fontaine@Bull.net.– **Club des Auditeurs de la Radio Mondiale,** 79 Ave.Paul Vaillant Couturier, 94400 Vitry sur Seine (F) – **QSL Club de France,** 40 rue de Haguenau, 67700 Saverne (F) – **Monde & Radiodiffusion,** 15, Cours du Midi, 84850 Camaret (F/E) —— **Radio DX Club d'Auvergne,** Centre Municipal P. et M. Curie, 2 bis, Rue du Clos Perret, 63100 Clermont-Ferrand. ☎ 📠 +04 73 37 08 46.

GERMANY Interessengemeinschaft Rundfunk Fernempfang e.V., Postfach 101311, 45813 Gelsenkirchen (G, exclusively for blind listeners) – **Arbeitsgemeinschaft DX,** Postfach 1107, 91001 Erlangen (G, has 10 member clubs, details for Rp.) *E-mail:* wwh@compuserve.com *Web:* http://ourworld.compuserve.com/homepages/wwh/agdx.html – **Assoziation Deutschsprachiger DXer e.V.,** Postfach 130124,

40551 Düsseldorf 13 (G) *E-mail:* 100644.575@compuserve. com *Web:* http://ourworld.compuserve.com/homepages/ADDX/ – **East and West Radio Club,** Banhofstr. 56, 50374 Erfstadt (G) – **Eastside DX,** c/o Jens Adolph, Nathusiusstr. 17a, 04129 Leipzig (G) – **Esperanto DX Club,** Kafkastr.48 5/M, 81737 München 83 (Esperanto) – **Kurzwellenfreunde Bremen,** Klaus Bergmann, Auf dem Berge 16a, 28759 Bremen (G) – **Kurzwellenfreunde Oldenburg,** Jan Lüschen, Stiekelkamp 2a, D-26125 Oldenburg (G) *E-mail:* jan_lueschen@ol.maus.de – **Kurzwellen-freunde Wuppertal,** Postfach 22 03 42, 42373 Wuppertal 22 (G) – **Kurzwellenclub Berlin e.V.,** Fregestr.59, 12159 Berlin 41 (G) – **Radiohörerclub International,** c/o Hans-Joachim Brustmann, Strasse am Park 16, 04209 Leipzig (G) —— **Rhein-Main-Radioclub e.V.,** Postfach 1168, D-65432 Flörsheim (G). *Web:* http://ourworld.compuserve. com/homepages/telstar901 —— **Klaus-Dieter Scholz,** Postfach 885, D-99017 Erfurt (G). *E-mail:* kdscholz@t-online.de — **UKW/TV-Arbeitskreis,** c/o H.J.Kuhlo, Wilhelm-Leuschner-Str. 293B, 64347 Griesheim (G) *Web:* http://ourworld. compuserve.com/homepages/wwh/ukwt-vak.htm – **Vereinigung Berliner Empfangsamateure,** Postfach 113, 13511 Berlin (G) – **Worldwide DX Club,** Postfach 1214, 61282 Bad Homburg 1. *E-mail:* wwdxc@compuserve.com *Web:* http://ourworld.compuserve.com/ homepages/wwdxc/ or http://www.european-media.com/wwdxc/

ITALY: Associazione Italiana Radioascolto, C.P 1338, 10100 Torino A.D. (I) *Web:* http://www.arpnet.it/~ard/ – **Coordinamento del Radioascolto,** c/o Marco Cerutti, C.P. 146, 13100 Vercelli (Umbrella organization of various Italian DX clubs) –**BCL Sicilia Club,** c/o Roberto Scaglione, C.P. 119 Succ.34, 90144 Palermo (I/E) – **Club Radio Ascolto,** c/o Emanuela Pelicioli, Via Battisti 11, 24010 Osio Sopra (BG) (I) – **Gruppo di Ascolto della Marca Trevigiana,** C.P. 3, Suc. 10, 31100 Treviso (I) –**Gruppo d'Ascolto Dei Due Mari,** c/o Carlo Pepe, via Battisti 27, 74100 Taranto (I/E) – **Gruppo d'Ascolto di Torino, c/o Riccardo Novarino,** C.P. 108, 10124 Moncalieri (TO) (I/E, speclaizes in difficult DX) – **Gruppod'Ascolto Lariano,** c/o Paolo Fumagalli, Viale Consolini 8, 22047 Molteno (CO) (I,only local activity) – **Gruppo d'Ascolto Radio dello Stretto,** c/o Giovanni Sergi, Via Crotone 33, 98149 Messina (Camaro Inferiore) (I) – **Gruppo d'Ascolto Radiotelevisivo della Sicila,** c/o Gioacchino Stallone, Via Sappusi 11 lotto 27 int. 3, 91025 Marsala (TP) – **Gruppo Radioascolto Liguria,** c/o Ricardo Storti, Via Mattei 25\1, 16010 Manesseno Sant'Olcese (GE) (I) – **Gruppo Utility Romagnolo,** c/o Fabrizio Magrone. Via Marengo 33, 47100 Forli (I, specialises in Utility DX) – **Mediterraneo Radio Club,** c/o Alberto Lo Passo, C.P. 172, 96100 Siracusa (I,only local activity) –**Play-DX,** c/o Dario Monferini, Via Davanzati 8, I-20158 Milano (I/E specialises in difficult DX) – **Ranch Radio Anchio,** c/o Sergio Citaglia, C.P. 38, 00040 Pavona Stazione (RM) (I) – **Victor Charlie Group sez. BCL,** C.P. 343, 30100 Venezia (I).

LITHUANIA: Lithuanian DX Club, P.O. Box 471, 235802 Klaipeda 2, Lithuania.

MOLDOVA: National Guild "DX Moldova", c/o Leonid Cutlculu, P.O. Box 3297, MD-2044, Chisinau (Umbrella organization of Moldovan DX Clubs. (Russian/Moldovan/E) — **FM-FANtasy,** c/o Evgeniy Alekseyev, str. Ialoveni 96/A-52, MD-2070 Chisinau (FM only, Russian/E) — **Pridnestrovian Radio Friends Society "Dniestr-X",** c/o Evgeniy veretennikov, per. Raevskogo 6-18, Tiraspol 27800 (Russian) — **Edinet Shortwave Fanciers,** c/o Vasiliy Spirin, str. Cirimpei 69, MD-4600 Edinet (SW only, Russian).

NETHERLANDS: Benelux DX Club, A.van Ostadestraat 106, 7944 XT Meppel (Dutch/E). *Web:* http://www-dx.deis.unibo.it/htdx/swls/ bdxc.html

NORWAY: DX Listeners' Club, P.O. Box 7080, Vestheiene, 4602 Kristiansand (Norwegian). *E-mail:* dxn@sn.no *Web:* http://www.dxlc.com/

RUSSIA: Club of DX-ers, c/o Vadim Alexeew, P.O. Box 65, Moscow A-581, 125581 (Russian/English/German). E-mail: gusev@itep.ru *Web:* http://limon.itep.ru/~gusev/RADIO/CLUBS/mdxc.html — **Irkutsk DX Club:** c/o Feodor Brazhnikov, P.O.Box 3036, Irkutsk-59, 664059 Russia *E-mail:* radio@tugr.irkutsk.su *Web:* http://www.icc.ru/radio/ — **Russian DX League,** ul. Tvardovskogo 23 - 365, 123458 Moscow (E/Russian) — **Sankt-Peterburg DX Club,** Aleksey Osipov, P.O. Box 46, 195213 Sankt-Peterburg. *E-mail:* mfactum@mail.dux.ru

SPAIN: Asociación DX Barcelona, Apartado 335, 08080 Barcelona (Sp). **Web:** http://www.redestb.es/adxb/ – **Asociación Española de Radioescucha (AER),** P.O. Box 4031, 28080 Madrid (Sp). *Web:* http://www.lander.es/~sedano/ – **Mediterranean DX Group,** P.O. Box 4212. 41080 Sevilla *E-mail:* jgarzon@arrakis.es *Web:* http://www.geocities.com/SiliconValley/4847/ — **MAF Ediciones,** c/o Manuel Castro Vinagre, P.O. Box 21126, 08080 Barcelona (FM/TV, Sp)

SWEDEN: Sveriges DX Förbund, Box 3108, 103 62 Stockholm. (Sw, umbrella organization of over 30 clubs). *E-mail:* Erik.Johansson@hks.se *Web:* http://www.lls.se/jal/sdxf/ – **Arctic Radio Club,** Box 5050, 350 05 Växjö (MW, Sw/E) – **Sveriges Radioklubb,** Pilvagen 86 nb, 191 42 Stockholm (Sw).

UNITED KINGDOM: British DX Club, 126 Bargery Rd, Catford, London SE6 2LR *E-mail:* 100437.173@compuserve.com *Web:* http://www.co.umist.ac.uk/BDXC/ – **The British FM and TV Circle,** 15 Boarhill Grove, Ashfield Park, Sutton-in-Ashfield, Nottinghamshire NG17 1HF. *E-mail:* skywaves.dx@virgin.net – **International Shortwave League,** 3 Bromyard Drive, Chellaston, Derby DE73 1PF (Also for amateur radio enthusiasts). *E-mail:* srj5@aber.ac.uk *Web:* http://freespace.virgin.net/nigel.dyche – **Medium Wave Circle,** 137a Hampton Road, Southport PR8 5DY (LW/MW only) *E-mail:* steve.whitt@zoo.bt.co.uk *Web:* http://www.geocities.com/ Hollywood/5613/mwc.html – **World DX Club,** 17 Motspur Drive, Northampton NN2 6LY. *E-mail:* mark@dxradio.demon.co.uk

AFRICA

CÔTE D'IVOIRE: DX-Ivoire, c/o Mr.Tolo Bakary, 10 B.P. 1300, Abidjan 10 (F).

NIGERIA: Africa DX Association, c/o Mr. Friday I. Okoloise, NITEL, P.M.B. 23, Lafia, Plateau State.

TUNISIA: Club des Auditeurs et de l'Amitié, c/o De Riadh Sakka, Route de Gremda Merkez Sahnoun, 3012 Sfax (F).

SOUTH AFRICA: South African DX Club, P.O. Box 18008, Hillbrow 2038.

TOGO: Club Inter Amitié Radio, CCF B.P. 2090, Lomé (F) – **Groupe Endoc,** B.P. 2667, Lomé (F)

UGANDA: The International DX Club of East Africa, c/o Ouma Samuel, PB 565, Iganga.

ASIA

BANGLADESH: International Radio Listeners Club, Konabari, P.O.Nilnagor, Gazipur, Dhaka (E/Bengali) — **'Rose' DW**

Listeners Club, c/o Ashik Eqbal "Tokon", Luximpur Greater Rd, Side of GPO, Rajshahi-6000.

INDIA: DX Network Club, P.O. Box 44, Golaghat 785621. **E-mail:** dxnet@apo.com *or* dxnet@sandh.com. **Web:** http://www.cyberspace.org/~babul/radio.html — **Foreign Radio Listeners' Club,** c/o Prasenjit Bhakat, P.O. Jhargram-721507, We. Bengal – **Indo-Universal DXing Society,** Peoples' House, Bahudarada, Bhadrak, Orissa 756100 – **North India DX-ers Association,** c/o Mr. Kumar Radha Raman Kanth, P.O. Banmankhi. Dist-Purnea, Bihar 854 202 – **Paribar Bandhu SWL Club,** c/o Mr. Anand Moham Bain, AT/Post. Pallari. Dist. Raipur, M.P. India. PIN.493228 – **Universal DX League,** C/o Kanwarjit Sandhu, P.O. Box: 1128, Chandigarh-160 015. *E-mail:* udxl@hotmail.com – **World DX Club & Library,** c/o Baidyanath Upadhyaya, Al Khairabarigaon, P.O. Khawrang, Udalguri 784 509.

INDONESIA: Indonesian DX Club, P.O. Box 50, Kutoarjo 54201, Jawa Tengah (E/Indonesian) – **Radio Listeners Club Indonesia,** P.O. Box 1983, Surabaya 60019, Ea. Java.

JAPAN: Japanese Association of DX-ers, P.O. Box 1766, Tokyo 100-91 (J/E) – **Asian Broadcasting Institute,** C.P.O. Box 1334, Tokyo 100-91 – **Japan BCL Federation,** 4F, Matsuokadudan Bldg.,2-8 Kudanminami 2-chome, Chiyoda-ku, Tokyo 102 (J) – **Japan Short Wave Club,** P.O. Box 29, Sendai Central 980-91 (J/E) – **Indonesian DX Circle,** c/o Satoshi Wakisaka, 6-25-21 Senriyama Nishi, Suita City, Osaka 565 – **Nagoya DXers Circle,** c/o Shigenori Aoki, 2-51 Kasumori-cho, Nakamura-ku, Nagoya 453 – **Radio Nuevo Mundo,** c/o Tetsuyu Hirahara, 5-6-6 Nukuikita, Koganei-shi, Tokyo 184.

NEPAL: Listeners' Club of Nepal, (registered no.144), P.O Box 126, Biratnagar-4.

PAKISTAN: National Society of Of Pakistani DXers, E-161/1, Iqbal Park, opposite Adil Hospital, Defence Housing Society Rd, Lahore Cantt.

SRI LANKA: Union of Asian DXers, c/o Victor Goonetilleke, "Shangri-La" 298 Kolamunne, Piliyandala.

PACIFIC

South Pacific Association of Radio Clubs **E-mail :** clarkb@sparc.org.nz (Umbrella organization of most Australian and NZ DX Clubs).

AUSTRALIA: Australian Radio DX Club Incorporated, 51 Bambil Road, Berowra 2081 NSW. *E-mail:* dxer@fl.net.au *Web:* http://www.aaa.com.au/dx/ – **DX'ers Calling Inc,** P.O. Box 285, Mt. Waverley, Vic. 3149 – **Southern Cross DX Club Incorporated,** G.P.O. Box 1487, Adelaide, S.A. 5001. *E-mail:* scdxc@tolstoi.saccii.net.au *Web:* http://tolstoi.saccii.net.au/~stephenn – **South Pacific Union Of Dxers Inc.,** P.O. Box 293, Coburg, Victoria 3058. *E-mail:* ddiamond@melbpc.org.au *Web:* http://www.celestial.com.au/~spud-pres/

NEW ZEALAND: New Zealand Radio DX League, P.O. Box 3011, Auckland. *Web:* http://navigator.co.nz/nzrdxl/ – **New Zealand DX Radio Association,** c/o R.Dickson, 88 Cockerell Str, Brockville, Dunedin.

NORTH AMERICA

Association of North American Radio Clubs (ANARC), 2216 Burkey Drive, Wyomissing, PA 19610-1553. Umbrella organization for most North American DX Clubs. Send 2 IRC's for more information about the association and its member clubs. *Web:* http://www.anarc.org/

CANADA: Canadian International DX Club, 79 Kipps Street, Greenfield Park, PQ J4V 3B1. *Web:* http://www.anarc.org/cidx/ – **Club Ondes Cortes du Quebec,** 5120 35ème rue, Grand-Mère, PQ G9T 3N6 (F). *E-mail:* dduplessis@infoteck.qc.ca *Web:* http://www.infobahnos.com /~pedro — **Ontario DX Association,** P.O. Box 161, Station 'A', Willowdale, ON M2N 5S8. *E-mail:* 73737.3453@compuserve.com *Web:* http://www.grove.net/~odxa/ – **Signal Surfer DX Club,** c/o Darcy Jabs, RR2, Burns Lake, BC V0J 1E0. *E-mail:* djabs@awinc.com *Web:* http://www2.awinc.com/users/ djabs/

USA: All Ohio Scanner Club, 20 Philip Drive, New Carlisle, OH 45344-9108 (non-broadcast Public Service Bands only). *Web:* http://aosc.rpmdp.com/ – **American Shortwave Listeners Club,** 16182 Ballad Lane, Huntington Beach, CA 92649-2204. *Web:* http://www.anarc.org/aswlc/ – **Association of Clandestine Enthusiasts,** P.O. Box 11201, Shawnee Mission, KS 66207-0201. *Web:* http://www.frn.net/ace/ – **Chicago Area DX Club,** c/o Edward G.Stroh, 53 Arrowhead Drive, Thornton, IL 60476 – **DecaloMania,** 9705 Mary NW, Seattle. WA 98117 (for collectors of station promotional items and airchecks). *E-mail:* bytheway@atk.com. *Web:* http://www.anarc.org/decal/ – **DX Audio Service,** P.O. Box 164, Mannsville, NY 13661-0164 (for sight-impaired listeners). *E-mail:* gnbc@wcoil.com *Web:* http://wcoil.com/~gnbc/ – **DX South Florida,** 1673 Palace Drive, Clearwater, FL 34616-1833 – **Fine Tuning,** 779 Galiliea Ct., Blue Springs, MO 64014 – **International Radio Club of America,** P.O Box 1831, Perris, CA 92572-1831 (mediumwave only) *E-mail:* Lynn Hollerman *Web:* http://fly.hiwaay.net/ ~waholler/irca.htm – **Longwave Club of America,** 45 Wildflower Road, Levittown, PA 19057 *Web:* http://users.aol.com/lwcanews/ – **Memphis Area Shortwave Hobbyists,** P.O. Box 3888, Memphis, TN 38173 – **Miami Valley DX Club (MVDXC),** Box 292132, Columbus, OH 43229 *E-mail:* dhammer@freenet.columbus.oh.us. *Web:* http://www.anarc.org/ mvdxc/ – **Michigan Area Radio Enthusiasts,** P.O. Box 530933, Livonia, MI 48153-0933 *E-mail:* xx024@detroit.freenet.org *Web:* http://detroit.freenet.org/sigs/l-radio/ – **Minnesota DX Club (MDXC),** 16330 Germane Ct W Rosemount, MN 55068. *Web:* http://www.anarc.org/mdxc/ – **National Radio Club,** P.O. Box 164, Mannsville, NY 13661-0164 (mediumwave only.) *E-mail:* gnbc@wcoil.com *Web:* http://wcoil.com/~gnbc/ – **North American Shortwave Association,** 45 Wildflower Road, Levittown, PA 19057 (SW only) *E-mail:* NASWA1@aol.com *Web:* http://www.anarc.org/ naswa – **Pacific Northwest, British Columbia DX Club,** c/o Bruce Portzer, 6546 19th Ave NE, Seattle WA 98115. *E-mail:* bytheway@atk.com *Web:* http://www.anarc.org/pnbcdxc/ – **Pirate Pages,** P.O. Box 109, Blue Ridge Summit, PA 17214 (pirate radio). *E-mail:* ayoder@cvn.net *Web:* http://www.frn.net/rfi/ – **Radio Communications Monitoring Association,** P.O. Box 542, Silverado, CA 92676 (utility stations) – **Southern California Area DXers,** 6398 Pheasant Drive, Buena Park, CA 90620-1056. *Web:* http://www.anarc.org/scads/ – **Washington Area DX Association,** 606 Forest Glen Rd, Silver Spring, MD 20901 – **Worldwide TV-FM DX Association,** P.O. Box 514, Buffalo NY 14205-0514 (FM/TV/Satellites). *E-mail:* 76255.2171@compuserve.com *Web:* http://www.users.sccoast.net/daustin/wtfda.html

MEXICO: Audio Pico DX Club, c/o César Granillo, Apartado 309, 94301 Orizaba, Ver – **Club DX Miguel Auza,** c/o Luis Antero, Apartado Postal 38, 98330 Miguel Auza, Zac – **Consultorio DX,** c/o Miguel Angel Rocha Gámez, Ap. Postal 31, 31820 Ascensión, Chih – **Nayarit DX Club,** c/o Iván López Alegria, Apartado Postal 62, 63001 Tepic, Nayarit – **Sociedad de Ingenieros**

Radioescuchas, c/o Cesar Fernandez de Lara Garcia, Ap. Postal 203, Admon. No.1, C.P. 91701 Veracruz, Ver.

ARGENTINA: Grupo Radioescucha Platense, Casilla 465, 1900 La Plata (Sp) *E-mail:* grplatense@usa.net

BRAZIL: Clube DX-ista do Para, a/c Djaci Franklin Soares da Silva, Tv. Angustura 1961, apto. 1205 - Pedreira, 66087-710 Belem - PA. *E-mail:* dx-brazil@bigfoot.com *Web:* http://www.amazon.with.br/~djaci or http://www.geocities.com/CapeCanaveral/6731 – **DX Clube Paulista,** C.P. 384, 09701-970 São Bernardo do Campo (P) *Web:* http://www.svn.com.br/dxclube/ – **Globo DX,** C.P. 21.429, São Paulo 04698-970 (P) – **Viamão DX Clube,** C.P. 101, Viamão, RS 94400-970 (P) – **Juventude DX Clube,** C.P. 08, Guabiruba, SC 88360-000 (P. For beginners).

CHILE: Amigos Radioescuchas de Santiago, Casilla 183, La Cisterna, Santiago 14 (Sp). *E-mail:* ars@mailcity.com / hlopez@interaccess.cl – **Departamento de Radioescucha,** c/o Carlos Toledo Verdugo, Carlos Gautier Lizana, Casilla 296, San Fernando, VI Region (Sp).

SURINAME: Suriname DX Club International, Bechaniestraat 58, Paramaribo.

URUGUAY: Asociación Diexman Uruguay, P.O. Box 6008, Montevideo 11000 (Sp). *E-mail:* cx4ban@adinet.com.uy or rialv@network.bbs.com.uy

VENEZUELA: Asociación Diexista de Venezuela, P.O. Box 65657, Caracas 1066-A (Sp) – **Club DXistas de la Amistad,** Ap.202, Barinas 5201-A (Sp).

RELIGIOUS BROADCASTING ORGANISATIONS

AWR - ADVENTIST WORLD RADIO
Worldwide HQ: 12501 Old Columbia Pike, Silver Spring, MD 20904-6600. (+1 (301) 680 6304. 2 +1 (301) 680 6303.
Web: http://www.awr.org/
L.P: Pres: Donald G. Jacobsen. Development Dir, North America: Mark Driskill. Contr: Richard Green. Board Chmn: Philip Follett. Board Vice Chmn: Matthew Bediako. Board Treasurer: Dennis Keith. Coordinator, Int. Rel: Adrian Peterson.
Regional + AWR-Africa, 08 B.P. 1751, Abidjan 08, Côte d'Ivoire. (+225 440097. 2 +225 442341— **AWR-Asia,** P.O. Box 7468, Agat, Guam 96928, USA. (+671 565 2000. 2 +671 565 2983 — **AWR Europe:** c/o Newbold College, Binfield, Bracknell, Berks RG42 4AN, England. (+44 (1344) 304204. 2 +44 (1344) 304169 — **AWR-PanAmerica:** P.O.Box 1177, 4050 Alajuela, Costa Rica. 2 +506 443 0966. 2 +506 441 1282 — **AWR-Russia,** Box 170, Tula Centre 300000, Russia.

CAM INTERNATIONAL
+ 8625 La Prada Drive, Dallas, TX 75228-5098, USA. (+1 (214) 327 8206. 2 +1 (214) 327 8201.
Web: http://www.bible.org/cam/
Operates TGNA Guatemala City, TGBA Barillas and TGMI San Sebastian, Guatemala (see World Radio section for details).

CEEH/MEBSH
+ c/o MFI Haiti, P.O. Box 15665, West Palm Beach, FL 33416, USA.
E-mail: radiolumiere@acn.com — **L.P:** CEEH Exec. Secr: Rigaud A. Antoine. MEBSH Pres: Raphael Lozama.
MEBSH is an association of Baptist Churches that operates the Radio Lumière network of five AM and four FM stations in Haiti ministering on behalf of the Evangelical churches of Haiti associated with CEEH, Concile des Eglises Evangéliques d'Haïti.

EVANGELIUMS-RUNDFUNK
+ Postfach 1444, D-35573 Wetzlar, Germany. (+49 (6441) 9570. 2 +49 (6441) 957122. **E-mail:** sino@dom.de **Web:** http://www.erf.de
ERF Switzerland: Web: http://www.erf.ch/ **E-mail:** erf@erf.ch
ERF So. Africa: Web: http://www3.fast.co.za/~graywolf/erf/erf.htm **E-mail:** twr-sa@iafrica.com
L.P: Chmn: Gerhard Hörster. Exec. Dir: Jürgen Werth.
ERF is the German-speaking partner of Trans World Radio.
Prgrs: German to Europe via TWR Monte Carlo, to So.America via TWR Bonaire, to the southern part of South America via RTM Uruguay, and to Ea.& So.Africa over TWR Swaziland (see World Radio section). Also via shortwave & mediuwave in Germany and via local and regional st's in Austria, Germany, Italy & Switzerland. Prgrs in other languages via TWR Monte Carlo.
New: prgrs via satellite — **V.** by QSL-card and letter.

FEBC RADIO INTERNATIONAL
+ Box 1, La Mirada, CA 90637-0001, USA. (+1 (310) 947 4651. 2 +1 (310) 943 0160. **E-mail:** 335091@mcimail.com
Web: http://www.febc.org
a   n   d
FEBA Radio
+ Ivy Arch Road, Worthing, West Sussex BN14 8BX, England. (+44 (1903) 237281. 2 +44 (1903) 205294. **E-mail:** user@febaradio.co.uk
Web: http://www.feba.org.uk/
An international group of private broadcasting facilities, broadcasting in over 150 languages for a total of approx. 9000 hours per month using 30 transmitters, and operating Christian radio sta-

tions in Korea, The Philippines, Saipan, and The Seychelles. Additional offices and/or studios in 18 countries.

GOOD NEWS BROADCASTING ASSOCIATION INC. BACK TO THE BIBLE INTERNATIONAL
+ Box 82808, Lincoln, NE 68501, USA. (+1 (402) 464 7200. 2 +1 (402) 464 7474. **E-mail:** markb@backtothe bible.org
Web: http://www.gospelcom.net/bttb/
L.P: Dir: Woodrow Kroll. Exec. Dir: Brian Erickson.
Produces three daily prgrs of evangelical preaching in the English language (Back to the Bible, Gateway to Joy, Confident Living), heard on over 600 stations in No. America. Another nine branch offices are scattered throughout the world, and through these offices programs are produced in 16 additional languages. Besides a radio ministry, Back to the Bible also has an extensive literature ministry in multiple languages — **PUB:** Free lists of st's carrying English prgrs as well as English literature catalog.

HIGH ADVENTURE MINISTRIES
+ Box 100. Simi Valley, CA 93062, USA. (+1 (805) 520-9460. 2 +1 (805) 520-7823. **E-mail:** voh@broadcast.net
Web: http://www.highadventure.org/
L.P: CEO: George Otis. Pres: John Tayloe.
High Adventure operates shortwave transmitters in the Middle East, Central Asia, the South Pacific and the USA to provide global coverage of the Christian message in over a dozen languages.

IBRA RADIO AB
+ IBRA Radio, S-105 36 Stockholm, Sweden.
HQ: Gammelgårdsvägen 38-42, Stora Essingen, Sweden. (+46 (8) 6192540. 2 +46 (8) 6192539. **E-mail:** hq@ibra.se
Web: http://www.ibra.se/ **BBS:** +46 (8) 6513010.
Prgrs: The broadcasts of IBRA can be heard in over 100 countries in approx 60 languages. On shortwave, IBRA prgrs.are carried by High Adventure (Lebanon), TWR Monaco, FEBA Seychelles, FEBC Manila & Saipan, Voice of Russia. About 200 local st's in Europe, Africa, Asia and Latin America are also used.

INTERNATIONAL CHRISTIAN MEDIA COMMISSION (ICMC)
+ P.O. Box 2165, Des Plaines IL 60017-2165 USA. (+1 847 699 1970. 2 +1 847 699 1570 — + P.O. Box 154 Keighley, BD20 6UU United Kingdom. (+44 1535 612100. 2 +44 1535 612101. — P.O. Box 100, Witney, OX8 7TD, United Kingdom. (+44 1993 776249. 2 +44 1993 776259.
E-mail: ICMC@XC.ORG **Web:** http://www.ICMC.Org/
The International Christian Media Commission is an association of Christian media professionals from around the world. Started in 1986, its membership encompasses 121 countries on six continents.

INTERNATIONAL LUTHERAN LAYMEN'S LEAGUE
+ 2185 Hampton Avenue, St.Louis, MO 63139-2983, USA. (+1 (314) 647 4900. 2 +1 (314) 647 6923. **E-mail:** lh_min@lhm.org
Web: http://www.lhm.org.
L.P: Exec. Dir: Larry E.Lumpe. Dir. International Ministries: Kenneth M.Peterson. Dir. Domestic Ministries: Rick Krueger. Dir. Marketing & Communication: James Telle.
Prgrs: The Lutheran Hour, Bringing Christ to the Nations, Woman-to-Woman Radio (weekly), This is the Life-TV, & more than 80 international prgrs in 33 languages from 26 production centers.

LUTHERAN WORLD FEDERATION DEPARTMENT FOR MISSION & DEVELOPMENT
Head Office: 150 route de Ferney (P. O. Box 2100), CH-1211

Geneva 2, Switzerland. (+41 (22) 791 6111. 2 +41 (22)798 8616. **E-mail:** dpj@wcc-coe.org **Web:** http://www.wcc-coe.org/lwf/ **L.P:** Gen. Secr: Dr. Ishmael Noko, Zimbabwe. Chairperson of Program Committee: Dean Karsten Nissen (Denmark). Secretary for Communication Consultancy: Rev. Bernhardur Gudmundsson, Iceland.

NATIONAL RELIGIOUS BROADCASTERS

+ 7839 Ashton Avenue, Manassas, VA 22110, USA. (+1 (703) 330 7000. 2 +1 (703) 330 7100. **Web:** http://www.nrb.com **L.P:** Pres: Dr. E.Brandt Gustavson. Chmn: Robert Straton, Walter Bennett Communications, Ft. Washington, PA, USA.
NRB is an association of over 800 religious national and international broadcasting organizations in radio, TV and related services. Founded 1944.
PUB: "Religious Broadcasting" magazine (monthly), "Directory of Religious Media" (annual). Conventions: Sponsors annual convention in Washington DC and other cities alternating years, late Jan/Feb, drawing delegates from around the world; also regional conventions in the US and Puerto Rico.

NOREA RADIO (Nordic Radio Evangelistic Assoc.)

+ Gimlekollen Mediesenter, Box 4087, N-4602 Kristiansand, Norway. (+47 38 093000. 2 +47 38 094100. **E-mail:** ged@gms.va.no
L.P: Exec. Dir: S.Pettersen. Dir. Prgr: L.Taule. Tech. Dept: J.Lien.
Prgrs in Norwegian on local radio st's in Norway and in 15 other languages over TWR Monaco, TWR Cyprus, TWR Swaziland, TWR Sri Lanka, TWR Guam, FEBA Seychelles and st's in Japan, Taiwan, Indonesia, Bolivia & Peru — **V.** by QSL-card & letter.

OMS INTERNATIONAL, Inc.

+ Box A, Greenwood, IN 46142, USA. (+1 (317) 881 6751.
Web: http://www.omsinternational.org/
OMS International is an evangelical, undenominational, faith mission. It operates 4VEH, the Evangelistic Voice of Haiti (details under 'Haiti').

PONTIFICAL COUNCIL FOR SOCIAL COMMUNICATIONS

+ Palazzo S.Carlo, 00120 Vatican City State. (+39 (6) 698 83197. 2 +39 (6) 698 85373.) 2019 PCCS VA — Pres: Archbishop John P. Foley. Secr: Bishop Pierfranco Pastore. Under-Secr: Hans-Peter Röthlin. Tech. Consultant: Rev. Antonio Stefanizzi.

SIM INTERNATIONAL

+P.O. Box 7900, Charlotte, NC 28241, USA. **E-Mail:** Radio@sim.org. **Web:** http://www.sim.org/radio/
L.P: Gen. Dir: Dr. Jim Plueddemann. Gen. Dir. Emeritus: Dr. Ian M. Hay. Deputy Gen. Dir's: Dr. Howard Brant, Eldon Howard, Gordon Stanley, Rev. Ron Wiebe.

TRANS WORLD RADIO INTERNATIONAL HEADQUARTERS

+ P.O.Box 8700, Cary, NC 27512-8700, USA. (+1 (919) 460 3700. 2 +1 (919) 460 3702. **E-mail:** info2@twr.org
Web: http://www.gospelcom.net/twr/
L.P: Pres: Thomas J. Lowell.
TWR broadcasts the Gospel in 100 languages via satellite and from nine transmitting locations: Albania, Monaco, Netherlands Antilles, Swaziland, Cyprus, Russia, Sri Lanka, Guam and Uruguay. **Trans World Radio - Europe:** Postfach 141, A-1235 Wien, Austria. (+43 (1) 863120. 2 +43 (1) 8631220 — European Regional Dir: Werner Kroemer.
PUB: TWR Magazine 4 times a year, TWReport 4 times a year.

UNDA - Association Catholique Internationale pour la Radio et la Télévision (International Catholic Association for Radio and Television).

UNDA General Secretariate: 12 rue de l'Orme, B-1040 Brussels, Belgium. (+32 (2) 734 97 08. 2 +32 (2) 734 70 18. **Cable:** Undabrussels.) 051-91 8023 GEONET G.
E-Mail: Zukowski@trinity.udayton.edu
Web: http://www.catholic.org/orgs/unda-int/
L.P: Secr. Gen: Victor Sunderaj.
Membership: National Associations in 139 countries and territories, plus 26 international or internationally-oriented organizations.
Pub: "Unda News" bulletin (6 per year) in English & French. "Educommunication News" bulletin in English & French (4 per year) on Media Education, plus various newsletters published at continental and national levels.

UNITED CHRISTIAN BROADCASTERS

+ P.O. Box 255, Stoke-on-Trent, Staffs ST4 8YY, England. (+44 (1782) 642000. 2 +44 (1782) 641121.**E-mail:** ucb@ucb.co.uk **Web:** http://www.ucb.co.uk/
L.P: MD: Gareth Littler. Broadc. Mgr: Ann Haccius. Tech. Mgr: Graeme Wilson. P.R. & Mkt: Patricia Hargreves.
Prgrs: Christian radio across Europe via the Astra satellite, rel. on AM/FM via tr's in Ireland. Contemporary Christian music with teaching, spoken features and interviews etc. Freely available for terrestrial re-broadcast.
PUB: Monthly newsletter, programme guides etc. — **V.** by letter.
F.PL: Additional terrestrial tr's. Two separate sces: one on MW for under 30's, one on LW/SW for older listeners.

WORLD ASSOCIATION FOR CHRISTIAN COMMUNICATION (WACC)

+ 357 Kennington Lane, London SE11 5QY, U.K. (+44 (171) 582 9139. 2 +44 (171) 735 0340.) 8812669 WACC G. **Cable:** WACC London SE11. **E-mail:** wacc@gn.apc.org
Web: http://www.oneworld.org/wacc/
L.P: Gen. Secr: Rev. Carlos A.Valle. Pres: Rev. Randy Naylor (Canada). Vice Pres: Ms. Manoushag Boyadjian (Lebanon). Secr: Rev. Andre Moussanga Epee (Togo). Treasurer: Rev. Kevin Engel (Australia).
WACC is a fellowship of organizations and individuals committed to Christian communication & media development.
PUB: ACTION newsletter (10 times a year). Media Development (quarterly). Occasional conference and research reports.

WORLD CHRISTIAN BROADCASTING CORP.

+ 605 Bradley Court, Franklin, TN 37064, USA. (+1 (615) 371 8707. 2 +1 (615) 371 8791. **E-mail:** KNLS@aol.com
Web: http://www.knls.org
L.P: Pres: Charles Caudill.
Operates SW-st. KNLS, Alaska. The broadcast format is diversified including informational and cultural prgrs. of general interest. (Further details in World Radio Section under "Alaska").

WORLD RADIO BIBLE BROADCASTS

+ P.O.Box 2000, West Monroe, LA 71294-2000, USA.
A non-denominational Christian broadcast, heard throughout the world in 16 languages.

WORLD RADIO MISSIONARY FELLOWSHIP, INC.

+ PO Box 39800, Colorado Springs, CO 80949-9800, USA. (+1 (719) 590 9800. 2 +1 (719) 590 9801. **E-mail:** 3417450@mic-mail.com
L.P: Pres: Ronald A. Cline. Exec. Dir: Scott Higgins.
Operates HCJB, La Voz de los Andes, Quito, Ecuador. Affiliated with KVMV, KOIR, KBNJ, KBNL, KBNR, KRIO (USA).

WORLD TIME TABLE

The differences marked + indicate the number of hours ahead of UTC. Differences marked – indicate the number of hours behind UTC. Variations from standard time during part of the year (in some countries referred to as Summer Time) are decided annually and may vary from year to year. N=Normal Time; S=Summer Time.

Place	N	S
Afghanistan	+4½	+4½
Alaska	−9	−8
	−10	−9
Albania	+1	+2
Algeria	+1	+1
Andorra	+1	+2
Angola	+1	+1
Anguilla	−4	−4
Antigua	−4	−4
Argentina (Ea.)	−3	−2
Argentina (rest)	−3	−3
Armenia	+4	+4
Aruba	−4	−4
Ascension Isl.	UTC	UTC
Australia		
Victoria & NSW	+10	+11
Queensland	+10	+10
Tasmania	+10	+11
N. Territory	+9½	+9½
S. Australia	+9½	+10½
(we. part)	+9	+10
W. Australia	+8	+8
Austria	+1	+2
Azerbaijan	+3	+4
Azores	−1	UTC
Bahamas	−5	−4
Bahrain	+3	+3
Bangladesh	+6	+6
Barbados	−4	−4
Belarus	+2	+3
Belgium	+1	+2
Belize	−6	−6
Benin	+1	+1
Bermuda	−4	−3
Bhutan	+6	+6
Bolivia	−4	−4
Bosnia/Hercegovina	+1	+2
Botswana	+2	+2
Brazil		
a) Oceanic Isl	−2	−2
b) Ea & Coastal	−3	−2
c) Manaos	−4	−3
d) Acre	−5	−4
Brunei	+8	+8
Bulgaria	+2	+3
Burkina Faso	UTC	UTC
Burundi	+2	+2
Cameroon	+1	+1
Canada		
a) NF, Labrador (So. Ea.)	−3½	−2½
b) Labrador (rest), NS, NB, PEI	−4	−3
c) ON, PQ	−5	−4
d) MB	−6	−5
e) AB, NWT	−7	−6
f) BC, YT	−8	−7
Cambodia	+7	+7
Canary Isl.	UTC	+1
Cape Verde Isl.	−1	−1
Cayman Isl.	−5	−4
Ce. African Rep.	+1	+1
Chad	+1	+1
Chile	−4	−3
China (P.R.)		
Beijing	+8	+9
Urumqi	+6	+7
Christmas Isl.	+7	+7
Cocos Isl.	+6½	+6½
Colombia	−5	−5
Comoro Rep.	+3	+3
Congo	+1	+1

Place	N	S
Congo (Dem. Rep)		
Kinshasa	+1	+1
Lubumbashi	+2	+2
Cook Isl.	−10	−9½
Costa Rica	−6	−5
Côte d'Ivoire	UTC	UTC
Croatia	+1	+2
Cuba	−5	−4
Cyprus	+2	+3
Czech Rep.	+1	+2
Denmark	+1	+2
Diego Garcia	+5	+5
Djibouti	+3	+3
Dominica	−4	−4
Dom. Rep.	−4	−4
Easter Isl.	−6	−5
Ecuador	−5	−5
Egypt	+2	+2
El Salvador	−6	−6
Equatorial Guinea	+1	+1
Estonia	+2	+3
Ethiopia	+3	+3
Falkland Isl.	−4	−4
(Port Stanley)	−4	−3
Faroe Isl.	UTC	+1
Fiji	+12	+12
Finland	+2	+3
France	+1	+2
Gabon	+1	+1
Galapagos Isl.	−6	−6
Gambia	UTC	UTC
Georgia*	+4	+5
(* exc. Abkhasia	+3	+4)
Germany	+1	+2
Ghana	UTC	UTC
Gibraltar	+1	+2
Greece	+2	+3
Greenland		
Scoresbysund	−1	UTC
Thule area	−3	−3
Other areas	−3	−2
Grenada	−4	−4
Guadeloupe	−4	−4
Guam	+10	+10
Guatemala	−6	−5
Guiana (French)	−3	−3
Guinea (Rep.)	UTC	UTC
Guinea Bissau	UTC	UTC
Guyana (Rep.)	−3	−3
Haiti	−5	−4
Hawaii	−10	−10
Honduras (Rep.)	−6	−6
Hong Kong	+8	+8
Hungary	+1	+2
Iceland	UTC	UTC
India	+5½	+5½
Indonesia		
a) Java, Bali, Sumatra	+7	+7
b) Kalimantan, Sulawesi Timor	+8	+8
c) Moluccas, We. Irian	+9	+9
Iran	+3½	+4½
Iraq	+3	+4
Ireland	UTC	+1
Israel	+2	+3
Italy	+1	+2
Jamaica	−5	−4
Japan	+9	+9

Place	N	S
Jordan	+2	+3
Kenya	+3	+3
Kazakhstan	+6	+7
Kiribati	+12	+12
Korea (Rep.)	+9	+10
Korea (D.P.R.)	+9	+9
Kuwait	+3	+3
Kyrgyzstan	+5	+6
Laos	+7	+7
Latvia	+2	+3
Lebanon	+2	+3
Lesotho	+2	+2
Liberia	UTC	UTC
Libya	+1	+2
Lithuania	+2	+3
Lord Howe Isl.	+10½	+11
Luxembourg	+1	+2
Macau	+8	+8
Macedonia	+1	+2
Madagascar	+3	+3
Madeira	UTC	UTC
Malawi	+2	+2
Malaysia	+8	+8
Maldive Isl.	+5	+5
Mali	UTC	UTC
Malta	+1	+2
Marshall Isl.	+12	+12
Martinique	−4	−4
Mauritania	UTC	UTC
Mauritius	+4	+4
Mayotte	+3	+3
Mexico	−6	−6
(see country section for exceptions)		
Micronesia		
Truk, Yap	+10	+10
Pohnpei	+11	+11
Midway Isl.	−11	−11
Moldova	+2	+3
Monaco	+1	+2
Mongolia	+8	+9
Montserrat	−4	−4
Morocco	UTC	UTC
Mozambique	+2	+2
Myanmar	+6½	+6½
Namibia	+1	+2
Nauru	+11	+11
Nepal	+5.45	+5.45
Netherlands	+1	+2
Neth. Antilles	−4	−4
New Caledonia	+11	+11
New Zealand	+12	+13
Nicaragua	−6	−6
Niger	+1	+1
Nigeria	+1	+1
Niue	−11	−11
Norfolk Isl.	+11½	+11½
N. Marianas	+10	+10
Norway	+1	+2
Oman	+4	+4
Pakistan	+5	+5
Palau	+9	+9
Panama	−5	−5
Papua N. Guinea	+10	+10
Paraguay	−4	−3
Peru	−5	−4
Philippines	+8	+8
Poland	+1	+2
Polynesia (Fr.)	−10	−10
Portugal	UTC	+1
Puerto Rico	−4	−4

Place	N	S
Qatar	+3	+3
Reunion	+4	+4
Romania	+2	+3
Russia		
Moscow	+3	+4
Novosibirsk	+7	+8
Khabarovsk	+10	+11
Petropavlovsk	+12	+13
Rwanda	+2	+2
Samoa Isl.	−11	−11
S. Tomé	UTC	UTC
Saudi Arabia	+3	+3
Senegal	UTC	UTC
Seychelles	+4	+4
Sierra Leone	UTC	UTC
Singapore	+8	+8
Slovakia	+1	+2
Slovenia	+1	+2
Solomon Isl.	+11	+11
Somalia	+3	+3
S. Africa	+2	+2
Spain	+1	+2
Sri Lanka	+6	+6
St. Helena	UTC	UTC
St. Kitts–Nevis	−4	−4
St. Lucia	−4	−4
St. Pierre	−3	−2
St. Vincent	−4	−4
Sudan	+2	+2
Suriname	−3	−3
Swaziland	+2	+2
Sweden	+1	+2
Switzerland	+1	+2
Syria	+2	+3
Tajikistan	+5	+5
Taiwan	+8	+8
Tanzania	+3	+3
Thailand	+7	+7
Togo	UTC	UTC
Tonga	+13	+13
Transkei	+2	+2
Trinidad	−4	−4
Tristan da Cunha	UTC	UTC
Tunisia	+1	+2
Turks & Caicos	−4	−4
Turkey	+2	+3
Turkmenistan	+5	+5
Tuvalu	+12	+12
Uganda	+3	+3
Ukraine	+2	+3
United Arab Em.	+4	+4
United Kingdom	UTC	+1
Uruguay	−3	−2
USA		
a) Eastern*	−5	−4
*) Indiana	−5	−5
b) Central	−6	−5
c) Mountain*	−7	−6
*) Arizona	−7	−7
d) Pacific	−8	−7
Uzbekistan	+5	+5
Vanuatu	+11	+12
Venezuela	−4	−4
Vietnam	+7	+7
Virgin Isl.	−4	−4
Wake Isl.	+12	+12
Wallis & Futuna	+12	+12
Yemen	+3	+3
Yugoslavia	+1	+2
Zambia	+2	+2
Zimbabwe	+2	+2

ALPHABETICAL LIST OF ITU COUNTRY CODES

*) not an ITU member.

	D = Germany	KWT = Kuwait	ROU = Romania
	DGA* = Diego Garcia	LAO = Lao P.D.R.	RRW = Rwanda
ABW* = Aruba	DJI = Djibouti	LBN = Lebanon	RUS = Russia
AFG = Afghanistan	DMA* = Dominica	LBR = Liberia	S = Sweden
AFS = South Africa	DNK = Denmark	LBY = Libya	SCN* = St. Kitts & Nevis
AGL = Angola	DOM = Dominican Rep.	LCA* = St. Lucia	SDN = Sudan
AIA* = Anguilla	E = Spain	LIE = Liechtenstein	SEN = Senegal
ALB = Albania	EGY = Egypt	LSO = Lesotho	SEY* = Seychelles
ALG = Algeria	EQA = Ecuador	LTU = Lithuania	SHN* = Saint Helena
AND = Andorra	ERI = Eritrea	LUX = Luxembourg	SLM = Solomon Islands
ARG = Argentina	EST = Estonia	LVA = Latvia	SLV = El Salvador
ARM = Armenia	ETH = Ethiopia	MAC* = Macao	SMA* = American Samoa
ARS = Saudi Arabia	F = France	MAU = Mauritius	SMO = Western Samoa
ASC* = Ascension	FIN = Finland	MCO = Monaco	SMR = San Mari*
ATG = Antigua and Barbuda	FJI = Fiji	MDA = Moldova	SNG = Singapore
ATN* = Netherlands Antilles	FLK* = Falkland Islands	MDG = Madagascar	SOM = Somalia
AUS = Australia	FRO* = Faroe Islands	MDR* = Madeira	SPM* = St. Pierre & Miquelon
AUT = Austria	FSM = Micronesia	MDW* = Midway Islands	SRL = Sierra Leone
AZE = Azerbaijan	G = United Kingdom	MEX = Mexico	STP = Sao Tome e Principe
AZR* = Azores	GAB = Gabon	MHL* = Marshall Islands	SUI = Switzerland
B = Brazil	GDL* = Guadeloupe	MKD = Macedonia	SUR = Suriname
BAH = Bahamas	GEO = Georgia	MLA = Malaysia	SVK = Slovakia
BDI = Burundi	GHA = Ghana	MLD = Maldives	SVN = Slovenia
BEL = Belgium	GIB* = Gibraltar	MLI = Mali	SWZ = Swaziland
BEN = Benin	GMB = Gambia	MLT = Malta	SYR = Syria
BER* = Bermuda	GNB = Guinea-Bissau	MNG = Mongolia	TCA* = Turks and Caicos
BFA = Burkina Faso	GNE = Equatorial Guinea	MOZ = Mozambique	TCD = Chad
BGD = Bangladesh	GRC = Greece	MRA* = Northern Marianas	TGO = Togo
BHR = Bahrain	GRD = Grenada	MRC = Morocco	THA = Thailand
BIH = Bosnia-Hercegovina	GRL* = Greenland	MRT* = Martinique	TJK* = Tajikistan
BLR = Belarus	GTM = Guatemala	MSR* = Montserrat	TKL* = Tokelau
BLZ = Belize	GUF* = Guiana	MTN = Mauritania	TKM = Turkmenistan
BOL = Bolivia	GUI = Guinea	MWI = Malawi	TON = Tonga
BOT = Botswana	GUM* = Guam	MYT* = Mayotte	TRC* = Tristan da Cunha
BRB = Barbados	GUY = Guyana	NCG = Nicaragua	TRD = Trinidad and Tobago
BRM = Myanmar	HKG* = Hongkong	NCL* = New Caledonia	TUN = Tunisia
BRU = Brunei Darussalam	HND = Honduras	NFK* = Norfolk Island	TUR = Turkey
BTN = Bhutan	HNG = Hungary	NGR = Niger	TUV* = Tuvalu
BUL = Bulgaria	HOL = Netherlands	NIG = Nigeria	TZA = Tanzania
CAF = Central African Rep.	HRV = Croatia	NIU* = Niue	UAE = United Arab Emirates
CAN = Canada	HTI = Haiti	NMB = Namibia	UGA = Uganda
CBG = Cambodia	I = Italy	NOR = Norway	UKR = Ukraine
CHL = Chile	ICO* = Cocos (Keeling) Islands	NPL = Nepal	URG = Uruguay
CHN = China	IND = India	NRU = Nauru	USA = United States
CHR* = Christmas Island	INS = Indonesia	NZL = New Zealand	UZB = Uzbekistan
CKH* = Cook Islands	IRL = Ireland	OCE* = French Polynesia	VCT = St. Vincent
CLM = Colombia	IRN = Iran (Islamic Rep. of)	OMA = Oman	VEN = Venezuela
CLN = Sri Lanka	IRQ = Iraq	PAK = Pakistan	VIR* = Virgin Islands (Am.)
CME = Cameroon	ISL = Iceland	PAQ* = Easter Island	VRG* = Virgin Islands (Br.)
CNR = Canary Isl.	ISR = Israel	PHL = Philippines	VTN = Vietnam
COG = Congo	J = Japan	PLW* = Palau	VUT = Vanuatu
COM = Comoros	JMC = Jamaica	PNG = Papua New Guinea	WAK* = Wake Island
CPV = Cape Verde	JON* = Johnston Island	PNR = Panama	WAL* = Wallis and Futuna
CTI = Côte d'Ivoire	JOR = Jordan	POL = Poland	YEM = Yemen
CTR = Costa Rica	KAZ = Kazakhstan	POR = Portugal	YUG = Yugoslavia
CUB = Cuba	KEN = Kenya	PRG = Paraguay	ZAI = Congo (Dem. Rep.)
CVA = Vatican	KGZ = Kyrgyzstan	PRU = Peru	ZMB = Zambia
CYM* = Cayman Islands	KIR = Kiribati	PTR* = Puerto Rico	ZWE = Zimbabwe
CYP = Cyprus	KOR = Korea (Rep. of)	QAT = Qatar	
CZE = Czech Rep.	KRE = Korea (D.P.R.)	REU* = Reunion	

ABBREVIATIONS & SYMBOLS USED
IN THIS HANDBOOK

✉ Mailing Address
☎ Telephone
🖹 Telefax
☽ Telex

a/c = alternating current
acc. = accepted
Addr. = Address
Admin. = Administrative
Ad = advertising
Af. = Africa
Ag. = Acting
alt. = alternate, alternative
AM = Amplitude
  Modulation
Am = America
Ann. = Announcement
approx. = approximately
Ap. = Apartado
As = Asia
Assoc. = Association
Asst. = Assistant
Aug. = August
Av., Ave = Avenida, Avenue
BBC = British Broad-
  casting Corporation
BBCWS = BBC World
  Service
B'casters = Broadcasters
Bldg. = Building
B.P. = Boite Postale
Broadc. = Broadcast(ing)
Cad. = Cadena
Carib = Caribbean
Cas. = Casilla
Cd. = Ciudad
Ce. = Central
Cel., Cnel. = Coronel
CE = Chief Engineer
cf. = refer to
Ch. = Channel
Chmn. = Chairman
Cl. = Club(e)
Co. = Company
Comm. = Commercial
Contr. = Controller
Corp. = Corporation
C.P. = Case/Caixa Postal
Cra = Carrera
Cult. = Cultura, Cultural
D = Daily
d = directional antenna
d/c = direct current
Dec. = December
Dem. = Democratic
Dep. = Deputy
Dept. = Department
Depto. = Departamento
Desp. = Despacho
Dif. = Difusora, Difusão

Diff = Diffusion
Dir. = Director
Div. = Division
DX = Long Distance
est = estimated
Ea = East
E.C = Electric Current
Edif. = Edificio
Edo. = Estado
Educ. = Education, Educa-
  tional, Educadora
Em. = Emis(s)ora
Eng. = Engineer, Engineer-
  ing
ERP = Effective Radiated
  Power
Esq. = esquina
Eu = Europe
Exc. = except
Exec. = Executive
Ext. = External
FE = Far East
Feb. = February
Fed. = Federal
FM = Frequency modulation
For.Rel. = Foreign Relations
F.Pl. = Future plans
freq. = frequency
Fri. = Friday
Ft. = Fort
G.C. = Geographical coor-
  dinates given in
  degrees and minutes
Gen. = General
GM = General Manager
Gov. = Government,
  Governmental
Gte = Manager (Sp.)
H = Horizontal polarization
h = hour
HQ = Headquarters
ID = (Station) Identification
i.e. = that is
Inc. = Incorporated
incl. = including
Inf. = Information
Int. = International
IRC = International Reply
  Coupon
irr. = irregular
Isl. = Island(s)
IS = Interval Signal
kHz = kiloHertz
kW = kiloWatts
Langs = Languages
L.P. = Leading Personnel
L.T. = Local Time
Ltd. = Limited
LV = La Voz
LW = Longwave

m. = metres
m.b. = metre band
Man. = Managing
max = maximum
ME = Middle East
MF = Mon-Fri
Mgr. = Manager
MHz = MegaHertz
Mil. = Military
Min. = Ministry, Ministerio
mins = minutes
Mon. = Monday
Mpal. = Municipal (Sp.)
Mpo. = Municipio
Mt. = Mount
MW = Mediumwave
N. = News, Network
Nal. = Nacional
Nat. = National
nd = nondirectional
  antenna
NE = North East
Netw. = Network
n.f. = nominal frequency
No. = North
Nom. = nominal
Nov. = November
Nte. = Norte
occ. = occasionally
Oce = Oceania
Oct. = October
Op's = Operations
Org. = Organization
Ote. = Oeste
Own = Owner
Pac. = Pacific
P = Programme
PD = Programme Director
Pol. = Polarization
Pop. = Population
P.O. = Post Office
PR = Public Relations
Pr. = Private
Pres. = President,
  Presidencia, Presidente
Prgr. = Programme
Pr.L.   =   Principal
  Language(s)
Prod. = Production
Prov. = Province, Provincial
Pte. = Presidente, Poniente,
  Puente
Pto. = Puerto
Pty = Party
Pub. = Publication(s)
QSL = confirmation of
  reception
R. = Radio
r = reported
Radiodiff. = Radiodiffusion

Re. = (reception) reports
Rec. = Recordings
Reg. = Regional
Rel. = Relay(ing), Relations
Rep. = Republic
Rlg. = Religious
Rp. = Return postage
  required
S. = San
Sat. = Saturday
Sce. = Service
Sched. = Schedule
Secr. = Secretary
Sen. = Senior
Sept. = September
SM = Station Manager
S/N, s/n = sin(sem)
  número
s/off = sign-off
s/on = sign-on
So. = South
Soc. = Sociedad(e)
Sp. = Spanish
SS = Sat/Sun
St. = Station, Saint
Sta. = Santa
Str. = Street
Su. = Summer
Sun. = Sunday
Superv. = Supervisor
Syst. = System
SW = Shortwave
TD = Technical Director
Te. = Telephone
Tech. = Technical
Thurs. = Thursday
Tlx. = Telex
Tr. = transmitter,
  transmission
Tues. = Tuesday
TV = Television
u.c. = under construction
UHF = Ultra High Frequency
Univ. = University,
  Universidad(e)
UTC = Coordinated
  Universal time
V. = Verification, Vertical
  polarization, Volts
VP = Vice-President
v. = varying
VHF = Very High Frequency
W. = Weekdays, Watt
We. = West
Web = World Wide Web
Wed. = Wednesday
Wi. = Winter
Wrp. = Weather report

WHERE TO OBTAIN THE WRTH

Australia
Dick Smith Electronics (Pty) Ltd., P.O. Box 321, North Ryde, NSW 2113
Bookwise International, 54 Crittenden Rd, Findon 5023, South Australia

Austria
ERB Handels GmbH, Amerlingstr. 1/11,
A-1120 Vienna

Denmark
GAD Import Booksellers, Siljangade 6-8,
DK-2300 Copenhagen

Finland
Tietoteos Publishing Co, POB 22, Oikkalantie 31,
FIN-02881 Veikkola

France
Brentano's, 37 Avenue de l'Opera, F-75002 Paris

Germany
Gert Wohlfarth GmbH, Stresemannstrasse 20-22, 47051, Duisburg 1

Greece
Librairie Cacoulides, Bld. Panepistimiou 25-29 GR 105 64 Athens

Hong Kong
Swindon Book Co. Ltd., 13-15 Lock Road, Tsim Sha Tsui, Kowloon

Israel
Steimatzky Ltd., P.O. Box 1444, Bnei Brak 51114

Italy
A.C. Distribuzione, Via Kramer 31, 20129 Milano

Netherlands
De Muiderkring B.V., P.O. Box 313, 1380 AH Weesp

New Zealand
Burnet Pollard Books, P.O. Box 149, Otaki 6471

Norway
BBV-Kjop, P.O. Box 88, N-1851 Mysen

South Africa
Technical Books (Pty) Ltd., P.O. 2866, Cape Town 8000
Verbatim Distributors, P.O. Box 190, Steenburg 7947

Spain
Marcambo S.A.,
Gran Vie de les Corts Catalanes 594,
E-08007 Barcelona
Diaz de Santos S.A., Books Department,
Legasca 95, E-28006 Madrid
Edinorte
Rúa da Muiña, 60
E-15890 Santiago de Compostela

Sweden
RADEX, Box 726, S-251 07 Helsingborg

Switzerland
Thali A.G., Industriestr. 14, CH-6285 Hitzkirch

Canada
General Publishing, 30 Lesmill Road, Don Mills, Ontario M3B 2T6

U.K. & Ireland
Windsor Books International, The Boundary, Wheatley Road
Garsington, Oxford OX44 9EJ

United States
Billboard Books, 1695 Oak Street, Lakewood, NJ 08701

Ask your local bookseller or electronic shop to order a copy for you.

WORLD RADIO BY COUNTRY

	Home	Ext.	TV		Home	Ext.	TV
Afghanistan	164	569	407	Czech Republic	65	577	379
Alaska	245	569	423	Denmark	67	579	379
Albania	57	569	376	Djibouti	148		402
Algeria	142	569	401	Dominica	282		426
Andorra	57		376	Dominican Republic	282		427
Angola	142	569	401	Easter Island	236		421
Anguilla	276	569		Ecuador	350	579	434
Antarctica	375		436	Egypt	149	580	402
Antigua & Barbuda	276	570	426	El Salvador	284		427
Argentina	299	570	430	Equatorial Guinea	150	581	403
Armenia	164	570		Eritrea	150		
Aruba	278		426	Estonia	68	581	380
Ascension Island	142	570		Ethiopia	150	581	403
Australia	229	570	419	Falkland Islands	356		434
Austria	57	571	376	Faroe Islands	69		380
Azerbaijan	164	571	408	Fiji	236		421
Azores	60		376	Finland	69	581	380
Bahamas	278		426	France	72	582	381
Bahrain	166		408	Gabon	150	583	403
Bangladesh	174	572	411	Galapagos Islands			421
Barbados			426	Gambia	150		403
Belarus	61	572	376	Georgia	166	583	413
Belgium	61	572	377	Germany	74	583	383
Belize	278	572	426	Ghana	151		403
Benin	143		401	Gibraltar	82		385
Bermuda	247		424	Greece	82	585	385
Bhutan	174			Greenland	253		425
Bolivia	305		431	Grenada	285		427
Bosnia-Hercegovina	62		377	Guadeloupe	285		427
Botswana	143		401	Guam	236	585	421
Brazil	310	572	431	Guatemala	286		427
British Indian Ocean				Guiana (French)	356	586	434
Territory	174		412	Guinea (Rep.)	151		403
Brunei Darussalam	174		412	Guinea-Bissau	151		
Bulgaria	63	573	377	Guyana (Rep.)	356		434
Burkina Faso	143		401	Haiti	288		427
Burundi			401	Hawaii	237	586	421
Cambodia	174	573	412	Honduras	289		427
Cameroon	146		401	Hong Kong	195		413
Canada	247	573	424	Hungary	83	587	386
Canary Islands	146		401	Iceland	85	587	386
Cape Verde	147			India	194	587	413
Cayman Islands	278			Indonesia	197	588	414
Central African Rep.	147		402	Iran	167	588	409
Chad	147		402	Iraq	167	589	409
Chile	339		433	Ireland	85	589	386
China (People's Rep.)	174	574	412	Israel	168	589	409
China (Rep. of)	189	576	413	Italy	87	590	387
Christmas Island	193			Jamaica	292		428
Cocos Islands	193			Japan	207	591	414
Colombia	342		433	Jordan	168	593	409
Comoros	148			Kazakstan	211	593	415
Congo (Dem. Rep.)	148		402	Kenya	151		403
Congo (Rep. of)	148		402	Kiribati	238		421
Cook Islands	236		421	Korea (Dem. Peoples Rep.)	212	593	416
Costa Rica	279	576	426	Korea (Rep.)	213	594	416
Côte d'Ivoire	148		402	Kuwait	169	594	409
Croatia	64	577	377	Kyrgyzstan	217		417
Cuba	280	577	426	Laos	217	595	417
Cyprus	166	577	408	Latvia	98	595	388

Country	Home	Ext.	TV
Lebanon	169	595	409
Lesotho	152		403
Liberia	152	595	404
Libya	152	595	404
Liechtenstein	90		
Lithuania	90	595	388
Lord Howe Island	238		
Luxembourg	92		389
Macau	217		417
Macedonia	92		389
Madagascar	152	596	404
Madeira	153		404
Malawi	153		
Malaysia	217	596	417
Maldives	218		417
Mali	153		404
Malta	92	596	390
Marshall Islands	238		422
Martinique	292		428
Mauritania	153		404
Mauritius	154		404
Mayotte	154		404
México	254	596	428
Micronesia	238		422
Midway Islands	238		
Moldova	92	596	390
Monaco	94	596	390
Mongolia	218	597	
Montserrat	292		428
Morocco	154	597	404
Mozambique	155	597	404
Myanmar	219		
Namibia	155		405
Nauru	238		422
Nepal	219		
Netherlands	94	598	390
Netherlands Antilles	293	598	428
New Caledonia	238		422
New Zealand	239	599	422
Nicaragua	293		429
Niger	156		405
Nigeria	156	599	405
Niue Island	242		422
Norfolk Island	242		422
Northern Marianas	242	599	422
Norway	95	600	390
Oman	170	600	410
Pakistan	219	600	417
Palau	243	600	423
Panama	295		429
Papua New Guinea	243		423
Paraguay	356		434
Peru	357		434
Philippines	220	600	418
Poland	96	601	391
Polynesia (French)	243		423
Portugal	102	602	392
Puerto Rico	296		429
Qatar	170	602	410
Réunion	157		405
Romania	103	602	392
Russia	106	603	393

Country	Home	Ext.	TV
Rwanda	157		
Samoa	244		423
Samoa (American)	244		423
San Marino	118		393
São Tomé	158	604	405
Saudi Arabia	170	604	410
Senegal	158		405
Seychelles	158	605	405
Sierra Leone	158		405
Singapore	223	606	418
Slovakia	118	606	393
Slovenia	119		394
Solomon Islands	244		
Somalia	158		406
South Africa	159	606	406
Spain	120	606	394
Sri Lanka	223	607	418
St. Helena	161		
St. Kitts & Nevis	297		429
St. Lucia	297		429
St. Pierre & Miquelon	266		425
St. Vincent	298		429
Sudan	161	607	407
Suriname	368		435
Swaziland	161	608	407
Sweden	127	608	397
Switzerland	128	609	398
Syrian Arab Rep.	171	609	410
Tajikistan	224	610	419
Tanzania	161	610	407
Thailand	224	610	419
Togo	162		407
Tonga	244		423
Trinidad & Tobago	298		429
Tristan da Cunha	162		
Tunisia	162		407
Turkey	171	610	411
Turkmenistan	227	611	419
Turks & Caicos	298		
Tuvalu	244		
Uganda	162		407
Ukraine	130	611	399
United Arab Emirates	173	611	411
United Kingdom	132	612	399
United States of America	267	616	425
Uruguay	369		435
Uzbekistan	227	621	419
Vanuatu	244		423
Vatican City State	140	621	
Venezuela	371	622	435
Vietnam	228	623	419
Virgin Isl. (American)	298		429
Virgin Isl. (British)	298		430
Wallis & Futuna	244		423
Yemen	173		411
Yugoslavia	140	623	400
Zambia	163		407
Zimbabwe	163		407

DIRECTORY OF ADVERTISERS

ADVERTISING OFFICES

Worldwide
World Radio TV Handbook
P.O. Box 7373
Milton Keynes MK12 5ZL
UK
Fax: +44 (0) 1908 321030

USA & Canada
World Radio TV Handbook
2698 Green Cove Roda
Brasstown, NC 28902
USA
Fax: +1 828 389 4007